Life

Life

Beginnings of Life
Animal Life
Plant Life
Evolution of Life
Behavior and Ecology of Life

Ricki Lewis
State University of New York at Albany

Contributing Authors

Animal Biology, Behavior and Ecology
Judith Goodenough
University of Massachusetts at Amherst

Plant Biology
Randall C. Moore
Wright State University

Wm. C. Brown Publishers

Book Team

Editor *Kevin Kane*
Developmental Editor *Margaret J. Manders*
Production Editor *Sherry Padden*
Visuals/Design Consultant *Marilyn Phelps*
Designer *Mark Elliot Christianson*
Art Editor *Janice M. Roerig*
Photo Editor *Carol Smith*
Permissions Editor *Vicki Krug*
Visuals Processor *Joseph P. O'Connell*

Wm. C. Brown Publishers

President *G. Franklin Lewis*
Vice President, Publisher *George Wm. Bergquist*
Vice President, Operations and Production *Beverly Kolz*
National Sales Manager *Virginia S. Moffat*
Group Sales Manager *Vince DiBlasi*
Vice President, Editor in Chief *Edward G. Jaffe*
Marketing Manager *Craig S. Marty*
Managing Editor, Production *Colleen A. Yonda*
Manager of Visuals and Design *Faye M. Schilling*
Production Editorial Manager *Julie A. Kennedy*
Production Editorial Manager *Ann Fuerste*
Publishing Services Manager *Karen J. Slaght*

WCB Group

President and Chief Executive Officer *Mark C. Falb*
Chairman of the Board *Wm. C. Brown*

Life
Front cover photo by © Robert Hernandez/Allstock

Part 1: Beginnings of Life
Front cover illustration by Mark Elliot Christianson based on photographs by © Lloyd M. Beidler/Science Photo Library/Photo researcher, Inc.

Part 2: Animal Life
Front cover photo by © Erwin & Peggy Bauer

Part 3: Plant Life
Front cover photo by © Michael Fogden/Oxford Scientific Films

Part 4: Evolution of Life
Front cover photo by © Henry Ausloos/Animals Animals

Part 5: Behavior and Ecology of Life
Front cover photo by © Larry Lefever/Grant Heilman Photography

The credits section for this book begins on page C-1, and is considered an extension of the copyright page.

Library of Congress Catalog Card Number: **Life**: 91-70426

ISBN **Life** Casebound, recycled interior stock: 0-697-05392-X
ISBN **Life** Paper binding, recycled interior stock: 0-697-14187-X
ISBN **Part 1: Beginnings of Life** Paper binding, recycled interior stock: 0-697-14193-4
ISBN **Part 2: Animal Life** Paper binding, recycled interior stock: 0-697-14195-0
ISBN **Part 3: Plant Life** Paper binding, recycled interior stock: 0-697-14197-7
ISBN **Part 4: Evolution of Life** Paper binding, recycled interior stock: 0-697-14199-3
ISBN **Part 5: Behavior and Ecology of Life** Paper binding, recycled interior stock: 0-697-14201-9
ISBN **Life** Boxed set, recycled interior stock: 0-697-14189-6

Printed in the United States of America by Wm. C. Brown Publishers, 2460 Kerper Boulevard, Dubuque, IA 52001

10 9 8 7 6 5 4 3 2

Publisher's Note to the Instructor

Recycled Paper

Life—in all its numerous binding options (listed here)—is printed on ***recycled paper stock.*** All of its ancillaries, as well as all advertising pieces for *Life*, will also be printed on recycled paper, subject to market availability.

Our goal in offering the text and its ancillary package on ***recycled paper*** is to take an important first step toward minimizing the environmental impact of our products. If you have any questions about recycled paper use, *Life*, its package, any of its binding options, or any of our other biology texts, feel free to call us at 1-800-331-2111. Thank you.

Kevin Kane
Senior Editor
Biology

Binding Option	***Description***	***ISBN***
Life, casebound	The full-length text (chapters 1-40), with hardcover binding.	0-697-05392-X
Life, paperbound	The full-length text, paperback covered and available at a significantly reduced price, when compared with the casebound version.	0-697-14187-X
Part 1 *Beginnings of Life*, paperbound	Part 1 features the first 4 units or 15 chapters of the text, covering the scientific method, the unity and diversity of life, basic chemistry, cell biology, reproduction and development, and genetics. This paperback option is available at a significantly reduced price when compared with both the full-length casebound and paperbound versions.	0-697-14193-4
Part 2 *Animal Life*	Part 2 features chapters 16-27 on the anatomy and physiology of animals—invertebrate, vertebrate, and human. This paperback is also available at a significantly reduced price when compared with the full-length versions of the text.	0-697-14195-0
Part 3 *Plant Life*	Part 3 features chapters 28-32 on plant form and function, with popular applications chapters on "Plants Through History" (28) and Plant Biotechnology (32). Paperback bound, it is available for a fraction of the full-length casebound or paperbound prices.	0-697-14197-7
Part 4 *Evolution of Life*	Part 4 features chapters 33-35 on evolution. Paperback bound, it is also available for a fraction of the full-length casebound or paperbound prices.	0-697-14199-3
Part 5 *Behavior and Ecology of Life*	Part 5 features chapters 36-40 on behavior and ecology. Paperback bound, it also sells for a fraction of the full-length book price.	0-697-14201-9
Life, the Boxed Set	The entire text, offered in an attractive, boxed set of all five paperback "splits." It is available at the same price as the full-length casebound text.	0-697-14189-6

Brief Contents

Contents

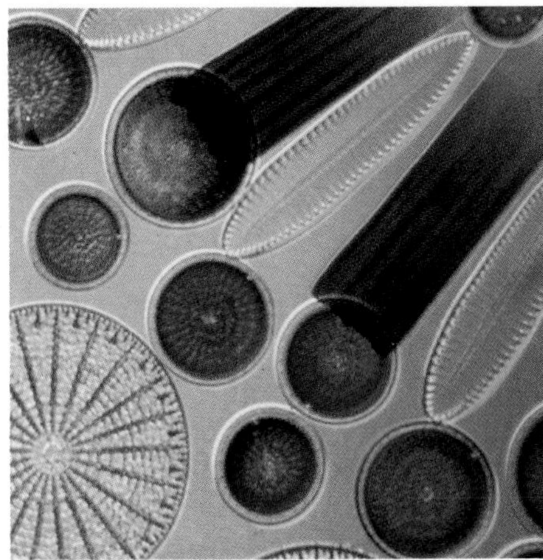

UNIT

1

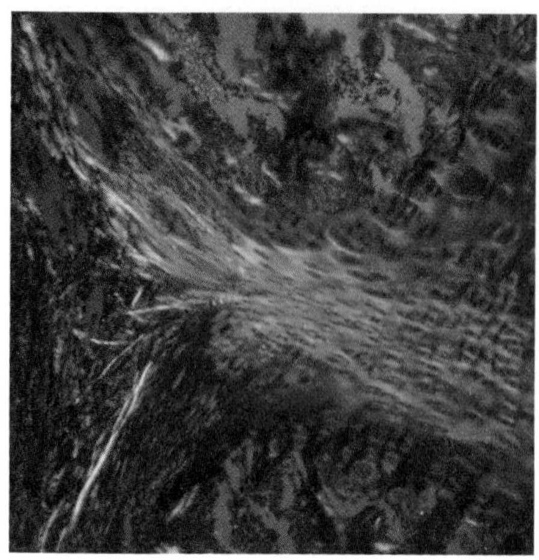

UNIT 2

Cell Biology 65

Chapter 4

Cells and Tissues 66

Chapter 5

Cellular Architecture 91

Chapter 6

Biological Energy 108

Chapter 7

Mitosis 130

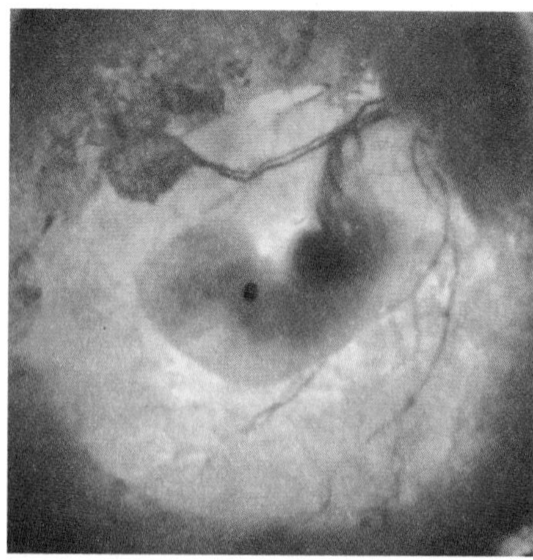

UNIT 3

Reproduction and Development 153

Chapter 8

Human Meiosis and Reproduction 154

Chapter 14

Chapter 15

UNIT 5

Chapter 16

Chapter 17

The Nervous System 330

Chapter 18

The Senses 348

Chapter 19

The Endocrine System 367

Chapter 20

The Skeletal System 389

Chapter 21

The Muscular System 404

Chapter 22

The Circulatory System 420

UNIT

6

Plant Biology 537

Chapter 28

Plants Through History 538

Chapter 29

Plant Form and Function 551

Chapter 30

Plant Life Cycles 570

Chapter 31

Plant Responses to Stimuli 587

Chapter 32

Plant Biotechnology 602

UNIT 7

Evolution 617

Chapter 33

Darwin's View of Evolution 618

Chapter 34

The Forces of Evolutionary Change 630

Chapter 35

Evidence for Evolution 647

UNIT 8

Behavior and Ecology 665

Chapter 36

The Behavior of Individuals 666

The *Life* Learning System

Chapter Outlines

Each chapter begins with an outline. These will allow students to tell at a glance how the chapter is organized and what major topics have been included in the chapter. The outlines include the first and second level heads for the chapter.

C H A P T E R

3

The Chemistry and Origin of Life

Chapter Outline

Learning Objectives

By the chapter's end, you should be able to answer these questions:

1. What characteristics distinguish living things from nonliving things?
2. What are the simplest forms of life?
3. What chemical components constitute living things?
4. What chemical compounds are important to human health?
5. How might living matter have evolved from nonliving chemicals?

38

Learning Objectives

Each chapter begins with a list of concepts stressed in the chapter. This listing introduces the student to the chapter by organizing its content into a few meaningful sentences. The concepts provide a framework for the content of each chapter.

Key Concepts

At the ends of major sections within each chapter, summaries briefly highlight key concepts in the section, helping students focus their study efforts on the basics.

68 *Cell Biology*

KEY CONCEPTS

The structural unit of an organism is the cell. Organisms are unicellular or multicellular. Complex cells contain specialized structures called organelles. All cells have structures in common to carry out basic life processes, but different numbers of organelles give cells distinct characteristics.

Viruses—Simpler Than Cells

The simplest form of life is a unicellular organism with no organelles, such as a bacterium. However, in chapter 3 we encountered several types of "infectious agents" that appear to be living while they are infecting cells but otherwise seem to be nonliving chemicals. Before describing how we examine cells and their contents, it is interesting to take a comparative look at the viruses, both to point out their noncellular organization and because they exert very noticeable effects on human health, causing such minor ills as colds and influenza and such deadly ones as AIDS. Reading 4.1 describes effects of the herpes simplex virus.

A *virus* consists of a nucleic acid (DNA or RNA) surrounded by protein. Figure 4.2 illustrates the human immunodeficiency virus (HIV), which causes AIDS. A virus must be within a cell to reproduce, and hence it is called an obligate parasite. Many viruses, such as HIV, cannot survive outside of a living cell. Some other viruses are afforded protection from the physical environment by their protein coverings. A virus reproduces by injecting its DNA or RNA into the host cell, where it situates itself within the host's DNA. In fact, viral DNA sequences can probably be found within your own chromosomes. (An RNA virus, such as HIV, is called a retrovirus and must first make a replica of its RNA in DNA form.)

Once viral DNA integrates into the host's DNA, it can either remain there and be replicated along with the host's DNA whenever the cell divides, but not cause harm, or the viral DNA can actively take over the cell, leading eventually to the cell's death. To do this, some of the virus's genes direct the host cell to replicate viral DNA rather than the host DNA. As viral DNA accumulates in the cell, some of it is used to manufacture proteins. (Recall from chapter 3 that the function of DNA is to provide information from which the cell constructs proteins.) Within hours or days, the infected cell fills with viral DNA and protein. Some of the proteins wrap around the DNA to form new viral particles. Finally, a viral enzyme is produced that cuts through the host cell's outer membrane. The cell bursts, releasing new viruses.

Viruses are known to infect all kinds of organisms, including animals, plants, and bacteria. A particular type of virus, however, infects only certain species, which constitute its *host range*. (Refer back to figure 3.6 for an illustration of a tomato infected by the tomato bunchy top virus.) Figure 4.3 illustrates what happens to a moth infected by a type of virus called a baculovirus.

Figure 4.2
A virus is a nucleic acid coated with protein. The human immunodeficiency virus (HIV), which causes AIDS, consists of RNA surrounded by several layers of proteins. Once inside a human cell (usually a T cell, part of the immune system), the virus uses an enzyme to convert its RNA to DNA, which then inserts into the host DNA. HIV damages the human body's protection against disease by killing T cells and by using these cells to make more of itself.

Dramatic Visuals Program

Colorful, informative photographs and illustrations enhance the learning program of the text as well as spark interest and discussion of important topics.

76 *Cell Biology*

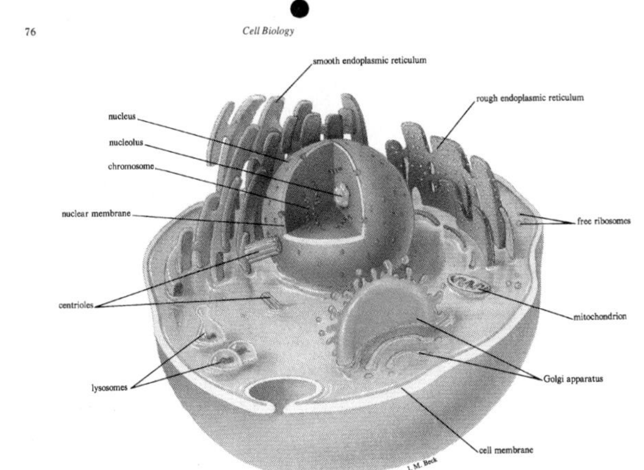

Figure 4.10
An animal cell is a eukaryotic cell. Note the appearances of the different organelles when viewed in cross section.

The organelles are described following in the context of their coordinated function in a familiar cellular activity—secretion. Table 4.4 lists organelle structures and functions with photographs for easy reference and review.

Organelles in Action—Secretion

The activities of the different organelles within a cell are coordinated to provide basic life functions and also to sculpt the distinguishing characteristics of a particular cell type. Consider a glandular cell in the breast of a human female. Dormant most of the time, the cell increases its metabolic activity during pregnancy, and then it undergoes a burst of productivity shortly after the baby is born. The ability of individual cells to manufacture the remarkably complex milk is made possible by the interaction of organelles, which function together to form a *secretory network*.

A new mother hears her infant cry, and within a minute she may feel her breasts swell in response. When the infant suckles, the milk that it receives is a mixture of cells and biochemicals, a highly nutritious food tailored specifically to meet the needs of a human child. Deep within the mother's breasts, glandular cells rapidly produce the proteins, fats, and sugars that are combined with immune system cells and biochemicals in specific proportions to form the milk.

Secretion begins in the nucleus, where, in humans, 23 pairs of rod-shaped **chromosomes** contain information that other parts of the cell use to construct proteins. Each chromosome consists of millions of DNA building blocks (nucleotides), long sequences of which comprise genes. The nucleotide sequence of a gene is transcribed, or rewritten, into another type of nucleic acid, **messenger RNA** (mRNA). (The mechanism for transcribing DNA into RNA is discussed in depth in chapter 13.) RNA building blocks are available in the substance surrounding the chromosomes, called the *nucleoplasm*, and are stored in a structure in the nucleus called the **nucleolus.** Messenger RNA exits the nucleus by passing through holes, called *nuclear pores*, in the two-layered *nuclear envelope* that surrounds the nucleus. In the manufacture of milk, for example, the gene

74 *Cell Biology*

Unlike bacteria, cyanobacteria contain internal membranes that are outgrowths of the cell membrane. These membranes, however, are not extensive enough to subdivide the cell into compartments, as membranes do in more complex cells. The cyanobacterium's membranes are studded with pigment molecules that absorb and extract energy from sunlight.

Cells require relatively large surface areas through which they can interact with the environment. Nutrients, water, oxygen, carbon dioxide, and waste products must enter or leave a cell through its surfaces. As a cell grows, its volume increases at a faster rate than does its surface area, a phenomenon that you can easily calculate (fig. 4.8). To put it another way, much of the interior of a large cell is far away from the cell's surface. The situation can be compared to the plight of a seafood lover living in the middle of a large country with an inadequate transportation system. To satisfy this food craving, this individual would be better off living either on a small island or in a country with a modern, rapid transportation system. Likewise, a cell's insides must either be close to the outside or have ready access to it.

A large cell lacking the means to bring in required chemicals from the environment or to eliminate wastes might die. As the earth became more populated and cells grew larger, they could survive only if they could somehow increase their surface areas relative to their increasing volumes. One strategy was for a large cell to divide in two, restoring a biologically suitable surface/volume ratio. This solution—cell division—occurs today in many of our own cells, as well as in single-celled organisms that have grown too large.

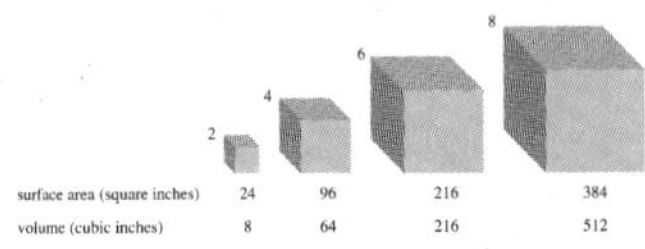

Figure 4.8
The important relationship between surface area and volume. As a cell grows larger, the amount of material inside it, its volume, increases faster than the area of the cell's surface. The surface of a cell is vital to its functioning because this is the site of communication between the cell's interior and the extracellular environment. The chemical reactions of life are more readily carried out when surface area is maximized. Imagine that a cell is a simple cube. Compare the surface area and volume of four increasingly larger cubes. (Surface area equals the area of each face multiplied by the number of faces; volume equals the length of a side cubed.) Can you see that volume increases faster than surface area?

KEY CONCEPTS

Cells exhibit the characteristics of life. Prokaryotes include the bacteria and cyanobacteria and are ancient and simple single-celled organisms. A prokaryotic cell is bounded by a cell wall and a plasmalemma. Its DNA is free of protein and located in the nucleoid area. Ribosomes are present, as are membranes derived from the plasmalemma.

Table 4.3
Comparison of Prokaryotic and Eukaryotic Cells

Characteristic	*Prokaryotic Cells*	*Eukaryotic Cells*
Organisms	Bacteria (including cyanobacteria)	Protists, fungi, plants, animals
Cell size	1-10 μm across	10-100 μm across
Oxygen required	By some	By all
Membrane-bound organelles	No	Yes
Ribosomes	Yes	Yes
DNA form	Circular	Coiled linear strands, complexed with protein
DNA location	In cytoplasm	In nucleus
DNA length	Short	Long
Protein synthesis	RNA and protein synthesis are not spatially separated	RNA and protein synthesis are spatially separated
Membranes	Some	Many
Cytoskeleton	No	Yes
Cellular organization	Single cells or colonies	Some single-celled, most multicellular with differentiation of cell function

Source: From Bruce Alberts, et al., *Molecular Biology of the Cell*. Copyright © 1983 Garland Publishing Company, New York, NY.

The Eukaryotic Cell

Another cellular solution to the problem of increasing size is to divide into compartments, or organelles, much like a growing store is subdivided into departments. Organelles are established by biological *membranes*, which are barriers composed of lipids and proteins. Organelles have access to the environment outside the cell and to each other by networks of bubblelike structures called *vesicles* (or *vacuoles*) that bud off from the membranes. Cells that have organelles are termed *eukaryotic*. The typical eukaryotic cell, which is roughly 1,000 times the volume of a typical prokaryotic cell, would not be able to function efficiently were it not for the division of

Boldfaced Words

New terms appear in boldface print as they are introduced within the text and are immediately defined in context. If any of these terms are reintroduced in later chapters, they are italicized. Key terms are defined in the text glossary with appropriate page reference.

Tables

Numerous strategically-placed tables list and summarize important information, making it readily accessible for efficient study.

Reading 5.2 *Liposomes—New Drug Delivers*

In 1961, English investigator Alec Bangham poured water into a flask containing a film of phospholipid molecules as part of his research on blood clotting. He was surprised to see that the lipid turned milky. Looking at the material under a microscope, Bangham saw that the phospholipid film had broken into thousands of tiny bubbles, each surrounding some of the water. The bubbles ranged in diameter from 250 Angstroms (A) to several micrometers. They were named liposomes, meaning "bodies of lipid."

What Bangham had discovered were microscopic spheres made of a simple lipid bilayer, identical to the structure that forms the basis of cell membranes. Some of the lipids even had more than one lipid bilayer coat, built a little like an onion skin. Throughout the 1960s, liposomes were used by cell biologists as models of cell membranes. In 1981, a young biologist named Marc Ostro realized the potential of liposomes as drug carriers. Drugs that are soluble in water can be packaged in the watery interior of liposomes; drugs soluble in fat can be lodged within the lipid bilayer. The advantages of packing a drug in a liposome are twofold—the drug can be released slowly from the liposome, and if a way to direct the liposome to diseased cells can be found, drug delivery can be targeted. This would solve a major drawback of conventional drug treatment, that is, getting enough drug to the site of disease to be effective yet keeping it away from healthy cells, where it can cause side effects.

The natural reaction of the human immune system to liposomes provides a way to move them to sites of disease. Large scavenger cells of the immune system, called macrophages, are attracted to liposomes and rapidly engulf them. The "swallowed" liposome is sent to a lysosome, where it is taken apart by enzymes. If the liposome contains a drug, some of it seeps into the cytoplasm and perhaps eventually out of the cell. The success of liposome-carried drugs lies in the function of macrophages, which normally congregate in parts of the liver, spleen, bone marrow, and lymph nodes and move to sites of inflammation or infection. Liposomes containing antibiotic drugs are engulfed by macrophages and transported to the infection site—precisely where they are needed. Liposomes harboring antiinflammatory drugs are taken, via macrophages, to inflamed arthritic joints. Liposomes are also useful for delivering cancer drugs, which when given in "free" form cause severe side effects because they kill healthy cells as well as cancer cells. When such toxic drugs are encapsulated in liposomes and injected, they accumulate in macrophage-laden places, where they leak out slowly and steadily enough to destroy the cancer cells but not the healthy ones.

Liposome-enclosed drugs to treat infection, inflammation, and cancer are injected because they are not absorbed well in the digestive tract. The tiny bubbles are useful, however, in some topical applications. "Artificial tears," for example, are liposomes packed with tear components (water, salts, lipid, and a mucuslike substance) and sooth the irritating condition called dry eye. Another liposome-enclosed drug treats fungal infections in the female reproductive tract, and a hair growing drug encased in liposomes is used to treat pattern baldness. Because of their slow release, liposome-enclosed topical drugs are needed less often and are therefore less irritating.

Figure 1
Liposomes are microscopic bubbles composed of lipid bilayers. They form spontaneously when certain concentrations of fatty molecules are mixed with water. The diameters of these liposomes are about 0.00015 millimeters.

Membrane proteins are diverse. Some lie completely within the lipid bilayer, and others traverse it to extend out of one or both sides. In animal cells, membrane proteins are often attached to branchlike sugar molecules to form **glycoproteins**, which protrude from the membrane's outer surface (fig. 5.6). The proteins and glycoproteins that jut from the cell membrane create the surface characteristics that are so important to a cell's interactions with other cells. Proteins within the oily lipid sandwich can move about, like slow-moving ships at sea. The protein-lipid bilayer is often called a **fluid mosaic** because the proteins can move and are not regularly arranged, as are the lipid molecules. A pigment protein found in the retina of a human eye, for example, moves to different depths within the lipid bilayer of the cell membrane depending upon the intensity of the incoming light. Figure 5.7 illustrates the membrane characteristics of another specialized cell, the red blood cell.

Readings

Throughout *Life*, selected readings both elaborate and entertain. Some describe experiments, some provide health information, and others are closer looks at specific topics. All readings are written by the author.

Chapter Summaries

At the end of each chapter is a summary. This should help students more easily identify important concepts and better facilitate their learning of chapter concepts.

There is a little bit of the scientist in each of us. We can all observe, form hypotheses, plan and carry out experiments, and attempt to interpret and use the results. However, your observations and ideas concerning the living world will be more meaningful if you have more extensive experience and learning on which to build. This book and your biology course will provide you with some of this valuable background. With it, you will be better equipped to apply the systematic approach of the biologist to your own observations of the living world. Indeed, you will soon know just what is in shampoo, what might have happened if you had not been immunized against various diseases, what is in polluted creek water and why it is harmful to human health, why heart disease can be caused by life-style habits, and what life before birth is like.

S U M M A R Y

Understanding the unity and diversity of living things through the study of biology can help us to appreciate our position in the living world. Biologists study life by applying the *scientific method*, in which observation and reasoning are used to make an educated guess, or *hypothesis*. One or more experiments are conducted to test the validity of the hypothesis, including *experimental controls* to ensure that only one phenomenon is being examined at a time. The experiment is usually repeated to test its accuracy, and then conclusions are drawn. Conclusions add to knowledge, but they also usually lead to more questions, and then the cycle of scientific inquiry begins anew.

Application of the scientific method does not always yield a complete answer to a question. A scientific investigation may be based on existing information, as in an *epidemiological study*, or it may be based on information generated specifically for the investigation, as in studies using experimental organisms. Each method has its advantages and disadvantages. Many answers to scientific questions are unusual or unexpected, and some are stumbled upon accidentally. Overall, the scientific method provides a systematic approach to exploring the living world. In addition, we often practice the method in our everyday lives without being aware of it.

Q U E S T I O N S

1. Read the following passage and identify the steps of the scientific method:

In 1953, a graduate student named Stanley Miller combined various chemicals in the presence of electrical sparks and heat to simulate conditions on earth before life existed. From the brew formed complex chemicals known to be important in the chemistry of living things.

In 1981, geologists discovered openings in the ocean floor where the water is extremely hot compared to the surrounding water. They found simple chemicals there, which they knew from Miller's experiment could react to yield the complex chemicals of living things. Several researchers suggested that life could have originated at these deep sea vents, because the chemical building blocks of life were present, plus the ocean would have protected the delicate precursors to life from the lightning, ultraviolet rays, and meteors that hit the earth's surface.

In 1988, Stanley Miller, now a chemist at the University of California at San Diego, challenged the popular idea that life began in these oceanic hot spots. Based on his knowledge of chemistry, Miller thought that the complex chemicals that might form under the suggested conditions would be unstable at the very high temperatures and decompose too rapidly to allow life to form. To test his idea, Miller conducted an experiment similar to what he had done as a graduate student. In the laboratory he combined the simple chemicals found in the deep sea vents and heated them to the temperatures found there. As Miller predicted, more complex chemicals formed, but they were dismantled by the heat in minutes or seconds. Therefore, deep sea vents are not likely to have been a place where life on earth started. Might the first life forms have arisen at cooler deep sea vents near the Galapagos Islands?

2. The most informative type of experiment is placebo controlled and double-blind.

- **a.** How does a placebo serve as an experimental control?
- **b.** How does withholding information from researchers and subjects ensure more reliable results?
- **c.** Design a double-blind study to test the effectiveness of a new antacid.
- **d.** DMSO (dimethylsulfoxide) is a chemical believed by some people to relieve a variety of ills, but its safety and effectiveness have not yet been proven by experiment. An unpleasant side effect of rubbing DMSO on the skin is that it produces a garlicky taste as it enters the body. How might this peculiarity complicate adequately controlled testing of DMSO for safety and effectiveness?

3. A chemical manufacturing plant closes down because a toxic substance, which was buried on the plant property a decade earlier, is now leaking from its containers. To examine the effects of the contamination, a survey is distributed among workers and town residents, inquiring about their health problems and asking them whether or not they believe that they were exposed to the chemical. What can be learned from this type of study? What are its limitations?

4. Animal models often provide an effective means of testing substances and procedures that are not yet well enough understood to be used on humans. Many people object that pain is inflicted on the animals used for such purposes. However, in what has been called the "Bambi factor," such objections tend to focus on "cute" organisms, such as rabbits and dogs, rather than the fruit flies, leeches, rats, and other less appealing organisms that are also used experimentally. What limits, if any, do you think should be placed on animal experimentation? What technique mentioned briefly in the chapter might be considered a compromise?

5. Scientists traditionally report their research findings in journals. Go to your science library, choose a biological or medical journal, and try to find an article that you can follow well enough to identify the steps of the scientific method.

T O T H I N K A B O U T

Many of our decisions about the foods we eat, the medicines we take, and the activities we participate in are based, in part, upon health information. Some of this information comes from well-executed studies that adhere to the scientific method, but some comes from flawed investigations. For each of the following real-life examples, state whether or not the scientific method was followed. If it was not, indicate which of the following specific faults was present:

- **a.** Conclusions not supported by experimental evidence
- **b.** Inadequate controls
- **c.** Biased sampling
- **d.** Inappropriate extrapolation of the experimental group to the general population
- **e.** Sample size too small
- **f.** Other problem

1. "I ran 4 miles (approximately 6 kilometers) every morning when I was pregnant with my first child," the woman told her physician, "and Jamie weighed only 3 pounds (approximately 1 kilogram) at birth. This time, I didn't exercise at all, and Jamie's sister weighed 8 pounds (approximately 3 kilograms). Therefore, doctor, running during pregnancy must cause low birth weight."

2. Eating foods high in cholesterol was found to be dangerous for a large sample of individuals with hypercholesterolemia, a disorder of the heart and blood vessels. It was concluded from this study that all persons should limit dietary cholesterol intake.

3. A dumping site in the small town of Love Canal, New York, was used for the disposal of industrial chemical wastes for several years. Later, toxic substances were found in the basements of schools and homes. A preliminary examination of the genetic material of several residents of Love Canal showed abnormalities. Exposure to the chemicals, then, must have caused the problems.

4. Each year, 500,000 American males join the 10 million who have already had a vasectomy, a simple sterilization procedure. In the early 1970s, animal studies suggested that heart disease might be a complication of vasectomy. To test this hypothesis, a study involving more than 20,000 men was begun. In this study, each of 10,590 vasectomized men was paired with a nonvasectomized man of the same age, race, and marital status and living in the same neighborhood at the time of the surgery. Each man was extensively questioned about medical conditions (including the 54 conditions statistically most likely to be associated with vasectomy) and health and life-style habits, such as smoking, drinking, diet, exercise, and frequency of checkups. Medical records were carefully evaluated. The results of this study showed that vasectomized men had no significantly higher rates for any of the health conditions examined. Conclusion: Vasectomy is safe.

5. The National Cancer Institute (NCI) sponsors clinical trials at its Maryland headquarters and in several research hospitals around the country. New treatments are carried out on human cancer patients and the results evaluated. Standard cancer treatments are used as controls. There are many more cancer patients seeking to participate in these trials than there are spaces available for them. The policy of the NCI is to select the healthiest, strongest patients possible—what one cancer researcher calls "the perfect patient." The goal is to give the new treatment the maximum chance of working by testing it on a patient who has cancer but is otherwise young, strong, and healthy. Whenever results of such trials seem promising, television news shows excitedly announce them as "breakthroughs."

6. A former NCI researcher was bothered by the practice of turning down large numbers of cancer patients who wanted to participate in clinical trials. His solution was to leave the institute and start his own company, which offers to do research on individuals' tumors. This research could lead to personalized experimental treatments. However, this researcher's company is not supported by government funds, so patients have to pay for the research service—about $60,000 a year. The company publishes the results of its research.

S U G G E S T E D R E A D I N G S

Dagani, Ron. November 12, 1984. In-vitro methods may offer alternatives to animal testing. *Chemical and Engineering News*. Traditional product tests using live, whole animals are being replaced with tests using animal cells and tissues.

Kanigel, Robert. January/February 1987. Specimen no. 1913. *The Sciences*. A rat's brief life in the service of science—from its point of view. An informative and absolutely riveting account of the use of animals in experiments.

Larkin, Tim. June 1985. Evidence vs. nonsense: A guide to the scientific method. *FDA Consumer*. The scientific method is fine in theory; in practice, it is sometimes difficult to implement.

Lecos, Chris W. February 1985. Sweetness minus calories = controversy. *FDA Consumer*. The natural sweetener aspartame and the artificial sweeteners cyclamate and saccharin have been put through many, many tests.

Lewis, Ricki. September 1984. Dioxin danger (?) *Biology Digest*. Just how harmful to human health is dioxin, a chemical contaminant of the herbicide Agent Orange? Different types of studies give different answers.

Segal, Marian. June 1990. Is it worth the worry? Determing risk. *FDA Consumer*. The scientific method is a major part of risk assessment.

Thompson, Richard C. December 1986/January 1987. Protecting human guinea pigs. *FDA Consumer*. How people participate in testing new drugs or medical devices.

Weisburd, Stefi. May 23, 1987. AIDS vaccines: The problems of human testing. *Science News*. Testing a vaccine is always controversial because healthy people must be the recipients. In the case of deadly AIDS, the situation is even more serious.

Young, Frank E. June 1987. Experimental drugs for the desperately ill. *FDA Consumer*. The former head of the FDA speaks out on loosening the scientific method in cases of hopeless illness.

Zumwalt, Elmo, III. June 1987. A war with hope. *Health*. An anecdotal look at the possible dangers to health of dioxin by a Vietnam veteran suffering from several types of cancer.

To Think About

Located at the end of each chapter, these questions are springboards for class discussions and term paper topics.

Questions

The end-of-chapter questions often continue the storytelling style of the chapter, using anecdotes and experiments from the chapter to illustrate and apply concepts.

Suggested Readings

A list of readings at the end of each chapter suggests references that can be used for further study of topics covered in the chapter. The items listed in this section were carefully chosen for readability and accessibility.

List of Readings

Foreword

Seasoned journalists freely admit—the very best stories write themselves. Each detail leads seamlessly to the next, so logically that the writer often seems a mere bystander as the words come pouring out. So it is with the stories of life. But it takes an extraordinary person to tell the tales of biology to an introductory audience effectively, particularly the unfamiliar events that unfold on a microscopic level. The technical words of a scientist, teamed with a science writer's polished adjectives, may not always suffice. While one author cannot experience, the other cannot always express. Beyond that, neither may have the teaching experience to know which explanations clarify and which confuse.

With Ricki Lewis, you have that "one-in-a-million" textbook author who effortlessly combines the skills of accomplished teacher, scientist, and writer. While most scientists have seldom experienced the autocatalytic writing the author does daily—writing for publications like *Discover, BioScience, The FDA Consumer, the Scientist,* or *The Journal of NIH Research*—most science writers have never directly experienced the joy of discovery or jolt of serendipitous finding that the author has in her genetics research. It is even more unlikely that either will have refined the intuitive skills of a "thirtysomething" pop-culture critic with more than 10 years teaching experience. In other words, neither may be as quick to notice when students' eyes have glazed over and be judicious enough to try a different way of explaining the confounding phenomena.

Furthermore, few textbook authors can bring to their efforts the cutting-edge, clinical perspective the author does as she applies her years of work as a genetic counselor. For example, in most textbooks, orphan diseases are not even mentioned. But here, they are presented through the very real example of Bradley, a boy whose genetic handicap is not seen in any other person east of the Mississippi. Counseling has also taught the author to be sensitive to the differences among us. For example, people with AIDS or cancer are never called victims. Diseases are described in terms of their everyday effects on real people.

The greatest difference between this text and others is also the least obvious and most difficult to explain—namely, the writing style. To evaluate it, you cannot just review the illustrations or the learning aids. The book, or at least passages from it, must be read. And when you do, you will be pleasantly surprised. Missing are such phrases as "in the last chapter we learned that" or "now we will examine"—annoying and constant reminders to students that this is not supposed to be fun, it's a textbook. To the contrary, cross-references are unobtrusively woven in, so as not to break the threads of learning. Here, the stories of *Life* actually live—transporting the reader from facts and concepts to illustrations and applications, then back again.

And the stories here are many. In chapter 4, "Cells and Tissues," the organelles are not merely a list of strange new terms, but a dynamic food assembly line responding to the wail of a hungry infant. In chapter 5, "Cellular Architecture," components of the cell surface, cell membrane, and cytoskeleton interact as an egg is fertilized. Chapter 11, "Mendel's Laws," shows how even scientists can be fooled. In the tale of the mysterious disease kuru in 1960s New Guinea, a viral infection spread by a cannibalistic ritual is mistaken for a pattern of inheritance. In the evolution chapters (33-35), the history of life itself is told with analogies to subjects as precise as algebra and as simple as Aesop's fables.

The point of this foreword and the point of this book is this: with the explosion of knowledge taking place in countless areas of research, biology has become the most dynamic discipline alive today. Its impact on the way we live our lives will become even more dramatic than the way computer science has already affected them. Such a discipline demands a textbook that will excite, cajole, and charm introductory students—whether they be future biologists or concerned citizens.

Catalogues of scientific fact (i.e., traditional textbooks), even when expertly crafted, cannot impart the sense of precision, wonder, and serendipity that biologists feel "at the height of their game." Only a marvelous book by a very gifted person will do. I sincerely believe that you have that here.

I invite you to read and enjoy.

Preface

Life was written with the nonbiology major in mind, but contains enough information to be suitable for a majors' course too.

Diversity in Action

While human examples and applications are emphasized, *Life*'s diversity is treated early in a separate chapter, later in an appendix on taxonomy, and is logically integrated into all chapters. The animal biology chapters, for example, explore a deep-sea shrimp's vision, an insect's exoskeleton, a cow's digestion, and much more. The behavior and ecology chapters are filled with glimpses into the lives of a variety of organisms, from aardwolves to fire ants to naked mole rats. The reader of *Life* will learn many new things, but also encounter familiar territory. The science of biology will not seem foreign—it will be fun and make sense.

Discovery and Evolution

Two conceptual threads weave their way through *Life*. The book opens with the first theme, discovery. The story of how the sweetener aspartame was discovered takes the student through the scientific method and experimental design, yet points out how the initial detection of the food additive was very much a surprise.

In chapter 2, "The Diversity of Life," taxonomy is alive and vibrant in the treetops of a Peruvian wildlife preserve, where biologists catalog the abundance of insect life; and in such an unlikely place as an urban fish market. A pair of children playing with spectacles led to the development of the compound microscope, as described in chapter 4. In chapter 6, "Biological Energy," the student can be the discoverer by using the reactions of photosynthesis to develop a photograph on a leaf. The inborn errors of metabolism, PKU (chapter 15, "Genetic Disease: Diagnosis and Treatment") was discovered thanks to a mother's alertness of her infant's odd-smelling diapers. And a simple treatment for newborn jaundice (chapter 24, "The Digestive System") was discovered by an observant English nurse changing "nappies" in the sunlight. Chapter 15 also tells the story of how a seemingly drunken sailor and his 5,000 living descendants helped provide the first genetic marker.

Not all discovery is accidental. The look at "Molecular Genetics" in chapter 13 is liberally sprinkled with descriptions of the most elegant experiments ever performed. The scientific method is reviewed in chapter 36, "The Behavior of Individuals," as students at the University of Miami track singing birds, and in chapter 38 "Populations," through ecologists conducting wildlife surveys. The creation of an artificial mini-biosphere, described in chapter 39 "Ecosystems," is an exciting view of scientific investigation—whether it works or not.

The second conceptual thread, evolution, accustoms the reader to continually wonder, "How did all of this happen?" How did a duo of protein and nucleic acid join forces long ago to form the first cell? How could random mutations in those early cells build the metabolic pathways of today? How did eukaryotic cells come by their highly successful "bags within a bag" organization? How do species arise, change, become extinct? How have our ideas about evolution themselves evolved?

Humor, History, and Human Values

An occasional foray into humor can help students learn. Consider the example of epistasis in chapter 11, borrowed from the soap opera "General Hospital," or the opening to chapter 34 "The Forces of Evolutionary Change," a love story between a moose and a dairy cow.

Historical references add interest and chronicle the evolution of ideas. The confusing multiple phenotypes of the blood disorder porphyria, for example, may have led the "mad king" George III to provoke the American Revolution. The study of genetics begins with early agricultural efforts nearly 10,000 years ago. How different were Edward Jenner's problems with how best to test his smallpox vaccine (chapter 28, "Plants Through History") from today's scientists' attempts to test AIDS vaccines? The state of the American temperate forest today reflects pioneer activity over the past centuries. Recent history brings the ecology chapters alive, from Mt. St. Helens to the Yellowstone fires to the nuclear explosion at Chernobyl.

Examining human values teaches the student to develop informed opinions and judgments about biologically relevant issues—a skill that will last long after the steps of glycolysis or the parts of the cell are forgotten. Should a pregnant woman who smokes or drinks alcohol be responsible for the health effects on her fetus? Should an employer be told the results of an employee's genetic marker test for Alzheimer's disease? Should we take extraordinary measures to save extremely premature babies if they will be handicapped after (or by) the treatment? Should we even attempt to clean animals drenched in oil from tanker spills? Should we limit reproduction? These disturbing queries are most often found in the "To Think About" sections at the chapters' ends, both so that they will not distract from learning major facts and concepts and so that the student is left thinking.

Integrating Technology

Technology has given new, exciting meaning to some difficult subjects. Discussing the development of extraembryonic structures segues into a peek at chorionic villus sampling. Liposomes are but an extension of cell membrane structure and function. Teaching DNA replication is no longer a hurdle, now that we have the polymerase chain reaction to demonstrate elegantly the power of the process. Filling in the details of food webs no longer requires being on the scene of a meal, thanks to stable isotope tracing (chapter 39, "Ecosystems").

The chapters on plant anatomy and physiology are bracketed by two unique applications chapters—chapter 28, "Plants Through History," chronicles our harvesting of the major crop plants, and chapter 32, "Plant Biotechnology," looks at how molecular and cellular techniques are likely to continue that harvest, via the genetic alteration of plant life.

Finally, Appendix A, "Microscopy", provides a closer look at the technology that really breathed life into biology, from the first crude lenses to today's powerful confocal microscopes. Yet the very technology that has taught us so much and made our lives so comfortable can get out of control, upsetting the delicate balance of life. Chapter 40 "Environmental Concerns," describes these problems, but emphasizes natural resiliency, leaving the reader, ultimately, with a sense of hope and purpose:

> *"This book has shown you the wonder that is life, from its constituent chemicals, to its cells, tissues, and organs, and all the way up to the biosphere. Do nothing to harm life—and do whatever you can to preserve its precious diversity. For in diversity lies resiliency, and the future of life on earth."*

Pedagogy

A great deal of creative energy has gone into the pedagogical aids, and some are quite different from those in the run-of-the-mill textbook. (For a visual walkthrough of these aids, examine the *Life* Learning System preview in this book's frontmatter.) The end-of-chapter "Questions" often continue the storytelling style of the chapter, using anecdotes and experiments from the literature to illustrate and apply concepts. The "To Think About" questions are springboards for class discussions and term paper topics. "Suggested Readings" go far beyond *Scientific American* and other textbooks, including sources such as *Science News, FDA Consumer* and the *New York Times*—sources that students are more likely to read, understand, and appreciate. "Learning Objectives," which open the chapters, "Key Concepts" following major sections, and end-of-chapter summaries reinforce main points.

"Readings" throughout the chapters both elaborate and entertain. Some describe experiments: "Enticing Cells to Divide in the Laboratory," "Recipes for Starting Life—Simulating Early Earth Conditions," "Tracking Development in Different Organisms;" some provide health information, "Cardiovascular Spare Parts," "Jon and Linda—The Plight of an Infertile Couple," "Our Overdrugged Elderly," "Steroids and Athletes—An Unhealthy Combination," "The War on Cancer;" others are closer looks, "A Closer Look at an Organelle—The Lysosome," "Tumor Necrosis Factor," "Odd Human Traits," or "The Herpes Simplex Virus." Some are practical, "Nutrition and the Athlete," "Food Inhalation and the Heimlich Maneuver" and many highlight diversity "Falling Felines," "Rumbles, Roars, Screeches, and Squeals—Animal Communication," or "Sexual Seasons."

Ancillaries

Instructor's Manual/Test Item File

Prepared by Heather McKean and James Hanegan of Eastern Washington University, the instructor's manual offers helpful suggestions for course outlines and developing daily lectures. Each chapter provides key concepts, key terms, chapter outlines, learning objectives, answers to the text's end-of-chapter questions, and suggested audiovisual materials. There are also 25 to 50 objective questions in a *Test Item File* in the back of the manual. (ISBN 0-697-10181-9)

Laboratory Manual

Written by Alice Jacklet, a colleague of mine at SUNY-Albany, the *Laboratory Manual* strongly emphasizes and guides students through *the process of scientific inquiry*. Beautifully illustrated in full-color, it features 20 self-contained exercises that can easily be reorganized to suit individual course needs. (ISBN 0-697-05637-6)

Laboratory Resource Guide

This helpful prep guide offers instructions for assembling lab materials and preparing reagents, as well as suggestions for using the Lab Manual in different kinds of lab settings. (ISBN 0-697-10178-9)

Customized Laboratory Manual

Inexpensive, one-color separates of each lab in the Laboratory Manual are available for individual use, for combination with labs of local origination, or for combination with labs from other Wm. C. Brown manuals. All materials will be custom, spiral-bound for your convenience. Contact your local Wm. C. Brown sales representative for more details.

Readings in Biology

A compilation of original journal and magazine articles by Ricki Lewis is also available to students at a nominal price. The readings, which correlate closely with the sequence of topics in the text, present additional high-interest information on cell biology, genetics, reproduction, and animal biology. (ISBN-0-697-12059-7)

Student Study Guide

Also written by Heather McKean and James Hanegan, the study guide offers students a variety of exercises and keys for testing their comprehension of basic as well as difficult concepts. (ISBN 0-697-05636-8)

TestPak

This computerized classroom management system/service includes a data base of objective test questions, copyable student self-quizzes, and a grade-recording program. Disks are available for IBM, Apple, and MacIntosh PC computers and require no programming experience. If a computer is not available, instructors can choose questions from the *Test Item File* and phone or FAX in their request for a printed exam, which will be returned within 48 hours.

Transparencies and Slides

More than 200 overhead *transparencies* or a comparable *slide set* is available for free to all adopters, on request. The acetates and slides feature key illustrations from the text that, in most cases, have images and labels that have been significantly enlarged for more effective classroom display. (Transparencies: 0-697-10179-7; Slides: ISBN 0-697-10167-3)

Customized Transparency Service

For those adopters interested in receiving acetates of text figures not included in the standard transparency package, a select number of acetates will be custom-made upon request. Contact your local Wm. C. Brown sales representative for more details.

Extended Lecture Outline Software

This instructor software features extensive outlines of each text chapter with a brief synopsis of each subtopic to assist in lecture preparation. Written in ASCII files for maximum utility, it is available in IBM, Apple, or Mac formats. It is free to all adopters, upon request.

You Can Make a Difference
by Judith Getis

This short, inexpensive supplement offers students practical guidelines for recycling, conserving energy, disposing of hazardous wastes, and other pollution controls. It can be shrink wrapped with the text, at minimal additional cost. (ISBN 0-697-13923-9)

How to Study Science
by Fred Drewes, Suffolk County Community College

This excellent new workbook offers students helpful suggestions for meeting the considerable challenges of a science course. It offers tips on how to take notes; how to get the most out of laboratories; as well as on how to overcome science anxiety. The book's unique design helps to stir critical thinking skills, while facilitating careful note-taking on the part of the student. (ISBN 0-697-14474-7)

The Life Science Lexicon
by William N. Marchuk, Red Deer College

This portable, inexpensive reference helps introductory-level students quickly master the vocabulary of the life sciences. Not a dictionary, it carefully explains the rules of word construction and derivation, while giving complete definitions of all important terms. (ISBN 0-697-12133-X)

Biology Study Cards
by Kent Van De Graaff, R. Ward Rhees, and Christopher H. Creek, Brigham Young University

This boxed set of 300, two-sided study cards provides a quick, yet thorough visual synopsis of all key biological terms and concepts in the general biology curriculum. Each card features a masterful illustration, pronunciation guide, definition and description in context. (ISBN 0-697-03069-5)

Special Software and Multi-Media Ancillaries

Life on Earth Videotapes

This critically acclaimed, twin-cassette package by David Attenborough, features thirteen programs, each about 25 minutes in duration, on Life's Diversity. Each cassette also features "Chapter Search," an on-screen numerical code for quick-scan access to each of the cassettes' thirteen programs and subtopics. The *Life on Earth* videotapes are available for free to all adopters of the text, upon request. (ISBN 0-697-14631-6)

Program Summary

1. THE INFINITE VARIETY
Nature's secrets found in ancient places.
2. BUILDING BODIES
First signs of life in the seas.
3. THE FIRST FORESTS
The world of plants, primitive and grand.
4. THE SWARMING HORDES
The ingenious adaptability of insects.
5. CONQUEST OF THE WATERS
Complexities of the great groups of fish.
6. INVASION OF THE LAND
The emergence of amphibian creatures.
7. VICTORS OF THE DRY LAND
Reptiles and the dinosaur dynasty.
8. LORDS OF THE AIR
Feathers, wings and birds in flight.
9. THE RISE OF THE MAMMALS
Where dinosaurs failed, mammals succeeded.
10. THEME AND VARIATIONS
The extremes of mammal evolution.
11. THE HUNTS AND THE HUNTED
Patterns of behavior in the animal kingdom.
12. LIFE IN THE TREES
Spotlighting monkeys and their relatives.
13. THE COMPULSIVE COMMUNICATORS
The development and achievements of humans.

Bio Sci II Videodisk

This critically acclaimed laser disk, produced by Videodiscovery for Wm. C. Brown, features more than 12,000 still and moving images, with a complete, bar-coded directory. Contact your Wm. C. Brown sales representative for more details. (ISBN 0-697-12121-6)

Mac-Hypercard and IBM Linkway Biostacks

These easy-to-use MacIntosh and IBM disks allow instructors to access the Bio Sci II laserdisk through a series of programmed lecture sequences. Contact your Wm. C. Brown representative for more details. (Mac Hypercard: 0-697-13273-1; IBM Linkway Biostacks, 3.5: 0-697-13275-7; IBM Linkway Biostacks, 5.2: 0-697-13274-9)

The Gundy-Weber Knowledge Map of the Human Body
by G. Craig Gundy, Weber State University

This thirteen disk, Mac-Hypercard program is for use by instructors and students alike. It features masterfully prepared computer graphics, animations, labeling exercises, self-tests and practice questions to help students examine the systems of the human body. Contact your local Wm. C. Brown representative or call 1-800-351-7671.

The Knowledge Map Diagrams

1. Introduction, Tissues, Integument System (0-697-13255-2)
2. Viruses, Bacteria, Eukaryotic Cells (0-697-13257-9)
3. Skeletal System (0-697-13258-7)
4. Muscle System (0-697-13259-5)
5. Nervous System (0-697-13260-9)
6. Special Senses (0-697-13261-7)
7. Endocrine System (0-697-13262-5)
8. Blood and the Lymphatic System (0-697-13263-3)
9. Cardiovascular System (0-697-13264-1)
10. Respiratory System (0-697-13265-X)
11. Digestive System (0-697-13266-8)
12. Urinary System (0-697-13267-6)
13. Reproductive System (0-697-13268-4)

Demo - (0-697-13256-0)
Complete Package - (0-697-13269-2)

GenPak: A Computer Assisted Guide to Genetics
by Tully Turney, Hampden-Sydney College

This Mac-Hypercard program features numerous, interactive/tutorial (problem-solving) exercises in Mendelian, molecular, and population genetics at the introductory level. (ISBN 0-697-13760-0)

Acknowledgments

Most of the credit for this book goes to the stories of life themselves. But thanks must also go to the scores of magazine editors who have shown me how to explain concepts clearly and concisely, yet retain a distinctive style; to the manuscript reviewers who corrected my errors and contributed so many valuable insights; to Gail Marsella, Randy Moore, Tom Gregg, Tom Wissing, and Judy Goodenough for assistance with selected chapters; to a fantastic bookteam; to my editor, Kevin Kane, and my developmental editor, Marge Manders, at Wm. C. Brown, who managed to keep me going at those times when the automatic pilot faltered; to my parents, who encouraged a little girl who brought home all sorts of creatures and to my parents-in-law who never lost faith; to my three daughters, whom I gestated along with this book; and most of all to my husband, Larry, who faithfully photocopied zillions of pages, listened to countless reviews, and never tired of hearing, yet one more time, "I've only got one more sentence left!" This really is the last sentence.

Reviewers

Stephen T. Abedon
University of Arizona

Holly Ahern
State University of New York at Albany

Sharon Antonelli
San Jose City College

Brenda C. Blackwelder
Central Piedmont Community College

James L. Botsford
New Mexico State University

Clyde S. Bottrell
Tarrant County Junior College

J. D. Brammer
North Dakota State University

William Caire
Central State University

Marcella D. Carabelli
Broward Community College

Galen E. Clothier
Sonoma State University

David J. Cotter
Georgia College

William P. Cunningham
University of Minnesota

Elizabeth A. Desy
Southwest State University

Durwood Foote
Tarrant County Junior College

Sally K. Frost
University of Kansas

Michael S. Gaines
University of Kansas

Judith Goodenough
University of Massachusetts at Amherst

Thomas G. Gregg
Miami University-Oxford

Thomas M. Haggerty
University of North Alabama

Madeline M. Hall
Cleveland State University

John P. Harley
Eastern Kentucky University

James Hanegan
Eastern Washington University

Ron Hoham
Colgate University

Alice Jacklet
State University of New York at Albany

Norma G. Johnson
University of North Carolina

Alan R. Journet
Southeast Missouri State University

William H. Leonard
Clemson University

Roger M. Lloyd
Florida Community College-Jacksonville

James O. Luken
Northern Kentucky University

Heather R. McKean
Eastern Washington University

Stephen A. Miller
The College of the Ozarks

Randall C. Moore
Wright State University

Patricia M. O'Mahoney-Damon
University of Southern Maine

Chris E. Peterson
College of DuPage

James E. Platt
University of Denver

Robert A. Ross
Linn-Benton Community College

Ron M. Ruppert
Cuesta College

Jon Sperling
Queens College-SUNY

Gerald Summers
University of Missouri

Philip Sze
Georgetown University

Sheila Tobias
University of California, San Diego

David L. Walkington
California State University-Fullerton

Larry G. Williams
Kansas State University

Geoffrey Zubay
Columbia University

About the Author

Ricki Lewis is a Lecturer at the State University of New York at Albany where she has taught human genetics, general biology, bioethics, and nutrition. An accomplished science journalist, she has had more than 500 articles published since 1982 in such diverse periodicals as *Discover*, *High Technology*, *Issues in Science and Technology*, *Self*, *Women's Sports*, *Science Health*, *BioScience*, *Science Digest*, *Biology Digest* and the Wm. C. Brown *Biology Newsletter*. She is also a genetic counselor and a reviewer for the *New York Times Review of Books*. She received her Ph.D. in genetics from Indiana University in 1980.

About the Contributing Authors

Randall C. Moore, Plant Biology

Randy Moore is Professor and Departmental Chair at Wright State University in Dayton, OH. He has taught general biology, general botany, and science writing. He is the Editor-in-Chief of the *American Biology Teacher*, Editor of the Wm. C. Brown *Biology Newsletter*, a Fulbright scholar, and the recipient of numerous writing and teaching awards. Moore has received numerous research grants and in 1986, directed a NASA research project carried aboard the space shuttle, *Columbia*. He received his Ph.D. in biology from the University of California, Los Angeles, in 1980.

Judith Goodenough; Animal Biology, Behavior and Ecology

Judy Goodenough is a Lecturer and Staff Associate at the University of Massachusetts at Amherst where she coordinates and teaches the introductory Animal Biology course. The recipient of a distinguished teaching award, she is a frequent reviewer of introductory textbooks and the co-author of an *Animal Behavior* text soon to be published. She received her Ph.D. in biology from New York University.

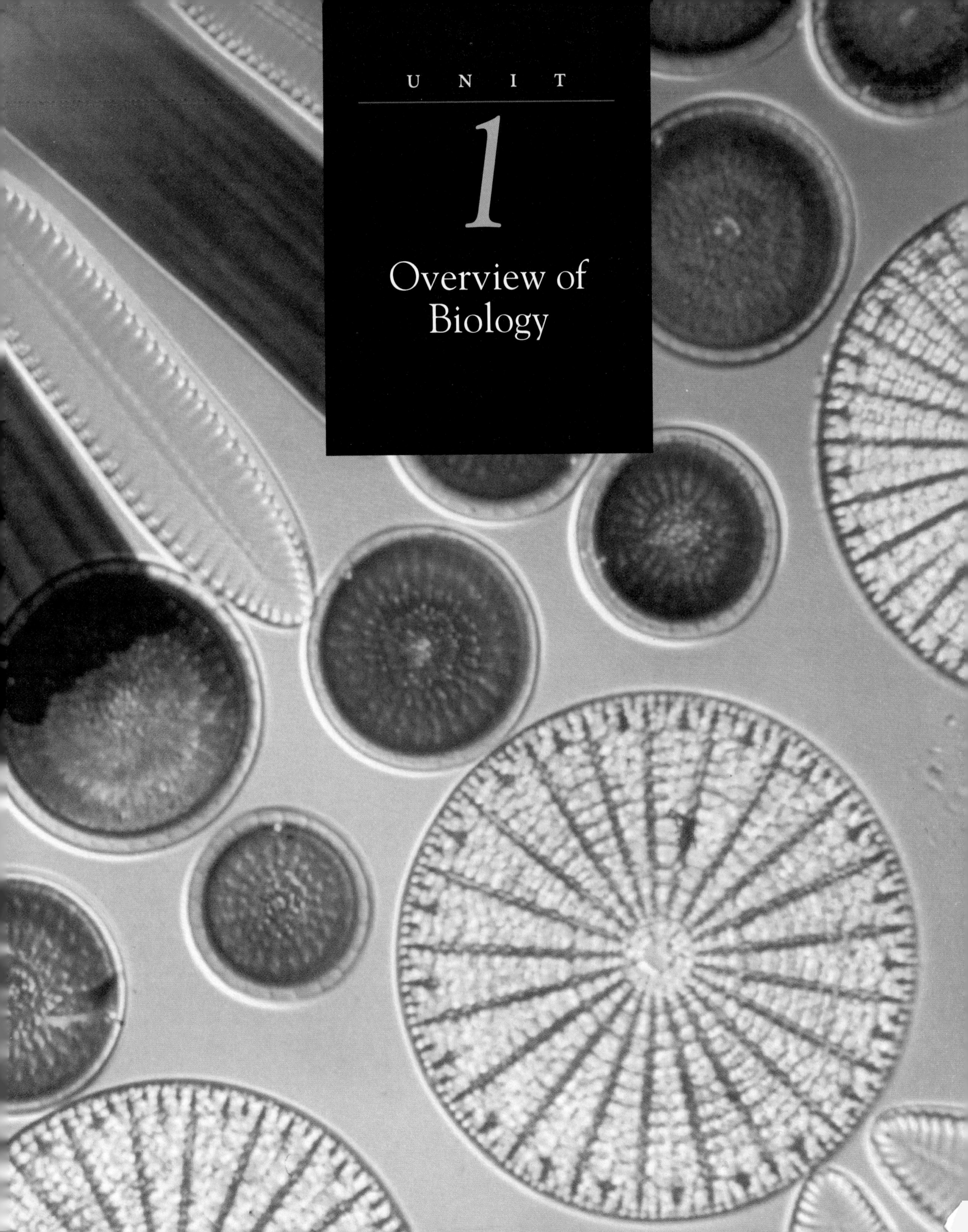

U N I T

1

Overview of Biology

C H A P T E R

1

Thinking Scientifically

Chapter Outline

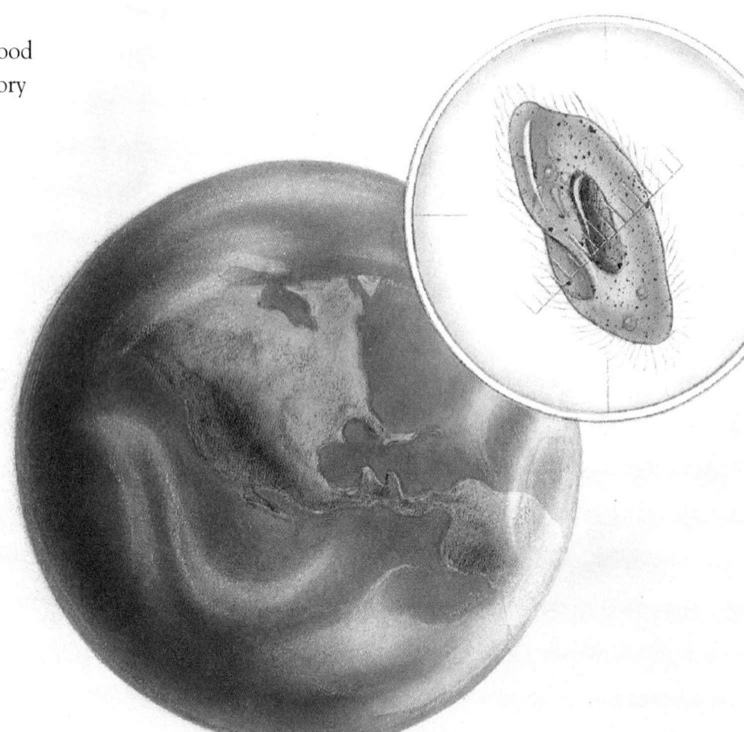

Learning Objectives

By the chapter's end, you should be able to answer these questions:

1. What are some of the ways in which biology affects you?
2. How do scientists use the scientific method to explain biological phenomena?
3. How are scientific experiments designed?
4. What are the limitations of the scientific method?
5. How do you use the scientific method in your everyday life?

You want to wash your hair, so you reach for a bottle of nonalkaline, high-protein, 100% organic, enzyme-powered, collagen-containing shampoo—with amino acids, glycoprotein, and DNA. What are these ingredients, where do they come from, and what do they do to your hair?

Years ago, people had a harder time staying healthy than they do today. They were susceptible to several diseases that we are now immunized against, including polio and mumps. Some of the diseases they feared, such as smallpox, do not even occur anymore, and others, such as measles and rubella, appear to be on their way out. However, seemingly new biological foes have arisen to take their place, including Reye's syndrome, toxic shock syndrome, Lyme disease, and acquired immune deficiency syndrome (AIDS). Are we really making any headway against illness?

A chemical company is dumping waste into the creek that runs through the college campus, a creek that many people use for swimming. How can we find out what effects this could have on the community?

A middle-aged man has chest pains and enters the hospital for a checkup. His family is alarmed to learn that he has atherosclerosis and needs immediate triple-bypass surgery. The doctor says that the man's life-style—smoking, overeating, and inactivity—has probably contributed to the problem. How have these behaviors contributed to his condition? What can you do to avoid a similar fate?

A woman undergoes an ultrasound examination to discover why, in only the fifth month of her pregnancy, she already looks and feels so big. The happy answer: twins!

Biology and You

Many of our common experiences are based on the form or function of the human body, from such simple decisions as what to eat or which shampoo to use to such complex questions as whether or not to undergo an operation. Raising a child is a constant exercise in human biology. Even the embarrassing observation that your stomach growls during a morning biology class when you have skipped breakfast, but not when you have eaten an omelet, can be a lesson in how the human body functions.

Humans, however, are only a small part of the living world. Our everyday experiences sometimes give us a limited glimpse of the diversity of living things and their relationships to one another. Bathing an itchy dog, for example, interferes with a relationship between two types of organisms—Fido and his fleas. Preparing the soil to plant a vegetable garden disrupts the biological relationships between the plants ("weeds") that would grow there naturally in the absence of human intervention. Sometimes you may be particularly aware of the diversity of life on earth. A New York City resident on a first visit to Florida is taken aback by the lush subtropical plant life. This individual cannot help but compare the swaying palms to the somehow less interesting maple and oak trees that he or she barely notices back home and is fascinated to see oranges actually growing on trees, rather than filling a box at the supermarket.

Much of biological diversity is not readily apparent to human eyes, perhaps because many life forms are in the realm of the microscopic, or because they dwell in habitats not populated by humans (fig. 1.1). However, all forms of life, from the structurally and functionally complex human down to a tiny bacterium, are built on a shared chemical theme, and all carry out the same repertoire of responses that qualifies them as living. *Biology* is the study of the unity and diversity of life. Understanding the similarities and differences between types of organisms can provide a broader perspective on human experience—whether that experience is choosing a shampoo or making a life-or-death medical decision.

A Biologist's View of the Living World

The scientist and nonscientist often share an almost insatiable curiosity about life and its diverse but related forms. The nonscientist may seek a view of life in a garden or zoo; the scientist may do so in a more exotic locale, perhaps perched in the treetops of a tropical rain forest or ensconced in a laboratory among cages of organisms specially bred for experimentation. The scientist follows a systematic approach to cataloging and interpreting observations of life, called the **scientific method.** In actuality, the scientific method, which consists of observing, reasoning, predicting, testing, concluding, and interpreting, is not very different from the way in which many nonscientists think about this fascinating thing we call life.

The Scientific Method

The scientific method begins with *observations*. In 1965, Jim Schlatter, a young researcher at a pharmaceutical company, was trying to purify a chemical needed to make a new ulcer drug. Jim's hand brushed against the rim of the flask he was working with, and a tiny amount of a fine white powder deposited on the rim got onto his fingertips. When Jim went to turn the page of his laboratory notebook, he licked his finger and was startled by the incredibly sweet taste. What had caused it?

His curiosity sparked, Schlatter retraced his steps to the flask and on close inspection spotted the white powder on the rim. He collected a sample and performed several standard chemical tests to identify it. He found that it was a surprisingly simple substance—just two building blocks (amino acids) of protein linked together. Because protein is a normal constituent of the diet, Schlatter knew that what he had found was probably safe to eat—and after all, he had already done so himself, quite by accident.

The next step in the scientific method is to devise an educated guess, or **hypothesis.** Jim Schlatter could hardly contain his excitement when he formulated his hypothesis. Could his "sweet-tasting discovery" be safe to use as a sweetener in or on foods? His guess was "educated" because he already knew, from his taste test and preliminary chemical analysis, that the substance was a normal nutrient.

Devising an *experiment* is the next step in the scientific method. Schlatter and by now many other intrigued researchers set up the first of many rigorous tests for the new sweetener, named aspartame after its two amino acid components, aspartic acid and phenylalanine. The first goal was to prove its safety, and for this the researchers enlisted the aid of large numbers of mice.

But they were not ordinary mice—the experimental subjects were genetically identical and raised in identical environments. Their bedding, cages, food, drink, and amount and type of exercise were the same, as was the air temperature and humidity of their quarters. Periods of light and darkness were the same, and the mice even had the same types of microorganisms in their bodies. By making sure that the mice had as much in common as possible, the researchers could then feed aspartame to half the mice, and any changes in their health or behavior could be attributed to the sweetener. The mice would be sacrificed at varying times in their development and 48 different tissues and organs examined for microscopic changes.

The scientists used several approaches to make their experimental data as meaningful as possible. For example, hundreds of mice were used because such a large *sample size* ensures that results are statistically significant. Conclusions cannot accurately be based on the reactions of fewer than 30 animals. Half the animals were fed aspartame, and the other half were fed a **placebo,** a substance similar in taste and appearance to the substance being evaluated but whose effects are already known. Using a placebo ensures that the mice taking aspartame are not simply reacting to having something added to their

a.

b.

Figure 1.1
Life on earth is amazingly resilient. *a.* Organisms struggle to survive in areas where humans have altered the environment, such as the tree that bursts forth between slabs of sidewalk or fish swimming in an inner city pond-turned-garbage dump. *b.* In the summer of 1988, 1 million acres of Yellowstone National Park burned, leaving a mosaic of charred and spared areas. Yet already, life is returning to the devastated area and new plant growth has begun. Animals driven from their home, such as this majestic elk, are returning.

food. Placebos are particularly valuable in studies on people, because some of us tend to feel better (or worse) if we know we are taking something.

A placebo is an example of an **experimental control,** which is an extra test that does not directly address the hypothesis but helps to ensure that a single factor, or variable, is responsible for any observed effects. Another safeguard is to make the experiment **double-blind,** which means that neither the researchers nor the subjects know who has been given the substance being evaluated and who has been given the placebo. This information is revealed only after the next step in the scientific method is completed—interpreting data and forming conclusions. Finally, the experimental results must be repeatable by other investigators (fig. 1.2).

The initial studies on animals fed aspartame confirmed what Jim Schlatter had hypothesized. He had indeed found a new and apparently safe natural sweetener. Those studies laid the groundwork for more than 100 others, some involving feeding different animals enormous doses of the substance to be certain that it is safe (Reading 1.1).

Science Is a Cycle of Inquiry

Even though a conclusion marks an end to the scientific method for a particular experiment, it is not the end of the process of scientific inquiry. A conclusion must be placed into perspective with existing knowledge.

After aspartame's safety was demonstrated in animals and humans of all ages, other questions arose, and these were answered with further experiments. Just how powerful was this sweetener? (Two hundred times as sweet as an equal amount of table sugar.) Was aspartame low in calories? (Yes.) Could people with diabetes, who must control their sugar intake, use aspartame safely as a sugar substitute? (Yes.) Could people with the inherited disorder phenylketonuria (PKU), in which the amino acid phenylalanine must be avoided, use aspartame safely? (No.) To

(*a. top*) Source: Betty Smith, *A Tree Grows in Brooklyn*, Viking, Penguin, a division of Penguin Books USA Inc., New York, NY. (*a. bottom*) From "Ponds" from THE MEDUSSA AND THE SNAIL by Lewis Thomas. Copyright © 1978 by Lewis Thomas. Reprinted by permission of Viking Penguin, a division of Penguin Books USA Inc.

Figure 1.2
Many advertisements attempt to simulate parts of the scientific method to make their products seem "clinically tested," "new and improved," or otherwise superior to the competition. In this ad, the pile of towels on the left is the control group, and the pile on the right, the experimental group.
© The Procter & Gamble Company. Used with permission.

Reading 1.1 *The Long Life of a Food Additive—The Aspartame Story*

TODAY YOU CAN FIND ASPARTAME IN A WIDE VARIETY OF FOODS AND BEVERAGES, IN DRUGS, AND AS A TABLETOP SWEETENER. But the journey of this tiny protein from discovery to availability has been long, relying repeatedly on the scrutiny of the scientific method along the way. The story of aspartame is not unlike that of many foods and drugs used by humans—it must be rigorously tested to assure its safety and effectiveness.

1965
Jim Schlatter accidentally makes a "sweet-tasting discovery" in his search for an ulcer drug.

1966
Schlatter files a patent on his discovery.

1967
The first of more than 100 large-scale, double-blind, controlled tests of the safety and efficacy (sweetness) of aspartame begins.

1970
Schlatter's patent is issued (it has since been extended).

March 1973
Schlatter's employer, G. D. Searle, seeks FDA approval of aspartame in dry foods and as a tabletop sweetener (to be sold in packets to limit consumption rather than as a bulk sweetener sold as loose powder).

July 1974
FDA approves aspartame for use in dry foods and as a tabletop sweetener.

1974–1977
Various consumer groups challenge the validity of Searle's studies indicating the safety of aspartame for its intended uses. Marketing is held up; FDA board of inquiry is formed to reevaluate studies.

December 1978
Panel of pathologists reviews and approves aspartame studies and informs FDA board of inquiry.

October 1980
FDA board of inquiry finds no evidence of damage to human brain or endocrine system from aspartame but requests longer-term animal tests for brain tumors before marketing approval is granted.

July 1981
FDA again approves aspartame for use in dry foods and as a tabletop sweetener; finds animal brain tumor studies unnecessary but requires Searle to monitor and report consumption levels of aspartame to FDA.

July 1983
FDA approves aspartame use in carbonated drinks.

December 1983
FDA approves aspartame use as an inactive ingredient in drugs for humans.

1983–1984
More challenges to aspartame's safety. Consumer groups point out that heat causes aspartame to break down into methanol, which can react to form formaldehyde. In large amounts in animals, methanol can cause blindness and formaldehyde can cause nasal tumors. FDA and Searle counter that methanol levels from aspartame breakdown are far below toxic levels, and that 12 ounces of tomato juice contains three times the methanol as a 12-ounce (approximately 355-milliliter) glass of diet soda sweetened with aspartame. Searle conducts experiments on "typical storage periods," storing aspartame-sweetened beverages for 0 to 52 weeks, at 41°F to 131°F (5°C to 55°C), and measuring methanol content. After heating and storing, methanol levels are still below toxic levels, but sweetness declines.

In the first half of 1983, FDA receives 108 complaints from individual consumers of aspartame causing headache, dizziness, and a wide range of other symptoms. In the second half of 1983, they receive 248 reports.

February 1984
FDA asks the Centers for Disease Control (CDC) to evaluate consumer complaints. CDC interviews 517 of 592 complainants.

November 1984
CDC concludes that the data "do not provide evidence for the existence of serious, widespread, adverse health consequences attendant to the use of aspartame." Few complaints were serious enough for the person to consult a physician, but some people may have unusual sensitivity to the sweetener, the CDC concludes.

Future
G. D. Searle continues to find new foods that can be sweetened with aspartame and to overcome the limitation of the product's breakdown with heat.

which foods could aspartame be added? (Drugs, dry foods, and beverages but not to foods that are heated.) How does aspartame compare in taste to the artificial sweeteners saccharin and cyclamate? (It has no aftertaste, as the others do.) After more than a decade of testing, aspartame became "one of the most thoroughly tested and studied food additives," according to the Food and Drug Administration, which assures the safety and effectiveness of foods, food additives, and drugs in the United States. Still, experiments are ongoing to expand uses of aspartame and to discover any side effects.

An underlying characteristic of scientific thinking is that the sequence of observing, reasoning, predicting, testing, concluding, and interpreting is a cycle, with new ideas spawned at every step (fig. 1.3). To the curious, scientific mind, a conclusion is never really the final answer. There is always something further to investigate.

KEY CONCEPTS

The scientific method consists of making observations, formulating a hypothesis, designing an experiment, collecting and interpreting data, and reaching conclusions. A good experiment has a large sample, it is controlled so that only one variable is assessed, it is placebo controlled, and it is conducted in a double-blind manner.

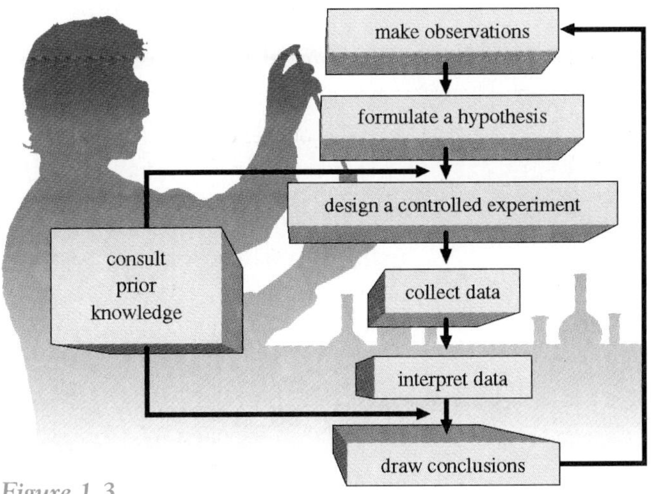

Figure 1.3
Steps of the scientific method.

Figure 1.4
On the way out: using whole animals to test products. Experimental animals can provide information on product safety, but many people object that the animals are subjected to unnecessary pain. An experimental step closer to humans than using standard bacterial tests is to use specially bred, genetically identical mice.

Designing Experiments

Superficially, the science of biology is a collection of facts that describes and explains the workings of the living world. The excitement of biology lies in intriguing observations and carefully planned experiments to explain them, as Jim Schlatter learned with his discovery of a new sweetener.

The science of biology is filled with exciting and often surprising experiments. Imagine the wonder felt by the early microscopists as they peered through their crude lenses and discovered a tiny living world (chapter 4). How would Gregor Mendel, the monk who laid out the principles of heredity a century ago, feel if he knew that the patterns of traits he observed in his pea plants would one day be used to explain how two perfectly healthy people could have a child with a crippling inherited illness (chapters 14 and 15)? It took keen observation, hypothesizing, experimentation, and reasoning to discover that a drop in the worldwide anchovy population reflected shifting winds over South America; that a mild and supposedly harmless tranquilizer called thalidomide would cause 10,000 European children to be born without arms and legs; and that a tiny virus causes the profound immunity breakdown that occurs with AIDS.

Experimental design can be as varied as the biological phenomena it probes. Many studies expose nonhuman organisms to treatments and situations that might provoke ethical objections if they were conducted on humans. The results of such studies are often extrapolated to humans. Potentially irritating cosmetics, for example, were not too long ago tested on the eyes and ears of rabbits (fig. 1.4). Bacteria are used extensively to test whether or not certain chemicals are likely to cause cancer. The jogging pig in figure 1.5*a* exercises to a point that even a human Olympian could not readily attain. The pig's heart and blood vessels are monitored for damage caused by the excessive exercise, and the results are used as guidelines in setting human limits. Similarly, a calf was used to test an artificial heart before the first human recipient, Barney Clark, had his own failing heart replaced with such a device (fig. 1.5*b*). The pig and calf, with cardiovascular systems remarkably like our own, serve as *animal models*, or experimental stand-ins, for humans.

Animal models are also used to examine biological phenomena simply for the sake of expanding our knowledge, with no immediate intention of extending the results to humans. Some animal experiments are used to assess the value of continuing a particular line of investigation on humans. Whatever the ultimate applications, using such animals as mice and fruit flies enables us to breed large numbers of nearly identical individuals so that experiments can be carefully controlled. In many experiments, the use of whole animals or plants is abandoned altogether, and parts of organisms grown in the laboratory are used instead (fig. 1.6).

In contrast to studies conducted on uniformly bred and raised experimental organisms, from which it is relatively straightforward to draw conclusions, are *epidemiological studies*, which are often used in the field of public health. **Epidemiology** is the analysis of data derived from real-life situations, ranging from one-person accounts (anecdotal evidence) to large-scale studies involving thousands of individuals. Such extensive studies may involve evaluating health effects of exposure to a certain chemical from industrial accidents, or they may be planned investigations conducted by researchers.

One type of epidemiological investigation used often in the social and biological sciences compares observations recorded in surveys. In general, two groups of people differing in one factor, usually a particular experience, are compared. In the best of such studies, each person from one group is matched to a person from the other group who is the same in as many characteristics as possible, such as sex, race, age, diet, and occupation. For example, the incidence and types of birth defects occurring among children of Taiwanese women exposed to cooking oil contaminated with polychlorinated biphenyls (PCBs) are compared to those of Taiwanese women not so exposed. A significant number of the children exposed to PCBs were shorter and lighter than the controls and had abnormal gums, skin, teeth, nails, and lungs. On the basis of this study, these health problems were attributed to PCB exposure before

a.

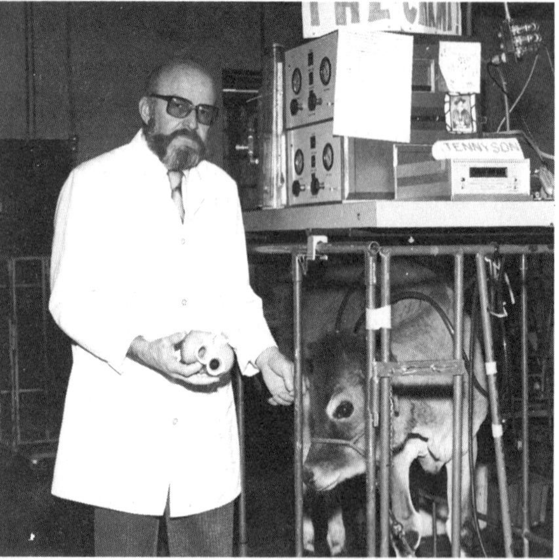

b.

Figure 1.5
Moe the jogging pig (*a*) and Tennyson the plastic-hearted calf (*b*) provide animal models of the human cardiovascular system.

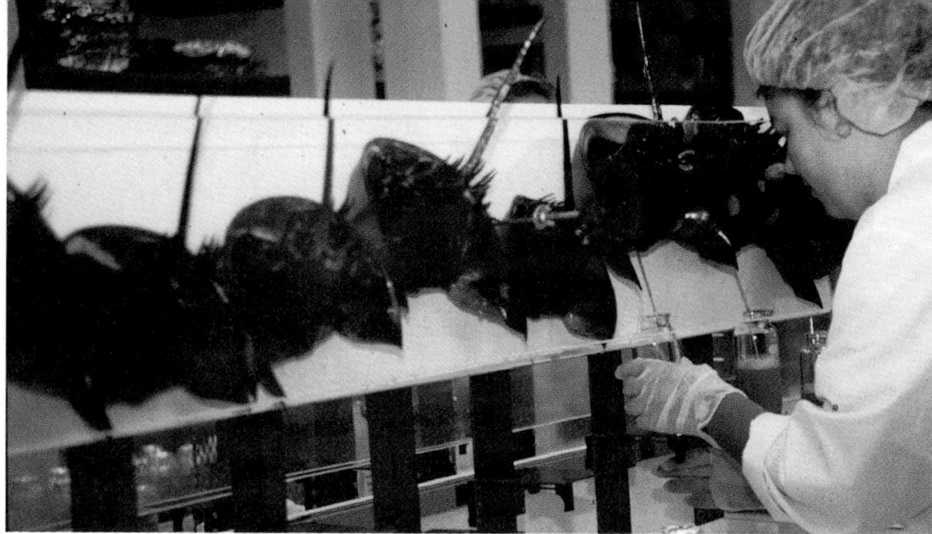

Figure 1.6
On the way in: animal parts replace whole animals in research. Cells from a horseshoe crab's blood form a gel in the presence of a common bacterial toxin. These cells are used as a standard test of laboratory and hospital equipment for bacterial contamination. Horseshoe crabs are "milked" for their blood in a facility on Cape Cod, then returned, apparently unharmed, to the sea. This test replaces a more costly and controversial test in which drugs thought to be contaminated were injected into rabbits. If the drug was contaminated, the rabbit ran a fever or died.

birth because this was the only difference between the two groups of children.

In some situations, though, epidemiological information can only demonstrate a correlation, and not a cause-and-effect relationship, between two observations. The information that runners are skinny may indicate that running is a good way to lose weight—or it may indicate only that those who are already thin tend to take up running. Further work must be done to determine what causes what in such cases.

An epidemiological study may seem simple in concept, but it is often quite difficult to carry out. Wading through masses of information in an attempt to set up experimental controls for all possible contributing factors is a formidable task. Additionally, biases often affect the reporting of information about events, especially events that occurred a decade or two earlier. There are no such problems with experimental mice. The To Think About section at the end of this chapter presents examples of interesting epidemiological and laboratory studies.

KEY CONCEPTS

Experiments typically utilize animal models or laboratory grown parts of organisms. Epidemiological studies utilize groups of people that, whenever possible, differ by only the variable being investigated.

Limitations of Applying the Scientific Method

Interpreting Results

The scientific method is not foolproof, and it is sometimes difficult to implement. Obtaining a sufficient number of experimental subjects, for example, might be possible when ordering mice from a biological supply catalog, but it is quite another story when a rare human disease is under scrutiny. Experimental evidence may lead to an unexpected or multiple conclusion, and even the most carefully designed experiment can fail to provide a definitive answer. For example, animals fed large doses of vitamin E live significantly longer than similar animals not fed the vitamin in excess. One could conclude that vitamin E retards aging, and indeed many people believe this. However, the excess vitamin E causes the animals to lose weight, and other experiments have associated weight loss with longevity. Do the vitamin E supplements extend life, or does the weight loss? The experiment of feeding animals large doses of vitamin E does not distinguish between these two possibilities. Can you think of further experiments to clarify whether vitamin E or weight loss extends life?

Observations or experimental results can be misinterpreted. In chapter 3 we will see how many investigations "proved" that life spontaneously arises from the nonliving by exposing decaying meat to the air and observing what grew on it. Also discussed in chapter 3 is an elegantly designed experiment whose conclusion is most difficult to validate. The Miller experiment sought to recreate the chemical reactions that might, on the primitive earth, have formed the chemicals found in living things today. Although the experiment had interesting results and inspired much further work, we cannot really know whether or not it successfully recreated the conditions and the events on the early earth—short of someone inventing a time machine.

Sometimes it is difficult to maintain the objectivity that the scientific method requires in drawing conclusions. In the 1960s, studies on the genetic material of residents in Scottish mental and penal institutions showed that a disproportionate number of these men had an extra Y chromosome when compared to men in the general population. It was widely concluded both by scientists and by the press that these studies showed that an extra Y chromosome predisposes a man to criminal behavior (chapter 14). However, the men with an extra Y chromosome also had in common greater than average height. Could their physical appearance have influenced how people reacted to them, which in turn might have affected their behavior? Today, geneticists concur that the only trait that can be associated with having an extra Y chromosome is greater than average height. It would be interesting to look at the occurrence of extra Y chromosomes among the members of basketball teams, who are notoriously tall.

When Implementing the Scientific Method Is Unethical

It was 16 years between the time that Jim Schlatter accidentally took the first taste of aspartame and when it became available as an alternative sweetener. Drugs also take several years to be thoroughly researched and approved, involving many placebo-controlled, double-blind tests. When treatments already exist for a particular disorder, studies compare the new drug to a standard drug whose effects are well known. What happens, though, when a new drug is the only treatment for a devastating illness? Can a researcher ethically give a very sick person a placebo, when the drug under study shows promise? The FDA was faced with this dilemma in the case of AIDS. In 1986, when a drug called azidothymidine began noticeably to extend the lives of certain AIDS patients participating in a clinical trial, the experimental protocol that called for giving a placebo to half of the volunteers was abandoned and 4,000 additional patients were put on the drug. On the other hand, two other potential AIDS treatments showed early promise but proved to be more harmful than the disease they were developed to combat. However, in light of the desperation of AIDS patients and others with terminal, untreatable illnesses, the FDA has ruled that the usual adherence to the scientific method can be suspended in these circumstances.

The ethics of the scientific method also come into question when testing vaccines for safety and effectiveness. This situation is even more controversial than testing drugs with the desperately ill, because a vaccine is given to healthy individuals to protect them against a specific disease. The problem is that to test the effectiveness of the vaccine, the person must be exposed to the disease-causing agent. The vaccine could actually cause the illness or fail to protect the individual.

In 1796, Edward Jenner took a great risk in testing his smallpox vaccine on boys. The first polio vaccines were given to young people who had recovered from polio, and the boost in immune system activity following immunization was taken as a sign that the vaccine worked. The FDA and many researchers are now wrestling with the dilemma of how to test an AIDS vaccine so that results are scientifically meaningful. To whom should the vaccine be given? Should vaccinated individuals be intentionally exposed to the virus to see whether the vaccine works or be educated on how to avoid AIDS for their own benefit? Who is responsible if the vaccine does not work or has side effects? The challenges of implementing the scientific method are intensified when human lives are involved.

Expect the Unexpected

Ideally, an investigator should attempt to keep an open mind toward observations, not allowing biases or expectations to cloud the interpretation of results. The scientist should expect the unexpected. But it is human nature to be cautious in accepting an observation that does not quite "fit" existing knowledge. The careful demonstration that life does not arise from decaying meat came as a surprise to many people who believed that mice were created by mud, that flies came from rotted beef, and that beetles sprouted from cow dung (chapter 3). Earlier this century, researchers were so enamored of the much-studied proteins that they were hesitant to follow up suggestions that the less-understood nucleic acid DNA was really the genetic material. The definitive identification of the stuff of heredity, described in chapter 13, not only demonstrated that DNA is the genetic material but also proved that protein is not.

Perhaps no one knows better the conflicting feelings of joy and frustration that accompany a scientist's discovery of a quirk of nature than Barbara McClintock. In the 1940s, McClintock studied inheritance of kernel color in corn—as she put it, "Asking the maize (corn) plant to solve specific problems and then watching it respond." While watching her carefully bred plants respond, McClintock noticed that some kernels had a peculiar pattern of spotting. After many repetitions of experiments, McClintock concluded that the kernel spots indicated that the units of inheritance (genes) jumped around inside corn cells. This idea seemed preposterous at the time, because genes were thought to be immovable parts of larger structures called chromosomes.

Despite McClintock's evidence and the repeated publication of her findings, the scientific community simply did not believe her. The idea of genes moving was just too unusual. Today, after a decade of additional discoveries of moving genes in many types of organisms, Nobel Prize winner Barbara McClintock has finally been recognized as the perceptive and talented researcher that she has always been. Her observation of roving genes in corn plants four decades ago may ultimately help to explain roving genes that cause certain cancers in humans.

Scientific Discovery by "Accident"

Scientific discovery is not always as well planned as the scientific method may seem to suggest. Sometimes new knowledge comes simply from being in the right place at the right time or from being particularly aware of the unusual, as were Barbara McClintock and Jim Schlatter. The creative side of science is making mental connections to take advantage of these accidental observations. Consider the experi-

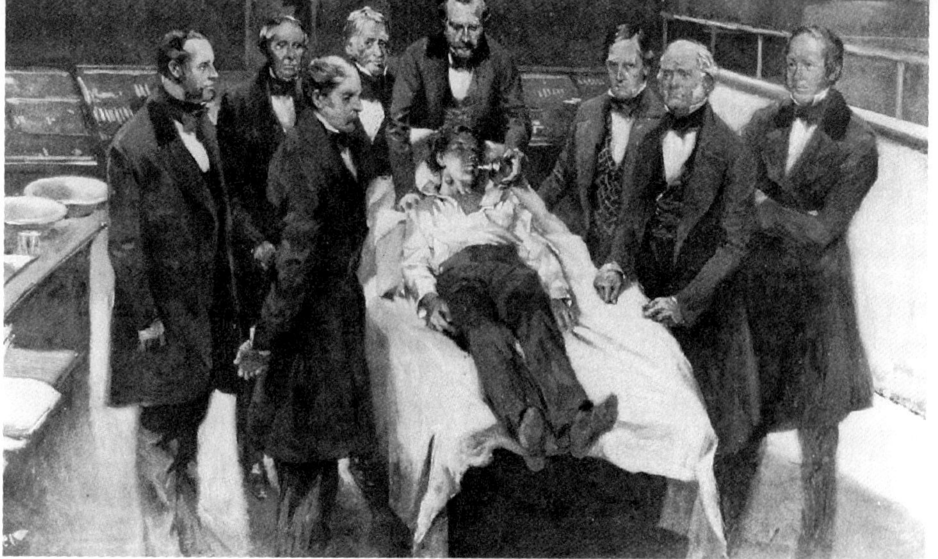

Figure 1.7
The discovery of ether's value as an anesthetic stems from a student's observation at a party that friends who were using the chemical to "get high" did not feel pain.

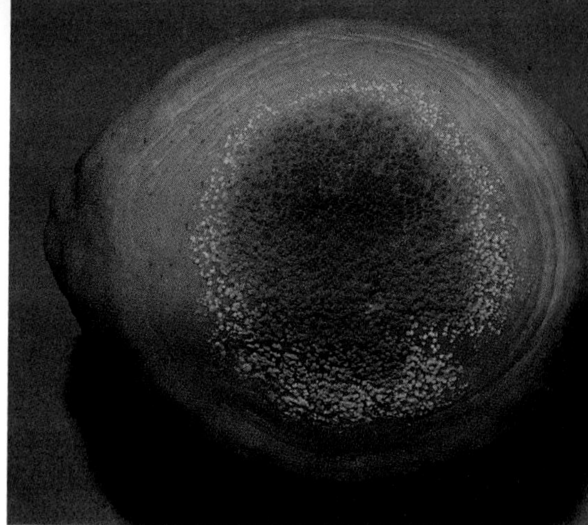

Figure 1.8
Sir Alexander Fleming discovered the bacteria-killing effects of penicillin by accident when a common mold grew on an uncovered culture of bacteria in his laboratory.

ence of Dr. Crawford Long, who was a medical student in the middle of the nineteenth century. While attending an "ether party," the observant student noticed that his friends who were intoxicated by the chemical appeared not to feel pain. After he obtained his degree, Long found that he could remove tumors painlessly by placing a towel soaked in ether over his patients' faces before starting surgery (fig. 1.7).

Some scientific advances occur by accident. In 1928, microbiologist Alexander Fleming left a culture dish of disease-causing *Staphylococcus* bacteria uncovered, and therefore it was exposed to contamination by other microorganisms. Just before tossing the contaminated dish in the garbage, Fleming noticed odd clear patches on the plate where the bacteria were not growing. Some contaminant from the air had apparently stopped the bacterial growth on the plate. Although he attempted to isolate and characterize the substance responsible for destroying the bacteria, Fleming was a poor chemist, and the work of others was ultimately required to identify penicillin, a product of a mold, whose effects he had inadvertently discovered (fig. 1.8).

A few years after Fleming's landmark discovery, a mother of two retarded children in Oslo, Norway, made another accidental discovery of medical value. She noticed that her children's urine had a peculiar smell, and she mentioned this to a relative who was a chemist. On analyzing the children's urine, the chemist detected a buildup of a certain body chemical that indicated the biochemical cause of the children's retardation. Because of this observation, today those who inherit the disease phenylketonuria (PKU) can grow up with normal intelligence if they follow a special diet during the first few years of their lives, when the brain is developing at its fastest.

KEY CONCEPTS

The value of the scientific method is limited by our ability to interpret experimental results. In some situations, use of the scientific method is unethical if it withholds effective treatment from human volunteers. As useful as the scientific method is, some discoveries are made completely by accident.

We All Think Like Scientists at Times

The cycle of questioning and answering that is the scientific method is second nature to a practicing biologist; it is more a way of life than a set of rules. This process probably played an important role in interpreting natural phenomena long before scientists were educated at universities. In fact, the steps of the scientific method are followed by many a nonscientist. Consider the question of whether or not a food is edible. Our ancestors probably consumed new foods in a trial-and-error fashion to determine if they were safe to eat, perhaps even assigning one person to this task. In 1820, a New Jersey colonel bravely ascended the steps of a county courthouse and consumed a tomato, to the amazement of townspeople who thought the fruit to be poisonous. His survival after this experiment provided valuable information. Today, new parents typically introduce different foods to their infants one at a time, in order to identify allergy-causing foods. In all of these examples, an experiment—tasting the food in question—was conducted to determine whether or not a food is safe to eat.

You probably use the scientific method without even realizing it. Consider the following situation: You notice that you have heartburn at night fairly frequently (*observation*), so you start to keep track of your diet. It does not take long to realize that your painful nights correspond to those when you consume pizza with mushrooms, sausage, and peppers (*further observations*). Perhaps the heartburn is due to the sausage (*hypothesis*), because eating mushrooms and peppers in other foods has not bothered you in the past (*prior knowledge*). To test this educated guess, you eat pizza with each topping alone (*experiment*), and maybe even with no topping or with pairs of toppings (*controls*), just to be sure that you know the cause. You try each variety several times (*sample size*). It appears that, for you, sausage does indeed cause heartburn (*conclusion*). On the basis of this experiment, you confine your pizza consumption to those without sausage, and perhaps you sample other foods containing sausage to see if they cause heartburn also (*further experiments*).

There is a little bit of the scientist in each of us. We can all observe, form hypotheses, plan and carry out experiments, and attempt to interpret and use the results. However, your observations and ideas concerning the living world will be more meaningful if you have more extensive experience and learning on which to build. This book and your biology course will provide you with some of this valuable background. With it, you will be better equipped to apply the systematic approach of the biologist to your own observations of the living world. Indeed, you will soon know just what is in shampoo, what might have happened if you had not been immunized against various diseases, what is in polluted creek water and why it is harmful to human health, why heart disease can be caused by life-style habits, and what life before birth is like.

SUMMARY

Understanding the unity and diversity of living things through the study of biology can help us to appreciate our position in the living world. Biologists study life by applying the *scientific method*, in which observation and reasoning are used to make an educated guess, or *hypothesis*. One or more experiments are conducted to test the validity of the hypothesis, including *experimental controls* to ensure that only one phenomenon is being examined at a time. The experiment is usually repeated to test its accuracy, and then conclusions are drawn. Conclusions add to knowledge, but they also usually lead to more questions, and then the cycle of scientific inquiry begins anew.

Application of the scientific method does not always yield a complete answer to a question. A scientific investigation may be based on existing information, as in an *epidemiological study*, or it may be based on information generated specifically for the investigation, as in studies using experimental organisms. Each method has its advantages and disadvantages. Many answers to scientific questions are unusual or unexpected, and some are stumbled upon accidentally. Overall, the scientific method provides a systematic approach to exploring the living world. In addition, we often practice the method in our everyday lives without being aware of it.

QUESTIONS

1. Read the following passage and identify the steps of the scientific method:

In 1953, a graduate student named Stanley Miller combined various chemicals in the presence of electrical sparks and heat to simulate conditions on earth before life existed. From the brew formed complex chemicals known to be important in the chemistry of living things.

In 1981, geologists discovered openings in the ocean floor where the water is extremely hot compared to the surrounding water. They found simple chemicals there, which they knew from Miller's experiment could react to yield the complex chemicals of living things. Several researchers suggested that life could have originated at these deep sea vents, because the chemical building blocks of life were present, plus the ocean would have protected the delicate precursors to life from the lightning, ultraviolet rays, and meteors that hit the earth's surface.

In 1988, Stanley Miller, now a chemist at the University of California at San Diego, challenged the popular idea that life began in these oceanic hot spots. Based on his knowledge of chemistry, Miller thought that the complex chemicals that might form under the suggested conditions would be unstable at the very high temperatures and decompose too rapidly to allow life to form. To test his idea, Miller conducted an experiment similar to what he had done as a graduate student. In the laboratory he combined the simple chemicals found in the deep sea vents and heated them to the temperatures found there. As Miller predicted, more complex chemicals formed, but they were dismantled by the heat in minutes or seconds. Therefore, deep sea vents are not likely to have been a place where life on earth started. Might the first life forms have arisen at cooler deep sea vents near the Galapagos Islands?

2. The most informative type of experiment is placebo controlled and double-blind.

a. How does a placebo serve as an experimental control?

b. How does withholding information from researchers and subjects ensure more reliable results?

c. Design a double-blind study to test the effectiveness of a new antacid.

d. DMSO (dimethylsulfoxide) is a chemical believed by some people to relieve a variety of ills, but its safety and effectiveness have not yet been proven by experiment. An unpleasant side effect of rubbing DMSO on the skin is that it produces a garlicky taste as it enters the body. How might this peculiarity complicate adequately controlled testing of DMSO for safety and effectiveness?

3. A chemical manufacturing plant closes down because a toxic substance, which was buried on the plant property a decade earlier, is now leaking from its containers. To examine the effects of the contamination, a survey is distributed among workers and town residents, inquiring about their health problems and asking them whether or not they believe that they were exposed to the chemical. What can be learned from this type of study? What are its limitations?

4. Animal models often provide an effective means of testing substances and procedures that are not yet well enough understood to be used on humans. Many people object that pain is inflicted on the animals used for such purposes. However, in what has been called the "Bambi factor," such objections tend to focus on "cute" organisms, such as rabbits and dogs, rather than the fruit flies, leeches, rats, and other less appealing organisms that are also used experimentally. What limits, if any, do you think should be placed on animal experimentation? What technique mentioned briefly in the chapter might be considered a compromise?

5. Scientists traditionally report their research findings in journals. Go to your science library, choose a biological or medical journal, and try to find an article that you can follow well enough to identify the steps of the scientific method.

TO THINK ABOUT

Many of our decisions about the foods we eat, the medicines we take, and the activities we participate in are based, in part, upon health information. Some of this information comes from well-executed studies that adhere to the scientific method, but some comes from flawed investigations. For each of the following real-life examples, state whether or not the scientific method was followed. If it was not, indicate which of the following specific faults was present:

- **a.** Conclusions not supported by experimental evidence
- **b.** Inadequate controls
- **c.** Biased sampling
- **d.** Inappropriate extrapolation of the experimental group to the general population
- **e.** Sample size too small
- **f.** Other problem

1. "I ran 4 miles (approximately 6 kilometers) every morning when I was pregnant with my first child," the woman told her physician, "and Jamie weighed only 3 pounds (approximately 1 kilogram) at birth. This time, I didn't exercise at all, and Jamie's sister weighed 8 pounds (approximately 3 kilograms). Therefore, doctor, running during pregnancy must cause low birth weight."

2. Eating foods high in cholesterol was found to be dangerous for a large sample of individuals with hypercholesterolemia, a disorder of the heart and blood vessels. It was concluded from this study that all persons should limit dietary cholesterol intake.

3. A dumping site in the small town of Love Canal, New York, was used for the disposal of industrial chemical wastes for several years. Later, toxic substances were found in the basements of schools and homes. A preliminary examination of the genetic material of several residents of Love Canal showed abnormalities. Exposure to the chemicals, then, must have caused the problems.

4. Each year, 500,000 American males join the 10 million who have already had a vasectomy, a simple sterilization procedure. In the early 1970s, animal studies suggested that heart disease might be a complication of vasectomy. To test this hypothesis, a study involving more than 20,000 men was begun. In this study, each of 10,590 vasectomized men was paired with a nonvasectomized man of the same age, race, and marital status and living in the same neighborhood at the time of the surgery. Each man was extensively questioned about medical conditions (including the 54 conditions statistically most likely to be associated with vasectomy) and health and life-style habits, such as smoking, drinking, diet, exercise, and frequency of checkups. Medical records were carefully evaluated. The results of this study showed that vasectomized men had no significantly higher rates for any of the health conditions examined. Conclusion: Vasectomy is safe.

5. The National Cancer Institute (NCI) sponsors clinical trials at its Maryland headquarters and in several research hospitals around the country. New treatments are carried out on human cancer patients and the results evaluated. Standard cancer treatments are used as controls. There are many more cancer patients seeking to participate in these trials than there are spaces available for them. The policy of the NCI is to select the healthiest, strongest patients possible—what one cancer researcher calls "the perfect patient." The goal is to give the new treatment the maximum chance of working by testing it on a patient who has cancer but is otherwise young, strong, and healthy. Whenever results of such trials seem promising, television news shows excitedly announce them as "breakthroughs."

6. A former NCI researcher was bothered by the practice of turning down large numbers of cancer patients who wanted to participate in clinical trials. His solution was to leave the institute and start his own company, which offers to do research on individuals' tumors. This research could lead to personalized experimental treatments. However, this researcher's company is not supported by government funds, so patients have to pay for the research service—about $60,000 a year. The company publishes the results of its research.

SUGGESTED READINGS

Dagani, Ron. November 12, 1984. In-vitro methods may offer alternatives to animal testing. *Chemical and Engineering News*. Traditional product tests using live, whole animals are being replaced with tests using animal cells and tissues.

Kanigel, Robert. January/February 1987. Specimen no. 1913. *The Sciences*. A rat's brief life in the service of science—from its point of view. An informative and absolutely riveting account of the use of animals in experiments.

Larkin, Tim. June 1985. Evidence vs. nonsense: A guide to the scientific method. *FDA Consumer*. The scientific method is fine in theory; in practice, it is sometimes difficult to implement.

Lecos, Chris W. February 1985. Sweetness minus calories = controversy. *FDA Consumer*. The natural sweetener aspartame and the artificial sweeteners cyclamate and saccharin have been put through many, many tests.

Lewis, Ricki. September 1984. Dioxin danger (?) *Biology Digest*. Just how harmful to human health is dioxin, a chemical contaminant of the herbicide Agent Orange? Different types of studies give different answers.

Segal, Marian. June 1990. Is it worth the worry? Determing risk. *FDA Consumer*. The scientific method is a major part of risk assessment.

Thompson, Richard C. December 1986/January 1987. Protecting human guinea pigs. *FDA Consumer*. How people participate in testing new drugs or medical devices.

Weisburd, Stefi. May 23, 1987. AIDS vaccines: The problems of human testing. *Science News*. Testing a vaccine is always controversial because healthy people must be the recipients. In the case of deadly AIDS, the situation is even more serious.

Young, Frank E. June 1987. Experimental drugs for the desperately ill. *FDA Consumer*. The former head of the FDA speaks out on loosening the scientific method in cases of hopeless illness.

Zumwalt, Elmo, III. June 1987. A war with hope. *Health*. An anecdotal look at the possible dangers to health of dioxin by a Vietnam veteran suffering from several types of cancer.

C H A P T E R

2

The Diversity of Life

Chapter Outline

Journey in the Present
A Living World of Contrasts
Life Within Life
Life Affects Life
Reading 2.1 Fungus + Alga = Lichen
Journey Through the Past
The Science of Biological Classification—Taxonomy
Characteristics to Consider
Problems in Taxonomy
The Linnaean System of Biological Classification
A Look at the Kingdoms
Monera
Protista
Fungi
Plantae
Animalia
Reading 2.2 Taxonomy in Action

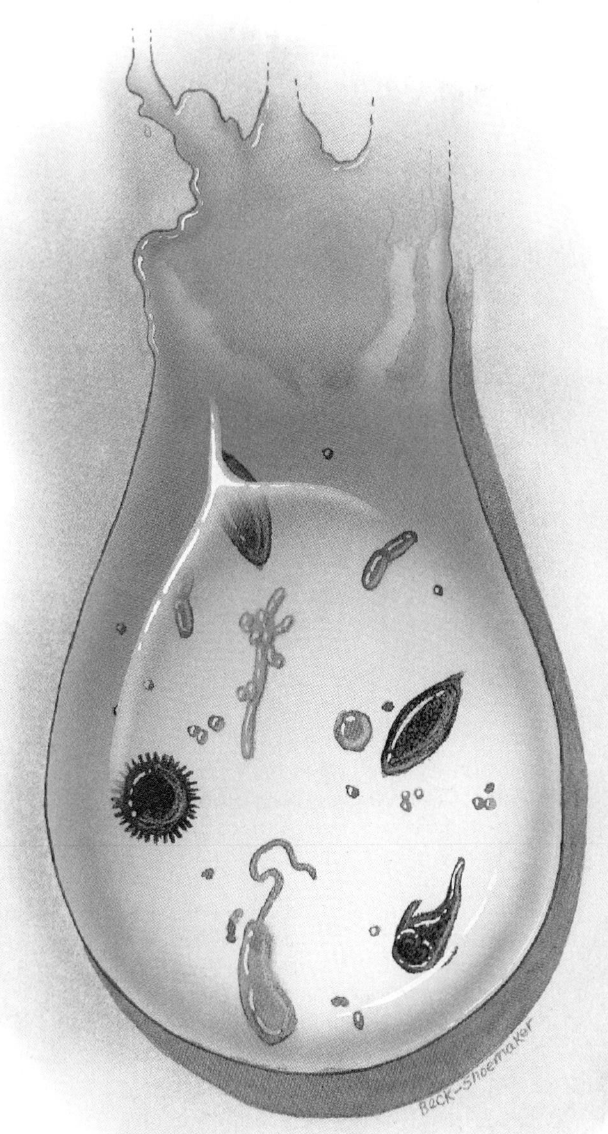

Learning Objectives

By the chapter's end, you should be able to answer these questions:

1. Where on earth are living organisms found?
2. What types of living organisms populated the earth in the distant past?
3. What criteria do biologists use to classify organisms?
4. What are the major groups of organisms that are living today?
5. How does biological classification reflect evolutionary relationships?
6. What is the full biological classification of humans?

The scene: a 2-square-mile (approximately 5-square-kilometer) wildlife preserve in Peru, home to perhaps the most varied assemblage of organisms ever to grace so small an area. In the treetops of the preserve's seven forest varieties live 40 kinds of parrots, 300 kinds of hummingbirds, and literally countless kinds of ants, beetles, and butterflies. Of the 2 million insect specimens collected by researchers each year in these forests, 80% are types previously unknown to science.

The scene: Hamelin Pool in western Australia. This very salty body of water is covered with green mats consisting of the same type of microscopic life, cyanobacteria (formerly called blue-green algae), that abounded on the very young earth. The only living organisms in evidence, besides the carpet of cyanobacteria, are a few fish. So harsh are conditions in the pool, at twice the salt concentration of the ocean, one wonders how any organisms manage to survive there at all. The few that can withstand the salinity of the pool are largely free of predators. Thus, the cyanobacteria flourish, much as their ancestors did a billion years ago. Hamelin Pool, off limits to most other life forms, is practically the cyanobacteria's own private turf (fig. 2.1).

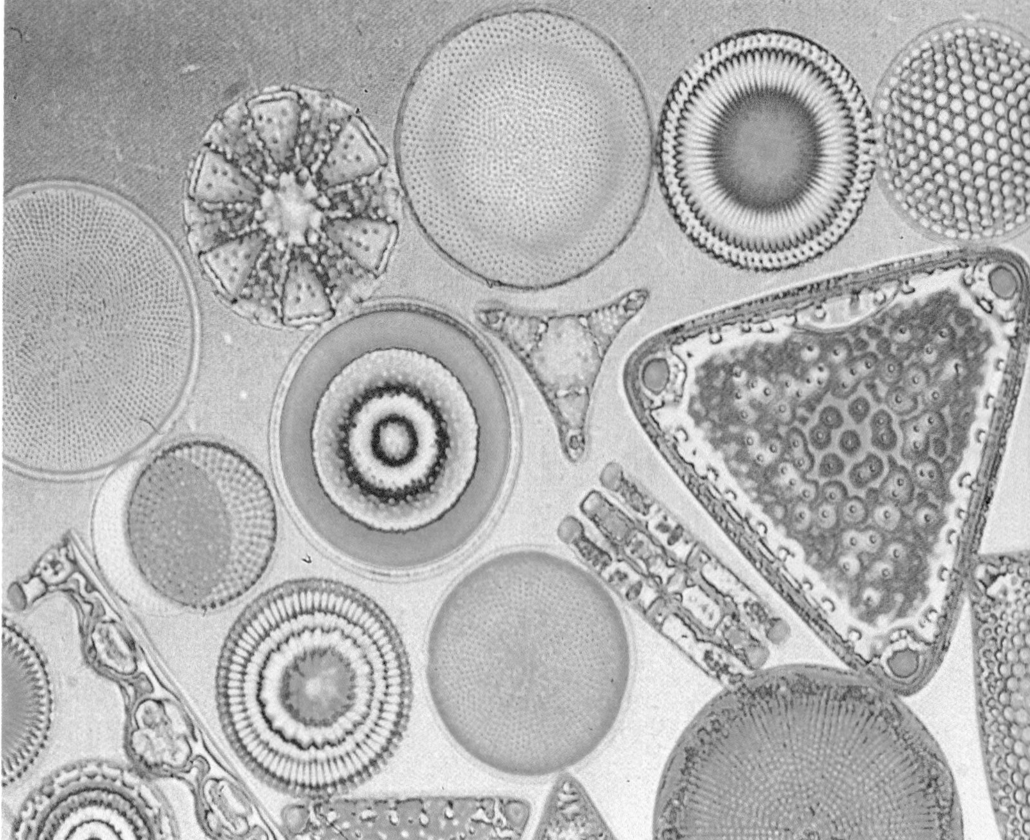

Figure 2.1
Life began in the oceans, and it continues to thrive there today. Even a single drop of seawater contains numerous organisms, and perhaps even the fossilized remains of microorganisms, such as these diatoms, a type of algae. The exquisite forms of diatoms and other microorganisms have inspired the works of many artists.

Journey in the Present

A Living World of Contrasts

The sparsely populated Australian pool contrasts sharply with the lush life of Peru's tropical rain forest. However, Hamelin Pool is just one of many seemingly inhospitable places that are in fact home to various organisms. Consider the cracks in the ocean floor known as thermal vents. These unlikely habitats are 1.5 miles (nearly 2.5 kilometers) beneath the ocean's surface, with local temperatures of 600°F (more than 300°C) and pressures as much as 265 times greater than the pressure at the surface. The murky water in such places is home to giant clams, worms, crabs, and mussels, as well as to a multitude of microorganisms. On close inspection, many other seemingly unlikely places for life reveal the presence of living organisms. Even hot desert rocks, which are almost devoid of life-supporting nutrients and water, have surfaces and crevices that are lined with microscopic forms of life (fig. 2.2). This chapter offers a view of the diversity of life on earth, introducing concepts that will be returned to in later chapters.

Life Within Life

Life appears to be nearly everywhere. A country corner left undusted soon houses a spider's web. Food forgotten in the back of the refrigerator soon wears a coat of fuzzy fungus. So abundant is life on earth that it often exists on or even in other organisms, forming a relationship called **symbiosis.** When organisms live in the intimacy of symbiosis, one of the pair benefits, and the other can either benefit also, be unaffected, or be harmed. In the type of symbiotic relationship called **mutualism,** both partners benefit, and, in some cases, one might not be able to survive without the other. A good example is the acacia tree that grows in Mexico. These trees are covered with ants, which eat sweet nectar secreted by the leaf tips. The ants, in turn, attack and drive off other insects that approach the tree and remove the tendrils of vines before they can establish a foothold on the trunk or branches. Without their ant defenders, the acacia trees are eaten by destructive insects or overwhelmed by surrounding vegetation; without the nectar provided by the acacias, the ants starve (fig. 2.3).

In **commensalism,** one member of the symbiotic pair benefits without affecting the other member. The human body is host to many microorganisms that do not threaten our health; in fact, some of them may actually help to maintain it by occupying areas that might otherwise be a home for a disease-causing organism. An interesting commensal relationship is that between the clownfish and the sea anemone. The stinging tentacles of the sea anemone protect the clownfish from predators, but the anemone seems to derive no benefit from the relationship.

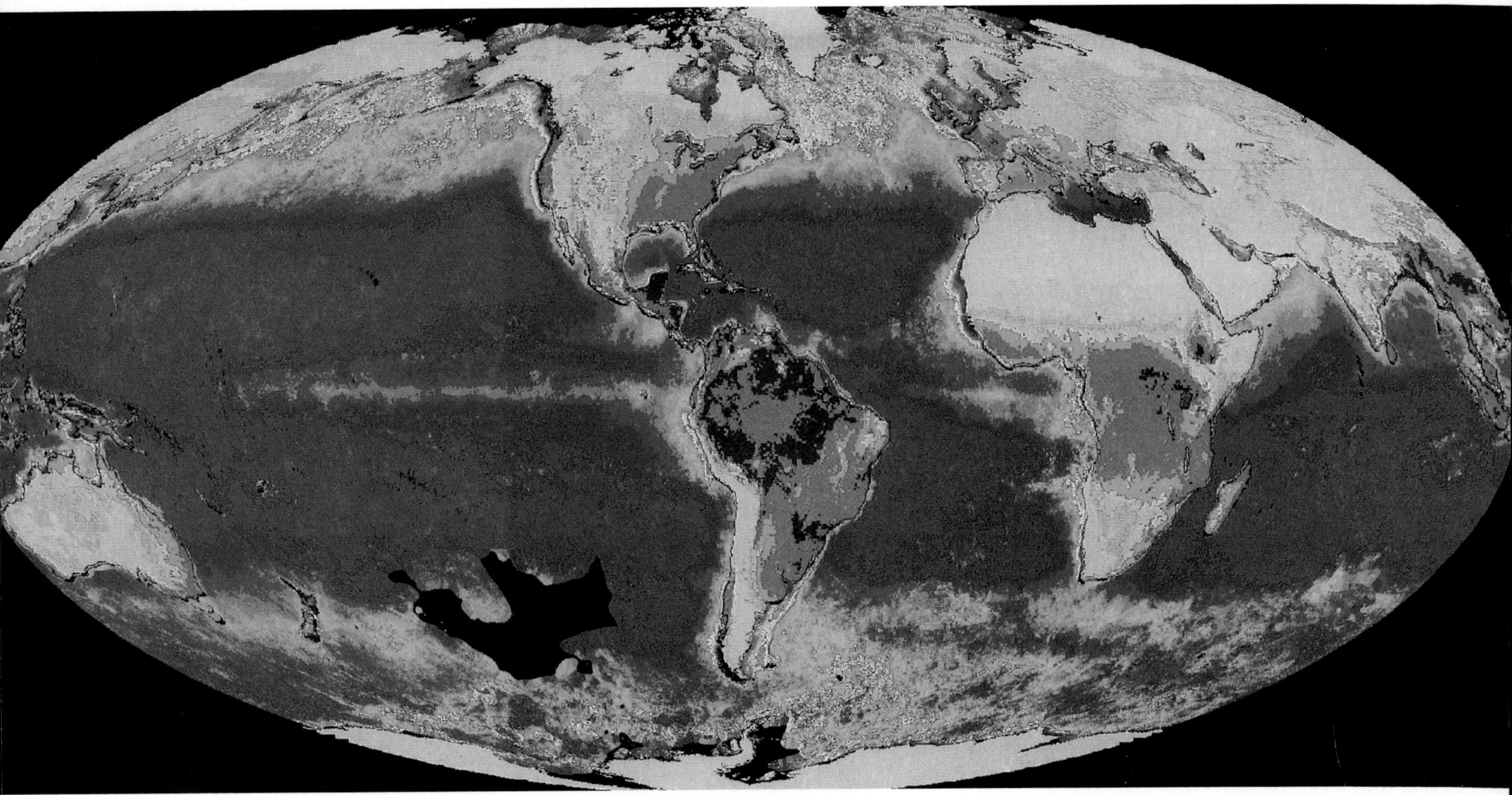

Figure 2.2
The first truly global view of the earth's biosphere is seen here. The composite image of the ocean chlorophyll concentration was produced from 31,352, scenes from November 1978, through June 1981, each of which corresponds to approximately 2 million square kilometers of ocean surface. Nearly 400 billion pieces of raw data located on more than 12,000, 9-track computer tapes were used to make this image. Land vegetation patterns are derived from 3 years of daily images from the NOAA-7's satellite's visible and near infrared sensors.

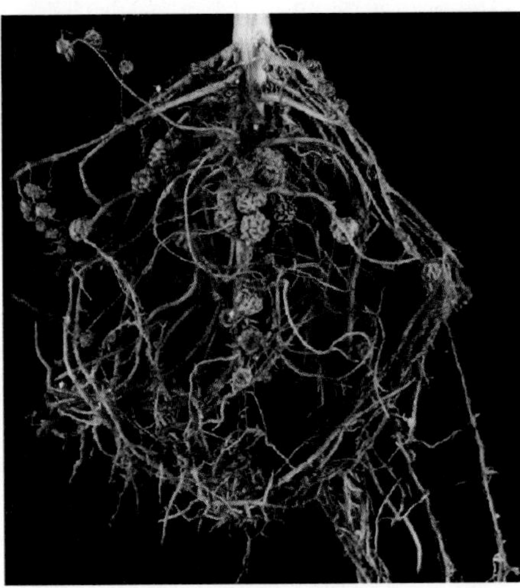

Figure 2.3
Bacteria in root nodules of plants chemically convert nitrogen gas into ammonia, which the plant can use to build some of the chemicals that it needs to survive. The bacteria, in turn, take sugars from the plant to use for energy.

Parasitic relationships, known as **parasitism,** are responsible for many human diseases. In this form of symbiosis, the parasite (usually small by comparison) derives benefit from the host, while harming but not usually killing it. Parasitic organisms cause such common problems as acne, athlete's foot, and strep throat, but they also cause such exotic illnesses as malaria and sleeping sickness. Sometimes a microorganism that is usually commensalistic becomes parasitic when the host's immune system is damaged and cannot regulate the size of the microorganism population. The result is an "opportunistic infection." Such infections are an early sign of AIDS.

The theme of life within life is striking in the termite, an insect that feeds on wood, often destroying wooden houses in the process. The true house wreckers, though, are the microorganisms that live in the termite's digestive system, for, without them, the termite could not digest wood. Not only are there protozoans (single-celled organisms) living symbiotically in the termite's gut, but an assortment of bacteria live symbiotically upon and within the protozoans. Together, these microorganisms break down cellulose, a large molecule in the wood, to produce smaller molecules that can be digested both by the microorganisms and by their termite host.

Life Affects Life

The earth has been, is, and (we hope) will continue to be home to far more species than we can catalog. One consequence of the rich parade of life on earth, or perhaps a factor contributing to it, is that organisms depend upon one another. All life is connected by complex chains and webs of "who eats whom." The importance of relationships between organisms becomes

Reading 2.1 *Fungus + Alga = Lichen*

One of the most fascinating biological living arrangements is that of certain fungi and green algae. The cohabitation of these representatives of two kingdoms of organisms produces a third type of organism, a **lichen**. The lichen often looks quite different from its constituents, suggesting that the fungus and the alga carry out some biochemical activities when they are associated that they do not carry out when they are apart. For example, some green algae produce certain types of alcohols only when they are part of a lichen. In the world of lichens, the whole is indeed more than the sum of its parts.

The algal constituent of a lichen contributes the ability to harness the sun's energy to manufacture various useful chemicals, the process of photosynthesis. Sometimes cyanobacteria are present, either as a stand-in for an alga or in addition to one. It is not known what, if anything, the alga gains from the fungus. Whatever the true nature of this symbiotic relationship, the lichen is a "plant pioneer" because it breaks down rock into biologically useful compounds. This is a first step on the road to soil formation, and, where there is soil, plants will soon follow.

Although uses for lichens are few today, the curious organisms have left their mark on history. The "manna lichen" *Lecanora esculenta* blows in the wind in the Near East, sometimes accumulating in drifts in sufficient quantities that it can be gathered and made into bread. In eastern Turkey, this lichen was called "wonder grain," and among the Kurds, "bread from heaven." Might the wind-borne lichen also have been the "manna from heaven" (Exodus 16) that sustained the starving Israelites in the desert in biblical times?

Lichens have been put to good use on this continent too. The Dakota Indians used *Usnea* lichen to dye porcupine quills, and the Wylackie tribe used it to tan leather. The Achomawi Indians soaked arrowheads in bright green "wolf moss" lichen for a year, rendering them quite lethal. The dark brown *Bryoria* lichen that resembles hair has been especially versatile. In western Canada, Indians baked it into a bitter-tasting black paste, which was eaten in times of famine. The Nez Perce Indians consumed the same lichen to treat stomach problems, and the Okanagan applied it to the navels of newborns to guard against infection. The Thompson Indians twisted strands of lichen together to fashion clothing.

Today, as in the past, lichens occupy a wide range of habitats. They commonly grow on trees and rocks, but they also exist in the minute crevices between the mineral grains in rocks, on the backs of beetles, in the driest deserts, and in the wettest tropical rain forests, and they spring up as if by magic in the damp soil of a fresh road cut. Lichens have even been found growing on stained-glass windows of cathedrals in Europe. Given lichen's environmental versatility, it is interesting to note one kind of habitat that is devoid of them—polluted areas. Lichens in such areas absorb toxic substances, such as sulfur dioxide from rainfall, but cannot excrete them. The buildup of toxins hampers photosynthesis, and the lichen dies. The disappearance of native lichens in an area is thus a sign that encroaching pollution is disturbing the environment. This is why lichens are seen growing on the bark of trees in the countryside but are not found on the trees in urban parks or along city streets.

a.

b.

Figure 1
Two shrubby lichens: (*a*) British soldiers *Cladonia cristatella;* and (*b*) reindeer moss *Cladonia subtenuis*.

apparent when disaster strikes. Consider what happens when, once or twice in a decade, a slack in the southeasterly trade winds over the ocean west of South America, an event known as El Nino, prevents the normal mixing of the nutrient-rich bottom layers of the ocean's water with the water above. Populations of microscopic organisms called phytoplankton decline sharply due to the lack of nutrients. As a result, anchovies, and other fish that normally feed on the phytoplankton, die of starvation. In the United States, farmers cannot obtain anchovy meal to feed their chickens and must turn to more expensive feed grains instead. Consumers in the United States, thousands of miles from El Nino, feel its effect in rising poultry prices.

KEY CONCEPTS

Diverse forms of life can be found all over the earth, even on or in other organisms in symbiotic relationships. In mutualism both types of organisms benefit; in commensalism one member benefits and the other is unaffected; in parasitism, one partner benefits at the other's expense. Different types of organisms interact with one another, especially in the complex chains and webs of "who eats whom."

Journey Through the Past

The great diversity of life on earth today is just the tip of an iceberg, the culmination of a parade of millions of types of organisms that no longer exist. Uncovering the links between past and present forms of life is one of the most fascinating aspects of the science of biology. Although obviously none of us was present to witness firsthand the evolution of life on earth, some tantalizing clues left behind from ancient organisms have helped modern-day evolutionary biologists paint a picture of life long ago on our planet. These glimpses may provide insight into our future as well.

On the very young earth, more than 4 billion years ago, the land was dotted with lava-belching volcanoes, bathed in an unearthly glow, and surrounded by bubbling pools of molten rock (fig. 2.4). The thin atmosphere, far different chemically from the atmosphere today, allowed high levels of ultraviolet light to reach the surface of the planet. Although life was not yet present, the complex chemicals that eventually gave rise to life were already combining in important new ways in the ocean depths.

Jump ahead to 3 billion years ago. The great geological disturbances of the past calmed. Nearly every part of the planet was inhabited by microscopic organisms, some of them long and thin, others spherical. This was the age of bacteria. The ability of these simple organisms to adapt to all parts of the ancient globe, still quite inhospitable to humans, allowed them not only to survive but to thrive almost anywhere, as they continue to do today.

Two billion years later (about 1 billion years ago), microorganisms still abounded, but some of them were more complex in organization, containing specific structures called **organelles** that carry out specific functions. These new, relatively complex microorganisms were the **eukaryotic cells.** Some of them aggregated to form colonies, which were the forerunners, perhaps, of the soft-bodied multicellular organisms that soon drifted near the sea's bottom. This early spurt of biological diversification signaled the explosion of new life forms soon to come.

A mere half a billion years later (500 million years ago), profound biological changes were apparent in the seas. Distant ancestors of modern-day clams cluttered the ocean bottom, providing footholds for the colorful, flowerlike crinoids (actually animals related to the sea stars) that winded their way upwards in the midst of lacy corals. Giant forerunners of modern squids drifted by. Other residents of the deep, the trilobites, looked a bit like stingrays when stretched out and like pill bugs when rolled up. They ranged from pea-sized to pumpkin-sized and were the ancestors of insects. This was the biggest explosion of biological diversity that the earth had yet experienced. It would continue for about 10 million years, introducing every basic animal body form present today. However, life was still confined to the seas (table 2.1).

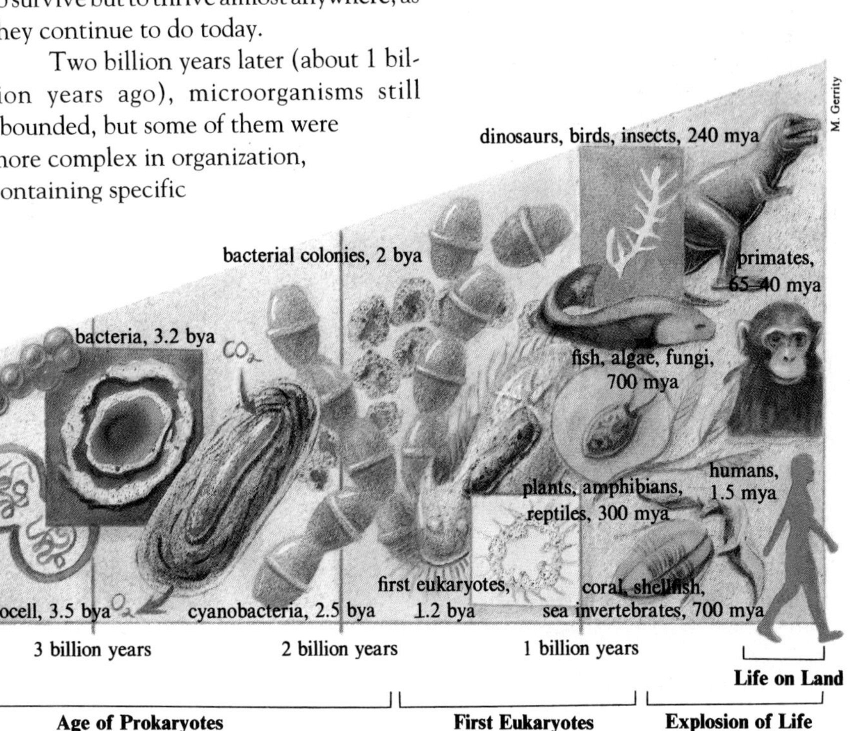

Figure 2.4
A panorama of life on earth over time.

Table 2.1
A Time Frame for Life on Earth
To put into perspective the vast lengths of time it has taken life to evolve into present forms, imagine that the entire history of life on earth is compressed into one week.

Day	Events
Sunday	
Morning	Earth forms First atmosphere and oceans, volcanoes, lightning
Evening	First organic (carbon-containing) compounds
Monday	
Morning	First prokaryotic microorganisms
Evening	Colonies of prokaryotes
Tuesday	Cyanobacteria Atmosphere changes (oxygen added)
Wednesday	Bacteria flourish and diversify
Thursday	First eukaryotic cells
Friday	First invertebrates (jellyfish, sponges, worms)
Saturday	Shellfish, fish, land plants, insects, forests, amphibians, dinosaurs, first mammals, birds, primates
Four seconds before midnight	Humans

By 300 million years ago, life had appeared on land. From space, the earth at this time would have appeared to have just one large, green landmass surrounded by a vast ocean. The waters were still brimming with life, and the land was packed with swamps ringed by lush forests of ferns and primitive trees. The air, which was warm year-round, was alive with the sounds of dragonflies, grasshoppers, and crickets, all giant versions of their present-day relatives. Various organisms squirmed through the slimy ooze. Many of these early earth inhabitants left parts of themselves behind as calling cards of their existence in the form of coal or sediments.

By 240 million years ago, a biological revolution was evident. Some organisms were able to live and breed solely on the land, a feat made possible by the development of watertight skin and eggs. In the continuing pattern of evolution, descendants gradually accumulated new traits. The individuals better adapted to a particular environment were more likely to survive to produce offspring, ensuring that their life-saving traits persisted in succeeding generations.

Gradually, the age of the dinosaurs, or "terrible lizards," dawned. Again, diversity was apparent. Some huge dinosaurs lived in the water, which helped to support their massive bodies. Others lived on land, browsing and grazing on the tough fronds of ferns that abounded in the forests. Some of these gentle giants fell prey to smaller meat-eating dinosaurs, like the familiar *Tyrannosaurus*.

Yet not all of the living things of this time were on the same scale as the giant dinosaurs. A different world existed on the forest floor. Insects scurried to avoid becoming the next meal of small, furry animals. These hairy newcomers, the earliest mammals, spent 135 million years underfoot of the reptilian monsters. The age of mammals was just around the corner. The skies, too, were populated with winged insects and the ancestors of birds.

By 180 million years ago, the living things on earth presented quite a contrast to those of the dinosaur age. The giant reptiles vanished, the temperature plummeted, and the small, furry insect-eating mammals were everywhere. By 30 million years ago, the great plains and forests were full of life. The treetops were home to the ancestors of modern monkeys, agile and nimble-fingered as they swung from branch to branch gathering food. Occasionally they encountered great, slumbering sloths that hung upside down from large branches. Beneath and all about the branches, insects buzzed about, and above, the sky became increasingly populated with birds and yet more insects.

As modern time approached, the trees still harbored their graceful primate occupants. However, some of these treetop residents ventured onto the grassy plains. They walked on two legs in a slightly hunched-over stance, and they had less hair than their cousins in the treetops. They are your ancestors.

KEY CONCEPTS

The organisms living today are only a small fraction of all those that have lived on earth. From the complex chemicals on the inhospitable earth of 4 billion years ago, the parade of life evolved. First came simple prokaryotic cells, then more complex eukaryotic cells, then simple soft-bodied multicellular organisms confined to the seas. Plants invaded the land, insects the skies, and gradually amphibians, reptiles, birds, and finally mammals flourished and diversified.

The Science of Biological Classification—Taxonomy

The planet earth, past and present, has been home to an impressive array of different organisms. How do biologists catalog the various kinds of living things? As with any large collection, subgroups can be arranged according to certain criteria. The problem of classifying life can be compared to that of constructing an efficient, useful list of foods. Arranging the foods in different groups could serve different purposes.

Suppose you were asked to organize the following foods into meaningful groups: bran cereal, bread, coffee, cake, donuts, lettuce, milk, orange juice, potatoes, steak, tomatoes, and tuna. You might divide the foods into those typically eaten at breakfast, lunch, or dinner. Such a classification might be useful to a restaurant manager.

Or, the foods could be grouped by biological source—plant or animal. This information could be valuable to a vegetarian. Separating them into main-meal foods and snack foods might help a dieter, and classifying them as solid or liquid might be useful to someone who has just had a tooth extracted. Of course, organizing the foods would be more complicated if each of them had synonyms. You would have to recognize "spuds" as potatoes or "chicken of the sea" as tuna. Indeed, organizing these dozen items into a system that would be understandable and useful to all could be very difficult.

Now imagine that you were given 500 million items to classify. How would you arrange them, and name them, in a meaningful manner? This is precisely the problem in sorting out the huge number of different organisms on earth. Were you to include the additional millions that have become extinct, the task of classification would become enormously complex.

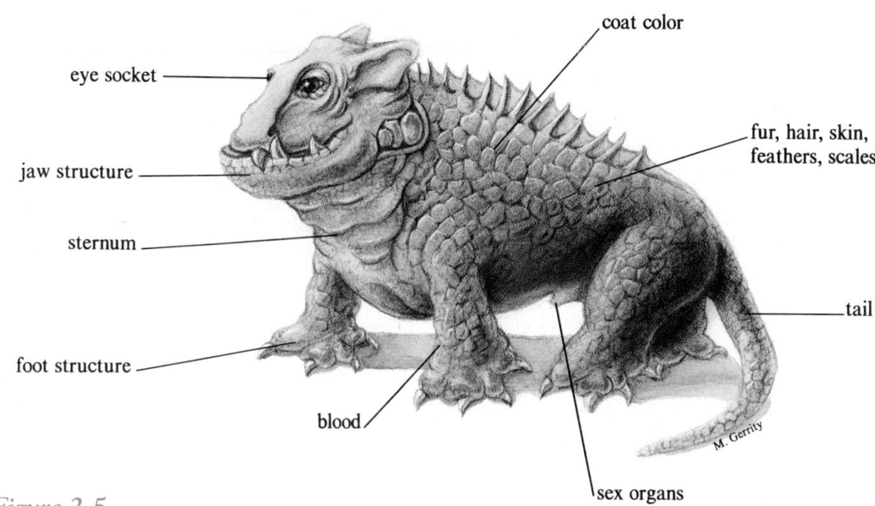

Figure 2.5
What is it? Taxonomists examine a variety of characteristics to identify and classify an organism, from the more obvious traits shown here to molecular sequences and embryonic forms.

Characteristics to Consider

Can life forms be grouped in a way that gives biologists information about the organisms living today as well as the ancestors that gave rise to them? The branch of biology dealing with classification, called **taxonomy,** attempts to do this by assigning a different name to each kind of organism. A taxonomic classification is based upon a set of distinguishing characteristics that reflects an organism's evolutionary descent from an ancestor. A taxonomic "name," then, provides information about an organism's evolutionary background.

Criteria for classification include structural, biochemical, and behavioral similarities as well as habitat preferences. To a taxonomist, the mating behavior of an organism, the way in which it obtains energy, and even the shape of its eye sockets are packed with information about its place in the system of classification (fig. 2.5). Each kind of organism described by a taxonomic name is a **species,** which may be defined as a group of similar organisms that interbreed in nature and are reproductively isolated from all other such groups. This definition of species works very well for animals, but it is not perfectly applicable to all organisms. Many organisms do not fit the definition because they reproduce asexually. Many single-celled organisms, for example, reproduce simply by dividing in two. Despite these apparent exceptions, the term "species" is useful, and it is applied even to those organisms that do not reproduce sexually.

Comparing similar characteristics among different species can be misleading, because it can be difficult to tell whether those species inherited the characteristic from a common ancestor or if they developed the trait independently in response to a shared environmental challenge. Does the similarity between the eyes of a human and those of an octopus reflect a shared relative back in the distant past? In this case, it is most likely that each species' eyes represent an independent response to a requirement for light-sensing organs, a common adaptation to a common problem.

One way to tell if similar structures reflect shared ancestors is to compare the sequences of chemical building blocks of biological molecules in different species. A new breed of taxonomist has evolved to do this: molecular taxonomists have given up the scattered skulls, teeth, and stuffed animals of traditional taxonomy for machines that sequence the long molecules of life.

Proteins are long molecules built of units (called **amino acids**) arranged in a specific sequence. This fact makes proteins ideal for the molecular approach to constructing biological family trees. The sequence of amino acids in a given protein molecule, for example, may differ slightly from one species to another. The more similar those sequences are between two species, the more recent was those species' divergence from a common ancestor. Because genes control protein production, changes in the sequence of amino acids in a protein reflect changes in the organism's genetic material. Because genetic changes occur very rarely, their frequency can be correlated to the passage of time. Thus, the fewer differences between a given protein's sequence in two species, the fewer genetic changes have occurred, which means less time has passed since the species diverged from a common ancestor.

In humans, for example, the amino acid sequence of hemoglobin, the blood's oxygen-carrying molecule, is identical to that of chimpanzees, and it differs from that of gorillas by only 2 amino acids out of about 300. This suggests that our closest relative is the chimp, followed by the gorilla. In another example, the sequence of amino acids in the egg-white protein albumin was compared between 2,000 pairs of bird species. The degrees of difference were then translated into "tree" diagrams, which made it possible to estimate the times at which those species diverged. The relationships and time estimates that result from comparing such sequences are called a "molecular clock."

Taxonomists using more traditional methods consider many characteristics in determining the relatedness of, say, humans and apes (table 2.2). Among the structural characteristics, for example, are the rates at which various bones harden in the young organism and the shapes and patterns of the teeth. Sexual similarities and differences are noted, including the relative distance between the mammary glands, the gestation period, the duration and frequency of intercourse, and the structure of the genitals. Sometimes new species are found in unlikely places (fig. 2.6).

a.

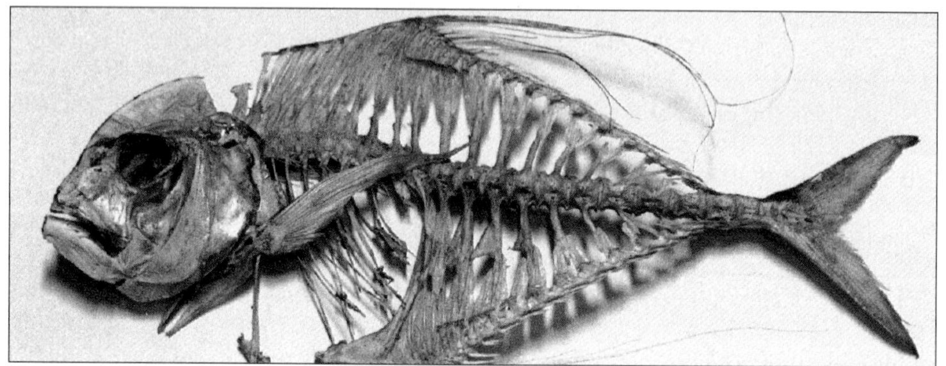

b.

Figure 2.6
a. For a select few of the half-million fish displayed in the Fulton Fish Market each day, the final destination is not someone's dinner plate but a display case at the American Museum of Natural History, a short ride uptown. The fish come to the market from the Gulf Coast, the West Coast, South America, Africa, the North Atlantic, the Caribbean, the Mediterranean, and the Pacific. Among the familiar and perhaps not-so-familiar animals, a sharp-eyed ichthyologist can spot a rare or even unknown species. The language, however, can be confusing. What a taxonomist recognizes as the red snapper (*Lutjanus campechanus*), for example, is a red porgy to the fish wholesaler, a pink snapper to a restaurant owner, and a white snapper to the person who finally eats it. *b.* The skeleton of an African pompany (*Alectix ciliaris*), caught by commercial fishermen.

Problems in Taxonomy

Taxonomy is a subjective science, and disagreement sometimes arises over classification schemes. Investigators who perceive many differences among related kinds of organisms and thus assign them to many taxonomic groups are termed "splitters," and those who see few important differences and assign them to fewer groups are called "lumpers." In the food-classification example, a splitter might classify the dozen items into four groups: breakfast (bran cereal, milk, and orange juice), lunch (tuna, bread, lettuce, and tomato), dinner (potato and steak), and snacks (coffee, cake, and donut). A lumper might see two groups (plant-derived and animal-derived foods) or perhaps only one (foods). Similarly, a lumper might classify all species of finches in a single group, whereas a splitter might discern many subgroups among them.

Another problem in biological classification is whether or not all characteristics should be given equal weight in determining relatedness between species. Even more generally, which characteristics should taxonomists consider at all? The problem of choosing those characteristics that reflect descent from a common ancestor is encountered in deciding which of the apes is most closely related to humans. If similarities in gene and protein sequences are the only criteria evaluated, our closest relative seems to be the chimpanzee. If the pattern of bone hardening is taken as the sole criterion for classification, we seem to be equally related to chimpanzees, gorillas, orangutans, and gibbons. But, an analysis of sexual structures and behaviors places us closest to the orangutans.

The molecular approach to taxonomy is regarded by many biologists as more

Table 2.2
Some Characteristics That Have Been Used to Classify Primates

Characteristic	*Human*	*Orangutan*	*Chimpanzee*	*Gorilla*	*Gibbon*
Short snout	Yes	Yes	Yes	Yes	Yes
Trapezoid-shaped nostrils	Yes	Yes	Yes	Yes	No
Thick molar enamel	Yes	Yes	No	No	No
Long hair	Yes	Yes	No	No	No
Wide-set nipples	Yes	Yes	No	No	No
Intercourse throughout menstrual cycle	Yes	Yes	No	No	No
Knuckle-walking	No	No	Yes	Yes	No

precise than comparing structural characteristics. This is because sequences of long molecules generally do not vary consistently within species (as structural characteristics often do), but they do vary consistently between species. Classification according to structural characteristics can also be confounded by organisms that look different at different stages of their life cycles. A caterpillar and the butterfly it will become certainly do not look like members of the same species, let alone like the same individual. Just recently, two protozoans that had been classified as different species, based partly upon differences in structure and partly upon the fact that they parasitize different organisms, were found to be the same species. The organism that had been called *Myxosporea* causes whirling disease in trout, in which the fish chase their own tails. The organism that had been called *Actinosporea* parasitizes the gut of a worm. However, it is now known that these two organisms represent different phases of the life cycle of a single species. The young inhabit the worms and are released to infect trout. When the trout die of whirling disease, immature parasites of the next generation are released to seek the worm host.

The Linnaean System of Biological Classification

The modern system of classifying life was introduced in the eighteenth century by the Swedish botanist Carolus Linnaeus. In this taxonomic scheme, a series of organizational levels are used to describe the ancestry and characteristics of organisms. These levels are increasingly restrictive as the most stringent one, species, is approached (table 2.3). The levels proceed from the most general, *kingdom*, to *division* (in plants) or *phylum* (in all other organisms), to *class*, *order*, *family*, *genus*, and finally *species*. Further taxonomic subdivisions are sometimes added to describe organisms more fully; the prefix "super" extends the limits of the designation, and the prefix "sub" restricts the limits. Commonly used supplementary designations include subphylum, superclass and subclass, superorder and suborder, superfamily and subfamily, subgenus, and subspecies. A helpful way to envision Linnaean classification is as a tree, with a kingdom as the trunk, phyla (or divisions) as limbs, classes as branches, and orders, families, genera, and species as finer and finer twigs.

Because stating an organism's full taxonomic position is rather cumbersome, scientists usually abbreviate it to just the genus and species. Thus, the common grass frog is called *Rana pipiens*, instead of Animalia Chordata Amphibia Salientia Ranidae *Rana pipiens*. The use of a double name is important, because neither the name of the genus nor the species alone furnishes enough information, just as neither "Ben" nor "Franklin" is specific enough to identify a particular person, but "Ben Franklin" is. The more closely related two species are, the more taxonomic levels, beginning with kingdom, they share. Table 2.4 compares humans and three related species.

Table 2.3
The Taxonomic Groups to Which Humans Belong

Category	*Name*	*Description*
Kingdom	Animalia	Complex cells; multicellular; nervous system
Phylum	Chordata	Body consisting of head, trunk, and tail; highly developed organ systems; three-layered embryo; cavity inside body
Class	Mammalia	Hair; mammary glands; internal fertilization; large skull; warm-blooded
Order	Primates	Complex brain; flexible toes and fingers; excellent vision
Family	Hominidae	Upright posture; small face; large brain
Genus	*Homo*	Large brain; relatively short arms; lightweight jaws; small teeth; large thumbs
Species	*Homo sapiens*	Only living species of genus *Homo*

Note: From kingdom to species, taxonomic designations become more precise, with more stringent requirements.

Table 2.4
Biological Classifications of Four Animals

Category	*Human*	*Gorilla*	*Squirrel*	*Katydid*
Kingdom	Animalia	Animalia	Animalia	Animalia
Phylum	Chordata	Chordata	Chordata	Arthropoda
Class	Mammalia	Mammalia	Mammalia	Insecta
Order	Primates	Primates	Rodentia	Orthoptera
Family	Hominidae	Pongidae	Sciuridae	Tettigoniidae
Species	*Homo sapiens*	*Gorilla gorilla*	*Sciurus carolinensis*	*Scudderia furcata*

KEY CONCEPTS

The biological science of taxonomy classifies organisms to reflect probable descent from common ancestors and therefore evolutionary relationships. A taxonomic name denotes a species, which is a group of organisms that are reproductively isolated from all others. Criteria for taxonomic classification include structural, behavioral, and biochemical similarities. The more similar the sequences of certain proteins and genes in two types of organisms, the more closely related they are. The accuracy of other criteria for taxonomic classification is limited by human subjectivity.

A Look at the Kingdoms

Different taxonomic schemes have been utilized since Linnaeus's time, and even today, taxonomists disagree on the number of kingdoms and on which types of organisms belong in which kingdom. Earlier this century, only two kingdoms were recognized—plants and animals. More recently, three-, four-, five-, and even six-kingdom classifications have been used. The members of each kingdom demonstrate a slightly different strategy in satisfying the same life requirements.

A basis for modern classification schemes is whether an organism's cells are of the structurally simple *prokaryotic* variety or the more complex *eukaryotic* type. **Eukaryotic cells** are distinguished by their membrane-bound **organelles,** which are the structures that specialize in certain subcellular functions, such as energy production and secretion. Such a cell takes the form of "bags within a bag." The genetic material of eukaryotic cells is enclosed in a membrane, and the resulting structure is called the **nucleus. Prokaryotic cells** lack organelles, and their genetic material is not enclosed in a membrane.

Modern taxonomists generally agree that the differences between prokaryotes and eukaryotes are greater than those between plants and animals. The two-kingdom plant-animal system did not allow this distinction based on cellular complexity. The five-kingdom system provides detailed descriptions, and we shall follow it here. The kingdoms are **Monera, Protista, Fungi, Plantae,** and **Animalia** (fig. 2.7). Organisms are assigned to their kingdoms on the basis of the complexity of their cells, whether they are unicellular or multicellular, and the ways in which they obtain nourishment.

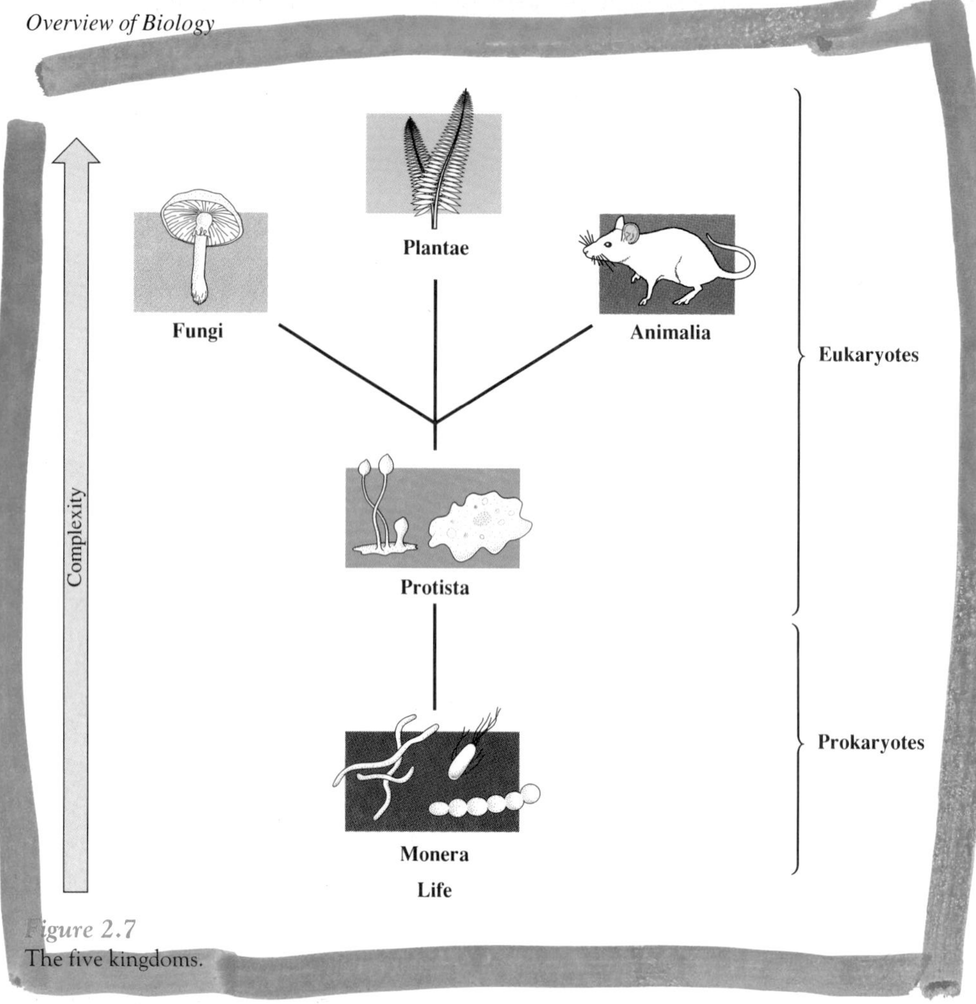

Figure 2.7
The five kingdoms.

Monera

The kingdom Monera consists of the **bacteria** and the **cyanobacteria.** Bacteria were probably among the first of the earth's inhabitants, and today they continue to flourish. Bacteria are single-celled prokaryotic organisms. They are a very successful biological group, occupying most parts of the planet, from deep sea thermal vents to the rocks of Antarctica. Bacteria can be found in your digestive tract, on your skin, in the linings of your nose and throat, and between your teeth. In very harsh environments, some bacteria can grow thick walls, condense their contents, and convert themselves into structures called *spores* that are resistant to extremes of temperature and moisture. Most bacteria require oxygen in the environment, but a few types do not.

Bacteria vary in their approaches to obtaining energy. **Autotrophic** bacteria convert inorganic compounds obtained from the environment into energy-rich organic compounds. Some autotrophic bacteria use the sun's energy to make organic compounds by a process quite similar to the photosynthetic process used by green plants. **Heterotrophic** bacteria obtain organic compounds from dead organisms or from the living. These bacteria then use enzymes (a type of protein) to break down these organic compounds to extract the energy held in the attractive forces (chemical bonds) that hold the compounds together. Using dead organic matter as a food source, heterotrophic bacteria function as decomposers, ultimately returning the chemicals in the organic matter to the environment. When they use a living organism as a food source, heterotrophic bacteria are parasites, often causing disease in the host organism.

Several criteria are used to classify the 5,000 or so recognized species of bacteria. Shape is one distinguishing characteristic. Rod-shaped bacteria are called *bacilli*, round-shaped bacteria are *cocci*, spiral-shaped ones are *spirilla*, and curved-shaped rods are *vibrios* (fig. 2.8). Bacteria are also categorized by certain chemical characteristics. For example, "gram-positive" bacteria have cell walls that stain purple in the presence of a stain developed by Danish physician Hans Christian Gram, whereas the cell walls of "gram-negative" bacteria become pale pink when the stain is applied. A bacterium's source of energy and whether it lives in an oxygenated (aerobic) or nonoxygenated (anaerobic) environment are also identifying characteristics in some classification schemes (fig. 2.9).

Bacteria reproduce by doubling their contents, then splitting in two. This type of asexual reproduction without the union of an egg and a sperm is termed **binary fission**—literally, "dividing in two." Some

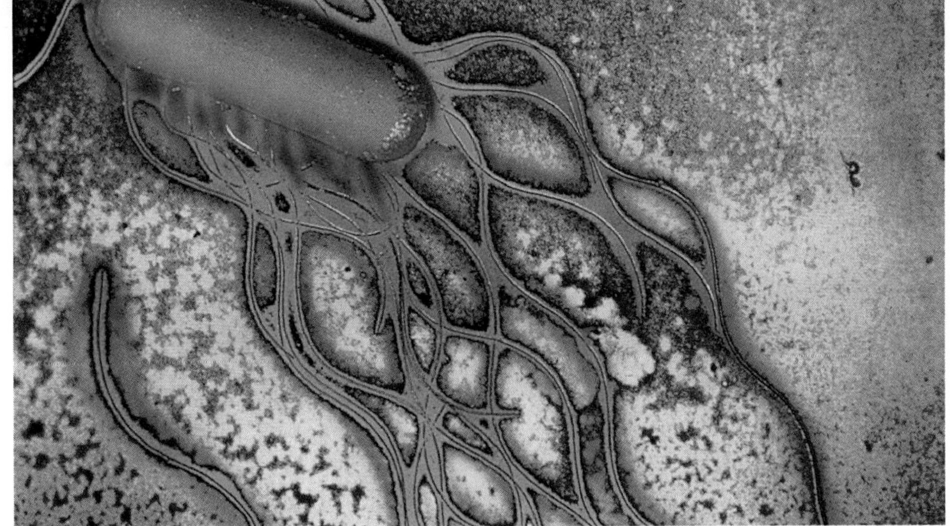

Figure 2.8
Salmonella poisoning. Each year in the United States rod-shaped bacteria called *Salmonella enteritidis* send 4 million people to the bathroom with days of unrelenting diarrhea—35,000 are hospitalized, 120,000 develop long-lasting joint pain, and 1,000 die.

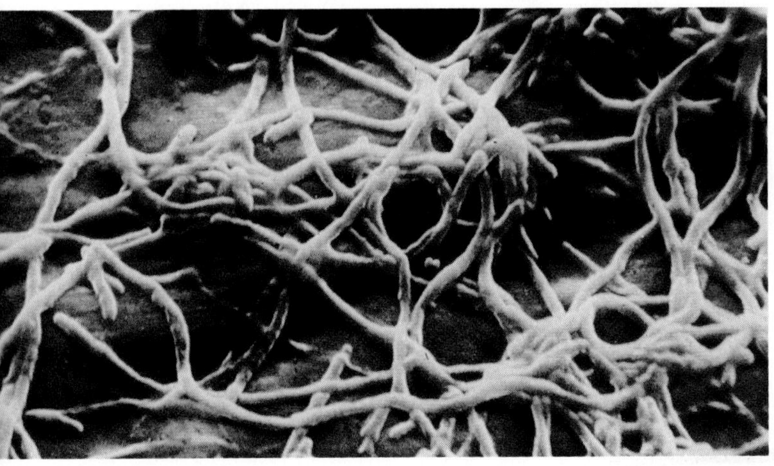

a.

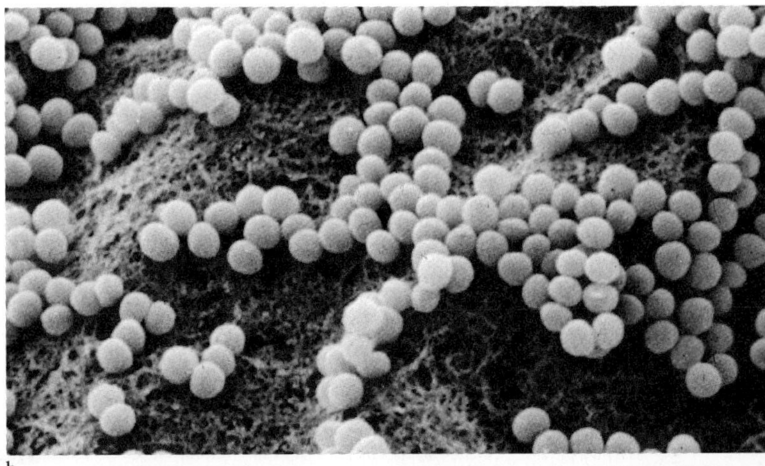

b.

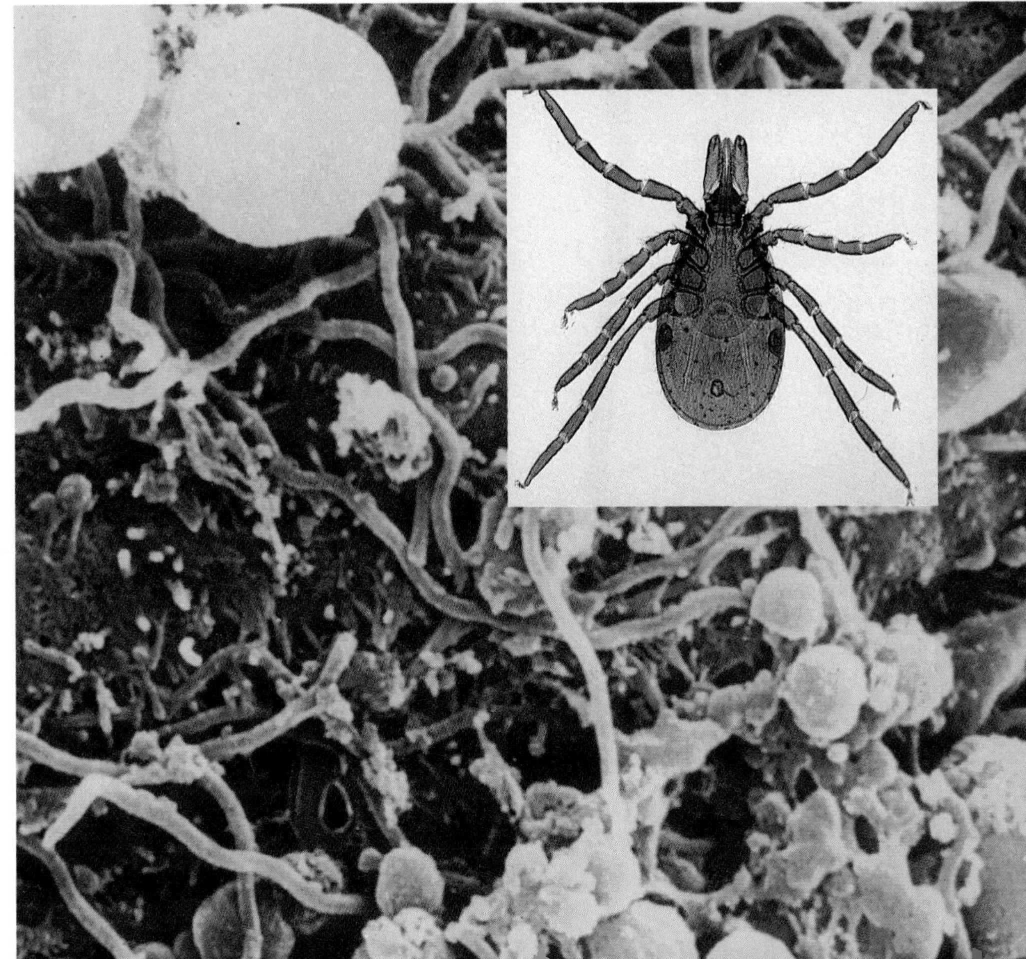

c.

Figure 2.9
A trio of "new" bacterial diseases.
a. Legionnaire's disease is a form of pneumonia caused by the bacterium *Legionella pneumophila.* Although this organism has probably been around for a very long time, it was not identified and associated with human illness until the 1970s, when it struck American Legion conventioneers in Philadelphia. *b.* Toxic shock syndrome first appeared in the early 1980s predominantly in women who used a certain brand of tampon. A material used in the manufacture of the tampons entrapped the common bacterium *Staphylococcus aureus,* provoking it to manufacture an apparently new toxic chemical, which caused severe symptoms (vomiting, diarrhea, plummeting blood pressure, and skin peeling) and was responsible for several deaths. *c.* Lyme disease produces a rash, aches and pains, and sometimes arthritis and nervous system complications. It is caused by a spiral-shaped bacterium called *Borrelia burgdorferi* and is transmitted to humans by a bite of the deer tick. Lyme disease was first noted in Lyme, Connecticut, in 1975, but in recent years it has affected thousands of people, particularly on the East Coast.

bacteria can also pass genetic material from one individual to another by means of a projection from the donor cell called a **pilus.** This transfer of genetic material is a type of sexual reproduction.

Human civilization has been influenced by bacteria in many ways. We have subjected these microscopic organisms to freezing, drying, salting, pressure cooking, high acidity, X rays, and chemical compounds galore in order to keep them from "spoiling" our food. Yet certain other "friendly" bacteria enable us to produce cheese and yogurt, and others are used in the production of antibiotic drugs and industrial chemicals.

Like the bacteria, the cyanobacteria are long-term earth residents. These one-celled organisms were once called blue-green algae because of their pigmentation and the fact that their internal chemistry is similar to that of certain eukaryotic organisms known as algae. Structurally, however, they are bacteria that can **photosynthesize** (use solar energy to produce nutrients)—hence the new name. Different species of cyanobacteria are distinguished by the particular patterns in which the individual cells aggregate.

Conspicuous growths of cyano- bacteria are most likely to be found in fresh water and moist soils. Cyanobacteria are famous for growing where nothing else will, as we saw at the beginning of this chapter. These colorful organisms even grow within the hairs of polar bears in zoos, turning their fur green. The cyanobacteria enter the hairs when the bears rub against the walls of their cages.

Cyanobacteria were pivotal in the course of life's history. About 2 billion years ago, as the cyanobacteria formed mats on pool bottoms much as they do today, they extracted hydrogen from water during photosynthesis. In the process, they released oxygen gas. Over many millions of years, this energy-obtaining strategy pumped more and more oxygen into the atmosphere. This great environmental change set the stage for oxygen-using life forms to follow, while at the same time ensuring that life would not begin again on earth in exactly the same way as it probably had the first time.

Protista

The kingdom Protista includes the simplest eukaryotic organisms: the protozoans, the algae, the water molds, the slime molds, and various other groups of relatively unfamiliar organisms. Most of the members of the 27 protistan phyla are unicellular, but some form many-celled colonies and some are truly multicellular (the brown algae, the red algae, and some of the green algae). This kingdom is believed to have given rise to the other three kingdoms of eukaryotes. Their structure, chemistry, and early developmental stages distinguish them from fungi, plants, and animals. Some protists are responsible for human diseases, including malaria, dysentery, and African sleeping sickness. Other protists are beautiful and fascinating to observe, such as the laboratory favorites the amoeba and the paramecium.

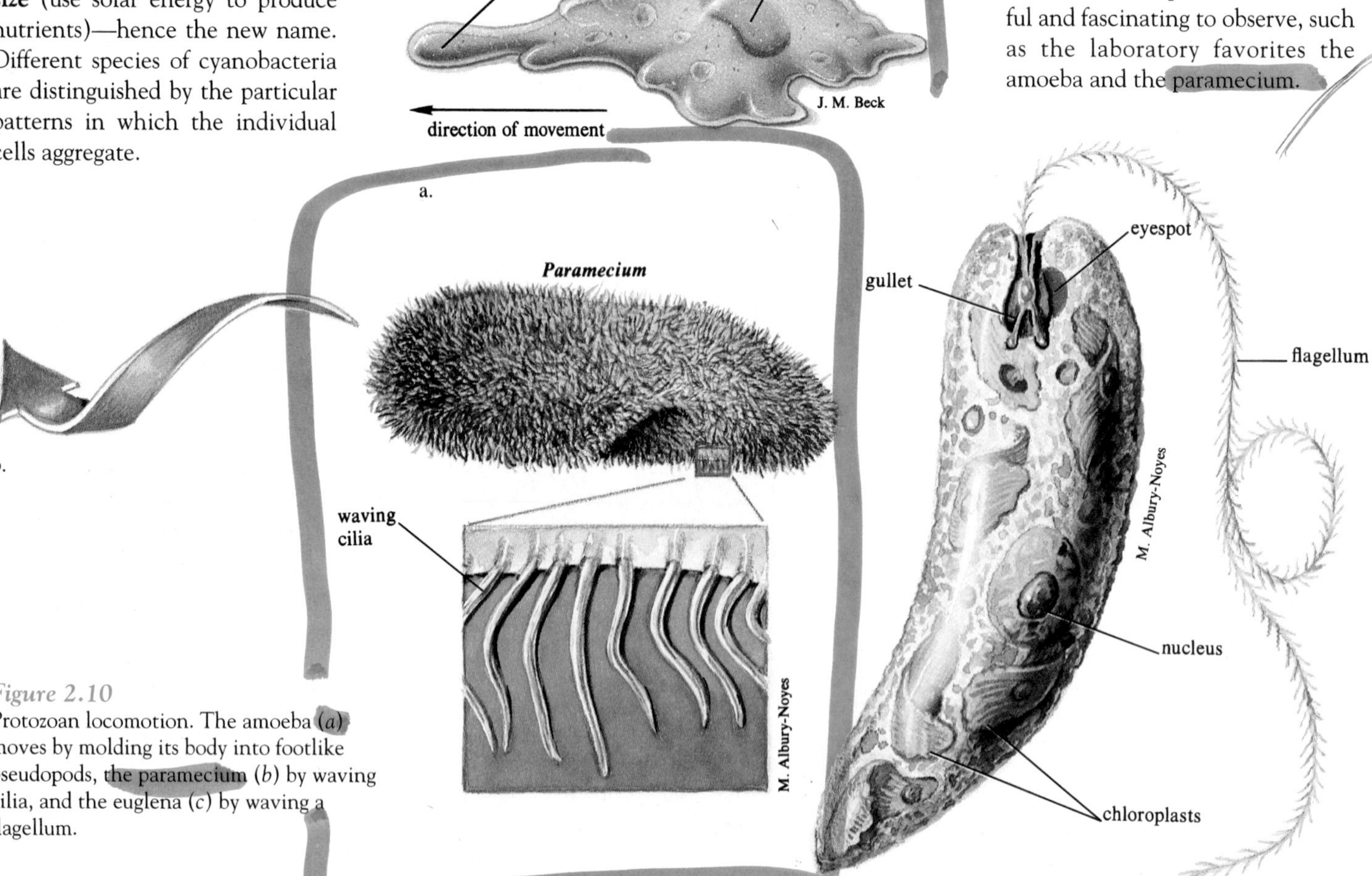

Figure 2.10
Protozoan locomotion. The amoeba (*a*) moves by molding its body into footlike pseudopods, the paramecium (*b*) by waving cilia, and the euglena (*c*) by waving a flagellum.

The **protozoans** make up seven phyla of the kingdom Protista. These "first animals" are single-celled, and most live in fresh or salt water or in very moist terrestrial habitats. Some protozoans live symbiotically with other organisms, and many are exquisite, resembling elaborate spaceships and Christmas tree ornaments. The first protozoans made their debut on earth about 1,200 million years ago. Today, about 64,000 species are recognized. They are sometimes classified by their method of movement. The *euglena* waves a taillike flagellum; the *amoeba* molds itself into footlike projections called pseudopods (fake feet); the *paramecium* glides by waving its hairlike cilia (fig. 2.10).

There are six phyla of **algae.** Like cyanobacteria and plants, the algae photosynthesize. The members of three phyla, the *euglenoids*, the *dinoflagellates*, and the *diatoms*, are single-celled. Of the other three phyla, two (the *red algae* and *brown algae*) consist entirely of multicellular species and the other (*green algae*) has both single-celled and multicellular members. In some classification schemes, the multicellular algae are included among the plants. Algae are classified by their color and internal chemistry. Familiar algae include the seaweeds belonging to the phylum of brown algae, of which there are about 1,500 species. Some of these seaweeds cling to rocky shores or skim the ocean's surface in huge, floating mats. Others live in shallow tidepools or at depths of 100 feet (about 30 meters). One, the fast-growing giant bladder kelp, is harvested off the coast of southern California's San Clemente Island. This algae is under study for use as food for humans and cattle, as fertilizer, and as a source of industrial chemicals (fig. 2.11).

Green algae of possible evolutionary note are members of the genus *Volvox*. Up to 50,000 single-celled individuals join to form a hollow ball that is just barely visible. The cells wave their flagella in a coordinated fashion, moving the entire sphere. Might an ancient *Volvox* colony have been a bridge of sorts between the single-celled and many-celled way of life? An opposing view is that *Volvox* is an "evolutionary dead end," because no other group of organisms seems to be directly descended from it. Today, the way that *Volvox* cells aggregate in the

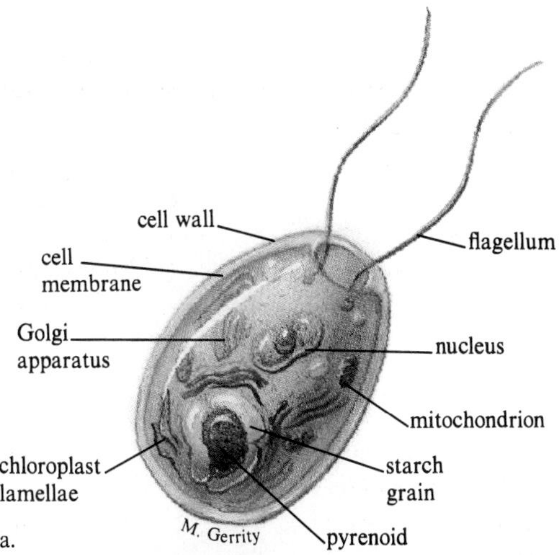

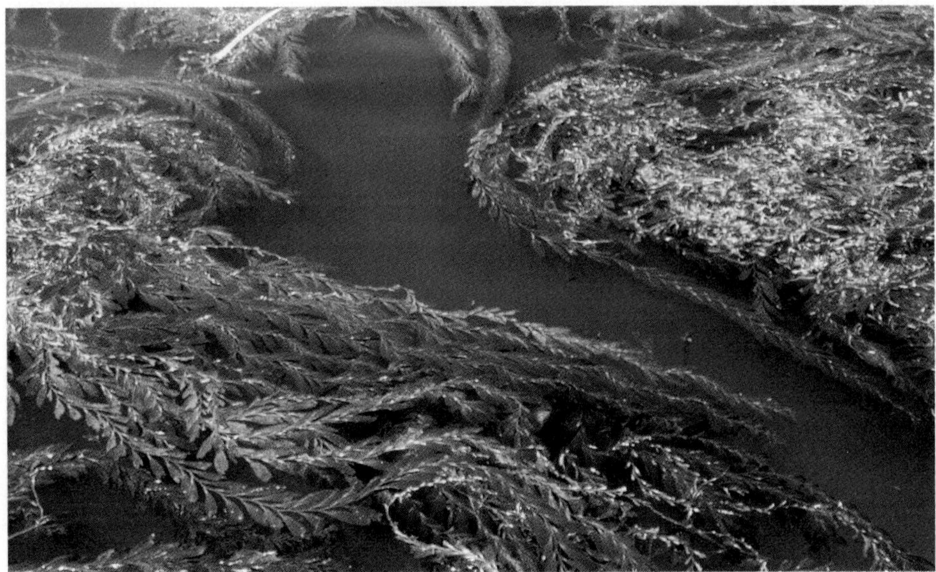

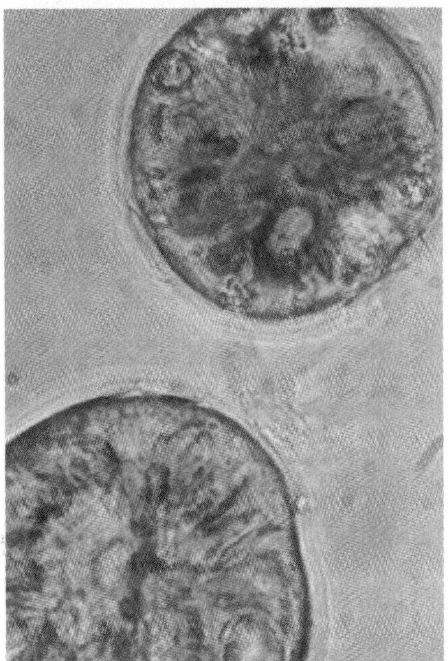

Figure 2.11
Algae. *a.* Green algae of the genus *Chlamydomonas* are single-celled, with a characteristic pair of flagella. *b.* Off the coast of southern California, a giant bladder kelp of the genus *Macrocystis* is harvested for use as livestock feed, fuel, and a source of industrial chemicals. *c. Gonyaulax tamarensis* is a type of alga called a dinoflagellate. The individual organisms are microscopic, but, when they are present in large numbers, these colorful dinoflagellates give the waters of their native Pacific Ocean a reddish color, called a "red tide." The dinoflagellates produce toxins that are eaten by mussels and other shellfish. Humans who eat the tainted shellfish develop nervous system poisoning that can lead to paralysis and death.

presence of toxic chemicals is being looked at as a way of testing for substances that are toxic to the human fetus.

Some protists, including the water molds, chytrids, and slime molds, are classified more for what they are not than for what they are. They do not photosynthesize, so they cannot be classed as algae or plants, but neither are they animals, protozoans, or fungi.

The *water molds* (phylum Oomycota) are found in fresh water or moist soil. Some of the 475 species of water molds are single-celled with two flagella, and others form long strands containing many nuclei. Water molds have cell walls built of cellulose, a biochemical found in plants. The best-known water molds are those that have thwarted agriculture. Perhaps the most notorious is *Phytophthora infestans*, which caused the potato famine in Ireland in which over a million people starved. Although the famine lasted from 1845 to 1847, the mold wiped out most of the potatoes in just a week (fig. 2.12).

The 750 species of *chytrids* (phylum Chytridiomycota) are quite varied. They are distinguished by their single flagellum and by cell walls built of *chitin*, a biochemical found also in the cell walls of fungi and the exoskeletons of insects. Some chytrids are parasites, feeding on water molds, algae, or plants.

The major difference between the two phyla of *slime molds* (Myxomycota and Acrasiomycota) is reflected in their popular names—acellular and cellular. Both organisms exist as single cells in one phase of the life cycle, but these cells aggregate into a mass that behaves as if it were a single organism. In the acellular slime molds, this mass (called a *plasmodium*) migrates in soil or along the forest floor consuming bacteria as it travels. The plasmodium contains many nuclei, but it is not divided into separate cells. In contrast, the individual cells of a cellular slime mold retain their separate identities when they collect into a mass (called a *pseudoplasmodium*). The slime molds are fascinating because they can switch body form in response to environmental changes (fig. 2.13). When the environment is moist and their bacterial food is plentiful, acellular slime molds swarm as huge plasmodia, and in times of drought or famine, they halt and form reproductive structures that ultimately release single cells, starting the cycle anew. Cellular slime molds feed mainly in the single-celled stage, aggregating into pseudoplasmodia only shortly before they form their reproductive structures.

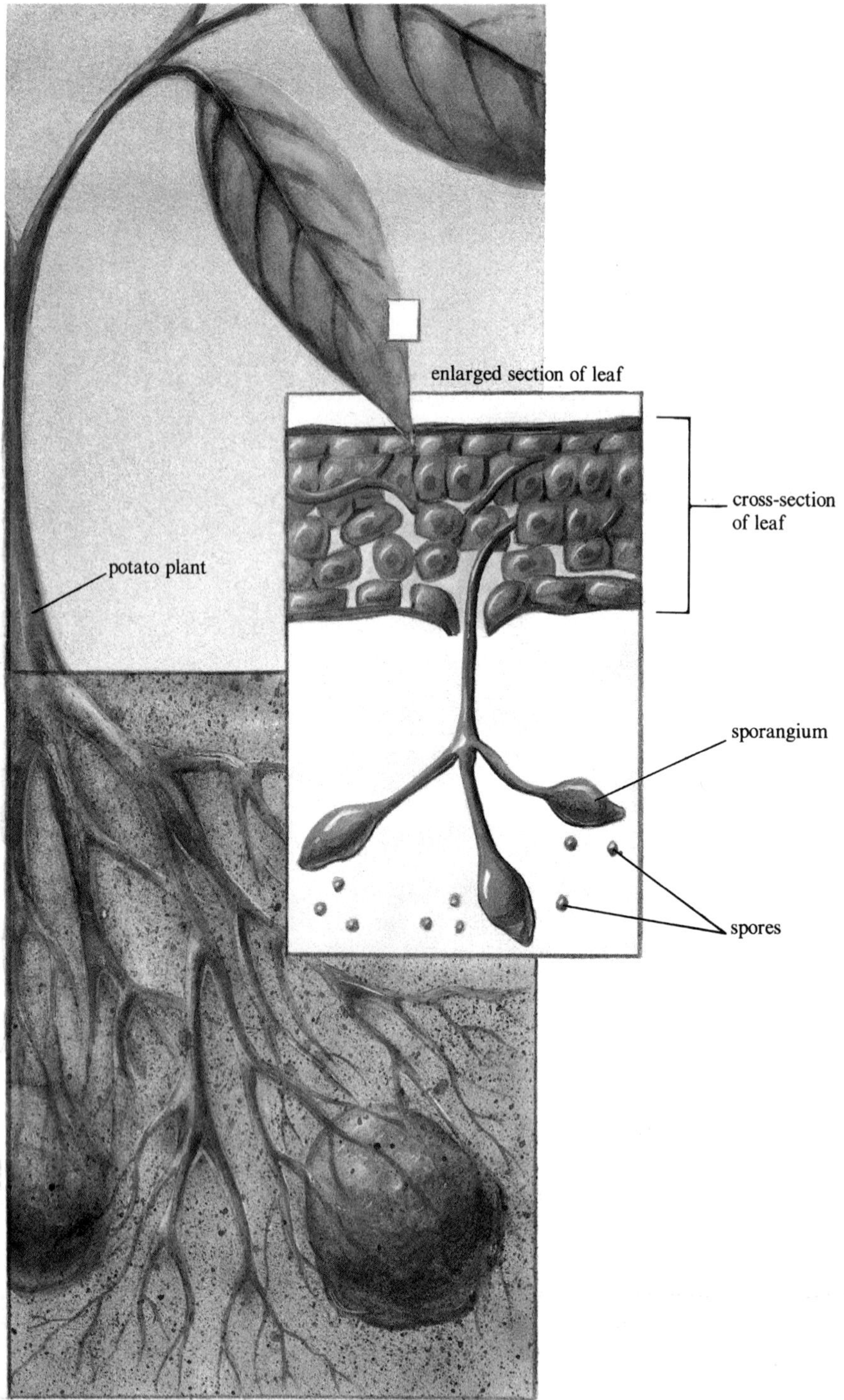

Figure 2.12
The water mold *Phytophthora infestans* entangles itself in potatoes. It caused the Irish potato famine of 1845 through 1847.

Fungi

The **fungi** have been an enigma to many taxonomists. Although fungi have been classified both with protists and with plants,

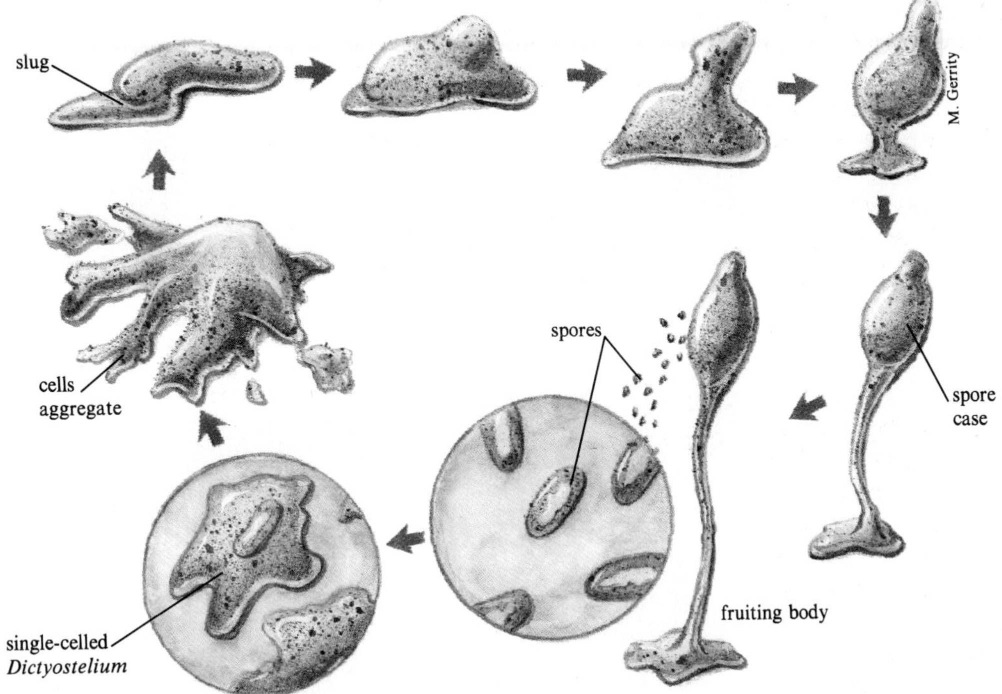

Figure 2.13
When food is lacking, the cellular slime mold *Dictyostelium discoideum* forms a multicellular slug and moves until it locates more of its bacterial food and conditions become moister. Then the mass forms a structure called a fruiting body, from which single-celled *Dictyostelium* form.

their unusual characteristics have prompted most taxonomists to place them in a separate kingdom, Fungi.

Most fungi are land-dwelling organisms, but a few forms are marine. They obtain energy by breaking down organic matter and absorbing small molecules. Some fungi feed on dead matter, and others are parasites. The fungi are important to life in general because they are decomposers, recycling the chemicals in once-living matter to the environment, where they can be used by other organisms. Many fungi are particularly helpful to plants when they form symbiotic relationships with their roots. The fungus-infested roots, called *mycorrhizae,* assist the plant in obtaining minerals from the soil.

Fungal cells are eukaryotic, with cell walls built of chitin. They are generally multicellular, with the exception of the *yeasts*, which are unicellular. Fungi do not have nervous systems and they do not photosynthesize. Except for the yeasts, the basic structural unit of a fungus is a threadlike filament called a **hypha.** An individual fungus consists of an assemblage of hyphae called a **mycelium.** Mushrooms, which are reproductive structures of certain fungi, consist of many tightly packed hyphae (fig. 2.14).

The three phyla of fungi are distinguished by their reproductive structures (fig. 2.15). The 600 species of **zygomycetes** (phylum Zygomycota) have sexual spores called *zygospores.* Most zygomycetes feed on decaying plant and animal matter, but some are parasitic. The black mold *Rhizopus stolonifer* that often grows on bread is a familiar zygomycete.

The phylum Ascomycota (the **ascomycetes**) includes many familiar species among its 30,000 members. These fungi have saclike sexual reproductive structures called *asci*. Many of the colorful growths on spoiled food are ascomycetes, as are the fungi responsible for Dutch elm disease, powdery mildew of plants, athlete's foot, ringworm, and the mouth infection called thrush. The ascomycete *Claviceps purpurea* parasitizes rye and other grasses, causing a disease called ergot. People who eat bread made from infected grain develop a serious condition known as ergotism, which causes convulsions, gangrene, and psychotic delusions. An extract of the fungus used in recent times as a recreational drug is lysergic acid diethylamide, also known as LSD.

A useful ascomycete is the yeast *Saccharomyces cerevisiae*. Its ability to ferment carbohydrates chemically to produce alcohol and carbon dioxide has been of great value to the brewing and baking industries. Yeast is used to make wine, cider, beer, and bread. Morels and truffles, considered to be great delicacies, are also ascomycetes. *Penicillium roquefortii* and *Penicillium camembertii* are used to ripen the cheeses named after them. Certain other species of the genus *Penicillium* produce the well-known antibiotics called the penicillins. Other ascomycetes are used in the manufacture of soy sauce and tofu (bean curd).

Figure 2.14
Mushrooms are the reproductive portions of fungi, specialized to produce spores, from which new fungal bodies, or mycelia, grow. Very valuable for its unique taste is the morel mushroom, which is found under dying elm trees, among other places.

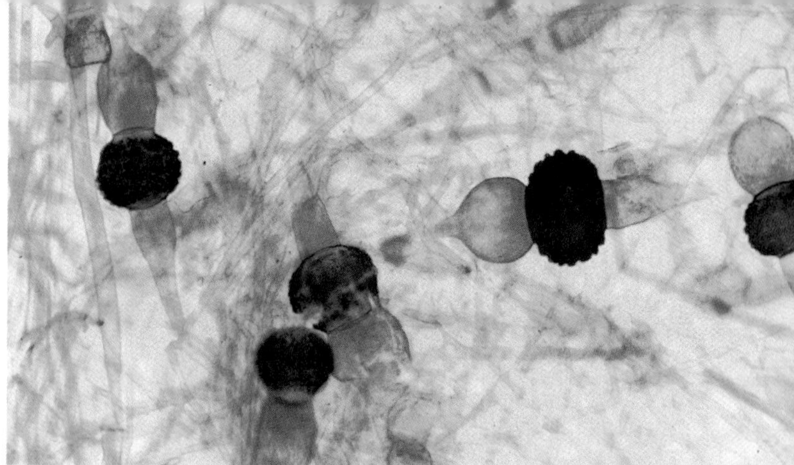

a.

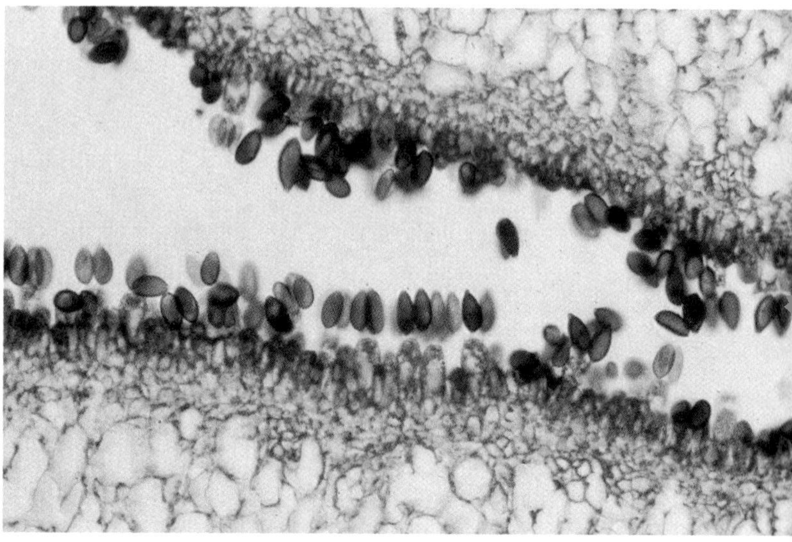

b.

Figure 2.15
Fungi are classified by reproductive structures. *a.* The zygomycetes are distinguished by sexual resting spores as in this bread mold. *b.* The basidiomycetes have spore-bearing structures called basidia as seen in this slice through the cap of a common mushroom. *c.* The ascomycetes carry sexual spores in saclike asci.

The third phylum of fungi, the **basidiomycetes** (phylum Basidiomycota), includes 25,000 species. Familiar members are the mushrooms, toadstools, puffballs, stinkhorns, shelf fungi, rusts, and smuts. They have characteristic spore-containing structures called *basidia*. Some mushrooms are edible, some are deadly, and others are hallucinogenic. The Aztecs regarded the mind-altering mushroom *Psilocybe mexicana* as sacred. Animals have been known to become intoxicated from eating hallucinogenic mushrooms.

Some fungi (mainly ascomycetes) can combine with a representative of another kingdom, usually a green alga, to form a third type of organism, a *lichen* (Reading 2.1).

Plantae

The kingdom Plantae consists of the plants—land-dwelling, multicellular organisms that extract energy from sunlight by using the pigment **chlorophyll** contained in their cells. The cells of a plant are eukaryotic, and they are enclosed in cell walls built of cellulose. Plant cells are assembled into specialized tissues and organs. A plant cannot move about as an animal can, but it can bend and grow under the influence of chemicals called hormones.

The life cycle of a plant is divided into two reproductive phases. In the **gametophyte** stage, the plant produces sex cells, known as **gametes.** These are two types: sperm cells and egg cells. A gamete must combine with a gamete of the opposite type to form a new individual, known as the sporophyte. In the **sporophyte** phase, a plant produces spores that can develop into a new individual known as the gametophyte, without fusing with another cell. This dual reproductive strategy of gamete-producing and spore-producing phases is termed **alternation of generations.** Members of different plant divisions spend characteristic proportions of their life cycles in each phase. In the more primitive plants, such as the mosses, the gametophyte phase is dominant. In the flowering plants, the sporophyte phase is dominant.

Adaptations of plants to life on land suggest that their ancestors evolved in the

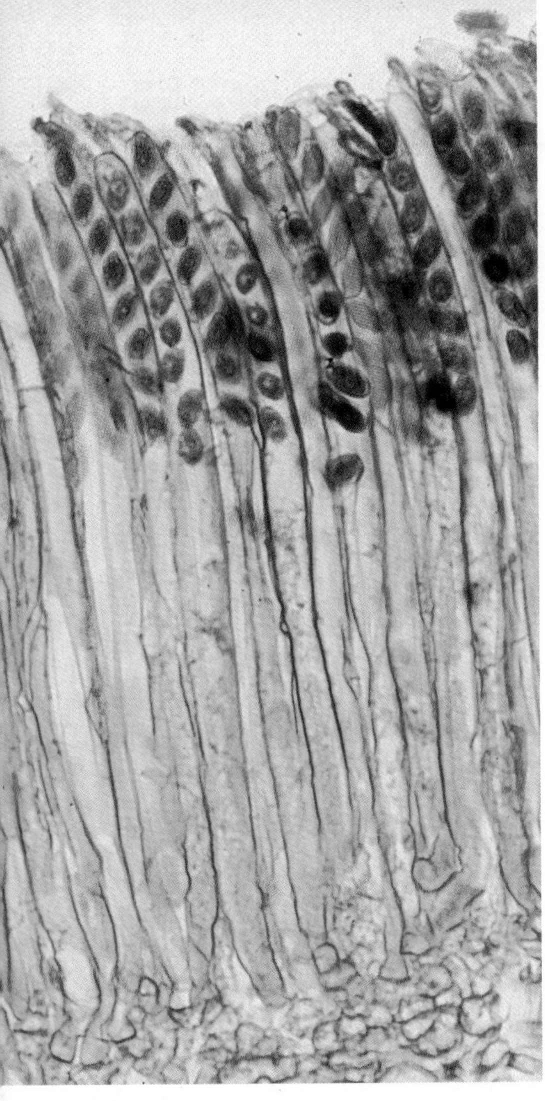

c.

water and then gradually developed the means to conquer drier habitats. Plants may have arisen, for example, from an aquatic green alga. Some species of anatomically simple modern plants echo an earlier aquatic habitat in that their sperm cells must travel in water to reach and fertilize egg cells. Adaptations of plants to land include various provisions to prevent water loss, such as wax coatings on the entire plant body, including the organs that contain the gametes. Openings in leaves and stems called *stomata* allow carbon dioxide to enter the plant from the atmosphere and oxygen to exit into the air. Plants lose, or transpire, vast quantities of water through their stomata.

The 10 divisions of the kingdom Plantae can be separated into two categories, the **bryophytes** (or nonvascular plants) and the **tracheophytes** (or vascular plants). The bryophytes, which are the most primitive plants, all belong to the single division Bryophyta. These plants are termed "nonvascular" because they lack specialized tissues to conduct water and nutrients. The bryophytes include the liverworts, hornworts, and mosses, and they lack true roots, stems, and leaves. A bryophyte spends most of its existence in the phase that eventually produces gametes. The most familiar of the bryophytes are the mosses, which are often found covering rocks, soil, or tree trunks with a carpetlike growth. Acids produced by the metabolic reactions of mosses help to fragment rock, initiating the slow process of soil building. Thus, like the lichens, mosses pave the way for other types of plant life.

The tracheophytes include the other nine divisions of the kingdom Plantae. These organisms have systems of internal tubes that conduct water and nutrients. **Xylem** tissue transports water and mineral nutrients from roots to leaves, and the **phloem** tissue conducts sugars from "sources" (such as photosynthesizing leaves) to "sinks" (such as growing shoot tips and fruits). In the vascular plants, the part of the life cycle in which spores are produced is larger, longer-lived, and more obvious to the human observer.

The most primitive species of vascular plants, like the bryophytes, require a film of water in which their sperm cells can swim. These species include the ferns (division Pterophyta), the whisk ferns (division Psilophyta), the lycophytes or club mosses (division Lycophyta), and the horsetails (division Spenophyta). The inefficient mode of sperm travel found in these organisms was altered with the evolution of the *seed plants*, in which sperm and egg are enclosed in protective structures. Cells that will develop into sperm cells are housed in *pollen* grains, and egg cells are enclosed in *ovules*. Sperm cells are transported to eggs by wind or by animals (usually insects). After the egg and sperm fuse at fertilization, the resulting embryo and a supply of nutrients are packaged into a *seed*. The seed plants, with the reproductive advantage of protecting the next generation, evolved about 225 million years ago.

Five divisions of seed plants are recognized. Four of these are **gymnosperms,** or "naked-seed plants." The gymnosperms include the conifers (division Coniferophyta), with such familiar members as the spruce, cedar, pine, fir, and redwood, as well as three less familiar divisions: Cycadophyta (cycads), Ginkgophyta (ginkgos), and Gnetophyta (gnetophytes). Gymnosperms have leaves that are specialized to form male and female cones, which produce pollen grains and ovules, respectively.

The **angiosperms** (seeds in a vessel), or flowering plants (division Anthophyta), include about 235,000 species. The flower is built of concentric arrangements of highly specialized leaflike appendages, and it can house either male or female reproductive organs, or both. The bright colors of parts of some flowers attract insects that inadvertently pick up sticky pollen and disperse it to the flowers of other plants, thereby aiding fertilization. After fertilization, angiosperm seeds are encased in a fruit (the "vessel"), which helps to protect the seeds from damage and may aid in their dispersal (table 2.5).

The tracheophytes, particularly the angiosperms, have had a large impact on human culture. Plant-based insecticides were used in ancient Rome, and aspirin was probably obtained from willow-bark tea long before recorded history. Today, many gums, waxes, tannins, oils, resins, dyes, rubbers, flavorings, drugs, and pesticides are commercially extracted from plants (table 2.6). Plant chemicals that are rare or difficult to extract are sometimes used as "models" from which chemists synthesize similar compounds. The widely used drugs Darvon and Demerol are laboratory-manufactured versions of plant-derived chemical cousins. Another interesting example of a plant model is the fruit of the common cocklebur, *Xanthium strumarium* (an angiosperm), which clings annoyingly to the pant legs or socks of passing humans. Swiss inventor George deMestral, after examining under the microscope some cocklebur fruits that had become attached to his clothing, was inspired to invent the clothing fastener called Velcro.

Animalia

The kingdom Animalia includes 32 phyla of living animals, which include many familiar species as well as many not-so-familiar ones. Animals have a combination of characteristics that distinguishes them from the members of other kingdoms: they are built of eukaryotic cells, derive energy from food, and have nervous systems (although in many cases these are rudimentary). Animal phyla are distinguished from one another by basic differences in body form, which can be considered in order of increasing complexity. This presumably reflects the sequence of animal evolution and

Table 2.5
Major Taxonomic Groups of a Flowering Plant—Corn (*Zea mays*)

Category	Name	Description
Kingdom	Plantae	Land-dwelling, multicellular, eukaryotic organisms with cell walls made of cellulose; capable of photosynthesis utilizing the pigment molecules chlorophyll *a* and *b*
Division	Anthophyta	Vascular plants with seeds and flowers; ovules enclosed in an ovary and mature seeds in fruits
Class	Monocotyledones	One seed-leaf
Order	Commelinales	Fibrous leaves
Family	Poaceae	Grasses
Genus	*Zea*	Separate male and female flowers
Species	*Zea mays*	Corn

Table 2.6
Some Drugs Extracted from Whole Plants

Drug	Medical Use	Plant Source
Cocaine	Local anesthetic	Coca
Codeine, morphine	Painkiller	Opium poppy
Colchicine	Gout treatment	Autumn crocus
Digitalis	Heart-disease treatment	Foxglove
Quinine	Malaria treatment	Trees and shrubs of genus *Cinchona*
Steroids	Oral contraceptives, muscle builders	Mexican yam
Vincristine	Cancer treatment	Madagascar periwinkle

Table 2.7
Some Animal Phyla

Phylum	Characteristics	Examples	Number of Species
Porifera	Simple multicellular body; feed by filtering water and nutrients through individual cells	Sponges	5,000
Cnidaria	Hollow two-layered body; interior jellylike; radial symmetry	Jellyfish, hydras, corals, sea anemones	9,000+
Platyhelminthes	Three-layered body; organs; bilateral symmetry	Flatworms	13,000
Nemertina	Three-layered body; one-way digestive tract; long muscular "nose" to catch prey	Ribbonworms	650
Nematoda	Three-layered body; one-way digestive tract; unsegmented; thick cuticle; nervous system	Roundworms	10,000+
Annelida	Three-layered body; one-way digestive tract; body cavity; segmented; circulatory, excretory, and nervous systems; sensory cells	Segmented worms	9,000
Mollusca	Soft body; hard shell; three body regions; muscular foot; circulatory, excretory, and nervous systems	Clams, oysters, snails, squids, octopuses	50,000
Arthropoda	Three body regions; segmented; outer jointed skeleton that is molted; blood in body cavities; complex nervous system	Insects, crabs, lobsters, spiders	1,000,000
Echinodermata	Five-part body plan; radial symmetry; spiny outer covering	Sea urchins, sand dollars, starfish, sea lilies	6,000
Chordata	Notochord (rod down back); hollow nerve cord above notochord; gill slits	Fishes, amphibians, reptiles, birds, mammals	45,000

diversification from a common ancestor. Detailed descriptions of the animal phyla are provided in table 2.7.

Three major groups of animals—Mesozoa, Parazoa, and Eumetazoa—branched from an ancestral **metazoan** (many-celled animal). The first animals were probably similar to modern-day mesozoans, which are simple wormlike parasites of soft-bodied marine organisms. The *mesozoans* (phylum Mesozoa) consist of only 20 to 30 cells organized into two layers. Slightly more complex are the animals in the group Parazoa. These include the members of the phylum Placozoa, which are built of two cell layers with fluid in between, and the sponges (phylum Porifera). The sponges are the simplest animals with specialized cell types, and they have a canal system that transports nutrients in and wastes out (fig. 2.16). The mesozoans and parazoans have a cellular level of organization, in contrast with more complex animals, which exhibit tissue or organ levels of organization (specialized groupings of cells).

The remaining 29 animal phyla are known as Eumetazoa, which are divided into two subgroups, Radiata and Bilateria. Radiata includes the phylum Cnidaria (hydroids, sea anemones, jellyfish, horny corals, and hard corals) and the phylum Ctenophora (sea walnuts and comb jellies). These organisms all have *radial symmetry*, in which the body parts are arranged around a central axis, like the spokes of a wheel. The cnidarians and ctenophorans have saclike bodies composed of two or three cell layers. They have diffuse networks of nerve cells (called *nerve nets*) and sense organs and therefore are said to have a tissue level of organization.

The other 27 animal phyla are placed in the group Bilateria because they have a bilateral (two-sided) body plan. The bilateral animals are further subdivided into the groups Protostomia and Deuterostomia based upon certain characteristics of the embryo (such as the origin of the mouth and anus and the pattern with which the first cells of the embryo divide). Protostomia is the more primitive group, and it is subdivided into three groups that are defined by the type of central body cavity (or **coelom**) that is present.

Figure 2.16
A barrel sponge, *Xestospongia muta*, grows like a giant vase on a reef off the Cayman Islands. Sponges of this type are being tested for use as potential drugs.

The most primitive protostomes are the *acoelomates* (no coelom), and these include three phyla—the flatworms (phylum Platyhelminthes), ribbonworms (phylum Nemertina), and jawworms (phylum Gnathostomidula). They lack a coelom, but they have a third layer added to the body plan, providing more material from which organs can develop. The beginnings of muscular and circulatory systems, and the anterior (forward) swelling of nerve tissue that is a forerunner of a brain, become apparent in the acoelomates. These animals are said to exhibit an organ level of organization.

On the next rung up the ladder of complexity are the *pseudocoelomates* (fake coelom), which have a coelom that lacks a lining seen in more advanced animals. This "pseudocoelom" enables these wormlike animals to move more freely than more primitive forms, and it provides more space for organs. The body cavity can also inflate, a property responsible for the name "bladder

Reading 2.2 *Taxonomy in Action*

THE PERUVIAN WILDLIFE PRESERVE CALLED TAMBOPATA IS ONLY 2 SQUARE MILES (APPROXIMATELY 5 SQUARE KILOMETERS) IN AREA, YET IT IS HOME TO PERHAPS THE MOST DIVERSE ASSORTMENT OF ORGANISMS ANYWHERE ON EARTH. Its seven types of forests contain a huge number of different species of plants and birds and far more insect species than can be cataloged in a lifetime. Yet Tambopata represents only a slice of the diversity of life in the tropical rain forest, a part of the world where half of the known living species of plants and animals now reside, including 3 million species that populate the treetops alone.

The Peruvian forest canopy teems with insects. Observing the six-legged organisms are two-legged entomologists, biologists specializing in insects, who are there to describe and classify the millions of previously unseen species. Areas 13 yards (approximately 12 meters) square are roped off and draped with cone-shaped nets. Then an insecticide fog is released into the area, and the insects fall into the nets and are funneled into collecting bottles. The fog kills only insects, and these are replaced in about 10 days by insects from neighboring areas. The "rain" of insects from a single fogging usually numbers about 10,000 specimens. They are transported to laboratories and museums, where they are meticulously examined. With the aid of a computer, the specimens are described and cataloged.

On the basis of Peru's insect diversity, represented by the Tambopata preserve, entomologists speculate that perhaps 50 million insect species currently occupy earth—that is roughly 50 times as many species as have been described so far.

Figure 1
Insects can be found nearly everywhere—even an amateur collector can find an unusual specimen.

worms" used to describe five of the seven pseudocoelomate phyla. One phylum, Nematoda, the nematode worms, is of particular interest because some of its members are parasites of humans, causing such conditions as trichinosis, hookworm disease, and elephantiasis.

The remaining protostomes are the *eucoelomates* (true coelom), which possess a lined coelom. Outgrowths of this peritoneum lining hold organs in place to form complex organ systems, thus introducing a new level of organization. There are 13 eucoelomate phyla, and the 3 major ones are Mollusca (including the snails, clams, oysters, squids, and octopuses), Annelida (the segmented worms, including the familiar earthworms), and Arthropoda (including the spiders, scorpions, ticks, mites, crustaceans, millipedes, centipedes, and insects). Arthropoda is a huge phylum, and its insect members alone are numerous enough to keep taxonomists busy for generations to come (Reading 2.2).

The phyla in the group Deuterostomia consist of bilateral animals with more complex embryos. The most primitive phylum of this group, Echinodermata (the echinoderms), includes the sea stars, brittle stars, sea urchins, sea cucumbers, and sea lilies. They are a curious mixture of the old and the new. The adult echinoderm has a five-part radial symmetry and is unsegmented with no obvious head region, a characteristic reminiscent of more primitive body plans. Yet the echinoderm embryo is clearly bilateral in its symmetry, and the nervous system of the adult is well developed, as it is in the other deuterostomes.

The most advanced animal phylum is Chordata, and it is named for the distinguishing **notochord,** a semirigid rod that runs down the length of the body. The notochord provides support and points of attachment for muscles. Chordates also have a nerve cord that runs on top of (dorsal to) the digestive tract (it runs beneath it in other phyla), gill slits in the throat region that function as feeding or respiratory organs, and a tail. Sometimes these chordate features (such as the gill slits and tail) are present only in the embryo. The chordates retain from their ancestors a three-layered embryo, bilateral symmetry, a coelom, and a prominent head region.

We will look at the phylum Chordata in greater detail than the others because it is the phylum to which humans belong. The chordates are subdivided into three groups. Two of them, the sea squirts (subphylum Urochordata) and the lancelets (subphylum Cephalochordata), are small groups of primitive chordates. The third group, Vertebrata (the vertebrates), is characterized by bone or cartilage around the spinal cord. The jawless fishes constitute the superclass Agnatha, and the jawed fishes and all four-limbed animals make up a second superclass, Gnathostomata. The six classes of Gnathostomata are familiar—the sharks and rays (class Chondrichthyes), the bony fishes (Osteichthyes), amphibians (Amphibia), reptiles (Reptilia), birds (Aves), and mammals (Mammalia). Humans are members of the class Mammalia.

The first mammals evolved about 150 million years ago from a reptilian ancestor, and they exploded in diversity between 70 million and 7 million years ago (fig. 2.17). Today, they remain an extremely successful group, due largely to their adaptations to many types of environments and for obtaining food. The name "mammal" comes from one identifying characteristic of the group—the presence of milk-secreting mammary glands. These are well developed in the female but only rudimentary in the male. Mammals have hair on their bodies, possess a large braincase, have fertilization within the female's body, and maintain a constant body temperature (warm-bloodedness).

The class Mammalia is subdivided into subclasses and infraclasses based upon how the young are carried and born. Subclass Prototheria includes the egg-laying mammals, the monotremes, such as the duck-billed platypus and the spiny anteater. Subclass Theria is large and includes the marsupials and the placental mammals. The marsupials (infraclass Metatheria) are the pouched mammals, such as the kangaroo and the opossum, in which the embryo is born when it is still very small and climbs along the mother's fur to her pouch, where it completes its early development. Placental mammals (infraclass Eutheria) nourish their unborn offspring within the mother's body through a specialized organ called the **placenta,** which develops between the mother and the fetus during pregnancy.

The 15 orders of placental mammals include many familiar animals, such as the dogs, cats, whales, and elephants. The members of the various orders are distinguished from each other by their dental patterns, specializations of their limbs, toes, claws, and hooves, and the complexity of their nervous systems. The order Primates includes the lemurs, monkeys, apes, and humans. Primates have larger brains than other animals, five digits (fingers or toes) on each limb, and the ability to grasp because the thumb can fold against the other fingers. Members of suborder Prosimii, small squirrellike primates such as the lemurs, tarsiers, and lorises, are primitive tree dwellers. The other living primates constitute suborder Anthropoidea, which means "resembling man." Superfamilies within this suborder are distinguished by nose structure, thumb position, tail use, and other characteristics. The New World monkeys (superfamily Ceboidea) include the spider and howler monkeys, and the Old World monkeys (superfamily Cercopithecoidea) include the mandrill and rhesus monkeys. The third superfamily of anthropoid primates, Hominoidea, consists of two families: Pongidae, which includes the gibbons, orangutans, chimpanzees, and gorillas; and Hominidae, which includes only humans, *Homo sapiens*. Humans are distinguished from the other primates by their more erect posture, reduced eyebrow ridges, lighter jaws, [illegible] front teeth, increased use of thumbs, [illegible] arms, complex speech, and higher intelligence. Some taxonomists have proposed that chimpanzees (and perhaps even gorillas) be classified in the family Hominidae also, based upon their biochemical similarities to humans.

The animal that we call *Homo sapiens*—the human animal—is the culmination of billions of years of evolution (fig. 2.18). We reflect a heritage of accumulating characteristics that have shaped life on earth from the simple single-celled organisms of the ancient seas to the diverse and highly complex organisms that live on earth today. But taxonomy is a science that does not end with today. What types of organisms will occupy earth in the future? What role (if any) will *Homo sapiens* play in determining the spectrum of life on earth 10,000 or a million or a billion years hence? Will there even be an earth then? Knowledge of the diversity of life on earth yesterday and today will enable us to speculate intelligently upon, and perhaps even to influence, the living earth of tomorrow.

KEY CONCEPTS

Types of organisms are classified into kingdoms according to the complexity of their cells and their strategies for meeting life's challenges. The kingdom Monera consists of the bacteria and the cyanobacteria, which are single-celled prokaryotes. They are classified by how they obtain energy, their shape, and their pigments. The kingdom Protista includes the simplest eukaryotes, such as the protozoa, algae, water molds, and slime molds. The fungi are eukaryotes that are mostly found on land, and they obtain energy by breaking down organic matter. Their cell walls contain chitin, and except for yeast, fungi are multicellular. The plants are land-dwelling multicellular eukaryotes that extract energy from sunlight and have a life cycle characterized by alternation of generations. Plant cell walls are built of cellulose. Members of the kingdom Animalia are multicellular eukaryotes, derive their energy from food, and have nervous systems. The different groups of animals are distinguished by body form.

a.

b.

c.

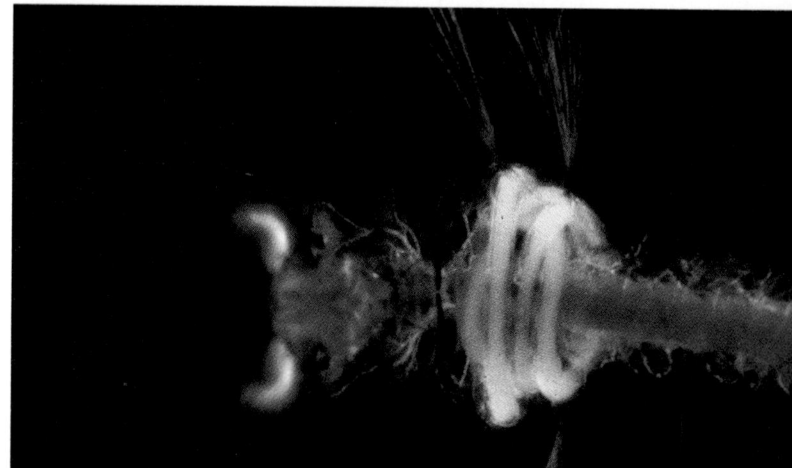

d.

e.

f.

Figure 2.17

Representatives of various animal phyla. *a.* The sponges have specialized cells but are considered to be an ancient offshoot of the animal kingdom that is an evolutionary dead end. *b.* The radially symmetrical hydra lives in fresh water, is built of two cell layers, and has sensory and nerve cells. *c.* The ribbonworm *Tubulanus polymorphus* is one of the most primitive bilateral organisms. *d.* This nematode worm is wrapped inside a mosquito larva. Other types of nematodes are digestive parasites of humans. *e.* The octopus is a complex mollusk. *f.* The earthworm is an annelid displaying segmentation. *g.* The trend of a segmented body is elaborated in the arthropods, including the insects. *h.* The "water bear" *Echiniscus spiniger* is a tardigrade and is only 1/100 to 2/100 inch (300 to 500 micrometers) long. In times of drought, it enters a state of suspended animation, reviving when moisture returns. *i.* The sea urchin is an echinoderm. *j.* The hippopotomus is a chordate, as are most familiar animals.

g.

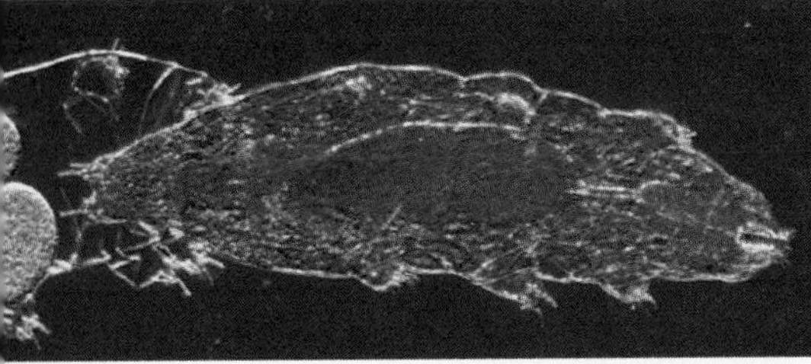

h.

j.

i.

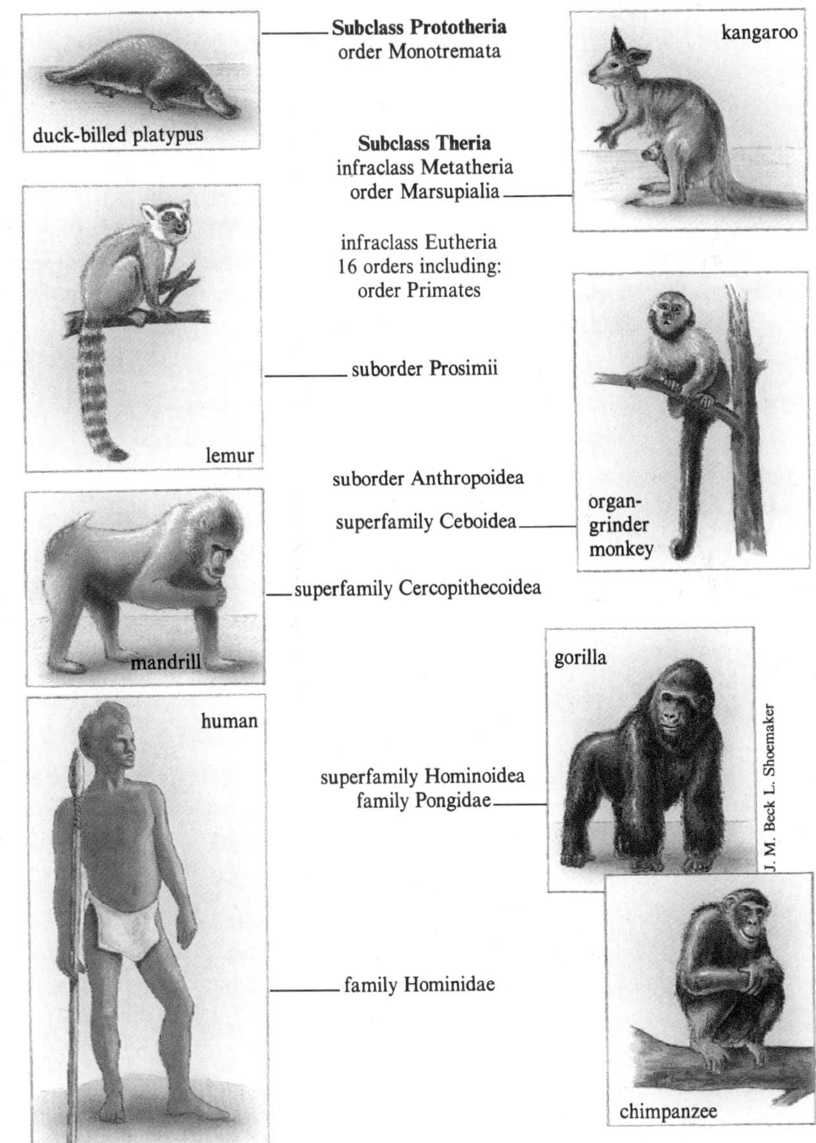

Figure 2.18
Classification of humans (*Homo sapiens*).

SUMMARY

Nearly every part of the earth's surface is occupied by life. Some regions, such as the tropical rain forests, contain very diverse collections of organisms. Other areas, such as salt pools and desert rocks, are home to only a few types of organisms. Life forms often exist within or on other life forms. Organisms directly affect each other through complex chains and webs of "who eats whom."

Biological beginnings date back some 5 billion years to novel chemical combinations in ancient seas that gave rise to the earliest forms of life. The first living things were probably single, structurally simple cells (*prokaryotes*). By 3 billion years ago, the seas were alive with many kinds of microscopic prokaryotes. By 1 billion years ago or so, more complex, organelle-containing cells (*eukaryotes*) had evolved, but there were, as yet, no multicellular organisms. By about 500 million years ago, the seas were densely populated by a great variety of multicellular organisms. Gradually, life adapted to existence on land, and biological diversification continued, as it does today.

The biological science of *taxonomy* classifies and names groups of closely related organisms, or *species*, that can successfully reproduce only among themselves. A taxonomic description consists of several levels of classification that distinguish species as well as indicate their ancestry. Criteria for classification include structural, metabolic, behavioral, and molecular characteristics. Problems in taxonomy include subjective differences among taxonomists; which traits to consider and how much importance to assign to each one; dealing with similar traits between two species that do not reflect descent from a common ancestor; and classifying organisms with different characteristics at different stages of their life cycles.

Linnaean levels of classification begin with the broadest, *kingdom*, and progress through *division* or *phylum*, *class*, *order*, *family*, *genus*, and *species*. Further subdivisions are sometimes used. The kingdom Monera includes the bacteria and cyanobacteria, which are single-celled prokaryotes. Bacteria are classified by their mode of obtaining energy, their shape, their cell wall composition, and whether or not they require oxygen. *Cyanobacteria* are pigmented and can photosynthesize. The kingdom Protista includes the *protozoans*, certain *algae*, the *water molds*, *chytrids*, and *slime molds*. The protists are the simplest eukaryotes. Members of the kingdom Fungi are usually terrestrial and multicellular and obtain energy by decomposing organic matter. They are eukaryotes with cell walls made of *chitin*. The plants (kingdom Plantae) are land-dwelling, multicellular eukaryotes that photosynthesize, have cell walls built of cellulose, and exhibit alternation of generations. The nonvascular *bryophytes* include the more primitive species, and the vascular *tracheophytes* introduce systems of internal conducting networks for water and nutrients. The species of the kingdom Animalia, the animals, are multicellular eukaryotes with nervous systems. Animals derive energy from food. Body form and symmetry in both embryonic and adult stages are important in distinguishing animal groups from each other. Humans, called *Homo sapiens*, belong to the phylum Chordata, subphylum Vertebrata, superclass Gnathostomata, class Mammalia, order Primates, suborder Anthropoidea, superfamily Hominoidea, and family Hominidae.

QUESTIONS

1. Why is it highly unlikely that life could begin again today in the same way that it did 5 billion years ago?

2. Give examples of criteria used by taxonomists to classify organisms at the kingdom, phylum or division, class, order, and family levels.

3. In a classic science fiction film (*Village of the Damned*), organisms from a distant planet arrive on earth and impregnate female humans. The women give birth to organisms that have characteristics of both parents. According to the classification criteria used by earthling taxonomists, how closely related are these aliens to us?

4. What biological trend exhibited in the types of organisms present on the earth from the start of life to today is also apparent in the five-kingdom classification scheme?

5. Some molecular taxonomists claim that the genetic material of humans and chimpanzees is so similar that the chimpanzee should be reclassified into our genus, *Homo*. Do you think the molecular correspondence between these two types of primates is sufficient to change the chimpanzee's taxonomic name to reflect a closer evolutionary relationship to *Homo sapiens?* What other characteristics are considered when comparing primates?

6. Cite two difficulties encountered in classifying organisms based on descent from a common ancestor. Suggest another framework for classifying life forms.

7. In recent years, genetic researchers have constructed novel life forms by combining genetic material from two or even three different species. What criteria should be met for such an engineered organism to be given a new taxonomic designation?

8. Name a human disease caused by bacterium, a protozoan, a fungus, a plant, and an animal.

9. The end of this chapter takes a closer look at one phylum, Chordata. Consult an appropriate advanced textbook to describe the taxonomic criteria used at each level of classification for another phylum or division.

TO THINK ABOUT

1. Researchers can combine genetic material of different species, increasing biological diversity. Other humans decrease diversity by destroying natural habitats, which sometimes leads to the extinction of species. What effects do you think human intervention in natural biological diversity will have on future life on this planet?

2. What objection might a Creationist (someone who believes in a biblical interpretation of the origin of life on earth) have to Linnaean classification?

3. What do you think is (or has been) the most successful group of organisms on earth? What is the basis for your answer?

4. Most textbooks list a distinguishing characteristic of humans to be our high intelligence. Is this an objective assessment? Why or why not?

5. Genetic material from species that are no longer living has been obtained and analyzed. If scientists can someday reconstruct an individual of a long-gone species, what do you think the consequences could be for natural biological diversity?

6. Taxonomy is ideally a system of biological classification used and understood by all biologists. However, this chapter has mentioned several disagreements among taxonomists: the placement of chimpanzees and gorillas in the family Hominidae; the classification of cyanobacteria; the classification of algae as protists or plants; even the number of kingdoms. Do you think it is possible to develop a taxonomic scheme that all biologists can agree on? Can you think of a way to minimize the confusion?

SUGGESTED READINGS

Attenborough, David. 1979. *Life on earth*. Boston: Little, Brown. A fascinating and remarkably photographed tour of the planet's natural history.

Calvin, William H. 1986. *The river that flows uphill*. New York: Macmillan. A weaving of diversity past and present along the Colorado River.

Diamond, Jared. April 1990. The search for life on earth. *Natural History*. A look at how taxonomy is regarded by modern biologists.

Duran Sharnoff, Sylvia. April 1984. Lowly lichens offer beauty—and food, drugs and perfume. *Smithsonian*. Members of two different taxonomic kingdoms join to form a third, and fascinating, type of organism.

Eldridge, Niles, and Ian Tattersall. March 1983. Future people. *Science 83*. What will our distant descendants be like?

Green, Lilias. March 1987. There's a protozoan in that painting. *BioScience*. The great variety of protozoan forms has inspired several artists.

Gurin, Joel. July/August 1980. In the beginning. *Science 80*. We humans have been a part of natural history for a very short time.

Lewis, Ricki. April 1988. 20,000 drugs under the sea. *Discover*. The great diversity of life in the sea offers new possible pharmaceuticals.

Margulis, Lynn, and Karlene V. Schwartz. 1982. *Five kingdoms*. San Francisco: W. H. Freeman. A detailed look at modern taxonomy.

May, Robert M. September 16, 1988. How many species are there? *Science*. Nobody knows the answer to this compelling question, but this article illustrates some ways that biologists attempt to answer it.

Playford, Phillip E. October 1980. Australia's stromatolite stronghold. *Natural History*. A look at harsh Hamelin Pool and its ageless inhabitants.

Stutz, Bruce. August 1986. Fish market taxonomy. *Natural History*. Taxonomists are not confined to jungle floors or museum cases. Fascinating fish species can be discovered amongst the bustle of the Fulton Fish Market.

Voelker, William. 1986. *The natural history of living mammals*. Medford, N.J.: Plexus Publishing. A look at modern diversity among the mammals.

Webster, Bayard. February 14, 1982. Classification is more than a matter of fish or fowl. *New York Times*. An amusing but informative look at the peculiarities of the science of taxonomy.

CHAPTER

3

The Chemistry and Origin of Life

Chapter Outline

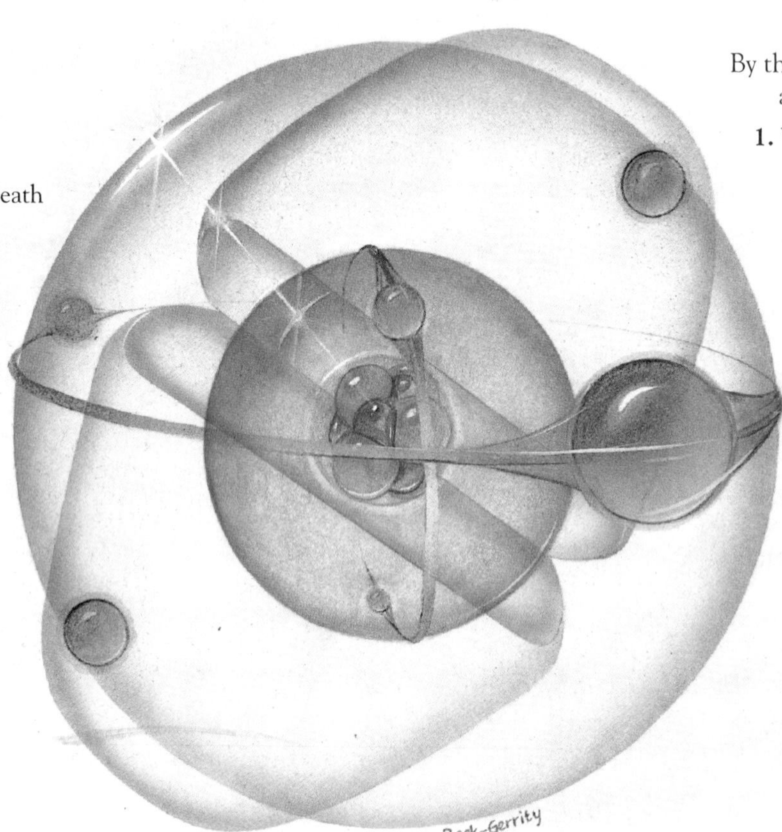

Learning Objectives

By the chapter's end, you should be able to answer these questions:

1. What characteristics distinguish living things from nonliving things?
2. What are the simplest forms of life?
3. What chemical components constitute living things?
4. What chemical compounds are important to human health?
5. How might living matter have evolved from nonliving chemicals?

At first glance, the blue-footed boobies of Hood Island (one of the Galápagos Islands off the coast of Ecuador) do not seem at all similar to the humans who watch them with obvious enjoyment. These birds are famous for their characteristic mating rituals. The male points his beak skyward and lifts his wings, revealing his normally hidden underside. But it is the female who evokes laughter from tourists as she prances in slow motion, lifting each bright blue foot in a gingerly fashion to tantalize her mate. To the onlooking humans, the birds seem strange indeed (fig. 3.1).

The blue-footed booby and the human, however different they may appear, actually share quite a few characteristics. Both organisms are built with systems of organs constructed of units called cells, which contain similar chemicals that react with one another in similar ways. The elaborate mating dances of the boobies are not really so unlike human interactions between the sexes: Both species engage in certain behaviors that can culminate in mating and the conception and birth of young. Booby and human respond similarly to some of the same environmental stimuli. If you threaten a human child, the parent protests. If you step within the circle of guano (bird excrement) that demarcates the territory of a young booby family, the adult birds strut and squawk their objections.

Some responses to environmental constraints have become a permanent part of being a booby or a human. In times of famine, a female booby is likely to lay only a single egg. If two or three eggs are laid, the firstborn hatchling throws the others outside the family guano ring, where the parents ignore them and, ultimately, they starve. Human females who have extremely low levels of body fat sometimes temporarily lose their ability to become pregnant. Both the boobies' response to famine and the temporary infertility of human females that is induced by semistarvation are biological responses to a situation in which starting a new generation is likely to be unsuccessful.

In short, what the blue-footed booby and the human have in common is that they are alive. Despite the great diversity of life forms discussed in the last chapter, and despite the obvious differences between the two organisms just described, all living things share certain characteristics. This chapter will explore these characteristics of life, the common chemical plan upon which all life is built, and the way in which this basic unity of life suggests that all species are descended from a common ancestor.

Figure 3.1
On the surface, humans and blue-footed boobies may seem to have very little in common. However, as complex forms of life, they share many characteristics, from the architecture of their bodies to certain behaviors.

The Characteristics of Life

Organization

Living matter consists of structures arranged in a particular three-dimensional relationship, often following a pattern of structures within structures within structures (fig. 3.2). Consider the human body. It is built of systems of **organs** that function together to perform specific activities, such as digestion, excretion, or circulation. The organs are constructed of functional parts, also. The kidney is not a solid mass but a highly organized collection of tiny tubes. Similarly, the heart is a system of pumps, chambers, valves, and tubes. These organ parts, in turn, are built of organized structures called **tissues,** which are made up of even smaller living units, **cells.** The parts of the cell are organized too, as we shall see in chapter 4. Ultimately, all life consists of chemicals.

At all levels, and in all organisms, structural organization is closely tied to function. Disrupt the structural plan, and function ceases. An injured organ does not work properly; if a fertilized hen's egg is shaken briskly, it will fail to develop. The reverse is true also—if its function is disrupted, a structure will eventually break down. Muscles that are not used, for example, soon begin to atrophy (waste away), a phenomenon that is familiar to anyone who has ever had a broken leg immobilized for weeks by a cast. Biological function and form, then, are interdependent. Some organisms can maintain their vital organization even when exposed to life-threatening drought by drying into a state of suspended animation (fig. 3.3). They can be revived later with a few drops of water.

Biological organization is apparent in all life. The noble frog, whose insides are laid open in biology labs, reveals a bodily organization quite similar to that of the student who is dissecting it. Plants are also organized into specialized organ systems, organs, tissues, and cells. Single-celled bacteria cannot boast such complexities as organ systems, yet even they are composed of intricately organized structures.

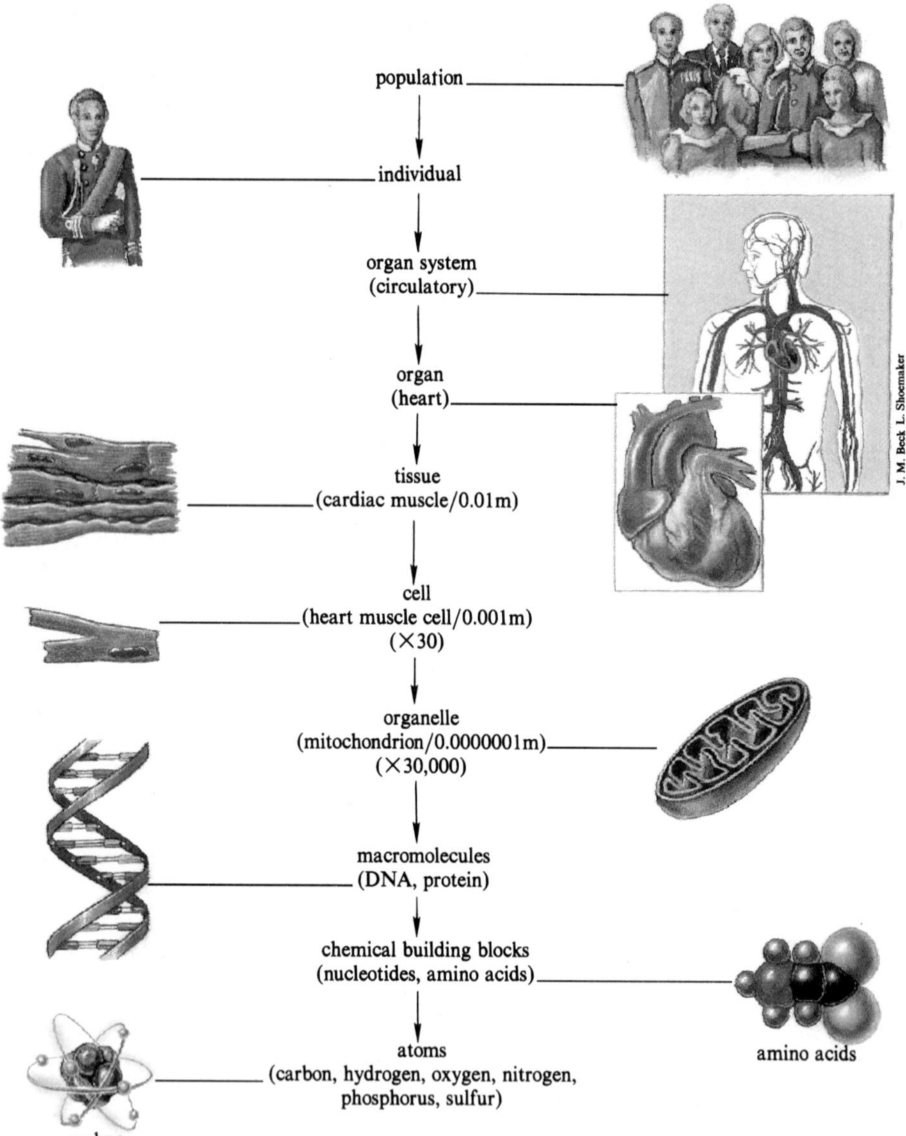

Figure 3.2
Levels of biological organization.

Metabolism

Important as organization is to a living system, organization alone is not enough to qualify something as "alive." The precisely organized structures of life require energy to power them. A living system must be able to acquire and use energy to build new structures, to repair or to break down old ones, and to reproduce. Plants tap the sun's energy, but animals derive their energy from food. The sum total of the chemical reactions that direct this energy of life, both building up and breaking down, is termed **metabolism.**

Irritability and Adaptation

Living organisms sense and respond to certain environmental stimuli while ignoring others. An immediate response to a stimulus is termed **irritability,** and it can be essential for survival. A man leaning on a hot stove yanks his hand away; a dog lifts its ears in response to a whistle; a plant grows toward the sun. Irritability may indeed be a characteristic that is unique to living things.

Whereas irritability is an immediate response to a stimulus, an **adaptation** is an inherited characteristic or behavior that enables an organism to survive a specific environmental challenge. Adaptations are often more complex than irritability, and they are likely to differ from one species to another. Both a human and a dog pull an extremity away from something hot (irritability), but, on a hot summer day, the human sweats and the dog pants (adaptations).

a.

b.

Figure 3.3
Life on hold. *a.* "Sea monkeys" are a child's delight—dried shreds that pop to life when placed in water. They are actually tiny brine shrimp that can suspend some of life's characteristics while in a dehydrated state.

b. The desert plant known as the scale-leaf spike moss can lose 97% of its water and still be capable of "returning" to life when given a few drops of water. This "resurrection plant" is a popular souvenir in the American Southwest.

Reading 3.1 *The Definition of Death*

THIRTY YEARS AGO, THE DEFINITION OF DEATH WAS OBVIOUS AND NONCONTROVERSIAL—LIFE CEASED WHEN THE HEART STOPPED BEATING AND THE LUNGS STOPPED INFLATING AND DEFLATING. In 1968, this definition was made obsolete by the development of heart transplants. How could a heart donor be considered dead if his or her heart continued to beat when it was removed for transplantation into another body? Medical technology further complicated the problem of defining life's end point with the invention of the respirator, a machine that could artificially maintain the heart and lungs. The moment of death went from a cessation of heart and lung function to a flick of a respirator switch.

Today, death is defined as the cessation of brain activity. Specifically, the criteria for "brain death" include the lack of observable response to any external stimulus; no movement or breathing; no reflexes (such as blinking or pupil constriction when a bright light is shone into the eye); no swallowing, yawning, or vocalizations; and a flat record of brain-wave patterns. Furthermore, the lack of movement and breathing without the aid of a respirator must be observed by physicians for at least an hour, and the flat brain-wave pattern must be observed a second time a day later.

It is becoming more and more commonplace for life's functions to be extended by means of donated or artificial organs or by machines. In recent years, many gravely ill people have defied death—some for a short time only, others somewhat longer—with the help of medical technology. Consider Barney Clark, who lived for several months with a mechanical heart in place of his own diseased one, and newborn Baby Fae, who lived for 20 days with a baboon heart implanted in her chest.

The case of Jamie Fiske provides another example of the extension of life through medical technology. Eleven-month-old Jamie was hours from death in a Minnesota hospital, her liver destroyed by a birth defect. Her only chance of survival was a liver transplant. As Jamie lay dying, a thousand miles away, in a small Utah town, 10-month-old Jessie Bellon had just been rushed to a hospital after the car he had been riding in had been struck by a train. Although his brain-wave pattern was flat, Jessie's organs were kept functioning with the aid of a respirator. Jessie's parents had heard Jamie's father's well-publicized plea for a liver donor, and they decided to help. Jessie's respirator was turned off, and his liver was prepared for shipment to Minnesota and transplantation into the dying Jamie. Thus Jessie's death gave new life to Jamie.

Consider, finally, the case of Jimmy Tontlewicz. On a bright winter day a few years ago, young Jimmy chased his sled onto frozen Lake Michigan and plunged through the thin ice into the frigid water below. After 20 minutes, he was rescued. He appeared to be dead, with no detectable heartbeat, pulse, or breathing. However, emergency care was administered as he was rushed to the hospital, and, within an hour, the characteristics of life had reappeared in the boy. Apparently, the initial blast of cold water had shocked Jimmy's body in such a way that, although his heart and lungs stopped temporarily, his other vital organs continued to function at a minimal level. Fortunately, this was enough to keep him alive until his heartbeat and breathing could be restored.

a.

b.

c.

Figure 3.4
The biological adaptations illustrated in these photographs arose from generations of selective pressure on inherited characteristics that were particularly well-suited to a specific environmental challenge. *a.* The viceroy butterfly is a master of mimicry. Because it looks just like the foul-tasting monarch butterfly (*b*), it is avoided by insect-eating birds that might actually find it quite palatable. *c.* This aggressive tabby fluffs out her fur, arches her back and bares her teeth, appearing more intimidating to the troublesome kittens than she actually is.

Adaptations develop because certain individuals of a species inherit qualities that enable them to survive and reproduce while others in the same environment do not. Thus, inherited traits of the better-equipped individuals are more likely to be passed on to offspring. Over many generations, those characteristics that provide adaptive advantages in the particular environment become more common in the population. However, if the environment suddenly changes, an inherited trait that was formerly adaptive can become a hindrance. This limited value of an adaptation may explain why less than 1% of the life forms that have ever existed on earth are alive today. Figures 3.4 and 3.5 show some interesting adaptations.

Reproduction

Reproduction can be as simple as the splitting of a single-celled organism into two or as complex as the conception and birth of a human baby. Reproduction is necessary if a population of organisms is to survive for more than one generation.

On the level of the individual organism, reproduction is the passing on of biochemical instructions (genes) to carry the characteristics of the organism from one generation to the next. Reproduction can be either asexual or sexual. In **asexual reproduction** in single-celled organisms, cellular contents are doubled, and the cell then splits in two. Some many-celled organisms can reproduce asexually, most notably the plants. In the potato plant, for example, a potato formed on an underground stem can sprout leaves and roots to form an entirely new potato plant. In some simple animals, such as sponges and sea anemones, asexual reproduction occurs when a fragment of the parent animal is detached and develops into a new individual. In **sexual reproduction,** genetic material from two individuals combines to begin the life of a third individual.

What Is the Simplest Form of Life?

In listing life's characteristics, an interesting question arises. At what point does a chemical, or group of chemicals, become a simple life form?

The tiniest candidates that we know of exhibit some characteristics of life, but they cannot function and reproduce without the help of more complex and obviously living things. The **virus** that causes

a.

b.

Figure 3.5
Adaptation in the peppered moth. *a.* At one time, virtually all of the peppered moths in England were light in color. Dark-colored moths of this species were extremely rare. Because pale lichens grew on the bark of the trees where the moths spent the daylight hours resting, the light-colored individuals would blend into the background, whereas any darker-colored ones would be readily seen and eaten by birds. *b.* Then came the Industrial Revolution. Smoke and soot from factories killed the lichens and darkened the bark of trees in and around cities. In such areas, the numbers of dark-colored moths increased dramatically and the light ones all but disappeared. In the unpolluted countryside, however, where lichens continued to grow on the bark of trees, the light-colored moths continued to flourish and the dark ones remained rare.

Figure 3.6
Plants can fall victim to an infectious (and possibly living) particle, the viroid. Shown here are (*left*) a tomato with "tomato bunchy top," caused by a viroid, and (*right*) a healthy tomato plant.

the crippling disease poliomyelitis is an example. The polio virus is a tiny piece of genetic material wrapped in another chemical, a protein. This structural simplicity is misleading, though, as anyone who has had polio can attest. The virus commandeers the components of the cells it invades and uses them to mass produce copies of itself. The tiny virus can metabolize and reproduce but only within the cells of a complex organism. Is it alive? Scientists disagree on this question. However, even viruses may not be the simplest life form.

Viroids are even more streamlined than viruses. They consist only of highly wound genetic material that, if it were untwisted and stretched out, would be about 3 feet (1 meter) long. Viroids cause several exotic-sounding plant diseases, such as "avocado sun blotch," "coconut cadang cadang" and "tomato bunchy top" (fig. 3.6). They nearly destroyed the chrysanthemum industry in the United States in the early 1950s.

Like viruses, viroids infect and take over the cellular apparatus of their host. However, even viroids may not represent the lower limit of living things. Tinier yet are **prions,** which are only 1/100 to 1/1,000 the size of the smallest known virus and are composed only of protein. They too invade cells and take over their chemical machinery. Prions are believed to cause the sheep disease scrapie and possibly the human ailments multiple sclerosis and Lou Gehrig's disease. Prions are a puzzlement because they do not contain genetic material.

KEY CONCEPTS

Living things are structurally organized. The human body is built of organ systems, organs, tissues, and cells. Biological functions depend upon structure. Living things display several characteristics. Organisms obtain and utilize energy through the chemical reactions of metabolism and deal with the environment through irritability and adaptation. Species are perpetuated by the asexual or sexual reproduction of individuals. Viruses, viroids, and prions display some, but not all, of the characteristics of life.

Chemistry Basics

Despite the tremendous variation among organisms, astounding similarities appear at the submicroscopic level of biological organization, the chemical level. All forms of life, from the single-celled to the trillion-celled, are built of the same types of chemical components. Not only is the human body constructed of chemicals (about $5 worth), but many chemicals can have profound effects on our health. In short, biology is based upon the same chemical principles that govern all matter.

The Atom

Subatomic Particles

All chemicals, including those that make up organisms, can be broken down into units called **atoms.** Atoms react with one another, but they cannot be further broken down by chemical means. An atom has a central region, the **nucleus,** which contains particles called **protons** and **neutrons** (except an atom of hydrogen, which contains only a proton). Surrounding the nucleus in "shells" of various diameters are subatomic particles of a third type, called **electrons.** An electron's distance from the atom's center reflects its energy level. The more energy an electron has, the closer it is to the nucleus. The electrons of the highest energy level are closest to the nucleus, and there can be one or two electrons in the innermost shell. The other shells can contain more electrons. The outermost shell is called the valence shell, and it is important in determining how atoms react with one another.

Electrons move about the nucleus at the speed of light, and so it is impossible to represent their position accurately in a diagram. Often electrons are illustrated as dots in concentric circles, much like planets orbiting the sun (fig. 3.7). Although these planetary models, known as Bohr models, permit visualization of interactions between atoms, they do not accurately capture the cloudlike distribution of constantly moving electrons about the nucleus. Another way to describe subatomic structure is to represent the movement of electrons, rather than depict them in static positions. The term *electron orbital* refers to the volume of space in which a particular electron can be found 90% of the time. The different electron shells are described in terms of orbitals (fig. 3.8). The shell closest to the nucleus consists of a single orbital containing up to two electrons. The second and third shells each consist of four orbitals, and each of these orbitals can contain two electrons. The electrons of an atom occupy their shells sequentially, so that the innermost shells are filled first. Electron arrangement, as we shall soon see, determines the chemical properties of an atom.

Subatomic particles are distinguished by mass and electrical charge. A proton has a mass of one unit and a positive electrical charge. A neutron also has a mass of one unit, but is has no charge. An electron has a negligible mass and a negative charge. The net charge and mass of an atom is the sum of the charges and masses of its constituent particles. Atomic weight, a measure of the mass of an atom, is calculated by adding the number of protons and neutrons. Another descriptive term, atomic number, equals the number of protons in an atom. An atom with six protons, six neutrons, and six electrons, for example, has an atomic weight of 12 and an atomic number of 6. It is electrically neutral (uncharged) because the number of positively charged protons balances the number of negatively charged electrons. This atom has two electron shells. The innermost shell contains two electrons; the outermost, four.

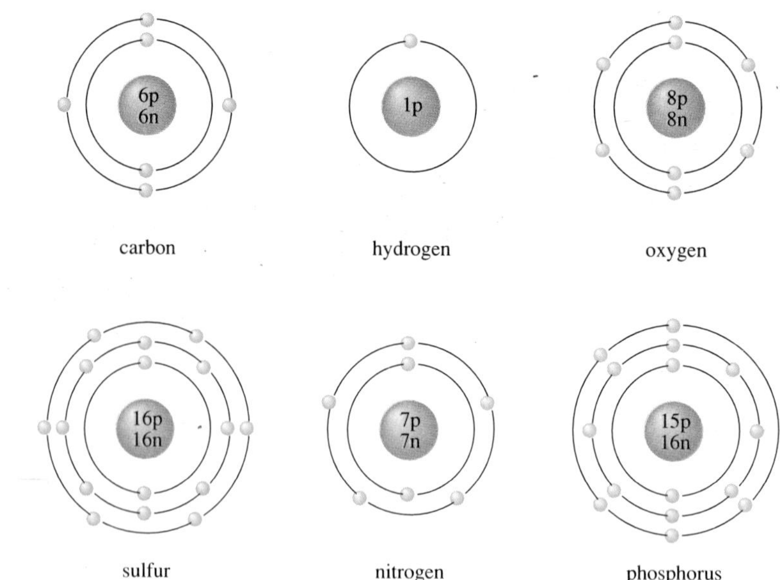

Figure 3.7
Structures of the atoms of life. Shown here are Bohr models of the six most common atoms of which living things consist.

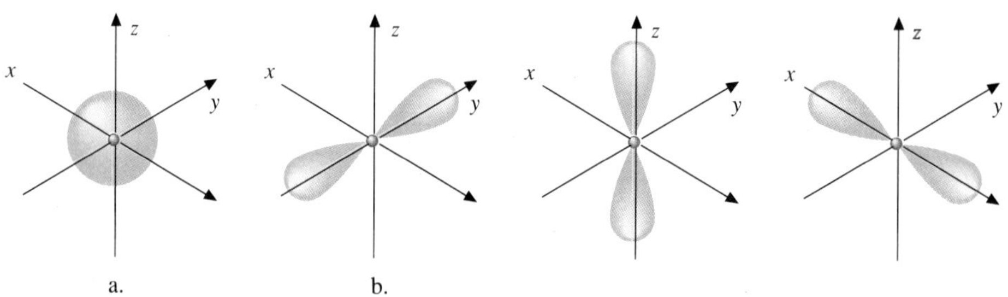

Figure 3.8
Electron orbitals. The ever-changing location of an electron is more accurately represented by an orbital, which is the volume of space in which an electron is likely to be found 90% of the time. *a.* The first energy level consists of one spherical orbital containing up to two electrons. The second energy level has four orbitals, each describing the distribution of up to two electrons. One of the orbitals of the second energy level is spherical; the other three are dumbbell-shaped and arranged perpendicular to one another, as the axis lines indicate (*b*). The nucleus is at the center, where the axes intersect.

Elements

All atoms are built of the same types of subatomic particles, but atoms are distinguished and named according to their numbers of protons. Atoms of a single type, each with the same number of protons, constitute an **element.** Each of the 106 known elements has been given a name and a corresponding symbol. The atom just described, containing six protons, is an

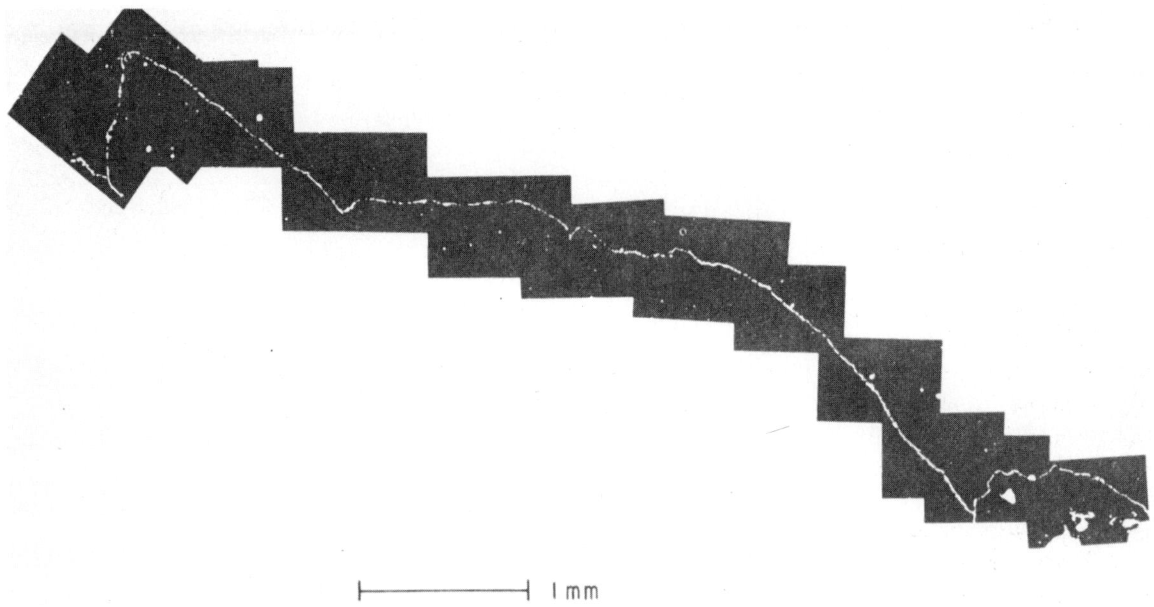

Figure 3.9
A radioactive form of hydrogen, called tritium, labels newly formed genetic material in a fruit fly. Energy emitted from the tritium exposes photographic film, just as the energy from light does.

atom of the element carbon, symbolized "C." It plays a very central role in the chemistry of life. Atoms of different elements vary remarkably in size. An atom of the element uranium, for example, contains 92 protons, 146 neutrons, and 92 electrons.

The elements can be arranged in a chart, the periodic table of the elements, which provides information about their characteristics. The elements are ordered from left to right by atomic number (number of protons). Elements in the same vertical column have similar characteristics because they have the same number of electrons in the outermost shell. Elements in the left-most column, for example, have one electron in their valence shells. The elements in the adjacent column have two electrons in this shell.

Atomic weights in the periodic table are actually averages. The atoms of a particular element can vary in the number of neutrons they contain and thus can have slightly different masses (and therefore different atomic weights). Each form of an element, that is, each different weight, is termed an **isotope.** Carbon's most prevalent isotope has six neutrons and an atomic weight of 12. Other common carbon isotopes have weights of 13 or 14. Sometimes an isotope is unstable, which means that it can change. When the isotope changes, it releases energy (radiation) and forms a more stable isotope. This property of emitting radiation is used widely in research as a "marker" of chemical activity in organisms (fig. 3.9). For example, a plant is exposed to carbon dioxide (CO_2) "labeled" with the radioactive isotope carbon-14 (designated ^{14}C). The carbon dioxide enters the plant. By tracing which parts of the plant become radioactive, the path of the radioactive carbon atoms can be followed. By extracting and identifying labeled chemicals from the plant, it is possible to see which ones are built from carbon dioxide. Investigators can thus learn how plants allocate their chemical energy to developing seeds, fruits, and other structures.

Atoms Meeting Atoms

Molecules

Atoms can combine to form **molecules.** A **compound** is a molecule containing different kinds of atoms. (In a few cases, two atoms of the same element join to form a molecule. Hydrogen gas and the oxygen and nitrogen in the air we breathe each consist of such "diatomic" molecules.) A compound has characteristics that are different from those of the elements it contains. Consider table salt, which is the compound sodium chloride. A molecule of salt contains one atom of sodium (abbreviated Na, from the Latin *natrium*) and one atom of chlorine (Cl). Sodium is a silvery, highly reactive metal that is a solid at room temperature, while chlorine is a yellow corrosive gas. When equal numbers of these two types of atoms combine, the resulting compound is a white crystalline solid, the familiar table salt.

A biologically important characteristic of a compound is whether it is a solid, liquid, or gas at particular temperatures and pressures. Water would not be very effective as the principal component of blood if, at body temperature, it took the form of steam or ice. Nor would bone mineral provide structural support in the skeleton if it were not a solid.

Molecules are described by writing the symbols of the constituent elements and indicating the numbers of atoms of each element in the molecule as subscripts. The sugar molecule glucose is represented as $C_6H_{12}O_6$, which shows that it contains 6 atoms of carbon, 12 of hydrogen, and 6 of oxygen. The number of molecules is indicated by a coefficient. Thus, 6 molecules of glucose is written $6C_6H_{12}O_6$.

Atoms and molecules react with one another by gaining, losing, or sharing electrons to produce new types of molecules. Such an interaction is called a *chemical reaction*. For example, if hydrochloric acid (HCl) is added to sodium hydroxide (NaOH), sodium chloride (NaCl) and water (H_2O) are produced. A chemical reaction entails the making and breaking of attractive forces, called bonds, between atoms (table 3.1). The type of chemical bond that forms between atoms depends upon the number of electrons the atoms have in their outermost shells.

Table 3.1
Chemical Bonds

Type	*Chemical Basis*	*Strength*	*Example*
Ionic	Attraction between oppositely charged ions	Strong	Sodium chloride
Covalent	Sharing of electron pairs between atoms	Strong	Carbon–carbon bonds
Hydrogen	Attraction of a hydrogen with a partial positive charge to negatively charged atoms in neighboring molecules	Weak	Cohesiveness of water

Ions and Ionic Bonds—Opposites Attract

Atoms of some of the elements most prevalent in living things (carbon, nitrogen, oxygen, phosphorus, and sulfur) are more chemically stable when they have eight electrons in their valence shells. This chemical tendency to fill the valence shell is called the *octet rule*. One way to accomplish this is for an atom with one, two, or three electrons in this shell to lose them to an atom with, correspondingly, seven, six, or five electrons in the valence shell. It is a kind of chemical give-and-take. A sodium atom, for example, has one electron in its outermost shell. When it donates this electron to an atom of chlorine, which has seven electrons in its outer shell, the two atoms are physically bonded together to form NaCl.

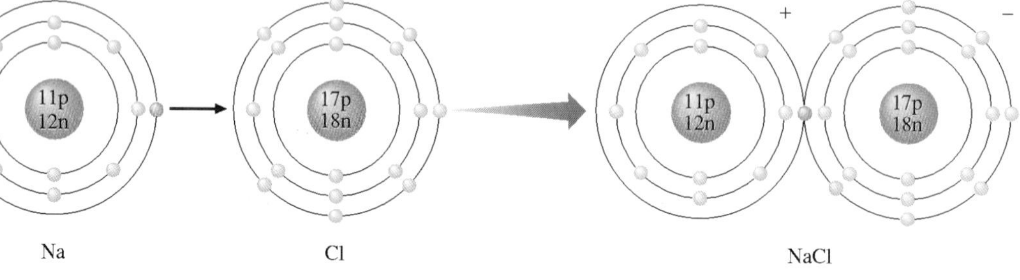

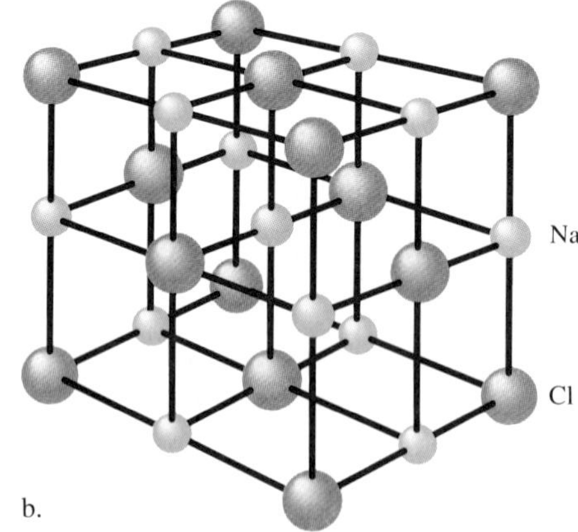

Figure 3.10
An ionically bonded molecule—table salt (NaCl). *a.* A sodium atom (Na) can donate the one electron in its valence shell to a chlorine atom (Cl), which has seven electrons in its outermost shell. The resulting ions (Na^+ and Cl^-) bond to form the compound sodium chloride (NaCl), better known as table salt. The octet rule has been satisfied. *b.* The ions that constitute NaCl are arranged in a repeating pattern.

Once an atom loses or gains electrons, it has an electric charge, and it is then called an **ion** instead of an atom. Atoms that lose electrons lose negative charges and thus carry a positive charge. Atoms that gain electrons carry negative charges. It is the attraction between oppositely charged ions that results in an **ionic bond,** such as the one that holds NaCl together (table 3.1). The oppositely charged ions Na^+ and Cl^- attract each other in such an ordered manner that a crystal results (fig. 3.10).

Ions are important in many biological functions. Nerve transmission depends upon the passage of sodium (Na^+) and potassium (K^+) ions in and out of cells. Muscle contraction depends upon calcium (Ca^{2+}) ions. Doctors debate the cause of high blood pressure—is it due to too much Na^+ or too little Ca^{2+}? Even the start of a new life is ionically controlled. In many types of organisms, calcium ions released by egg cells clear a path for sperm entry.

Acids, Bases, and Salts

Ionically bonded molecules tend to break up to form ions in the presence of water. A molecule that releases hydrogen (H^+) ions into water is an **acid.** Hydrochloric acid (HCl) and sulfuric acid (H_2SO_4) are familiar acids. A molecule that releases hydroxide (OH) ions into water is a **base,** such as sodium hydroxide (NaOH), commonly known as lye. Molecules that release ions other than H^+ and OH are *salts*, such as sodium chloride (NaCl). Water containing ions is known as an *electrolyte solution* and can carry an electrical charge. Athletes who have lost valuable salts by sweating and individuals who have lost them through diarrhea can drink solutions of the appropriate salts to restore the normal electrolyte balance of the body's fluids.

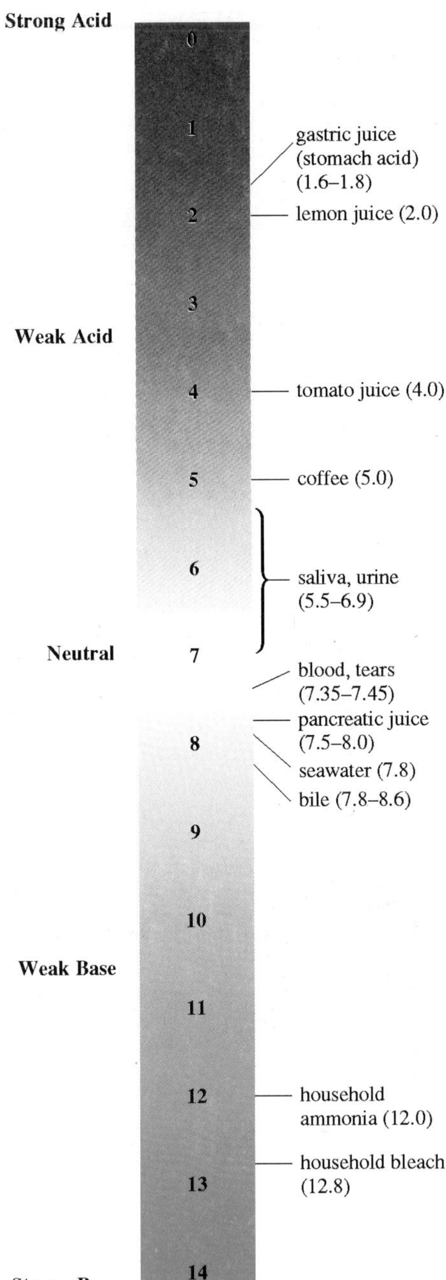

Figure 3.11
The pH scale is commonly used to measure the strength of acids and bases.

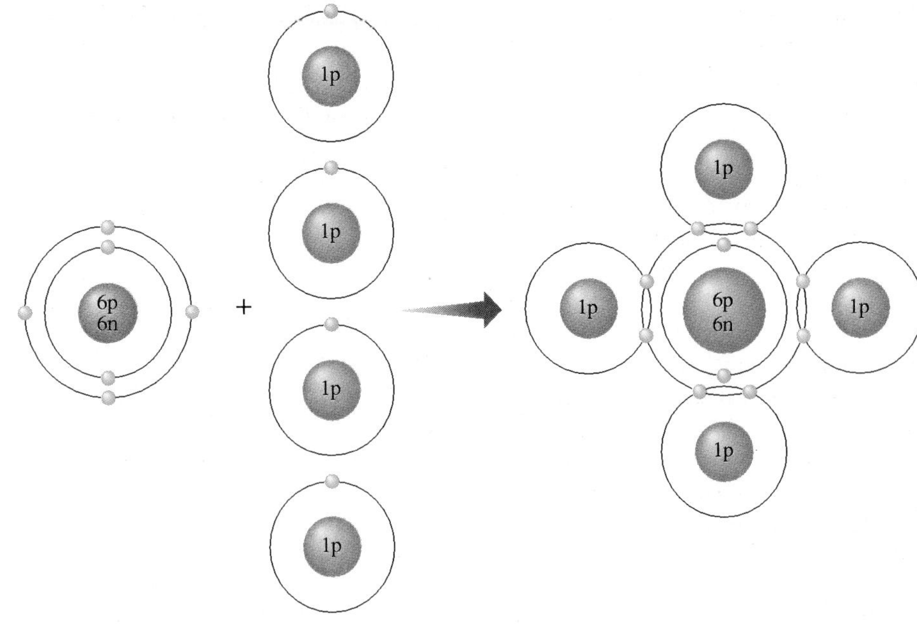

Figure 3.12
A covalently bonded molecule—methane (CH_4). By sharing electrons, one carbon and four hydrogen atoms complete their outermost shells. Note that the first electron shell is complete with two electrons.

In order for vital chemical reactions to occur in an organism, many biological fluids (water with other chemicals dissolved in it) must have a specific acidity or basicity. A solution's acidity or basicity influences its interactions with other molecules. A system of measurement called the **pH scale** gauges how acidic or basic a solution is in terms of its concentration of hydrogen ions. The pH scale ranges from 0 to 14, with the 0 representing strong acidity and the 14 representing strong basicity. A neutral solution has a pH of 7.

Many fluids in the human body function within a narrow pH range (fig. 3.11). The pH of blood ranges from 7.35 to 7.45, and disruptions in its acid-base balance result in illness. For example, changes in blood pH cause the "altitude sickness" sometimes felt by athletes who are accustomed to working out at sea level when they exercise at higher altitudes. Because there is less oxygen in the air at high altitudes, the athletes must breathe harder (hyperventilate) to get enough oxygen. They exhale so much carbon dioxide that the pH of their blood increases, which makes them feel sick. On a larger biological scale, acid (low-pH) precipitation has devastating effects on life in lakes.

Covalent Bonds—Sharing Electron Pairs

Atoms that have three, four, or five electrons in their valence shells are more likely to share electrons in a **covalent bond** than they are to swap them in the electron give-and-take of an ionic bond. Carbon, with four electrons in its outer shell, is a good example. It can attain the stable eight-electron configuration in its outer shell by sharing electrons with four hydrogen atoms, each of which has one electron in its only shell. The resulting compound is methane, CH_4, a gas given off in swamps (fig. 3.12). Methane can be represented graphically in several ways (fig. 3.13).

Two or three electron pairs can also be shared in covalent bonds, which are termed, respectively, double and triple bonds. Carbon atoms can form all three types of covalent bonds with other carbon atoms, building the frameworks of a biologically important class of compounds called **hydrocarbons.** These carbon-based rings and chains are also covalently bonded to hydrogen atoms. Bonding of additional chemical groups produces various other biological molecules.

Hydrogen Bonds

Sometimes the sharing of electrons in a covalent bond is not equal. In a water molecule (H_2O), for example, the single oxygen atom has a stronger hold on the shared electrons than do the two hydrogen atoms. As a result, the shared electrons are closer most of the time to the oxygen atom. Because of this unequal (or "polar") sharing, the hydrogen atoms have a partial

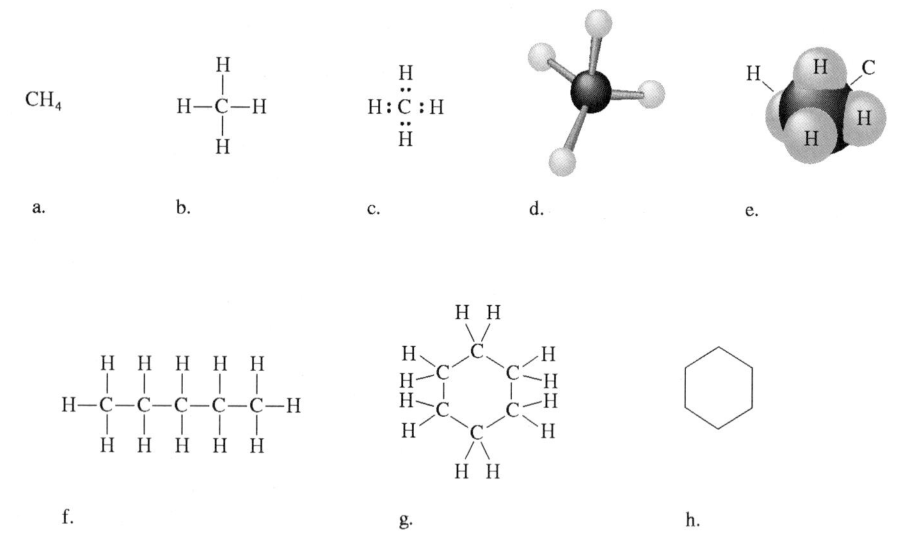

CH_4

a. b. c. d. e. f. g. h.

Figure 3.13
Different types of diagrams are used to represent molecules. Consider methane, a gas present on the early earth that may have played a pivotal role in the debut of life. *a.* The molecular formula indicates that methane consists of one carbon atom bonded to four hydrogen atoms. *b.* The structural formula shows those bonds as single. *c.* The electron dot diagram shows the number and arrangement of shared electron pairs. *d.* A ball-and-stick model gives further information—the angles of the bonds between the hydrogens and the carbon. *e.* A space-filling model shows bond relationships as well as the overall shape of a molecule. The organic compounds of life are built of carbon chains (*f*) and rings (*g*), with attached hydrogens, oxygens, and sometimes other elements such as nitrogen, phosphorus, and sulfur. *h.* The carbons of ring-shaped molecules are sometimes indicated only by the vertices of the bond lines, and the hydrogen atoms in such structures are usually not explicitly shown.

positive charge and are therefore attracted to other, negatively charged atoms nearby. This attraction, which is weak compared with ionic and covalent bonds, is a **hydrogen bond.** It is represented graphically by a dashed line (fig. 3.14). Hydrogen bonds form between water molecules, and they are also important in the structure of genetic material, as we shall see in chapter 13.

Chemical Bonding in Biology

The fact that a carbon atom forms four covalent bonds to satisfy the octet rule allows this element to assemble into long chains and intricately branching and ring-shaped *organic* (meaning "carbon-containing") molecules. Organic molecules are found extensively in living things of all kinds, as well as in some nonliving things. Biological molecules (those organic molecules found in living things) are also rich in the elements hydrogen, oxygen, nitrogen, phosphorus, and sulfur. Many organic molecules in living systems are quite large and are termed *macromolecules*. The plant pigment chlorophyll, for example, contains 55 carbon atoms, 68 hydrogens, 5 oxygens, 4 nitrogens, and 1 magnesium. The blood pigment hemoglobin is about the same size. *Molecular weight*, which is a measure of a molecule's size, is calculated by adding the atomic weights of the constituent atoms. Unlike organic molecules, most inorganic molecules (molecules that do not contain carbon) are small and are ionically bonded.

The molecules of life react with one another in many ways, creating the reaction cycles and pathways that enable an organism to obtain and use energy. These reactions constitute metabolism. Two frequently encountered types of metabolic reactions are **oxidation** and **reduction,** and the two typically occur together as oxidation-reduction ("redox") reactions. In general, oxidation entails the loss of electrons, and reduction entails the gain of electrons. Among organic molecules, these reactions often occur through the gain or loss of hydrogen atoms, each of which contains a single electron. In organisms, sequences of redox reactions form electron-transport chains, in which an electron is lost by one molecule and gained by another, and so on for several steps. Energy is transferred as a result. An electron-transport chain is part of photosynthesis, the process by which plants harness energy from sunlight. Chapter 6 explores certain metabolic pathways more closely.

KEY CONCEPTS

Chemicals are built of units called atoms, which in turn consist of protons and neutrons in a central nucleus, plus electrons that orbit the nucleus. Electron arrangement determines the properties of an atom, and the number of protons distinguishes pure substances, or elements. Atoms with the same number of protons but different numbers of neutrons are isotopes of the same element and can be radioactive. Atoms combine to form molecules, and a molecule built of more than one element is a compound. Compounds have different properties than their constituent atoms.

Atoms and molecules chemically react with one another by forming and breaking attractive forces called bonds. Ionic bonds form when atoms gain or lose electrons to complete their outermost, or valence, electron orbitals. An atom that gains or loses an electron is called an ion. A molecule that releases H^+ into water is an acid, and one that releases OH^- is a base; others are salts. The pH scale describes acidity and basicity. Atoms with three, four, or five electrons in their outermost shells tend to share electrons to form covalent bonds. A weaker type of bond is the hydrogen bond. Hydrocarbons are organic molecules that contain many hydrogen and carbon atoms and are important in life. In oxidation reactions, electrons are lost; in reduction reactions, electrons are gained.

Life's Chemical Components

About 99% of any living thing, from a single-celled bacterium to a thousand-pound hippopotamus, is composed of only six elements: carbon (C), hydrogen (H), nitrogen (N), oxygen (O), phosphorus (P), and sulfur (S). Living things are built almost entirely of organic compounds containing these elements, plus a substance with a unique collection of chemical properties that accounts

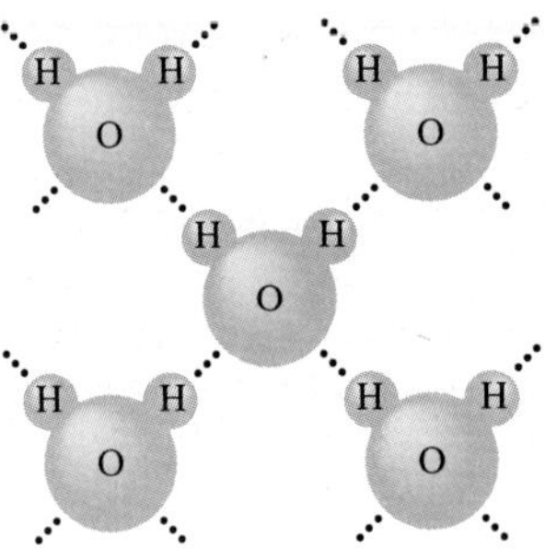

Figure 3.14
Hydrogen bonds hold water together. Hydrogen bonding (indicated by dotted lines) is responsible for water's cohesiveness. Each water molecule forms hydrogen bonds with four others.

for its importance in life—water (H_2O). In fact, water accounts for more than 50% of all living matter and more than 90% of the living matter of most plants.

Characteristics of Water

Molecules of water have a tendency to stick to each other. This is due to hydrogen bonding, which, in turn, is due to the fact that an individual water molecule is positively charged around its hydrogen atoms and negatively charged near its oxygen atom (fig. 3.14). Attraction of a substance to itself is called *cohesiveness*. In addition to bonding readily to itself (cohesion), water bonds to many other compounds (*adhesion*), making it an ideal basis for body fluids that carry vital dissolved substances. The movement of water from a plant's roots to its highest leaves depends both upon the cohesion of water within the plant's water-conducting tubes and upon the adhesion of water to the walls of those tubes. The adhesiveness of water also accounts for *imbibition*, which is the tendency of water to be absorbed by certain substances, causing them to swell. Rapidly imbibed water swells a seed so that it bursts through the seed coat, releasing the embryo within. Imbibition is why macaroni swells when it is placed in boiling water.

Anyone who has exercised outdoors on a hot summer day knows how important water is to temperature regulation. Water is valuable in biological temperature control because a great deal of heat is needed to raise its temperature. Thus an organism can be exposed to considerable heat before its water-based body fluids become dangerously warm. High *heat capacity* is a term used to describe this characteristic. A related function of water is its high *heat of vaporization*, which means that a lot of heat is required to make water evaporate. When enough heat is generated by exercise to produce sweat, its evaporation from the skin carries heat away from the body (table 3.2).

Table 3.2
Properties of Water

Property	*Example of Importance to Life*
Adhesiveness	Forms the basis of many biological fluids that carry dissolved substances
Imbibition	Embryo within seed absorbs water to swell and burst through seed coat
High heat capacity	Maintenance of constant internal body temperature
High heat of vaporization	Sweating (cooling mechanism)

Water is essential to life on a global scale. Because water expands upon freezing (unlike most other fluids), and therefore becomes lighter, ice floats on the surface of liquid water. When the air temperature drops sufficiently, a small body of water freezes at the surface, forming a solid cap of ice that retains the heat in the water below, protecting the organisms that live in the water from freezing. Many, many organisms are able to survive very cold weather in such shielded lakes and ponds. If water were to become denser upon freezing, ice would sink to the bottom, and the body of water would gradually turn to ice from the bottom up. Any organisms living in the water would be frozen in the process.

Water in the Human Body

Water does much more in the human body than just provide the basic material for blood, sweat, and tears. Water is in saliva, intestinal juice, and cerebrospinal fluid (which bathes and protects the brain and spinal cord). Lymph, a fluid that flows in vessels throughout the human body, consists of liquid that has seeped out into the tissues through the walls of the tiniest blood vessels. Lymph is 92% water. The amniotic fluid that cushions a developing fetus in the womb is also mostly water. Even the start of a new human life depends upon this remarkable substance. At the time of month when a woman is most likely to conceive, the mucus in her reproductive tract becomes more watery, easing the sperm's journey to the egg. On a microscopic level, water accounts for a large part of each cell. On a macroscopic level, about 67% of the human body is water.

Too much or too little water can affect health in obvious ways. When the intestines absorb too much water from the feces, constipation can result. When they absorb too little, diarrhea is the unpleasant consequence. The kidneys control the concentration of urine by removing or adding water. A breast-feeding woman is especially aware of the importance of water in her body, because she needs to consume extra water in order to manufacture milk. In fact, her body has a built-in mechanism to make sure that she does so: When the baby begins to suck, a nursing mother typically feels an immediate and intense thirst.

Organic Compounds of Life

Four types of organic compounds are especially important to the functioning of an organism: carbohydrates, fats, proteins, and nucleic acids. Vitamins are another group of biologically important organic molecules.

Carbohydrates

Carbohydrates include the sugars and starches and contain the elements carbon, hydrogen, and oxygen, with twice as many hydrogens as oxygens (fig. 3.15). They are classified by size. The **monosaccharides** (single sugars) contain five or six carbons.

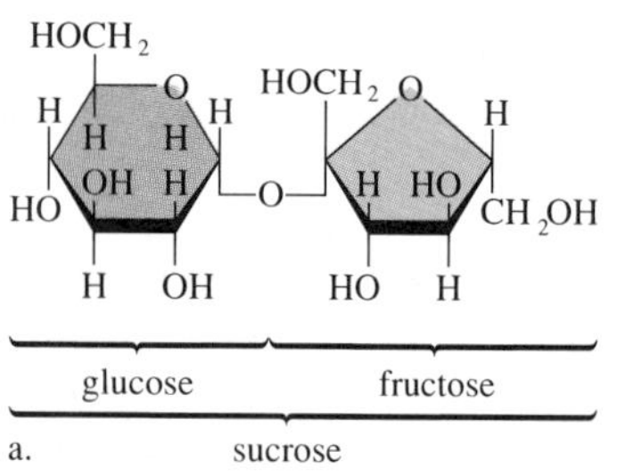

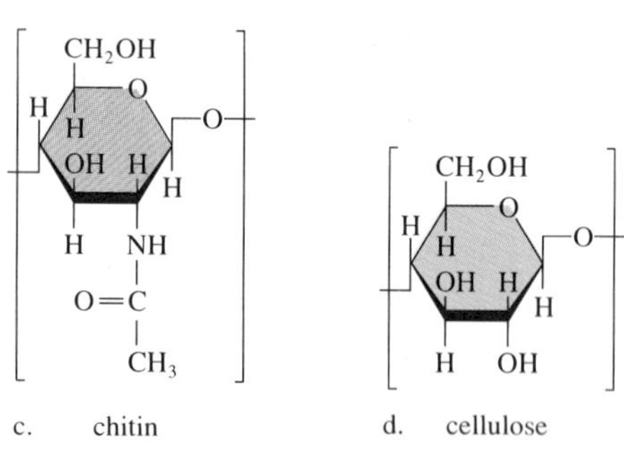

Figure 3.15

Carbohydrates—sugars and starches. *a.* Sucrose, or table sugar, is a disaccharide (double sugar) composed of the monosaccharides (single sugars) glucose and fructose. *b.* The strong jaws of insects are constructed of a complex carbohydrate called chitin. They can crunch through the complex carbohydrate of plants called cellulose. Although chitin and cellulose are both polysaccharides, the repeating single sugar of chitin (*c*), a six-carbon sugar with a nitrogen-containing chemical group bonded to it, is chemically different from that of cellulose (*d*).

Three six-carbon sugars with the same molecular formula, but different structures, are *glucose* (blood sugar), *galactose*, and *fructose* (fruit sugar).

Disaccharides (double sugars) form when two monosaccharides join and, in the process, lose a molecule of water (H_2O). This is an example of a type of chemical reaction that we will encounter again, called **dehydration synthesis** (making by losing water). In the opposite reaction, **hydrolysis** (breaking with water), a disaccharide molecule and a molecule of water react to form two monosaccharide molecules. Dehydration synthesis builds molecules and hydrolysis breaks them down into smaller molecules.

Table sugar, *sucrose*, is a disaccharide formed from one glucose and one fructose molecule. Its molecular formula is $C_{12}H_{22}O_{11}$, and its structure is shown in figure 3.15. Sucrose is found in sugar cane, beets, and many other vegetables and fruits. *Maltose*, a disaccharide formed from two glucose molecules, is found in sprouting seeds and is used to make beer. *Lactose*, or milk sugar, is a disaccharide formed from glucose and galactose.

Monosaccharides and disaccharides are together termed **simple carbohydrates.** These biochemicals provide energy. In the human body, simple carbohydrates are all broken down to, or converted into, glucose before energy is extracted from their chemical bonds. This means that any one type of sugar is probably no better or worse for your health than any other. In fact, many sugars sold as "health foods" are rather high in sucrose, the very substance that health-food advocates often tell us we should not eat (table 3.3).

Complex carbohydrates are polysaccharides (many sugars), which are chains of simple sugars linked by dehydration synthesis (fig. 3.16). Such long molecules built of similar smaller ones, like links of a chain, are widespread in the chemistry of life. The long molecule is termed a **polymer** (many units), and each piece is a *monomer* (one unit). The complex carbohydrates discussed following are all glucose-based polymers that are broken down in the digestive system by hydrolysis to their sugar subunits. You can test this by keeping a piece of bread in your mouth for a few minutes. The bread begins to taste sweet as a chemical in your saliva, salivary amylase, breaks the starch down into molecules of maltose, a simple sugar.

Starch, which is found in plants, and *glycogen*, its animal equivalent, are energy-storing complex carbohydrates. In the human body, glycogen accumulates mainly in the liver and muscles. During strenuous endurance sports, such as marathon running, glycogen is broken down to release energy-rich sugar molecules. In fact, the total exhaustion ("hitting the wall") that runners often experience at about the 20-mile (30-kilometer) mark in the 26.2-mile (42.16 kilometer) marathon marks the point at which glycogen reserves are depleted and the body turns to fat reserves for energy.

Cellulose is a complex carbohydrate that provides support in plants. It is the most abundant polymer in nature. Cellulose is the major component of dietary fiber, which was called roughage a generation ago. Foods that are rich in fiber, such as bran, speed the movement of feces through the intestines. This hastening action aids both in the digestion of nutrients

Table 3.3
The Composition of Various Sweeteners

Sweetener	*Composition*		
	Sucrose (%)	*Glucose (%)*	*Fructose (%)*
Table sugar	100		
Brown sugar	98		
Molasses	54	9	6
Honey (clover)	trace	33	39
Corn syrup (light)		19	2
Fructose syrup		6	72
Fructose powder		trace	96-100

cellulose

starch

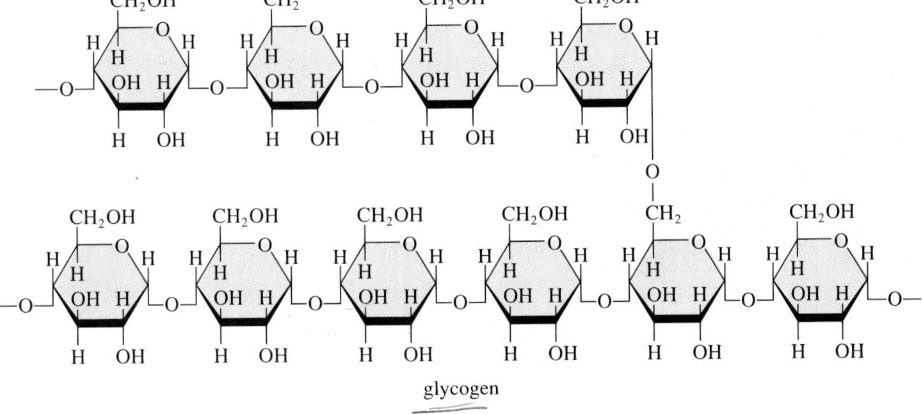

Figure 3.16
The complex carbohydrates cellulose, starch, and glycogen are all chains of glucose subunits, but they differ in structure. The glucose subunits of cellulose (found in plants) are linked together differently than the glucose subunits of starch (also found in plants). Glycogen is also a glucose polymer, found in animals, but it is more highly branched than either cellulose or starch.

and the rapid elimination of toxins from the body, and it may even prevent disease by reducing the time that disease-causing microorganisms stay in the intestinal tract. Cellulose is used in the food industry to keep ice cream, salad dressing, and whipped toppings creamy, to hold fish cakes together, and to thicken milkshakes.

Chitin, another complex carbohydrate, is the second most abundant natural polymer. It forms the outer coverings of many organisms, including insects, crabs, and lobsters. In Japan, 300 tons of chitin are processed each year for water purification and waste treatment. Research in harnessing this biological resource is growing, and chitin will soon be found in surgical thread, burn dressings, contact lenses, animal feed, and drugs.

Fats

Fats contain the same elements as carbohydrates but with proportionately less oxygen. They are part of a broader class of organic compounds, the **lipids,** which are characterized by their insolubility in water. A fat is built of two other kinds of organic molecules: a small molecule of *glycerol*, which serves as a backbone from which extend three *fatty acids* (fig. 3.17). **Adipose cells,** which make up the fatty tissues of the body, specialize in storing fat and contain little else.

Fats have many biological functions. They add flavor to food, and they slow digestion, thereby delaying the return of hunger. Even the strictest diet must contain some fat, for this much-maligned substance is vital for proper growth and the utilization of some vitamins. Fats are excellent energy sources, providing more than twice as much energy as equal amounts (by weight) of carbohydrate or protein. Lipids are the major components of membranes, the selective barriers that surround a cell and divide it into compartments. They are the basis of the steroid hormones (such as testosterone) that are important in reproduction, as well as a chemical precursor of these hormones, cholesterol. Although a certain amount of cholesterol is necessary to manufacture these hormones, excess cholesterol collects on the inner walls of blood vessels, sometimes causing heart and circulatory disease.

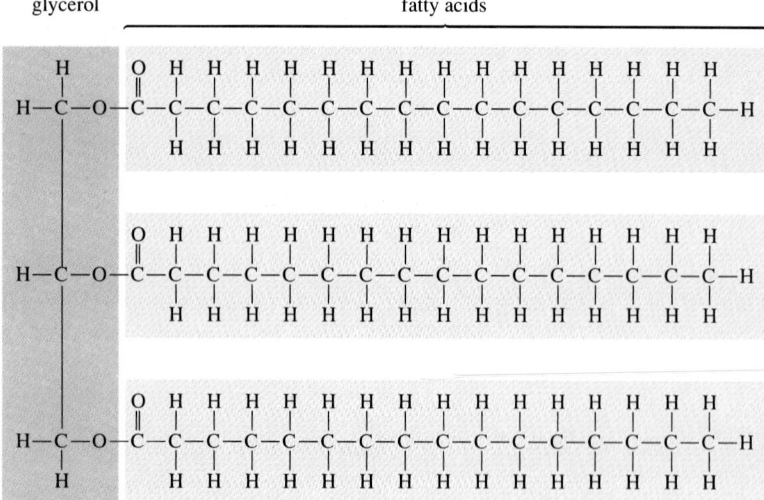

Figure 3.17
Recipe for a fat: one glycerol, three fatty acids. This is tripalmitin, a type of fat called a triglyceride.

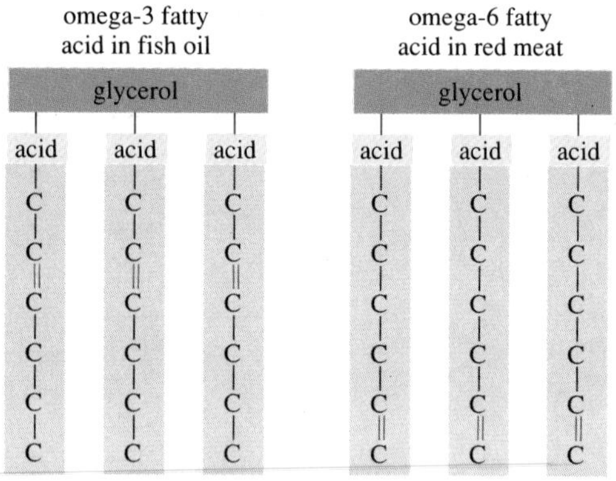

omega-3 fatty acid in fish oil

omega-6 fatty acid in red meat

glycerol

Figure 3.18
Omega-3 fatty acids in the diet may lower risk of heart disease. Very low heart disease rates among Greenland Eskimos and Dutch men who eat lots of fish suggest that the unsaturated omega-3 fatty acids in fish oil might be healthier for the human heart than the unsaturated omega-6 fatty acids found in red meat.

The fatty acids distinguish different fats from one another, and their particular structures seem to have a great deal to do with their effects in the human body. Fat from beef, for example, has fatty acid "tails" 14 carbons long, whereas the fat in butter has 4 carbons in its tails. A fat is **saturated** when its fatty acids contain all the hydrogens that they possibly can—that is, it has all single bonds between its carbons. A fat is **unsaturated** if it has one double bond between carbons and *polyunsaturated* if it has more than one double bond. Vegetable oils tend to be more unsaturated than fats in meat. It is the saturated type of oil that is thought to cause heart disease in humans.

The site of saturation (the double bond) also seems to have an impact on health. These sites are numbered beginning with the carbon closest to the acid group (COOH), which attaches the fatty acid tail to the glycerol backbone. Most fats in a typical, heavy on red meat Western diet are classified as "omega minus 6," which means that the single double bond is six carbons down from the one next to the COOH, which is called the omega carbon (fig. 3.18). (Omega is the last letter in the Greek alphabet, corresponding to the COOH end of the fatty acid.) Greenland Eskimos, who have very low incidences of heart disease but tend to bleed easily, eat mostly omega-3 fatty acids in the seal, whale, and fish that make up most of their diet. Further evidence that omega-3 fatty acids promote a healthy heart comes from studies of Dutch men who ate a very fishy diet for 20 years. They had significantly less heart disease than Dutch men who did not eat much fish. The blood-thinning effects of omega-3 fatty acids are thought to be caused by still unknown influences on body chemicals called prostaglandins.

Some fats make animal fur and bird feathers water repellent, and others cushion organs and help to retain body heat by providing insulation. Fat cells are often found together as adipose tissue, of which there are two types. *White adipose tissue* accounts for most of the fat in human adults. It is packed with energy-rich lipids. *Brown adipose tissue* is rare in adults, but it is found in layers around the neck and shoulders and along the spine of newborns.

Brown adipose tissue was originally identified about 40 years ago in small hibernating mammals, where it was thought to provide insulation during cold weather or hibernation. Today we know that brown fat warms the body by converting the energy contained in its lipids directly into heat, rather than storing it as chemical energy as white adipose tissue does. In this process of thermogenesis (making heat), a brown fat cell releases up to 80 times the amount of heat as other cells. Thermogenesis is triggered by a drop in temperature, and it is controlled by hormones. Obesity in humans may in some cases be caused by brown adipose tissue that does not respond normally to the temperature and hormonal cues that should cause it to begin producing heat, resulting in a buildup of lipids.

Proteins

Proteins have varied uses in organisms (table 3.4). Hemoglobin shuttles oxygen from the lungs to the rest of the body. Collagen and elastin are part of the connective tissue that literally holds the human body together. Proteins turn the genetic material on and off at appropriate times, orchestrating the development of a complex organism from a fertilized egg to an adult. Tiny peptides called enkephalins, which consist of only five amino acids each, influence the perception of pain and pleasure.

The functions of proteins are intimately tied to their structures. Proteins are polymers built of one or more chains of **amino acids.** An amino acid contains a central carbon atom bonded to the following:

1. A hydrogen atom.
2. Another carbon atom that is double-bonded to an oxygen and single-bonded to a *hydroxyl group* (OH). This is a *carboxyl group*, and it is characteristic of all organic acids.

Table 3.4
Some Proteins and How They Function in the Human Body

Function	*Protein*	*Where Produced in Body*
Digestion	Amylase Pepsin	Mouth and pancreas Stomach
Transport of oxygen	Hemoglobin Myoglobin	Blood Muscle
Storage of iron	Ferritin	Liver
Motion	Actin, myosin	Muscle
Support	Collagen, elastin	Connective tissue, cartilage, bone, skin, blood vessels
Protection against infection	Antibodies	Blood plasma
Regulation of blood-sugar level	Insulin	Pancreas

amino group — acid group

a.

peptide bond

b.

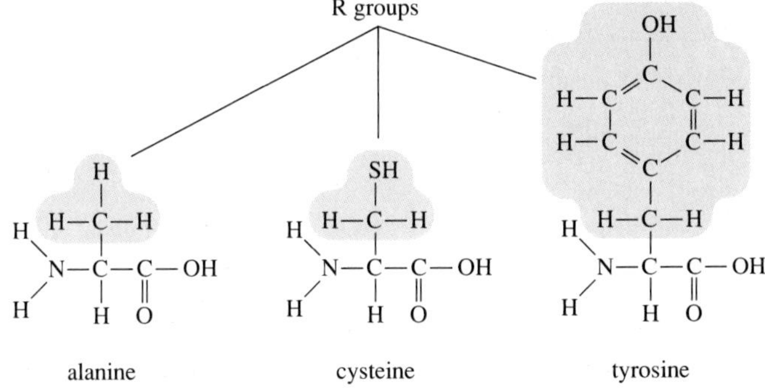

c.

Figure 3.19
Amino acids, the building blocks of protein. *a.* An amino acid consists of a central carbon atom covalently bonded to an amino group, an acid (carboxyl) group, a hydrogen, and a side group designated "R." *b.* By the removal of the elements of water (H_2O) from the amino group of one amino acid and the acid group of another, two amino acids bond to form a dipeptide. This is the synthesis of a peptide bond. *c.* Different amino acids are distinguished by their R groups. Twenty amino acids are important in life, three of which are shown here.

3. A nitrogen atom that is single-bonded to two hydrogen atoms. This is an *amino group*.
4. A side chain, or *R group*, that can be any of several other chemical groups.

The nature of the R group distinguishes the 20 different amino acids commonly found in living things (fig. 3.19). Therefore, proteins differ from carbohydrates and fats in that, in addition to carbon, hydrogen, and oxygen, they contain nitrogen and sometimes sulfur in the R groups of their amino acids.

Amino acid chains are held together by several types of chemical bonds. Two adjacent amino acids are joined by dehydration synthesis (losing a water molecule), which links the acid-group carbon of one amino acid to the nitrogen of the other. This is a **peptide bond** (fig. 3.19). Each chain is generally at least 100 amino acids in length, built up piece by piece from a *dipeptide* (2 amino acids), to a *tripeptide* (3 amino acids), to a larger *peptide*, and finally to a long chain, termed a *polypeptide*. A protein breaks down into its constituent amino acids by hydrolysis (reacting with water).

Hydrogen bonds between the R groups of different amino acids help to determine a protein's overall shape. Another type of bond, called a *disulfide bond,* occurs between two sulfur atoms, each of which is a part of the R group of a particular amino acid, cysteine. Disulfide bonds form bridges between parts of an amino acid chain. For example, some proteins contain regions called zinc fingers, which consist of loops of amino acids that wrap around zinc atoms. Strategically located cysteines, which are attracted to one another, fold the protein into loops. Such zinc finger proteins may be involved in activating certain hormones. Disulfide bonds are also found in keratin, the protein that forms hair.

Proteins are usually not just straight chains of amino acids but complex three-dimensional molecules. This shape, or **conformation,** is due largely to attractive forces between the amino acids that are linked to form a polypeptide chain. The amino acid sequence of a polypeptide is called its **primary** (1°) **structure.** Chemical attractions between amino acids that are close together in the 1° structure fold the polypeptide chain into its **secondary** (2°) **structure.** For example, the amino acid chain in figure 3.20 is wound by secondary attractions into a coil. Chemical attractions between amino acids that are far apart in the amino acid chain provide larger loops, known as the molecule's **tertiary** (3°) **structure.** Some proteins are built of more than one polypeptide chain. The arrangement of these chains constitutes the protein's **quaternary** (4°) **structure.** For example, the protein in blood that carries oxygen, hemoglobin, has four constituent polypeptide chains. The liver

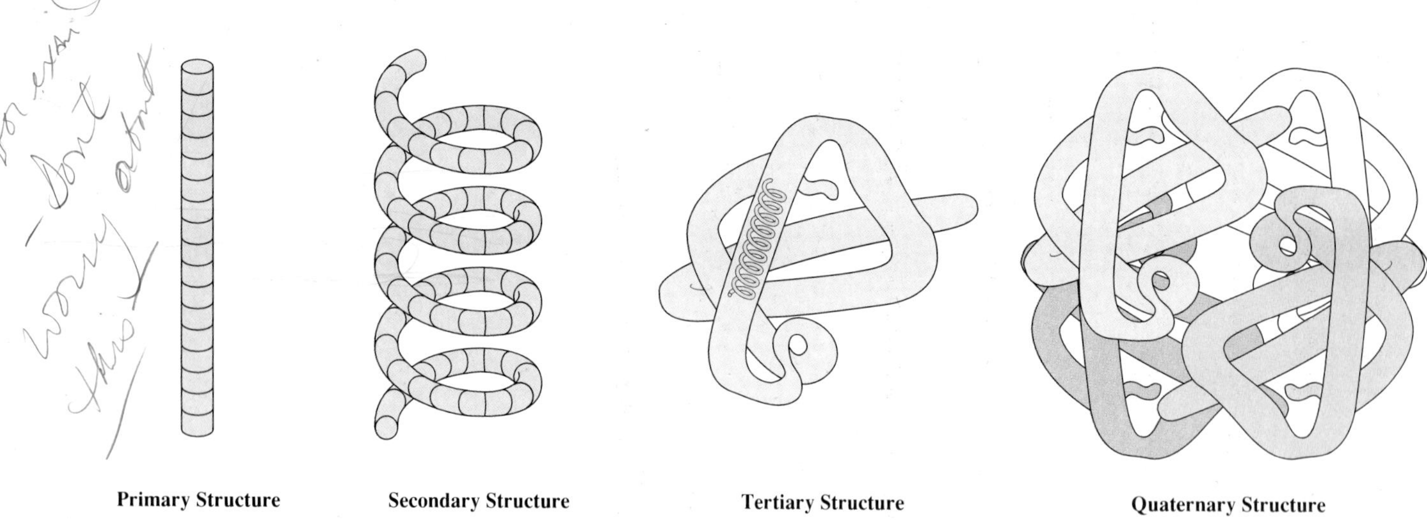

Figure 3.20
Protein structure determines protein function. A protein has a characteristic three-dimensional form, or conformation. It is determined by the amino acid sequence (primary structure), attractive forces between amino acids (secondary and tertiary structures), and, in the case of a protein consisting of more than one polypeptide, arrangement of polypeptide subunits (quaternary structure).

protein ferritin is built of 20 identical polypeptide subunits of 200 amino acids each. In contrast is the muscle protein myoglobin, which is a single polypeptide chain. By mimicking common shapes called motifs, which are seen in natural proteins, "protein engineers" are trying to invent new types of proteins (fig. 3.21).

Proteins are vital to human health, and they are an important part of our diet. Some of the amino acids the human body needs for building its own proteins can be manufactured by recycling other body chemicals. The others, though, must come from food. The proteins in food are broken down into amino acids by digestion, travel in the circulatory system to cells throughout the body, and are then built up again into human proteins. Many peptides can be used as drugs, but they cannot be taken orally because the digestive system dismantles them. For example, the peptide insulin, which helps to regulate blood-sugar level, is given by injection to people who have diabetes.

Different foods contain different mixtures of amino acids, and a food with a good balance of what the body needs is said to have "complete protein." Animal products supply complete protein, but many plant foods do not. This is why a vegetarian must mix foods with complementary amino acid components to obtain a balanced diet. Macaroni and cheese or beans and rice are combinations that provide a fairly good balance of amino acids.

A type of protein of great importance to living things is the **enzyme.** Enzymes are large protein molecules that alter the rates of chemical reactions without being used up in the process, a phenomenon called *catalysis*. A little bit of an enzyme goes a long way. Some enzymes can increase reaction rates a billionfold. Without these proteins, many important chemical reactions in organisms would proceed far too slowly to support life. Enzymes are essential to biological growth, repair, and waste disposal. They function under specific conditions of pH and temperature, and they are usually also specific to particular chemical reactions.

Consider, for example, the enzyme carbonic anhydrase. In human tissues, it speeds the reaction of carbon dioxide (CO_2) with water (H_2O) to form bicarbonate ion (HCO_3^-) and hydrogen ion (H^+) by a factor of 10 million. This reaction allows carbon dioxide, which is produced abundantly in the tissues as a by-product of energy use, to be transported as the easily dissolved bicarbonate ion to the lungs, where it is released as carbon dioxide gas.

In humans, enzymes are essential to digestion, blood clotting, nerve transmission, and a complex assortment of metabolic reactions. A missing or defective enzyme can be devastating. A child who has Tay-Sachs disease is missing the enzyme hexoseaminidase, resulting in fat buildup on nerve cells that leads to gradual deafness, blindness, and paralysis and death by the age of three or four years. If the enzyme whose absence causes an illness is known, its measurement can serve as a diagnostic tool. Thus, determination of hexoseaminidase levels in blood can identify adults likely to have a Tay-Sachs child. (Such adults have less of the enzyme than most people, but enough that their own health is not impaired.) Determination of hexoseaminidase levels in cells taken from a fetus can reveal whether or not the child will be born with the disorder.

Some enzymes are used directly as medicines, such as streptokinase and tissue plasminogen activator, which dissolve blood clots in the heart. Such therapeutic enzymes can come from cadavers, animals, or genetically engineered bacteria, or they can be synthesized in a laboratory. An enzyme treatment can even be discovered by accident. An 11-year-old girl who suffered abdominal pain and diarrhea after eating sucrose found that her symptoms did not develop when she ate fresh baker's yeast too. The girl's love for fresh bread led to a treatment that worked for others with the same condition. Apparently an enzyme in yeast—sucrase—was able to replace the sucrase that the girl's body lacked.

Figure 3.21
Protein engineering. Protein structures are enormously variable because they are composed of 20 types of building blocks that can be ordered in any possible sequence. While some researchers attempt to decipher the "rules" of protein folding, others try to imitate the helices, sheets, loops, kinks, turns, bundles, folds, and barrels of natural proteins to create new molecules. One of the first human-made proteins is a ring of four helices, modeled after natural proteins that monitor passage of ions across cell membranes.

The key to an enzyme's specific action lies in a region of its surface called the **active site** (fig. 3.22), to which a very few molecules, called substrates, are chemically attracted. Substrates fit into the active site to form a short-lived partnership, the *enzyme-substrate complex* (fig. 3.23). An enzyme can hold two substrate molecules that react to form one product molecule, or it can hold a single substrate molecule that splits to yield two product molecules. In either case, once the enzyme and the substrate (or substrates) have formed the complex, the reaction takes place very rapidly. Then the complex breaks down to release the products (or product) of the reaction. The enzyme is unchanged, and its active site is once again empty and ready to pick up more substrate.

The specificity of enzymes and the speed with which they work make them essential to life. The same qualities also give enzymes great industrial value. About 90% of the enzymes that are important to industry are derived from microorganisms, such as bacteria of the genus *Bacillus* and the fungus *Aspergillus*, and are used for cheese making, detergent manufacture, food processing, chemical synthesis, and waste treatment. Many companies maintain their own supplies of these microorganisms in huge fermenting vats, to which appropriate nutrients and substrates are added, creating brews that stimulate the organisms to produce more of the desired enzymes. After a time, the microorganisms are harvested, and the desired enzymes are separated, purified, dried, prepared, and packaged.

A growing trend in the industrial use of enzymes is to engineer their production genetically. This is done by extracting from a complex organism the genetic material that governs the production of a particular enzyme and inserting it into bacterial cells, which then churn out

Figure 3.22
The enzyme shown in green, HIV-1 protease, has a molecular stranglehold on the AIDS virus, preventing its replication.

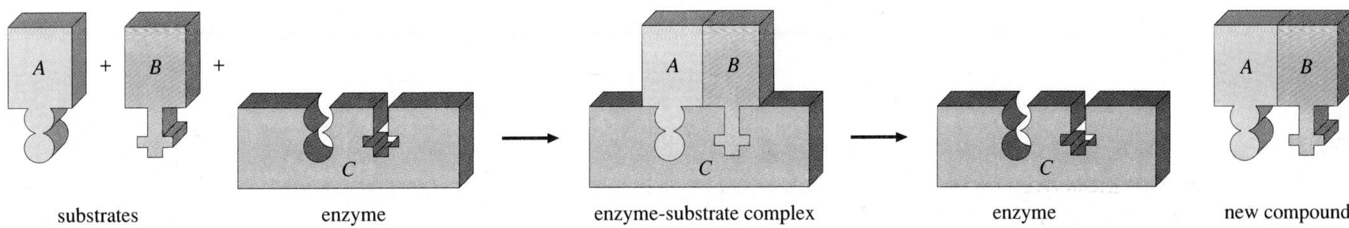

Figure 3.23
Enzyme action is like a "lock and key" fit. Substrate molecules A and B fit into the active site of enzyme C. An enzyme-substrate complex is formed, and A and B react. A new compound, AB, is released, and the enzyme is recycled. Enzyme-catalyzed reactions can break down as well as build up substrate molecules.

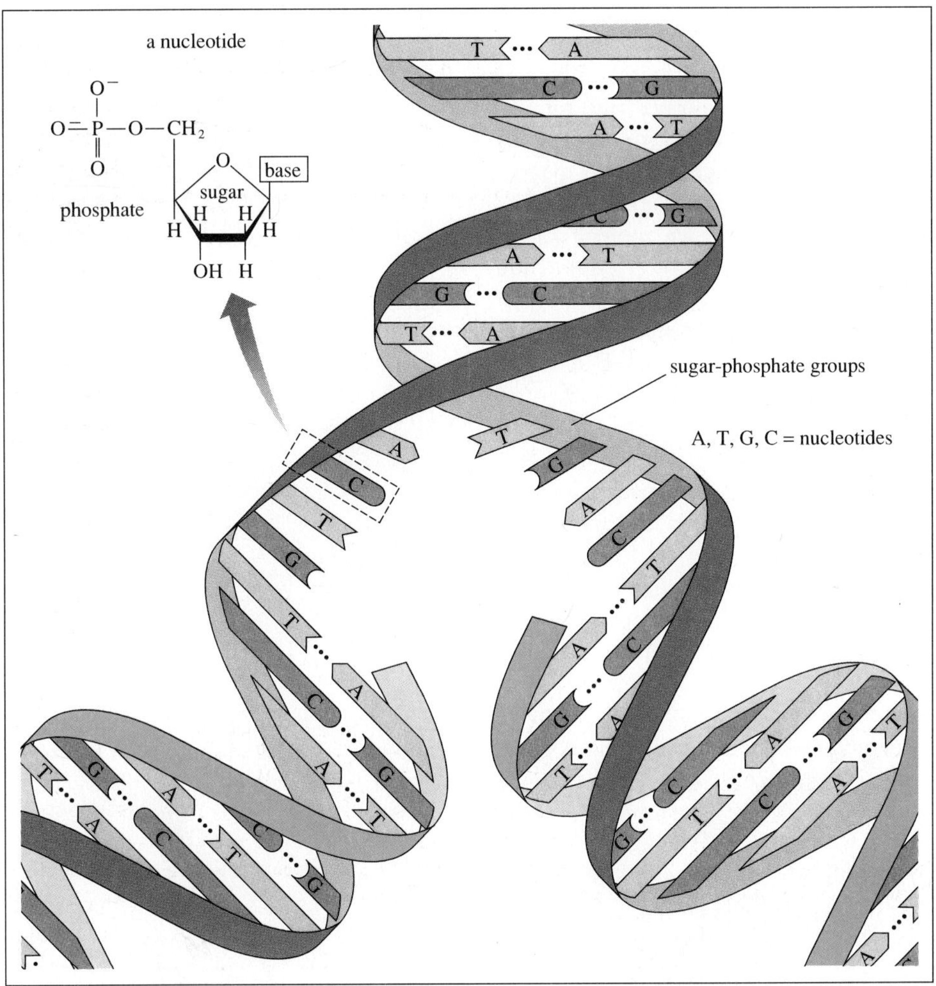

Figure 3.24
DNA, the hereditary molecule. A DNA molecule is constructed of building blocks called nucleotides, each of which consists of a five-carbon sugar, a phosphate group, and one of four different types of nitrogen-containing bases. The molecule resembles a spiral staircase and the steps are specific pairs of nucleotide bases and the sides are chains of sugar-phosphate groups. The double helix shape of the molecule depends upon the fact that the steps in the staircase are built of only two kinds of base pairs: A-T (or T-A) and G-C (or C-G). The molecule replicates by locally unwinding so that the base pairs split apart. Each exposed base then chemically attracts a free-floating nucleotide of the complementary type, building two identical DNA double helices from the original one.

the enzyme in large quantities. Research chemists are now trying to improve upon nature by constructing nonprotein chemicals that mimic the active sites of known enzymes. They hope to build catalysts that work efficiently but are not restricted to the narrow pH and temperature range of a living organism, as true enzymes are.

Nucleic Acids

The human digestive system breaks food proteins down into amino acids. How are these amino acids built up once again into the specific protein molecules that have so many important functions in the body? The answer lies in another polymer, a *nucleic acid*. The chemical building blocks of nucleic acids are called **nucleotides.** Each contains a five-carbon sugar, a *phosphate* group (PO_4), and one of four types of nitrogen-containing ring compounds, or *nitrogenous bases* (fig. 3.24). The nitrogenous bases are **adenine** (A), **guanine** (G), **thymine** (T), and **cytosine** (C). Long sequences of these nucleotides provide information that guides other cellular components in assembling free-floating amino acids into polypeptide chains. This process, which is much like translating one language (nucleic acid) into another (polypeptide), will be discussed in detail in chapter 13.

The nucleic acid **deoxyribonucleic acid** (DNA) is the genetic material of all species and many viruses, and it stores information that specifies the construction of proteins. A sequence of DNA that specifies a particular polypeptide is a **gene.** Built into the structure of a DNA molecule is a mechanism for the molecule to construct a replica of itself. The DNA molecule resembles a spiral staircase, with the sugar and phosphate groups of the nucleotides forming the sides of the staircase and the nitrogenous bases pairing to form the "steps." Pairs form only between adenine and thymine and between guanine and cytosine, forming a symmetrical double helix. When the DNA molecule replicates, the spiral staircase splits down the center and unwinds, separating each nucleotide pair. Each half of the pair now chemically attracts a free-floating nucleotide of the proper complementary type, and little by little the spaces created by the splitting of the original molecule are filled in. The final result is two new DNA molecules, each identical to the original. The ability of a molecule to replicate itself was an important requirement for the origin of life, as we shall soon see.

Vitamins

Vitamins are organic chemicals that are essential in small amounts for the normal growth and function of an organism but that cannot be synthesized by it. Vitamins are a chemically diverse group. A chemical that is a vitamin for one type of organism may not be a vitamin for another. For example, ascorbic acid (vitamin C) is required in the diets of humans, guinea pigs, monkeys, and the Indian fruit bat, but most other animals can manufacture all the ascorbic acid they need from glucose. For most animals, then, ascorbic acid is not considered a vitamin.

Although vitamins are needed in much smaller amounts than carbohydrates, fats, and proteins, they are no less essential to good health. Vitamin deficiencies (and, in some cases, vitamin excesses) can lead to illness. Vitamin C and the B-complex vitamins (including thiamine, riboflavin, pantothenic acid, niacin, pyridoxine, folic acid, biotin, and cyanocobalamin) are soluble in water. Excesses of these vitamins are excreted in the urine. Most of these water-soluble vitamins are essential to the proper functioning of enzymes. Vitamins A, D, E, and K dissolve in fat rather than water. Because they are not water soluble, they are not easily excreted, and excesses of these vitamins tend to accumulate in the body. Consuming too much of any of the fat-soluble vitamins can result in illness.

Inorganic Compounds in Life—Minerals

More than 90% of a living organism is composed of water and various organic compounds. The inorganic materials that make up the remainder are also essential to life. In fact, without these *minerals*, our muscles would not contract, our bones would not support our weight, our nerves would not relay messages, and many enzymes would be stopped cold, bringing life to a sudden halt. Minerals are master regulators, controlling blood clotting, heartbeat, oxygen transport, and the pressures of body fluids. Inorganic compounds required in large amounts (100 milligrams per day or more) are termed *bulk minerals* and include calcium, phosphorus, potassium, sulfur, sodium, chloride, and magnesium. The *trace elements*, including zinc, iron, manganese, copper, iodine, cobalt, fluoride, chromium, and selenium, are needed only in very small amounts. *Ultratrace elements* are needed in even smaller amounts. These include some substances that are dangerous if present in anything other than very tiny amounts, such as cadmium, lead, and arsenic.

A certain amount of each mineral is optimal for good health, and, as for vitamins, deficiencies and excesses can be dangerous. Excesses are particularly harmful because they can disrupt the balances of other minerals. Many children are admitted to emergency rooms because they have swallowed iron supplements. Complications of mineral excesses can include anemia, broken bones, nervous disorders, and even birth defects. Mineral balance is especially important for the very young, the elderly, pregnant and breast-feeding women, and the chronically ill.

Some elements not needed in the diet can be quite harmful if ingested. Symptoms of lead poisoning, including lethargy, muscle weakness, abdominal pain, and behavioral changes, were noted as much as 2,000 years ago. Today, the problem is most likely to show up as learning disabilities in children who have eaten chips of lead-based paint. Pregnant women who eat fish heavily contaminated with mercury risk giving birth to babies with the deformities and mental retardation of Minamata disease. Cadmium poisoning disrupts the body's proper utilization of iron, copper, and calcium, resulting in anemia, kidney disease, and bone destruction.

Minerals are found throughout the living world. Each molecule of the plant pigment chlorophyll, which harnesses the sun's energy to make carbohydrates, is built around a magnesium atom. The iron-containing compound magnetite functions as a built-in magnet in bacteria, birds, and bees, where it is thought to aid in navigation. The elements phosphorus, potassium, magnesium, iron, calcium, chlorine, zinc, and copper are vital to all organisms.

KEY CONCEPTS

Living things are built almost entirely of organic compounds containing carbon, hydrogen, nitrogen, oxygen, phosphorus, sulfur, and water. Water is vital to life because it forms the basis of many body fluids, it is imbibed by structures such as seeds, and it controls temperature regulation. Carbohydrates include the sugars and starches and are broken down in the body to release the energy held in their covalent bonds. Like carbohydrates, fats (a type of lipid) contain carbon, hydrogen, and oxygen but with less oxygen. Fats provide energy, they slow digestion, they enable certain vitamins to function, and they form cell membranes, cholesterol, and some hormones. Proteins are long chains of amino acids. The amino acids interact to give the protein its three-dimensional conformation, which is important to its function. Proteins have a wide variety of functions in organisms, and one vital one is as enzymes. Protein sequences are specified by DNA sequences, another organic molecule whose building blocks are nucleotides. DNA is the genetic material. Vitamins are organic compounds and minerals are inorganic compounds, both of which are important to living things.

The Origin of Life on Earth

So far, we have considered life from a whole organism viewpoint and in terms of its chemical constituents. Attempting to define or describe life leads to one overriding and perhaps unanswerable question: How did all the amazingly efficient living organisms on this planet come into existence?

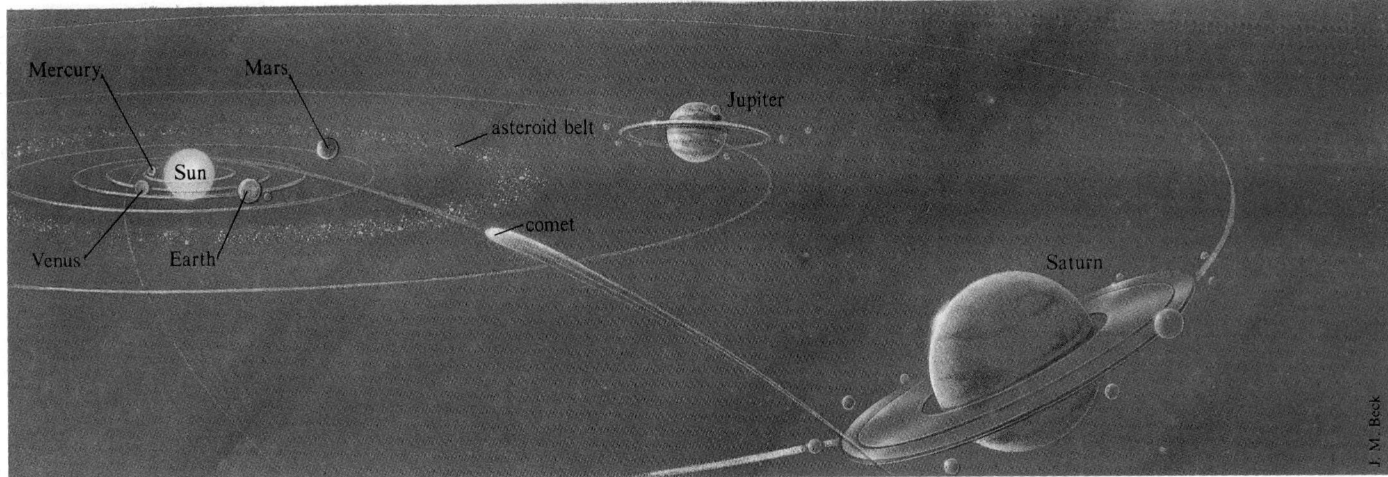

Figure 3.25
Did life on earth come from space? Comets, meteors, and interstellar dust are rich in the same organic molecules found in cells. Some scientists interpret this to mean that life on earth was seeded from preexisting life in space. Can you think of an alternative hypothesis to how life originated?

Many theories have been proposed to answer this question. They represent different perspectives, ranging from an unquestioning belief in religious scriptures to a scientific search for clues to past life. Attempts have even been made to recreate the conditions that existed when life first appeared on earth.

Spontaneous Generation

In the seventeenth and eighteenth centuries, a popular theory of life's origins was **spontaneous generation,** the appearance of the living from the nonliving. "Evidence" abounded: Beetles and wasps seemed to sprout from cow dung. Egyptian mice seemed to arise from the mud of the Nile. A recipe for making bees was to kill and bury a young bull in an upright position so that its horns protruded from the earth. After 1 month, a swarm of bees was supposed to fly out of the corpse of the bull. Even such noted scientists as Newton, Harvey, and Descartes did not question the theory that life could arise from practically any dirty mess.

There were some doubters, though. In the middle of the seventeenth century, Francesco Redi, an Italian physician and poet, conducted a simple but effective experiment. He filled two jars with meat, leaving one open and covering the other one lightly with cloth. He saw flies enter the open jar. Shortly afterwards, maggots appeared, and then new flies. The covered jar produced no flies. He had shown that new flies come from old flies, and not from decaying meat, as was previously thought. Soon John Needham, an English naturalist, showed that microorganisms flourished in various soups that had been exposed to the air. A few years later, Lazzaro Spallanzani, an Italian biologist, boiled soup for 1 hour in flasks that were then sealed by melting the mouths of the flasks shut. When the soup in those flasks was examined some days later, no microorganisms could be found. However, soup in sealed flasks that had been boiled for only a few minutes, as well as thoroughly boiled soup in flasks that had been sealed only with corks, did reveal the presence of microorganisms. Spallanzani's conclusion was that the microorganisms in the flasks of spoiled soup had entered from the air. Life begat life. In the next century, Louis Pasteur conducted several experiments with yeast soup in a variety of containers, and he too disproved the once-popular theory of spontaneous generation.

Life from Space

Life drifting to the earth from space has been the theme of science fiction tales, scientific hypotheses, and even some scientific experiments. The basis for serious thought on this matter is the fact that such extraterrestrial materials as asteroids, meteorites, comets, and interstellar dust contain organic compounds of the sort found in life on earth (fig. 3.25). These are the very chemicals that might have led to the first living things. How might this have happened?

One scenario of extraterrestrial chemicals "seeding" life on earth envisions simple organic molecules in interstellar dust clouds forming complex organic compounds in comets. Heat released by exploding stars could have melted ice in the comets to provide the water for these chemicals to brew into a living soup. When pieces of those comets crashed to earth eons ago, they may have seeded the planet with the molecules that led to life, or possibly even with simple forms of life itself. However, the presence of organic compounds in extraterrestrial matter does not necessarily signify life. The very fact that these chemicals have been found in many unearthly rocks and that they have even been synthesized in laboratories by simulating conditions in outer space suggests that they may be simply the result of common chemical phenomena.

Common Ancestry

Returning to earth, the chemical similarities of diverse organisms are astounding.

All life forms use nucleic acids as their genetic material, they use the same genetic "code" to translate the nucleic acids into proteins, and they use the same energy-generating molecules. All organisms use the same 20 amino acids to build proteins, even though many more varieties are chemically possible. The common chemical plan that is the basis of all life on earth has been interpreted by many scientists to mean that all life has descended from a common ancestor.

What was this first organism like? How did it arise from the molecules that were present on the ancient earth? At what point did a collection of chemicals develop the capacity to replicate, elevating itself from the nonliving to the living? Scientists from many disciplines are trying to answer these questions (Reading 3.2).

The *Voyager* missions to Jupiter and Saturn have provided information about the physical and chemical environments on those giant planets and their moons. If all of the planets in our Solar System formed at the same time, then knowing the conditions on the other planets may provide some clues about the atmosphere of the prebiotic (before life) earth. Chemists, mathematicians, and computer scientists have delineated possible schemes by which likely chemical combinations might have been brewed into life. Paleontologists (scientists who study past life) and geologists have rooted these possibilities in reality by providing a time frame. Evidence of microscopic life 3.5 billion years old has been found—only a billion years or so after the formation of the planet itself.

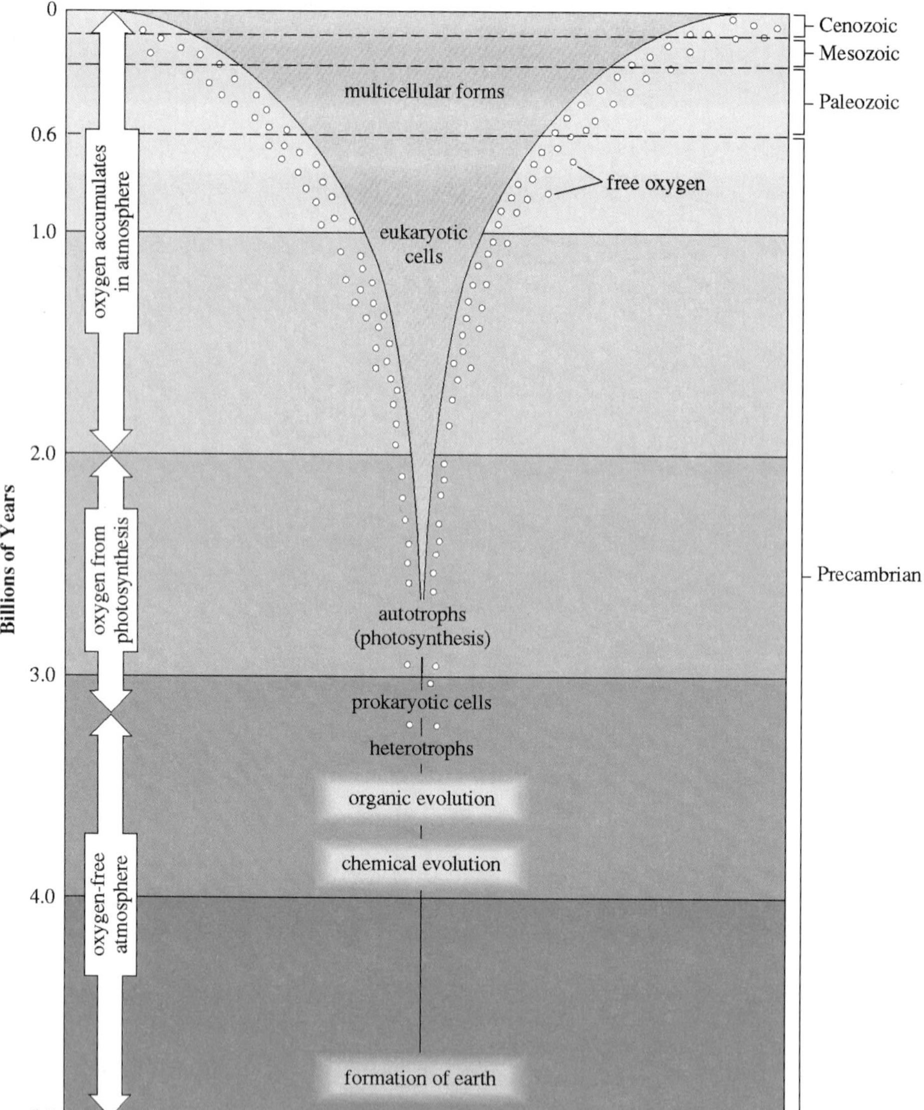

Figure 3.26
Chemical evolution led to biological evolution. About a billion years after the earth formed, in a way that we do not fully understand, complex organic chemicals became the first and simplest living things. These first earth inhabitants were heterotrophic, or dependent upon the environment for food. Later forms could extract energy to produce food molecules from the sun (photosynthesis). A by-product of photosynthesis, oxygen, altered the atmosphere in a way that made it hospitable for more complex life forms.

Chemical Evolution

Whatever happened during those first billion years paved the way for the diversity of life on the earth today. Before life (or even molecules suggestive of life) appeared, certain chemical changes had to occur. In the first stage of this prebiotic chemical evolution, energy must have been available to provoke small molecules to react and produce the small organic monomers that form polymers in living organisms. These earliest molecules on the road to life might have been the gases ammonia (NH_3), hydrogen (H_2), methane (CH_4), carbon dioxide (CO_2), and water vapor (H_2O). The energy may have come from sunlight, lightning, or volcanoes.

The second stage of chemical evolution could be called the period of polymerization, as chemical building blocks bonded to form long chains. The final and least understood phase of this chemical prelude to life was the transition of these polymers from random chains of monomers to information-containing molecules capable of self-reproduction (fig. 3.26).

The prime candidates for life's chemical forerunners are proteins and nucleic acids. Which polymer appeared first? Proponents of the "protein first" view, or **proteinoid theory,** point to the observation that, in the laboratory, certain combinations of amino acids, other molecules,

Reading 3.2 *Recipes for Starting Life—Simulating Early Earth Conditions*

A MAJOR PROBLEM IN STUDYING THE ORIGIN OF LIFE ON EARTH IS SIMPLY THAT NONE OF US WAS THERE TO WITNESS IT. However, chemicals thought to have been present on the young planet can be combined in an attempt to see if they will react to produce the more complex molecules of life. This idea was suggested by Soviet biochemist Alexander Oparin in 1927. He hypothesized that random physical and chemical processes acting upon such simple compounds as methane (CH_4) and ammonia (NH_3) in the presence of sunlight might have set the stage for the natural synthesis of larger organic molecules. In 1953, a graduate student named Stanley Miller picked up where Oparin had left off.

Miller set up a simple but clever apparatus. In a large glass container he mixed the gases Oparin had suggested, plus hydrogen (H_2) and water vapor (H_2O). For an energy source, he exposed the gases to electric sparks that simulated lightning. Next, the gases were condensed in a narrow tube and passed over an electric heater, a laboratory version of a volcano. This prebiotic soup brewed for a week. Then, using chemical techniques to analyze what had formed in this concoction, Miller found that he had indeed cooked up four amino acids! Because these chemical building blocks link to form proteins, and proteins are such an important part of life, Miller and others hypothesized that once amino acids formed, the other chemical reactions of life could also have taken place.

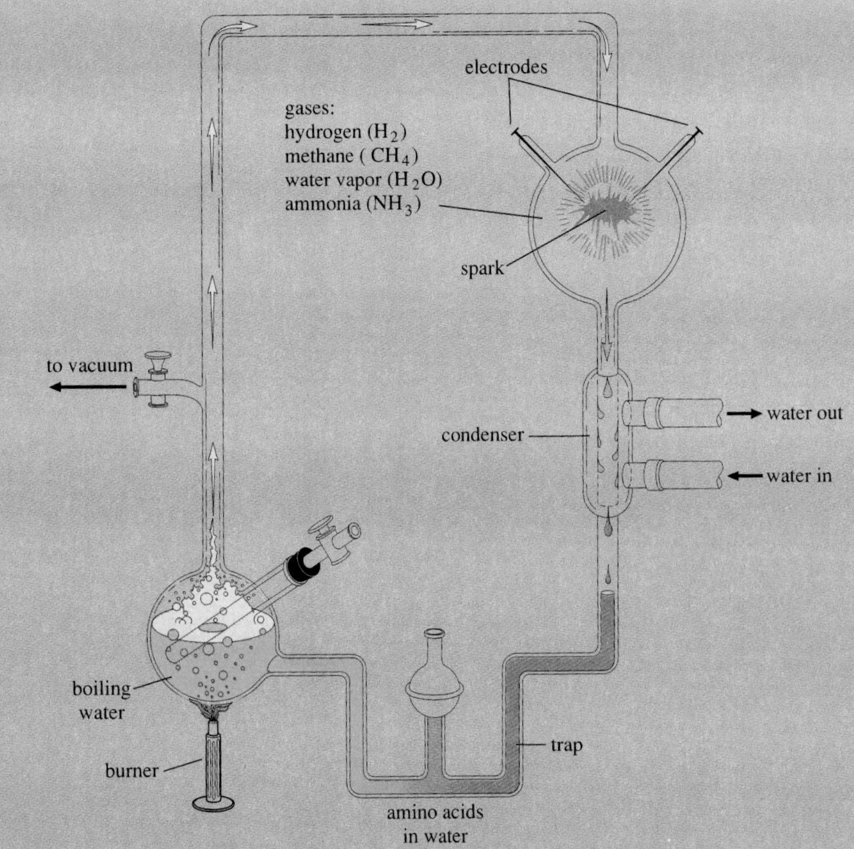

Figure 1
When Stanley Miller passed an electrical spark through heated gases, he generated amino acids, a more complex chemical that may have played a role in the origin of life.
From *A View of Life*, by Luria, Gould and Singer. Copyright © 1981, Benjamin/Cummings Publishing Company, Menlo Park, Calif. Reprinted by permission.

Many researchers successfully repeated the Miller experiment, and it is even duplicated as an exercise in some undergraduate chemistry courses. However, recent variations on Miller's original recipe have raised other intriguing possibilities. For example, Clifford Matthews of the University of Illinois has suggested that ammonia and methane in the ancient atmosphere formed clouds of hydrogen cyanide, which formed long polymers when exposed to the ultraviolet radiation from the sun. When these hydrogen cyanide chains fell into the oceans, they reacted with water to yield amino acids. When Matthews combined his ingredients in laboratory glassware, leaving out the water, he obtained a yellow-brown sticky substance containing six different amino acids. Still other variations on Miller's recipe for prebiotic soup produce not only amino acids but nucleotides too, the building blocks of genes. These multiple successes suggest that formation of these organic compounds reflects a chemical tendency but not necessarily a first step toward life. The discovery of nucleic acids in meteorites would seem to confirm this.

If life indeed began as a series of such random chemical combinations, did it do so only once? If so, what prevented life from starting again and again? One possibility is that once life began, it very gradually changed the atmosphere by adding oxygen to it. This new atmosphere destroyed the chemicals in the prebiotic soup, ensuring that, once life was on its way, it could not begin again—at least not in the same fashion.

and heat produce tiny spheres that have some characteristics of living cells. Supporters of the "nucleic acid first" view, or the **naked-gene hypothesis,** mention that DNA is the only molecule that can replicate and that DNA controls protein synthesis, not the other way around. It is also possible that both types of polymers formed simultaneously, perhaps in moist clays that were rich in chemical building blocks. The ordered mineral structure of clays may have provided a model for the formation of polymers such as proteins and nucleic acids.

It is not clear just how the first protein-nucleic acid partnership formed, whether a proteinoid sphere "swallowed" a wandering strand of nucleic acid or a naked gene donned a protein coat. What is important, though, is that this marriage did occur. Somehow, this "protocell" developed a way to manufacture and put together its own amino acids so that it was no longer dependent upon random chemical reactions among the prebiotic molecules in its environment. It also developed a way to generate energy.

These important steps could not have occurred overnight. Many different combinations of proteins and nucleic acids must have formed spontaneously over millions of years. With the occurrence of so many chance experiments of nature, it seems almost inevitable that a true protocell, with the ability to replicate provided by the nucleic acid, would, over time, persist and multiply. Eventually, as its internal chemistry grew more complex, it would become what we call a living thing.

KEY CONCEPTS

The once-popular theory that life springs from nonliving matter is called spontaneous generation, and several investigators have disproved it. The complex organic molecules that may have preceded life on earth could have come from extraterrestrial sources. Such chemicals, in the presence of energy sources from lightning and perhaps volcanic activity, were probably "brewed" into the earliest living things.

SUMMARY

Living things are highly organized systems of matter that can utilize energy in *metabolic reactions*. Life responds to the environment, both on the individual level as *irritability* and over generations in populations as *adaptation*. Organisms also grow and reproduce. It is debatable what the simplest form of life is, but *viruses*, *viroids*, and *prions* are possibilities.

All forms of life are composed of, and are affected by, the same types of chemicals. The basic chemical unit is the *atom*, which contains a centrally located *nucleus* consisting *of protons* and *neutrons* and *electrons* that circle the nucleus at different energy levels. Different types of atoms, distinguished by the number of their protons, constitute elements. Atoms chemically bond by losing, gaining, or sharing electrons to form molecules.

Living things consist mostly of the *elements* carbon, hydrogen, nitrogen, oxygen, phosphorus, and sulfur. Water is vital to life because of its cohesiveness and high heat capacity and heat of vaporization. Four main classes of organic compounds are found in organisms. *Carbohydrates* include sugars and starches, which release energy when their chemical bonds are broken. *Fats* also store and provide energy, as well as form parts of membranes, retain heat, and enable certain vitamins to be utilized in organisms. *Proteins* are built of amino acids and have many different biological activities. *Enzymes* are a particularly important type of protein in that they alter the rates of chemical reactions. *Nucleic acids* constitute the genetic material. Complex carbohydrates, proteins, and nucleic acids are examples of *polymers*, which are long molecules composed of linked chemical building blocks.

A central question in biology is how life originated. Although many theories have been advanced, the most popular is that simple chemicals reacted in the presence of energy to form more complex compounds until a self-replicating "protocell" appeared. Such a protocell is presumed to have been the common ancestor of the diverse forms of life on earth today.

QUESTIONS

1. In an episode of the television show "Star Trek," three members of the crew of the starship USS *Enterprise*, while exploring a strange planet, meet an enemy who dehydrates each one into a small box. He then rearranges the boxes and crushes one of them. He later attempts to restore them to life, but only two of them reappear. What biological principle is illustrated here?

2. Give three reasons why an automobile does not qualify as a living thing.

3. In another "Star Trek" episode, life forms based on the element silicon (Si) are encountered by the crew of the *Enterprise*. Considering the fact that life on earth is based upon carbon, why is silicon also a logical choice? Consult the periodic table for your answer.

4. The vitamin biotin contains 10 atoms of carbon, 16 of hydrogen, 3 of oxygen, 2 of nitrogen, and 1 of sulfur. What is its molecular formula?

5. A medical journal article concludes, "The consumption of as little as one or two fish dishes per week may be of preventive value in relation to coronary heart disease." To what substance in fish does this statement refer?

6. A magazine article describes the industrial use of a biological molecule to make cheese. What does the following sentence from the article mean?: "The enzyme, derived from the microorganism *Aspergillus oryzae*, catalyzes the hydrolytic conversion of lactose to galactose and glucose."

7. What are the sources of the carbon, hydrogen, oxygen, and nitrogen in the amino acids that Stanley Miller brewed in his prebiotic soup?

TO THINK ABOUT

1. Some people believe that human life begins with the union of sperm and egg, while others believe that it begins at birth. Some even believe that it has no start or finish but is a continuum, because we all possess cells that have the potential, when each is joined with another of the appropriate sort, to create the next generation. What do you think?

2. The "reductionist" approach to the definition of life is that the whole organism is the sum of its parts. In contrast, the "holist" approach states that the whole organism comprises its parts plus something else, a "vital force" perhaps. Which do you agree with? Can you think of other explanations for this thing we call life?

3. Humans have already (a) engineered bacteria to use genes of other organisms, (b) bred plants not found in nature, (c) designed functional plastic human organs, even a heart, (d) grown muscle cells in laboratory glassware, and (e) fertilized human eggs in a dish. Do you think we will ever create life in the laboratory? If so, how do you imagine it might be accomplished?

4. One of the first studies to suggest the value of fish oil in preventing heart disease examined the effects of the oil on 20 people who had hypertrigylceridemia, a condition in which the blood has from 15 to 30 times the normal amount of triglycerides. Blood triglyceride and cholesterol levels fell significantly in every participant when he or she was on a fish oil diet, but not when on a low-fat, low-cholesterol diet lacking in fish oils. What further studies would you suggest be conducted before this information is extrapolated to the general, healthy population?

5. The atmosphere of Titan, one of Jupiter's moons, is similar to that hypothesized for the prebiotic Earth. What do you think this might indicate about the origin of life on Earth and elsewhere?

6. What organisms or organismlike structures present on earth today are chemically similar to what the first life forms may have been like?

7. One modern view of the mechanism of life's origins is that life arose from nonliving matter only once and under very special conditions. How does this approach compare to the "spontaneous generation" view of the origin of life? Do you think any of the concepts concerning the origin of life that were discussed in the chapter are untestable? Which ones, and why?

SUGGESTED READINGS

Ballentine, Carol. May 1985. The essential guide to amino acids. *FDA Consumer*. How the body uses amino acids in the diet.

Blakeslee, Sandra. September 6, 1988. NASA to probe heavens for clues to life's origins on earth. *New York Times*. The chemicals that were somehow "brewed" into life may have come from beyond the earth.

Easton, Thomas. May 1985. Sublife mysteries. *Biology Digest*. At what point does a complex chemical combination become a living thing?

Freundlich, Naomi J. March 1986. Scientists stir up new recipes for life. *Industrial Chemical News*. The Miller experiment has prompted many other investigators to suggest other ways that the molecules of life could have formed on the early earth.

Morse, Gardiner. August 11, 1984. Viroids: Nature's littlest killers. *Science News*. Are viroids alive?

Poole, Robert. June 29, 1990. Closing the gap between proteins and DNA. *Science*. Chemists are synthesizing self-replicating molecules that may have been precursors to life.

Raloff, J. September 30, 1986. Is there a cosmic chemistry of life? *Science News*. Do amino acids and nucleic acids form beyond the earth—and could they lead to life there?

Thomsen, D. E. February 7, 1987. A periodic table for molecules. *Science News*. Molecules can be organized by properties into a chart.

Vogel, Shawna. October 1988. The shape of proteins to come. *Discover*. Researchers are adapting nature's rules to build new types of proteins.

Yellin, Arthur K. April 1986. Enzymes—the movers and shakers of our body chemistry. *FDA Consumer*. Enzymes are absolutely vital to our health—we could not live without them.

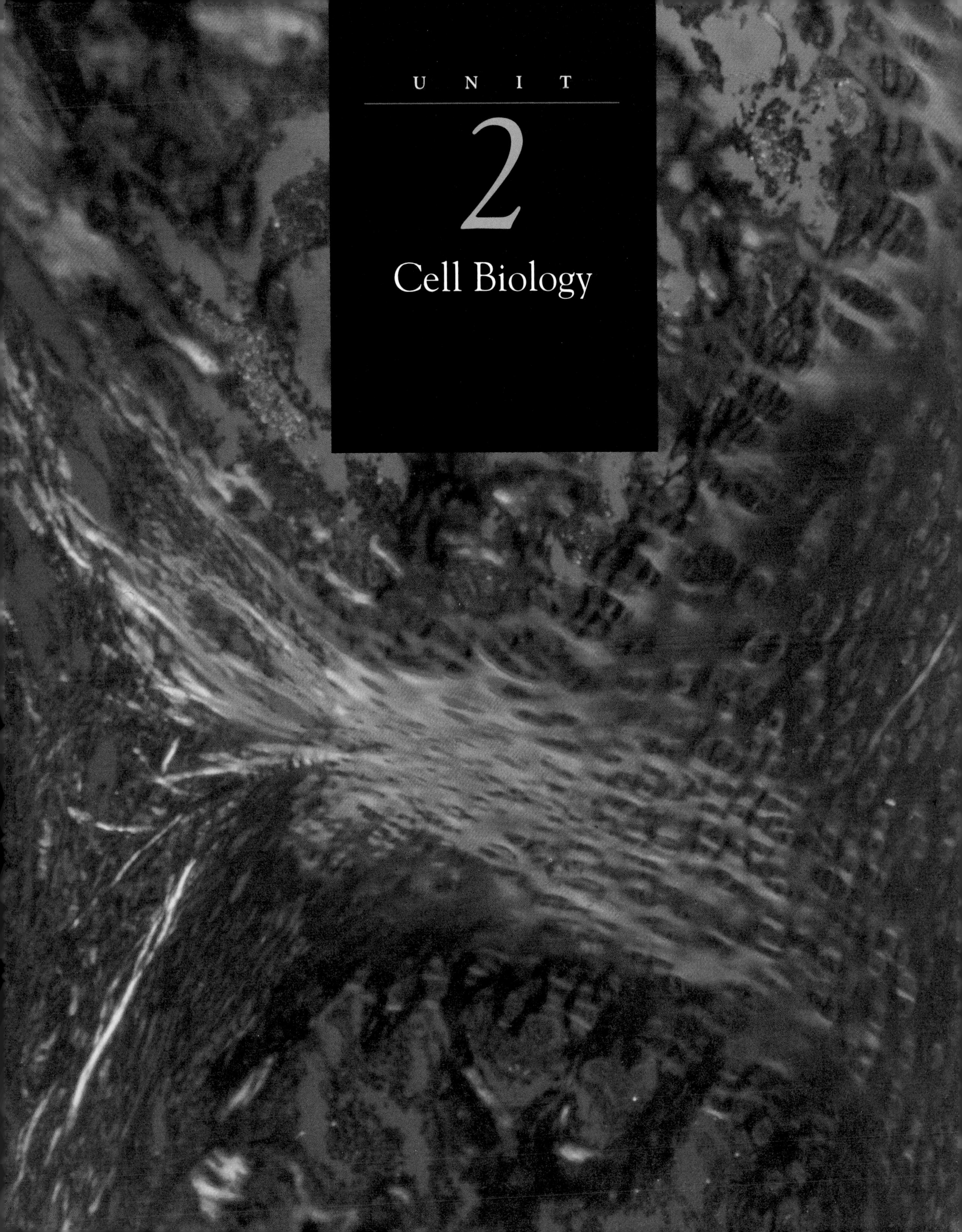

UNIT

2

Cell Biology

CHAPTER

4

Cells and Tissues

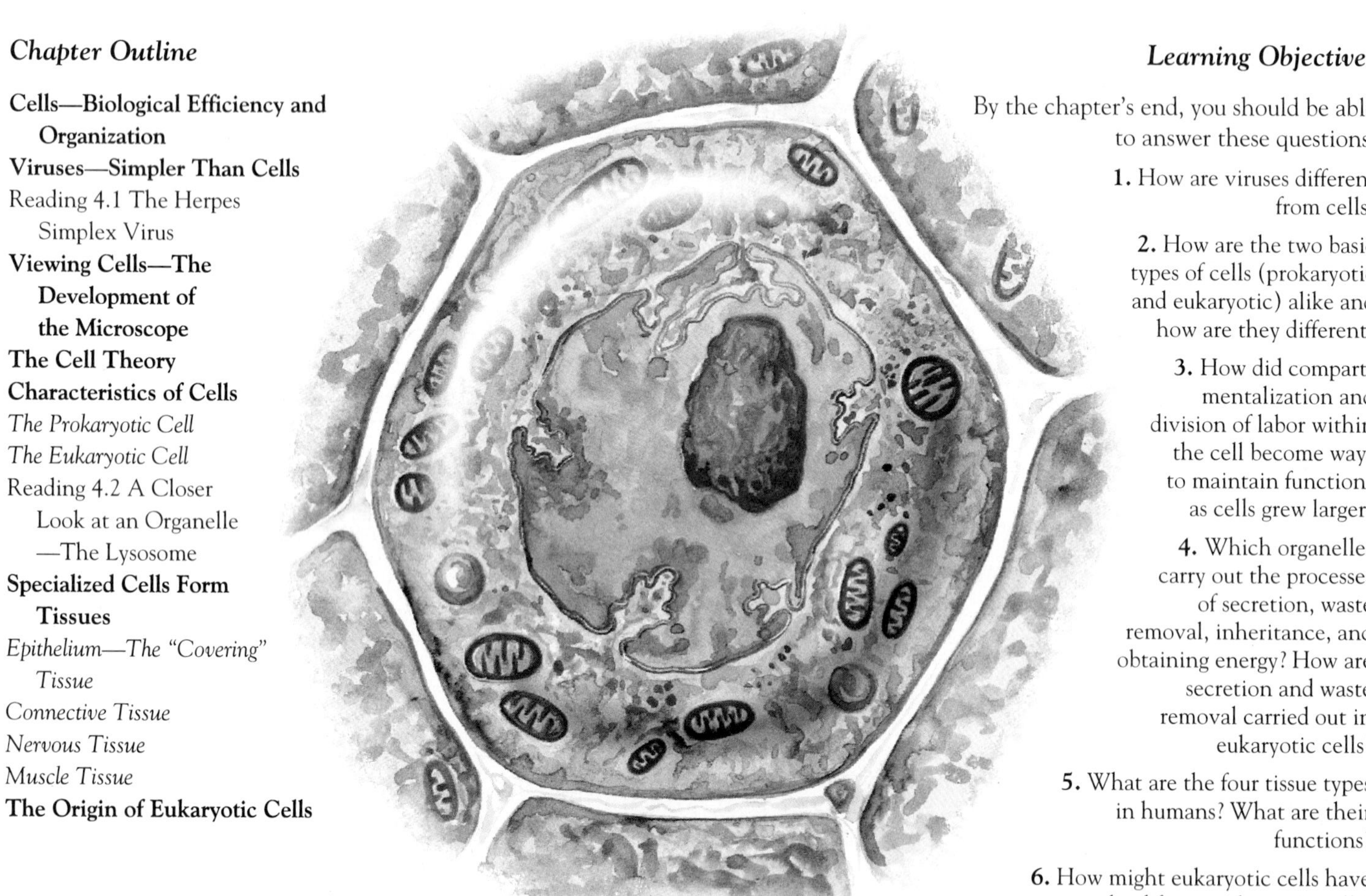

Chapter Outline

Learning Objectives

By the chapter's end, you should be able to answer these questions:

1. How are viruses different from cells?
2. How are the two basic types of cells (prokaryotic and eukaryotic) alike and how are they different?
3. How did compartmentalization and division of labor within the cell become ways to maintain functions as cells grew larger?
4. Which organelles carry out the processes of secretion, waste removal, inheritance, and obtaining energy? How are secretion and waste removal carried out in eukaryotic cells?
5. What are the four tissue types in humans? What are their functions?
6. How might eukaryotic cells have evolved from prokaryotic cells?

Department stores are marvels of efficiency and organization. Within the confines of a single building, nearly everything necessary to make life comfortable can be found. The interior is organized into departments, so a particular item can rapidly be located. Departments are often placed near each other if their contents are typically used together—lawn furniture might be next to barbecue equipment, housewares near linens.

A living organism is also a model of organization and efficiency, as evidenced by the organ systems of your own body, which enable you to breathe, move, resist disease, and basically carry out all of the activities that qualify you as a living thing. This biological organization continues on a microscopic level, within and between **cells,** which are the basic structural units of all organisms.

Cells—Biological Efficiency and Organization

A cell is defined by its outer membrane, just as the merchandise that comprises a department store is defined by the building it occupies. *Unicellular* organisms consist of a single cell, such as bacteria and protists. *Multicellular* organisms, such as ourselves, are built of many cells. Structures within the cells of multicellular organisms and within some more complex unicellular organisms, called *organelles* (little organs), carry out specific functions. In multicellular organisms different numbers of particular types of organelles endow some cells with specialized functions, such as skin, muscle, and bone cells and specialized cells such as those shown in figure 4.1. Yet even the structurally simple cells of bacteria, lacking organelles, are organized and efficient. All cells have some structures in common that allow them to perform the basic life functions of reproduction, growth, response to stimuli, and energy conversion.

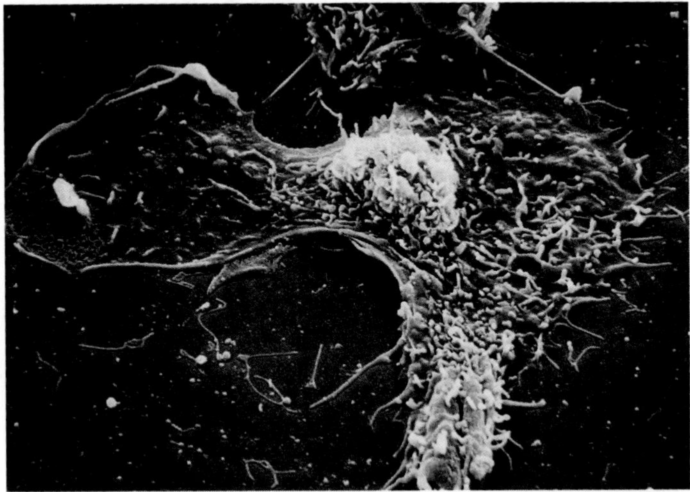

a.

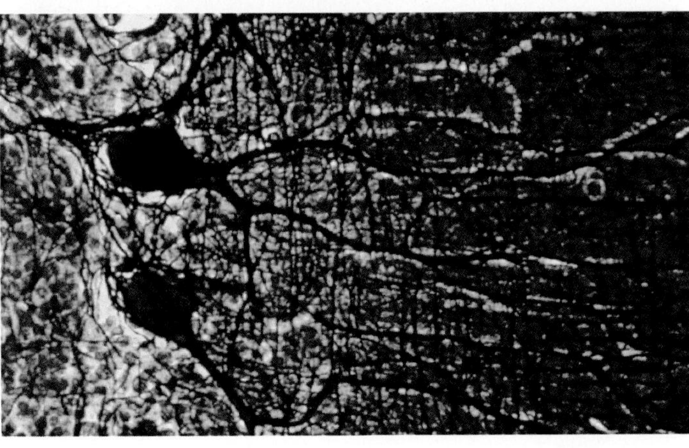

b.

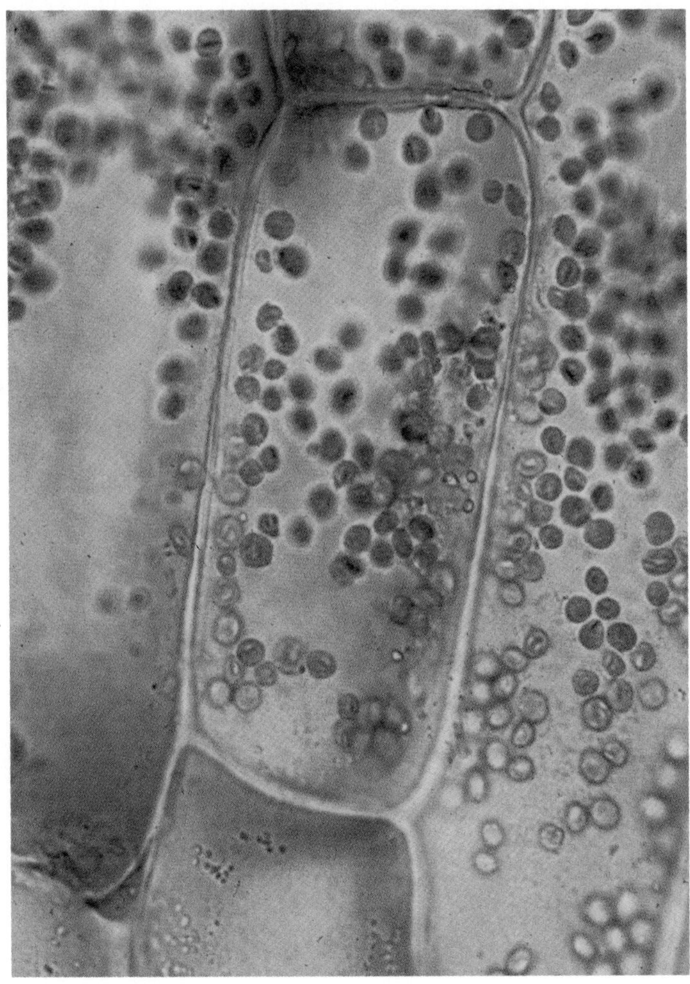

c.

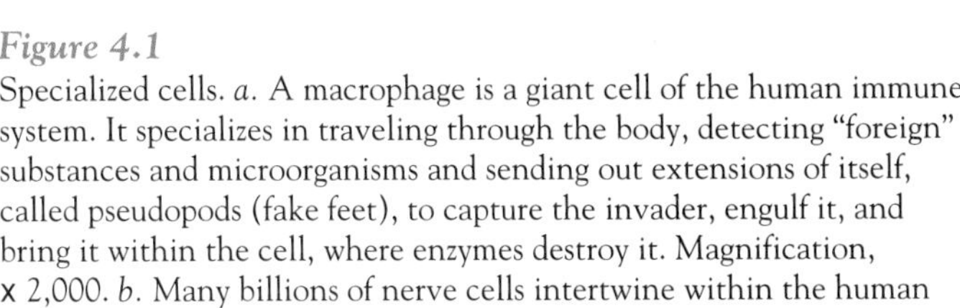

Figure 4.1
Specialized cells. *a.* A macrophage is a giant cell of the human immune system. It specializes in traveling through the body, detecting "foreign" substances and microorganisms and sending out extensions of itself, called pseudopods (fake feet), to capture the invader, engulf it, and bring it within the cell, where enzymes destroy it. Magnification, x 2,000. *b.* Many billions of nerve cells intertwine within the human brain. Note the roundish cell bodies and long, tangled extensions of these nerve cells, or neurons. *c.* The green discs seen in these *Elodea* cells are chloroplasts, which give the cells of green plants their color. Chloroplasts contain structures and biochemicals that permit the cell to utilize the energy from sunlight to manufacture nutrients.

KEY CONCEPTS

The structural unit of an organism is the cell. Organisms are unicellular or multicellular. Complex cells contain specialized structures called organelles. All cells have structures in common to carry out basic life processes, but different numbers of organelles give cells distinct characteristics.

Viruses—Simpler Than Cells

The simplest form of life is a unicellular organism with no organelles, such as a bacterium. However, in chapter 3 we encountered several types of "infectious agents" that appear to be living while they are infecting cells but otherwise seem to be nonliving chemicals. Before describing how we examine cells and their contents, it is interesting to take a comparative look at the viruses, both to point out their noncellular organization and because they exert very noticeable effects on human health, causing such minor ills as colds and influenza and such deadly ones as AIDS. Reading 4.1 describes effects of the herpes simplex virus.

A *virus* consists of a nucleic acid (DNA or RNA) surrounded by protein. Figure 4.2 illustrates the human immunodeficiency virus (HIV), which causes AIDS. A virus must be within a cell to reproduce, and hence it is called an obligate parasite. Many viruses, such as HIV, cannot survive outside of a living cell. Some other viruses are afforded protection from the physical environment by their protein coverings. A virus reproduces by injecting its DNA or RNA into the host cell, where it situates itself within the host's DNA. In fact, viral DNA sequences can probably be found within your own chromosomes. (An RNA virus, such as HIV, is called a retrovirus and must first make a replica of its RNA in DNA form.)

Once viral DNA integrates into the host's DNA, it can either remain there and be replicated along with the host's DNA whenever the cell divides, but not cause harm, or the viral DNA can actively take over the cell, leading eventually to the cell's death. To do this, some of the virus's genes direct the host cell to replicate viral DNA rather than the host DNA. As viral DNA accumulates in the cell, some of it is used to manufacture proteins. (Recall from chapter 3 that the function of DNA is to provide information from which the cell constructs proteins.) Within hours or days, the infected cell fills with viral DNA and protein. Some of the proteins wrap around the DNA to form new viral particles. Finally, a viral enzyme is produced that cuts through the host cell's outer membrane. The cell bursts, releasing new viruses.

Viruses are known to infect all kinds of organisms, including animals, plants, and bacteria. A particular type of virus, however, infects only certain species, which constitute its *host range*. (Refer back to figure 3.6 for an illustration of a tomato infected by the tomato bunchy top virus.) Figure 4.3 illustrates what happens to a moth infected by a type of virus called a baculovirus.

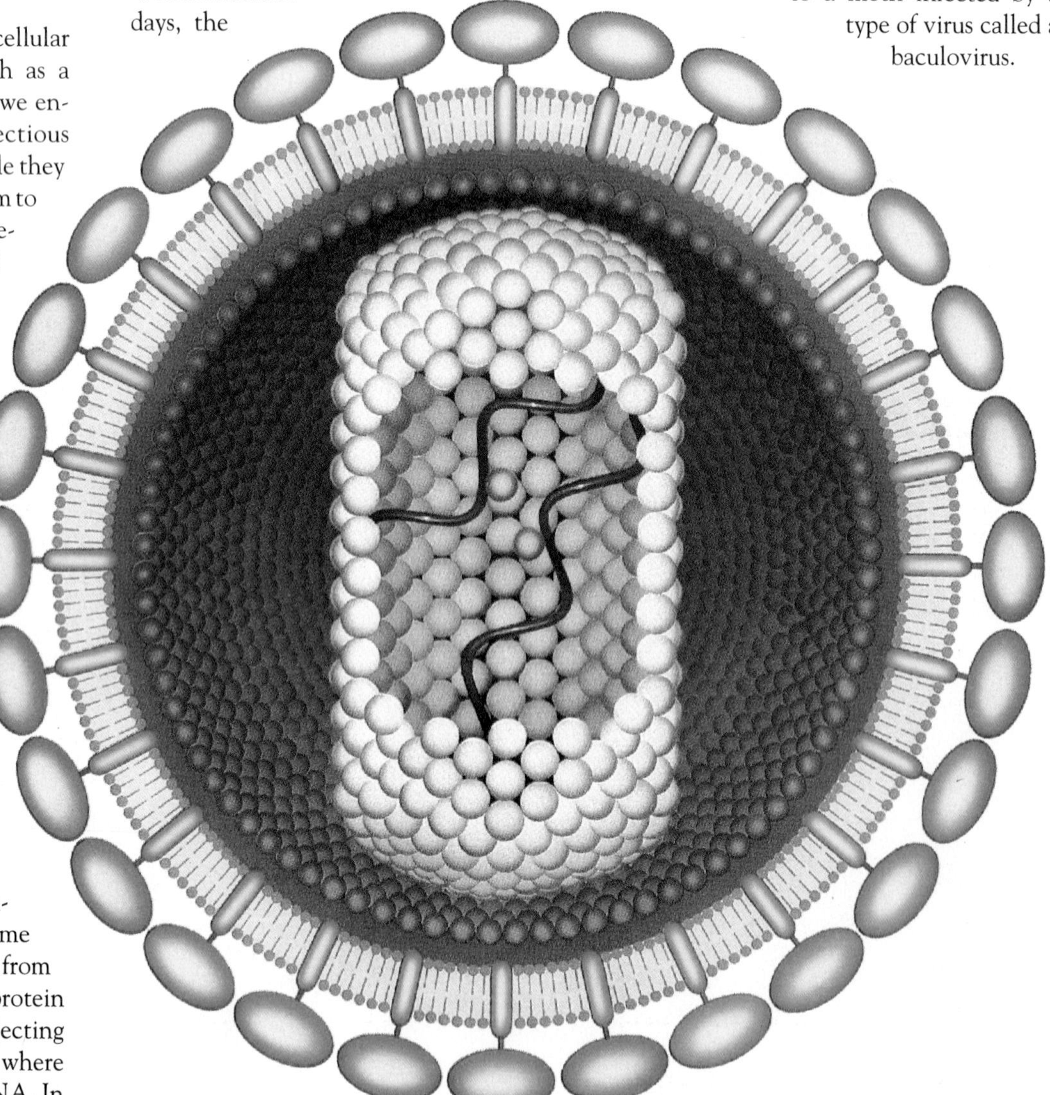

Figure 4.2
A virus is a nucleic acid coated with protein. The human immunodeficiency virus (HIV), which causes AIDS, consists of RNA surrounded by several layers of proteins. Once inside a human cell (usually a T cell, part of the immune system), the virus uses an enzyme to convert its RNA to DNA, which then inserts into the host DNA. HIV damages the human body's protection against disease by killing T cells and by using these cells to make more of itself.
From "AIDS viron" (January 1987 cover painting), copyright © 1987, by Scientific American, Inc., George V. Kelvin, all rights reserved.

Reading 4.1 The Herpes Simplex Virus

THE HERPES SIMPLEX VIRUS IS FAR SIMPLER IN STRUCTURE THAN EVEN THE SMALLEST PROKARYOTIC CELL. Like all viruses, it is built only of nucleic acid surrounded by a protein coat. Viruses are too simple to be considered organisms, but whether or not they should be considered to be alive is a matter of debate. Yet, despite its organization and its questionable biological status, it is clear that the herpes simplex virus can have drastic effects on human health.

Herpes simplex virus Type 1 (HSV-1) usually produces cold sores on the mouth and HSV-2 usually causes genital blisters. However, each of the two viral types has been known to cause infections of the body part usually infected by the other type. Once HSV-1 or HSV-2 enters human skin, it multiplies quickly. The person initially feels tingling or itching. Within 2 weeks, the area of tingling has turned into mouth sores or a rash of painful, clear blisters on the genitals. The eruptions last for about 3 weeks, during which time the virus can be passed on to others by skin contact.

Once the outbreak disappears though, the person is not free of the virus. Both viral types enter nerve cell branches in the skin, and then the viruses follow these cells further into the nervous system. HSV-1 retreats to the brain and HSV-2 to the spinal cord. Within the nervous system, the virus is protected from attack by the person's immune system. The virus may resurface to cause a new eruption a week, a month, or a year later, or it may never appear again. Recurrences, sufferers claim, seem to be sparked by stress.

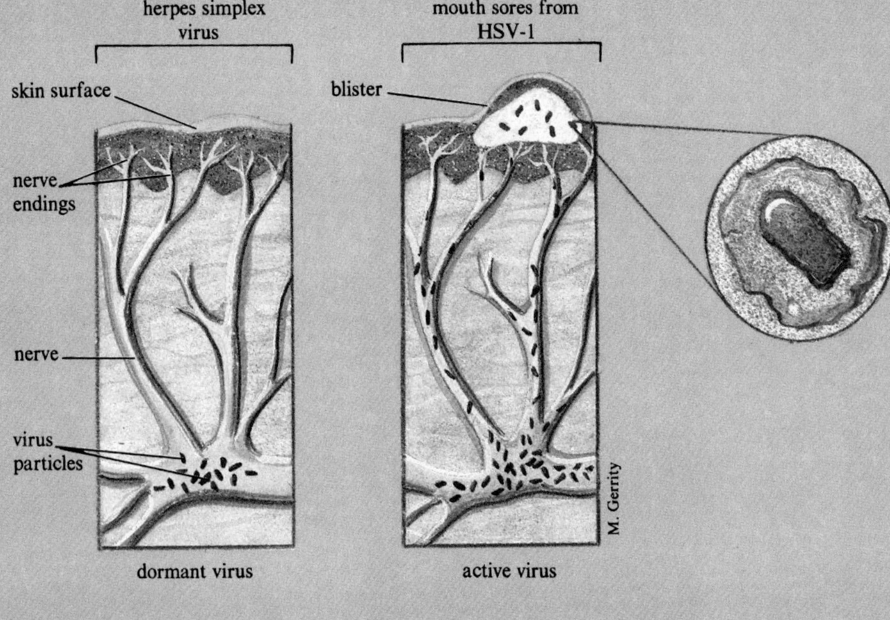

Figure 1
Herpes simplex lodges in nerve cells near the skin's surface.

Herpes simplex infections are commonly spread by sexual contact. Genital herpes can be acquired by having sexual intercourse with someone who has open sores. Oral sex can spread the infection from the mouth of one partner to the genitals of the other, or vice versa. Although it is true that herpes cannot be spread unless the virus is "shed" from open sores, it is possible for a person's symptoms to be so mild (particularly if the attack is not the first one, or in a male), that he or she does not even know that contagion is possible. Another complication is that females can sometimes have internal blisters and be unaware of the attack.

The most seriously affected host of the herpes simplex virus is a newborn who comes in contact with open sores during the journey of birth. Because the infant's immune system is not yet completely functional, a herpes infection is especially devastating. Forty percent of babies exposed to active vaginal lesions become infected, and half of these infants die from it. Of those infants who are infected but survive, 25% have severe nervous system damage, and another 25% have widespread skin sores. To prevent exposure of newborns to herpes, a woman with a history of the infection can have her vaginal secretions checked periodically during pregnancy for evidence of an outbreak. If she has sores at the time of delivery, a cesarean section is performed to protect the child.

At the present time, herpes simplex infections cannot be cured completely. The drug acyclovir, however, inhibits the formation of new sores, decreases the healing time of sores, and shortens the time during which live virus is "shed" from a sore. The drug appears to help both new sufferers and those who suffer from recurrent attacks.

KEY CONCEPTS

A virus consists of a nucleic acid surrounded by protein and must invade a cell to reproduce. Viral DNA inserted into the host cell remains dormant or is replicated using the cell's organelles. Viruses attack specific species.

Viewing Cells—The Development of the Microscope

Our understanding of the structures inside cells depends upon technology, because most cells are too small to be seen by the unaided human eye. Today, sophisticated equipment allows us to probe the inner workings of cells by greatly magnifying their contents. However, the ability to make objects appear larger than they are probably dates back to ancient times, when people noticed that pieces of glass or very smooth, clear pebbles could magnify small

a.

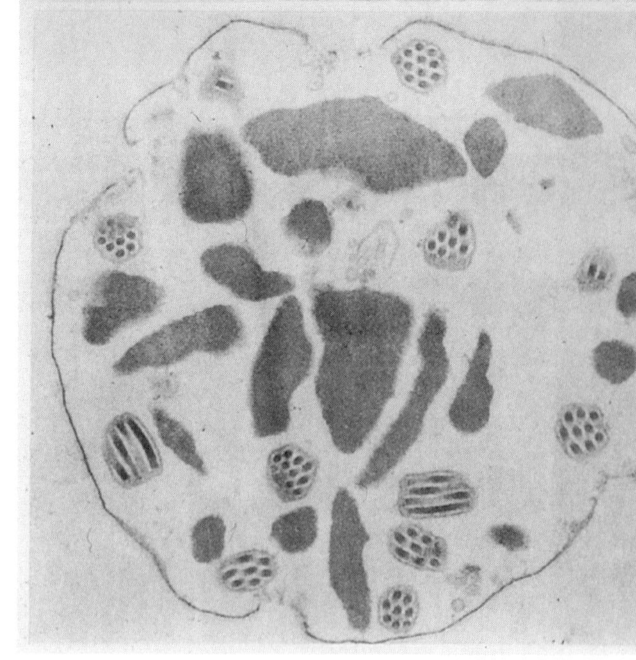

b.

Figure 4.3
Baculovirus infects cells of moths and butterflies. *a.* The gypsy moth caterpillar crawls out onto a leaf to eat, and in addition to its leafy meal, ingests polyhedron-shaped proteins containing baculovirus. Once inside the caterpillar's intestines, alkaline digestive juices dissolve the polyhedra, releasing viruses that quickly insert their DNA into the insect's intestinal cells. Six to 10 hours after the caterpillar eats the protein-encased viruses, its gut cells are already mass-producing viral DNA. *b.* By 12 hours after ingestion, the caterpillar's cells are manufacturing polyhedra. By 18 to 24 hours, fully assembled virus-containing polyhedra can be seen under a microscope. Some viruses remain "naked," without the protective polyhedra, and these continue to invade more insect cells. Four or 5 days after the original leaf is consumed, the caterpillar dies, its cells bursting with baculovirus. As the caterpillar decomposes, it leaves behind on the vegetation baculovirus that is shielded from the environment by the protein polyhedra—waiting for the next hungry victim.

or distant objects. By the thirteenth century, the value of such "lenses" in aiding people with poor vision was widely recognized in the Western World.

It was more than three centuries until the effect of using lenses in pairs was first noted. The origin of a double-lens **compound microscope** can be traced to two Dutch spectacle makers, Johann and Zacharius Janssen. Actually, their children were unwittingly responsible for this important discovery. One day in 1590, a Janssen youngster was playing with two lenses, looking through them at distant objects. Suddenly he screamed—the church spire looked as if it was coming toward him! Looking though both pieces of glass, as the elder Janssens quickly did, the faraway spire did indeed look as if it was approaching. One lens had magnified the spire, and the other lens had further enlarged the magnified image. The first compound optical device, a telescope, was thus invented. Soon, similar double-lens systems were constructed to focus on objects too small to be seen by the naked human eye, and the compound microscope was born.

It was not long before such lenses were turned towards objects of nature. By 1660, an inquisitive and imaginative English physicist, Robert Hooke, melted together strands of spun glass to create lenses that were optically superior to any that had been available before. Hooke focused his lenses on many objects, including bee stingers, fish scales, fly legs, feathers, and any type of insect he could hold still long enough to study. He was particularly fascinated by cork, which is actually bark from a type of oak tree. Under the lens, the cork appeared to be divided into little boxes. Hooke called these units "cells," because they looked like the cubicles (cellae) in which monks studied and prayed. Although he did not realize the significance of his observation, Hooke was the first human to see cells, the fundamental structural units of life.

Lenses were soon improved again, at the hands of a Dutchman, Anton van Leeuwenhoek. He used only a single lens, but it was more effective at magnifying and produced a clearer image than most two-lens microscopes then available. One of his first objects of study was tartar scraped from his own teeth, and his words best describe what he saw there:

> *"To my great surprise, I found that it contained many very small animalcules, the motions of which were very pleasing to behold. The motion of these little creatures, one among another, may be likened to that of a great number of gnats or flies disporting in the air."*

Leeuwenhoek discovered bacteria and protozoa, and in so doing opened up a vast new world to the human eye and mind (figs. 4.4 and 4.5). He spent much of the rest of his life describing with remarkable accuracy microorganisms and microscopic parts of larger organisms.

KEY CONCEPTS

The magnifying ability of lenses was discovered in the late 1500s, and Robert Hooke first magnified biological objects using lenses in 1660. Leeuwenhoek improved lenses and described many microbes.

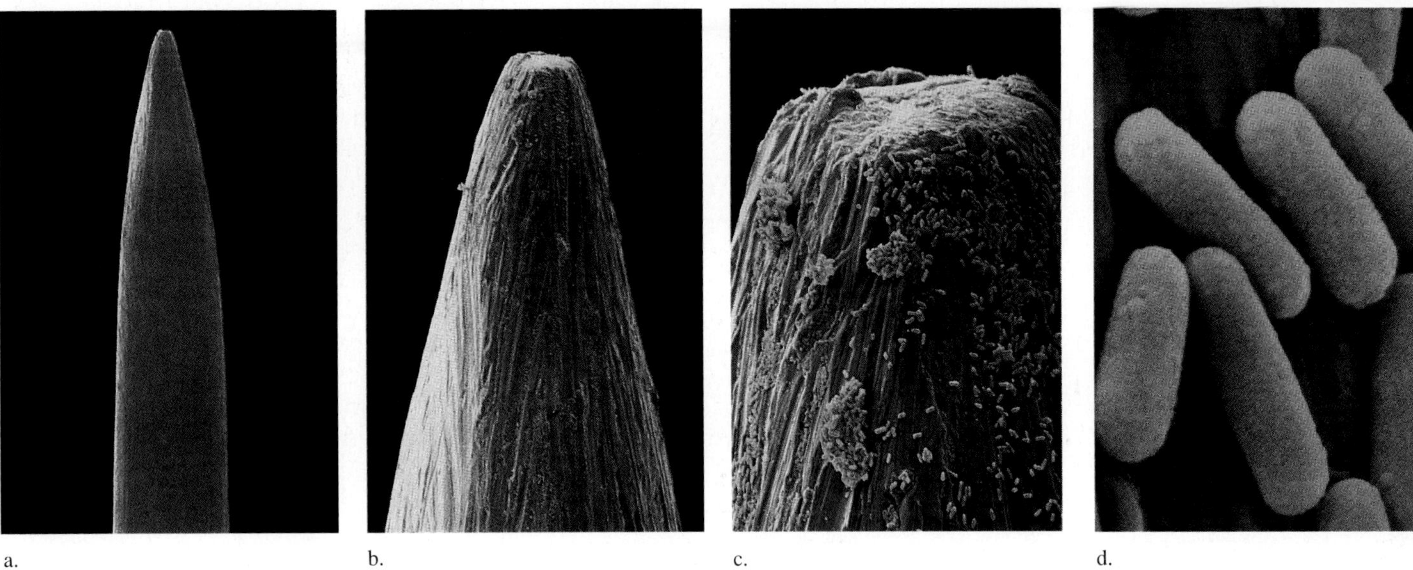

Figure 4.4
The living world beyond our vision. *a.* To the naked eye, this pin does not appear to be a likely site for bacterial growth. However, when the pin is examined under the scanning electron microscope at increasing magnifications (*b, c,* and *d*), rod-shaped bacteria are revealed. Magnification, *a,* x7; *b,* x35; *c,* x178; *d,* x4,375.

Figure 4.5
The microscope reveals another part of our living world. "Egad, I thought it was tea, but I see I've been drinking a blooming micro-zoo!" says this horrified, proper nineteenth-century London lady, when she turns her microscope to her tea. People must have been shocked to learn that there is a very active living world too small for us to see.

The Cell Theory

By the early nineteenth century, the optics of the compound microscope were well understood. The gaze of the early microscopists now focused more intensely on cells. Over the next century, the **cell theory** was developed. In 1839, the first part of the theory was contributed by German biologists Matthias J. Schleiden and Theodor Schwann, who stated that all living matter is composed of cells, and that cells are the basic structural and functional units of life. Shortly afterwards, German physiologist Rudolph Virchow added that all cells come from preexisting cells. Virchow also suggested that human disease results from changes taking place on the cellular level. For example, the loss of nerve and muscle function experienced by people with multiple sclerosis is caused by a lack of lipid within the membranes of cells that surround nerve cells. Reading 4.2 describes other disorders based upon a defect at the cellular level.

The second half of the nineteenth century saw a flurry of activity in the field of microscopy, as cellular contents were rapidly described. By the twentieth century, subcellular function and the microscopic aspects of disease were well under study. Soon, the need became clear for a device that could reveal structures even smaller than those that could be seen with the compound light microscope. In the 1940s, the invention of the *electron microscope* answered this need (fig. 4.6).

KEY CONCEPTS

The cell theory states that all living matter is composed of cells; the cell is the structural and functional unit of life; and that all cells come from preexisting cells. Human disease often reflects abnormality at the cellular level.

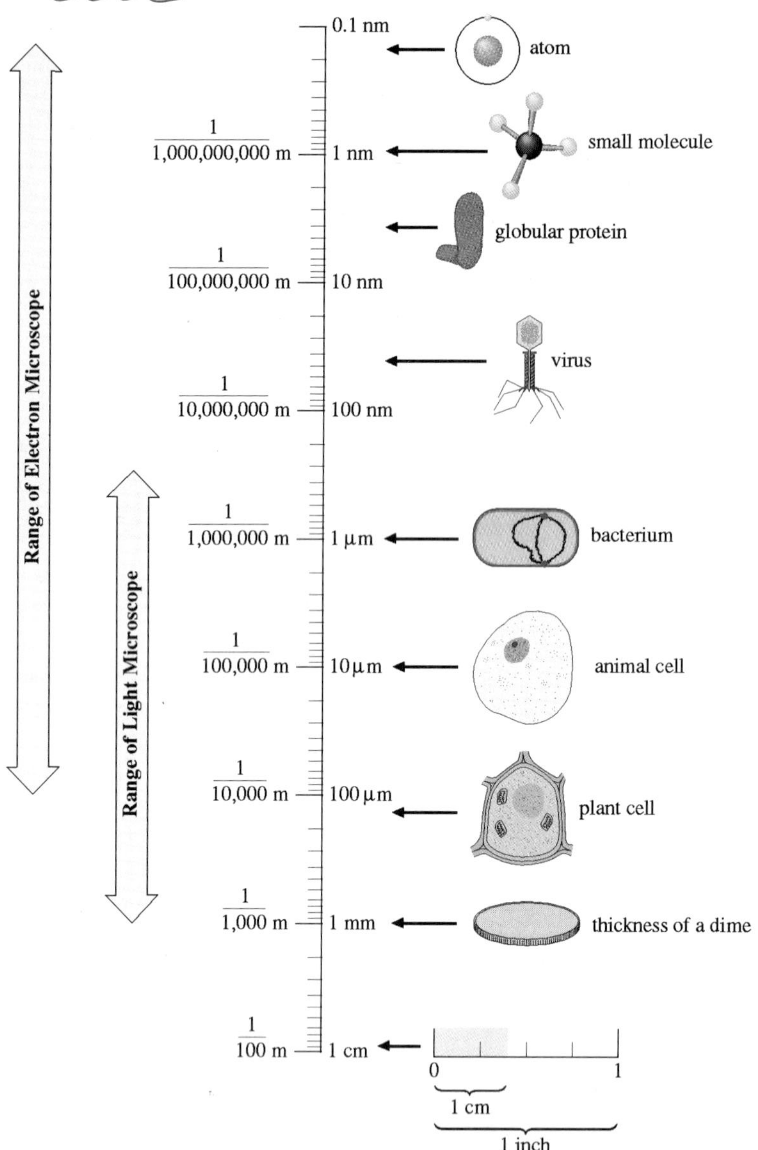

Figure 4.6

How biologists measure size. Biologists usually use the metric system to measure size. The basic unit of length is the meter (m), which equals 39.37 inches (slightly more than a yard). Smaller metric units are used to measure many biological and chemical structures. A centimeter (cm) is 0.01 meter (about 2/5 of an inch); a millimeter (mm) is 0.001 meter; a micrometer (μm) is 0.000001 meter; a nanometer (nm) is 0.000000001 meter; and an angstrom unit (A) is 0.0000000001 meter (or 1/10 of a nanometer). Although each segment of the scale is the same length, each represents only 1/10 the length of the segment beneath it. The sizes of some familiar structures are indicated next to the scale.

Characteristics of Cells

Cells are the basic units of life and, as such, they exhibit all of the characteristics of life. A cell requires energy, genetic information to direct biochemical activities, and structures to carry out these activities. Movement occurs within living cells, and some cells, such as the swimming sperm cell, can move about in the environment. In short, the functions of cells are similar to the functions of whole organisms.

The Prokaryotic Cell

The oldest cells to have left fossil evidence were smaller and simpler than the cells that make up the human body. These were single-celled *prokaryotes*—organisms including bacteria and *cyanobacteria*, which are also called blue-green algae. The fact that prokaryotes flourish today, comprising the majority of living cells on earth, illustrates the success of this rather streamlined type of cell. The prokaryotic cell has a much less specialized internal organization than the cells of multicellular plants and animals, but it is nevertheless organized and efficient enough to support the biochemical reactions of life. Prokaryotes affect humans in many ways, both for better and for worse (tables 4.1 and 4.2).

Most prokaryotic cells are surrounded by a rigid **cell wall** that is built of peptidoglycans (peptide-sugars). Many antibiotic drugs, including the penicillins, cephalosporins, vancomycin, and bacitracin, halt infection by interfering with the bacterium's ability to build its cell wall. The shape and staining properties of cell walls are used to classify bacteria as round, rod-shaped, curved, or spiral, as we saw in chapter 2. Species whose cell walls turn purple in the presence of the Gram stain are termed *gram positive;* those whose cell walls turn pink are called *gram negative*. Cell wall characteristics are used together with metabolic and biochemical criteria in distinguishing bacterial species.

Beneath the prokaryote's cell wall is a cell membrane, or *plasmalemma*. The cell membrane pinches inward in places, which may indicate sites at which the cell can divide in two. Embedded in the cell membrane are enzymes that speed the rates of certain biochemical reactions, enabling the cell to obtain and utilize energy. In some prokaryotes, taillike appendages called **flagella**, which enable the cell to move, are anchored in the cell wall and underlying cell membrane.

The genetic material of a prokaryote is a single circle of DNA (fig. 4.7). The DNA is described as "naked" because it stands alone, not complexed with protein or surrounded by a membrane as it is in more complex cells. The part of a prokaryotic cell in which the DNA is located is called the **nucleoid** (nucleuslike), and it sometimes appears fibrous under a microscope. Nearby are molecules of RNA and spherical structures built of RNA and protein, called **ribosomes.** Ribosomes enable the cell to utilize DNA sequence information to direct the manufacture of proteins. Because the DNA, RNA, and ribosomes in prokaryotic cells are in close contact with one another, protein synthesis in prokaryotes is rapid when compared to the process in more complex cells, where these cellular components are separated.

Table 4.1
Prokaryotic Foes

Organism	Illness
Treponema pallidum	Syphilis
Neisseria gonorrhoeae	Gonorrhea
Salmonella typhosa	Typhoid fever
Vibrio cholerae	Cholera
Yersinia pestis	Bubonic plague
Staphylococcus aureus	Toxic shock syndrome, food poisoning
Clostridium tetani	Tetanus
Clostridium botulinum	Botulism
Legionella pneumophila	Legionnaire's disease
Salmonella species	Food poisoning
Clostridium perfringens	Food poisoning
Streptococcus species	"Strep" throat, rheumatic fever

Table 4.2
Prokaryotic Friends

Organism	Use
Streptococcus lactis, *S. cremoris*	Production of buttermilk, American cheese
Leuconostoc bulgaricus *Streptococcus thermophilus*	Production of yogurt
Lactobacillus bulgaricus *Propionibacterium shermanii*	Production of Swiss cheese
Escherichia coli	Production of genetically engineered insulin and other drugs and genetically engineered enzyme (rennin) used in cheddar-cheese production
Acetobacter suboxydans	Production of food-grade acetic acid (vinegar) from corn alcohol
Pseudomonas species	Conversion of sulfur in crude oil into water-soluble compounds
Clostridium acetobutylicum	Extraction of petroleum
Rhizobium species	Fixation of atmospheric nitrogen into compounds useful to plants

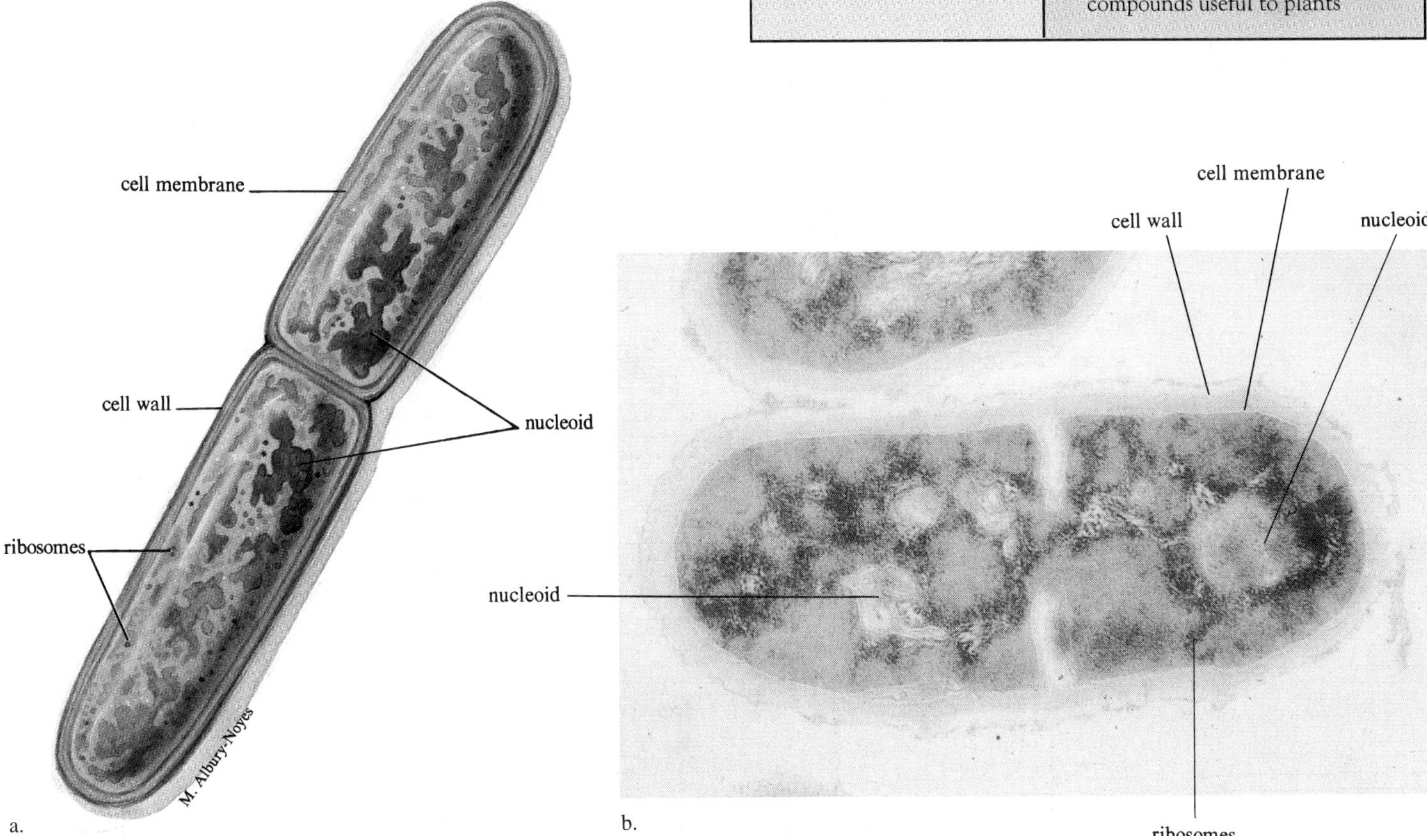

Figure 4.7
A generalized prokaryotic cell. The single DNA molecule of a prokaryotic cell resides in a dense-appearing region called the nucleoid, but there is no true membrane-bound nucleus to house the DNA, as there is in more complex cells. The prokaryotic cell contains many ribosomes, as well as fats, proteins, carbohydrates, and pigments. The cell is surrounded by a cell membrane, and most are also surrounded by a cell wall.

Unlike bacteria, cyanobacteria contain internal membranes that are outgrowths of the cell membrane. These membranes, however, are not extensive enough to subdivide the cell into compartments, as membranes do in more complex cells. The cyanobacterium's membranes are studded with pigment molecules that absorb and extract energy from sunlight.

Cells require relatively large surface areas through which they can interact with the environment. Nutrients, water, oxygen, carbon dioxide, and waste products must enter or leave a cell through its surfaces. As a cell grows, its volume increases at a faster rate than does its surface area, a phenomenon that you can easily calculate (fig. 4.8). To put it another way, much of the interior of a large cell is far away from the cell's surface. The situation can be compared to the plight of a seafood lover living in the middle of a large country with an inadequate transportation system. To satisfy this food craving, this individual would be better off living either on a small island or in a country with a modern, rapid transportation system. Likewise, a cell's insides must either be close to the outside or have ready access to it.

A large cell lacking the means to bring in required chemicals from the environment or to eliminate wastes might die. As the earth became more populated and cells grew larger, they could survive only if they could somehow increase their surface areas relative to their increasing volumes. One strategy was for a large cell to divide in two, restoring a biologically suitable surface/volume ratio. This solution—cell division—occurs today in many of our own cells, as well as in single-celled organisms that have grown too large.

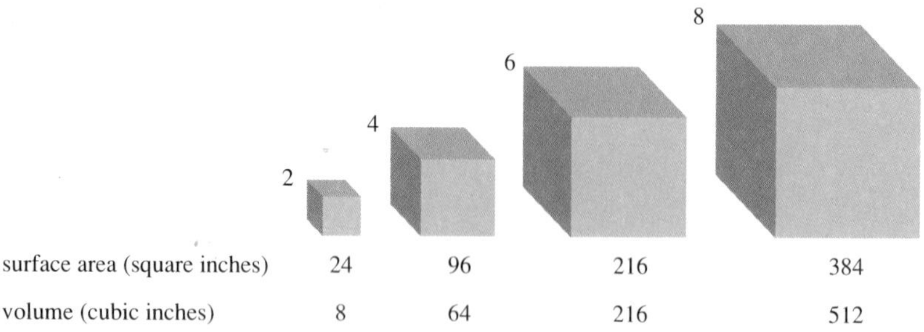

Figure 4.8
The important relationship between surface area and volume. As a cell grows larger, the amount of material inside it, its volume, increases faster than the area of the cell's surface. The surface of a cell is vital to its functioning because this is the site of communication between the cell's interior and the extracellular environment. The chemical reactions of life are more readily carried out when surface area is maximized. Imagine that a cell is a simple cube. Compare the surface area and volume of four increasingly larger cubes. (Surface area equals the area of each face multiplied by the number of faces; volume equals the length of a side cubed.) Can you see that volume increases faster than surface area?

KEY CONCEPTS

Cells exhibit the characteristics of life. Prokaryotes include the bacteria and cyanobacteria and are ancient and simple single-celled organisms. A prokaryotic cell is bounded by a cell wall and a plasmalemma. Its DNA is free of protein and located in the nucleoid area. Ribosomes are present, as are membranes derived from the plasmalemma.

The Eukaryotic Cell

Another cellular solution to the problem of increasing size is to divide into compartments, or organelles, much like a growing store is subdivided into departments. Organelles are established by biological *membranes*, which are barriers composed of lipids and proteins. Organelles have access to the environment outside the cell and to each other by networks of bubblelike structures called *vesicles* (or *vacuoles*) that bud off from the membranes. Cells that have organelles are termed *eukaryotic*. The typical eukaryotic cell, which is roughly 1,000 times the volume of a typical prokaryotic cell, would not be able to function efficiently were it not for the division of

Table 4.3
Comparison of Prokaryotic and Eukaryotic Cells

Characteristic	*Prokaryotic Cells*	*Eukaryotic Cells*
Organisms	Bacteria (including cyanobacteria)	Protists, fungi, plants, animals
Cell size	1-10 μm across	10-100 μm across
Oxygen required	By some	By all
Membrane-bound organelles	No	Yes
Ribosomes	Yes	Yes
DNA form	Circular	Coiled linear strands, complexed with protein
DNA location	In cytoplasm	In nucleus
DNA length	Short	Long
Protein synthesis	RNA and protein synthesis are not spatially separated	RNA and protein synthesis are spatially separated
Membranes	Some	Many
Cytoskeleton	No	Yes
Cellular organization	Single cells or colonies	Some single-celled, most multicellular with differentiation of cell function

Source: From Bruce Alberts, et al., *Molecular Biology of the Cell*. Copyright © 1983 Garland Publishing Company, New York, NY.

labor afforded by organelles. The plants and animals with which you are familiar, including your own body cells, are eukaryotic, as are all microorganisms other than the bacteria and cyanobacteria. Prokaryotic and eukaryotic cells are compared and contrasted in table 4.3.

The extensive membrane networks formed by the organelles and vesicles of eukaryotic cells have many functions. Membranes surround organelles, keeping within them chemical reactions whose products might harm other parts of the cell. Some organelles are constructed of membranes that are studded with enzymes, allowing certain chemical reactions to occur on their surfaces. On some membranes, different enzymes are organized according to the sequences in which they participate in biochemical reactions. In general, then, organelles keep related biochemicals and structures together to make their functioning more efficient. The structure and function of biological membranes are discussed in detail in the next chapter.

The organization of a eukaryotic cell can be described as "bags within a bag" (figs. 4.9, 4.10, and 4.11). The most prominent organelle is the *nucleus*, which contains the genetic material (DNA). The remainder of the cell consists of other organelles and a jellylike fluid called **cytoplasm.** About half of the volume of an animal cell consists of organelles; a plant cell contains large volumes of water. The cytoplasm and the organelles are considered the living parts of the cell, called *protoplasm*. Nonliving cellular components include stored proteins, fats, and carbohydrates, pigment molecules, and various inorganic chemicals. Arrays of protein rods within an animal cell form a framework called the **cytoskeleton,** which helps to give the cell its shape. Protein rods and tubules also form cellular appendages that enable certain cells to move, and they comprise structures involved in cell division. These cell components built of protein rods and tubules are discussed in chapters 5 and 7.

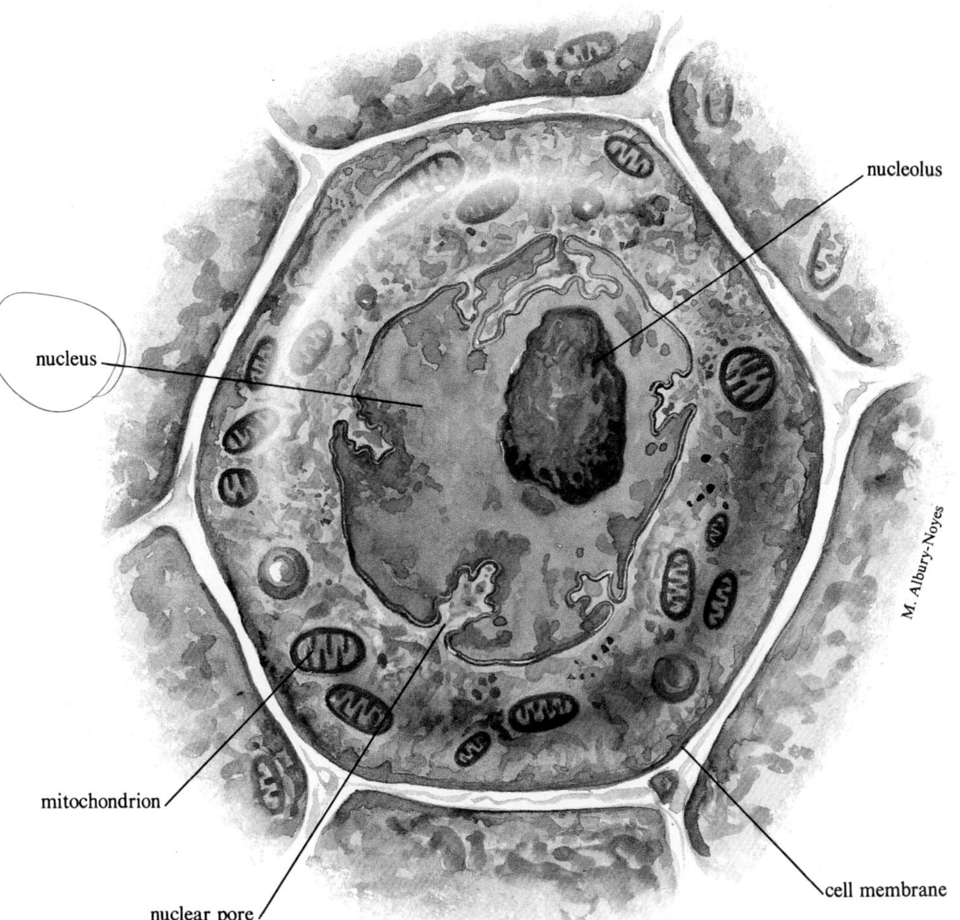

a.

mitochondrion
nuclear membrane
nucleus
nucleolus

b.

Figure 4.9
Eukaryotic cells. A eukaryotic cell contains numerous organelles, the most prominent of which is the nucleus. See table 4.4 for details about individual organelles.

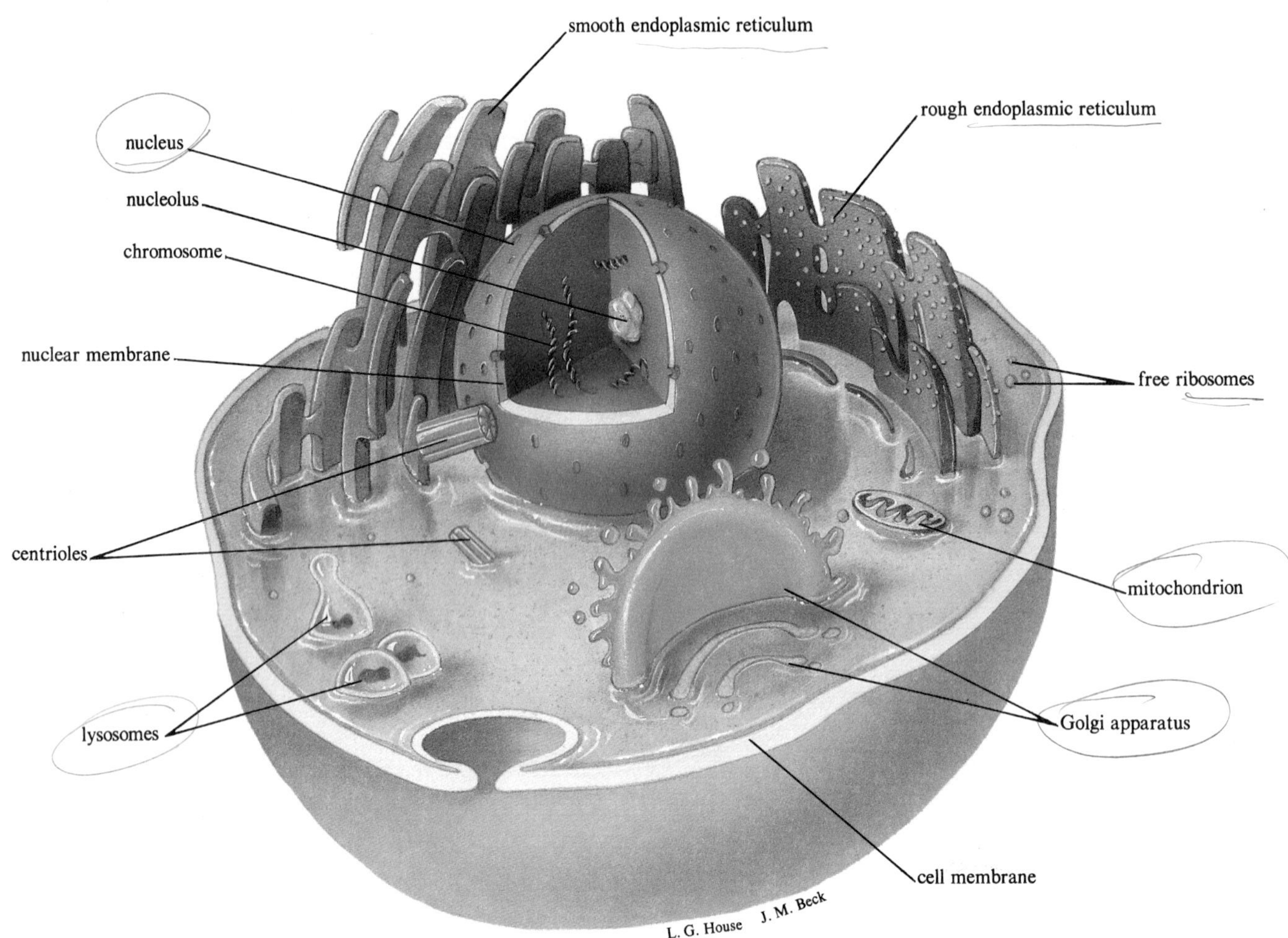

Figure 4.10
An animal cell is a eukaryotic cell. Note the appearances of the different organelles when viewed in cross section.

The organelles are described following in the context of their coordinated function in a familiar cellular activity—secretion. Table 4.4 lists organelle structures and functions with photographs for easy reference and review.

Organelles in Action—Secretion

The activities of the different organelles within a cell are coordinated to provide basic life functions and also to sculpt the distinguishing characteristics of a particular cell type. Consider a glandular cell in the breast of a human female. Dormant most of the time, the cell increases its metabolic activity during pregnancy, and then it undergoes a burst of productivity shortly after the baby is born. The ability of individual cells to manufacture the remarkably complex milk is made possible by the interaction of organelles, which function together to form a *secretory network*.

A new mother hears her infant cry, and within a minute she may feel her breasts swell in response. When the infant suckles, the milk that it receives is a mixture of cells and biochemicals, a highly nutritious food tailored specifically to meet the needs of a human child. Deep within the mother's breasts, glandular cells rapidly produce the proteins, fats, and sugars that are combined with immune system cells and biochemicals in specific proportions to form the milk.

Secretion begins in the nucleus, where, in humans, 23 pairs of rod-shaped **chromosomes** contain information that other parts of the cell use to construct proteins. Each chromosome consists of millions of DNA building blocks (nucleotides), long sequences of which comprise genes. The nucleotide sequence of a gene is transcribed, or rewritten, into another type of nucleic acid, **messenger RNA** (mRNA). (The mechanism for transcribing DNA into RNA is discussed in depth in chapter 13.) RNA building blocks are available in the substance surrounding the chromosomes, called the *nucleoplasm*, and are stored in a structure in the nucleus called the **nucleolus.** Messenger RNA exits the nucleus by passing through holes, called *nuclear pores*, in the two-layered *nuclear envelope* that surrounds the nucleus. In the manufacture of milk, for example, the gene

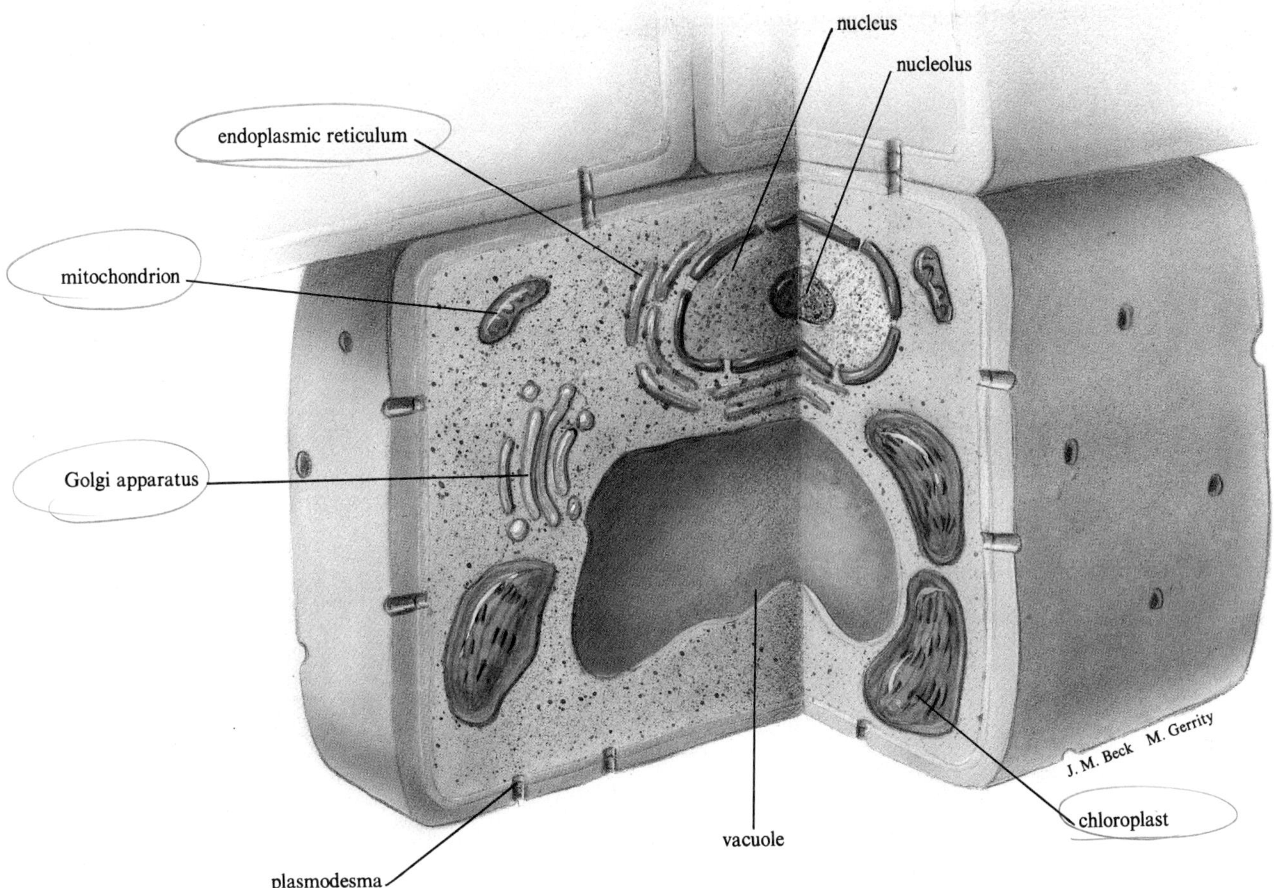

Figure 4.11
A plant cell is eukaryotic but is distinguished from animal cells in that it also has a cell wall and chloroplasts.

containing the "code" for a protein, casein, is transcribed into messenger RNA, which then leaves the nucleus through the nuclear pores (fig. 4.12).

Once in the cytoplasm, messenger RNA encounters a maze of interconnected membranous tubules and sacs that winds from the nuclear envelope to the cell membrane. This labyrinth is the **endoplasmic reticulum** (ER). The portion of this membranous system near the nucleus is flattened and studded with ribosomes, and this region is called *rough ER* because of its appearance in the electron microscope. Messenger RNA attaches to the ribosomes. It is here, extending from the rough ER, that free-floating amino acids are strung together to build proteins, following the instructions encoded in the sequence of messenger RNA building blocks. Some proteins are released into the cytoplasm, where they serve some function. Other proteins—such as the casein destined to become part of breast milk—enter the tubules of the rough ER and begin their journey out of the cell.

As the rough ER winds its way outwards towards the cell membrane, the ribosomes become fewer, and the diameter of the tubules widens, forming a section called *smooth ER*. Lipids are synthesized here, and they are added to the proteins that are transported from the rough ER. The lipids and proteins travel along, and the tubules of the smooth ER eventually narrow and end. The proteins and fats accumulated in the smooth ER exit in vesicles that pinch off of the tubular endings of the membrane. A loaded vesicle takes its contents to the next stop in the eukaryotic production line, a structure called the **Golgi apparatus.** The Golgi apparatus is a system of flat, stacked, membrane-enclosed sacs. Here sugars bond to one another to form starches, or they bond to proteins to form glycoproteins or to lipids to form glycolipids. The components of secretions such as breast milk are temporarily stored in the Golgi apparatus.

Table 4.4
Structure and Function of Major Organelles

Single Membrane

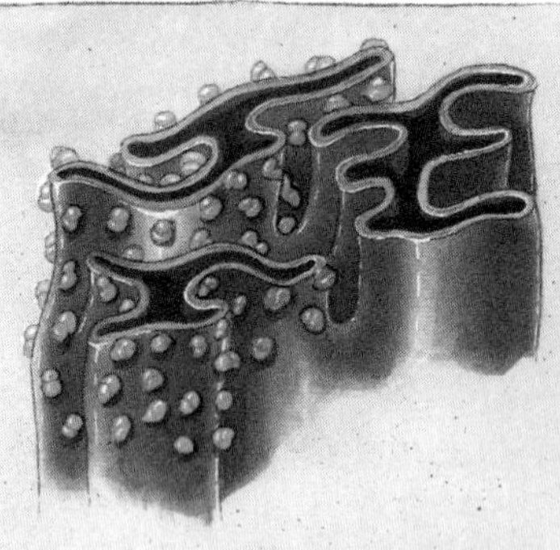

Endoplasmic Reticulum (ER)

Membrane network: rough ER has ribosomes, smooth ER does not

Protein and lipid synthesis

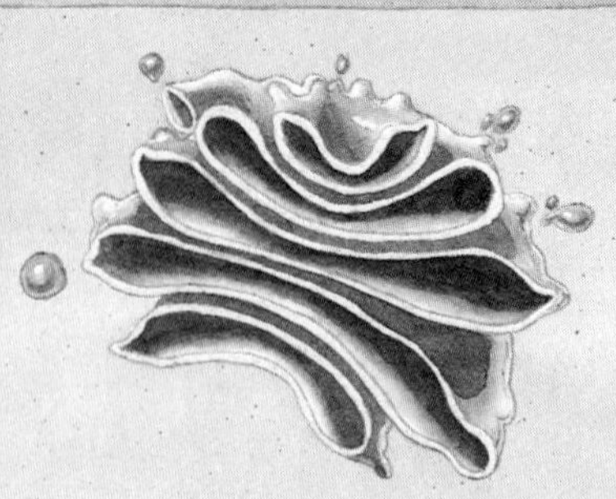

Golgi Apparatus

Stacks of membrane-enclosed sacs

Assembly, storage, packaging, and transport of several types of organic molecules

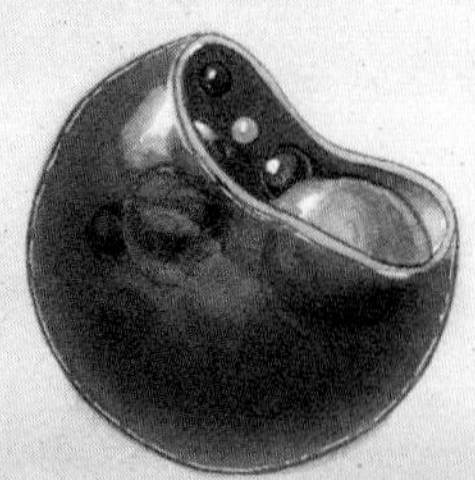

Lysosome

Sac containing digestive enzymes

Degradation of intracellular debris, recycling of cell components

Double Membrane

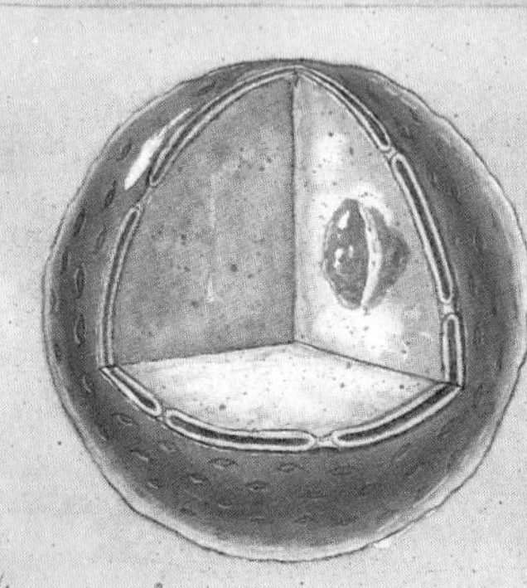

Nucleus

Cell compartment containing DNA with pores in surrounding membrane

Separation of genetic material from rest of cell

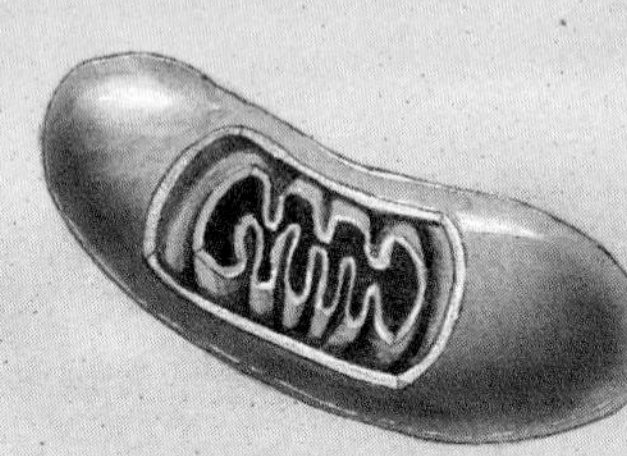

Mitochondrion

Inner membrane highly folded and studded with enzymes

Cellular respiration

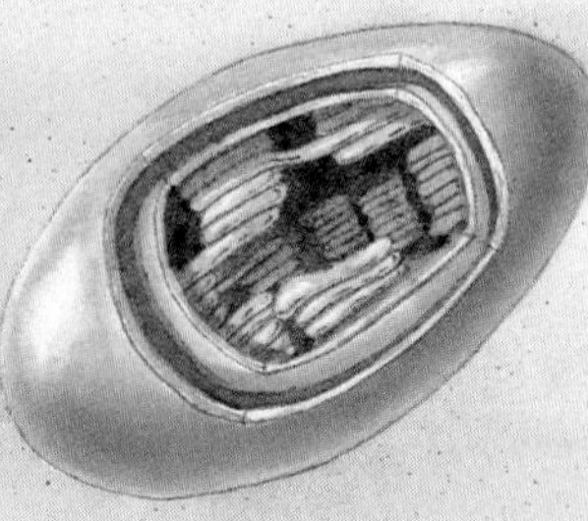

Chloroplast

Plastid built of stacks of flattened sacs containing pigments

Photosynthesis

L. Shoemaker

Milk proteins bud off of the Golgi apparatus in vesicles that travel outwards to the plasmalemma. The protein-carrying vesicles fleetingly become part of the plasmalemma and then open out facing the exterior of the cell. This process, called **exocytosis,** releases free proteins outside the cell. Fat droplets retain a layer of surrounding membrane when they leave the cell. The milk constituents are stimulated to exit the cell when the baby sucks. A hormone released in the mother's system when she feels the sucking action causes muscle cells surrounding balls of glandular cells to contract, which squeezes milk from them. The milk is released into ducts that lead to the nipple. It is interesting that the composition of milk is different in different species. Human milk, for example, is rich in lipids, which suits the rapid development of the child's brain. Cow's milk, in contrast, is rich in protein, which is suited to the calf's need to build muscle rapidly.

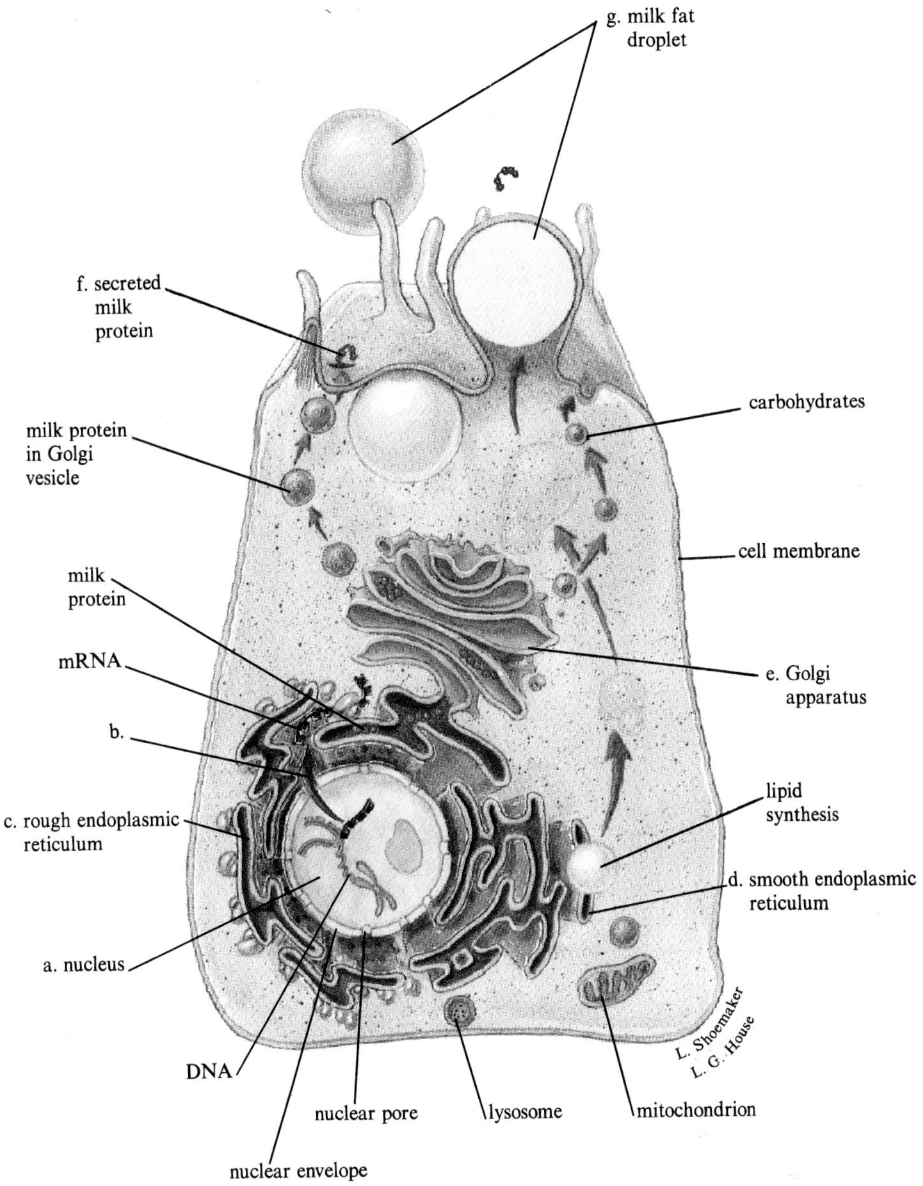

Figure 4.12
Milk secretion illustrates how organelles interact to synthesize, transport, store, and export biological molecules. Secretion begins in the nucleus (*a*), where genes provide information for the manufacture of proteins. Messenger RNA copies of the DNA information for production of milk protein are made and carried through a nuclear pore in the nuclear envelope to the cytoplasm (*b*). Most proteins are synthesized on the membranes of the rough endoplasmic reticulum (ER) (*c*), using amino acids that float free in the cytoplasm. The smooth endoplasmic reticulum (ER) (*d*) is the site of lipid synthesis, and other organic molecules are assembled and stored in the Golgi apparatus (*e*). In an active cell in a mammary gland, milk proteins are released from vesicles that bud off of the Golgi apparatus (*f*), and fat droplets are surrounded by a layer of lipid from the cell membrane and exit the cell (*g*). When the baby sucks, he or she receives a chemically complex secretion—human milk.

KEY CONCEPTS

Organelles are membrane-bound specialized structures of eukaryotic cells that partition off biochemicals that function together. The nucleus contains DNA, and the rest of the cell is cytoplasm. Protein rods form the cytoskeleton. The process of secretion illustrates the coordinated functioning of organelles. Following hereditary instructions, ribosomes in the cytoplasm use the information in a sequence of mRNA to build a protein. Lipids are added where the endoplasmic reticulum becomes free of ribosomes, and sugars are added at the Golgi apparatus. The products bud off in vesicles and exit the cell by exocytosis.

Other Organelles and Structures

The activities of secretion, as well as the many chemical reactions taking place in the cytoplasm, require a steady supply of energy. Cellular energy is provided by organelles called **mitochondria,** within which the energy-generating reactions of *cellular respiration* occur. A typical cell has about 1,700 mitochondria, although cells with high energy requirements, such as muscle cells, may have many more.

A mitochondrion has an outer membrane similar to those of the ER and Golgi apparatus and an intricately folded inner membrane. The folds of the inner membrane are called **cristae,** and they contain many of the enzymes that take part in cellular respiration. The mitochondrion is especially interesting because it contains genetic material. Another unique characteristic of mitochondria is that they are

Reading 4.2 A Closer Look at an Organelle—The Lysosome

A lysosome is a membrane-bound sac of enzymes that buds off from a Golgi apparatus. The key to the functioning of a lysosome can be summed up in one word—balance. The lysosomes in a particular type of cell contain a balanced mix of enzymes that is appropriate to the function of that cell. If that balance is disrupted, the cell's function can be altered—sometimes drastically.

More than 40 lysosomal enzymes are known, and most break down fats and carbohydrates. This enzymatic digestion is applied to worn-out organelles and membranes within the cell, as well as to particles engulfed by the cell by inward budding of the cell membrane. Material to be digested is carried to the lysosome in a vesicle. The membranes of the lysosome and the vesicle fuse, and the appropriate enzymes go to work. Lysosomes are sometimes called "suicide sacs" because if they rupture and release their enzymes, the entire cell is digested from within and dies. Lysosomes may actually play a role in aging by destroying cells in this way.

The absence or malfunction of just one lysosomal enzyme can be devastating to health, creating a lysosomal storage disease. In these inherited disorders, the molecule that is normally degraded by a missing or abnormal lysosomal enzyme accumulates in the lysosome. The lysosome swells with the excess waste, crowding organelles and interfering with the cell's biochemical activities. Usually only the cells that constitute a particular tissue are affected. Because there are a number of different lysosomal enzymes, several types of lysosomal storage diseases are known. Symptoms reflect the tissue whose cells are affected.

Tay-Sachs disease is a lysosomal storage disease that results from a missing enzyme that normally breaks down lipids in nerve cells. Without the enzyme, cells of the nervous system gradually accumulate abnormal amounts of lipids. Symptoms are usually noted at about 6 months of age, when an infant begins to lag behind in the acquisition of motor skills. On a cellular level, however, signs of Tay-Sachs disease are present even earlier as enlarged lysosomes. Children who have inherited Tay-Sachs disease soon lose their vision and hearing, and they are paralyzed by the time they die, usually before they are 4 years old. A less severe form of Tay-Sachs affecting adults is also known.

Pompe disease is another lysosomal storage disease of childhood. An enzyme that breaks down the complex carbohydrate glycogen into simple sugars is missing. Glycogen builds up in muscle and liver cells. The young patients usually die of heart failure, because the muscle cells of the heart swell so greatly that they can no longer function properly. Another lysosomal storage disease, *Hurler's disease*, causes bone deformities. Affected bone cells are revealed by the electron microscope to contain huge lysosomes, swollen with mucuslike substances called mucopolysaccharides.

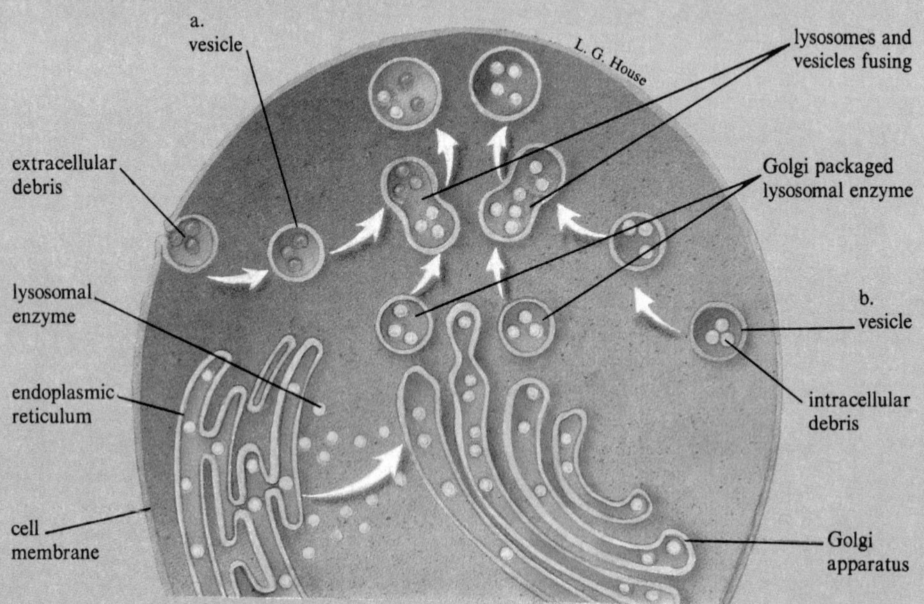

Figure 1
Lysosomal enzymes are synthesized in the endoplasmic reticulum and transported to the Golgi apparatus. In a poorly understood process, the Golgi apparatus detects and pulls out those enzymes destined for lysosomes (perhaps by recognizing a particular sugar that is attached to these enzymes) and packages them into vesicles that ultimately become lysosomes. Lysosomes fuse with vesicles carrying debris from the outside or from within the cell. Lysosomal enzymes degrade the debris.

inherited from one's mother only. This is because mitochondria are found in the tails of sperm cells but not in the head region, which is the portion that actually enters the egg to fertilize it. A class of inherited diseases is caused by abnormal mitochondria, and these disorders are always passed from mother to offspring. These mitochondrial illnesses usually produce symptoms of extreme muscle weakness, because muscle is a highly active tissue dependent upon the functioning of many mitochondria. We will return to this fascinating organelle shortly, and it is discussed in greater detail in chapter 6.

Eukaryotic cells break down molecules and structures as well as produce them. Organelles called **lysosomes** are the cell's "garbage disposals," chemically dismantling captured bacteria, worn out organelles, and other debris. The lysosome is a sac that buds off of the ER or Golgi apparatus. In humans each lysosome contains more than 40 different digestive enzymes. These enzymes can work only in a very acidic environment (pH 5), and the compartmentalization provided by the lysosome membrane provides a highly acidic region for the enzymes, without harming other cellular constituents. Material is brought to a lysosome by a vesicle, either from within the cell or from the plasmalemma, which can bud inward to entrap a particle in a process called **endocytosis.** In humans lysosomes are particularly abundant in liver cells, perhaps because cells of this gland dismantle toxins. The correct balance of enzymes within a lysosome is important to human health (Reading 4.2). Lysosomes are less prevalent in plant cells than they are in animal cells.

Another organelle that is a single membrane-bound sac containing enzymes is the **peroxisome,** which buds from the smoother ER, either remaining attached to it by a thin stalk of membrane or floating nearby like a small bubble. Peroxisomes house enzymes important in oxygen utilization. It is believed that peroxisomes may be very ancient structures that participated more heavily in using oxygen before mitochondria evolved.

The organelles and structures discussed so far—the nucleus, ribosomes, endoplasmic reticulum, Golgi apparatus, mitochondria, lysosomes, and peroxisomes—are present in nearly all eukaryotic cells, although some are more abundant in certain specialized cell types than in others. A few structures are peculiar to animal or plant cells. **Centrioles** are oblong structures built of protein rods called **microtubules,** and they are found in pairs in animal cells, oriented at right angles to one another near the nucleus. Centrioles appear to play a role in organizing other microtubules to pull replicated chromosomes into two groups during cell division.

An organelle unique to plants and algae, the **chloroplast,** gives these organisms their green color. Chloroplasts house the chemical reactions of **photosynthesis,** by which the cells capture solar energy and use it to manufacture organic molecules. The inner membrane of a chloroplast is studded with enzymes necessary for photosynthesis. It is organized into stacks, called **grana,** of flattened membranous disks, called **thylakoids.** Like mitochondria, chloroplasts contain genetic material.

The green pigment **chlorophyll** within the chloroplasts harnesses the solar energy. An actively photosynthesizing plant cell, such as a cell on the side of a leaf that faces the sun, may contain more than 50 chloroplasts (fig. 4.1*c*). Chloroplasts are the most abundant of a general class of pigment-containing organelles, called *plastids*, found in plant cells.

Mature plant cells may also be distinguished from other eukaryotic cells by the presence of a large, centrally located vacuole. Water stored in a plant cell's vacuole can amount to 90% of the cell's total volume. In addition, plant cells are surrounded by a rigid cell wall built of the carbohydrate *cellulose*. The cell wall helps to support the cell and to protect its contents. Animal cells do not have cell walls.

KEY CONCEPTS

Mitochondria are built of a double membrane. Along the folds of the inner membrane enzymes carry out biochemical reactions that harness energy in food molecules. Lysosomes contain biochemicals that degrade cellular debris. Peroxisomes contain enzymes to assist the cell to utilize oxygen. Centrioles are peculiar to animal cells. Plastids, including chloroplasts, are found only in plant cells.

Specialized Cells Form Tissues

A human life begins with a single cell, the fertilized ovum. It is a very large cell, containing an abundance of all organelle types found in human cells, plus stores of nutrient molecules. Nine months later, that organism has grown to about 20 inches (50 centimeters) in length and weighs 5 to 11 pounds (2 to 4 kilograms). During the 9 months that follow, the progeny of the original single cell divide many times, molding the fetus into a perfectly formed miniature version of an adult human, its component parts built of 200 different types of specialized cells. Although all cells contain the structures and organelles necessary for survival, the combinations of these components give many cells specialized, or *differentiated*, characteristics.

The differentiated cells of humans and other multicellular organisms are grouped into *tissues*. Some tissues also contain nonliving materials in which cells are embedded or suspended. Tissues are grouped to form *organs*, and functionally related organs are grouped to form *organ systems*. For example, the respiratory system (an organ system) includes the lungs (organs), which are built of tissues. The cells constituting different tissues are specialized in structure, in function, and in the kinds of molecules they manufacture in large quantities. Four basic tissue types are recognized in humans: **epithelial tissue,** which forms coverings and linings; **connective tissue,** which provides support; **nervous tissue,** which offers rapid communication networks between cells; and **muscular tissue,** which provides motion (fig. 4.13). Human tissues are introduced here.

Epithelium—The "Covering" Tissue

The human body has many surfaces, from the most obvious one, the skin, to the numerous surfaces of internal organs. The tissue lining the insides of the blood vessels in a single human body, if spread out, would cover many square miles. Such lining tissue is called **epithelium,** and it consists of closely aggregated cells with very little extracellular material between them.

Different types of epithelia are classified by the shape of the cells and the number of layers that they form (table 4.5). A layer of epithelium that is one cell thick is called *simple epithelium*, and layers two or more cells thick are called *stratified epithelium*. A single layer of cells whose nuclei are at different levels gives the illusion of stratification, and it is called *pseudostratified epithelium*. Flat epithelial cells are called *squamous*. Squamous epithelium in the top layer of the skin accumulates a hard protein, called keratin, and becomes so thin that the cells flake off (fig. 4.14). Squamous

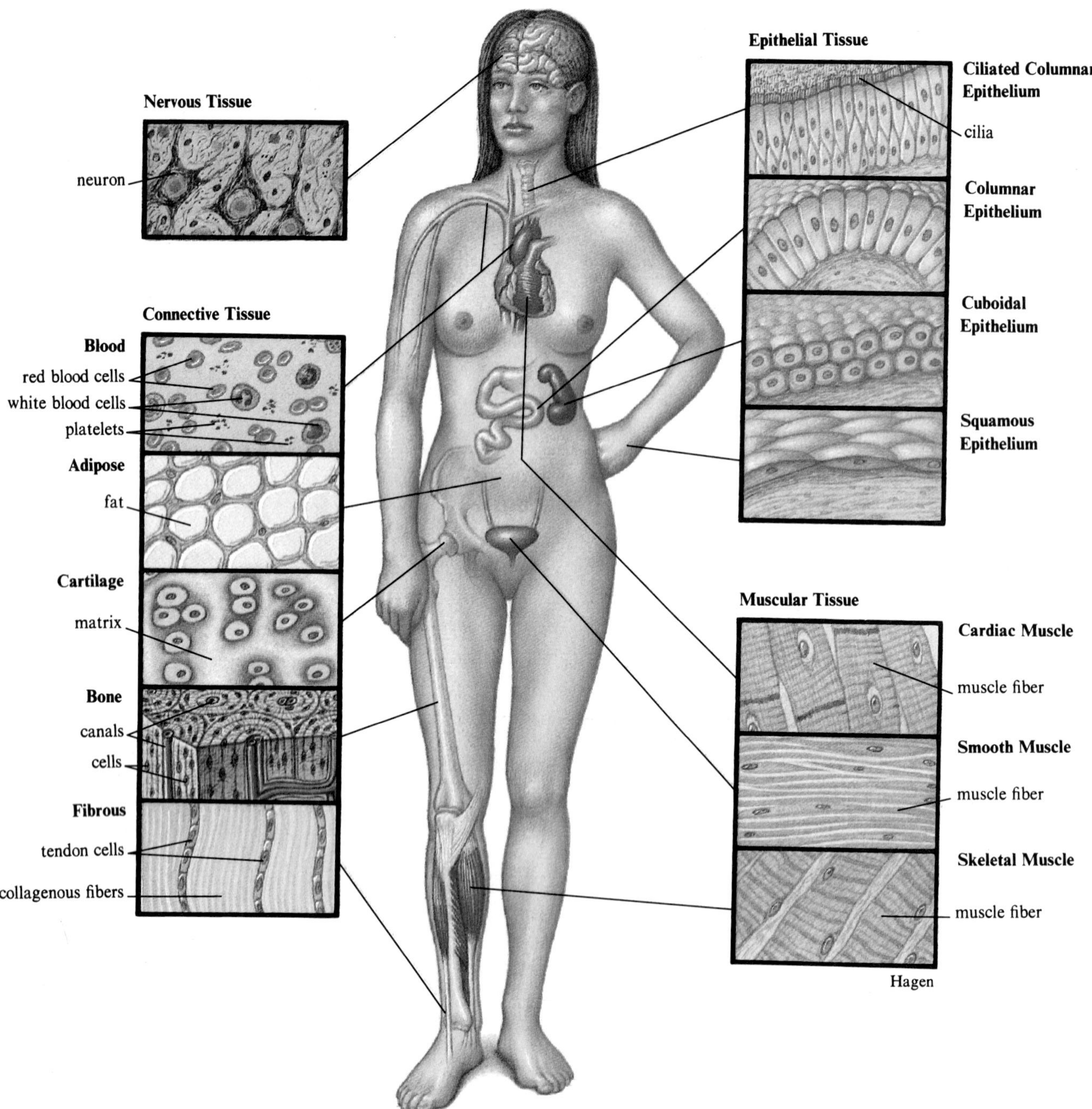

Figure 4.13

Human cell types. The organs in the human body are built of four basic tissue types: epithelial tissue, which lines organs and other structures; connective tissue, which provides form and support and includes loose and fibrous connective tissue, adipose tissue, blood, cartilage, and bone; nervous tissue, which provides cell-to-cell communication; and three types of muscle cells, which contract, providing the ability to move.

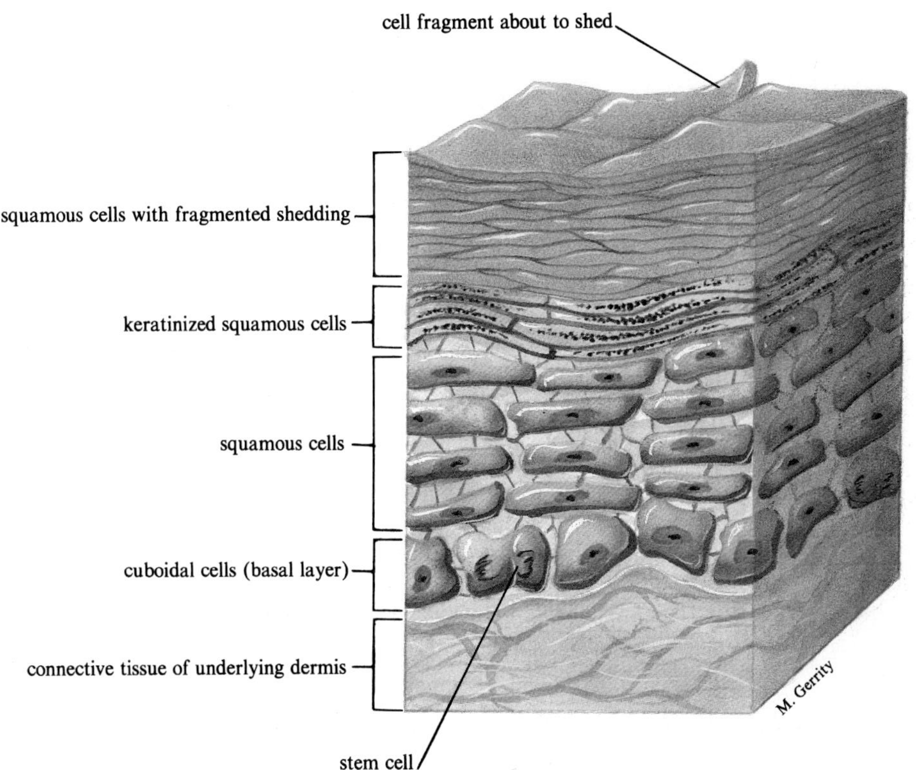

Figure 4.14
Skin contains different types of epithelium. The topmost layer of human skin consists of squamous cells that flake off. Beneath the squamous layer, the cells gradually assume a more cuboidal shape. Some of the cells in the bottommost, or basal, layer divide to replenish the skin cell population above.

cells lining the mouth, esophagus, and vagina, however, do not produce keratin. Epithelium can also be cube-shaped (*cuboidal*) or tall (*columnar*).

Epithelium has many functions. It protects the inner and outer surfaces of organs, and the epithelial linings of blood vessels and the digestive cavities participate in the absorption and transport of nutrients. The amount of space between epithelial cells determines how readily substances can cross the lining they form. The cells that form the lining of the microscopic blood vessels in the brain, for example, are packed so closely together that they constitute a "blood brain barrier," which excludes certain chemicals from the delicate nervous tissue of the brain.

Epithelium surrounds groups of secretory cells, forming glands that specialize in manufacturing certain substances, such as mucus or digestive enzymes. Epithelial cells lining parts of the respiratory system are fringed with waving protein projections called **cilia,** which move dust particles up and out of the body. The epithelium forming the outer layer of the skin, called the epidermis, surrounds specialized structures such as hair and nails, as well as sensory receptors for pain, heat, and touch.

Connective Tissue

A large part of the human body is built of connective tissue, which fills in spaces, attaches epithelium to other tissues, protects and cushions organs, and provides mechanical support. In short, connective tissue gives the human body form and strength. Connective tissue follows an anatomical plan of cells embedded in a nonliving substance called a *matrix*. Although all connective tissue is built of cells in a matrix, the tissue varies in different parts of the body in the proportions of cells to matrix and in the chemical components of the matrix.

A very abundant type of connective tissue cell is the **fibroblast** (fiber builder). It manufactures two types of protein fibers that are part of the matrix—*collagen*, a flexible white protein that resists stretching, and *elastin*, a yellowish protein that stretches readily, much like a rubber band (table 4.6). The matrix also consists of a thin gel made of *proteoglycans*, which are complex carbohydrates linked to proteins.

The major types of connective tissues in the human body are loose and fibrous connective tissues, blood, cartilage, and bone (fig. 4.15). Each type is identified by the composition of its matrix and the specializations of its cells. *Loose connective tissue* is the "glue" of the body, consisting of widely spaced fibroblasts and a few adipose (fat) cells surrounded by a meshwork of collagen and elastin fibers. In contrast, *fibrous connective tissue* is built of dense tracts of collagen, and it forms ligaments, which bind bones to each other, and tendons, which connect muscles to bones. Much of the middle, or dermis, layer of the skin is composed of fibrous connective tissue.

Blood is a complex mixture of different cell types suspended in a matrix called **plasma.** Blood cells are said to be "wandering" connective tissue cells because they circulate through the body; other connective tissue cells are termed "fixed," because they do not move. An adult has about 10 pints (5 liters) of blood, comprising 7% of body weight. **Red blood cells** transport oxygen and constitute the bulk of the cells in the blood. A typical adult male has about 5.4 million red blood cells for each milliliter of blood; a typical adult female has about 4.8 million. **White blood cells** are less numerous than red blood cells, with only about 8,000 per milliliter of blood, but they come in more varieties. White blood cells protect against infection and help to clear the body of its own cells that have worn out or become abnormal. Blood also contains cell fragments called **platelets,** which release chemicals that promote blood clotting. About 250,000 platelets are found in every milliliter of blood. The blood plasma is about 92% water, and it carries cells, dissolved salts and gases, proteins, nutrients, and waste products. In a healthy individual, the different types of blood cells are present in specific proportions. Alterations in blood

Table 4.5
Types of Epithelium

Type	Examples	
Simple squamous	Lining of the heart	
Simple cuboidal	Lining of the kidney tubles	
Simple columnar	Lining of the oviduct	
Stratified squamous	Lining of the vagina	
Pseudostratified	Lining of the trachea	

composition can signify illness. For example, too few red blood cells indicates anemia; too many white blood cells most often means infection but can be a sign of leukemia, a cancer in which certain white blood cells lose their specific characteristics and divide more frequently than healthy white blood cells.

Cartilage is a connective tissue that cushions organs and forms a structural framework to keep tubular organs from collapsing, such as in the ear, the nose, and in the respiratory passages. In joints, cartilage can sustain weight while allowing bones to move against one another. Cartilage also forms the skeleton in the embryo, and it is gradually replaced with bone, which is a much harder tissue.

a.

Table 4.6
Types of Connective Tissue

Type	Cells	Matrix	Cell/Matrix Ratio	Site
Loose connective tissue	Fibroblasts, adipose cells, white blood cells	Loose collagen and elastin networks	High	Beneath skin
Fibrous connective tissue	Fibroblasts	Dense collagen and elastin networks	Low	Tendons, ligaments
Blood	Red blood cells, white blood cells, platelets	Plasma	High	
Cartilage	Chondrocytes	Collagen	Low	Ears, bone tips, joints, respiratory passages, embryonic skeleton
Bone	Osteoclasts, osteoblasts, osteocytes, osteoprogenitor cells	Collagen, calcium, salts	Low	Skeleton

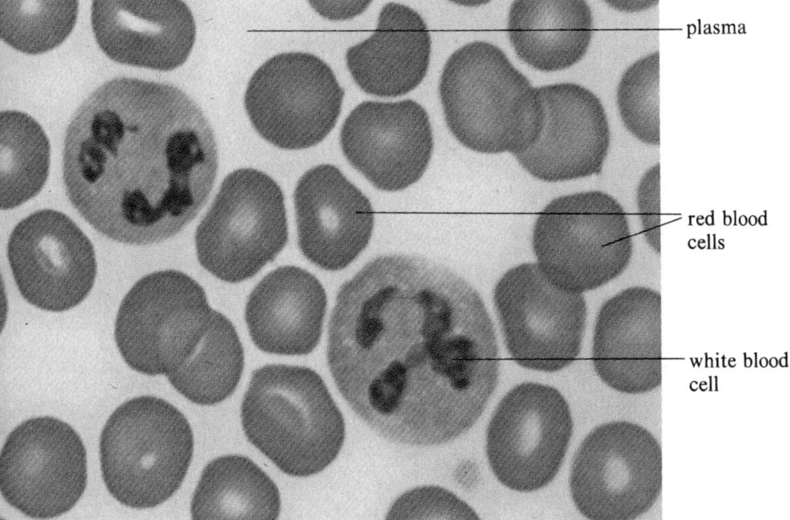

b.

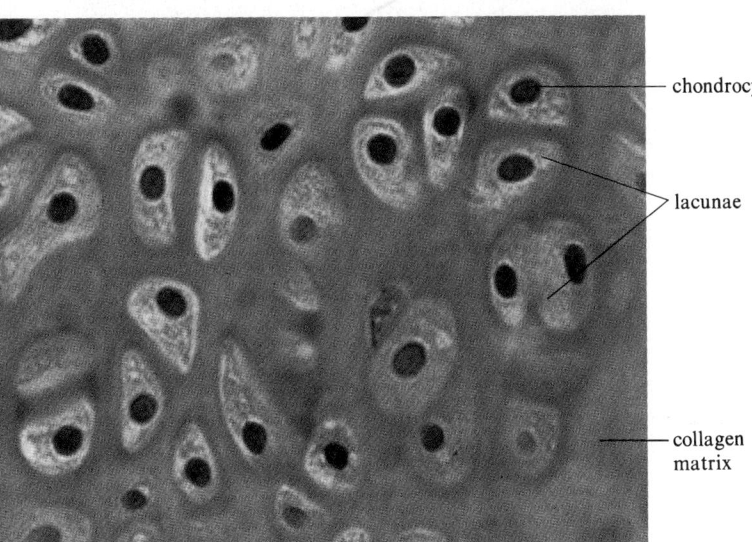

c.

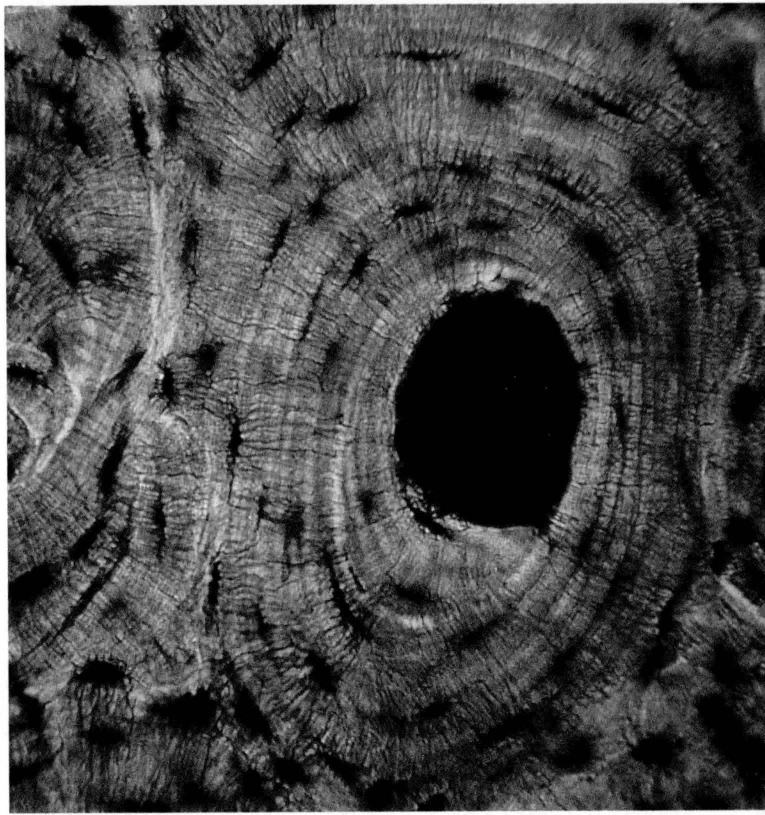

d.

Figure 4.15
a. Fibrous connective tissue is more organized than loose connective tissue, as can be seen in this close-up of connective tissue beneath a person's skin. *b.* Blood is a fluid connective tissue consisting of red blood cells, white blood cells, and platelets (not shown), suspended in a watery matrix called plasma. *c.* Cartilage is a flexible, resilient connective tissue consisting of chondrocytes housed within spaces called lacunae, which are embedded in a collagen matrix. *d.* Bone consists of four types of cells surrounded by complex tunnel systems of matrix hardened by minerals.

Cartilage is the simplest connective tissue because it has a single cell type, the *chondrocyte*, lodged within oblong spaces called *lacunae* embedded in a collagen matrix. Some cartilage also contains the protein elastin. The tissue grows as chondrocytes secrete collagen. Cartilage cells have large nuclei and extensive endoplasmic reticulum, a sign of their high protein output.

The strong networks of collagen and elastin fibers give cartilage great flexibility yet also the ability to distribute and support weight. Cartilage matrix also contains proteoglycans consisting of protein chains bonded to long chains of the disaccharide *hyaluronic acid*. The proteoglycans are negatively charged, which attracts tremendous amounts of water. The water entrapped within cartilage provides support as well as resiliency, which is the ability of the tissue to return to its original shape after being deformed.

Cartilage is covered with a thick and tough shell of collagen, and unlike most other tissues, it lacks nerves and blood vessels. Because blood is lacking, the nutrients required by the chondrocytes can enter only by diffusing slowly in. Once in the tissue, nutrients are distributed by the movement of the abundant water that is trapped within the proteoglycan framework. The lack of a blood supply is why injuries to cartilage take a long time to heal.

Bone is a familiar type of connective tissue. Bone is supportive, providing maximum strength with minimum weight, and it protects other tissues and organs. The long bones of the arms and legs shield the delicate bone marrow within; the skull protects the brain; and the ribs guard the lungs and heart. Bones also provide points of attachment for muscles.

Bone follows the connective tissue organization of widely separated cells within a matrix but with an important addition—mineral salts. The mineral *hydroxyapatite*, containing calcium and phosphate, constitutes most of the *mineral phase* of bone. The hard mineral is deposited to form a labyrinth of tunnels. A slice of bone tissue has many concentric rings, which are the various canal systems seen in cross section.

The *organic phase* of bone consists almost entirely of collagen. Cells called *osteocytes* occupy spaces, called lacunae, in the bone. Long narrow passageways called *canaliculi* connect the lacunae, and osteocytes send out extensions that touch each other through the canaliculi. The canaliculi and lacunae are arranged around larger passageways called *Haversian canals*, which surround blood vessels. Still other canals connect the inner ones to the outer surface of the bone and to the marrow cavity within, where blood cells are manufactured.

Bone is an unusual tissue in that it is continually being reconstructed by its several cell types. *Osteoblasts* secrete bone matrix, while large cells with many nuclei, called *osteoclasts*, degrade bone matrix. Osteocytes are osteoblasts that have become trapped in their lacunae by the collagen that they secrete. Other bone cells, called *osteoprogenitor* cells, line the passageways of bone and serve as a reserve supply of cells that can transform into osteoblasts or osteoclasts in the event of growth or injury.

The precise balance between destruction and reconstruction of bone matrix molds bones during development, enabling them to enlarge yet maintain their characteristic shapes throughout life. The dynamic nature of the tissue allows bones to respond to stress. This is why a weight lifter develops stronger and more massive bones. Imbalance in the breaking down and building up of bone can result in bones that are too brittle or too soft. In the condition osteoporosis, for example, loss of bone mineral results in easily fractured bones. Interestingly, the balance between bone building and bone destruction is altered during space travel. Calcium salts detected in the urine of astronauts following a flight indicate that bone destruction occurs faster than bone building while they are away from the force of gravity. Fortunately, the normal process of bone remodeling returns when the travelers come back to earth.

Nervous Tissue

Nerve cells, called **neurons,** and the cells that structurally support them, called *neuroglia*, constitute nervous tissue. Neurons convey information, and they can be quite long and branching. A typical neuron consists of an enlarged portion called the *cell body* (which contains the nucleus), a thick branch called the *axon*, and several thinner branches called *dendrites* (fig. 4.16). Most dendrites receive information in the form of chemicals called **neurotransmitters,** which are released from the axon of another neuron or from direct energy stimulation such as light, heat, or pressure. The arrival of this neurotransmitter or sensory stimulation alters the plasmalemma of the receiving cell's dendrite so that different types of ions can enter and leave the cell. This membrane change alters the electrical potential of the receiving cell, and the electrochemical change is sent along the neuron's cell membrane. When it reaches the end of the axon, the electrochemical "wave" triggers the release of a neurotransmitter. The information is thereby passed on to another cell, usually a neuron but sometimes a muscle (which contracts in response) or a gland cell (which secretes in response). Nervous information is therefore passed by a combination of electrochemical and neurotransmitter signals.

Neurons impinge upon one another to build intricate nerve networks. A single neuron in the brain, for example, might receive thousands of incoming messages from other neurons at any one time. Several types of supportive cells called neuroglia, as well as connective tissue, are found around and between neurons. One abundant type of neuroglia, Schwann cells, have very fatty cell membranes that wrap around axons, forming an insulating sheath called myelin, which aids in the conduction of nerve impulses. Other neuroglia provide a structural scaffolding on which highly branched neurons rest.

Muscle Tissue

Muscle tissue can contract, which enables the human body to move either under voluntary control, such as in walking, or involuntarily, such as the heart's regular pumping and the rhythmic inflation and deflation of the lungs. Muscles contract when two types of protein filaments (*actin* and *myosin*) slide past one another, shortening their total length. Other, less abundant proteins are involved in muscle contraction too, such as dystrophin, which is lacking in people with muscular dystrophy. Muscle cells have many mitochondria that provide the energy for contraction. Four types of contractile cells are recognized:

skeletal, cardiac, and smooth muscle cells and myoepithelial cells, which have characteristics of both epithelium and muscle.

A fiber of **skeletal muscle** consists of one huge cell with many nuclei that appears striped, or striated, when viewed under a microscope. These striations are caused by the arrangement of proteins within the fibers. A skeletal muscle fiber can be several centimeters long and can contain more than a hundred nuclei. Skeletal muscle makes possible voluntary movements. **Cardiac muscle,** so named because it is found only in the heart, is also striated, but it is built of cells with single nuclei. Cardiac muscle cells are joined to each other by disclike structures. **Smooth muscle** is not striated, and its involuntary contractions are slow when compared with those of other contractile cells. Smooth muscle cells are responsible for the pulsations along the digestive tract that help to move food along and for erecting hairs at the back of the neck in response to fright. Myoepithelial cells, which are not striated, are found in epithelium, where they contract to expel secretory products, such as milk, saliva, and sweat, from the glands that produce them.

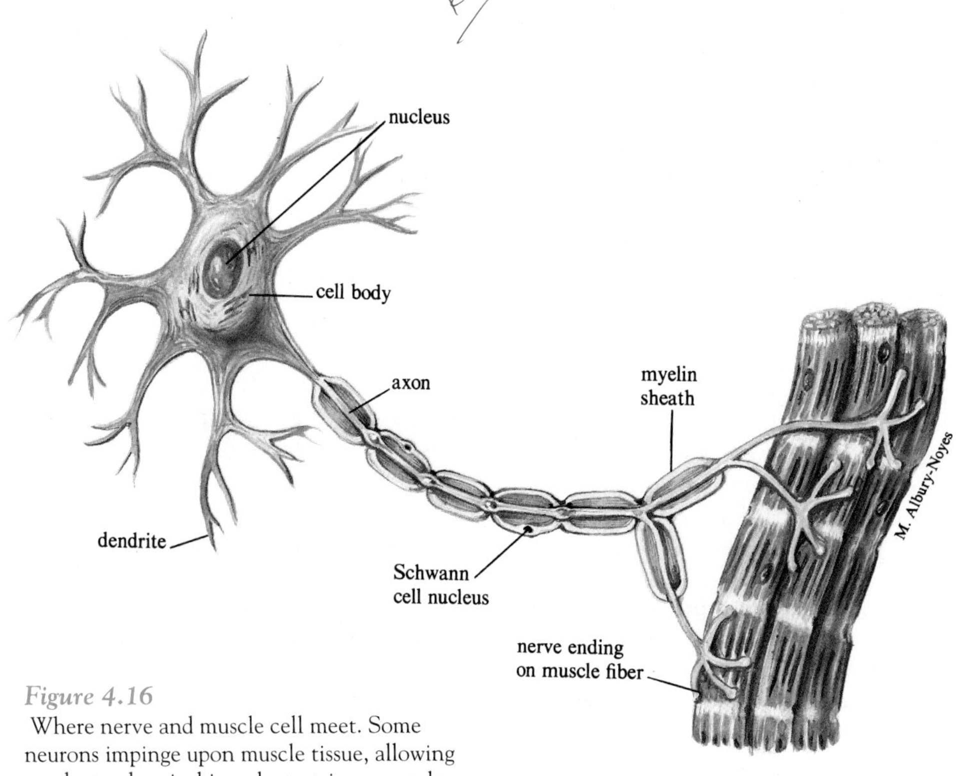

Figure 4.16
Where nerve and muscle cell meet. Some neurons impinge upon muscle tissue, allowing an electrochemical impulse to trigger muscle contraction.

KEY CONCEPTS

Differentiated cells are arranged into tissues, and tissues are arranged into organs and organ systems. Epithelial tissue forms surfaces and coverings and is described by number of layers and cell shapes. Connective tissue consists of cells embedded in a matrix and includes fibroblasts, blood, cartilage, and bone. Nervous tissue consists of neurons and their supporting cells. Muscle cells contain actin and myosin, which contract when they slide past each other.

The Origin of Eukaryotic Cells

The 200 types of specialized cells in the human body are incredibly complex when compared to the typical prokaryotic cell. How did the sophisticated eukaryotic cell arise on a planet populated by the far simpler prokaryotes? Many biologists favor the **endosymbiont theory,** which states that eukaryotic cells formed from large prokaryotic cells that incorporated smaller and simpler prokaryotic cells. (An endosymbiont is an organism that can live only inside another organism.) The compelling evidence in support of the endosymbiont theory is that the mitochondria and chloroplasts found in eukaryotic cells bear striking resemblances to prokaryotic cells.

Mitochondria and chloroplasts resemble bacteria in size, shape, and membrane structure. Each organelle reproduces by splitting in two and contains its own DNA. In addition, the DNA, messenger RNA, and ribosomes within chloroplasts and mitochondria function in close association with each other, which is the way in which prokaryotes use their genes to make proteins. In contrast, DNA in a eukaryotic cell's nucleus is physically separated from the RNA and ribosomes. There are other similarities. Pigments in the chloroplasts of eukaryotic red algae are similar to pigments used by cyanobacteria to carry out photosynthesis. The projections that propel sperm cells may be descendants of ancient spiral-shaped bacteria. The endosymbiont theory specifically proposes that mitochondria descended from aerobic (oxygen-using) bacteria, that the chloroplasts of red algae descended from cyanobacteria, and that the chloroplasts of green plants descended from yet another type of photosynthetic microorganism. This hypothesized merger of organisms might have happened as follows.

Picture a mat of bacteria and cyanobacteria, thriving in a pond some 2.5 billion years ago. The flourishing cyanobacteria pumped oxygen into the atmosphere as a by-product of photosynthesis. The oxygen content of the atmosphere increased, and only those organisms that could tolerate the presence of free oxygen survived. Free oxygen tends to react with molecules found in living things (such as nucleotides, amino acids, and sugars), turning these chemicals into oxides that can no longer carry out biological functions. One way for a large cell to solve the problem of living in an oxygen-rich environment would be to engulf an aerobic bacterium in an inward-budding vesicle of its cell membrane. Eventually, the membrane of this vesicle became the outer membrane of the mitochondrion. The outer membrane of the engulfed aerobic bacte-

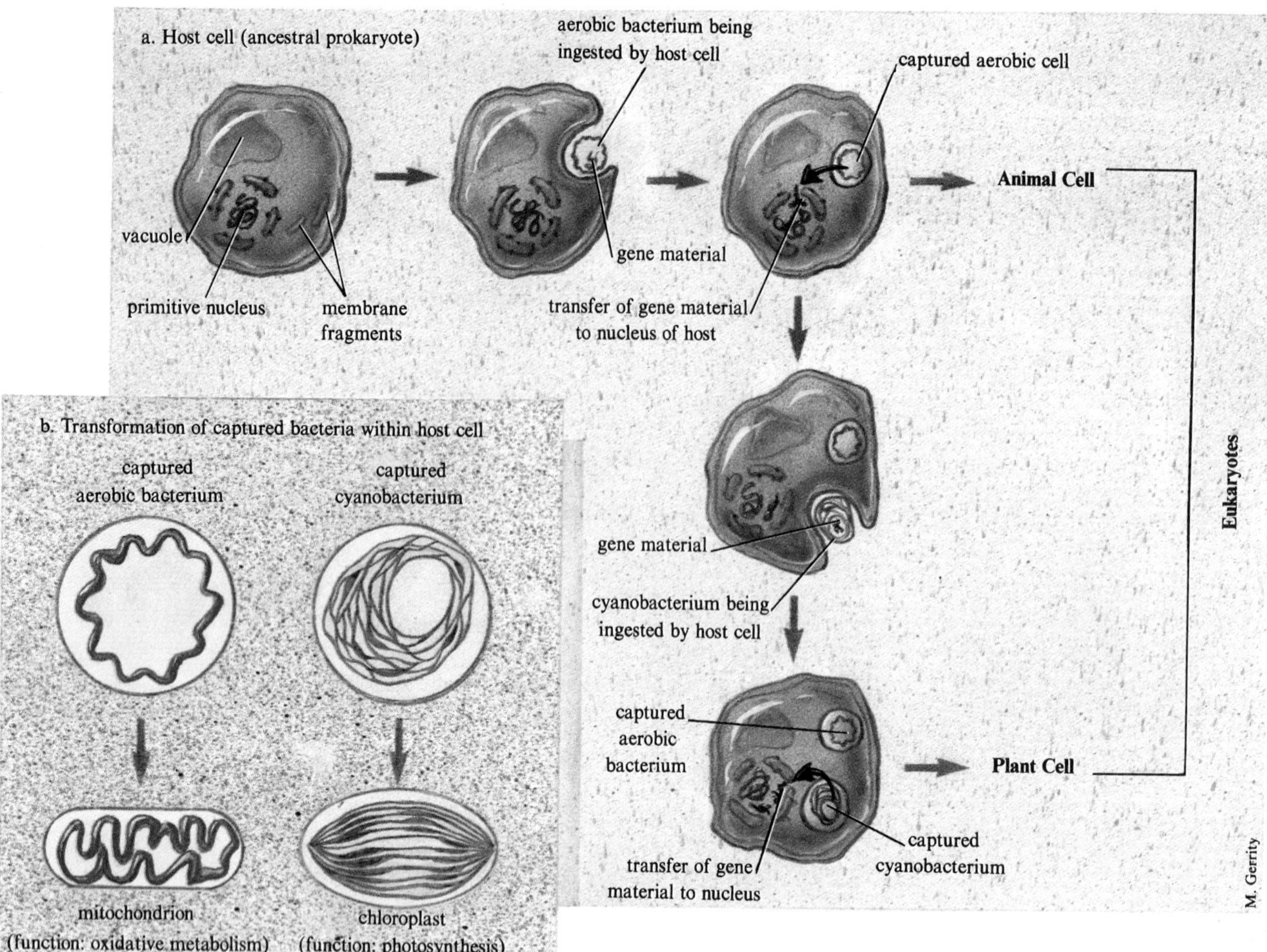

Figure 4.17
The endosymbiont theory. About 2.5 billion years ago, earth was home to a variety of single-celled organisms. Some of these early cells were probably larger than the others and may have already contained genetic material in a structure resembling a nucleus. Pieces of membrane may have floated in the cell. *a.* The large cells surrounded and engulfed some of their smaller contemporaries—aerobic bacteria and photosynthetic microorganisms. *b.* The captured cells eventually lost some of their genes to the nuclei of the larger cells, relinquishing their status as free-living organisms and becoming endosymbionts of the host organisms. The photosynthetic microorganisms evolved membranes, becoming the chloroplasts of some eukaryotic cells. Similarly, the swallowed aerobic bacteria developed into mitochondria.

rium became the inner membrane system of the mitochondrion, complete with respiratory enzymes (fig. 4.17). The smaller bacterium found a new home in the larger cell; in return, the host cell could survive in the newly oxygenated atmosphere.

Similarly, large cells that picked up a cyanobacterium or some other small cell capable of photosynthesis obtained the forerunners of chloroplasts, becoming the early ancestors of red algae or of green plants. Once such ancient cells had acquired their endosymbiont organelles, genetic changes occurred that impaired the ability of the captured prokaryotes to live on their own outside of the host cell. The result of this biological interdependency is the compartmentalized cells of modern eukaryotes, including our own.

KEY CONCEPTS

The endosymbiont theory proposes that complex cells descended from large prokaryotes that incorporated smaller and simpler prokaryotes.

SUMMARY

The cells that constitute organisms are organized and efficient, especially as compared to viruses, which are built only of nucleic acid and protein and must exist within cells. Cells were not observed until the late seventeenth century, when Robert Hooke turned his crude microscope to a piece of cork. Subsequent microscopists amassed the observations that led to the development of the *cell theory*, which states that all life is composed of cells, that cells are the functional units of life, and that all cells come from preexisting cells. Today, a variety of compound light microscopes and electron microscopes greatly aid our ability to view the microscopic living world.

Subcellular structures must communicate with the extracellular environment in order to obtain the materials necessary to carry out the biochemical activities of life and to eliminate wastes. As a cell grows larger, its surface area cannot keep pace with expanding volume. One solution to the problem of diminishing surface area is for a cell to split in two. Another biological strategy is the evolution of compartments within cells, called *organelles*, that specialize in certain functions and communicate with the environment by means of membranes and vesicles. Small, structurally simple cells that lack organelles are called *prokaryotic*. Bacteria, including cyanobacteria, are prokaryotes, and these unicellular organisms contain genetic material, ribosomes (structures used to manufacture proteins), enzymes needed for obtaining energy, and various biochemicals. The more complex *eukaryotic* cells house their genetic material in a membrane-bound *nucleus;* synthesize, store, transport, and release biological molecules along a network of organelles (*endoplasmic reticulum*, *Golgi apparatus*, *vesicles*); degrade wastes in *lysosomes;* extract energy from biological compounds in *mitochondria;* and in certain organisms, such as plants, extract energy from sunlight in *chloroplasts*. A cell membrane, the *plasmalemma*, surrounds both prokaryotic and eukaryotic cells.

Many eukaryotic cells have *differentiated* characteristics resulting from the presence of different numbers of the different types of organelles. Differentiated cells are grouped to form *tissues*, tissues are grouped to form organs, and organs are grouped to form organ systems. There are four basic types of human tissues. *Epithelial tissue* forms linings. Individual cells are squamous, cuboidal, or columnar and can form layers one or several cells thick. *Connective tissues* consist of cells embedded in a nonliving matrix material and include loose and fibrous connective tissue, blood, cartilage, and bone. *Loose connective tissue* forms the "glue" of the body, and *dense connective tissue* forms ligaments, tendons, and part of the skin. *Blood* consists of cells and cell fragments suspended in a watery plasma matrix. *Cartilage* consists of *chondrocytes* embedded in collagen, and it supplies support and flexibility in joints and other parts of the body. *Bone* is a mineralized connective tissue consisting of four cell types embedded in an intricate tunnel system. *Nervous tissue* consists of *neurons* and their supportive *neuroglia*. Neurons are long, branching cells that transmit chemical and electrochemical signals. *Muscle tissue* consists of four types of contractile cells that move by the sliding action of their protein components.

Much thought has been given to the origin of eukaryotic cells. The *endosymbiont theory* suggests that chloroplasts and mitochondria evolved from once free-living prokaryotes that were swallowed by larger prokaryotes. According to the theory, chloroplasts evolved from cyanobacteria and mitochondria evolved from aerobic bacteria. The evidence for the endosymbiont theory is that mitochondria and chloroplasts resemble small aerobic bacteria in size, shape, membrane structure, reproduction, and the ways in which their DNA, RNA, and ribosomes interact to manufacture proteins.

QUESTIONS

1. How do the structures of a virus, bacterium, and human cell differ?

2. A fibroblast cell is shaped like a slightly irregular tile, and it is part of the connective tissue that cements the human body together. A nerve cell can be up to a meter in length, with many branches. A muscle cell is spindle-shaped. A white blood cell looks like a blob, and it can move about and engulf particles. Each of these cell types has a nucleus, a nucleolus, mitochondria, ribosomes, endoplasmic reticulum, one or more Golgi apparatuses, vesicles, and lysosomes. Yet each cell has a characteristic structure and function. How can these cells contain the same components, yet be so different from one another?

3. As a cube increases in size from 3 cm on a side, to 5 cm on a side, to 7 cm on a side, how does its surface/volume ratio change?

4. A type of liver cell called a hepatocyte has a volume of 5,000 μm^3. Its total membrane area, including the inner membranes that are part of organelles as well as the outer cell membrane, provides an area of 110,000 μm^2. For a cell in the pancreas that manufactures digestive enzymes, the volume is 1,000 μm^3, and the total membrane area is 13,000 μm^2. Which cell is probably more efficient in carrying out activities that require extensive membrane surfaces? State the reason for your answer.

5. Name three structures or activities found in both prokaryotic and eukaryotic cells. List eight differences between the two cell types.

6. Cells of the green alga *Chlamydomonas rheinardi* are grown for a few hours in a medium that contains amino acids that have been "labeled" with radioactive hydrogen, called tritium, and then returned to a nonradioactive medium. At various times, the cells are applied to photographic film (which radiation from the tritium exposes, producing dark silver grams), and the film is then examined under an electron microscope. After 3 minutes, the film shows radioactivity in the rough ER; after 20 minutes, in the smooth ER; after 45 minutes, in the Golgi apparatus; after 90 minutes, in vesicles near one end of the cells; after 2 hours, there is no radioactive label. What cellular process has this experiment traced? How might a similar technique be used to follow a lysosome's activity?

7. The defects that underly some disorders can be observed at the organelle level. For the following medical problems, indicate which organelles might be malfunctioning, and state whether the organelle is underactive or overactive.

a. A man is infertile because his sperm cannot "swim."
b. A child experiences kidney failure due to Fabry's disease. The cells lining the inside of her kidney tubules accumulate abnormally large amounts of a glycolipid (a molecule consisting of a sugar and a fat) that is usually degraded by an enzyme.
c. In a child with cystic fibrosis, certain cells in the lungs greatly overproduce mucus. The child's parents must vigorously rub his back several times a day to dislodge the mucus and enable him to breathe.
d. A woman dies in minutes after ingesting cyanide because the energy-generating molecule adenosine triphosphate (ATP) is no longer synthesized in her cells.
e. A single cell in a smoker's lung has turned cancerous. It doubles its DNA and splits in two much faster than a normal lung cell.

8. Which human tissues are the following substances found in?

a. Keratin
b. Collagen
c. Elastin
d. Proteoglycans
e. Hyaluronic acid
f. Hydroxyapatite
g. Neurotransmitters
h. Myelin

9. List specialized cells that make up each of the four tissue types of the human.

10. State the cell theory and the endosymbiont theory. What is the evidence for each?

TO THINK ABOUT

1. Do you think that a virus is alive? Cite a reason for your answer.

2. In terms of structure, prokaryotic organisms are far simpler than organisms built of eukaryotic cells. Yet the "simpler" prokaryotes have dwelt on the earth far longer than we comparatively complex eukaryotes, and prokaryotes occupy a far more diverse range of environments than eukaryotes. Can you explain this seeming contradiction?

3. What advantages does compartmentalization confer on a large cell?

4. Why are tough and fibrous connective tissues, blood, cartilage, and bone all considered to be connective tissues?

5. The amoeba *Pelomyxa palustris* is a single-celled eukaryote with no mitochondria, but it contains symbiotic bacteria that can live in the presence of oxygen. How does this observation support or argue against the endosymbiont theory of the origin of eukaryotic cells?

6. Some of the genetic material contained in the nuclei of some species normally jumps from one chromosome to another. Do you think that such movable DNA could account for the presence of DNA in mitochondria and chloroplasts? If so, what subcellular structures do you think might participate in the transfer of genetic material from the nucleus to mitochondria and chloroplasts?

SUGGESTED READINGS

de Kruif, Paul. 1966. *Microbe hunters*. New York: Harcourt Brace Jovanovich. This engrossing historical account of major discoveries in the field of microbiology highlights the effects that prokaryotic organisms can have on human health.

Gurin, Joel. July/August 1980. In the beginning. *Science 80*. The endosymbiont theory.

Larkin, Tim. May 1985. "Friendly" microbes: A world of miniature workaholics. *FDA Consumer*. Microorganisms, including prokaryotes, have greatly influenced human civilization.

Lewis, Ricki. June 1987. Harvesting the cell. *High Technology*. Pharmaceutical companies are turning to huge collections of active cells as sources of useful biochemicals.

Payne, Claire M. January 1985. The ultrastructural pathology of cell organelles. *Biology Digest*. Many disorders can be traced to abnormal organelles.

Raff, Rudolf A., and Henry R. Mahler. August 18, 1972. The nonsymbiotic origin of mitochondria. *Science*, vol. 177. An alternative to the endosymbiont theory.

Thomas, Lewis. 1974. *The lives of a cell*. New York: Viking Press. Essays entitled "The Lives of a Cell" and "Organelles as Organisms" discuss the nature of organelles and the endosymbiont theory.

C H A P T E R

5

Cellular Architecture

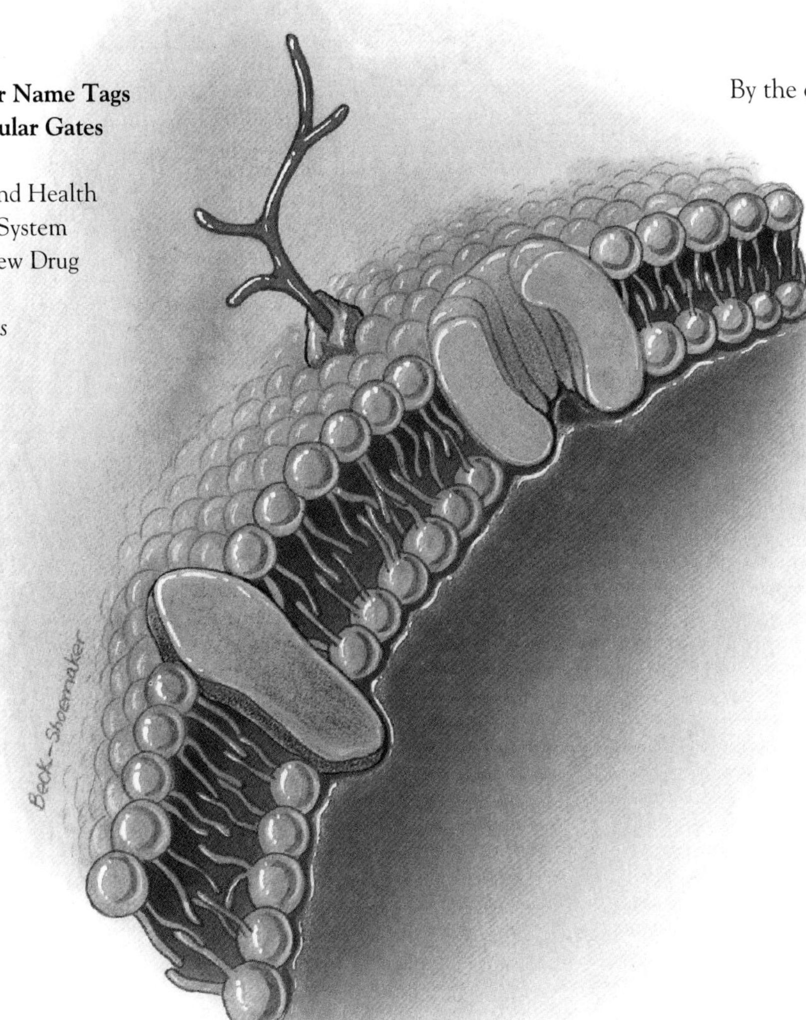

Chapter Outline

Learning Objectives

By the chapter's end, you should be able to answer these questions:

1. What structures comprise the cellular architecture?
2. How are a cell's surface molecules and their arrangements important in biological functioning?
3. What are the components of a cell membrane and how are they organized?
4. How do substances cross a cell membrane?
5. What are the functions of the cytoskeleton, and what are its components?

The 2-year-old's health already appeared to be returning, mere hours after the transplant. Her new liver, needed to replace her degenerated one, came from a child just killed in an automobile accident. Her skin was losing its usual sickly yellowish pallor, and she was becoming much more alert. Yet already, just as the new organ was taking over the jobs that her old one had abandoned, cells of her immune system had detected the new organ and, interpreting it as an infection, were beginning to produce chemicals that would attack it. Even though the donor's liver had been carefully "matched" to the little girl, meaning that the molecules on the surfaces of its cells appeared to be very similar to those on the cells of her own liver, the match was not perfect. A rejection reaction was in progress, and the transplanted organ would soon cease to function. The little girl would need another transplant to survive.

The rejection of a transplanted organ illustrates the importance of cell surfaces in the coordinated functioning of a multicellular organism. The cell surface is one component of the cellular architecture—structures that give a cell its particular three-dimensional shape and topography (fig. 5.1). The surface molecules of a cell are anchored in the **cell membrane,** the outer covering of a cell. Just beneath the cell membrane are protein fibers that are part of the cell's interior scaffolding, or **cytoskeleton**. Together, the cell surface, cell membrane, and cytoskeleton form a structural framework that helps to distinguish cells from one another.

The importance of a cell's architecture can be appreciated by considering what can happen when it is disrupted. A cancer cell's architecture, for example, is profoundly altered, noticeable in its unusual shape and interactions with other cells. Often immune system cells detect this architectural difference and destroy the cancer cells (fig. 5.2). A cancerous growth in a person's large intestine or colon starts out as a single, odd-looking cell, rounder than the surrounding tilelike cells and standing somewhat apart from them. The cell membrane of this tumor cell is oilier than those of its neighbors, and its surface is dotted with a different array of molecules. The cell is not alone for long. It divides in two, and then again and again at more frequent intervals than do the surrounding healthy intestinal cells. A breakdown among the cytoskeletal fibers that pull one cell into two is partly responsible for this runaway cell division.

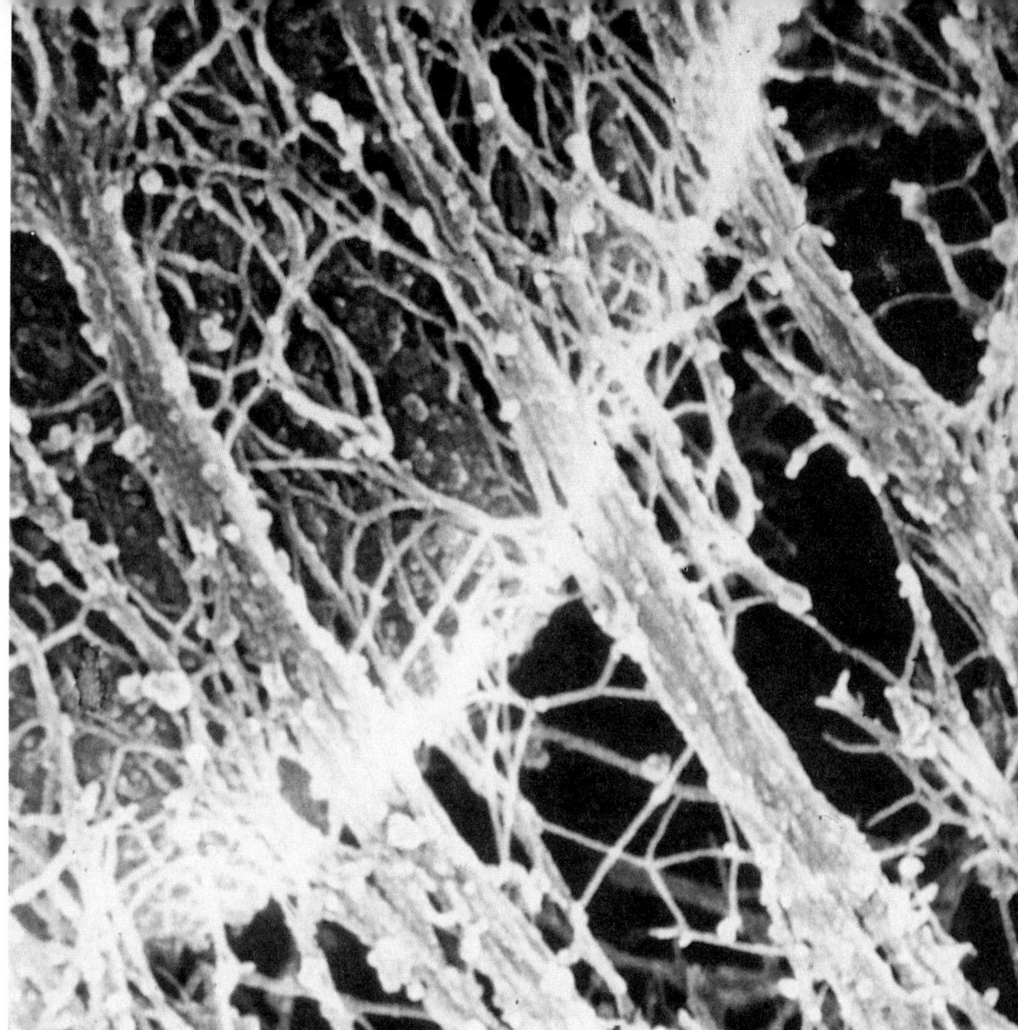

Figure 5.1
Cellular architecture. When the early microscopists first turned their lenses to cells, what they saw was crude compared to the complex three-dimensional structures revealed today in the scanning electron microscope. A cell's overall form is molded by the cellular architecture, including surface molecules, membranes, and the inner framework, or cytoskeleton. The central bulging over the nucleus, and the flattened edges pushing outwards of a typical fibroblast, a connective tissue cell, result from an inner scaffolding built largely of microfilaments of the protein actin.

Soon, perhaps in a matter of weeks, the original peculiar cell is a part of a tiny, glistening lump, jutting from the intestine's wall as the cells grow all over one another and invade surrounding healthy tissue. Eventually the intestine is blocked so that it bleeds as undigested food squeezes through. This rectal bleeding is often the first sign of illness. Colon cancer is highly curable by surgery if caught early. Understanding the disrupted cellular architecture of the cancer cell may provide clues to stopping its unchecked growth earlier.

We will look at a cell's architecture from the outside in, beginning with the surface, moving to the cell membrane, and then to the cytoskeleton within.

The Cell Surface—Cellular Name Tags

At a conference where most of the participants do not know one another, name tags are often used to establish identities. Cells also have name tags, in the form of sugars and proteins that protrude from their surfaces. These cellular name tags are found on all cells, from the single cells of simple bacteria to the trillion cells of the human body. Some surface molecules distinguish cells of one species from cells of another, like company affiliations on name tags.

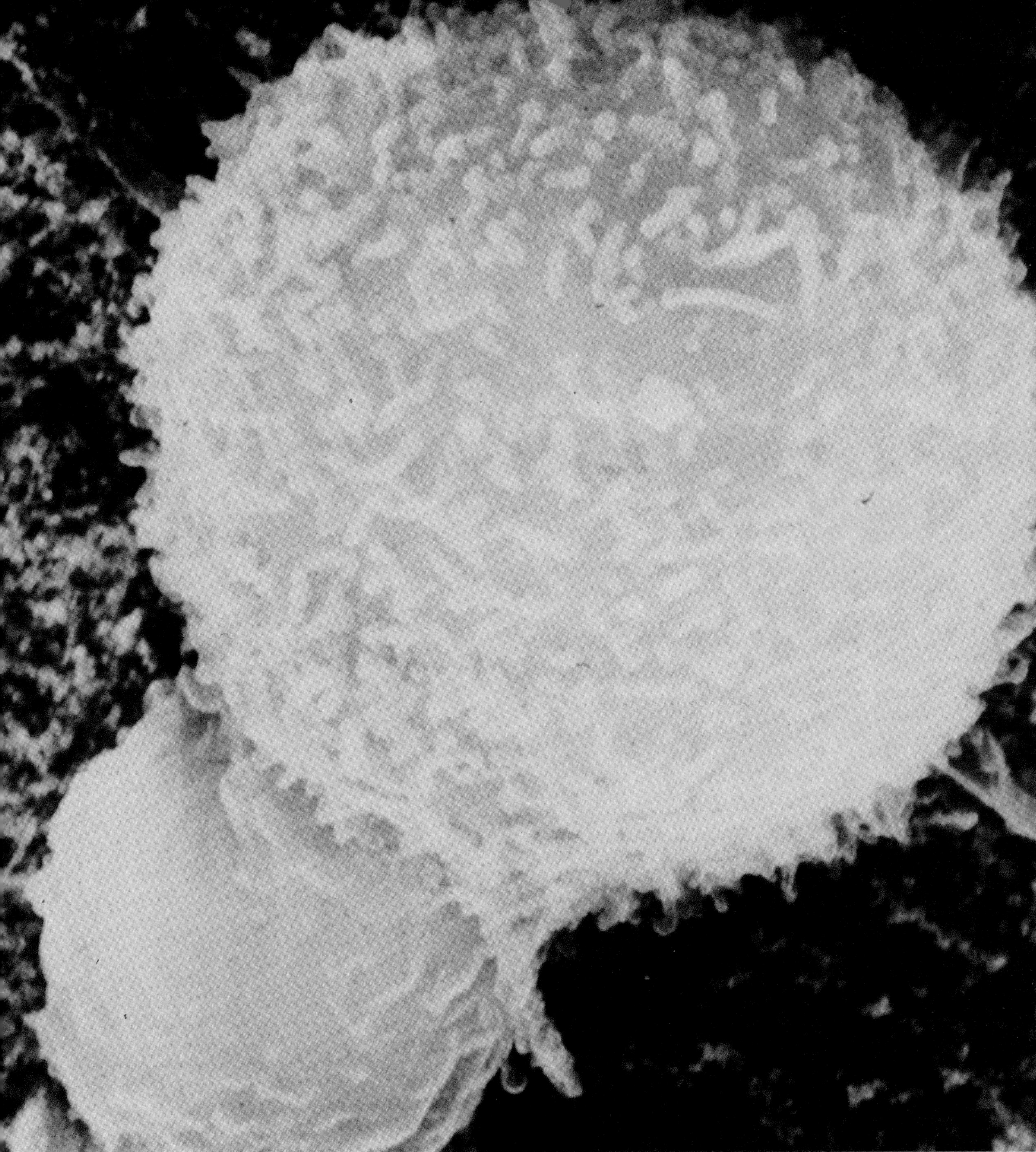

Figure 5.2
The smaller cell is a T cell, which homes in on the surface topography of the large cancer cell above it. The T cell will literally break the cancer cell apart, leaving nothing behind but scattered fibers.

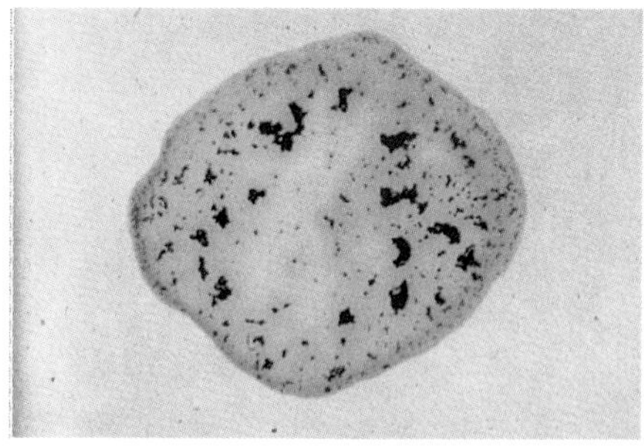

a.

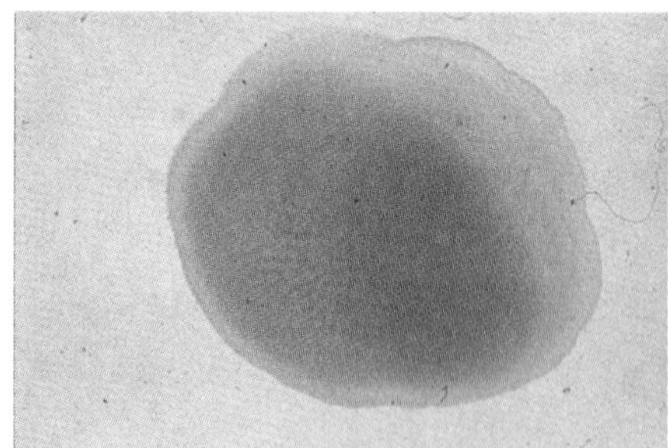

b.

Figure 5.3
Human blood types are based on cell surface differences. *a.* If type A blood is mixed with type B blood, which contains anti-A antibody, clumping occurs. *b.* If type A blood is mixed with type A blood, which contains anti-B antibody, clumping does not occur.

Other surface structures distinguish individuals within a species from one another. A multicellular organism must have a way to determine which cells are part of the body and which cells are not. Within your body, a huge collection of white blood cells and the biochemicals they produce form the immune system, which recognizes the surfaces of your cells as "self" and all other cell surfaces as "nonself." When the immune system encounters the nonself surfaces of some bacteria that have entered the body, it launches an attack, calling forth other cells and biochemicals to squelch the infection.

Surface structures also distinctively mark cells of different tissues within an individual, so that a bone cell's surface is different from that of a nerve cell or a muscle cell. The human body has more than 200 cell types, and each has its own personal landscape. These surface differences between cell types are particularly important during the development of the embryo, when different cells sort themselves out to grow into distinctive tissues and organs. Cell surfaces also change over time, with the number of surface molecules declining as a cell ages. It is possible that cancer cells have reverted to an embryonic form, acquiring both the cell surface characteristics of cells in the embryo as well as their ability to divide very frequently.

The fact that people have unique patterns of sugars and proteins on their cell surfaces is used in medical practice (fig. 5.3). The closer the match between two persons' cell surfaces, the more likely that the immune system of one person will recognize the cells of the other person as self. When an organ is available for transplanting, its cell surface is determined and the information entered into a computer, which then searches information on cell surface characteristics of people all over the country who are awaiting organs. When a good "match" is found—such as in the case of the little girl described at the start of the chapter—the organ is shipped to the patient, and the transplant is performed. Drugs are given to suppress the recipient's immune response. If all goes well, the immune system is "fooled" into accepting the healthy organ, and a life is prolonged or saved.

A person's particular pattern of cell surface features may one day be used to predict the likelihood of developing certain diseases. The *human leukocyte antigens* (HLA) are cell surface molecules that appear in different patterns in different people. (The name comes from the fact that these cell surface molecules, or antigens, were first studied in leukocytes, which are white blood cells.) A person who has a certain assortment of HLA cell surface molecules may have a much greater chance of developing a certain disease than a person with a different HLA makeup. In the future, an HLA profile taken at birth may be used to predict disease susceptibilities. People may then choose to alter their life-styles, such as diet and exercise habits, to try to prevent or delay certain illnesses (Reading 5.1).

KEY CONCEPTS

Cells within and between organisms can be distinguished by the pattern and types of molecules on their surfaces. The human leukocyte antigens predict the likelihood of development of certain diseases.

The Cell Membrane—Cellular Gates

Just as the character of a community is molded by the people who enter and leave it, the special characteristics of different cell types are shaped in part by the substances that enter and leave them. The spindly, pulsating muscle cells of the heart have different types of molecules entering and leaving them than do the highly branched nerve cells whose endings approach each other to exchange chemical and electrical signals. The movement of molecules into and out of a cell is monitored by the cell membrane, which is a selective barrier that completely surrounds the cell.

In eukaryotes, membranes are also found within cells, where they compartmentalize and protect structures. Membranes covering organelles guard them from chemicals in the cytoplasm that might harm them or be harmed by them. For example, the nuclear membrane shields the delicate genetic material from gene-cutting enzymes in the cytoplasm. The membrane surrounding a lysosome sac keeps this organelle's harsh digestive enzymes from chewing up the cell's contents. Some organelles are composed entirely of membranes, such as the endoplasmic reticulum and the Golgi apparatus of the secretory network, and the mitochondria, which generate chemical energy. Specific biochemical reactions take place on the ribbonlike twists and turns of these intracellular membrane systems.

The Protein-Lipid Bilayer

How does a membrane determine which substances it will allow to cross and which to keep out? The chemical characteristics and the arrangement of the molecules that build a membrane determine the activity of the membrane.

The structure of a biological membrane is possible because of a chemical property of the **phospholipid** molecules that comprise it—one end of such a molecule is attracted to water, while the other end is repelled by water. Phospholipids are lipid molecules with attached phosphate groups (PO_4, which represents a phosphorus atom bonded to four oxygen atoms). The phosphate end of the molecule, which seeks

Reading 5.1 Cell Surfaces and Health Predictions—The HLA System

WOULD YOU WANT TO KNOW WHICH MEDICAL CONDITIONS YOU ARE LIKELY TO SUFFER FROM LATER IN LIFE, AND PERHAPS EVEN DIE FROM? In just a few years, a blood test performed at birth, coupled with computer analysis of disease-incidence statistics, might make it possible to construct at least a partial health-prediction profile for any newborn baby. All of this may be possible because of a group of proteins called the human leukocyte antigens (HLA), which dot the surfaces of human white blood cells (leukocytes).

Individuals are distinguished by different combinations of HLA molecules. Certain HLA combinations are statistically associated with a higher-than-normal probability of developing a particular disease. So far, about 50 medical conditions have been linked to specific HLA types. As Table 1 shows, the HLA-associated disorders produce symptoms in a range of organ systems, but all of these disorders seem to stem from immune system abnormalities. HLA-linked diseases also tend to run in families, but not in any predictable manner.

An example of an HLA-linked disease is ankylosing spondylitis, a disease characterized by inflammation and deformation of the vertebrae. If a person has an antigen called B27 on his or her white blood cells, then the chances of developing the condition are 100 times greater than those of a person who lacks the B27 antigen. This HLA prediction is based on the observation that more than 90% of the people who suffer from ankylosing spondylitis possess the B27 antigen, although this antigen is present in only 5% of the general population. Of course, these statistics suggest that predictions of disease from HLA tests should be approached with caution: As the statistics show, 10% of people who have ankylosing spondylitis do *not* have the B27 antigen, and some people who have the antigen never develop the disease.

So far, the precise nature of the relationship between HLA types and human health is not well understood. A particular HLA type may indicate a genetic susceptibility to a disease or perhaps a sensitivity to certain viruses. An editorial in the *New England Journal of Medicine* sums up the current state of knowledge, "Thus, the disease-associated HLA antigen may be thought of as an innocent but visible witness to a health 'crime' rather than the actual culprit." But even though our present knowledge of the HLA system is not complete, the statistical associations may indeed prove clinically useful in the form of warnings of high risk for certain conditions. Chapter 15 discusses another way to predict disease–tracing DNA sequences called genetic markers.

Table 1
Medical Problems Linked to Specific HLA Types in Humans

Condition	*Description*
Ankylosing spondylitis	Inflammation and deformation of vertebrae
Reiter's syndrome	Inflammation of joints, eyes, and urinary tract
Rheumatoid arthritis	Inflammation of joints
Psoriasis	Scaly skin lesions
Dermatitis herpetiformia	Burning, itchy skin lesions
Systemic lupus erythematosis	Rash on face; destruction of heart, brain, and kidney cells; very high, persistent fever
Addison's disease	Malfunction of adrenal glands producing anemia, discolored skin, diarrhea, low blood pressure, and stomachache
Grave's disease	Malfunction of thyroid gland, producing goiter
Juvenile-onset diabetes	Defect in beta cells of pancreas, disrupting sugar metabolism
Multiple sclerosis	Degenerative disease of brain or spinal cord producing weakness and poor coordination
Myasthenia gravis	Progressive paralysis
Celiac disease	Childhood diarrhea
Gluten-sensitive enteropathy	Sensitivity of intestine to wheat
Chronic active hepatitis	Inflammation of liver

water, is said to be **hydrophilic** (water-loving), while the other end, consisting of two fatty acid chains, moves away from water and is said to be **hydrophobic** (water-hating). Because of these water preferences, phospholipid molecules in water spontaneously arrange into a **lipid bilayer.** This is a sandwichlike structure whose outer surfaces—exposed to the watery medium outside and inside the cell—are hydrophilic and whose unexposed interior is hydrophobic (figs. 5.4 and 5.5). Lipid bilayers are being used in the pharmaceutical industry to construct microscopic bubbles, called liposomes, which are used to encapsulate drugs (Reading 5.2).

The lipid bilayer forms the structural backbone of a biological membrane, its hydrophobic interior a barrier to most substances dissolved in water. However, passageways for water-soluble molecules and ions are formed by proteins that are embedded throughout the lipid bilayer. The membranes of living cells, then, consist of lipid bilayers and the proteins within and extending out of them.

hydrophilic group

choline

a.

hydrophobic fatty acid "tails"

Figure 5.5

Membrane structure is studied using a technique called freeze fracture. A cell is frozen and splits easily down the lipid bilayer of its cell membrane when fractured with a special sharp tool. The bumps revealed in the scanning electron micrograph of this cell membrane represent proteins that protrude from the lipid bilayer.

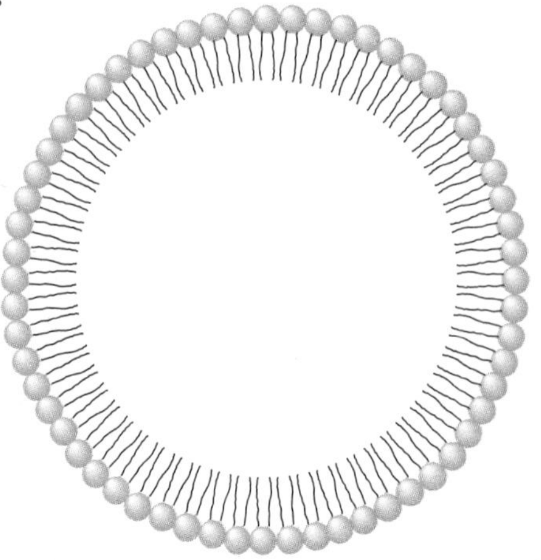

b.

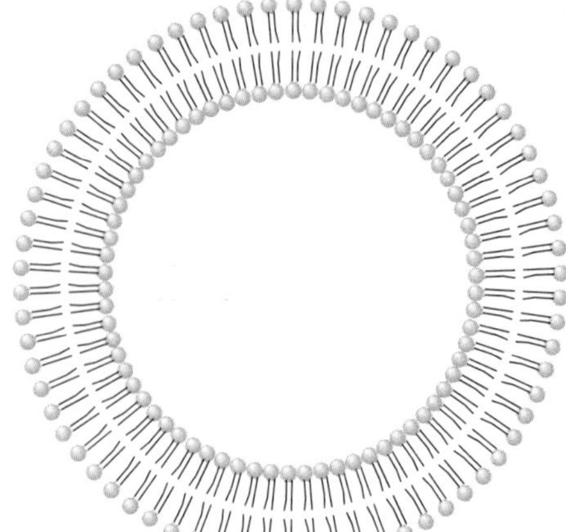

c.

Figure 5.4

The two faces of membrane lipids. *a.* A phospholipid is literally a two-faced molecule, with one end attracted to water (hydrophilic, or water-loving) and the other repelled by it (hydrophobic, or water-hating). *b.* Phospholipid molecules arrange themselves to satisfy the water preferences of their parts, such as in this simple structure, called a micelle. *c.* Phospholipid molecules form lipid bilayers, which are the structural frameworks of biological membranes.

(c) "Adopted from 'Liposomes'," by Marc J. Ostro.

Reading 5.2 *Liposomes—New Drug Delivers*

In 1961, English investigator Alec Bangham poured water into a flask containing a film of phospholipid molecules as part of his research on blood clotting. He was surprised to see that the lipid turned milky. Looking at the material under a microscope, Bangham saw that the phospholipid film had broken into thousands of tiny bubbles, each surrounding some of the water. The bubbles ranged in diameter from 250 Angstroms (A) to several micrometers. They were named liposomes, meaning "bodies of lipid."

What Bangham had discovered were microscopic spheres made of a simple lipid bilayer, identical to the structure that forms the basis of cell membranes. Some of the lipids even had more than one lipid bilayer coat, built a little like an onion skin. Throughout the 1960s, liposomes were used by cell biologists as models of cell membranes. In 1981, a young biologist named Marc Ostro realized the potential of liposomes as drug carriers. Drugs that are soluble in water can be packaged in the watery interior of liposomes; drugs soluble in fat can be lodged within the lipid bilayer. The advantages of packing a drug in a liposome are twofold—the drug can be released slowly from the liposome, and if a way to direct the liposome to diseased cells can be found, drug delivery can be targeted. This would solve a major drawback of conventional drug treatment, that is, getting enough drug to the site of disease to be effective yet keeping it away from healthy cells, where it can cause side effects.

The natural reaction of the human immune system to liposomes provides a way to move them to sites of disease. Large scavenger cells of the immune system, called macrophages, are attracted to liposomes and rapidly engulf them. The "swallowed" liposome is sent to a lysosome, where it is taken apart by enzymes. If the liposome contains a drug, some of it seeps into the cytoplasm and perhaps eventually out of the cell. The success of liposome-carried drugs lies in the function of macrophages, which normally congregate in parts of the liver, spleen, bone marrow, and lymph nodes and move to sites of inflammation or infection. Liposomes containing antibiotic drugs are engulfed by macrophages and transported to the infection site—precisely where they are needed. Liposomes harboring antiinflammatory drugs are taken, via macrophages, to inflamed arthritic joints. Liposomes are also useful for delivering cancer drugs, which when given in "free" form cause severe side effects because they kill healthy cells as well as cancer cells. When such toxic drugs are encapsulated in liposomes and injected, they accumulate in macrophage-laden places, where they leak out slowly and steadily enough to destroy the cancer cells but not the healthy ones.

Liposome-enclosed drugs to treat infection, inflammation, and cancer are injected because they are not absorbed well in the digestive tract. The tiny bubbles are useful, however, in some topical applications. "Artificial tears," for example, are liposomes packed with tear components (water, salts, lipid, and a mucuslike substance) and sooth the irritating condition called dry eye. Another liposome-enclosed drug treats fungal infections in the female reproductive tract, and a hair growing drug encased in liposomes is used to treat pattern baldness. Because of their slow release, liposome-enclosed topical drugs are needed less often and are therefore less irritating.

Figure 1
Liposomes are microscopic bubbles composed of lipid bilayers. They form spontaneously when certain concentrations of fatty molecules are mixed with water. The diameters of these liposomes are about 0.00015 millimeters.

Membrane proteins are diverse. Some lie completely within the lipid bilayer, and others traverse it to extend out of one or both sides. In animal cells, membrane proteins are often attached to branchlike sugar molecules to form **glycoproteins**, which protrude from the membrane's outer surface (fig. 5.6). The proteins and glycoproteins that jut from the cell membrane create the surface characteristics that are so important to a cell's interactions with other cells. Proteins within the oily lipid sandwich can move about, like slow-moving ships at sea. The protein-lipid bilayer is often called a **fluid mosaic** because the proteins can move and are not regularly arranged, as are the lipid molecules. A pigment protein found in the retina of a human eye, for example, moves to different depths within the lipid bilayer of the cell membrane depending upon the intensity of the incoming light. Figure 5.7 illustrates the membrane characteristics of another specialized cell, the red blood cell.

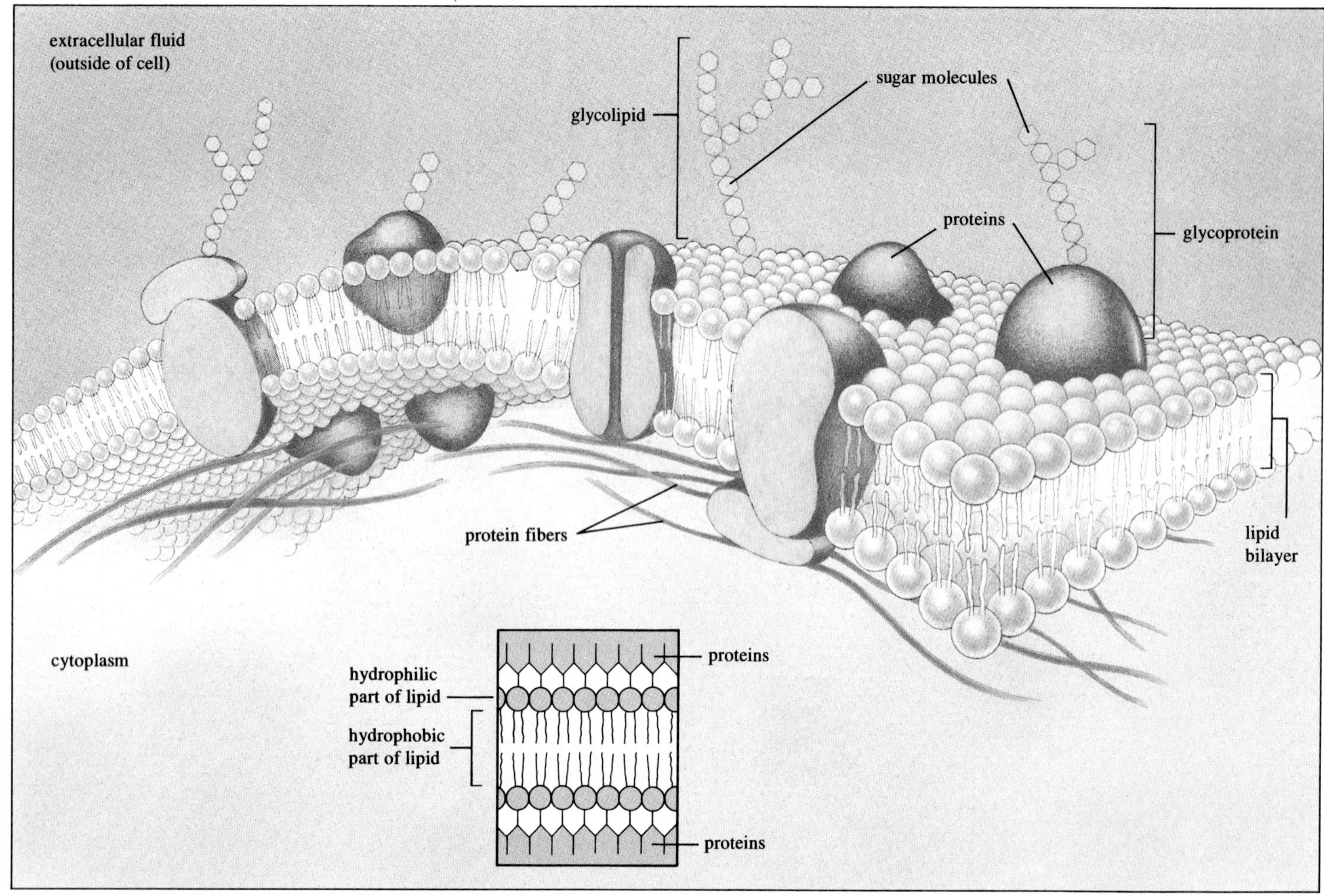

Figure 5.6
The structure of a cell membrane was first described in 1935 as the highly ordered, rigid protein-lipid bilayer shown in the inset. Improved microscopic techniques in recent years have shown, though, that mobile proteins are embedded throughout the lipid bilayer to produce a structure that is somewhat fluid in consistency. The cell membrane is supported by an underlying meshwork of protein fibers. Jutting from the membrane's outer face are sugar molecules attached to proteins (glycoproteins) and sugars attached to lipids (glycolipids). Because a membrane is built of different components (lipids and proteins) and proteins can move within the lipid bilayer, the structural organization of a membrane is described as a fluid mosaic.

The number or distribution of cell membrane proteins distinguish different cell types. Many cell membranes are about half protein and half lipid. The cell membrane of a nerve cell, however, is about 80% protein. In contrast, the cell membrane of a fatty Schwann cell wrapped around that nerve cell is 80% lipid, which enables the Schwann cell to insulate the nerve cell. If Schwann cell membranes contain too much or too little lipid, illness can result. In multiple sclerosis, the lipids in Schwann cell membranes are destroyed, leading to symptoms of visual impairment, numbness, tremor, and difficulty moving. In Tay-Sachs disease, Schwann cell membranes accumulate too much lipid, and the nerve cells beneath them cannot transmit messages to muscle cells. Paralysis and loss of sight and hearing result.

Movement Across Membranes

A muscle cell can contract and a nerve cell conduct a message only if certain molecules and ions are maintained at certain levels inside and outside the cell. In all cells, the cell membrane oversees these vital concentration differences. Before considering how cells control which substances enter and leave them, it is helpful to define some terms.

A **solution** is a homogeneous mixture of a substance (the *solute*) dissolved in water (the *solvent*). Concentration refers to the relative number of one kind of molecule compared to the total number of molecules present, and it is usually given in terms of the solute. When solute concentration is high, the proportion of solvent (water) present is low, and the solution is concentrated. When solute concentration is low, solvent is proportionately high, and the solution is dilute. Lemonade made from a powdered mix illustrates the relationship between solute and solvent: the solvent is water, the solute is the powdered mix, and the solution is lemonade.

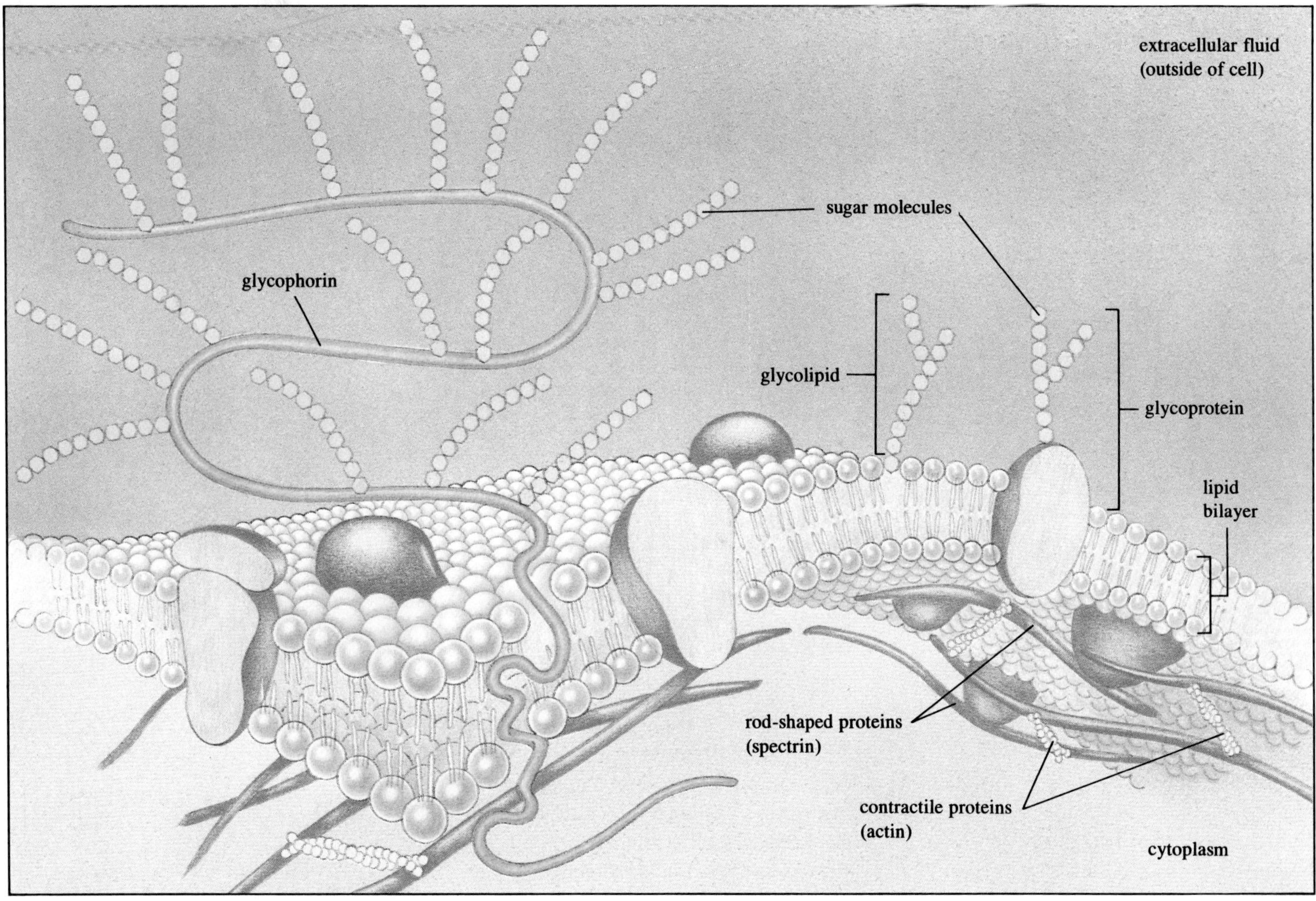

Figure 5.7

The cell membrane of a human red blood cell. A red blood cell travels a rough road in the human body, squeezing through narrow vessels and repeatedly withstanding the force of the heart's contractions. The red blood cell's outer membrane is especially tough. It is easy to study because it is the only membrane in the cell and is therefore easy to separate from other cell parts. A strong underlying network of rod-shaped proteins linked to contractile actin proteins provides flexibility and support. On the outside face of the membrane is a glycoprotein called glycophorin. Part of glycophorin's protein portion is exposed on the cell's surface, and extending from it are 16 spokes of 10 sugar molecules each. The remainder of the protein part of the molecule extends through the lipid bilayer and is exposed to the cytoplasm on the inner face of the membrane. Because a large majority of the red blood cell's surface is glycophorin, this membrane glycoprotein influences the interactions of red blood cells with their environment.

Passive Diffusion

The cell membrane is selectively permeable—that is, some molecules are able to pass freely through the membrane (either between the molecules of the lipid bilayer or through special protein-lined channels), while others are not. Molecules of oxygen (O_2), carbon dioxide (CO_2), and water (H_2O) are among those that freely cross through biological membranes, and they do so by the simple process of diffusion.

Diffusion is the movement of a substance from a region where it is very concentrated to a region where it is not very concentrated. It is a familiar phenomenon demonstrated by a tea bag placed in a cup of hot water. Compounds in the tea leaves dissolve gradually and diffuse throughout the cup. The tea is at first concentrated in the vicinity of the bag, but the brownish color eventually spreads to give a uniform brew. The natural tendency of a substance to move from where it is highly concentrated to where it is not is called "moving down" or "following" its concentration gradient.

Movement of molecules through a membrane responding only to this natural tendency to travel from high concentration to low concentration is called **passive diffusion,** because it requires no input of energy. Passive diffusion reaches a point at which the concentration of the substance is the same on both sides of the membrane. After this point, random movements of molecules of the substance continue but at the same rate back and forth across the membrane, so that the concentration remains the same on both sides.

A Special Case of Diffusion—The Movement of Water

The fluids that continually bathe our cells consist of molecules dissolved in water. Because cells are constantly exposed to water, it is important to understand how the entry of water into a cell is regulated. If too much water enters a cell, it swells; if water leaves, it shrinks. In either case, the cell's functions may be hampered. Water moves across biological membranes by a form of passive diffusion called **osmosis,** which is influenced by the concentration of dissolved substances inside and outside the cell (fig. 5.8).

When the solute concentrations on each side of a membrane differ, water moves across the membrane. According to the general tendency of diffusion, water moves in a direction that dilutes the solute. That is, water moves to where the solute is more concentrated. Most cells are *isotonic* with reference to their surrounding fluid—that is, solute concentrations are the same within and outside the cell, so that there is no net flow of water.

If a cell's isotonic state is disrupted, its shape changes as water rushes in or leaks out. If a cell is placed in a solution in which the concentration of solute is lower than it is inside the cell (a *hypotonic* solution), water enters the cell to dilute the higher solute concentration there. The cell swells. In the opposite situation, if a cell is placed in a solution in which the solute concentration is higher than it is inside the cell (a *hypertonic* solution), water leaves the cell to dilute the higher solute concentration outside. This cell shrinks. (Hypotonic and hypertonic are relative terms in that they can be applied both to the surrounding solution and to the solution inside the cell. It may help to remember that hyper means "more than" and hypo means "less than.") The effects of immersing a cell in a hypertonic or hypotonic solution can be demonstrated with a human red blood cell, which is normally suspended in an isotonic solution called plasma (fig. 5.9).

Living things have evolved interesting strategies of regulating osmosis in a way that maintains cell shapes. Some single-celled inhabitants of the ocean are isotonic with their salty environment, so their shapes are unaltered. The paramecium, a single-

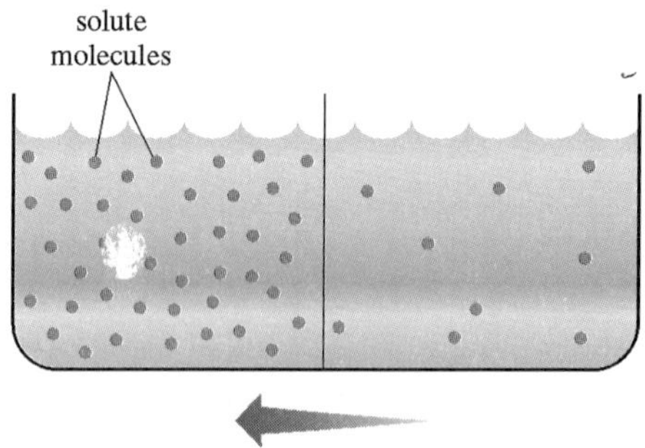

Figure 5.8
Osmosis—a biologically important type of diffusion. Either a synthetic nonliving material (called an artificial membrane) or a biological membrane can be used to demonstrate osmosis, which is the movement of water down its concentration gradient. Water molecules diffuse from a region where its concentration is high to where its concentration is low. In other words, water moves through the membrane to the side where the solute (the dissolved chemical) is more concentrated.

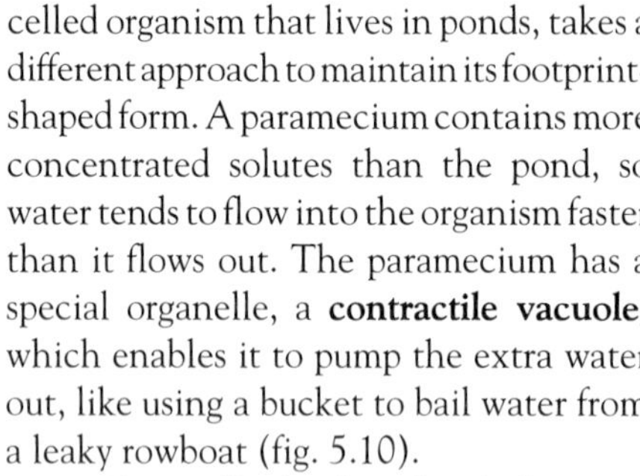

celled organism that lives in ponds, takes a different approach to maintain its footprint-shaped form. A paramecium contains more concentrated solutes than the pond, so water tends to flow into the organism faster than it flows out. The paramecium has a special organelle, a **contractile vacuole,** which enables it to pump the extra water out, like using a bucket to bail water from a leaky rowboat (fig. 5.10).

Plant cells also face the challenge of a concentrated interior. Instead of expelling the extra water that rushes in, like the paramecium does, plant cells expand until their cell membranes are restrained by their cell walls. The resulting rigidity, caused by the force of water against the cell wall, is called **turgor pressure** (fig. 5.11). This can be demonstrated with a piece of wilted lettuce. When it is placed in water, it becomes crisp, as the individual cells expand like inflated balloons.

The frequency with which we urinate is influenced by osmosis. Kidney tubules have very active membranes that return valuable substances to the blood and excrete in the urine substances not needed in the blood. One chemical that is recycled in the kidney is water, and the amount returned to the blood is influenced by intake of

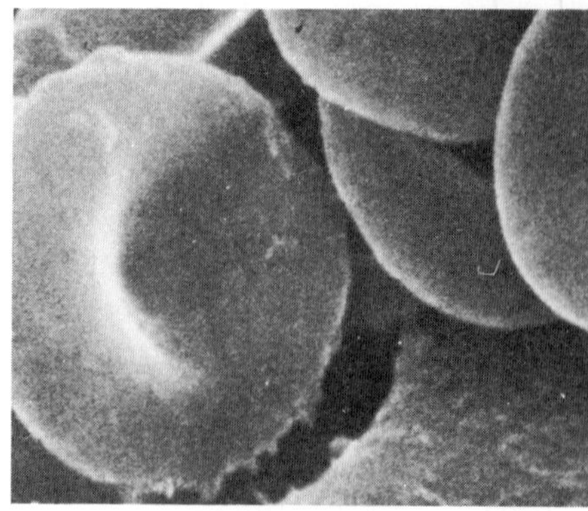

a.

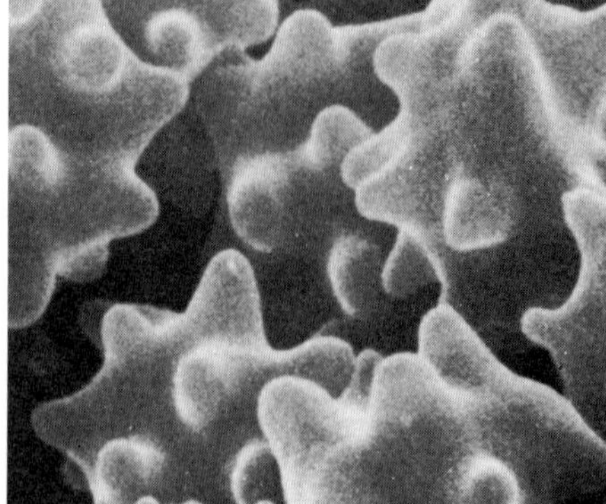

b.

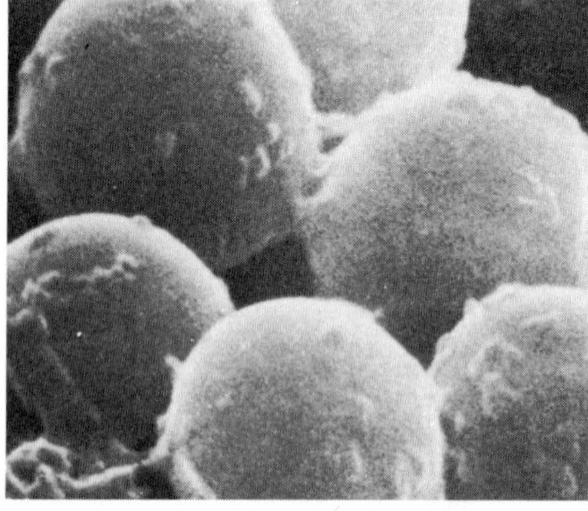

c.

Figure 5.9
A red blood cell changes shape in response to changing plasma solute concentrations. *a.* A human red blood cell is normally a biconcave disc and isotonic to the surrounding blood plasma. This means that water enters and leaves the cells at the same rate, so their shape is maintained. *b.* When the salt concentration of the plasma increases, becoming hypertonic to the cells, water leaves the cells to dilute the outside solute faster than water enters the cells. The cells shrink. *c.* When the salt concentration of the plasma decreases, becoming hypotonic to the cells, water flows into the cells faster than it leaves. The cells swell and may even burst.

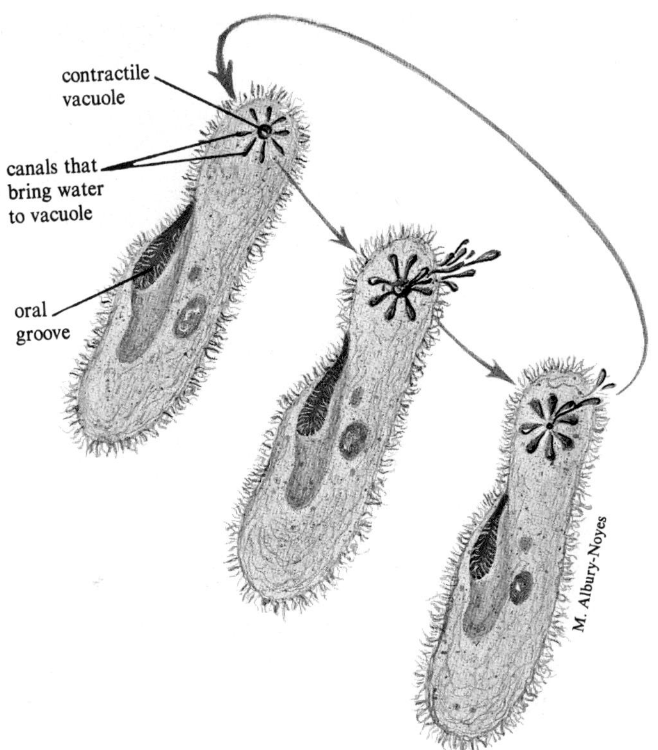

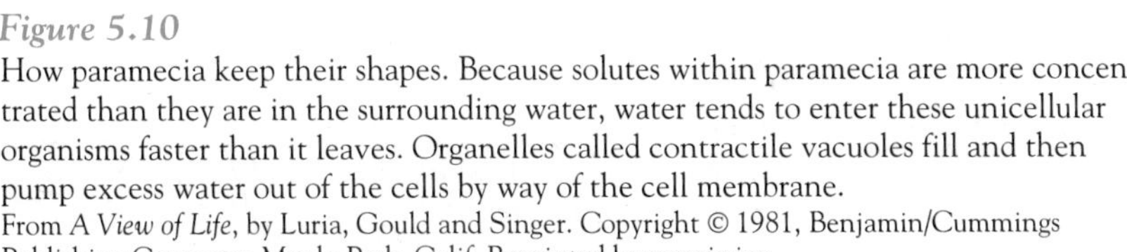

Figure 5.10
How paramecia keep their shapes. Because solutes within paramecia are more concentrated than they are in the surrounding water, water tends to enter these unicellular organisms faster than it leaves. Organelles called contractile vacuoles fill and then pump excess water out of the cells by way of the cell membrane.
From *A View of Life*, by Luria, Gould and Singer. Copyright © 1981, Benjamin/Cummings Publishing Company, Menlo Park, Calif. Reprinted by permission.

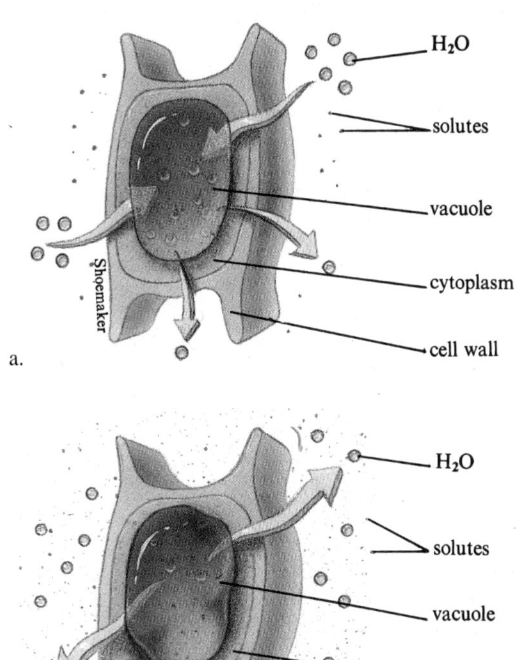

Figure 5.11
How plant cells keep their shapes. Like paramecia, plant cells usually have more concentrated solutes than their surroundings, prompting water to flow in. *a.* Vacuoles collect this water and the cells swell against their rigid, restraining cell walls, generating turgor pressure. *b.* When a plant cell is placed in a hypertonic environment (where solutes are more concentrated outside the cells), water flows out of the vacuole, and the cell shrinks.

water as well as by other chemicals, such as caffeine and alcohol. Have you ever drunk a huge mug of coffee or tea, and then spent much of the next few hours running to the bathroom? The caffeine in coffee and tea is a diuretic, causing frequent passage of watery urine by decreasing the movement of water from the kidney tubules back into the blood. Beer has a similar effect, caused both by its high water content and by the diuretic action of alcohol. In the rare disorder diabetes insipidus, a hormone that controls water resorption in the kidneys is absent. People with this disorder urinate profusely, up to 7 gallons (26 liters) a day! In the opposite situation, in which water returns to the blood too readily, urine volume is low and the solutes in the urine are concentrated.

Facilitated Diffusion

A molecule can passively diffuse through a membrane only if a pathway is open for it, either between lipid molecules or through spaces in membrane proteins (fig. 5.12). However, some molecules that are too big to slip through a custom-fit channel in the membrane can nevertheless cross it. They do this with the aid of a *carrier protein,* which functions like a revolving door, picking up its molecular cargo on one side, then turning to release it on the other side, then swiveling again to pick up another load. A substance's crossing a membrane down its concentration gradient (that is, from a region where it is highly concentrated to where it is less concentrated) with aid from a carrier protein is called **facilitated diffusion.** Glucose molecules, for example, enter red blood cells by this mechanism.

Active Transport

Sometimes a cell accumulates a particular substance at higher concentrations than are present outside the cell. For example, liver cells take up glucose molecules and string them into the starch glycogen. To do this, glucose must enter the already glucose-packed liver cells from the tissue fluid, where glucose may not be highly concentrated. The glucose is able to make this journey, opposite its concentration gradient, with the aid of both a carrier protein and energy provided by a molecule called **adenosine triphosphate** (ATP). The carrier protein delivers the glucose to the opposite side of the membrane, and the simultaneous release of the phosphate group from an ATP molecule provides a pulse of energy. It is a little like giving a revolving door an extra push. Movement of a molecule through a membrane against its concentration gradient using a carrier protein and energy is called **active transport.**

So effective is active transport that it can produce a concentration of a substance on one side of a membrane that is as much as 50 times as great as its concentration on the other side. All animal cells use active transport to maintain a higher concentration of sodium ions outside the cell than inside and a higher concentration of potassium ions inside the cell than outside. The energy-driven "sodium/potassium pump" helps to control a cell's volume by setting up solute concentrations on either side of the cell membrane, which affect osmosis. Active transport is also vital in the kidney tubules, where ATP and carrier proteins help return useful salts to the bloodstream.

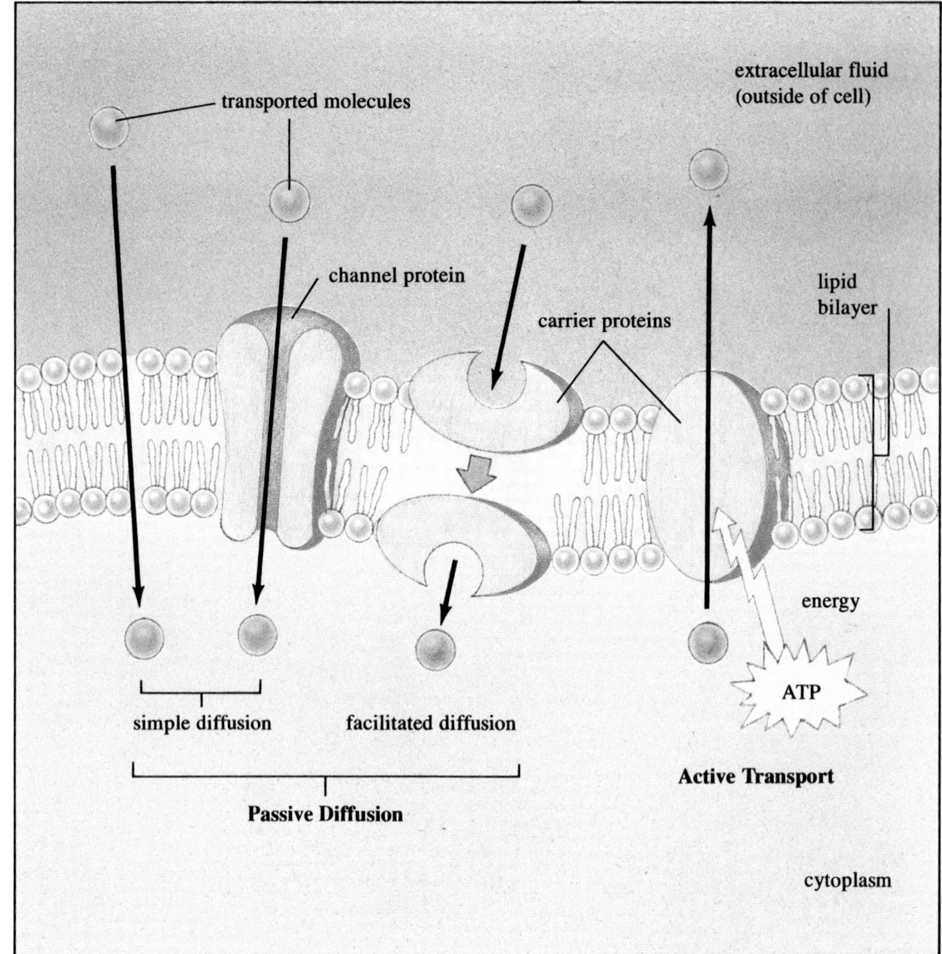

Figure 5.12
Passive diffusion and active transport. Molecules cross cell membranes from a region where they are highly concentrated to a region where they are not highly concentrated, either through spaces between the lipid or protein components of the membrane or with the aid of a carrier protein. Such diffusion does not require energy. Movement of a molecule against its concentration gradient, however, requires energy, which is provided by ATP.

Endocytosis

Most molecules dissolved in water are small, and they can easily cross cell membranes by passive diffusion, facilitated diffusion, or active transport. Large molecules (and even bacteria) can also get into and out of cells, with the help of the cell membrane.

Large particles enter cells by **endocytosis,** in which the cell membrane in a localized region moves outward to surround and enclose the particle (fig. 5.13). The pocket of membrane then pinches off from the interior of the membrane, producing a vesicle (a bubblelike structure) containing the particle. The vesicle is released into the cytoplasm, where it eventually fuses with a lysosome. (See Reading 4.2 on lysosomes and Reading 5.2.) Digestive enzymes within the lysosome dismantle the foreign particle. Endocytosis enables a white blood cell to engulf a bacterium, which it then destroys with lysosomal enzymes. Some substances that enter cells by endocytosis are used, rather than destroyed. This is true for some of the chemical messengers of the nervous system.

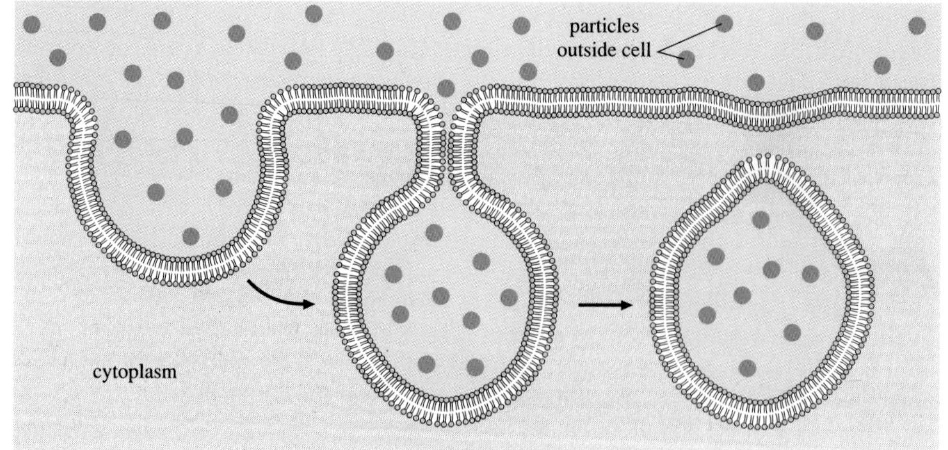

Figure 5.13
Endocytosis—into the cell. Large particles and even bacteria can enter a cell through endocytosis, in which a small portion of the cell membrane buds inward, entrapping the particles. A vacuole forms, bringing the substances into the cell.

Exocytosis

Large molecules leave cells by **exocytosis.** Inside the cell, a vesicle made of a lipid bilayer surrounds a structure that is to be transported out of the cell—a droplet of a secretion such as mucus, for example. The vesicle moves to the cell membrane, joins with it, and the transported molecule is released on the outside of the membrane (fig. 5.14). Nerve transmission relies on the release of chemical messengers by exocytosis. Table 5.1 summarizes the mechanisms of transport of substances across membranes.

KEY CONCEPTS

Membranes control which substances enter and leave a cell. Membranes also form compartments within cells. A membrane is built of a lipid bilayer embedded with moveable proteins. The number and arrangement of proteins differs in different cell types. Molecules cross membranes by various mechanisms. In passive diffusion, a substance moves down its concentration gradient. Passive diffusion of water is called osmosis, and control of osmosis influences a cell's shape. In facilitated diffusion, a substance moves down its concentration gradient with the aid of a carrier protein. In active transport a substance moves against its concentration gradient using energy from split ATP. In endocytosis a membrane surrounds a substance, bringing it into the cell, and in exocytosis, a bit of membrane surrounds a substance and joins a larger membrane, transporting its contents.

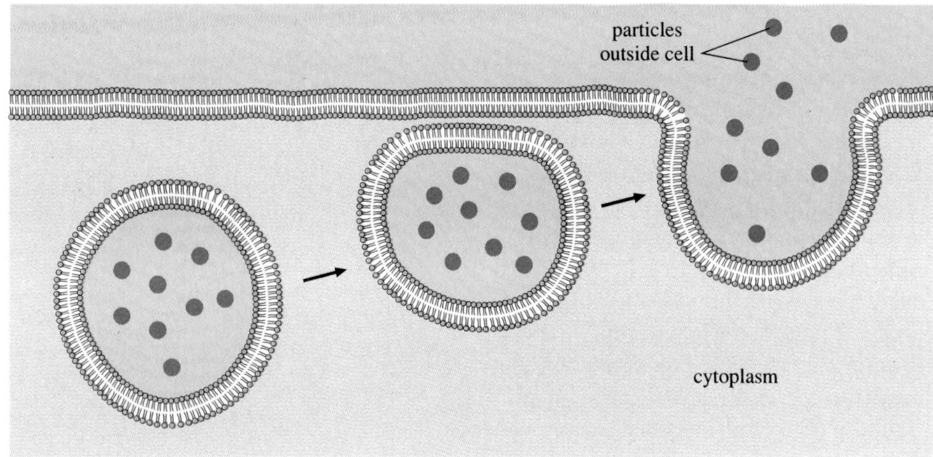

Figure 5.14
Exocytosis—out of the cell. Biochemicals or particles can exit cells by exocytosis. The structures to be exported are surrounded by lipid bilayer spheres within the cell, travel to the cell membrane, and merge with it. The particles are then released to the outside.

Table 5.1
Movement Across Membranes

Mechanism	*Characteristics*	*Example*
Passive diffusion	Follows concentration gradient	Diffusion of oxygen from lung into capillaries
Osmosis	Passive diffusion of water	Reabsorption of water from kidney tubules
Facilitated diffusion	Follows concentration gradient but is assisted by carrier protein	Diffusion of glucose into red blood cells
Active transport	Moves against concentration gradient, assisted by carrier protein and energy from ATP	Reabsorption of salts from kidney tubules
Endocytosis	Membrane engulfs substance and draws it into cell in membrane-bound vesicle	Ingestion of bacterium by white blood cells
Exocytosis	Membrane-bound vesicle fuses with cell membrane, releasing its contents outside of cell	Release of neurotransmitters by nerve cells

The Cytoskeleton—Cellular Support

Until about 20 years ago, eukaryotic cells were thought to be little more than sacs containing various structures that floated randomly in the jellylike cytoplasm. Today, thanks to modern methods of viewing cells, we know that the units of life are highly organized. Much of this organization is provided by the cell's skeleton, or cytoskeleton, which is a meshlike network of fine rods and tiny tubes of protein. Like a human body without bones, a cell without its cytoskeleton would be more like a blob than a solid structure with a distinctive form. The remnants of the cancer cell in figure 5.2 left after attack by immune system cells vividly displays the cytoskeleton.

The cytoskeleton gives a cell its shape by supporting its outer membrane and defining spaces inside where particular organelles lodge. The different cell shapes shown in chapter 4 are made possible by distinctive cytoskeletons—the macrophage with its tentaclelike extensions (fig. 4.1), the tilelike squamous epithelia (fig. 4.14), whose shapes are maintained by extensive internal frameworks of protein rods, and the muscle cell (fig. 4.16), which moves when its protein filaments slide past one another. Another function of the cytoskeleton is the control of cell movements. The meanderings of macrophages throughout the body and their movements to surround foreign particles are made possible by cytoskeletal elements that contract like muscles. The fringe that waves back and forth on cells of the respiratory system is built of moveable cytoskeletal parts. Within a cell in the process of division, duplicated chromosomes are distributed into two cells by the cytoskeleton. The cytoskeleton builds the cell wall in plants cells, their outermost boundary.

Microtubules

All eukaryotic cells have a supply of long, hollow tubes called **microtubules,** which are largely responsible for cellular movements. Each microtubule is only about a millionth of an inch (about 25 nanometers) in diameter and is a long chain of a protein called *tubulin*. Cells contain both formed microtubules and individual tubulin molecules. When the cell requires microtubules to carry out a specific function—dividing into two, for example—the free tubulin "self-assembles" into more of the tiny tubes. After the cell has divided, some of the microtubules fall apart into individual tubulin molecules, replenishing the cell's supply of building blocks (fig. 5.15). Cells, then, are in a perpetual state of flux, building up and breaking down microtubules to carry out particular functions. Some drugs used to fight cancer (in which certain cells divide too frequently) work by dismantling the microtubules that pull a cell's duplicated chromosomes into two, halting cell division.

Microtubules enable sperm cells to swim, which is essential for them to travel in the female's reproductive tract to meet with an egg cell. A sperm's "tail," or **flagellum,** contains microtubules that can slide past one another, generating movement. If a man's sperm cells cannot move their tails, he will be unable to father a child.

Many types of eukaryotic cells are fringed with **cilia,** hairlike structures built of microtubules that move in a coordinated fashion to produce a wavelike motion. An individual cilium is constructed of nine microtubule pairs surrounding a central pair, and the entire structure is studded with another type of protein (fig. 5.16). Cilia move using energy obtained from ATP, the biological energy molecule. ATP is generated in mitochondria from molecules derived from food nutrients—the subject of the next chapter.

In animals, beating cilia move liquids over cell surfaces. The cilia of the cells lining the upper respiratory tract in humans sweep mucus, inhaled dust, and other foreign matter up to the throat, where it is eliminated by swallowing or coughing. Smoking cigarettes destroys these vital cilia.

Figure 5.15
Microtubules self-assemble. Microtubules are built by adding tubulin molecules at one end and torn down by losing them from the other.

Similarly, the cilia on the cells that line the reproductive tract in the human female wave an egg cell down the tract toward a possible meeting with a sperm cell. Cilia can also be found on the sperm cells of primitive plants (such as the mosses, ferns, and liverworts) in which fertilization occurs in a film of water. Here, the cilia help propel the sperm towards the egg cells. In single-celled organisms, tracts of waving cilia provide locomotion. Some such organisms are completely covered with cilia—one type of paramecium has about 17,000 individual cilia that help propel it through the water.

Microfilaments

Another building block of the cytoskeleton is the **microfilament,** a tiny rod made of the protein *actin* (fig. 5.17). Microfilaments are about one-third as thick as microtubules, and they are not hollow. In muscles, actin microfilaments are interspersed with rods of another type of protein, myosin. When the actin microfilaments and larger myosin rods slide past one another, muscle contraction occurs. Microfilaments are also important in endocytosis by white blood cells. They align beneath animal cell membranes and propel portions of the membrane outwards to entrap particles. You can actually see microfilaments in action by watching the blood clot that forms on a scraped knee. The retraction of the clot as the wounded area is replaced with new skin is carried out by microfilaments and other proteins.

KEY CONCEPTS

A cell's shape is largely determined by its cytoskeleton, which is a network of protein rods and tubes. Microtubules are built of tubulin and are involved in cell division and form cilia. Microfilaments are built of actin and form part of muscle.

Coordination of Cellular Architecture—The Meeting of Sperm and Egg

A vivid example of the coordinated functioning of the cellular architecture is the first step to a new life—the meeting of sperm and egg, or fertilization. Much of what we know about fertilization comes from studying sea urchins that live in the Pacific Ocean off the California coast.

A sea urchin is a brightly colored animal that clings to the rocky ocean bottom. It is about the size of a human fist and looks a little like a curled up porcupine. The animals do not mate but release their sex cells directly into the water—the female depositing millions of eggs at a time, and the male, billions of sperm. The large numbers of sex cells help ensure that at least some of them will meet to start the next generation. Researchers can easily collect the cells from the water and observe their interactions (fig. 5.17).

The waters off of California teem with sea urchin sex cells, the sperm swimming under the power of their whiplike

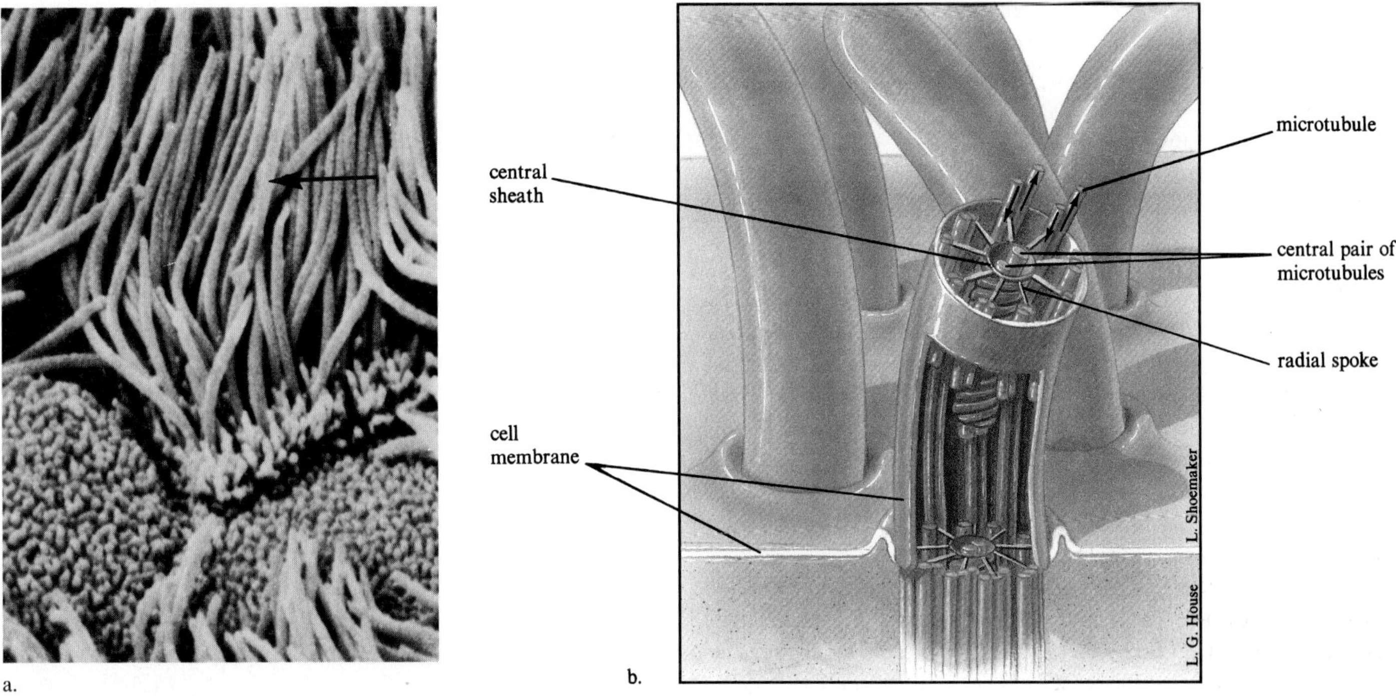

Figure 5.16
Cilia are hairlike appendages found on many eukaryotic cells. *a.* On cells lining the human respiratory tubes, such as these, beating cilia move dust particles upwards, so they can be exhaled. Cigarette smoking damages lung tissue by destroying respiratory cilia. *b.* A cilium is built of nine pairs of microtubules surrounding a central pair. The microtubules bend when ATP molecules are broken down, and the bending action causes the cilium to wave.

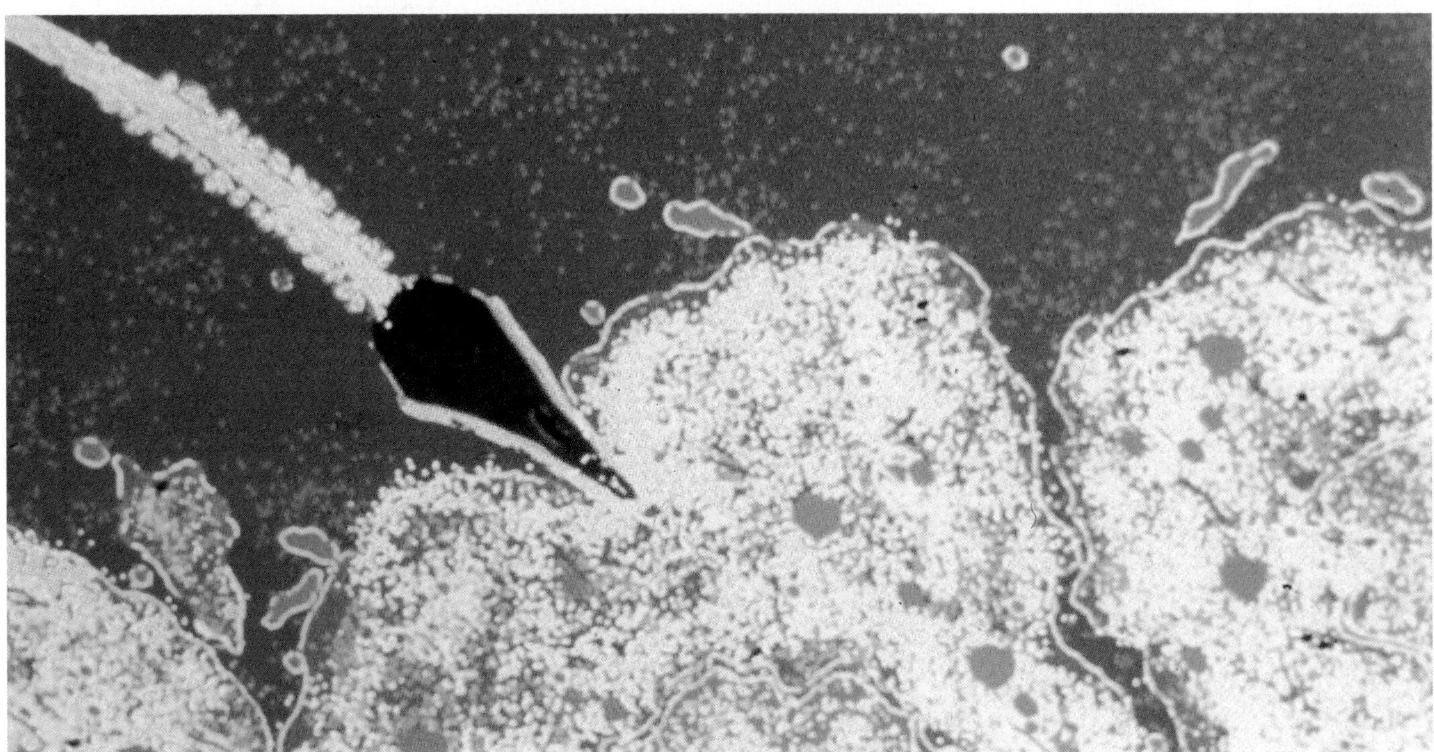

Figure 5.17
Fertilization—the coordination of cellular architecture. In a matter of minutes, components of the cellular architecture of two very different appearing cells, the sperm and the egg, interact so that the cells can meet and merge to start a new life on its way.

microtubule tails and the eggs tossed about by the waves. The cells are from several different species, but a sperm and egg can join to produce a new individual only if they are from the same species. The key to a correct match-up lies in cell surfaces. The pattern of proteins and microfilaments that dot an egg's surface attract sperm only of the same species.

Within seconds of the release of sex cells from the spiny sea urchins, each egg is surrounded by hundreds of sperm cells. Only one sperm will succeed in entering the egg, perhaps the first one to arrive or one that approaches at a particular angle. When the successful sperm makes contact with the egg's surface, it rapidly strings together actin molecules in its front tip, forming a projection that pokes at the egg. In an instantaneous response, microfilaments that protrude from the egg's surface organize to form a "fertilization cone," which rises upward to engulf the sperm head. The head, packed with genetic material from a male sea urchin, is drawn into the egg by endocytosis, leaving its microtubule tail behind.

As the sperm's nucleus is drawn into the egg cell, the surface of the egg immediately changes. Vesicles beneath the egg's cell membrane release phospholipids into the protein-lipid bilayer, swelling the membrane to more than twice its normal surface area. To accommodate the sudden superabundance of membrane material, the egg's 130,000 microfilaments near the surface extend to more than three times their normal length. Between 20 and 30 seconds after the sperm nucleus is engulfed, the egg cell releases enzymes by exocytosis that dissolve its surface proteins and attract sperm. Between 30 and 40 seconds later, other vesicles release chemicals that further protect the egg from additional sperm entry, and a change in the electrical properties of the cell membrane complete the female cell's defenses. Thus, within little more than a minute of the entry of the sperm nucleus, the cell membrane of the egg is effectively closed to entry by other sperm. These changes are vital to the new life being fashioned—if a sperm were to enter an already fertilized egg, development would not occur because the excess genetic material would upset the delicate chemical balance of the cell.

All of the other sperm cells clinging to the egg cell drop away. Inside the egg cell, the single engulfed sperm nucleus is pulled by microfilaments towards the egg cell's nucleus. When the nuclei of egg and sperm touch, the microfilament guides disassemble, and the genetic packages of the two cells unite as the nuclear membranes fuse to form a single nucleus. The genetic material of the parents mingles. The first cell of a new life is formed, thanks to the coordinated functioning of the cellular architecture.

SUMMARY

Many cellular functions depend upon a cell's *surface*, *membranes*, and *cytoskeleton*, which together can be thought of as the *cellular architecture*. Different cell types can be distinguished by the characteristics of their cellular architecture, and abnormalities in the cellular architecture are often associated with disease.

The features of a cell's surface identify it as belonging to a particular species, to a particular individual, and to a particular tissue within that individual. The surface consists of intricate patterns of sugars, proteins, and other molecules jutting from the cell's membrane. The human immune system recognizes the surfaces of body cells as "self" and the surfaces of other cells or particles as "nonself." Cell surfaces are particularly important in the sorting out of embryonic cells to form tissues and organs.

Cell membranes control which substances enter or leave a cell, and this, in turn, affects the cell's functions. A biological membrane is constructed of a *lipid bilayer* in which moveable proteins are embedded. The percentage and distribution of proteins varies in the outer membranes of different cell types and in organelles. Substances pass through cell membranes in several ways. A molecule that passes through openings in a membrane by following its concentration gradient moves by *passive diffusion*. *Osmosis* is the passive diffusion of water. Movement of molecules through a membrane down the concentration gradient aided by a carrier protein is termed *facilitated diffusion*. Movement of molecules through a membrane opposing the concentration gradient requiring both a carrier protein and energy supplied by ATP is called *active transport*. Vesicles inside the cell can carry molecules and bacteria to the cell membrane, where they fuse with the membrane and release the cargo outside the cell. This process is called *exocytosis*. Vesicles derived from the cell membrane envelop and transport molecules into the cell by *endocytosis*.

The cytoskeleton is a network of rods and tubes that provides cells with form, support and the ability to move. *Microtubules* self-assemble from hollow tubulin subunits to become such cellular constituents as *flagella*, *cilia*, and the spindle fibers that separate one cell into two in cell division. *Microfilaments* are smaller than microtubules, they are solid, they are composed of the protein actin, and they provide contractile motion.

The steps involved in fertilization of a sea urchin egg by a sperm illustrate interaction of the components of the cellular architecture.

QUESTIONS

1. What role does the cell surface play in the following processes?

- **a.** The body's reaction to transplanted tissue
- **b.** Bacterial infection
- **c.** Tissue and organ formation in the embryo

2. How would you test the hypothesis that a particular medical condition is associated with a particular HLA type?

3. How is a liposome structurally similar to a cell membrane?

4. Some cells have huge, multibranched protein molecules jutting out from their surfaces. Many of these cell surface markers are manufactured within the cell that they sit upon. What cellular structures and activities might assist these proteins in getting from their sites of synthesis inside the cell to the cell's surface?

5. How is ATP utilized in membranes? How is it utilized in the cytoskeleton?

6. How do the cytoskeleton and surface features of a cancer cell differ from those of a normal cell?

7. In the film version of Isaac Asimov's *Fantastic Voyage*, a group of scientists and a ship are shrunken and injected into the body of an important person to treat his brain tumor from within. At one point, the scientists wander from their miniature ship, and one of them is approached by a white blood cell. The cell attempts to swallow her. What parts of the cell surface and cytoskeleton participate in this attack?

TO THINK ABOUT

1. In many species, including our own, sperm cells have surface proteins that bind to receptor molecules on the surface of the egg cell. Use this information to suggest two possible methods of birth control.

2. In rabbits, two proteins made in the uterus early in pregnancy attach to the cell surfaces of the embryo and mask its surface proteins. What function might these two uterine proteins serve? What might happen to the pregnant rabbit and its developing offspring if one of the uterine proteins were absent?

3. What do you think should be the biological criteria for deciding who should get an available organ?

4. The cell membranes of many animal cells contain approximately equal amounts of lipid and protein. However, the inner mitochondrial membrane is roughly 80% protein and 20% lipid, and the outer membranes of Schwann cells, which insulate nerve cells, are roughly 80% lipid and 20% protein. How do these differences in the membrane composition of specialized cell types and organelles argue against the original model of a biological membrane as a rigid protein-lipid bilayer, as depicted in figure 5.6?

5. At the tip of a human sperm cell is a small elevation called the acrosome, which contains enzymes that allow the sperm cell to enter an egg cell. The acrosome contains many vesicles beneath its membrane. How do you suppose the enzymes are released from the tip of the sperm cell?

6. A drop of a 5% salt (NaCl) solution is added to a leaf of the aquatic plant *Elodea*. When the leaf is viewed under a microscope, colorless regions appear at the edges of each cell as the cell membranes shrink away from the cell walls. What is happening to these cells?

SUGGESTED READINGS

Edelson, Edward. July/August 1981. Scaffold of the cell. *Mosaic*, vol. 12, no. 4. An overview of the cytoskeleton.

Elgsaeter, Arnljot, et al. December 5, 1986. The molecular basis of erythrocyte shape. *Science*, vol. 234, no. 4781. The regular arrangement of proteins on the inner surface of the red blood cell's outer membrane allows this highly specialized cell to squeeze through the narrow passageways of the circulatory system.

Galloway, John. April 17, 1986. Transmembrane molecule helps to unravel proteins. *New Scientist*. The position of a protein embedded in a membrane provides information about the arrangement of its hydrophilic and hydrophobic amino acids.

Leff, David. July/August 1981. The mystery of cell movement. *Mosaic*, vol. 12, no. 4. The cytoskeleton lies behind the mystery.

Wassarman, Paul M. January 30, 1987. The biology and chemistry of fertilization. *Science*, vol. 235, no. 4788. The events of fertilization are like a dance, with specific steps carried out in a specific sequence by elements of the cellular architecture.

Weiner, Jonathon. April 1982. Inner tubes. *Science*, vol. 22, no. 4. Microtubules are the bones and muscles of the cell.

CHAPTER

6

Biological Energy

Chapter Outline

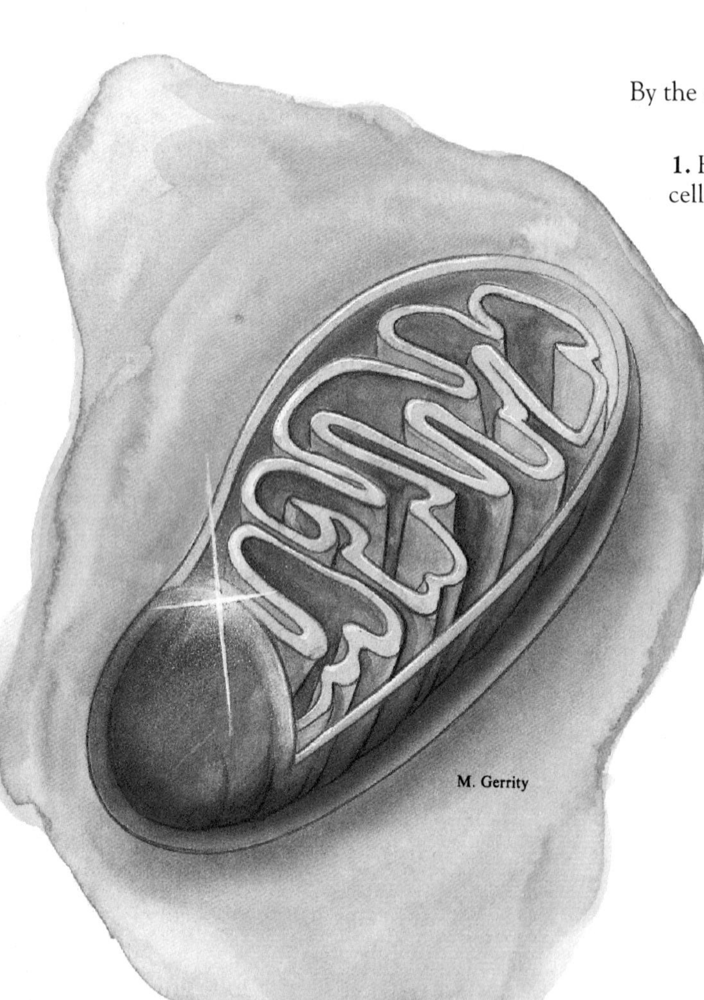

Learning Objectives

By the chapter's end, you should be able to answer these questions:

1. How is energy essential to life, on a cellular, whole-body, and global level?

2. Where does the energy used by living organisms ultimately originate?

3. How do the reactions of photosynthesis capture energy from sunlight and use it to manufacture glucose?

4. How do the reactions and pathways of cellular respiration extract energy from a glucose molecule?

5. How are the pathways of energy metabolism interrelated?

6. How might the pathways of energy metabolism have evolved?

It is 10:30 in the morning. You have just listened to 2 hours of lectures, and it has been a long time since breakfast. You feel very tired, and the morning is only half over. During a break, you eat a granola bar. By the time the next lecture begins, you are relieved to feel a burst of energy that enables you to pay close attention to the lecturer.

We all experience ebbs and flows of energy during a day's activities. We know that food provides "fuel" that seems to infuse us with a sense of overall alertness. However, energy is continually used in the human body on a more pervasive, though less noticeable, level. Each cell is the site of literally thousands of chemical reactions, some that require energy and some that release it. Our cells take in energy from the environment, store it, use it, perhaps convert it to another form of energy, and ultimately lose it to the environment. All forms of life on earth are interrelated by the very similar ways in which they are able to obtain energy for cellular functions. This chapter will explore both the whole-body sense of energy, with which we are all familiar, and the energy produced and consumed within cells, which keeps us alive and reveals our fundamental relationships to all other living things.

Energy in Living Systems

Energy can be defined as the ability to do work, and it is evidenced by movement. The turning of a windmill as it is powered by moving air demonstrates energy at work, as does the kicking of an animal's legs as they are moved by muscles. Energy cannot be created or destroyed, but it can change form. For example, electrical energy entering a kitchen range is converted to heat energy; similarly, energy stored in the chemical bonds of food molecules powers the activities of life.

The energy of motion is called *kinetic energy*. Energy contained in the structure or position of matter is called *potential energy*. A boulder on a hilltop illustrates the difference between kinetic and potential energy: When at rest on top of the hill, the boulder has potential energy. When it crashes down the slope, the potential energy is converted to the kinetic energy of motion and is released as heat and sound, both of which result from the movement of molecules.

adenine

phosphate groups

ribose sugar

adenosine monophosphate (AMP)	(1 phosphate)
adenosine diphosphate (ADP)	(2 phosphates)
adenosine triphosphate (ATP)	(3 phosphates)

Figure 6.1
ATP—energy currency of the cell. Energy is required to form a bond between negatively charged phosphate (PO_4) groups. The squiggly lines that are used to represent these bonds indicate that a great deal of energy is released when they are broken.

ATP—Biological Energy Currency

In living organisms, potential energy is stored in the chemical bonds of organic molecules, such as proteins, carbohydrates, and fats. It takes energy to make bonds, and energy is released when bonds of these molecules are broken. Much of this released energy of life is stored temporarily in the covalent bonds of the molecule **adenosine triphosphate (ATP).** ATP is a nucleotide, composed of the nitrogen-containing base adenine, a sugar group (ribose), and three phosphate groups (a phosphorus atom bonded to four oxygen atoms) (fig. 6.1). The covalent bonds that connect the phosphate groups to each other can break, releasing a large amount (7 kilocalories) of energy. Because of their ability to release so much energy, these phosphate bonds are sometimes called "high-energy" bonds, and they are represented graphically by squiggly lines. When the endmost phosphate group of an ATP molecule detaches, the disruption of its bond releases energy and the molecule becomes adenosine diphosphate (ADP). Yet another phosphate bond can be broken to yield adenosine monophosphate (AMP) and another release of energy.

The energy contained in the phosphate bonds of ATP can be tapped by another chemical reaction, if that reaction occurs at the same time as the splitting, or hydrolysis, of ATP. Indeed, hundreds of a cell's chemical reactions are "coupled" to the release of energy from ATP. These reactions use the energy in ATP's bonds to synthesize molecules, to break down large molecules into smaller ones, to power the active transport of molecules across membranes, and to move such structures as cilia. For example, the synthesis of the amino acid glutamine requires that another amino acid, glutamic acid, react with ammonia. The hydrolysis of ATP to ADP provides the energy that drives this reaction. The stringing together of the building blocks of DNA is another reaction that is coupled to ATP hydrolysis, but this reaction requires so much energy that, once ATP is split to yield ADP, the ADP is also split to yield AMP. On a larger scale, ATP is responsible for muscular motion and the transmission of nerve impulses. The human body uses an estimated 2 billion ATP molecules each minute merely to stay alive. ATP serves as a universal energy currency of life on earth.

Metabolic Pathways—Energy on a Cellular Level

The reactions that take place within living cells are collectively termed **metabolism,** which is from the Greek word for "change."

The general functions of metabolism are to build structures and to convert stored energy into forms that can be directly used to power such biological functions as synthesis, motion, and transport. The reactions of metabolism are powered by enzymes. Recall that an enzyme is a protein that greatly increases the rate of a particular chemical reaction. In general, the reactions of metabolism extract chemical energy from the bonds of nutrient molecules (carbohydrates, fats, and proteins) to form phosphate bonds in ATP. This temporarily stored energy is then used to build up or break down a wide variety of molecules that are essential to life.

Metabolic reactions are organized into pathways that consist of several chemical reactions linked sequentially so that a product of one reaction becomes a reactant (starting material) of another. These chains of biological reactions can extend to 10 or 15 (or more) individual steps, each catalyzed by an enzyme. When an end product of a reaction pathway is also the starting material for the first reaction in the chain, a cycle is formed (fig. 6.2). Metabolism consists of many such chains and cycles. Because the same molecule can participate in different pathways, the chains and cycles of metabolism often have steps in common. When the many metabolic reactions of a human cell are written out to show where they intersect, the resulting diagram is very complicated.

The energy-requiring pathways that build large molecules from small ones are called **anabolism,** or biosynthesis. In anabolism, a small number of chemical building blocks join in different ways to produce many different large molecules. Anabolism is said to diverge, because a few types of precursor molecules combine to yield many different types of products. For example, the 20 different types of amino acids are linked together in different sequences to synthesize hundreds of different proteins.

Metabolic pathways that break down large molecules are known as **catabolism,** or degradation. These pathways release energy. Catabolic pathways converge, in that many different types of large molecules degrade to yield fewer types of small molecules. For example, you eat many, many different types of foods, consisting of large

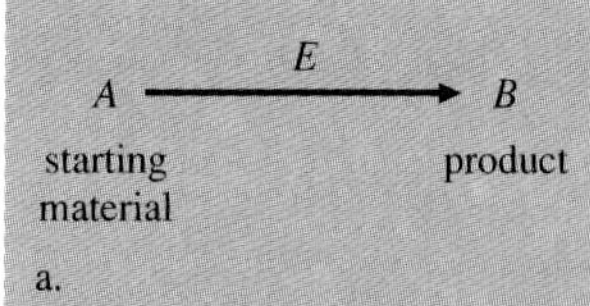

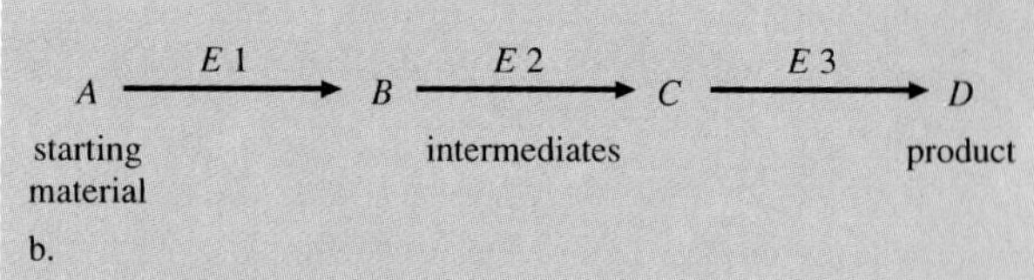

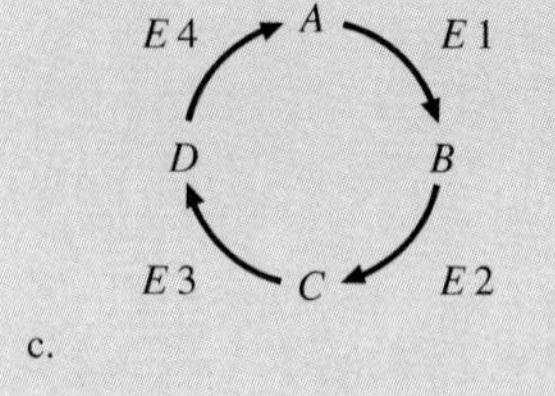

Figure 6.2
Metabolic pathways—chains and cycles. An enzyme-catalyzed chemical reaction (*a*), biochemical pathway (*b*), and biochemical cycle (*c*), in which the product of the last reaction is the starting material of the first. Cycles release by-products important in other biochemical pathways.

Table 6.1
Differences Between Catabolism and Anabolism

Catabolism	*Anabolism*
Breakdown of large molecules	Synthesis of large molecules
Energy is released	Energy is required
Reactions converge	Reactions diverge

and diverse molecules of carbohydrates, fats, and proteins. These major nutrients are broken down into glucose, glycerol and fatty acids, and amino acids, respectively. These smaller components of food molecules are in turn catabolically converted to even smaller organic molecules. Electrons are released in catabolic reactions, and their energy is used to manufacture ATP.

Anabolism and catabolism appear to be opposites, in that one set of pathways builds molecules and the other breaks molecules apart. Chemically, however, the situation is not this straightforward, because the reactions of anabolism and catabolism are not usually simple reverses of one another. Converting substance A to substance B often requires a different enzyme than the conversion of B to A. The result is a separate enzymatic control of synthesis and degradation. The reaction that proceeds in one direction may go much faster or slower than the reaction that proceeds in the opposite direction. In addition, the two reactions may take place in different parts of the cell. The two "directions" of metabolism—anabolism and catabolism—can be compared to building a car and taking it apart. Different skills and tools are needed for each task. (The basic differences between anabolism and catabolism are summarized in table 6.1.)

Control of Metabolism

The overall result of metabolism is that the cell is in a constant state of flux, with some molecules being torn apart while others are being built up. If the balance between the building up and breaking down of molecules is disturbed, the life of the cell is threatened. Several biological mechanisms regulate cellular metabolism and preserve this delicate balance.

The key to metabolic regulation lies in the pacesetters of these reactions, the enzymes. In a metabolic pathway, the enzyme whose reaction proceeds the slowest

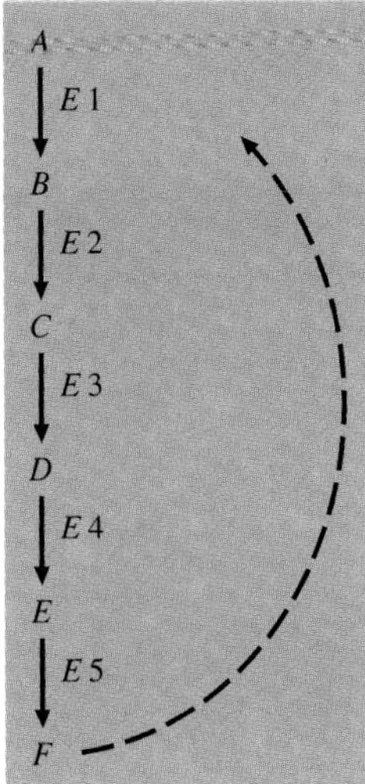

Figure 6.3
Negative feedback loops regulate metabolism. In this diagram of a metabolic pathway, *A* is the starting material, *B*, *C*, *D*, and *E* are intermediates, *F* is the final product, and *E1* through *E5* are the enzymes that catalyze the various reactions. When more than enough of the product *F* has accumulated for the cell to function, the buildup signals the regulatory enzyme *E1* to cease its activity temporarily.

controls the pathway's productivity, because each subsequent reaction requires the product of the preceding reaction to continue. The reaction catalyzed by this enzyme is called the *rate-limiting step*, and the enzyme is a *regulatory enzyme*. Often a regulatory enzyme is highly sensitive to a chemical cue. For example, when the end product of the metabolic pathway begins to accumulate, the regulatory enzyme responds by temporarily ceasing to function. When the level of the end product has fallen back to normal, the regulatory enzyme switches on again.

When an enzyme is turned off by the accumulation of a product, just as a building's heating system is turned off by its thermostat when the inside temperature reaches a certain point, it is responding to **negative feedback,** also called feedback inhibition (fig. 6.3). This occurs, for example, in the biosynthesis of amino acids. Through such feedback loops, a cell can maintain optimal levels of specific amino acids and other biological molecules. Sometimes the accumulation of a particular intermediate in a metabolic pathway under the regulatory enzyme's control signals that the pathway is not active enough. In response, the regulatory enzyme steps up its activity. This is *positive feedback*.

Metabolic balance is also maintained by **hormones.** In plants and animals, these chemical messengers are manufactured in one tissue and carried to other tissues, where they stimulate or inhibit metabolic reactions. An example of hormonal control of metabolism in humans is the relationship between levels of the pancreatic hormone insulin and glucose in the blood (blood sugar). When you eat a candy bar, the blood-glucose level rises and, in response, the pancreas secretes more insulin. The insulin binds to muscle and liver cells, enabling them to take up the extra glucose. Once inside the muscle and liver cells, the glucose is either broken down to provide energy for cellular activity or polymerized and stored as glycogen. As the glucose enters these cells, its level in the blood falls, and, in response, so does insulin secretion.

The Evolution of Metabolic Pathways

The reactions of metabolism in diverse species are remarkably similar, with many pathways virtually identical. The similarities may reflect the evolution of species from a single ancestral type, in which the foundations of metabolism were established. How might metabolic pathways have evolved? It seems highly unlikely that the first organism suddenly appeared fully formed, complete with 2,000 distinct, enzymatically orchestrated chemical reactions conducting the biochemical business of life. As with any scientific explanation of the origin and evolution of life on earth, events probably proceeded from the simple to the complex.

Return for a moment to the "protocell" introduced in chapter 3, the nucleic acid-protein partnership that floated in a primordial sea. It must have fed on a molecule (let's call it nutrient A) that was present in its environment. As the protocells divided, producing more of themselves, their single food source must have dwindled. However, because of the capacity for change that is inherent in all genetic material (even in the nucleic acids of protocells), it is likely that the ancient seas were home to more than one genetic variety of protocell. One type might have evolved that made an enzyme that could convert some other nutrient (let's call it B) into the original one (A). (Recall that genetic material contains the instructions for manufacturing enzymes.) The new variety of protocell would have a nutritional advantage, because it could extract energy from two different food sources—B and A.

Soon the first type of organism, totally dependent on nutrient A, would die out as its food supply was consumed. The variant form that could convert nutrient B to A would flourish, but, in time, nutrient B would also become scarce. Then, a variety of organism with an additional enzyme that converted a nutrient C to B (and then, using the first enzyme, B to A) would arise and flourish, and so on. Over time, an enzyme-catalyzed biochemical pathway, D—C—B—A, evolved (fig. 6.4). More pathways evolved. Intermediates of one pathway became starting points of another, and metabolism gradually arose.

KEY CONCEPTS

Energy is the ability to do work. The energy of motion is called kinetic, and the energy held in a structure or position is potential. Energy held in the phosphate bonds of ATP is used to power many biochemical reactions. The reactions in cells constitute metabolism, and they form pathways and cycles. Anabolic reactions build molecules and require energy. Catabolic reactions break down molecules and release energy. Metabolic reactions are regulated by enzymes, positive and negative feedback loops, and hormones. Diverse species have metabolic reactions in common. Metabolic pathways may have arisen as ancient protocells evolved new ways to use nutrients.

Whole-Body Metabolism—Energy on an Organismal Level

Metabolic reactions supply cells with the energy to survive, as well as the specialized functions that enable the cell to be a valuable part of a multicelled organism. However, we are most familiar with biological energy on a larger scale—the moving muscles, flowing blood, and other bodily functions that constitute *whole-body metabolism*.

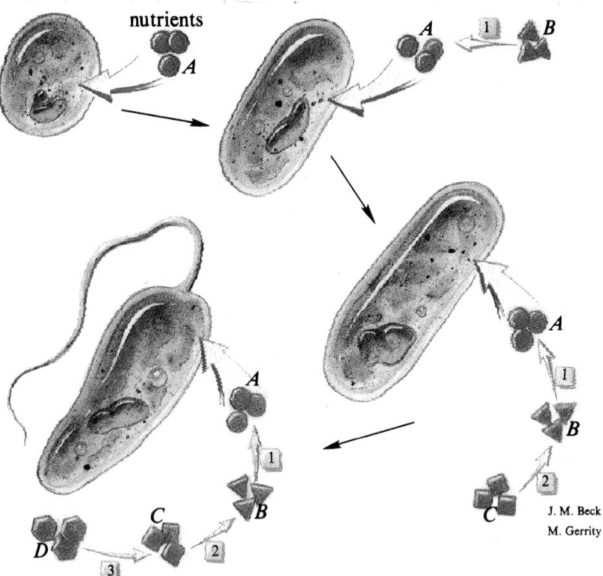

Figure 6.4
The evolution of new enzymes probably expanded the number of different nutrients that could be used by certain of the earth's first organisms, thus giving them a survival advantage over other organisms competing for limited food. *1*, *2*, and *3* are enzymes; A through *D* are nutrients.

Energy enters the human body in the form of food molecules and exits in the form of heat or activity, either in bodily movements or in the numerous energy-requiring functions entailed in merely staying alive. Usually, energy intake and output are in balance. When they are not, the result can be underweight or obesity (Reading 6.1). Nutritionists measure energy in units called **Calories** (with a capital C). A single Calorie equals the amount of energy needed to raise the temperature of 1 kilogram (slightly more than a quart) of water by 1°C. Because the nutritional Calorie is 1,000 times as large as a chemist's calorie (with a lowercase c), it is preferable to refer to the nutritional Calorie as a kilocalorie, in order to avoid confusion.

The energy required by an organism simply to stay alive is described as the **basal metabolic rate** (BMR), which measures the kilocalories needed for heartbeat, breathing, the functioning of nerves, kidneys, and glands, and the maintenance of body temperature when the subject is awake, physically and mentally relaxed, and has not eaten anything for 12 hours. BMR does not include the energy needed to digest food or for physical activity. For an adult human male, the average BMR is 1,750 kilocalories in 24 hours; for a female, the average BMR is 1,450 kilocalories. However, most people are not usually completely at rest and fasting, so the number of kilocalories actually needed to get through a day generally exceeds the average basal requirement.

Several factors can influence one's BMR, including age, sex, weight, and body proportions. Even one's regular level of activity can affect the BMR. A sedentary college student's BMR may be 1,700 kilocalories in 24 hours. His roommate, who is on the swim team and runs 5 miles each morning, may burn 5,000 kilocalories per day and have a BMR of 2,500. When the two students are asleep, their metabolic rates are likely to fall below the basal level, but the swimmer's rate will still be higher than his roommate's, because regular exercise leads to elevated metabolism even during rest. This is why regular exercise—for at least 20 minutes three or more times a week—can accelerate weight loss if kilocalorie intake is maintained or decreased.

The basal metabolic rate rises from birth to about age five, then declines until the teen years, when it peaks again. As we age, BMR drops in parallel to declining energy needs. BMR increases during the course of some illnesses and whenever body temperature is elevated. It also increases in pregnant or breast-feeding women. BMR falls during starvation as the body attempts to conserve energy. BMR is also related to body composition. For two individuals of the same weight, age, and sex, the one with a greater proportion of "lean tissue" (muscle, nerve, liver, and kidney) to fat tissue will have a higher BMR because lean tissue is active and consumes more energy than relatively inactive fat tissue.

The **thyroid gland,** located in the neck, is one of several glands that affects basal metabolism. The thyroid manufactures the hormone *thyroxine,* which increases energy expenditure. Thyroid function is often evaluated in hospitalized patients as an indicator of metabolic activity. When too much thyroxine is produced, the BMR may nearly double, resulting in the body's excessive consumption of kilocalories and a consequent loss of weight. If too little thyroxine is available, BMR slows, fewer kilocalories are used, and weight is gained, even with a relatively modest food intake.

It is interesting to look at basal metabolism in species other than our own. In general, the smaller the organism, the higher its metabolic rate. A shrew has a higher metabolic rate than a cat, which has a higher rate than a dog, which, in turn, has a higher rate than a human. This is because smaller organisms have higher surface-to-volume ratios and therefore lose more heat to the environment than larger organisms. They must burn relatively more fuel to maintain a nearly constant body temperature, and hence they have higher rates of metabolism than larger organisms. Within a species, the smaller individuals have higher metabolic rates.

Different species have evolved different mechanisms of altering metabolic rate to cope with environmental change. The energy generated by basal metabolism is sufficient to keep naked humans warm outdoors if the air temperature is about 74°F (23°C). If the temperature falls much below this level, unclothed humans can raise their metabolic rates by as much as a factor of five by shivering. This action contracts muscles that play against one another, using the energy of metabolism to generate potentially lifesaving heat.

Reading 6.1 The Facts About Fat

MAINTAINING A CERTAIN BODY WEIGHT IS A MATHEMATICAL MATTER: IF THE NUMBER OF KILOCALORIES CONSUMED EXCEEDS THOSE USED IN BASAL METABOLISM, DIGESTION, AND PHYSICAL ACTIVITY OR DISCARDED BY THE BODY AS WASTES, THE EXCESS FOOD ENERGY IS STORED AS FAT. It takes 3,500 kilocalories to produce a pound of fat. The average adult in North America consumes about 3,000 kilocalories per day, getting 130 kilocalories from each ounce of protein or carbohydrate in the diet and 270 from each ounce of fat. A meal at a typical fast-food restaurant may give you a whopping 1,500 kilocalories or more.

Despite our society's preoccupation with thinness, 25% to 30% of the adults in the United States and 10% of the children are obese. An obese person is someone whose body contains a significantly higher proportion of fat tissue than normal. For example, an 18-year-old female, whose ideal proportion of fat tissue is 18% to 24%, is considered obese if more than 30% of her body weight is fat. However, an 18-year-old male, whose body should ideally contain only 12% to 18% fat, is considered obese if his proportion of fat tissue is greater than 20%. A thin marathon runner is likely to be technically underweight, with a body fat measurement of only 4% or 5%. Not all heavy people are obese. Some have a higher proportion of muscle tissue and normal proportions of fat tissue. Such people do not face as great a health risk from their excess weight as do those whose extra poundage is due to fat.

How can you tell if you are overweight? Sometimes all it takes is looking in a mirror or trying to get into tight-fitting clothing that used to fit you perfectly. A technical way to determine the percentage of fat tissue in the body is to use a clamplike device called a caliper on the skin at the back of the arm. The proportion of fat in this tissue (which is determined by its thickness) is taken as an estimate of the percentage in the entire body. Alternatively, a variety of charts can be consulted that indicate a range of "desirable" or "recommended" weights for certain heights and for each sex. Yet another way to determine if your weight is excessive is to multiply your weight by 703, divide this value by your height in inches, and divide the resulting figure again by your height. A "normal" value is between 20 and 25.

Today, "fat" is often regarded as a dirty word. In the face of seasonal famine, however, the storage of extra energy-rich fat can be a lifesaver. As a result, some researchers have suggested that a survival mechanism has evolved in humans—a metabolic control called weight setpoint, in which the brain alters the metabolic rate of a hungry person. This effective survival ploy has become an obstacle to dieters, because it can sabotage an attempt to lose weight. When a person diets, hunger pangs signal the body's craving for food energy. When food intake does not follow the pangs, the weight-setpoint mechanism compensates by slowing the metabolic rate—making more, so to speak, of a limited resource. It is possible, however, to elevate one's setpoint by increasing physical activity on a regular basis.

Some cases of obesity are due to energy metabolism that is too efficient. In individuals of normal weight, a fraction of the kilocalories in food is stored in the bonds of ATP and the remainder is released as heat energy. When an overly efficient metabolism converts more than the normal amount of food energy to ATP, less food is needed to produce an adequate supply of ATP. The person, however, may still consume a normal amount of food. The result: an obese person who may actually eat no more than a person of normal weight.

For most of us, gaining weight is simply a matter of food intake exceeding exercise output. Normal weight is best maintained by eating a balanced diet in small portions and exercising regularly.

Although extreme cold can kill, some organisms have evolved interesting adaptations to keep warm. Plants detect the diminishing hours of daylight that herald the arrival of cold weather and drop their leaves and slow metabolism in their protected inner tissues and roots. Many animals hibernate (fig. 6.5), slowing metabolism and sleeping at a time when food is scarce and the environment is harsh. Hibernating animals can metabolize the fat in their brown adipose tissue to release heat, as we saw in chapter 3. In hibernating animals, a dangerous dip in body temperature stimulates the brain to arouse the animal and prompts fur fluffing and shivering, creating more efficient insulation and generating temporary warmth to help the animal survive.

KEY CONCEPTS

Energy that we are aware of is termed whole-body metabolism. The basal metabolic rate measures the energy needed to stay alive. BMR is influenced by age, sex, and activity level, but it varies greatly from individual to individual. The thyroid gland regulates basal metabolism. Different species have diverse mechanisms for altering metabolism to cope with environmental change. In organisms energy is stored in the bonds of nutrient molecules and is converted to more available energy in the phosphate bonds of ATP.

A Global View of Biological Energy

What is the source of the energy that powers life? For humans, the immediate source is the chemical bonds of the carbohydrate, fat, and protein molecules in food. Ultimately, however, the energy that makes the human body a moving, breathing, thinking machine comes from the sun.

Solar energy is harnessed by green plants, algae, and certain bacteria in a series of metabolic pathways called **photosynthesis** (making from light). In photosynthesis, carbon dioxide and water react, in the presence of sunlight and certain pigment molecules, to produce oxygen and energy-rich organic molecules. Energy from

Figure 6.5
The black bear is a hibernator, although its body temperature does not drop appreciably in winter. The bear's metabolic rate slows to half that of normal, which conserves energy. One interesting metabolic adaptation is the rerouting of the body chemical urea to form amino acids rather than urine, because heat is normally lost in urination. The bear thus builds proteins from amino acids while asleep and does not urinate. Its warm pelt and relatively low surface-to-volume ratio (a consequence of large size) also help it to conserve heat.

the sun is thus stored in the chemical bonds of these molecules, which are passed along to other organisms when the photosynthetic organisms are eaten. Intricate *food webs* of "who eats whom" are actually routes of energy transfer in the living world. Each energy transfer is only about 10% efficient, however. At each level of a food web, most of the energy consumed is lost as heat in the metabolic processes of the consuming organism. Compared with meat eaters, strict vegetarians are conserving biological energy. The foods they eat have obtained their energy directly from the sun, and the captured solar energy is transferred only once.

Humans and other organisms that obtain energy from other living (or once-living) sources are called **heterotrophs** (feeding on others). Organisms that obtain energy from nonliving sources, such as the sun, are called **autotrophs** (self-feeding). Some autotrophs are not dependent on the sun but are able to extract energy from inorganic chemicals in their environments. For example, bacteria that live around thermal vents on the ocean floor can extract energy from the bonds of hydrogen sulfide (H_2S) and then combine this energy with carbon dioxide to synthesize organic compounds. Such organisms are called *chemoautotrophs*.

Like photosynthetic organisms, chemoautotrophic bacteria form the basis of food webs. The organic compounds that the bacteria synthesize with the aid of the energy derived from hydrogen sulfide are used to power their own life functions. Microscopic animals (zooplankton) that drift in the ocean obtain the organic compounds by consuming the bacteria. The bacteria also provide nourishment to some of the large residents of thermal vents by actually living within their bodies. For example, *Riftia pachyptila* is a giant tube worm that has no mouth or digestive tract. It is nourished by colonies of chemoautotrophic bacteria that live in its tissues. The blood in the worm's circulatory system brings a constant supply of hydrogen sulfide from the outside environment to the bacteria. The diverse living communities of thermal vents are made possible by the chemoautotrophic bacteria, just as the lives of animals elsewhere on earth are made possible by algae and plants.

The discovery of deep sea thermal vents and their diverse and thriving residents came as a great surprise. Other biological energy sources may await discovery. For example, soil-dwelling microorganisms have been found that derive energy from chemicals that are toxic to us, such as polychlorinated biphenyls (PCBs). Researchers are trying to devise ways to encourage growth of these organisms to clean up toxic waste.

Once chemical energy is obtained by an organism, either directly or by consuming other organisms, it must be extracted in order to be utilized in life functions. The long-term storage of energy in chemical bonds of nutrient molecules is converted to short-term storage in the bonds of ATP by various metabolic pathways in cells, and energy extracted from the bonds of ATP is used to power cellular activities. Thus, the global view of energy in biological systems leads directly back to the cellular level.

KEY CONCEPTS

Energy enters biological systems as autotrophs (plants) produce organic molecules using solar energy, and other organisms (heterotrophs) eat the autotrophs or each other. Chemoautotrophs extract energy from inorganic chemicals in their environments.

Photosynthesis

The journey of biological energy transfer begins with the sun. In the reactions of photosynthesis, plants capture the energy in sunlight and convert it to stable chemical energy. This is a complex task, accomplished in a series of steps by "exciting" certain electrons with light. The excited electrons then become stabilized by losing energy in increments. This released energy is used by the cell to first synthesize molecules that contain the energy in their bonds for short periods and ultimately to synthesize organic molecules, primarily the simple sugar glucose, which store the energy indefinitely. The cells then draw on this long-term energy supply to carry out the functions of life.

Overall, the reactions of photosynthesis convert 6 molecules of carbon dioxide and 6 molecules of water to 1 molecule of glucose and 6 molecules of oxygen. The reaction is written as follows:

$$6CO_2 + 6H_2O \xrightarrow[\text{pigment}]{\text{light}} C_6H_{12}O_6 + 6O_2$$

Glucose ($C_6H_{12}O_6$) is an excellent energy-storage molecule found in all organisms. As it is dismantled to carbon dioxide and water, the energy released from its bonds can be captured to make 36 "high-energy" ATP bonds. In a sense, then, 1 glucose molecule can be "exchanged" for 36 molecules of ATP. Glucose and compounds derived from it are participants in many metabolic pathways.

Light

Light powers the photosynthetic reactions, so it is important to understand the nature of light energy. Visible light (or white light) is a small part of a continuous spectrum of radiation called the *electromagnetic spectrum* (fig. 6.6). Visible light is actually composed of the six colors of the visible spectrum. Electromagnetic energy consists of tiny packets of energy called *photons* that travel in waves. These waves are of different wavelengths in different portions of the electromagnetic spectrum. Invisible regions of the spectrum may also be familiar, such as microwaves, infrared radiation that provides heat, television and radio waves, and the ultraviolet rays that tan skin.

The visible portion of the electromagnetic spectrum is vital to life because of its role in photosynthesis. Unlike radiation of shorter wavelengths, such as X rays and ultraviolet radiation, which can break chemical bonds, visible light excites molecules. When a photon impinges upon a molecule, it is absorbed, causing electrons close to its atom's nucleus to jump to a higher energy level. These electrons are said to be in an excited state. Photon-boosted electrons are like the boulder on the hilltop, in terms of their potential energy. Like the boulder rolling downhill, when excited electrons fall back to their original positions near the nucleus, they release energy.

Chlorophyll and Chloroplasts

The plant molecules that typically capture light energy in their excitable electrons are pigments. **Chlorophyll** *a* is a large pigment molecule that plays a major role in photosynthesis in many plants. It absorbs wavelengths corresponding to red, orange, blue, and violet, and it reflects and transmits green wavelengths. This is why leaves, in whose cells photosynthesis takes place, are

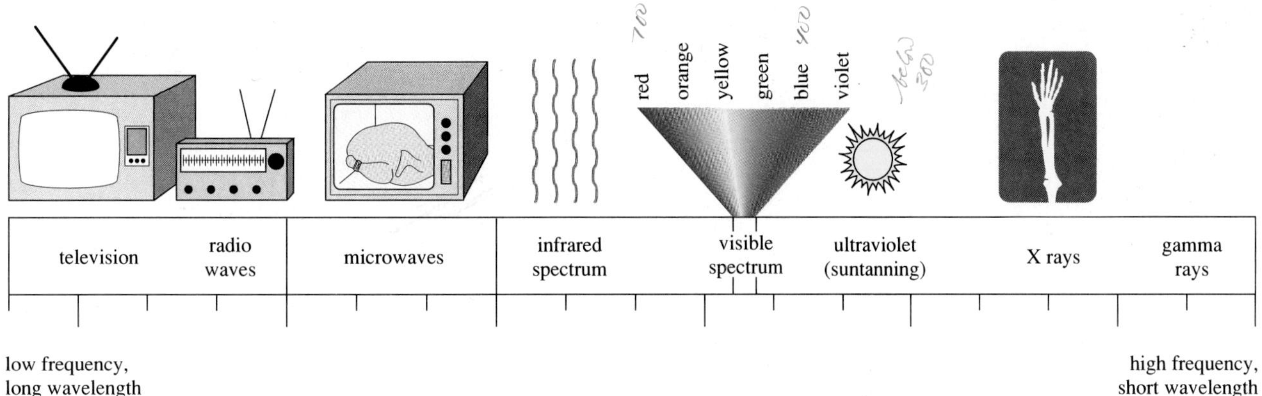

Figure 6.6
Visibile light, microwaves, and the electromagnetic radiation that makes television and radio transmission possible are parts of a continuous spectrum of radiation termed the electromagnetic spectrum.

green. Other pigment molecules (carotenoids and xanthophylls) absorb light energy of other visible wavelengths and transfer it to chlorophyll *a* (fig. 6.7).

Several different pigment molecules cluster together to form **photosystems.** The majority of molecules in a photosystem absorb energy, and they transfer the energy to a few special molecules that actually harness it in excited electrons for further use in photosynthesis. The entire photosynthetic process consists of two photosystems that are said to be "linked" because one system passes electrons to the other.

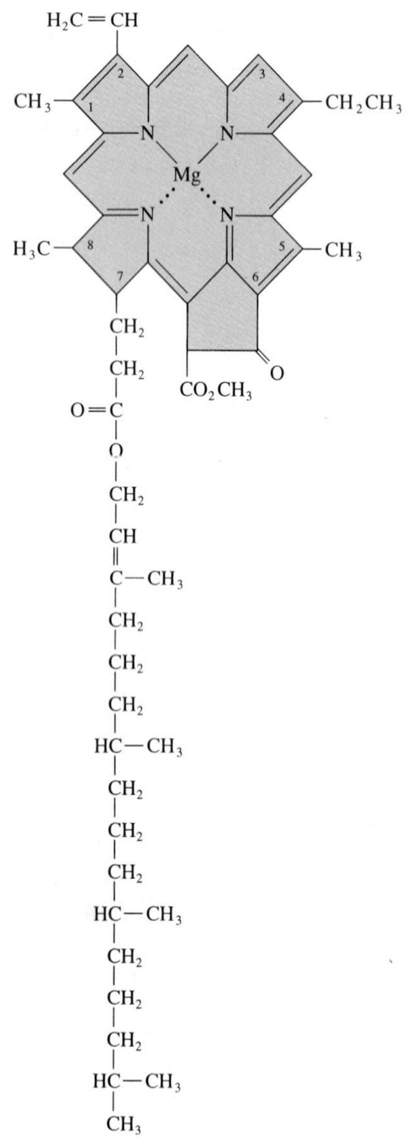

Figure 6.7
The green pigment chlorophyll *a* is the dominant pigment in the photosynthesizing cells of plants. Each of the large molecules of this pigment consists of a central magnesium (Mg) atom surrounded by four nitrogens (N) and various organic rings. A long, hydrophobic (water-hating) tail anchors the molecule into the chloroplast membranes.

Photosynthesis in eukaryotes occurs in a specialized organelle, the **chloroplast.** A chloroplast is constructed of two outer membranes surrounding a highly folded third membrane (fig. 6.8). The nonmembranous inner region is called the **stroma,** and it contains several types of proteins. Inner chloroplast membranes that are loosely packed are called **stroma lamellae,** and these contain the pigment molecules. Much of the inner membrane is organized into flat, interconnected sacs called **thylakoids,** which are oriented toward the sun. Stacks of thylakoids are called **grana.** Photosynthesis in organisms that lack chloroplasts, such as cyanobacteria, takes place on membranes that are not as intricately organized as those of eukaryotic chloroplasts.

KEY CONCEPTS

In the reactions of photosynthesis, the energy from sunlight powers the conversion of six molecules of carbon dioxide and six molecules of water into one molecule of glucose and six of oxygen. Energy is released as ATP when glucose is broken down to carbon dioxide and water. Participants in photosynthesis include photons and chlorophyll and other pigments clustered into photosystems. The reactions take place in chloroplasts, which are plant organelles built of membranes (thylakoids and stroma lamellae) that contain pigments.

The Chemical Reactions of Photosynthesis

Photosynthesis is accomplished by several types of chemical reactions. Among those that occur early in the process are oxidation-reduction reactions linked to form **electron-transport chains,** which are series of molecules that readily accept and then lose electrons. (Recall from chapter 3 that electrons are lost in an oxidation reaction and gained in a reduction reaction.) There are two such chains in photosynthesis, one of which links the two photosystems. In the first electron-transport chain, an electron from a molecule of chlorophyll *a* that has been excited by light energy is passed along a series of electron-carrier molecules, each of which briefly holds the electron at a slightly lower energy level than the one before. As the electron moves "downhill" from one carrier molecule to another, some of the energy that is released is packaged in the form of ATP. An electron-transport chain ends with a final electron-accepting molecule.

A type of chemical reaction that occurs early in photosynthesis is **photophosphorylation** (adding a phosphate using light). This is the reaction in which the energy released by the electron-transport chain linking the two photosystems is stored in the phosphate bonds of ATP. In a third type of photosynthetic reaction, called **photolysis** (splitting by light), electrons are stripped from water molecules (to replace electrons lost by molecules of chlorophyll *a*), which split to yield oxygen gas and protons (H^+).

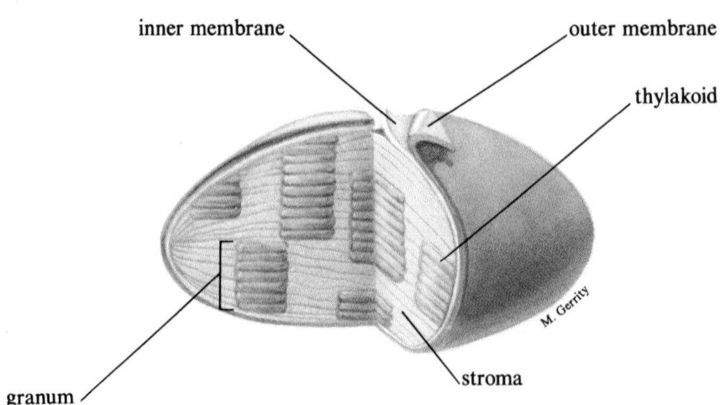

Figure 6.8
The chloroplast is the site of photosynthesis in plant and algal cells. This double-membraned organelle contains flat, interconnected sacs called thylakoids, which are organized into stacks called grana.

Two sets of reactions constitute photosynthesis. In the **light reactions** (which include the reactions just mentioned), light is required and water is split. Along the way, ATP is produced and a molecule called *nicotinamide adenine dinucleotide phosphate* ($NADP^+$) is reduced (picks up electrons) to form NADPH (which, like ATP, is used to store energy temporarily). In the **dark reactions** of photosynthesis, so-called because they do not require light, the products of the light reactions are used to produce glucose from carbon dioxide.

KEY CONCEPTS

The types of chemical reactions in photosynthesis include electron-transport chains (series of molecules that accept and release electrons); photophosphorylation (energy is stored in phosphate bonds of ATP); and photolysis (electrons are released from splitting water molecules).

The Light Reactions

Photosynthesis begins in the cluster of pigment molecules that constitute photosystem II. (The two photosystems were named by the order of their discovery, and not the order of their functioning.) The energy from incoming photons is absorbed by the pigment molecules in photosystem II. The energy is transferred from one pigment molecule to another in the photosystem until it reaches a particularly reactive molecule of chlorophyll *a*, one of whose electrons is excited to a higher energy level. Such a chlorophyll *a* molecule is thought to be more reactive than the others because it is physically associated with certain membrane proteins.

The excited electron leaves the reactive chlorophyll *a* molecule and is accepted by the first electron-carrier molecule of the electron-transport chain that links the two photosystems (fig. 6.9).

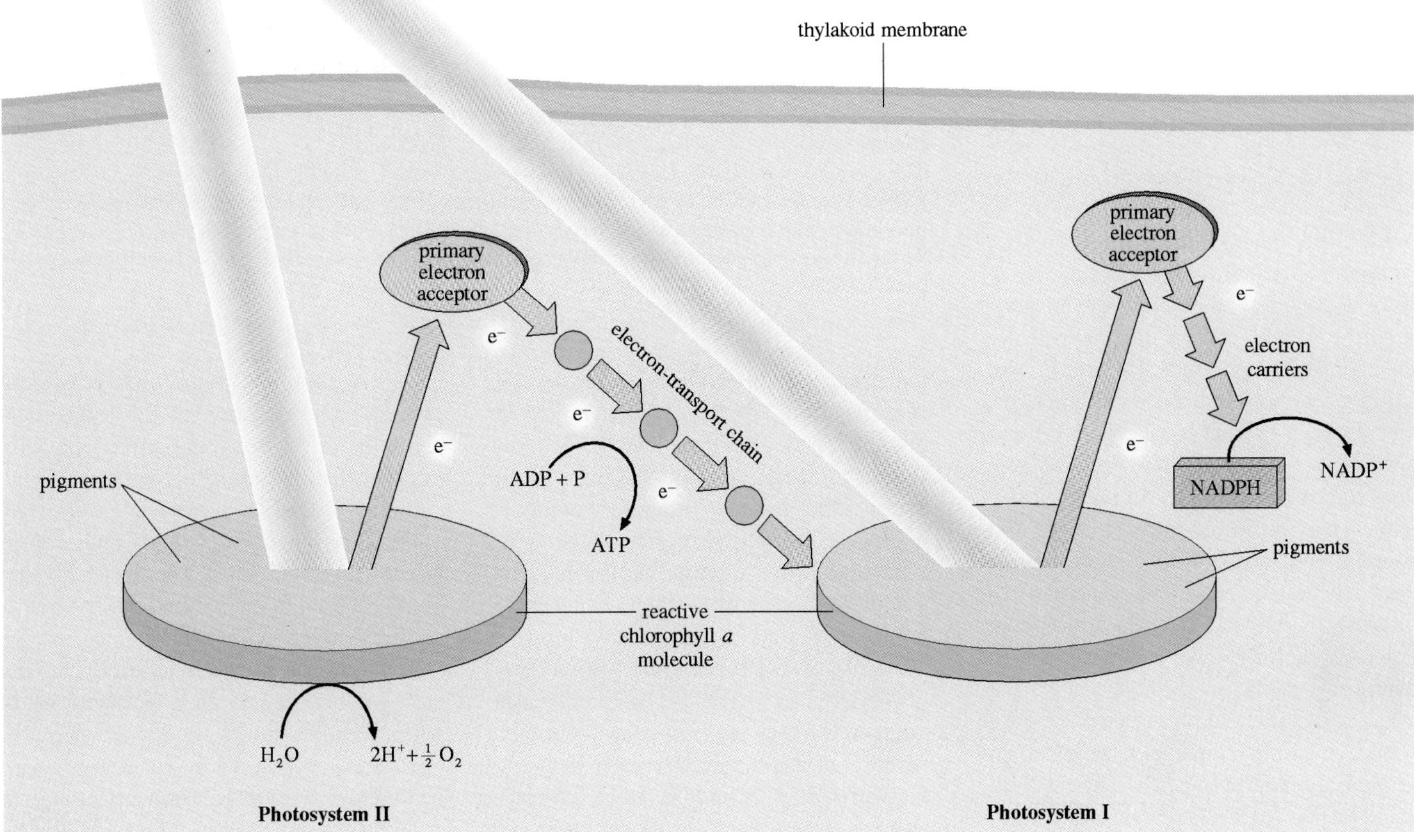

Figure 6.9
The light reactions of photosynthesis. Energy from the sun propels electrons, from reactive molecules of chlorophyll *a*, through a series of electron-carrier molecules. In the process, the energy sources for the dark reactions—ATP and NADPH—are formed.

Meanwhile, the reactive chlorophyll *a* molecule replaces its lost electron with an electron released from the splitting (photolysis) of a molecule of water into oxygen gas and protons (H^+). When the electrons that have been boosted from the reactive chlorophyll *a* molecule pass through the electron-transport chain, they lose energy, which is captured and used to add a phosphate group to an ADP molecule so that it becomes ATP. Energy is stored in the phosphate bonds of ATP.

As in photosystem II, the energy of photons propels electrons from the reactive chlorophyll *a* of photosystem I to the first electron-carrier molecule in a second electron-transport chain. The boosted electrons from the reactive chlorophyll *a* molecule of photosystem I are replaced by the electrons passed down the first electron-transport chain from photosystem II. Finally, the transported electrons of photosystem I reduce a molecule of $NADP^+$ to NADPH. This NADPH, plus the ATP generated in photosystem II, are the sources of energy used for the dark reactions to follow. The overall products of the light reactions are oxygen, ATP, and NADPH.

The Dark Reactions

The end result of photosynthesis is the incorporation of carbon from carbon dioxide into glucose and other organic compounds that a living cell can use to store energy. The dark reactions of photosynthesis "fix" carbon into biologically useful organic compounds. These reactions, which occur both in darkness and in light, are powered by the products (ATP and NADPH) of the light reactions. Where does the carbon dioxide in biological systems originate? Photosynthetic algae and bacteria absorb dissolved carbon dioxide from the water that surrounds them. In plants, carbon dioxide from the atmosphere enters through openings, called stomata, in leaves and stems.

Some of the dark reactions of photosynthesis form a metabolic cycle known as the Calvin cycle (fig. 6.10). The Calvin cycle begins with the reaction of carbon dioxide and a five-carbon molecule to form a six-carbon molecule, which splits to yield two three-carbon products. After several more reactions, the three-carbon molecule glyceraldehyde phosphate is produced. This is the immediate product of the Calvin cycle. Other dark reactions convert molecules of glyceraldehyde phosphate to glucose and other organic molecules. To complete photosynthesis, the Calvin cycle makes two complete turns, converting 18 molecules of ATP to ADP and 12 molecules of NADPH to $NADP^+$. These molecules are recycled back to photosystems I and II, where they again pick up energy by being converted to ATP and NADPH. The fact that photosynthesis ultimately produces carbohydrates can be used to form a photograph on a leaf (Reading 6.2).

In summary, photosynthesis is about energy. The function of these chemical reactions is to convert light energy into stable chemical energy. However, this is a difficult task and requires many steps. The first set of reactions converts light energy to unstable chemical energy (ATP and $NADPH_2$). This unstable chemical energy is then converted to stable chemical energy, which can be stored indefinitely, in the form of glucose, via the dark reactions. In photosystems I and II, the excited electrons become stabilized by losing energy in increments. Eventually, enough energy is released to synthesize ATP and NADPH.

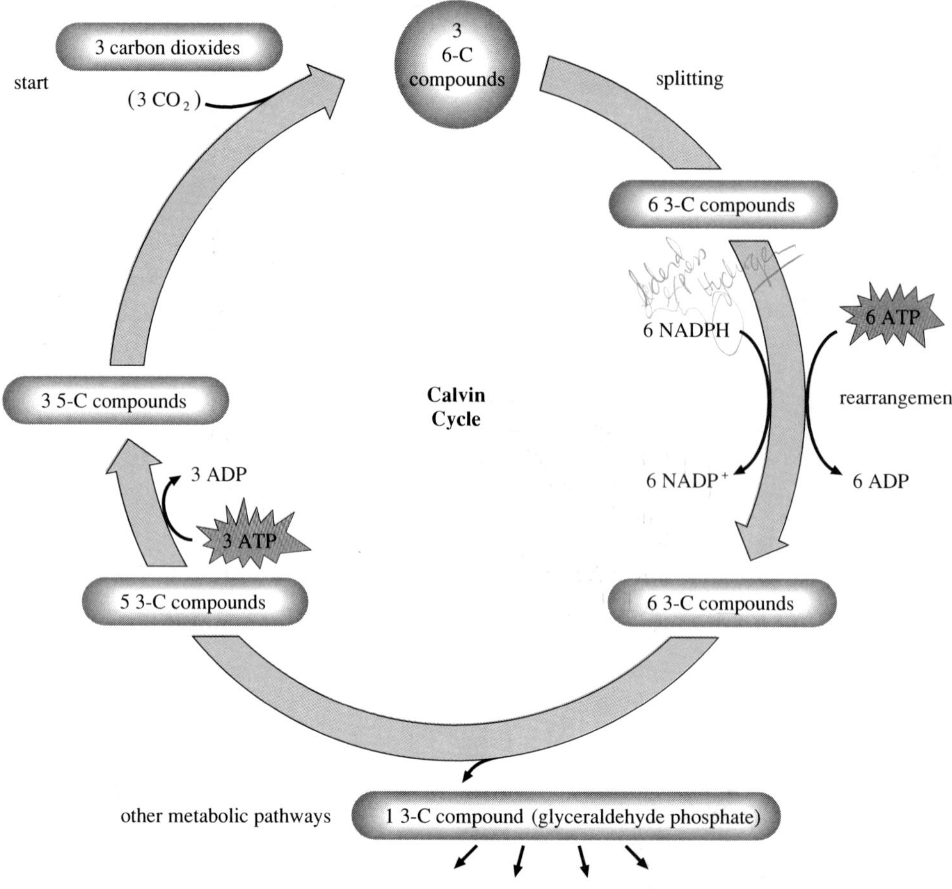

Figure 6.10
The products of the light reactions of photosynthesis—ATP and NADPH—drive the Calvin cycle (shown here in simplified form). The cycle spawns a three-carbon molecule, glyceraldehyde phosphate, which reacts in additional dark reactions to produce glucose and other carbohydrates. This pathway operates in most photosynthesizing plants, but some have evolved significant modifications.

Reading 6.2 Photosynthetic Photography

THE REACTIONS OF PHOTOSYNTHESIS ARE VIVIDLY DISPLAYED IN FIGURE 1, WHICH WAS DEVELOPED ON A GERANIUM LEAF. Such photosynthetic photography relies on the fact that photosynthesis produces carbohydrates, and carbohydrates stain a dark color in the presence of iodine.

To make a photosynthetic photograph, you will need a leaf, a piece of black cloth, a solution of baking soda (sodium bicarbonate), two sheets of glass, a cup of water, a slide projector, a negative, a solution of 86% alcohol, iodine, and potassium iodide.

First, place the leaf in the dark for 2 days. During this time it will use up available carbohydrates in metabolism. Moisten the piece of black cloth with the sodium bicarbonate. This provides carbon dioxide for photosynthesis. Take the leaf off of its stalk and place it on the black cloth. Then sandwich the leaf on its black cloth backing between the two sheets of glass. Hold it upright, so that the piece of stalk at the bottom sits in the cup of water. Place the negative of the desired photograph in the slide projector, and focus the image onto the leaf for 1 hour.

Figure 1
A photograph can be "developed" on a leaf by using the staining reaction of carbohydrate with iodine.

The final step, as with any photograph, is to develop the image. First dip the leaf into a solution of boiling 86% alcohol to leach out the pigments. Next, place the blanched leaf into a dish containing iodine and potassium iodide. Can you see how the image forms on the leaf? Photosynthesis occurs most readily in portions of the negative where the most light comes through. Dark areas do not prompt photosynthesis; transparent areas do. Watch as the iodine turns darkest wherever photosynthesis, and therefore the light, was the most intense.

Photosynthetic photography was invented by Austrian plant physiologist Hans Molisch in 1914 and has been adapted for use in biology classes by Howard Gest at Indiana University in Bloomington. By the way, the man whose image appears on the leaf here is Dr. Jan Ingen-Housz, a Dutch physician who discovered photosynthesis in the eighteenth century.

KEY CONCEPTS

In the light reactions, photon energy is absorbed by pigment molecules of photosystem II and sent to a reactive chlorophyll a *molecule. An excited electron in this chlorophyll is boosted to an electron-transport chain linking the two photosystems. Photolysis replaces the lost electron in the chlorophyll* a. *The energy released through the electron-transport chain converts ADP to ATP. The electrons from photosystem II allow boosted electrons from a reactive chlorophyll* a *in photosystem I to enter a second electron-transport chain. The electrons of photosystem I reduce $NADP^+$ to NADPH. This NADPH plus the ATP from photosystem II are used as energy sources in the dark reactions, which ultimately produce organic molecules via the Calvin cycle and other pathways.*

Energy Extraction—From Glucose to ATP

The capturing of the sun's energy in the bonds of carbohydrate molecules provides plants and other photosynthetic organisms (autotrophs) with a long-term energy supply that they can draw on even when sunlight is not available. Plants store these carbohydrates in their tissues, and some plants have structures that specialize in stockpiling the chemical energy derived from photosynthesis (e.g., the tubers of the potato plant). Other organisms (heterotrophs) benefit from the energy produced in photosynthesis by consuming the energy-rich molecules of photosynthetic organisms. After glucose has been obtained—either directly from photosynthesis or by digesting the carbohydrates of plants—all organisms face the task of breaking the sugar down to obtain and utilize the energy stored in its bonds.

In most organisms, energy is retrieved from glucose in a catabolic process that occurs in two stages. The first, **glycolysis,** takes place in the cytoplasm. The second stage, **cellular respiration,** is a series of reactions that occur in the mitochondrion. Plants carry out both photosynthesis and cellular respiration (fig. 6.11). Overall, in the presence of oxygen, glucose is oxidized to yield carbon dioxide, water, and energy. The energy is transferred to the phosphate bonds of ATP. Glycolysis and cellular respiration are quite efficient, with 1 glucose molecule ultimately yielding 36 ATP molecules. The situation is like converting a $360 savings account (a glucose molecule) into 36 $10 bills (ATP molecules). The ATP, like currency, can then be used in several ways.

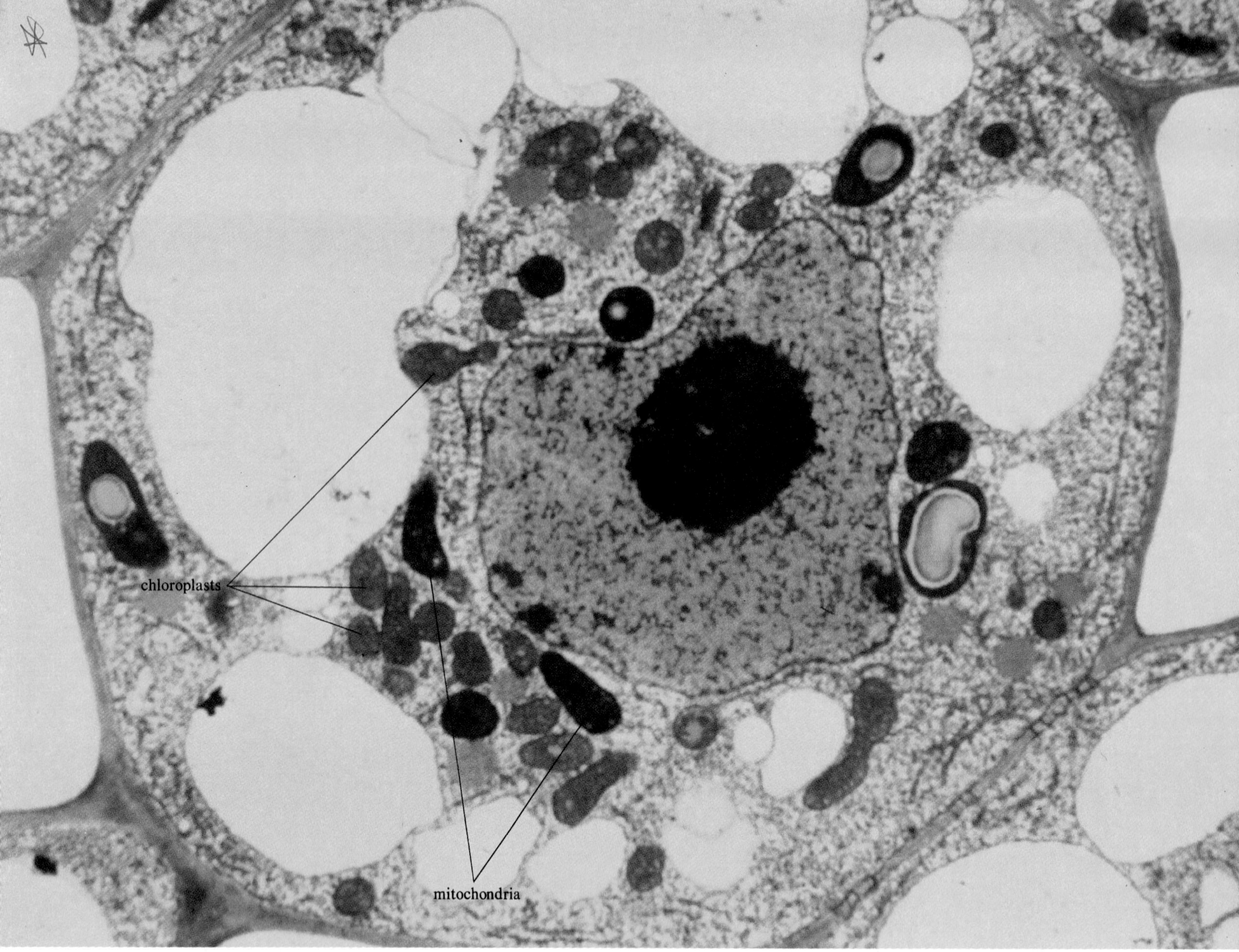

Figure 6.11
Both photosynthesis and cellular respiration occur in plants. Plant cells harness solar energy to manufacture organic molecules in which energy is stored, but they must also use this energy (after extracting it in the form of ATP). Thus, plant cells carry out both photosynthesis and cellular respiration. In this micrograph of a plant cell, the organelles responsible for photosynthesis (chloroplasts) and cellular respiration (mitochondria) can both be seen.

Glycolysis

Glycolysis means "glucose splitting," and this is precisely what happens to a glucose molecule at this stage of metabolism. This is the first step in breaking down the glucose molecule to release the energy in its bonds. In nine enzyme-catalyzed steps, one glucose molecule is rearranged and split to yield two three-carbon molecules of a compound called *pyruvic acid.* Two molecules of one of the intermediate organic compounds formed along the pathway lose hydrogens to molecules of *nicotinamide adenine dinucleotide* (NAD^+), which are thereby reduced to NADH. Like ATP and NADPH, NADH is used to store energy temporarily. A gain of hydrogens is a reduction reaction because electrons are gained; losing hydrogens is an oxidation reaction because electrons are lost. Each round of glycolysis produces two molecules of pyruvic acid, two molecules of NADH, and two molecules of ATP. Glycolysis occurs in all organisms. The steps of the pathway are illustrated schematically in figure 6.12.

KEY CONCEPTS

Glycolysis and cellular respiration retrieve energy stored in the bonds of the organic products of photosynthesis. In the nine steps of glycolysis, one glucose molecule is split and rearranged to yield two molecules of pyruvic acid. Two NAD^+ molecules are reduced to NADH and two ATPs are also produced. Glycolysis occurs in the cytoplasm of all cells.

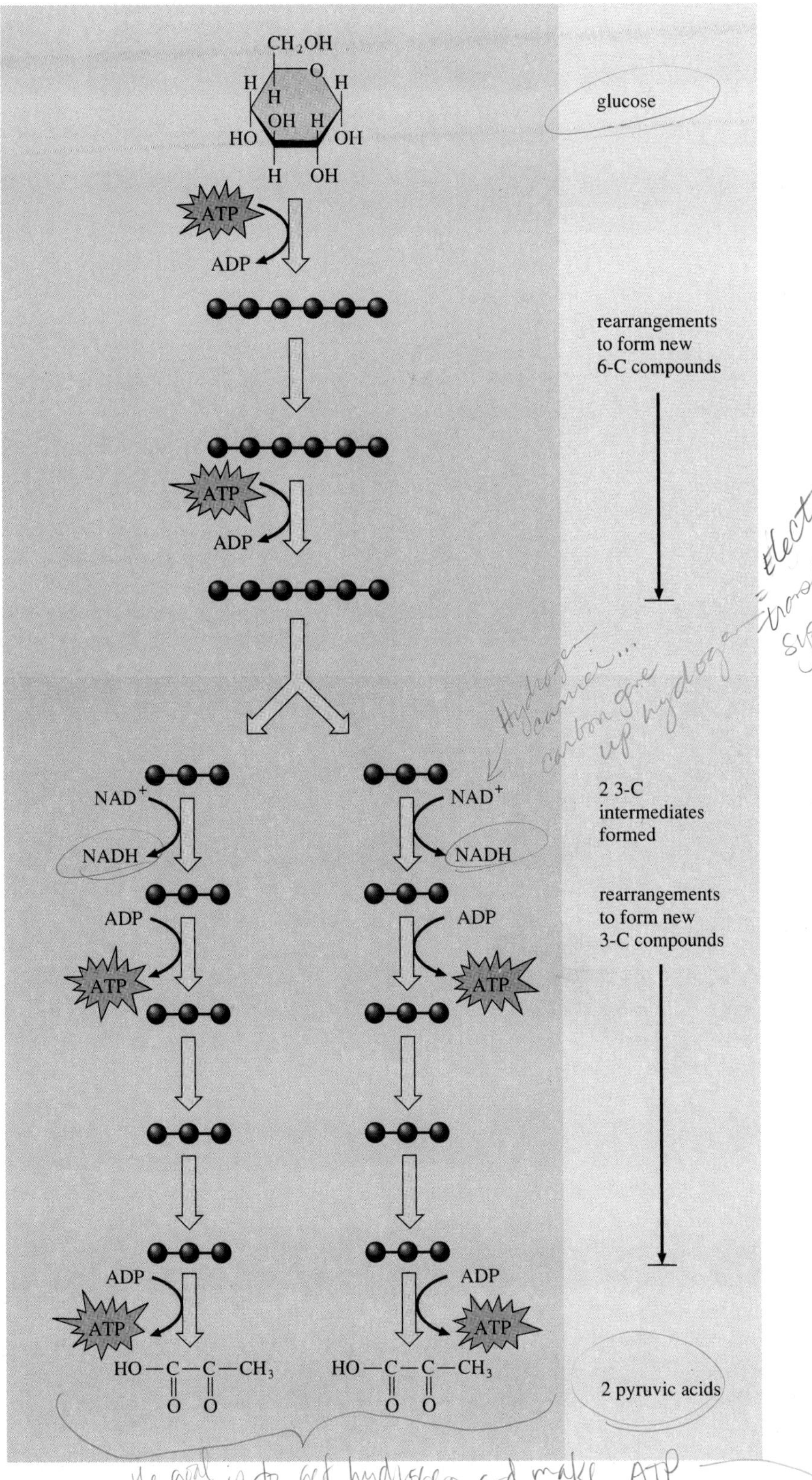

Figure 6.12
In the reactions of glycolysis, glucose is rearranged, split into two three-carbon intermediates, then rearranged further to yield eventually two molecules of pyruvic acid. Along the way, four ATPs and two NADHs are produced.

Fermentation—In the Absence of Oxygen

The metabolic fate of the product of glycolysis, pyruvic acid, depends upon the environment of the organism in which the reactions are taking place. Species that live in environments lacking oxygen are called *anaerobes*, and after glycolysis they utilize a short metabolic pathway called *fermentation* or *anaerobic respiration*. Organisms that live in oxygen-rich environments are called *aerobes*, in which glycolysis is followed by metabolic pathways that are more complex than fermentation. Chemoautotrophic bacteria that dwell in deep sea thermal vents cannot use oxygen at all and are called *strict anaerobes*. Some species that can function with or without oxygen, such as yeast, are called *facultative anaerobes*.

The reactions that follow glycolysis, whether anaerobic or aerobic, remove hydrogens from NADH to produce NAD^+. (The NAD^+ is recycled to glycolysis, where it is reduced again to NADH.) Two types of fermentation pathways accomplish this, and both occur in the cytoplasm. Yeast cells convert pyruvic acid to ethanol (an alcohol) and carbon dioxide in **alcoholic fermentation,** converting NADH to NAD^+ in the process (fig. 6.13). Yeast fermentation is used industrially to manufacture baked goods and alcoholic beverages. Depending upon the nature of the material being fermented, the variety of yeast used, and whether or not carbon dioxide is allowed to escape during the process, yeast fermentation can be used to produce wine or champagne (from grapes) or beer (from grain).

Some anaerobic bacteria (and animal cells that are temporarily deprived of oxygen) convert the pyruvic acid of glycolysis to the three-carbon compound lactic acid in a single step, converting NADH and NAD^+. This **lactic acid fermentation** occurs in human muscle cells when they are exercising so strenuously that their consumption of pyruvic acid exceeds their supply of oxygen. In this condition of "oxygen debt," the muscle cells revert to anaerobic respiration. The accumulation of lactic acid is felt as muscle fatigue and cramps (fig. 6.14). Later, when the oxygen supply is once again adequate, lactic acid is converted back to pyruvic acid in the liver,

Alcoholic Fermentation

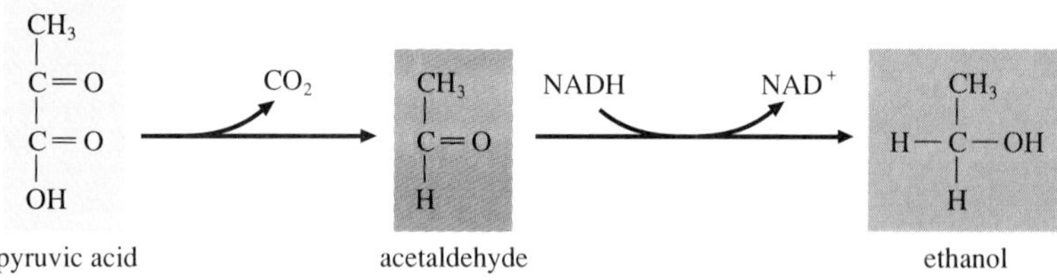

Lactic Acid Fermentation

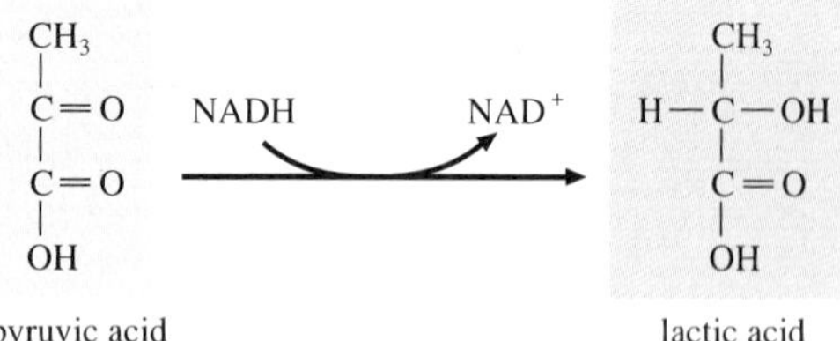

Figure 6.13
Alcoholic fermentation in yeast is used commercially to manufacture baked goods and alcoholic beverages. Anaerobic bacteria as well as animal cells deprived of oxygen carry out lactic acid fermentation.

Figure 6.14
Exercise and anaerobic respiration. Former President Jimmy Carter near the end of a 10-kilometer race. His muscle cells have been using oxygen so rapidly that they switch to lactic acid fermentation. The lactic acid buildup causes muscle fatigue and cramps, as is evidenced in Mr. Carter's grimace.

from which the body can extract more energy. Athletic coaches sometimes measure lactic acid levels in the blood to assess the physical condition of swimmers and sprinters.

KEY CONCEPTS

In the absence of oxygen, the product of glycolysis, pyruvic acid, undergoes anaerobic respiration (fermentation). Yeast cells convert pyruvic acid to ethanol and carbon dioxide, and some anaerobic bacteria and animal cells deprived of oxygen convert pyruvic acid to lactic acid. In both forms of fermentation, NADH is converted to NAD^+, which is recycled to glycolysis.

Aerobic Respiration—In the Presence of Oxygen

The pyruvic acid formed in glycolysis contains a lot of energy that could be used to produce considerably more ATP than is released by anaerobic respiration. For earth's first residents, however, glycolysis was probably the only means of extracting useful cellular energy. Glycolysis must have evolved at a time when there was not enough oxygen in the atmosphere to be used in aerobic respiration. As oxygen was pumped into the atmosphere by the first photosynthetic organisms, it became possible for aerobic respiration to evolve. Aerobic respiration is a far more efficient pathway than glycolysis or anaerobic respiration, because it allows the dismantling of more of the bonds in glucose.

Aerobic respiration occurs in the mitochondrion, an organelle constructed of an outer membrane and a highly folded inner membrane (fig. 6.15). The folds of the inner membrane, called **cristae,** contain enzymes and electron-carrier molecules that participate in aerobic respiration. In aerobic prokaryotes, which lack mitochondria, respiratory enzymes are embedded in the cell membrane.

The role of oxygen in aerobic respiration is to accept electrons that have been passed through a series of electron-accepting molecules. The energy that is released along the way is harnessed to convert ADP to ATP. Aerobic respiration starts where glycolysis ends, with a reaction called **acetyl CoA formation.** In the presence of oxygen, the pyruvic acid produced in glycolysis in the cytoplasm crosses both mitochondrial membranes to enter the organelle. Once inside the mitochondrion, each molecule of pyruvic acid loses a carbon dioxide, converts NAD^+ to NADH, and attaches to an enzyme called coenzyme A to form a new molecule, acetyl CoA (fig. 6.16). The 2-carbon acetyl CoA is an important molecule in metabolism. In addition to its being produced from pyruvic acid, acetyl CoA is part of the catabolic pathways of fats and amino acids. In energy metabolism, the formation of acetyl CoA is a bridge between glycolysis and aerobic respiration. At this point in metabolism, a molecule of glucose has been converted to 2 molecules of pyruvic acid, each of which has then been converted to a molecule of acetyl CoA. Two molecules of carbon dioxide have been released.

Next, each acetyl CoA enters a cycle of reactions called the Krebs cycle (fig. 6.17). Energy is released as hydrogen atoms and carbon dioxide are removed from the intermediates of the cycle. For each acetyl CoA that enters the Krebs cycle, 1 ATP molecule and 3 NADH molecules are produced, and 1 molecule of another electron-carrier molecule, *flavin adenine dinucleotide (FAD)*, is reduced to $FADH_2$. At this point, the energy yield from each glucose molecule is twice this, however, because each glucose molecule sends 2 acetyl CoA molecules through the Krebs cycle.

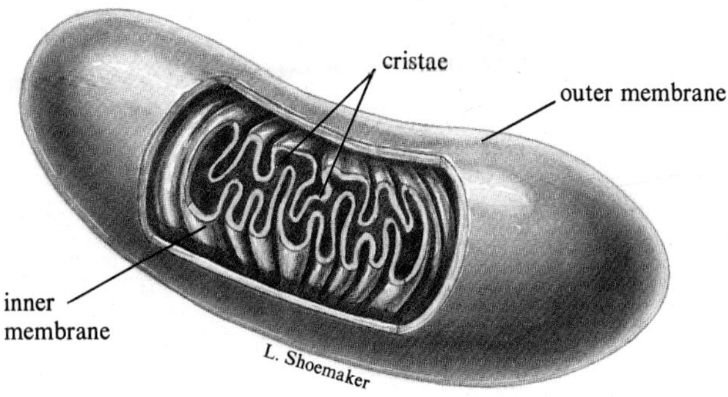

Figure 6.15
The mitochondrion's interior is a complex folded arrangement of membrane that is studded with enzymes and electron-carrier molecules that are important in aerobic respiration. An outer membrane surrounds the organelle.

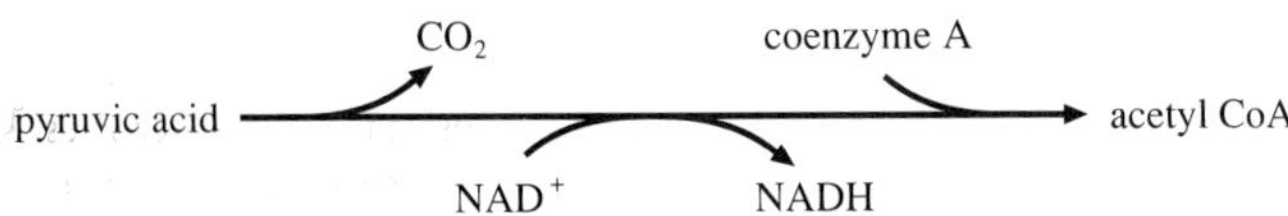

Figure 6.16
Acetyl CoA formation is a bridge between glycolysis and aerobic respiration. After pyruvic acid enters the mitochondrion, crossing both membranes, it loses CO_2, converts NAD^+ to NADH, and combines with coenzyme A to yield acetyl CoA. The next step is the Krebs cycle.

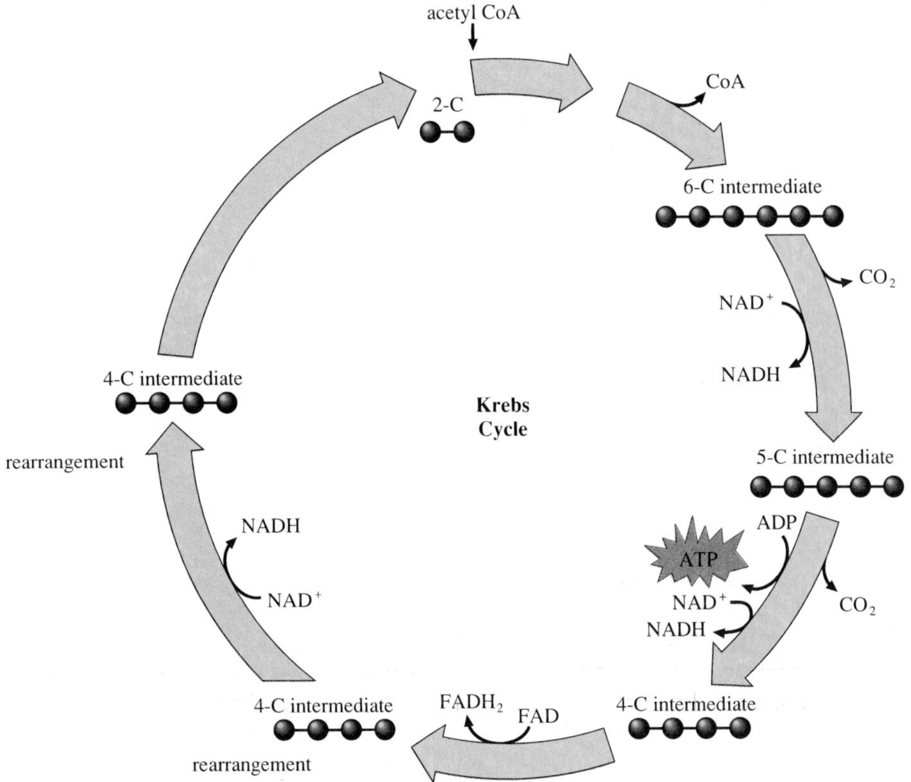

Figure 6.17
Each turn of the Krebs cycle generates one molecule of ATP, three molecules of NADH, one molecule of $FADH_2$, and two molecules of CO_2. One glucose molecule yields two molecules of acetyl CoA. Therefore, one glucose molecule is associated with two turns of the Krebs cycle.

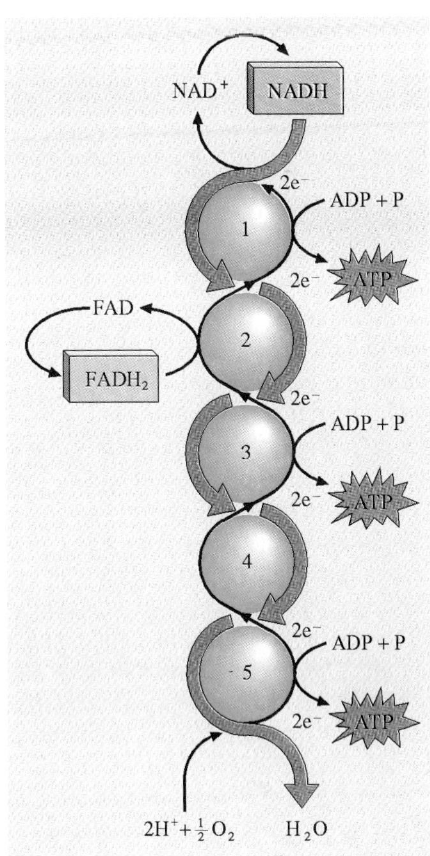

Figure 6.18
The respiratory chain. Electrons from NADH and $FADH_2$ generated in glycolysis and the Krebs cycle are passed along electron-carrier molecules (numbered circles) in the respiratory chain. The energy released along the way is captured in the phosphate bonds of ATP. In each chain, an oxygen accepts the electrons and combines with hydrogens to produce water.

In the final stage of extracting energy from glucose, the energy contained in the NADH and $FADH_2$ molecules generated so far is used to convert ADP into ATP. This occurs on a series of electron-accepting enzymes that are embedded in the inner mitochondrial membrane, forming a **respiratory chain.** (This is similar to the electron-transport chains of photosynthesis.) Electrons that enter glucose catabolism as parts of hydrogen atoms in the intermediates of glycolysis and the Krebs cycle are now transferred from NADH and $FADH_2$ to electron-carrier molecules in the respiratory chain, losing energy as they pass from one carrier molecule to the next (fig. 6.18). The released energy is used to transport protons (H^+) from the inner mitochondrial compartment (the matrix) across its membrane. Energy is required to do this because the membrane is impermeable to ions. The electron-fueled movement of protons establishes an electrochemical gradient across the membrane. The energy traveling through the membrane can be used by other proteins residing there. Most of it is used to activate the enzyme ATP synthetase, which catalyzes the addition of a phosphate to ADP, regenerating ATP. (Recall that a synthesis reaction requires an input of energy). The ATP then exits the inner mitochondrial membrane by facilitated diffusion (fig. 5.12), being escorted by a protein carrier through a protein-lined channel. This link between the chemical pathways that ultimately release high energy electrons (those of photosynthesis as well as aerobic respiration) and movement of protons (H^+) across a membrane is termed **chemiosmosis**, literally meaning "a chemical push." Some of this released energy is captured to manufacture ATP. There are "exchange rates" between the electron-carrier molecules and ATP formation: 1 molecule of NADH is used to convert 3 molecules of ADP to ATP, and the energy from 1 molecule of $FADH_2$ is used to convert 2 molecules of ADP to ATP. Oxygen is the final electron acceptor, combining with hydrogen to form water.

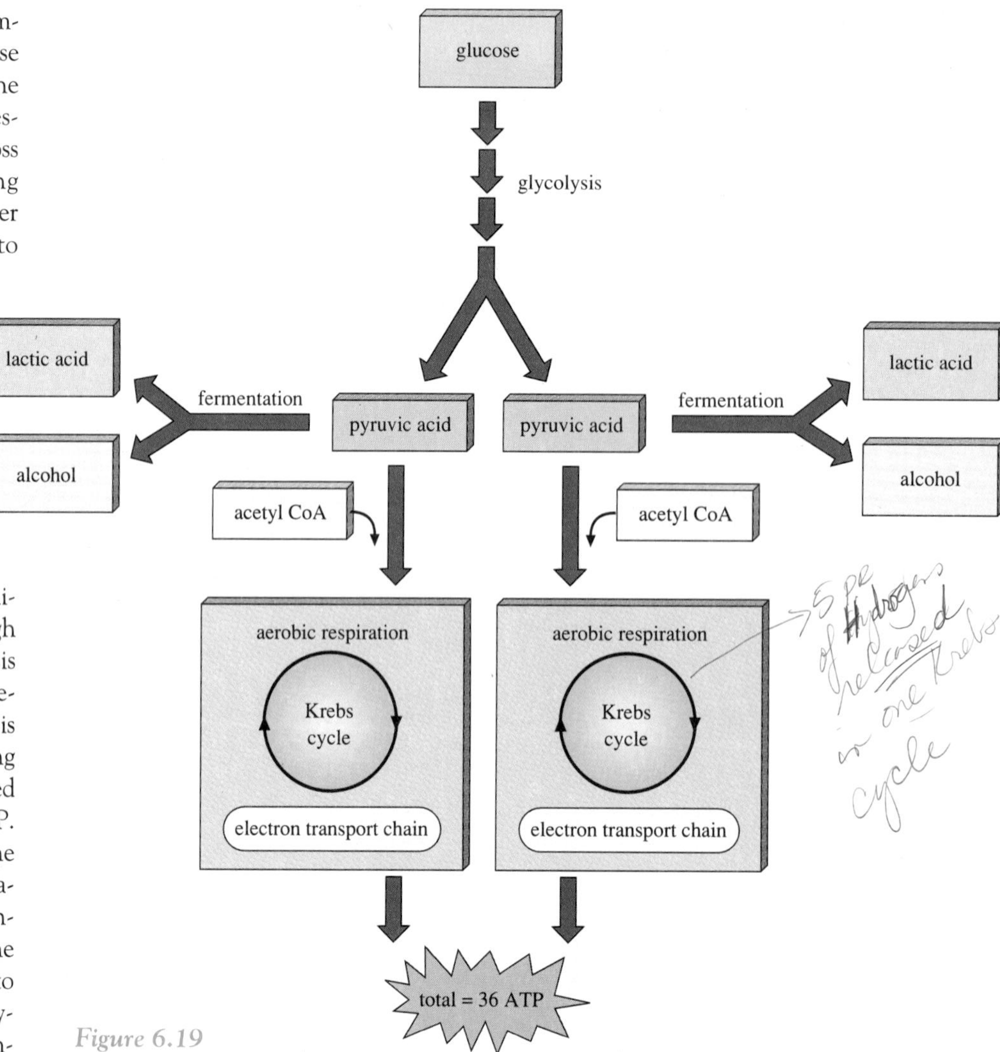

Figure 6.19
Extracting energy from glucose.

We can now look back over the pathways of glycolysis and aerobic respiration and consider the ATP "exchange rates" of NADH and $FADH_2$, plus the ATP released directly, to calculate the total energy yield from a molecule of glucose. Glycolysis generates 2 NADH molecules, each of which powers the formation of 3 ATP molecules from 3 ADP molecules. Two ATP molecules are produced directly in glycolysis, for a total of 8 ATP molecules. However, 2 ATPs are consumed in transporting the NADH from glycolysis into the mitochondrion. The net energy yield after the NADH molecules from glycolysis have passed through the respiratory chain is 6.

Next, the 2 NADH molecules produced from a single glucose molecule in acetyl CoA formation contribute another 6 ATP molecules, after the NADHs are used to manufacture ATP via the respiratory chain. The net ATP production from acetyl CoA formation, then, is also 6. The two turns of the Krebs cycle generate 2 ATP molecules directly, plus 6 NADH molecules (used to produce 18 ATP molecules along the respiratory chain), plus 2 $FADH_2$ molecules (used to produce 4 ATP molecules along the respiratory chain). The ATP output from the Krebs cycle is therefore 2 + 18 + 4, or 24. A total ATP accounting (6 from glycolysis, 6 from acetyl CoA production, and 24 from the Krebs cycle) indicates that 36 ATP molecules are generated from the complete breakdown of 1 glucose molecule (fig. 6.19 and table 6.2). This is very efficient energy transfer—ATP formation captures 40% of the energy stored in the bonds of glucose. In comparison, energy production from a coal-fired steam turbine is only 20% to 30% efficient.

Figure 6.20 shows at which points in the energy pathway the major nutrients enter.

KEY CONCEPTS

Aerobic respiration is far more efficient that anaerobic respiration. These reactions take place along the enzyme-dotted folds of the inner mitochondrial membrane. In the first step, pyruvic acid enters the mitochondrion, loses a carbon dioxide, converts NAD^+ to NADH, and attaches to coenzyme A to form acetyl CoA. Acetyl CoA enters the Krebs cycle, where 1 ATP, 3 NADHs and 1 $FADH_2$ are produced as the molecules lose carbon dioxide and rearrange. Along a respiratory chain, the energy in NADH and $FADH_2$ is used to convert ADP to ATP. Using the conversion factors of 1 NADH yielding 3 ATPs and 1 $FADH_2$ yielding 2 ATPs, we can trace back over all the pathways and arrive at a total of 36 ATPs from 1 molecule of glucose.

Table 6.2 **ATP Accounting**	
Exchange rates: 1 NADH → 3 ATPs 1 $FADH_2$ → 2 ATPs 1 glucose → 2 turns of Krebs cycle	
Pathway	***ATP Yield***
Glycolysis	
2 ATP	2 ATP
2 NADH	6 ATP
Acetyl CoA formation	
2 NADH	6 ATP
Krebs cycle (2 turns)	
1 ATP x 2 turns	2 ATP
3 NADH x 2 turns x 3 ATP/NADH	18 ATP
1 $FADH_2$ x 2 turns x 2 ATP/NADH	4 ATP
Subtotal	38 ATP
Energy expended to enter mitochondrion	-2 ATP
Total	36 ATPs /glucose

How Did the Energy Pathways Evolve?

The energy-generating metabolic pathways are complicated, each including several enzyme-catalyzed steps that release or store energy gradually. How might these connected collections of biochemical reactions have evolved? It is likely that glycolysis was the first of the energy pathways to form, because it is common to nearly all cells. In contrast to glycolysis, photosynthesis occurs only in green plants, algae, and cyanobacteria; fermentation is restricted to certain species; and aerobic respiration occurs only in cells that utilize oxygen. Hence, photosynthesis, fermentation, and aerobic respiration are more specialized metabolic processes. The simplest and most common pathway, glycolysis, must have appeared first (fig. 6.21).

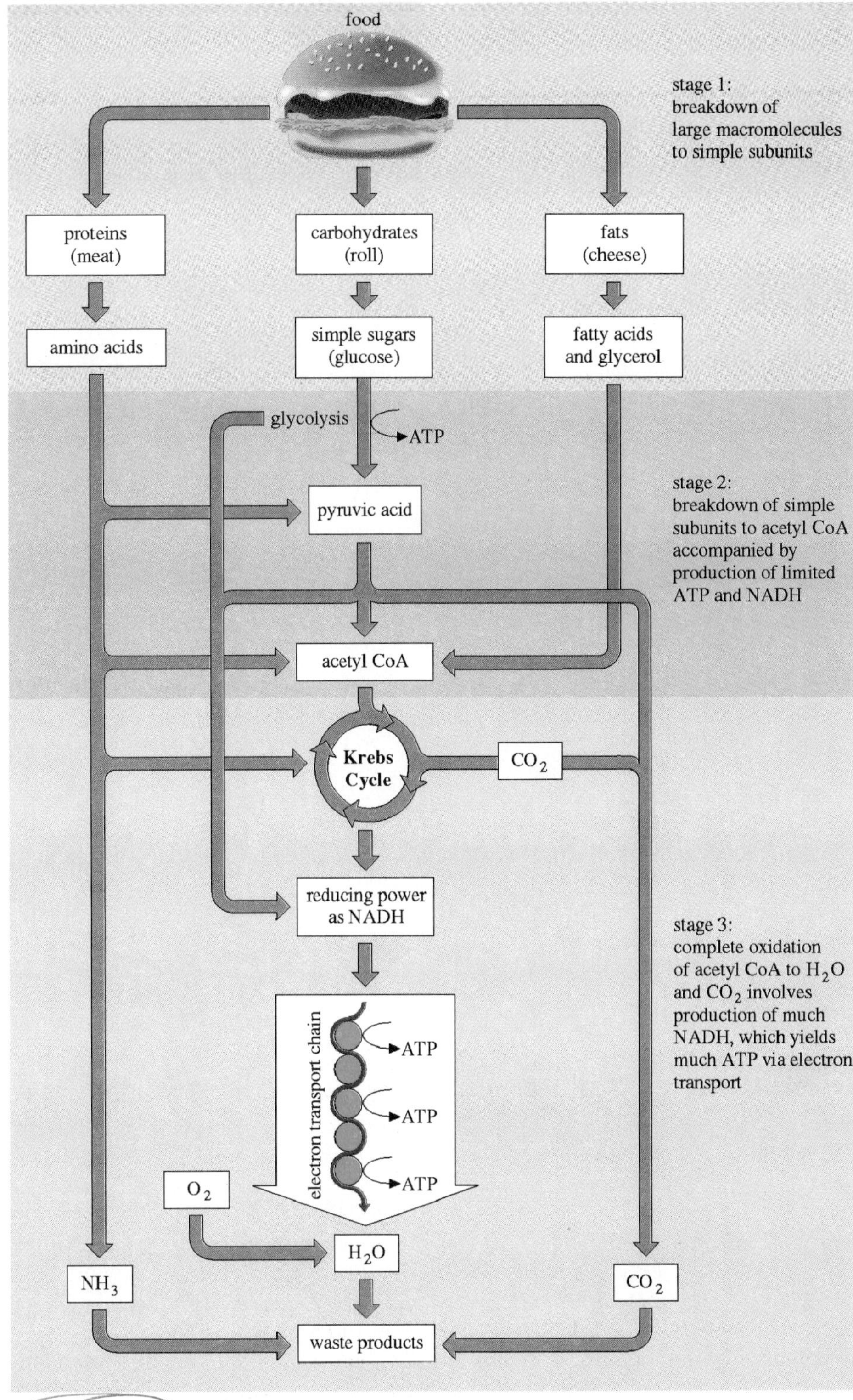

Figure 6.20 NOTE
The energy pathways extract energy from food. Complex nutrient molecules of proteins, carbohydrates (polysaccharides), and fats are broken down in the human digestive system into amino acids, simple sugars, and fatty acids and glycerol, respectively. The sugars enter glycolysis as glucose or fructose; ammonia (NH_3) is removed from amino acids in liver cells, and the remaining organic portion of the dismantled amino acids can enter the pathways of catabolism at the end of glycolysis (pyruvic acid formation), acetyl CoA formation, or the Krebs cycle. Fatty acids enter the energy pathways at acetyl CoA formation. The ATP produced in these reactions is used in other energy-requiring cellular reactions, such as the synthesis of biomolecules.

Glycolysis probably evolved when the earth's atmosphere lacked oxygen. These reactions enabled the earliest organisms to extract energy from simple organic compounds present in the nonliving environment. Photosynthesis may have evolved from glycolysis because some of the reactions of the Calvin cycle are the reverse of some of the reactions of glycolysis. In a more general sense, glycolysis takes a six-carbon compound and converts it to two three-carbon compounds; some of the dark reactions of photosynthesis do just the reverse. By some still unknown chemical rearrangement, the reactions of photosynthesis arose. When they did, the earth and life on it were changed forever.

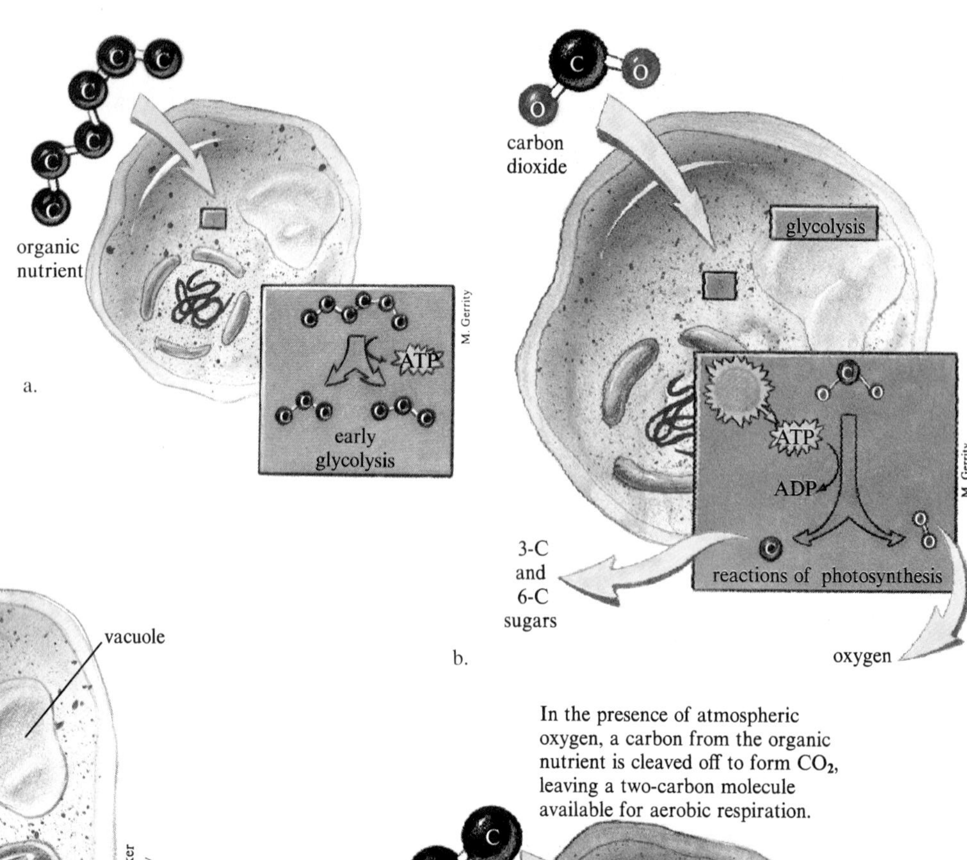

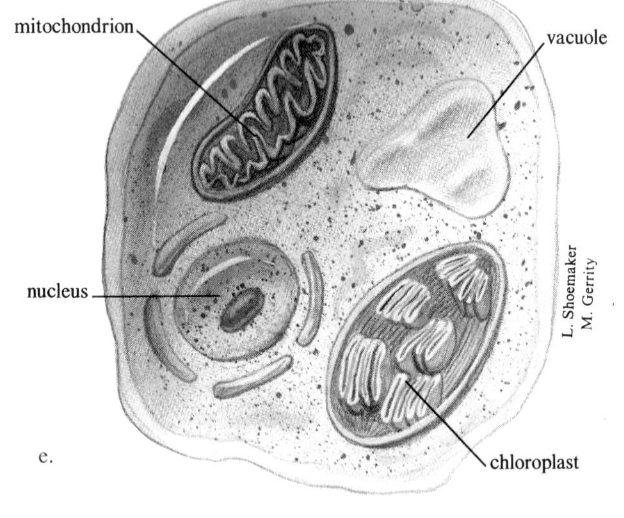

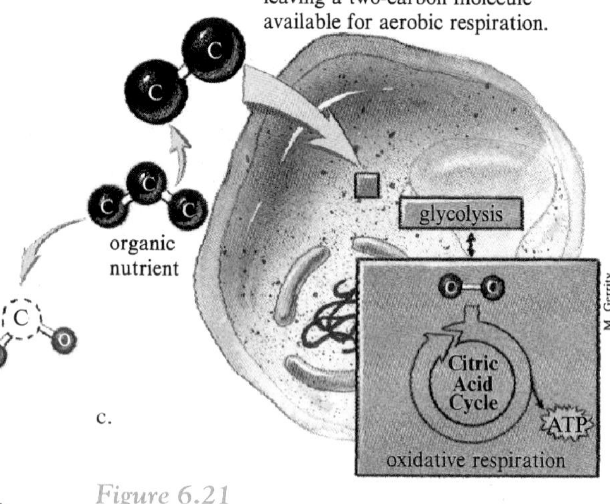

respiring (aerobic) prokaryote is engulfed

glycolysis

citric acid cycle

vacuole

nucleus

photosynthesis

photosynthesizing prokaryote is engulfed

M. Gerrity

d.

Large, Primitive Eukaryote

Figure 6.21

A scenario for the evolution of energy pathways. The very first prokaryotic organisms probably fed on organic compounds of nonbiological origin that were present in the oxygen-free ancient environment. *a.* A pathway similar to present-day glycolysis enabled these organisms to extract energy from these molecules. *b.* Somehow, perhaps from a reversal of some of the reactions of glycolysis, the ability to harness the sun's energy to produce simple carbohydrates—photosynthesis—evolved in some of these organisms. The first photosynthetic prokaryotes most likely carried out photosynthesis on internal membranes. *c.* A by-product of the process, oxygen, was to change the atmosphere and, ultimately, to change life on earth profoundly. Oxygen in the air led to a more energy-efficient extension of glycolysis. Respiration had arrived. *d.* Eventually, a large cell in the primordial ooze swallowed a photosynthesizing prokaryote and a respiring one. *e.* The result, according to the endosymbiont theory (chapter 4) was the first photosynthesizing eukaryote, which may have been the ancestor of modern plants. The photosynthesizing prokaryote (*b*) gave rise to the chloroplast; the respiring one (*c*) gave rise to the mitochondrion.

The evolution of photosynthesis, over time, pumped oxygen into the primitive atmosphere. No longer were organic compounds present in the environment required for nutrition; a mechanism now existed to produce organic compounds constantly. The appearance of the photosynthetic pathways paved the way for an explosion of new life forms capable of utilizing this new atmospheric component, oxygen. An added bonus to life was that the splitting of water in the atmosphere by electrical storms released single oxygen atoms that joined with diatomic oxygen (O_2) to produce ozone (O_3). Ozone tends to accumulate high in the atmosphere, where it blocks harmful ultraviolet radiation from reaching the planet's surface, where it could damage genetic material.

Photosynthesis could not have made its initial appearance in a plant cell, however, because such advanced organisms were not present on the early earth. An early photosynthetic organism was a prokaryotic cell, probably an anaerobic bacterium that used hydrogen sulfide (H_2S) instead of the water (H_2O) used by plants. These first photosynthetic bacteria would have released sulfur, rather than oxygen, into their environment. Eventually, pigments evolved that allowed some bacteria to use water instead of hydrogen sulfide. If a large "protoeukaryote" engulfed an ancient bacterium of this sort (recall the endosymbiont theory discussed in chapter 4), the result would have been a eukaryotic cell that may have been the ancestor of modern plants. Respiration might have evolved in a similar way, with bacteria capable of utilizing oxygen being "swallowed" by protoeukaryotes.

The present interrelationships of the reactions of photosynthesis and cellular respiration demonstrate a unifying theme of biology: All types of organisms are related at the biochemical level. Table 6.3 summarizes the energy pathway, and figure 6.22 shows how the pathways are interrelated.

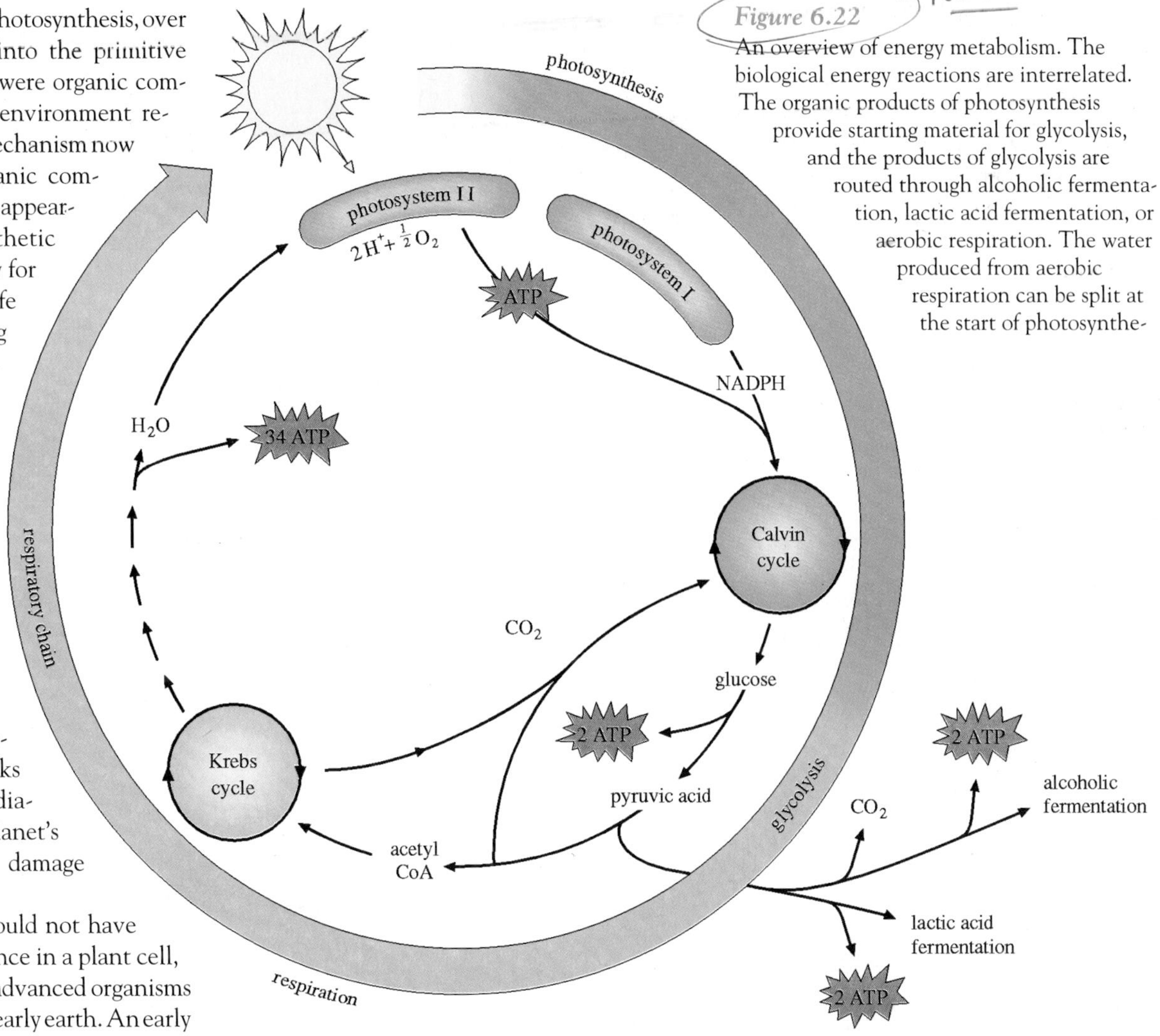

Figure 6.22
An overview of energy metabolism. The biological energy reactions are interrelated. The organic products of photosynthesis provide starting material for glycolysis, and the products of glycolysis are routed through alcoholic fermentation, lactic acid fermentation, or aerobic respiration. The water produced from aerobic respiration can be split at the start of photosynthe-

Table 6.3
Summary of Energy Metabolism

Form	*Pathway*	*Location*	*Reactants*	*Products*
Photosynthesis	Light	Thylakoids	$6CO_2 + 6H_2O$	Oxygen gas, ATP, NADPH
	Dark	Stroma	ATP, NADPH	Glucose, H_2O
Glycolysis		Cytoplasm	Glucose	2 pyruvic acid, 6 ATP if oxygen is present
Fermentation	Alcoholic	Cytoplasm	Pyruvic acid	Ethanol, CO_2
	Lactic acid	Cytoplasm	Pyruvic acid	Lactic acid
Aerobic respiration	Acetyl CoA formation	Mitochondria	Pyruvic acid	NADH, CO_2, acetyl CoA
	Krebs cycle (2 turns)	Mitochondria	Acetyl CoA	2 ATP, 6 NADH, 4 CO_2, 2 $FADH_2$
	Respiratory chain	Inner mitochondrial membrane	10 NADH, 2 $FADH_2$	34 ATP

KEY CONCEPTS

Glycolysis is probably the most ancient metabolic pathway because it is present in all cells. It may have evolved when oxygen was scarce, allowing early organisms to extract energy from bonds of organic compounds in the environment. The reactions of photosynthesis may have evolved as a reverse of part of glycolysis. Through photosynthesis, anaerobic bacteria added oxygen to the atmosphere and provided a source of organic molecules. Cellular respiration might have evolved in primitive aerobic bacteria. These bacteria, "swallowed" by protoeukaryotes, could have been the forerunners of plants and animals.

SUMMARY

The living cell requires energy to power its thousands of chemical reactions. The ultimate energy source for life is the sun, whose energy is harnessed by *photosynthetic* organisms. Sulfur-based compounds supply energy to *chemoautotrophic* bacteria in deep sea thermal vents. All other organisms obtain nourishment from photosynthetic or chemosynthetic products. The energy of life is stored in the chemical bonds of such molecules as *adenosine triphosphate* (ATP), which functions as a short-term energy reserve. The chemical reactions that utilize and release energy in an organism are termed *metabolism*. *Catabolic* reactions degrade molecules, and *anabolic* reactions synthesize molecules. Metabolic reactions are enzyme-catalyzed and are organized into chains and cycles. *Whole-body metabolism* refers to energy expenditure on an organismal level. Energy enters nonphotosynthetic organisms as food, which is metabolized to keep the organism alive (*basal metabolism*) and to power activities beyond this requirement. The metabolic rate of a human depends upon age, sex, activity level, weight, body composition, and thyroid function.

In photosynthesis, six molecules each of carbon dioxide and water are converted with the aid of light and pigment molecules into a molecule of glucose and six molecules of oxygen. Photosynthesis in eukaryotes begins in the *thylakoid membranes* of the *chloroplast*. Many molecules of the pigment *chlorophyll a* are associated to form two *photosystems*, which are linked by a series of molecules that pass electrons from one photosystem to the other.

The *light reactions* of photosynthesis must occur in the light. As the energy level of an electron from a particularly reactive chlorophyll *a* molecule is boosted in photosystem II, water is split to yield oxygen. The electron hits an electron-accepting molecule, and energy released by the electron as it is passed down an *electron transport chain* is used to convert ADP to ATP in a process called *photophosphorylation*. This electron then replaces one boosted from a reactive chlorophyll *a* molecule of photosystem I. After meeting other electron-accepting molecules, the electrons reduce *NADP*$^+$ to *NADPH*. The products of the light reactions—ATP and NADPH—power the *dark reactions* of photosynthesis. In these reactions, carbon dioxide enters the *Calvin cycle*. Additional reactions lead to the synthesis of glucose and other organic compounds of use to the cell.

Other metabolic pathways extract energy from the glucose produced in photosynthesis. *Glycolysis*, which occurs in the cytoplasm of nearly all cells, splits and rearranges glucose to yield 2 molecules of *pyruvic acid*. In the absence of oxygen, pyruvic acid can either lose carbon dioxide and convert to ethanol (*alcoholic fermentation*) or react in a single step to produce lactic acid (*lactic acid fermentation*). In the presence of oxygen, the process of *aerobic respiration* takes place: Pyruvic acid enters the mitochondrion, loses carbon dioxide, converts NAD$^+$ to NADH, and bonds to *coenzyme A* to produce *acetyl CoA*. This molecule enters the *Krebs cycle*, which yields 1 molecule of ATP, 3 of *NADH*, and 1 of *FADH*$_2$. Finally, an electron-transport chain (the *respiratory chain*) converts the NADH and FADH$_2$ to ATP. Overall, 1 molecule of glucose generates 36 molecules of ATP.

QUESTIONS

1. List three similarities and three differences between photosynthesis and cellular respiration.

2. A type of inherited disease called an "inborn error of metabolism" is characterized by a buildup of a certain substance, often to such a high level that it becomes toxic to the body. Suggest a metabolic basis for such an accumulation. In some inborn errors, several substances are present in greater-than-normal amounts. How might this occur?

3. When a cell contains more than enough ATP to carry out its functions, the excess ATP signals a slowing down of the respiratory chain by decreasing production of acetyl CoA. When ATP levels fall below what is required for cellular functions, the increase in ADP and AMP signals increased breakdown of glycogen to glucose and speeds up the respiratory chain. What general metabolic phenomenon is demonstrated by these relationships?

4. How are humans biochemically dependent upon plants for life?

5. Some photosynthetic bacteria utilize hydrogen sulfide (H_2S) instead of water in photosynthesis. They produce sulfur instead of oxygen as a photosynthetic product. How do these bacteria demonstrate that the source of oxygen in photosynthesis is water rather than carbon dioxide?

6. The growth of vegetables and flowers in greenhouses in the winter greatly increases when the level of carbon dioxide is raised to two or three times the level in the natural environment. What is the biological basis for the increased rate of growth?

7. Is photosynthesis a catabolic or anabolic process? Is glycolysis catabolic or anabolic? Is cellular respiration a reversal of photosynthesis? Why or why not?

8. Where does each of the 36 ATP molecules derived from a single glucose molecule come from in energy metabolism?

TO THINK ABOUT

1. Fructose intolerance is an inherited disorder in which a missing enzyme makes a person unable to utilize fructose, a simple sugar found in fruit. Infants with the condition display abnormally low mental and motor functioning. In later childhood, pronounced lethargy and mild mental retardation are evident. An adult with fructose intolerance has a deteriorating nervous system, which eventually causes mental illness and death. Molecules that are derived from fructose are intermediates in the first few reactions of glycolysis, and the enzyme missing in people with fructose intolerance would normally catalyze these reactions. Considering this information about the whole-body and biochemical effects of fructose intolerance, suggest what might be happening on a cellular level.

2. When heavy smokers stop smoking, they often put on weight. How might this weight gain be explained in terms of metabolism?

3. When a sea slug eats green algae, chloroplasts enter some of its cells, where they continue to photosynthesize. Similarly, experiments have shown that chloroplasts injected into mouse cells in laboratory cultures can continue to photosynthesize, even though they are in cells from a nonphotosynthesizing organism. Based on this information, some scientists foresee a time when actively photosynthesizing chloroplasts could be transplanted into the cells of human skin. What do you think would be the advantages and disadvantages of carrying chloroplasts in your skin?

4. A student is accustomed to running 3 miles each afternoon, at a slow, leisurely pace. One day, she runs a mile for time, to see how fast she can do it. Afterwards, she is winded, with pains in her chest and leg muscles. She is upset at her aches and pains, because she thought that she was in good physical condition. What has she experienced, in terms of energy metabolism?

5. A cancer cell buried deep within a tumor often resists radiation and drug treatment simply because it is hard to reach. Such hidden cancer cells can also evade the body's own defense because they are surrounded by a locally acidic environment, which drives off the immune system's cancer-fighting white blood cells and antibody molecules. The cancer cells become surrounded by acid because their energy metabolism adapts to the fact that they are buried deep within the tumor, far from oxygen. What metabolic pathway discussed in this chapter is probably utilized by these cancer cells?

6. Diverse organisms, such as humans and yeast, have certain of the same enzymes and even share entire metabolic pathways. What are the implications of these similarities in terms of evolution?

SUGGESTED READINGS

Hendry, George. May 1990. Making, breaking, and remaking chlorophyll. *Natural History*. Photosynthesis is vital for nearly all life on earth.

Levi, Primo. October 1984. Travels with C. *The Sciences*. An eloquent travelogue of how a single atom of carbon circulates through the living world.

Lewis, Ricki. February 1984. Blood sugar ups and downs. *Biology Digest*. Hormones can control metabolism.

Lewis, Ricki. October 1989. Mitochondria–eclectic organelles. *Biology Digest*. Energy reactions occur in the mitochondria.

McCammon, Robert R. 1987. *Swan song*. New York: Pocket Books. A chilling look at what happens in the years after a nuclear holocaust—when the sky is so blackened with debris that plants cannot photosynthesize.

Moskowitz, Harold. May 1982. Biochemistry is beautiful. *Science 82*. The chart of all of the metabolic pathways is overwhelming but fascinating.

Patrusky, Ben. May–June 1982. Photosynthesis: From the sparkplug. . . . *Mosaic*. A good review of the basics of photosynthesis.

Roberts, Leslie. August 28, 1987. Discovering microbes with a taste for PCBs. *Science*. Some microorganisms derive energy from toxic wastes.

Whitney, Eleanor Noss, and Eva May Hamilton. 1987. *Understanding nutrition*, 5th ed. St. Paul: West Publishing. Chapter 7 presents an excellent summary of the pathways and cycles of metabolism.

C H A P T E R

7

Mitosis

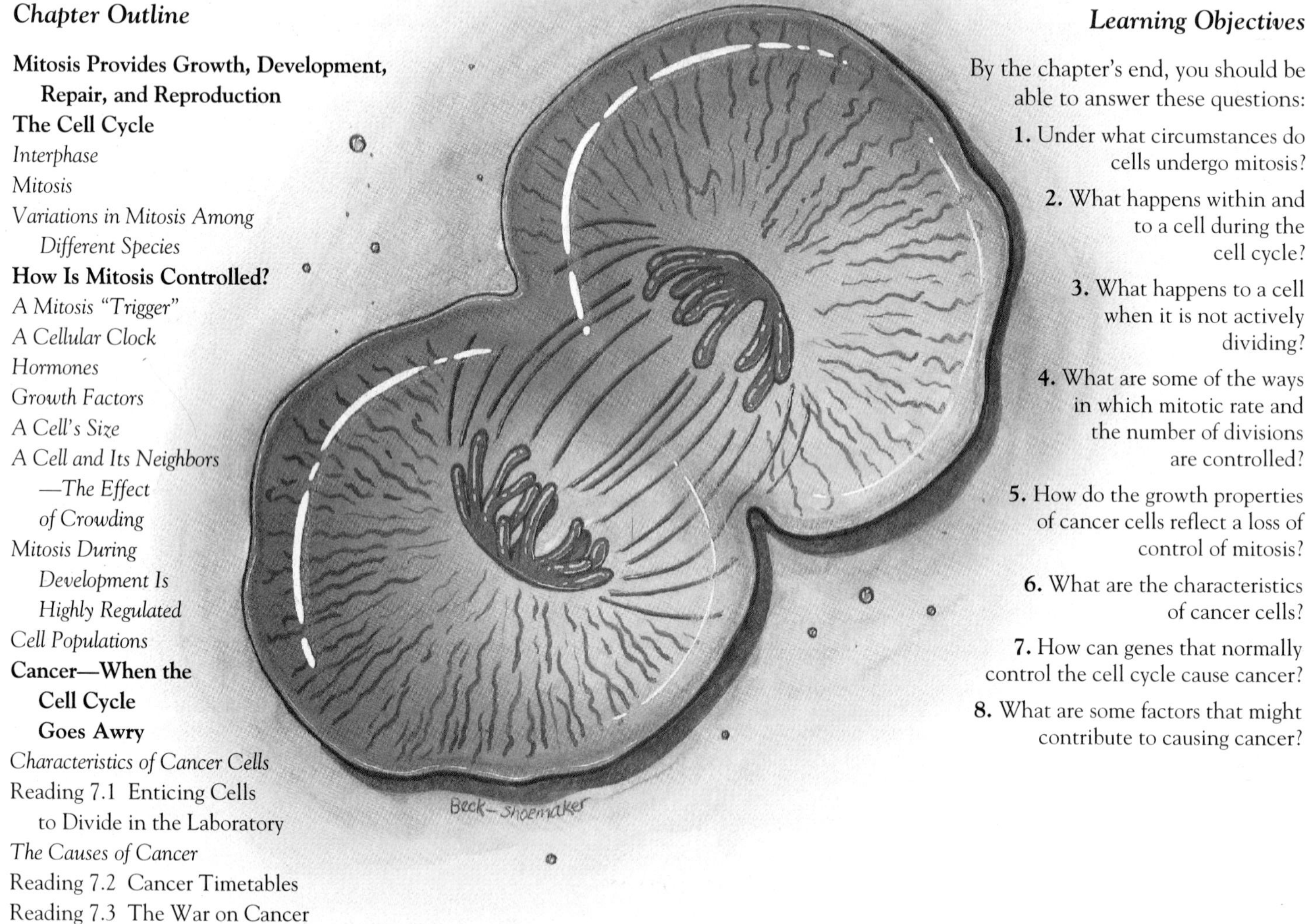

Chapter Outline

Learning Objectives

By the chapter's end, you should be able to answer these questions:

1. Under what circumstances do cells undergo mitosis?
2. What happens within and to a cell during the cell cycle?
3. What happens to a cell when it is not actively dividing?
4. What are some of the ways in which mitotic rate and the number of divisions are controlled?
5. How do the growth properties of cancer cells reflect a loss of control of mitosis?
6. What are the characteristics of cancer cells?
7. How can genes that normally control the cell cycle cause cancer?
8. What are some factors that might contribute to causing cancer?

It is late spring, a few days after the first seeds were planted in the vegetable garden. Already tiny shoots peek through the soil, where yesterday there was no sign of growth at all. New cells are being added to these pole bean plants so rapidly that they seem to grow inches in a single day. Much activity is taking place below the surface too, as roots extend downwards and outwards between the fine particles of soil.

Ninety minutes after a sperm and an egg cell fuse to begin the development of a new sea urchin, the first cell division occurs. What was one huge cell is now two smaller ones. Within 5 hours of the fusion of sperm and egg, four such divisions have taken place, and the young organism consists of 16 cells. By 12 hours, the embryo consists of more than 200 cells. By 18 hours, it is a hollow ball of more than 400 cells, fringed with cilia that form a tuft at one end. By 70 hours, the animal is a free-swimming larva of more than 1,500 cells. The young sea urchin continues to grow by increasing its cell number, and within 6 weeks of the meeting of sperm and egg, it consists of more than 50,000 cells.

A major artery in a man's heart is so blocked with fatty deposits that blood can no longer pass through it, robbing a portion of the heart muscle of its vital oxygen supply. The man suffers a heart attack as cells in this part of his heart muscle die. Somehow, the death of these muscle cells triggers the proliferation of cells of another type, fibroblasts, which secrete a network of protein fibers. The protein patch fills in the devastated area, and a scar is born.

The bacterium *Escherichia coli* divides every 20 minutes. This means that, three times every hour, a population of these tiny organisms can double. If a single *E. coli* cell had unlimited nutrients and space, and divided every 20 minutes, it would produce a colony of approximately 5,000,000,000,000,000,000,000 (5 trillion billion) cells in just a day. After a day and a half, that first cell would have produced enough offspring to cover the entire planet in a 1-foot-thick coating.

Mitosis Provides Growth, Development, Repair, and Reproduction

Cells of the rapidly growing bean seedling, the developing sea urchin embryo, and the self-healing human heart tissue, are all undergoing **mitosis,** a form of cell division in which two identical cells are generated from one (fig. 7.1). The bacteria undergo **binary fission,** a form of asexual reproduction that is very similar to the mitosis seen in higher organisms. In mitosis and binary fission, the original cell first doubles its genetic material and then distributes it equally among the two "daughter" cells and other cellular constituents, such as organelles and macromolecules, that have been replenished throughout the original cell's existence.

Mitosis occurs in **somatic** (body) **cells,** which include cells other than sperm and egg cells. A second form of cell division, called **meiosis,** halves the amount of genetic material to fashion the sperm and egg, which are sex cells. The halving of genetic material in meiosis ensures that the required amount is passed on to succeeding generations. That is, without meiosis, the number of chromosomes would double with each generation. Meiosis is discussed in chapter 8.

Mitosis is a highly regulated process, and it maintains the numbers and arrangements of specialized cells that build tissues and organs throughout a multicellular organism's life. The cells in the pole bean divide at a rate and in a pattern such that the roots extend down far enough to obtain nutrients and water and the shoots grow tall enough to capture sunlight, but not so fast that they become too thin and collapse. Mitosis in the bean plant adds cells to foster growth and development. In the sea urchin, the pace and extent of mitosis oversees how the embryo grows and develops. In the man's heart, the fibroblasts divide sufficiently to replace damaged tissue, but not so extensively that they form an abnormal growth. Mitosis in this instance provides repair. For the bacterium, cell division is the mechanism of reproduction. Mitosis, then, is a form of asexual cell division that provides growth, development, repair, and reproduction.

Mitosis is important throughout the human life span. The fast growth of a developing human embryo, for example, depends upon rapid division of somatic cells. By the time of birth, however, the pace of mitosis slows dramatically. Why might this tight control be necessary? In an adult, cells divide according to a precise plan that replaces the many cells lost daily through both normal wear and tear as well as those lost through injury.

Regulation of mitosis in the human body is quite a task. The average adult is built of about 10 trillion cells, and nearly 1/10 of these are replaced every day, adding up to some 10 quadrillion mitoses in a lifetime. These divisions do not occur at random. If there is too little mitosis, an injury may go unrepaired. If there is too much, an abnormal growth may develop.

This chapter will first examine the process of normal mitosis and then look at a consequence of mitosis that has gone out of control—cancer.

The Cell Cycle

The life of a cell, in terms of whether it is dividing or not, is described by a series of events called the **cell cycle.** The cell cycle is divided into two major stages: mitosis, when the cell is actively dividing, and **interphase,** when the cell is not actively dividing. Mitosis entails both a division of the genetic material, called **karyokinesis,** and distribution of cytoplasm, macromolecules, and organelles into two daughter cells, called **cytokinesis.**

Interphase

Interphase was mistakenly described by biologists 30 years ago as a time of cellular rest, because the rod-shaped chromosomes that contain the genes are not visibly dividing when viewed under a microscope at this time. The DNA that comprises the chromosomes is so unwound during interphase that many stains cannot bind to it, and as a result chromosomes cannot be generally visualized

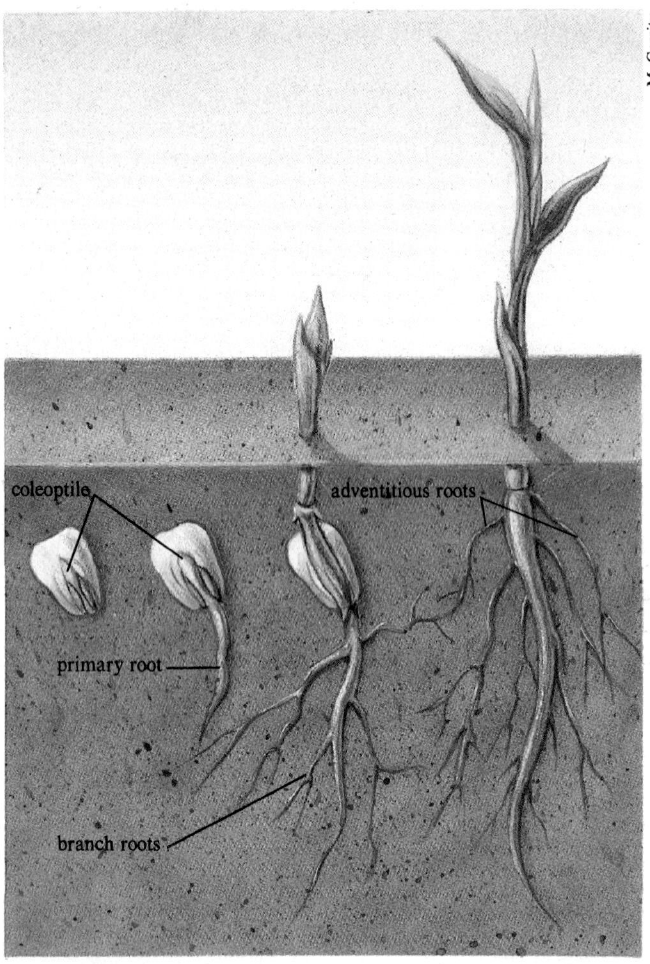

a.

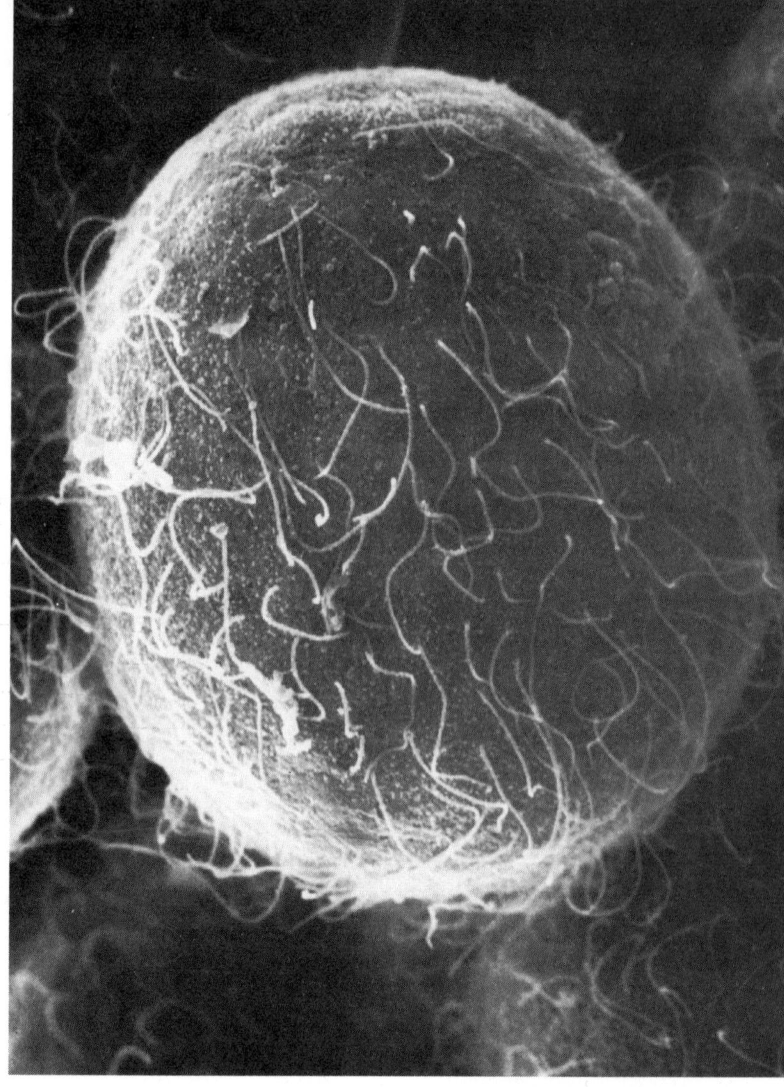

b.

c.

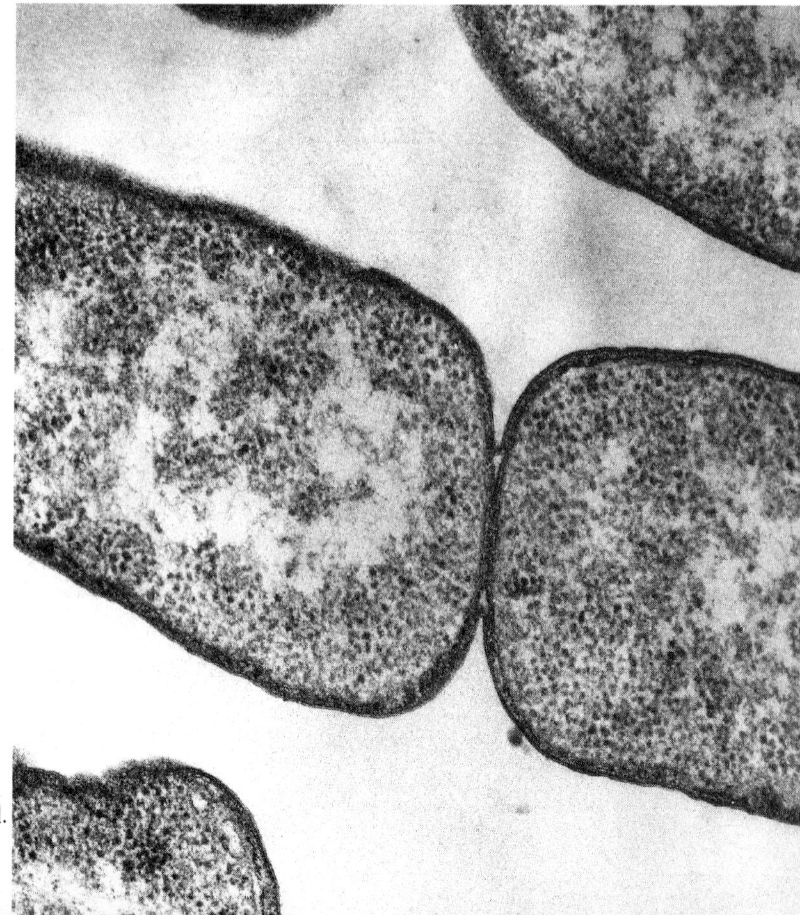

d.

Figure 7.1
Mitotic cell division is part of growth, development, repair, and reproduction. *a.* The rapid growth of a corn seedling and (*b*) in a sea urchin embryo illustrates the importance of mitosis in growth and development of a young organism. Magnification, x4,000. *c.* A child's scraped knee is well on the way to recovery. Beneath the scab, fibroblasts secrete collagen, endothelial cells divide to build new blood vessels, and epidermal cells divide, forming new skin. *d. E. coli* cells to divide every 20 minutes. Magnification, x132,000.

(special, recently-developed vitamin-based stains allow some interphase chromosomes to be seen). However, interphase is actually a very active time. The cell not only continues the basic biochemical functions of life but also replicates its genetic material and other subcellular structures in preparation for splitting into two daughter cells.

Interphase is divided into phases of *gap* (designated "G") and *synthesis* ("S") (fig. 7.2). (Some texts refer to the "G" phases as "growth.") During the first gap phase, the **G_1 phase,** proteins, lipids, and carbohydrates are synthesized. These molecules will be utilized to surround the two cells that form from one. G_1 is the period of the cell cycle that varies most in duration among different types of specialized cells. Slow-growing cells, such as those of the human liver, may remain in this first gap phase for years. In contrast, the fast-growing cells of human bone marrow speed through G_1 in 16 to 24 hours, and early embryonic cells may skip G_1 entirely. Interestingly, the total time spent in the other stages of the cell cycle is nearly the same in all cells of multicellular organisms.

The next period of interphase, the S phase, is a time of great synthetic activity as the immense job of replicating the genetic material is undertaken. In most human cells, this takes from 8 to 10 hours, and it entails the replication of billions of DNA building blocks. (DNA replication is discussed further in chapter 13.) Many proteins are manufactured during the S phase, including those that are part of the chromosomes and proteins that coordinate the many events taking place in the nucleus and cytoplasm.

Also during the S phase, microtubules are synthesized, which will be assembled into a **spindle apparatus** early in the mitotic process. The spindle apparatus will pull the replicated chromosomes apart into two separate sets of chromosomes, each containing a complete set of genetic instructions.

In the second gap phase, the **G_2 phase,** more proteins are synthesized. Membranes are assembled and stored as small empty vesicles beneath the cell membrane, and they will be used to provide enough membrane material to enclose two cells rather than one. The end of G_2 is signaled by the DNA winding more tightly around its associated proteins. The genetic material is now so condensed that it is visible under a microscope when certain stains are applied to it. The start of chromosome condensation at the end of G_2 signals impending mitosis. Interphase has ended.

KEY CONCEPTS

Mitotic cell division is responsible for binary fission in single-celled organisms and for growth, development, and repair in somatic cells of multicellular organisms. The cell cycle describes a cell's state of division (mitosis) or nondivision (interphase). Division of genetic material is called karyokinesis, and division of organelles and macromolecules into daughter cells is called cytokinesis. During interphase, biochemicals are synthesized. Proteins, lipids, and carbohydrates are made during G_1; DNA and other proteins are manufactured during the S phase; and still more proteins are produced during G_2.

Mitosis

During mitosis, or M phase, the replicated genetic material divides. Mitosis is a continuous process, but for ease of study, it is considered as several stages.

A cell's chromosomes are replicated during the **S phase.** A replicated chromosome consists of two strands of identical chromosomal material, called **chromatids,** which are held together by a constriction called a **centromere** (fig. 7.3). At a certain point during mitosis, the centromeres split

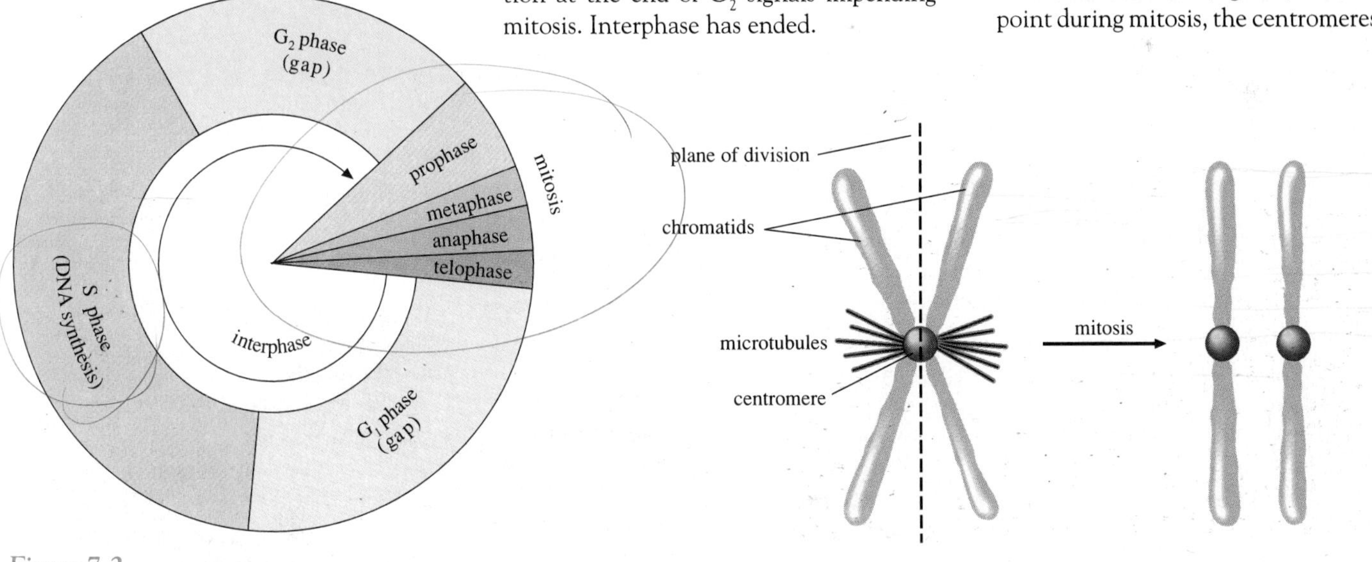

Figure 7.2
The cell cycle is divided into interphase, when cellular components are replicated, and mitosis, when the cell splits in two, distributing its contents into two daughter cells. Interphase is divided into two gap phases (G_1 and G_2), when specific molecules and structures are duplicated, and a synthesis phase (S), when the genetic material is replicated. During mitosis the genetic material divides (karyokinesis) and then the cytoplasm, organelles, and molecules are partitioned into two cells (cytokinesis).

Figure 7.3
Replicated and unreplicated chromosomes. Chromosomes are replicated during the S phase, before mitosis begins. The two genetically identical chromatids of a replicated chromosome are attached at the centromere. In anaphase, the centromeres split, and each chromosome consists of only one chromatid.

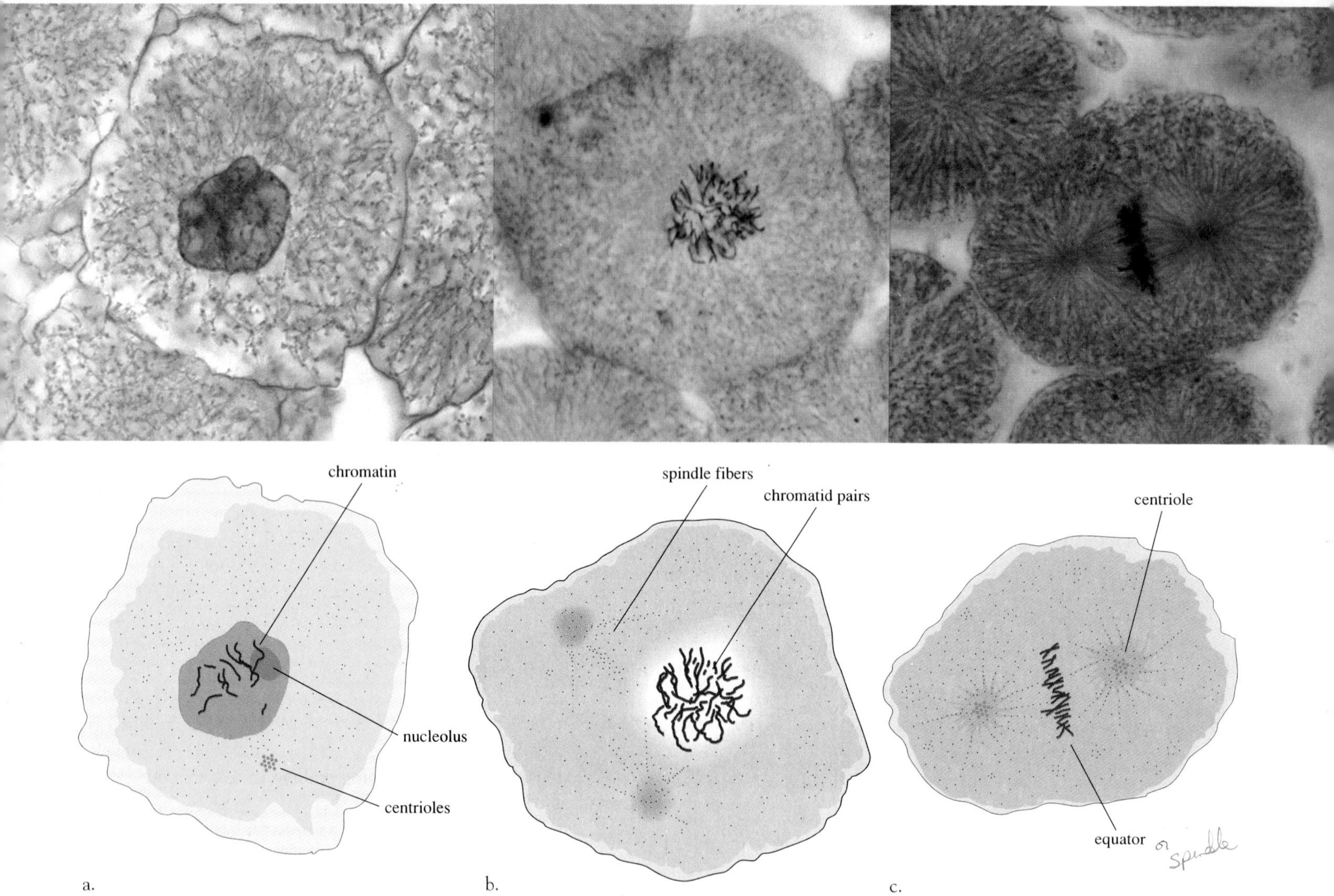

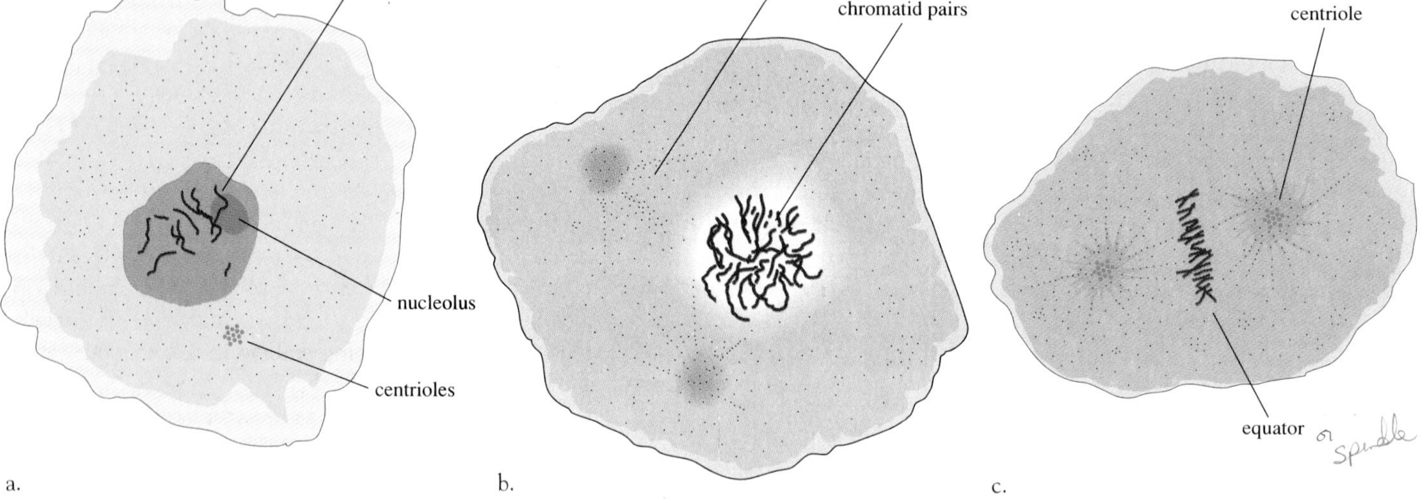

Figure 7.4
Mitosis in a human cell. *a.* Chromosomes are not yet condensed, and therefore not usually visible, in a cell in interphase. The large, dark-staining structure is the nucleus. *b.* In prophase the chromosomes are condensed and easily visible when stained, the spindle apparatus is assembled from microtubules, centrioles appear at opposite poles of the cell, and the nuclear membrane disappears. *c.* In metaphase, the chromosomes align along the plane of division of the spindle, which is also called the equator.

and each chromosome then consists of only one chromatid. Thus, chromosomes may have one or two chromatids, depending upon the stage of mitosis they are in. To follow the number of chromosomes in a cell, count the number of centromeres. The number of chromosomes remains the same in a cell before it has undergone mitosis and in its daughter cells after mitosis. In contrast, meiosis halves the chromosome number in cells to form the sperm and egg.

Prophase, the first stage of mitosis, is defined as the time when chromosomes are considered and, when stained, become visible under a light microscope. The DNA is coiled very tightly around the chromosomal proteins, which shorten and thicken the chromosomes (fig. 7.4). Chromosomes can separate into two groups more easily when they are condensed in this manner, rather than when they are long, tangled strands. It is like separating several tightly-wound packages of yarn from each other compared to untangling long strands of loose yarn.

The mitotic spindle forms during prophase from microtubules. Toward the end of prophase, the membrane surrounding the nucleus breaks down, and a dark spot that has been visible in the nucleus called the **nucleolus** can no longer be seen. In the nucleolus a type of the nucleic acid RNA is manufactured, which participates in protein synthesis.

Metaphase follows prophase. At this stage the spindle microtubules attach to the free-floating chromosomes at their centromeres and align them in the center of the cell. Because of this arrangement, when the centromeres eventually split, each daughter cell will receive a complete set of genetic information (i.e., one chromatid from each chromatid pair). Metaphase chromosomes are under a great deal of tension. They appear motionless because they are pulled with equal force from both sides, a little like the rope in a game of tug-of-war.

In the next stage of mitosis, **anaphase,** the centromeres split, relieving the tension and sending one chromatid from each pair

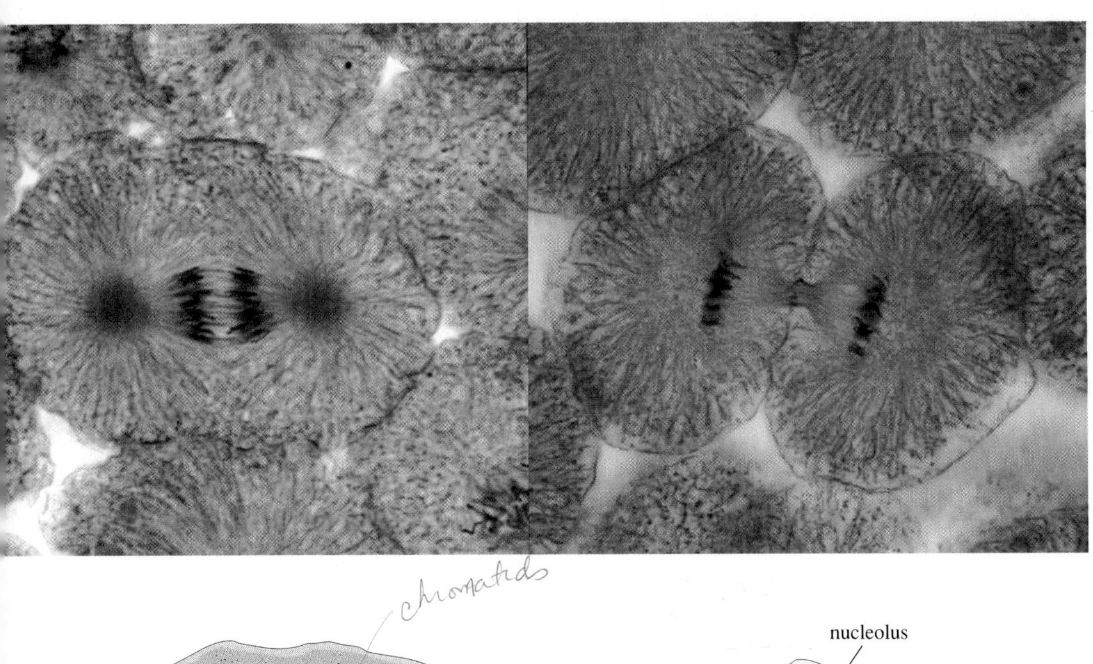

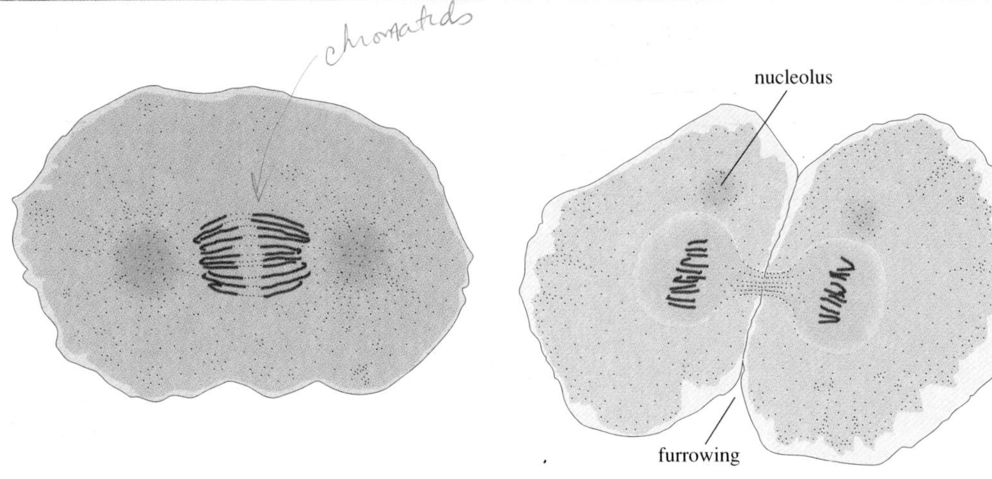

d. e.

d. In anaphase, the centromeres split and the chromatids separate. *e.* In telophase, the spindle apparatus disassembles and the nuclear membrane reforms. In cytokinesis, which begins during telophase, the cytoplasm and other cellular structures are pinched off into two daughter cells.

to opposite ends of the cell, called poles. It is a little like the tug-of-war rope breaking in the middle and the participants falling into two groups. As the chromatids separate, some microtubules in the spindle shorten and some lengthen in a way that actually moves the poles farther apart, stretching the dividing cell. During the very brief time of anaphase, a cell contains twice the normal number of chromosomes because each chromatid is now an independently moving chromosome.

Finally, in **telophase,** an animal cell looks like a dumbbell with a set of chromosomes at each end. The spindle is disassembled, and nucleoli and nuclear membranes reform at each end of the stretched-out cell. Division of the genetic material, karyokinesis, is now complete. During or just after karyokinesis, the other cellular contents, including organelles and macromolecules, are distributed among the two forming daughter cells. This is cytokinesis. Finally, the daughter cells physically separate. In most cell types, the daughter cells are about equal in size.

Animal and plant cells have evolved different strategies for cytokinesis. During anaphase in an animal cell, a slight indentation forms around the cell, aligned with the points at which the metaphase chromosomes line up. This indentation is the first sign of the formation of a band of microfilaments located beneath the cell membrane. The microfilament band then contracts like a drawstring to separate the newly formed daughter cells.

Plant cells complete an additional step when they divide—building a new cell wall that separates the two daughter cells. To do this, microscopic fibers made of cellulose are laid down along with other polysaccharides and proteins on the cell membranes of the daughter cells. This arrangement of cellulose fibers embedded in surrounding material is very strong and rigid, resembling reinforced concrete. Figure 7.5 illustrates the square shape of plant cells due to the rigid cell walls, as well as what cells in different stages of mitosis look like. The longer in duration a particular stage, the more cells can be seen demonstrating it.

In some instances mitosis occurs without cytokinesis, resulting in large cells that have many nuclei each. For example, plant endosperm tissue, which nourishes the developing embryo in a seed, consists of cells that have many nuclei sharing a common cytoplasm. Some algae and fungi undergo mitosis without immediate cytokinesis, producing tremendous multinucleated cells that can extend a meter or more in length.

Variations in Mitosis Among Different Species

Mitosis is quite similar among all species in that cellular parts are duplicated and then divided. It is interesting, though, that the process is more complex in more complex organisms. For example, bacteria have just one chromosome. Unlike the chromosomes of eukaryotes, a lone bacterial chromosome is not wrapped in protein, and it is not enclosed in a nucleus. The replicated DNA molecules of bacteria are attached directly to the cell membrane. The cell membrane and the cell wall elongate between the attached DNA molecules so that the two copies of the genetic material are pulled apart. Bacterial cell division is technically

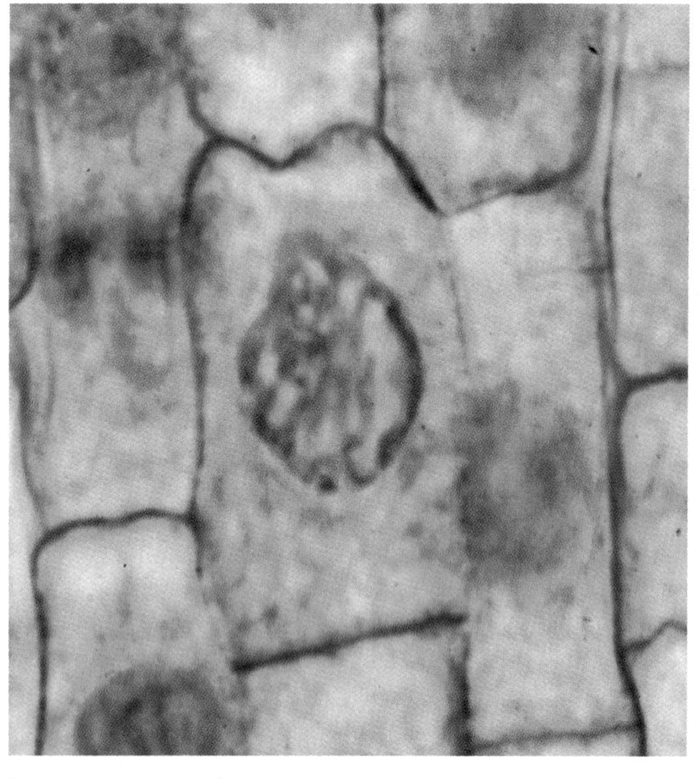

a.

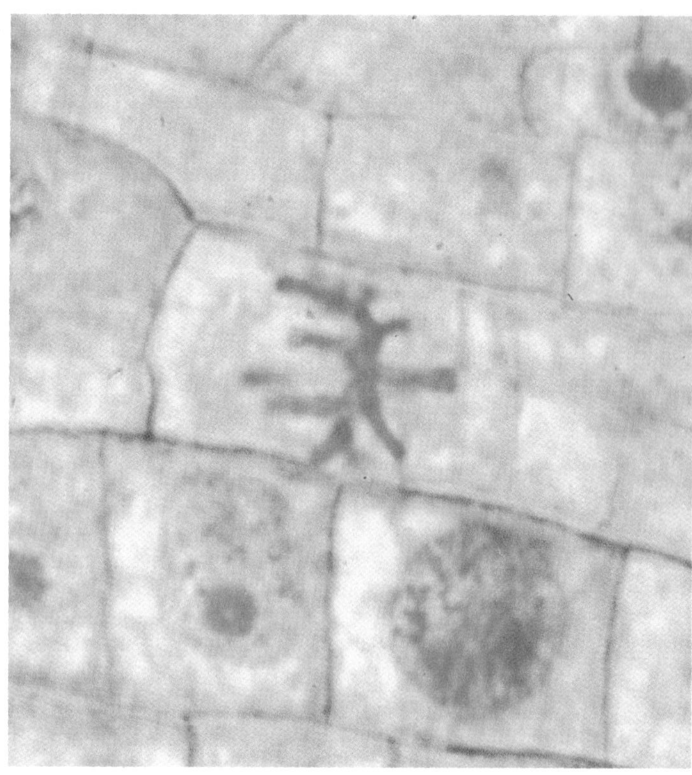

b.

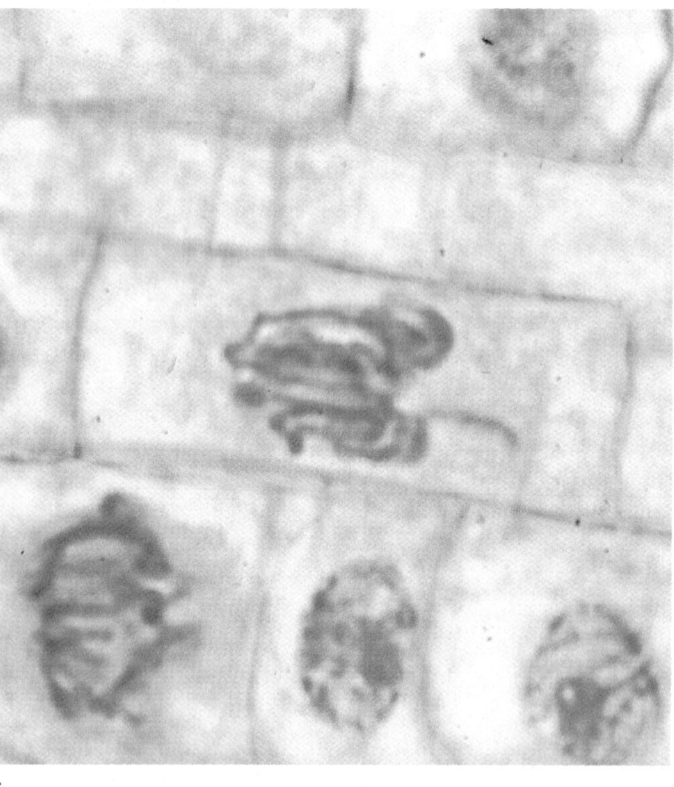

c.

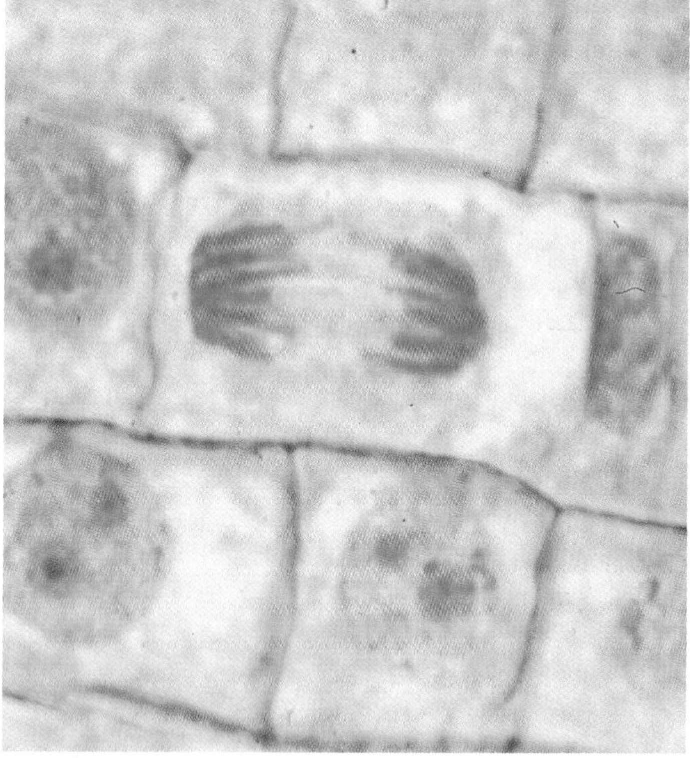

d.

Figure 7.5
Mitosis in a plant cell. Plant cells in telophase do not develop indentations as animal cells do, but they have cell plates, which are soft boundaries between daughter cells. When new cell membranes and cell walls form between the daughter cells, mitosis is complete. Note the square shape that the cell wall imparts to plant cells. Can you identify the different stages of mitosis in these dividing cells in an onion root tip? Magnification, x 1,000.

not considered to be mitosis because bacteria lack nuclei, chromosomal proteins, and microtubules.

Mitosis in some dinoflagellates, which are single-celled eukaryotes, is more sophisticated than that of bacteria, yet not as complex as in our own cells. Dinoflagellate chromosomes are free of associated proteins (like those of bacteria) but are enclosed in a nucleus. Microtubules press on the nuclear membrane to guide the movements of chromosomes from outside the nucleus. In higher organisms, the nuclear membrane actually breaks down, permitting direct interaction of chromosomes and the microtubules that pull them into two daughter cells. The differences in these details of mitosis in modern-day species may provide clues to how the process of mitosis evolved as life forms increased in complexity from single-celled prokaryotes, to single-celled eukaryotes, to multicellular eukaryotes such as ourselves. Can you envision how this might have happened over many millions of years?

KEY CONCEPTS

Chromosomes about to divide are replicated, each consisting of two chromatids attached to one centromere. In prophase the replicated chromosomes become visible as they condense. The spindle forms and the nuclear membrane breaks down. During metaphase, the chromosomes align down the center of the cell, held in place by the spindle. In anaphase, centromeres split, and one set of chromosomes, now in unreplicated form, is pulled to each end of the cell. In telophase, the cell pinches down the middle, as karyokinesis completes. As the cells separate, cytokinesis is completed. In animal cells a band of microfilaments assists cytokinesis. In plant cells, new cell walls form around the daughter cells. A process similar to mitosis occurs in bacteria.

How Is Mitosis Controlled?

A Mitosis "Trigger"

What determines when a somatic cell divides? Mechanisms for controlling the rate of mitosis operate at the molecular, cellular, and tissue levels. These vital mechanisms are not yet well understood, and they constitute a major focus of current biological research. On a practical note, understanding how mitosis is controlled may provide clues to how abnormalities of the cell cycle, such as cancer, arise.

One theory of the control of mitosis is that an individual cell may be induced to divide by the presence of a still-undiscovered "trigger molecule." Experimental evidence supports this idea. When two cells in different stages of the cell cycle are fused in the laboratory to form a large cell with two nuclei, the "younger" nucleus (the one in an earlier stage of the cell cycle) quickly "catches up" to the stage of the older nucleus. For example, when one nucleus is in the S phase and the other is in G_1, the G_1 nucleus begins to replicate its DNA ahead of schedule, as if it has been pushed to the S phase. Perhaps some substance in the S nucleus, which causes it to begin DNA synthesis, leaks into the shared cytoplasm and stimulates the G_1 nucleus to start DNA synthesis also. Levels of this triggering molecule are hypothesized to rise and fall during the cell cycle, perhaps being synthesized rapidly during G_1 and, by the beginning of the S phase, accumulating to the point that DNA synthesis is initiated. Once DNA synthesis begins, mitosis is inevitable. When prophase begins, the level of the triggering molecule drops, and the cycle begins anew.

A Cellular Clock

Mammalian cells grown in a laboratory seem to obey an internal "clock," which allows them to divide a maximum number of times. This rule is called the *Hayflick limit*, after its 1960s discoverer, Leonard Hayflick. A fibroblast taken from a human fetus, for example, divides from 35 to 63 times, the average being about 50 times. However, a fibroblast taken from an adult divides only 14 to 29 times, with a greater number of divisions in cells taken from younger individuals. It is as if the cells "know" how long they have existed and how much longer they have to exist.

Within an organism, however, different cell types undergo mitosis at different characteristic rates that may exceed the division limit seen in cells grown in the laboratory—or not even come close to it. A cell lining a person's small intestine might divide throughout life; a cell in the brain may never divide; a cell in the deepest skin layer of a 90-year-old could divide a dozen or more times if the person lives long enough. In fact, by the time a very elderly person dies, many cells may not even have begun to tap their proliferative potential, that is, their built-in mitotic clock.

Hormones

Certain cells divide frequently at some times, yet infrequently or not at all at others, due to the influence of biochemicals called **hormones.** A hormone is manufactured in a gland and travels in the bloodstream to another part of the body, where it exerts an effect. For example, at a particular time in the monthly hormonal cycle in the human female, peak levels of the hormone estrogen stimulate the cells lining the uterus (the womb) to divide, building tissue in which a fertilized egg can implant. If an egg is not fertilized, another hormonal shift breaks the lining down, resulting in menstruation. Hormones also control the cell proliferation necessary to rapidly convert a fatty breast into an active milk-producing gland.

Growth Factors

When an organism is wounded, different cells in the damaged area must begin or increase the frequency of mitosis in order to build new tissue. Such wound healing is mediated by proteins called **growth factors.** Unlike hormones, growth factors are not carried in the bloodstream but act more locally. Little is known about growth factors—we do not yet know how many there are, where they are made, and how they act and interact.

One of the best studied growth factors is *epidermal growth factor* (EGF), which stimulates epithelium (lining tissue) to undergo mitosis. An example of EGF action is the filling in of new skin underneath the scab of a skinned knee. EGF is also produced in the salivary glands. Can you see how this site of synthesis might explain why an animal's licking a wound aids healing? EGF also helps mend ulcers of the digestive system when it is swallowed.

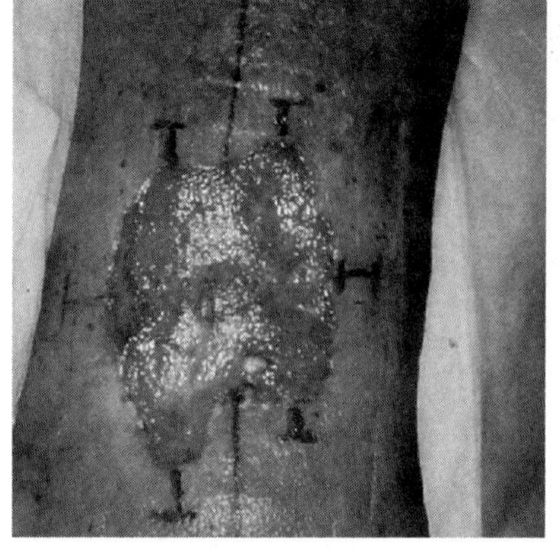

a.

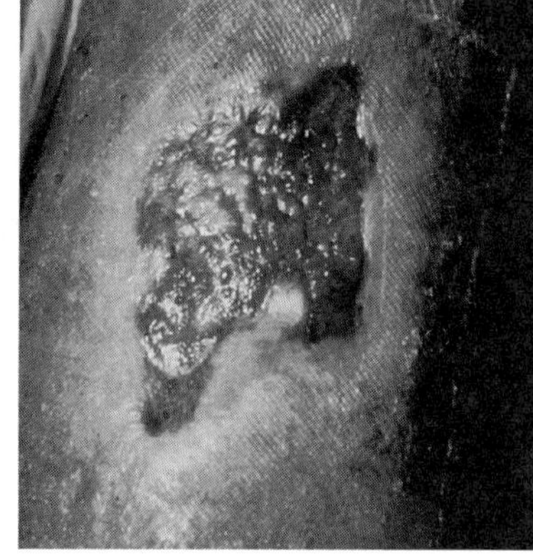

b.

Figure 7.6
Leg ulcer caused by chronic venous stasis dermatitis present for many years in an elderly man. *a.* Pre-treatment. *b.* One month after beginning bi-weekly applications to the left half of the ulcer of a mixture growth factors from fluid from a cow's blood platelets.

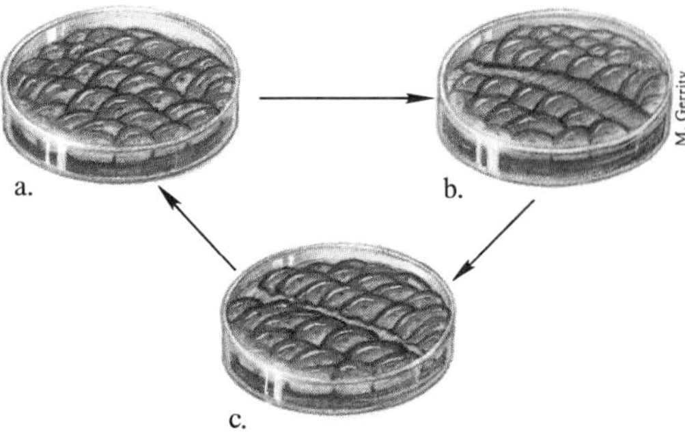

Figure 7.7
a. Normal cells in culture divide until they line their container in a one-cell-thick sheet (called a monolayer). If the monolayer is torn (*b*), the cells at the wound site grow and divide (*c*) to fill in the gap.

Fibroblast growth factor (FGF) participates in another component of the wound healing process. It stimulates division of endothelial cells that form the one-cell-thick walls of the tiniest blood vessels (capillaries) and the inner linings of the larger blood vessels. FGF also provokes mitosis in fibroblasts, which secrete collagen, a protein that also helps build blood vessels. The formation of new blood vessels, called *angiogenesis*, is important in wound healing because the increased blood supply brings oxygen and nutrients to nourish the new tissue.

Yet another growth factor, *platelet-derived growth factor* (PDGF), is synthesized by the large cells that give rise to blood platelets. Platelets break apart at the site of a wound, where they release biochemicals that cause the blood to clot locally. PDGF provokes mitosis of fibroblasts in the area of a wound, which, like EGF, helps to fill in the damaged area beneath the clot.

Growth factors can be produced in the laboratory using genetic engineering techniques (discussed in chapter 13), and use of the growth factors as drugs is being investigated. EGF, for example, can speed healing of a corneal transplant. The cornea is a one-cell-thick layer covering the eye, and normally these cells do not divide. However, a torn or transplanted cornea treated with EGF will divide to restore a complete cell layer. EGF is also being tested as a treatment for skin grafts and for nonhealing skin ulcers, which occur as a frequent complication of diabetes (fig. 7.6). FGF may help heal surgical incisions by filling in fibroblasts and repair heart tissue damaged by a heart attack by building a new blood vessel network.

A Cell's Size

A cell may divide when its surface-to-volume ratio becomes too small for the cell to obtain enough nutrients and to excrete sufficient wastes. (This concept of a cell growing too large to carry out the functions of life efficiently was discussed in chapter 4.) Experiments with the single-celled amoeba demonstrate the role of cell size. Consider two amoebas at the same stage of the cell cycle, one of them much larger than the other. The large one always divides first. If some of the larger amoeba's cytoplasm is removed just before the cell would normally divide, mitosis is delayed until the cell grows and the cytoplasm once again accumulates to the threshold level.

A Cell and Its Neighbors—The Effect of Crowding

Crowding can slow or even halt mitosis. Normal cells growing in culture stop dividing when they form a one-cell-thick layer (a monolayer) lining their container. If the layer is torn, the remaining cells that border the tear will grow and divide to fill in the gap but stop dividing once the space is filled in (fig. 7.7). This control of mitosis in culture by the presence of adjacent cells is called **contact inhibition.** Similarly, many cells in the adult human are inhibited from dividing by surrounding cells. In some tissues, only cells in certain positions can divide, such as those deep in the skin. In plants, cells in special regions called *meristems* at the tips of roots and shoots divide frequently.

Division of bacterial cells is also limited by the proximity of other cells, which compete for space and nutrients. Because a bacterium is a one-celled organism, each division produces a new generation. The duration of a bacterial cell division is called its *doubling time*.

The doubling time for pathogenic (disease-causing) bacteria influences the course of the illness. Bacteria that cause food poisoning, in which symptoms occur within hours or days of infection, have short doubling times. *Salmonella typhi*, for example, doubles in 25 minutes, and symptoms appear rapidly. Bacteria with longer doubling times produce symptoms weeks after infection. Tuberculosis, for example, is caused by *Mycobacterium tuberculosis*, which doubles every 14 hours. The doubling time of pathogenic bacteria is considered when a physician prescribes a course of treatment with an antibiotic drug.

Bacterial cell division that is not limited proceeds in an exponential or logarithmic pattern. That is, after the first division, there are 2 cells; after the second division, when those 2 divide, there are 4 cells; then 8; then 16; and so on. Bacteria

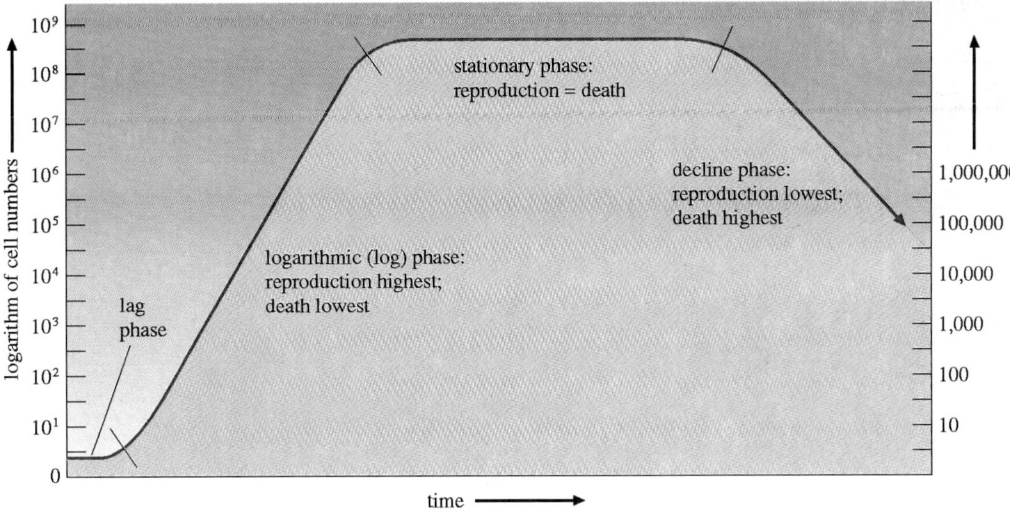

Figure 7.8

Bacterial growth curve. A newly founded bacterial population passes first through a lag phase, when the number of dividing cells is balanced by the number of cells that die in adjusting to the new environment. During the lag phase, division occurs at an exponential rate. During the stationary phase, cells begin to die again as nutrients and space are depleted. Finally, in the decline phase, dying cells overtake dividing cells.

From *Fundamentals of Microbiology*, Second Edition, by I. Edward Alcamo. Copyright © 1987, Benjamin/Cummings Publishing Company, Menlo Park, Calif. Reprinted by permission.

growing in a contained space (a flask or within an organism) form a population of cells. The rate of division of those cells depends upon nutrient and space availability and waste buildup and is described in a characteristic **population growth curve** (fig. 7.8).

The growth curve has four phases. In the *lag phase*, the population starts as bacterial cells adjust to their new environment. There is some cell division, but it is balanced by the approximately equal number of cells that die. Some bacteria growing in a flask might be killed as a student swishes the flask around; bacteria growing in a vegetable might die as the vegetable is cooked; bacteria in a person's bloodstream might be killed by the immune system's white blood cells. In lag phase, the population does not increase appreciably in size.

Once the bacterial population becomes accustomed to its surroundings, cell division rate reaches a maximum while death rate reaches a minimum, and for a while, the population increases exponentially. This is the *log phase*. At this stage a bacterial population is easiest to study because many cells are present. It is also the stage at which a bacterial infection produces symptoms in the unfortunate host organism.

The log phase ends when nutrients and growth factors are depleted, or perhaps when antibiotic drugs or immune system biochemicals begin to kill more bacteria. This is the *stationary phase*, when cell division rate about equals cell death rate. Finally, wastes accumulate, resources are exhausted, and as a result cell death rate overtakes cell division rate. The population is now in its *decline phase*. A few cells may survive this final stage, living off of the debris from the dying and dead cells.

Bacterial population growth curves are important in the dairy industry—and whenever a container of milk is taken out of the refrigerator. Pasteurization removes most pathogenic bacteria from dairy products before they appear on store shelves. Bacteria remaining in the milk at the grocery, in a refrigerator, or in a container that has been opened only once is in the lag phase. Each time the milk is taken out of the refrigerator, the bacterial population comes closer to entering the log phase. Returning milk to the refrigerator promptly can delay the beginning of exponential growth and therefore extend shelf life. Once the log phase begins, the milk spoils rapidly. The expiration date stamped on dairy products is determined by estimating when the log phase will begin.

KEY CONCEPTS

Control of mitotic rate is important in growth and development. Cells may be induced to divide by a trigger molecule, a cellular "clock" mechanism, hormones, growth factors, the cell's size, and the number of neighboring cells. Bacterial growth is also controlled and is described by a growth curve. In the lag phase, bacterial cell division is balanced by cell death. In the log phase the cell population increases exponentially as division is maximized and death is minimal. As nutrients are depleted, the bacterial population enters a stationary phase of equal cell division and death. In the final decline phase, wastes accumulate, resources are depleted, and cells die faster than new cells arise.

Mitosis During Development Is Highly Regulated

Precise control of mitosis is necessary for normal growth and development. The livers of a newborn human and an adult differ dramatically in size, but each is recognizable as a liver. The balance of cell death and cell reproduction maintains the organization of tissues forming organs in a growing individual. In many tissues and organs, cells are continually dying and being replaced by mitosis.

Many tissues have cells that divide often, termed **stem cells.** For example, a stem cell in the skin divides to give rise to two daughter cells—one that will become specialized and not divide again and the other a stem cell. In this way, the specialized nature of a tissue can be maintained in addition to its ability to generate new cells (fig. 7.9).

Cell Populations

Groups of cells can be described as **cell populations** to indicate the percentage of cells in particular stages of the cell cycle. In a *renewal cell population*, the cells are actively dividing. Renewal cell populations maintain linings within animal bodies that are continually shed and rebuilt from dividing cells, such as the cell layers that line the inside of digestive tracts. In the human body, renewal cell populations replace many millions of cells each day.

In an *expanding cell population,* in which up to 3% of the cells are dividing. The remaining cells of the expanding population are not actively dividing, but they can enter mitosis when a tissue is injured and new cells are required to repair it. The fast-growing tissues of young organisms, as well as adult tissues of the kidney, liver, pancreas, and bone marrow, consist of expanding cell populations.

Cells that are highly specialized and no longer divide comprise *static cell populations*. A mature red blood cell, for example, circulates in the blood for 120 days before it dies. It does not divide and does not even have a nucleus. Red blood cells form from bone marrow cells that can divide. Nerve and muscle cells also form static cell populations, remaining in the first gap phase (G_1). These cells grow by cell enlargement rather than by mitosis. A single nerve cell may grow to a meter in length, but it cannot divide.

KEY CONCEPTS

In multicellular organisms, stem cells divide often. Cell populations are groups of cells in which a certain proportion are in a particular stage of the cell cycle. Renewal cell populations are actively dividing. Expanding cell populations have up to 3% of cells dividing. Static cell populations are inactive and do not contain dividing cells.

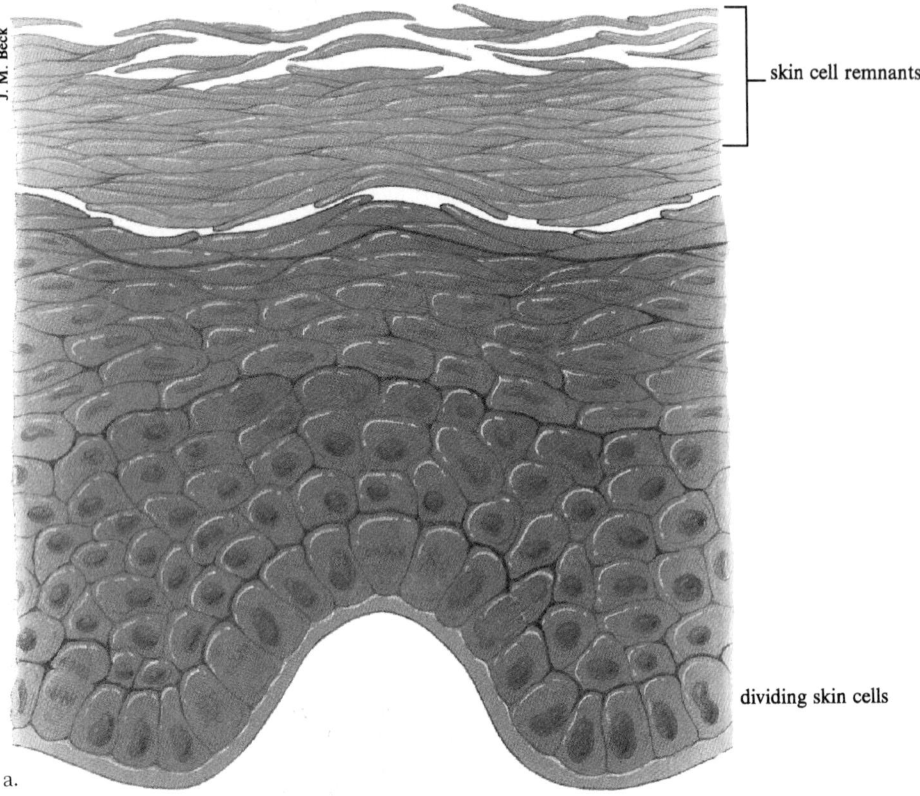

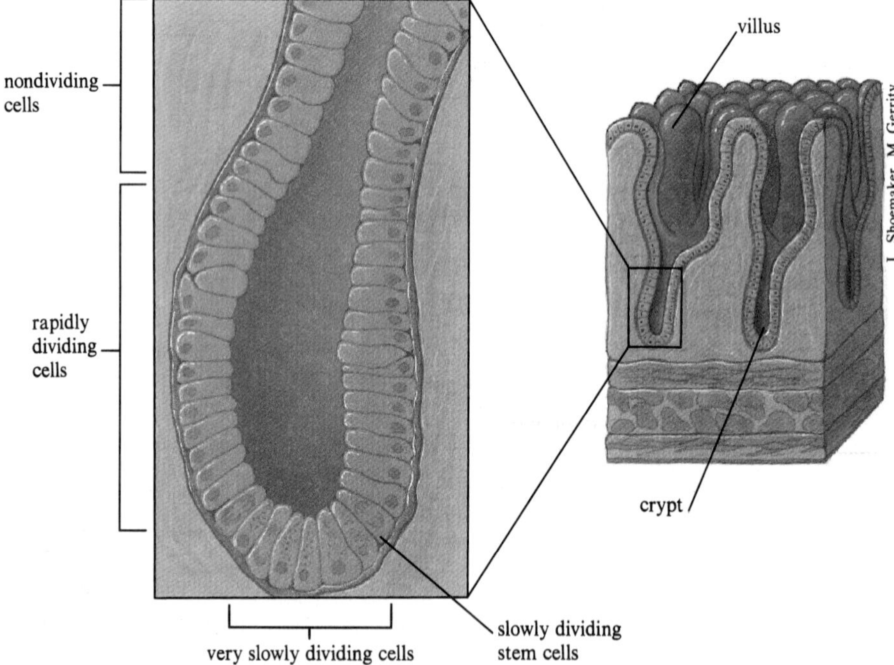

Figure 7.9

In some tissues, only cells in certain positions divide. *a.* The outer layer of human skin, the epidermis, has actively dividing stem cells in its deepest level that push most of their daughter cells upward yet maintain a certain number of stem cells. The cells pushed upward accumulate so much of the protein keratin that eventually the nuclei and organelles are squeezed aside and degenerate. The cell remnant is flattened and shed from the skin's outer surface. *b.* The lining of the small intestine also has deep cells that divide and push some of the daughter cells toward the surface. Rapidly dividing cells, such as those in the skin and small intestine, are most easily damaged by some forms of radiation. This is why radiation from bombs or cancer therapy can cause skin problems and digestive distress.

Figure 7.10
When a plant is wounded, a common soil bacterium, *Agrobacterium tumefaciens*, can infect the wound and produce crown galls, which are plant tumors with many of the characteristics of animal cancers. The tumors consist of disorganized masses of cells that do not develop into specialized tissues, such as roots and shoots; their cells divide rapidly; the cancerous qualities of their cells are inherited from one cell generation to the next; and cells transplanted from a crown gall tumor to a healthy plant produce new tumors. The source of the cancer characteristics is actually a small circle of DNA, called a plasmid, that contains genes that transform cells into a cancerlike state. Once the infecting bacteria have introduced the plasmid into the plant cells, the crown gall tumor continues to grow, even if the bacteria are removed. This crown gall is growing on a tobacco plant.

Cancer—When the Cell Cycle Goes Awry

One out of three of us will develop **cancer,** a group of disorders in which certain cells lose normal control over both mitotic rate and the number of divisions they undergo. Cancer begins with a single cell, which divides to produce others like itself, growing into a mass called a cancerous or malignant *tumor* or traveling in the blood. Cancer cells probably arise from time to time in everyone, because mitosis occurs so frequently that an occasional cell is bound to escape from the mechanisms that normally control the process. Fortunately, most cancer cells are destroyed by the immune system. Humans are not the only organisms to develop cancer. Cats and dogs develop cancer, and even plants have an abnormal sort of growth that is similar to cancer (fig. 7.10).

Characteristics of Cancer Cells

Cancer cells can divide uncontrollably and eternally, given sufficient nutrients and space. These characteristics are vividly illustrated by the cervical cancer cells of a woman named Henrietta Lacks, who died in 1951. Her cells persist today as standard cultures in many research laboratories (Reading 7.1).

Although it is said that cancer cells divide frequently, their rate of mitosis is actually a relative matter. Some cells normally divide frequently, others rarely. Even the fastest-dividing cancer cells, which complete mitosis every 18 to 24 hours, do not divide as often as some normal cells of the human embryo (table 7.1). Therefore, it is more accurate to say that a cancer cell has a rate of mitosis that is faster than the rate for the normal cell type from which it arose, or that it divides at the normal rate but does not "know" when to stop dividing. Once a tumor forms, it may grow faster than surrounding tissue simply because a larger proportion of its cells are actively undergoing mitosis.

Some cancers grow at an alarmingly fast rate. The smallest detectable fast-growing tumor is half a centimeter in diameter and can contain a billion cells, dividing at a rate that produces a million or so new cells in an hour. If 99% of the tumor's cells are destroyed, a million would still be left to proliferate. Other cancers are very slow to develop and may not even be noticed for several years (Reading 7.2). Lung cancer may take 3 to 4 decades to develop. However, the rate of a tumor's growth is slower at first, because there are fewer cells to divide. By the time the tumor is the size of a pea—when it is usually detectable—billions of cells are actively dividing.

When a cell becomes cancerous, its abnormal division characteristics are passed on to its descendants. That is, when a cancer cell divides, both daughter cells are also cancerous, having inherited the altered cell cycle control. Therefore, cancer is said to be *heritable*, in the sense that it is passed from cell to daughter cell. (Very few cancers are inherited, that is, passed from parents to offspring.)

A cancer cell is also *transplantable*. That is, if a cancer cell is injected into a healthy animal, the disease spreads as more cancerous cells divide from the original one. A cancer cell also looks different from a normal cell (fig. 7.11). It is rounder, possibly because it is less adhesive to surrounding normal cells than usual, and because the cell membrane is more fluid and allows different substances across it than normal.

A cancer cell is somewhat *dedifferentiated*, exhibiting less specialization than the normal cell type from which it derives. A skin cancer cell, for example, is rounder and softer than the flattened, scaly healthy skin cells above it in the epidermis. Cancer cells also act much differently than normal cells. Whereas normal cells placed in a container divide to form a monolayer, cancer cells pile up on one another. In an organism, this pileup would produce a tumor. Cancer cells growing all over one another therefore lack contact inhibition.

Reading 7.1 *Enticing Cells to Divide in the Laboratory*

BECAUSE OBSERVING CELLS DIVIDE IN A LIVING ORGANISM IS DIFFICULT, MUCH OF OUR KNOWLEDGE OF MITOSIS COMES FROM CELLS GROWING IN GLASS CONTAINERS, A TECHNIQUE CALLED *CELL CULTURE*. Often cultured cells multiply into very large numbers, and their secreted products can be extracted and used as drugs. Cell culturing is as much an art as it is a science. Some types of cells, such as blood cells, float freely in containers of liquid. Most cell types, though, adhere in single layers to surfaces. Researchers have devised glassware with intricate nooks and crannies to encourage maximal cell growth in minimal space. For example, a 4-liter vessel containing extensive surface area in its bumpy interior holds 40 billion cells—a number that formerly required 1,300 smooth-surfaced bottles! Another help in growing cells is to sculpt labware surfaces to resemble the contours of naturally occurring proteins. This mimicking of nature apparently creates an environment more like that on which a cell would normally grow, and the cultures flourish.

A major hurdle in early attempts to culture cells was the fact that most cells of vertebrate animals divide in culture only 50 times and then die. The reason for this natural time limit on mitosis in the laboratory is not known, but it is startling to observe. For example, if a human cell that has divided 20 times is frozen for a few years and then thawed, it will usually resume dividing until it has reached 50 divisions. To be of use in biomedical experiments that follow cellular changes over long periods of time, cell cultures need to remain mitotically active much longer. A way to lift the 50-division limit is to enlist the aid of cancer cells, which divide well beyond 50 times. Cancer cells are obtained from a tumor in an animal, or by exposing normal cells in culture to cancer-causing viruses. Cellular "immortality" is a boon to cell culture.

Perhaps the greatest help to cell culture technology has come from a seemingly unlikely source—a woman named Henrietta Lacks who died of cervical cancer in 1951. Before she died, she donated a few of her cancerous cells to a research laboratory at Johns Hopkins University. Although the cancer killed Henrietta in 8 months, her cells lived on. These "HeLa" cells, named (by abbreviation) for their donor, were the first human cells to be cultured successfully, and they remain in widespread use in biological research projects today. Ironically, the very characteristics that have made HeLa cells so valuable to research—their unrelenting mitosis and their seeming immortality—have also created a problem. The cells grow so well, dividing nearly once a day in any environment, that they rapidly proliferate in any culture of non-HeLa cells that they actually come into direct contact with. Within a few days, such a HeLa-contaminated culture contains nearly all HeLa cells. So far, some 90 cell types thought to be of non-HeLa origin have been found to be HeLa cells!

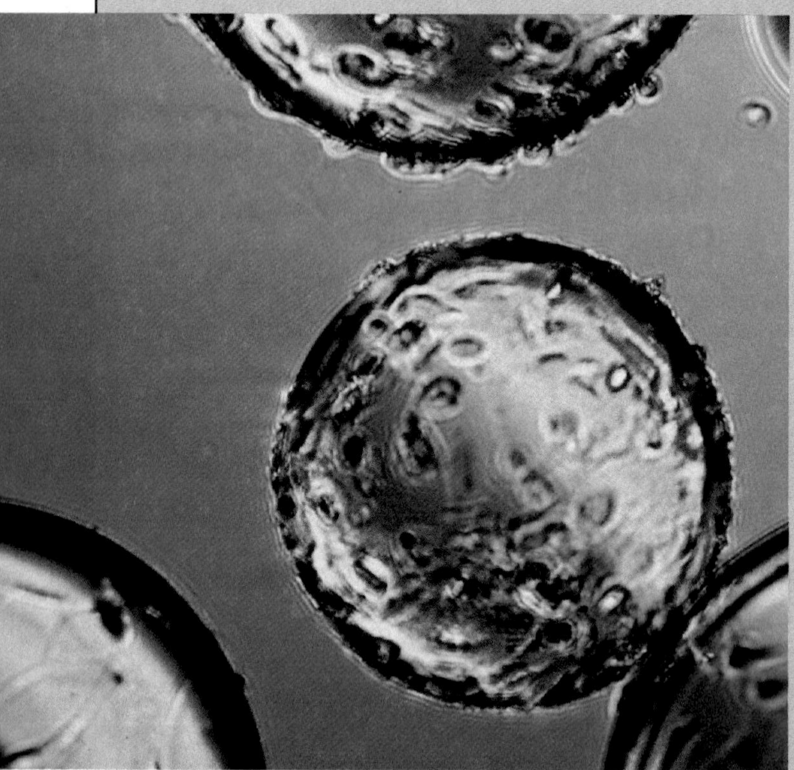

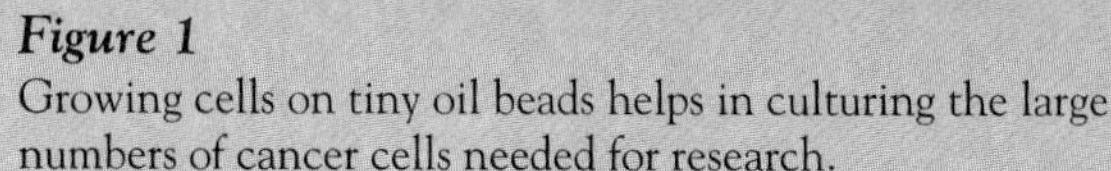

Figure 1
Growing cells on tiny oil beads helps in culturing the large numbers of cancer cells needed for research.

Figure 2
Culture dishes whose surfaces are modified to resemble biological proteins increase the variety of animal cell types that can be grown.

Table 7.1 Time to Complete One Cell Cycle in Some Human Cells	
Cell Type	*Hours*
Normal cells:	
Bone marrow precursor cells	18
Lining cells of large intestine	39
Lining cells of rectum	48
Fertilized ovum	36–60
Cancer cells:	
Stomach	72
Acute myeloblastic leukemia	80–84
Chronic myeloid leukemia	120
Lung (bronchus carcinoma)	196–260

Not all tumors are cancerous. A malignant tumor differs from a *benign* (gentle) tumor. A benign tumor is usually round and appears distinct from surrounding tissues. If a benign tumor remains small, shrinks, or is surgically removed, it does not threaten health. A growing benign tumor, though, can damage healthy organs by crowding them. However, a benign tumor does not travel to other locations in the body. Tumors called *fibroids,* which are often found in the uterus, are an example of a type of benign tumor. Fibroids can be painless, they sometimes shrink on their own or do not enlarge, or they can be surgically removed if they are painful.

Cancerous cells have surface structures that enable them to squeeze into any available space, a property known as *invasiveness.* They anchor themselves to tissue boundaries, called *basement membranes,* where they secrete chemicals that cut paths through healthy tissue (fig. 7.12). An invasive malignant tumor, in contrast to a benign tumor, grows irregularly, sending tentacles of itself in all directions (fig. 7.13). In fact, the word "cancer," which means "crab" in Greek, comes from the resemblance of malignant tumors to crabs, first noticed by the fifth-century B.C. Greek physician Hippocrates.

Eventually, the malignant cells of a tumor reach the bloodstream, in which they travel to other parts of the body and

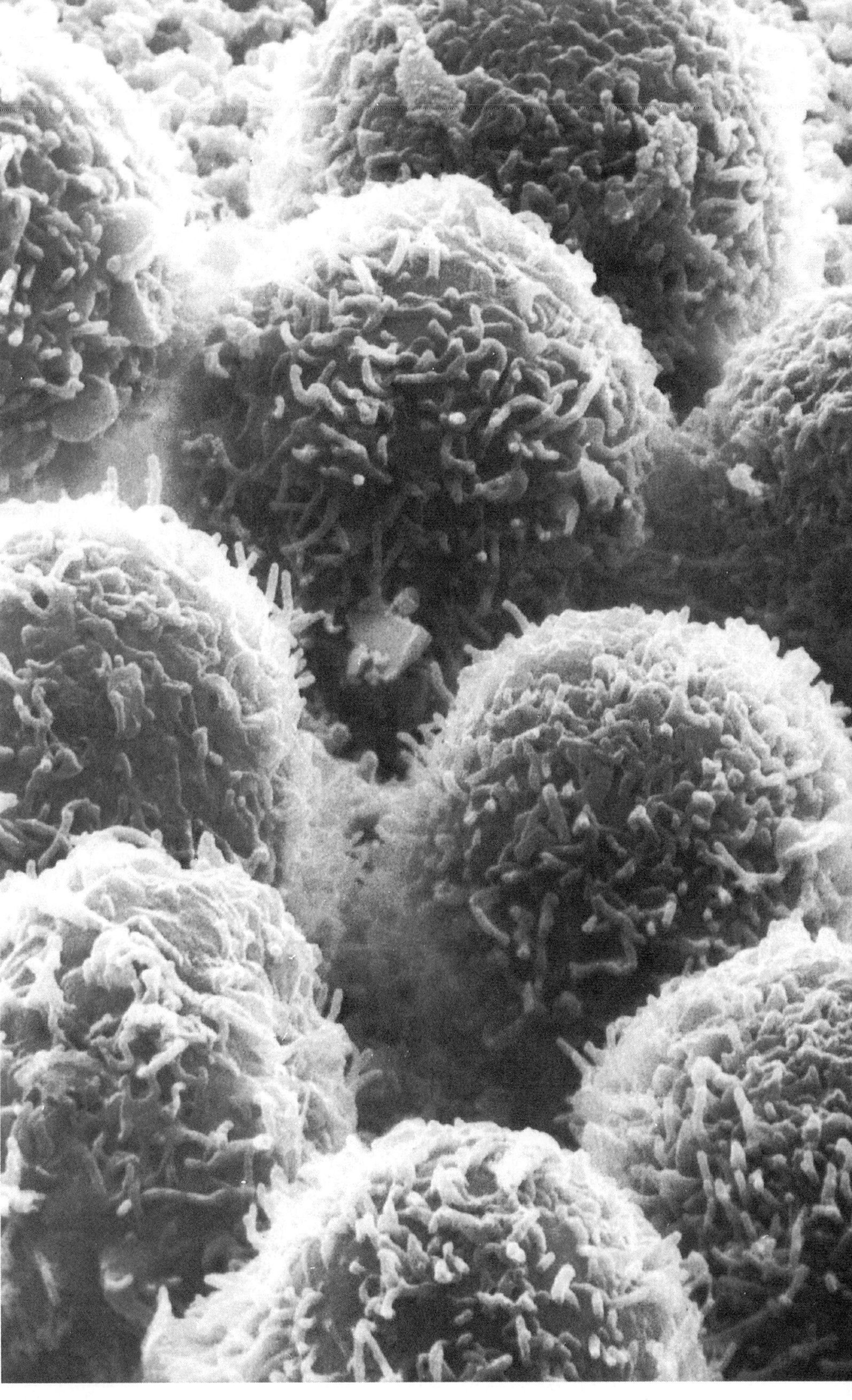

Figure 7.11
These breast cancer cells have already invaded the layer of extracellular material beneath them. They will soon advance through the healthy tissue below.

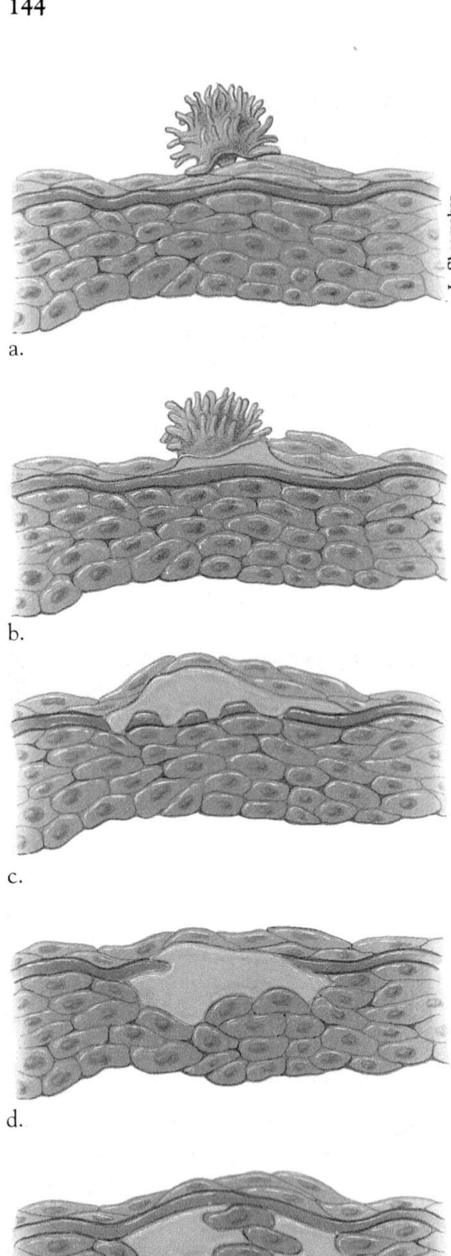

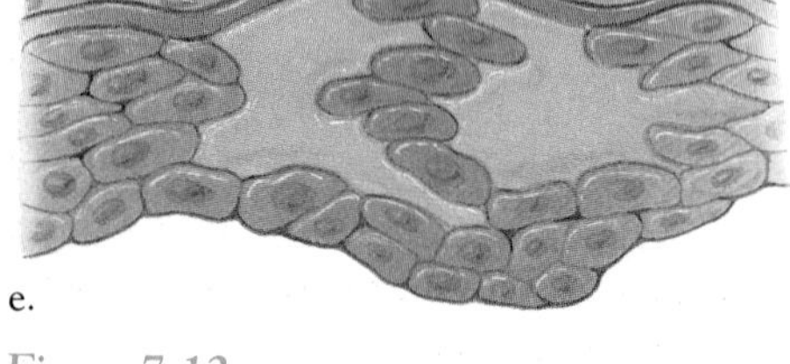

Figure 7.12
A cancer's spread takes many steps. *a.* A cancer cell adheres to normal cells that lie next to a basement membrane that separates two sections of tissue. *b.* The cancer cell secretes substances that cause the neighboring normal cells to move away, so that the cancer cell can now attach directly to the basement membrane. *c.* Next, the cancer cell secretes enzymes that allow it to penetrate the basement membrane and (*d*) to invade the tissue in the adjacent compartment. *e.* The cancer cell continues its migration and divides, starting a new tumor.

Figure 7.13
Note the spread of this solid malignant tumor, derived from fibroblasts. Magnification, x 200.

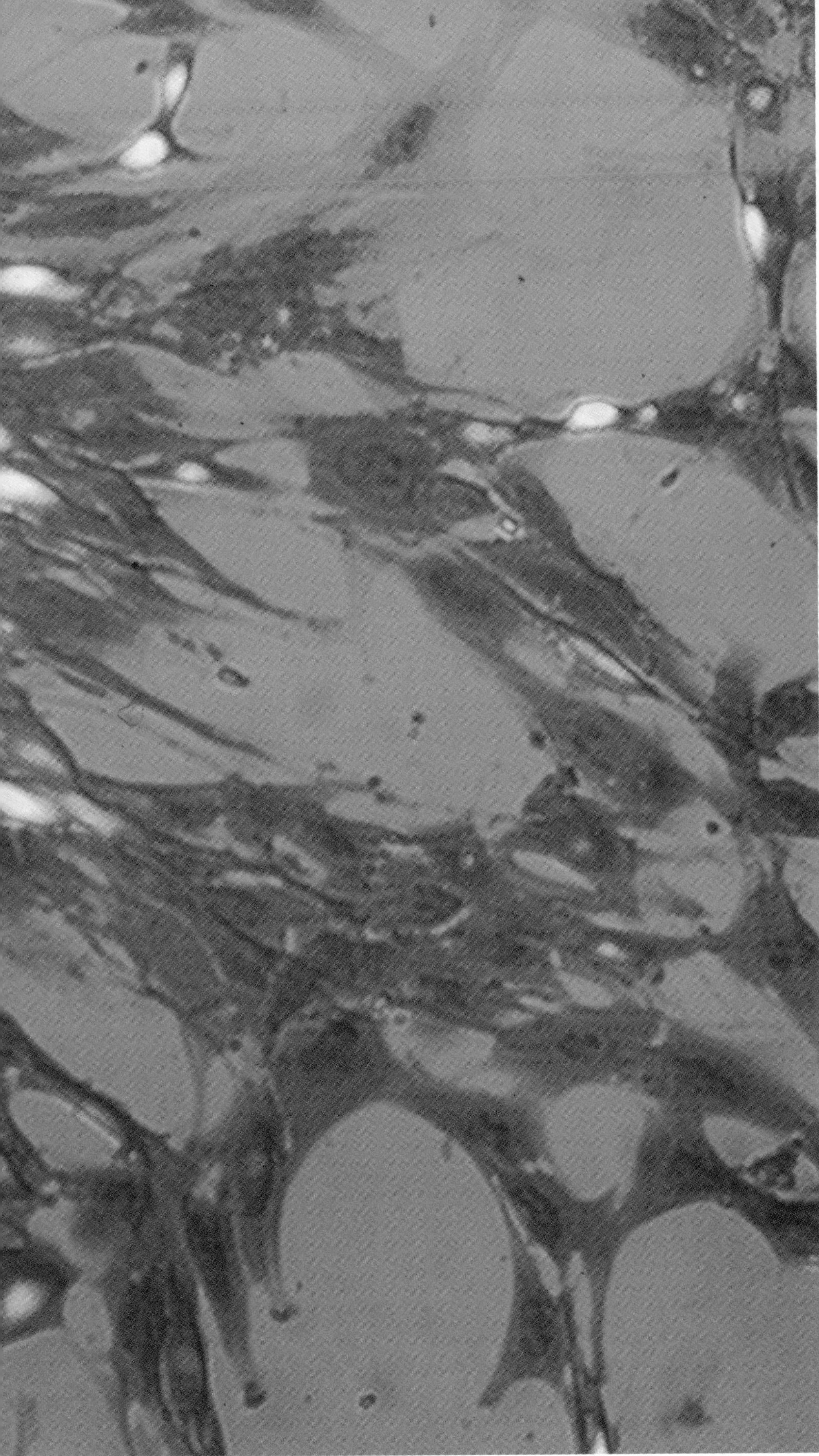

establish new or secondary tumors. The traveling cells secrete chemicals in their new location that stimulate the production of tiny blood vessels to nourish the rapidly accumulating cells. The cancer has spread, or **metastasized** (from the Greek for "beyond standing still"). Once a cancer spreads, it becomes very difficult to treat because the cells of secondary tumors often undergo genetic changes, making them different from the original tumor cells. To the patient, this means that a drug that shrinks a malignant stomach tumor may have no effect at all on a secondary tumor in the liver. However, for some cancers, the type of genetic change in a secondary tumor provides clues as to which treatment is likely to work.

KEY CONCEPTS

Cancer occurs when cells divide faster or more times than they normally would. Cancer cells are heritable, transplantable, dedifferentiated, and lack contact inhibition. Benign tumors can crowd healthy tissue but do not invade surrounding tissue as malignant tumors do. Cancer cells cut through basement membranes and eventually metastasize to other tissues.

The Causes of Cancer

Cancer can be viewed as a normal process—mitosis—that is mistimed or misplaced. For example, if cells in an adult's liver divide at the rate or to the extent that cells in an embryo's liver divide, the resulting overgrowth may lead to liver cancer. Because the pace of mitosis is in part controlled by proteins, specifically by growth factors, and proteins are constructed using genetic information, genes play a role in causing cancer. Genes that normally control cell division, but whose ill-timed or ill-placed activation leads to cancer, are called **oncogenes** ("onco-" means cancer). For example, one human oncogene is normally activated at the site of a wound, where it stimulates production of growth factors, which prompt mitosis to heal the damage. When the oncogene is activated at a site other than a wound, it still hikes growth factor production and therefore stimulates

Reading 7.2 Cancer Timetables

Most cancers are many years in the making (table 1). Cancer of the small intestinal lining develops very slowly because of the highly folded arrangement of cells. A cancer cell arising in a "crypt," where rapidly dividing cells are located, is shed from the body in a bowel movement or pushed down by neighboring cells streaming into the crypt. The earliest symptoms of this type of cancer, bleeding and discomfort in the area of the tumor, may not even be noticed until 16 years or so after the first cancer cell has formed.

Lung cancer accounts for 25% of all cancers in the United States, and 85% of all lung cancers are directly attributable to smoking. The precise chemical causes of smoking-induced lung cancer are hard to identify, because the chemicals in cigarette smoke are metabolized into more than 4,000 different compounds in the body. A person who begins smoking cigarettes at age 18 and sticks to the habit may die of lung cancer by age 45 (table 2). Once a disease primarily of men, lung cancer is being seen more and more in women, paralleling an increase in women's smoking.

Cancer of the uterine cervix also takes many years to develop. The cervix is the area of tissue lying between the uterus (womb) and the vagina. The cells there change abruptly from the tall lining cells in the uterus to flat cells in the vagina, and the boundary between the two cell types is especially prone to cancer. Cervical cancer has no early warning signs, but fortunately, a simple test called a Pap smear identifies the cellular changes that can precede the only symptom of cervical cancer—bleeding between menstrual periods—by 14 years (table 3). Cervical cancer in its early stages is successfully treated in over 95% of cases by removing the unhealthy cells.

Table 1
The Development of a Tumor

Time	*Number of Cells*	*Tumor Diameter*
0	1	Microscopic
6 months	2	Microscopic
4 years	500	Microscopic
10 years	500,000	1 mm
13 years	2 billion	2 mm

Table 2
The Progression of Lung Cancer

Age (years)	*Condition*
18	Smoking starts; lungs healthy, lined with tall, ciliated cells
19	Ciliated lining cells destroyed and replaced with flat, scaly cells; can regenerate if smoking stops
23	Flat, scaly cells accumulate on air passageways
26	First cancer cell forms
28–38	Small cancers develop but are contained by healthy tissue
38–40	Continued smoking damages tissue containing small cancers; cancers "escape" and spread within lung, then to liver, brain, bone marrow, or elsewhere
40	First symptoms (chest pain)
45	Death

Table 3
The Stages of Cervical Cancer

Stage	*Description of Tissue*
Metaplasia	Tall lining cells found in the cervix, where they should be flat; these are normal cells in an abnormal place
Dysplasia	Cells are large and blobby, with large nuclei, and are less specialized than normal lining cells; this is a precancerous state
Carcinoma in situ	"Surface cancer"; cancer cells are restricted to the outermost layer of the cervix; symptoms of bleeding between periods first noticed; cancer still 95% curable by freezing, laser treatment, or surgery
Invasive cancer	Cancer cells have crossed the basement membrane beneath the cervix lining cells and entered the bloodstream

mitosis, but, because there is not damaged tissue to replace, the new cells form a tumor instead.

We still do not understand many aspects of oncogene function. The new technology of **transgenic organisms,** in which a multicellular organism is engineered to contain a particular oncogene in every cell, is being used to study the development of cancer (fig. 7.14).

How are oncogenes "turned on" to cause cancer? Researchers think it happens when an oncogene is placed next to a gene that it is not normally next to, which boosts its expression. A virus infecting a cell, for example, may insert its genetic material next to an oncogene. When the viral DNA begins to reproduce itself rapidly, as viruses do, it also stimulates unusually rapid expression of the oncogene next to it. The heightened activation of the oncogene leads to increased production of the growth factor or other protein it controls, which promotes inappropriate mitosis.

Oncogenes can also be activated when they are moved from their normal location on a particular chromosome and placed next to a gene that is normally very active. This appears to be the case in *Burkitt's lymphoma,* a cancer of the white

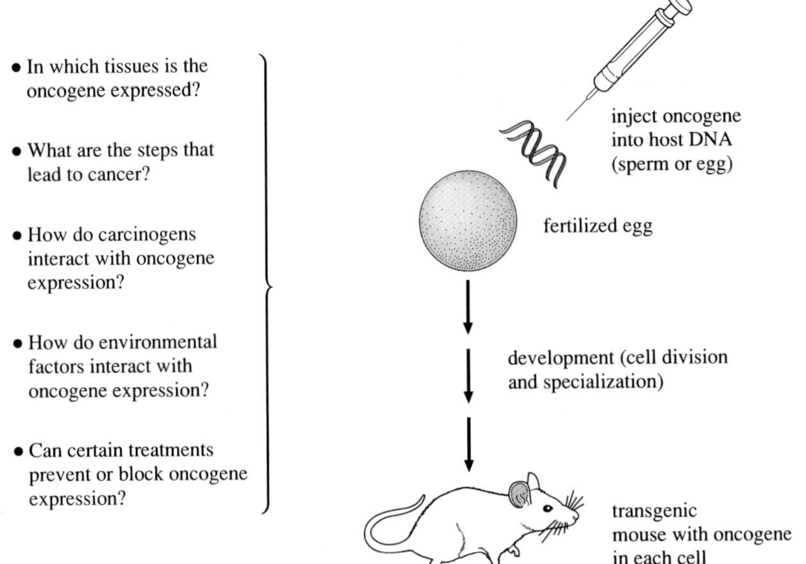

Figure 7.14
Transgenic animals permit study of cancer-causing genes (oncogenes). To create a transgenic animal, a gene of interest is injected into the nucleus of a fertilized egg. As the egg develops, the injected gene is perpetuated in each new cell. An injected oncogene will be expressed in certain tissues at a certain time in development. Researchers at the Massachusetts Institute of Technology and DuPont have developed a transgenic mouse containing a human breast cancer gene.

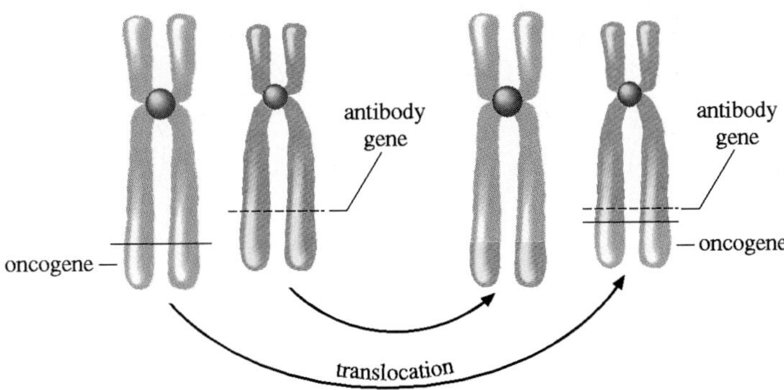

Figure 7.15
The cause of Burkitt's lymphoma appears to be the movement of an oncogene on the 8th largest human chromosome to a specific site on the 14th largest chromosome. The oncogene is placed near an immune system gene that is normally highly expressed. Overexpression of the oncogene in certain cells sets into motion the biochemical changes of cancer.

blood cells, which manufacture immune system chemicals called antibodies. It is thought that in the cancer, a virus triggers chromosome breakage, which places an oncogene next to an antibody gene. In this abnormal location, the oncogene is more highly expressed than normal. Indeed, in Africa, where Burkitt's lymphoma is common, patients exhibit a specific chromosome breakage pattern in which an oncogene is placed near an antibody gene (fig. 7.15). These people are also infected with the Epstein-Barr virus. Infection with the Epstein-Barr virus appears to trigger chromosome breakage, which places an oncogene in a position where it is overexpressed. Such a chromosomal repositioning of an oncogene near a highly expressed immune system gene is also behind a form of childhood leukemia, a cancer of the blood.

Unlike cancers caused by the activation of oncogenes, some cancers result from loss of a gene whose normal function is to suppress tumor formation. For example, the childhood kidney cancer called Wilms' tumor is caused by the absence of a gene that normally halts mitosis in the rapidly developing kidney tubules in the fetus. With the gene missing, mitosis does not cease on schedule, and the child's kidney retains pockets of cells dividing as frequently as if they were still in the fetus. In the child these cells form tumors. Other cancers are even more complicated. Colon cancer, for example, takes several stages to develop, including tumor suppressor gene losses as well as oncogene activation.

Other possible causes of cancer include certain chemicals, nutrient deficiencies, and radiation. Associations between these factors and cancer are often based upon population data. For example, people who smoke cigarettes are 40 times more likely to develop lung cancer than people who do not smoke. Therefore, something in cigarette smoke causes cancer.

Chemical carcinogens were recognized as long ago as 1775, when British physician Sir Percival Potts suggested that the high rate of skin cancer in the scrotum among chimney sweeps was due to their exposure to a chemical in soot (table 7.2). Today, carcinogens are identified both by epidemiological studies and on the Ames test, in which the ability of a chemical to cause genetic change in bacteria is taken as a strong indication that it may cause cancer in higher organisms. A substance that tests positively on the Ames test is typically tested further on whole animals.

Three general classes of chemical carcinogens are recognized. *Direct carcinogens* cause cancer when they are applied to standard fibroblast cells growing in culture. An example of a direct carcinogen is benzene, a solvent used in chemical laboratories and formerly in the dry-cleaning industry. *Procarcinogens* are safe outside the body, but once inside they are metabolized to produce intermediate compounds that cause cancer. They include certain organic dyes and cigarette tars and the nitrites and nitrates used to preserve processed meats. *Promoters* are chemicals that make other carcinogens more powerful. They include alcohol, certain hormones, and various chemicals found in cigarette smoke.

Numerous studies on cells in culture, animals, and human populations have linked dietary habits to increased risk of developing certain cancers (table 7.3). All of these associations are under investigation at the National Cancer Institute, where researchers are attempting to devise nutritional ways to prevent cancer. By identifying possible cancer-causing factors, we can learn more about these disorders and also take direct action to prevent them (Reading 7.3).

KEY CONCEPTS

If a gene that normally controls cell division is activated inappropriately, cancer may result. Such oncogenes may be activated by being placed near a viral sequence or a very active gene. Some cancers are caused by the absence of tumor-suppressing genes. Chemicals, nutrient deficiencies, and radiation can also cause cancer.

The adult human body is built of trillions of cells, of 200 different types, assembled into the tissues and systems of organs that enable us to eat, breathe, move, feel, and reproduce. Maintaining the specific structures and functions of the body's organs requires precise control over the rates at which different cells undergo mitosis. Considering the enormity of this task, it is amazing that disruption of the cell cycle—leading to cancer—does not occur even more frequently than it does.

Table 7.2
Suspected Carcinogens

	Agent	*Site of Cancer*
Foods and drugs:	Caffeine	Breast
	Diethylstilbestrol (DES)	Vagina and cervix
	Nitrites and nitrates (processed meat)	Stomach
	Saccharin (artificial sweetener)	Bladder
	Sequential birth-control pill, estrogens	Uterine lining
	Vegetable oils (reheated)	Colon and stomach
	Drugs to prevent organ transplant rejection	Blood
	Cigarette smoke	Lung, mouth, throat
	Smokeless tobacco	Mouth and throat
Environment:	Soot and fly ash	Lung
	Ultraviolet radiation (sun)	Skin
	X rays	Skin, thyroid, tonsils, thymus, blood
	Radon (radioactive gas formed from decay of uranium in soil and rock)	Lung
Industrial:	Asbestos (insulation material)	Digestive system, lung, epithelium
	Chromium, gold, silver, steel	Lung
	Ethylene dibromide (a pesticide)	Tumors in experimental organisms
	Nickel	Nasal cavity, sinuses, lung
	Solvents and dyes	Bladder, urinary tract, blood
	Tars and waxes	Skin
Agricultural:	Herbicides (phenoxy variety)	Blood
	Insecticides	Immature red blood cells

Table 7.3
Cancer-Diet Connections

Dietary Factor	*Site of Increased Cancer Risk*
Obesity	Colon, breast, prostate, gallbladder, ovary, uterus
Riboflavin deficiency	Esophagus
Fiber deficiency, fat excess	Colon, rectum
Vitamin C deficiency	Stomach, cervix
Vitamin A deficiency	Lung, bladder, larynx, cervix
Alcohol excess	Mouth, throat, liver
Folic acid deficiency	Leukemia, cervix
Selenium deficiency	Colon, rectum, breast

Reading 7.3 The War on Cancer

Prevention

THERE IS A GREAT DEAL THAT YOU CAN DO TO PREVENT CANCER. The National Cancer Institute claims that 20,000 cancer deaths could be prevented each year if Americans were to follow these suggestions:

1. Avoid obesity and eat less fat.
2. Eat more fruits and vegetables, especially those rich in vitamins A and C. Cabbage and cauliflower are also valuable because they contain minerals that stimulate production of enzymes that break down carcinogens.
3. Limit consumption of foods containing mutagens or carcinogens, such as smoked, nitrated, or charred meats.
4. Limit alcohol intake.

Cancer may also be prevented by not smoking cigarettes, following health and safety rules in the workplace, avoiding X rays whenever possible, and protecting the skin from ultraviolet radiation (from the sun or artificial tanning facilities).

Learn to recognize and report to a physician the warning signs of cancer:

1. A change in bowel or bladder habits
2. A sore that does not heal
3. Unusual bleeding or discharge
4. A lump
5. Persistent indigestion
6. Difficulty in swallowing
7. A change in the appearance of a wart or mole
8. Chronic cough or hoarseness

Diagnosis

The key to treating cancer successfully is early detection, so that the deadly spread of cancer cells can be stopped. Increasing emphasis is being placed on simple examinations and tests that can be performed at home. Women are encouraged, beginning at age 20, to examine their breasts at the same time each month for lumps that could be tumors. A breast exam takes about 10 minutes and consists of palpating each breast in widening concentric circles so that all of the tissue is felt. Indeed, 56% of all breast cancers are initially detected by women themselves. Males between the ages of 15 and 35 should periodically examine their testicles for lumps.

Colon cancer can be spotted early with an at-home *hemoccult* (hidden blood) test, in which a person sends a stool sample to a laboratory, where the presence of blood—a sign of colorectal cancer 50% of the time—is detected. Adult males in particular should do a hemoccult every year, especially if they have family members who have had intestinal growths called polyps.

Other low-cost cancer tests are performed routinely by doctors to screen for certain cancers in a healthy population. Most women have a yearly *Pap smear* to detect cervical cancer. Many women over the age of 40 undergo yearly *mammography*, an X-ray technique that detects breast tumors still too small to be felt. For women in this age group, the risk of developing cancer from the X rays in the test is less than that of developing breast cancer.

Some diagnostic equipment is still too costly to be implemented in large-scale screening of healthy populations, and its use is restricted to people who already suspect that they might have cancer. A standard X ray can pinpoint the shadow of a large tumor in a lung. *Computerized axial tomography*, more commonly known as a CAT scan, is a three-dimensional X ray that can quickly detect small tumors that have not yet spread to an extent detectable by X rays. An even more sensitive technology, *magnetic resonance imaging*, uses radio waves and a magnetic field to spot tiny tumors.

A blood test to diagnose cancer in early stages takes advantage of molecules found in larger numbers on cancer cells than on healthy cells. Highly pure preparations of immune system proteins called *monoclonal antibodies* have specialized binding sites that fit certain cancer cell surface molecules like a key fits a lock. Monoclonal antibodies are mixed with a sample of blood, and if a growing tumor somewhere has shed molecules from its cell surfaces, the monoclonal antibodies bind to them and can be detected. Monoclonal antibody tests are very valuable in detecting a cancer's spread or tumors too small to be felt. They are particularly useful in spotting recurrences of ovarian cancer, for which explorative surgery was previously the only diagnostic option.

Cancer Treatment

Cancer has traditionally been treated by any of three approaches. Surgery is effective in removing solid tumors that are confined to a particular area but are not close to vital structures. Breast cancer is usually treated surgically. Localized tumors that are unreachable by surgery are often treated with radiation. In chemotherapy, drugs are used alone or in combination to attack different characteristics of cancer cells. A drug that halts mitosis might be used along with one that prevents a tumor from establishing its nourishing blood supply. Chemotherapy is typically used to treat blood cancers. A patient may undergo all three types of treatment, or perhaps combine a standard therapy with an experimental one. For example, it has been found that heating the inside of a tumor with ultrasound, microwaves, or radiowaves, an approach called *hyperthermia*, is a powerful partner to radiation therapy, which sometimes kills only the cells in the outermost portion of a tumor. Stray cancer cells remaining after surgery on the liver can be killed by application of extreme cold with an instrument called a *cryoprobe*.

Radiation and chemotherapy kill dividing healthy cells along with cancerous ones, producing such side effects as hair loss and nausea, which involve rapidly dividing healthy tissue. A fourth major approach to cancer treatment is *biotherapy*, the use of immune system cells and biochemicals to boost a person's natural defenses against cancer cells. For example, monoclonal antibodies chemically linked to cancer-fighting drugs are attracted to tumor-specific surface molecules, where they release their cargo. Another immune system booster is interferon, used to treat the rare blood cancer, hairy cell leukemia.

SUMMARY

Mitosis, which is the division of somatic cells, is responsible for growth, development, reproduction, and repair of damaged tissue. Cells reproduce by first duplicating their contents, including the genetic material, and then dividing into two "daughter" cells. The *cell cycle* is a sequence of events that describes whether or not a particular cell is dividing or preparing to do so. The cell cycle consists of *interphase*, when genetic material, macromolecules, and organelles are duplicated, and mitosis, when the cellular constituents are distributed into two daughter cells.

Interphase is further broken down into two *gap* periods, during which proteins are synthesized (G_1 and G_2), and a *synthesis* period (S) when chromosomal constituents are manufactured. Each such replicated chromosome consists of two complete sets of genetic information, called *chromatids*, that are attached by a constriction called a *centromere*.

Mitosis consists of four stages. In *prophase*, the chromosomes condense and become visible when stained, the nuclear membrane disassembles, and the *spindle apparatus* is built from microtubules. In *metaphase*, the replicated chromosomes attach to spindle fibers that align the chromosomes down the center of the cell. In *anaphase*, the chromatids of each replicated chromosome separate, such that a complete copy of the genetic material segregates to each end of the cell. In *telophase*, the spindle breaks down and two nuclear membranes form around each of the two sets of chromosomes. The distribution of the genetic material into daughter cells is called *karyokinesis*. Also during telophase *cytokinesis* begins, which is the distribution of cytoplasm, organelles, and macromolecules into daughter cells.

The rate, timing, and number of mitotic divisions among the various specialized cells of a multicellular organism are highly regulated. Although cell cycle control is not fully understood, some regulatory factors appear to be the action of hormones, *growth factors*, a cell's size, and its proximity to other cells. Bacterial cell divisions are called *binary fission* and are a form of asexual reproduction. The growth of a population of bacteria depends upon availability of space and nutrients. The population growth curve for bacteria typically proceeds through *lag*, *log*, *stationary*, and *decline* phases. Healthy human cells generally do not divide more than 50 times when grown in culture, and in the body their division is also limited by the presence of other cells. Different *cell populations* are defined by their proportion of cells that are in different stages of the cell cycle.

A loss of cell cycle control can result in growth of a *malignant* or cancerous *tumor*. A tumor cell divides more frequently or more times than cells surrounding it, has altered surface properties, has lost the specializations of the cell type from which it arose, and produces daughter cells like itself. A *benign* tumor is rounded in shape and localized, whereas a malignant tumor infiltrates nearby tissues. A cancerous tumor can spread, or *metastasize*, by attaching to *basement membranes* that separate tissue compartments from each other and secreting enzymes that penetrate tissues and open a route to the bloodstream. From here a cancer cell can travel and exit to establish secondary tumors.

Some cancers appear to be caused by *oncogenes*, which are genes whose protein products control mitotic rate in certain cells at certain times in development, possibly by influencing growth factors. An oncogene activated in the wrong place or at the wrong time can alter mitotic rate or extent in a way that causes cancer. Oncogenes can be activated by being moved to a region of DNA that is very active, possibly stimulated to do so by the presence of a virus. Loss of tumor suppressor genes can also cause cancer. Risk factors that contribute to the likelihood of some cancers include smoking cigarettes, deficiencies of certain nutrients, and exposure to chemical carcinogens.

QUESTIONS

1. Describe the events that take place during the process of mitosis.

2. What is a cell doing during interphase?

3. Give an example of how mitosis that is too frequent or extensive can harm health and an example of how too infrequent or too little mitosis can harm health.

4. A cell is taken from a newborn human, allowed to divide 19 times, and then frozen for 10 years. Upon thawing, what is the cell likely to do in terms of mitosis?

5. If a layer of cells is torn, mitosis sometimes fills in the missing cells. Cite an example of this phenomenon observed in the laboratory and in the human body.

6. In what ways do cancer cells differ from normal cells?

7. A tumor is removed from a mouse and broken up into cells. Each cell is injected into a different mouse. Although all of the mice used in the experiment are genetically identical and raised in the same environment, the animals develop cancers with different rates of metastasis. Some mice die quickly, some linger, others recover. What do these results indicate about the characteristics of the cells that comprised the original tumor?

8. How do biologists think a virus can cause cancer?

9. A young boy had a tumor in his stomach. Because the doctors said the tumor was benign, his parents refused to allow it to be removed, thinking it was not dangerous. The boy died. How did this happen?

10. How can one cancer be caused by activation of a gene, yet another be caused by absence of a gene?

TO THINK ABOUT

1. A researcher classifies the cells of a tissue in a mouse according to what stage of the cell cycle they are in. She finds that 3% of the cells are dividing. Of the cells in interphase, 50% are in the G_1 phase, 40% are in the S phase, and 10% are in the G_2 phase. Based on this information, the investigator concludes that, for the cells in this tissue, the G_1 phase is of the longest duration, followed by the S phase. The cells spend the least amount of time in G_2. What is the basis for this interpretation?

2. A device called a "fluorescence-activated cell analyzer" measures the DNA content in cells of a large cell population. The device distinguishes three groups of cells. Group A is large, and its cells have a certain amount of DNA. Group B is also large, and its cells have twice as much DNA as the cells in group A. The third group, C, is very small, and each of its cells has four times as much DNA as a cell from group A. What stage of the cell cycle corresponds to cells in groups A, B, and C?

3. Many of the human victims of the atomic bombs dropped over Japan in World War II were not immediately killed but suffered slow, agonizing deaths from radiation sickness. The massive, sudden doses of radiation that they received affected the cells in their bodies that proliferate at the highest rates. What types of cells and tissues were affected?

4. Tumor cells can often be grown in culture, where the responses of the cells to drugs can more easily be observed. How might such a procedure benefit a cancer patient?

SUGGESTED READINGS

Alberts, Bruce, et al. 1989. *Molecular biology of the cell*. New York: Garland Publishing. Chapter 11, "Cell Growth and Division," provides a clear but detailed explanation of the cell cycle.

Angier, Natalie. 1988. *Natural obsessions*. New York: Warner Books. How the secrets of oncogenes were unlocked.

Carlson, Elof Axel. 1984. *Human genetics*. Lexington, Mass.: D. C. Heath. In chapter 22, the author writes from personal experience about the links between smoking and lung cancer.

DeYoung, Garrett. 1986. *The cell builders*. Garden City, New York: Doubleday. A personalized account of one of the first recipients of immune system chemicals to fight cancer.

Division of Cancer Communication, National Cancer Institute, National Institutes of Health, Bethesda, Maryland, 1-800-4-CANCER. Excellent information on all aspects of cancer can be obtained by calling or writing this organization.

Gallagher, Gayl Lohse. March 1990. Evolutions: the mitotic spindle. *The Journal of NIH Research*. A short history of what we know about the mitotic spindle, with beautiful illustrations.

Graham, Jory. 1982. *In the company of others*. New York: Harcourt Brace Jovanovich. A patient's gripping account of her battle with breast cancer.

Lewis, Ricki. June 1990. Wilms' tumor—The genetic plot thickens. *The Journal of NIH Research*. Pockets of embryonic tissue persisting in the kidney lead to cancer.

Lewis, Ricki. October 1990. Neurofibromatosis I revealed. *The Journal of NIH Research*. Benign tumors can result from loss of tumor suppressor genes too.

Marx, Jean. December 15, 1989. Many gene changes found in cancer. *Science*. The route to colon cancer entails oncogene activation and chromosome loss.

Marx, Jean. August 3, 1990. New clue to cancer metastasis is found. *Science*. Animals diverse as slime molds, fruit flies, and humans have cellular mechanisms to suppress tumor formation.

Miller, Julie Ann. September 1990. Genes that protect against cancer. *BioScience*. Researchers are discovering more and more genes whose normal role is to prevent cancer.

Pelech, Steven. July/August 1990. When cells divide. *The Sciences*. The history of our knowledge of the mitotic process, and current thoughts on its pacemaker.

Rawls, Rebecca L. February 25, 1985. In the search to control cancer, understanding metastasis is crucial. *Chemical and Engineering News*. A step-by-step look at the biochemical changes that precede metastasis.

Thomas, Lewis. 1982. MSKCC: The Memorial Sloan-Kettering Cancer Center. *The youngest science: Notes of a medicine-watcher*. New York: Viking Press. A humane look at a devastating collection of illnesses.

VanBrunt, Jennifer, and Arthur Klausner. January 1988. Growth factors speed wound healing. *Biotechnology*. Natural growth factors can function as drugs, helping healing.

UNIT 3

Reproduction and Development

C H A P T E R

8

Human Meiosis and Reproduction

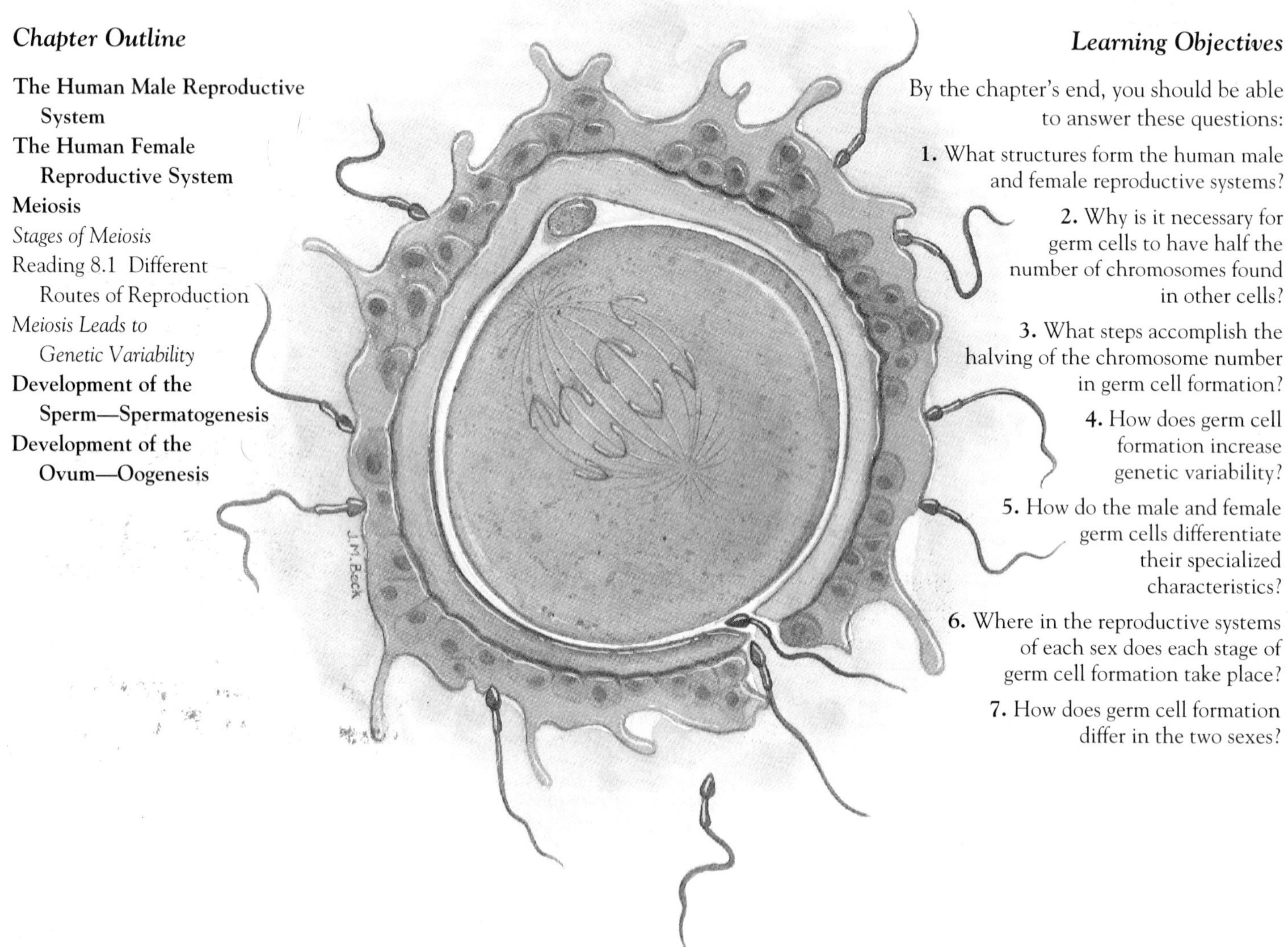

Chapter Outline

Learning Objectives

By the chapter's end, you should be able to answer these questions:

1. What structures form the human male and female reproductive systems?
2. Why is it necessary for germ cells to have half the number of chromosomes found in other cells?
3. What steps accomplish the halving of the chromosome number in germ cell formation?
4. How does germ cell formation increase genetic variability?
5. How do the male and female germ cells differentiate their specialized characteristics?
6. Where in the reproductive systems of each sex does each stage of germ cell formation take place?
7. How does germ cell formation differ in the two sexes?

As the two cells approach one another, their differences become more apparent. The smaller of the two is a tiny, streamlined cell, anatomically specialized as a powerful swimmer. The front end of this **sperm** cell is swollen with a package of genetic material representing the male individual from whom it came. The rest of the cell is a whiplike tail powered by many mitochondria.

The other cell, an egg cell, or **oocyte,** is huge by comparison, containing 90,000 times the volume of the smaller cell. It is packed with organelles, membranes, vesicles, and nutrients, plus its own genetic cargo, which represents the female individual from whom it came. The oocyte carries a large supply of nearly everything needed to launch a new individual on the developmental journey of life. When the two cells meet and their nuclei merge, that is precisely what happens—a new human being is conceived.

Sometime between 36 and 48 hours after the genetic material of the sperm and oocyte (now called an *ovum*) fuse, rapid cell division begins. The new individual is, for a while, a small ball of cells that gradually becomes hollow. Within 2 weeks, an elongated collection of cells on the interior surface of the hollow ball folds and twists to form a three-layered structure, the **embryo,** which will be protected and nourished by surrounding membranes and sacs that will develop. In the weeks to follow, swellings and elongations come and go as the person-to-be begins to resemble a human, complete with functioning, specialized cells. After many months of growth and elaboration of the structures that were laid down in the early weeks, an organism weighing about 8 pounds (3.6 kilograms), and built of a trillion or so cells, leaves its watery, membranous environment and enters the world. A new person is born. The next three chapters explore this most fascinating of biological journeys, the development of a human being.

Although the formation of a new individual begins with two cells, the sperm and the ovum, the story of development actually opens with the construction of these two very important cell types in the bodies of the parents. Not only does sperm and egg formation provide a mechanism for starting a new life, but it also mixes up genetic contributions from past generations, so that each new individual has a unique combination of traits. It is this variability that can allow a species to survive environmental upheaval. In the face of extreme temperatures or a new illness, for example, individuals with certain combinations of traits will survive to reproduce, passing on those traits and strengthening the species under those conditions.

Examining the production of sperm and ovum takes us to the organ and organ system levels of biological organization—specifically, to the reproductive system. The reproductive systems of the human male and female are similarly organized. Each system has paired structures in which the sperm and ova are manufactured, a network of tubes to transport these cells, and various hormones and glandular secretions that control the entire process.

The Human Male Reproductive System

Sperm cells are manufactured within a 125-meter-long network of tubes called *seminiferous tubules,* which are packed into paired, oval organs called **testes** (sometimes called testicles) (fig. 8.1). The testes lie outside the abdomen within a sac called the *scrotum*. Their location outside of the abdominal cavity allows the testes to maintain a temperature lower than that of the rest of the body, which is necessary for the sperm cells to develop properly. Leading from each testis is a tightly coiled tube, the **epididymis,** in which sperm cells mature and are stored. Each epididymis continues into another tube, the **vas deferens.** Each vas deferens bends behind the bladder and joins the *urethra*, the tube that also carries urine out through the *penis*.

Along the sperm's path, three glands contribute secretions. The vas deferentia pass through the prostate gland, which produces a thin, milky, alkaline fluid that activates the sperm to swim. Opening into the vas deferens is a duct from the **seminal vesicles,** which add the sugar fructose for energy, plus hormonelike prostaglandins, which may stimulate contractions in the female reproductive tract that help the meeting of sperm and ovum. The *bulbourethral glands* (also called Cowper's glands), each about the size of a pea, are attached to the urethra where it passes through the body wall. They secrete alkaline mucus, which coats the urethra before sperm are released. All of these secretions combine to form the *seminal fluid*, in which the sperm cells travel.

During sexual arousal, the penis becomes erect so that it can penetrate the vagina and deposit sperm in the female reproductive tract. At the peak of sexual stimulation, a pleasureable sensation called *orgasm* occurs, accompanied by rhythmic muscular contractions that eject the sperm from the vas deferens through the urethra and out of the penis. The discharge of sperm from the penis is called *ejaculation*. One ejaculation in the human typically delivers 400 million sperm cells.

KEY CONCEPTS

The human male and female reproductive systems each have paired structures housing reproductive cells, tubes for transporting these cells, and glands whose secretions enable the cells to function. Sperm develop in the seminiferous tubules, which are wound into the testes and reside in the saclike scrotum. Sperm cells mature and are stored in the epididymis, which leads from the testes and continues into the vas deferentia. Each vas deferens joins the urethra in the penis. Three glands contribute secretions to the semen. The prostate gland adds an alkaline fluid, the seminal vesicles add fructose and prostaglandins, and the bulbourethral glands secrete mucus. About 400 million mature sperm cells are ejaculated during orgasm.

The Human Female Reproductive System

The female sex cells develop within paired organs in the abdomen, the **ovaries** (fig. 8.2). Within each ovary of a newborn female are about a million oocytes, which are immature egg cells. An individual oocyte is surrounded by **follicle cells,** which nourish it. In the adult, the ovary houses oocytes in different stages of development. Once a month, the most mature oocyte from one ovary is released and swept by a current

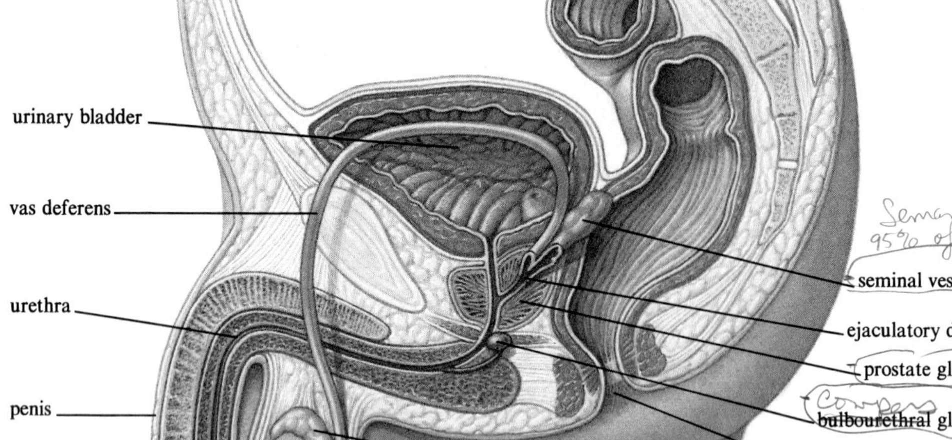

Figure 8.1
The human male reproductive system. Sperm cells are manufactured within the seminiferous tubules, which are tightly wound within the testes, which descend into the scrotum. Sperm mature and are stored in the epididymis and exit through the vas deferens. The paired vas deferens join in the urethra, from which they exit the body. Secretions are added to the sperm cells from the prostate gland, the seminal vesicles, and the bulbourethral gland.

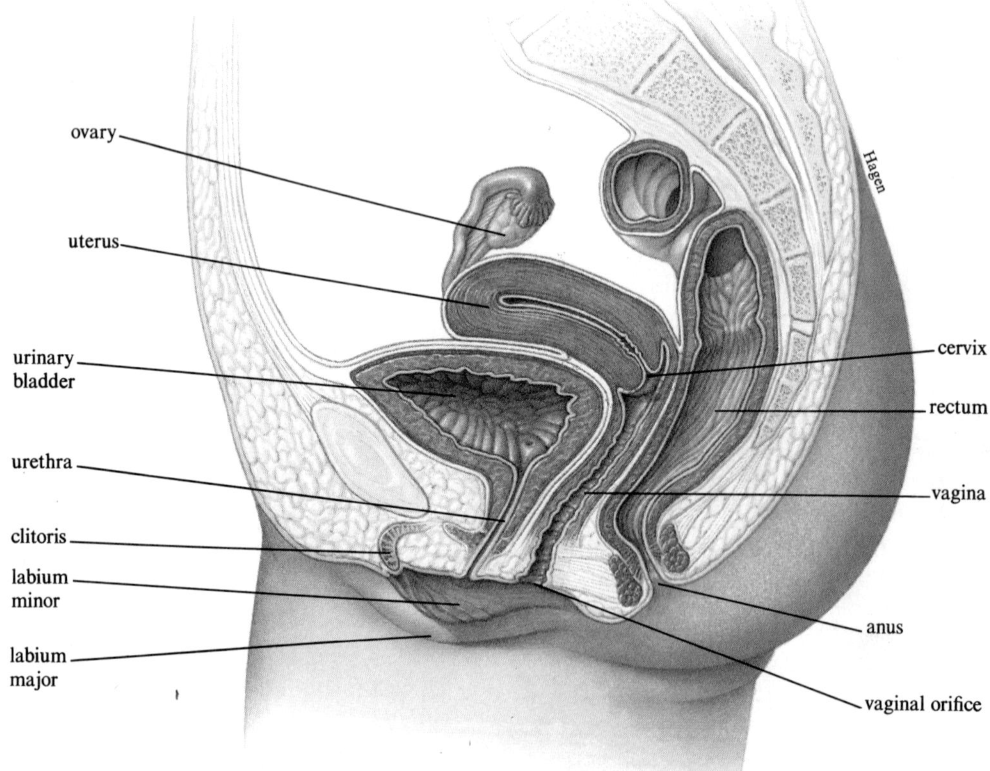

Figure 8.2
The human female reproductive system. Immature egg cells are packed into the paired ovaries. Once a month, one oocyte is released from an ovary and is drawn into the fingerlike projections of a nearby fallopian tube by ciliary movement. If the oocyte is fertilized by a sperm cell in the fallopian tube, it continues into the uterus, where built-up it is nurtured for 9 months as it develops into a new individual. If the ovum is not fertilized, it is expelled along with the built-up uterine lining through the cervix and then through the vagina. The external genitalia consist of inner and outer labia and the clitoris.

generated by beating cilia on the fingerlike projections of the nearest of the paired **fallopian tubes.** The tube carries the oocyte into a muscular saclike organ, the **uterus,** or womb.

Once released from the ovary, an oocyte can live for about 72 hours, but it can be fertilized only during the first 24 hours of this period. Recent research suggests that this "window" of fertilization is only 6 hours. If the oocyte encounters a sperm cell in the fallopian tube and the cells combine and their nuclei fuse, the oocyte completes its development and is called a *fertilized ovum*. It travels into the uterus and becomes implanted in the thick, blood-rich uterine lining, called the **endometrium,** and develops there, over the next 9 months, into a new human being. If the oocyte is not fertilized, it and the endometrium (which accumulated over the preceding few weeks to nurture any oocyte that might be fertilized) are expelled as the *menstrual flow*.

The lower end of the uterus narrows to form the **cervix,** which opens into the *vagina*. The vaginal opening is protected on the outside by two pairs of fleshy folds called the *labia majora* (major lips), which protect thinner, inner flaps of tissue called the *labia minora* (minor lips). At the upper juncture of both pairs of labia is a 2-centimeter-long structure called the *clitoris*, which is anatomically similar to the penis. Rubbing the clitoris stimulates females to experience orgasm, and this can happen during intercourse. The cycle of egg maturation and the preparation of the uterus to nurture it are controlled by rising and falling levels of hormones.

KEY CONCEPTS

Oocytes develop in the ovaries. Each month after puberty, one oocyte is released from an ovary and is captured by fingerlike projections from a fallopian tube. Each of the paired tubes leads to the uterus. If the oocyte is fertilized by a sperm, it nestles into the endometrium to develop. Otherwise, it exits the body with the monthly menstrual flow, passing through the cervix and vagina.

Meiosis

The sperm and ovum are often referred to by other, more general names: **gametes,** *germ cells*, or *sex cells*. These cells are different from all others in the human body in that they contain only half the usual amount of genetic material. Each germ cell contains 23 different chromosomes, whereas other cells, called **somatic** (body) **cells,** contain two of each type of chromosome, for a total of 46 chromosomes, or 23 pairs. Germ cells are termed **haploid** (or n) to indicate that they have only one of each type of chromosome, and somatic cells are **diploid** ($2n$), signifying their double chromosomal load. Can you see why this halving of the chromosome number in germ cells is necessary? If the human sperm and ovum each contained 46 chromosomes, then, when they joined to form the first cell of a new individual, it would contain twice the normal number of chromosomes. Such a genetically overloaded cell would not develop normally, if at all.

Germ cells are formed from certain somatic cells, called *germ-line cells*, by a special form of cell division called **meiosis,** in which the chromosome number is halved. A further step, *maturation*, sculpts the distinctive characteristics of the sperm and egg. Meiosis plus maturation constitute gametogenesis (making gametes). Meiosis occurs in sexually reproducing species (Reading 8.1).

Stages of Meiosis

Meiosis entails two divisions of the genetic material. The first division is called **reduction division** (or meiosis I) because it reduces the number of chromosomes in humans from 46 to 23. The second division, called the **equational division** (or meiosis II), is much like a mitotic division, producing four cells from the two cells formed in the first division (fig. 8.3).

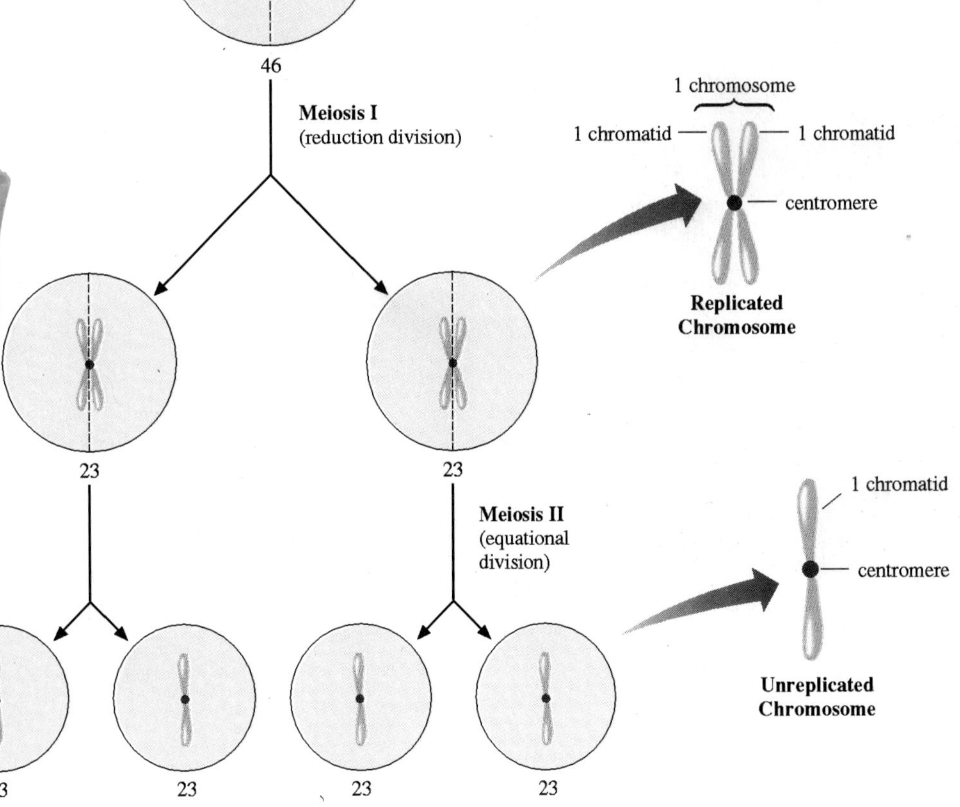

Figure 8.3
Meiosis is a special form of cell division in which certain cells are set aside to give rise to haploid germ cells. In humans the first meiotic division reduces the number of chromosomes to 23, all in the replicated form (inset). In the second meiotic division, each cell from the first division essentially undergoes mitosis. The result of the two divisions of meiosis is four haploid cells.

Reading 8.1 *Different Routes of Reproduction*

AN OBVIOUS REQUIREMENT FOR THE CONTINUED EXISTENCE OF LIFE ON EARTH IS FOR ORGANISMS TO REPRODUCE. Reproduction generally follows two routes. In **asexual reproduction,** a single cell divides mitotically to produce two genetically identical daughter cells. The growth of a new plant from a "cutting" is also a form of asexual reproduction, because the new plant is genetically identical to the original plant and is derived from somatic cells. In **sexual reproduction,** a cell containing a new combination of genes is produced by the fusion of two genetically different haploid parent cells formed by meiosis. A fertilized egg develops into a unique individual.

The single-celled *Amoeba proteus* (fig. 1) follows an asexual life-style by splitting in two (binary fission). The bacterium *Escherichia coli* usually reproduces asexually by binary fission, but it can also transfer genetic material to another bacterium by way of a projection from its cell membrane called a sex pilus (fig. 2). This transfer of DNA between bacterial cells is similar to sexual reproduction in that a new combination of genes results. The yeast *Saccharomyces cerevisiae* is a single-celled organism that can reproduce either asexually or sexually. A single diploid cell can "bud" mitotically to produce genetically identical diploid daughter cells (fig. 3). Alternatively, yeast can form a meiotic structure called an ascus, which gives rise to haploid germ cells that can then fuse to produce diploid cells. Plants have an alternation of haploid and diploid generations that allows individuals with new gene combinations to form.

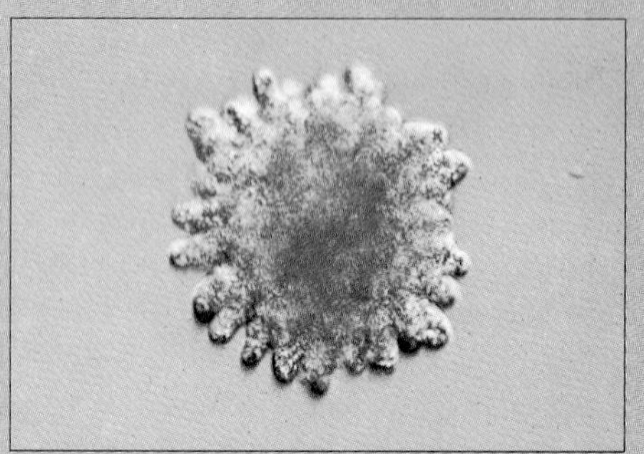

a. 0 minutes

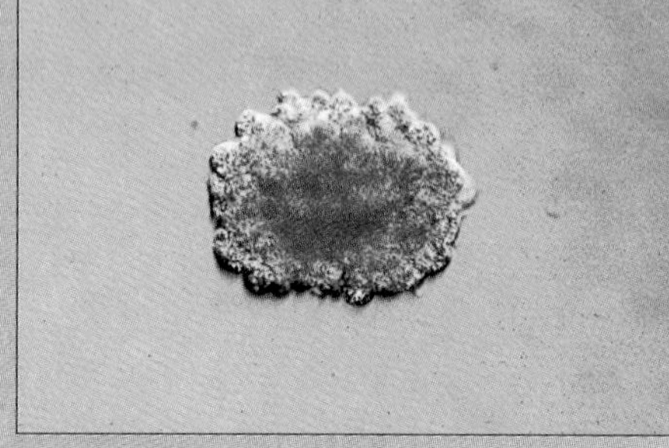

b. + 6 minutes

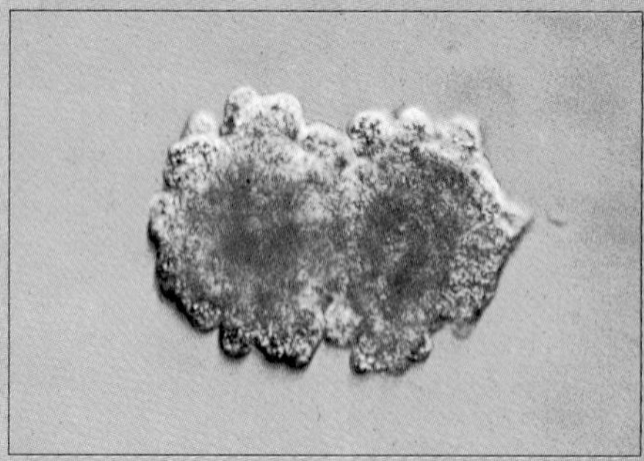

c. + 8 minutes

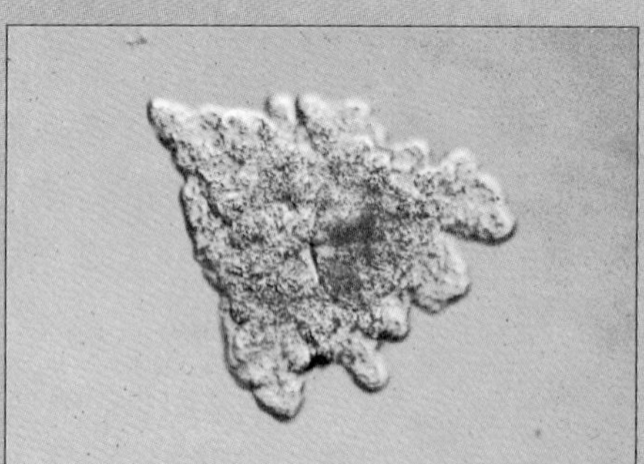

d. + 13 minutes

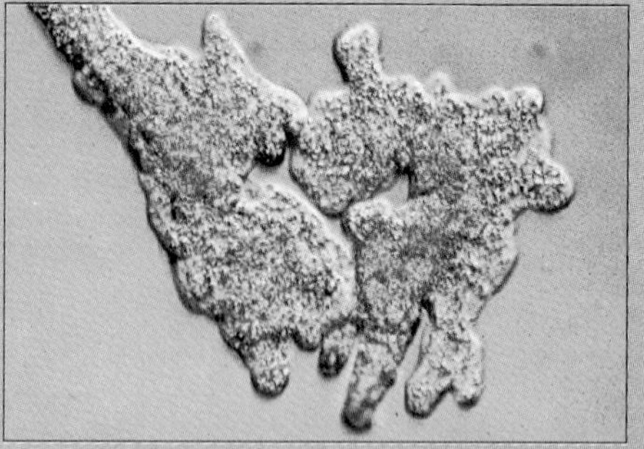

e. + 18 minutes

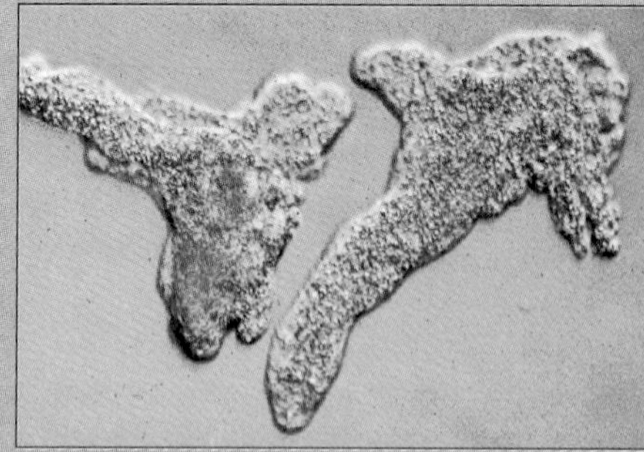

f. + 21 minutes

Figure 1
Amoeba proteus undergoing binary fission.

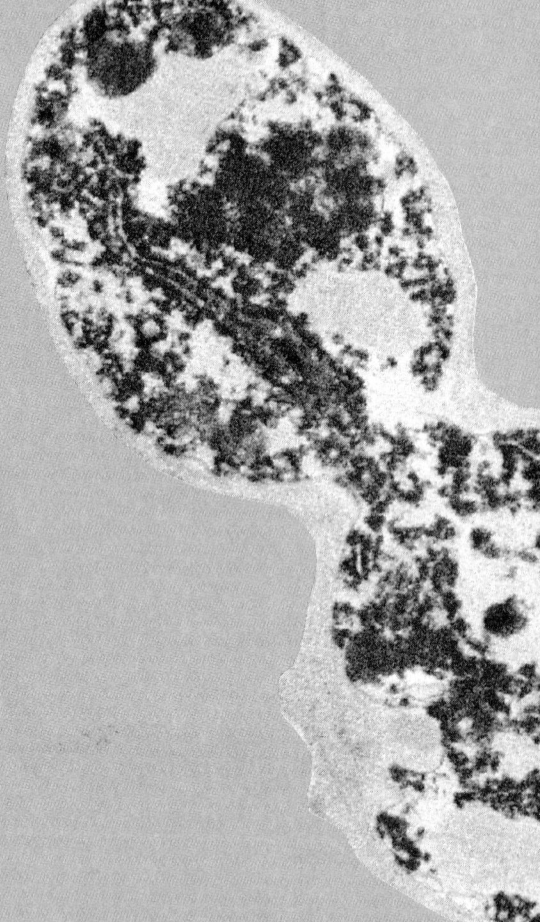

Figure 3
Yeast budding.

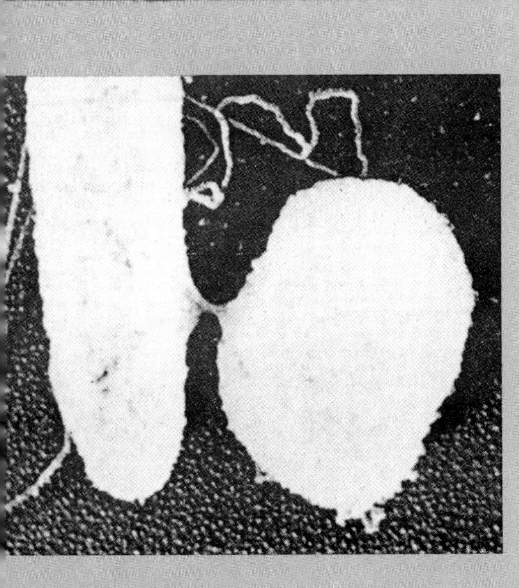

Figure 2
E. coli transferring genetic material to another cell using a sex pilus.

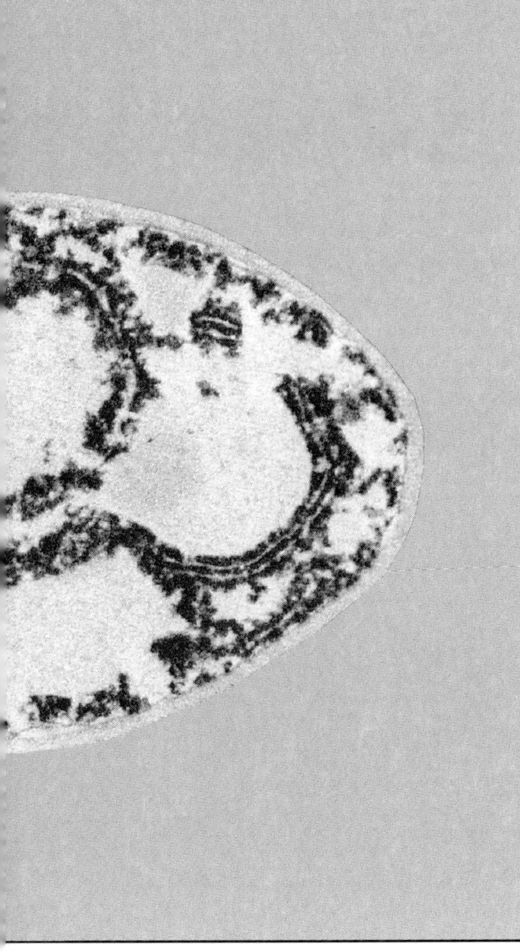

Table 8.1
Comparison of Mitosis and Meiosis

Mitosis	*Meiosis*
One division	Two divisions
Two daughter cells per cycle	Four daughter cells per cycle
Daughter cells genetically identical	Daughter cells genetically different
Chromosome number of daughter cells same as that of parent cell (2*n*)	Chromosome number of daughter cells half that of parent cell (*n*)
Occurs in somatic cells	Occurs in primordial germ cells
Occurs throughout life cycle	In humans, completed only after sexual maturity
Used for growth, repair, and asexual reproduction	Used for sexual reproduction, in which new gene combinations arise

As in mitosis, meiosis is preceded by an interphase period during which the genetic material replicates (table 8.1). The germ-line cell in which meiosis begins has two of each type of chromosome (that is, it is diploid). These chromosome pairs are called **homologous pairs,** or just homologs for short. Homologs look alike and carry the genes for the same traits in the same sequence. One of each homologous pair comes from the person's mother, and the other from the father. When meiosis begins, each homolog is replicated to form two chromatids joined by a centromere (fig. 8.3). The chromosomes are not yet condensed enough to be visible.

Interphase is followed by *prophase I* (so called because it is the prophase of meiosis I). Early in prophase I, replicated chromosomes condense and become visible (fig. 8.4*a*). Towards the middle of prophase I, the homologs line up next to one another, gene by gene, a phenomenon called **synapsis**. The paired chromosomes are held together by a "glue" made of RNA and protein. Toward the end of prophase I, the synapsed chromosomes pull apart, but they remain attached at a few points along their lengths (fig. 8.4*b*). These points of contact are the sites at which the two homologs exchange parts in a process called **crossing over** (fig. 8.5).

Crossing over is one reason why siblings are never exactly alike, unless they are identical twins. This exchange of genetic material between the two chromosomes of a pair provides variety by producing chromosomes with new combinations of genes. In any diploid cell of a sexually reproducing organism, one member of each chromosome pair is derived from each parent. After crossing over, each chromosome in the pair contains genes from each parent. New gene combinations arise from crossing over when the parents carry different forms of the same gene, called **alleles**—blue or brown alleles of an eye-color gene, for example.

Consider a simplified example. Suppose that the two chromosomes of a pair carry genes for hair color, eye color, and finger length. One of the chromosomes has alleles of those genes for blond hair, blue eyes, and short fingers. Its homolog carries alleles of the same genes but for black hair, brown eyes, and long fingers. After crossing over, one of the chromosomes might bear alleles for blond hair, brown eyes, and long fingers, the other having alleles for black hair, blue eyes, and short fingers. Thus, each of the chromosomes carries a new combination of alleles.

Meiosis continues in *metaphase I,* when the pairs of chromosomes line up at a region called the metaphase plate (fig. 8.4*c*). A spindle forms, and each chromosome of a homologous pair attaches to a spindle fiber that goes to one pole of the cell, such that each member is anchored to an opposite end of the cell.

The precise pattern in which the chromosomes line up during metaphase is important in generating genetic diversity. Recall that one chromosome of each homologous pair comes from the mother of the organism undergoing meiosis, and the other comes from the father. When the chromosome pairs align down the center of the spindle of the metaphase I cell, whether a maternally or paternally derived chromosome goes to one pole or another occurs at random. That is, all of the maternally derived chromosomes do not necessarily attach to the spindle fibers emanating from one pole and the paternally derived chromosomes to the other side.

The greater number of chromosomes an organism has, the more genetic diversity is possible. If an organism has two pairs of homologs, then four (2^2) different metaphase configurations are possible (fig. 8.6). For three pairs of homologs, eight (2^3) different configurations are possible. Using the formula 2^n, where n equals the number of homologous pairs, we can calculate that the 23 chromosome pairs of a human can line up in 2^{23}—or 8,388,608—different ways. This random arrangement of the members of homologous pairs in the metaphase cell is called **independent assortment,** and is discussed in chapter 11. It accounts for a basic law of inheritance.

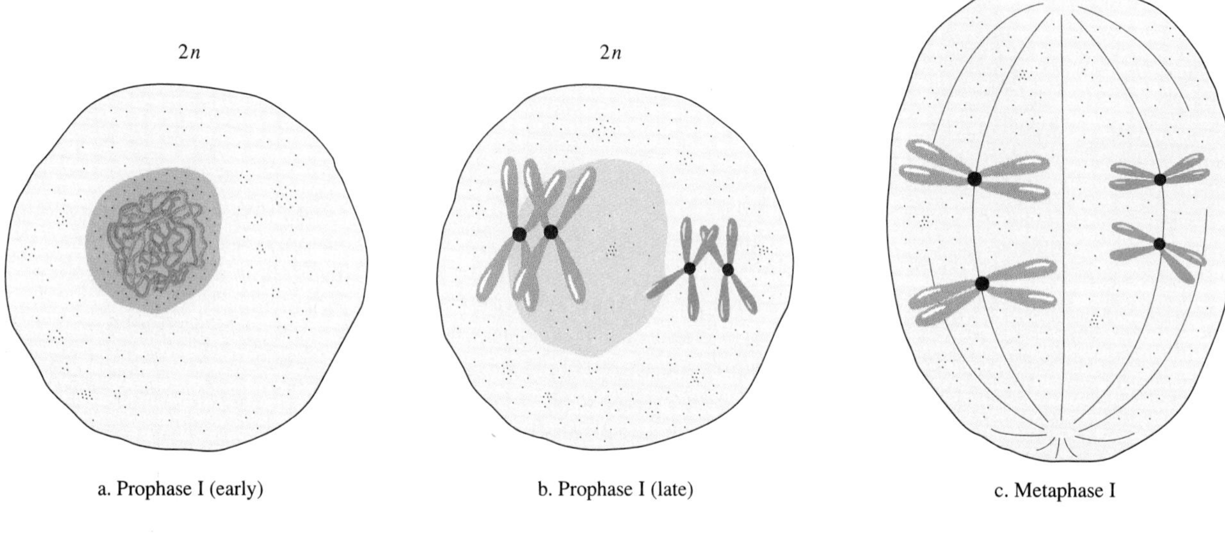

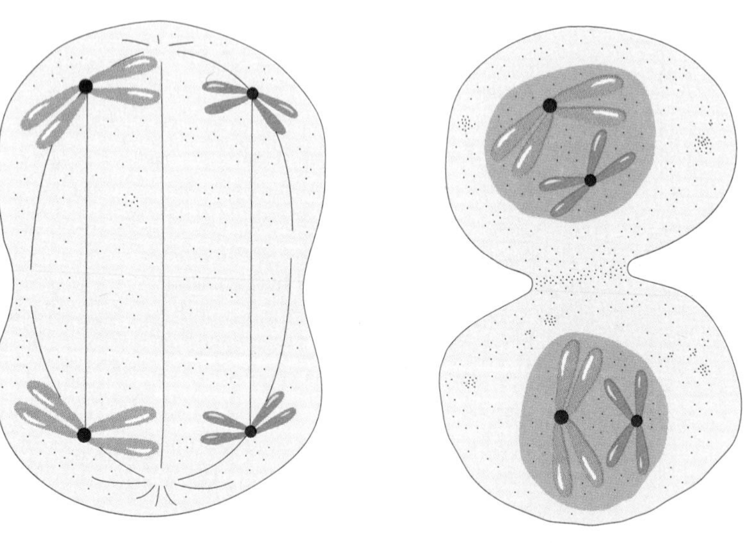

Figure 8.4
a. In early prophase I, replicated chromosomes condense and become visible as a tangle within the nucleus. *b.* By late prophase I, the pairs are aligned and the homologs cross over. *c.* In metaphase I, spindle fibers align the homologs, and (*d*) in anaphase I, the homologs move to opposite pole. *e.* In telophase I, the genetic material is partitioned into two daughter nuclei, each containing only one homolog from each pair. In most species cytokinesis occurs between the meiotic divisions, forming two cells after telophase I.

Homologs separate in *anaphase I*, and they complete their movement to opposite poles in *telophase I* (figs. 8.4*d* and 8.4*e*). Nuclear envelopes may form at this time or in the ensuing interphase that precedes the second meiotic division in some species. During this second interphase, the chromosomes once again become invisible, spreading into very thin threads. Proteins may be manufactured, but the genetic material is not replicated a second time. It is the single DNA replication, followed by the double division of meiosis, that halves the chromosome number.

Prophase II marks the start of the second meiotic division. The chromosomes are again condensed and visible (fig. 8.7*a*). In *metaphase II* (fig. 8.7*b*) the chromosomes (each consisting of a pair of chromatids) align down the center of the spindle. In *anaphase II* (fig. 8.7*c*), the centromeres split, each chromatid pair divides in two, and the resulting single-chromatid chromosomes are pulled to opposite poles. In *telophase II* (fig. 8.7*d*) nuclear envelopes form around the four nuclei (from each of the two cells produced by meiosis I).

Cytokinesis completes the process of meiosis by separating the four nuclei into individual cells (fig. 8.7*e*). In most species cytokinesis occurs between the first and second meiotic divisions, forming first two cells, and then four. In some species, cytokinesis occurs only following telophase II, separating the four nuclei enclosed in one large cell into four smaller haploid cells. The net result of meiosis: four haploid cells, each carrying a new assortment of genes and chromosomes.

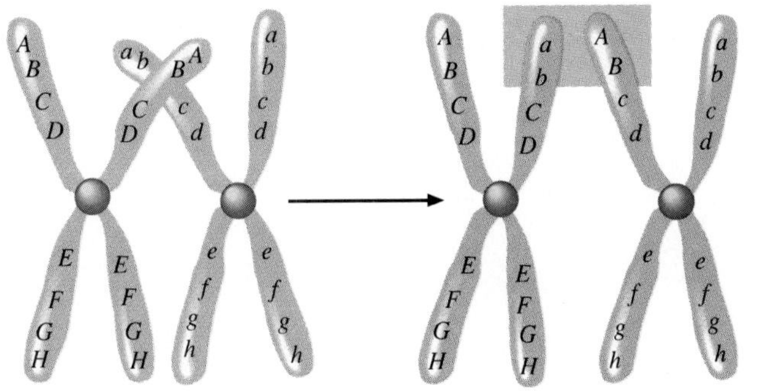

Meiosis Leads to Genetic Variability

Meiosis provides a mechanism for generating astounding genetic variety. Within homologs, crossing over mixes up the genetic contributions from the previous generation. Independent assortment introduces

Figure 8.5
Crossing over recombines genes. The capital and lowercase forms of the same letter represent different forms (alleles) of the same gene.

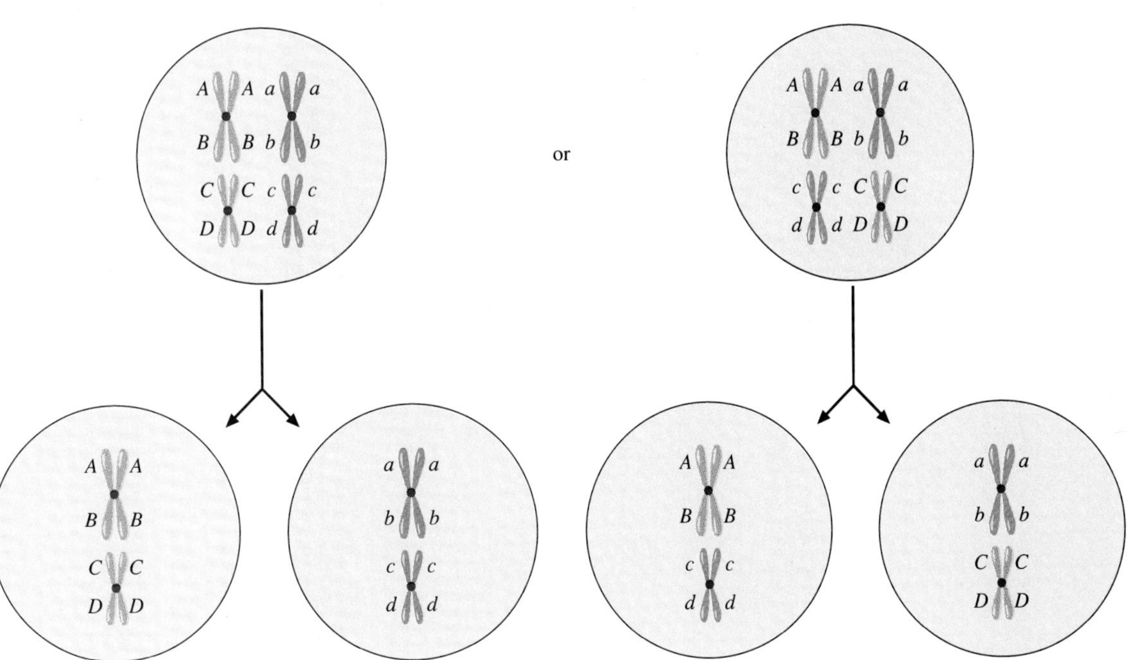

Figure 8.6
Independent assortment. The pattern in which homologs align during metaphase I determines the combination of chromosomes in the daughter cells. This illustration follows two chromosome pairs with different alleles of the same gene indicated by capital and lowercase forms of the same letter. Two pairs of chromosomes can align in two different ways to produce four different possibilities in the daughter cells. The potential variability generated by meiosis skyrockets when all 23 chromosome pairs and the effects of crossing over are considered.

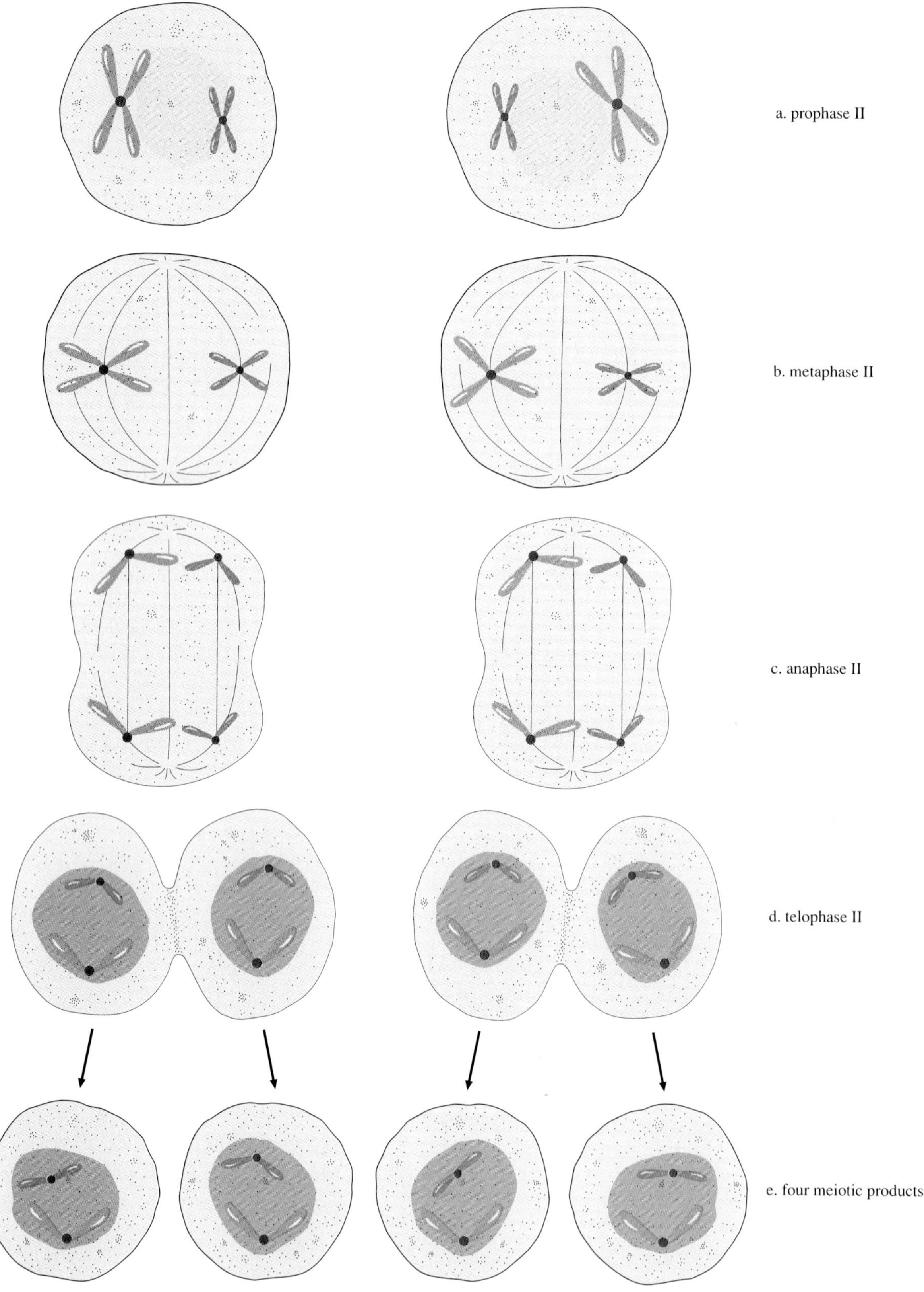

Figure 8.7
Meiosis II. The second meiotic division is very similar to mitosis. *a.* In prophase II, the chromosomes are visible. *b.* In metaphase II, the spindle apparatus aligns the chromosomes. *c.* In anaphase II, the centromeres split, and each chromatid pair divides into two chromosomes, which are pulled toward opposite poles. *d.* In telophase II in some species, each of the two separated sets of chromosomes is enclosed in a nuclear membrane, and (*e*) in cytokinesis they are partitioned into individual haploid cells. The net yield from the entire process of meiosis: four haploid daughter cells.

yet more variability in the way that chromosomes are packaged into daughter cells. Independent assortment provides more than 8 million different ways for human chromosomes to distribute themselves. An additional source of variability is that any one of a person's more than 8 million possible combinations of chromosomes can combine with any one of the more than 8 million combinations of his or her mate. This raises the possible variability to more than 70 trillion (8,388,608 multiplied by itself) genetically unique individuals possible for the next generation. When crossing over is considered, the number of genetic combinations that can result from the meeting of a single sperm and a single ovum is staggering.

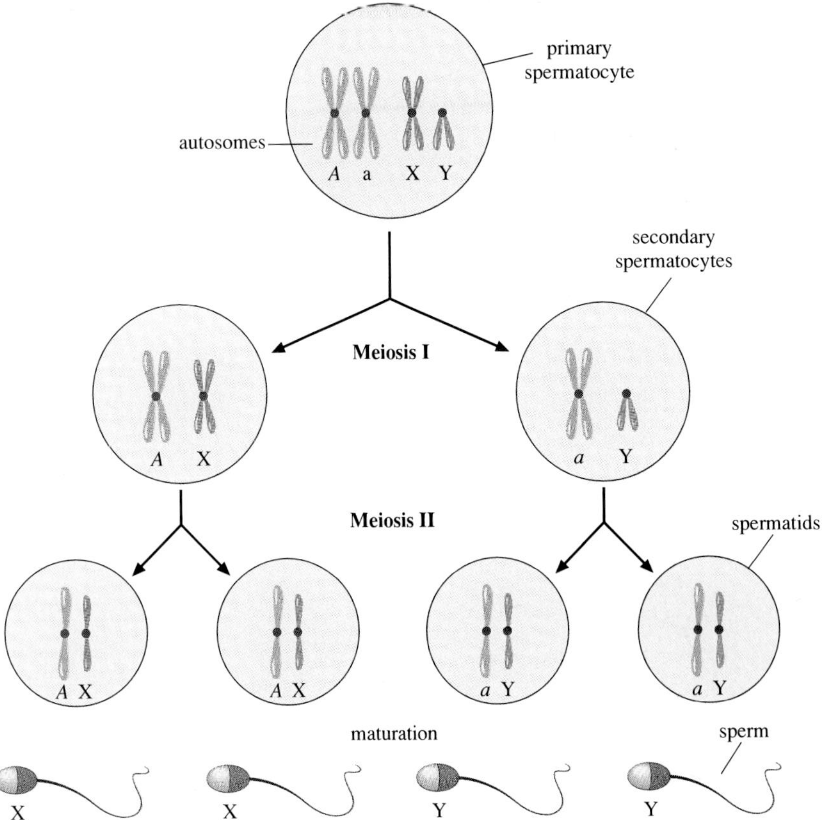

Figure 8.8
Sperm formation. In humans, primary spermatocytes have the normal diploid number of 23 chromosome pairs. The yellow pair of chromosomes represents a pair of autosomes, and the pink pair represents the sex chromosomes.

KEY CONCEPTS

Sperm and oocytes are haploid and are derived from special diploid germ-line cells by meiosis and maturation. Meiosis is important because it enables a species' chromosome number to remain constant over generations, and because it generates novel combinations of genes. In the first meiotic, or reduction, division, the number of chromosomes is halved. Reduction division is followed by the second meiotic, or equational, division, in which each of two cells from the first meiotic division divides mitotically, yielding four cells from the original one. Each meiotic division proceeds through the stages of prophase, metaphase, anaphase, and telophase, similar to mitosis. The chromosome number is halved because there are two divisions but only one round of DNA replication. Genetic diversity is generated by crossing over and by the independent assortment of homologs. Cytokinesis completes meiosis.

Development of the Sperm—Spermatogenesis

Although meiosis occurs in males and females, differences in the process between the sexes ensure that each type of germ cell is structurally suited for its particular function. Thus, the product of male meiosis is the motile, lightweight sperm, which must travel to meet its mate; the result of female meiosis is the relatively large ovum, packed with nutrients and organelles. We will look first at meiosis in the male.

The differentiation of sperm cells is called **spermatogenesis** (sperm making) (fig. 8.8). A diploid cell destined to produce sperm cells is called a *spermatogonium*. Several spermatogonia are attached to each other by bridges of cytoplasm, and they undergo meiosis together. The spermatogonia accumulate cytoplasm and replicate their genetic material, becoming *primary spermatocytes*. During reduction division (meiosis I), each primary spermatocyte divides to form two equal-sized haploid cells called *secondary spermatocytes*. In meiosis II, each secondary spermatocyte divides to yield two equal-sized *spermatids*. Each spermatid then specializes, developing the characteristic sperm "tail," or flagellum. The tail has many mitochondria and ATP molecules, and this energy system enables the sperm to swim once it is inside the female reproductive tract.

Up until this point, the products of meiosis are still attached by a bridge of cytoplasm. After spermatid formation, some of the cytoplasm is stripped away, leaving mature, tadpole-shaped *spermatozoa*, or sperm cells for short (fig. 8.9). Each sperm cell consists of a tail, body, and head region and has a small protrusion on its front end called the **acrosome**. This bump contains enzymes that will help the cell penetrate the ovum's outer membrane. By the time meiosis and maturation are completed, each spermatogonium has yielded four sperm cells.

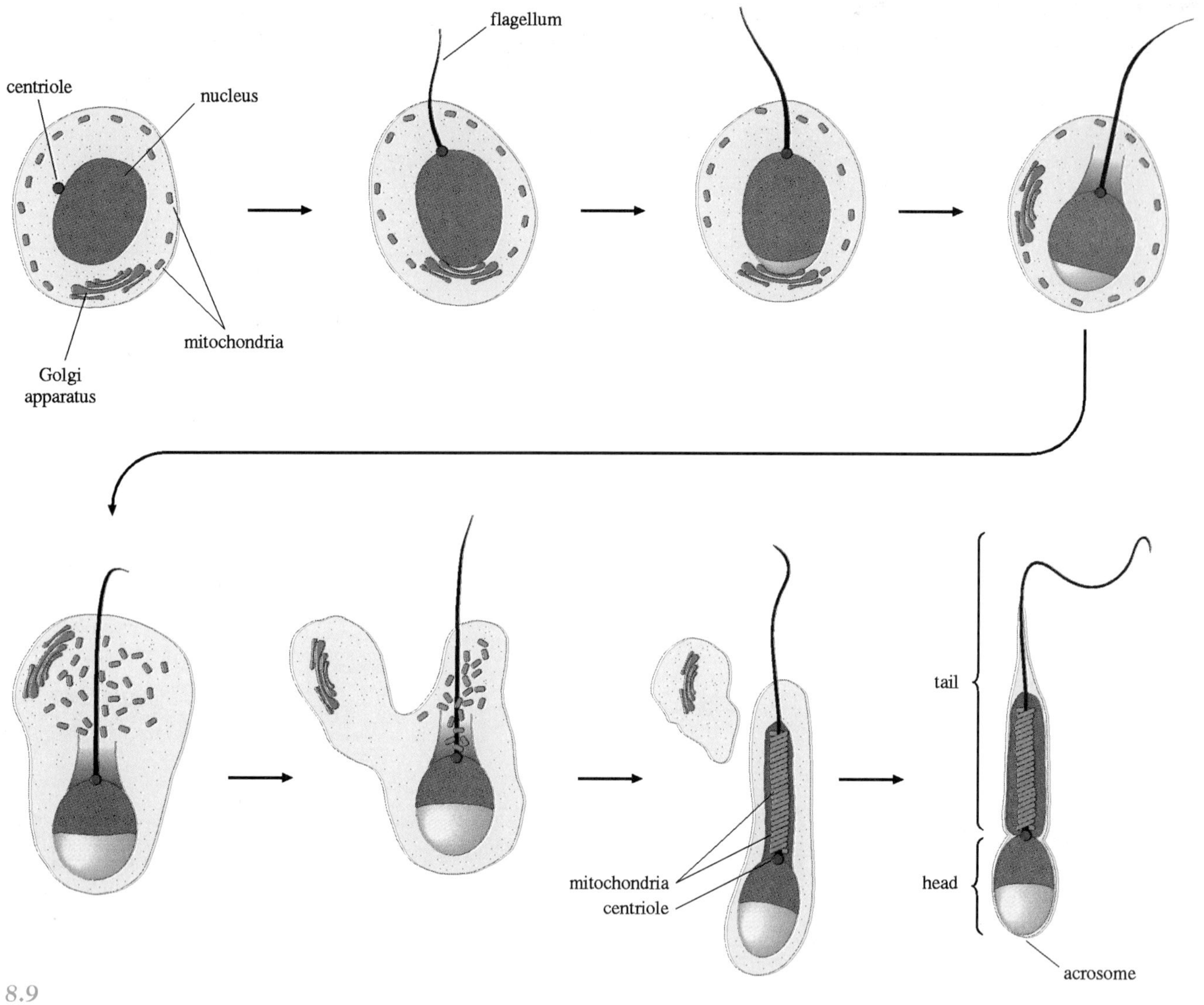

flagellum
centriole
nucleus
mitochondria
Golgi apparatus
mitochondria
centriole
tail
head
acrosome

Figure 8.9
From spermatogonium to spermatozoan. The head of the mature sperm consists mostly of the cell's nucleus. The tail elongates as excess cytoplasm is stripped away.

Male meiosis begins in the seminiferous tubules. Spermatogonia reside at the side of the tubule that is farthest from the lumen (the central cavity). When a spermatogonium divides, one daughter cell moves towards the lumen and accumulates cytoplasm, becoming a primary spermatocyte. The other daughter cell remains in the tubule wall. It is a type of stem cell, because it is unspecialized, but continually gives rise to cells that specialize.

The developing sperm cells travel towards the lumen of the seminiferous tubule, and by the time they are released into it, they are spermatids. The spermatids are stored in the epididymis, where they complete their differentiation into spermatozoa. The entire process, from spermatogonium to spermatocyte, takes about 2 months. A human male manufactures trillions of sperm in his lifetime (fig. 8.10).

When the epididymis contracts during ejaculation, the sperm cells are propelled into the vas deferens, which passes through the **prostate** gland and joins the urethra. The sperm, along with the secretions from the accessory glands, form semen, which exits the body through the penis. Then the challenge begins—the sperm, a mere 0.0023 inch (0.06 millimeter) long must swim about 7 inches (180 millimeters) to reach the product of female meiosis.

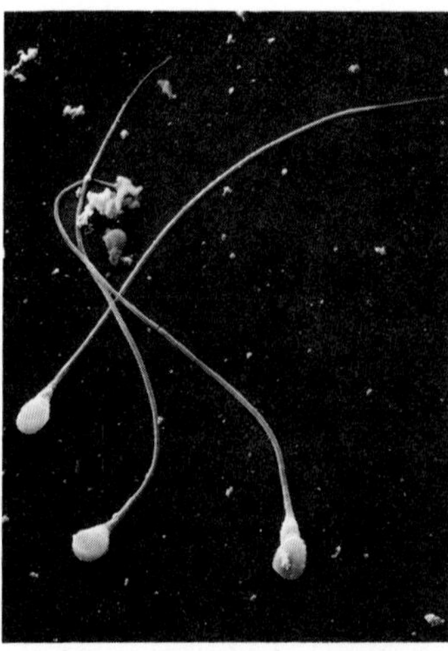

Figure 8.10
Scanning electron micrograph of human sperm cells.

KEY CONCEPTS

Sperm development begins with spermatogonia, which are located deep within the walls of the seminiferous tubules. A diploid spermatogonium divides mitotically, yielding one stem cell and one cell that accumulates cytoplasm, becoming a primary spermatocyte as it moves toward the lumen of the tubule. In meiosis I, each spermatogonium halves its genetic material to form two secondary spermatocytes. In meiosis II, each secondary spermatocyte divides to yield two equal-sized spermatids. In the epididymis, the spermatids further specialize, developing their characteristic mitochondria-laden flagella. These developing sperm cells are attached to each other by bridges of cytoplasm. When the spermatids mature, the cells separate and excess cytoplasm is shed. The mature spermatocyte has a tail, body, and head region, with a bump on the head, the acrosome, which contains enzymes.

Development of the Ovum—Oogenesis

Meiosis in the female is called **oogenesis** (egg-making), and it begins, like spermatogenesis, with a diploid cell. This cell is called an *oogonium*. Unlike the male cells, oogonia are not attached to each other but are surrounded by a layer of follicle cells. Each oogonium grows, accumulating cytoplasm and replicating its chromosomes, becoming a *primary oocyte*. The ensuing meiotic division in oogenesis, unlike those in spermatogenesis, produces cells of different sizes.

In meiosis I, the primary oocyte divides to become a small cell with very little cytoplasm, called a **polar body,** and a much larger cell called a *secondary oocyte* (fig. 8.11). Each is haploid. In meiosis II, the tiny polar body may divide to yield two polar bodies of equal size, or it may simply decompose. The secondary oocyte, however, divides unequally in meiosis II to produce a small polar body and the mature egg cell or ovum, which contains a large amount of cytoplasm. In other words, most of the cytoplasm among the four meiotic products is concentrated in only one of them, the ovum. Polar bodies are absorbed by the woman's body and normally play no further role in development.

The timetable for ovum formation differs greatly from that of sperm formation. Six months before a human female is born, her ovaries contain 2 million or more primary oocytes. From then on, the number of primary oocytes declines. A million are present by birth, and about 400,000 remain by the time of puberty. At birth, the oocytes are arrested in prophase I. After puberty, meiosis I is completed in only one or a few oocytes each month. These oocytes stop meiosis again, this time at metaphase II. At specific hormonal cues each month, one such secondary oocyte is released from an ovary, or *ovulated*. If the oocyte membrane is penetrated by a sperm, then meiosis is completed, and a fertilized ovum forms. Therefore, female meiosis does not even finish unless the process of fertilization has already begun. If the secondary oocyte is not fertilized, it degenerates and leaves the body in the menstrual flow.

A female will only ovulate about 400 oocytes between puberty and menopause. Only a few of these are likely to be entered by a sperm cell. Only 1 in 3 of those oocytes that do meet and merge with a sperm cell will continue growing, dividing, and specializing to form a new human life.

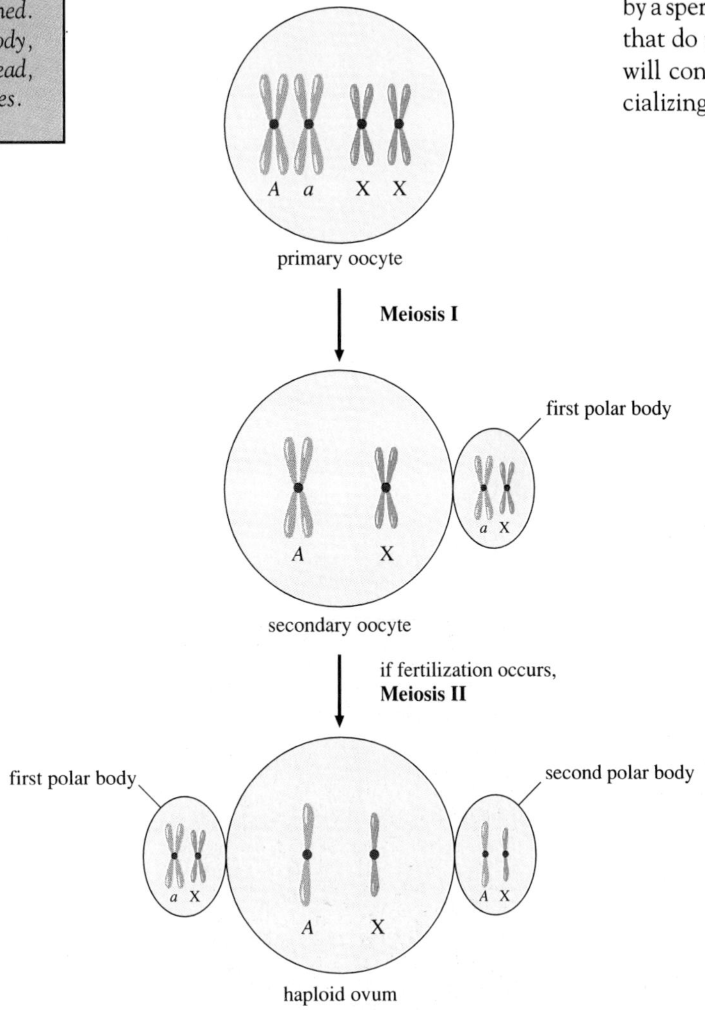

Figure 8.11
Ovum formation. In humans, primary oocytes have the normal diploid number of 23 chromosome pairs. The yellow pair of chromosomes represents a pair of autosomes, and the pink pair represents the sex chromosomes.

KEY CONCEPTS

Ovum development begins in the ovary in an oogonium. An oogonium accumulates cytoplasm and replicates its chromosomes, becoming a primary oocyte. In meiosis I, the primary ooctye divides to form a small polar body and a large, haploid secondary oocyte. In meiosis II, the secondary ooctye divides to yield another small polar body and a mature ovum. The million oocytes that a female is born with are arrested at prophase I. One oocyte continues meiosis as it is ovulated, and meiosis is completed only when a secondary oocyte is fertilized.

SUMMARY

The cells that combine to form a new human are produced in the male and female reproductive systems. These systems are built of paired structures in which sperm and ova are manufactured, networks of tubes, and glands. Male *germ cells (spermatozoa)* originate in *seminiferous tubules* within the paired *testes* and pass through a series of tubes, the *epididymis* and *vas deferentia*, where they mature and are stored before exiting the body through the *urethra* during sexual intercourse. The *prostate gland*, the *seminal vesicles*, and the *bulbourethral glands* add secretions to the sperm. Female germ cells originate in the *ovary*, and each month after puberty, one cell is released into a *fallopian tube* and travels to the uterus.

Gametogenesis produces the male sperm and female ovum that fuse to form the first cell of a new individual, the fertilized ovum or *zygote*. *Meiosis* is a special form of cell division that halves the chromosome number in germ cells from that in somatic cells. As a result, the chromosome number of a species stays the same. This constancy is achieved by two cell divisions of meiosis, with only one round of DNA replication. Meiosis ensures genetic variability by partitioning different combinations of genes into germ cells, as a result of *crossing over* and *independent assortment*. Maturation completes the manufacture of the gametes.

Spermatogenesis begins with *spermatogonia*, which accumulate cytoplasm and replicate their DNA, becoming *primary spermatocytes*. After *reduction division* (meiosis I), the cells are haploid *secondary spermatocytes*. In meiosis II, an *equational division*, the secondary spermatocytes divide to each yield two spermatids, which then differentiate as they travel along the reproductive tract of the male.

In *oogenesis*, *oogonia* grow and replicate their DNA, becoming *primary oocytes*. In meiosis I, the primary oocyte divides with an unequal apportionment of cytoplasm to yield one large *secondary oocyte* and a much smaller *polar body*. In meiosis II, the secondary oocyte divides to yield the large ovum and another small polar body. The development of a sperm cell takes about 2 months. The development of an ovum takes at least 13 years, and female meiosis is completed only at fertilization.

QUESTIONS

1. A dog has 39 pairs of chromosomes. Considering only independent assortment (the lining up at random of maternally and paternally derived chromosomes), how many genetically different puppies are possible from the mating of two dogs? Is this number an underestimate or overestimate? Why?

2. In a birth-control technique called "coitus interruptus," the penis is withdrawn from the vagina just before ejaculation. However, a small drop of semen often escapes the penis just before ejaculation, sometimes before the man can withdraw. How can a single drop of semen released into the vagina render this form of birth control ineffective?

3. Define the following terms having to do with meiosis.

a. Crossing over
b. Gamete
c. Haploid
d. Homolog
e. Synapsis

4. In what ways are the male and female reproductive tracts similar?

5. How many sets of chromosomes are present in each of the following cell types?

a. An oogonium
b. A primary spermatocyte
c. A spermatid
d. A cell during anaphase of meiosis I, from either sex
e. A cell during anaphase of meiosis II, from either sex
f. A secondary oocyte
g. A polar body derived from a primary oocyte

TO THINK ABOUT

1. Why is the halving of the chromosome number accomplished in meiosis necessary for sexual reproduction?

2. Why is it extremely unlikely that a child will be genetically identical to a parent?

3. Many veterans of the Vietnam War who believe they were exposed to the herbicide Agent Orange claim that their children—born years after the exposure—had birth defects caused by a contaminant of the herbicide called dioxin. What types of cells in these men would have to have been affected by the chemical to have caused birth defects years later? Explain your answer.

4. A woman who is about 4 weeks pregnant suddenly begins to bleed and pass some tissue through her vagina. After a physician examines the material, he explains to her that a sperm fertilized a polar body instead of an ovum, and an embryo could not develop. What has happened? Why do you think a polar body cannot support the development of an embryo, whereas an ovum, which is genetically identical to it, can?

5. How do the structures of the male and female human germ cells aid in their functions?

SUGGESTED READINGS

Alberts, B. et al. 1989. *The molecular biology of the cell*. New York: Garland Publishing. Chapter 14 presents an in-depth look at meiosis and germ cell formation.

Gould, Stephen Jay. 1980. Dr. Down's syndrome. *The Panda's Thumb*, New York: W.W. Norton. An entertaining essay on the importance of meiosis, as evidenced by what can happen when it goes awry.

Keller, Sylvia Fox. 1983. *A feeling for the organism: The life and work of Barbara McClintock*. San Francisco: W.H. Freeman. Chapter 3 clearly explains meiosis.

Levine, Richard J. et al. July 5, 1990. Differences in the quality of semen in outdoor workers during summer and winter. *The New England Journal of Medicine*. Lower birth rates in the spring may reflect sensitivity of sperm to the heat of the previous summer.

C H A P T E R

9

A Human Life—Development Through Aging

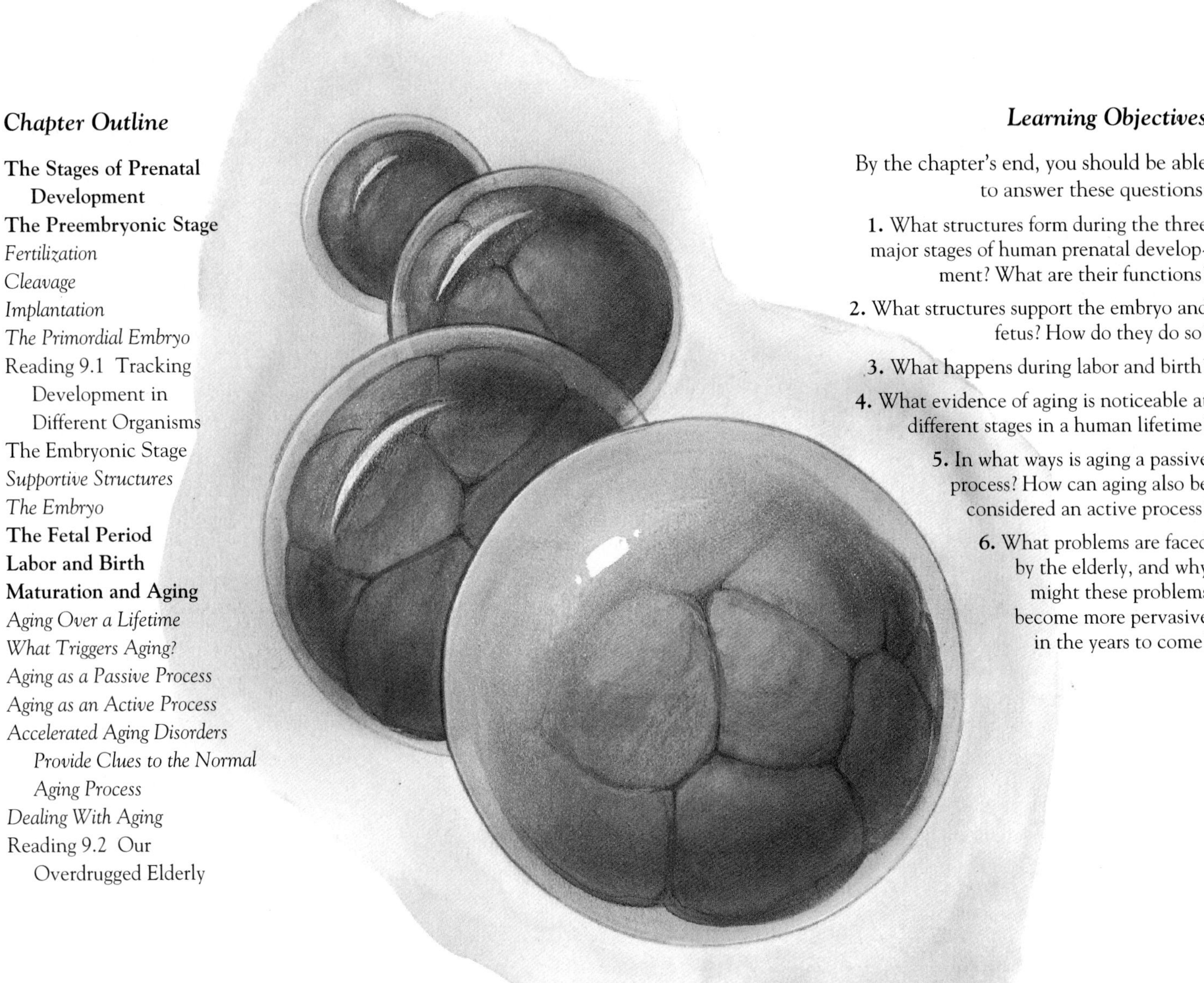

Chapter Outline

Learning Objectives

By the chapter's end, you should be able to answer these questions:

1. What structures form during the three major stages of human prenatal development? What are their functions?
2. What structures support the embryo and fetus? How do they do so?
3. What happens during labor and birth?
4. What evidence of aging is noticeable at different stages in a human lifetime?
5. In what ways is aging a passive process? How can aging also be considered an active process?
6. What problems are faced by the elderly, and why might these problems become more pervasive in the years to come?

The young woman did not feel quite right. She was so exhausted that getting out of bed in the morning was difficult, and she was more tired than usual at work. She had little appetite. At night she would curl up with a book right after dinner and fall asleep. When her fatigue did not lift in the time it would take a viral infection to run its course, she consulted her doctor.

It did not take the doctor long to determine the source of the woman's symptoms, and a blood test confirmed her suspicions in a few minutes by detecting a hormone manufactured by a human embryo but not by an adult. The woman was pregnant. Her menstrual periods were irregular, and when the most recent two periods were scant, she had not taken notice.

From feeling the size of the uterus, the doctor concluded that the woman was already about 7 weeks pregnant. To be certain when the woman would deliver, the doctor performed an ultrasound examination. The patient drank several glasses of water, then lay on a table. A small instrument was passed over her abdomen, painlessly bouncing sound waves off of her uterus. On a screen appeared a shadowy bean-shaped structure, with a pulsating blip in the center. This, the doctor said, was her baby-to-be.

The structure was a human embryo, and it already had arms, legs, toes, and fingers. The pulsating blip was its heart, already beating and sending blood through tiny vessels (fig. 9.1). The developing organism was in fact approaching a developmental milestone, the end of the embryonic stage of its *prenatal* (before birth) development. By 8 weeks, with all organs present in rudimentary form, it would be considered a fetus. The structures laid down during embryonic existence would continue to grow and specialize during the fetal period. Although prenatal development appears to be most extensive when the pregnant woman is obviously "with child," from a biological standpoint the most vital events take place in the first few months—sometimes before a woman even realizes that she is pregnant.

The Stages of Prenatal Development

During the 38 weeks of prenatal development a most amazing biological process unfolds. Through an enormous number of cell divisions and precise cell specializations and interactions, a single cell gradually gives rise to a newborn baby. The newborn is built of some 200 million cells and can move, breathe, digest, excrete, think, sense, and respond to the environment. The series of events that mold the newborn from the fertilized ovum is called *morphogenesis*.

Three stages of morphogenesis are recognized. The first 2 weeks are the *preembryonic stage*, and it includes fertilization, division of the fertilized ovum as it is transported through the woman's fallopian tube towards the uterus, implantation into the uterine wall, and formation of a structure with three distinct layers that will develop into the embryo.

The *embryonic stage* lasts from the start of the third week until the end of the eighth week. During this vital period the cells of the three layers laid down in the second week grow, specialize, and interact to form all of the body's organs. Structures that support the embryo also develop—the **placenta, umbilical cord,** and **extraembryonic membranes.**

The third stage of prenatal development is the *fetal period*, when organs begin to function and coordinate to form organ systems. Growth is very rapid. Prenatal development ends with *labor* and *parturition*, which is the birth of a baby. Table 9.1 lists these major stages in the prenatal development of humans and other animals, and Reading 9.1 highlights fascinating experiments that have contributed to our knowledge of animal development.

Let's take a closer look at the complex parade of events that culminate with the appearance of a new human being.

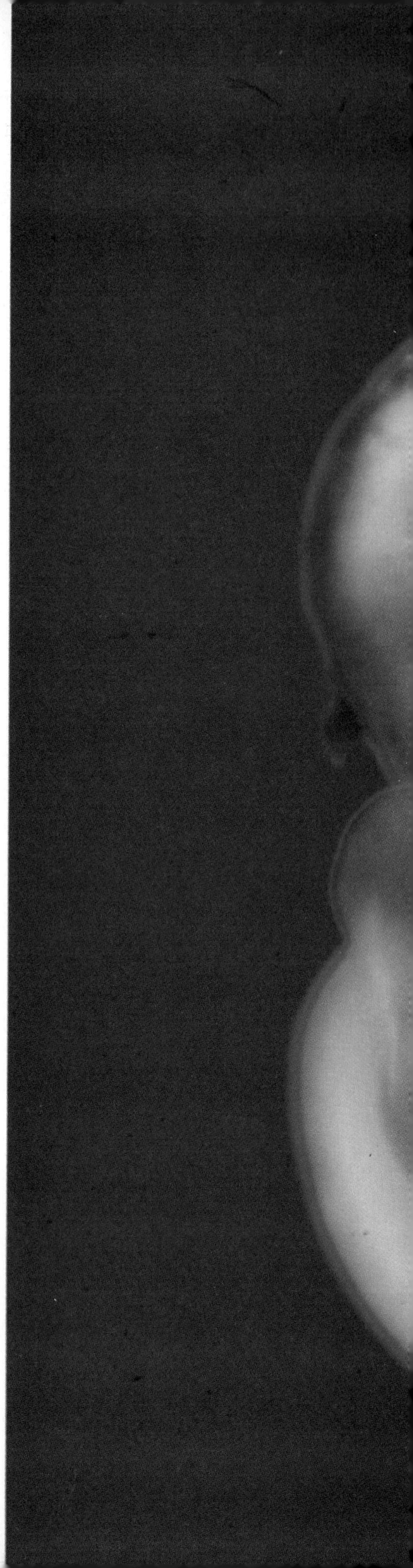

Figure 9.1
A human embryo at 6 weeks. Nearly all of the organs are well developed.

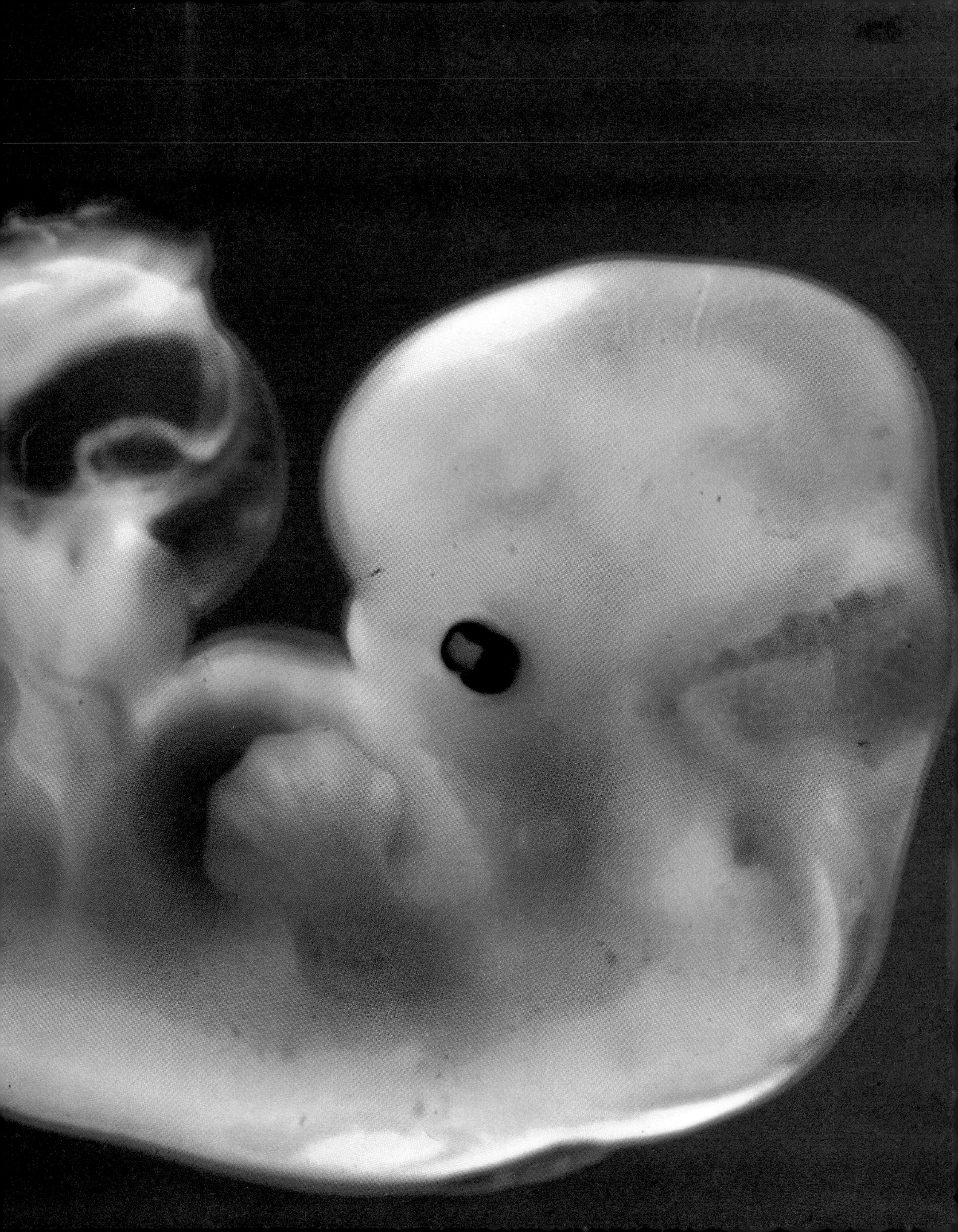

Table 9.1
Stages of Animal Development

Stage	*Events*	*Timing in Humans*
Gametogenesis	Manufacture of sperm and oocytes	In adults
Fertilization	Sperm and oocyte meet	Within 24 hours after ovulation
Cleavage	Rapid cell division	Days 1—3
Gastrulation	Germ layers form	Week 2
Organogenesis	Cells specialize, forming organs	Weeks 3 through 8
Further development	Growth and tissue specialization	Week 9 throughout life

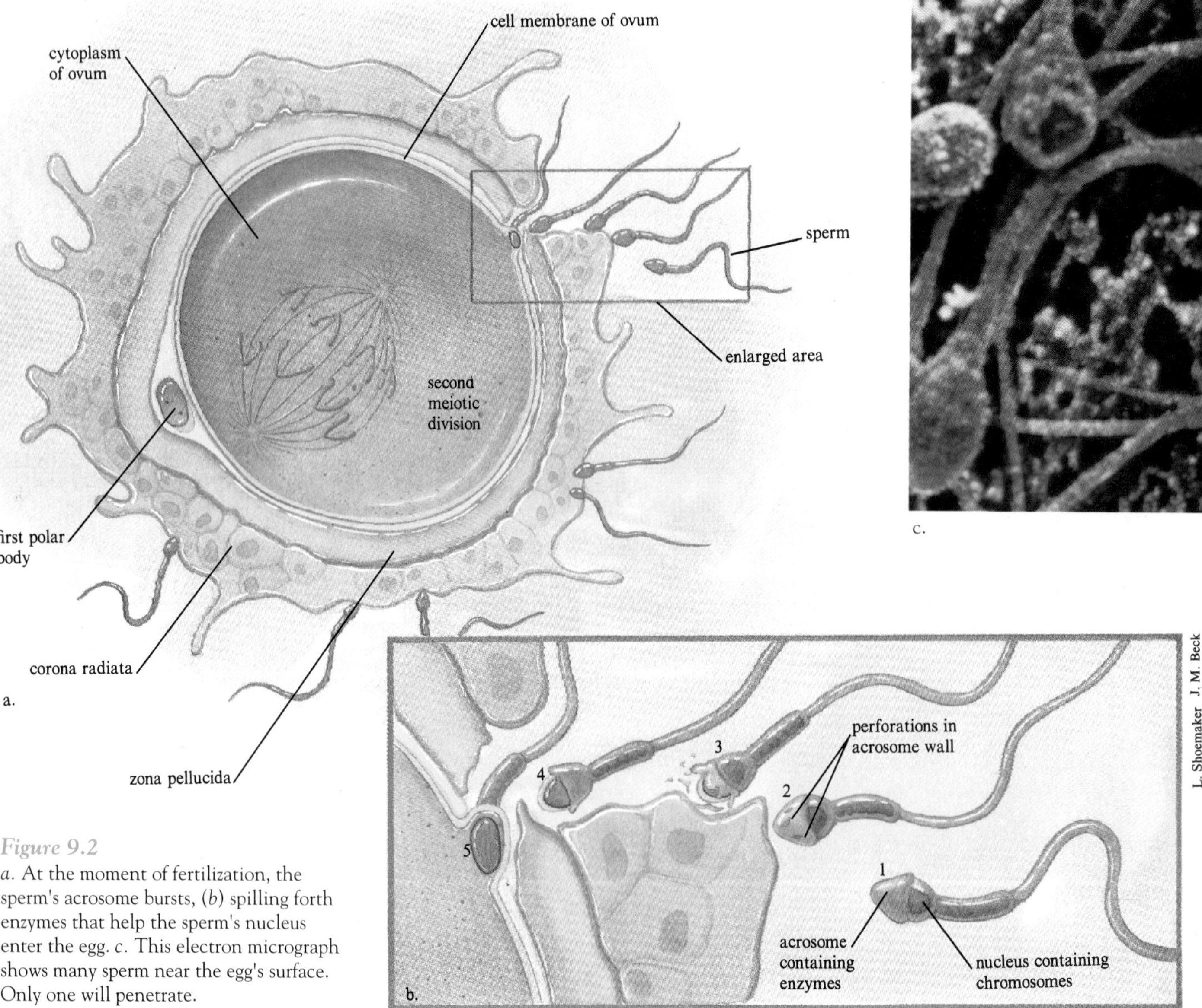

Figure 9.2
a. At the moment of fertilization, the sperm's acrosome bursts, (*b*) spilling forth enzymes that help the sperm's nucleus enter the egg. *c.* This electron micrograph shows many sperm near the egg's surface. Only one will penetrate.

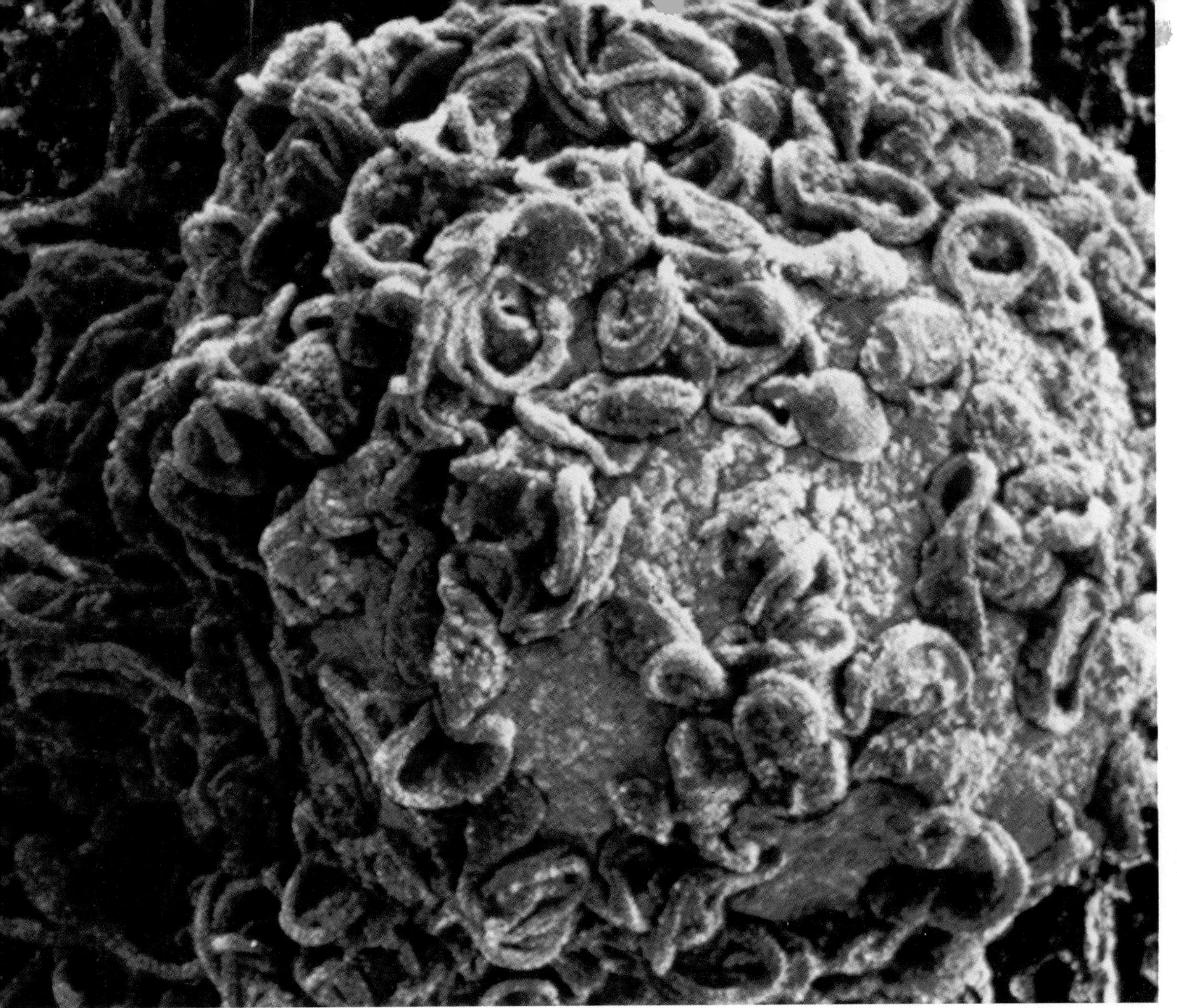

The Preembryonic Stage

Fertilization

The first step in prenatal development is the initial contact between the male's sperm cell and the female's secondary oocyte. Recall that the female cell is arrested in the metaphase of the second meiotic division. It is encased in a thin, clear layer of protein and sugars called the **zona pellucida,** which is further surrounded by a layer of cells called the **corona radiata** (fig. 9.2*a*).

Hundreds of millions of sperm cells are deposited in the vagina during sexual intercourse. Although a sperm cell can survive in the woman's body for up to 3 days, it can only fertilize the oocyte in the 12 to 24 hours following ovulation (the release of the oocyte from the ovary).

Reaching the female cell is quite a task—only 100 or so sperm cells swim far enough to encounter it. Before a sperm can fertilize an oocyte it must be activated by biochemicals in the female reproductive tract in a process called **capacitation,** which is not fully understood. After capacitation the sperm cells move upwards in the female reproductive tract. They are assisted in their journey by mucus that is swept into motion by cilia on the cells of the cervix and by contractions of the muscles in the walls of the vagina. When the sperm cells have traveled about two-thirds of the distance to the fallopian tube, their flagella help to move them along as well.

When a particular sperm makes contact with the secondary oocyte, its acrosome bursts, spilling its enzyme contents (fig. 9.2*b*). The enzymes then digest the zona pellucida and corona radiata protecting the oocyte. The moment of fertilization, also called *conception*, occurs when the outer membranes of the sperm and secondary oocyte meet. This is usually in the fallopian tube.

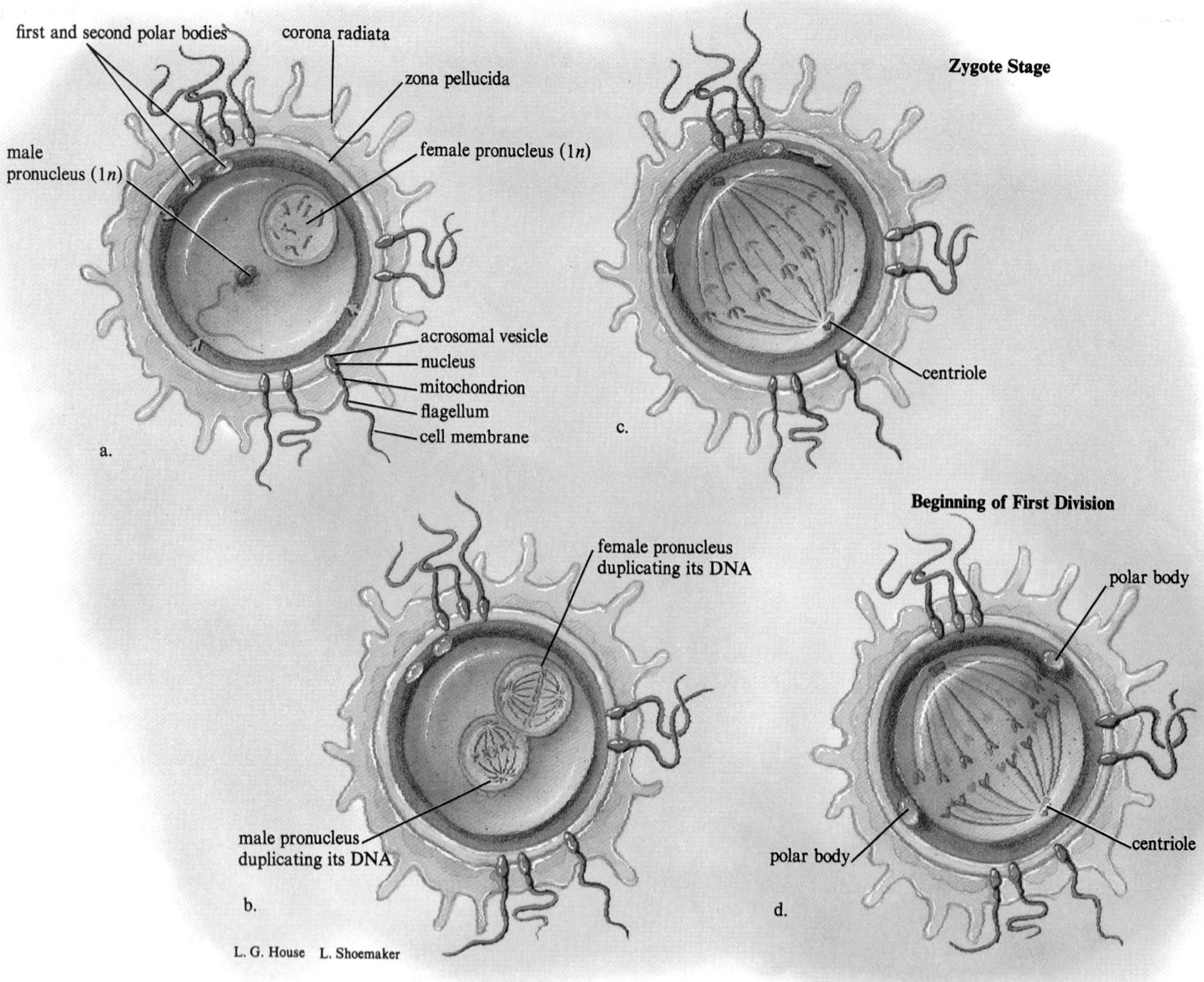

Figure 9.3
a. Once the sperm cell enters the egg, (*b*) its pronucleus approaches the egg's pronucleus, and (*c*) finally the 23 sets of chromosomes are together. *d.* The first cell of the new organism now exists.

At the site where the sperm touches the oocyte's surface, a wave of electricity is generated, spreading physical and chemical changes across the entire oocyte surface. These changes produce a "block to polyspermy," which means that they keep all other sperm out. If more than one sperm enters an oocyte, the resulting cell has too much genetic material to develop normally. Very rarely such individuals survive to be born, but they have defects in many organs and die within days. However, when two sperm fertilize two oocytes, fraternal twins result.

As the sperm enters the secondary oocyte, the female cell completes meiosis and becomes a fertilized ovum. Fertilization is not complete, however, until the genetic packages, or **pronuclei,** of the sperm and ovum meet. Within 12 hours of the sperm's penetration, the nuclear membrane of the ovum disappears, and the two sets of chromosomes mingle (fig. 9.3). The first cell of the new individual thus forms. It has 23 pairs of chromosomes and is called a **zygote** (Greek for "yolked together"). It is still within a fallopian tube.

Cleavage

By about 30 hours after fertilization, the zygote undergoes its first mitotic division. This starts a period of rapid cell division called **cleavage.** The resulting cells are called **blastomeres.** By 60 hours another division has occurred, and the zygote now consists of 4 cells. Cleavage continues until a solid ball of 16 or more cells forms. It is called a **morula,** which is Latin for "mulberry," which it resembles (fig. 9.4).

Three days have passed since fertilization, and the new organism is still within the fallopian tube, although it is drawing closer to the uterus. It is still about the same size as the fertilized ovum, because the initial cleavage divisions produce daughter cells that are about half the size of the parent cell. After the first few cleavage divisions, the cells have about as much cytoplasm as any somatic cell. During

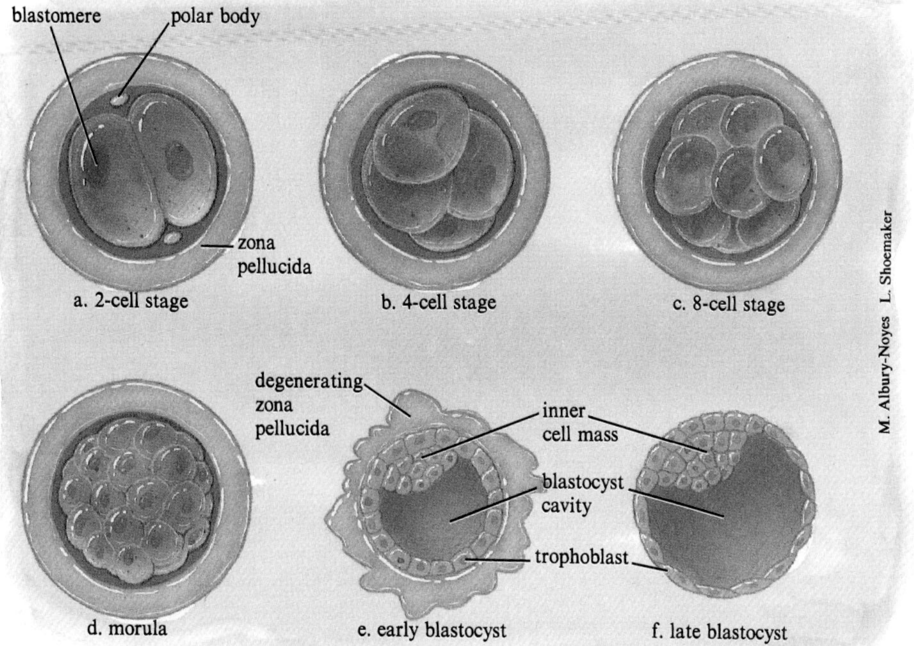

Figure 9.4
Cleavage divisions lead to formation of a mulberrylike structure, the morula, which then hollows out to form the blastocyst. The outer cell layer is the trophoblast, and the clump of cells on one side of the interior is the inner cell mass.

cleavage, cellular activity is still somewhat under the control of the organelles and molecules from the secondary oocyte's cytoplasm, but some of the developing organism's genes are active.

The morula travels down the fallopian tube and arrives in the uterus sometime between days 3 and 6. The ball of cells hollows out, its center filling with fluid that seeps in from the uterus. The organism is now called a **blastocyst,** Greek for "germ bag." The blastomeres are now located in one of two positions. Some blastomeres form an outer layer of single cells called the **trophoblast** (Greek for "nourishment of germ"). Certain trophoblast cells will develop into a membrane called the *chorion*, which eventually forms the fetal portion of the placenta, the special organ that brings oxygen and nutrients to the fetus and removes wastes from it.

A clump of blastomeres inside the blastocyst constitutes the **inner cell mass,** and these cells will develop into the embryo plus its supportive extraembryonic membranes. The fluid-filled center of the ball of cells is called the *blastocyst cavity* (fig. 9.5).

Implantation

Sometime between days 5 and 7, the blastocyst attaches to the uterine lining, which is called the **endometrium** (fig. 9.6). The portion of the blastocyst where the inner cell mass is located lies against the endometrium. Digestive enzymes produced by the trophoblast now break through the outer epithelial layer of the uterine lining, and the blastocyst becomes surrounded by ruptured blood vessels, which bathe it in nutrient-rich blood. This nestling of the blastocyst into the uterine lining is called *implantation*. The trophoblast layer directly beneath the inner cell mass thickens and sends out fingerlike projections into the endometrium at the site of implantation (fig. 9.7). These projections develop into the chorion.

The trophoblast cells now secrete a hormone, **human chorionic gonadotropin** (HCG), which prevents menstruation. In this way, the blastocyst helps to ensure its own survival, for if menstruation occurs, the new organism would be eliminated from the woman's body along with the buildup of tissue in the uterus. HCG continues to be produced for about 10 weeks.

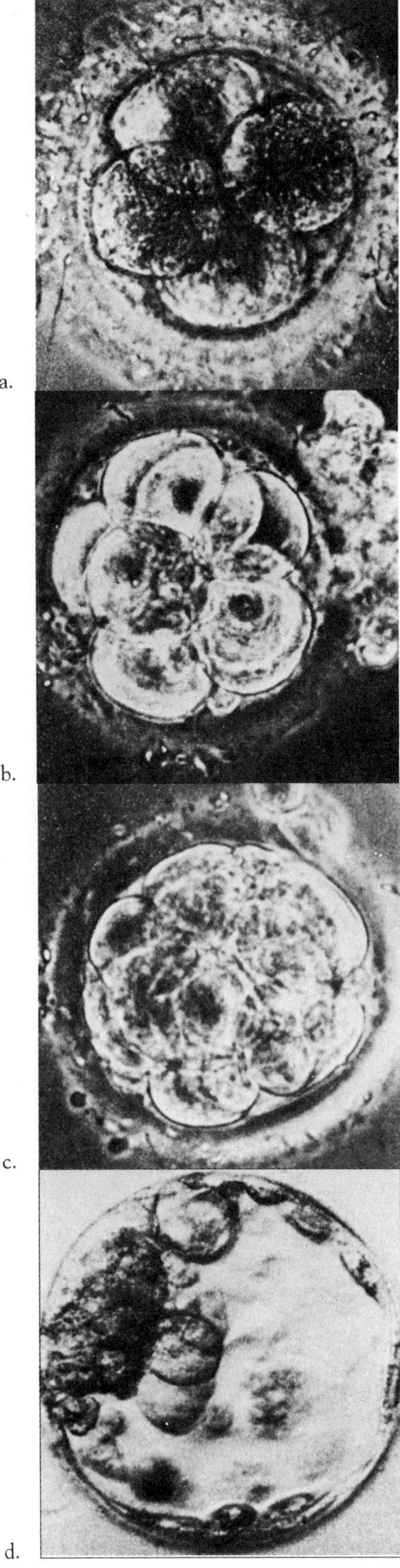

Figure 9.5
Electron micrographs of a human at the (*a*) 4-cell stage, (*b*) 16-cell stage, (*c*) morula stage, and (*d*) blastocyst stage.

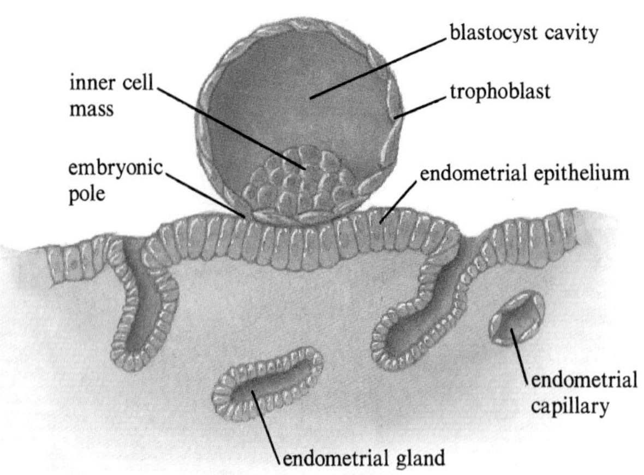

Figure 9.6
Implantation. On about day 6, the blastocyst implants in the uterine lining, the endometrium.

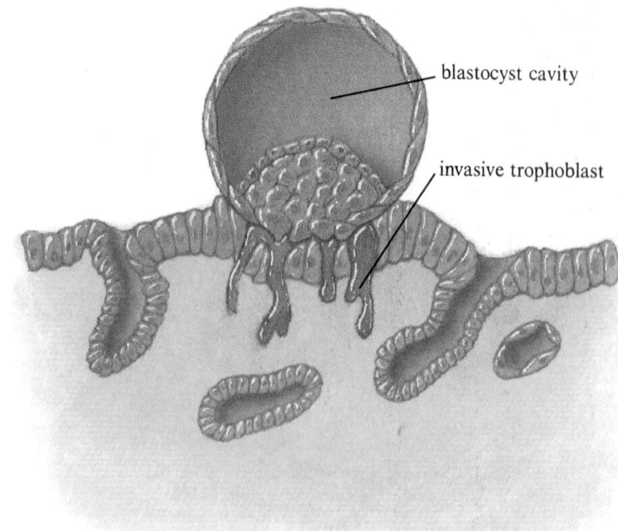

Figure 9.7
Invasion of the endometrium. Part of the trophoblast adjacent to the endometrium sends out projections, which will develop into the chorion.

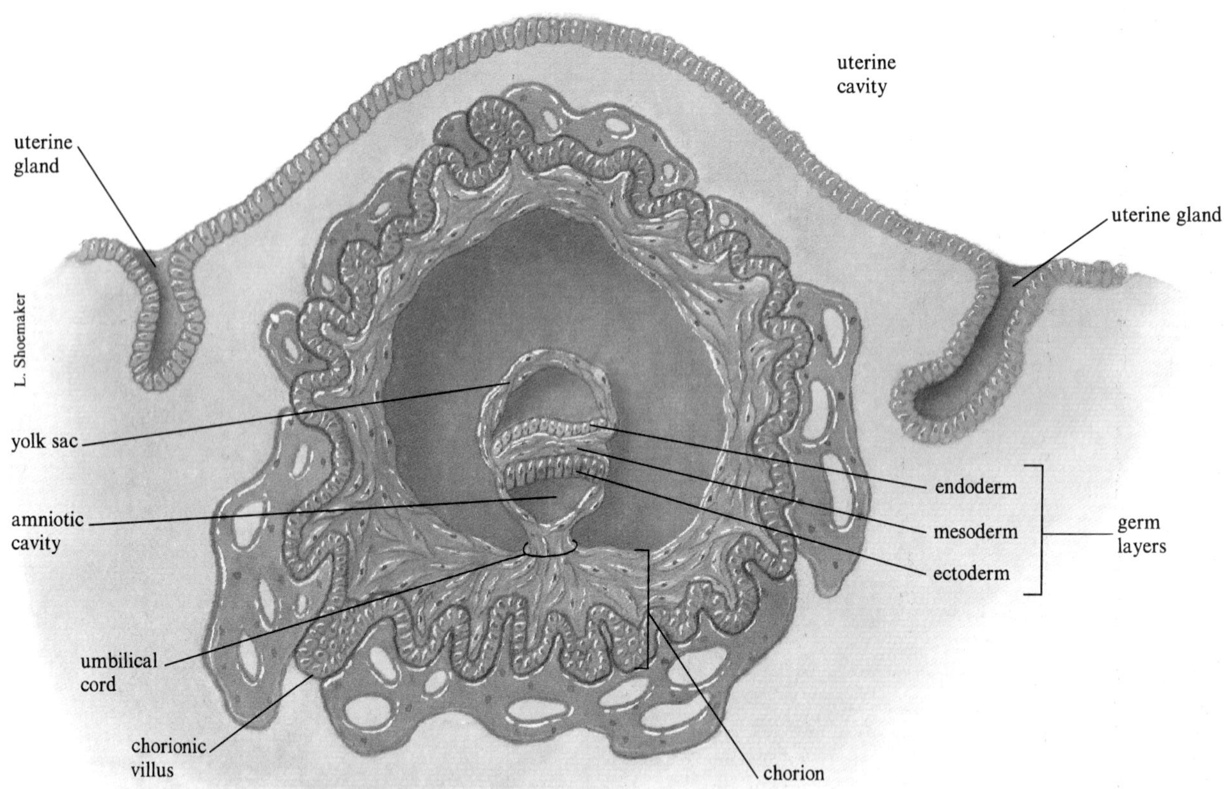

Figure 9.8
By 2 weeks, the primary germ layers have formed, and the embryonic stage of prenatal development begins.

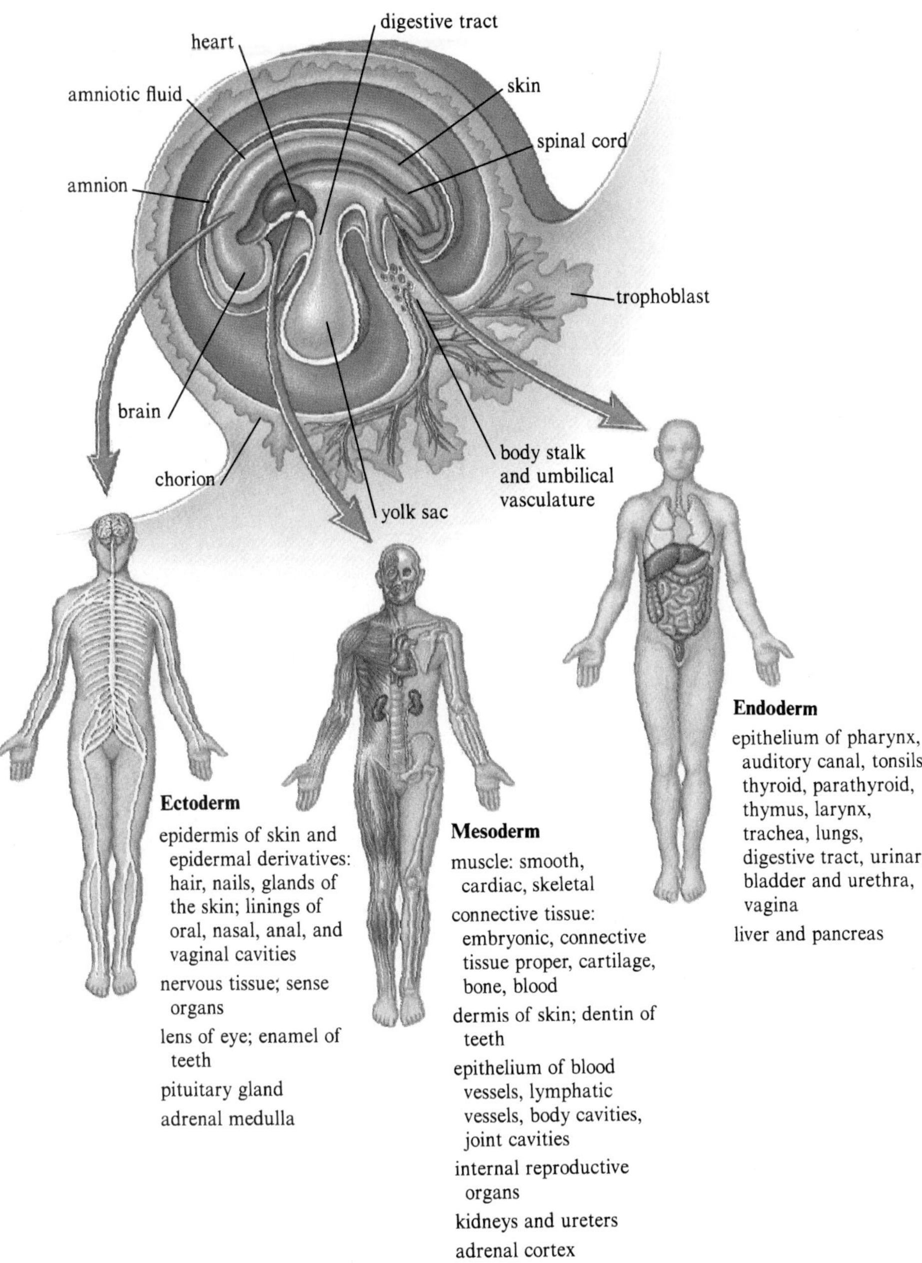

Figure 9.9
These organ systems arise from each of the primary germ layers.

The Primordial Embryo

During the second week of prenatal development, the blastocyst completes implantation, and the inner cell mass changes. A space, called the *amniotic cavity*, forms between the inner cell mass and the portion of the trophoblast that "invades" the endometrium. The inner cell mass then flattens and is called the *embryonic disc*.

The embryonic disc at first consists of two layers. The outer layer, nearest the amniotic cavity, is called the **ectoderm** (Greek for "outside skin"). The inner layer, closer to the blastocyst cavity, is the **endoderm** (Greek for "inside skin"). Shortly after, a third and middle layer forms, called the **mesoderm** (Greek for "middle skin"). This three-layered structure is the *primordial embryo* or *gastrula*. The process of forming the primordial embryo is called gastrulation, and the layers are called *germ layers* (fig. 9.8).

Gastrulation is an important process in prenatal development because a cell's fate is determined by which layer it is in (fig. 9.9). Cells of the ectoderm are destined to become part of the nervous system, sense organs, outer skin layer (epidermis), hair, nails, or skin glands. Mesoderm cells develop into bone, muscles, blood, inner skin layer (dermis), reproductive organs, or connective tissue. Cells of the endoderm eventually form the organs and linings of the digestive, respiratory, and urinary systems.

It is now the end of the second week of prenatal development (table 9.2). Although a woman has not yet "missed" her menstrual period, she might suspect something is happening because her breasts may be swollen and tender and she may be unusually tired. She may even be carrying twins (fig. 9.10). A highly sensitive pregnancy test performed on her blood or urine may already be able to detect small amounts of HCG. The pregnancy test that announced your arrival was less sophisticated. A sample of your mother's urine was injected into a female rabbit. The rabbit was then sacrificed and its ovaries examined. HCG present in the urine sample caused the rabbit to ovulate. The rabbit's swollen ovaries meant that you were on the way. (Thus, the expression "the rabbit died" indicated that a woman was pregnant.)

KEY CONCEPTS

The events of fertilization include capacitation of sperm; the action of acrosomal enzymes on the secondary oocyte and its surrounding zona pellucida and corona radiata; and the block to polyspermy as the sperm enters the oocyte. The chromosomes from the male and female cells meet to form the zygote. Cleavage cell divisions follow, forming the morula and then the blastocyst. The outer layer of cells and the cells that invade the endometrium are part of the trophoblast. The inner cell mass will develop into the embryo and the extraembryonic membranes. The blastocyst cavity is fluid-filled. Enzymes from the trophoblast help the blastocyst to implant in the endometrium between days 5 and 7. Trophoblast cells secrete HCG. During gastrulation in the second week, the germ layers of the primordial embryo form. Cells in a specific germ layer later become part of particular organ systems.

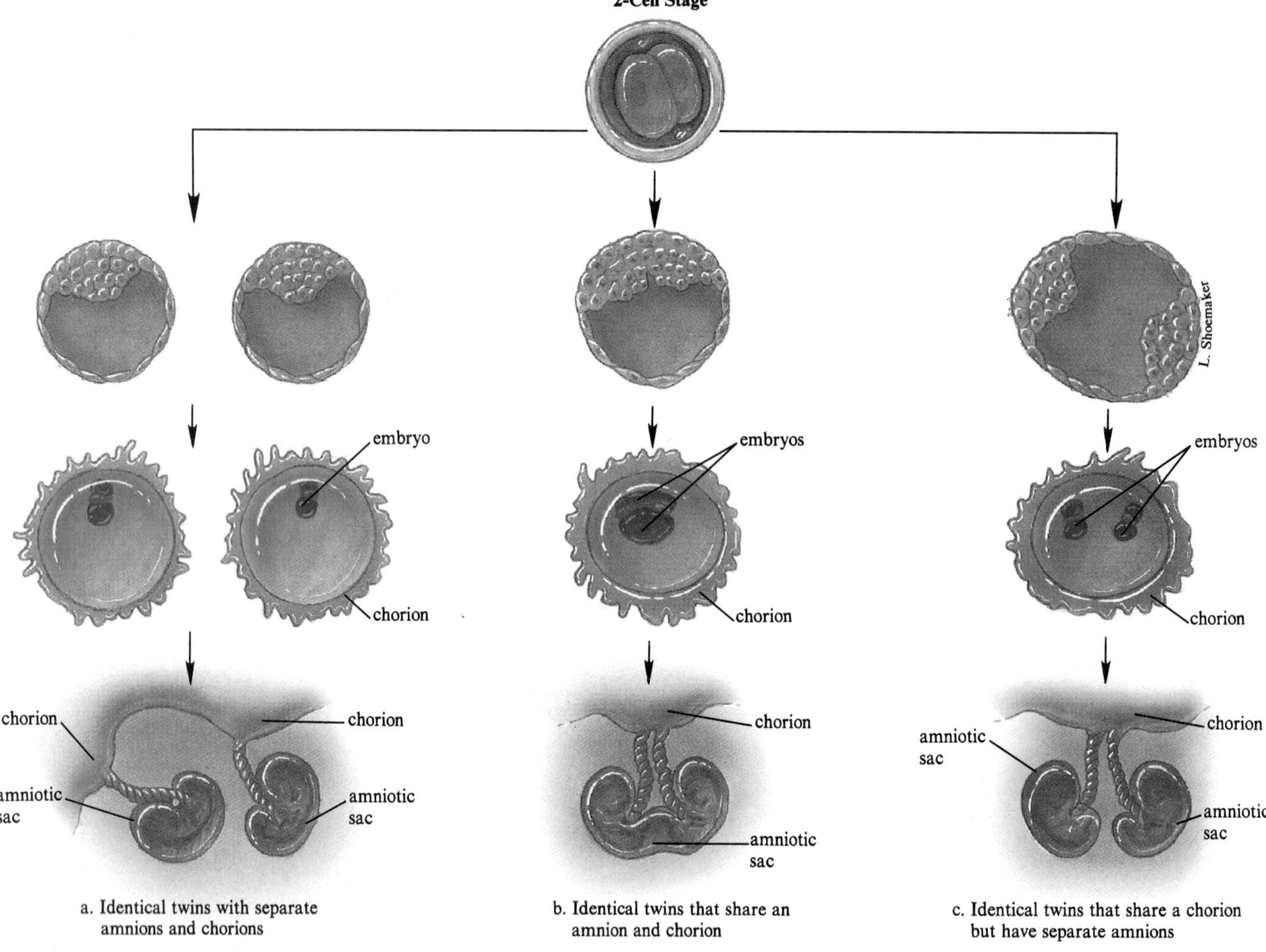

Figure 9.10
Facts about twins. One in 81 births in the United States produces twins. Identical twins originate at three points in development. *a.* In about one-third of identical twins, separation of cells into two groups occurs before the trophoblast forms on day 5. These twins have separate chorions and amnions. *b.* About 1% of identical twins share a single amnion and chorion, because the tissue splits into two groups after these structures have already formed. *c.* In about two-thirds of identical twins, the split occurs after day 5 but before day 9. These twins share a chorion but have separate amnions. Fraternal twins result from two sperm fertilizing two secondary oocytes. The twins develop their own amniotic sacs, yolk sacs, allantois, placentae, and umbilical cords. Fraternal twins can have two different fathers, if the mother had intercourse with two men within a short period of time and ovulated twice. Fraternal twins run in families. In about 55% of all twin conceptions, one twin dies before birth. Many times the parents do not even know that their baby was a twin!

Table 9.2
Morphogenetic Stages and Principal Events During Preembryonic Development

Stage	*Time Period*	*Principal Events*
Zygote	12–24 hours following ovulation	Egg is fertilized; zygote has 23 pairs of chromosomes (diploid) from haploid sperm and haploid egg; genetically unique
Cleavage	30 hours to third day	Mitotic divisions increase number of cells
Morula	Third to fourth day	Hollow ball-like structure forms; single layer thick
Blastocyst	Fifth day to end of second week	Inner cell mass and trophoblast form; implantation occurs; embryonic disc forms followed by primary germ layers

Reading 9.1 Tracking Development in Different Organisms

ALTHOUGH THE BASIC PHENOMENA OF CELL SPECIALIZATION AND TISSUE FORMATION ARE EXHIBITED IN ALL MULTICELLULAR ORGANISMS, DIFFERENT EVENTS ARE MORE STRIKINGLY SEEN IN SOME SPECIES THAN OTHERS. Many types of "model" organisms have assisted biologists in probing the intricacies of development. Following are descriptions of some organisms that are widely used by biologists and the fascinating aspects of their development that have been revealed to us.

Mice—A Cell's Developmental Options Narrow With Time

Cells of a mouse at the 4-cell stage are not yet specialized, and each has the potential to differentiate into any adult structure. If cells at this early stage are separated and implanted into a mouse's uterus, each cell develops into a normal mouse. If cells from a mouse embryo at the 1,000-cell stage are implanted, they do not develop into embryos because the cells no longer have the potential to develop into all of the specialized cell types needed to form a complete organism.

In a fascinating experiment, 4-cell embryos of female black, brown, and white mice were dismantled, and the cells mixed together. The result was a tricolored mouse (fig. 1). She has six parents—a pair for each of the three original embryos. The tricolored mouse was bred to a white mouse, and she gave birth to the three mice shown above her in the photograph. The fact that the tricolored mouse gave birth to solid-colored mice of the original three colors indicates that some of her germ cells descended from each of the original three donor embryos.

Nematode Worms—Tracing Cell Fates

Biologists have long sought to understand the complete development of an organism. A tiny soil-dwelling roundworm, *Caenorhabditis elegans* (fig. 2), is helping researchers do just that.

Like other multicellular organisms, *C. elegans* begins life as a single cell. The single cell divides several times, growing into an embryo within its egg shell, which is soon released to the environment by the parent worm. Next, all but 6 of the cells die. These 6 "founder cells" divide to produce all of the 558 cells that constitute the tiny worm that hatches from the egg just half a day after the first cell division. The hatchling has muscles, nerves, an outer body wall, and intestines. By the end of its 3-day life, after some cell divisions, it is an adult built of exactly 959 cells. Along the way, as a normal part of its developmental program, 113 cells die.

Figure 1
The tricolored mouse has six parents. The other three mice are her offspring.

C. elegans is a wonderful organism with which to study development for several reasons. The animal is transparent, so the locations of daughter cells as cells divide can actually be watched. The worm lives for only a short time, so entire life spans of many individuals can easily, but painstakingly, be observed. The worm is built of only a few cells, so following cell lineages—that is, which cells give rise to which cells—is not an overwhelming task. Compare this to studying the fates of the trillions of cells in a human!

Each cell of the worm is "determined" at an early stage to become a specific adult structure, and inductive interactions (influences of one group of cells on the specialization of an adjacent group) are few. Every individual undergoes the same number of cell divisions at the same times and in the same places. All of these

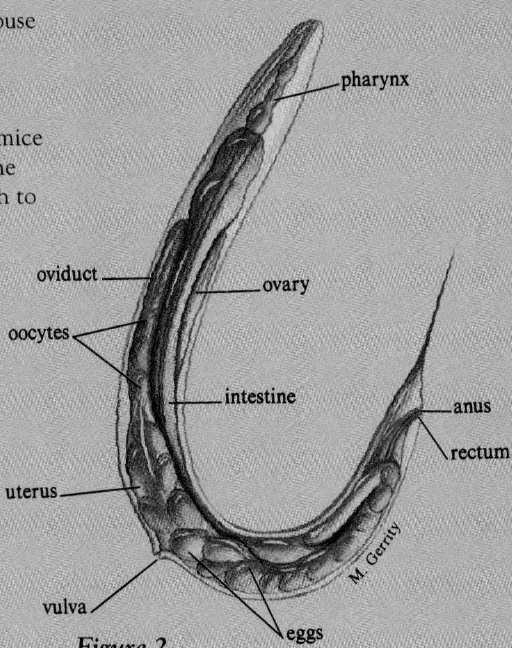

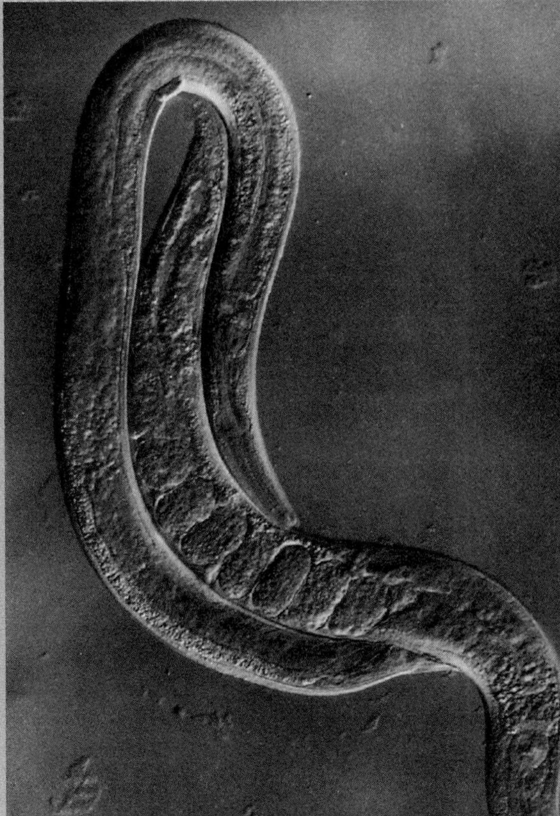

Figure 2
Cell fates are followed in the simple roundworm *C. elegans*. In the photo, the long, thin worm on the outside, a male, is attempting to mate with a shorter worm, a mutant called "dumpy". Magnification, x100.

Reading 9.1 Continued

characteristics add up to an organism whose cell-by-cell development, including differentiation, can easily be studied with the aid of a microscope.

Two biologists, Einhard Schierenberg of West Germany and John Sulston of England, studied these worms from 1975 to 1980, tracing the developmental fates of each and every cell. Thanks to these and other researchers, the tiny worm is now the best understood multicellular organism on earth.

Fruit Flies—Cell Packets

The fruit fly *Drosophila melanogaster* starts life as a fertilized egg and develops into a many-celled embryo within the eggshell. Soon, a small larva hatches. It sheds its outer covering twice before metamorphosing inside a cocoon to emerge later as an adult fly.

The larva contains several round, flat structures called *imaginal discs*, which are packets of cells that are already committed to develop into specific structures (fig. 3). Disc cells are inactive in the larva but are "awakened" by hormones once the larva wraps itself in its cocoon. In response to this chemical cue, the disc cells rapidly proliferate and develop into increasingly complex two- and then three-dimensional sheets of cells. Gradually, adult structures—eyes, antennae, mouth parts, legs, wings, and genitals—telescope out from the discs as the larval structures are degraded by enzymes. A particular disc always gives rise to a specific adult structure—a wing disc always produces a wing; an antennal disc, an antenna.

The fact that disc cells are committed to follow a path of development can be shown by removing a disc from one larva and injecting it into the abdomen of another larva that is about to metamorphose into an adult. When metamorphosis begins, the transplanted disc cells differentiate within the host's abdomen. After metamorphosis, an adult structure corresponding to the particular type of disc floats freely in the fly's abdomen! If a leg disc is transplanted, for example, then a perfectly formed leg can be seen within the translucent body of the adult fly recipient. Correspondingly, the donor fly does not develop a leg.

Sometimes development takes a wrong turn, and imaginal discs give rise to the wrong structures. The result is a fly with mixed-up body parts. These developmental detours are caused by abnormal genes called *homeotics*. One group of homeotic genes transforms structures at the front end of the fly into structures normally found in more posterior positions—for example, an antenna grows where a mouth part should be (see fig. 11.15) or a leg grows where an antenna should be. Another group of homeotic genes affects structures towards the rear of the fly, causing abdominal segments to develop the more centrally located wings, for example (fig. 11.17).

One explanation for the existence of these genetic variants that transform the ends of the animal into more "middle" segments is that they reflect a time when the ancestors of modern insects had two pairs of wings and were built of many identical body segments. Although homeotic mutations appear to be restricted to certain insect species, similar genes have been found in other organisms, including mice and humans. These sequences of genetic information—called *homeoboxes*—appear to participate in organizing cells and tissues in the developing embryo.

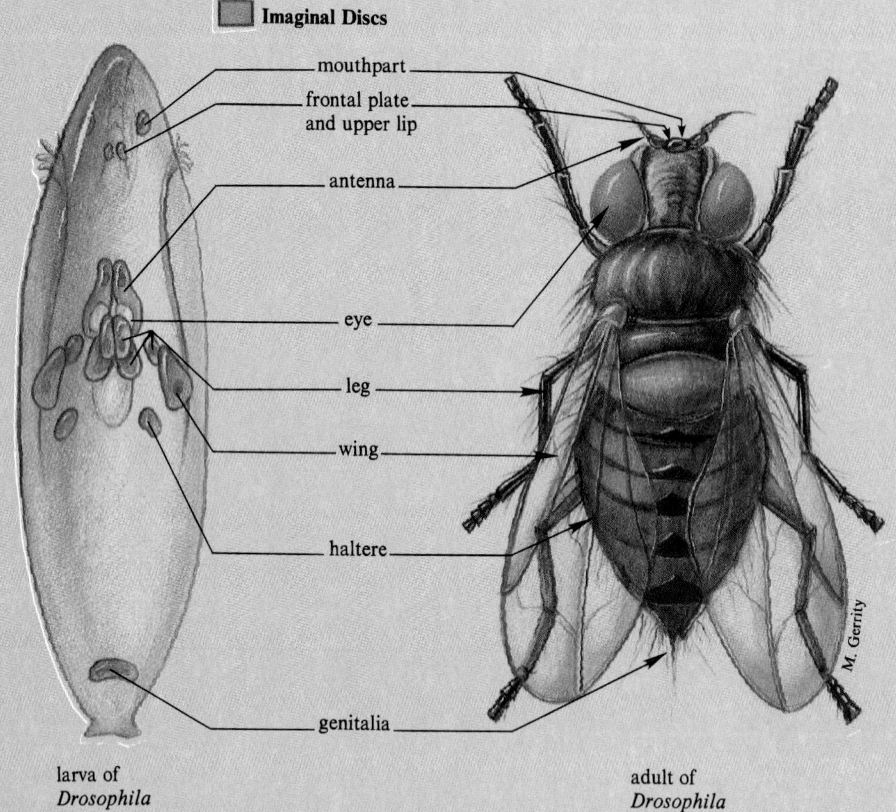

Figure 3
Imaginal discs in the fruit fly.

Toads—Differentiation Depends on Selective Gene Use

An elegant series of experiments on the African clawed toad, *Xenopus laevis*, conducted in the early 1960s demonstrated that differentiated cells contain a complete genetic package, but that they use only some of that information. British developmental biologist J. B. Gurdon designed a way to "turn on" the genes that a differentiated cell normally "turns off." He removed the nucleus from a specialized cell (one lining the small intestine) of a tadpole and injected it into an egg of the same species whose nucleus had previously been destroyed.

Can you deduce what happened? Some of the altered eggs developed into tadpoles, a few actually developing into perfectly normal adult toads (fig. 4). (The experiment did not work every time partly because it is physically difficult to manipulate single cells and nuclei without damaging them.) The fact that a nucleus from a differentiated cell (the intestinal cell) could support normal development from the single-cell stage proved that such a nucleus contains the complete set of genetic instructions for the species. Genes must therefore be inactivated—not lost as some investigators had thought—as time proceeds in the life of an individual multicellular organism. Genes are selectively turned off by proteins that physically block their activation. Gurdon's work has been repeated in other species, using different types of specialized cells. Today, his technique of *nuclear transplantation* is used to make genetic replicas, or clones, of crop plants, and someday it may be used to help preserve endangered animal species.

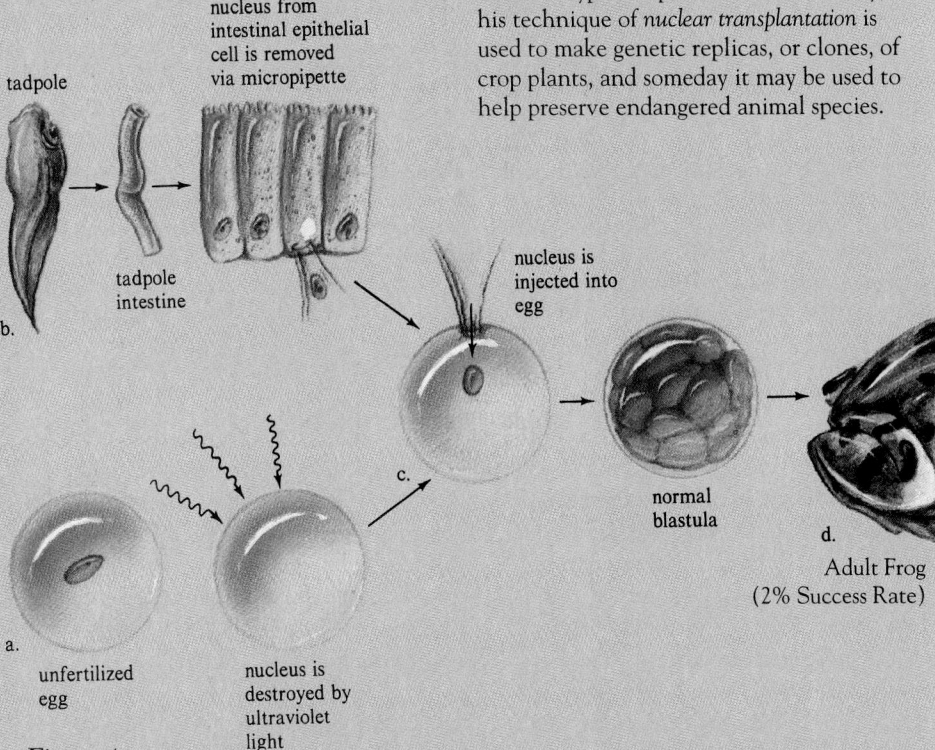

Figure 4
Nuclear transplantation shows that the nucleus of a differentiated cell can support complete development. *a.* In the first step of the procedure, a toad egg nucleus is inactivated with radiation. *b.* A nucleus is removed from a tadpole's differentiated cell. *c.* Next the nucleus from the differentiated cell is transferred to the enucleated egg. *d.* The egg, controlled by its transplanted nucleus, develops into a tadpole and then a toad.

The Embryonic Stage

During the third to eighth weeks of prenatal development, organs begin to develop and structures form that will nurture and protect the developing organism. The embryo begins to organize around a central axis, and towards the end of this period, development proceeds rapidly, with events unfolding in a precisely regulated sequence. The embryonic stage includes development of the embryo and formation of the extraembryonic membranes (the yolk sac, allantois, and amnion), the placenta, and the umbilical cord.

Supportive Structures

By about the third week after conception, as the embryo body is folding into its three-layered form, the fingerlike projections from the chorion, called **chorionic villi,** extend further into the uterine lining. These embryonic outgrowths reaching towards the mother's blood supply are the beginnings of the placenta. It is a unique organ because one side of it—the chorion tissue—is contributed from the new organism, and the other side consists of blood pools derived from the mother's circulatory system (fig. 9.11).

The blood systems of mother and embryo are separate, but they lie side by side. The chorionic villi extend between pools of maternal blood. This proximity of blood systems makes it possible for nutrients and oxygen to diffuse across the chorionic villi cells from the mother's circulation to that of the embryo and for wastes to leave the embryo's circulation and enter the mother's circulation, from where they are eventually excreted.

The placenta is completely developed in 10 weeks, establishing a vital link between mother and fetus that will last throughout pregnancy. When fully formed, the placenta is a reddish-brown disc about the size of a Frisbee. In addition to providing a lifeline to the embryo and fetus, the placenta secretes hormones that maintain the pregnancy. Placental hormones also alter the mother's glucose metabolism so that the fetus receives much of the energy-rich sugar.

A generation ago it was thought that the placenta filtered out any substances in the mother's circulation that could harm the embryo. Today we know that although the placenta can detoxify some substances, many chemicals pass readily from mother to unborn child. Nicotine from cigarette smoke crossing the placenta stunts fetal growth. Addictive drugs such as heroin and cocaine reach the fetus, causing addiction in the newborn. It is best for a pregnant woman to avoid all drugs, because some substances that do not harm her may damage the fetus. Viruses are also small enough to cross the placenta and may cause devastating effects in the fetus but only minor symptoms in the mother. Chapter 10 discusses some of these environmental factors that may cause birth defects.

The start of the embryonic period is also when the extraembryonic membranes form (fig. 9.12). The *yolk sac* begins to appear beneath the embryonic disc at the end of the second week. It manufactures blood cells until about the sixth week, when the liver takes over this function. At this time, the yolk sac starts to shrink. Parts of it eventually develop into the intestines and germ cells. Despite the name, the yolk sac in humans does not actually contain yolk. Similar structures in other animals do contain yolk, which provides nutrients.

By the third week, an outpouching of the yolk sac forms the *allantois*, another extraembryonic membrane. It too contributes blood cells and gives rise to the fetal umbilical arteries and vein. During the second month, most of the allantois degenerates, but part of it persists in the adult as a ligament supporting the urinary bladder.

While the yolk sac and allantois are developing, the amniotic cavity swells with amniotic fluid. This "bag of waters" cushions the embryo, maintains a constant temperature and pressure, protects the embryo if the mother falls, and allows development to proceed unhampered by the forces of gravity. The amniotic fluid is derived from the mother's blood, and it also contains fetal urine and cells from the amniotic sac, the placenta, and the fetus.

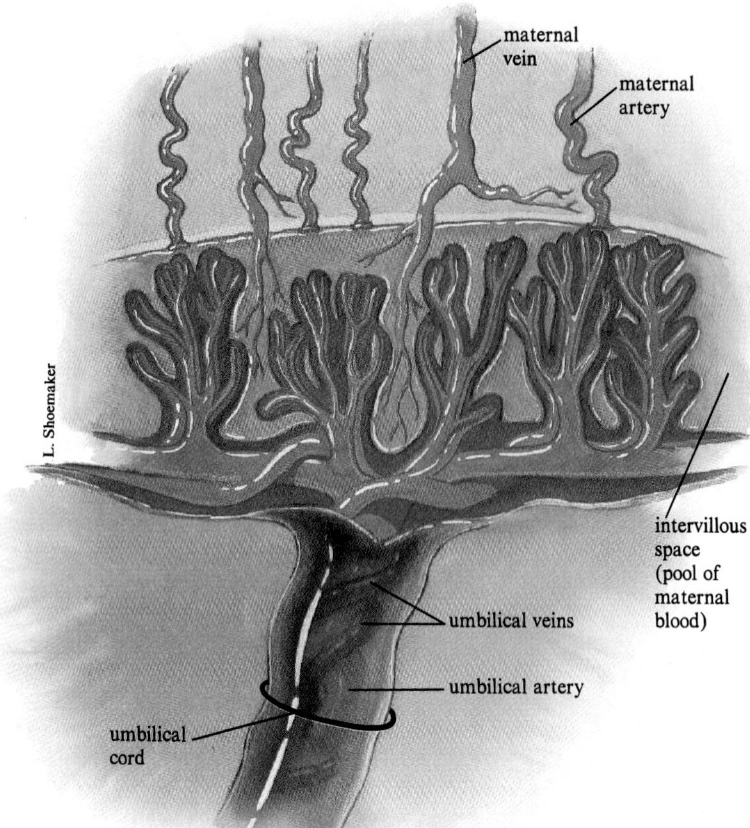

Figure 9.11
The circulations of mother and unborn child come into close contact at the site of the chorionic villi, but they do not actually mix. Branches of the mother's arteries in the wall of her uterus open into pools near the chorionic villi. Her nutrient and oxygen-rich blood flows across the villi cell membranes and enters fetal capillaries (tiny blood vessels) on the other side of the villi cells. The fetal capillaries lead into the umbilical vein, which is enclosed in the umbilical cord. From here the fresh blood circulates through the fetus's body. Blood that the fetus has depleted of nutrients and oxygen returns to the placenta in the umbilical arteries, which branch into capillaries, from which waste products diffuse to the maternal side.

The presence of fetal cells in amniotic fluid is the basis of the prenatal test called **amniocentesis.** The fluid is sampled when enough has accumulated so that a needle can remove some without harming the fetus. Amniocentesis is generally performed at 16 weeks gestation, but several researchers are experimenting with doing the test as early as 11 weeks (see fig. 15.7). An earlier prenatal test, *chorionic villus sampling*, examines the chromosomes of cells taken from the chorionic villi at 8 weeks (fig. 15.8). Can you see why an abnormality in the chorionic villus cell reflects the genetic health of the embryo?

Towards the end of the embryonic period, as the yolk sac shrinks and the amniotic sac swells, another structure forms. This is the *umbilical cord*. The cord is 2 feet (0.6 meter) long and 1 inch (2.5 centimeters) in diameter. It houses two umbilical arteries, which transport blood depleted of oxygen to the placenta, and one umbilical vein, which brings oxygen-rich blood to the embryo. The umbilical cord attaches to the center of the placenta. It is twisted because the umbilical vein is slightly longer than the umbilical arteries.

The Embryo

As the days and weeks of prenatal development proceed, different rates of cell division in different parts of the embryo lead to more intricate folding of its tissues. In a little understood process called **embryonic induction,** specialization of one group of cells causes adjacent groups of cells to specialize. Gradually, these changes mold the three primary germ layers into organs and

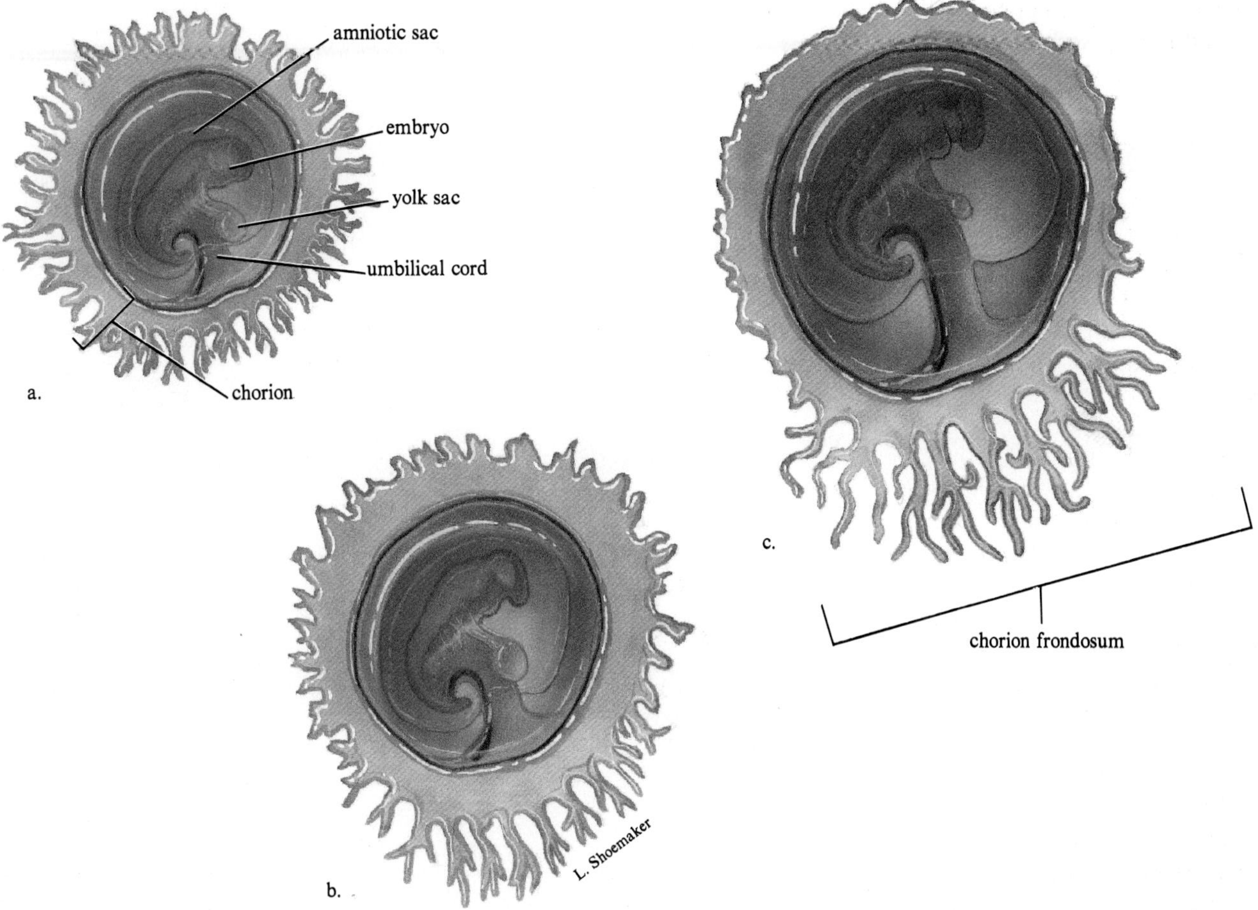

Figure 9.12
During the fourth week of embryonic existence, the extraembryonic membranes rapidly form.

organ systems. *Organogenesis* (making organs) is the term used to describe the transformation of the structurally simple three-layered embryo into an individual with distinct organs. It is during the weeks of organogenesis that the organism is most sensitive to environmental factors, such as chemicals and viruses.

During the third week of prenatal development, a band appears along the back of the embryonic disc. This is the **primitive streak,** and it gradually elongates to form an axis, an anatomical reference point around which other structures organize as they develop. The primitive streak eventually gives rise to connective tissue precursor cells and the **notochord,** a structure that forms the basic framework of the skeleton. A reddish bulge appears, which contains the heart. It begins to beat around day 18, and soon the central nervous system starts to take form. Figure 9.13 shows some early embryos.

The fourth week of embryonic existence is one of spectacularly rapid growth and differentiation. Blood cells begin to form and to occupy primitive blood vessels. Immature lungs and kidneys appear. A **neural tube,** which will one day be the central nervous system, runs along the length of the embryo by the 23d day. If this tube does not close normally at about day 28, a deformity called a *neural tube defect* results, in which parts of the brain or spinal cord protrude from the back. If this happens, a substance made in the fetal liver called alpha fetoprotein leaks abnormally quickly into the mother's circulation. If a maternal blood test at the 15th week of pregnancy detects high levels of alpha fetoprotein, the fetus may have a neural tube defect.

Also during the fourth week, small buds appear where the arms and legs will develop. In the early 1960s, the embryos of European women who took the mild tranquilizer thalidomide between days 28 and 42 experienced interference with fetal development during this period of limb development. About 10,000 children were born with flipperlike stumps where their arms and legs should have been. (see fig. 10.6)

The embryo has a distinct head and jaws now, and the eyes, ears, and nose begin to take shape. The inklings of a digestive system appear as a long, hollow tube, from which the intestines will eventually form. A woman carrying this embryo, which is now only 1/4 inch (0.6 centimeter) long, may begin to suspect that she is pregnant, because her menstrual period is about 2 weeks late.

By the fifth week, the embryo appears to be somewhat off balance, its head far too large for the rest of its body. Limbs extend from the body, but they end in platelike structures. Tiny ridges run down the plates, and by week 6 the ridges deepen, molding fingers and toes. The eyes are open, but

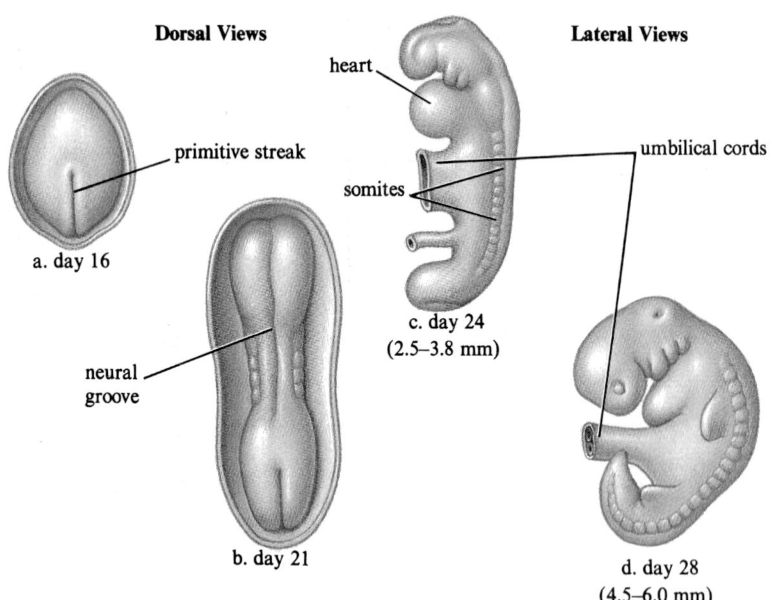

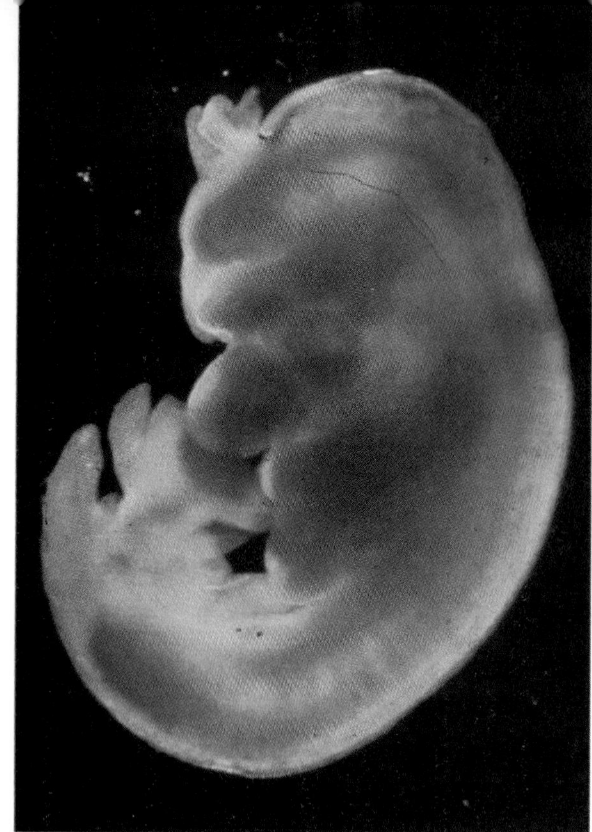
e.

Figure 9.13
Early embryos. It takes about a month for the embryo to look like a "typical" embryo. At first all that can be distinguished is the primitive streak (*a*), but soon the central nervous system begins to form (*b*). By the 24th day, the heart becomes prominent as a bulge (*c*), and by the 28th day, the organism is beginning to look human (*d* and *e*).

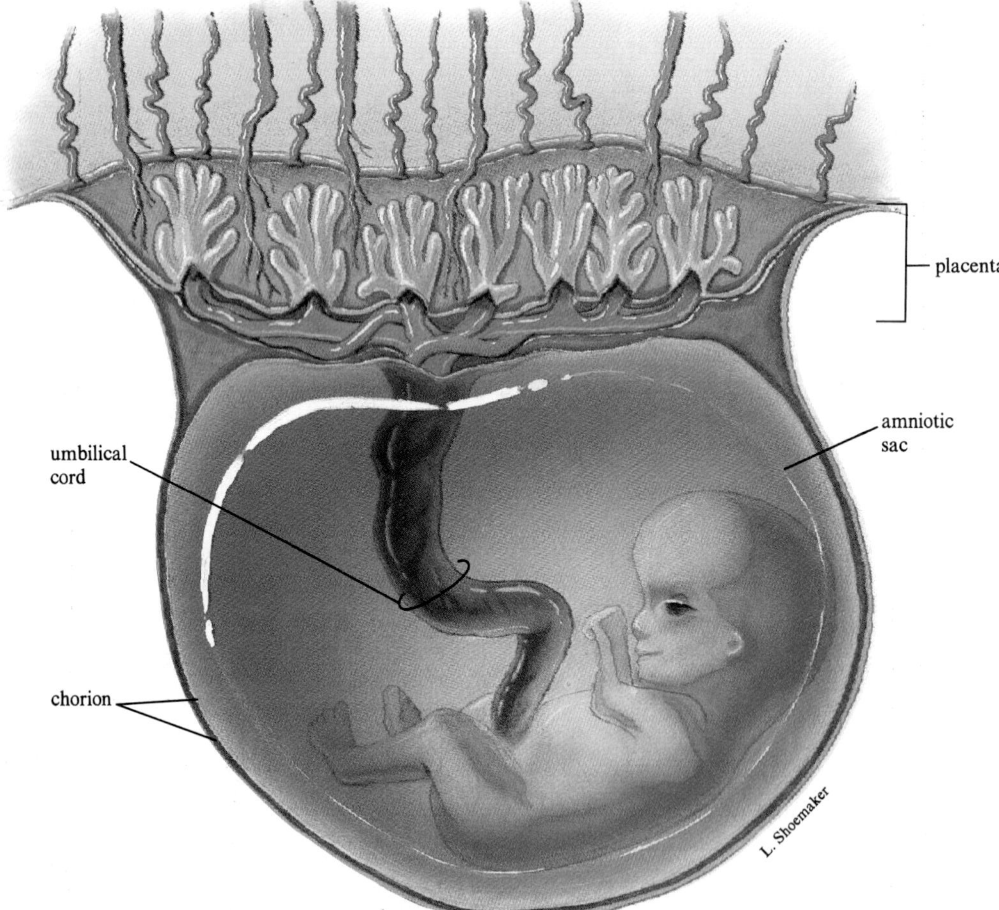

Figure 9.14
At 7 weeks of development, the chorionic villi and the tissue between them form the placenta. Blood from the embryo flows to and from the placenta in the umbilical vein and arteries.

they do not yet have eyelids or irises. The head still seems large, and cells in the brain are rapidly differentiating. The embryo is now about 1/2 inch (1.3 centimeters) from its head to its buttocks.

By the seventh and eighth weeks, a skeleton built of cartilage is present. The placenta is now almost fully formed and functional, secreting hormones that maintain the blood-rich uterine lining. To the outside world, the mother-to-be looks much the same as always; inside her, the embryo is now 1 inch (2.5 centimeters) long. It weighs 1/30 of an ounce (0.85 gram)—about the same weight as a paper clip. Its organs have all started to form, and the increase in detail from just a few weeks ago is astonishing. The eyes are now sealed shut and will stay that way until the seventh month. Nostrils are present, but they are closed. A neck appears as the head begins to comprise proportionately less of the embryo, and the abdomen has flattened somewhat. The 8-week-old unborn child, on the border of being an embryo and a fetus, already looks quite human (figs. 9.1 and 9.14).

KEY CONCEPTS

The embryonic period consists of weeks 3 through 8 of prenatal development. During the third week, chorionic villi extend towards the maternal circulation. The blood supplies of mother and embryo lie adjacent to one another. Nutrients and oxygen enter the embryonic circulation, and wastes are passed from the embryo into the maternal circulation. Many substances pass through the placenta to the embryo. The yolk sac forms early in the embryonic period and manufactures blood cells until the liver takes over at week 6. The allantois forms from part of the yolk sac and produces blood cells and eventually forms the umbilical arteries and vein. The amniotic cavity expands with fluid, as the yolk sac and allantois shrink. The umbilical cord forms late in the embryonic period. During the third week the primitive streak appears, and then, rapidly, the central nervous system, heart, notochord, neural tube, limbs, fingers, toes, eyes, ears, nose, and other organ rudiments form.

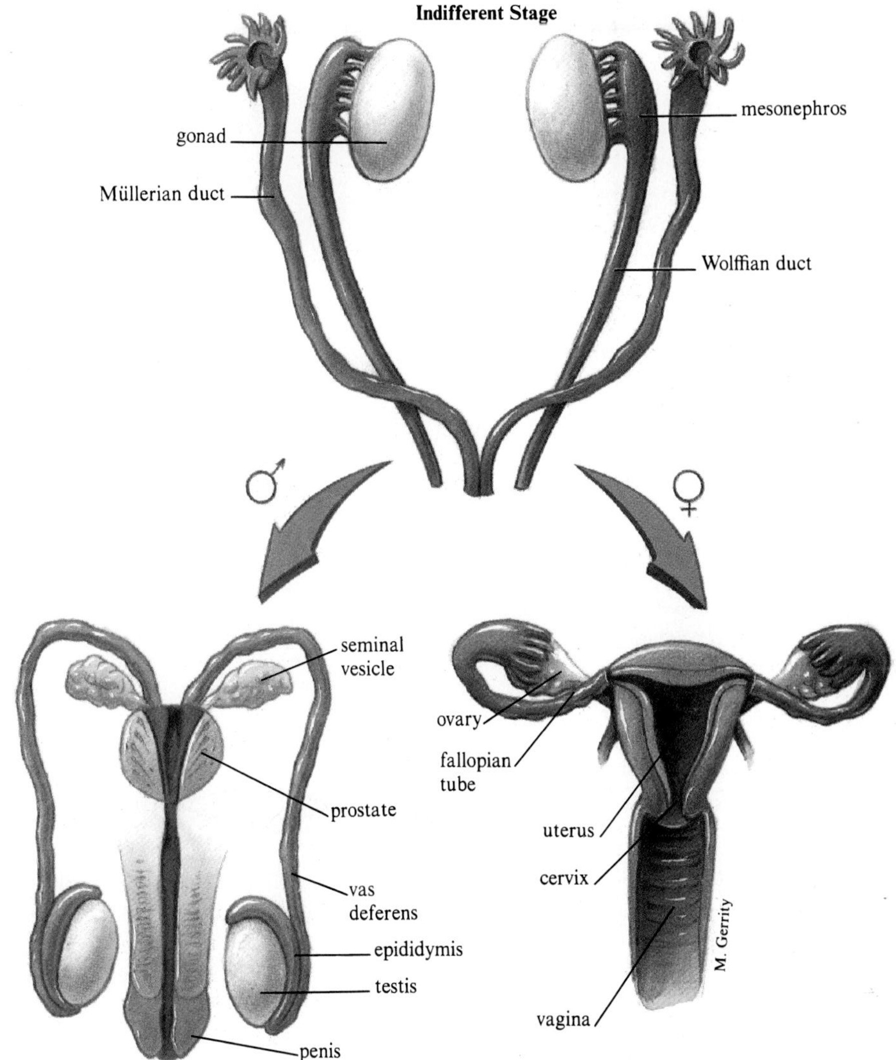

Figure 9.15
Until 6 weeks of development, all embryos have the same "indifferent" sexual structures. If male hormones are released after week 6, a male develops. If not, a female develops.

The Fetal Period

In the third month of pregnancy, a woman may have begun to outgrow her jeans, but she does not yet look obviously pregnant. In her expanding uterus, the body proportions of the 4-inch-long (10-centimeter), 1-ounce (28-gram) fetus appear a little more like those of a newborn. The ears lie low, and the eyes are widely spaced. In the centers of the cartilaginous bones appear bone tissue, which will grow until most of the cartilage is replaced by the harder tissue. Soon, as the nerves and muscles become coordinated, the fetus will be able to move its arms and legs.

Sex organs differentiate throughout the month, and whether the fetus is male or female is obvious by the 12th week, if one could peek into the uterus (fig. 9.15). The fetus sucks its thumb, kicks, makes fists and faces, and has the beginnings of baby teeth. It breathes amniotic fluid in and out, and urinates and defecates into it. The first *trimester* (three months) of pregnancy is over.

During the second trimester, the body proportions of the fetus become even more like those of a baby. By the fourth month, it has hair, eyebrows, lashes, nipples, and nails. Some fetuses even scratch themselves before birth. The cartilage skeleton continues to be replaced by bone. The fetus's muscle movements become stronger, and the woman may become aware of them as slight flutterings. By the end of the fourth month, the fetus is about 8 inches (20 centimeters) long and weighs about 6 ounces (170 grams).

During the fifth month, the fetus becomes covered with an oily substance called *vernix caseosa,* which looks like cottage cheese. It protects the developing skin, which is just beginning to grow. The vernix is held in place by white downy hair called *lanugo,* which may persist for a few weeks after birth. By 18 weeks the vocal cords have formed, but the fetus makes no sounds because air is not breathed. By the end of the fifth month, the fetus is curled into the classic head-to-knees position. It weighs about 1 pound (454 grams) and is 12 inches (30.5 centimeters) long.

During the sixth month the skin appears wrinkled because there is not much fat beneath it, and it turns pinkish as blood-filled capillaries extend. By the end of the second trimester, the mother feels distinct kicks and jabs and may even detect a fetal hiccup. Another person can feel movements by placing a hand over the woman's swollen abdomen. The fetus is now about 9 inches (23 centimeters) long.

The middle three months of pregnancy are usually the most pleasant for a mother. The queasiness and fatigue of the first trimester usually pass, and the fetus's movements are felt. Weight gain is not rapid until the last 3 months, when a woman may suffer from backaches and leg cramps from the excess poundage.

Fetal movements during the last 3 months are pronounced. Many obstetricians ask their patients to keep a chart of fetal movements during the last month, and if 24 hours go by without a kick, a problem may be indicated. The fetus may seem to become more active in response to certain noises or motions made by the mother. Some women even report that their fetuses respond to the music of certain performers, and one study found that youngsters respond to theme songs of soap operas that their mothers watched frequently while pregnant with them!

In the final trimester connections between fetal brain cells form at a fast pace. The major organs mature rapidly, and a layer of fat is laid down beneath the fetus's skin. The digestive and respiratory systems mature last, which is why infants born prematurely often have difficulty digesting milk and breathing. Figure 9.16 shows human fetuses at various stages.

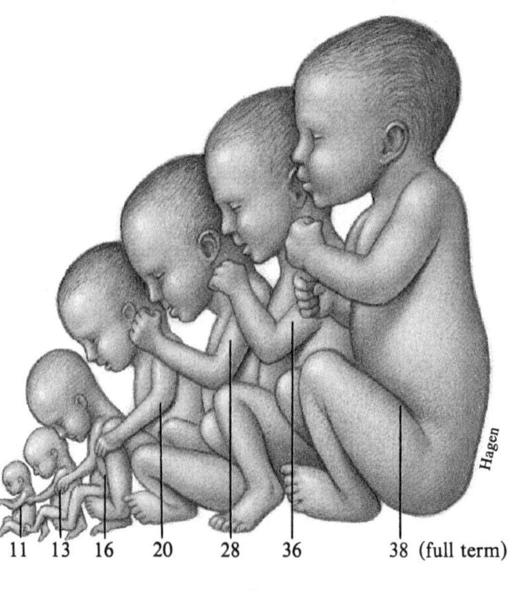

Figure 9.16
Note the changing proportions of the head to the rest of the body in a prenatal human.

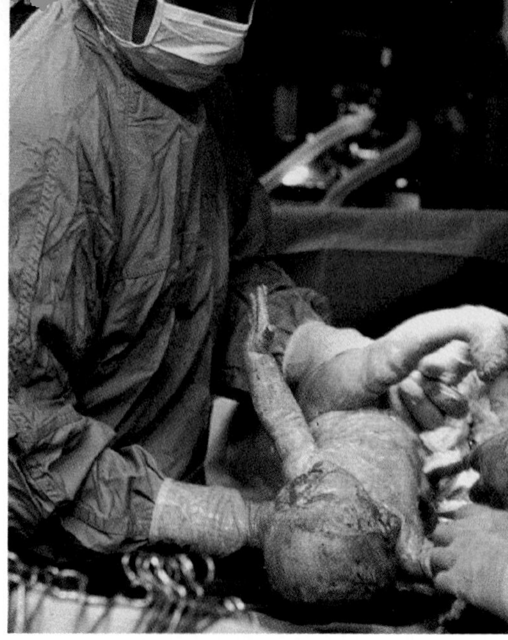

Figure 9.17
The moment of birth. After nearly a day of experiencing the contractions of labor, this woman finally pushes her baby into the outside world. The new father assists by helping her to control her breathing, which can make her less aware of the pain.

KEY CONCEPTS

During the fetal period, structures grow and increase in detail and begin to interact with one another. Bone replaces cartilage in the skeleton, the head comprises less of the body, and sex organs become more distinct. The fetus is active now. During the fifth month the fetus becomes coated with vernix caseosa, and downy hair called lanugo appears. In the final trimester, fetal movements are pronounced, fat fills out the skin, and growth is rapid.

Labor and Birth

Approximately 266 days after a single sperm burrowed its way into an oocyte, a baby is ready to be born. He or she is about 21 inches (53 centimeters) long and weighs from 6 to 10 pounds (2.7 to 4.5 kilograms). The skin is bluish pink in babies of all races because pigment is made only after sunlight is encountered. Usually the fetus settles in a head down, or vertex, position. Sometimes this is noticeable in the pregnant woman as the fetus "dropping."

The actual birth of a baby is called *parturition*, and the events preceding it constitute *labor*. Often labor begins when the woman feels a gush of fluid, which is the amniotic fluid spilling out of its sac as it is punctured by the downward movement of the fetus. Because the sac is broken, infection may set in unless the woman gives birth within 24 hours.

The onset of labor may not be as abrupt as the amniotic fluid sac breaking. A woman may detect a slight discharge from the vagina consisting of blood and mucus. This is called "bloody show" and signals the onset of labor within a day or two—or possibly much sooner. Or, a woman may begin to feel very mild contractions, or labor pains, in her abdomen every 20 minutes or so, whether or not the amniotic sac has broken or bloody show appears. Another sign of impending labor, which is more folklore than established fact, is a "nesting instinct," a flurry of activity by the mother. It is quite common for a pregnant woman to clean the house vigorously from top to bottom, then go into labor hours later.

Labor is different for each woman and for each pregnancy. The contractions may last for days or as little as an hour or two. Usually, though, labor for a first baby lasts about 24 hours and about 12 to 18 hours for a woman who has already had a baby.

Knowing what to expect in the labor process can bring a feeling of control over the situation that can actually lessen pain. One way to do this is to take a prepared childbirth class in which breathing and relaxation techniques are learned and practiced. The father or a friend learns how to be an effective "coach" and prepares to stay at the mother-to-be's side throughout the experience and literally cheer her on. A variety of drugs are also available to ease the pain of giving birth.

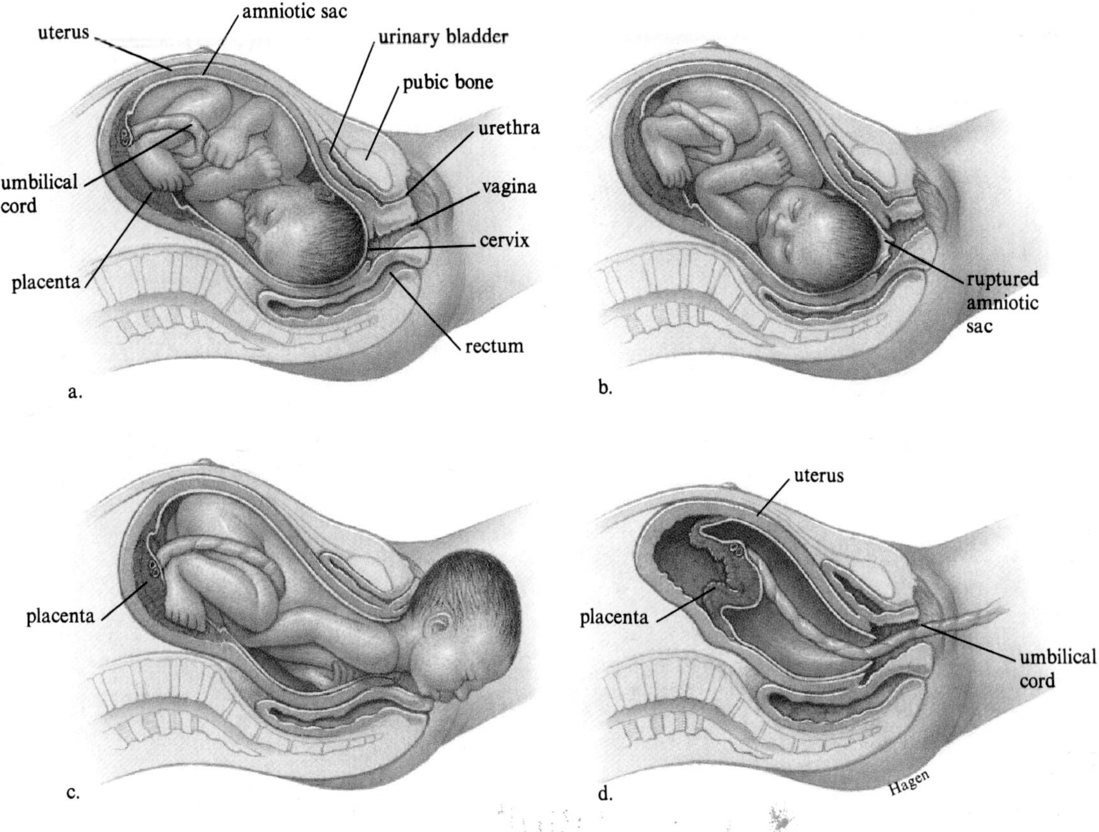

Figure 9.18
About 2 weeks before birth, the baby "drops" in the woman's pelvis and the cervix may begin to dilate. At the onset of labor, the amniotic sac may break. The baby and then the placenta and extraembryonic membranes are pushed out of the birth canal.

As labor proceeds, the uterine contractions, prompted by hormones, gradually increase in frequency and intensity. During the first stage of labor, the baby is pushed against the cervix by the contractions. The cervix dilates (opens) a little bit with each contraction. At the start of labor, the cervix is a thick, closed band of tissue. By the end of the first stage of labor, the cervix is stretched open to about 10 centimeters. The opening of the cervix can sometimes take several days, beginning well before a woman realizes that labor has started.

The second stage of labor is the actual delivery of the baby (figs. 9.17 and 9.18). It begins when the cervix is completely dilated to 10 centimeters. At this point, the woman feels a tremendous urge to push. Anywhere from a few minutes to 2 hours later, the baby descends through the cervix and vagina and is born. In the third and last stage of labor, the placenta and extraembryonic membranes leave the woman's body as the "afterbirth."

Sometimes the fetus cannot be pushed out, and a surgical procedure called a *cesarean section* is necessary. The baby is removed through an incision made in the mother's abdomen. About one in five births in the United States is currently performed by cesarean section for a variety of reasons. The baby may be too large to fit through the mother's pelvis; it may be positioned feet down (breech), and the head may become caught if it is delivered vaginally; the fetus may be side-to-side (transverse lie), refusing to budge. If a device called a fetal monitor detects a falling heartbeat, an emergency cesarean section is performed. Often in such situations, the umbilical cord is wrapped around the baby's neck. Cesarean section is also indicated if labor is prolonged but the cervix is not dilating. It takes only a few minutes to deliver a baby by cesarean section, but it takes much longer for the mother to recuperate than from a vaginal delivery.

Giving birth to a baby is perhaps the most exciting moment in a person's life, a magical time filled with overwhelming relief and unimaginable joy. As the new father and mother recover from the excitement and exhaustion of the birth process, they must now get ready for one of life's greatest challenges—parenthood.

The birth of a baby seems to be a miracle when the number of potential human lives that do not complete prenatal development is considered. Of every 100 secondary oocytes that are released from an ovary and exposed to sperm, 84 are fertilized. Of these, 69 are implanted in the uterine lining, 42 survive 1 week or longer, 37 survive 6 weeks or longer, and only 31 are born alive. The next chapter considers situations in which the carefully regulated process of prenatal development does not have a happy ending.

KEY CONCEPTS

There are several signs that labor and parturition are about to begin—the amniotic sac rupturing, a bloody mucus discharge, and mild labor contractions. During labor, contractions of the uterus push the baby down against the cervix, which gradually dilates. After the baby is born, the afterbirth is born. In some situations that threaten the mother or child, a cesarean section is needed.

Maturation and Aging

The process of development does not end when the fetus enters the world outside the uterus, or even when a child passes through adolescence and into adulthood. The structural and functional changes in the body that constitute aging occur throughout life, although they may not be particularly obvious until adulthood (fig. 9.19). These changes occur at different rates in different individuals, but some general bodily changes are noticed at certain ages by many people.

Aging Over a Lifetime

Aging actually begins even before fertilization, because the primary oocyte has been dormant in the ovary since the female reached puberty. The fact that children born to women over 35 have an increased likelihood of inheriting an abnormal number of chromosomes may be related to the greater age of the female germ cell at conception. An abnormal chromosome number is usually associated with miscarriage or severe birth defects. It is thought that "older" oocytes have had more time to accumulate genetic damage from long-term exposure to chemicals, viruses, and radiation.

Human cells begin to die even before an individual is born, but many more cells form to replace them in the rapidly growing embryo and fetus. Fingers and toes, for example, are gradually carved from weblike extremities by cells in the webbing that die. Cells die and cells form throughout human life. The life spans of cells are reflected in the waxing and waning of biological struc-

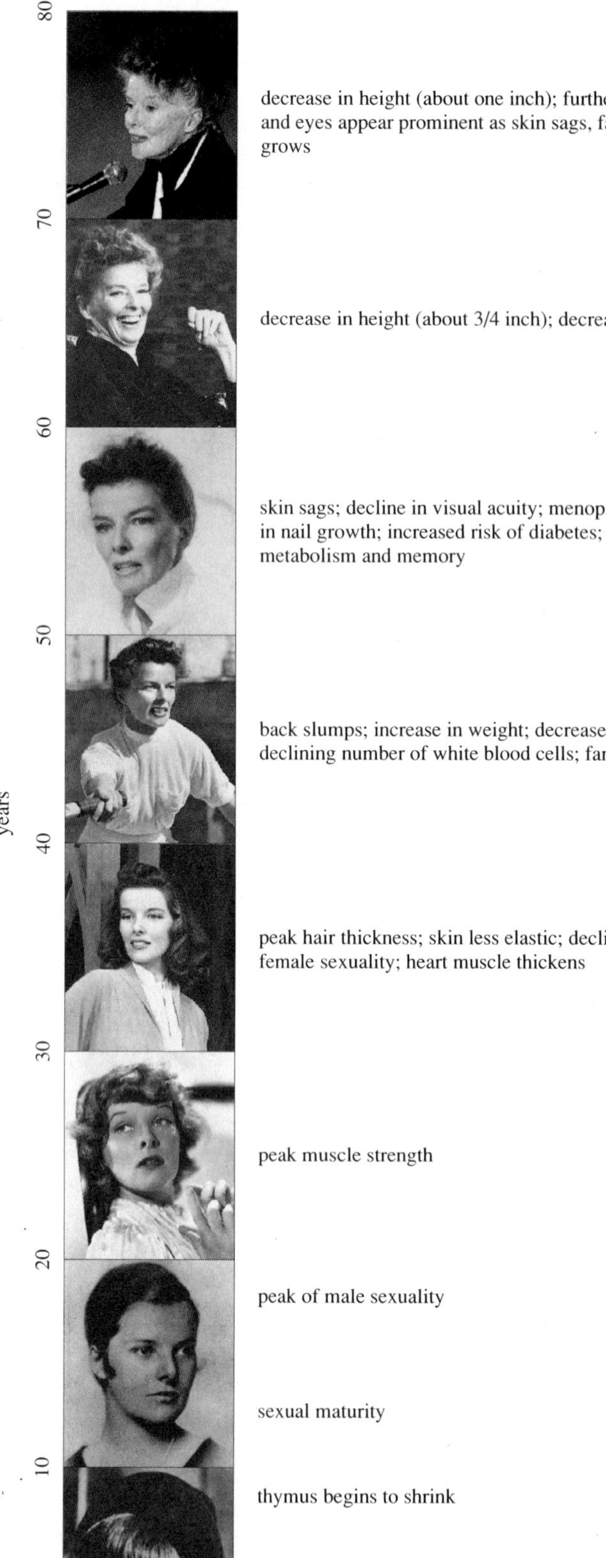

Figure 9.19
The aging person.

tures and functions, as different abilities attain a peak level and then decline at characteristic rates.

By the age of 2 years, for example, a child's brain cells are as numerous as they will ever be. The brain has been growing over the past couple of years at the same rate as it did in the last 6 months in the uterus. At the age of 10, a person's hearing is the best that it will ever be. Shortly before sexual maturity is reached (around age 12 for a girl, 14 for a boy), a small gland in the chest called the *thymus* reaches its greatest size (about the size of a walnut) and then begins to shrink gradually until it is almost microscopic by about the seventh decade of life. The thymus is part of the body's immune system, and the organ's declining activity may play a key role in the increasing susceptibility to certain illnesses seen with advancing age.

By age 18, the human male is producing the highest level of the sex hormone testosterone that he will ever have in his lifetime, and his sex drive is correspondingly at its strongest. In the 20s, muscle strength peaks in both sexes, and the hair is at its fullest, with each hair as thick as it will ever be. By the end of the third decade of life, obvious signs of the aging process may first appear as a loss in the elasticity of facial skin, producing small wrinkles around the mouth and eyes. Height is already starting to decrease, but this is not yet at a detectable level.

The age of 30 seems to be a developmental turning point, although the signs of aging vary tremendously among individuals. Hearing often becomes less acute. Heart muscle begins to thicken. The elasticity of the ligaments between the small bones in the back lessens, setting the stage for the slumping posture that becomes apparent in later years. Some researchers estimate that beginning roughly at age 30, the human body becomes functionally less efficient by about 0.8% every year. Yet growing older has its benefits—many women attain their greatest desire for sex during their 30s.

During their 40s, many people weigh 10 to 20 pounds (4.5 to 9 kilograms) more than they did at the age of 20, thanks to a slowing of metabolism and decrease in activity level that is often not accompanied by eating less. They may be 1/8 inch (0.3 centimeter) shorter, too. Their hair may be graying or falling out, and they may become farsighted or nearsighted. Certain of the immune system's white blood cells are less efficient now, making the body more prone to infection and cancer.

The early 50s bring further declines in the functioning of the human body. Nail growth slows, taste buds die, and the skin continues to lose its elasticity. For most people, the ability to see close objects clearly becomes impaired, but for the nearsighted, vision improves. Women stop menstruating, although this does not necessarily mean an end to or loss of interest in sex. Decreased activity of the pancreas may lead to diabetes. By the decade's end, muscle mass and weight begin to decrease. A male produces less semen but is still sexually active. His voice may become higher as his vocal cords degenerate. A man has half the strength in his arm muscles and half the lung function as he did at age 25, which is measured by the amount of air taken in a deep breath that can be exhaled. He is about 3/4 inch (2 centimeters) shorter.

The 60-year-old may experience minor memory losses. A few million of the person's trillion or so brain cells have been lost over his or her lifetime, but for the most part, intellect remains quite sharp. In fact, some researchers suggest that mental deterioration associated with aging is due largely to breakdown of the cells that physically support nerve cell networks, and that if this cellular scaffolding remained intact, the brain could function adequately for 150 to 200 years.

By age 70, height decreases a full inch (2.5 centimeters). Sagging skin and loss of connective tissue, combined with continued growth of cartilage, make the nose, ears, and eyes more prominent. For some people life ceases when they are in their 70s. But death is like giving birth in that it is different for everyone. Some people die suddenly of heart failure; others suffer debilitating illnesses such as strokes or cancer for years. Life-style habits can influence life span considerably. Although there is no guarantee of immortality by following a healthful diet, exercising regularly, and avoiding excesses of alcohol, unnecessary drugs, and cigarettes, these habits can probably make a person's last years more pleasant.

KEY CONCEPTS

Aging in a sense occurs throughout the human lifetime, including prenatal existence, because certain cells die at particular times. Although organs such as the brain and thymus cease cell division early in life, aging usually becomes apparent after the age of 30. Physical aging changes become more noticeable in one's 40s and 50s, and mental symptoms such as memory loss often begin in the 60s.

What Triggers Aging?

The aging process is difficult to analyze because of the intricate interactions of the body's organ systems. Breakdown of one structure ultimately affects the functioning of others, so that tracing the primary causes of the changes that occur in aging is complicated. The field of **gerontology** examines the biological changes of aging at the molecular, cellular, organismal, and population levels. Although we actually know very little about the mechanisms of aging, current theories center around two viewpoints—one, that aging is a passive process; the other, that it is an active one. There is evidence for both views, and it is likely that they each contribute to aging.

Aging as a Passive Process

Aging as a passive process involves breakdown of existing structures and a slowing of existing functions. At the molecular level, passive aging is seen in the degeneration of the elastin and collagen proteins of connective tissue, which literally holds the body together. With the progressive loss of these proteins, skin loses its elasticity and begins to sag, and muscle tissue becomes less firm. Molecular changes are thus experienced at the whole-body level.

During a long lifetime, biochemical abnormalities accumulate as time passes. For example, mistakes occur throughout life when DNA duplicates itself in dividing cells. Usually special "repair" enzymes correct this damage as soon as it occurs. But perhaps over many years of exposure to chemicals, viruses, and radiation, which can disrupt DNA replication, the error burden simply becomes too great to be fixed. The cell may die as a result of faulty genetic instructions.

Another sign of aging at the biochemical level is the breakdown of lipids, which make up biological membranes. As aging membranes that surround the cell and some of its organelles leak due to lipid degeneration, a fatty, brown pigment called *lipofuscin* forms. Lipofuscin does not cause aging but is a characteristic of old cells. Mitochondria also begin to break down in older cells, decreasing the supply of chemical energy to power the cell's functions. These cellular changes occur gradually and become especially noticeable in the cells of older people.

The cellular degradation associated with aging may be set into action by highly reactive by-products of metabolism called **free radicals.** A molecule that is a free radical has only one electron in its outermost valence shell, and it tends to grab electrons from other molecules to complete the outer shell. The electron-stealing action of free radicals makes the molecules that they steal from less stable, and a chain reaction of chemical instability begins that could kill the cell. Free radicals originate when stable molecules of nearly any kind are exposed to radiation or toxic chemicals (such as those introduced into the body by cigarette smoke). Enzymes that usually inactivate free radicals before cellular constituents can be damaged by them, diminish in number and activity in the later years. One such enzyme is *superoxide dismutase* (SOD). Some health-food stores promote it as an antiaging remedy. Even though the enzyme is a natural free radical fighter, there is no evidence that it stalls aging on a whole-body level.

Passive aging is also apparent at the organ system level. The thymus gland produces molecules called thymosins, which seem to control the longevity of the immune system. By age 70, the thymus is 1/10 the size it was at the age of 10. Declining efficiency of the immune system with age increases a person's susceptibility to infection and cancer.

Medical conditions common in the elderly may reflect the gradual shutting down of specific organs. The accumulation of fatty cholesterol plaques along the inner surfaces of arteries, perhaps as a consequence of a lifetime of eating fatty foods, can clog circulation, perhaps even producing blood clots that can disrupt the vital functions of the heart, brain, and lungs. Hearing loss is often due to the wearing away of sensory hairs in the inner ear as a result of repeated exposure over a lifetime to loud noise. Interestingly, leading causes of death reflect the ability of medical science to treat certain disorders (table 9.3).

Table 9.3
Five Leading Causes of Death: 1900 and 1986
Many of the main causes of death in 1900 have been conquered, presenting modern medicine with a new hierarchy of ills to overcome.

1900 Rank	*Percent of Total Deaths*	*1986 Rank*	*Percent of Total Deaths*
1. Pneumonia and influenza	11.8	1. Heart disease	30.0
2. Tuberculosis	11.3	2. Cancer	21.0
3. Diarrhea and enteritis	8.3	3. Stroke	14.0
4. Heart disease	8.0	4. Chronic respiratory disease	8.0
5. Stroke	6.2	5. Injuries	7.0

Source: Centers for Disease Control, Atlanta, GA.

Aging as an Active Process

Aging can entail the initiation of new activities or the appearance of new substances, as well as breakdown of structure and function. One such "aging substance" may be the lipofuscin granules that build up in aging muscle and nerve cells. Accumulation of lipofuscin with increasing age illustrates both passive and active aging. Lipofuscin actively builds up with age, but it does so because of the passive breakdown of lipids. Another example of active aging is autoimmunity, in which the immune system turns against the body, attacking its cells as if they were invading organisms.

Active aging begins before birth, as certain cells die as part of the developmental program encoded in the genes. This "programmed cell death" occurs regularly in the embryo, degrading certain structures to pave the way for new ones. In the adult, programmed cell death can be seen in the brain cells that die as we grow older and in blood and skin cells that die and are replaced at a high rate. Programmed cell death occurs dramatically in various species that undergo metamorphosis, in which adult structures replace larval ones. Cell death, then, is not a phenomenon that is restricted to the aged. It is a normal part of life.

Accelerated Aging Disorders Provide Clues to the Normal Aging Process

Programmed cell death suggests that aging is controlled at the cellular level. The role of genes in setting the pace of development is suggested by a class of inherited diseases in which the aging timetable is accelerated.

The most severe aging disorders are the **progerias.** In *Hutchinson-Gilford syndrome* (fig. 9.20), a child appears normal at birth but, by the first birthday, growth is obviously retarded. Within just a few years, the child ages with astounding rapidity, assuming a shocking appearance of wrinkles, baldness, and the prominent facial features characteristic of advanced age. The body is aging on the inside as well, as arteries clog with fatty deposits. The child usually dies of a heart attack or a stroke by the age of 12, although some patients live into their 20s. Only a few dozen cases of this syndrome have ever been reported. In an "adult" form of progeria called *Werner's syndrome,* which becomes apparent before the 20th birthday, death from old age usually occurs in the 40s.

Not surprisingly, the cells of progeria patients show aging-related changes. Recall that normal cells in culture divide only 50 times before dying. Cells from progeria patients die in culture after only 10 to 30 divisions, as if they were programmed to die prematurely. Certain structures seen in normal cultured cells as they near the 50-division limit (glycogen particles, lipofuscin granules, many lysosomes and vacuoles, and a few ribosomes) are seen in the cells of

Figure 9.20
Progeria causes rapid aging. Eight-year-old Fransie Geringer of South Africa thought he was the only person with progeria in the world. But in 1981, he met 9-year-old Mickey Hays of Texas. The two children enjoyed a day at Disneyland, where they were met by yet a third progeria patient. People from the hometowns of the three elderly looking children had raised money to grant them their wishes of visiting Disneyland.

progeria patients early on. Understanding the mechanisms that cause these diseased cells to move through the aging process at an abnormally rapid pace may help us to understand the biological aspects of normal aging.

A more prevalent aging disorder with a hereditary component is *Alzheimer's disease,* which, until recently, was simply referred to as "senility." German neurologist Alois Alzheimer first identified the condition in 1907 as affecting those in their 40s and 50s, and it became known as "presenile dementia." Similar symptoms of memory loss and inability to reason among persons over the age of 60 were regarded by the medical community as normal facets of aging that affect a subset of the elderly. Today, Alzheimer's disease is recognized as an abnormal condition that strikes some 2 million Americans each year, killing 100,000 people annually. It affects 5% of United States citizens over the age of 65 and 20% of those over 85. Some cases of Alzheimer's disease may be inherited, and it is possible, in some families, to use DNA-based "marker" tests to predict which family members have inherited the disorder even before symptoms appear. Marker tests are discussed in chapter 15.

A person with Alzheimer's disease may begin to experience memory loss and impairment of reasoning ability as early as the age of 50. Mental function declines steadily for 3 to 10 years after the first symptoms appear. In the beginning, the family may not understand what is happening. It is not uncommon for confused and forgetful Alzheimer's patients to wander away from family and friends and become lost. Finally, the patient cannot perform basic functions such as speaking or eating and must be cared for in a hospital or nursing home. Death is often due to infection, a common killer of bedridden people.

Alzheimer's disease can only be definitely diagnosed at autopsy, when characteristic brain lesions are noted (fig. 9.21). However, as more is learned about this prevalent disorder, descriptions of symptoms are becoming more precise, and an affected family can more readily recognize the stage of the illness in which a loved one may be.

KEY CONCEPTS

In passive aging, structures built of collagen, elastin, muscles, and membranes begin to break down, and DNA repair becomes less efficient. Free radicals accumulate, the thymus shrinks, and cholesterol deposits hamper circulation. Active aging may be evidenced by the accumulation of lipofuscin pigments in cells and programmed cell death. The progerias and Alzheimer's disease demonstrate the role of genes in aging. Aging is probably due to both passive and active events.

Dealing With Aging

In the age-old quest for longer life, people have sampled everything from turtle soup to owl meat to human blood. A Russian-French microbiologist, Ilya Mechnikov, believed that a life span of 150 years could be achieved with the help of a steady diet of milk cultured with bacteria. He thought that the bacteria would take up residence in the large intestine and somehow increase the human life span. (He died at 71.) Many people have ironically died in pursuit of a literal "fountain of youth." The exercises and dietary supplements recommended by Chinese alchemists in the second century B.C. to help people attain eternal life bear an eerie resemblance to the "life extension" products sold at health-food stores today.

A true fountain of youth has never been found. Aging is inevitable, and people must deal with the problems that aging creates (see Reading 9.2). The average age of the population of the United States is increasing. Each day 5,000 people turn 65. By the year 2000, 1 in 8 people in the United States will be over 65, and the "senior" population will total about 32 million (fig. 9.22).

The emerging field of gerontology is helping us to understand the process of aging, and with this information, we are learning how we can live our lives to increase the chances of our last years being enjoyable ones. We may be as unsuccessful as the ancient Chinese alchemists in achieving immortality, but perhaps a growing knowledge of the inevitable march of development and aging will help us to improve the quality of life—during all of our years.

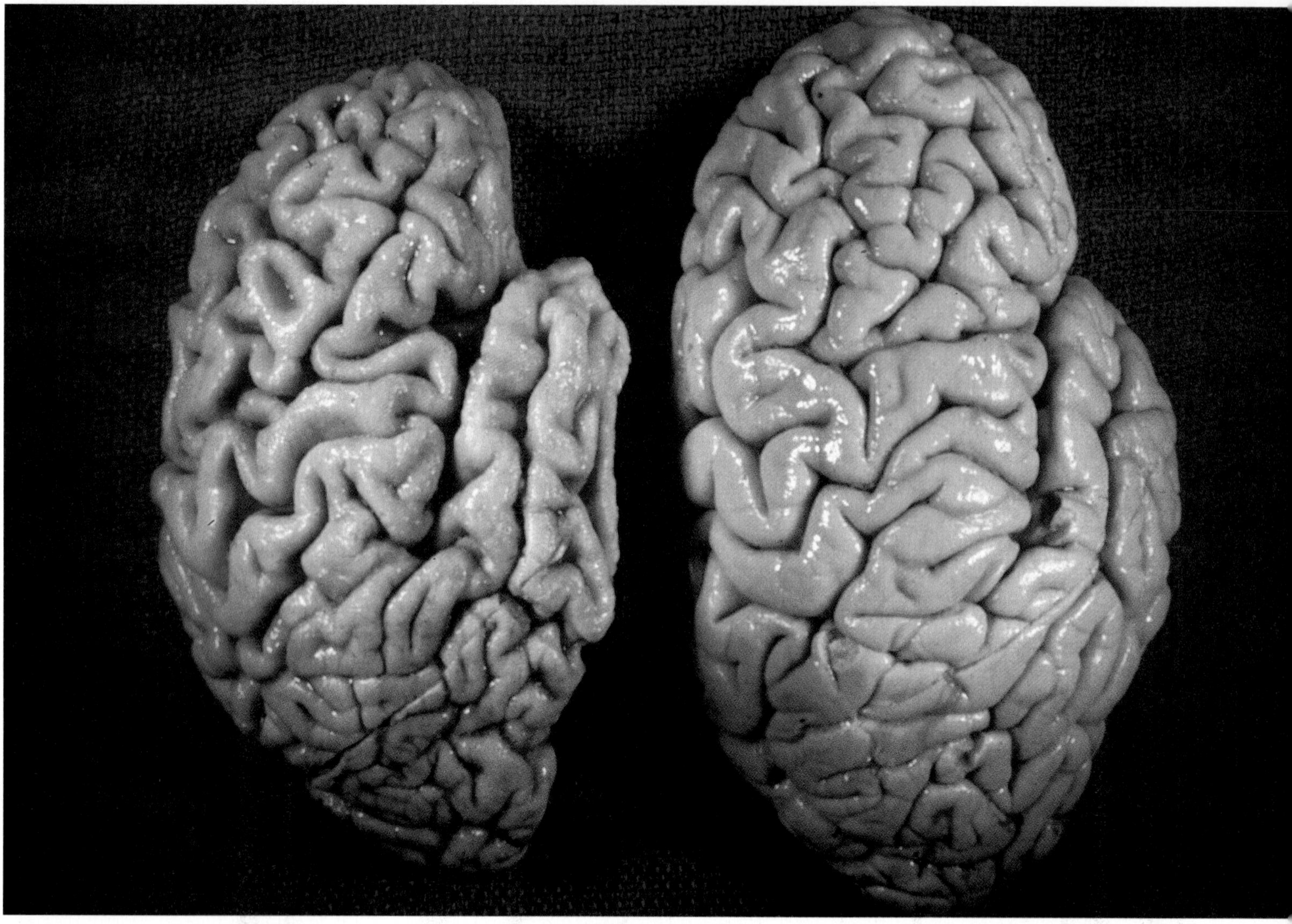

Figure 9.21

The Alzheimer's brain. Upon autopsy, a specimen of brain tissue from a person who died of Alzheimer's disease shows loss of nerve cells in regions that provide memory and reasoning ability. Many of the neurons present contain abnormal twisted fibers. Blood vessels are studded with gummy protein, and cellular debris is widespread. All of these changes are seen in healthy elderly people too but not in the abundance with which they occur in the Alzheimer's brain. Interestingly, adult patients who have Down syndrome (a chromosomal abnormality associated with mental retardation) show very similar brain lesions to those of Alzheimer's patients.

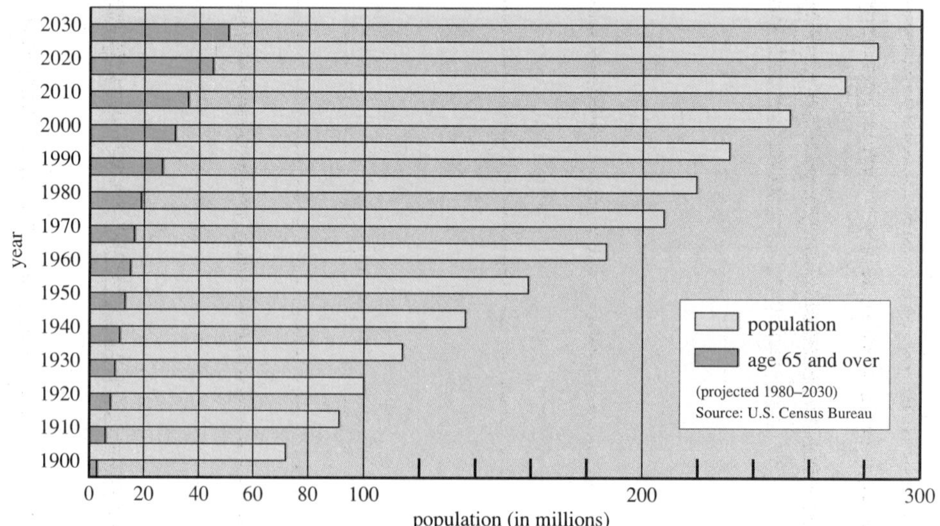

Figure 9.22

Our aging population. Bars show the number of persons age 65 and older compared with total population from 1900. The chart extends to year 2030.
Source: National Institute on Aging.

Reading 9.2 Our Overdrugged Elderly

A few years ago, 67-year-old Rose Zimny stood before a congressional inquiry committee and described her symptoms. She could not walk or control her bowels or bladder, and she suffered mental lapses that made her feel as helpless as a small child. She then demonstrated the cause of her problems by dramatically spilling out the contents of three shopping bags—all of the drugs she took for various ailments. Including both prescription and over-the-counter medications, Rose Zimny was downing some 40 pills a day!

Because of a lack of communication between Zimny's doctors and pharmacists, the mixture of her medications actually caused some of her problems—a common situation among the elderly, who do not often question a doctor's advice. But Rose Zimny was lucky. Caring physicians finally cut back her drug schedule to four types of pills a day to treat her original problems of a nervous disorder and asthma. The rest of her problems vanished.

Rose Zimny is one of thousands—maybe millions—of people over the age of 65 who take so many different drugs that they cannot keep track of them. A senior citizen may have half a dozen or more prescription medications, each of which must be taken several times a day. It is a schedule that would confuse anyone, but it is especially difficult for a person whose memory may be failing. Plus, some elderly people share their drugs. An estimated 30,000 older people die each year of drug interactions or unnecessary or inappropriate drugs.

Drugs meant to treat conditions can cause symptoms by affecting nutrition. Drugs can affect the rates at which nutrients are absorbed through the digestive tract and how quickly these nutrients are metabolized. For example, an older person taking diuretics to lose excess body fluid also loses potassium, calcium, and other vital minerals. Regular use of antacids and laxatives can deplete bone phosphorus. Aspirin causes gastrointestinal bleeding and consequent losses of iron, folic acid, and vitamin C. A simple snack of wine and cheese can dangerously raise the blood pressure of an older person taking certain antidepressant medications.

Help is on the way for the overmedicated elderly. Doctors are making more of an effort to educate older patients about drug interactions and side effects, and pharmacists use computers to keep closer watch over prescriptions. The Food and Drug Administration, the agency that certifies the safety and effectiveness of drugs in the United States, does not yet determine dosages specifically for the aged. Just as an "adult" dose of a painkiller may be excessive for a child, it may also be dangerously high for an elderly person. When we learn more about the special medical and dietary needs of the elderly, drug-induced aches and pains—and worse—should disappear.

The aging of our population presents problems at the societal level but also at the level of the individual family. How can an elderly person who has little money and cannot easily travel obtain adequate medical care in an age when doctors do not make house calls and their offices are often very crowded? How can such an elderly person shop for, afford, prepare, and clean up after three well-balanced meals a day and coordinate those meals with medication schedules? How should adult children deal with a parent rendered physically helpless by a stroke or made dangerously forgetful by Alzheimer's disease?

SUMMARY

Fertilization occurs after a sperm cell is *capacitated*, reaches a secondary oocyte, and burrows through the *zona pellucida* and *corona radiata* with *acrosomal enzymes* and when its *pronucleus* meets the oocyte's pronucleus. This is the *zygote*. *Polyspermy* is blocked. *Cleavage* ensues, and a 16-celled *morula* forms. Between days 3 and 6, the morula arrives at the uterus and hollows out to form the *blastocyst*. Individual cells are *blastomeres*. The *trophoblast* layer and *inner cell mass* form. Around day 6 or 7, the blastocyst implants in the *endometrium*, and trophoblast cells secrete *human chorionic gonadotropin*, which prevents menstruation.

During the second week, the *amniotic cavity* forms as the inner cell mass flattens, forming the *embryonic disc*. *Ectoderm* and *endoderm* form, and then *mesoderm* appears, establishing the germ layers of this *primordial embryo* or *gastrula*. Cells in a particular germ layer develop into parts of specific organ systems. During the third week, the *chorion* begins to develop into the *placenta*, and the *yolk sac*, *allantois*, and *umbilical cord* begin to form as the amniotic sac swells with fluid. *Organogenesis* occurs throughout the embryonic period. Gradually structures appear, including the *primitive streak*, the *notochord*, and *neural tube*, arm and leg buds, the heart, facial structures and skin specializations, and a skeleton.

The *fetal* period begins in the third month. The organ rudiments laid down in the embryonic period grow and specialize as the developing organism moves and reacts. Gradually the proportions of the body more closely resemble those of a baby. The mother feels the fetus move. The fetus is covered in *vernix caseosa* and *lanugo*. In the last trimester the brain develops rapidly and fat is deposited beneath the skin. The digestive and respiratory systems mature last. Signs of *labor* include rupture of the amniotic sac, labor pains, and a vaginal discharge. Strong uterine contractions gradually dilate the *cervix* to 10 centimeters, and the baby is then pushed out.

Aging begins in prenatal life in the sense that certain cells die. Throughout life different organs show signs of aging at characteristic times, such as the brain and thymus early in life. Minor physical signs of aging usually appear in one's 30s and become more noticeable in the 40s and 50s, and memory loss and other disorders may appear later than this.

Aging is both a passive and active process. In passive aging, structures such as collagen, elastin, muscles, and membranes gradually break down, and DNA repair is less efficient. Free radicals build up, the thymus shrinks, and cholesterol is deposited in blood vessels. Active aging is shown by accumulation of *lipofuscin* pigments in cells and *programmed cell death*. The *progerias* and *Alzheimer's disease* alter aging patterns.

QUESTIONS

1. Arrange these prenatal humans into chronological order, from youngest to oldest: morula, gastrula, zygote, fetus, blastocyst, and embryo.

2. What events must take place for fertilization to occur?

3. How do the developmental fates of a cell in the trophoblast and a cell in the inner cell mass differ? What about cells in the ectoderm, endoderm, and mesoderm?

4. Exposure to toxins usually causes more severe medical problems if it occurs during the first 8 weeks of pregnancy, compared to the later weeks. Why?

5. What is the biological basis of the following medical tests?
   a. A pregnancy test that detects HCG in maternal blood or urine
   b. Chorionic villus sampling
   c. Amniocentesis
   d. Maternal alpha fetoprotein detection

6. Why can't a fetus born in the fourth month survive?

7. What are the functions of the following?
   a. The allantois
   b. The amnion
   c. The placenta
   d. Vernix caseosa
   e. Lanugo

TO THINK ABOUT

1. When sperm is mechanically injected into an oocyte, without being in a female's reproductive tract, an embryo does not develop. Based on what you know about fertilization, why doesn't development ensue?

2. A woman has been pregnant for 41 weeks. Her doctor performs a cesarean section to deliver the child because biochemical tests indicate that the cells of the placenta are dying. Why is it vital to the child's survival to perform the cesarean section?

3. Rats raised in an "enriched environment" (provided with toys and given a lot of attention by their keepers) live a third of an average life span longer than normal, and their brains show cellular patterns similar to those of much younger rats. Other animal studies have correlated lower than normal body temperature and lower caloric intake with longevity. How might these findings be applied to the study of human aging?

4. When does aging begin? What structures exhibit aging long before a person appears to be elderly?

5. What factors do you think contribute to longevity?

6. What problems would be encountered in studying the inheritance of longevity?

7. In the year 2050, 1 in every 20 persons in the United States will be over the age of 85. What provisions would you like to see made for the elderly people of the future (especially if you will be one of them)? What can you do now to increase the probability that your final years will be healthy, enjoyable ones?

SUGGESTED READINGS

DeRobertis, Eddy M., Guillermo Oliver, and Christopher V. E. Wright. July 1990. Homeobox genes and the vertebrate body plan. *Scientific American*. A family of genes in frogs, flies, mice, and humans tells cells how to form organs.

Flieger, Ken. October 1988. Why do we age? *FDA Consumer*. Is aging a slowing down of existing function or the start of new activities?

Henig, Robin Marantz. 1985. *How a woman ages* and *How a man ages*. New York: Ballantine Books. A step-by-step look at the various changes associated with aging in the sexes.

Lewis, Ricki. January 1990. Prenatal peeks. *Biology Digest*. Thanks to technology, much of the mystery of prenatal existence has been lifted.

Lord, Thomas R. January 1989. Exploring the twinning phenomenon in *Homo sapiens*. *Biology Digest*. Conceiving twins is far more complex than was previously thought.

Milunsky, Aubrey. 1987. *How to have the healthiest baby you can*. New York: Simon & Schuster. There are things that a pregnant woman can do to ensure that her baby is as healthy as possible.

Wassarman, Paul M. January 30, 1987. The biology and chemistry of fertilization. *Science*, vol. 235. Fertilization is a complex series of interactions between germ cells.

Weiss, Rick. November 5, 1988. Forbidding fruits of fetal-cell research. *Science News*. Should fetal cells that would otherwise be discarded be used to treat disorders in adults?

CHAPTER

10

Human Reproductive and Developmental Problems

Chapter Outline

Learning Objectives

By the chapter's end, you should be able to answer these questions:

1. What are some of the physical causes of infertility in the human male and female? How can infertility be treated?
2. What are some possible causes of spontaneous abortion?
3. What are some of the medical problems faced by a baby born prematurely?
4. What are the effects of some toxic chemicals that interfere with prenatal development?
5. How can the reproductive technologies of artificial insemination, in vitro fertilization, and embryo transfer help infertile couples to have children?
6. What are some commonly used and experimental birth-control devices and methods and how do they work?
7. What are some sexually transmitted diseases?

They are each named Louise, and each, in her own special way, has contributed much to our knowledge of prenatal development. Louise Joy Brown was born in England in 1978, a normal little girl in every way except for her place of conception—a piece of laboratory glassware. "Adorable Louise" was born in 1985, a normal donkey in every way except for her place of prenatal development—the uterus of a horse.

Louise Brown was the first human to be conceived outside of the body and transferred as preembryo to the uterus of the egg donor, where Louise continued development. The techniques that led to Louise Brown's birth were perfected in nonhuman animals, and Louise's historic beginning has since guided the technology that has led to the births of more than a thousand "test-tube" babies.

Adorable Louise was conceived in a female donkey by a male donkey, but 8 days after fertilization, she was transferred by veterinary researchers at Cornell University to the uterus of a horse. Normally, the immune system of a female horse would abort the "foreign" donkey embryo. The Cornell researchers wanted to prevent this rejection by exposing the horse before the embryo transfer to white blood cells from Louise's natural donkey parents. It worked. The horse's immune system, "tricked" into recognizing donkey cells as its own, accepted the donkey embryo. A year later, the horse gave birth to Adorable Louise (fig. 10.1).

An approach similar to the Cornell donkey-in-a-horse feat is now offered at a few clinics to certain women who suffer repeat early pregnancy losses. The mothers' immune systems apparently do not "ignore" the embryos, as they normally would, and instead attack them as if they were foreign, due to an incompatibility with the father. The women receive injections of the husbands' white blood cells before they attempt pregnancy, which somehow tricks their immune systems into accepting the embryos.

The two Louises illustrate the trend in reproductive biology to transfer basic research and agricultural work to human health care. As medical science learns more about the complex biological interactions that build a newborn human from a fertilized egg, more ways are found to encourage or discourage conception, to prevent or treat birth defects, and to treat the fetus or newborn.

a.

b.

Figure 10.1
Two Louises. *a.* Louise Joy Brown was conceived when her father's sperm cell met her mother's egg cell in the laboratory. Here, at age 10, Louise, the oldest "test-tube baby," holds one of the youngest, Andrew Macheta. The children were among 600 test-tube babies attending a 10-year anniversary party at the clinic near London where they were conceived. *b.* Adorable Louise looks like a normal donkey, but a horse gave birth to her, following embryo transfer. The techniques used to allow Louise to gestate in a horse led to treatments that can prevent miscarriage in humans.

Reading 10.1 *What's New at the Zoo?*

The birth of a baby ox at the Bronx Zoo in 1981 caused quite a stir. Little Manhar was the first member of an endangered species to be born following embryo transfer. Manhar's genetic parents were guars, a rare species of ox found in India and Nepal. The female who carried him in her uterus, however, was a common Holstein dairy cow.

In the Cincinnati Zoo, dairy cows carry embryos of a near-extinct antelope, an embryo of a wild sheep nestles in the uterus of a domestic sheep, tiger embryos are carried by lions, and housecats are surrogate mothers to embryos of endangered feline relatives. At the Louisville Zoo, a horse gives birth to a zebra, and in the London Zoo, a donkey does the same. In other zoos, cow and horse embryos at the four-cell stage are treated with digestive enzymes that separate them into single cells, and each cell is implanted into a surrogate mother of the same species. Identical quadruplets are born from the four surrogate mothers. Still other techniques are used to create species not possible by the mating of members of two existing species. For example, embryos of a sheep and goat can be separated into single cells and reassembled to produce a hybrid embryo built of cells from both species (fig. 1). The goal is not to add something new to the zoo but to boost declining populations of endangered species.

But sophisticated technology is not always needed to encourage reproduction in rare species. Sometimes all that is required to beef up a falling zoo population is a little romance. Consider the cheetah, a magnificent cat so close to extinction that some zoos artificially inseminate the females. When the cheetahs in the San Diego Zoo were not mating in their very-visible home they were moved to a more private compound, and pregnancies soon followed. Lion-tailed macaques had nearly died out at one zoo because their birthrate was not keeping pace with their death rate. The females nursed their young for so long that they became pregnant only every 2 years. (Breast-feeding inhibits ovulation.) Zookeepers separated mothers from suckling babies and bottle-fed the young. Soon, the population rose. But how does a zookeeper matchmake 9-foot-long Komodo dragons, when the sexes cannot be readily distinguished by humans? The hormone levels in the animals' droppings are measured, a safe alternative to probing the private parts of these gigantic reptiles.

A major problem in maintaining or increasing the sizes of zoo populations is that when such groups are small, animals that are closely related to one another mate. When organisms inbreed in this manner, they can bring together identical forms of genes that they inherited from a common ancestor, causing frequent spontaneous abortions and birth defects. (This is the reason why first cousins are discouraged from marrying in some cultures.) If inbreeding continues for several generations, the animals become more and more alike genetically. The species is threatened as more individuals have difficulty reproducing successfully.

One way for zoos to increase the genetic diversity of their populations is to swap animals—or, better yet, germ cells of these animals. It is far simpler to mail frozen sperm than it is to ship a whole hippo, for example. Embryos of a species of antelope known as the bongo were recently transported transcontinentally in containers of nutrients that were held in the armpits of researchers, where they were kept warm. Many zoos use computers to catalog genetic information about their residents. Freezing somatic cells may also be valuable, because their nuclei could be used to clone new individuals.

Figure 1
This animal has the face and horns of a goat but the body proportions of a sheep. It is engineered by combining early embryonic cells of a goat and a sheep.

Infertility

For one out of every six couples in the United States, trying to conceive a child is not joyous as it is for most, but it is a period of anxiety and growing frustration as the couple realizes that they may be infertile. **Infertility** is the inability to conceive after 1 year of frequent intercourse without the use of contraceptives. A physical cause for infertility can be identified in 90% of all cases, and many of these problems can be corrected. But infertility in the United States seems to be on the rise.

Male Infertility

About 40% of the time, infertility is traced to the male. Another 40% of the time, it is traced to the female. In the remainder of cases, both partners contribute to the problem. Infertility in the male is easier to detect, but it is sometimes harder to treat than female infertility.

Some men have difficulty fathering a child because they produce fewer than the 400 million sperm found in a normal ejaculate (fig. 10.2). If this *low sperm count* is due to a hormonal imbalance, administering the appropriate hormones may boost sperm output. Sometimes the problem is *immune infertility*, in which a man's immune system interprets his own sperm as foreign, as if they were invading bacteria. His white blood cells produce antibodies, which cover up the molecules on sperm cell surfaces that enable them to bind to oocytes. Male infertility can also be due to a varicose vein in the scrotum. This enlarged vein brings too much heat to developing sperm, so they cannot mature. A scrotal varicose vein can be removed surgically.

For many men with low sperm counts, becoming a father is just a matter of time. If the ejaculate contains at least 60 million sperm cells, then fertilization is likely to happen eventually. A man with a low sperm count can donate several semen samples over a period of weeks to a fertility clinic. The samples are kept in cold storage until there are a sufficient number of sperm cells to make conception likely. The samples are pooled, and some of the seminal fluid is withdrawn to leave a sperm cell "concentrate," which is then transferred to the woman's reproductive tract (artificial insemination).

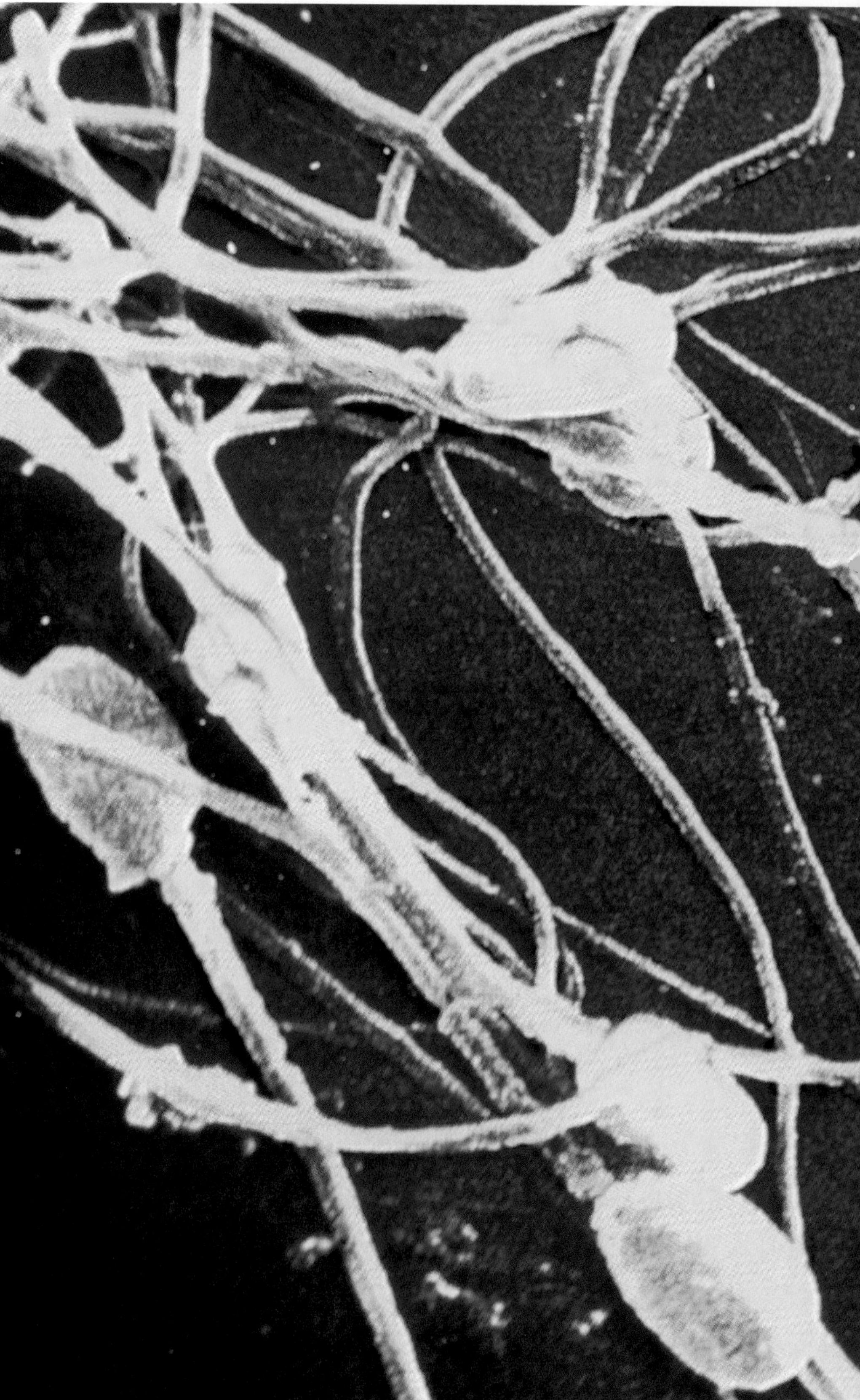

Figure 10.2
The man who donated these sperm cells became a father—his sperm are abundant, motile, and normal in shape. Magnification, x 2,900.

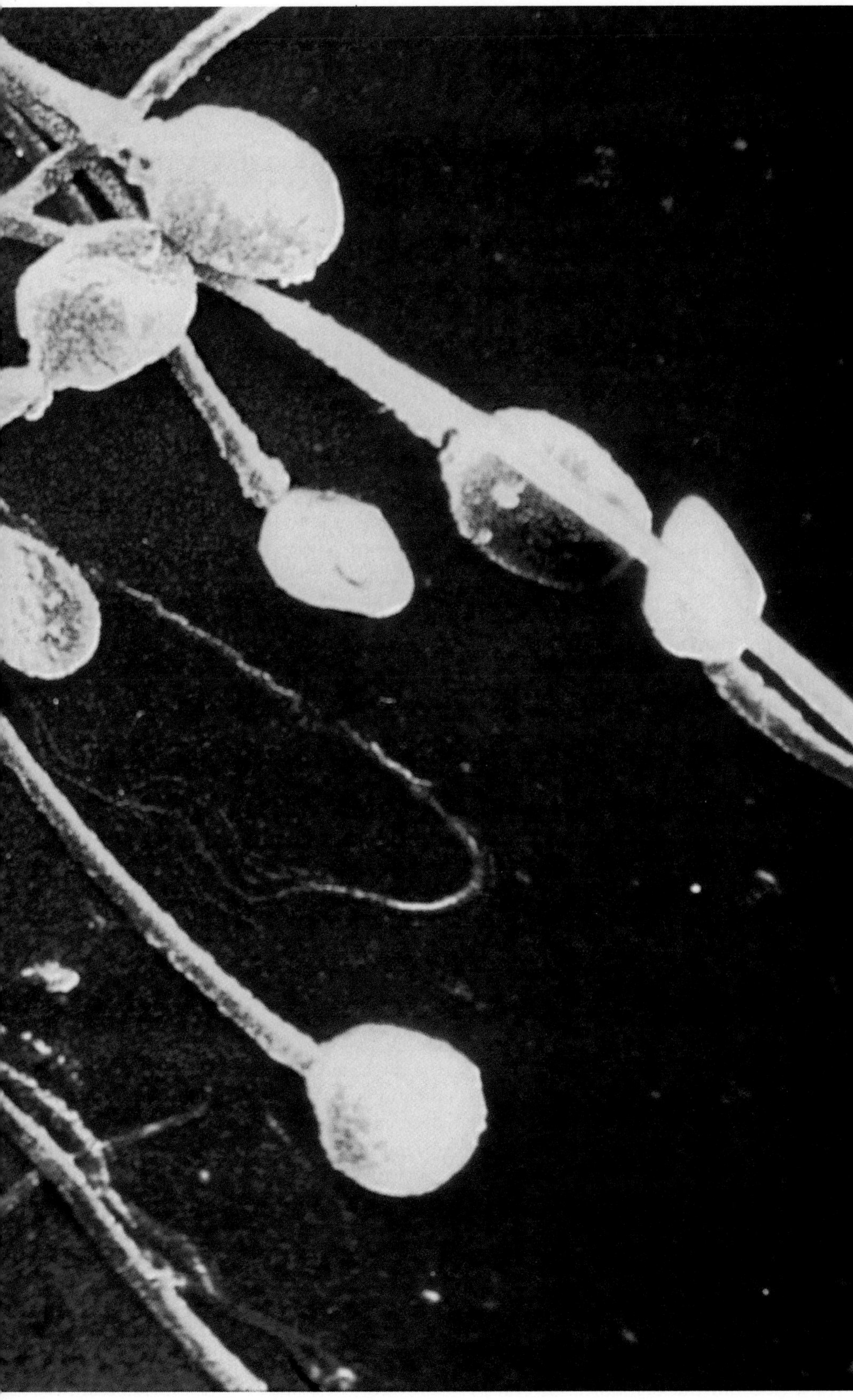

Sperm quality is even more important than its quantity. Sometimes a man manufactures plenty of sperm, but the cells are unable to move, and therefore cannot travel to an oocyte to fertilize it. If the lack of motility is due to a physical defect, such as misshapen or missing flagella, there is no treatment. If the cause is hormonal, however, administering the appropriate hormone can sometimes restore sperm motility. Sperm shape and the speed and direction of their movements can be tracked by computer.

KEY CONCEPTS

Infertility is the inability to conceive a child after a year of trying. It is traced to the male 40% of the time, the female 40% of the time, and to both 20% of the time. Male infertility can be caused by a low sperm count; an immune reaction by a man against his own sperm; a varicose vein in the scrotum; and malformed sperm.

Female Infertility

Female infertility can be due to abnormalities in any part of the reproductive system. Many infertile women have irregular menstrual cycles, making it difficult to pinpoint when conception is most likely. In an average menstrual cycle of 28 days, ovulation usually occurs around the 14th day after menstruation begins, and this is when a woman is most likely to conceive. For a woman under 30 years old not using birth control, pregnancy is likely to occur within 3 or 4 months. A woman with irregular menstrual periods can use an ovulation predictor test, which detects a peak in the level of a certain hormone that precedes ovulation by a few hours.

The hormonal imbalance that usually underlies irregular ovulation can have various sources—a tumor in the ovary or in the pituitary gland in the brain that hormonally controls the reproductive system; an underactive thyroid gland; or steroid-based drugs such as cortisone. Sometimes a woman's endocrine system produces too much prolactin, the hormone that normally promotes milk production and suppresses

Reading 10.2 *Jon and Linda—The Plight of an Infertile Couple*

WHAT SHOULD A COUPLE DO AFTER THEY HAVE TRIED UNSUCCESSFULLY FOR A YEAR OR MORE TO BECOME PARENTS? Consider the case of Jon and Linda Williams.

Jon and Linda met in law school, when they were both 25 years old. After graduation, they married and went to work for different law firms. Linda loved her job. After all those years in school, she was grateful to be out in the real world at last. Besides, she and Jon still had educational loans to repay. Starting a family was way off in the future.

Ten years later, when both were secure in their jobs, loans had been repaid, and they had a house of their own with an empty nursery, the future had finally arrived. Linda looked forward to a few years off to raise their baby, and Jon was eager for fatherhood. But, after a year of trying, Linda had still not become pregnant. Rather than going directly to one of the hospital-run fertility clinics in their area, the couple decided first to visit a urologist specializing in fertility problems to check Jon's sperm count. They made this decision because it is easier, less costly, and less painful to check the man first. Jon's sperm cells were normally active, but his sperm count was 350 million cells per ejaculate, slightly below the normal number. The doctor felt that, after a year of trying to conceive, some factor other than Jon's slightly low sperm count must be contributing to the problem.

The next step was to visit Linda's gynecologist, also a fertility expert. She first determined that Linda was ovulating by measuring the level of the hormone progesterone in her blood. Next she performed an ultrasound examination of Linda's pelvic region to see if all of her reproductive organs were present and normally formed and to determine if her uterine lining would support an embryo. All was well. Then the doctor scheduled a *postcoital test*. Immediately after intercourse at the time of ovulation, Linda's cervix was examined to see if her secretions were thin enough to allow sperm to swim through. Again, Linda passed.

The doctor told the anxious couple that, so far, everything looked normal. But she cautioned that Linda's age—36 years—might mean that it would take her longer to conceive, according to statistics. Her advice: try for another 6 months. She suggested that they use a specially designed thermometer to take Linda's temperature each morning. A slight rise (0.2 to 0.5°F) would indicate that she was ovulating—a good time to try for parenthood. Linda could also use an ovulation predictor test. If the couple had no luck, then more tests could be run, and possibly fertility drugs could be prescribed.

Linda got into the habit of checking her temperature at the same time each morning, and she and Jon had intercourse at her most fertile times. Three months later, she was pregnant! Apparently Linda's slightly irregular menstrual cycle and Jon's slightly low sperm count had combined to thwart their efforts to become parents.

Had the couple continued to have difficulty conceiving, more complicated tests would have been in store for Linda. The next step would have been an *endometrial biopsy*, in which a small piece of her uterine lining would be scraped off and examined under the microscope to see if it could support an embryo. Next, in a *hysterosalpingogram*, a dye would be injected into her fallopian tubes and its path followed with an optical device to see if the tubes were obstructed. *Laparoscopy* would be next: A small incision would be made near Linda's navel, and a tiny, illuminated optical device called a *laparoscope* would be inserted to detect small amounts of scar tissue, fibroid tumors, or endometriosis inside the fallopian tubes, an area not easily seen in ultrasound exams or readily accessible through the vagina. If the laparoscope detected a problem, a *laparotomy* would be performed, in which the abnormal tissue would be surgically removed.

In some cases of infertility, no cause can be found. Psychological factors may be at play, or it may simply be that the failure to conceive is the result of consistently poor timing. It is not at all unusual for an "infertile" couple to adopt a child, only to conceive one of their own shortly thereafter.

ovulation in new mothers. If prolactin is abundant in a nonpregnant woman, she will not ovulate and therefore cannot conceive.

Fertility drugs can be used to stimulate ovulation, but they also cause women to "superovulate," producing more than one egg each month. The result is twins, triplets, or even higher multiples. If a woman's ovaries are completely inactive, or if they are absent (due to a birth defect or to surgery), pregnancy is impossible. The only recourse is an ovary transplant (still experimental) or one of the reproductive alternatives discussed later in the chapter.

Female infertility can also be due to blocked fallopian tubes. Because fertilization usually occurs within these tubes, the blockage either prevents sperm cells from reaching the oocyte or keeps a fertilized egg from descending into the uterus. Fallopian tubes can be blocked as a birth defect or, more likely, from an infection. A woman may not know she has blocked tubes until she has difficulty conceiving and a doctor discovers the problem. Blocked tubes can sometimes be opened surgically.

Defects in the uterine lining may make it inhospitable to an embryo. One in five women develop noncancerous uterine tumors called *fibroids*. If fibroids do not grow too large and do not interfere with the functioning of an organ, they do not produce any symptoms. A woman can have several plum-sized fibroids and not even know it. However, fibroids growing within the uterine lining may prevent conception. Fibroids often disappear on their own, but if they cause pain they can be removed surgically.

Secretions in the vagina and cervix may be hostile to sperm, containing antibodies that attack sperm as if they were bacteria. If the mucus in the vagina and cervix is abnormally sticky, sperm cells become entrapped and they cannot move far enough up the reproductive tract to

Table 10.1
Causes of Human Infertility

Problem	Possible Causes	Diagnostic Test	Treatment
Males			
Low sperm count	Hormone imbalance, varicose vein in scrotum, excessive exposure to heat by wearing tight pants or taking hot showers	Examine sperm under microscope	Hormone therapy, surgery, avoiding excessive heat
Immobile sperm	Abnormal sperm shape	Examine sperm under microscope	None
	Infection	Examine semen for infective agents	Antibiotics
	Malfunctioning prostate	Test semen for chemical abnormalities	Hormones
Antibodies against sperm	Problem in immune system	Test semen for antibodies	Drugs
Females			
Ovulation problems	Pituitary or ovarian tumor	Measure hormone levels, take daily temperature readings, X rays	Surgery
	Underactive thyroid	Same as above	Drugs
Antisperm secretions	Unknown	Postcoital test	Acid or alkaline douche, estrogen therapy
Blocked fallopian tubes	Infection caused by IUD or abortion or by sexually transmitted disease	X rays, ultrasound examination, hysterosalpingogram, laparoscopy	Laparotomy, egg removed from ovary and placed in uterus
Endometriosis	Delaying parenthood until the 30s	Laparoscopy	Hormones, laparotomy

encounter an oocyte. Sometimes vaginal secretions are so acidic or alkaline that sperm cells are weakened or killed. Secretion problems can be treated with low doses of the hormone estrogen or by douching daily with an acidic solution such as acetic acid (vinegar) or an alkaline solution such as bicarbonate to alter the pH of the vagina so that it is more receptive to sperm cells. Table 10.1 provides a summary of the causes of infertility in males and females.

The Age Factor in Female Infertility

The older a woman is, the longer it may take her to conceive. In the United States and Europe, female fertility peaks in the middle 20s, declines until age 29 when it levels off, and again dips between ages 30 and 34.

One reason for difficulty in conceiving over the age of 30 is *endometriosis*. In a healthy woman, the lining of the uterus (the endometrium) builds up blood and tissue each month in preparation for the implantation of an embryo. If no embryo appears, the lining is shed as the menstrual flow. In endometriosis, this lining forms on the outside of the uterus, on or in the fallopian tubes, or even on a nearby organ such as the bladder. At the hormonal cue that conception has not occurred, this misplaced tissue breaks down and bleeds, just as it would inside the uterus. The tissue blocks implantation of a zygote (fig. 10.3).

Endometriosis is sometimes called the "career woman's disease" because it is more common in women who postpone childbearing, but it is not a new phenomenon. In the 1950s, it was known as the "schoolteacher's disease" and was recognized as far back as the nineteenth century. Today, the painful condition can sometimes be corrected well enough by surgery or hormone treatment to permit pregnancy to occur. Symptoms can be alleviated by using a nasal spray of the drug nafarelin acetate, which suppresses menstruation. Curiously, the birth of a child seems to prevent recurrence of endometriosis.

Age affects female fertility in other ways. Older women are more likely to produce oocytes with an abnormal chromosome number. The developmental consequences of a chromosomal imbalance are often so severe that the pregnancy ends before it is even detected. Losing very early embryos may be mistaken for infertility because the bleeding caused by the aborted embryos appears to be just the normal menstrual flow. The higher incidence of meiotic errors in older women may be because their oocytes are older and therefore have had more time to be exposed to harmful chemicals, viruses, and radiation.

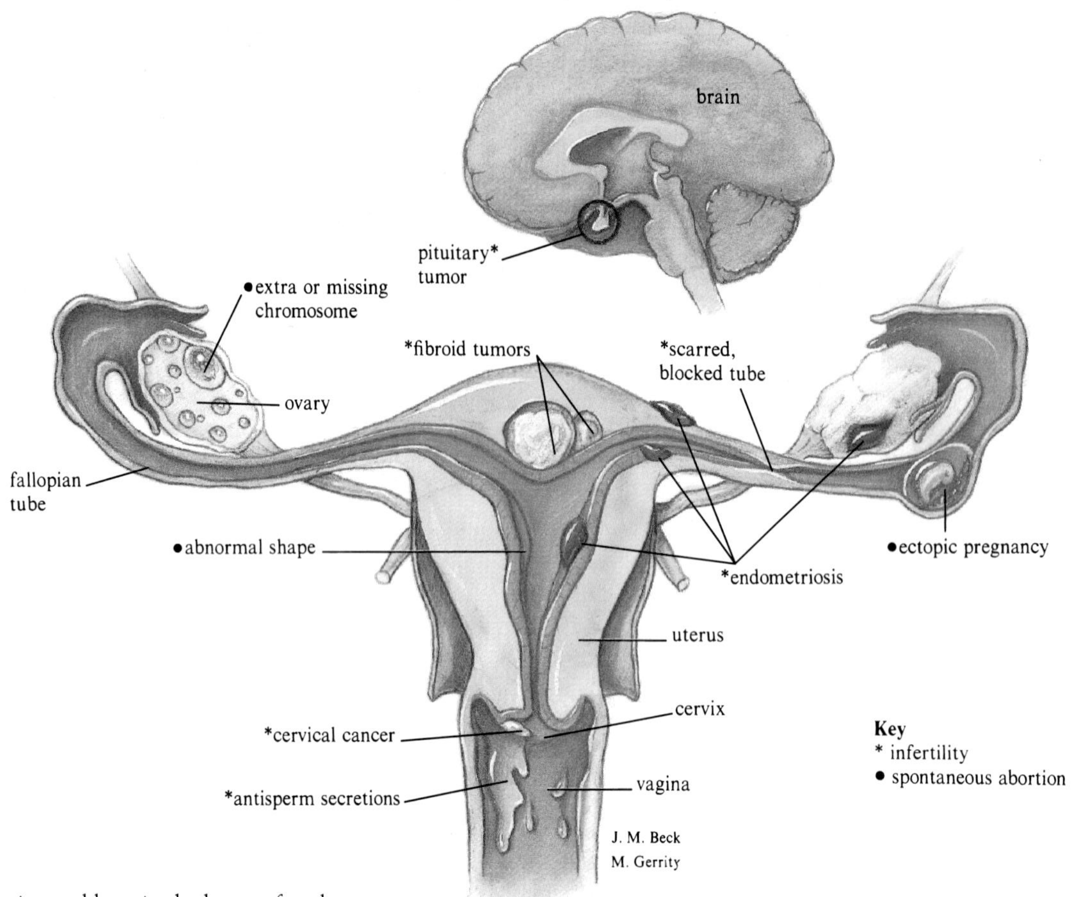

Figure 10.3
Sites of reproductive problems in the human female.

> **KEY CONCEPTS**
>
> *Female infertility can be due to an irregular menstrual cycle, which can be traced to a tumor in the ovary or pituitary gland, an underactive thyroid, or excess prolactin. Blocked fallopian tubes, which can result from infection or a birth defect, can cause infertility. Fibroid tumors in the uterus may prevent implantation, and secretions in the vagina and cervix may inactivate sperm. Women over 30 are more prone to endometriosis, which can prevent pregnancy. Older women are also more likely to produce oocytes with abnormal chromosome numbers.*

Spontaneous Abortion

Perhaps even more upsetting than infertility is *spontaneous abortion*, a pregnancy that ends naturally before the fetus is developed enough to survive in the outside world. Sometimes the embryo or fetus is deformed. Sometimes it is normal, but some problem in the mother's body prevents her from carrying the fetus to full term. Although nearly one in five women who know they are pregnant have spontaneous abortions, actually two in three conceptuses may not survive. Many losses occur before the woman realizes that she is pregnant.

A spontaneous abortion is similar to an unusually heavy menstrual period, with cramping and more copious bleeding. The more advanced the pregnancy, the more severe the symptoms. If the pregnancy is past the first trimester, the spontaneous abortion is called a miscarriage. If a woman in the second trimester notices that the fetus has not moved in 24 hours, it may be a sign that a miscarriage is imminent. A doctor will usually advise immediate bed rest, but most miscarriages cannot be prevented. In a pregnancy that is 6 months along or later, the woman must endure labor to give birth to a child that she knows will be born dead, or *stillborn*.

Why don't all embryos and fetuses complete development? Many embryos cannot survive for more than a few weeks because they have an abnormal number of chromosomes. Sometimes the zygote is normal, but the trophoblast cells that surround it do not produce enough HCG to suppress menstruation, and implantation is blocked. If the structure of the mother's uterus is abnormal, the fetus may be lost. Another maternal problem is an *incompetent cervix*, which opens long before the fetus is ready to be born. If an incompetent cervix is detected early, the pregnancy can be maintained by closing the cervix with sutures, which are removed when labor begins.

Fibroid tumors in the uterus may crowd the developing embryo or fetus, stunting its growth and possibly interfering with its nutrient supply. Spontaneous abortion may occur if the zygote implants in a fallopian tube rather than in the uterus. This dangerous and painful condition,

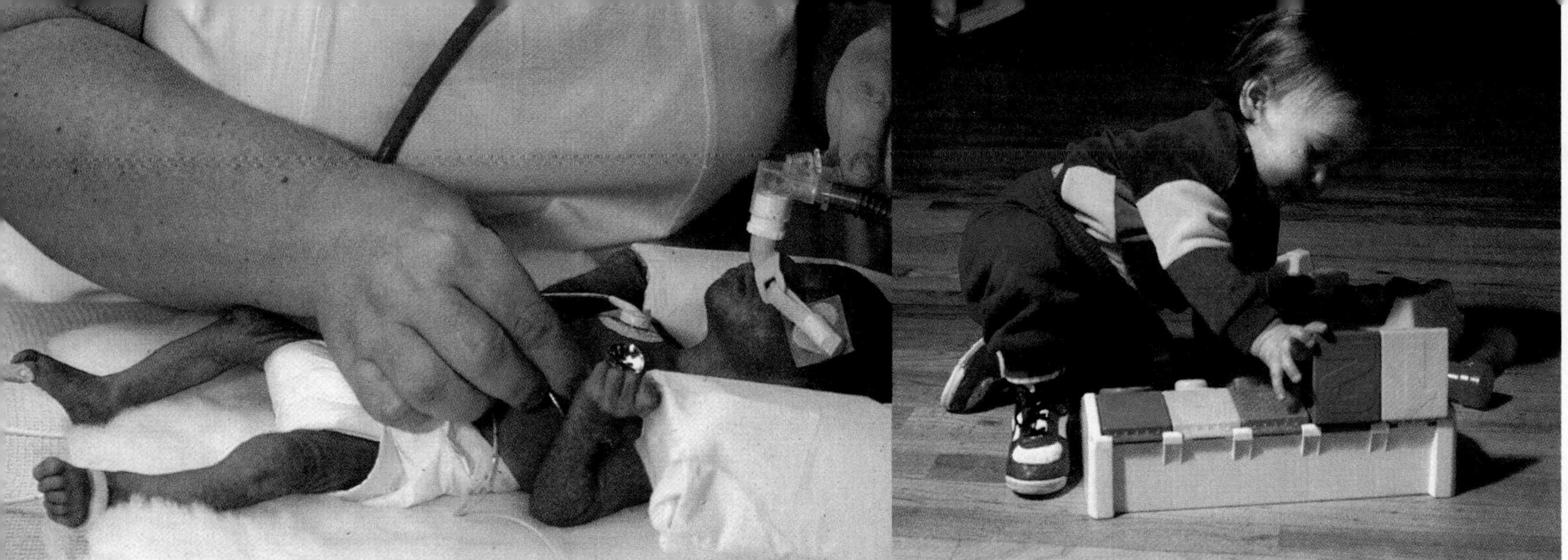

Figure 10.4
Babies born too soon must have special care at first, but by a year of age, most have caught up.

called an **ectopic pregnancy,** usually requires an immediate therapeutic abortion. Otherwise, the growing embryo will eventually burst the tube, and the woman can lose a great deal of blood and become seriously infected. Once rare, ectopic pregnancies are now on the rise. This increase may be related to the parallel increase in infections that damage the fallopian tubes and, curiously, to the ability to treat them. Before pelvic infections could be diagnosed early and treated with antibiotics, the tubes would completely close with scar tissue, so pregnancy could not occur. Today, antibiotic treatment can halt infection so that the tubes are only partially obstructed by scar tissue. This scar tissue, however, entraps fertilized ova, and the result is an ectopic pregnancy.

It is normal for a couple experiencing spontaneous abortion to grieve for their expected child. They often wonder, "Can this happen again?" or "Was it my fault?" For most couples, the chances of recurrence of a chromosomal abnormality is only 5% to 20%, so they are usually encouraged by their doctors to wait a few months and try again. If the cause of the spontaneous abortion was hormonal or due to certain malformations, the situation can sometimes be corrected by hormone therapy or surgery. However, some women spontaneously abort, for unknown reasons, at about the same point in several successive pregnancies. These women are advised to stay in bed during this critical time in their pregnancies.

KEY CONCEPTS

Spontaneous abortion is caused by chromosomal defects, too little HCG, an abnormally shaped uterus, an incompetent cervix, or fibroid tumors. In an ectopic pregnancy, the zygote implants in a fallopian tube and must be surgically removed.

Born too Soon—The Problem of Prematurity

Human prenatal development normally takes about 38 weeks. Babies born more than 4 weeks early, or weighing less than 5 pounds (2.3 kilograms), are considered *premature*. Their chances of survival increase with time spent in the uterus and with birth weight. Babies born more than 12 weeks prematurely usually live only a few days, but rapidly developing technology for treating tiny infants is greatly increasing the survival chances for these and even more immature infants. Each year about 10,000, or 8%, of the babies born in the United States are premature. Twins and triplets are more likely to be born prematurely because of their cramped prenatal environment. Teens and malnourished women also are more likely to deliver prematurely. However, a "preemie" can be born to even a well-nourished, healthy woman, for unknown reasons.

The most serious problem of premature infants is their immature reflexes. The baby born too soon is often not yet ready to breathe properly, because normally the placenta would still be providing oxygen and carrying away carbon dioxide. *Respiratory distress syndrome* develops when the baby's lungs are not sufficiently covered with a soapy substance called *surfactant*. The small air sacs in the lungs do not inflate without it. When the baby strains to draw air into its deflated lungs, the tissue is damaged.

A baby with insufficient surfactant can be assisted in breathing with a synthetic surfactant dripped into its windpipe. The infant is then placed on a respirator. Within minutes, the rosy blush of a functioning respiratory system is seen for the first time.

Often preemies cannot suck and must be fed either intravenously or through a stomach tube inserted through the nose. They are also more susceptible to injury during birth, and they are more prone to infection. Preemies must be kept warm in heated units called isolettes because the combination of small body size, low body fat, and the immaturity of the brain's temperature-control mechanism makes them lose heat rapidly (fig. 10.4).

Sometimes a fetus is carried full term but weighs less than 5 pounds (2.3 kilograms) at birth. Such an infant is termed "small for gestational age" and usually has few problems beyond the low birth weight that can cause it to rapidly lose body heat. Unlike the infant born too soon, this type of tiny tot has mature breathing and sucking reflexes. Women who are malnourished or who smoke during pregnancy often have babies born on time but with low birth weights. Abnormal development of the placenta is sometimes responsible for low birth weight in full-term infants.

KEY CONCEPTS

Babies born more than 4 weeks early or weighing less than 5 pounds (2.3 kilograms) often suffer health problems because of their small sizes and immature reflexes. Premature infants may have difficulty sucking, breathing, and staying warm. Respiratory distress syndrome can often be treated with surfactant.

Birth Defects

The prenatal period is a time of rapid growth and complex changes. It is perhaps not surprising, then, that a variety of developmental errors can occur. Some birth defects are caused by genetic abnormalities, and some are caused by exposure of the embryo or fetus to drugs, poisons, or viruses. Although many things can go wrong during prenatal development, it is reassuring to know that most babies—about 97%—are completely normal at birth.

The Critical Period

If development stops, a spontaneous abortion or stillbirth results. Alternatively, development can continue abnormally, and the child is born with a defect. The specific nature of the defect reflects the structures that were developing when the damage occurred. The time when a specific structure can be altered is called its **critical period** (fig. 10.5). Some body parts, such as fingers and toes, are sensitive for short periods of time. In contrast, brain development can be disrupted throughout prenatal development, as well as during the first 2 years of life. Because of the extensive critical period of the brain, many birth defects include mental retardation. The continuing sensitivity of the brain after birth is why toddlers who eat lead-based paint develop learning disabilities.

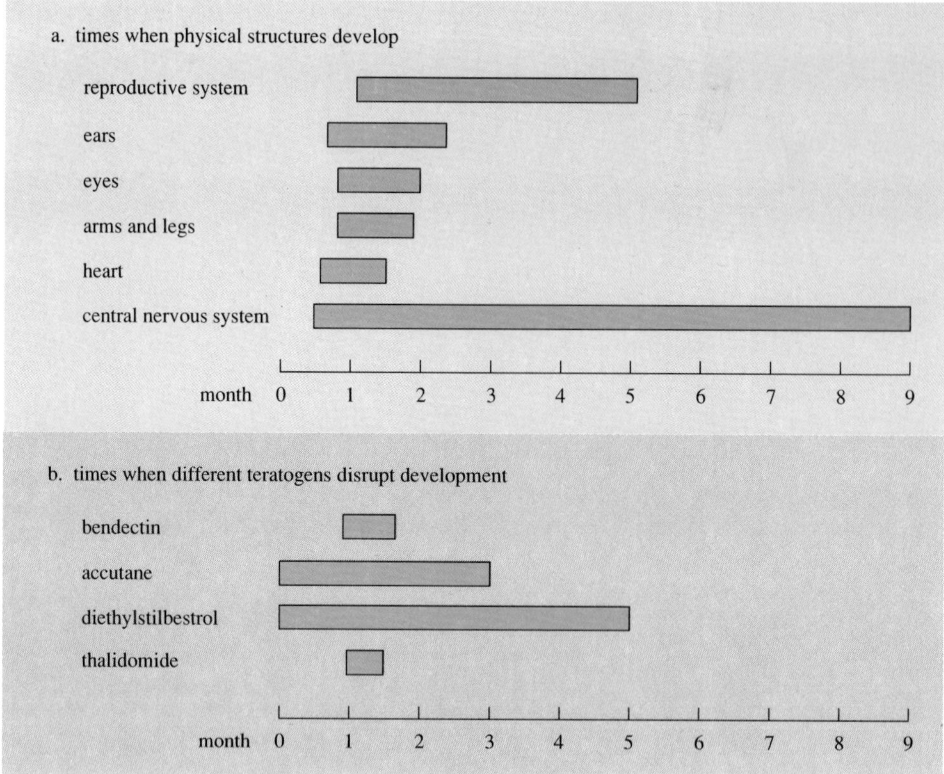

Figure 10.5
Critical periods of some structures (*a*) and times in development when certain drugs act (*b*).

About two-thirds of all birth defects stem from a disruption during the embryonic period. More subtle defects that become noticeable only after infancy, such as learning disabilities, are often caused by disruptions in the fetal period. For example, chemicals that disrupt brain development in the first trimester are likely to lead to a mentally retarded newborn. A substance that affects brain function in the seventh month of pregnancy, however, might be responsible for difficulty in learning to read.

A birth defect can be due to an abnormal gene that exerts its influence at a specific point in prenatal development. In the rare inherited condition called phocomelia (seal limbs), for example, an abnormal gene halts limb development from the third to fifth week of embryonic existence, causing the infant to be born with "flippers" where the arms and legs should be. A genetically caused birth defect has a predictable chance of recurrence in the offspring of the affected child.

Many birth defects, however, are caused by toxic substances ingested by the expectant mother, and these problems cannot be passed on to future generations by the affected child. Chemicals or other agents that cause birth defects are called **teratogens** (Greek for "monster causing"), and the study of birth defects is called teratology.

Teratogens

For many years it was believed that the unborn child is protected from harmful substances by the placenta. This idea was tragically disproven between 1957 and 1961, when 10,000 children were born in Europe with what seemed, at first, to be phocomelia. Realizing that this genetic disorder is very rare, doctors began to look for another cause. They soon identified a mild tranquilizer, *thalidomide*, which all of the mothers of the deformed infants had taken early in pregnancy, during the time of limb formation. The "thalidomide babies" were born with stumps or nothing at all where their legs and arms should have been (fig. 10.6). How does the thalidomide disaster illustrate the critical period concept?

At about the same time that the severity of the thalidomide crisis was being realized, another teratogen, this one a virus, was sweeping the United States. The teratogenic effects of the *rubella* virus that causes German measles were first noted in Australia in 1941, but they were not brought to public attention here until the early 1960s, when a rubella epidemic caused 20,000 birth defects and 30,000 stillbirths. Children of women who contracted the virus in the first trimester of pregnancy ran a high risk of being born with cataracts, deafness, and heart defects. The effects of rubella on fetuses exposed during the second or third trimesters of pregnancy showed up later as learning disabilities, speech and hearing problems, and juvenile-onset diabetes.

Alcohol is a teratogen. A pregnant woman who has just one or two drinks a day, or perhaps a large amount at a crucial time in prenatal development, risks *fetal alcohol syndrome* in her unborn child. Because the effects of small amounts of alcohol at different stages of pregnancy are not yet well understood, and because each woman metabolizes alcohol slightly differently, it is best to avoid drinking alcohol entirely when pregnant, or when trying to become pregnant.

A child with fetal alcohol syndrome has a characteristic small head, misshapen eyes, and a flat face and nose (fig. 10.7). He or she grows slowly before and after birth, and intellect is impaired, ranging from minor learning disabilities to mental retardation. One study found that if a mother-to-be consumed three mixed drinks a day her child lost five IQ points. Problems in children of alcoholic mothers were noted by Aristotle more than 23 centuries ago. In the United States today, fetal alcohol syndrome is the third most common cause of mental retardation in newborns, and 1 to 3 in every 1,000 infants has the syndrome—more than 40,000 born each year.

A newly recognized teratogen is cocaine, which can cause spontaneous abortion by inducing a form of stroke in the fetus. Infants who were exposed to cocaine in the uterus do not react normally to environmental stimuli and seem unable to make sense of their surroundings. The effects of cocaine on children are still being investigated, and we may see other health problems as these children grow.

Chemicals in cigarette smoke are teratogens. Carbon monoxide produced from cigarette smoke crosses the placenta and robs the fetus of oxygen. Other chemicals in smoke prevent nutrients from

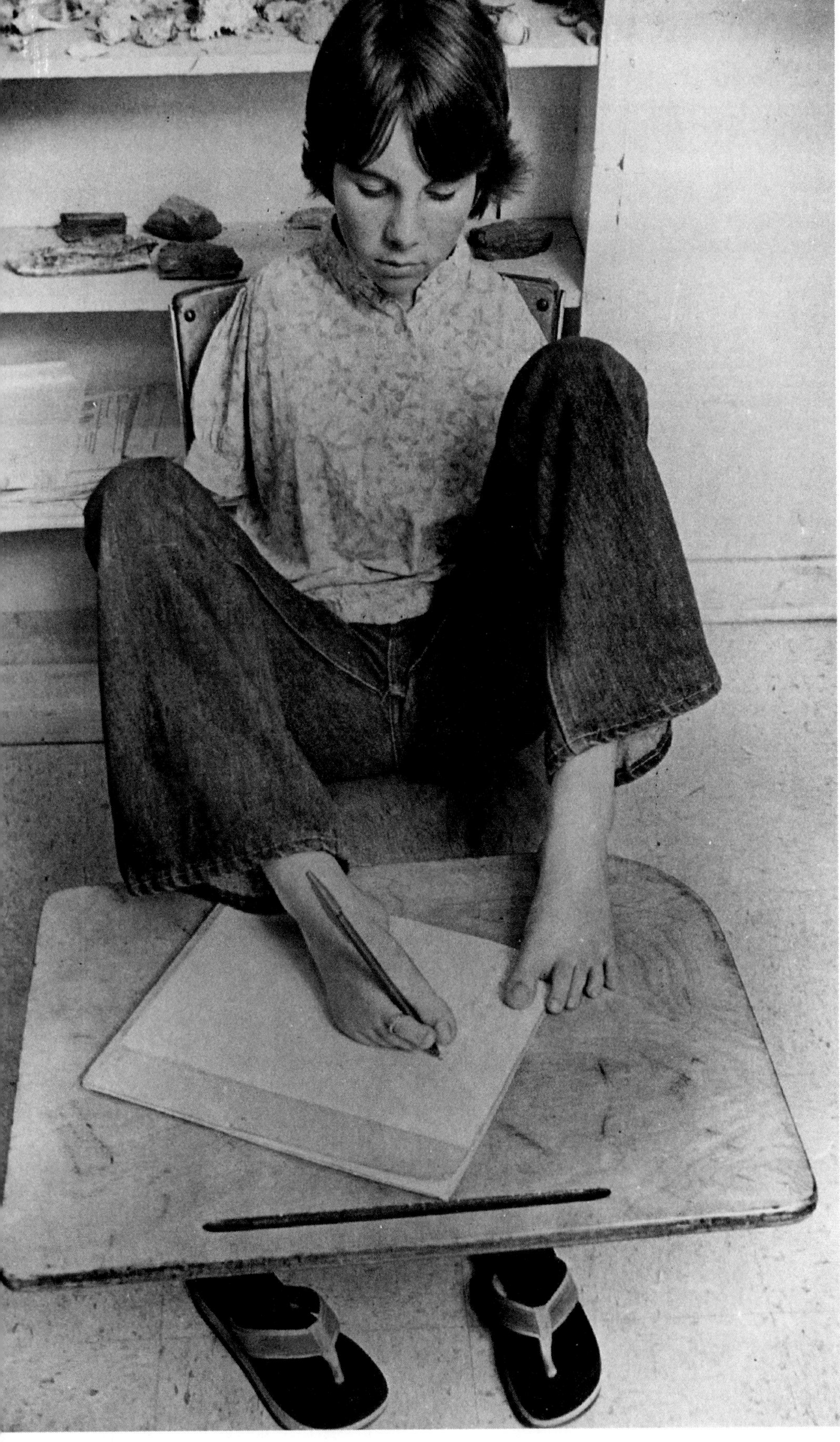

Figure 10.6
This child is 1 of about 10,000 children born between 1957 and 1961 in Europe to mothers who had taken the tranquilizer thalidomide early in pregnancy. The drug acted as a teratogen on developing limbs, resulting in infants with stumps in place of arms and/or legs.

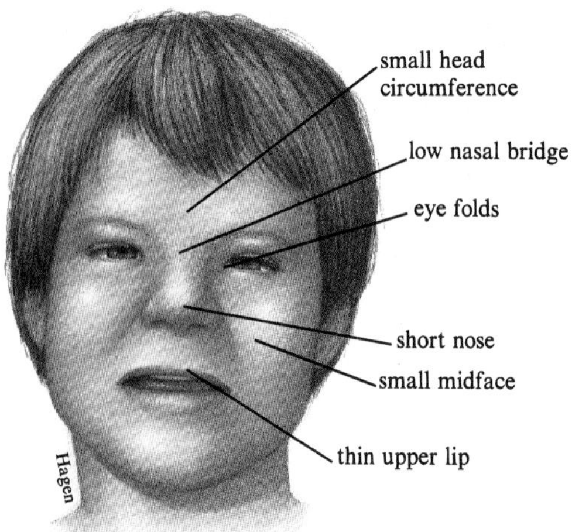

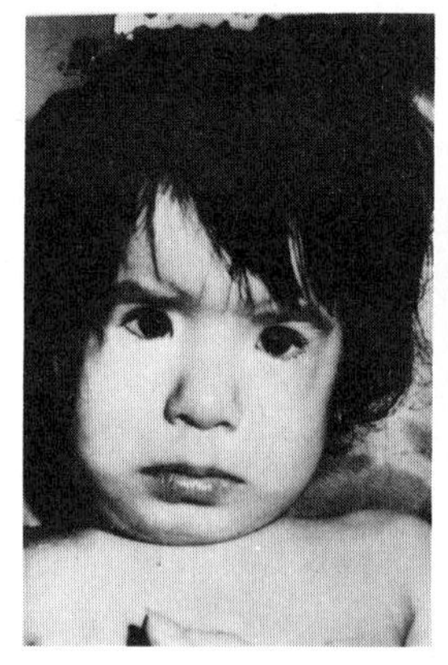

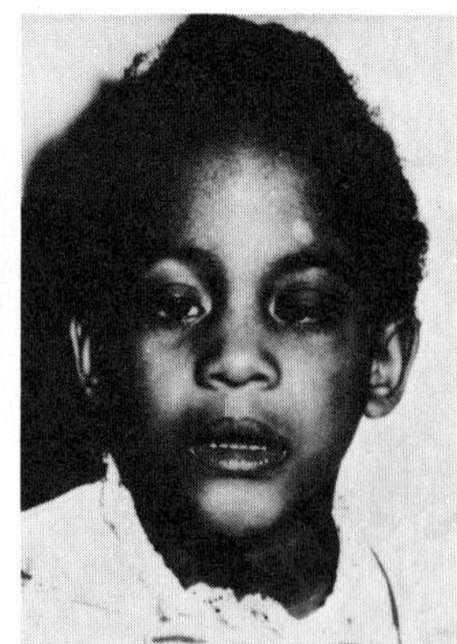

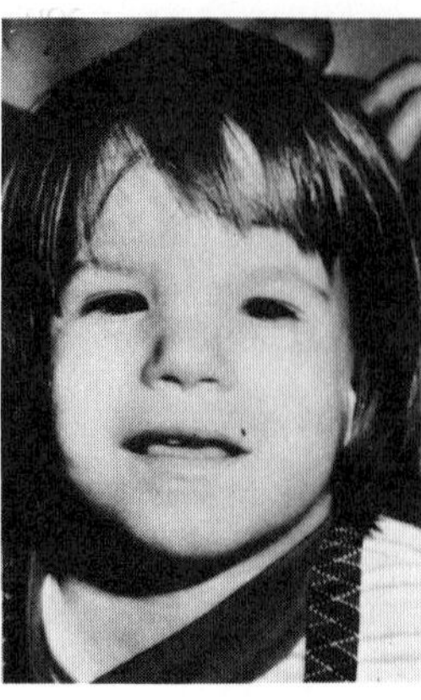

Figure 10.7
Fetal alcohol syndrome. Children whose mothers drank heavily during pregnancy have characteristic flat faces (*a*) that are strikingly similar in those of all races (*b*). They are often mentally retarded or learning disabled. Drinking 1 or 2 ounces of alcohol a day (equivalent to one or two mixed drinks) during pregnancy is associated with a 10% risk of fetal alcohol syndrome, and drinking 2 to 3 ounces a day is linked to 19% risk. Alcoholic mothers-to-be have a 30% to 45% chance of having a child with fetal alcohol syndrome.

reaching the fetus, and studies comparing placentas of smokers and nonsmokers show that smoke-exposed placentas lack important growth factors. The result is poor growth before and after birth. Cigarette smoking in pregnancy is linked to miscarriage, stillbirth, prematurity, and low birth weight.

Despite the climate of caution raised by the thalidomide disaster, drugs still occasionally cause birth defects. The acne medication *isotretinoin* (Accutane) is a derivative of vitamin A that causes spontaneous abortions and defects of the heart, nervous system, and face. The tragic effects of this drug were noted exactly 9 months after dermatologists began prescribing it to young women in the early 1980s. Today, the drug package bears prominent warnings, and it is never prescribed to pregnant women. A vitamin A–based drug used to treat psoriasis, as well as excesses of vitamin A itself, also cause birth defects. This is because some forms of vitamin A are stored in body fat for up to 3 years after ingestion.

Another nutrient that can harm a fetus when taken by the mother in excess is vitamin C. The fetus becomes accustomed to the large amounts that the mother takes, and after birth, when he or she gets far less of the vitamin, symptoms of vitamin C deficiency (scurvy) set in. The baby bruises easily and is prone to infection.

Malnutrition in a pregnant woman threatens the fetus. Obstetrical records of pregnant women before, during, and after World War II link inadequate nutrition early in pregnancy to an increase in spontaneous abortion incidence. The aborted fetuses had very little brain tissue. Poor nutrition later in pregnancy affects development of the placenta, and the infant has a low birth weight and is at high risk for short stature, tooth decay, delayed sexual development, learning disabilities, and possibly mental retardation.

Some teratogens are encountered in the workplace. Increased rates of spontaneous abortion and birth defects have been noted among women who work with textile dyes, lead, certain photographic chemicals, semiconductor materials, mercury, and cadmium. We do not know much about the role of the male in environmentally caused birth defects. One built-in protection is that if a sperm cell is severely damaged, it will not be able to move. However, men whose jobs expose them to sustained heat, such as smelter workers, glass manufacturers, and bakers, may produce sperm that can nonetheless move, fertilize an oocyte, and possibly lead to spontaneous abortion or a birth defect. It is also possible that a virus or a toxic chemical carried in semen can cause a birth defect.

KEY CONCEPTS

The time when a structure is sensitive to being damaged by a faulty gene or environmental insult is called its critical period. Most birth defects are caused during embryonic existence and are generally more severe than problems caused later in pregnancy. Agents that cause birth defects are called teratogens. Teratogens include viruses, recreational and therapeutic drugs, and malnutrition.

The Baby Doe Dilemma

Thirty years ago, if a woman went into labor after only 24 weeks of pregnancy, the result was almost always a spontaneous abortion or an infant that died within a few hours after birth. Today, nearly half of such pregnancies produce extremely small live infants. Thanks to the modern medical technology of **neonatology,** (study of the newborn), many of these infants survive. Some of them, however, are born with severe and sometimes multiple medical problems. Often new parents and physicians must make difficult decisions over whether it is kinder to subject a severely ill newborn to corrective surgery and drug treatments or to leave the child alone, letting nature take its course.

Consider the case of Andrew. Because the placenta detached early, Andrew was born only 24 weeks after he was conceived. After labor, his mother awoke, expecting to be comforted about her miscarriage, only to learn that she was the mother of a 1-pound 12-ounce (794-gram) baby boy. The doctors called him "marginally viable," giving him less than a 5% chance of survival. After several months of medical manipulation, Andrew died. In his brief life, Andrew suffered from dehydration, broken bones, collapsed lungs, seizures, and diseases of the blood, eyes, urinary tract, liver, and heart. Before his death, his mother summed up her confusion: "I'm afraid my baby is going to die. I'm afraid my baby is going to live."

The anguish felt by parents of severely ill newborns has been dubbed the "Baby Doe dilemma," after a baby who was born with a blocked esophagus and Down syndrome in Bloomington, Indiana, in 1982. Unlike Andrew's parents, Baby Doe's parents elected to withhold food, water, and medical help because even though the esophagus could be unblocked, the baby would still have Down syndrome. The baby died at a few months of age. In the years that followed, other Baby Does came to national attention, and the troubling questions of how to treat—or not treat—them are still handled on a case-by-case basis.

Most disturbing was the case of Baby Fae, who was born in 1984 with part of her heart undeveloped, a defect that could not be repaired. The infant was kept alive for a few additional weeks by means of a transplanted baboon heart because a human heart was not available. The controversy centered around the fact that the baboon heart transplant may have been done for experimental purposes, rather than realistically to save a child's life.

The Fetus as a Patient

A partial solution to the Baby Doe problem is to extend medical treatment to the fetus. Because fetus and mother function as a unit, some prenatal medical problems can be treated by administering drugs to the mother or by altering her diet. An abnormally small fetus can be given a nutritional boost by putting the mother on a high-protein diet. A fetus that cannot produce adequate amounts of a specific vitamin can sometimes overcome the deficiency by giving the mother large doses of the vitamin. It is also possible to treat prenatal medical problems directly: A tube inserted into the uterus can drain the dangerously swollen bladder of a fetus with a blocked urinary tract, providing relief until the problem can be surgically corrected at birth. A similar procedure can remove excess fluid from the brain of a hydrocephalic fetus. Drugs can be delivered to the fetus, bypassing the mother, through a tube placed in the umbilical cord.

Little Blake Schultz made medical history when he underwent major surgery 7 weeks before his birth. Ultrasound had revealed that his stomach, spleen, and intestines protruded through a hole in his diaphragm, the muscle sheet separating the abdomen from the chest. This defect would have suffocated him shortly after birth, were it not for pioneering surgery by Michael Harrison at the University of California at San Francisco. Through an incision in the mother, the surgical team exposed Blake's left side, gently tucked his organs in place, and patched the hole with Gore-Tex, a synthetic material used in clothing.

KEY CONCEPTS

Neonatology is the branch of medicine that deals with the gravely ill newborn. The degree of medical intervention in cases of extremely ill newborns is controversial. A few medical procedures can be performed on fetuses.

Reproductive Alternatives

Until recently, there was only one way to make a baby: sexual intercourse at about the time of ovulation. However, a growing number of couples with fertility problems are turning to "reproductive alternatives." Many of these techniques were first perfected in animals (table 10.2).

Donated Sperm—Artificial Insemination

The oldest reproductive alternative is **artificial insemination,** in which a woman whose partner is infertile receives the sperm of an anonymous donor in a laboratory setting. More than 250,000 babies have been born worldwide as a result of this procedure. A couple who chooses artificial insemination can select sperm from a donor whose personal characteristics (race, blood type, hair and eye color, build, even educational level and interests) match those of the man. Sperm is frozen and stored in *sperm banks,* which provide the cells to obstetricians, who perform the insemination. Artificial insemination is also used in cattle breeding, where sperm from a single bull is often used to conceive thousands of calves.

A Donated Uterus—Surrogate Motherhood

If a man produces healthy sperm but his mate's uterus is missing or cannot maintain a pregnancy, a *surrogate mother* may help. A surrogate mother is artificially inseminated with sperm donated by the man. When the child is born, the surrogate gives it up to the couple. The child is genetically that of the sperm donor and the surrogate mother. These situations are usually arranged by an attorney. The surrogate mother signs a statement signifying her intent to give up the baby, and she is paid $10,000 or more for her 9-month job.

Although problem surrogacy cases tend to make headlines, there are also happy outcomes. For example, one surrogate mother carried her own triplet grandchildren for her daughter, who could not become pregnant. Like artificial insemination, surrogate motherhood is also valuable in agriculture. Consider Mist, a 5-year-old Holstein cow who produces a whopping 9 gallons (41 liters) of milk a day. Using surrogate mothers, Mist will probably mother 100 calves in her lifetime—10 times the normal number.

In Vitro Fertilization

Louise Brown, the child described at the beginning of the chapter, was the first human product of *in vitro fertilization* (IVF), which literally means "fertilization in glass." A woman might elect to undergo IVF if her ovaries and uterus are functional but her fallopian tubes are blocked. To begin, the woman takes a superovulation drug that causes her ovaries to release more than one "ripe" oocyte at a time. A lit tube called a *laparoscope* is inserted through an incision made near her navel, and several of the largest oocytes are removed and transferred to a laboratory dish. Chemicals that mimic

CERTIFICATE OF LIFE

Commonwealth of California, Department of Health's Vital Records

Subject:	Baby Boy, Miller
Date of conception:	November 15, 2018, 12:15 P.M.
Place:	Comprehensive Fertility Institute, Beverly Hills, CA
Number of parents:	Three, including surrogate mother—mother donated egg, father sperm
Method of conception:	In vitro fertilization followed by embryo transfer. Mother's body had rejected her artificial fallopian tube. After 8 days on Pergonal, mother produced two eggs. Both were removed during routine laparoscopy and screened for possible defects. Eggs united with father's sperm. After 48 hours in incubator, embryos were removed from growth medium and placed in surrogate's womb. Only one embryo attached itself to uterine wall.
Prenatal care:	Ultrasound at 3 months. Fetal surgery performed at 5 months.
Date/time of birth:	Jason Lawrence Miller born July 20, 2019, 4:15 A.M.
Father:	Jason L. Miller, Sr..
Mothers:	Amy Wong (natural); Maribeth Rivers (surrogate)
Birth method:	Newly lifed in Morningstar Birthing Center, division of Humana Corporation. Natural delivery after 5-hour labor. Labor pains controlled through acupuncture. Therapeutic touch used for last hour of labor. Child's father, adopted sister, and natural mother attended the delivery.
Weight/length:	10 pounds.; 25 inches
Eye color:	Green
Projected life span:	82 years

Figure 10.8
A look ahead.

the female reproductive tract are added, and sperm donated by her mate are applied to the oocytes. A day or so later, some of the zygotes are transferred to her uterus. If human chorionic gonadotropin appears in the woman's blood a few days later, she is pregnant.

IVF costs from $3,000 to $10,000, and the success rate is only 10% to 15%. Usually several embryos are transferred, which can result in multiple births. An alternative is to freeze some of the embryos for later use by the couple or another couple. Freezing embryos can have curious consequences. One man sued his wife for possession of their frozen embryos as part of the divorce settlement. Identical twins were born 18 months apart because one was

Table 10.2
Some Landmarks in Reproductive Technology

	In Animals	*In Humans*
1782	Use of artificial insemination in dogs	
1799		Pregnancy reported from artificial insemination
1890s	Birth from embryo transplantation in rabbits	Artificial insemination by donor
1949	Use of cryoprotectant to successfully freeze and thaw animal sperm	
1951	First calf born after embryo transplantation	
1952	Live calf born after insemination with frozen sperm	
1953		First reported pregnancy after insemination with frozen sperm
1959	Live rabbit offspring produced from in vitro fertilization (IVF)	
1972	Live offspring from frozen mouse embryos	
1976		First commercial surrogate motherhood arrangement reported in the United States
1978	Transplantation of ovaries from one female to another in cattle	Baby born after IVF in United Kingdom
1980		Baby born after IVF in Australia
1981	Calf born after IVF	Baby born after IVF in United States
1982	Sexing of embryos in rabbits	
	Cattle embryos split to produce genetically identical twins	
1983		Embryo transfer after uterine lavage
1984		Baby born in Australia from embryo that was frozen and thawed
1985		Baby born after gamete intrafallopian transfer (GIFT)
		First gestational surrogacy arrangement reported in the United States
1986		Baby born in the United States from embryo that was frozen and thawed.

Source: Office of Technology Assessment, *Infertility: Medical and Social Choices* (Washington, D.C.: U.S. Congress, Government Printing Office, May 1988).

frozen as an embryo. In 1985, a plane crash killed the wealthy parents of two early embryos sitting in suspended animation at -320°F (-195°C) in a hospital in Melbourne, Australia. Adult children of the couple found themselves in the curious position of possibly having to share their estate with two eight-celled embryonic relatives. Figures 10.8 and 10.9 show some of the complexities of IVF, surrogate mothers, and artificial insemination.

Gamete Intrafallopian Transfer (GIFT)

A technique similar to IVF but less invasive is GIFT, or gamete intrafallopian transfer. After the woman takes a superovulation drug for a week, several of her largest oocytes are removed. The man submits a sperm sample, from which the most active cells are separated. The collected oocytes and sperm are deposited together in the woman's fallopian tube, past the obstruction so that fertilization may occur. GIFT is about 40% successful. Can you think of a reason why GIFT is more successful than IVF?

Embryo Adoption

In *embryo adoption*, a woman with malfunctioning ovaries but who has a healthy uterus and wants to experience pregnancy can carry an embryo that results from her mate's artificially inseminating a woman who produces healthy oocytes. If the woman conceives, the embryo is gently flushed out of her uterus when it is a week old and inserted through the vagina of the husband's mate into her uterus. The child is genetically that of the husband and the woman who carries it for the first week, but it is born from the woman who cannot produce healthy oocytes.

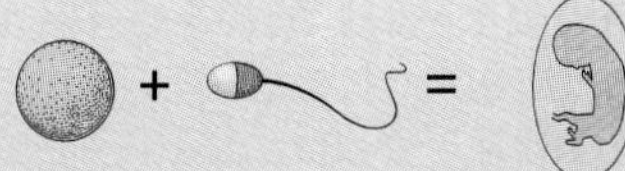

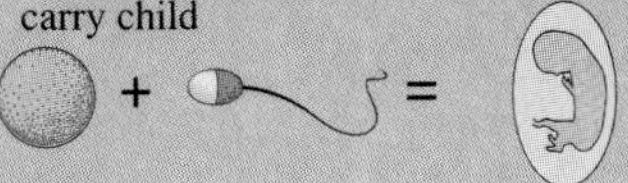
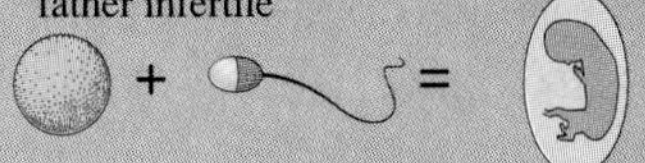

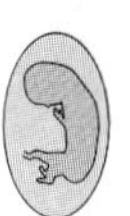
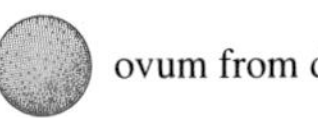

Figure 10.9
Reproductive alternatives. Artificial insemination, in vitro fertilization, surrogate mothering, and embryo freezing and transfer have provided many more ways to produce a baby than the "natural" way. These are possible parental combinations with artificial insemination by donor and in vitro fertilization.

Reading 10.3 *Ethical Questions Raised by Reproductive Alternatives*

- If more zygotes develop in the laboratory than the couple who donated them can use, should the extra be discarded? Experimented on? Donated to another couple?
- If a procedure is too successful and a woman is pregnant with more fetuses than can survive, should some of the fetuses be aborted, in order that some survive?
- For how long, and under what conditions, should embryos be frozen?
- How can procedures that involve a third parent—such as a sperm donor or surrogate mother—ensure that the donated biological material is not infected or will not transmit an inherited disease?
- What should be done when a surrogate mother, upon seeing the child she has carried for another couple, changes her mind about giving the baby to them?
- Will surrogate mothering be abused, with wealthy couples hiring poor women as surrogates, because the adoptive mother simply does not wish to be pregnant?
- Should a child conceived or nurtured using a reproductive alternative be told of his or her origin?

KEY CONCEPTS

Reproductive alternatives include artificial insemination, surrogate motherhood, in vitro fertilization, freezing embryos, gamete intrafallopian transfer, and embryo adoption. Many reproductive technologies were first perfected on animals.

Sex and Health

The reproductive system is unique among the body's organ systems in that it does not become active until sexual maturity, or puberty, is reached. Puberty itself brings health changes, and when sexual activity begins, preventing pregnancy and sexually transmitted diseases (STDs) become important concerns.

Birth Control

Contraception is the use of devices or practices that are "against conception." These birth-control methods physically block the meeting of sperm and oocyte or alter the environment of the female system so that it is hostile to sperm or to a zygote's implantation (table 10.3).

In 1960 the first contraceptive pill became available. The "combination pill" contains synthetic versions of the hormones estrogen and progesterone, which hinder conception in three ways: by suppressing ovulation, by altering the uterine lining so that a zygote cannot implant, and by thickening the mucus in the cervix so that sperm cells cannot get through.

Many women in more than 80 countries (not including the United States) use a birth-control injection of progesterone called Depo-Provera. This drug is given to a woman every 3 months, and it suppresses ovulation. American women may soon be able to choose from several new implants that deliver contraceptive hormones already used in birth-control pills—an injection in the hip every 3 months, rubber rings worn in the vagina, or a pellet or rod inserted beneath the skin in the arm every 1 to 5 years.

A promising candidate for male birth control is a substance called *gossypol*, which is derived from the seeds, stems, and roots of the cotton plant. Its dampening effects on sperm count were discovered in a rural community in China in the 1950s, where male infertility was traced to the use of cottonseed oil in cooking. Gossypol decreases levels of lactate dehydrogenase in the testicles, where it is needed for sperm production. Although gossypol has been used successfully as a male birth-control pill on thousands of Chinese, its side effects of appetite loss, weakness, heart rhythm changes, lowered sex drive, and occasional permanent infertility are problems. Gossypol is currently undergoing animal testing in the United States.

The idea of using the immune system to develop a vaccine against pregnancy stems from studies of women whose vaginal secretions contain substances that inactivate sperm cells. One experimental vaccine prompts a female's immune system to produce antibody molecules that attach to the zona pellucida, which surrounds an oocyte (table 10.4). The bound antibodies block a sperm's entry. Another experimental female birth-control vaccine is a piece of a molecule that is normally found on the surfaces of sperm cells. When injected into the woman's bloodstream, the fragment programs her immune system to recognize the molecule—and the sperm cells that carry it—as foreign. The sperm cells are destroyed by the woman's immune system. A single injection seems to be effective for a few months.

In the early 1970s, 45 million women in the United States used *intrauterine devices* (IUDs). These devices are placed inside the uterus by a physician, where they disrupt the uterine lining so that implantation of a zygote cannot occur. In 1974, however, the IUD was found to cause uterine infections so severe that 20 women died. The string that hung into the vagina from the IUD in the uterus provided a very effective conduit for bacteria. Users frequently experienced heavy and painful menstrual periods, pregnancies sometimes occurred with the device, and some babies were born accompanied by a dislodged IUD! Today, only one brand of IUD is available, and it is designed so that it is unlikely to cause infection.

Contraceptives that provide a physical and/or chemical barrier between sperm and oocyte are very popular because they prevent and protect against sexually transmitted diseases. A *condom* is a sheath worn over the erect penis. Spermicidal (sperm-killing) jellies, foams, creams, and suppositories are inserted into the vagina, and they can be used in conjunction with devices that block the cervix, such as the *diaphragm*, *contraceptive sponge*, or *cervical cap*. Most products now on the market use the spermicide nonoxynol-9.

Interestingly, the sponge and the cervical cap are both new to the United States but are actually quite ancient devices. A contraceptive sponge was mentioned in the Talmud, a religious document written in A.D. 800 that reflected practices over the 10 preceding centuries. Centuries ago Europeans used sponges soaked in lemon juice, which killed sperm. Cervical caps, which adhere to the cervix by suction, have been around for thousands of years and are constructed of materials such as beeswax, fibers from opium poppies, and paper. Casanova, an eighteenth-century Italian romantic figure, had his mates use a cervical cap and spermicide in the form of a half lemon.

"Natural" birth control or "fertility awareness" methods rely on signs that conception is not likely to occur. The *rhythm method* charts a woman's menstrual cycle. If a woman begins to menstruate every 28 days, she is most likely to conceive on the 12th to 16th days following the onset of bleeding. A more precise version of the rhythm method is for a woman to take her temperature every morning with a specially calibrated thermometer. Body temperature rises slightly just before ovulation, indicating the days on which intercourse should be avoided.

Observing the consistency and appearance of vaginal mucus provides clues to when a woman is likely to conceive. The secretion is thin, elastic, and clear during fertile periods and cloudy, sticky, or absent at other times. Some women can feel differences in the position and texture of the cervix at the time of ovulation.

Table 10.3
Birth-Control Methods

	Method	Mechanism	Advantages	Disadvantages	Pregnancies per Year per 100 Women*
	None				80
Barrier and Spermicidal	Condom	Worn over penis, keeps sperm out of vagina	Protection against sexually transmitted diseases	Disrupts spontaneity, can break, reduces sensation in male	3–10
	Condom and spermicide	Worn over penis, keeps sperm out of vagina, and kills sperm that escape	Protection against sexually transmitted diseases	Disrupts spontaneity, reduces sensation in male	2–5
	Diaphragm and spermicide	Kills sperm and blocks uterus	Inexpensive	Disrupts spontaneity, messy, needs to be fitted by doctor	3–17
	Cervical cap and spermicide	Kills sperm and blocks uterus	Inexpensive, can be left in 24 hours	May slip out of place, messy, needs to be fitted by doctor	5–20
	Spermicidal foam or jelly	Kills sperm and blocks vagina	Inexpensive	Messy	5–22
	Spermicidal suppository	Kills sperm and blocks vagina	Easy to use and carry	Irritates 25% of users, male and female	3–15
	Contraceptive sponge	Kills, blocks, and absorbs sperm in vagina	Can be left in for 24 hours	Expensive	5–16
Hormonal	Combination birth-control pill	Prevents ovulation and implantation, thickens cervical mucus	Does not interrupt spontaneity, lowers risk of some cancers, decreases menstrual flow	Raises risk of cardiovascular disease in some women, causes weight gain and breast tenderness	0–10
	Minipill	Blocks implantation, deactivates sperm, thickens cervical mucus	Fewer side effects	Weight gain	1–10
Behaviorial	Rhythm method	No intercourse during fertile times	No cost	Difficult to do, hard to predict timing	13–21
	Withdrawal	Removal of penis from vagina before ejaculation	No cost	Difficult to do	9–25
Surgical	Vasectomy	Sperm cells never reach penis	Permanent, does not interrupt spontaneity	Requires minor surgery	0
	Tubal ligation	Egg cells never reach uterus	Permanent, does not interrupt spontaneity	Requires surgery, entails some risk of infection	0
Other	Intrauterine device	Prevents implantation	Does not interrupt spontaneity	Severe menstrual cramps, increases risk of infection	1–5

*The lower figures apply when the contraceptive device or technique is used correctly. The higher figures take into account human error.

Table 10.4
Understanding the Events of Early Development Leads to New Contraceptive Approaches

Event	*Block*
Sperm loosely attach to gellike coat (zona pellucida) around ovum	Vaccine prompts woman to produce antibodies that bind to glycoproteins on zona
Sperm surface proteins attach to receptors on zona	Vaccine consists of antibodies that cover sperm surface molecules, preventing them from binding to zona
As sperm penetrates zona and ovum, electrical and chemical changes in ovum membrane block additional sperm from entering	Premature induction of these "blocks to polyspermy" prevents any sperm from entering zona or ovum
Fertilized ovum implants in uterus	Implantation prevention: a pill blocks progesterone receptors on uterine lining; a vaccine blocks human chorionic gonadotropin, a hormone produced by a fertilized ovum that maintains uterine lining

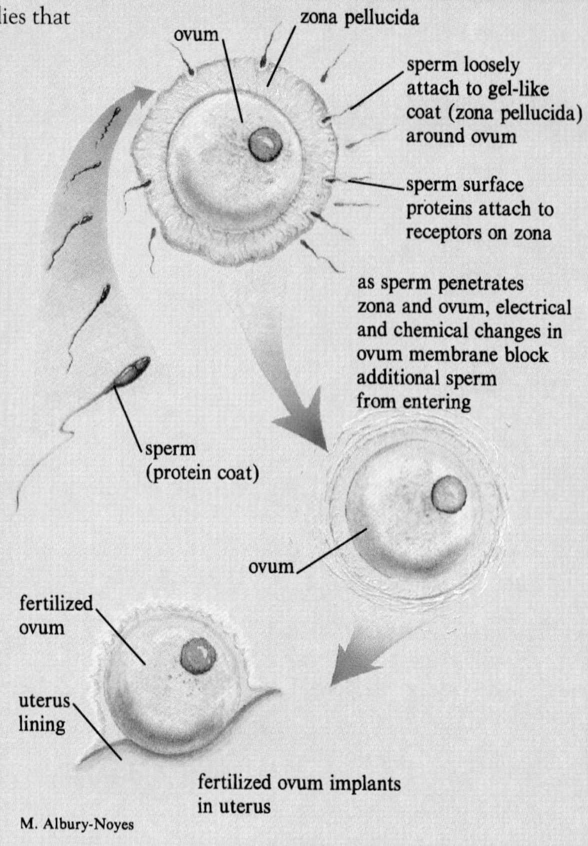

Sterilization provides permanent birth control. In males, the vas deferens leading from each testicle is cut or burned shut in a procedure called a *vasectomy*, so that sperm cannot leave the testicles. The germ cells degenerate in the man's body. Females can have their fallopian tubes tied off with a *tubal ligation* (fig. 10.10).

Terminating a Pregnancy

An unwanted pregnancy can be terminated by preventing the zygote's implantation in the uterus, or, later on, by removing an embryo or fetus from the uterus. First-trimester abortions use a sucking or scraping device, and later abortions utilize more intense scraping, injection of a salt solution into the amniotic sac, or prostaglandin suppositories, which trigger uterine contractions and expulsion of the fetus within 24 hours. A drug developed in France, RU-486, can induce abortion if taken during the first 7 weeks of pregnancy, but the ease of its use has made it very controversial.

Most people feel that abortion should not be considered a form of contraception along with such devices as the diaphragm, birth-control pill, or condom, all of which act earlier in the process. In the Soviet Union, however, two-thirds of all women have had at least one abortion because of the inavailability of contraceptives. Abortion is often used when a fetus is diagnosed with a devastating medical condition for which there is no treatment.

Sexually Transmitted Diseases

The 20 recognized sexually transmitted diseases (STDs) are often called "silent infections" because the early stages may not produce symptoms, especially in women (table 10.5). By the time symptoms appear, it is often too late to prevent complications or the spread of the infection to sexual partners. Because many STDs have similar symptoms, and some of the symptoms are also seen in diseases or allergies that are not sexually related, it is wise to consult a physician if one or a combination of these symptoms appears:

1. Burning sensation during urination
2. Pain in the lower abdomen

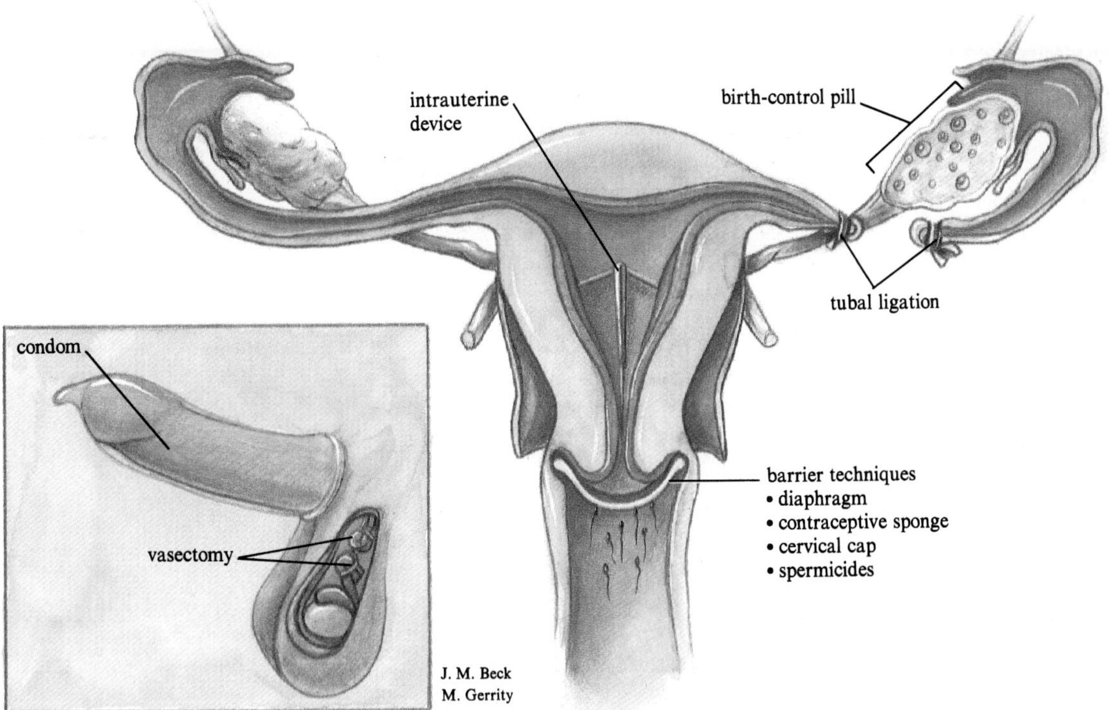

Figure 10.10
Sites of birth-control action. Birth-control pills prevent the release of egg cells from the ovary. In a tubal ligation the fallopian tubes are tied so that egg cells cannot pass through them to the uterus. An intrauterine device embeds in the uterine wall, preventing implantation of a zygote. The diaphragm, contraceptive sponge, and cervical cap block the cervix so that sperm cells cannot enter the uterus. Spermicidal jellies, creams, and suppositories enhance the effectiveness of barrier contraceptives. A male can prevent impregnating a woman by wearing a condom or by undergoing a vasectomy, which can now be done without surgery.

3. Fever or swollen glands in the neck
4. Discharge from the vagina or penis
5. Pain, itch, or inflammation in the genital or anal area
6. Pain during intercourse
7. Sores, blisters, bumps, or a rash anywhere on the body, particularly the mouth or genitals
8. Itchy, runny eyes

One possible complication of the STDs gonorrhea and chlamydia is *pelvic inflammatory disease*, in which bacteria enter the vagina and spread throughout the reproductive organs. The disease begins with intermittent cramps, followed by sudden fever, chills, weakness, and severe cramps. Hospitalization is usually necessary so that antibiotics can be given intravenously to stop the infection. The uterus and fallopian tubes are often scarred, resulting in infertility and an increased risk of ectopic pregnancy.

The most frightening sexually transmitted disease is *acquired immune deficiency syndrome* (AIDS), which is a steady deterioration of the body's immune defenses caused by a virus that attacks certain immune system cells. The body becomes overrun by infection and often cancer, diseases that are usually conquered by the cells and biochemicals of the immune system. The AIDS virus is passed from one person to another in body fluids such as semen, blood, milk, and tears and is most frequently transmitted during anal intercourse or by using a needle containing contaminated blood.

The human reproductive system is unique in that it does not begin to function until several years after birth. It is also unique in that an individual can survive without it. On the population level, however, the reproductive system is essential to the survival of the species. A healthy reproductive system can make life very fulfilling, both in providing the pleasures of sexuality and in starting new lives.

KEY CONCEPTS

Contraceptive methods include the birth-control pill; the IUD; barrier methods such as the condom, diaphragm, cervical cap, and contraceptive sponge; the spermicide nonoxynol-9; vasectomy and tubal ligation; and monitoring natural changes that indicate when a woman is most fertile. Experimental contraceptives include Depo-provera, gossypol, and vaccines. There are 20 known sexually transmitted diseases, and their symptoms can be confusing. Two of the more serious are pelvic inflammatory disease and AIDS.

Table 10.5
Some Sexually Transmitted Diseases

Disease	*Cause*	*Symptoms*	*Number of Reported Cases (U.S.)*	*Effects on Fetus*	*Treatment*	*Complications*
Acquired immune deficiency syndrome	Human immunodeficiency virus	Fever, weakness, infections, cancer	1–2 million (infected)	Exposure to AIDS virus and other infections	Drugs to treat or delay symptoms; no cure	Dementia and death
Chlamydia infection	Bacteria of genus *Chlamydia*	Painful urination and intercourse, mucus discharge from penis or vagina	3–10 million	Prematurity, blindness, pneumonia	Antibiotics	Pelvic inflammatory disease, infertility, arthritis, ectopic pregnancy
Genital herpes	Herpes simplex virus type I or II	Genital sores, fever	20 million	Brain damage, stillbirth	Antiviral drug (acyclivor)	Increased risk of cervical cancer
Genital warts	Human papilloma virus	Warts on genitals	1 million	None known	Chemical or surgical removal	Increased risk of cervical cancer
Gonorrhea	*Neisseria gonorrheae* bacteria	In women, usually none; in men, painful urination	2 million	Blindness, stillbirth	Antibiotics	Arthritis, rash, infertility, pelvic inflammatory disease
Syphilis	*Treponema pallidum* bacteria	Initial chancre sore usually on genitals or mouth; rash 6 months later; several years with no symptoms as infection spreads; finally damage to heart, liver, nerves, brain	90,000	Miscarriage, prematurity, birth defects, stillbirth	Antibiotics	Death

SUMMARY

Human developmental and reproductive problems are becoming more amenable to treatment as we learn more about the functioning of the reproductive system. A major area of concern is *infertility*, the inability of a couple to conceive a child after a year of trying. Causes of infertility in the male include a *low sperm count*, a malfunctioning immune system, a varicose vein in the scrotum, heat exposure, structural sperm defects, and abnormal hormone levels. Female infertility can be due to absent or irregular ovulation, blocked fallopian tubes, an inhospitable uterus, or antisperm secretions. Older women generally take longer to conceive than younger women.

An embryo or fetus *spontaneously aborts* in about one out of every five recognized pregnancies, usually due to an extra or missing chromosome or to a hormonal or structural abnormality in the woman's reproductive tract. Children born a month or more prematurely often experience a range of health problems, the most serious of which are immature lungs and breathing reflexes. Birth defects can be due to genetic anomalies or to a *teratogen* to which the woman is exposed at a *critical period* in prenatal development. The field of neonatology deals with the medical problems of newborns.

Infertile couples are increasingly turning to reproductive alternatives such as *artificial insemination*, *in vitro fertilization*, *embryo transfer*, *frozen embryos*, and *surrogate mothers*. While relatively new to humans, these techniques are standard practices in animal breeding. These reproductive technologies have spawned many fears.

Contraceptives prevent pregnancy by blocking the meeting of sperm cell and oocyte, or by making the uterus inhospitable to the implantation of a zygote. Birth-control devices in current use include hormone-containing pills or injections, barrier contraceptives that block the cervix, *spermicides*, *intrauterine devices*, and *condoms*. *Vasectomy* and *tubal ligation* are permanent means of birth control. Contraception in the future, for both sexes, may be in the form of vaccines developed by manipulating the immune response.

Once sexual activity begins, *sexually transmitted diseases* can become a threat to health. Prevalent STDs include *chlamydia infection*, *gonorrhea*, *syphilis*, *herpes simplex infection*, and *acquired immune deficiency syndrome*.

QUESTIONS

1. Causes of infertility sometimes lead to the development of new birth-control methods. Match the causes of infertility listed on the left with the contraceptive method listed on the right that is based on it.

Cause of infertility	*Contraceptive device*
a. Failure to ovulate due to hormonal imbalance	Intrauterine device
b. Fibroid tumor on the uterine lining	Tubal ligation
c. Fallopian tubes blocked by endometriosis tissue	Birth-control pill
d. Low sperm count	Birth-control vaccine
e. Immune infertility	Gossypol

2. A couple is having trouble conceiving a child. After several visits to fertility specialists, they find that their problem is twofold. The male has a sperm count of 50 million sperm per ejaculate, and the woman ovulates only three times a year. Suggest a way for the couple to conceive.

3. How can "fertility awareness" be used to conceive a child? How can it be used as a contraceptive measure?

4. Ultrasound detects twin embryos in a woman who is 6 weeks pregnant. Seven months later, only one baby is born. What has happened?

5. A few years ago, a doll manufacturer sold a "preemie" model. The doll was cute and pudgy. Is this an accurate representation of an infant born too soon?

6. In Aldous Huxley's book *Brave New World,* egg cells are fertilized in a laboratory and allowed to develop assembly line style. To render some of the embryos less intelligent than others, alcohol is given to them. What medical condition in humans is this scenario similar to?

7. Contraception literally means "against conception." According to this definition, is an intrauterine device a contraceptive? Why or why not?

TO THINK ABOUT

1. Intervention by lawyers and government regulatory agencies into biological matters has been mentioned several times in this chapter. What do you think the role of the law should be in the following procedures or situations?

- **a.** Treatment of multiply handicapped newborns
- **b.** What to do with "extra" fertilized ova or embryos growing in vitro
- **c.** Legal liability if an embryo or fetus conceived by a reproductive alternative is spontaneously aborted or is born with a birth defect
- **d.** A surrogate mother decides, after giving birth to the child that she has agreed to give to the adoptive couple, that she wants to keep the child
- **e.** Under what circumstances a woman can obtain an abortion
- **f.** Whether a minor's parents should be informed that he or she has sought birth control

2. What do you think are the advantages and/or disadvantages of a woman's waiting until her middle 30s to have her first child?

3. A man produces sperm that cannot swim, but his wife is fertile. They opt for artificial insemination, so they consult a sperm bank's listing of possible donors. They select a donor who, like the husband, has brown hair and brown eyes, a small frame, the same ethnic and religious background as the husband, plus he is studying to be an architect—the husband's profession. Why might they be surprised at the characteristics of their child?

4. A study performed on male college students who smoked marijuana found that they had lower-than-average sperm counts. However, the fact that some of these students drank alcohol, took other drugs, and ejaculated frequently was not taken into account. Nevertheless, some people interpreted the low sperm count finding to mean that marijuana smoking lowers the sperm count in all men. How could such a study be conducted in a more thorough manner?

5. Some potentially teratogenic substances encountered in everyday life—caffeine, cigarette smoke, alcohol—can harm an embryo before the woman even knows that she is pregnant. How can a woman who is trying to conceive make use of this information?

6. Adoption is one reproductive alternative for a couple with a fertility problem. Identity of the biological parents is usually confidential. Do you think this is a wise policy? Why or why not? Suggest another reproductive alternative in which confidentiality is an issue.

7. Do you see any inconsistencies in a society that treats some fetuses with medical problems in the uterus, undertakes "heroic measures" to prolong the life of a premature and possibly very ill infant, and also allows abortion of healthy fetuses?

SUGGESTED READINGS

Bower, B. February 4, 1989. Drinking while pregnant risks child's IQ. *Science News*. Just two or three alcoholic drinks a day can harm a fetus.

Djerassi, Carl. November 1984. The making of the pill. *Science 84*. The birth-control pill had its origins in Mexican yams in the 1930s.

Edwards, Robert, and Patrick Steptoe. 1980. *A matter of life*. New York: William Morrow. The story of the development of Louise Joy Brown—the first "test-tube baby."

Fleming, Anne Taylor. March 29, 1987. Our fascination with Baby M. *New York Times Magazine*. A detailed look at a famous surrogate mother case.

Lewis, Ricki. July/August 1987. Brave new contraceptives. *Massachusetts Medicine*. New injectable, implanted, and barrier contraceptives are on the way.

Lewis, Ricki. 1989. The vulnerable fetus: New risks examined. *Britannica Medical and Health Annual*, p. 401. The thalidomide disaster was the start of our new awareness of dangers to the fetus.

Office of Technology Assessment. May 1988. *Infertility: Medical and social choices*. Washington, D.C.: U.S. Congress, Government Printing Office. A very informative collection of reports on infertility causes and treatment.

Rothman, Barbara Katz. 1986. *The tentative pregnancy*. New York: Penguin Books. When a fetus is severely abnormal, difficult choices lie ahead.

Stinson, Robert, and Peggy Stinson. 1983. *The long dying of Baby Andrew*. Boston: Atlantic Monthly Press. Born months too soon, Baby Andrew was kept alive despite serious medical problems. His parents tell his story.

Wallis, Claudia. February 4, 1985. Chlamydia: The silent epidemic. *Time*. The most prevalent and fastest-spreading sexually transmitted disease is not the highly publicized and widely feared herpes or AIDS but Chlamydia infection.

Willis, Judith. October 1988. New warning about accutane and birth defects. *FDA Consumer*. Vitamin A derivatives can harm fetuses.

Zamula, Evelyn. April 1986. Syphilis and gonorrhea: old-fashioned VD still flourishing. *FDA Consumer*. AIDS may get most of the attention, but other sexually transmitted diseases still exist.

UNIT 4

Genetics

C H A P T E R

11

Mendel's Laws

Chapter Outline

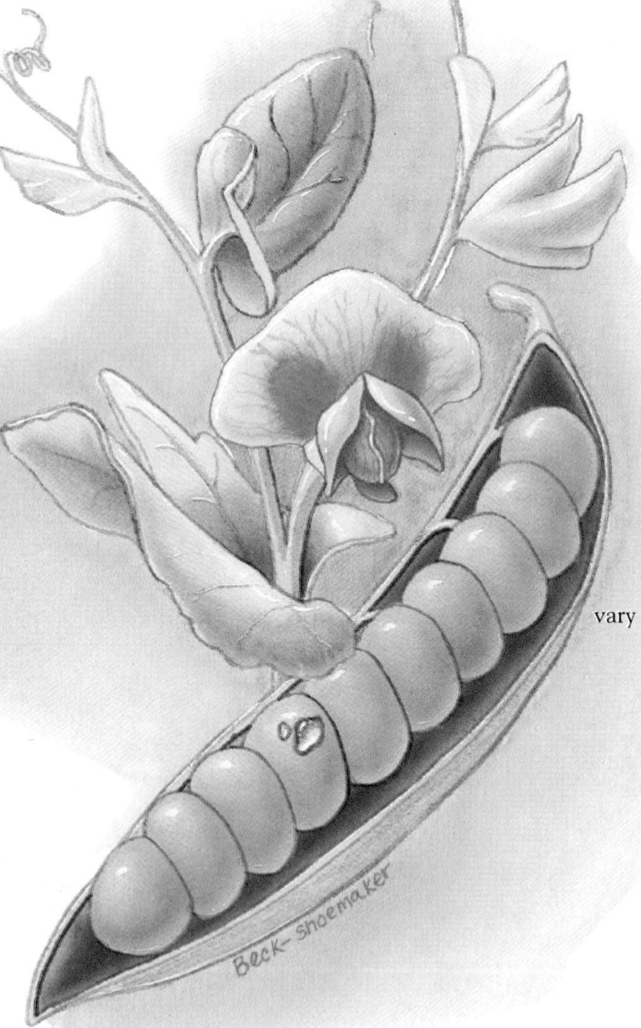

Learning Objectives

By the chapter's end, you should be able to answer these questions:

1. How and where did the modern science of genetics originate?
2. How did Gregor Mendel follow the inheritance of a single gene, and of two genes, by observing the offspring of crosses of pea plants?
3. How can Mendel's observations be explained in terms of meiotic events?
4. How can the outcomes of genetic crosses be predicted?
5. What are the terms that are commonly used in genetics?
6. How can different forms of a gene affect one another to alter the results of crosses that are expected according to Mendel's laws?
7. How can the expression of a gene vary among individuals?

"Oh my, she has your nose and mouth!"

"Look, she's got your widow's peak—and your funny ears."

"Well, maybe, but that yell is just like her brother's, and she already seems to prefer her right side, just like you."

"I'm so glad she's got blond hair, but look how pale her eyelashes are!" comments the new mother, already wondering whether her tiny daughter will want to wear eye makeup some day.

What these new parents are clucking over is the brand new person whom they have created together. To an outside observer visiting this nursery, one bundled-up baby looks much like another. But to these new parents, their newborn immediately stands out from the crowd. In her they see parts of themselves and other family members, but at the same time they are a little awed by the realization that they are responsible for a unique individual, a person with an entirely new combination of the traits of her relatives—possibly including a few characteristics that have not appeared for several generations. As the infant grows into a toddler, child, adolescent, and adult, the parents will continue to marvel at her similarities to some family members, as well as at the traits that seem to be hers alone. Still other characteristics—an athletic body, perhaps, or an intense interest in science—will develop as a result of some aspect of the environment. All of these traits, influenced by heredity and the environment, combine to form a unique individual.

Sometimes people do not think much about the science of genetics until they begin to consider passing their own traits on to a new generation. Then they might wonder about such obvious characteristics as hair color and eye color, height and weight, or more complex qualities such as intelligence and personality. If prospective parents have read about inherited illness, or perhaps have a relative with such a disorder, they might think about one of the many medical conditions rooted in the genes. If a relative has some special talent, they might hope that their child inherits it.

Just how does a child inherit a father's kinky hair, a mother's long legs, or a grandparent's color blindness? The units of inheritance, **genes,** are long chains of the molecule **deoxyribonucleic acid (DNA)** and are responsible for an amazing array of characteristics. Genes mold our physical traits, down to the pattern of the hair on our bodies, to our fingerprints, and even to such skills as the ability to wiggle our ears, smell the odor of a squashed skunk, or carry a tune.

On a microscopic level, genes direct the cell's synthesis of particular proteins, or they control the activities of other genes. It is sometimes difficult to envision how the tiny piece of DNA that is a gene causes, for example, the painful swollen joints of sickle cell disease. However, it is our increasing knowledge of the connections between gene structure and gene function that has brought the science of genetics into the forefront of human health care.

Mendel's Laws of Inheritance

Many human cultures have been intensely interested in the passage of traits from generation to generation. Farmers in Mexico 6,000 years ago used genetics when they carefully set aside seed from the hardiest plants of a wild grass each season and used it to start the next season's crop. In this way, over many plant generations, domesticated corn was bred. Similarly, Asian farmers 4,000 years ago bred horses and kept records of the animals' physical characteristics, suggesting that they knew something about inheritance. Assyrian priests living in the ninth century B.C. pollinated date palms by hand, controlling the traits of the next generation of plants (fig. 11.1). Ancient Hebrew writing describes the transmission of a blood-clotting disorder from a mother, who does not suffer from it, to her son, who does—a description of hemophilia.

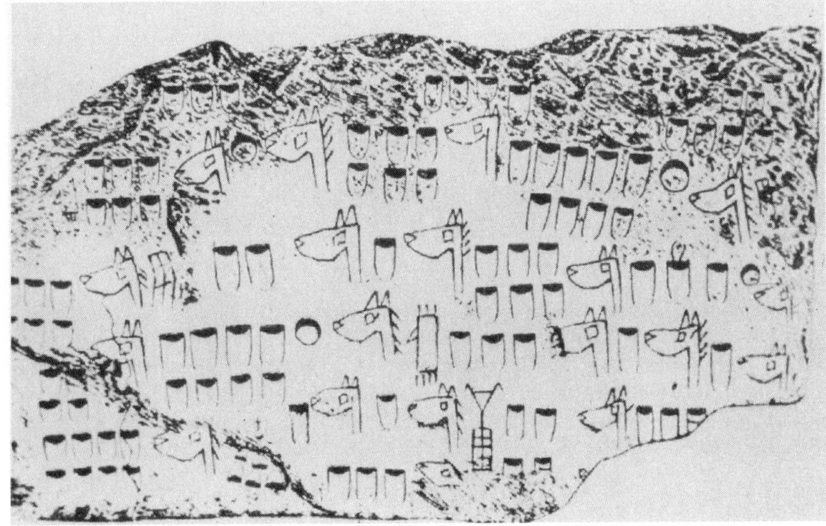

a.

b.

Figure 11.1
Ancient cultures recognized inheritance patterns. *a.* A horse breeder in Asia 4,000 years ago etched a record of his animal's physical characteristics in stone. *b.* In the ninth century B.C., Assyrian priests hand-pollinated date palms.

Genetics as a modern science was born in an Austrian abbey garden between 1857 and 1865, where an Augustinian monk named Gregor Mendel combined two interesting hobbies, mathematics and botany. Mendel followed the passage of easily observed traits such as seed shape, color, and texture over several generations in pea plants (fig. 11.2). He called the determinants of these traits "characters," although he had no idea of what their chemical nature might be. He did know about the cell theory, but **chromosomes,** the structures in the cell's nucleus that we now know to be composed of genes, had not yet been discovered.

Segregation—Following the Inheritance of One Gene at a Time

Mendel noted that short pea plants bred to other short pea plants were "true-breeding," that is, they consistently yielded seeds that gave rise only to more short plants. However, this was not always the case with tall pea plants. Some tall plants, when crossed with a short plant or another tall plant, produced only tall plants in the next generation, suggesting that their "tallness" masked "shortness." But when certain other tall plants were crossed with each other, about one-quarter of the plants in the next generation were short (fig. 11.3). Of the remaining three-quarters of the plants, one-third proved (by further crosses to short plants) to be "true-breeding tall," but the other two-thirds produced some short plants in the next generation.

Mendel was intrigued by these interactions between the so-called characters. He suggested that the gametes—the egg cells and pollen (or sperm cells in animals)—distributed characters, because these cells provide a physical link between generations. He hypothesized that characters separate from one another as a plant forms its gametes. When the gametes combine at fertilization to form a new generation, the characters are grouped into new combinations. If each character was packaged into a separate gamete, and if gametes combine at random, Mendel reasoned, then the different ratios of traits

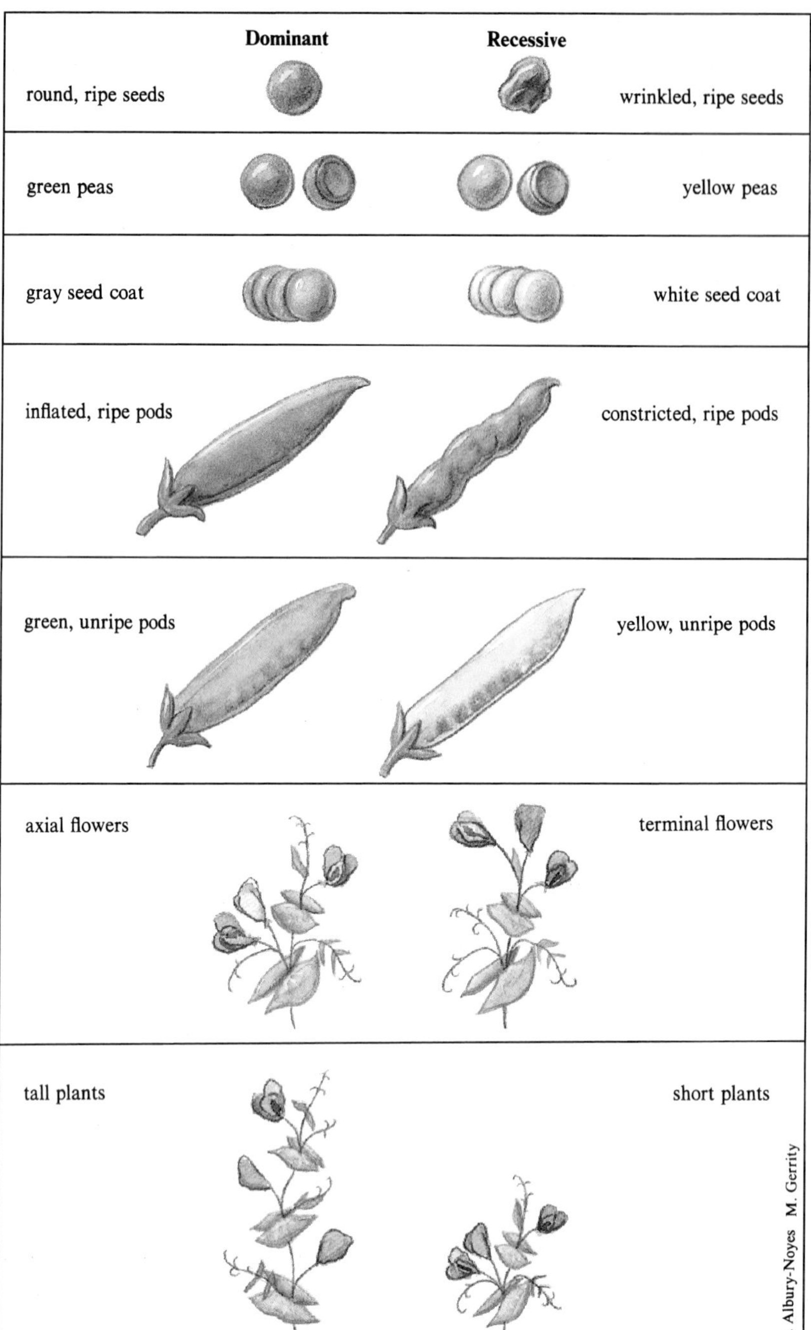

Figure 11.2
Gregor Mendel studied the transmission of seven traits in the pea plant. Each trait had two expressions, which were easily distinguished from each other. Seeds were either round or wrinkled; peas were either green or yellow; seed coats were either gray or white; ripe pods were either inflated or constricted; unripe pods were either green or yellow; flowers were either axial (arising sideways from the stems) or terminal (emanating only from the top of the plant); and plants were either tall (6 to 7 feet) or short (3/4 to 1 1/2 feet).

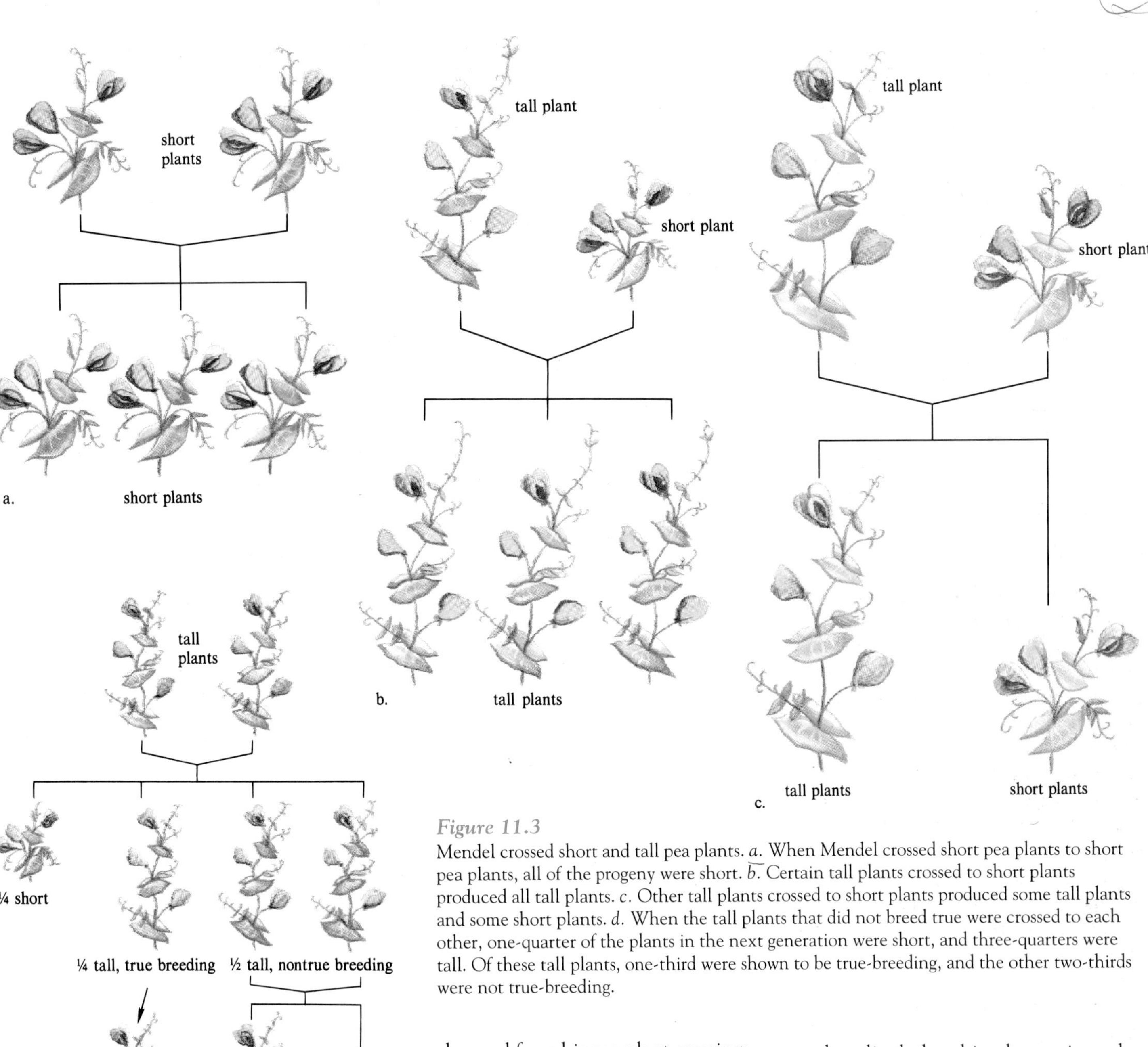

Figure 11.3

Mendel crossed short and tall pea plants. *a.* When Mendel crossed short pea plants to short pea plants, all of the progeny were short. *b.* Certain tall plants crossed to short plants produced all tall plants. *c.* Other tall plants crossed to short plants produced some tall plants and some short plants. *d.* When the tall plants that did not breed true were crossed to each other, one-quarter of the plants in the next generation were short, and three-quarters were tall. Of these tall plants, one-third were shown to be true-breeding, and the other two-thirds were not true-breeding.

observed from his pea plant crossings could be explained. Mendel's idea of separation of characters would later be called the law of **segregation.**

Mendel presented his experimental data on the inheritance of certain genes to a local natural history society, to limited response. In 1866, he published his results in a journal compiled by the society, but this "father of genetics" did not receive due credit for his work until after his death. In 1900, three other botanists, Hugo deVries of Holland, Carl Correns of Germany, and Erich von Tschermak of Austria, read Mendel's 1866 paper and realized that his observations, hypotheses, and conclusions explained data that they had gathered on crosses in other plant species.

Mendel's ideas were to become even more widely accepted in the years to follow, as the role of chromosomes in heredity was unraveled. In the early 1900s, several researchers noted that chromosomes behaved much like Mendel's characters. Specifically, paired characters and paired chromosomes separate at each generation, so that they are contributed, one from each parent, to the offspring. Mendel's characters were inherited in random combinations, as were chromosomes (fig. 11.4). It began to look like chromosomes provided a physical basis for what Mendel described but could not actually see. Mendel's "characters" were named "genes" (Greek for "give birth to") in 1908 by the English biologist William Bateson.

Using our current knowledge of the nature of the genetic material, we can clarify Mendel's observations further. His characters were actually segments of the DNA of chromosomes that control the expression of particular traits. All somatic cells of a diploid organism contain two copies of each type of chromosome and therefore also two copies of each gene. When gametes form (during meiosis), the two chromosomes of each type (and therefore the two copies of each gene) separate into different gametes. At fertilization, the diploid number of chromosomes is restored, and the nucleus once again contains two copies of each gene.

Mendel's observation of two different expressions of a trait—for example, "short" and "tall"—is explained by the fact that a gene can exist in alternate forms, called **alleles.** Mendel worked with "short" and "tall" alleles of a height-determining gene in pea plants. A plant or animal having two identical alleles for a particular gene is **homozygous** for that gene. A plant or animal having two different alleles is **heterozygous.** Although Mendel did not know anything about alleles, he noted that for some genes one allele could mask the expression of another. The allele that masks the expression of the other is completely **dominant,** and the masked allele is **recessive.** When Mendel crossed a true-breeding tall plant to a short plant, the "tall" allele was completely dominant to the "short" allele, and the plants of the next generation were all tall.

Mendel's laws can be applied to different crosses and to different organisms. To do this, abbreviations are used to represent alleles—capital letters for dominant alleles, and lowercase letters for recessive alleles. Therefore, a capital letter *T* denotes the dominant allele in peas that is responsible for "tallness," and a lowercase letter *t* signifies the "shortness" allele. If both of a pea plant's height genes are recessive, signified by *tt*, the plant is said to be homozygous recessive for this gene and is short. A plant with two copies of the dominant allele, designated *TT*, is called homozygous dominant. The other possible allele combination for this gene, written *Tt*, is a heterozygote. The heterozygote is tall, despite the presence of the "short" allele, because the *T* allele is dominant over the *t* allele.

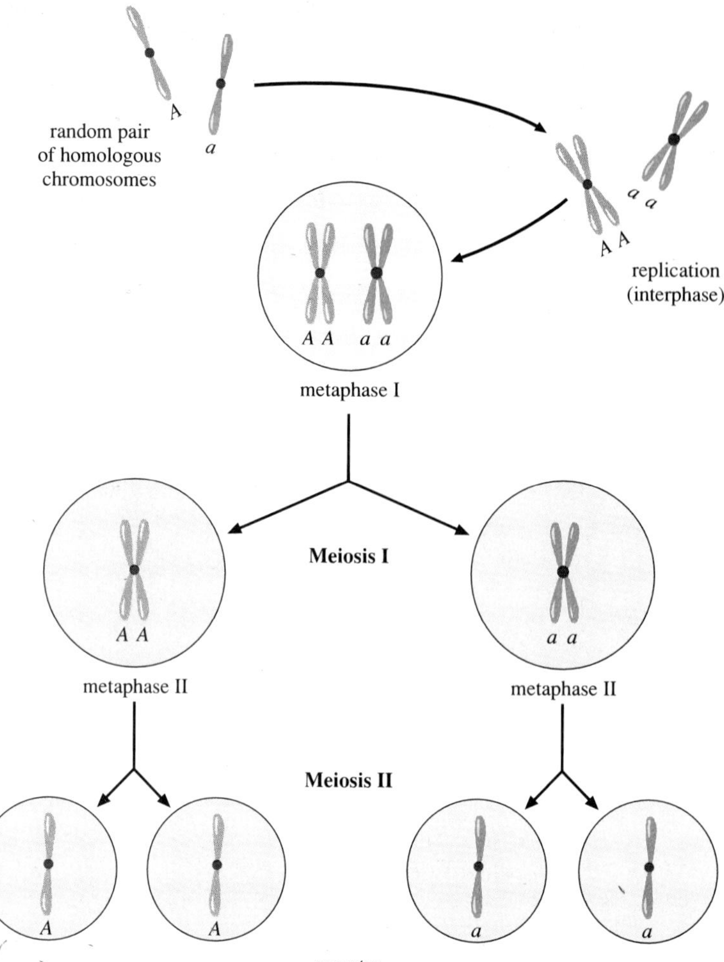

Figure 11.4
Mendel's first law—gene segregation. During meiosis, homologous pairs of chromosomes (and the genes that comprise them) separate from one another and are packaged into separate gametes. At fertilization, gametes combine at random to form the individuals of a new generation.

An organism's appearance does not always reveal which alleles of a particular gene are present. For example, both a *TT* and a *Tt* pea plant are tall, but one is a homozygote and the other a heterozygote. **Genotype** is the term that describes the alleles present for a particular gene, and **phenotype** describes the expression of an allelic combination. A pea plant may have a tall phenotype but be either genotype *TT* or *Tt*. A phenotype is often designated **wild type** if it is the most common expression of a particular gene in a population or **mutant** if it is not. Tall pea plants are wild type; short plants mutant.

In analyzing genetic crosses, the first generation considered is called the *parental generation*, or P_1; the second generation is the *first filial generation*, or F_1; the next generation is the *second filial generation*, or F_2, and so on. In your family, your grandparents might be considered the P_1 generation, your parents the F_1 generation, and you and your siblings the F_2 generation.

Mendel's observations on the inheritance of a single gene reflect the events of meiosis. When a gamete is produced, the two copies of a particular gene separate. In female gametes

	T	t
male gametes T	TT	Tt
t	Tt	tt

Figure 11.5
A Punnett square is helpful in following the transmission of traits. It is a diagram of gametes and how they can combine in a cross between two particular individuals. The different types of female gametes are listed along the top of the square; male gametes are listed on the left-hand side. Each compartment within the square contains the genotypes that result when gametes join from the individuals corresponding to that compartment. The Punnett square here describes Mendel's monohybrid cross of two tall pea plants. In the progeny, tall plants outnumber short plants 3:1. Can you determine the genotypic ratio?

Figure 11.6
Kernels on an ear of corn represent progeny of a single cross. When a plant, which has a dominant allele for purple kernels and a recessive allele for yellow kernels is self-crossed, the resulting ear has approximately three purple kernels for every yellow kernel.

a plant of genotype *Tt*, for example, gametes carrying either *T* or *t* are formed in equal numbers in anaphase of the first meiotic division. When gametes meet to start the next generation, they combine at random. That is, a *t*-bearing egg cell is not more or less "attractive" to a *t*-bearing sperm cell than it is to a *T*-bearing one. Equal gamete formation and random combinations of gametes underlie Mendel's laws.

Mendel crossed short pea plants (genotype *tt*) with true-breeding tall pea plants (genotype *TT*). The resulting seeds grew into F_1 plants that all had the same phenotype: tall (genotype *Tt*). Next he crossed these F_1 individuals with each other. This is called a **monohybrid** cross, because one trait is being followed by crossing two heterozygous, or hybrid, individuals. There are three possible outcomes of such a cross: *TT*, *tt*, and *Tt*. A *TT* individual results from the fertilization of a *T*-bearing egg cell by a *T*-bearing sperm cell; a *tt* individual results from the meeting of a *t*-bearing egg cell and a *t*-bearing sperm cell; and a *Tt* individual results from either one of two different combinations—a *t*-bearing egg cell fertilized by a *T*-bearing sperm cell, or a *T*-bearing egg cell fertilized by a *t*-bearing sperm cell.

Because there are twice as many ways to produce a heterozygote than there are to produce either homozygote, the *genotypic ratio* expected of a monohybrid cross is 1 *TT*:2 *Tt*:1 *tt*. The corresponding *phenotypic ratio* is three tall plants to one short plant, a 3:1 ratio. An easy way to keep track of the contributions of each parent and the resulting genotypes and phenotypes in the next generation is to construct a *Punnett square* (fig.11.5).

Mendel's data were in the form of numbers, not ratios. When he crossed *Tt* pea plants to other *Tt* pea plants, he catalogued 787 tall plants and 277 short plants in the next generation, which is very close to a 3:1 ratio. Mendel then bred the tall plants resulting from the monohybrid cross to *tt* individuals. If a tall plant produced tall plants and short plants when crossed to a *tt* plant, then it was genotype *Tt*. But if it produced only tall plants, then it must be *TT*. The technique of crossing an individual of unknown genotype to a homozygous recessive individual is called a **test cross.** It is based on the fact that the homozygous recessive is the only genotype that can be identified by observation of the phenotype—that is, a short plant can only be of genotype *tt*. The homozygous recessive serves as a "known" with which an individual of an unknown genotype can be crossed. Figures 11.6 and 11.7 show the results of a monohybrid cross in two other familiar species.

a.

b.

Figure 11.7
a. Albinism can result from a monohybrid cross in a variety of organisms. A heterozygote, or "carrier," for albinism has one allele that directs the synthesis of an enzyme needed to manufacture the skin pigment melanin and one allele that fails to make the enzyme. Each child of two carriers has a one in four chance of inheriting the two deficient alleles and being unable to manufacture melanin. He or she is an albino. *b.* An albino mouse with a normally pigmented sibling.

KEY CONCEPTS

Many physical characteristics and abilities are inherited. The unit of inheritance is the gene, a sequence of DNA of a chromosome. Gregor Mendel established the laws of heredity in breeding experiments with pea plants. From the observation that crosses of two tall plants sometimes produced short plants, Mendel deduced that the two "characters" for plant height segregate from one another during meiosis, then combine at random with those from the opposite sex cell at fertilization. This is the law of segregation.

Alternate forms of a gene are alleles. An organism is homozygous for a gene if it has two identical alleles and heterozygous if it has two different alleles. The allele that is expressed in the phenotype of a heterozygote is dominant; the allele not expressed is recessive. The alleles present in an organism constitute its genotype. The most common phenotype is called the wild type; others are mutant. A monohybrid cross yields a genotypic ratio of 1:2:1 and a phenotypic ratio of 3:1. A test cross is a cross to a homozygous recessive individual to reveal an "unknown" genotype.

Independent Assortment—Following the Inheritance of Two Genes at a Time

The law of segregation follows the inheritance of two alleles of a single gene. In a second set of experiments, Mendel examined the inheritance of two different traits, each of which had two different alleles. He looked at seed shape, which was either round or wrinkled (determined by the *R* gene), and seed color, which was either yellow or green (determined by the *Y* gene). When he crossed plants with round yellow seeds to plants with wrinkled green seeds, all individuals of the resulting generation were round and yellow. Therefore, round was completely dominant to wrinkled, and yellow was completely dominant to green. Next, he took F_1 plants (genotype *RrYy*) and crossed them to each other. This is a **dihybrid** cross, because individuals heterozygous for two genes are crossed. He found four types of plants in the F_2 generation: round and yellow (315 plants), wrinkled yellow (101 plants), round green (108 plants), and wrinkled green (32 plants). This is an approximate ratio of 9:3:3:1, which has come to be synonymous with a dihybrid cross and the demonstration of Mendel's second law (fig. 11.8).

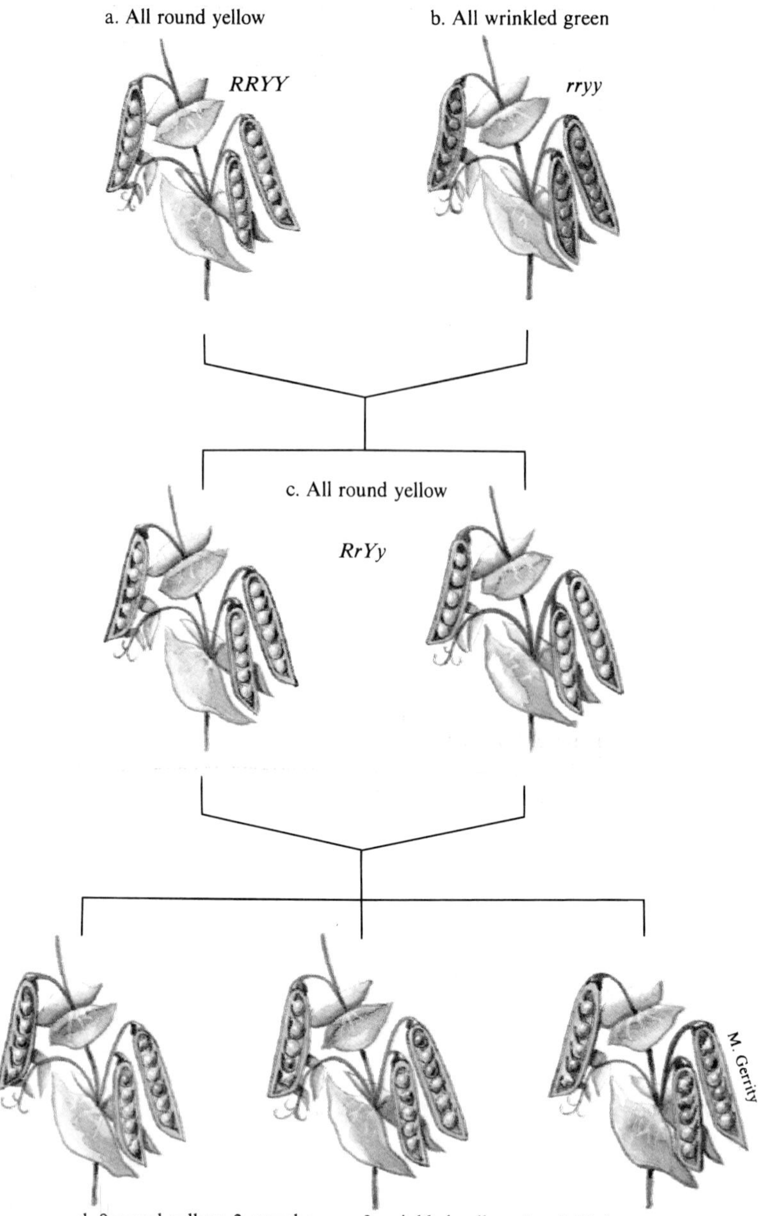

Figure 11.8
Mendel's crosses involving two genes. To study the inheritance pattern of two genes, Mendel crossed (*a*) a true-breeding plant with round yellow seeds (*RRYY*) to (*b*) a plant with wrinkled green seeds (*rryy*). *c*. The peas of the F_1 generation were all round and yellow (*RrYy*). *d*. When the F_1 dihybrid pea plants were crossed to each other, the F_2 generation exhibited the phenotypes round yellow (*RrYr*, *RRYy*, *RrYY*, and *RRYY*), round green (*Rryy* and *RRyy*), wrinkled yellow (*rrYY* and *rrYy*), and wrinkled green (*rryy*) in a 9:3:3:1 ratio.

Mendel took each individual from the F_2 generation and crossed it to wrinkled green (*rryy*) plants. These test crosses established whether each F_2 plant was true-breeding for both genes (*RRYY* or *rryy*), true-breeding for one but heterozygous for the other (*RRYy*, *RrYY*, *rrYy*, or *Rryy*), or heterozygous for both genes (*RrYy*). Based upon the results of the dihybrid cross, Mendel concluded that a gene for one trait does not influence the transmission of a gene for

Table 11.1
A Glossary of Genetic Terms

Term	Definition	Term	Definition
Allele	One of two or more alternate forms of any given gene.	Meiosis	A form of cell division that produces gametes (sperm cells or egg cells).
Autosome	Any chromosome other than a sex chromosome.	Monohybrid	An individual heterozygous for one particular gene.
Chromosome	A rod-shaped structure in a cell's nucleus composed of DNA and protein.	Mutant	A phenotype or allele that is not the most common for a certain gene in a population.
Dihybrid	An individual heterozygous for two particular genes.	Phenotype	The observable expression of a specific genetic constitution in a specific environment.
Dominant	An allele that masks the expression of another allele.	Recessive	An allele whose expression is masked by the activity of another allele.
Gene	A segment of a chromosome, composed of DNA, that directs the cell's synthesis of a particular protein or controls the activity of another gene.	Segregation	The separation of alleles of a gene into separate gametes during meiosis.
Genotype	The genetic constitution of an individual.	Sex chromosome	A chromosome that determines sex.
Heterozygous	Possessing different alleles of the same gene.	Sex linked	A gene located on the X chromosome or a trait that results from the activity of a gene on the X chromosome.
Homozygous	Possessing identical copies of an allele of the same gene.	Wild type	A phenotype or allele that is the most common for a certain gene in a population.
Independent assortment	Mendel's second law. A gene on one chromosome does not influence the inheritance of a gene on a different (nonhomologous) chromosome because chromosomes are packaged randomly into gametes during meiosis.		

another trait. This is Mendel's second law, **independent assortment,** and it is true only for genes located on different chromosomes. The seed shape and seed color genes that Mendel worked with fit this criterion.

With the idea of independent assortment, Mendel had again inferred a principle of inheritance that has its physical basis in meiosis. Independent assortment occurs because chromosomes derived from each parent combine in a random fashion (fig. 11.9). In Mendel's dihybrid cross, each parent produces equal numbers of gametes of four different types: *RY*, *Ry*, *rY*, and *ry*. (Note that each of these combinations has one gene for each trait.) When a Punnett square for this cross is worked out (fig. 11.10), we see that the four phenotypic classes—round yellow (*RRYY*, *RrYY*, *RRYy*, and *RrYy*); round green (*RRyy*, *Rryy*); wrinkled yellow (*rrYY*, *rrYy*); and wrinkled green (*rryy*)—are present in the ratio 9:3:3:1, just as Mendel found.

Using Probability to Analyze More Than Two Genes

A Punnett square can be cumbersome when tracing more than two genes at a time. For three genes, it would have 64 boxes; for four genes, 256 boxes. An easier way to make genetic predictions is to use the mathematical laws of probability. (Punnett squares are actually a way of using probability too.) To calculate the frequencies of traits controlled by many genes located on different chromosomes, simply predict the frequencies of inheriting one gene at a time, and multiply the frequencies.

Let's look at an example. Returning to Mendel's dihybrid cross, suppose you want to know the chance of obtaining a wrinkled green (*rryy*) plant from dihybrid (*RrYy* crossed to *RrYy*) parents. Consider the dihybrid one gene at a time, and construct a Punnett square for *Rr* crossed to *Rr*. This shows that the probability of *Rr* plants producing *rr* progeny is 25%, or 1/4. Similarly, the chance of two Yy plants producing a yy individual is also 25%, or 1/4. By multiplying these probabilities, we predict that the chance of dihybrid (*RrYy*) parents producing homozygous recessive (*rryy*) offspring is 1/4 multiplied by 1/4, or 1/16. Now consult the 16-box Punnett square for Mendel's dihybrid cross (fig. 11.10). Indeed, only 1 of the 16 boxes is of genotype *rryy*. Table 11.1 presents common genetic terms.

KEY CONCEPTS

Mendel's second law, that of independent assortment, considers genes transmitted on different chromosomes. In a dihybrid cross of heterozygotes for seed color and shape, Mendel demonstrated a phenotypic ratio of 9:3:3:1 and concluded that a gene on one chromosome does not influence the transmission of another gene (assumed to be on another chromosome). Segregation and independent assortment are explained by the behavior of chromosomes during meiosis. Punnett squares can be used to follow the independent assortment of two genes, but it is easier to multiply probabilities when three or more genes are considered.

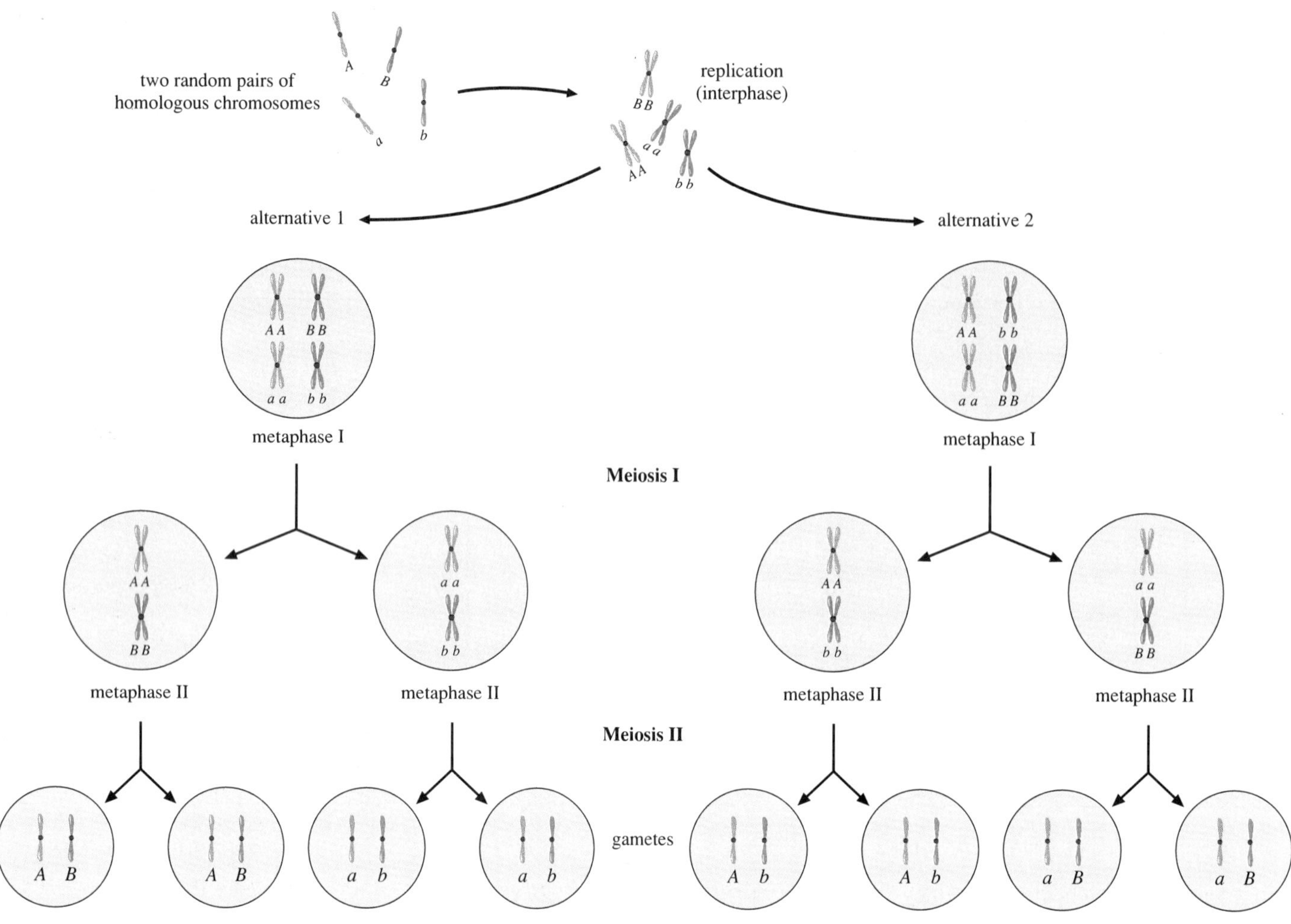

Figure 11.9
The independent assortment of genes carried on different chromosomes results from the random alignment of chromosome pairs during metaphase of meiosis I. An individual of genotype *AaBb*, for example, manufactures four types of gamates, containing the dominant alleles of both genes (*AB*), the recessive alleles of both genes (*ab*), and a dominant allele of one with a recessive allele of the other (*Ab* and *aB*). The allele combination depends upon which chromosomes are packaged together into the same gamate—and this happens at random.

Disruptions of Mendelian Ratios

Mendel's genetic crosses yielded offspring that were easily observed and distinguished from each other. A pea is either yellow or green, round or wrinkled; a plant is either tall or short. In many instances, however, what is seen is not what is expected by working out Punnett squares. Mendel's laws are in operation, but either the nature of the phenotype or influences from body chemicals, other genes, or the environment alter the observable phenotypic ratios.

Lethal Alleles

Some allele combinations are deadly to embryos, fetuses, or larvae, and therefore they are not observed as a phenotypic class of offspring. An allele that results in such early death is called a *lethal allele*. In humans, lethal alleles are responsible for some spontaneous abortions. Consider a man and a woman each heterozygous for a particular gene, possessing one dominant normal allele and one recessive lethal allele. Each person produces equal numbers of gametes containing each allele. When the man and woman try to have children, each fertilized egg has a 25% chance of inheriting both normal alleles, a 50% chance of inheriting one normal allele and one lethal allele, and a 25% chance of inheriting both lethal alleles. (Note that this is a 1:2:1 monohybrid cross ratio.) But if the couple's children are examined, all are healthy. The expected 3:1 phenotypic ratio of healthy to sick children is not seen because homozygous recessive embryos do not survive long enough to be born—they are miscarriages.

Inheritance of lethal alleles can be observed using experimental organisms. Two phenotypically normal fruit flies known to be heterozygous for a particular gene are placed together in a jar. The female lays 20 fertilized eggs, but only 15 of them emerge 10 days later as flies. An experimenter analyzing only those progeny that mature into flies might conclude that the two parents produce only 15 offspring, all normal; however, 5 have died and decomposed.

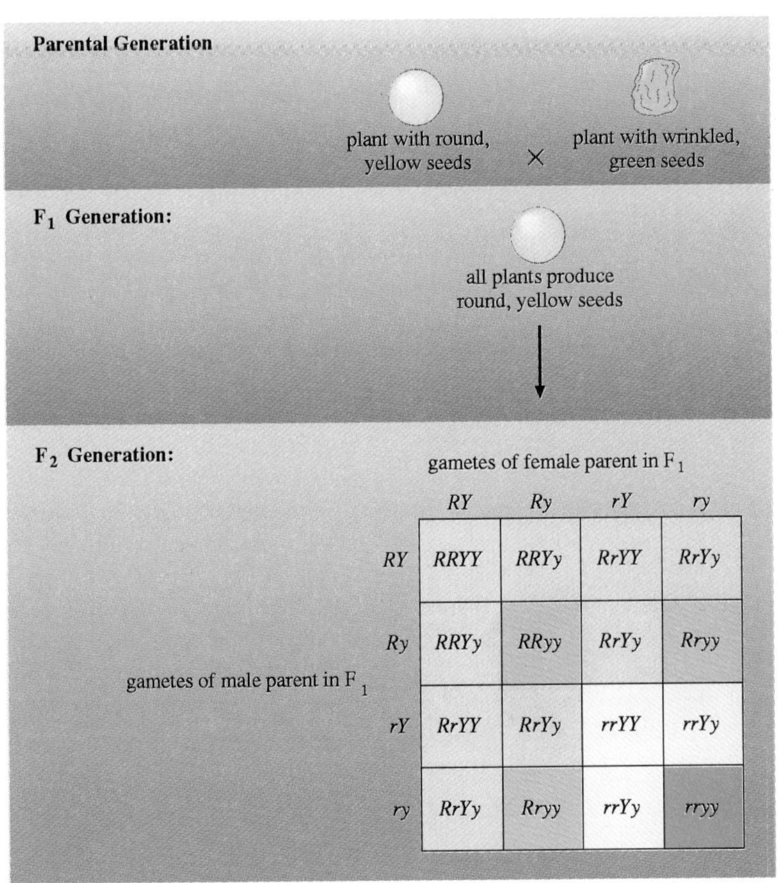

Figure 11.10
A Punnett square can be used to represent the random combinations of gametes produced by dihybrid individuals.

A dominant allele that produces a mutant phenotype sometimes behaves as a recessive lethal allele as well. That is, the heterozygote expresses the dominant phenotype, but the homozygous dominant organism dies before birth. For example, cattle inheriting a dominant allele of a gene called Dexter, plus a recessive allele, have abnormally short limbs. Cattle inheriting two copies of the dominant Dexter allele are severely deformed and generally do not survive long enough to be born.

The Influence of Gender—Of Breasts and Beards

An individual's sex can affect gene expression. The most important influence of sex on inheritance concerns genes carried on the **sex chromosomes,** which are designated X and Y in humans. These chromosomes determine a person's sex. (The remaining 22 pairs of chromosomes, called **autosomes,** are not involved in sex determination.) A human male has one X and one Y chromosome, and a human female has two X chromosomes. The Y chromosome carries only a few genes, and these are different from those carried on the X chromosome. Because a female has two alleles of genes on the X chromosome (because she has two X chromosomes) and a male only has one, an allele that is recessive in a female is dominant in a male. *Sex-linked traits* are discussed in the next chapter.

A **sex-limited trait** affects a part or function of the body that is present in one sex but not the other. Consider the following conversation:

> *Son: "Dad, how can I have such a thick beard, if your face is as smooth as a baby's rear?"*
>
> *Dad: "Don't ask me. You must have inherited it from your mother!"*

The father may not be far from the truth. The mother of course does not have a beard, because her hormones do not encourage the growth of facial hair. However, she may still pass on to her son genes controlling beard growth. The father would pass on such genes too, if they are located on an autosome, but in this case his particular alleles do not foster heavy beard growth. A similar situation may exist for a young woman with large breasts whose mother is flat-chested. Sex-limited inheritance is important in animal breeding. In cattle, for example, milk yield and horn development are traits that affect only one sex, but the genes controlling them can be transmitted by either parent (fig. 11.11).

In **sex-influenced inheritance,** an allele is dominant in one sex but recessive in the other. This difference in expression can be caused by hormonal differences between the sexes. For example, a gene for hair growth pattern has two alleles, one that produces hair all over the head and another that causes pattern baldness (fig. 11.12). The baldness allele is dominant in males but recessive in females, which is why more men than women are bald. A heterozygous male is bald, but a heterozygous female is not. What is the genotype of a bald woman?

Different Dominance Relationships

Mendel's tall pea plants of genotypes *Tt* and *TT* are indistinguishable from each other because the *T* allele is completely dominant to the *t* allele. Complete dominance is also seen in many inherited illnesses. In Tay-Sachs disease, for example, an affected child is homozygous recessive for a certain allele and lacks a vital enzyme. His heterozygous parents each have half the amount of the enzyme found in a homozygous dominant person, but this amount is enough to maintain their health.

In some genes, however, one allele may not completely overshadow another. In **incomplete dominance,** the phenotype of the heterozygote is intermediate between those of homozygotes. For example, crossing a snapdragon plant with red flowers (*RR*) to one with white flowers (*rr*) results in a plant with pink flowers (*Rr*) (fig. 11.13). Two alleles that are both expressed in a heterozygote are termed **codominant.** This is illustrated in a person who is of blood type AB. A person of blood type A has red blood cells with antigen A on their surfaces. Blood type B corresponds to red blood cells with antigen B. A person with the unusual blood type AB has red blood cells with both the A and B molecules, and the red cells of a person with type O blood has neither antigen (table 11.2).

An **overdominant** heterozygote is one that is more vigorous than either of the parent homozygotes. Over-dominance is often called hybrid vigor. A typical hybrid corn plant, for example, has a higher grain yield than either of the parent strains from which it was derived. The same is also said to be true of mixed-breed dogs: mongrels tend to be healthier than purebred dogs. The genetic basis of hybrid vigor is not completely understood.

Penetrance and Expressivity

Cystic fibrosis is a genetic disease in which large amounts of mucus clog the lungs and pancreas, making breathing and digestion difficult. Although cystic fibrosis patients all have the same gene affected, they may live very different lives—one may die of respiratory failure in infancy, another of infection at

Figure 11.11
Both male and female sheep of some breeds have genes that control horn development, but only the males have horns.

a.

b.

c.

d.

Figure 11.12
Pattern baldness is a sex-influenced trait and was a genetic trademark of the illustrious Adams family. *a.* John Adams (1735-1826) was the second president of the United States. He was the father of John Quincy Adams (1767-1848) (*b*), who was the sixth president. John Quincy was the father of Charles Francis Adams (1807-1886) (*c*), who was a diplomat and the father of Henry Adams (1838-1918) (*d*), who was a historian.

age 8, and yet another may survive into his or her 30s. In other words, identical genotypes do not always mean identical phenotypes. Environmental factors such as nutrition and exposure to toxins, other illnesses, and even other genes can all affect the expression of a gene. It is also possible that different parts of the same gene are affected in different individuals, resulting in differing severity of the illness. For some conditions, an individual may actually inherit a disease-causing genotype and for reasons unknown remain healthy.

A genotype is said to be *completely penetrant* if all individuals who have it express the associated phenotype. Cystic fibrosis, as well as the traits that Mendel followed in peas, are completely penetrant. A genotype is *incompletely penetrant* if some individuals do not express the phenotype. Having more than 10 fingers or toes, a condition called polydactyly (many digits), is incompletely penetrant. Some of the individuals who inherit the dominant allele have more than 5 digits on a hand or foot (fig. 11.14), yet others who are known to have the allele from tracing family background have the normal number of fingers and toes. The penetrance of a gene is described numerically. If, for example, 80 of 100 people who have inherited the dominant polydactyly allele have extra digits, the allele is 80% penetrant.

A phenotype is said to be *variably expressive* if it is expressed to different degrees among individuals. One person with polydactyly, for example, might have an extra digit on both hands and a foot; another might have two extra digits on both hands and both feet; a third person might have just one finger that has an extra fingertip. So while **penetrance** refers to the all-or-none expression of a genotype, **expressivity** refers to the severity of a phenotype.

Table 11.2
Codominance in Human ABO Blood Types

Whether your blood type is A, B, AB, or O reflects which alleles you have of a gene called I. This gene controls antigens on the surfaces of red blood cells—A antigen for type A, B antigen for type B, both antigens A and B for type AB, and neither antigen for type O. The I gene is a three-allele system, with allele I^A and I^B dominant to i but codominant to each other. The following table lists the phenotypes and genotypes associated with the I gene.

Phenotype (blood type)	Genotype
A	I^AI^A or I^Ai
B	I^BI^B or I^Bi
AB	I^AI^B
O	ii

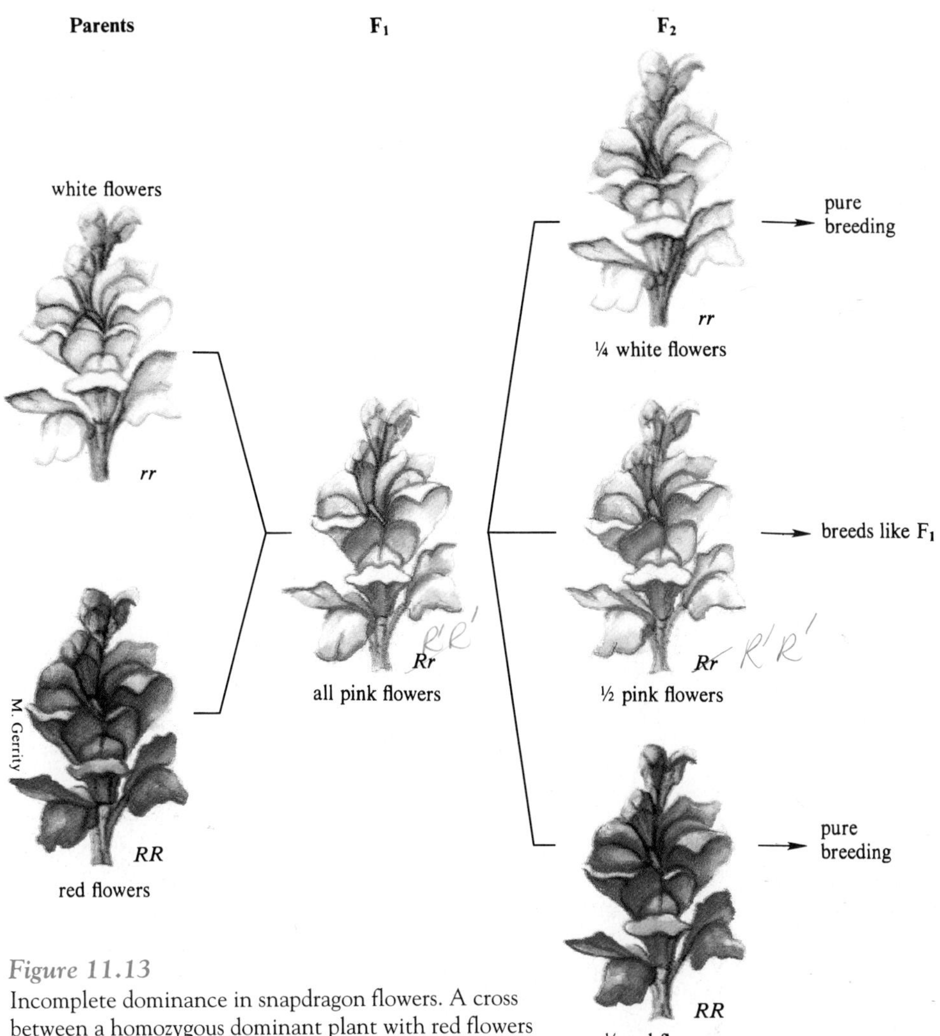

Figure 11.13
Incomplete dominance in snapdragon flowers. A cross between a homozygous dominant plant with red flowers (*RR*) and a homozygous recessive plant with white flowers (*rr*) produces a heterozygous plant with pink flowers (*Rr*). When *Rr* pollen fertilizes *Rr* egg cells, one-quarter of the progeny are red flowered (*RR*), one-half are pink flowered (*Rr*), and one-quarter are white flowered (*rr*). The phenotypic ratio of this monohybrid cross is 1:2:1 (instead of the 3:1 seen in cases of complete dominance) because the heterozygous class has a phenotype different from that of the homozygous dominant class.

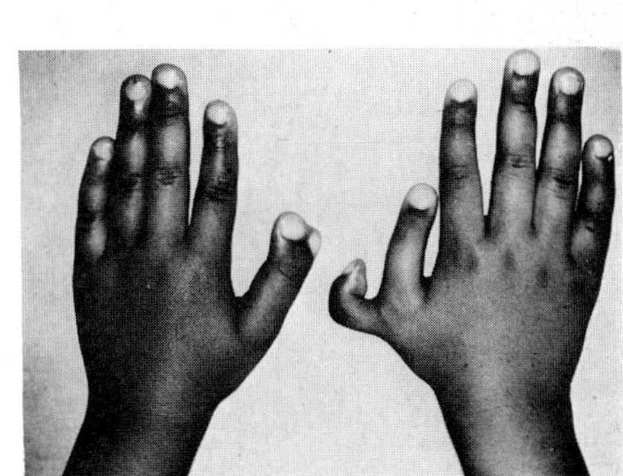

Figure 11.14
The dominant allele that causes polydactyly is incompletely penetrant and variably expressive. Some people who inherit the allele have the normal number of fingers and toes. Those in whom the allele is penetrant express the phenotype to differing degrees.

KEY CONCEPTS

In certain situations, Mendelian ratios are not what is expected. When an allele is lethal in homozygous form, this progeny class is not observed. Expression of sex-limited traits is influenced by sex hormones. A sex-influenced allele is dominant in one sex and recessive in the other. In incomplete dominance, the heterozygote's phenotype is intermediate between those of the homozygotes, and in overdominance, the heterozygote is more vigorous than either homozygote. Genotypes vary in their penetrance (percent of individuals affected) and expressivity (degree of expression).

Influences of the Environment

The expressions of some genes are exquisitely sensitive to the environment. A diet very low in the amino acid phenylalanine, for example, can prevent the mental retardation caused by the genetic disease phenylketonuria (PKU), in which phenylalanine builds up in the blood. In fact, diet is used to treat several genetic disorders (see Reading 15.2).

Temperature is another environmental influence on gene expression. Coat color in some mammals is affected by temperature, because the pigment molecules (whose synthesis is controlled by genes) are destroyed by extreme heat or cold. In Siamese cats, for example, the ears, nose, feet, and tail are darker than the rest of the body because the location of these anatomical parts makes them colder than the animal's abdomen. Temperature sensitivity of certain alleles is very striking in the fruit fly *Drosophila melanogaster*. At 82°F (approximately 28°C), the mouth of a fly of a certain genotype is transformed into a leg; at 65°F (approximately 18°C), the mouth in a fly of the same genotype develops as an antenna. At an intermediate temperature of 77°F (25°C), the mouth develops as a mixture of leg and antennal tissue (fig. 11.15).

Reading 11.1 *Odd Human Traits*

Do you have uncombable hair, misshapen toes or teeth, a pigmented tongue tip, or an inability to smell skunk? Do you lack teeth, eyebrows, eyelashes, nasal bones, thumbnails, or fingerprints? Can you roll your tongue or wiggle your ears? If so, you may find your unusual trait described in a book called *Mendelian Inheritance in Man*, compiled by Johns Hopkins University geneticist Victor McKusick, which catalogs more than 5,000 known human genetic variants. Most of the entries consist of family histories, clinical descriptions, and hypotheses of whether the trait is transmitted by a single gene, whether it is recessive or dominant, and whether it is inherited on a sex chromosome or an autosome. But amidst the medical terminology can be found some fascinating inherited traits in humans, from top to toes (fig. 1).

Genes control whether hair is blond, brown, or black, whether or not it has red highlights, and whether it is straight, curly, or kinky. Widow's peaks, cowlicks, a whorl in the eyebrow, and white forelocks run in families, as do hairs with triangular cross sections. Some people have multicolored hairs like cats, and others have hair in odd places, such as on the elbows, nosetip, knuckles, palms of the hands, or soles of the feet. Teeth can be missing or extra, protuberant or fused, present at birth, or "shovel-shaped" or "snow-capped." One can have a grooved tongue, duckbill lips, flared ears, egg-shaped pupils, three rows of eyelashes, spotted nails, or "broad thumbs and great toes." Extra breasts have been observed in humans and guinea pigs, and one family's claim to fame is a double nail on the littlest toes.

Unusual genetic variants can affect metabolism, sometimes resulting in disease or other times producing harmless yet quite noticeable effects. Members of some families experience "urinary excretion of odoriferous component of asparagus" or "urinary excretion of beet pigment" after eating the vegetable in question. In "blue diaper syndrome," an infant's inherited inability to break down an amino acid results in urine that turns blue on contact with air.

One father and son were plagued with a failure to open their mouths completely. Some families suffer from "dysmelodia," the inability to carry a tune. Those who have inherited "odor blindness" are unable to smell either musk, skunk, cyanide, or freesia flowers. Motion sickness, migraine headaches, and stuttering may be inherited. Uncontrollable sneezing may be due to inherited hayfever or to Achoo syndrome

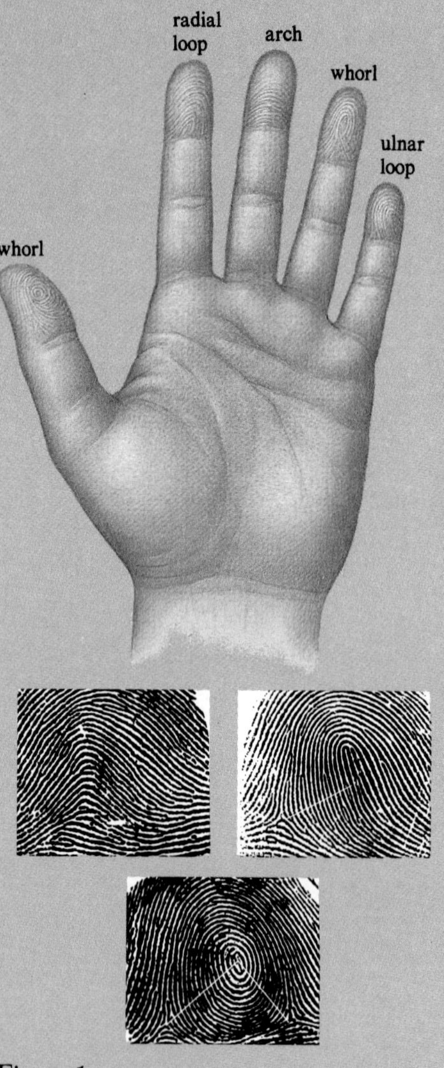

Figure 1
Fingerprint patterns form characteristic arches, loops, and whorls. The pattern is determined by the actions of several genes.

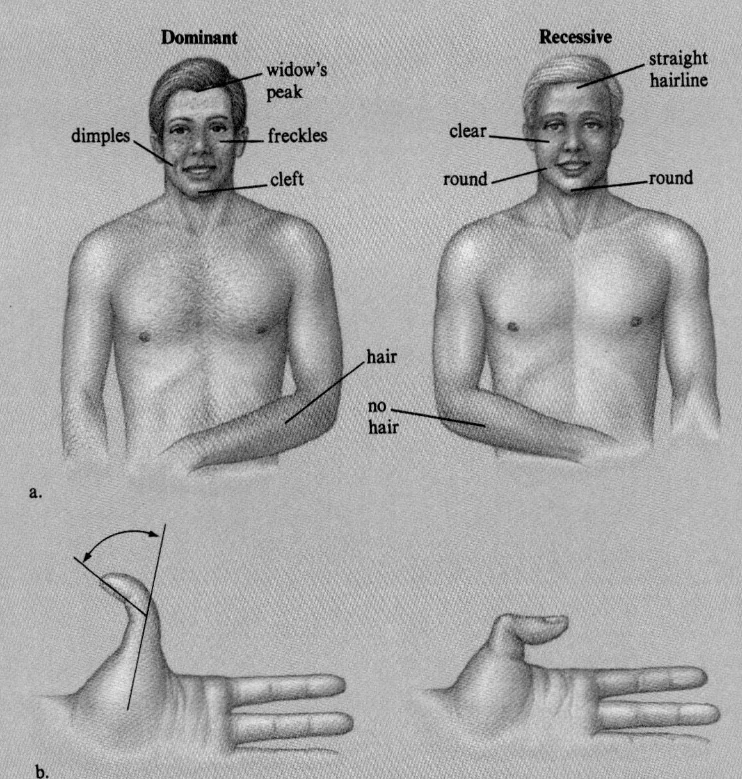

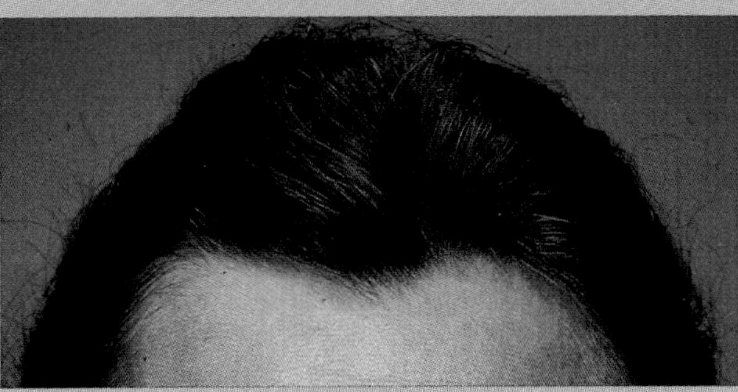

(an acronym for "autosomal dominant compelling helicophthalmic outburst syndrome").

Perhaps the most bizarre inherited illness is the "jumping Frenchmen of Maine syndrome." This is an exaggerated startle reflex that was first noted among French-Canadian lumbermen from the Moosehead Lake area of Maine, whose ancestors were from the Beauce region of Quebec. The disorder was first described at a meeting of the American Neurological Association in 1878 and was confirmed with the aid of videotape in 1980. McKusick describes the disorder as follows:

> *"If given a short, sudden, quick command, the affected person would respond with the appropriate action, often echoing the words of command . . . . For example, if one of them was abruptly asked to strike another, he would do so without hesitation, even if it was his mother and he had an ax in his hand."*

Some more common genetic quirks are illustrated in figure 2.

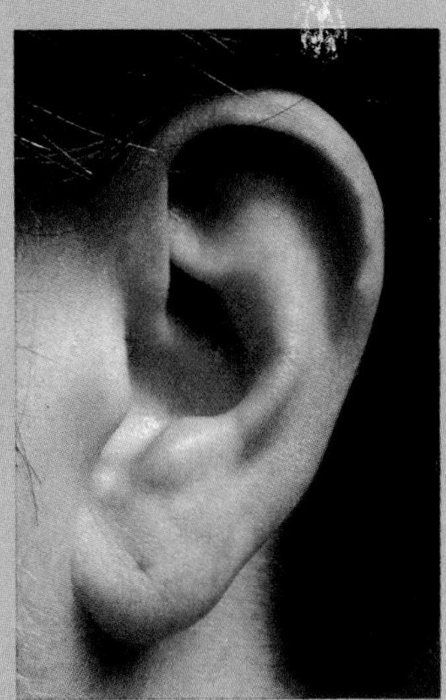

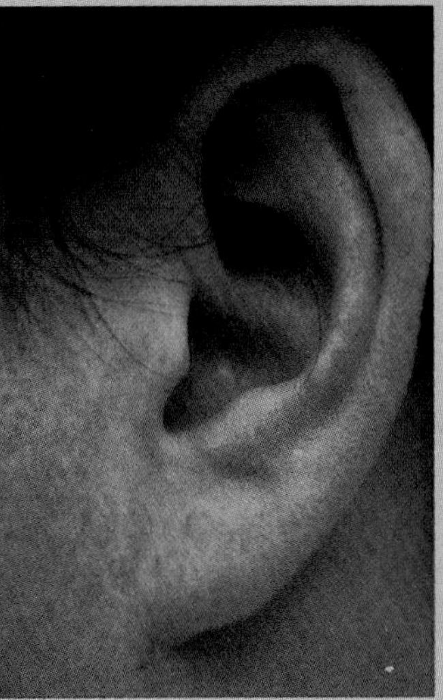

Figure 2
Inheritance of some common traits: (*a*) freckles, dimples, widow's peak, hairy elbows, and a cleft chin; (*b*) the ability to bend the thumb backwards or forwards; (*c*) widow's peak; and (*d*) attached or unattached earlobes.

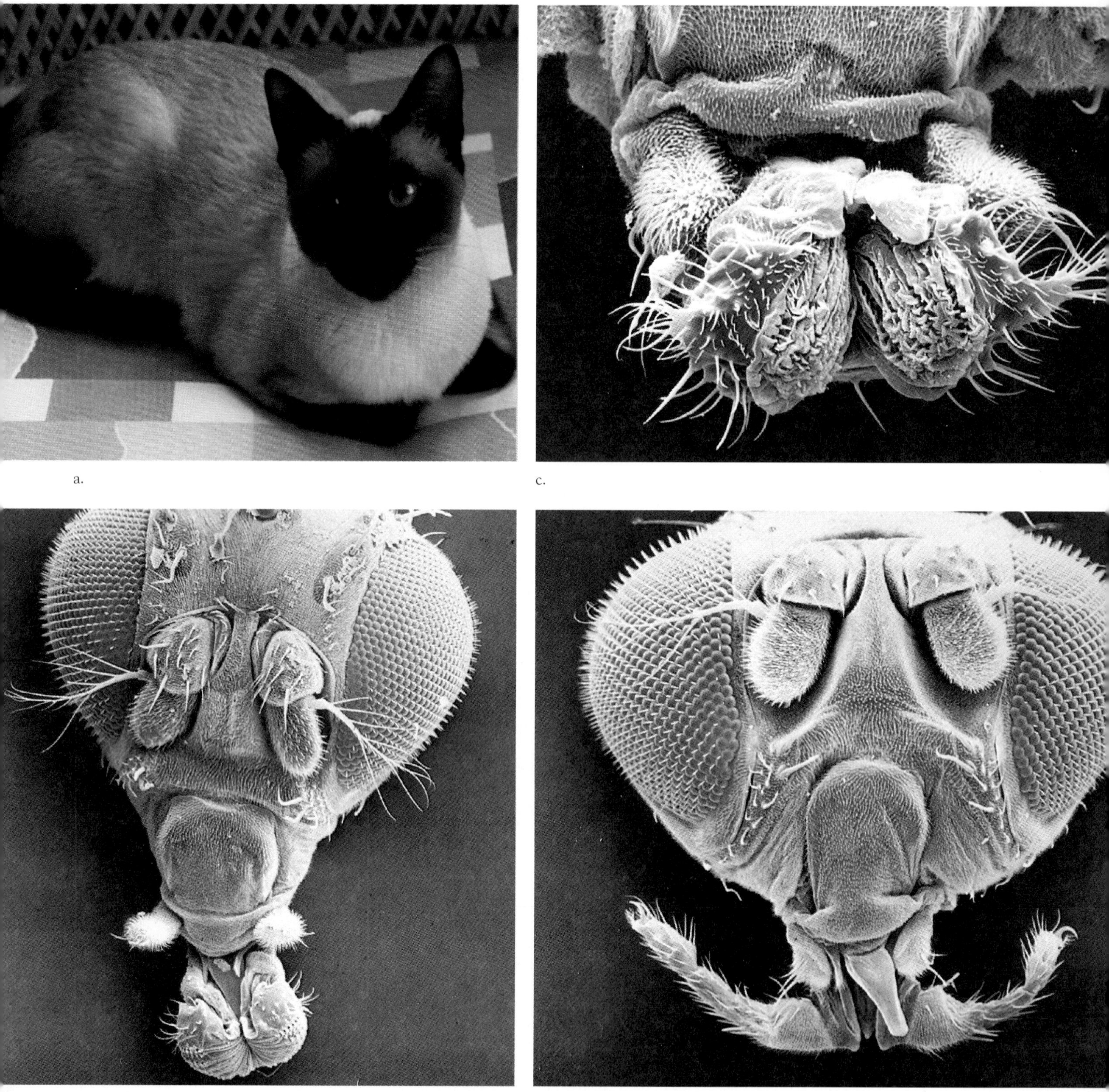

a.

c.

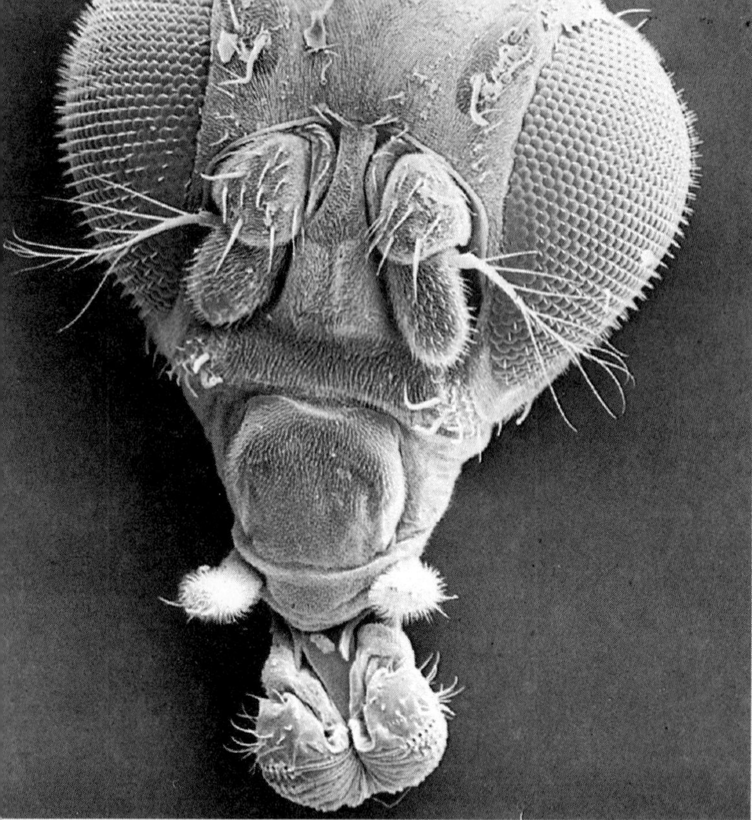

b.

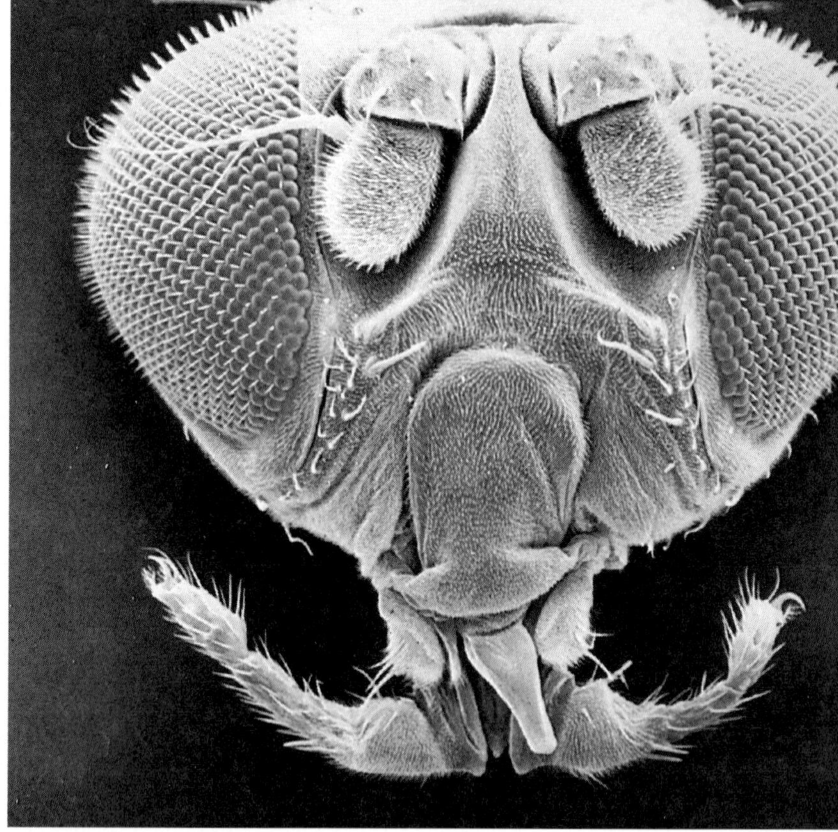

d.

Figure 11.15
An allele in the Siamese cat specifies an enzyme that produces pigment at cool temperatures. The enzyme is active in the colder portions of the animal, such as the nose, toes, ears, and tail. A wild type fruit fly's mouth is two rounded, fleshy lobes (*b*). Fruit flies homozygous for an allele of the proboscipedia gene have mouths transformed into an antenna at 65°F (approxmately 18°C) into (*c*) and into a leg at 82°F (approximately 28°C)(*d*).

Pleiotropy and King George III

Because living organisms are biochemically and anatomically complex, a protein that is abnormal due to a single defective gene can cause a variety of different symptoms, depending upon where and how in the body it is used. A gene that has multiple expressions is called **pleiotropic.** Consider a child who has sickle cell disease, in which red blood cells are sickle-shaped instead of the normal concave sphere. The sickled cells get caught in narrow blood vessels and break open, causing anemia. The heart, pumping harder than normal to move the oddly shaped red cells, may fail with the effort. The spleen becomes overtaxed in breaking down the red blood cells and may lose its ability to protect the body from infection. The sickle-shaped red blood cells thicken the blood with wide-ranging effects. Blood that clumps in the brain may lead to paralysis; in the muscles to rheumatism; and in the lungs to pneumonia. A child with sickle-cell disease may show any combination of these symptoms.

Another pleiotropic blood disease had interesting effects on the royal families of Europe. King George III ruled England during the time of the American Revolution. At the age of 50, he experienced his first terrifying bout with what would later be identified as porphyria. It began with abdominal pain and constipation, followed by the passage of dark red urine, weakness in the limbs, fever, hoarseness, and a fast pulse. Next, nervous system symptoms appeared, including insomnia, headache, visual problems, restlessness, delirium, convulsions, and stupor. Some of his reactions to physical discomfort were interpreted as signs of mental derangement. For example, the king would rip off his wig and clothing when at the peak of a fever. However, he also had the confused and racing thoughts characteristic of a severe mental disturbance. While Parliament and the king's baffled doctors spent 4 months debating whether or not the king's lunacy was permanent, George mysteriously recovered, just before he was about to be relieved of his royal duties.

But George's plight was far from over. He suffered a relapse 13 years later, then again 3 years after that. Always the symptoms appeared in the same order, beginning with abdominal pain, fever, and weakness and progressing to the nervous system symptoms. Finally, an attack in 1811 placed him in an apparently permanent stupor, and he was dethroned by the Prince of Wales. King George lived on for several years, experiencing further episodes of his bizarre illness. History books would later refer to him as the "mad king."

In George III's time, doctors were permitted to do very little to the royal body and made their diagnoses based on what the king told them. Twentieth-century researchers identified George's problem by the fact that his urine was dark red. This, they reasoned, reflected a metabolic error that routed part of the blood pigment hemoglobin, called a porphyrin ring, into the urine. When red blood cells die, porphyrin is normally broken down and metabolized by cells. In porphyria, an enzyme is missing that is necessary for the metabolism of porphyrin. The buildup of porphyrin attacks both the peripheral and central nervous systems. Examination of physicians' reports on George's relatives—easy to obtain for a royal family—showed that several of them had symptoms of porphyria as well. Different individuals exhibited different combinations of the symptoms and to differing degrees.

Genetic Heterogeneity

In a genetics clinic at a large hospital, two sets of parents discuss their experiences with their children. Each child visits the clinic for treatment of hemophilia, a disorder in which the blood cannot clot. The parents are surprised to learn that one of the boys inherited the responsible gene from his mother; but the other boy received hemophilia-causing genes from both parents. How could that be? Alleles of different genes can indeed produce the same phenotype if the genes' products act at different points in the same biochemical pathway. This **genetic heterogeneity** is seen in hemophilia, because 11 different genes specify 11 different "clotting factors" necessary for blood to clot. The most common form of hemophilia is caused by an allele on the X chromosome, but genes on autosomes can cause hemophilia as well.

Epistasis—Gene Masking at "General Hospital"

Another complication of Mendel's laws is that one gene can mask the effect of a different gene, a phenomenon called **epis-tasis.** (Do not confuse this with dominance relationships between alleles of the same gene.) An example of epistasis in humans was demonstrated in a most unlikely place—a television soap opera, "General Hospital" (fig. 11.16). It concerned the Bombay phenotype, a result of two interacting genes: the *I* gene, which is shown in table 11.2 directly controls ABO blood type, and another gene designated *H*.

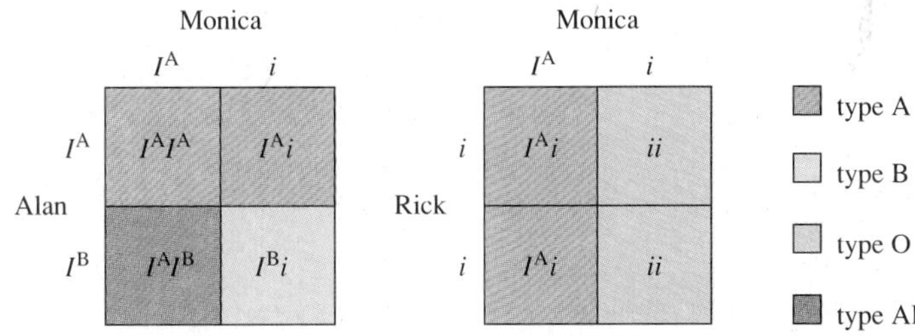

Figure 11.16
Gene masking at "General Hospital." Monica has just had a baby. But who is the father—her husband, Alan, or friend and fellow doctor, Rick? Monica's blood type is A, Alan's is AB, and Rick's is O. The child's blood type is O. At first glance, it looks as if Rick is the father. *a.* Whether Monica's genotype is I^Ai or I^AI^A, she cannot have a type O child with Alan, whose genotype is I^AI^B. *b.* If Monica's genotype is I^Ai, she could have a type O child with Rick, whose genotype is *ii*. Unless . . . nosy nurse-in-training Amy has just learned about the Bombay phenotype and suggests that the baby have a blood test to see if he manufactures the *H* protein, and that the adults look into their family histories to see if any other relatives' blood types are not what is predicated based on their parents' blood types. Sure enough, the baby is of genotype *hh*, and Alan's family has had past incidences in which the blood types of a child did not match what was predicted by parents' blood types. On further testing, both Monica and Alan are found to be *Hh*, but Rick is *HH*. Alan Jr. is *hh*. *He* has a type O phenotype, but his ABO genotype can be either I^AI^A, I^Ai, I^Bi, or I^AI^B. What he cannot be is Rick's son.

The relationship between the two genes affects the expression of the ABO blood type.

The product of the dominant allele *H* is an enzyme that inserts a sugar molecule onto a particular glycoprotein on the surfaces of red blood cells. The recessive allele *h* produces an inactive form of the enzyme, which cannot insert the sugar. The A and B antigens, which determine the ABO blood type, attach to the sugar molecule that is controlled by the *H* gene. As long as at least one *H* allele is present, the ABO blood type associated with the person's ABO genotype is expressed, because the A and B antigens have something to attach to. However, a person with genotype *hh* cannot attach the sugar to its glycoprotein base, and the A and B antigens have nothing to attach to on the red blood cells. The result is that an individual of genotype *hh* always has blood type O, although the ABO genotype can be any of the six possibilities.

Multiple Alleles

A diploid cell has a "place," or *locus*, for two alleles of a given gene, one on each of a pair of homologous chromosomes. The alleles can be identical copies (homozygous) or different (heterozygous). Even though an individual diploid organism has only two alleles for a particular gene, a gene can exist in more than two allelic forms within a population. This is because a change, or **mutation,** in one of the hundreds or thousands of subunits of a gene can lead to a different allele. In rabbits, coat color is determined by a series of alleles: c is recessive; c^h is dominant to c; c^{ch} is dominant to c^h and c; C is dominant to all alleles (table 11.3). In humans, the gene that causes cystic fibrosis can be altered in any of 70 or more different ways.

The more alleles a gene has, the more phenotypes and genotypes are associated with it. For a gene with two alleles, such as round (*R*) versus wrinkled (*r*) peas, three genotypes are possible—*RR*, *Rr*, and *rr*. For the blood type *I* gene, which has the three alleles I^A, I^B, and i, six genotypes form—I^AI^A, I^Ai, I^BI^B, I^Bi, I^AI^B, and ii.

Table 11.3
Phenotypes and Their Corresponding Genotypes for Coat Color in the Rabbit

Phenotype	Possible Genotypes
Gray	CC, Cc^{ch}, Cc^h, Cc
Chinchilla	$c^{ch}c^{ch}$
Light gray	$c^{ch}c^h$, $c^{ch}c$
Himalayan	c^hc^h, c^hc
Albino	cc

Phenocopies—When It's Not Really in the Genes

A condition caused by an environmental factor can sometimes resemble a known genetic disorder or mimic inheritance by occurring in certain relatives. An environmentally caused trait that appears to be inherited is called a **phenocopy** (fig. 11.17). The infant who picks up acquired immune deficiency syndrome (AIDS) from his or

Figure 11.17
Phenocopies mimic inherited characteristics. A condition caused by a substance in the environment can mimic one caused by a defective gene. A fruit fly exposed to ether at an early stage of development has two pairs of wings, just like a fly with the bithorax mutation.

her mother while passing through the birth canal has a medical problem in common with the mother, but the disease is infectious, not hereditary. Behaviors such as overeating and aggressiveness may also seem to be inherited when they are learned from relatives. Sometimes the reverse occurs—an inherited condition appears to be environmentally caused. For example, parents of an infant who had multiple fractures were wrongly accused of child abuse. The child had inherited osteogenesis imperfecta, in which abnormal collagen protein in the bones causes them to shatter very easily.

A birth defect can be a phenocopy. In the early 1960s, thousands of babies in Europe were born without limbs. A genetic disease called phocomelia was known to produce such "flipper babies," but how could this disease appear so suddenly in such large numbers? The problem was actually a drug, thalidomide, taken early in pregnancy to help women relax. Tragically, it also disrupted embryonic development at about the same time that the phocomelia gene exerts its effects. Because thalidomide and the phocomelia gene interfere with development at the same time, the drug-induced birth defect was identical to the hereditary variety. A "thalidomide baby," then, is a phenocopy of genetic phocomelia (see fig. 10.6).

Another phenocopy is a disease called kuru, a neurological disorder found prior to the 1970s only among the Fore people and their immediate neighbors, all primitive inhabitants of a mountainous rain forest in the interior of New Guinea (fig. 11.18). The Fore often battled their neighbors and honored their dead war heroes by eating their brains. In the 1950s, kuru accounted for nearly one-third of all Fore deaths. The disease began suddenly, with a loss of coordination and speaking ability, muscle spasms, poor balance, and eventually inability to swallow. The person rapidly became helpless, and death came within a year.

A look at the Fore kuru sufferers suggests that it is inherited. Three-quarters of those affected were women and the remaining one-quarter were children of both sexes over the age of 4 years. Men rarely succumbed. The 3:1 ratio suggested a Mendelian mode of inheritance—that is, that a single gene was responsible for kuru. Could kuru be due to an autosomal gene that is dominant in females and recessive in males? Homozygous dominant (*KuKu*) individuals of both sexes would experience symptoms early in life (the children); heterozygous (*Kuku*) females would develop symptoms as adults, but heterozygous males would be normal. Homozygous recessive (*kuku*) males and females would be normal.

Family data supported this explanation, as did other observations. For example, outsiders who moved to Fore settlements did not contract kuru, apparently ruling out the disorder as a communicable infection. Fore people who moved away still died from kuru, suggesting that they took it with them in their genes. But despite this convincing evidence, kuru, in fact, is not a genetic disorder. It is caused by a virus that is transmitted through the Fore cannibalism ritual.

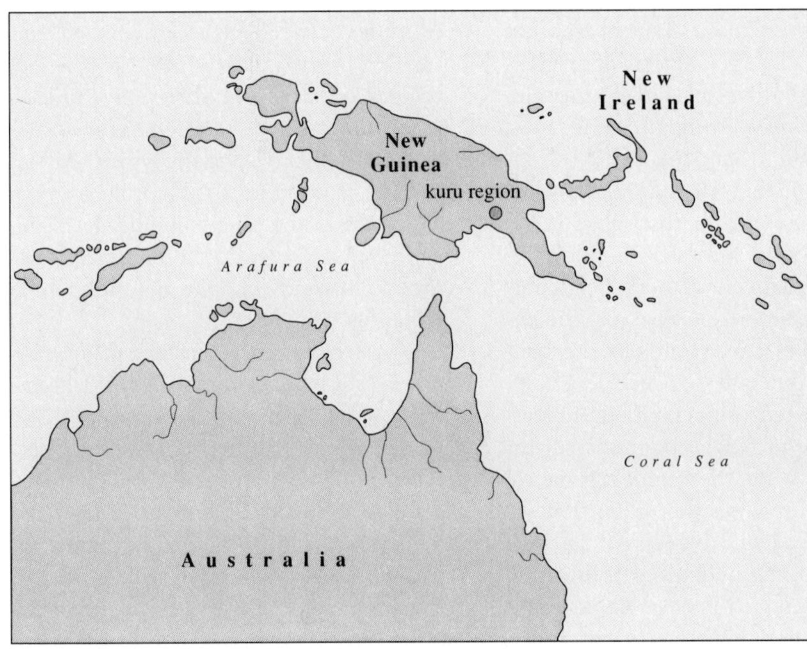

a.

b.

Figure 11.18

a. Kuru is a degenerative disease of the central nervous system that struck many women and children of the Fore people, who live in an isolated region of New Guinea. Although the pattern of the disease's occurrence in the population prior to 1970 suggested that it was inherited, the cause was actually a virus that was passed from person to person in a cannibalism ritual. *b.* The boy, who has kuru, could not stand, sit, or talk when this photo was taken and died a few months later.

The first clue to the origin of kuru came in the early 1960s, when a researcher from the National Institutes of Health, D. Carleton Gajdusek, noted that the symptoms of kuru were similar to those in sheep suffering from a viral disease called scrapie. Scrapie was caused by a "slow" virus, one that replicates very slowly, remaining in the host for many years before symptoms gradually appeared. Several degenerative diseases of the human nervous system are also caused by slow viruses. To investigate the possibility that kuru was caused by a slow virus, researchers injected monkeys with brain tissue from people who had died of kuru. A year and a half after the injection, the monkeys developed the disease, and they could pass it to other monkeys.

The next step was to take a closer look at the Fore way of life, particularly the custom of handling and consuming brain tissue. Indeed, it was the women and the children of both sexes over the age of four who prepared the brain tissue, cooked it, and ate it, often without washing themselves before or after the procedure. (These people could survive if there were no war dead.) Men merely watched the preparations. Kuru was caused by a slow virus and spread by exposure to uncooked infected brain tissue.

The lesson from kuru was that what seems to be a Mendelian ratio is not always so. From the kuru example, geneticists outlined criteria for an inherited trait: it should vary in frequency in different populations; it should not be transmitted to unrelated individuals who encounter the same environment as affected individuals; it should occur more frequently among relatives of an affected individual than in the general population; it should occur more frequently in identical twins, who share the same genes, than in fraternal twins; and it should produce symptoms at a particular age.

KEY CONCEPTS

Nutrition and temperature alter expression of some genes. A gene with more than one expression is pleiotropic. Genetic heterogeneity arises when more than one gene produces the same phenotype. Epistasis occurs when one gene masks the expression of another. A gene can have more than two alleles, which occur in pairs to produce several genotypes and phenotypes. A trait caused by the environment but resembling a known genetic characteristic or occurring in certain family members is called a phenocopy.

SUMMARY

For centuries societies have noticed and even attempted to manipulate the inheritance of certain traits in different organisms. Between 1857 and 1865, an Austrian monk, Gregor Mendel, was the first to investigate the nature of inheritance in a systematic way. He set up crosses between certain pea plants and followed the inheritance of one or two traits at a time. The *genes* for the traits that he studied were carried on different *chromosomes*, and each trait had two easily distinguished forms. From the raw data published by Mendel, geneticists a generation later (who knew then what chromosomes were) deduced the basic two laws of inheritance.

Mendel's first law, *segregation*, states that inherited "characters" separate at each generation, so that each new individual receives one copy of each factor from each parent. Mendel's second law, *independent assortment*, follows the transmission of two or more characters located on different chromosomes. A random assortment of maternally and paternally derived chromosomes (and the genes that they carry) in meiosis results in gametes that have different combinations of genes.

An *allele* is an alternate form of a gene. An allele whose expression masks that of another is *dominant;* an allele whose expression is masked by a dominant allele is *recessive*. A diploid individual who has two different alleles of a particular gene is *heterozygous* for that gene; if the two copies of a particular gene are identical, he or she is *homozygous* for that gene. If the two alleles are dominant, the individual is homozygous dominant for that gene; if the alleles are recessive, he or she is homozygous recessive for that gene. An allelic combination constitutes a *genotype*, and the expression of a particular genotype is its *phenotype*. The outcomes of genetic crosses can be predicted using Punnett squares, which use the principles of mathematical probability to follow the joining of gametes.

In some crosses the proportions of phenotypes predicted by Mendel's laws are not observed. Individuals who are homozygous recessive for *lethal alleles*, cease developing before birth and therefore are not detectable as a type of offspring. An individual's sex can influence phenotype. A *sex-limited trait* can be transmitted by either sex but is only expressed in one because of hormonal differences between the sexes. A *sex-influenced trait* is controlled by an allele that is dominant in one sex but recessive in the other.

Different types of dominance relationships between alleles influence the ratios of offspring classes. Heterozygotes of *incompletely dominant* alleles have phenotypes intermediate between those associated with the two homozygotes. Alleles that are *codominant* to each other are each expressed in the phenotype. In *overdominance* (hybrid vigor), the heterozygote is more fit than either homozygote from which it was derived.

An *incompletely penetrant* genotype is not expressed in every individual who inherits it. Phenotypes that vary in intensity among individuals are said to be *variably expressive*. The environment can influence gene expression. *Pleiotropic* genes have several expressions. A particular phenotype can be caused by several different genes. In *epistasis*, one gene masks the effect of another. Sometimes a characteristic that appears to be inherited is not genetic at all but caused by an environmental factor.

QUESTIONS

1. Bob and Joan know from a blood test that they are each heterozygous for the autosomal recessive gene that causes sickle cell disease. If their first three children are healthy, what is the probability that their fourth child will have the disease?

2. Mendel crossed pea plants heterozygous for the height gene (*Tt*) and obtained the monohybrid phenotypic ratio of 3:1 and the genotypic ratio of 1:2:1. Calculate the genotype and phenotypic ratios for the following crosses:

- **a.** Homozygous dominant crossed to homozygous recessive
- **b.** Homozygous dominant crossed to heterozygous
- **c.** Homozygous recessive crossed to heterozygous

3. A man and a woman each have dark eyes, dark hair, and freckles. The genes for these traits assort independently. The woman is heterozygous for each of these autosomal traits, but the man is homozygous. The dominance relationships of the alleles are as follows:

B = dark eyes *b* = blue eyes
H = dark hair *h* = blond hair
F = freckles *f* = no freckles

- **a.** What is the probability that their child will have the same phenotype as the parents?
- **b.** What is the probability that their child will have the same genotype as each parent?

Use probability or a Punnett square to obtain your answers. Which method is easier?

4. Which genetic principle does each of the following examples illustrate?

- **a.** All of the members of the Livingston family are vegetarians, and each has anemia (too little hemoglobin in the red blood cells). They think the condition is inherited because they all have it, but their physician says the anemia is caused by the lack of red meat in their diets.
- **b.** Marfan syndrome is a defect in connective tissue. Symptoms include a weakened aorta, nearsightedness, long and thin fingers, very long arms and legs, and deformity of the sternum (breastbone). Affected persons can express all or some of these symptoms.
- **c.** In the Smith family, the two sons are albinos, although the parents have normal skin pigmentation. The boys inherited the gene for albinism from the X chromosome donated by their mother. Their maternal grandfather was also an albino. In the Jones family, a boy and a girl are albinos and their parents too have normal skin pigmentation. The form of albinism in the Jones family is transmitted on an autosome, and the parents are carriers.
- **d.** Genes on two different chromosomes affect hearing in humans. The *D* gene controls development of the auditory nerve, with allele *d* impairing development. The *F* gene controls development of the cochlea, which is part of the inner ear, with allele *f* blocking that development. If a person is homozygous recessive for either gene (*dd* or *ff*), he or she is deaf, irrespective of the genotype at the other locus.
- **e.** An inherited illness causes the uterus to collapse and extend into the vagina. This is an autosomal recessive trait, which means that it is transmitted through both sides of the family.

TO THINK ABOUT

1. Do you think Mendel's observations would have differed if the traits he had chosen to study were part of the same chromosome?

2. If you wanted to confirm Mendel's law of independent assortment by recording the phenotypes and genotypes of the offspring of a dihybrid cross, what sorts of traits would you choose to examine, and in what organism?

3. A white woman with fair skin, blond hair, and blue eyes gives birth to fraternal twins with a black man with dark brown hair and eyes. One twin has blond hair, brown eyes and light skin, and the other has dark hair and dark skin. What Mendelian principle does this real-life case illustrate?

4. How can parents carry recessive lethal genes and not know it? Why do we have no evidence for the existence of dominant lethal mutations?

5. Why was it thought that kuru is an inherited illness? Can you think of a characteristic or behavior that might appear to be inherited but is actually caused by something in the environment?

SUGGESTED READINGS

Brown, William L. November 1984. Hybrid vim and vigor. *Science 84*. Studying inheritance in corn has shed much light on genetic principles, but it has also provided practical information of great value.

Corcos, Alain, and Floyd Monaghan. April 1985. Some myths about Mendel's experiments. *The American Biology Teacher*, vol. 47, no. 4. Many textbooks (but not this one) attribute conclusions to Mendel that he did not actually make.

Gajdusek, Carleton. Unconventional viruses and the origin and disappearance of kuru. 1977. *Science*, vol. 197, p. 943. Kuru looked like it was inherited—but it wasn't.

Macalpine, Ida, and Richard Hunter. July 1969. Porphyria and King George III. *Scientific American*. An unusually easy-to-read and fascinating account of a historical genetic problem.

Miller, Julie Ann. February 1984. Mendel's peas: A matter of genius or guile? *Science News*. Some say that Mendel's experimental results were too good to be true.

Weiss, Rick. May 20, 1989. A genetic gender gap. *Science News*, vol. 135. Is there an alternative explanation for Mendel's results?

C H A P T E R

12

Linkage

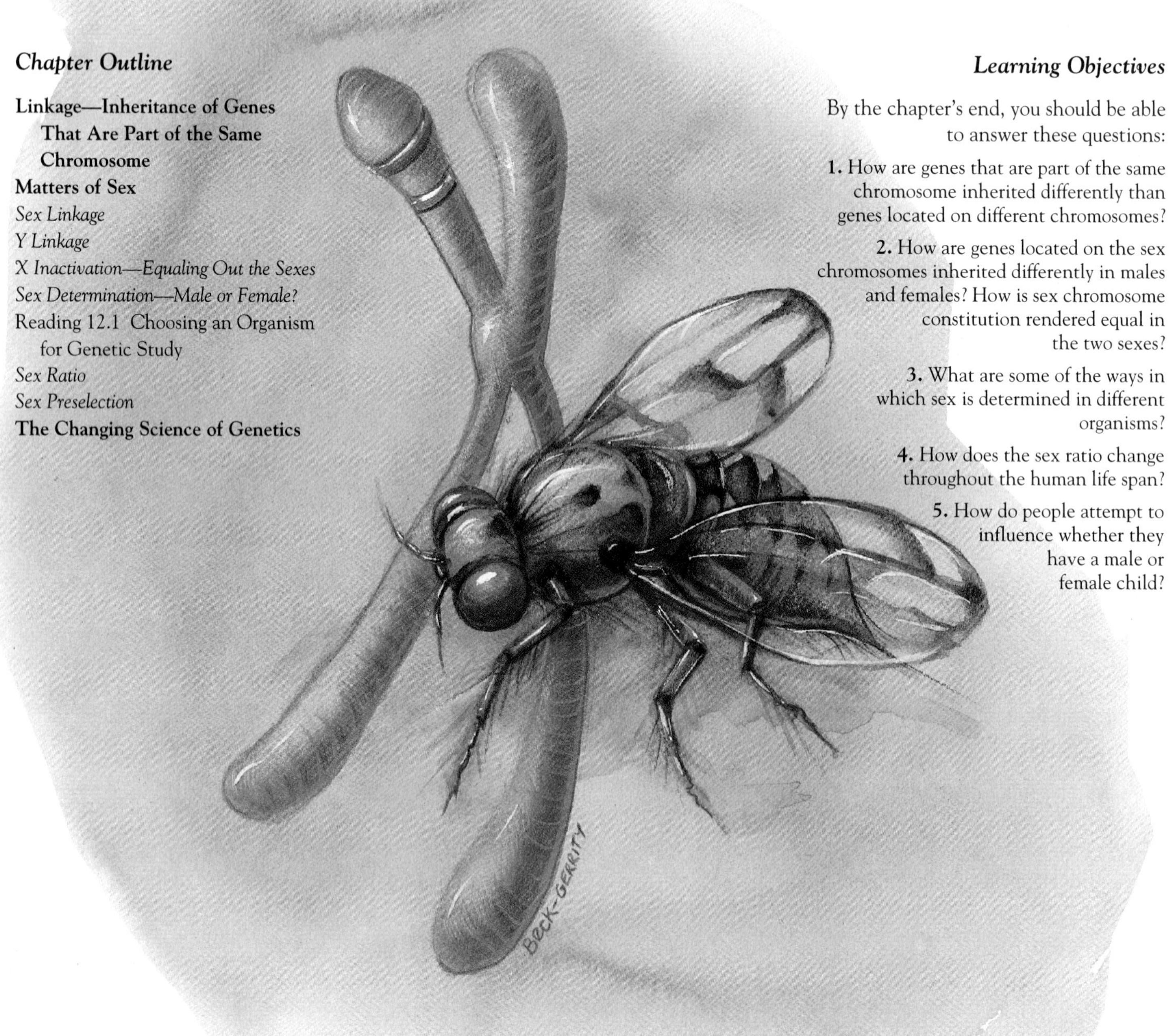

Chapter Outline

Learning Objectives

By the chapter's end, you should be able to answer these questions:

1. How are genes that are part of the same chromosome inherited differently than genes located on different chromosomes?
2. How are genes located on the sex chromosomes inherited differently in males and females? How is sex chromosome constitution rendered equal in the two sexes?
3. What are some of the ways in which sex is determined in different organisms?
4. How does the sex ratio change throughout the human life span?
5. How do people attempt to influence whether they have a male or female child?

When biologists discovered at the beginning of this century that genes are actually part of chromosomes, they realized that the number of known heritable traits in people was far greater than the number of chromosomes. That is, we have 23 pairs of chromosomes but far more than 23 inherited traits. This was confusing. Scientists had only recently rediscovered Mendel's laws and renamed his "characters" genes, and now another form of genetic unit—the chromosome—came along. Plus, there were too few chromosomes to account for the diversity of human inheritance. The first explanation was that during meiosis, chromosomes fall apart into gene-sized pieces. This not only accounted for the discrepancy between the number of inherited traits and the number of chromosomes, it also explained how genes behave as if they are physically separate when they assort into gametes. The hypothesis that chromosomes break into gene-sized pieces during meiosis indeed fits all the known facts—but it was not correct. The reason was that not all the facts were known. Genes located on the same chromosome are inherited together and do not assort independently. The traits that Mendel followed in his pea plants were specified by different chromosomes. Had these genes been carried near each other on the same chromosome, Mendel would have generated markedly different results in his dihybrid crosses.

Linkage—Inheritance of Genes That Are Part of the Same Chromosome

In the early 1900s, William Bateson and R. C. Punnett showed that the inheritance of two genes transmitted on the same chromosome is indeed different from inheritance of two genes each on different chromosomes. They set up a hybrid cross involving genes for flower color and pollen grain shape in peas, crossing plants that were homozygous dominant for both genes (*AABB*) to plants that were homozygous recessive (*aabb*) to generate F1 dihybrids of genotype *AaBb*. When they crossed the dihybrids to each other, instead of observing the 9:3:3:1 phenotypic ratio in the F2 generation that Mendel had seen for a dihybrid cross, they found that the progeny of the dihybrids were almost entirely of two genotypes—*AABB* and *aabb*. These are the genotypes of the parental (P1) generation, and the F2 individuals that resemble them are termed the *parental class*. The other F2 genotypes, which occurred less frequently than Mendel's rules predict, are called the *recombinant class* because they are different combinations of alleles than were present in the P1 generation.

The designation of parental class and recombinant class is relative, depending upon the genotypes of the P1 generation. In the preceding example, if the parental plants had been genotypes *AAbb* and *aaBB*, the F1 would still all be genotype AaBb, but the parental classes would be *AAbb* and *aaBB* and the recombinant classes *AABB* and *aabb*.

At about the same time that Bateson and Punnett were studying deviations from the famous 9:3:3:1 phenotypic ratio in peas, other geneticists were noting similar discrepancies in other organisms. It was as if the P1 combinations of alleles were sticking together through the generations. In the fruit fly, for example, four groups of traits appeared to be inherited together. Such groups of inherited traits are called **linkage** groups. Was it just a coincidence that the fruit fly has four different pairs of chromosomes? The answer is no. The fact that genes on the same chromosome tend to be inherited together explained the disruption of Mendelian ratios.

The rare recombinant offspring could also be explained by the behavior of chromosomes. During prophase of the first meiotic division, chromosomes of a pair can physically exchange material with one another, a process called **crossing over.** When this happens, alleles are exchanged from one chromosome to its mate. This alters the allelic combinations that are packaged into gametes. Most of the gametes produced by a P1 plant of genotype *AaBb* might carry a chromosome containing *AB* or *ab*. If crossing over occurs, however, a few of the gametes would carry *Ab* or *aB* (fig. 12.1).

As the inheritance patterns of more traits were studied in different organisms, it was found repeatedly that the number of linkage groups was the same as the number of chromosomes. It was also noticed that the farther apart two genes are on a chromosome, the more likely crossing over is to occur between them, simply because there is more room for chromosomes to twist about one another and exchange segments. The correlation between crossover frequency and the distance between genes is used to construct *linkage maps*, which are diagrams of the orders of genes on chromosomes and the relative distances between them. The frequency with which crossing over occurs between any two linked genes is inferred from the proportion of offspring that are recombinant. Genes that occupy different ends of the same chromosome would have crossovers between them often, generating a large recombinant class. But genes lying very close on the chromosome would rarely be separated by a crossover.

Linkage maps have been used for decades with the organisms that geneticists use to study heredity. Recently, linkage maps have begun to provide valuable medical information for the diagnosis of certain inherited illnesses for which little is known about the disease-causing allele. Researchers can detect a segment of DNA that is closely linked on the chromosome to the disease-causing gene. Because closely linked genes tend to be inherited together, detecting the piece of nearby DNA (the **genetic marker**) signals a high probability that the disease-causing gene has been inherited too. Blood tests can identify genetic markers for a growing number of human diseases, and they are discussed in greater depth in chapter 15.

KEY CONCEPTS

Genes that are carried on the same chromosome are said to be linked, and they are inherited in different patterns than the unlinked genes whose transmission Mendel observed. Crosses involving linked genes produce a large parental class and a small recombinant class, which is caused by crossing over. The farther apart two genes are on a chromosome, the more likely crossing over is to occur between them. The relationship between crossing over and distance is used to construct linkage maps.

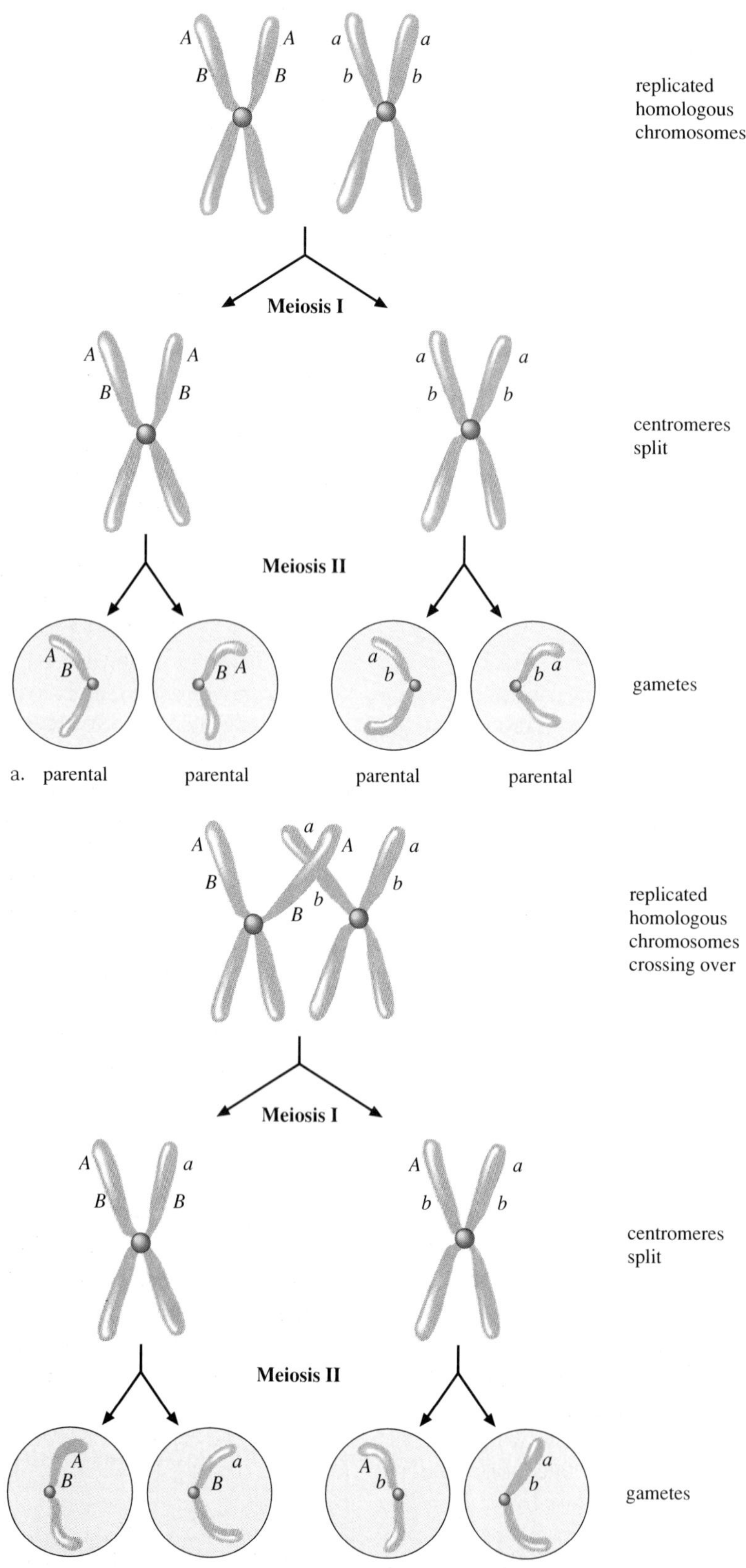

Figure 12.1

a. Genes that are linked closely to one another on the same chromosome are usually inherited together when that chromosome becomes packaged into a gamete. *b.* Linkage between two genes can be interrupted if the chromosome on which they are located crosses over with its homolog at a point between the two genes. Such crossing over packages recombinant arrangements of the genes into gametes.

Matters of Sex

Sex Linkage

Despite this age of sexual equality, it is a genetic fact of life that, in terms of our sex chromosome endowment, the sexes are not created equal. The female of the human species has two large X chromosomes, whereas the male has only one X chromosome plus a small Y chromosome. Each X chromosome has more than 1,000 genes, as do many of the autosomes. The Y chromosome, however, contains only a few genes, and most of these control male sexual characteristics. The X and Y chromosomes pair and separate in meiosis, like homologs, but they do not look alike or carry genes for the same traits.

Because the sexes have different sex chromosome pairs, genes on the X chromosome are inherited somewhat differently in males than they are in females. Any gene on the X chromosome of a male is expressed in his phenotype, because there is no second allele for that gene. An allele on an X chromosome in a female may or may not be expressed (i.e., it may be dominant or recessive) depending upon the nature of the allele for that gene contributed by the second X chromosome. Genes on the X chromosome are said to be **sex-linked,** and the human male is said to be **hemizygous** for sex-linked traits because he has half the number of genes as does the female.

A male always inherits his Y chromosome from his father and his X chromosome from his mother (fig. 12.2). A female inherits one X chromosome from each parent. If a mother is heterozygous for a particular sex-linked gene, her son has a 50% chance of inheriting either allele from her. Sex-linked genes are therefore passed from mother to son. Because a male does not receive an X chromosome from his father (he inherits the Y chromosome from his father), a sex-linked trait is not passed from father to son.

A sex-linked allele causes the most common form of hemophilia, a blood-clotting disorder caused by lack of a specific protein clotting factor. A female "carrier" for the disease has one hemophilia-causing allele, plus an allele that enables her to produce the normal clotting factor. She does not have the disease, but a son who inherits her hemophilia-causing allele suffers from the illness because he does not have a normal allele. Members of the European royal family suffered from hemophilia, beginning with Queen Victoria's children. The transmission of hemophilia in subsequent generations can be followed using Punnett squares and a pedigree chart (fig. 12.3). Hemophilia was actually described long before it appeared among English royalty. The Talmud, a Jewish book of law written around A.D. 500, discusses how the illness appears in a family:

> *If she circumcised her first son and he died, and she had the second son circumcised and he died, she must not circumcise her third son. It once happened with four sisters from Tzippori that the first had her son circumcised and he died, the second sister had her son circumcised and he died, the third sister had her son circumcised and he also died, and the fourth sister came before Rabbi Shimon ben Gamliel and he told her, "You must not circumcise your son."*

Table 12.1 lists other sex-linked human diseases.

KEY CONCEPTS

A human female has two X chromosomes, and males have one X and one Y chromosome. A male is hemizygous, expressing genes on his X chromosome, whereas a female expresses recessive alleles on the X chromosome only if she is homozygous for them. Sex-linked traits are passed from carrier (or affected) mother to son.

Y Linkage

In *Y-linked inheritance*, a trait is passed from father to son on the Y chromosome. This mode of inheritance is not often seen because the Y chromosome carries so few genes. A "hairy ear" trait seen in the men of some families in Ceylon, India, Israel, and aboriginal Australia is apparently passed along the Y chromosome because it never appears in females (fig. 12.4). Perhaps the most celebrated case of Y linkage occurred in three generations of an English family, beginning in 1718 with the birth of a boy

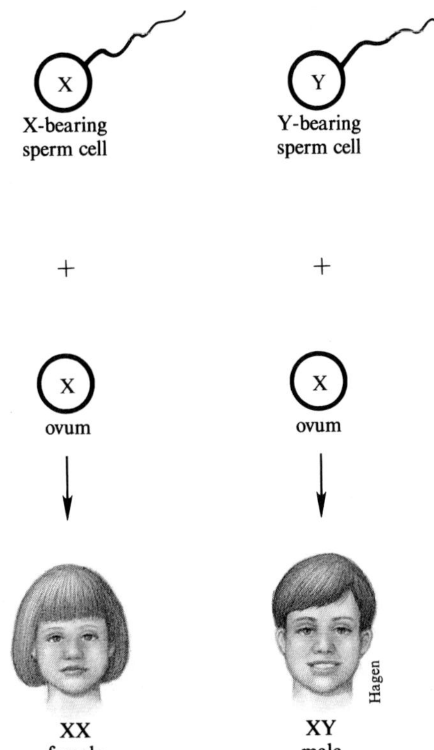

Figure 12.2
How sex is determined in humans. An egg cell typically contains a single X chromosome. A sperm cell contains either an X chromosome or a Y chromosome. If a Y-bearing sperm cell fertilizes an egg cell, the zygote is a male (XY). If an X-bearing sperm cell fertilizes an egg cell, then the zygote is a female (XX).

covered with scaly skin and long bristly hairs. He became known as the "porcupine man," and as an adult he was exhibited in circuses. Some records indicate that several males inherited the condition from him, but never females, suggesting Y linkage. In 1958, however, researchers found church records and correspondence from the family that showed there were indeed affected females, ruling out Y linkage. The family hid its affected female members. Today the porcupine man syndrome, called ichthyosis hystrix, is known to be caused by a dominant allele inherited on an autosome.

X Inactivation—Equaling Out the Sexes

Human females have two alleles for every gene on the X chromosome and males have only one. This inequality in the number of sex-linked genes between the sexes is balanced by a mechanism called **X inactivation,** which operates in all mammals. In females, early in development, one X chromosome in each cell in inactivated. Which X chromosome is turned off—the one she inherited from her mother or the one from her father—is a matter of chance. As a result, a female mammal expresses the X chromosome genes she inherited from her father in some cells and those from her mother in others.

X inactivation means that a female is a genetic mosaic of any genes on the X chromosome for which she is heterozygous, because some cells express one allele, and other cells express the other allele. However, her heterozygosity can still offer her protection from sex-linked disorders. If she inherits one allele that specifies a vital enzyme, and another allele that specifies an inactive version, she will probably still be healthy, because some of her cells manufacture the enzyme. A male who has the defective allele would not survive. Rarely, a female who is heterozygous for a sex-linked gene expresses the associated condition. This can happen in a female carrier of hemophilia. If the X chromosome carrying the normal allele for the clotting factor is, by chance, turned off in many of her immature blood platelet cells, then her blood will take longer than normal to clot—mild symptoms of hemophilia. A female carrier of a sex-linked trait who expresses the phenotype is called a *manifesting heterozygote*.

Once an X chromosome is inactivated in a cell, all of the cells that form when the cell divides have the same X chromosome turned off. Because the inactivation occurs early in development, the adult female has patches of tissue that are phenotypically different with respect to the expression of sex-linked genes. Now that each cell in her body has only one active X chromosome, she is genetically equivalent to the male.

X inactivation is easily observed in female cells. The turned-off X chromosome absorbs a stain much faster than does the active X chromosome. The nucleus of a female cell in interphase has one dark-staining X chromosome, called a **Barr body** after its discoverer Murray Barr, a Canadian

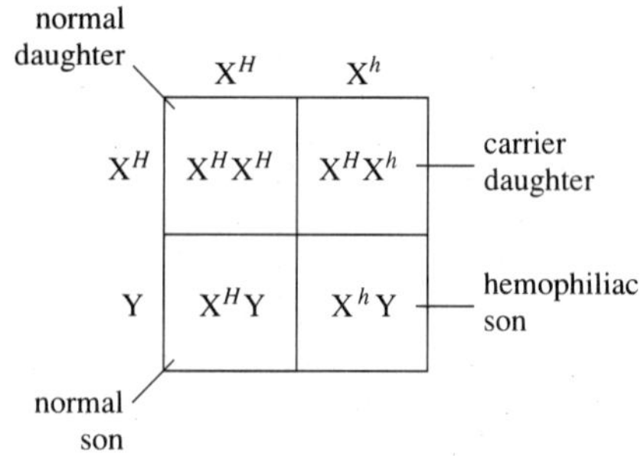

normal daughter
X^H
X^h
X^H
X^HX^H
X^HX^h
carrier daughter
Y
X^HY
X^hY
hemophiliac son
normal son

a.

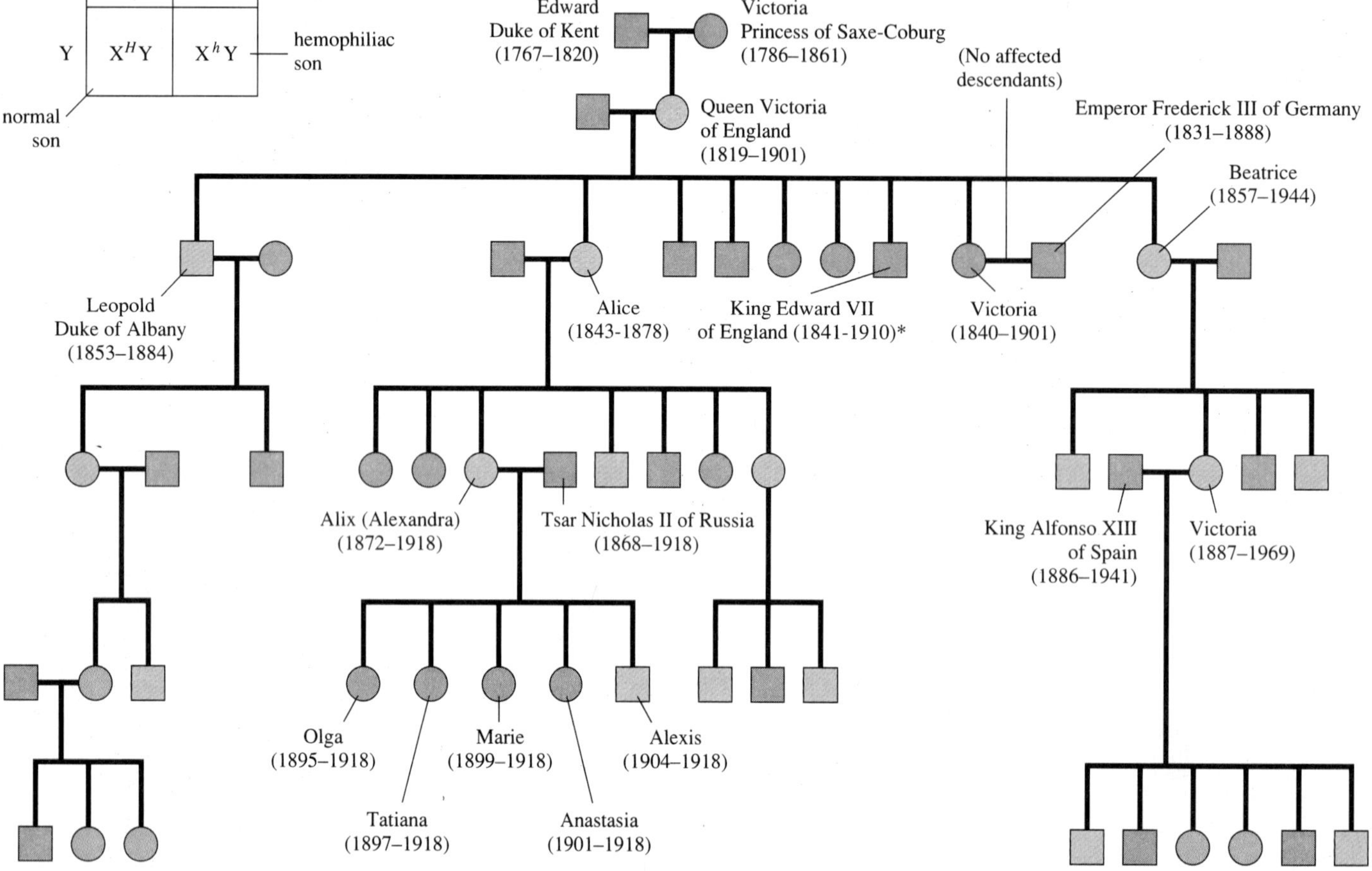

Edward
Duke of Kent
(1767–1820)
Victoria
Princess of Saxe-Coburg
(1786–1861)
(No affected descendants)
Queen Victoria
of England
(1819–1901)
Emperor Frederick III of Germany
(1831–1888)
Beatrice
(1857–1944)
Leopold
Duke of Albany
(1853–1884)
Alice
(1843-1878)
King Edward VII
of England (1841-1910)*
Victoria
(1840–1901)
Alix (Alexandra)
(1872–1918)
Tsar Nicholas II of Russia
(1868–1918)
King Alfonso XIII
of Spain
(1886–1941)
Victoria
(1887–1969)
Olga
(1895–1918)
Marie
(1899–1918)
Alexis
(1904–1918)
Tatiana
(1897–1918)
Anastasia
(1901–1918)

Key: noncarrier female; carrier (heterozygous) female; possible carrier female; normal male; affected male

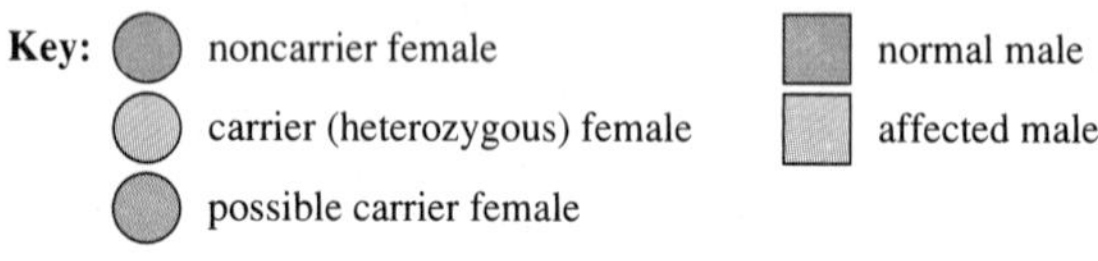

*(descendants include present British royal family)

b.

Figure 12.3

a. The sex-linked form of hemophilia is usually transmitted from a heterozygous woman (designated X^HX^h, where X^h is the hemophilia-causing allele) to heterozygous daughters or hemizygous sons. *b.* Male hemophiliacs generally do not live long enough or feel well enough to father children. The disorder has appeared in the royal families of England, Germany, and Russia. The mutant allele apparently arose in Queen Victoria. She was either a carrier or produced an egg cell that underwent a mutation to carry the allele. Queen Victoria passed the allele to daughters Alice and Beatrice, who were carriers, and to Leopold, who had a case mild enough so that he fathered children. In the third generation, Alexandra was a carrier and married Nicholas II, Tsar of Russia, passing the allele to that family. Also in the third generation, Irene married Prince Henry of Prussia, passing the allele to the German royal family. Note that hemophilia is not present or carried in the modern royal family in England.

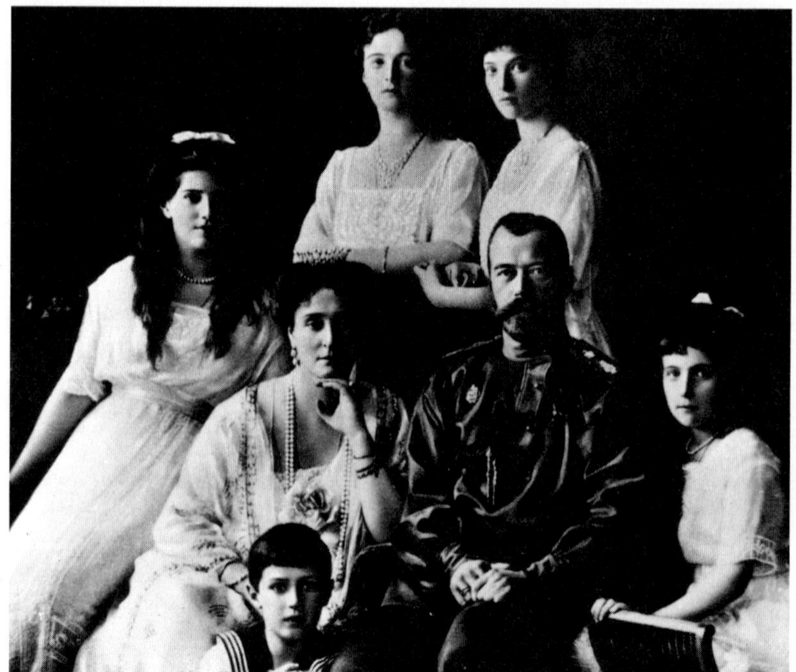

Table 12.1
Some Sex-Linked Human Diseases

Condition	*Description*
Cleft palate	Palate does not close during fetal development as it normally does
Diabetes insipidus	Inability of kidneys to respond to antidiuretic hormone that suppresses urination; autosomally inherited forms and noninherited forms are known
Duchenne muscular dystrophy	Muscle degeneration in childhood
Green color blindness	Absence of green color vision pigment in cone cells of retina
Hemophilia A	Absence or deficient function of blood-clotting factor VIII; blood clots very slowly or not at all
Ichthyosis	Rough, scaly skin on scalp, ears, neck, and front of abdomen and legs
Lesch-Nyhan syndrome	Deficiency of the enzyme HGPRT causes mental retardation, spastic cerebral palsy, urinary stones, and self-mutilative behavior
Red color blindness	Absence of red color vision pigment in cone cells of retina
Testicular feminization	Male embryo does not respond to male hormones, so female phenotype develops, although genotype is male (XY)

researcher who noticed the bodies in 1949 in nerve cells of female cats. A normal male cell has no Barr body because his one X chromosome remains active (fig. 12.5).

In 1961, English geneticist Mary Lyon proposed that the Barr body is the inactivated X chromosome, and that the turning off takes place early in development and is irreversible. She reasoned that for homozygous sex-linked genes, X inactivation would have no functional effect. No matter which X chromosome was turned off, the same allele would be expressed. But for heterozygous genes, X inactivation leads to expression of one allele or the other.

For most sex-linked genes, X inactivation is not obvious because the cells of the body work together, and the overall phenotype reflects the mixture of cell types. (An exception is the manifesting heterozygote.) But one strikingly obvious example of X inactivation is the coat color pattern of the calico cat. The shapes and positions of its characteristic orange and black patches are controlled by two sex-linked alleles, one specifying orange and one black. Cells in which the orange allele is inactivated develop into black patches, and cells in which the black allele is inactivated develop into orange patches. The earlier in development X inactivation occurs, the larger the patches are (fig. 12.6). The only way in which a male calico cat can arise is through a rare meiotic mishap resulting in an animal with the sex chromosome constitution XXY. Only 1 in 10,000 calico cats is a male.

X inactivation has a valuable medical application in detecting carriers for *Lesch-Nyhan syndrome*, a sex-linked disorder in which a child has cerebral palsy, bites his or her fingers and lips to the point of mutilation, is mentally retarded, and passes painful urinary stones. The mutant allele causes a deficiency of an enzyme called HGPRT. Lesch-Nyhan syndrome is so severe that many people who have the disease in their families would like to know the chances of a fetus inheriting the condition. A carrier mother can be detected by testing hairs taken from widely separated parts of her head and testing them for the presence of HGPRT. If some hairs contain the enzyme, but others do not, then the woman is a carrier for Lesch-Nyhan syndrome. The hair cells that lack the enzyme have the X chromosome that carries the normal allele turned off; the hair cells that manufacture the enzyme have the X chromosome that carries the disease-causing allele turned

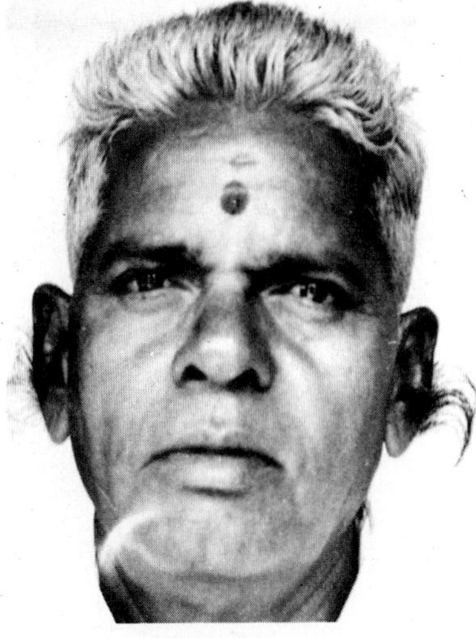

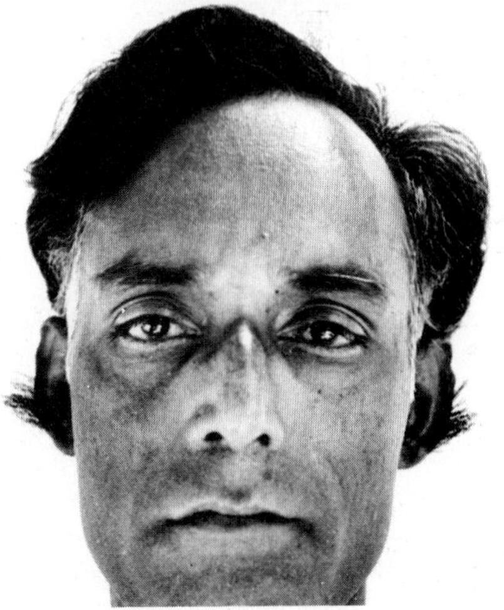

Figure 12.4
"Hairy ears" is an apparently Y-linked trait seen in some families in Ceylon, India, Israel, and aboriginal Australia.

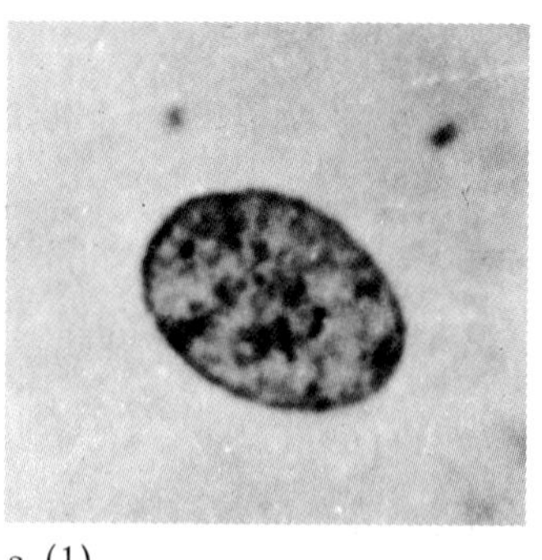
a. (1)

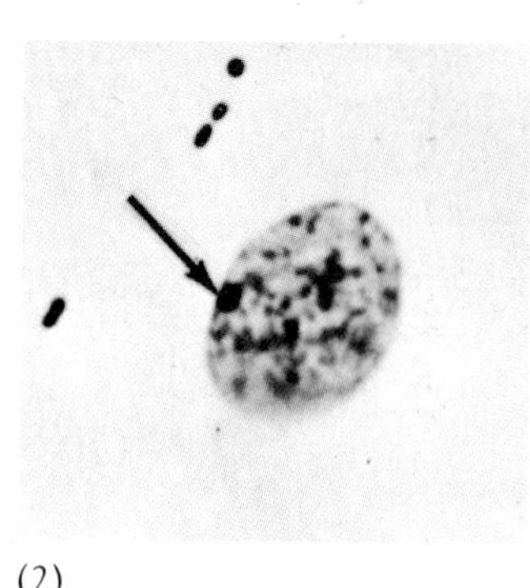
(2)

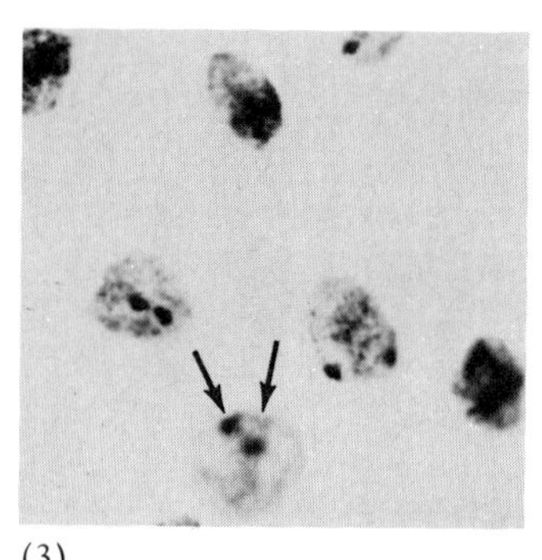
(3)

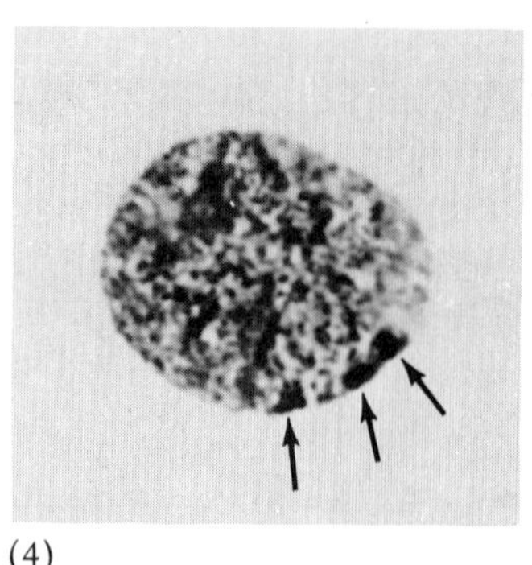
(4)

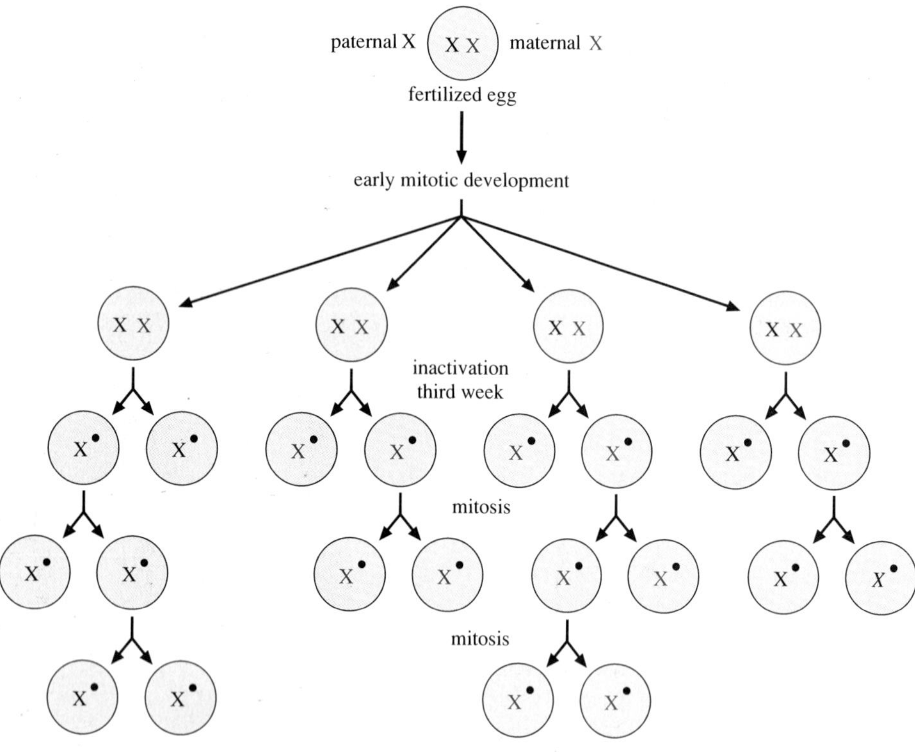

b.

Figure 12.5
a. A Barr body marks an inactivated X chromosome. One X chromosome is inactivated in the cells of a female mammal. The turned-off X chromosome absorbs a stain called Giemsa faster than the active X chromosome, forming a dark spot called the Barr body, which is visible at the edge of the nucleus. A normal male cell has no Barr body (*1*), and a normal female cell has one Barr body (*2*). Disorders in which there are abnormal numbers of sex chromosomes are diagnosed by observing symptoms and the number of Barr bodies in cells. A "superfemale" has two Barr bodies in her cells (*3*) because she has three X chromosomes. A "super-superfemale" has three Barr bodies in each cell (*4*). How many X chromosomes does she have? *b.* At about the third week of embryonic development in the human female, one X chromosome in each diploid cell is inactivated, and all daughter cells of these cells have the same X chromosome turned off.

off. If the woman is a carrier, and she is pregnant with a male fetus, the child has a 50/50 chance of inheriting the disease-causing allele and manifesting the condition.

KEY CONCEPTS

The Y chromosome contains very few genes. Y-linked traits are passed from father to son. In mammals, differences in chromosome constitution of the sexes is evened out by X inactivation. Early in development, one X chromosome in each cell of the female is turned off.

Sex Determination—Male or Female?

Today we know that sex is an inherited trait, determined by the sex chromosomes received at conception. Before this was realized in the early 1900s, interesting theories abounded concerning the age-old question of what makes a baby a boy or a girl.

Some ancient cultures thought sex was determined by the phase of the moon, the direction of the wind, or what the couple wore, ate, or said just prior to conception. The ancient Greeks thought sex was determined by which testicle sperm came from. This idea was perpetuated in some European royalty, who had their left testicles tied off or removed to guarantee conception of a son. Mendel was actually the first to suggest that sex is inherited.

In humans, sex is determined by the type of sperm—that bearing an X chromosome or a Y chromosome—that fertilizes an egg cell, which carries one X chromosome. But sex determination mechanisms are diverse among species. The sex

a. b.

Figure 12.6
A striking exhibition of X inactivation occurs in the calico cat. Each orange patch is made up of cells descended from a cell in which the X chromosome carrying the coat color allele for black was inactivated; each black patch is made of cells descended from a cell in which the X chromosome carrying the orange allele was turned off. *a.* X inactivation happened early in development, resulting in large patches. *b.* X inactivation occurred later in development, producing smaller patches.

Reading 12.1 *Choosing an Organism for Genetic Study*

THE BEAUTY OF THE SCIENCE OF HEREDITY IS THAT ITS PRINCIPLES CAN BE DEMONSTRATED IN ANY ORGANISM THAT CAN BE HANDLED IN A LABORATORY OR OBSERVED IN THE WILD. However, some species are easier to work with than others. A new bacterial generation can be seen every 20 to 30 minutes; in humans the generation time is 20 to 30 years. Obtaining chromosomes from a plant is simpler than obtaining them from a hippopotamus. A fly that produces 50 offspring at a time provides much more statistically significant genetic information than a cow that gives birth to a single calf. An experimental organism must also be amenable to having its environment and even its behavior manipulated by a researcher. Some "model organisms" are commonly found in genetics laboratories, but many other species are used as well.

Much of what we know about the structure and function of genes at the molecular level was worked out using bacteria and their viruses. Bacteria have DNA for their genetic material, but the DNA is not enclosed in a nucleus. The generation time of bacteria is very short, and they can easily be cultured and manipulated. Viruses contain either DNA or the related biochemical RNA as their genetic material. They have genes, but they must use structures found within living cells to replicate themselves. Figure 1 shows an *Escherichia coli* bacterium surrounded by viruses.

Paramecium aurelia is a protozoan that is of interest to geneticists because it can reproduce in several ways (fig. 2). The organism contains three nuclei: two diploid micronuclei and a macronucleus that has many copies of each chromosome. Paramecia can divide asexually, with each micronucleus undergoing mitosis and the macronucleus dividing in two. This happens about every 5 hours. Two paramecia can pair in an event called conjugation, and their micronuclei undergo meiosis, each ultimately producing gametes. A gamete from one organism fuses with a gamete from the other. In yet a third type of reproduction, micronuclei within a single organism undergo meiosis, and the resulting gametes fertilize each other. Geneticists can trace the inheritance of traits carried on genes in either a micronucleus or a macronucleus.

The bread mold *Neurospora crassa* is used to study meiosis, revealing linkage relationships between genes (fig. 3). Two haploid parents are crossed to produce a diploid zygote, which then undergoes meiosis. After the first meiotic division, the original diploid cell yields two haploid cells that are physically attached. After the second meiotic division, these two cells divide to yield four cells, but they are still attached, representing the order in which they formed. Next, the four meiotic products undergo a mitotic division, but they remain attached, forming eight cells called ascospores that are enclosed in a sac called an ascus. By following genes that produce different- colored ascospores, geneticists can determine when crossing over has occurred. Those asci with four dark ascospores in a row and four light ascospores in a row are parental types because the parents were dark and light. The ascospores that alternate two dark and then two light ascospores indicate that crossing over has occurred. By conducting many crosses, geneticists can count ascospores as parental or recombinant, just as they can the offspring of a cross of diploid organisms, and deduce the order and distances between genes linked on a chromosome. The life cycle of *Neurospora crassa* takes about 2 weeks.

The 10 pairs of chromosomes in corn (*Zea mays*) are easily observed in each stage of meiosis. Many traits of corn have been well studied, they are generally easy to observe, and pollinations are simple to make. The plant can be bred either in the field or in a greenhouse. The major drawback to working with corn is its relatively long generation time: the life cycle is 4 months.

The fruit fly *Drosophila melanogaster* is easy to raise in a laboratory and hundreds of its traits have been cataloged and mapped to one of its four chromosomes (fig. 4). The generation time is 2 weeks. Large chromosomes for study are obtained from cells of the salivary glands, in which an unusual form of mitosis occurs. The genetic material here replicates many times, but the cell does not divide. Salivary gland cells contain 1,024 copies of each chromosome, perfectly aligned gene by gene. The glands are dissected out, placed on a microscope slide, and a drop of stain added. A coverslip is placed on the material, and it is "squashed" by the investigator's pressing down with a thumb. The squashing action spreads the chromosomes out, and under a microscope, they appear intricately banded. The bands represent genes (fig. 5). For example, a fly missing several genes will also be missing certain bands.

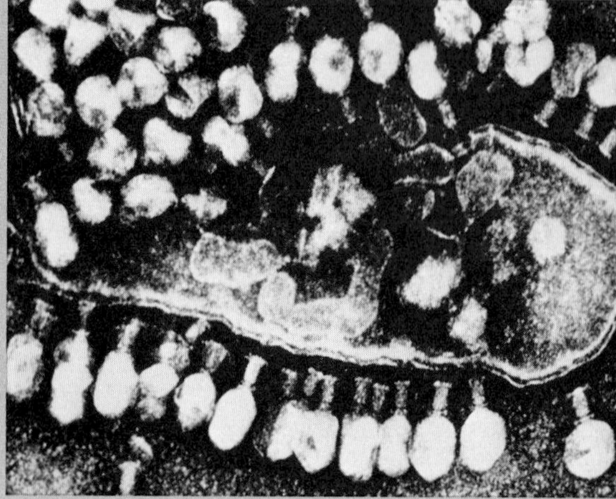

Figure 1
Bacteriophage adhere to an *E. coli* cell.

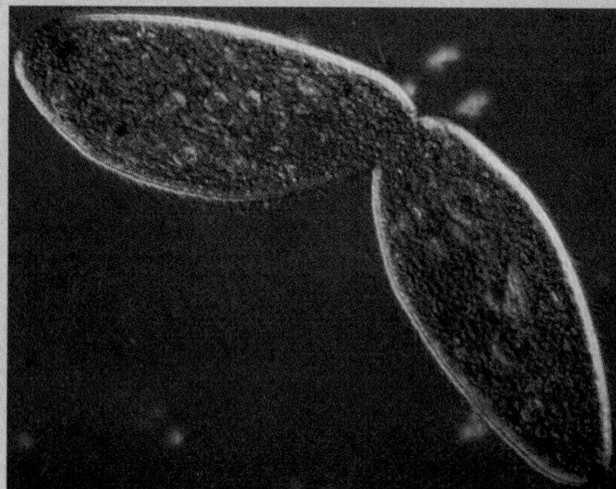

Figure 2
Paramecium aurelia reproducing.

determination in grasshoppers, for example, is termed XO. The insect has only one type of sex chromosome, designated X. If two sex chromosomes are present (XX), it is a female; if only one is present (XO), it is a male.

Sex determination of humans is XY, in which maleness depends upon the presence of a Y chromosome, and femaleness on the absence of a Y chromosome. The human female is called the **homogametic sex** (gamete same), because she has two copies of a single type of sex chromosome. The male is the **heterogametic sex** (gamete different), because he has two different sex chromosomes.

The human XY mechanism of sex determination was deduced from individuals with abnormal sex chromosome constitutions. A person whose cells have two X chromosomes and one Y chromosome is a male with *Klinefelter syndrome*. He has small testes, does not manufacture sperm, excretes sex hormones in excessive amounts, and has underdeveloped secondary sexual characteristics. A person whose cells have a single X chromosome and no Y chromosome is a female with *Turner syndrome*. She does not mature sexually, is short in stature, and has a characteristic flap of skin on the back of her neck. Sex chromosome anomalies are discussed in greater detail in chapter 14.

The fruit fly *Drosophila melanogaster* exhibits yet another type of sex determination, in which the X:autosome ratio is important. Individuals with a pair of X chromosomes and a pair of each autosome are normal females; those with one X chromosome and a pair of each autosome are normal males. Although fertile male fruit flies are XY, it is not the presence of the Y chromosome that determines maleness but the ratio of only one X chromosome to the paired autosomes. An XO fruit fly is a sterile male (recall that in humans an XO is female) and an XXY fruit fly is female (in humans XXY is male). X:autosome sex determination is also seen in species of higher plants in which individuals have either male or female flowers, such as willows and the American holly.

Birds, butterflies, moths, and some fish have two types of sex chromosomes, but the homogametic sex is the male, and the heterogametic sex is the female (opposite humans). Sex determination is more

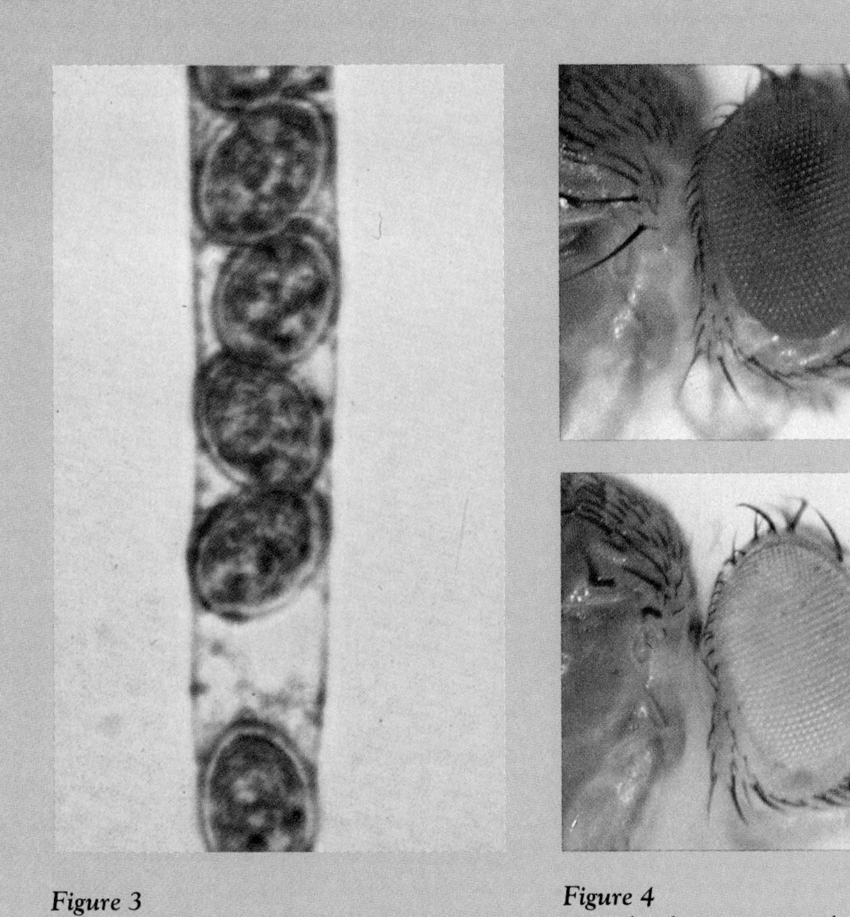

Figure 3
An ascus containing ascopores in the bread mold *Neurospora crassa*.

Figure 4
Normal and a mutant eye color in the fruit fly *Drosophila melanogaster*.

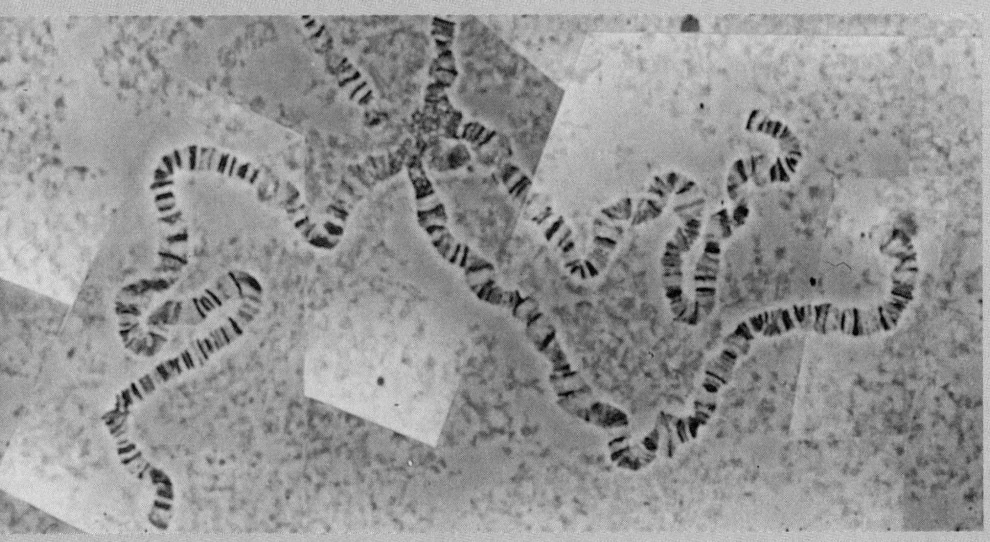

Figure 5
Stained and squashed chromosomes of the fruit fly coalesce at their centromeres (characteristically located constrictions in each chromosome).

complex in some other species. Certain beetles have 12 X chromosomes, 6 Y chromosomes, and 18 autosomes. In other species, the number of sets of chromosomes is the deciding factor in whether an individual develops as a male or a female. In bees, a fertilized egg becomes a diploid female; an unfertilized egg develops into a haploid male. The queen bee actually determines sex. The eggs that she fertilizes become females, and those that she does not become males.

Autosomal genes can influence sex determination. In fruit flies, a mutation called "transformer" gives an XX individual a male phenotype. Different alleles can also determine sex. In the bread mold *Neurospora crassa,* the two sexes are called mating types, and each is specified by a different allele of a particular gene. Two individuals mate only if they have different alleles for this gene. In one species of wasp, a gene with nine alleles controls sex determination. All heterozygotes are female, and all homozygotes are male.

In some species, the environment determines sexual fate. For some turtles, sex is determined by the temperature of the land on which the eggs are laid. In the sea-dwelling worm *Bonellia viridis,* larvae mature into females if no adult females are present. In the presence of an adult female, however, the same larvae would develop as males in response to hormones secreted by the female. Some marine bivalves and fish can even change their sex as adults (fig. 12.7).

KEY CONCEPTS

Sex is determined in various different ways in different species. Humans have XY sex determination, in which maleness depends upon the presence of a Y chromosome, and femaleness upon its absence. Fruit flies have X:autosome sex determination, in which a female has a pair of X chromosomes and a male has one X chromosome. Sex determination in birds, butterflies, and moths is opposite that in humans. Beetles have several X and Y chromosomes, and in bees, differences in the number of sets of chromosomes determines sex. The bread mold has mating types, and for turtles environmental temperature influences sex. For some marine organisms, sex is determined by how many of each sex are present in a population.

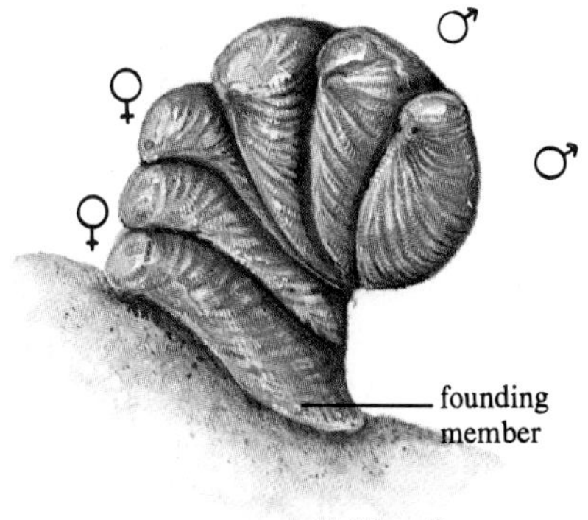

Figure 12.7
Sex reversal in the slipper limpet. To the slipper limpet *Crepidula fornicata,* position is literally everything in life. These organisms aggregate in groups anchored to rocks and shells on an otherwise muddy sea bottom. The founding member of a group is a small male. As he grows, he becomes a she, and a new, small male comes in at the top. When this second member grows large enough to become female, a third recruit enters. Thus, one's size and position in the pile determines sex. The slipper limpet shown here is in the intermediate stage.

Sex Ratio

According to Mendel's law of segregation, a male produces equal numbers of sperm cells carrying an X chromosome and a Y chromosome. This suggests that the numbers of males and females in a population should be about equal. In actuality, the proportion of males to females, called the **sex ratio,** differs throughout the human life cycle.

At the beginning, males clearly outnumber females, with 120 to 150 males conceived for every 100 females. This is called the *primary sex ratio.* Males outnumber females at conception because the lighter-weight Y-bearing sperm cells can travel faster in the female reproductive tract, therefore reaching the egg cell faster, than the X-bearing sperm cells. The *secondary sex ratio* compares numbers of males to females at birth and is 106 males to 100 females (fig. 12.8). What has happened to the "excess" male zygotes? They may have died from the action of lethal alleles inherited along with their single X chromosome. This may be why two to four times as many spontaneous abortions are male than females.

After birth, the *tertiary sex ratio* is measured at decade-long intervals. During the teen years, the sex ratio begins to approach equality. The slight decline in the number of males from birth to adulthood may represent sex-linked disorders that cause death during childhood and adolescence. There are about equal numbers of males and females between the ages of 20 and 40 years. For individuals in their 50s, the proportion of males gradually declines, so that by the 10th decade, females outnumber males five to one. One reason for the preponderance of females in the later years of life is that the so-called weaker sex is actually the healthier sex. Of 64 causes of death studied by the United States Census Bureau, 57 affected more males at any age than females. Five of the seven remaining conditions affect female reproductive structures.

The average shorter life span of males may in part be caused by life-style differences between the sexes in the United States. Males have traditionally been sent to the front lines in wars. More men work outside the home than do women and tend to occupy different sorts of jobs than do women. These occupational differences may introduce different stresses into their lives, which may ultimately affect health. It will be interesting to see if and how the sex ratio in the later years of life changes as more women enter careers held mostly by men in the past.

KEY CONCEPTS

The primary sex ratio in humans is 120 to 150 males for every 100 females at conception. By birth, the secondary sex ratio is 106 males to 100 females. After birth, the tertiary sex ratio gradually favors females as the population ages.

Sex Preselection

Two pregnant women meet at the water fountain in their office.

"Hi! I hear Stephen is going to have a sister!" says one to the other.

"Yes, and I understand you're painting the nursery blue," answers her coworker.

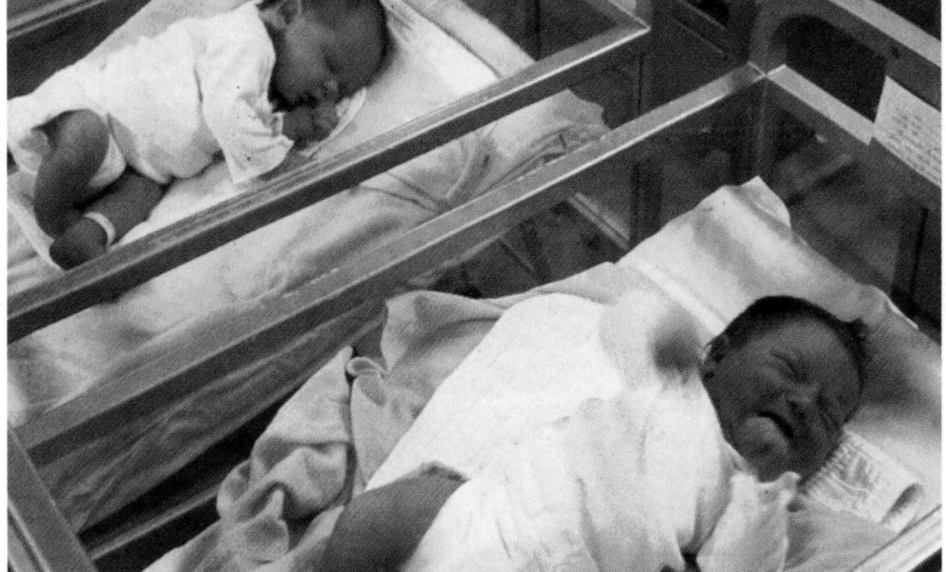

a.

b.

c.

Figure 12.8
The ratio of males to females differs for people of particular ages. There are 120 to 150 males conceived for every 100 females conceived, but at birth (*a*), there are only slightly more males than females. By adolescence (*b*), the numbers of each sex are about equal, but by the later years (*c*), females outnumber males.

What these women take for granted—knowledge of the genders of their babies-to-be—was an accepted mystery of pregnancy just a few years ago. Each of these women had undergone a procedure called chorionic villus sampling to see if certain disorders were present, and in addition to receiving clean bills of health for their babies, they learned their infants' sexes.

Although many couples learn the sexes of their children before birth as part of a prenatal test, others seek to influence the determination of sex. One way to do this is to alter behavior to enhance the probability that one sperm cell type (Y-bearing or X-bearing) reaches the egg cell before the other. Some couples attempt to do this by carefully scheduling the time of intercourse. If intercourse occurs within a few hours of ovulation, more of the lighter-weight Y-bearing sperm will reach the egg cell than X-bearing sperm cells. If intercourse takes place a day or two before ovulation, many of the Y-bearing sperm cells will already have died by the time the egg cell is released, and an X-bearing sperm is more likely to reach it, resulting in a female conception.

However, what works in theory does not always work in practice. It is difficult to know when ovulation occurs, and at any given time in the 48 hours following intercourse, both types of sperm are present in the female reproductive tract. Other factors influence the abilities of the different types of sperm cells to reach the egg cell. Y-bearing sperm cells swim faster in an alkaline environment, such as is present in the female reproductive tract at about the time of ovulation. X-bearing sperm "prefer" an acidic environment. Some women use an alkaline douche (e.g., a baking soda solution) to encourage conception of a boy or an acidic douche (a dilute vinegar solution) to try for a girl.

Timing of intercourse and altering the pH level in the vagina are not very reliable methods of determining the sex of a child. For people who want greater control over this process, sperm can be donated by the male and separated in a laboratory into X-bearing and Y-bearing enriched fractions. The two types of sperm cells are separated by a variety of physical techniques based upon their differing weights and densities. But this is a very involved and costly process—and hardly romantic. One couple, who had two sons and wanted a daughter, went to a "fertility institute" to have sperm separated and the woman artificially inseminated. The mother described the experience:

> *I had to keep track of my temperature using a basal thermometer for 3 months. They did an ultrasound, and looked at the eggs to see if I was ovulating or not. They took temperature readings, checked the cervical mucus in an internal exam, and determined the exact time of ovulation. They took my husband's sperm and put it in the special medium in which the X-bearing sperm stick to the bottom. A half hour later, I was inseminated.*

Nine months afterwards, a girl was born. Sperm separation has an 80% to 85% success rate in giving couples a child of the sex of their choice.

More controversial than attempting to manipulate conception is to induce abortion of an embryo or fetus because its sex is not what is desired. Abortion based upon sex can be medically justified if the mother carries a severe sex-linked disease, and the fetus is known (by chromosome analysis or Barr body identification) to be male.

In England, where manipulation of human pre-embryos is permitted, a stain that highlights the Y chromosome can identify males as early as the 8-stage cell. A woman who had several children with the devastating Lesch-Nyhan syndrome used this technique to select female zygotes growing in the laboratory following in vitro fertizatin for further development in her uterus.

Some people choose abortion solely to contol the sexes of their children. A study conducted in China in 1979 focused on 100 women who had the sexes of their embryos identified at 8 weeks of development. Twenty-nine women carrying girls chose abortion, whereas only one woman carrying a boy chose it.

What would be the effects on our society if sex selection, by whatever means, was practiced by everyone? Would it drastically alter the numbers of boys and girls—and a generation later, of men and women? Sociologists Charles F. Westoff and Ronald R. Rindfuss addressed this compelling question in 1974, when they worked at the Office of Population Research at Princeton University. Westoff and Rindfuss found that in a large sample of women who were not yet mothers, 63% preferred to have a boy. Based on surveys, the investigators estimated that if sex selection was available to couples desiring two children, the sex ratio would remain its natural 106 boys to every 100 girls born because most people would desire one child of each sex. (Interestingly, most people preferred the eldest child to be male.) But in one-child families, boys would be favored. Would this lead to a country overrun with males? It might for a few years, but probably not in the long run, according to gynecologist Joe Leigh Simpson of the Northwestern University Medical School. Like any rare commodity, as girls became fewer, their popularity would rise. Dr. Simpson wrote in an editorial in the *New England Journal of Medicine*, "Common sense and biologic selection would eventually regain what nature has evolved—a sex ratio nearly equal to one."

KEY CONCEPTS

People attempt to influence the sex of their children by having intercourse at certain times, by douching to alter the pH of the vagina, or by having sperm separated into X-bearing or Y-bearing enriched fractions, followed by artificial insemination. Some people use information from a prenatal test identifying sex to abort fetuses of the unwanted sex. If sex preselection was practiced on a large scale, the sex ratio would eventually revert to the natural near 1:1 ratio.

The Changing Science of Genetics

The study of inheritance had its beginnings in the field, the barnyard, the garden, and the research laboratory. While today new knowledge of hereditary mechanisms continues to come from these sources, genetics is beginning to affect us on a more personal level. Genetic information guides many of our choices about life-style and the bearing and rearing of children in particular. Human health care and nutrition are already being bettered by genetic engineering, in which the genes of microorganisms and cells growing in culture are manipulated to produce proteins of use in the pharmaceutical and agricultural industries. Many formerly incurable illnesses are becoming treatable thanks to techniques that probe the inner workings of Mendel's "characters," the fascinating biochemicals that we call genes. In the next three chapters, we will explore the molecular nature of inheritance and the impact of genetics on our lives.

SUMMARY

Inheritance patterns for genes that are linked on the same chromosome depart from Mendelian ratios because the genes do not assort independently. *Sex-linked* genes are inherited differently in males and in females because the male does not have alleles on a second X chromosome to interact with the corresponding alleles on his X chromosome. A male receives an X chromosome from his mother and a Y chromosome from his father. A female receives one X chromosome from each parent.

Sex-linked diseases are passed from mother to son on an X chromosome. Inequality of sex chromosome constitution is balanced by X *inactivation* in which, at an early stage of development in female mammals, one X chromosome in each cell is turned off. In humans sex is determined by the presence or absence of a Y chromosome, but different mechanisms of sex determination operate in different species.

The ratio of males to females, called the *sex ratio*, is 120 to 150 males for every 100 females at conception, 106 males to 100 females at birth, and gradually reaches equality by the end of adolescence. Starting in the fifth decade of life, females increasingly outnumber males. Medical technology is allowing people to preselect the sex of their children.

QUESTIONS

1. How are genes that are part of the same chromosome inherited differently than genes that are located on different chromosomes?

2. How are genes that are located on the sex chromosomes inherited differently in male and female humans?

3. What are some of the ways in which sex is determined in different organisms?

4. A normal-sighted woman whose mother has normal vision but whose father is color blind marries a color-blind man. The trait is sex-linked. What are the chances that a son will be color blind? A daughter?

5. The tribble is a cute, furry creature ideal for embryonic experimentation. Cells from tribble embryos genetically destined to become brightly colored adults can be transplanted into tribble embryos that would otherwise develop into albino (pure white) adults. The numbers and positions of cells descended from the original transplanted cells can be followed by observing the pattern of colored patches in the adult tribble.

A gene in the tribble confers coat color. The brown allele *B* is dominant to the orange allele *b*. A single cell from a female embryo at the 8-cell stage with genotype *Bb* is transplanted into an albino embryo. The resulting adult tribble has both brown and orange patches on an albino background. But when a single cell from an embryo at the 64-cell stage is transplanted into an albino embryo, only orange patches, or only brown patches, appear in the adult.

State the genetic principle demonstrated here. What is the difference between the two experiments?

6. Explain why an XO human is female but an XO fruit fly is male. Explain why an XXY human is male but an XXY fruit fly is female.

TO THINK ABOUT

1. Why doesn't the inheritance pattern of linked genes disprove Mendel's laws?

2. Offer a genetic explanation for why blond hair/blue eyes and black hair/brown eyes are more common phenotypic combinations than blond hair/brown eyes and black hair/blue eyes.

3. A healthy man and woman have a daughter who has hemophilia. The distraught husband, claiming that hemophilia is inherited as a sex-linked recessive trait, sues his wife for divorce on grounds of adultery. List three ways in which the husband's charge could be scientifically incorrect, using information from this and the preceding chapter.

4. Suggest a genetic explanation for why sex-linked hemophilia is rare in females.

5. Why are male calico cats rare?

6. Why are humans poorly suited organisms for genetic experimentation? If you were to visit another planet to study the genetics of its life forms, what characteristics would you look for in an experimental organism?

7. Do you think the technology to alter sex determination in humans should be developed? If so, under what circumstances?

8. Do you think that the changing role of women in our society could affect the male-to-female sex ratio, which currently declines with age? If so, in what way?

SUGGESTED READINGS

Gould, Stephen Jay. August 1983. Sex and size. *Natural History*. A look at some interesting variations on the sex determination theme.

Lewis, Ricki. January 1987. Your next child will be a (boy) (girl). *Health*. Medical technology is now allowing some couples to influence the sex of their children.

Westoff, Charles F., and Ronald R. Rindfuss. May 10, 1974. Sex preselection in the United States: Some implications. *Science*, vol. 184. In this oft-quoted paper, two sociologists look at the implications of choosing the sex of children.

C H A P T E R

13

Molecular Genetics

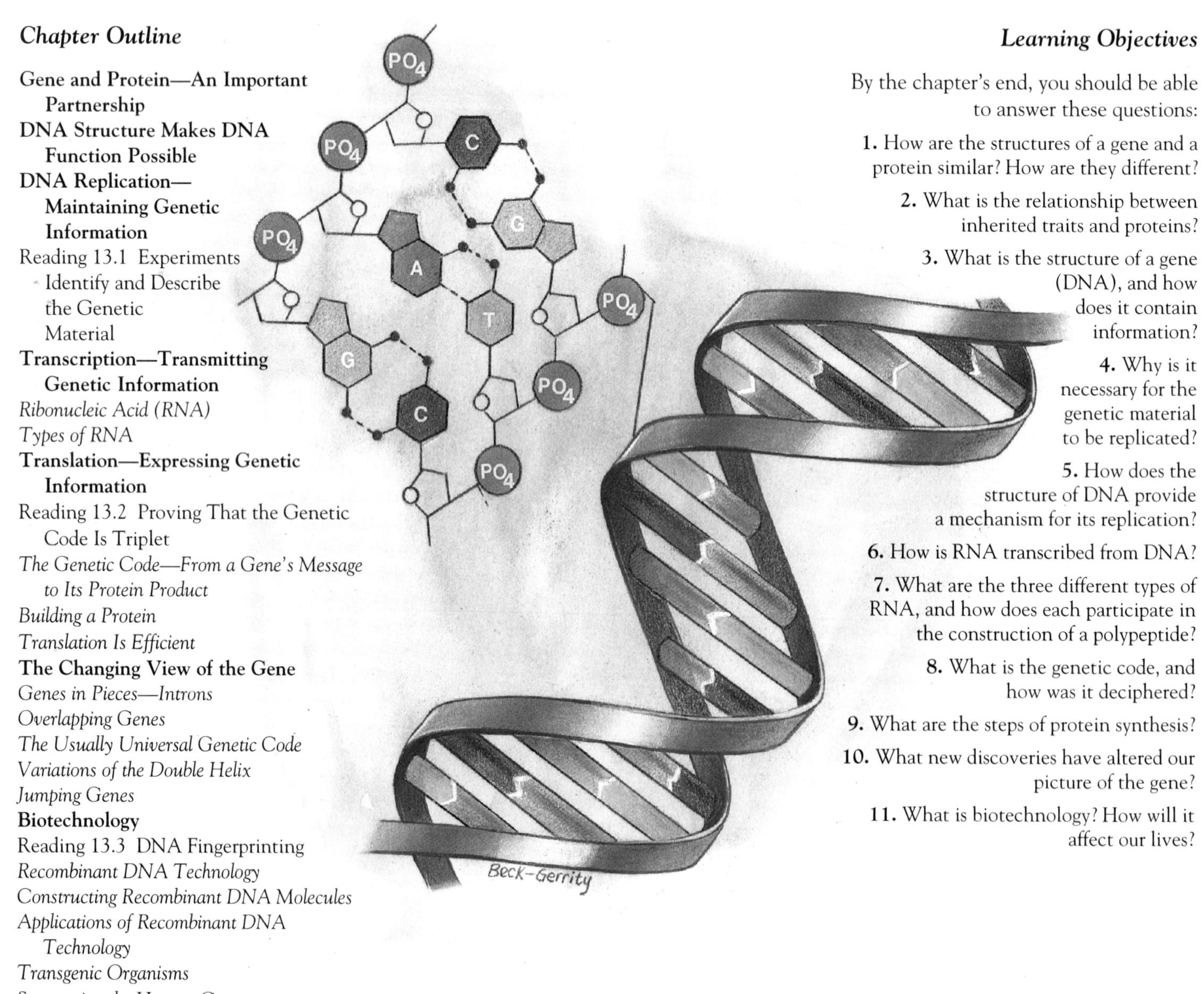

Chapter Outline

Learning Objectives

By the chapter's end, you should be able to answer these questions:

1. How are the structures of a gene and a protein similar? How are they different?

2. What is the relationship between inherited traits and proteins?

3. What is the structure of a gene (DNA), and how does it contain information?

4. Why is it necessary for the genetic material to be replicated?

5. How does the structure of DNA provide a mechanism for its replication?

6. How is RNA transcribed from DNA?

7. What are the three different types of RNA, and how does each participate in the construction of a polypeptide?

8. What is the genetic code, and how was it deciphered?

9. What are the steps of protein synthesis?

10. What new discoveries have altered our picture of the gene?

11. What is biotechnology? How will it affect our lives?

Before 1982, the insulin injected daily by 2 million of the nation's diabetics came from pancreases removed from cattle in slaughterhouses. Cattle insulin is so similar to the human variety, differing in only 2 of its 51 amino acids, that most diabetics get along just fine using it. However, about 1 in 20 diabetics is allergic to cow insulin because of the slight chemical difference. Until recently, the allergic patients had to use expensive combinations of insulin from a variety of other animals or from human cadavers.

In 1978, a way to obtain large amounts of pure human insulin became available because of a new process, recombinant DNA technology. The gene that instructs certain pancreas cells to produce insulin was snipped out of its chromosome and inserted into the genetic material of an *E. coli* bacterium. When the bacterium reproduced, the human insulin gene was duplicated along with its own genetic material. When the bacterial genes directed the synthesis of *E. coli* proteins, the transplanted human gene directed the manufacture of its protein product, human insulin. Today, *E. coli* bacteria engineered to manufacture human insulin are grown in huge vats at a major pharmaceutical company. The diabetic can now purchase genetically engineered human insulin at the local drugstore—and because it is the human variety, allergic reactions do not occur.

Improved insulin for diabetics is possible because of our understanding of the structure and function of the gene. From this knowledge has grown our increasing ability to manipulate genes; insulin is only the first of several "genetically engineered" drugs. Handling genetic material has spawned much controversy but has also, as the diabetic can attest, already improved the quality of life.

Much research and insight has built the science of genetics, from the rediscovery of Mendel's laws at the beginning of this century to the development of the genetic technology that is making practical inroads into human health care, veterinary science, agriculture, and industry. This chapter complements the preceding two by considering the molecular structure and function of Mendel's characters—genes—and leads directly into the next chapters, which examine the impact of genetics on human health (fig. 13.1).

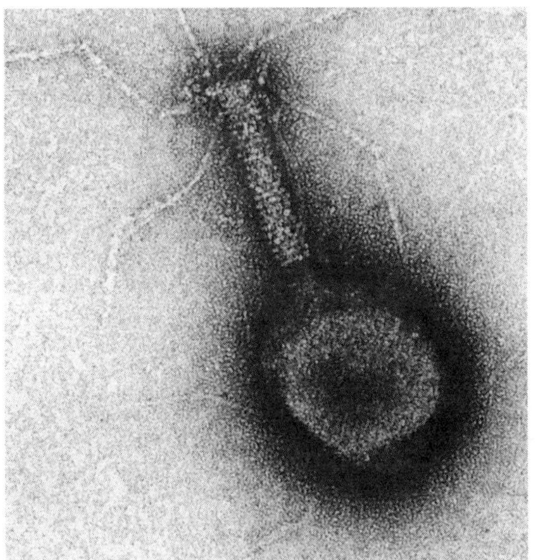

a.

b.

c.

Figure 13.1

The genetic material. Gregor Mendel perceptively described the transmission of "characters," but he did not know their chemical composition. Today we can actually see the genetic material—with the aid of powerful microscopes, of course. *a.* This T4 virus has burst its protein "head," releasing its genetic material, DNA. *b.* The tangled mass is the single DNA double helix of an *E. coli* bacterium, worn by this book's author. *c.* DNA structure is revealed in the scanning tunneling microscope. This new type of microscope allows complex molecules to be viewed in the conformations they assume when surrounded by water—just as they are in an organism. This is the DNA double helix, first imaged this way in 1987.

Gene and Protein—An Important Partnership

A **gene** is a long DNA molecule that specifies the production of another type of long molecule, a protein. The activity of the protein is responsible for the trait that is associated with the gene. More specifically, the information contained in the sequence of the **nucleotide** building blocks of DNA is used to specify the sequence of amino acid building blocks of a polypeptide chain. Proteins are built of one or more polypeptides. Both nucleic acids and proteins are polymers—that is, they are strings of building blocks. Nucleic acids have 4 varieties of nucleotides, and proteins have 20 varieties of amino acids. Different orderings of these building blocks in each molecule result in different genes and proteins. In contrast, a polymer of identical building blocks, such as the starch glycogen (in which all of the components are molecules of the sugar glucose), cannot encode any information.

The connection between inherited traits and the molecular structure and function of the gene is not always clear, but it probably lies in the many activities of proteins in living systems. Pea color may come from pigment proteins; plant height may be controlled by a protein hormone. Earlier chapters considered some of the varied roles of proteins in the human body. Enzymes catalyze the chemical reactions of metabolism; proteins such as collagen and elastin provide structural support in connective tissues; tubulin builds the cytoskeleton; hemoglobin transports oxygen in the blood; antibodies furnish immune protection. Malfunctioning or inactive proteins, reflecting genetic defects, can be devastating to health. Chapter 14 examines a class of inherited human disorders called inborn errors of metabolism, which are caused by specific inactive enzymes that lead to dangerous buildups of other biochemicals.

The amino acids that genes assemble into proteins ultimately come from the diet. Animals obtain proteins from foods and then digest them into the component amino acids, which are carried in the blood and enter cells, where they are built into new proteins. For example, the hemoglobin that you eat in a rare steak is degraded

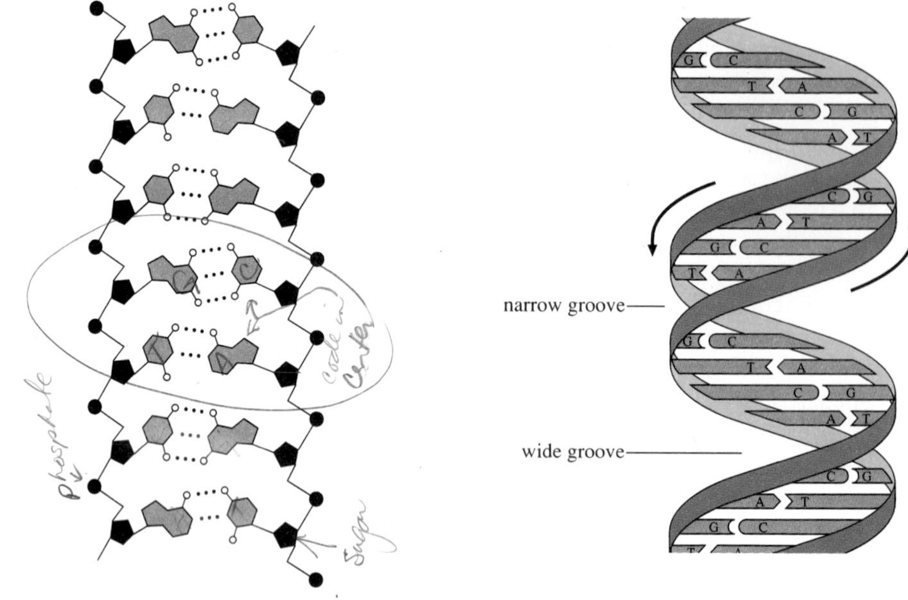

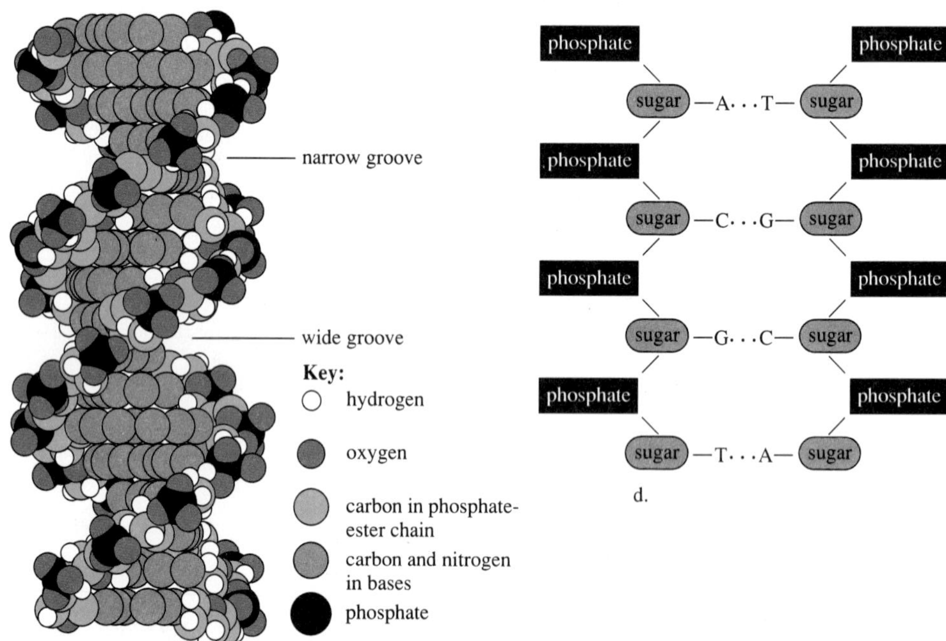

Figure 13.2

Different ways to represent the DNA double helix. *a*. The helix is unwound to show the base pairs in color and the sugar-phosphate backbone in black. *b*. The sugar-phosphate backbone is emphasized and the fact that the two strands run in opposite directions. *c*. This representation shows the relationships of all of the atoms. *d*. A schematic representation of an unwound section of the double helix outlines the relationship of the sugar-phosphate rails to the base pairs. The informational content of DNA lies in the sequence of bases. The sugar-phosphate rails are identical in all DNA molecules.

into its constituent amino acids by your digestive system and then built up into the amino acid sequence of a protein, possibly even the human version of hemoglobin, whose amino acid sequence is slightly different from that of the hemoglobin in the steak.

KEY CONCEPTS

A gene is a sequence of nucleotide building blocks that specifies a sequence of amino acids in a polypeptide. The function of the resulting protein in an organism provides an inherited trait.

DNA Structure Makes DNA Function Possible

DNA must transmit the information contained in its building block sequence in order to direct the synthesis of proteins. To do this, the information in DNA is copied into an intermediate form, the related polymer **ribonucleic acid (RNA).** This leaves the genetic material available so that protein synthesis can occur repeatedly and at a rate fast enough to support the activities of the cell. DNA must also be maintained so that its information can be transmitted to daughter cells through mitosis or meiosis. The structure of the DNA molecule makes possible both of these functions—maintaining and transmitting information.

The genetic material is actually two long molecules of DNA that are entwined about each other to form a double helix, resembling a twisted ladder (fig. 13.2). The rungs of the ladder are pairs of *nitrogenous* (nitrogen-containing) *bases*. DNA has four different types of nitrogenous bases. The ladder's rails consist of alternating units of *deoxyribose* (a five-carbon sugar) and *phosphate* (PO_4). Each DNA molecule makes up half of the ladder, consisting of a sugar-phosphate rail and many half-rungs (each one a single nitrogenous base). A single building block of DNA, a nucleotide, consists of one deoxyribose sugar, one phosphate, and one nitrogenous base.

The key to DNA's function as an information molecule lies in the sequence of nitrogenous bases. The four types are **adenine** (A), **guanine** (G), **cytosine** (C), and **thymine** (T). Adenine and guanine are **purines,** which have a double organic ring structure, and cytosine and thymine are **pyrimidines,** which have a single organic ring (fig. 13.3). The sleek, symmetric double helix of DNA is formed when nucleotides containing adenine pair with those containing thymine, and nucleotides containing guanine pair with those containing cytosine. These specific purine-pyrimidine couples, called **complementary** base pairs, are held together by hydrogen bonds (fig. 13.4). The discovery of the base-pairing relationships—A with T and G with C—was a major clue that led to the deciphering of DNA's structure in 1953 (Reading 13.1).

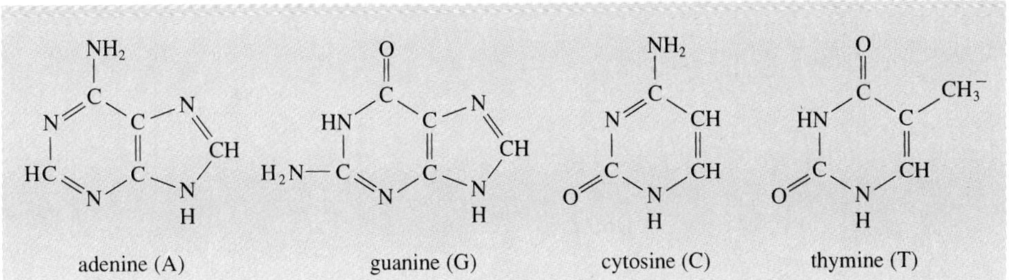

Figure 13.3
DNA bases. Adenine and guanine are purines, composed of a six-membered organic ring and a five-membered ring. Cytosine and thymine are pyrimidines, built of a single six-membered ring.

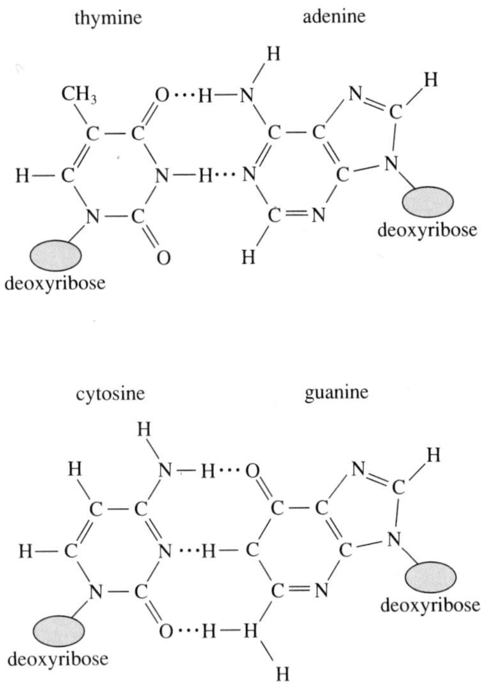

Figure 13.4
DNA base pairs. The key to the constant width of the DNA double helix is the pairing of purines with pyrimidines. Specifically, adenine pairs with thymine with two hydrogen bonds and cytosine pairs with guanine with three hydrogen bonds.

The DNA double helix fits within a cell, or within a eukaryotic cell's nucleus, because it is highly coiled (fig. 13.5). The single "chromosome" of DNA in the bacterium *E. coli*, for example, is 10 million base pairs long. If the bases were typed as A, C, T, and G, the genetic material of *E. coli* would fill 2,000 pages of 5,000 letters each. The human **genome**—all of a cell's genetic material—is about 3 billion base pairs long and would fill the equivalent of 4,000 books of 500 pages each!

KEY CONCEPTS

The structure of DNA enables it to remain in the nucleus, be replicated when the cell divides, and be copied into an intermediate molecule, RNA, which goes to the cytoplasm, where it carries out protein synthesis. The DNA double helix consists of a backbone of alternating deoxyribose and phosphate groups, with "rungs" formed by complementary pairs of the bases adenine with thymine and guanine with cytosine. Adenine and guanine are purines; thymine and cytosine are pyrimidines.

DNA Replication—Maintaining Genetic Information

New DNA molecules are constructed, or replicated, from nucleotides that float freely in the cell. These nucleotides are synthesized from precursor molecules available from digestion and metabolism. DNA

replication is the mechanism behind the chromosome duplication of mitosis. Because replication faithfully copies a DNA molecule's sequence, the process ensures that genetic information is passed to the next cell generation.

When DNA replicates, the two strands of the double helix twist apart. Free-floating nucleotides bond with their complementary bases in the exposed part of the double helix. Thus, DNA replication follows the base-pairing rules. If a region of the double helix opens exposing five guanines in a row along one strand, five cytosines bond with them. Next, the sugars and phosphates brought in with the nucleotides form covalent bonds with each other, creating a new second half of the ladder (fig. 13.6). Meanwhile, the same event occurs on the other strand of the original, or parental, helix. Here, the five exposed cytosines (which were originally bonded to the five guanines on the first strand) bond with five free-floating guanines. This process continues until all of the DNA in a chromosome is replicated. One helix has become two. When the cell

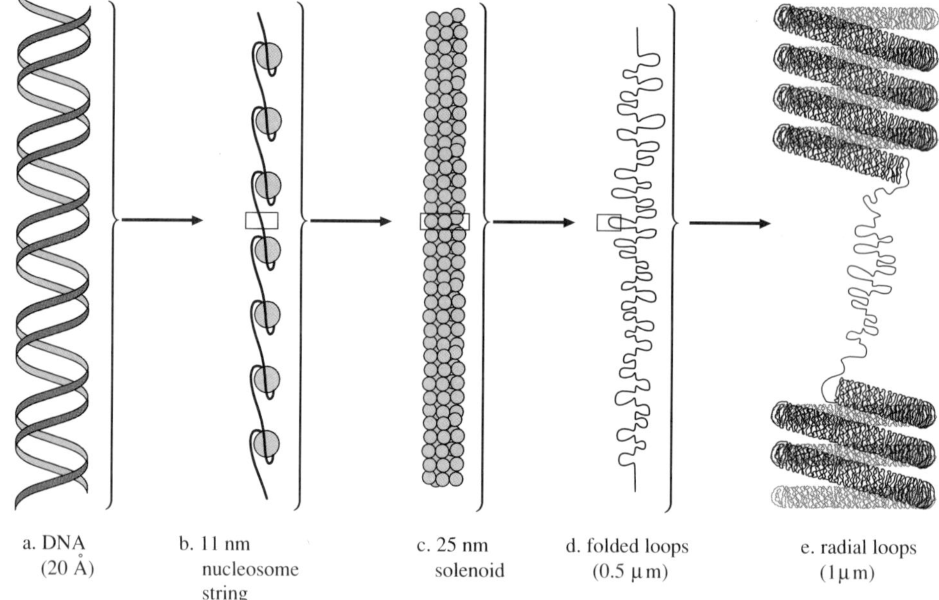

Figure 13.5
The DNA inside eukaryotic cells is highly coiled at several levels. *a.* The DNA double helix is wrapped around proteins, resembling beads on a string (*b*). Each bead contains approximately 146 nucleotides and is about 54 nucleotides from the next bead. *c.* The beads-on-a-string structure is folded into fibers 25 nanometers in diameter. *d.* The 25-nanometer fibers are folded into thicker loops. *e.* The loops are twisted into yet larger structures. Ultimately the DNA assumes the form of a chromosome.

Reading 13.1 *Experiments Identify and Describe the Genetic Material*

PEOPLE HAVE LONG BEEN FASCINATED BY INHERITANCE, BUT IT WAS NOT UNTIL THE NINETEENTH CENTURY THAT THE IDEA THAT CERTAIN STRUCTURES IN THE BODY WERE RESPONSIBLE FOR THE TRANSMISSION OF TRAITS BECAME POPULAR. What we now call genes then went by a variety of colorful names, including pangens, bioblasts, idioblasts, plasomes, plastidules, gemmules, and stirps. The first investigator to attach any sort of chemical significance to an hypothesized genetic material was Swiss biochemist Fredrich Miescher. In 1871, he isolated an acidic substance from the nuclei of white blood cells in pus, which he collected from soiled bandages, and correctly connected it with the substance of heredity. In 1909, the important link between inheritance and protein was first made by English physician Archibald Garrod, who noted that certain inherited disorders that he called "inborn errors of metabolism" were associated with missing enzymes. In later years, the relationship between inherited traits and enzymes would be made in other species: unusual eye color in fruit flies was found to be caused by missing enzymes, and variants of the bread mold *Neurospora crassa*, which needed nutritional supplements in order to grow, were missing key enzymes.

In 1928, English microbiologist Frederick Griffith made an observation that he could not explain but that eventually led to the identification of DNA as the genetic material. Griffith noticed that people with pneumonia harbored one of two types of *Diplococcus pneumoniae* bacteria. Type R bacteria had rough coats, and type S had smooth coats. Mice injected with type R bacteria did not develop pneumonia, but mice injected with type S died of pneumonia. When type S bacteria were heated, which killed them, they no longer could cause pneumonia in mice. However, when Griffith injected mice with a mixture of type R bacteria plus "heat-killed" type S bacteria—each of which alone did not kill the mice—the mice died of pneumonia. Their bodies contained live type S bacteria. What was happening?

The answer came in 1944, with the work of Rockefeller University physicians Oswald Avery, C. M. MacLeod, and Maclyn McCarty. They hypothesized that something in the heat-killed type S bacteria "transformed" the normally harmless type R strain into a killer. The "transforming principle" was DNA, which the researchers showed by isolating DNA from heat-killed type S bacteria and injecting it along with type R bacteria into mice. Again the mice died, and their bodies contained active type S bacteria. The conclusion: the material apparently passed from type S bacteria to type R bacteria was DNA.

But scientists were hesitant to accept DNA as the chemical of heredity. Protein, with its 20 different building blocks and known important functions in organisms, was thought to be more complex and therefore capable of carrying more information than the simpler nucleic acids, about which relatively little was known. Some researchers even suggested that the Rockefeller group's DNA preparations were contaminated with protein, which was responsible for transformation.

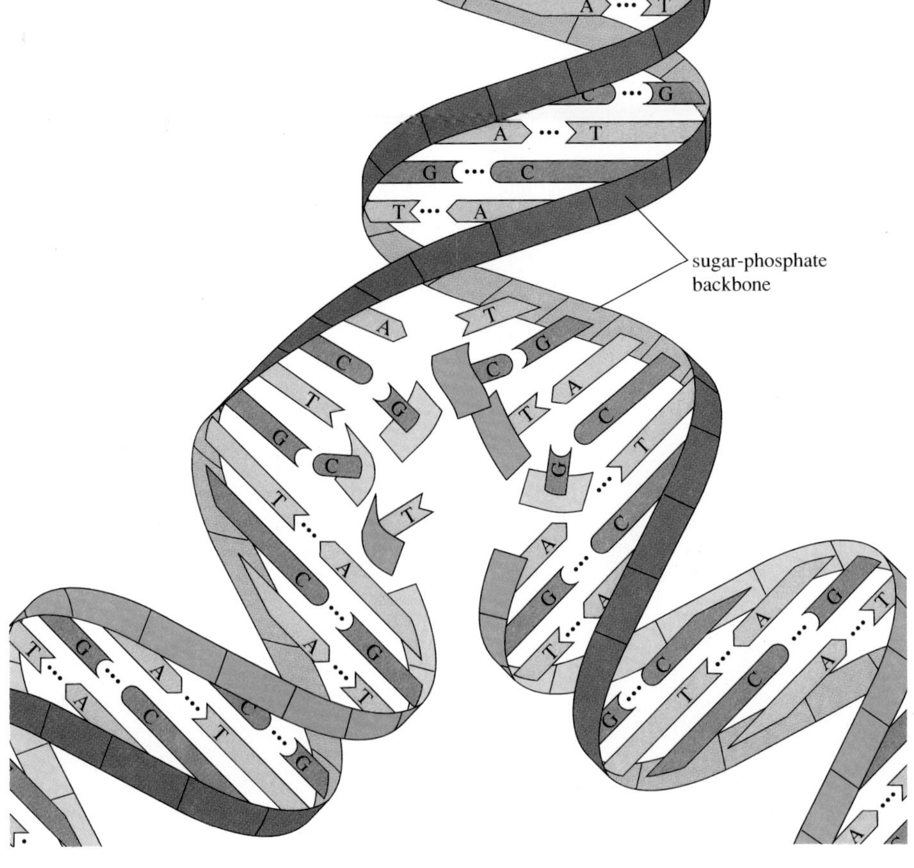

Figure 13.6
When DNA replicates, the double helix unwinds and the bases of each half attract free-floating nucleotides bearing complementary bases.

Figure 1
James Watson *left*, and Francis Crick.

Six years after Avery, MacLeod, and McCarty's experiments, American microbiologists Alfred Hershey and Martha Chase confirmed that DNA is the genetic material. They used *E. coli* bacteria infected with a virus consisting of a protein "head" surrounding nucleic acid. Hershey and Chase first showed that virus grown with radioactive sulfur incorporated the "label" into the protein coat; when they repeated the experiment with radioactive phosphorus, the label was incorporated into the nucleic acid. This showed that sulfur is found in protein but not in nucleic acid, and that phosphorus is found in nucleic acid but not in protein.

Next, the researchers labeled two batches of virus, one with radioactive sulfur, and the other with radioactive phosphorus. They infected two batches of bacteria, each with one type of labeled virus. After several minutes, during which the virus bound to the bacteria and injected their genetic material into them, each mixture was agitated in a blender. This action shook free the empty virus protein coats. The contents of each test tube were centrifuged, and the infected bacteria sank to the bottom of each tube because they were denser than the virus coats. In the tube containing virus labeled with sulfur, no radioactivity was found in the infected bacteria. But in the other tube, radioactive phosphorus was detected in the infected bacteria. This meant that the part of the virus that could enter bacteria and direct them to mass-produce more virus was the part that had incorporated the phosphorus label—the DNA. The genetic material, therefore, was not protein but DNA.

Although DNA was known to contain phosphorus, carbon, nitrogen, hydrogen, and oxygen, its precise structure had not yet been worked out. In the early 1950s, two lines of experimental evidence provided the direct clues that finally led to the discovery of the structure of DNA. Chemical analysis by Austrian-American biochemist Erwin Chargaff showed that DNA in several species contained equal amounts of adenine and thymine and equal amounts of guanine and cytosine. Next, English physicist Maurice Wilkins and English chemist Rosalind Franklin bombarded DNA with X rays, a technique called X ray crystallography, and then observed the pattern in which the X rays were deflected from it. This pattern revealed a structure of building blocks that repeated at regular intervals.

In 1953, American biochemist James Watson and English physicist Francis Crick, working together in England, built a replica of the DNA molecule using ball-and-stick models. Their model had equal amounts of adenine and thymine and equal amounts of guanine and cytosine, and it satisfied the symmetry shown in the X ray diffraction pattern. The result of their insight, based upon the experimental evidence of so many others, was the now familiar double helix.

physically divides, distributing one double helix and approximately half of the organelles into each of two daughter cells, cell division is complete (fig. 13.7).

Each newly replicated double helix, then, consists of one parental strand and one new daughter strand. This reproduction of the double helix is called *semiconservative* replication, because half of the parent molecule is always conserved in each of the new ones formed. In replication, then, each DNA strand serves as a template for the other. It is like a line of couples at a square dance, and an equal number of people on the side, waiting to dance. The dancing couples move apart and select new partners from the onlookers to create two new lines of couples. One person from each new couple is an "old" dancer, and one is "new."

The DNA double helix unwinds a little at a time so that correct complementary bases can be inserted without disrupting the overall structure of the molecule. To speed the process, replication

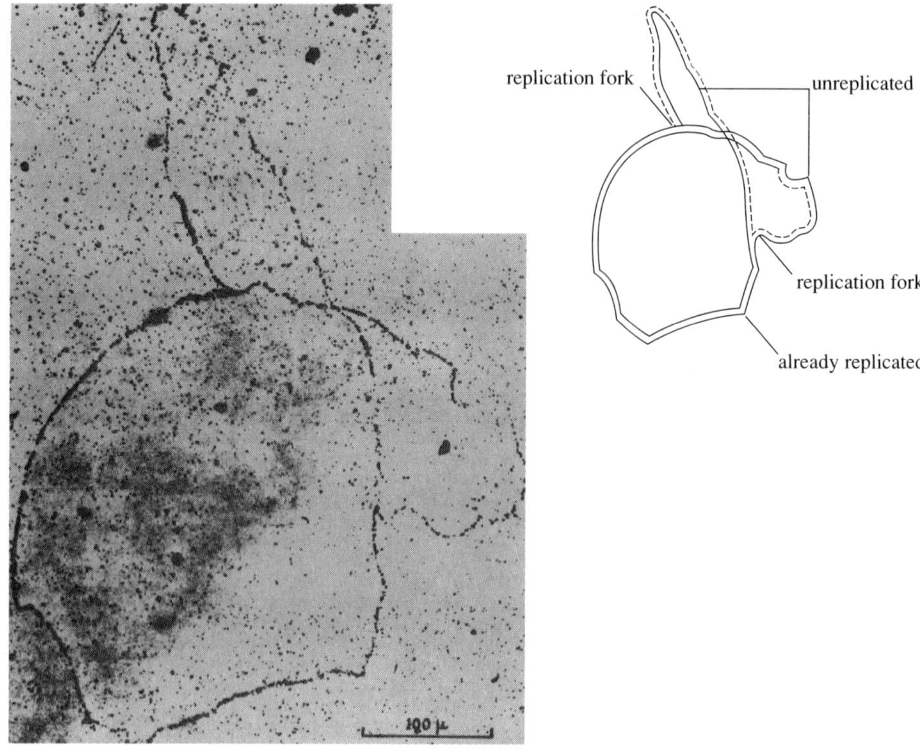

a.

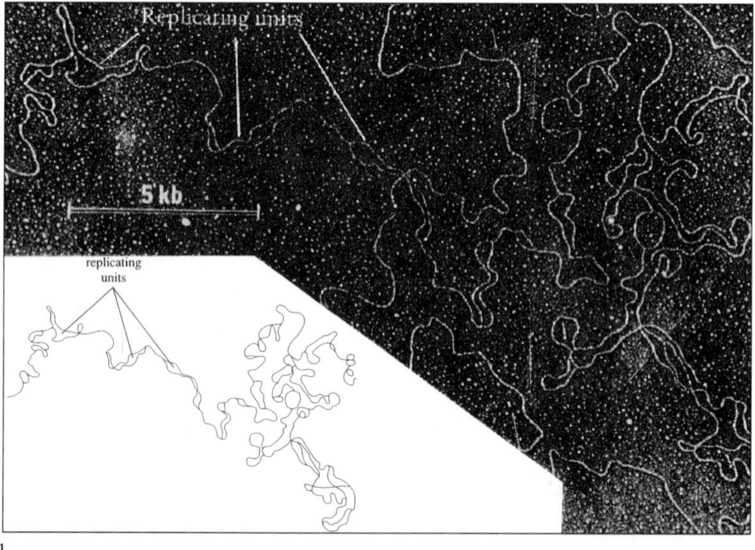

b.

Figure 13.8
DNA replication can be visualized by a process called autoradiography. Cells replicate their DNA in the presence of radioactive thymidine, which is a precursor of the nucleotide thymine. The daughter strands incorporate the radioactive thymine. After a certain amount of time elapses, the cells are broken open, and the contents collected on filter paper. The paper is then placed against a sheet of X ray film and kept in a dark drawer for 2 months. During this time, the DNA labeled with the thymine emits electrons, which cause dark grains to appear in the film. When the film is developed, the grains outline the position of the newly replicated DNA. The thickest lines represent parts of the DNA molecule that had already been replicated at the time when the cells were broken open; the thinner lines represent DNA that had not yet been replicated. An *E. coli* chromosome "caught" in the middle of DNA replication is depicted in (*a*). *b.* This DNA is from the fruit fly *Drosophila melanogaster*. Note the several sites of replication origin in the fly DNA, as compared to the single origin in the bacterial DNA.

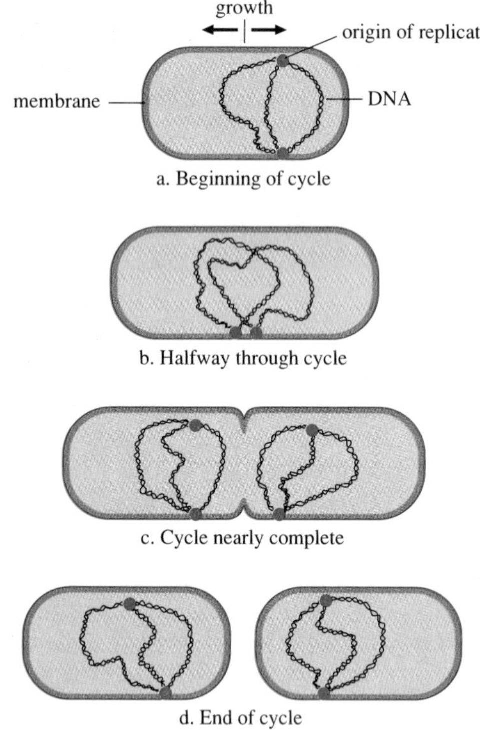

Figure 13.7
DNA replication is part of cell division. Cell division in the bacterium *E. coli* begins with DNA replication (*a*) and (*b*). When replication is complete, the cell wall and cell membrane pinch down the center of the cell (*c*), eventually partitioning the two DNA double helices into the two daughter cells (*d*).

beginning at any given point proceeds in two directions at a time. In eukaryotes, a chromosome can replicate bidirectionally from 100 or so local unwindings of the double helix (fig. 13.8). If DNA replication in human cells did not proceed in two directions from many sites of origin, it could not be replicated fast enough for the body to grow and replace worn-out cells.

Replicating the enormous DNA molecule is a little like separating two very long strands of spaghetti entwined about each other. To keep the DNA from forming a hopeless tangle, portions of the double helix undergoing replication, called *replication forks*, are held open by *unwinding proteins* with such colorful names as gyrase and swivelase. The energy that powers the unwinding comes from adenosine triphosphate (ATP). In the replication forks, enzymes called **DNA polymerases** add new nucleotides as well as detect and correct any noncomplementary base pairs that may have formed by accident. This *DNA repair* is so accurate that only 1 in 10 billion bases is inserted against a noncomplementary base. Finally, enzymes called *DNA ligases* (ligate means "to tie") link the sugars and the phosphate groups to form the rails of the new strand.

So fast and accurate is DNA replication that a new laboratory tool, called *gene amplification*, is based on it. In this procedure, DNA of interest—from human tissue left at the scene of a crime, for example—is mixed with free nucleotides, a short lab-made piece of DNA corresponding to a particular gene, and DNA polymerase. The short piece of synthetic DNA binds to its complementary sequence in the DNA sample (the crime-scene tissue), and the DNA polymerase brings the free nucleotides in to complete the new strand of DNA. Now this double helix is heated, which separates the strands. The DNA polymerase goes back to work, knitting a complete new double helix from each separated strand. After another round of heating, the two double helices separate, and the polymerase manufactures four double helices. After 20 such cycles, there are more than a million copies of the original targeted portion of the sample!

This cycling of DNA replications is called the *polymerase chain reaction*. Gene amplification is valuable when tiny amounts of DNA must be scaled up before other tests can be performed, such as on preserved DNA in woolly mammoths and mummies; in forensic samples as small as a single hair or sperm cell; on viral DNA in human tissues to diagnose infectious disease; and on fetal DNA for prenatal diagnoses. Can you think of other uses for gene amplification?

KEY CONCEPTS

DNA replication is semiconservative. A double helix unwinds locally, and free-floating bases are brought by enzymes to pair with their complementary mates. DNA replication starts at several points on a chromosome and proceeds in two directions. Errors in DNA replication are repaired as they occur. The polymerase chain reaction rapidly copies a DNA sequence.

Transcription—Transmitting Genetic Information

DNA sequences contain 4 types of nucleotides; protein sequences contain 20 types of amino acids. How does the 4-letter gene alphabet contain the information for the 20-letter protein alphabet? The information transfer from gene to protein is accomplished by ribonucleic acid (RNA).

The first step in constructing a protein is to make an RNA molecule that is complementary in sequence to one strand of the DNA double helix. RNA synthesis from DNA is called **transcription.** RNA is manufactured in the nucleus but then enters the cytoplasm, where it participates in building protein molecules in a process called **translation.** DNA, then, does not directly synthesize proteins but provides the instructions for the amino acid sequence of a protein by way of RNA molecules. Which genes are transcribed in which cells, and when, are very important in development. A bone cell transcribes collagen genes, whereas a muscle cell transcribes myocin and actin and other muscle protein genes.

Ribonucleic Acid (RNA)

RNA is a nucleic acid like DNA, but it differs from DNA in a few ways (fig. 13.9). In RNA, thymine is replaced by another pyrimidine, **uracil** (U), and deoxyribose is replaced by another sugar, *ribose*. RNA forms a single strand rather than a double helix. RNA is transcribed from DNA by a mechanism quite similar to that of DNA replication, based upon complementary base pairing. Enzymes unwind the double helix, and RNA bases bond with the exposed complementary bases on one strand of the DNA double helix (fig. 13.10). The enzyme *RNA polymerase* knits together the RNA nucleotides in the sequence specified by the DNA. For example, the DNA sequence GCGTATG is transcribed into the RNA sequence CGCAUAC.

For a particular gene, RNA is transcribed from only one strand of the double helix, which is called the **sense strand.** The strand that is not transcribed is called the **antisense strand.** However, the half of the double helix that serves as the sense strand for one gene may be the antisense strand for another gene. Researchers use synthetic antisense sequences as "gene silencers," which bind to a gene's RNA, preventing it from instructing the cell to string together a particular amino acid sequence. Antisense sequences can be used to silence DNA from disease-causing viruses, or can be directed against the gene for an enzyme that hastens ripening in the tomato, extending its shelf life.

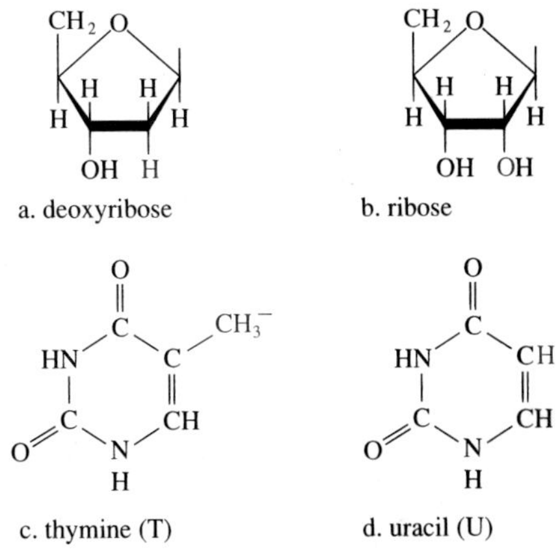

a. deoxyribose

b. ribose

c. thymine (T)

d. uracil (U)

Figure 13.9

DNA and RNA differ structurally from each other in three ways. DNA nucleotides have the sugar deoxyribose (*a*), whereas RNA contains ribose (*b*). DNA nulceotides include the pyrimidine thymine (*c*), whereas RNA has uracil (*d*). DNA is double-stranded, and RNA is generally single-stranded. The two types of nucleic acids have different function.

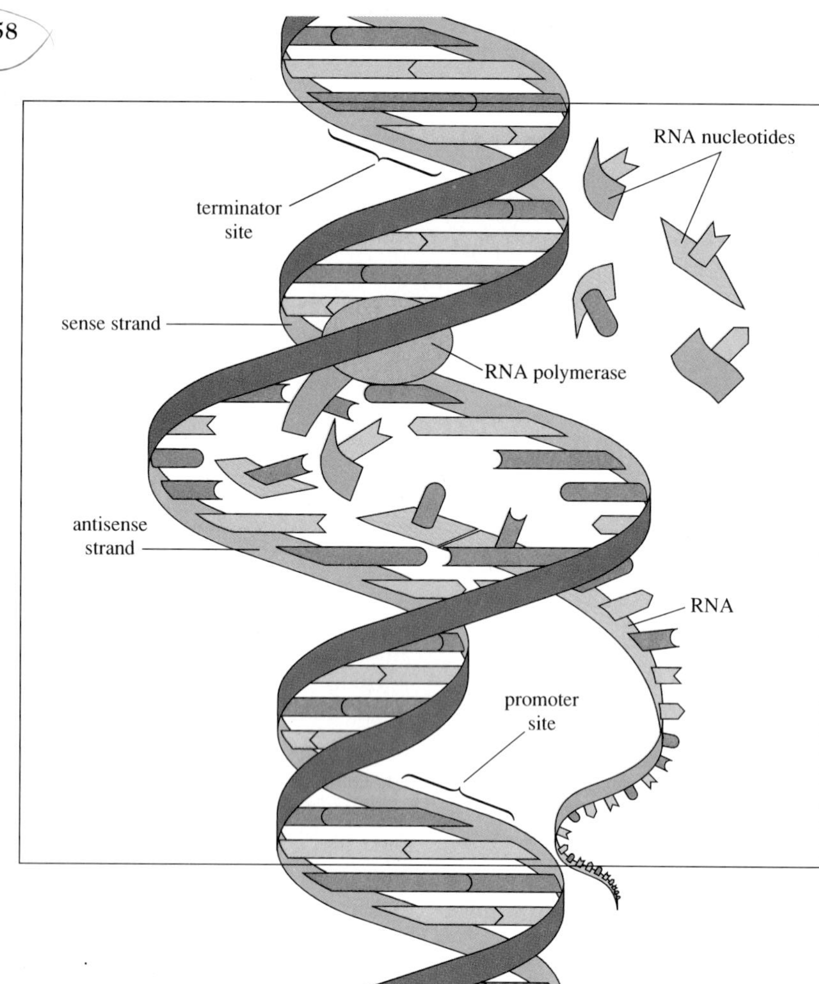

Figure 13.10
Transcription of RNA from DNA. RNA nucleotides hydrogen bond to complementary bases in the DNA sense strand. The DNA sequence itself signals enzymes where to start and end transcription: a promoter sequence marks the start of a gene to be transcribed, and a terminator sequence signals the end of a gene.

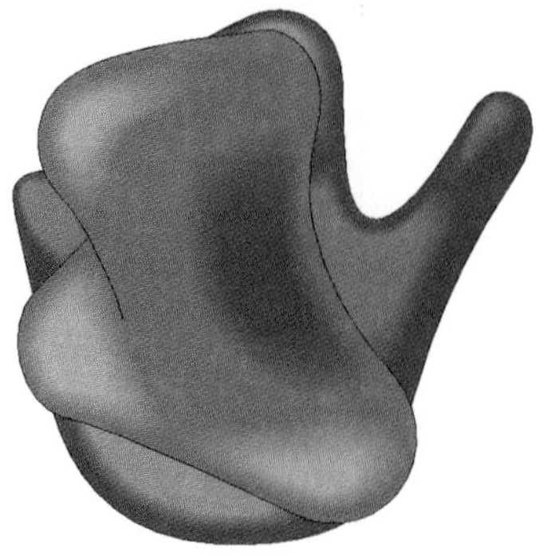

Figure 13.11
A ribosome consists of about half protein and half ribosomal RNA. In *E. coli,* the larger subunit is built of two rRNA molecules and 35 proteins, and the smaller subunit is one rRNA molecule and 21 proteins. A ribosome from a eukaryote also has two subunits, but they are each slightly larger than the corresponding prokaryotic subunits and contain 80 proteins and four rRNA molecules altogether.
From Lake, copyright © 1981, by Scientific American, Inc., George V. Kelvin, all rights reserved.

One double-stranded DNA molecule in a human carries many hundreds or even thousands of genes. As RNA is synthesized along its DNA template, it curls into characteristic shapes, or conformations, that are determined by complementary base pairing within the same RNA molecule. The conformations of RNA molecules are very important for their functioning.

Types of RNA

There are three major types of RNA, distinguished by their sizes and functions. **Messenger RNA (mRNA)** contains in its sequence of nucleotides information that is translated into the sequence of amino acids of a particular protein, the gene's product. Each three mRNA bases in a row form a genetic code word, or **codon,** that specifies a particular amino acid. A different mRNA represents each gene, and therefore mRNA molecules can vary greatly in length. Most mRNAs, however, are between 500 and 1,000 nucleotides long.

Ribosomal RNA (rRNA) ranges from 100 to nearly 3,000 nucleotides long and joins with certain proteins to form a structure called a **ribosome,** which is a structural support for protein synthesis (fig. 13.11). A ribosome has two subunits that float separately in the cytoplasm but come together to play their role in building a protein.

Transfer RNA (tRNA) molecules are "connectors," binding to an mRNA codon at one end and to a specific amino acid at the other. A tRNA molecule is small, only 75 to 80 nucleotides long. Although like all RNA it is single-stranded, certain of its bases hydrogen bond with each other, folding the tRNA into a characteristic "cloverleaf" shape. Figures 13.12 and 13.13 show how small RNA folds. One loop of the tRNA molecule has three bases in a row, called the **anticodon,** which is complementary to an mRNA codon. The end of the tRNA opposite the anticodon covalently bonds to a specific amino acid. A tRNA with a particular anticodon always carries the same amino acid. A tRNA with the anticodon sequence AAG, for example, is bonded to the amino acid phenylalanine. Special enzymes attach amino acids to tRNA bearing the appropriate anticodon sequences, one enzyme type for each anticodon–amino acid match.

KEY CONCEPTS

RNA mediates the transfer of information encoded in the DNA sequence into a protein's amino acid sequence. RNA contains uracil but not thymine, has the sugar ribose, and is single-stranded. Messenger RNA transmits the information in a gene's sense strand. Each three mRNA bases in a row is a codon, specifying a particular amino acid. Ribosomal RNA, along with ribosomal proteins, form ribosomes, which physically support the other components of protein synthesis. Transfer RNAs are connector molecules. They are shaped like cloverleaves, with a base sequence complementary to a particular mRNA codon at one end and an amino acid bound at the other end.

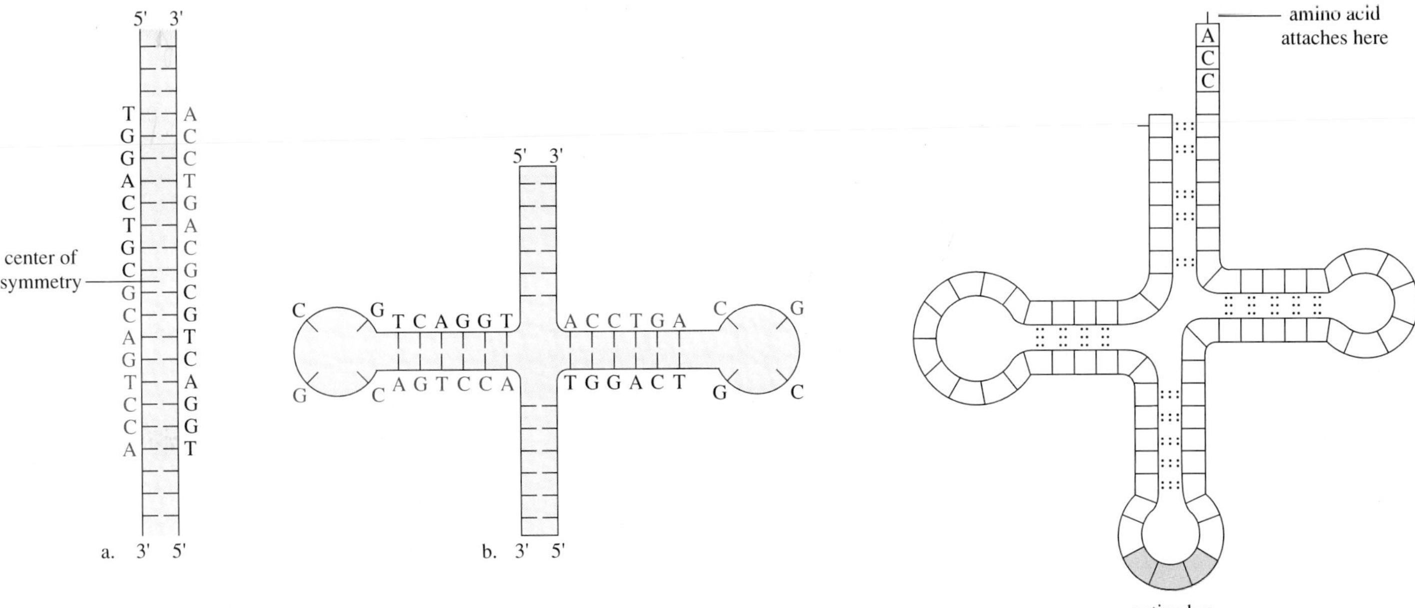

Figure 13.12

Palindromes are phrases or sentences in the English language that read the same backwards or frontwards, such as the statement "Madam, I'm Adam." Many nucleic acids have palindromes in their sequences. *a.* The red bases and black bases illustrate the symmetry about an axis—each segment is a "reverse repeat" of the other segment in that color. *b.* If the strands separate, as they would in DNA replication or RNA transcription, hydrogen bonds that form between bases within the same molecule produce a cross shape. Palindromes help confer the three-dimensional shapes to long nucleic acid molecules that are closely tied to their functions. For example, the enzymes thatregulate replication, transcription, and translation bind to palindromes in DNA. Genes in many species contain palindromes. Mouse DNA folds into about 40,000 tiny loops, and the rRNA of paramecium is a single long palindrome. A gene in the bacteriophage MS2 folds into a flower-like configuration, and the gene for human globin (the protein portion of hemoglobin) contains a palindrome 60 bases long.

Figure 13.13

Transfer RNA. Certain nucleotide bases within a tRNA molecule hydrogen bond with each other to give the molecule a conformation that can be respresented in two dimensions. The dotted lines indicate hydrogen bonds, and the filled-in bases at the bottom form the anticodon. Each tRNA terminates with the sequence CCA, and a particular amino acid covalently bonds with the RNA at this end.

Translation—Expressing Genetic Information

Transcription copies the information encoded in a gene's base sequence into the complementary language of mRNA. The next step is translation of this "message" to the specified sequence of amino acids. Particular mRNA codons (three bases in a row) correspond to particular amino acids (fig. 13.14). This correspondence between the languages of mRNA and protein is called the **genetic code.** Which mRNA codons correspond to which amino acids was deciphered in the 1960s by several researchers, including Francis Crick, the codiscoverer of the structure of DNA. The genetic code was "cracked" using both logic and experimental evidence. Certain questions had to be answered to decipher the genetic code.

The Genetic Code—From a Gene's Message to Its Protein Product

Question 1—How many RNA bases specify one amino acid? Because the number of different protein building blocks (20) exceeds the number of different mRNA building blocks (4), each codon must contain more than 1 mRNA base. In other words, if a codon consisted of only 1 mRNA base, then only 4 different amino acids could be specified, 1 corresponding to each of the 4 bases, A, C, G, and U. If a codon consisted of 2 bases, then 16 different amino acids could be specified, 1 corresponding to each of the 16 possible orders of 2 RNA bases (AA, CC, GG, UU, AC, CA, AU, UA, CG, GC, GU, UG, GA, AG, UC, and CU). If a codon consisted of 3 bases, then as many as 64 different amino acids could be specified. Because 20 different amino acids must be indicated by at least 20 different codons, the minimum number of bases in a codon is 3. Experiments that confirm the triplet nature of the genetic code are described in Reading 13.2.

Question 2—Is the genetic code overlapping? Consider the hypothetical mRNA sequence AUCAGUCUA. If the genetic code is triplet and nonoverlapping (i.e., each three bases in a row form a codon, but any one base is part of only one codon), then this sequence contains only three codons: AUC, AGU, and CUA. If the code is overlapping, the sequence contains seven codons: AUC, UCA, CAG, AGU, GUC, UCU, and CUA. An overlapping code seems economical in that it packs maximal information into a limited number of bases. However, an overlapping code constrains protein structure because certain amino acids would always be followed by certain others. For example, the amino acid specified by the first codon listed, AUC, would always be followed by an amino acid whose codon begins with UC.

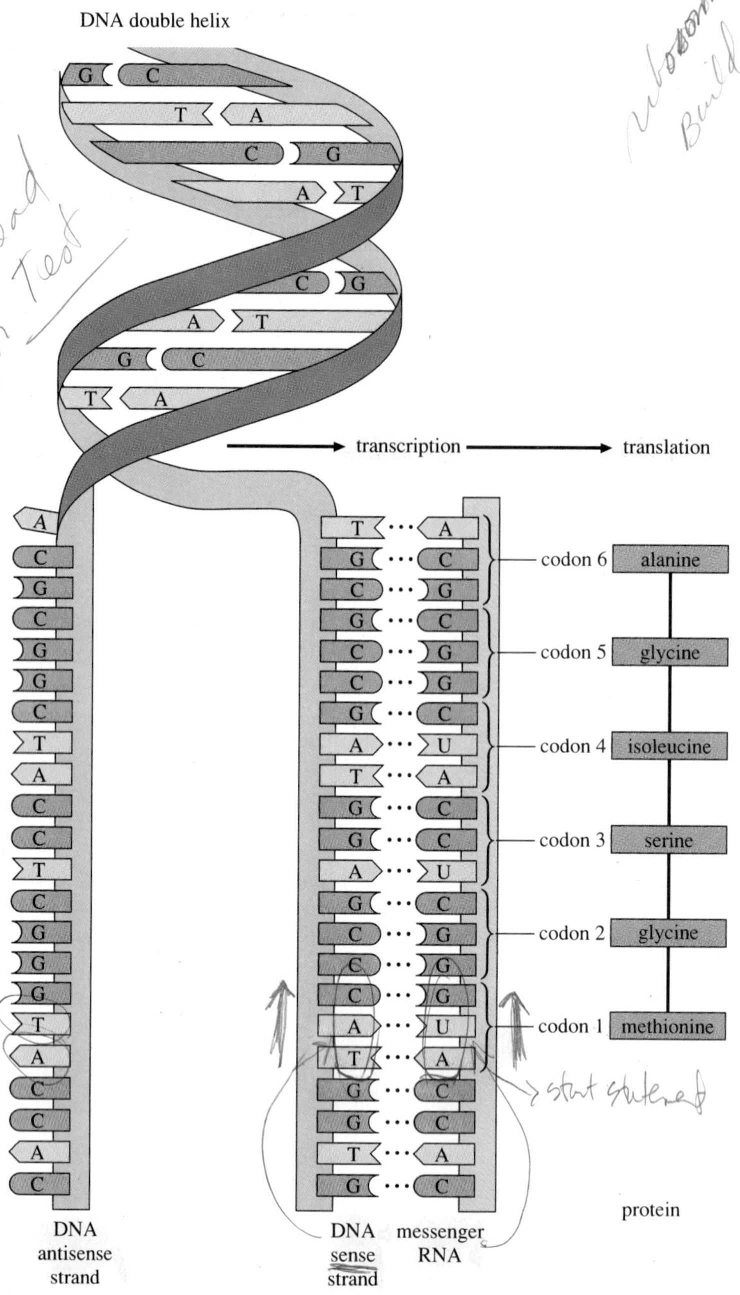

Experiments in which proteins are taken apart amino acid by amino acid in sequence demonstrate that a specific amino acid is not always followed by another. Any amino acid can be followed by any other amino acid.

Question 3—Is the genetic code punctuated? Do parts of an mRNA molecule have functions other than coding for amino acids? For example, is there a chemical group or even a nucleotide sequence that separates adjacent codons from one another, as words are sometimes separated by commas? Chemical analysis of mRNA shows that nucleotides are recognized one right after the other, and that no additional chemical groups designate the start or the end of a codon. However, the genetic code does contain directions for starting and stopping the translation of a protein. The codon AUG signals "start," and the codons UGA, UAA, and UAG each signify "stop." Another form of "punctuation" is a short sequence of bases at the start of each mRNA, called the *leader sequence,* which enables the mRNA to form hydrogen bonds with part of the rRNA in a ribosome.

Question 4—Do all organisms use the same genetic code? Because biochemical evidence indicates that all life on earth evolved from a common ancestor, it is logical that all species should use the same mRNA codons to specify the same amino acids. For the most part this is true, and the genetic code is therefore said to be universal. The fact that mRNA from one species can be translated in a cell of another has made genetic engineering possible (fig. 13.15).

Question 5—Which codons specify which amino acids? The details of the genetic code—that is, which codons specify which amino acids—began to be deciphered by biochemist Marshall Nirenberg and his co-workers at the National Institutes of Health in 1961. They synthesized mRNA molecules in the laboratory and added them to test tubes containing all of the chemicals and structures needed for translation, which were extracted from *E. coli* cells.

The first synthetic mRNA tested had the sequence UUUUUUUU. . . . In the test tube, this was translated into a polypeptide consisting entirely of the amino acid phenylalanine. The first entry in the genetic code dictionary was in: the codon UUU specifies the amino acid phenylalanine. Also, the fact that the number of phenylalanines was always equal to one-third of the number of mRNA bases confirmed that the genetic code is triplet and nonoverlapping. In the next three experiments, AAA was found to code for the amino acid lysine, GGG for glycine, and CCC for proline.

The next laboratory-made mRNA had the sequence AUAUAUAUA. . . . This introduced the codons AUA and UAU, and it produced an amino acid sequence of alternating isoleucines and tyrosines. The next mRNA manufactured had the sequence UUUAUAUUUAUA . . . . This produced an amino acid sequence of alternating phenylalanines and isoleucines. Because it was known from the first experiment that UUU codes for phenylalanine, then AUA must code for isoleucine. If AUA codes for isoleucine, then UAU must code for tyrosine. Two more puzzle pieces had been found.

Figure 13.14
From DNA to RNA to protein. Messenger RNA is transcribed from a locally unwound portion of DNA. In translation, transfer RNA matches up mRNA codons with amino acids.

Reading 13.2 Proving That the Genetic Code Is Triplet

THE TRIPLET NATURE OF THE GENETIC CODE WAS EXPERIMENTALLY DEMONSTRATED BY FRANCIS CRICK AND THREE OTHER INVESTIGATORS USING A WELL-STUDIED GENE, CALLED *rII*, IN THE VIRUS T4. The gene controls which strains of *E. coli* the virus will attack and the pattern of clear areas that arise on a plate of bacteria when the virus infects them. The researchers exposed T4 DNA to a compound called proflavin, which inserts into DNA between the bases. When DNA exposed to proflavin replicates, the inserted proflavin causes one or more bases in a row to be added or deleted. The effects of added or deleted bases were observed on the expression of the *rII* gene, for which the normal base sequence was known.

The researchers found that adding or deleting one or two bases "disrupted the reading frame"—that is, the resulting protein did not have the expected amino acids from a point in the amino acid sequence corresponding to the site of the inserted or deleted bases in the DNA. However, if three bases in a row were added or deleted, the reading frame was disrupted only locally, and a near-normal *rII* protein resulted.

It is easiest to follow the T4 experiment with a sentence analogy in which an mRNA codon (three bases in a row) is represented by a three-letter word:

THE BOY AND THE DOG RAN . . .

Adding one letter to the sentence destroys the meaning.

THE BOY ANQ DTH EDO GRA N . . .

Deleting one letter from the sentence does the same:

THE BOY ANT HED OGR AN . . .

Adding or deleting two bases also interferes with the meaning:

THE BOY ANQ QDT HED OGR AN . . .

THE BOY ANH EDO GRA N . . .

However, adding or deleting three letters together causes localized interference, but the general meaning of the sentence can still be deciphered:

THE BOY AND QQQ THE DOG RAN . . .

THE BOY AND DOG RAN . . .

Because adding or deleting three letters in a row does not appreciably disrupt the meaning of the sentence, but adding or deleting one or two letters does, the size of a word—a unit of meaning—must be three letters. Analogously, the size of a codon—a genetic unit of meaning—must be three bases.

By the end of the 1960s, the entire genetic code had been "cracked," with many research groups contributing entries to the genetic code dictionary (table 13.1). Each step of the way was built on knowledge gained from previous experiments. Some amino acids were specified by more than 1 codon. In fact, 60 of the possible 64 codons were found to specify 1 of the 20 amino acids, the other codons indicating "stop" or "start." For example, phenylalanine is coded for by UUU as well as by UUC. Different codons that specify the same amino acid are called **degenerate,** and they often differ from one another by the base in the third position. This pattern of degeneracy may protect against genetic change, or **mutation.** A change in the third base will often result in a codon specifying the same amino acid. Thus, even though a genetic change has occurred, the amino acid sequence is unchanged, and therefore the trait associated with the gene appears normal.

KEY CONCEPTS

The genetic code is triplet, nonoverlapping, continuous, universal, and degenerate.

a.

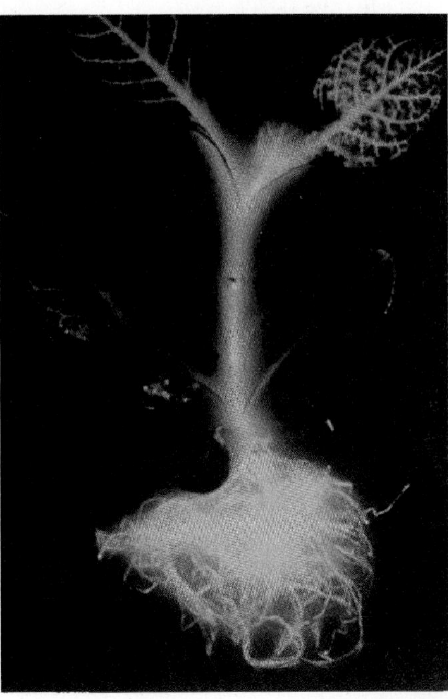

b.

Figure 13.15
The genetic code is universal. Recombinant DNA technology is built on the fact that all organisms utilize the same DNA codons to specify the same amino acids. A striking illustration of the universality of the code is shown in this tobacco plant (*a*), which has been genetically engineered to contain genes from the firefly that specifies the "glow" enzyme luciferase. The plant is bathed in a chemical that allows the enzyme to be expressed (*b*), causing the plant to glow.

Table 13.1
The Genetic Code

First Letter	Second Letter: U	Second Letter: C	Second Letter: A	Second Letter: G	Third Letter
U	UUU } phenylalanine (phe)	UCU } serine (ser)	UAU } tyrosine (tyr)	UGU } cysteine (cys)	U
	UUC	UCC	UAC	UGC	C
	UUA } leucine (leu)	UCA	UAA	UGA	A
	UUG	UCG	UAG	UGG tryptophan (try)	G
C	CUU } leucine (leu)	CCU } proline (pro)	CAU } histidine (his)	CGU } arginine (arg)	U
	CUC	CCC	CAC	CGC	C
	CUA	CCA	CAA } glutamine (gln)	CGA	A
	CUG	CCG	CAG	CGG	G
A	AUU } isoleucine (ilu)	ACU } threonine (thr)	AAU } asparagine (asn)	AGU } serine (ser)	U
	AUC	ACC	AAC	AGC	C
	AUA	ACA	AAA } lysine (lys)	AGA } arginine (arg)	A
	AUG[+] methionine (met)	ACG	AAG	AGG	G
G	GUU } valine (val)	GCU } alanine (ala)	GAU } aspartic acid (asp)	GGU } glycine (gly)	U
	GUC	GCC	GAC	GGC	C
	GUA	GCA	GAA } glutamic acid (glu)	GGA	A
	GUG	GCG	GAG	GGG	G

Building a Protein

Initiation

Protein synthesis requires mRNA, tRNA molecules carrying amino acids, ribosomes, energy-storing molecules such as ATP, and various protein factors. These pieces come together at the beginning of translation, called *initiation*. First, the mRNA leader sequence hydrogen bonds with a short sequence of rRNA in a small ribosomal subunit. The first mRNA codon to specify an amino acid is always AUG, which attracts an initiator tRNA that carries a special form of the amino acid methionine, called *formylated methionine* (abbreviated fmet), because it has an organic formyl group (CHO) attached to it (fig. 13.16). Fmet signifies the start of a polypeptide. The small ribosomal subunit, the mRNA bonded to it, and the initiator tRNA and its attached fmet form the *initiation complex*.

Elongation

In the next stage of translation, called *elongation*, a large ribosomal subunit attaches to the initiation complex. The codon adjacent to the initiating codon (AUG), which is GGA in figure 13.16, then bonds to its complementary anticodon, which is part of a free-floating tRNA that carries the amino acid glycine. The two amino acids (fmet and glycine in the example), still attached to the tRNA molecules that connect them to the mRNA, align with each other and, with the help of an enzyme, join by forming a **peptide bond** between them. Once the two amino acids have bonded to each other, the first tRNA is released. It will pick up another amino acid and be used elsewhere. The ribosome and its attached mRNA are now bound to a single tRNA, with two amino acids dangling from it. This is the beginning of the polypeptide chain.

Next, the ribosome moves down the mRNA by one codon. A third RNA enters, carrying its amino acid (cysteine in figure 13.16). This third amino acid aligns with the other two and forms a peptide bond to the second amino acid in the growing chain. The tRNA attached to the second amino acid in the chain is released and recycled. The polypeptide is constructed one amino acid at a time, each piece brought in by a tRNA whose anticodon corresponds to an mRNA codon.

Termination

Elongation halts when one of the mRNA "stop" codons (UGA, UAG, or UAA) is reached because there are no tRNA molecules that correspond to those codons. The last tRNA is released from the ribosome, the ribosomal subunits separate from each other and are recycled, and the new polypeptide floats away. Recall that the three-dimensional conformation of a polypeptide is determined by attractive forces between its amino acids. These attractions occur as the polypeptide is being synthesized. By the time that translation is terminated, the polypeptide has assumed the conformation that is essential to its function.

After a polypeptide is manufactured, it sometimes undergoes some structural modification before it becomes a functional protein. Some polypeptide chains must be shortened by enzymes to become active. The polypeptide insulin, which is 51 amino acids long, for example, is initially translated as the polypeptide proinsulin, which is 80 amino acids long. Some polypeptides must join with others to form larger protein molecules. The blood protein hemoglobin, for example, consists of four polypeptide chains.

Translation Is Efficient

Protein synthesis is an economical biological process, enabling a cell to produce large amounts of a particular protein with just one or two copies of a gene. A plasma cell in the human immune system, for example, can produce 2,000 identical antibody molecules per second. To mass-produce on this scale, the molecules and structures of protein synthesis are used and reused quite efficiently. Many mRNA transcripts can be made from a single gene. In addition, an mRNA transcript can be translated by several ribosomes simultaneously, each at a different point along the message (fig. 13.17). RNA, ribosomes, enzymes, and factors are recycled. Bacteria go one step further in conserving resources: they begin translation of an mRNA before its transcription is completed (fig. 13.18). This is not possible in eukaryotic cells because the nucleus physically separates transcription from translation.

Many antibiotic drugs work by interfering with protein synthesis in the bacteria that cause infections. Some antibiotics "trick" tRNA molecules into misreading codons and inserting the wrong amino acids, resulting in malfunctioning polypeptides that kill the microorganism. Other antibiotics release fmet from the initiation complex, dismantling the protein synthesis machinery before polypeptide construction can get under way. Other

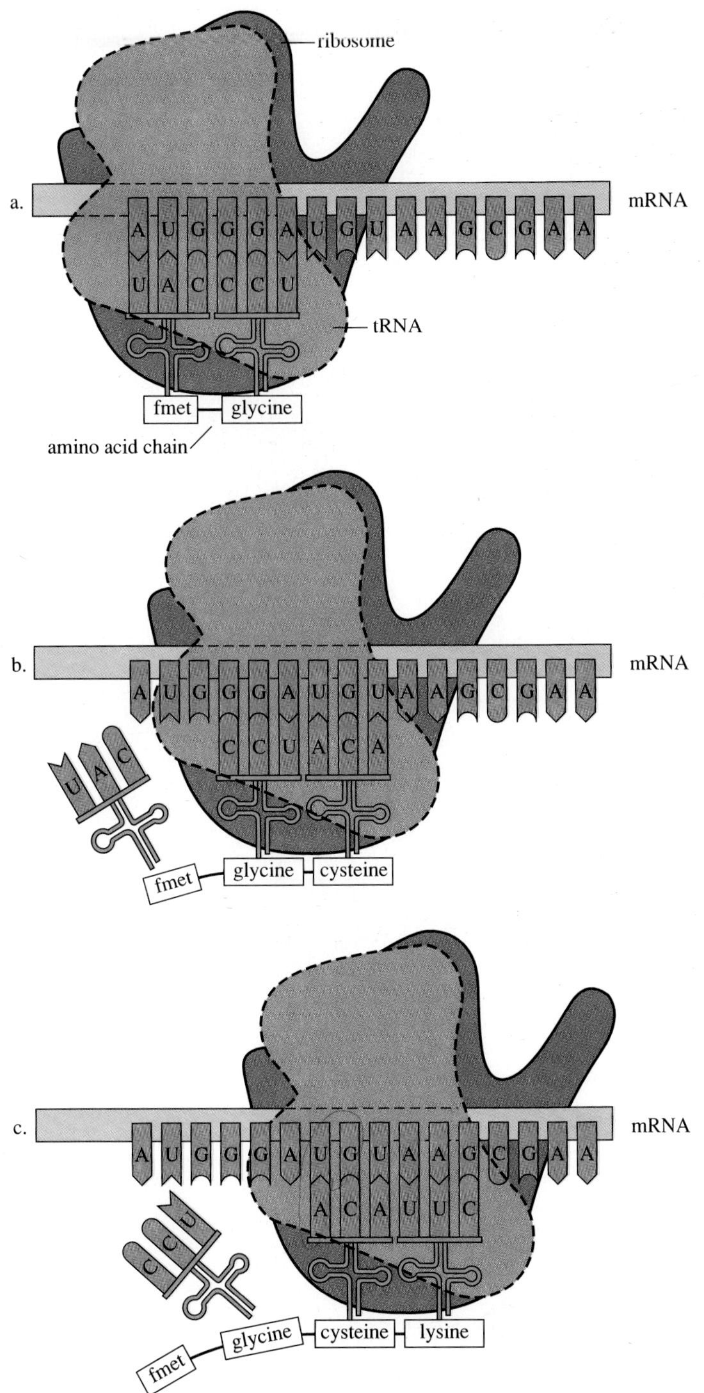

Figure 13.16

Translating a polypeptide. Translation begins when a mRNA molecule binds to a segment of rRNA that is part of a small ribosomal subunit. The anticodon of a tRNA bearing formylated methionine (fmet) hydrogen bonds to the initiation codon (AUG) on the mRNA. These bound structures form the initiation complex. Next, a large ribosomal subunit binds to the complex, and a tRNA bearing a second amino acid (glycine, in this example) forms hydrogen bonds between its anticodon and the second mRNA's codon (*a*). The fmet brought in by the first tRNA forms a dipeptide bond with the amino acid brought in by the second tRNA, and the first tRNA detaches and floats away. The ribosome moves down the mRNA by one codon, and a third tRNA arrives, carrying the amino acid cysteine in this example (*b*). A fourth amino acid is linked to the growing polypeptide chain (*c*), and the process continues until a termination codon is reached.

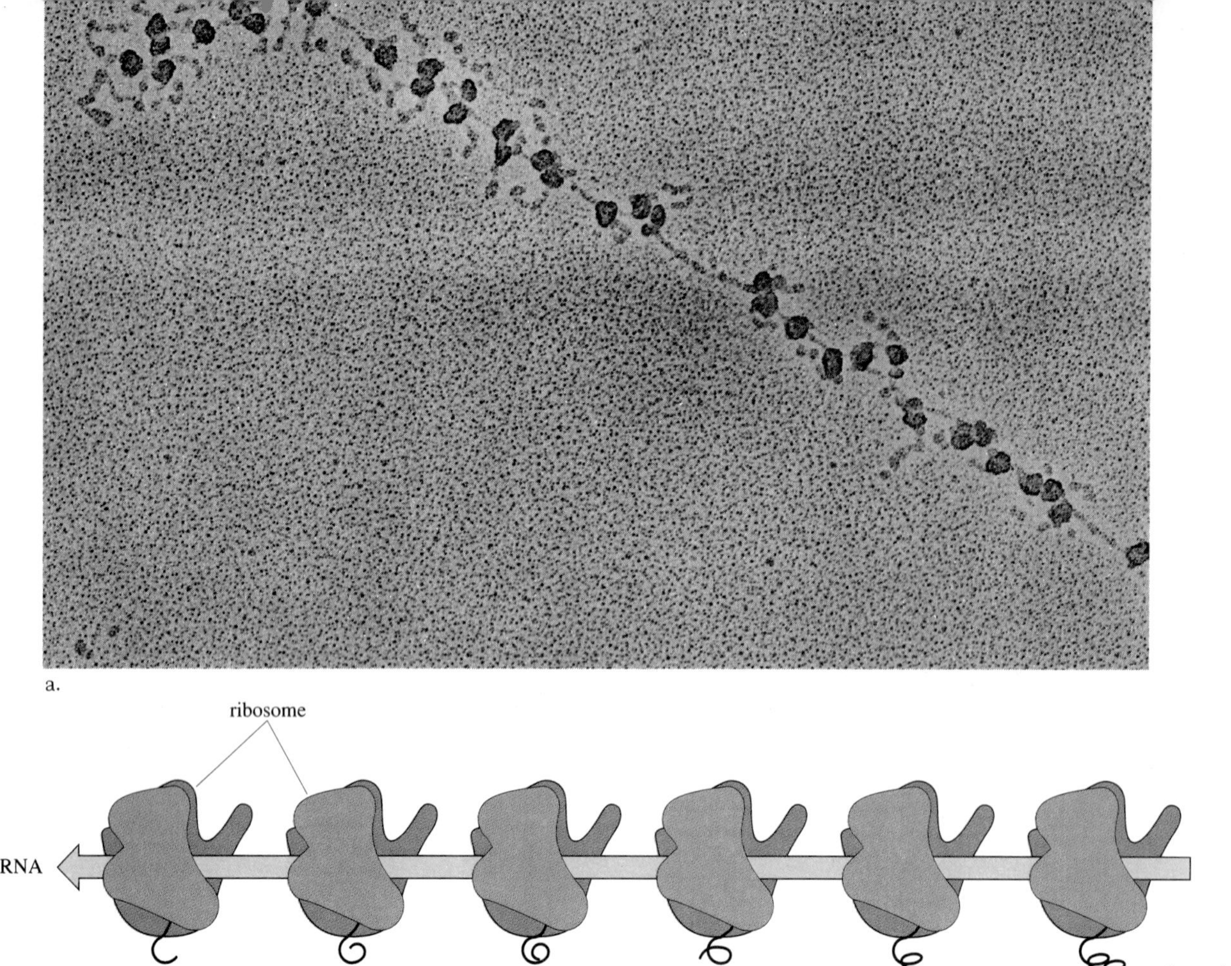

Figure 13.17
Many polypeptides can be manufactured simultaneously from an mRNA molecule. A single mRNA molecule can be translated by several ribosomes at one time. The ribosomes have different-sized polypeptides dangling from them—the closer a ribosome is to the end of a gene, the longer its polypeptide chain. *a.* The line is an mRNA molecule, the dark round structures are ribosomes, and the short chains extending from the ribosomes are polypeptides. *b.* A schematic illustration of polyribosomes.

start of mRNA
polypeptide chains
end of mRNA

Figure 13.18
Transcription and translation occur simultaneously in prokaryotic cells because they do not have nuclei to separate DNA from ribosomes. The concurrent processes can be visualized in an electron microscope as a long thin strand (the DNA) from which increasingly long strands (mRNA) extend. The mRNA transcripts are dotted with globular ribosomes, which indicates that translation is taking place even before an mRNA transcript is completed.

antibiotics prevent tRNA molecules from binding to the ribosome, and still others block the movement of ribosomes down mRNA.

KEY CONCEPTS

In translation initiation, mRNA, tRNA with bound amino acids, ribosomes, energy molecules, and protein factors assemble. First, the mRNA leader sequence binds to a portion of rRNA in the small subunit of a ribosome, and the first codon attracts a tRNA bearing formylated methionine. Next, in elongation, the large ribosomal subunit attaches, and then the appropriate anticodon portions of tRNAs bind to successive codons in the mRNA. As the amino acids attached to the aligned tRNA molecules form peptide bonds, a protein is built. When the termination codon is reached, protein synthesis ceases. Protein synthesis is very efficient.

The Changing View of the Gene

Until the middle 1970s, the gene was envisioned as a continuous sequence of chemical units that were read off three at a time in a nonoverlapping manner to specify a particular sequence of amino acids. It was believed that each gene was confined to a particular location on its chromosome, and that all species utilized exactly the same genetic code. But increasingly powerful technology has demonstrated some intriguing exceptions to what we previously thought was the final answer to the question, What is a gene?

Genes in Pieces—Introns

Some genes of eukaryotic species contain nucleotide sequences that do not appear in the mRNA that is translated into protein. A gene might be 10,000 bases long, for example, but its corresponding mRNA only 6,000 bases long. These "extra" noncoding regions of a gene are called **introns**; the stretches of DNA bases that *do* specify amino acids are called **exons** (fig. 13.19). Using a sentence analogy, a gene containing an intron might read:

THE BOY AND XXX XXX XXX THE DOG RAN

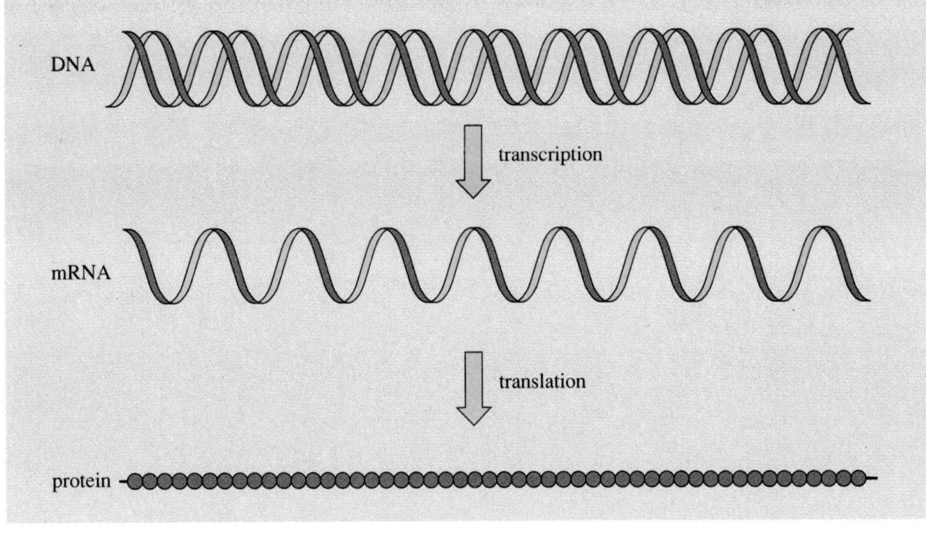

a.

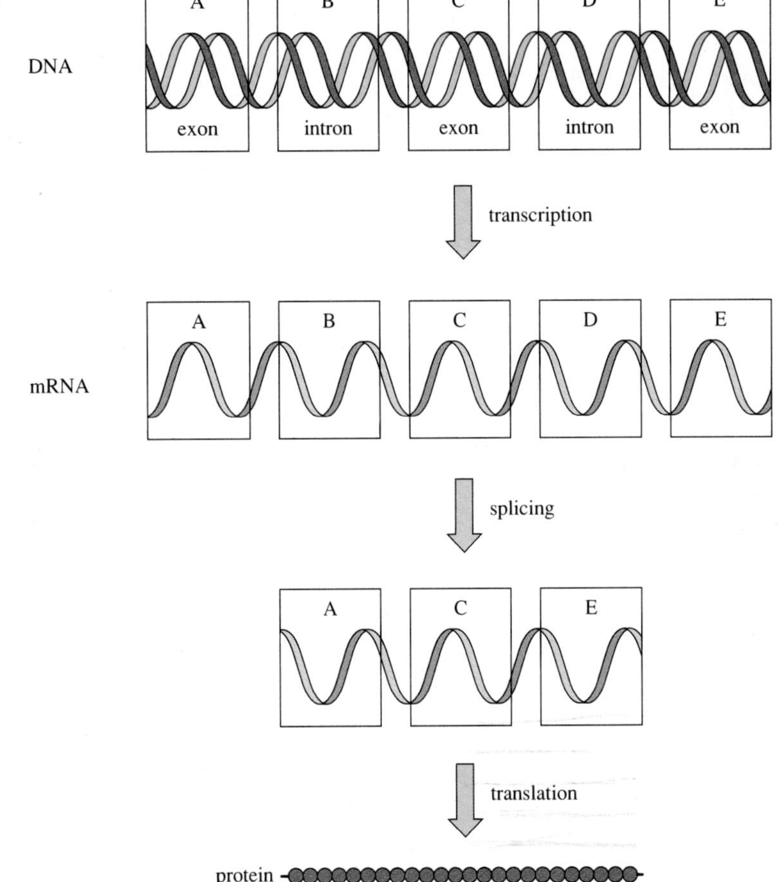

b.

Figure 13.19

Introns interrupt some genes. Most of the early advances in molecular genetics were made in prokaryotes, in which mRNA molecules contain the same number of bases as the genes from which they are transcribed. In 1977 a gene in the mouse was found to contain a stretch of 600 bases that did not correspond to amino acids in the gene's polypeptide product, beta globin. Soon after, segments of DNA that were not represented in the corresponding gene products were discovered in several eukaryotes, including humans. Apparently the intron sequences are transcribed into the initial mRNA and are then spliced out by enzymes before the RNA is translated. Introns are also found in tRNA and rRNA molecules. *b.* Eukaryotic "split" genes.

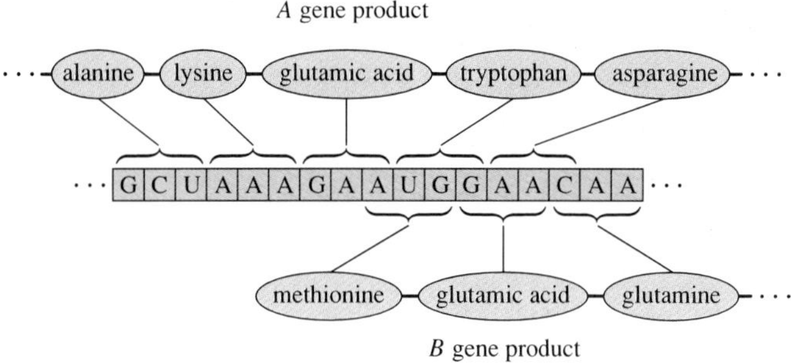

Figure 13.20
These 5,400 base pairs are able to code for 5,700 amino acids because one gene begins before another ends. The initiation codon for the second gene is actually part of two adjacent codons in the first gene. Because the reading frames are different for the two genes, the same sequence of bases codes for different amino acids. Overlapping genes have been found in other viruses, and this economic genetic organization may be possible because of the small amount of genetic material present.

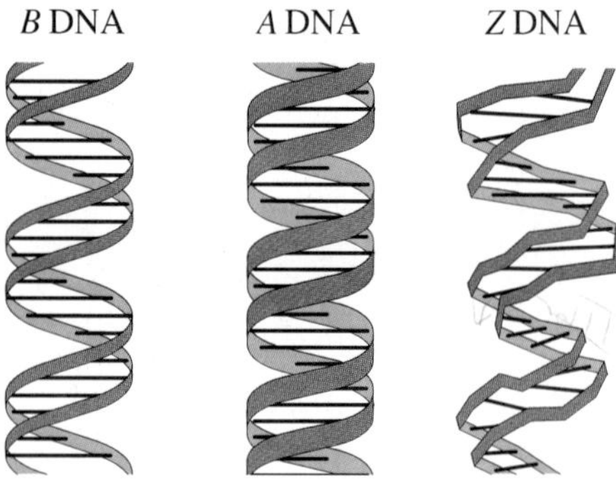

Figure 13.21
Alternate forms of DNA. Watson and Crick described DNA in the *B* form, in which base pairs are perpendicular to the axis of the helix, and each turn of the helix includes 10 base pairs. A DNA has 11 base pairs for each turn of the helix. Z DNA is the only left-handed helix. Its sugar-phosphate rails form a zigzag pattern.

The intron is transcribed, but it is removed from the mRNA before it is translated. The mRNA prior to intron removal is called *pre-mRNA*. Introns are removed by yet another type of RNA, called *small nuclear RNA* (snRNA), which complexes with a protein to form a small nuclear ribonucleoprotein (snRNP), or "snurps," as they are whimsically known. Several snurps work together to form a spliceosome, which actually cuts the introns out and knits the exons together to form the mature mRNA, which exits the nucleus.

Introns can take up a lot of genetic space. While the average exon is 100 to 300 bases long, the average intron is about 1,000 bases long. Introns range from 65 bases to 100,000 bases! Many genes are riddled with introns—the human collagen gene, for example, contains 50 of them. The number, size, and arrangement of introns varies from gene to gene.

The functions of introns (if they indeed have any) are not understood. Introns may somehow control transcription and translation, but why, then, do they not occur in all genes? Another hypothesis is that they are base sequences without a function but from which future genes might somehow be carved. Perhaps they are very ancient genes that have lost their original function. Introns have been called "junk" or even "parasitic" DNA, suggesting that they might be remnants of the genetic material of viruses that once infected the cell.

Overlapping Genes

The tiny virus uX174 contains 5,400 bases that code for a number of polypeptides that altogether contain 5,700 amino acids. Where do the extra 300 amino acids come from? Sequence analysis shows that this virus is a genetic economist: two of its genes actually contain second genes that are read from the same bases but in a different reading frame (fig. 13.20).

The Usually Universal Genetic Code

Mitochondria are organelles that house the chemical reactions of cellular respiration. They also carry a limited amount of genetic information in the form of their own DNA and ribosomes, and they synthesize some of their own proteins. However, mitochondria require products of genes in the nucleus, so they are intimately tied to the genetic headquarters of the cell. Mitochondrial genes in some species use certain codon–amino acid "assignments" that are different from those in the nucleus. In humans, for example, the mitochondrial DNA is 15,569 bases long and encodes mRNA, tRNA, and rRNA. During protein synthesis in mitochondria, UGA codons, which normally encode "stop," specify the amino acid tryptophan. The mRNA codons AGA and AGC, which normally code for the amino acid arginine, are instead read as stop codons. An AUA codon codes for methionine, rather than the normal isoleucine. These differences in the genetic code in mitochondria may support the endosymbiont theory, which suggests that mitochondria were once free-living organisms that became engulfed by primitive eukaryotic cells. Over many millions of years, according to the hypothesis, the mitochondria underwent genetic changes that made them dependent upon their eukaryotic host cells for survival. Perhaps mitochondria, then, are descended from a form of life that used a slightly different genetic code.

Exceptions to codon—amino acid assignments are also seen in the nuclear DNA of four species of ciliated protozoa. All genes in ciliates end in the stop codon UGA. The codons UAA and UAG, which spell "stop" in all other species, specify the amino acid glutamine in these four species.

Variations of the Double Helix

Watson and Crick were correct that DNA is a double helix, but recent improvements in X-ray diffraction of short, laboratory-synthesized DNA molecules demonstrate that the molecule can assume not one but five types of double helices (fig. 13.21). The most common form, which Watson and Crick described, is B DNA. It is long and thin, with the bases perpendicular to the sugar-phosphate rails. Types A, C, and *D* are right-handed like *B*, but the angles between the bases and the rails differ. Z DNA is a thin, left-handed double helix

Figure 13.22
The mitochondrial DNA of the single-celled organism *Crithidia fasciculata* forms an elaborate pattern of loops, rather than a double helix. The significance, if any, of this DNA shape is not known.

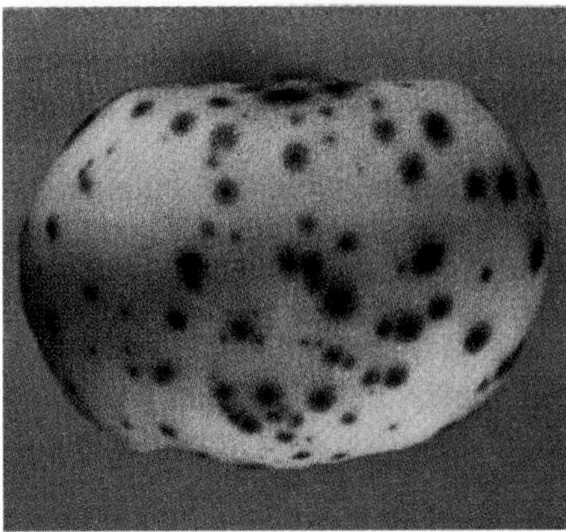

Figure 13.23
Corn kernels are splotched because their cells contain jumping genes. Normally, a dominant gene inhibits the production of pigment in the outer layer of the kernel. But when a jumping gene inserts into the dominant pigment-inhibiting gene, normally turned-off pigment genes are turned on. That cell and all of its descendant cells produce a pigment, until the jumping gene jumps out. The result is a speckled kernel. The earlier in development the jumping gene turns pigment production on, the larger the colored section resulting from the event.

with zigzagging rails. (Comparing a right-handed double helix to a left-handed one is like comparing spiral staircases that wind in opposite directions.) DNA may assume alternate helical forms in regions where genes are actively being transcribed. For example, the base pairs in Z DNA are located closer to the outside of the molecule than they are in the *B* conformation. This may make the bases more accessible to transcription enzymes. Sometimes DNA does not form a double helix at all (fig. 13.22).

Jumping Genes

In the 1940s, many scientists did not believe corn geneticist Barbara McClintock, who insisted that certain genes jump from chromosome to chromosome, disturbing the expression of the genes into which they insert (fig. 13.23). But in the 1970s, when the base sequences of genes could be determined, McClintock was vindicated. Genes that move from one chromosomal location to another have since been identified in bacteria, yeast, fruit flies, and humans. Just as McClintock suggested, these "jumping genes" (she called them transposable elements) can block gene expression by inserting into a gene and preventing its transcription.

Our changing view of the gene does not really alter the basic facts: DNA is the genetic material, it can replicate itself, and it can be transcribed into RNA, which guides protein synthesis. However, exceptions to our ideas about the gene make a philosophical statement that applies to all scientific investigation and thought: The answers are never final. Exceptions arise in the most unexpected places, and they can be detected only by an openly questioning mind.

KEY CONCEPTS

There are exceptions to our view of the gene. Some genes are interrupted by introns. In a few viruses, some genes overlap, such that more than one protein is encoded in a single DNA sequence. Mitochondrial genes and the genes of some protozoa utilize some different codon—amino acid assignments. DNA can assume forms different from the particular double helix envisioned by Watson and Crick, and some genes move.

Biotechnology

A man feels a shooting pain in his arm, then a thundering in his chest. Realizing that he is in the throes of a heart attack, he reaches for his self-injector of tPA (tissue plasminogen activator) and quickly injects himself. The tPA begins to break apart the blood clots that are blocking his heart's circulation. This lifesaving protein is naturally found in the human body in tiny amounts. The man's tPA drug, although identical to his own, was manufactured in bacteria.

Pigs raised for food are usually fed steroid hormones and antibiotics to ensure that their meat is plentiful and infection free. But the use of antibiotics can provoke the rise of microorganisms that are resistant to the drugs, because these are the only ones that can survive in the animals. An alternative to antibiotics in pig feed is porcine growth hormone (PGH)—it is cheaper than antibiotics, it does not encourage growth of antibiotic-resistant microbes, and it produces animals with very lean meat. Like tPA, PGH is naturally rather hard to come by. But if the PGH gene is transferred into and used by bacteria, the resulting hormone is so abundant that it costs only 5¢ to 11¢ per dose.

The fertilized egg of a cow is injected with a human gene coding for the protein collagen, plus a piece of DNA that controls in which tissues the gene will function. The calf descended from this altered cell has the human collagen gene quietly tucked away in each of its cells. The control sequence turns the gene on in breast tissue after the grown-up calf has herself given birth—and her milk is rich in collagen, a protein not normally found in milk. The collagen is separated from the milk and used as an ingredient in skin-care products.

These three scenarios are all possible because of our knowledge of the chemistry of the gene. The quest to understand how DNA transmits hereditary information began in the 1950s and 1960s with purely academic goals—discovering the steps of a fascinating biological process. Beginning in the 1970s and 1980s, knowledge of the molecular workings of the gene were put to practical use in human and veterinary health care, agriculture, food processing, and forensics (Reading 13.3).

The alteration of cells or biological molecules with specific applications in mind is broadly defined as **biotechnology.** One application of biotechnology is the construction of immortal immune system cells, called hybridomas, that secrete large quantities of pure, or *monoclonal*, antibodies, which are then used to treat cancer (discussed in chapter 7). Growing enormous numbers of cells and then extracting valuable biochemicals from them is another application of biotechnology, called *large-scale cell culture* (discussed in chapter 4).

Reading 13.3 DNA Fingerprinting

TOMMIE LEE ANDREWS WAS A VERY METICULOUS RAPIST. He picked out his victims months before he attacked, watching them so that he knew exactly when they would be home alone. On a balmy Sunday night in May 1986, Andrews lay in wait for Nancy Hodge, a young computer operator at nearby Disney World in Orlando, Florida. The burly man surprised her when she was in the bathroom removing her contact lenses, then raped and brutalized her repeatedly.

Andrews was very careful not to leave fingerprints, threads, hairs, or any other indication that he had ever been in Hodge's home. But he had not counted on the then-fledgling technology of DNA fingerprinting. In this test, differences in DNA sequence between individuals are used to establish, or disprove, identity. If the DNA fingerprint of a forensic sample such as sperm, blood, or hair matches the DNA fingerprint of a suspect's cells, the person's presence at the crime scene is established.

Thanks to Nancy Hodge's quick arrival at the hospital following the attack, where a "rape kit" including a vaginal secretion sample containing sperm cells was taken, and two district attorneys who had read about DNA fingerprinting, some of Tommie Lee Andrews's sperm ended up at Lifecodes, a Westchester, New York, biotechnology company. There, the DNA was extracted from the sperm cells and cut with restriction enzymes. The DNA pieces were then mixed with laboratory-synthesized genes, called probes, that were labeled with radioactive isotopes. The probes bound to those segments of Andrews's DNA containing the complementary base sequence.

The same procedure of extracting, cutting, and probing the DNA was done on white blood cells taken from Nancy Hodge and Tommie Lee Andrews, who had been apprehended on March 1, 1987, and held as a suspect in several assaults. When the radioactive DNA pieces from each sample were separated out and displayed according to their sizes, the resulting pattern of bands—the DNA fingerprint—matched exactly for the sperm sample and Andrews's blood but was different for Nancy Hodge. There was little doubt that he was the attacker.

The sizes of the DNA pieces in a DNA fingerprint vary from person to person because of differences in DNA sequence in the regions surrounding the probed genes. The power of the technology stems from the fact that there are far more ways for the 3 billion bases of the human genome to vary than there are people on earth. Figure 1 shows the steps of DNA fingerprinting.

Figure 1 (facing page)
DNA fingerprinting. A blood sample (*1*) is collected from the suspect. White blood cells containing DNA are extracted and burst open (*2*), releasing the DNA strands (*3*). The strands are snipped into fragments (*4*), using scissorlike restriction enzymes. Electrophoresis is used to align the DNA pieces by size—the longest pieces at one end, shortest pieces at the other—in a groove on a sheet of gel (*5*). Next the resulting pattern of DNA fragments is transferred to a nylon sheet (*6*). It is exposed to radioactively tagged probes (*7*) that home in on the DNA areas that are used to establish identity. When the nylon sheet is placed against a piece of X-ray film (*8*) and processed, black bands appear where the probes had stuck (*9*). This pattern of black bands in a white column constitutes a DNA print (*10*).

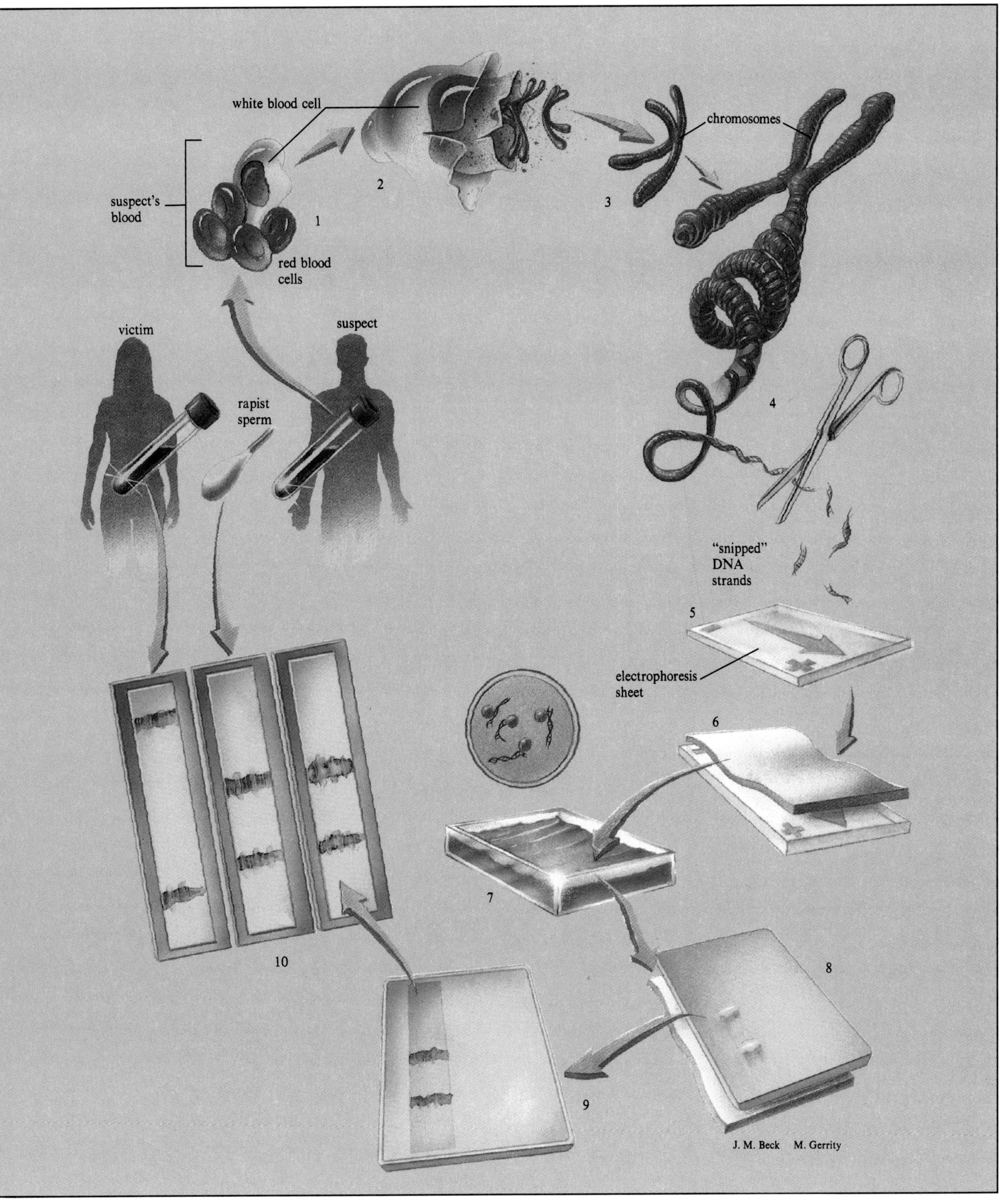
white blood cell
suspect's blood
red blood cells
1
2
chromosomes
3
4
victim
suspect
rapist sperm
"snipped" DNA strands
5
electrophoresis sheet
6
7
8
9
10
J. M. Beck M. Gerrity

The biotechnology that is often called genetic engineering refers to manipulations of genetic material. This includes altering the DNA of an organism to suppress or enhance the activities of its own genes and combining genetic material of different species, which is called **recombinant DNA technology.** Recombinant DNA allows bacteria to produce large quantities of desired proteins, or when introduced into multicellular organisms, endows them with novel traits.

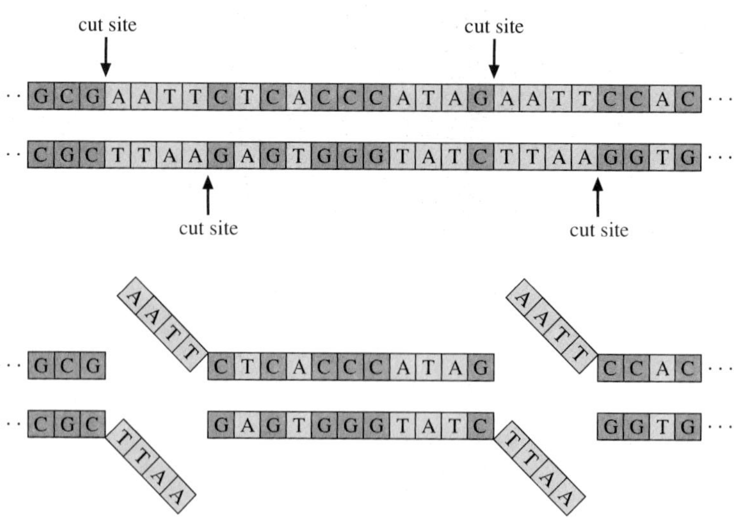

Figure 13.24
A restriction enzyme makes "sticky ends" in DNA and cuts it at specific sequences. *a.* The enzyme *EcoR1* cuts the sequence GAATTC between the G and the A. *b.* This staggered cutting pattern produces "sticky ends" of sequence AATT.

Recombinant DNA Technology

In February 1975, 140 molecular biologists convened at Asilomar, a seaside conference center on California's Monterey Peninsula, about 90 miles south of San Francisco. The scientists met to discuss the safety and implications of a new type of experiment that had been developed over the preceding two years. Investigators had found a simple way to combine the genes of two species, the technique of *recombinant DNA synthesis*. The scientific community was initially concerned because certain proposed experiments required the use of a cancer-causing virus, but there was general concern about where the field was headed. Said Paul Berg, the biologist who called for the meeting, "We are placed in an area of biology with many unknowns; indeed, the greatest risk may well be our ignorance, and it is this ignorance which compels us to pause, reflect, and assess the magnitude of the potential risks associated with this line of research."

The scientists discussed safety—what restrictions could and should be placed on the sorts of organisms used in recombinant DNA research, and what could be done to prevent "escape" of a recombinant organism from the laboratory. The guidelines drawn up at the Asilomar meeting were adopted by the National Institutes of Health, which funds much of the research conducted in university laboratories. These guidelines outlined measures of physical containment, such as specialized hoods and airflow systems that would keep the recombinant organisms inside the laboratory, and approaches to biological containment, such as the use of organisms that were weakened so that they could not survive outside of the laboratory.

A decade after the Asilomar meeting, many members of the original group reconvened at the meeting site to assess progress in the field. Nearly all agreed on two things: recombinant DNA technology had proven to be safer than most had predicted, and the technology had spread from the research laboratory to industry far faster and in more diverse ways than anyone had imagined. Today, thousands of small companies are devoted to recombinant DNA work and other biotechnologies, and many major chemical and pharmaceutical firms have growing biotechnology components.

Constructing Recombinant DNA Molecules

The manufacture of recombinant DNA molecules uses scissorlike biochemicals called **restriction enzymes** to cut a gene from its normal location, insert it into a circular piece of DNA, and then transfer the circle of DNA into cells of another species. Restriction enzymes are found only in bacteria, and each of the 200 or so types described so far cuts DNA at a specific base sequence (fig. 13.24). The natural function of restriction enzymes is to protect bacteria by cutting and thereby inactivating the DNA of infecting viruses. The bacterium's own DNA is shielded from its restriction enzymes by protective proteins. Geneticists isolate restriction enzymes from bacteria and use them to cut DNA at specific base sequences.

Another natural "tool" used in recombinant DNA technology is a *vector*, which is any piece of DNA to which DNA from one type of organism can be attached and transferred into a cell of another organism. A commonly used type of vector is a **plasmid,** which is a small circle of double-stranded DNA found in some bacteria, yeasts, plant cells, and other types of organisms (fig. 13.25). Plasmids are not living, but they can transfer traits between their host cells because they carry genes.

In putting together a recombinant DNA molecule, DNA is isolated from a donor species (e.g., the gene for insulin might be isolated from a human cell) and cut by a particular restriction enzyme. An enzyme is chosen that cuts at sequences known to bracket the gene of interest. The enzyme leaves single-stranded ends dangling from the cut DNA, each bearing a characteristic base sequence (fig. 13.26). Next, a plasmid is isolated and cut with the same restriction enzyme that was used to cut the donor DNA. Because both the donor DNA and the plasmid DNA are cut by the same restriction enzyme, the same single-stranded base sequences extend from the cut ends of each. When the plasmid

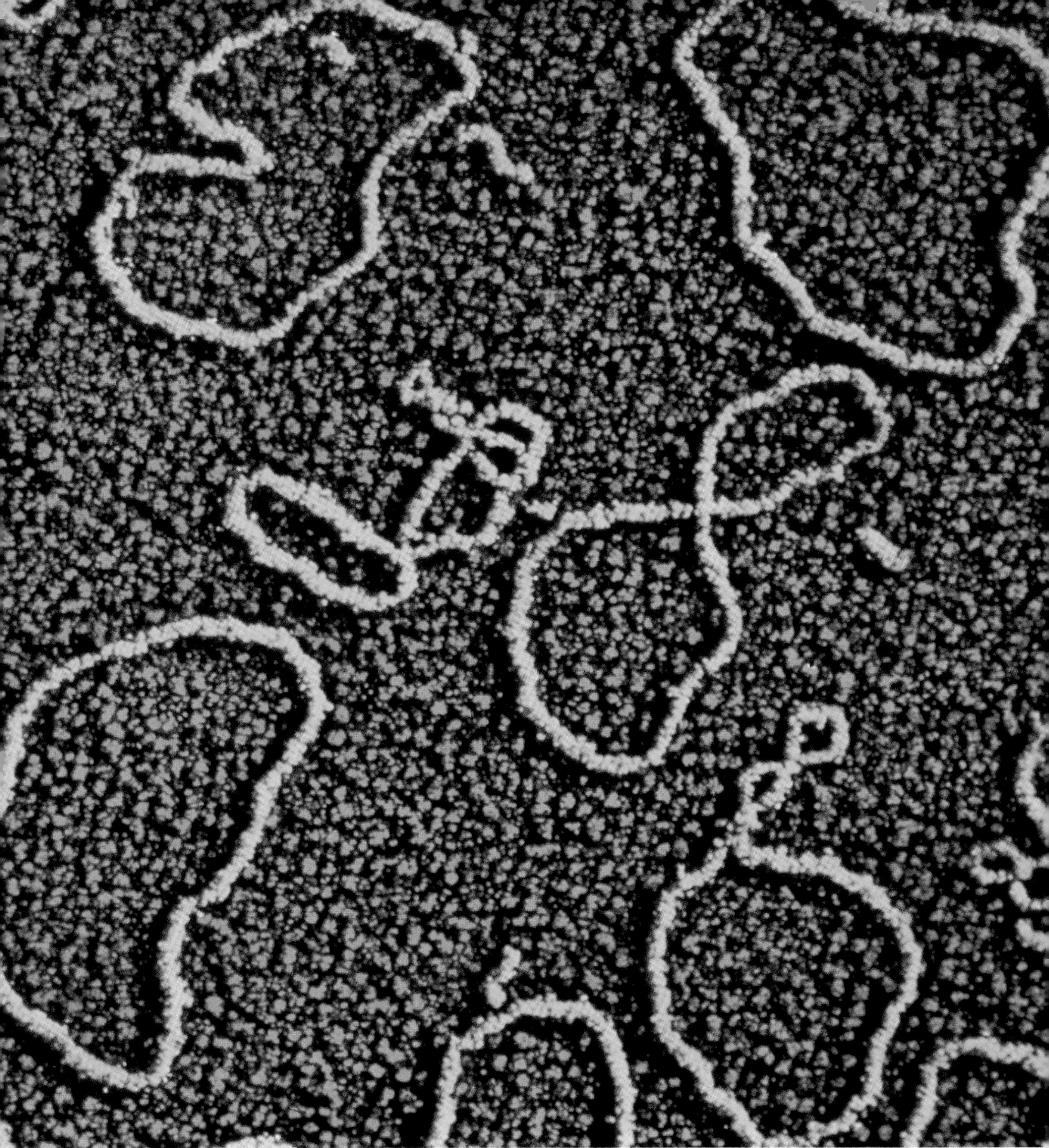

Figure 13.25
Plasmids are small circles of DNA found naturally in the cells of some organisms. A plasmid can replicate itself as well as any other DNA that is inserted into it. For this reason, plasmids make excellent "cloning vectors"—structures that carry DNA from one species into the cells of another.

and the donor DNA are mixed together, the single-stranded "sticky ends" of some plasmids base pair with the sticky ends of the donor DNA. (The ends are called "sticky" because their complementary sequences attract.) The result—a recombinant DNA molecule, such as a plasmid carrying the human insulin gene. The plasmid and its stowaway human gene can now be transferred to a cell from an individual of another species.

But first, the recombinant molecules must be separated from molecules consisting of just donor DNA or just plasmid DNA. This can be taken care of ahead of time by setting up a "selection system" in the particular DNA molecule used. To do this, a plasmid is used containing a gene that makes it able to grow in the presence of a particular antibiotic. A piece of donor DNA is used including a gene that enables it to grow in the presence of a different antibiotic. After the plasmid and donor DNAs are mixed, they are exposed to both antibiotics at the same time. Only the recombinant plasmids can grow in the presence of both.

Once selected, the recombinant plasmid is inserted into a bacterial cell. As the bacterium divides, so does the plasmid, just as a normal plasmid would. Within hours, the original bacterium has given rise to a culture of bacterial cells containing the recombinant plasmid. Because the enzymes, ribosomes, energy molecules, and factors necessary for protein synthesis are present in the bacterial cells, the plasmid DNA and its stowaway human gene are transcribed and translated. The bacterial culture produces its human protein.

Applications of Recombinant DNA Technology

Recombinant DNA technology provides a way to isolate individual genes from complex organisms and observe their functions on the molecular level—that is, which proteins they produce. If the protein product of a particular gene is useful as a drug, then recombinant DNA technology can mass-produce the drug. Often such a genetically engineered drug is less expensive to produce and safer to use than drugs extracted directly from organisms.

The human insulin described at the beginning of this chapter is a good example of a drug improved by recombinant DNA technology, as is the clotting factor missing in hemophiliacs. A few boys with hemophilia have contracted AIDS from the clotting factor obtained from donated blood. Human growth hormone is another drug produced with recombinant DNA technology and is used to treat pituitary dwarfism in children. The hormone was formerly collected from pituitary glands of human cadavers, a costly source because

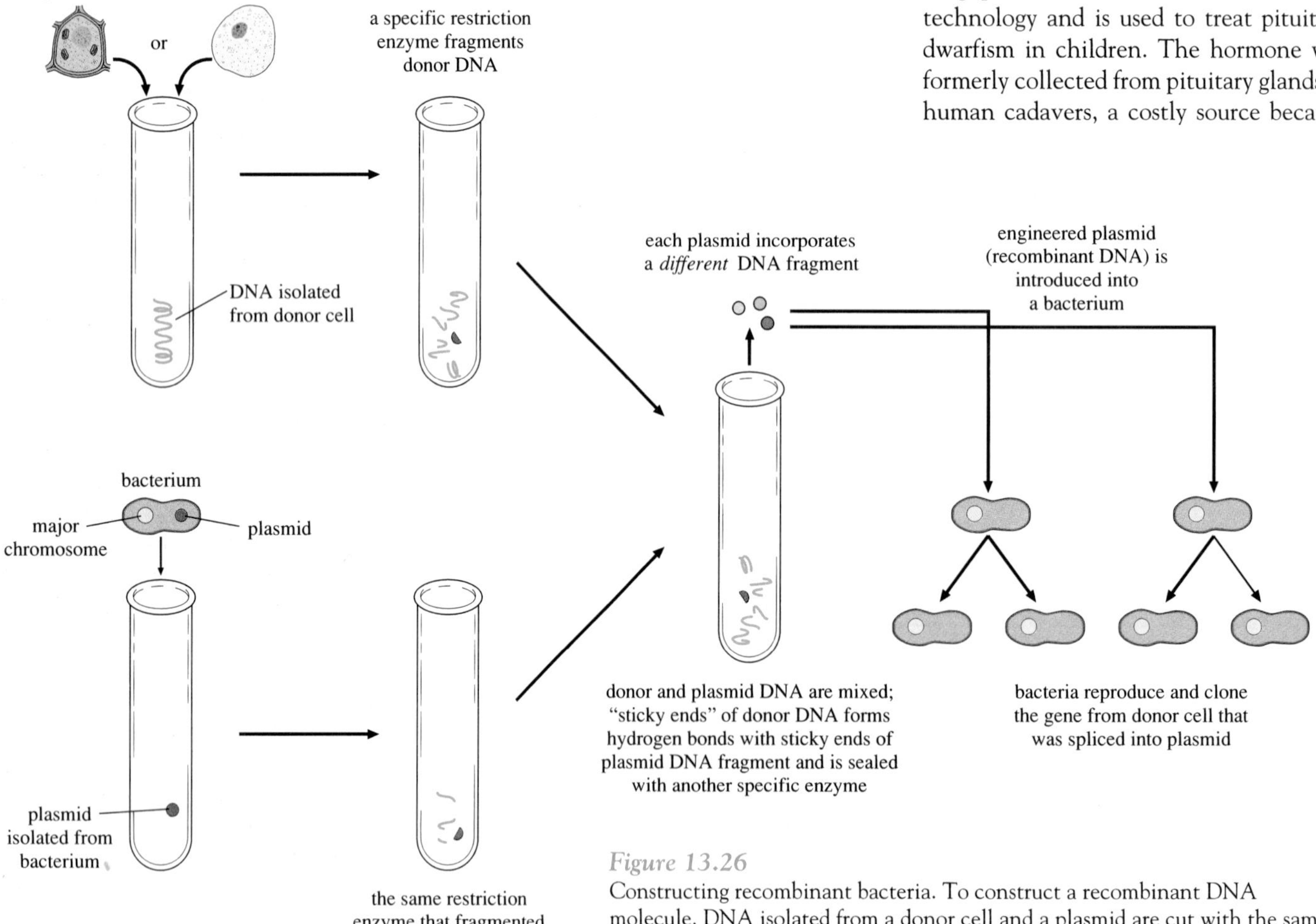

Figure 13.26
Constructing recombinant bacteria. To construct a recombinant DNA molecule, DNA isolated from a donor cell and a plasmid are cut with the same restriction enzyme and mixed. Some of the resulting "sticky ends" from the donor DNA hydrogen bond with the sticky ends of the plasmid DNA, forming recombinant DNA molecules. When such an engineered plasmid is introduced into a bacterium, it is mass-produced as the bacterium divides.

very little of the substance is found in each gland. Children sometimes acquired viral infections of the central nervous system through human growth hormone obtained from cadavers. The hormone produced in genetically engineered *E. coli*, however, is not contaminated, and large amounts can easily be obtained. *E. coli* bacteria can also be genetically engineered to produce human antibodies, which they do more efficiently than monoclonal antibodies. So already one biotechnology is replacing another.

Products of recombinant DNA technology are also used in the food industry. The enzyme rennin, for example, is normally produced in calves' stomachs and is used in cheese making. The gene coding for the enzyme is inserted into plasmids and transferred to bacteria, which are mass-cultured and produce large quantities of pure rennin. Other once-scarce substances now plentiful because of recombinant organisms are cited in table 13.2.

Recombinant DNA technology can recover genes from preserved remains of extinct or mummified organisms, insert this DNA into plasmids and then into bacteria, and discover the genes' products. This has been done with DNA taken from a salt-preserved hide of a quagga, an extinct relative of the zebra that roamed the South African plains more than 100 years ago. Mammoths frozen in the ice in Arctic regions have also had their DNA studied. Deciphering gene sequences from extinct organisms and comparing them to sequences in their modern-day relatives provides information on evolutionary and taxonomic relationships. If a preserved sample yields only a few cells, gene amplification can be used to increase the amount of genetic material for study. Because such DNA is rare, however, it is unlikely that recombinant DNA technology will bring back the dinosaurs.

Transgenic Organisms

Recombinant DNA technology was pioneered in bacteria and is now being applied to more complex organisms. When a cell of a multicellular organism receives a foreign gene, and then an individual develops from the engineered cell, the resulting organism is termed *transgenic*. The altered cell can be from the somatic tissue of a plant or from an oocyte or fertilized egg of a sexually reproducing animal.

Transgenic technology permits the rapid introduction of new traits. For example, a gene that confers an agriculturally useful characteristic—such as the ability to withstand a particular pesticide—is isolated from one species, inserted into a vector, and the recombinant vector placed into single plant cells whose cell walls have been removed. If a whole plant is regenerated from the genetically engineered cell, it has the gene for the transferred trait in all of its cells. Frequently used plant vectors include the Ti plasmid (for "tumor-inducing"), which occurs naturally in the bacterium *Agrobacterium tumefaciens*, and viruses found in plant cells. For example, a gene from the bacterium *Bacillus thuringiensis* specifies a protein that destroys the stomach linings of insects. When the gene is engineered into tobacco cells and the cells regenerated into plants, the resulting tobacco plants produce their own insecticide.

Transgenic animals may revolutionize the pharmaceutical industry, because they can provide vast quantities of valuable drugs. Consider Ethel, a sheep who is the matriarch of a very special flock in Edinborough, Scotland. When she was a fertilized egg, she received a human gene for factor VIII, the protein needed to clot the blood of most hemophiliacs. Another piece of DNA instructed the gene to be activated in the sheep's milk. The valuable substance need only be separated from the milk. Now in its third generation, the flock can provide enough clotting factor to supply the world's hemophiliacs. Transgenic mice have been similarly engineered to secrete human tPA in their milk. These genetically altered animals are also valuable research tools. One strain of transgenic mouse harbors a human gene that causes breast cancer, enabling researchers to study the very early development of this disease. A variation of transgene technology is gene targeting, in which a multicellular organism has a particular gene inactivated in each of its cells. By observing these organisms, researchers can pinpoint the roles of particular genes.

Table 13.2
Drugs Produced Using Recombinant DNA Technology

Drug	Use
Atrial natriuretic factor	Dilates blood vessels, promotes urination
Epidermal growth factor	Accelerates healing of wounds and burns; treatment of gastric ulcers
Erythropoietin	Stimulates production of red blood cells in treatment of anemia
Factor VIII	Promotes blood clotting in treatment of hemophilia
Fertility hormones (follicle stimulating hormone, leutinizing hormone, human chorionic gonadotropin)	Treatment of infertility
Human growth hormone	Promotes growth of muscle and bone in pituitary dwarfs
Insulin	Allows cells to take up glucose in treatment of diabetes
Interferon	Destroys some cancer cells and some viruses
Lung surfactant protein	Helps alveoli in lungs to inflate in infants with respiratory distress syndrome
Renin inhibitor	Lowers blood pressure
Somatostatin	Decreases growth in muscle and bone in pituitary giants
Superoxide dismutase	Prevents further damage to heart muscle after heart attack
Tissue plasminogen activator	Dissolves blood clots in treatment of heart attacks and arterial blockages

KEY CONCEPTS

Biotechnology provides products based on altered cells or biological molecules. Recombinant DNA technology uses restriction enzymes to cut out genes from one organism and introduce them, via plasmids or other vectors, into other organisms, where they are expressed. Products of recombinant DNA technology, from engineered bacteria as well as transgenic organisms, are used in such varied fields as health care, food technology, agriculture, and forensics.

Sequencing the Human Genome

In the summer of 1986, another historic meeting in the field of molecular genetics took place, this time at the opposite end of the country from Asilomar, on Long Island. The Cold Spring Harbor meeting, convened by DNA-discoverer James Watson, had a goal so mindboggling that it could hardly have been imagined when many of those present had originally met at Asilomar a decade earlier. The goal—to determine how to decipher the complete genetic instructions of humans, our genome. Just 2 years after that first meeting, a worldwide effort was underway to sequence the genome, a little at a time. The big questions were: In what order should the mass-sequencing proceed? Should disease-causing genes be looked at first, and if so, which ones? Whose genome should we sequence? What should we do with the resulting avalanche of information?

Sequencing the human genome became possible when chemical DNA sequencing methods developed in the 1970s were automated in the mid-1980s. The sequencing project uses **gene libraries,** which are collections of recombinant bacterial cultures, each cell containing a piece of the human genome. Altogether, the cells of a library harbor the entire human genome of 3 billion bases. A gene library is constructed by cutting all of the DNA in a human cell with several restriction enzymes and inserting each resulting piece into plasmids and then into bacteria. Different libraries can be obtained by using different combinations of restriction enzymes. Researchers contribute their human gene libraries to centralized human DNA repositories, from which investigators can "withdraw" and "deposit" particular bits of DNA.

It is now only a matter of time before we know the sequences of all 100,000 or so human genes and what their products are. Molecular biologist Walter Gilbert, who was present at Asilomar, codeveloped a DNA sequencing method in the 1970s, and is part of the human genome sequencing effort, calls the goal "an incomparable tool for the investigation of every aspect of human function." It is certainly true that within the sequence of 3 billion nucleotides lie the secrets to what makes us human—our strengths and weaknesses, our similarities and differences. The technology is here, a direct outgrowth of our probings into the molecular nature of the gene over the past four decades. We must now rely on our wisdom to effectively use the secrets unlocked from our genes.

SUMMARY

DNA, the genetic material, contains information in its nucleotide sequence that is used to synthesize polypeptides. A DNA *nucleotide* consists of a sugar (*deoxyribose*), a phosphate group, and a nitrogenous base, of which there are two types, *purines* and *pyrimidines*. The DNA purine bases *adenine* (A) and *guanine* (G) contain two carbon rings each, and the DNA pyrimidine bases *cytosine* (C) and *thymine* (T) contain one carbon ring each. Two DNA molecules align and twist about each other, A pairing with T and G with C, to form a double helix. This complementary base pairing provides a mechanism for DNA replication in which the double helix unwinds and free-floating nucleotides hydrogen bond to the exposed bases in each parental strand.

Transcription is the enzyme-assisted manufacture of *ribonucleic acid* (RNA). RNA differs from DNA in that it is single-stranded and that it contains the pyrimidine base *uracil* instead of thymine and the sugar *ribose* instead of deoxyribose. *Messenger RNA* (mRNA) is complementary to the *sense strand* of a DNA molecule, which specifies the amino acid sequence of a particular polypeptide. Each group of three consecutive mRNA bases is a *codon*, and it specifies a particular amino acid. This correspondence between mRNA codons and amino acids is called the *genetic code*. Of the 64 different codons, 61 specify amino acids and 3 signal the termination of polypeptide synthesis. There are 20 different amino acids, so many of them are specified by more than one degenerate codon. The genetic code is nonoverlapping, triplet, and identical in nearly all genes of all species. The code was deciphered by constructing synthetic RNA molecules, exposing them to the contents of *E. coli* cells, and seeing which amino acids were strung into peptides. Other experiments demonstrated the triplet nature of the genetic code.

Translation of a polypeptide from an mRNA molecule requires *transfer RNA* (tRNA), *ribosomes* (which are composed of ribosomal RNA and proteins organized into small and large subunits), and certain energy-storage molecules, enzymes, and factors. Transfer RNA has a specific amino acid covalently bonded to one of its ends and a three-base sequence, the anticodon, in the middle of the molecule. The *anticodon* is complementary to a specific mRNA codon.

In translation initiation, mRNA representing a gene, a small ribosomal subunit, and a tRNA carrying formylated methionine (fmet) come together to form an initiation complex. In translation elongation, a large ribosomal subunit joins the small one already in place. A second tRNA binds by its anticodon to the next mRNA codon, and its amino acid forms a dipeptide bond with the fmet brought in by the first tRNA. Transfer RNAs whose anticodons are complementary to each successive codon bring in their amino acids, and a polypeptide chain is built as each new amino acid bonds to the existing chain. The ribosome moves down the mRNA as the polypeptide chain grows. When a "stop" codon is reached, the ribosome falls apart into its subunits and is released, and the new polypeptide breaks free. After translation, some polypeptides are cleaved and some aggregate to form multisubunit proteins. The proteins are either used in the cell or secreted.

Recent discoveries have altered our picture of the gene. Some genes of eukaryotes contain sequences, called *introns*, that are not represented as amino acids in the gene's polypeptide product. Some viral genes overlap, and some eukaryotic genes are nested. Mitochondria and some ciliated protozoa utilize slightly different genetic codes. Some genes can jump from one site on a chromosome to another.

Biotechnology refers to the manipulation of cells or biochemicals. In *recombinant DNA technology*, DNA from a donor organism and DNA from a vector (such as a *plasmid*) are cut with the same *restriction enzymes* and joined to form recombinant DNA molecules, which are inserted into bacteria. The recombinant DNA replicates along with the bacterial DNA and is transcribed and translated. DNA stored in gene libraries and repositories is being used to sequence the human genome.

QUESTIONS

1. Refer to figure 1 to answer the following questions.

- **a.** Label the mRNA and the tRNA molecules, and draw in the ribosomes.
- **b.** What are the next four amino acids to be added to peptide (*b*)? (Consult table 13.1 for your answer.)
- **c.** Fill in the correct codons in the mRNA opposite the sense strand (*c*).
- **d.** What is the sequence of the DNA antisense strand (as much of it as can be determined from the figure)?
- **e.** Is the end of the peptide encoded by this gene indicated in the figure? How can you tell?
- **f.** What might happen to peptide (*b*) after it is terminated and released from the ribosome?

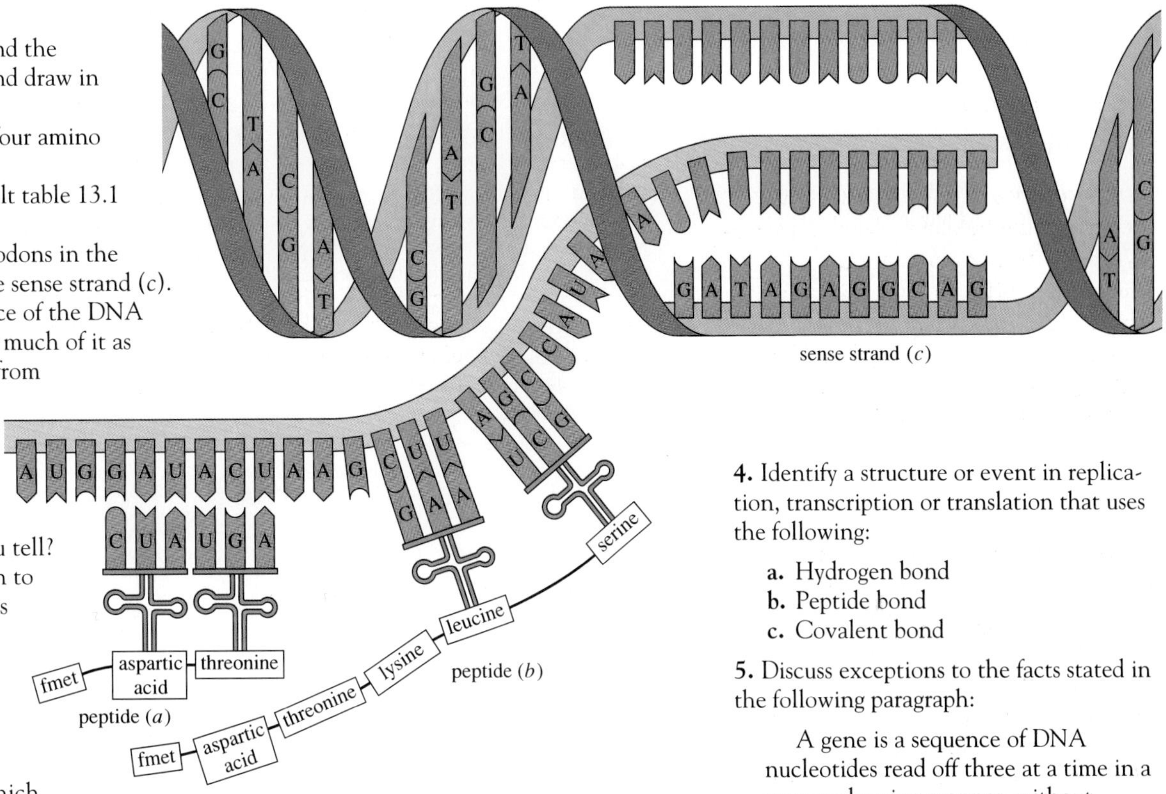

2. To answer the following questions, refer to this DNA sequence, which is part of a sense strand:

GCAAAACCGCGATTATCATGCTTC

- **a.** What is the sequence of the antisense strand?
- **b.** What is the sequence of the mRNA transcribed from the sense strand?
- **c.** What is the amino acid sequence specified by the sense strand?
- **d.** What would be the amino acid sequence if the genetic code were completely overlapping, that is, if a new codon began with each consecutive base?
- **e.** If, by mutation, the 15th base were changed to thymine, what would be the consequence to the amino acid chain?

3. How is complementary base pairing responsible for

- **a.** the structure of the DNA double helix?
- **b.** replication of DNA?
- **c.** transcription of RNA from DNA?
- **d.** the attachment of mRNA to a ribosome?
- **e.** codon/anticodon pairing?
- **f.** tRNA conformation?
- **g.** the use of DNA probes to diagnose certain infections and inherited disorders?

4. Identify a structure or event in replication, transcription or translation that uses the following:

- **a.** Hydrogen bond
- **b.** Peptide bond
- **c.** Covalent bond

5. Discuss exceptions to the facts stated in the following paragraph:

> A gene is a sequence of DNA nucleotides read off three at a time in a nonoverlapping manner, without punctuation, to specify a particular polypeptide. Three contiguous nucleotides constitute a codon and specify a particular amino acid, and the codon—amino acid assignments are the same in all organisms. A gene resides at a certain site on a chromosome.

6. List four useful products of recombinant DNA technology.

TO THINK ABOUT

1. Many articles on genetics in the popular press state, "We are now beginning to crack the genetic code," or "Everyone has his or her own unique genetic code." What is wrong with such statements?

2. How can a mutation alter the sequence of DNA bases in a gene but not produce a noticeable change in the gene's polypeptide product?

3. Why do different cell types have different rates of transcription and translation?

4. How do proteins assist in their own synthesis?

5. Human cells contain far more DNA than is apparently needed to code for polypeptides. The human body contains about 10 times as many different proteins as an *E. coli* cell, but each of our cells contains roughly 1,000 times as much DNA. What functions might be attributed to this seemingly excess genetic material?

6. What restraints, if any, do you think should be placed on recombinant DNA technology? Do you think regulations should differ for research conducted at universities and research conducted in private industries? Do you think that people living near laboratories in which recombinant DNA research is conducted should have any input into how the research is done? Why or why not?

7. How might recombinant DNA technology affect you personally?

8. Do you think researchers should attempt to sequence the human genome? If so, how do you think they should go about this monumental task? What sorts of problems do you predict that they will encounter?

SUGGESTED READINGS

Ahern, Holly. May 1989. DNA repair. *Biology Digest*. DNA has a built-in mechanism to maintain itself.

Baltimore, David. November 1984. The brain of a cell. *Science 84*. Watson's and Crick's work put into historical perspective by a leading molecular geneticist.

Hall, Stephen S. February 1990. Biologists zero in on life's very essence. *Smithsonian*. Biologists worldwide are cooperating to sequence the human genome.

Lewis, Ricki. February 1987. History of recombinant DNA technology. *Biology Digest*. From ancient Mexican Indians to today's scientific meetings, how we have learned to manipulate the genetic material—and why.

Lewis, Ricki. June 1988. DNA fingerprints: Witness for the prosecution. *Discover*. At the DNA sequence level, we are all different, a fact that has been used to fashion a powerful forensic tool.

Lewis, Ricki. October 1989. Making sense out of antisense. *BioScience*, vol. 39, no. 9. Antisense sequences can be used to silence genes selectively.

Lewis, Ricki. November 1990. Beyond PCR. *Genetic Engineering News*. Alternatives to the polymerase chain reaction use different DNA replication enzymes.

Mullis, Kary B. April 1990. The unusual origin of the polymerase chain reaction. *Scientific American*. How the multipurpose technique of gene amplification arose from a brainstorm.

Steitz, Joan Argetsinger. June 1988. "Snurps." *Scientific American*. How the nonsense is deleted from genetic messages.

Watson, James D. 1968. *The double helix*. New York: New American Library. An exciting, personal account of the discovery of the structure of DNA.

Watson, J. D., and F. H. C. Crick. April 25, 1953. Molecular structure of nucleic acids: A structure for deoxyribose nucleic acid. *Nature*, vol. 171, no. 4356. The original paper describing the structure of DNA.

Weaver, Robert F. December 1984. Changing life's genetic blueprint. *National Geographic*. A clearly illustrated look at DNA structure and function.

C H A P T E R

14

Human Genetics

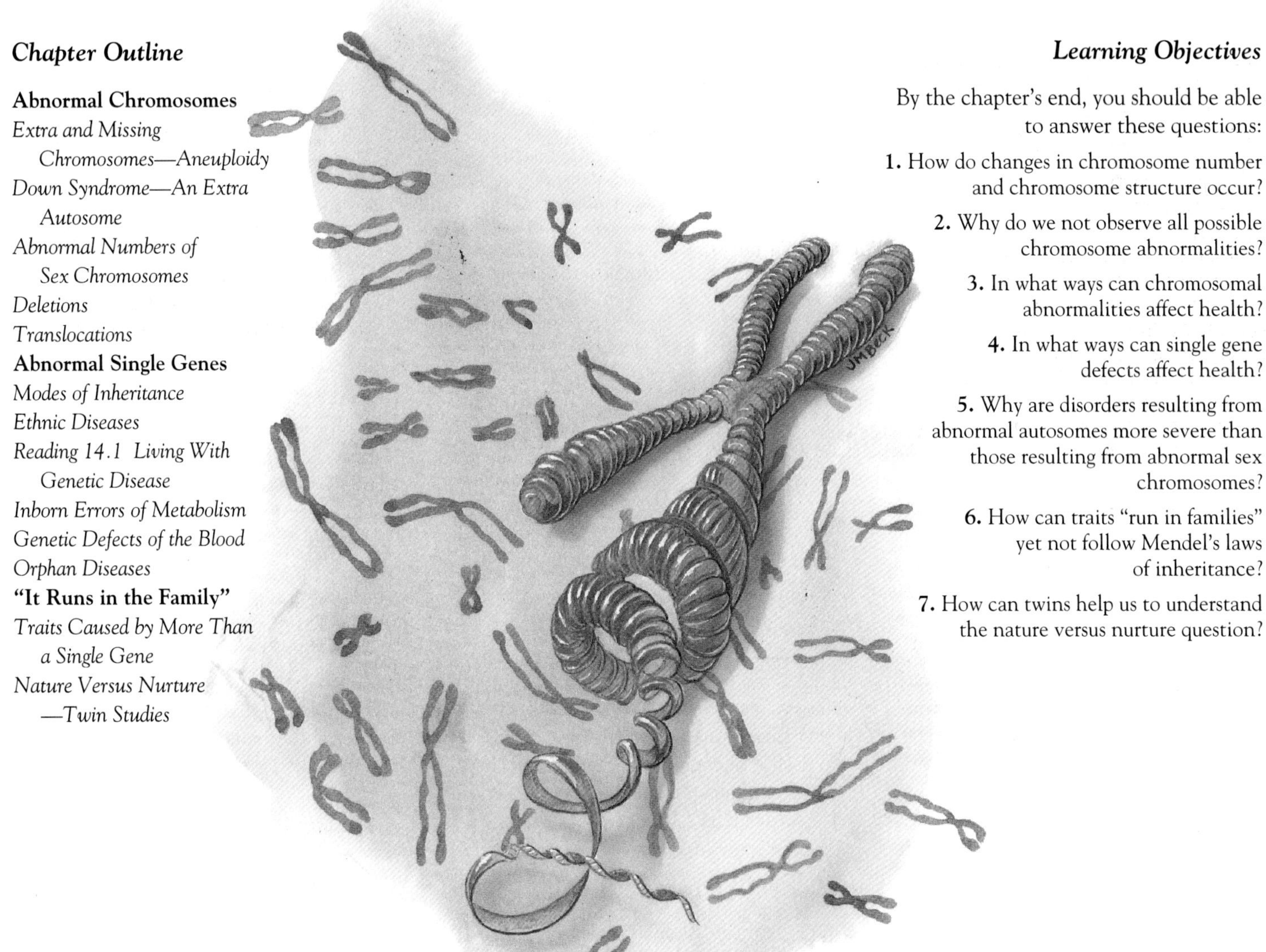

Chapter Outline

Abnormal Chromosomes

Extra and Missing Chromosomes—Aneuploidy

Down Syndrome—An Extra Autosome

Abnormal Numbers of Sex Chromosomes

Deletions

Translocations

Abnormal Single Genes

Modes of Inheritance

Ethnic Diseases

Reading 14.1 Living With Genetic Disease

Inborn Errors of Metabolism

Genetic Defects of the Blood

Orphan Diseases

"It Runs in the Family"

Traits Caused by More Than a Single Gene

Nature Versus Nurture—Twin Studies

Learning Objectives

By the chapter's end, you should be able to answer these questions:

1. How do changes in chromosome number and chromosome structure occur?
2. Why do we not observe all possible chromosome abnormalities?
3. In what ways can chromosomal abnormalities affect health?
4. In what ways can single gene defects affect health?
5. Why are disorders resulting from abnormal autosomes more severe than those resulting from abnormal sex chromosomes?
6. How can traits "run in families" yet not follow Mendel's laws of inheritance?
7. How can twins help us to understand the nature versus nurture question?

The new mother and father were quite surprised at the appearance of their daughter. Although they had expected her to be blond because they were both fair haired, they were astonished at the absolute whiteness of her fuzzy damp hair and her pink eyes. The doctor explained that the child was an albino, and that although she might have some visual problems, she would probably live a perfectly normal life. Her blond hair might one day even be the envy of her friends. The young couple wanted more information, and so they consulted a genetic counselor. As the fair-haired infant slept in her mother's arms, the counselor asked each of the parents about their families and sent them home armed with questions to ask their elder relatives. Sure enough, they found albinos in both branches of their family trees. When the couple's second child was born, a boy with brown hair and bright blue eyes, they were a little disappointed that they did not have two special children.

The woman was only 10 weeks pregnant, yet she already knew that her daughter had Down syndrome. Although a test of fetal cells had shown the extra chromosome that causes the disorder, it could not show how severely affected the child would be—only mildly retarded, or severely retarded, or perhaps also with a serious heart defect (fig. 14.1). Now the woman was meeting with her genetic counselor for a second time, to discuss Down syndrome in more detail and to try to come to a decision about whether to end the pregnancy or raise a handicapped child. Her initial meeting with the counselor had been because this was her first pregnancy, and she was 38 years old. Her age put her at a higher risk of having a child with Down syndrome than a younger woman would have. The genetic counselor had suggested the test that had led to the diagnosis of Down syndrome. Now the hard part began.

It was not until the frail child was a year old that doctors thought to perform the blood test that showed the abnormal sickle shape of his red blood cells, a finding that explained his lethargy and frequent infections. A genetic counselor told the boy's parents that the abnormal red blood cells of his inherited sickle cell disease slowed his circulation, causing pain wherever the flow of blood to a vital organ was blocked. With daily antibiotics the child could avoid the severe infections associated with sickle cell disease, but once past childhood he might suffer from pneumonia, muscle pain, and maybe even heart failure, the counselor gently told them. The distraught and confused parents wanted to know how two healthy people like themselves could have passed on such a disorder to their child. If they have a second child, they asked the counselor, what are the chances that he or she would also inherit sickle cell disease?

Figure 14.1
Medical genetics can present difficult choices. Down syndrome is usually caused by an extra chromosome 21. The symptoms vary greatly from individual to individual. About half of Down syndrome patients die before 1 year of age, often due to malformations of the heart or kidneys. Other patients must undergo heart surgery or suffer very often from common illnesses. However, some Down syndrome patients are happy, loving children and young adults. Unfortunately, the test that diagnoses Down syndrome in a fetus cannot detect how severely affected the child will be. The decision of whether to terminate such a pregnancy or raise a child with Down syndrome is very difficult.

These real-life examples illustrate just a few ways in which the inheritance of an unusual characteristic can affect a family. As researchers have learned more about the transmission of traits (chapter 11, and chapter 13 on the molecular nature of the gene), the field of medical genetics has emerged to focus on understanding, and eventually treating or even preventing, inherited disorders.

A normal human somatic cell contains 23 pairs of chromosomes and is said to be euploid (good set). These 46 chromosomes contain enough DNA to account for 50,000 to 100,000 genes, each of which provides the information for the assembly of a particular polypeptide. So far, the polypeptide products of about 3,550 human genes have been identified. Genetic disease can result from a problem as great as an abnormal number of chromosomes, or as specific as inheriting two disease-causing recessive alleles of a gene, or by inheriting a single dominant disease-causing allele. Because even a single gene product can participate in several biochemical reactions, many genetic diseases have several symptoms and therefore constitute a syndrome. The particular symptoms depend upon the functions of the polypeptide whose structure is abnormal. Expression of a genetic disease is also influenced by the environment and sometimes by the actions of other genes.

Abnormal Chromosomes

Chromosomes can be abnormal in a variety of ways. This can be shown on charts called **karyotypes** (fig. 14.2). Sometimes an error in meiosis produces a sperm cell or an egg cell that has an extra complete set of chromosomes, a condition called **polyploidy** (many sets). For example, if a sperm that has the normal one copy of each chromosome fertilizes an egg with two copies of

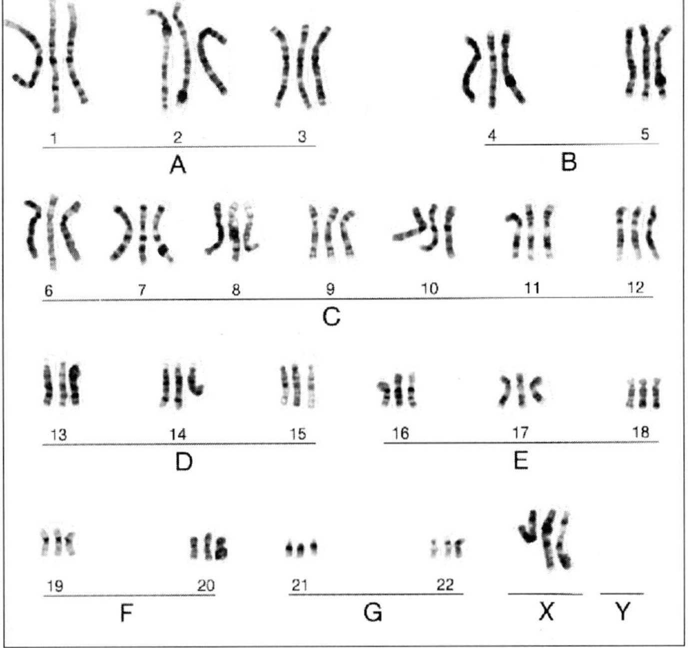

a.

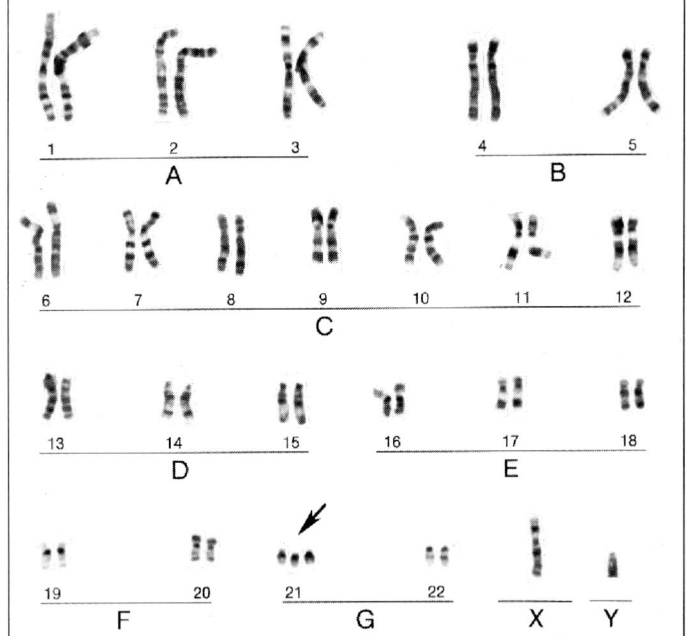

b.

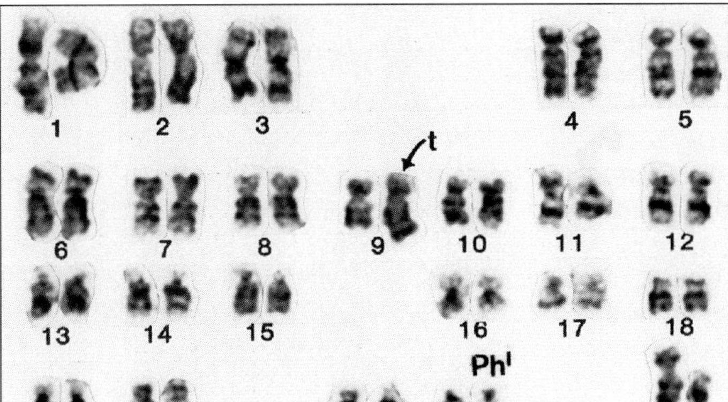

c.

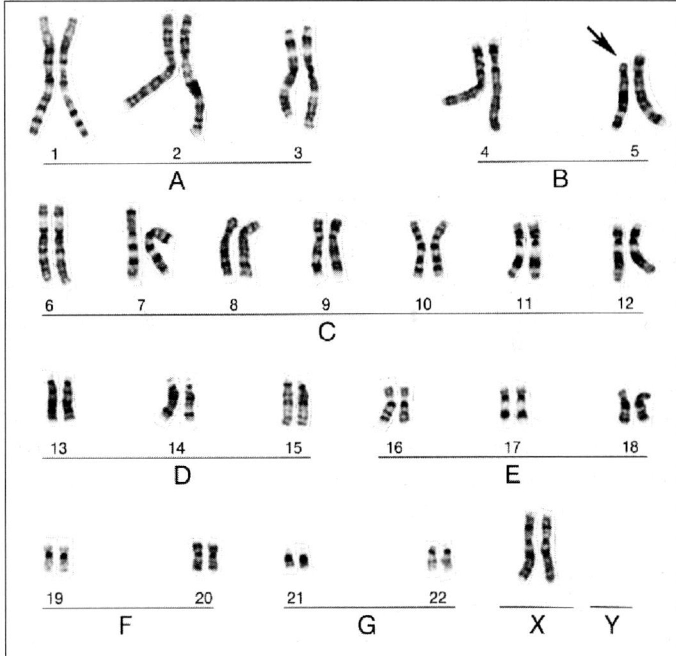

d.

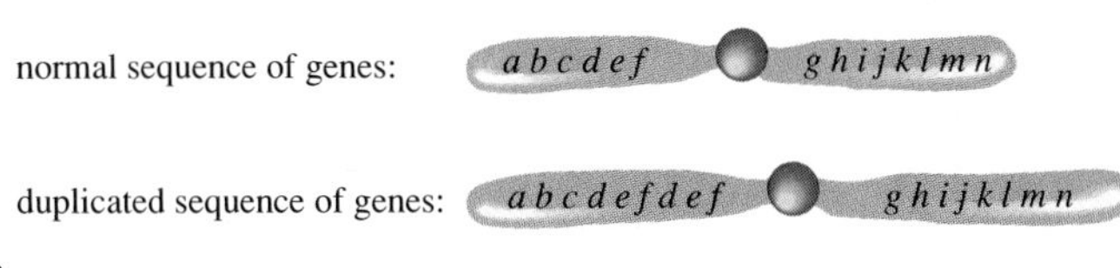

e.

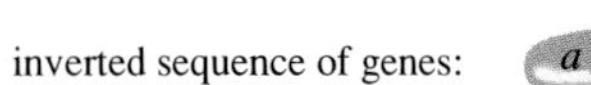

f.

Figure 14.2

Types of chromosomal defects. A normal fertilized human egg has 23 pairs of chromosomes. Variations in this number, or in the arrangement of genes in chromosomes, can affect health. *a.* Persons with three copies of each chromosome (triploids) account for 17% of all spontaneous abortions and 3% of stillbirths and newborn deaths. *b.* An extra or missing single chromosome is frequently lethal before birth and is usually associated with major defects if the individual survives birth. Most Down syndrome patients have an extra chromosome 21. *c.* Most children with chronic myeloid leukemia have a translocation, or "mixed-up" chromosome, in which the tip of chromosome 22 is attached to chromosome 9. *d.* A child born with cri-du-chat syndrome is missing a piece of chromosome 5. He or she has a cry like that of a cat, has an unusually small head, and is mentally retarded. *e.* In a duplication, genes are repeated within a chromosome, but this usually has no effect on health. *f.* In an inversion, a sequence of genes is turned around. This does not affect health unless a gene is physically disrupted.

each chromosome, the resulting zygote will have three copies of each chromosome, a type of polyploidy called *triploidy*. Most human polyploids die as embryos or fetuses, but occasionally one will survive for a few days after birth, with defects in nearly all organs. However, polyploidy is tolerated well in about 30% of flowering plant species. Many agriculturally important plant variants are derived from polyploids.

An individual whose cells have an extra or missing chromosome is called an **aneuploid** (not good set). Symptoms depend upon which chromosomes are missing or extra. Other abnormalities affect chromosome structure but do not alter chromosome number. In an **inversion,** for example, a chromosome is broken in two places so that a piece is momentarily removed, flipped around by 180°, and rejoined into the chromosome. If a gene is cut in two by an inversion, it cannot provide the complete information for the synthesis of its polypeptide product, and symptoms may result. A chromosome can contain extra copies of a gene (a *duplication*) or lack some genes (a *deletion*). Duplications rarely cause medical problems, but deletions can be devastating to health. In a **translocation,** two different chromosomes break apart and then come together so that they have exchanged parts. If a gene is broken during a translocation event, a medical problem may result. Let's take a closer look at abnormal chromosomes.

KEY CONCEPTS

Somatic cells of healthy humans contain 22 pairs of autosomes and a pair of sex chromosomes. Of the 100,000 or so genes, the functions of only 3,550 are known. Chromosome aberrations include polyploidy, aneuploidy, inversions, duplications, deletions, and translocations.

Extra and Missing Chromosomes—Aneuploidy

A person with an extra or missing chromosome is often mentally retarded. This is because most of the chromosomes carry many genes that contribute to brain function, so that nearly any type of chromosome irregularity disrupts at least one of them. So profound is the effect of just one extra or missing chromosome that individuals with more than one almost never survive birth.

Extra genetic material is less dangerous than missing material, and this is why most children born with the wrong number of chromosomes have an extra one (called a **trisomy**) rather than a missing one (called a **monosomy**). Trisomies and monosomies are named according to the particular chromosome that is involved. The autosomes are numbered from 1 to 22, from largest to smallest. A fetus with Down syndrome has "trisomy 21," or an extra copy of chromosome 21. The absence of a chromosome (a monosomy) is usually lethal to an embryo or fetus. Aneuploidy of the autosomes is generally more severe than aneuploidy of the sex chromosomes. This may be because the sex chromosomes do not contain as many genes needed for vital functions as do the autosomes.

Aneuploidy is due to a meiotic mishap called **nondisjunction** (fig. 14.3). In normal meiosis, pairs of homologous chromosomes separate so that each of the resulting sperm or egg cells contains only one member of each pair. In nondisjunction, a chromosome pair fails to separate, either at the first or at the second meiotic division, resulting in a sperm or egg that has two copies of a particular chromosome or none, rather than the normal one copy. When such a gamete fuses with its mate at fertilization, the resulting zygote has either 45 or 47 chromosomes instead of the normal 46. Because there are 23 types of human chromosomes, there are 49 ways in which one chromosome can be missing or extra—an extra or missing copy of each of the 22 autosomes, plus the five abnormal types of sex chromosome combinations (YO, XO, XXX, XXY, XYY). However, only 9 of these 49 possibilities are seen in newborns because most halt development before birth. In fact, about 50% of spontaneous abortions are caused by extra or missing chromosomes.

Aneuploidy and polyploidy can also arise during mitosis, producing groups of somatic cells with the particular chromosomal aberration. If only a few cells are altered, health may not be affected. For example, the human liver often contains patches of polyploid cells, but these have no obvious effect on the organ's activity. However, if such a mitotic abnormality occurs in a very early embryo, so that many cells descend from the original defective one, a serious problem can result. Some people who have mild Down syndrome, for example, are really *chromosomal mosaics*—that is, some of their cells carry the extra chromosome 21, and some are normal.

Down Syndrome—An Extra Autosome

Five types of autosomal aneuploids are known to survive birth. Edward's syndrome (trisomy 18) and Patau's syndrome (trisomy 13) produce multiple defects, and are lethal in infancy or earlier. The most common type of autosomal aneuploidy is trisomy 21, or *Down syndrome*. The characteristic slanted eyes and flat face of the Down syndrome patient prompted Sir John Langley Down to coin the inaccurate term "mongolism" when he described the syndrome in the 1880s. Down was the medical superintendent of a facility for the profoundly mentally retarded, and he noted that about 10% of his patients resembled individuals of the Mongolian race. In his paper "Observations on an Ethnic Classification of Idiots," he wrote: "The great mongolian family has numerous representatives, and it is to this division I wish in this paper to call special attention. A very large number of congenital idiots are typical Mongols. So marked is this that when placed side to side, it is difficult to believe that the specimens compared are not children of the same parents."

It is just a coincidence that the faces of Down syndrome patients resemble those of people of the Mongolian race. Characteristic facial features are associated with many inherited disorders. Males and females of all races can have Down syndrome. The association of the problem with an extra chromosome was made in 1959, just 3 years after the normal number of human chromosomes was deduced.

A person with Down syndrome is short, with straight, sparse hair and a tongue protruding through thick lips. The hands have an abnormal pattern of creases, the joints are loose, and reflexes and muscle tone are poor, resulting in a "floppy" appearance. Development is slow, and it may

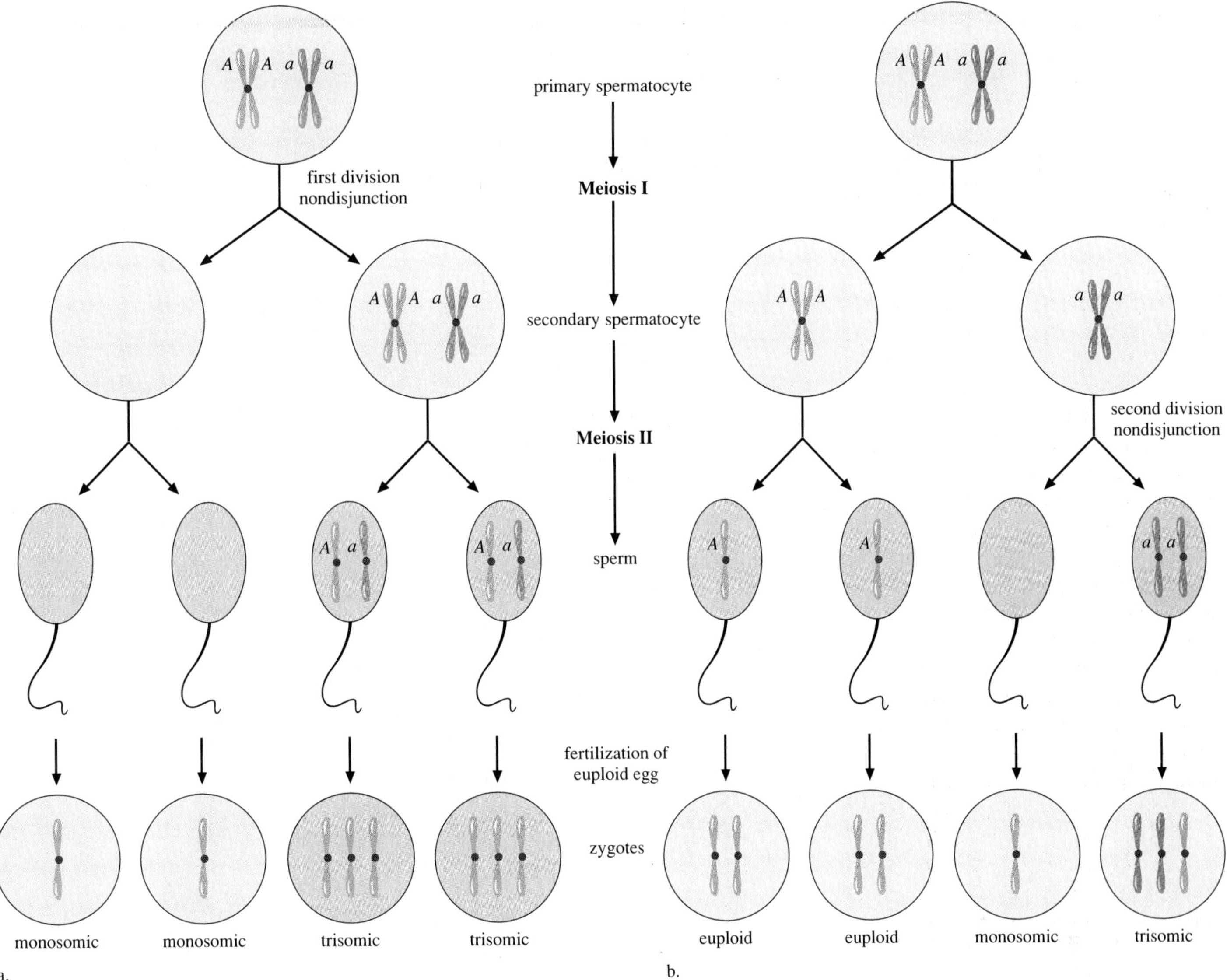

Figure 14.3
Extra and missing chromosome—aneuploidy. Unequal division of chromosome pairs into sperm and egg cells can occur at either the first or the second meiotic division. *a.* A single pair of chromosomes unevenly partitioned into the two cells arising from the first division of meiosis in a male. The result: two sperm cells that have two copies of the chromosome, and two sperm cells that have no copies of that chromosome. When a sperm cell with two copies of the chromosome fertilizes a normal egg cell, the zygote produced is trisomic for that chromosome; when a sperm cell lacking the chromosome fertilizes a normal egg cell, the zygote is monosomic for that chromosome. Symptoms depend upon which chromosome is involved. *b.* This nondisjunction occurs at the second meiotic division. Because the two products of the first division are unaffected, two of the mature sperm are normal and two aneuploid. Egg cells can undergo nondisjunction as well, leading to zygotes with extra or missing chromosomes when they are fertilized by normal sperm cells.

take several years, for example, for a Down syndrome child to be toilet trained. Intelligence is subnormal, but some patients can follow simple directions and even learn to read. One young man with Down syndrome graduated from a junior college in California, and another starred in a television series. These special people tend to have warm, loving personalities and enjoy art and music.

Down syndrome patients often die young of complicating conditions, 50% before their first birthdays. Heart and kidney defects are common. Many patients have suppressed immune systems and may die from a prolonged cold or influenza. A Down syndrome child is 15 times as likely to develop leukemia (a cancer of the white blood cells) than a healthy child. Down syndrome patients rarely live beyond age 40. Those who reach adulthood develop black tangles of protein called *amyloid* in their brains—the same problem seen in the brains of people who have died from *Alzheimer's disease*. Alzheimer's disease strikes 2 million American adults, causing severe memory loss and impairment of reasoning. Sometimes Alzheimer's disease involves a gene on chromosome 21, the same chromosome that in excess causes Down syndrome. Although the exact connection between the two disorders is not yet fully understood, both seem to involve accelerated aging of parts of the brain and accumulation of a sticky protein called amyloid.

The likelihood of giving birth to a Down syndrome child increases dramatically with the age of the mother. For women under 30, the chances of conceiving such a child are 1 in 3,000. By age 48, the number jumps to 1 in 9 (fig. 14.4). The age factor in Down syndrome may be due to the fact that meiosis in the female is completed after conception. The older a woman is, the longer her egg cells have been arrested on the brink of completing meiosis. During this time, the eggs may have been exposed to chemicals or radiation that can damage chromosomes.

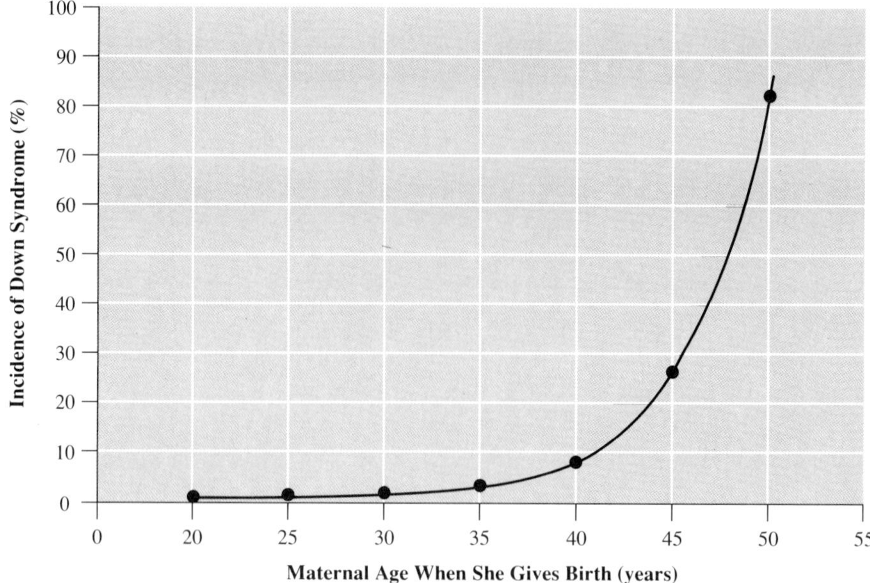

a.

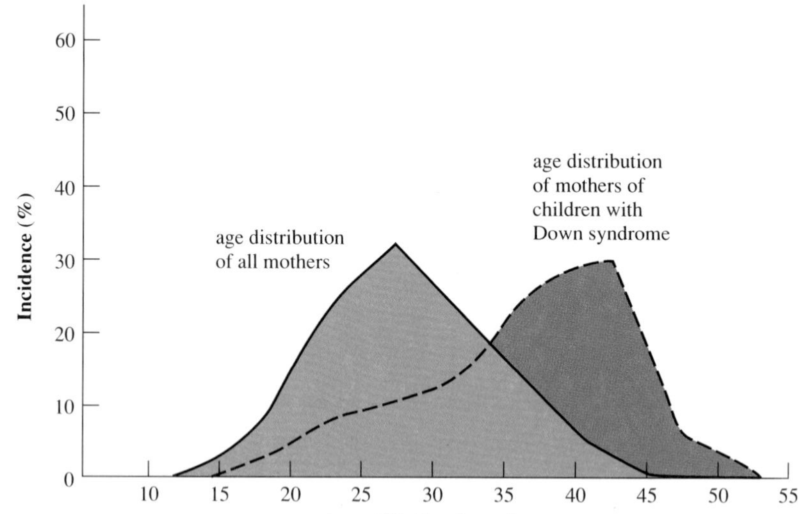

b.

Figure 14.4
Down syndrome and age of the mother. The incidence of Down syndrome in the general population is about 1 in every 770 births. *a.* Among women over the age of 35 years, however, the incidence of delivering a child with Down syndrome increases. *b.* The correlation between maternal and Down syndrome risk is striking when the age distribution for all mothers is compared to that of mothers of Down syndrome children.

Abnormal Numbers of Sex Chromosomes

Too many or too few sex chromosomes affects the development of characteristics that distinguish the sexes. This is far less devastating than extra or missing autosomes.

Recall that a normal human female has two X chromosomes; a normal male, an X and a Y chromosome. A female lacking an X chromosome (XO) has *Turner syndrome*. In childhood, the only signs of the condition are wide-set nipples and extra skin that forms flaps on the back of the neck. At sexual maturity, sparse body hair develops, but the child does not ovulate or menstruate, and she has underdeveloped breasts and genitals. Her ovaries, fallopian tubes, and uterus are very small and immature. Turner syndrome patients are sometimes mentally below normal. The syndrome appears in 1 out of 3,000 female births but is found more frequently among aborted fetuses. These individuals probably have complications affecting vital organ functions.

About 1 of every 1,000 females is born with an extra X chromosome in each of her cells, a condition called *triplo-X*. The only symptom seems to be menstrual irregularities.

Males with an extra X chromosome have *Klinefelter syndrome*. They have underdeveloped testes and prostate glands, no pubic and facial hair, very long arms and legs, and large hands and feet, and they may be mentally impaired. Klinefelter syndrome occurs in 1 out of 500 male births.

One male in 1,000 has an extra Y chromosome. In the 1960s, studies conducted in Swedish mental hospitals and Scottish maximum security prisons and hospitals for the criminally insane found a disproportionate incidence of XYY men in their facilities compared to the general population (1 in 7 compared to 1 in 1,000). Typically, these "extra-Y" men were very tall, prone to acne, aggressive and antisocial, and of lower-than-average intelligence. Many geneticists as well as the press first attributed these traits to the extra Y chromosome, despite the fact that many of the studies were not properly controlled (e.g., they did not also assess the characteristics in XY males in the same institutions). In fact, in 1968 two murderers in France and Australia pleaded innocent on the grounds

Table 14.1
Is an Extra Y Chromosome Associated With a Syndrome?

Population Studied		*Largest Pooled Study*
Newborn males		1 in 1,080 (37,779)
Young boys		1 in 1,300 (3,901)
Males in general population	Any height	1 in 1,000 (11,000)
	Tall	1 in 325 (7,798)
Males in mental institutions	Any height	1 in 316 (2,526)
	Tall	1 in 72 (649)
Males in penal institutions	Any height	1 in 223 (5,805)
	Tall	1 in 56 (1,683)
Males in mental-penal institutions	Any height	1 in 51 (4,010)
	Tall	1 in 29 (2,233)

that their extra Y chromosomes made them not responsible for their behavior. The question of whether or not an extra Y chromosome causes a syndrome is still unanswered. Although several studies have supported the higher incidence of XYY males in mental and penal institutions than in the general population, most geneticists believe that the only trait associated with the XYY condition is above-average height (table 14.1). Perhaps XYY boys are treated differently because of their older appearance, and this makes them become aggressive.

A person with one Y chromosome and no X chromosomes has never been observed. Probably too much genetic material is missing to sustain development beyond a few cell divisions.

Deletions

Sometimes only part of a chromosome is missing (a deletion) or extra (a duplication). For example, *cri-du-chat syndrome* (French for "cat's cry") is associated with the deletion of a piece of chromosome 5 (fig. 14.2*d*). Children with this syndrome have pinched faces and a peculiar cry that sounds like the mewing of a cat and are profoundly mentally retarded.

Translocations

Deletions and duplications can be caused by *translocations*, in which different (nonhomologous) chromosomes exchange parts. These chromosome exchanges can be caused by certain viruses (e.g., the mumps virus), drugs (anticancer drugs), and radiation (medical X rays or ultraviolet radiation). There are two types of translocations—nonreciprocal and reciprocal.

In a *nonreciprocal translocation*, a piece of one chromosome breaks off and attaches to another chromosome. In 97% of patients with chronic myeloid leukemia, for example, the tip of one copy of chromosome 22 is attached to chromosome 9 (fig. 14.2*c*). In a *reciprocal translocation*, two different chromosomes exchange parts. If this swapping does not break any genes, then the person may be perfectly healthy and is called a *translocation carrier*. He or she has the normal amount of genetic material, but it is rearranged. However, a translocation carrier can produce "unbalanced gametes," which are sperm or eggs that have deletions or duplications of genes in the translocated chromosomes. This occurs when the sperm or egg receives one reciprocally translocated chromosome but not the other, resulting in a genetic imbalance. A common reciprocal translocation is between chromosomes 14 and 21; figure 14.5 shows the chromosomal abnormalities associated with it.

A physician or genetic counselor would suspect a translocation carrier in a family with many instances of birth defects and spontaneous abortions. The effects of a translocation can also be seen in plants. An ear of corn resulting from a cross in which one parent is a translocation carrier has many tiny, undeveloped kernels. These are embryos that had too much extra or missing genetic material to complete development.

KEY CONCEPTS

Nondisjunction during meiosis leads to aneuploidy. Trisomics are more likely to survive than monosomics. Sex chromosome aneuploidy is less severe than autosomal aneuploidy. Mitotic nondisjunction can result in chromosomal mosaics. Down syndrome (trisomy 21) is the most common aneuploid. Sex chromosome aneuploids include Turner syndrome (XO), Klinefelter syndrome (XXY), triplo-X females, and XYY males. In a nonreciprocal translocation, a piece of one chromosome breaks off and attaches to another chromosome. In a reciprocal translocation, two chromosomes exchange parts. If a translocation leads to a deletion or duplication, symptoms may result.

Abnormal Single Genes

When large blocks of genetic material are lost, gained, or rearranged, health can be seriously affected, as we have just seen. Change in a single DNA building block can also affect health. An extra or missing nucleotide, for example, can disrupt the reading frame of a gene, resulting in the construction of a polypeptide that does not function because it is built of the wrong sequence of amino acids. A substitution of one nucleotide for another can alter the folding pattern of a polypeptide that is essential to its activity. Most of the known inherited diseases are caused by defects in single genes.

Modes of Inheritance

Disorders caused by abnormalities in a single gene are classified by whether they are carried on an autosome or a sex chromosome and by whether one or two defective alleles are necessary to cause symptoms. Most known single-gene defects are inherited as *autosomal recessives*. That is, each parent passes one copy of a disease-causing allele on a non–sex chromosome to a child who inherits the disorder. Autosomal recessive conditions affect the sexes with equal frequency. When two healthy parents have a

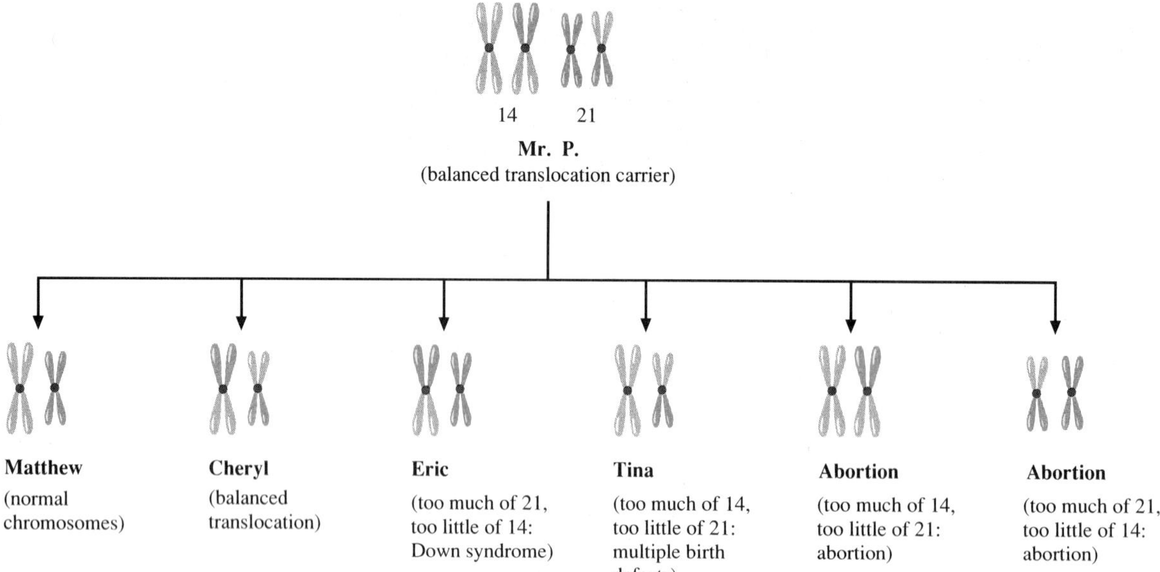

Figure 14.5

Families with many birth defects may have a translocation. After two spontaneous abortions, Mr. and Mrs. P. were overjoyed to have two healthy children, Matthew and Cheryl. Their third child, Eric, had Down syndrome. When the doctor told them it probably could not happen again, they tried for a fourth child. Tina was born with multiple medical problems. Chromosome studies revealed that Mr. P. was a translocation carrier. His 14th largest and 21st largest chromosomes had exchanged parts. This had no effect on him, because his cells still had all of the genes normally found in a human cell. But whenever one mixed-up chromosome was packaged into a sperm cell without the other, and that sperm fertilized a normal egg, an abnormal child was conceived. Tina inherited extra material from chromosome 14 and too little from chromosome 21. Eric inherited the opposite condition, resulting in Down syndrome. Cheryl too had abnormal chromosomes, but hers were balanced, so she was a translocation carrier like her father. Matthew had normal chromosomes. Mrs. P.'s two spontaneous abortions were also probably caused by her husband's translocation—the embryos lacked too much genetic material to develop.

child with an inherited illness, such as the child with sickle cell disease mentioned at the beginning of this chapter, it is usually due to autosomal recessive inheritance. The parents, heterozygotes for the disease-causing allele, are called *carriers*.

Sex-linked recessive conditions are transmitted on the X chromosome. They are more likely to be expressed in males than in females, because a female's second X chromosome can carry a normal allele of the gene, masking the effect of the abnormal one. Sex-linked recessive disorders are passed from carrier mothers to sons, who have the disorder. A sex-linked condition that is not life-threatening, such as color blindness, may be passed from a father who has the condition to a daughter, who is a carrier of it.

Autosomal dominant conditions can be passed to offspring even if only one parent is affected (fig. 14.6), because a single "dose" of a dominant allele is sufficient to cause symptoms. Each child of a couple in which one partner is affected has a 50% chance of developing the condition. These disorders often do not produce symptoms until adulthood. Whereas autosomal recessive and sex-linked recessive conditions can skip generations because they are carried in heterozygotes, an autosomal dominant trait appears in every generation. If by chance no one in a given generation inherits the defective allele, its transmission in the family ends.

Sometimes it is difficult to tell if a medical condition is inherited because it affects different family members to different degrees. A disorder is considered to be hereditary if it meets at least one of the following three criteria: the disorder appears in other family members in a pattern that can be explained by Mendel's laws; a chromosomal abnormality is evident; or a biochemical defect is detected that is caused by a mutant allele. Reading 14.1 describes some of the more prevalent hereditary disorders caused by defects in single genes.

KEY CONCEPTS

Most genetic diseases are inherited as autosomal recessives, affecting males and females equally and being expressed when two deleterious alleles are inherited from carrier parents. Sex-linked recessive traits and illnesses are transmitted on the X chromosome and are therefore passed from carrier or affected mothers to sons. Autosomal dominant characteristics are inherited as a single gene from an affected parent.

Ethnic Diseases

Some genetic diseases are found more frequently among specific population groups. *Sickle cell disease* affects 1 in 500 black babies but only 1 in 160,000 nonblacks. *Cystic fibrosis* affects 1 in 2,000 white babies but only 1 in 250,000 nonwhite babies. One in 3,600 babies born to Jewish people of central or Eastern European descent has *Tay-Sachs disease* but only 1 in 600,000 babies among other population groups has the condition.

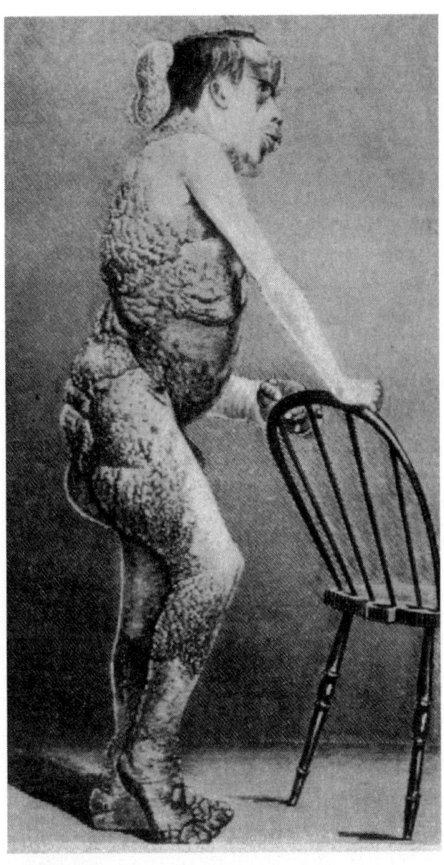

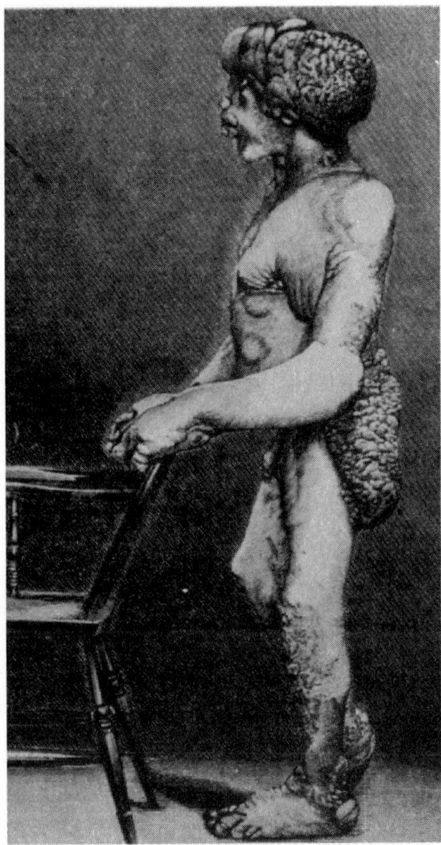

a.

b.

Figure 14.6
Autosomal dominant conditions—when one allele is enough to cause disease. *a.* In neurofibromatosis, the skin is covered with tumors, some of which are cancerous. The condition is often mild, with as few as six light brown pigment spots on the skin, but it can be grossly disfiguring, as it was for this man depicted in the *British Medical Journal* in 1886. *b.* Woolly hair is inherited as an autosomal dominant trait in some Norwegian families.

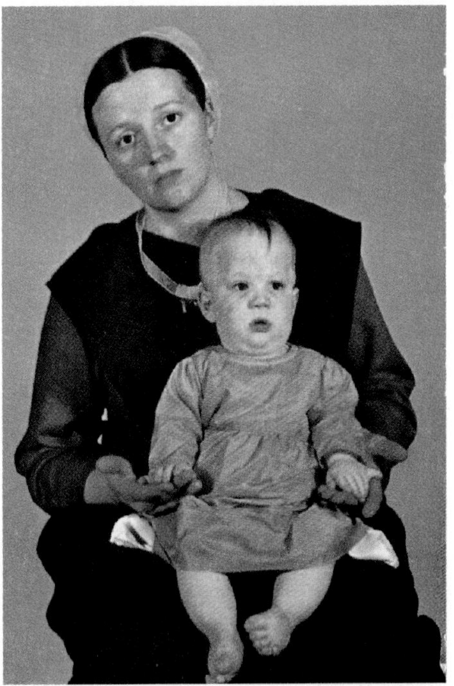

Figure 14.7
Marriage customs can influence the prevalence of genetic disorders. Six-fingered dwarfs, such as this child, are more prevalent among the Old Order Amish of Lancaster County, Pennsylvania, than in the general population because the Amish tend to marry among themselves.

The predominance of certain inherited diseases in particular populations often is a result of marriage practices in which people tend to choose mates of the same ethnic background as themselves, keeping any disease-causing alleles common to that group restricted to it. An extreme example of the effect of marriage customs on gene distribution is seen among the Old Order Amish of Lancaster County, Pennsylvania. In this very tightly knit community of 12,500 people, marriage to outsiders is forbidden. Most husbands and wives are related, and are often both carriers of alleles brought in with the original 60 settlers who came from Europe in the early 1700s. Six-fingered dwarfism, for example, is far more common among the Amish than in the general population, found today in 33 Amish families (fig. 14.7).

The Amish community has provided a gold mine of information to geneticists, because their families are large, they keep detailed family histories, and their behavior is constant. For example, an inherited form of manic depression was traced through

Reading 14.1 Living With Genetic Disease

Figure 1
Parents usually first suspect that their child has Tay-Sachs disease when he or she cannot sit or stand at an age when other children can. Another clue to Tay-Sachs disease is a "cherry red spot" on the retina not seen in normal eyes or in any other disease.

Autosomal Recessive Diseases

MICHAEL WAS A CUTE, SEEMINGLY NORMAL INFANT. But at a year of age, he still could not crawl, while others his age were already taking their first wobbly steps. By 18 months, Michael was blind, his muscles becoming more and more rigid each day as his nervous system gradually deteriorated. He died at the age of 4 years of a respiratory infection that his weakened immune system could not overcome.

Michael had inherited *Tay-Sachs disease*, particularly common among Jewish people of Eastern European descent (fig. 1). A missing enzyme, hexosaminidase, causes lipid to accumulate on nerve cells, triggering slow paralysis. Although carriers can be detected and fetuses diagnosed, there is presently no treatment. A milder, adult form is also known.

Cystic fibrosis is the most prevalent genetic disease among white Americans, affecting 1 in 2,000 newborns each year. A child with cystic fibrosis produces abnormally large amounts of thick mucus, which blocks the lungs and pancreas, causing respiratory distress and inability of the intestines to absorb fats because of the plugged-up pancreas. The detective gene specifies an abnormal membrane protein, which entraps chloride ions inside cells, drawing water into them, which thickens mucus in affected tissues.

The child with cystic fibrosis fails to gain weight and has very frequent respiratory and ear infections. An initial diagnosis is often "failure to thrive," with cystic fibrosis confirmed after a characteristic high salt content is detected in the sweat of the youngster, and detecting the responsible mutant gene, which was discovered in 1989. An early reference to cystic fibrosis was a 17th century English rhyme, which says, "A child that is salty to taste will die shortly after birth."

Before cystic fibrosis was clinically described in 1938, affected children usually died in infancy. Today, thanks to antibiotics and physical therapy performed by the parents, many children survive into their thirties and a few have reached age 50. Alex Deford was not so lucky. She died at the age of 8 years, and her father describes her battle in a book, *Alex, the Life of a Child*. The fight began anew each day:

> *Alex's day would start with an inhalation treatment that took several minutes. This was a powerful decongestant mist that she drew in from an inhaler to loosen the mucus that had settled in her lungs. Then, for a half hour or more, we would give her postural drainage treatment to accomplish the same ends physically. It is quite primitive, really, but all we had, the most effective weapon against the disease. Alex had to endure 11 different positions, each corresponding to a section of the lung, and Carol or I would pound away at her, thumping her chest, her back, her sides, palms cupped to better "catch" the mucus. Then, after each position, we would press hard about the lungs with our fingers, rolling them as we pushed on her in ways that were often more uncomfortable than the pounding.*
>
> *Some positions Alex could do sitting up, others laying flat on our laps. But a full 4 of the 11 she had to endure nearly upside down, the blood rushing to her head, as I banged away on her little chest, pounding her, rattling her, trying somehow to shake loose that vile mucus that was trying to take her life away. One of her first full sentences was, "No, not the down ones now, Daddy."*
>
> *Only slowly did the recognition come that she was singled out for these things. Then she began to grope for the implications. One spring day when she was four, Alex came into my office and said she had a question. Just one was all she would bother me with. All right, I asked, what was it. And Alex said, "I won't have to do therapy when I'm a lady, will I?"*
>
> *It was a leading question; she knew exactly where she was taking me.*
>
> *As directly as I could I said, "No, Alex"—not because I would lie outright about it, but because I knew the score by then. I knew that she would not grow up to be a lady unless a cure was found.* *

Sex-Linked Recessive Disease

Eric has *hemophilia*. He is in near-constant pain because of bleeding in the joints of his knees, elbows, ankles, and hips. A minor cut could be fatal. His brother Steve also has the disease, but no one knew it until he lost a tooth at age 6 and had trouble stopping the bleeding. Until recently, 75% of hemophiliacs died of their illness before the age of 25. Today, many hemophiliacs can receive their missing clotting factor and live a near-normal life. However, the treatment is expensive, costing about $10,000 a year. About 1 in 10,000 males has hemophilia, but only 1 in 100 million females has it. This is largely because an affected female would have to have both an affected father and a carrier mother, an unlikely combination because both are so rare. As more males with hemophilia live longer and healthier lives, the incidence of the disorder may rise.

Autosomal Dominant Diseases

Mary looked like any other 5-year-old, except for the soft yellow lumps between her fingers and behind her knees. One day, while skipping rope, she very suddenly collapsed and died of a heart attack. Joe's case was more typical. He died of a heart attack at age 42. Mary and Joe had *familial hypercholesterolemia*, an autosomal dominant condition. Mary was a homozygote, possessing two copies of the defective allele and therefore suffering from a more severe form of the illness. Joe was a heterozygote, and having inherited only one defective gene, he had a milder case. Mary was literally 1 in a million, but Joe's plight may affect 1 in 500 people, accounting for a considerable proportion of those who die of heart attacks in middle age.

Mary and Joe had excess cholesterol in their blood, which accumulated as plaques on the interior walls of the arteries, obstructing blood flow and leading to their fatal heart attacks. The cholesterol buildup is due to a decrease in the number of receptor molecules that normally allow cholesterol molecules to pass into cells from the bloodstream. Heterozygotes like Joe have half the normal number of receptors; homozygotes like Mary have none. Heterozygotes can live longer by carefully monitoring their diet, exercising, and avoiding stress and taking drugs that lower blood pressure can also help. Homozygotes like Mary can have their blood removed and cleansed in a machine, but nearly all die of heart failure during childhood.

At age 37, Suzanne began to experience peculiar mood changes, lashing out at her family for no apparent reason. She became clumsy, and would lose her balance, walking as if she was intoxicated. When she fell in the street one day and children taunted her for being drunk, she went to her physician. When the nurse took her family history, Suzanne mentioned that her father had also become clumsy before he died in a car accident at age 42, and that her father's mother was presently in a mental institution. On the basis of Suzanne's symptoms, her family history, and a DNA test, the doctor concluded that she was showing the early symptoms of Huntington disease (fig. 2), an autosomal dominant condition that she had inherited from her father, and he from his mother. Suzanne would continue to decline mentally and physically, until she would need constant care. Her only consolation was that her children could be tested to see if they too would develop symptoms later in life.

*From *Alex: The Life of a Child* by Frank Deford. Copyright © 1983 by Frank Deford. Reprinted by permission of the publisher, Viking Penguin, a division of Penguin Books USA Inc., and Sterling Lord Agency, New York, NY.

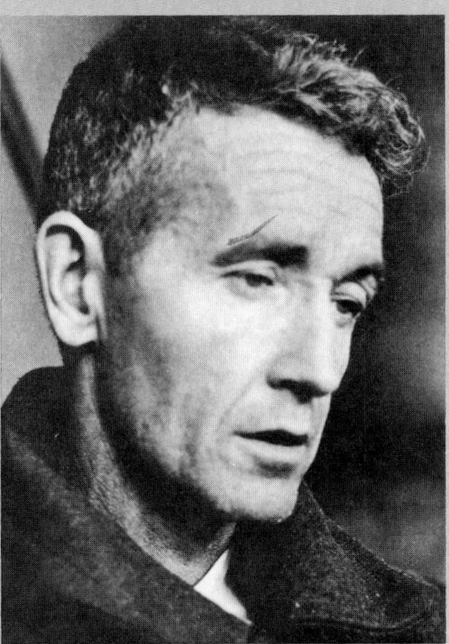

Figure 2
Folksinger Woody Guthrie lingered for more than a decade with the mental and physical degeneration of Huntington disease. The top photo was taken before symptoms arose; the bottom photo shows how this condition makes one prematurely age.

a large Amish family. The symptoms of wild mood swings were easy to observe amongst the backdrop of the quiet, conservative Amish life-style. The Amish manic depression appears to be inherited as an autosomal dominant condition.

KEY CONCEPTS

Some genetic diseases are more common in certain ethnic, religious, or racial groups, reflecting the fact that people often select mates like themselves.

Inborn Errors of Metabolism

Many genes control the production of enzymes, the proteins that speed the rates of certain chemical reactions. Conditions caused by missing or inactive enzymes are called **inborn errors of metabolism,** or "nature's perfect experiments in biochemistry" (fig. 14.8). In these disorders, the chemical reaction that the enzyme normally catalyzes cannot proceed fast enough to be of use to the body. The reactants accumulate, and products do not form, either of which causes the symptoms. Most inborn errors of metabolism are inherited as autosomal recessive disorders. People who have two copies of the disease-causing allele have the associated syndrome. Heterozygotes are usually healthy, but examination of the tissue directly affected by the enzyme involved may reveal half the normal amount of that enzyme. For example, carriers of Tay-Sachs disease are detected by their half-normal levels of hexosaminidase.

Inborn errors of metabolism were first described in 1902, when English physician and biochemist Archibald Garrod became interested in newborns who had a curious symptom—their urine turned black on exposure to air. The chemical culprit was homogentisic acid, a metabolic product that is normally broken down in the body but turns black in the presence of oxygen. Garrod suggested that this disorder, called *alkaptonuria,* was caused by a missing enzyme, which blocked the normal breakdown of homogentisic acid (then called alkapton, hence the name of the disorder.) He further proposed that alkaptonuria was inherited, based upon his observation of the disorder among siblings, particularly when the unaffected parents were first cousins. Archibald Garrod's identification of the first inborn error of metabolism was the first direct application of Mendelian principles to human health.

KEY CONCEPTS

Inborn errors of metabolism are caused by defects in genes that code for essential enzymes. These disorders are generally inherited as autosomal recessives.

Genetic Defects of the Blood

A change in a single nucleotide building block of DNA is sometimes enough to cause serious illness. In fact, much of what we know about these single nucleotide changes—called *point mutations*—comes from studies on the blood protein *hemoglobin*.

Red blood cells are filled with hemoglobin, which transports oxygen. A hemoglobin molecule is built of four tangled polypeptides, called *globin chains*, each wrapped around a small iron-containing group called a *heme* group (fig. 14.9). Two of these chains, designated *alpha* (∂) chains, are coded for by genes on chromosome 16. The other two, called *beta* (ß) globin chains, are coded for by genes on chromosome 11. Different types of *anemia* are caused by point mutations within the globin chains or by hemoglobin molecules built of the wrong types or numbers of globin chains. Anemia is felt as extreme fatigue.

More than 300 hemoglobin variants are known, but not all of them are harmful to health (fig. 14.10). The most common inherited anemia is sickle-cell disease, which is transmitted as an autosomal recessive. The hemoglobin of sickle cell disease is called hemoglobin S, and it has the amino acid valine in the sixth position in its beta globin chains, instead of the normal glutamic acid (fig. 14.11). A substitution of the amino acid lysine at the same position produces hemoglobin C, which also causes anemia. Hemoglobins S and C bond within themselves when oxygen levels are low, causing the hemoglobin to crystallize, which bends the red blood cells into a sickle

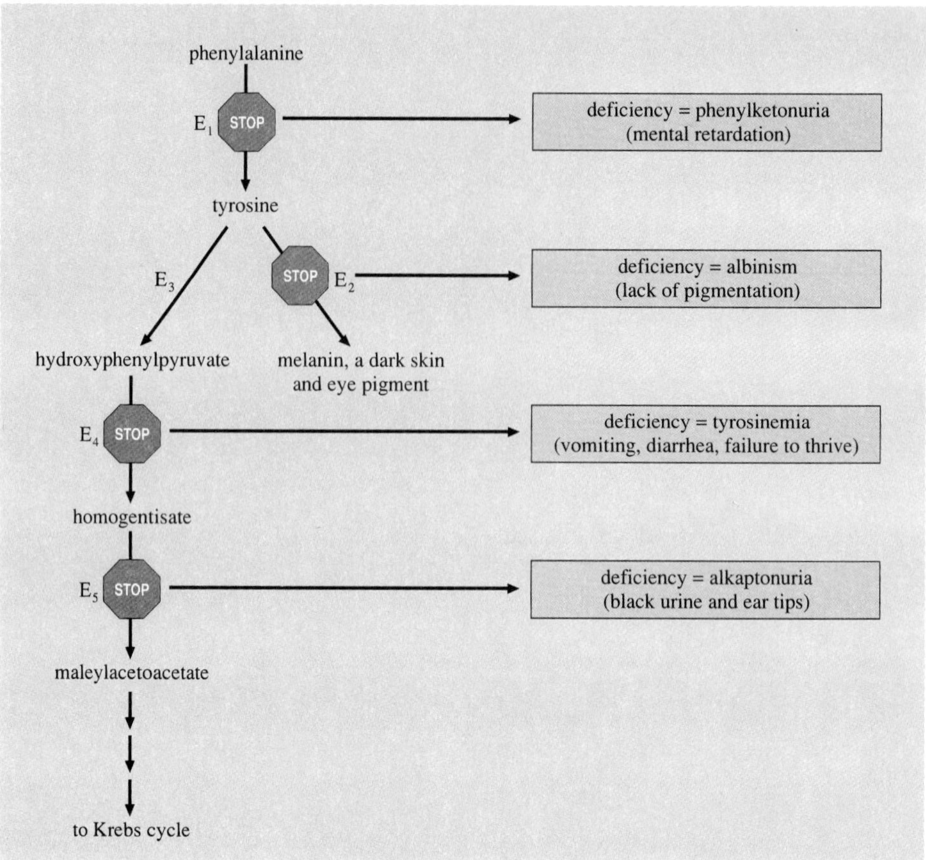

Figure 14.8
Nature's experiments in biochemistry—inborn errors of metabolism. Four inborn errors of metabolism are caused by inherited blocks in the same biochemical pathway, which leads to the breakdown of the amino acid phenylalanine.

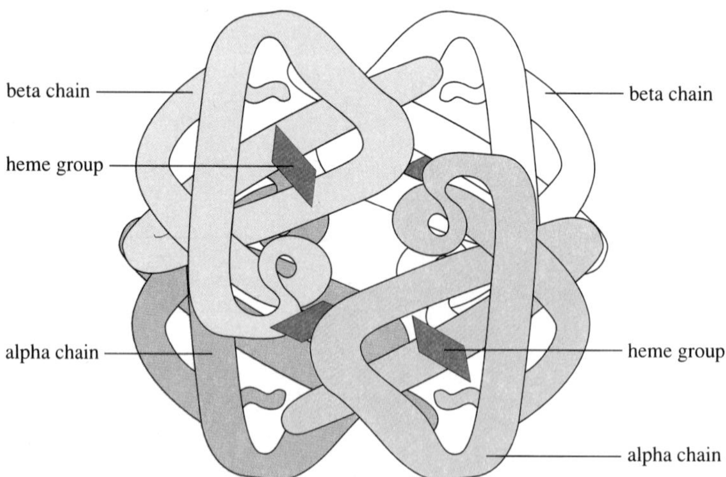

Figure 14.9
Defects in the beta globin genes. More than 300 variants of the human beta globin gene (Hb) are known, each caused by a different mutation. Mutations can disrupt the binding of the globin chains to each other or to the iron groups. Some mutations change the gene's stop codon to a codon specifying an amino acid, resulting in a beta globin chain that is longer than normal. Many mutations are "silent," having no effect on health because they change a codon into one specifying the same amino acid, or they substitute an amino acid in a part of the beta globin chain that is not essential to the molecule's function. A particular mutation is often named for the town or family in which it was first detected. (Refer back to table 13.1 for a listing of the abbreviations of amino acids.)

alpha chain												
		5	15	16	22	30	47	54	58	68	87	116
normal	val	ala	gly	lys	gly	glu	asp	gln	his	asn	his	glu
Hb J Toronto		*										
Hb J Oxford			*									
Hb I				*								
Hb J Medellin					*							
Hb G Chinese						*						
Hb L Ferrara							*					
Hb Mexico								*				
Hb Shimonoseki								*				
Hb M Boston									*			
Hb G Philadelphia										*		
Hb M Iwate											*	
Hb O Indonesia												*

beta chain																		
	1 2	6	7	16	22	26	43	46	61	63	67	70	79	87	95	121	132	143
normal	val, his	glu,	glu	gly	glu	glu	glu	gly	lys	his	val	ala	asp	thr	lys	glu	lys	his
Hb Tokuchi	*																	
Hb C		*																
Hb S		*																
Hb San Jose			*															
Hb Siriraj			*															
Hb Baltimore				*														
Hb G Coushatta					*													
Hb E						*												
Hb G Galveston							*											
Hb K Ibadan								*										
Hb Hikari									*									
Hb M Saskatoon										*								
Hb Zurich										*								
Hb Milwaukee											*							
Hb Seattle												*						
Hb G Accra													*					
Hb D Ibadan														*				
Hb N															*			
Hb O Arab																*		
Hb K Woolwich																	*	
Hb Kenwood																		*

Figure 14.10
Defect in the beta globin genes. More than 300 variants of the human beta globin gene (tlb) are known, each caused by a different mutation. This chart illustrates the different sites of mutation in various mutant hemoglobins. The numbers indicate the locations of particular amino acids in the primary sequence of beta globin.

shape. Some red cells are destroyed, and others clump in blood vessels, causing muscle pain, pneumonia, or brain damage, depending upon where they lodge.

The hemoglobin molecule is especially interesting because it changes at different times in development to suit the differing oxygen needs of the embryo, fetus, and newborn. Our alpha globin chains remain the same throughout these stages, but the embryo has two epsilon chains and the fetus has two gamma chains in place of the two beta chains, which gradually appear in the first 6 months after birth.

The genes for embryonic and fetal hemoglobin are normally turned off in the adult. In some disorders, however, they are expressed. For example, fetal hemoglobin is present in adults who have "hereditary persistence of fetal hemoglobin." The fact that these people are healthy has led to a treatment for a type of anemia called *beta-thalassemia,* in which the beta globin chains are absent. Drugs are used to activate the normally quiet fetal globin genes, and the fetal hemoglobin is produced, correcting the anemia. Fetal hemoglobin manufactured after birth, however, may not always be harmless. One theory of the cause of sudden infant death syndrome is that these infants, who die in their sleep for no apparent reason, may have embryonic or fetal hemoglobin.

Anemia can also be caused by abnormal numbers of alpha globin genes and the polypeptides they control. A normal person has four genes that specify alpha globin chains, two next to each other on each copy of chromosome 16. If the two copies of chromosome 16 misalign during meiosis, a sperm or egg cell can form that has zero, one, three, or four alpha globin genes instead of the normal two (fig. 14.12). Fertilization results in a zygote with an abnormal number of these genes. A person with at least three alpha globin genes produces enough hemoglobin to be healthy, and someone with only two is only mildly anemic. A person with a single alpha globin gene is severely anemic, however, and a fetus with no alpha globin genes dies before birth.

a.

b.

Figure 14.11
Hemoglobin S causes sickle cell disease. *a.* A normal red blood cell is a concave disc containing about 200 million molecules of the protein hemoglobin. *b.* A point mutation in the beta globin gene results in abnormal hemoglobin that crystallizes when oxygen tension is low, bending the red blood cells into sickle shapes. These abnormally shaped cells obstruct circulation, causing pain and the loss of function of various organs.

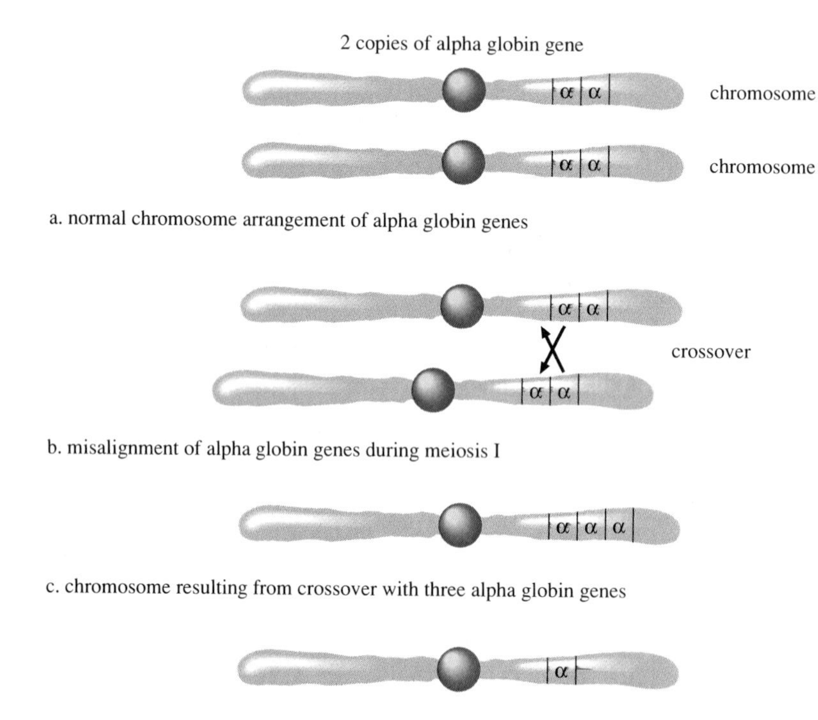

Figure 14.12
Defects in the alpha globin genes. *a.* A normal human body cell contains four alpha globin genes, a pair on each chromosome 16. A normal sperm or egg cell contains two copies of the gene. However, sperm or eggs with zero, one, three, or four alpha globin genes can form if the two chromosome 16s misalign during prophase of meiosis I, and then crossing over occurs between the alpha globin genes. *b.* This happens when the first alpha gene on one chromosome lies directly opposite the other on the homolog, a phenomenon often seen in repeated genes such as these. After the crossover, one chromosome has three alpha globin genes, and the other has one. If the chromosome on (*c*) is fertilized by a normal gamete, an individual develops who has five alpha globin genes in each body cell; if the chromosome in (*d*) is fertilized by a normal gamete, the resulting individual has three alpha globin genes in each body cell. Neither of these people is anemic. However, if two people produce gametes carrying only a single alpha globin gene and these gametes join, the child has only two alpha globin genes in each body cell and has severe anemia.

KEY CONCEPTS

Point mutations in the globin genes can cause anemias, including sickle cell disease and hemoglobin C. Different types of hemoglobin are present in the human embryo and fetus and after birth. Missing beta globin chains produces beta-thalassemia, and missing alpha globin chains produces alpha-thalassemia.

Orphan Diseases

Any inherited illness is devastating to a family. Those who have a prevalent genetic disorder, such as sickle cell disease, cystic fibrosis, or Tay-Sachs disease, can take some comfort in the knowledge that research into treating their conditions and perhaps even preventing them is proceeding. These families can participate in organizations such as the Cystic Fibrosis Foundation, helping others and learning to cope with their illness on a daily basis. However, a family whose members have an extremely rare inherited illness may not have a support group to turn to and may have difficulty obtaining information about coping with the symptoms of the illness. An added dilemma is that drug development for these rare *orphan diseases* has been slow due to lack of financial incentives. (Very rare diseases that are not inherited are also termed orphan diseases.)

Consider the case of Bradley. At the age of 2 years, he lost the strength in his arms. Soon after, his legs grew weaker. Bradley's mother took him to several doctors before finding one who was able to make a firm diagnosis: her son was suffering from the autosomal recessive Hallervorden-Spatz disease. Bradley would grow more rigid and would gradually lose his mental abilities, the doctor said. He would slowly become unable to make voluntary movements, and chewing and swallowing would become nearly impossible. He would probably die before the age of 30 years. Bradley was the only current case, the doctor said, east of the Mississippi.

Sometimes an inherited illness may actually be prevalent but appear to be rare because affected people do not know that other people also suffer from the illness, or they are reluctant to admit that they have it because of embarrassing symptoms. Such was the case for Tourette's syndrome, which is inherited as an autosomal dominant in 10% of the 1 in 100 people in the United States who have it. Tourette's syndrome first becomes noticeable in childhood as involuntary tics of the face and limbs and progresses to uncontrollable grunting and barking, echoing of others' speech, and foul language. A Tourette's patient in school might suddenly repeat what the teacher has just said and then curse. It is easy to see why people would be reluctant to discuss such symptoms and also why these behaviors might not be recognized as part of a medical condition. Tourette's syndrome was brought to public attention when a popular television program highlighted it in an episode dealing with the problem of drug development for orphan diseases. Another example of an orphan genetic disease in which those affected became aware of one another only after media exposure is progeria, the accelerated aging disorder discussed in chapter 9.

"It Runs in the Family"

Traits Caused by More Than a Single Gene

The medical conditions mentioned so far are caused by single genes that are inherited according to Mendel's laws. For disorders in which the mode of inheritance is known, geneticists can predict the probability that a certain family member will inherit the condition. Some diseases, though, seem to "run in families," appearing in a few relatives with no apparent pattern. Sometimes a single gene problem may not appear to follow Mendel's laws because of extreme variability in the severity and combination of symptoms in individuals, as we saw in chapter 11. In other cases, though, failure to adhere to a Mendelian ratio can mean that more than one gene contributes to the problem, or that the environment influences gene expression.

An example of a trait that runs in families but is not inherited as a single-gene defect is breast cancer. A woman whose mother or sister has breast cancer is nearly twice as likely to develop the disease than a woman whose close female relatives do not have it, but the likelihood is less than would be predicted for a classic recessive or dominant condition. Neural tube defects, the most common birth defect in the United States, also run in families, but not according to Mendelian ratios.

A *neural tube defect* forms when the neural tube, part of the embryo's nervous system, fails to close at 1 month gestation. The child is born with nervous tissue bulging through a portion of the spine (spina bifida), which is often accompanied by water accumulation around the brain (hydrocephaly) or a lack of brain tissue (anencephaly). Many children with these disorders are paralyzed from the waist down, but about 20% can be helped by surgery.

Incidence of neural tube defects is highest in the world in Wales and Northern Ireland, where 4 to 8 of every 1,000 newborns are affected; in the United States and Canada, incidence is only 1 or 2 out of every 1,000 infants. But in any country, the risk goes up if an affected child has already been born. A couple who has had one child with a neural tube defect has a 5% chance of having another, and if they have had two such children, the risk of having a third is 12%. It is not known precisely what causes neural tube defects, but it is probably some combination of genes and the environment.

Two more familiar traits that are influenced by heredity and the environment are height and intelligence. These traits tend to have characteristic expressions within families but can vary greatly. Traits such as these that show great variability among individuals are said to be continuously varying (fig. 14.13). Their inheritance is usually controlled by more than one gene.

KEY CONCEPTS

A trait determined by the environment or by more than one gene may recur in a family but may not adhere to the ratios predicted by Mendel's laws.

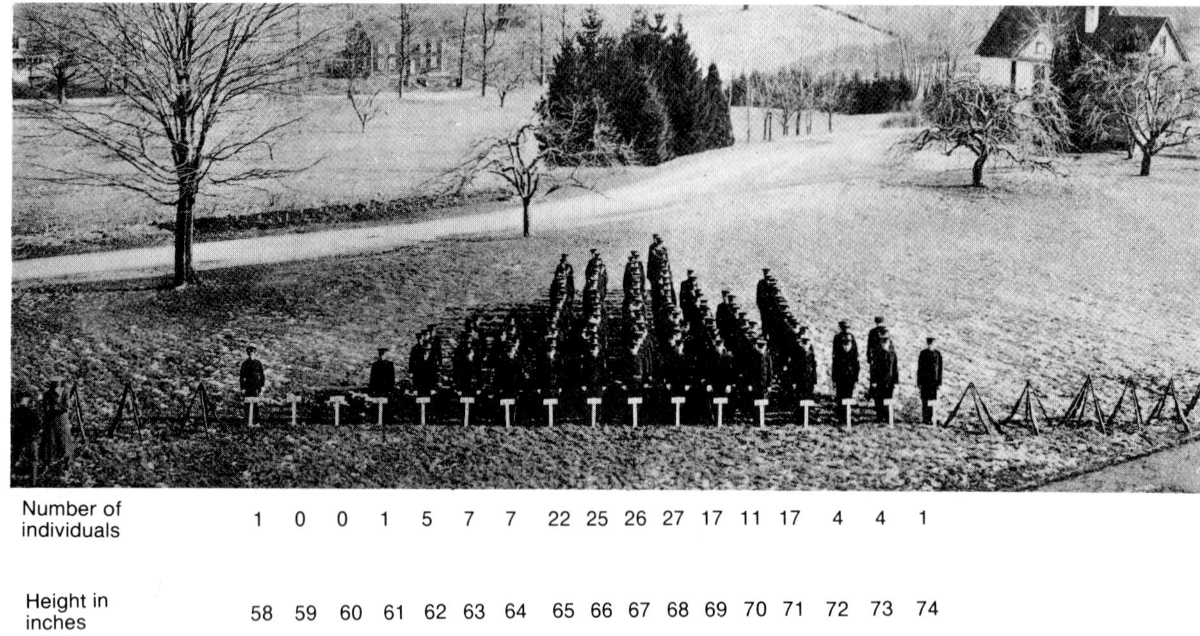

a.

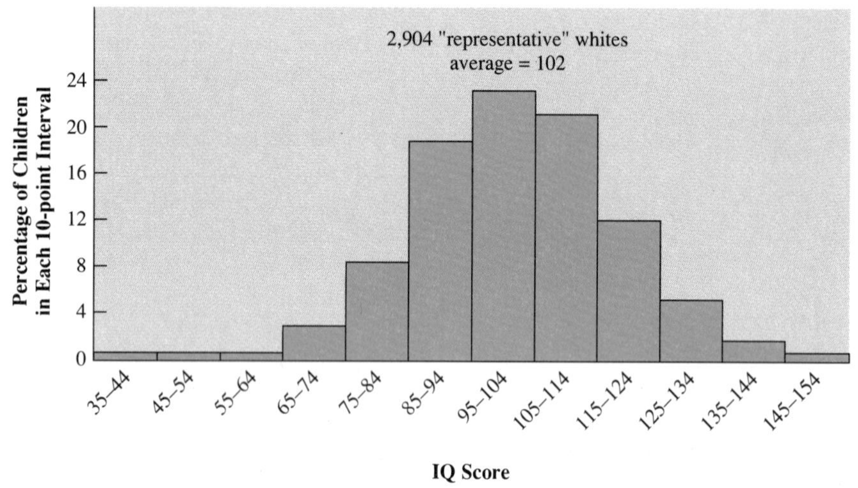

b.

Figure 14.13
Continuously varying traits. *a.* One hundred seventy-five soldiers were asked to line up according to their heights. The heights form a continuous distribution. Height is thought to be influenced by more than one gene and also by environmental factors such as nutrition. *b.* This frequency distribution represents scores on intelligence tests achieved by 2,904 American-born white children. These scores also form a continuous distribution, suggesting that intelligence is influenced by a variety of factors.

Nature Versus Nurture —Twin Studies

A woman who is a prolific writer has a daughter who grows up to become a gifted novelist. An overweight man and woman have obese children. A man whose father was an alcoholic is himself an alcoholic. Are these characteristics of writing talent, obesity, and alcoholism inherited, or are they the result of imitating the habits of close relatives? Geneticists have wondered for many years to what degree certain traits are determined by "nature" (the genes) or "nurture" (the environment). One way to determine this is by looking at twins, who represent a kind of natural experiment.

Twins occur in 1 out of 80 births. Identical, or *monozygotic* (MZ), twins result from the splitting of a single fertilized egg, and therefore the twins are always of the same sex and have identical genes. Fraternal, or *dizygotic* (DZ), twins arise from two distinct fertilized eggs, and therefore they are no more similar genetically than any other two siblings, although they have shared the same prenatal environment.

If a trait tends to occur more frequently in both members of identical twin pairs than it does in both members of fraternal twin pairs, then it is at least partially controlled by heredity. Geneticists describe the inherited component of a trait with a value called **concordance,** calculated for both MZ and DZ twins as the number of pairs in which both twins express the trait divided by the number of pairs in which at least one twin expresses the trait. Diseases caused by single genes, whether dominant or recessive, always show 100% concordance in MZ twins—that is, if one twin has it, the other does too. However, among DZ twins, concordance for a dominant single gene trait is 50%, and for a recessive trait, 25%. Note that these are the same values as apply to any two siblings. For a trait determined by several genes, concordance values for MZ twins are significantly greater than for DZ twins. Finally, a trait largely molded by the environment has a similar concordance value for both types of twins. Several twin studies conducted in the 1930s attempted to determine the hereditary component of such poorly-defined conditions as "criminality" and "feeblemindedness." Table 14.2 lists concordance values for some interesting characteristics.

A limitation of comparing identical and fraternal twins to assess the genetic component of a trait is that this method assumes that both types of twins are treated similarly. In fact, identical twins are often in closer contact with each other than fraternal twins, particularly if fraternal twins are of the opposite sex. This discrepancy led to some misleading results in twin studies conducted in the 1940s. One study concluded that tuberculosis is inherited because concordance among identical twins was higher than among fraternal twins. In actuality, the infectious disease was more readily passed between identical twins because their parents tended to keep them in close physical contact.

Another type of twin study overcomes the problem of how twins are raised. At the University of Minnesota, a worldwide effort is underway to bring together identical twins who were parted early in life. The premise is simple—because the twins have the same genes, but were raised apart, any differences between them must have been induced by the environment.

In the Minnesota study, each pair of twins undergoes a 6-day battery of tests that measures both physical and behavioral traits, including 24 different blood types, handedness, the direction of hair growth, fingerprint pattern, height, weight, the functioning of all organ systems, intelligence, allergies, and dental patterns. Facial expressions and body movements in different circumstances are videotaped, and fears, vocational interests, and superstitions probed.

So far, the researchers have found that identical twins separated at birth and reunited later are remarkably similar, even when their adoptive families are quite different (fig. 14.14). Idiosyncrasies are particularly striking. A pair of identical male twins raised in different countries in different religions, for example, were astounded when they were reunited as adults to find that they both laugh when someone sneezes and flush the toilet before using it. A pair of girls reunited in late adolescence

Table 14.2
Is a Trait Caused by Heredity or the Environment? Ask Twins!

Observed Disease or Behavior	*Percentage Concordance*	
	MZ Twins	*DZ Twins*
Tuberculosis	54	16
Cancer at the same site	7	3
Clubfoot	32	3
Measles	95	87
Scarlet fever	64	47
Rickets	88	22
Arterial hypertension	25	7
Manic-depressive syndrome	67	5
Death from infection	8	9
Rheumatoid arthritis	34	7
Schizophrenia (1930s)	68	11
Criminality (1930s)	72	34
Feeble-mindedness (1930s)	94	50

Source: From *Heredity Evolution and Society*, 2d ed., by I. Michael Lerner and William J. Libby. W. H. Freeman and Co. Copyright © 1976.

Separated at birth, the Mallifert twins meet accidentally.

Figure 14.14
Drawing by Chas. Addams; © 1981 The New Yorker Magazine, Inc.

found they were both afraid of swimming. Twins who met for the first time in their 30s paused for 30 seconds after being asked a question, and each then rotated a gold necklace she was wearing three times, then answered.

The "twins reared apart" approach to separating nature from nurture is not a perfect experiment. Identical twins share an environment in the uterus and possibly in early infancy that may affect later development. Also, adoption agencies often search for families of similar socioeconomic or religious backgrounds to adopt separated twins, so that environment may not be as different as it might be for two unrelated adoptees. However, at the present time, the "twins reared apart" method is providing some very interesting insights into the number of small body movement patterns and psychological quirks that seem to be rooted in our genes.

KEY CONCEPTS

Concordance is the proportion of twins in which both individuals display a particular trait and indicates the hereditary component of that trait. For a dominant or recessive genetic disease, MZ twins show 100% concordance, but for DZ twins concordance is the same as for any siblings. Studies of twins reared apart can indicate effects of the environment.

SUMMARY

Imbalances of the genetic material can affect health. When chromosome pairs fail to separate in meiosis, unbalanced gametes form that, when fertilized, produce *aneuploid* individuals with missing or extra chromosomes. *Down syndrome* is the most common autosomal aneuploid—most others are lethal before birth, and those that survive have widespread defects. Sex chromosome aneuploids are less serious and include *Turner syndrome*, *Klinefelter syndrome*, *XYY syndrome*, and *triplo-X*. Some syndromes are caused by deleted chromosomes. A person who is a *translocation carrier* can produce gametes with abnormal amounts of genetic material, leading to birth defects or spontaneous abortions in the next generation.

Some of the most prevalent genetic disorders are caused by single-gene defects, as are many of the rare orphan diseases. Single-gene disorders are classified by whether they are recessive or dominant, and whether they are transmitted on autosomes or sex chromosomes. Much of our knowledge about genetics comes from studies on blood disorders involving the protein hemoglobin. Not all traits follow Mendelian patterns of inheritance because they can be caused by more than a single gene and can be influenced by the environment. Comparing the occurrence of traits among pairs of identical and fraternal twins can give a rough measure, called the *concordance*, of the relative influences of heredity and the environment on a trait. Traits in identical twins reared apart also provide valuable information on the "nature" versus "nurture" question.

QUESTIONS

1. For each of the following medical conditions, indicate the underlying genetic defect and the gene product that is involved, where known.

- **a.** Crie-du-chat syndrome
- **b.** Tay-Sachs disease
- **c.** Sickle cell disease
- **d.** Edward's syndrome
- **e.** Huntington disease
- **f.** Down syndrome
- **g.** Alkaptonuria
- **h.** Tourette's syndrome
- **i.** Alpha thalassemia
- **j.** Klinefelter syndrome
- **k.** Hemophilia
- **l.** Familial hypercholesterolemia

2. For an exercise in a college genetics laboratory course, a healthy student takes a drop of her blood, separates out the white cells, stains her chromosomes, and constructs her own chromosome chart. She finds only one chromosome 3 and one chromosome 21, plus two unusual chromosomes that do not seem to have matching partners.

- **a.** What type of chromosomal abnormality do you think she has?
- **b.** Why doesn't she have any symptoms?
- **c.** Would you expect any of her relatives to have any particular medical problems? If so, which ones?

3. In Texas there is a summer camp run for children and teenagers who have sickle cell disease. It is not unusual for campers to fall in love with each other and marry. What is the chance that a child born to such a couple inherits sickle cell disease?

4. A screening program to detect carriers of Tay-Sachs disease is conducted at a college with a large Jewish population. A young woman who plans to participate says to her husband, "I'll have the test. You don't need to be tested unless I have a positive result." Why did she say this?

5. At the Center for Germinal Choice in California, male Nobel Prize winners donate sperm, and women of high intellect are artificially inseminated with this sperm to try to produce children of extreme intelligence. Why is this idea incorrect?

TO THINK ABOUT

1. Based on your knowledge of genetics from reading the past few chapters, why do you think that an excess of genetic material (such as an extra chromosome or a gene duplication) is generally less harmful to health than a deficit of genetic material (a missing chromosome or gene deletion)?

2. A fetus dies in the uterus. Several of its cells are examined for their chromosomal content. Approximately 75% of the cells are diploid, and 25% are tetraploid (4 copies of each chromosome). What do you think has happened, and when in development did it probably occur?

3. In 1965, a man with sex chromosome complement XYY stood trial for killing a prostitute. His attorneys claimed that he was unfit to stand trial because of his genetic abnormality, which was responsible for his behavior. Do you agree with his attorneys' claims? Why or why not?

4. Do you think that parents can influence their children's intelligence? How might they do this?

5. The doctor of a child born 2 months prematurely and weighing 3 pounds says, "Because of her poor nutrition while in the uterus, she will never reach her genetic potential for height." What does this statement mean about the inheritance of height?

6. Design an experiment to determine to what extent obesity is inherited.

7. Cells are taken from a fetus and an inversion is detected in a particular chromosome. What additional piece of evidence would demonstrate that the inversion does not cause symptoms?

SUGGESTED READINGS

Bouchard, T.J., Jr., D.T. Lykken, M. McGue, N.L. Segal, A. Tellegen. October 12, 1990. Sources of human psychological differences: the Minnesota study of twins reared apart. *Science*, vol. 250 no. 4978, p. 223. Identical twins separated at birth offer clues to the hereditary and environmental influences on behavior.

Deford, Frank. 1983. *Alex, the life of a child*. New York: Viking Press. An in-depth and very moving account of living with cystic fibrosis.

Gelehrter, Thomas D. and Francis S. Collins. 1990. *Principles of Medical Genetics*. Baltimore: Williams and Wilkins. An exciting look at the molecular secrets behind many human diseases.

Hook, Ernest B. January 12, 1973. Behavioral implications of the human XYY genotype. *Science*, vol. 179. Are XYY males really prone to criminal behavior?

Lewis, Ricki. October 1990. A glimpse of neurofibromatosis 1 protein function. *The Journal of NIH Research*. The abnormal gene behind neurofibromatosis alters a protein involved in a cell's reception of incoming signals.

Loupe, Diane E. December 2, 1989. Breaking the sickle cycle. *Science News*. New treatments are on the horizon for this well-understood genetic disease.

McKusick, Victor A. 1990. *Mendelian inheritance in man*. Baltimore: Johns Hopkins University Press. This compendium of all known inherited human traits is a technical book, but a browse through it makes fascinating reading.

Pierce, Benjamin A. 1990. *The family genetic source book*. New York: John Wiley & Sons. Seven chapters explain genetic principles, followed by a catalog of fascinating human genetic traits.

C H A P T E R

15

Genetic Disease—Diagnosis and Treatment

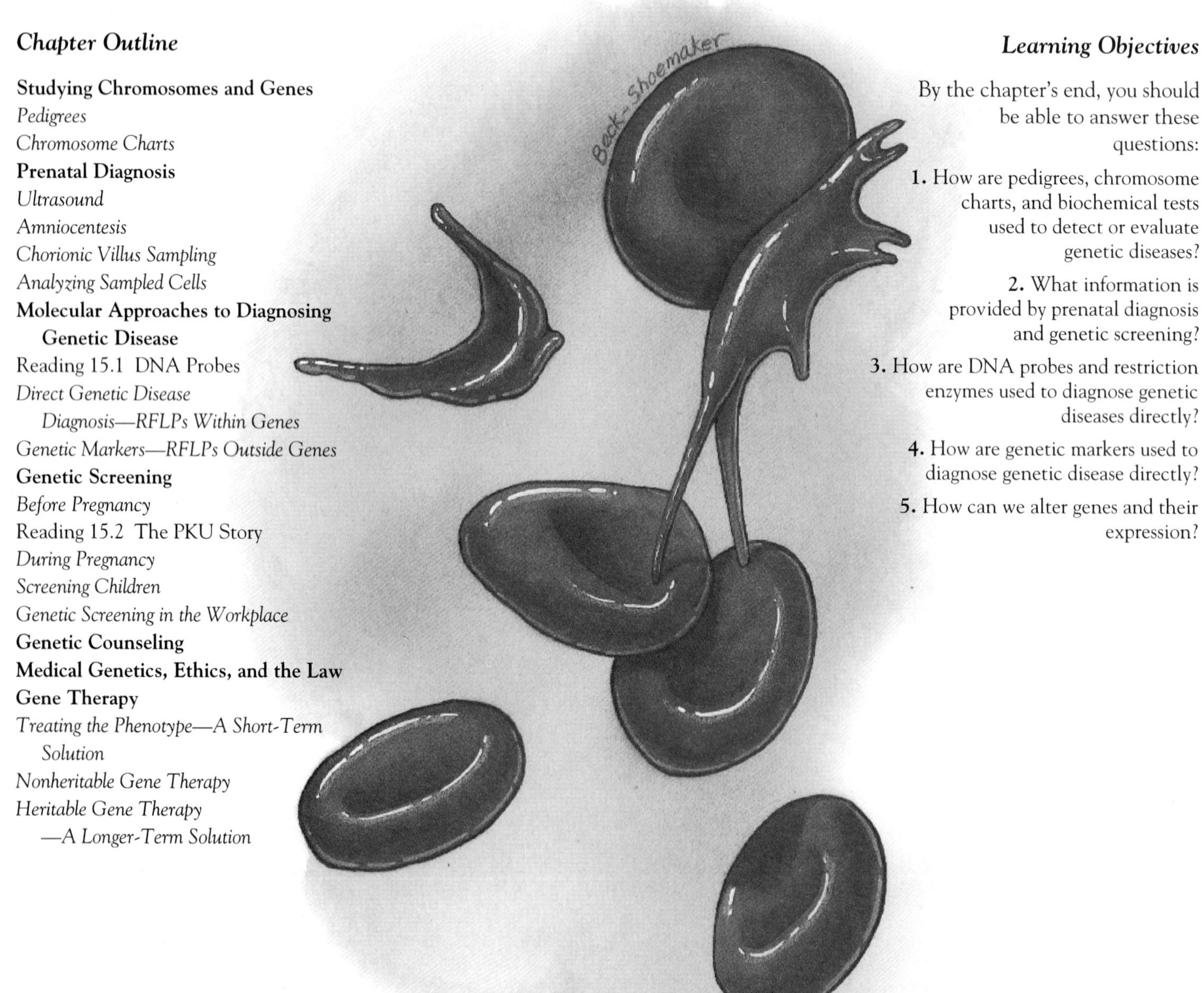

Chapter Outline

Learning Objectives

By the chapter's end, you should be able to answer these questions:

1. How are pedigrees, chromosome charts, and biochemical tests used to detect or evaluate genetic diseases?
2. What information is provided by prenatal diagnosis and genetic screening?
3. How are DNA probes and restriction enzymes used to diagnose genetic diseases directly?
4. How are genetic markers used to diagnose genetic disease directly?
5. How can we alter genes and their expression?

Moments after the baby lets out her first angry bellow as she enters the world, a nurse collects material containing immature blood cells from the freshly cut umbilical cord. As the new mother and father become acquainted with their daughter, the sampled blood cells are sent to a laboratory. Some will be frozen for use later in life to replenish bone marrow if the new person ever needs a transplant. The DNA fingerprint is run to establish identity, to prevent a hospital baby mix-up, and also in case the child is ever lost or abducted. Finally, a series of predictive genetic tests is run. Telltale clues in the sequences of A, T, G, and C in the newborn's DNA tell her parents that she will one day probably develop diabetes, hypertension, and colon cancer.

All of the procedures just mentioned are already possible, although a newborn has yet to undergo all of them.

Studying Chromosomes and Genes

Pedigrees

Despite spectacular recent advances in genetic technology, much information about heredity still comes from where it always has—observation of the passage of traits over generations, be it in peas or people. Information on how family members are related and who among them have certain characteristics is compiled into a chart called a **pedigree** (figs. 15.1 and 15.2), which is much like the pedigrees used to trace traits in race horses and purebred dogs and cats. In people, pedigrees can reveal how a trait is inherited.

Pedigrees can be difficult to construct and interpret for several reasons. People sometimes hesitate to supply needed information because they are embarrassed by symptoms affecting behavior or mental stability. Tracing family relationships can be complicated by adoption, children born out of wedlock, and serial marriages and the resulting "blended" families. Reproductive alternatives such as artificial insemination and surrogate mothering also complicate pedigrees. In our nation of immigrants, many people cannot trace their families back more than three or four generations, which may not provide sufficient evidence to reach a conclusion about a mode of inheritance.

The nature of a disorder can make it hard to trace. Recall the inherited blood disorder porphyria, described in chapter 11. It has so many different symptoms that it appeared to be several disorders in the

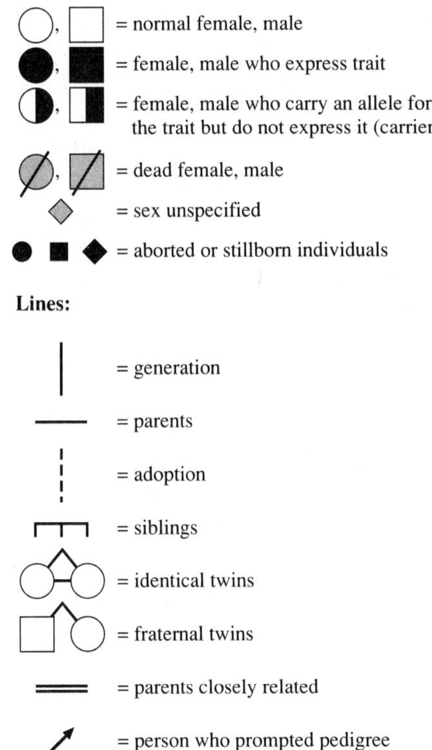

Figure 15.1
Symbols used in pedigree construction are connected to form a pedigree chart, which displays the inheritance patterns of particular traits.

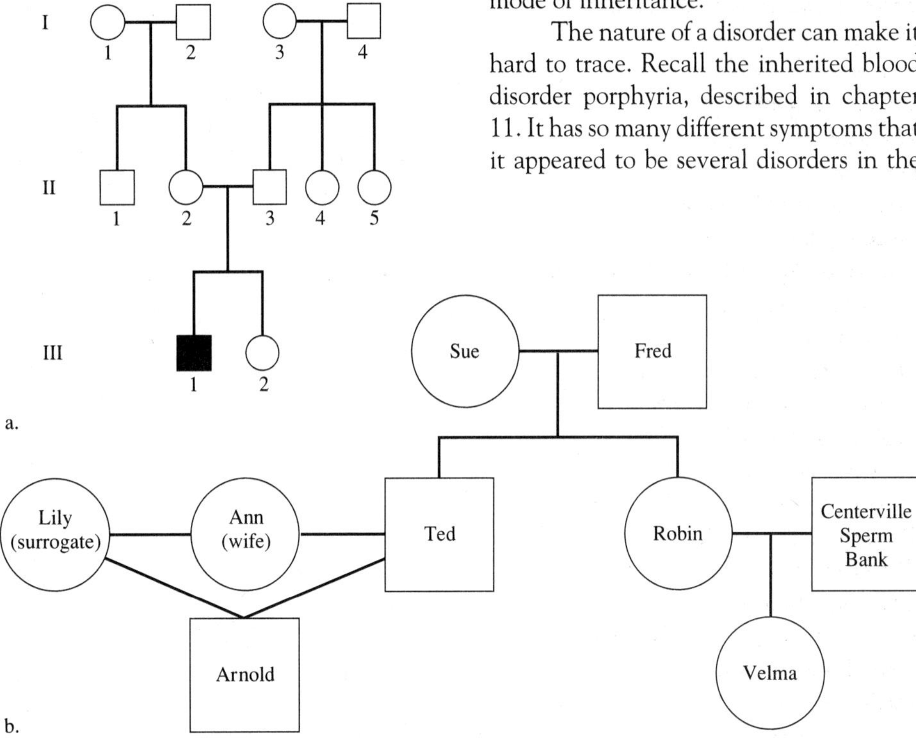

Figure 15.2
Pedigrees are not always conclusive. a. In generation I, one healthy couple has a healthy son and daughter, and another healthy couple has a healthy son and two healthy daughters. The daughter of the first couple (individual II-2) has two children with the son of the second couple (individual II-3). The firstborn child (individual III-1) is a boy who has an illness known to be hereditary. His sister (individual III-2) is healthy. What is the mode of inheritance for this disorder? It cannot be dominant, because it is not expressed in generations I or II, but it appears in generation III. The disorder could be inherited as a sex-linked recessive, with individual II-2 a carrier who passes her disease-causing allele on the X chromosome to her son, who has the illness. The disorder could also be inherited as an autosomal recessive, with individuals II-2 and II-3 being carriers, each passing on a defective allele to their son. b. Compiling human pedigrees is complicated by such social factors as adoptions, multiple marriages, and children born to unmarried parents. A new complication may be alternate ways of conceiving children—as shown in this fanciful pedigree.

royal family that suffered from it. In late-onset disorders such as Huntington disease, a person who has inherited the disease-causing gene may die of another cause before the genetic disorder produces symptoms. Also confusing in pedigree construction are conditions that run in families but do not follow Mendel's laws, such as some forms of diabetes and breast cancer.

Despite these drawbacks, a pedigree can indicate the possibility of a certain genetic disorder, suggesting further tests on which a diagnosis can be based.

Chromosome Charts

Whenever possible, a pedigree is supplemented with photographs showing the chromosome abnormality thought to be the cause of the patient's symptoms. A chart of chromosomes arranged in pairs and in decreasing size order is called a **karyotype.** It can indicate, for example, whether Down syndrome is caused by an extra or a translocated chromosome. Figure 14.2 shows several karyotypes.

To prepare a karyotype, a few drops of blood are taken from a person's finger, and the white blood cells separated out and treated with the chemical colchicine, which halts the movement of chromosomes in dividing cells. Chromosomes "caught" dividing are condensed enough to be visible when viewed under a microscope. Next, the cells are placed in a hypotonic salt solution, where water enters, swelling them. The swollen cells are then drawn up into a pipette and dropped onto a microscope slide on which stain has been placed. As the cells splash onto the slide, they burst, and the tangled mass of chromosomes in each nucleus spreads apart. The investigator locates a cell in which the chromosomes are nicely displayed, and then uses a camera attached to the microscope to photograph it. After the photograph is developed and enlarged, the individual chromosomes are cut out and arranged in pairs from smallest to largest.

The first karyotypes, constructed about 30 years ago, were rather crude because the ability of DNA to absorb stains was not yet fully appreciated. Many of the chromosome pairs looked the same, and they were arranged into seven groups, A through G, in decreasing size order. Some chromosome pairs were further distinguished by the site of their centromere, the characteristic constriction that is either in the center, off center, or at an end of a chromosome. In 1958, a mentally retarded child with a deletion in chromosome 2 would have been diagnosed as having an "A group chromosome" disorder.

Today, researchers use several different stains to identify which parts of which chromosomes are associated with a person's symptoms. The chromosomes are numbered from 1 to 22 in decreasing size order and the sex chromosomes are designated as X and Y. The different stains zero in on different parts of the DNA molecule or the protein surrounding it, creating intricate banding patterns unique to each chromosome (fig. 15.3). More than 2,000 bands are detectable on the 46 human chromosomes, each band corresponding to about 20 genes. Abnormal chromosomes can have disrupted banding patterns. A karyotype of the man described in figure 14.5, for example, would have an unusual chromosome with bands characteristic of chromosomes 14 and 21—this is the translocated chromosome causing so much heartbreak in the family. Similarly, an inversion is recognized by a reversed banding patten.

Constructing a chromosome chart in this "photograph-cut-and-paste" way takes hours or even days. Computerized karyotyping devices produce a chart in minutes. Researchers program the device to scan the burst cells in a drop of fluid and select the one in which the chromosomes are the most visible. The user can then outline each chromosome and place it in its proper sized place on a video screen, or "teach" the computer to recognize the band patterns of each stained chromosome pair and automatically sort the structures into a size-order chart. Push a button, and the device prints out the finished karyotype. If an abnormal band pattern is recognized, a data base pulls out identical or similar karyotypes from previous patients, providing information on possible symptoms.

Karyotypes are useful at several levels. A chromosome chart can be used to diagnose a chromosomal "accident" such as Down syndrome, which occurs in a child of chromosomally normal parents. Karyotypes can also diagnose the second most common form of mental retardation, *fragile X syndrome*. Males inherit this disorder from carrier mothers, and it is detected by growing chromosomes in a special medium low in folic acid. The affected X chromosomes appear to have characteristic knobs (fig. 15.4).

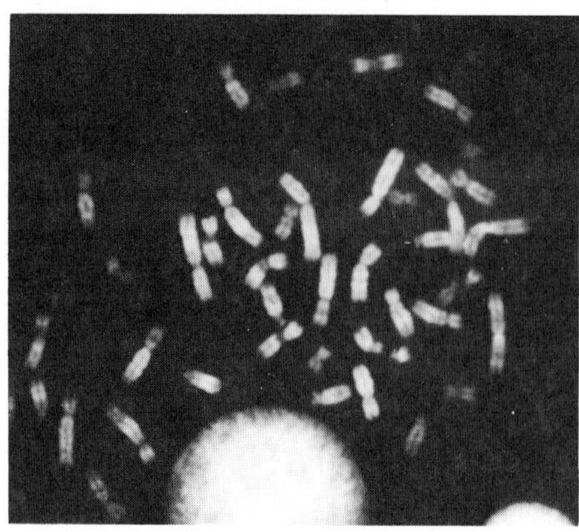

Figure 15.3
Chromosome banding is based upon chemical attractions between stains and DNA. In Q banding, a stain called quinicrine binds to DNA regions that are rich in adenine and thymine. Fluorescent bands are seen through a microscope under ultraviolet light. The tip of the human Y chromosome glows prominently when stained with quinicrine, making this a useful stain for determining fetal sex. Other stains highlight particular autosomes. What would three highlighted chromosome number 21 indicate?

Sometimes recognizing an abnormal chromosome can be lifesaving. In one New England family, several adult members died from a very rare form of kidney cancer. Because inheriting the cancer was so unusual, the family's health was studied closely, and karyotypes showed a translocation between chromosomes 3 and 8—in every relative who had cancer. When two healthy young family members were found to have the translocation, their kidneys were examined, and indeed they had very early stages of the cancer! They were successfully operated on.

Comparing karyotypes from individuals within different populations can sometimes identify environmental toxins, if abnormalities appear only in a group exposed to a particular contaminant. Because chemicals and radiation that can cause cancer and birth defects often also break chromosomes into fragments or rings

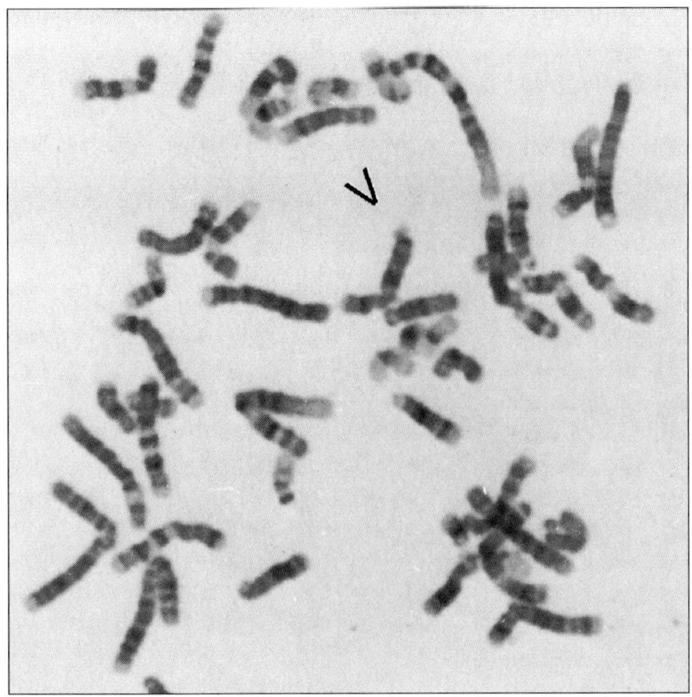

Figure 15.4
Fragile X syndrome. An X chromosome that has a knob attached by a narrow strand of material to the rest of the chromosome is seen in specially treated cells of 20% to 30% of moderately mentally retarded men. These men also have large ears, jaws, and testicles and flat cheeks. One-third of females who are heterozygous for the fragile X chromosome are slightly mentally retarded. Pregnant women who have several retarded relatives can now have their blood tested for the "fragile X" chromosome. If it is present and the fetus is male, he has a 50% chance of inheriting the abnormal chromosome and of being mentally retarded. In the general population, 1 in 1,350 males has fragile X syndrome, and 1 in 677 females is a carrier.

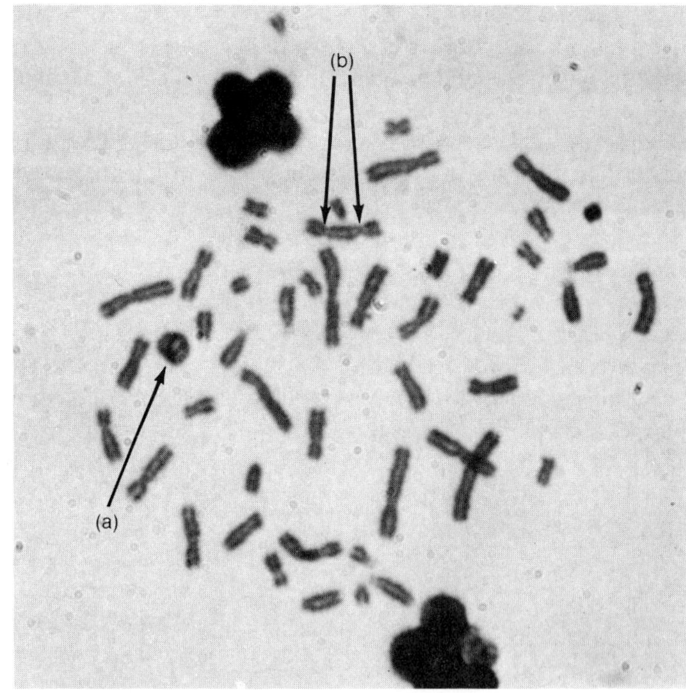

Figure 15.5
Ring chromosomes. Exposure to toxic chemicals and radiation can break the tips off of chromosomes. *a.* The tipless chromosomes sometimes attract one another and fuse, forming rings. The appearance of ring chromosomes or chromosome fragments in a karyotype is a sign of possible exposure to a toxin. The ring chromosome in this photograph arose from exposure to X rays. *b.* This chromosome is derived from two chromosomes that fused, as evidenced by its two centromeres.

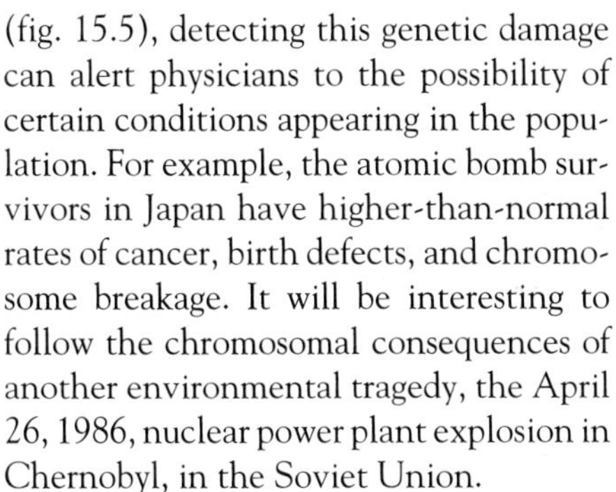

(fig. 15.5), detecting this genetic damage can alert physicians to the possibility of certain conditions appearing in the population. For example, the atomic bomb survivors in Japan have higher-than-normal rates of cancer, birth defects, and chromosome breakage. It will be interesting to follow the chromosomal consequences of another environmental tragedy, the April 26, 1986, nuclear power plant explosion in Chernobyl, in the Soviet Union.

Karyotypes can be compared between species to clarify evolutionary relationships. The more recent the divergence of two species from a common ancestor, the more closely related they are, and the more alike their karyotypes should be. Indeed, the bands of human chromosomes are strikingly similar to those of our closest relative, the chimpanzee.

KEY CONCEPTS

Pedigrees are charts that display the relationships of individuals within a family and indicate who among them expresses a particular inherited trait. A karyotype is a chart of chromosomes arranged into homologous pairs from largest to smallest. By staining chromosomes, certain abnormalities can be detected as disruptions in banding patterns. Karyotypes provide information on genetic disease, exposure to environmental toxins, and evolutionary relationships between species.

Prenatal Diagnosis

The techniques of *prenatal diagnosis* provide glimpses into the world of the human embryo and fetus. In many instances doctors can predict if a child will be born with a particular medical problem—one that is inherited (a genetic birth defect) or caused by something in the environment or a noninherited error in development (a congenital birth defect). Although prenatal diagnosis cannot reveal all medical problems, it is taking much of the mystery out of life before birth.

Ultrasound

Mrs. F. is 38 years old and is 15 weeks pregnant with her first child. She mentions to her doctor that her sister just gave birth to a child with Down syndrome. Should she be concerned about her own child? The doctor is alerted because of her age and family background and suggests a few tests, beginning with an *ultrasound exam*. In this exam, sound waves are bounced off of the fetus and converted into an image on a screen.

The next day a nervous Mr. and Mrs. F arrive for the ultrasound. After Mrs. F drinks several glasses of water, a nurse spreads a cold jelly on her belly, then presses a hand-held device called a transducer across her slightly rounded abdomen. It does not hurt, and the parents-to-be become so excited at seeing fingers and toes that they almost forget that the doctor is looking for the organ abnormalities that sometimes occur with Down syndrome. All looks well. The beating heart is clear, and the lengths of the tiny arms and legs are just about right for a 15-week fetus (fig. 15.6). Body parts that are too large or too small for a particular point in gestation indicates a problem.

Amniocentesis

The patient's relief following the ultrasound exam is short-lived, however. The doctor advises her to have an **amniocentesis** a week later. In this procedure, fetal cells and fluids are removed from the uterus with a needle (fig. 15.7). The cells are cultured in the laboratory and karyotypes constructed from a sample of 20 cells. The fluid is examined for telltale biochemicals that could indicate a metabolic disorder. With amniocentesis, 400 or so of the more than 5,000 known chromosomal and biochemical problems can be detected. Ultrasound is used to guide the needle so that the fetus is not harmed, and the procedure only hurts the mother for a minute. Amniocentesis causes spontaneous abortion in about 1 in 200 cases. Results come back in about 10 days. In the case of Mr. and Mrs. F, the cytogeneticist who examines their sample will look specifically for the extra chromosome 21 that indicates Down syndrome.

Although apprehensive about having a needle stuck in her middle, Mrs. F. is most concerned that the results of the test will not arrive until she is 18 weeks pregnant—she knows that a fetus can survive at just 6 weeks past that point.

Chorionic Villus Sampling

The time factor is a problem in amniocentesis, the doctor concurs. Researchers are experimenting with amniocentesis as early as 12 weeks. Had the doctor known of the Down syndrome in the family by the ninth week of pregnancy, Mrs. F. could have had an experimental procedure, *chorionic villus sampling,* that snips cells from the finger-like projections that develop into the placenta (fig. 15.8).

In chorionic villus sampling, the cells are removed through the vagina. A karyotype is prepared directly from the collected cells, rather than first culturing them, as is the case with amniocentesis. Results are ready in days. Because chorionic villus cells are descended from the fertilized egg, their chromosomes are identical to those of the embryo. Occasionally, though, chromosomal mosaicism occurs, which means that the karyotype of a villus cell differs from that of an actual embryonic cell. This is why the procedure is slightly less accurate than amniocentesis,

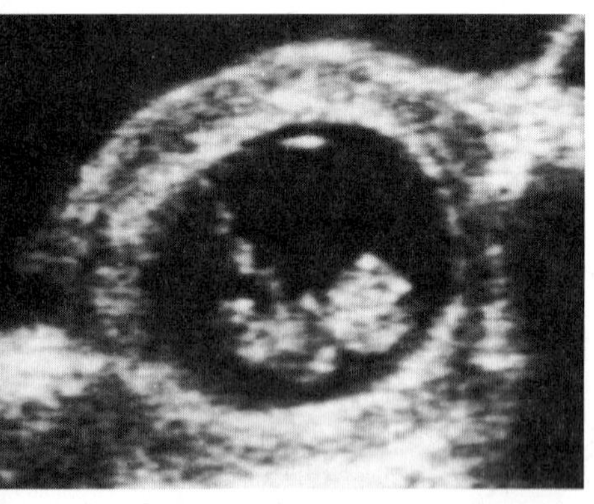

a.

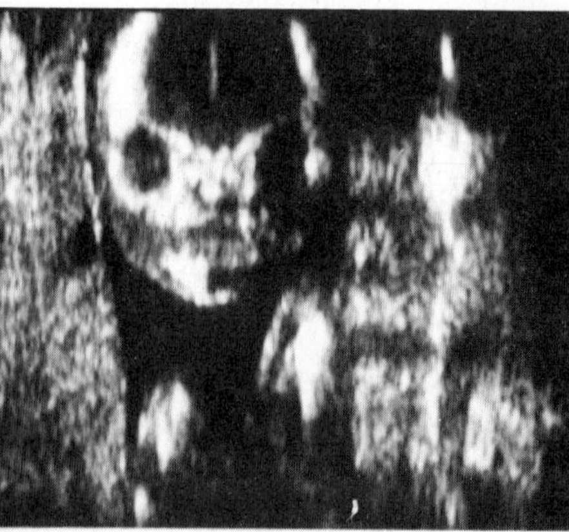

b.

Figure 15.6
In an ultrasound exam, sound waves are bounced off of the embryo or fetus, and the pattern of deflected sound waves is converted into an image. Not very much detail is visible in the 6-week embryo (*a*), but by 13 weeks (*b*), the face can be discerned.

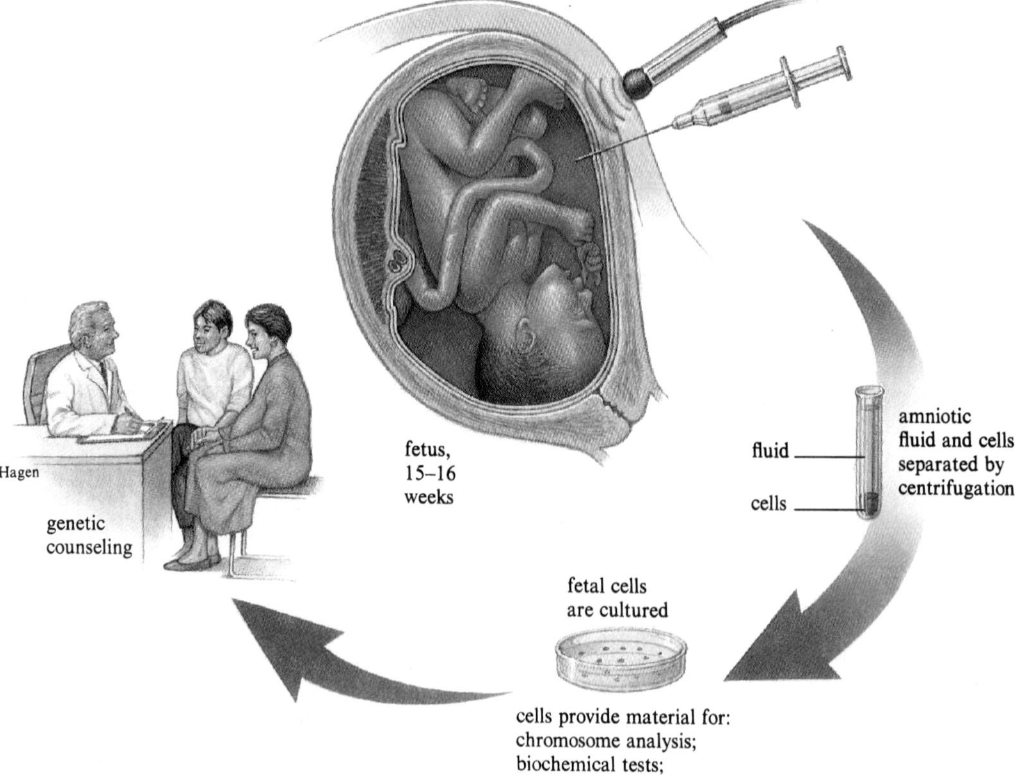

Figure 15.7
In amniocentesis, fetal cells are removed from a sample of amniotic fluid taken from the woman's abdomen with a needle at the 16th week of pregnancy. It takes another 10 days for enough fetal cells to grow for a karyotype to be constructed.

which probes fetal cells. Also, the sampling procedure does not include amniotic fluid, so the biochemical tests that are part of amniocentesis cannot be done.

Analyzing Sampled Cells

The few drops of tissue recovered from amniocentesis or chorionic villus sampling can reveal much about a fetus's health. The fetus is a boy if one Y and one X chromosome appear and a girl if there are two X chromosomes. Determining gender is valuable if the mother is a carrier of a sex-linked disorder, because a male fetus has a 50% chance of inheriting the condition. Chromosome abnormalities are also revealed—fortunately, Mr. and Mrs. F's unborn daughter did not have the extra chromosome 21 that caused Down syndrome in her cousin.

Biochemicals released by fetal cells also provide information on health. A very low level of the enzyme hexosaminidase, for example, signals Tay-Sachs disease. Too much of a substance called alpha-fetoprotein may mean that the fetus has an opening in its spinal cord through which the chemical leaks, a neural tube defect that appears as spina bifida or hydrocephaly (water on the brain) in the newborn. About 100 conditions can be prenatally diagnosed by such clues in fluids.

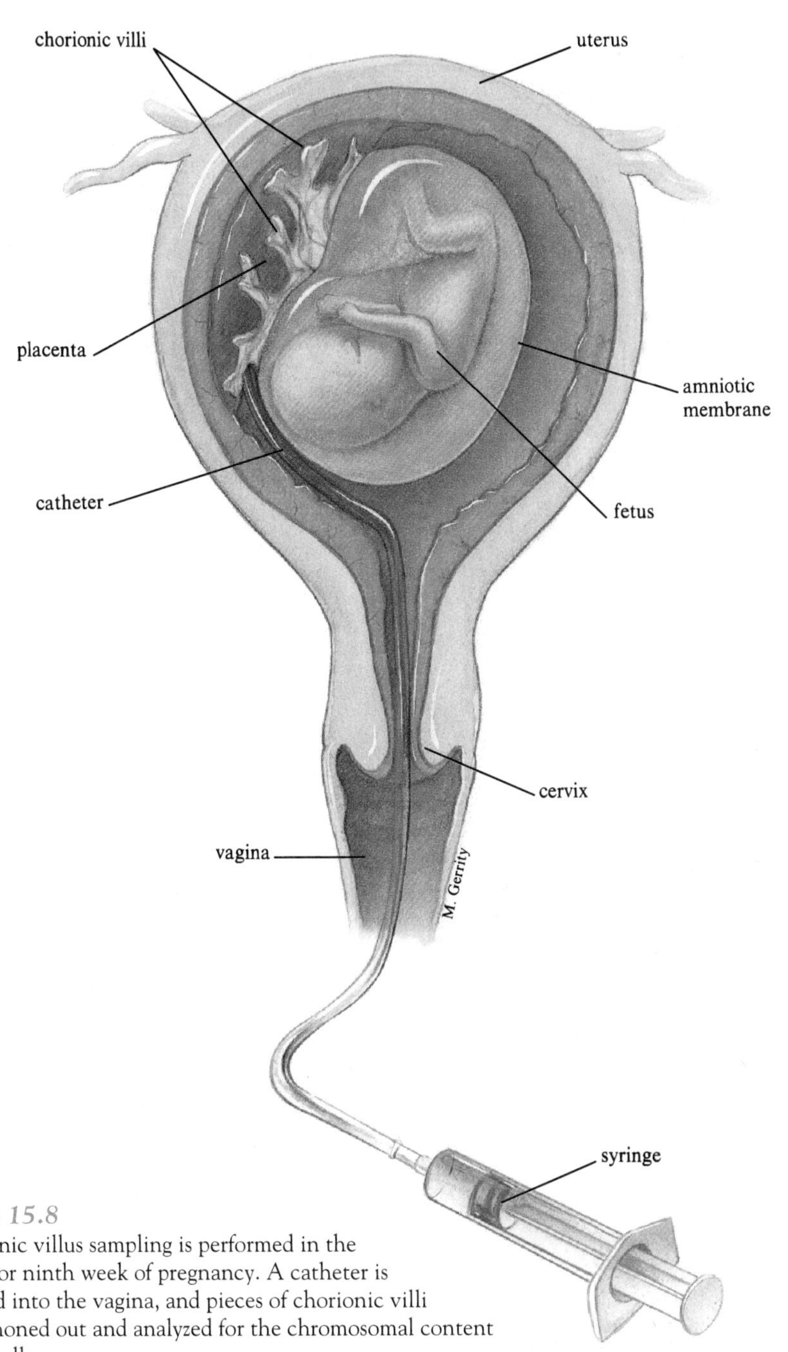

Figure 15.8
Chorionic villus sampling is performed in the eighth or ninth week of pregnancy. A catheter is inserted into the vagina, and pieces of chorionic villi are siphoned out and analyzed for the chromosomal content of the cells.

KEY CONCEPTS

Ultrasound bounces sound waves off of the uterus to image the developing fetus, highlighting normal as well as abnormal structures and guiding for amniocentesis. In amniocentesis, a sample of amniotic fluid containing fetal cells is withdrawn at 16 weeks gestation. The fetal cells divide in the laboratory, and the chromosomes are examined. Biochemical tests are performed on the fluid. Chorionic villus sampling examines cells surrounding the early fetus that develop from the fertilized egg.

Molecular Approaches to Diagnosing Genetic Disease

Examining the DNA of sampled cells to diagnose genetic disorders is based on the fact that all cells of an individual, from conception on, contain the full set of genetic instructions. For example, even though cystic fibrosis is expressed as an overproduction of thick mucus in the pancreas and lungs, the mutation that causes it is present in all cells of an affected individual. Unusual chromosome structures or DNA sequences can confirm a diagnosis based on symptoms and family history. Returning to the cystic fibrosis example, detecting the causative gene mutation in a newborn confirms a more traditional test that detects excess salt in sweat. Genetic tests can also spot carriers of recessive conditions, and diagnose affected individuals in whom symptoms have not yet appeared.

The techniques of molecular biology allow us closer peeks at disease-causing genes than are possible with karyotyping stained chromosomes. However, it is often a chromosomal abnormality that tells geneticists where among the 23 chromosome pairs to begin the search for a gene. For example, the gene for Duchenne's muscular dystrophy was localized to its specific

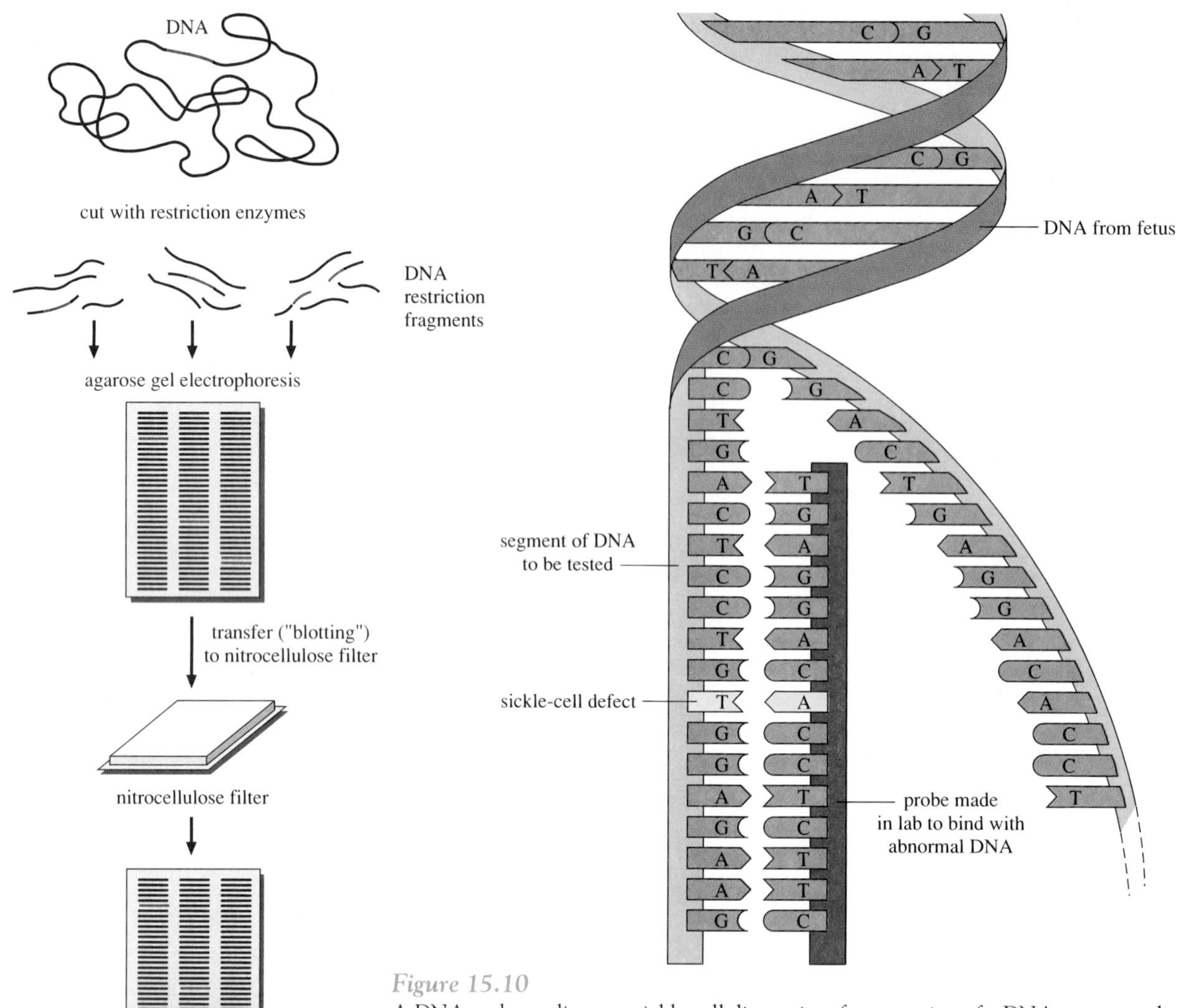

Figure 15.9
Southern blotting is used to display DNA fragments. In this procedure, DNA pieces that have been separated by size are blotted onto a nitrocellulose filter (which looks like a square of paper) and exposed to radioactively tagged pieces of DNA, or probes. When the probed filter is placed next to X-ray film, the sites at which the probes bind to the DNA fragments expose the film. The completed Southern blot shows the probed DNA fragments in size order. (An actual Southern blot looks more like a line of smears than the neat lines shown in the illustration.)
Source: National Institute of General Medical Sciences, "The New Human Genetics." NIH Publication No. 84-662, September 1984, page 29.

Figure 15.10
A DNA probe to diagnose sickle cell disease in a fetus consists of a DNA sequence that is complementary to the part of the sickle cell allele containing the single base difference that causes the disorder. Fetal cells are collected from a sample of amniotic fluid. The DNA is separated into single strands and unwound, and the radioactively labeled probe is applied. DNA from a fetus who has inherited two copies of the disease-causing allele (and will therefore have sickle cell disease) shows twice as much fluorescence as DNA from a fetus who has only one disease-causing allele (a carrier).

site on the X chromosome thanks to young patients whose chromosomes bore deletions in the region. The two tools that make possible this new molecular genetic medicine are *restriction enzymes* and *DNA probes*.

Recall from chapter 13 that restriction enzymes are bacterial biochemicals that are used as molecular scissors to cut DNA at specific sequences. DNA probes are pieces of DNA about 1,000 to 6,000 bases long that are complementary to part or all of a particular gene or any portion of a chromosome's DNA. Probes are synthesized in the laboratory or are obtained from "libraries" consisting of all of an organism's DNA cut into pieces and cloned in recombinant bacteria. Probes are allowed to replicate in the presence of radioactive DNA precursors so that they are labeled. In a procedure called **Southern blotting,** probes pick out their complementary sequences in sample DNA and are visualized by their radioactivity, which exposes photographic film (fig. 15.9).

A DNA probe consisting of precisely the part of a gene that is abnormal and causes a disorder can be used directly to diagnose that disorder. In sickle cell disease, for example, a probe containing the mutant sequence binds only when a person's cells contain the mutant gene (fig. 15.10).

It is unusual, though, to have enough information about a gene's product or sequence to manufacture such an explicit

Reading 15.1 DNA Probes

THE CHEMICAL ATTRACTION BETWEEN COMPLEMENTARY STRANDS OF DNA IS THE BASIS OF DNA PROBE TECHNOLOGY. A DNA probe is a short piece of DNA, about 1,000 to 6,000 bases long, that is complementary to a segment of DNA in an infectious organism or virus or to part of the sequence of a gene that causes an inherited trait.

Consider a DNA probe used to diagnose hepatitis B, a virus whose genetic material consists of 3,200 base pairs. A single strand of DNA that is complementary to one strand of the viral DNA is manufactured synthetically and attached to a molecule of a fluorescent dye. This chemical combination, the DNA probe, is added to a sample of blood from a person with symptoms of hepatitis. Next, the sample is treated with a chemical or heat so that its DNA separates into single strands. If the hepatitis B virus is present, the DNA probe hydrogen bonds to it (hybridizes), forming a double helix. Next, hybridized (double) strands are separated from unhybridized (single) strands. If the hybridized DNA fluoresces under lighting that detects the dye, then the hepatitis B virus is present. A diagnosis has been made. The probe can be mass-produced using recombinant DNA technology.

Many types of infectious diseases can be diagnosed using DNA probes. A dentist swabs a patient's gums with a paper point, then sends it to a lab where DNA probes detect the three microorganisms that cause periodontal disease, the major cause of tooth loss. Herpes simplex infections and stomachaches caused by *Shigella* bacteria can be diagnosed with DNA probes. "Kits" containing DNA probes are used by homeowners to detect fungal infections in turf grasses. A very valuable DNA probe tests for the presence of *Salmonella* bacteria in food samples. The test takes 2 days, compared to the week that it takes to detect the bacteria using conventional microbiological techniques to grow the organism. Had the *Salmonella* probe been available in the spring of 1985, some of the 17,000 people in the Midwest who suffered days of intestinal pain following ingestion of contaminated milk might have been spared their experience.

Inherited diseases can be diagnosed before birth using DNA probes. A DNA probe used to diagnose sickle cell disease in fetuses, for example, is complementary to 1,000 bases of the allele that causes sickle cell disease. In this recessive disorder, abnormal hemoglobin bends red blood cells into a sickle shape, which obstructs blood flow, causing pain and the destruction of organs. A probe can also detect the cystic fibrosis gene.

DNA probes are also used to study genes even if practical applications are not immediately apparent. For example, the three genes in the human that provide color vision were identified by DNA probe technology. Genes for red and green color vision were localized to the X chromosome, as was expected from the sex-linked inheritance pattern of red-green color blindness. A third gene, which provides blue color vision, was found on the seventh largest chromosome. Now that the human genes for color vision have been identified, the molecular basis of color vision—and of color blindness and other visual abnormalities—can be studied.

probe. More often, probes attract large fragments of DNA that contain the complementary sequence. The sizes of these fragments depend upon where the restriction enzymes used to make them cut, which in turn depends upon the sequence of the sample DNA. Differences in restriction enzyme cutting sites between individuals are called **restriction fragment length polymorphisms,** or RFLPs (pronounced "riflips").

Direct Genetic Disease Diagnosis—RFLPs Within Genes

If an allele that causes a disorder is disrupted in a way that alters the base sequence at a site where a particular restriction enzyme normally cuts, or if the disease-causing allele contains a cutting site not usually present, that particular restriction enzyme can be used to distinguish between the defective and normal alleles. That is, the normal allele yields a different pattern of fragment sizes than the disease-causing allele, if each is cut with the same restriction enzyme.

A childhood form of Gaucher's disease is an example of a condition that can be probed directly with a restriction enzyme. In this disorder, a biochemical that usually breaks down fat in various tissues is abnormal, causing an enlarged liver and spleen and neurological impairment that usually kills the child by age three. The DNA base change that causes the disease falls at a site that is recognized by a certain restriction enzyme in the normal allele. When that restriction enzyme is applied to the mutant allele, it does not cut at the site, as it normally would. This is observed as longer than normal DNA pieces resulting from cutting the chromosome with the restriction enzyme and singling them out with appropriate probes. The unusually long gene pieces signal Gaucher's disease in a fetus. Sickle cell disease can also be detected by direct use of restriction enzymes, because cutting sites occur within the gene (fig. 15.11).

Genetic Markers—RFLPs Outside Genes

For disorders in which the gene has not been identified, restriction enzymes can be used for diagnosis if a piece of DNA located near the disease-causing gene has a characteristic extra or missing restriction enzyme cutting site. This indirect approach uses a **genetic marker,** a sequence of DNA bases that is linked to the disease-causing gene (i.e., it lies nearby on the same chromosome.) When the marker is detected, the disease-causing allele's presence is inferred.

To be useful as a genetic marker, the linked DNA sequence must always be present in ill family members but never in healthy relatives. In a family with Huntington disease, for example, the marker might be present on chromosome 4 of an elderly grandmother institutionalized with the late, incapacitating stage of the illness, in her son who is beginning to feel early personality changes and loss of coordination, and perhaps also in some children, although they are too young to have symptoms.

On a molecular level, this means that the marker must always reside on the same homolog as the disease-causing gene. Can you see why an RFLP on the homolog that does not contain the disease-causing gene could not serve as a marker? In a direct RFLP test, such as that for Gaucher's disease, anybody can be tested. But for a genetic marker test, entire families must be tested to ensure that the RFLP travels only with the defective gene.

The first genetic marker was discovered in 1983, and the story behind this medical milestone illustrates the interaction of different subdisciplines of genetics and a good helping of luck. The search for a genetic marker for Huntington disease began in a remote village on the shores of Lake Maracaibo, Venezuela. Seven generations ago, in the 1800s, a local woman married a visiting Portuguese sailor who, the folklore goes, walked as if he was intoxicated. Like most couples in the poor fishing village, the woman and her sailor had many children, and some of them grew up to walk in the same peculiar way as their father. Of the couple's nearly 5,000 descendants, 250 of them today have Huntington disease (fig. 15.12). Another 727 have at least one affected parent, and 39 individuals have two affected parents, giving geneticists the unusual opportunity to study people inheriting two dominant disease-causing alleles.

The idea of using RFLPs as genetic markers arose in the late 1970s, and the huge Venezuelan clan and their disease seemed a perfect starting point. In 1981, Columbia University psychologist Nancy Wexler began making yearly trips to the Lake Maracaibo residents, who still lived in huts perched on stilts much as their ancestors had. The

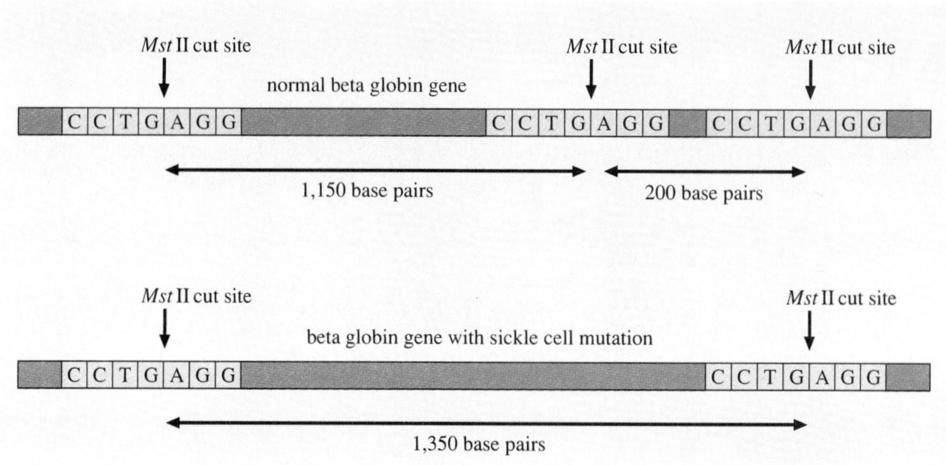

Figure 15.11
Restriction enzymes can be used to diagnose sickle-cell disease. The normal beta globin gene contains a cutting site for the restriction enzyme *Mst*II that is lacking in the allele that causes sickle cell disease. Therefore, cells of a fetus with sickle cell disease have two unusual segments of DNA that are 1,350 bases long. A fetus who has inherited two normal copies of the gene has pieces for that gene of 1,150 bases long and 200 bases long. What size pieces would you expect for a carrier of sickle cell disease?

Figure 15.12
A large Venezuelan family helped locate the first genetic marker. Psychologist Nancy Wexler and her co-workers frequently consulted this huge pedigree of thousands of individuals while hunting for a marker for Huntington disease. Wexler is "at-risk" for the disease.

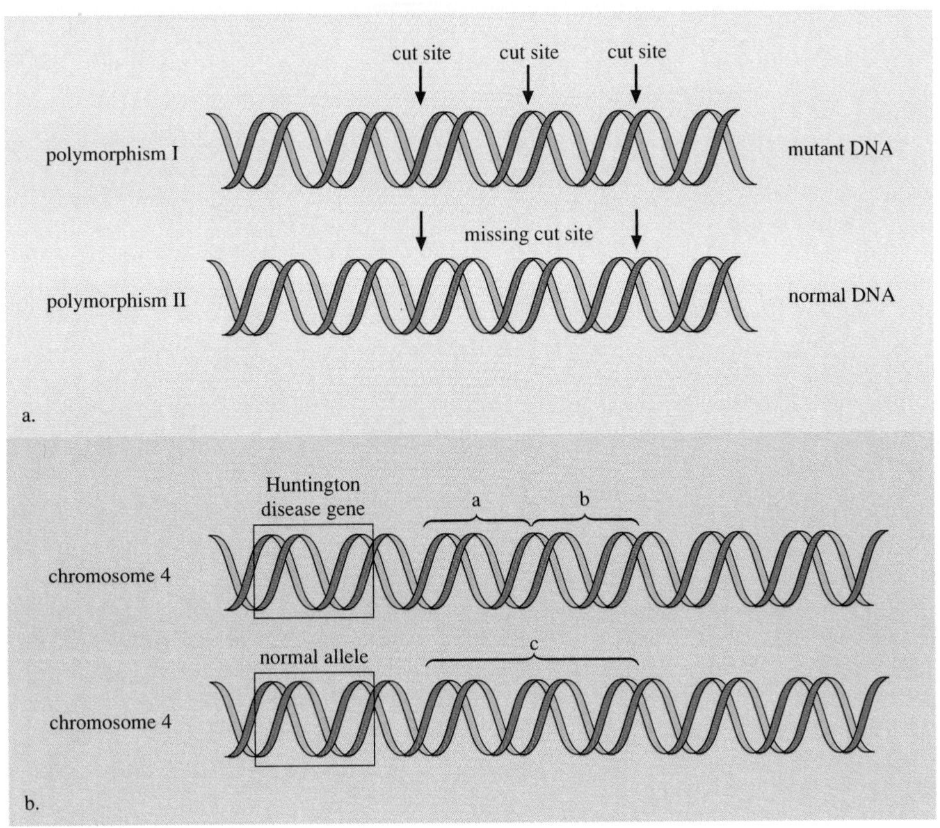

Figure 15.13

An RFLP serves as a marker for Huntington disease. *a*. For a marker to be informative, the person being tested must be heterozygous for the RFLP—that is, the DNA sequence at a particular point differs on the two homologs. *b*. The RFLP must also be on the same homolog as the disease-causing gene. In a family with this gene arrangement, cutting an individual's DNA with the appropriate restriction enzyme yields pieces *a* and *b* if the person has inherited the disease.

people grew to look forward to Wexler's visits and would permit skin biopsies and blood samples to be taken in exchange for blue jeans and M&M candies.

Meanwhile, similar tissue samples were being collected from a family in Iowa consisting of 41 members, 22 of whom had Huntington disease. James Gusella and his co-workers at Massachusetts General Hospital began the laborious task of extracting DNA from the samples, cutting it with restriction enzymes, and testing random DNA probes to see if any probe always appeared in sick individuals but never in healthy ones. Gusella had several hundred probes, and the task was expected to take years, maybe even a decade.

Yet a third research group, led by P. Michael Conneally at Indiana University, tried to identify probes that served as markers within specific pedigrees. One warm May night in 1983, the computer being fed all of this information came up with a match. By sheer luck, the 12th probe Gusella had chosen, called G8, worked—within both the Venezuelan and Iowa families, the presence of a distinctly sized piece of DNA pulled out by the G8 probe indicates Huntington disease.

The G8 probe is about 96% accurate—that is, in these families 4% of the people who have the DNA sequence have not inherited the Huntington gene, and vice versa. The marker is not perfect because the RFLP it recognizes lies several thousand bases away from the actual disease-causing gene, which means that crossing over between the gene and the marker can occur, separating the RFLP from the gene. However, newer Huntington markers are closer to the gene, increasing accuracy to about 99% (fig. 15.13). By the time you read this, the Huntington disease gene itself may be known.

Presymptomatic tests for late-onset disorders lacking treatments, such as Huntington disease and inherited Alzheimer's disease, present difficult ethical and psychological problems. The decision to take the test is very upsetting, and those who receive bad news often become profoundly depressed. On the other hand, knowing one's medical future permits informed life decisions. In one study, people at risk for developing Huntington disease were asked, prior to 1983, whether they would take a presymptomatic test. Two-thirds said yes. Asked again when they knew that the marker test was a reality, most of them said that they would not take it.

Even though many people debate the usefulness of markers, finding these signposts of health-related genes can lead researchers to the disease-causing genes themselves, and this might lead to effective treatments. To do this, researchers cut the chromosomal region containing the gene with several restriction enzymes and sequence the resulting pieces. By lining up the sequences where they overlap, investigators "walk" closer and closer to the gene. If a RFLP is very far from a disease-causing gene, restriction enzymes can be used to loop out some of the intervening material, and the investigator is then said to "jump" closer to the gene. The cystic fibrosis gene was the first to be identified by such chromosome jumping.

KEY CONCEPTS

DNA can be cut with restriction enzymes and the fragments probed with radioactively labeled DNA probes. Direct diagnosis of inherited disease is possible when the probe is complementary to the part of the disease-causing allele that is abnormal. Diagnosis is also possible if a RFLP falls within a disease-causing allele and by using genetic markers if a RFLP is closely linked to such an allele. Genetic markers must be traced through families.

Genetic Screening

Genetic screening tests diagnose certain inherited illnesses early enough so that treatment can be of maximum benefit, or so that situations that aggravate or provoke

Reading 15.2 The PKU Story

In 1934, a mother in Oslo, Norway, noticed that the urine of her two retarded children had a peculiar odor. She mentioned this to a relative who was a chemist, and he analyzed the urine. He found that the urine's odd smell was caused by the buildup of a certain biochemical, which in turn was caused by a missing enzyme. The enzyme was missing because of a defective gene. The chemical accumulations resulting from this "inborn error of metabolism" caused the mental retardation. The observant young mother and her relative laid the groundwork for today's successful treatment of phenylketonuria, or PKU.

In the United States, PKU strikes 1 in every 14,000 white newborns. In Ireland and Scotland, it is much more prevalent, affecting 1 in 5,000 infants. Among blacks it is extremely rare, affecting only 1 in 300,000 people. The mental retardation caused by the buildup of the amino acid phenylalanine in the blood can be avoided by following a diet very low in phenylalanine from birth until age eight, when most brain development is complete. All newborns are required by law to be tested at birth for PKU, which is done on a small blood sample taken from the heel. The reason for the required test is not so much a matter of technology as economics.

Newborn screening for PKU is a classic example of an economically feasible genetic screening program. That is, it is cheaper to screen the entire population and treat detected individuals for PKU (at a cost of $3.3 million) than it is not to screen and to institutionalize patients who would otherwise be helped by the diet (at a cost of $189 million).

The ability to treat the mental retardation (the phenotype) of PKU does not correct the underlying genetic defect (the genotype), and this limitation has led to an unexpected problem. Most of the thousands of PKU children treated over the last 3 decades have grown into adults who are now transmitting the mutant gene to their own offspring. All children of mothers with PKU, whether or not they have inherited the disease, are born severely retarded. This is because the mother, who is no longer on the restrictive diet of her childhood, produces excess phenylalanine, even though the high level no longer affects her brain. To protect the fetus, the PKU mother must return to the low phenylalanine diet during pregnancy. About 3,000 such women in the United States are currently of childbearing age, and public health officials are trying to locate and help them. Some geneticists are now advising that people with PKU remain on low phenylalanine diets throughout their lives. They are particularly advised to avoid foods sweetened with aspartame, which consists of the amino acids aspartic acid and phenylalanine.

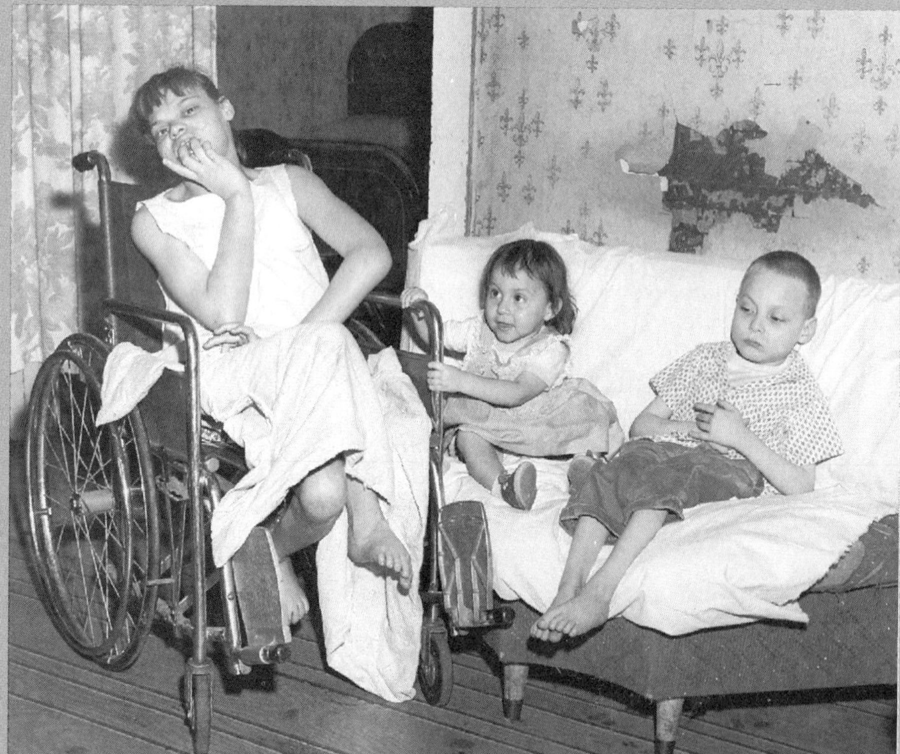

Figure 1
These three siblings have each inherited PKU. The older two are mentally retarded, but the youngest child is of normal intelligence, thanks to a special diet followed since birth that counteracted the effects of the mutant gene.

Figure 2
Those with PKU should avoid aspartame, which contains large amounts of phenylalanine. Researchers do not yet know whether aspartame has dangerous effects on carriers of PKU.

symptoms can be avoided. Genetic disease diagnosis of a fetus gives parents the option of abortion if the symptoms are particularly devastating or untreatable, or it allows them time to prepare for dealing with the problem.

Genetic screens are largely a matter of economics and treatment availability. A test must save money by identifying affected individuals early enough so that they can be helped, making more expensive care later on unnecessary. Detection and treatment of an inherited form of mental retardation called phenylketonuria is an excellent example of an effective genetic screening program, because the symptoms can be avoided by following a special diet from birth (Reading 15.2). Genetic screens are useful at different points in life.

Before Pregnancy

Michael and Sharon are college juniors planning to marry when they graduate. Because they hope to have children one day, they participate in a program on campus offering blood tests to identify carriers of Tay-Sachs disease—the neurological childhood killer 100 times more common among Eastern European Jewish people than others. Neither Michael nor Sharon has a relative with Tay-Sachs disease, but each has grandparents who came to the United States from the Soviet Union. Even though they know that 1 in 30 people from their ethnic background is a carrier, they are astounded to learn that both of them fall into this category. They are distressed but in a way relieved. Now when Sharon becomes pregnant, she will know to ask for a prenatal test.

During Pregnancy

Julie hardly noticed when the nurse took yet another sample of her blood, a test for some unpronounceable condition that her child could not possibly have. Unfortunately, Julie turned out to be one of the 50 pregnant women out of every 1,000 whose blood contains unusually high amounts of a substance called alpha-fetoprotein. For 2 of these 50 women, further tests (a repeat blood test, ultrasound, amniocentesis) bring bad news—the fetus has a neural tube defect. But for the other 48, the elevated protein is explained by a miscalculated due date (the amount in the blood normally rises during pregnancy), or, as in Julie's case—twins! Too low levels of alpha-fetoprotein can signal Down syndrome.

Screening Children

When the 6-month-old infant developed a fever of 102°F (approximately 39°C), the nervous mother called her doctor. A nurse told her to call back in the morning if the fever had not dissipated. Most babies run high fevers from time to time, the nurse assured the anxious parent. But by morning, the baby was dead.

The nurse was right—most babies do run very high fevers occasionally. But this child was not the run-of-the-mill infant. He had sickle-cell disease, which had never been diagnosed. Although carrier tests and prenatal diagnosis for this disorder have been available for years, they were mandatory in only a few states because it was thought that there was no treatment. But early identification of these children can warn parents to watch out for the normally mild infections that, for reasons still unknown, kill 15% of children with the disease. Screening of all newborns—not just blacks—is now available in all states. At 22¢ a test, it is both a bargain and a lifesaver.

Genetic Screening in the Workplace

Another site of genetic screening is the workplace. Some chemical companies administer blood tests to identify workers who are genetically predisposed to become ill from exposure to certain chemicals. One company screens employees to spot carriers of sickle cell disease and enters this information in their health records. Another company examines workers' chromosomes before employment begins and again after the workers have been in contact with potentially dangerous chemicals. This is how the tendency of benzene to damage chromosomes was discovered.

KEY CONCEPTS

Genetic screening tests are usually cost-effective programs that detect an inherited disease, carrier status, or chromosomal damage. Such tests are performed at various stages of the human life cycle.

Genetic Counseling

Diagnosis of a genetic disease can be confusing and terrifying, often requiring life-and-death decision making at a time of intense emotional turmoil. The genetic counselor is a new breed of medical professional who helps families understand their problem and who evaluates pedigrees and test results to predict who in a family is likely to be affected by the condition. The genetic counselor supplies the information necessary for a family to make a rational, informed choice but cannot ethically tell a family which choice to make.

The 38-year-old pregnant woman described earlier in this chapter, Mrs. F, illustrates a very common reason for seeking the help of a genetic counselor. Mrs. F might be alarmed over the prenatal tests her doctor advises and may not understand their comparative risks and benefits or why they are recommended. The counselor calms her fears, explaining that amniocentesis is routine for women over 35 because of their slightly increased risk of carrying a child with a chromosomal abnormality. The counselor also stresses that the test can rule out certain conditions, but it is not a guarantee of a healthy baby. The patient might check back with the counselor after the test results are in.

A genetic counselor helps patients put medical information into perspective, calming panic as well as ensuring that patients adequately understand risks. Parents of a child with PKU need to be told how the dietary treatment works. The young couple who are carriers for Tay-Sachs disease must make decisions about becoming pregnant or adopting a child. The woman with the "false positive" alpha-fetoprotein test needs to be helped through her anxiety when undergoing further tests. The parent of the child with sickle cell disease might want to learn about new treatments or be put in touch with a patient support group.

Many genetic counseling sessions deal with the disturbing issue of abortion. Personal views range widely, from those opposed to abortion for any reason, to those who would choose abortion for any abnormality, to those who make the decision based upon the expected quality of life for a child born with a particular problem. One of the many difficulties encountered in making such a decision is that for many disorders, a prenatal diagnosis does not indicate how severely affected the child will be. A couple may know that their fetus has the extra chromosome 21 of Down syndrome but not whether he or she will be a happy, playful youngster who is a slow learner but otherwise healthy or a severely retarded child with an associated heart defect requiring frequent surgery.

Even when a child is born with a serious birth defect, medical science cannot always predict the child's future. In one striking example, a boy born in the 1950s with a severe neural tube defect was sent home from the hospital with instructions from the doctors to his parents to just wait for him to die. Today, that boy has grown into an active man who has earned a Ph.D. in social work and is a counselor for disabled children. He was one of the 5% of neural tube defect patients who spontaneously recovers. Others are helped by surgery performed at birth. Prenatal detection of a neural tube defect can alert medical specialists to be present at the birth to provide immediate treatment.

KEY CONCEPTS

Genetic counseling helps patients understand genetic tests and diagnoses and the risks of particular family members' developing certain conditions.

Medical Genetics, Ethics, and the Law

With the increasing power of medical genetics lies the potential for abuse. Doctors have grown concerned about the growing number of couples who choose to abort a healthy fetus because amniocentesis revealed it to be a girl and they had wanted a boy, or vice versa. Some people worry that, in the future, decisions about whether to have an abortion might be made on such criteria as physical attractiveness or intelligence.

Similarly, genetic screening by industry presents an ethical dilemma. On the positive side, testing employees to detect inherited susceptibilities or sensitivities may provide information that can be used to prevent illness. On the negative side, companies might attempt to prevent chromosomal abnormalities not by making the workplace safer but by refusing to hire genetically susceptible people. Because some genetic conditions affect certain ethnic or racial groups much more than others, the use of genetic tests to keep some workers from areas that use chemicals that they might be particularly sensitive to could be construed as discrimination rather than preventive medicine.

For example, until recently carriers of sickle cell disease were kept out of the Air Force Academy and were required to pay higher premiums by health insurance companies, even though they suffer no ill effects as a result of their genotype unless they undergo excessive physical exertion. As it becomes possible to identify carriers of more genetic diseases, and as more inherited susceptibilities become detectable, will more groups be perhaps unfairly limited in their life-styles and livelihoods?

Because of ethical concerns, it is not surprising that the new genetics has entered the courtroom. Consider these cases:

—A couple had two children with the disfiguring genetic disorder known as neurofibromatosis. The husband had a vasectomy, but the wife became pregnant anyway. The third child also had the disease. The Pennsylvania State Supreme Court ruled that the parents could sue their obstetrician for the "wrongful birth" of the child and for the physical inconvenience and mental stress of raising her.

—A 35-year-old woman requested, and was denied, amniocentesis at a church-affiliated hospital in Ohio. Her son was born with Down syndrome, severe heart defects, and other problems. The parents sued the hospital.

—A court in California ruled that a child with a genetic disease whose birth could have been prevented can sue her parents for allowing her to be born. Other states have passed legislation banning such laws.

KEY CONCEPTS

Genetic tests can be misused or abused, causing discrimination. Not offering available genetic tests can lead to lawsuits against medical practitioners.

Gene Therapy

The practical goal of medical genetics is to relieve suffering caused by defective genes. Treating inherited disease follows two general approaches—altering the phenotype or altering the genotype.

Treating the Phenotype—A Short-Term Solution

Treating the symptoms of an inherited disease alleviates pain in the individual but does not alter the genetic misinformation that could be passed to the next generation. These treatments may increase the frequency of disease-causing alleles in a population by allowing individuals to become parents who otherwise would not be healthy enough to do so.

The phenotypes of some genetic disorders can already be altered. A child with cystic fibrosis, for example, sprinkles a powder consisting of cow digestive enzymes onto a serving of applesauce, which she eats before each meal to replace the enzymes that her mucus-plugged pancreas cannot secrete. A boy with hemophilia receives a clotting factor. The biochemical buildups caused by several inborn errors of metabolism are counteracted by following restrictive diets. Even wearing eyeglasses is a way of altering the expression of one's inheritance.

Nonheritable Gene Therapy

Correcting any inherited problem at its source, the gene, is a complex challenge in a many-celled organism. Correcting only the affected somatic cells, called *nonheritable gene therapy*, will be implemented sooner than the more complex *heritable gene therapy*, in which the genetic material of a germ cell or fertilized egg is altered so that all cells of the individual harbor the change.

One very useful tissue to alter in nonheritable gene therapy is endothelium, the tilelike cells that form the capillaries, which are microscopic blood vessels. Engineering endothelium allows a needed substance to be deposited right into the bloodstream. In a diabetic, endothelium might be altered to secrete the needed insulin; for a hemophiliac, it would ooze the missing clotting factor. Skin grafts can also be engineered to secrete a needed biochemical, such as growth hormone.

Cystic fibrosis is correctable in cells from the human respiratory tract grown in culture, by inserting a normal copy of the gene. A therapy might consist of an aerosol spray to deliver the gene to the cells of the mucus-clogged airways of a cystic fibrosis sufferer.

The first nonheritable gene therapies will be for diseases that affect the precursors of blood cells in the bone marrow, because these conditions are well understood and the tissue involved can be manipulated. The cells are removed from a patient with a hypodermic needle, grown in the laboratory, and exposed there to viruses that contain the normal alleles of the defective genes. Cells that incorporate the replacement alleles are then reinjected into the patient, where they repopulate the marrow and divide to form healthy blood cells. This procedure is used to treat severe hereditary combined immunodeficiency, in which lack of the enzyme adenosine deaminase kills the immune system's T and B cells, which are white blood cells that fight infection. Patients usually die of massive infection before they reach 10 years of age.

Another target of nonheritable gene therapy in humans is Lesch-Nyhan syndrome, a sex-linked recessive disorder in which a missing enzyme in the brain causes mental retardation, cerebral palsy, and self-destructive behavior in which the patient uncontrollably chews his lips, fingers, toes, and shoulders, causing severe injuries. The normal allele can restore enzyme activity in cultured cells taken from a Lesch-Nyhan patient, but therapy requires knowing where to replace the gene on the brain. Treating Lesch-Nyhan syndrome is not as clear-cut as severe hereditary combined immunodeficiency, in which the abnormal cells can easily be isolated and then returned to the body.

Nonheritable gene therapy is probably far in the future for the many genetic diseases whose sites of action in the body are still unknown. In Tay-Sachs disease, for example, we know which enzyme is missing and the consequence of the deficit, but we do not know which nerve cells become buried in fat. Even when we know which tissues are affected, delivering healthy genes is a challenge. Transplants of healthy cells into muscular dystrophy patients, for example, only restore muscle activity presisely where they are implanted. Any correction provided by this localized form of gene therapy helps the individual, but not his or her children, because the defective gene in the sex cells is not altered.

a.

b.

Figure 15.14
Gene therapy—building a better mouse. *a.* The huge size of the mouse on the left compared to her litter mate on the right is due to a rat growth hormone gene that she acquired by gene therapy when she was just a fertilized egg. The gene is activated by traces of heavy metal in the diet. Because the transplanted gene is present in her sex cells, her unusual size will be passed on to her offspring. *b.* The newborn mouse on the left has also been altered by gene therapy. Her sibling on the right suffers from beta-thalessemia caused by a deletion in the beta globin gene. Note its pale, small body. The healthy pink newborn on the left has had its thalessemia deletion "patched" with a human beta globin gene.

Heritable Gene Therapy—A Longer-Term Solution

Heritable gene therapy is accomplished by creating transgenic animals, in which a gene is altered or replaced at the germ cell or fertilized egg stage. In fruit flies, immature egg cells are removed from the animal and an allele conferring a particular eye color replaced with an allele conferring a different color. When the egg cell is transferred back to the animal and fertilized, the new eye color is passed to subsequent fly generations and is expressed in the appropriate tissues at the correct time in development (the adult eye).

Heritable gene therapy has been used to create a "supermouse" (fig. 15.14). A rat's growth hormone gene is implanted into the fertilized egg of a mouse. The rat gene is attached to a DNA sequence that is activated in the presence of a heavy metal. After the fertilized egg develops into a mouse, and the mouse is fed trace amounts of a heavy metal, the rat growth hormone gene turns on. The result is a mouse nearly twice the normal size.

Even more dramatic heritable gene therapy in the mouse is the "curing" of beta thalassemia. This inherited blood disorder is caused by a deletion in the beta globin gene, and its human version is often fatal during childhood. Researchers injected fertilized mouse eggs lacking the beta globin genes with normal human beta globin genes. The resulting mice were healthy and robust. In addition, the transgenic mouse has a new type of hemoglobin—a mixture of mouse and human globin chains. In another approach called gene targeting, a transgenic organism lacks a particular gene.

Yet another type of gene therapy is "gene silencing." Rather than endowing cells or organisms with new genes, "gene silencing" shuts off expression of certain genes. The technique uses laboratory-made pieces of RNA that are the same sequence as the antisense strand of a particular gene—that is, the RNA is complementary to the gene. When the antisense RNA enters the cell, it seeks out and binds to the selected gene, preventing it from being transcribed. Gene silencing might be useful, for example, in turning off oncogenes, which cause cancer.

KEY CONCEPTS

Treating the symptoms of genetic disease is treating the phenotype. Nonheritable gene therapy corrects the genetic defect in the somatic tissue in which the defect is expressed. In heritable gene therapy, defective genes in germ cells or fertilized ova are replaced with normal alleles. In gene silencing and gene targeting, certain genes whose overproduction causes disease are blocked from being expressed.

Now that we know the biochemical nature of the gene, understand what genes do, and are beginning to be able to manipulate phenotypes and even genotypes, we face the most difficult challenge of all—how to use this information responsibly. We will face this challenge on several levels.

Individuals may find relief from diseases such as diabetes and hemophilia without worry of infection by using genetically engineered drugs. A couple may face tough decisions after receiving a prenatal diagnosis of their fetus, or choose not to have a family at all if they are both carriers of an untreatable disease. Multigenerational families will be asked to cooperate to trace genetic markers through their members, perhaps leading to presymptomatic diagnoses. Communities may have a say in what sorts of research are conducted at facilities in their area. Ethnic groups will increasingly try to benefit from screening tests for diseases they are more likely to get, while preventing discrimination based on this knowledge. As the international effort to sequence the human genome gains momentum in the years to come, we will be screened for more and more of the genetically influenced disorders that affect us all.

Genetics is certain to be a part of your future.

SUMMARY

Several analytical methods are used in medical genetics. *Pedigrees* provide information on which family members are affected by a particular trait. *Karyotypes* are charts of the chromosomes stained and arranged in size order and indicate chromosomal abnormalities. Conditions caused by chromosomal aberrations or single-gene defects can sometimes be diagnosed prenatally using *ultrasound examination*, *chorionic villus sampling*, and *amniocentesis*. DNA probes are used to detect abnormalities in individual genes directly. Restriction enzymes are used to diagnose certain genetic diseases directly or to locate markers that indicate the presence of a disease-causing gene. A genetic marker must be traced through many family members.

Genetic screening applies the technologies of prenatal diagnosis and carrier detection to identify individuals within a population who can transmit disease-causing genes. It is conducted before and during pregnancy, on newborns, and on people who have inherited disease but do not yet experience symptoms. Genetic screening is valuable if the knowledge gained is useful, if the condition detected is treatable, and if the test is economically sound. A genetic counselor can help people understand the results of genetic tests. The growth of medical genetics has introduced many ethical and legal problems.

Gene therapy is a major goal of medical genetics. Methods to treat phenotypes already exist. In *nonheritable gene therapy*, only the tissues affected by a disease-causing gene are genetically altered, and subsequent generations are not affected by the treatment. In *heritable gene therapy*, which is directed at germ cells or fertilized eggs, the correction is passed on to subsequent generations.

QUESTIONS

1. Copy the pedigrees in figure 15.2 twice. Then complete the pedigrees by indicating which individuals must be carriers and which might be carriers (using the symbols in Figure 15.1) for cystic fibrosis and for hemophilia. Why would this pedigree probably not represent a family with Huntington disease?

2. A woman is concerned about the chances of her children inheriting "kinky hair disease," a rare genetic disorder that killed her two brothers when they were infants. The boys had kinky white hair, and they died before they were 2 years old from brain degeneration. The woman's parents are normal, but she had a male cousin on her mother's side who died of the same condition. All other relatives are healthy.

- **a.** What is the likely mode of inheritance of this condition?
- **b.** How might chorionic villus sampling or amniocentesis help this woman?

3. A 41-year-old woman receives the following report of the results of her chorionic villus sampling: "The fetus is XX, with an apparently normal chromosome complement. Alpha-fetoprotein level is within the normal range." The woman calls her husband. "We can relax now. Our little boy is guaranteed to be healthy!" In what two ways is this woman's assessment incorrect?

4. Compare and contrast the advantages and drawbacks of chorionic villus sampling and amniocentesis.

5. How might you distinguish between mental retardation caused by trisomy 21 and mental retardation caused by fragile X syndrome?

6. How can DNA probes be used to diagnose inherited diseases as well as infectious diseases?

7. If heritable gene therapy was possible on humans, why could it not help someone already suffering from a genetic disease?

TO THINK ABOUT

1. If you were a genetic counselor, what would you say to each of the patients described at the beginning of chapter 14?

2. Symptoms of an inherited form of Alzheimer's disease usually appear in the fifth or sixth decade of life as increasingly severe forgetfulness and confusion. Genetic markers have made it possible for some younger people who have not yet developed symptoms to learn that they have indeed inherited the dominant disease-causing gene. What are the advantages and disadvantages of having such knowledge?

3. Do you think that screening to detect the one or two neural tube defects among every 1,000 pregnancies is justified, considering the stress that must be endured by the 50 or so couples who have positive results on the initial blood test?

4. The medical community is preparing to institute population screening for cystic fibrosis, whose gene was discovered in 1989. Past screening attempts for other genetic diseases have had different results. Tay-Sachs disease screening was highly successful, preventing the births of many affected children, whereas sickle cell disease screening led to discrimination. How did the natures of Tay-Sachs disease and sickle cell disease probably affect the outcomes of the screening programs? What might we learn from the experiences to help make cystic fibrosis screening valuable?

5. Why is heritable gene therapy more difficult to accomplish than nonheritable gene therapy?

SUGGESTED READINGS

Katz-Rothman, Barbara. 1986. *The tentative pregnancy: Prenatal diagnosis and the future of motherhood*. New York: Penguin Books. Amniocentesis puts pregnant couples through an anxious few weeks of waiting. When the news is bad, agonizing choices lay ahead.

Lewis, Ricki. Fall 1987. Genetic marker testing—are we ready for it? *Issues in Science and Technology*. We can prenatally and presymptomatically detect several genetic diseases. But should we?

Lewis Ricki. December 1990. Genetic screening: fetal signposts on a journey of discovery. *FDA Consumer*. Soon, genetic screening will affect us all.

Milunsky, Aubrey. 1987. *How to have the healthiest baby you can*. New York: Simon & Schuster. Chapters 8 and 9 discuss prenatal diagnosis and genetic disease.

Nash, J. Madeline. September 17, 1990. Tracking down killer genes. *Time*. A riveting interview with Francis Collins, one of the discoverers of the genes behind cystic fibrosis and neurofibromatosis.

Oakey, R. E. April 1987. Following up a natural mistake. *Chemistry in Britain*. Tracing an inborn error of metabolism is biochemical detective work.

Suzuki, David, and Peter Knudtson. 1989. *Genethics: The clash between the new genetics and human values*. Cambridge: Harvard University Press. Understanding the new genetic technologies must be accompanied by consideration of their effects on our lives.

UNIT 5

Animal Biology

C H A P T E R

16

Neurons

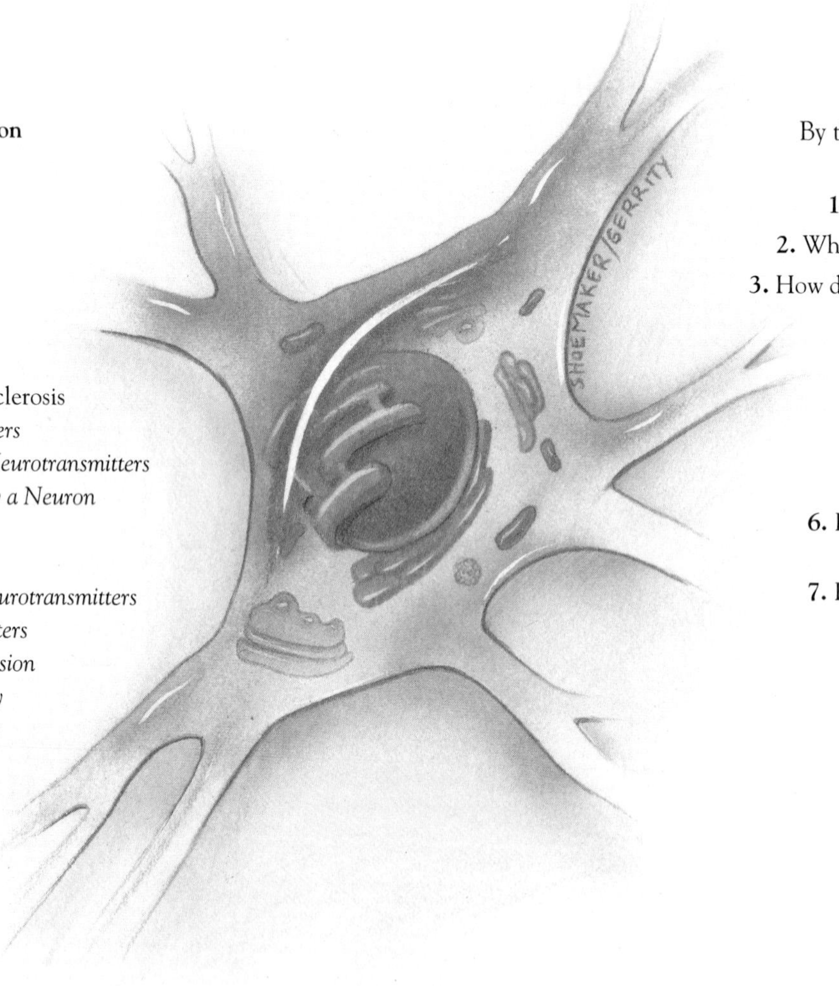

Chapter Outline

The Anatomy of a Neuron
Types of Neurons
A Neuron's Message
The Resting Potential
The Action Potential
The Myelin Sheath and Saltatory Conduction
Synaptic Transmission
Reading 16.1 Multiple Sclerosis
Disposal of Neurotransmitters
Excitatory and Inhibitory Neurotransmitters
Synaptic Integration—How a Neuron Interprets Its Messages
Neurotransmitters
Psychoactive Drugs and Neurotransmitters
Disease and Neurotransmitters
The Biochemistry of Depression
Opiates in the Human Body
Reading 16.2 Parkinson's Disease

Learning Objectives

By the chapter's end, you should be able to answer these questions:

1. What are the parts of a neuron?
2. What are the three types of neurons?
3. How does a resting neuron differ from a neuron firing a nerve impulse?
4. What happens during a nerve impulse?
5. How does the myelin sheath increase the speed of nerve transmission?
6. How do nerve cells communicate across a synapse?
7. How do nerve cells communicate with each other to integrate their functions?

The crowd roars with the crack of the bat. It is a high fly to the outfield, where the center fielder sees the ball and begins to move forward. He anticipates the path of flight and quickly moves into position to catch the ball. His arm extends his glove upward to intercept it. He feels the ball strike the glove, quickly retrieves it, and fires it towards home plate.

The ability to complete this complex task involves many systems of the body, and it is all orchestrated by the nervous system. The ball player's respiratory system brings oxygen to the circulatory system, which delivers it along with nutrients to all of the cells. Wastes produced by the cells are carried by the circulatory system to the kidneys for excretion. The skin keeps him cool by dissipating sweat. The muscles acting on the levers of the skeleton move the player at his will. The digestive system provides the basic nutrients for energy conversion and work done by the various cells. The endocrine system secretes adrenalin, which sharpens his awareness and facilitates the many systems for rapid action. All the while, the immune system is on alert for infection through the skin or respiratory passages.

The act of catching a ball demonstrates the major roles of the nervous system. Information arrives in the form of visual images (sensory). The player compares this information with past experiences of catching a ball and decides how to act (decision). The motor systems coordinate the muscles of the body to move the player into position to catch the ball (motor). These events are initiated at the conscious level of the brain. Meanwhile the heart beats, breathing continues, and balance is maintained by sensory, decision-making, and motor output at the unconscious level.

The nervous system, like all other systems of the body, is composed of single cells that act together to perform a complex function. These cells are termed **neurons,** or nerve cells. Neurons make possible a variety of sensations, actions, emotions, and experiences. Networks of these interacting cells control mood, appetite, blood pressure, coordination, and perception of pain and pleasure. The unique ability of neurons to communicate rapidly with each other enables us to not only be aware of the environment and to react to it but also to screen out unimportant stimuli, to form memories, and to learn. Yet despite these diverse functions, all neurons communicate in a similar manner, using a "language" of electrical and chemical changes that are passed along and between cells.

The Anatomy of a Neuron

Describing a typical neuron is like describing a typical house. Just as all houses share characteristics, neurons have common features. However, although a cabin and a castle both have a roof, walls, and a door, there are significant differences between them. Likewise, all neurons have the same basic parts, but they vary considerably in shape and size.

A neuron's shape, a rounded central portion from which many long, fine extensions emanate, is ideally suited for its job of receiving, integrating, and conducting messages over long distances (fig. 16.1). A motor neuron, for example, carries a message from the central nervous system (the brain or spinal cord) directing a muscle to contract. The central portion of the neuron, called the **cell body,** does most of the neuron's metabolic work. It contains the usual assortment of cellular organelles: a nucleus, extensive endoplasmic reticula, mitochondria to produce the ATP needed for maintaining a neuron in a state of readiness to send a message, and ribosomes to manufacture the proteins needed to convey a message to another neuron.

The extensions from the cell are of two types. In a motor neuron, the shorter, branched, and more numerous extensions are called **dendrites.** Dendrites usually receive information from other neurons and transmit it toward the cell body. The many branching dendrites allow a neuron to receive input from a tremendous number of other neurons. The second type of extension from the cell body is an **axon.** Typically, an axon conducts the message away from the cell body and transmits it to another cell. Because a nerve's message may have to be transmitted to a cell quite far away, an axon is usually longer than a dendrite, sometimes surprisingly so. An axon permitting you to wiggle your big toe, for example, extends from the base of your spinal cord to the toe, a distance that may be 3 feet (0.9 meter) or more. Axons and dendrites differ in other ways too. An axon is usually thicker than a dendrite, and there is usually only one axon per neuron. Axons are sometimes called **nerve fibers.**

To picture the size relationship among the parts of a motor neuron, imagine that the cell body is the size of a tennis ball. The axon might then be 1 mile (1.6 kilometers) long and a half an inch (1.27 centimeters) thick. The dendrites would fill an average-size living room.

Types of Neurons

Neurons are classified into three groups according to their general function: sensory neurons, motor neurons, and interneurons (fig. 16.2). The shape of each type reflects its function. A neuron that brings information toward the central nervous system, called a **sensory** (or **afferent**) **neuron,** has long dendrites that carry its message from a body part, such as from the skin, toward the cell body, which is located just outside the spinal cord. The axon of the sensory neuron is relatively short, because it delivers the message to another neuron whose dendrites are located within the spinal cord.

On the other hand, a **motor** (or **efferent**) **neuron** conducts its message outward, from the central nervous system toward a muscle or gland. Therefore, it has a long axon to reach the **effector** (the muscle or gland) and short dendrites. A motor neuron stimulating a muscle cell causes it to contract, and when a gland is stimulated it secretes. A third type of neuron, an **interneuron,** connects one neuron to another to integrate information from many sources and coordinate responses. Interneurons are the most variable in shape. In the brain and spinal cord, large, complex networks of interneurons receive information from sensory neurons, process and store this information, and generate the messages carried by the motor neurons to the effector organs.

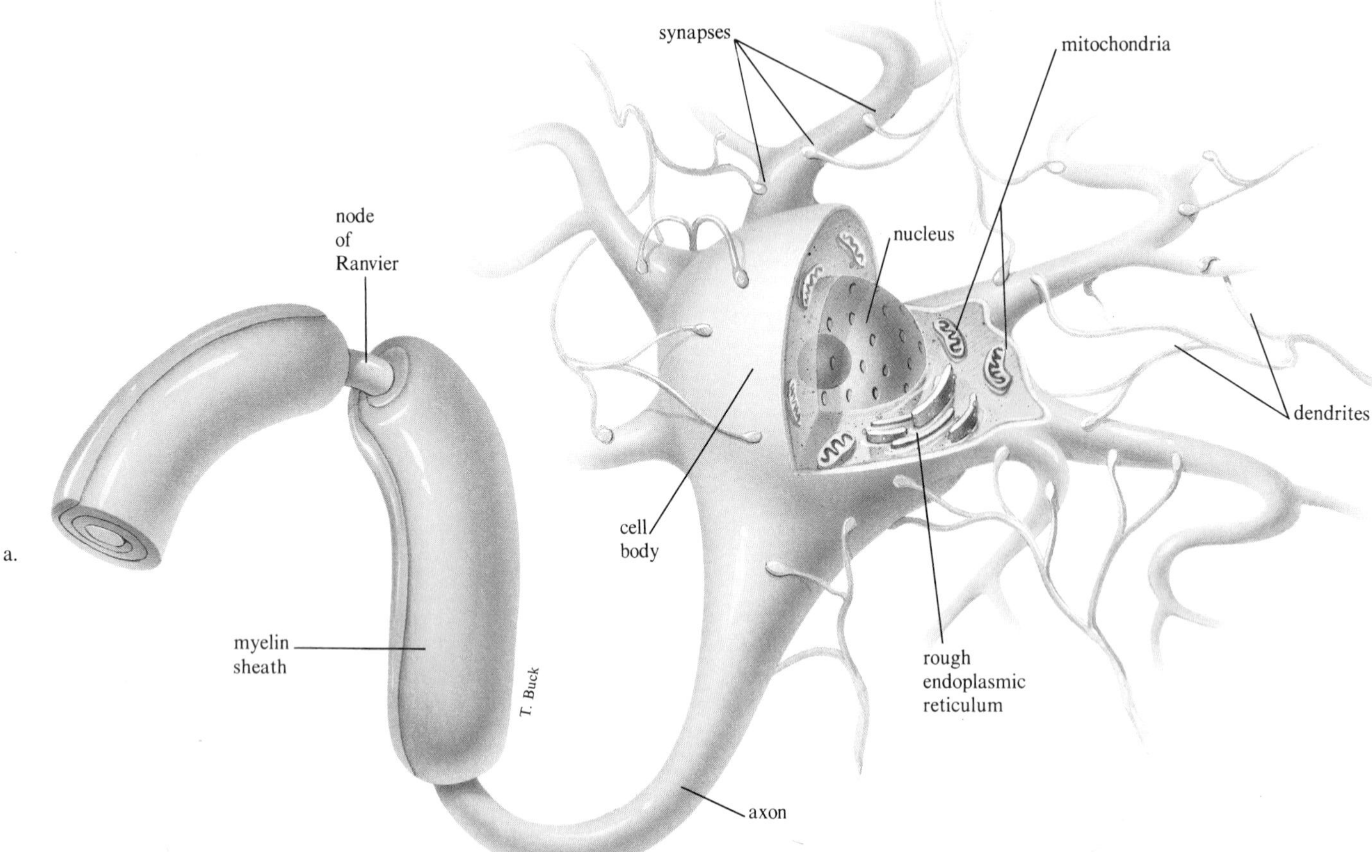

Figure 16.1
a. A neuron consists of a rounded cell body, "receiving" branches called dendrites, and "sending" branches called axons. The space between the axon terminals of a neuron and the dendrites of an adjacent neuron is called the synapse. Many axons are encased in myelin sheaths that are formed from the cell membranes of Schwann cells. Unmyelinated regions between adjacent Schwann cells are called nodes of Ranvier. *b.* These neurons from the cortex of a human brain are magnified 500 times. Note their entangled axons and dendrites.

b.

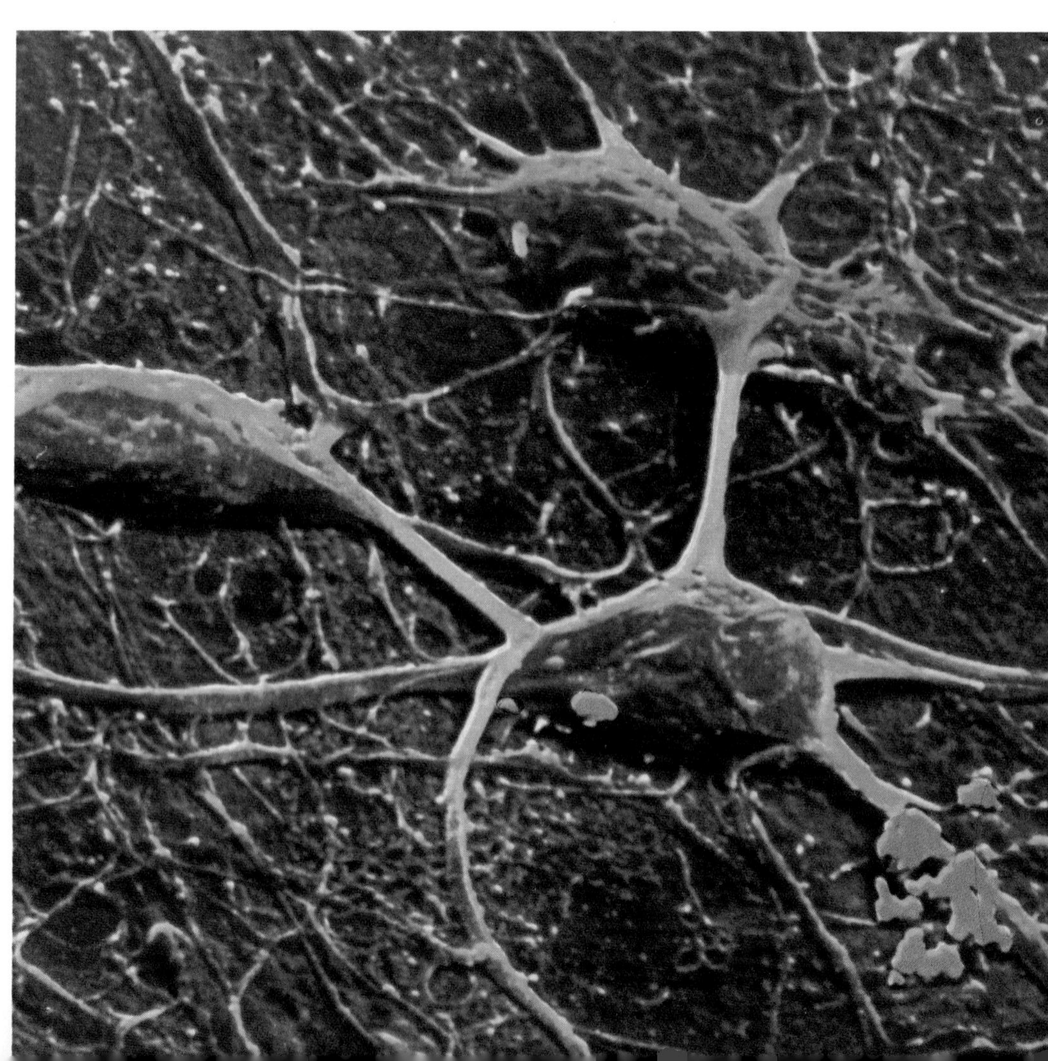

KEY CONCEPTS

A neuron consists of a cell body, which carries out most metabolic functions, dendrites, which receive information, and axons, which transmit information. Neurons usually have one thick axon and numerous thin dendrites. Sensory, or afferent, neurons carry information towards the central nervous system. Motor, or efferent, neurons carry messages from the central nervous system to muscles or glands. Interneurons connect other neurons.

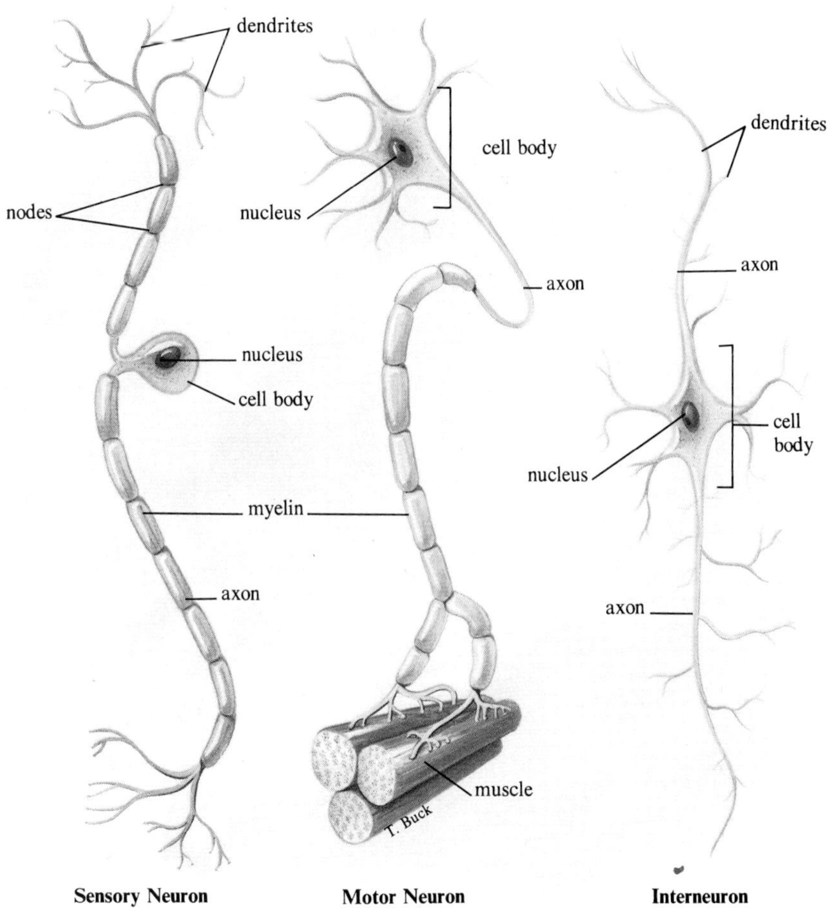

Figure 16.2
Types of neurons. Sensory neurons transmit information from the environment to the central nervous system. They have long dendrites and short axons. Motor neurons send information from the central nervous system to muscles or glands and have long axons and short dendrites. Interneurons connect other neurons.

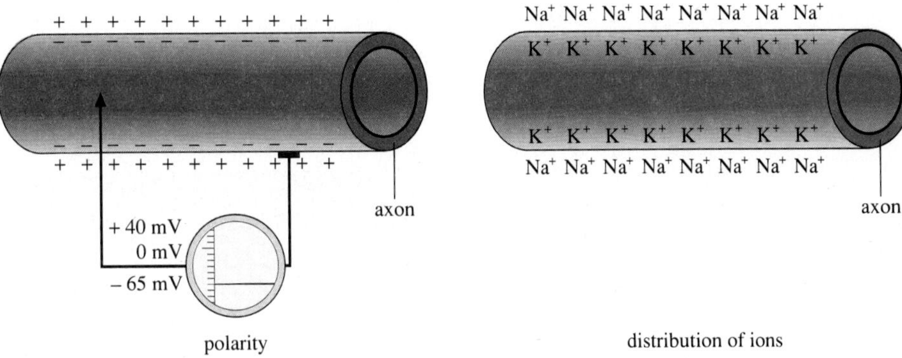

Figure 16.3
The resting potential. An oscilloscope measures the difference in electrical potential between two electrodes. When one electrode is placed inside an axon at rest and one is placed outside, the electrical potential inside the cell is -65 millivolts relative to the outside. In an axon at rest, Na^+ is more concentrated outside the cell and K^+ is more concentrated inside.

A Neuron's Message

The message that a neuron conducts is called a nerve impulse, which is actually an electrochemical change caused by ions moving across the cell membrane. This change is described by a measurement called an **action potential.** To understand how and why these ions move, it helps to be familiar with the state of the neuron when it is not conducting an impulse.

The Resting Potential

The inside of a resting neuron (one not conducting an impulse) has an electrical potential that is slightly negative with respect to the outside. This is called the **resting potential,** and it measures -65 millivolts.

A membrane in this condition is described as being **polarized**, because the inside carries a different electric charge from the outside. The charge differences across the membrane result from the unequal distribution of sodium ions (Na^+) and potassium ions (K^+). The concentration of K^+ is 30 times greater inside the cell than outside, and the concentration of Na^+ is 10 times greater outside than inside (fig. 16.3).

How is this unequal distribution of ions established and maintained? The properties of the cell membrane provide part of the answer. The cell membrane is **selectively permeable,** meaning that it admits some substances but not others. Ions move through the membrane through small pores called channels. Some channels are always open, but others are opened or closed by the position of gates, which are proteins that change shape to block or clear the channel (fig. 16.4). Whether a gate opens or closes a channel depends upon the membrane potential, so that such gates are termed voltage regulated. Some of these membrane channels are specific for Na^+ and others are specific for K^+.

Another property of the membrane important in understanding how ions come to be distributed as they are is the **sodium-potassium pump,** a mechanism that uses cellular energy (ATP) to transport Na^+ out of the cell and K^+ into the cell. The sodium-potassium pump uses active transport, moving Na^+ and K^+ against their concentration gradients.

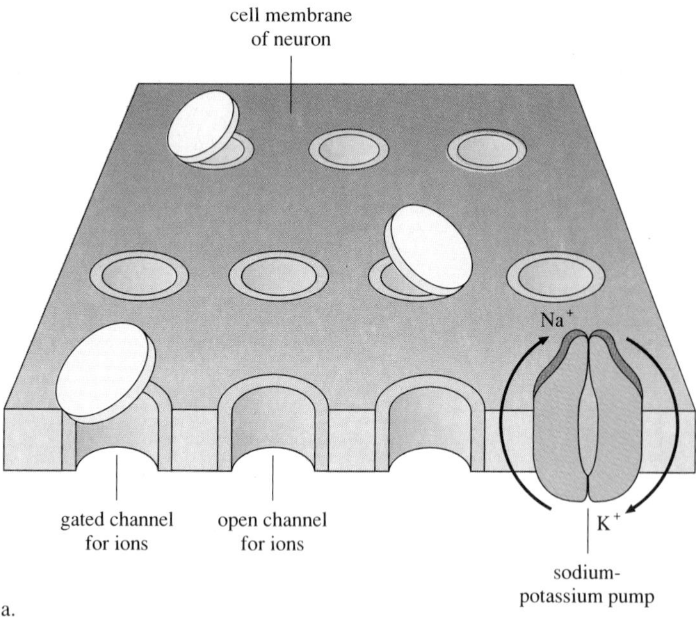

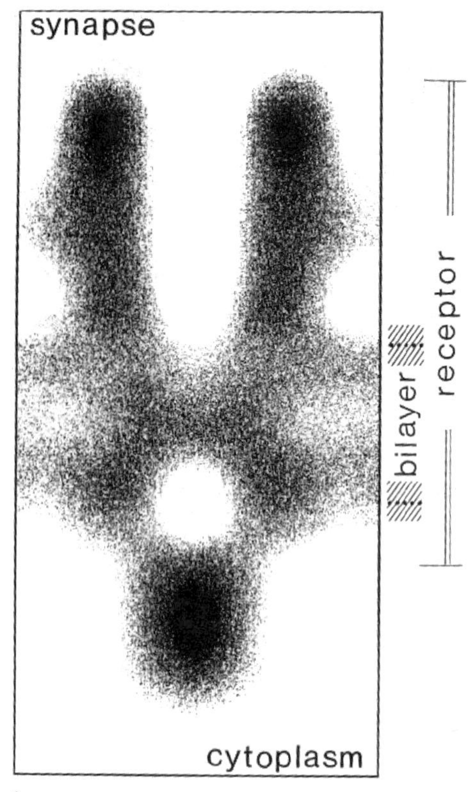

Figure 16.4
Ion pathways across the cell membrane. *a.* Recall that a cell membrane is a fluid mosaic of proteins embedded in a lipid bilayer. Some of the proteins form pathways for ions. *b.* In 1988, researchers at the Medical Research Council in England used electron microscopy to image this profile of the "gatekeeper" protein complex that forms an ion channel in a neuron's cell membrane. The structure is built of five proteins.

Limited by these membrane properties, ions distribute themselves in response to two forces. First, ions follow an **electrical gradient.** Like charges (negative and negative; positive and positive) tend to repel one another. Unlike charges (negative and positive) attract one another. Second, ion distribution is affected by a **concentration gradient.** Ions passively diffuse from an area in which they are highly concentrated toward an area of lower concentration. So a particular ion in the region of a channel specific for it will enter or exit the cell, depending upon which side is more concentrated.

How does all this explain the resting potential? Three basic mechanisms are at work. First, the sodium-potassium pump, using ATP for energy, concentrates K^+ inside the cell and Na^+ outside. Approximately 20 to 30 times more K^+ is inside than outside the cell, and 10 times more Na^+ is outside the cell than inside. The pump ejects three Na^+ while pumping in two K^+. Second, large negatively charged proteins and other negative ions are trapped inside the cell because the cell membrane is not permeable to them. Third, the membrane is 40 times more permeable to K^+ than to Na^+ in the resting state.

Because of the concentration gradient and the high permeability, K^+ is able to diffuse out of the cell. As K^+ moves through the membrane to the outside of the cell, it carries a positive charge with it, leaving behind the large negatively charged molecules. A charge or potential is therefore established across the membrane; positive on the outside and negative on the inside. The magnitude of the charge is determined by the balance of opposing forces acting on K^+. The concentration gradient drives K^+ outward, and the negative charge inside the cell tends to hold K^+ in. When these two opposing forces are equal there is no net movement of K^+ and the cell is in equilibrium. At this time, the cell is uniformly charged over its entire surface.

The importance of the sodium-potassium pump in maintaining the resting potential can be seen when the pump is stopped by a metabolic poison such as cyanide. K^+ slowly diffuses out and Na^+ in, destroying the concentration gradients. Nerve transmission is impossible, because there is no longer a charge across the membrane. Death occurs in minutes.

It is curious that the neuron uses more energy while resting than it does while conducting an impulse. Presumably, expending energy to maintain a resting potential allows the neuron to respond more quickly than it could if it had to generate a potential difference across the membrane each time it receives a stimulus. This is analogous to pulling back the string on a bow and arrow to be ready to shoot.

KEY CONCEPTS

In the resting state, the inside of the membrane is negative compared to the outside due to the unequal distribution of ions. The membrane holds some large negatively charged proteins within the cell and allows K^+ to flow outward because it is 30 times more concentrated inside than outside. Not only is Na^+ held out of the membrane, but it is also actively pumped out so that Na^+ is 10 times more concentrated outside than inside. The sodium-potassium pump maintains the concentration gradient so that the neuron is always charged and ready to respond to stimuli.

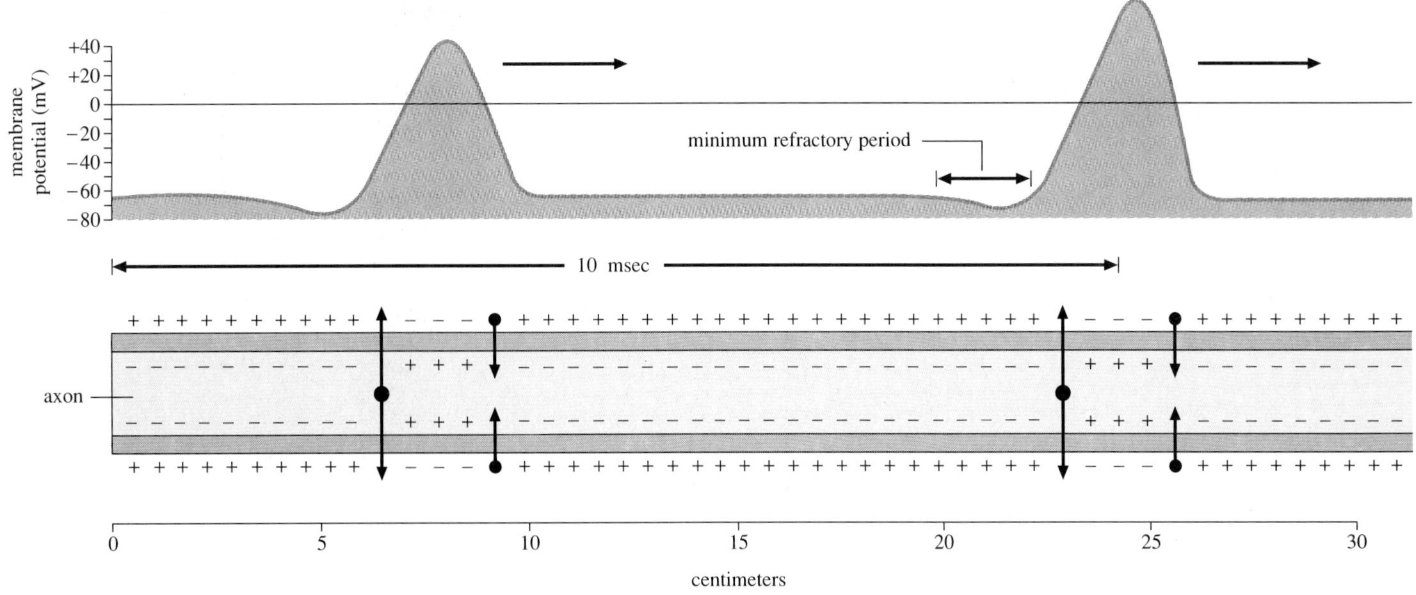

a.

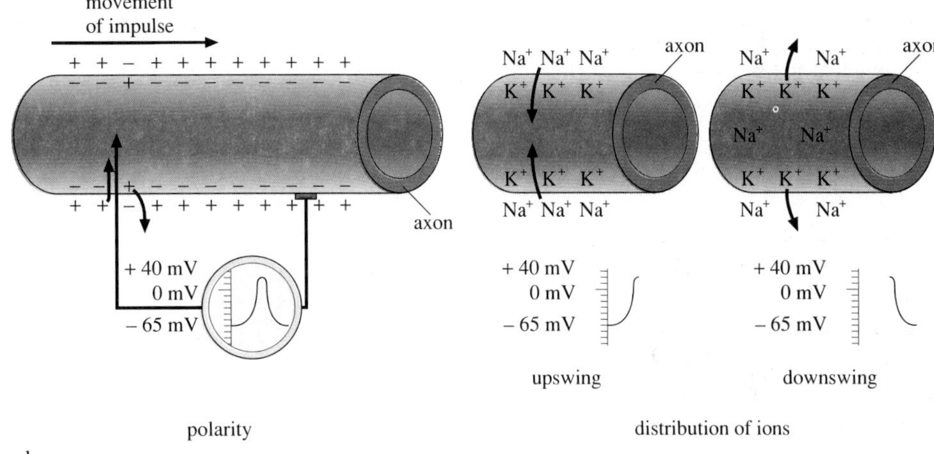

b.

Figure 16.5
a. An action potential occurs when a stimulus alters the neuron's cell membrane permeability so the Na^+ flows in, and the K^+ flows out, until the electrical potential inside the cell reaches +40 millivolts. *b.* An action potential is a wave of localized depolarization that travels down the axon.
From *A View of Life*, by Luria, Gould and Singer. Copyright ©1981, Benjamin/Cummings Publishing Company, Menlo Park, Calif.

The Action Potential

During an action potential (the evidence of a nerve impulse and therefore somewhat synonymous to it) the Na^+ and K^+ quickly redistribute themselves across a small patch of the cell membrane, creating an electrochemical change that moves like a wave along the nerve fiber. An action potential is usually initiated by a chemical stimulus from another neuron, but other stimuli include a change in pH, an electric shock, or pressure applied to the neuron.

An action potential begins when a stimulus (a change in pH, a touch, or a signal from another neuron) changes the permeability of the membrane so that some Na^+ begins to leak into the cell (fig. 16.5). As Na^+ enters the neuron, the interior becomes less negative. The membrane is said to become **depolarized** by the influx of Na^+. When enough Na^+ enters to depolarize the membrane to a certain point (the "threshold"), the sodium gates in that area of the membrane open, making the membrane even more permeable to Na^+. Driven by both the electrical gradient and the concentration gradient, Na^+ floods the inside of the cell so that the interior becomes positively charged. Still driven inward by the concentration gradient, Na^+ enters until a peak positive charge is reached.

At this peak of the action potential, membrane permeability changes again. Permeability to Na^+ halts by the closing of sodium gates, but permeability to K^+ suddenly increases by the opening of potassium gates. As a result, Na^+ is prevented from entering in large numbers. However, a mass exodus of K^+ now begins, driven by both electrical and concentration gradients. K^+ flows outward because it is more concentrated inside than outside and because the inside of the membrane is now positively charged due to the influx of Na^+.

The loss of positively charged K^+ now restores the negative charge to the interior of the cell. Therefore, the outward flow of K^+ repolarizes the cell membrane. In fact, the electrical potential fleetingly drops below the resting value, because the K^+ gates stay open slightly longer than the Na^+ gates. This slight increase in negative charge compared to the resting state is corrected by the sodium-potassium pump.

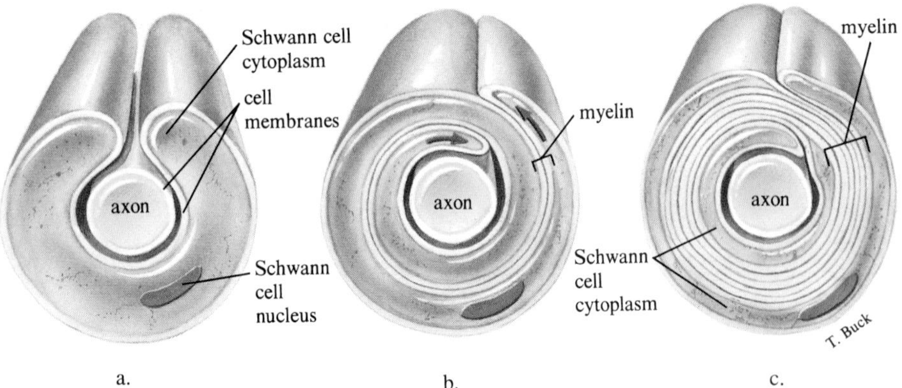

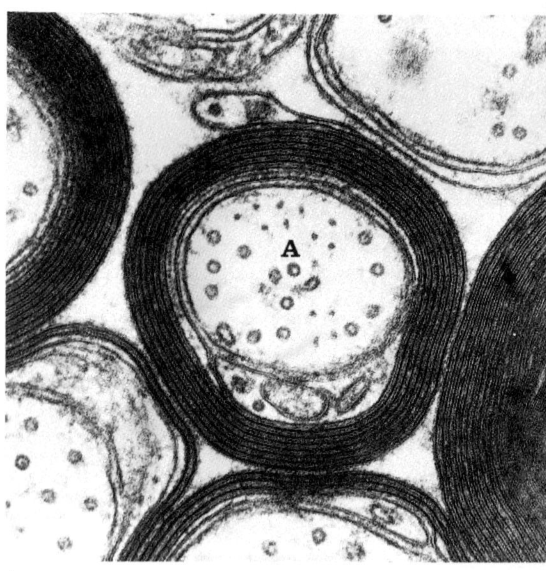

d.

e.

Figure 16.6
Many axons of vertebrate nerve cells are encased in a sheath of fatty myelin. This covering forms when a Schwann cell (*a*) winds around an axon in a spiral fashion (*b*), such that several layers of its lipid-rich cell membrane surround the axon. *c.* The Schwann cell's cytoplasm and nucleus often are squeezed into the periphery of the sheath. *d.* Some healthy neurons outside the central nervous system are coated with myelin. *e.* In multiple sclerosis, some of these axons lose their myelin coats. See Reading 16.1 for a closer look at this disorder.

While the Na^+ gates and then the K^+ gates are open, a second action potential cannot be initiated. Still, an action potential takes only 1 to 5 milliseconds. This rapid transmission is why the nervous system is an effective communication network.

The characteristic changes in membrane permeability that constitute the action potential travel along the neuron, usually from dendrite to cell body and down an axon. The action potential spreads because some of the Na^+ rushing into the cell at a particular point moves to the neighboring part of the neuron. The appearance of a positive charge in the nearby region triggers an influx of Na^+ there, producing an action potential there. The action potential moves along the neuron just as surely as a line of properly spaced dominoes will fall when the one on the end is bumped. Action potentials are thus said to be self-perpetuating. Furthermore, if enough Na^+ leaks in to open the sodium gates, a threshold is reached, and a full-blown action potential is generated. An action potential is therefore an all-or-none phenomenon, that is, there are no degrees of action potentials—they either happen or they do not.

If all action potentials are of equal magnitude, exhibiting the same changes in electrical potential, then how is the intensity of a stimulus encoded? That is, how do we feel different degrees of stimulation? Sensitivity is transmitted by the frequency of action potentials. Whereas a light touch to nerve endings in the skin might produce 10 impulses in a given time period, a hard hit might generate 100 impulses, intensifying the sensation. Our neurons also discern the type of stimulation. We can distinguish light from sound because the neurons stimulated by light end in a different place in the brain than those stimulated by sound.

KEY CONCEPTS

An action potential begins when a stimulus (chemical, electrical, or change in pressure) changes the permeability of the neuron membrane so that Na^+ begins to leak through. When enough Na^+ enters the cell the sodium gates open and these ions rush inside, making the interior of the cell positive. The positive charge inside the membrane triggers the closing of the sodium gates and increases permeability to K^+, which then rushes out of the cell, eventually restoring the original inside negative resting potential. The shift in ions in one area triggers the same changes in the next area. As a result, a wave of depolarization and repolarization (the action potential) travels along the axon. All action potentials are of the same magnitude. Differences in the intensity of stimuli are communicated by the frequency of action potentials.

The Myelin Sheath and Saltatory Conduction

Not all nerve fibers conduct impulses at the same speed. The speed of conduction is determined by certain characteristics of the fiber. The diameter of the fiber is important; the greater the diameter, the faster it conducts an action potential. However, vertebrate nerve fibers can remain thin and conduct impulses very rapidly when coated by a fatty material called a **myelin sheath.**

Outside the brain and spinal cord, myelin sheaths are formed by **Schwann cells,** which contain enormous amounts of lipid. A Schwann cell wraps around an

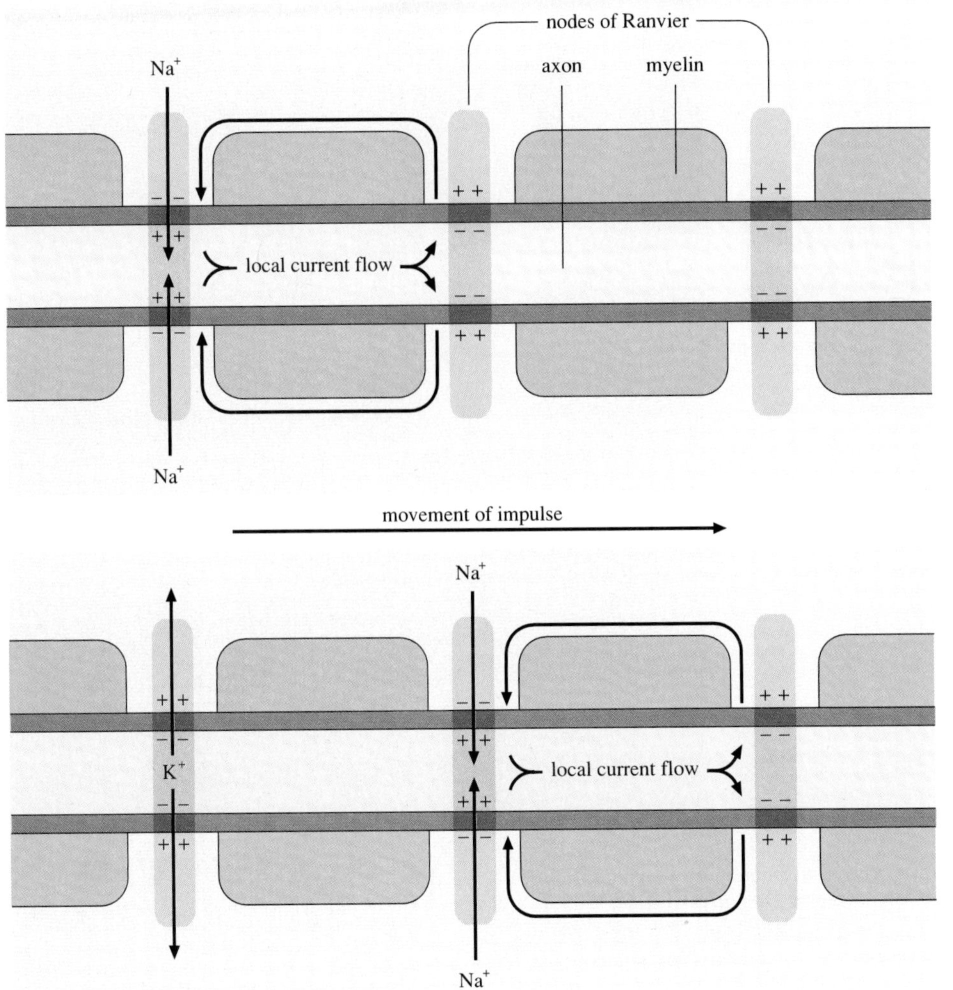

Figure 16.7
Saltatory conduction. In myelinated axons, action potentials "jump" from one node of Ranvier to the next one.

axon many times, forming a whitish coating (fig. 16.6). Many Schwann cells form the sheath, each wrapping a small segment of the axon. Between each Schwann cell is a short region of exposed axon called a **node of Ranvier.** Some neurons in the brain and spinal cord are wrapped in myelin produced by cells called oligodendrocytes.

When an impulse is conducted along a myelinated axon, the action potential "jumps" from one node to the next, a type of transmission called **saltatory conduction** (fig. 16.7). The impulse leaps from node to node because the myelin insulation prevents ion flow, but a small electric current spreads instantly between nodes. In addition, the voltage-regulated ion channels are unevenly distributed. Na^+ channels, for example, are found at a density of 10,000 per square micrometer at the nodes but are almost nonexistent in the areas in between that are wrapped in myelin.

Just as tossing a ball from one person to another in a line would cause it to travel faster than one could run with it, the jumping of the action potential from node to node increases the speed of transmission. In fact, myelinated axons may conduct impulses 100 times faster than unmyelinated axons, at speeds of up to 120 meters per second (an astounding 270 miles an hour)!

Myelinated fibers, which impart their coloration to the **white matter** of the nervous system, are found in pathways that transmit impulses over long distances. The **gray matter** of the nervous system consists of cell bodies and interneurons that lack myelin. These areas usually specialize in integration.

KEY CONCEPTS

Neurons outside the brain and spinal cord are wrapped in myelin sheaths, formed by fatty Schwann cells. Action potentials jump between nodes of Ranvier, which are interruptions in the myelin sheath. The saltatory conduction from node to node permits fast nerve impulse transmission.

Synaptic Transmission

In order to form a communication network, a neuron's message, the action potential, must be conveyed to another neuron (or a muscle or gland). Neurons do not actually touch each other, so the action potential cannot be passed directly from cell to cell. Instead, the action potential is converted into a chemical, which travels from a "sending" cell to a "receiving" cell across a tiny space between the cells. Once across this space, the **neurotransmitter** chemical alters the permeability of the receiving cell's membrane either to provoke an action potential or prevent one.

The space between neurons is called a **synapse.** The end of an axon has tiny branches that enlarge at the tips to form **synaptic knobs.** Within a synaptic knob are many small sacs, called **synaptic vesicles,** that contain neurotransmitter molecules. An action potential passes down the axon of the cell sending the message, which is called the **presynaptic cell.** When the action potential reaches the membrane near the space or **synaptic cleft,** the permeability of the membrane changes and calcium ions enter the cell, driven by their concentration gradient. The calcium ions cause the vesicles containing neurotransmitter molecules to move toward the synaptic membrane, fuse with it, and dump their contents into the synaptic cleft by exocytosis (fig. 16.8).

Neurotransmitter molecules diffuse across the cleft and attach to protein **receptors** on the membrane of the receiving neuron (the **postsynaptic cell**) (fig. 16.9). A particular neurotransmitter fits only into a specific receptor type, like a lock and key. When the neurotransmitter attaches to the receptor, the conformation (three-dimensional shape) of the receptor protein

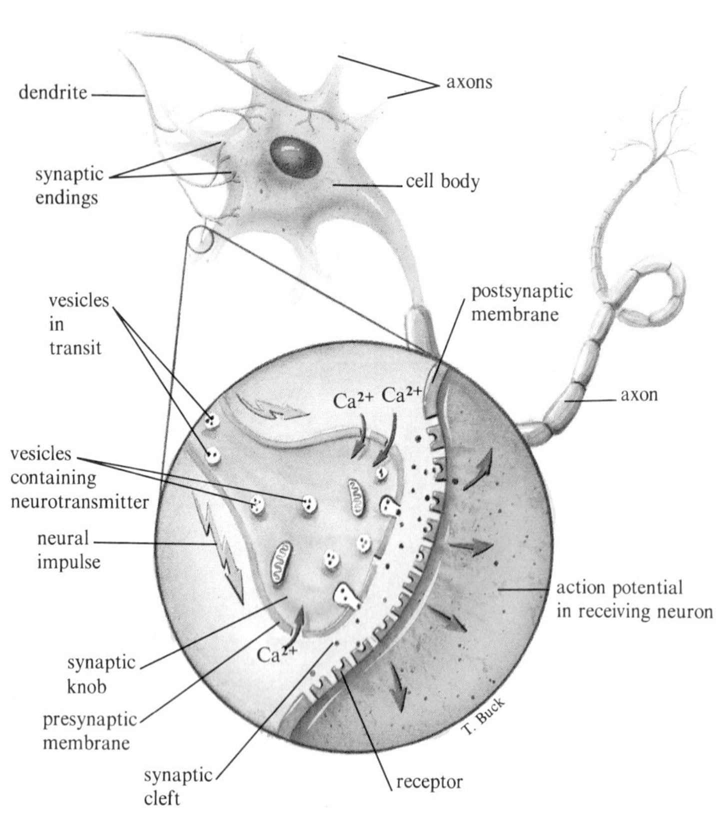

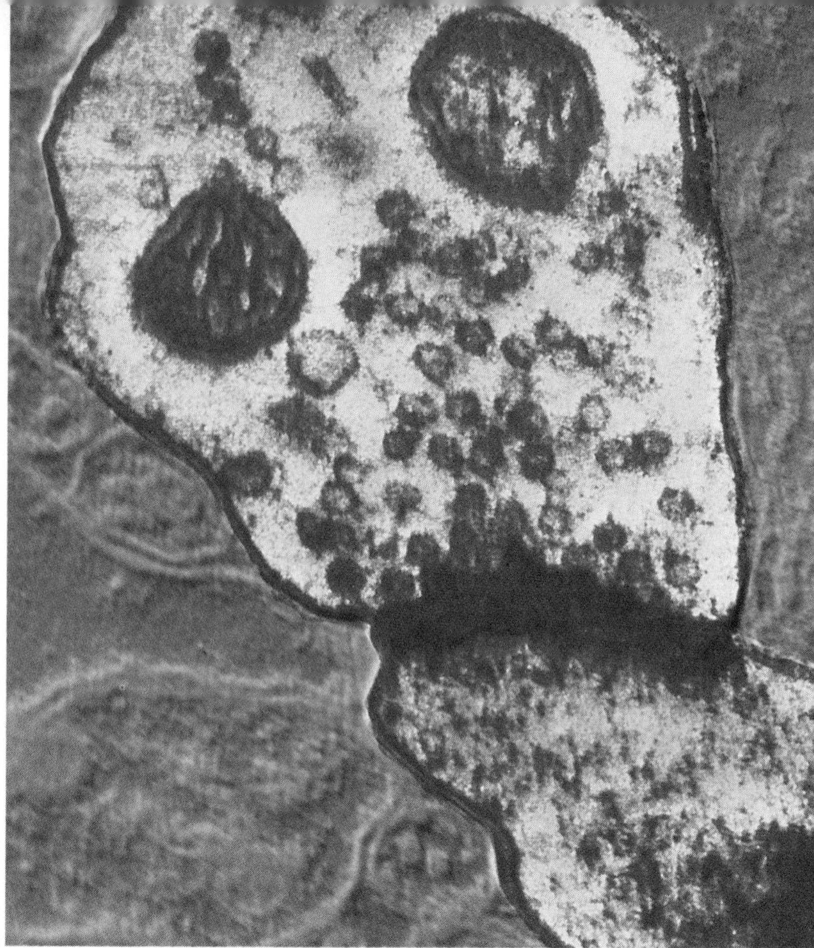

Figure 16.8

The synapse—where neurons meet. A neuron may form synapses with hundreds or even thousands of other cells. A neurotransmitter released from the presynaptic cell in response to an action potential traverses the synaptic cleft and binds to receptors on the postsynaptic cell's membrane. If sufficient excitatory neurotransmitters bind to the postsynaptic cell, it will "fire" an action potential. Magnification, x17,600.

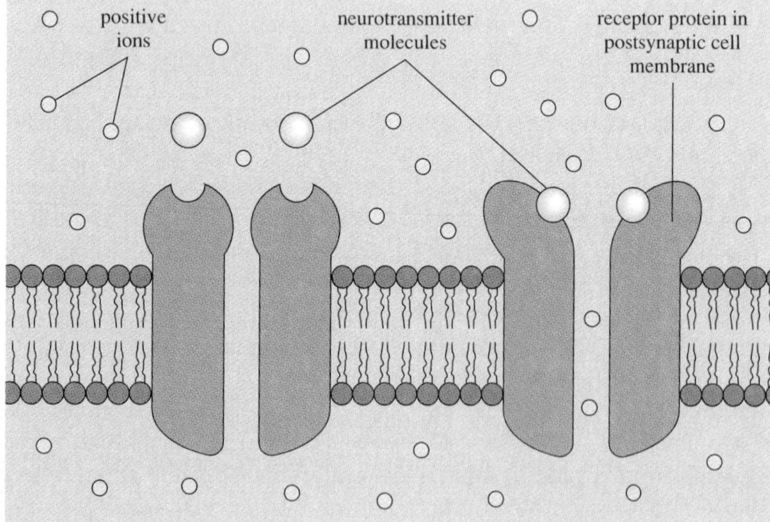

Figure 16.9

Receptors on the postsynaptic membrane. The neurotransmitter acetylcholine is synthesized in the presynaptic cell and is packaged into vesicles there. When an action potential arrives, acetylcholine is released, crosses the synaptic cleft, and binds to receptor proteins on the highly folded postsynaptic membrane—the membrane of a muscle cell, for example. The binding opens channels in the membrane, permitting entry of sodium and other positively charged ions. Sufficient ions enter the postsynaptic cell to trigger an action potential there. The neuron's "message" has been sent.

changes so that channels open in the postsynaptic membrane, allowing specific ions to flow through and changing the probability that an action potential will be generated.

If synaptic transmission is blocked, the consequences can be severe, possibly even fatal. In the deadly food poisoning known as botulism, a bacterial toxin physically blocks vesicles in the presynaptic cell from releasing the neurotransmitter acetylcholine into the synapses to muscle cells. Lacking stimulation, the muscle cells cannot contract. If an antidote is not given before the muscles responsible for breathing become paralyzed, the person will suffocate. Another such neurotoxin is tetrodotoxin, which is made by bacteria in puffer fish—a delicacy in Japan and used to produce "zombies" in Haiti (fig. 16.10). Tetrodotoxin causes paralysis by blocking Na^+ channels. Sometimes blocking synaptic

Reading 16.1 Multiple Sclerosis

IN 1964, AT AGE 20, SKIER JIMMIE HUEGA WON THE OLYMPIC BRONZE MEDAL IN THE SLALOM. In 1967, his vision became blurry, and a few months later his legs became slightly numb. These symptoms were intermittent, and Jimmie ignored them. After a while, they disappeared. Three years later, however, they reappeared, and this time Jimmie made an appointment with a doctor to have his eyes examined. The doctor then referred him to a nerve specialist, who diagnosed his problem as multiple sclerosis (MS). Jimmie learned that multiple sclerosis usually strikes between the ages of 20 and 40, with precisely the symptoms that he had ignored for so long: attacks of fatigue and visual problems that come and go. The patient can actually function quite normally during the symptom-free periods. But for many of the 2 million sufferers worldwide, the progressive deterioration eventually leaves them permanently paralyzed. So far, Jimmie has been lucky. He is still able to ski and cycle, and he does situps and pushups regularly, but he never knows when the on-again off-again disease will strike.

MS is felt in the muscles as weakness, tremors, or loss of coordination or in the sensory system as blurred vision, tingling, or numbness, but the defect is actually in the central nervous system. The symptoms are caused by the destruction of myelin, the insulating material around nerves. In various sites throughout the brain and spinal cord, the myelin coating forms hard scars called scleroses. The neurons that they surround can no longer transmit messages. Muscles that no longer receive input from motor neurons stop contracting. Without stimulation, muscles atrophy.

The symptoms depend upon which neurons are affected. Short circuiting in one part of the brain may affect fine coordination in one hand; if another brain part is affected, vision may be altered. What might be responsible for the destruction of myelin in MS? A virus may cause the body's immune system to attack the cells producing myelin. A virus is suspected for a few reasons—viral infections are known to strip neurons of their myelin sheaths; viral infections can cause repeated bouts of symptoms; and most compelling, MS is far more common in some geographical regions (the temperate zones of Europe, South America, and North America), suggesting a pattern of infection. Can you think of an alternate explanation for why MS is more prevalent in some areas than others?

A possible culprit is the virus that causes measles. The series of events that might explain the misdirected attack by the immune system is as follows: When today's MS patients were children, the vaccine for measles had not yet been developed, and nearly all children had the measles. For most of these children, the virus produced a rash and a fever for a week or two and then was gone, squelched by the immune system.

For some children though, the virus invaded the usually protected myelin-producing cells of the central nervous system. The virus used these cells' organelles to manufacture more of its own genetic material, as all viruses do. But the virus borrowed something more. To package itself into a protein coat, it took some proteins from the outer membranes of the fatty cells.

A decade or so after the initial measles attack, something provoked the virus to spread throughout the nervous system. The immune system tried to attack the new viruses. Because the viruses bore proteins on their coats that they had "borrowed" from cells, the immune system also attacked the insulating cells wrapped around neurons, recognizing them as viral invaders. This misplaced attack is multiple sclerosis. MS is an example of an autoimmune disease, in which the body's immune system attacks the body's own cells.

The hypothesis that a past viral infection lurks behind MS has been popular for almost a century. Still, the responsible virus has not been identified, the precise immune system failure is not known, and the uneven geographical distribution of MS sufferers remains a mystery. But researchers are confident that soon the key to the devastating attack on the nervous system will be understood—just as such viral villains as polio and smallpox were conquered.

Figure 16.10
Neurotoxins block synaptic transmission. Japanese Shinto priests preside at a festival in honor of fugu or puffer fish. Fugu, prized as a culinary delicacy in Japan, contains the poison tetrodotoxin. Specially trained chefs remove the ovaries, liver, and intestines, which contain most of the poison, but traces left behind cause a tingling in the mouth when the fish is eaten. Each year a few diners die from improperly prepared puffer fish. In Haiti, a powder containing dried puffer fish toxin was rubbed into the skin of some unfortunate people, inducing a deathlike trance. These "zombies" were taken away as dead and enslaved when they awoke.

transmission can be beneficial. Some anesthetics may work by binding to receptors, preventing nerve transmission and thereby temporarily blocking pain.

Disposal of Neurotransmitters

A neurotransmitter does not linger in the synapse. If it did, its effect on the receiving cell would be continuous, perhaps causing it to fire unceasingly. Life would become a meaningless barrage of stimuli, with uncontrolled movements and seizures. Such chemical chaos is avoided because soon after a neurotransmitter is released, it is either destroyed by an enzyme or taken back into the axon that released it. For example, *acetylcholine* is broken down by the enzyme *acetylcholinesterase* into its component parts, acetate and choline, which are then absorbed by the axon and used to resynthesize acetylcholine. A molecule of acetylcholine must work with lightning speed, for it probably has no more than 1/500 of a second to act before its destruction. Other neurotransmitters, such as adrenaline, nonadrenaline, and serotonin, are reabsorbed by the presynaptic cell (a process called active reuptake).

If a neurotransmitter is not quickly destroyed, there may be dire consequences. For example, nerve gas and certain insecticides block the breakdown of acetylcholine by inhibiting acetylcholinesterase. As a result, acetylcholine is not destroyed and stays active in the synapse. These nerve endings stimulate skeletal muscle to contract continuously, and the victim convulses and dies. This can be seen in the twitching legs of a cockroach sprayed with insecticide.

Excitatory and Inhibitory Neurotransmitters

Synapses in the nervous system are of two types. **Excitatory synapses** depolarize the postsynaptic membrane, and **inhibitory synapses** increase the polarization. A neurotransmitter acting at an excitatory synapse increases the probability that an action potential will be generated in the second neuron by slightly depolarizing it. For example, when acetylcholine binds to the receptors at an excitatory synapse, channels open that admit Na^+ into the postsynaptic cell. In just a millisecond, these channels let in half a million sodium ions. If enough Na^+ enters to reach a threshold level of depolarization, an action potential is triggered in the postsynaptic cell.

On the other hand, a neurotransmitter may make it more difficult for an impulse to be generated in the postsynaptic cell. Such a neurotransmitter is said to be inhibitory. In this case, the binding of the transmitter to its receptor causes different channels to open. These channels allow negatively charged chloride ions (Cl^-) to enter the cell, driven by their concentration gradient. The presence of negative chloride ions makes the cell's interior more negative than the usual resting potential; that is, the inhibitory neurotransmitter **hyperpolarizes** the postsynaptic membrane. Now, extra Na^+ has to enter before the membrane becomes depolarized enough for an action potential to be generated.

Synaptic Integration—How a Neuron Interprets Its Messages

What is the point of having both excitatory and inhibitory synapses? Presumably it gives finer control over a neuron's activities, just as having both an accelerator and a brake allows for finer control over the motion of a car.

A single neuron in the nervous system may receive input from thousands of other neurons, some excitatory and others inhibitory. Nearly half of its receiving surface forms synapses. Whether that neuron transmits an action potential depends on the sum of the excitatory and inhibitory impulses that it receives. If excitatory impulses are in the majority, the postsynaptic cell is stimulated; if inhibitory messages predominate, it is not. It is almost as if the transmitters exciting the postsynaptic cell by depolarizing it are having a chemical tug-of-war with the inhibitory transmitters that are hyperpolarizing the same cell. Only when the postsynaptic membrane is depolarized to threshold level is an action potential generated. A neuron's evaluation of impinging nerve messages, which determines whether an action potential is "fired" or not, is termed **neural** or **synaptic integration.**

In the human body this interplay among transmitters allows us to filter stimuli. Consider sleep. We sleep when the activity of certain cells in a part of the brain called the reticular activating system is inhibited by the release of the neurotransmitter serotonin by other neurons. While we slumber, unimportant stimuli, such as a roommate's snoring, may be filtered out. In other words, the excitation generated by irrelevant stimuli is not sufficient to overcome the inhibition by serotonin. However, a relevant stimulus, such as someone whispering your name or warning that the building is on fire, will generate excitatory impulses strong enough to overwhelm the inhibitory input, and you wake up.

In a more general sense, synapses markedly increase the informational content of the nervous system. The brain has 10 billion neurons, which can be looked at as bits of information. But if a synapse is considered to be a unit of information, then the informational capacity of the brain increases 1,000-fold, because a typical brain neuron has synaptic connections to 1,000 other neurons.

KEY CONCEPTS

The space between two adjacent neurons is a synapse. Chemical neurotransmitters are released from the presynaptic cell in response to an action potential, and they cross the synapse. By binding to receptors on the postsynaptic cell, transmitters alter membrane permeability so that the cell is either depolarized (an excitatory event) or hyperpolarized (an inhibitory event). Neurotransmitters are either enzymatically dismantled in the synaptic cleft or they reenter the presynaptic cell. A neuron typically receives input from many other neurons, and whether or not it transmits an action potential depends upon the sum of the excitatory and inhibitory impulses it receives. Synapses greatly increase the informational capacity of the nervous system.

Neurotransmitters

The peripheral nervous system (the part outside the brain and spinal cord) uses three neurotransmitters: acetylcholine, noradrenaline, and adrenaline. The central nervous system (the brain and spinal cord) has many additional transmitters. Among the 40 or more neurotransmitters found in the central nervous system are dopamine, serotonin, the inhibitory transmitter

GABA (gamma amino butyric acid), and the fascinating internal opiates, endorphins. It was once believed that a single neuron could produce only one neurotransmitter. However, we now know that some neurons produce more than one transmitter, releasing the same combination of them at each of its synapses. Different neurotransmitters seem to be associated with particular behaviors or responses.

Neurotransmitter levels are modulated by chemicals called **second messengers.** One second messenger is cyclic adenosine monophosphate (cAMP). When a neuron is stimulated repeatedly, it produces cAMP, which activates an enzyme called phosphokinase. This enzyme attaches phosphate groups (PO_4) to specific proteins, thereby altering their structure and function. In neurons, the phosphokinase alters the proteins that form K^+ channels so that the ions cannot rush out of the cells as fast as they otherwise would. This makes the action potential last longer.

Table 16.1
Mechanisms of Drug Action

Drug	*Neurotransmitter Affected*	*Mechanism of Action*	*Effect*
Tryptophan	Serotonin	Stimulates neurotransmitter synthesis	Sleepiness
Reserpine	Noradrenaline	Packaging neurotransmitter into vesicles	Limb tremors
Curare	Acetylcholine	Decreases neurotransmitter in synaptic cleft	Muscle paralysis
Valium	GABA	Enhances receptor binding	Decreases anxiety
Nicotine	Acetylcholine	Stimulates synthesis of enzyme that degrades neurotransmitter	Increases alertness
Cocaine	Noradrenaline	Blocks reuptake	Euphoria
Tricyclic antidepressants	Noradrenaline	Blocks reuptake	Mood elevation
Monoamine oxidase inhibitors	Noradrenaline	Blocks enzymatic degradation of transmitter in presynaptic cell	Mood elevation

Psychoactive Drugs and Neurotransmitters

Understanding how neurotransmitters fit receptors can explain the actions of certain drugs (table 16.1). When a drug alters the activity of a neurotransmitter, synaptic transmission is either halted or enhanced. A drug that binds to a receptor, blocking a neurotransmitter from binding there, is called an **antagonist.** A drug that activates the receptor, triggering an action potential, or helps a neurotransmitter to bind is called an **agonist.** The effect of a drug depends upon whether it is an antagonist or an agonist and on the particular behaviors regulated by the affected neurotransmitter.

Pathways using noradrenaline as a transmitter control arousal, dreaming, and mood. Amphetamine and cocaine enhance noradrenaline activity, thereby heightening alertness and mood. Amphetamine's structure is so similar to that of noradrenaline that it binds to the noradrenaline receptors and triggers the same changes in the postsynaptic membrane. Cocaine, on the other hand, blocks the receptors on the membrane of the presynaptic cell that normally function as a reuptake gateway for noradrenaline. Because noradrenaline cannot reenter the presynaptic cell, it accumulates in the synaptic cleft. The resulting overabundance of noradrenaline in the synapses produces the feelings of a cocaine high.

GABA is an inhibitory neurotransmitter used in a third of the brain's synapses. The drug valium causes relaxation and inhibits seizures and anxiety by helping GABA bind to receptors on postsynaptic neurons. It is therefore a GABA agonist.

Disease and Neurotransmitters

Disturbances of neurotransmitter balance are thought to be behind a variety of medical problems, including epilepsy, insomnia, and sudden infant death syndrome. However, it is often difficult to tell if the neurotransmitter abnormality is a result of a disorder or a cause of it. Using drugs to alter neurotransmitter levels is one approach to treating such conditions. Table 16.2 describes several illnesses that are associated with unusual neurotransmitter levels or distributions. Reading 16.2 takes a closer look at the relationship between Parkinson's disease and neurotransmitters.

Understanding the functioning of neurotransmitters has been particularly beneficial in treating certain mental disorders. It is sometimes comforting for patients and their families to realize that there is a biochemical explanation for symptoms. Depression is a good example of an illness that can result from a neurotransmitter imbalance.

The Biochemistry of Depression

Marsha is profoundly and inexplicably sad. At age 46, she has an exciting career, a solid marriage, and her two children are healthy and happy, yet she cannot eat or sleep, has lost interest in sex, and begins and ends each day in tears. She does not know where to turn for help.

Marsha suffers from **endogenous depression**, along with 10 million others in the United States. Unlike situational depression, in which a particular event such as a death or divorce triggers the low mood, endogenous depression can often be traced to a biochemical abnormality in the brain. If Marsha does not get help soon, she may be 1 of the 1.5 million depressives who attempt suicide in this country each year.

Thirty years ago, Marsha would probably have ended up on a locked ward in a psychiatric hospital, where therapists would have probed her confused thoughts, seeking past events that might have triggered her depressed mood. Today a depressed patient like Marsha is likely to have her body fluids tapped as well as her thoughts. Telltale biochemicals in urine, blood, and cerebrospinal fluid (the fluid surrounding the brain and spinal cord) can guide physicians to rapid and accurate diagnoses of biochemical abnormalities that express themselves as mental illnesses, permitting quick identification of the most effective drug. The concentration of a breakdown product of serotonin in cerebrospinal fluid can even indicate whether or not a depressed patient is likely to attempt suicide.

Drugs are now standard treatment of mental illness because many disorders of the "mind" are known to be associated with neurotransmitter imbalances. Marsha's depression was linked to a deficiency of noradrenaline. The evidence was that her cerebrospinal fluid contained abnormally low amounts of the breakdown products of noradrenalin. Marsha began taking a drug known as a tricyclic antidepressant, which works by blocking the reuptake of noradrenaline by presynaptic cells in the brain. Because of the drug's action, noradrenaline accumulated in Marsha's synaptic clefts, and, within 8 weeks, she began to feel better. Some depressed people have abnormal levels of serotonin or dopamine. A new antidepressive drug, Prozac, elevates mood almost immediately, but may cause other mental disturbances as side effects.

Table 16.2
Disorders Associated With Neurotransmitter Imbalances

Condition	Symptoms	Imbalance of Neurotransmitter in Brain
Alzheimer's disease	Memory loss, depression, disorientation, dementia, hallucinations, death	Deficient acetylcholine
Endogenous depression	Debilitating, inexplicable sadness	Deficient noradrenaline
Epilepsy	Seizures, loss of consciousness	Excess GABA leads to excess noradrenaline and dopamine
Huntington disease	Personality changes, loss of coordination, uncontrollable dancelike movements, death	Deficient GABA
Hypersomnia	Excessive sleeping	Excess serotonin
Insomnia	Inability to sleep	Deficient serotonin
Mania	Elation, irritability, overtalkativeness, increased movements	Excess noradrenaline
Myasthenia gravis	Progressive muscular weakness	Deficient acetylcholine at nerve-muscle junctions
Parkinson's disease	Tremors of hands, slowed movements, muscle rigidity	Deficient dopamine
Schizophrenia	Inappropriate emotional responses, hallucinations	Deficient GABA leads to excess dopamine
Sudden infant death syndrome ("crib death")	Baby stops breathing, dies if unassisted	Excess dopamine
Tardive dyskinesia	Uncontrollable movements of facial muscles	Deficient dopamine

Opiates in the Human Body

Opiate drugs, such as morphine, heroin, codeine, and opium, are potent painkillers derived from the poppy plant. In addition to altering one's perception of pain, thus making it easier to tolerate, these drugs elevate mood. The effects of opiates are enjoyable, and animals can become physically addicted to these drugs.

In the 1970s the human body was found to produce its own opiates. These biochemicals, called **endorphins** (for "endogenous morphine") are peptides considered to be a class of neurotransmitter. Like the poppy-derived opiates that they structurally resemble, endorphins influence mood and perception of pain.

The discovery of endorphins began in 1971 in research laboratories at Stanford University and the Johns Hopkins School of Medicine, where researchers exposed pieces of brain tissue from experimental mammals to morphine. The morphine was radioactively labeled (some of the atoms were radioactive isotopes) so that its destination in the brain could be followed.

The researchers found that the morphine indeed bound to receptors on the membranes of certain nerve cells, particularly in the neurons that transmit pain. Why, the investigators wondered, would an animal's brain contain receptors for a chemical made by a poppy? Could it be that the mammalian body manufactures its own opiates? The opiate receptor, then, would normally bind the body's own opiates (the endorphins) but would also be able to bind the chemically similar compounds made by the poppy.

By 1975, researchers had used chemical extraction methods on different parts of the human brain (from cadavers) and identified the first endorphin, a peptide consisting of 5 amino acids named *enkephalin* (Latin for "in the head"). A few months later, *beta-endorphin*, 30 amino acids long, was discovered. Several other endorphins have been identified since then. One particularly powerful endorphin is *dynorphin*. It was discovered in 1979 and is 200 times as potent as morphine.

Endorphins have several effects on the human body. Elevated endorphin levels in the brain are associated with the feeling of well-being reported by runners after a vigorous workout (popularly known as "runner's high"). High endorphin levels are found both in the blood of a woman in labor and in the placenta, indicating that these chemicals provide pain relief to mother and child alike during childbirth.

Reading 16.2 *Parkinson's Disease*

PARKINSON'S DISEASE IS A NERVOUS SYSTEM DISORDER CAUSED BY A DEFICIT IN A PARTICULAR NEUROTRANSMITTER. The condition usually appears after age 60 as slowed movements, muscle rigidity, and tremors that are described as "shaking palsy." An odd symptom of Parkinson's disease is in the sufferer's handwriting—at the end of a sentence the writing often becomes smaller and descends below the level of the rest of the writing. This is caused by the loss of fine coordination of the hand muscles.

Autopsies on Parkinson's disease patients performed during the 1930s revealed that a small dark area at the base of the brain, called the *substantia nigra* (dark substance), was far smaller than normal. The dopamine-producing cells in this brain region are crucial in controlling complex movements; the symptoms of Parkinson's disease are apparently caused by a deficiency of dopamine in this area of the brain. Symptoms appear when dopamine levels in the substantia nigra fall 80% below normal.

Since 1970, many patients have been helped with a drug called L-dopa, which enables the cells of the substantia nigra to produce greater amounts of dopamine. Dopamine is not administered directly because it cannot pass into the fluid around the brain due to the specialized capillaries there that form a "blood-brain barrier." L-dopa can pass through this barrier and once it enters the neurons of the substantia nigra, it is chemically converted to dopamine. L-dopa does not work for all Parkinson's patients. Furthermore, the large doses that are needed in the later stages of the disease produce symptoms of schizophrenia, a mental disorder characterized by inappropriate emotional responses. Interestingly, schizophrenia is caused by too much dopamine, and it is possible that while L-dopa elevates dopamine in certain areas of the brain to counteract Parkinson's, it also causes an excess of the neurotransmitter in areas that lead to symptoms of schizophrenia.

Recently, some Parkinson's patients have been successfully treated by surgically grafting dopamine-producing tissues from their own adrenal glands (located atop the kidneys) into the brain. A controversial experimental therapy is to graft fetal brain cells from the substantia nigra into the brains of Parkinson's patients. This raises the disturbing possibility of women intentionally becoming pregnant, then having an abortion to provide a treatment for a relative with Parkinson's disease.

Researchers are seeking the precise cause of the dopamine shortage in the Parkinson's brain. An answer may come from a most unexpected source—young heroin addicts who inject synthetic drugs manufactured by amateur chemists. Shortly after injecting "designer drugs" (chemicals slightly different from heroin) severe Parkinson's symptoms begin. This phenomenon captured medical attention because the disease is usually gradual in onset.

This apparent new form of Parkinson's disease first appeared in San Jose, California, in 1982. Dr. J. William Langston, director of Parkinson's disease research at the Institute for Medical Research in San Jose, was alerted to the cases of two young heroin abusers whose muscles had suddenly become extremely rigid. Shortly afterward, four similar patients were found. Another neurologist noted that the puzzling new cases were similar in their sudden, early appearance to a few cases seen among chemists who had worked with a chemical compound called MPTP. Could MPTP be the chemical used as a heroin substitute? Indeed it was.

The fact that a chemical can cause Parkinson's disease symptoms in young people may provide a clue to the cause of classic Parkinson's disease. The key seems to be in the rate of aging of the part of the brain from which the symptoms arise. Autopsies show that in normal aging, the substantia nigra loses 5% to 7% of its neurons every 10 years. Most people simply do not live long enough to lose the amount of dopamine here that would bring on Parkinson's symptoms. In Parkinson's disease, the rate of dopamine decrease is accelerated. By age 50 or 60, more than 80% of the dopamine is depleted, and symptoms appear.

For some people, Langston suggests, this accelerated aging of the substantia nigra may be triggered by exposure to chemicals such as MPTP. But heroin addicts and chemists may not be the only people at risk for chemically caused Parkinson's disease.

In the brain, MPTP is enzymatically converted into an even more toxic compound—one that is chemically similar to certain potent herbicides, such as paraquat. Epidemiological studies are currently underway to see if Parkinson's disease correlates in prevalence to herbicide use in particular geographical regions. It is possible that the familiar Parkinson's disease of middle age may indeed be the result of a lifetime of exposure to herbicides.

Endorphins may also be responsible for the pain relief provided by the medical art of *acupuncture,* in which needles inserted at certain points in the body relieve pain in other areas. Studies on cats and mice have shown that cells in the spinal cord that normally fire rapid action potentials in the presence of a painful stimulus have a slower rate of firing when acupuncture is used. Pain relief lasts for a while after the needle is removed. Researchers suggest that rather than acting directly on the cells of the spinal cord, acupuncture causes the pituitary gland (attached to the base of the brain) to release endorphins, which then bind to receptors in the spinal cord. This idea is based on the observation that acupuncture does not relieve pain in rats whose pituitaries have been removed. In humans, acupuncture is used to treat back pain, migraine headaches, arthritis, joint pain, and menstrual cramps.

The existence of endorphins explains why some people who are addicted to opiate drugs such as heroin experience withdrawal pain when they stop taking the drug. Initially, the body interprets the frequent binding of heroin to its endorphin receptors as an excess of endorphins. To bring the level down, the body slows its own production of endorphins. Then, when the addict stops taking the heroin, the body is caught short of opiates (heroin and endorphins). The result is pain.

How can a single molecule produce such varied effects as pain relief, euphoria, and addiction? Endorphins—as well as other neurotransmitters—bind to several types of receptors. Pharmaceutical researchers try to develop drugs that bind to only those receptors that alleviate pain. The active ingredient in marijuana, for example, is known to stop the nausea caused by some cancer treatments. So far, however, researchers have been unable to separate this property from the drug's effect on mood.

Neurons form intricate connections with each other, sending action potentials and neurotransmitters in pathways to enable you to do something as seemingly simple as lifting a fork, remembering a tune, driving a car, or catching a high fly ball to the outfield. Although we can understand the biological activity of a single neuron, it is how these cells interact that provides us with the ability to carry out an amazing variety of tasks. The next two chapters explore what we know about this fascinating organ system.

KEY CONCEPTS

Neurons utilize several different transmitters. Certain neurotransmitters are associated with specific behaviors. Some drugs act by influencing neurotransmitter function, and certain diseases are associated with altered neurotransmitter levels. The body produces endorphins, which are neurotransmitters that mediate pain perception.

SUMMARY

A *neuron* has a *cell body*, which contains organelles; *dendrites*, which receive impulses and transmit them toward the cell body; and *axons*, which conduct impulses toward another cell and transmit information across the synapse by releasing a chemical neurotransmitter. A *sensory neuron* carries information about stimuli towards the brain and spinal cord. A *motor neuron* carries information away from the brain and spinal cord and causes an effector, a muscle or gland, to respond. An *interneuron* conducts information between two other neurons.

The neuron's message is a nerve impulse or an *action potential*. In a section of the neuron at rest, the concentration of K^+ is 30 times greater inside the cell than outside, and the concentration of Na^+ is 10 times greater outside than inside. In addition, negatively charged proteins within the cell give the interior a negative charge. When an action potential is initiated, membrane channels open and allow Na^+ to enter. The positively charged Na^+ depolarizes the membrane. At the peak of depolarization, Na^+ channels close. Repolarization occurs as K^+ leaves the cell, restoring the resting potential. The action potential spreads along the nerve fiber because the depolarization in one region of the membrane triggers depolarization in the adjacent region.

Myelination increases the speed of nerve transmission. A *myelin sheath* is formed around some nerve fibers by fatty cells that wrap around the fiber many times. The gaps between these insulating cells are called *nodes of Ranvier*. In a myelinated fiber, the nerve impulse jumps from one node of Ranvier to the next, increasing the rate of conduction about 100 times. Transmission of an impulse along a myelinated fiber is called *saltatory conduction*.

When an impulse reaches the end of an axon, it causes vesicles in the terminal portion of the *presynaptic cell* to move toward the membrane and release *neurotransmitters* into the *synaptic cleft*. These chemicals diffuse across the cleft and fit receptors on the surface of the *postsynaptic cell*. If the neurotransmitter is *excitatory*, the postsynaptic membrane is slightly depolarized, making it more likely that an action potential will begin. If the transmitter is *inhibitory*, the membrane is hyperpolarized, and it is less likely that an action potential will begin. The summing of excitatory and inhibitory messages, called *neural integration*, allows finer control over the activity of a neuron. *Secondary messengers* such as cAMP modulate neurotransmitter action. A neurotransmitter is quickly removed from the synapse by either enzymatic destruction or active reuptake.

Neurotransmitter balance is important to maintain health, and many drugs and diseases produce their effects by altering the levels of particular transmitters. One class of neurotransmitters, the *endorphins*, reduces the perception of pain.

QUESTIONS

1. Sketch a neuron and label the cell body, dendrites, and axon.

2. What are the differences in the jobs of sensory, motor, and interneurons?

3. What is the distribution of charges in the membrane of a resting neuron?

4. What ionic events are responsible for the wave of depolarization and repolarization called an action potential?

5. How does the presence of myelin alter the conduction of an impulse along a nerve fiber?

6. Describe the differences between transmission of an action potential and synaptic transmission.

7. Sketch a synapse including the following structures: presynaptic membrane, vesicles, receptors, postsynaptic membrane.

8. Which part of the nervous system is affected in each of the following conditions?

- **a.** Parkinson's disease
- **b.** Schizophrenia
- **c.** Epilepsy
- **d.** Botulism
- **e.** Heroin addiction
- **f.** Depression
- **g.** Multiple sclerosis

TO THINK ABOUT

1. Do you think that severely depressed people should be routinely tested for a chemical that might indicate likelihood of committing suicide? What should be done after a depressed person is found to have the chemical indicator for suicidal tendency?

2. What do you think the next steps should be in continuing Dr. Langston's pioneering work on linking Parkinson's disease to herbicides?

3. What are the possible advantages and disadvantages of a psychiatrist's telling a depressed patient that his or her mental illness is actually the result of a biochemical imbalance?

4. The depressed person described in the text, Marsha, was helped by a tricyclic antidepressant drug, which blocks reuptake of noradrenaline. Another class of antidepressants, called monoamine oxidase inhibitors, works on noradrenaline by a different mechanism. What might it be?

SUGGESTED READINGS

Barinaga, Marcia. November 16, 1990. Technical advances power neuroscience. *Science*. Imaging cell culture and molecular manipulations have given us new views of the brain in action.

Franklin, Deborah. November/December 1990. Hooked/not hooked. Why do some cocaine users become addicted and some do not?

Gold, Philip W. et al. August 11, 1988. Clinical and biochemical manifestations of depression. *The New England Journal of Medicine*. Depression is a biological illness.

Goldstein, Avram. March 1978. Endorphins. *The Sciences*. One of the first and best analyses of these fascinating body chemicals.

Heppenheimer, T. A. February 1988. Nerves of silicon. *Discover*. Computers are now built to imitate neural networks.

Kanigel, Robert. Fall 1987. Learning at the subneural level. *Mosaic*. The plasticity of the human brain is due to the great number of synapses.

Langston, J. William. January/February 1985. The case of the tainted heroin. *The Sciences*. How a strange muscle paralysis in heroin addicts shed light on understanding the biochemistry of Parkinson's disease.

Lewis, Ricki. October 1984. The biochemistry of mental illness. *Biology Digest*. Biochemical imbalances are behind schizophrenia, depression, and manic-depression.

Peele, Stanton. July/August 1989. Ain't misbehavin'. *The Sciences*. Addiction often has a biological basis.

Van Brunt, Jennifer. November 1988. Neurobiotech: From spiders to Alzheimer's. *Biotechnology*. Understanding how neurotransmitters bind to receptors gives us information to "rationally" design drugs.

Wilbur, Robert. March 1986. A drug to fight cocaine. *Science 86*. Antidepressant drugs may be useful in combating cocaine addiction.

C H A P T E R

17

The Nervous System

Chapter Outline

Organization of the Vertebrate Nervous System
The Central Nervous System
The Spinal Cord
Reading 17.1 Nervous System Diversity
The Brain
Reading 17.2 Mysterious Sleep
Memory
Reading 17.3 PET Peeks at the Brain
Protection of the Central Nervous System
The Peripheral Nervous System
The Somatic Nervous System
The Autonomic Nervous System
When Nervous Tissue Is Damaged
Neuronal Cell Culture

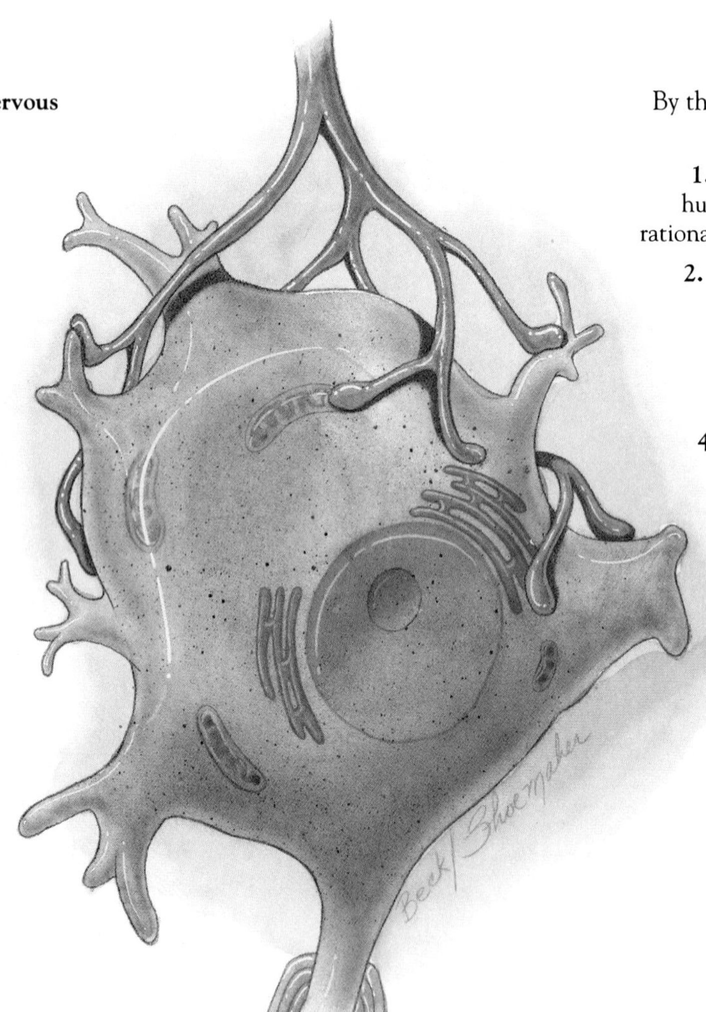

Learning Objectives

By the chapter's end, you should be able to answer these questions:

1. What are the major divisions of the human nervous system, and what is the rationale for this organizational framework?
2. What are the functions of the spinal cord?
3. What functions are provided by each of the major structures of the human brain?
4. What is thought to be the physical basis for memory?
5. How does the nervous system respond to a threat and then recover afterwards?

The dedicated parent was becoming awfully tired of reading *The Fuzzy Duckling.* Each night he would plop his young daughter on his lap and read the story, as he had done since she was a few weeks old, thinking that she somehow understood. Now he had his doubts. At 18 months, she seemed more interested in wriggling away.

Then one day father and daughter visited a lake. All of a sudden, the child dropped the father's hand and toddled over to the shore. He actually heard her quacking before he saw the white, fluffy ducks and their fuzzy ducklings. Not only did she quack and imitate the ducks' waddle (perhaps unintentionally), but she uttered her versions of several words from the book—duck, baby, go, water. The thrilled father could almost see the proverbial light bulb over his child's head. She jumped up and down, quacking and babbling in excitement—and her dad did too. Although we all use our brains daily to recall, think, associate, and reason, at no time, perhaps, do these skills seem quite as miraculous as when a child first displays them—and is aware that she has.

The responses shown by the child seeing her first real ducks are orchestrated by her nervous system. This vast labyrinth of neurons also floods her with emotion, enables her to tune out distractions temporarily, while keeping her heart and lungs functioning. In the animal kingdom, collections of neurons and their intricate connections to each other and sometimes to muscle and gland cells oversee an incredible array of skills—from the simple detection of a stimulus to a child's ability to connect a word with a fuzzy duckling to a father's recollection of his own childhood.

Organization of the Vertebrate Nervous System

With its billions of component cells and their trillions of interactions, the human nervous system is enormously complex. Deciphering the physical basis—that is, the specific neuronal connections—underlying behavior is one of the most intensely pursued avenues of biological research. Understanding this complex organ system is helped by considering subdivisions of it—organized groups of neurons—that are associated with specific functions.

The most general division of the vertebrate nervous system is into the **central nervous system** (CNS), which consists of the brain and spinal cord, and the **peripheral nervous system** (PNS), which consists of nerves and **ganglia** (collections of cell bodies) that carry information to and from the CNS. The peripheral nervous system is further subdivided and will be discussed later in the chapter (fig. 17.1).

The Central Nervous System

The Spinal Cord

Encased in the bony armor of the vertebral column (backbone), the **spinal cord,** a tube of neural tissue, extends about 17 inches (43 cm) from the base of the brain to an inch or so below the last rib. It carries impulses to and from the brain and is a site of interaction between neurons involved in spinal reflex actions. A **reflex** is a rapid,

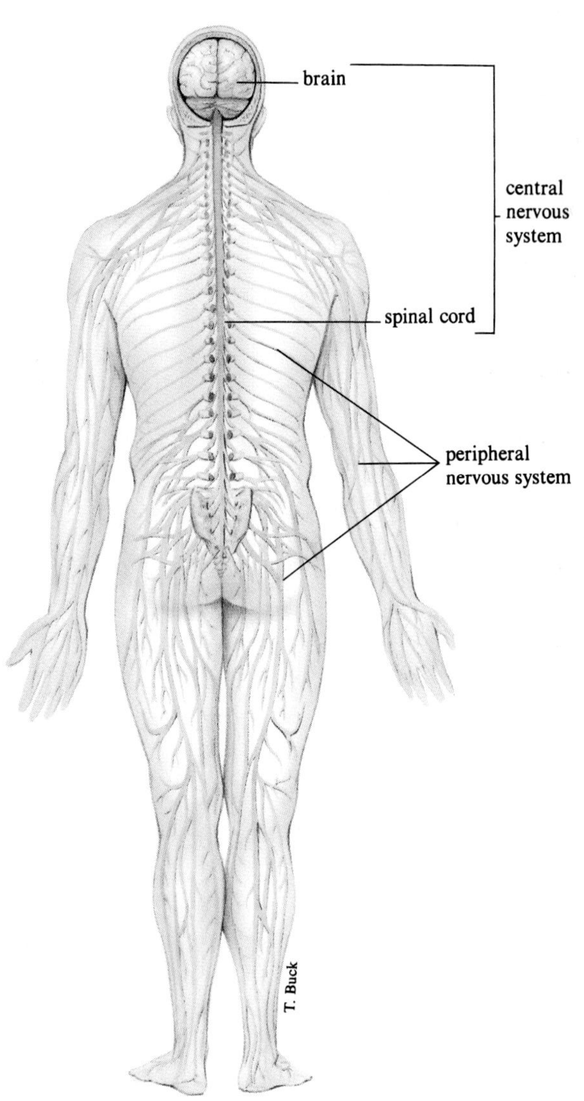

Figure 17.1
Major divisions of the nervous system. The central nervous system (CNS) consists of the brain and the spinal cord, which integrate and interpret incoming information. All other nervous tissue is part of the peripheral nervous system, which sends impulses from effectors to the CNS and delivers impulses from the CNS to effectors.

involuntary response to a stimulus either from within the body or from the outside environment.

The spinal cord communicates with the rest of the body through spinal nerves. On each side of the cord is a pair of nerves; sensory information is delivered to the rear of the spinal cord and instructions for activity of skeletal muscles pass outward from the front of the cord. The sensory and motor nerve fibers form a single cable, which passes through the opening between vertebrae (fig. 17.2).

In a developing human embryo, the spinal nerves leave the spine through a nearby opening. However, as a person grows, the backbone grows faster than the nerve cord. As a result, the spinal nerves of an adult often pass out of the vertebral column below the tip of the spinal cord. Therefore the vertebral column of the lower back contains only spinal nerves and fluid. Because the rubbery spinal nerves of this area slip out of the way of a needle, spinal anesthetics are often administered in the lower back (fig. 17.3). Spinal anesthetics temporarily deaden sensation of the neurons leading into the lower portion of the spinal cord and are sometimes given to ease childbirth pain. Samples of the cerebrospinal fluid around the spinal cord may be taken, and they can provide clues helpful in the diagnosis of some cancers, infections, and mental disorders.

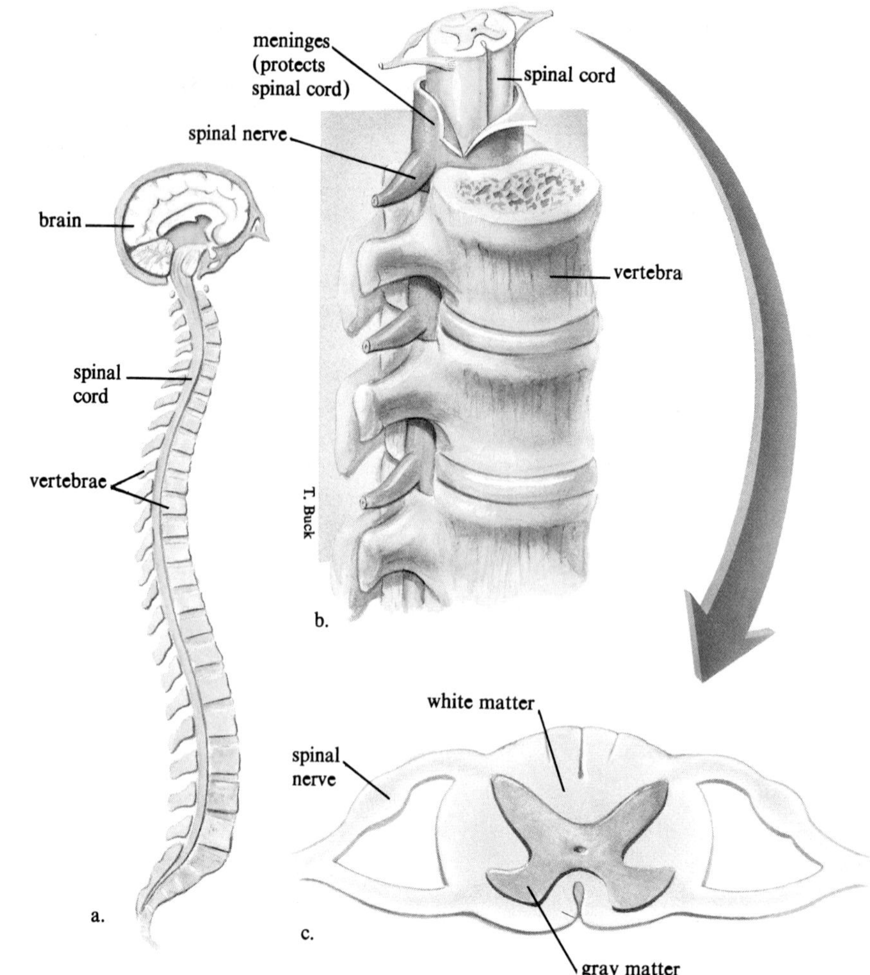

Figure 17.2
The spinal cord. *a*. The CNS consists of the brain, which is encased in the bony skull, and the spinal cord, which is protected by the bony vertebrae. *b*. The spinal cord is surrounded by the subarachnoid space and then by three membranes called the meninges. *c*. Like the brain, the spinal cord contains both gray and white matter. The gray central region, which resembles the letter *H*, consists of short interneurons, motor neuron cell bodies, and glial cells. The white matter surrounding the H consists of myelinated nerve tracts.

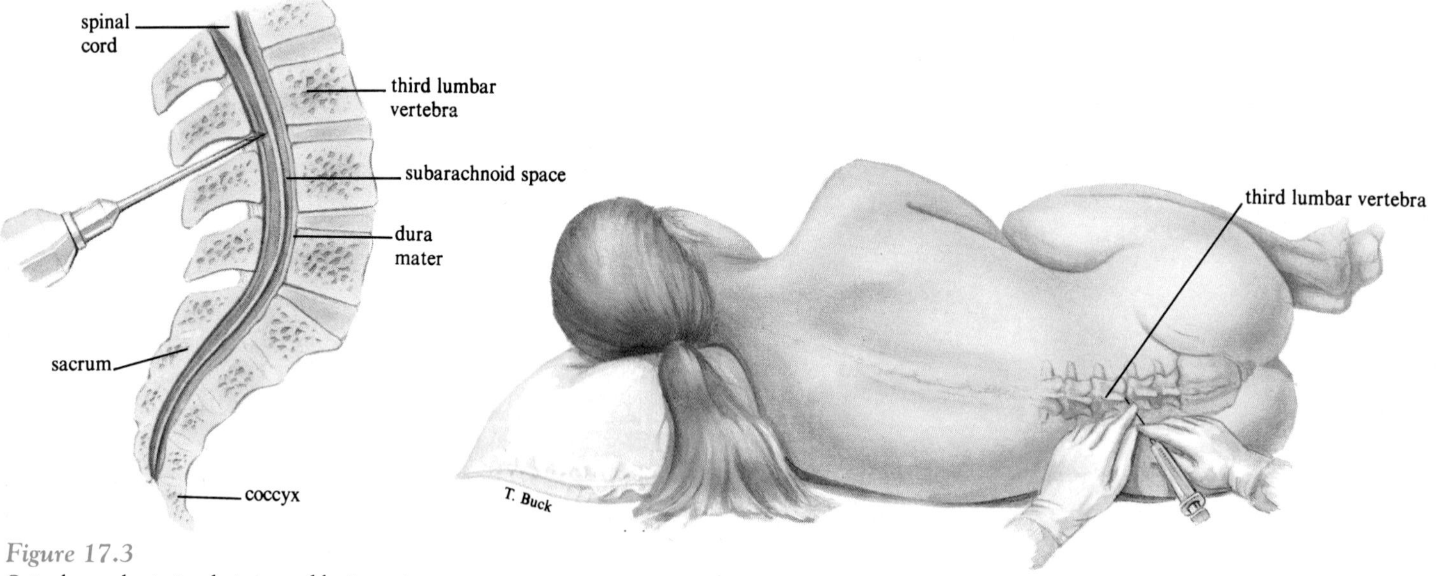

Figure 17.3
Spinal anesthesia is administered by inserting a needle into the space at the tip of the spinal cord.

The spinal cord conducts information to and from the brain via myelinated fibers, which form the white matter at the periphery of the cord. Specific types of information are carried in particular tracts (axons and dendrites) of the white matter. **Ascending tracts** carry sensory information to the brain, and **descending tracts** carry motor information from the brain to muscles and glands. The interior of the spinal cord, shaped a little like the letter H in cross section, consists of gray matter, which is motor neuron cell bodies, interneurons, and glial cells.

The spinal cord can also handle some situations without interacting with the brain. Such a **spinal reflex** usually involves several neurons but may require only two. Even a simple neural connection of a sensory neuron to interneuron to motor neuron allows a more rapid response than would be possible if the information had to travel to the brain. Because of their speed, reflexes are often protective.

Consider a **reflex arc,** a neural pathway linking a sensory receptor and an effector such as a muscle. A reflex arc is formed by the three neurons that cause you to pull your hand away from a painful stimulus such as a thorn (fig. 17.4). A dendrite of a sensory neuron in the skin is specialized as a **sensory receptor,** detecting the sharp thorn. An action potential is generated along the axon and is transmitted to the dorsal (back) side of the spinal cord. Within the spinal cord, the sensory neuron's axon synapses with an interneuron. Next, within the gray matter, the interneuron synapses with a motor neuron. The motor neuron's axon exits the spinal cord on the ventral side, and its action potential stimulates a skeletal muscle cell

Reading 17.1 *Nervous System Diversity*

HOW DID THE HUMAN BRAIN EVOLVE? The structure and organization of nervous systems of modern organisms that have changed little from their long-ago ancestors provide clues to the evolution of the human brain. The simplest nervous systems are diffuse networks of neurons located within the body walls of members of the phylum Cnidaria (fig. 1*a*). The scattered neurons synapse with muscle cells near the body surface, enabling the organism to swim and to move its tentacles. This diffuse arrangement and the ability of the neurons to conduct impulses in both directions allow the animal to react to stimuli approaching from any direction. The *Hydra,* which commonly sits on the bottom of a pond where danger and food may approach from any direction, is ready to react to a change in the environment.

Nervous tissue is organized in the flatworms (phylum Platyhelminthes) into a brain consisting of two cerebral ganglia and two nerve cords extending down the body, which are connected to each other to form a "ladder" type of nervous system (fig. 1*b*). Motor structures and the neurons that control them are paired, allowing coordinated forward motion. Paired symmetric sense organs at the head end allow the planarium to detect a stimulus and crawl toward it. Bilaterally paired receptors allow the animal to determine the direction of stimulation by comparing the intensity of the stimulus at each receptor. The receptor responding most strongly is closest to the stimulus.

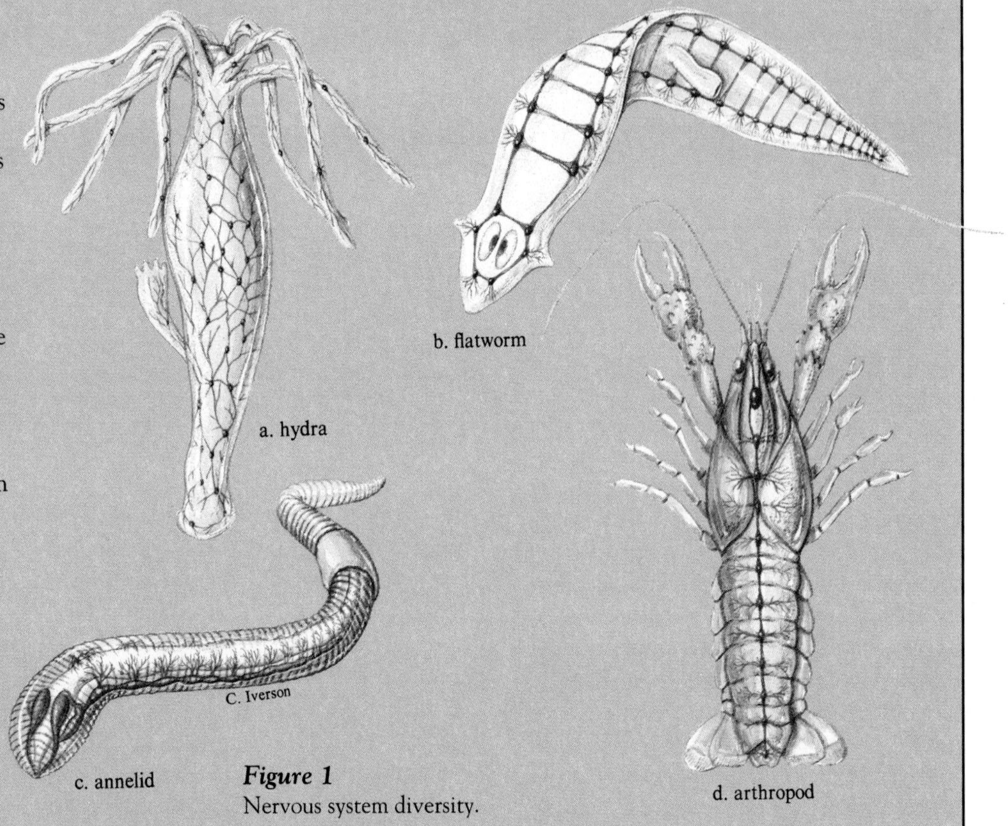

Figure 1
Nervous system diversity.

The segmented worms (phylum Annelida) have a yet more elaborate ladder arrangement, with a larger brain (fig. 1*c*). Two nerve cords extend down the body beneath the digestive tract and the cords are solid and fused to each other. Peripheral nerves branch off from the central fused ladder. Like the nervous systems of flatworms and annelids, that of the arthropods consists of a brain and a solid fused nerve cord located ventrally (fig. 1*d*). In addition, the arthropod nervous system has highly developed sensory organs for touch, smell, hearing, chemical reception, and balancing, and these animals are capable of complex behaviors and even rudimentary learning.

to contract. When a sufficient number of muscle fibers contract, the hand is pulled away from the thorn. The original sensory neuron synapses with other interneurons too. Some of these go to the brain, perhaps prompting you to yell, "Ouch!" This protective mechanism prevents further injury before you even realize you are hurt. A reflex can be carried out even if the brain is not functional. Chicken farmers are familiar with the phenomenon of a chicken continuing to run around the barnyard even after its head has been cut off.

KEY CONCEPTS

The brain and spinal cord form the central nervous system. The spinal cord carries impulses to and from the brain and can bypass the brain to produce spinal reflexes. The cord is a tube of neural tissue with paired nerves on the sides. The outside of the cord consists of white matter, which consists of tracts of axons and dendrites. The inside is gray matter, which comes from interneurons, glia, and cell bodies.

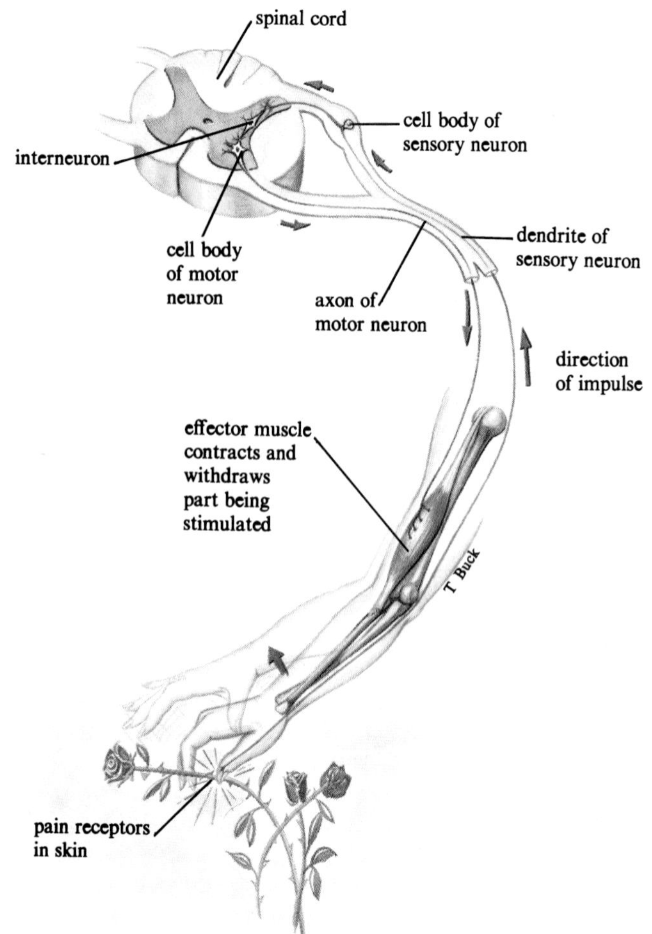

Figure 17.4
Neuron types within a reflex arc. A reflex arc is formed when a sensory neuron, responding to an environmental stimulus, synapses in the spinal cord with an interneuron, which relays the impulse to a motor neuron, which, in turn, stimulates a muscle cell to contract. Motor neurons may also activate other neurons or gland cells.

The Brain

The human brain weighs a mere 3 pounds (1.36 kilograms) and has the appearance and consistency of grayish pudding, yet it is the seat of our individuality. To oversee the functioning of our organs and also to provide the qualities of "mind"—learning, reasoning, and memory—the brain requires a large and constant energy supply. At any given time, the activity of the brain accounts for 20% of the body's consumption of oxygen and 15% of its consumption of blood glucose. Permanent brain damage sets in after just 5 minutes of oxygen deprivation. In people who suffer from hypoglycemia, low levels of glucose in the blood deprive the brain of glucose, causing mental confusion.

The human brain is built of three major regions and a few smaller structures. The three largest regions are the *brain stem*, the *cerebellum*, and the *cerebrum*. The brain stem and the cerebellum are considered to be more primitive brain structures, because they are present in species that are less complex than humans. They are sometimes referred to as "lower" brain structures, both for the evolutionary reason and because they are physically located closer to the spinal cord than the cerebrum (fig. 17.5). The cerebrum is most highly developed in the primates, especially our own species.

The Brain Stem

Anatomically, the brain stem is a continuation of the spinal cord within the skull. It is aptly named for its appearance. Like the stem on a head of cauliflower, it looks like a stem supporting the brain. But it does much more. The brain stem controls vital physiological functions.

The section of brain stem closest to the spinal cord is the **medulla,** which regulates physiological processes that are essential to life. The medulla maintains breathing and heartbeat while other activities are carried out. It regulates blood pressure by altering the diameter of blood vessels. The medulla adjusts these activities to suit the body's varying needs, increasing heart and breathing rates and blood pressure during vigorous activity and slowing them during restful periods. In addition, the medulla contains reflex centers for vomiting, coughing, sneezing, urinating, defecating, swallowing, and hiccoughing.

All messages entering or leaving the brain must pass through the medulla. As the neurons carrying sensory or motor messages pass through this region most of them cross from one side of the body to the other, so, curiously, the left side of the brain receives impressions from and directs the motor activity of the right side of the body, and vice versa. No one knows why this twist exists.

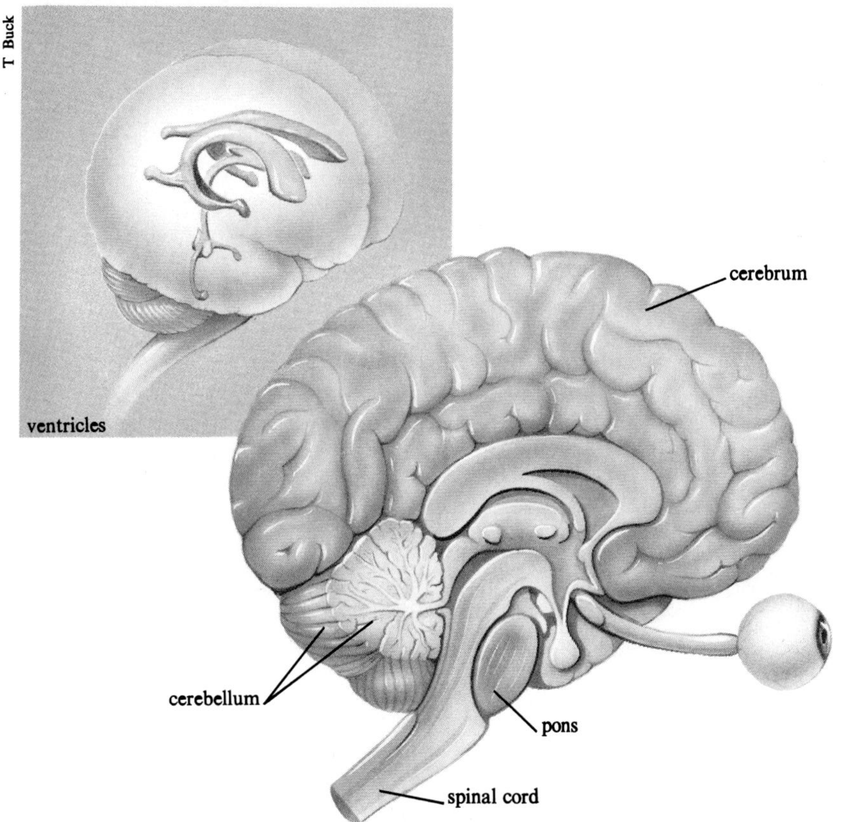

Figure 17.5
Major structures of the human brain.

The section above the medulla is called the **pons,** which means "bridge." This is a suitable name for this oval mass because white matter tracts in the pons connect the medulla and higher brain structures. In addition, gray matter in the pons controls some aspects of respiration. A narrow region above the pons, called the **midbrain,** is also part of the brain stem. Gray matter in the midbrain contributes to seeing and hearing, and white matter connects the region to the cerebrum and other structures above.

The Cerebellum

In back of the brain stem but connected to the cerebrum above is the grooved, easily recognized **cerebellum** (little brain). The cerebellum is gray on the outside, white with gray patches on the inside, and divided into two hemispheres. It is the second largest structure in the human brain, accounting for almost one-eighth of the brain's mass.

The neurons of the cerebellum refine motor messages, resulting in well-coordinated muscular movements. This action is not conscious. The cerebellum receives sensory input from the cerebral cortex (the outer portion) and the peripheral nervous system. It then compares the action the cerebrum intended with the actual movement and makes the corrections necessary to make the two agree. For example, the cerebellum is active when we try to bring the tips of the index fingers together without looking. The initial miss is quickly recognized and the motion corrected.

Many of our conscious activities involve subconscious components governed by the cerebellum. If it were necessary to plan every movement in catching a ball consciously, for instance, you would probably not be quick enough to prevent it from striking your head. The automatic program acquired by training and practice that is responsible for the precision movements of athletes and dancers resides in the cerebellum.

The structures of the brain can also be grouped according to their location. In this organization, the midbrain is part of the brain stem, the hindbrain is the cerebellum, and the forebrain includes the *thalamus* and *hypothalamus*, part of the *reticular activating system*, and the *cerebrum*. The structures of the forebrain are important in complex behaviors, such as learning and memory, language utilization, motivation, and emotion.

The Thalamus

The **thalamus** is a gray, tight package of nerve cell bodies and their associated glial cells located beneath the cerebrum. It is a relay station for sensory input, processing incoming information and sending it to the appropriate part of the cerebrum.

The Hypothalamus

The **hypothalamus,** situated just beneath the thalamus, is more important than its size suggests. Weighing less than 4 grams and occupying less than 1% of the brain volume, the hypothalamus regulates a surprising number of vital functions. It helps to maintain homeostasis by regulating body temperature, heartbeat, water balance, and blood pressure. Groups of nerve cell bodies in the hypothalamus control hunger, thirst, sexual arousal, and feelings of pain, pleasure, anger, and fear. The hypothalamus also regulates the production of hormones in the pituitary gland at the base of the brain. Thus, the hypothalamus is an important link between the nervous system and the endocrine system, the two communication systems of the body.

The Reticular Activating System

The **reticular activating system (RAS)** is a diffuse network of cell bodies and nerve tracts extending through the brain stem and into the thalamus. Reticular means "little net," alluding to the RAS's role in filtering or screening sensory information so that only certain impulses reach the cerebrum. If the RAS did not filter input from the environment, our senses would be overwhelmed with stimulation. Imagine sitting in a lecture class. Without your RAS, you would be acutely aware of every cough, sneeze, and rustle of paper from those around you, and you would even be bothered by the touch of your clothing against your skin. Concentrating on the lecture would be quite difficult.

Reading 17.2 Mysterious Sleep

THE PERSON TO GO THE LONGEST WITHOUT SLEEP WAS A 17-YEAR-OLD STUDENT WHO STAYED AWAKE, IN A SLEEP LABORATORY UNDER MEDICAL SUPERVISION, FOR 264 HOURS. Fortunately he suffered no ill effects, but during his ordeal he was irritable and had blurred vision, slurred speech, and memory lapses. Toward the end, he seemed confused about his identity.

Rats experimentally deprived of sleep do not fare as well. In a study conducted at the University of Chicago, rats kept awake by a moving floor developed skin sores and hormonal and metabolic changes, dying within 11 to 32 days. Control rats allowed to nap survived.

It is clear from these and other experiments that we need sleep to maintain health, but exactly why we need it is not known. About the best we can do is to describe this state of consciousness in which we are not very responsive to outside stimuli and are at the mercy of sometimes vivid dreams.

We know that sleep is physiologically distinct from the waking state because the electrical activity of the brain—the brain waves detectable in an electroencephalogram (EEG) tracing—differs when we are awake or in a stage of sleep. In 1953 it was discovered that we experience two types of sleep—REM sleep, named for the rapid eye movements visible through closed lids, and nonREM sleep, which in turn is subdivided into four levels of increasingly deep slumber.

During REM sleep, the nervous system is curiously quite active. The autonomic nervous system is on alert, with heart rate and respiration elevated. Brain temperature rises as the brain's blood flow and oxygen consumption increase. Action potentials zip about the brain more rapidly than during nonREM sleep, or even during wakefulness. REM sleep is a time of vivid dreaming, although dreaming occurs at other times as well. NonREM sleep proceeds through four continuous stages, from light sleep to deep sleep, and the EEG wave tracings become less frequent but higher in amplitude (size).

Sleep follows a predictable pattern of 70 to 90 minutes of nonREM followed by 5 to 15 minutes of REM, repeated many times a night. If a person is deprived of REM sleep and its dreams, the next night's pattern compensates for it, with more of each. Sleep may vary a great deal with age. A newborn sleeps 16 hours a day, a toddler 9 to 12 hours, adults 7 or 8 hours, and the elderly often less than 6 hours.

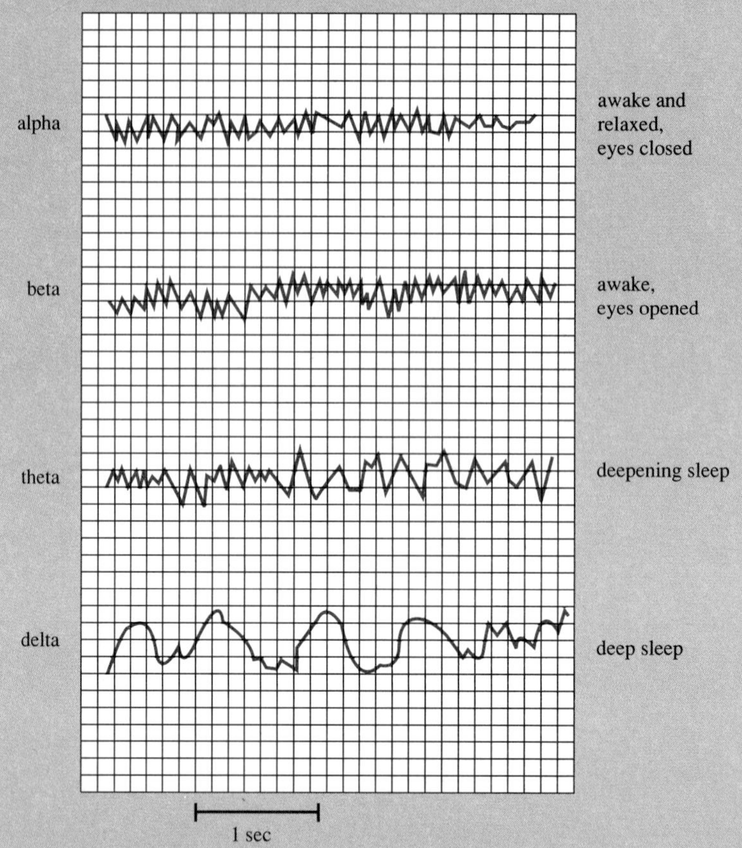

Figure 1
Different states of consciousness are associated with distinct characteristic brain-wave patterns.

As its name implies, the reticular activating system is also important in overall activation and arousal. When certain neurons within the RAS are active, you are awake. When they are inhibited by other neurons, you sleep. Reading 17.2 looks more closely at sleep.

KEY CONCEPTS

The brain stem looks like a stem supporting the brain, and it includes the medulla, pons, and midbrain. The medulla controls breathing, heartbeat, blood pressure, and several reflexes. Neurons passing through the medulla from one side of the brain cross to the other side of the body. The pons connects the medulla to higher brain structures and helps regulate respiration. The midbrain participates in hearing and seeing. Subconscious refinement of movement is controlled by the cerebellum. In the forebrain, the thalamus serves as a relay station, and the hypothalamus helps to regulate a variety of vital functions, including heartbeat, body temperature, blood pressure, water balance, and certain drives and feelings.

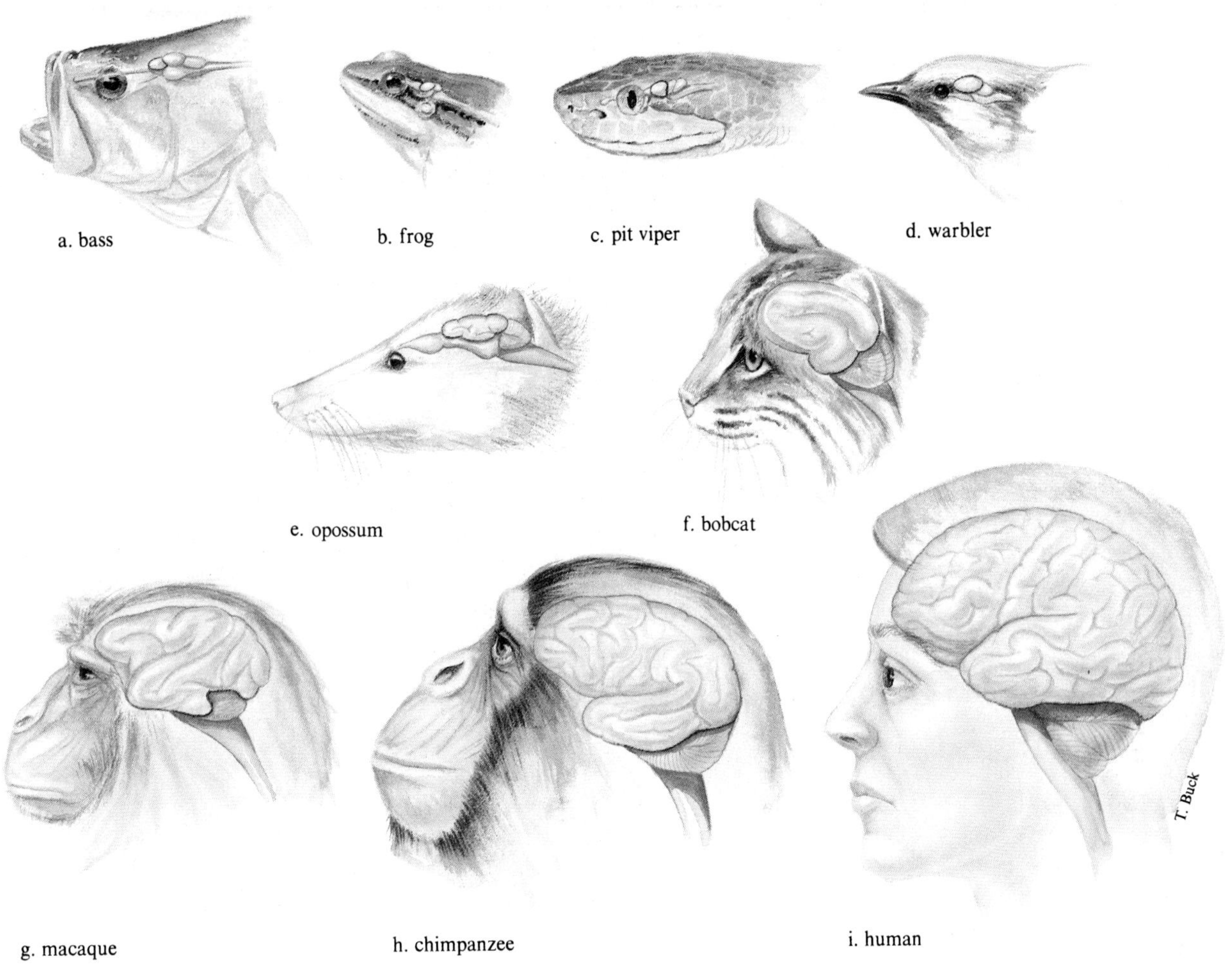

Figure 17.6
The cerebrum in different vertebrate species. The cerebrum increases in both size and in the complexity of its neural connections in more advanced species, as indicated by the dark portion of each animal's brain. Fishes (*a*) and amphibians (*b*) lack cerebral cortexes, whereas reptiles (*c*) and birds (*d*) have a small amount of gray matter covering their cerebrums. More primitive mammals, such as the opossum (*e*), have smooth cortexes. Carnivores such as the cat (*f*) have larger cerebrums with a cortex with a few convolutions. In the primates (*g* and *h*), the cerebrum is much increased relative to other brain structures, and the cortex is more convoluted. The human cerebrum (*i*) obviously dominates the brain, and it is highly convoluted.

The Cerebrum

The **cerebrum** controls the qualities of "mind," including intelligence, learning, perception, and emotion. The two hemispheres of the cerebrum comprise 80% of total brain volume. Ten billion neurons are in the cortex alone—the entire brain has 100 billion neurons.

The outer layer of the cerebrum, the cortex, consists of gray matter that integrates or makes sense of incoming information. During development the cortex grows faster than the underlying white matter. As a result, the cortex folds into convolutions. Nonhuman primates have fewer convolutions in the cerebral cortices than do humans (fig. 17.6).

The Cerebral Cortex

The cerebral cortex is divided into sensory, motor, and association areas. **Sensory areas** receive and interpret messages from sense organs about temperature, body movement, pain, touch, taste, smell, sight, and sound. **Motor areas** send impulses to skeletal muscles. **Association areas,** so named because they were once thought to connect the sensory and motor pathways, do not appear to be either sensory or motor. These little-understood parts of the cortex are the seats of learning and creative abilities.

A band of cerebral cortex extending from ear to ear across the top of the head controls voluntary muscles. The region is called the **primary motor cortex.** Just behind it is the **primary sensory cortex,** which receives input from the skin (fig. 17.7). Nearly every part of the body is represented in both the sensory and motor cortex. The surface area devoted to a particular body part is proportional to the degree of sensitivity and motor activity of the area. For example, the hands, tongue, and face are extensively represented in both the sensory and motor regions of the cortex (fig. 17.8). Other senses are represented in other areas of the cortex.

Maps of the cerebral cortex that link certain areas to certain behaviors have been constructed by intentionally damaging parts of the cortex in experimental animals or by studying loss of function in brain-damaged people (fig. 17.9). Experimental evidence suggests that the correspondence between sensory ability and cortical representation is highly specific. For example, the five rows of whisker follicles on a mouse's snout correspond to five rows of cell clusters similarly arranged in the cortex, each cluster containing roughly 2,000 neurons.

Specializations of the Cerebral Hemispheres

Each cerebral hemisphere has some specific functions, yet the hemispheres work together. In most people, parts of the left hemisphere are associated with speech, linguistic skills, mathematical ability, and reasoning, while the right hemisphere specializes in spatial, intuitive, musical, and artistic abilities.

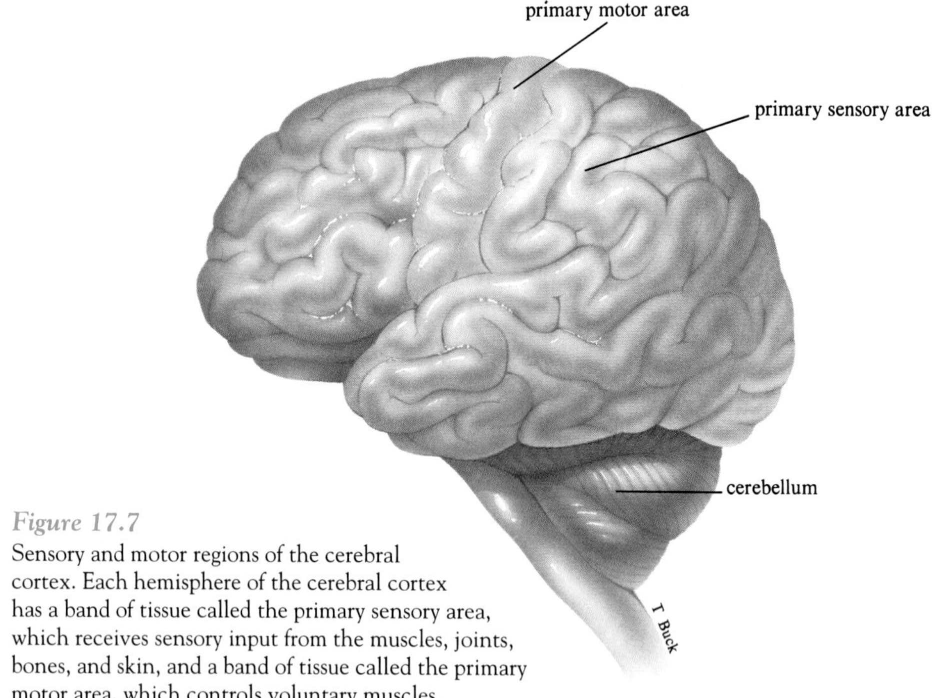

Figure 17.7
Sensory and motor regions of the cerebral cortex. Each hemisphere of the cerebral cortex has a band of tissue called the primary sensory area, which receives sensory input from the muscles, joints, bones, and skin, and a band of tissue called the primary motor area, which controls voluntary muscles.

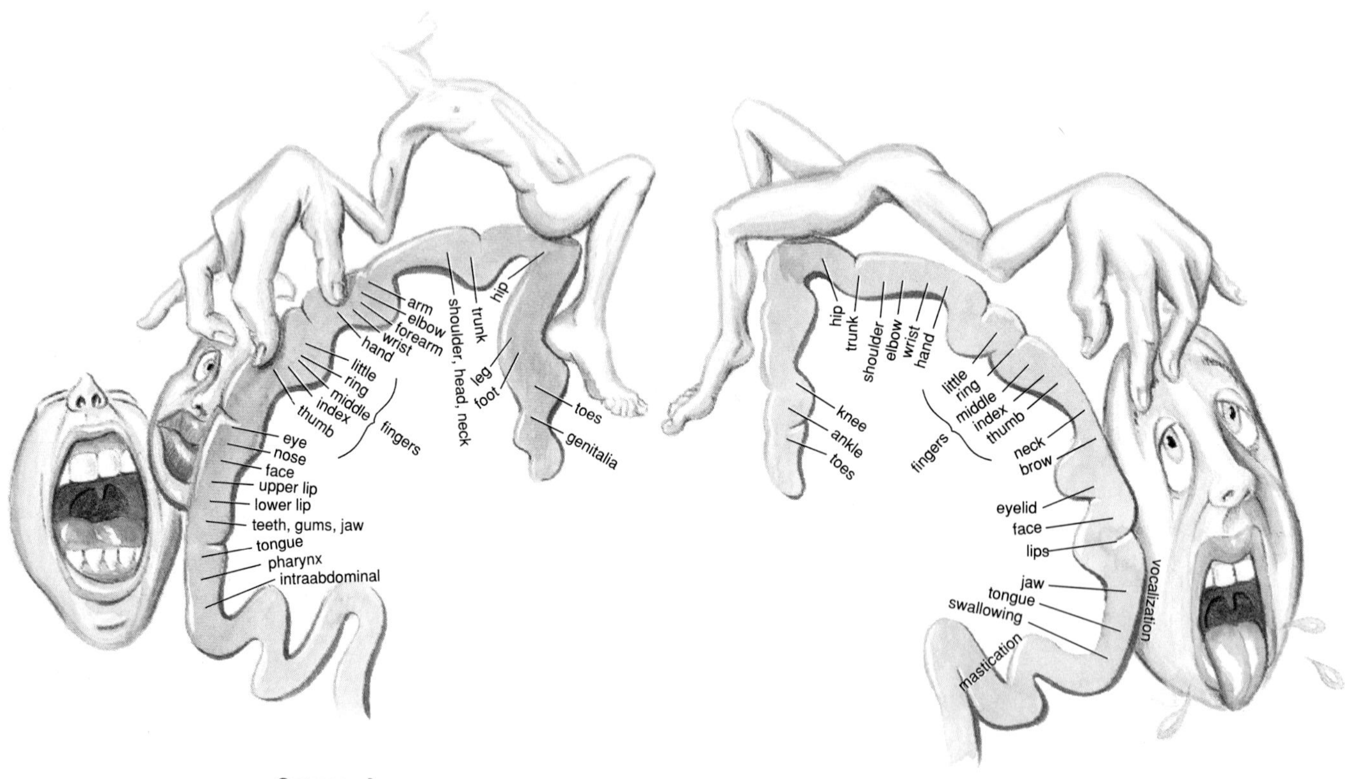

Figure 17.8
Different body parts are innervated to different degrees. The amount of each area devoted to a particular body part depends not upon the size of the part but upon how precisely it must be controlled by the nervous system. A distorted human figure superimposed on a diagram of the primary sensory and motor areas indicates those areas that are finely innervated, such as the face and the hands.

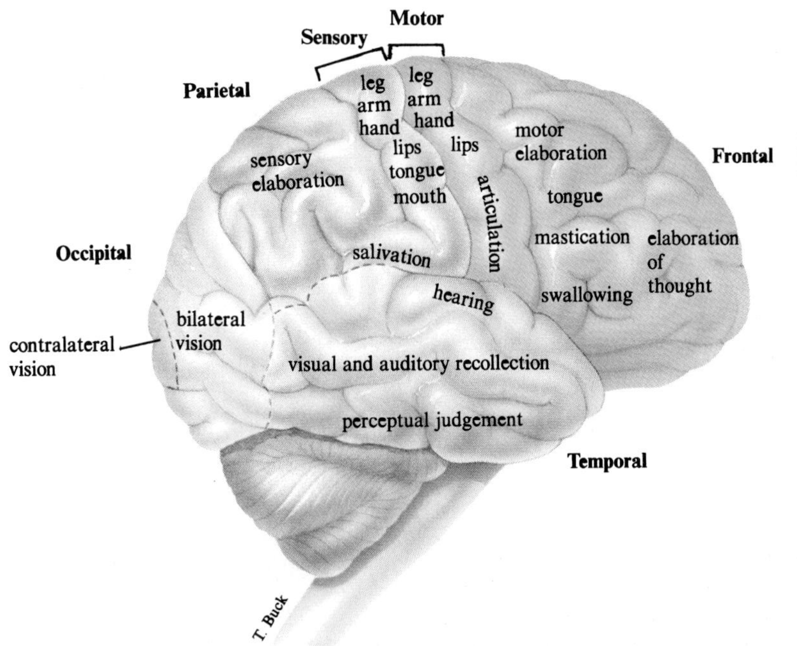

Figure 17.9
Cerebral specialization. The cerebrum can be subdivided into four sections—the frontal, temporal, occipital, and parietal lobes. Experiments have shown that different parts of the cerebrum provide different specific functions.

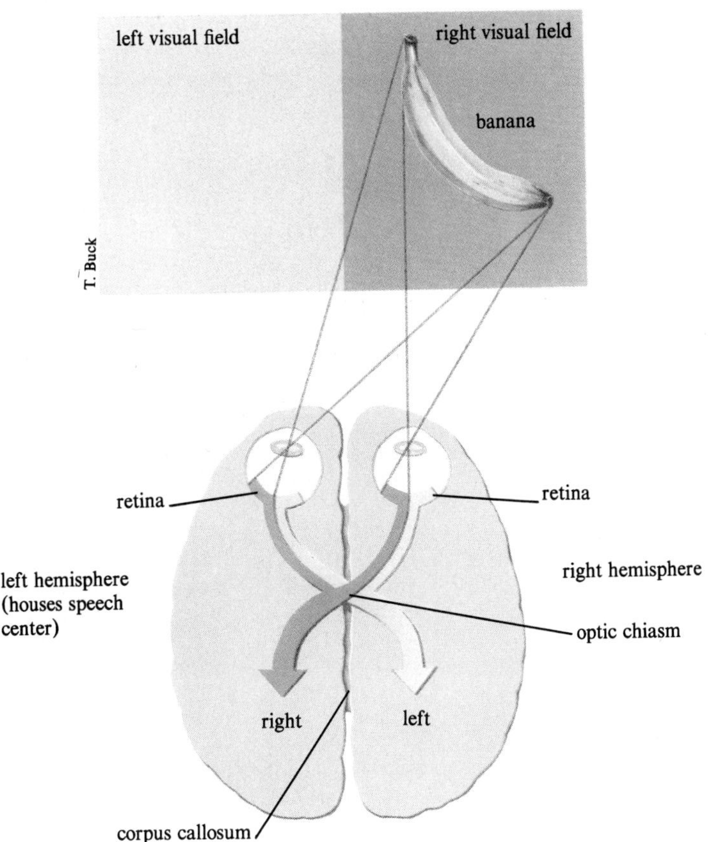

Figure 17.10
A split-brain person can name a familiar object placed in the right visual field because the stimulus goes to the left cerebral hemisphere, which houses the speech center. An object placed in the left visual field cannot be named, but the split-brain person can point to it.

The specializations of the hemispheres have been studied in individuals who have severed corpus callosums. The corpus callosum is a thick band of hundreds of millions of axons, running between the cerebral hemispheres, and enabling them to share information. When this structure is cut, information delivered to only one hemisphere cannot pass to the opposite side. This "split-brain" phenomenon has been studied in children born without corpus callosums, in laboratory animals whose corpus callosums are intentionally cut, and most commonly in people who have severe epilepsy and whose corpus callosums have been cut.

Epilepsy is an electrical disturbance in the brain resulting in seizures, loss of consciousness, and sensory disturbances. When drugs fail to help these individuals, cutting the corpus callosum reduces the severity and frequency of seizures. It is thought that severing the connection between the cerebral hemispheres allows one of the hemispheres to function normally. People (and other animals) with split brains retain their intelligence and personalities and have little difficulty functioning in everyday life.

Under normal conditions, both cerebral hemispheres gather information simultaneously. However, in the laboratory, input can be restricted to only one side of the cerebrum so that specializations of the hemispheres can be studied. If a split-brain person is shown a banana in only his or her right visual field, the information goes to the left cerebral hemisphere. This hemisphere houses the speech center, and the subject quickly says "banana" (fig. 17.10). If the banana is shown to the person's left visual field, the information goes to the right cerebral hemisphere. Because there is no speech center here, the subject does not say "banana." But the right hemisphere recognizes the banana in other ways. If shown several different fruits in the left visual field and asked to point to the banana, the person can indeed do so. The right hemisphere cannot process language, but it is better than its left counterpart in comprehending spatial relationships. Which hemisphere do you think would predominate in putting together a puzzle?

Reading 17.3 *PET Peeks at the Brain*

Many mental illnesses and neurological disorders, as well as normal thoughts, feelings, and behaviors, are caused by fluctuations in the chemistry of the brain. A technique that allows researchers and clinicians to observe the chemical functioning of the human brain is **positron emission tomography,** or **PET.** Already the new tool is creating multicolored views of the brains of patients who have Huntington disease, epilepsy, or Alzheimer's disease—and those engaged in such mundane activities as reading and speaking (fig. 1).

A PET scan is painless, although it does emit low-level radiation in the part of the body being examined. First, the patient is injected with or inhales a chemical that contains a radioactive isotope, or "label." This chemical might be a radioactive form of glucose (a normal brain metabolite) or perhaps a labeled form of a drug used to treat the patient's particular problem. As the chemical circulates in the patient's body, it gives off positrons, which are positively charged subatomic particles. When a positron collides with a negatively charged electron, given off by a cell, the two particles "annihilate" one another, an event that releases two gamma rays. These rays hit a detector system that encircles the patient, and crystals in the detector convert the gamma rays into light rays. A photomultiplier next converts the light rays into electrical energy, which is then amplified. A computer records the annihilations, at a rate of 150,000 per second, for the entire 5 to 60 minutes that a typical PET scan takes.

Finally, the computer correlates the rate of annihilations coming from a particular part of the brain (or whatever organ is being examined) with a color, which tells researchers how metabolically active the region is. Bright red and white indicate rapid circulation and metabolism, which occurs in a tumor or the site of the abnormality in a patient with epilepsy. Blue and purple signal slowed circulation and metabolism, such as in a portion of the brain that has died as a result of a stroke. Green and yellow represent normal brain metabolism.

In the PET scan in figure 1, a patient with a brain tumor has been given a labeled amino acid, which is converted to a red area where it concentrates in tumor that is left following surgery. PET scans can also show whether epilepsy is operable or a stroke reversible, and sometimes they indicate the cause of dementia.

The white matter beneath the cortex consists of tracts of myelinated nerve fibers. Some of these tracts connect regions within a hemisphere, some form a bridge between the hemispheres called the **corpus callosum,** and others connect the cerebrum to lower brain structures and the spinal cord.

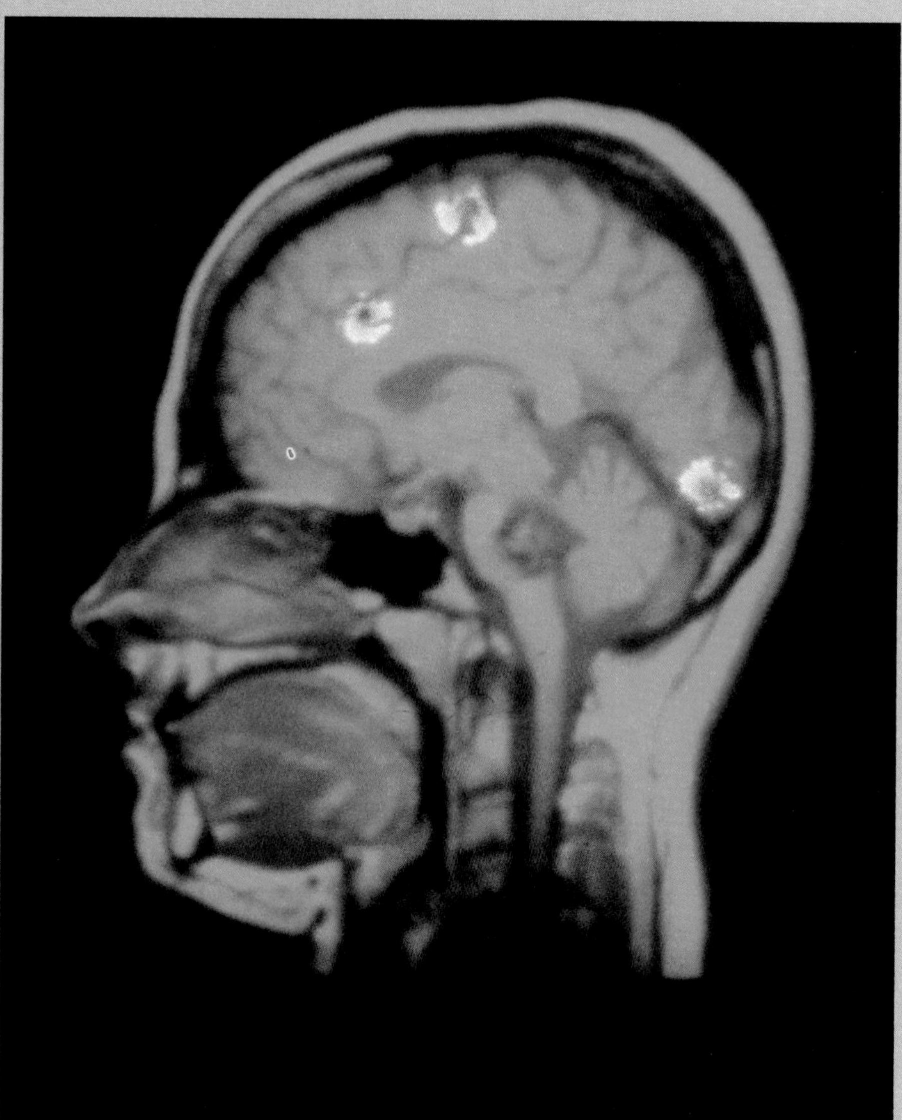

Figure 1
PET scans superimposed upon a purely anatomical view of the brain provided by magnetic resonance imaging (MRI) highlight sites of activity during language use. The frontmost highlighted region deciphers a word's meaning. The middle spot is activated during speech. The back area is active during reading.

KEY CONCEPTS

The cerebrum is a massive structure in humans, consisting of an outer cortex of gray matter covering white nerve tracts that communicate with other brain regions. The cortex has sensory, motor, and association areas. The primary motor cortex controls voluntary muscles, and the primary sensory cortex receives input from the skin. The cerebral hemispheres are connected by a band of axons, the corpus callosum. The left hemisphere specializes in speech, language skills, mathematical ability, and reasoning. The right hemisphere is linked to spatial, intuitive, musical, and artistic abilities. If the corpus callosum is cut, the hemispheres cannot interact.

Memory

It is easy to remember exactly where you were and what you were doing when you learned that the space shuttle *Challenger* exploded shortly after launching, but you cannot recall where you put your car keys this morning. You cannot remember what your biology professor said just an hour ago, but you can easily recite the lyrics to a song that you enjoyed when it was popular years ago but have not heard very much since.

Two types of memory, *short term* and *long term*, have been recognized for many years, and researchers are now beginning to realize that they differ in characteristics other than duration. Understanding how neurons in different parts of the brain encode memories, and how short-term memories are converted to long-term memories (a process called consolidation) is at the forefront of research into the functioning of the human brain.

Short-term memories are thought to be electrical in nature. Neurons may be connected in a circuit so that the last in the series stimulates the first. Once an impulse begins, around and around it goes for seconds or hours. As long as the pattern of stimulation continues, you remember the thought. When the reverberation ceases, so does the memory—unless it enters long-term memory.

Long-term memory probably involves some change in the structure or function of neurons that enhances synaptic transmission, perhaps by establishing certain patterns of synaptic connections. This idea was first suggested by Italian psychiatrist Eugenio Tanzi a century ago. Today it is widely accepted, because synaptic patterns fulfill two requirements of long-term memory. First, there are enough of them to encode an almost limitless number of memories—each of the 10 billion neurons in the cortex can make tens of thousands of synaptic connections to dendrites of other neurons. Second, a certain pattern of synapses can remain unchanged for years.

Indeed, structural changes such as an increase in the number of synapses and changes in the shape of neurons in rat brains have been noted following repeated electrical stimulation. If such stimulation is consistently associated with the animals' mastering a task, such as remembering how to run through a maze, learning based on memory may be possible. What do you think happens in your own brain cells as you read this chapter and attempt to retain the information in it?

One theory of how new facts become lasting memories is called **long-term synaptic potentiation.** It proposes that in an area of the cerebral cortex called the **hippocampus,** frequent and repeated stimulation of the same neurons strengthens their synaptic connections. This strengthening could be in the form of a greater electrical change in the action potentials triggered in postsynaptic cells in response to the repeated stimuli. For example, certain neurons are consistently stimulated when a 2-year-old first learns what an apple is. His senses tell his brain that the object is round, red, shiny, and hard, and that it smells and tastes sweet. With repetition, the stimulation of the same neurons communicating different aspects of "apple" become, somehow, associated into a single memory. Soon, just the smell of an apple or a drawing of one summons up a mental image—the memory—of the real thing.

There seem to be two types of long-term memory—one for skills or habits, for instance solving a jigsaw puzzle, and another for factual information. Skills and facts are encoded in different parts of the nervous system and appear to depend on different biochemical mechanisms.

Skill memories form in the cerebellum and in masses of nerve cell bodies within the cerebrum called the **basal ganglia.** The neural connections that enable us to remember how to perform a particular task seem to involve the synthesis of proteins. Inhibitors of protein synthesis tend to block this type of learning. For example, rats whose protein synthesis was blocked learned shock avoidance tasks more slowly than did rats in whom protein synthesis was not blocked. We do not yet know precisely how protein synthesis is involved in forming a memory.

Factual memory is encoded in the hippocampus and in a nearby area called the **amygdala.** One theory about how it is encoded points to calcium ions as regulators of the process. A rapid rate of action potentials opens calcium channels on postsynaptic membranes, and the influx of Ca^{2+} activates an enzyme that in turn alters proteins of the cytoskeleton. The altered proteins change the shape of the neuron, enabling it to make new, and specific, synaptic connections—this is the memory.

Like much of our knowledge about the human brain, we know about the role of the hippocampus and amygdala by observing the unusual behaviors and skills of experimental animals or people in whom these structures have been damaged (fig. 17.11). In 1953, a surgeon removed parts of the hippocampus and amygdala of a young man called H. M., thinking this drastic action might relieve his severe epilepsy. His seizures indeed became less frequent, but H. M. suffered a profound loss in the ability to consolidate short-term memories into long-term ones. As a result, events in H. M.'s life fade from memory as quickly as they occur. He is unable to recall any events that took place since surgery, living today as if it was the 1950s. He can read the same magazine article repeatedly with renewed interest each time. With practice, he improves skills that require procedural memory, such as puzzle solving. But, since factual memory is impossible, he insists that he has never seen the puzzle before.

KEY CONCEPTS

Short-term memories are thought to be temporary and electrical, whereas the long-term memories that they may be consolidated into are probably encoded as patterns of particular synaptic connections. Long-term memory of skills form in the cerebellum and basal ganglia and may involve protein synthesis. Factual long-term memories form in the hippocampus and amygdala, possibly by an influx of Ca^{2+}, which leads to a series of biochemical reactions that change the cell's shape, permitting new synaptic connections.

Protection of the Central Nervous System

The functioning of the brain and spinal cord are essential to health. The central nervous system is protected from environmental insults in several ways. First, the bones of the skull and vertebral column shield the delicate nervous tissue from bumps and blows. Additional protection comes from trilayered membranes called **meninges** that jacket the central nervous system. Furthermore, a **blood-brain barrier** is created by specialized brain capillaries. The tilelike cells that form these tiny blood vessels fit so tightly against one another that only certain substances can cross into the **cerebrospinal fluid** that bathes and cushions the brain and spinal cord. The fluid is made by cells that line spaces in the brain called **ventricles,** and it is also found in the hollow, innermost section of the spinal cord. Cerebrospinal fluid is similar in composition to plasma and tissue fluid but with fewer cells and proteins.

The Peripheral Nervous System

All nervous tissue other than the brain and spinal cord belongs to the peripheral nervous system. The PNS is subdivided into **sensory pathways,** which transmit impulses from a stimulus to the CNS, and **motor pathways,** which carry impulses from the CNS to effectors such as muscle or gland cells. The motor pathways are in turn subdivided into the somatic nervous system, which leads to skeletal muscles, and the autonomic nervous system, which goes to smooth muscle, cardiac muscle, and glands. Finally, the autonomic nervous system consists of the sympathetic division, which mobilizes the body to respond to environmental stimuli, and the parasympathetic system, which carries out more mundane functions such as respiration and heart rate at rest (fig. 17.12).

The Somatic Nervous System

The nerves of the **somatic nervous system** send impulses to the muscles, the sense organs, and the sensory receptors in the skin, resulting in sensations such as light, sound, pain, body position, and contracting voluntary muscles. A motor neuron of the somatic nervous system originates in the CNS and ends at an effector (muscle or gland).

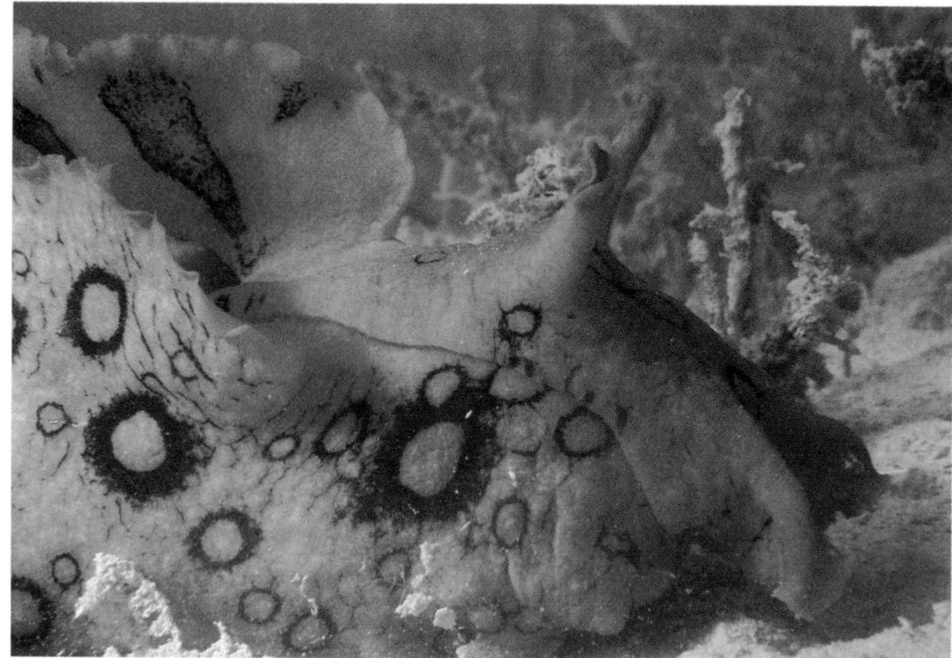

a.

Figure 17.11
Learning about learning from the sea snail *Aplysia californica*. *a.* It is almost impossible to study memory and learning in the human brain, with its billions of nerve cells. The sea snail *Aplysia* is an excellent model of learning however, because it has only 20,000 neurons, many of which are large and easily isolated. If *Aplysia* is struck on the head or tail, it withdraws its gill, which is a protective action. It can be trained to withdraw its gill even in response to a light tap. Particular sensory neurons, after this learning, release more neurotransmitter and cAMP—amplifying their messages. *b. (facing page)* A single neuron dissociated from the central nervous system of *Aplysia* shows many fibers sprouting from the cell body. This neuron had been living and growing in culture for 13 days when the photograph was taken. The projections would contact other neurons if they were present.

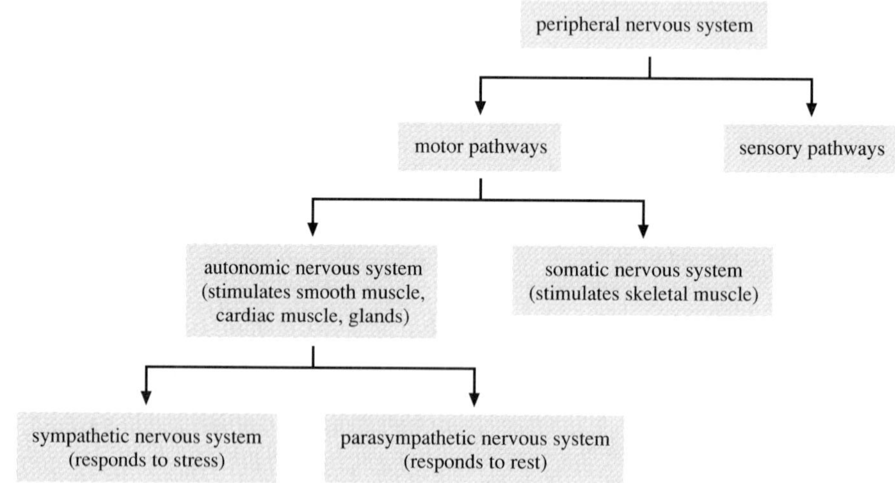

Figure 17.12
The peripheral nervous system consists of all nervous tissue that is not part of the central nervous system. Most generally, it can be divided into motor and sensory pathways. The motor pathways are further subdivided into the somatic nervous system, which innervates skeletal (voluntary) muscle, and the autonomic nervous system, which stimulates effectors in an involuntary matter. The autonomic nervous system is further subdivided into the sympathetic nervous system, which predominates in emergency situations, and the parasympathetic nervous system, which predominates in restful circumstances.

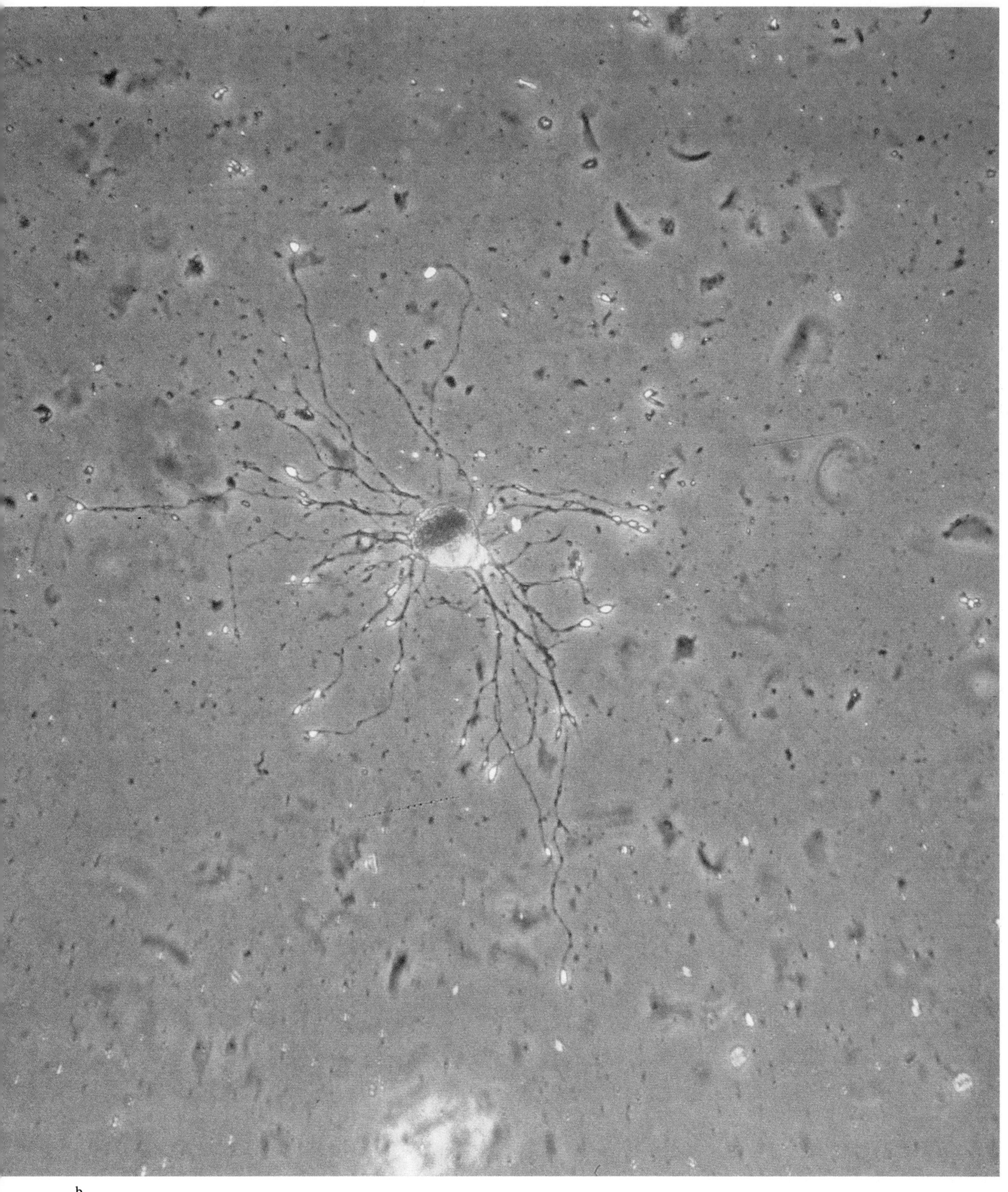

b.

Somatic nerves are classified by their site of origin. Twelve pairs of **cranial nerves** arise from the brain. Thirty-one pairs of **spinal nerves** exit the spinal cord, to emerge between the vertebrae. Eleven of the cranial nerve pairs innervate portions of the head or neck. The exception, the **vagus nerve,** leads to the internal organs. Each pair of spinal nerves innervates a section of the body near its point of departure from the spinal cord. Loss of feeling in a particular patch of skin sometimes indicates damage to the spinal nerves that lead to that area.

The Autonomic Nervous System

The **autonomic nervous system** is the "automatic pilot" that keeps internal organs functioning properly without conscious awareness. The nerves transmit impulses to smooth muscle, cardiac muscle, and glands.

The autonomic nervous system is subdivided into the **sympathetic nervous system** and **parasympathetic nervous system.** Under stressful situations the sympathetic division dominates, and during relaxing times, the parasympathetic system reigns. Although both the sympathetic and parasympathetic nervous systems often innervate the same organ, they usually have opposite effects on it (fig. 17.13).

The sympathetic nervous system prepares the body to face emergencies—accelerating heart rate and breathing rate; shunting blood to the places that need it most, such as the heart, brain, and the skeletal muscle necessary for "fight or flight"; and dilating the airways so gas exchange can take place more easily. All synapses in the sympathetic nervous system depend on the neurotransmitter noradrenalin.

The parasympathetic nervous system is active while the body is at rest. After an emergency, heart rate and respiration return to normal and digestion resumes. Different organs return to normalcy at different rates, because the arrangement of the parasympathetic nerves permits independent control of organs. Synapses in the parasympathetic nervous system use the neurotransmitter acetylcholine.

The opposing actions of the sympathetic and parasympathetic nervous systems are illustrated in the following scenario: Chatting with a friend, you leisurely stroll across the street to class. Without warning a truck zooms toward you. The sympathetic nervous system prepares the body for a "fight or flight" by increasing breathing rate and the volume of blood flowing to skeletal muscles. The heart pounds. Fleet feet carry you safely to curbside. Once safely in class, your parasympathetic nervous system becomes active. Breathing and heart rate slow, and much of the blood that had rushed to the muscles to prime them for action is rerouted to help digest lunch (fig. 17.14).

KEY CONCEPTS

The PNS consists of all nonCNS nervous tissue, and it has several subdivisions. The PNS has sensory pathways (stimulus to CNS) and motor pathways (CNS to muscles or glands). Motor pathways of the somatic system provide sensations and stimulate voluntary muscles. Motor pathways of the autonomic nervous system control involuntary life-sustaining impulses to smooth and cardiac muscle and glands. Within the autonomic nervous system, parasympathetic nerves call forth physical responses to threatening situations, and sympathetic nerves maintain normal functioning in nonthreatening situations.

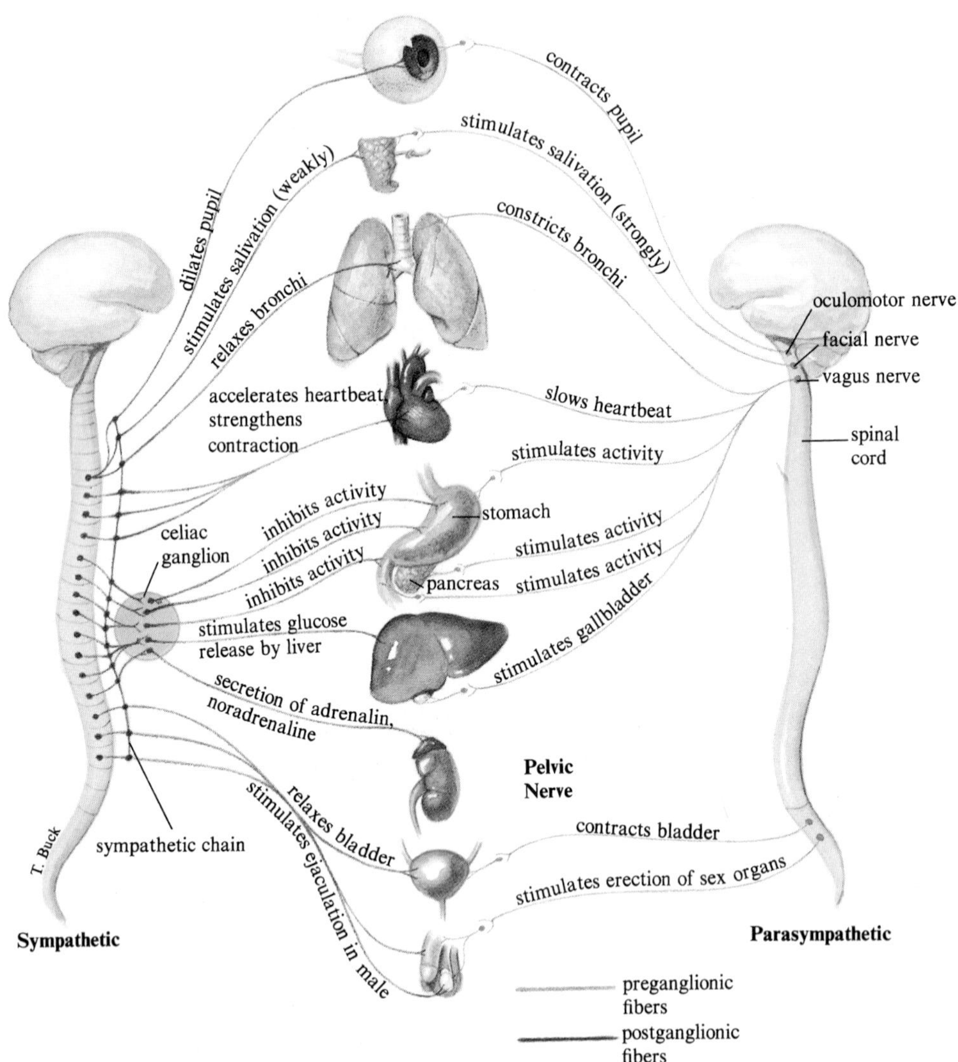

Figure 17.13

The autonomic nervous system. The internal organs are innervated by neurons from the parasympathetic and sympathetic divisions of the autonomic nervous system. Note that in the parasympathetic division, the nerves emerging from the central nervous system are long and those near the organ are short. The reverse is true in the sympathetic nervous system. Both divisions have opposite effects on the same organs.

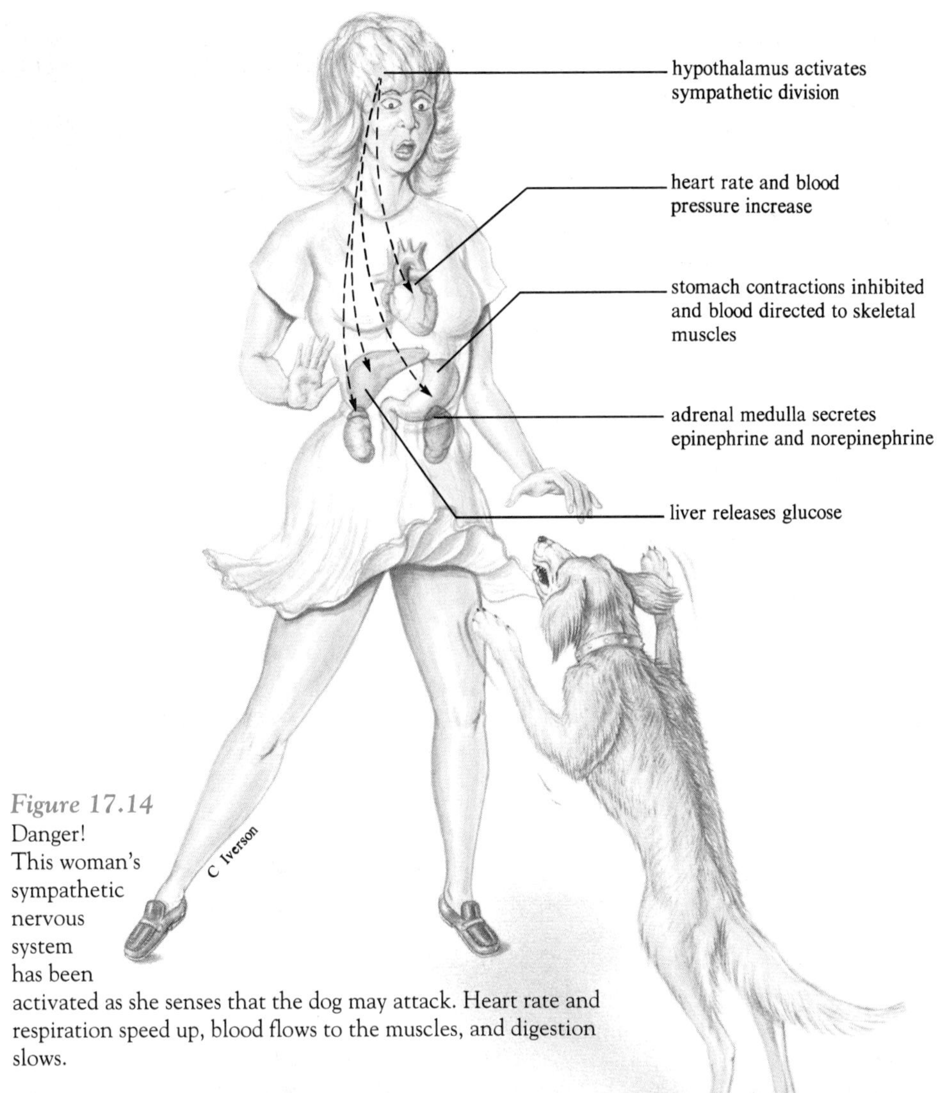

Figure 17.14
Danger! This woman's sympathetic nervous system has been activated as she senses that the dog may attack. Heart rate and respiration speed up, blood flows to the muscles, and digestion slows.

When Nervous Tissue Is Damaged

From all of the functions attributed to the nervous system in this chapter, it is clear that injury to any of the structures could be devastating to health (table 17.1). Many nervous system problems are particularly difficult to treat because, unlike other specialized tissue, mature neurons usually cannot divide to repair damaged tissue. How, then, can the nervous system compensate for damage?

Neurons of the PNS can actually regenerate. Schwann cells at the tip of a damaged peripheral nerve cell grow out beyond the underlying neuron, providing a pathway and physical support for new growth. If a peripheral nerve is crushed, it can grow back, making the same synaptic connections as it originally did.

In the mammalian central nervous system, however, heavily myelinated and damaged neurons cannot regenerate at all. Damaged unmyelinated neurons can sprout new axons, but these do not extend as far as the old ones. Unlike the crushed peripheral nerve, the damaged spinal nerve is replaced not with correctly connected neurons but with a scar consisting of glial cells, connective tissue, and random axons.

Table 17.1
Common Nervous System Problems

Problem	*Symptoms*
Amnesia	Disorientation and memory loss
Cerebral palsy	Damage to motor cortex, basal ganglia, and/or cerebellum, either before birth, during, or shortly after; paralysis and absence of muscular control
Cerebrovascular accident	A blood clot in an artery of the brain (stroke), hemorrhage in the brain, hardening or plaque in cerebral arteries, or aneurysm (ballooning out of artery) in brain
Coma	A state of unconsciousness from which a person cannot be aroused
Concussion	Brief unconsciousness followed by confusion, caused by sudden movement of brain due to violent blow to head
Dyslexia	A disorder in language skills in which a person cannot easily recognize sequences of letters as meaningful words
Epilepsy	Seizures caused by electrical disturbances in the brain
Fainting	A brief loss of consciousness caused by blow to head, standing rapidly, or viewing something shocking
Headache	Often caused by dilation of blood vessels in meninges (membranes surrounding brain)
Meningitis	Viral or bacterial infection of meninges around brain or spinal cord
Polio	Viral infection of nerve cell bodies in anterior horn of spinal cord; symptoms progress from fever, headache, and pain to loss of reflexes and muscle paralysis and atrophy
Schizophrenia	Withdrawal into a fantasy world; person makes inappropriate emotional responses to stimuli

Even though neurons in an adult mammal's CNS cannot regenerate appreciably after injury, plasticity among surviving neurons can help to maintain the functions of the system. When a neuron dies, neighboring undamaged neurons extend new terminals to the cells that previously synapsed with the destroyed cell, like office workers assuming the duties of a departed co-worker.

Neuronal Cell Culture

Researchers can now grow human brain cells in laboratory culture. In the past, efforts to view interactions of neurons outside the body have failed, because these cells could not be coaxed to divide. Using rapidly dividing brain cancer cells was inaccurate, because these neurons are highly abnormal. However, brain cells from an 18-month-old girl with a rare disease that makes them divide rapidly, but that appear normal, yielded the first neuronal cell culture in 1990, and others have since been started using other approaches. Nerve cell cultures have already revealed how some embryonic cells "decide" to specialize as particular neurons, and how some fetal neurons form synaptic connections and five action potentials in unison to establish pathways of neural communications. Neuronal cultures offer a safer and more accurate way to test new drugs and might someday provide material for transplants.

Despite its plasticity, the brain, like other organs, loses efficiency over time. In the aging brain, some synaptic connections deteriorate, some postsynaptic receptors become less sensitive to their particular neurotransmitters, and levels of some neurotransmitters fall. By age 60, the brain has decreased in weight by about 3 ounces (85 grams). Although memory may be less reliable and learning may be slowed, a person can continue to be mentally alert and capable of reasoning and enjoying the sensations, feelings, and perceptions made possible by the nervous system for many decades.

SUMMARY

The vertebrate nervous system can be divided into the *central nervous system*, consisting of the brain and spinal cord, and the *peripheral nervous system*, consisting of all other neurons.

The *spinal cord* is a tube of neural tissue within the vertebral canal. The white matter on the periphery of the cord conducts impulses to and from the brain. The spinal cord also serves as a reflex center. A *reflex* is a quick, automatic, protective response that depends on a *reflex arc*. The reflex arc usually consists of a sensory neuron, a spinal interneuron, and a motor neuron.

The three largest regions of the human brain are the *brain stem*, *cerebellum*, and *cerebrum*. The brain stem includes the *medulla*, *pons*, and *midbrain*. The medulla controls many vital life functions such as breathing, heart rate, blood pressure, urination, and defecation. The pons forms a bridge between the medulla and higher brain regions. In addition to serving as a means of communication between lower and higher brain regions, the midbrain processes visual and auditory sensory information. The cerebellum is important in sensory motor coordination.

The forebrain consists of the *thalamus*, a relay station between lower and higher brain regions; the *hypothalamus*, which regulates many vital physiological processes, is involved in many emotional states, and regulates the level of some pituitary hormones; the *reticular activating system*, which filters sensory input and is important for arousal; and the cerebrum, which integrates sensory input and directs motor responses, intelligence, learning, and many emotions.

The cerebrum has two layers. The lower layer of white matter is covered by a convoluted layer of gray matter. The cerebrum is also divided into two hemispheres. The left cerebral hemisphere receives sensory input from and directs the motor responses of the right side of the body, and vice versa. The hemispheres are specialized for different functions. Whereas the left hemisphere is usually specialized for language and analytical reasoning, the right hemisphere is specialized for abilities involving comprehension of spatial relationships.

Short-term memory is thought to depend on activity within neuronal circuits. Short-term memories are later consolidated into *long-term memories*, which depend on chemical or structural changes in the brain.

The central nervous system is protected by the bones of the skull and vertebrae, the *cerebrospinal fluid*, the *blood-brain barrier*, and the *meninges*.

The peripheral nervous system is composed of the *somatic nervous system* and the *autonomic nervous system*. The *spinal* and *cranial nerves* comprise the somatic nervous system. The autonomic nervous system has two branches, the *sympathetic nervous system*, which prepares the body to face an emergency, and the *parasympathetic nervous system*, which dominates during restful periods. Neurons of the PNS can regenerate, but those in the CNS cannot. Undamaged neurons can sometimes take over functions of damaged neurons. Neurons can be grown in culture.

QUESTIONS

1. List 10 skills you use in everyday life that depend upon some part of your nervous system. Where possible, indicate what part of the nervous system provides a particular skill.

2. The human nervous system is incredibly complex. Suggest three ways discussed in the chapter in which researchers can learn about the functions of our nervous system.

3. What functions might be lost if the spinal cord is severed in an accident?

4. Does the spinal cord always act in concert with the brain? Cite evidence for your answer.

5. In a condition called anencephaly, a fetus fails to develop "higher" brain structures. Often such a fetus survives long enough to be born, but then only lives a day or two. It has a face and a brain stem but nothing else within the skull. What few functions does such a newborn have? What nervous system functions does it lack? Why is it possible to use some of an anencephalic baby's organs for transplantation?

6. What role does the cerebellum play in athletics?

7. What symptoms might arise from a tumor growing in the hypothalamus?

8. A person who has narcolepsy (uncontrollably falling asleep) probably has a defect in what part of the brain?

9. In what part of the brain do the qualities of "mind" lie?

10. How does the structure of the cerebral cortex provide for different degrees of sensitivity in different body parts?

11. A person suffering from amnesia can recall very recent events but cannot remember long-ago events. A psychiatric patient treated with electroconvulsive therapy can recall long-ago events but not events surrounding the treatment, even if that treatment was performed just days ago. What do these observations indicate about our capacity for memory?

12. Suggest the events that might take place as a person learns to recognize a particular person. How might this process be disturbed in H. M., the epilepsy patient who had parts of his amygdala and hippocampus removed?

13. You are walking home from an evening class when you hear footsteps approaching from behind and heavy breathing, and then you see a tall shadow behind you. What is your nervous system likely to do?

TO THINK ABOUT

1. Studying mitosis and meiosis in stages enabled us to understand these events but does not really reflect the continuous nature of cell division. Do you think that the subdivisions of the human nervous system are similarly devices to help us understand its functioning, or do they more closely reflect the relationship between structure and function in an organ system?

2. If a student gets only 3 hours of sleep for several days because of studying, how might her sleeping pattern differ from normal on her next full night of sleep?

3. All human brains are about the same size, contain the same major structures, and function in the same ways. How, then, does each of us develop a distinct personality?

4. Smoking marijuana kills some brain neurons, which cannot regenerate. Why do individuals apparently not suffer noticeable brain damage from smoking marijuana?

5. Now that you know a little about the physical basis of memory, can you think of some ways to make learning biology a little easier?

6. A controversial new use of PET scans is to detect subtle changes in brain chemistry that precede the onset of symptoms for such conditions as Huntington disease and schizophrenia. What do you think the advantages and disadvantages of such a use might be?

7. Fossilized skulls of humans and our immediate ancestors reveal that the brain has increased in capacity over 2 million years from 440 cubic centimeters to the present-day range of 1,350 to 1,400 cubic centimeters. How might you test the hypothesis, "brain size increases as intellectual skills grow"?

SUGGESTED READINGS

Angier, Natalie. May 1990. Storming the wall. *Discover*. The blood brain barrier is so effective that even helpful drugs sometimes cannot traverse it.

Bower, Bruce, November 17, 1990. Gone but not forgotten. *Science News*, vol. 138. Unconscious thought affect memory and recall.

Bower, Bruce. November 26, 1988. The brain in the machine. *Science News*, vol. 134. Can computer neural networks simulate—and maybe teach us about—the human brain?

Brown, Thomas H., et al. November 4, 1988. Long-term synaptic potentiation. *Science*, vol. 242, p. 724. Researchers are trying to decipher the synaptic connections that encode memories.

Lewis, Thomas A. August/September 1990. Squid: The great communicator. *National Wildlife*. Squid have a complex nervous system—perhaps even language.

Montgomery, Geoffrey. March 1989. The mind in motion. *Discover*. Imaging techniques such as PET are providing peeks at the functioning of the normal and diseased brain.

Montgomery, Geoffrey. December 1989. Molecules of memory. *Discover*. Memory has a chemical basis.

Ornstein, Robert, and David Sobel. 1988. *The healing brain*. New York: Simon & Schuster. What role does the brain play in maintaining health?

Sacks, Oliver. 1985. *The man who mistook his wife for a hat*. New York: Harper & Row. Fascinating tales of human brain malfunctions.

Small, Meredith F. March/April 1990. Political animal. *The Sciences*. How do our brains enable us to function in complex societies?

Weiss, R. May 5, 1990. Human brain neurons grown in culture. *Science News*. A long-sought goal has been achieved—growing human nerve cells in the laboratory.

C H A P T E R

18

The Senses

Chapter Outline

Learning Objectives

By the chapter's end, you should be able to answer these questions:

1. What is the difference between sensation and perception?
2. What components of the nervous system participate in sensory perception?
3. How can all of the senses be based on transmission of action potentials yet provide different perceptions?
4. How do we perceive different intensities of stimuli?
5. How do receptor cells in the different sense organs transmit information to the brain, which enables us to sense and perceive the environment?

Every animal inhabits a world of its own, a world whose character is determined by the animal's sense organs. The sensory abilities of some animals are vastly different, and sometimes amazingly more acute, than ours. If you and your dog were to walk through the New England woods in the fall, your sensations would be primarily visual. You would note the leaves aglow in hues of red, orange, and yellow contrasted against the brilliant blue sky. Unimpressed by the colors in nature's paintbox, your dog would more likely be overwhelmed by odors that paint a picture missed by your eyes—fox urine left as a calling card near an oak and the scent in a shrub revealing the site of a deer's bed.

If this is an evening stroll, you may encounter an animal with yet another view of the scene, the little brown bat. The bat's *eye* view of the world is not an eye view at all, but an *ear* view. Bats emit high-frequency pulses of sound and analyze the resulting echo to form a sound picture of the world around them. We cannot experience the bat's world because our ears cannot hear such high-frequency sound. Although a bat emits sound pulses that are about as loud as a passing train, to our ears these are silent screams.

Why do animals have such different sensory worlds? Each species is sensitive to the range of stimuli most critical to its survival. Cells in a frog's eyes respond selectively to small moving objects, thereby helping to locate dinner. Male and female butterflies that look alike to us appear quite different to the butterflies (fig. 18.1). The ears of female cricket frogs respond to the frequencies in the croaks of males of the species but are deaf to other frequencies.

a.

b.

Figure 18.1

To us, the male and female "dogface" butterflies *Colias cesonia* look alike, both yellow and black (*a*). The insects, however, can see in the ultraviolet, and to them the female appears dark blue, and the male has reflectant patches on his wings (*b*). This color difference is important to sex recognition in courtship behavior.

General Principles of Sensory Reception

Specialized neurons called sensory receptors detect certain environmental stimuli and pass the information on to sensory neurons, which in turn deliver nerve impulses to the brain. Sensory receptors are sometimes gathered into sense organs, such as the eye or ear. Once the brain receives sensory input, it integrates the information, perhaps even interpreting it by consulting memories, to form perceptions, which are the individual's particular view of environmental phenomena. The sensory receptors, then, are portals through which our nervous systems experience the world.

Sight is different than sound, and yet the action potentials on the sensory neurons are all alike. This is because sensory receptors are selective. They absorb a particular type of energy, that of light for example, and convert it into the electrochemical energy of an action potential. The action potentials generated by a specific type of receptor are carried to the brain over a specific sensory pathway. Different qualities of sensation such as light, sound, or pressure are the brain's interpretation of the input from different pathways.

In addition to information on the type of stimulus present, it is often helpful to know the strength of the stimulus. One way that the intensity of the stimulus is signaled to the brain is in the rate of nerve impulses. Sensory receptors respond to environmental change with a **receptor potential.** A receptor potential is a change in membrane potential caused by the redistribution of ions. In this respect it is similar to a nerve impulse. However, unlike an action potential, the magnitude of a receptor potential varies with the strength of the stimulus. For example, louder sound may cause greater depolarization resulting in a larger receptor potential.

The change in the receptor potential in turn influences the likelihood that an action potential will be generated. In some receptors, the strength and duration of the receptor potential determine the rate at which action potentials are generated and how long they last. The brain interprets this as an increase in stimulus intensity. A second way information about stimulus strength is sent to the brain is by variations in the number of sensory neurons carrying the message. As the stimulus becomes stronger, more receptors are stimulated, and these activate more sensory neurons.

Many sensory receptors detect changes in input. Once a sensation is noted, if it remains constant, it becomes less noticeable. This mechanism, called **sensory adaptation,** keeps the nervous system from becoming too sensitive. The strong smell of a fish market, for example, may be overpowering to anyone who has just entered, but to the people working there, it is hardly noticeable. Without sensory adaptation, you would be distinctly aware of the touch of your clothing and every sight and sound in your presence, almost to the point of pain. Concentrating on a single stimulus, such as a person talking to you, would be difficult.

Humans have multiple receptors, making ours a very rich reality indeed. A person's skin, for example, contains 4,000,000 pain receptors, 500,000 pressure receptors, 150,000 receptors specific for cold, and 16,000 specific for heat. Specialized receptors in the joints, tendons, ligaments, and muscles provide information that, when combined with information from sense organs such as the eyes and ears, gives a feeling of where certain body parts are in relation to the rest of the body. Although the five most familiar senses are vision, hearing, taste, smell, and touch, experiencing the world around us blends these into a far larger number of sensations.

Reading 18.1 *Mixed-up Senses—Synesthesia*

"His name was purple."
"The song was full of shimmering green triangles."
"The paint smelled red."
"The sunset was salty."

TO 1 IN 500,000 OF US, SENSES BECOME UNUSUALLY ASSOCIATED IN A LITTLE-UNDERSTOOD CONDITION CALLED SYNESTHESIA. A sensation involving one sense is perceived in terms of another, so that, most commonly, visions take on characteristic smells or sounds are associated with particular colors. Although synesthetic people are unlikely to share the same weird associations, they do share some aspects of their talents—their particular quirks are involuntary, remain constant over their lifetimes (a certain name or song is always a certain color), and they are absolutely convinced that what they perceive is reality, not something in their "mind's eye."

We do not know what causes synesthesia, although it does seem to be inherited. The condition has been sheepishly reported to psychologists and physicians for at least 200 years. Various theories (all unproven) have attributed the condition to an immature nervous system that cannot sort out sensory stimuli; altered brain circuitry; and simply an exaggerated use of metaphor—taking such descriptions as a sharp flavor, warm color, and sweet person too literally.

In 1980, measurements of cerebral blood flow linked synesthesia to a brain abnormality. A person who tasted in geometric shapes inhaled a harmless radioactive chemical, xenon-133, which was absorbed by her tissues. The rate at which the chemical left certain parts of her brain indicated how metabolically active that part was. When brain activity was so assessed during a synesthetic experience, it was found that blood flow in her left hemisphere, particularly in the temporal lobe, plummeted 18%—a decrease seen only when tissue dies as a result of a stroke. But the woman was perfectly healthy. The left hemisphere is the site of the language center. Another clue is that people with temporal lobe epilepsy are often synesthetic. Researchers believe that synesthesia reflects a breakdown in the translation of a perception into language—but we have much to learn about this fascinating mixing of the senses.

KEY CONCEPTS

Animals perceive the world in species-specific ways because of differing sensory abilities. Sensory receptors detect stimuli in the form of energy of specific wavelengths and reach a receptor potential. This causes an action potential in sensory neurons, which transmits the information to the brain, which in turn integrates and interprets the sensations to form perceptions.

Chemoreception

A flavor combines sensations of smell and taste, plus texture, temperature, pain, touch, and even appearance. Taste is intimately tied to smell, as anyone whose tasting ability has been dulled by a stuffy nose can attest. For the 3 million people in the United States who cannot smell normally, eating is a tasteless experience. This is because about 75% to 80% of flavor comes from smell (olfaction).

Smell

Smell is the detection of certain molecules by specialized **olfactory receptor cells** in a patch of tissue approximately the size of a quarter located high in the nasal cavity (fig. 18.2). To be smelled, a substance must combine with receptor sites on the cilia of the receptor cells in this patch of tissue, which is called **olfactory epithelium.** The location of the receptors makes it necessary that the chemical be a gas or consist of small particles suspended in the air passed over the olfactory epithelium during breathing. When you are particularly interested in an odor, you assist the delivery of these particles by sniffing. To stimulate the receptor cells, a substance must also dissolve in the mucous layer covering the cilia. Once in the nose, molecules are probably picked up by a molecule that ferries them to the receptors. Such an *odorant binding protein* has been identified in rats and probably operates in humans too.

Part of each olfactory receptor cell passes through the skull and synapses with neurons in the **olfactory bulb** of the brain, which relays the message to other regions of the brain. In the **cerebral cortex,** the message is interpreted as an odor and identified. Exactly how this occurs is still a mystery. How, for example, do the myriad of molecules stimulating olfactory receptors become identified as chicken soup?

One theory to explain how a limited number of receptors can detect a nearly limitless number of odors is that each odor stimulates a distinct subset of receptor subtypes. The brain then recognizes the combination of receptors as an *olfactory code*. For example, perhaps there are 10 types of odor receptors. Banana might stimulate receptors 2, 4, and 7; garlic receptors 1, 5, and 9.

Sensory information from olfactory receptors is also conducted to the limbic system, a brain center for memory and emotions. This is why we may become nostalgic over a scent from the past. A

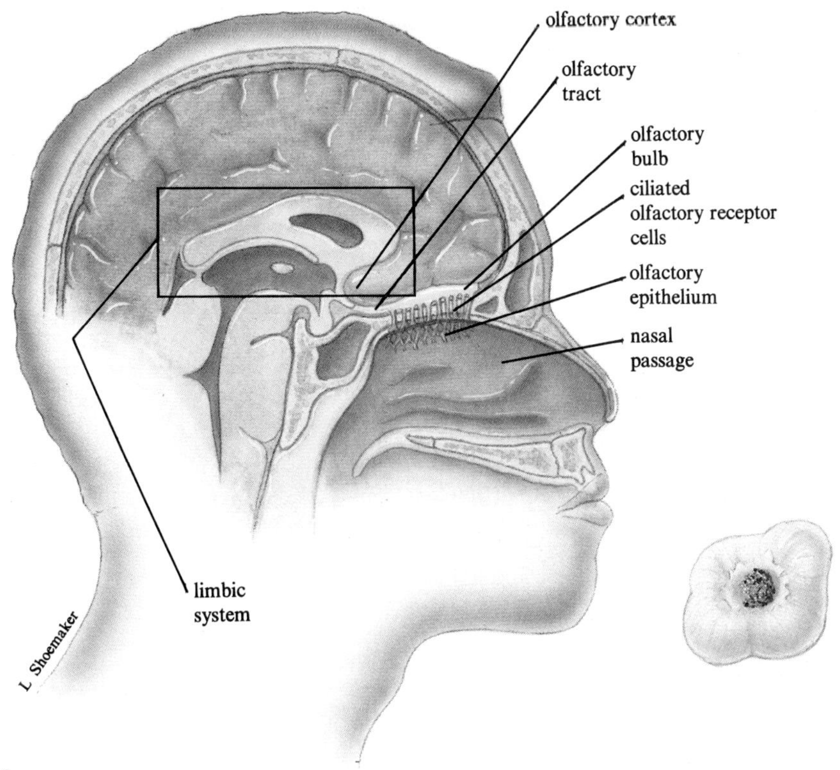

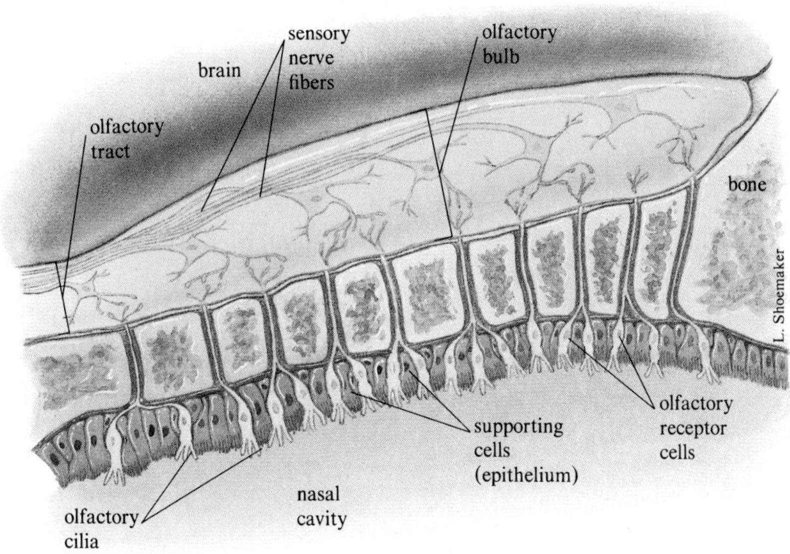

Figure 18.2
How smells signal the brain. *a.* The powerful sense of smell derives from an inch-square section of sensitive olfactory receptor cells in each nostril. *b.* An olfactory receptor cell physically binds a particular odoriferous molecule at its ciliated end and transmits an impulse to fibers of the olfactory nerve, which leads to the brain, where the sensation is interpreted as a particular smell. Some impulses go to the limbic system, the brain's seat of memory and emotion. This is why a particular food may evoke powerful feelings.

whiff of the perfume that grandma used to wear may bring back a flood of memories. The input to the limbic system also explains why odors can alter mood so easily. For example, the scent of new-mown hay or rain on a summer's morning generally makes us feel good.

Although the sense of smell in other animals may be much more acute than ours, the human sense of smell is better than most people think it is. We have about 12 million receptor cells, each with 10 to 20 cilia, which increase the surface area for receiving smells. We can detect a single molecule of the chemical that gives a green pepper its odor in 3 trillion molecules of air! Yet a bloodhound has 4 billion receptor cells and is used in forensics to "sniff" out individuals.

Taste

Chemicals are also detected by **taste receptors,** which are generally found in the parts of the body that contact food. A lobster not only tastes good, it tastes well, using its walking legs and antennules to detect food sources. Insects also taste with their legs. A blowfly, for instance, will land on its food and taste it while walking around. Before placing a morsel in its mouth, the octopus samples food with taste receptors on its tentacles. In humans, taste is centered primarily in the mouth. Although taste buds are lightly scattered around all the places that food and drink are likely to touch, they are concentrated on the tongue in the grooves around certain of the bumps, or papillae, on the tongue's upper surface (fig. 18.3). Each of the 10,000 taste buds contains 60 to 100 receptor cells, which transmit impulses to the brain when molecules bind to them. The receptor cells live for only a week to 10 days and are then replaced by new ones.

For many years it has been claimed that there are four primary taste sensations—sweet, sour, salty, and bitter. More recently it has been argued that there are many more tastes than just these four. According to this view, our sensation of taste depends upon the pattern of activity across all the taste neurons. We do know that most taste buds are sensitive to all four of the "primary" tastes, but each responds most strongly to one or two of these. The

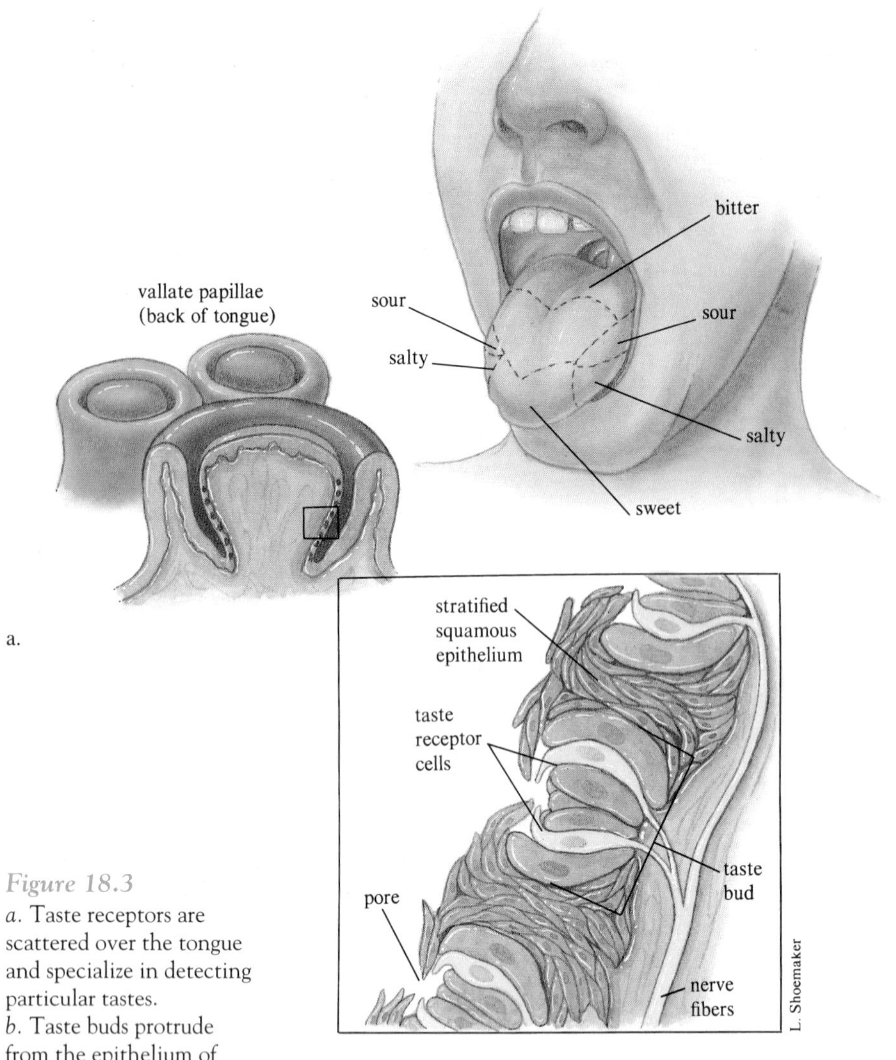

Figure 18.3
a. Taste receptors are scattered over the tongue and specialize in detecting particular tastes.
b. Taste buds protrude from the epithelium of the tongue.

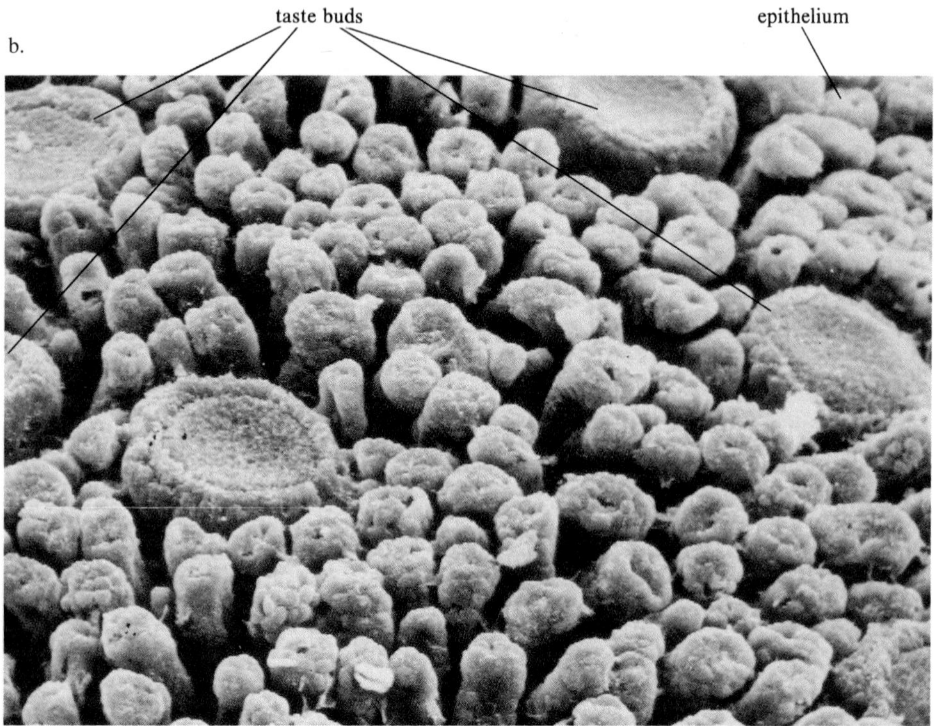

receptors that respond most strongly to sugars are concentrated on the tip of the tongue. (Perhaps this explains why we usually lick a lollipop rather than chew it.) Sour receptors are most common along the sides of the tongue. The receptors along the rim of the tongue respond most strongly to salts. At the back of the tongue are the receptors that are most sensitive to bitter substances.

On the basis of information from the taste receptors, we decide whether to swallow substances in the mouth. Sweet tastes are generally perceived as pleasant and are quickly consumed. Bitter tastes, on the other hand, are often refused. Rejection of bitter substances can be lifesaving. For example, the poisonous alkaloids in wild herbs often taste bitter. Figure 18.4 shows how taste and smell contribute to our enjoyment of food.

KEY CONCEPTS

Chemoreception includes the senses of smell and taste. A substance that we smell binds to receptor sites on cilia of receptor cells in olfactory epithelium. These cells stimulate neurons in the olfactory bulb of the brain, which relays the message to other brain areas that interpret the messages as smells. Taste receptors in the tongue detect specific types of tastes and pass the information to the brain.

Photoreception

Can you see the light? This feat is only possible for animals with **photoreceptors.** A photoreceptor contains a pigment that is associated with a membrane. When the pigment absorbs light, its structure is altered in a way that changes the charge across the membrane. In higher organisms, these membrane changes are eventually converted to action potentials.

The compound eye of insects and crustaceans does form an image, but it may provide a picture of the world that is quite different from ours. The compound eye consists of many visual units called **ommatidia** (fig. 18.5). Because each ommatidium contains a lens that transmits light to its own photoreceptor cells or those that are nearby, each receives its own tiny view of the world. The information from individual

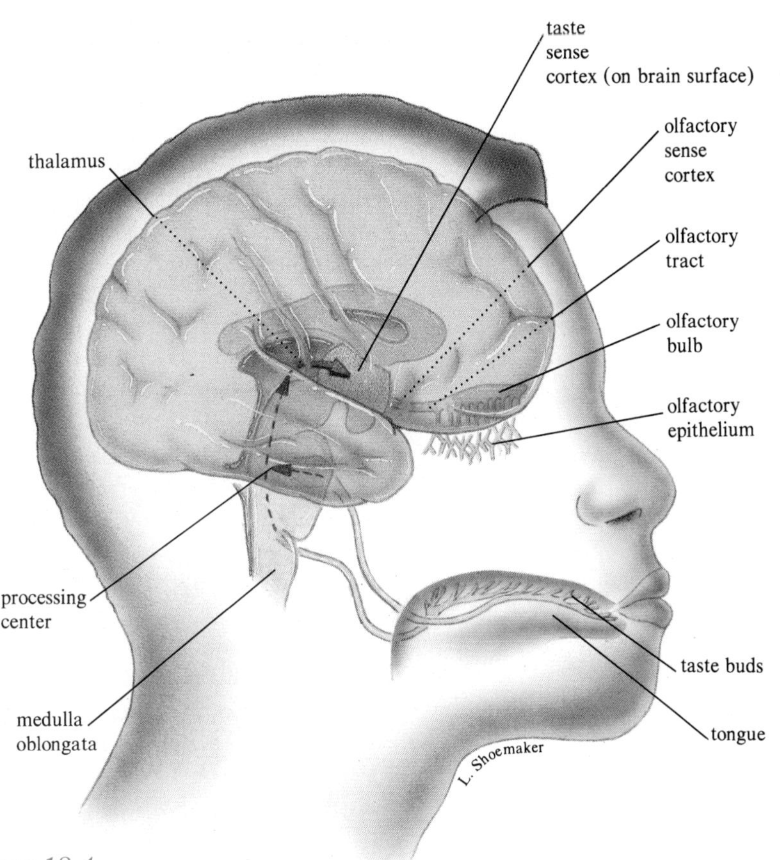

Figure 18.4
Neural connections for taste and smell.

Figure 18.5
The invertebrate eye. The compound eyes of the southern aeshna dragonfly have more ommatidia than those of any other insect—about 30,000 each.

ommatidia is integrated by the nervous system. Although it was once believed that all images from compound eyes were coarse mosaics, many vision scientists now suspect that, in those animals whose compound eyes have many ommatidia, images may be as clear as ours.

The Human Visual System

Our eyes, like those of all vertebrates, work the way a camera does. Both a camera and the eye have a **lens** system in the front that focuses incoming light into an image on a surface specialized for photoreceptors (light detection). This surface in a camera is the film; in the eye it is a sheet of photoreceptors called the **retina.** Various other structures support and nourish the photoreceptive area.

The human eyeball is a fluid-filled sphere built of three distinct layers. The white of the eye is the outermost layer, the **sclera,** which protects the inner structures of the eye (fig. 18.6). Towards the front of the eye, the sclera is modified into the **cornea,** which is a transparent curved win-

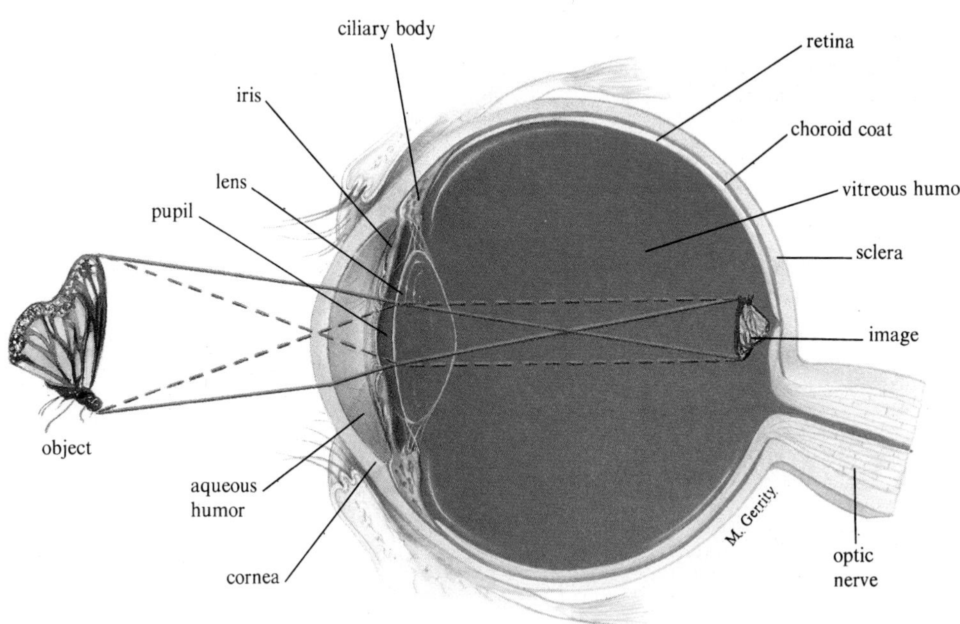

Figure 18.6
Parts of the eye. Note that the image on the retina is upside down. The brain enables us to see the image right side up.

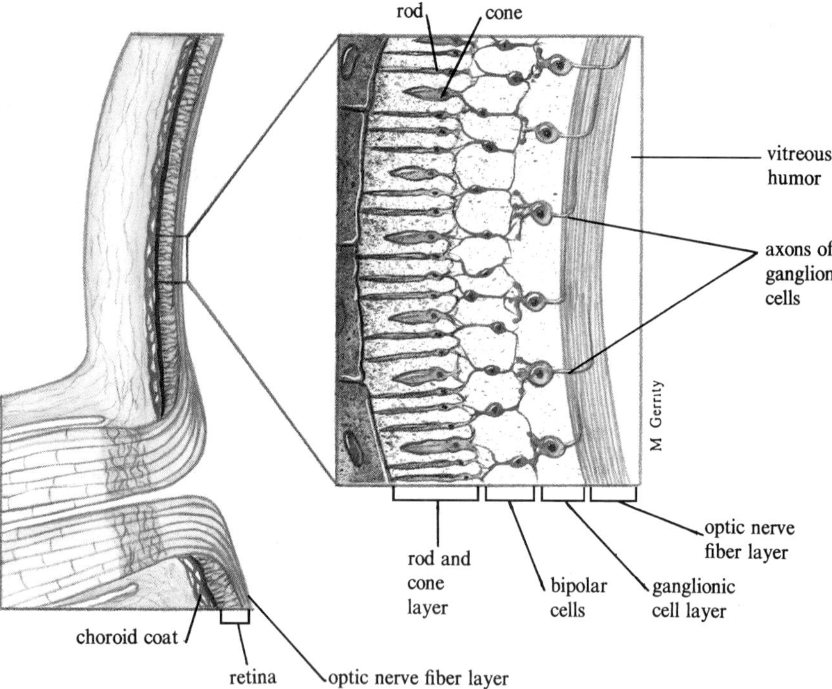

Figure 18.7
The three layers of the retina. The outermost layer of the retina contains the rods and cones, which are the photoreceptor cells providing black-and-white and color vision, respectively. The rods and cones synapse onto bipolar neurons, which form the middle layer of the retina. The bipolar neurons then synapse with neurons in the ganglionic cell layer, whose axons form the optic nerve.

dow through which light enters. This curve bends incoming light rays, helping to focus them on the photoreceptor cells.

The middle layer of the eyeball, the **choroid coat,** is rich in blood vessels that nourish the eye. Because the choroid coat contains a dark-colored pigment, it absorbs light and prevents it from reflecting off the retina, much like the black paint inside a camera. Like the sclera, the choroid coat is specialized into different structures at the front of the eyeball. The choroid thickens into a highly folded structure called the **ciliary body,** which houses the **ciliary muscle,** whose function is to alter the shape of the lens to adjust the focus of the image during near and far vision. Adjacent to the ciliary body, the choroid becomes the thin opaque **iris,** which is responsible for eye color. Like the diaphragm in a camera, the iris regulates the amount of light entering the eye. An opening in the iris, the **pupil,** admits light. In bright light, the pupil constricts, shielding the retina from excessive stimulation. In dim light, the pupil

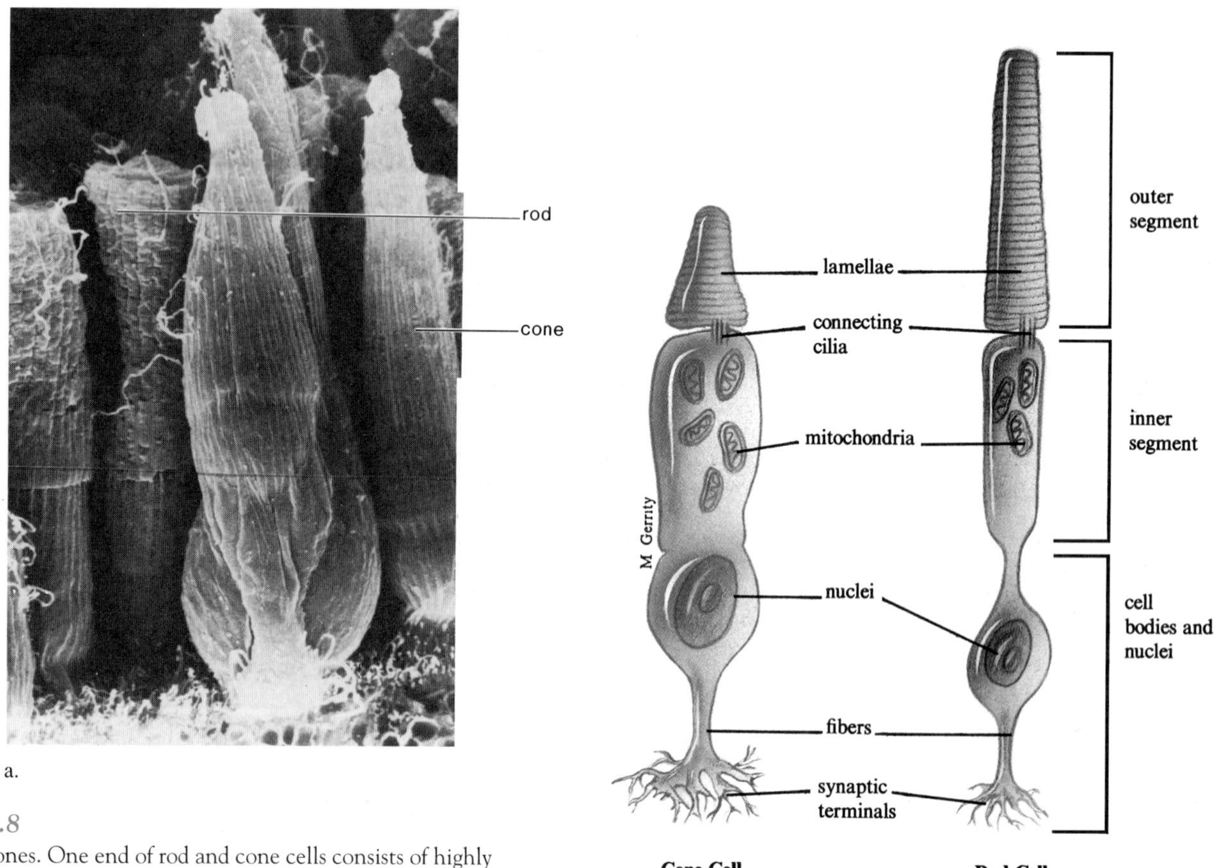

Figure 18.8
Rods and cones. One end of rod and cone cells consists of highly folded membrane, which contains rhodopsin. The other end of the cells contains the nuclei and organelles.

opens up, letting more light strike the retina. Behind the iris, the lens focuses the light into an image at the back of the eyeball onto the retina.

The third and innermost layer of the eyeball, the **retina,** is itself built of three cell layers (fig. 18.7). The outermost layer contains the photoreceptor cells, called **rods** and **cones,** named for their shapes (fig. 18.8). The human eye contains about 125 million rod cells and 7 million cone cells. Rod cells are concentrated around the edges of the retina, and they provide black-and-white vision in dim light. The rods enable us to see at night. Cone cells detect color and are concentrated more centrally in the retina. An indentation directly opposite the lens, called the **fovea centralis,** contains only cones. The number of cones here is a direct measure of the acuity of an animal's sight. Humans have about 150,000 cones per square millimeter, whereas some predatory birds have a million. Readings 18.2 and 18.3 explore visual diversity further.

The distribution of rods and cones explains some of our visual experiences. To see detail in bright light, it is best to look directly at the object because this focuses light on the central area dense with cones.

Reading 18.2 A Shrimp's Eye View

AN ORGANISM WITH VERY UNUSUAL VISUAL ORGANS IS THE SHRIMP *RIMICARIS EXOCULATA* (SHRIMP WITHOUT EYES) DISCOVERED IN 1985 LIVING AROUND DEEP SEA THERMAL VENTS. They lack the stalklike eyes of other shrimp but have a pair of curious light-reflecting patches on their backs (fig. 1). When graduate student Cindy VanDover dissected the patches, she found that they were attached by nerves to the animal's brain. Could they be eyes? She called in other researchers to see.

Steven Chamberlain from Syracuse University compared the anatomy of the structures of various invertebrates' eyes but could not determine whether it was a visual organ or a gland. Next, Ete Szuts at the Marine Biological Laboratory in Woods Hole mashed up several of the "eyes" and indeed found that they were packed with rhodopsin, the pigment molecule that evokes nerve impulses when it is split by light (fig. 2).

Excited by the discovery of rhodopsin, anatomist Chamberlain examined the structures more extensively and found their surfaces to be covered with about 1,500 clusters of photoreceptor cells each. Lacking lenses, the "eyes" cannot form images, but the large amount of rhodopsin allows them to sense the direction, strength, and size of a light source.

But in a murky, deep sea thermal vent some 2 miles (3.5 kilometers) beneath the mid-Atlantic Ridge, just what are these animals seeing? We do not yet know.

Figure 1
The bright spots on these shrimp are apparently eyes—but what do they see 2 miles beneath the ocean's surface?

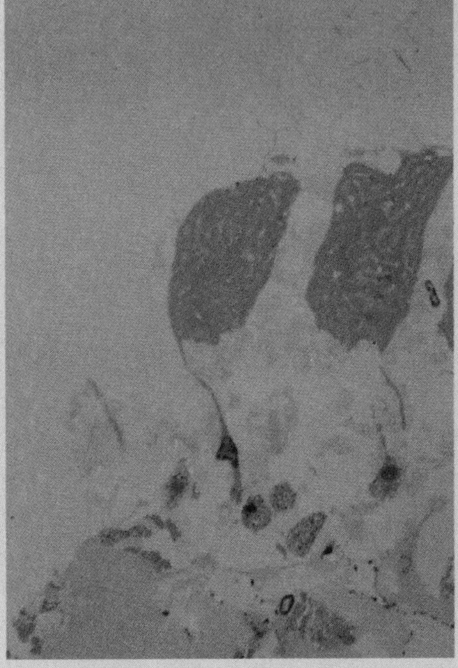

Figure 2
The lower portion of this photoreceptor cell contains the nucleus and other organelles. The huge upper portion is packed with rhodopsin. Magnification, x25,000.

However, at night an object is seen more clearly from the corner of the eye. From this perspective, the object will stimulate the region of the eye rich with rods. Only rods are sensitive enough to be stimulated in dim light.

The rods and cones synapse with bipolar neurons, which form the middle layer of the retina. The bipolar neurons synapse upon cells of the third retinal layer, consisting of **ganglion cells.** The fibers of the ganglion cells form the **optic nerve,** which leads to the visual cortex in the brain.

The eyeball is filled with fluid. Behind the lens, making up most of the volume of the eyeball, is the jellylike **vitreous humor.** Between the cornea and the lens is the watery **aqueous humor.** The aqueous humor nourishes the cornea and lens, bends light rays to help focus them on the retina, and creates pressure against the sclera that maintains the shape of the eyeball. The aqueous humor is continuously secreted and absorbed. With a fresh supply of aqueous humor every four hours, impurities in this fluid rarely disrupt vision. A jellylike fluid also bathes the rods and cones. Its function is being intensely studied.

KEY CONCEPTS

Photoreceptors are cells that contain pigments, which are chemically altered by light in a way that eventually produces an action potential. Diverse species have photoreceptors but see in different ways. The human eye is a fluid-filled sphere with supportive structures, a lens, and a photoreceptive layer, the retina. The sclera is the outermost layer, and at its front is the transparent cornea, which bends incoming light rays. The middle layer is the choroid coat, which includes the iris and pupil in front. The retina has three layers: the photoreceptive rods and cones, bipolar neurons, and ganglion cells that form the optic nerve. The vitreous humor fills the eyeball behind the lens, and the aqueous humor occupies the region in front of the eyeball.

Focusing the Light

Seeing begins when light rays pass through the cornea and lens and are focused on the retina. The light rays must bend to hit the retina, because most objects that we see are considerably larger than our retina, on which the object's image forms. The bending of light rays is accomplished by the cornea, the lens, and the humors of the eye.

The lens changes its shape to focus light from distant or close objects onto the retina. This molding of the lens to suit the distance of the object being viewed is called **accommodation.** To focus on a very close object, the ciliary muscle contracts, rounding the lens so that it can bend incoming light rays at sharper angles, enabling the close-range light rays to form an image on the retina. Close work often causes eyestrain

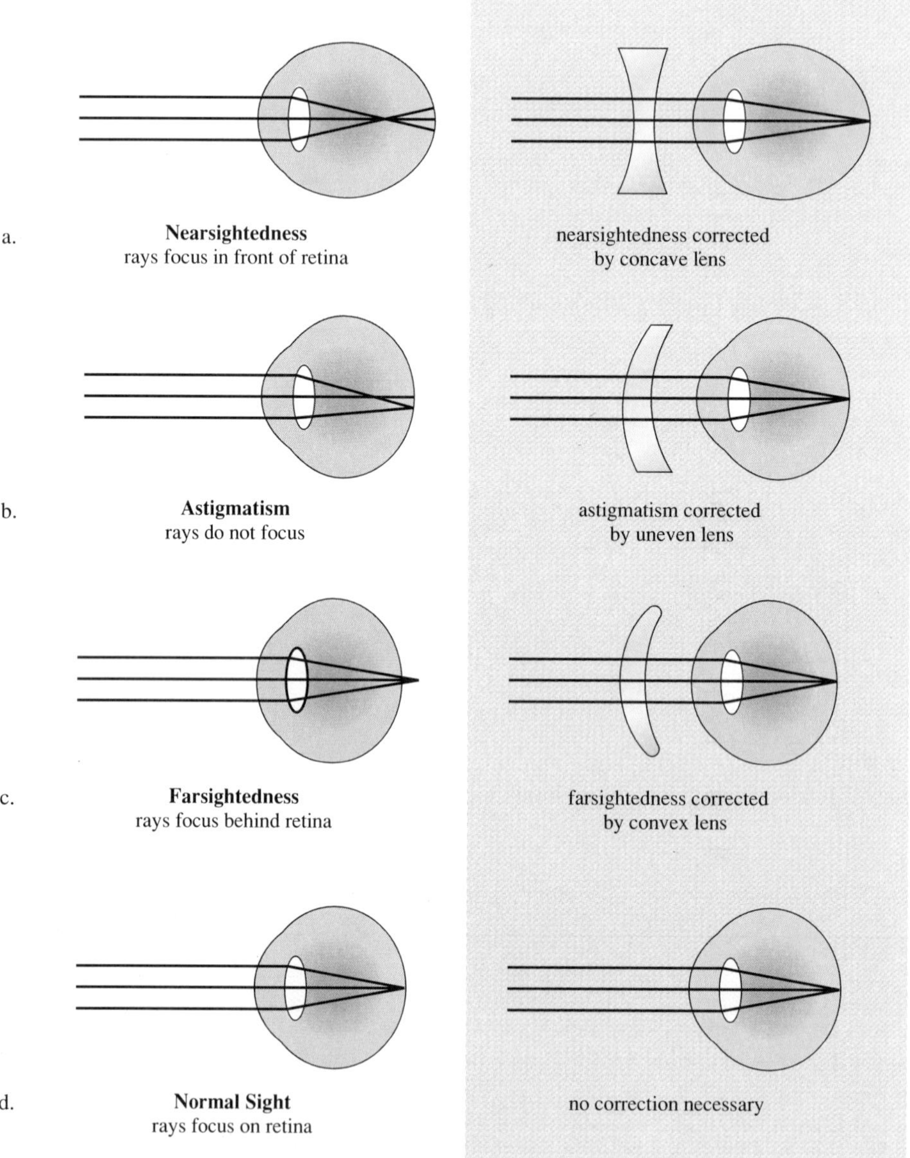

Figure 18.9

Corrective lenses. *a.* If the eyeball is unusually elongated, the incoming light rays converge before the retina, resulting in the inability to see distant objects (nearsightedness). A concave lens can alter the point of focus to the retina in nearsightedness. *b.* Sometimes the eyeball is shaped so unusually that incoming light rays do not focus at all. In such cases of astigmatism, an irregularly shaped lens can be fashioned that compensates for the eyeball's abnormal shape, allowing the light rays to focus on the retina. *c.* If the eyeball is too short, light rays focus beyond the retina, and the person has difficulty seeing close objects (farsightedness). A convex lens can correct this problem. *d.* The shape of the eyeball plays an important role in the ability of the cornea and lens to focus light rays into an image on the retina.

because the ciliary muscle must be contracted to view nearby objects. When you are viewing a faraway object, the lens is flattened by relaxation of the ciliary muscle. That is why eyestrain can be relieved by gazing into the distance.

When an image is not correctly focused on the retina, it appears blurry. In nearsighted individuals, the eye is too long or the lens curves too much so that the image of a distant object is focused in front of the retina. Therefore, a nearsighted person has trouble seeing remote objects. Farsighted persons, on the other hand, have difficulty seeing things that are close by but can see well at a distance. This is because the image is focused behind the retina due to a lens that does not curve enough or an eyeball that is too short. Our eyesight often begins to fail as we get older because the lens becomes less flexible and can no longer change shape to focus on objects at various distances. Eyeglasses or contact lenses can often be used to correct focusing errors so that vision becomes clear (fig. 18.9).

The image of the object that is projected onto the retina is actually upside down. In some way that is not yet understood, the brain processes this information so that we perceive a right-side-up world.

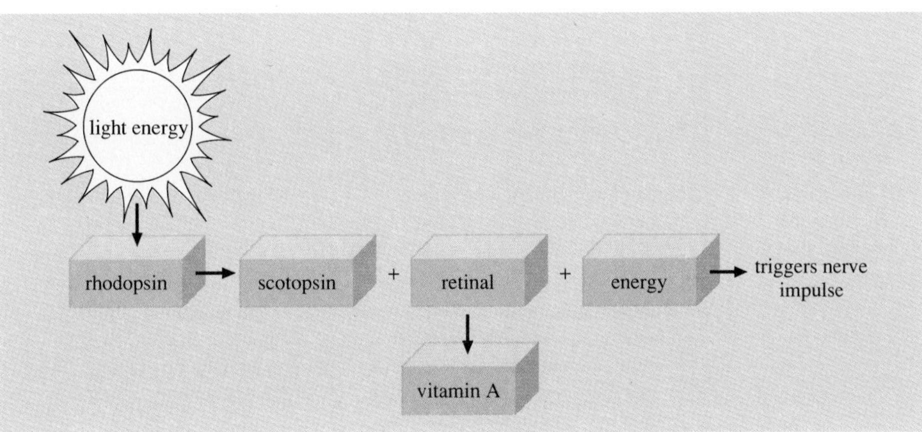

a.

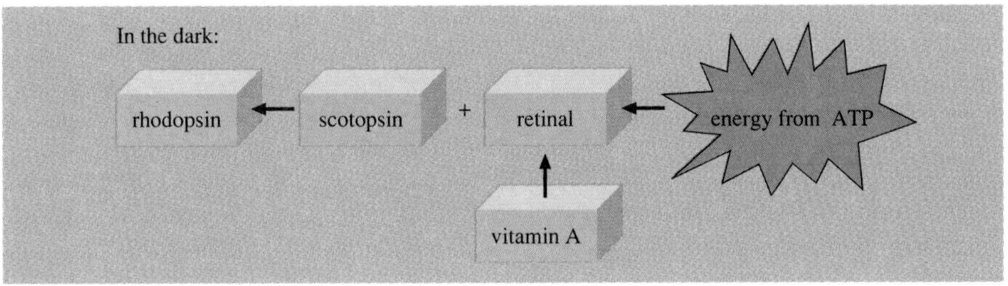

b.

Figure 18.10
Breaking and making rhodopsin. Light splits rhodopsin, which causes depolarization of the rod cell containing it. In the dark, rhodopsin is reformed.

Converting Light Energy to Neural Messages

Visual information is converted to receptor potentials by changes in pigment molecules within the rods and cones. In the rods, molecules of the pigment **rhodopsin** are stored in a highly folded membrane at one end of the cell. Light impinging on a rod cell splits rhodopsin to yield a protein, called **scotopsin,** and a pigment molecule called **retinal,** which is chemically derived from vitamin A (fig. 18.10). The splitting of rhodopsin depolarizes the rod cell, and it releases a neurotransmitter, passing the nerve impulse, via a bipolar neuron, to the optic nerve. In the dark, rhodopsin is resynthesized from scotopsin and retinal. It takes several seconds for rhodopsin to be reformed, and this is why it takes our eyes a short while to make out shapes in a very dark room. Many types of organisms use rhodopsin to detect light (Reading 18.3).

Reading 18.3 *Visual Diversity*

VISION IS INTERESTING FROM AN EVOLUTIONARY STANDPOINT BECAUSE THE PAIRING OF PHOTORECEPTOR CELLS AND PIGMENT MOLECULES APPEARS TO BE QUITE ANCIENT, YET THE VISUAL ABILITIES OF DIFFERENT SPECIES ARE ADAPTED TO PARTICULAR WAYS OF LIFE. Nearly all multicellular species with photoreceptor cells utilize the pigment molecule rhodopsin, splitting it in response to light to provoke a nerve impulse. Even some unicellular organisms use rhodopsin, such as the alga *Chlamydomonas*. A photosynthetic bacterium, *Halobacterium halobium*, uses a rhodopsin variant in place of chlorophyll to harness light energy from the sun.

The turbellarian flatworm has photoreceptor cells gathered into two eyespots, which provide sufficient sensation of light for the animal to orient itself. The starfish has cells at the tip of each arm that detect differences in light intensity. The horseshoe crab, a primitive arthropod, has two compound eyes built of many ommatidia plus two simple eyes. The bee, a complex arthropod, has two compound eyes, each of which has 15,000 ommatidia. The bee's brain receives many separate images, which it merges into a single picture. The bee cannot see fine detail, but it can detect movement.

The eyes of amphibians are adapted to life on the land, with tear glands to moisten the eyes and eyelids to protect them. The frog's retina has many rods and cones, providing color vision acute enough for the animal to catch insect prey on its sticky tongue. Nocturnal mammals such as the bat have only rods in their retinas, producing black-and-white night vision. In contrast, the gray squirrel is active only during the day. Its retinas have only cones, and the animal is virtually blind at night. We humans are lucky—unlike most mammals so

The photochemistry of rhodopsin explains many common visual experiences. For example, rhodopsin is sensitive to very small amounts of light. As a result, we can see the face of a loved one by dim candlelight or moonlight. In bright light, however, rods are of little use because rhodopsin is split faster than it can be resynthesized. We cannot see colors at night because the cones, which are responsible for color vision, are not as sensitive as the rods and do not respond in dim light.

Color vision is possible because the cone cells contain different pigments, each of which absorbs light of different wavelengths (colors). One pigment absorbs the wavelengths around red best, another green, and the third blue. Although an individual cone cell has only one type of pigment, the wavelengths absorbed by each cone type overlap quite a bit.

When light is absorbed by a cone, its pigment molecule is chemically changed and a receptor potential results. The ratio of activity among the cone types is interpreted by the brain as light of a particular color. If one or more of the cone types is missing, an individual cannot distinguish among all colors and is **color blind.** This and other visual problems are presented in table 18.1. Many traditional drug or surgical treatments for eye problems are being replaced by laser methods.

Color vision in humans falls within the 400 to 700 nanometer wavelength of light range, but ants, bees, spiders, and goldfish can detect wavelengths down to 360 nanometers, which means that they can detect ultraviolet light. (Refer back to fig. 6.6 for a review of the electromagnetic spectrum.)

Because our eyes are located close to one another on the front of our heads, the visual fields of each eye overlap. An object directly in front of you can be seen when either eye is closed, although each eye sees it from a slightly different angle. When both eyes function, the visual cortex integrates and interprets the information from each eye to produce a three-dimensional perception. Such depth perception probably evolved in some of our primate ancestors, and was selected for because it enabled them to see branches in three dimensions so that they could swing among them without falling.

Table 18.1
Visual Problems

Problem	*Cause*	*Treatment*
Astigmatism	Asymmetrical cornea or lens; light rays do not focus to form sharp images	Uneven corrective lens focuses on retina
Cataracts	Lens is opaque due to birth defect, injury that perforates cornea, complication of diabetes or glaucoma, or part of aging	Surgical removal of cataracts
Red-green color blindness	Red or green cones absent	None
Farsightedness (cannot see close objects)	Light focuses in back of retina because eyeball is too compact	Corrective convex lens focuses light on retina
Glaucoma	Aqueous humor is blocked from draining from ducts near the iris, as it should; buildup of fluid presses capillaries that nourish nerve fibers in retina, damaging optic nerve, causing blindness	Drugs to inhibit synthesis of aqueous humor; drugs that thin iris, increasing drainage of fluid Surgical removal of piece of iris
Nearsightedness (cannot see distant objects)	Light focuses in front of retina because eyeball is too elongated	Corrective concave lens focuses light on retina

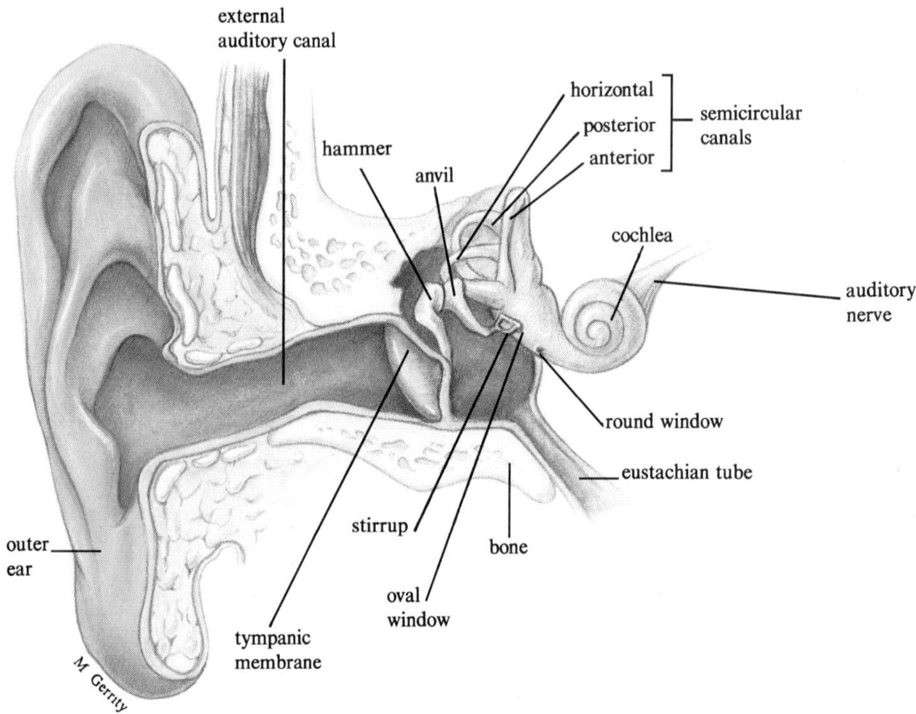

Figure 18.11
Anatomy of the ear. Sound enters the outer ear and impinges upon the tympanic membrane, which vibrates upon the three bones (hammer, anvil, and stirrup) of the middle ear. The vibrations next hit the oval window, which is the opening into the inner ear, where hair cells in the cochlea convert the sound waves into action potentials, which are passed along the auditory nerve to the brain. Hair cells in the semicircular canals and in the vestibule provide a sense of balance. The eustachian tube connects the middle ear to the throat, allowing air pressure to be equalized.

> **KEY CONCEPTS**
>
> *Light entering the eye is bent by the cornea, lens, and humors and is focused on the retina. The ciliary muscle alters the shape of the lens to focus on near or far objects. The image on the retina is upside down, but the brain provides right-side-up perceptions. Rod cells contain rhodopsin, which provokes a nerve impulse when it is split by light. This stimulation is interpreted as black-and-white vision. Perception of color depends upon the types of cone pigments that are stimulated. The location of our eyes on the front of our heads provides three-dimensional visual perception.*

Mechanoreception

Mechanoreceptors convert mechanical energy to action potentials. This usually occurs when the receptor is bent or deformed in some way. Specialized mechanoreceptors function in our senses of hearing, balance, and touch.

Hearing

The clatter of a train, the sounds of a symphony, a child's wail—what do they all have in common? All sounds, regardless of the source, originate when something vibrates. The vibrating object creates repeating pressure waves in the surrounding medium, which, for humans, is usually air. The size and energy of these pressure waves determine the intensity or loudness of the sound. The number of waves (cycles) per second determines the frequency or pitch of the sound. The more cycles per second, the higher the pitch.

Specialized organs for hearing are found primarily in insects and vertebrates. Insects usually hear with their **tympanal organ,** a thin region of the outer body covering that vibrates in response to sound and stimulates special receptor cells as it is displaced. Different insects have these "ears" in a variety of locations that often seem odd to us. The tympanal organs of a noctuid moth, for example, are under the wings, and in crickets they are on the legs.

Among vertebrates, the hearing organ is the ear. In humans, the fleshy outer ear traps sound waves and funnels them into a wax- and hair-lined ear canal to impinge upon the **tympanic membrane,** or eardrum (figs. 18.11 and 18.12). The sound pressure waves cause the eardrum to vibrate, which moves three small bones, the **hammer, anvil,** and **stirrup,** located in the **middle ear**. These bones transmit the incoming sound and amplify it 20 times. At the end of the middle ear, the vibrations hit the **oval window,** which is a membrane that opens into the **inner ear.**

The inner ear is a fluid-filled chamber that houses both balance and hearing structures. The first two parts of the inner ear, the semicircular canals and the vestibule, are concerned with balance. The hindmost section of the inner ear is the snail-shaped **cochlea.** If the spirals of the cochlea are unwound, three fluid-filled canals are revealed: the vestibular, cochlear, and tympanic canals. The vestibular and tympanic canals are actually a continuous U-shaped tube (fig. 18.13). Between them is the cochlear canal. When the last bone of the middle ear, the stirrup, moves, it pushes on a membrane, the oval window. This action transfers the vibration to the fluid in the vestibular canal. The pressure of each vibration is dissipated by the movement of the round window at the

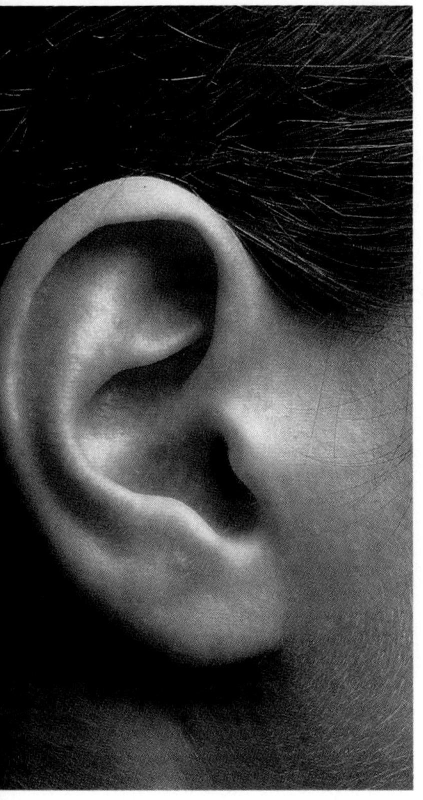

Figure 18.12
The outer ear.

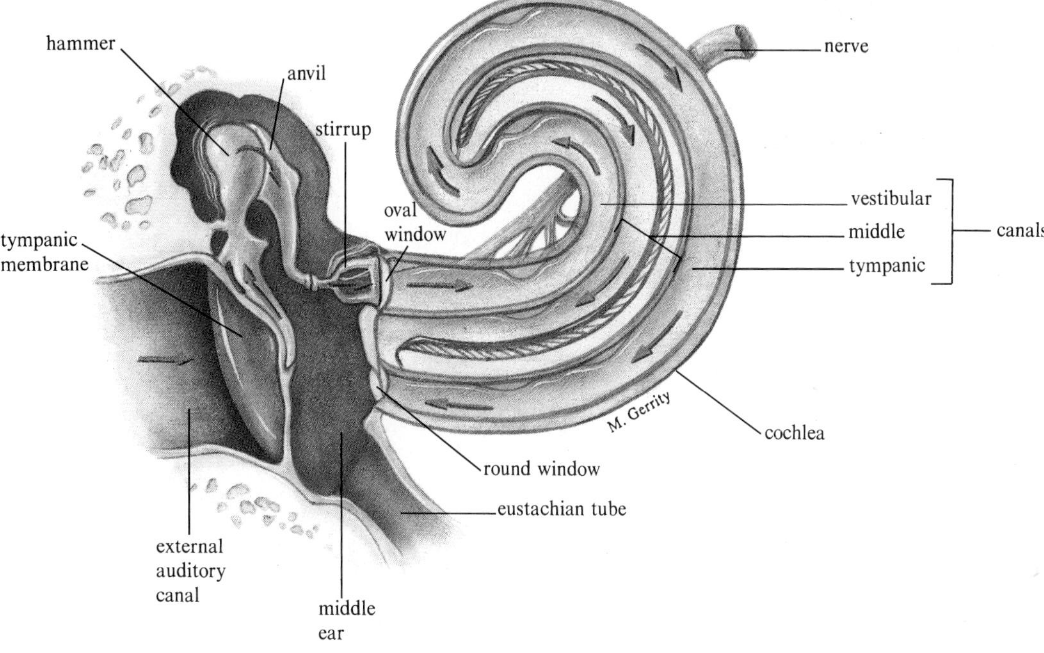

Figure 18.13
The cochlea is a spiral-shaped structure consisting of three fluid-filled canals. When the stirrup moves against the oval window, the fluid in the vestibular canal vibrates. This vibration eventually converts the sound wave to an action potential.

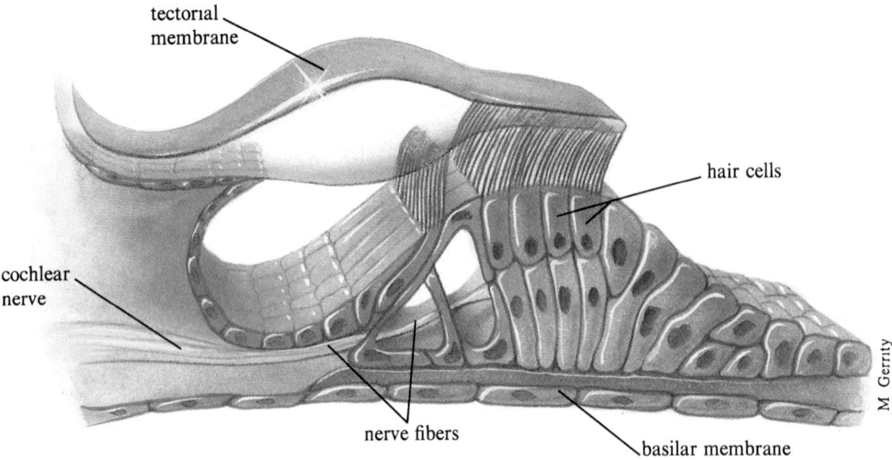

Figure 18.14
From vibration to nerve impulse. Hair cells lie between the tectorial membrane above and the basilar membrane below. Vibrations along particular portions of the basilar membrane correspond to different sound frequencies. The vibration exerts pressure on the hair cells, which in turn push the tectorial membrane, which leads to an action potential in the auditory nerve.

other end of the U. The vibration of the fluid also moves the basilar membrane and initiates the change of mechanical energy to receptor potentials.

How is sound translated into the universal language of the nervous system? Within the cochlea, specialized **hair cells** (the mechanoreceptors) lie between the **basilar membrane** below and the **tectorial membrane** above (figs. 18.14 and 18.15). The basilar membrane is narrow and rigid at the base of the cochlea and widens and becomes more flexible nearer the tip. Because of the variation in width and flexibility, different places along the basilar membrane vibrate more intensely when exposed to different frequencies. The high-pitch tinkle of a bell stimulates the narrow region of the basilar membrane at the base of the cochlea, while the low-pitch tones of a tugboat whistle stimulate the wide end.

When a region of the basilar membrane vibrates, the hair cells there are pushed against the tectorial membrane, and action potentials are spiked in fibers of the **auditory nerve.** This nerve carries the impulses to the brain, where the input from different regions is interpreted as sounds of different pitches (fig. 18.16). The louder the sound, the greater the vibration of the basilar membrane, and consequently, the more hair cells are stimulated. The resulting increase in the rate of firing and the number of neurons firing is interpreted as an increase in amplitude, or loudness.

Hearing Loss

A sound is heard only if it is conducted from outside the ear to the cochlea and neural messages generated in the inner ear are transmitted to the brain. Consequently, there are two types of hearing loss. One type, **conductive deafness,** occurs when the transmission of sound through the middle ear is impaired. It affects all pitches of sound equally. This type of loss might be caused by an infection of the middle ear, a buildup of earwax, or damage to the eardrum. The buildup of fluid in the middle ear that sometimes accompanies a cold can temporarily affect hearing because it hinders conduction.

The other type of hearing loss—**sensory,** or **neural, deafness**—is due to the inability to generate action potentials in the cochlea, blocked communication between the cochlea and the brain, or the brain's inability to make sense of the sensory message. An individual with this type of deafness may hear some frequencies better than others. Sensory deafness may accompany aging because a progressive degeneration of the hair cells at the base of the cochlea (those that respond to high frequencies) begins at the age of 40. This type of deafness is commonly caused by exposure to loud noise.

The human ear is simply too sensitive for many of the sounds of our times, such as a subway's roar or the clanking of farm

a.

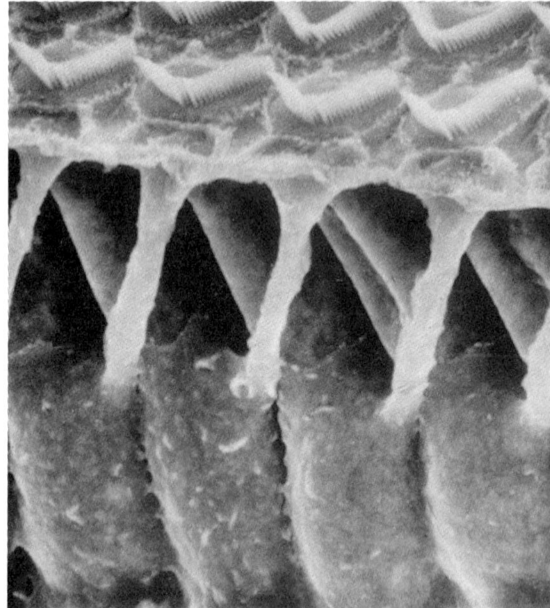

b.

Figure 18.15
A closer look at the cochlea. *a.* The cochlea is the snail-shaped part of the inner ear on which sound waves are converted into action potentials. *b.* This is accomplished by tiny hair cells, each of which responds to a particular wavelength of sound.

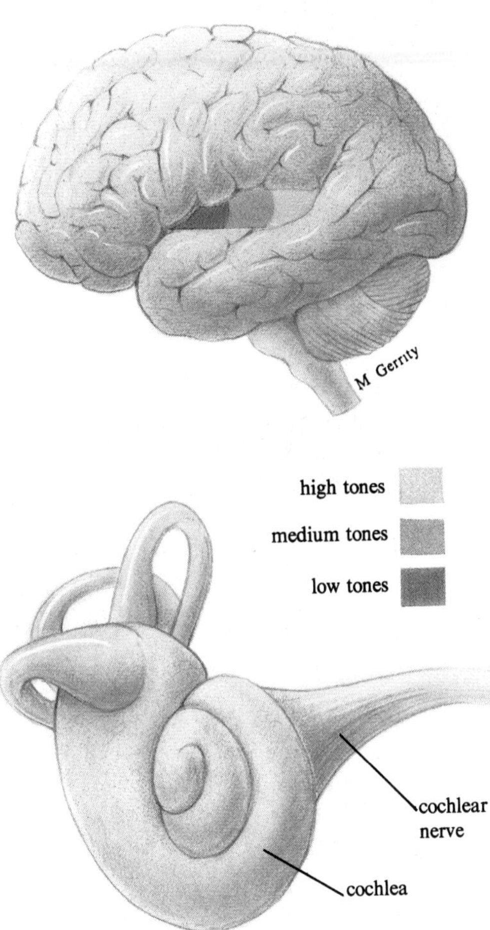

Figure 18.16
Correspondence between cochlea and cortex. Sounds of different frequencies (pitches) excite different sensory neurons in the cochlea. These in turn send their input to different regions of the auditory cortex.

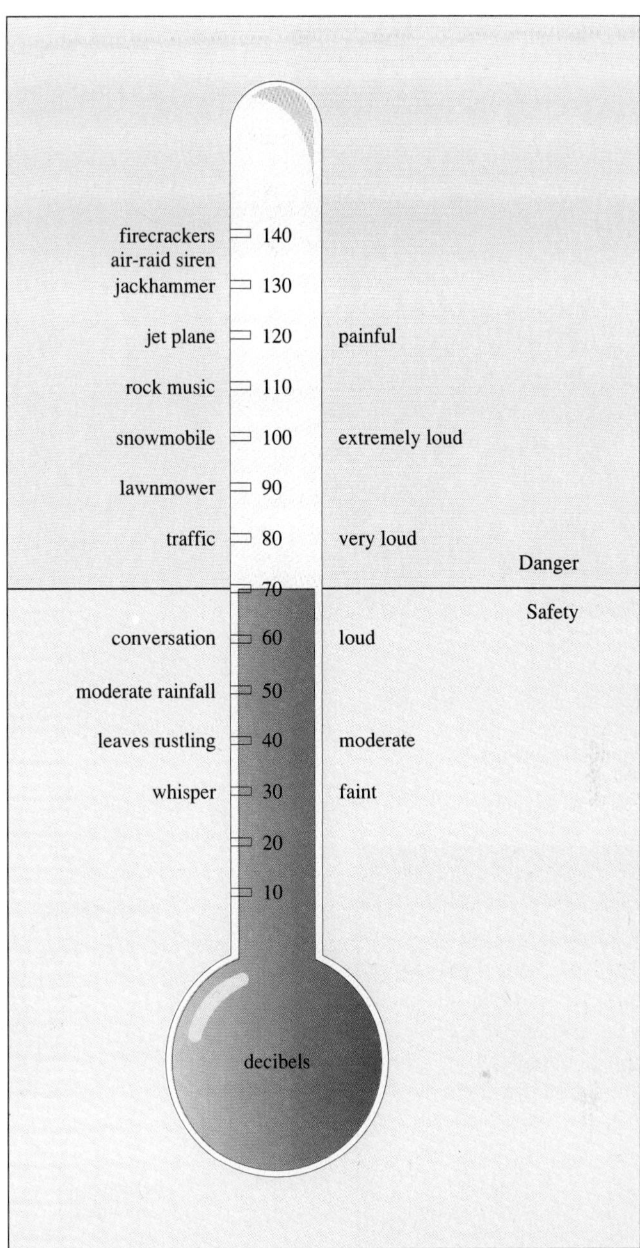

Figure 18.17
Turn down that stereo! Sound intensity is commonly measured in decibels. Each increase in 10 decibels represents a 10-fold increase in sound intensity. Damage to hair cells begins to occur at about 80 decibels. The degree of damage depends upon both decibel level and duration of exposure to the sound. As the figure shows, many commonplace sounds are well above this safety level.

machinery. More than 16 million Americans have a hearing loss caused by excessive noise. Although exposure to loud noise is sometimes unavoidable, it is sometimes elective. For example, the amplitude of music at rock concerts, and sometimes in homes, is sufficient to cause permanent damage to the hair cells (fig. 18.17). Many rock stars of the 1960s are going deaf in the 1990s! The damage begins on the hair cells as bulges similar to blisters that eventually pop. The tissue beneath the hair cells swells and softens until, eventually, the hair cells and sometimes even the neurons leaving the cochlea degenerate, becoming blanketed with scar tissue (fig. 18.18).

A common result of slightly damaged hair cells in the inner ear is **tinnitus,** or ringing in the ears. It can range from a barely perceptible distant ringing to a constant blaring screech. In 5% of the cases, tinnitus stems from an ear infection, drug reaction (such as aspirin), brain tumor, or wax buildup on the eardrum. The cause of the other 95% is loud noise.

Extensively damaged hair cells can cause profound deafness, which affects 2 million Americans. Help for several thousand of them may be provided by a **cochlear implant,** a device that delivers an electronic stimulus directly to the auditory nerve, bypassing the function of the hair cells. For the 15 million Americans with partial conductive hearing loss, hearing aids, special amplified telephones, the use of sign language, and hearing-aid dogs (the counterpart of the seeing eye dog) can help.

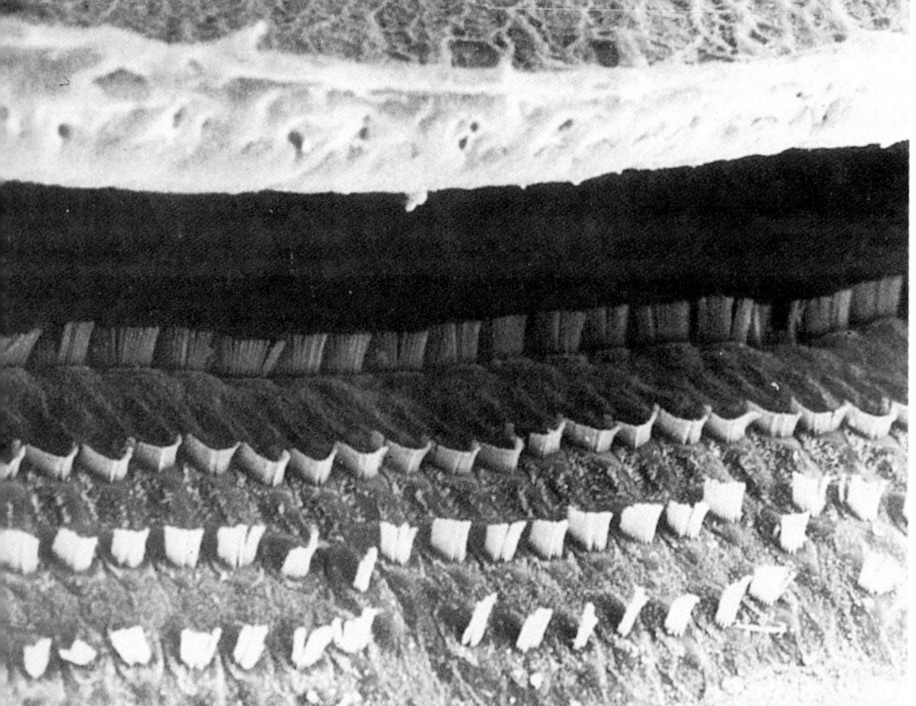

a.

b.

Figure 18.18
After 24 hours of exposure to noise at the level of loud rock music, normal hair cells (*a*) have become scarred and lost (*b*).

KEY CONCEPTS

The sense of hearing results from the vibration of sound waves stimulating receptor cells, which then produce action potentials in a nerve leading to the brain. In humans, sound waves impinge upon the tympanic membrane, which vibrates to move the hammer, anvil, and stirrup bones of the middle ear. After being amplified, the vibrations hit the oval window and are transferred to the inner ear. There, within the cochlea, the vibration causes specific hair cells to push against the tectorial membrane, stimulating the auditory nerve, which leads to the brain, where the vibration is perceived as a particular sound. Different types of hearing loss depend upon which structures within the ear are damaged.

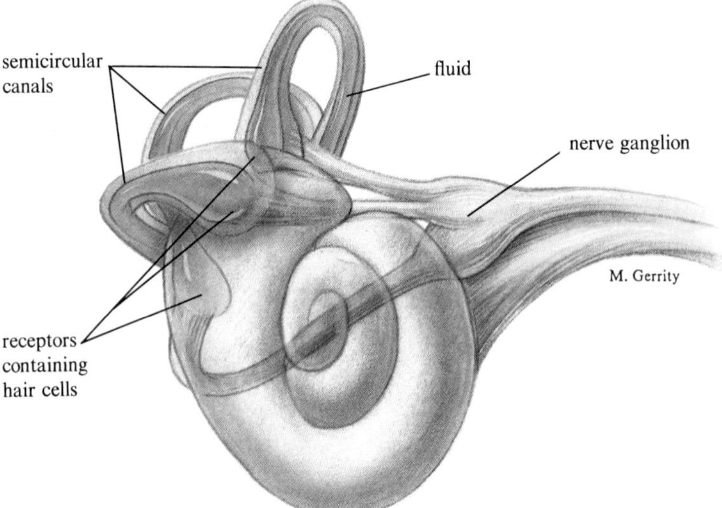

Figure 18.19
Changes in the position of the head are detected by pressure exerted on the receptors in the semicircular canals.

Balance

The semicircular canals and the vestibule of the inner ear are concerned with our sense of balance (fig. 18.19). The **semicircular canals** tell us when the head is rotating and help us maintain the position of the head in response to sudden movement. The enlarged bases of the semicircular canals, called the **ampullae,** are lined with small ciliated hair cells. The semicircular canals are perpendicular to each other, and the fluid that fills them swishes back and forth in response to a person's movements. This fluid motion bends the cilia on the hair cells in the ampullae, which in turn stimulates those cells to generate action potentials in a nearby cranial nerve. The perception of body position results from the brain's interpretation of these impulses.

Information from the **vestibule** (fig. 18.13) tells us the position of the head with respect to gravity. In addition, it senses changes in velocity when traveling in a straight line. When riding in a car, for example, we can sense acceleration and deceleration. The vestibule functions in a similar way to the semicircular canals. It contains two pouches, the **utricle** and the

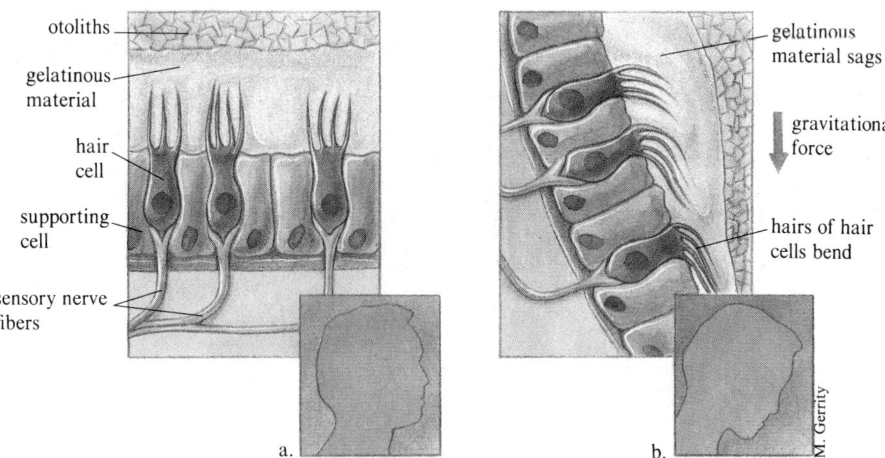

Figure 18.20
The sense of balance. Gravitational forces on the head move the otoliths. *a.* When the head is upright, the otoliths balance atop the cilia of hair cells. *b.* When the head bends forward, the otoliths move, bending the cilia, which provokes a nerve impulse.

saccule, which are filled with jellylike fluid and lined with ciliated hair cells (fig. 18.20). Granules of calcium carbonate, called **otoliths,** rest within the fluid. The granules in the utricle move in response to vertical body movements. The motion of the granules bends the cilia on the hair cells. When the cilia are bent in one direction, the rate of sensory impulses to the brain increases. However, a shift in the opposite direction inhibits the sensory neuron. The brain interprets this information as a change in velocity.

Motion sickness is a disturbance of the inner ear's sensation of balance. Nine out of 10 people have experienced this nausea and vomiting, usually when riding in a car or on a boat. A form of motion sickness called space adaptation syndrome has been reported by astronauts since the Apollo 8 moon voyage in 1968, when spacecraft began to be made roomy enough for astronauts to move about while in flight.

Although the cause of motion sickness is not known, one theory is that it results when visual information contradicts the inner ear's sensation that one is motionless. Consider a woman riding in a car. Her inner ears tell her that she is not moving, but the passing scenery tells her eyes that she is moving. The problem is compounded if she tries to read. The brain reacts to these seemingly contradictory sensations by signaling a "vomiting center" in the medulla oblongata in the brain.

Touch

Our sensitivity to touch comes from several types of receptors in the skin—**Pacinian** and **Meissner's corpuscles** and **free nerve endings** (fig. 18.21). Pacinian corpuscles are stimulated by firm pressure—a bear hug for instance. On the other hand, Meissner's corpuscles are stimulated by light touch, particularly if the object is moving lightly over the skin, as in a gentle caress. The free ends of sensory neurons found in epithelial surfaces including the skin are sensitive to touch, pressure, and pain. In signaling pain, they may respond to pressure, temperature, or chemicals associated with tissue damage.

KEY CONCEPTS

Movement of fluid in the semicircular canals against ciliated hair cells provides information on body position. Information on head position and velocity change comes from the movement of otoliths in the utricle and saccule, two pouches in the vestibule. Pacinian and Meissner's corpuscles and free nerve endings detect touch.

Thermoreception

We know very little about thermoreception. In humans, the thermoreceptors may simply be two types of specialized free nerve endings within the skin, heat receptors and cold receptors. Cold receptors are stimulated by temperatures between 50°F (10°C) and 68°F (20°C). Heat receptors are active as the temperature climbs to 77°F (25°C), but become inactive at temperatures above 113°F (45°C). The sensation of hot or cold fades rapidly. For example, the water in a hot tub bath may feel as if it is scalding at first, but it very quickly feels comfortably warm. This is because the thermoreceptors adapt rapidly.

Detection of Magnetic Fields—A Sixth Sense?

A magnetic mineral called magnetite (an iron oxide) is found in a wide variety of organisms, including certain bacteria, pigeons, bees, dolphins, and in the bones forming the sinuses of some humans. These built-in magnets are thought to play a role in orientation. Certain bottom-dwelling bacteria, for instance, are thought to use the earth's magnetic field to determine which way is down. Migrating birds and homing pigeons may also use the earth's magnetic field as a navigation cue. Homing behavior is discussed further in chapter 36.

Could we, too, have a magnetic "sixth sense"? In an interesting British study, blindfolded students were taken on a circuitous bus trip, then asked to point either in the direction of home or in a specific compass direction. The students performed consistently worse when they wore magnets atop their heads. This was interpreted to

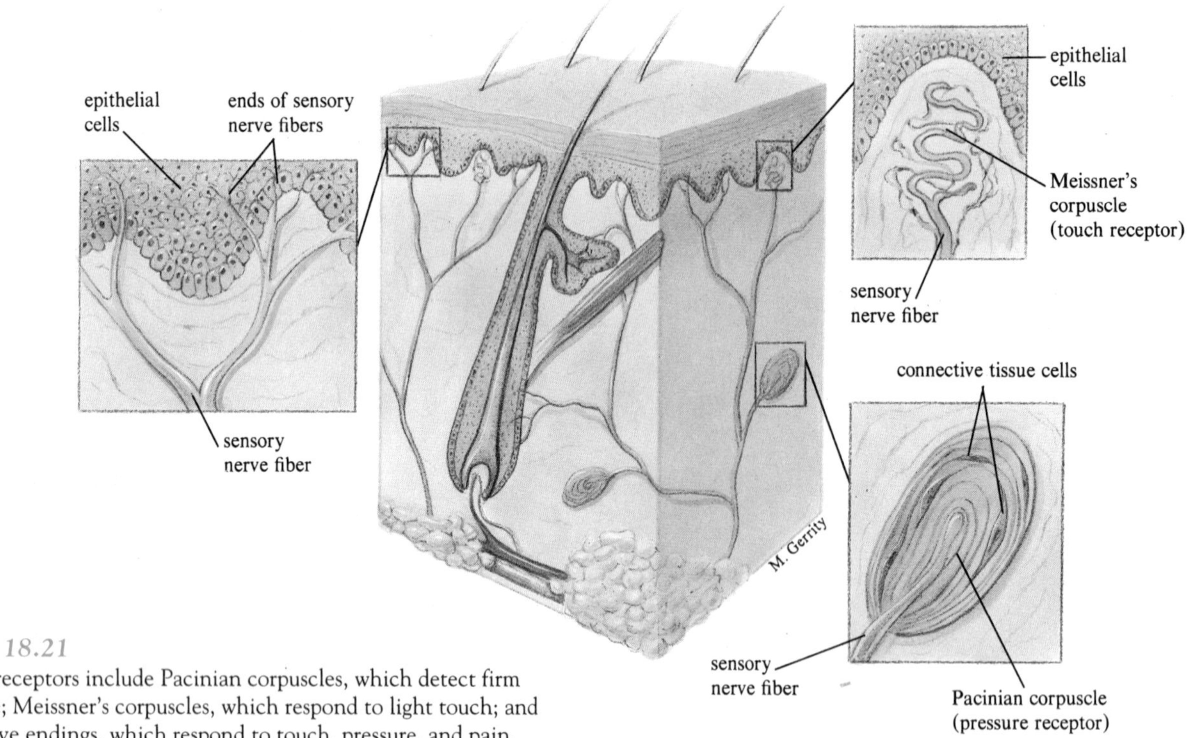

Figure 18.21
Touch receptors include Pacinian corpuscles, which detect firm pressure; Meissner's corpuscles, which respond to light touch; and free nerve endings, which respond to touch, pressure, and pain.

Table 18.2
Information Flow From the Environment Through the Nervous System

Information Flow	*Smell*	*Taste*	*Sight*	*Hearing*
Sense recepters ↓	Olfactory cells in nose ↓	Taste bud receptor cells ↓	Rods and cones in retina ↓	Hair cells in cochlea ↓
Stimulation of nerve ↓	Olfactory nerve ↓	Sensory fiber in taste bud ↓	Optic nerve ↓	Auditory nerve ↓
Impulse transmission to CNS ↓	Cortex ↓	Cortex ↓	Midbrain and visual cortex ↓	Midbrain and cortex ↓
Sensation (memory, experience) ↓	A pleasant smell ↓	A sweet taste ↓	A small, round, red object ↓	A crunching sound ↓
Perception	The smell of an apple	The taste of an apple	The sight of an apple	Biting into an apple

mean that the magnets somehow interfered with a natural magnetic sense of direction, as they did in similar experiments with homing pigeons (fig. 36.13). However, other scientists who have attempted to repeat this experiment are not enthusiastic about their results. We will have to wait for future investigation to resolve the controversy.

How we experience the world is largely due to input from our senses and its interpretation by the brain into perceptions. The path taken from an environmental stimulus to perception is similar for different senses, as Table 18.2 demonstrates. Specialized sensory receptor cells are stimulated and relay information to the CNS, where it is interpreted as a sensation. The sensation is modified by experience and memory to produce meaningful information, a perception.

SUMMARY

A sensory receptor is selectively responsive to a single form of energy, converting it to the energy of nerve impulses. The immediate response of a receptor is the generation of a *receptor potential*, which is a change in membrane potential that varies in magnitude with the strength of the stimulus. If a stimulus remains constant, a sensory receptor generally ceases to respond within a few minutes and we are no longer aware of the stimulus. This process is called *sensory adaptation*.

Chemoreception includes the senses of smell *(olfaction)* and taste. Olfaction occurs when chemicals excite receptors in the *olfactory epithelium* of the nasal passages. Taste occurs when chemicals stimulate the receptors within taste buds.

Photoreceptors vary in complexity, but all have a light-absorbing pigment associated with a membrane. Some invertebrate eyes do not form images, but the compound eyes of insects and crustaceans do form images. The compound eye consists of many image-forming visual units called *ommatidia*.

The human eye contains three layers. The outer layer, the *sclera*, is protective. It forms the transparent cornea in the front of the eyeball. The next layer, the *choroid coat*, is pigmented toward the rear of the eye and absorbs light. It also contains the blood vessels nourishing the eye. In the front of the eye, the choroid coat is modified in several ways. It forms the *ciliary body*, which controls the shape of the lens to focus light on the photoreceptors, and the opaque *iris*. The size of the opening in the center of the iris, the *pupil*, is adjusted to regulate the amount of light entering the eye.

The innermost layer is the multilayered retina. The outermost layer contains the photoreceptors: *rods* for black-and-white vision in dim light and *cones* for color vision in brighter light. These synapse with bipolar cells that form the middle retinal layer. The bipolar cells synapse with ganglion cells that leave the retina as the optic nerve, carrying the neural messages to the brain for interpretation. The pigment in a rod or a cone is split in two by absorbed light. This generates a receptor potential. There are three types of cones, each containing a pigment that absorbs maximally light of particular wavelengths. The ratio of the activity of the three cone types is interpreted as a color. Mechanoreceptors typically respond when the receptor is bent or deformed. They function in our senses of hearing, balance, and touch. In human hearing, sound is funneled into the auditory canal and causes vibrations in the eardrum *(tympanic membrane)*. These vibrations are transmitted through the middle ear and amplified by three tiny bones, the *hammer*, *anvil*, and *stirrup*. The movement of these bones causes pressure changes in fluid within the *cochlea*, which in turn causes the *basilar membrane* to vibrate. As the basilar membrane moves, it pushes *hair cells* against the *tectorial membrane*, which signals the brain to perceive the pitch of the sound by the location of the hair cells that respond. The loudness of the sound is encoded in the frequency of action potentials and the number of hair cells that are stimulated. Body position and movement are sensed by the semicircular canals and the vestibule. The movement of fluid within these areas stimulates sensory hair cells, and this information is interpreted by the brain, giving us a sense of equilibrium.

Pacinian corpuscles, *Meissner's corpuscles*, and *free nerve endings* act as mechanoreceptors involved with the sense of touch. Humans have heat receptors and cold receptors. These are thought to be specialized nerve endings in the skin. They adapt rapidly. Some animals are able to orient in magnetic fields. Deposits of a magnetic material, magnetite, have been located in these animals and are thought to serve as magnetoreceptors.

QUESTIONS

1. How do the senses of smell and taste enable us to enjoy eating?

2. Why doesn't burning your tongue on a piping hot slice of pizza permanently damage your sense of taste?

3. How does your vision differ from that of an insect? From a shrimp living 2 miles beneath the ocean's surface?

4. How do we see colors other than red, blue, or green?

5. How do the three major structures of the inner ear (cochlea, semicircular canals, and vestibule) function similarly?

6. Sensation is generally the result of an anatomical pattern in which many sensory cells transmit impulses to a nerve leading to a specific part of the brain. Describe neuronal arrangements in three senses that fit this pattern of funneling incoming information from many sensory neurons to a nerve fiber leading to the brain.

TO THINK ABOUT

1. The major symptoms of Alzheimer's disease and Parkinson's disease are often preceded by loss of the sense of smell. How do you think this information might be used? What additional information is needed to make medical use of this association?

2. People can inherit an inability to smell certain substances, such as freesia flowers, jasmine, and hydrogen cyanide. How can the theory that odors are encoded in a combination of stimulated receptors explain these conditions?

3. Why is vitamin A in the diet important for good eyesight?

4. Cite two examples of how people lacking one sense can compensate by relying more heavily on another.

5. Why do you think people who are deaf due to damage to the cochlea do not suffer from motion sickness?

6. We have relatively few sensory systems. How, then, do we experience such a huge and diverse number of sensory perceptions?

SUGGESTED READINGS

Appel, Camille E. July 1985. Taste = flavor. *ChemTech*. The flavor of food involves more than just the senses of smell and taste.

Blakeslee, Sandra. October 4, 1988. Pinpointing the pathway of smell. *New York Times*. Specialized proteins ferry odoriferous molecules to the olfactory epithelium in the nose, enabling us to smell even very faint odors.

Gillis, Anna Maria. November 1990. What are birds hearing? *BioScience*. Birds can distinguish a remarkable number of different sounds.

Hechinger, Nancy. March 1981. Seeing without eyes. *Science 81*. Individuals lacking one sense often compensate by relying more on other senses.

Lewis, Ricki. September 1986. New help for the hearing impaired. *Biology Digest*. A variety of new devices can help restore hearing.

Marantz Henig, Robin. March 26, 1989. The aging eye. *New York Times Magazine*. Visual acuity declines as we age.

Miller, J.A. February 15, 1986. Sensory surprises in platypus, mantis. *Science News*. The platypus detects electrical fields with its bill, and the praying mantis detects ultrasound waves with a single hearing organ in its throat.

Montgomery, Geoffrey. December 1988. Seeing with the brain. *Discover*. The brain is essential to seeing in color.

Patlak, Margie. March 1990. Pursuing 20/20 at 40-plus. *FDA Consumer*. New types of glasses and contact lenses offer help to those whose vision falters as they age.

Weinstock, Cheryl Platzman. February 1990. Hearing aids: A link to the world. *FDA Consumer*. Many people with hearing loss can be helped.

Weisburd, S. December 22/29, 1984. Whales and dolphins use magnetic "roads." *Science News*. Several species can detect magnetic fields.

Weiss, Peter L. September 15, 1990. Eye diving. *Science News*. A thin sheet of gel next to the rods and cones may be crucial to vision.

CHAPTER

19

The Endocrine System

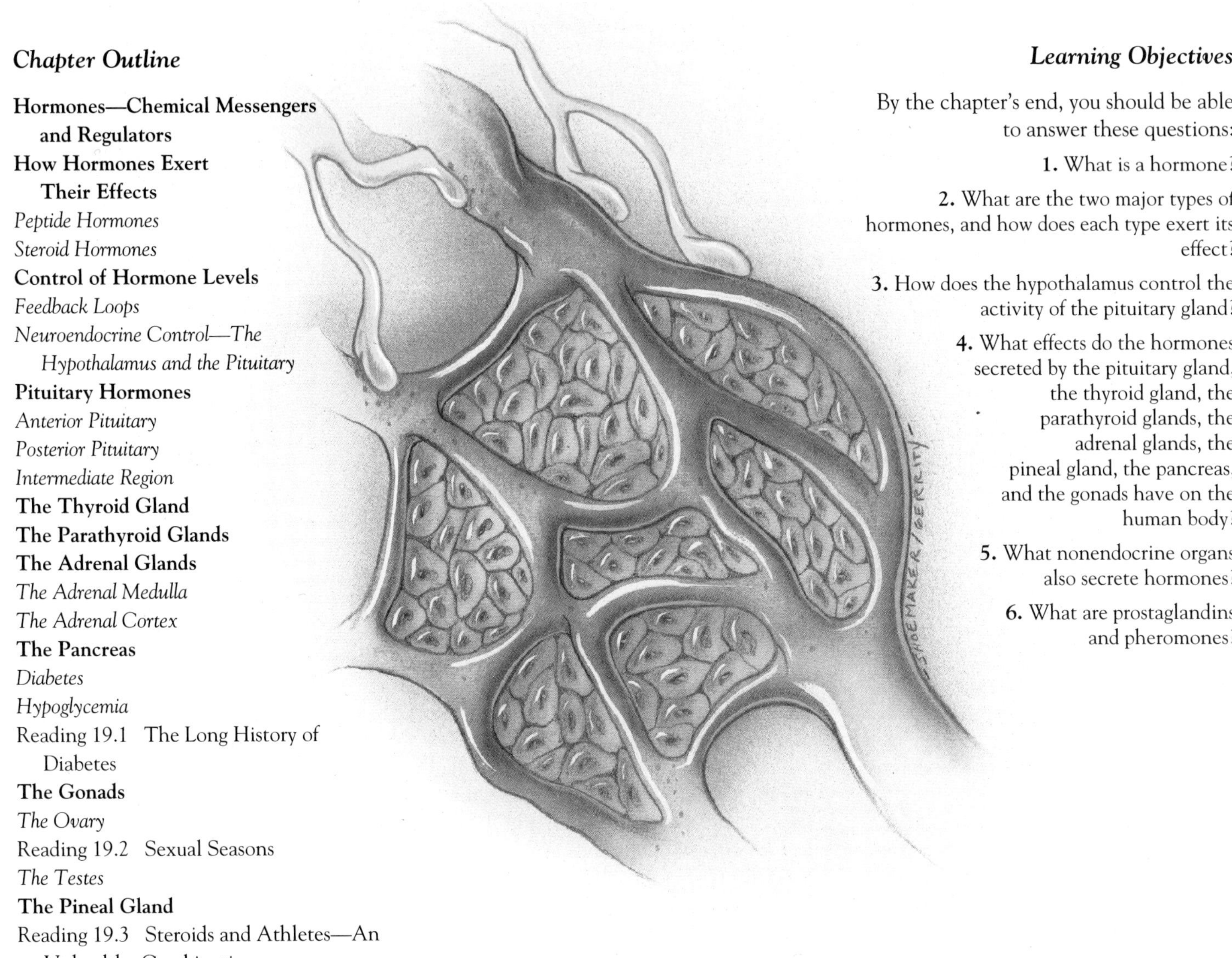

Chapter Outline

Hormones—Chemical Messengers and Regulators
How Hormones Exert Their Effects
Peptide Hormones
Steroid Hormones
Control of Hormone Levels
Feedback Loops
Neuroendocrine Control—The Hypothalamus and the Pituitary
Pituitary Hormones
Anterior Pituitary
Posterior Pituitary
Intermediate Region
The Thyroid Gland
The Parathyroid Glands
The Adrenal Glands
The Adrenal Medulla
The Adrenal Cortex
The Pancreas
Diabetes
Hypoglycemia
Reading 19.1 The Long History of Diabetes
The Gonads
The Ovary
Reading 19.2 Sexual Seasons
The Testes
The Pineal Gland
Reading 19.3 Steroids and Athletes—An Unhealthy Combination
Hormones Not Associated With Endocrine Glands
Hormonelike Molecules
Prostaglandins
Pheromones

Learning Objectives

By the chapter's end, you should be able to answer these questions:

1. What is a hormone?
2. What are the two major types of hormones, and how does each type exert its effect?
3. How does the hypothalamus control the activity of the pituitary gland?
4. What effects do the hormones secreted by the pituitary gland, the thyroid gland, the parathyroid glands, the adrenal glands, the pineal gland, the pancreas, and the gonads have on the human body?
5. What nonendocrine organs also secrete hormones?
6. What are prostaglandins and pheromones?

Puberty is a time that most of us would like to forget. The change that swept through our bodies seemed simultaneously too sudden and too slow. We can thank hormones for our transfiguration. At puberty, the hypothalamus in the brain began sending chemical messages to a pea-sized gland in the brain known as the pituitary. In turn, the pituitary's hormonal messages began to stimulate the production of sex hormones by the ovaries or testes.

Soon reproductive structures began to enlarge and mature. Hair appeared in new places. Males found that new muscle growth broadened their shoulders, while thickening vocal cords and an enlarging larynx deepened their voices. Females broadened in the hips and developed breasts. Other changes probably escaped notice—alteration in bone growth, calcium metabolism, and red blood cell formation to name a few. Behavior was also modified with new feelings and attitudes, especially toward members of the opposite sex. As if in protest of all this change, and just when appearance seemed so critical, our skin erupted with acne. Indeed, puberty is probably the time when we were most aware of hormones. The pervasive effects of the sex hormones are typical of hormones in general.

Figure 19.1
This vibrant photograph shows crystals of estrogen, a hormone that exerts several effects on the female reproductive system.

Hormones—Chemical Messengers and Regulators

Communication within the body is essential and results from the coordinated functioning of the body's two information systems—the nervous system and the endocrine system. The endocrine system produces chemical messengers called hormones, which regulate cellular activities. Although the endocrine system responds more slowly than the nervous system, its effects are generally more prolonged. These two systems work together to produce a variety of responses, from sexual and reproductive behavior, to control of growth and development, to adjusting the delicate chemical balance of body fluids.

A **hormone** is a chemical released by cells into the bloodstream, which carries it to some distant part of the body where it alters cellular activity (fig. 19.1). A baby sucking on its mother's nipple stimulates nerve cells that cause a hormone called oxytocin to be released from the pituitary gland in the brain. The oxytocin enters the mother's circulation and travels to musclelike cells surrounding milk-filled cells in the breast. The oxytocin causes the musclelike cells to contract, which squirt milk out of the 16 or so small holes in the nipple, and thus the baby is fed.

Hormones are produced by **endocrine glands,** which secrete their hormones into the circulatory system directly (fig. 19.2). In contrast, exocrine glands, such as sweat and salivary glands, secrete their products into ducts (fig. 19.3).

Dozens of hormones circulate to all the cells of the human body, but only certain cells respond to a particular hormone. This specificity exists because cells have molecules called receptors on their surfaces that bind to certain hormones. A cell that binds a particular hormone is called its **target cell.** A target cell's receptor physically fits the shape of the hormone, and the hormone-receptor complex actually initiates the cell's response.

A cell that lacks the appropriate receptors cannot respond to the hormone. For example, certain men lack receptors for the sex hormone testosterone, so the body behaves as if the hormone is not being produced. These men never develop male characteristics even though they produce adequate amounts of the necessary hormone.

KEY CONCEPTS

Hormones are secreted into the bloodstream by cells of the glands of the endocrine system. Endocrine responses are slower and longer lasting than those of the nervous system. Hormones bind to receptors on their target cells.

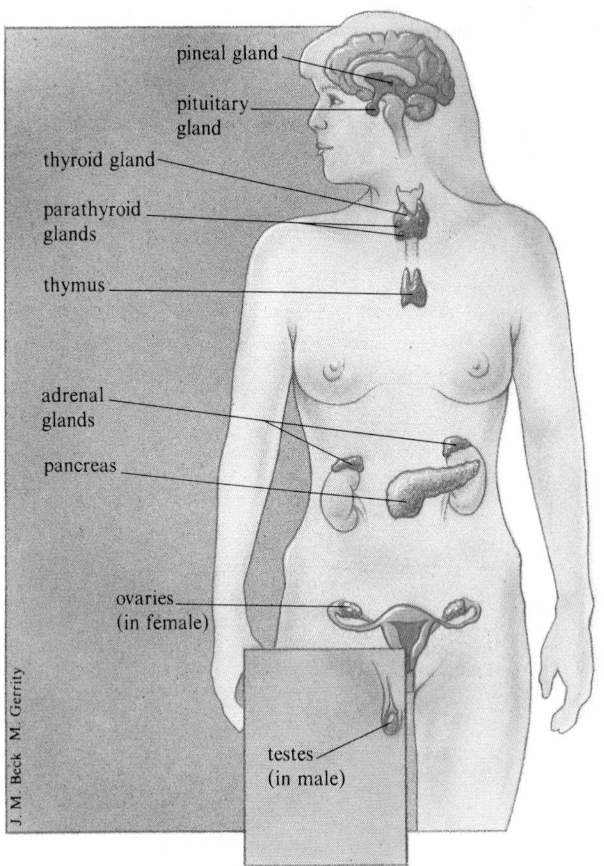

Figure 19.2
Location of the endocrine glands. The endocrine system includes several glands that contain specialized cells that secrete hormones, as well as cells scattered among some of the other organ systems that also secrete hormones.

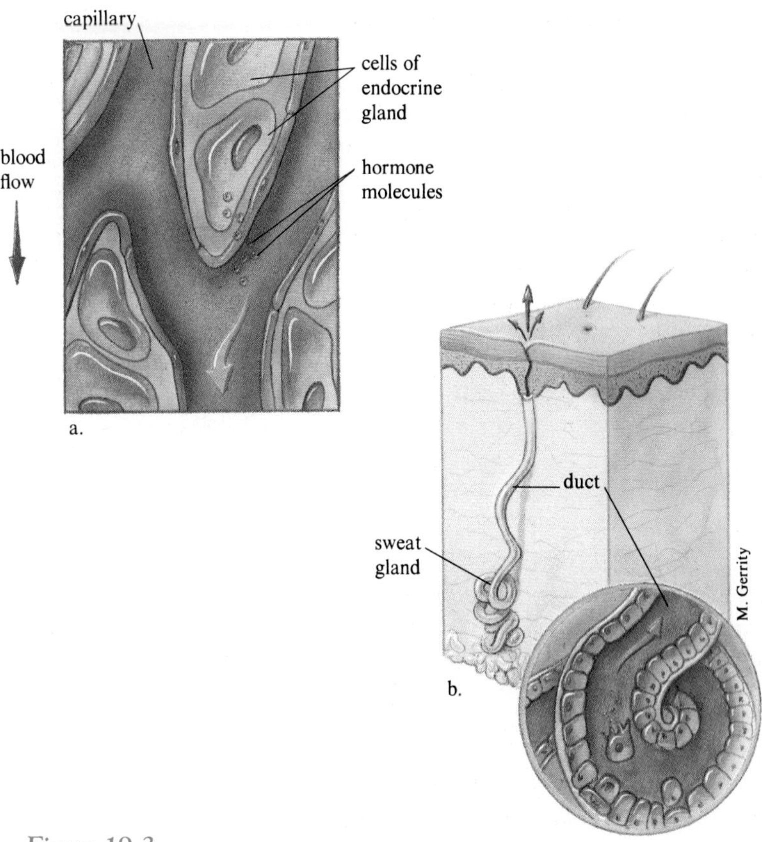

Figure 19.3
Endocrine glands (*a*) release their hormones directly into the circulatory system, but exocrine glands (*b*) release their secretions through ducts.

How Hormones Exert Their Effects

Most known hormones are either **peptide hormones,** which are short chains of amino acids, or **steroid hormones,** which are lipids. These two general types of hormones have different mechanisms of action.

Peptide Hormones

Peptide hormones are water soluble, so they are easily carried in the watery plasma of the bloodstream to virtually all of the body's cells. However, because the peptide hormones are lipid insoluble, the fatty plasma membranes of cells present a barrier to them. A peptide hormone itself never enters the target cell but binds to one of the hundreds or even thousands of receptors on the target cells that fit it like a key fits a lock. A peptide hormone binding to its receptor initiates the target cell's response, which might be the opening of specific ion channels in the membrane. As a result, the concentration of ions within the cell is altered, and this may change the rate of certain cellular activities.

Most commonly the formation of a complex between a hormone and a membrane receptor activates another substance within the cell that then triggers the cellular responses. This intracellular signaling of the hormone's effects is called a *second messenger* because it actually exerts the hormone's effects by activating other enzymes in the cell, which then produce the effects associated with the hormone.

A common second messenger is cyclic adenosine monophosphate, abbreviated cyclic AMP, or cAMP. When the hormone binds to the receptor, an enzyme called adenyl cyclase is activated on the inner face of the membrane (fig. 19.4). Adenyl cyclase then catalyzes the conversion of ATP into cyclic AMP (cAMP). Recall that ATP is the energy currency of the cell. In turn, cAMP activates specific enzymes within the cell. These enzymes catalyze the reactions within the cell that characterize the response to the hormone. Different hormones can stimulate cAMP formation, but it is the sequence of specific biochemical events following cAMP formation that produces the characteristic effects of different peptide hormones.

Parathyroid hormone is an example of a peptide hormone. It is synthesized in cells of the parathyroid glands in the throat. The hormone exits these cells and enters the bloodstream, eventually binding to receptors on specific kidney cells, where it converts ATP to cAMP. The cAMP sets into motion a series of biochemical reactions that cause the kidney tubules to reabsorb calcium.

Steroid Hormones

Unlike the peptide hormones, steroid hormones are small and lipid soluble, and therefore they can easily pass through the membranes of target cells. Once inside a cell, the steroid binds to a receptor in the cytoplasm, and the hormone-receptor complex enters the cell's nucleus. There, a particular gene is stimulated to direct the manufacture of a particular protein. The steroid hormone testosterone, for example, activates genes in muscle cells that direct the synthesis of muscle proteins (fig. 19.5). Steroid hormones are synthesized from cholesterol, which is one reason why some cholesterol is essential for health.

KEY CONCEPTS

Peptide hormones bind to target cell receptors but do not enter the cells. They trigger a second messenger, usually cAMP, which sets into motion the biochemical reactions that produce the cell's response. Steroid hormones are lipid soluble and can therefore cross cell membranes to enter target cells, where they activate genes whose products carry out the cell's response.

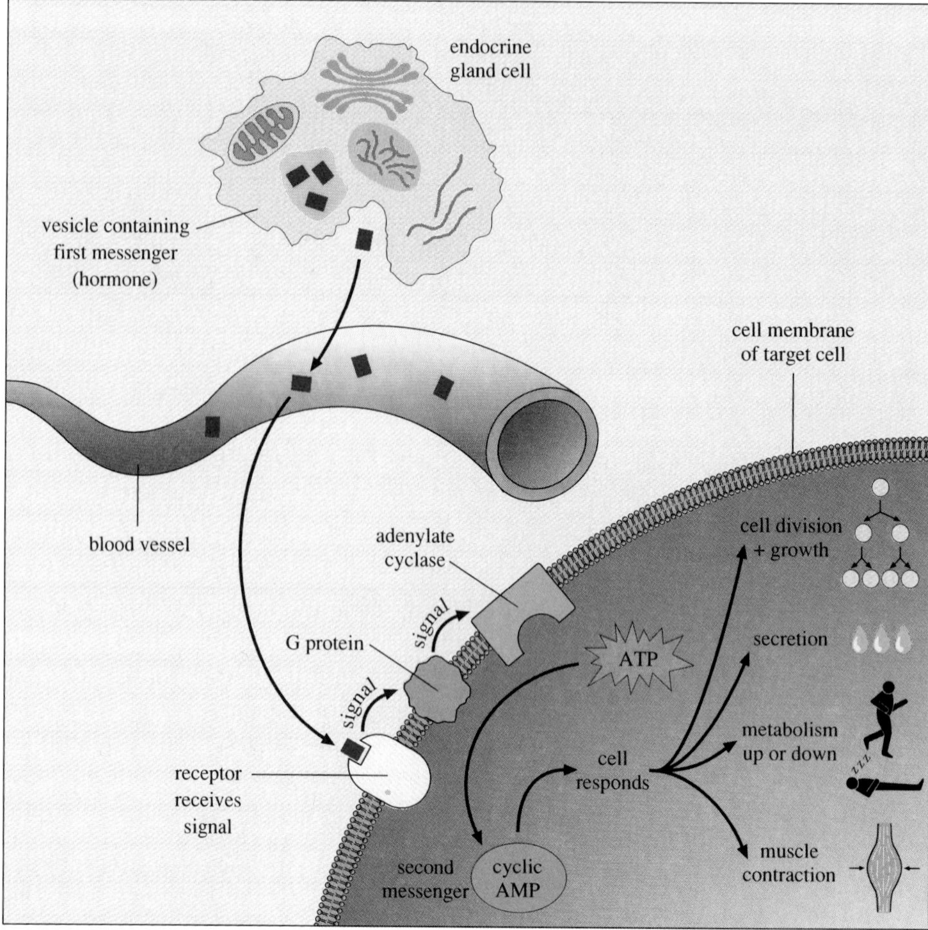

Figure 19.4
Mechanism of action of peptide hormones. A peptide hormone is synthesized under the direction of a gene in a cell of the endocrine system. After leaving the cell, the peptide hormone enters a capillary and travels in the bloodstream, eventually exiting into the tissue fluid. The hormone binds to a cell that has receptors specific for it on its surface. When the hormone binds to the receptor, the enzymatic conversion of ATP to cAMP is triggered on the inner face of the target cell's membrane. This reaction sets into motion other chemical reactions, which produce the physiological effect associated with the particular hormone.

Control of Hormone Levels

Hormones must be present at certain levels to work effectively. Hormone levels are altered by changes in the level of ions or nutrients in the cellular environment, instructions from the nervous system, and directives from other hormones.

Feedback Loops

To maintain a stable internal chemical environment in the body (homeostasis), the levels of specific hormones in the bloodstream at a given time are strictly regulated. The level of a particular hormone in the blood can often maintain itself by means of a complex interaction between the hormone and its precursors called a **feedback loop.**

Negative feedback loops are the most common. In such a control system, when a certain biochemical accumulates to above-normal levels, its synthesis slows or temporarily halts. This keeps the level of the biochemical relatively constant. Negative feedback operates much like a thermostat—when heat builds up and a room's temperature rises, the thermostat turns off the heat source until it is needed again. Hormones controlled by negative feedback act in the same way.

A negative feedback loop, for example, keeps blood-glucose levels within limits. After you eat a meal rich in carbohydrates, the digestive system breaks down the complex carbohydrates into the simpler carbohydrate glucose, which enters the circulation in the walls of the small intestine. The resulting rise in blood-sugar level stimulates certain cells of the pancreas to secrete the hormone insulin. Insulin stimulates target cells to admit glucose, which is then utilized in the cell to generate energy or is stored as the polymer glycogen. As glucose leaves the bloodstream to enter cells, the rate of insulin secretion from the pancreas decreases. Rising insulin levels create the situation that eventually turns off insulin production as the hormone's "job" is done (fig. 19.6).

In the rare **positive feedback loop,** an accumulating biochemical increases its own production. For example, at the onset of labor, the uterus contracts, which releases the hormone oxytocin and hormonelike substances called prostaglandins. These substances intensify the uterine contractions. The contractions further stimulate

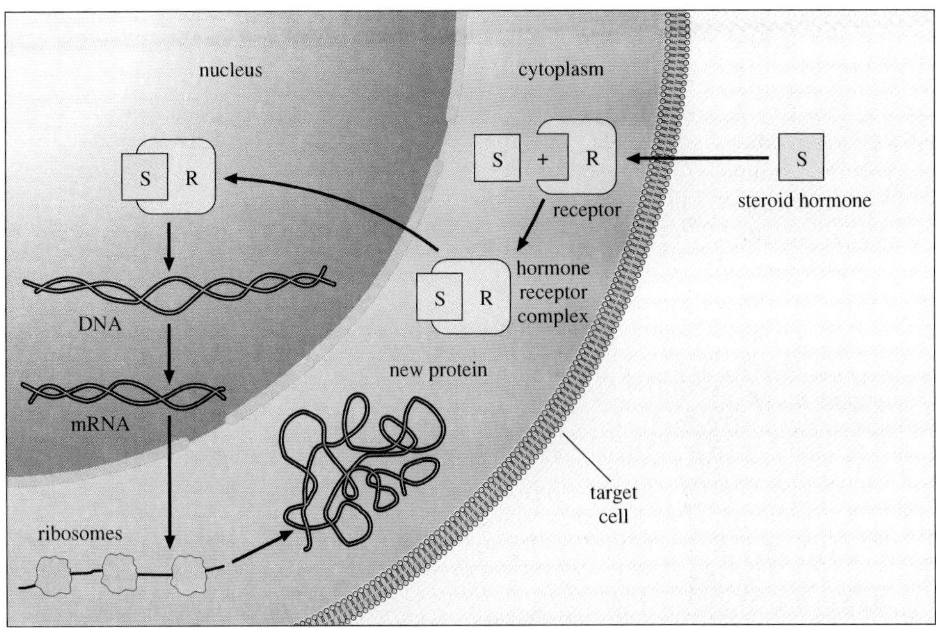

Figure 19.5
Mechanism of action of steroid hormones. A steroid hormone is synthesized in certain cells of the endocrine system. The hormone exits the cell and is carried in the bloodstream to target tissues. Once at a target cell, the steroid hormone crosses the cell membrane and binds to an intracellular protein receptor. The hormone-receptor complex enters the target cell's nucleus, where it activates a particular gene to produce its corresponding protein. It is the action of this protein that produces the effect associated with the specific steroid hormone.

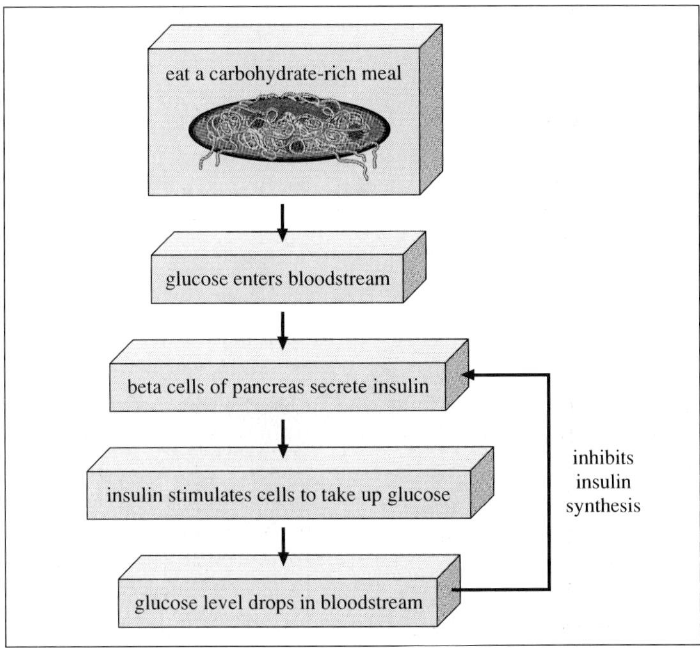

Figure 19.6
Control of blood-glucose level by negative feedback. The digestion of carbohydrates sends glucose into the bloodstream, which triggers insulin secretion by the pancreas. As the insulin stimulates cells to take up glucose, the blood-glucose level drops, signaling insulin output to decrease.

the production of oxytocin and prostaglandins. The cycle of increase stops abruptly when the baby is born due to other hormonal changes.

Neuroendocrine Control—The Hypothalamus and the Pituitary

Hormone levels are regulated by the nervous system and by other hormones through the actions of two glands in the brain: the **hypothalamus** and the **pituitary.** The pituitary gland, a small endocrine gland in the brain, was once called the "master gland" because it produces several hormones that regulate the production of many other hormones. Indeed, the pituitary plays a critical role in coordinating hormone synthesis. Like the conductor of a band, the pituitary makes many other endocrine glands work together harmoniously rather than allowing each to march to the beat of a different drummer.

The pituitary, however, takes its orders from the hypothalamus, a region of the brain that, in addition to releasing its own hormones in response to nerve stimulation, produces other hormones that control the output of several pituitary hormones. There is, therefore, a chain of command in the production of many hormones: the hypothalamus directs the pituitary output, which controls the secretion of hormones by other endocrine glands. A hormone produced by one gland that influences the secretion of a hormone by another gland is called a **tropic hormone,** or if its source is the hypothalamus, it is called a **releasing hormone.**

The pituitary gland dangles by a stalk from the brain's hypothalamus (fig. 19.7). Structurally and functionally, the pituitary gland is actually two glands, with each half releasing a different set of hormones and regulated by the hypothalamus in a different manner. Between the anterior and posterior lobes of the pituitary is an intermediate region known to produce at least one hormone.

The posterior hypothalamus develops from the central nervous system, so its close relationship with the brain is not surprising. In fact, all parts of the brain are touched by nerve fibers extending from the hypothalamus, and so the hypothalamus is

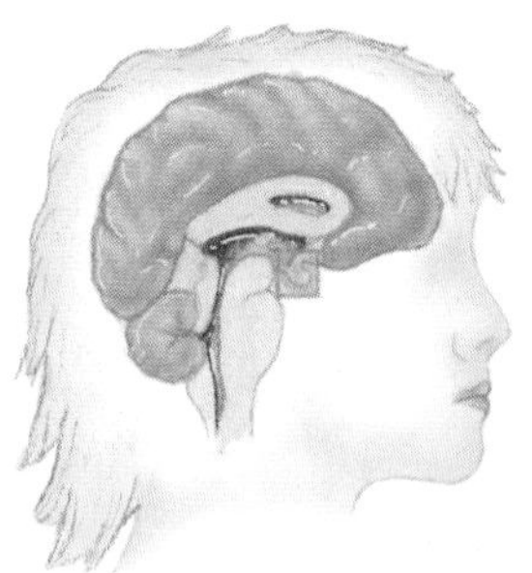

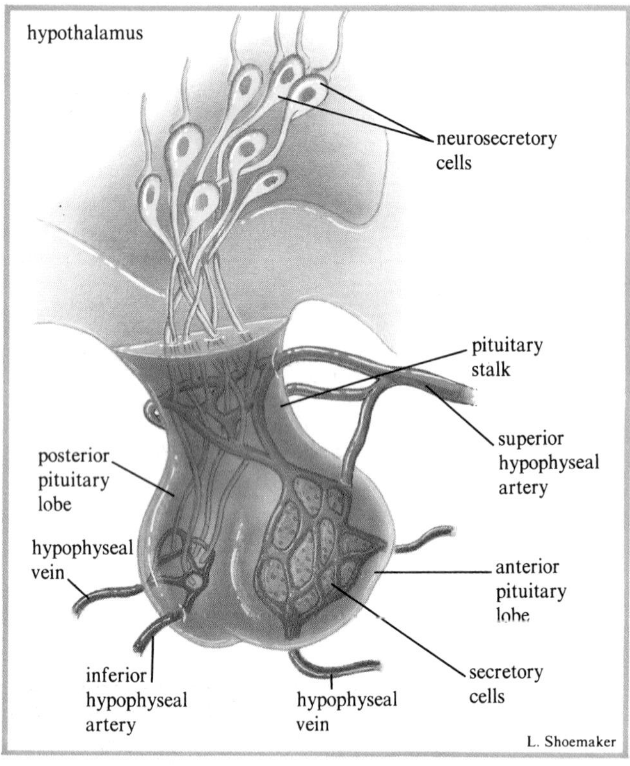

a.

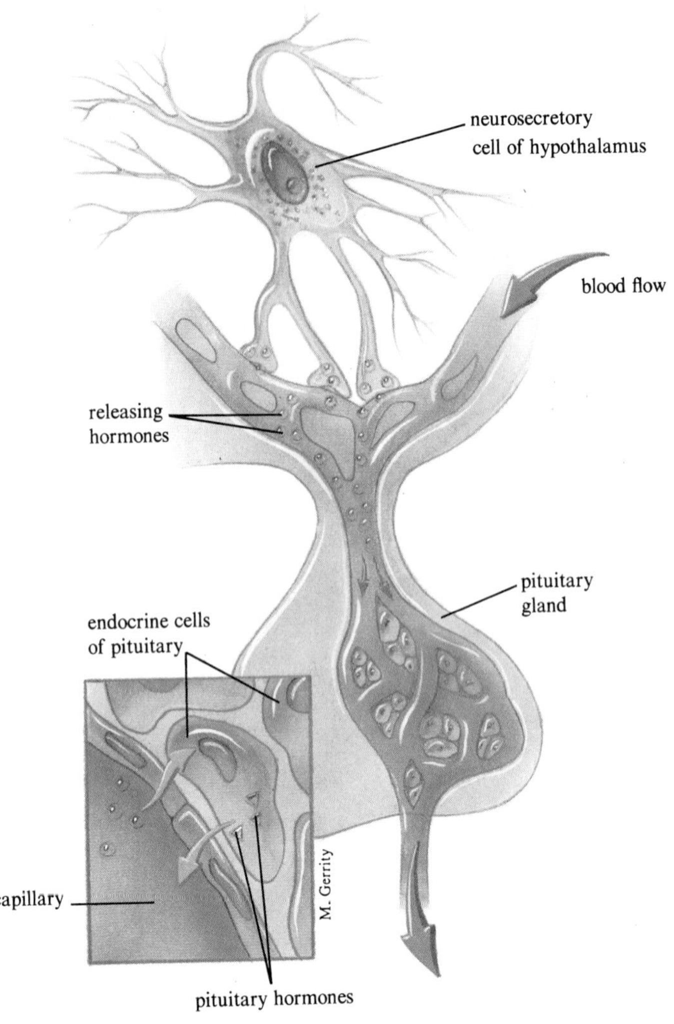

b.

Figure 19.7
The hypothalamic pituitary connections. *a.* Neurosecretory cells in the hypothalamus secrete hormones into a portal system of capillaries. These hormones travel to the anterior pituitary, where they control the release of its hormones. Other neurosecretory cells in the hypothalamus produce the hormones oxytoxin and ADH. These hormones are stored in and released from the posterior pituitary gland. *b.* This close-up diagram shows the secretion of releasing hormones from neurosecretory cells in the hypothalamus. These hormones are carried to the anterior pituitary gland and affect its activity.

a link between the nervous and endocrine systems. The hormones released from the posterior pituitary, ADH and oxytocin, are actually synthesized by **neurosecretory cells** in the hypothalamus. These cells function like neurons at one end, receiving neurotransmitters from other nerve cells and generating action potentials, but like endocrine cells at the other end, secreting a hormone. ADH and oxytocin are transported through the stalk along the axons of nerve cells that produced them to the posterior lobe of the pituitary, where they are stored within the nerve cell endings. When the neurosecretory cells are stimulated by neural activity in the brain, they release their hormones.

The anterior lobe of the pituitary is also controlled by the hypothalamus but in a different manner. Other neurosecretory cells in the hypothalamus secrete hormones that either stimulate or inhibit the production of anterior pituitary hormones. These hypothalamic hormones are delivered to the anterior pituitary through a specialized system of vessels.

Following is a closer look at the specific hormones.

KEY CONCEPTS

Hormone levels are influenced by feedback loops, the nervous system, and other hormones. In a negative feedback loop, a product's accumulation temporarily shuts off its synthesis. In a positive feedback loop, a product's accumulation triggers further synthesis. Under the direction of the hypothalamus, the pituitary integrates hormone function. The posterior pituitary releases ADH and oxytocin, which are synthesized by neurosecretory cells in the hypothalamus. The anterior pituitary secretes several hormones under the influence of other hypothalamic neurosecretory cells.

Pituitary Hormones

Anterior Pituitary

The anterior pituitary gland produces hormones that affect body parts directly and hormones that affect other endocrine glands (fig. 19.8). Growth hormone (GH) promotes growth and development of all tissues by increasing rates of protein synthesis and cell division. GH stimulates cells to take up more amino acids and to mobilize fat, and glucose is released from the liver to supply energy. GH increases cell division rate in cartilage and bone, thereby promoting height.

Levels of GH peak in the preteen years and cause the growth spurts of adolescence. When the levels of GH during the growing years are unusually high, the individual grows very tall. A deficiency of the hormone during childhood leads to **pituitary dwarfism** (fig. 19.9). Until recently, pituitary dwarfs were treated with growth hormone extracted from pituitary glands removed from cadavers. This treatment was costly and dangerous, because a serious viral infection of the brain was sometimes passed to patients receiving growth hormone pooled from many donors.

Human growth hormone can now be manufactured using recombinant DNA technology (chapter 13), and so it is more plentiful and pure. But it is sometimes put to questionable uses. For example, some people are requesting it to treat normal children who are of short stature. GH given to healthy, elderly men builds up their muscles and reduces fat. Some dairy cows are given injections of bovine growth hormone (the cow version of human growth hormone), which bolsters their natural supplies, causing them to manufacture more milk. The hormone does not affect milk quality or human health.

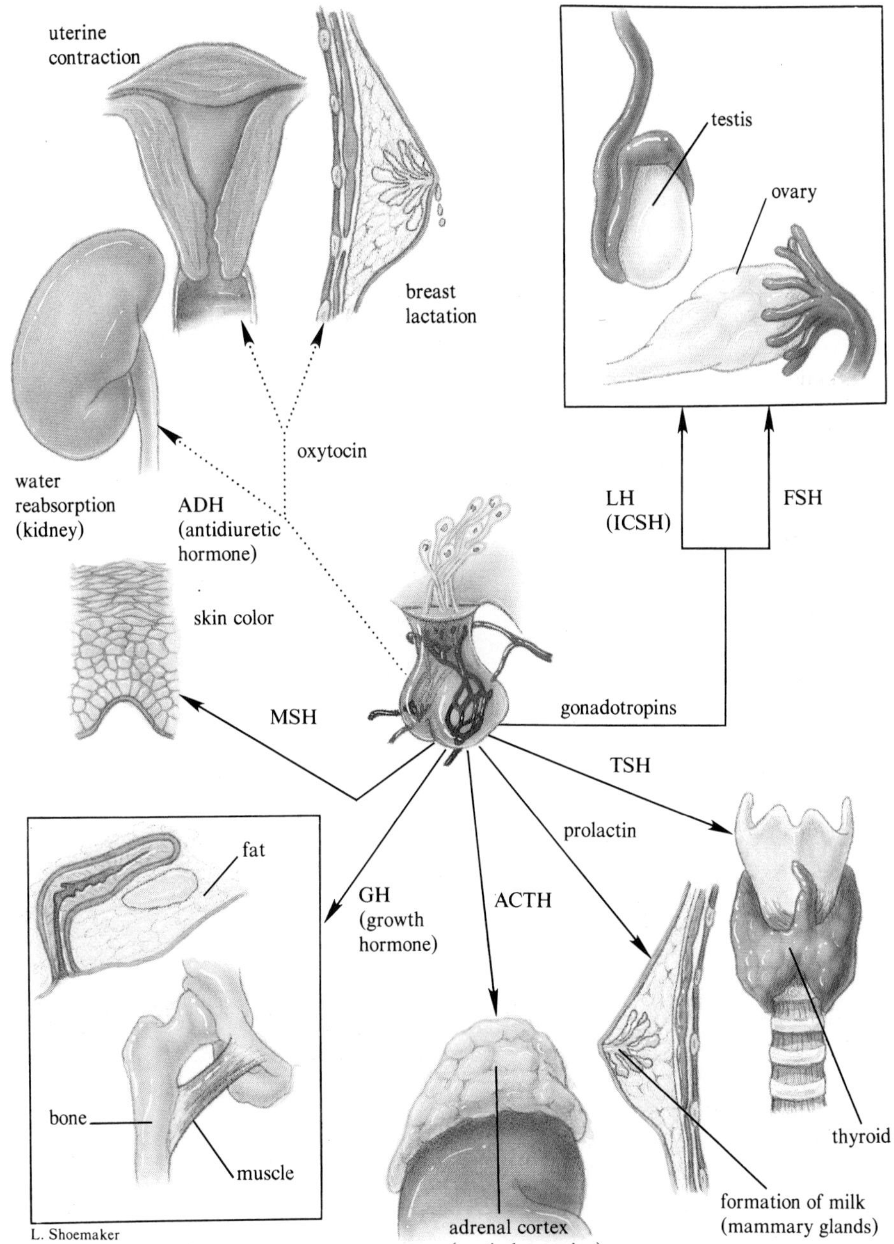

uterine contraction
breast lactation
testis
ovary
oxytocin
water reabsorption (kidney)
ADH (antidiuretic hormone)
LH (ICSH)
FSH
skin color
MSH
gonadotropins
TSH
prolactin
fat
GH (growth hormone)
ACTH
bone
muscle
thyroid
formation of milk (mammary glands)
adrenal cortex (cortisol secretion)
L. Shoemaker

Figure 19.8
The hormones of the pituitary gland have many functions. The anterior lobe of the pituitary gland secretes six hormones—thyroid stimulating hormone (TSH), prolactin (PRL), follicle stimulating hormone (FSH), luteinizing hormone (LH) (called interstitial cell stimulating hormone (ICSH) in males), adrenocorticotropic hormone (ACTH), and growth hormone (GH). The posterior lobe of the pituitary gland secretes oxytocin and antidiuretic hormone (ADH), which is also called vasopressin. The intermediate lobe produces melanocyte stimulating hormone (MSH).

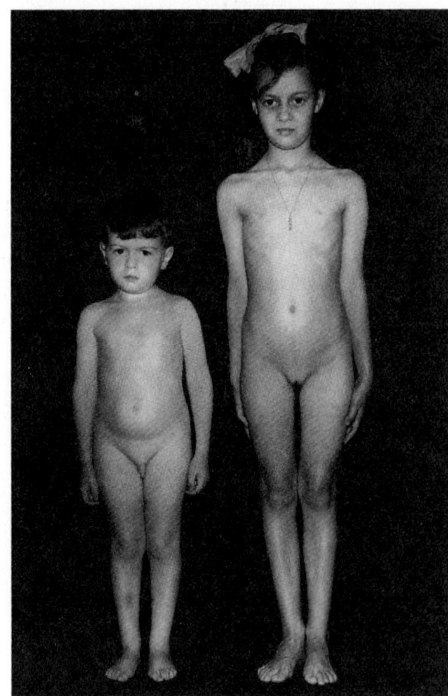

Figure 19.9
Body height is influenced by the amount of growth hormone present during adolescence. These children are twins, but the child on the left has a deficiency of growth hormone, which is responsible for his short stature.

Figure 19.10
Acromegaly is caused by an overproduction of growth hormone beginning in adulthood. The limb bones are no longer responsive to growth hormone in adulthood, however, the bones of the hands and face are responsive. These bones increase in size considerably, as can be seen by the changes in the facial features of this woman as she ages: (*a*) 9 years, (*b*) 16 years, (*c*) 33 years, (*d*) 52 years.

Excess growth hormone affects health. A tumor of the pituitary gland including cells that secrete GH causes overgrowth of many tissues. In a child, such a tumor produces a **pituitary giant.** In an adult, overall height is not greatly affected by excess GH because the growth regions in the long bones are no longer active. Instead, excess GH results in **acromegaly,** a thickening of the bones that are capable of responding, most noticeably in the hands and face (fig. 19.10).

Several anterior pituitary hormones affect reproduction. **Prolactin** stimulates milk production in a woman's breasts after she gives birth. In males and in women who are not breast-feeding, a prolactin-inhibiting chemical produced in the hypothalamus suppresses prolactin synthesis in the anterior pituitary. The inhibition is prevented in nursing mothers by nerve impulses generated when the infant sucks on the nipples.

The remaining hormones of the anterior pituitary influence the level of hormone secretion by other endocrine glands (fig. 19.11). The **gonadotropic hormones** produced in the anterior pituitary affect the gonads, which are the ovaries in the female and the testes in the male. In the human female, **follicle stimulating hormone (FSH)** leads to the development of ovarian follicles (the cells surrounding developing oocytes), the maturation of oocytes, and the release of the hormone estrogen from the follicles. In the male, FSH promotes the development of

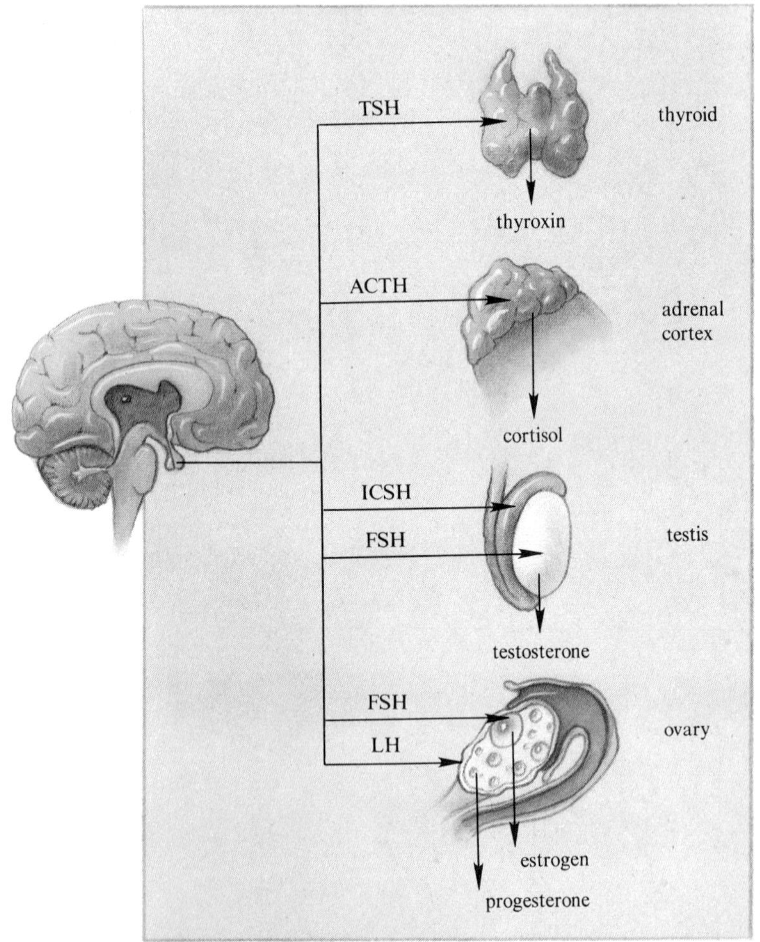

Figure 19.11
Several of the hormones of the anterior pituitary regulate the production of hormones by other endocrine glands.

the testes and the manufacture of sperm cells. In the female, **luteinizing hormone (LH)** causes release of an oocyte from an ovary each month. In the male, LH is known as **interstitial cell stimulating hormone (ICSH),** and it prompts the testes to produce the hormone testosterone.

Thyroid stimulating hormone (TSH) causes the thyroid gland in the throat to release its two hormones. **Adrenocorticotropic hormone (ACTH)** prompts the release of the glucocorticoid hormones from the outer portion (cortex) of the adrenal glands, which are located on top of the kidneys.

KEY CONCEPTS

The anterior pituitary produces growth hormone, which increases the rates of protein synthesis and cell division in all tissues; prolactin, which stimulates milk production; the gonadotropic hormones FSH, LH, and ICSH, which affect the gonads; thyroid stimulating hormone; and adrenocorticotropic hormone, which releases hormones from the adrenal cortex.

Posterior Pituitary

Two hormones are synthesized in the hypothalamus and stored in and released from the posterior pituitary. One of these hormones has two different names, **vasopressin** and **antidiuretic hormone (ADH).** This hormone has been called vasopressin because it causes the smooth muscle cells lining the blood vessels to contract, thereby raising blood pressure. It is called antidiuretic hormone because it makes the collecting ducts of the kidneys more permeable to water so that more water can be reabsorbed by the body.

Control of ADH secretion helps maintain the chemical balance of body fluids. Specialized cells in the hypothalamus called **osmoreceptors** sense the concentration of water in the blood. If the blood is too concentrated, the osmoreceptor cells stimulate the posterior pituitary to release more ADH. As ADH causes the blood to become more diluted, the osmoreceptor cells signal the cells in the hypothalamus that manufacture ADH to decrease their activity. Disruption of the synthesis or release of ADH can produce **diabetes insipidus,** in which a person drinks excessively and passes large volumes of very watery urine.

Oxytocin is the other hormone produced in the hypothalamus and stored and released from the posterior pituitary. This hormone contracts both the myoepithelial cells in the breasts, causing milk to be released when a baby nurses, and in the uterine muscles, causing the force that delivers the newborn during labor. A synthetic form of oxytocin is often given to induce labor or to accelerate labor contractions in a woman who is becoming exhausted from the effort of giving birth.

Intermediate Region

Melanocyte stimulating hormone (MSH) is produced in a region of the pituitary between the anterior and posterior lobes. MSH is responsible for color changes observed in many vertebrates. This hormone binds to receptors on pigment cells in the skin called melanocytes. The bound MSH causes pigment granules to disperse, darkening the skin (fig. 19.12).

KEY CONCEPTS

ADH raises blood pressure by contracting the smooth muscles in blood vessel walls and enables the kidneys to reabsorb more water. Oxytocin contracts myoepithelial cells in the breasts and the uterine muscles. MSH darkens skin.

a.

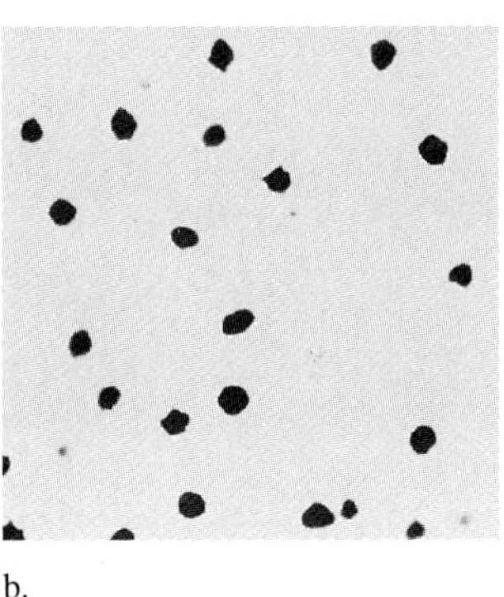

b.

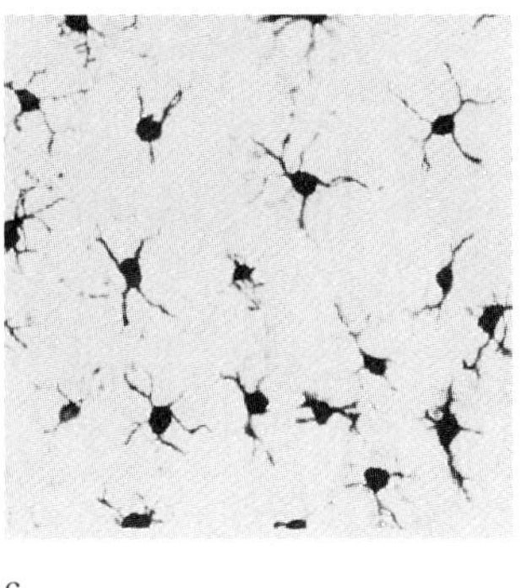

c.

Figure 19.12
Effect of melanocyte stimulating hormone (MSH) on skin coloration. MSH causes the dispersion of pigment molecules in pigment-containing structures of cells. This darkens skin color. The skin color of the amphibians (*a*) and the distribution of pigment in the organelles (*b*) are shown. MSH is best studied in fish, amphibians, reptiles, and some mammals. The effect in humans, if any, is unknown.

The Thyroid Gland

The thyroid is a two-lobed gland located at the front of the larynx and trachea in the throat (fig. 19.13). Two hormones from the thyroid, **thyroxine** and **triiodothyronine,** increase the rate of cellular metabolism. Different cells within the thyroid produce a third hormone, **calcitonin,** which decreases blood calcium levels under certain conditions. Iodine is required in the diet to make thyroxine and triiodothyronine. The gland concentrates iodine, so levels there are 25 times higher than levels in the blood.

The thyroid hormones bind to many different cell types, where their effect is to speed metabolism by increasing the number of enzymes that take part in cellular respiration. The specific nature of this activity depends upon cell type. Under thyroid stimulation, gas exchange in the lungs occurs faster, the small intestine absorbs nutrients more readily, and the fat levels in cells and in blood plasma are lowered.

Abnormal levels of thyroid hormones influence health. The effects of an underactive thyroid gland are termed **hypothyroidism.** The metabolic rate slows and, because fewer calories are burned, the person gains a great deal of weight despite a poor appetite. Heartbeat slows, and blood pressure and body temperature are lower than normal. Hypothyroidism beginning at birth is called **cretinism,** and the child is physically and mentally retarded. Fortunately, a blood test at birth can detect cretinism, and if thyroxine is given before 3 months of age, the child develops normal intelligence. Hypothyroidism beginning in adulthood is termed **myxedema** and is characterized by lethargy, a puffy face, and dry, sparse hair.

Hypothyroidism due to a lack of iodine in the diet causes the gland to swell and become visible as a lump in the neck called a **goiter** (fig. 19.14). Ancient Egyptian doctors found that this type of goiter could be cured by giving the patient seaweed, although they did not realize that it was the iodine in the seaweed that was reversing the condition by allowing more thyroid hormone to be produced. Today goiter due to iodine deficiency is rare in the United States and Canada because iodine is added to table salt.

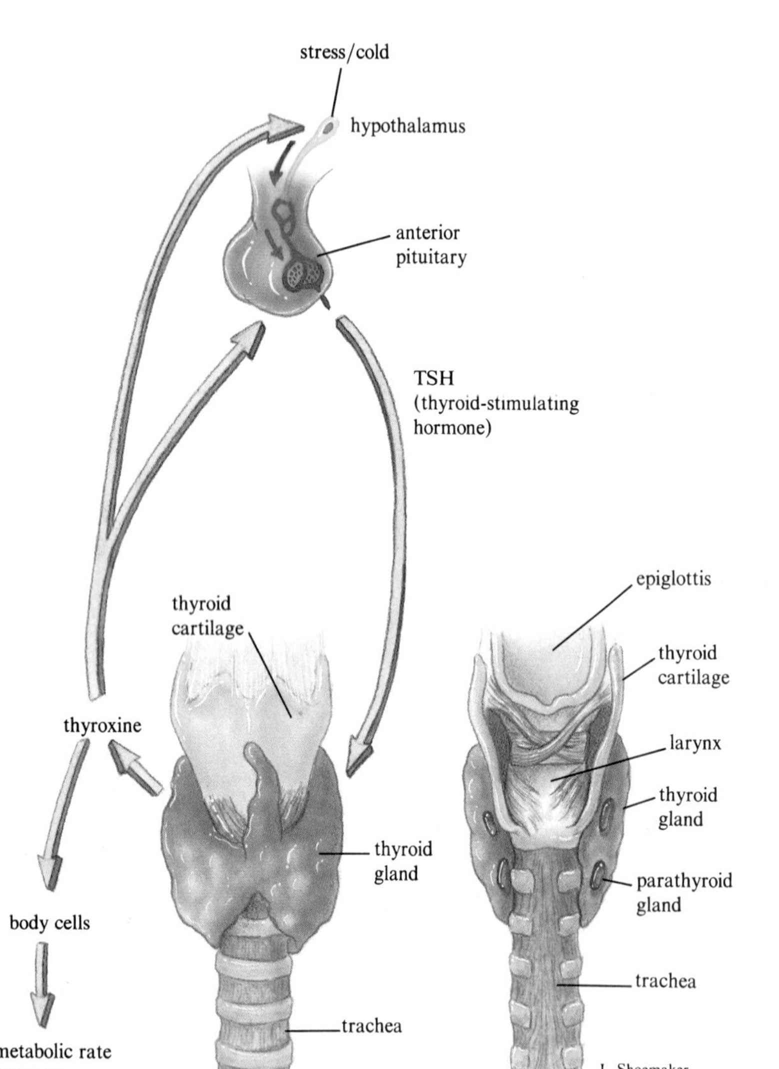

Figure 19.13
The thyroid gland surrounds the front of the voice box. It produces hormones that increase the metabolic rate of most cells. The production of thyroid hormones is regulated by a negative feedback loop involving thyroid stimulating hormone from the anterior pituitary.

a.

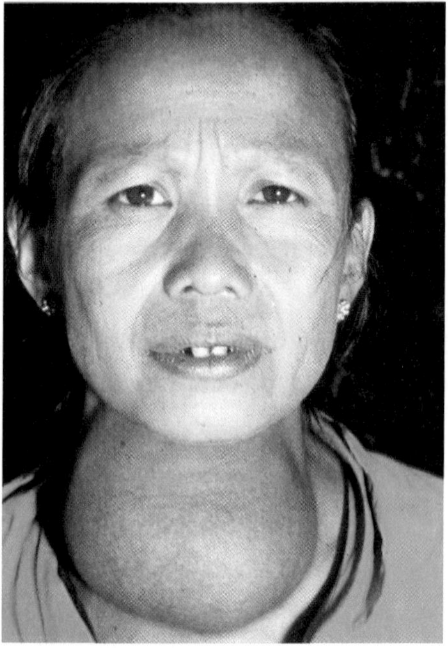

b.

Figure 19.14
a. Leonardo da Vinci drew this likeness of a man with a goiter in the fifteenth century. *b.* A woman with a goiter.

An overactive thyroid, **hyperthyroidism,** produces a swelling in the neck called a toxic goiter. Other symptoms of hyperthyroidism reflect the dangerously accelerated metabolism caused by the abnormally high levels of the thyroid hormones. The person has a very short attention span and is irritable and hyperactive. Heart rate, blood pressure, and temperature are elevated. Appetite is great, but the high metabolic rate keeps the person thin. In some cases, the person's eyes protrude, perhaps as a result of swelling behind the eyeball. Drugs called goitrogens are used to inhibit thyroid function, but in severe cases part of the gland must be removed or destroyed by radioactive iodine.

KEY CONCEPTS

The thyroid gland in the throat produces thyroxine, triiodothyronine, and calcitonin. Thyroxine and triiodothyronine require iodine, and they speed metabolism. Hypothyroidism results in a slowed metabolism, with symptoms of low blood pressure and body temperature, slow heartbeat, and goiter. Hyperthyroidism dangerously raises metabolic rate.

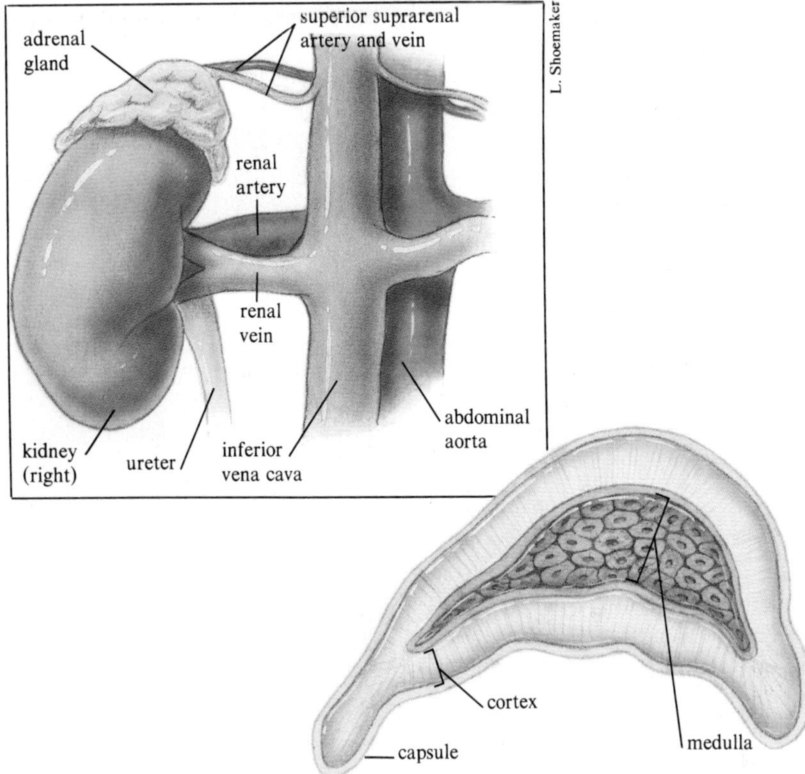

Figure 19.15
The adrenal glands are located on top of the kidneys. Each gland is built of two structurally, functionally, and developmentally distinct parts, the cortex and the medulla.

The Parathyroid Glands

Embedded in the thyroid gland are four small groups of cells called the **parathyroid glands** (fig. 19.13). **Parathyroid hormone** maintains levels of calcium in blood and tissue fluid by releasing calcium from bones and by enhancing calcium absorption through the digestive tract and kidney. Note that the action of parathyroid hormone is opposite that of calcitonin. Parathyroid hormone also inhibits reabsorption of phosphate by the kidneys.

Calcium is vital to many biological functions—muscle contraction, nerve impulse conduction, blood clotting, bone formation, and the activities of many enzymes. Because of calcium's importance in human physiology abnormal levels of it can drastically affect health. In fact, underactivity of the parathyroids can be swiftly fatal. Excess parathyroid hormone (hyperparathyroidism) causes calcium to be removed from bones faster than it is deposited, resulting in easily broken bones. This condition, called **osteoporosis,** is most common in women who have reached menopause (the cessation of menstrual periods). The decrease in the hormone estrogen that accompanies menopause is believed to make bone-forming cells more sensitive to parathyroid hormone, which results in the depletion of bone mass.

KEY CONCEPTS

The parathyroid glands are four cell clusters embedded in the thyroid gland. Parathyroid hormone controls the level of calcium in the bones and blood.

The Adrenal Glands

The paired **adrenal glands** ("ad" means near, "renal" means kidney) sit on top of the kidneys (fig. 19.15). Each adrenal gland is built of two parts. The inner portion of each gland is called the **adrenal medulla,** and the outer portion is called the **adrenal cortex.** The adrenal medulla and cortex differ in structure, function, and embryonic origin and can really be considered separate, but physically attached, glands.

The Adrenal Medulla

The hormones of the adrenal medulla, **epinephrine** (also known as adrenaline) and **norepinephrine** (also called noradrenaline), help the body deal with an emergency. These hormones are called **catecholamines.** Under the control of the sympathetic nervous system, which provides a quick response in the face of imminent danger, the adrenal medulla supplies a support system for the nervous system during the initial stages of a stress reaction. Recall the discussion in chapter 17 describing the "fight or flight" response. The responses to the hormones of the adrenal medulla are the same. Heart and breathing rates increase. Air passageways open, speeding oxygen delivery to the cells. Blood is shunted to the skeletal muscles and brain, regions of the body that will require more oxygen to fuel a swift and alert response. Glucose is mobilized from the liver to provide energy.

The Adrenal Cortex

The adrenal cortex secretes three types of steroid hormones: the mineralocorticoids, glucocorticoids, and sex hormones. The mineralocorticoids and glucocorticoids function together in times of prolonged stress to mobilize energy reserves while keeping blood volume and blood composition constant.

The **mineralocorticoids** maintain blood volume and electrolyte balance by stimulating the kidney to return sodium ions and water to the blood but to excrete potassium ions. The major mineralocorticoid, **aldosterone,** maintains the level of sodium ions in the blood by altering the amount reabsorbed in the kidneys. Since water reabsorption follows sodium ion reabsorption, aldosterone also affects blood volume. This action becomes particularly important in compensating for fluid loss in severe bleeding.

The **glucocorticoids,** the most important of which is **cortisol,** are essential components of the body's response to prolonged stress. Whereas the adrenal medulla braces the body against an immediate danger, the glucocorticoids take over for the long haul, for instance during an infection. Glucocorticoids affect carbohydrate, protein, and lipid metabolism. Specifically, they break down proteins into amino acids and then stimulate the liver to synthesize glucose from these freed amino acids. This glucose production provides a source of energy for healing after the immediate supply of glucose has been used up during stressful situations. If bleeding is excessive, the constriction of blood vessels indirectly caused by glucocorticoids will offset the drop in blood pressure caused by blood loss. In addition, the anti-inflammatory action of the glucocorticoids keeps the body's response to injury from becoming disruptive rather than protective.

A deficiency of hormones from the adrenal cortex causes Addison's disease. Lack of mineralocorticoids causes ion imbalance, which lowers water retention so that the person has low blood pressure and is dehydrated. Symptoms include weight loss, mental fatigue, weakness, and impaired resistance to stress. Lack of glucocorticoids causes a darkening of the skin. Fortunately, Addison's disease can be treated by providing hormones. Former president John F. Kennedy had the condition.

Excess adrenal cortical hormones causes Cushing's syndrome, in which body fat redistributes. Although the legs may be thin, an unusual amount of fat is deposited in the face, behind the shoulder blades, and in the abdomen. In addition, the person bruises easily and wounds heal slowly (fig. 19.16).

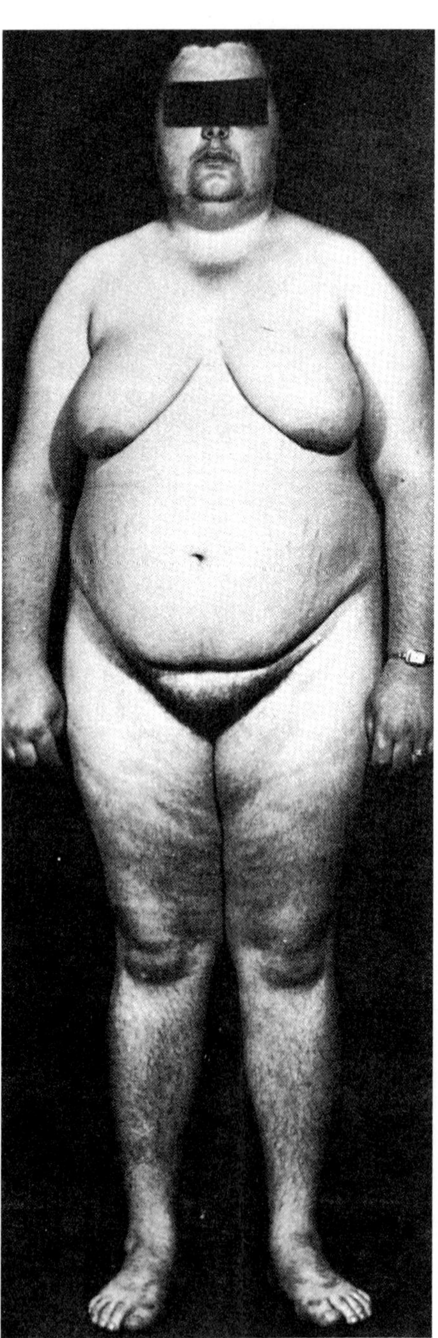

Figure 19.16
Cushing's syndrome. Excess adrenal cortex secretion causes an extreme redistribution of fat.

The cortex also secretes small amounts of **sex hormones.** Although both male and female sex hormones are produced, more male sex hormone (testosterone) is produced than female sex hormones (estrogen and progesterone).

KEY CONCEPTS

The adrenals are paired glands sitting on top of the kidneys. Each adrenal gland has two distinct parts. The adrenal medulla produces epinephrine and norepinephrine, which help the body respond to a threat. The adrenal cortex, the outer portion, secretes mineralocorticoids, which maintain blood volume and electrolyte balance; and glucocorticoids, which break down proteins into amino acids, which the liver converts to glucose. The cortex also secretes some sex hormones.

The Pancreas

The pancreas is a large gland located between the spleen and the small intestine (fig. 19.17*a*). It is multifunctional, producing digestive enzymes as well as hormones. Randomly located among the cells that produce the digestive enzymes are clusters of cells called **islets of Langerhans** (fig. 19.17*b*). These islets constitute the endocrine portion of the pancreas, and the hormones they secrete are polypeptides that regulate the body's utilization of nutrients.

Three hormones are produced by the islets, each by a different type of cell: **insulin** (Latin for "island"), **glucagon,** and **somatostatin.** The actions of insulin and glucagon oppose one another in regulating blood-glucose levels. Secreted when glucose levels are high, such as after a meal, insulin lowers the blood-sugar level by stimulating most of the body's cells to take up glucose from the blood and by stimulating reactions inside cells that metabolize glucose or store it as glycogen. In contrast, glucagon breaks down glycogen into glucose, thereby raising blood-glucose levels and providing energy between meals. Somatostatin controls the rate of nutrient absorption into the bloodstream.

Diabetes

Lack of insulin, or an inability of the body to utilize it, results in **diabetes mellitus** in which glucose cannot be used or stored properly, and is therefore present in the urine after fasting. Blood tests reveal other signs of diabetes. In the glucose tolerance test, a dose of glucose is given orally. A nondiabetic's blood glucose peaks after 2 hours at about 145 milligrams per 100 milliliters of blood. A diabetic's blood-glucose level sustains a peak of 300 milligrams or more for 4 hours or longer (fig. 19.18).

Nearly 11 million Americans suffer from diabetes mellitus. Of these, 15% have **insulin-dependent diabetes,** which is also called juvenile-onset diabetes because symptoms often first appear in childhood. In this more serious form of the disorder, insulin is not manufactured in sufficient quantity, so the person must inject insulin daily. Without it, the individual is very thirsty, has blurred vision, and is weak, tired, irritable, nauseated, and underweight. Untreated, a severe insulin deficiency can result in a diabetic coma that is ultimately lethal. Insulin-dependent diabetes is thought to be an autoimmune disease, in which the body perceives some of its own cells—in this case, those that produce insulin—to be foreign and therefore destroys them.

The more common form of the disease, **insulin-independent diabetes,** also known as adult-onset diabetes, reflects an inability to use insulin. The source of this problem may be defective insulin receptors on cell surfaces, starving the cells for glucose, or an abnormality in a recently-discovered pancreatic protein called amylin, which is thought to regulate insulin's activity. Symptoms include fatigue, itchy skin, blurred vision, slow wound healing, and poor circulation. Insulin-independent diabetes strikes twice as many women as men, more blacks than whites, and is more often seen in overweight individuals. Diet and exercise can help about 90% of these diabetics, but the remainder must receive drugs that enable the body to more effectively use available insulin. Reading 19.1 explores diabetes further.

Hypoglycemia

Hypoglycemia is a low level of glucose in the blood. When blood sugar levels drop below 45 milligrams per milliliter of blood, cells do not receive enough glucose, and a person feels weak, sweaty, anxious, and shaky. Hypoglycemia is caused by excess insulin. This is most likely to happen to a diabetic who injects too much insulin or to someone with a tumor of the insulin-producing cells of the pancreas. A healthy person might experience transient hypoglycemia following very strenuous exercise. Hypoglycemia is a rare condition that can often be relieved by following a diet of frequent, small meals that are low in carbohydrates and high in protein, which prevents surges of insulin that lower the blood-sugar level. The symptoms of hypoglycemia are also caused by many other conditions, as well as by stress.

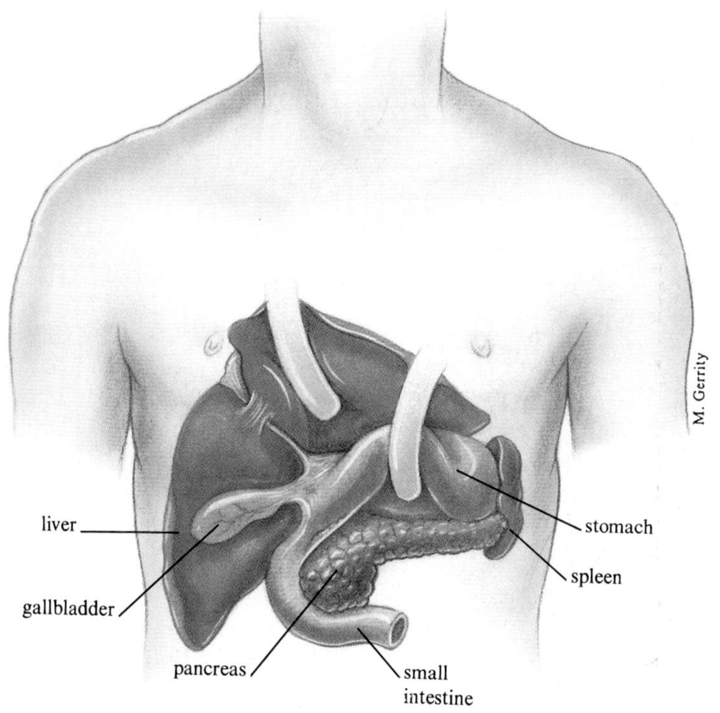

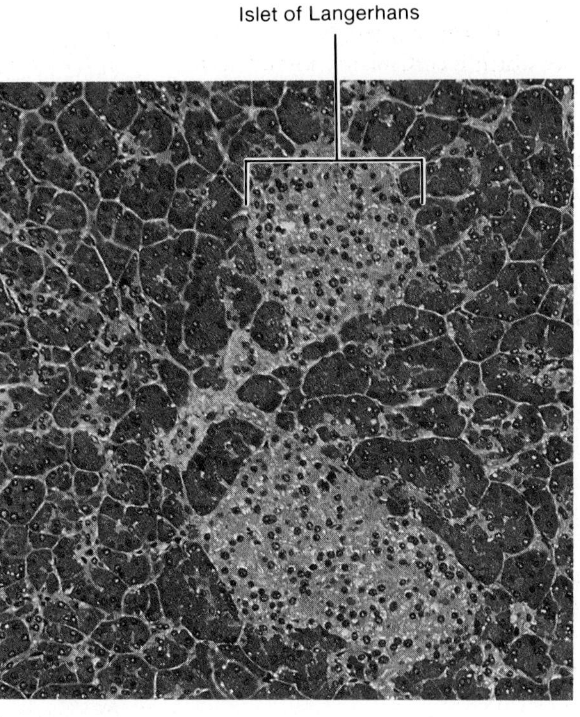

Figure 19.17
The pancreas. *a.* This structure is located above the small intestine and beneath the stomach. It is built of many lobes and produces digestive enzymes and hormones. *b.* Light micrograph of an islet of Langerhans within the pancreas.

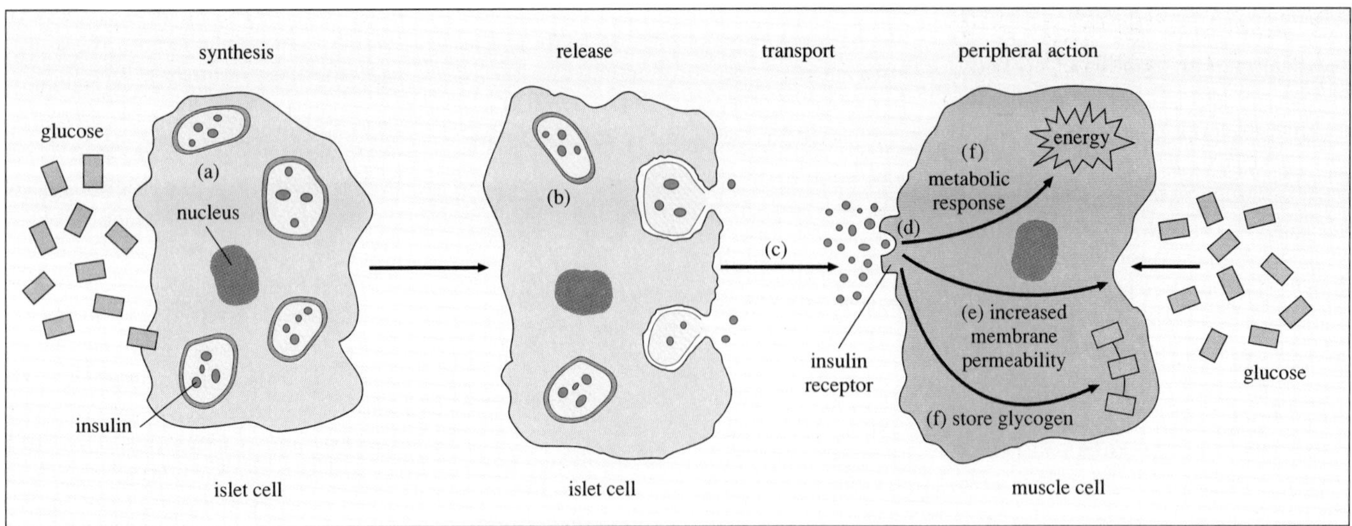

Figure 19.18
Diabetes is the result of abnormal insulin function. Diabetes can result from a defect at any point in the synthesis or functioning of insulin. This may occur in the pancreatic cells (*a*), where insulin is manufactured in response to increasing blood-glucose level. If the insulin-producing cells are absent or cannot synthesize or secrete normal insulin (*b*), diabetes can result. Once released from the cells, insulin must travel to receptor cells (*c*), but it can be destroyed en route by antibodies produced in an autoimmune attack. Once insulin reaches a target cell, it must bind to receptors there (*d*). Diabetes can result from deficient or structurally abnormal receptors. When insulin binds to receptors, the cell membrane allows glucose to enter by facilitated transport (*e*), and then the glucose is either metabolized to generate energy or is stored as the polymer glycogen (*f*). Defects in glucose uptake and utilization can also produce the symptoms of diabetes.

Reading 19.1 The Long History of Diabetes

The sweet-smelling urine that is the hallmark of diabetes mellitus was noted as far back as an Egyptian papyrus from 1500 B.C. In A.D. 96 in Greece, Aretaeus of Cappadocia described the condition as a "melting down of limbs and flesh into urine." By 1674, not much more about diabetes had been learned. At this time English physician Thomas Willis wrote, "The urine was wonderfully sweet as if imbued with honey or sugar." In the late 1700s and early 1800s, researchers probed the cause of the unusual sweet-smelling urine of the diabetic. In 1815, French chemist Michel Chevreul identified the sugar in the urine as the same found in grapes—glucose. Later in the nineteenth century, several investigators traced diabetes to the pancreas by removing the gland from animals and observing the rapid onset of diabetes.

Despite this progress, medical science still could not do much to help the diabetic of a century ago. In the 1880s, diabetes was thought to be caused by too much alcohol, too much sex, or too much of anything that "tends to impoverish the blood." Children who wet their beds because of their illness were punished. Treatments included avoiding carbohydrates, wearing warm clothing, and taking long sea voyages. Salt-water sponge baths were prescribed for the less affluent.

In 1921, a link was made between diabetes symptoms, sugary urine, and the pancreatic hormone insulin by Canadian physiologists Sir Frederick Grant Banting and Charles Herbert Best. To isolate insulin, they worked with two groups of dogs. The dogs of one group had their pancreases removed so that they developed severe diabetes. The dogs of the other group had the ducts that carry digestive enzymes from the pancreas tied off, which led to degeneration of the nonendocrine portion of the organ. After several days, the remaining pancreas consisted only of islets. Banting and Best then made an extract of the remaining pancreatic tissue and injected it into the dogs whose pancreases had been removed. The extract "cured" the induced diabetes.

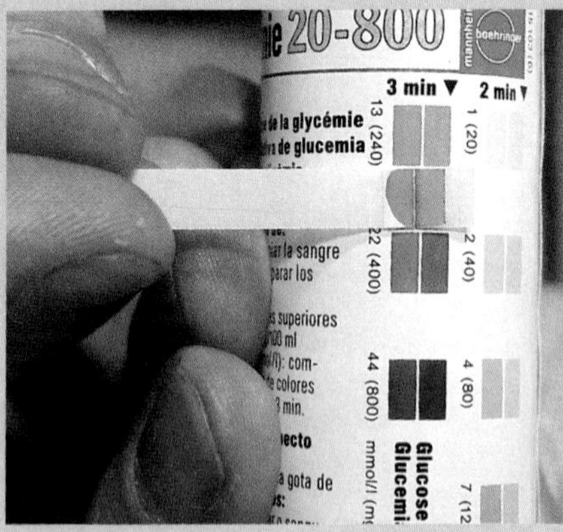

Figure 1
Today diabetes mellitus is a fairly well controlled and understood condition. Diabetics use a testing paper that can be purchased at drug stores to monitor their urine or blood-glucose levels, which can tell a particular patient when the next dose of insulin is needed or when the diet needs to be altered.

The association of diabetes with insulin insufficiency has saved millions of human lives, because insulin from animals such as pigs and cows is so chemically similar to the human molecule that most diabetics could inject animal insulin to relieve their symptoms. In 1982, the gene for human insulin was genetically engineered into bacteria, so that pure human insulin is available to all. The beginning of chapter 13 describes this landmark in medical history. Research efforts into diabetes today focus on delivering insulin in a way that closely mimics its normal secretion from the pancreas.

KEY CONCEPTS

The islets of Langerhans produce insulin and glucagon, which oppose each other in regulating blood-glucose level, and somatostatin, which controls nutrient absorption. In insulin-dependent diabetes, insulin production is insufficient. In insulin-independent diabetes, insulin is synthesized but not utilized normally. Diabetes results in glucose backing up into the blood. Hypoglycemia is low blood sugar, producing weakness and shakiness.

The Gonads

The Ovary

In a woman's monthly menstrual cycle, the hypothalamus stimulates the pituitary to release hormones, which in turn stimulate the ovaries to produce the steroid hormones **estrogen** and **progesterone.** Estrogen increases cell division rate in the vagina, uterus, and breasts and triggers a buildup of fat beneath a female's skin. Progesterone controls secretion patterns associated with reproductive function.

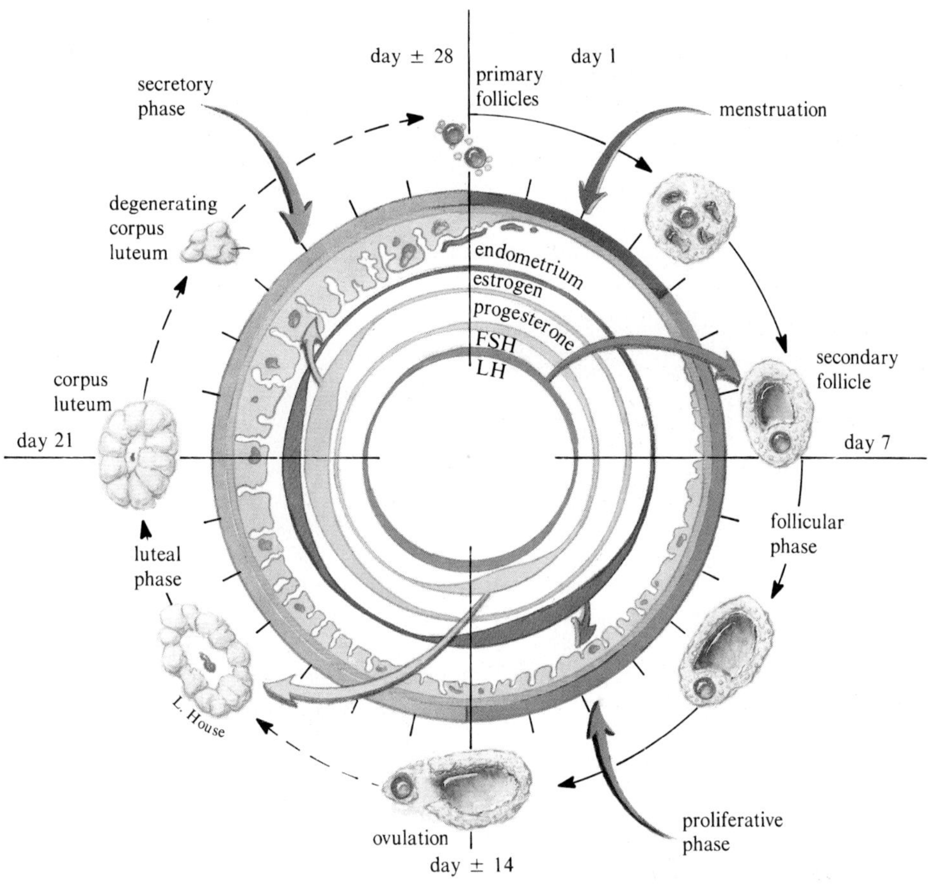

Figure 19.19
Three types of biological changes associated with the menstrual cycle are illustrated—the development of the oocyte within its follicle, the thickening of the uterine lining (the endometrium), and the changing levels of the hormones FSH, LH, estrogen, and progesterone. On the first day of the cycle, the oocyte is within its follicle, the uterine lining is actively being shed, and levels of all four hormones are low. The low levels signal the hypothalamus to release LHRH, which increases the anterior pituitary's output of FSH and then LH. By midcycle, LH level surges, which prompts the oocyte to burst out of its follicle. Meanwhile, the follicle has been producing estrogen. After ovulation, if conception does not occur, the ruptured follicle produces estrogen and progesterone. The uterine lining continues to build as the corpus luteum (the follicle without its oocyte) first enlarges and then degenerates. Finally, by the end of the cycle, the decreasing levels of estrogen and progesterone cause the padding of blood and tissue in the uterine lining to be shed.

On the first day of menstrual bleeding, blood levels of estrogen and progesterone are low (fig. 19.19). The hypothalamus detects these low levels and sends **luteinizing hormone releasing hormone (LHRH)** to the anterior pituitary, stimulating it to release large amounts of follicle stimulating hormone (FSH) and small amounts of luteinizing hormone (LH). The increase in FSH prompts the largest follicle in the ovary to produce estrogen. The ensuing increase in blood-estrogen level stimulates growth of the lining of the uterus in preparation for pregnancy. The rising estrogen level also lowers the level of FSH in the blood, which stimulates the hypothalamus to produce more LHRH. The rising LHRH level dramatically increases production of LH in the anterior pituitary. On day 14 of the cycle, blood levels of LH peak, causing the largest ovarian follicle to release an oocyte. This is **ovulation,** which is when a woman is at her most fertile.

After ovulation, the follicle cells that had surrounded the released oocyte send biochemical signals to lower blood-estrogen level. The declining estrogen supply lowers levels of LHRH and therefore lowers levels of LH too. The follicle now enlarges to form a gland, the **corpus luteum,** which produces estrogen and progesterone. The increasing levels of estrogen and progesterone inhibit production of LH and FSH and stimulate the buildup of the uterine lining.

It is now 9 days since ovulation. If a fertilized ovum has implanted in the uterine lining, a new hormone, **human chorionic gonadotropin (HCG),** is secreted by the fertilized ovum into the woman's blood and urine. Detection of HCG forms the basis for pregnancy tests. If conception does not occur, the progesterone level in the blood continues to rise, which inhibits production of LHRH. As a result, estrogen and progesterone levels in the blood fall, leading to the breakdown of the thick uterine lining. This tissue passes out of the vagina as the menstrual flow. Estrogen and progesterone levels in the blood are low, and the cycle begins anew. The menstrual cycle usually begins at about age 12 and ceases around age 45. Reproductive cycles differ in different species (Reading 19.2).

Reading 19.2 Sexual Seasons

THE HUMAN FEMALE IS UNUSUAL AMONG ANIMALS IN THAT SHE IS SEXUALLY RECEPTIVE AT ALL TIMES. Because of her monthly cycle, she is referred to as a **cyclic ovulator.** Many women may be aware of shifting moods and shifting sexual feelings throughout the month, but most women cannot pinpoint when they ovulate.

Most female mammals are **seasonal ovulators** and go through periods of estrus (heat) when they are both sexually receptive and ovulating. The coinciding of sexual behavior and peak fertility may be an adaptation to maximize the chances of conceiving. During estrus, the seasonal ovulator's surging estrogen levels make her frantically seek sex, and her need is quite apparent to potential mates in her scent, appearance, posture, and behavior. When not in estrus, she is unreceptive to mating behavior of any sort. In some species, females are actually physically incapable of copulation except when they are in estrus. The African bush baby, for example, grows a covering of skin over her vagina in the sexual off-season. The male adapts to the female pattern. In some species, males manufacture sperm only when females are in estrus.

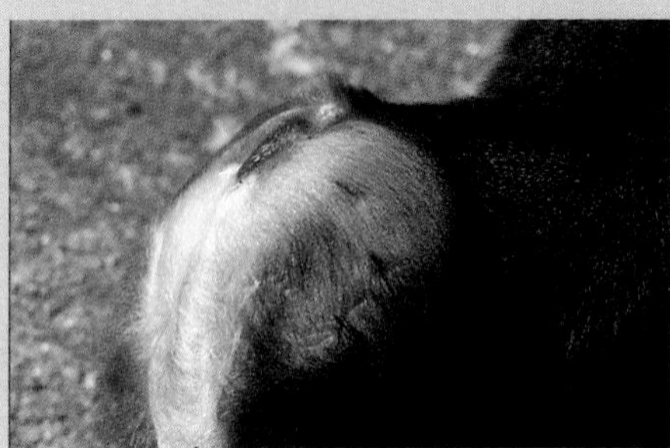

Figure 1
The female mandrillus wears a badge proclaiming her approaching fertility—her bottom turns bright red.

Even greater reproductive economy is seen among the **induced ovulators.** In sheep and goats, for example, ovulation is induced by the mere presence of a male. In rabbits, copulation induces ovulation. The female mink ovulates when her mate bites her on the back of the neck.

In mice, a male can disrupt a female's hormones even after she becomes pregnant. If a female is less than 4 days pregnant when she meets a "new" male (a male other than the father), his presence causes her to abort her embryo and to become sexually receptive to him. To trigger this "pregnancy block," the male emits a smelly substance called a pheromone that causes the female's hypothalamus to stimulate her pituitary to release FSH and LH. Estrus and ovulation are induced, and blood levels of the hormone prolactin plummet. This prevents the embryo from implanting in the uterus.

The Testes

In the human male, hormone levels are constant rather than fluctuating. LHRH from the hypothalamus stimulates the anterior pituitary to release FSH and interstitial cell stimulating hormone (ICSH, which is chemically identical to LH). FSH and ICSH travel in the bloodstream to the testes. There, FSH stimulates the early stages of sperm formation, and ICSH completes sperm production and stimulates the testes to synthesize the male hormone testosterone. Testosterone promotes the development of male secondary sexual characteristics, including facial hair, deepening of the voice, and increased muscle growth. Some athletes take steroid hormones to increase muscular strength, but this practice is dangerous (Reading 19.3).

KEY CONCEPTS

The menstrual cycle involves hormones from the hypothalamus, pituitary, and ovaries. When bleeding begins, low levels of estrogen and progesterone signal the hypothalamus to release LHRH, which prompts the anterior pituitary to release FSH and LH. The FSH triggers the ovary to manufacture estrogen, which builds uterine tissue. Rising estrogen lowers blood FSH, which stimulates the hypothalamus to manufacture more LHRH, which in turn increases LH. Midcycle, the LH peak triggers ovulation.

After ovulation the follicle cells lower blood-estrogen level, which lowers LHRH and then LH levels. The corpus luteum forms and produces estrogen and progesterone, which lowers LH and FSH and builds the uterine lining. If a fertilized ovum implants in the uterus, HCG is secreted. If there is no pregnancy, blood progesterone rises, inhibiting LHRH synthesis. Estrogen and progesterone levels fall and menstruation begins.

In human males, LHRH stimulates release of FSH, which stimulates the early stages of sperm formation, and ICSH, which completes sperm production and stimulates the testes to manufacture testosterone. Testosterone controls development of male secondary sexual characteristics.

The Pineal Gland

The **pineal gland** is a small oval structure located in the brain near the hypothalamus. It produces **melatonin,** a hormone that seems to help regulate reproduction in certain mammals, possibly even humans, by inhibiting the anterior pituitary hormones that regulate the activities of the gonads.

Melatonin production depends upon the pattern of lightness and darkness to which an individual is exposed. Light inhibits melatonin synthesis; darkness stimulates it. Because melatonin synthesis depends upon the absence of light, levels of the hormone may signal an organism to prepare for changing seasons, such as by growing a heavier coat or by initiating mating behavior.

Experiments have shown that melatonin influences the functioning of other endocrine glands. When rodents are given injections of melatonin, levels of thyroid hormone, melanocyte stimulating hormone, and sex hormones decrease.

The role of the pineal gland in humans is not well understood, possibly because we alter natural light-dark cycles

Reading 19.3 Steroids and Athletes—An Unhealthy Combination

CANADIAN TRACK STAR BEN JOHNSON FLEW PAST HIS COMPETITORS IN THE 100-METER RUN AT THE 1988 SUMMER OLYMPICS IN SEOUL. But 72 hours later, the gold medal awarded for his record-breaking time of 9.79 seconds was rescinded after traces of the drug stanozolol were detected in his urine.

Stanozolol is one of several synthetic versions of the steroid hormone testosterone. Like testosterone, these drugs promote signs of masculinity (their androgenic effect) and increased synthesis of muscle proteins (their anabolic effect). Used in the past to treat a handful of medical conditions—anemia and breast cancer among them—steroids are used by professional and amateur athletes alike, all looking to build muscle tissue easily.

Steroid users may improve their performances and physiques in the short term, but in the long run they may suffer for it. Steroids hasten adulthood, stunting height and causing early baldness, and in males lead to breast development and in females to a deepened voice, hairiness, and a male physique. The kidney, liver, and heart may be damaged, and atherosclerosis may develop because steroids raise LDL and lower HDL—the opposite of a healthy cholesterol profile. In males, the body mistakes the synthetic steroids for the natural hormone and lowers its own production of testosterone. The price of athletic prowess today may be infertility later.

Figure 1
Canadian track star, Ben Johnson, ran away with the gold medal in the 100-meter race at the 1988 Summer Olympics—he then had to return the award when traces of a steroid drug showed up in his urine.

Steroid use began in Nazi Germany, where the drugs were used by Hitler to fashion his "super race." Ironically, steroids were used shortly after to build up the emaciated bodies of concentration camp survivors. In the 1950s Soviet athletes began using steroids in the Olympics, and a decade later, United States athletes did the same. In 1976 the International Olympic Committee banned the use of steroids by athletes and required urine tests.

Ben Johnson was caught in his tracks by such a test, refined by 1988 so that part-per-billion traces of synthetic steroids can be detected even weeks after they are taken. Even though Johnson at first claimed the stanozolol in his urine was the result of a spiked drink of an approved anti-inflammatory drug used on his ankle, a test of his natural testosterone showed it to be only 15% of normal—a sure sign that this athlete had been taking steroids for a long time.

with artificial lighting. Some researchers hypothesize that mood swings in people are linked to abnormal melatonin secretion patterns, particularly a form of depression called seasonal affective disorder (SAD). Preliminary studies have shown that exposing such individuals to additional hours of daylight seems to elevate their moods, although clearly more work is needed to determine that the extra light is indeed the cause of the mood change.

KEY CONCEPTS

The pineal gland in the brain produces melatonin. This hormone's function, which is not well understood in humans, seems to be linked to patterns of light and dark.

Hormones Not Associated With Endocrine Glands

In recent years, the field of endocrinology has broadened beyond the study of distinct glands devoted in whole or in part to the synthesis and release of hormones to cells scattered throughout other organ systems that secrete hormones.

The mucous-rich lining of the stomach and intestines contains scattered hormone-secreting cells. These "gut hormones" include secretin, gastrin, cholecystokinin, and several others currently under investigation. Most of the gastrointestinal hormones regulate movement of specific parts of the digestive system and control the secretion rates of other digestive biochemicals.

The heart is another organ that contains endocrine tissue (fig. 19.20). Cells called cardiocytes in the atria of mammals, including humans, secrete **atrial natriuretic factor (ANF).** This hormone is released in response to the stretching of musclelike filaments in the cardiocytes. ANF exerts complex effects on several organs and endocrine glands, regulating blood pressure, blood volume, and the excretion of sodium ions, potassium ions, and water by binding to certain areas of the brain and kidney and by blocking the release of ADH from the hypothalamus and of aldosterone from the adrenal glands. The heart hormone also promotes relaxation of blood vessel walls. ANF binds to the ciliary body of the eye, the lungs, and the small intestines, with as yet unknown effects. ANF is being tested for use as a drug to treat hypertension (high blood pressure) and congestive heart failure.

Other organs, including the kidney and the placenta, contain hormone-secreting cells. As technology enables endocrinologists to detect more widely dispersed hormone-producing cells, it is likely that the list of recognized components of the endocrine system will grow.

KEY CONCEPTS

Organs not traditionally considered to be part of the endocrine system secrete hormones, including the stomach, intestines, heart, kidneys, and placenta.

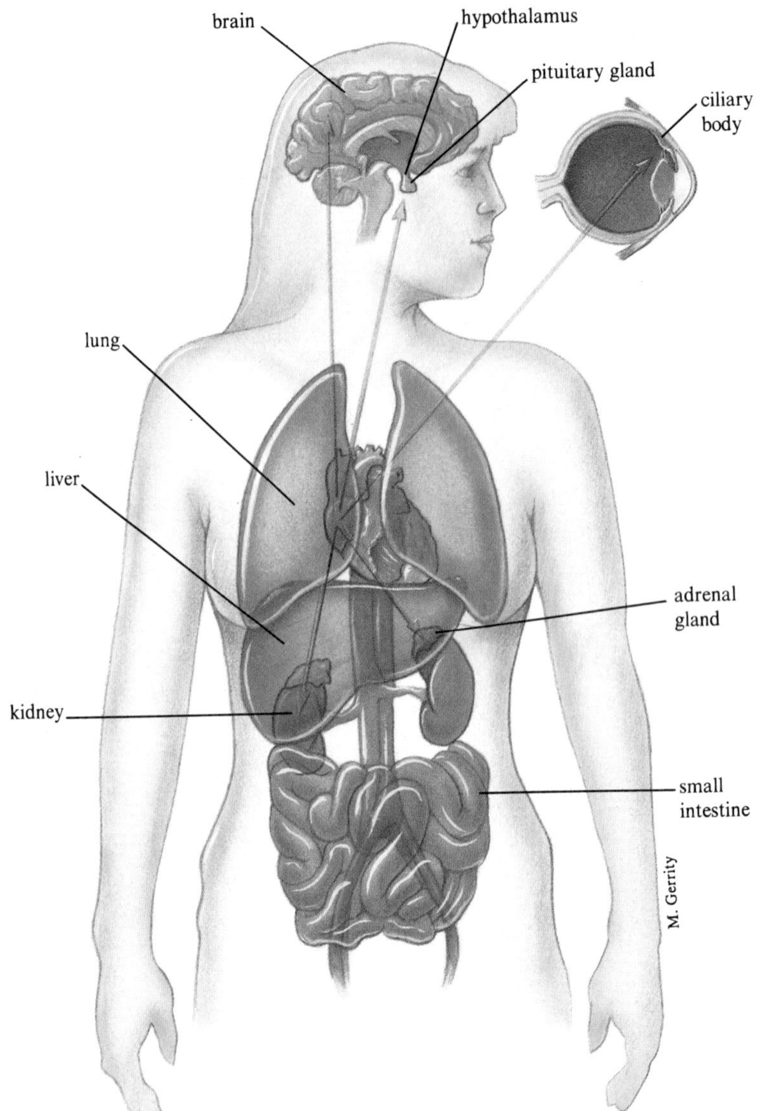

Figure 19.20
A heart hormone. Cells in the atria of the human heart secrete a hormone, atrial natriuretic factor (ANF), which has profound effects on the body.

Hormonelike Molecules

Prostaglandins

Prostaglandins are little-understood molecules that exert profound effects on various tissues and organs of the human body, often by altering hormone levels. Prostaglandins are lipids that appear locally and transiently when cells are disturbed. When a cell membrane is disrupted by an injury, the binding of a hormone, or an attack by the immune system, certain fatty acids are released from the damaged membrane into the cytoplasm. These fatty acids stimulate the synthesis of enzymes, which catalyze the formation of the different prostaglandins.

Like hormones, prostaglandins are chemical regulators, but they do not fit the classical definition of a hormone for several reasons. Prostaglandins function at the site of their synthesis, rather than traveling in the bloodstream to a target tissue as hormones do. In addition, prostaglandins are released by different types of stimuli than are hormones. Also, prostaglandins have been found in every mammalian tissue type thus far examined, whereas hormones are produced by specialized cells of the endocrine system.

Prostaglandins have a variety of functions: they affect smooth muscle contraction, secretion, blood flow, reproduction, blood clotting, respiration, transmission of nerve impulses, fat metabolism, the immune response, and inflammation. Consider the varied effects of prostaglandins on smooth muscle. Two types stimulate the smooth muscle in blood vessel walls to adjust blood flow to regions of the body. Other prostaglandins affect smooth muscle in the walls of airways within the lungs, where they alter the ease of oxygen delivery. Still other prostaglandins contract uterine muscles. Depending on the circumstance, these uterine contractions may assist sperm in their journey to the ovum or a fetus out of the womb, or they may cause menstrual cramps.

Understanding how different prostaglandins exert their specific effects on the body has explained the mechanisms of certain drugs and has led to some interesting new medical applications. For example, some prostaglandins promote inflammation, which is the immune system's attempt to fight infection at the site of a wound by increasing fluid accumulation and sending in white blood cells. Aspirin relieves the pain of inflammation by inactivating an enzyme needed for the synthesis of several prostaglandins. Aspirin also blocks synthesis of prostaglandins that stimulate pain receptors in the nervous system.

About 30% to 50% of all women in the United States suffer from painful menstrual cramps, a condition called dysmenorrhea. The discovery that these women have elevated levels of two prostaglandins in their uterine fluid led to the development of drugs that reduce levels of these prostaglandins, thus eliminating the cramps, nausea, diarrhea, and nervousness of the condition. Uterine prostaglandins also stimulate the contractions of labor. This knowledge has already had medical applications. An intrauterine device is now being tested that releases prostaglandins into the uterus to prevent implantation of a fertilized ovum by causing the uterus to contract. Administered after pregnancy has been established in the uterus, prostaglandins induce abortion. Drugs that lower levels of uterine prostaglandins can sometimes prevent miscarriage or premature birth.

The role of prostaglandins in blood clotting and controlling blood vessel diameter may play a part in heart attacks. Daily doses of aspirin may reduce the risk of heart attack by altering prostaglandin activity. Prostaglandins that widen blood vessels may be used as a treatment for hypertension, which results from constricted blood vessels.

KEY CONCEPTS

Prostaglandins are lipids that are synthesized transiently in all tissues and, unlike hormones, act locally. They have a wide variety of effects, including control of smooth muscle contraction, secretion, nerve impulse transmission, the immune response, fat metabolism, and inflammation.

Pheromones

Another type of communication biochemical that has some characteristics in common with hormones is the **pheromone.** Pheromones are substances secreted by an organism that stimulate a physiological or behavioral response in another individual of the same species. They are similar to hormones in that they affect reproduction and behavior, but they differ in that they transmit information between individuals, rather than within an organism.

Also unlike hormones, pheromones are species-specific. A pheromone secreted by a female cockroach attracts a male of the same species but has no effect on other insects. Pheromones reach their "target" individuals through the air (olfactory), by being eaten, or by being absorbed through the skin (fig. 19.21).

a.

b.

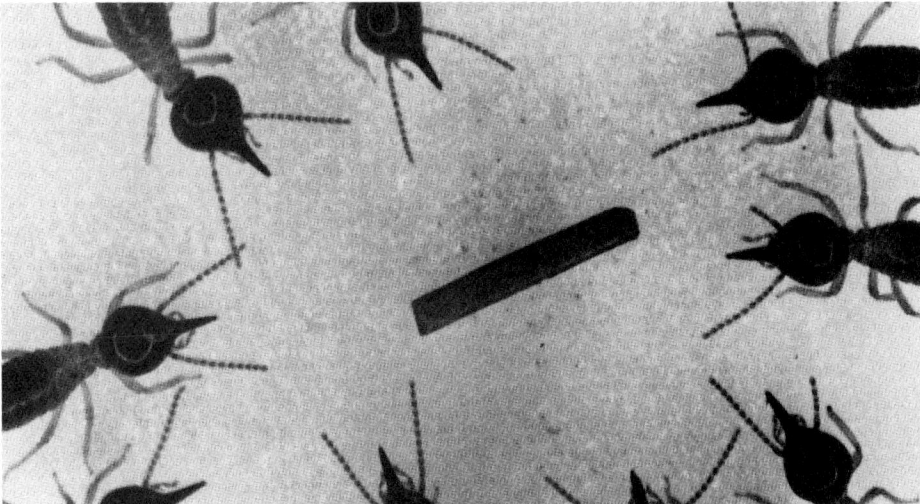

c.

Figure 19.21

Pheromones. *a.* This clover cabbage looper moth releases sex pheromones from a gland under its tail end. *b.* Pheromones from glands in their chins are used by rabbits to mark objects with their scent. *c.* The small metal bar in the center has been sprayed with a secretion from a termite soldier. The pheromone signals "alarm" to the other termites, who rush to the scene.

Pheromones were discovered in 1959, and the first ones studied were sex attractants. Since that time, the chemicals have been found in birds, fish, and mammals, although a human pheromone has yet to be characterized.

The endocrine system exerts pervasive effects on the human body, from such obvious physical characteristics as height and weight to control of body fluid composition, calcium levels in the blood and bones, blood pressure, and metabolic rate (tables 19.1 and 19.2). Hormones drastically influence how we feel, affecting sexual desire, mood, and energy level. We can overcome many threatening situations by the quick action afforded by the interplay between the nervous system and the endocrine system. Finally, and perhaps most important from an evolutionary standpoint, the endocrine system enables us to become parents, thus ensuring that our species continues.

Table 19.1
Functions of the Endocrine Glands

Gland	*General Functions*
Pituitary	Hormones of the anterior lobe increase growth and development of all tissues (growth hormone); concentrate pigment granules in melanocytes (melanocyte stimulating hormone); stimulate secretion from the thyroid (thyroid stimulating hormone), the adrenal cortex (adrenocorticotropic hormone) and the gonads (follicle stimulating hormone, luteinizing hormone, and interstitial cell stimulating hormone); and initiate milk production (prolactin) Hormones manufactured in the hypothalamus but released from the posterior pituitary raise blood pressure (antidiuretic hormone) and contract uterine muscles and milk-secreting cells (oxytocin)
Thyroid	Regulates metabolic rate and body temperature (thyroxine and triiodothyronine); lowers blood-calcium level (calcitonin)
Parathyroids	Raises blood-calcium level (parathyroid hormone)
Adrenal cortex	Mobilizes energy reserves while maintaining blood volume and composition (mineralocorticoids and glucocorticoids)
Adrenal medulla	Prepares body to cope with emergency situations (adrenaline and noradrenaline)
Pineal gland	Regulates effects of light-dark cycle on other glands (melatonin)
Pancreas	Regulates body's utilization of nutrients (insulin, glucagon, somatostatin)
Ovary	Maturation of oocyte and ovulation (follicle stimulating hormone, luteinizing hormone, estrogen, progesterone)
Testis	Sperm maturation, secondary sexual characteristics (follicle stimulating hormone, interstitial cell stimulating hormone, testosterone)

SUMMARY

The *endocrine system* includes several glands and tissues and the *hormones* that they secrete into the bloodstream. A hormone exerts a specific physiological effect on *target cells*, which have receptors specific for it. *Peptide* hormones bind to surface receptors of target cells, stimulating the enzyme-catalyzed conversion of ATP to cAMP on the inner face of the cell membrane. The cAMP second messenger sets into motion a chemical cascade ending in a specific metabolic effect. *Steroid* hormones cross target cell membranes, bind to cytoplasmic receptors, and then activate genes to direct the synthesis of particular proteins.

The levels of many hormones are regulated by *negative feedback loops*, in which an excess of the hormone or the product of a hormone-induced response suppresses the synthesis or release of that hormone until levels return to normal. In a *positive feedback loop*, the hormone causes an event or process that increases the hormone's production. Feedback loops may be controlled by the level of particular ions or nutrients near the endocrine cells, input from the nervous system, or other hormones.

Endocrine system function is hierarchical. The *hypothalamus* produces *releasing hormones*, which travel in neurosecretory cells to the anterior lobe of the *pituitary* gland, where they control the release of its hormones: *growth hormone*, which stimulates cell division, protein synthesis, and growth in all cells; *thyroid stimulating hormone*, which prompts the thyroid gland in the throat to release *thyroxine* and *triiodothyronine*, which regulate metabolism; *adrenocorticotropic hormone*, which stimulates the adrenal cortex to release hormones that enable the body to cope with a serious threat; *prolactin*, which stimulates milk production; and the *gonadotropic hormones*, which control sex cell development. The hypothalamus also manufactures two hormones that are stored in and released from the posterior lobe of the pituitary gland—*antidiuretic hormone*, which regulates body fluid composition, and *oxytocin*, which contracts the uterus and milk ducts. In many vertebrate species, the region between the anterior and posterior lobes of the pituitary secretes *melanocyte stimulating hormone*, which causes changes in skin color.

The parathyroid glands secrete parathyroid hormone, which increases the calcium level in the blood by releasing it from bone and increasing calcium absorption by the gastrointestinal tract and kidney. Another hormone affecting calcium metabolism is *calcitonin*, secreted by the thyroid gland. Calcitonin lowers the level of calcium in the blood by increasing the rate of its entry into bone.

The adrenal *cortex* secretes *mineralocorticoids* and *glucocorticoids*, which mobilizes energy reserves during times of stress and maintain blood volume and blood composition. The adrenal *medulla* secretes

Table 19.2
Summary of Major Hormone Functions in the Human Endocrine System

Hormone	*Site of Synthesis*	*Target*	*Effects*
Releasing hormones	Hypothalamus	Anterior pituitary	Stimulates release of hormones from anterior pituitary
Growth hormone (GH)	Anterior pituitary	All cells	Increases cell division, protein synthesis
Thyroid stimulating hormone (TSH)	Anterior pituitary	Thyroid gland	Stimulates secretion of thyroxine and triiodothyronine
Adrenocorticotropic hormone (ACTH)	Anterior pituitary	Adrenal cortex	Stimulates secretion of glucocorticoids and mineralocorticoids
Prolactin	Anterior pituitary	Myoepithelial cells in mammary glands	Stimulates milk secretion
Melanocyte stimulating hormone (MSH)	Anterior pituitary	Skin	Increases skin pigmentation
Follicle stimulating hormone	Anterior pituitary	Ovaries	Stimulates follicle development, maturation of oocytes, release of estrogen
Luteinizing or interstitial cell stimulating hormone (LH or ICSH)	Anterior pituitary	Ovaries, testes	Promotes ovulation; stimulates late development of sperm cells, synthesis of testosterone
Antidiuretic hormone (ADH)	Hypothalamus (released from posterior pituitary)	Kidneys, smooth muscle cells of blood vessels	Helps to maintain composition of body fluids
Oxytocin	Hypothalamus (released from posterior pituitary)	Uterus, myoepithelial cells of mammary glands	Stimulates muscle contraction
Thyroxine, triiodothyronine	Thyroid gland	All cells	Increases metabolic rate
Calcitonin	Thyroid gland	Bone	Increases rate of Ca^{2+} deposition in bone
Parathyroid hormone	Parathyroid glands	Bone, digestive	Releases Ca^{2+} from bone, increases Ca^{2+} absorption in digestive tract and kidneys
Glucocorticoids	Adrenal cortex	All cells	Increases glucose levels in blood and brain
Mineralocorticoids	Adrenal cortex	Kidneys	Maintains blood volume and electrolyte balance
Catecholamines	Adrenal medulla	Blood vessels	Raises blood pressure, constricts blood vessels, slows digestion
Melatonin	Pineal gland	Other endocrine glands	Regulates effects of light-dark cycle on other glands
Insulin	Pancreas	All cells	Increases glucose uptake by cells and conversion of glucose to glycogen
Glucagon	Pancreas	All cells	Stimulates conversion of glycogen to glucose
Somatostatin	Pancreas	Small intestine	Regulates rate of nutrient absorption
Progesterone	Ovaries	Uterine lining	Controls monthly secretion patterns
Estrogen	Ovaries	Uterine lining	Increases rate of mitosis
Human chorionic gonadotropin (HCG)	Fertilized ovum	Uterine lining	Interrupts menstrual cycle and maintains pregnancy
Testosterone	Testes	Skin, muscles	Maintains secondary sexual characteristics

epinephrine and norepinephrine, which ready the body to cope with an emergency.

The endocrine portion of the pancreas secretes *insulin*, which stimulates cells to take up glucose; *glucagon*, which breaks down glycogen to yield glucose; and *somatostatin*, which regulates the production of insulin and glucagon. Lack of insulin or the inability to utilize it leads to *diabetes mellitus*.

The female gonads, the *ovaries*, secrete *estrogen* and *progesterone*. The hormones stimulate the development of female sexual characteristics, and together with LHRH, FSH, and LH control the menstrual cycle. The male gonads, the *testes*, secrete *testosterone*, which controls the development of secondary sexual characteristics.

The *pineal gland* may regulate the responses of other glands to light-dark cycles through the production of its hormone, *melatonin*.

The gastrointestinal tract secretes several hormones that regulate digestion. In mammals, the atria of the heart secrete *atrial natriuetic factor*, which regulates blood pressure, blood volume, and the excretion of sodium ions, potassium ions, and water. Other biochemicals share some characteristics with hormones. The prostaglandins are lipids that are formed enzymatically when certain other lipids are released from disturbed cell membranes. Prostaglandins function at the site of their release, and the different types exert different effects on the body. *Pheromones* are odoriferous compounds released from one individual of a species that stimulate a behavioral or physiological response in another member of the same species.

QUESTIONS

1. Why doesn't a hormone exert effects on all body cells?

2. How do peptide hormones and steroid hormones differ in the ways they affect cells?

3. How do hormones regulate their own levels in the body? Describe the regulation of a specific hormone.

4. What bodily effects might result from a tumor on the anterior lobe of the pituitary gland? The posterior lobe of the pituitary gland? The hypothalamus?

5. List two hormones that exert opposite effects.

6. Why would someone with an insufficient supply of thyroid hormones feel tired all the time?

7. Describe how insulin and glucagon interact to regulate blood-glucose levels.

8. A queen honeybee secretes a substance from a gland in her jaw that inhibits the development of ovaries in worker bees. Is this substance most likely a hormone, prostaglandin, or pheromone? Cite a reason for your answer.

TO THINK ABOUT

1. Imagine that you are a researcher who has just found a small glandlike structure in a human. How would you determine whether it is an endocrine gland? If it is, how might you go about characterizing the role of the gland in the endocrine system?

2. Do you think there is any danger in a person's attributing any signs of nervousness to hypoglycemia without confirmation from a physician? Why or why not?

3. Female college students living in the same section of the dorm and sharing a bathroom tend, after several months, to get their menstrual periods at about the same time each month. Offer a biological explanation for this phenomenon.

4. It is estimated that 5% to 10% of women in the United States suffer monthly premenstrual syndrome (PMS) and experience a combination of backache, headache, bloating, clumsiness, painful breasts, irritability, crying, or depression. Symptoms begin a few days before ovulation and persist until the onset of menstruation. Fifty percent of all women may experience such symptoms occasionally. PMS is thought to be due to a decrease in endorphins, a deficiency of progesterone, or an increase in prostaglandins. Do you feel that women should not hold positions of authority because of the possibility that they may suffer from premenstrual syndrome? Should the condition be considered a valid legal defense against charges of murder or other violent crimes?

SUGGESTED READINGS

Bower, B. May 26, 1990. Bright light therapy expands its horizons. *Science News*. Could depression result from malfunctioning melatonin.

Cantin, Marc, and Genest, Jacques. February 1986. The heart as an endocrine gland. *Scientific American*. In mammals, the atria of heart secretes a hormone that exerts powerful effects on the body.

Corie, Beverly, D.V.M. April 1990. Bovine growth hormone: Harmless for humans. *FDA Consumer*. Bovine growth hormone injected into cows ups milk production.

Dusheck, Jennie. October 19, 1985. Fish, fatty acids and physiology. *Science News*. Prostaglandins may explain why some people suffer from cardiovascular disease and others do not.

Marshall, E. October 14, 1988. The drug of champions: Athletes and body builders support a $100-million black market in steroids, while medical science has been slow to see why. *Science*, vol. 242. The short-term effects of steroids may seem beneficial, but the long-term effects are dangerous.

Paxton, Mary Jean W. 1986. Endocrinology: Biological and medical perspectives. Dubuque, Iowa: William C. Brown. A detailed look at each of the endocrine system glands and hormones, with fascinating historical anecdotes and medical applications.

Rudman, D. July 5, 1990. Effects of human growth hormone in men over 60 years old. *The New England Journal of Medicine*. Can growth hormone reverse signs of aging?

Snider, Sharon. December 1990. The pill: 30 years of safety concerns. *The FDA Consumer*. The birth control pill "fools" the endocrine system.

Snyder, S.H. October 1986. The molecular basis of communication between cells. *Scientific American*. The cell-to-cell communication afforded by the nervous and endocrine systems is essential to our lives.

Stehlin, Dori. April 1990. Lactation suppression—Safer without drugs. *FDA Consumer*. If a new mother chooses not to breast-feed, her hormones can make her breasts very painful for a few days.

C H A P T E R

20

The Skeletal System

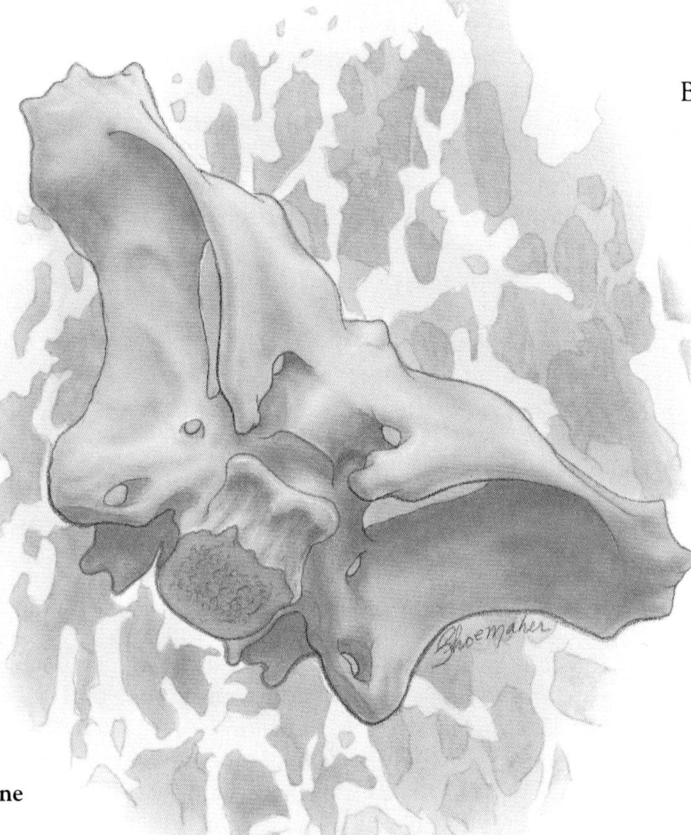

Chapter Outline

Learning Objectives

By the chapter's end, you should be able to answer these questions:

1. What are the functions of skeletal systems?
2. What are some of the forms of skeletal systems in different species?
3. What types of molecules and cells make up the human skeletal system, and how are they arranged in bone tissue?
4. How does the human skeleton develop and grow in the embryo, fetus, and child?
5. How does the human body repair bone fractures?
6. How is the mineral composition of bone regulated?
7. How are bones classified?
8. How are bones arranged in the human skeleton?
9. What is the structure of the synovial joint that connects some bones to each other?

The dancers are lined up nervously, each one perfectly still, waiting for the signal from the choreographer. At the cue, they suddenly come alive, jumping and turning in unison, their graceful movements precisely timed to the music (fig. 20.1).

The elderly man's arthritis makes his hands and wrists very painful. He takes the aspirin from the medicine chest, but his pain-wracked hands, swollen at the joints from arthritis, are no match for the cap on the child-resistant, tamperproof (and sadly older-person-proof) aspirin bottle.

These examples illustrate the human skeletal and muscular systems at their best and worst. Although most of us are not professional dancers or arthritis sufferers, we can still appreciate how smoothly these two organ systems function together just by considering normal daily activities. Our ability to move about depends upon interconnected structures of bone and muscle.

Skeletal Diversity

A skeleton is a supporting structure or framework. It gives the body its shape, it protects internal organs, and it provides something firm for the muscles to contract against, permitting movement. Skeltons worn on the outside also offer protection.

Hydrostatic Skeletons

The simplest type of skeleton is a hydrostatic skeleton ("hydro" means water), which consists of liquid surrounded by a layer of flexible tissue. **Hydrostatic skeletons** are found in squid, sea anemones, slugs, and certain worms, including annelids such as the familiar earthworm (fig. 20.2). The tension of the constrained fluid provides support and helps to determine shape, as it does in a water-filled balloon.

To see how a hydrostatic skeleton permits movement consider earthworm locomotion. The earthworm has two layers of muscle in the body wall. One layer encircles the body and the other runs lengthwise. When the circular muscles contract against the fluid in the body cavity, the worm lengthens, just as a water-filled balloon would if it were squeezed at one end. The worm then sticks its bristles into the ground as an anchor. Contraction of the longitudinal muscles shortens the animal, pulling it forward.

Exoskeletons

A more complex skeleton is a **braced framework,** which consists of solid structural components that are strong enough to resist pressure without collapsing. Muscles can attach to the surfaces of the skeleton, facilitating movement.

An **exoskeleton** ("exo" means outside) is a braced framework that protects an organism from the outside, much like a suit of armor. Exoskeletons are found in many groups of invertebrates, including arthropods (such as lobsters and insects) and molluscs (such as snails). Fossil evidence indicates that exoskeletons became prevalent rather suddenly some 570 million years ago. Muscles attach to the inner surface of an exoskeleton and contract against it to bring about movement. Clams and snails use readily available minerals to produce hard calcium-containing shells. These rigid shells and spikelike structures thicken as the organism grows, remaining with it for life.

The jointed suits of armor of a cockroach and a lobster are built of the polysaccharide chitin. However, these exoskeletons have a shortcoming in common with armor—they can be outgrown. The growing animals must shed, or "molt," the outgrown exoskeleton and grow a new one that is slightly larger. Until the new exoskeleton has formed and hardened, the animal is in greater danger of being eaten by predators (fig. 20.3). Seafood lovers may be familiar with the results of molting: A soft-shelled crab has just begun the process of growing a new exoskeleton, and a hard-shelled crab has already completed the process.

Endoskeletons

An **endoskeleton** ("endo" means inner) is an internal scaffolding found in vertebrates (animals with backbones). An endoskeleton offers several advantages over an exoskeleton—it does not restrain growth and, because it consumes less of an organism's total body mass than an exoskeleton does, it increases mobility.

Figure 20.1
The graceful movements of dancers are possible because muscles contract against bones that function as a lever system.

A few animals, such as cartilaginous fishes including the shark, have endoskeletons made of cartilage. Most vertebrates, including humans, have endoskeletons composed primarily of bone. A cat's endoskeleton is particularly interesting, because the animal can manipulate itself to survive falls (Reading 20.1).

KEY CONCEPTS

A hydroskeleton consists of liquid constrained by flexible tissue and is found in certain worms, squid, sea anemones, and slugs. Braced framework skeletons include the exoskeleton, which is similar to a suit of armor (such as in insects and molluscs), and the endoskeleton, which is an internal supportive structure found in vertebrates.

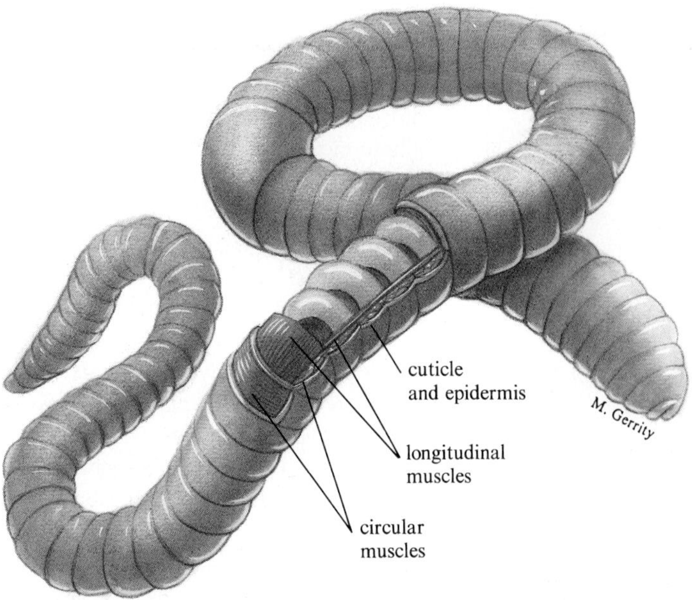

Figure 20.2
The earthworm has a hydrostatic skeleton.

Figure 20.3
The lobster molts its exoskeleton. Before the new exoskeleton hardens, the animal grows.

The Human Skeletal System

Skeletal Functions

The human skeletal system has several functions:

1. *Support*—The skeleton is a framework that supports the body against gravity. To a large extent it is responsible for the body's characteristic shape. Indeed, without skeletal support, a human would resemble a jellyfish.
2. *Lever system for movement*—The vertebrate skeleton is also a system of muscle-operated levers. Typically the two ends of a skeletal muscle are attached to different bones that are connected by a joint. When the muscle contracts, one bone is moved.
3. *Protection of internal structures*—Some bones form cages protecting the soft and delicate internal organs. The backbone surrounds and shields the sensitive spinal cord, the skull protects the brain, and the ribs protect the heart and lungs.
4. *Production of blood cells*—Many bones, such as the long bones of the human arm and leg, contain and protect a tissue called red marrow in which red blood cells, white blood cells, and platelets are produced.
5. *Mineral storage*—The skeleton also provides a storehouse for calcium and phosphorus, minerals that are required for a number of important activities. The minerals in bone are continually being withdrawn for use elsewhere in the body but are later replaced.

Composition

The skeletal system is built primarily of cartilage and bone. These two types of connective tissue are specialized to serve as an internal framework.

Cartilage

Cartilage is an important, but often forgotten, component of the skeletal system. During embryonic development the skeleton first forms in cartilage, providing a mold for the bone that will later replace it. Furthermore, cartilage is widely dispersed in the adult body. An elastic form of the tissue is found in the external ear and in the nose. The flexibility of elastic cartilage permits the expansion of the rib cage when the lungs inflate. Another type of cartilage, found in the pads between the bones comprising the backbone, has the ability to withstand compression. In the skeleton, smooth cartilage coverings allow the adjoining bones of the limbs to slide past one another.

Cartilage is firm yet flexible, thanks to its structure. It follows the general connective tissue pattern of cells within a matrix (fig. 4.15c). The cells are sparsely distributed within the matrix, where they occupy small spaces called lacunae. The large nuclei and extensive endoplasmic reticula of cartilage cells reflect their function of producing large amounts of protein.

The proteins produced by cartilage cells give the tissue its properties. Because the cartilage proteins, collagen and elastin are pliable, the tissue is flexible. The collagen also allows cartilage to bear great weight. This is possible because collagen forms strong networks of fibers that distribute weight, thereby enabling the tissue to resist breakage and stretching. The supportive collagen network of cartilage is similar in function to the steel framework of a building.

Reading 20.1 *Falling Felines*

A FALL FROM A FIFTH STORY WINDOW ONTO CONCRETE IS ALMOST ALWAYS FATAL—IF YOU ARE A HUMAN OR A DOG. Cats, however, can fall from the same distance and come away with only minor injuries. How can this be if humans, dogs, and cats all have vertebrate skeletons? The reason is that the lightness of a cat's skeleton and the way it holds itself during a fall helps protect it.

Veterinarians at the Animal Medical Center in New York City evaluated 115 accidental cat falls from an average of 5 stories. Only 10% of the felines died, and although the others broke bones, most were treated and survived. One survivor fell 32 stories and only chipped a tooth!

Do cats really have nine lives, or is there a physical explanation for their ability to withstand falls? The researchers cite several contributing factors:

Cats are light, and no matter how far they fall, they do not exceed a velocity of 60 miles per hour. Humans fall at 120 miles per hour.

Falling felines accelerate for only 100 feet, then continue their fall at 60 miles per hour. After this point, they seem to relax, spreading out their limbs in a way that distributes the impact over the entire body. This may reduce the severity of injury.

Cats are pros at tumbling and instinctively contort their bodies in flight so that they land on all fours.

Studying how cats use their skeletons to minimize injury might provide clues to prevent the 13,000 human falling deaths in the United States each year.

Figure 1
The cat's lightweight frame and the way it holds itself help it to survive falls.

The firmness and resilience (the ability to return to the original shape after deformation) of cartilage is due to other proteins in the matrix called proteoglycans. These molecules look like bottle brushes; the central core is protein and the radiating "bristles" are sugars. The proteoglycans adhere to long chains of a disaccharide (hyaluronic acid), creating enormous structures that interact with collagen (fig. 20.4) and carry a strong overall negative charge. These negative charges attract tremendous amounts of water. The overall result is that cartilage consists mostly of water trapped in a protein framework. The water provides support and firmness, reminiscent perhaps of the more primitive hydrostatic skeleton. The high water content of the matrix makes cartilage an excellent shock absorber, which is its function between many bones. When cartilage is compressed the water is squeezed out of the protein mesh but returns to restore the original shape when the pressure is released. The motion of water within cartilage that accompanies normal movement is what cleanses the bloodless tissue and nourishes it with dissolved nutrients.

Bone

Bone is a wonder of biological engineering, packaging maximal strength into a lightweight form. Among nature's strongest materials, 1 cubic inch of bone can bear loads of 19,000 pounds (approximately 8,618 kilograms). The strength of bone rivals that of light steel, but a steel skeleton would be four or five times heavier than ours.

Bone's strength is due to both minerals and collagen in the matrix. The properties of each component enhance those of the other—strength from collagen and hardness from minerals. Collagen fibers are amazingly strong. Although a collagen strand is a mere millimeter thick, a 20-pound (9-kilogram) weight could dangle from it without causing the fiber to snap. Collagen's strength is an asset to bone, but by itself it is flexible and elastic. The hardness and rigidity of bone is due to the minerals, primarily calcium and phosphate, which precipitate out of body fluids coating the collagen fibers. If the minerals were removed from bone, it could be tied in a

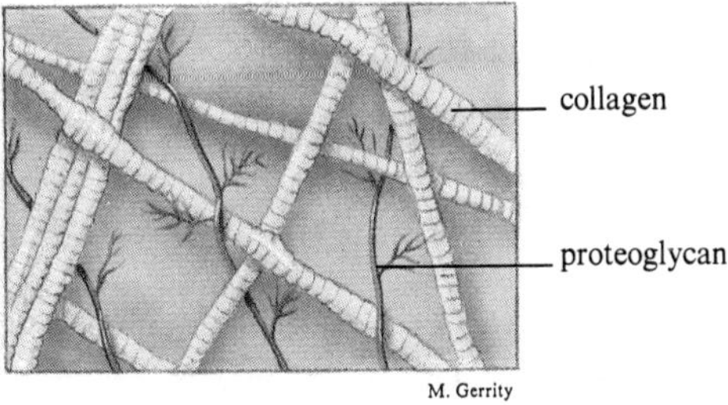

Figure 20.4
Proteoglycans and hyaluronic acid molecules within cartilage form a network that entraps enormous amounts of water.

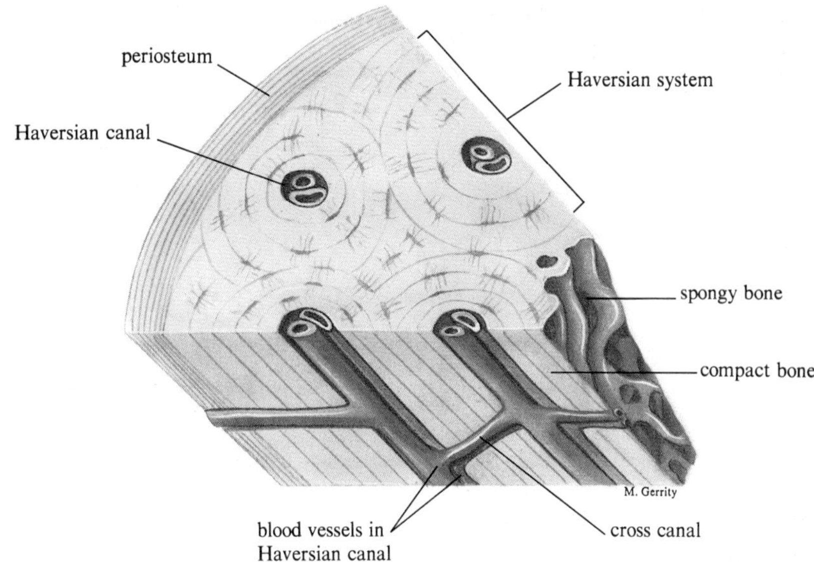

Figure 20.5
Bone cells are arranged in concentric circles around a central passageway, the Haversian canal, that contains the blood supply.

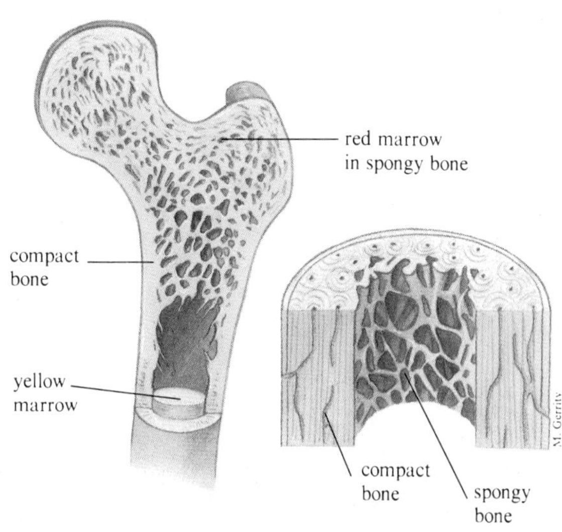

Figure 20.6
The structure of a long bone. The shaft of a bone is hollow and contains yellow marrow. Next is a layer of spongy bone. The outer hard coat consists of compact bone. The knobby ends of the bone are spongy bone coated with compact bone.

knot like a garden hose. This is why chicken bones boiled in soup easily fall apart. However, removal of the collagen would make the bone brittle and prone to crumble.

Embedded in the matrix are living bone cells that are responsible for growth and repair (fig. 4.15*d*). Nutrients and wastes cannot diffuse through the matrix; instead, blood is piped to the cells through small canals that penetrate the matrix (fig. 20.5). The most solid bone tissue is called compact bone. Here, the cells, called **osteocytes,** are located in spaces (lacunae) that are arranged in concentric rings, called Haversian systems, around a central portal containing the blood supply, the **Haversian canal.** Cross canals link adjacent Haversian canals. Narrow passageways called **canaliculi** connect the lacunae. Extensions from osteocytes extend through the canaliculi, keeping widely separated osteocytes in physical contact. Materials can be passed from the blood, and from osteocyte to osteocyte, through a "bucket brigade" consisting of a chain of up to 15 cells. Still other canals connect the entire labyrinth to the outer surface of the bone and to the marrow cavity within. The entire structure is surrounded by a layer of connective tissue called periosteum.

Part of the wonder of bone is that it is lightweight as well as strong. This is because bones are not solid. Most of the irregularly shaped flat bones and almost all of the bulbous tips of long bones are made of **spongy bone.** As its name suggests, this bone resembles a sponge. It has many large spaces between a web of bony struts. The struts are aligned with the lines of stress, a feature that increases the bones' strength. The spaces within spongy bone are not wasted; they are filled with **red marrow,** a nursery for blood cells and platelets that spews out more than a million cells a minute. Spongy bone is covered with a layer of more solid hard **compact bone.** The weight of long bones, such as those in the arms or legs, is further reduced by a cavity in the shaft, the **marrow cavity,** that contains fatty yellow marrow consisting of fat cells and a few blood cells (fig. 20.6). The strength of long bones is actually increased by this cavity. According to physical principles, a hollow tube is more resistant to certain types of stress than is a solid rod of equal size.

KEY CONCEPTS

The human skeleton provides support, a lever system for movement, protection of internal structures, and a place for blood cell production and mineral storage. Cartilage forms the embryonic skeleton and smooth coatings on bones so that they can slide. Cartilage is firm and flexible. Its proteins give the tissue a negative charge that attracts water, making cartilage a good shock absorber. Bone is strong yet lightweight due to collagen and minerals. In compact bone, osteocytes reside in lacunae, which surround Haversian canals. The lacunae are connected by canaliculi, through which osteocytes stay in contact. Flat bones and bone tips contain spongy bone, which is filled with red marrow. Yellow marrow is found in long spaces within limb bones.

Bone Growth and Development

Before Birth

Cartilage and bone interact throughout development to sculpt the everchanging skeleton. Most of the skeleton originates in the embryo as a cartilage model that is gradually replaced by bone tissue in a process called **ossification** (fig. 20.7). Each cartilage model has a shape that is similar to that of the bone it will become, and it is surrounded by a layer of connective tissue called the **perichondrium** (surrounding cartilage).

When the embryo is about 4 weeks old, cells just beneath the perichondrium secrete a "collar" of compact bone around the central shaft of the developing bone. Meanwhile, cartilage cells and their lacunae within the shaft enlarge, squeezing the cartilage matrix into thin spicules. The compressed cartilage matrix becomes

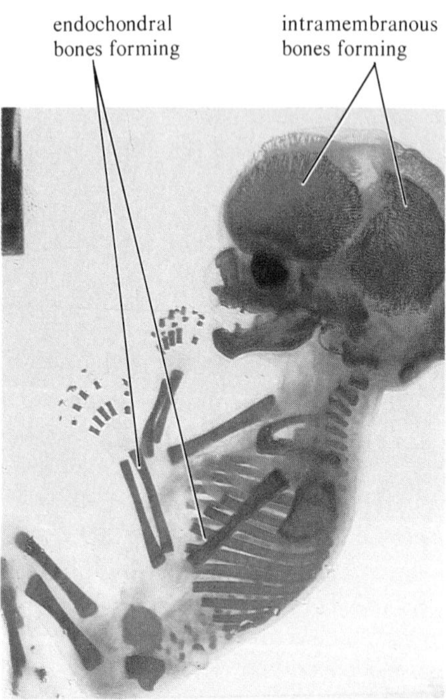

Figure 20.7
The cartilage model of a fetal skeleton. During fetal development, the model will be reformed in bone.

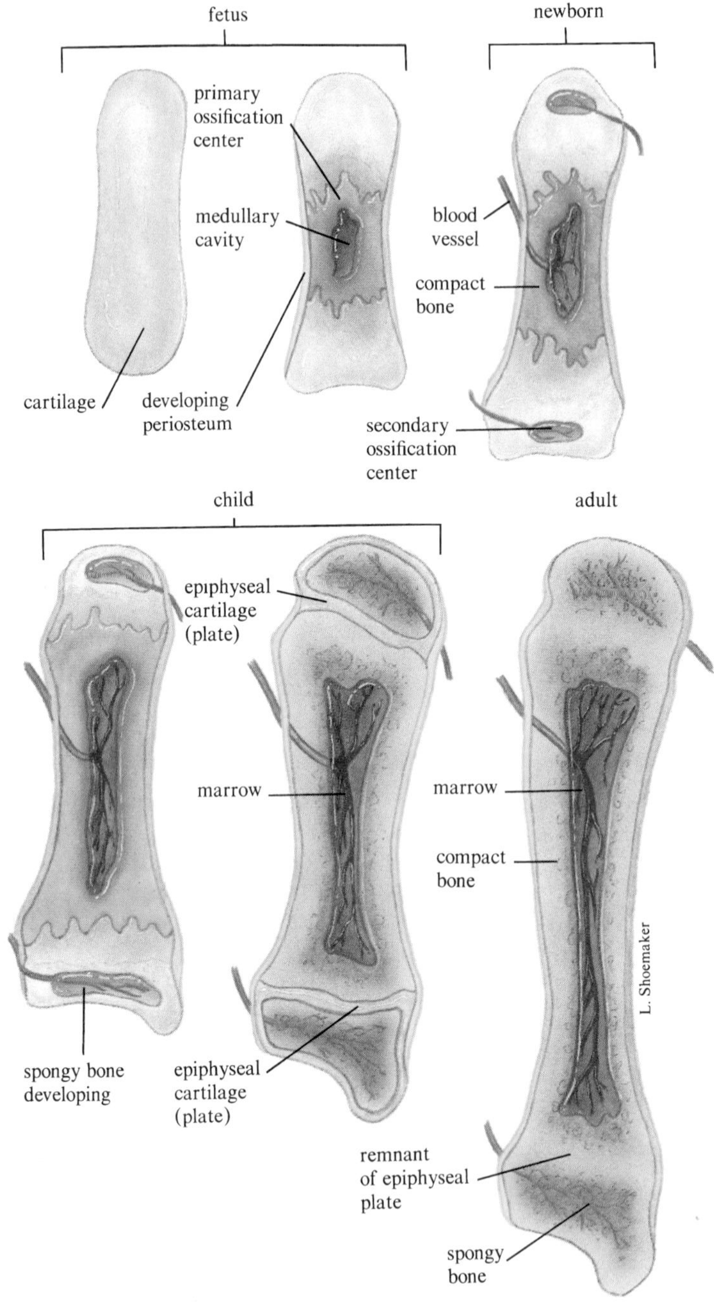

Figure 20.8
Bone growth. During fetal development, the cartilage skeleton is replaced by bone. Bone forms first around the central shaft as cartilage cells die within a matrix that is calcifying. Later, bones elongate from cartilage disks, the epiphyseal plates, located between the shaft and the knob at the end of the bone.

hardened with calcium salts. Since cartilage lacks a blood supply and relies on the diffusion of materials through its watery matrix for nourishment and waste removal, this loss of fluidity kills the cartilage cells. The hardened cartilage matrix degenerates. Blood capillaries invade the pockets left by the rapidly retreating cartilage. The bone cells enter and secrete bone matrix to establish the first internal region of new bone (fig. 20.8).

Bone Elongation During Childhood

The skeleton of the newborn is still built largely of cartilage. Within a few months of birth, bone growth becomes centered near the ends of the long bones in thin disks of cartilage (the **epiphyseal plates**). These cartilage disks enable bones to elongate as the child grows. The cartilage cells in the plate divide, pushing daughter cells towards the shaft, where they calcify. The dividing cartilage cells keep the disk relatively constant in thickness, but the calcified daughter cells become part of the shaft, elongating it. This growth continues until the late teens, when the cartilage plates begin to be replaced by bone tissue. By the early 20s, bone growth is complete, and only a line remains to mark the position of a cartilage disk. Reading 20.2 describes a new technique for coaxing bones to grow in very short individuals.

Repair of Fractures

Cartilage is called into action again when a bone breaks. The immediate reaction to a fracture is bleeding, followed rapidly by blood-clot formation at the site of the break (fig. 20.9). The blood clot is gradually replaced with dense connective tissue fibers that are secreted by invading connective tissue cells. Next, cartilage cells enter the dense connective tissue to build a fibrous "callus" that fills the gap left by the injury. Bone cells begin to manufacture new bone tissue. As cartilage is destroyed and replaced by spongy bone, the gap is closed. When the injured bone is exercised, compressive forces stimulate bone cells in the region of the break to secrete more collagen, which compacts the newly formed bone. This is why inactivity can actually impede a fracture's healing.

KEY CONCEPTS

Ossification is the replacement of the cartilage model of the skeleton before birth with bone tissue. Cartilage cells are compressed and die, capillaries invade, and bone cells arrive and build new tissue. During childhood, new bone forms at the epiphyseal plates of long bones, and elongation continues through adolescence. When a bone breaks, a blood clot forms and is replaced with dense connective tissue, then cartilage, then bone.

Bone as a Mineral Store

The nonliving appearance of bone is misleading, for bone is an active, metabolizing tissue. Bone tissue is continuously renewed as its structural material is broken down and built up again. The activity of bone tissue is highly attuned to the environment, responding to external influences, such as gravity, as well as to internal ones, such as nutrition and hormones (fig. 20.10).

Even when bone is not developing or healing, it is constantly being produced and destroyed. The continual renovation

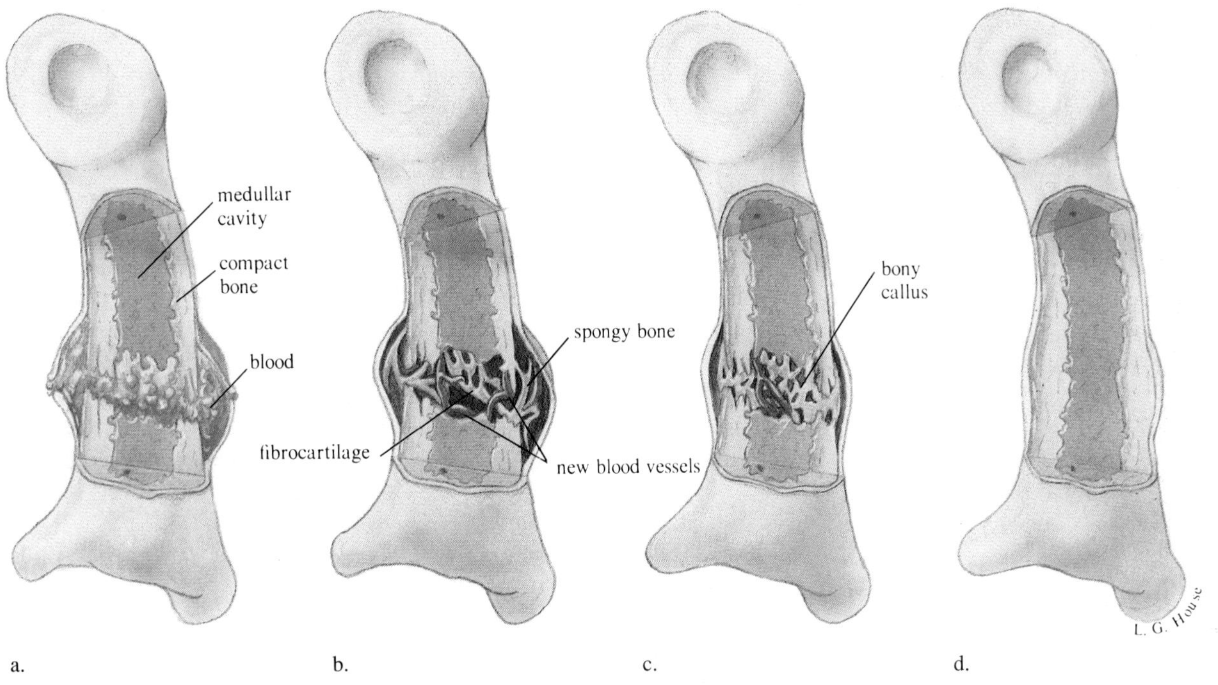

Figure 20.9
A fracture is healed.

Reading 20.2 *Forcing Bones to Grow*

Reza Garakani was tired of being a dwarf. He was born with achondroplasia, a hereditary form of dwarfism, and by age 14 was only 40 inches (101.6 cm) tall. He could not reach doorknobs, faucets, and elevator buttons, and he yearned to grow more so that he could lead a normal life. His goal—reaching 4 feet 4 inches (132.1 cm)—would enable him to drive a car.

In April 1988, Reza began a brutally painful treatment that would literally stretch his frame—bone lengthening. He had heard about 12-year-old Anthony Terravachio, who had begun treatment in August 1985. In his legs the tibias had been stretched 7 inches (17.8 cm) and the femurs 5 inches (12.7 cm), and work was ongoing on his arms, which had already been stretched 6 inches (15.2 cm). "I can fit my hands into my pockets now!" Anthony had said excitedly on a television program. So Reza bravely underwent the procedure. The outer shells of compact bone in his legs were fractured, leaving the blood vessels and marrow undisturbed. Then pins were inserted so that they projected from the broken bone ends, like the spokes of a bicycle wheel. That was the easy part. After the surgery, Reza's mother would turn screws attached to the pins four times a day, moving the pieces of bone apart 0.039 inches (1/4 mm) each time. The separated area between the pieces would fill in with bone tissue and calcify.

It hurt terribly, but it was worth it, Reza says. For by September—just 6 months after the surgery—his thigh bones had grown 3 1/2 inches (8.9 cm) and his legs 7 inches (17.8 cm). Reza just may get that coveted driver's license by his 16th birthday.

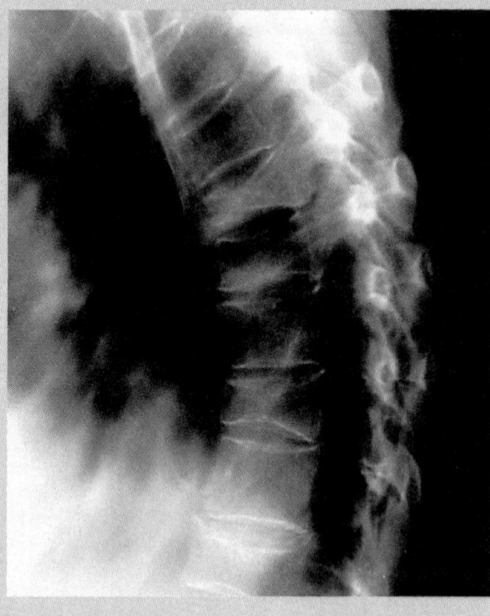

Figure 1
Forcing bones to grow.

Figure 20.10
Rickets. Lack of vitamin D has prevented the leg bones in this child from hardening normally. The bones bend under stress.

is under precise biological control. This is why a particular leg bone has the same characteristic shape, but a different size, in the fetus, the child, and the adult. Regulation of bone structure is an example of biological balance, called **homeostasis.** The balance that maintains bone form also controls its function as a storehouse of biologically important minerals, particularly calcium and phosphorus.

About 99% of the calcium in the human body is stored in the bones. Of the remainder, some is carried in the blood plasma. Calcium is also present in and around cells and is vital for many body functions, including the activity of some enzymes, muscle contraction, blood clotting, cell cohesion, and cell membrane permeability. Calcium is constantly being shuttled between blood and bone in a highly regulated fashion to maintain the blood concentrations of calcium necessary to these body functions.

The mineral content of bone is controlled by hormones, (chapter 19). Parathyroid hormone and calcitonin work together to regulate blood-calcium level. **Parathyroid hormone,** manufactured in the parathyroid glands in the throat, raises the level of calcium in the blood in three ways: it causes bone to release calcium into the blood; it converts vitamin D into its active form, which increases absorption of dietary calcium from the intestine; and it decreases excretion of calcium from the kidneys. Conversely, a hormone produced by the thyroid gland (also in the throat), **calcitonin,** lowers blood calcium by halting the release of calcium from bone.

Another hormone that affects calcium content in the skeleton is the sex hormone **estrogen.** Its decline in post-menopausal women can worsen the loss of bone mass, a condition called osteoporosis (Reading 20.3). Decline in bone mass is also seen in astronauts who have spent prolonged periods in a microgravity environment.

KEY CONCEPTS

Bone is constantly being broken down and built up, and this highly regulated process is influenced by gravity, nutrition, and hormones. Bone is a storehouse for calcium. The level of calcium is controlled by parathyroid hormone, calcitonin, and estrogen.

Reading 20.3 Osteoporosis

DECREASING BONE MASS DUE TO LOSS OF CALCIUM SALTS IS A NORMAL PART OF AGING. This decreasing bone mass can lead to osteoporosis (holes in the bones), in which bones become highly susceptible to breakage. Ten percent of women have severe enough osteoporosis by the age of 50 to suffer from fractures. Nearly 25% of all women develop osteoporosis by age 60, and by age 90, one-third of all women and one-sixth of all men have it. Osteoporosis is responsible for the shrinking stature and chronic back pain suffered by some of the elderly, their frequent fractures, and the disfiguring "dowager's hump."

Females are more likely to suffer from osteoporosis than males for a variety of reasons. The longer life span of women contributes to their overrepresentation among the ranks of osteoporosis sufferers. The bone mass of the average woman is about 30% less than that of a man, making her bones more easily depleted of calcium stores. A decline in the level of certain female hormones also makes a woman more prone to osteoporosis.

Three lines of evidence implicate declining estrogen levels in the development of osteoporosis. Young women who have had their ovaries removed make very little estrogen and become prone to bone mass loss. Women who suffer from anorexia nervosa, a condition characterized by self-starvation, sometimes cease to menstruate because of their low body fat. Their estrogen levels subsequently fall. Although their frequent exercise seems to prevent overt symptoms of osteoporosis, microscopic examination of their bone tissue often shows depletion of calcium salts. Women who have passed menopause, particularly those who are underweight, have sharply declining estrogen levels and comprise the population group at highest risk for osteoporosis. For this reason, doctors now advise all women to take 1,000 to 1,500 milligrams of calcium daily and to exercise regularly.

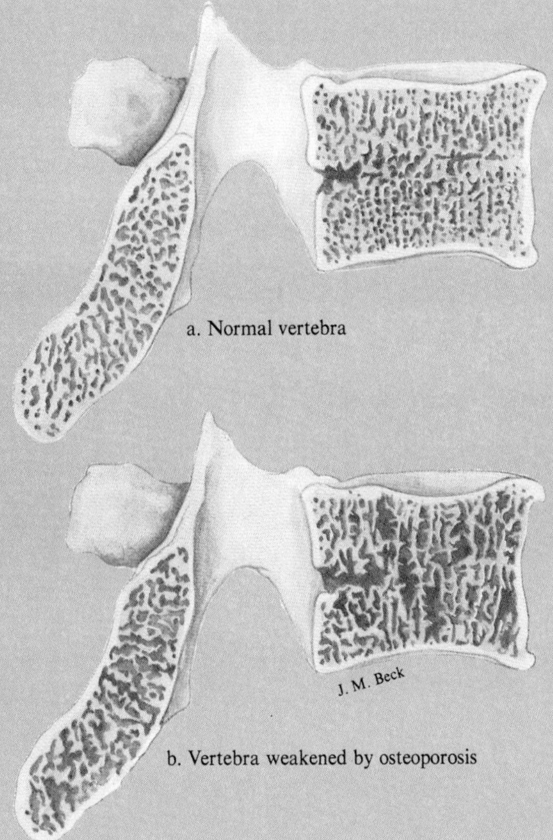

Figure 1
Osteoporosis is a condition in which bone is weakened by the loss of calcium. *a.* Normal vertebra and (*b*) vertebra weaken by osteoporosis..

Skeletal Organization

The 206 bones of the human skeleton are often grouped into two parts. The **axial skeleton,** so named because it is located in the longitudinal central axis of the body, consists of the skull, vertebral column, ribs, and breastbone. The **appendicular skeleton** consists of the limbs and bones that support them (the collarbones and shoulder blades, which support the forelimbs, and the pelvis, which supports the hind limbs). Both parts of the skeleton are constructed in such a way that they are able to absorb the tremendous shocks generated in locomotion (fig. 20.11).

Axial Skeleton

The axial skeleton is specialized for its role in shielding soft body parts. The **skull** protects the brain and many of the sense organs and is built of the hardest, densest bones in the human body (fig. 20.12). These bones fit together like puzzle pieces at immovable boundaries called sutures. While holding the pieces of the skull together, the sutures help dissipate the shock of a blow to the head. Inside the skull are nooks and crannies sculpted so that the brain and eyes fit snugly. A hole called the foramen magnum at the base of the skull allows the nervous tissue of the brain to continue into that of the spinal cord. All of the head bones are joined with immovable joints except for those of the lower jaw and the middle ear. Movement of these bones is necessary for chewing, speech, and hearing. Spaces within the bones of the head, called sinuses, help to lighten the skull and are a resonating chamber for the voice.

A baby's skull bones still contain dense connective tissue at the sutures, and the regions between the bones, called "soft spots" or "fontanels" (from a French word meaning "little fountain"), pulsate due to the underlying blood vessels. Because the skull bones of the newborn are not fused, the infant's head can squeeze through the birth canal, and the head is often temporarily misshapen in the process (fig. 20.13). The soft head of the newborn also permits growth of the skull as the brain enlarges. The fontanels usually are completely filled in by bone tissue by 18 months of age.

The vertebrae protect the spinal cord down to the lower back. These bones form the **vertebral column,** also known as the spine or backbone (fig. 20.14). The position of the vertebral column and its multipart construction distributes weight so that locomotion is possible. Consider, for example, the 33 vertebrae of humans. The 7 **cervical vertebrae** in the neck are the smallest and allow the widest range of motion so that we can turn our heads in many directions. In the upper back are 12 heavier **thoracic vertebrae**. These are specialized with indentations on each side that help anchor the ribs in place. Next are the 5 **lumbar vertebrae** in the small of the back. These are the largest of the bones of the spine, presumably because they support most of the body's weight. The 5 fused pelvic vertebrae are called the **sacrum.** These form a wedge between the two hipbones.

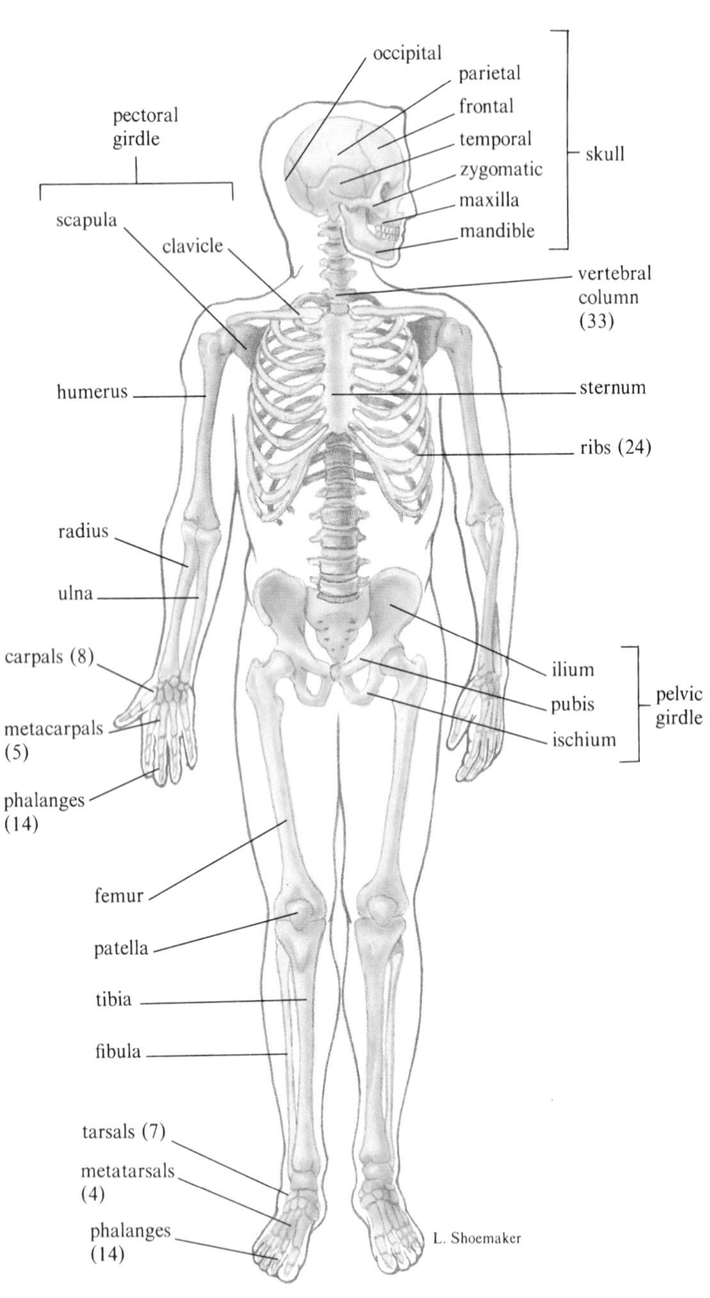

Figure 20.11
The human skeleton is divided into the axial skeleton (indicated in the lighter color) and the appendicular skeleton (indicated in the darker areas). The axial skeleton in the human includes the skull, vertebral column, and rib cage. The bones of the limbs and those that support them comprise the appendicular skeleton.

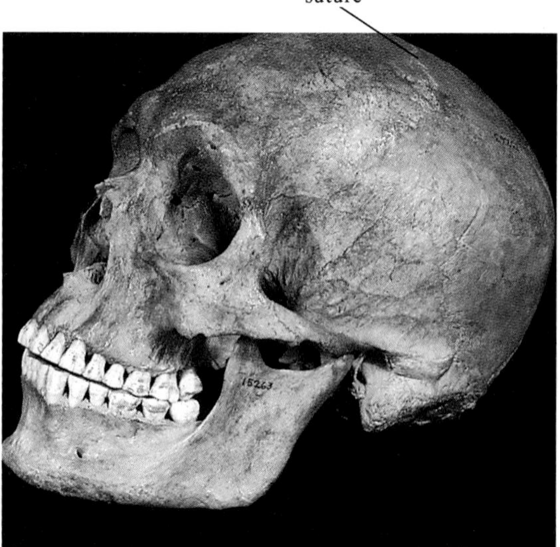

Figure 20.12
The skull is a protective case for the brain and eyes. The brain case is composed of several bones that are held together by immovable joints called sutures.

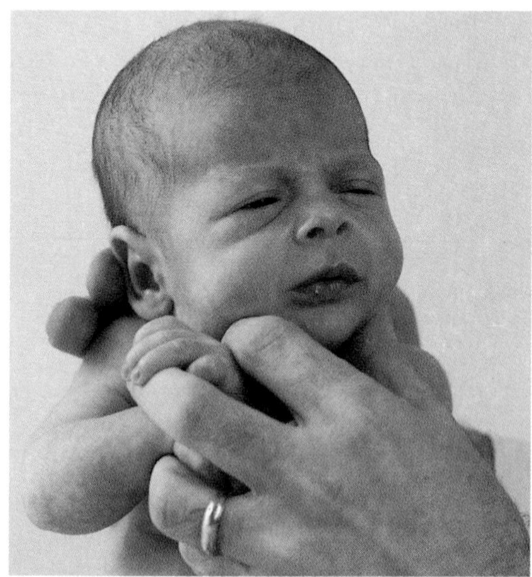

Figure 20.13
A skeletal adaptation for birth. The molded shape of this newborn's head is a souvenir of her recent squeeze through the birth canal. The skull bones can actually overlap slightly to ease the passage. Her head will assume a more rounded shape in a few weeks.

The final 4 vertebrae are fused to form the **coccyx,** or tailbone, which does not seem to have a function.

A normal spine extends from the neck to the hips, curving outwards near the shoulders and inwards in the lower back. About 5% to 10% of all adolescents have **scoliosis,** a type of abnormal spinal curvature (fig. 20.15). In scoliosis, the vertebrae shift sideways, so that the spine deforms into an S or C shape. If the curvature is not stopped—usually by wearing a brace or undergoing surgery—the vertebrae and their attached ribs twist, crowding the lungs and heart and disrupting their functions.

An abnormal forward curvature of the spine called **lordosis** is sometimes seen in long-distance runners who do no other form of exercise. The movements of running strengthen the back muscles but not the abdominal muscles in front. The disproportionate muscle use misaligns the spine. Fortunately lordosis caused by intense exercise limited to certain muscle groups can be stopped by adding other forms of exercise, such as weight training or situps.

The **rib cage** is also part of the axial skeleton. It protects the heart and lungs, resembling a packing crate. In the human, it is built of 10 pairs of ribs that are attached to the sternum, or breastbone, and 2 additional pairs not attached that are said to "float." All of the ribs are also attached to the vertebral column. The flexibility of the cartilage connections between the ribs and other bones allows muscles to elevate the ribs. This movement is essential for breathing.

Figure 20.14
The vertebral column consists of 33 vertebrae that provide a flexible protective case for the spinal column. The vertebrae have become specialized for different functions in various regions of the spinal column.

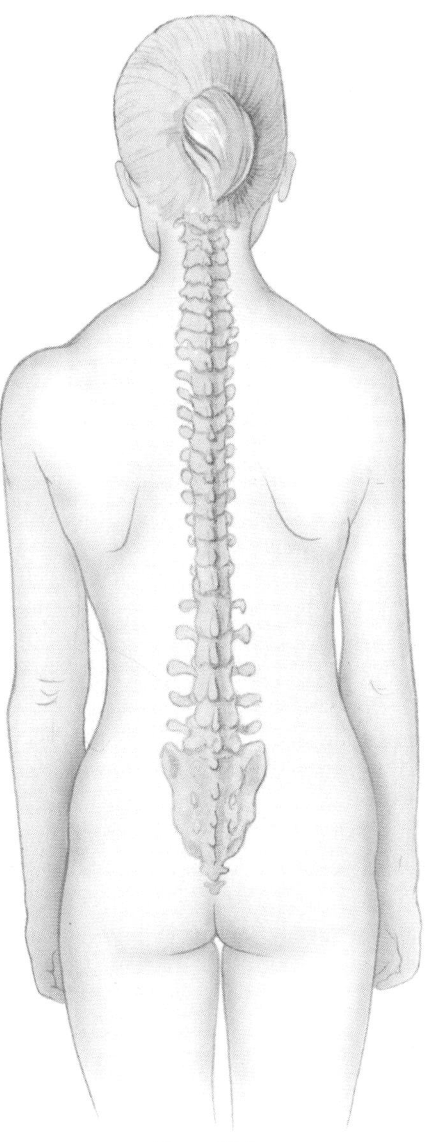

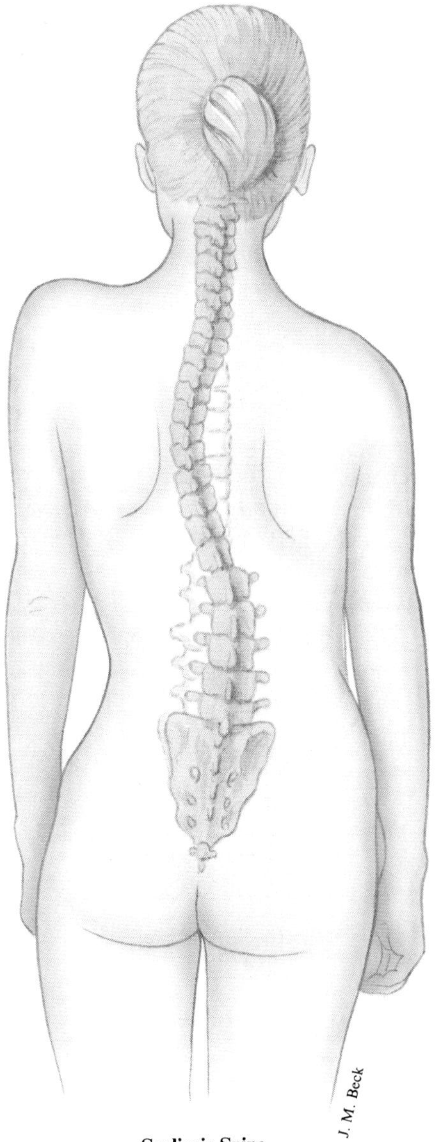

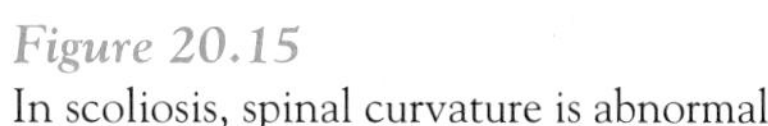

Figure 20.15
In scoliosis, spinal curvature is abnormal.

KEY CONCEPTS

The axial skeleton includes the skull, vertebral column, ribs, and breastbone. These bones shield delicate body parts. The bones of the skull fit tightly together. In the head, only the lower jaw and middle ear bones move. The vertebrae are specialized to allow locomotion. Scoliosis and lordosis are types of abnormal spinal curvatures. The ribs protect the heart and lungs.

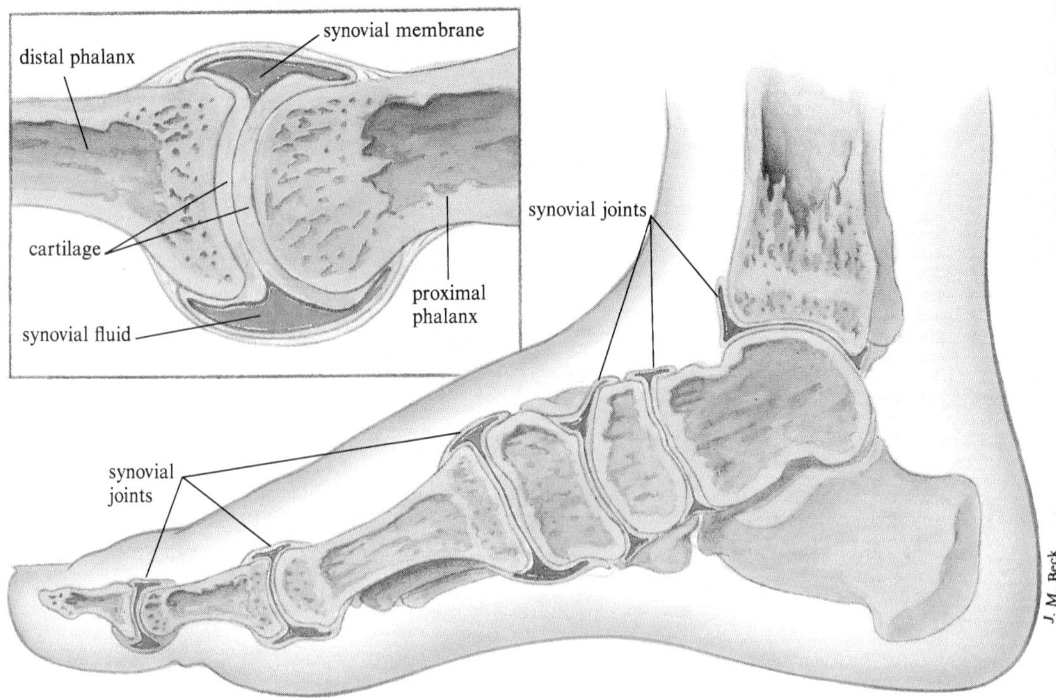

Figure 20.16
The synovial joint—where bone meets bone. A synovial joint is a capsule of fibrous connective tissue ligaments that is filled with fluid. This illustration shows the synovial joints in the foot and ankle.

Appendicular Skeleton

The appendicular skeleton ("appendicular" refers to appendages, or limbs) permits movement. It includes two structures, called limb girdles, to which the limb bones attach. The long and short bones of the appendicular skeleton function as lever systems, which, when powered by muscles, cause movement. As the number of bones connected by movable joints increases, finer and more variable movement is possible.

The **pectoral girdle,** which forms the shoulders, consists of two clavicles (collarbones) and two scapulae (shoulder blades) (fig. 20.11). Connected to a scapula on each side is an upper arm bone, called a **humerus,** which in turn is jointed to the lower arm bones, the **radius** and the **ulna.** Next, the wrist bones, called **carpels,** are joined to the **metacarpal** bones of the hand. Connected to these are the finger bones, or **phalanges.** Because the wrist and the hand are composed of many small bones, we can grip objects and make delicate hand movements such as those needed for writing, sewing, or even performing surgery.

The leg bones are attached to the **pelvic girdle.** This girdle is built of the two hipbones, each of which is actually three separate bones that are fused. The hipbones join the sacrum in the rear and meet each other in front, creating a bowllike pelvic cavity. (The term "pelvis" is Latin for "basin.") The bony pelvis protects the lower digestive organs, the bladder, and some of the reproductive structures (especially in the female). The front of the female pelvis is broader than the corresponding region in the male, and it is larger and has a wider bottom opening. These differences in the female pelvis are adaptations for childbirth. In the later months of pregnancy, hormones loosen the ligaments holding the pelvic bones together, allowing them greater flexibility when the infant passes through.

The hipbone is connected to the thighbone (**femur**). The shin, or **tibia,** is the larger of the two bones of the lower leg, and the **fibula** is the smaller. The fibula is used as a source of bone graft material to replace infected or damaged bone tissue elsewhere in a person's body. The bony part of the kneecap is the **patella.** The ankle bones are called **tarsals.** The arch and ball of the foot allow body weight to be distributed over a triangular area much larger than the base of the leg bones. Although they are constructed in a basically similar manner, the toe bones, or phalanges, are shorter than the finger bones and consequently most of us cannot use our toes for grasping or for fine motor tasks. However, the toes, and the entire complicated ensemble of bones in the foot, are important for balancing the body while standing. If you stand on one foot for a moment and notice the subtle movements of the many bones of the ankle and foot as your weight shifts, you will appreciate their role in maintaining an upright posture, in walking, and in running.

KEY CONCEPTS

The appendicular skeleton includes the limb bones and limb girdles, which provide movement when powered by muscles. The pectoral girdle consists of the collarbones and shoulder blades. The bones of the arms, wrists, and hands extend from the shoulder blades. The two hipbones form the pelvic girdle, and the leg bones extend from here. The female pelvis is specialized for childbirth. The bones of the feet are arranged to support the body's weight.

Joints—Where Bone Meets Bone

The skeleton is constructed of bones attached to other bones. The attachments between specific bones and the degrees of movement that they permit have much to do with the locomotive function of the skeletal system. Various parts of the skeleton have different degrees of mobility. The bones of the skull, for example, interlock like the pieces of a jigsaw puzzle and join each other only by a thin layer of

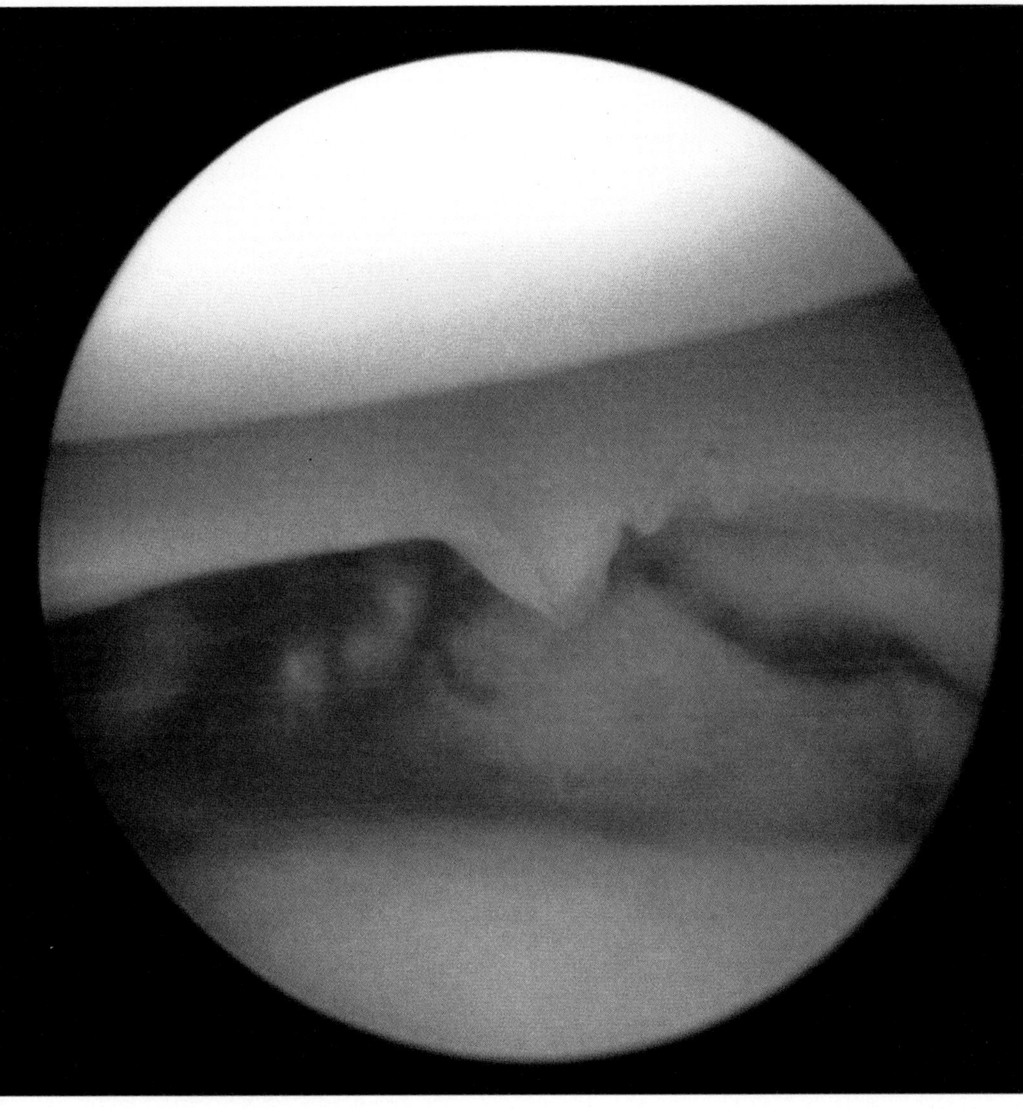

Figure 20.17
Joints are frequently the sites of athletic injuries. The arthroscope is a device that enables a medical specialist to look into a joint without using extensive surgery. A small illuminated tube is inserted into a joint, usually in the knee or shoulder, where it provides a magnified view of the structures within. Arthroscopes are used to diagnose joint problems, to provide a route for the insertion of instruments used to remove damaged cartilage, and even to perform surgery. The procedure requires only local anesthesia and a stitch or two. A notable supporter of the use of the arthroscope is Joan Benoit Samuelson, who won first place in the 1984 Olympic Marathon trial just 17 days after having arthroscopy performed on her knee.

fibrous connective tissue. After 18 months of age, these bones are immobile. The vertebrae, which are attached to each other by dense connective tissue and fibrous cartilage in the back, move only slightly.

Most of the rest of the skeleton is built of freely movable bones that are separated from each other by cartilage but joined by a capsule of fibrous connective tissue called a **synovial joint** (figs. 20.16 and 20.17). Such a joint is filled with fluid, which, along with the slipperiness of the cartilage on the bone ends, allows bones to move against each other in a nearly friction-free environment. Indeed, the friction of cartilage on cartilage in a healthy synovial joint is only 20% that of ice on ice.

The joint capsule is formed by tough bands of fibrous connective tissue called **ligaments.** Lining the interior of the cavity is the **synovial membrane,** which secretes the lubricating synovial fluid. The amount of synovial fluid in joints decreases with age, which is one reason why older people typically have stiffer joints than younger people. Within a synovial joint, small membrane-lined packets, called **bursae,** store synovial fluid and help to reduce friction between bones and other nearby structures such as skin and muscle. Calcium deposits in the bursae cause inflammation, known as **bursitis**. Tennis elbow is a familiar example of bursitis.

Freely movable joints are a prime target for disease and injury. Advancing age makes these places where bone meets bone especially vulnerable. More than 75 diseases of joints are recognized. The most common type of joint problem is **arthritis,** a term that refers to joint inflammation.

The most serious form of arthritis, rheumatoid arthritis, is an inflammation of the synovial membrane, usually of the small joints of the hands and feet. Rheumatoid arthritis may be due to a faulty immune system that attacks the synovial membranes.

Osteoarthritis is more common than rheumatoid arthritis. It is sometimes called the "wear-and-tear" disease because it seems to result from normal use of the joints over many years. The cartilage in the joints begins to wear away and, as the bone is exposed, small bumps of new bone begin to form. Osteoarthritis usually becomes noticeable after age 40 as stiffness and soreness in certain joints.

Along with the muscular and nervous systems, the skeletal system endows us with the capacity to carry out an enormous variety of bodily motions. To ensure that this masterpiece of natural engineering continues to serve its functions throughout the life span, we must use it regularly and provide it with the nutrients necessary to sustain the continual, microscopic ebb and flow of its components.

KEY CONCEPTS

Synovial joints are capsules of fibrous connective tissue that join movable bones. The fluid in the joints plus slippery cartilage on bone ends allow bones to move against each other with very little friction. The joint capsule is made of ligaments, and it is lined by the synovial membrane, which has small fluid-filled areas called bursae. Arthritis is an inflammation of the joints.

SUMMARY

An animal's skeleton provides support for its body and something firm against which muscles can contract. A *hydroskeleton* results from the pressure of a constrained fluid. A *braced framework*, which has solid components, can be worn on the organism's exterior as an *exoskeleton* or within the body as an *endoskeleton*. Exoskeletons include shells that thicken and grow throughout life and jointed suits of armor that are periodically shed and replaced. An endoskeleton grows along with the organism.

The human skeleton provides support, protection, points of attachment for other tissues, and a storehouse for minerals that are important to many body functions. Skeletal components are living, metabolizing structures built of bone, cartilage, and connective tissue. Cartilage forms the embryonic skeleton, participates in bone repair in the adult, and is present on the ends of bones. In cartilage, cells occupy lacunae within a bloodless matrix of collagen and water-trapping *proteoglycans*. Bone is a connective tissue with a rigid matrix. Its strength is due to collagen and its hardness comes from minerals (calcium and phosphate). Calcium balance is maintained between the bones and blood plasma by hormones.

In compact bone, living bone cells *(osteocytes)* are arranged in Haversian systems. The central channel for blood supply is a *Haversian canal*. Tiny channels, *canaliculi*, connect the spaces housing bone cells (lacunae) and the Haversian canal. Processes from bone cells extend through the canaliculi and materials are transported from cell to cell.

The sites of initial bone formation *(ossification)* in the embryo occur in the middle of the shaft of a long bone. Bone elongation during childhood occurs in disks of cartilage between the shaft of a long bone and the knobby ends.

In a vertebrate, the *axial skeleton* consists of the skull, vertebral column, breastbone, and ribs. The *appendicular skeleton* is composed of the limbs and the limb girdles (pectoral and pelvic) that support them. Bones are attached to one another by joints. Some joints, such as those holding the skull bones in place, are immovable. The free-moving connections between bones are *synovial joints*, which are built of cartilage and connective tissue ligaments and contain lubricating synovial fluid.

QUESTIONS

1. Distinguish between a hydrostatic skeleton, exoskeleton, and endoskeleton. Give an example of an animal with each type of skeleton.

2. What advantages and disadvantages does a jointed exoskeleton have over a shell? What advantage does an endoskeleton have over a jointed exoskeleton?

3. List five functions of the human skeleton.

4. What role does cartilage play in the skeletal system?

5. What are the two major components of bone matrix? How do these work together to give bone its characteristics?

6. How is a bone cell nourished? Relate this process to the structure of compact bone.

7. Describe each of the following and indicate where each is found:

- **a.** Spongy bone
- **b.** Compact bone
- **c.** Red marrow
- **d.** Yellow marrow

8. How does the structure of a long bone increase its strength?

9. Describe the process of bone growth that occurs during fetal development.

10. How do bones elongate as we grow throughout childhood?

11. Describe how a broken bone heals.

12. What is osteoporosis? Why is it more common among women than men?

TO THINK ABOUT

1. Skeletal remains of our humanlike ancestors are uncovered in Ethiopia. Examination of some characteristics of the bones of several individuals suggests that there are four types of humans represented by the remains. Specifically, microscopic examination shows that the bone masses of some skeletons are about 30% less than those of the other type, and the skeletons with the lower bone mass have front pelvic bones that are broader than those of the heavier bone mass group. Within the two groups defined by bone mass and pelvic bone shape, smaller skeletons have bones with evidence of epiphyseal plates, but larger bones have only a thin line where the epiphyseal plates should be. What might the four types of individuals in this find be?

2. Cartilage cells manufacture a chemical that suppresses the growth of blood vessels. A tumor needs a large blood supply to grow and spread. How might the cartilage chemical be used to fight cancer? Why would a shark be a better source of cartilage to use to search for this substance than a rat?

3. If a bedridden hospital patient is not moved for a long time, the patient's cartilage begins to degenerate. What might be the molecular basis for this phenomenon?

4. What role does fluid play in hydroskeletons, cartilage, and synovial joints? How might the functions of fluid in modern skeletons be a reflection of the evolution of skeletal systems in general?

5. Spondylitis is a painful condition in which ligaments in the backbone harden. How can this impair the functioning of the backbone? What might symptoms be?

6. Supersaurus is the largest known dinosaur. Bones 6 feet (1.83 meters) long suggest that the animal was at least 120 feet (36.58 meters) long. The pelvic bones are hollow. What functions might hollow bones have served in this dinosaur?

SUGGESTED READINGS

Caplan, Arnold I. November 1984. Cartilage. *Scientific American*. A fascinating consideration of how the molecular structure of cartilage provides the tissue with the properties that make it essential for skeletal structure and function.

Harris, Edward D. Jr. M.D. May 3, 1990. Rheumatoid arthritis. *The New England Journal of Medicine*. In this painful and debilitating condition, the body's immune system attacks the joints.

Hecht, Annabel. December 1984-January 1985. The sinuses are obsolete troublemakers. *FDA Consumer*. These holes in the head can cause quite a bit of pain.

Stolzenburg, William. August 25, 1990. When life got hard. *Science News*. Evidence of ancient injuries suggests that skeletons may have evolved to provide protection from predators.

Stehlin, Dori. May 1988. The silent epidemic of hip fractures. *FDA Consumer*. Hip fractures are common among the elderly, but there are ways to prevent and treat them.

Wallis, Claudia. January 14, 1985. Making bones as good as new. *Time*. The fibula can be used as a bone graft elsewhere in a person's body.

Zamula, Evelyn. April 1989. Back talk: Advice for suffering spines. *FDA Consumer*. Nearly 7 million Americans suffer from back pain.

C H A P T E R

21

The Muscular System

Chapter Outline

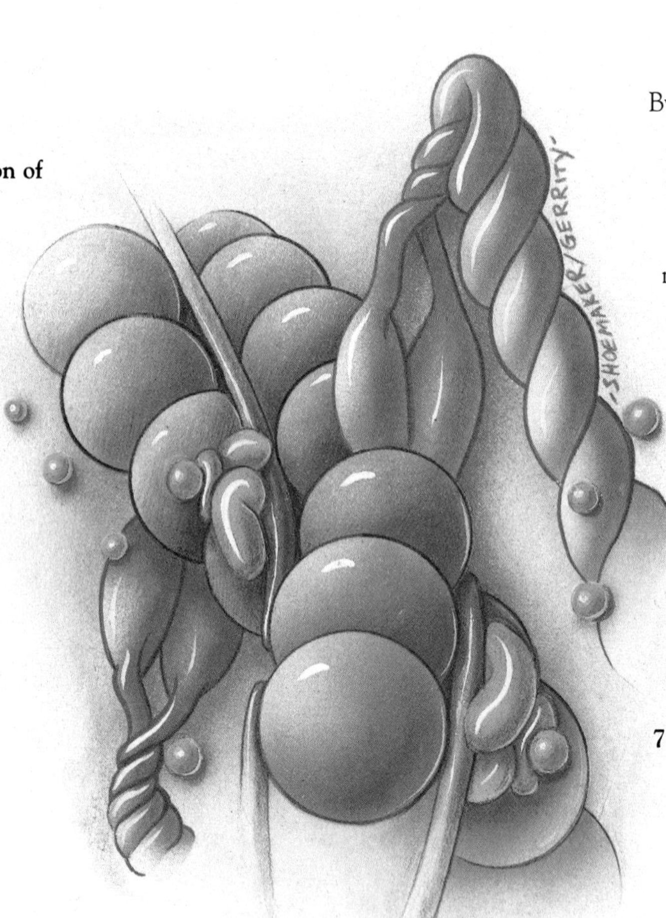

Learning Objectives

By the chapter's end, you should be able to answer these questions:

1. How are the three types of contractile cells alike and how are they different?
2. What structures compose skeletal muscle, both macroscopically and microscopically?
3. How do thick and thin protein rods interact to cause the sliding action that underlies muscle contraction?
4. What factors contribute to the number, distribution, speed of contraction, and resistance to fatigue of muscle fibers within a skeletal muscle?
5. How do nerves affect a muscle's activity?
6. How do muscles, joints, and bones function together in lever systems?
7. How does exercise affect muscle tissue?

The sprinters line up. As the gun goes off, the runners shoot forward almost instantaneously. Deep within their working leg muscles, nerves signal muscle cells to contract. The shortening muscle fibers pull on ropelike tendons, which move the bones to which they are attached. Within the long, thin cells of the contracting muscles, the relative movements of countless tiny protein filaments add up to power the athlete's limbs.

One function of the muscular system—locomotion—is dramatically displayed in athletic competition. Muscle tissue also provides less obvious, but vital, functions: the sustained beating of the heart, the waves of contractions that move food along the digestive tract, and breathing. Muscle also shapes the human form. Roughly half of the body mass is composed of muscle tissue.

Muscle Cell Types

Several types of muscle cells are recognized (fig. 21.1). All forms of muscle contract by the controlled sliding of tiny protein rods within the cells.

Smooth muscle cells are long and tapered. Each cell has a single nucleus. Because smooth muscle cells function without conscious control, they form what are called involuntary muscles. Smooth muscle is found in many parts of the body. In the walls of certain blood vessels, smooth muscle contracts to regulate blood pressure and direct blood flow to the appropriate regions of the body. Waves of contraction of smooth muscle in the wall of the intestines push food through the digestive system. Smooth muscle forms linings in several hollow organs.

Cardiac muscle cells are found in the heart, where their contraction creates the force that propels blood around the body. Cardiac muscle cells are striped, or striated, and their control is involuntary. They have one nucleus per cell. Cardiac muscle cells are branched to form a netlike arrangement, and they are joined to one another by tight foldings in the membrane called **intercalated disks.**

Probably most familiar are the skeletal muscle cells, which are attached to bones and are under conscious, or voluntary, control. Skeletal muscle cells are very

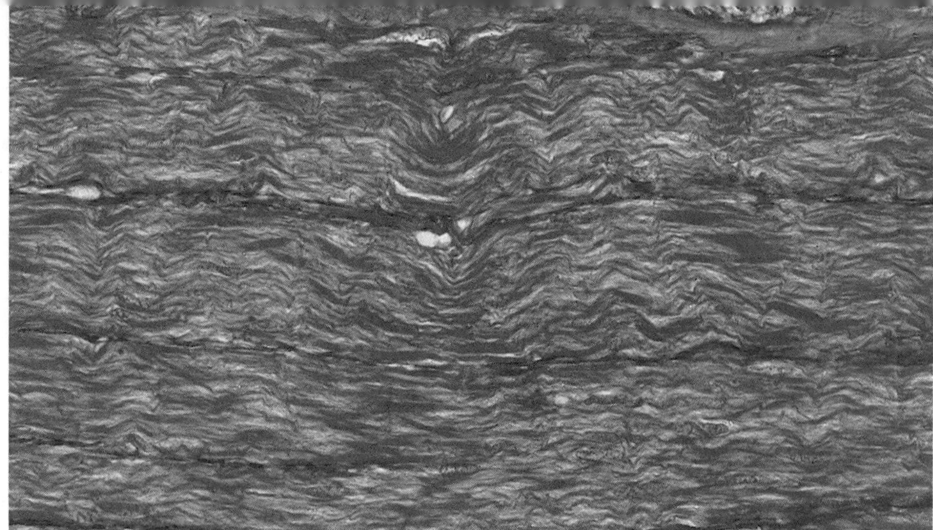

a.

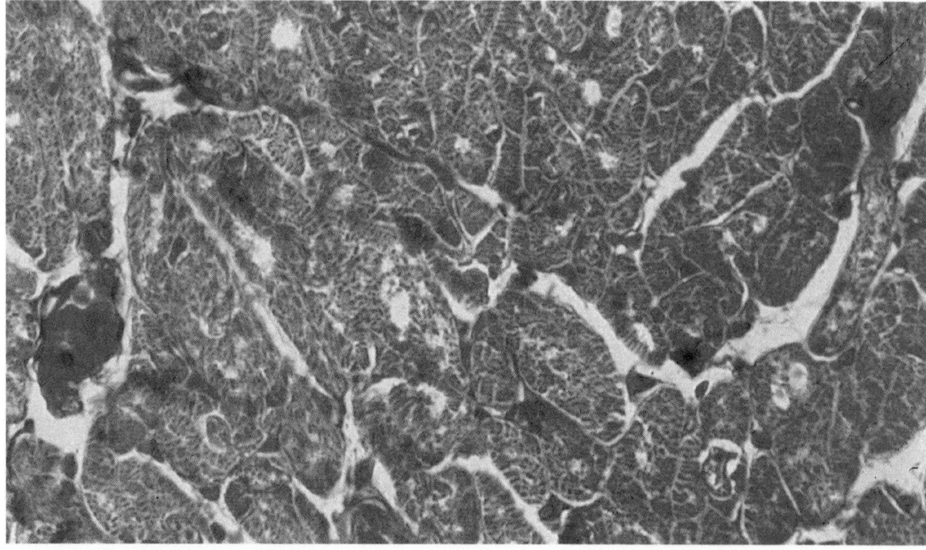

b.

c.

Figure 21.1

There are three types of muscle tissue. *a.* Smooth muscle is composed of spindle-shaped cells that often form layers. We have no conscious control over the contraction of smooth muscle. *b.* Cardiac muscle is unique to the heart. Its striated cells are joined at tight connections called intercalated disks. The cells form a branching network, and their contraction is involuntary. *c.* Skeletal muscle is under voluntary control. The cells are striated due to the orderly arrangement of contractile proteins.

long and appear striated under the microscope due to the orderly arrangement of contractile filaments within them. A single cell has several nuclei, which are pushed to the outer surface. We will take a closer look at the familiar skeletal muscles.

> **KEY CONCEPTS**
>
> *Smooth muscle cells are long, with a single nucleus, and contract involuntarily. Smooth muscle cells line some blood vessels and hollow organs. Cardiac muscle cells are in the heart. They are striated, have one nucleus, and are under involuntary control. Cardiac muscle cells are joined by intercalated disks. Skeletal muscle cells are striated and voluntary and form the muscles that move bones.*

Skeletal Muscle Organization

The powerful leg muscles that propel a sprinter can be viewed at several levels (fig. 21.2). The most familiar level is the whole muscle. Each muscle lies along the length of a bone with opposite ends attached to different bones by a heavy band of fibrous connective tissue called a **tendon.** The word "muscle" comes from the Latin word for mouse, because the shape of a typical skeletal muscle and its attached tendons resembles the shape of a mouse. A typical skeletal muscle has a connective tissue sheath surrounding many bundles, called **muscle fasciculi,** of very long cells called muscle fibers. Each muscle fiber itself has a connective tissue covering.

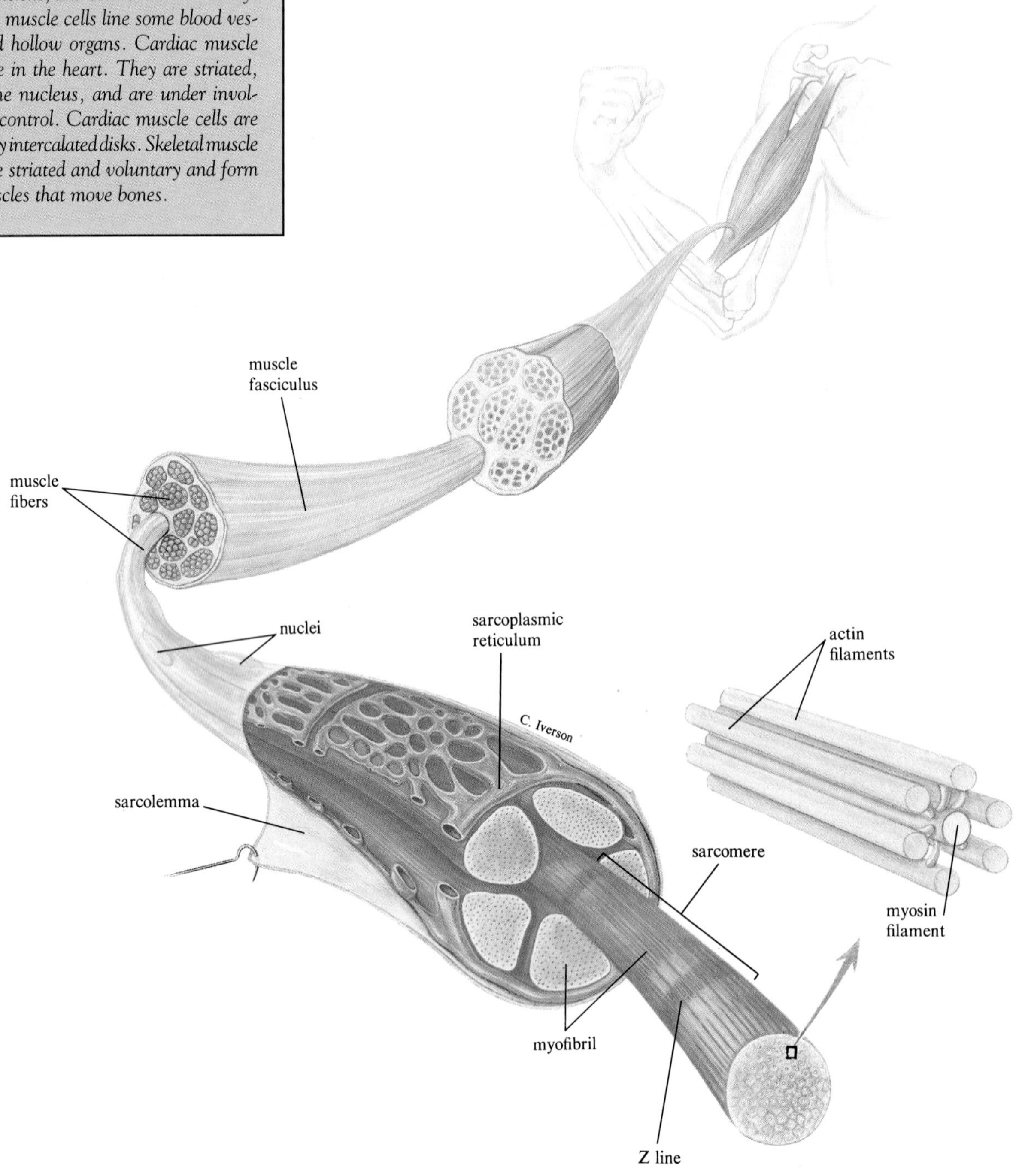

Figure 21.2
Levels of skeletal organization. A skeletal muscle is composed of bundles of muscle fibers. Each muscle fiber, or muscle cell, contains many myofilaments.

Between the bundles of muscle fibers are blood vessels and nerves. A rich blood supply is needed so that every muscle cell receives the nutrients and oxygen needed to power contraction and so that cellular wastes can be removed. The nerves trigger and control muscle contraction.

Microscopic Structure and Function of Skeletal Muscle

Muscle fibers—that is, individual muscle cells—can be more than 1 foot (.3 meter) long. Most of the volume of a muscle fiber consists of hundreds of thousands of cylindrical subunits called **myofibrils.** The rest of the cell contains the usual cellular components. The cell membrane and endoplasmic reticulum (ER) of the fibers are extensive and have special names—**sarcolemma** for the cell membrane, and **sarcoplasmic reticulum** for the ER. Both membrane networks are folded against one another at many points. The parts of the outer membrane that jut into the inner membrane are called the **transverse,** or **T, tubules** (fig. 21.3).

The myofibrils contain even finer "strings," the **myofilaments.** The protein molecules that constitute the myofilaments are the actual site of muscle contraction. Myofilaments are of two types—thin and thick. Each thin myofilament is composed primarily of the protein **actin.** Actin molecules are almost spherical but join together in a chain that resembles a string of beads. A thin myofilament is a twisted double strand of the protein actin (fig. 21.4*a*). The thin myofilaments also contain two proteins that are important in controlling muscle contraction, **troponin** and **tropomyosin.**

The thick myofilaments are composed of the protein **myosin.** Each myosin molecule is shaped like a golf club (fig. 21.4*b*). The "shafts" of myosin molecules adhere to one another, forming a bundle that may consist of several hundred molecules. The "heads" protrude from both ends of the bundle in a spiral pattern. The myofilament assemblage is three-dimensional, so each myosin myofilament is surrounded by six thin actin myofilaments (fig. 21.4*c*). This arrangement allows cross-bridges between actin and myosin to form, break apart, and form again. These interactions between the two main muscle proteins are essential for muscle contraction.

Skeletal muscle appears striped when viewed under a microscope because of the arrangement of thick and thin myofilaments. The striping occurs in a pattern of repeated units, which are called sarcomeres (muscle units). Figure 21.5 shows a microscopic view of a sarcomere and a diagrammatic representation of how the myofilaments interact to produce the various shadings. Different areas within a sarcomere have specific designations.

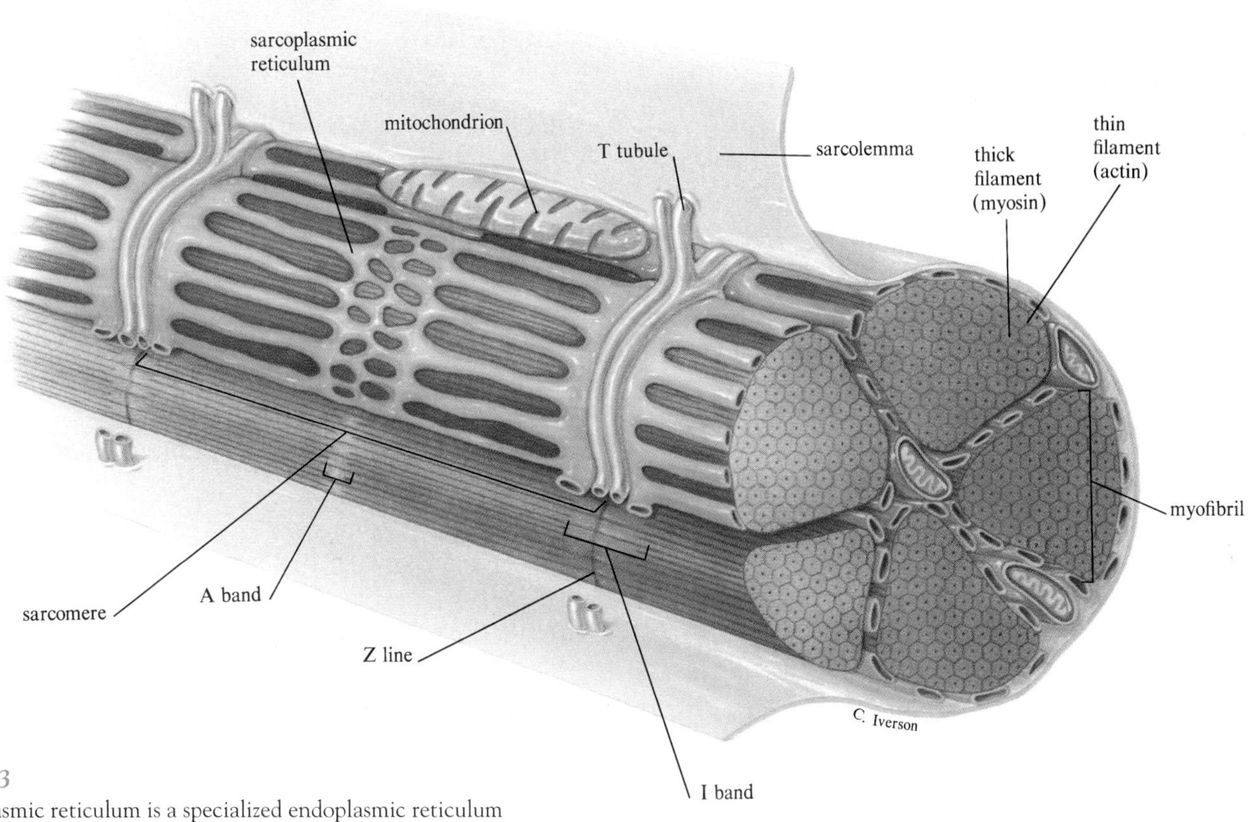

Figure 21.3
The sarcoplasmic reticulum is a specialized endoplasmic reticulum in a muscle cell. It is highly folded and serves as a reservoir for the calcium ions that regulate muscle contraction. Transverse tubules (T tubules) are membranes that fold inward between contractile units of the muscle cell. They conduct the electrical changes caused by a nerve impulse to the interior of the cell. The Z line, I band, and A band reflect the arrangement of muscle protein filaments, and a closer look is provided in figure 21.5.

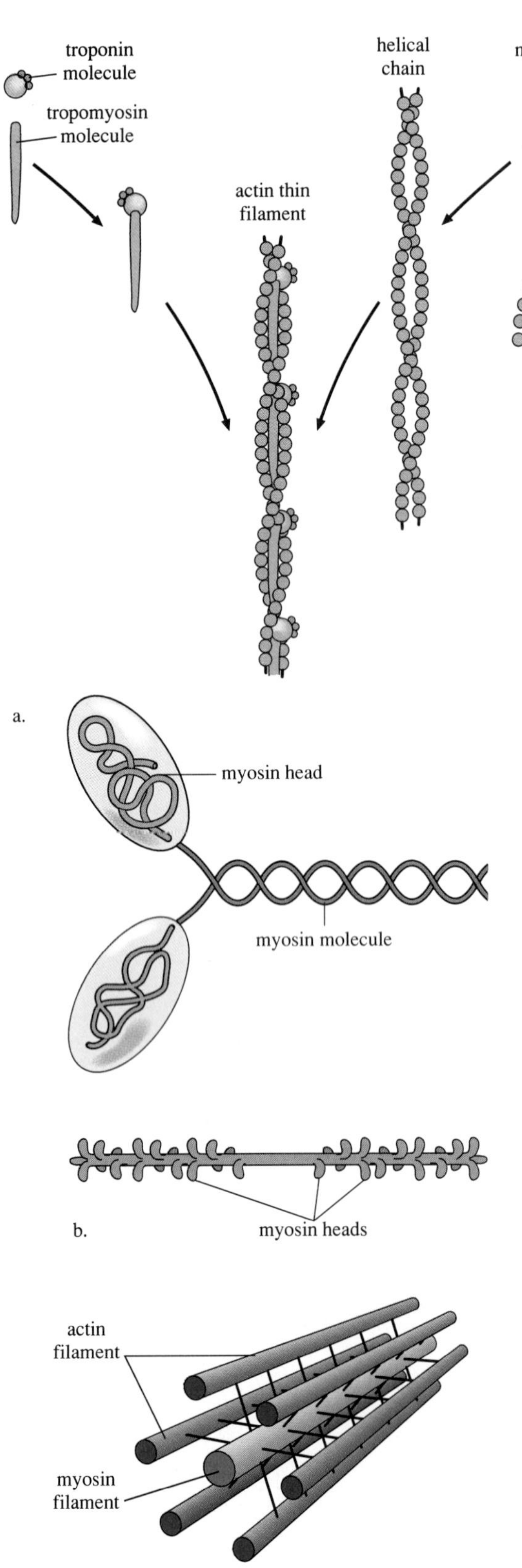

Figure 21.4
a. A muscle's thin myofilaments are composed of actin strands wound around troponin and tropomyosin. *b.* Thick myofilaments are composed of the protein myosin. Individual molecules aggregate to form a linear, spiral structure with protruding myosin heads. *c.* Myofilament arrangement. Six actin myofilaments surround each myosin myofilament.

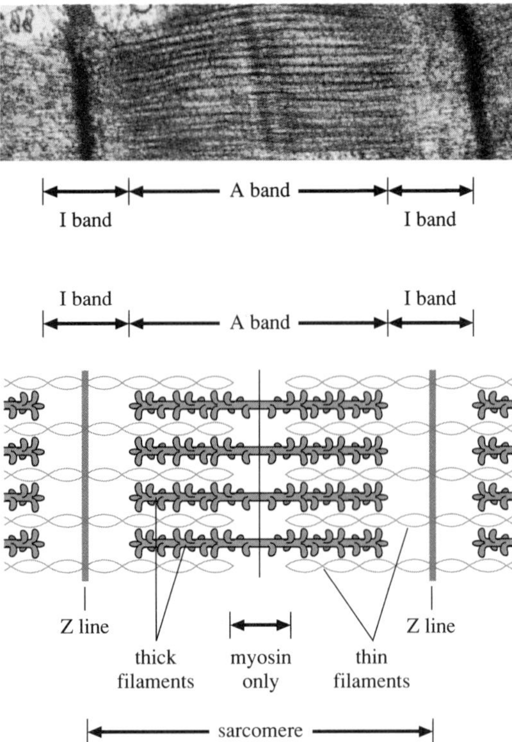

Figure 21.5
The boundaries of a sarcomere are formed by membranes called the Z lines, to which the actin myofilaments are attached. The area on either side of a Z line occupied only by actin myofilaments is called the I band. It appears light in color when viewed under a light microscope. The entire area containing myosin is the A band. The outer borders of the A band look dark under a light microscope because these are the regions where the thick and thin myofilaments overlap. The midregion of the A band is occupied by myosin alone and is light in color. The light and dark bands formed by the alignment of myofilaments give skeletal muscle its striated appearance.

KEY CONCEPTS

Muscle can be studied at many levels. A whole muscle lies along a bone and is attached to other bones by tendons. Within a muscle, connective tissue surrounds bundles of muscle fibers. Blood vessels and nerves run between the bundles. A muscle fiber is a cell, and it contains myofibrils, which in turn contain myofilaments. Thin myofilaments are built of actin, troponin, and tropomyosin, and thick myofilaments consist of myosin. Muscles contract when the thin and thick myofilaments slide past each other.

How Skeletal Muscle Contracts

Muscle contracts when the thin myofilaments slide between the thick ones. This motion shortens the sarcomere without shortening its protein components. You can illustrate this using your hands as a model: Hold your hands in front of you so that you are looking at your palms and your fingertips are touching, then slide the fingers of one hand between those of the other. The distance between your palms shortens even though the length of your fingers remains the same (fig. 21.6). The movement of protein myofilaments to shorten skeletal muscle cells is called the **sliding filament model** of muscle contraction.

A key to understanding how actin and myosin are propelled past one another is in the myosin heads. The club-shaped heads are attached to the shafts with hingelike connections that allow the head to rock back and forth relative to the shaft as the actin and myosin filaments move. This swiveling motion is similar to the path of an oar as a stroke propels a boat through the water. The movement causes actin to slide past myosin in the same manner that the oar's motion moves a boat.

Myosin heads can bind both actin and the energy-laden molecule, ATP. The splitting of this ATP provides the energy that powers muscle contraction. In addition to ATP and the four types of proteins comprising the thin and thick myofilaments, muscle contraction requires calcium ions (Ca^{2+}). It is a highly regulated, multistep process.

Actin and myosin must be in physical contact for muscle contraction to occur. The connection of a myosin head against an actin molecule is called a cross-bridge. In a muscle at rest, troponin holds tropomyosin in such a way that it occupies a groove between the two entwined actin strands, blocking actin's binding sites to myosin. Troponin and tropomyosin thus prevent the formation of the cross-bridges, holding the muscle at rest.

The directive to contract begins when a message from a motor nerve cell that extends from the brain or spinal cord to a muscle cell releases a chemical messenger, acetylcholine, at a special junction of nerve and muscle called a neuromuscular junction (fig. 21.7). This messenger binds to the outer membrane of the muscle cell, inducing an electrical wave that is quickly sent to storage sacs within the sarcoplasmic reticulum. The sacs release calcium ions, which flood the cell's interior. A recently discovered protein, dystrophin, may relay the nerve's signal to the calcium storage sacs. A lack of dystrophin results in muscular dystrophy, a genetic disease in which muscle wastes away (Reading 21.1).

When the cellular concentration of calcium ions increases, they bind to the troponin molecules attached to the actin myofilaments (fig. 21.8). Troponin moves to place the attached tropomyosin molecules in the interior of the thin myofilament, a location where they no longer block actin from contacting the myosin of the surrounding thick filaments. The movement of tropomyosin activates the myosin heads to touch the exposed actin.

Meanwhile, ATP slips into its site on the myosin head (fig. 21.9). An enzyme here splits ATP into ADP and inorganic phosphate, releasing energy that activates myosin. This step is somewhat analogous to cocking a pistol. Activated myosin then forms a cross-bridge to actin. When this occurs, the energy stored in myosin is released, and the myosin head forcefully

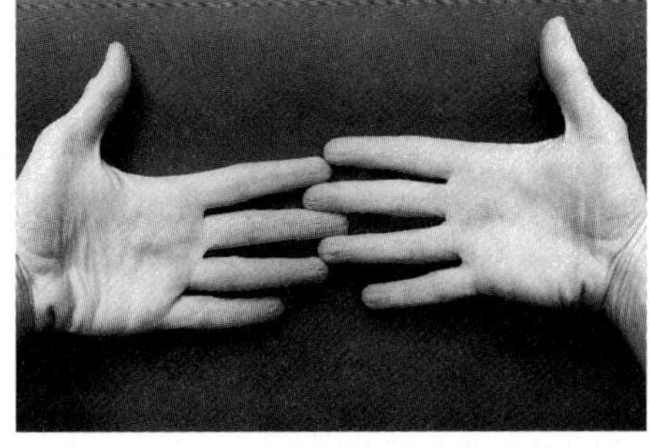
a.

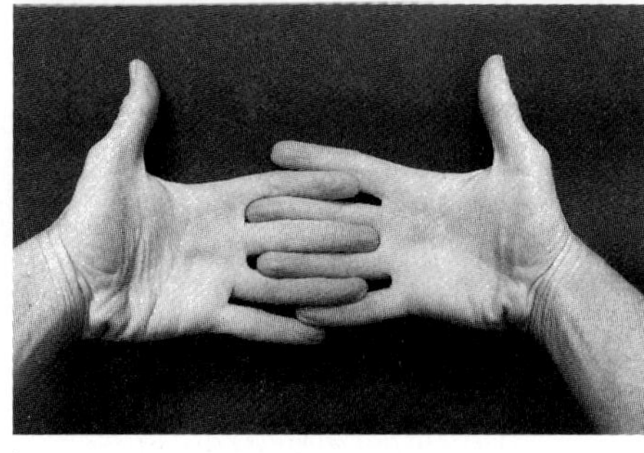
b.

Figure 21.6
Demonstrating the sliding filament model using the movements of fingers. As fingers slide between one another, the distance between the palms shortens. This is analogous to the relative movements of thick and thin myofilaments during muscle contraction.

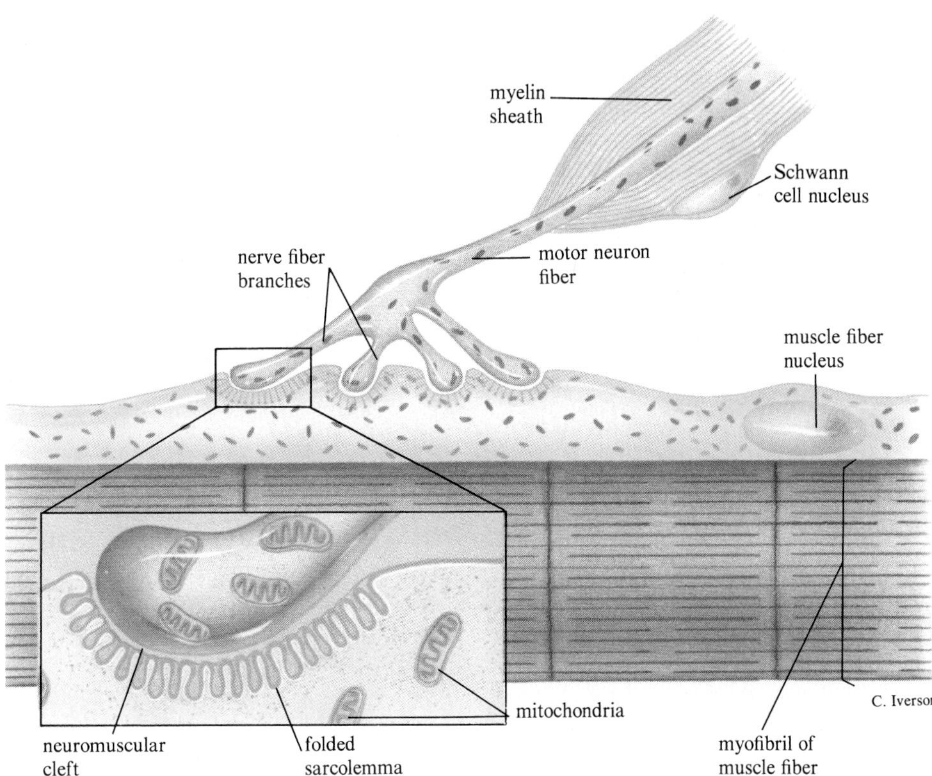

Figure 21.7
A neuromuscular junction. A motor axon branches forming specialized junctions with muscle fibers. The axon releases a chemical that triggers electrical charges in the muscle cell membrane, which cause the release of calcium ions and the initiation of muscle contraction.

swings back to its original position, causing the filaments to move past one another. This step is like pulling the trigger on a gun. Immediately after this power stroke, ADP and the inorganic phosphate pop off the myosin head, leaving it free to bind a new ATP, and the cross-bridge between actin and myosin is broken. The myosin head flips back to its original position and binds a new actin further along on the chain.

This process of filament sliding repeats about five times a second on each of the hundreds of myosin molecules of a thick filament. Although the movement of filaments due to the interactions of single actin and myosin filaments would be no more astounding than the forward progress of a boat with a single person rowing, a skeletal muscle contracts quickly and forcefully due to the efforts of many thousands of "rowers."

A muscle cell does not normally remain contracted. Myofilament sliding stops when calcium ions are no longer available. Shortly after calcium ions are released as a result of motor nerve stimulation, they return to the sarcoplasmic reticulum by active transport. Tropomyosin goes back to its original position blocking actin from interacting with myosin heads, so the sarcomere relaxes.

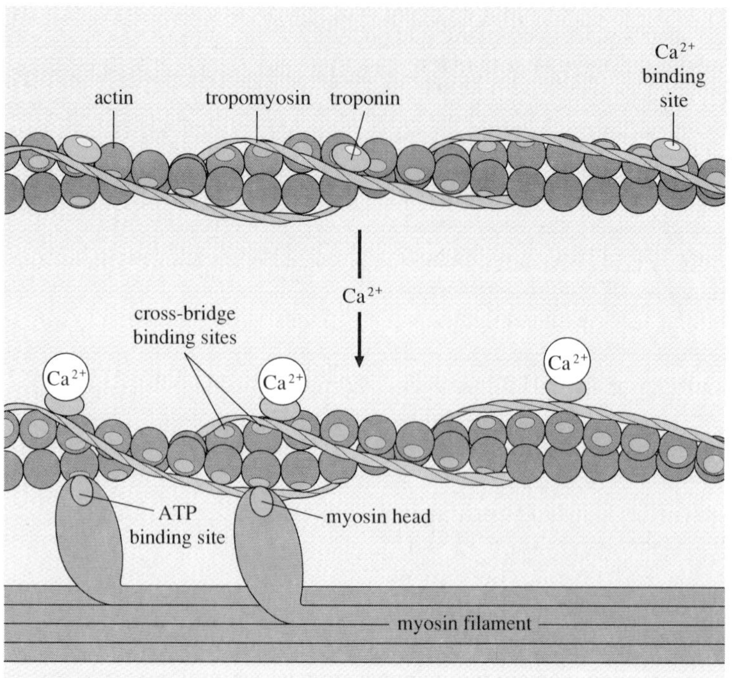

Figure 21.8
The regulation of muscle contraction by calcium ions. In the absence of calcium ions, troponin holds tropomyosin over the sites on myosin that bind to actin. When calcium ions are released after a nerve impulse, they bind to troponin. As a result, tropomyosin moves away from the actin binding sites and cross-bridges can form.
From *Biology*, by Campbell. Copyright ©1987, Benjamin/Cummings Publishing Company. Menlo Park, Calif.

KEY CONCEPTS

When thin myofilaments slide between thick myofilaments, the sarcomere shortens, although the lengths of the filaments are unchanged. This sliding of filaments is possible because the myosin heads can swivel. Muscle contraction requires four major proteins, calcium ions, and ATP for energy. In a muscle at rest, troponin and tropomyosin block contact between actin and myosin. Contraction begins when a nerve message signals sacs in the sarcoplasmic reticulum to release calcium ions, which remove troponin and tropomyosin from actin. Myosin touches actin, ATP splits, and the myosin head moves, causing the filaments to slide. ADP and inorganic phosphate then leave the myosin head, which flips back, binds a new ATP, and forms a cross-bridge to a new actin molecule.

Energy for Muscle Contraction

The contraction of skeletal muscle requires huge amounts of ATP, both to power the return of Ca^{2+} to the sarcoplasmic reticulum and to break the connection between actin and myosin, allowing a new cross-bridge to form. After death, muscles run out of ATP. The cross-bridges cannot be broken, and the muscles remain in position. This stiffening is called rigor mortis.

The body has several ways to generate ATP. When muscle activity begins, ATP can be replenished as rapidly as it is depleted thanks to the molecule **creatine phosphate,** which is stored in muscle fibers. Creatine phosphate can donate its high-energy phosphate to ADP to regenerate ATP. However, after the supply of creatine phosphate is diminished, ATP must be generated from other sources. As long as enough oxygen can be supplied to the muscle tissue, ATP is formed by aerobic respiration in the mitochondria (see chapter 6). The fuel sources are fatty acids and carbohydrates. During more intense exercise, the circulatory system cannot meet the demands of the muscle tissue for fuel, and increasing amounts of oxygen become necessary for aerobic respiration.

When not enough ATP is provided by aerobic respiration and creatine phosphate, the muscle cell obtains ATP from anaerobic respiration, a less efficient metabolic route. The muscle stores glycogen, a polymer of glucose, which can be broken down to provide a ready supply of glucose for metabolism. However, recall that a product of anaerobic respiration is lactic acid. As this biochemical accumulates, it causes muscle fatigue and cramping. To metabolize the lactic acid, the exhausted exerciser continues to pant after the physical exertion, taking in oxygen that is then used to reconvert the lactic acid back to pyruvate and then to metabolize it to carbon dioxide and water. This need for oxygen to complete the metabolism of lactic acid is termed **oxygen debt.**

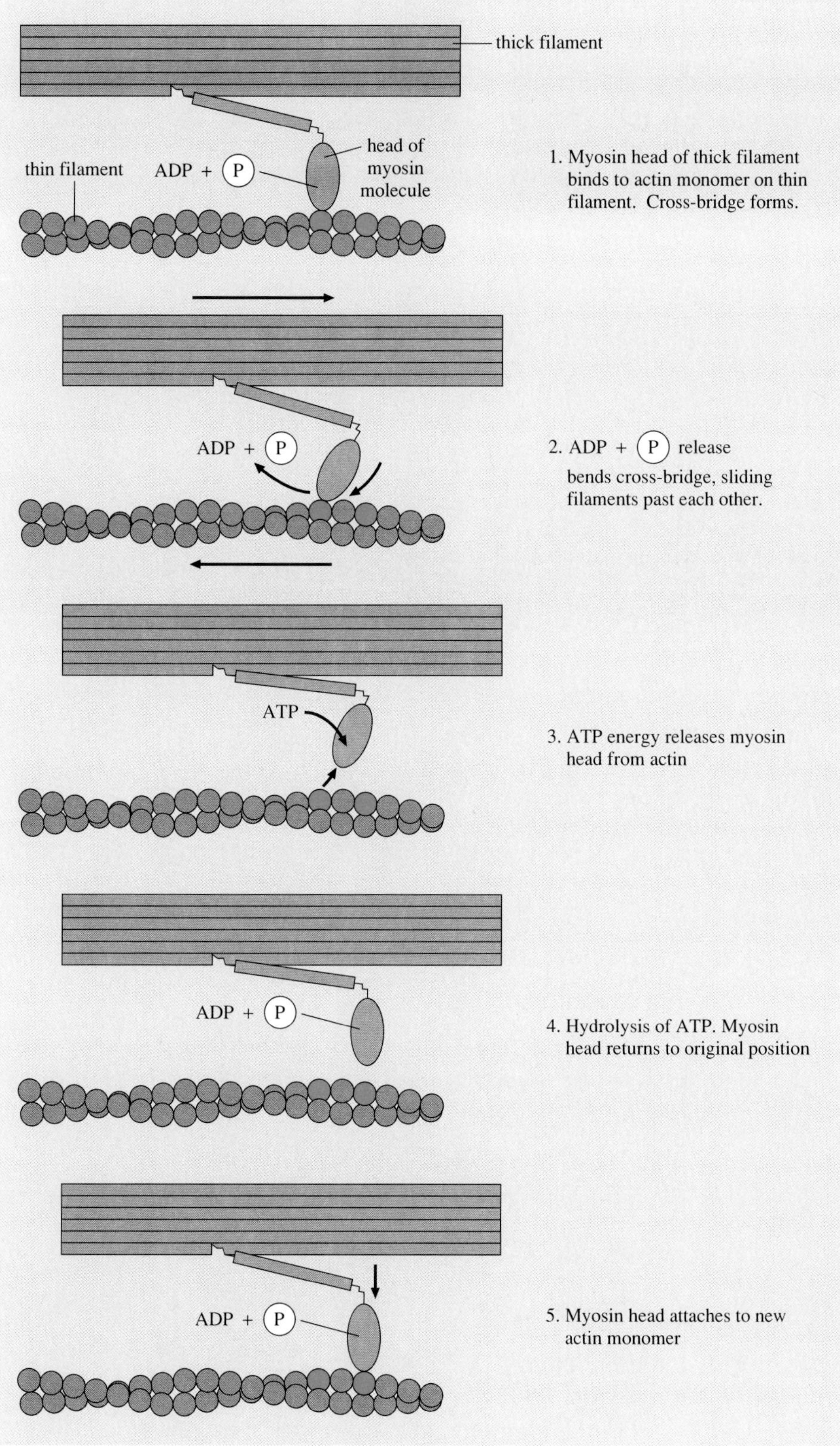

Figure 21.9

The sequence of molecular interactions involved in the sliding filament model of muscle contraction.

From *Biology*, by Campbell. Copyright ©1987, Benjamin/Cummings Publishing Company. Menlo Park, Calif.

KEY CONCEPTS

ATP supplies the energy to return Ca^{2+} to the sarcoplasmic reticulum and to break the contact between actin and myosin. ATP is supplied by creatine phosphate, from aerobic respiration, or from anaerobic respiration.

Macroscopic Structure and Function of Skeletal Muscle

The components of an individual muscle fiber work well as a unit. If nerve stimulation releases calcium within the fiber, then all of its sarcomeres contract and the entire fiber shortens. Therefore, the muscle fiber is said to respond in an all-or-none fashion. If many of the muscle fibers encased in the same connective tissue sheath contract, then the sheath pulls on the attached tendon, which in turn pulls on the bone.

The strength of contraction of a muscle depends both on how many individual fibers contract and on the distribution of the contracting fibers among the many sheathed bundles that constitute the muscle. Thus, the muscle as a whole does not respond in an all-or-none manner, because not all of the fibers are contracted at one time.

A nerve cell and all of the muscle fibers it contacts are called a **motor unit** (fig. 21.10). Depending on how many muscle fibers are touched by the endings of a single motor nerve cell, the neuron may exert either a localized effect or a large effect on a muscle. A motor nerve cell in the eye, for example, may only impinge upon a few muscle fibers, producing the fine small-scale movements of this sense organ. In contrast, a single nerve cell in the upper arm may have hundreds of endings, each in contact with a different muscle cell. As a result, when a single neuron fires, many muscle cells in the upper arm respond, providing the strength to lift a heavy weight.

As can be seen in figure 21.11, the rate of stimulation influences the pattern of muscle response. A single stimulation causes the muscle to contract quickly and

Reading 21.1 *Dystrophin—Unraveling the Defect in Muscular Dystrophy*

ON A BRIGHT APRIL MORNING IN 1990, 9-YEAR-OLD SAM LOOPER WIGGLED THE BIG TOE ON HIS LEFT FOOT, AND MADE MEDICAL HISTORY. He was the first human recipient of dystrophin, the protein that his body lacks, causing his muscular dystrophy. Sam had received a transplant of his father's healthy, dystrophin-producing muscle cells in his toe, so that if side effects arose, they would harm only a small part of his body.

It took researchers until 1986 to identify the dystrophin deficiency behind muscular dystrophy, because this protein accounts for only 0.002% of the total protein in a skeletal muscle cell. Lack of dystrophin means that skeletal muscle fibers cannot relay a nerve's message to the calcium storage sacs within the muscle tissue. As a result, Ca^{2+} is not released, and muscle contraction is halted at the very first step. Not only is contraction impaired, but the abnormal Ca^{2+} levels trigger release of an enzyme that destroys actin and myosin. The body attempts to repair the damaged muscle by hardening it, but this only blocks the blood supply, destroying muscle function further. It is a deadly sequence of effects (fig. 1).

Duchenne's muscular dystrophy, the major form of the illness, is inherited on the X chromosome from a carrier mother, and therefore most sufferers are male. Typically signs of muscle weakness do not begin until about age 3, when the boy stands and walks oddly to compensate for weak leg muscles. He falls easily and then cannot right himself. By age 6 he may exhibit "gower's sign," in which he rises from a sitting position by pushing off with his hands, because his thigh muscles are weakened. By age 12, he is wheelchair bound. He is likely to die by his early 20s of pneumonia as his respiratory muscles weaken.

The future is very bright, though, for those with muscular dystrophy. All that remains is for researchers to develop the most effective way to deliver **dystrophin** where it is needed. Soon, Sam Looper and the 200,000 others with Duchenne's or Becker's muscular dystrophy, may be able to wiggle more than their big toes.

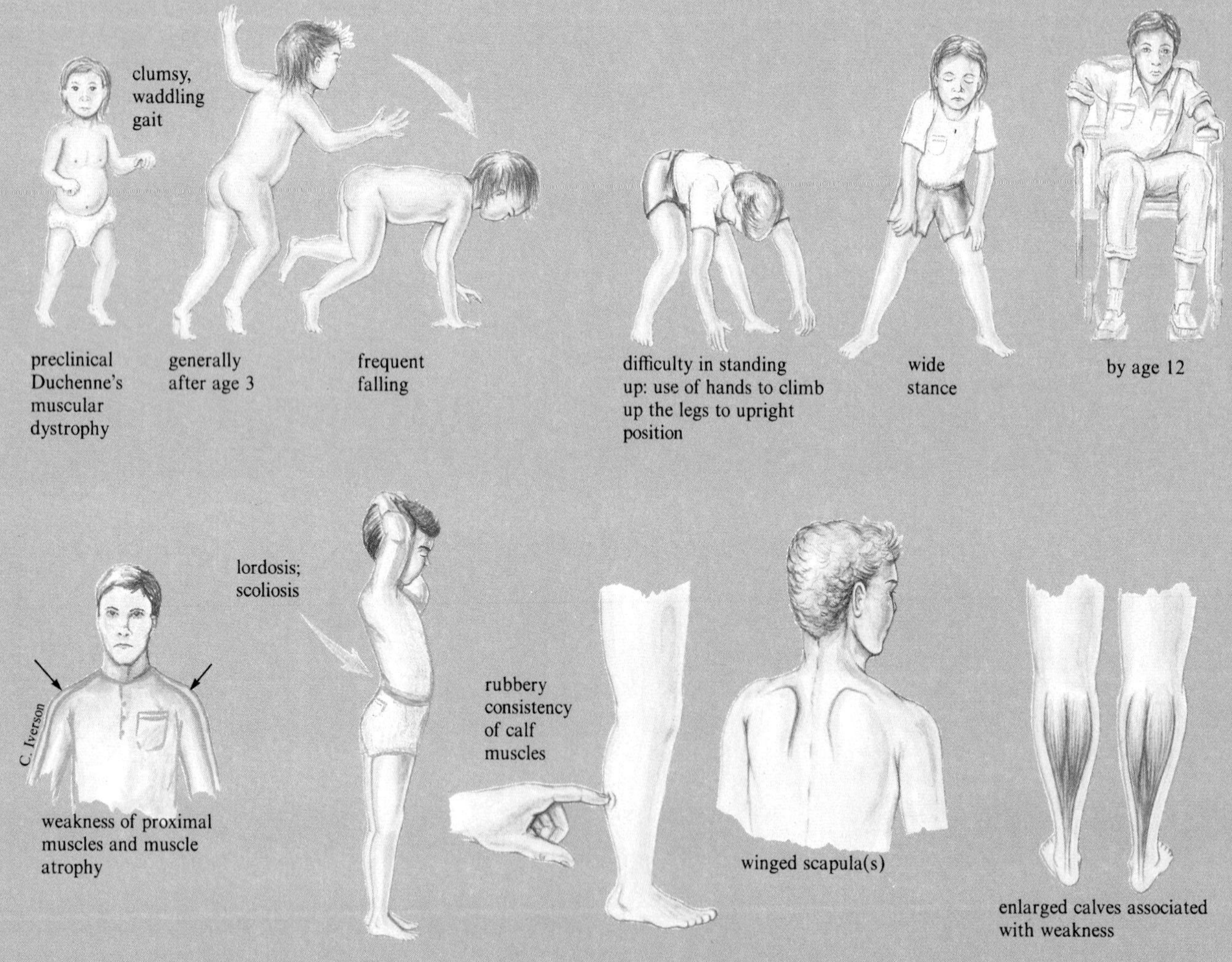

Figure 1
The young sufferers of Duchenne's muscular dystrophy have an increasingly difficult time moving around as their muscles weaken. It is all due to an abnormal gene and its missing protein, dystrophin.

then relax. This response is called a **twitch.** It is followed by a refractory period of 2/100 of a second during which the muscle fiber cannot respond to further stimulation. If the muscle is stimulated again, after the refractory period but before it is fully relaxed, the second contraction is added to the remains of the first one. Therefore, the strength of contraction increases. This phenomenon is called **summation.** If a muscle receives repeated strong stimulation without time to relax, the strength of muscle contraction builds until a smooth and continuous contraction called a **tetanus** occurs. Tetanus ceases when the muscle uses up its chemical energy reserves, and it is no longer able to contract, even when further stimulation is applied. (Muscle tetanus is not the same thing as the infectious disease called tetanus, which is also known as "lockjaw." The disease was so named because of the spasms of the jaw and back associated with it.)

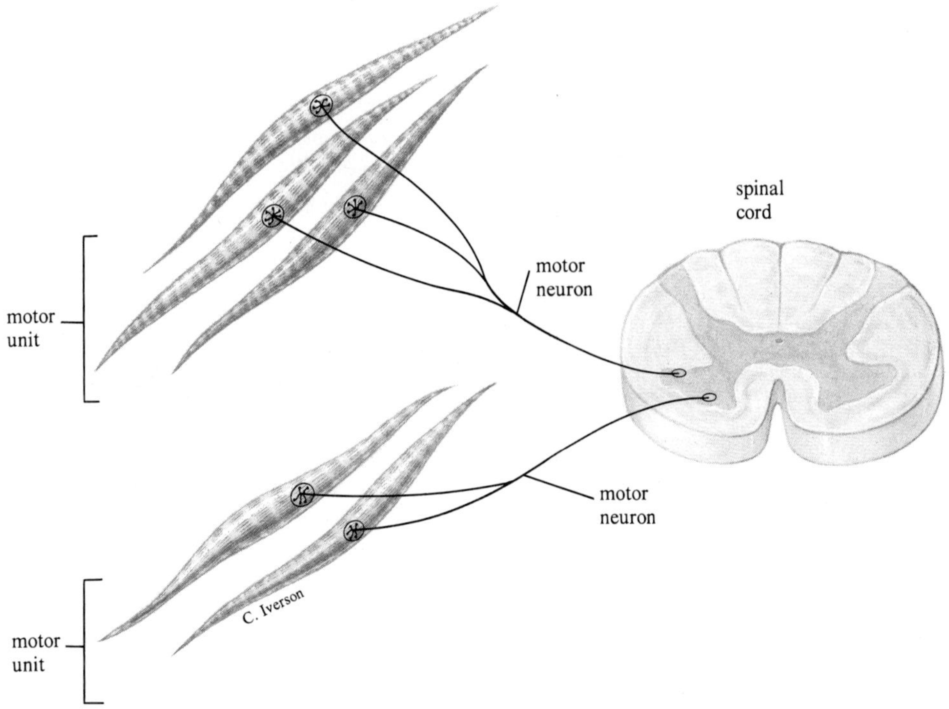

Figure 21.10
A motor unit consists of a motor nerve cell axon and the muscle fibers it innervates. If one nerve cell stimulates many muscle cells, they all contract at once, providing strength. If only a few muscle cells are stimulated by one nerve cell, finer movements are generated.

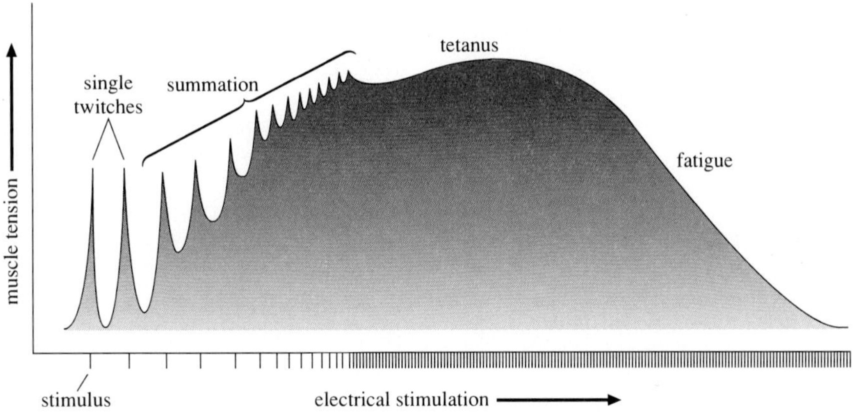

Figure 21.11
Responses of a muscle to stimulation. When stimulated once, a muscle cell quickly contracts and relaxes, a response called a twitch. If the muscle is stimulated a second time before it has completely relaxed, the second contraction is longer than the first. This increase in the strength of contraction is called summation. If the stimulation is rapid, the muscle reaches a state of sustained contraction called tetanus.

KEY CONCEPTS

Muscle fibers respond in an all-or-none manner, but whether or not a muscle moves a bone depends upon how many of its muscle fibers contract and their distribution. A nerve cell and the muscle fibers it contacts form a motor unit. The strength of muscle contraction depends upon how many muscle fibers a motor nerve cell stimulates. A twitch is a muscle's response to a single stimulation. Further stimulation leads to summation and then a tetanus.

Inborn Athletic Ability and Muscle Fiber Types

Is there such a thing as a "natural athlete?" Are some individuals born with certain body features that give them a competitive edge? One way this might occur is by inheriting specific types of muscle fibers.

Most skeletal muscles contain fibers of three different **twitch types,** distinguished by how quickly they contract and tire (fig. 21.12). **Slow twitch–fatigue resistant fibers** contract slowly because the myosin heads split ATP slowly. However, they resist fatigue because they are well supplied with oxygen that is bound to a large number of the oxygen-carrying molecule myoglobin or delivered there by the extensive blood supply serving these fibers. Myoglobin is similar to hemoglobin, the carrier for oxygen in the blood supply, and is also reddish. Therefore these slow twitch, slow fatiguing muscles are also called red or dark fibers. They are found in muscles specialized for endurance. For example, the breast meat (flight muscles) of ducks and geese, which need muscle endurance for long, migratory flights, is reddish "dark meat."

Fast twitch–fatigue resistant fibers also have abundant oxygen, and therefore do not tire easily, but their myosin heads split ATP quickly, resulting in a twitch

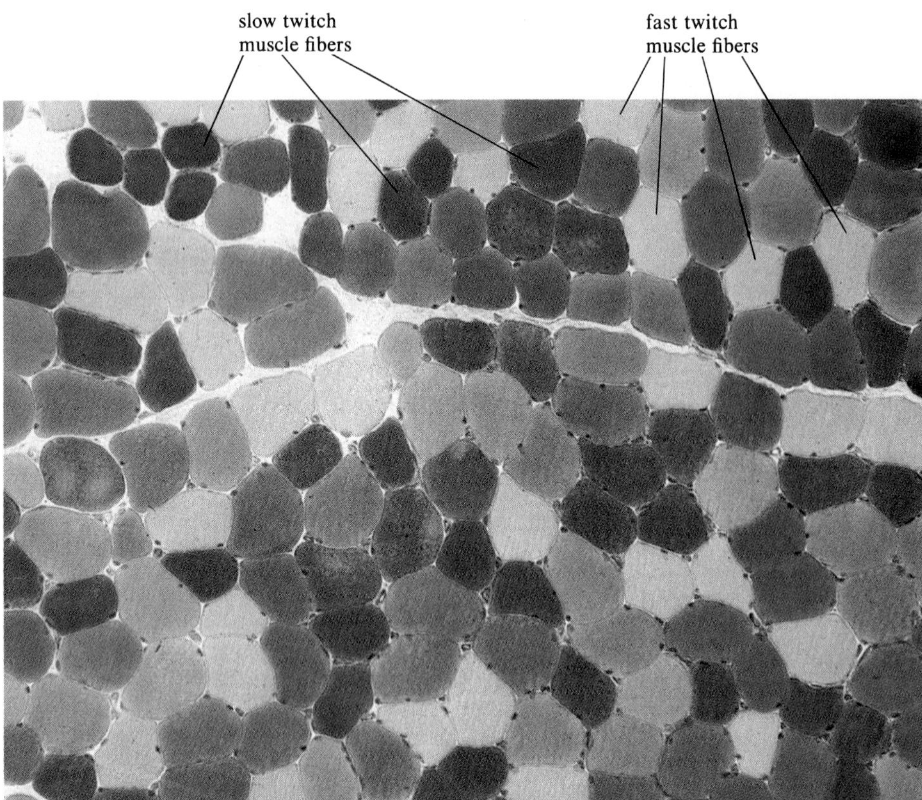

Figure 21.12
Types of muscle fibers. The darker fibers are the slow twitch fibers, which are specialized for endurance. The dark color comes from myoglobin, a molecule that binds oxygen. The lighter fibers are fast twitch fibers. These tire more quickly than the dark fibers.

response that may be 10 times as fast as that of a slow twitch fiber. The third type of muscle fiber, **fast twitch–fatigable fibers,** split ATP quickly but do not have much of an oxygen supply because there are few capillaries and little myoglobin. They are white because they lack myoglobin and a rich blood supply. These muscle fibers are specialized for short bouts of rapid contraction. The very white breast muscle of the domesticated chicken, for example, can power barnyard flapping of short duration but cannot support sustained flight over long distances.

The proportion of fast twitch to slow twitch fibers within particular muscles affects performance. Most people have about equal numbers of fast and slow twitch muscle fibers. However, those who have a higher proportion of slow twitch fibers tend to excel at endurance sports, such as long-distance biking, running, and swimming. Consider Alberto Salazar, who holds many first-place finishes in the marathon but does not do nearly as well in the 400-meter dash. Salazar's leg muscles consist of about 93% slow twitch fibers. When the muscle fiber balance is in favor of the fast twitch variety, short, fast events (such as sprinting, weight lifting, and shot-putting) are easier. In some European nations, measurement of the ratio of fast twitch to slow twitch fibers is used as a predictor of athletic success in certain events. Small samples of muscle tissue from athletes are taken to do this.

KEY CONCEPTS

The proportion of muscle fiber types within a muscle may affect performance. Slow twitch–fatigue resistant fibers contract slowly but have a large supply of oxygen and can contract for a long time. Fast twitch–fatigue resistant fibers split ATP quickly and have large oxygen supplies. Fast twitch–fatigable fibers split ATP quickly but also deplete oxygen quickly.

Muscles Working Together

Lever Systems

The interactions of muscles, joints, and bones to generate movement form biological lever systems. A lever is a structure that can pivot around a point, called the fulcrum, when a force is applied to it. In the body, the bones function as levers, the joints are fulcrums, and the force that moves the levers is supplied by skeletal muscles. Force generated in the muscles is passed to the bones by way of the attached tendons.

A skeletal muscle is attached to one end of each of two bones by its tendons. The muscle's **origin** is the end on the bone that does not move, and the muscle's **insertion** is the end on the movable bone (fig. 21.13).

In any lever system, the effort needed to overcome resistance depends on the length of the lever. In the body, the length of the lever depends on where the muscle is inserted. Interestingly, small differences in the placement of a muscle's insertion can make a big difference in strength. Consider the hypothetical situation depicted in figure 21.14. Except for the insertion, all things are equal—the bone of the forearm, the weight to be lifted, and the strength of the muscles. If the muscle is inserted 2 inches (5.08 centimeters) from the elbow, the effort needed to lift the 35-pound (16-kilogram) weight is 187 pounds (85 kilograms). However, with the muscle inserted 1 inch (2.54 centimeters) from the elbow, 565 pounds (256 kilograms) of force is required!

Antagonistic Pairs

Each movable bone is flanked by two muscles or muscle groups that move it in opposite directions and are therefore termed **antagonistic muscles.** The muscles of the upper arm, for example, form an antagonistic pair. Contraction of the biceps (the muscle that bulges when you "make a muscle") bends the arm at the elbow. The arm is straightened by the contraction of the triceps, the muscle at the back of the upper arm (fig. 21.15).

Typically, one member of an antagonistic pair is stronger than the other. For example, the muscles that close the jaw

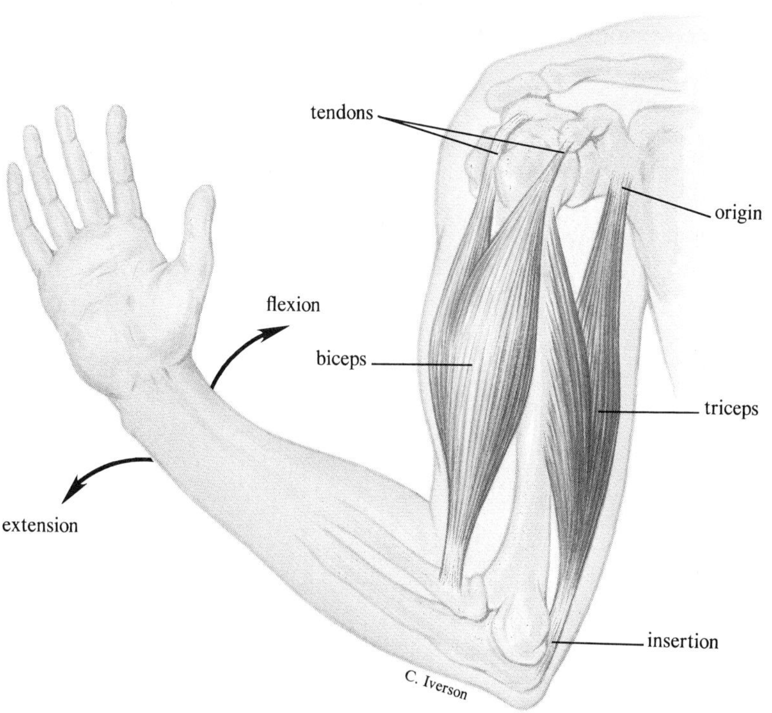

Figure 21.13
The origin and insertion of a skeletal muscle. A muscle is attached to two bones. The attachment to the stationary bone is called the origin. The attachment to the bone that moves when the muscle contracts is called the insertion. The bone can be flexed or extended.

Figure 21.14
A small difference in the insertion of a muscle can make a large difference in one's strength.

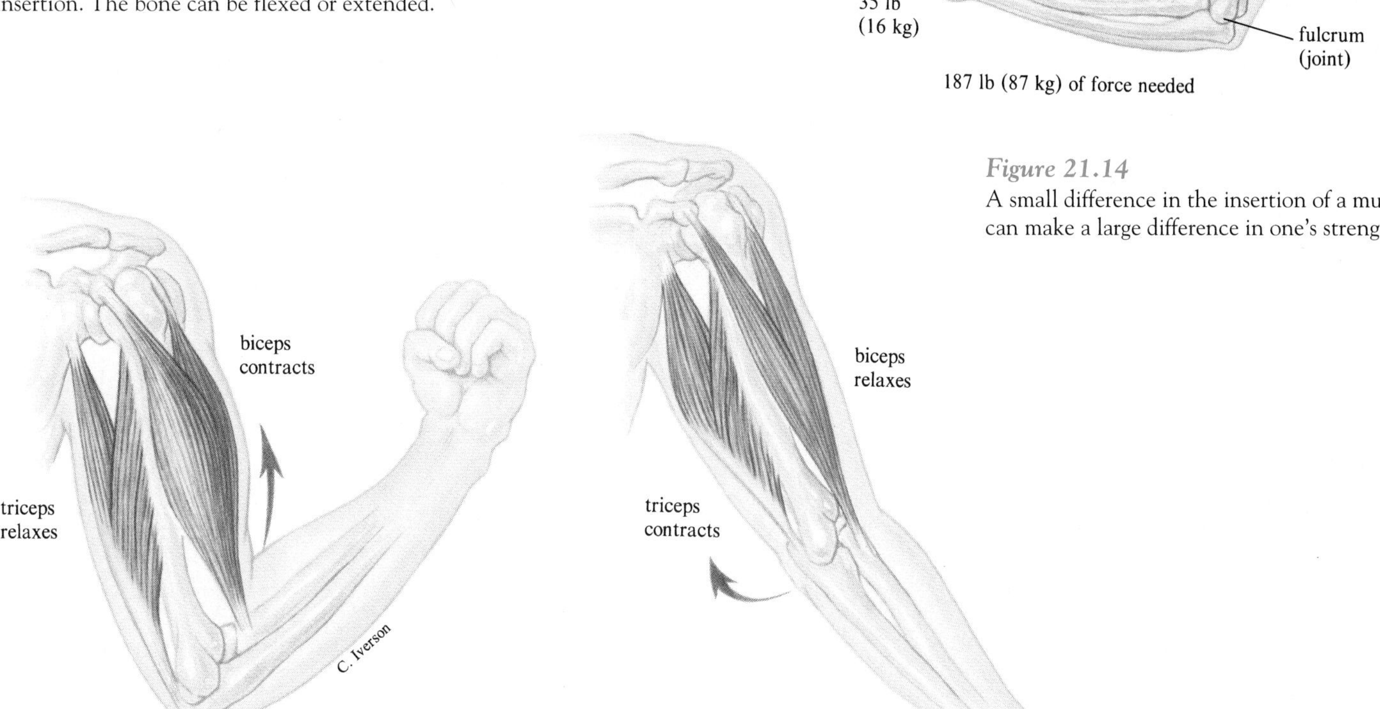

Figure 21.15
Antagonistic pairs of muscles. Contraction of one muscle of an antagonistic pair moves a bone in one direction, while the contraction of the other member of the pair pulls the bone in the opposite direction.

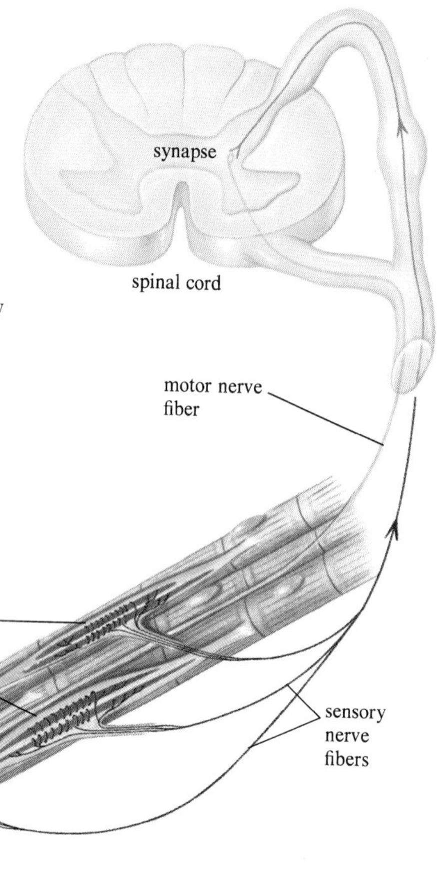

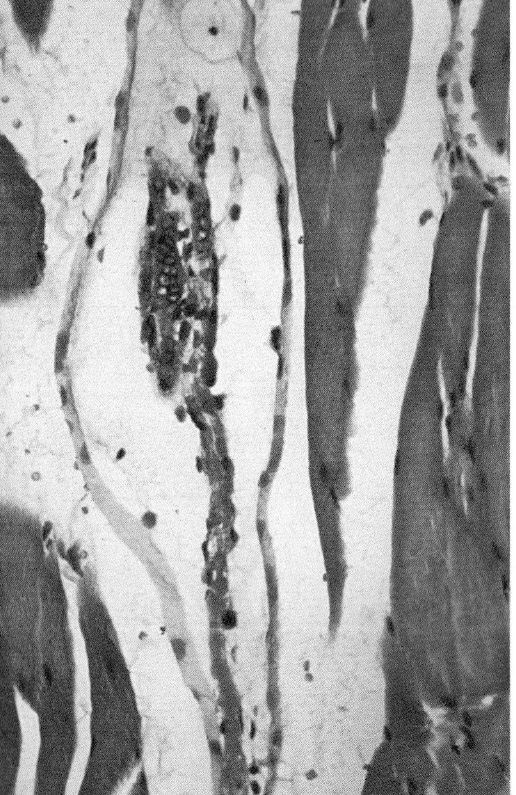

Figure 21.16
Muscle tone results from the contraction of some muscle fibers within a muscle. Sensory receptors called muscle spindles detect the stretch of a muscle and send the information to the central nervous system so that muscle tone can be adjusted. *a.* The reflex pathway responsible for muscle tone. *b.* A micrograph of a muscle spindle. Muscles involved in very fine movements have many spindles.

for chewing are more powerful than those that open the mouth. (If you ever have to wrestle an alligator, remember that it is easier to hold its mouth closed than to keep it open.) In the arm, the biceps, which is used for lifting, is generally stronger than the triceps. Likewise, the calf muscle, which must lift the weight of the body with every step, is more powerful than the muscles near the shin, which lift only the foot.

The pairing of antagonistic limb muscles can set the stage for injury when one member of a pair is exercised more than the other over a long period of time. Many sports-related injuries are the direct result of the extensive use of the calves in the backs of the legs at the expense of the muscles in front. Because the lower legs support the body, the result of overuse of the calf muscles can be Achilles tendinitis, back pain, shin splints (shooting pain in the front of the lower leg), thigh pain, and other muscle pains.

Muscle Tone

Just being able to function in the presence of gravity requires muscle contraction. At any given time some fibers in skeletal muscle are contracted, a property of muscle called **muscle tone.** Receptors called **muscle spindles** monitor the degree of tone in a muscle. Sensory information from the spindles is conducted to the brain so that muscle tone can be adjusted (fig. 21.16).

If we lost this constant interplay between muscle spindles and the nervous system, we would simply collapse. This became obvious to Russian cosmonauts who spent 211 days in space at zero gravity. Despite daily exercise, the absence of gravity meant that their muscle spindles did not function to the degree that they normally do on earth. When the cosmonauts returned to earth, their movements were wobbly due to sluggish muscle spindles, and it took a week for them to be able to walk normally. However, experiments conducted on rats sent repeatedly into space show that muscle degeneration worsens on each flight.

KEY CONCEPTS

Muscles and bones form lever systems, with bones as levers, joints as fulcrums, and force supplied by skeletal muscles. A muscle's origin attaches to the end of the bone that does not move, and its insertion attaches at the end of a movable bone. The placement of the insertion greatly influences strength. Movable bones are flanked by antagonistic muscle pairs, one of which is usually stronger than the other. Muscle spindles are receptors in skeletal muscles that keep some fibers contracted in response to gravity, providing muscle tone.

Effects of Exercise on Muscle

Regular exercise strengthens the muscular system and enables it to use available energy more efficiently. Intense research efforts throughout the world are probing the function of the healthy human body during athletic performance. Athletes are monitored on treadmills and other machines, and their tissues are examined to shed light on the anatomy and physiology of the exercising human body.

Consider what has been learned about someone who runs six times a week and logs 6 miles (9.7 kilometers) each time. When running, muscles consume more than 90% of the total energy generated in the body. During the few months after the runner begins training, gradually increasing mileage to 6 miles, leg muscles increase in size noticeably. This exercise-induced increase in muscle mass is called **hypertrophy,** and it is due to an increase in the size of individual skeletal muscle cells rather than to an increase in their number.

Examination of a sample of muscle tissue from a trained runner shows that the enzymes within the muscle fibers are both more active and more numerous, and that the mitochondria are larger and more abundant than the corresponding structures in a muscle fiber of a sedentary person. When the arms of a runner and a sedentary person are examined in a nuclear magnetic resonance scanner, which detects phosphorus, the runner scores higher—indicating that the muscles have more ATP than those of the nonrunner. As a result, the runner's muscles can withstand far more exertion than can the untrained person's before they respire anaerobically (fig. 6.14). This means that a conditioned athlete is less likely to suffer the pain and fatigue of lactic acid buildup than a sedentary person who suddenly decides to go for a run. The athlete's muscles contain more blood and store more glycogen than do those of an untrained person.

Exercise-induced hypertrophy is even more pronounced in the weight lifter, because the muscles are greatly stressed by the resistance of the weights. The opposite condition of muscle degeneration resulting from the lack of use or immobilization in a cast is called **atrophy.**

The changes in muscles due to regular exercise disappear quickly if activity is discontinued. Just 2 days after regular workouts stop, mitochondrial enzyme activity drops in skeletal muscle cells. After a week without the accustomed exercise, the efficiency of aerobic respiration falls by 50%. The number of small blood vessels surrounding muscle fibers declines, lowering the body's ability to deliver oxygen to the muscle. Metabolism of lactic acid becomes less efficient, and glycogen reserves fall. After 2 or 3 months of inactivity, the benefits that were provided by regular exercise have all but disappeared.

Injuries to the Muscular System

Injuries to tendons are fairly common. A tendon transmits the force generated within a muscle to bone, but because the tendon is smaller in diameter than the muscle, and unlike the muscle it cannot contract, it is under more strain. One very common sports injury is Achilles tendinitis, an inflammation of the Achilles tendon, which attaches the calf muscle to the heel bone. Although the Achilles tendon is one of the strongest tendons in the body, it is subject to extensive stress in many sports because of its location.

When a muscle is subjected to more stress than it can bear, muscle fibers actually tear. The site of a muscle tear depends upon the activity that stressed the muscle. Sprinters tend to tear the hamstring muscles in the back of the thigh, swimmers the deltoid muscles in the shoulder, and jockeys the inner thigh muscles.

For first aid for suspected injuries of muscles, ligaments, joints, and bones, the RICE plan, suggested by the American College of Sports Medicine, can be followed until a medical professional can be consulted:

Rest to prevent further damage.
Ice to contract blood vessels in the damaged area to stop bleeding and promote healing.
Compression to decrease swelling and promote healing.
Elevation of the legs above the level of the heart so gravity drains excess fluid.

The next time you reach back to catch a ball, bend to lift a pile of books, or simply take a breath of fresh air, you can thank the sliding protein filaments that have provided locomotion in living systems probably since ancient bacteria used them to move. The microscopic protein filaments within muscle cells work together to build a system that powers the stride of an Olympic athlete, the fluid motions of the dancer, and the rhythmic contractions of respiration, digestion, and circulation that keep us alive.

KEY CONCEPTS

Regular exercise leads to muscle hypertrophy, in which muscle cells increase in size, with enzymes and mitochondria more active and more numerous. Likewise, muscle degenerates when exercise ceases or muscles are not used at all. Tendons are frequent sites of injury, and muscles can tear.

SUMMARY

Smooth muscle cells are spindle-shaped, and their contraction is involuntary. Smooth muscle lines many organs. It pushes food through the digestive system and regulates blood flow and pressure. *Cardiac muscle cells*, found only in the heart, are striated and not under conscious control. The cells are separated by *intercalated disks* and form a branching pattern. *Skeletal muscle cells* are voluntary and move bones.

An intact whole skeletal muscle is attached to bones by *tendons*. Within the muscle are bundles of muscle fibers. Each muscle fiber is a long cylindrical cell composed primarily of *myofilaments*. The thick myofilaments consist of bundles of *myosin* molecules. Each myosin molecule has a head attached to a shaft. The heads project outward from each end of a myosin bundle. The thin myofilaments are composed of a double strand of spherical *actin* molecules and the two regulatory proteins *troponin* and *tropomyosin*.

A muscle fiber, or cell, is a chain of many contractile units called sarcomeres. Myofilaments are arranged within a sarcomere in an orderly manner, giving the tissue its striated appearance. According to the *sliding filament model*, muscle contraction or shortening occurs when the thick and thin myofilaments slide past one another so that the region of overlap increases.

Muscle contraction is regulated by the availability of calcium ions. When a motor nerve cell stimulates a muscle fiber, the chemical acetylcholine is released at a neuromuscular junction. The electrical changes on the muscle cell membrane cause calcium ions to be released from the *sarcoplasmic reticulum*, a specialized endoplasmic reticulum. The calcium ions bind to the troponin molecules perched on the actin myofilaments. As a result, troponin is moved from its position blocking the binding of actin and myosin. This allows cross-bridges to form. After the nerve impulse, calcium ions are actively pumped back into the sarcoplasmic reticulum and tropomyosin prevents actin and myosin interactions. When ATP attached to the myosin head splits, the head moves, causing the actin myofilament to slide past the myosin myofilament. ADP and inorganic phosphate are released. A new ATP is bound to the myosin head and the cross-bridge to actin is broken. The myosin head returns to its original position and a new cross-bridge is formed further along the actin myofilament.

The energy for muscle contraction comes first from stored ATP. This energy supply is quickly exhausted. For a short time, new ATP can be formed from the high-energy phosphate bond donated by *creatine phosphate* stored in the muscle cells. When creatine phosphate is used up, new ATP is generated by aerobic respiration. If the use of ATP exceeds the ability of muscle cells to produce ATP aerobically, the cells switch to anaerobic respiration and generate lactic acid as a by-product. Later, while the muscles rest, enough oxygen is supplied to convert the lactic acid to pyruvic acid. The need for oxygen to convert lactic acid to pyruvic acid is called *oxygen debt*. An important source of energy for muscle contraction is glycogen, a polymer of glucose.

When stimulated, a muscle cell responds in an all-or-none fashion. However, not all of the many muscle cells in a whole muscle contract. When a muscle is stimulated once, it contracts and relaxes, a response called a *twitch*. If the response rate increases, the muscle cell does not completely relax between pulses, and the strength of response increases. This is called a *summation*. At a high rate of stimulation, the muscle cells reach a sustained state of maximal contraction, called *tetanus*.

Most voluntary muscles are attached to bones, forming biological lever systems. When a muscle contracts, a bone is moved. The muscle attachment to the stationary bone is called its *origin*. The end attached to the bone moved by the contraction of that muscle is called the *insertion*.

Muscles are arranged in *antagonistic* pairs. The contraction of one member of the pair causes a bone to move in one direction. The contraction of the other muscle causes the opposite movement of the bone.

All muscles are always partially contracted, a condition known as *muscle tone*. Muscle tone is controlled by receptors in muscles *(muscle spindles)* that send impulses to the brain or spinal cord when a muscle is stretched.

When a muscle is exercised regularly, it increases in size *(hypertrophy)* because each muscle cell gets thicker. No new muscle cells develop. When the muscle is not used, it gets smaller *(atrophy)*. Regular exercise causes changes in muscle cells that enable them to use available energy more efficiently.

QUESTIONS

1. How do the three types of muscle differ, and how are they the same?

2. When skeletal muscle contracts, what happens to the following:
- **a.** Sarcomeres
- **b.** Muscle fibers
- **c.** Muscle fiber bundles
- **d.** Whole skeletal muscle
- **e.** Tendon
- **f.** Attached bone

3. How does the muscular system interact with the following:
- **a.** The nervous system
- **b.** The skeletal system
- **c.** The circulatory system (blood)
- **d.** Connective tissue

4. What are five proteins involved in muscle contraction, and what does each one do?

5. What is the source of the stripes seen in skeletal muscle cells under the microscope?

6. Why are tendons particularly prone to injury?

7. How is ATP supplied to a skeletal muscle cell?

8. How do antagonistic muscle pairs move bones? Give an example of such a pair.

9. A man begins a regular running program. He is surprised that his legs actually seem thicker, even after months of running, because he thought the exercise would make him lose weight. What has happened?

TO THINK ABOUT

1. The aerobics instructor chants, "Just concentrate and feel your muscles expand and contract." Is her statement an accurate description of muscle motion? Why or why not?

2. Researchers at the University of Texas Southwestern Medical School are studying adult muscle growth in cats as a model of the phenomenon in humans. Cats are taught to exercise only one forelimb by lifting weights to obtain a food reward. After a year or more of this training, the exercised forelimb shows signs of hypertrophy. On microscopic examination, the hypertrophied forelimb has more muscle fibers than the unexercised forelimb. Is this observation consistent with what has been found in the skeletal muscles of human athletes? Why or why not?

3. A novice cross-country skier takes a bad plunge and feels a sharp pain in his calf. What should he do until he receives medical care?

4. A father takes his son to a sports medicine specialist and requests that the boy's leg muscles be tested to determine the percentage of slow twitch and fast twitch muscle fibers. He wants to know if the healthy boy should try out for soccer or cross-country (long-distance) running. Do you think testing muscle tissue should be used for this purpose? Why or why not?

5. Some runners, cross-country skiers, and swimmers attempt to boost athletic performance by a technique called blood doping. About 6 weeks before an athletic event, a pint of blood is removed from the athlete and frozen. A few days before an athletic event, when the athlete's blood level has returned to normal, the stored blood is thawed and reinfused. How might extra blood increase athletic performance?

6. A nuclear magnetic resonance scan shows that a runner's wrist muscles have a large supply of ATP, even though she does not exercise her wrists. What does this result suggest about the inborn nature of athletic stamina?

SUGGESTED READINGS

Alexander, R. M. July-August 1984. Walking and running. *American Scientist*. Getting around requires much coordination between muscles and bones.

Cohn, Jeffrey P. March 1988. Making a stand against leg cramps. *FDA Consumer*. What causes painful leg cramps?

Karkowsky, Nancy. May 1989. Exercise with care—Fitness is not risk-free. *FDA Consumer*. The RICE plan can help treat an injury.

Lewis, Ricki. October 1989. Mitochondria: eclectic organelles. *Biology Digest*. Abnormal mitochondria can cause diseases of the muscles.

Malacinski, George M., et al. May 1989. Developmental biology in outer space. *BioScience*. Low gravity has profound effects on organisms—particularly the muscular and skeletal systems.

Weisburd, Stefi. March 26, 1988. The muscular machinery of tentacles, trunks and tongues. *Science News*. Not all muscles generate movement in the same way.

Weiss, R. January 2, 1988. Muscular dystrophy protein identified. *Science News*. The mystery behind a major muscular disease is solved.

Wickelgren, I. December 17, 1988. Endurance superstars may be "born to run." *Science News*. All skeletal muscles in elite athletes conserve ATP.

CHAPTER

22

The Circulatory System

Chapter Outline

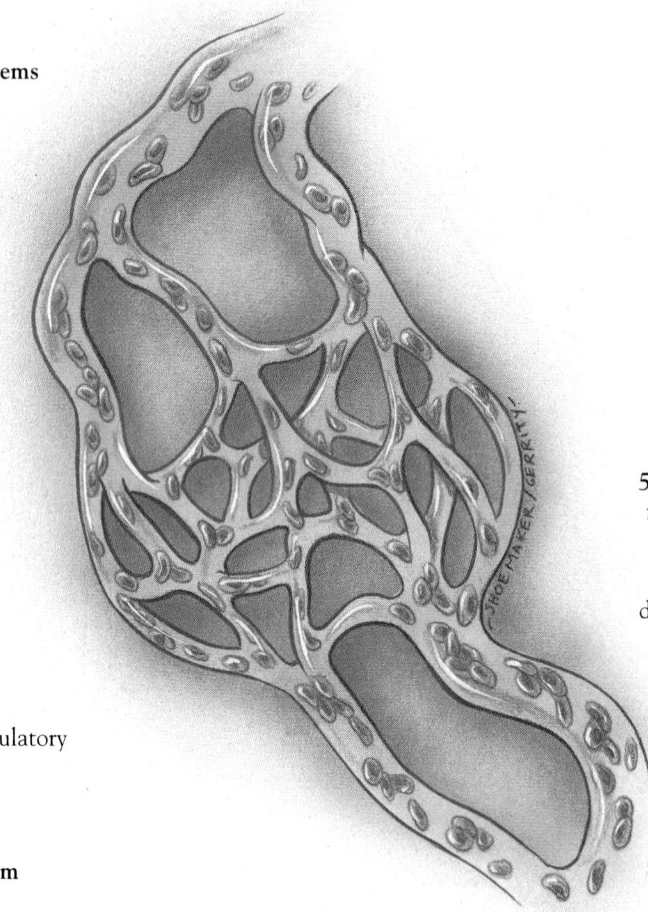

Learning Objectives

By the chapter's end, you should be able to answer these questions:

1. How do circulatory systems in diverse organisms work?
2. How do the structures of the human circulatory system provide oxygen and nutrients to the body's cells and remove wastes from them?
3. What are the functions of each of the blood's components?
4. What is the structure of the blood vessels and heart?
5. What is the pathway of the blood to and from the lungs and to and from other body tissues?
6. At what level of the circulatory system does exchange of oxygen and nutrients and wastes occur?
7. What factors determine and control heartbeat?
8. What effects does exercise have on the circulatory system?
9. What are the structures and functions of the lymphatic system?

Oxford is a small city nestled in the quiet farmland of western Ohio, about an hour's drive from the much larger cities of Cincinnati and Dayton. Yet, despite Oxford's relative isolation, the thousands of students who attend its Miami University can still order seafood at a restaurant, even though the closest ocean is nearly 800 miles away. Students can come and go by car, bus, train, or even small private plane. Transport systems allow this small town deep in the interior of a huge country to survive, while retaining its own charm.

So it is in a multicellular organism. Cells and tissues buried deep within the body must be in contact with the outside to survive. Nutrients must be brought in and waste products removed, while at the same time the internal environment must remain nearly constant in order to sustain life. These two related functions—communication with the outside and maintenance of the inside—are accomplished in multicellular organisms by several organ systems that deliver nutrients and vital gases and remove wastes. The circulatory system—the heart and a vast network of vessels—is one such "transport" system. Within it, blood circulates, sustaining life by nourishing and cleansing each cell.

Diversity Among Circulatory Systems

Nutrient delivery and waste removal are essential for life, and different organisms meet these needs in different ways. Diffusion, the movement of materials from an area of high concentration to an area of lower concentration, is a reliable means of distributing materials, but it is exceedingly slow. Nonetheless, for single-celled organisms and multicellular animals whose cells are close to the environment, diffusion across the body surface is speedy enough to deliver raw materials and remove wastes (fig. 22.1). In a protozoan, molecules enter and leave the cell simply by diffusing across the cell's membrane, and they move about inside the cell by cytoplasmic streaming. Not all the cells of a sponge (phylum Porifera) have equal exposure to the environment. However, canals lead from the surface to a sponge's inner body mass, carrying nutrient-laden water to the organism's inner cells. In flatworms (phylum Platyhelminthes), a central cavity branches so that all of the organism's cells lie close enough to a branch or to the body surface for diffusion to meet the cell's metabolic demands.

In animals more active than a flatworm, the cells require greater supplies of raw materials and oxygen and generate larger amounts of waste. An increase in an animal's size decreases its surface-to-volume ratio, making diffusion an inefficient means of distributing materials within the organism. Therefore, large active animals require a circulatory system to deliver vital materials to the cells and to cleanse the cells of the wastes they continuously produce. A circulatory system is composed of a fluid **(blood)**, a pump, or **heart,** to force the blood around the body, and **blood vessels** to serve as a network of conduits in which the blood is delivered to and removed from tissues.

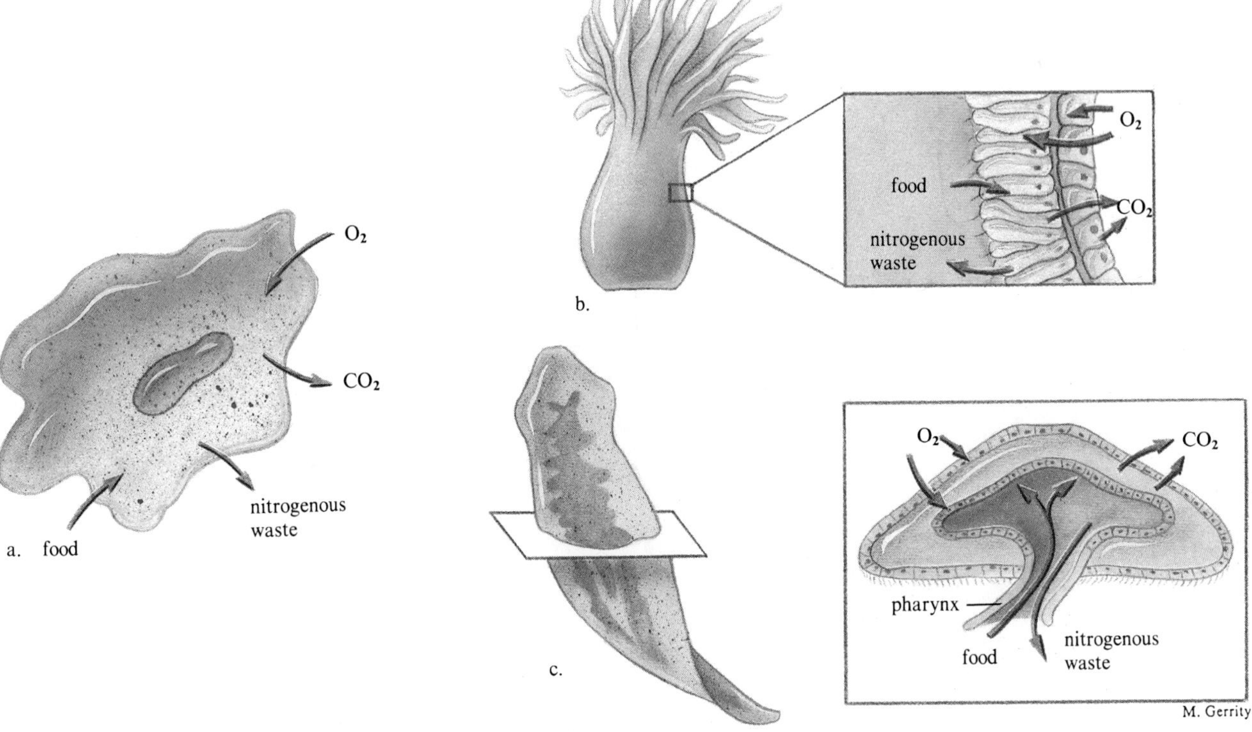

Figure 22.1
Exchange of material by diffusion across the body surface. The cells of small organisms such as protozoans (*a*), cnidarians (*b*), and flatworms (*c*) maintain close enough contact with the environment that they have no need for a circulatory system.

In an **open circulatory system** the blood is not always contained in blood vessels. For example, the open circulatory systems of most mollusks (e.g., clams and snails) and arthropods (e.g., insects, spiders, and crayfish) consist of a heart and blood vessels that lead to spaces, called sinuses, in which blood directly bathes cells before returning to the heart (fig. 22.2).

In a **closed circulatory system,** the blood remains within blood vessels. Large vessels conduct blood away from the heart and form smaller vessels and finally divide into a network of very tiny thin vessels. Materials move between the cells and the blood across the walls of these smallest vessels. Blood is then collected into slightly large vessels, which unite to form larger vessels, which carry blood back to the heart.

At the level of the vertebrates, the circulatory system has a highly specialized heart, blood with red and white cells in a "closed" system, and hemoglobin that is carried in cells.

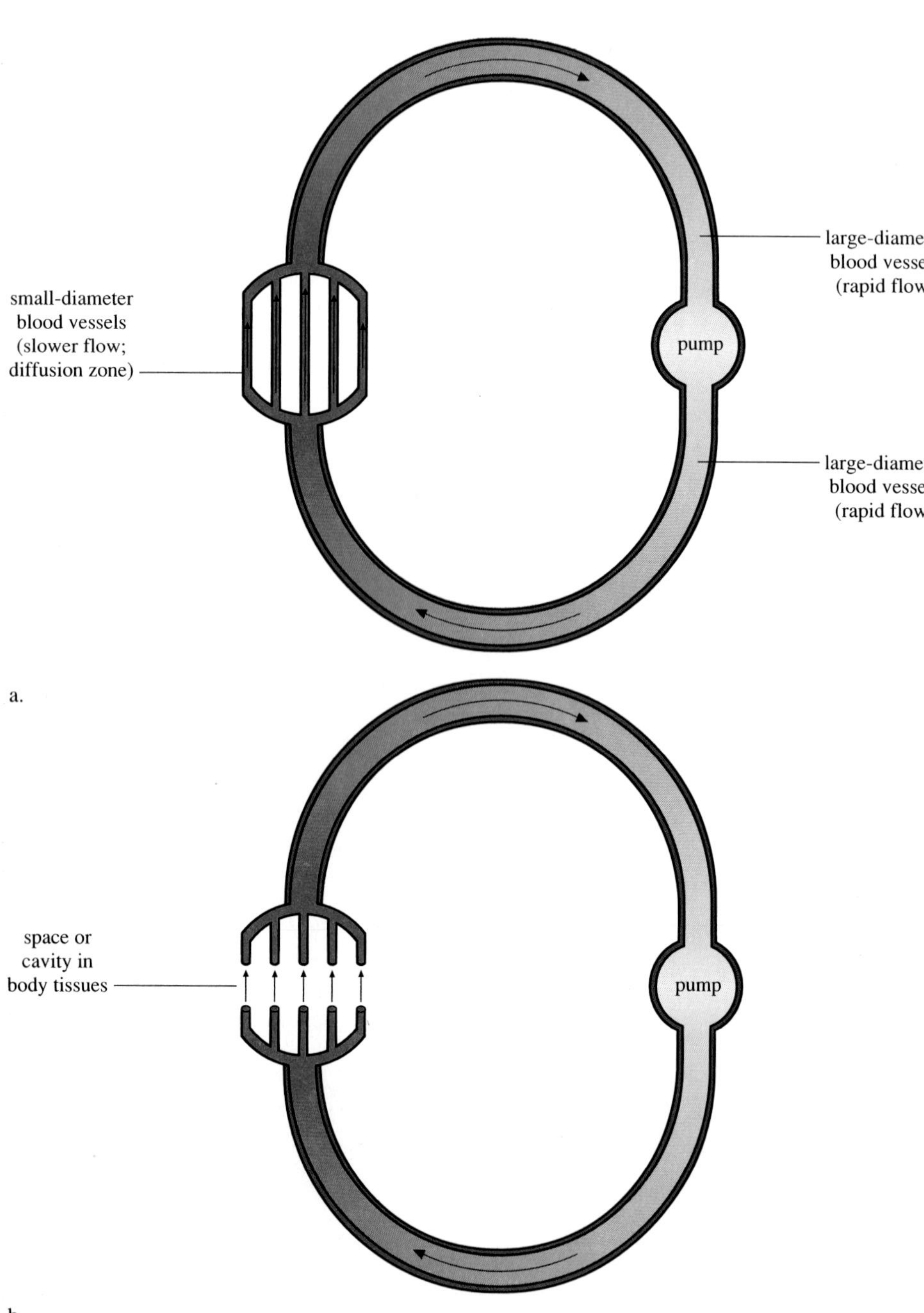

Figure 22.2
General scheme of closed and open circulatory systems. *a.* In a closed circulatory system, blood is kept within vessels. *b.* In an open circulatory system, blood bathes the cells directly.

KEY CONCEPTS

Circulatory systems transport nutrients to and remove wastes from cells. Diffusion effectively, if slowly, does this in single-celled animals and in simple animals whose cells are in close contact with the environment. In more complex animals, circulation is carried out by a system of a pump, blood, and vessels. In an open circulatory system, blood in spaces bathes cells. Closed circulatory systems contain blood in networks of vessels.

The Human Circulatory System

The human circulatory system consists of a central pump, the heart, and an attached continuous network of tubes, the blood vessels. The system is also called the cardiovascular system (cardio refers to the heart, vascular to the vessels). The circulatory system transports blood, a complex fluid that nourishes cells, removes their waste products, and helps to protect against infection. So extensive is the circulatory system that no cell is more than a few cell layers away from one of the system's smallest branches. So vital is the system that medical researchers have built working replicas of its major parts (Reading 22.1).

Functions of Blood

An obvious function of the blood is to keep cells and the abundant fluid around them "fresh." To do this, blood brings oxygen and nutrient molecules to cells and removes their wastes. Carbon dioxide (CO_2), a product of cellular respiration, is taken by the blood primarily in the form of bicarbonate ion (HCO_3^-) to the lungs, from which it is exhaled as CO_2. Urea, a product of

Reading 22.1 *Cardiovascular Spare Parts*

ON THE SURFACE, THE HUMAN CIRCULATORY SYSTEM APPEARS TO BE RATHER SIMPLE—OXYGEN-CARRYING MOLECULES TRAVELING THROUGH NUMEROUS CONDUITS, PROPELLED BY THE CONTRACTION OF A POWERFUL PUMP. But when we try to duplicate any of these components—blood, blood vessels, and the heart—it becomes apparent just how complex these structures are.

Artificial Blood

Blood has several functions, because it is a mixture of various types of formed elements and molecules. Synthetic bloods—all still experimental—try to duplicate blood's oxygen-carrying ability. They are variations on chemicals that bind oxygen, mimicking the hemoglobin carried in red blood cells.

The first such synthetic blood was silicone oil. When in 1965 Leland Clark at the University of Cincinnati submersed a mouse into a beaker of silicone oil, its lungs quickly filled with the liquid, but the animal was able to "breathe" the oxygen in the oil (fig. 1). Next researchers tried related chemicals containing fluorine and carbon, which bind oxygen similarly to silicone oil. It was not until 1990 that the first red blood cell substitute was approved for use to maintain localized blood flow during a procedure called balloon angioplasty, in which a balloon is inflated against deposits blocking an artery's lumen. The product, Fluosol, consists of two flourine compounds, a mild detergent, and lipid from egg yolk.

In another approach, hemoglobin molecules are linked together and administered alone, rather than in cells as occurs in the circulatory system. Yet another type of artificial blood consists of hemoglobin packaged into synthetic microcapsules, which are tiny fat bubbles currently used in carbonless carbon paper and paper that releases an odor when scratched. However, these "neohemocytes" so far provoke an immune attack from the animals in which they have been tested. Artificial bloods based on hemoglobin do not offer the freedom from contamination possible with a completely synthetic product.

Blood Vessel Equivalent

A purely synthetic blood vessel attracts bacteria, and transplanted blood vessels prompt the immune system to attack. But a company called Organogenesis in Cambridge, Massachusetts, has combined the best of each of these approaches into a "living blood vessel equivalent." These flexible yet strong tubes are indeed quite like the real thing, built of an inner layer of endothelium, a middle layer of smooth muscle cells in collagen, and an outer layer of connective tissue. However, cell surface molecules likely to cause an immune reaction are removed, and an internal, synthetic mesh is woven into the structure to lend strength (fig. 2).

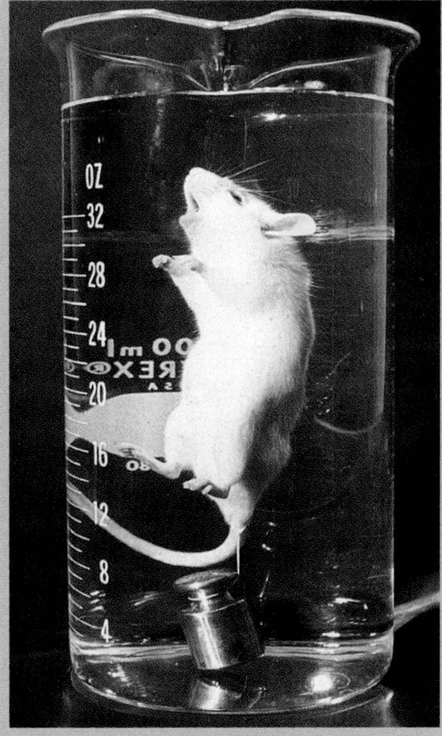

Figure 1
This mouse isn't drowning—it is "breathing" a fluorocarbon liquid that is saturated with oxygen.

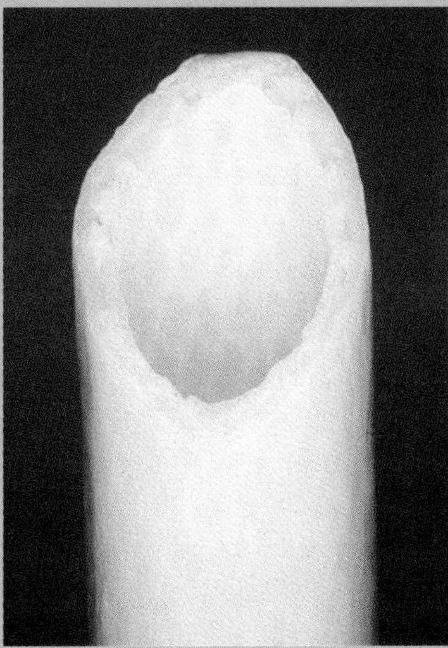

Figure 2
The living blood vessel made on a cylindrical mold.

The blood vessel equivalent is remarkably versatile. It can be fashioned to any length or width, and it withstands pressures comparable to that of the blood racing through the circulatory system. The structure can be stitched next to natural blood vessels so smoothly that blood clots are not encouraged to form as they are with synthetic grafts. The living blood vessel equivalent, is being developed for use in cardiac bypass surgery, to replace damaged arteries in the legs, and to replace brain arteries damaged by a stroke.

Artificial Heart

In late 1982, dentist Barney Clark was given a short new lease on life—112 days—by the first artificial heart to be implanted in a human. Previously, an identical device had sustained a calf named Tennyson (fig. 1.5*b*) for 268 days. The first artificial heart was developed at the University of Utah Division of Artificial Organs by Dr. Robert Jarvik. The Jarvik heart is connected to the pulmonary artery, the aorta, the atria, and an external power supply. An artificial heart is really just two artificial ventricles and their associated valves and is built of metal valves, a plastic body, and velcro attachments. Since Barney Clark's experience, a few other people have joined the ranks of the plastic-hearted but unfortunately, none survived very long. The Jarvik heart apparently causes deadly blood clots to form, and the valves are breeding grounds for antibiotic-resistant bacteria. Another drawback is that the wearer needs 375 pounds of equipment to keep the spare part running. Because of these problems, the Jarvik heart was retired in 1990, but a new generation of replacement hearts is in development. These new ventricular assist devices are the size of a softball and replace just the left ventricle.

protein metabolism, travels in the circulatory system from the liver, where it is produced, to the kidneys, where it is sent to the bladder and excreted from the body. In addition, blood carries hormones from the glands that produce them to the cells that use them.

The composition of blood strongly influences immediate cellular environments, because the fluid that bathes cells is derived from blood. Blood helps to maintain a fairly stable surrounding for the cells in several ways. One way is by regulating its own pH. Blood's pH of 7.4 is maintained by protein molecules in the blood, each of which can release a hydrogen ion (H^+) from one end of its amino acid chain and can absorb a hydrogen ion at the other end.

Blood also maintains biological constancy, or homeostasis, by regulating the water content of the cells. Dissolved sodium ions and certain blood proteins, principally albumin, maintain the osmotic pressure of the blood. The osmotic pressure results from the movement of water from a region of high concentration to one of low concentration, and it is this pressure that keeps the liquid part of the blood from seeping out of the circulatory system's vessels. Albumin controls the movement of water between the blood's liquid part and the tissue fluid surrounding the blood vessels. Finally, the blood helps maintain a stable body temperature in birds and mammals because it absorbs heat and carries it to the body surface, where it can be dissipated.

The blood provides protection from injury in two ways—clotting, which involves cell fragments called platelets, and inflammation, a swelling induced by white blood cells. Furthermore, the white blood cells defend the body against foreign invaders, which are usually disease-causing microbes or the toxins they produce.

KEY CONCEPTS

In the human circulatory system, blood brings oxygen and nutrients to cells and removes CO_2 and urea. Blood helps to maintain homeostasis by regulating its pH and the water content of cells. Blood protects by clotting and by fighting infection.

Blood Composition

Blood follows the general connective tissue scheme of cells within a matrix (see chapter 4). Unlike the matrix of any of the other connective tissues, however, the matrix of blood is a liquid, called **plasma,** in which many molecules are either suspended or dissolved. Blood cells or cell fragments are called **formed elements.** The three varieties of formed elements are the **red blood cells** (erythrocytes), the **white blood cells** (leukocytes), and the **platelets** (fig. 22.3). A cubic millimeter of blood normally contains about 5 million red blood cells, 7,000 white blood cells, and 250,000 platelets.

Plasma

Blood plasma, which represents slightly more than half of the blood's volume, is 90% to 92% water. It contains various dissolved molecules, such as salts, nutrients, hormones, metabolic wastes, and gases. Although the concentration of some of these substances is low, their presence remains critical (fig. 22.4). For example, the blood is usually about 0.1% glucose, but if glucose levels fall 0.04%, convulsions begin. In addition, the plasma is about 7% to 8% dissolved proteins. There are over 70 different plasma proteins.

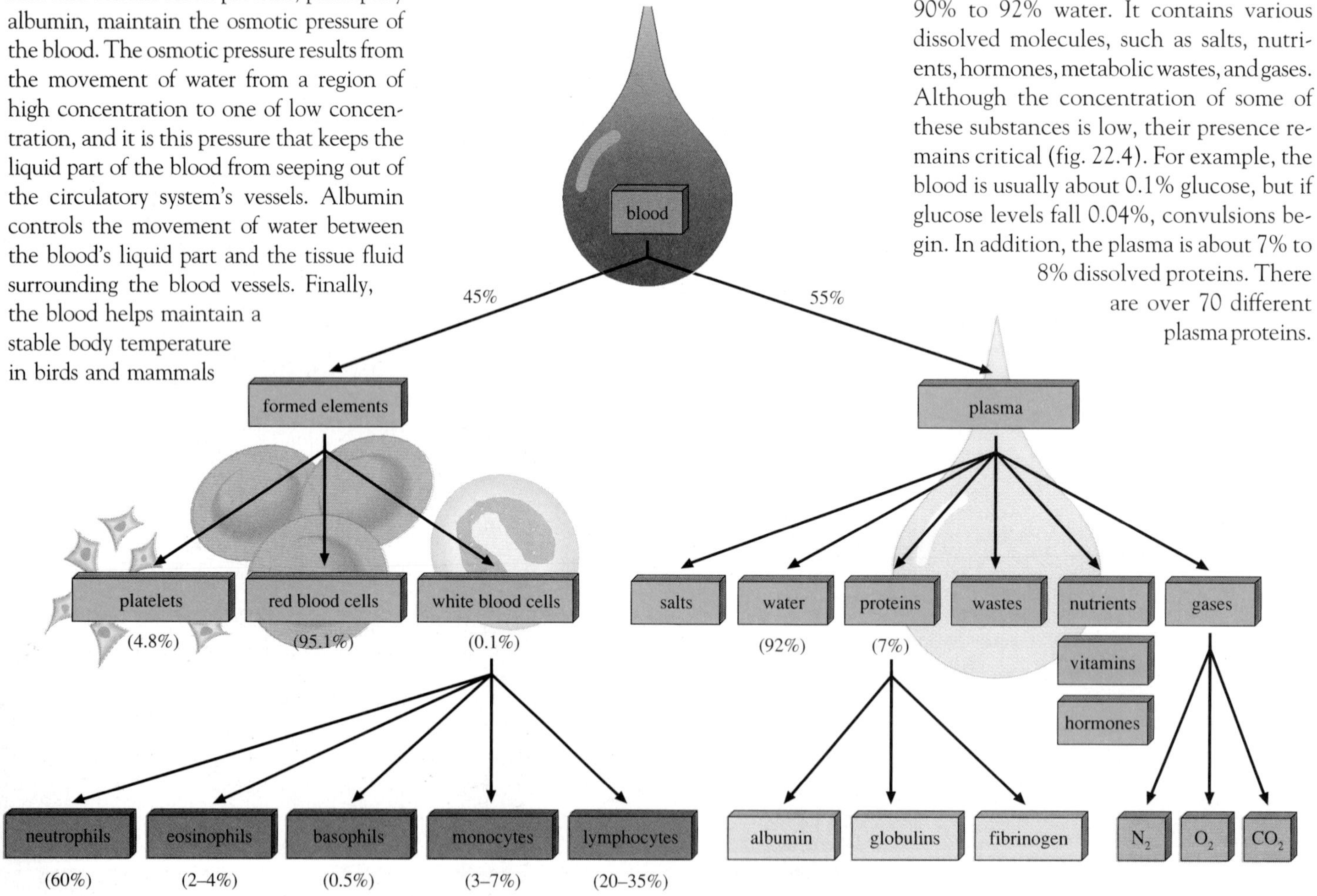

Figure 22.3
Blood composition. Human blood is a complex mixture of formed elements (platelets, red blood cells, and white blood cells) suspended in a liquid plasma that is 90% to 92% water, 7% proteins, and 1% salts, wastes, nutrients, hormones, and dissolved gases. Alterations in blood composition may be a sign of illness.

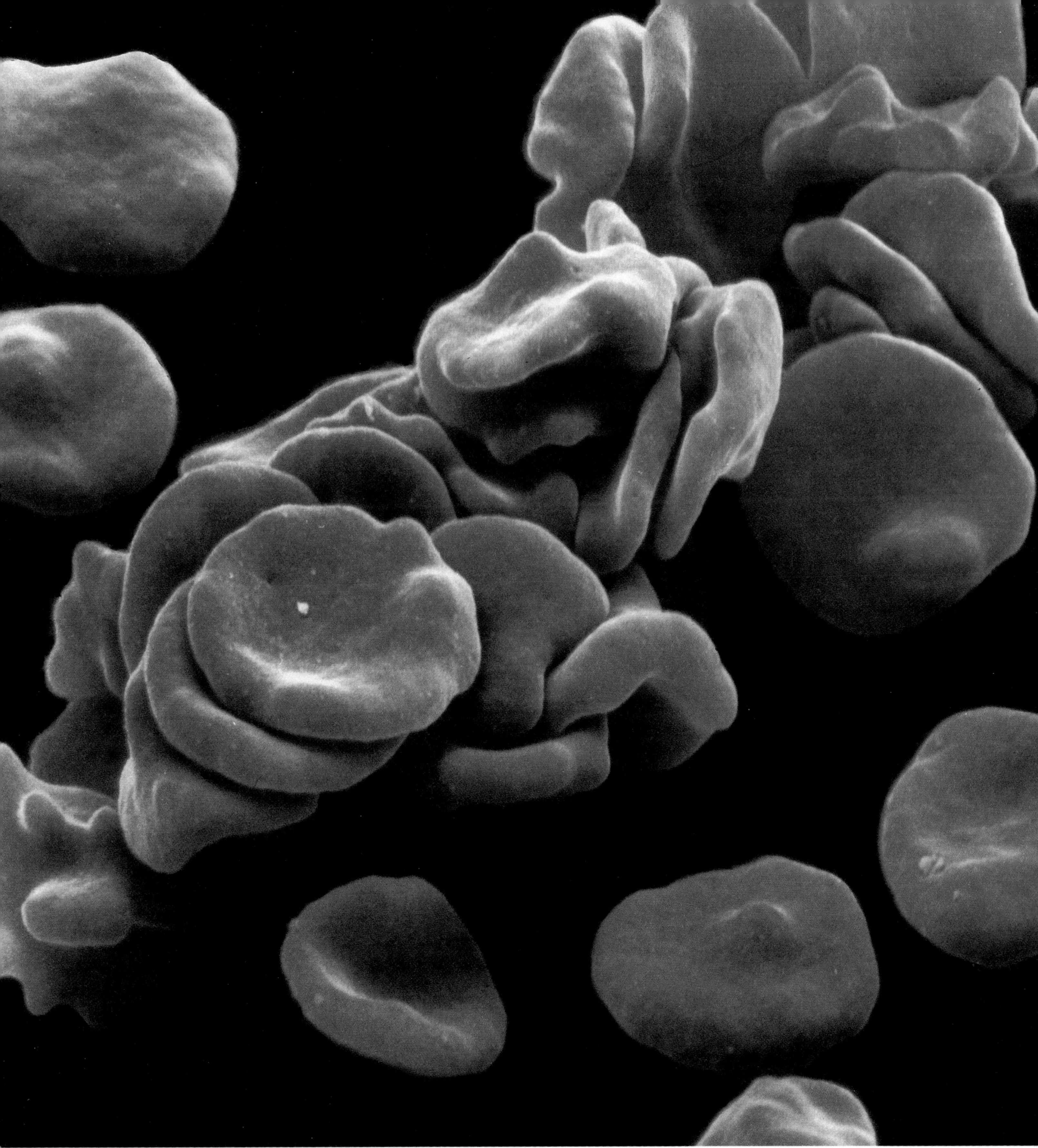

Figure 22.4
A scanning electron micrograph of red blood cells in a less than normal salt solution. The cells are swollen with absorbed water. In the body, the salt concentration of the plasma is kept constant, so these changes do not occur.

Some, such as the albumins, are important in maintaining osmotic pressure. Globulins take part in immune responses, and some transport lipids or metals, including zinc, iron, and copper. The third major plasma protein, fibrinogen, is needed for blood clotting.

Red Blood Cells

Red blood cells are by far the most numerous of the formed elements, which is why blood is red. Formed within the red bone marrow at the rate of 2 to 3 million every second, mature red blood cells are biconcave discs virtually packed with the oxygen-carrying molecule hemoglobin. During development they lose their nuclei, ribosomes, and mitochondria. Therefore, they are unable to reproduce or to carry out cellular metabolism. During a red blood cell's 4-month life span, it journeys through the body some 75,000 times, being pounded against artery walls and squeezed through tiny capillaries. Finally, the red blood cell is destroyed in the liver or spleen. Most of its components are recycled for future use.

The red blood cell's shape and content ideally suit it to transport oxygen. The thin, easily bent cell can squeeze through narrow passageways with ease, and its biconcave shape increases its surface area for gas exchange (fig. 22.5). The total surface area of all the red blood cells in the human body is roughly 2,000 times as great as the body's exterior surface, providing ample space for gas exchange to occur. About 280 million hemoglobin molecules pack into a red blood cell.

The hemoglobin molecule is built of four polypeptide chains known as **globins,** each of which is bound to an iron-containing complex called a **heme** group—hence the name hemoglobin. Hemoglobin exists in two functional forms. **Oxyhemoglobin** is bright red due to the oxygen it has picked up in the lungs. This oxygen is bound to the iron in the molecule's heme groups. When the oxyhemoglobin gives up its oxygen to the cells, it is transformed into the deep red **deoxyhemoglobin,** the reduced form of the molecule. Deoxyhemoglobin can also pick up carbon dioxide and transport it to the lungs, where it is exhaled. If all of the oxygen-binding sites of the body's hemoglobin molecules are blocked, death follows swiftly. Carbon monoxide (CO) poisoning occurs because this molecule binds more strongly to heme groups than oxygen does. Fluorine-based organic molecules that bind oxygen are being looked at as red blood cell substitutes (Reading 22.1).

Adequate oxygenation of the body's tissues depends on a sufficient number of red blood cells carrying enough hemoglobin. Fortunately, the proportion of red cells in the blood is attuned to the environment.

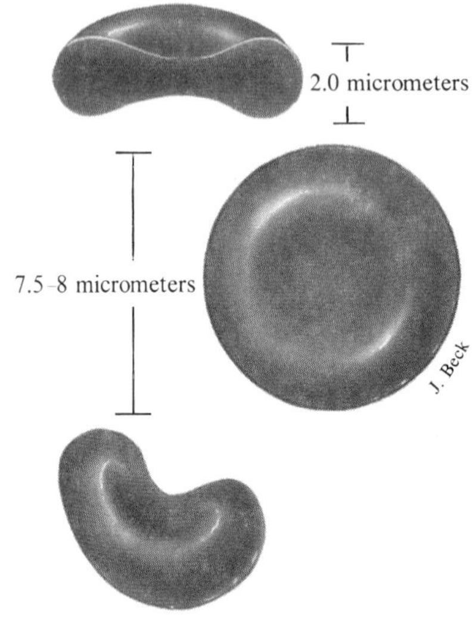

Figure 22.5
The shape of the red blood cell maximizes its surface area for gas exchange and allows it to be easily bent as it squeezes through capillaries.

Both the maturation of red blood cells and the synthesis of hemoglobin respond to the availability of oxygen. When certain cells in the kidney are not receiving enough oxygen, they produce a substance that combines with a plasma protein to form the hormone **erythropoietin.** This hormone stimulates red blood cell production in the red bone marrow. The gene coding for human erythropoietin has been engineered into bacterial cells (see chapter 13), and the hormone can be used to boost red blood supplies in people who have lost blood during surgery.

At high altitudes, where the oxygen content of air is low, the production of red blood cells increases to maintain an adequate supply of oxygen to the tissues. A similar adaptation occurs as a result of regular aerobic exercise, when utilization of oxygen by muscles increases. Conversely, a sedentary individual may have fewer than the average number of red blood cells.

Anemia is a reduction in red blood cells or in hemoglobin, which limits the body's ability to deliver oxygen to the cells. Because oxygen is necessary to maximize the energy extracted from food molecules and stored in the energy-laden ATP molecule, it is not surprising that the symptoms of anemia are fatigue and lack of tolerance to cold. Anemia can be due to red blood cells that are too small or too few, that contain too little hemoglobin, or that are manufactured too slowly or die too quickly. Several types of anemia are recognized, and these are described in table 22.1. The most

Table 22.1
Types of Anemia

Type	Cause	Defect
Aplastic anemia	Toxic chemicals, radiation	Damaged bone marrow
Hemolytic anemia	Toxic chemicals	Red blood cells destroyed
Iron deficiency anemia	Dietary lack of iron	Hemoglobin deficiency
Pernicious anemia	Inability to absorb vitamin B_{12}	Excess of immature cells
Sickle-cell disease	Defective gene	Red blood cells abnormally shaped
Thalassemia	Defective gene	Hemoglobin deficiency; red blood cells short-lived

common cause of anemia is iron deficiency, which can be corrected by eating more iron-rich foods or taking iron supplements.

Molecules called antigens lie on the surfaces of red blood cells. Antigens are the basis of blood types, such as the ABO blood types and Rh blood types. When blood from incompatible types is mixed, a clump forms, disrupting circulation (fig. 5.3). Furthermore, the red blood cells in this clump are likely to break open and release their hemoglobin. The free hemoglobin clogs the microscopic tubules in the kidney, damaging this vital organ.

White Blood Cells

The five varieties of white blood cells—neutrophils, eosinophils, basophils, lymphocytes, and monocytes—protect against toxins, invading organisms, and to some extent, even cancer (fig. 5.2). White blood cells are larger than red blood cells and retain their nuclei. They originate in the bone marrow and typically live about 1 year, of which only 3 or 4 days is spent in the blood. The white blood cell is a wanderer and can often be found among a variety of tissue types. The types of white cells are distinguished by size, life span, location in the body, shape of the nucleus, number of granules in the cytoplasm, and staining properties (fig. 22.6).

White blood cells defend the body against any unrecognized invader. The different types of leukocytes cause the warmth and swelling of inflammation, which occurs at the site of an injury or infection. Some leukocytes surround microbes and destroy them, and others produce proteins called antibodies that escort invaders to white blood cells, which destroy them. The specific functions and interactions of white blood cells and their products are discussed in chapter 27.

White blood cells are a window on health, because elevated or diminished numbers of specific types of white blood cells may be a clue to the type of infection or disease present. Detection of altered blood composition is used to confirm a diagnosis based on more obvious symptoms. For example, in tuberculosis and whooping cough, the proportion of lymphocytes in the blood is elevated. Unusually high levels of eosinophils are seen in hookworm and tapeworm infections, in a newly-described condition called eosinophilia-myalgia syndrome caused by a contaminant in an amino acid supplement manufactured in Japan, and excess monocytes are found in the blood of typhoid and malaria sufferers. In mononucleosis, monocytes are abnormally prevalent and larger than usual.

The leukemias are cancers of the white blood cells, in which one type of cell greatly predominates because it divides more often and continuously compared to other blood cells. The flow of these cells competes for space in the circulatory system with red blood cells, so that the patient's "white cell count" is elevated and the "red cell count" is diminished. Thus, leukemia is also a cause of anemia.

Having too few white blood cells is even more devastating than having too many. In acquired immune deficiency syndrome (AIDS), for example, a virus kills lymphocytes called helper T cells so that the immune system can no longer fight viral infections. In leukopenia, bone marrow is severely damaged by acute exposure to radiation or toxic chemicals. Without immediate replacement of the white cells, death occurs in a day or two from rampant infection of the lungs and digestive tract.

neutrophil

basophil

J. Beck

eosinophil

lymphocyte

monocyte

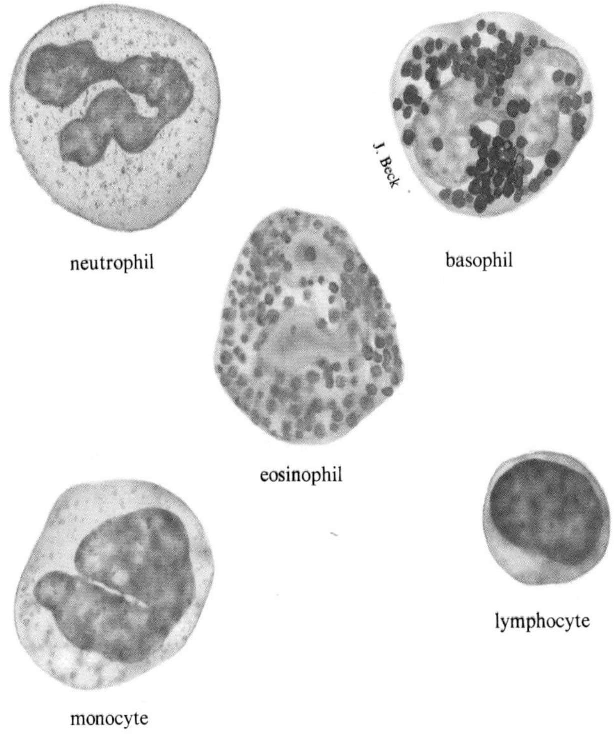

Figure 22.6
Types of white blood cells. Neutrophils surround, engulf, and destroy foreign invaders. Eosinophils attack parasites that the neutrophils do not affect. Basophils release chemicals that are part of the inflammatory response. Monocytes attack organisms that are resistant to neutrophils. Lymphocytes produce antibodies.

Platelets

Platelets are small, colorless cell fragments that live about 1 week and are responsible for initiating blood clotting, or **coagulation.** A platelet originates as part of a huge bone marrow cell called a **megakaryocyte** (fig. 22.7). This giant cell has organized rows of vesicles that divide the cytoplasm into distinct regions, like a sheet of stamps, but in three dimensions. The vesicles enlarge and coalesce to form *platelet demarcation channels*, which "shed" the cell fragments that are the platelets. A platelet is bounded by a membrane, contains mostly cytoplasm, secretory granules, and a few organelles, and lacks a nucleus. The flattened shape of the platelet is maintained by a ring of microtubules.

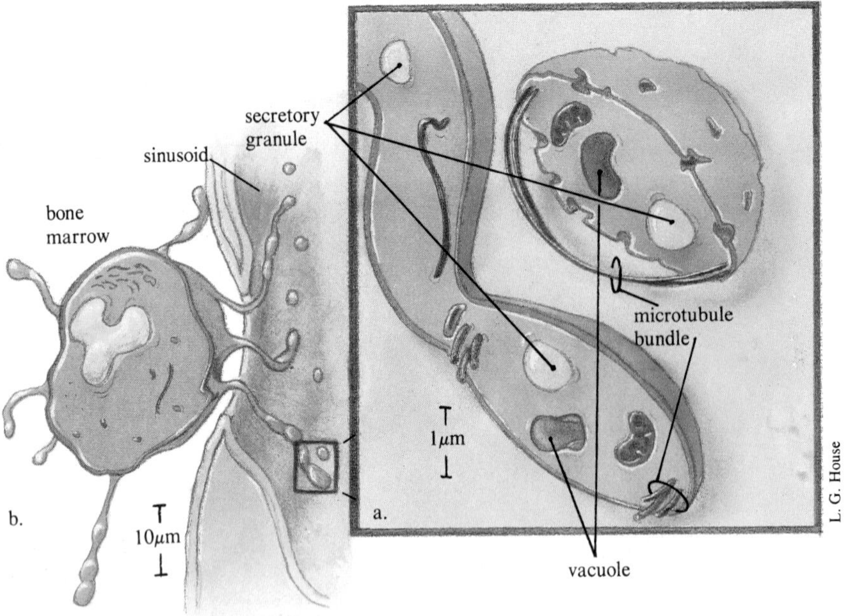

Figure 22.7
Platelet formation. *a.* A platelet is a plate-shaped structure lacking a nucleus that breaks off of a large cell called a megakaryocyte. *b.* The megakaryocyte is perforated by rows of vesicles called platelet demarcation channels, which join together to shed platelets. The platelets contain the biochemicals needed to start the process of blood clotting.

In a healthy circulatory system, platelets travel freely within the vessels with the help of heparin, a chemical produced by white blood cells called basophils. (Heparin is sometimes given as a drug to prevent abnormally fast clotting.) But when a wound nicks a blood vessel, or a blood vessel's normally smooth inner lining becomes obstructed, platelets snag on the bumps. The platelets shatter, releasing biochemicals from their secretory granules, which, along with plasma proteins, start a complex series of reactions that result in blood clotting.

Consider a familiar scenario—a scraped knee. A child falls and cuts his leg on a stone. A small blood vessel near the skin's surface is damaged, and as a result, the wound bleeds. As the child runs crying to his parent for aid, his body is already healing itself. Immediately, the injured vessel constricts to slow the flow of blood. Within seconds, platelets collect at the site. Meanwhile, cells from the blood vessel wall release a protein, **thromboplastin,** which, in the presence of calcium ions, converts the blood protein prothrombin into thrombin (fig. 22.8). Thrombin, in turn, initiates a series of chemical reactions that ultimately convert the plasma protein fibrinogen into fibrin. As its name suggests, fibrin molecules are threadlike, and they form a meshwork of fibrils that entrap red blood cells and additional platelets. The platelets trapped in the clot release more thromboplastin, which keeps the clotting cycle going. Clot platelets also polymerize actin and myosin into a contractile apparatus, similar to that of skeletal muscle, that contracts the clot as new tissue forms to fill in the wound.

If clotting occurs too slowly or too rapidly, health is seriously affected. The inherited bleeding disorder hemophilia, discussed in Reading 14.1, is caused by absence of factor VIII, one of the proteins essential for blood to clot. Deficiencies of vitamins C or K can also dangerously slow blood clotting.

Blood that clots too readily is extremely dangerous. One of the hazards of atherosclerosis, a condition in which fatty deposits form within arteries, is the possibility of clot formation as platelets snag on rough spots in the lining. A clot that blocks a blood vessel or the heart is called a **thrombus.** It can cut off circulation, causing death. The blockage of a coronary artery (a vessel that brings blood to the heart muscle) by such a clot is called a coronary thrombosis. A clot that travels in the bloodstream to another location is called an **embolus.** An embolus that obstructs a blood vessel in one of the lungs is termed a pulmonary embolism, and it too can kill.

KEY CONCEPTS

Blood is a connective tissue consisting of formed elements suspended in plasma. Plasma is mostly water but also contains proteins, salts, nutrients, hormones, gases, and metabolic wastes. Red blood cells are the most abundant of the formed elements, and they are biconcave discs lacking nuclei and some organelles but filled with hemoglobin. The large surface area of the cell is ideal for gas exchange. Hemoglobin binds and releases oxygen. White blood cells are larger than red blood cells, are less numerous, and have nuclei. These cells protect against injury, infection, and cancer by surrounding foreign agents or releasing biochemicals that destroy them. Platelets are small, pale cell fragments. When bumped by a vessel irregularity, platelets release biochemicals, which start a cascade of reactions that clot blood in the area. The vessel constricts, releasing thromboplastin, which converts prothrombin to thrombin. Thrombin starts a reaction series that eventually converts fibrinogen to fibrin, which forms the clot. Abnormal proportions of a formed element affect health.

Blood Vessels

The arrangement of blood vessels can be compared to an arrangement of roads. To get to a particular destination by car, you might first travel on a large superhighway, then on a smaller highway, then an avenue, and finally on a small street. Returning home, you might reverse the pattern but take a different set of roads to avoid traffic. In the circulatory system, large elastic blood vessels called **arteries** leave the heart and branch into smaller arteries, which in turn branch into even smaller **arterioles,** finally forming the tiniest vessels of all, the **capillaries.** On the blood's return voyage, capillaries flow into small **venules,** then small **veins,** and finally large veins that return the blood to the heart (fig. 22.9).

Arteries

The arteries, which lead away from the heart, have walls so thick that the largest arteries contain smaller blood vessels in their walls to nourish the vessel's cells. Even arteries that are not carrying blood remain open because their thick walls do not allow them to collapse.

The outermost layer of an artery consists of connective tissue, the middle layer is built of elastic tissue and smooth muscle, and the inner layer is a smooth one-cell-thick lining called endothelium (fig. 22.10). The middle layer is extremely important because it makes the arteries elastic and contractile. When the heart muscle contracts, a tidal wave of blood pounds against the arterial walls. The arteries expand to accommodate this surge. Then, while the heart relaxes, the arteries recoil, driving the blood along.

As arteries branch farther from the heart, their thickness diminishes and the outermost layer may be lost. In arterioles, the smooth muscle layer is predominant, a change in structure that reflects the gradual change in function. Because of the smooth muscle, these small vessels can easily constrict or widen, thereby regulating blood pressure as well as the distribution of blood in the body.

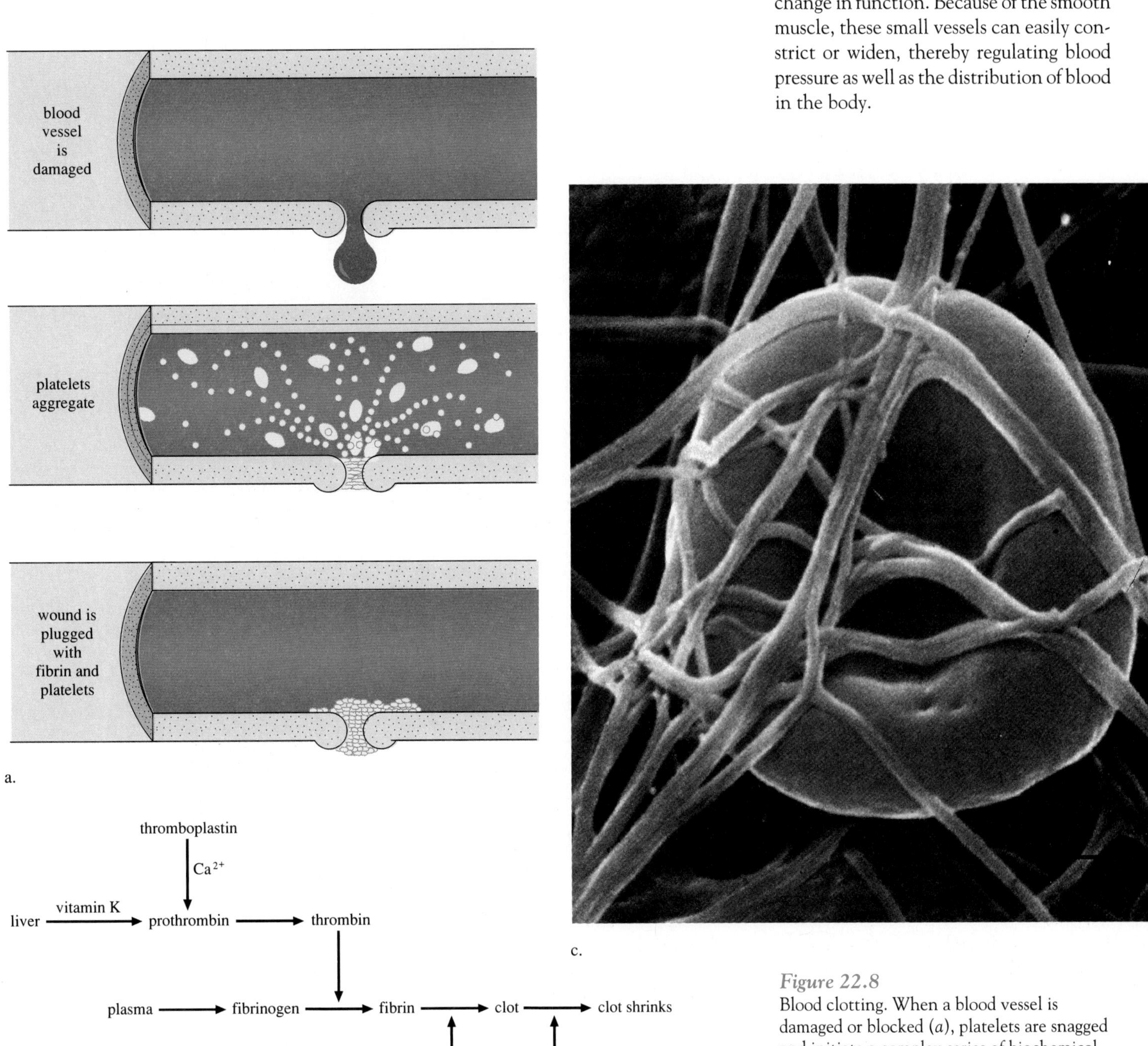

Figure 22.8
Blood clotting. When a blood vessel is damaged or blocked (*a*), platelets are snagged and initiate a complex series of biochemical reactions that lead to clot formation (*b*).
c. Close-up of a red blood cell caught in the fibrin meshwork of a clot.
(*a*) Source: *FDA Consumer*, Washington, DC.

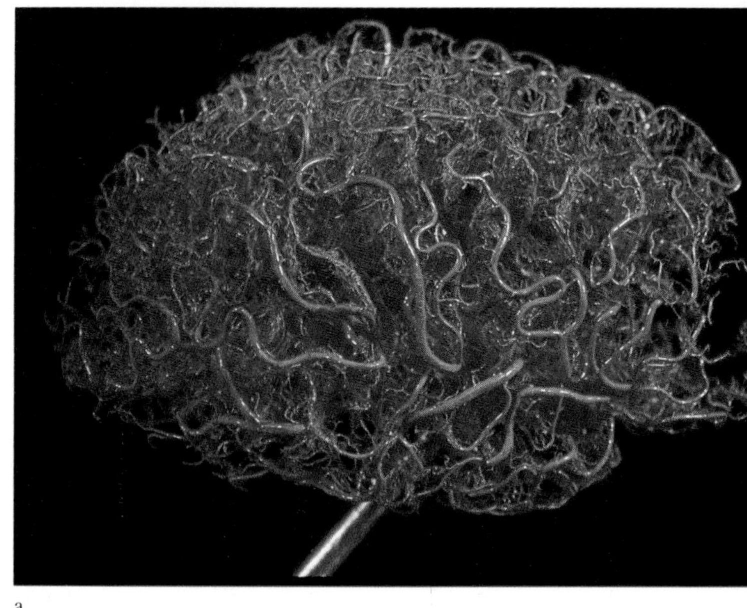

a.

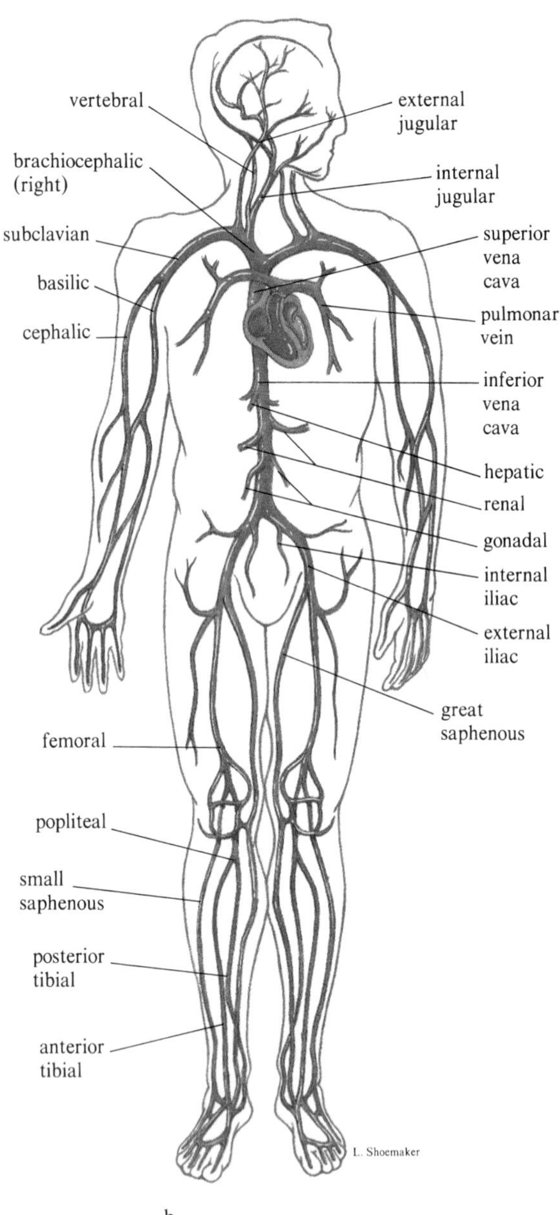

b.

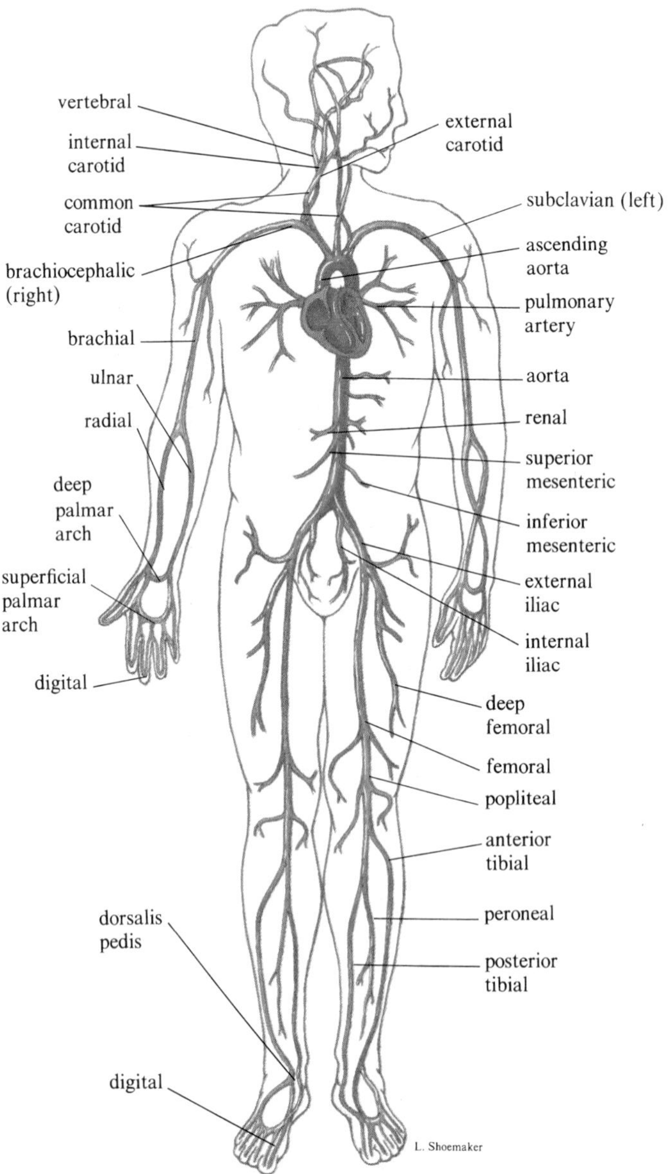

c.

Figure 22.9
a. This resin cast of the blood vessels in the human brain illustrates how numerous the vessels are. No cell is far from a capillary. *b.* The venous system. *c.* The arterial system.

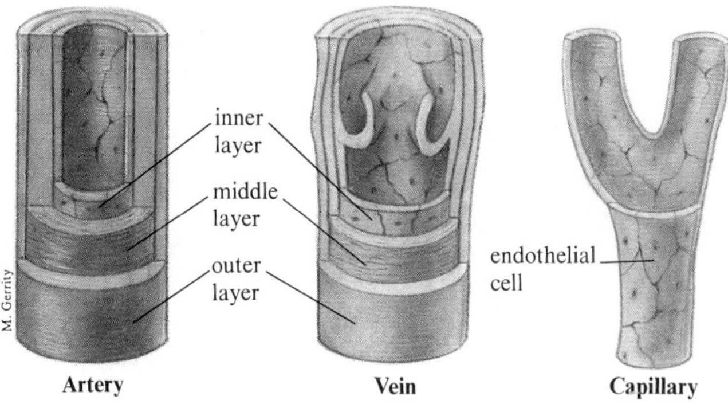

Figure 22.10
Cross sections of blood vessels. The walls of arteries and veins have three layers: the outermost layer is connective tissue; the middle layer is elastic and muscular; the inner layer is smooth endothelium. The walls of arteries are much thicker than those of veins. The middle layer is greatly reduced in veins.

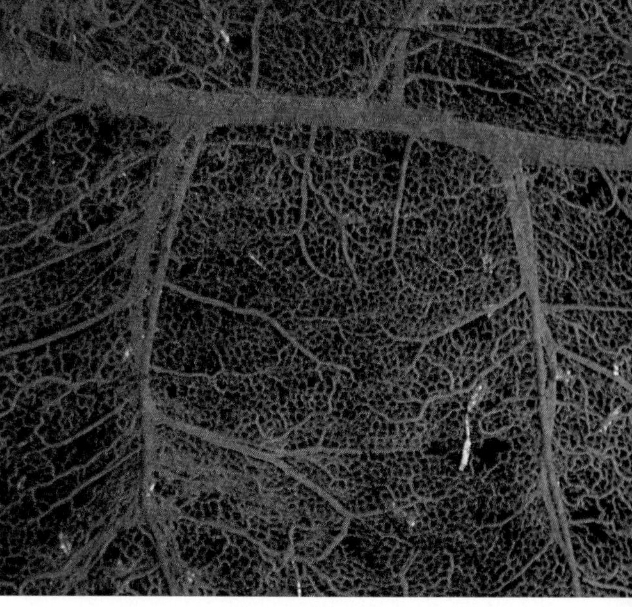

Figure 22.11
Capillaries are tiny, one-cell-thick vessels that bring the blood close to every cell of the body.

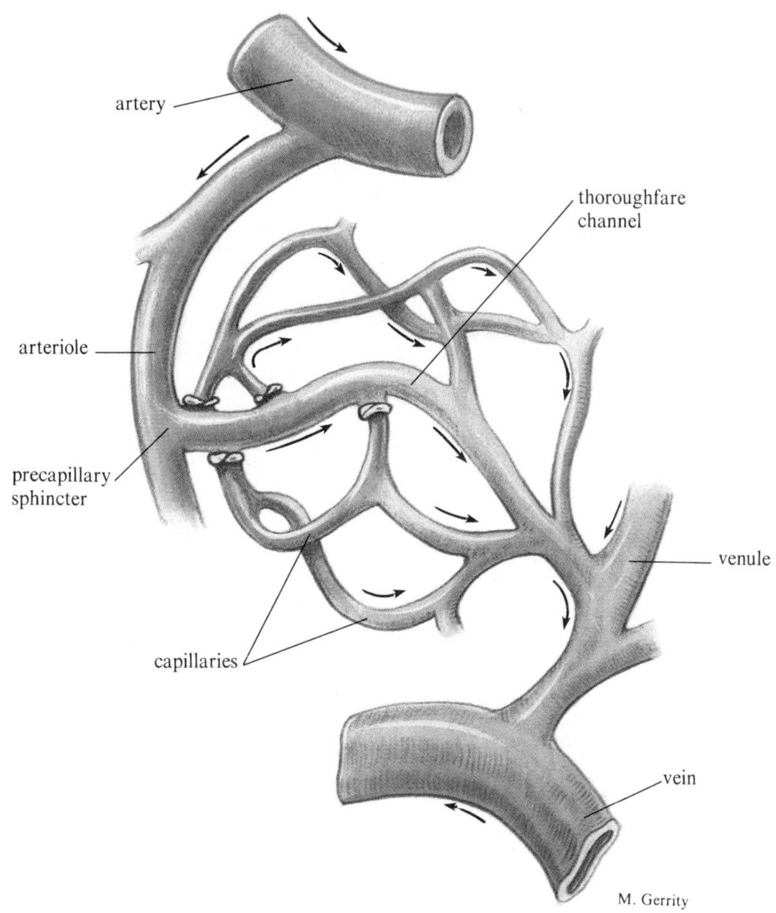

Figure 22.12
A capillary bed is a network of tiny vessels between an arteriole and a venule. Blood does not flow through all a person's capillaries at one time. To divert blood away from some capillaries, perhaps from a tissue that is not active at the moment, sphincters located in the arterioles can close off the blood flow to all but one capillary in a bed. Blood flow is maintained through one capillary, which is called a thoroughfare channel.

Capillaries

Capillaries are built only of endothelium. Imagine a sheet of linoleum, folded into a tube. It is this streamlined construction that enables capillaries to function as the sites for exchange of oxygen, carbon dioxide, nutrients, and waste (fig. 22.11). Because capillaries carry out this vital chemical exchange in nearly all tissues, they are numerous, providing extensive surface area. If all of the capillaries in the human body were laid end to end, they would circle the earth twice, with enough left over to make two round trips between Los Angeles and New York City. To gain an appreciation for the extensiveness of the capillaries serving an area, push down on your fingernail and watch the underlying tissue pale as the capillaries are squeezed closed.

Capillaries are arranged in networks, called capillary beds, that extend between an arteriole and a venule (fig. 22.12). Not all capillary beds in the body are open at one time. There simply is not enough blood to fill them all. The smooth muscle in an arteriole can contract and reduce the blood flow to the capillary bed it feeds or relax and open the floodgates to that area.

The circulatory needs of individuals vary with their activity and with environmental conditions. The pattern of dilation and constriction of arterioles provides adaptation to exercise, when the blood supply to the muscles increases at the expense of the supply to organs not in immediate use, such as those of the digestive tract. Runners, for example, need 20 times as much blood in their calf muscles when they are running as they do when they are sleeping. Blood must be redistributed for other reasons too. A cold-water drowning victim may survive until rescue comes if the blood is shunted from the extremities to the heart and brain. In very hot weather, arterioles at the skin's surface dilate, bringing warm blood to the surface to be cooled. Arterioles monitor temperature regulation, and they redirect blood to the heart and lungs from the extremities in emergency situations. In short, blood vessels open and close in a coordinated fashion that ideally meets the localized metabolic needs of tissues.

Veins

Veins are thinner and less elastic than arteries. Thickness and elasticity are not necessary in a vessel conducting blood with so little pressure. The middle layer of a vein is much reduced or even absent. When they are not filled with blood, veins collapse.

Because the blood pressure in the veins is low, some medium and large veins, particularly in the lower legs, where blood must move against gravity to return to the heart, have flaps called **venous valves** that allow blood to flow in only one direction. As skeletal muscles contract, they shorten. This squeezes the veins and propels the blood in the only direction it can move, toward the heart (fig. 22.13).

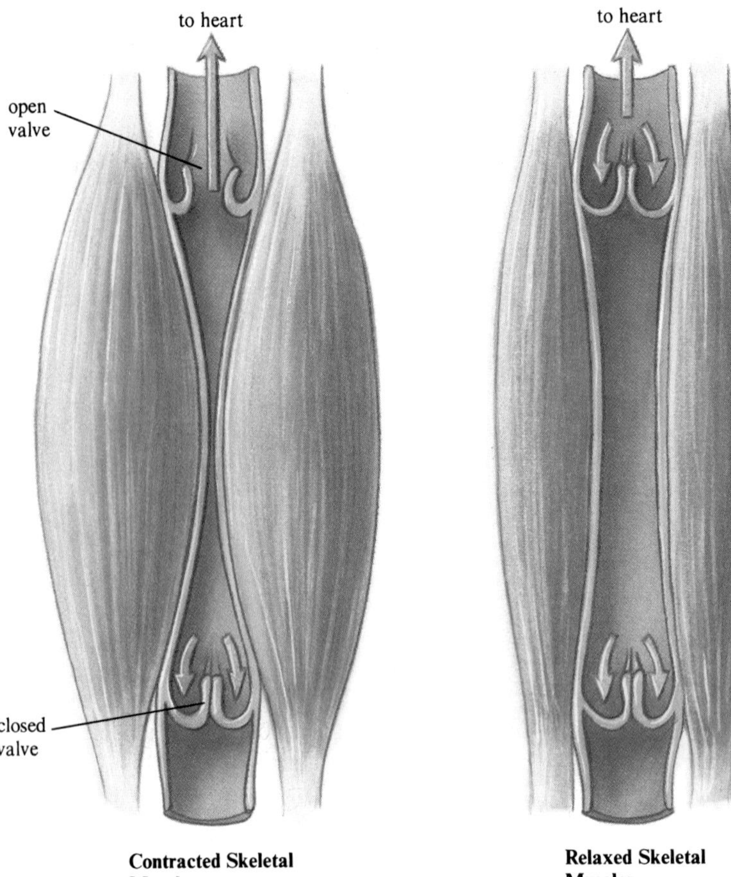

Figure 22.13
Valves in the veins assist the return of blood to the heart. When skeletal muscles contract they get shorter and push against the veins, propelling the blood onward. While skeletal muscles relax, the blood is prevented from flowing backward by valves in the veins.

Systemic Circulation

How are blood vessels that guide the blood's journey through the body arranged? Blood leaves the heart to reach the tissues through a large artery, the **aorta** (fig. 22.14). Among the first branches of the aorta are the vessels that nourish the heart itself. The remaining branches of the aorta diverge. About a quarter of its blood travels to the kidneys, and another quarter takes oxygen and nutrients to the muscles. About 15% of the blood goes to the organs of the abdomen, 10% to the liver, 8% to the brain, and the rest to other parts of the body.

In the liver, circulating blood detours from the usual pattern of capillary to venule to vein. The **hepatic portal system** (hepatic means liver) is a special division of the circulatory system that helps to harness the chemical energy in digested food quickly. Here, capillaries from the stomach, small intestine, pancreas, and spleen continue into four veins, which converge into the **hepatic portal vein,** which leads to the liver. Once in the liver, the hepatic portal vein diverges into venules and then capillaries, and these capillaries converge into the hepatic veins, which empty into the **inferior vena cava** and finally back to the heart. Can you see how this arrangement is slightly different from that elsewhere in the circulatory system?

The unusual position of the hepatic portal vein between two beds of capillaries allows dissolved nutrients that enter the circulation at the digestive organs to reach the liver rapidly, where they are metabolized to meet the body's energy needs. The liver also removes toxins and microbes from the material picked up from the digestive organs.

Blood Pressure

The same force that drives blood through vessels is also exerted outward on the vessel walls. This **blood pressure** results from the heart's pumping action. Blood pressure reaches a maximum when heart chambers called the **ventricles** contract, and this peak is called the **systolic pressure.** Blood pressure is at a minimum when the ventricles relax, and this low point is called the **diastolic pressure.** Blood pressure is generally measured by determining the millimeters of mercury displaced in a pressure gauge called a **sphygmomanometer** (fig. 22.15). A "normal" reading for blood pressure is expressed as 120 millimeters of mercury for the systolic pressure and 80 millimeters of mercury for the diastolic pressure, giving the expression "120 over 80" (written 120/80).

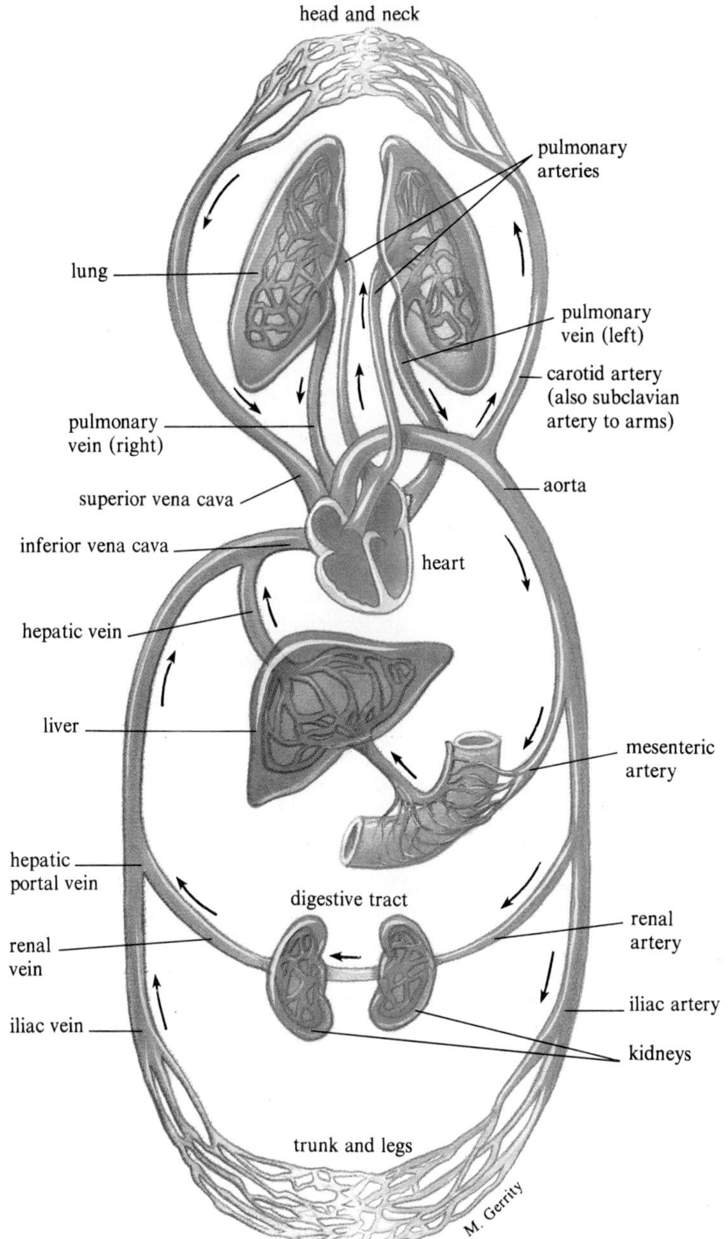

Figure 22.14
Pulmonary and systemic circulation—the journey of blood.

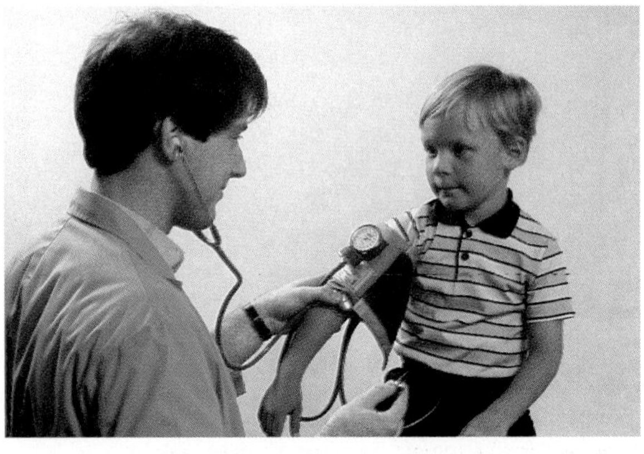

a.

b.

Figure 22.15
Blood pressure is measured with a sphygmomanometer, which consists of an inflatable cuff attached to a pressure gauge. The cuff is wrapped around the upper arm and inflated until no pulse can be felt in the wrist, which signifies that the circulation to the lower arm has been temporarily stopped. A stethoscope is placed on the arm just below the cuff, so the sound of blood flow returning can be heard when the cuff is slowly deflated. When a thumping is heard through the stethoscope, the listener notes the pressure on the gauge. This sound is the blood rushing through the arteries past the deflating cuff as the pressure peaks, due to ventricular contraction. The value on the gauge is the systolic blood pressure. The sound fades until it is muffled, and the pressure reading at this point is the diastolic blood pressure.

Blood pressure actually varies in different parts of the body. The value 120/80 reflects the blood pressure in arteries near the heart. Pressure drops with increasing distance from the heart, so that it is highest in the arteries and arterioles but drops dramatically at the capillaries. The decrease in blood pressure with distance from the heart is a physical consequence of the increasing cross-sectional area of the blood vessels as the number of branches from the aorta increases. The situation is similar to traffic on a crowded highway (a large artery) dissipating after many of the cars exit onto diverging roads (smaller arteries and arterioles). By the time the capillaries are reached, the cross-sectional area is 600 to 800 times that of the aorta. This results in a slowing of blood flow through the capillaries, which facilitates nutrient and waste exchange there. Once in the veins, blood is helped to move by contraction of surrounding skeletal muscles and by the venous valves that prevent the blood flow from reversing direction.

Blood pressure rises when blood vessels narrow **(vasoconstriction)** and lowers when they open **(vasodilation)**. The number of capillaries that are opened or closed depends on the blood volume in the arterioles and indirectly on several other factors, including temperature, blood chemistry, and the level of activity of the individual. Such emotions as anger and fear can also alter blood pressure. Blood pressure that is significantly lower than normal, termed **hypotension,** may produce fainting. Blood pressure that is higher than normal is termed **hypertension,** and it deserves its reputation as the "silent killer" (Reading 22.2).

KEY CONCEPTS

Arteries lead away from the heart and remain open because their walls are very thick. They eventually branch into arterioles, which are thinner. Arterioles lead into one-cell-thick capillaries, which are the sites of exchange of oxygen, carbon dioxide, nutrients, and waste. Capillaries provide very extensive surface area. Muscle contraction controls which capillary beds are open. Veins lead toward the heart and have collapsible walls that are thinner than those of arteries. Valves keep blood flow unidirectional. In systemic circulation, blood leaves the heart through the aorta and branches to service the tissues. The hepatic portal system is a special "stop" for nutrients from the digestive organs. The heart's pumping produces blood pressure. Maximal blood pressure is called systolic and is measured when the heart contracts. Minimal blood pressure, called diastolic, corresponds to when the heart muscle relaxes.

The Heart

The heart is an incredibly active pump. Each day it sends more than 7,000 liters of blood through the body, contracting more than 2.5 billion times in a lifetime. This fist-sized muscular pump is specialized not only to create the force that propels the blood through the body but also to ensure that blood flows in one direction through a loop of vessels. The structure of the heart is well suited to these functions.

The Structure of the Pump

The force propelling blood comes from the contraction of cardiac muscle, which comprises most of the walls of the heart. Cardiac muscle cells are striated because they are packed with neatly organized contractile proteins. The cells branch, forming an almost netlike arrangement. As a result, a cardiac muscle cell contracts in two dimensions, wringing the blood out of the heart (fig. 22.16). The heart is enclosed in a tough connective tissue sac, the **pericardium** (around the heart), that protects the heart while giving it freedom to move even during vigorous beating.

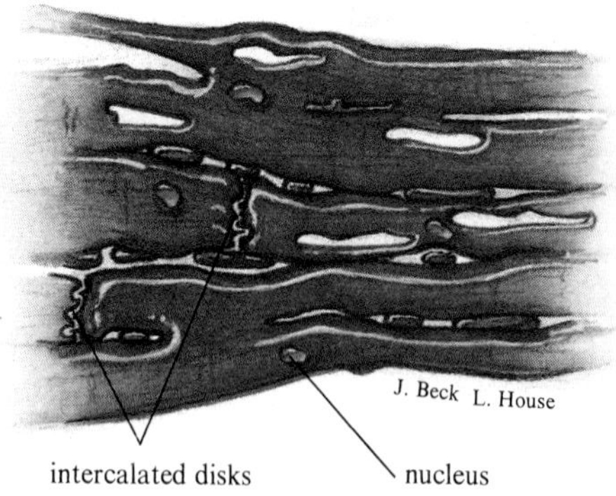

Figure 22.16
Cardiac muscle. The structure of cardiac muscle is well suited to its function. The striations reveal the orderly arrangement of contractile proteins. The branching network results in shortening of muscle in two dimensions, not just one. The intercalated disks are tight connections that allow the electrical activity to spread from cell to cell.

The heart consists of four chambers, which pump blood to and from the lungs, where it picks up oxygen, and to and from the rest of the body (fig. 22.17). The two **atria** receive the blood returning to the heart and pump it to the ventricles. The two ventricles are much larger than the atria and have thick muscular walls, which are necessary to generate enough force to push blood to even the most remote body parts.

Heart Valves

Heart valves are structures that keep blood flowing in a single direction through each closed loop of vessels. Two pairs of valves in the heart prevent the backflow of blood. The **atrioventricular valves** (AV valves) are located between each atrium and ventricle. These are thin flaps (cusps) of tissue that move in response to the pressure changes accompanying the contraction of the ventricle. The AV valve on the right side of the heart is the tricuspid valve; the AV valve on the left side is the bicuspid, or mitral, valve. When the ventricle contracts, pressure increases, pushing the flaps upward and causing them to spread out and prevent blood from moving backward into the atrium. The valves cannot be pushed completely into the atria because they are anchored by strings of connective tissue to muscles extending from the wall of the ventricle.

The **semilunar valves** consist of three pocketlike flaps sitting in a ring in the arteries just outside each ventricle. Blood leaving the ventricle moves past these valves easily. However, blood moving backward toward the ventricle is blocked as it fills these pockets and causes the valves to balloon outward, just as a parachute does when it fills with air (fig. 22.18).

The "lub-dub" sound of the heartbeat is caused by two sets of heart valves closing. The "lub" corresponds to valves closing between atria and ventricles, and the "dub" is the sound of blood sloshing against and closing the valves that lead to the arteries. Some variation on the "lub-dub" sound often reflects abnormally functioning valves and is called a **heart murmur.** It could mean a dangerous backflow of blood, forcing the heart to work harder, or it could be completely harmless.

Each year in the United States about 75,000 people receive replacement heart valves, made of synthetic materials such as a metal and a ceramic or of tissue from a cow or pig.

About 10% of the population has a condition called mitral valve prolapse, in which the mitral valve is stretched and balloons back into the atrium when the ventricle contracts. This can cause palpitations or a heart murmur or occasionally chest pain, but there are often no symptoms at all.

The Journey of Blood

The heart is sometimes considered as two pumps that beat in unison because it pumps the blood through two closed loops of blood vessels—the right side sending blood to the lungs and the left side powering the vast systemic circulation. Events on both sides of the heart are basically the same, but the blood from each side has a different destination.

Dark red, oxygen-poor blood from the systemic circulation enters the heart at the right atrium. The blood is delivered to the heart by the two largest veins in the body, the **superior vena cava** and the **inferior vena cava.** Another smaller vein entering the right atrium, the coronary sinus, returns blood that has been circulating within the muscle of the heart itself. The blood passes from the right atrium through an AV valve into the

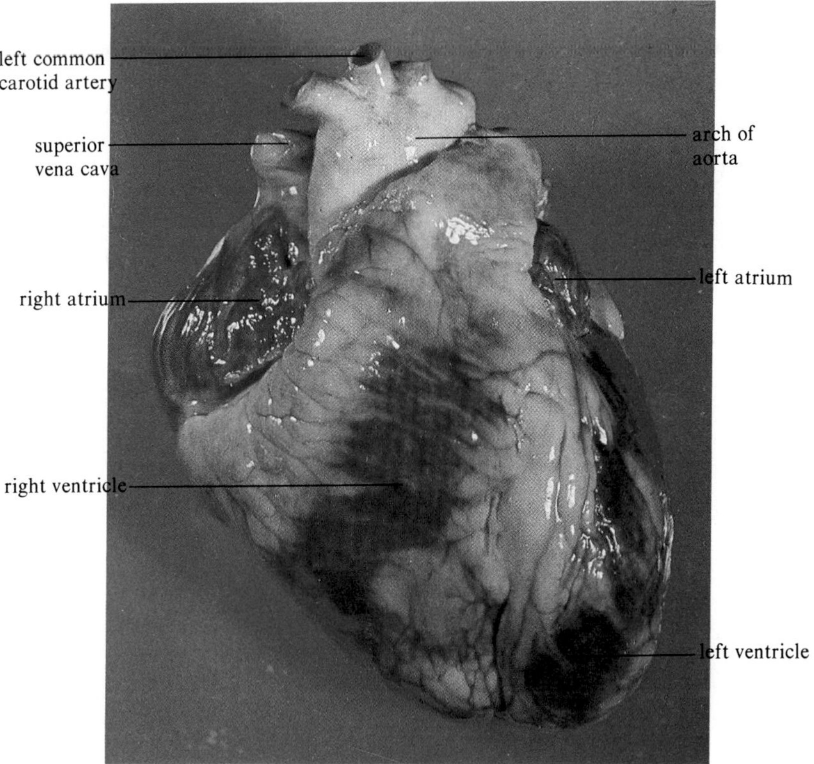

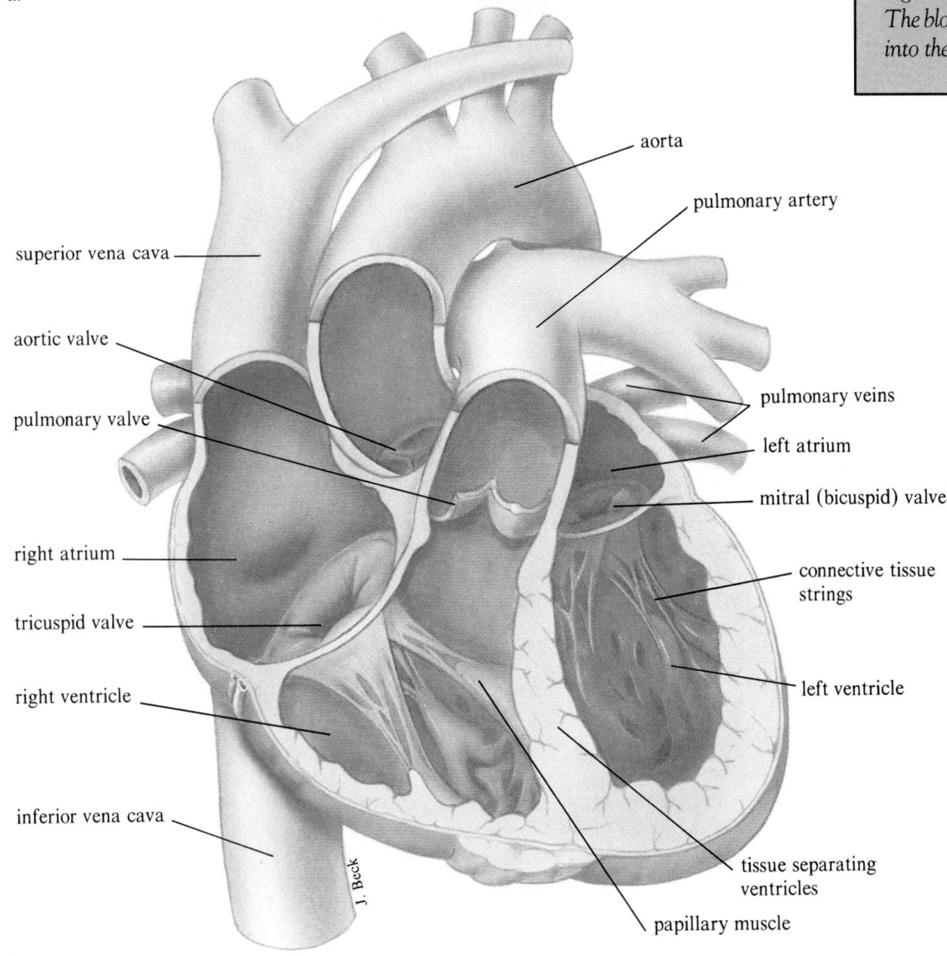

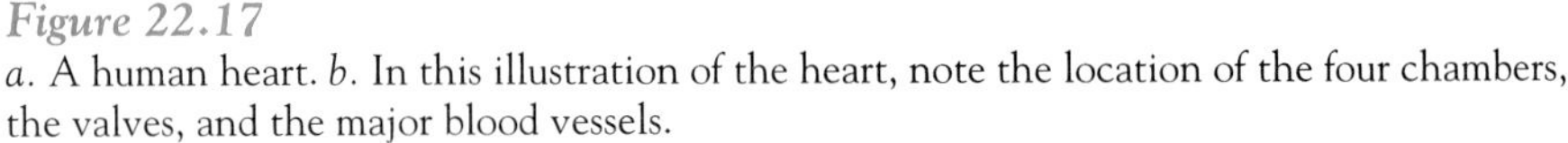

Figure 22.17
a. A human heart. *b.* In this illustration of the heart, note the location of the four chambers, the valves, and the major blood vessels.

right ventricle. Next, the blood flows out of the right ventricle, past the **pulmonary semilunar valve,** and through the **pulmonary artery** to the lungs, where it picks up oxygen. The oxygen-rich blood, bright red in color, now returns through four **pulmonary veins** to the heart, where it enters the left atrium. The oxygenated blood then flows from the left atrium through the **bicuspid valve** and into the left ventricle. The massive force of the left ventricle, the most powerful of the heart chambers, sends the blood through the **aortic semilunar valve** and into the **aorta,** the largest artery in the body.

KEY CONCEPTS

The branching arrangement of cardiac muscle cells allows them to contract in two dimensions, generating great power. The heart has four chambers: two atria, which receive blood, and two ventricles, which pump the blood out. The pericardium surrounds the heart. The atrioventricular valves just outside each ventricle keep blood flowing in one direction. Blood flows through two closed loops, to the lungs and to the rest of the body. Blood from the tissues enters the right atrium in the venae cavae, and crosses a valve to the right ventricle. Next it goes to the lungs, returning to the heart at the left atrium. The blood passes to the left ventricle, from where it is pumped into the aorta and distributed throughout the body.

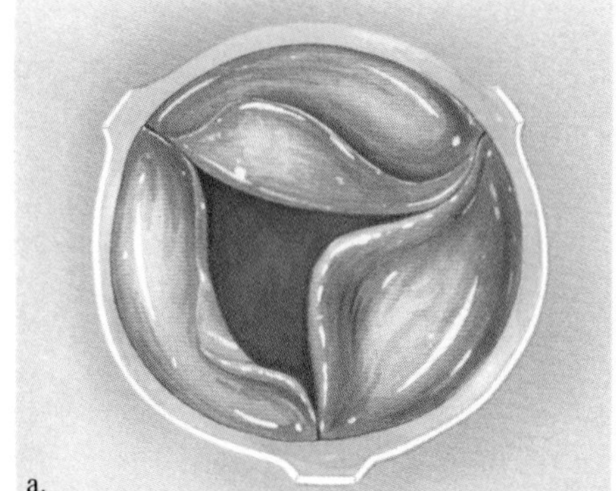

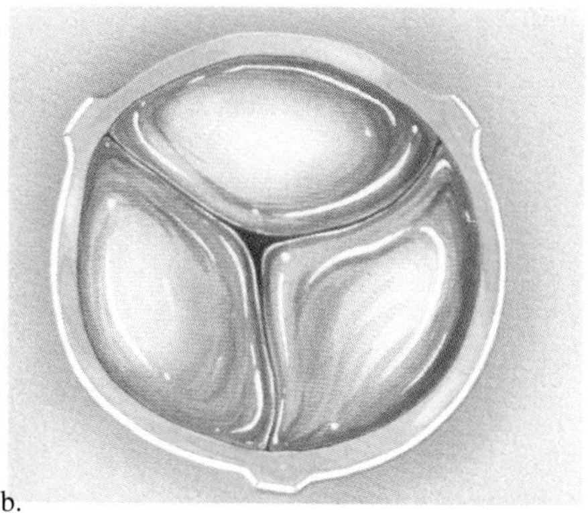

Figure 22.18
Heart valves. *a.* Semilunar valve open. *b.* Valve nearly closed.

Reading 22.2 The Unhealthy Circulatory System

Coronary Heart Disease

DISEASES OF THE HEART AND BLOOD VESSELS ACCOUNT FOR 53% OF ALL DEATHS IN THE UNITED STATES EACH YEAR. Nearly 87% of these deaths are due to **coronary heart disease,** which affects the coronary arteries that nourish the heart. A first step in coronary heart disease is often **atherosclerosis,** in which fatty plaques accumulate inside the walls of the coronary arteries and reduce blood flow to the heart muscle. In some people this accumulation is due to an inherited inability of cells to remove cholesterol from the bloodstream, facilitating a buildup of the fatty molecule in the blood vessels. In others, a lifetime of eating fatty food simply overwhelms the cell surface receptors that transport cholesterol into cells, and the excess backs up into the blood. In either case, just as a blocked roadway causes traffic problems for miles around, a blocked coronary artery can be devastating to the body.

Although high levels of cholesterol in the blood are related to the incidence of atherosclerosis, measuring the total cholesterol level does not tell the whole story. Some types of cholesterol may not be harmful and might even be beneficial. Because cholesterol, like any fat, is insoluble in water, it must be transported by a carrier protein. One carrier, the low-density lipoprotein (LDL), functions like a delivery truck. LDLs transport cholesterol to artery walls, where it is deposited. The "good" carriers are high-density lipoproteins (HDLs), which serve as garbage trucks that haul cholesterol to the liver for elimination. HDL levels can be raised by exercise and by substituting polyunsaturated fats for saturated fats in the diet. Avoiding obesity, high blood pressure, and cigarette smoking are also important in maintaining heart health.

Consequences of atherosclerosis include heart attack, angina pectoris, arrhythmia, and congestive heart failure.

Heart Attack

When blood flow through a coronary artery is blocked, the muscle in the region dies. This is a heart attack, and it is often caused by a blood clot triggered by vessels narrowed by coronary heart disease. The dying portion of the heart develops altered electrical activity, which upsets the heartbeat, leading to a wild twitching of the heart muscle **(fibrillation)**. The person may initially feel pressure, fullness, or a squeezing sensation in the chest that spreads to the shoulder, neck, and arms and lasts for at least 2 minutes. He or she may also feel sweaty, short of breath, dizzy, or nauseated.

A heart attack can come on quite suddenly, and the outcome depends to some extent on the availability of help. About two-thirds of all heart attack patients die within minutes or hours of the attack. Some of these people could be saved if someone trained in cardiopulmonary resuscitation (CPR) is present and can perform mouth-to-mouth resuscitation and closed-chest heart massage until an emergency medical technician or a doctor can take over. Once help arrives, a drug is given to relax the patient and to slow the erratic heartbeat, and oxygen is administered to keep as much of the heart muscle functioning as possible, given the impaired coronary circulation. In the ambulance, an electrocardiogram locates the damage and assesses its extent while monitoring the heartbeat. If the heart continues to beat erratically, a device called a **defibrillator** sends an electric shock to the heart to restore a regular beat.

At the hospital, the patient is rushed to the coronary care unit, where heart rate and blood pressure are carefully monitored. Dye is injected into the coronary arteries to see which ones are blocked, blood chemistry is evaluated to see if clots form too quickly, and enzyme measurements tell the extent of heart muscle damage. Because this damage will continue over the next 72 hours, drug therapy is begun to minimize the damage and to ward off a second heart attack. A shot of a "clot-bursting" drug, such as tissue plasminogen activator (t-PA) or streptokinase, given within 4 hours of a heart attack can be lifesaving and eliminate further treatment.

If a person survives the few weeks following the initial heart attack, and is perhaps given drugs to ward off a second attack, coronary circulation can be restored as other arteries deliver blood to the damaged part of the heart. This natural healing is called **collateral circulation.** The patient may even begin light exercise in a few weeks.

Angina Pectoris

Blood flow in the heart need not be completely blocked to be dangerous. Decreased oxygen flow in the heart due to a partially blocked coronary blood vessel can lead to a gripping, viselike pain called **angina pectoris.** This is a confusing condition because it is often difficult to tell whether a given instance of such pain is a warning of impending heart attack, an isolated event, or simply indigestion. The circumstances of the attack provide clues to its origin. *Angina at rest* occurs during sleep, and it is due to a calcium imbalance or a blood clot rather than atherosclerosis. *Angina of exertion* follows sudden activity in a sedentary person and is also not usually a result of atherosclerosis. *Unstable angina pectoris* is the dangerous variety, producing chest pain either when the person is at rest (but awake) or is engaged in minimal exertion, such as housecleaning or brisk walking. A cardiologist (heart specialist) who detects this condition may suggest bypass surgery to provide the blood with a detour around clogged arteries, thereby preventing the attack.

Arrhythmia

An arrhythmia is an abnormal heartbeat. Some arrhythmias may be transient flutters or a racing feeling that lasts only a few seconds. These arrhythmias usually are centered in the atria. They may reflect a malfunctioning pacemaker, which slows ventricular pumping so that blood flow is too slow to support life. Electronic pacemakers can supplement a faulty pacemaker. In the sudden and deadly **ventricular fibrillation,** pacemaker cells go completely out of control, causing the heart muscle to twitch wildly. In this state the heart is an ineffective pump. The plummeting of the blood pressure to 0/0 causes sudden death. This often occurs soon after a heart attack, and it is responsible for 77% of in-hospital deaths following a heart attack.

Congestive Heart Failure

In **congestive heart failure,** the heart is too weak to pump sufficiently to maintain circulation. As a result, fluid accumulates in tissues causing swelling. The 5 million people who have congestive heart failure in

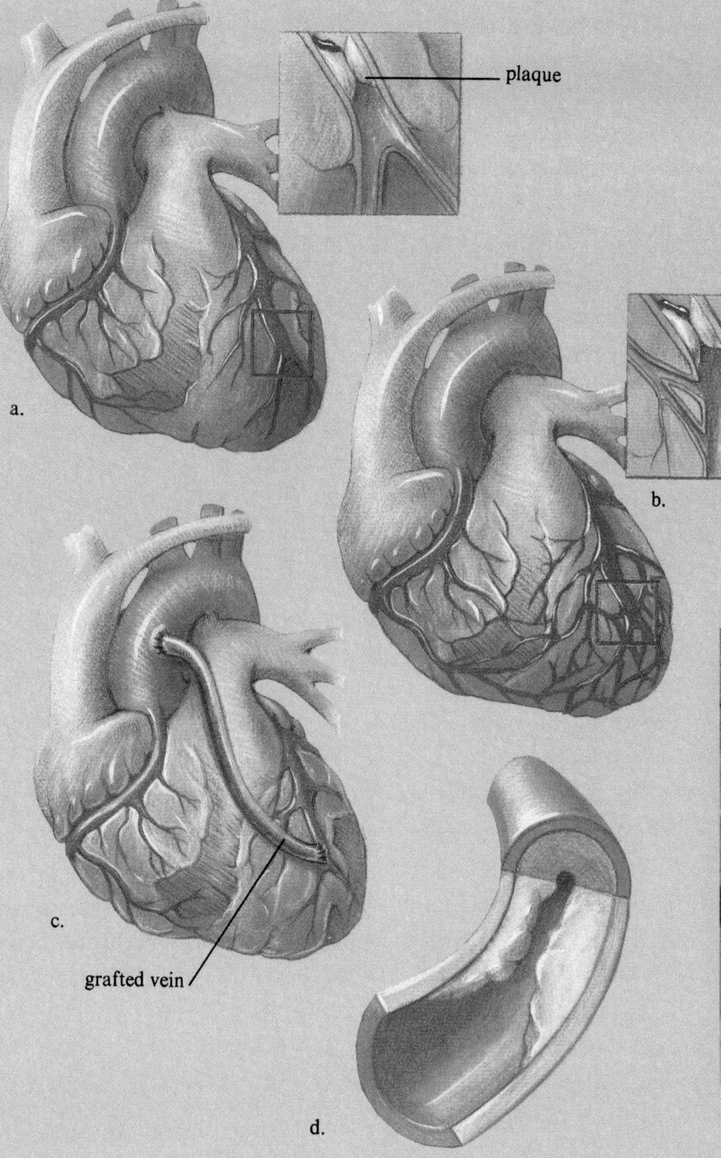

Figure 1
Bypassing a blocked coronary artery. *a.* A heart attack has occurred—an atherosclerotic plaque has blocked a coronary artery, and the tissue normally served by that artery, robbed of nutrients and oxygen, dies. *b.* If the attack is mild and the person recovers, other arteries may grow and attach to nonblocked arteries, establishing a collateral circulation to the damaged area. *c.* A coronary bypass operation uses a vein removed from the person's leg to fashion an alternative pathway for blood past the obstruction. *d.* This artery is nearly completely occluded by a plaque. *e.* Close-up of a post-mortem specimen of a human heart, centered on an artery blocked (occluded) by deposition of plaque. Degeneration of arterial walls due to formation of fatty plaques and scar tissue is a major cause of coronary heart disease.

the United States also experience fatigue, weakness, and shortness of breath, and 20,000 of them die from the condition each year. Congestive heart failure can be caused by atherosclerosis or defective heart valves. Treatment includes drugs—digitalis to strengthen the heart muscle and diuretics to rid the body of its excess fluid.

Hypertension

Consistently elevated blood pressure, called hypertension, affects 23% of the population of the United States. Blood pressure of 140/90 is considered high and values of 200/100

Reading 22.2 Continued

are dangerously high. Hypertension strains the heart, raising the risk of heart attack or failure. The excessive pressure may cause a stroke, in which the flow of blood to the brain is disrupted due to a blood clot in a blood vessel or the rupture of a vessel. Strokes can result in paralysis and memory loss, depending upon which parts of the brain are robbed of oxygen and nutrients. Often, high blood pressure damages the delicate filtering system in the kidney. Hypertension promotes the formation of lipid deposits in the arteries (atherosclerosis), elevating blood pressure further. Hypertension is a "silent killer," because it often has no symptoms and can only be diagnosed by a high blood pressure reading.

What causes hypertension? About 10% of the time a malfunctioning organ is at fault. Narrowed or hardened arteries account for many cases. In rare cases, a faulty gene is responsible for arterioles that are surrounded by too many nerves, causing them to constrict too much and thus elevating blood pressure. The kidneys, pituitary, thyroid, or adrenal cortex can also raise blood pressure.

Some people are more likely to develop hypertension than others. Risk factors that cannot be controlled include age, sex, and race. Incidence of hypertension rises with age and is greater among males and blacks. Fortunately, some risk factors can be controlled, such as diet, body weight, and smoking.

Excess salt in the diet, for example, can lead to fluid retention, which strains the circulatory system. Eating many fatty foods raises blood pressure when cholesterol is deposited in blood vessels, impeding blood flow. Deficiencies of calcium, potassium, and vitamins A and C are also linked to hypertension.

Excessive weight increases the risk of high blood pressure because the circulatory system has to serve more tissue than is present in a person of normal weight. The nicotine in cigarettes constricts blood vessels, elevating blood pressure. Another component of cigarette smoke, carbon monoxide, occupies sites on hemoglobin molecules that would otherwise carry oxygen, thereby decreasing the efficiency of oxygen transport to the tissues.

Rheumatic Heart Disease

A complication of a common "strep throat" can damage heart valves. Most people who get strep throat are sick for only a few days. But 1 in 100 become ill again a few weeks later with a more serious disorder, rheumatic fever. In some people, for reasons unknown, the antibodies that the immune system manufactures to attack this second bout of strep infection also attack heart valve cells, whose surfaces are strikingly similar to the surfaces of streptococcus bacteria. As the body fights the infection, it also fights a part of itself. Scar tissue forms on the damaged heart valves, and they begin to malfunction as a result. This is rheumatic heart disease. It can be prevented by treating a strep throat with antibiotics so that the body does not manufacture the misguided antibodies.

Varicose Veins

Veins near the surface of the body do not have much support from the surrounding tissue. Because vein walls are thin, they stretch easily. The tendency to stretch coupled with a weak valve allows blood to seep backward until it reaches a good valve. The blood pools there, increasing the pressure and damaging that valve. As a result, the superficial veins may become unusually distended, a condition called **varicose veins.**

Twenty percent of the people in the United States have varicose veins. Many suffer no discomfort at all, while others may feel dull aches or suffer from night cramps in the calf muscles. Rarely, varicose veins are a warning of serious complications—**phlebitis,** in which the vein wall is inflamed, or **thrombophlebitis,** in which inflammation of the vein wall is complicated by the formation of blood clots.

To some extent varicose veins are hereditary, but obesity, a low-fiber diet that leads to constipation, excessive sitting or standing, and pregnancy can contribute to their development. Treatments range from support stockings to surgical removal of the varicose veins.

Cardiac Cycle

Although the heart seems to have two distinct sides, the actions of these two pumps are highly synchronized. Each beat of the heart, called a **cardiac cycle,** consists of a sequence of contraction **(systole)** and relaxation **(diastole)**. Most of the blood pumped with each heartbeat flows right through the atria and begins to fill the ventricles while all four heart chambers relax. When the ventricles are about 70% full, both atria contract and fill the still-relaxed ventricles. Next, while the atria relax, the ventricles contract. A torrent of blood gushes into the arteries.

The thudding of the heart continues inexorably, beginning just a few weeks after conception and continuing until it is silenced by death. What causes the rhythmic contractions that accomplish this herculean task?

The muscular contraction of the heartbeat originates in the cardiac muscle cells themselves. These cells can contract on their own, without nervous stimulation. When grown outside the body, isolated heart cells beat at different rates. However, as two cardiac muscle cells make contact, they begin to beat synchronously. The spread of excitation is made possible by unusually tight connections within junctions between cells called **intercalated disks.** Because so many heart cells contract in unison, the force is great enough to drive the powerful ejection of blood.

The contraction is coordinated by specialized cardiac muscle cells that initiate each beat and ensure that excitation reaches each region of the heart in a pattern that ensures a forceful contraction (fig. 22.19). The beat begins within specialized cells in the **sinoatrial node (SA node)** located in the wall of the right atrium slightly below the opening of the superior vena cava. Because the SA node sets the tempo of beating, it is called the **pacemaker.** Once kindled, the impulses triggering contraction race across the atrial wall to another region of specialized muscle cells, the **atrioventricular (AV) node.** The AV node branches into specialized cardiac muscle fibers called **Purkinje fibers,** which conduct electrical stimulation six times faster than do other parts of the heart muscle. It

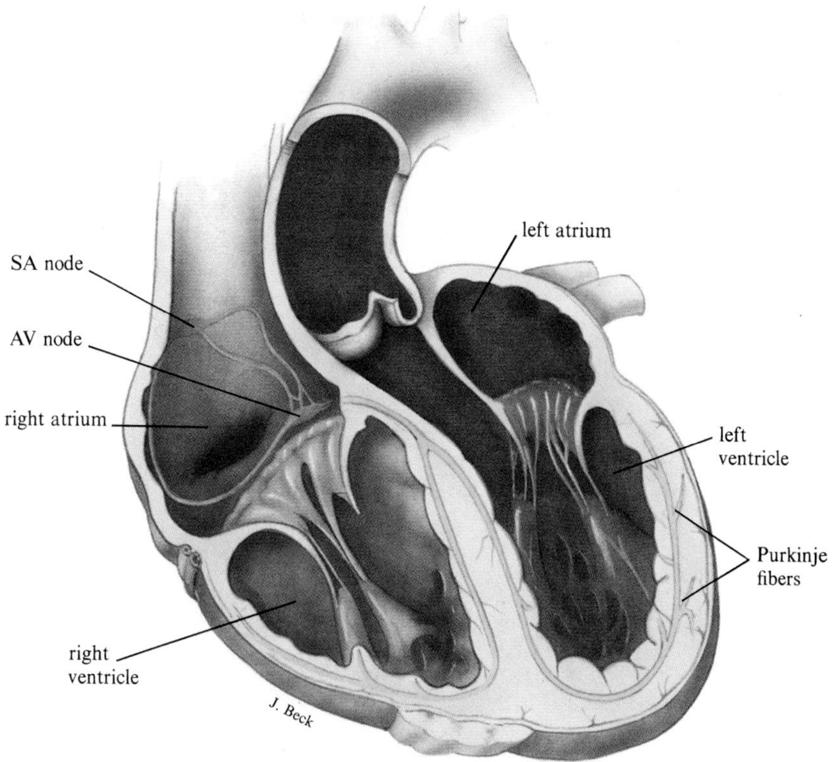

Figure 22.19
Specialized muscle cells form a conduction network through the heart. Contraction starts in the SA node and travels through the atrial wall to the AV node. From there it is conducted to the ventricle walls through the Purkinje fibers.

takes only 8/10 of a second from the time the SA node produces its electrical stimulation until the ventricles contract. The SA node, AV node, and Purkinje fibers collectively form the **Purkinje system,** the network of muscle fibers permeating the heart that triggers contraction of the ventricles.

Like all biological functions, heart rate is regulated. Two branches of the **autonomic** (involuntary) **nervous system** work opposite each other to adjust the rate of heartbeat to the needs of the body. During quiet times, the **parasympathetic nervous system** (one division of the autonomic nervous system) is dominant, inhibiting the SA node and slowing the heart. When life is threatened, the **sympathetic nervous system** is dominant, stimulating the SA node and causing the heart to beat faster. We can observe the electrical changes accompanying the contraction of the heart using a device called an electrocardiograph (EKG) (fig. 22.20). An irregular heartbeat is called an arrhythmia (Reading 22.2).

KEY CONCEPTS

A heartbeat, or cardiac cycle, consists of a contraction and relaxation. The heartbeat spreads because of the tight association and synchronous contractions of cardiac muscle cells. Specialized cells in the SA node originate the heartbeat in the right atrium. The beat goes to the AV node and then to the Purkinje fibers. Heartbeat is controlled by the nervous system. Its pattern of electrical charge is traced with an EKG. An arrhythmia is an irregular heartbeat.

Coronary Circulation

Heart tissue is not nourished by the blood pumped through the heart. Instead, it is nourished in the same manner as all tissues, by exchange across capillary walls. Shortly after leaving the heart, the aorta branches, sending arteries to all parts of the body. Two branches are the left and right **coronary arteries,** which diverge from the aorta above the semilunar valve and branch to surround and enter the heart muscle itself, carrying about 4% of the blood leaving the heart. The heart is infused with an extensive network of blood vessels that nourish the active muscle cells, and this **coronary circulation** returns in the coronary vein to the right atrium. Note the blood vessels visible in the heart shown in figure 22.17*a*.

Blood flow to the heart and heart rate are influenced by the **vasomotor center,** which is located in the lower part of the brain. It has four areas that affect circulation, mostly by sending nerve impulses through the spinal cord to the sympathetic nervous system and ultimately to the heart and blood vessels. The **vasoconstriction area** and the **cardioaccelerator area** stimulate circulation by constricting blood flow and speeding the heart, respectively. These areas are activated when blood pressure is low, blood is lost, or the blood contains too much carbon dioxide. The **vasodilation area** and the **cardioinhibition area** dilate vessels and slow the heart. These areas are activated when blood pressure is high.

Exercise and the Circulatory System

Regular exercise offers many benefits for the cardiovascular system, benefits that reduce the risk of heart attack. Some of the changes that result from regular exercise reduce the heart's workload. The heart of a dedicated exerciser does its job more efficiently than that of a sedentary person, pumping at the rate of 50 beats per minute the same amount of blood that the heart of a sedentary person pumps at the rate of 75 beats per minute (table 22.2). The lowest heart rate ever recorded in an athlete was

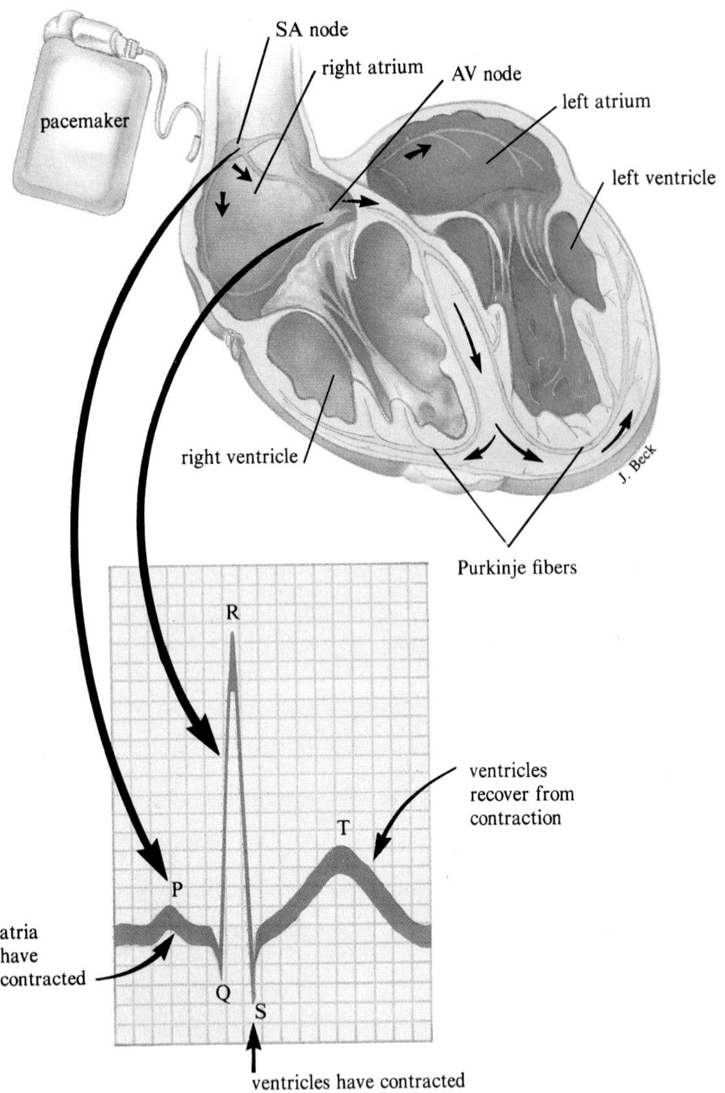

Figure 22.20
An electrocardiogram (EKG) is a printout of the electrical changes on the body's surface that reflect electrical changes as the heart contracts. An EKG has a characteristic pattern in a healthy heart. The P wave indicates atrial excitation. The QRS complex charts the spread of depolarization through the ventricles. The T wave results from ventricular repolarization.

Table 22.2
The Athletic Heart Versus the Average Heart

	Athletic Heart (50 beats per minute)	*Average Heart (75 beats per minute)*	*Difference*
1 minute	50	75	25
1 hour	3,000	4,500	1,500
1 day	72,000	108,000	36,000
1 month	2,160,000	3,240,000	1,080,000
1 year	25,920,000	38,880,000	12,960,000
72 years	1,866,240,000	2,799,360,000	933,120,000

Source: Date from Gabe Mirkin and Marshall Hoffman, *The Sports Medicine Book*. Copyright © 1978, Little, Brown & Company, Boston, MA.

25 beats per minute! At the same time, the number of red blood cells increases, and these cells are packed with a greater quantity of hemoglobin, allowing more oxygen to be delivered to tissues.

Exercise brings about changes that reduce the chance of developing atherosclerosis (Reading 22.2). High blood pressure and atherosclerosis go hand in hand. Exercise can lower blood pressure significantly. In addition, exercise elevates the level of high-density lipoproteins (HDLs), the cholesterol carriers that pick up cholesterol and transport it to the liver for elimination. HDLs may even remove cholesterol from the arteries.

Another benefit of exercise is the development of collateral circulation, or extra blood vessels. The additional blood vessels may prevent a heart attack by providing alternative pathways for blood flow to the heart muscle should a major coronary artery be blocked.

How much exercise is enough to benefit the circulatory system? To achieve the benefits of exercise, the heart rate must be elevated to 70% to 85% of its "theoretical maximum" for at least half an hour three times a week. You can calculate your theoretical maximum by subtracting your age from 220. If you are 18 years old, your theoretical maximum is 202 beats per minute. Seventy to 85% of this value is 141 to 172 beats per minute. Some good activities for raising the heart rate are tennis, skating, skiing, handball, vigorous dancing, hockey, basketball, biking, and fast walking.

It is wise to consult a physician before starting an exercise program. People over the age of 30 are advised to have a **stress test,** which is an electrocardiogram that is taken while the subject is exercising. (The standard electrocardiogram is taken while the subject is at rest.) An arrhythmia that appears only during exercise may indicate heart disease that has not yet produced symptoms.

KEY CONCEPTS

Regular exercise can lower heart rate and blood pressure, improve oxygen delivery, raise HDL level, and develop collateral circulation. These changes improve heart health.

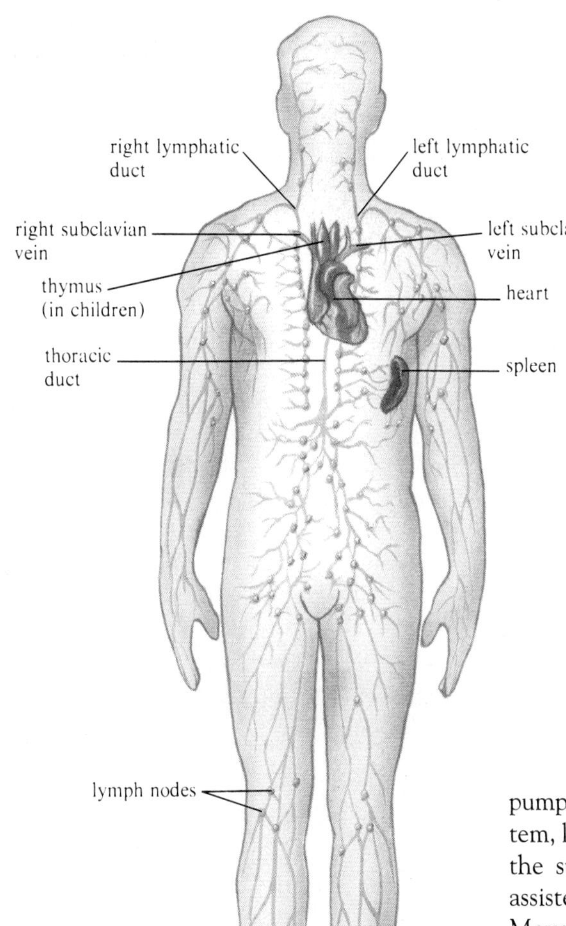

Figure 22.21
The lymphatic system is a network of vessels that collect excess fluid and proteins that leak from the blood capillaries and return them to the blood. At places along the lymph vessels are enlargements, known as lymph nodes, that filter the lymph.

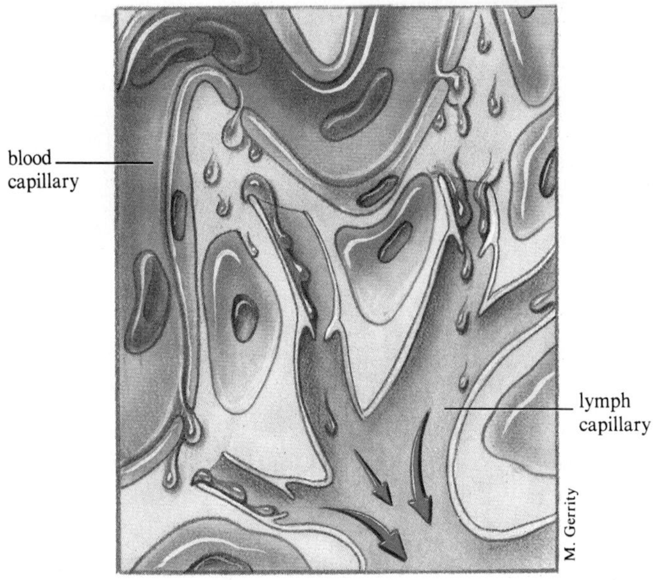

Figure 22.22
The lymph capillaries permeate tissues. They absorb excess tissue fluid, preventing the tissues from swelling.

The Lymphatic System

The **lymphatic system** is another transport system (fig. 22.21). The fluid contained in these capillaries and vessels, **lymph,** is made up of blood plasma minus certain large proteins that are too large to diffuse out of the blood capillaries. Lymphatic vessels begin as **lymph capillaries** that are dead-ended (fig. 22.22). They gradually converge to form larger **lymph vessels** that eventually empty into veins. Lacking the strong pumping action of the cardiovascular system, lymph is propelled by contractions of the surrounding skeletal muscles and is assisted by valves in the lymph vessels. Movement of lymph is sluggish. The lymphatic system normally returns to the veins less than 1 ounce (28 grams) of fluid per minute, in contrast to the 4 to 5 quarts (5 or 6 liters) pumped through the circulatory system in the same period of time. Tissue fluid surrounding the cells enters lymphatic vessels and flows toward the chest region, where it is returned to the blood. Special lymph capillaries in the small intestine, called lacteals, absorb dietary fats.

Lymph capillaries are highly permeable to permit the entry of fluid that has seeped out of blood capillaries and into tissue spaces. A breakdown in lymphatic functioning is quickly apparent because the body swells with the excess fluid, a condition called **edema.** The importance of compensating for the slow leakage of protein and fluid from the blood circulatory system is seen when the process is prevented, as it is in elephantiasis (fig. 22.23). In this disease parasitic worms reside in lymph vessels, blocking the flow of lymph and causing grotesque swelling.

In addition to vessels and fluid, the lymphatic system has structures called **lymph nodes,** which contain white blood cells that protect against infection. The kidney-shaped lymph nodes are located primarily in the abdomen, where the limbs meet the trunk, and in the neck (fig. 22.24). Lymph nodes range in size from microscopic to larger masses, such as those that are easily felt in a sore throat as swollen glands.

Figure 22.23
Elephantiasis is a disease in which the lymph vessels are blocked by a parasitic worm. As a result, the excess tissue fluid cannot be returned to the bloodstream. Instead it accumulates in the tissues, causing extreme swelling.

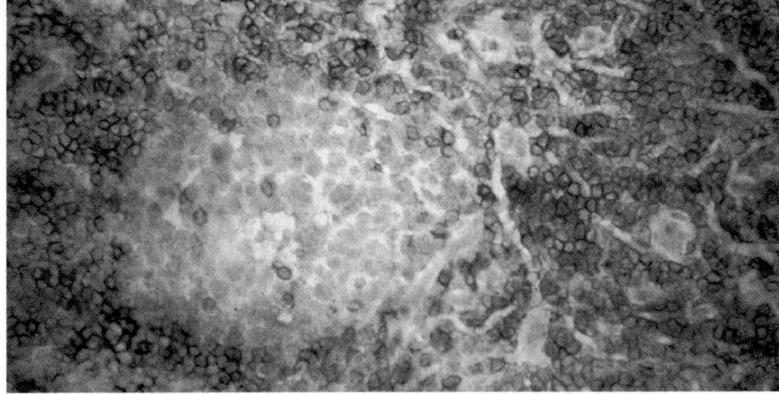

Figure 22.24
A slice through a lymph node reveals collections of white blood cells. In this lymph node, T cells are stained dark, and B cells stained light.

A lymph node is built of fibrous tissue with many pockets, each filled with millions of white blood cells that filter out cellular debris and bacteria from the lymph flow. The lymph nodes are very efficient. The fluid that enters them is often packed with infectious particles, 99.9% of which is destroyed in the node. Lymph nodes can detect and destroy cancer cells too. However, if a cancer originates in or spreads to a lymph node and is not destroyed, it spreads quickly to the rest of the body in the lymph fluid. The lymph nodes continually add lymphocytes to the lymph, from which they are transported to the blood.

As their name implies, lymphocytes are manufactured by parts of the lymphatic system—the spleen, the thymus, and possibly the tonsils. The largest component of the lymphatic system is the **spleen,** which is located to the left of the stomach and somewhat behind it. In addition to producing and storing lymphocytes, the spleen contains reserve supplies of red blood cells, which can be added to the systemic circulation in case of an injury when blood is lost. The **thymus** is a lymphatic organ located in the upper chest near the neck that is prominent in children but begins to degenerate in early childhood. The thymus is important in "educating" certain lymphocytes in the fetus to distinguish body cells from foreign cells. **Tonsils** are collections of lymphatic tissue high in the throat. Tonsils do not filter lymph but are thought to help protect against infection. Years ago, infected tonsils in children were routinely removed. Today, tonsils are left in whenever possible and the infection is treated with antibiotics, because we are not yet sure what the tonsils do.

The blood circulatory system and the lymphatic system work in concert, the blood carrying a continuous supply of nutrients and oxygen to cells and removing wastes from them, and the lymphatic system screening fluid that has leaked from blood capillaries for bacteria, debris, and cancer cells and returning the filtered fluid to the blood. Like the transport systems that link relatively isolated Oxford, Ohio, to the rest of the large country that contains it, the blood circulatory system and the lymphatic system nurture even those cells that are far removed from the body's surface, ensuring that the trillions of cells of the human body can carry out the functions of life.

KEY CONCEPTS

Blind-ended lymph capillaries and lymph vessels carry plasma minus some of its large proteins and pick up lymphocytes at lymph nodes. These lymphocytes filter bacteria, cellular debris, and cancer cells from the circulation. Lymphocytes are manufactured in the spleen, the thymus, and possibly the tonsils.

SUMMARY

The extensive branches of the circulatory system carry *blood* to nearly all tissues, delivering nutrients, gases, and infection-fighting cells and removing metabolic wastes. Blood is a mixture of cells and cell fragments (collectively called *formed elements*), proteins, and molecules that are dissolved or suspended in a liquid matrix, the *plasma*. *Red blood cells* (erythrocytes) are packed with the protein hemoglobin, which transports oxygen. Red cells develop in the bone marrow, and the mature cell is a biconcave disc lacking a nucleus. Having too few red blood cells or too little hemoglobin results in anemia.

There are five types of *white blood cells*, or leukocytes. Neutrophils, the most abundant type of white cell, surround and degrade bacteria at the site of inflammation. Eosinophils affect parasites that the neutrophils do not affect, and basophils release certain chemicals that are part of the inflammatory response. Monocytes attack bacteria that are resistant to neutrophils. Lymphocytes produce antibodies that protect against infection. Certain diseases are associated with white cell imbalances.

Platelets initiate clotting by being broken and collecting near a wound, where they release chemicals that trigger a chemical cascade resulting in clot formation. Poor clotting can lead to hemorrhage; abnormally rapid clotting can lead to a *thrombus* or an *embolus*.

The circulatory system consists of the heart and a system of vessels that leads to and from the lungs in the pulmonary circulation and to and from the rest of the body in the systemic circulation. Blood leaves the heart in the *aorta*, and it travels in increasingly narrower *arteries* and *arterioles* to the *capillaries*, where nutrient exchange and waste exchange occur. Blood flows from capillaries to *venules* and then to *veins*, and it reenters the heart through the *venae cavae*. Arteries have thicker walls than veins.

The human heart is a muscular pump that is divided into four chambers. The upper two chambers are called *atria*, and the lower two, which are more massive, are called *ventricles*. Heart valves are located between the right atrium and right ventricle, between the left atrium and left ventricle, and where the pulmonary and systemic blood pathways leave the heart. The heart valves ensure one-way blood flow.

Heart rate is set by the pacemaker or *sinoatrial node*, a collection of specialized cardiac muscle cells in the right atrium. From there heartbeat spreads to the AV node and then along the *Purkinje fibers* through the ventricles. The force exerted by circulating blood against blood vessel walls generates blood pressure.

Circulatory system function is controlled from within by the volume of blood in the arterioles, the ability of the heart to monitor its own activity, and capillary control of fluid balance. The *parasympathetic* and *sympathetic* branches of the *autonomic nervous system*, under the influence of the brain's vasomotor center, speed or slow heart action and dilate or constrict blood vessels in response to environmental cues. The kidneys influence cardiovascular function by controlling urine volume.

The *lymphatic system* is a network of vessels that drain fluid from the body's tissues, purify it, and return it to the blood. It also contains lymph nodes which filter lymph, and such structures as the *spleen* and *thymus*.

QUESTIONS

1. Why do large, active animals need circulatory systems?

2. How do open and closed circulatory systems differ?

3. Maintaining the proper portions of formed elements in the blood is essential for health. What can happen when there are too few or too many red blood cells, white blood cells, or platelets?

4. How is a red blood cell adapted for its function?

5. Why is it difficult to fashion a synthetic blood?

6. The usual time for prothrombin to be converted to thrombin and thrombin to cause the formation of fibrin is 13 seconds. When one's diet lacks vitamin K, however, these chemical reactions take 30 seconds or longer. What function of the circulatory system is impaired? What symptoms might a dietary deficiency of vitamin K cause?

7. A young man suffers from the genetic disease hemophilia. He contracts acquired immune deficiency syndrome from injecting a contaminated sample of donated factor VIII, which he needs to control his hemophilia. What abnormalities are present in this man's blood?

8. What is meant by blood pressure?

9. What causes the heartbeat? Why is the heart sometimes referred to as "two hearts that beat in unison"?

10. A man feels "flutters" in his chest, so he has a checkup. The doctor says, "Your diastolic pressure is elevated. You seem to have angina of exertion, but the stress test does not show an arrhythmia or a blocked coronary artery. Your leukocyte count and clotting time are normal, but your red count is a bit low." What is—and is not—wrong with the man's cardiovascular system?

11. Where does lymph originate? What causes it to move through the lymphatic vessels?

TO THINK ABOUT

1. A bacterial infection damages the inside wall of a blood vessel. What events are likely to follow?

2. In "white coat hypertension," a person's blood pressure skyrockets when entering a doctor's office because he or she is uneasy in a medical setting. How can a person with this problem obtain a more accurate blood pressure reading?

3. Why might a person consider undergoing coronary bypass surgery?

4. Dr. George Sheehan, a noted running expert, once showed heart specialists electrocardiograms of the New York Giants football team. The doctors found that many of the EKGs were abnormal, showing arrhythmias. But when Dr. Sheehan told his audience that the patients were professional athletes, the doctors changed all of their diagnoses to normal. Why do you think they had this change of heart?

5. Athletes tend to be slim and strong, to have low blood pressure, to abstain from smoking, and to alleviate stress through exercise. How might these characteristics complicate a study to assess the effects of exercise on the cardiovascular system?

SUGGESTED READINGS

Epstein, Franklin H. July 5, 1990. Regulatory functions of the vascular endothelium. *The New England Journal of Medicine*. vol. 323, no. 1. Endothelium is much more than just a lining.

Houston, Tim P., et al. July 4, 1985. The athletic heart syndrome. *The New England Journal of Medicine*, vol. 313, no. 1. The heart of an athlete differs in several ways from the heart of a sedentary person.

Lewis, Ricki. May 1990. Cardiovascular spare parts. *FDA Consumer*. Artificial hearts, blood, and blood vessels can be teamed with preventive measures to build a better cardiovascular system.

Noonan, David. February 1990. Dr. Doolittle's question. *Discover*. A complex cascade of chemical reactions lies behind a blood clot.

Pool, Robert. January 4, 1991. Making 3-D movies of the heart. *Science*. Magnetic resonance imaging reveals weakened parts of the heart.

Stone, John. May 21, 1989. Tempo of the heart. *New York Times Magazine*. Artificial pacemakers have come a long way since 1962.

Thompson, Richard C. April 1988. Who donates blood better for you than you? *FDA Consumer*. Being your own blood donor can protect against infection.

Zamula, Evelyn. February 1990. Getting a leg up on varicose veins. *FDA Consumer*. Varicose veins can have no symptoms or be quite bothersome.

C H A P T E R

23

The Respiratory System

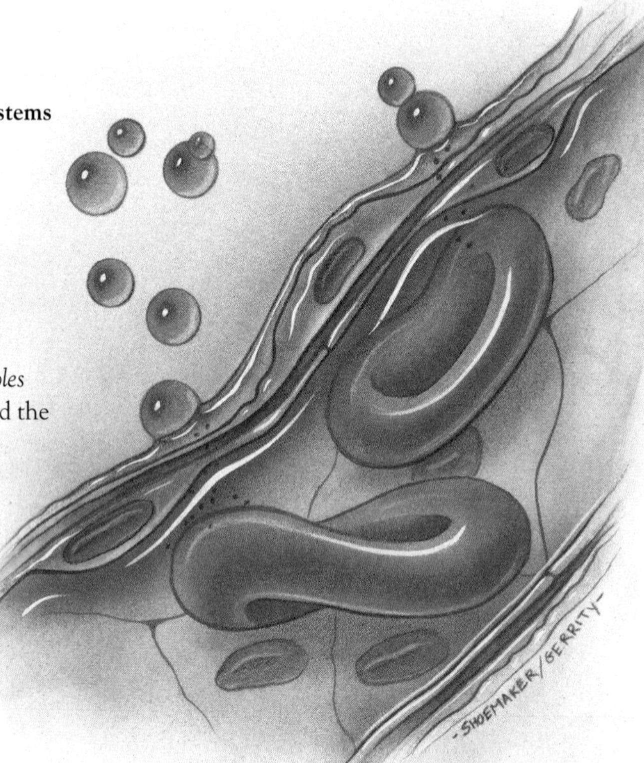

Chapter Outline

Learning Objectives

By the chapter's end, you should be able to answer these questions:

1. Why do we require oxygen?
2. What kinds of structures have different species evolved to respire?
3. What structures form the human respiratory system, and what does each do?
4. How do we breathe?
5. How are gases exchanged between the respiratory system, the circulatory system, and the body's tissues?
6. How is breathing rate controlled?
7. What are some disorders of the respiratory system?

The first breath of life is the toughest. A baby often emerges from the birth canal with a bluish color that may be frightening to the new parents who do not realize this is normal. Soon, as the baby musters up 15 to 20 times the strength she will need for subsequent breaths, the millions of tiny sacs in her lungs, each only partially inflated, balloon out with air for the very first time. As the oxygen rapidly diffuses into her bloodstream and is delivered to her tissues, the infant miraculously turns a robust pink and lets out a yowl. The insides of her lungs are a healthy pink (fig. 23.1).

The lungs of the man who has spent his 50 years breathing polluted air, and many years smoking cigarettes, are quite different. The passageways leading down to his lungs are dotted with bare patches where dense cilia once waved, moving particles up and out of his respiratory tract. Deep within his lungs, the pattern of bare patches continues, with sections of air sacs deflated or gone altogether. Unlike the pure pink of the newborn's pristine lung linings, the smoker's are a sooty black. The tissues that capture life-giving oxygen for distribution to the man's tissues have been ravaged beyond repair. He is well aware of the damage with each hacking cough.

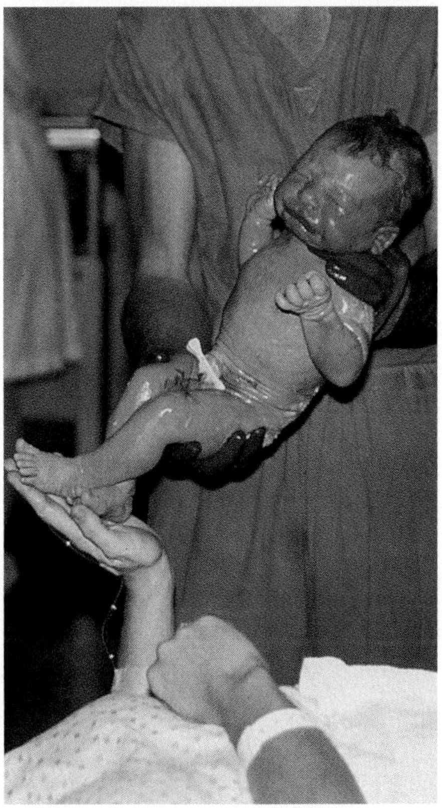

Figure 23.1
The first breath of life is the hardest.

Why We Need Oxygen

Our intense need for oxygen becomes quickly apparent when we intentionally hold our breaths. Less obvious is the requirement of oxygen by individual cells. Oxygen is used by cells to extract the maximal amount of energy from nutrient molecules and to store that energy in the bonds of ATP. The energy in ATP can then be tapped for a variety of biological processes—maintaining the potential difference across a nerve cell membrane, muscle contraction, active transport across membranes, and hormone synthesis. For many cells a steady oxygen supply is absolutely crucial. A nerve cell deprived of oxygen for just a few minutes dies. This is why even brief choking or strangulation can result in brain damage.

In the presence of oxygen, cells produce ATP much more efficiently. When cells lack oxygen, they can produce only 2 to 3 ATP molecules from a single glucose molecule. In contrast, the total metabolism of the same molecule of glucose in the presence of oxygen can yield at least 36 ATP molecules. Recall from chapter 6 that energy is slowly liberated from food molecules by stripping off electrons and channeling them through a series of electron carriers, each at a lower energy level than the previous one. The energy released at each step is used to form ATP. Oxygen is the final acceptor of the low-energy electrons at the end of the chain of acceptors. If oxygen is not present, the electrons cannot pass along this series of acceptors, so the breakdown of organic molecules stops at a much earlier step and a great deal of the energy remains locked in the food molecule.

Carbon dioxide is a by-product of the Krebs cycle. In the process of removing electrons, carbons are cleaved from food molecules. Oxygen is added after electron transport, to release carbon dioxide, which is ultimately exhaled.

Diversity Among Respiratory Systems

In all organisms, adequate amounts of oxygen and carbon dioxide must be exchanged by diffusion between the cells and the environment across a moist membrane, and exchange of gases is called respiration. This is a physical process, compared to the chemical respiration in the mitochondrion. Respiratory systems must fulfill several requirements.

The respiratory membrane must provide enough surface area for gas exchange to meet the metabolic needs of the organism. Larger and more active animals use more oxygen and produce more carbon dioxide than smaller animals, and consequently they need a greater respiratory surface. Because diffusion can only effectively distribute substances over a short distance, about 0.5 mm, large active animals also require a circulatory system to transport the gases between the respiratory membrane and the cells.

Animal species that live in an aquatic environment have evolved different respiratory structures than species that dwell on the land, because these environments differ in oxygen availability. Oxygen is not very soluble in water. Water saturated with oxygen contains only about 1/30 of the oxygen present in an equal volume of air, and the warmer or saltier the water, the less oxygen is present. Furthermore, oxygen takes about 300,000 times longer to diffuse through water than through air, so it takes longer to replenish oxygen used for respiration. This means that to obtain the same quantity of oxygen, the amount of water that an aquatic organism must move over its respiratory surface is greater than the volume of air that a land dweller must move across its respiratory surface. In addition, because water is more dense and more viscous than air, an animal in an aquatic environment must use more energy to keep oxygen-rich water flowing over its respiratory system than a land-dwelling animal does moving air over its respiratory surface. As a result, an aquatic organism requires a larger respiratory surface than a terrestrial animal of similar size and energy requirements. However, land dwellers have

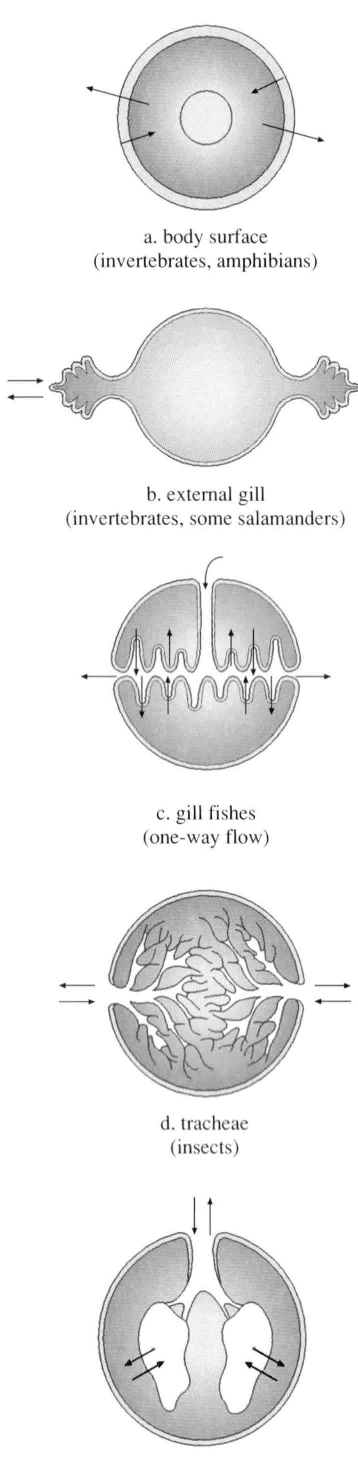

Figure 23.2
Diversity in respiratory surfaces. *a.* The sea anemone has the simplest type of respiratory surface—diffusion across a thin layer. *b.* A mud puppy uses an external gill, and (*c*) a fish uses an internalized gill. *d.* Land-dwelling arthropods exchange gases using an extensive network of tracheae. *e.* Terrestrial vertebrates use lungs.

their own problem: keeping the respiratory membrane from drying out. Therefore, the respiratory surface of terrestrial animals is inside the body, where it can be kept moist.

Body Surface

In the simplest mechanism for gas exchange, all cells are in close enough contact with the environment that gases can be exchanged across the cell membrane and distributed by diffusion. A single-celled organism, for instance, can use this method because its cell membrane provides enough surface area to transport sufficient oxygen into the cell.

Among multicellular organisms different strategies for gas exchange have evolved (fig. 23.2). One way to keep cells in contact with the oxygen-rich environment is a flat body form. For example, cnidarians, such as *Hydra,* have only two cell layers, so all their cells are in contact with the environment. Flatworms (*phylum Platyhelminthes*) are somewhat thicker, but they are thin enough for both gas exchange and the distribution of gases to occur by diffusion.

In larger animals, gases cannot be effectively distributed by diffusion. A circulatory system transports gases between body cells and the respiratory membrane. Some organisms, such as earthworms (*phylum Annelida*) and amphibians, still use the body surface for gas exchange but employ a circulatory system to transport the gases between the cells and the environment. The relationship between the respiratory surface and the circulatory system is intimate and extensive; just beneath the skin surface is a myriad of tiny blood vessels. Gases diffuse across the skin into or out of the circulatory system.

Tracheal Systems

Terrestrial arthropods including insects, centipedes, millipedes, and some spiders have cells that are close to the environment, but the respiratory surface is protected and water loss is prevented by bringing the air into the body in close contact with every cell. These organisms have tough, waterproof exoskeletons that are indented to form an extensively branching system of tubules, called **tracheae,** that bring the outside environment in close enough contact with every cell for gases to be exchanged by diffusion. The openings to the tracheae are guarded by valves that close them off in very dry environments, keeping the cells from drying out. In small arthropods the tracheae are short enough for diffusion to deliver oxygen to the cells effectively. Large active arthropods, however, must move air in and out of the tracheal system to refresh the oxygen supply. They use their abdominal muscles to do this. A foraging honeybee, for example, pumps segments of her abdomen back and forth.

Gills

Large, active organisms require a surface for gas exchange that is greater than the skin surface. In aquatic organisms, the vast respiratory surface usually takes the form of gills, which are extensions of the body wall that are so highly folded and branched that they may superficially resemble feathers. Within the gill is a dense network of capillaries. Oxygen diffuses from the water across the delicate gill membrane and the thin capillary wall into the blood, which delivers it to the cells. Carbon dioxide diffuses in the opposite direction.

The complex gills of fishes create the extensive surface area necessary to extract maximal oxygen from water and deliver it to cells (fig. 23.3). Gases are exchanged across a very thin respiratory membrane and the single layer of cells of the capillary and transported by the circulatory system. These delicate feathery gills would be easily damaged if they were not protected by a flap called the **operculum.** Maximal gas exchange requires a continuous supply of fresh water flowing from the mouth over the gills. This is more energy-efficient than moving water into a gill chamber only to reverse the direction of flow and push it out again.

The flow of blood through the capillaries is in the opposite direction of the flow of water over the gill membrane. This arrangement, which is called **countercurrent flow,** maximizes the amount of oxygen that can be extracted. Recall that substances diffuse from an area of higher concentration to an area of lower concentration. The aim then is to keep the level of oxygen in the water passing over the gills higher than that of the blood. Because of the countercurrent flow, the blood leaving the gills, which has a relatively high oxygen concentration, encounters water entering the gills, which has the highest available oxygen concentration. Since there is still a gradient, oxygen diffuses into the blood and the oxygen load of the blood is topped off as it leaves the gills. The oxygen level in the water drops as it continues to flow over the gills. However, the blood vessels it encounters have less oxygen, so there is always a gradient for diffusion.

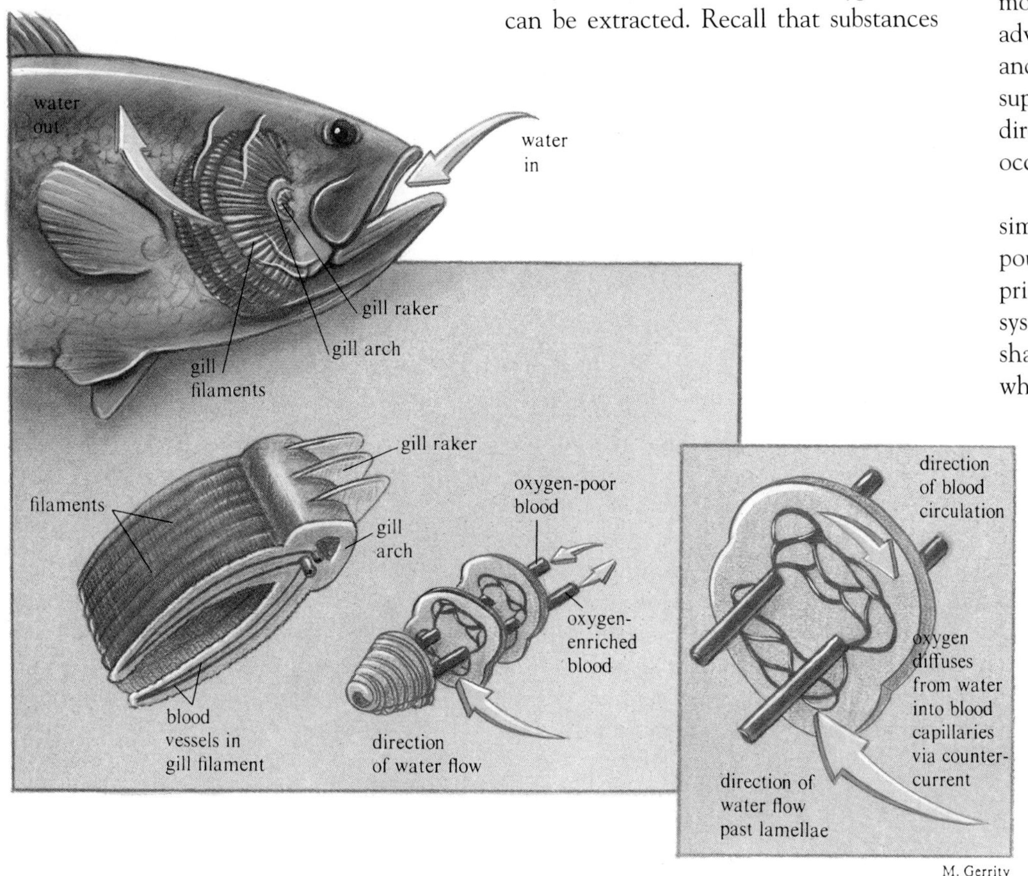

Figure 23.3
Gills. When the protective operculum is removed, the feathery gills of this fish are visible. Each side of the head has four gill arches, and each arch consists of many filaments. A filament houses capillaries within lamellae. Note that the direction of water flow opposes that of blood flow. This countercurrent flow allows the maximal amount of oxygen to be extracted from the water.

Lungs

A lung is a respiratory surface within the body that is usually in close association with the circulatory system. An advantage of lungs, particularly for terrestrial animals, is that the respiratory surface can be kept moist, a necessity for gas exchange. A disadvantage is that air must be moved into and out of the lung to renew the oxygen supply, rather than simply flowing in one direction over the respiratory membrane as occurs in the gills of fishes.

The complexity of lungs varies. The simplest lungs are little more than air-filled pouches used by some present-day but primitive fishes to supplement their gill systems. The bichir, for example, lives in shallow, oxygen-poor waters in Africa, where it occasionally swims to the surface to gulp air. Oxygen diffuses into the lush network of capillaries that invest its pouchlike lung.

The lungs of amphibians and reptiles have some infoldings, which increase surface area for gas exchange. This maximization of internal respiratory surface area is spectacular in the mammals, whose lungs are subdivided into sections and contain millions of microscopic, capillary-surrounded air sacs. If the human lung surface was completely spread out, it would take up 50 times the surface area of the skin. An elephant seal can obtain oxygen to support frequent diving because of an abundant blood supply (fig. 23.4).

Figure 23.4
Elephant seals store extra oxygen in their blood. An elephant seal can spend 85% of its time diving, with only short recovery periods. It can do this because of its tremendous blood supply (two and one-half times that of humans, for its weight), which is packed with oxygen-carrying red blood cells. Beneath 130 feet the lungs actually collapse.

KEY CONCEPTS

Oxygen is needed by our cells to extract maximal energy from nutrient molecules. Carbon dioxide is produced as a by-product. Respiration is the exchange of gases between cells and the environment across a moist membrane. Respiratory surfaces must be extensive enough to meet an animal's needs and have a way to transfer oxygen to the cells. Aquatic and terrestrial organisms have evolved different respiratory structures because oxygen is less plentiful in water than air. Single-celled organisms respire by diffusion across their cell membranes. In simple multicellular organisms, a flat body form keeps all cells in direct contact with the environment. More complex organisms use circulatory systems to transport oxygen. Terrestrial arthropods have tracheal systems, fishes have gills, and terrestrial vertebrates have lungs.

The Human Respiratory System

The Nose

Air entering the respiratory system must be processed before oxygen is extracted from it in the lungs. About 150,000 airborne bacteria enter the nose each day, along with many dust particles. Furthermore, winter temperatures in many parts of the world would freeze lung tissues and exposure to dry air would kill lung cells.

Cells lining the nose purify and warm incoming air. These cells are plentiful, because the nose is subdivided into compartments. Each of the two nasal cavities is partitioned into channels by three shelflike bones, the **nasal conchae.** The entire surface is covered with a mucous membrane containing many blood vessels (fig. 23.5).

Within the nose, air is cleaned by hairs, mucus, and cilia. The incoming air is first filtered by hairs at the entrance of the nasal cavities. Most bacteria and particles that slip past the initial sieve are caught in sticky mucus produced by the mucous membrane, and if inhaled further, they are trapped by mucus lower in the respiratory tract and are swept back out by waving cilia. A large particle that is inhaled is likely to trigger a sensory cell in the nose. This cell signals the brain to orchestrate a sneeze, which forcefully ejects the particle. Particles ejected from the nose in a sneeze can travel up to 100 miles per hour. That's 169 kilometers per hour if you happen to sneeze in Canada.

The nose protects the lung tissue by adjusting the temperature and humidity of incoming air. Air is warmed by the blood within the many vessels permeating the mucous membrane and is moisturized by the mucus. These vital functions of the nose are disrupted by cigarette smoking.

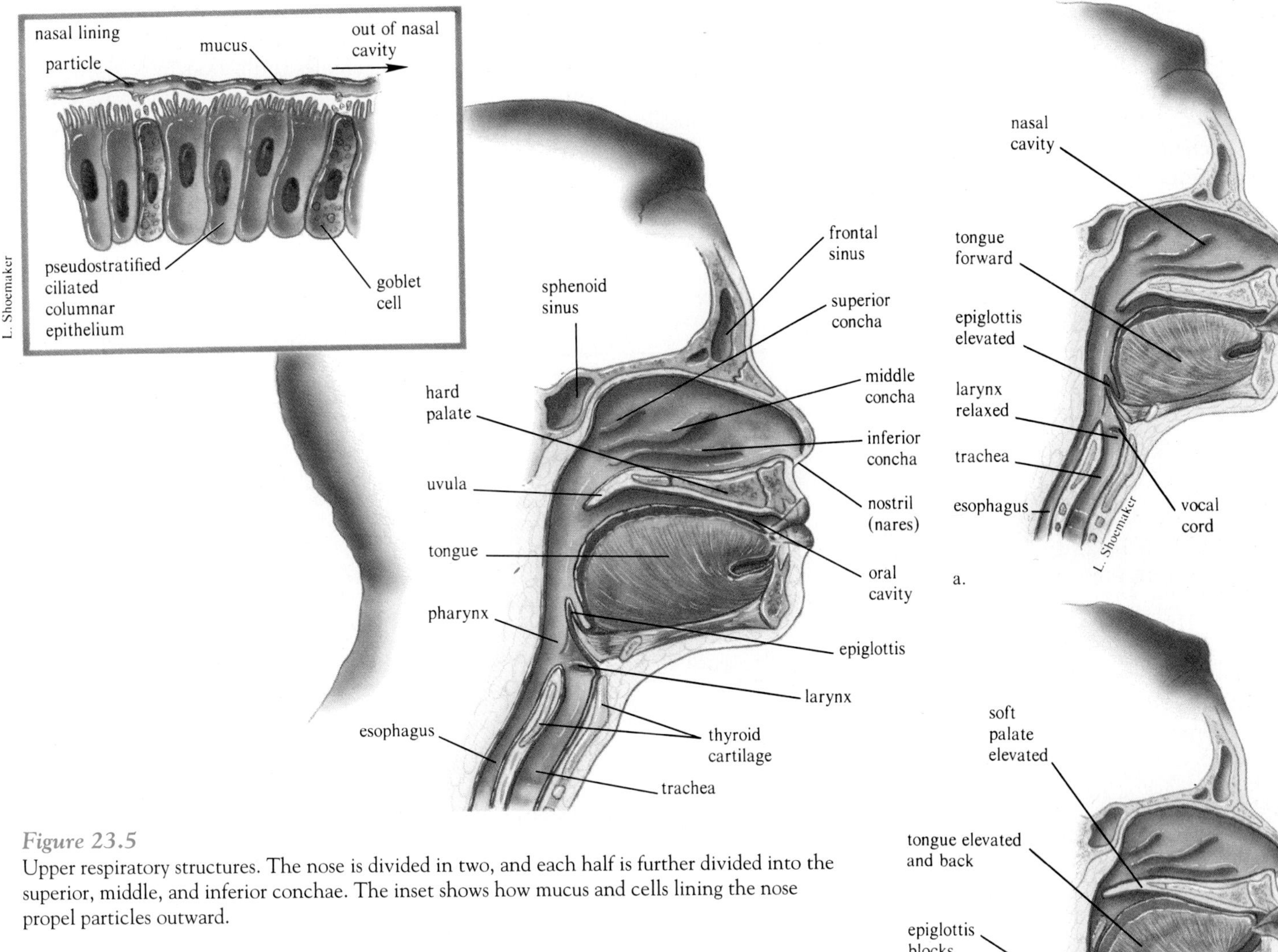

Figure 23.5
Upper respiratory structures. The nose is divided in two, and each half is further divided into the superior, middle, and inferior conchae. The inset shows how mucus and cells lining the nose propel particles outward.

Figure 23.6
Separating air from food. *a.* When we breathe, air passes through the glottis, down the trachea. *b.* When we swallow, the epiglottis blocks entry to the trachea, routing food to the digestive system.

The Pharynx and the Larynx

The back of the nose leads into the **pharynx,** or throat, a 4.5-inch (12-centimeter) tube that conducts food and air. The **larynx,** or Adam's apple, is a boxlike structure in front of the pharynx that directs passing materials. It is also responsible for the voice. A reflex action directs air to the lungs but steers food and fluid toward the digestive system. Inhaled air passes through the opening to the larynx, called the **glottis.** However, swallowing moves the larynx upward, and a piece of cartilage called the **epiglottis** flips down and covers the glottis like a trapdoor (fig. 23.6). Food that enters the respiratory tract can be deadly (Reading 23.1).

Stretched over the glottis are two elastic bands of tissue, the **vocal cords,** that vibrate when air rushes past them, producing sounds that can be molded into speech. The sound resonates in the pharynx, the nasal cavities, and the mouth and is molded by variations in the position of the tongue and lips. The deepening voice of a male undergoing puberty is caused by expansion of the Adam's apple, creating a larger space.

The Trachea, Bronchi, and Bronchioles

The larynx sits atop the **trachea,** a 5-inch (11-centimeter) tube about 1 inch (roughly 2.5 centimeters) in diameter that is also known as the windpipe. Horseshoe-shaped rings of cartilage hold the trachea open in spite of the negative pressure created by exhalation (fig. 23.7). You can feel these rings in the lower portion of your neck. The inside surface of the trachea is ciliated and mucus-secreting. It contributes to the purifying, warming, and moistening of incoming air.

As the trachea approaches the lungs, it branches into two **bronchi** (singular, bronchus), which continue the tracheal organization of C-shaped cartilage rings and an inner ciliated layer and outer layer of smooth muscle. The bronchi branch repeatedly, each branch decreasing in diameter until it consists of only a thin layer of smooth muscle and elastic tissue. At this microscopic level, the tubes are called **bronchioles** (little bronchi).

The respiratory passageways within the lung are called the "bronchial tree" (fig. 23.8), because their branching pattern resembles that of an upside-down tree. The 10 million cilia lining the bronchial tree,

Reading 23.1 *Food Inhalation and the Heimlich Maneuver*

ALTHOUGH THE REFLEX MOVEMENTS OF THE LARYNX NORMALLY PREVENT FOOD FROM BLOCKING THE AIR PASSAGEWAYS IN THE RESPIRATORY SYSTEM, ACCIDENTS OCCUR MORE FREQUENTLY THAN MANY OF US REALIZE. Approximately 8,000 to 10,000 people choke to death in the United States each year. This is more likely to happen when the individual talks with food in the mouth, hiccups, or is intoxicated. A person who is choking may look like he or she is having a heart attack because of agitation. A choking person is unable to speak.

What should you do if you witness a choking incident? First, try to calm the person and determine whether the air passages are completely blocked. Most choking incidents do not involve complete blockage of the air passageways. Therefore, if the person remains calm, it may be possible to move enough air in and out to keep the individual alive until he or she can be brought to a hospital. If breathing is possible, do not act. However, if the person is turning blue, it means that the air passages are completely blocked and immediate action is necessary.

The best way to assist an upright, severely choking person is to use the **Heimlich maneuver** (fig. 1), pushing inward and upward under the rib cage from the center of the abdomen with clenched fists. This creates a sudden burst of pressure as air is pushed out of the lungs, hopefully with enough force to dislodge the blockage in the air passage. In a person who is lying on the ground, push inward and upward in the anterior part of the abdomen. If you are alone and you begin to choke, you can "Heimlich" yourself by throwing the upper abdominal region against the rounded corner of a table or other stationary object.

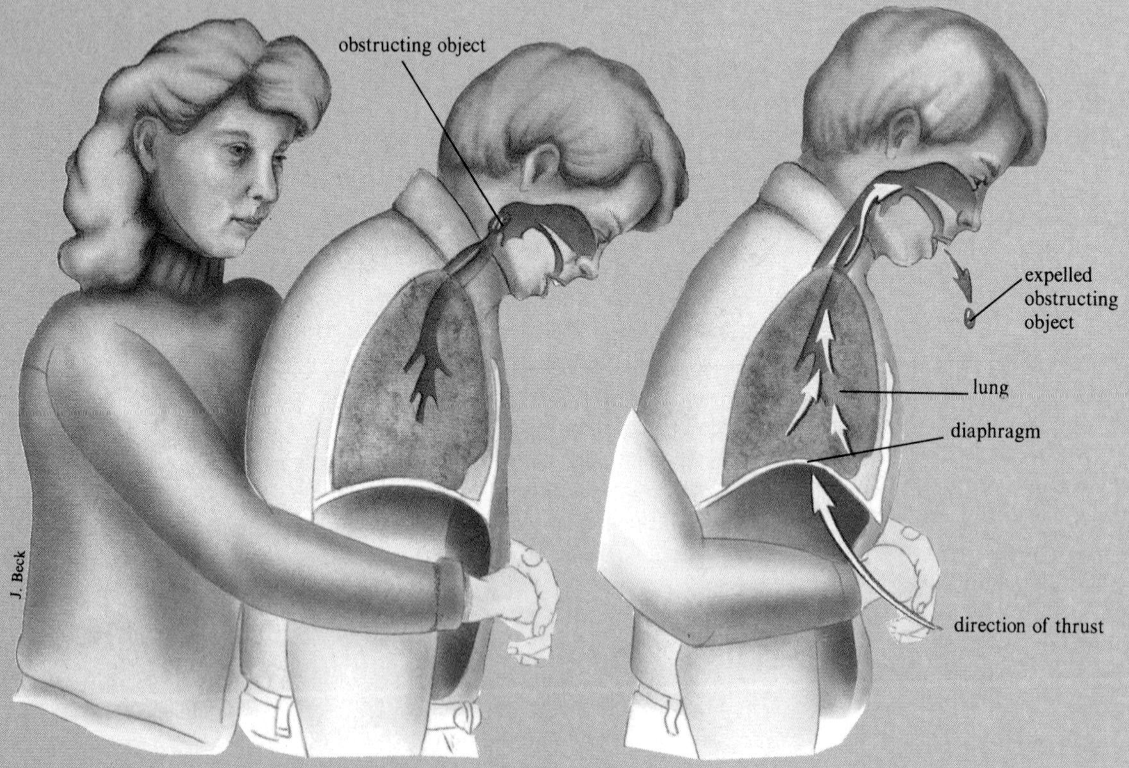

Figure 1
The Heimlich maneuver is used to expel food stuck in the trachea.

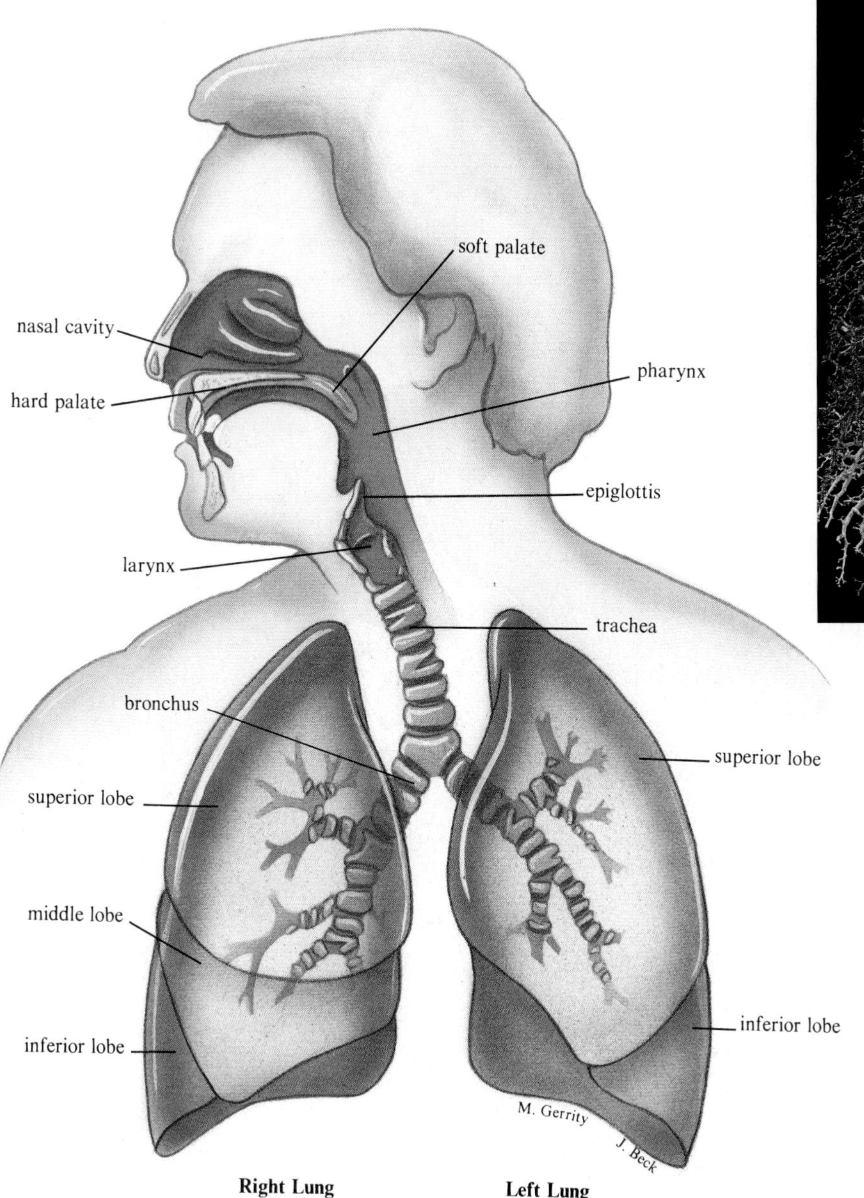

Figure 23.7
The trachea leads from the upper respiratory structures to the lungs.

Figure 23.8
The bronchial tree.

beating hundreds of times a minute, work constantly to remove inhaled particles.

The bronchioles have no cartilage, but their walls do contain smooth muscle. The contraction of these muscles is controlled by the autonomic nervous system so that air flow is adjusted to suit metabolic needs. During a stressful or emergency situation, certain muscle cells may require more oxygen to produce the ATP that fuels the response to the emergency. Under these conditions, the sympathetic nervous system causes the bronchioles to constrict, reducing their diameter and then returning them to normal.

The contraction of the muscle in bronchial walls is usually well adjusted to the body's needs. Sometimes, however, as in **asthma,** the bronchial muscles go into spasm, making air flow exceedingly difficult (fig. 23.9). Asthma is a chronic condition characterized by recurring attacks of wheezing and difficulty breathing. Most asthma attacks are triggered by an allergy to pollen, dog or cat dander (skin particles), or tiny mites in house dust (see fig. 27.16). However, an attack can also be caused by infections, certain drugs, inhaling irritating substances, vigorous exercise, and psychological stress. Some attacks start for no apparent reason. The inhalants often prescribed to treat asthma attacks usually work by relaxing the bronchial muscles. Some aerosol inhalers spray epinephrine onto the bronchial walls. Since epinephrine is the neurotransmitter released by the sympathetic nervous system, these inhalants mimic the effect of the sympathetic nervous system and dilate the bronchioles.

The Alveoli

Each bronchiole narrows into several **alveolar ducts,** each of which opens into a grapelike cluster of **alveoli** (singular alveolus) (fig. 23.10). Each alveolus is a tiny sac with a wall that is only one cell thick. Most of the lung is composed of alveoli, some 300 million of them, which makes the lung structure similar to that of foam rubber. If the total surface area created by alveoli were measured, it would cover about half a tennis court. Surrounding each cluster of alveoli is a vast network of capillaries. Gas exchange occurs through the thin walls of the alveoli and those of the neighboring

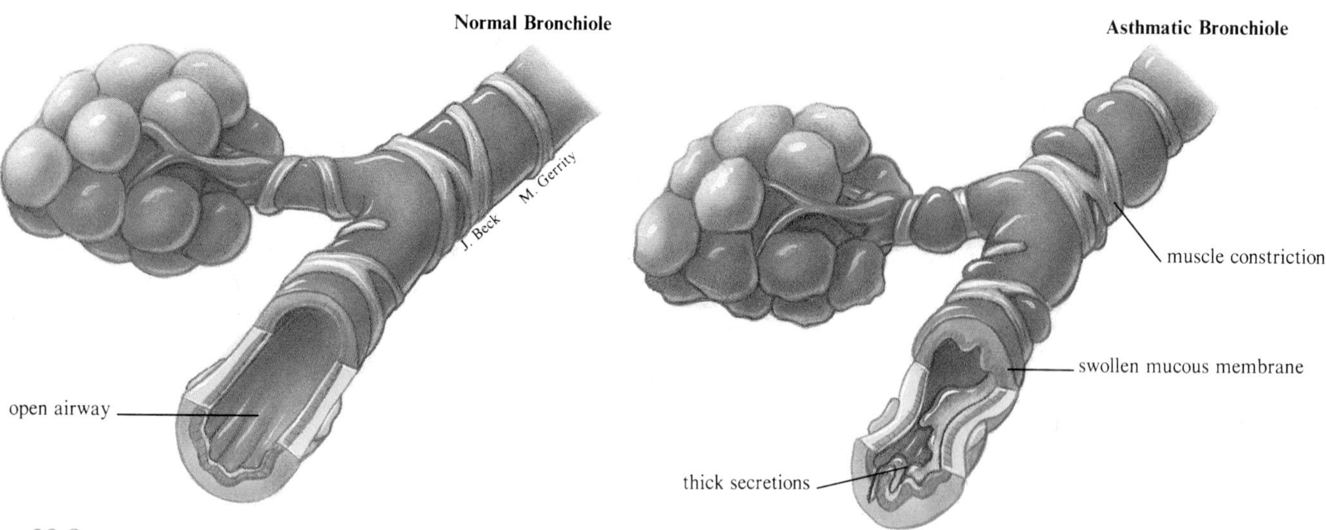

Figure 23.9

During an asthma attack, temporarily obstructed bronchioles make breathing difficult. Dr. Stanley Reichman describes an attack in his book *Breathe Easy*: "Shoulders hunched, head thrust forward, chest bellowed out, desperation fills you as you struggle to push air out of your lungs and to acquire new air. Often, within moments, sweat pours from your body, trickles over your scalp. Your back becomes hard as wood. Your chest feels totally constricted. If you are wearing anything close-fitting around your neck . . . you want to tear it off. You wheeze, gasp, even moan."

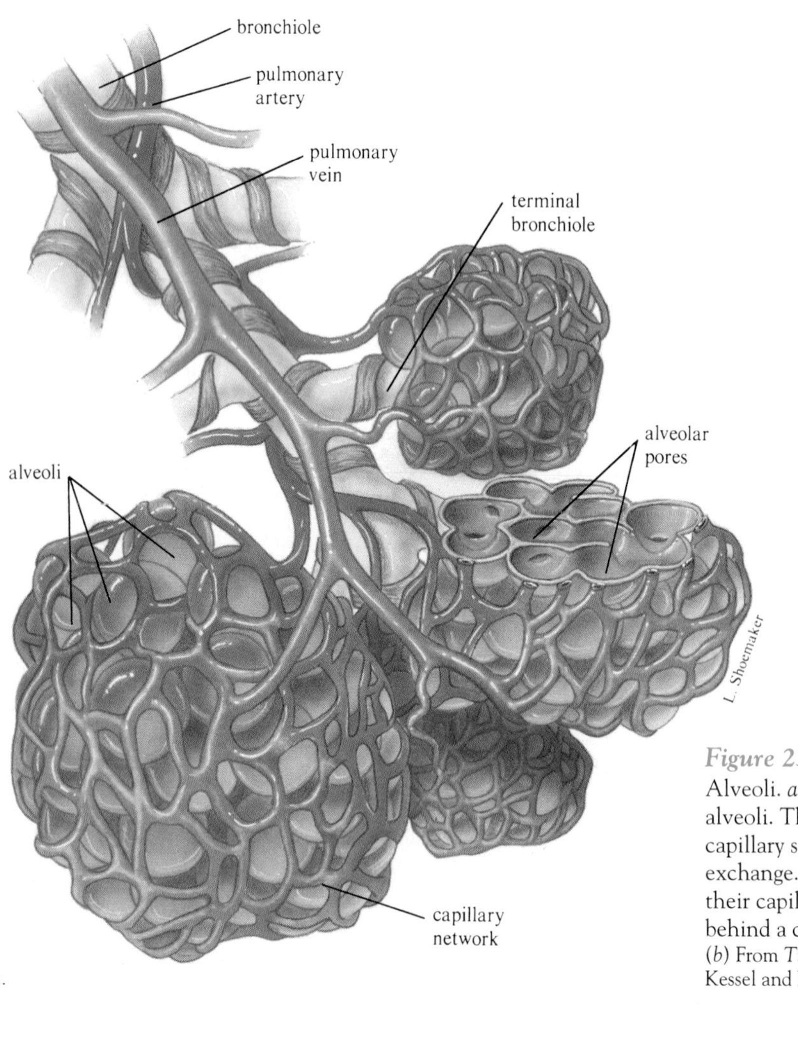

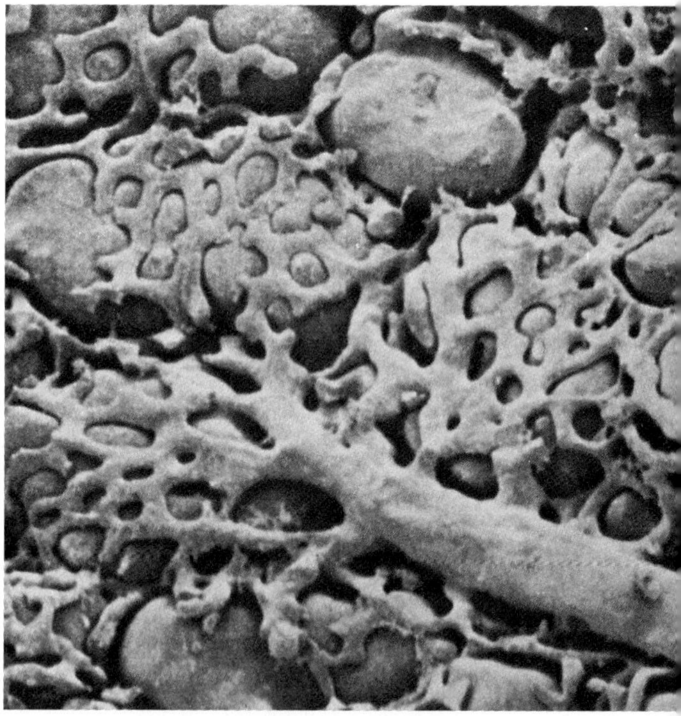

Figure 23.10

Alveoli. *a.* Bronchioles narrow to form alveolar ducts, which lead to alveoli. The alveoli resemble grape clusters. Each is surrounded by a lush capillary supply. Alveoli greatly increase the available surface area for gas exchange. *b.* This electron micrograph was made by filling alveoli and their capillaries with resin. When the tissue decomposed, the resin left behind a cast of the respiratory structures.

(*b*) From *Tissues and Organs: A Text-Atlas of Scanning Electron Microscopy* by R. G. Kessel and R. H. Kardon, ©1979 W. H. Freeman.

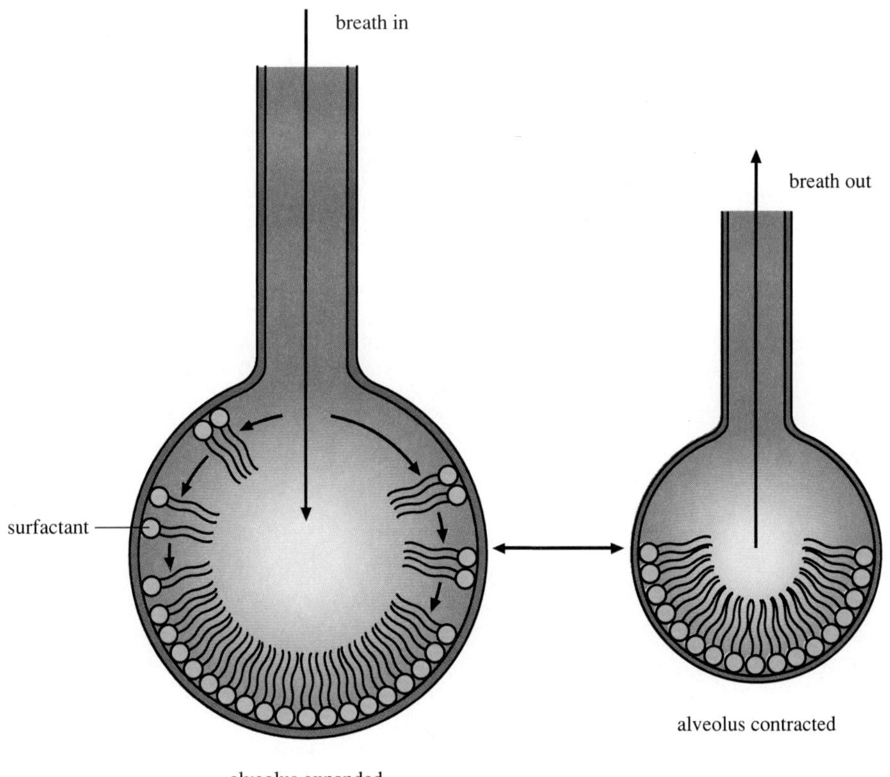

Figure 23.11
Lung surfactant inflates the alveoli. Human lung surfactant is a mixture of phospholipid molecules. They insert into the interface between liquid and air on the inside of alveoli. This reduces the surface tension of the liquid lining, resulting in an expansion of the alveoli. When a breath is taken, more surfactant is forced into the alveoli, inflating them.

capillaries by diffusion. Here, oxygen enters the blood and carbon dioxide leaves the blood.

The alveoli must be inflated for their extensive surface area to be efficiently utilized for gas exchange. This is accomplished by a mixture of phospholipid molecules called **human lung surfactant.** This substance keeps the alveoli open by occupying the space between the watery film that lines the inner surface of the alveoli and the air within the alveoli. Without surfactant, the watery lining exerts surface tension, a property of liquid surfaces that is a consequence of cohesive forces between molecules (see chapter 3). This surface tension generates a contractile force that would collapse the alveoli. The surfactant counters this force, so the alveoli remain open (fig. 23.11).

Human lung surfactant is especially important at the moment of birth, when a baby's lungs must inflate so that the first of a lifetime of breaths can be taken. In order to fill a collapsed alveolus, the infant must overcome the force of surface tension, which is 15 to 20 times more difficult than ventilating expanded alveoli. However, once the alveoli have been expanded by the first breath, surfactant, which has been collecting for the preceding few weeks, keeps them open and subsequent breaths are much easier.

In many babies born prematurely, however, not enough surfactant has yet been produced, and alveoli collapse after each breath. This lack of surfactant in the newborn is called infantile respiratory distress syndrome. Such a newborn must fight as hard for every breath as a normal newborn does for the first one. Chapter 10 describes some new ways to supply the vital surfactant.

The Lungs

The bronchial tree and the alveoli are housed within the paired **lungs.** Each lung weighs about 1 pound (454 grams) and is spongy in consistency due to the numerous alveoli. The lung tissues of a newborn are characteristically pink, but they darken with age as the person is exposed to environmental pollutants, particularly cigarette smoke (Reading 23.2). The left lung is divided into two regions, called "lobes," and the right is divided into three. Between the lungs is the heart.

KEY CONCEPTS

The hairs, mucus, and cilia in the nose filter bacteria and particles from incoming air and warm and moisturize it. The pharynx is the throat, and the larynx lies in front of it. Movement of the larynx and of the epiglottis routes air to the trachea and food to the digestive tract. The trachea branches into bronchi and then bronchioles. These tubes are muscular and lined with cilia that continually sweep particles out. The bronchioles lead to alveolar ducts, which end at alveoli. The alveoli are inflated when surfactant counters surface tension. These tiny sacs are surrounded by capillaries, through which gas exchange occurs. The lungs are spongy organs that house the bronchial tree.

Mechanism of Breathing

Air moves between the atmosphere and the lungs in response to pressure gradients. Air moves in when the pressure in the lungs is lower than the air pressure outside the body, and it moves out when the pressure in the lungs is greater than the pressure in the atmosphere.

The anatomy of the thoracic (chest) cavity explains how the pressure changes responsible for pulmonary ventilation are generated. The thoracic cavity has no direct connection to the outside of the body. Furthermore, it is separated from the abdominal cavity by a broad sheet of muscle, the **diaphragm.** Membranes coat all the surfaces in the thoracic cavity, which is essentially filled by the lungs. Changing the size of the thoracic cavity creates pressure changes that move air in and out of the lungs.

Air is drawn into the lungs when the size of the thoracic cavity increases due to the contraction of the muscles of the rib cage and the diaphragm. This process is called **inspiration,** or inhalation. When the

Reading 23.2 *The Effects of Cigarette Smoking on the Respiratory System*

DAMAGE TO THE RESPIRATORY SYSTEM FROM CIGARETTE SMOKING IS SLOW, PROGRESSIVE, AND DEADLY. A healthy respiratory system is continuously cleansed. The mucus produced by the respiratory tubules traps dirt and disease organisms and cilia sweep it all toward the mouth, where it can be eliminated. This housekeeping is severely impaired by cigarette smoke. With the very first drag, the beating of the cilia slows. With time, the cilia become paralyzed and, eventually, disappear altogether. The loss of cilia leads to the development of smoker's cough. The cilia are no longer effective in removing mucus, so the individual must cough it up. Coughing is usually worse in the morning because mucus has accumulated during sleep.

To make matters worse, excess mucus is produced and accumulates, clogging the air passageways. Disease organisms that are normally removed now have easier access to the respiratory surfaces and the resulting lung congestion favors their growth. This is why smokers are sick more often than nonsmokers. In addition, a lethal chain reaction begins. Smoker's cough leads to chronic bronchitis. Mucus production increases and the lining of the bronchioles thickens, making breathing difficult. The bronchioles lose elasticity and are no longer able to absorb the pressure changes accompanying coughing. As a result, a cough can increase the air pressure within the alveoli enough to rupture the delicate alveolar walls; this condition is the hallmark of smoking-induced emphysema. Emphysema is 15 times more common among individuals who smoke a pack of cigarettes a day than among nonsmokers.

It pays to quit. Much of the damage to the respiratory system can be repaired. Cilia are restored and the thickening of alveolar walls due to emphysema can be reversed. But ruptured alveoli are gone forever.

Simultaneous with the structural changes progressing to emphysema may be cellular changes leading to lung cancer. First, cells in the outer border of the bronchial lining begin to divide more rapidly than usual. Eventually, these displace the ciliated cells. Their nuclei begin to resemble those of cancerous cells—large and distorted with irregular numbers of chromosomes. Once again, quitting is rewarded. Up to this point, the damage can be repaired. However, if smoking continues, these cells may eventually break through the basement membrane and begin dividing within the lung tissue, forming a tumor with the potential of spreading throughout lung tissue (fig. 1). Eighty percent of lung cancer cases are due to cigarette smoking. Unfortunately, only 13% of lung cancer patients live as long as 5 years after the initial diagnosis. The world has lost many people in their primes due to lung cancer. In the field of entertainment, Steve McQueen, Humphrey Bogart, Yul Brynner, John Wayne, Robert Taylor, and Gary Cooper all died from the disease. See chapter 7, Reading 7.2, "Cancer Timetables," for more information on the relationship of smoking and lung cancer.

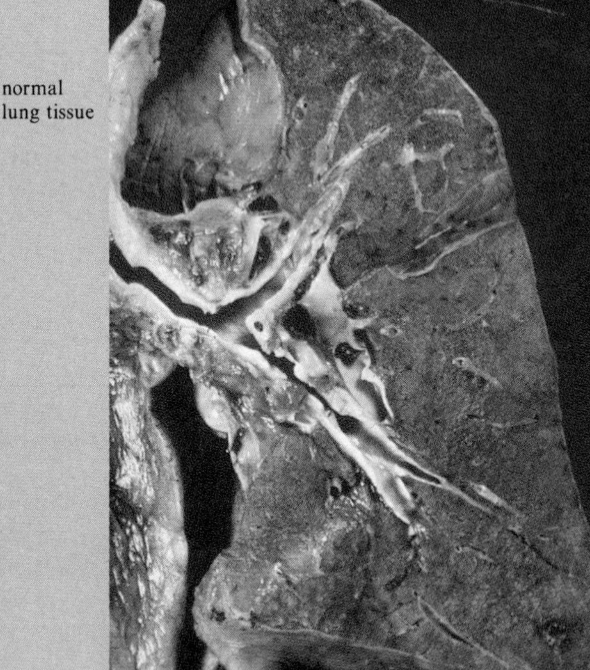

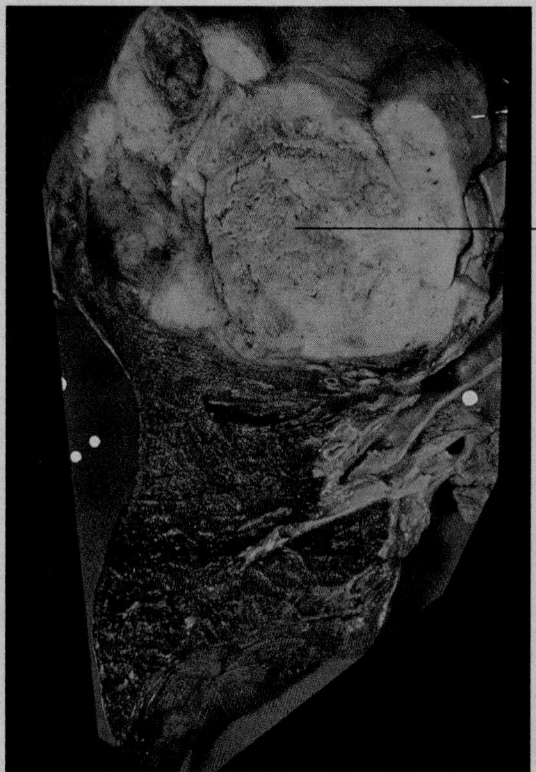

Figure 1
The lung on the left is healthy. Compare its appearance to that of the lung on the right, in which a cancerous tumor fills approximately the top half of the organ. What symptoms do you think this might produce?

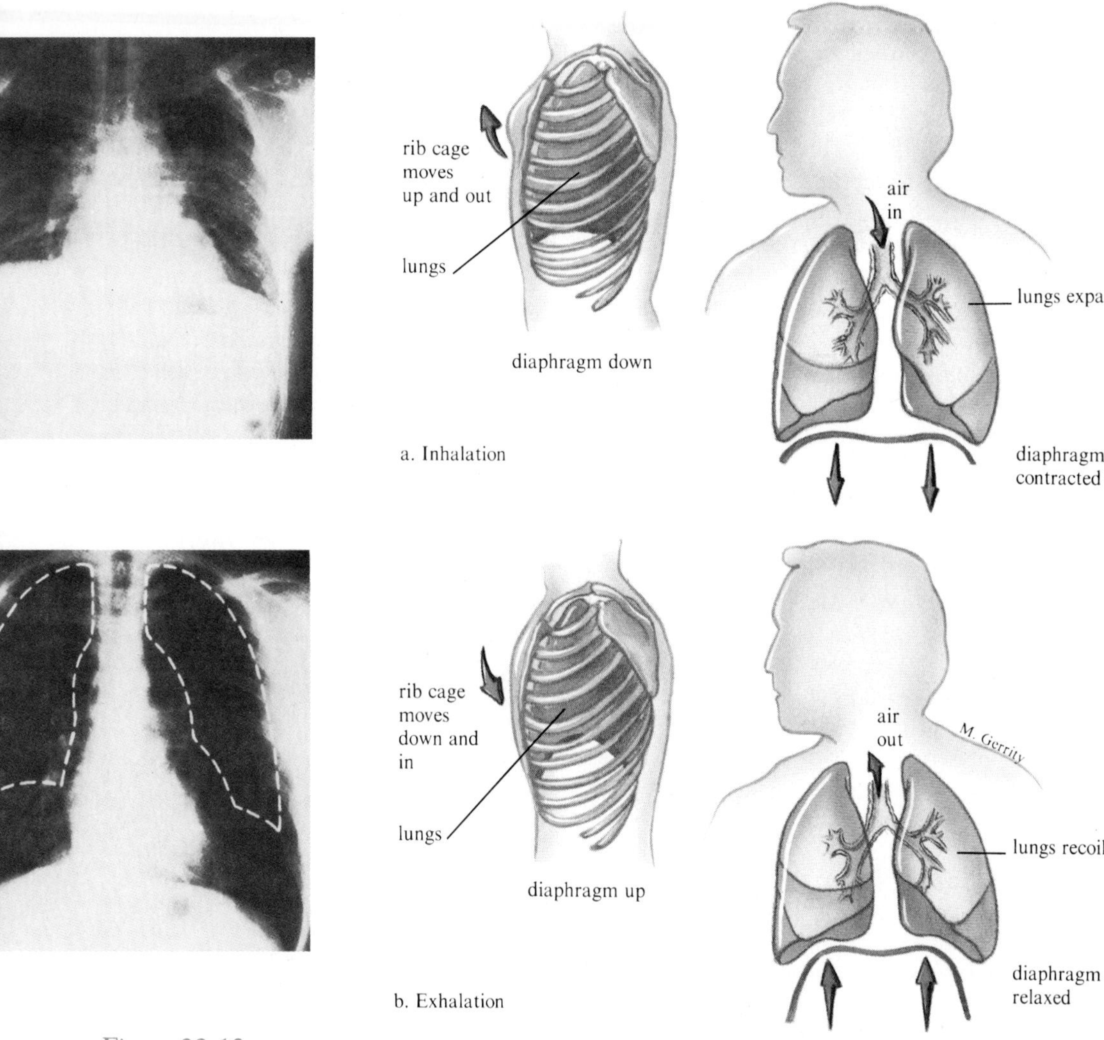

Figure 23.12
How we breathe. When we inhale, the diaphragm lowers and the rib cage rises, expanding the lungs. When we exhale, the diaphragm rises and the rib cage lowers. X-rays reveal the change in lung volume when we breathe in and out.

rib muscles contract, they pull the rib cage upward and forward. If you place your hands on your rib cage as you breathe, you can feel it move this way—up and out. The elevation of the rib cage increases the size of the thoracic cavity from the front to the back. Meanwhile, the contraction of the diaphragm elongates the thoracic cavity. This is because when the diaphragm is relaxed, it is shaped like an inverted bowl and its convex side projects into the thoracic cavity. Contraction of the diaphragm causes it to flatten, thereby lengthening the thoracic cavity from top to bottom. The increase in the size of the thoracic cavity causes a drop in the pressure between the lungs and the outer wall of the cavity (this area is called the pleural cavity), and air is drawn in (fig. 23.12).

When the muscles of the rib cage and the diaphragm relax, the thoracic cavity is decreased in size, causing **expiration,** or exhalation. Now, the rib cage falls back to its former lower position, and the diaphragm bulges into the thoracic cavity again. The elastic tissues of the lung recoil, deflating the lungs. As the volume of the lungs decreases, the pressure within them increases. When the pressure within the lungs exceeds atmospheric pressure, air is pushed out.

Not all of the respiratory tract actively participates in gas exchange. The part that is not used in gas exchange—the air in the pharynx, trachea, and the upper third of the lung—is called **dead space.** The air in the bottom third of the lungs is not exchanged with each breath; because most of this air remains in the lungs it is called **residual air.**

These limitations on air flow within the respiratory system have some consequences for the amount of air exchanged with each breath. While breathing quietly you move roughly 1 pint (about 0.5 liter) of air in and out with each breath. The amount of air inhaled or exhaled during a normal breath is called the **tidal volume.** If you were to take the deepest breath possible and exhale until you could not force any more air from your lungs, you could measure your **vital capacity,** the maximal amount of air that can be moved in and out of the lungs during forceful breathing.

KEY CONCEPTS

Air moves in and out of the respiratory system because of pressure differences between the atmosphere and in the thoracic cavity. Inspiration occurs when the diaphragm contracts and moves down and the rib cage moves up and out. Expiration occurs when the diaphragm relaxes and the rib cage moves down and in. Dead space air is not used in gas exchange, and residual air remains at the bottom of the lungs during breathing. Tidal volume is the amount of air breathed in and out, and vital capacity is the maximum amount one can breathe.

Transport of Gases

Oxygen

From the alveoli, oxygen is carried by the blood throughout the body, even to the most distant cells. It is quite an efficient process, for about 97% of the oxygen that reaches the cells is carried there bound to hemoglobin in the red blood cells. The remaining 3% of the oxygen delivered to the cells is dissolved in the plasma. Hemoglobin carries oxygen more readily if it is ferried in red blood cells than if it floats free in the plasma. The biconcave shape of the red blood cell increases surface area (fig. 22.5). A red blood cell spending only a second or two in the alveolar capillaries becomes almost completely saturated with oxygen.

Each hemoglobin molecule is composed of four subunits. Each of the subunits has a protein chain, the globin, and a heme group (fig. 14.9). The heme group includes an iron atom, which actually binds to the oxygen. All four of the protein chains have a heme group and, therefore, each hemoglobin molecule can carry four molecules of oxygen. Defects in the structure of the hemoglobin molecule can cause certain genetic diseases of the blood (chapter 14).

The compound formed when hemoglobin binds with oxygen is called oxyhemoglobin. The most important factor determining whether hemoglobin will bind to oxygen is the partial pressure of oxygen, which is the portion of the total air pressure that is due to oxygen. The partial pressure of a gas in a mixture is directly related to the concentration of that gas. Hemoglobin associates with oxygen when the partial pressure is high, as it is in the lungs, and releases it when the partial pressure is low, as occurs near the cells (fig. 23.13). About 97% of the hemoglobin leaving the lungs is saturated with oxygen. However, as cells use oxygen, they lower the partial pressure of oxygen in the blood there. Under these conditions, oxygen is released from the hemoglobin. In the vicinity of normally active cells, only 70% of the hemoglobin is saturated with oxygen. In other words, about 27% of the hemoglobin molecules release their oxygen to the cells.

Oxygen delivery is amazingly responsive to the requirements of the cells. The amount of oxygen delivered to the cells can be increased more than threefold without increasing the rate of blood flow. Vigorous exercise, for example, causes muscle cells to require more oxygen than usual for the production of ATP. Under these metabolic conditions, hemoglobin releases more of its oxygen load to the cells. Hemoglobin does this in response to a rise in acidity in the blood, which results from the exercise. Strenuous exercise can prompt muscle cells to respire anaerobically, releasing lactic acid (chapter 6). In addition, a very active muscle cell produces more carbon dioxide as a waste product of energy production. The carbon dioxide forms carbonic acid when it dissolves in the water of the tissue fluid and blood, a phenomenon called the Bohr effect. Can you see why the Bohr effect is adaptive?

Active tissues also produce heat as a by-product of energy metabolism. Blood flowing through warmer parts of the body releases more of its oxygen. Once again, more oxygen is released to cells that are using oxygen at a higher rate (table 23.1).

Carbon Dioxide

Besides delivering oxygen, the blood must also remove the carbon dioxide (CO_2) that the cells produce. Carbon dioxide transport is more complicated than oxygen transport in that it occurs in three ways. A small amount of carbon dioxide, about 7%, is transported dissolved in plasma. When it reaches the lungs, it simply diffuses into the alveoli. Slightly less than a quarter (23%) of the transported carbon dioxide is carried by hemoglobin molecules. Unlike oxygen, however, the carbon dioxide binds with the protein portion of the hemoglobin molecule (rather than the heme portion as oxygen does).

About 70% of the carbon dioxide is transported as a bicarbonate ion dissolved in the plasma. The bicarbonate ion forms when carbon dioxide produced by the cells diffuses into the blood and into the red blood cells. The carbon dioxide forms carbonic acid, which dissociates to form hydrogen ions (H^+) and bicarbonate ions (HCO_3^-). Within the

Table 23.1
Volume of Air Inhaled Depends Upon Activity

Activity	Liters Inhaled per Minute
Bed rest	8
Sitting	16
Walking	24
Running	50

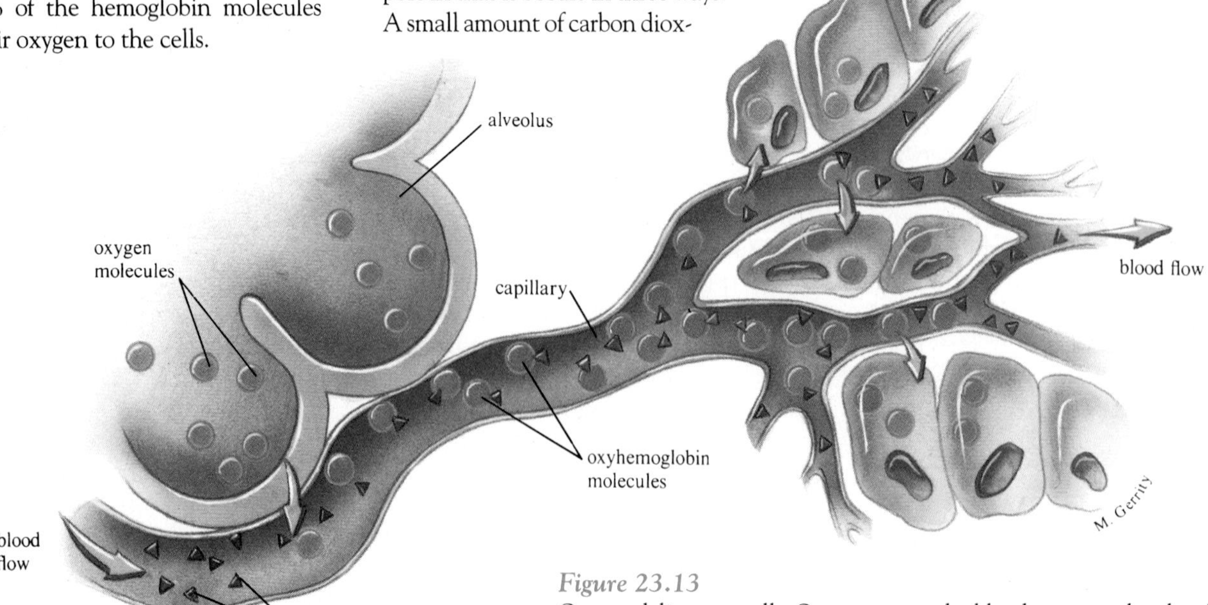

Figure 23.13
Oxygen delivery to cells. Oxygen enters the bloodstream at the alveolus, binds to hemoglobin, and is transported to tissues. Here oxygen is released from hemoglobin and taken up by cells.

red blood cell is an enzyme, **carbonic anhydrase,** that speeds up this reaction. The H^+ then combines with hemoglobin, keeping the pH of the blood fairly stable. The bicarbonate ions diffuse out of the red blood cell into the plasma. In the lungs, the whole process is reversed. In the presence of carbonic anhydrase within the red blood cell, carbonic acid reforms CO_2 and H_2O. The carbon dioxide then diffuses into the alveolar air.

Thus in the tissues, carbon dioxide enters the bloodstream, where its partial pressure is lower, but in the lungs, carbon dioxide moves from the bloodstream to the alveoli because the partial pressure is lower in the alveoli.

Carbon Monoxide Poisoning

Carbon monoxide (CO) is a colorless, odorless gas that is a by-product of combustion. It is released by campfire smoke, car exhaust, cigarette smoke, kerosene heaters, woodstoves, and home furnaces and can be particularly dangerous in poorly ventilated areas. Carbon monoxide is poisonous because it binds to hemoglobin more readily than does oxygen. Plus, it is less likely to leave the hemoglobin molecule. As a result, carbon monoxide releases oxygen, and the unsuspecting person begins to experience symptoms of oxygen deprivation.

When 20% of one's hemoglobin molecules carry carbon monoxide instead of oxygen, the person feels tired, nauseous, and may have a bad headache. A heavy smoker may experience these symptoms at lower exposures to carbon monoxide, because his or her blood always harbors some of it. When 30% of the hemoglobin molecules carry carbon monoxide, the person loses consciousness and may go into a coma. If the person survives, he or she may be left with permanent confusion, memory loss, or coordination problems.

KEY CONCEPTS

Oxygen binds to hemoglobin molecules in red blood cells where its partial pressure is high, as in the lungs. Oxygen is then released near the cells. Metabolic changes associated with exercise increase the acidity of the blood, which prompts hemoglobin to release more oxygen. Partial pressure also controls the transport of carbon dioxide from the tissues to the lungs. About 7% of CO_2 dissolves in plasma; 23% is carried by hemoglobin; 70% is converted to bicarbonate ion, which diffuses into the plasma. In the lungs reformed CO_2 diffuses from the capillaries into the alveolar air. Carbon monoxide is poisonous because it binds to hemoglobin more strongly than does oxygen.

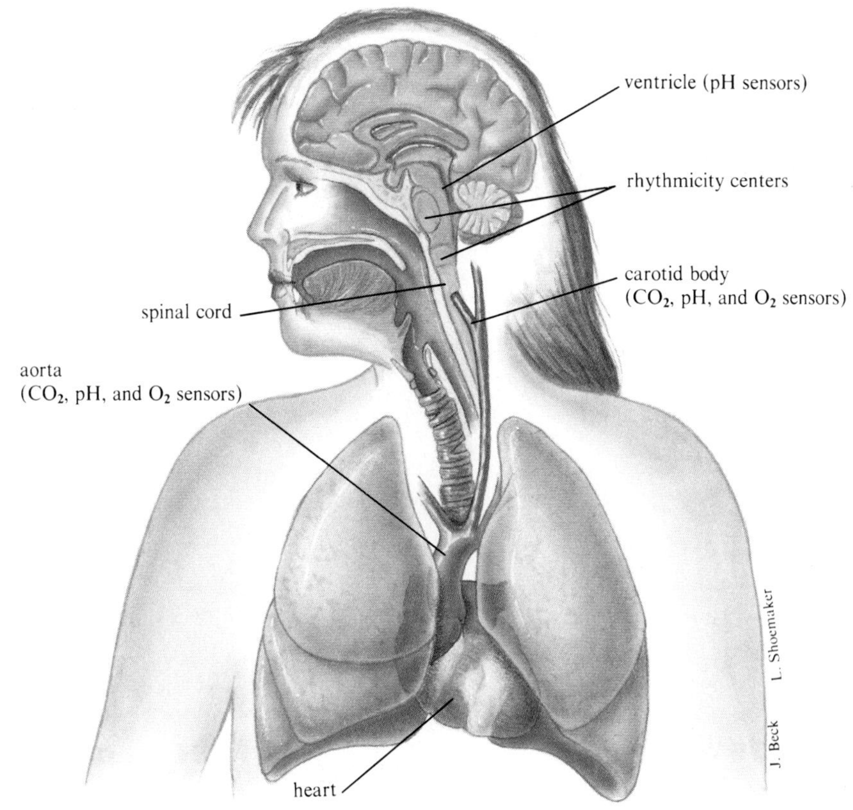

Figure 23.14
Control of breathing. Levels of oxygen, carbon dioxide, and hydrogen ions in the bloodstream are detected by receptors in the aorta and carotid arteries, and the information is passed to the rhythmicity center in the brain's medulla.

Control of Respiration

Awake or asleep, without thinking, you breathe an average of 12 times a minute. This steady breathing is controlled by a rhythmicity center in the brain's medulla (fig. 23.14). During quiet breathing, part of this region called the inspiratory center is in control, with its neurons very active for about 2 seconds, followed by inactivity for 3 seconds. The impulses sent by these neurons contract muscles used in inhalation, increasing the size of the chest cavity and drawing air into the lungs. At the end of 2 seconds, the neurons in the inspiratory center rest for 3 seconds, and you passively exhale. During strenuous activity, the respiratory cycle is changed so that exhalation is no longer a passive event, occurring when the inspiratory muscles relax, but is instead an active process. During heavy breathing, the inspiratory center is active as it is in quiet breathing, but it also sends impulses to a part of the medulla's rhythmicity center called the expiratory center. This contracts muscles that pull the rib cage down and the sternum inward, and the abdominal muscles, causing the abdominal organs to push the diaphragm further into the thorax than usual. Air is quickly pushed out of the lungs.

Other brain areas may modify the basic breathing pattern. The **apneustic center,** for example, allows the forceful inspiration of a deep breath. Furthermore, we can voluntarily alter our pattern of breathing through the impulses originating in the **cerebral cortex** (the "thinking" part of the brain). We can control breathing for speech, and we can voluntarily pant like a dog, sigh, and hold our breath. Holding our breath while swimming underwater is obviously adaptive. Such voluntary control can also protect us from inhaling smoke or irritating gases. But you cannot kill yourself by holding your breath, even though children often threaten to do so. The buildup of carbon dioxide in the blood overcomes this effort and forces you to breathe. Reading 3.1 describes a diving reflex that enables some children to survive drowning by temporarily stopping breathing.

The most important factor in regulating breathing rate is the blood CO_2 level, or more precisely, the number of H^+ formed when the CO_2 goes into solution and forms carbonic acid ($CO_2 + H_2O \rightarrow H_2CO_3 \rightarrow H^+ + HCO_3^-$). Since, under most circumstances, CO_2 is produced as a by-product of the energy-extracting reactions that require oxygen, monitoring blood CO_2 levels is a way to determine how quickly the cells are using oxygen.

Chemoreceptors in the medulla, and to some extent those in the aortic bodies (part of the main blood vessel delivering blood to the body) and carotid bodies (part of the blood vessels that deliver blood to the brain), monitor the blood CO_2 levels. Although all these chemoreceptors are involved in the body's response to CO_2, the key areas are in the medulla. These receptors are in an ideal location—near the surface of the medulla, where they are bathed in cerebrospinal fluid. Since the cerebrospinal fluid lacks the blood's ability to buffer changes in pH, dissolved CO_2 changes the pH of cerebrospinal fluid more than that of blood.

When the blood CO_2 level increases, thereby raising the H^+ concentration, chemosensitive areas are stimulated and send messages to the inspiratory center in the medulla. As a result, breathing rate is increased. The increased breathing rate, called **hyperventilation,** decreases the level of CO_2. Prolonged hyperventilation (except during exercise) will elevate the pH and one will become dizzy and perhaps faint. This can happen when we are severely frightened or at extremely high altitudes (Reading 23.3).

Although the concentration of CO_2 in the air we breath is low, small changes in its concentration can trigger a tremendous increase in the rate of breathing. For example, a 5% increase in CO_2 in the air inhaled causes us to breath 10 times faster.

Since oxygen, not carbon dioxide, is so critical to survival, it is surprising that the oxygen level is less important than the carbon dioxide level in regulating breathing. In fact, oxygen is not involved in regulating breathing rate unless its level falls dangerously low. Activation of the inspiratory center due to low oxygen levels is a last-ditch effort to increase breathing rate and restore the normal oxygen levels. Sudden infant death syndrome (SIDS), in which a baby dies while asleep, may be caused by a failure of the carotid bodies and/or the brain's respiratory centers to respond to low oxygen levels in arterial blood (fig. 23.15). This "forgetting" to breathe is called apnea, and it also causes normal snoring.

KEY CONCEPTS

Breathing is controlled by the rhythmicity center in the medulla, which has an inspiratory center whose neurons are active during quiet breathing and also during heavy breathing, when an expiratory center is also active. The apneustic center and cerebral cortex also control breathing. Blood CO_2 level is the most important factor in regulating breathing. Chemoreceptors in the medulla, aorta, and carotid arteries detect H^+ formed from CO_2 and alert the inspiratory center so breathing rate is increased.

Reading 23.3 *Breathing at High Altitudes*

THE HUMAN RESPIRATORY SYSTEM FUNCTIONS BEST AT THE 21% OXYGEN CONCENTRATION NEAR SEA LEVEL. At high elevations, the percentage of oxygen in the air falls gradually, so that at an elevation of almost 10,000 feet (about 3,000 meters) above sea level, a third less oxygen is inhaled with each breath. Each year, 100,000 mountain climbers and high-altitude exercisers experience varying degrees of altitude sickness. This occurs when the body's efforts to get more oxygen—including increased breathing and heart rate and stepped-up production of red blood cells and hemoglobin—cannot keep pace with the declining concentration of oxygen. Altitude sickness can be cured by descending slowly at the first appearance of symptoms. Table 1 indicates the severity of symptoms associated with increasing altitude.

Table 1
The Effects of High Altitudes

Condition	Altitude	Symptoms
Acute mountain sickness	5,900 ft (1,800 m)	Headache, weakness, nausea, poor sleep, shortness of breath
High-altitude pulmonary edema	9,000 ft (2,700 m)	Severe shortness of breath, cough, gurgle in chest, stupor, weakness; person can drown in accumulated fluid in lungs
Cerebral edema	13,000 ft (4,000 m)	Brain swells, causing severe headache, vomiting, loss of coordination, hallucinations, coma, and death

The Unhealthy Respiratory System

The Common Cold

The most common respiratory problem is the viral infection known as the cold. The nasal congestion and scratchy throat of a cold can be due to any one of a family of viruses, the **rhinoviruses,** whose members number more than 80 (fig. 23.16). The great number of cold-causing viruses is one reason why finding a cure for the cold is difficult. Another reason is that many of these viruses can alter their arrangement of surface molecules, which hinders development of a vaccine that relies on the immune system's recognition of viral surfaces.

Cold remedies fight the symptoms of colds: **decongestants** shrink nasal membranes and ease breathing, **antihistamines** decrease mucus secretion and combat watery eyes and sneezing, and **analgesics** relieve pain and discomfort. One popular cold remedy is good old-fashioned chicken soup, which has been scientifically tested and shown to clear a stuffy nose. It may be just the heat of the soup that temporarily clears nasal passages, because inhaling vapor from hot water has the same effect as eating chicken soup.

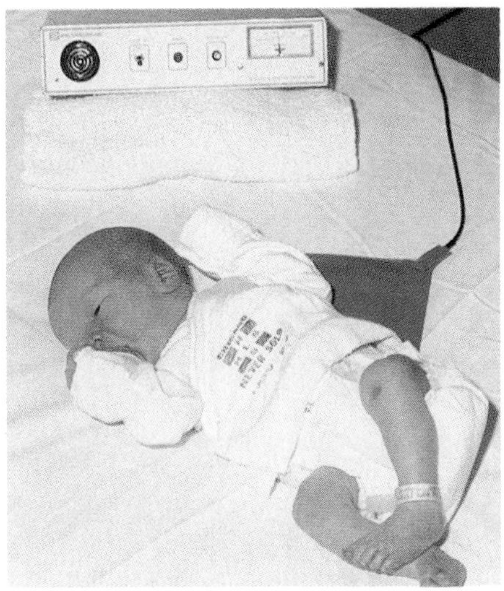

Figure 23.15
Sleep apnea. Each year, 10,000 babies in the United States die because they "forget" to breathe. This child has apnea, the tendency to stop breathing while asleep, and must be attached to a monitor, which sounds an alarm when breathing ceases. The parents are alerted and can then attempt to revive the child.

Influenza

Another common respiratory problem is influenza, often called the flu. Although the flu may start out with a stuffy nose and therefore feel like a cold, the onset of fever, joint aches, fatigue, and weakness indicates the flu. Bed rest is usually necessary for only a few days, but the fatigue may persist for a few weeks. Flu is viral in origin, and the causative viruses change their surface characteristics so frequently that vaccines cannot be developed fast enough to keep up with them. This is why a flu vaccine that effectively prevents the disease one winter may not do so the next. Flu epidemics occur every few years. In 1918, such an epidemic killed 20 million people worldwide. Like cold symptoms, flu symptoms can be treated, but a cure for influenza has not yet been found.

Bronchitis

Bronchitis is an inflammation of the mucous membrane of the bronchi. It can be caused by viruses or bacteria. The main symptom of the disease is a deep cough that produces gray or greenish yellow phlegm. In addition, the patient is often breathless, wheezy, and feverish.

Bronchitis can be acute or chronic. Acute bronchitis often follows a cold. The cold virus can cause the inflammation itself or it may lower the body's resistance to bacteria, which then invade the trachea and bronchi. If the cause is bacterial, an antibiotic will hasten recovery.

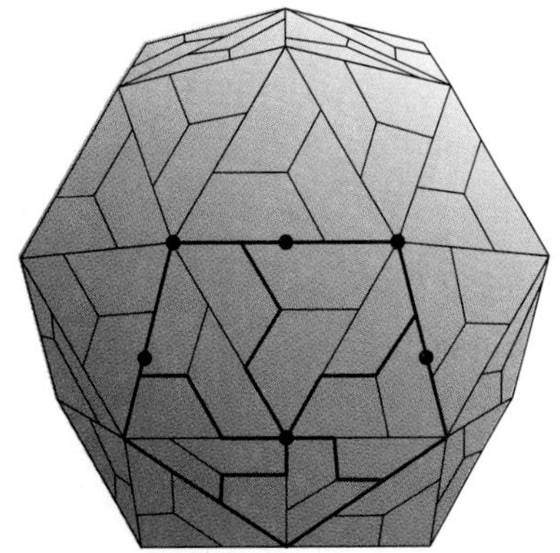

Figure 23.16
The structure of a cold virus. More than 80 different viruses can cause the group of symptoms we call the "common cold." The structure of this cold virus was determined by X-ray crystallography, in which the paths of X-rays passed through the virus are converted into an image. This cold virus is composed of 20 triangular faces, each made of protein, with DNA inside.

Chronic bronchitis is a more serious long-term or recurring condition with symptoms similar to acute bronchitis. It is associated with cigarette smoking and air pollution because these add to the irritation of the bronchial tree. The repeated infections of the bronchi that occur in chronic bronchitis cause the linings of the air tubules to thicken, narrowing the passageway for air. The air tubules are further obstructed by the contraction of muscles in their walls and the secretion of mucus. In addition to causing labored breathing, these degenerative changes in the lining of the air tubules make drainage of mucus and pus more difficult.

Chronic bronchitis can have serious consequences. The poor drainage predisposes the patient to lung infections such as pneumonia. Famed muppet-maker Jim Henson died from a very rapid-acting form of pneumonia.

Emphysema

In **emphysema,** the alveoli become overinflated and burst. As the alveolar walls break down, neighboring alveoli merge, creating fewer and larger alveoli (fig. 23.17). This reduces the surface area available for gas exchange and increases the volume of residual or "dead" air in the lungs. A loss of elasticity of the lungs also accompanies the rupture of alveoli. Recall that exhalation is a passive process and that the elasticity of lung tissue helps push air out of the lungs. As the dead air space increases, more forceful inspiration is necessary to adequately ventilate the lungs and this ruptures additional alveolar walls, increasing the dead air space. Although the lungs increase in size as the residual volume of air becomes greater, creating a characteristic barrel chest, gas exchange becomes more difficult. To experience what poor lung ventilation due to increased dead air space feels like, take a deep breath. Now exhale only slightly. Continue taking very shallow breaths, leaving the lungs almost completely filled with air.

The decreased surface area for gas exchange and the increased dead air space produce the main symptom of emphysema—shortness of breath. As the disease progresses, alveolar walls thicken with fibrous connective tissue, making gas exchange even more difficult. The oxygen that does make it to the alveoli has difficulty crossing the connective tissue to enter the blood, as does carbon dioxide, which builds up in the blood. The increase in acidity of the blood caused by the carbonic acid formed when carbon dioxide dissolves in the blood kills brain cells in the inspiratory area so that respiration rate decreases. Some cases of emphysema are caused by an inherited enzyme deficiency, but most cases are caused by irritating the lungs, such as by smoking.

Cystic Fibrosis

Reading 14.1 describes the daily tortures that were endured by Alex Deford as her parents beat her chest to loosen the mucus clogging her lungs. Alex had **cystic fibrosis**, and she died at the age of 8, before researchers began to understand the molecular basis of her genetic disease.

In 1989 the gene for cystic fibrosis was identified and found to specify a very large protein that forms an ion channel in epithelial cells lining the lungs, pancreas, and sweat glands. The protein in people with cystic fibrosis lacks only one amino acid, but this defect removes an ATP binding site. As a result the ion channel cannot regulate passage of chloride ions (Cl^-) and water. In the lungs, mucus is not sufficiently mixed with water, causing lung congestion that encourages infection. Similarly, in the pancreas, thick mucus blocks release of digestive enzymes and the person becomes very thin. In the sweat glands, Cl^- leaves but cannot be reabsorbed, resulting in the salty sweat that many parents notice as the first sign of an abnormality. Now that researchers have found the abnormal protein behind cystic fibrosis, the next step is to get it into patients where they need it. Can you think of a way to do this?

Tuberculosis

Tuberculosis, or TB, is caused by Mycobacterium tuberculosis bacteria inhaled into the lungs. An infected person sends bacteria-laden droplets into the air when coughing or sneezing. The disease is highly contagious, perhaps because the bacteria can survive for very long periods in air. Because the bacteria are inhaled, the lungs are usually the first site attacked. However, the bacteria can also infect other parts of the body, especially the brain, kidneys, or bone.

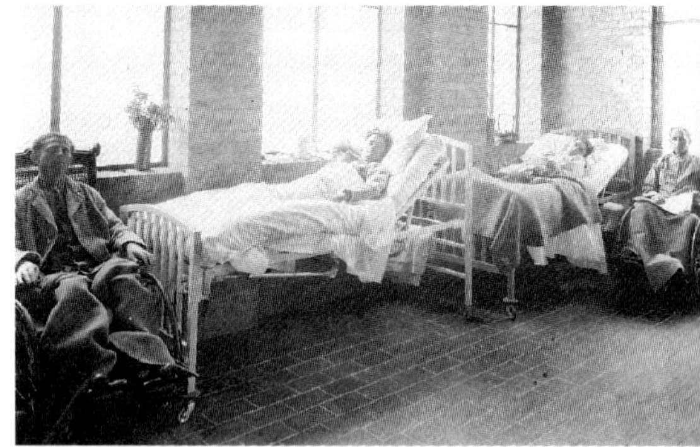

Figure 23.18
Tuberculosis. At the turn of the century, the best treatment for tuberculosis was to ship patients to sanatoriums in the country for fresh air and rest. Today patients are treated with drugs and stay in the hospital for a short time.

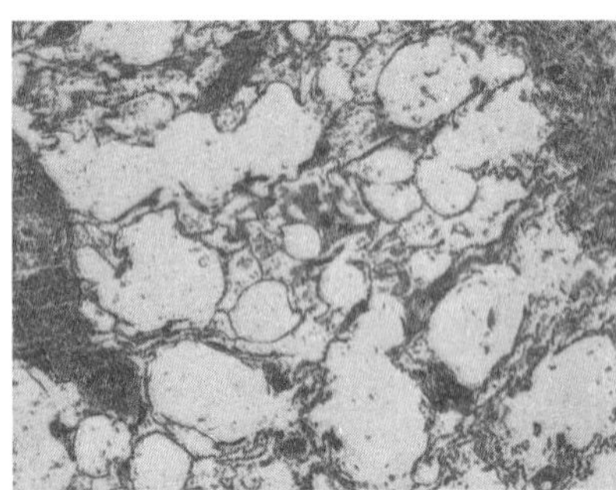

a.

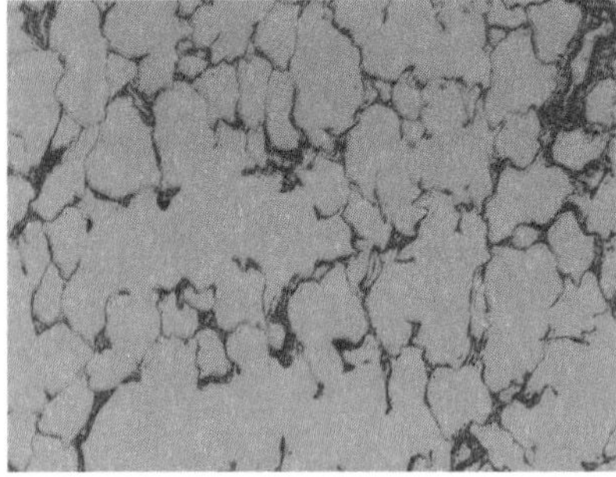

b.

Figure 23.17
In emphysema, exposure to pollutants or a genetic defect causes alveoli to coalesce and bronchioles to collapse: (*a*) normal alveoli; (*b*) emphysema alveoli. The decreased surface area for gas exchange greatly diminishes lung capacity.

Table 23.2
The Unhealthy Respiratory System

Condition	*Cause*	*Symptoms*
Bronchitis	Cilia destroyed by bacterial or viral infection or irritation	Persistent heavy cough
Cystic fibrosis	Genetic disease causes mucus to accumulate	Severe lung congestion, impaired digestion
Emphysema	Loss of elastic fibers around alveoli causes their overinflation	Cough, fatigue, wheezing, difficulty breathing
Lung cancer	Inhalation of asbestos, cigarette smoking	Fever, bloody cough, hoarseness, wheezing
Pleurisy	Inflammation of pleural linings due to infection	Chest pain, fever, dry cough
Pneumonia	Inflammation of bronchioles and alveoli due to bacterial or viral infection	Fatigue, fever, bloody cough, chest pain, difficulty breathing
Respiratory distress syndrome (infantile)	Deficiency of surfactant in lungs of newborn	Newborn cannot breathe
Streptococcal sore throat (strep throat)	Bacterial infection of throat	Painful swallowing, fever
Sudden infant death syndrome (crib death)	Unknown	Breathing of infant suddenly stops (apnea)
Pulmonary tuberculosis	Bacterial infection of lungs	Fever, emaciation, cough

As the bacteria multiply, they cause a small area of inflammation. During the primary stage of infection, the body resists the bacteria by killing them or sealing them in a fibrous connective tissue capsule. The walled off section is called a **tubercle,** which is the source of the name of the disease. The formation of tubercles slows the spread of the disease, but it does not actually kill the bacteria. When the immune system cannot kill all the bacteria, the disease may progress to the secondary stage. After many symptomless years, pockets of bacteria throughout the body may become active again. Cavities form in the lungs, impairing their function. Bacteria can escape from the tubercles and be carried by the bloodstream to other parts of the body. The disease may flare up whenever the person becomes weak, ill, or poorly nourished.

Tuberculosis was a very serious public health problem before 1944, when the antibiotic streptomycin was introduced. Evidence of the disease has been found in Egyptian mummies and pre-Colombian skeletons and was described by ancient Greek and Roman doctors. From the mid-1850s to the early twentieth century, sufferers of "consumption," one of several names for tuberculosis, rested for long periods in sanatoriums (fig. 23.18). Today many antibiotics can control tuberculosis, but some 22,000 people in the United States contract it each year, and because of acquired immune deficiency syndrome (AIDS), the number is actually rising slightly. Tuberculosis is also discussed in chapter 34.

Because lungs are so sensitive to the environment, workers in certain industries are especially prone to respiratory problems. Miners and cereal grain workers develop sensitivities to the dust they breathe. Cotton workers develop "brown-lung disease," glass blowers suffer from their own form of emphysema, and fire fighters suffer from coughs, shortness of breath, and wheezing caused by frequent inhalation of smoke. Table 23.2 lists some common respiratory disorders.

The process of breathing provides an elegant example of the interaction and coordination of organ systems. The respiratory, circulatory, nervous, skeletal, endocrine, and muscular systems all function together to enable you to take the 7 million or so breaths you have taken each year since that first, most difficult of all, breath.

KEY CONCEPTS

Colds are caused by rhinoviruses. Cold symptoms can be relieved, but the infection cannot be cured. Flu is a viral illness more severe than a cold. Bronchitis is inflammation of the bronchial mucous membrane, can be caused by a virus or bacterium, and can be acute or chronic. In emphysema, alveoli burst and coalesce and the lungs lose elasticity, hampering gas exchange. The thick mucus in the lungs of the cystic fibrosis patient is caused by ion channels in lung epithelium that prevent water from diluting the mucus. Tuberculosis is a bacterial infection of the lungs.

SUMMARY

Oxygen is necessary to increase the amount of energy in food that can be converted to ATP, an energy-laden molecule. In producing ATP, carbon dioxide is formed and must be eliminated from the body. Oxygen and carbon dioxide are exchanged by diffusion across a moist membrane. Several factors, including an animal's size, metabolic requirements, and habitat, have influenced the evolution of respiratory systems. Simple organisms exchange oxygen and carbon dioxide across the body surface. Larger or more active animals using skin exchange employ a circulatory system to distribute the gases between the cells and the environment. Terrestrial arthropods bring the environment into contact with almost every cell through a highly branched system of tubules called *tracheae*. Complex aquatic animals typically exchange gases across the membranes of gills, extensions of the body surface specialized for gas exchange. The gills of bony fishes are specialized to maximize gas exchange. They are thin, highly folded membranes in close association with the circulatory system over which water flows in the opposite direction of blood flow. Vertebrate lungs create a highly vascularized internal surface for gas exchange that is sheltered from extreme water loss.

In humans, air is usually inhaled through the nose, which purifies, warms, and moisturizes the air. The air then flows through the *pharynx,* a common passageway for food and air, and on through the *larynx* (voice box) to the *trachea.* The trachea, or windpipe, is supported open with C-shaped rings of cartilage. It divides to form smaller tubules called bronchi that deliver the air to the lungs. The bronchi branch extensively to form tinier air tubules called *bronchioles*. The bronchioles end blindly, usually with a cluster of thin-walled, cup-shaped structures called alveoli. Enveloping the alveoli are many capillaries. Oxygen diffuses into the blood from the alveolar air while carbon dioxide diffuses from the blood into the alveolar air. The alveoli are kept open by *surfactant,* a chemical that reduces the surface tension that would cause alveoli to collapse.

Breathing brings a fresh supply of oxygen-rich air into the lungs and removes the air high in carbon dioxide. *Inspiration* (inhalation) occurs when the thoracic cavity increases in size due to the contraction of the diaphragm and the muscles that elevate the rib cage. This creates a negative pressure that pulls air into the lungs. *Exhalation* results when these muscles relax and the size of the thoracic cavity decreases.

Dead air, that in the pharynx, trachea, and upper third of the lung, is the air that is not used in gas exchange. *Residual air* is that which remains in the lungs after exhalation. The volume of air moved in and out of the lung during a normal breath is called the *tidal volume*. The amount of air moved in and out of the lungs during forced breathing is the *vital capacity*.

Almost all of the oxygen transported to the cells is bound to hemoglobin within the red blood cells. A small amount is dissolved in the plasma. An increase in the acidity of the blood, due primarily to an increase in carbon dioxide level, and elevated temperature due to metabolism increase the amount of oxygen that is delivered to the cells.

Most of the carbon dioxide in the blood is transported as bicarbonate ion, but some is carried bound to hemoglobin or dissolved in plasma. The bicarbonate ion forms when carbonic acid from carbon dioxide ionizes, generating hydrogen ions and bicarbonate ions. An enzyme within the red blood cells speeds up this reaction. The process is reversed in the lungs.

During quiet breathing, the inspiratory center within the medullary rhythmicity center is spontaneously active for about 2 seconds and generates impulses that result in inhalation. Then it rests for approximately 3 seconds. During heavy breathing the inspiratory center becomes active as usual and then the expiratory center becomes active, causing a more forceful exhalation. The basic breathing pattern may be altered by other brain regions, notably the *apneustic center* and the cerebrum.

Breathing rate is adjusted to the needs of the body by chemical factors. When carbon dioxide levels rise, the rise in the acidity of the blood is sensed by chemoreceptors in the medulla, the aortic bodies, and carotid bodies. This triggers an increase in breathing rate. Changes in blood oxygen are also sensed. However, oxygen levels do not change breathing rate unless the levels are critically low.

The common cold is an upper respiratory infection caused by one of many *rhinoviruses*. Other viruses cause a more severe infection, influenza. *Bronchitis* is an inflammation of the bronchi during which mucus production is increased and the individual coughs frequently and has difficulty breathing. In *emphysema*, gas exchange is made difficult because many alveolar walls break down and remaining ones thicken. In *cystic fibrosis*, mucus lining the lungs becomes very thick due to abnormal ion channel proteins that prevent water from diluting it. *Tuberculosis* is a bacterial infection of the lungs. The body defends itself against these bacteria by walling them off into connective tissue capsules called *tubercles*.

QUESTIONS

1. What color might the insides of prehistoric human lungs have been? Why?

2. How does an animal's size, activity level, and environment influence the structure and function of its respiratory system?

3. Trace the path of an oxygen molecule from a person's nose to the red blood cell.

4. How is air cleaned before it reaches the lungs?

5. Describe the process of inhalation. What muscles are involved?

6. How is oxygen transported to the cells?

7. What mechanisms cause more oxygen to be delivered to the most metabolically active cells?

8. What is the primary means of transporting carbon dioxide? What enzyme facilitates this? What other ways is carbon dioxide transported?

9. How is the basic breathing rhythm established by the brain?

10. What chemical change in the blood is most important in altering the breathing rate?

TO THINK ABOUT

1. It is below 0°F outside, but the dedicated runner bundles up and hits the roads anyway. "You're crazy," shouts a neighbor. "Your lungs will freeze." Why is the well-meaning neighbor wrong?

2. Why does breathing through the mouth instead of the nose dry out the throat?

3. On December 13, 1799, George Washington spent the day walking on his estate in a freezing rain. The next day, he had trouble breathing and swallowing. Several doctors were called in. One suggested a tracheostomy, in which a hole would be cut in the throat so that the president could breathe. He was voted down. The other physicians suggested bleeding the patient, plastering his throat with bran and honey, and placing blister beetles on his legs to produce blisters. Within a few hours, Washington's voice became muffled, breathing was more labored, and he was restless. For a short time he seemed euphoric, and then he died.

Today, Washington's problem is recognized as *epiglottitis*, in which the epiglottis swells to 10 times its normal size. How does this diagnosis explain Washington's symptoms? Which suggested treatment might have worked?

4. Why can't you commit suicide by holding your breath?

5. When a woman is very close to delivering a baby, she often breathes very rapidly, hyperventilating. This is not the controlled breathing that she may have learned in a childbirth preparation class but a panicky, uncontrolled fast breathing. Her childbirth coach can help by having her breathe into a paper bag. How does this action return her breathing to normal?

6. Asbestos is a substance known to cause respiratory system cancers. Many workers are exposed to it. Do you think that efforts should be made to improve masks worn by workers or to rid the workplace of the asbestos threat altogether?

7. Why do you think the times of endurance events at the 1968 Olympics, held in 2,200-meter-high Mexico City, were rather slow?

8. Why might it be dangerous for a heavy smoker to use a cough suppressant?

SUGGESTED READINGS

Cohn, Jeffrey P., and Bill Rados. November 1988. Here come the bugs: Cold and flu season's back . . . and here's help: Modern pharmacy's answers to chicken soup. *FDA Consumer*. We still cannot cure the common cold—but we can ease the symptoms.

Davies, Kevin A. March 1990. The race to discover the gene for cystic fibrosis. *Biology Digest*. A new-found gene explains the oldest known symptom of cystic fibrosis.

Hecht, Annabel. December 1986–January 1987. TB: Curable, preventable, but still a killer. *FDA Consumer*. TB should have disappeared by now, but it is actually on the rise.

LeBoeuf, Burney J. February 1989. Incredible diving machines. *Natural History*. The elephant seal has adaptations for respiring aerobically for up to an hour underwater.

Lewis, Ricki. March 1990. Neutrophils in emphysema: The smoking gun? *The Journal of NIH Research*. A complex enzymatic interplay lies behind the lung ravaging of emphysema.

Zamula, Evelyn. July/August 1990. More than snuffles: childhood asthma. *FDA Consumer*. Asthma in children requires prompt treatment.

C H A P T E R

24

The Digestive System

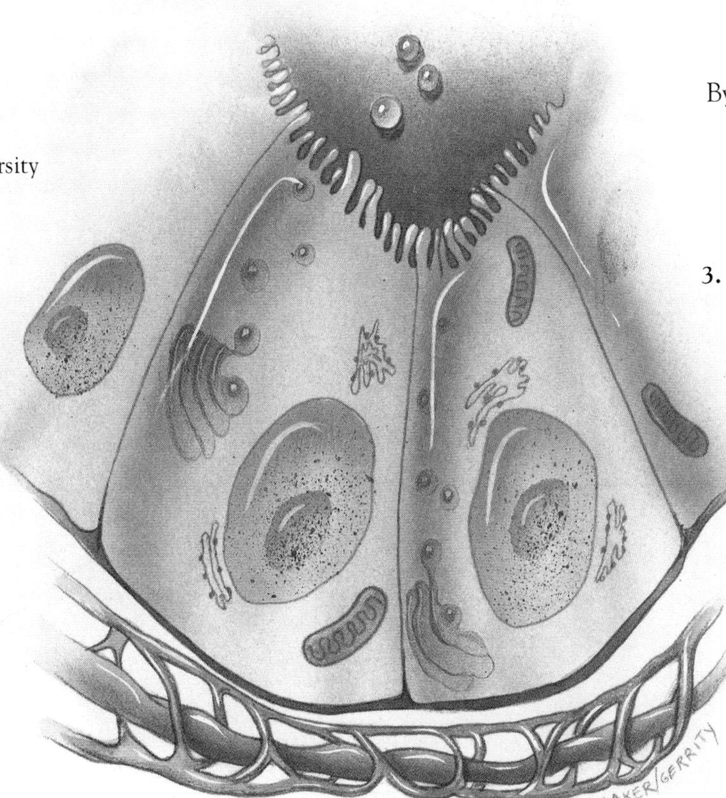

Chapter Outline

Learning Objectives

By the chapter's end, you should be able to answer these questions:

1. Why do animals need to eat?
2. What are some different types of digestive systems?
3. How do the structures and biochemicals of the human digestive system mechanically and chemically degrade food into nutrient molecules small enough to be absorbed and enter the bloodstream?
4. What structures aid digestion?
5. How are the activities of the human digestive system regulated?
6. What are the building blocks into which proteins, fats, and carbohydrates are broken down?

Tossed green salad
1/4 pound chopped steak
Baked potato
Green beans
Lime sherbet
Orange juice

This well-balanced dinner is eaten with much enjoyment. The diner savors the crunchy greens, succulent meat, fluffy potato, tasty beans, and especially the sweet ice cream. He may sit back afterwards, full and satisfied, but probably pays no further thought to the meal unless he feels ill.

Meanwhile, that meal presents quite a challenge to the organs, glands, and biochemicals of the man's digestive system. Consider the chopped steak. It consists of actin and myosin that were once a cow's muscle and probably quite a bit of fat. The actin and myosin are long chains of amino acids, and the fats are triglycerides and cholesterol, which are also large molecules. The salad greens and beans contain complex carbohydrates, one of which, cellulose, cannot be digested by the human digestive system at all. The orange juice and sherbet, mostly sugars, are easier to digest.

Eating and Digesting

For the man in the preceding example to use the proteins, carbohydrates, and fats in that delicious dinner, his body must dismantle these nutrient molecules into smaller molecules—proteins to amino acids, complex carbohydrates (starch) to simple carbohydrates (sugar), and fats to fatty acids and glycerol. Only such small molecules can enter cells lining the digestive tract, from which they enter the circulation. Once inside cells, nutrient molecules are further broken down to release the energy in their chemical bonds, or they are built up into human versions of macromolecules. For example, the cow myosin that is part of the steak is degraded to its amino acids, and then those amino acids are strung together to make a human protein—perhaps even myosin. Nonnutritive parts of food are eliminated from the body as feces.

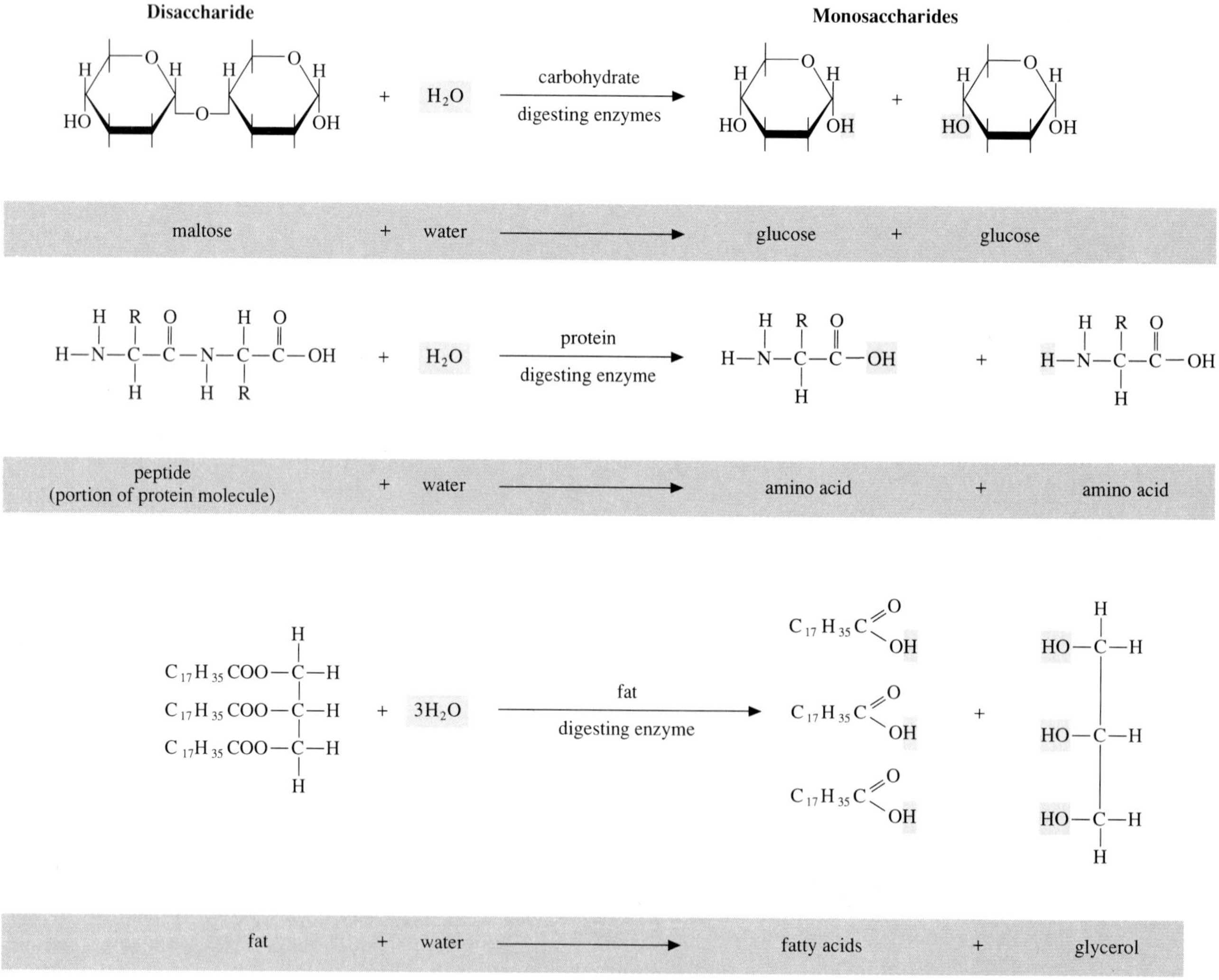

Disaccharide + H_2O —carbohydrate digesting enzymes→ Monosaccharides +

maltose + water → glucose + glucose

$$\mathrm{H{-}N(H){-}C(R)(H){-}C(=O){-}N(H){-}C(H)(R){-}C(=O){-}OH} + H_2O \xrightarrow[\text{digesting enzyme}]{\text{protein}} \mathrm{H{-}N(H){-}C(R)(H){-}C(=O){-}OH} + \mathrm{H{-}N(H){-}C(R)(H){-}C(=O){-}OH}$$

peptide (portion of protein molecule) + water → amino acid + amino acid

$$\begin{matrix} & \mathrm{H} \\ C_{17}H_{35}COO{-} & \mathrm{C{-}H} \\ C_{17}H_{35}COO{-} & \mathrm{C{-}H} \\ C_{17}H_{35}COO{-} & \mathrm{C{-}H} \\ & \mathrm{H}\end{matrix} + 3H_2O \xrightarrow[\text{digesting enzyme}]{\text{fat}} 3\ C_{17}H_{35}C(=O)OH + \begin{matrix} \mathrm{H} \\ \mathrm{HO{-}C{-}H} \\ \mathrm{HO{-}C{-}H} \\ \mathrm{HO{-}C{-}H} \\ \mathrm{H}\end{matrix}$$

fat + water → fatty acids + glycerol

Figure 24.1
Chemical digestion. Hydrolytic enzymes break complex carbohydrates down to maltose, and then maltose to glucose. Proteins are broken down to peptides, and then the peptides to their constituent amino acids. Fats are digested to fatty acids and glycerol.

Animals eat to replace energy expended in the activities of life and to provide the raw material for the structural components that are needed for growth, repair, and maintenance of the body. Although different types of animals eat different foods, obtain those foods in diverse ways, and have characteristic ways of digesting, all use the same sorts of nutrient molecules. Digestion is chemically carried out by hydrolytic (water-splitting) enzymes, which enable cells to add water molecules between the building blocks of large nutrient molecules, thereby splitting them apart (fig. 24.1).

Humans obtain food from the supermarket or a restaurant or may grow some of it. A humpback whale, in contrast, eats by opening its mouth and gulping down a few thousand fish and crustaceans. Cilia help to move the food into and along its digestive tract, and it is not particularly choosy about what enters its great mouth. A protozoan "eats" by surrounding microscopic plankton, drawing the microbes into itself, and surrounding them by a bubble of membrane that forms a food vacuole. Enzymes in the vacuole complete digestion. Many familiar animals use teeth or other chewing mouthparts to grab food, and some, such as snakes and lizards, swallow small mammal meals whole (fig. 24.2). Yet others drink their food—as anyone knows who has ever watched a mosquito alight on human skin and take a "blood meal." Some organisms digest their food alive, such as a spider consuming an insect or a starfish dining on a clam.

Types of Digestive Systems

Since living bodies are composed of the same general types of substances as are nutrient molecules, digestive enzymes could just as easily attack the body as food molecules. To prevent this, digestion commonly occurs within a specialized compartment.

Lower invertebrates, such as rotifers, brachiopods, and simple sea dwellers, separate digestion from other activities by sequestering food in food vacuoles. These membrane-bound sacs fuse with other sacs containing digestive enzymes, which dismantle large nutrient molecules into a usable size. Digested nutrients exit the food vacuole and are used in other parts of the cell. Waste is extruded from the cell. Such digestion within food vacuoles in cells is called **intracellular digestion.** Parts of our own cells, the lysosomes, carry out intracellular digestion to break down fats and carbohydrates of worn-out cell parts. Reading 4.2 takes a closer look at lysosomes.

Intracellular digestion can handle only very small food particles. In **extracellular digestion** larger food particles are dismantled by hydrolytic enzymes in a cavity. When nutrient molecules are small enough, they enter cells lining the cavity. The digestive space is considered to be outside the body. Extracellular digestion eases waste removal and is more efficient and specialized than intracellular digestion. The cavity in which extracellular digestion occurs constitutes a digestive system.

Some animals, such as the cnidarian hydra and the flatworm planaria, have digestive systems with a single opening. Food that cannot be digested exits through the same opening it entered. Consequently, food must be digested and the residue expelled before the next meal can begin. This two-way traffic makes impossible any specialized compartments for storage, digestion, or absorption within the digestive cavity. In these organisms, the digestive cavity doubles as a circulatory system, distributing the products of digestion to the body cells for use.

a.

b.

Figure 24.2
Having a meal and being a meal. *a.* The chipmunk furtively stuffs its cheeks with seeds. To sustain itself over the winter, it stores at least a half bushel of seeds in its underground home. *b.* A corn snake swallows its house mouse meal whole.

Reading 24.1 *Digestive Diversity*

Animals obtain and digest food in diverse ways. Birds metabolize nutrients very rapidly, and they must eat continually and digest food quickly to fuel their lightweight bodies. Their mouths have adapted to their diets. The stork's bill scoops fish (fig. 1); the strong, short beak of the finch cracks and removes the hulls from seeds; the vulture's hooked beak rips carrion into manageable chunks. A bird's esophagus has an enlargement called a crop, which temporarily stores food. The stomach is divided into two sections. The first, called the proventriculus, secretes gastric juice that chemically digests food. The second part of the stomach, the gizzard, is a muscular organ lined with ridges. It mechanically digests food, with the aid of sand and small pebbles swallowed by the bird.

Many mammals have specializations of the digestive system that enable them to break down the cellulose in the cell walls of plant foods. The cow has a four-sectioned stomach (fig. 2). From the esophagus, grass enters the first section, the rumen, where it is broken down into balls of cud by bacteria. The cud is regurgitated back up to the cow's mouth, where it is mechanically broken down by chewing. The cow's teeth can mash the cud in a forward and backward movement as well as up and down and sideways. The chewed cud is finally swallowed again, but this time it bypasses the rumen, continuing digestion in the other three sections of the stomach. Sheep, deer, buffalo, and elephants also have quadruple stomachs.

Elephants have a particularly difficult time digesting their meals of leaves and twigs. Because they have only tusks and molars, a lot of wood lands in the massive stomach. There it stays for the better part of 2 1/2 days, churning about amidst digestive juices and cellulose-degrading bacteria. Still, elephant dung contains many undigested twigs.

Figure 1
The stork captures a fishy meal in its beak.

Rabbits have their own variation on getting the most out of a meal. Like cows and elephants, rabbits eat leaves that churn about in the stomach. Instead of regurgitating the partially digested plant matter, rabbits excrete it in a moist pellet. Then they eat the moist pellets, which continue digestion in the stomach and intestines. This time around, the rabbit excretes a dry pellet, which it does not eat.

The panda bear is in an unfortunate situation: it has the short digestive tract characteristic of meat-eaters but lives surrounded by bamboo, its primary food source (fig. 3). The panda also lacks bacteria that degrade cellulose and an organ to store bulky plant matter. To obtain adequate nutrition, the panda spends virtually all of its waking time (15.4 hours a day) eating—and about as much time defecating. Even though the animal consumes 6% of its body weight in bamboo each day (compared to 2% for most plant-eating mammals), it just barely meets its basic nutritional needs.

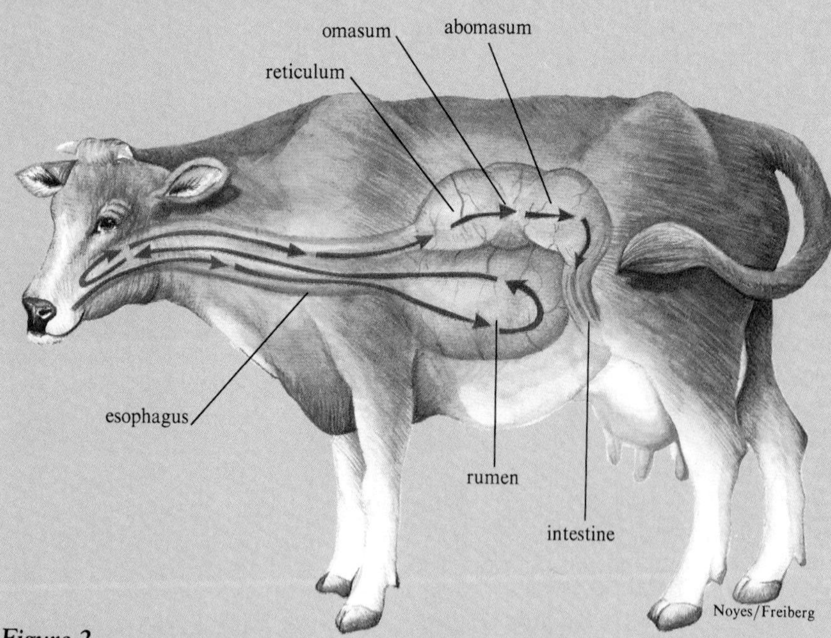

Figure 2
A cow's meal goes first to the rumen and is then sent back up to the mouth for more chewing. Next it bypasses the rumen, undergoing further digestion in the other three stomachs—the reticulum, omasum, and abomasum.

Figure 3
The panda consumes enormous amounts of bamboo.

The digestive systems of annelids, mollusks, arthropods, echinoderms, and chordates have two openings, an entrance and an exit. Hydrolytic enzymes are released into this tube, and the products of digestion are absorbed and delivered to the cells via the circulatory system. Substances that cannot be digested remain in the tube and leave the body with the feces. A major advantage of a two-opening digestive system is that regions of the tube can become specialized for different functions: breaking food into smaller particles, storage, chemical digestion, and absorption.

KEY CONCEPTS

Carbohydrates, proteins, and fats in foods must be broken down by hydrolytic enzymes into their chemical building blocks in order to enter cells, where they are used for their energy or built up into macromolecules. Carbohydrates are digested to glucose, proteins to amino acids, and fats to fatty acids and glycerol. Nonnutrients are eliminated. Animals eat to supply energy and raw materials for growth, repair, and maintenance of tissues. Different species obtain food in different ways but carry out the same sorts of reactions to utilize it. Lower invertebrates use intracellular digestion, whereas more complex organisms use extracellular digestion in digestive systems.

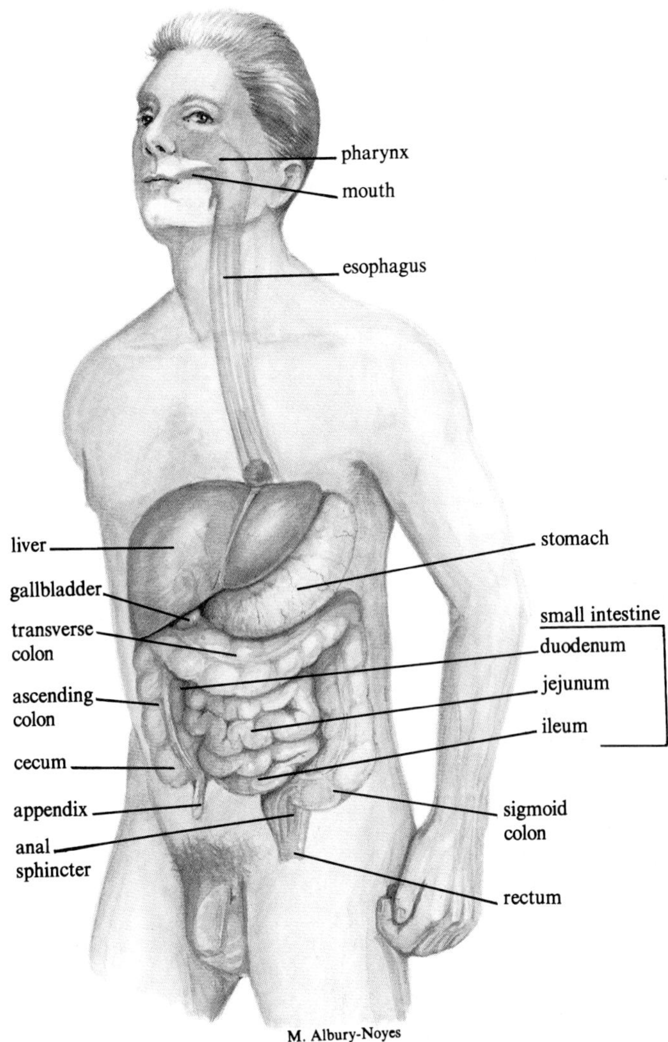

Figure 24.3
The human digestive system.

An Overview of the Human Digestive System

The human digestive system consists of a continuous tube called the **gastrointestinal** (stomach-intestine) **tract** (including the mouth, esophagus, stomach, small intestine, and large intestine) and accessory structures. These structures aid the breakdown of food either mechanically (the teeth and tongue) or chemically by secreting digestive enzymes (the salivary glands and pancreas). Other accessory structures, the liver and gallbladder, produce and store bile, which assists the digestion of fats (fig. 24.3). The digestive organs and glands are supported by the *mesentery,* which is an epithelial sheet reinforced by connective tissue.

The digestive lining begins at the mouth (you can feel it inside your cheeks) and ends at the anus. The moist innermost layer secretes the mucus that helps lubricate the tube so that food slips through easily. The mucus lining protects underlying cells from rough materials in food and from digestive enzymes. In some regions of the digestive system, cells in the innermost lining also secrete digestive enzymes. Beneath this layer is a layer of connective tissue containing blood vessels and nerves. The blood nourishes the cells of the digestive system and in some regions picks up and transports digested nutrients. The nerves coordinate the contractions of the next two layers, which are muscular.

The muscles of the inner layer circle the tube, constricting it when they contract, and the muscles in the outer layer run lengthwise, shortening the tube when they contract (fig. 24.4). This muscle arrangement churns food until it is liquefied, mixing

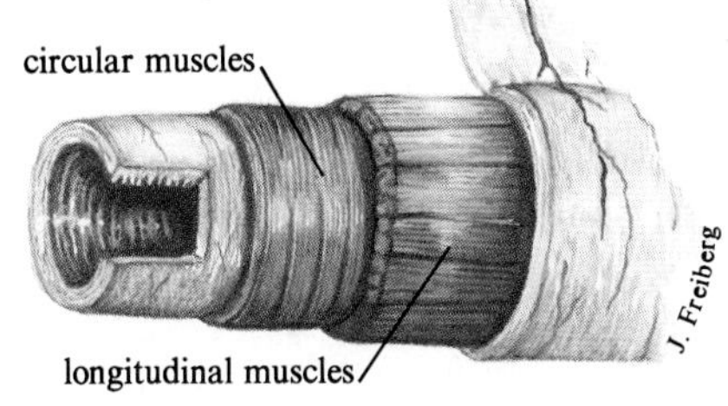

Figure 24.4
Muscles move food. Coordinated contraction of the two muscle layers of the digestive tract move food along in one direction. These contractions occur about three times per minute, even between meals. Note that the muscle layers are arranged perpendicular to each other.

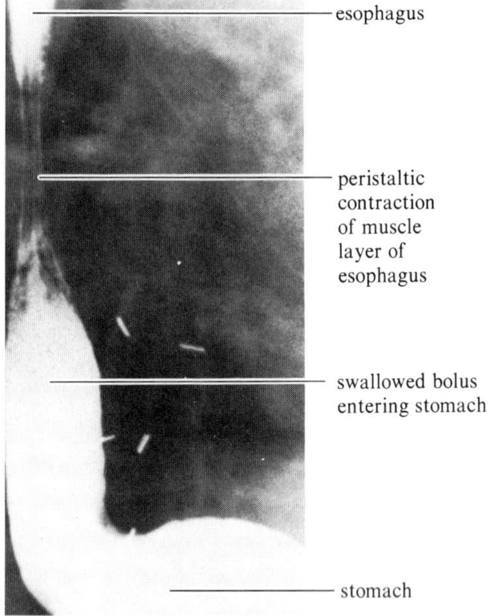

a.

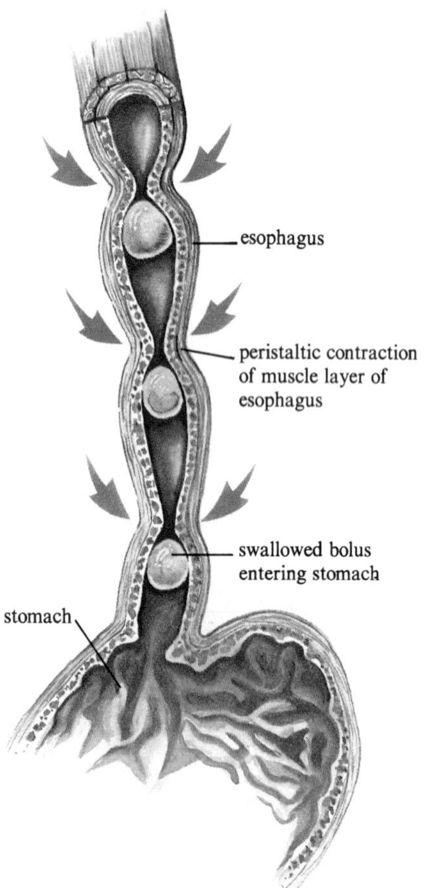

b.

Figure 24.5
Peristalsis. The muscular contractions of peristalsis move food along the digestive tract.

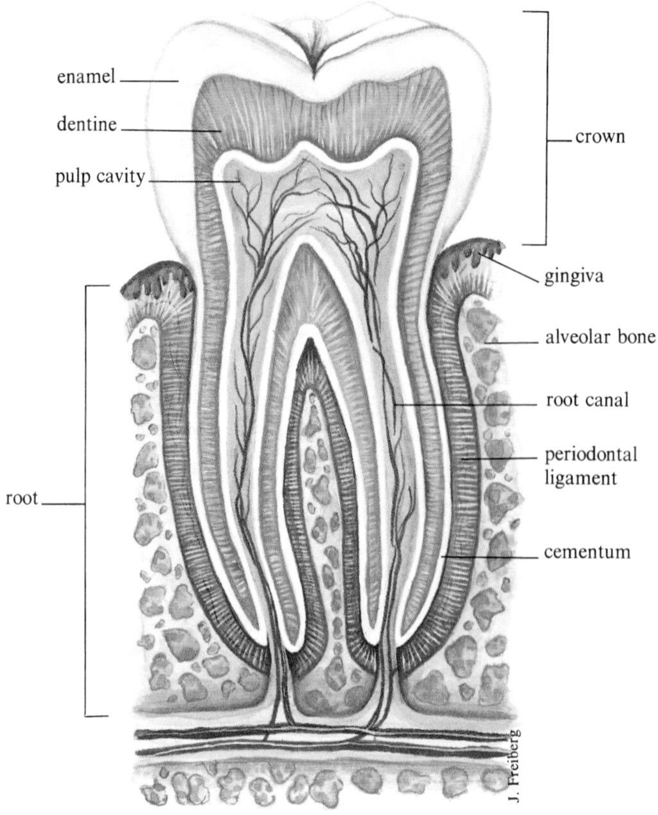

Figure 24.6
Anatomy of a tooth.

it with enzymes. Waves of contraction called **peristalsis** propel the food through the digestive system. When the walls of the tube are distended by food, the circular muscles immediately behind the food mass contract, squeezing the food forward (fig. 24.5).

> **KEY CONCEPTS**
>
> *The human digestive system consists of the organs of the gastrointestinal tract and accessory structures. The tract has four layers: an innermost mucus layer; then a layer of connective tissue, nerves, and blood vessels; and two muscle layers, the inner one circular and the outer one longitudinal. The muscles mix food and propel it in waves of peristalsis.*

Structures of the Human Digestive System

The Mouth and the Esophagus

Digestion begins when the thought or smell of food triggers secretion of **saliva** from three pairs of **salivary glands** near the mouth. Chemical digestion begins as the enzyme **salivary amylase** breaks down starch (complex carbohydrate) molecules into molecules of the disaccharide maltose. You can test the action of salivary amylase by holding a piece of bread in your mouth. Within a minute, it begins to taste sweet as the starch is converted to sugar.

Water and mucus in saliva aid the teeth in the first stage of mechanical digestion, chewing (fig. 24.6). Teeth tear food into small pieces, increasing the surface area upon which chemical digestion can begin. The thick **enamel** that covers a tooth is the hardest substance in the human body. Beneath the enamel is the bonelike **dentine,** and beneath that, the soft inner **pulp,** which contains connective tissue, blood vessels, and nerves. Two layers on the outside of the tooth, the **periodontal membrane** and the **cementum,** anchor the tooth to the gum and jawbone. The visible part of a tooth is the **crown,** and the part below the surface is the **root.** Teeth, like bones, are hardened by calcium compounds. Taste buds in the tongue relay nerve messages to the brain, and as a result, we experience the flavor of food.

The tongue rolls chewed food into a lump called a **bolus,** which is pushed to the back of the mouth. As the bolus of food is swallowed, it passes first through the **pharynx** (the throat) and then to the **esophagus,** a muscular tube leading to the stomach. During swallowing, the epiglottis covers the passageway to the lungs, so that food is routed along the digestive tract and away from the respiratory system.

Food does not merely slide down the esophagus due to gravity—it is actively pushed along by contracting esophageal muscles in a wave of peristalsis. This is why it is possible to swallow while standing on your head or even on the moon. Although an astronaut's dinner may float off the table due to the absence of gravity, peristalsis ensures that the swallowed material will move along the digestive system.

KEY CONCEPTS

Mechanical digestion begins as the teeth tear food, increasing surface area. Saliva contains water and mucus, which aid the teeth, and salivary amylase, which begins starch digestion. The tongue rolls food into a bolus, which is swallowed and moved down the esophagus by peristalsis.

The Stomach

The stomach is a J-shaped bag about 12 inches (30 centimeters) long and 6 inches (15 centimeters) wide with three important functions: storage, some digestion, and regulation of the flow of food into the small intestine (fig. 24.7).

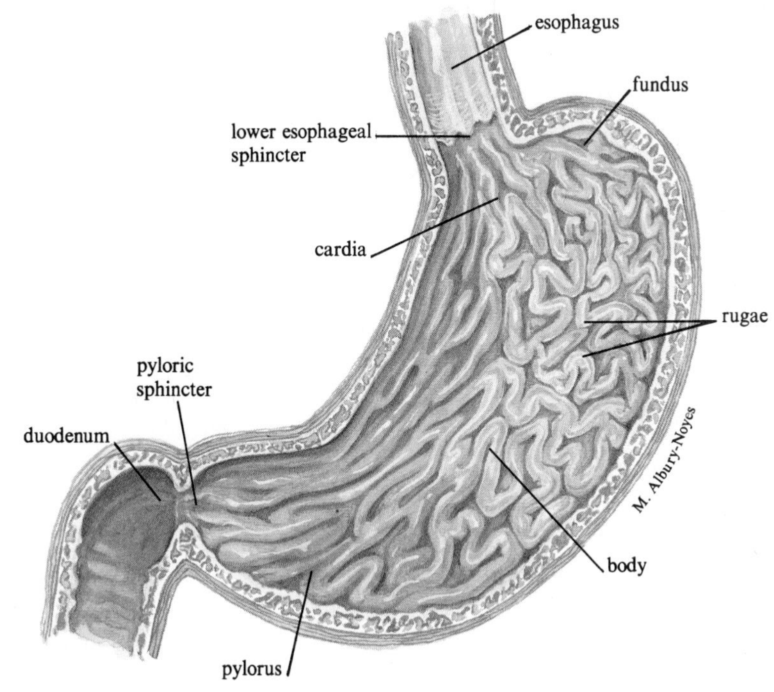

a.

fundus

pyloric sphincter

rugae

duodenum

body

pylorus

b.

Figure 24.7
The stomach is a J-shaped bag that stores food until it is fluid enough to move on to the small intestine. *a.* The stomach has four regions, two sphincters, and folds called rugae that increase its capacity. An X-ray of a stomach is shown in (*b*).

Although it is the size of a large sausage when empty, the stomach can expand to hold a fairly large volume of food—as much as 3 or 4 quarts (3 or 4 liters). Folds in the stomach's mucosa, called **rugae,** can unfold like the pleats of an accordion as the stomach is stretched by a large meal. Entry to and exit from the stomach are controlled by muscular rings called **sphincters,** which function like rubber bands, pinching off the openings to the stomach so the contents are held for treatment. The stomach has four regions. The entry to (or "neck" of) the stomach is called the **cardia.** The domelike top is called the **fundus,** the midsection is the **body,** and the bottom is the **pylorus.**

Food usually remains in the stomach for several hours, fats from 3 to 6 hours, proteins for up to 3 hours, and carbohydrates only 1 or 2 hours. What does this suggest about the best type of meal to eat before athletic competition? What might be the best type of food to eat if it will be a long time before you eat again?

Both mechanical and chemical digestion occur in the stomach. The waves of peristalsis cause mechanical digestion. When the sphincter at the bottom of the stomach closes, peristalsis pushes the food against the stomach bottom, and the food sloshes backwards somewhat. This churning action mixes the food with **gastric juice** secreted by stomach cells to produce a semifluid mass called **chyme.**

Gastric juice is responsible for the chemical digestion that occurs in the stomach. About 40 million cells lining the stomach's interior secrete 2 to 3 quarts (2 to 3 liters) of gastric juice per day (fig. 24.8). These stomach secretions consist of water, mucus, salts, hydrochloric acid, and enzymes. The **hydrochloric acid** creates a highly acidic environment, which activates an enzyme called **pepsin** from its precursor, **pepsinogen** (fig. 24.9). Pepsin breaks down proteins to yield polypeptides, a first step in protein digestion. Another stomach enzyme, **gastric lipase,** splits butterfat molecules found in milk. However, little fat is digested in the stomach because, unlike pepsin, gastric lipase works better in a less acidic environment.

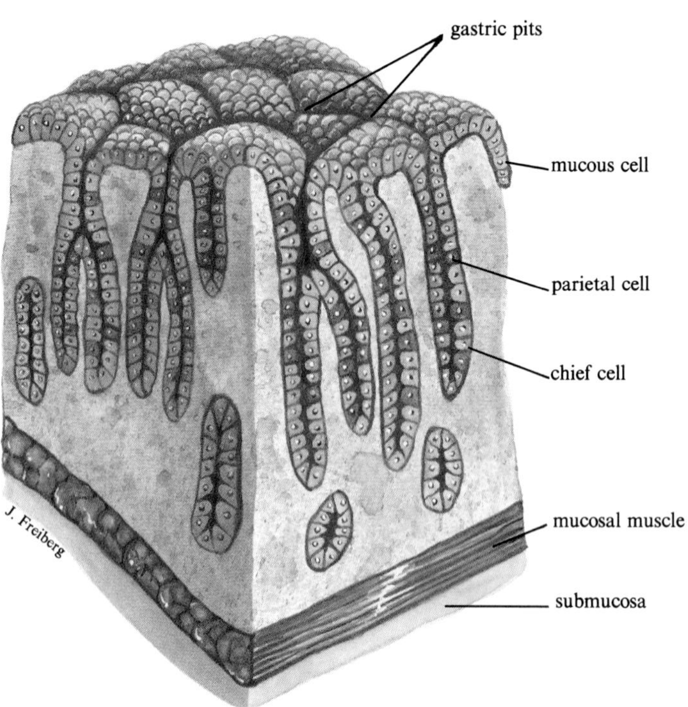

Figure 24.8
The lining of the stomach contains indentations, called gastric pits, where chief cells secrete pepsinogen and parietal cells secrete hydrochloric acid. Still other lining cells secrete the abundant mucus that coats the stomach lining, preventing it from digesting itself.

Secretion of gastric juice is regulated by nerves and hormones. Anticipation of eating or the presence of food in the mouth initiates nerve impulses that cause secretion of between 6 and 7 ounces (about 200 milliliters) of gastric juice. This prepares the stomach to receive food. When the food is actually present in the stomach, gastric cells release a hormone, **gastrin,** that causes another 20 ounces (600 milliliters) of gastric juice to be secreted. The ability of stress to induce gastric juice secretion has been directly observed in people with fistulas, openings that reveal stomach activity (fig. 24.10).

Why doesn't the stomach digest the protein of its own cells along with that in the food? The organ actually has built-in triple protection. First, most gastric juice is not secreted until food is present for it to work on. Secondly, some of the stomach cells secrete mucus, which coats and protects the stomach lining from the corrosive action of gastric juice. Finally, pepsin is produced in an inactive form, pepsinogen, and cannot digest protein until hydrochloric acid is present.

The stomach's automatic protection, however, sometimes fails. In this case, excess hydrochloric acid, possibly caused by stress, or a deficiency of protective mucus can result in an **ulcer,** a raw craterlike sore in the stomach lining (fig. 24.11). Approximately 1 in 5 men and 1 in 10 women in the United States develops an ulcer at some time.

Very few nutrients are absorbed from the stomach; most food has not yet been digested into component building blocks. However, some water and salts, a few drugs (such as aspirin), and alcohol can be absorbed. This is why aspirin can irritate the stomach lining as it breaks down and alcohol's intoxicating effects are felt quickly.

Contraction of the sphincter muscles at either end of the stomach holds most of the chyme within the sac. Eventually, the sphincter guarding the exit to the stomach (the pyloric sphincter) is directed to relax by neural messages triggered by stretching or by gastrin. Then the stomach contents can enter the small intestine.

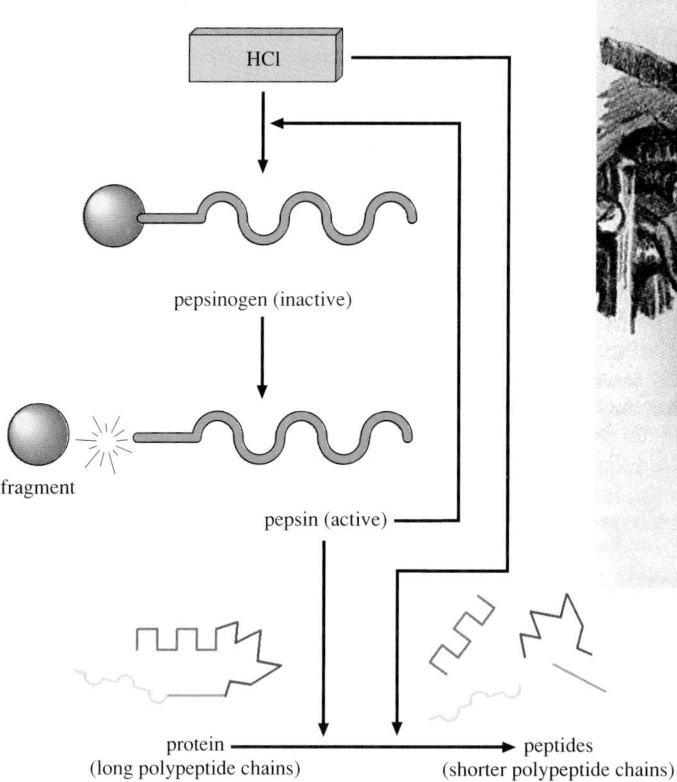

Figure 24.9
Pepsin activation. The hydrochloric acid in the stomach enables pepsinogen to split, yielding pepsin. Pepsin then splits more pepsinogen. Pepsin digests proteins into peptides.

Figure 24.10
Collecting gastric juice. Much of what we know about the stomach's functioning comes from a French-Canadian explorer, Alexis St. Martin, who in 1822 accidentally shot himself in the abdomen. His extensive injuries eventually healed, but a hole, called a fistula, was left, allowing observers to look at his stomach in action. A United States Army surgeon, William Beaumont, spent eight years watching food digesting in the stomach, noting how the stomach lining changed in response to stress.

Chyme is transferred from the stomach to the small intestine in squirts. When the upper part of the small intestine is full of chyme, receptor cells on the outside of the intestine are stimulated, which in turn temporarily shut the pyloric sphincter. Once pressure on the small intestine lessens, the sphincter opens and more chyme enters.

Sometimes the sphincters may not function perfectly. For example, when the sphincter at the upper end of the stomach is unable to hold back the stomach contents, acidic chyme will squeeze into the esophagus. This creates the burning sensation of **heartburn.** A more serious condition is **pyloric stenosis,** which affects premature infants. These children are born with an enlarged circular muscle in the pyloric sphincter, which as a result cannot open properly. The defect must be surgically corrected or the infant will starve.

KEY CONCEPTS

The stomach is a muscular sac that stores food, partially digests it, and passes it on to the small intestine. The stomach has four regions, it has a sphincter on both ends, and it is folded to increase surface area. Gastric juice contains water, mucus, salts, hydrochloric acid, and enzymes. In the presence of hydrochloric acid, pepsinogen is split to yield pepsin, which begins protein digestion. Gastric lipase begins digestion of butterfat. Secretion of gastric juice is controlled by nerves and hormones. The stomach is protected from digesting itself. Chyme is squirted into the small intestine.

The Small Intestine

The small intestine is a 23-foot-long (7-meter-long) tubular organ where digestion is completed and the products are absorbed. The first 10 inches (25 centimeters) are

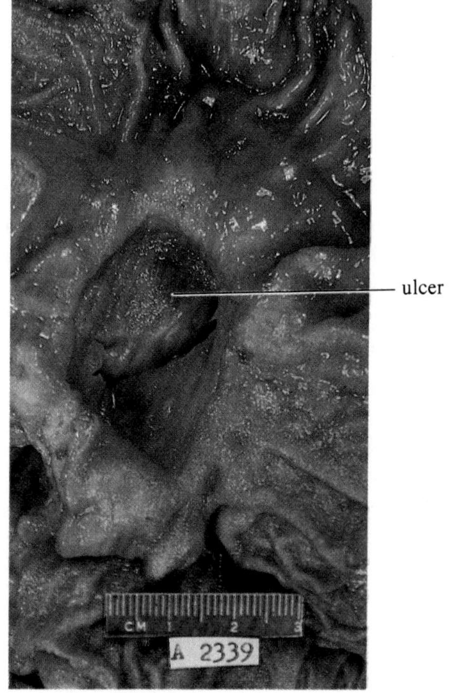

Figure 24.11
A peptic ulcer in the stomach. Stress can provoke the stomach to produce too much gastric juice, which eats away at the stomach lining, causing an ulcer.

called the **duodenum,** the next two-thirds the **jejunum,** and the remainder the **ileum.** Mechanical digestion in this organ results primarily from localized muscle contractions called **segmentation** (fig. 24.12). During this process the chyme is squished back and forth between segments of the small intestine that are created by the temporary contraction of bands of circular muscle. Food is moved along the intestine by peristalsis.

Chemical digestion in the small intestine acts on all three major types of food molecules. Digestive secretions come from the small intestine itself as well as from two glands, the liver and the pancreas. The intestinal secretions contain mucus, water, and various enzymes. The intestinal enzymes are attached to the cell membranes of the epithelial lining cells. This arrangement may help protect the cells from digestion by their own enzymes.

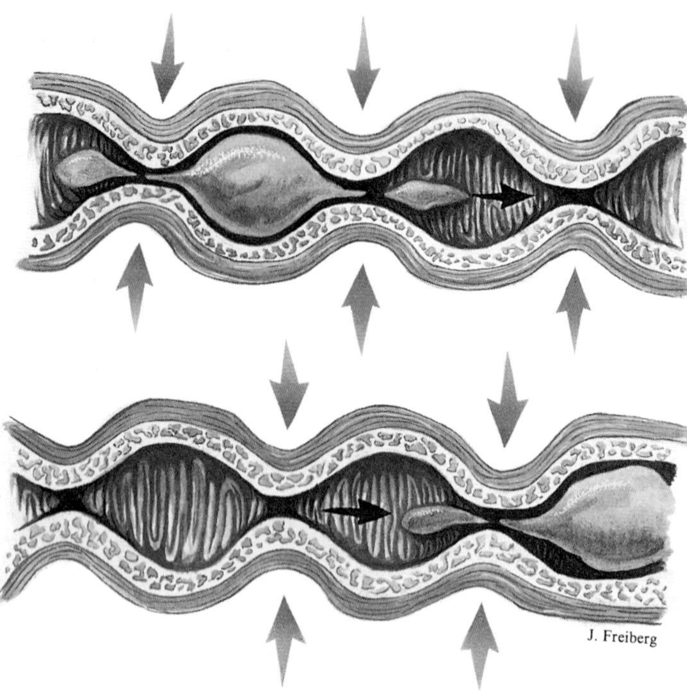

Figure 24.12
Segmentation. Coordinated muscular contraction of different parts of the small intestine at the same time mix the chyme with mucus and digestive enzymes.

The small intestine completes the protein digestion initiated in the stomach. This is accomplished by **peptidase** enzymes manufactured by intestinal cells and by the enzymes **trypsin** and **chymotrypsin,** which are made in the pancreas.

Fats present an interesting challenge to the digestive system. **Lipases,** the enzymes that chemically digest fats, are water soluble, but fats are not. Therefore, lipase can only act at the surface of a fat droplet. The total surface area of many small droplets is greater than the surface area of a single large one. **Bile,** produced by the liver, is needed to emulsify fat, breaking it into droplets small enough to remain in suspension. Bile works as a detergent in much the same way as the dish detergent that cleans grease off dinner plates (fig. 24.13). With the fats emulsified, pancreatic lipases have a much greater surface area from which they can chemically digest fats. Most fats in our diet are triglycerides, which are digested into fatty acids and glycerol.

Some starch has been digested to maltose by salivary amylase in the mouth. The pancreas sends more amylase to the small intestine to digest starches missed in the mouth. The small intestine produces **carbohydrase** enzymes, which chemically break down certain disaccharides into specific monosaccharides. Sucrase, for example, is a carbohydrase that breaks down the disaccharide sucrose into the monosaccharides glucose and fructose.

Deficiency of a particular carbohydrase can cause digestive distress when certain foods are eaten. For example, **lactose intolerance** results from the absence of the

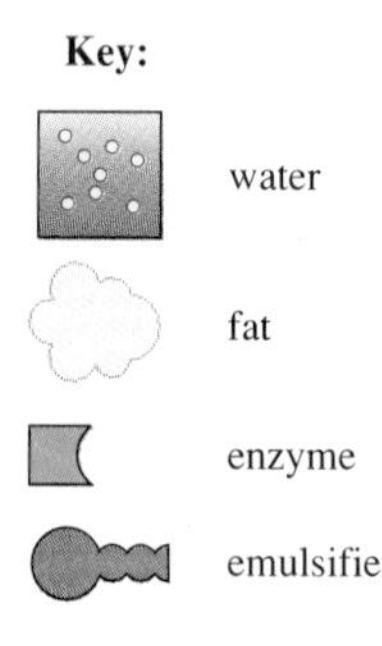

Figure 24.13
Bile emulsifies fats. A familiar type of emulsifier is a detergent, which works by breaking up grease spots on dishes or clothing. Once the spots are small enough, the water washes them away.

enzyme *lactase* in the small intestine. This enzyme breaks down the disaccharide lactose (milk sugar) found in cow's milk into the monosaccharides glucose and galactose. Although lactase is plentiful in infants, it is lacking in some adults. When lactose remains undigested, as it does in lactose intolerance, the sugar is fermented by bacteria in the large intestine. Abdominal pain, gas, diarrhea, bloating, and cramps occur after milk or other dairy products are consumed. These symptoms can be avoided by replacing fresh milk with fermented dairy products such as yogurt, buttermilk, and cheese, in which the lactose has already been broken down. Adding lactase to milk also prevents discomfort.

Hormones coordinate the various digestive activities of the small intestine. When the contents of a meal require a particular digestive fluid, the message is relayed to the appropriate structure via hormones produced by the small intestine. When the stomach squirts acidic chyme into the small intestine, the hormone **secretin** is released by intestinal cells. Secretin triggers the release of bicarbonate from the pancreas, which quickly neutralizes the acidity. In addition, this hormone stimulates the liver to secrete bile. When the chyme contains fats, the intestinal cells secrete another hormone, **cholecystokinin (CCK),** which signals the release of substances needed for fat digestion. This hormone stimulates the liver to continue bile secretion and triggers contraction of the gallbladder, causing the bile stored within to squirt into the small intestine. CCK also releases pancreatic enzymes, including lipase, into the small intestine.

Hormonal regulation of digestive agents helps protect the intestinal cells, because substances are produced when they have something to digest. As an additional safeguard, other glands in the small intestine secrete mucus, which protects the intestinal wall from digestive juices and assists in neutralizing stomach acid. However, the protection offered by mucus is limited. Many intestinal lining cells succumb to the caustic contents. The epithelial lining is replaced every 36 hours, and nearly one-quarter of the bulk of feces consists of dead epithelial cells from the small intestine.

At long last, carbohydrates are digested to monosaccharides, proteins to amino acids, and fats to fatty acids and glycerol (fig. 24.14). These products now travel to the cells lining the small intestine, where they are absorbed. From here they enter the circulation.

Absorption of nutrients must take place at a surface, and that of the small intestine provides an elegant example of biological maximization of surface area (fig. 24.15). The organ is very long, fitting into the abdomen because it is highly coiled. In addition, the innermost layer is corrugated with circular ridges that are almost half an inch high. The surface of every hill and valley of the lining looks velvety due to additional folds; about 6 million tiny projections called **villi** (fig. 24.16). The tall epithelial cells on the surface of each villus bristle with projections of their own called **microvilli.** Together the villi and the 500 microvilli on each cell lining them increase the surface area of the small intestine at least 600 times, creating an area for absorption of 200 to 300 square meters—about the size of a tennis court!

Within each villus is a capillary network into which amino acids, monosaccharides, and water, as well as some vitamins and minerals, pass into the circulatory system for distribution to the cells. The capillary network surrounds a lymph vessel, called a **lacteal,** that absorbs fat molecules. The ways that digested nutrients enter intestinal lining cells illustrate the transport mechanisms discussed in chapter 5.

Monosaccharides are absorbed by facilitated diffusion and active transport. Amino acids are taken up by active transport. The products of fat digestion move into cells by passive diffusion, which is possible because they are soluble in the lipids of the intestinal cell membranes. Within the epithelial cells, the fatty acids and glycerol are reassembled into triglyc-

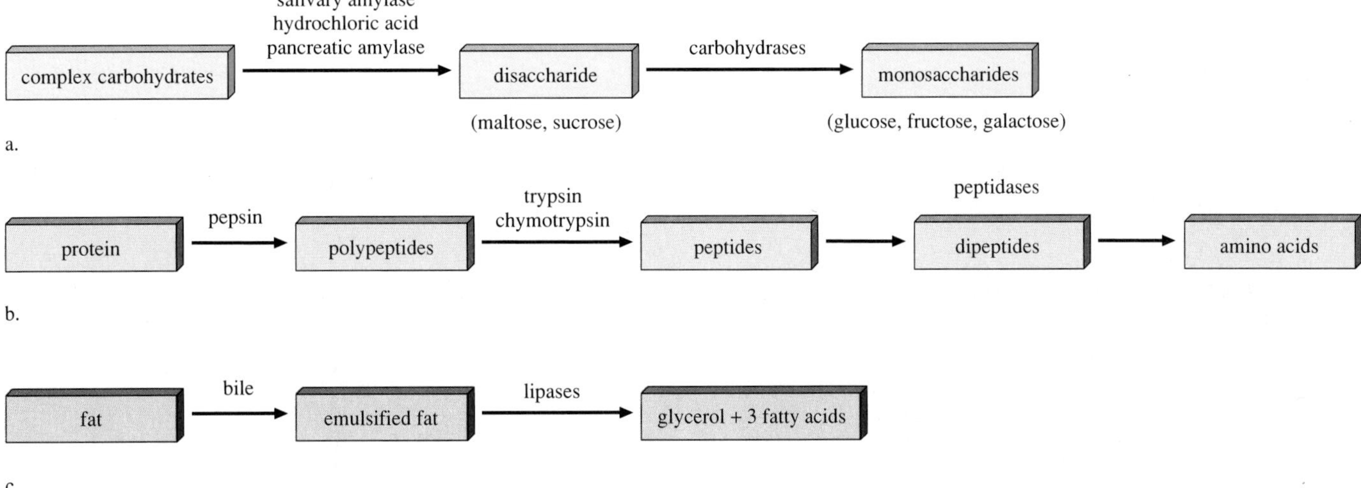

Figure 24.14
Digestion is a gradual breakdown of macromolecules. *a.* The initial stages of carbohydrate digestion are carried out by salivary amylase, hydrochloric acid, and pancreatic amylase. Disaccharides are degraded to monosaccharides by carbohydrases. *b.* Protein digestion begins in the stomach, where pepsin breaks down proteins into polypeptides. In the small intestine, trypsin, chymotrypsin, and peptidases continue the chemical degradation, ultimately yielding amino acids. *c.* Fats are emulsified by bile and then degraded to glycerol and fatty acids by lipases.

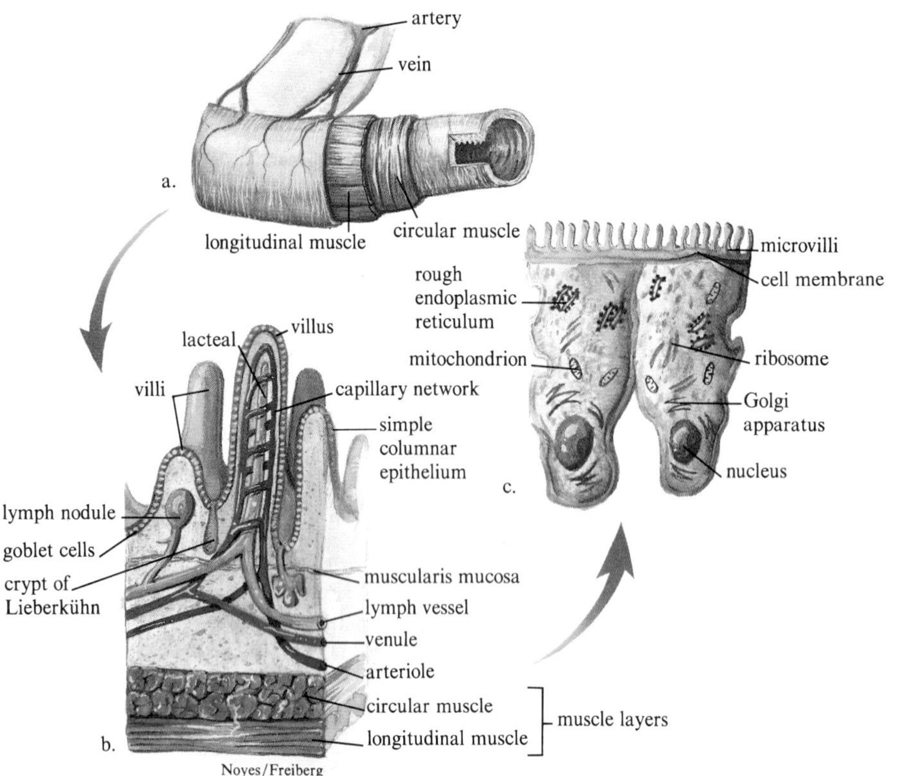

Figure 24.15

The small intestine maximizes surface area. The small intestine's 23 feet (7 meters) are wound into the abdomen. Its surface has ridges, and the inner lining is folded into villi. Each villus cell, in turn, has extensive surface area because its cell membrane is folded into microvilli.

erides, which are coated with a carrier protein to make them soluble before entering the lymphatic system. The fats enter the blood when the lymphatic system joins the circulatory system (fig. 24.17).

KEY CONCEPTS

Digestion is completed in the small intestine. Segmentation movements mechanically digest chyme. Secretions that continue protein digestion to amino acids include peptidases made in the small intestine and trypsin and chymotrypsin supplied by the pancreas. Fats are emulsified by bile, which is made in the liver and stored in the gallbladder, and chemically digested to fatty acids and glycerol by pancreatic lipases. Pancreatic amylase and the small intestine's carbohydrases continue carbohydrate digestion to monosaccharides. Secretin and cholecystokinin are hormones that regulate digestion in the small intestine. Hormonal regulation, mucus, and rapid replacement of cells protect the small intestine from digesting itself. The structure of the small intestine provides ample surface area for absorption of digested nutrients. This absorption takes place through capillaries and lacteals in intestinal villi, using a variety of transport mechanisms.

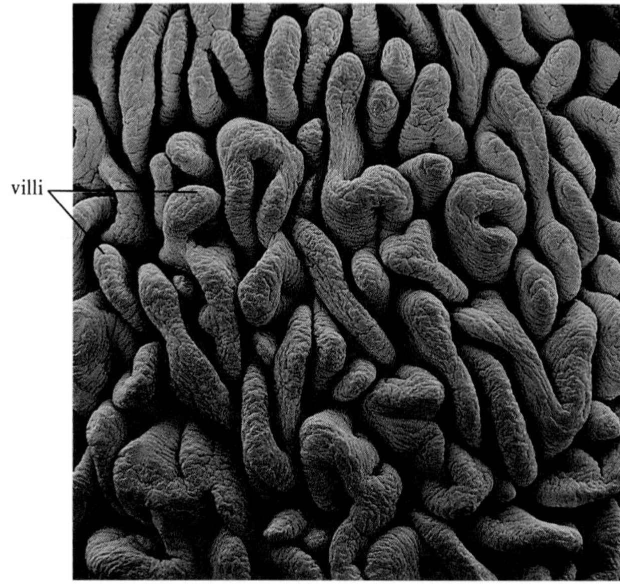

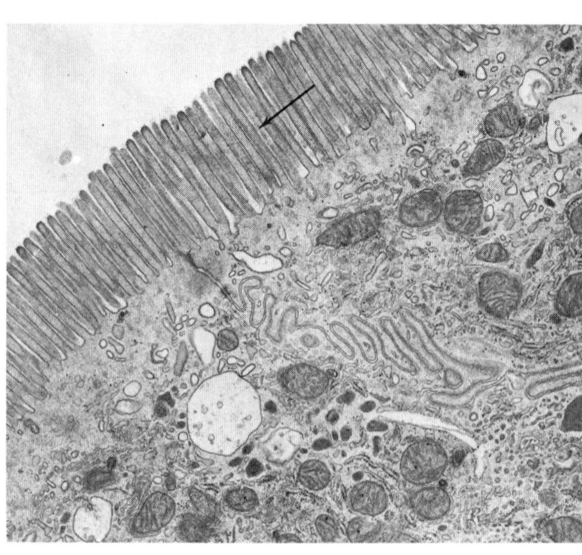

Figure 24.16

Small intestines make spectacular photographs. *a.* The lining of the small intestine is folded into many villi. *b.* The cell membrane of villus cells is folded into hundreds of microvilli.

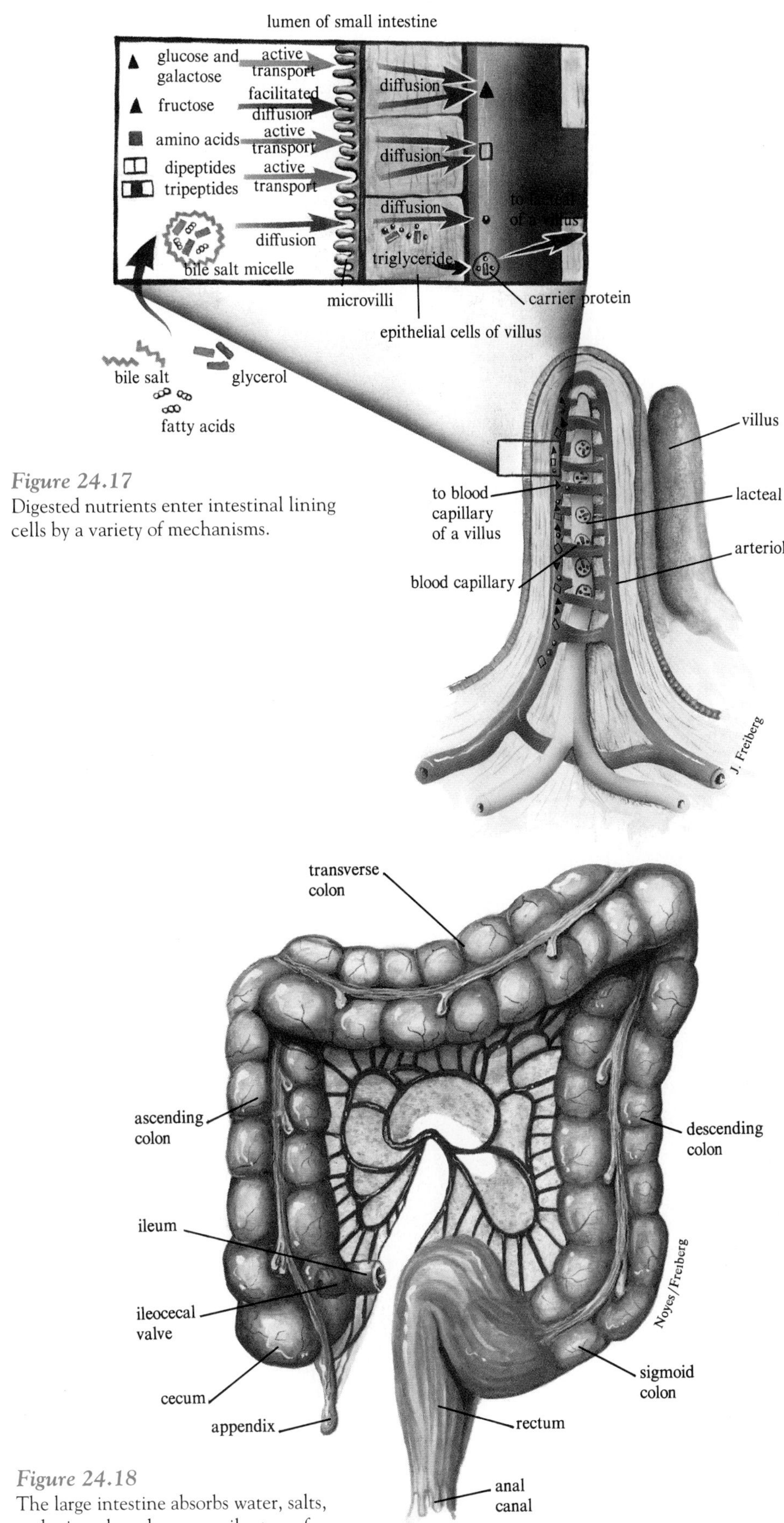

Figure 24.17
Digested nutrients enter intestinal lining cells by a variety of mechanisms.

Figure 24.18
The large intestine absorbs water, salts, and minerals and temporarily stores feces.

The Large Intestine

The material remaining in the small intestine after the nutrients have been absorbed then enters the large intestine, or **colon.** The large intestine is shorter than the small intestine; it is called large because its 2.5-inch (6.5-centimeter) diameter is greater than that of the small intestine. The 5-foot (1.5-meter) tube is arranged around the convoluted mass of the small intestine in the shape of a question mark. A pouch at the start of the large intestine is called the **cecum** (fig. 24.18). Dangling from the cecum is the **appendix,** a thin tube resembling a worm.

In our nonhuman primate ancestors the appendix may have helped digest fibrous plant matter, but today it has no apparent function and can actually be a hindrance. If bacteria or undigested food become trapped in the appendix, the area can become irritated and inflamed and infection can set in, producing severe abdominal pain. Immediate surgical removal is required so that the appendix does not burst and spill its contents into the abdominal cavity, which would produce a severe infection.

The large intestine absorbs most of the remaining water, plus salts and some minerals, from chyme, producing a solid or semisolid mass known as feces. Also present are billions of bacteria, representing about 500 different species that are normal inhabitants of the healthy human large intestine. These "intestinal flora" decompose any nutrients in chyme that escaped absorption in the small intestine; they produce vitamins B_1, B_2, B_6, B_{12}, K, folic acid, and biotin; and they break down bile and foreign chemicals such as those in drugs.

Sometimes the actions of intestinal bacteria can be embarrassing—they produce foul-smelling compounds and gases that cause the characteristic odors of intestinal gas and feces. Intestinal flora also help prevent disease-causing microorganisms from causing an intestinal infection. Antibiotics often kill the normal bacteria, allowing other microbes to grow. This bacterial alteration causes the diarrhea that is sometimes a side effect of using antibiotics.

The remnants of digestion—cellulose, bacteria, bile, and intestinal cells—collect in the final section of the digestive

Reading 24.2 *Disorders Involving the Large Intestine*

THE LARGE INTESTINE IS THE SOURCE OF MANY MEDICAL PROBLEMS, FROM SUCH FAMILIAR DIGESTIVE DISCOMFORTS AS GAS TO MORE SERIOUS DISORDERS.

Intestinal Gas

People do not often talk about intestinal gas, but this common evidence of digestion is a source of pain and sometimes embarrassment to many. What exactly is in intestinal gas? Most of it is nitrogen and oxygen gulped in while breathing and eating. Undigested food fermented by bacteria contributes methane (CH_4), carbon dioxide (CO_2), and hydrogen. These five gases account for 99% of intestinal gas. The other 1% comes from compounds also produced by intestinal bacteria, and these impart foul odors. Intestinal gas can be minimized by eating slowly, avoiding milk if you are lactose intolerant, and not eating gas-inducing foods—beans, bagels, bran, broccoli, brussels sprouts, cabbage, cauliflower, and onions.

Diarrhea

Because the large intestine absorbs water from material within it, the rate of movement through it determines the consistency of feces. **Diarrhea,** the frequent and too-rapid passage of loose feces, results when material moves along so quickly that too little water is absorbed. The condition may reflect poisoning, infection, a diet too high in fiber, or nervousness. When the large intestine is the site of an infection or chemical irritation, diarrhea protects by flushing toxins out of the body. Diarrhea is also caused by nervousness or eating too much fiber.

Constipation

Constipation, the infrequent passage of hard feces, is caused by abnormally slow movement of fecal matter through the large intestine. Because the feces remain in the large intestine longer than usual, excess water is absorbed. Constipation can be caused by a failure of the sensory cells in the rectum to signal the spinal cord to defecate or by the conscious suppression of defecation, both of which can be a result of emotional stress. A diet low in fiber can also cause constipation by slowing fecal movement through the large intestine. Eating foods high in fiber, drinking at least eight 8-ounce glasses of water a day, and regular exercise can prevent constipation.

More Serious Disorders

In **diverticulosis,** parts of the intestinal wall weaken, and the inner mucous membrane protrudes through (fig. 1). Many times there are no symptoms, but if the outpouching becomes blocked with chyme and then infected (a condition called

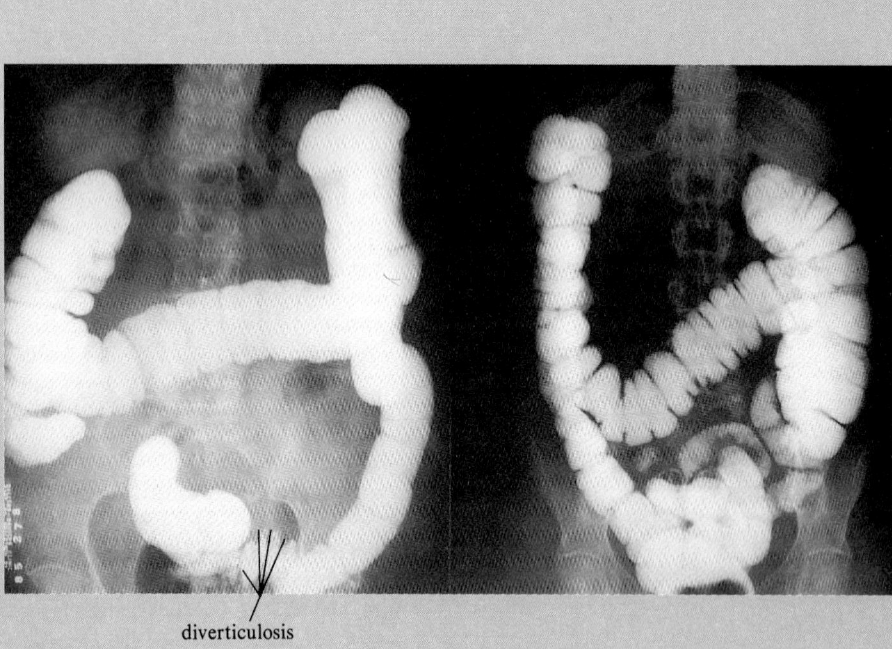

Figure 1
Diverticulosis of the large intestine. The large intestine on the right is healthy. The organ on the left shows diverticula on the left side.

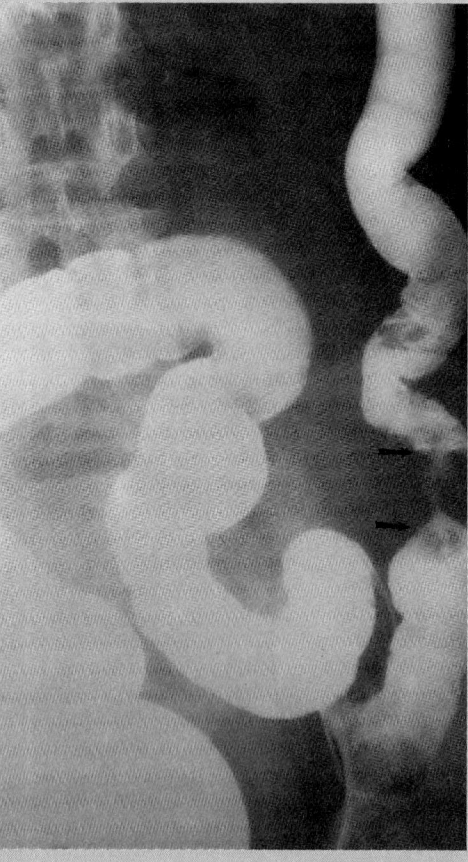

Figure 2
To detect colon cancer, a patient receives an enema containing barium. The barium highlights the lower digestive tract, revealing an obstruction caused by a tumor.

diverticulitis), antibiotics or surgery may be required. More than half of all Americans develop diverticulosis by age 60. It is not known exactly what causes diverticulosis, but the fact that it is nonexistent in populations whose diets are high in fiber and that it was not known in the United States before refined foods were introduced earlier this century suggests that lack of dietary fiber may be a cause. Fiber speeds the movement of material through the digestive system, hastening exit of toxic substances.

In **ulcerative colitis,** inflammation of the inner lining of the colon and rectum produces abdominal and rectal pain, bloody diarrhea, and weight loss. Drugs can often treat the symptoms, but sometimes removal of the colon is necessary. Again, the cause of this disorder is not known.

Cancer of the large intestine and rectum, known as **colorectal cancer,** is the fourth most prevalent cancer in the United States, with 113,000 new cases and nearly 60,000 deaths yearly. Symptoms include a change in the frequency or consistency of bowel movements, bloody feces, and abdominal pain. A home test kit called a **hemoccult** (hidden blood) **test** can detect intestinal bleeding that may signal the presence of a cancer. Blood in feces is often black and not detectable by sight. Follow-up at a doctor's office entails use of a fiber-optic colonoscope that searches and samples colorectal tissue for cancer. Like diverticulosis, colorectal cancer is linked to lack of fiber in the diet.

Cancer or other severe maladies may require that the large intestine be removed. A new opening for feces to exit the body is then needed. The free end of the intestine can be surgically attached to an opening created through the skin of the abdomen, and a bag attached to the opening to collect the fecal matter. This procedure is called a **colostomy.**

tract as feces. This 6- to 8-inch- (15- to 20-centimeter-) long region is called the **rectum.** The digestive system ends in the **anal canal,** which is 1 inch (2.5 centimeters) long. The opening to the anal canal, the **anus,** is usually closed by two sets of sphincters, an inner one of smooth muscle, which is under involuntary control, and an outer one of skeletal muscle, which is voluntary. When the rectum is full, receptor cells trigger a reflex that eliminates the feces. Other nerves can override the reflex. During childhood, voluntary control of the defecation reflex is acquired, much to the relief of parents. Figure 24.19 summarizes the steps of digestion in humans.

KEY CONCEPTS

In the large intestine, water, salts, and minerals are absorbed, and bacteria decompose leftover nutrients and foreign chemicals and produce several vitamins. Feces are stored briefly in the rectum and pass through two sphincters to exit the digestive tract at the anus.

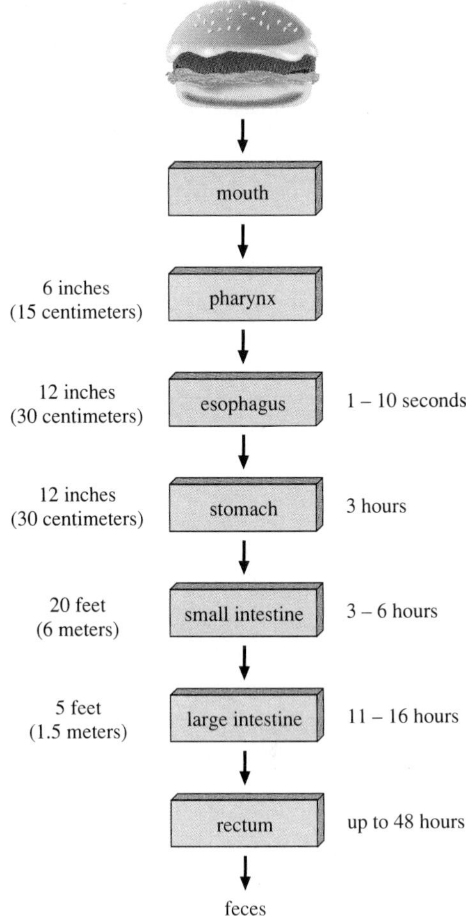

Figure 24.19
The journey of food. Food enters the human body at the mouth, then moves through a system of tubes and hollow organs. Along this pathway, food is mechanically and chemically degraded. Digested nutrients are absorbed into the blood or lymphatic systems at the small intestine. Material remaining after further absorption in the large intestine is stored in the rectum and eliminated from the body as feces.

Associated Glands and Organs

The **pancreas** is a multifunctional structure associated with the digestive tract (fig. 24.20). This gland sends about a liter of fluid to the duodenum each day, including trypsin and chymotrypsin to digest polypeptides, pancreatic amylase to digest carbohydrates, pancreatic lipase to further break down emulsified fats, and nucleases to degrade DNA and RNA. Pancreatic "juice" also contains sodium bicarbonate to neutralize the acidity of chyme caused by hydrochloric acid added in the stomach. Recall from chapter 19 that the pancreas also functions as an endocrine gland, regulating blood-sugar level.

Weighing about 3 pounds (1.4 kilograms), the **liver** is the largest solid organ in the body. It has many different functions, including detoxification of harmful substances in the blood, storage of glycogen and fat-soluble vitamins, synthesis of blood proteins, and monitoring the blood-sugar level. The liver's contribution to digestion is the production of greenish yellow **bile.** Bile is stored in the **gallbladder** until fatty chyme in the small intestine triggers its release. The cholesterol in bile sometimes crystallizes, forming one or more gallstones that partially or completely block the duct to the small intestine (fig. 24.21).

Bile consists of pigments derived from the breakdown products of hemoglobin, salts, and cholesterol. This colorful substance is responsible for the brown color of feces, the pale yellow of blood plasma and urine, and the abnormal yellow complexions of people with **jaundice,** in which the bile pigment bilirubin is overproduced and is deposited in the skin (Reading 24.3).

Jaundice can be a symptom of liver inflammation, which is called **hepatitis.** At least four different viruses are responsible for hepatitis, which occurs in forms known as A, B, C, and D. Symptoms of each include overwhelming fatigue, jaundice, dark urine, and pale-colored feces. It usually takes several weeks of bed rest to recover.

Type A hepatitis is usually transmitted through food. Restaurant workers who carry the virus and do not wash thoroughly before handling food pass on many cases of hepatitis A. One taco maker in a New Jersey restaurant, for example, passed the disease to 55 patrons who ate the lettuce he had shredded. Hepatitis A can also be picked up from eating clams or oysters from waters contaminated with the virus in sewage.

The hepatitis B virus is usually transmitted when a body fluid (blood, semen, or saliva) from an infected person contacts a break in the skin of an uninfected person. Drug abusers often contract hepatitis B by sharing needles. Fortunately the hepatitis B virus has been nearly eliminated from the blood supply for transfusions, and a highly effective vaccine is available. In addition, pregnant women are routinely screened for evidence of hepatitis B infection so that their babies, who are very likely to be infected, can be treated shortly after birth with the vaccine, which seems to prevent hepatitis-associated medical problems. The newly recognized D form of hepatitis can occur only if hepatitis B is already present. It is detected by additional viral proteins in the person's blood. Cases of hepatitis that are not associated with the A, B, or D virus are called non-A non-B hepatitis and include at least two additional viruses, one of which is called C. About 3% to 7% of people receiving blood transfusions contract non-A non-B hepatitis, and many of these go on to develop complications, including cirrhosis (in which scar tissue replaces liver tissue) and liver cancer.

You now have a much better idea of how your body handles a meal. The next chapter will show you how to choose the best possible meals. It contains information not usually found in a biology text but that you will probably find very useful.

KEY CONCEPTS

The pancreas is a gland that supplies amylase, trypsin, chymotrypsin, lipase, and nucleases to digest various nutrients and sodium bicarbonate to neutralize stomach acid. The liver supplies bile, which emulsifies fats, and the gallbladder stores bile. Jaundice results from excess bile pigments. Hepatitis is inflammation of the liver.

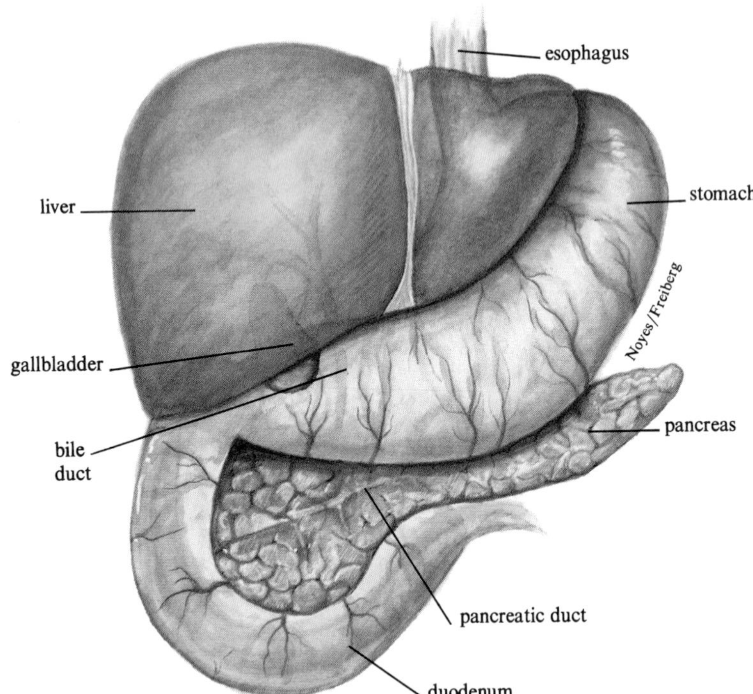

Figure 24.20
Associated structures of the digestive system include the liver, gallbladder, and pancreas.

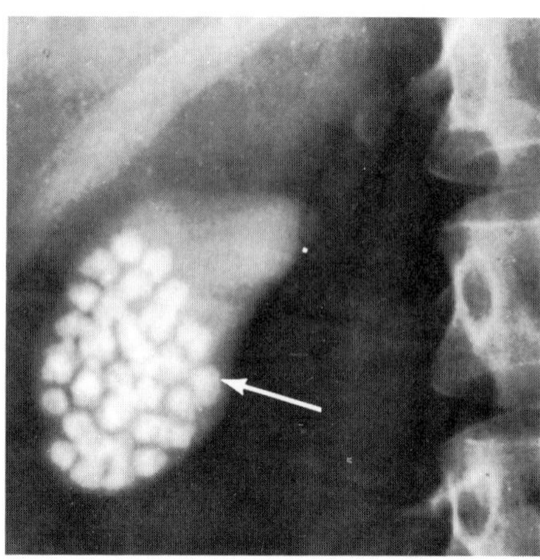

a.

b.

Figure 24.21
a. This X-ray shows gallstones in a gallbladder. *b.* The gallbladder has been opened to reveal the stones within. The dime indicates the size of the stones.Gallstones are easily removed by using laser "band-aid" surgery.

Reading 24.3 A Symptom With Many Meanings—Jaundice

THE YELLOWISH SKIN AND EYE WHITES OF JAUNDICE CAN REFLECT ABNORMALITIES IN THE PRODUCTION, DELIVERY, OR BREAKDOWN OF BILE.

In hemolytic anemia abnormally high numbers of red blood cells break apart, overloading the liver with hemoglobin. This leads to overproduction of bilirubin, the bile pigment responsible for the characteristic yellow color. Jaundice might also be caused by a blocked bile duct. This prevents the release of bile from the gallbladder to the small intestine, increasing the amount of bile absorbed into the bloodstream. Jaundice can also signify that the liver cells are malfunctioning, perhaps due to infection or inflammation (hepatitis).

Jaundice is common in hospital newborn nurseries. While in the uterus the fetus has more red blood cells than it needs after birth, and these cells are broken down during the first week of life. A fetus depends on its mother to eliminate excess bilirubin; infants' enzymes to do this may not be plentiful enough. Usually a newborn's ability to excrete bilirubin is activated by the 10th day of life, but by this time the accumulated bilirubin may have caused brain damage.

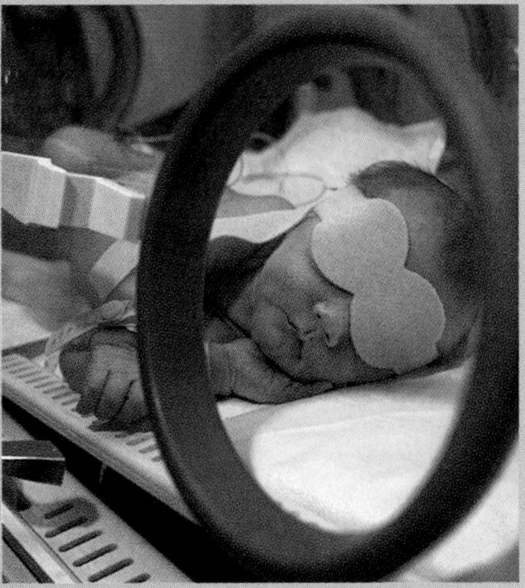

Figure 1
An infant undergoing phototherapy for jaundice wears only a diaper and goggles, which protect the eyes from the ultraviolet radiation. The "bili lights" help the infant's immature liver to break down bilirubin.

Fortunately, an effective treatment for newborn jaundice was found by a nurse working with premature babies in a British hospital in 1958. The nurse liked to take her tiny charges out in the sun, and she noticed that a child whose skin had a yellow pallor developed normal pigmentation when he lay in the sun. However, the part of the child's body that was covered by a diaper, and therefore had not been exposed to the sun, remained yellow. Further investigation showed that sunlight enables the body to break down bilirubin. Today, thousands of jaundiced newborns spend the first few days of their lives under artificial "bili lights" (fig. 1).

SUMMARY

The digestive system breaks food molecules into their component building blocks and absorbs the products. Mechanical digestion breaks food into smaller pieces that are more easily chemically digested by hydrolytic enzymes. Proteins are digested to amino acids, fats (triglycerides) are broken down to fatty acids and glycerol, and starches are broken down to monosaccharides. The circulatory system can then distribute the products to the cells, where they are either oxidized to provide energy for cellular activity, used to make new cellular materials, or stored.

In simple organisms such as protists and sponges, digestion is *intracellular,* but in other organisms it is *extracellular* and occurs in a special cavity outside the cells. Digestive cavities can have one opening or two. In a system with two openings such as ours, food enters through the mouth and is digested and absorbed, and undigested material leaves through a second opening, the anus.

In humans, food is processed first in the mouth. Teeth break the food into smaller pieces. The *salivary glands* produce *saliva,* which moistens food and begins the digestion of starch. The tongue moves food, mixes it with saliva, and tastes it. Food is swallowed and moves through the *esophagus* to the stomach. The movement of food along the digestive system occurs by a wave of contraction called *peristalsis.*

In the stomach, food is stored, churned until it is liquefied, and mixed with gastric juice. Hydrochloric acid in the gastric juice activates another component, *pepsinogen,* forming the protein-splitting enzyme *pepsin.* In the small intestine digestion is completed and the products are absorbed, using enzymes produced by both the small intestine and the pancreas. *Bile,* made by the liver and stored in the gallbladder, emulsifies fats in the small intestine to facilitate digestion by lipases. The surface area for absorption is tremendous due to the length of the small intestine, the circular folds, the *villi,* and *microvilli.* In the villus, amino acids and monosaccharides are absorbed into the capillaries, and the products of fat digestion enter the lacteal.

The large intestine absorbs water, minerals, and salts, thereby adjusting the consistency of the feces. Many bacteria digest remaining nutrients and produce useful vitamins.

The release of digestive secretions is regulated both by the nervous and the endocrine systems. Reflexes trigger the release of enzymes when food is present in the mouth or gut. Produced by the stomach when food is present, the hormone *gastrin* causes the release of gastric juice. The release of *secretin* and *cholecystokinin* by cells of the small intestine coordinates digestive secretion by the small intestine, liver, and pancreas.

QUESTIONS

1. How do the circulatory, muscular, nervous, and endocrine systems take part in digestion?

2. What are the products of digestion of starches, proteins, and of fats?

3. What happens in the digestive tract to cause the following:

a. Heartburn **b.** Diarrhea
c. Hepatitis **d.** Ulcers

4. What is the difference between mechanical digestion and chemical digestion? How does mechanical digestion facilitate chemical digestion?

5. What structures or mechanisms ensure that the stomach and small intestine do not digest themselves?

6. How is surface area maximized in the stomach and in the small intestine? Why is it necessary for a digestive system to have extensive surface area?

7. Is the presence of bacteria in the large intestine a sign of infection? Why or why not?

TO THINK ABOUT

1. How can people consume vastly different diets, yet all obtain adequate nourishment?

2. A few years ago, health-food stores marketed a protein derived from kidney beans as a "starch blocker." It worked by inhibiting the action of amylase. Supposedly, one could consume huge amounts of starch without digesting it. After a summer in which millions of people took the pills and packed away plates of pasta, studies found that the starch blocker did not work. Many people developed constipation, diarrhea, gas, nausea, and vomiting. Why do you think the starch blocker failed? What could have caused the side effects?

3. Why can't nutritional benefit be gained by finely grinding meat and injecting it into the bloodstream?

4. In recent years, hepatitis has been seen increasingly among families whose children attend the same day-care centers. What type of hepatitis is this likely to be? Suggest a mechanism for the spread of hepatitis in a day-care center.

5. If you had to pick one organ or gland from the gastrointestinal tract and engineer an artificial version of it, which would be the most difficult to replicate? Why?

SUGGESTED READINGS

Farley, Dixie. April 1988. Living with inflammatory bowel disease. *FDA Consumer.* Nearly 2 million Americans suffer from cramps and diarrhea caused by two disorders of the intestines—Crohn's disease and ulcerative colitis.

Fox, Stuart Ira. 1990. *Human Physiology*, Dubuque, Iowa: Wm. C. Brown Publishers. Chapter 18 presents a detailed look at the functioning of the human digestive system.

Lewis, Ricki. May 1991. The gallbladder, an organ you can live without. *FDA Consumer.* Gallstones can be very painful, but are easily treated.

Miller, Roger W. April 1987. The digestive discomfort of gas. *FDA Consumer.* The cause of gas lies in the large intestine.

Murphy, Frederick A. February 2, 1990. Protection against viral hepatitis. *Morbidity and Mortality Weekly Report.* A group of at least four unrelated viruses causes inflammation of the liver and its symptoms.

Noble, Robert C. Spring 1990. Death on the half shell: The health hazards of eating shellfish. *Perspectives in Biology and Medicine.* Several digestive problems can result from eating unclean shellfish.

Turbak, Gary. February–March 1991. The cat survives by a hare. *National Wildlife.* The lynx can survive on a very limited diet.

Weck, Egon. July–August 1987. New hope for those with diverticular disease. *FDA Consumer.* One consequence of a low-fiber diet could be diverticulosis.

C H A P T E R

25

Nutrition

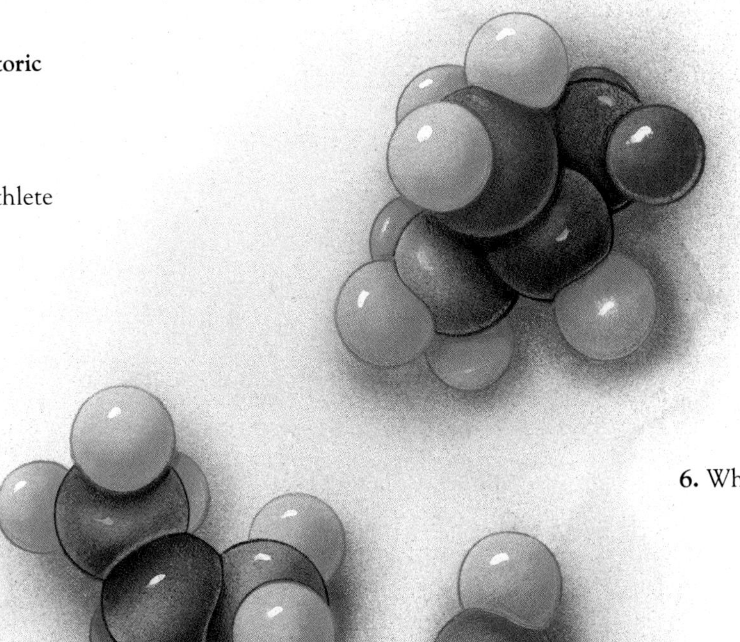

Chapter Outline

Human Nutrition—From a Prehistoric Meal to Fast Food
The Nutrients
Planning a Balanced Diet
Reading 25.1 Nutrition and the Athlete
The Food Group Plan
Dietary Guidelines
The Exchange System
Nutrient Deficiencies
Reading 25.2 Designing a Personal Diet Plan Using the Exchange System
Starvation
Marasmus and Kwashiorkor
Anorexia Nervosa
Bulimia
Overweight
Gaining Weight
Reading 25.3 Human Milk—The Perfect Food for Human Babies

Learning Objectives

By the chapter's end, you should be able to answer these questions:

1. What are the nutrients for humans?
2. How is the energy content of foods measured?
3. What are some ways to plan a diet?
4. What forms of starvation are seen in developing nations?
5. What are eating disorders?
6. Why is being overweight dangerous?

Hunters of long ago never knew where their next meal would come from. With luck, they might kill a large deer, providing many meals. The cooked deer meat was stringy, dry, and rather tough, but 82% of its calories came from protein, an important nutrient for the strongly muscled hunters. The other 18% of the calories came from fat, half of it polyunsaturated and therefore not likely to build up in the hunters' hearts and blood vessels.

Many animals migrated, so meat meals were infrequent. Fortunately grains, fruits, and vegetables were abundant, even though none were yet planted intentionally. However, plant foods were so naturally varied that these Cro-Magnon people regularly consumed adequate levels of all important vitamins and minerals. The large amounts of green vegetables eaten provided sufficient calcium for an adult diet.

The back teeth of these long-ago people wore down from the extensive chewing that their high-fiber diet required, so that 30,000 years later, anthropologists would discover their teeth and infer that they had eaten predominantly plants. Those teeth were free of cavities, because except for an occasional lucky find of honey, sweets were the sugars in fruits, which also contained fiber to brush the stickiness from tooth surfaces.

All in all, these Cro-Magnons were quite a healthy lot. Their remains leave little evidence of the heart disease, cancer, stroke, and diabetes that plague their modern-day descendants, although these conditions may not have arisen simply because the life span was shorter. Nor were these ancient people obese, for exercise was essential to their survival. Their preserved bones have large, dense areas that are points where muscles once attached, attesting to their great muscular strength. The Cro-Magnon tended to die from infection, accidents, or environmental extremes. Their prehistoric diet, high in protein and complex carbohydrates and low in fat, was quite similar to the type of diet recommended today to reduce risk of heart disease.

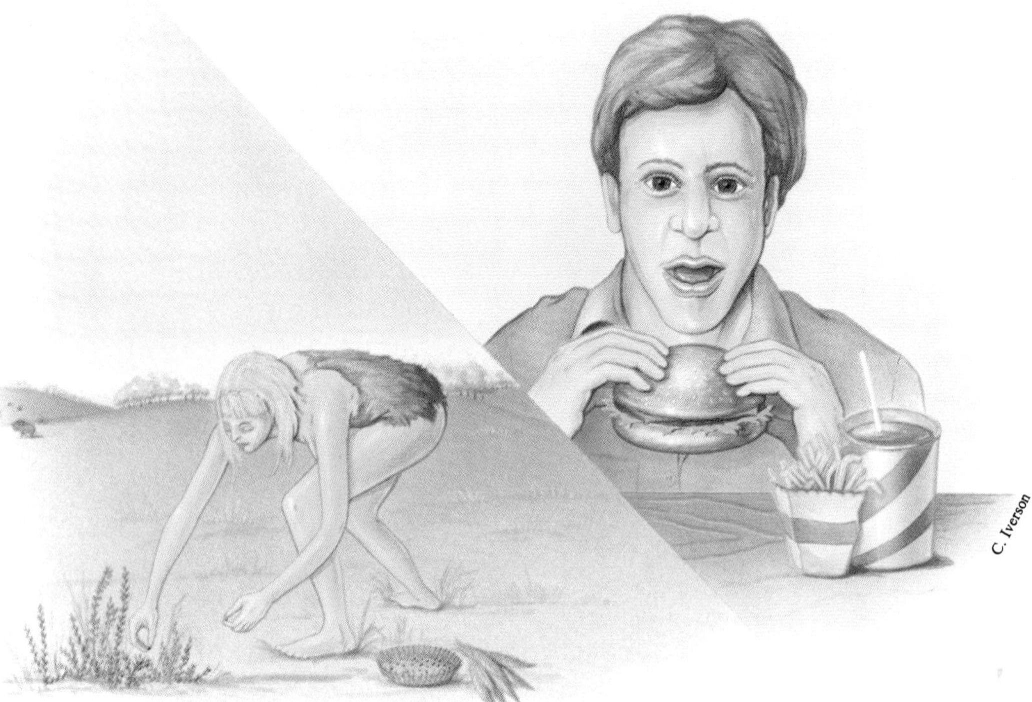

Figure 25.1
The Cro-Magnon dines on fruit and buffalo; the 1990s human wolfs down a hamburger.

Human Nutrition—From a Prehistoric Meal to Fast Food

A thousand generations after the Cro-Magnons dwelled in European caves, diets have changed dramatically. Lunch at a typical fast-food restaurant adds up to far more calories than a meal eaten in a Cro-Magnon cave. Unlike the muscular animal that provided the Cro-Magnon's rare meat meal, the cow that became the fast-food hamburger was artificially fattened to the point that its meat yields 16% of its calories from protein and 84% of its calories from fat—and only 5% of that fat is the healthier polyunsaturated variety (fig. 25.1).

Although the Cro-Magnon hunter and junk-food fan eat vastly different foods, each eats for the same reasons. A substance that is obtained from food and used in the body to promote growth, maintenance, and repair is a **nutrient.** Nutrition is the study of nutrients and their fate in the body, including their ingestion, digestion, absorption, transport, metabolism, interaction, storage, and excretion.

Many aspects of nutrition have already been discussed—chemistry of the nutrients in chapter 3, how they are metabolized in chapter 6, and how they are digested in chapter 24. This chapter will explore how nutritional information can be used to plan a healthy diet—something the Cro-Magnons of 30,000 years ago seemed to know how to do instinctively.

KEY CONCEPTS

Animals consume nutrients to promote growth, maintenance, and repair of body tissues. The field of nutrition examines the nutrients and what happens to them in the human body.

The Nutrients

The nutrients include carbohydrates, proteins, and fats, which are called **macronutrients** because they are needed in large amounts. They are also called **energy nutrients** because they supply energy. Vitamins and minerals are needed in very small amounts and are called **micronutrients.** Water is also a vital part of the diet.

Table 25.1
Vitamins and Health

Vitamin	*Function*	*Food Sources*	*Deficiency Symptoms*
Water-Soluble Vitamins			
Thiamine (vitamin B_1)	Growth, fertility, digestion, nerve cell function, milk production	Pork, beans, peas, nuts, whole grains	Beriberi (neurological disorder), loss of appetite, swelling, poor growth, heart problems
Riboflavin (vitamin B_2)	Energy use	Liver, leafy vegetables, dairy products, whole grains	Hypersensitivity of eyes to light, lip sores, oily dermatitis
Pantothenic acid	Growth, cell maintenance	Liver, eggs, peas, potatoes, peanuts	Headache, fatigue, poor muscle control, nausea, cramps*
Niacin	Growth	Liver, meat, peas, beans, whole grains, fish	Dark rough skin, diarrhea, mouth sores, mental confusion (pellagra)
Pyridoxine (vitamin B_6)	Protein use	Red meat, liver, corn, potatoes, whole grains, green vegetables	Mouth sores, dizziness, nausea, weight loss, neurological disorders*
Folic acid	Manufacture of red blood cells, metabolism	Liver, navy beans, dark green vegetables	Anemia, cancer
Biotin	Metabolism	Meat, milk, eggs	Skin disorders, muscle pain, insomnia, depression*
Cyanocobalamin (vitamin B_{12})	Manufacture of red blood cells, growth, cell maintenance	Meat, organ meats, fish, shellfish, milk	Pernicious anemia
Ascorbic acid (vitamin C)	Growth, tissue repair, bone and cartilage formation	Citrus fruits, tomatoes, peppers, strawberries, cabbage	Weakness, gum bleeding, weight loss (scurvy)
Fat-Soluble Vitamins			
Retinol (vitamin A)	Night vision, new cell growth	Liver, dairy products, egg yolk, vegetables, fruit	Night blindness, rough dry skin
Cholecalciferol (vitamin D)	Bone formation	Fish-liver oil, milk, egg yolk	Skeletal deformation (rickets)
Tocopherol (vitamin E)	Prevents certain compounds from being oxidized	Vegetable oil, nuts, beans	Anemia in premature infants*
Vitamin K	Blood clotting	Liver, egg yolk, green vegetables	Bleeding, liver problems*

*Deficiencies of these vitamins are rare in humans, but they have been observed in experimental animals.

Carbohydrates, which include the sugars and starches, are the major source of energy, which cells extract from the covalent bonds of glucose. Complex carbohydrates and disaccharides are broken down to glucose during digestion, and the glucose is metabolized to yield energy in the form of ATP. Energy can also be derived from proteins and fats under certain conditions, but these nutrients have other major functions. Tables 25.1 through 25.4 provide details on the specific vitamins and minerals. Recall that vitamins function as coenzymes, and minerals are essential to many biochemical activities in the body.

Table 25.2
Vitamin Fallacies and Facts

Fallacy	*Fact*
Vitamins from foods are superior to those in tablet form	A vitamin molecule is chemically the same, whether it comes from a bean or a bottle
The more vitamins, the better	Too much of a water-soluble vitamin results in excretion of the vitamin through urination; too much of a fat-soluble vitamin can harm one's health
A varied diet provides all needed vitamins	Many people do need vitamin supplements, particularly pregnant and breast-feeding women
Vitamins provide energy	Vitamins do not directly supply energy; they aid in the release of energy from carbohydrates, fats, and proteins

Table 25.3
Bulk Minerals in the Human Diet

Mineral	*Food Sources*	*Functions in the Human Body*
Calcium	Milk products, green leafy vegetables	Bone and tooth structure, blood clotting, hormone release, nerve transmission, muscle contraction
Chloride	Table salt, meat, fish, eggs, poultry, milk	Digestion in stomach
Magnesium	Green leafy vegetables, beans, fruits, peanuts, whole grains	Muscle contraction, nucleic acid synthesis, enzyme activity
Phosphorus	Meat, fish, eggs, poultry, whole grains	Bone and tooth structure
Potassium	Fruits, potatoes, meat, fish, eggs, poultry, milk	Body-fluid balance, nerve transmission, muscle contraction, nucleic acid synthesis
Sodium	Table salt, meat, fish, eggs, poultry, milk	Body-fluid balance, nerve transmission, muscle contraction
Sulfur	Meat, fish, eggs, poultry	Hair, skin, and nail structure, blood clotting, energy transfer, detoxification

Table 25.4
Trace Minerals Important to Human Health

Mineral	*Food Sources*	*Functions in the Human Body*
Chromium	Yeast, pork kidneys	Regulates glucose utilization
Cobalt	Meat, eggs, dairy products	Part of vitamin B_{12}
Copper	Organ meats, nuts, shellfish, beans	Part of many enzymes, storage and release of iron in red blood cells
Fluorine	Water in some areas	Maintains dental health
Iodine	Seafood, iodized salt	Part of thyroid hormone
Iron	Meat, liver, fish, shellfish, egg yolk, peas, beans, dried fruit, whole grains	Transport and utilization of oxygen (as part of hemoglobin and myoglobin), part of certain enzymes
Manganese	Bran, coffee, tea, nuts, peas, beans	Part of certain enzymes, bone and tendon structure
Selenium	Meat, milk, grains, onions	Part of certain enzymes, heart function
Zinc	Meat, fish, egg yolk, milk, nuts, some whole grains	Part of certain enzymes, nucleic acid synthesis

Essential nutrients are those that must be obtained from the diet because the body cannot synthesize them. They are listed in table 25.5. **Nonessential nutrients,** such as 11 of the 20 amino acids, are found in foods, but they are also synthesized in the body.

The amount of energy that a nutrient provides is measured in units called **kilocalories** (kcal)—these are the calories with which we are familiar. Caloric content of a food is determined by burning it in a **bomb calorimeter,** which is a chamber surrounded by water (fig. 25.2). Burning food placed in the chamber raises the temperature of the water, and this temperature change is converted into kilocalories. One kilocalorie is the energy needed to raise 1 gram of water 1 degree Celsius. Bomb calorimetry studies have shown that 1 gram of carbohydrate yields 4 kilocalories; 1 gram of protein yields 4 kilocalories; and 1 gram of fat yields 9 kilocalories. To picture these amounts, a teaspoon of dried food weighs approximately 5 grams.

Good nutrition is a matter of balance. Taking in more kilocalories than are expended in basal metabolism (the energy used to stay alive when at rest when not digesting food) and a lack of exercise lead to overweight; not consuming sufficient kilocalories to support the activities of life leads to underweight and, if taken to extreme, starvation. Balancing the vitamins and minerals is important to health too.

KEY CONCEPTS

Macronutrients are the energy nutrients—carbohydrates, proteins, and fats. Micronutrients include the vitamins and minerals. Water is also a nutrient. The energy obtained from nutrients is measured in kilocalories. One gram of carbohydrate and 1 gram of protein each yields 4 kilocalories, and 1 gram of fat yields 9 kilocalories.

Table 25.5
Essential Nutrients in the Human Diet

Carbohydrate	*Fat*	*Protein*	*Vitamins*	*Minerals*	*Water*
Glucose	Linoleic acid	Histidine	Vitamin A	Calcium	Water
	Linolenic acid	Isoleucine	Vitamin D	Chloride	
		Leucine	Vitamin E	Chromium	
		Lysine	Vitamin K	Copper	
		Methionine	Thiamin	Fluoride	
		Phenylalanine	Riboflavin	Iodine	
		Threonine	Niacin	Iron	
		Tryptophan	Pantothenic acid	Magnesium	
		Valine	Biotin	Manganese	
			Vitamin B_6	Molybdenum	
			Vitamin B_{12}	Nickel	
			Folate	Phosphorus	
			Vitamin C	Potassium	
				Selenium	
				Silicon	
				Sodium	
				Sulfur	
				Vanadium	
				Zinc	

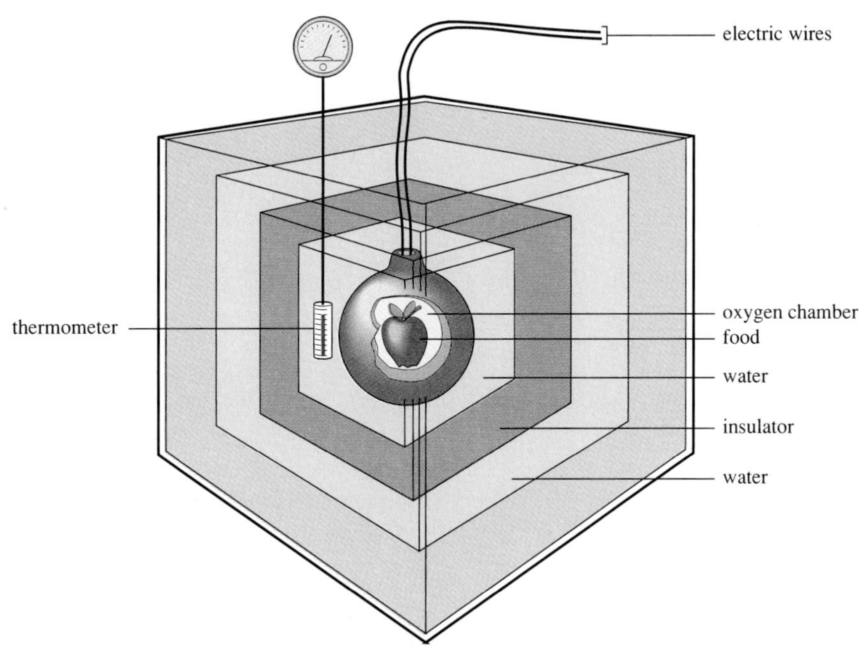

Figure 25.2
The number of kilocalories provided by a particular food is determined by burning the food in a bomb calorimeter. Each degree celsius that 1 gram of water is raised by the burning food is equivalent to 1 kilocalorie of energy. The apple in the diagram holds about 80 kilocalories of energy.

Planning a Balanced Diet

Consuming optimal amounts of all the nutrients can be complicated. To help people follow adequate diets, the United States government releases **Recommended Dietary Allowances (RDAs)** every 5 years. The **energy RDA** covers carbohydrates, proteins, and fats and lists optimal kilocalorie intake for each sex, at various ages. It is based on the mean kilocalories needed in the population. For example, an 18-year-old male needs about 2,800 kilocalories per day and an 18-year-old female requires 2,100 kilocalories. However, individuals vary greatly, with most 18-year-old males eating between 2,100 and 3,900 kilocalories per day and most 18-year-old females eating between 1,200 and 3,000 kilocalories per day.

RDAs for vitamins and minerals are set so that if they are followed, 97% of the healthy adult population will receive adequate amounts. To set vitamin and mineral RDAs, a panel of scientists considers

Reading 25.1 Nutrition and the Athlete

CAN A MARATHONER, CROSS-COUNTRY SKIER, WEIGHT LIFTER, OR COMPETITIVE SWIMMER EAT TO WIN? A diet of 60% or more carbohydrate, 18% protein, and 22% fat should be adequate to support frequent, strenuous activity.

Carbohydrates

As the source of immediate energy, carbohydrates are the athlete's best friend. Athletes should get the bulk of their carbohydrates from vegetables and grains in frequent meals, because the muscles can store only 1,800 kilocalories worth of glycogen.

A dietary regimen called carbohydrate loading is popular among marathoners, who deplete their muscle glycogen at about the 20-mile mark of their 26-mile, 385-yard event—an experience aptly called "hitting the wall." The safest form of carbohydrate loading is to decrease exercise for a week before a marathon and increase the proportion of complex carbohydrates in the diet, while not increasing total kilocalories. The goal is to pack the muscles with as much glycogen as they can hold. Carbohydrate loading should not be attempted by teenagers or people with heart disease, diabetes, or high levels of blood lipids. (An older version of carbohydrate loading, in which an all-out "depletion" run is followed by 3 days of avoiding carbohydrates and then 4 days of a high carbohydrate intake, should be avoided. It causes nausea, weakness, and depression.)

Protein

Many people erroneously believe that an athlete needs protein supplements. One survey found that two-thirds of high school coaches in Texas believe this to be true. Excess dietary protein is turned into fat, not muscle. The kidneys are strained in ridding the body of the excess nitrogen, and as a result the athlete may become dehydrated as more water is used in urine. The American Dietetic Association suggests that athletes eat 1 gram of protein per kilogram of weight per day, compared to 0.8 grams for nonathletes. Do not rely solely on meat for protein, because these foods can be high in fat. Supplements are only necessary for young athletes at the start of training, under a doctor's supervision. Too little protein in an athlete is linked to "sports anemia," in which hemoglobin levels decline and blood may appear in the urine.

Fat

The body stores 140,000 kilocalories of fat, so it is clear why we do not need to replenish that constantly with fatty foods. Use low-fat milk and meats.

Water

A sedentary person loses a quart of water a day as sweat; an athlete may lose 2 to 4 quarts of water an hour! To stay hydrated, drink 3 cups of cold water 2 hours before an event, then 2 more cups 15 minutes before the event, and small amounts every 15 minutes during the event. Drink afterwards too. Another way to determine water needs is to weigh yourself before and after training. For each pound lost, drink a pint of water. Also avoid sugary fluids, which slow water's trip through the digestive system, and alcohol, which increases fluid loss.

Vitamins and Minerals

If an athlete eats an adequate, balanced diet, vitamin supplements are not needed. Supplements of sodium and potassium are usually not needed either, because the active body naturally conserves these nutrients. To be certain you have enough sodium, put a little salt on your food; to get enough potassium, include bananas, dates, apricots, oranges, or raisins in your diet.

A healthy pregame meal should be eaten 2 to 5 hours before the game, provide 500 to 1,500 kilocalories, include 4 or 5 cups of fluid, and be high in carbohydrates, which taste good, provide energy, and are easy to digest.

for each nutrient conditions caused by its deficiency and excess, how the body stores the nutrient, the rate at which the stores are depleted, and the rate of intake and excretion of the nutrient.

The RDAs that appear on food packages are the United States Recommended *Daily* Allowances, and these are set at the upper range of the Recommended Dietary Allowances. The United States RDAs are directed at the "generalized human adult," which does not include children, the elderly, and women who are pregnant or breastfeeding.

The Food Group Plan

RDAs are not very easy to fit into an active life-style—it is difficult to keep track of kilocalories consumed and vitamin and mineral contents of individual foods. To make nutrient watching easier, the United States government in the 1940s introduced the **food group plan.** The groups include meat and meat substitutes, milk and milk products, fruits and vegetables, and grains. Some nutritionists include an additional group, which includes foods that are almost entirely fat, sugar, or alcohol.

A modified food group plan corrects deficiencies in vitamins B_6 and E and the minerals magnesium, zinc, and iron, which can develop from the standard four or five food group plan. The new plan adds two servings of legumes or nuts. Vegetarians can follow modified food group plans.

Dietary Guidelines

Since the first food group plan was introduced, much has been learned about the many connections between nutritional status and specific diseases. Adequate fiber, for example, prevents constipation, diverticulosis, and possibly colon cancer, and moderating fat intake reduces chances of developing heart disease and possibly

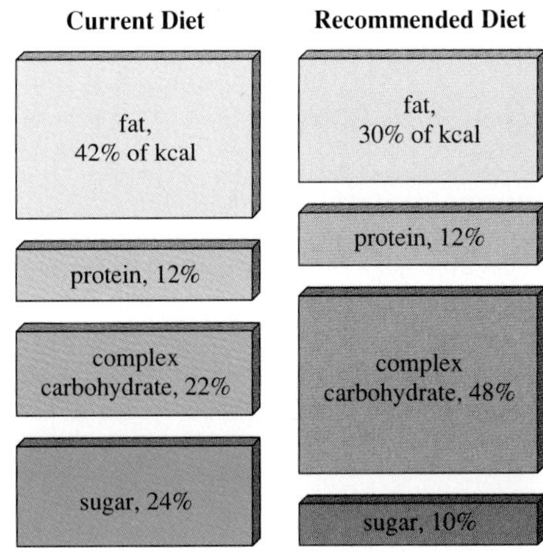

Figure 25.3
United States dietary goals.
Source: U. S. Government Printing Office, Washington, DC.

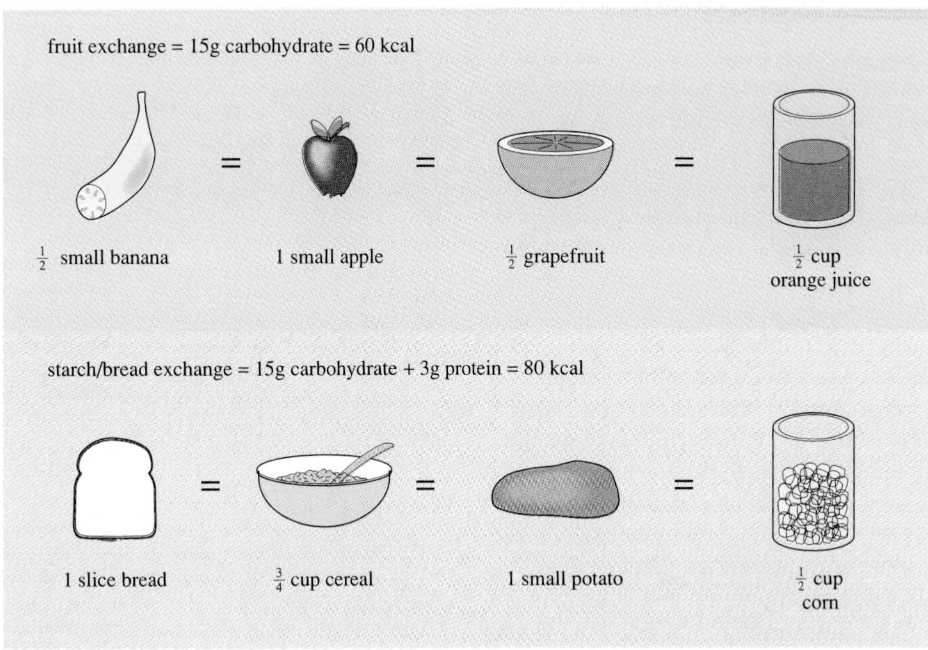

Figure 25.4
Some examples of food exchanges. Detailed lists of food exchanges can be found in nutrition textbooks or by consulting a registered dietitian.

breast cancer. In recognition of these diet–disease links, the United States Department of Agriculture and the Department of Health and Human Services suggest the following:

1. Eat a variety of foods.
2. Maintain a desirable weight.
3. Cut down on fats and cholesterol. To do this, eat more fish and beans, trim the fat from meat, broil or bake foods instead of frying them, and eat less oil and butter.
4. Increase intake of fiber and starch by eating more fruits and vegetables, grains, beans, peas, and nuts.
5. Reduce sugar intake, because sugar contributes to tooth decay.
6. Lower intake of sodium, which can contribute to the development of hypertension (high blood pressure). Sodium is found in salt and in many processed foods.

Figure 25.3 compares dietary goals to the actual breakdown of nutrient types in the American diet.

The Exchange System

A more specific way than the food group plan to select nutritious foods is the **exchange system.** It is based on a diet plan developed in the 1950s for diabetics, who must monitor and control their diets to maintain blood-sugar levels.

The exchange system groups foods by their proportions of carbohydrate, protein, and fat. Six lists are consulted—starch/bread, meat (lean, medium fat, and high fat), vegetables, fruits, milk, and fats. Each list consists of several foods, with the number of kilocalories, portion size, and proportions of the energy nutrients shown for each. For example, equivalent foods from the starch/bread list include one cob of corn, half a bagel, one slice of bread, and a half cup of pasta. Each provides 15 grams of carbohydrate, 3 grams of protein, and no fat, totaling 72 kilocalories (fig. 25.4).

The food exchange lists include only foods that are **nutrient dense**—that is, foods that offer a maximum amount of nutrients with a minimum number of kilocalories. Because foods in the exchange system are grouped by their carbohydrate/protein/fat composition, a particular food is not always listed with the foods it is listed within the food group plan. For example, an avocado is considered a fat in the exchange system. Using the food exchange system can be complicated for multifood dishes such as stews and casseroles. Reading 25.2 shows how to use the system to devise an indvidualized diet plan.

KEY CONCEPTS

There are several ways to plan a balanced diet. Recommended Dietary Allowances indicate the mean number of kilocalories consumed daily (the energy RDA) and amounts of vitamins and minerals that are sufficient for most people. Food group plans can be used to devise meals that contain a variety of foods. The 1985 Dietary Guidelines help us eat to prevent certain diseases. The exchange system allows diets to be planned for individuals based on certain proportions of the energy nutrients.

Nutrient Deficiencies

Obtaining an adequate number of kilocalories each day from a variety of foods is sometimes easier said than done. A student at exam time may eat nothing but pizza for a week; a busy working person may skip meals or eat on the run; many individuals may not be able to afford enough milk or meat. These are all situations that may lead to **primary nutrient deficiencies,** which are those caused by diet.

Secondary nutrient deficiencies, in contrast, result from an inborn metabolic condition in which a particular nutrient is not absorbed sufficiently in the small or large intestine, is excreted too readily, or is destroyed. Consider Menkes disease, also known as kinky hair syndrome. In this rare inherited disorder, copper is not adequately absorbed from food in the small intestine. The connection between the copper deficiency and the symptoms of twisted white hair, failure to grow, and profound nervous system destruction was made with the help of sheep. In 1971 an Australian researcher noted that a young patient with Menkes disease had hair very similar to that of sheep suffering from copper deficiency. Indeed, a sample of the child's hair revealed copper deficiency.

Micronutrient deficiencies develop slowly, with subtle changes occurring before health is noticeably affected. For example, a vegetarian who does not eat many iron-containing foods may have a blood test that shows an abnormally low number of red blood cells, or cells that are small with little hemoglobin (fig. 25.5). The person feels fine, temporarily, as the body uses its iron stores. At this point the nutrient deficiency is termed **subclinical,** because there are no symptoms. After several weeks, however, the signs of anemia become apparent—weakness, fatigue, frequent headaches, and a pale complexion.

A compulsive disorder that may result from mineral deficiency is **pica,** in which people consume huge amounts of nondietary substances such as ice chips, dirt, sand, laundry starch, clay and plaster, and even such strange things as hair, toilet paper, matchheads, inner tubes, mothballs, and charcoal. The condition is named for the magpie bird, *Pica pica*, which eats a range of odd things. Pica affects people of all cultures and has been noted as early as 40 B.C. The connection to dietary deficiency stems from the observation that slaves suffering from pica in colonial America recovered when their diets improved, particularly when they were given iron supplements. But pica is largely a medical mystery—trace mineral deficiencies apparently both cause it and result from it.

Reading 25.2 *Designing a Personal Diet Plan Using the Exchange System*

THE MOST EFFECTIVE DIET PLAN SUITS A PERSON'S PARTICULAR NEEDS AND FOOD PREFERENCES. The exchange system can help to do this.

The first step is to decide how many kilocalories are needed each day to gain, lose, or maintain weight. Consult the energy RDA, or keep track of how much you eat and exercise to maintain your weight. Next, decide what proportion of your daily caloric intake should be carbohydrate, protein, and fat, and use these percentages to calculate how many kilocalories of each energy nutrient are needed (fig. 1).

Translate kilocalories into food amounts by converting them to grams. For carbohydrates and proteins, divide kilocalories by 4 kilocalories/gram; for fats divide by 9 kilocalories/gram. The 2,000-kilocalorie diet of 60% carbohydrate, 15% protein, and 25% fat in figure 1 equals 300 grams of carbohydrate, 75 grams of protein, and 55.5 grams of fat. The next step is the most challenging—select the number of exchanges from each of the six lists (found in any nutrition text), so that food preferences and the desired amounts of the energy nutrients are met. Consult table 1 for the gram equivalents of the various food exchanges. Finally, assign particular foods to a day's meals to satisfy the proportions of carbohydrate, protein, and fat desired.

Table 1
Gram Equivalents of Various Food Exchanges

Exchange List	*Carbohydrate (grams)*	*Protein (grams)*	*Fat (grams)*	*Calories*
Starch/Bread	15	3	trace	80
Meat				
Lean	–	7	3	55
Medium fat	–	7	5	75
High fat	–	7	8	100
Vegetable	5	2	–	25
Fruit	15	–	–	60
Milk				
Skim	12	8	trace	90
Low fat	12	8	5	120
Whole	12	8	8	150
Fat	–	–	5	45

a. number of kcal = 2,000

b. percentage of energy nutrients

60% carbohydrate = 1,200 kcal

15% protein = 300 kcal

25% fat = 500 kcal

c. convert kcal to grams

$$\frac{1{,}200 \text{ kcal carbohydrate}}{4 \text{ kcal/gram}} = 300 \text{ grams carbohydrate}$$

$$\frac{300 \text{ kcal protein}}{4 \text{ kcal/gram}} = 75 \text{ grams protein}$$

$$\frac{500 \text{ kcal fat}}{9 \text{ kcal/gram}} = 55.5 \text{ grams fat}$$

d. number of exchanges to satisfy kcal-energy nutrient distribution

	# Exchanges	Carbohydrate (grams)	Protein (grams)	Fat (grams)
starch/bread	10	150	30	—
meats (medium fat)	3		21	15
vegetables	3	15	6	—
fruit	7	105	—	—
whole milk	2	24	16	16
fats	5	—	—	25
		294	73	56

e. convert exchanges into a day's meals ( ) = number of exchanges

Exchanges	Breakfast	Lunch	Dinner	Snacks
starch/bread (10)	1 cup bran flakes (2)	1 large potato (2)	1 cob corn (1)	3 cups popcorn (1)
	1 bagel (2)		$\frac{1}{2}$ cup peas (1)	8 animal crackers (1)
meats (medium fat) (3)		2 oz hamburger (2)	1 oz veal cutlet (1)	
vegetables (3)		1 tomato (1)	$\frac{1}{2}$ artichoke (1)	
		collard greens (1)		
fruit (7)	1 banana (2)		1 kiwi fruit (1)	apple (1)
	1 cup orange juice (2)			peach (1)
whole milk (2)	1 cup milk (1)			1 cup milk (1)
fats (5)	1 tsp margarine (1)	1/4 avocado (2)		
		2 tbsp salad dressing (2)		

Figure 1
The food exchange system.

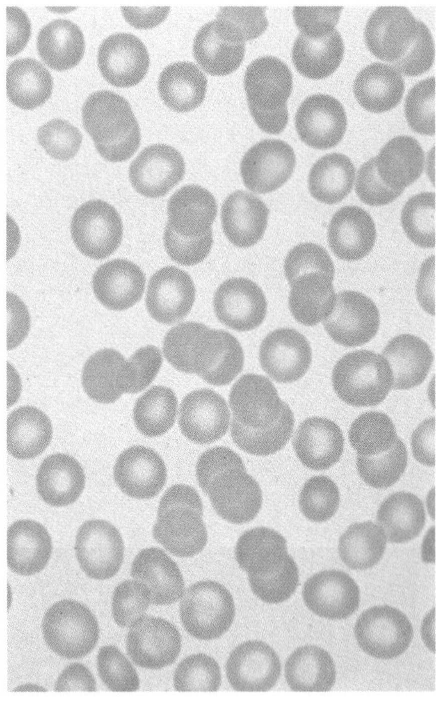

a.

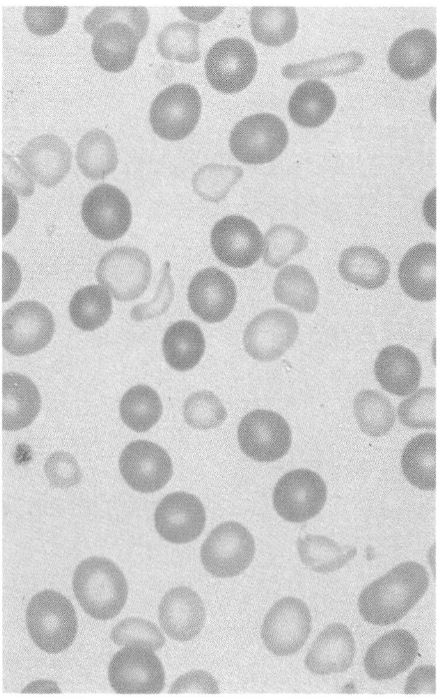

b.

Figure 25.5
Iron deficiency anemia. *a.* These red blood cells are normal. *b.* These, however, are small and pale. They contain too little hemoglobin, because iron is lacking in the diet.

KEY CONCEPTS

Primary nutrient deficiencies are caused by diet; secondary nutrient deficiencies are caused by metabolic disorders. Deficiencies develop slowly, with stores being used before symptoms appear.

Starvation

A healthy human can stay alive for 50 to 70 days without food. In prehistoric times, this margin allowed survival during seasonal famines. In some areas of Africa today, famine is not a seasonal event but a constant condition, and many millions have starved to death. Starvation is also seen in hunger strikers, in inmates of concentration camps, and in sufferers of psychological eating disorders such as anorexia nervosa and bulimia.

Whatever the cause, the starving human body begins to digest itself. After only 1 day without eating, the body's reserves of sugar and starch are gone. Next, the body extracts energy from fat and then from muscle protein. By the third day, hunger is no longer felt as the body uses energy from fat reserves. Gradually, metabolism slows to conserve energy, blood pressure drops, the pulse slows, and chills set in. Skin becomes dry and hair falls out as the proteins in these structures are broken down to release amino acids that are used for the more vital functioning of the brain, heart, and lungs. When the immune system's antibody proteins are dismantled for their amino acids, protection against infection is lost. Mouth sores and anemia develop, the heart beats irregularly, and bone begins to degenerate. After several weeks without food, coordination is gradually lost. Near the end, the starving human is blind, deaf, and emaciated.

Marasmus and Kwashiorkor

In the town of Abyei, just south of the war-torn Sudan, only 1,100 children were left in the fall of 1988 from tens of thousands of the Dinka tribe alive a year before that. Of those 1,100, 500 were too weak to stand up and stagger the few hundred yards to the feeding center, where they could receive lifesaving protein biscuits and a mixture of sugar, flour, and milk. For the "lucky" 600 other children, walking was an agony and even speaking an effort. The children all are suffering from the profound nutrient deficiency called **marasmus.** All nutrients are severely lacking, and as a result, the children resemble living skeletons (fig. 25.6*b*). Saddest of all is that no children under the age of two are left—all have died of measles, their immune systems too weakened to fight off this normally mild viral illness.

Some starving children do not look skeletal but have protruding bellies (fig. 25.6*a*). These youngsters suffer from a form of protein starvation called **kwashiorkor,** which in the language of Ghana means "the evil spirit which infects the first child when the second child is born." Kwashiorkor typically appears in a child who has recently been weaned from the breast, usually because of the birth of a sibling. The switch from protein-rich breast milk to the protein-poor gruel that is the staple of many developing nations is the source of this protein deficiency. The children's bellies swell with fluid, which builds up when protein is lacking, and their skin may develop lesions. Infections overwhelm the body as the immune system becomes depleted of its protective antibodies.

Anorexia Nervosa

In **anorexia nervosa,** starvation is self-imposed. The sufferer, typically a well-behaved adolescent girl from an affluent family, perceives herself to be overweight and eats barely enough to survive (figs. 25.7 and 25.8). The anorectic is terrified of any weight gain and usually loses 25% of her original body weight. In addition to eating only small amounts of low-calorie foods, she further loses weight by vomiting, by taking laxatives and diuretics, or by intense exercise. She develops low blood pressure, a slowed or irregular heartbeat, constipation, and constant chilliness, and she stops menstruating as her body fat level plunges. Like any starving person, the anorectic's hair becomes brittle and her skin dries out. Curiously, she may develop soft, pale, fine body hair called lanugo, which is also seen on the developing fetus. Lanugo preserves body heat.

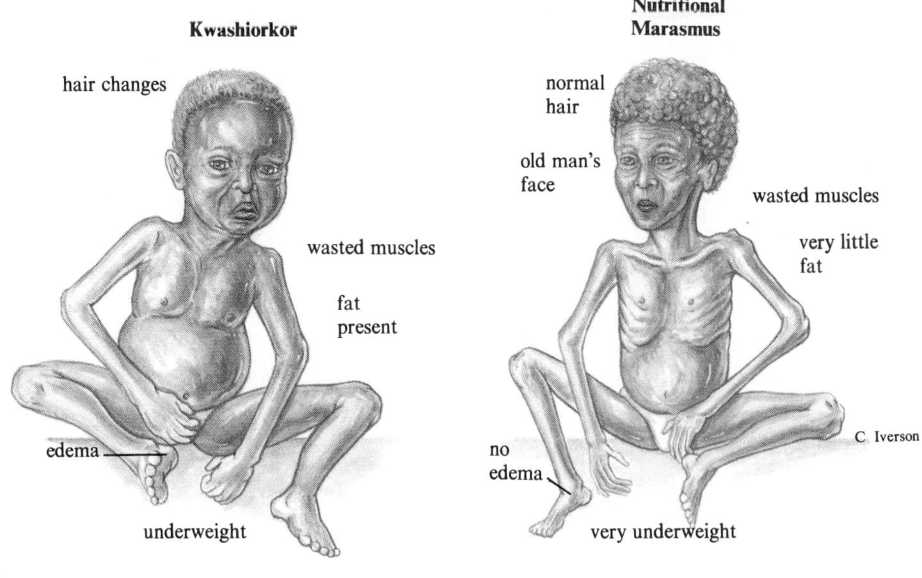

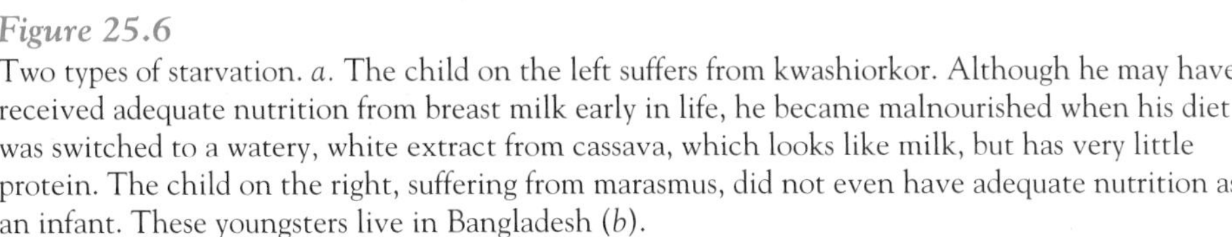

a.

b.

Figure 25.6
Two types of starvation. *a*. The child on the left suffers from kwashiorkor. Although he may have received adequate nutrition from breast milk early in life, he became malnourished when his diet was switched to a watery, white extract from cassava, which looks like milk, but has very little protein. The child on the right, suffering from marasmus, did not even have adequate nutrition as an infant. These youngsters live in Bangladesh (*b*).

The anorectic's eating behavior is often ritualized. She may meticulously arrange her meager meal on her plate or consume only a few foods. One anorectic, for example, ate only plain tuna fish washed down with diet cola. Some anorectics cook elaborate meals for their families, then sit back and watch them eat while nibbling on a carrot or celery stick. An anoretic uses devious means to avoid eating. One 18-year-old anorectic explained, "I'd go out for dinner with my friends and not eat, telling them I'd already eaten at home. Then I'd go home and skip dinner, telling my parents I'd eaten out with my friends."

When the anorectic reaches an obviously emaciated state, her parents usually have her hospitalized, where she is fed intravenously so that she does not starve to death or die suddenly of heart failure due to a mineral imbalance. She is also given psychotherapy and nutritional counseling. Despite these efforts, 15% to 21% of anorectics die. The condition is reported to affect 1 out of 250 adolescents, and 95% of them are female.

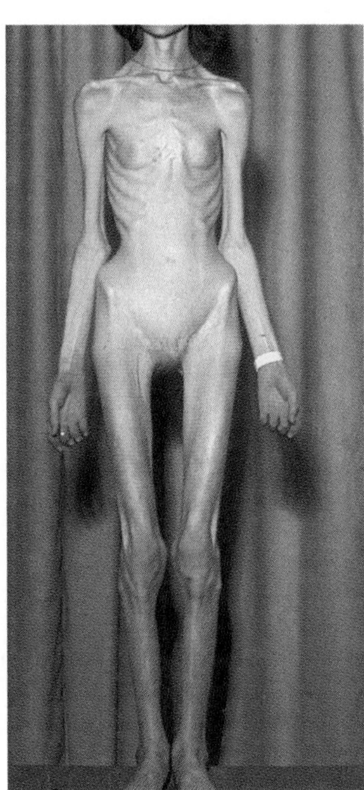

Figure 25.7
This woman is emaciated from the self-starvation of anorexia nervosa.

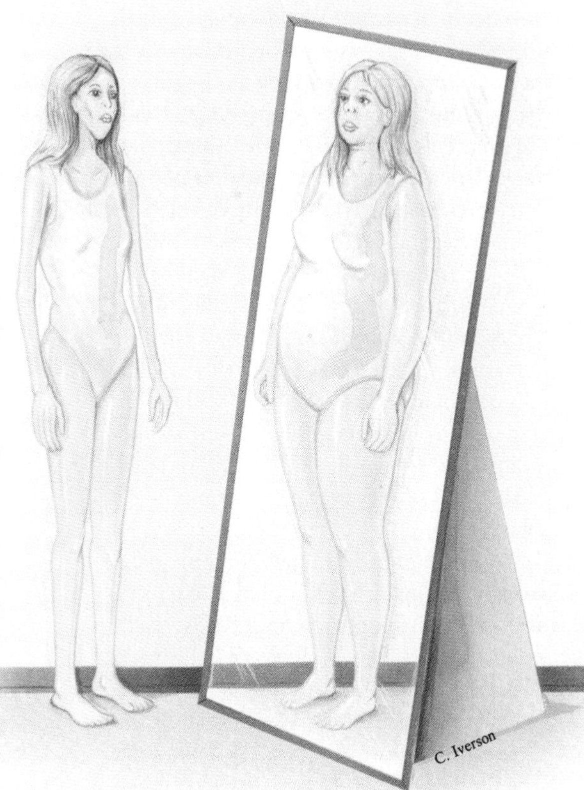

Figure 25.8
The anorectic typically perceives herself as obese.

Anorexia nervosa has no known physical cause. One suggestion is that the anorectic is rebelling against her approaching womanhood. Indeed, her body is astonishingly childlike, with a flat chest and slim hips, and she has often ceased to menstruate. She typically has low self-esteem and believes that others, particularly her parents, are controlling her life. Her weight is something that she can control. Anorexia can be a one-time, short-term experience—little more than a weight-loss diet gone out of control—or a lifelong obsession.

Bulimia

Unlike the anorectic, a person suffering from the eating disorder **bulimia** is often of normal weight. She eats whatever she wants, often in huge amounts, but she then rids her body of the thousands of extra kilocalories by vomiting, taking laxatives, or exercising frantically. For an estimated one in five college students, the great majority of them female, "binging and purging" appears to be a way of coping with stress.

The bulimic knows that her eating habits are not normal, and so often she binges in privacy, eating well beyond the point of pain. Bulimics tend to eat soft foods that can be consumed in large amounts quickly with minimal chewing. Reveals one bulimic, "For me a binge consists of a pound of cottage cheese, a head of lettuce, a steak, a loaf of Italian bread, a 10-ounce serving of broccoli, spinach or a head of cabbage, a cake, an 18-ounce pie, with a quart or half gallon of ice cream. When my disease is at its worst, I eat raw oatmeal with butter, laden with mounds of sugar, or a loaf of white bread with butter and syrup poured over it." This young woman follows her "typical" 20,000-kilocalorie binge with hours of bicycle riding, running, and swimming.

Sometimes a bulimic's dentist is the first to spot her problem by observing teeth decayed from frequent vomiting. The backs of her hands may bear telltale scratches from efforts to induce vomiting. Her throat is raw and her stomach lining ulcerated from the stomach acid forced forward by vomiting. The binge and purge cycle is very hard to break, even with psychotherapy and nutritional counseling. A famous recovered bulimic is actress Jane Fonda, who binged and purged up to 20 times a day from age 12 until age 35.

KEY CONCEPTS

Starvation occurs under a variety of conditions, but the effects on the body are the same—first carbohydrates are used for energy, then fat, then protein. As the proteins of the immune system are used, infection sets in. Marasmus is a form of starvation in which all nutrients are lacking, and the sufferer looks like a skeleton. Kwashiorkor is protein starvation seen in children weaned from breast milk to protein-poor gruel and is characterized by a swollen belly. In anorexia nervosa a person starves herself. In bulimia a person binges and purges.

Overweight

The National Institute of Health considers obesity a "killer disease," and for good reason. A person who is **obese**—defined as 20% above "ideal" weight based on population statistics considering age, sex, and build—is at higher risk for diabetes, digestive disorders, heart disease, kidney failure, hypertension, stroke, and cancers of the female reproductive organs and the gallbladder. The body is enormously strained to support the extra weight—miles of blood vessels are needed to nourish the additional pounds.

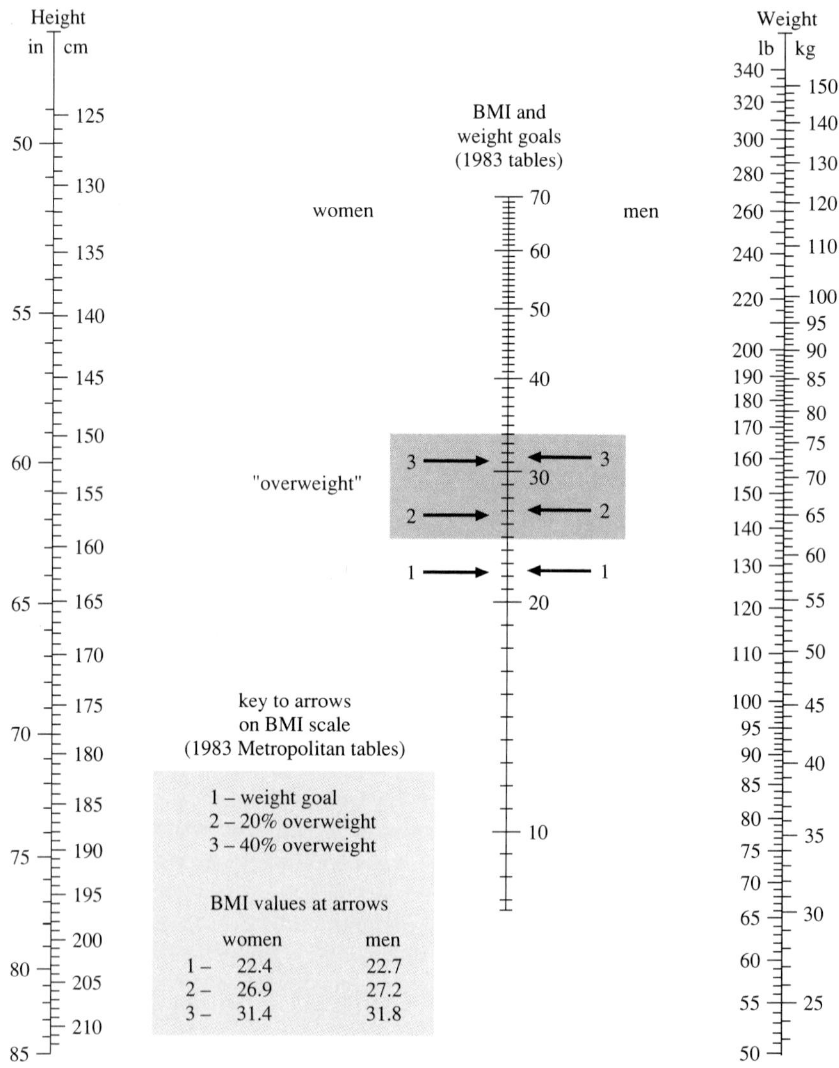

Figure 25.9
Are you obese? Body mass index (BMI) = weight (kg)/height ^{2} (m) and provides an estimate of obesity. To calculate your BMI, draw a line from your height on the left side of the figure to your weight on the right side. The point where the line crosses the scale in the middle is your BMI. A value above 27.2 for men and 26.9 for women indicates obesity. A value above 30 is cause for concern; a value above 35 is life threatening.

Obesity refers specifically to extra pounds of fat. A weight lifter may be overweight but not obese, because the additional body weight is primarily muscle, bone, connective tissue, and water, known as **lean tissue.** The proportion of fat in a human body ranges from 5% to more than 50%, with "normal" for males falling between 12% and 23% and for females between 16% and 28%. An elite athlete may have a body fat level considerably lower—basketball player Magic Johnson's body fat level is 4%. Moreover, the distribution of body fat affects health. Excess poundage above the waist is linked to increased risk of heart disease, diabetes, hypertension, and lipid disorders. Figure 25.9 shows a do-it-yourself way to estimate obesity using a measurement called body mass index.

Obesity is caused by both heredity and the environment. We inherit genes that control metabolism, but the fact that identical twins reared in different households can grow into adults of vastly different weights indicates that environment influences weight too. Studies comparing body mass index between adult identical twins reared apart indicate that weight is about 70% influenced by genes and 30% by the environment. Do you think that obese children of obese parents are overweight because of their genes or because they have learned bad eating habits—eating quickly, eating when not hungry, eating high-fat foods—from the parents?

A safe goal for weight loss is 1 pound of fat per week. A pound of fat contains 3,500 kilocalories of energy, so that pound can be shed by an appropriate combination of food restriction and exercise. This might mean eating 500 kilocalories less per day or exercising off 500 kilocalories each day. Actually more than a pound of weight will drop because water is lost as well as fat.

Calorie cutting should apply to the energy nutrients but never to the vitamins and minerals. A rule of thumb is to leave the proportion of protein kilocalories about the same or slightly increased, cut fat kilocalories in half, and cut carbohydrates by a third. Choose foods that you like, and distribute them into three or four balanced meals of 250 to 500 kilocalories each.

Diet plans abound (see the first "To Think About" question). Many are simply variations on the preceding suggestions, with a gimmick added, such as a cup of chili or an ice cream cone each day. Other diets emphasize a particular food (the banana, pineapple, or rice diet). Avoid diets that are based on erroneous but impressive-sounding scientific principles, such as the "immune power" diet that predicts that the effects of certain food extracts on isolated white blood cells will be echoed in the body, or diets that suggest that the order in which foods are eaten influences how many kilocalories are absorbed. Also avoid diets very low in carbohydrates (these deplete energy), diets high in protein (these strain the kidneys), and diets with less than 1,200 kilocalories per day, which is starvation level. Figure 25.11 shows some drastic measures used for weight reduction.

KEY CONCEPTS

Obesity means being 20% or more above ideal weight. Elevated disease risks associated with obesity depend upon percent body fat and fat distribution. A weight-loss plan can be devised by considering that a pound of body fat equals 3,500 kilocalories of energy and working out a diet and exercise program to decrease daily kilocalorie intake.

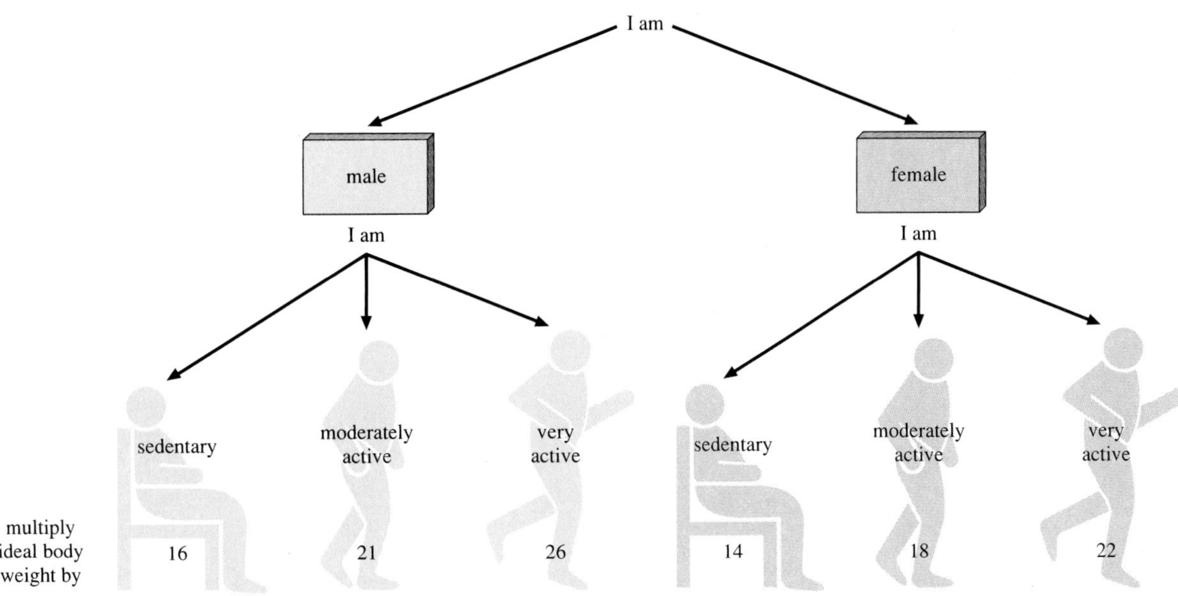

Figure 25.10
How many kilocalories are needed to maintain weight?

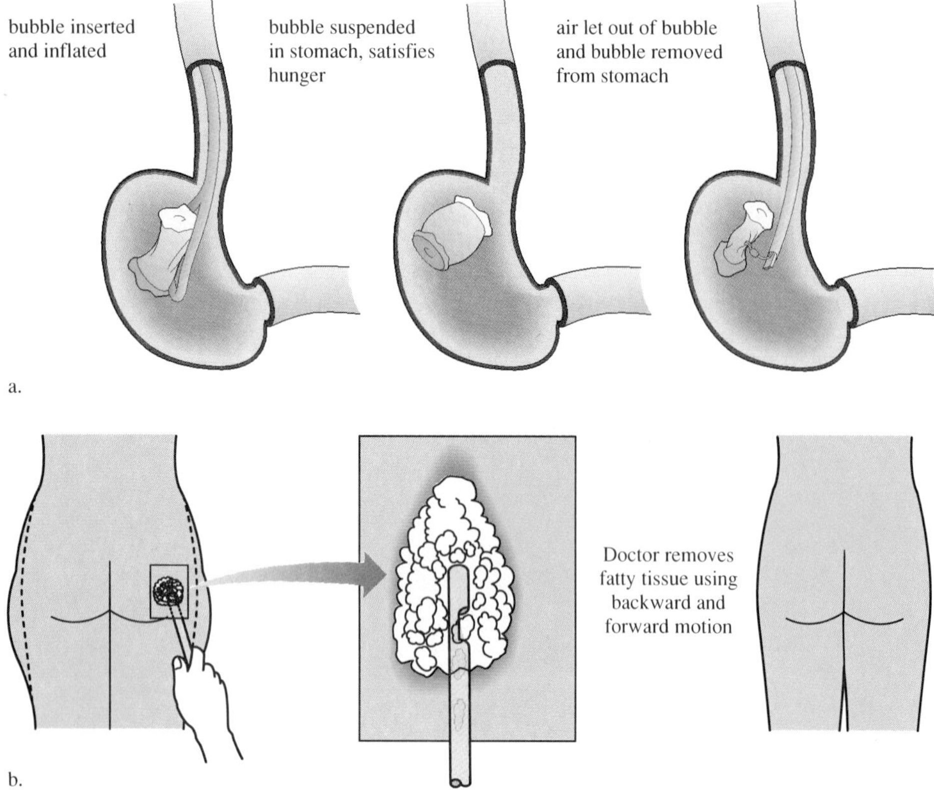

Figure 25.11
Two drastic ways to lose weight. *a.* The stomach bubble is based on the observation that long-haired cats are thinner than short-haired cats. Long-haired cats regularly swallow fur balls, which provide a sense of fullness because they take up space in the stomach. The stomach bubble works in a similar manner. *b.* In liposuction, fat is literally sucked out.

Gaining Weight

Diet can be altered to gain weight, which may be desired by hospital patients who have lost weight, those who are naturally thin, and certain athletes. One weight-gain diet for a convalescent patient offers 3,000 to 3,500 kilocalories per day, with approximately 14% of kilocalories as protein, 24% as fat, and 62% as carbohydrate. National body-building champion Gary Strydom follows a different recipe for weight gain, stressing the protein needed for muscle development. He consumes 5,500 to 7,000 kilocalories per day, with 40% protein, 50% carbohydrates, and 10% fat. An infant also needs to gain weight rapidly. The ideal food for a human infant is human milk (Reading 25.3).

Whether you are a body builder, a hospital patient, or just not happy with your weight, the nutritional key to good health is to eat a variety of foods. Even our ancestors 30,000 years ago seemed to know that.

SUMMARY

Nutrients promote growth, maintenance, and repair of body tissues. The field of nutrition considers how nutrients are handled in the body. *Macronutrients*, or *energy nutrients*, include carbohydrates, proteins, and fats. *Micronutrients* include vitamins and minerals. Water is also vital. Energy provided by food is measured in *kilocalories*, using a *bomb calorimeter*. One gram of carbohydrate or protein yields 4 kilocalories, and 1 gram of fat yields 9 kilocalories.

Recommended Dietary Allowances, Recommended Daily Allowances, *food group plans*, dietary guidelines, and the *exchange system* are used to devise an adequate, *nutrient-dense* and varied diet. A *primary nutrient deficiency* is caused by diet. A *secondary nutrient deficiency* is caused by a metabolic defect.

During starvation, the body uses its nutrient stores to maintain the functions of vital organs, which sacrifices immunity. *Marasmus* is severe lack of all nutrients. *Kwashiorkor* is extreme protein deficiency. *Anorexia nervosa* is self-starvation due to an underlying psychological obsession with being overweight. A person with *bulimia* binges and purges. Being *obese*—20% above ideal weight—is associated with a variety of increased disease risks.

QUESTIONS

1. What is a nutrient?

2. What units do nutritionists use to measure the energy content of foods? To what is this unit equivalent?

3. List three ways that you can reduce the fat and cholesterol in your diet.

4. List three ways in which human breast milk differs nutritionally from infant formula.

5. A woman is 5 feet 8 inches and weighs 135 pounds. According to her body mass index, is she overweight?

6. Zinc deficiency is characterized by an enlarged liver and spleen, poor growth, delayed sexual development, poor wound healing, and loss of the ability to taste and smell. A teenager following a poor diet has zinc deficiency symptoms. Another person has a zinc deficiency due to an inborn inability to absorb this micronutrient. Which of these people suffers from a primary nutrient deficiency and which from a secondary nutrient deficiency?

7. A diet plan boasts, "Follow our regimen and you will lose 10 pounds of fat in a week." Is this possible? Why or why not?

Reading 25.3 *Human Milk—The Perfect Food for Human Babies*

The female human body manufactures milk that is a perfect food for a human newborn in several ways (table 1). Human milk is rich in the lipid needed for rapid brain growth, and it is low in protein. Cow's milk is the reverse, with three times as much protein as human milk. Much of this protein is casein, which is fine to spur a calf's rapid muscle growth but forms hard-to-digest curds in a human baby's stomach. The protein in human milk has a better balance of essential amino acids than does the protein in cow's milk.

Human milk protects a newborn from many infections. For the first few days after giving birth, a new mother's breasts produce a clear fluid called colostrum, which has less sugar and fat than mature milk but has more protein. Colostrum is rich in antibodies, which are biochemicals produced by the immune system of the mother. The antibodies transfer protection to the baby from such infections as Salmonella poisoning and polio. When the milk matures by a week to 10 days, it has antibodies, enzymes, and white blood cells from the mother that continue the infection protection introduced in the colostrum. A milk protein called lactoferrin binds iron, making it unavailable to microorganisms that might use it to thrive in the newborn's digestive tract. Another biochemical in human milk, bifidus factor, encourages the growth of the bacteria *Lactobacillus bifidus*, which manufactures acids in the baby's digestive system that kill harmful bacteria.

A breast-fed baby typically nurses until he or she is full, not until a certain number of ounces have been drunk, which may explain why breast-fed babies are less likely to be obese than bottle-fed infants. Babies nurtured on human milk are also less likely to develop allergies to cow's milk.

But breast-feeding is not the choice for all women. It may be impossible to be present for each feeding or to provide milk. Also, many drugs taken by the mother enter breast milk and can affect the baby. A nursing mother must eat about 500 kilocalories per day more than usual to meet the energy requirements of milk production—but she also loses weight faster than a mother who bottle-feeds, because the fat reserves set aside during pregnancy are used to manufacture the milk. Another disadvantage of breast-feeding is that the father cannot do it.

An alternative to breast-feeding is infant formula, which is usually cow's milk plus fats, proteins, carbohydrates, vitamins, and minerals added to make it as much like breast milk as possible. Although infant formula is nutritionally sound, the foul-smelling and bulkier bowel movements of the bottle-fed baby compared to the odorless, loose, and less abundant feces of a breast-fed baby indicate that breast milk is a more digestible first food than infant formula. Last, but certainly not least, is the fact that breast-feeding continues the biological link between mother and child that has existed for the previous 9 months. This closeness has emotional benefits for both parties.

Table 1
Human Milk—The Perfect Food for Human Babies

Nutrient	*Human Milk*	*Cow's Milk*	*Infant Formula (Similac)*
Protein (g/100 ml)	1.1	3.5	1.6
Fat (g/100 ml)	4.5	3.6	3.6
Carbohydrate (g/100 ml)	6.8	4.9	7.2
Calcium (mg/100 ml)	34	118	100
Phosphorus (mg/100 ml)	14	92	77

TO THINK ABOUT

1. For each of the following diets, indicate how the diet is nutritionally unsound (if it is) and why you would or would not be able to follow it.
   a. The bikini diet consists of 500 kilocalories per day, including 45% to 55% protein, 40% to 50% carbohydrate, and 4% to 6% fat.
   b. The Cambridge diet is a powder mixed with water and drunk three times a day. One day's intake equals 33 grams of protein, 40 grams of carbohydrate, and 3 grams of fat.
   c. For the first 10 days of the Beverly Hills diet, nothing but fruit is eaten. On day 10 you can eat a bagel and butter, and then it is back to fruit only until day 19, when you can eat steak or lobster. The cycle repeats, adding more meat. This diet is based on "conscious combining"—the idea that eating certain combinations of foods leads to weight loss.
   d. The Weight Loss Clinic diet consists of 800 kilocalories per day, with 46.1% protein, 35.2% carbohydrate, and 18.7% fat.
   e. The macrobiotic diet includes 10% to 20% protein, 70% carbohydrate, and 10% fat, with a half hour of walking each day. Most familiar foods are forbidden, but you can eat many unusual foods—such as rice cakes, seaweed, barley stew, pumpkin soup, rice gruel, kasha and onions, millet balls, wheat berries, and parsnip chips.
   f. The No Aging diet maintains that eating foods rich in nucleic acids (RNA and DNA) can prolong life, since these are the genetic materials. Recommended foods include sardines, salmon, calves liver, lentils, and beets.

2. Why are the RDAs that appear on food packages not very useful for a family consisting of a 4-year-old, an infant, a breast-feeding mother, and an elderly grandmother?

3. A young man takes several vitamin supplements each day, claiming that they give him energy. Is he correct? Why or why not?

4. A soccer coach advises his players to eat a hamburger and french fried potatoes about 2 hours before a game. Suggest a more sensible pregame meal.

5. Anorexia nervosa is a form of starvation. If it is a nutritional problem, then why should treatment include psychotherapy?

6. Why do starving children often die of infections that are usually mild in well-nourished children?

SUGGESTED READINGS

Bouchard, Claude. May 24, 1990. The response to long-term overfeeding in identical twins. *The New England Journal of Medicine*. A tendency towards obesity is in the genes—but a lot depends on how and what a person eats.

Bower, B. October 20, 1990. Tooth analysis may decipher prehistoric diets. *Science News*. Plant remains preserved on teeth of primate forerunners to humans provide clues to ancient diets.

Crump, Martha L. February 1991. You eat what you are. *Natural History*. More than 1300 species obtain precisely the right nutrient balance—by eating their own kind.

Farley, Dixie. May 1986. Eating disorders: When thinness becomes an obsession. *FDA Consumer*. Anorexia nervosa can be deadly.

Graf, Joan Stephenson. November 1981. Death by fasting. *Science 81*. How the body deals with lack of food.

Henderson, Doug. May 1987. Nutrition and the athlete. *FDA Consumer*. Athletes should stress carbos, eat a moderate amount of protein, and avoid fats.

Lecos, Chris. May 1988. We're getting the message about diet-disease links. *FDA Consumer*. A balanced diet not only controls weight but may also prevent disease.

Lewis, Ricki. March 1988. Essential to health—trace elements. *Biology Digest*. A look at human health problems that result from too little or too much of a trace element.

Madsen, David B. July 1989. A grasshopper in every pot. *Natural History*. For some human cultures, dining on insects was quite common.

Segal, Marian. May 1988. Fruit: Something good that's not illegal, immoral or fattening. *FDA Consumer*. A healthful diet includes lots of fruits.

Stehlin, Dori. December 1986. Good nutrition for breastfeeding mothers. *FDA Consumer*. A woman who is nursing a baby must be extra careful about what she eats.

Stehlin, Dori. January–February 1991. Women and nutrition: A menu of special needs. *FDA Consumer*. Diet can influence "female" problems, such as osteoporosis, breast cancer, and iron deficiency.

C H A P T E R

26

Homeostasis

Chapter Outline

Temperature Regulation
Reading 26.1 A Day in the Life of a Marine Iguana
Ectothermy and Endothermy
Temperature Regulation in Humans
Regulation of Body Fluids
Nitrogenous Waste Removal
Osmoregulation
The Human Excretory System—An Overview
The Kidney
Activities Along the Nephron
Bowman's Capsule
Reading 26.2 Urinalysis—Clues to Health
The Proximal Convoluted Tubule
The Loop of Henle
The Distal Convoluted Tubule
The Collecting Duct
Control of Kidney Function
The Unhealthy Excretory System
Urinary Tract Infections
Kidney Stones
Kidney Failure

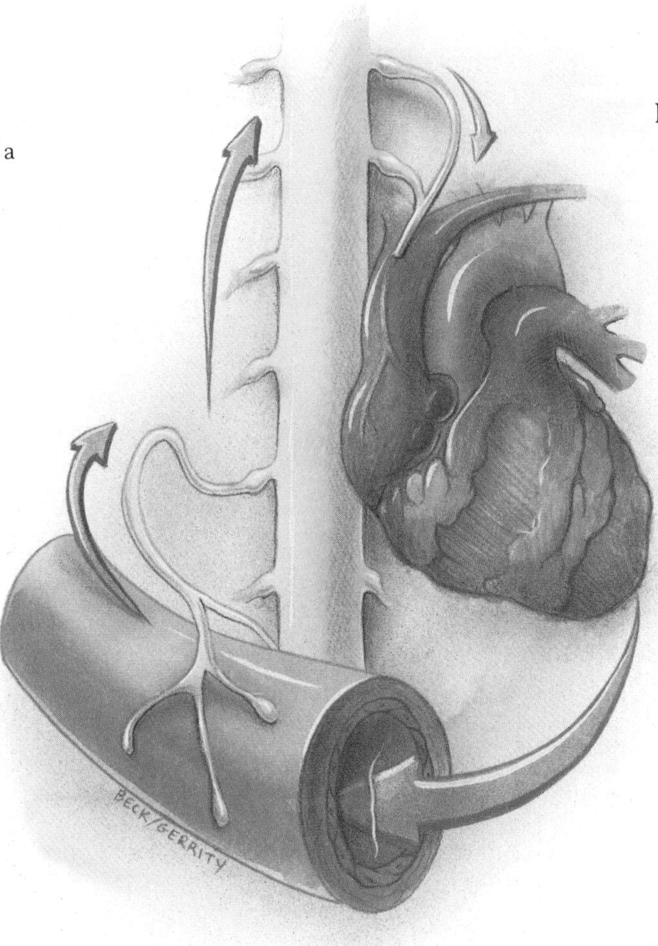

Learning Objectives

By the chapter's end, you should be able to answer these questions:

1. Why is temperature regulation of an organism important to its functioning?
2. How do organisms use their behaviors and metabolisms to regulate their temperatures?
3. Where do nitrogen-containing wastes come from, and why are they toxic to animals?
4. What happens if sufficient water levels are not maintained in living tissues?
5. What structures comprise the human excretory system?
6. How do the microscopic structures within the kidneys filter certain molecules from the blood, and then recycle some of them to the bloodstream, and excrete others in the urine?
7. What biochemicals control kidney function?
8. What are some disorders of the excretory system?

The summer of 1988 was blisteringly hot, with day after day of temperatures topping 100°F (37.8°C) in many parts of the United States. Those living in these areas sweated, and swam, and did what they could to stay cool. Despite the unrelenting heat and lack of rain, people coped. Body temperatures still hovered about the 98.6°F (37°C) mark, tissues had sufficient water to function, the pH of blood stayed within its normal range, and the distribution and concentrations of various dissolved biochemicals in blood remained as always. If we had made these measurements back in the unusually frigid winter of 1977, the results would have been pretty much the same.

Coping with environmental extremes while keeping a relatively stable internal environment is critical to life. This ability to maintain constancy of body temperature, fluid balance, and chemistry is termed **homeostasis.** Although there are many aspects of homeostasis, this chapter focuses on how humans and some other animals maintain body temperature within tolerable limits, regulate water and salt balance within the body fluids, and excrete nitrogen-containing metabolic waste.

Temperature Regulation

Why is it important that an animal's body temperature remain within limits? One reason is that temperature influences the rate of the chemical reactions that sustain life. Recall from chapter 3 that enzymes work most efficiently around specific temperatures. An animal's enzymes work best at its customary body temperature. Should cellular temperatures vary from this optimum, the enzymes function less efficiently, so vital biochemical reactions occur more slowly. If enzymatic steps in a sequence of reactions are affected differently by temperature change, their precisely coordinated timing may be disrupted.

Extreme temperatures alter biological molecules. For example, a drastic increase in temperature destroys a protein's three-dimensional shape. If the misshapen

Reading 26.1 *A Day in the Life of a Marine Iguana*

IGUANA LIZARDS SPEND MUCH OF THEIR TIME REGULATING THEIR BODY TEMPERATURES, WHICH ARE ABOUT THE SAME, OR SLIGHTLY HIGHER, THAN OUR OWN.

Marine iguanas on the Gálapagos Islands begin their days by basking in the rising sun. They drape themselves on boulders by the sea and on hardened lava, sunning their backs and sides. After about an hour, they turn and aim their undersides at the sun, almost like a human sunbather rolling over to get an even tan.

By midmorning the air temperature is rising rapidly. The iguanas cannot sweat; instead they escape the blazing sun. They lift their bodies by extending their short legs, removing their bellies from the hot rocks, and allowing breezes to fan them. But by noon these pushups are insufficient to stay cool. The iguanas retreat to the shade of crevices in the rocks (fig. 1).

By midday, the animals are hungry. The iguanas dive into the ocean, which is too cold for them to stay in for more than a few minutes. They eat green algae on the ocean's floor or nibble seaweed by the shore, hanging off of rocks to reach it. The water is so cold that the lizards' body temperatures rapidly drop. Constriction of arteries near their body surfaces helps to conserve heat.

After feeding, the iguanas stretch out on the rocks again as they did at the start of the day, warming sufficiently to digest their meal. They continue basking as the day ends, absorbing enough heat to sustain them until a new day begins.

Figure 1
Marine iguanas retreat to a crevice to escape the heat of the noonday sun.

protein is an enzyme, it can no longer function. An extreme drop in temperature solidifies lipids. Since lipids are an integral component of biological membranes, such a change would impede many functions, especially those involving movement of substances across membranes.

Ectothermy and Endothermy

Animals have evolved some interesting ways to regulate their body temperatures. Animals called **ectotherms** lose or gain heat to their surroundings by moving into areas where the temperature is suitable. A snake or a lizard warms itself by basking in the sun on a rock. By careful choice of a sunny spot and proper positioning, an ectotherm can absorb enough solar heat to increase its body temperature by 33.8°F (1°C) a minute, even on chilly mornings. When it becomes too warm, the animal moves out of the sun's rays by finding shade or by burrowing. During the nighttime or on overcast days, an ectotherm cannot warm itself. Activity slows. Because defense reactions are slowed, the animal seeks shelter offering some safety from predation. An ectotherm thus regulates its body temperature by its behavior. The marine iguana lizard is a master at moving about to maintain a relatively stable body temperature (Reading 26.1).

In contrast to the ectotherm's external heat source, an **endotherm's** source of heat is internal, that is, the source is its own metabolic heat. The metabolic rate of an endotherm is generally five times that of an ectotherm of similar size and body temperature. Layers of insulation, feathers, or fur help retain body heat.

Ectothermic and endothermic ways of life each have advantages and disadvantages. The ectotherm is dependent upon a continuous ability to escape or take advantage of the environment. An injured iguana that could not squeeze into a crevice to escape the broiling noon sun would cook to death. In contrast, an endotherm's heat is generated internally, enabling it to maintain body heat even in the middle of the night. However, this internal constancy comes at a cost. The endotherm must eat large amounts of food compared to the ectotherm, and 80% of the energy from that food is used to maintain body temperature.

Perhaps these pluses and minuses of ectothermy and endothermy are why many organisms utilize both strategies (fig. 26.1). European honeybees exhibit a fascinating combination of behavior and physiology to survive freezing temperatures. The bees align themselves, and then their flight muscles shiver (fig. 26.2). About a fifth of each insect's body mass is flight muscle, and when these muscles are active, their metabolic rate is 25 times as great as when they are at rest. Within these muscles, chemical energy is converted to mechanical energy and heat. The numerous hairs over the animals' chests, woven to form an almost continuous fabric by the bee's alignment, retain the heat given off by the muscles beneath. The overall result is that the bees can survive for hundreds of hours

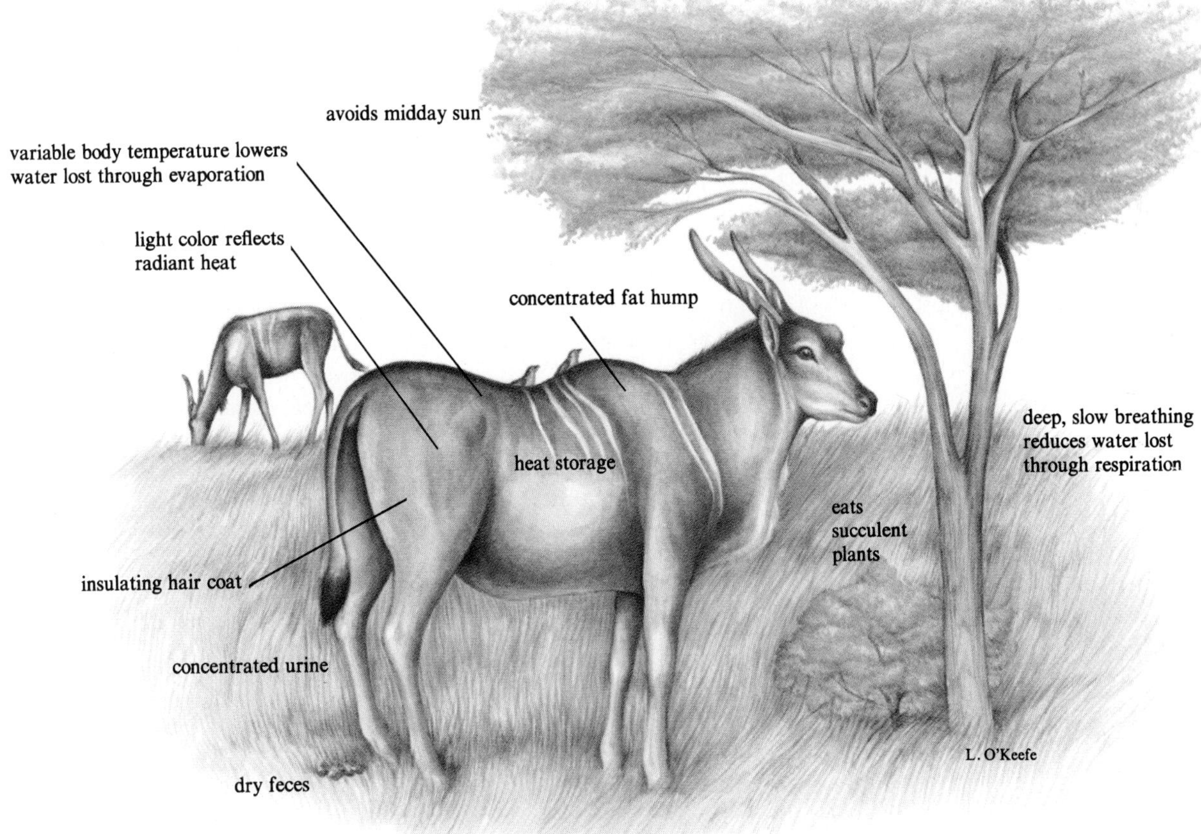

Figure 26.1
The eland dwells on the hot plains of central Africa. It has evolved a variety of adaptations to maintain its body temperature in the heat.

in temperatures as low as -112°F (-80°C). Alone, a bee would die in an hour of continuous exposure below 28.4°F (-2°C).

Animals differ not only in how they warm their bodies but also in how stable they keep their body temperatures. In some species, body temperature fluctuates with the temperature of their surroundings, peaking when the animal is active on a hot sunny day but dropping quite low on a cold night. In contrast, the body temperature of a human remains fairly constant regardless of environmental conditions. Other animals seem to compromise, maintaining a constant body temperature most of the time but sometimes varying it or doing so only in a part of the body.

A hummingbird, for example, flutters its wings while foraging to generate heat. However, such intense activity is a tremendous drain on energy resources and heat is lost quickly from its tiny body. At night, when food is no longer being gathered, the bird enters a sleeplike state called **torpor** in which the body temperature falls toward that of the surroundings.

Some large fish such as the mako shark and tuna conserve energy by heating only parts of the body. These fish keep their core body temperatures 41°F to 59°F (5°C to 15°C) warmer than the frigid water around them, trapping heat from their deeper body muscles in a network of blood vessels. The heat from blood leaving the active muscles is transferred to cooler blood returning from the skin instead of traveling to the body surface, where it would be lost to the surroundings.

Figure 26.2
To conserve heat in their hive, European honeybees align in groups of 2,000 or more individuals. Their chest hairs and shivering flight muscles generate and maintain sufficient heat to allow bees in the interior to survive for many hours.

Temperature Regulation in Humans

We consciously control our temperature by dressing appropriately for weather conditions, seeking comfortable locations, or altering the environment by heating or cooling the air. We also have physiological responses that adjust body temperature.

The most noticeable human response to cold is shivering, when certain skeletal muscles contract from 10 to 20 times per second, up to 5 times faster than they do when at rest. Blood flowing through shivering muscles is heated and carries the warmth throughout the body. Other musclelike cells in the skin also contract in the cold, forming goose bumps and making our hair "stand on end." This is not very effective in conserving our body heat because we do not have enough hair to form a good layer of insulation. However, the same physiological response is very helpful to other mammals and to birds. Raising fur or feathers, a process called **piloerection,** traps a layer of air near the body. Air insulates because it is a poor heat conductor.

We and other vertebrates tend to feel cold most in our noses, hands, and feet. Recall from figure 11.15*a* that temperature-sensitive genes cause dark pigmentation of the coldest body parts in Siamese cats. A heat-conserving adaptation called vasoconstriction narrows the capillaries near the body's surface, reducing blood flow in the extremities (fig. 26.3). Finally, when body temperature begins to drop, the hypothalamus in the brain triggers the release of hormones that increase metabolic rate, thereby increasing heat production.

A very dramatic response to sudden, extreme, and prolonged exposure to cold (hypothermia) is seen in young drowning victims (see Reading 3.1, "The Definition of Death"). Under these drastic conditions, blood flow is temporarily rerouted so that vital functions are maintained for a short time.

The human body cools itself by perspiring. When the outside temperature reaches 95°F (35°C), about 95% of the heat lost from the body is due to evaporation of sweat. Sweat is a secretion of sweat glands in the skin, and it consists of water and salts (fig. 26.4). Although we sweat all of the time, it is not obvious unless humidity prevents its evaporation or a person exercises very vigorously. During strenuous exercise on a hot, humid day, as much as 4 quarts (4 liters) of sweat can be lost in an hour. Because prolonged, profuse sweating can upset the balance of water and salts, it is essential to drink water before, during, and after exercising. Table 26.1 lists common heat-related illnesses.

Figure 26.3 (Facing Page)
Vasoconstriction controls heat loss from different body parts. The thermographs of a man, woman, and boy are white in hotter areas and blue in colder areas. Note the hot red bald spot in the man—much heat is lost from the head, which is why wearing a hat in winter is important for conserving heat.

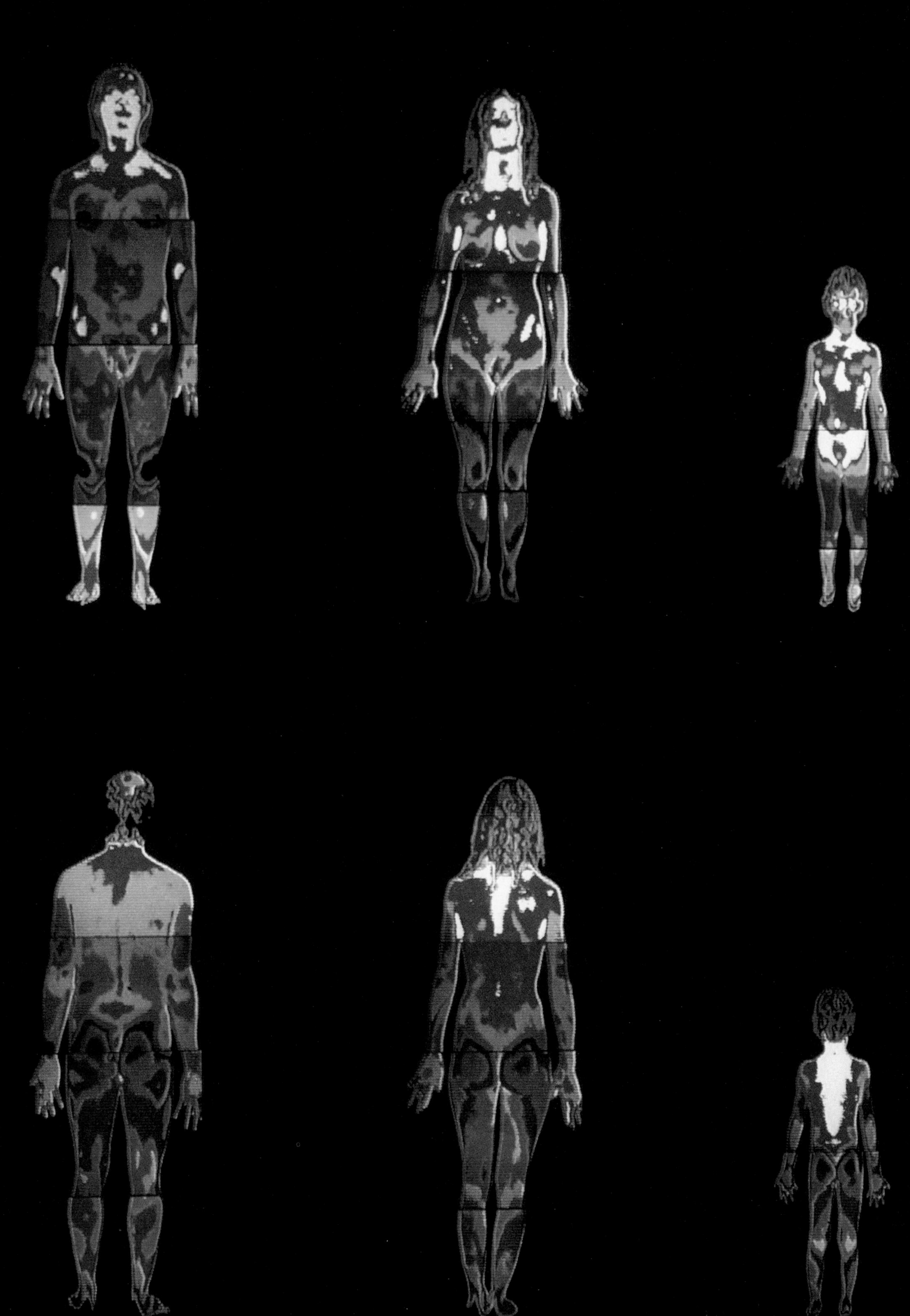

The "normal" human body temperature of 98.6°F (37°C) is controlled by the hypothalamus, which responds to temperature-sensitive receptors in the skin. Human body temperature varies throughout the day, being lowest at night and peaking in the afternoon before dipping once again. Therefore, a temperature of 99°F (37.2°C) is of little significance if it is recorded in the afternoon but may indicate a fever if taken early in the morning.

KEY CONCEPTS

Homeostasis is the ability to maintain constant conditions in the body, such as temperature, fluid balance, and chemistry. Biochemicals work optimally at certain temperatures. Ectotherms regulate body temperature primarily by their behaviors, whereas endotherms do so by utilizing metabolic energy to generate heat. Many organisms use both ectothermic and endothermic strategies. Some species maintain a constant body temperature at all times, some alter their body temperatures at certain times, and some can alter the temperature only in certain body parts. Humans retain heat by shivering and vasoconstriction, and the body's "thermostat" is set by the hypothalamus. Evaporation of sweat cools the body. Normal human body temperature varies throughout the day.

Figure 26.4
Sweat helps the human body cool off.

Table 26.1
Heat-Related Illnesses

	Condition	*Symptoms*
Least severe	Heat cramps	Cramps in calf muscles
	Heat exhaustion	Excessive sweating
		Rapid pulse
		Headache
		Dizziness
Most severe	Heat stroke	Skin is dry and feels hot
		Body temperature near 106°F (41.1°C)

Regulation of Body Fluids

Each cell is bathed in a fluid derived from the blood. If the composition of that fluid varies too much, it spells certain death for the cells. Two important ways in which the composition of body fluids is regulated are removal of nitrogenous wastes and osmoregulation (water balance).

Nitrogenous Waste Removal

As blood courses through the body, it delivers vital substances, including nutrients and hormones, and picks up and removes metabolic wastes. Nitrogen-containing wastes resulting from the breakdown of protein and genetic material are exceedingly toxic and would kill cells if they accumulated within the body.

Three types of nitrogenous wastes are generated by protein destruction: ammonia, urea, and uric acid. The first step in dismantling an amino acid is the removal of the amino group (-NH_2). Each amino group then picks up a hydrogen ion and becomes **ammonia,** NH_3. Using ammonia as the primary nitrogenous waste product is energetically efficient because it requires only one reaction (fig. 26.5). However, ammonia is very toxic and must be excreted in a very dilute solution.

Some animals that live in fresh water can take in water from the environment to dilute the ammonia. Land-dwelling animals, who must conserve water, use energy to convert ammonia to less toxic substances. Adult amphibians and mammals, including humans, convert ammonia to **urea** as their primary nitrogenous waste. Birds, reptiles, and insects convert much of their

$H_2N—C—NH_2$

a. urea

b. uric acid

Figure 26.5
Nitrogenous wastes. *a.* Mammals and sharks excrete nitrogen-containing wastes primarily in the form of urea, a water-soluble compound that forms in the liver, travels in the bloodstream, and is excreted from the body in the urinary tract. *b.* Birds, reptiles, and insects excrete much of their nitrogenous waste as uric acid.

Figure 26.6
The islands off the coast of South America are covered with guano, which are bird droppings rich in solid uric acid. These "gooney" birds nest on the islands, which they build up with their waste. Guano deposits are used as fertilizer. Removing it from these islands was an important trade in the late 1800s.

nitrogenous wastes to **uric acid,** which can be excreted in an almost solid form. In birds uric acid is mixed with undigested food in a storage organ called the cloaca to form a pasty substance called guano—the familiar "bird dropping" (fig. 26.6).

Humans produce very small amounts of uric acid. Our uric acid has a different source than that of birds and reptiles. Instead of being generated as a breakdown product of proteins, our uric acid comes from the breakdown of the purine bases of DNA and RNA. In the metabolic disorder **gout,** too much uric acid is produced. Crystals of uric acid accumulate in the joints, causing pain.

Osmoregulation

Osmoregulation refers to the control of water and salt balance in the body. Water loss can have rapid and drastic effects on human health (fig. 26.7). When water loss (dehydration) reaches 1% of body weight, the mouth becomes very dry and thirst prompts drinking to replace lost water. If the loss reaches 10%, the person becomes deaf, delirious, and can no longer feel pain. Water loss of 12% or more of the body weight is usually fatal because muscles, including those that control swallowing, cease to function. A person can die of thirst in only 3 days without water from food and drink.

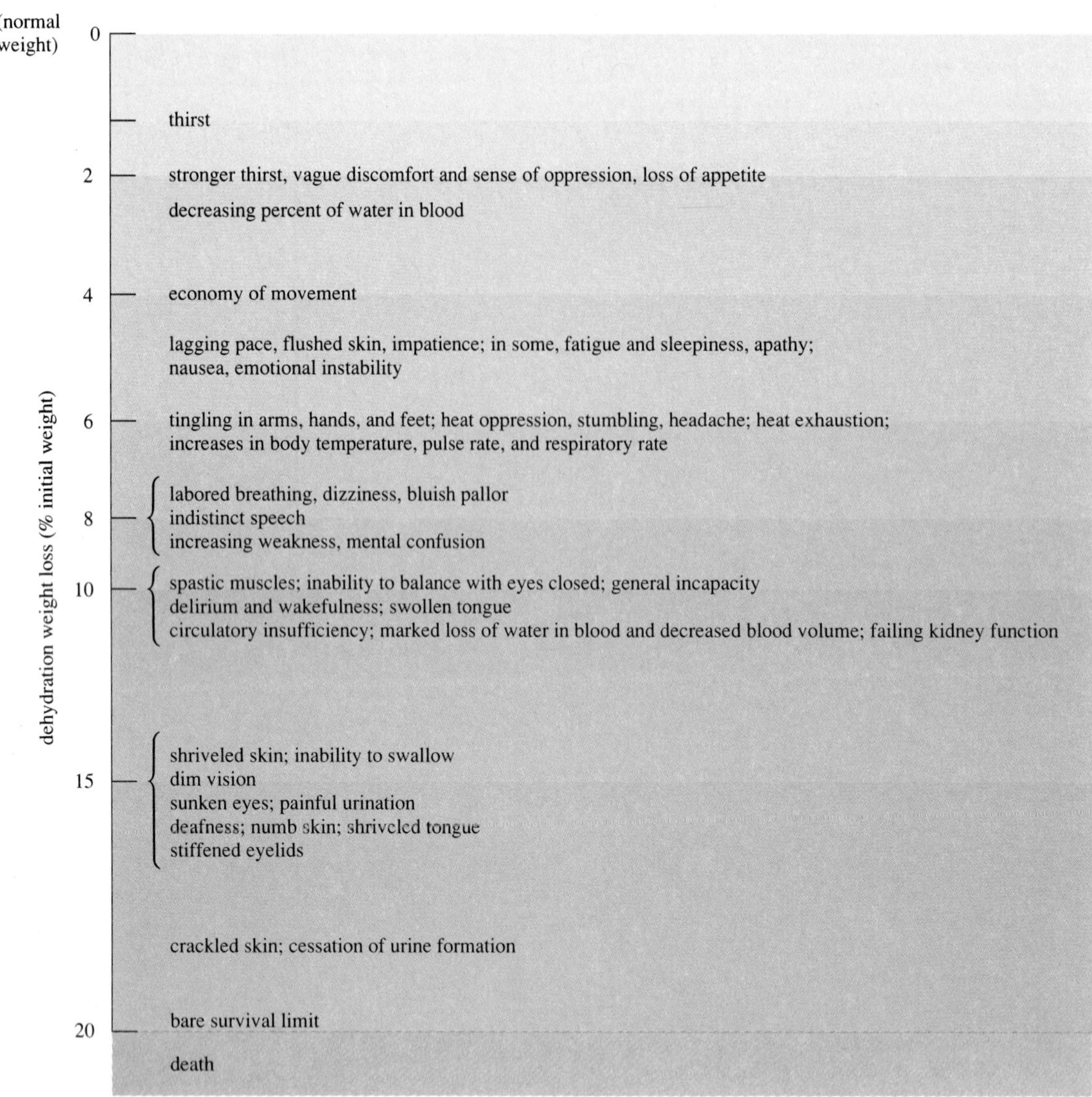

Figure 26.7
When a person stops eating and drinking, symptoms of dehydration set in rapidly.
Source: E. M. Roth, *Compendium of Human Responses to the Aerospace Environment*, Volume 3, section 15, NASA Contract No. NASr-115, 1967.

The challenge then is to balance water gains and losses. Most water is gained from food and drink, but some results from the oxidation of food molecules occurring during cellular respiration. By far the most important process in regulating water loss is excretion. However, smaller amounts of water are lost from the skin by evaporation, from the lungs during breathing, and with the feces.

KEY CONCEPTS

Nitrogenous wastes form from the metabolic breakdown of proteins and nucleic acids. The amino groups of amino acids gain a hydrogen ion to form ammonia, which is excreted in dilute form by some aquatic animals but converted to urea by amphibians and mammals and to uric acid by birds, reptiles, and insects. Water gain must be balanced against water loss to keep tissues sufficiently hydrated.

The Human Excretory System—An Overview

The paired **kidneys** are the major organs responsible for both excretion and osmoregulation (fig. 26.8). Located against the wall of the upper back within the abdominal cavity, each is about the size of an adult fist and weighs about half a pound. Urine is formed within each kidney and drains into a muscular tube about 11 inches (28 centimeters) called a **ureter.** The two ureters (one from each kidney) lie at an angle and are lined with valves, keeping urine flow one way. A back flow could cause infection

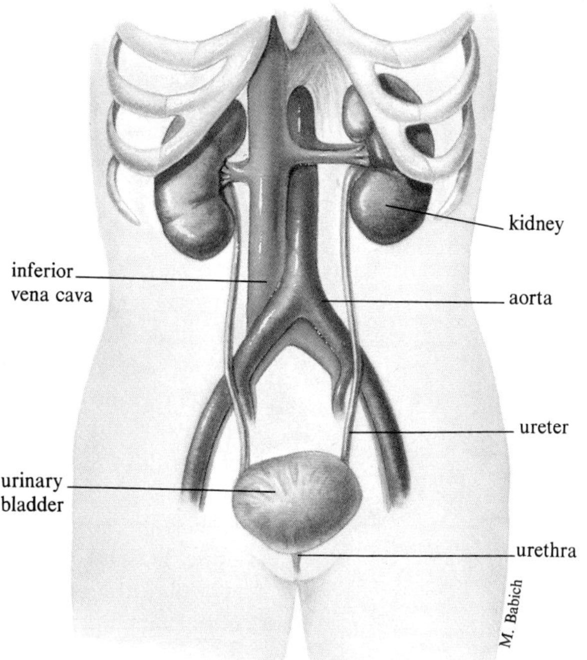

Figure 26.8
The human excretory system includes the kidneys, ureters, urinary bladder, and urethra.

if microorganisms are picked up from outside the body. Waves of muscle contraction squeeze the urine along the ureters and squirt it into a saclike muscular **urinary bladder.** Urine drains from the bladder and exits the body through the **urethra.** In the female, this tube is about an inch (2 to 3 centimeters) long and opens between the clitoris and vagina. In the male, the urethra is about 8 inches (20 centimeters) long and extends the length of the penis. Can you suggest why infections of the urinary tract are more frequent among women than men?

The exit from the bladder is guarded by two rings of muscle, called sphincters, both of which must relax before urine can leave the bladder. The innermost sphincter is involuntarily controlled by a spinal reflex. At about 2 years of age, we learn to consciously control the relaxation of the outer sphincter.

Although the adult bladder can hold about 20 ounces (600 milliliters) of urine, stretch receptors in the bladder wall are stimulated when only 10 ounces (300 milliliters) of urine accumulate. The receptors send impulses to the spinal cord, which stimulates nerves that contract the bladder muscles, generating a strong urge to urinate. The urge can be suppressed for a short time by the external sphincter. The cerebral cortex directs the sphincters to relax and the contractions of bladder muscles squirt urine out of the body.

The Kidney

The kidneys form an incredibly efficient blood-cleansing mechanism by simultaneously adjusting the composition, pH, and volume of the blood. As the blood is processed, urine forms.

Each kidney is packed with 1.3 million microscopic tubules called **nephrons.** An individual nephron is built of a continuous **renal tubule** ("renal" means "of the kidney"), plus a blood supply, the **peritubular capillaries,** that entwines around it. A renal tubule stretched out would be about a half an inch (12 millimeters) long.

The same regions of each nephron lie at similar positions within the kidney (fig. 26.9). The outermost part of the kidney is called the **cortex,** and it is grainy in appearance. The middle section, called the **medulla,** looks like it is composed of many aligned strings, and this region corresponds to the long sections of the renal tubules. The innermost portion of the kidney, the **pelvis,** collects the urine produced from each nephron before it leaves the kidney to enter a ureter. The kidney illustrates the biological plan of extensive surface area (afforded by the renal tubules) packed into a relatively small volume (the kidneys). Specific portions of the nephrons carry out specific functions.

The body's entire blood supply courses through the kidney's blood vessels every 5 minutes. At that rate, the equivalent of 425 to 525 gallons (1,600 to 2,000 liters) of blood per day passes through the kidneys. Yet a person excretes only about .4 gallons (1.5 liters) of urine daily. Most of the material that enters the renal tubules from the blood is reabsorbed back into the blood rather than excreted in the urine.

The kidney retains and recycles important dissolved chemicals and water by three processes. It filters wastes, nutrients, and water from the blood into the renal tubule, while large structures such as proteins and blood cells remain in the blood. It reabsorbs salts and nutrients back into peritubular capillaries and returns water to the bloodstream. Finally, various toxic substances are secreted from the capillaries into the renal tubule. Overall, the nephrons extract wastes and recycle valuable nutrients and salts to the bloodstream.

KEY CONCEPTS

The human excretory system consists of the kidneys, ureters, urinary bladder, and urethra. Sphincters and the nervous system control the bladder. A kidney is built of microscopic nephrons. A nephron consists of a renal tubule and its surrounding capillaries. The outer portion of the kidney is the cortex, and the inner portion, the medulla. Urine formed along the nephrons drains into the innermost kidney region, the pelvis. The kidney extracts wastes and toxins from the blood and recycles valuable substances by filtering, reabsorbing, and then secreting.

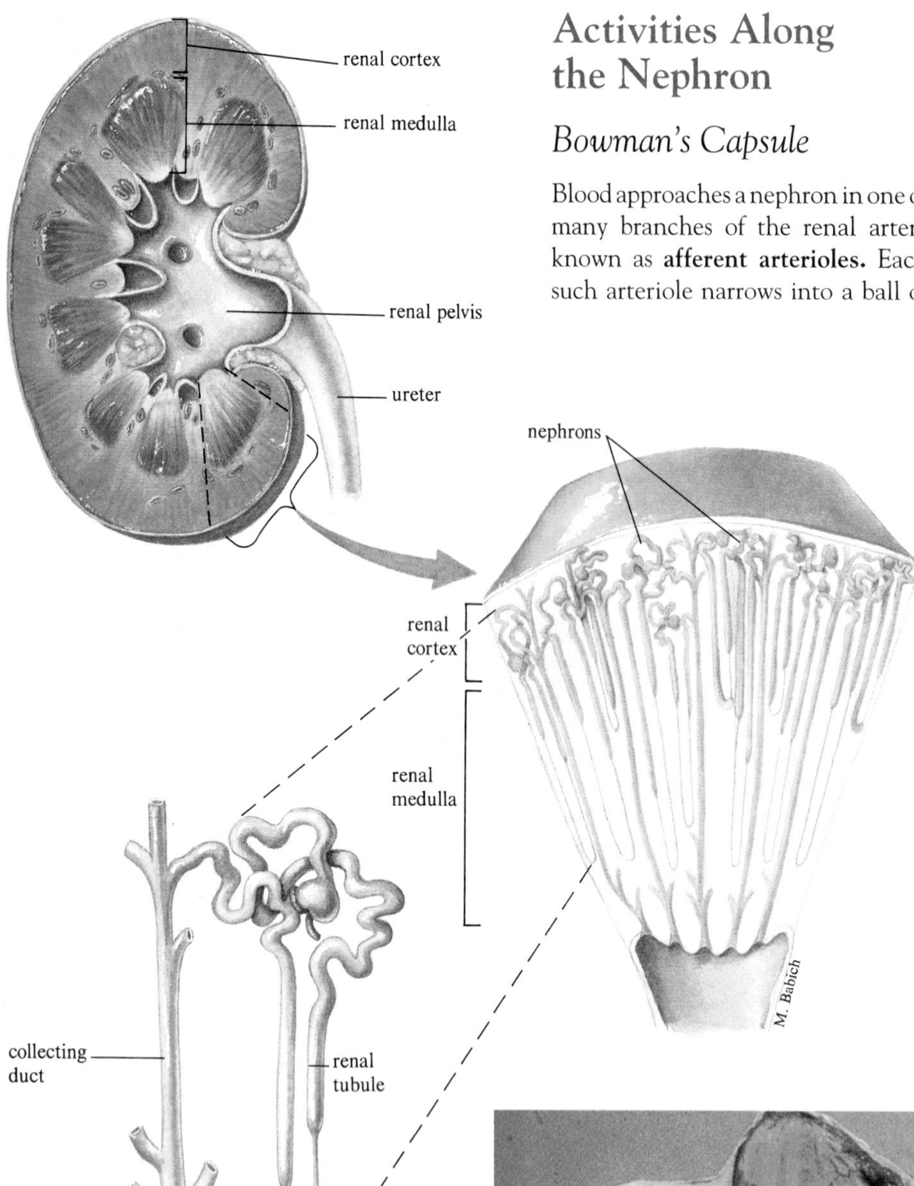

Activities Along the Nephron

Bowman's Capsule

Blood approaches a nephron in one of many branches of the renal artery known as **afferent arterioles.** Each such arteriole narrows into a ball of capillaries called the **glomerulus** (Latin for "tiny ball"). These capillaries then come together to form an **efferent arteriole.** The glomerulus is surrounded by the cup-shaped end of the renal tubule, which is known as **Bowman's capsule** (figs. 26.10 and 26.11). Bowman's capsules lie in the cortex of the kidney. To envision the structure of Bowman's capsule, picture the end of the finger of a glove pushed inward to form a two-layered cup.

Blood is filtered by the glomerular capillaries into Bowman's capsule. Anything that fits through the pores in the glomerulus and that is not repelled by the charge on the membrane passes into Bowman's capsule. Like a catcher's mitt firmly grasping a ball, Bowman's capsule surrounds the glomerulus and captures all of this material.

What creates the pressure driving substances out of the glomerulus? The diameter of the afferent arteriole is greater than that of the efferent arteriole. The pressure generated by this difference forces fluid and small dissolved molecules across the capillary walls through tiny spaces between the cells of the Bowman's capsule. Large molecules (such as plasma proteins) and formed elements of the blood (cells

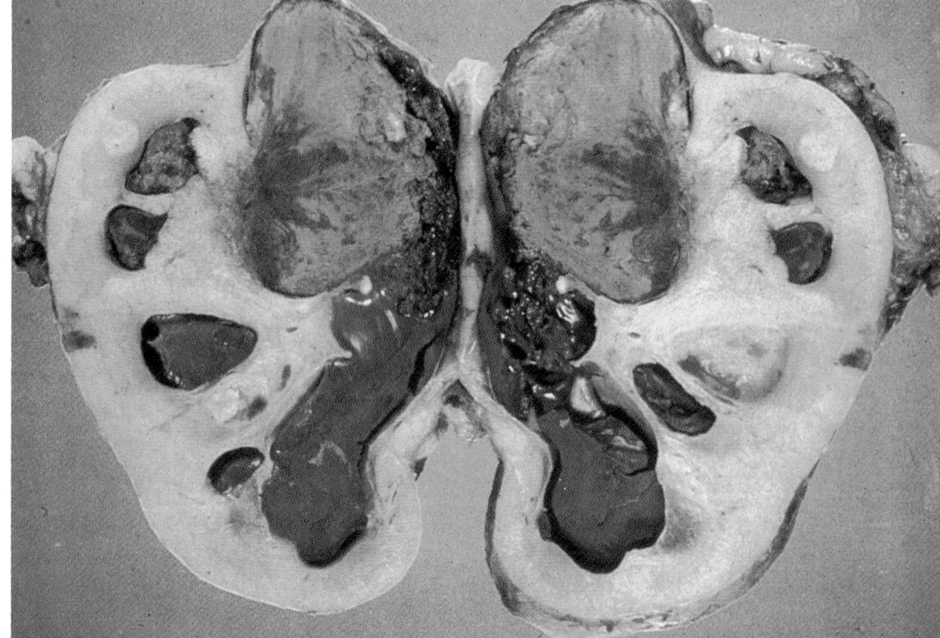

Figure 26.9
Anatomy of a kidney. *a.* A kidney sliced down the middle reveals an outer area, the cortex, and an inner section, the medulla. The nephrons are aligned and, as a result the same parts are found in the same region of the kidney. *b.* A calf kidney.

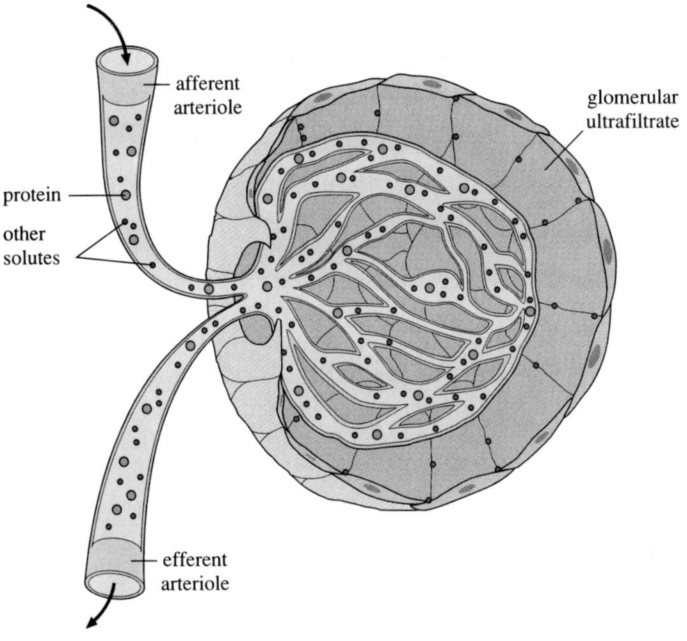

Figure 26.10
The glomerulus and Bowman's capsule. Small molecules dissolved in blood plasma cross the capillary walls and squeeze between the cells lining the Bowman's capsule. Larger proteins, blood cells, and platelets are left behind.

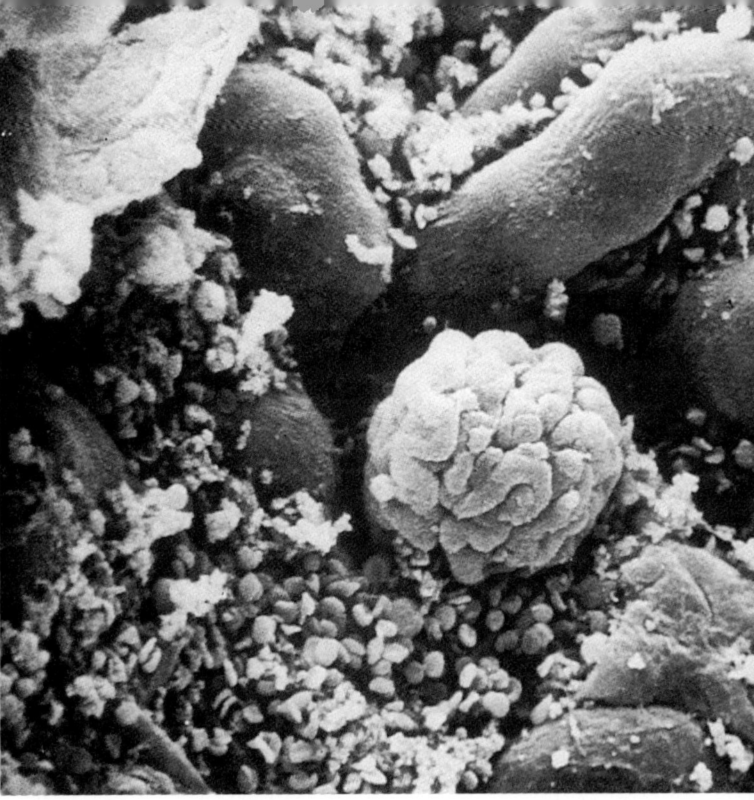

Figure 26.11
Scanning electron micrograph of a glomerulus. Magnification, x300.

Reading 26.2 *Urinalysis—Clues to Health*

LIKE THE BLOOD, URINE IS FASCINATING TO MEDICAL MINDS. Urine has been a part of many folk remedies, variously used as a mouthwash, a toothache treatment, and a cure for sore eyes. Hippocrates (460–377 B.C.) was the first to observe that the condition of the urine can reflect health, noting that frothy urine denoted kidney disease. During the Middle Ages, health practitioners frequently consulted charts in which certain urine colors were matched to certain diseases. In the seventeenth century, British physicians diagnosed diabetes by tasting sugar in the urine.

Today, urine composition is still used as an indicator of health. The urine of a healthy individual contains water, urea, creatinine, uric acid, ammonia, amino acids, and several salts (table 1). It is pale yellow, with a pH of 4.8 to 8.0. These characteristics may fluctuate slightly due to diet. Urine is also routinely tested for traces of biochemicals that indicate a person has taken an illegal drug.

What can a urine specimen reveal about the health of the donor? Blood in the urine indicates bleeding in the urinary tract. More than a trace of glucose may be a sign of diabetes mellitus. Carrier molecules actively transport glucose out of the fluid, or filtrate, in the kidney tubules and into the blood. If there are more glucose molecules than carrier molecules, the excess glucose will remain in the filtrate and enter a ureter. In diabetes mellitus, the pancreas produces too little insulin or the insulin produced cannot be used by the body. In either event, insulin is prevented from doing its job, which is to lower the level of glucose in the blood. Therefore, individuals with diabetes mellitus have a high blood level of glucose and consequently so much in the filtrate that glucose appears in the urine.

Table 1
The Composition of Urine

Component	*Grams/24 Hours (1,200 ml)*
Glucose	< 0.05
Amino acids	0.80
Ammonia	0.80
Urea	25.0
Creatinine	1.5
Uric acid	0.7
Hydrogen (H^+)	Enough to maintain pH 5–8
Sodium (Na^+)	3.0
Potassium (K^+)	1.7
Calcium (Ca^{2+})	0.2
Magnesium (Mg^{2+})	0.15
Chloride (Cl^-)	6.3
Phosphate (PO_4^{2-})	1.2
Sulfate (SO_4^{3-})	1.4
Carbonate (HCO_3^-)	0–3

There are other reasons why glucose in the blood may spill over into the urine—such as a high carbohydrate diet or stress. Stress causes secretion of excess epinephrine, which stimulates the liver to break down more than the usual amount of glycogen into glucose.

Albumin in the urine may be a sign of damaged nephrons, because this plasma protein is too large to fit through the pores at the entrance to the kidneys' microscopic tubules. The presence of pus along with complete absence of glucose indicates infection. The pus consists of white blood cells, which fight infection, and traces of glucose are consumed by the infecting bacteria.

and platelets) remain behind in the bloodstream and exit the nephron in the efferent arteriole, which ultimately leads into the capillary network surrounding the remaining tubules of the nephron. These capillaries then empty into a venule, which runs into the renal vein, which finally joins the inferior vena cava.

The chemical composition of the material in Bowman's capsule is similar to that of blood plasma. Because only some substances from the blood enter the nephron here, the material passing into the Bowman's capsule is called the **glomerular filtrate.** It is the product of the first step in the manufacture of urine. About 425 to 525 gallons (1,600 to 2,000 liters) of blood per day pass through the kidneys, producing approximately 45 gallons (180 liters) of glomerular filtrate.

The remainder of the renal tubule consists of a winding passageway, which can be considered as four functional regions: the proximal convoluted tubule, the loop of Henle, the distal convoluted tubule, and the collecting duct (fig. 26.12).

The Proximal Convoluted Tubule

The glomerular filtrate passes from the Bowman's capsule into the **proximal convoluted tubule.** This is an important site for selective reabsorption, a process that returns useful components of the glomerular filtrate to the blood. The wall of the proximal convoluted tubule is highly folded, which increases its surface area, and its cells have many mitochondria to power the active transport of molecules of dissolved nutrients back into the blood. Here, specialized cells actively reabsorb all the glucose and vitamins and about 75% of the amino acids. Important ions (also called electrolytes because they can conduct electricity when in solution) are also reabsorbed. Some ions are actively transported and others tag along passively, attracted by the electrical charge on the transported ion. With so many substances being removed from the filtrate, it now contains fewer solutes than the blood.

The Loop of Henle

After the proximal convoluted tubule, the renal tubule of a mammal forms the **loop of Henle,** where water is conserved and the urine concentrated. The loop of Henle consists of a **descending limb** and an **ascending limb,** and it dips into the medulla region of the kidney.

Differing permeabilities of the descending and ascending limbs create a situation in which the bottom of the loop sits in a salty brew, which forces water to diffuse out of the renal tubule and into the surrounding capillaries (fig. 26.13). The cells that form the ascending limb are impermeable to water, and they actively transport sodium ions (Na^+) into the space in the medulla of the kidney not occupied by nephrons. Because of the movement of Na^+, the medulla builds up a higher concentration of Na^+ than does the interior of the descending limb. Unlike the situation in the ascending limb, the cells of the descending limb are permeable to water, which passively diffuses into the medulla, following its concentration gradient. In this way, the descending limb conserves water by sending it into the medulla, where it can diffuse into the peritubular capillaries and back into the circulation.

As water leaves the descending limb, the Na^+ exiting the ascending limb passively diffuses into the descending limb.

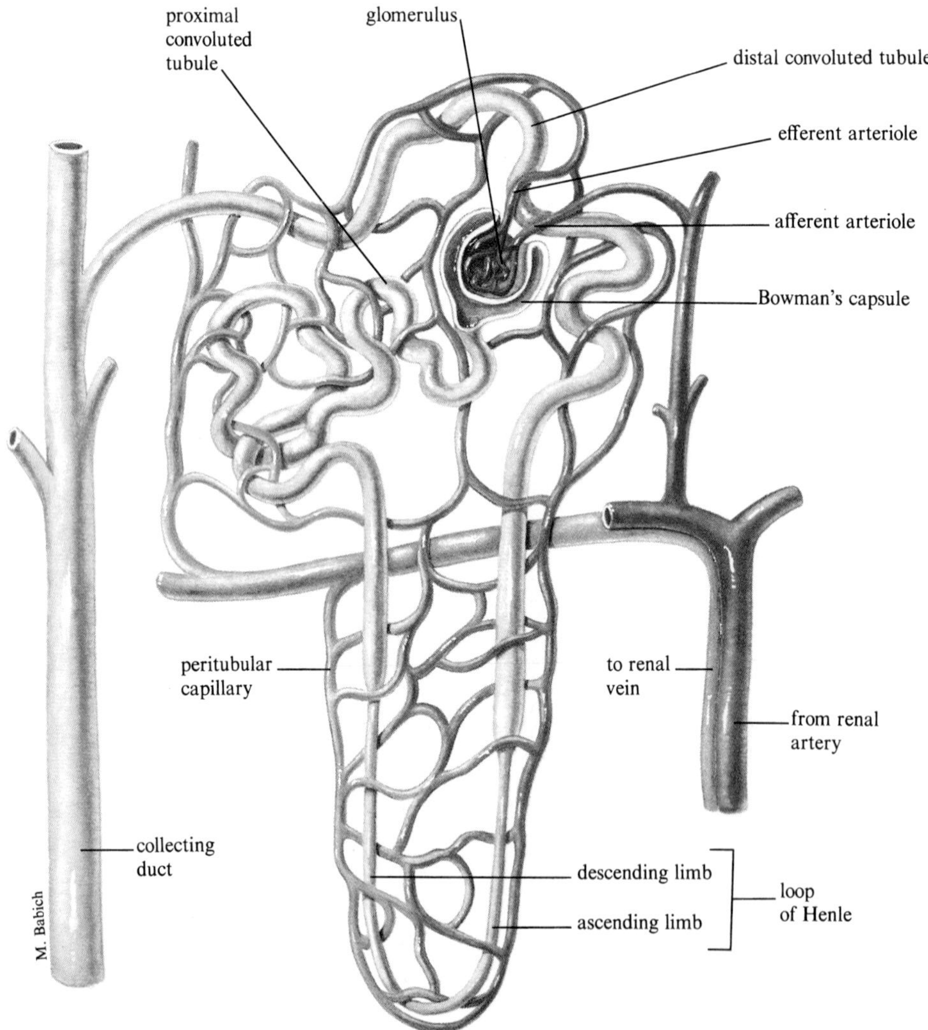

Figure 26.12
The nephron.

This is because there is now a greater concentration of Na^+ in the medulla space than inside the descending limb. Once inside the descending limb, the Na^+ is transported along the loop until it reaches the ascending limb, from which it is actively transported. Can you see how a cycle is set up?

The movement of Na^+ and water between the limbs of the loop of Henle and the medullary space is called a **counter-current multiplier system.** The "counter-current" refers to the fact that fluid movement direction is opposite in the two limbs (down and then up). The effect is multiple because of both the active extrusion of Na^+ from the ascending limb and the passage of Na^+ into the descending limb.

The loop of Henle is an adaptation of the mammalian nephron to conserve water. It is interesting to see how the loop's relative length differs among mammalian species. The longer the loop, the more water is reabsorbed into the bloodstream and the more concentrated is the urine. The beaver and the Australian hopping mouse are at two extremes. The beaver lives surrounded by water and so does not have to conserve it. This animal has a short loop of Henle and excretes a watery urine that is twice as concentrated as its blood plasma. The Australian hopping mouse is a desert dweller with the longest loop of Henle of all mammals in relation to its body size. Its urine is 22 times as concentrated as its blood plasma. In comparison, human urine is 4.2 times as concentrated as blood plasma.

The Distal Convoluted Tubule

Once the glomerular filtrate has passed the loop of Henle, it is considerably more concentrated because of the water returned to the circulation in the descending limb. The next region of the renal tubule is the **distal convoluted tubule,** located in the kidney's cortex. In the distal convoluted tubule, sodium is reabsorbed into the peritubular capillaries. This is accomplished by active transport, because the medulla already has a high concentration of Na^+ due to the active transport of this ion from the ascending limb of the loop of Henle. The accumulation of Na^+ outside the distal convoluted tubule stimulates water to diffuse passively out of the tubule into the capillaries.

The distal convoluted tubule also helps to maintain the pH of the blood between 6 and 8. Any variance from this range is deadly. The blood is more likely to become too acidic than too alkaline, largely because most metabolic wastes are acids. If the pH of the blood falls too low, the distal convoluted tubules can raise it by secreting H^+ into the urine. A blood pH that is too high can be lowered by inhibiting the secretion of hydrogen ions by the tubules.

The distal convoluted tubule also secretes waste molecules, such as creatinine, a breakdown product of phosphocreatine, which supplies high energy phosphate groups to muscle, and drugs such as penicillin.

The Collecting Duct

The next portion of the renal tubule, the **collecting duct**, descends into the medulla, as did the loop of Henle. Several renal tubules drain into a common collecting duct. Urea diffuses out of the collecting duct as it passes through the medulla, drawing water out of the collecting ducts as well as from the bottom region of the descending limb of the loop of Henle.

After adjustment by reabsorption and secretion, the filtrate is urine. From the collecting duct, urine accumulates in the pelvis of the kidney before draining through the ureter to the urinary bladder and, finally, moving out of the body through the urethra. Table 26.2 summarizes the functions of the parts of the nephron.

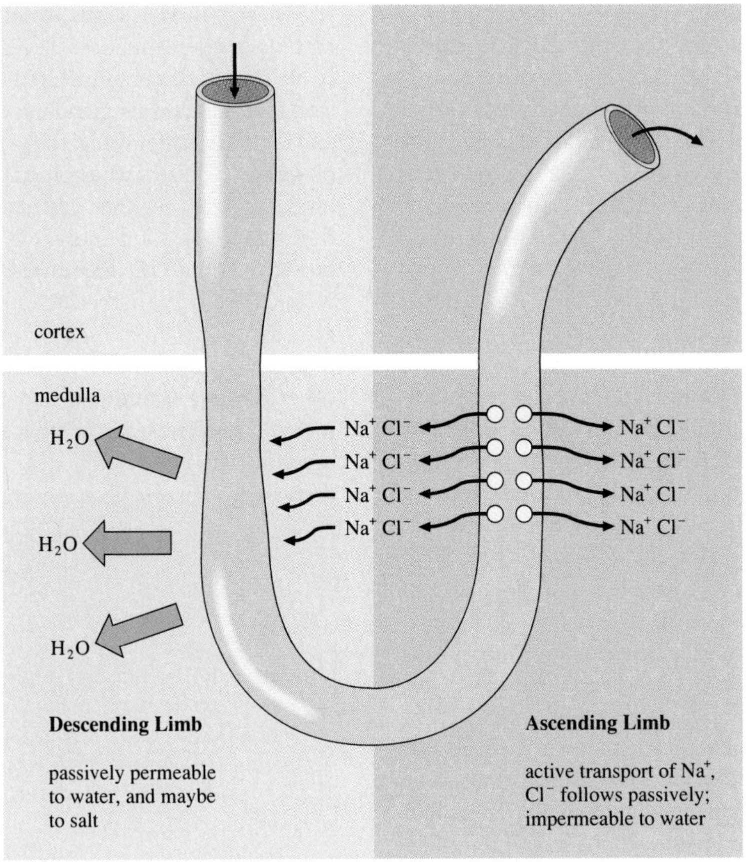

Figure 26.13
The loop of Henle conserves water. The active transport of Na^+ from the ascending limb and the passive entry of Na^+ into the descending limb sets up a situation in which water diffuses out of the loop and eventually enters the surrounding capillaries.

KEY CONCEPTS

Urine formation begins as fluid and small molecules cross from the glomerulus into the Bowman's capsule. In the proximal convoluted tubules, glucose, ions, vitamins, and amino acids are selectively reabsorbed back into the peritubular capillaries. In the loop of Henle, the cycling of Na^+ from the ascending limb to the descending limb forces water to leave the renal tubules. In the distal convoluted tubule, Na^+ is sent out, larger wastes and toxins are taken from the blood, and blood pH is regulated. In the collecting duct some urea diffuses out to the medulla, increasing water reabsorption back at the bottom of the loop of Henle. What is left in the renal tubules is finally urine.

Table 26.2
Summary of Functions of Parts of the Renal Tubule

Part	Site in Kidney	Filtration
Bowman's capsule	Cortex	Filtration
Proximal convoluted tubule	Cortex	Selective reabsorption of water, amino acids, electrolytes
Loop of Henle	Medulla	Countercurrent exchange
Distal convoluted tubule	Cortex	Secretion of large molecules
Collecting duct	Medulla	Further reabsorption of water

Control of Kidney Function

It is important to the health of each cell that the pH, volume, and composition of the blood not deviate too much from optimum values. Since the kidney plays a key role in regulating these factors, its activities are continuously monitored and adjusted to meet the body's needs.

The amount of water reabsorbed from the filtrate influences two important characteristics of blood, osmotic pressure and volume. The osmotic pressure of the blood, that is, how concentrated the plasma solutes are, affects many cellular activities, particularly the exchange of materials between the cells and the blood. The volume of blood is a factor determining blood pressure and, therefore, cardiovascular health.

The solute concentration of the blood is kept constant in spite of variations in the amount of water we consume in food or drink. When excess water is consumed, our kidneys allow more to pass into the urine. However, if water is scarce, our kidneys conserve it by producing concentrated urine. Osmoreceptor cells within the hypothalamus in the brain determine how much water should be retained. When the blood plasma becomes too concentrated, increasing osmotic pressure, the osmoreceptor cells send impulses to the posterior pituitary gland in the brain, which then secretes **antidiuretic hormone (ADH).** ADH increases the water permeability of both the distal convoluted tubule and the collecting duct. This increases the reabsorption of water into the blood.

Conversely, if the blood plasma is too dilute, that is, if its osmotic pressure is too low, the same cells in the hypothalamus detect this and respond by signaling the posterior pituitary gland to stop production of ADH. As a result, the distal convoluted tubule and collecting ducts become less permeable to water. More water is retained and excreted and the osmotic pressure of the plasma and tissue fluids is decreased. In a disease called diabetes insipidus, ADH activity is insufficient, and the person urinates from 1 to 2 gallons (5 to 10 liters) per day. Thirst is intense, yet the person has trouble drinking enough to compensate for this abnormal water loss from the kidneys.

The ethyl alcohol in alcoholic beverages such as beer is a diuretic, meaning that it increases the volume of urine. It stimulates urine production by decreasing production of ADH, thereby increasing the permeability of the tubules to water. By increasing water loss to urine, an alcoholic beverage actually intensifies thirst. The dehydration of body tissues that results from drinking too much alcohol is responsible for some of the discomfort of a hangover. The caffeine in coffee is also a diuretic, exerting its effects at the Bowman's capsule by increasing the amount of water that enters the renal tubule. Diuretic drugs (water pills) are used to increase urine volume by decreasing reabsorption of Na^+ in the proximal tubules. Diuretics lower blood pressure and relieve edema (painful tissue swelling).

Aldosterone is a hormone synthesized in the adrenal glands, which sit on top of the kidneys. It enhances the reabsorption of Na^+ in the distal convoluted tubules, as well as in the salivary glands, sweat glands, and large intestine. When the sodium level of the blood falls, or when blood pressure or blood volume declines, aldosterone synthesis increases and more Na^+ is transported back into the bloodstream. Groups of specialized cells in the afferent arterioles also sense lowered blood volume and pressure. They respond by releasing another hormone, **renin,** which initiates a series of chemical reactions that eventually boost aldosterone levels.

KEY CONCEPTS

The blood's solute concentration and volume are regulated by osmoreceptor cells in the hypothalamus, which control (via the pituitary) secretion of ADH. ADH increases the permeability to water of the distal convoluted tubule and the collecting duct, increasing reabsorption of water into the blood. Aldosterone increases reabsorption of Na^+ in the distal convoluted tubules, thereby controlling blood-sodium level. Renin, a hormone released by special cells in the afferent arterioles, controls aldosterone secretion.

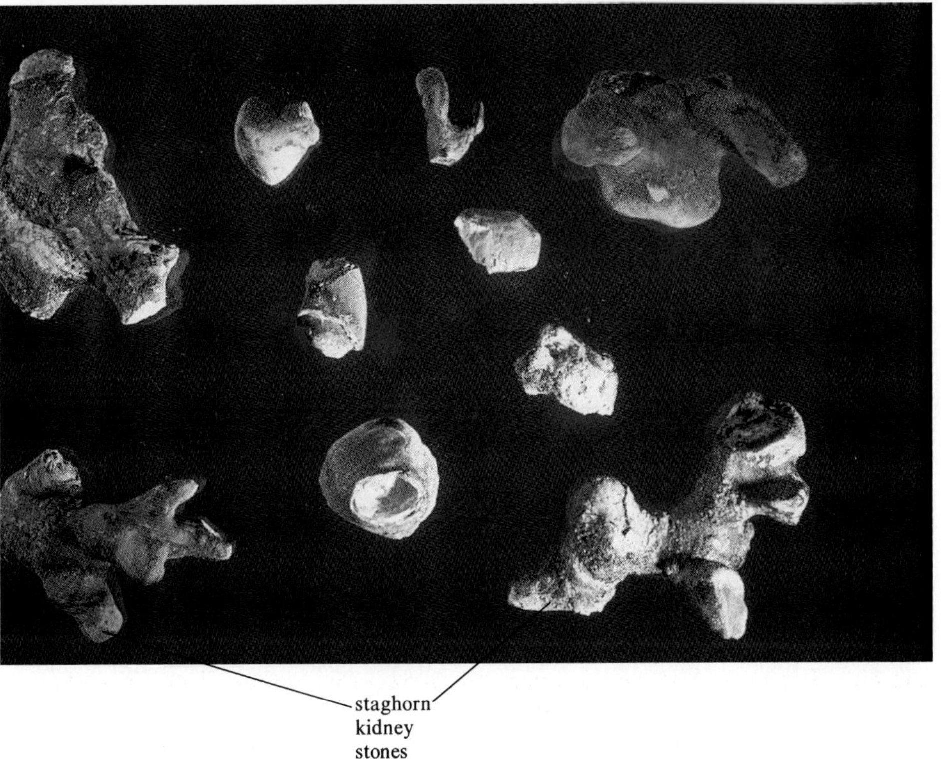

Figure 26.14
A large kidney stone assumes the shape of the region where it formed, where the collecting ducts empty into the renal pelvis. Such a stone is called a "staghorn" because it resembles a deer's antlers.

The Unhealthy Excretory System

Urinary Tract Infections

Urinary tract infections are a prevalent excretory system problem. "UTIs" most often affect women, because the placement of the urethra near the vaginal and anal openings provides an open pathway for bacteria to enter. Older men sometimes develop UTIs when the prostate gland swells and blocks urination. A UTI is often first experienced as frequent, painful urination, and it may be accompanied by fever and lower abdominal pain. Typically, the urine is cloudy and has a foul odor. Although a UTI usually clears up on its own, antibiotics are used to prevent the spread of the infection to the kidneys.

Kidney Stones

Kidney stones form when salts, usually calcium salts or uric acid, precipitate out of the newly formed urine and accumulate in the kidney tubules and pelvis (fig. 26.14). Kidney stones can be caused by excessive milk consumption, infection, dehydration, hormonal problems, or the inherited metabolic disorder gout, in which excess uric acid is present in the blood, tissues, and renal tubules.

Kidney stones cause fever, chills, frequent urination, and kidney pain. If the stones are passed down through the ureters, they can cause excruciating pain. Treatment for large stones has traditionally been surgery, but today there are several other techniques. The most exciting new treatment is extracorporeal shock wave lithotripsy (ESWL). At first this treatment required that the patient be submerged in a bath of warm water. However, now the bath can be replaced by a water cushion coupled to the patient's body. In either case, shock waves are focused on the stone and passed through the water and the patient's body to shatter it. The much smaller fragments are then passed in the urine.

Kidney Failure

Kidney failure is the result of damaged renal tubules. One theory attributes kidney failure to elevated blood pressure in the glomerulus, which in itself could have a number of causes. The high blood pressure strains the filtering capabilities of the glomerular capillaries. As this part of the nephron becomes overworked, it hardens and eventually cannot function. As individual nephrons progressively shut down, remaining nephrons "adapt" and attempt to compensate for the loss by working harder. Gradually, these adaptive nephrons become overworked too. A cycle is set into motion that culminates with the failure of the kidney as a whole. Although all of the details of this self-perpetuating mechanism of kidney failure have yet to be worked out, one contributing factor appears to be high-protein diets, which can overwork the kidneys.

Malfunctioning kidneys lead to a rapid buildup of toxins in the blood, altered ion concentrations in the tissues, and water retention that causes painful swelling of tissues (edema). A replacement for the kidney must be instituted quickly to preserve life.

For about 20,000 Americans, a dialysis machine (or "artificial kidney") takes over kidney function for a few hours twice a week. The patient's blood is passed through a tube that is separated from a balanced salt solution by an artificial semipermeable membrane. Following concentration gradients, substances that are more concentrated in the salt solution diffuse into the blood. The result is "cleansed" blood.

Newer, more convenient and time-saving forms of dialysis use the patient's own membrane, specifically the peritoneal membrane lining the body cavity, instead of an artificial membrane. The balanced salt solution is placed into the patient's abdominal cavity. Waste products and excess water diffuse from the person's blood into the fluid. After about half an hour, the fluid is suctioned out. A similar method of dialysis interferes even less with the patient's activity. A balanced salt solution is put into the abdominal cavity and left there for several days before replacement.

A more convenient but harder-to-find alternative to dialysis is a kidney transplant. Because the body usually rejects tissues that it does not recognize as "self," this procedure is most successful when the donor is a blood relative (a person can live with only one kidney). When no suitable relative is available, computers are utilized to find unrelated donors whose cell surfaces are similar enough to those of the potential recipient for a transplant to be successful. Since 1983, success rates of kidney and other transplants have soared thanks to the use of a new drug, cyclosporin, which suppresses the body's rejection of transplanted tissue—a topic explored further in the next chapter.

KEY CONCEPTS

Urinary tract infections develop when bacteria enter the urethra. Kidney stones form when salts precipitate out of urine in the renal tubules and pelvis. Kidney failure has many causes and is not completely understood, but it happens when renal tubules are damaged. Kidney failure leads to buildup of toxins in the blood, altered chemistry in the tissues, and water retention. Dialysis techniques can artificially cleanse the blood.

SUMMARY

Maintaining a stable internal environment is called *homeostasis*. The enzymatic reactions critical to life depend on a regulated temperature. Biochemicals such as lipids in membranes are altered by heat, disrupting their function.

Animals differ in the ways they warm their bodies. *Ectotherms* use solar energy and regulate their body temperature behaviorally by choosing an environment with the appropriate temperature. *Endotherms* use metabolism to generate heat. Some animals maintain a constant body temperature, and others have temperatures that fluctuate with that of their surroundings. In yet others, body temperature changes sometimes, or parts of the body change temperature.

When an organism that maintains its body temperature metabolically is chilled, the body increases heat production by shivering and hormonally increasing metabolic rate. At the same time, heat is conserved by erecting feathers or fur to trap a layer of air around the body for insulation and by constricting the blood vessels delivering blood to the surface of the body. Perspiring helps rid the body of excess heat.

The composition of the fluids bathing each cell must also be regulated. Nitrogenous wastes must be removed, and the amount of water and salts must be adjusted (osmoregulation). Most nitrogenous wastes come from the breakdown of protein, and include ammonia, urea, and uric acid. *Ammonia* uses the least energy to produce but requires the most water for excretion. *Urea* uses more energy but conserves some water. *Uric acid* requires the most energy to produce but can be excreted in an almost solid form.

The human urinary system excretes nitrogenous waste (mostly urea) and regulates water. Urine is produced in the *kidneys*, each of which drains into a *ureter*, which leads to the *urinary bladder* for storage. Urine leaves the body through the *urethra*.

The functional unit of the kidney is the *nephron*. The blood is filtered from a tuft of capillaries called the *glomerulus* into a cuplike structure called *Bowman's capsule*. This filtrate is adjusted by reabsorption of important materials from the filtrate back to the blood and by secretion of other substances into the urine. The adjustment begins in the part of the renal tubule directly connected to Bowman's capsule, the *proximal convoluted tubule*. The filtrate then moves into a long tubule that dips into and then out of the center, or *medulla*, of the kidney. This section of the tubule is called the *loop of Henle*. Its most important function is the concentration of the urine to conserve water. Water is drawn out of the filtrate because an osmotic gradient is established in the fluid around the loop of Henle. Na^+ and urea are cycled between tubules to establish this gradient. In the next region of the nephron, the *distal convoluted tubule*, more reabsorption and secretion occur. The secretion of H^+ here helps regulate the pH of the blood. The filtrate then moves on to the collecting duct and finally to the central cavity of the kidney, the *pelvis*, before draining through the ureter.

The amount of water and salt reabsorbed from the distal convoluted tubule is regulated by *antidiuretic hormone* (ADH) from the posterior pituitary gland of the brain in response to signals from the hypothalamus, which senses the osmotic pressure of the blood. ADH increases the permeability of the distal convoluted tubule and the collecting duct so more water is reabsorbed and the urine is concentrated. *Aldosterone*, another hormone, increases the retention of salt by the kidneys. Aldosterone is released by the adrenal glands in response to either low sodium concentration in the plasma or low blood pressure. Water reabsorption is a consequence of salt retention.

Possible problems within the excretory system are *urinary tract infections*, *kidney stones*, and *kidney failure*. If the kidneys fail, they must be replaced by a kidney transplant or their function must be taken over by dialysis.

QUESTIONS

1. What is homeostasis?

2. How do humans and iguana lizards differ in how they regulate their body temperatures?

3. Why is it important to maintain a certain level of water in the human body?

4. What are the three types of nitrogenous waste, and where do they come from?

5. Draw a nephron and label the parts. Indicate which regions of the renal tubule are specialized for each of the three processes involved in urine formation.

6. How is urine concentrated?

7. How can what you eat and drink affect the volume and concentration of your urine?

8. What are some medical conditions that might be detected by examining the urine?

9. How do ADH and aldosterone control kidney function?

10. Why can't you live without at least one kidney, unless you undergo dialysis?

TO THINK ABOUT

1. Imagine that you are adrift at sea. Why would you dehydrate more quickly if you drank seawater to quench your thirst?

2. An infant suffers from severe and frequent vomiting and diarrhea for 24 hours. When her father takes her to the doctor, he is surprised when the doctor hospitalizes the child and orders that she be given fluids intravenously. Why do you think this drastic action was taken?

3. Urinary tract infections frequently accompany sexually transmitted diseases. Why?

4. Would an excess or deficiency of renin be likely to cause hypertension (high blood pressure)? Cite a reason for your answer.

5. Why is protein in the urine a sign of kidney damage? What structures in the kidney are probably affected?

6. Why is coffee not a good treatment to ease a hangover?

7. Why are people following high-protein diets advised to drink large quantities of water?

8. How could very low blood pressure impair kidney function?

SUGGESTED READINGS

Bainbridge, J. J. Jr. January 1991. Five weeks on a magnet. *Audobon*. Aves Island, in the Caribbean, is built up by guano—bird droppings.

Flieger, Ken. March 1990. Kidney disease: When those fabulous filters are foiled. *FDA Consumer*. Drugs, devices, and transplants can restore vital kidney function.

Harrison, George H. February–March 1991. Little dynamo. *National Wildlife*. The chickadee eats, and eats, and eats, to stay warm in winter.

Lewis, Ricki. June 1990. Wilms' tumor: The genetic plot thickens. *The Journal of NIH Research*. Unraveling a childhood kidney cancer reveals how this complex organ develops.

Southwick, Edward E., and Gierhard Heldmaier. June 1989. Temperature control in honey bee colonies. *Bioscience*. European honeybees rely on social behavior as well as metabolism to regulate body temperature.

Underwood, Benjamin A. December 1990. Bee Cool. *Natural History*. Bees have interesting adaptations to retain heat.

Zamula, Evelyn. March 1985. Urinary infection: The unwelcome night visitor. *FDA Consumer*. A urinary infection is annoying and temporarily painful but, with antibiotic treatment, is rarely serious.

For information on kidneys and health, write:

The National Kidney Foundation, Inc.
30 East 33 Street
New York, NY 10016

C H A P T E R

27

The Immune System

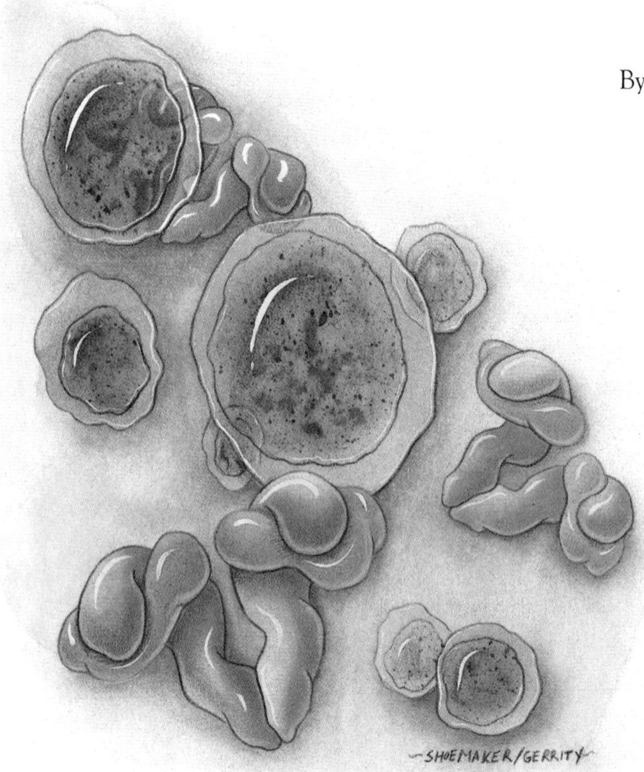

Chapter Outline

Learning Objectives

By the chapter's end, you should be able to answer these questions:

1. What is the overall function of the immune system?
2. What are some ways that the body protects itself nonspecifically?
3. How is the immune response fast, specific, and diverse?
4. How does the immune system remember?
5. How do the cells and biochemicals of the immune system interact with each other to protect against specific invaders?
6. How does immune function change over a lifetime?
7. How is immune function altered by AIDS, autoimmune disorders, and allergies?
8. How can the immune system be altered to prevent, diagnose, and treat disease?

In 1974, a Danish physician working in a poor village in Zaire, Africa, came down with the first inklings of a mystifying disease. For the first 2 years, the illness manifested itself mostly as inexplicable, profound fatigue, weight loss, and persistent diarrhea. But by the time the woman returned to Denmark in 1977, her body was wracked by infections that ravaged her mouth, bloodstream, and lungs. After dozens of medical tests proved fruitless, just before Christmas, Margrethe Rask became one of the first to die of what we know today as acquired immune deficiency syndrome (AIDS). She and thousands of others since then have experienced the gradual breakdown of a most vital body system—the immune system.

Picture the human body as a castle surrounded by a moat and patrolled inside by an army. The moat is the skin, a physical-chemical barrier against the outside environment. Backing up the skin are secretions in and around natural openings (such as the nose and mouth) that also physically and chemically prevent microorganisms from entering the body. The internal army protecting the body is an enormous contingent of cells and biochemicals that distinguishes among the substances that should or should not enter the body. Some lines of defense seek out and destroy a particular organism, and others are more or less equally effective against all cells or organisms that are not recognized as belonging in the body.

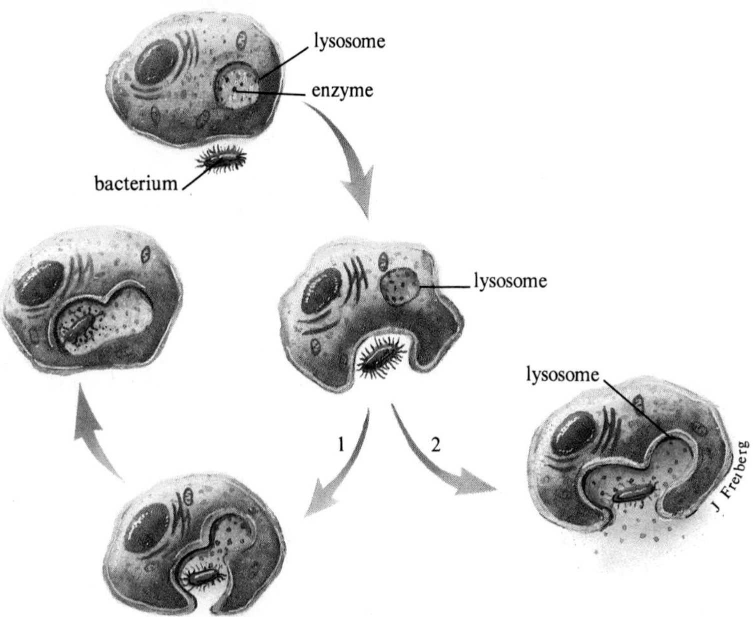

Figure 27.1
Phagocytosis. Neutrophils and macrophages surround and engulf bacteria much as an amoeba captures its food. Sometimes the vacuole formed around the bacterium fuses with a lysosome entirely within the cell (1), and sometimes the lysosomal enzymes spill out of the cell (2), contributing to inflammation.

Nonspecific Defenses

Barriers nonspecifically prevent microbes from entering the body. Unpunctured skin is the most pervasive and obvious of these walls, but there are others. Mucus in the nose traps inhaled dust particles; tears wash chemical irritants from the eyes and contain lysozyme, a biochemical that kills bacteria by rupturing their cell walls; wax traps dust particles in the ears. The microbes that pass these barriers and enter the stomach die in a vat of acidic stomach secretions. Those entering the respiratory system may meet a similar demise as they are swept out of the airways by cilia and then swallowed.

Phagocytosis

Intruders in the body are met by scavenger cells, called **phagocytes,** that engulf and digest them (fig. 27.1). Some phagocytes are anchored in particular tissues, but others, such as **neutrophils** and **macrophages,** roam about. Like soldiers on the front lines, these cells often die in battle. A neutrophil can engulf only about 20 bacteria before it dies, but a macrophage lives longer and can engulf up to 100 bacteria. Dead phagocytes are a component of pus, a sure sign of infection.

Inflammation

When the skin is punctured, the wounded area is quickly infiltrated by protective phagocytes that attack entering bacteria. Plasma accumulates at the wound site, which dilutes toxins secreted by bacteria. Increased blood flow warms the area, turning it swollen and red. This is **inflammation,** which creates an environment at the site of injury that is hostile to microbes (fig. 27.2).

Antimicrobial Substances

The **interferons** are polypeptides produced by a cell infected with a virus. Interferon diffuses to healthy neighboring cells and stimulates them to produce biochemicals that block viral replication. When these cells become infected, the viruses are unable to take over the protein synthetic machinery to manufacture more of themselves. The spread of the infection is halted.

The **complement system** is a group of proteins that assist, or complement, several of the body's other defense mechanisms. Some complement proteins trigger a chain reaction that punctures the cell membranes of microbes, bursting them. Other complement proteins accentuate inflammation by causing **mast cells** to release a biochemical called **histamine.** Histamine, in turn, widens blood vessels, easing entry of white blood cells to the injured area. Still other complement proteins attract phagocytes to the area.

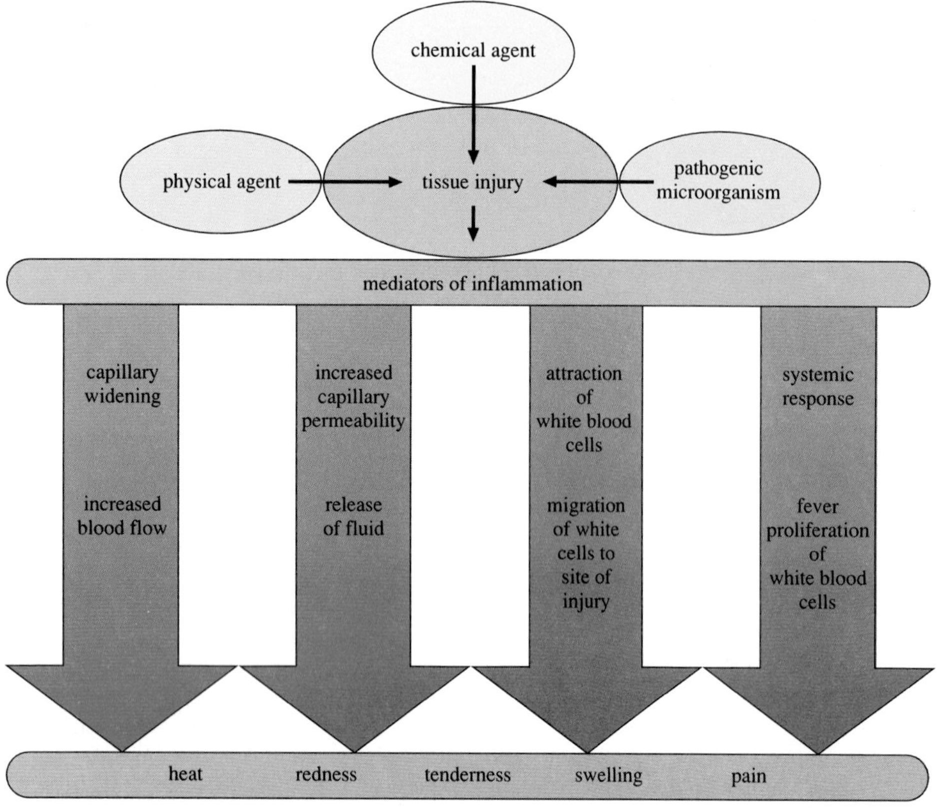

a.

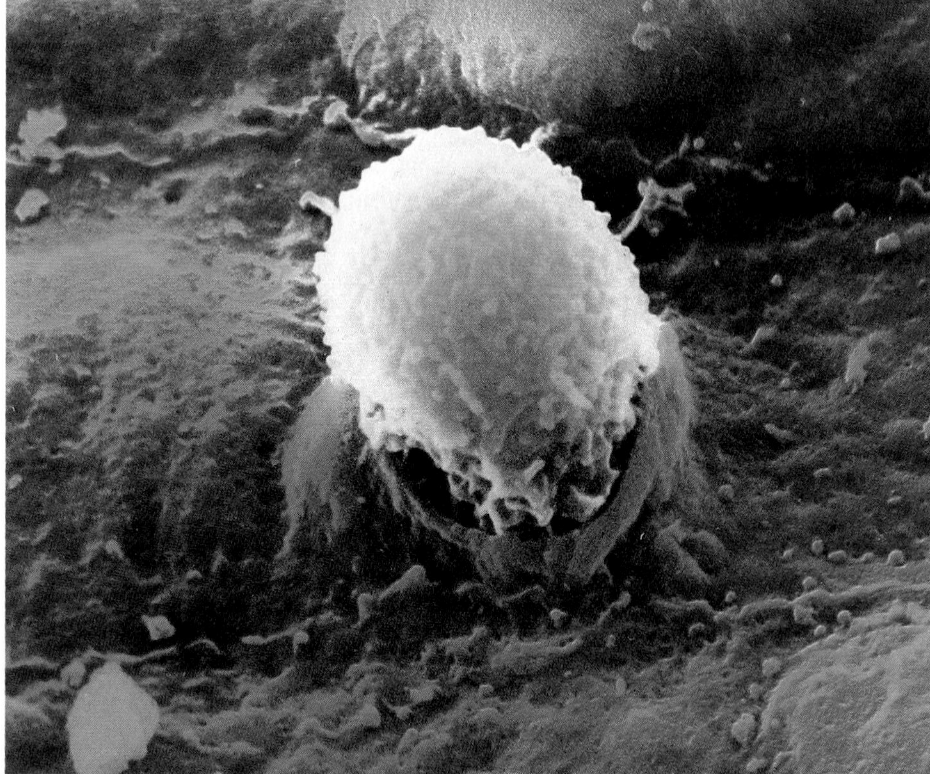

b.

Figure 27.2
Inflammation. *a.* An injury with the threat of infection sets into motion the several steps of the inflammatory response. *b.* This white blood cell is migrating through the endothelium lining of a vein, perhaps on its way to the site of an injury.

Fever

A fever is a powerful protective strategy. When a viral or bacterial infection sets in, white blood cells stimulated to proliferate, giving rise to cells that secrete a protein called **endogenous pyrogen** (fire maker from within). Endogenous pyrogen resets the thermoregulatory center in the brain's hypothalamus to maintain a higher body temperature.

A fever-reducing medication may actually counteract an effective biological defense. In such diverse animal species as fish, lizards, and rabbits, suppressing fever increases the likelihood of dying from infection. One study on humans suggests that fever may help the body fight viral diseases. In the study, participants all had a common cold. Half were treated with aspirin to reduce fever and the other half given a placebo (a substance that resembles the drug being tested but that lacks the action of the drug). Individuals whose fevers were reduced by aspirin shed more viruses from their noses and throats than did those in the feverish placebo group. Other studies show that interferon more efficiently prevents viral replication at higher body temperature.

Microbial infections are also stemmed by fever because higher body temperature reduces the level of iron in the blood. Since bacteria and fungi need more iron as the temperature rises, their growth is squelched in a fever-ridden body. Plus, phagocytes attack microorganisms more vigorously when the temperature rises.

KEY CONCEPTS

Viruses and bacteria entering the human body encounter a variety of nonspecific defenses. They are kept out by barriers such as the skin, mucus, tears, ear wax, stomach acid, and respiratory cilia. Phagocytes engulf and destroy the invaders. When skin is punctured, the inflammatory response dilutes toxins and sends in white blood cells to stem infection. Interferons inhibit viral replication. The complement system bores holes in microbes' cell membranes, bursting them. Fever fights viral infections by boosting interferon action and stems bacterial infections by reducing blood-iron levels.

Specific Defenses

In addition to nonspecific defenses, the human immune response includes protection against specific "foreign" structures or entities. It distinguishes "self" (the body's cells) from "nonself" surfaces (e.g., a disease-causing microorganism) by recognizing the particular arrangement of cell surface molecules that is unique to cells of an individual, except identical twins. This recognition of "self" cell surfaces was important in the recovery of the young liver transplant recipient described at the start of chapter 5.

The immune system attacks nonself cells and chemicals with an army of lymphocytes plus specialized biochemicals that some of these cells synthesize and release. The specific parts of cells or chemicals that elicit a response from the immune system are called **antigens.** An antigen can be as small as a few amino acids linked together on the surface of a microorganism.

The immune response is swift and specific. Many infections are halted before symptoms even arise. Attack by this built-in defense is specific, inactivating pathogenic organisms and some toxic chemicals but not disturbing the microorganisms that normally inhabit the human body, such as the bacteria that colonize the large intestine.

The immune system is also incredibly diverse, fighting such agents as a cold virus, a microbe infecting a cut, a fungal infection of the foot, food poisoning caused by bacteria growing in milk, and even cancer. The components of the immune system carry out different battle strategies. Some cells act directly in a kind of hand-to-hand combat, while others are specialized for chemical warfare. The immune system tailors an attack specifically aimed at every invader that is not recognized as belonging in the body, even on the first encounter. After that first meeting, the immune system "remembers." Measles, for example, is usually contracted only once, because the immune system recognizes the virus when it returns.

KEY CONCEPTS

The immune system distinguishes between self and nonself antigens with specialized white blood cells and the biochemicals they produce. The immune response is fast, highly specific, and very diverse, and it has memory.

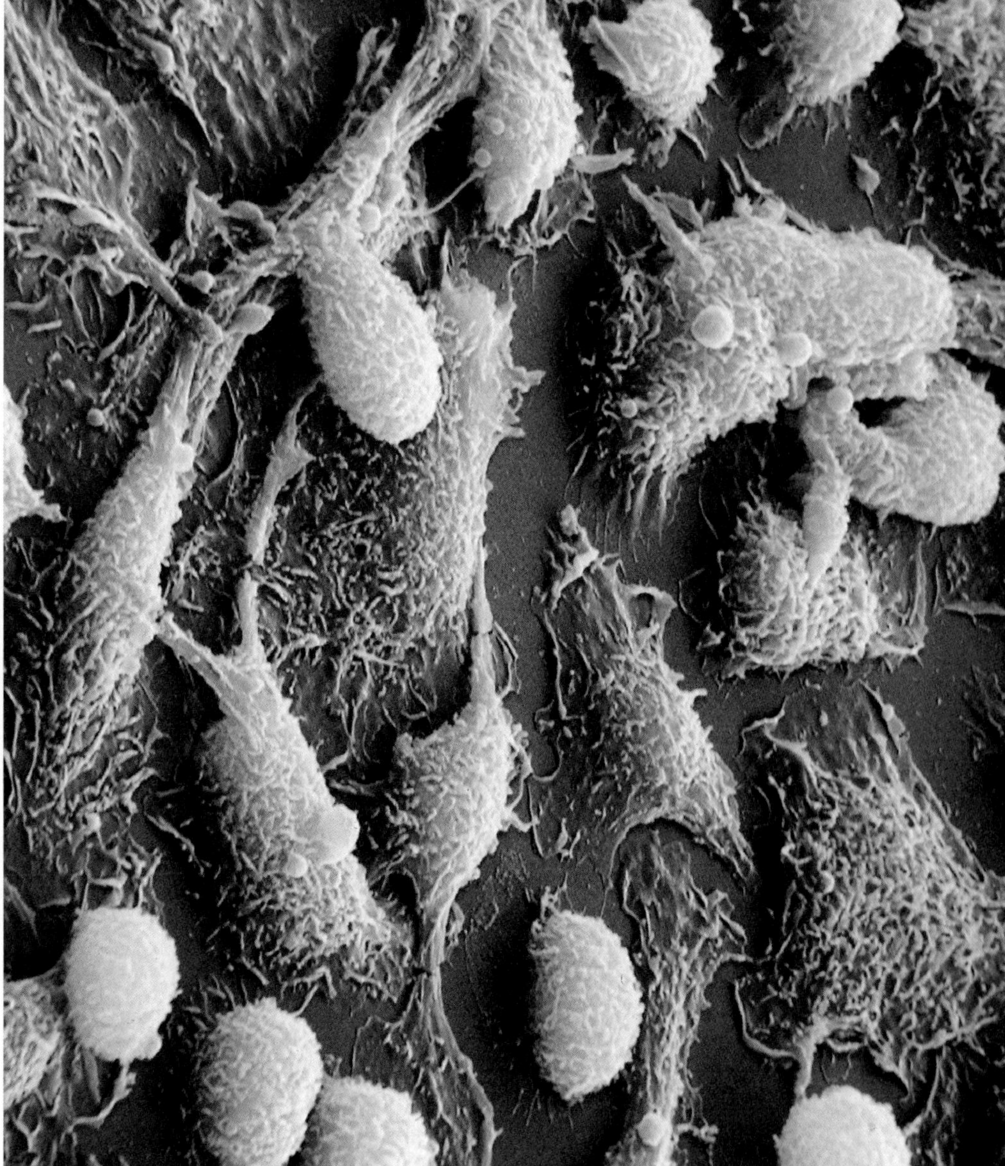

Figure 27.3
Macrophages bind to lymphocytes. When the body is infected, macrophages bind to lymphocytes (helper T cells), activating them to trigger other immune defenses. In this photograph, the round cells are lymphocytes and the cells bearing projections are macrophages. Magnification, x1,040.

Cells and Chemicals of the Immune System

The immune system is a grouping of components of the lymphatic system and the blood circulatory system. Specifically, immune protection is provided by some 2 trillion **lymphocytes** (white blood cells), which weigh about 2 pounds altogether. Lymphocytes produce a variety of biochemicals called **lymphokines,** which are essential to the immune response. Lymphocytes are made in the bone marrow and then sent to the lymph nodes, spleen, tonsils, and thymus gland, and also they circulate in the blood and tissue fluid. The skin is a vital outpost for lymphocytes, for this is where many infectious agents enter the body. Other specialized immune cells in the outermost skin layers signal lymphocytes, which then coordinate an immune defense.

Macrophages

One of the first cell types to respond when the body is infected is the macrophage. These scavengers engulf foreign matter and cellular debris and activate lymphocytes. Figure 27.3 shows a macrophage interacting with a lymphocyte.

Macrophages alert lymphocytes by displaying a bit of protein, an antigen, from the engulfed invader. The antigen is held to the surface of the macrophage by a protein badge that labels the macrophage as part of the body. This protein badge, specified by genes called the **major histocompatibility complex (MHC),** is shared by all body cells. Displaying the foreign antigen like a flag, the macrophage travels to the nearest lymph node, where it is likely to encounter several varieties of lymphocytes. When both the macrophage's flag and its MHC "self" label are recognized by a lymphocyte called a helper T cell, other immune reactions are set into motion. Simultaneously, the macrophage secretes a lymphokine called interleukin-1, which stimulates the T cell to begin replicating and causes a fever, which may slow microbial activity. The action of lymphocytes to protect the body is both intricate and highly coordinated (fig. 27.4).

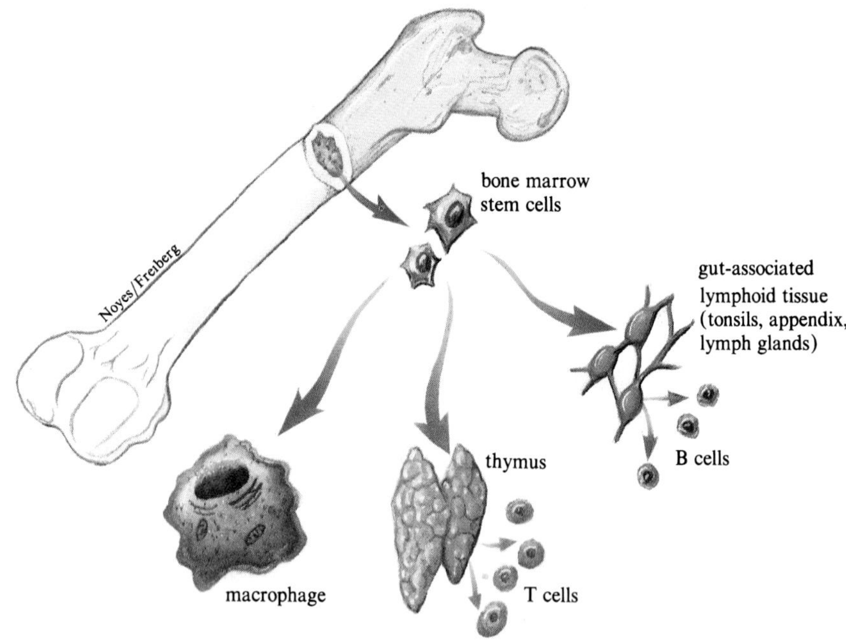

Figure 27.4
Three major types of immune system cells. Macrophages, T cells, and B cells interact to generate immune responses. All three cell types originate in the bone marrow and migrate into the blood. Macrophages engulf bacteria and also stimulate helper T cells to proliferate and activate B cells. T cells mature in the thymus gland and in the skin, and some of them manufacture lymphokines (including interferon and interleukin-2), which are released during viral infections. T cells also attack cancer cells and transplanted tissues. T cell function constitutes cellular immunity. B cells are released from lymphoid tissues such as the appendix, tonsils, and lymph nodes. B cells secrete antibodies, constituting the humoral immune response.

The Humoral Immune Response—B Cells Produce Antibodies

Lymphocytes called **B cells** destroy antigens by secreting proteins called **antibodies** into the bloodstream. The release of antibodies is called the **humoral immune response** ("humor" means fluid). Several cells interact to manufacture antibodies. An infecting influenza virus, for example, is first detected by a macrophage, which recognizes part of the virus' coat as nonself. The alerted macrophage then activates a helper T cell, which stimulates a B cell to produce antibodies specific for that antigen. The activated B cell divides, producing a clone (a group of identical cells) of B cells that can identify the particular antigen that stimulated its production.

The stimulated B cells develop into **plasma cells** and **memory cells.** Plasma B cells are antibody factories, each cell spewing out antibodies specific for the detected antigen at a rate of 2,000 per second, continuously, until the cell dies in 4 to 5 days. The memory B cells linger on, prepared to respond to the antigen more quickly and forcefully should it be encountered again (fig. 27.5).

The immune system's reaction to its first meeting with a foreign antigen is termed the **primary immune response,** and it takes a few days. During the delay, B cells divide and mature into plasma cells and produce the appropriate antibodies. By the 10th day, there are enough antibodies to be detected in a blood test. Five to 10 days is also about the time it takes for symptoms of the "flu" to abate—a sign that the immune system is working.

The body's reaction to any subsequent encounters with an antigen is called a **secondary immune response.** This defense is quicker and more intense than the primary immune response, so the person may not even be aware of the threat. The fast response is possible because the memory B cells are ready and waiting with instructions for producing the appropriate antibodies.

Antibody Structure

Antibodies are large, complex molecules. The simplest individual antibody molecule is built of four polypeptide chains connected to one another by disulfide bonds to form the shape of the letter Y (fig. 27.6). The two larger polypeptide chains are called **heavy chains,** and the other two are called **light chains.** The lower portion of each antibody polypeptide chain consists of a sequence of amino acids that tends to be the same or very similar in all antibody molecules. These regions are called **constant regions.** The amino acid sequence of the upper portions of each polypeptide chain can vary a great deal between individual antibody molecules, and these are called **variable regions.** Thus, an antibody molecule has two heavy chains and two light chains, and each chain is partly constant and partly variable in its amino acid sequence.

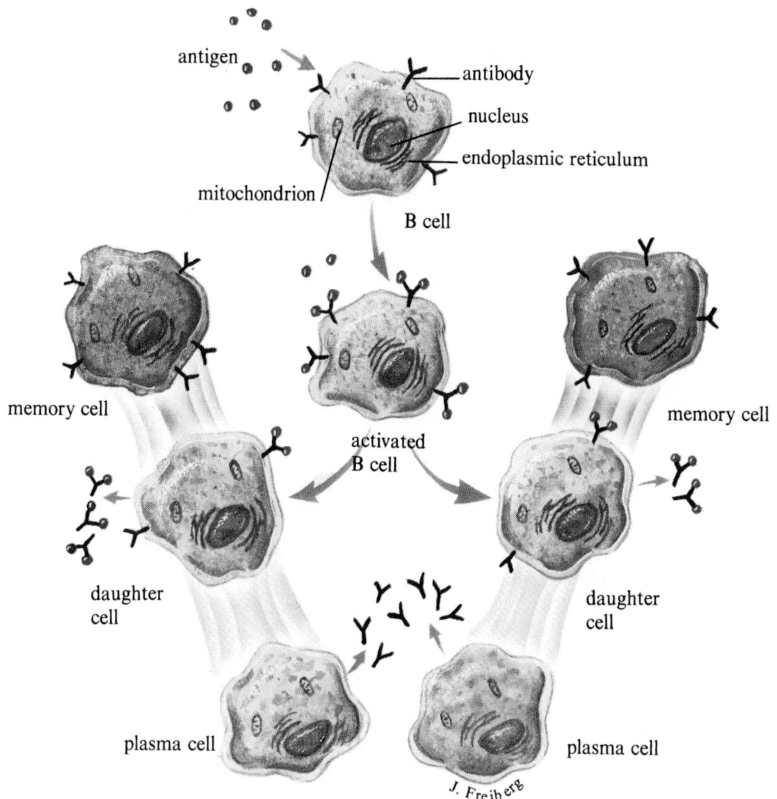

Figure 27.5
B cells mature into plasma cells, which produce antibodies. When antibodies on a B cell's surface bind antigen, the cell divides. The daughter cells specialize into plasma cells, which are antibody factories, and memory cells. Note the extensive endoplasmic reticulum and Golgi apparati in the plasma cells. How does this structure reflect the cell's function?

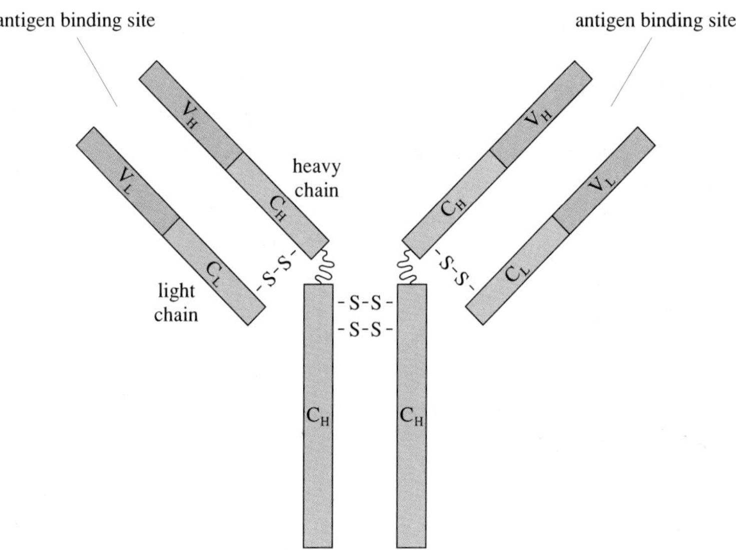

Figure 27.6
Antibody structure. The simplest antibody molecule is built of four polypeptide chains, two heavy and two light, which are held together by disulfide bonds that form between sulfur-containing amino acids. Approximately half of each polypeptide chain has a constant sequence of amino acids, and the other half has a variable sequence. The tops of the Y-shaped molecules form antigen binding sites. More complex antibodies are built of several Y-shaped units.

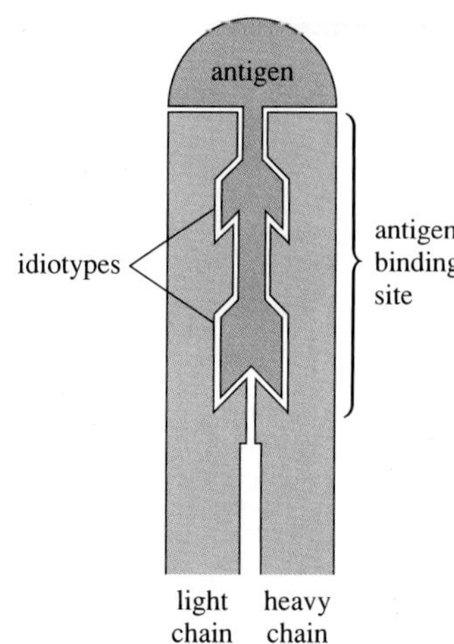

Figure 27.7
The antigen binding site. Antigens bind to antibody molecules in the custom-fit cleft at the top part of the Y-shaped molecule.

Antibody molecules can bind to certain antigens because of the three-dimensional shape of the variable regions. Parts of these areas fit structures on particular antigens like a key fits into a lock. These specialized ends of the antibody molecule are called **antigen binding sites,** and the particular parts that actually bind the antigen are called **idiotypes** (fig. 27.7).

The binding of an antibody to an antigen inactivates the microbe or neutralizes the toxins it produces. Sometimes antibodies cause pathogens to clump, making them more visible to macrophages, which then destroy them in a frenzy of phagocytosis. Antibodies also activate the complement system, which triggers a chain reaction that is fatal for the microbe.

Antibody molecules can aggregate in groups of two or five to form antibody complexes (fig. 27.8). The bound antibody-antigen complexes are destroyed by macrophages or the complement system. Five classes of antibodies are distinguished according to their locations in the body and their functions (table 27.1). Antibodies are also called immunoglobulins (Igs), which means that they are globular proteins that provide immunity.

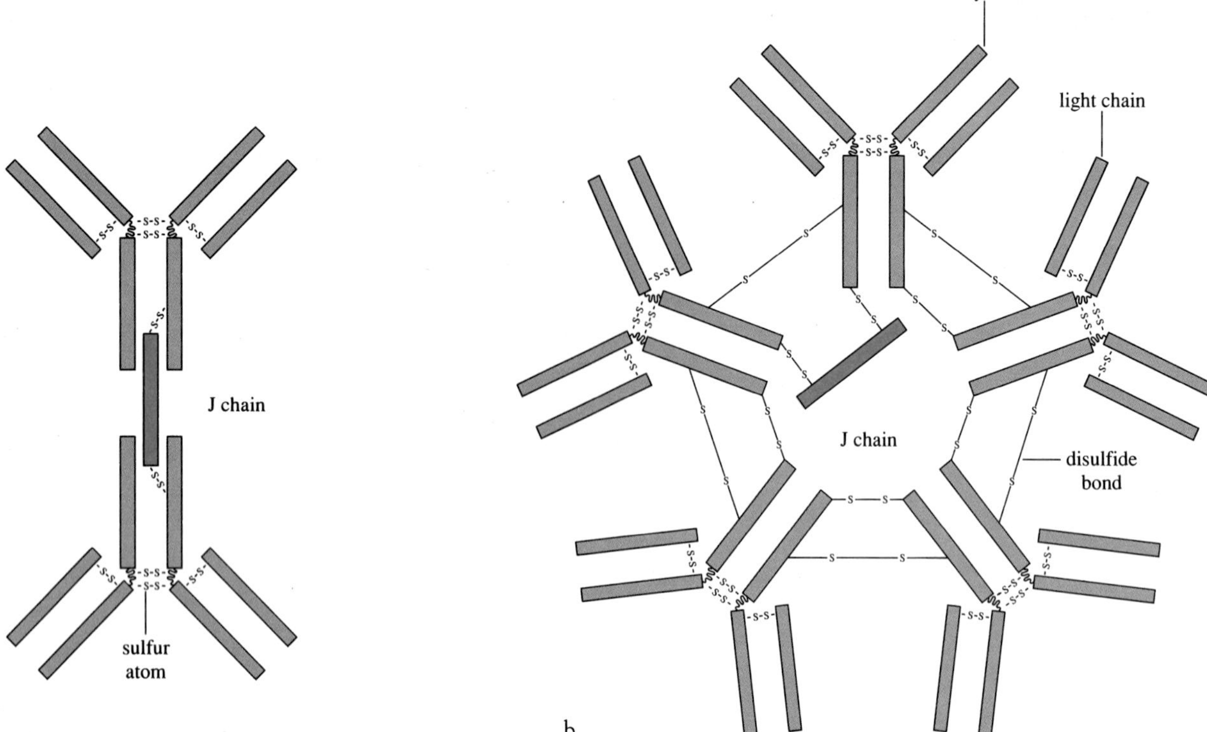

Figure 27.8
Antibody complexes. Antibodies of class IgA and IgM are built of multiple four-polypeptide antibody molecules held together by short peptides called J (for "joining") chains. *a.* IgA, found in several secretions, is composed of two such subunits. *b.* IgM, secreted into the blood during a primary immune response, is built of five subunits. IgD, IgE, and IgG each consist of only one four-polypeptide molecule.

The humoral immune response displays amazing biological economy, because an apparently limitless number of different antibody proteins can be manufactured from a limited number of antibody specifying genes. If each possible antibody was represented by a different gene, there would not be enough genes in any animal's genome to code for their millions of different antibodies. How can this be?

The tremendous diversity of antibody types originates by shuffling sections of genetic material. During the early development of B cells, sections of their antibody genes are randomly moved to other locations among the chromosomes, creating new gene sequences and, consequently, instructions for producing different antibodies. It is a little like using the limited number of words in a language to compose an infinite variety of stories.

Table 27.1
Types of Antibodies

*Type**	*Location*	*Functions*
IgA	Milk, saliva, and tears; respiratory and digestive secretions	Protect against microorganisms at points of entry into body
IgD	B cells in blood	Stimulates B cells to make other antibodies (little is known about IgD)
IgE	Mast cells in tissues	Receptors for antigens that cause mast cells to secrete allergy mediators
IgG	White blood cells and macrophages in blood	Bind to bacteria and macrophages at same time, assisting macrophage in engulfing bacteria; activate complement; abundant in secondary immune response; passed from mother to fetus
IgM	B cells in blood	Activate complement in primary immune response

*The letters A, D, E, G, and M refer to the specific conformations of heavy chains characteristic of each class of antibody.

> **KEY CONCEPTS**
>
> *The immune system consists of trillions of macrophages, B and T cells, which are made in the bone marrow and collect in lymphatic tissue, the bloodstream, tissue fluid, and the skin. Macrophages engulf invaders and alert helper T cells to start other defenses. The helper T cell stimulates a B cell to divide and differentiate into a plasma cell, which secretes huge amounts of a specific antibody, and a memory B cell. Producing antibodies constitutes the humoral immune response. A primary immune response takes 5 to 10 days, and a secondary response is faster because of memory cells. Antibodies are built of Y-shaped four-polypeptide subunits, each of which has two light chains and two heavy chains, with two constant and two variable regions. The shape of the tips of the Y forms the antigen binding site. Antibody binding to antigen inactivates or detoxifies invaders, makes them more visible to macrophages, or stimulates the complement system. Antibody diversity is generated by shuffling gene pieces.*

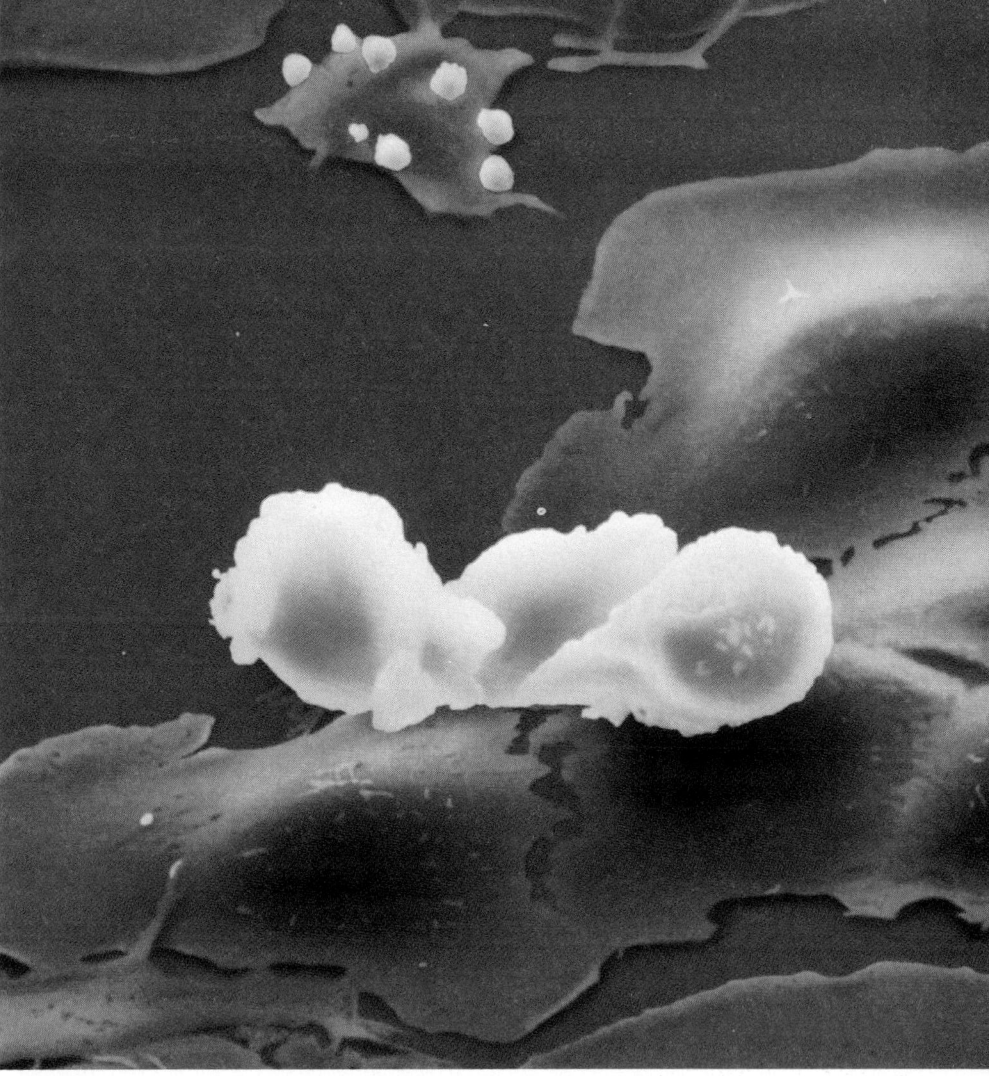

Figure 27.9
T cells are "educated" in the thymus. This color-enhanced scanning electron micrograph shows T cells (blue) attached to epithelial cells (red) in the thymus gland.

The Cellular Immune Response—T Cells

The cellular immune response is carried out by lymphocytes called **T cells.** This facet of immunity is called cellular because the T cells must actually travel to where they are needed, unlike B cells, which secrete antibodies to carry out the defense. T cells originate in the bone marrow and then pass through the thymus, a gland located in the chest. The "T" indicates the dependence on the thymus gland for processing. T cells are said to "mature" or be "educated" in the thymus, where they acquire the ability to recognize particular nonself cell surfaces and molecules (fig. 27.9).

Several types of T cells are distinguished by their functions. **Helper T cells** stimulate B cells to produce antibodies, secrete lymphokines, and activate another type of T cell, called a killer.

Killer T cells attack nonself cells by attaching to them and releasing chemicals. They do this using two surface peptides among many hundreds, which are linked to form receptors for foreign antigens. When a killer T cell encounters a nonself cell, the T cell receptors draw the two cells into physical contact. The killer cell then releases a chemical called cytolysin (cell-cutter), which drills holes in the foreign cell's membrane. The holes disrupt the flow of chemicals in and out of the foreign cell, and it dies. Killer T cells are also attracted by their surface receptors to body cells that are covered with certain viruses. The T cells destroy such cells before the viruses on them can enter and replicate themselves to spread the infection. Killer T cells also dismantle cancer cells, as can be seen in figure 5.2. **Suppressor T cells** inhibit the response of all lymphocytes to foreign antigens, shutting off the immune response when an infection is controlled. Figure 27.10 summarizes the relationships of the lymphocytes.

> **KEY CONCEPTS**
>
> *The cellular immune response is carried out by T cells, which are made in the bone marrow and then pass through the thymus. Helper T cells stimulate B cells and killer T cells and secrete lymphokines. Killer T cells bind to nonself cells and burst them. Suppressor T cells inhibit all lymphocytes.*

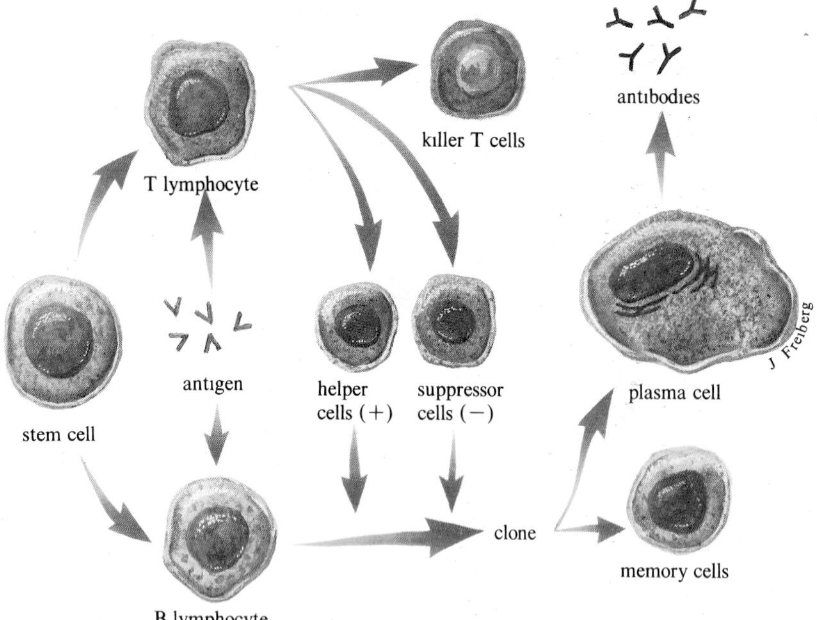

Figure 27.10
All lymphocytes descend from stem cells in the bone marrow. T lymphocytes (T cells) spend time in the thymus and account for 65% to 85% of all lymphocytes. B cells do not pass through the thymus. Upon stimulation by a foreign antigen, B cells divide and differentiate into plasma cells, which secrete antibodies, and memory cells. Helper T cells stimulate B cells. Suppressor T cells inhibit B cells. Killer T cells attack nonself cells.

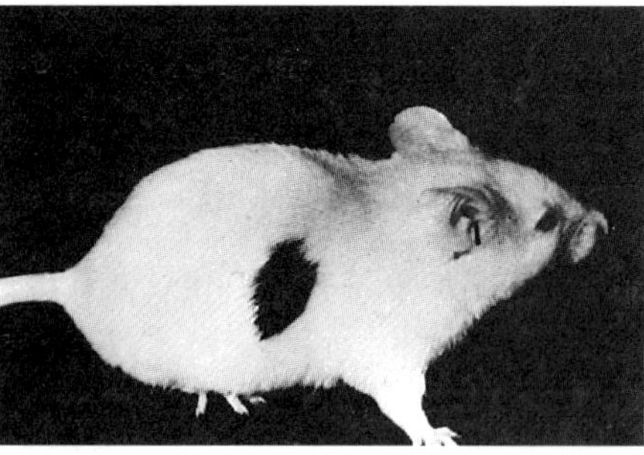

Figure 27.11
The immune system learns to distinguish self from nonself during fetal existence. A white mouse exposed to the blood of a brown mouse while in the uterus will accept a transplant of skin from the brown mouse after birth.

Development of the Immune System

The immune system does not become fully functional until about the 18th month of life. The fetus cannot mount its own immune defense, termed **active immunity.** If it did, it would attack cells and biochemicals from the mother that crossed the placenta and entered the fetal circulation. The fetus somehow "catalogs" its own cells, so its immune system can learn to distinguish self from nonself. Experimental evidence supports this idea. Mice exposed to a chemical while in the uterus do not recognize that chemical as a foreign antigen when exposed to it after birth but perceive it as self. Similarly, a mouse exposed in the uterus to the skin of an unrelated mouse can later in life receive a transplant from that mouse (fig. 27.11).

At birth the newborn's immune system is not yet active, however, the newborn received some antibodies from the mother (termed **passive immunity**) during fetal existence, which continue to provide temporary immunity. The newborn receives an extra boost of protection if breast-fed. Shortly before and after a woman gives birth, her mammary glands secrete a yellow substance called colostrum, which is rich in antibodies that protect the newborn against certain digestive and respiratory infections. Within a few days, the colostrum is replaced by mature milk, which contains antibodies that protect against several intestinal parasites.

A newborn begins to manufacture antibodies soon after birth, and the first of these antibodies are likely to be synthesized in response to bacteria and viruses transmitted by whoever is in closest physical contact with the baby. Likewise, the mother's system makes antibodies to the bacteria and viruses on her baby's skin (faster than the baby can), and these antibodies are passed to the infant in the mother's milk. This gradual exposure to foreign antigens seems to be part of the normal development of the immune system. By 6 months of age, a baby can produce enough of a variety of his or her own antibodies so that most infections are easily overcome. Certain infections, such as those caused by the herpes simplex family of viruses, cannot be fought off until the second year of life.

The immune system begins to lose effectiveness early in life. The thymus gland reaches its maximal size in early adolescence and then slowly degenerates. By age 70, the thymus is 1/10 the size it was at the age of 10, and the immune system is only 25% as powerful. The declining strength of the immune response is why elderly persons often succumb to infections that they might have fought off at an earlier age.

KEY CONCEPTS

The fetal immune system learns to distinguish self from nonself while not rejecting maternal cells and biochemicals. For the first few months until active immunity begins, the infant is protected by antibodies transmitted first through the placenta and then through breast milk. Gradually the baby makes his or her own antibodies. The immune system begins to decline early in life.

When Immunity Breaks Down

AIDS

In late 1981 and early 1982, physicians from large cities in the United States began reporting to the Centers for Disease Control cases of formerly rare infections in otherwise healthy young men. Some of the infections were fairly prevalent in the general population, such as herpes simplex and cytomegalovirus, but the cases in these young men were unusually severe. Some of the infections were caused by organisms known to infect only nonhuman animals, such as life-threatening diarrhea caused by protozoan parasites of the genus Cryptosporidium. Some diseases, for instance a form of pneumonia caused by a protozoan called *Pneumocystis carinii* and a form of cancer, Kaposi's sarcoma, were known only in individuals whose immune systems were suppressed (fig. 27.12). It was this type of pneumonia that killed the Danish doctor in Zaire described at the start of the chapter. On autopsy, her lungs were filled with the organisms.

The bodies of the sick young men had become nesting places for all types of infectious agents, including viruses, bacteria, protozoans, and fungi. The infections were opportunistic, meaning that they seemed to take advantage of a weakened immune system. The appearance of Kaposi's sarcoma, and then other formerly rare cancers, confirmed the growing belief that malfunctioning immunity was a common link in the rise of infections being seen across the nation.

AIDS generally starts with recurrent fever, weakness, and weight loss. If the first infection is fought off, a second one soon sets in. The syndrome is usually lethal within 2 years of symptom onset, but the causative virus can be present for a decade or longer before the person feels ill. Some people develop a mild form of the condition termed **AIDS-related complex (ARC),** characterized by weakness, swollen glands in the neck, and frequent fever. Still other people carry the virus but have apparently normal immunity. It is not clear whether these three types of responses to the AIDS virus are distinct clinical entities or different stages of the same condition.

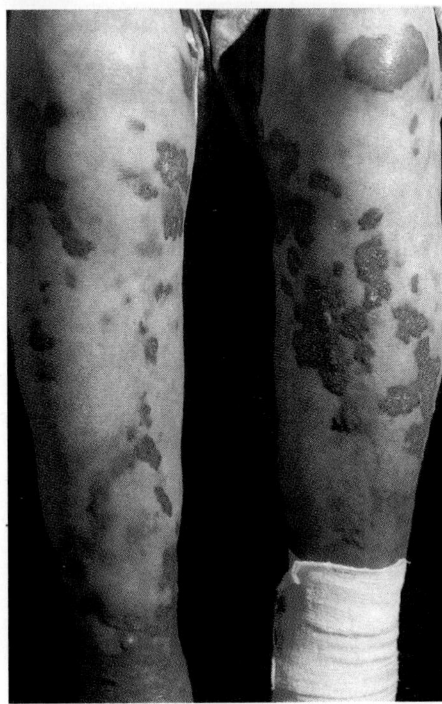

Figure 27.12
Prior to the appearance of AIDS, Kaposi's sarcoma was a rare cancer seen only in elderly Jewish and Italian men and in people whose immune systems are suppressed. In these groups it produces purplish patches on the legs, but in AIDS patients, Kaposi's sarcoma patches appear all over the body and sometimes internally too. These legs display characteristic lesions.

As the scientific and medical communities joined forces to identify the source of the unusual outbreak of infections, attention first turned to those who were affected. What did they have in common? All had had blood contact with another AIDS patient. Many were homosexual men who transferred the infection through fragile rectal tissues during anal intercourse. AIDS can also be transferred during vaginal intercourse between males and females if any break in the tissue permits blood contact. Intravenous drug users spread AIDS when they share needles. AIDS was also passed in blood transfusions prior to 1985, when safer blood banking precautions were instituted. Those with hemophilia were particularly hard hit, because they received blood products pooled from many donors. A growing AIDS population includes children who acquired the disease during birth from infected mothers.

By late 1986, it was known that AIDS is caused by the human immune deficiency virus, or HIV, that attacks the immune system (fig. 4.2). HIV is found in blood, sperm, and to a lesser extent in milk, tears, and saliva. It is a retrovirus, which means that its genetic material is ribonucleic acid (RNA). The virus attaches to a receptor called CD4 on helper T cells and sends in its RNA, including instructions to manufacture an enzyme, reverse transcriptase. The stowaway reverse transcriptase builds a DNA strand complementary to the viral RNA, which then replicates to form a DNA double helix representing the viral genetic material (fig. 27.13). The viral DNA then enters the T cell's nucleus. Can you see why one avenue of research into preventing AIDS is to find a compound that plugs up CD4 receptors?

The viral gene that controls HIV's replication has a very similar sequence to two genes activated in the helper T cell to fight viral infections—those for interferon and interleukin-2. So when some other virus activates a helper T cell, it also activates HIV. The viral DNA is transcribed and translated, and the helper T cell fills with HIV RNA and proteins. The commandeered helper T cell can no longer release lymphokines or stimulate B cells to manufacture antibodies as it normally would. Finally, as the helper T cell dies, it bursts and unleashes many new HIV particles. After several weeks of helper T cell destruction, the suppressor T cells dominate, and the effect is felt on the bodily level as a suppression of immunity. Opportunistic infections set in.

Thousands of people have died from AIDS, and millions are infected with HIV (table 27.2). Treatment so far has taken three approaches: inhibiting the ability of the virus to infect and replicate, boosting the functions of the immune system, and treating the various opportunistic infections. The major goal is to develop a vaccine, but this effort has been hampered because of the virus's changeable surface. Still, several vaccines are being tested in humans.

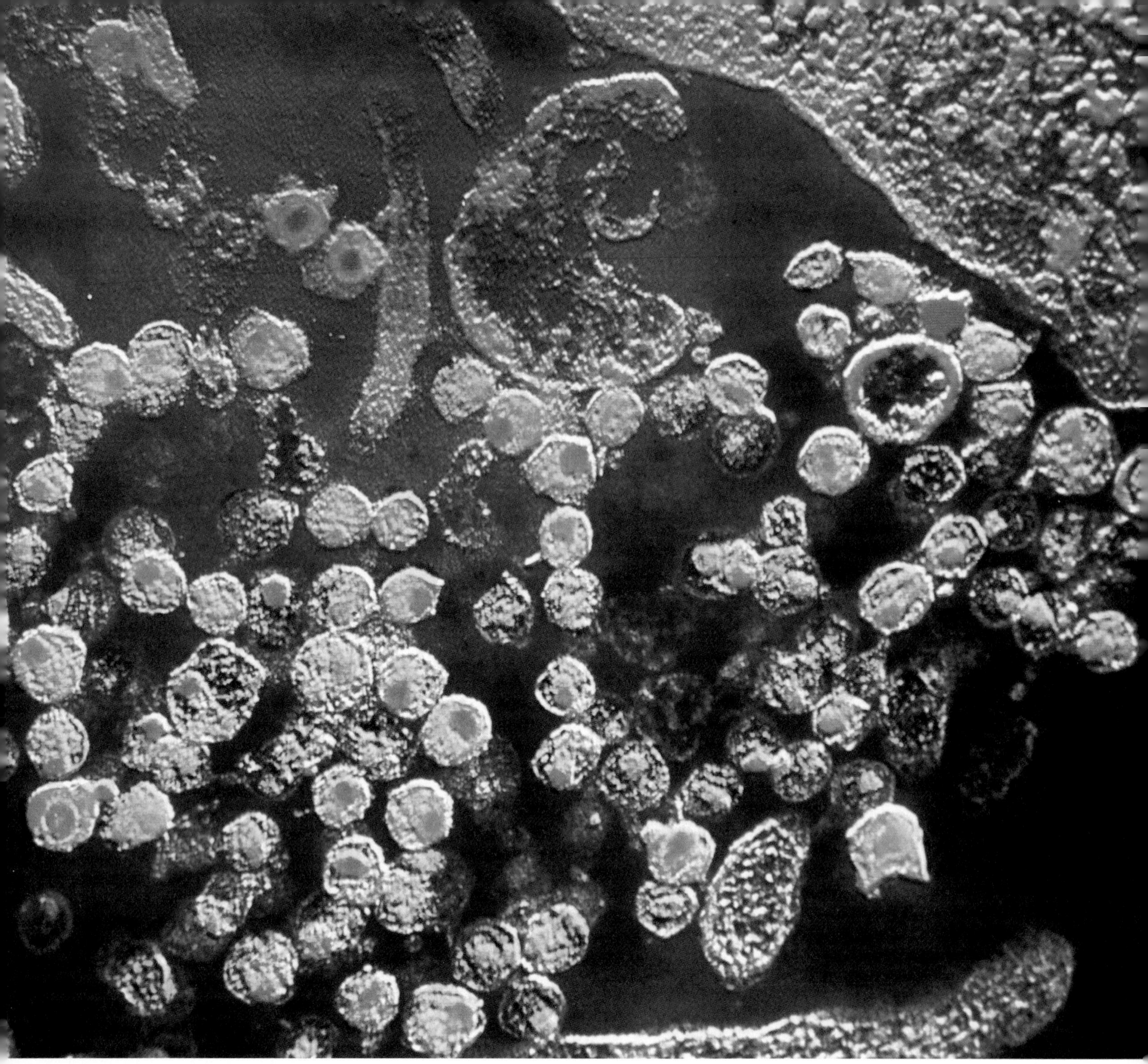

a.

Figure 27.13
How HIV infects a helper T cell. *a.* Helper T cells are not the only ones taken over by HIV. This killer T cell harbors many viral particles, seen in this transmission electron micrograph as red circles. The red is RNA. Magnification, x8,800. *b.* HIV sends its RNA and reverse transcriptase into a helper T cell by docking at a CD4 receptor. Once inside, reverse transcriptase builds a complementary DNA chain, and after another round of replication, the viral genetic information, in the form of DNA, enters the host cell's nucleus. When infection by another type of virus stimulates the T cell to divide and manufacture lymphokines, the HIV genes are activated too. Instead of assisting B cells and secreting lymphokines, the helper T cell becomes an HIV factory.

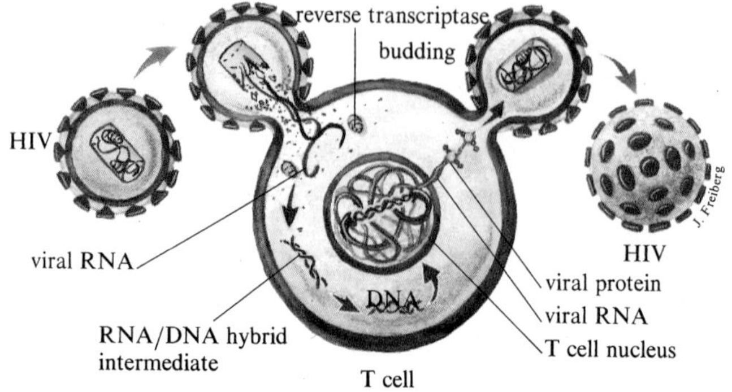

b.

Table 27.2
Numbers of Living AIDS Patients in the United States

Year	Number
1989	92,000–98,000
1990	100,000–122,000
1991	127,000–153,000
1992	139,000–225,000
1993	151,000–225,000

Source: Centers for Disease Control, 1990.

Figure 27.14
David, the "bubble boy," was born without a thymus gland. Because his T cells could not mature, he was virtually defenseless against infection.

Severe Combined Immune Deficiency

Each year, a few children are born defenseless against infection due to **severe combined immune deficiency,** in which neither T nor B cells function. "David" was one such youngster (fig. 27.14). Born in Texas in 1971, he had no thymus gland and spent the 12 years of his life in a vinyl bubble, awaiting a treatment that never came. As David reached adolescence, he wanted to leave his bubble. An experimental bone marrow transplant was performed, but soon afterwards David began vomiting and developed diarrhea, signs of an infection. David was allowed to leave the bubble for his last few days of life, before a massive infection killed him.

Today children with inborn immune deficiences can be helped by a bone marrow transplant, or with adenosine deaminase (ADA), an enzyme whose absence inactivates T and B cells. One youngster with this ADA deficiency is receiving infusions of her own white blood cells genetically engineered to produce ADA—the first attempt at human gene therapy.

Chronic Fatigue Syndrome

Some 2 to 5 million people in the United States suffer from a poorly-understood immune system imbalance called chronic fatigue syndrome. The condition begins suddenly, producing fatigue so great that getting out of bed is an effort. Chills, fever, sore throat, swollen glands, muscle and joint pain, and headaches are also symptoms.

The various disabling aches and pains reflect an overactive immune system. These people have up to 40 times the normal amount of interleukin 2, and excess killer T cells, yet too little interferon. It is as if the immune system mounts a defense, and then doesn't know when to shut it off. The culprit behind chronic fatigue syndrome seems to be a virus, but none has yet been indentified that is common to all sufferers.

KEY CONCEPTS

AIDS is an assault on helper T cells and other cells by a retrovirus, HIV. Symptoms include profound fatigue, weight loss, opportunistic infections, and cancer. HIV inserts its genetic material and an enzyme to convert it to DNA into a host cell, where its RNA is converted to DNA by reverse transcriptase and viral particles are produced when the host cell divides and manufactures proteins to fight another viral infection. Severe combined immune deficiency is an inborn form of immunity breakdown. Chronic fatigue syndrome reflects an overactive immune system.

Autoimmunity

Sometimes the immune system backfires, manufacturing antibodies that attack the body's own cells. These misdirected molecules are called **autoantibodies.** One theory of the source of this autoimmunity (immunity to self) is that a virus, while replicating within a human cell, "borrows" proteins from the host cell's surface and incorporates them onto its own surface. When the immune system "learns" the surface of the virus to destroy it, it also learns to attack the human cells that normally bear the particular protein.

The specific nature of an autoimmune disease depends upon the cell types that are the target of the immune attack. In myasthenia gravis, neuromuscular junctions are destroyed, resulting in muscular weakness in the arms and legs. In multiple sclerosis, the myelin coat around neurons is attacked. In rheumatoid arthritis, the synovial membranes of the joints become inflamed. An immune system attack on pancreatic beta cells can produce insulin-dependent diabetes. Male infertility may be due to antibodies manufactured against a man's own sperm cells. In systemic lupus erythematosus, autoantibodies attack DNA and RNA of blood cells, affecting many tissues.

Allergies

The immune system can be too sensitive, attacking substances that are not a threat to health. Such an inappropriate response is an **allergy.** In an allergic reaction, the offending substances, called **allergens,** activate antibodies of type IgE. These antibodies in turn prompt mast cells to release substances called **allergy mediators** explosively, which include histamine and heparin (fig. 27.15). Allergy mediators cause symptoms by increasing the

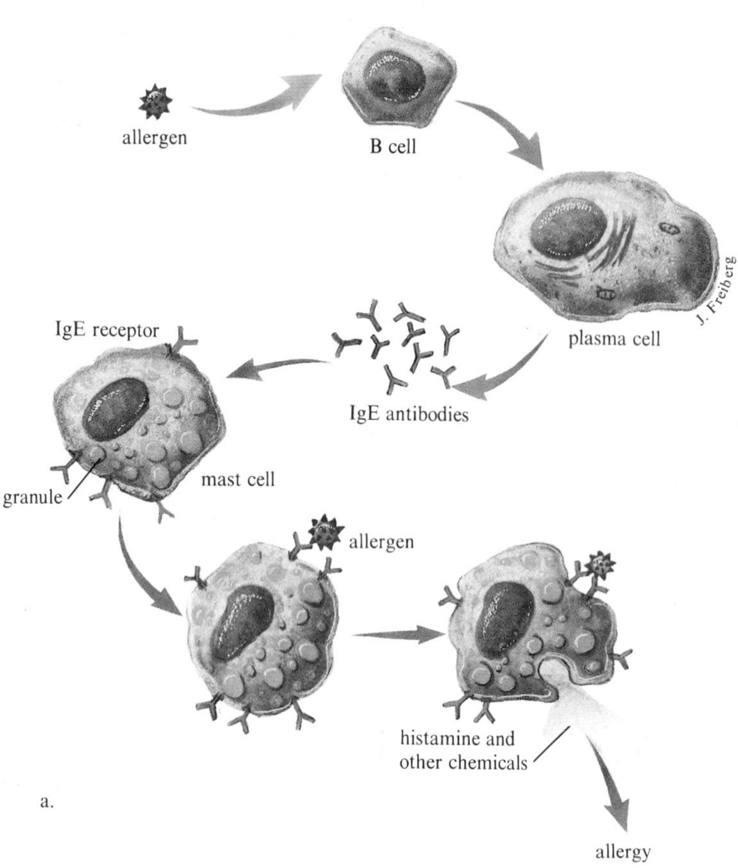

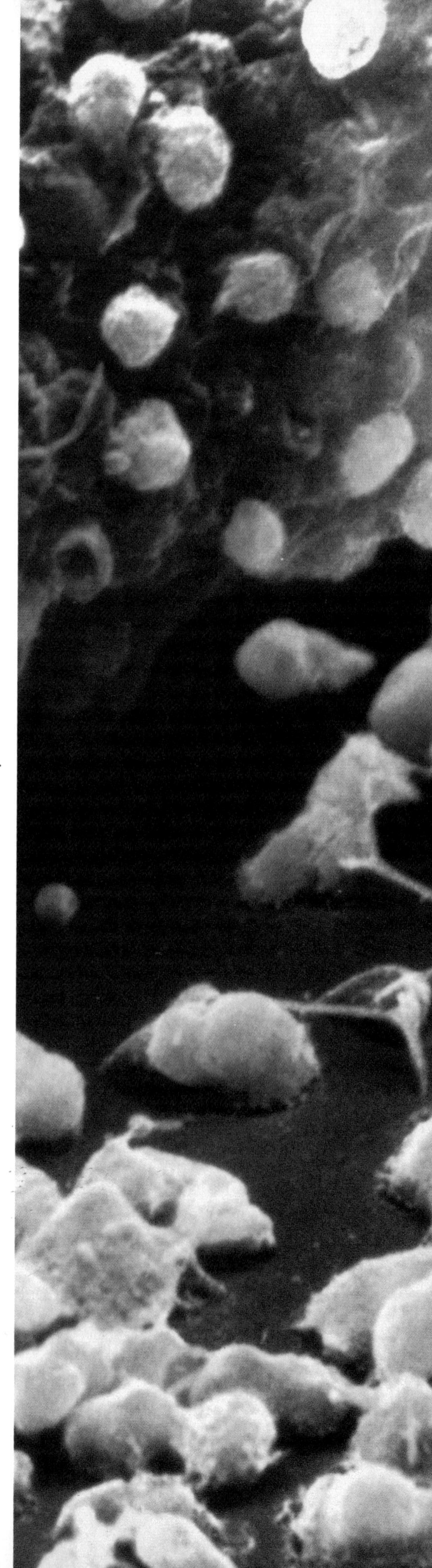

Figure 27.15
Allergy symptoms result from exploding mast cells. *a.* A bout of hay fever is triggered when allergens (pollen) activate B cells to secrete antibodies of class IgE into the tissues. The antibodies bind to receptors on mast cells. The allergens then bind to adjacent bound antibodies, pulling them in such a way that the mast cell releases its supply of histamine by exocytosis. *b.* A mast cell releases histamine granules.

permeability of capillaries and venules, contracting smooth muscle, and stimulating mucous glands. Because histamine is responsible for many allergy symptoms, a remedy is a dose of antihistamine.

The symptoms of a specific allergic response depend upon the site in the body where mast cells release their mediators. Because many mast cells are found in the skin, respiratory passages, and digestive tract, allergies tend to affect these organs. This is why hives, runny nose and eyes, asthma, and nausea, vomiting, and diarrhea are common allergic reactions.

Among the most common allergens are foods, dust mites, pollen, and fur (fig. 27.16). Many people are allergic to more than one substance and must severely restrict their diets and activities. Allergies run in families. The mind can also influence an allergic reaction. A person allergic to tulips, for example, may experience allergy symptoms when presented with an artificial tulip!

To find the cause of an allergy, a physician injects extracts of suspected allergens beneath the skin on the lower arm or on the back of a sufferer. If a red bump develops, the substance injected at that point is an allergen for that person. Can you think of a way that an allergy sufferer can help identify an allergen?

Hay fever and asthma sufferers can sometimes be helped by **desensitization,** in which small amounts of the allergen are periodically injected under the skin. The doses are not enough to stimulate production of IgE, but they do stimulate production of IgG. When the person next encounters the allergen in a natural setting—pollen during a camping trip, for example—IgG binds to the allergen before IgE can. Because only IgE can stimulate mast cells to explode, the allergy attack is prevented.

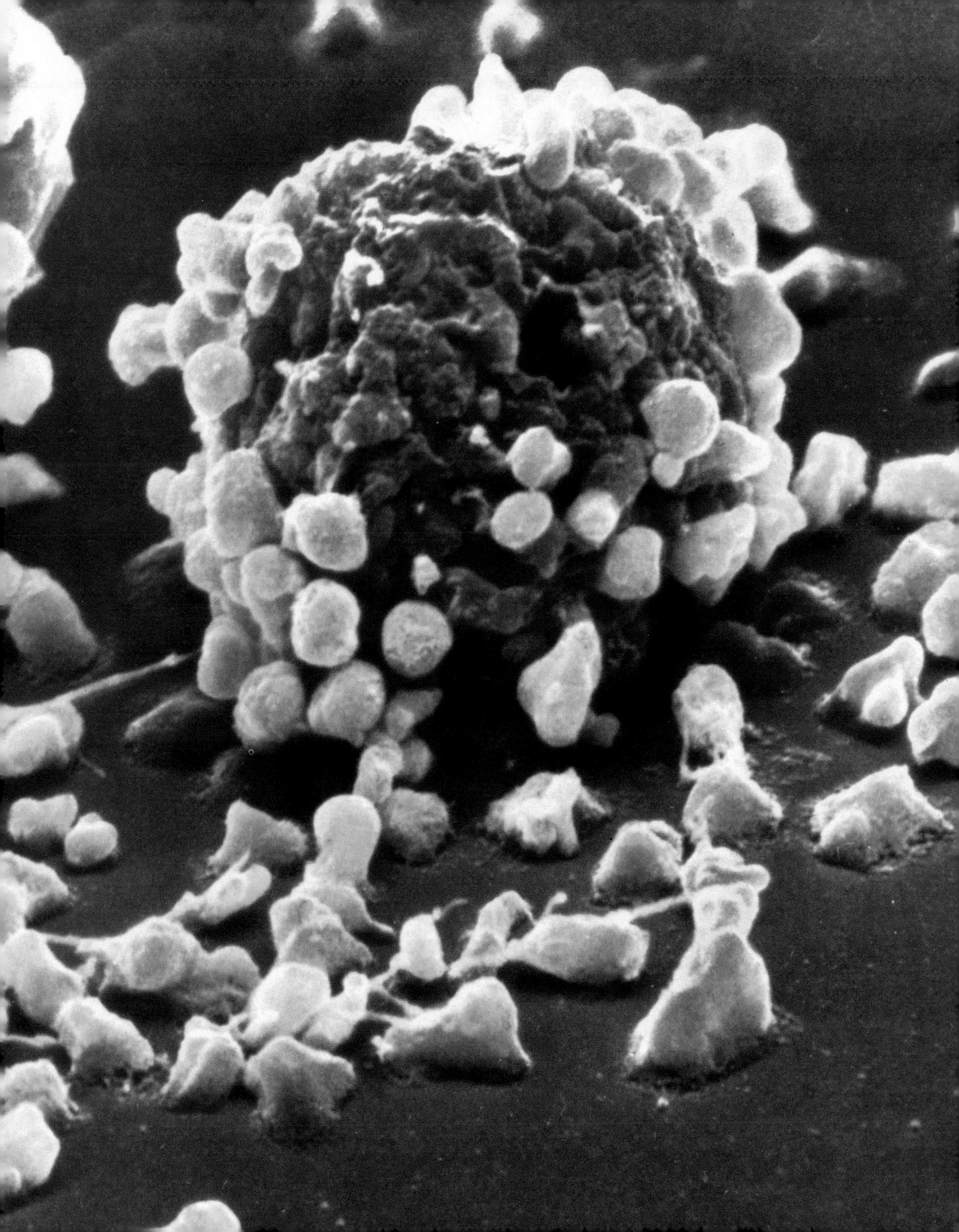

a.

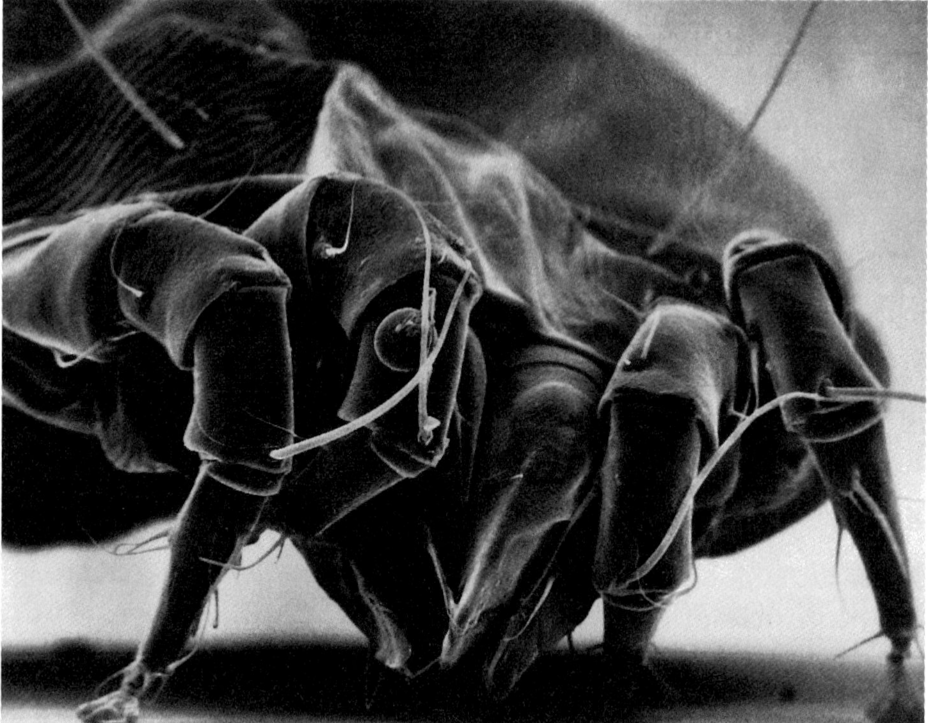

b.

Figure 27.16
Causes of two common allergies: (*a*) ragweed pollen causes hay fever and (*b*) dust mites cause year-long runny nose and eyes and asthma. (*a*. Magnification, x325.)

Not all allergic reactions are as benign (although uncomfortable) as hives and watery eyes. Some individuals may react to certain stimuli with a frightening and potentially life-threatening reaction called **anaphylactic shock,** in which mast cells release mediators throughout the body. The person may at first feel an inexplicable apprehension, and then suddenly the entire body itches and breaks out in fiery red hives. He or she may vomit and have diarrhea. The face, tongue, and larynx begin to swell, and breathing becomes difficult. Unless the person receives an injection of epinephrine (adrenaline) and sometimes a tracheotomy (an incision into the windpipe so that breathing is restored), he or she will lose consciousness and die within 5 minutes. Anaphylactic shock most often results from an allergy to penicillin or insect stings. Fortunately, thanks to prompt medical attention and people who know they have allergies avoiding the allergens, fewer than 100 people a year actually die of anaphylactic shock.

Anaphylactic shock seems to be a severe response to something as innocuous as an insect bite or penicillin. One theory of the origin of allergies, particularly anaphylactic shock, is that they evolved at a time when insect bites and the natural substances from which antibiotics such as penicillin are made threatened human survival. The observation that IgE protects against roundworm and flatworm infections, in addition to taking part in allergic reactions, supports the idea that this antibody class is a holdover from times past, when immunity requirements might have been different from what they are today.

KEY CONCEPTS

In autoimmune disorders, autoantibodies attack a person's own tissues. In allergies, allergens binding to IgE on mast cells release allergy mediators in response to nonthreatening substances, causing common symptoms in the skin and respiratory and digestive tracts and anaphylactic shock. In desensitization, periodic injections of the allergen prompt production of IgG, which binds the allergen before IgE can, preventing the response. Allergies may be an adaptation to once life-threatening substances.

Altering Immune Function

Vaccines—Augmenting Immunity

An army is more effective fighting a familiar enemy than fighting an unknown force. So it is with the immune system. When the human immune system meets a disease-causing virus or bacterium for the second time, it is already armed with a circulating collection of memory cells that manufacture the appropriate antibodies. A shortcut to evoking such a secondary immune response is provided by a **vaccine,** which is a killed or weakened form of a virus or bacterium or merely the part of the infectious agent that starts the immune response. The vaccine does not cause illness, but it introduces the immune system to the enemy. Memory cells are then placed on guard, ready to stimulate antibody production when the agent is encountered again.

Vaccines are either live or killed by heat or a toxic chemical. Killed vaccines are damaged just enough so that they cannot cause disease, but their surfaces are preserved so that an immune response is kindled. A problem with live vaccines is that occasionally a virus or bacterium mutates so that it causes disease. In rare instances a killed vaccine can cause illness when a virus or bacterium somehow survives a killing treatment and mutates so that it becomes pathogenic.

A safer vaccine uses only the precise part of the virus's or bacterium's surface that elicits an immune response. A hepatitis B vaccine, for example, consists of a single peptide, and a vaccine against hoof-and-mouth disease, which strikes cattle, is only 20 amino acids long, taken from a viral surface containing 240 proteins.

Another new way to make vaccines echoes back to the field's beginnings, when Edward Jenner developed the smallpox vaccine, which is a virus called vaccinia that causes cowpox (Reading 27.1). A "super vaccine" can be constructed by inserting genes for surface proteins from various disease-causing viruses into the well-studied vaccinia virus. One such vaccine protects against herpes simplex type I, hepatitis B, influenza, and of course smallpox. The vaccinia virus has room for a dozen or more genes to be inserted, and it is possible to stitch "cassettes" of genes into it to tailor multiple vaccines to particular populations or even individuals.

Thanks to vaccines many formerly prevalent diseases have been nearly eliminated. The once nearly unavoidable "childhood diseases" of diphtheria, pertussis (whooping cough), and mumps are becoming rare. Vaccines against chickenpox and strep throat will soon be available. The potential power of vaccines to eradicate illness is realized when the incidence of a disease diminishes greatly thanks to vaccine usage and then rises as people who were not vaccinated or vaccinated ineffectively contract the disease. This has happened with measles.

Before the measles vaccine became available in 1963, 500,000 cases occurred in the United States each year. By 1983, incidence of measles had reached its lowest annual level ever—1,497 reported cases. But in 1984, the number of cases rose to 2,534, and in 1985, to 2,704. By 1989, measles was reported in 45 states, with incidence up 350% from just a year earlier. New outbreaks occurred in 1990. The reason for this resurgence of measles is twofold—many inner-city children are not vaccinated, and vaccines used before 1980 were not powerful enough. The Centers for Disease Control is urging the 100 million people in the United States who have not had measles or the vaccine to be vaccinated. Many colleges and universities now require evidence of measles immunity for admission. With cooperation by the public, measles will soon be eradicated.

KEY CONCEPTS

A vaccine is a portion of a virus or bacterium or an inactivated virus or bacterium that evokes a secondary immune response when the disease-causing agent is encountered.

Organ Transplants—Suppressing Immunity

For the body to accept tissue from another individual, the immune system must recognize the grafted material as "self." Doctors and scientists have made tremendous strides in transplantation in recent years.

One of the first organ transplants was performed in 1905, when the cornea of an 11-year-old boy who lost his eye in an accident was transplanted into a man whose cornea had been destroyed by a splash of a caustic chemical. An important step in successful transplantation also came early in the century, when surgeon Alexis Carrel deciphered the intricate blood vessel connections necessary to support various organs removed from the body.

In 1954, the first kidney transplant was performed between identical twins. Can you see why it was successful? Later in the 1950s, researchers learned how to use blood types to predict how successful a transplant would be between two particular individuals. Because blood types are determined by surface characteristics of blood cells, the more blood types two people share, the more likely a transplant is to "take" between them.

In 1967, South African surgeon Christiaan Barnard performed the first heart transplant. The patient lived for 18 days, and a flurry of transplants followed. However, by the mid-1970s, heart transplants became a rarity, because life was extended only a few months. Rejection by the immune system, surgical complications, and failure to alleviate the underlying medical condition all contributed to the decreasing popularity of transplants. Drugs that suppressed immunity sufficiently to prevent rejection also dampened defenses against infection and cancer. What good was receiving a new organ if the patient died soon afterward from an infection? Some medical centers banned transplants altogether.

By the 1980s, technological developments including improved surgical techniques, better biochemical methods to match donor to recipient, a way to strip antigens from donated tissue, and safer immunity-suppressing drugs made transplants popular again. Most important has been the drug **cyclosporin,** a fungal product originally discovered in a soil sample from Switzerland picked up routinely by a pharmaceutical researcher. Cyclosporin suppresses the T cells that reject transplanted tissue but not those that attack cancer cells or stimulate B cells to produce antibodies. Therefore, infection and cancer can be minimized in the transplant recipient (fig. 27.17).

Reading 27.1 *The Development of Vaccines*

The Smallpox Vaccine

THE FIRST SUCCESSFUL VACCINE WAS DEVELOPED IN 1796 BY EDWARD JENNER, AN ENGLISH PHYSICIAN, AGAINST SMALLPOX, AN ILLNESS THAT ONCE KILLED A THIRD OF THE PEOPLE WHO CONTRACTED IT. Smallpox often led to terrible scarring. The luckiest people were those who developed mild cases, because they were left with relatively smooth skin and could not contract the disease a second time. With this observation in mind, people in medieval China and Europe attempted to immunize themselves (although they knew nothing about the immune system) by stuffing the crusted matter of smallpox lesions into their nostrils. Unfortunately, this hastened the spread of the disease.

Jenner investigated the common belief that people who worked around cows did not develop smallpox. Milkmaids whose faces had the characteristic lesions of the mild disease cowpox did not get smallpox, while others in the area who did not work with animals did. We know today that the cowpox virus is so similar to the smallpox virus that the immune system exposed to cowpox virus develops immunity to smallpox as well.

Jenner found a young milkmaid with active cowpox, took fluid from a blister, and injected it under the skin of a healthy boy. The boy contracted cowpox. Two months later, Jenner gave him fluid from a smallpox blister. The boy did not develop smallpox—he had become immune to it. It took Jenner 2 years to find another milkmaid with active cowpox and another young volunteer. He repeated his experiment and published his results. Although Jenner was ridiculed, his vaccine, which he named from the Latin for "cow," eventually freed the world from smallpox (fig. 1).

For the first few years, the smallpox vaccine was administered in an arm-to-arm fashion—a person recently vaccinated would rub the resulting skin lesion against the arm of another person at a site where a skin scratch had been made. But this crude method spread other infectious diseases. In 1860, the smallpox vaccine began to be manufactured in calf lymph. In 1950, a freeze-dried variety was developed that was very easy to transport, bringing an end to smallpox in many developing nations. In 1959 the World Health Organization began its smallpox eradication campaign, and the disease was virtually gone from the United States and Europe in 1971, from Asia in 1975, and from Africa in 1977. Today, smallpox virus exists only in laboratories, and the vaccine is no longer necessary.

The Polio Vaccine

In the early 1950s, microbiologist Jonas Salk extended Jenner's idea to conquer another disease. He developed the first vaccine against the crippling disease poliomyelitis (commonly called polio) using a killed form of the causative virus. His first batch of polio vaccine was tried in 1952 on children who had already recovered from the disease. The vaccine raised the children's antibody levels, indicating that Salk was on the right track. The next step—a daring one—was to inject healthy children with the vaccine. Would they develop polio when exposed to the virus? They did not, proving that the vaccine worked. By 1954, youngsters all over the United States were lining up for the first polio shots.

A few years later, another microbiologist, Albert Sabin, developed a live oral polio vaccine. Sabin tried his vaccine first on laboratory animals, then on himself, and then on volunteer prison inmates. It too worked. The oral vaccine is safer than its predecessor because it goes to the small intestine, where the polio virus first establishes itself before it travels in the bloodstream to nerves, where it causes paralysis. In contrast, the killed polio vaccine halts the virus in the bloodstream, allowing the virus still to colonize the small intestine. Thus, a recipient of the killed vaccine cannot contract polio but can harbor the virus in the small intestine. If that virus is infectious, it could be spread through a bowel movement to other people.

Although the use of polio vaccines has drastically reduced the incidence of polio in many nations, about 250,000 cases still occur each year, mostly in the developing countries. But that is expected to change, thanks to renewed efforts to disseminate the vaccine. The World Health Organization has declared that by the year 2000, polio, like smallpox before it, will be eradicated.

Figure 1
People were so afraid of being vaccinated against smallpox that it was commonly thought that cow parts would grow out of those receiving the vaccinia virus as a smallpox vaccine.

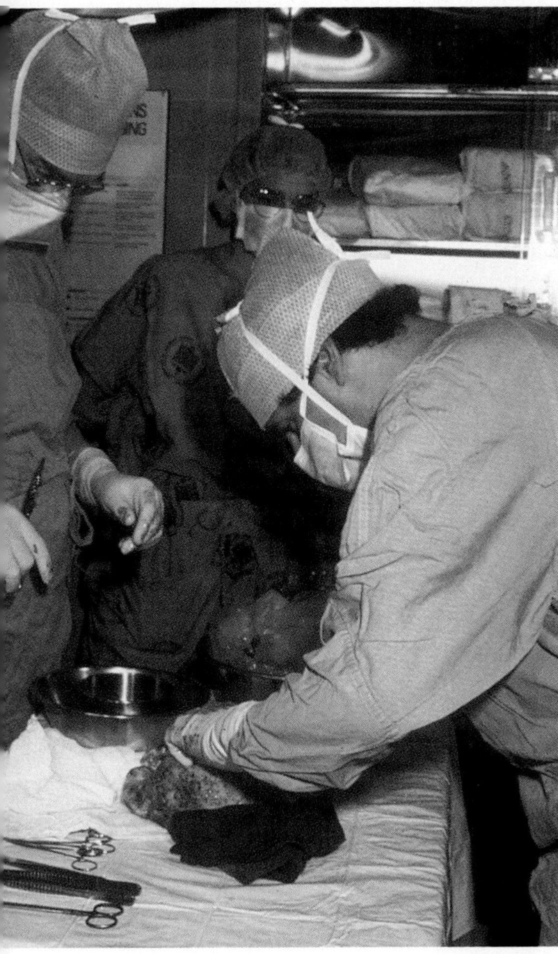

Figure 27.17
Both lungs and the heart can be transplanted to correct congenital abnormalities. Here, lungs are prepared to be removed from a deceased donor.

Continuing challenges to successful transplants are the lack of donor organs and rejection of transplants by the recipient's immune system. Many creative ways around these obstacles are being developed. A person can donate part of his or her liver to a relative in need of this vital organ. A form of autologous (from oneself) transplantation is the practice of storing one's own blood for use in a future transfusion. Stem cells taken from a newborn's umbilical cord can be stored and then cultured later to provide healthy bone marrow, in case the person needs a bone marrow transplant to treat leukemia, for example.

KEY CONCEPTS

The key to a successful transplant is to suppress immunity against foreign tissue but to retain immunity against infection and cancer. Cyclosporin does this.

Monoclonal Antibodies—Targeting Immunity

A single plasma cell (an activated B cell) secretes up to 2,000 identical molecules of antibody per second, all directed against a certain antigen. Isolating a single such cell would provide a rich source of a pure, natural substance that could be used to fight infections and perhaps even cancer. But isolating a B cell is very difficult, because an organism manufactures several types of antibodies at once, in response to different antigens on the surface of the infectious agent. Even if a single B cell could be isolated, it does not live long enough in culture to obtain enough antibodies with which to work. An ingenious way to capture the antibody-making capacity of the B cell was developed in 1975 by British researchers Cesar Milstein and Georges Köhler, who received the Nobel Prize in medicine in 1984 for their contribution of monoclonal antibody (MAb) technology.

Monoclonal antibodies are antibodies that descend from a single B cell. Milstein and Köhler injected a mouse with a foreign antigen, red blood cells from a sheep (fig. 27.18). They then isolated a single B cell from the mouse's spleen and fused it with a cancerous white blood cell called a myeloma. The resulting fused cell is called a **hybridoma,** and it has a valuable pair of talents: like the B cell it produces large amounts of a single type of antibody; like the cancer cell, it divides continuously.

Although hybridoma technology can supply pure and abundant specific antibodies, the procedure is inefficient and laborious. An alternate approach, called combinatorial libraries, produces MAbs much faster than the hybridoma route by engineering bacteria to harbor antibody genes from a higher organism, such as a mouse or human. These genes behave as they would in their natural cellular backgrounds, combining to specify virtually any type of antibody molecules—but only one type per bacterial cell.

Monoclonal antibodies have found applications in basic research as well as human and veterinary health care and agriculture. Cell biologists can use pure antibodies to localize and isolate each of the thousands of proteins in a cell. On the clinical front, more than 100 diagnostic MAb "kits" detect minute amounts of a single molecule. Most kits consist of a paper strip impregnated with a MAb. The appropriate body fluid is then applied to the paper. For example, a woman who suspects she is pregnant places drops of her urine on the paper. A color change ensues if the MAb binds to the hormone (human chorionic gonadotropin) that is present only in the urine of a pregnant woman. Similarly, in tests that predict the time of ovulation, a MAb detects the surge of luteinizing hormone that immediately precedes this time of peak fertility. A MAb kit to diagnose turf grass disease is used to maintain golf courses.

To diagnose viral or bacterial diseases, MAbs bind to the infectious agent or to a molecule that it produces. Because MAbs can detect very small amounts of specific antigens, they can be used to diagnose cancer before a tumor can be felt or detected by other means. Typically the MAb is chemically bonded to a radioisotope (a form of an atom that is radioactive), which is then detected when the MAb binds to the antigen.

MAbs can be used to treat cancer as well as detect it, and they have the advantage over existing cancer treatments in that they affect cancer cells and not healthy ones. Several hundred patients in the United States who have cancer of the B cells are currently being treated with MAbs that bind to molecules protruding from the surfaces of the cancerous B cells. When the MAbs bind to the cancer cells, macrophages are attracted, and they destroy the MAb–cancer cell complex.

MAbs can ferry conventional cancer treatments to where they are needed. Chemotherapy drugs or radioisotopes are attached to MAbs that are attracted to particular types of cancer cells. When injected into a patient, the MAb and its cargo are engulfed by the

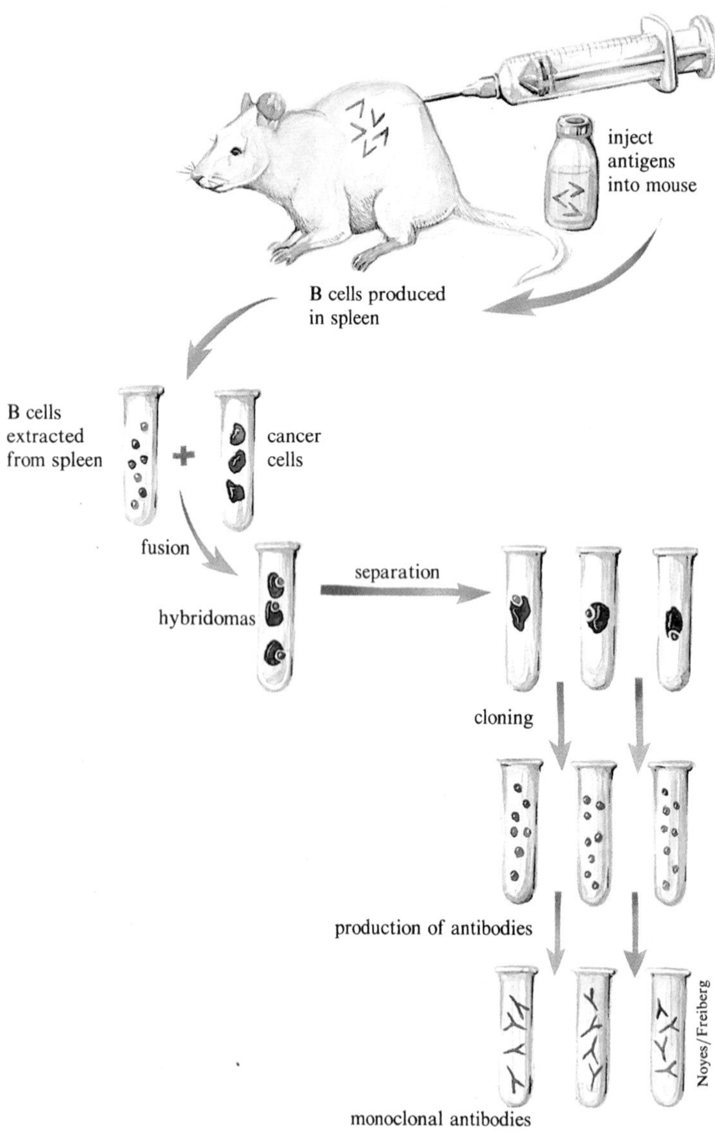

a.

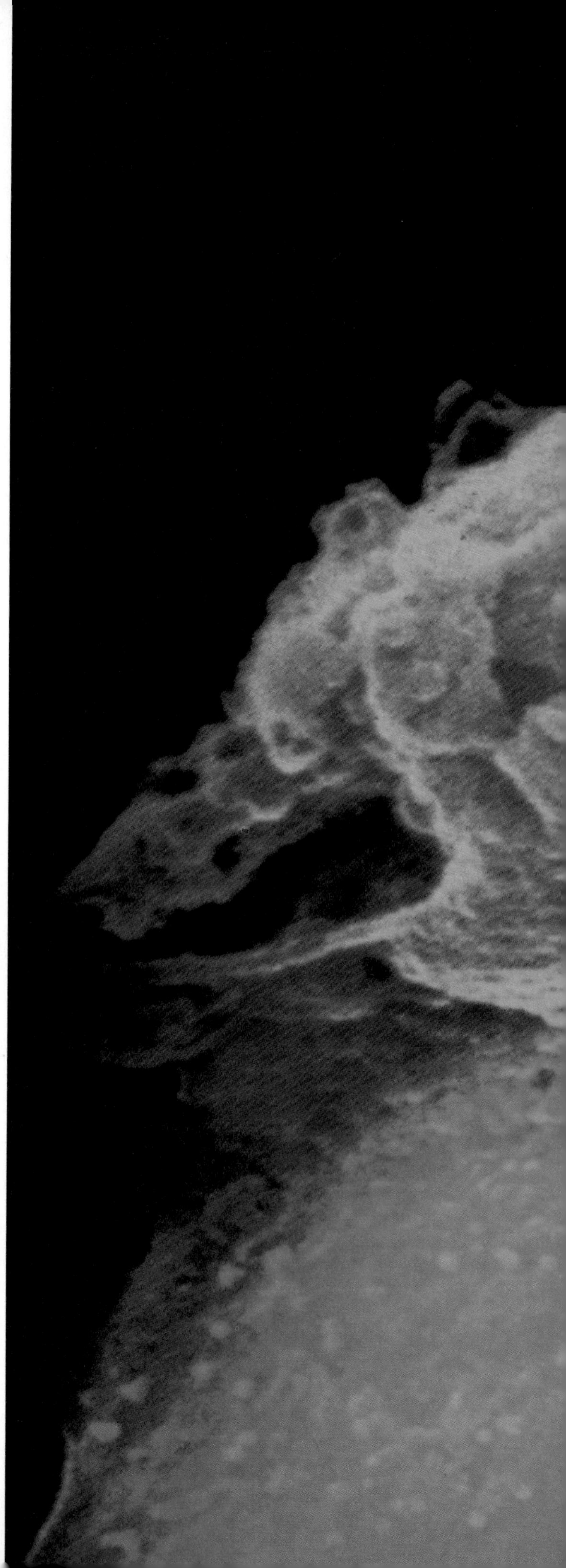

b.

Figure 27.18
a. The steps in making monoclonal antibodies. *b.* B cell + cancer cell = hybridoma. A hybridoma monoclonal antibody factory is created by fusing a cancer cell (the flat cell in blue) with a B cell (the rounded green cell).

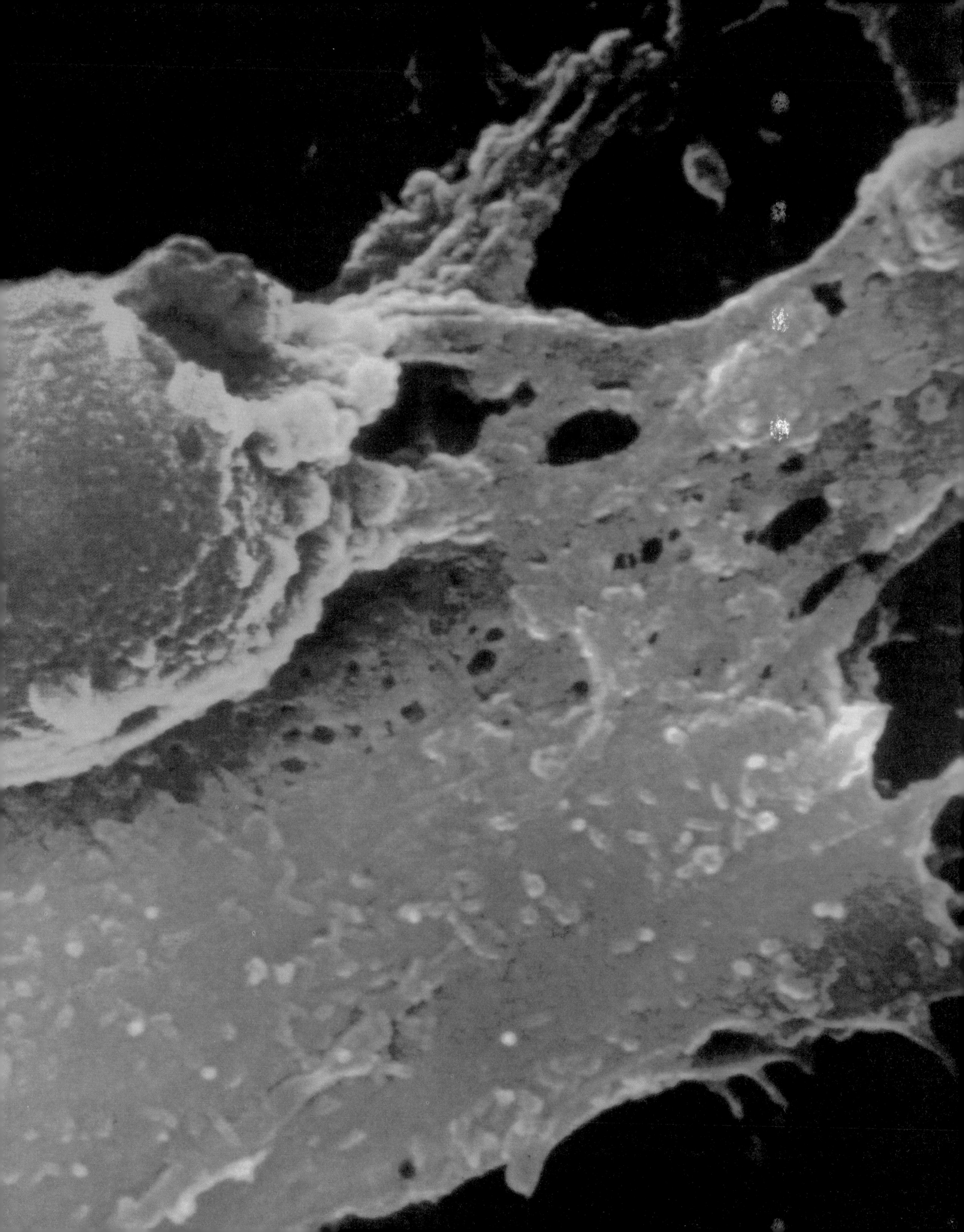

Reading 27.2 Tumor Necrosis Factor—A Closer Look

TUMOR NECROSIS FACTOR (TNF) IS A MOLECULAR JACK-OF-ALL-TRADES, OVERSEEING AN ARRAY OF RESPONSES INVOLVED IN INFECTION, INFLAMMATION, AND CANCER. In small amounts it releases growth factors; in large amounts it causes septic shock (blood poisoning) and the aches, pains, and fever associated with bacterial infection. TNF also clots blood, degrades bone, and stimulates lymphocytes.

In the late nineteenth century, doctors noted that when patients spontaneously recovered from cancer, they also often had infections. Could the body's response to invading bacteria also squelch cancer? In 1893, New York surgeon William Coley tested the theory by intentionally infecting cancer patients with streptococcus bacteria. But in this preantibiotic era, an infection could be life-threatening, so Coley killed his bacteria before administering them. It worked. A solid tumor would bleed inward, turn black, and dry up. So successful was the approach that in 1934, the American Medical Association declared "Coley's toxins" the "only systemic cancer therapy."

But Coley's toxins were replaced by radiation therapy and chemotherapy in the treatment of cancer. Some research continued, though, and the agent that attacked cancer was identified. First the culprit was thought to be endotoxin, the bacterial biochemical that is present when infection symptoms arise. But it turned out that endotoxin stimulates a human biochemical, which actually destroys tumors. In 1973, the substance was named tumor necrosis factor (necrosis means rot). TNF is made by macrophages.

Meanwhile, researchers at Rockefeller University and the Southwest Texas Medical Center were on the trail of another intriguing biochemical. They were interested in a condition called cachexia—a profound wasting seen in animals with cancer or persistent infection. In the late 1970s, the researchers isolated a protein from rabbits suffering from sleeping sickness, a parasitic wasting disease. They named the substance cachectin. Would blocking its action stem the fever, weight loss, and low blood pressure caused by infection? The investigators showed that it did in several animal species. Also as predicted, cachectin alone was enough to induce septic shock. In 1984, the two roads of inquiry—Coley's toxins (TNF) and cachectin—merged as both proteins and their genes were sequenced. They were one and the same.

Could this molecule with its hodgepodge of effects be of medical value? Indeed it kills or slows several human cancer cell types growing in culture, apparently distinguishing between normal cells and their cancerous counterparts. In people, however, TNF alone causes inflammation and flulike symptoms. The trick may be to team TNF with other substances. TNF plus interleukin-2 or interferon, for example, more effectively fights cancer than either alone, at least in the test tube.

TNF is responsible for inflammation due to autoimmunity, infection, or injury and may be useful in treating the joint pain and knotty-looking bones of rheumatoid arthritis. Some commonly used antiinflammatory drugs, such as cortisone and ibuprofen, work by blocking synthesis of TNF. Fish oil may quell inflammation by blocking TNF as well.

cancer cells. The cells are then destroyed by the drug or radiation. Cancers of the breast, lung, and prostate are being treated with MAbs bound to anticancer drugs, and liver cancer is being treated with MAbs bound to radioisotopes.

In yet another approach, MAbs are being developed to cleanse cancerous bone marrow. Some bone marrow is removed and infused with millions of magnetic beads, each coated with MAbs specific to cancer cells. When the marrow is passed through a magnet, the cancer cells are pulled out. The clean marrow can then be returned to the patient.

KEY CONCEPTS

Monoclonal antibodies are produced from hybridomas, which are artificial cells resulting from the fusion of a single B cell and a cancer cell. MAbs can be used to diagnose a wide range of infections, pinpoint hormonal events, and detect and possibly treat cancer.

Biotherapy—Using Immunity to Treat Cancer

MAb cancer therapy is part of an approach to treating cancer called **biotherapy,** which uses substances naturally made in the body, such as immune system cells and biochemicals.

In the 1960s biotherapy was first conducted on experimental animals. Tumor cells were injected into mice, and then various immune system chemicals were injected to treat the cancer that developed. Although the biotherapy worked in some cases, this early animal work did not adequately approximate human cancer because the disease was induced. In addition, it was treated at a stage much earlier than could be detected in humans.

Biotherapy trials begun in humans in the late 1960s were difficult to conduct because the needed biochemicals could only be obtained in small and impure amounts from cadavers. In the 1970s, however, two powerful technological tools catapulted biotherapy into practical use. Monoclonal antibodies could be used to deliver the biochemicals selectively, and recombinant DNA technology could provide vast amounts of pure proteins.

Interferon was the first substance to be tested on a large scale once *E. coli* bacteria were genetically engineered to manufacture it. Although interferon did not live up to early expectations of being a wonder drug, it has been proven effective against a few types of cancer, including hairy cell leukemia (a rare blood cancer), melanoma (skin cancer), and some types of lymphoma and kidney cancer.

In another form of biotherapy, a patient is given interleukin-2, and then killer T cells are removed from samples of tumors. The T cells are incubated in the laboratory in interleukin-2 as the patient receives even more of it. Finally, the activated killer T cells, along with more interleukin-2, are injected into the patient. Although these "lymphokine activated killer cells" have dramatically shrunk some tumors, a few patients have died from the treatment. Clearly more work needs to be done.

Cancer treatment in the future is likely to consist of a combination of immune system cells and biochemicals, as well as standard treatments. For example, tumor necrosis factor (another T cell product) can alter the immune system so that a patient can withstand higher doses of a conventional drug (Reading 27.2). Or, a MAb tagged with a radioisotope might kill 95% of the cells in a tumor, and then a dose of interferon or interleukin-2 could be used to destroy the remaining cancer cells. To treat tumors that have evaded conventional treatment because they are encased in tough connective tissue, substances that induce inflammation can be used to free the tumor. The potential practical uses of the immune system will be limited only by our imaginations.

SUMMARY

Nonspecific defenses prevent infectious agents from entering the body. These include barriers such as skin, mucous membranes, secretions, and cilia. If the barriers are penetrated, *phagocytes* engulf and digest the microbes. The inflammatory response, antimicrobial substances, and fever create an environment hostile to pathogens.

The cells and biochemicals of the immune system differentiate between self and nonself, providing internal protection against many disease-causing agents and cancer cells. An infectious agent encountered again triggers an immune response, which was initiated during the first encounter.

Macrophages quickly phagocytize invading microbes. The macrophage displays a piece of the microbe's surface on its own surface. This and other molecules on the macrophage surface *(major histocompatibility markers)* allow the macrophage to alert and activate helper T cells, which in turn activate *B cells*, which carry out the *humoral immune response* by maturing into plasma cells and secreting antibodies.

An *antibody* is a Y-shaped protein built of units of four polypeptide chains, two *heavy* and two *light*. Each chain has a region of *constant* amino acid sequence and a region of *variable* sequence. The tips of the Y-shaped antibody form an antigen binding site, where a specific part of an infectious agent, the *antigen*, binds. Antibodies bind antigens to form immune complexes, which are large enough to be detected and destroyed by other immune system components. The human body can produce a far greater variety of antibody molecules than there are genes to direct antibody manufacture. This can be explained by the shuffling of DNA segments during early B cell development.

The cellular immune response is carried out by *T cells*, lymphocytes that pass through the thymus gland. Killer T cells release biochemicals that bore through bacterial cell walls and membranes, killing them. Killer T cells also destroy body cells that are covered with viruses. Helper T cells activate killer T cells and also activate B cells. *Suppressor T cells* stop an immune response.

During fetal existence, the developing human immune system learns to distinguish self from nonself. *Passive immunity* is obtained via the placenta and breast milk. The immune system does not mature sufficiently to protect against many illnesses until the second year of life. The immune system begins to decline in function early in life.

In AIDS, HIV prevents helper T cells from activating B cells and causes the T cells to burst, releasing many viruses. The result is a suppression of immunity, and the person develops severe opportunistic infections. HIV has a sequence similar to that for two *lymphokines*, so that when a helper T cell is activated to defend against another virus, HIV replication is set into motion.

In *severe combined immune deficiency*, a person is born lacking T and B cells. Chronic fatigue syndrome reflects an overactive immune system. In an autoimmune disease, the body manufactures antibodies against its own tissues. In an *allergy*, an immune attack is mounted against an *allergen*. When a person comes into contact with an allergen, IgE binds to mast cells, which release *allergy mediators* that cause the symptoms of a particular allergy. A systemic, serious allergic reaction is *anaphylactic shock*.

Vaccines are killed or weakened strains of bacteria or viruses or the parts of them that stimulate antibody production, but they do not cause disease. Transplants succeed only if immune attack against nonself tissue is shut off. Individual activated B cells fused with cancer cells form hybridomas, which secrete monoclonal antibodies. MAbs can diagnose and treat a variety of conditions. Biotherapy directs immune system components against cancer cells.

QUESTIONS

1. Which parts of the immune system are affected in each of the following conditions?

a. AIDS

b. Myasthenia gravis

c. Hay fever

d. Anaphylactic shock

2. How and with which other immune system cells does each of the following interact?

a. Mast cells

b. Macrophages

c. Killer T cells

d. Helper T cells

e. B cells

3. How can a vaccine cause the illness that it is intended to prevent?

4. A young man eats gooseberry pie and soon begins to feel uneasy and warm. Then he itches as hives pop out on his skin. The itching intensifies, and suddenly he heads for the bathroom, where he vomits and has diarrhea. Then his throat swells, and he seeks help. What is happening to him?

5. Explain three ways in which HIV disrupts immune function. How might the virus have come into existence?

TO THINK ABOUT

1. Edward Jenner, Jonas Salk, and Albert Sabin encountered the problem of finding individuals to test their new vaccines. How do you think AIDS vaccines should be tested?

2. Exposure to ultraviolet light in tanning booths increases the body's proportion of suppressor T cells to helper T cells. Would the use of such machines help or harm people with AIDS? State a reason for your answer.

3. A man and a woman intentionally conceived a child so that she could serve, after birth, as a bone marrow donor for her teenage sister, who suffers from leukemia. Do you think that this action is ethical? Why or why not?

4. T cells "learn" to distinguish self from nonself during prenatal development. How could this learning process be altered to prevent allergies? To enable a person to accept a transplant?

5. HIV can be detected in a person's blood long before symptoms of AIDS arise. Do you think healthy people who test positively should be told the results? Should their families be told? Should these people be restricted in their activities in any way to prevent the spread of AIDS?

6. One out of every 310,000 children who receives the vaccine for pertussis (whooping cough) develops permanent brain damage. The risk of suffering such damage from pertussis is about 1 in 30,000. Some parents are refusing to vaccinate their children because of the few reported adverse cases. What are the dangers, both to the individual and to the population, of parents refusing to allow their children to be vaccinated against pertussis?

7. Soon after a heart attack, small amounts of the muscle protein myosin are released from damaged cardiac muscle tissue into the surrounding tissue fluid. Considering this information, devise a way for monoclonal antibodies to be used to assess the extent of damage from a heart attack.

SUGGESTED READINGS

Bayer, Peter Brandon. April 2, 1989. A life in limbo. *New York Times Magazine*. The author tells what it is like to have contracted AIDS by being treated for hemophilia.

Cowley, Geoffrey. November 12, 1990. Chronic fatigue syndrome. *Newsweek*. This disabling illness reflects immune system imbalance.

Jaroff, Leon. May 23, 1988. Stop that germ! *Time*. An excellent review of immune system function.

Kilbourne, Edwin D. July 4, 1990. New viral disease—a real and potential problem without boundaries. *The Journal of the American Medical Association*. Vol. 264, no. 1. The human immune system is challenged to keep pace with ever-changing viruses.

Lewis, Ricki. July/August 1990. Antibody applications get a boost. *BioScience*. Production of monoclonal antibodies from hybridomas was a brilliant invention—but a genetic engineering approach may be even better.

Lewis, Ricki. July 1991. Joint attack: arthritis. *FDA Consumer*. Rheumatoid arthritis is an autoimmune disorder.

Moses, Phyllis B. March 1986. Vaccinia virus: Reinventing the wheel. *BioScience*. Some of the newest vaccines use the oldest vaccine.

Old, Lloyd J. May 1988. Tumor necrosis factor. *Scientific American*. TNF is a multifunctional immune system protein.

Rosenberg, Steven A. May 1990. Adoptive immunotherapy for cancer. *Scientific American*. Immune system cells and biochemicals can be harnessed to fight cancer.

Scientific American. 1989. *The Science of AIDS*. New York: W. H. Freeman. This collection of articles from the October 1988 issue of *Scientific American* is a fascinating and detailed look at AIDS.

Segal, Marian. May 1989. Anaphylaxis: An allergic reaction that can kill. *FDA Consumer*. Mast cells releasing allergy mediators throughout the body leads to a systemic reaction, anaphylactic shock.

Winter, Greg and Cesar Milstein. January 24, 1991. Man-made antibodies. *Nature*. Vol. 349, no. 6307. Genetic engineering can provide antibodies not seen in nature.

U N I T

6

Plant Biology

C H A P T E R

28

Plants Through History

Chapter Outline

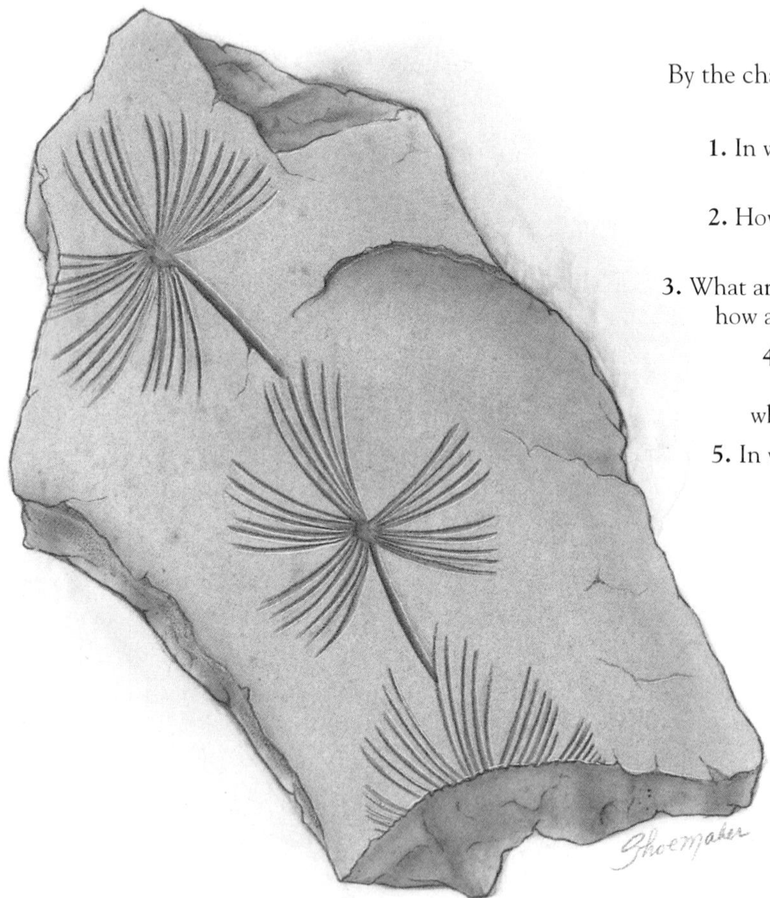

Learning Objectives

By the chapter's end, you should be able to answer these questions:

1. In what ways do we use and depend upon plants?
2. How might agriculture have arisen? Where and when did it arise?
3. What are the major parts of a grain, and how are they used in the human diet?
4. From what types of plants is it thought that the modern bread wheats and modern corn evolved?
5. In what major way does the history of rice differ from that of wheat and corn?
6. How might we use the plant amaranth?
7. What are some examples of plant biochemicals used as medicines?
8. How have plants helped to spread malaria, as well as to treat it?

It is difficult to name an area of human existence in which plants are not important or even vital. Grains, fruits, and vegetables feed us. Fibrous stems and leaves clothe us. Plants generate the oxygen we breathe. Trees provide us with building materials, paper to write on, and welcome shade on hot days. Many plants manufacture chemicals that have important medicinal properties or synthesize brilliant dyes, industrial chemicals, or useful oils (see table 2.6). The colors and fragrances of flowers and foliage satisfy our aesthetic senses. Plant-derived spices enhance our enjoyment of food. Yet plants are not entirely beneficial to people. Plants produce some of the most potent poisons known. Overgrowth of some plants clogs streams and water pipes. Weeds harm crops, and some plants can cause severe allergic reactions in people.

For many centuries, perhaps even longer, humans have learned to recognize those plants that help us and those that harm us. As a result, plants have become an integral part of human civilization (fig. 28.1).

a.

b.

Figure 28.1
The banana and the coconut—staples of the tropics. *a.* The banana has been cultivated in tropical Africa for the past 2,000 years. The starchy bananas called plantains are a major part of the human diet in this part of the world. Sweeter varieties are popular in the United States. *b.* The 50 to 100 coconuts that grow on a palm tree each year provide proteins, oils, and carbohydrates, while the trunks, leaves, and coconut shells make excellent building materials.

Plants as Food

Obtaining a wide variety of plant foods today is as simple as a trip to the supermarket or a visit to the garden or farm stand. But for our ancestors living 12,000 years ago, finding a meal was quite a challenge. Tribes of people were constantly on the move in search of food. In what is now the United States, some of our forebearers hunted the large plant-eating mammals that roamed the great plains, while others collected seeds, nuts, edible roots, fruits, and grasses. In this **hunter-gatherer** lifestyle, people were at the mercy of the environment. If there was no food in a given area, they had no choice but to move on or starve (fig. 28.2).

a.

b.

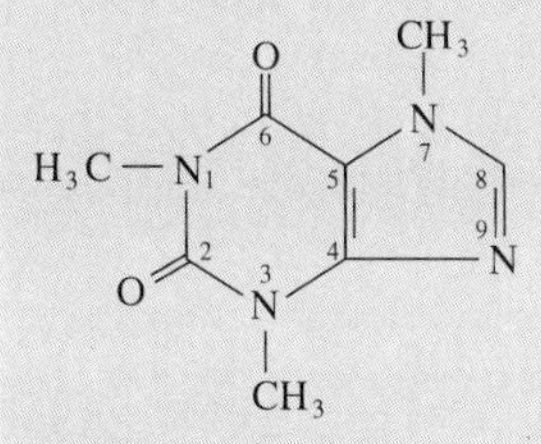

c.

Figure 28.2
Coffee and tea. *a.* Coffee was originally cultivated in the mountains of Ethiopia and was spread elsewhere by Arab traders. Today, *Coffee arabica* is grown for export in 50 nations. *b.* Tea comes from *Camellia sinensis* and originated in the mountains of subtropical Asia. *c.* Both coffee and tea contain caffeine, which has a stimulating effect on the human central nervous system.

From Hunter-Gatherers to Farmers—The Dawn of Agriculture

The dependence upon the environment for food changed dramatically between 12,000 and 10,000 years ago. Several factors contributed to this first spark of civilization. The last ice age was ending. As the great ice sheets receded, the land that was revealed became inhabited by a vast assemblage of animals and flowering plants. Humans no longer had to follow herds or gather berries to eat, because food could be found in more areas. The people could stay in one place, eating readily available small game, fish, and wild plants. By about 10,000 years ago, humans learned to take care of certain animals, such as wild sheep and goats, so that they could rely on a continual supply of milk and meat. The wild weeds that sustained many tribes also became subject to human interference. The transition from weed to crop may have happened almost by accident, perhaps as follows: A wandering band of humans found some tasty grain growing in a sparsely vegetated area. They stopped, set up camp, and ate the grain. The next day they moved on, leaving some of the grain on the ground where they had eaten it. A few months or even years later, the same people came back to the campsite, and, to their initial surprise, found the same type of grain growing in abundance. Eventually, the connection between the discarded grain of one season and the bountiful crop of the next was discovered. The people learned to leave seeds in the ground intentionally to ensure a future food supply.

Gradually, these early farmers discovered more about plant reproduction and used their knowledge to improve their food supply. They deduced by trial and error how to plant the correct seeds in the right soil and to see that they got enough water to sprout. They learned to recognize when the plants had developed sufficiently to harvest. The people regularly saved the seeds from one season's most useful plants to sow the next year's crop, thereby encouraging certain combinations of traits (a practice called **artificial selection**) that might not have predominated in the wild (fig. 28.3).

Later on, people learned to preserve their harvested food by drying it, reducing even further their dependence on the unpredictable weather and climate. The domestication of animals and the intentional planting and cultivation of crops marked the birth of **agriculture.** Farming probably arose independently in many places around the world. Archaeological evidence indicates domestication of sheep and goats and cultivation of grassy wheats and barley in the Fertile Crescent region of what is now

Figure 28.3
Jojoba—an old plant finds new uses. The earliest farmers fashioned crops by taking robust specimens and using them for the next season's crop. This technique also works well today. Before 1980, the jojoba plant was an obscure grayish green bush growing in the southwest United States. Then some innovative people realized the potential for the light golden oil from jojoba's seeds as a lubricant, a cosmetic ingredient, and even as a no-cholesterol cooking oil. In 1980, seeds from wild-growing jojoba were collected and used to start the first cultivated plantings. The next season, researchers took cuttings from the healthiest bushes and used them to start the next crop. Today, the United States' jojoba yield grows by at least 50% a year.

the Middle East about 11,000 years ago. Gradually, agriculture spread to Eastern Europe around 8,000 years ago, to the Western Mediterranean and Central Europe about 7,000 years ago, and to the British empire about 4,000 years ago. Egyptian and Assyrian tombs, mummy wrappings, paintings, and hieroglyphics from 4,000 years ago depict a rich agricultural society, which cultivated pomegranates, olives, grapes, figs, dates, and cereals. A similar spread of agriculture occurred in the Americas, based upon native corn, tomatoes, and chili peppers. Yams and cassava were early crops in Africa.

With the growth of towns and then cities, the human influence on cultivated plants spread. The distribution of crops across the planet today indicates that in many cases, when humans colonized new lands, they brought their native plants with them and likewise carried new plant varieties from the new land back to the old. Tracing the origins of crop plants is difficult because archaeological evidence is often destroyed over time. The available evidence, however, suggests that very different crop plants were domesticated in the Old and New Worlds. Today, thanks to efficient transportation, many different kinds of plants are grown quite far from where they originated (table 28.1).

Today, more than 80% of the total kilocalories consumed by humans comes from crop plants, mostly from wheat, rice, corn, potatoes, sweet potatoes, and manioc (the source of tapioca). The cereals provide much dietary carbohydrate, and the seeds of legumes (beans, lentils, peas, peanuts, and soybeans) are rich in protein. Cereals have different types of amino acids than legumes, so eating these two foods together provides a good protein balance. The best plant-derived nutrition combines a cereal (a rich carbohydrate source) with a legume (a protein source) with a green leafy vegetable (rich in vitamins and minerals) and perhaps small amounts of sunflower oil, avocado, or olives (which provide fats). Plants also yield spices, which have been used to flavor food since at least the time of the Roman Empire. Spices can also preserve foods, which made possible the colonization of the New World. The biochemicals that we use as spices offer a defense against herbivores to the plant.

Table 28.1
Agricultural Origins

Time (years ago)	*Location*	*Crops*
15,000	Yellow River, China	Millet, rice
12,000	Nile Valley, Egypt	Barley, wheat, lentils, chickpeas
	Thailand	Rice
10,000	Fertile Crescent: Iran, Iraq, Syria, Lebanon	Barley, wheat, lentils, broad beans, olives, dates, grapes, pomegranates
9,000–8,000	Mexico, Peru	Corn, kidney beans, lima beans, peanuts, chili peppers, tomatoes, tobacco, cocoa, pumpkins, avocados, squashes
5,000	Africa	Sorghum, millet, okra, coffee, cotton
4,000	Andes (South America)	Potatoes
3,000	India	Cotton

KEY CONCEPTS

About 12,000 years ago, humans were hunter-gatherers, moving to find food. As the ice age ended, ample food was revealed, so gradually people stayed in one place and raised animals and crops for food. Agriculture dawned as people artificially selected crops and learned to store and preserve plant foods and domesticate animals. Agriculture began at different times in different parts of the world. Agriculture changed as people moved to start new settlements and took their crops and animals with them. Today 80% of kilocalories consumed by humans comes from plant foods.

Cereals—Staples of the Human Diet

In the Orient, the principal food is rice. Corn is a major part of the human diet in South America. In the United States, many of our foods are based on wheat. Rice, corn, and wheat, along with oats, rye, barley, and a few others, are all **cereals,** defined as members of the grass family (Poaceae), which have seeds that can be stored for long periods of time. The seed is the kernel or grain, and it consists of the embryo, a large starch supply called the **endosperm,** and outer protective layers called the **aleurone** and the **pericarp,** which are rich in protein, lipid, and vitamins. Figure 28.4 shows the parts of a wheat grain and their dietary uses. Although wheat, corn, and rice feed much of the human population, other edible cereals await discovery.

Wheat

Modern wheat comes in thousands of varieties. It grows in a range of climates, from the Arctic to the equator, from below sea level to 10,000 feet (3,048 meters) above it, and from areas with less than 2 inches (5 centimeters) of annual rainfall to places with more than 70 inches (178 centimeters) of rain per year. Most wheat is used for food. The grain of the "hard" bread wheats contains 11% to 15% protein, mostly of a type called gluten, and when ground and mixed with water it forms an elastic dough that is excellent for making bread. The grain of the "soft" bread wheats contains 8% to 10% protein and is best for making cakes, cookies, crackers, and pastries.

Today's wheats fall into three categories, based upon their number of chromosomes—14 (diploid), 28 (tetraploid), or 42 (hexaploid) (table 28.2). The bread wheats are hexaploid; durum wheat used to make macaroni is tetraploid. Geneticists study the number of chromosomes in modern wheats to reconstruct the evolution of these important crop plants. It is a fascinating and complex story. Wheat as we know it apparently evolved from an accidental merger between a distant

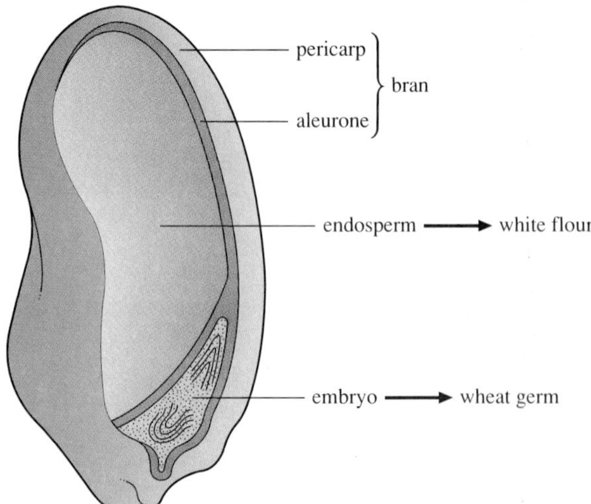

Figure 28.4
Anatomy of a wheat grain. Different components of a wheat grain give us different food products. The endosperm comprises 80% of the grain, and when separated and mashed, it yields a white flour. If the embryo, or wheat germ, is included along with the endosperm, the resulting flour is more nutritious than pure white flour because of vitamins in the germ, but it does not last as long, because of the high fat content of the germ. Bran, also vitamin-rich, consists of the pericarp and aleurone layers, which surround the endosperm. Flour made from ground whole wheat grains is called graham.

Table 28.2
Types of Wheat

Type	Number of Chromosomes	Location	Uses
Einkorn	14 (diploid)	Hills of southeast Europe, southwest Asia	Cattle and horse feed
Emmer	28 (tetraploid)	Europe, United States	Stock feed, macaroni
Bread wheat	42 (hexaploid)	Throughout the world	Bread

wheat ancestor and a weedy grass. It may have happened as follows: About 10,000 years ago, wandering peoples in what is now Jericho, in Israel, came upon a rich oasis, a spring in the desert ringed by hills covered with wild grass. In looking about for food, the people discovered that the grain held in the grass, when ground, made a fine flour. For many years, they did not know how to encourage the grass's growth intentionally, so each season they would simply forage for whatever nature provided.

Then about 8,000 years ago, something happened that was to have profound effects on agriculture and human civilization. The grass that the people had grown fond of, really an ancient wheat called Einkorn, crossbred with another type of grass that was not very good to eat (fig. 28.5). The hybrid grass then underwent a genetic accident (nondisjunction) that prevented the separation of chromosomes

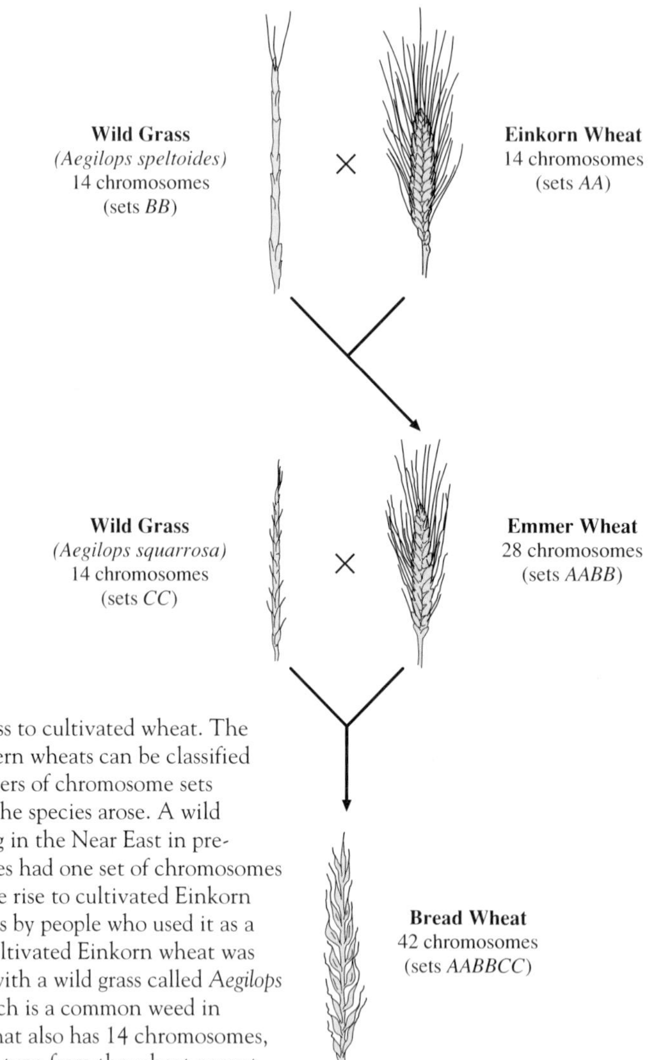

Figure 28.5
From wild grass to cultivated wheat. The fact that modern wheats can be classified by their numbers of chromosome sets suggests how the species arose. A wild wheat growing in the Near East in pre-Neolithic times had one set of chromosomes (AA) and gave rise to cultivated Einkorn wheat, perhaps by people who used it as a cereal. The cultivated Einkorn wheat was then crossed with a wild grass called *Aegilops speltoides*, which is a common weed in wheat fields that also has 14 chromosomes, but 7 of the *A* type from the wheat parent and 7 of the *B* type from the grass parent. An "accident" in which the chromosome number doubled then led to the appearance of Emmer wheat, which has 28 chromosomes (*AABB*). The bread wheats arose in the Bronze Age, when Emmer wheat crossed with another grassy weed, *Aegilops squarrosa,* a pest also known as goat grass, which had 14 chromosomes of yet a third type, CC. The initial hybrid of Emmer wheat with goat grass had 21 chromosomes (7 *A*, 7 *B*, and 7 C), which accidentally doubled to yield a wheat with 42 chromosomes (*AABBCC*), which is modern bread wheat.

Figure 28.6
A popular grain hybrid is triticale, which is a combination of wheat (genus *Triticum*) and rye (genus *Secale*). The cells of triticale are polyploid, containing a complete set of chromosomes from wheat as well as a set from rye. The plant has the high yield of wheat and the ability to cope with harsh environments of rye.

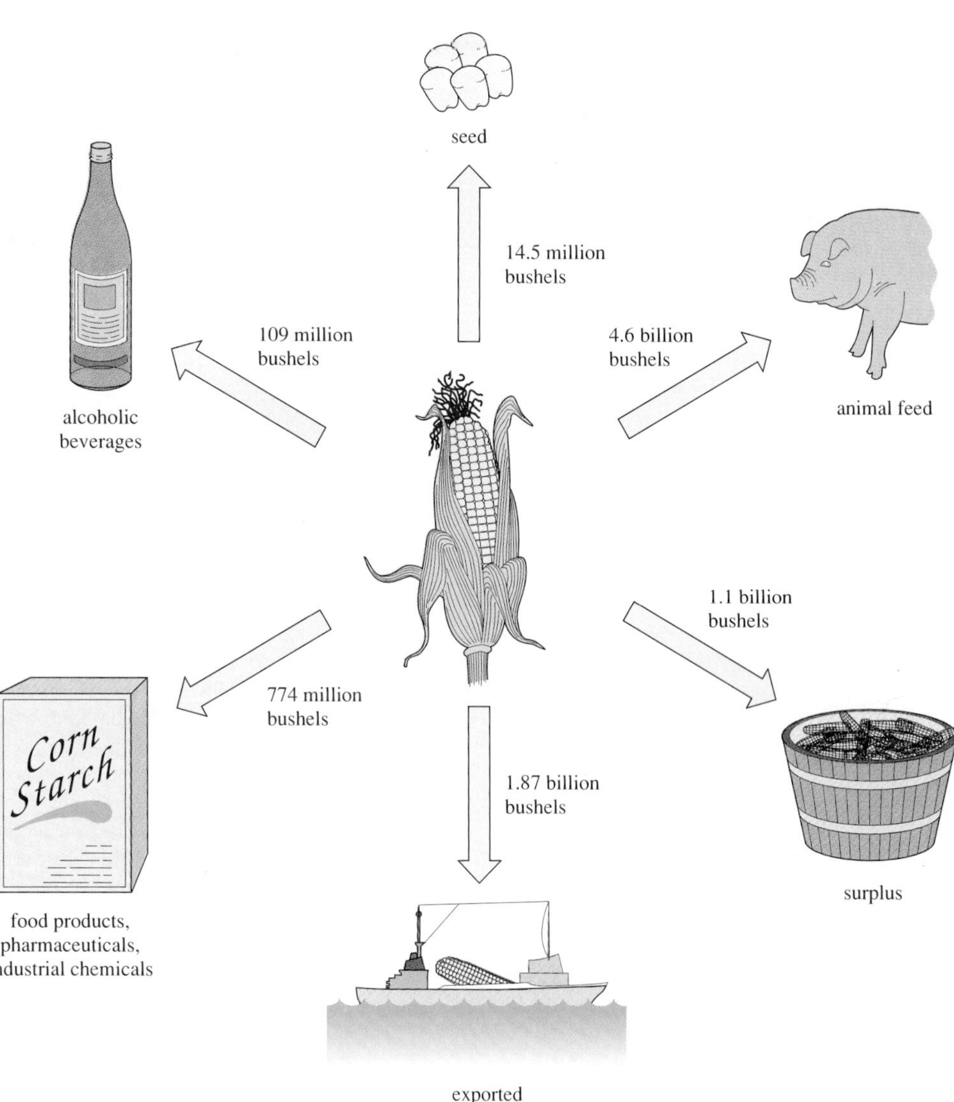

Figure 28.7
The corn crop in the United States in 1982 hit a then all-time high of 8.4 billion bushels.

in some of the developing germ cells. The result was a new type of plant that had twice the number of chromosomes as either parent plant, or 28 chromosomes total. This plant was called Emmer wheat, and it was a far better source of food than the parent wheat whose grains had become a staple. The doubled number of chromosomes in each cell resulted in a plant with larger grains. Plus, the grains were attached to the plant in such a way that they could be easily loosened and spread by the wind, so that the new wheat was soon plentiful. It was probably about this time that early farmers learned to select seed from the most robust plants to start the crops of the next season.

Then about 6,000 years ago, another mistake of nature further improved the quality of wheat. Emmer wheat crossed with another weed, goat grass, and after another fortuitous "accident" of chromosome doubling, led to bread wheat, which has 42 chromosomes. Bread wheat has even larger grains that the Emmer wheat that gave rise to it, but at a cost. Its ears are so compact that, on its own, the grain cannot be released. However, with the help of farmers the rich seed was collected each season for food, and a certain percentage held to be planted the next season. The interdependence of humans and crops that is the basis of modern agriculture was thus born. Today, interesting hybrids add variety to our diets, such as triticale, a combination of wheat and rye (fig. 28.6).

Corn

The tasty, sweet, or starchy kernels of the corn plant *Zea mays* have sustained the Incas, Aztecs, and Mayan Indians of South America and the pilgrims of colonial Massachusetts, and today they continue to be a dietary staple from Chile to Canada. The corn crop in the United States presently exceeds 9 billion bushels a year and is used to manufacture food products, drugs, and industrial chemicals, as well as to feed humans and animals directly (fig. 28.7).

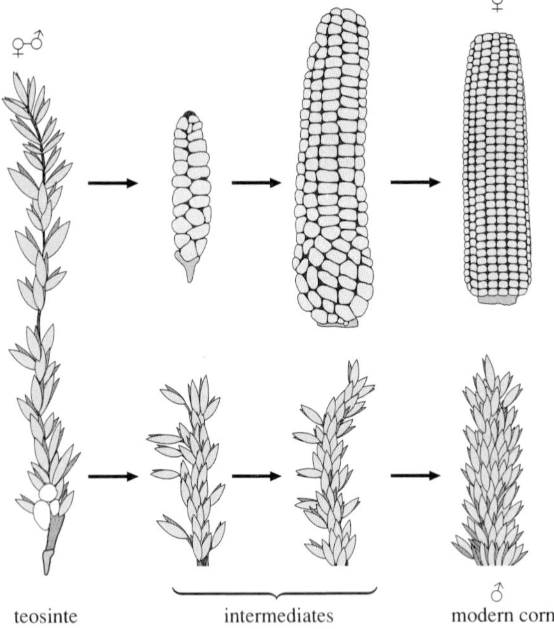

Figure 28.8
Corn from teosinte. The wild grass teosinte (*Zea mexicana*) bears structures that may have given rise to modern corn (*Zea mays*). The small hard seeds of teosinte evolved into corn kernels, and the tassels eventually became the familiar tassels of corn.

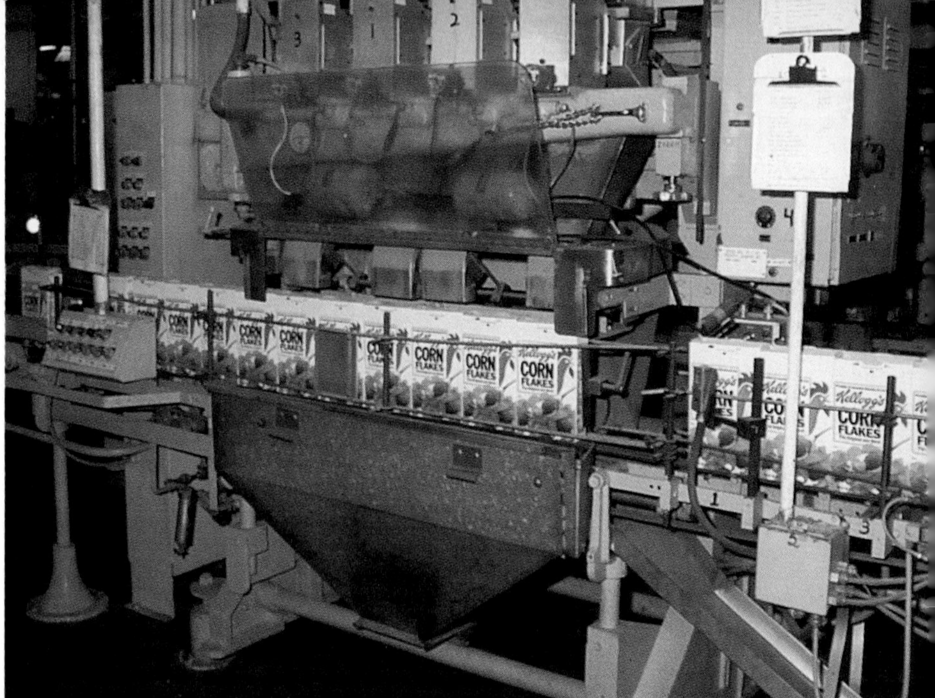

Figure 28.9
The making of a cornflake. The invention of flaked cereals took place accidentally at the Battle Creek Sanitarium for the Seventh Day Adventist Church in 1894. The sanitarium was run by a physician, Dr. John Harvey Kellogg, and his business manager brother, Keith. The Kellogg brothers were health-food advocates somewhat ahead of their time, and they regularly put wheat dough through rollers to form wheat sheets, which when ground up, became a sort of cereal.

At first, the patients didn't like the consistency, so the brothers worked to perfect their invention. One night, the brothers were called away on an emergency and left some wheat soaking for many hours longer than usual. When they tried to make their wheat sheets the next day, they found instead that the moisture had equalized in each wheat berry. When run through the rollers, each wheat berry formed its own flake. The sanitarium had a new food favorite, wheat flakes.

Four years later, the Kellogg brothers tried their approach with corn, added some malt flavoring, and the cornflake was born. One of their patients paid particularly close attention to the flaking process. His name was C. W. Post, and he would become one of the Kellogg brothers' chief competitors.

Like wheat, modern corn probably arose from a naturally occurring wild grass that initially produced a grasslike type of corn with small ears. Fossil evidence exists of such a primitive corn. A clue to the origin of modern corn comes from a grass called **teosinte** *(Zea mexicana)*, which grows in Mexico today and probably has for many thousands of years. The Indians call teosinte "madre de maize," which means "mother of corn." Corn and teosinte have a few interesting similarities. Teosinte is more grasslike than corn, but it does have small ears that each bear a single row of small, hard seeds. Both species have 20 chromosomes and produce very fertile hybrids when they are crossed. Sometimes wild teosinte grows on the outskirts of cultivated corn fields.

Evolutionary biologists have constructed a scenario of corn's origins, based upon the similarities between modern-day corn and teosinte (fig. 28.8). About 7,500 years ago, a population of wild teosinte faced an environmental stress that was overcome only by individual plants whose tassels included some female structures. Farmers noticed that the unusually hardy plants had larger and better-tasting kernels, and they chose those plants to cultivate. Part of the teosinte tassel might have enlarged to evolve into the corn tassel. Over the centuries, modern corn, with female ears and male tassels, was artificially selected by farmers seeking plants with plump and tasty kernels.

Charles Darwin, the father of evolution, was interested in the origin of corn, and in the late nineteenth century, he bred the plants in greenhouses. He found that corn plants that were continually self-fertilized produced increasingly weaker offspring—that is, inbred plants were more prone to disease and produced ears with fewer rows of kernels. Darwin noted, however, that when he crossed unrelated corn plants to each other, the offspring were strong and healthy, with many rows of plump kernels. Although Darwin knew little about the genetic discoveries of his contemporary Gregor Mendel, what he had demonstrated was the genetic phenomenon of hybrid vigor. Plants with unrelated parents were more vigorous than self-fertilized plants because they had inherited new combinations of genes.

In the first decade of the twentieth century, George Shull, at the Station for Experimental Evolution in Cold Spring Harbor, New York, carried Darwin's greenhouse experiments one step farther, thanks to the additional insight provided by the rediscovery of Mendel's laws of inheritance. Shull continually bred plants from the same ears of corn, artificially selecting highly inbred lines. The inbred corn plants were sickly looking, with very small ears. But when Shull crossed two

Figure 28.10
Planting several varieties of rice encourages success. This rice paddy is planted with several varieties of rice. Should disease strike, some may survive.

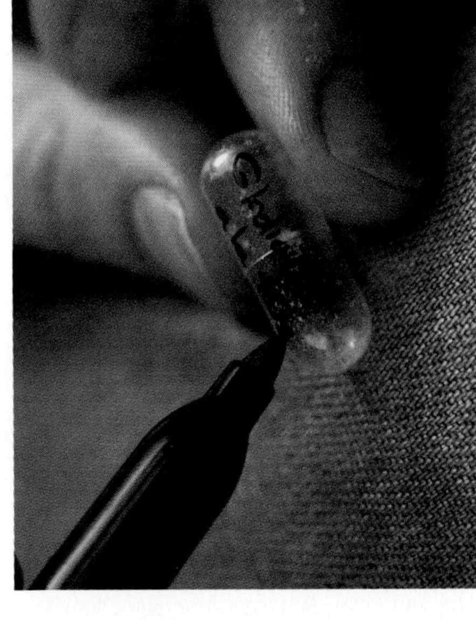

Figure 28.11
Banking plant genes. At the cryogenic gene bank at the University of California at Irvine, pollen is frozen in small plastic ampules.

different inbred lines to one another, the offspring—hybrids—were exceptionally vigorous. Shull had developed hybrid corn, which revolutionized corn output worldwide. United States production of corn has jumped from 21.9 bushels per acre in 1930 to 95.1 bushels per acre in 1979 and well beyond that today, thanks largely to hybrid corn. Many of us start our days with this tasty grain (fig. 28.9).

Rice

In Taiwan, a hungry person is not asked, "Are you hungry?" or "What would like to eat?" but "Would you like some rice?" In much of the world, like in Taiwan, rice is so much a dietary staple that it is synonymous with "food." For 2 billion people in Asia, rice provides 80% of their calories. The 400 million metric tons of rice produced each year there are grown in many types of environments, ranging from 53° north latitude to 40° south latitude.

Today's rices include 20 wild species plus 2 cultivars, which are domesticated species produced by agriculture rather than by the evolutionary forces of natural selection. Cultivars do not grow in the wild. The oldest species of rice for which we have fossil evidence, *Oryza sativa*, originated about 130 million years ago in parts of South America, Africa, India, and Australia, which were then joined into one landmass. So unlike wheat and corn, rice is a very ancient plant that evolved in a tropical, semiaquatic environment.

Human cultivation of rice began about 15,000 years ago in the area bordering China, Burma, and India. By 7,000 years ago, efforts to control rice growth had spread to China and India; by 2,300 years ago, the crop was growing in the high altitudes of Japan. By 300 B.C., rice was a staple throughout Asia. It has only been over the past six centuries that rice has been eaten in West Africa, Australia, and North America.

As migrating peoples took their native rices with them to new lands, the plants adapted to a wide range of environments, from deep salty water to the driest of drylands. From the 1930s to the 1950s many nations collected hundreds of native rice varieties, growing small amounts of each type every season just to keep the collections going (fig. 28.10).

In the 1960s, highly productive newcomers, the **semidwarf rices,** became very popular with farmers and soon accounted for nearly all of China's crop and almost half of the rices of many other Asian nations. To offset a potential disaster due to reliance on a few types of rice, the International Rice Research Institute (IRRI) was founded in 1961 in the Philippines. It soon became a clearinghouse for the world's rice varieties, cold storing the seeds of 12,000 natural variants by 1970 and of more than 70,000 by 1983. Representatives of other important crops are being banked as well. Potato cells are stored at the International Potato Center in Sturgeon Bay, Wisconsin, and wheat cells are banked at the Kansas Agricultural Experimental Station. Pollen and seeds from 250 endangered species of flowering plants have been frozen at a plant gene bank at the University of California at Irvine (fig. 28.11).

Plant banks offer three priceless services to humanity: a source of variants in case a major crop is felled by disease or an environmental disaster; the return of endangered or extinct varieties to their native lands; and, perhaps most important, a supply of genetic material from which researchers can fashion useful plants in the years to come, even after the species represented in the bank have become extinct.

Establishing a seed or pollen bank is one solution to a growing problem in traditional plant agriculture—the decrease in genetic diversity that makes a crop vulnerable to disease or a natural disaster because it does not have resistant variants. Farmers felt

Figure 28.12
The sunflower—a native American plant. The sunflower was first cultivated in what is now the United States by Mexican Indians. Today it remains a favorite among home gardeners but it is also an important cash crop because of its oil. The Soviet Union produces more than one-half of the world's sunflower crop, and the plant is being grown for its oil in the Mediterranean nations and Spain as well.

Figure 28.13
Amaranth—a "new" grain food. This regal-looking plant is amaranth, which may one day be a major food source in the United States, as it was in years past to many Central and South American peoples.

this short-sightedness painfully in 1970, when the southern corn leaf blight fungus destroyed 15% of the United States corn crop, at a cost of a billion dollars. Genetic uniformity of the potato crop in Ireland contributed to the extent of the famine suffered from 1846 to 1848, when the mold *Phytophthora infestans* decimated the crop. Two million people died and an equal number left the country. Today in Ireland, more than 60 species of potatoes are actively farmed to prevent crop vulnerability.

Another strategy to increase our supply of plant foods is to look for new crops among the many naturally occurring plants. Fewer than 30 of the 240,000 species of flowering plants provide more than 90% of plant-based foods eaten by people (fig. 28.12).

Amaranth

A plant with great promise as a food source is the majestic-looking **amaranth,** which stands 8 feet (2.4 meters) tall and has broad purplish green leaves and massive seed heads (fig. 28.13). Each plant harbors a half million of the mild, nutty-tasting seeds, which are each the size of a grain of sand. The flowers are a vivid purple, orange, red, or gold. Amaranth was cultivated extensively in Mexico and Central America until the arrival of the Spanish conquistadors in the early 1500s, who banned the plant from use as a crop because of its importance in the Aztec religion.

Amaranth is being grown experimentally at the Rodale Research Center in Kutztown, Pennsylvania, and is available from commercial seed catalogs for home garden use. It grows fast, adapts to a wide range of environmental conditions (high salt, high acid, high alkalinity), yields many seeds, and comes in many varieties. Problems with cultivating amaranth include combating weeds and pests and difficulty in harvesting the tiny seeds.

Nutritionally, amaranth is superb. Its seeds contain 18% protein, compared to 14% or less for wheat, corn, and rice. Amaranth is rich in amino acids that are poorly represented in the other major cereals. The seeds can be used as a cereal, a popcornlike snack, and a flour to make graham crackers, pasta, cookies, and bread. The germ and bran together are 50% protein, making them ideal to add to prepared foods and animal feeds. The broad leaves of amaranth are rich in vitamins A and C and the B vitamins riboflavin and folic acid. They can be cooked like spinach or eaten raw in salad.

KEY CONCEPTS

Most human diets are rich in cereals. Seeds of these grasses consist of the embryo, the starchy endosperm, and the surrounding aleurone and pericarp. Wheats are classified by their chromosome numbers, which provide evolutionary clues. Polyploid edible wheats evolved from hybridization of wheats with fewer chromosome sets. Modern corn probably evolved from a similar wild grass, teosinte. Charles Darwin and later George Shull discovered that hybrid corn varieties were more vigorous than pure strains. Rice is much more ancient than wheat or corn. Overreliance on a few very productive species is dangerous. Pollen and seed banks help to preserve species. We can also use less common grains, such as amaranth.

Nature's Botanical Medicine Cabinet

Several years ago, I was on a plant collecting trip near Serengeti National Park in East Africa. I was searching for the rare medicinal plant Kigeria africana. After 3 days of combing the Serengeti plain, I came upon an outcrop of K. africana trees. Feeling very lucky indeed, I climbed up onto one of the tree's branches to collect the sausage-shaped fruit.

But as I busily went about my collecting, I realized I was being watched. Two menacing eyes from an adjacent tree stared at me through the branches. Those eyes belonged to a rather large leopard. I tried with all my might to keep from falling from my perch as my knees knocked and sweat poured from my body. This brief encounter ended abruptly, however, when the leopard decided I was not a suitable meal and disappeared into the bushes. Well, I thought, another day in the life of a natural products chemist.

Isao Kubo, Professor of Natural Products Chemistry, University of California at Berkeley, *From Medicine Men to Natural Products Chemists* (1985)

Figure 28.14
The lemon—a multipurpose plant to humans. The tart fruit of the lemon has had various uses in human health care. The juice promotes urination and perspiration and has been used to stimulate these body functions in the treatment of conditions as commonplace as the common cold and as serious as malaria. The astringent nature (skin pore closing) of lemon juice makes it a soothing treatment for sunburn, an effective gargle for a sore throat, and, it has been claimed, a cure for hiccups. Lemon juice kills bacteria on teeth and gums, and chewing the rind can reduce tartar buildup. The high vitamin C content of the fruit makes it an excellent defense against scurvy, a condition characterized by weakness, anemia, and hemorrhage caused by severe vitamin C deficiency. The leaf of the lemon plant is useful too—it makes a soothing, sleep-promoting tea.

Natural products chemists search through the bounty of chemicals manufactured by organisms for substances that may treat illness in people. A chemical that oozes from a tree's bark to discourage hungry caterpillars from eating it may also be an effective drug. For example, a chemical in the bark of the Indian neem tree keeps desert locusts off the tree. The people of Serengeti National Park in east Africa chew the twigs of the Indian neem tree to prevent tooth decay. Other, more familiar plants have healing properties as well (fig. 28.14). Few natural products chemists are as adventurous as Isao Kubo. Most work in laboratories, trying to imitate nature by synthesizing compounds that plants normally make.

Today's natural products chemist is a modern version of the "medicine man" (or woman) who traditionally explored the healing powers of plants. Herbal medical practices may have begun in prehistoric times, when some individuals became botanical experts by sampling plants themselves. Clay tablets carved 4,000 years ago in Sumaria list several plant-based medicines, as do records from ancient Egypt and China. Roman philosopher Pliny the Elder wrote in the first century A.D., "If remedies were sought in the kitchen garden, none of the arts would become cheaper than the art of medicine." Modern-day medicine men called the Bwana mgana practice herbal medicine in the region of East Africa explored by Isao Kubo.

A good example of medicines derived from plants used today are the chemicals called **alkaloids,** which come from the periwinkle plant and other species. Alkaloids have helped revolutionize the treatment of some leukemias (blood cancers), and alkaloid narcotics derived from the opium poppy, including morphine, are excellent painkillers. Today, nearly half of all prescription drugs contain chemicals manufactured by plants or bacteria, and many other drugs contain compounds that were synthesized in a laboratory but were modeled after plant-derived substances. The many medicines "borrowed" from the plant kingdom provide one compelling reason why we must halt the present rapid destruction of the world's tropical rain forests. Here, plant life is so abundant and diverse that all of the species have not even been cataloged by taxonomists.

One disease whose spread has been greatly influenced by plants is malaria, which kills more people worldwide each year than any other disease. Malaria starts with chills and violent trembling and progresses to an extremely high fever accompanied by delirium. Finally, the person sweats profusely, is completely exhausted, and has a dangerously enlarged spleen. The disease strikes in a relentless cycle, with symptoms returning every 2 to 4 days. Within weeks, the sufferer either dies of circulatory system collapse or manages to marshall the body's immune defenses against the invading parasite that causes the disease.

Malaria is an interesting example of the relationship between plants and human disease, because plants both contribute to

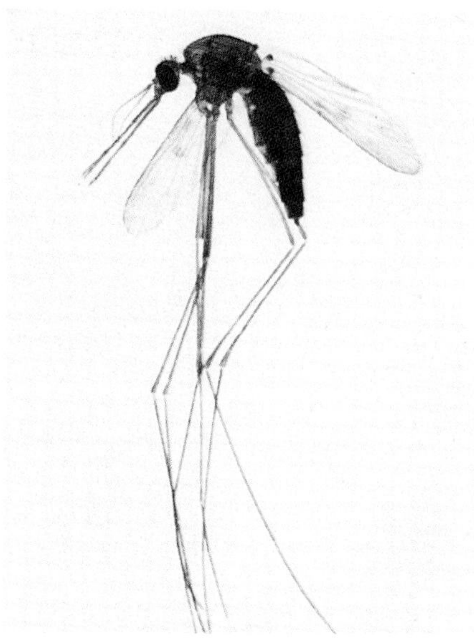
a.

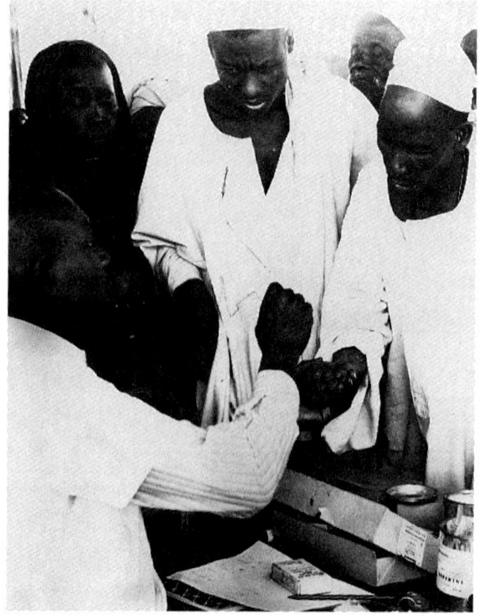
b.

c.

Figure 28.15
Plants and malaria. In an estuary community near the Demerava River in Guyana, malaria was almost unheard of before the 1960s, when the natural vegetation was destroyed and replaced with the "cash crop" rice. *a.* The *Anopheles aquasalis* mosquito thrived in the new, damp environment. *b.* So did malarial parasites within the mosquitoes, and soon the people were ill with the fever. Here, malaria sufferers receive a drug treatment. *c.* A potent antimalarial drug called artemisinin, new to the Western world but a folklore treatment in China for centuries, is derived from the dried leaves and flowers of *Artemisia annua,* which grows near Washington, D.C.

the disease's spread and to its cure. Like many parasitic diseases, the malaria parasite (*Plasmodium falciparum, vivax, malariae,* or *ovale*) must spend part of its life cycle within an intermediate host organism—mosquitoes of the genus *Anopheles.* This insect must bite humans for the disease to be spread. The type of vegetation in an area determines whether or not the mosquito, and the parasite it carries, will thrive. Unfortunately, agriculture often ushers in malaria by replacing dense forests with damp rice fields that are a haven for the mosquitoes.

However, some plant products can kill the malaria parasite. In the sixteenth century, natives of Peru gave Jesuit missionaries who were on their way to Europe their secret malaria remedy—the bark of the cinchona tree. It was not until 1834, though, that French chemist Pierre Joseph Pelletier extracted the active ingredient from cinchona bark, which was called quinine. This substance reigned as the standard treatment for malaria until the 1930s, when chemists began to synthesize substitutes because the malaria parasites were developing resistance to the bark-derived treatment. Malaria spread through the developing nations as more and more land was cleared for farming. New natural antimalarial drugs were sought, and one very promising one has come from ancient Chinese folk medicine.

In 1967, researchers in the People's Republic of China began a systematic study of all plants known to have medicinal properties in search of a new drug to fight malaria. In a document entitled "Recipes for 52 Kinds of Diseases" unearthed from a Mawangdui Han dynasty tomb from 168 B.C., a plant called qinghao was described as a treatment for hemorrhoids. A reference from A.D. 340 cited the same plants (also known as *Artemisia annua* or sweet wormwood, a relative of tarragon and sagebrush) as a treatment for fever. In 1596, a Chinese herbalist prescribed qinghao to combat malaria.

In the 1970s, chemists isolated the active ingredient from sweet wormwood and called it artemisinin (fig. 28.15). By 1979 it had been tested on more than 2,000 malaria patients, in whom it cured the fever in 72 hours and rid the blood of parasites within 120 hours. The drug is more than 90% effective in treating cerebral malaria, the most severe form of the disease. In animal tests, artemisinin is proving effective against other parasitic diseases as well, including the worm infection schistosomiasis.

Plants are essential to our comforts and to our very existence. Nature has certainly provided a bountiful harvest from which we can choose. The following chapters probe into the structures and functions of plants and conclude with new types of plants made possible by biotechnology.

KEY CONCEPTS

Natural products chemists explore biochemicals with healing properties, many of which come from plants. This is an ancient art as well as a modern science. Many drugs are derived from plant chemicals or intentionally resemble them. Plants both contribute to the spread of malaria and provide a treatment.

SUMMARY

Plants help provide us with food, clothing, shelter, and medicine. About 12,000 to 10,000 years ago, groups of people gradually changed from a *hunter-gatherer* life-style to an *agricultural* way of life, intentionally saving and planting seeds from the strongest individual plants of crops that could be used as food. Encouraging the propagation of certain individuals *artificially selected* particular traits, and in this way, humans have influenced plant evolution.

Cereals are members of the grass family, whose edible seeds (grains) can be stored for long periods. The *endosperm* is the starchy food supply in the seed, the *germ* is the embryo, and outer protective layers are the *aleurone* and *pericarp*. Modern wheats contain two, four, or six sets of chromosomes, and they probably evolved from a natural cross between ancient Einkorn wheat (a diploid) and a wild grass (also a diploid), followed by chromosome doubling in some germ cells (nondisjunction), about 8,000 years ago. This event yielded tetraploid Emmer wheat. Emmer wheat too crossed with a wild grass to produce the first hexaploid wheats, which today are used to manufacture bread and other baked goods.

Corn probably evolved from an ancient grass relative, *teosinte*. Corn and teosinte each have 20 chromosomes, they sometimes grow in the same fields, and they can crossbreed. About 7,500 years ago, an environmental stress may have selected teosinte plants with large kernels, and early farmers may then have cultivated these individual plants. Breeding experiments by Charles Darwin and later by George Shull led to the development of *hybrid corn*, which results from crosses between separate inbred lines to yield bountiful crops.

Rice is perhaps the most widely consumed modern cereal crop, and fossils of the plants date back some 130 million years. Today, rice grows in a wide range of environments. In 1961, a rice *seed bank* was started in the Philippines to preserve different varieties and prevent reliance on a few types. Seed and pollen banks have since been founded for many other valuable plant species. Amaranth is a new crop. Its abundant seeds are highly nutritious.

Plants have been used for their medicinal properties probably since the beginning of human existence. Today, *natural product chemists* use a combination of laboratory techniques and information from folklore and herbal medicine to synthesize compounds with similar activities to plant-derived compounds, in the search for new and more effective drugs. Malaria is one disease whose spread has been greatly influenced by plants. The clearing of land for agricultural purposes in many tropical regions has encouraged the spread of the malaria-carrying *Anopheles* mosquito. On the other hand, several plant products have been used to treat the fever of malaria.

QUESTIONS

1. What is a cereal?
2. What is agriculture?
3. How might agriculture have arisen?
4. Describe the ways in which scientists hypothesize that modern varieties of wheat, corn, and rice arose.
5. Why is reliance on only a few varieties of a crop plant dangerous? Cite two examples of when such reliance led to devastation.
6. In what two ways can biochemicals from plants help treat human disease?

TO THINK ABOUT

1. Artificial selection has sculpted a modern corn plant that has tasty, nutritious kernels. However, modern corn cannot reproduce successfully without the help of people. The kernels are so tightly protected within the ears that a human or animal must disperse the kernels to start the next year's crop. Do you think that this intervention with a natural process is justified? Why or why not?

2. Devise an experiment to demonstrate that the scenario for the evolution of bread wheat from tetraploid and diploid relatives breeding with wild grasses is possible.

3. Devise an experiment to demonstrate that the scenario of corn evolving from teosinte is possible.

4. Would you eat, or attempt to grow, amaranth? Why or why not?

5. You are planning your vegetable garden and intend to purchase five packets of carrot seeds. The seed catalog describes a new variety of carrot that is resistant to nearly every known garden pest and produces long, highly nutritious carrots in a variety of climates. It sounds too good to be true, but the company that publishes the catalog has had a lot of experience in plant breeding. If you want to obtain as many carrots as possible, would you be better off buying five packets of the new variety or five different types of carrot seeds? Give a reason for your answer.

6. In an effort to conserve and perhaps increase genetic diversity of crop plants, many nations are cooperating by donating seeds to plant gene banks. Sometimes, however, a developing nation will donate to a bank and that genetic material is then used in more industrialized countries to breed or engineer new plant varieties. The new plants—based on the genetic material from the poor nation but the technology of the more wealthy nation—are then sold to the developing nation. Can you think of a more equitable way for nations to cooperate in the development of new plant varieties?

7. Many people claim that certain herbs have specific healing powers. How would you design a scientifically sound experiment to test whether or not a particular plant product alleviates symptoms of a specific human ailment?

SUGGESTED READINGS

Brown, William L. November 1984. Hybrid corn. *Science 84*. The invention of hybrid corn revolutionized the crop worldwide.

Boucher, Douglas H. February 1991. Cocaine and the coca plant. *BioScience*. The coca plants *Erythroxylon coca* and *E. novogaratense* are rich in alkaloids, which have been used, medicinally and recreationally, for centuries.

Doebley, John. June 1990. Molecular evidence for gene flow among *Zea mays*. *BioScience*. Corn evolved from teosinte: now genes engineered into corn are being transferred to teosinte.

Feldman, Moshe and Ernest R. Sears. February 1981. The wild gene resources of wheat. *Scientific American*. Modern crops arose from wild plants. we may have to return to wild varieties to regain vigor.

Kubo, Isao. April 1985. The sometimes dangerous search for plant chemicals. *Industrial Chemical News*. Natural product chemists can sometimes be found in the laboratory and are called modern-day "medicine men."

Plucknett, Donald L., and Nigel J.H. Smith. January 1989. Quarantine and the exchange of crop genetic diseases. *BioScience*. Saving seeds today may be vital to our future.

Shulman, Seth. November 1986. Seeds of controversy. *BioScience*. Who owns and should have access to stored plant genetic material?

Tucker, Jonathan B. January 1986. Amaranth: The once and future crop. *BioScience*. Amaranth is an ancient crop with much future promise.

Vaughan, Duncan A. and Lesley A. Sitch. January 1991. Gene flow from the jungle to farmers. *BioScience*. How can we best use wild rice genes?

CHAPTER

29

Plant Form and Function

Chapter Outline

Learning Objectives

By the chapter's end, you should be able to answer these questions:

1. What are the main tissues and structures of the plant body?
2. What functions are provided by a plant's tissues and structures?
3. How does a plant transport water and nutrients?
4. How does a plant elongate?
5. How does a plant increase in girth?

A summer garden offers a spectacular display of plant life. Towering and flowering green bean plants curl about each other, as the tiniest tomatoes peek out from where vibrant yellow flowers stood a few days earlier. Melon and cucumber vines snake along the ground, and cabbages are just starting to build their tight knots of leaves. Feathery green tufts in one corner herald growing carrots beneath the surface, as robust leafy plants indicate potatoes forming below. In the evergreens that ring the garden, birds gather, eager to attack the succulent ears of corn wrapped in their tight leafy jackets.

Although these plants look very different from one another, each is built of similar cells, tissues, and organs. These structural building blocks, like those of other types of organisms, are evolutionary adaptations that provide such basic requirements as protection, structural support, transport and storage of water and nutrients, and waste removal. This chapter explores how plant cells and tissues interact to carry out the activities of life.

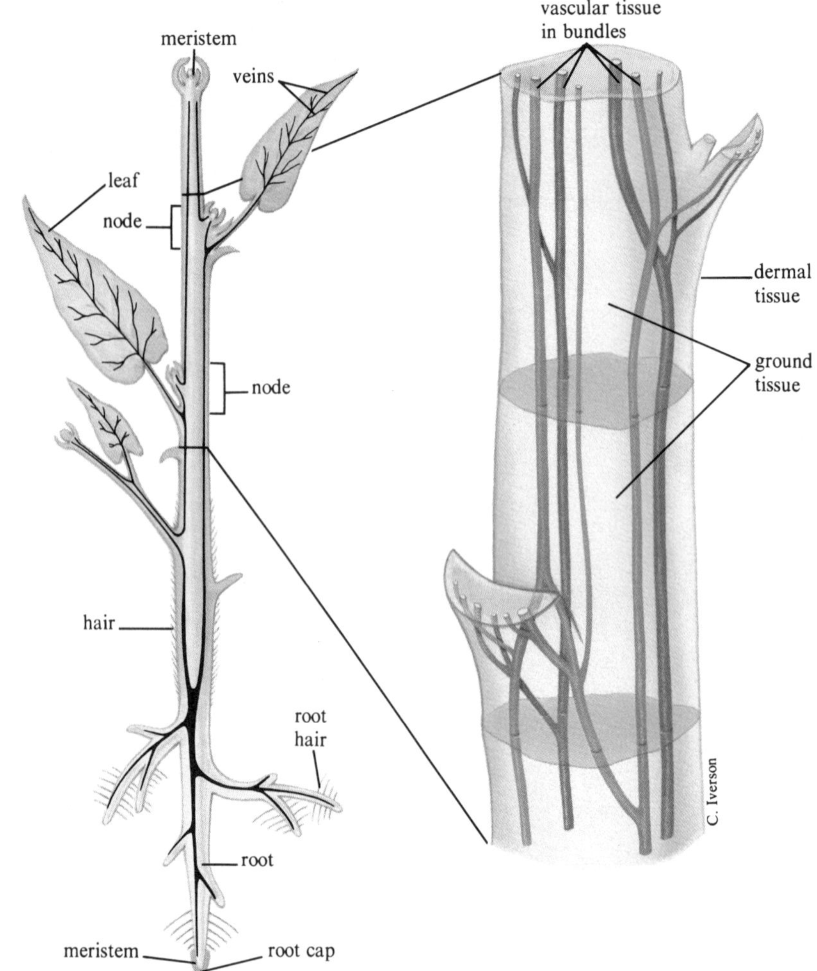

Figure 29.1
Parts and tissues of higher plants.

Primary Tissues

The **primary body** of a plant is an axis consisting of a root (usually below ground) and a shoot (above ground). Shoot and root systems are extensive in many plants and increase available surface area on which the chemical reactions of life can occur. Shoots and roots support one another. Shoots, through photosynthesis, provide carbohydrate nourishment to the roots below. Roots gather water and minerals, which are sent to the shoots.

The **primary tissues** of the plant body, like tissues in general, are groups of cells with a common function. Four tissues comprise the primary plant body: meristems, ground tissue, dermal tissue, and vascular tissue (fig. 29.1). Each of these is made of distinctive cell types and provides a specific function.

Meristems

Meristems are localized regions of cell division, consisting of undifferentiated plant tissue from which new cells arise. They are the ultimate source of all the cells of a plant, although some cell types descended from meristems can themselves divide. The cells farthest away from meristems are the most mature and differentiated, and they may be many times larger than the meristematic cells from which they arose. Meristems function throughout a plant's life, and because of them, plants never stop growing.

Apical meristems are found near the tips of roots and shoots. Cells in these regions are small and unspecialized. When these meristematic cells divide and elongate, the plant lengthens. This extension is called **primary growth.** Apical meristems give rise to three primary meristems, which form the other three primary tissues.

Lateral meristems grow outward, thickening the plant, which is called **secondary growth.** Wood formation is due to secondary growth. Unlike primary growth, secondary growth does not occur in all plants.

Intercalary meristem is an unusual type of dividing tissue found in grasses between mature regions of stem. If the tip of a stem or leaf is torn off—by a lawnmower or hungry cow, for example—intercalary meristems reform the structure of the plant. This is why grass blades grow back so quickly after a lawn is mowed.

Ground Tissue

Ground tissue comprises most of the primary body of a plant, filling much of the interior of roots, stems, and leaves. Cells of ground tissue have many functions, including storage, support, and basic metabolism. Ground tissue consists of three cell types: parenchyma, collenchyma, and sclerenchyma.

Parenchyma cells are the most abundant cells in plants. They are relatively unspecialized and often lack distinctive features. More specialized cell types probably evolved from these versatile parenchyma cells. At maturity, parenchyma cells are alive and capable of dividing. This is an important characteristic, because it enables them to specialize in response to a plant's changing situation. For example, parenchyma cells can help a plant respond to wounding or adapt to a new environmental condition.

The substances that give plants their familiar edible parts, such as the carbohydrates in a potato, an ear of corn, or a coffee bean, are stored in parenchyma cells. Fragrant oils, salts, pigments, and organic acids are also stored in parenchyma cells. Oranges and lemons, for example, store citric acid, which gives them their tart taste. Parenchyma cells are also important sites of life-support functions such as photosynthesis, cellular respiration, and protein synthesis. **Chlorenchyma** are chloroplast-containing parenchyma cells that take part in photosynthesis, imparting to leaves their green color (fig. 29.2).

Collenchyma cells are elongated, living cells that differentiate from parenchyma and support the growing regions of shoots. Collenchyma cells have unevenly thickened primary (outer) cell walls that can stretch, enabling the cells to elongate. As a result, they provide support without interfering with the growth of young stems or expanding leaves. Collenchyma strands are the "strings" in celery that often lodge between your teeth.

Sclerenchyma cells are long with thick, nonstretchable secondary cell walls (a trilayered structure beneath the outer cell wall). These cells support regions of plants that are no longer growing and are usually dead at maturity. Two types of sclerenchyma form from parenchyma: sclereids and fibers.

Sclereids have many shapes and occur singly or in small groups. The gritty texture of a pear is due to small groups of sclereids. Occasionally, sclereids form hard layers, such as the hulls of peanuts. Fibers are elongated cells that usually occur in strands varying in length from a few to a few hundred millimeters. Many sclerenchyma

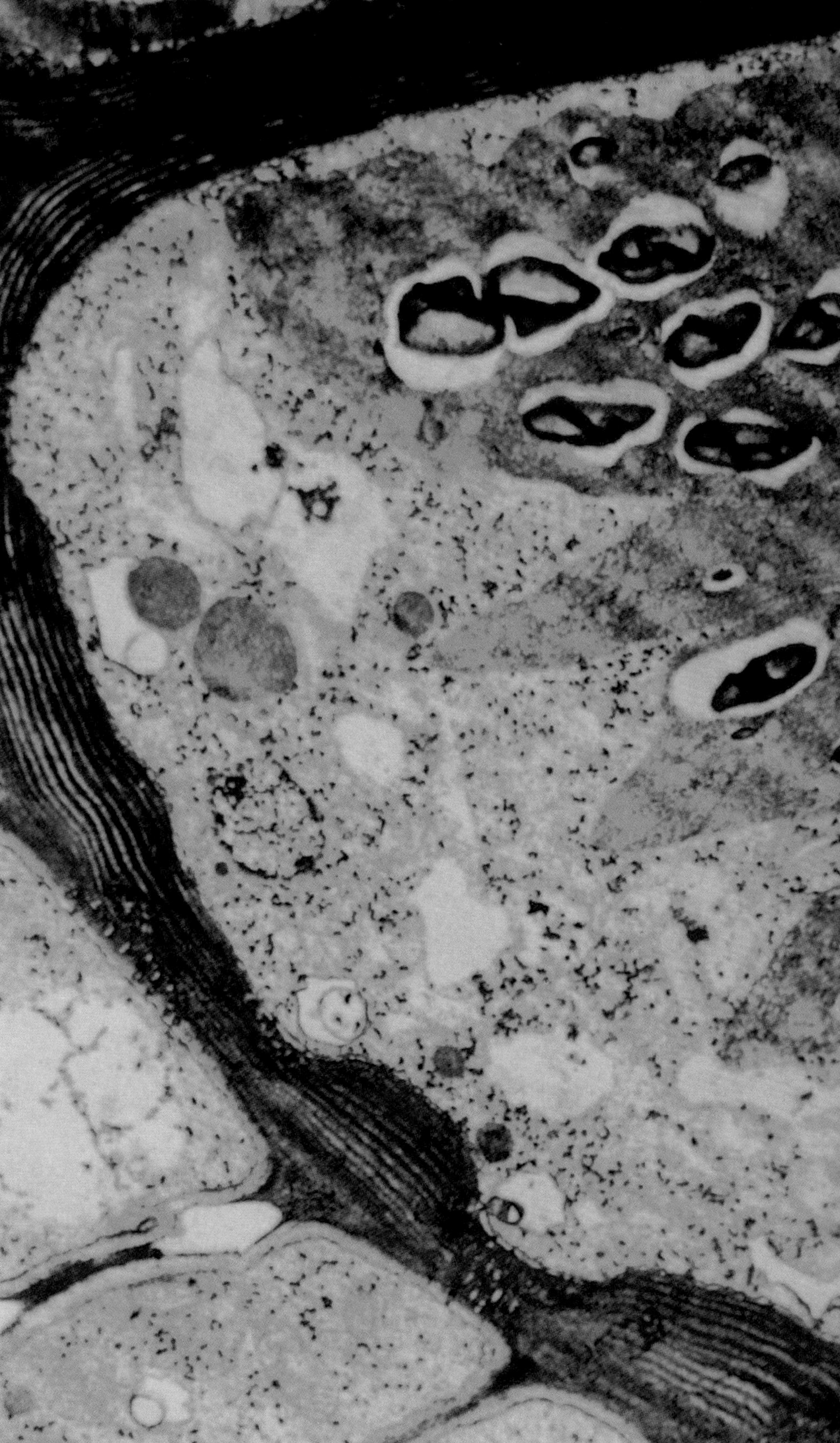

Figure 29.2
Chlorenchyma in a grass leaf.

fibers are used to produce textiles. Humans currently cultivate more than 40 families of plants for fibers and have fashioned cords from fibers since 8000 B.C. The fibers of *Agave sisalana*, commonly known as sisal or the century plant, are used to make brooms, brushes, and twines (see fig. 31.13). Linen comes from the fibers of *Linum usitassimum*, or flax. Even the notorious *Cannabis sativa*, from which marijuana comes, is used to make twine and rope.

KEY CONCEPTS

Plants face similar life-support challenges as animals and meet them through organizations of cells and tissues. Four primary tissues form the root and shoot of a plant—meristems, ground tissue, dermal tissue, and vascular tissue. Meristems are localized regions of mitosis. Apical meristems near root and shoot tips provide primary growth, and lateral meristems thicken the plant (secondary growth). Most of the plant's primary body is ground tissue, which has three cell types. Parenchyma cells are abundant, capable of division, and have diverse functions such as response to stress, nutrient storage, and metabolism. Collenchyma cells derive from parenchyma and support growing parts of shoots because they can stretch. Sclerenchyma cells are more rigid and support mature parts of plants. Sclerenchyma cells form sclereids (groups of cells) and fibers.

Dermal Tissue

Dermal tissue covers the plant body. The **epidermis** covers the primary plant body and is usually only one cell thick. Epidermal cells are flat, transparent, and tightly packed. Special features of the epidermis such as the cuticle, stomata, and trichomes provide diverse functions, including protection, gas exchange, secretion, and digestion.

The **cuticle** is a covering over all but the roots of a plant that protects and keeps out water. The cuticle consists primarily of **cutin,** a fatty material produced by epidermal cells. This covering prevents desiccation by retaining water. As a result, the plant can maintain a watery internal environment—a prerequisite to survival on dry land. Also, the cuticle and underlying epidermal layer act as a plant's first line of defense against predators and infectious agents. In many plants a smooth, whitish layer of wax covers the cuticle and when thick can be seen on leaves and fruits. The layer on the undersides of wax palm leaves may be more than 5 millimeters thick. It is harvested and used to manufacture polishes, record albums, and lipstick.

Since the tightly packed epidermal cells are covered with an impermeable cuticle, how do plants exchange water and gases with the atmosphere? This problem is solved with pores called **stomata. Guard cells** control the opening and closing of the pore, regulating gas and water exchange. The movements of stomata regulates the amount of carbon dioxide diffusing into a leaf for photosynthesis, and loss of water from leaves due to evaporation (transpiration). Stomata may be very numerous. The underside of a black oak leaf, for example, has 100,000 or so stomata per square centimeter! Because stomata help plants conserve water, they are an essential adaptation for life on land (fig. 29.3).

Trichomes are outgrowths of epidermal cells found in almost all plants. Single celled or multicellular, trichomes deter predators in interesting ways. Hook-shaped trichomes may impale marauding animals. Other times, predators inadvertently break off the tips of trichomes that release a sticky substance that traps the invading animals. Trichomes of stinging nettle have spherical tips that break off and penetrate a predator's body, where the poisonous contents are injected into the wounds. Trichomes of carnivorous plants such as the Venus's-flytrap secrete enzymes to digest trapped animals. These trichomes then absorb the digested prey.

Trichomes often reflect light, which helps to prevent overheating. Since cooler plants lose less water via evaporation, trichomes prevent excessive water loss. **Root hairs** are trichomes that appear near root tips and absorb water and minerals from soil.

Many trichomes are economically important. Cotton fibers, for example, are trichomes produced by the epidermis of cotton seeds. Menthol comes from peppermint trichomes, and hashish, a powerful narcotic, is purified resin from *Cannabis* trichomes.

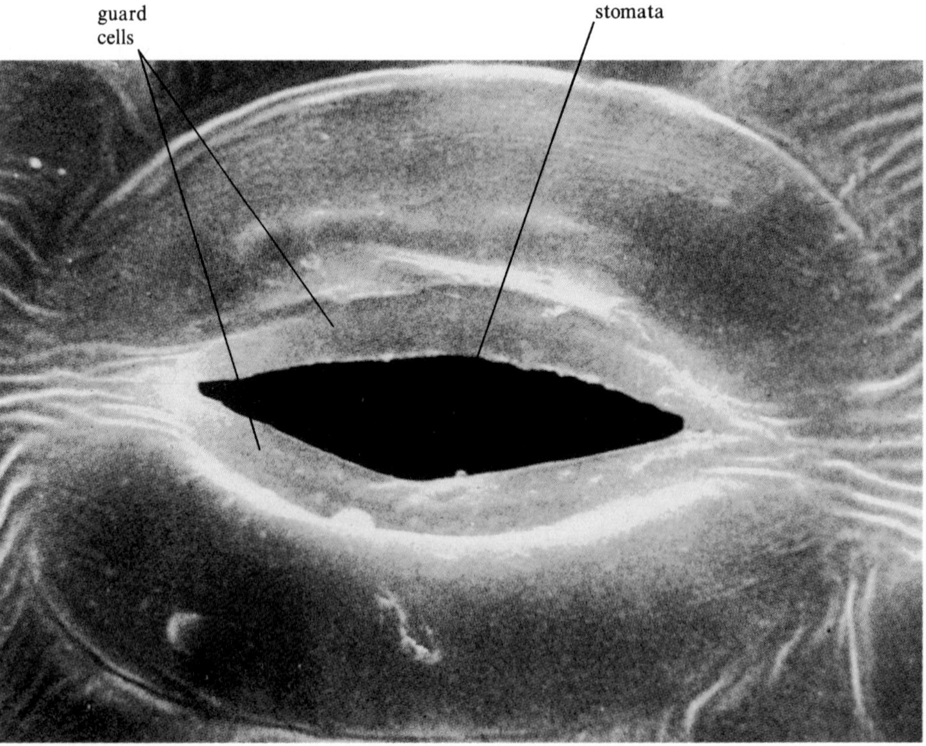

Figure 29.3
Stomata are found on leaf hairs of *Mascagnia macroptera*. Guard cells hug the stomatal pore and closely packed epidermal cells lie under the cuticle.

Vascular Tissues

Vascular tissues are specialized conducting tissues, and form the veins in leaves. Vascular tissues transport water, food, and other materials to all parts of a plant. Two kinds of vascular tissues occur in plants: xylem and phloem. Each is formed during primary and secondary growth.

Xylem forms a continuous system that transports water with dissolved nutrients from roots to all parts of a plant, only in an upward direction. The two kinds of conducting cells in xylem are called **tracheids** and **vessel elements.** Both are elongate, are dead at maturity, and have thick walls. Thick walls are essential for these conducting cells, because water is pulled through plants under negative pressure (suction). Without thick cell walls, tracheids and vessel elements might collapse.

Tracheids are the least specialized conducting cells. These cells are long and narrow, and their tapered ends overlap one another. Water moves from tracheid to tracheid through thin areas in cell walls called pits. Vessel elements are more specialized cells that evolved from tracheids. Unlike tracheids, vessel elements are short, wide cells shaped like barrels. Vessel elements are stacked end to end, and their end walls usually dissolve, forming hollow tubules, or vessels, for efficient transport of water (fig. 29.4).

Phloem transports water and food materials, primarily dissolved sugars, throughout a plant. Unlike xylem, which transports water upward under negative pressure (suction), phloem transports substances under positive pressure, which is like blowing up a balloon. Thus, water and dissolved sugars can move through phloem in all directions. Also unlike xylem, the conducting cells of phloem are alive at maturity. Their cell walls have thin areas perforated by many **sieve pores,** through which solutes move from cell to cell. Phloem has two kinds of conducting cells: sieve cells and sieve tube members.

Sieve cells are the more primitive (that is, less specialized) conducting cells in phloem. They are elongate cells with tapered, overlapping ends. Sieve pores permeate all walls of sieve cells. **Sieve tube members** are more complex, shorter, and wider than sieve cells. Sieve tube members are arranged end to end to form long sieve tubes. The pores of individual cells are concentrated on contacting end walls, forming an area called a **sieve plate.** This organization allows faster, more efficient transport of nutrients (fig. 29.5). Mature sieve tube members lack nuclei but contain living protoplasm—unique among plant cells. Can you think of a functioning type of animal cell that also lacks a nucleus? Table 29.1 reviews the major cell types in plants.

KEY CONCEPTS

The epidermis is a one-cell-thick layer that covers the plant and is itself covered by a cuticle. These coverings help to retain water and keep out invaders. Stomata are pores that allow exchange of water and gases with the atmosphere. Guard cells regulate the opening and closing of the pores. Trichomes are epidermal projections that provide protection, reflect light, and increase absorptive surface area. Xylem is a vascular tissue that transports water and dissolved nutrients upwards. Tracheids are less specialized xylem and vessel elements are more specialized. Tracheid cells overlap, and vessel elements form hollow tubules. Phloem transports water and nutrients in all directions. Phloem includes the more primitive sieve cells and the more complex and organized sieve tube members. Xylem and phloem form during primary and secondary growth.

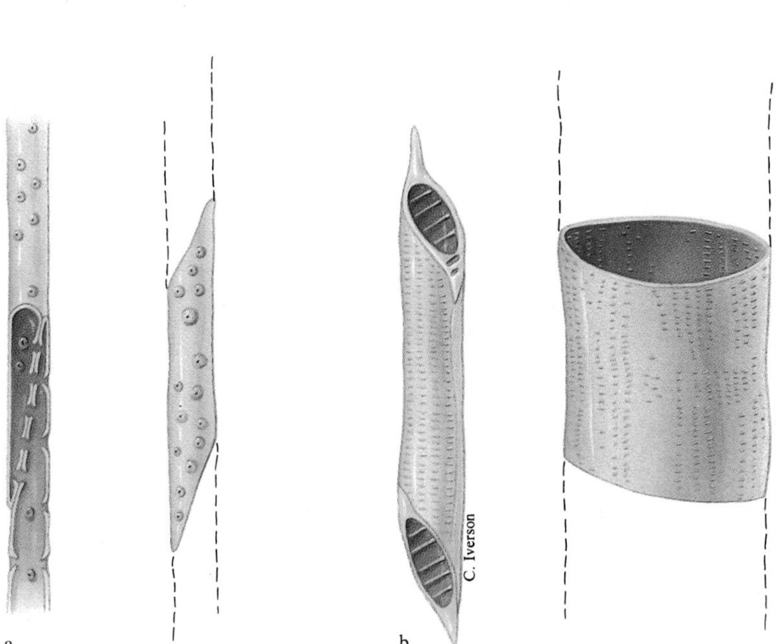

Figure 29.4
Xylem: (*a*) tracheids and (*b*) vessel elements.

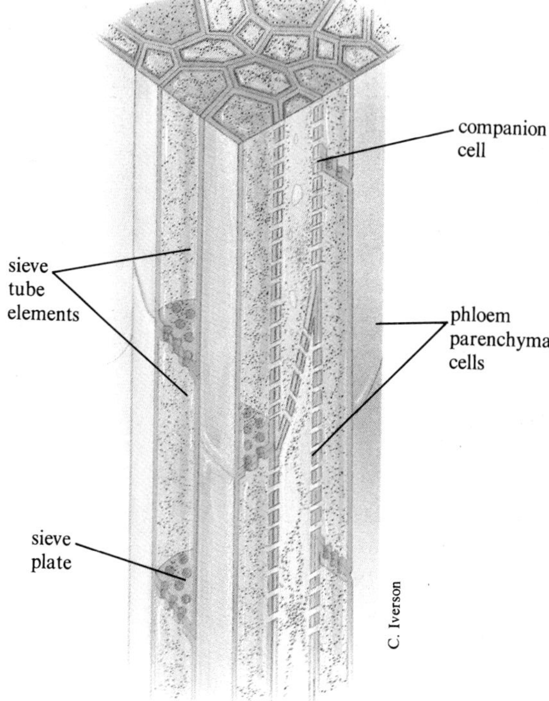

Figure 29.5
A longitudinal view of phloem in a tobacco stem.

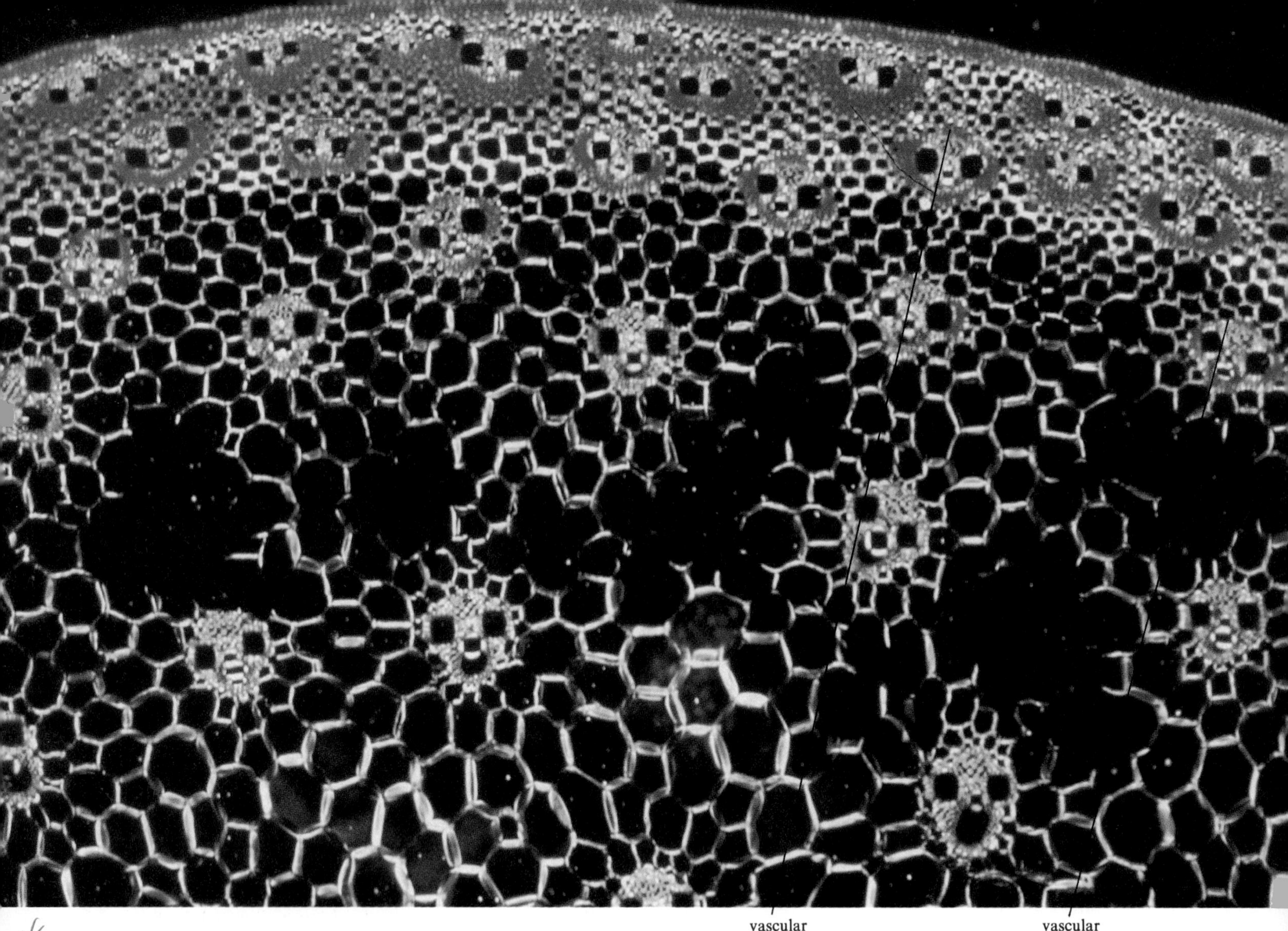

Figure 29.6
Cross sections of (*a*) monocot (*Zea mays*) and (*b*) dicot (*Helianthus*) stems. Notice the scattered vascular bundles in the monocot and the ring of vascular bundles in the dicot.

Table 29.1
Major Cell Types in Plants

Cell	*Tissue Type*	*Function*
Parenchyma	Ground	Storage, division, metabolism
Collenchyma	Ground	Support
Sclerenchyma	Ground	Support
Tracheids	Vascular—xylem	Water transport
Vessel elements	Vascular—xylem	Water transport
Sieve cells	Vascular—phloem	Water and nutrient transport
Sieve tube members	Vascular—phloem	Water and nutrient transport

Parts of the Plant Body

The stems, leaves, and roots of a plant are all built of the same meristematic, dermal, ground, and vascular tissues. Yet plants exhibit a wide variety of forms—contrast a prickly cactus to a lush vine to the majestic sunflower. The different arrangements of plant tissues are sculpted by the forces of evolution, because certain forms are better adapted to particular environments than others.

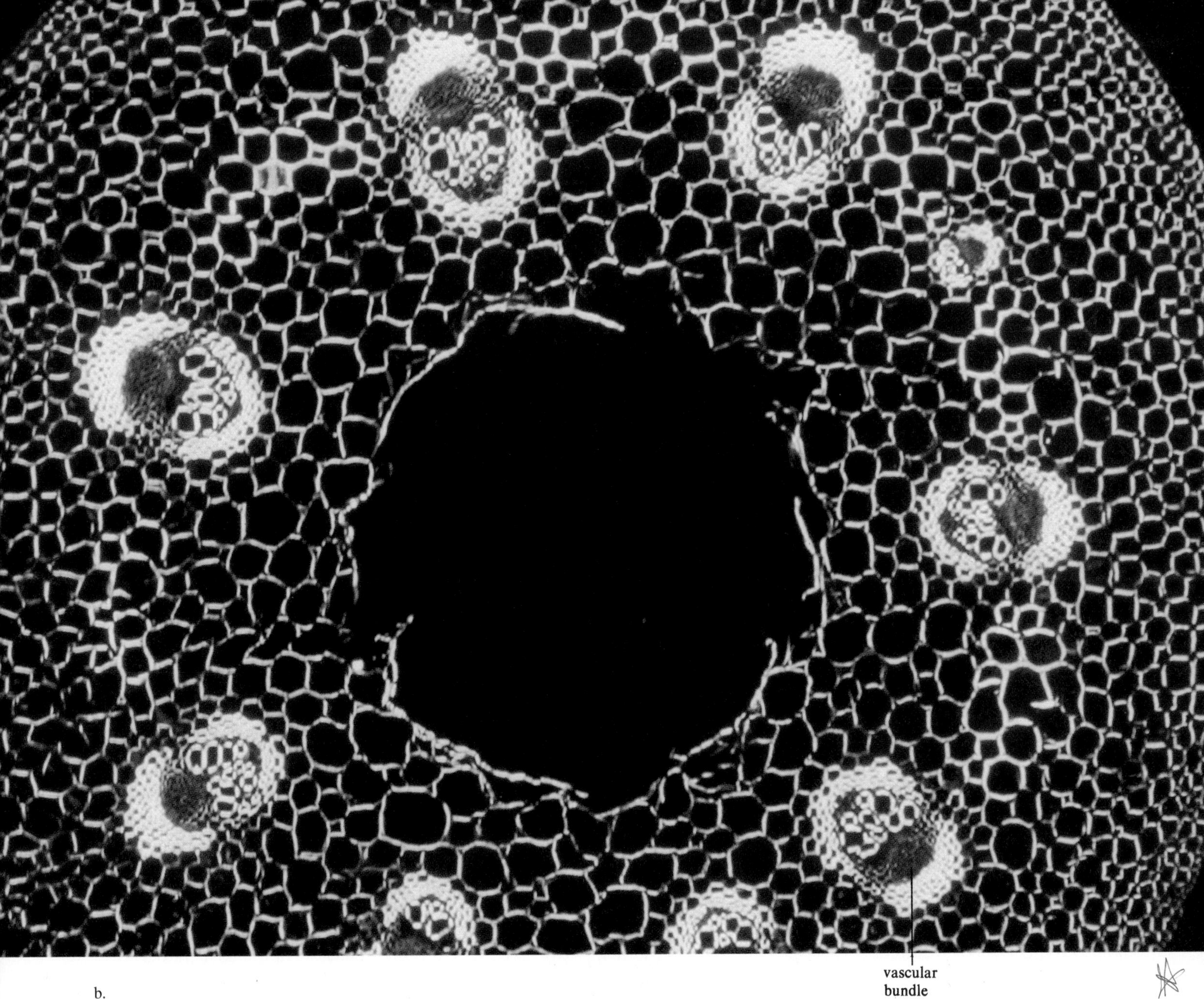

b.

Stems

The central axis of a shoot system is the stem, which is a collection of nodes and internodes. **Nodes** are areas of leaf attachment, and **internodes** are portions of stem between nodes (see fig. 29.1). The region between a leaf stalk and stem is a leaf **axil.** Axillary buds are undeveloped shoots that form in leaf axils. Although axillary buds may elongate to form a branch or flower, many remain small and dormant.

Normally, stem elongation occurs in the internodal regions. However, some plants have stems called **rosettes** that do not elongate. Rosettes have short internodes and overlapping leaves. A banana tree is a rosette—its "trunk" is made of large, tightly packed leaves.

The epidermis surrounding a stem is a transparent, unicellular layer. It contains stomata—fewer than the epidermis of a leaf, however. The epidermis of a stem also may possess protective trichomes.

Vascular tissues in stems are organized into **vascular bundles,** which branch into leaves at nodes. Food-conducting phloem forms on the outside of a bundle, whereas water-conducting xylem forms on the inside of a bundle. Often, thick-walled sclerenchyma fibers are associated with vascular bundles. These fibers strengthen the vascular tissue.

Vascular bundles are arranged differently in different types of plants. Consider the familiar flowering plants, which are divided into two classes: monocotyledons (**monocots** for short), which have one first, or "seed," leaf; and dicotyledons **(dicots),** which have two seed leaves. Monocots such as corn have vascular bundles scattered throughout their ground tissue, whereas dicots such as sunflower have a single ring of vascular bundles (fig. 29.6). In contrast to flowering plants, pines have an outer

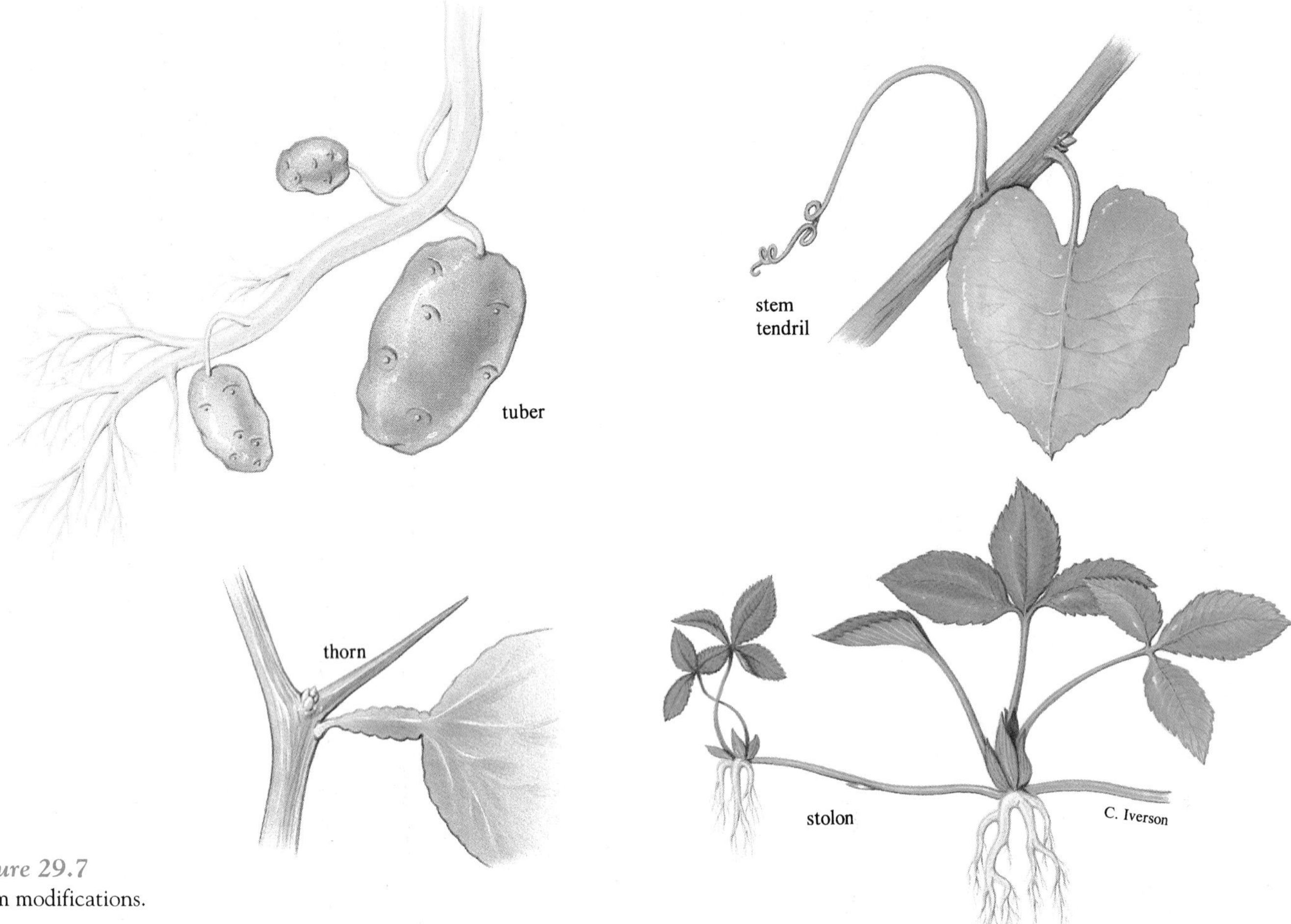

Figure 29.7
Stem modifications.

cylinder of phloem surrounding an inner cylinder of xylem. In all plants, a single layer of cells between the xylem and phloem remains meristematic. In dicots and pines, this layer becomes a lateral meristem.

The ground tissue that fills the area between the epidermis and vascular tissue in stems is called the **cortex.** Although a few collenchyma strands may help support the cortex, most cortical cells are parenchyma. Some cortical cells are green and photosynthetic and store starch. In plants having concentric cylinders of xylem and phloem—pine, for example—the ground storage tissue in the center of the stem is called **pith.**

Stems support leaves, produce and store food, and transport food and water between roots and leaves. Asparagus is an example of an edible stem. Many plants modify their stems for special functions such as reproduction, climbing, protection, and storage (fig. 29.7):

Stolons, or runners, are stems that grow along the soil surface. New plants form from their nodes. Strawberry plants have stolons after they flower, and several plants can arise from the original one.

Thorns often are stems modified for protection, such as thorns on a rosebush.

Succulent stems of plants such as cacti are fleshy and store large amounts of water.

Tendrils are shoots that support plants by coiling around objects, sometimes attaching by their adhesive tips. The stem tendrils of green bean plants readily wrap themselves around posts put in the ground by a helpful gardener. If the garden is planted too densely, the green bean plants will wrap around its neighboring plants. It can form a new anchorage in just minutes!

Tubers are swollen regions of stems that store nutrients. Potatoes are tubers produced on burrowing stolons.

KEY CONCEPTS

The stem forms the central axis of the plant body. Leaves attach at nodes. Elongation occurs at internodes. Rosettes are nonelongated stems. Stem vascular bundles have phloem on the outside and xylem on the inside. Bundles are scattered in monocots and form a ring in dicots. Between the epidermis and vascular tissue of stems lies ground tissue called the cortex. In general, stems provide support, food storage, and transport of nutrients and water, but they can be specialized for reproduction, climbing, protection, and storage.

a.

b.

c.

Figure 29.8
The diversity of leaves: (*a*) the leaves of *Costus*, on a spiral stem; (*b*) the succulent leaves of a stone plant (*Lithops*); and (*c*) *Setaria*, an African grass. Magnification, x10.

Leaves

In addition to the stem, a shoot system consists of leaves. Leaves are the primary photosynthetic organs of most plants. Like stems, leaves are made of epidermal, vascular, and ground tissues. Leaves are the most diverse of all plant organs. The fronds of tropical palms may be 65 feet (20 meters) long, whereas the leaves of *Wolffia*, an aquatic plant, are no larger than a pinhead. A mature American elm has several million leaves, whereas a desert plant called *Welwitschia mirabilis* produces only two leaves during its entire lifetime. Leaves may be needlelike, feathery, waxy, or smooth. The leaves on no two species are identical. Botanists classify leaves according to their basic forms as well as their arrangements on stems (fig. 29.8).

Most leaves consist of a flattened **blade** and a supporting, stalklike **petiole.** There are four basic kinds of leaves. *Simple* leaves have flat, undivided blades. The blades of *compound* leaves are divided into leaflets. *Pinnate* leaflets are paired along a central line, whereas *palmate* leaflets are all attached to one point at the top of the petiole.

Leaves can be arranged in different patterns on stems, which maximizes sun exposure. Plants with one leaf per node have *alternate*, or spiral, arrangements. Plants with two leaves per node have *opposite* arrangements, and plants with three or more leaves per node have *whorled* arrangements (fig. 29.9).

The leaf epidermis consists of tightly packed transparent cells. The epidermis is usually nonphotosynthetic and has many stomata—more than 11 million in a cabbage leaf, for example. Although plants use stomata for gas exchange, they also lose

Figure 29.9
Botanists use a dizzying variety of terms to describe leaves: (*a*) a palmately compound leaf of buckeye; (*b*) pinnately compound leaf of black walnut; (*c*) a tulip tree's simple alternate leaves; (*d*) simple leaves of dogwood; (*e*) palmately veined leaf of maple; (*f*) succulent leaves of string-of-pearls; (*g*) pinnately veined oak leaf; (*h*) leaf of grass; (*i*) whorled leaves of bedstraw; (*j*) linear leaves of yew; (*k*) fan-shaped leaf of ginkgo.

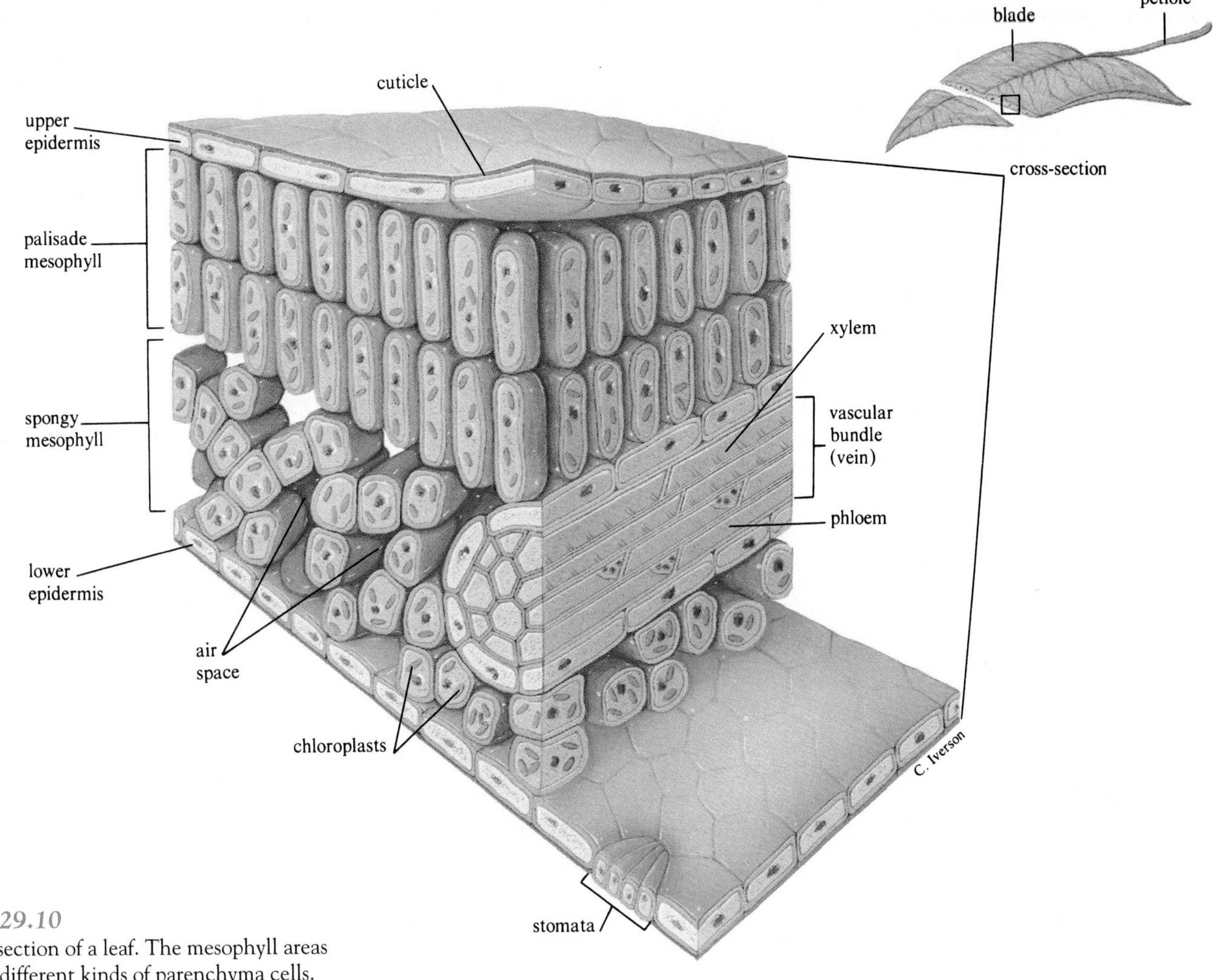

Figure 29.10
A cross section of a leaf. The mesophyll areas contain different kinds of parenchyma cells.

large amounts of water through stomata in transpiration. In fact, in Amazon rain forests, transpiration accounts for half of the moisture in rainfall. However, this water loss is minimized because stomata are concentrated on the protected lower side of leaves. How do you think stomata are distributed on vertically oriented leaves?

Vascular tissues in leaves occur in strands called **veins.** Xylem forms on the upper side of a vein, and phloem forms on the lower side. Leaf veins often are associated with thick-walled sclerenchyma fibers for support. Also, a layer of parenchyma cells surrounds and supports leaf veins. Leaf veins may be *net,* with minor veins branching off from larger, prominent midveins, or *parallel,* with several major parallel veins connected by smaller minor veins. Most dicots have netted veins, and many monocots have parallel veins. *Vein endings* are the blind ends of minor veins, where water and solutes move into and out of veins.

The ground tissue of leaves is called mesophyll, which is parenchyma. Here chlorenchyma cells produce sugars via photosynthesis. Horizontally oriented leaves have two types of chlorenchyma. The long, columnar cells along the upper side of a leaf are called **palisade mesophyll** cells, and they are specialized for light absorption. Located below the palisade layer are **spongy mesophyll** cells, which are irregularly shaped chlorenchyma cells separated by large intercellular spaces. These cells are specialized for gas exchange. In contrast to horizontally oriented leaves, vertically oriented leaves have uniformly shaped chlorenchyma cells and are called uniform mesophyll cells (fig. 29.10).

In addition to their photosynthetic role, leaves provide support, protection, and nutrient procurement and storage with the following specializations:

Tendrils are modified leaves that wrap around nearby objects to support climbing plants. Pea plants growing in a garden will hold to a fence with leaf tendrils. (Both leaves and stems can be modified into tendrils.)

Spines of plants such as cacti are leaves modified to protect plants from predators and excessive sunlight.

Bracts are floral leaves that protect developing flowers. They are colorful in some plants, such as poinsettia.

Figure 29.11
Leaf modification for trapping insects: the leaves of the Venus's-flytrap snap shut to catch prey. This fascinating plant is found only in North and South Carolina.

Storage leaves are fleshy and store food. Onion bulbs consist of the bases of such leaves.

Insect-trapping leaves are found in about 200 types of carnivorous plants and are adapted for attracting, capturing, and digesting prey. Some leaves have sticky "flypaper" surfaces, whereas others form water-filled chambers in which insects drown. Trigger hairs of Venus's-flytrap respond to movements of a visiting insect by stimulating the two leaves to snap together. The insect is trapped, as the leaves secrete digestive enzymes that destroy it (fig. 29.12).

Cotyledons are embryonic leaves found in flowering plants that often store energy used for germination.

Anyone who has ever raked leaves is well aware that leaves have a limited life span. **Leaf abscission** is the normal process by which a plant sheds its leaves. **Deciduous trees** shed their leaves at the end of a growing season. Evergreens have leaves throughout the year but gradually shed a few leaves at a time.

Leaves are shed from an **abscission zone,** a region at the base of the petiole. In response to environmental cues such as shortening days or climatic changes, a separation layer forms in the abscission zone, isolating the dying leaf from the stem. Eventually, wind, rain or some other disturbance, such as a scurrying squirrel, breaks the dead leaf from the stem.

KEY CONCEPTS

Leaves are photosynthetic organs. They are very diverse and are classified by blade structure, attachment pattern, organization around nodes, and leaf venation. Leaf ground tissue, or mesophyll, is specialized as palisade, spongy, or uniform cells. Like stems, leaves are modified into a variety of structures.

Roots

Although plants are immobile, much biological activity takes place underground. Roots are so indispensable to plant growth and photosynthesis that annual production of roots often consumes more than half of a plant's energy and may account for

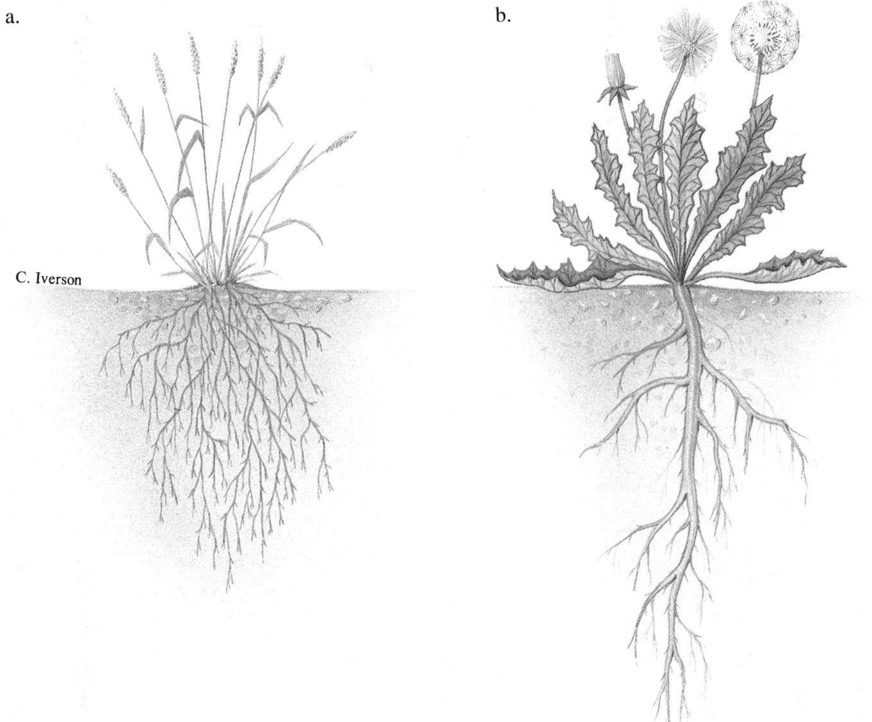

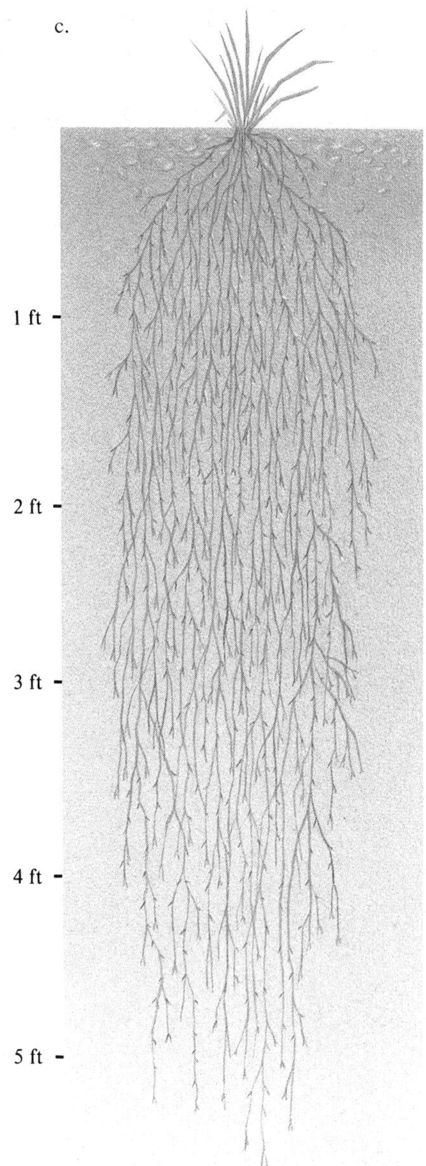

Figure 29.12
Two main root types: (*a*) the fibrous root system of barley and (*b*) the taproot of a dandelion. c. A mature root system may be very complex, as is this one of winter wheat.

a substantial portion of its body. Roots provide anchorage and absorb, transport, and store water and nutrients. They absorb oxygen from between soil particles. Roots pushing through very firmly packed soil may die from lack of oxygen.

The first root to emerge from a seed is the **radicle.** In a **taproot system,** the radicle enlarges to form a major root that persists throughout the life of the plant. Taproots grow fast and deep, maximizing support and enabling a plant to use material located deep in the soil. Engineers once found a mesquite root 174 feet (53 meters) below the earth's surface! Most dicots develop taproot systems.

In a **fibrous root system,** the radicle is short-lived. **Adventitious roots,** which are roots that form on stems or leaves, replace the radicle. The result is an extensive system of similarly sized roots. Fibrous root systems are relatively shallow, rapidly absorbing materials from near the soil surface and preventing soil erosion. Most monocots form fibrous root systems (fig. 29.12).

As rapidly growing roots push through the soil, as many as 10,000 cells per day are lost. Cell replacement and protection are provided by the **root cap,** a thimble-shaped structure covering the tips of roots. A root cap has a rapidly dividing meristem that continually pushes cells forward. Eventually, these cells are pushed to the outside of the root, where they are sloughed off as the root grows (fig. 29.13). The root cap also produces **mucigel,** a slimy substance that protects root tips from desiccation. Mucigel lubricates the root tips as they force their way between soil particles. Mucigel even helps roots absorb water and nutrients from the soil.

Just behind the root cap is a cluster of seemingly inactive cells called the **quiescent center.** This region functions as a reservoir to replace damaged cells of the adjacent meristem. The portion of the root above the tip anchors the plant in the ground. The region immediately behind the root cap is called the **subapical region.** It is loosely divided into three zones. The **zone of cellular division** is meristematic. Cells here divide as rapidly as every 12 to 36 hours. The meristem of a root surrounds the quiescent center. The **zone of cellular elongation** lies behind the zone of cellular division. Cells here elongate by as much as 150-fold as their vacuoles fill with water. This action pushes the root rapidly through the soil.

Cells mature and differentiate in the **zone of cellular maturation.** This is also called the root hair zone, because it is here that tiny root hairs protrude from epidermal cells (fig. 29.14). Root hairs are plentiful, a

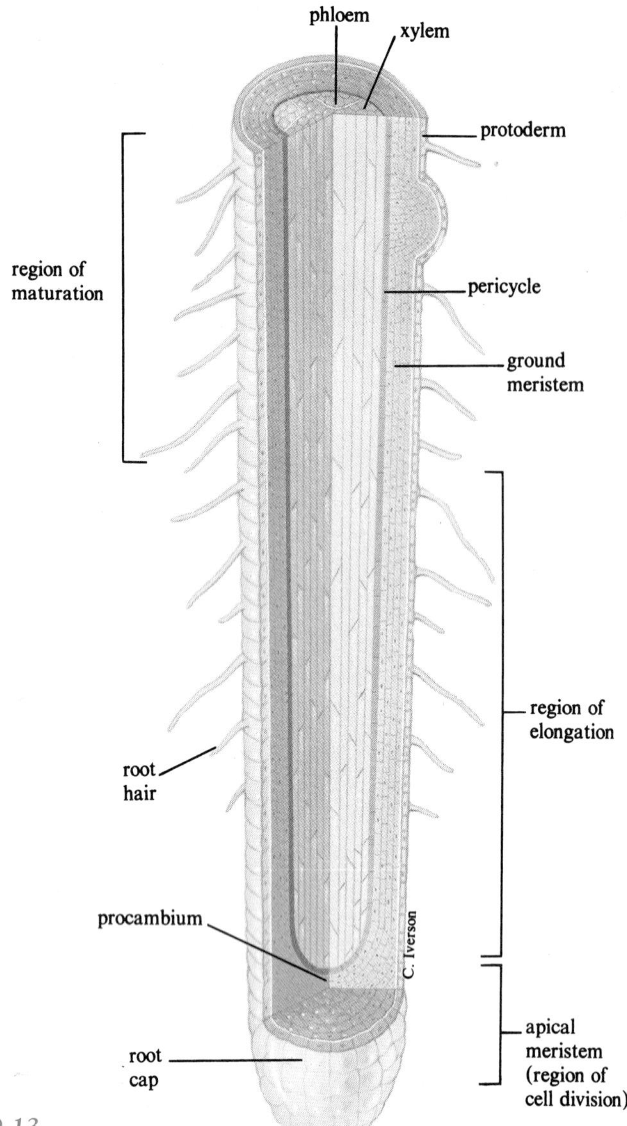

Figure 29.13
A dicot root tip has distinct regions.

single plant growing several billion of them, and they greatly increase the surface area of the root through which water is absorbed. The root hairs give this portion of the root a downy, fuzzy appearance. When a garden plant is transplanted, many root hairs are destroyed. It is only after they begin to grow back that the plant regains its vigor. Can you see why root hairs grow in the zone of cellular maturation and not at the root tip?

The zone of cellular maturation is also where primary tissues such as the epidermis and cortex develop. The epidermis surrounds all of the root except the root cap. Epidermal cells either lack or have a very thin cuticle and are thus well adapted for absorbing water and minerals. Root epidermal cells usually lack stomata.

The epidermis surrounds all the cortex, which consists of three layers: hypodermis, storage parenchyma cells, and endodermis. The **hypodermis** is the outermost, protective layer of the cortex. Loosely spaced storage parenchyma cells make up most of the cortex. These cells form a vast collecting system that absorbs water and minerals moving through the epidermis and stores these nutrients for future growth.

The **endodermis** is the innermost ring of the cortex. It consists of a single layer of tightly packed cells, called a **Casparian strip,** which contains a waxy, waterproof material called **suberin.** The endodermal cells resemble bricks in a wall, and the Casparian strip is like the mortar surrounding each brick. Due to this arrangement, the endodermis can regulate the movement of nutrients and water into and out of the central vascular tissue of the root. Figure 29.15 shows a monocot root in cross section.

Inside the cortex, the **pericycle** is the ring of parenchyma cells that produces branch roots that burst through the cortex and epidermis and finally into the soil. The root's vascular tissues are interior to the pericycle, where bundles of xylem alternate with bundles of phloem. Roots absorb water and minerals from the soil and transport them to shoots via the xylem. In return, roots receive organic nutrients from shoots via the phloem. Often, these organic nutrients are stored in cortex cells for later use.

Like stems and leaves, roots are modified for special functions (fig. 29.16):

Storage is a familiar root specialization. Beet and carrot roots store starch, sweet potato roots store sugar, and desert plant roots store large amounts of water.

Pneumatophores are specialized roots of plants growing in oxygen-poor environments. These roots grow up into the air, and oxygen diffuses into the plant body.

Adventitious roots that form and grow in the air are called **aerial roots.** They are important modifications for mangroves, trees that typically grow in low-oxygen environments. Mistletoe and orchids are two familiar plants that have aerial roots.

Many roots form mutualistic associations called **mycorrhizae** with beneficial fungi. The fungi absorb nutrients from the soil, while the host plants provide the fungi with vitamins or other needed substances.

Roots of legume plants, such as peas, are often infected with bacteria of the genus *Rhizobium*. The roots form nodules in response to the infection (fig. 2.3). The bacteria function as built-in fertilizer, providing the plant with nitrogen "fixed" into compounds that it can use. Genetic engineering to bring the benefits of this nitrogen fixation to plants not normally inhabited by *Rhizobium* is discussed in chapter 32.

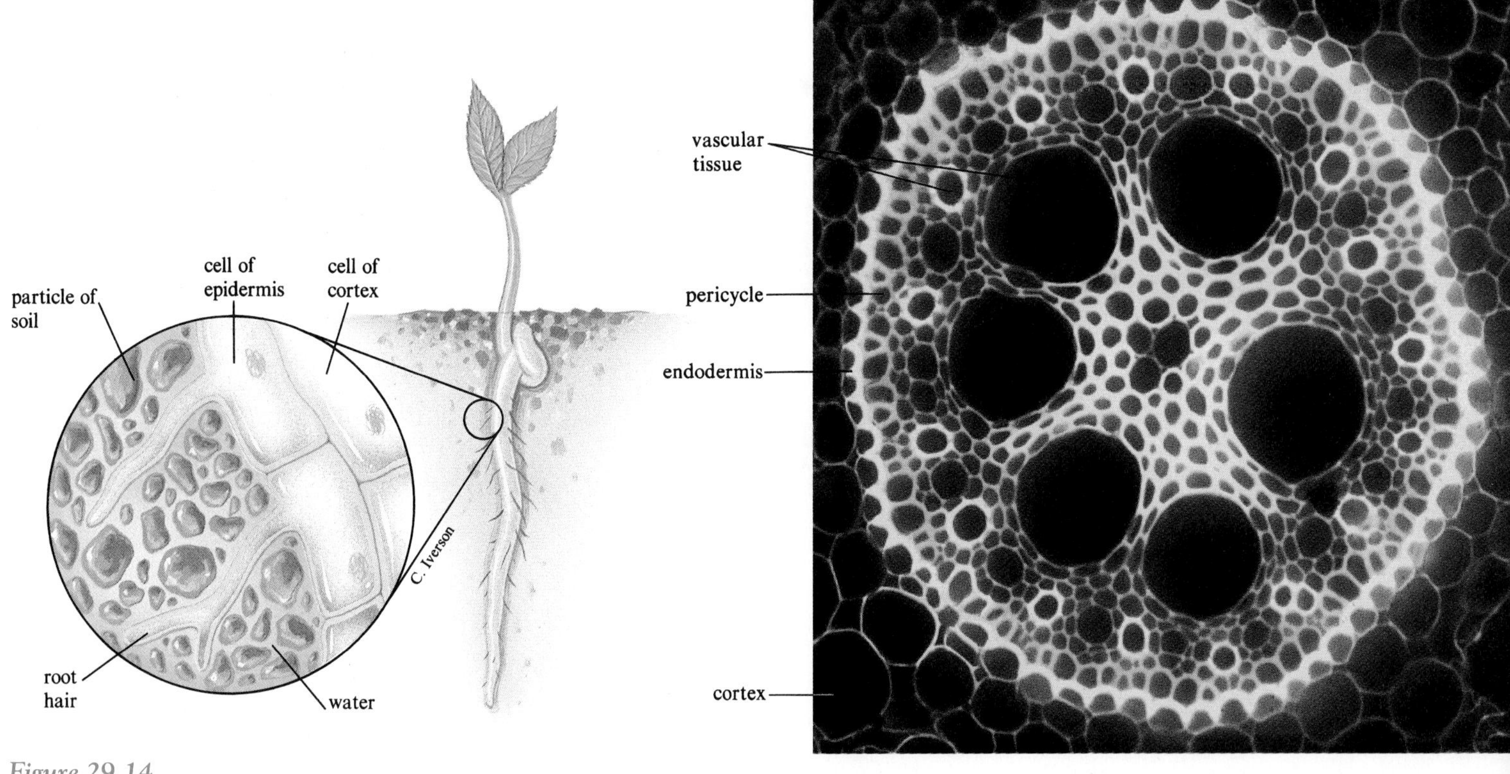

Figure 29.14
Root hairs. These outgrowths of epidermal cells extend through the soil, greatly increasing absorptive surface area.

Figure 29.15
Cross section of a monocot (corn) root.

a.

b.

c.

Figure 29.16
Root modifications. *a.* Yams are fleshy roots that store carbohydrate. *b.* The banyan tree has aerial roots growing out of its branches. *c.* This tropical fig tree has roots that are so enormous that they resemble a trunk.

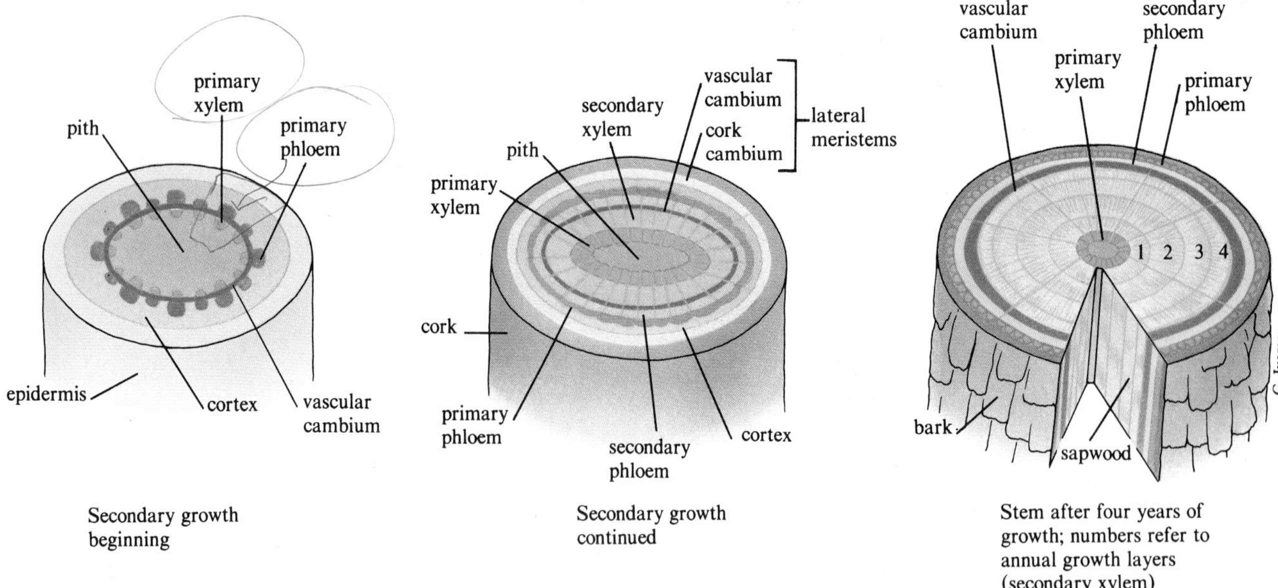

Figure 29.17
The secondary growth of a woody stem involves the activities of two lateral meristems—vascular cambium and cork cambium. Primary xylem and phloem are produced by the apical meristem, whereas secondary xylem and phloem are produced by the vascular cambium.

KEY CONCEPTS

Roots anchor plants and absorb, transport, and store nutrients. Taproots are large, deep, and fast-growing and last a lifetime. Fibrous roots are smaller, shallow, and short-lived. The root cap provides new cells and protects the root tip. The region behind the root cap includes zones of division, elongation, and maturation. In the maturation zone, numerous root hairs greatly increase absorptive surface area. Here also is found epidermis and the three-layered cortex. The hypodermis protects, parenchyma stores nutrients, and the endodermis regulates entry of substances into vascular tissue lying interior to the pericycle. Roots can be modified.

Secondary Growth

Plants must compete with each other to intercept light. One way they maximize light absorption is by growing taller than neighboring plants. Continued elongation presents a problem for plants, however, because primary tissues cannot support them very well. Thus, an improved support system has evolved: the formation of lateral meristems that increase the girth of stems and roots. These meristems, called the vascular cambium and cork cambium, produce the secondary plant body.

The results of secondary growth can be impressive. A 2,000-year-old tule tree in Oaxaca, Mexico, is 148 feet (45 meters) in circumference and only 131 feet (40 meters) tall. A 328-foot-tall (100-meter) giant Sequoia in northern California is more than 23 feet (7 meters) in diameter.

The **vascular cambium** produces most of the secondary plant body and is a thin cylinder of meristematic tissue found in roots and stems. Generally, the vascular cambium forms only in plants that exhibit secondary growth—primarily dicots and gymnosperms (conifers). Meristematic cells produce secondary xylem on the inner side of the vascular cambium and secondary phloem on the other side. Overall, the vascular cambium produces much more secondary xylem than secondary phloem. Secondary xylem is more commonly known as wood (fig. 29.17).

The vascular cambium produces wood during the spring and summer. During the moist days of spring, the vascular cambium produces *spring wood* made of large, thin-walled cells. Spring wood, also called early wood, is specialized for conduction. During the drier days of summer, the vascular cambium produces *summer*, or late, *wood*. Summer wood consists of small, thick-walled cells and is specialized for support. Because of the differences between spring and summer wood, wood often displays visible demarcations called **growth rings.** Reading 29.1 explains how growth rings can be "read" to reveal climatic conditions of years past.

Hardwoods are woods of dicots such as oak, maple, and ash. **Softwoods** are woods of gymnosperms such as pine, spruce, and fir. Recall that xylem has two kinds of conducting cells, tracheids and vessel elements. Hardwoods contain tracheids, vessels, and supportive fibers, whereas softwoods are 90% tracheids. As a result, hardwoods are stronger and denser than softwoods. In fact, hardwoods often are too hard to penetrate with nails and as a result are not used for construction.

Different regions of wood are specialized for different functions. **Heartwood,** wood in the center of a tree, collects a plant's waste products. It forms a solid,

Reading 29.1 Tree Rings Reveal the Past

ASK A DENDROCHRONOLOGIST WHAT THE TEMPERATURE AND RAINFALL WERE IN THE AMERICAN SOUTHWEST FROM A.D. 341 TO 832 AND HE OR SHE MAY HAVE AN ANSWER. With the help of slices from the interiors of trees, the tree ring specialist can reconstruct "weather reports" back thousands of years (fig. 1).

The pattern of secondary xylem indicates the passage of time because the cells of wood laid down in the spring are large, with thin walls (appearing light colored), and wood cells laid down in the summer are smaller, with proportionately thicker walls. The contrast between the latest wood of one season and the earliest of the next creates the characteristic annual ring. The most recent ring is next to the vascular cambium, which is microscopic. On the outside of the cambium is the phloem and then the bark. Can you see why tropical species, which have secondary growth all year long, do not have rings?

Tree rings provide other information. The larger the ring, the more plentiful was rainfall that year. A fire leaves behind a charred "burn scar." A season when voracious caterpillars or locusts chomped away the early leaves is represented by two rings very close together, because nutrients for growth were in short supply with photosynthesis inhibited. Dendrochronologists can also read information on light availability, altitude, temperature, and length of the growing season in the patterns of tree rings.

But not all temperate species can reveal the climatic past. Meaningful rings have one predominating environmental factor, such as temperature in Alaska and precipitation in the American Southwest, which must change from year to year yet have similar effects on trees in a wide geographical area. Trees growing in the best conditions—high groundwater and low run-off—produce uniformly wide rings, despite changing rainfall patterns. They are very healthy but provide no information. Trees in flat areas, such as roadsides, lakesides, and river valleys, produce these noninformative, or "complacent" rings. In contrast, useful, or "sensitive," rings are found on rocky hills, where drainage is good and there is no underground water for the roots to tap into when rainfall is low.

An occasional missing ring can be a problem. This happens because the tree rings are actually cones in three dimensions, with uneven widths throughout the tree. By chance, the section of the tree viewed may be missing one ring. Dendrochronologists date a tree by removing a slice of wood through a tiny hole that they bore. A disinfectant and a bandage protects the tree.

The most revealing use of tree ring dating is to align common ring patterns of several trees that lived at different times but with some overlap, including dead trees. As long as one of the trees has a ring next to the vascular cambium so that the most recent year is known, such series can yield a very accurate "master chronology." For example, the oldest known living tree is a bristle cone pine (*Pinus longaeva*) growing in the White Mountains of California. It is 4,906 years old! Combining its tree ring information with that from other trees in the area has provided rainfall data going back 8,200 years!

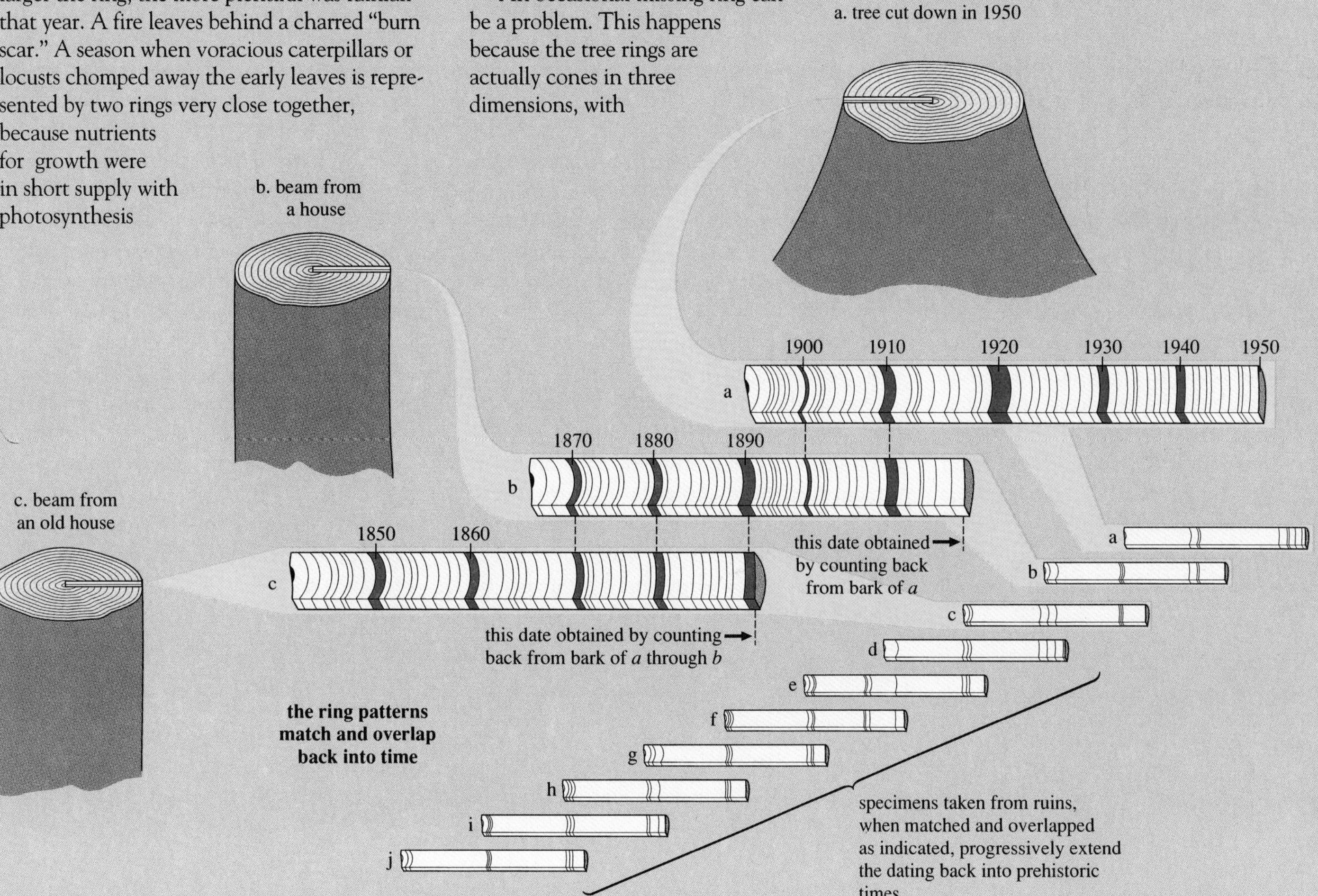

rodlike center that adds support to the organism. **Sapwood,** wood located nearest the vascular cambium, transports water and dissolved nutrients within a plant.

Bark includes all of the tissues outside of the vascular cambium. Secondary phloem forms the inner layer of bark, but only the innermost secondary phloem functions in transport. The **periderm** portion of the bark is the outer protective covering on mature stems and roots. It forms after secondary growth breaks through the epidermis. The periderm consists of the cork cambium, cork, and phelloderm.

The **cork cambium** is the lateral meristem that produces the periderm. The cork cambium produces cork cells to the outside; it produces phelloderm to the inside. **Cork cells** are waxy, densely packed cells covering the surfaces of mature stems and roots. They are dead at maturity and form waterproof, insulating layers that protect plants. Areas of loosely packed cells penetrate cork layers and enable gas exchange to occur. **Phelloderm** consists of living parenchyma cells, which may be photosynthetic and store nutrients. A familiar example of cork is the "skin" of a potato. The cork used to stopper wine bottles comes from a cork oak tree that grows in the Mediterranean. Every 10 years, harvesters remove the cork cambium and cork, which grows back.

Ninety percent of a typical tree is secondary xylem. Secondary xylem plays important roles in our lives. United States forests produce 18.6 million cubic meters of secondary growth per day. That may not seem like so much when you consider that each year the average American uses about 500 pounds of paper! We use 250,000 tons of napkins and 2 million tons of newsprint and writing paper each year. We can thank the secondary growth of plants for such diverse products as rubber, chewing gum, turpentine, cardboard, rayon, synthetic cattle food, and ice cream fillers.

KEY CONCEPTS

Vascular cambium and cork cambium are lateral meristems that increase a plant's girth. Vascular cambium forms secondary xylem (wood) on its inner side and secondary phloem on its outer side. The cells of spring wood are large, and of summer wood small, so that rings appear. Hardwoods and softwoods have different proportions of xylem components. Bark includes secondary phloem and the outer periderm, which is produced by the cork cambium.

SUMMARY

Meristematic tissue actively divides. *Apical meristems* are located at the plant's tips, and *lateral meristems* add girth. *Ground tissue* is abundant and multifunctional. It includes *parenchyma* cells, which can divide, store substances, and carry out photosynthesis, respiration, and protein synthesis; *collenchyma* cells, which are supportive; and *sclerenchyma* cells, which support nonliving plant parts. *Dermal tissue* includes the *epidermis* and the *cuticle*. Gas and water exchange can occur through pores called *stomata*. *Trichomes* are epidermal outgrowths. Vascular tissue consists of *xylem*, which transports water upward, and *phloem*, which transports water and nutrients in all directions. Xylem cells include *tracheids* and the more specialized *vessel elements*. Phloem includes *sieve cells* and the more specialized *sieve tube members*.

A stem is the central axis of the shoot and consists of nodes, where leaves attach, and internodes. *Vascular bundles* in stems contain xylem and phloem, which are scattered in monocots and form a ring in dicots. Between a stem's epidermis and vascular tissue lies the *cortex*, made of ground tissue. *Pith* is storage ground tissue in the center of a stem. Stem modifications include stolons, thorns, tendrils, bulbs, and tubers.

Leaves are photosynthetic organs. *Simple* leaves have undivided blades, and *compound* leaves form leaflets. *Pinnate* leaves have a central axis, and *palmate* leaves extend from a common point. Leaf arrangement is *alternate*, *opposite*, or *whorled*. Leaf epidermis is tightly packed, transparent, and nonphotosynthetic. Vascular tissue is found in veins, which may be in *net* or *parallel* formation. Leaf ground tissue, *mesophyll*, is organized into palisade and spongy layers. Leaf modifications include tendrils, spines, and bracts. Leaves are shed from an *abscission zone* in response to environmental cues.

A plant's first root is the *radicle*. *Taproot systems* have a large, persisting major root, whereas *fibrous root systems* are shallow and shorter-lived. A meristem in the *root cap* replaces cells lost during a root's rapid extension. The *subapical* region behind the root cap is divided into zones of cellular division, elongation, and maturation. This last zone includes root hairs, epidermis, and cortex, which itself has three layers. Some roots are specialized for storage or adapted to low-oxygen environments. Some roots form symbiotic relationships with fungi or bacteria. *Vascular cambium* and *cork cambium* produce secondary xylem and phloem, which increase a plant's girth.

QUESTIONS

1. Which plant tissue is responsible for the fact that plants never stop growing?

2. Which tissue is the most abundant in plants?

3. What are the functions of the following substances?
   a. Cutin
   b. Mucigel
   c. Suberin

4. Is the trunk of a banana plant a stem or leaves? How might you demonstrate which type of structure it is?

5. What are some of the ways in which leaves are classified?

6. Cite a stem and a leaf specialization that provide protection.

7. Corn is a monocot and cucumber a dicot. How do these plants differ in the following?
   a. Stem structure
   b. Leaf venation
   c. Root organization

TO THINK ABOUT

1. If some stems and leaves of a tomato plant are torn off by an overanxious tomato picker, the plant regrows these parts in a few weeks. Which tissue type is responsible for this regrowth?

2. Why are stomata necessary?

3. Why can phloem transport materials in all directions, whereas xylem can transport materials upward only?

4. How does a tree "know" when to shed its leaves?

5. How can roots grow in compact soils, such as clays?

6. How are root hairs similar to human intestinal villi?

7. What would happen (or not happen) if a plant's quiescent centers were destroyed?

SUGGESTED READINGS

Briffa, K. R. et al. August 2, 1990. A 1,400-year tree-ring record of summer temperatures in Fennoscandia. *Nature*. Tree-ring data indicate that there was a "little Ice Age" between 1570 and 1650.

Feldman, Lewis, J. October 1988. The habits of roots. *BioScience*. Roots can be meticulously separated from the soil and their complex structures revealed.

Gower, Stith T. and James H. Richards. December 1990. Larches: deciduous conifers in an evergreen world. *BioScience*. Larches shed their leaves like deciduous trees, but the leaves are shaped and water is conducted in manner similar to evergreens.

Ross, Gary N. December 1986. Night of the radishes. *Natural History*. The root of the radish provides the plant with stored nutrients. It also serves as sculpture material for Mexican families.

Stern, Kingsley R. 1991. *Introductory plant biology*. Dubuque, Iowa: William C. Brown. A more detailed yet easy-to-read look at the basic parts of the plant body.

C H A P T E R

30

Plant Life Cycles

Chapter Outline

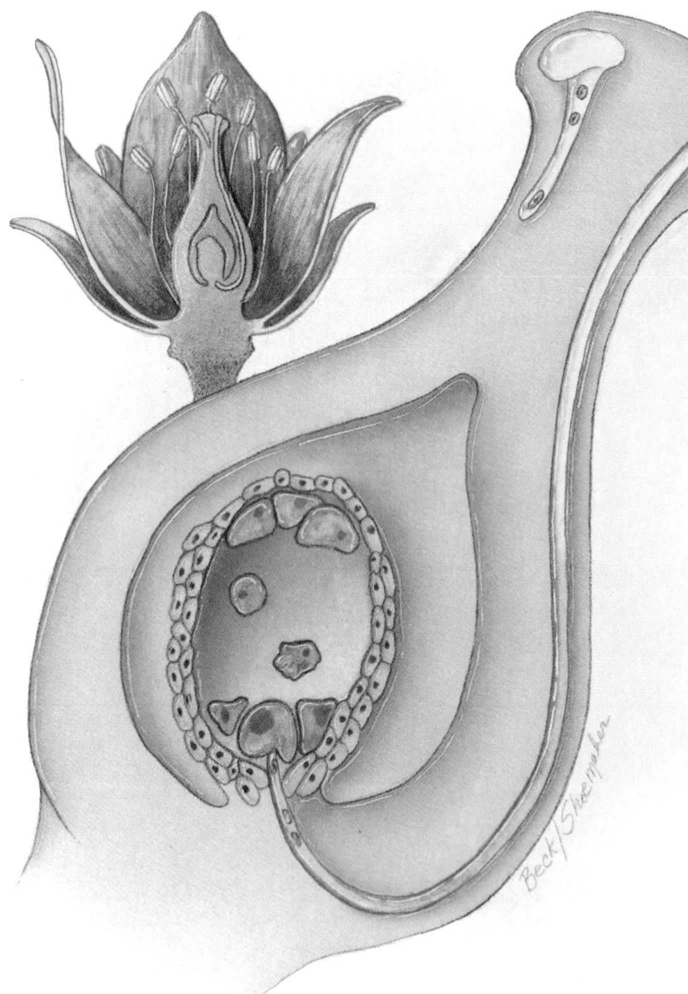

Learning Objectives

By the chapter's end, you should be able to answer these questions:

1. What are the diploid and haploid phases of the life cycle of a sexually reproducing plant?
2. What are the functions of the parts of a flower?
3. How do the gametes of flowering plants come together?
4. What events follow fertilization to form and nourish the plant embryo?
5. What is a fruit, and how does it assist a plant's reproduction?
6. What factors provoke a seed to germinate?
7. How does sexual reproduction in pines differ from that in flowering plants?
8. How do plants reproduce asexually?

Newly emerged from his cocoon, the male wasp is already eager to locate a female. The females, though, will not emerge for another week. Meanwhile, the male wasp flies among orchid plants of the genus *Ophrys*. Suddenly he becomes intensely excited. He smells a female! The unmistakable fragrance of a sexually receptive female wasp wafts from the orchid flowers. Sexually stimulated, the male wasp approaches the blooms more closely. To his eyes, which can discern ultraviolet light that human eyes cannot, one petal of each flower looks incredibly like a female wasp, seeming to sport eyes, antennae, and wings.

The now-frantic male alights on the provocative petal, which not only looks and smells like a female but whose fuzzy surface feels like the touch of a potential mate. Completely fooled, he begins the motions of copulation wasp style, and in the process, small packets of pollen are deposited on his confused head. Despite his sexual movements, the male does not achieve sexual satisfaction, so, still highly agitated, he moves on to the next flower. He tries again to mate with a beguiling petal, and in doing so, he leaves behind some of the pollen clinging to his head from the first flower (fig. 30.1).

The behavior of the male wasp seeking a mate and finding a flower may seem to be a waste of biological energy, but it is an amazingly efficient mechanism for pollen—the male sex cells of flowering plants—to be delivered from one plant to another. The sexual reproduction of orchids and other flowering plants is tied intimately to the behavior of certain animals. Rather than being just pretty appendages of plants, flowers are sophisticated, highly evolved structures. They facilitate the recombination of genetic material that provides the genetic diversity essential to the survival and success of a species.

Plants evolved from a multicellular green alga that invaded land more than 430 million years ago. Although life on land was harsh compared with life in water, plants adapted to the terrestrial environment by developing waxy coverings to contain moisture and mechanisms to combine sex cells without relying on delivery via water. Today tracheophytes and bryophytes are the two main groups of plants. The

Figure 30.1
A male wasp finds this orchid quite enticing.

tracheophytes, or vascular plants, dominate on land. They possess vascular tissues, which transport water and dissolved nutrients throughout a plant. Bryophytes lack vascular tissue and are smaller than vascular plants.

Alternation of Generations

The life cycle of a plant differs from that of an animal. In mammals special haploid cells are set aside that combine from two individuals to form an offspring. In plants,

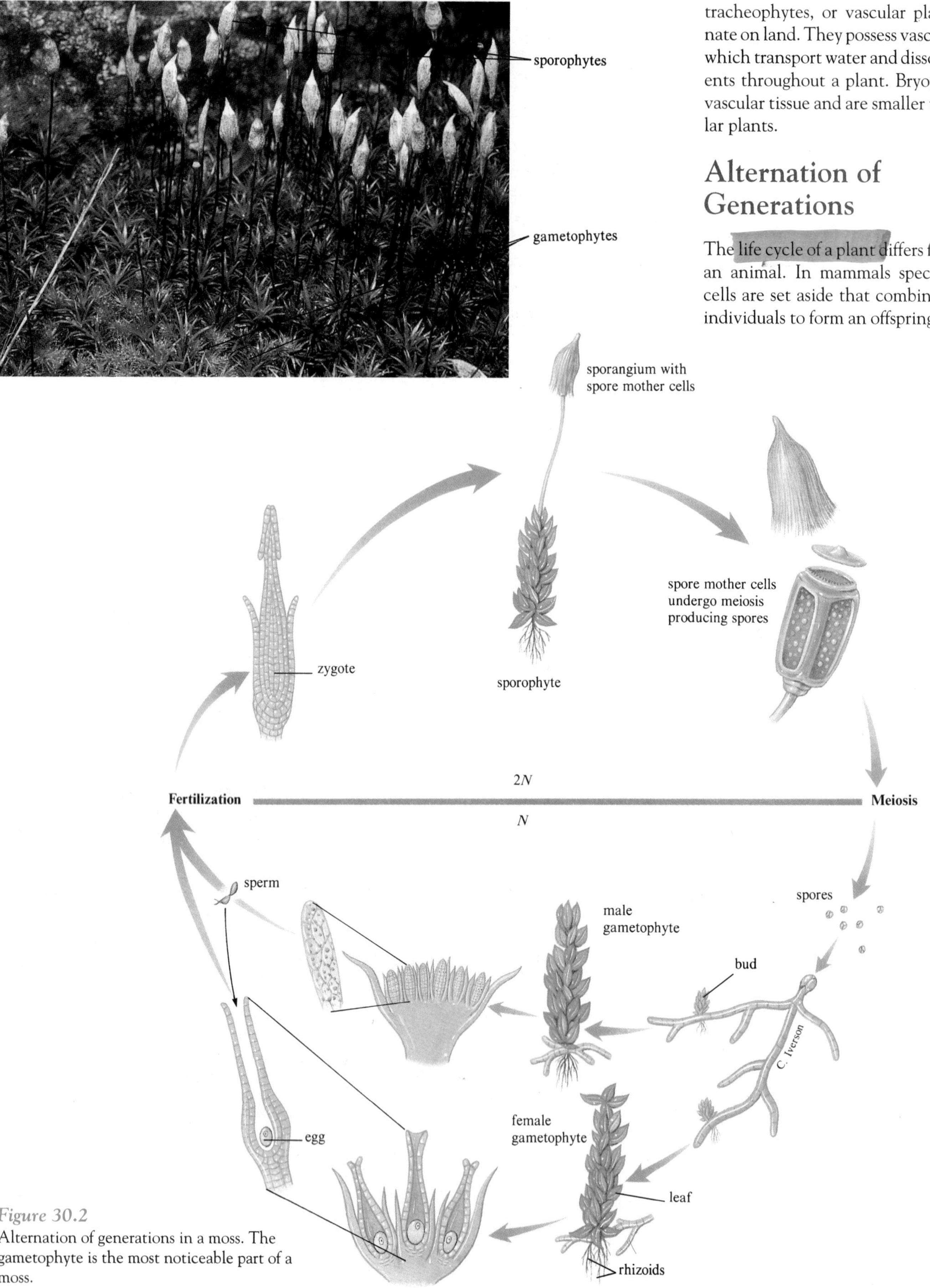

Figure 30.2
Alternation of generations in a moss. The gametophyte is the most noticeable part of a moss.

however, a complete life cycle of an individual includes both a diploid stage and a haploid stage. These stages are called **generations,** and because they alternate within a life cycle, sexually reproducing plants are said to undergo **alternation of generations.**

The diploid generation, or **sporophyte,** produces haploid spores through meiosis. Haploid spores divide mitotically to produce a multicellular haploid individual, the **gametophyte.** Eventually, the gametophyte produces haploid gametes—eggs and sperm—which fuse to form a **zygote.** The zygote grows to become a sporophyte, and the cycle begins anew.

In most vascular plants, gametophytes produce either eggs or sperm, but not both. Because often the female egg-producing gametophytes are larger than the male sperm-producing gametophytes, the female gametophytes are called **megagametophytes,** and the male gametophytes are called **microgametophytes.** Male and female gametophytes arise from two different types of spores called **megaspores** and **microspores.** Not surprisingly, these spores form in two different types of structures, **megasporangia** and **microsporangia** ("mega" simply means "big," and "micro" means "small").

Although alternation of generations is characteristic of all plants, the gametophyte stage dominates some plant life cycles, while the sporophyte stage dominates others (figs. 30.2 and 30.3). For example, the gametophyte stage is the more obvious phase in bryophytes such as the mosses. In contrast, vascular plants have a reduced gametophyte phase and a dominant sporophyte phase. Familiar vascular plant forms such as a fern, a tree, and grass are sporophytes.

KEY CONCEPTS

Plants exhibit alternation of generations. A diploid sporophyte phase produces haploid spores, which divide to form the haploid gametophyte. The gametophyte produces haploid gametes, which unite to form a zygote, which grows into a sporophyte. In some plants the sporophyte dominates, and in some the gametophyte dominates.

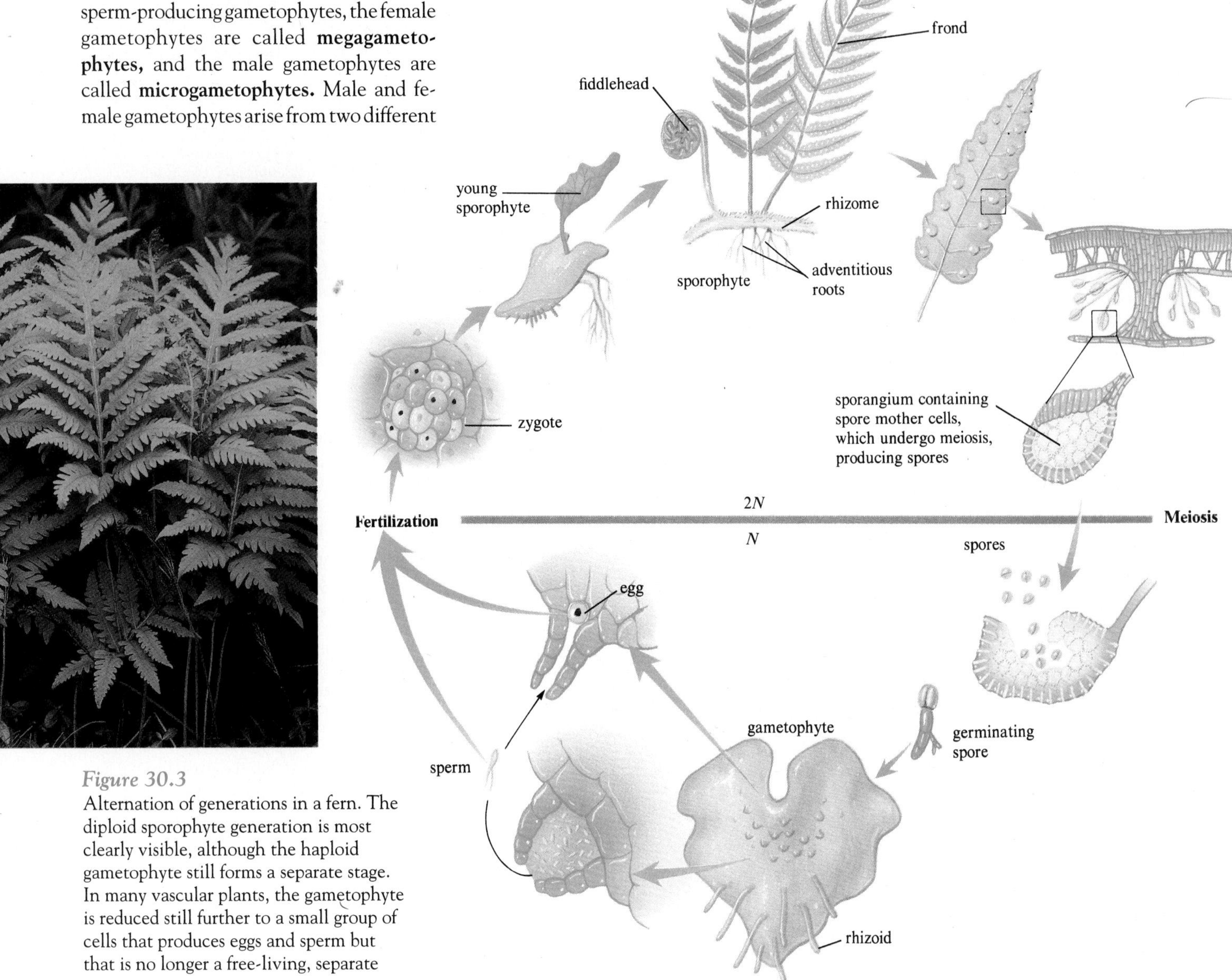

Figure 30.3
Alternation of generations in a fern. The diploid sporophyte generation is most clearly visible, although the haploid gametophyte still forms a separate stage. In many vascular plants, the gametophyte is reduced still further to a small group of cells that produces eggs and sperm but that is no longer a free-living, separate stage.

Flowering Plant Life Cycle

Flowering plants, or **angiosperms,** are the dominant group of vascular plants. The approximately 235,000 species of angiosperms include such familiar plants as shrubs, grasses, vegetables, and grains. The reproductive structures of angiosperms are **flowers.**

Structure of the Flower

Complete flowers have four types of organs, all of which are modified leaves. Each type of organ occurs in a whorl, or circle, about the end of the flower stalk (fig. 30.4). The calyx is the outermost whorl of a flower. It is made up of green, leaflike **sepals** that enclose and protect the inner floral parts. Inside the calyx is the corolla, a whorl made up of **petals.** Petals often are large and colorful, especially when they are important in attracting pollinators to the flower. The **calyx** and the **corolla** do not play a direct role in sexual reproduction and are therefore referred to as accessory parts.

The two innermost whorls of a flower, the **androecium** and the **gynoecium,** are essential for sexual reproduction. The whorl to the inside of the corolla, the androecium, consists of the male reproductive structures. These are the **stamens,** which are built of stalklike filaments bearing pollen-producing oval bodies called **anthers** at their tips. The whorl at the center of a flower, the gynoecium, consists of the female reproductive structures, or **pistil.** The gynoecium is formed from one or more **carpels,** leaflike structures enclosing the **ovules.** The carpels and their enclosed ovules are referred to as the **ovary.** The ovary gives rise to a stalklike **style** bearing a **stigma** at its tip. The stigma receives pollen.

Formation of Gametes

Microspores form in the anthers. Each anther contains four microsporangia called **pollen sacs.** Pollen sacs contain **microspore mother cells** that divide meiotically to produce four haploid microspores. Each microspore then divides mitotically, producing a haploid **generative nucleus** and a haploid **tube nucleus.** A thick, resistant wall forms around each two-celled structure, forming male microgametophytes, more familiarly known as **pollen grains.** The generative cell divides to form two sperm cells, either before or after the pollen grains are shed. Millions of pollen grains may be released when mature pollen sacs burst open.

Megaspores form in the ovary of the flower. Each ovule within the ovary is a megasporangium containing a **megaspore mother cell,** which divides meiotically to produce four haploid cells. Three of these cells quickly disintegrate, leaving one large haploid megaspore. The megaspore undergoes three mitotic divisions, forming a female megagametophyte having eight nuclei but only seven cells—one large cell has two nuclei called **polar nuclei.** In a mature megagametophyte, also called an **embryo sac,** one of the cells with a single nucleus is the egg. The top part of figure 30.5 illustrates formation of the pollen grains and embryo sac.

KEY CONCEPTS

Flowers are reproductive structures. Accessory parts include the calyx, the outermost whorl made up of sepals, and inside it the corolla, made of petals. Inside the corolla are the male structures (stamens built of filaments and anthers) and the female structures (ovaries built of carpels and ovules). The style and stigma arise from the ovary. Microspore mother cells are in pollen sacs, and they undergo meiosis to yield four haploid microspores. Each microspore divides mitotically to yield a generative nucleus and a tube nucleus, which, surrounded by a thick wall, form a pollen grain. An ovule contains a megaspore mother cell, which yields four haploid cells, three of which disintegrate. The remaining megaspore mitotically divides three times, forming the eight cells of the embryo sac.

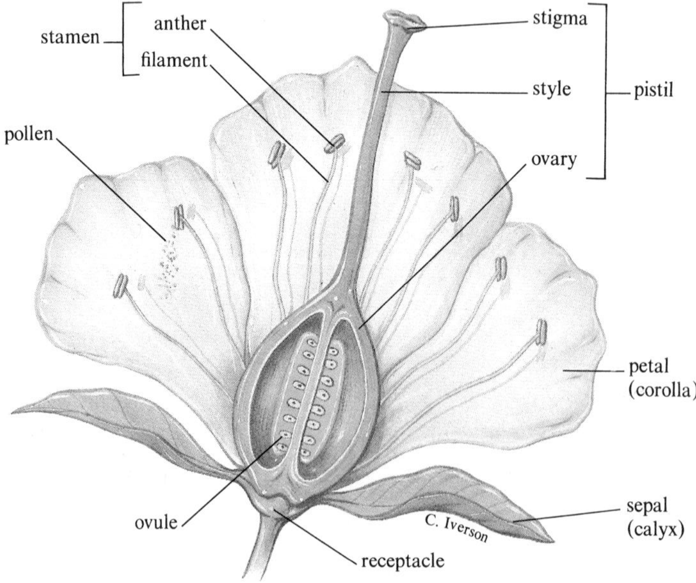

Figure 30.4
Flower anatomy. The stamen comprises the androecium, and the pistil comprises the gynoecium. Flowers can be diverse from one another yet still contain the same basic parts.

Pollination

Pollination is the transfer of pollen from an anther to a receptive stigma. Some angiosperms are self-pollinating, meaning that pollen grains are transferred from the anther of one flower to the stigma of the same flower. Other angiosperms are **outcrossing** species in which pollen grains from one flower are carried to the stigma of another flower.

Self-fertilization leads to reduced genetic variability and is common in temperate regions where environmental conditions are relatively uniform. Outcrossing is crucial in a changing environment because it produces a genetically variable population that can adapt to changing conditions. Interestingly, some angiosperms actually promote outcrossing by producing physically isolated anthers and stigmas. Others lack the androecium or gynoecium, making self-fertilization impossible. The magnolia, one of the first flowering

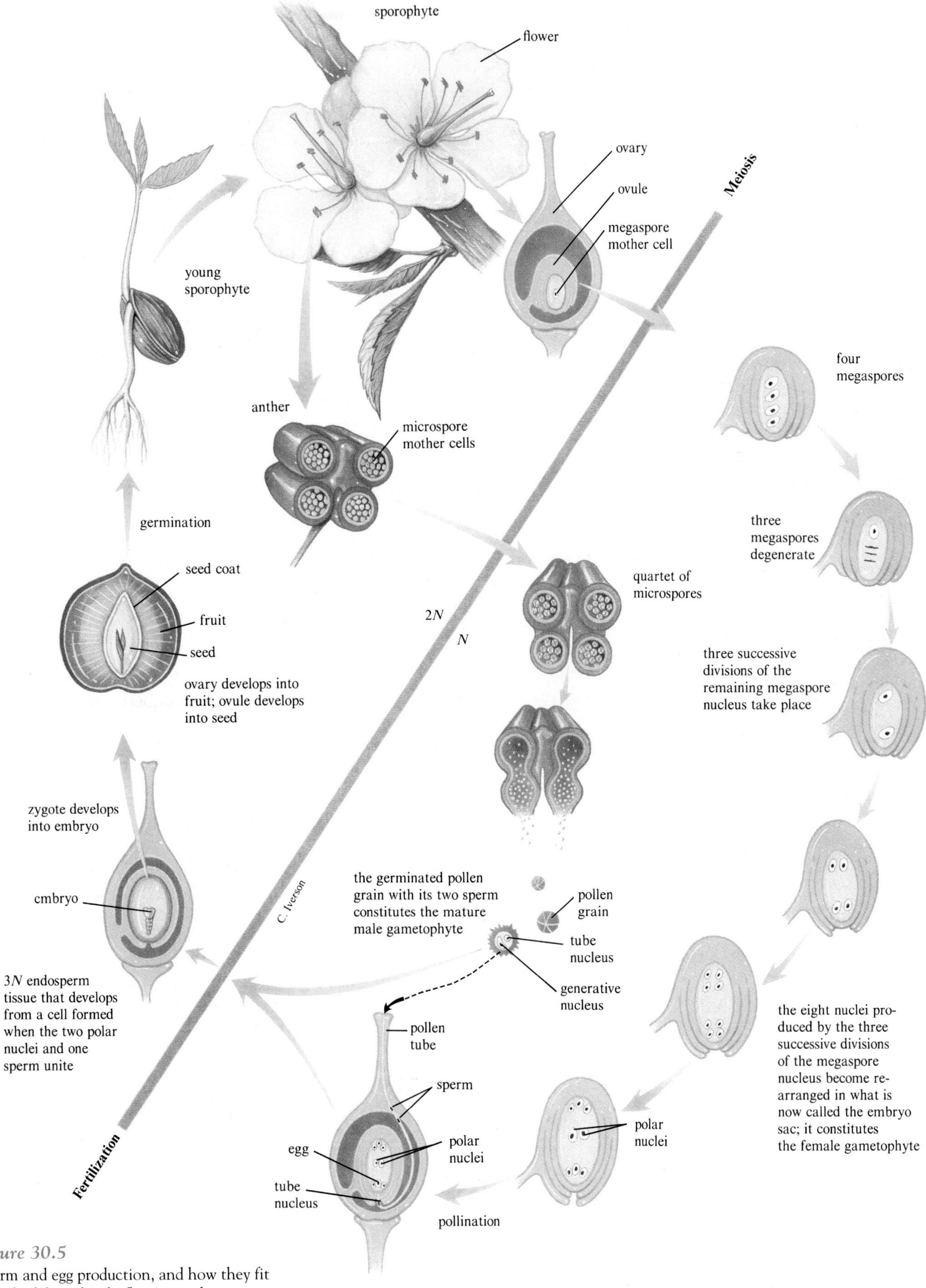

Figure 30.5
Sperm and egg production, and how they fit into the life cycle of a flowering plant.

plants, ensures cross pollination by releasing its pollen only after its eggs have been fertilized by pollen from other individuals.

Pollen is often assisted in traveling from plant to plant. Animals, particularly insects, play an important role in pollination (fig. 30.6). In animal-pollinated angiosperms, floral characteristics such as morphology, color, shape, and odor often attract particular animals to particular flowers, as the wasp in the orchid described in the chapter's opening illustrates. (See Reading 30.1 for further examples of orchid adaptations to attract pollinators.) Some flowers produce heat, which volatizes their aromatic molecules, releasing their characteristic fragrances. Pollinating animals visit flowers in search of food, such as sugary nectar, and end up carrying pollen from flower to flower. Nectar is a sweet-tasting substance that has no known function in the plant other than attracting pollinators. Myrtle plants take another approach to luring insects. They make two types of pollen—one to entice visiting insects, and the other type are the actual sex cells.

Bees, the most common pollinators of flowers, initially locate a food source by its fragrance and then by its color. Because ultraviolet light is highly visible to bees, bee-pollinated flowers often have conspicuous ultraviolet markings (fig. 30.7).

Different types of insects have characteristic floral preferences. In contrast to bees, which are partial to blue or yellow sweet-smelling blooms, beetles prefer spicier scents with dull-colored flowers.

Birds are attracted to red flowers, a color that insects cannot distinguish. The structure of the flower almost ensures that a visiting bird will transport pollen. A hummingbird flitting about a California fuchsia plant, for example, sticks its long, thin beak into the flower. The bird catches its bill in a network of fine threads, on which pollen grains descend. Carrying the pollen on its bill, the hummingbird ferries the cells to the next plant it visits.

Some flowers are pollinated by butterflies and moths. These flowers often are white or yellow and are heavily scented—characteristics that make them easy to locate at night, when these insects are the most active. The flowers may even have flat surfaces on which their pollinators can land. Bats are important pollinators in the tropics, where flowers are open at night.

a.

b.

Figure 30.6

Pollinators. Bumblebees (*a*) and hummingbirds (*b*) are efficient pollinators of many angiosperm species. Animal pollinators range from ants to butterflies to bats. When wind is the principal pollinator, modifications may be necessary. Corn, for example, must be planted in blocks to allow wind pollination.

Reading 30.1 Orchids Entice Pollinators

ORCHIDS ARE AN EXTREMELY SUCCESSFUL TYPE OF FLOWERING PLANT, WITH 35,000 SPECIES COMPRISING 10% OF KNOWN ANGIOSPERMS. The unusually large number of species, and the fact that naturally occurring orchid hybrids are very rare, reflects what is called "flower fidelity," the very specific match between what orchids display and what pollinating animals seek.

Evolution has molded traditional floral structures into traps that attract an insect and force it to transport pollen. Consider the 650 species of orchid growing in the American tropics. Their flowers release an intoxicating nectar or oil that quickly makes a visiting bee lose its coordination and stagger about. To make its way out of the labyrinth of flower parts, the tottering insect passes structures that shower it with pollen. The effects of the intoxication subside just as the insect nears an exit, and the insect flies on, carrying pollen sticking to its body to the next flower.

Orchids of genus *Epipactus* growing in marshes sport a large petal, called a labellum, that forms a runway of sorts leading to a pool of nectar at the base of the flower. But once the insect alights on the petal, the runway clamps up, trapping him. Like a circus visitor winding his way through a funhouse with only one exit, the insect wanders until it locates the only way out—past structures that trigger a pollen shower before he leaves.

Pollinating insects are very much attuned to the details of flower construction. They recognize distinctive scents, spatial relationships between the stigma and anthers, how far apart plants of its favorite species grow from one another, and the precise timing of blooming. Although the benefit to the plants of these relationships seems obvious, it is not so clear just what insects derive from them. Some insect pollinators are rewarded with a tasty treat of nectar, yet others seem attracted simply by a whiff of a fragrant oil. Whatever the subtle attractions between plant and pollinator, it is an amazingly efficient adaptation for maximizing reproductive potential.

Figure 30.7
Insects can "see" ultraviolet light, which makes these black-eyed Susans appear purple. To us they are yellow.

Not all floral scents are pleasing to the human nose. In South Africa, the "carrion flowers" of stapelia plants smell remarkably like rotting flesh, a highly attractive scent to flies. As if the stench were not sufficient to beckon the insects, stapelia's leaves are wrinkled and brown, resembling decayed meat.

Wind-pollinated angiosperms such as oaks, cottonwoods, ragweed, and grasses shed large quantities of pollen that are dispersed on breezes. The wind, however, does not carry pollen very far. This is why wind-pollinated plants grow closely together. Wind pollination is far less precise than animal pollination. Even though enormous amounts of pollen are released, the wind does not provide a mechanism to deliver it precisely to other plants of the same species, as animals do. Many people are allergic to the pollen of wind-pollinated plants such as ragweed (see fig. 27.16). The flowers of wind-pollinated angiosperms are small, greenish, and odorless. Rather than invest energy in flowers, it is more efficient for these plants to invest energy in pollen production.

Fertilization

After a pollen grain lands on a stigma (pollination), the pollen grain gives rise to a growing *pollen tube*. The pollen grain's two sperm cells enter the pollen tube as it

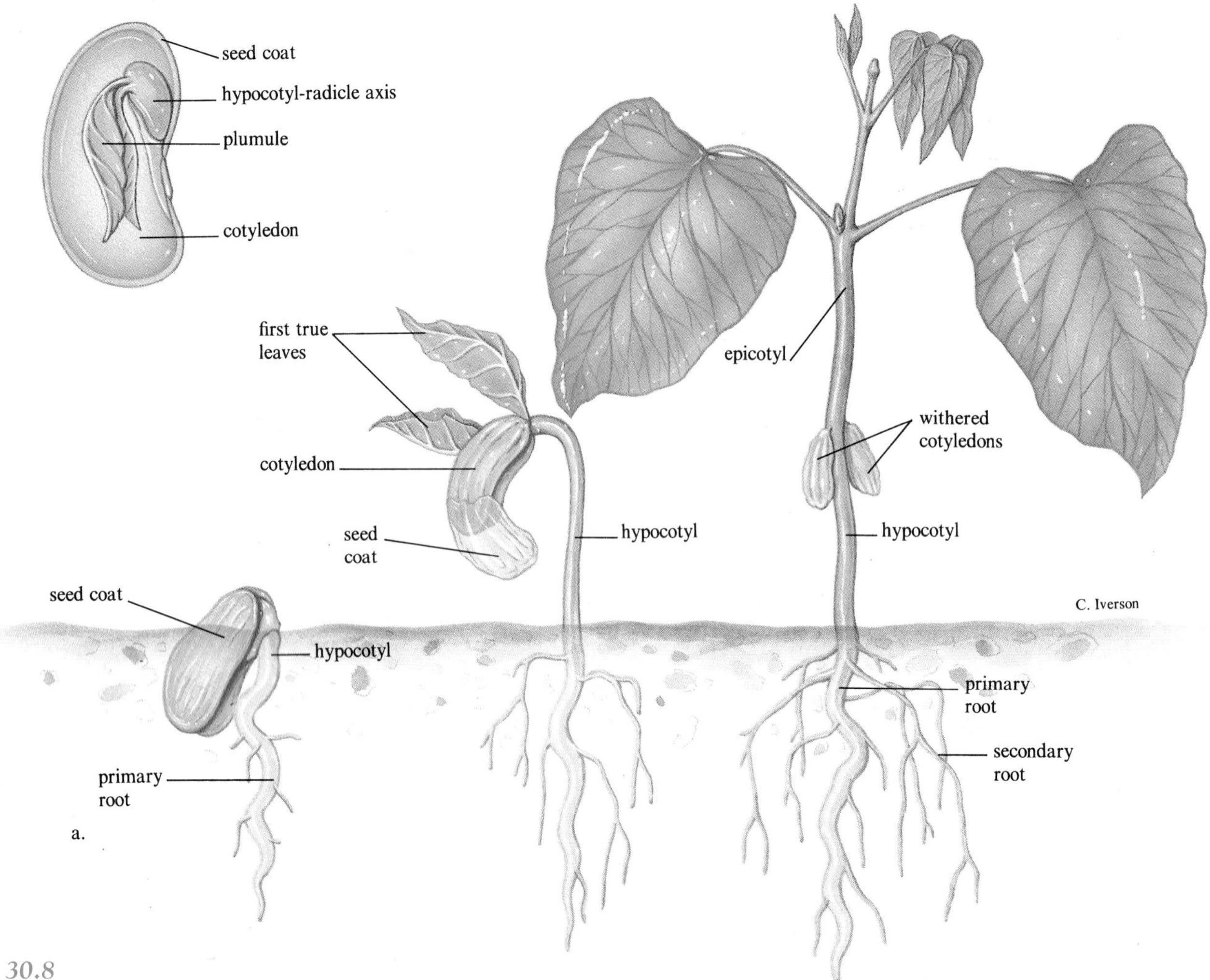

Figure 30.8
Seed structure, germination, and development in (*a*) a dicot (green bean).

grows down through the tissues of the stigma and the style towards the ovary. When the pollen tube reaches an ovule, it discharges its two sperm cells into the embryo sac. One sperm cell fuses with the egg, forming a diploid **zygote.** After a series of cell divisions, the zygote will become the **plant embryo.** The second sperm cell fuses with the two polar nuclei, forming a triploid nucleus that will divide to form a nutritive tissue called **endosperm.** Endosperm is stored food for the developing embryo. Familiar endosperms are coconut milk and the fleshy part of a kernel of corn. Notice that the egg and the polar nuclei both are fertilized, a process termed **double fertilization.** The bottom of figure 30.5 depicts fertilization and subsequent development of the seed and fruit.

Seed Development

A **seed** is a temporarily dormant sporophyte individual surrounded by a tough protective coat. Immediately after fertilization, the ovule contains an embryo sac with a diploid zygote and a triploid endosperm nucleus, both of which are surrounded by several layers of protective maternal tissue. Initially, the endosperm nucleus divides more rapidly than the zygote, forming a large mass of nutritive endosperm. The developing embryo forms **cotyledons,** or seed leaves. Angiosperms that have one cotyledon are called **monocots,** while those having two cotyledons are **dicots** (fig. 30.8).

Further development in monocot and dicot seeds differs. In many dicots, the cotyledons become thick and fleshy as they absorb the endosperm. In monocots the cotyledon does not absorb the endosperm, but absorbs and transfers food from the endosperm to the embryo.

The apical meristems form early in embryonic development. Recall that these are regions of cellular division that will remain active throughout the entire life of a plant. The **shoot apical meristem** forms at the tip of the **epicotyl,** which is the stemlike region above the cotyledons. The **root apical meristem** differentiates near the tip of the embryonic root, or **radicle.** When one or more embryonic leaves form on the epicotyl, the epicotyl plus its young leaves is called a **plumule.** The stemlike region below the cotyledons is the **hypocotyl.**

In monocots, a sheathlike structure called the **coleoptile** covers the plumule. Also, the ovary wall remains attached to a monocot seed, forming a fruit. A cereal grain such as a corn kernel actually is a hard fruit.

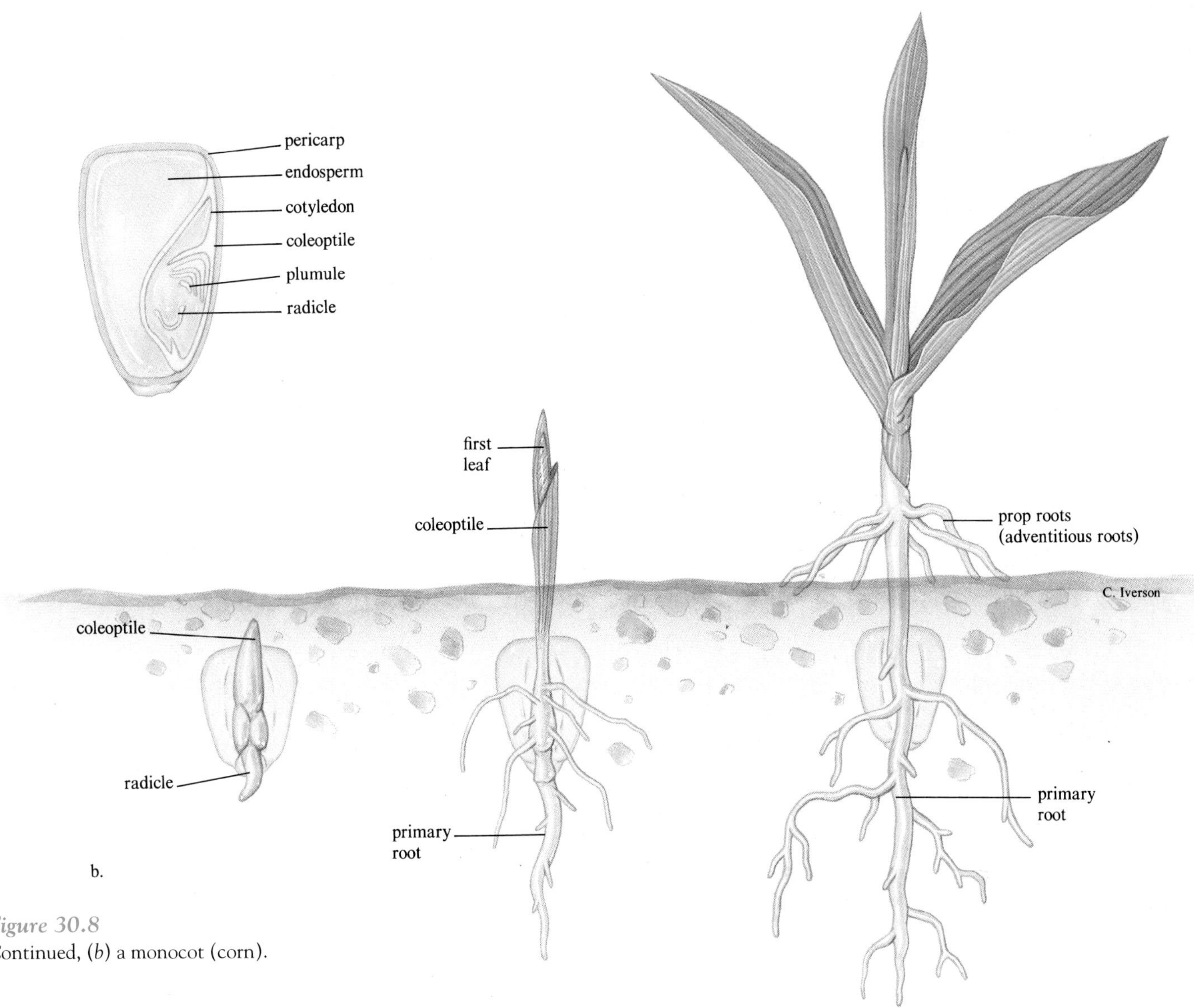

Figure 30.8
Continued, (*b*) a monocot (corn).

KEY CONCEPTS

Pollen is transferred from anthers to stigmas by animals or the wind. Here the pollen grain forms a pollen tube, which grows through the stigma and style to the ovary, where its two nuclei enter the embryo sac. One sperm nucleus fuses with the egg to form the zygote, and the other fuses with two polar nuclei to form the endosperm. A dicot embryo forms two seed leaves, and a monocot embryo forms one. Dicot cotyledons grow thick by absorbing endosperm, but in monocots the cotyledon transfers food from endosperm to embryo. The epicotyl develops above the cotyledons, with an apical meristem. Another meristem develops at the embryonic root, or radicle.

Seed Dormancy

At a certain point in embryonic development, cell division and growth stop, and the embryo becomes dormant. The dormant plant embryo and its food supply are protected by a tough outer layer called the **seed coat.** Together, the plant embryo, stored food, and seed coat comprise the seed.

Why should a plant embryo simply stop growing? Seed dormancy is a crucial adaptation enabling seeds to postpone development when the environment is unfavorable such as during a drought. Seeds of some plants can delay development for hundreds of years. Favorable conditions trigger growth to resume when young plants are more likely to survive. Dormant seeds are also more likely to disperse into new environments.

Fruit Formation

A flower begins to change during seed formation. When a pollen tube begins growing, the stigma produces large amounts of **ethylene,** a simple organic molecule that is a plant hormone. (Recall that a hormone is a biochemical produced in one part of an organism that exerts a physiological effect on some other part of the organism.) Ethylene triggers senescence of the flower. Floral parts that are no longer needed—all parts except the ovary—wither and fall to the ground. Sometimes the ovary swells

a.

b.

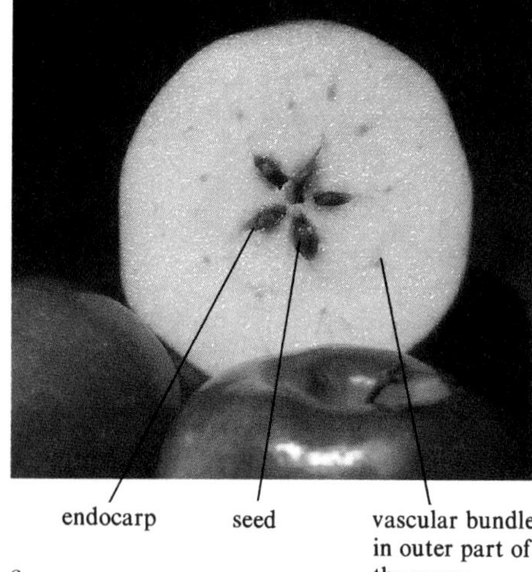

c.

Figure 30.9
Fruits. *a.* A drupe is a fleshy fruit with a hard pit surrounding a single seed. A peach is a drupe. *b.* A berry is fleshy and contains many seeds. Tomatoes, peppers, and eggplants are berries, although we mistakenly think of them as vegetables. Ironically, blackberries, raspberries, and strawberries are not true berries. They are termed aggregate fruits and consist of several small fruits clustered together. *c.* An apple is an example of a pome, a fleshy fruit that develops mostly from the tissues surrounding the ovary. The outermost layer feels like paper.

and develops into a **fruit,** which is a ripened ovary enclosing a seed. Different types of fruits are shown in figure 30.9.

Plant hormones continue to influence fruit development. Seeds within fruits synthesize the hormone **auxin,** which stimulates fruit growth. Ethylene hastens fruit ripening in many species, including tomatoes, apples, and pears. Fruit falls from the plant when auxin levels drop and ethylene levels rise. (Plant hormones are discussed in the next chapter.)

Fruit and Seed Dispersal

In addition to protecting vulnerable seeds from desiccation, fruits facilitate seed dispersal. Attractive fruits such as shiny berries are eaten by birds and other animals. The animals carry the ingested seeds to new locations where they are released in the animals' feces. Surprisingly, some seeds will germinate only after passing through the intestines of birds or mammals. The beginning of chapter 33 describes the relationship between the now-extinct dodo bird and the hard fruits of the calvaria tree. The plant could not germinate unless the hard coats of its fruits were dismantled in the digestive tracts of the large birds.

Mammals spread seeds from place to place when fruits bearing hooked spines become attached to their fur. The fruit of the burdock plant, for example, has barbed hooks that cling to a passing deer or the jeans of a hiker. The inspiration for velcro, a fuzzy fabric that sticks to other fabrics, came from the annoyingly strong attachment of burdock fruits (fig. 30.10).

Wind-dispersed fruits such as those of dandelions and maples have wings or other structures that enable them to ride far from their places of origin on air currents. Coconuts are water-dispersed fruits that travel long distances before colonizing distant islands.

Seed Germination

Germination, the resumption of growth and development, is a sort of reawakening for a seed. Germination usually requires water, oxygen, and a source of energy. The first step is **imbibition,** which is the absorption of water by a seed (fig. 30.11). In some seeds imbibition causes the embryo to release hormones that stimulate the breakdown of the endosperm or stored food reserves. Starch is converted to sugar that the embryo uses for energy. Imbibition also swells a seed, eventually rupturing the seed coat, and exposing the plant embryo to oxygen. At this point a seed may resume growth. However, seeds of many plants normally germinate only after some additional requirement is met—exposure to light of a certain intensity or a series of cold days, for example.

Plant Development

After the growing embryo bursts out of the seed coat, further growth and development depend upon the root and shoot apical meristems. The hypocotyl, with its attached radicle, emerges first from the seed. In response to gravity, the radicle grows downward and anchors the plant in the soil. Root systems develop rapidly because plants require a constant water supply for continued growth.

The ways in which the shoot emerges from the soil differ from species to species. In most dicots, the elongating hypocotyl forms an arch that breaks through the soil and straightens in response to light, pulling the cotyledons and epicotyl out of the soil. In most monocots the epicotyl begins growing upward shortly after the initiation of root growth, protected by its coleoptile as it pushes through the soil. The single cotyledon remains underground.

A plant is producing its first chloroplasts by the time its shoot emerges from the soil. When embryonic food reserves are exhausted, plants can produce their own food photosynthetically.

a.

b.

c.

d.

e.

Figure 30.10
Seed dispersal. *a.* Velcro resembles the burdock fruit. *b.* This cedar waxwing helps disperse the seeds of winterberry, a type of holly. *c.* These tumbleweed plants look calm and firmly rooted, but if a brisk wind comes along, an entire plant can blow away. *d.* Seeds of the dandelion have fluff that enables them to float on a breeze. *e.* Some seed pods have their own explosive methods of scattering seeds. Beans pop out of this garbanzo bean pod.

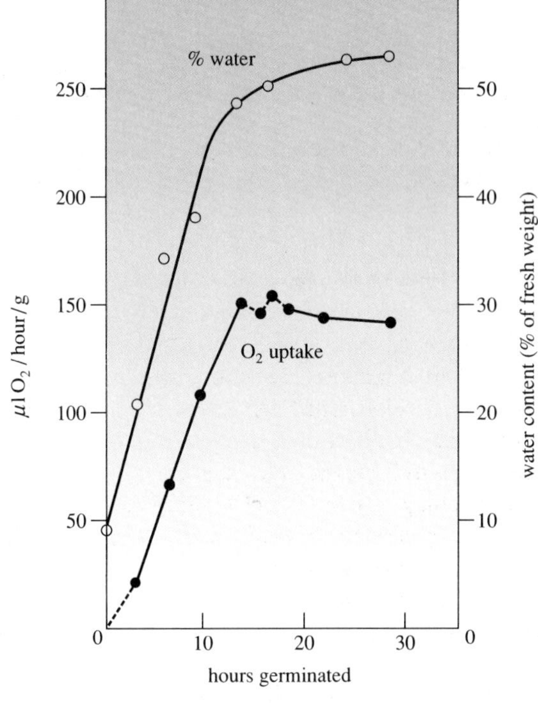

Figure 30.11
Seed germination. The imbibition of water is necessary for a seed to germinate. Notice that as the percentage of water increases, the oxygen uptake increases too, signaling the onset of cellular respiration.

KEY CONCEPTS

Plant embryos become dormant, as seeds, enabling them to avoid environmental extremes and to be dispersed. When seeds form, most parts of the flower fall off, but the ovary enlarges to become a fruit. The fruit protects the seed and disperses it. Germination begins with imbibition of water. Oxygen and energy and sometimes environmental cues must also be present. Nutrient stores are broken down to yield energy. The embryo bursts from the seed coat and primary growth ensues. As the shoot emerges, photosynthesis begins.

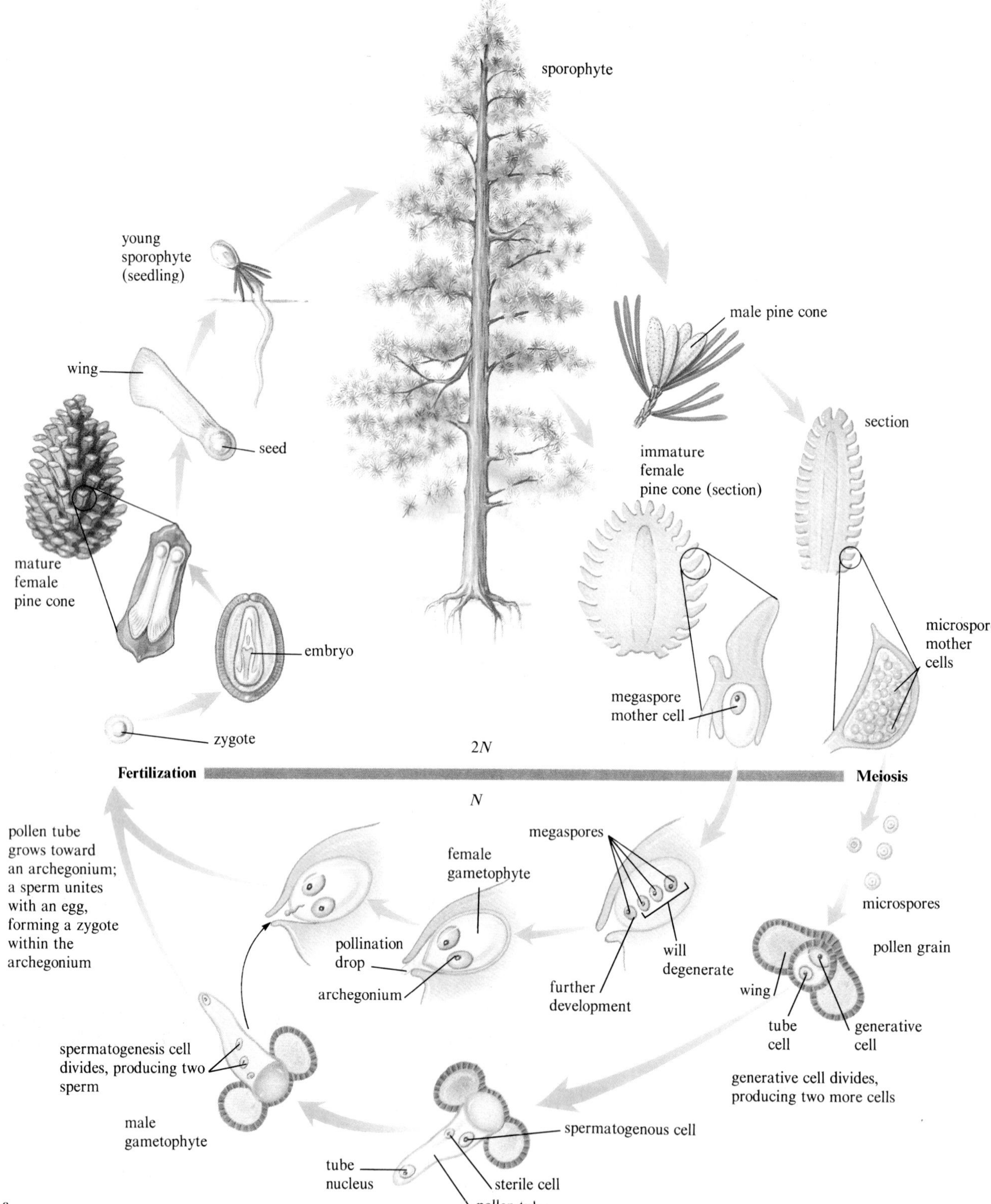

Figure 30.12
a. Pine life cycle. Notice once again the alternation of generations. Pines have no flowers, but they do possess male and female cones, which produce pollen and eggs. Notice also that the sporophyte stage is entirely dominant, and the gametophyte consists of only a few cells. *b.* Pollen can be seen blowing from the male cones of this tree.

b.

Pollination occurs when airborne pollen grains drift down between the scales of female cones and adhere to drops of a sticky secretion. Afterwards the cone scales grow together, and the pollen tube begins growing through the ovule toward the egg. But the pollen is not yet mature—before the pollen tube reaches the egg, the pollen grain must undergo two more cell divisions to become a mature, six-celled microgametophyte. Two of six cells become active sperm cells, one of which fertilizes the egg cell. The whole process happens so slowly that fertilization occurs about 15 months after pollination!

Within the ovule, the developing embryo is nourished by the haploid tissue of the megagametophyte. Following a period of metabolic activity, the embryo becomes dormant and the ovule develops a tough, protective seed coat. It may remain in this state for another year. Eventually, the ovule is shed as a seed. If conditions are favorable, the seed germinates, giving rise to a new tree.

The gymnosperm life cycle differs from the angiosperm life cycle in several ways:

- The reproductive structures are cones instead of flowers.
- Ovules lie bare on female reproductive structures rather than embedded in their tissues.
- Gymnosperms have no fruit.
- Gymnosperms have single fertilization rather than double fertilization.
- The haploid tissue of the female gametophyte rather than a triploid endosperm nourishes the gymnosperm embryo.

The Pine Life Cycle

Angiosperms are the only vascular plants that bear seeds in protective structures. Vascular plants whose seeds are not protected are the gymnosperms, a term that literally means "naked seed." Conifers are the best-known group of gymnosperms, including such familiar species as pines, spruces, firs, and redwoods.

Cones are the reproductive structures of pines. Large, female cones bear two ovules (megasporangia) on the upper surface of each **scale.** Through meiosis, each ovule produces four haploid megaspores, three of which degenerate and one of which continues to develop into a female gametophyte. Over many months the female gametophyte undergoes mitosis (fig. 30.12*a*). Finally, two to six structures called archegonia form, each housing an egg that is ready to be fertilized.

Small male cones have pairs of microsporangia borne on thin, delicate scales. Microsporangia produce microspores through meiosis, and each microspore eventually becomes a four-celled pollen grain (microgametophytes). Millions of winged pollen grains are released when microsporangia burst. If you tap a male cone in the spring, you can see a halo of pollen released (fig. 30.12*b*). Most pollen grains, however, never reach a female cone.

KEY CONCEPTS

Gymnosperm seeds are unprotected. In pines, two ovules occupy each scale of a female cone. In meiosis, the ovule produces three haploid cells that degenerate and one that eventually produces eggs. Pairs of microsporangia on male cones yield four-celled pollen grains, which are released to the air. Pollen lands directly on female cones, and the pollen tubes grow toward the egg. The pollen divides two more times. Of the six cells that result, two are sperm and one of them fertilizes the egg. The embryo metabolizes briefly, then lies dormant for a year, after which it may germinate.

a.

b.

c.

Figure 30.13
Vegetative propagation comes in many forms. New shoots can come from underground, as in bamboo (*a*), or new plantlets can form on leaves, as in a fern (*b*). In the laboratory, greenhouse, or home kitchen window, cuttings from many plants can be coaxed to grow roots and even flower in water or in a growth medium, like this pussy willow in water (*c*).

Asexual Reproduction

Many plants reproduce by asexually forming new individuals by mitotic cell division. Compared with sexual reproduction, it is a simple process—there is no meiosis, no gamete production, and no fertilization. Instead, a parent simply gives rise to genetically identical individuals called **clones.**

Asexual reproduction is particularly advantageous when environmental conditions are stable and plants are well-adapted to their surroundings. Recall that sexual reproduction introduces genetic variation into a population. Why risk losing a favorable combination of genes when asexual reproduction will produce clones well suited to the environment?

Plants often reproduce or propagate, vegetatively (asexually) by forming new plants from portions of their roots, stems, or leaves (fig. 30.13). For example, adventitious buds form on roots of cherry, pear, apple, and black locust plants, and when they sprout, aerial shoots grow upward. These shoots are called "suckers" because they draw on materials from the parent plant. If separated from the parent plant, suckers become new individuals.

A few plants use their leaves to reproduce asexually. When the leaves of walking fern or maternity plant lie on a moist surface, roots and shoots develop at their edges. These plantlets become new individuals when they are shed from the parents' leaves.

Asexual reproduction also occurs from modified stems. Stolons, also called runners, grow along the soil surface. Roots and shoots form at intermittent nodes and eventually form new plants some distance from the parent plant. Stolons are common in plants such as strawberry, Boston fern, spider plant, and crabgrass.

Tubers are swollen regions of stems that grow below ground. They produce nodes that can grow roots and shoots. A potato is a swollen stem, and the eyes are buds from which new potato plants arise asexually.

The reproductive structures of plants add a great deal of joy to our lives. Imagine a celebration without flowers! How wonderful that these beautiful and fragrant biological structures also provide an efficient mechanism for the mixing of genetic material, which is the backbone of evolutionary success in these organisms.

KEY CONCEPTS

Asexual reproduction in plants is much simpler than sexual reproduction. Vegetative propagation occurs from adventitious buds, leaves, and modified stems.

SUMMARY

Sexually reproducing plants exhibit *alternation of generations*, in which the diploid *sporophyte* generation produces spores through meiosis. The spores divide mitotically to produce the haploid *gametophyte*, which in turn produces haploid gametes. In different species, either the sporophyte or gametophyte generation dominates.

Flowers are reproductive structures of angiosperms. The *calyx*, made of *sepals*, and the *corolla*, made of *petals*, are accessory structures. Inside the corolla is the *androecium*, which consists of the male *stamens* and their pollen-containing *anthers*. At the center of the flower is the *gynoecium*, enclosing the *ovary*. The *stigma* extends from the ovary and captures pollen.

In the anther, four *pollen sacs* contain *microspore mother cells*, each of which divides by meiosis to yield four haploid *microspores*, which divide mitotically to yield a haploid *generative nucleus* and a haploid *tube nucleus*. These two cells and their covering constitute a *pollen grain*. Sperm cells arise from the generative cells. In the ovary, *megaspore mother cells* divide meiotically to yield four haploid cells, one of which persists as a haploid *megaspore*, which then divides mitotically three times. The resulting megagametophyte, or *embryo sac*, contains several cells, one of which is the egg.

Pollen is transferred from an anther to a stigma, either within the same flower or plant or between plants. Flower structures and odors have evolved to appeal to pollinators or to utilize wind dispersal. Once on a stigma, a pollen grain grows a *pollen tube*, in which the sperm approaches the ovary. In the embryo sac, one sperm fertilizes the egg to form the zygote, and the second sperm fertilizes the *polar nuclei*, forming the *endosperm*, which nourishes the embryo. *Cotyledons* develop, absorbing the endosperm in dicots and transferring nutrients in monocots. Apical meristems promote growth of the shoot and root ends of the embryo.

After fertilization, no longer needed floral parts fall off, and the ovary may develop into a fruit, under the influence of hormones. A seed germinates in response to an environmental cue, and it requires oxygen, energy, and water. When the embryo bursts from the seed coat, primary growth ensues.

In pines, large female cones bear two ovules per scale, which yield three haploid degenerate cells and one haploid megaspore through meiosis. Microsporangia grow on the small scales of males cones. Through meiosis, microspores are produced, which become pollen grains. Pollen lands on sticky material on female cones, and the pollen tube grows towards the egg. Here the pollen matures by dividing twice. Two of the resulting cells are sperm, one of which fertilizes the egg. The embryo is nourished by the megagametophyte and enters dormancy, protected by a tough seed coat.

In asexual reproduction, a parent plant gives rise to clones, which can develop from roots, stems, or leaves.

QUESTIONS

1. Describe the steps in alternation of generations.

2. What is the function of flowers? Which floral structures participate directly in this function? Which structures participate indirectly?

3. What floral structures might a male wasp encounter when pollinating orchids?

4. What are three adaptations that have evolved in plants to encourage outcrossing?

5. What type of pollinator might be attracted to a plant with the following:
   a. Red, sweet-smelling flowers
   b. Dull-colored, spicy-smelling flowers
   c. Yellow, heavily scented flowers

6. Why do wind-pollinated species produce more pollen than animal-pollinated plants?

7. Name a tissue or cell in an angiosperm that is haploid, diploid, and triploid.

8. Gardeners know to plant many more seeds than should suffice for the number of plants desired, because not all of them will germinate. What are some reasons why seeds might fail to germinate?

9. How is the life cycle of a corn plant and a spruce pine similar and different?

TO THINK ABOUT

1. Chefs consider a plant food a fruit or a vegetable according to when it is eaten and how sweet it tastes. How does this differ from the biological definition of a fruit?

2. Why is it important for a particular insect species to pollinate only one species of flowering plant?

3. How could you tell if a portion of a plant is part of the sporophyte generation or the gametophyte generation?

4. Under what type of environmental condition might asexual reproduction be of most benefit to a species, and when would sexual reproduction be the most beneficial?

5. Humans and plants each have diploid as well as haploid cells. How do the life cycles of humans and flowering plants differ?

6. How are petals, which are considered accessory structures, nevertheless necessary for reproduction?

7. In the Midwest, children can get summer jobs "detasselling" corn, which is removing the male parts. Why might a farmer want this done to the crop?

8. Why might a flower smell like putrefied meat?

SUGGESTED READINGS

Beattie, Andrew J. February 1990. Ant plantation. *Natural History*. Ants help disperse seeds.

Darwin, Charles. 1862. *On the various contrivances by which British and foreign orchids are fertilized*. The father of evolution was particularly fascinated by the adaptations of flowers to their pollinators' tastes.

Goldberg, Robert B. June 10, 1988. Plants: Novel developmental processes. *Science*, vol. 240. Structurally plants are simpler than animals, but on the molecular level they too are complex.

Huntly, Brian, and I. Colin Prentice. August 5, 1988. July temperatures in Europe from pollen data, 6,000 years before present. *Science*, vol. 241. The structures of fossilized pollen hold clues to long-ago environments.

Pennisi, Elizabeth. April-May 1990. Planting the seeds of a nation. *National Wildlife*. Native wildflowers make spectacular garden displays.

Raab, Mandy M., and Ross E. Konig. November 1988. How is floral expansion regulated? *BioScience*, vol. 38, no. 10. For successful pollination, a flower must unfold rapidly in a specific way.

Robackes, David C., et al. June 1988. Floral aroma. *BioScience*, vol. 38, no. 6. The fragrances of flowers are closely tied to the olfactory preferences of their pollinators.

C H A P T E R

31

Plant Responses to Stimuli

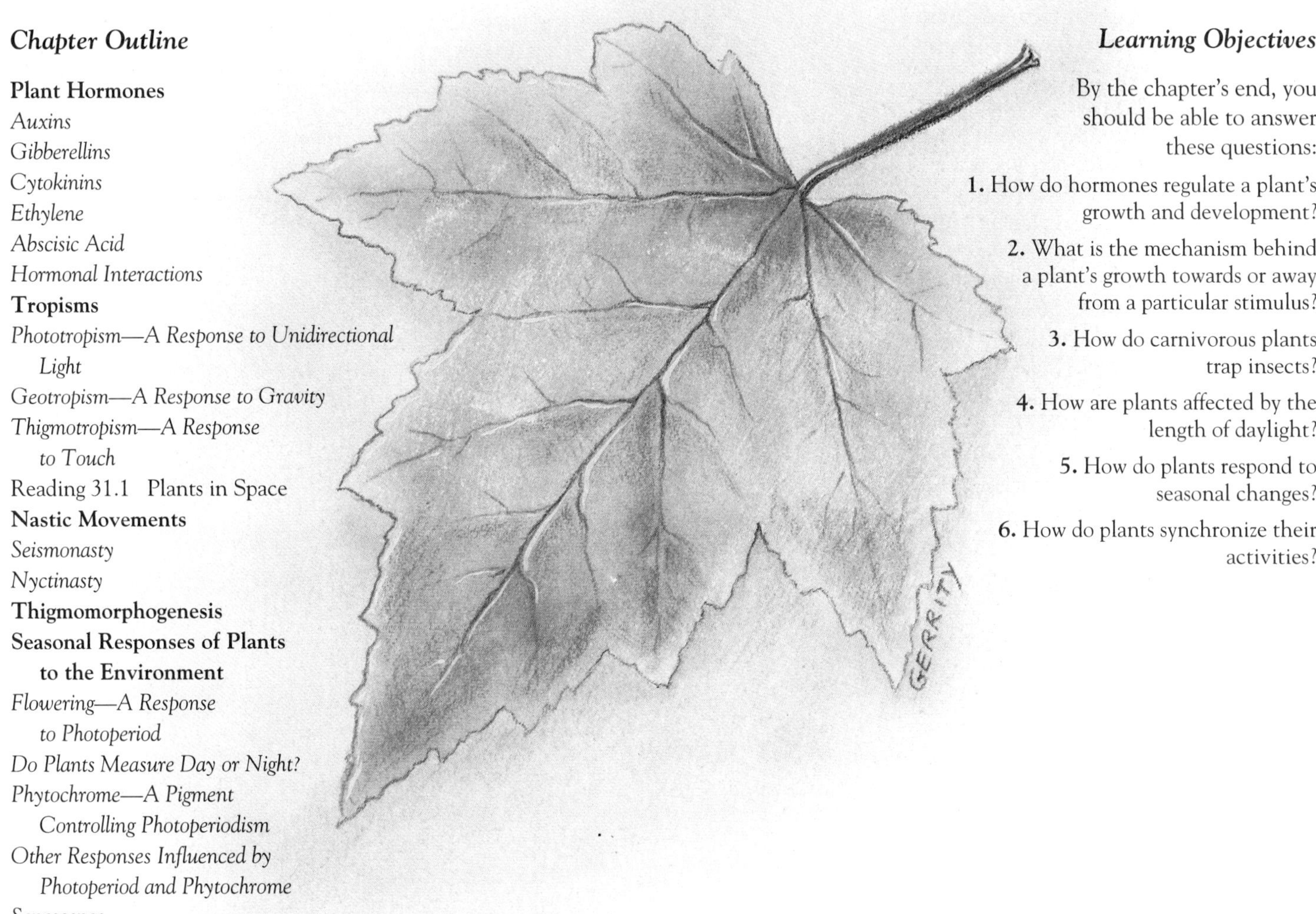

Chapter Outline

Learning Objectives

By the chapter's end, you should be able to answer these questions:

1. How do hormones regulate a plant's growth and development?
2. What is the mechanism behind a plant's growth towards or away from a particular stimulus?
3. How do carnivorous plants trap insects?
4. How are plants affected by the length of daylight?
5. How do plants respond to seasonal changes?
6. How do plants synchronize their activities?

To a human observer, the sundew plant is a magnificent member of a swamp community. To an insect, however, its beckoning club-shaped leaves, bearing nectar-covered tiny tentacles, are treacherous. Once the insect alights on the plant and begins to enjoy its sweet, sticky meal, the surrounding hairlike tentacles begin to move. Gradually they fold inward, entrapping the helpless visitor, forcing it down toward the leaf's center. Here, powerful digestive enzymes go to work, dismantling the insect's body, releasing its component nutrients. After 18 hours, the leaves open. All that remains of the previous day's six-legged guest is a few bits of indigestible matter (fig. 31.1).

The response of the sundew plant to an insect touching structures on its leaves dramatically illustrates a plant's response to its immediate environment. Other plant responses may not appear as exciting but are also the culmination of a complex interplay of biochemicals and little-understood biological activities.

Unlike animals, most plants cannot escape unfavorable conditions by moving away. Plants adapt to the environment by growing, which is influenced by both internal and external factors. A plant's DNA contains the blueprints for organizing specialized cells into tissues and organs, and these genetic instructions can supply several options. External factors, such as light intensity or temperature, may determine which option is expressed. The result can be a finely tuned growth response that enables the organism to adapt to its surroundings and survive.

Plant Hormones

Many aspects of plant growth are regulated by hormones, which are biochemicals produced in small quantities. Like animal hormones, plant hormones are synthesized in one part of the organism and are transported to another part, where they stimulate or inhibit growth. Yet unlike animal hormones, plant hormones are not produced in tissues specialized for hormone production, and they do not have definite target areas.

Five major classes of plant hormones are known: **auxins, gibberellins, cytokinins, ethylene,** and **abscisic acid** (table 31.1

Figure 31.1
This sundew plant (*Drosera rotundifolia*) grows in the swamps of upstate New York. Its attractive leaves are actually insect traps.

Table 31.1
Major Classes of Plant Hormones

Class	Principal Actions
Auxins	Cell elongation in seedlings, shoot tips, embryos, leaves
Gibberellins	Cell elongation and division in seeds, roots, shoots, young leaves
Cytokinins	Stimulate cytokinesis in seeds, roots, young leaves, fruits
Ethylene	Hastens fruit ripening
Abscisic acid	Inhibits growth

and fig. 31.2). Plant hormones have both characteristic and unpredictable effects. Their influences depend upon which other hormones are present, and the sensitivity of the tissue to the hormones. As a result, a single hormone can elicit numerous responses.

Auxins

Auxin was the first plant hormone to be described. In the late 1870s, decades before the chemical identification of plant hormones, Charles Darwin and his son Francis learned that a plant-produced "influence" caused growth towards light (fig. 31.3). They were describing auxin, a plant hormone that stimulates cell elongation in grass seedlings and herbs. Auxin apparently coaxes cells to elongate by stretching their cell walls. Auxin acts rapidly, spurring noticeable growth in a grass seedling in minutes.

Three naturally occurring auxins in plants are known. The most active auxin is **indoleacetic acid,** or **IAA.** It is produced in shoot tips, embryos, young leaves, flowers, fruits, and pollen. Synthetic compounds having auxin-like effects are important commercially, such as 2,4-D (2,4-dichlorophenoxyacetic acid), which is used extensively as an herbicide. When applied in concentrations higher than auxin would normally be present, 2,4-D causes weeds to elongate rapidly and literally grow to death.

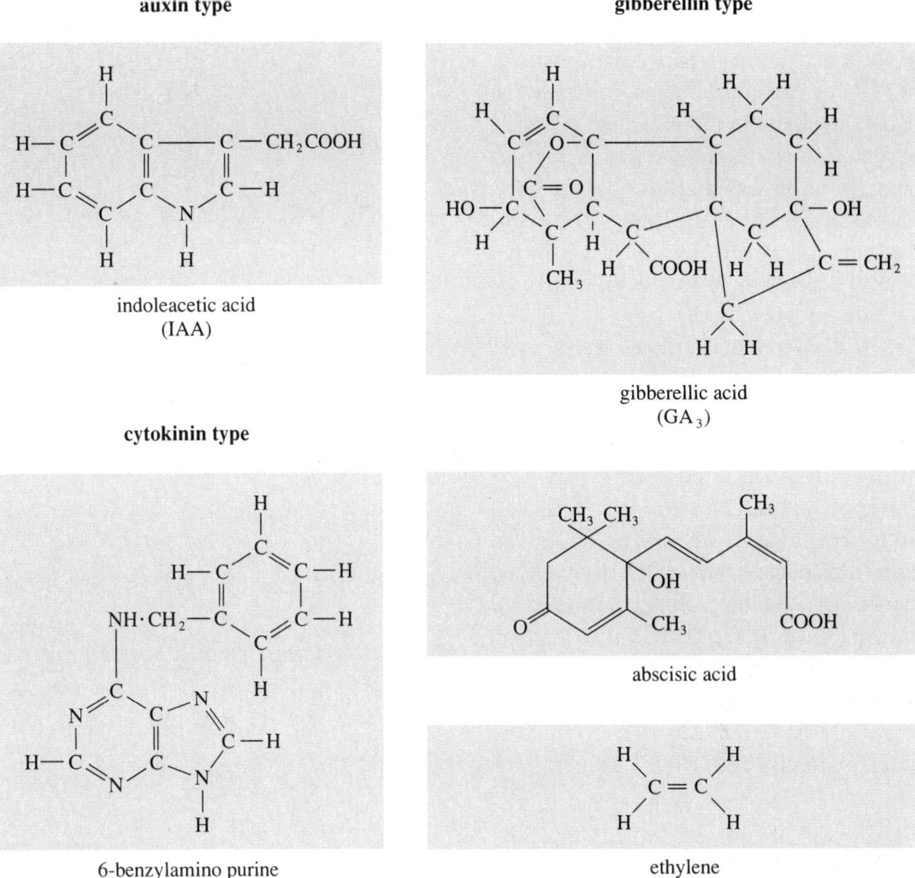

Figure 31.2
The major groups of plant hormones. All hormones produce varying effects when the plant uses them in combination.

Figure 31.3
Effects of auxin. The pea plants on the left were not treated with auxin; those on the right were.

Gibberellins

In the late 1920s, Japanese botanists studying "foolish seedling disease" in rice discovered gibberellin, another plant hormone involved in shoot elongation. Plants suffering from foolish seedling disease grow rapidly because they are infected by a fungus that produces gibberellin. The infected plants become so spindly that they fall over and die (fig. 31.4).

Gibberellin, abbreviated GA, stimulates shoot elongation in mature regions of trees, shrubs, and a few grasses and is also found in immature seeds, apices of roots and shoots, and young leaves of flowering plants. GA induces both cell division and elongation. GA-induced elongation occurs after about a 1-hour delay, which is much slower than auxin-induced growth. More than 65 naturally occurring gibberellins are known.

Figure 31.4
Effects of gibberellin. Note the size difference between these two California poppy plants. Can you tell which one was treated with gibberellin?

Cytokinins

As early as 1913, scientists knew that some compound stimulates cell division in plants. It was not until 1964 that the first naturally occurring cytokinin was discovered in corn kernels. Since then, researchers have isolated other cytokinins and have synthesized several artificial ones.

Cytokinins are so-called because they stimulate cytokinesis (the division of the cell after the genetic material has replicated and separated) by pushing cells into mitosis. Cytokinins do not work alone—auxin must be present before mitosis begins. The effects of cytokinins are similar to some of the effects of auxin and gibberellin—they promote cell division and growth and participate in development, differentiation, and senescence.

In flowering plants, most cytokinins are found in roots and actively developing organs such as seeds, fruits, and young leaves. Synthetic cytokinin-like compounds have a variety of uses, including keeping wheat stalks short so they are not blown over; shortening shrubs to a manageable level; and extending the shelf-life of lettuce and mushrooms.

Ethylene

The effects of ethylene were known long before its discovery. In 1910, scientists in Japan observed that bananas ripened prematurely when stored with oranges. Apparently, the oranges produced something that induced rapid ripening. By 1934, scientists realized it was the simple gas ethylene that hastened ripening.

Ethylene ripens fruit in many species. Ripening is a complex process involving pigment synthesis, fruit softening, and breakdown of starches to sugars. A tomato picked from the garden when it is hard and green with streaks of pale orange will turn, aided by ethylene, into a soft, red, succulent tomato.

Ethylene also helps ensure that a plant will survive injury or infection. The hormone is produced when the plant is damaged, and it hastens aging of the affected part so that it can be shed before the problem spreads to other regions of the plant.

Ethylene is synthesized in all parts of flowering plants, but large amounts form in roots, the shoot apical meristem, nodes, and ripening fruits. The dark spots on a ripening banana peel are concentrated pockets of ethylene. Because ethylene is a gas, its effects can be contagious. The expression "one bad apple spoils the batch" refers to the ability of the ethylene released from one apple to hasten the ripening, and spoiling, of others nearby (fig. 31.5).

Although the chemical structure of ethylene is simple, its production by a plant is actually the culmination of several biochemical steps. Genetic engineers have cloned the gene that specifies the enzyme necessary for the very last step in ethylene production. By blocking this gene, they can therefore block ethylene production. What commercial benefit might blocking a plant's ethylene production serve?

Abscisic Acid

Could plants manufacture growth-inhibiting substances too? By the 1940s, botanists thought so. Twenty years later, researchers isolated such a compound.

Abscisic acid, or ABA, inhibits the growth-stimulating effects of many other hormones. ABA is used commercially to inhibit growth of nursery plants so that they are less likely to be damaged during shipping.

Hormonal Interactions

Plant hormones seldom function alone. Several plant hormones, for example, influence abscission, which is the shedding of leaves or fruit. Abscission is preceded by senescence, the aging and death of a plant or plant part. Recall from chapter 29 that during senescence, an abscission zone forms at the base of the organ that will be shed. Eventually, the leaf or fruit separates from the plant at this abscission zone.

Senescence is normally retarded by auxin produced by actively growing leaves and fruit, along with cytokinin and gibberellin produced by roots. Senescence begins when auxin production drops in response to environmental stimuli, such as injury or the shorter days of autumn. During senescence, cells in the abscission zone begin producing

Figure 31.5
As ethylene accumulates, these tomatoes ripen.

ethylene, which swells the cells and prompts them to produce biochemicals that digest cell walls. As a result, the cells separate and the leaf or fruit drops.

If decreasing levels of auxin lead to abscission, then the application of synthetic auxins should prevent abscission. Indeed, synthetic auxins often are used to prevent preharvest fruit drop in orchards, to hold berries on holly, and to coordinate abscission of fruits at harvest time.

The coordinated production of several hormones permits a plant to survive the changing seasons as well as extreme weather conditions. This enables the plant to develop its reproductive organs and structures, ensuring the perpetuation of the species.

KEY CONCEPTS

Plant hormones coordinate growth and development. Auxins provoke rapid cell elongation in various growing plant parts. Gibberellins promote cell division and elongation in roots, shoots, leaves, and immature seeds of flowering plants and in mature parts of trees and shrubs. Cytokinins stimulate cells to enter mitosis and are found in rapidly dividing plant parts. Ethylene is a gaseous hormone that hastens ripening. Abscisic acid inhibits the growth-stimulating effects of other hormones.

Tropisms

Despite their obvious immobility plants are acutely responsive to some environmental signals. The glorious heads of sunflowers turn towards the sun, and roots grow downwards in response to gravity. Some plant responses are rather short-term. A Venus's-flytrap closes in less than a second, and the curving of a stem towards light usually takes only a few hours. Other behaviors, such as flowering, are long-term responses associated with changing seasons. All of these responses to environmental stimuli are a result of growth.

The term **tropism** refers to plant growth toward or away from environmental stimuli, such as light or gravity. Each of the many types of tropisms is named for the stimulus eliciting the response. Phototropism, for example, is a growth response to unidirectional light, and geotropism is a growth response to gravity.

Tropisms result from differential growth, in which one side of the responding organ grows faster than the other, bending the part. Curving of an organ towards the stimulus is called a *positive tropism*, such as stems growing toward light. Curvature of an organ away from a stimulus is a *negative tropism*. Roots are negatively phototropic.

Phototropism—A Response to Unidirectional Light

During **phototropism,** cells on the shaded side of a stem elongate faster than cells on the lighted side of the stem. The rapid elongation of cells along the shaded side of coleoptiles is controlled by auxin coming from the apex.

Precisely how auxin controls phototropism was discovered in the 1950s by Winslow Briggs and his colleagues. First, they determined that the amount of auxin produced by coleoptiles grown in the light is the same as that of coleoptiles grown in the dark—that is, light does not destroy auxin. They then discovered that more auxin could be collected from the shaded side of coleoptiles than from the lighted side, suggesting that light causes auxin to migrate to the shaded side of the stem. More recent experiments support this finding. Auxin labeled with radioactive carbon (^{14}C) and exposed to unidirectional light moves to the shaded side of coleoptiles. Cells in the shade elongate more rapidly than cells in the light, which curves the coleoptile towards the light (fig. 31.6).

How does a plant sense unidirectional light? Only blue light with a wavelength less than 500 nanometers effectively induces phototropism. The yellow pigment **flavin** is probably the photoreceptor molecule for phototropism. Flavin alters transport of auxin to the shaded side of the stem or coleoptile.

Figure 31.6
Positive phototropism. This kidney bean was grown under a hood (shown on the right) for 5 days. Note how it has grown toward the light.

Geotropism—A Response to Gravity

Charles Darwin and his son Francis also studied **geotropism,** which is a growth response to gravity. How would roots grow if their caps were removed, they wondered. They found that decapped roots grew, but not downwards, in response to gravity. Therefore, the root cap is necessary for geotropism. It is interesting to examine root growth in the absence of gravity, which is possible by growing plants aboard orbiting spacecraft, where gravity is minimal (Reading 31.1). The nation's schoolchildren are evaluating the growth of tomatoes that spent 5 1/2 years in space as seeds. NASA has supplied "space" seeds and earthbound controls to interested classes, and the students record the growth and conduct other experiments.

A plant's shoot is negatively geotropic, because it grows upwards, and roots are positively geotropic, extending downward along with the gravitational force. Curiously, accumulation of auxin seems to provoke each of these opposite responses. In a horizontal stem or shoot, auxin accumulates on the lower side, stimulating differential growth there. As a result, the structure bends upward (fig. 31.7). However, if a root is held horizontally, accumulating auxin in the lower regions causes downward growth. How can this be? One hypothesis is that root tissue is far more sensitive to auxin than shoot tissue, and the amount of auxin accumulating is so great that it actually inhibits cell elongation. With cells in the upper portion of a horizontal root growing faster, the root bends downward.

Figure 31.7
Formerly horizontal branches of this fallen tree now grow upward, due to geotropism.

Thigmotropism—A Response to Touch

The coiling tendrils of twining plants such as morning glory and bindweed display **thigmotropism,** a response to touch. When hanging free, tendrils often grow in a spiral fashion, which increases their chances of contacting an object to which they can cling (fig. 31.8). Contact with an object is detected by specialized epidermal cells, which induce differential growth of the tendril. The tendril completely encircles the object in only 5 or 10 minutes. Thigmotropism is often long-lasting. Stroking a

Reading 31.1 *Plants in Space*

PLANTS WILL BE A KEY PART OF SPACE COLONIZATION, PROVIDING FOOD AND OXYGEN. But how will organisms that have evolved under constant gravity function in its near-absence beyond the earth?

Researchers are studying the effects of microgravity on a variety of species by sending them on space shuttle voyages. Such experiments can more realistically assess plant growth and development in space than previous earthbound simulations, which used a rotating device called a clinostat to diminish gravitational force.

So far, it appears that plants are profoundly affected by a lack of gravity, from subcellular structural organization to whole organismal functioning. These responses will present interesting challenges to space farmers of the future.

Subcellular Responses to Microgravity

Plant cells grown in space have fewer starch grains and more abundant lipid-containing bodies, indicating a change in energy balance. This may mean that space farmers may have plenty of vegetable oil but no french fries to cook in it.

Organelle organization is grossly altered in the absence of gravity. Endoplasmic reticula are bunched together and arranged randomly, and mitochondria swell when freed from the constraints of gravity. Nuclei enlarge, and chromosomes break, perhaps due to greater exposure to cosmic radiation. Chloroplasts have enlarged thylakoid membranes and small grana. Most interesting are amyloplasts, which are starch-containing granules in specialized gravity-sensing cells in roots. On earth, amyloplasts aggregate at the bottoms of these cells, but in space, they randomly float about the cell. As a result, a root tip cannot elongate downward in response to gravity (fig. 1).

Cell Division

Mitosis is halted by microgravity, usually at telophase, resulting in some cells with more than one nucleus. Oat seedlings germinated in space had only 1/10 as many dividing cells as seedlings germinated on earth. The spindle apparatus also seems to be disrupted by microgravity. Cell walls formed in space are considerably thinner than their terrestrial counterparts, with less cellulose and lignin. Regeneration is also inhibited in space. A decapped root will regenerate in 2 to 3 days on earth, but it will not do so at all in space. Distribution of dividing cells is altered. Lettuce roots, for example, have a shortened elongating zone when grown in space.

Growth and Development

Germination is less likely to occur in space than on earth because of chromosome damage, but it does happen. Early growth success seems to depend upon the particular species—bean, oat, and pine seedlings grow more slowly than on earth, and lettuce, garden cress, and cucumbers grow faster. Many species, including wheat and peas, cease growing and die before they flower. However, in 1982 Arabidopsis plants (mustard weed) successfully completed a life cycle in space—indicating that human space colonies with plant companions may indeed be possible (see fig. 32.2).

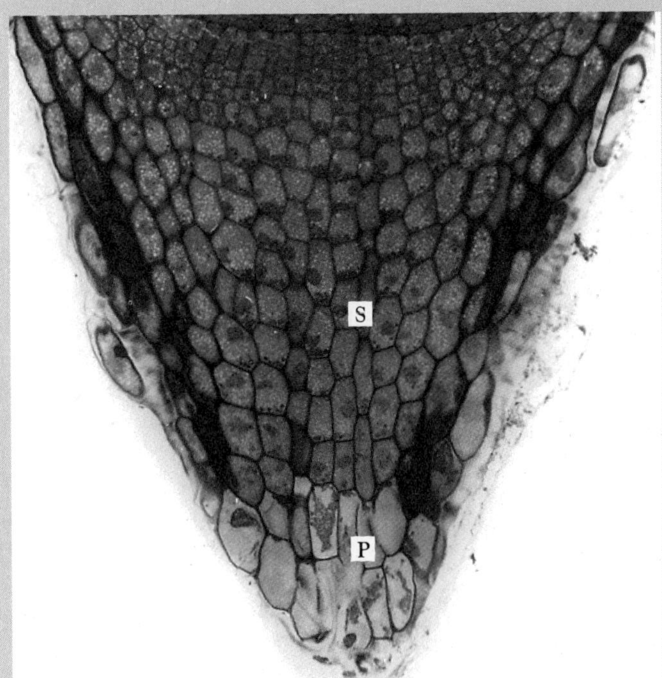

a.

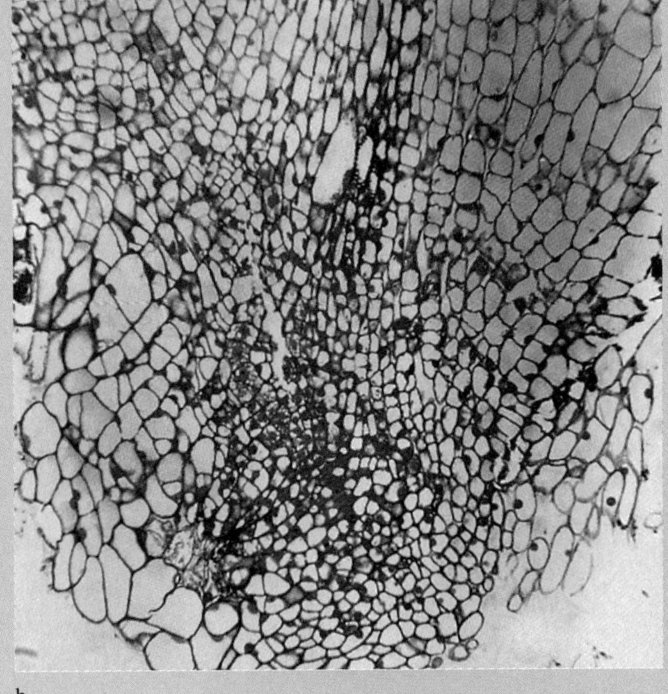
b.

Figure 1
a. On earth, a root whose cap has been removed regenerates an organized, functional root cap. *b.* In the microgravity of space, however, regrowth of a decapped root is disorganized. Clearly, gravitational cues are needed to direct normal regeneration.

Figure 31.8
Thigmotropism. The twining of tendrils is a rapid response to touch. This tendril of a passion vine is wrapped around a blackberry stem.

tendril of garden pea for only a couple of minutes can induce a curling response that lasts for several days. Auxin and ethylene control thigmotropism.

Interestingly, tendrils seem to remember touch. Tendrils growing and touched in the dark do not respond until they are illuminated. It seems that sensory information is stored by tendrils in the dark, but a response does not occur until light is present.

KEY CONCEPTS

Plant growth toward or away from a stimulus is called a positive or negative tropism. Tropisms result from differential growth, which is controlled by hormone distributions. Phototropism is response to light; geotropism is response to gravity; thigmotropism is a coiling response to touch.

Table 31.2
Plant Responses

Stimulus	*Tropism*	*Nastic Movement*
Light	Phototropism	Photonasty
Dark		Nyctinasty
Gravity	Geotropism	
Touch	Thigmotropism	Thigmonasty
Temperature	Thermotropism	Thermonasty
Chemical	Chemotropism	Chemonasty
Water	Hydrotropism	Hydronasty

Nastic Movements

Nondirectional plant motions, called **nastic movements,** include some of the most fascinating responses seen in the plant kingdom (table 31.2).

Seismonasty

A nastic movement resulting from contact or mechanical disturbance is **seismonasty.** Seismonastic movements depend upon a plant's ability to transmit rapidly a stimulus from touch-sensitive cells in one part of the plant to responding cells located elsewhere. Consider leaf movement in the "sensitive plant" mimosa. When leaves of this plant are touched, the leaflets fold and the petiole droops. Touching the leaf elicits a reaction that causes the cell membrane of motor cells at the base of the leaflet to become more permeable to K^+ and other ions. As ions move out of the motor cells, the osmotic potential in the surrounding area decreases, moving water out of the motor cells by osmosis. This loss of water shrinks the motor cells and seismonastic movement occurs. Reversal of this process causes the leaves to unfold in approximately 15 to 30 minutes (fig. 31.9).

What is the adaptive significance of seismonastic movements in plants? Seismonastic movements probably decrease a leaf's chances of being eaten, because folded leaves are more difficult for an animal to see. These movements may defend sensitive plants in other ways as well. For example, sharp prickles located along a leaf axis are exposed when the leaflets close and motor cells secrete noxious substances called tannins, both of which discourage hungry animals.

The Venus's-flytrap is famous for its dramatic seismonastic response (see fig. 29.11). Unlike the sensitive plant in which seismonastic movements result from reversible changes in turgor pressure, movements of the Venus's-flytrap result from irreversible increases in cellular size, which can be initiated by acidifying the cell walls to pH 4.5 and below.

The leafy traps are built of two lobes, each of which has three sensitive "trigger" hairs overlying motor cells. When a meandering animal touches these hairs, signals are sent to the plant's motor cells, which then initiate transport of H^+ to the walls of epidermal cells along the outer surface of the trap. The resulting acidification of these cell walls expands the outer epidermal cells along the central portion of the leaf. Since epidermal cells along the inner surface of the leaf do not change volume, the flytrap snaps shut.

Closure of the Venus's-flytrap takes 1 to 2 seconds and requires a large expenditure of ATP—motor cells use almost one-third of their ATP to pump the H^+ that acidifies the cell walls and closes the trap. The trapping mechanism is quite sophisticated. A trap will not close unless two of its trigger hairs are touched in succession or one hair is touched twice. By responding to two stimuli instead of one, the plant can distinguish potential prey from other objects, such as falling leaves.

Figure 31.9
The mimosa (sensitive) plant on the top has not been touched; the one on the bottom has. This response illustrates seismonasty.

Figure 31.10
The prayer plant exhibits nyctinasty. When the sun goes down, the prayer plant's leaves turn inward.

When the trap closes, the captive animal is pressed against glands located on the inner surface of the trap that secrete digestive enzymes. Digestion of the unlucky animal may take 1 or 2 days. Opening of empty traps usually requires 8 to 12 hours and results from expansion of epidermal cells along the midrib of the inner surface of the leaf. The sundew in figure 31.1 is another carnivorous plant, like the Venus's-flytrap.

Nyctinasty

The nastic response caused by daily rhythms of light and dark is known as **nyctinasty** (from the Greek nyktos, meaning "night," and nastos, meaning "pressed together"), or "sleep movement." The prayer plant maranta is an ornamental houseplant exhibiting nyctinastic responses. Leaves of prayer plants orient themselves horizontally during the day, which maximizes their interception of sunlight. At night, the leaves fold vertically into a configuration resembling a pair of hands in prayer (fig. 31.10).

The movement of the prayer plant's leaves in response to light and dark occurs by changes in turgor pressure of motor cells at the base of each leaf. In the dark, K^+ moves out of cells along the upper side and into cells along the lower side of a leaf base. This ion flux moves water, via osmosis, into cells along the lower side of the leaf base, swelling them. Meanwhile, cells along the upper side lose water and shrink. Overall, the leaf stands vertically due to the changes in cellular volume. At sunrise, the process is reversed and the leaf again lies horizontally. Changes in leaf position can decrease loss of water and heat.

Sorrel and legumes such as beans have similar sleep movements occurring at the same time each day. A clever use of these regular movements was made by Carl Linnaeus, a famous Swedish botanist. Linnaeus filled wedge-shaped portions of a circular garden with plants having sleep movements occurring at different times.

By seeing which plants of his so-called horologium florae (flower clock) were "asleep," Linnaeus could then tell the time of day.

Thigmomorphogenesis

Plants are extremely sensitive to mechanical disturbances such as rain, hail, wind, animals, and falling objects. In response, plants typically inhibit cellular elongation, remaining short and stocky, and produce large amounts of thick-walled supportive tissue (collenchyma and sclerenchyma fibers). For example, spraying tomato plants with water for only 10 seconds per day reduces their growth by 40%. This response to mechanical disturbances is called **thigmomorphogenesis,** and it is controlled by ethylene.

KEY CONCEPTS

Nondirectional plant movements are called nastic movements. Nastic response to contact is seismonasty; to daily patterns of light and dark is nyctinasty; to touch is thigmomorphogenesis.

Seasonal Responses of Plants to the Environment

Seasonal changes affect plant responses. Autumn brings cooler nights and shorter days, which produce beautifully colored leaves, dormant buds, and decreased growth in preparation for winter. In the spring, buds resume growth and rapidly transform a barren forest into a dynamic, photosynthetic community.

Flowering—A Response to Photoperiod

Flowering reflects seasonal change. Many plants flower only during certain times of the year. Clover and iris flower during the long days of summer, and poinsettias and asters bloom in the short days of early spring or fall.

Studies of how flowering is controlled by seasonal changes began in the early 1900s. W.W. Garner and H.A. Allard at a United States Department of Agriculture research center in Maryland were studying tobacco, which flowers during late summer in Maryland. One group of tobacco mutants did not flower as did the rest of the crop but continued to grow vegetatively into autumn. They became quite large, leading Garner and Allard to name them Maryland Mammoth. Since these oversized mutants had the potential for increasing yield of tobacco crops, Garner and Allard moved their Mammoth plants into the greenhouse to protect them from winter's cold and continued to observe their growth. To their surprise, the mutants finally flowered in December!

Could the plants somehow measure day length? To test this hypothesis, Garner and Allard set up several experimental plots of soybeans, each planted approximately a week apart. All of the plants flowered at the same time, despite the fact that the different planting times resulted in plants of different ages and sizes. From these experiments, Garner and Allard established the term **photoperiodism,** which is a plant's ability to measure seasonal changes by the length of day and night.

The adaptive significance of the ability of plants to anticipate and plan for seasonal changes in climate is obvious. But why would plants measure and respond to day length rather than other climatic factors such as rainfall or temperature? The answer is that weather is unpredictable, whereas day length is consistent due to the position of the earth as it travels around the sun.

Plants are classified into one of four groups, depending upon their responses to photoperiod (duration of daylight). **Day-neutral plants** do not rely on photoperiod to stimulate flowering, and include roses, snapdragons, cotton, carnations, dandelions, sunflowers, tomatoes, cucumbers, and many weeds. **Short-day plants** require light periods that are shorter than some critical length. These plants usually flower in late summer or fall. For example, ragweed plants flower only when exposed to 14 hours or less of light per day. Asters, strawberries, poinsettias, potatoes, soybeans, and goldenrods are short-day plants.

Long-day plants flower when light periods are longer than a critical length, usually 9 to 16 hours. These plants usually bloom in the spring or early summer and include lettuce, spinach, beets, clover, corn, and iris. **Intermediate-day plants** flower only when exposed to days of intermediate length, growing vegetatively at other times. They include sugarcane and purple nutsedge (fig. 31.11).

During which season a plant flowers depends upon whether the photoperiod is longer or shorter than some species-specific critical length. For example, ragweed and spinach both flower if exposed to 14 hours of daylight, yet ragweed is a short-day plant while spinach is a long-day plant. Thus, spinach flowers in the long days of summer, while ragweed blooms in the short days of fall.

Geographical distribution of plants is greatly influenced by their flowering response to photoperiod. For example, many short-day plants do not grow in the tropics where daylight is always too long to induce flowering. The measuring system in many plants is remarkably sensitive. Henbane, a long-day plant, flowers when exposed to light periods of 10.3 hours but not when the light period is 10.0 hours. Photoperiod is sensed by leaves; plants whose leaves are removed do not respond to photoperiod changes.

Some plants will not bloom unless they are exposed to the correct photoperiod and are said to exhibit **obligate photoperiodism.** For plants such as soybeans, the requirement for an inductive photoperiod is absolute—these short-day plants will not flower unless exposed to long nights. Conversely, other plants, including marijuana and Christmas cactus, will eventually flower even without an inductive photoperiod. For them, an inductive photoperiod merely hastens flowering. This response is called **facultative photoperiodism.** In some plants, the photoperiodic requirement for flowering can be influenced by other factors. For example, poinsettias are short-day plants at high temperatures and long-day plants at low temperatures.

Do Plants Measure Day or Night?

Plant physiologists Karl Hamner and James Bonner continued Garner's and Allard's work by studying the photoperiodism of the cocklebur, a short-day plant requiring 15 or fewer hours of light to flower. Controlled-environment growth chambers were used to manipulate photoperiods. The researchers were startled to discover that plants responded to the length of the dark period rather than the light period. The cocklebur plants flowered only when the dark period exceeded 9 hours.

Hamner and Bonner also discovered that flowering did not occur if the dark period was interrupted by a 1-minute flash of light, even if the regular light period remained less than 15 hours. Similar experiments in which the light period was interrupted with darkness had no effect on flowering. Furthermore, a long-day plant flowering on a photoperiod of 16 hours light to 8 hours dark will also flower on a photoperiod of 8 hours light to 16 hours dark if the dark period is interrupted by a 1-minute exposure to light. Other experiments with long- and short-day plants confirmed that flowering requires a specific period of uninterrupted dark rather than uninterrupted light. Thus, short-day plants are more accurately described as long-night plants, because they flower only if their uninterrupted dark period exceeds a critical length. Similarly, long-day plants are more accurately termed short-night plants.

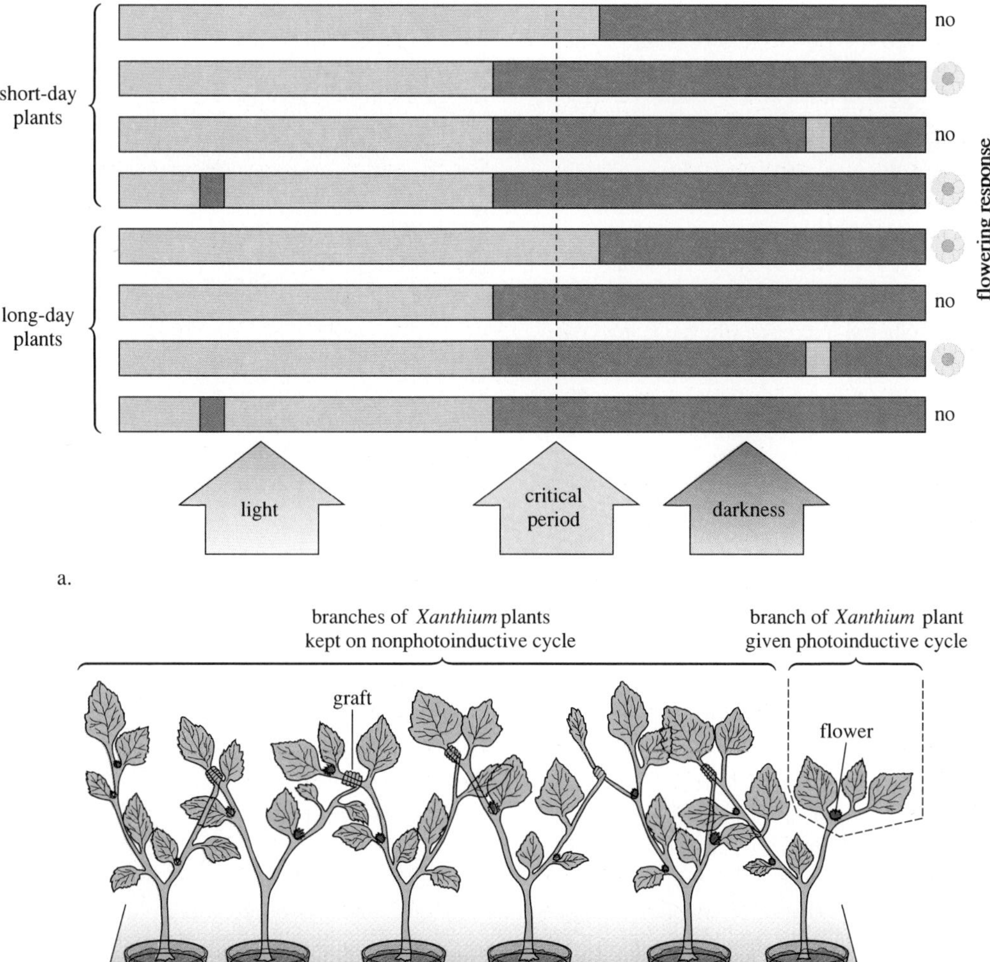

Figure 31.11
Photoperiod. *a*. Flowering response in plants requires not only the proper amount of light but also a critical length of darkness. If either period is interrupted, even for a short time, the plant will not flower. *b*. That the induction of flowering requires some chemical substance can be shown by this experiment. All the plants flowered, even though only one branch of one plant had the right amount of light and darkness, because all the plants were grafted together and could exchange sap.

KEY CONCEPTS

Plants flower in response to a certain ratio of light to dark in a day, which reflects the season. This response is species-specific and is called photoperiodism. Day-neutral plants do not show photoperiodism; short-day plants require light shorter than a certain duration; long-day plants require light longer than a certain duration; intermediate-day plants require daylight of intermediate length. Photoperiod requirements can be quite specific. Plants may respond to the duration of darkness rather than to the duration of light.

Phytochrome—A Pigment Controlling Photoperiodism

Because photoperiodism is a response to light, botanists suspected that it might be carried out by a pigment molecule whose structure is altered by absorbing light of a particular wavelength. The existence of such a pigment was suggested by the observation that red light inhibits flowering when it is used to interrupt the dark period. This inhibition can be reversed if the interruption of red light is immediately followed by far-red light, a form of red light corresponding to a wavelength at the edge of the electromagnetic spectrum (fig. 6.6). From these observations, botanists concluded that the pigment existed in two forms, one of which absorbed red light and the other far-red light.

In 1959, a pale blue pigment was isolated and identified as **phytochrome.** As hypothesized, phytochrome exists in two interconvertible forms, P_r and P_{fr}. The inactive form of phytochrome is synthesized as P_r. When P_r absorbs red light, it is converted to P_{fr}, which is the active form of phytochrome in flowering. P_{fr} promotes flowering of long-day plants and inhibits flowering of short-day plants. P_{fr} is converted to P_r when it absorbs far-red light (fig. 31.12).

How does the ratio of P_r to P_{fr} provide information on photoperiod? Sunlight has proportionally more red light than far-red light. Therefore, during the day, P_r is converted to P_{fr}. This abundance of P_{fr} could

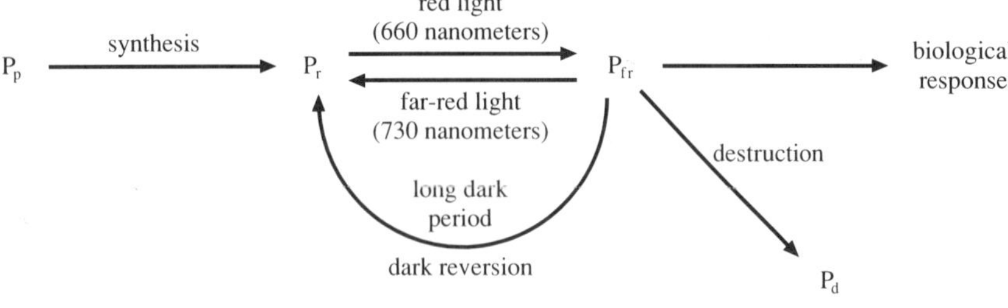

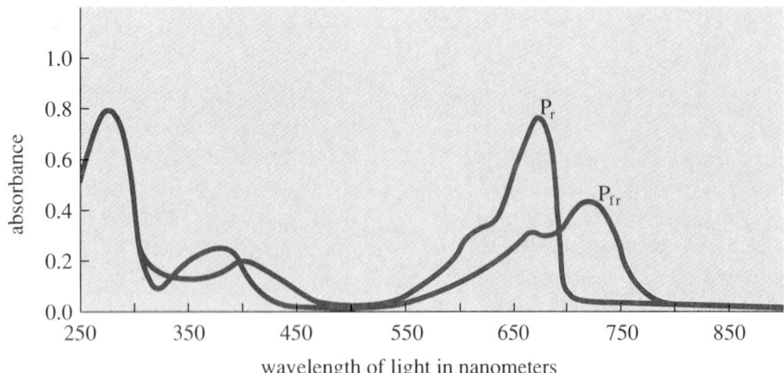

Figure 31.12
Two forms of phytochrome are P_r and P_{fr}. *a.* P_{fr} stimulates biological responses, whereas P_r does not. *b.* The two different forms of phytochrome absorb light at different wavelengths.

Figure 31.13
The century plant. These plants flower only once in several decades and then die.

signal the plant that it is in light. When the plant is placed in darkness, P_{fr} is slowly converted to P_r. This dark conversion of phytochrome to P_r was originally believed to initiate a set of reactions enabling the plant to measure the length of darkness relative to the length of light. Indeed, this reasoning would explain why an uninterrupted period of dark (rather than light) controls photoperiodism. However, the dark reversion of P_r to P_{fr} requires only 3 to 4 hours, and thus it cannot fully explain the light-dark sensing mechanism that controls flowering. Some internal clocklike mechanism probably also influences flowering.

A plant can respond to a photoperiodic initiation of flowering only if it is reproductively mature. Reaching reproductive maturity can take only a few days (as in the Japanese morning glory) or weeks (as in annuals) to several years. Some species of the century plant require decades before flowering (fig. 31.13). The giant bamboo of Asia flowers only about every 33 or 66 years! The delay in flowering until reproductive maturity ensures that the plant has stored enough food to completely form and maintain its flowers. Moisture, soil conditions, nearby plants and temperature may also influence flowering.

Other Responses Influenced by Photoperiod and Phytochrome

Seed germination is also affected by phytochrome. In seeds of lettuce and many weeds, red light stimulates germination, and far-red light inhibits germination. Seeds alternately exposed to red and far-red light are affected only by the last exposure. Thus, germination occurs after exposure to red/far-red/red/far-red/red light as it does after a single exposure to red light. Treatments with far-red and red/far-red/red/far-red/red/far-red light result in no germination. Therefore, the phytochrome system can inform a seed that sunlight is nearby for photosynthesis and thus promote germination. If seeds are buried too deeply in the soil, P_{fr} is absent (due to no sunlight), and germination does not occur.

Phytochrome also controls the early growth of seedlings. Seedlings grown in the dark have abnormally elongated stems, small roots and leaves, a pale color and a spindly appearance. This condition is termed **etiolated** (fig. 31.14). Bean sprouts used in Chinese cooking are etiolated.

By rapidly elongating, etiolated plants reach the light before exhausting their food reserves. Etiolated growth is replaced by normal growth once plants are exposed to the light. Red light controls transformation from etiolated to normal growth. Etiolated plants are very sensitive to red light. Normal growth ensues after exposure to only 1 minute of red light. If red light is followed immediately by an exposure to far-red light, P_{fr} is converted to P_r, and etiolated growth continues.

Phytochrome may also help direct shoot phototropism. Light coming from one direction would presumably create a gradient of P_r and P_{fr} across the stem. P_{fr} would be most abundant on the illuminated side of the stem and P_r most abundant on the shaded side of the stem. This gradient of phytochrome could bend a shoot as P_r promotes stem elongation and P_{fr} inhibits it.

Yet another function of phytochrome is to provide information about shading by overhead plants. Chlorophyll in a plant canopy absorbs much of the red light of sunlight. By the time light reaches underlying plants, such as those on a forest floor, there is less red light, and therefore less P_r is converted to P_{fr}. Since P_r promotes stem elongation (as in etiolated plants), information provided by the phytochrome system can help a plant reach sunlight more rapidly.

Horticulturists use photoperiodism to produce flowers when they are wanted for sale. For example, chrysanthemums (a short-day plant) can be made available year-round by using shades to create an inductive, short-day photoperiod. Similarly, poinsettias can be induced to flower near Christmas if kept in darkness for at least 14 hours a day for at least a month.

Figure 31.14
Etiolated seedlings, *left*, have had insufficient light for normal growth.

KEY CONCEPTS

Photoperiodism is carried out by phytochrome, a blue pigment that interconverts between two forms. Inactive P_r absorbs red light and is converted to active P_{fr}, which absorbs far-red light. P_{fr} causes long-day plants to flower and inhibits short-day plants from doing so. The state of phytochrome imparts information about day length because sunlight has more red light than far-red light. A plant must be mature to be sensitive to photoperiod. Phytochrome and photoperiodism are also involved in early seedling growth, providing information about light conditions, and seed germination.

Senescence

Senescence, or aging, is also a seasonal response of plants. Aging occurs at different rates in different species. Flowers of plants such as wood sorrel and heron's bill shrivel and die only a few hours after being formed. Slower senescence is seen in the colorful turning of leaves in autumn.

Whatever its duration, senescence is not merely a gradual cessation of growth but an energy-requiring process brought about by new metabolic activities. Leaf senescence begins during the shortening days of summer, as nutrients are mobilized and proteins broken down. By the time a leaf is shed, most of its nutrients have long since been transported to the roots for storage. The fallen leaves that so beautifully litter the ground in autumn contain little more than cell walls and remnants of nutrient-depleted protoplasm.

The destruction of chlorophyll in leaves is part of senescence. In autumn, the yellow and orange **carotenoid pigments,** which were previously masked by chlorophyll, become visible. Senescing cells also produce pigments called **anthocyanins.** The loss of chlorophyll, the visibility of the carotenoids, and the production of anthocyanins are responsible for the spectacular colors of autumn's leaves (fig. 31.15).

Dormancy

Before the onset of harsh environmental conditions such as cold or drought, plants often become **dormant,** a state of decreased metabolism. Like leaf senescence, dormancy involves structural and chemical changes. Cells synthesize sugars and amino acids, which function as antifreeze, preventing or minimizing cold damage. Growth inhibitors accumulate in buds, transforming them into winter buds covered by thick, protective scales. These changes in preparation for winter are called **acclimation.**

Growth resumes in the spring as a response to changes in photoperiod and/or temperature. Lengthening days release dormancy in birch and red oak, whereas fruit trees such as apple and cherry resume growth only after exposure to winter's cold. Apple and cherry trees transplanted in warm climates are late bloomers in the spring. The exact mechanism by which photoperiod or cold breaks dormancy is unknown, although hormonal changes are probably involved.

In some plants, dormancy is triggered by factors other than photoperiod or temperature. In many desert plants, rainfall alone releases dormancy, while potato requires a dry period before renewing growth.

> **KEY CONCEPTS**
>
> *Plants age at different rates. Autumn leaf colors result from destruction of chlorophyll and unmasking of carotenoids and anthocyanins. Dormancy protects plants against cold damage via such acclimation events as antifreeze synthesis and growth inhibitors. In the spring, growth resumes in response to photoperiod, temperature, or moisture.*

Figure 31.15
Fall leaves. Senescent leaves lose their chlorophyll and then other pigments become visible.

Circadian Rhythms

Some rhythmic responses in plants are not seasonal. Consider the common four-o'clock, which opens its flowers only in late afternoon, whereas the yellow flowers of evening primrose open only at nightfall. Similarly, nyctinastic movements of prayer plants occur at the same time every day, and some dinoflagellates glow in warm ocean waters within a few minutes of midnight each evening. These regular, daily rhythms are called **circadian rhythms** (from the Latin circa, meaning "about," and dies, meaning "day"). Other plant activities that occur in daily rhythms include cell division, stomatal opening, protein synthesis, secretion of nectar, and synthesis of growth regulators. Many eukaryotic organisms, including humans, have circadian rhythms (chapter 36).

How do plants measure a day? Is the passage of a day controlled internally by a plant or externally by environmental factors? Several experiments have shown that circadian rhythms in many species do not coincide with a 24-hour day. That is, they may be a few hours longer or shorter than 24 hours. In addition, circadian rhythms often continue even in constant environmental conditions. Thus, circadian rhythms are probably controlled internally, by a little-understood mechanism called a biological clock, rather than externally.

A plant's circadian rhythm may be altered by environmental factors, such as a change in photoperiod, regardless of the rhythm's internal control. This resynchronization of the biological clock by the environment is called **entrainment.** However, entrainment to a new environment is limited. If the new photoperiod is too different from a plant's biological clock, the plant reverts to its own internal rhythms. Also, a plant maintained in a modified photoperiod over a long period of time reverts to its natural rhythms when placed in constant light.

Biological clocks allow plants to synchronize their activities. In this way flowers of a particular species open when they are most likely to be visited by pollinators. For some plants this timing is quite precise. Flowers of genus *Cereus* are pollinated by bats and must therefore open at night when bats are active. Furthermore, different individuals must flower within a few days of each other, because *Cereus* flowers persist for only a week.

The next chapter describes some interesting ways in which we can alter plants to our specifications. Although these methods are part of the relatively new field of biotechnology, they are actually a continuation of the agricultural approaches discussed at the start of this unit.

> **KEY CONCEPTS**
>
> *Circadian rhythms are internally controlled daily responses characteristic of a species that coordinate activities. The internal timing mechanism of a circadian rhythm is called a biological clock. Modification of a biological clock by the environment is called entrainment.*

SUMMARY

Plants respond to the environment by growing. Plant hormones have characteristic effects on growth, yet they interact in complex ways. *Auxins* stimulate cell elongation in shoot tips, embryos, young leaves, flowers, fruits, and pollen. *Gibberellins* stimulate cell division and elongation but act slower than auxins. *Cytokinins* stimulate mitosis and are found in actively developing plant parts. *Ethylene* speeds ripening. *Abscisic acid* inhibits the growth-inducing effects of other hormones.

A *tropism* is a growth response toward (positive) or away from (negative) an environmental stimulus usually caused by different rates of growth in different parts of an organ or structure. In *phototropism*, light sends auxin to the shaded portion of the plant, causing growth there and bending the plant towards the light. Shoot growth is a *negative geotropism*, and root growth is a *positive geotropism*. Auxin accumulation can cause these opposite responses because of different sensitivities of the responding tissues. *Thigmotropism* is response to touch and is evidenced by clinging tendrils.

Nastic movements are nondirectional. *Seismonasty* is response to contact, such as a carnivorous plant's entrapping an insect in its leaves. Nastic response to light and dark is *nyctinasty* and can be caused by osmotic changes that differentially alter cell volume. *Thigmomorphogenesis* is a growth-inhibiting response to mechanical disturbance.

Plants are very sensitive to seasonal changes. *Photoperiodism* is the ability of a plant to measure length of day and night. Flowering can depend upon photoperiodism. Short-day plants flower only when the duration of light is less than some critical length, whereas long-day plants require a light period longer than some critical length. Intermediate-day plants need days of intermediate length to flower. The type of plant determines the season during which it flowers. Plants requiring a precise photoperiod display *obligate photoperiodism*, whereas plants whose flowering is merely hastened by a certain photoperiod display *facultative photoperiodism*. Plants may actually respond to length of darkness rather than length of daylight.

A plant's response to light is controlled by a pigment called *phytochrome*. The inactive form, P_r, absorbs red light to become P_{fr}, the active form. P_{fr} promotes flowering of long-day plants and inhibits flowering of short-day plants. P_{fr} reconverts to P_r by absorbing far-red light. Relative abundances of the two forms of this molecule provide information because sunlight has more red than far-red light. Phytochrome also controls early seedling growth, provides information about shading, and directs shoot phototropism and seed germination.

Senescence is both an active and passive cessation of growth. Growth becomes dormant during cold or dry times and resumes when environmental conditions are more favorable. Daily responses are called *circadian rhythms* and are controlled by internal biological clocks. These clocks can be altered, or *entrained*, by environmental change.

QUESTIONS

1. How is plant growth different from and similar to animal growth?

2. How does the action of gibberellin differ from that of auxin?

3. A tendril's pattern of coiling can be due to thigmotropism or seismonastic coiling. What is the difference between the two responses?

4. For three plant species, describe how movement of ions influences movement of leaves.

5. Give examples of how a tropism, nastic response, and flowering response to photoperiod are adaptive.

6. What is the experimental evidence that a short-day plant is more accurately described as a long-night plant?

7. How are auxin and phytochrome involved in shoot phototropism?

TO THINK ABOUT

1. Several times in the scientific investigation of plant responses, the existence of a substance was suspected before it was actually observed. How is the scientific method used in such a situation? Provide an example of how the scientific method was used to explain a plant response.

2. How can different plant hormones, or their synthetic equivalents, be used in agriculture?

3. You want to make a fruit salad for a barbecue tomorrow, but the bananas are not ripe yet. How might you hasten their ripening? Which plant biochemical would you use?

4. Why won't rain cause the leaves of a Venus's-flytrap to close?

5. Although spinach and ragweed each requires 14 hours of sunlight to flower, spinach flowers in the summer and ragweed (as hay fever sufferers can attest) flowers in the fall. Explain the seasonal difference in flowering.

6. How is the function of phytochrome similar to that of rhodopsin, the pigment that functions in human vision?

SUGGESTED READINGS

Evans, Michael L., Randy Moore, and Karl Hasenstein. December 1986. How roots respond to gravity. *Scientific American*, pp. 112–19. A plant's roots are quite in tune to gravitational forces.

Lewis, Ricki. May 28, 1990. Scientists take to the classroom to inspire youngsters. *The Scientist*. The author is one of many teachers leading schoolchildren in observing the growth of plants from space seeds.

Marsella, Gail. November 1984. Plant assassins. *Biology Digest*. Some plants protect themselves by producing biochemicals that ward off attackers.

Moore, Randy. January 1988. How gravity affects plant growth and development. *Biology Digest*. How will plants fare when the gravity under which they have evolved is absent?

Raab, Mandy M., and Ross E. Koning. November 1988. How is floral expansion regulated? *BioScience*, vol. 38, no. 10. The timing of flowering is intimately tied to a plant's environment.

CHAPTER

32

Plant Biotechnology

Chapter Outline

Beck/Shoemaker

Learning Objectives

By the chapter's end, you should be able to answer these questions:

1. What is biotechnology?
2. What are the similarities and differences between plant biotechnology and traditional plant breeding?
3. How can protoplast fusion produce new plant varieties?
4. How can plant cells be grown in culture to produce identical offspring or variant offspring?
5. How can novel combinations of plant cell nuclei, cytoplasms, and organelles be developed?
6. How can recombinant DNA technology introduce new traits into plants?
7. How do dicots and monocots differ in their abilities to be manipulated by specific biotechnologies?
8. What are some of the environmental questions that must be addressed when plants altered by biotechnology are field tested?

If you were asked to design a new type of plant that would offer more to human consumers than anything available in nature, what might it be? Popcorn that tastes so buttery by itself that it does not need butter, or perhaps a plant that sprouts tomatoes above ground and potatoes below? A farmer might desire plants that "fix" nitrogen from the atmosphere into a biologically usable form without costly fertilizer or crops that manufacture their own herbicide and insecticide and resist disease. Why not mix up nutritional characteristics into healthful new combinations—corn that is rich in the amino acid tryptophan, which it normally lacks, or a hardy new rice derived from cultivated and wild strains?

All of these interesting types of plants have indeed been developed in recent years, thanks to techniques that either manipulate cells or delve into the genetic material. These new variants have been produced far faster than is possible using conventional plant breeding methods. It is estimated that by 1995, 5% of agricultural output in the United States will come from technologies that manipulate plants at the cellular and subcellular levels.

Figure 32.1
Opium, ancient product of biotechnology. The opium poppy is the source of powerful painkillers called morphine alkaloids. Evidence of medicinal use of poppy extracts dates back at least 3,500 years. Morphine comes from a green ovary capsule, which protects the growing seeds. The capsule can be seen 80 days after planting, when the flower petals fall off. The active alkaloid is contained in a whitish fluid within the capsule. This juice is collected and air dried to form a black semisolid, which is the fresh opium product from which drugs are derived.

The Challenge of Agricultural Biotechnology

Biotechnology is the use of organisms or their components to provide goods or services. It includes such age-old practices as fermenting wine from grapes and "folk" medicines derived from living things (fig. 32.1); cell manipulations developed throughout the twentieth century; and the genetic alterations performed and perfected since the mid-1970s. The first products of modern biotechnology served the medical, microbiological, and veterinary markets. Another major area of biotechnology is agriculture. Biotechnology applied to crop plants may become very important in expanding food supplies to serve a growing human population. Agricultural biotechnology can introduce qualities that appeal to farmers, food processors, and consumers alike, improving such agronomic characteristics as adaptability, yield, and resistance to disease and pesticides; making anatomical alterations that ease harvesting; increasing nutritional content; and fine-tuning sweetness, acidity, texture, flavor, size, and shape to consumer preferences.

Although agricultural biotechnology may ultimately have the greatest impact upon human existence of all the applications of biotechnology, it may be the most difficult to achieve. Manipulating the genes and cells of plants has lagged behind similar work on animals and bacteria because the arrangement of genetic material in plants is more complex than in organisms. Plant DNA, for example, often contains vast regions of repeated nucleotide sequences of unknown function. Recessive traits are difficult to study in polyploids. What plant geneticists have needed for some time has been a genetically simple plant. The mustard weed *Arabidopsis thaliana* fits the bill (fig. 32.2).

Understanding plant ecology is also important. "Engineered" plants must eventually be grown in the environment, where they can interact with other organisms. A genetically engineered vaccine or drug, in contrast, would be confined to a laboratory facility.

A first set of hurdles in plant biotechnology has already been surmounted for a few species—that is, growing a mature plant from an initial cell that has been manipulated, demonstrating the new trait in the regenerated plant, and showing that the regenerated altered plant passes on the characteristic to its progeny. A second hurdle will be to understand which combinations of individual altered traits contribute to desired characteristics. Although most of the biotechnologies manipulate one inherited trait at a time, traditional plant breeding work has shown that valuable characteristics are often the product of particular combinations of inherited traits.

Traditional Plant Breeding Versus Biotechnology

The steps in traditional breeding and biotechnology are similar. First, an interesting trait is identified and bred, or engineered, into a plant whose other characteristics comprise a valuable package. For example, larger fruit size might be desired in a plant

Figure 32.2
A model organism for unraveling plant genetics—the mustard weed. With only five chromosomes, just 1% of the DNA content of wheat, it seems to say with a few genes what other plants do with many. For example, the mustard weed codes for the protein portion of chlorophyll with 3 genes, whereas the petunia does so with 16 genes. The weed also has little repetitive DNA. It is a small plant that produces lots of seeds and is easy to grow, and its life cycle is only 5 weeks. Botanists hope to unravel molecular details in this weed and then apply what they learn to more complex plants.

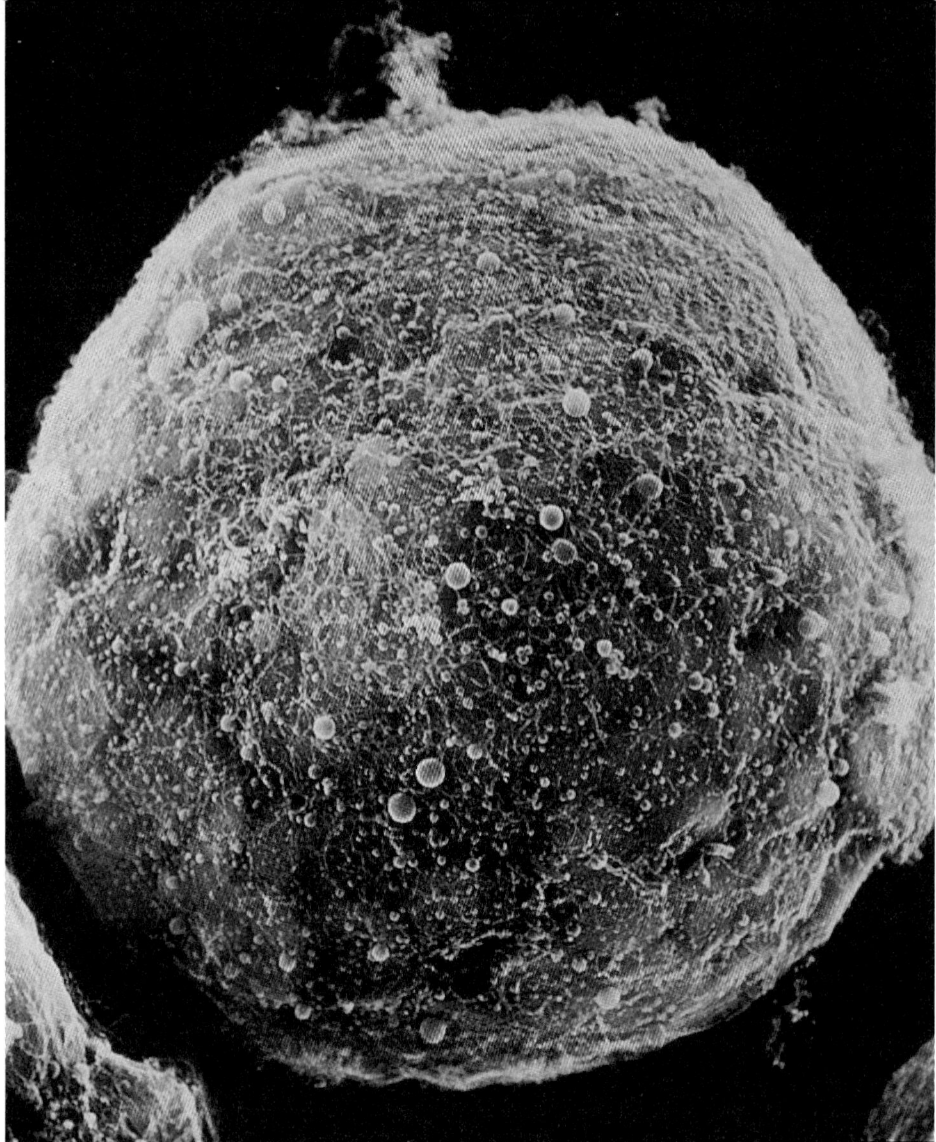

Figure 32.3
Protoplasts are denuded plant cells. Protoplasts can be fused to create hybrid cells with interesting new mixtures of traits. Magnification, x512.

that can already withstand temperature extremes and resist pests. The new variety is then tested in several different habitats and during different seasons to determine the conditions under which it grows best. Finally, seeds are distributed to growers.

Traditional plant breeding introduces new varieties by a sexual route. Pollen carrying the genes for traits in male sex cells fertilizes ovules that bear egg cells carrying a different set of traits. Because each sex cell brings with it a different combination of the parent plants' traits, offspring from a single cross are not genetically uniform. This is the reason that human siblings are not identical; similarly, some plants may be taller or more robust or produce smaller seeds than others.

Rather than beginning with sperm and egg, which have a half set of genetic instructions each, most biotechnologies begin with somatic (body) cells, which form nonsexual parts of the plant such as leaves and stems, or may come from embryos. Somatic cells have a complete set of genetic instructions. Plants that are regenerated from somatic cells do not have the unpredictable mixture of characteristics found in plants derived from sexual reproduction, because the somatic cells are usually genetic replicas, or **clones,** of each other. Biotechnology, then, offers a degree of precision, plus it can assure consistency in crop quality from season to season.

Biotechnology can alter plant structure or function at any of several levels of cellular organization: the individual cell, its organelles, and the genes within its nucleus.

Protoplast Fusion—The Best of Two Cells

In **protoplast fusion,** new types of plants are created by combining cells from different species and then regenerating a mature plant hybrid from the fused cell. A protoplast is simply a plant cell whose cell wall has been removed by treatment with digestive enzymes. Two protoplasts may join on their own, or they can be stimulated to do so

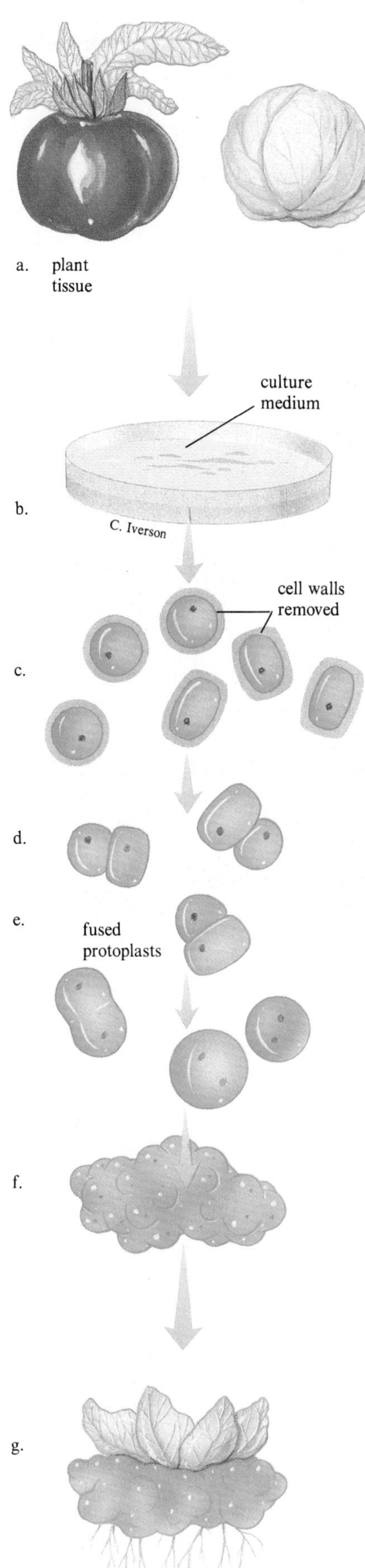

by exposure to polyethylene glycol (an antifreeze), a brief jolt of electricity, or being hit by a laser beam (figs. 32.3 and 32.4).

A single gram (about 1/28 of an ounce) of plant tissue can yield as many as 4 million protoplasts, each of which is a potential new plant, either by itself or when fused with another protoplast. If the protoplasts from two different species look different or can be distinguished biochemically, then hybrids can be separated. A plant regenerated from a protoplast fusion of two plant types is called a **somatic hybrid.**

Not all protoplast fusions yield mature plants, and of those that do, not all are useful. Consider the "pomato" plant, which is the result of a tomato protoplast fused with a potato protoplast. The regenerated plant produces both types of vegetables in the proper part of the plant, but the tomatoes and potatoes are small and seed quality is poor. The interesting hybrid cannot be propagated. A plant derived from a protoplast fusion between parsley and carrot also produces the desirable parts—carrot roots and parsley leaves—but they are small, and the plant has tiny seeds. Even less useful is a fusion of radish and cabbage. The not-very-tasty plant grows radish leaves and cabbage roots! Protoplast fusion is more successful when the parent cells come from closely related species. For example, when a protoplast from a potato plant that is normally killed by the herbicide triazine is fused with a protoplast from the wild black nightshade, a relative that is naturally resistant to the herbicide, the resulting potato hybrid grows well in soil treated with triazine to control weeds.

Protoplast fusion has limitations. Bringing together quantities of two cell types gives mixed and unpredictable results: single cells, fused cells of the same species, and the sought-after fused cells of two different types. Not all species form protoplasts that can be coaxed to fuse. Of those that do, only some of the fusion products go on to divide and develop into plants. Of these, there is no guarantee that they will have useful traits. Many dicots can be regenerated from protoplasts taken from leaf mesophyll tissue, but monocots such as corn, rice, and sugarcane grow best from protoplasts derived from less specialized embryonic cells (fig. 32.5). Monocot somatic hybrids have been constructed between pearl millet and Einkorn wheat, between pearl millet and sugarcane, and between cultivated and wild rice.

Figure 32.4
Protoplast fusion—one cell from two. Protoplast fusion can yield plants with traits from two species. Tissues from different plants (*a*) are placed in tissue culture (*b*), and the cell walls are dissolved with enzymes or pierced by electricity or a laser (*c*). The resulting protoplasts are mixed and encouraged to fuse (*d*). Among the fusion products may be hybrid protoplasts (*e*), which are selected and grown into undifferentiated tissue called callus (*f*). Sometimes, plants can be regenerated from the callus, yielding interesting new combinations of traits (*g*).

KEY CONCEPTS

Agricultural biotechnology can ease harvesting, improve nutritional content, and alter taste. Plants are more difficult to manipulate genetically than are animals, and altered plants will ultimately have to grow among other species. Many plant biotechnologies use somatic cells. In protoplast fusion, plant cells that have had their cell walls removed are fused, and a hybrid plant can develop. Such hybrids, however, have unpredictable combinations of traits and may be difficult to propagate.

Cell Culture

A fascinating thing happens when protoplasts or tiny pieces of plant tissue, called **explants,** are nurtured in a dish with nutrients and plant hormones. After a few days, the cells lose the special characteristics of the tissues from which they were taken and form a white lump called a **callus.** The lump grows, its cells dividing, for a few weeks. Then certain cells of the callus grow into either a tiny plantlet with shoots and roots or a tiny embryo. An embryo grown from callus is called a **somatic embryo** because it derives from somatic, rather than sexual, tissue.

Researchers are not sure how or why calli of some species give rise to somatic embryos, some give rise to plantlets, and others never develop beyond a lump of tissue (fig. 32.6). Callus growth of calli is apparently a phenomenon unique to plants. In humans, it would be the equivalent of a cultured skin cell, for example, multiplying into a blob of unspecialized tissue and then sprouting tiny humans or human embryos!

Callus cells are unspecialized, even though they derive from differentiated tissue in which only some genes are expressed. All of the genes of a callus cell are capable of expression, much like those of a fertilized egg. Most of the time, embryos or plantlets grown from a single callus are genetically identical to each other. Sometimes, however, the embryos or plantlets differ from each other because certain of the callus cells undergo genetic change (mutation). Biotechnologists can to some extent control whether or not growths from a callus are identical by altering the nutrients and hormones in the callus culture. This ability to control callus growth is valuable, because under some circumstances agriculture may benefit from a uniform crop, and at other times, new varieties may be sought.

Cell Culture for Uniformity

Somatic Embryos as Artificial Seeds

A natural seed is a plant embryo and its food supply, packaged in a protective shell. An **artificial seed** can be fashioned by suspending a somatic embryo in a transparent polysaccharide gel containing nutrients and hormones and providing protection and shape with an outer biodegradable polymer coat. For example, alfalfa embryos are encapsulated in a calcium alginate gel to form artificial seeds (fig. 32.7). To develop artificial seeds, the callus culture is manipulated so that the somatic embryos that they yield are genetically identical. Biotechnologists can actually improve upon nature by packaging somatic embryos with pesticides, fertilizer, nitrogen-fixing bacteria, and even microscopic parasite-destroying worms.

The advantage of artificial seeds for the farmer is their guarantee of a uniform crop because they are genetically identical. Plants that mature at the same rate, for example, can make harvesting simpler and cheaper. In contrast, the embryos of seeds obtained from sexual reproduction contain unpredictable mixtures of parental traits.

Artificial seed technology so far has been most successful for dicots such as celery, lettuce, cotton, and alfalfa, but embryos of the monocots corn, rice, and sugarcane can grow from calli if certain synthetic plant hormones and herbicides are added. Again monocot embryos grow only from calli derived

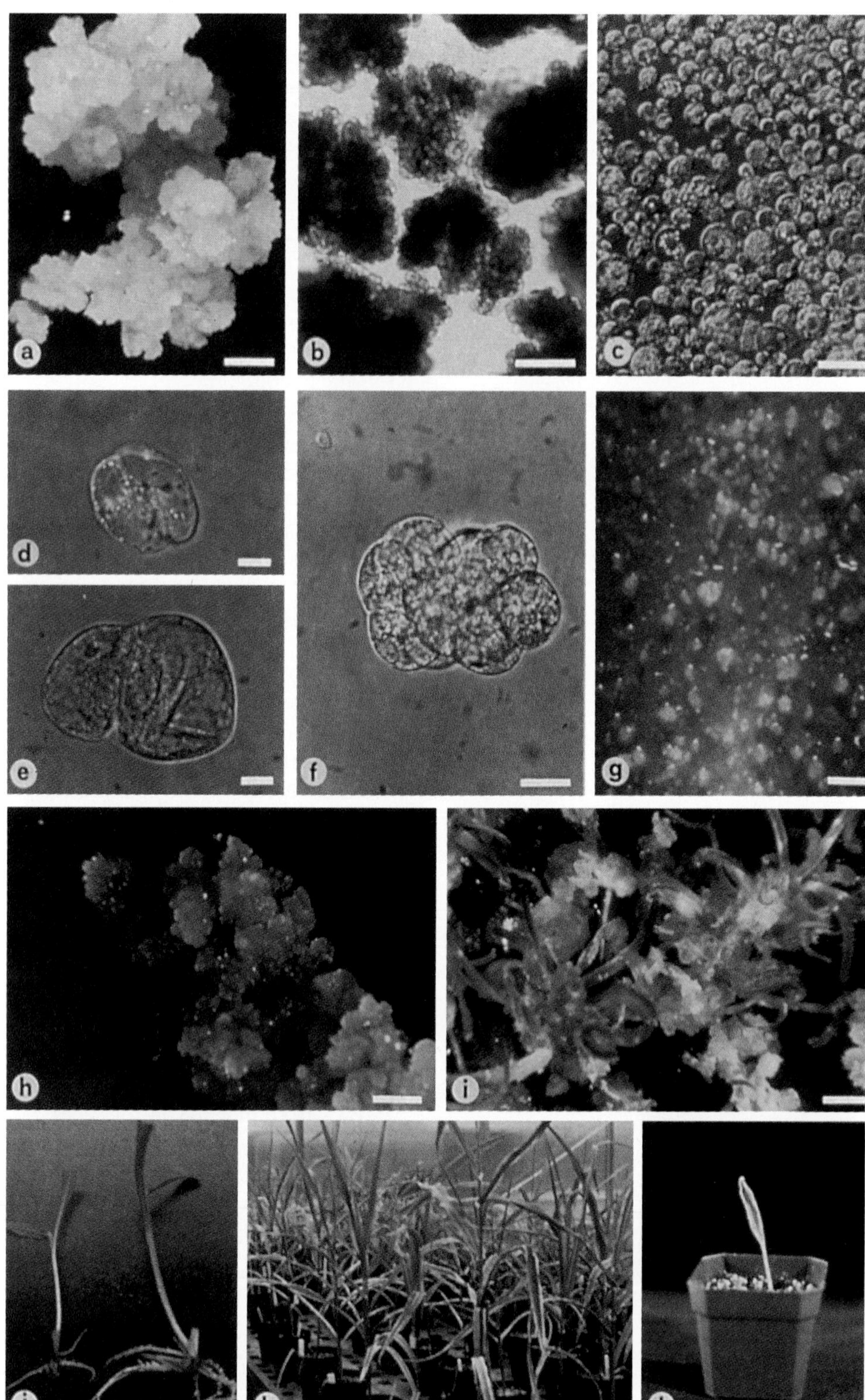

Figure 32.5

Regenerating corn plants from protoplasts, a long-sought goal of biotechnologists, was accomplished in 1988. The process begins with a corn callus developed from embryo tissue (*a*). The callus cells are gently suspended in culture (*b*) and then purified (*c*). The protoplast cells divide (*d* and *e*). After 2 weeks, a colony has formed (*f*), which gives rise to calli (*g* and *h*). When certain nutrients and hormones are added, tiny plantlets form (*i*). After some growth, the plantlets are ready to be placed in soil (*j*) and then are raised in a greenhouse (*k*). These plants grow to maturity and give rise to their own offspring (*l*). In 1990, wheat was successfully grown from protoplasts.

from embryonic or unspecialized cells. Artificial seed technology for the cereals is not economically feasible, however, because these plants provide abundant seed in their natural form. In addition, growing calli takes up a great deal of laboratory space.

Clonal Propagation

Uniform plants are also produced by **clonal propagation,** in which cells or protoplasts are cultured in the laboratory and then grown into genetically identical plants (clones) (fig. 32.8). As long ago as the 1930s, some plants were grown this way in standard laboratory glassware. Today, large and intricately designed tanks called bioreactors are used to house plant cells growing in culture.

Clonal propagation offers several advantages. It can speed growth of plants that are difficult to grow vegetatively and of slow-growing valuable trees, such as oil palm, redwoods, and chestnuts. Plants that are clonally propagated are grown in disease-free conditions and can be grown year round. The ornamental flower industry clonally propagates orchids, for example, because the flowers are uniform in appearance. So far clonal propagation has been used mostly for tropical root and tuber plants, such as bananas, plantains, cassava, potato, sweet potato, and yam, but also for asparagus, carrot, potato, and strawberry. Clonally propagated oil palm, orchids, and potatoes are grown commercially. Like artificial seeds, clonally propagated plants are often too costly to be practical.

Cell Culture for Variety

Somaclonal Variation

Embryos and plantlets have been grown from callus since the 1950s. When researchers attempted to grow uniform embryos and plantlets, they were sometimes confounded by the appearance of new variants. A callus covered with tiny green tomato plantlets might sprout one that is darker or lighter than the others. This occasional variability,

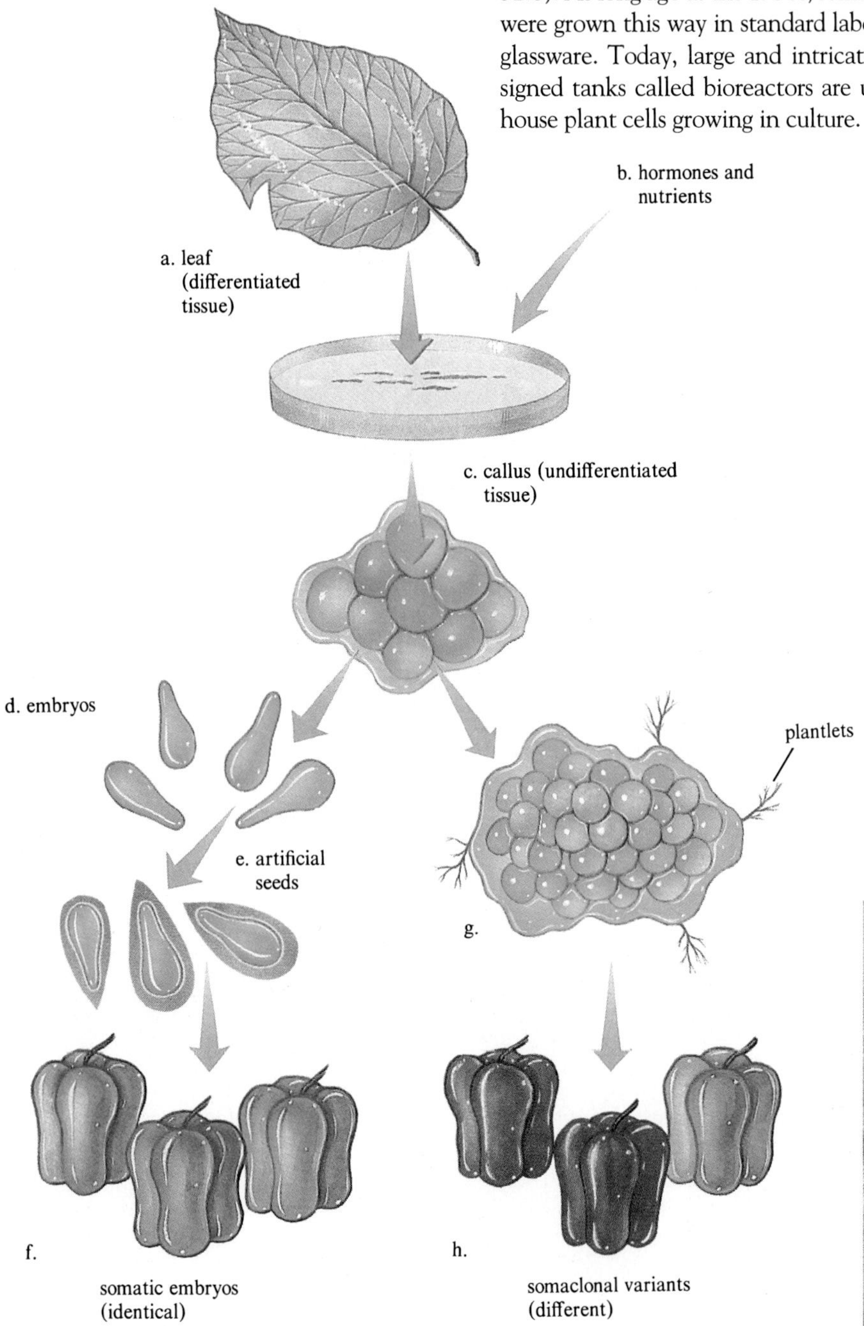

a. leaf (differentiated tissue)
b. hormones and nutrients
c. callus (undifferentiated tissue)
d. embryos
plantlets
e. artificial seeds
g.
f.
h.
somatic embryos (identical)
somaclonal variants (different)

Figure 32.6
The fate of a callus—embryos or plantlets. New plants grown from cell culture are valuable either for their uniformity or their variation, depending upon the goal of the researcher. In somatic embryogenesis, plant cells are cultured into callus (*a-c*), which is then exposed to nutrients and hormones to give rise to genetically identical plant embryos (*d*). The embryos, when packaged in a protective shell to form artificial seeds (*e*), guarantee uniform, consistent crop (*f*). In somaclonal variation, the culture medium encourages variety. Plants are regenerated from the resulting callus (*g*), some of which display new traits (*h*).

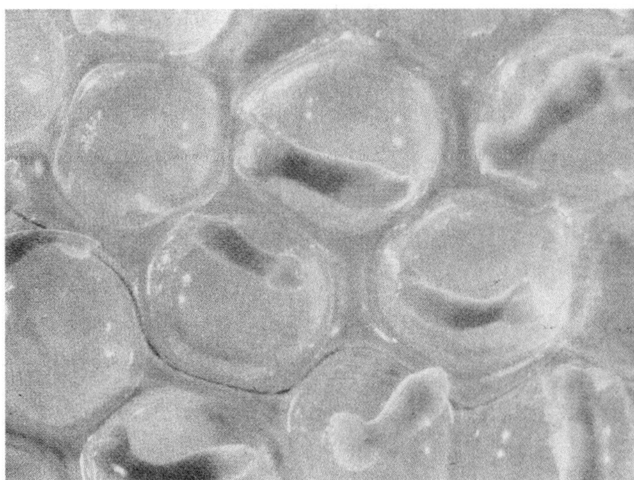

Figure 32.7
These artificial seeds consist of somatically derived alfalfa embryos encapsulated in a calcium alginate gel. As supplied to growers, the gel package would also include nutrients, hormones, pesticides, and other components needed for embryo-to-plant development. Artificial seeds, however, are more expensive than natural seeds.

a.

b.

c.

d.

Figure 32.8
Identical (cloned) carrots can be grown from callus. *a.* To clone carrots, sections of root are cored, and thin slices are placed in a culture dish containing nutrients and hormones to stimulate callus formation. *b.* After a few days, the carrot root cells lose their specific characteristics and form a callus (shown at 5 weeks). *c.* After about 3 months, the callus begins producing differentiated tissue, which is transferred to another medium to encourage growth of shoots and roots. *d.* Eventually, the new carrot plantlets are strong enough to be moved from their sterile medium into regular potted soil.

once regarded as a liability, has been turned into an entirely new technology in its own right, called **somaclonal variation.** Unusual plants arising from protoplast or cell culture are called somaclonal variants because they are derived from a single somatic cell (the one that gave rise to the callus) and are therefore clones of it and each other.

In 1981, researchers in Australia suggested that somaclonal variants could be agriculturally useful. Normally new plant varieties arise literally one in a million by spontaneous mutation, but somaclonal variants arise much more frequently. In 1983, researchers at DNA Plant Technologies, a New Jersey company, decided to intentionally look for variants from callus. They took a normal, medium-sized red tomato plant, chopped up bits of leaf tissue, grew the bits into callus, and regenerated 230 plantlets. After growth to maturity, 13 of the 230 plants were markedly different from the original tomato plant. One variant had tangerine-colored fruits; two others lacked a joint between the stem and the tomato, making harvesting easier; and two other variants had a high solids content (fig. 32.9).

One would have to look through millions of naturally grown tomato plants to find 13 genetic variants. How does the culture process uncover genetic variation? Possibly a genetic variant not noticeable if it exists only in one leaf cell among thousands becomes obvious when the cell that contains it gives rise to an entire new plant. (If this could happen in humans, imagine that a skin cell in a Caucasian underwent a mutation giving it a dark color. It would not be noticeable in the person, but if that mutated cell was used to regenerate a new person, he or she would

a.

b.

Figure 32.9
Somaclonal variation produces interesting new tomatoes. *a.* A high-solids tomato, the product of somaclonal variation, is of interest to the canned-food industry because less boiling is required to produce tomato paste, sauce, and soup. *b.* The same technique has also yielded larger than normal and differently pigmented varieties.

Figure 32.10
Rice plants are grown from cultured sex cells in gametoclonal variation. The inset shows various stages of the process—pollen-bearing anthers cut from rice plants are placed in culture, *left*, where the pollen grains form callus, *center*, and eventually produce plantlets, *right*. When they are strong enough, the plantlets are moved to pots and examined for beneficial variants.

have dark skin.) Alternatively, there is evidence that culturing alters the pattern of protective chemical groups on chromosomes, which may lead to cells with new traits.

Somaclonal variation technology makes it possible to alter or add traits one at a time to an existing genetic background. High-solids tomatoes derived from somaclonal variants go quicker from field to soup can, because less water needs to be boiled off. Crunchier carrots and stringless celery from somaclonal variants are sold at supermarkets and fast-food restaurants. Another healthful somaclonal variant is popcorn with built-in buttery taste.

Gametoclonal Variation

Whereas somaclonal variation uses somatic cells, sex cells can also be used to grow callus, from which plantlets can be coaxed to form. Genetically variant plantlets that grow from callus initiated by sex cells are said to arise from **gametoclonal variation** (fig. 32.10). Because such a callus consists of mass-produced sex cells, each cell has half the number of chromosomes found in somatic cells of that particular species. A plant regenerated from such a gamete-derived callus cannot itself form gametes, and so it cannot reproduce. To get around this drawback, gametoclonally derived plantlets are exposed to the drug colchicine, which duplicates the chromosomes, creating a polyploid. Obtaining such a homozygous plant by conventional breeding would take at least 5 or 6 years; using cell culture, it takes 1 or 2 years.

Mutant Selection

Researchers can choose specific characteristics of new plant variants arising in cell cultures by exposing cells or protoplasts to noxious substances and then selecting only those cells that survive. The surviving cells possess a gene (or genes) that enables them to manufacture a biochemical providing resistance to the substance. Looking for genetic variants that offer a desired characteristic is called **mutant selection.** If a plant can be regenerated from a cell that has been mutant selected, then that plant and its progeny may also be resistant.

A practical application of mutant selection is to tailor seeds to be resistant to particular herbicides, so that a seed company can sell an irresistible package—seed along with an herbicide that is biologically guaranteed not to harm it (fig. 32.11).

Altering Organelles

Combinations of nuclei, cytoplasms, and organelles not known in nature can be devised to yield interesting new variants. Chloroplasts and mitochondria are good candidates for such **organelle transfer** because they contain their own genes, some of which confer such traits as male sterility (important in setting up crosses), herbicide resistance, and increased efficiency in obtaining and using energy.

Cybridization is a technique that produces a plant cell having cytoplasm derived from two cells but containing a single nucleus. A cybrid is created by fusing two protoplasts, then destroying the nucleus of one with radiation. Researchers then select fused cells that contain one nucleus and the desired combination of organelles from the original cells. In another approach, individual chloroplasts or mitochondria are isolated and encapsulated in a fatty bubble called a liposome (see Reading 5.2). The liposome can transport its contents across the cell membrane into a selected cell. Introducing a chloroplast in this manner creates a cell called a **chlybrid**; sending in a mitochondrion produces a **mibrid** (figs. 32.12 and 32.13).

KEY CONCEPTS

Cultured protoplasts or explants can be cultured to give rise to calli, from which somatic embryos or plantlets may grow. Genetically identical somatic embryos can be encapsulated to form artificial seeds, and genetically identical plants are grown from calli by clonal propagation. Plantlets grown from calli can develop from somatically mutated cells, yielding somaclonal variants that may have interesting new qualities. Gametoclonal variation is seen in plantlets grown from calli initiated with sex cells. Use of colchicine in such plantlets produces homozygous polyploids. Mutant selection identifies cells and plants resistant to certain harmful substances. New varieties of plant cells can be generated by introducing new combinations of mitochondria and chloroplasts.

Within the Nucleus—Recombinant DNA Technology

In **recombinant DNA technology,** single genes are transferred from a cell of one type of organism to a cell of another. The first recombinant organisms were bacteria engineered to carry and express the genes of higher organisms, such as the bacterium *E. coli* altered to produce human insulin. Multicellular organisms that contain a "foreign" gene in each of their cells, resulting from foreign DNA introduced at the fertilized egg stage, are called **transgenic.** Figure 13.15 shows a transgenic tobacco plant, engineered to display a firefly's "glow." Recombinant DNA technology—on the simplest bacteria or in complex plants and animals—works because all species use the same genetic code. That is, all organisms use the same sequences of DNA to order the cell to manufacture the same amino acids. For example, a gene from a bean plant placed in

Figure 32.11
Mutant selection yields herbicide-resistant corn. Corn developed by mutant selection to resist the herbicide imidazolinone, *left,* is compared with a nonresistant strain after treatment of the field with herbicide.

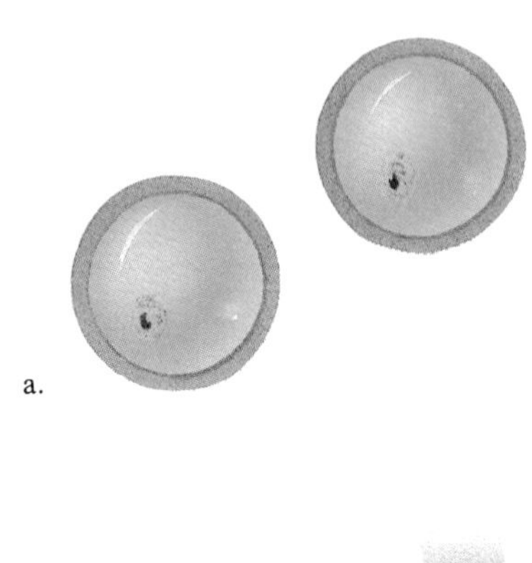

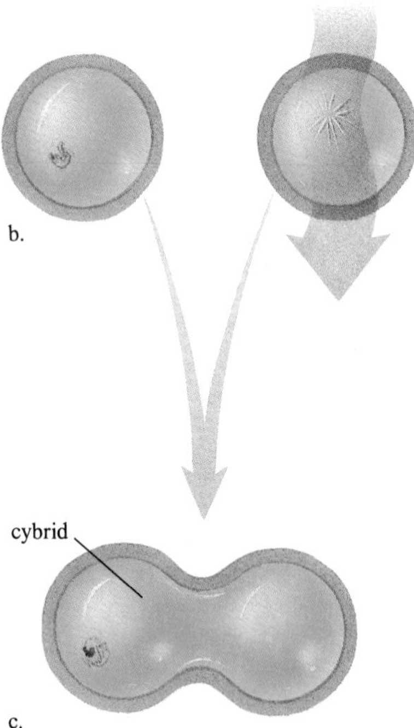

Figure 32.12
Cybridization, a technique similar to protoplast fusion, produces a plant cell having cytoplasm and organelles derived from two cells but containing only one nucleus. Just before two protoplasts are fused (*a*), the nucleus of one is destroyed by radiation (*b*). The fusion product (*c*) is called a cybrid.

a sunflower cell instructs the cell to produce the highly nutritious bean proteins, in a sunflower, a feat that has improved the protein quality of sunflower seeds.

Recombinant DNA technology in plants begins by identifying and isolating an interesting gene. Bacteria are often the source of such genes, which can confer on plants built-in resistances to disease, insecticides, herbicides, and environmental extremes. The donor DNA as well as the "vector" DNA, which transports it into the plant cell, are then cut with the same restriction enzyme so that they can attach to each other at the ends to form a recombinant molecule. The vector and its cargo gene are then sent into the plant cell. Chapter 13 discusses the steps of recombinant DNA technology in greater detail.

In dicots, foreign genes are introduced in a **Ti plasmid** (which stands for "tumor inducing"), a ring of DNA found in the microorganism *Agrobacterium tumefaciens* (fig. 32.14). In its natural state the Ti plasmid invades plant cells and causes a cancerlike growth called crown gall disease (fig. 7.10). However, the tumor-causing genes of the plasmid can be chemically removed without impairing the plasmid's ability to enter a plant cell's nucleus. The "disabled" Ti plasmid brings into the plant cell whatever foreign genes are stitched into it.

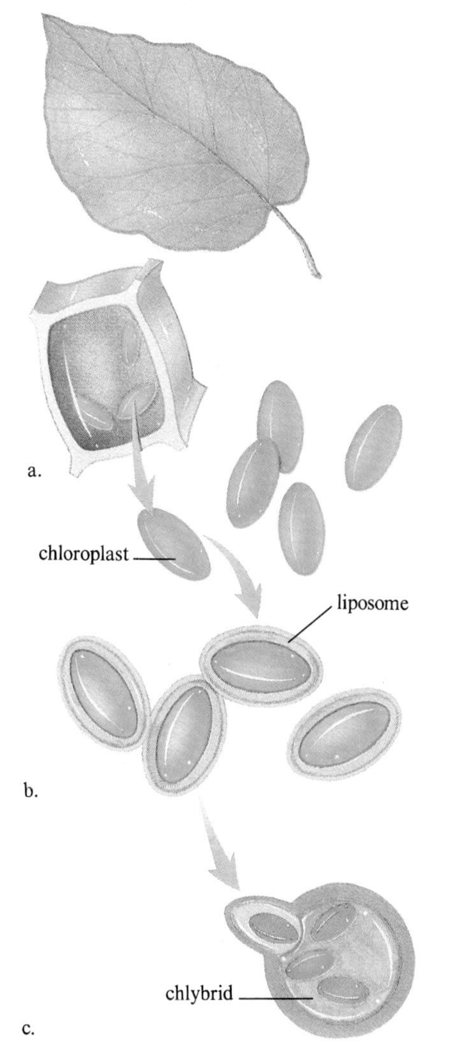

a.
chloroplast
liposome
b.
chlybrid
c.

Figure 32.13
Chlybrids and mibrids. Chloroplasts can be isolated from cells of a parent plant (*a*) and encased in liposomes, which are microscopic spheres of lipid (*b*). When mixed with protoplasts of selected cells, the liposomes fuse with the cell membrane (*c*), delivering their contents to the cellular interior and creating "chlybrid" cells. Mitochondria introduced in this way produce "mibrid" cells.

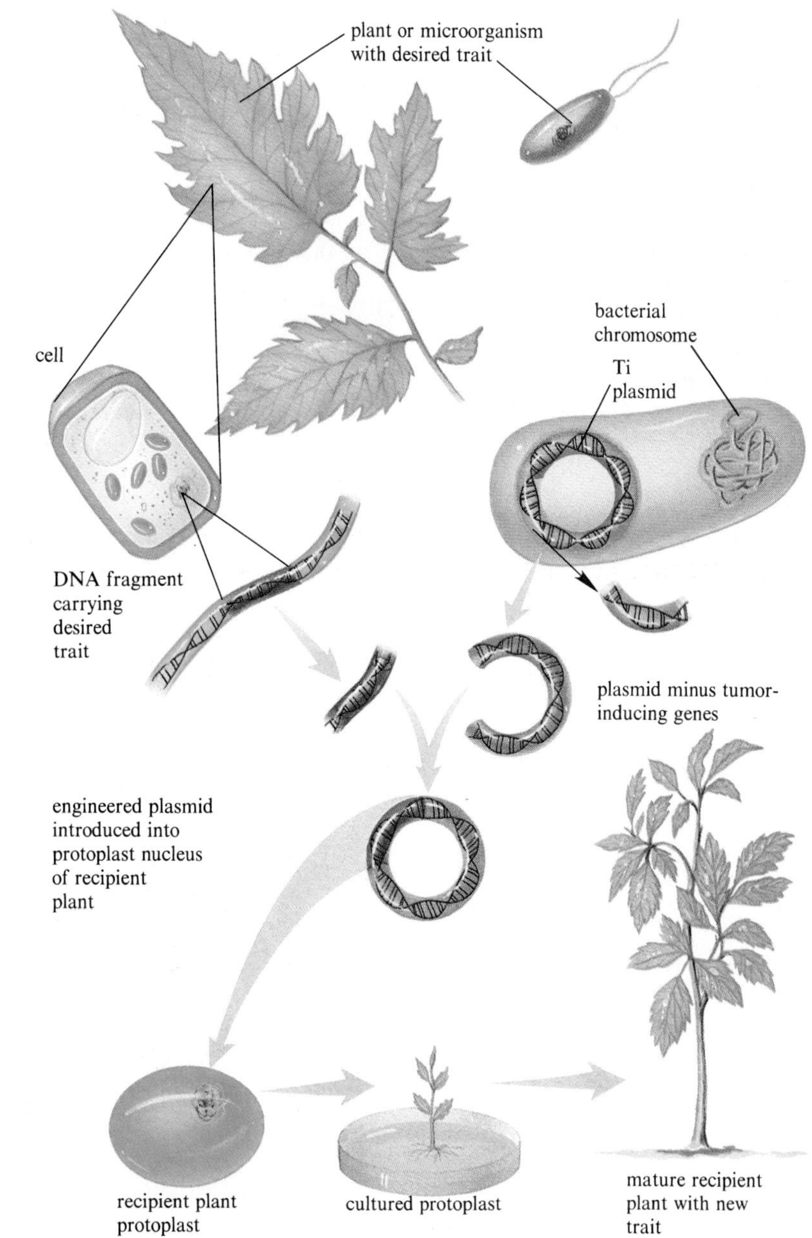

Figure 32.14
Making a transgenic plant. A fragment of DNA carrying the desired gene—conferring resistance to an herbicide, for example—is isolated from its natural source and spliced into a Ti plasmid from which the tumor-inducing genes have been removed. The plasmid incorporating the foreign DNA is then allowed to invade a protoplast of the recipient plant, where it enters the nucleus and integrates into the plant's DNA. Finally, by means of cell culture, the protoplast is regenerated into a mature, transgenic plant that expresses the desired trait and passes it on to its progeny.

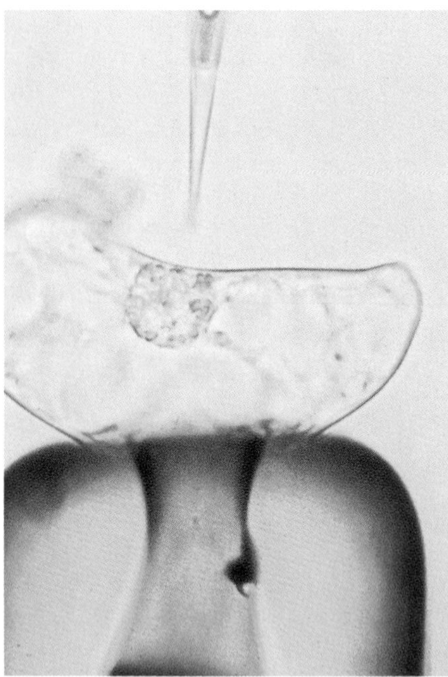

Figure 32.15
Injecting monocot cells with foreign DNA. One way to move foreign DNA into a plant cell nucleus is by direct injection with a microscopic glass needle.

Genetic engineering of the monocots requires more creative approaches than that of the dicots, because many naturally occurring plasmids do not enter monocot cells. One solution is to use monocot protoplasts, because removing the cell wall makes the cell membrane more likely to admit foreign DNA. In a technique called **electroporation,** a brief jolt of electricity opens up transient holes in the cell membrane of monocots that may permit entry of foreign DNA. Genetic material can also be injected into monocot cells using microscopic needles (fig. 32.15) or sent across the cell membrane within liposomes.

Another way to introduce DNA into a plant cell is with **electric discharge particle acceleration,** also known as a "gene gun." A gunlike device shoots tiny metal particles, usually gold, that have been coated with the foreign DNA. Some of the projectiles enter the target cells. This method has been used to "shoot" soybean seed meristem cells with a gene from *E. coli* that stains cells expressing it a vibrant blue, allowing the gene transfer to be detected (fig. 32.16).

Once foreign DNA is introduced into a target cell, it must enter the nucleus and then be replicated along with the cell's own

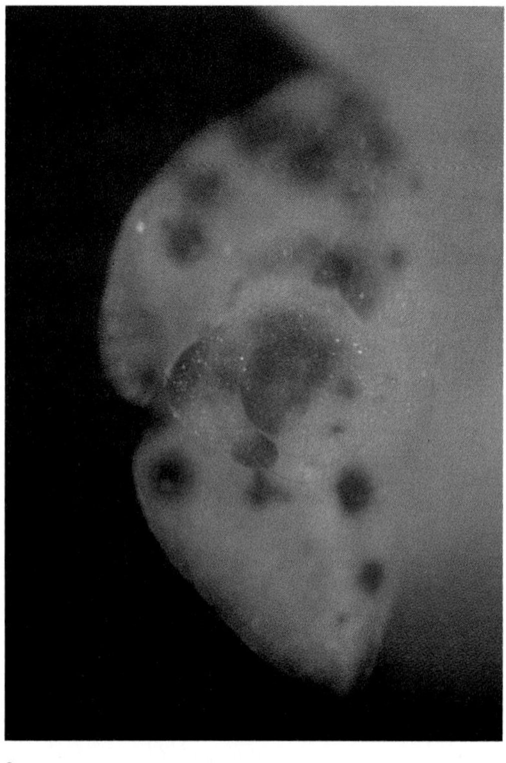

a.

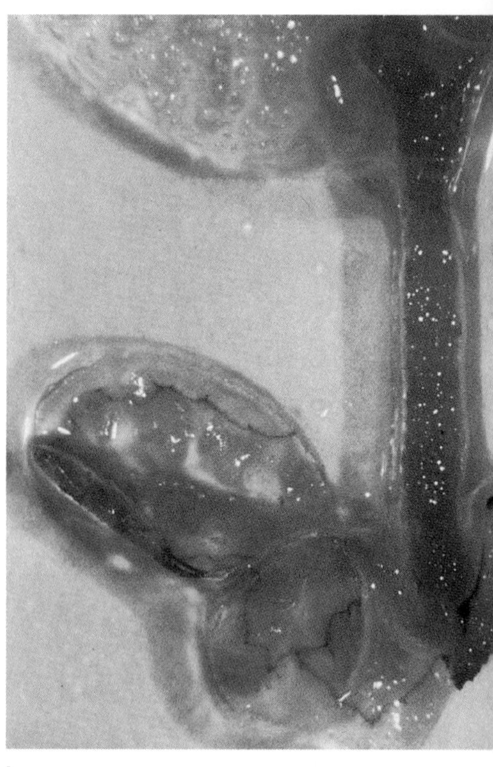

b.

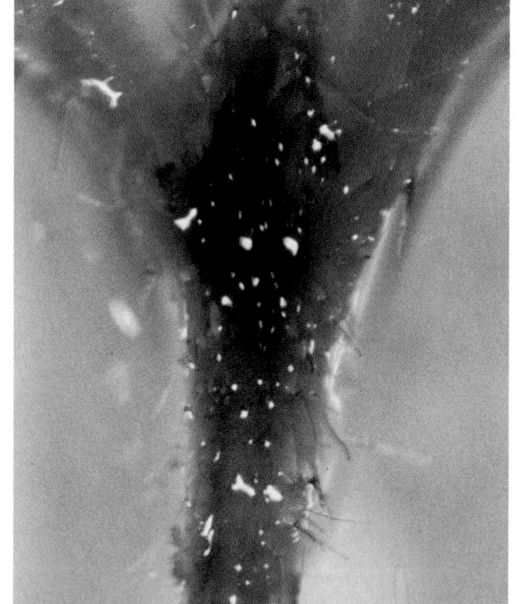

c.

d.

Figure 32.16
Transgenic soybeans created using a gene gun. A gene gun was used to send the *E. coli* gene for beta-glucuronidase into cells of soybean seed meristems. *a.* The blue spots on the meristem indicate cells that have taken up the foreign gene because this enzyme produces a blue color in the presence of its substrate. *b.* Nonengineered soybean meristem is shown for comparison. *c.* The plantlet has some engineered cells, and (*d*) this plantlet has taken up the bacterial gene in nearly all of its cells, as indicated by the dark blue stain.

a. b.

Figure 32.17
Two routes to herbicide resistance. Petunias and tobacco were both genetically engineered to resist the herbicide glyphosate but in different ways. Glyphosate kills plants by suppressing the activity of an essential enzyme, EPSP synthase. *a.* The petunias in the back row were given a viral gene that allows them to overproduce the enzyme sufficiently to counteract the effect of a commercial glyphosate spray. In contrast, the unmodified petunias in the front row show no resistance to the spray. *b.* The glyphosate-sprayed tomato plants in the center row carry a bacterial gene for a form of EPSP synthase that functions even in the presence of glyphosate. In the left row are engineered tomato plants that were not sprayed; in the right row are unmodified tomato plants that have succumbed to the glyphosate spray.

Table 32.1
Recombinant DNA Solutions to Agricultural Challenges

Challenge	*Possible Solution*
Crops damaged by frost	Spray crops with bacteria genetically engineered to lack surface proteins that promote ice crystallization. Bacteria can also be engineered to encourage ice crystallization; used to increase snow buildup in winter sports facilities.
Crops damaged by herbicides and pesticides	Isolate resistance genes from an organism that is not affected by the chemical and engineer the gene into crop plant.
Crops need costly nitrogen fertilizer because atmospheric nitrogen is not biologically usable	Short term: genetically engineer nitrogen-fixing *Rhizobia* bacteria to overproduce enzymes that convert atmospheric nitrogen to a biologically usable form in root nodules of legumes. Alter *Rhizobia* to colonize a wider variety of plants. Long term: transfer *Rhizobia* nitrogen-fixation genes into plant cells and regenerate transgenic plants.
A plant food is low in a particular amino acid	Transfer gene from another species that controls production of a protein rich in the amino acid normally lacking in the crop plant.
A crop plant is killed by a virus	Genetically engineer crop plant to manufacture a protein on its cell surface normally found on the virus's surface. Plant becomes "immune" to virus.
Public concern about the safety of synthetic pesticides	Engineer *Bacillus thuringiensis* to overproduce its natural pesticide, which destroys insects' stomach linings. Transfer *B. thuringiensis* bioinsecticide gene to crop plant.
Fruits and vegetables ripen quickly once picked	Suppress or slow production of ripening enzymes such as cellulase.

DNA and transmitted when the cell divides. Finally, a mature plant must be regenerated from the engineered cell and express the desired trait in the appropriate tissues and at the right time in development. Then the plant must pass the characteristic on to the next generation. That's a tall order! A quicker way to use genetic engineering to endow a plant with a new capability is to manipulate bacteria that normally live on a crop plant's roots (fig. 32.17). Table 32.1 lists some ways that recombinant DNA technology can be used to solve agricultural challenges.

KEY CONCEPTS

Transgenic plants are generated by introducing foreign DNA into a cell and regenerating a plant from that cell. The plant must express the foreign gene appropriately and transmit it to future generations. Methods of gene transfer include the Ti plasmid, electroporation, microinjection, liposomes, and gene guns.

Biotechnology Provides Different Routes to Solving a Problem

Different biotechnological approaches can address a single problem. For example, how might you devise a crop that cannot be damaged by the herbicide used to protect it from weeds? The traditional approach is to find a weed that is resistant to the herbicide and related to the crop plant. The hardy weed and its domesticated relative are then crossbred until a variant arises that retains the desired qualities of each parent plant—resistance to the herbicide, plus the characteristics that make the plant a valuable crop. In the past, if an herbicide-resistant crop could not be bred, then the herbicide was simply not used.

Today, instead of changing herbicides to fit crops (called the "spray and pray" approach), biotechnologists are altering crops to fit herbicides. In mutant selection, resistant cells cultured in the presence of the herbicide are isolated and used to regenerate resistant plants. A callus grown in the presence of the herbicide would yield embryos or plantlets that are resistant. The recombinant

Reading 32.1 *Plant Biotechnology Moves from the Laboratory to the Land*

It is a frigid evening in January 1987, and several angry residents of Tulelake, California, are meeting in a town hall to discuss what to them is a frightening prospect. Researchers at the nearby University of California at Berkeley want to paint potato plants, on town land, with genetically altered bacteria to see whether the bacteria will render the potatoes frost resistant. The residents have been given a very complete, 500-page environmental impact statement by the researchers. While it looks like the proposed experiment is relatively safe and has been checked out in the laboratory and greenhouse extensively, the people are nonetheless disturbed. Words like "mutant" and "recombinant DNA" invoke fear; and many are worried by the uncertainty of the scientists' language and the fact that they cannot guarantee that the bacteria will not spread beyond the test plot, harm native plants, or cause disease in them or their animals.

The story of the "ice-minus" bacteria of Tulelake actually begins back in 1977, when a graduate student at the University of Wisconsin at Madison, Steven Lindow, found that ice forms on potato leaves by gathering around "ice-nucleating proteins." The proteins are part of the surface of the bacterium *Pseudomonas syringae*. What would happen to the potato plants, Lindow wondered, if the bacteria could not produce the ice-forming protein? He found a naturally occurring *P. syringae* mutant that made fewer than normal ice-nucleating proteins; he made the mutant more severe by deleting about a third of the responsible gene's base pairs. In greenhouse experiments, Lindow validated his hypothesis—potato leaves coated with unaltered, wild type *P. syringae* froze at 28°F (-2.2°C); leaves coated with the genetically engineered ice-minus bacteria did not freeze until the mercury dipped to 23°F (-5°C). By 1982, Lindow wanted to test his bacteria in the field, thinking ahead that frost-resistant bacteria could cut into the $1.5 billion in damage caused by frozen crops in the United States each year.

Lindow's plans for the logical next step—the field test—were sidelined for 4 years by a lawsuit brought by the Foundation on Economic Trends in Washington, D.C. Meanwhile, Lindow continued his work—he brought native Tulelake plants into the greenhouse, where he exposed them to ice-minus bacteria; he studied local wind patterns, consulted wildlife experts, and talked with Tulelake townspeople, who did not, at first, seem to object to what he planned to do. The Environmental Protection Agency and the National Institutes of Health applauded Lindow's efforts to ensure safety.

But in 1985, the young scientist hit another snag. A company called Advanced Genetic Sciences injected similarly altered bacteria, called Frostban, into trees on the roof of their Oakland facility—without approval of the EPA. The Frostban experiment hit the press, and a wave of apprehension spread northwards to Tulelake. The town meetings began.

Finally, in the spring of 1987, the first "deliberate release" of genetically engineered bacteria was allowed. A Sacramento County Superior Court Judge declared the Frostban bacteria safe on April 23, and the next day, 2,400 strawberry plants were sprayed with it in Brentwood. But vandals ripped out the plants overnight. On April 19, Steven Lindow planted in Tulelake 3,000 potato seedlings that had been coated as seeds with ice-minus bacteria. Again, vandals struck, removing half of the plants. Researchers replanted most of them the next day. Then, on May 28, Lindow sprayed ice-minus bacteria on the seedlings.

The ice-minus bacteria indeed protected the potato plants. Elaborate monitoring experiments showed that, even many months later, none of the 6 trillion released bacteria had wandered beyond the 30 meters of bare soil surrounding the test plot. And so began the era of deliberate release of organisms altered by humans.

Scientists are developing methods to counter some possible dangers of releasing genetically altered organisms to the environment. For example, bacteria that are manipulated to contain genes providing a valuable resistance factor to crop seeds can also be given a gene that makes them glow or change color in the presence of a particular substance. Recombinant *Pseudomonas fluorescens* bacteria coated onto winter wheat are altered to cause a color change on standard bacterial growth plates when exposed to the sugar lactose. Using this tracing system, the bacteria were shown to have traveled 7 inches (17.8 centimeters) horizontally and 12 inches (30.5 centimeters) vertically in 10 weeks along the roots of the wheat plants.

A way to control the spread of a genetically engineered organism is the "suicide plasmid," which causes altered bacteria coated on crop plants to self-destruct once they have completed their job. Recall that a plasmid is a small, circular piece of DNA found naturally in some bacterial cells (see fig. 13.25). A plasmid can be engineered to contain a gene that turns on production of a DNA-cutting enzyme when a particular biochemical is not present. Self-destruct plasmids are inserted into the engineered bacteria, which are coated onto seeds of crop plants. (For example, consider bacteria that produce an insecticide and are coated onto corn seeds.) A chemical that suppresses the DNA-cutting function of the plasmid is applied along with the bacteria. Once the chemical is degraded or washed away, the plasmid activates its DNA-cutting enzymes, and the bacteria are destroyed. The corn seeds have by this time germinated, insect free, and the bacteria can no longer pass on their engineered resistance to nearby weeds.

While geneticists alter crop plants one gene at a time and devise ways to introduce these plants into the environment as safely as possible, other biologists are concerned about the consequences of these efforts because they deviate from the sorts of genetic changes that occur naturally. For example, pests are known to evolve resistances to pesticides, usually in the form of resistance genes carried on plasmids. Such natural resistance usually proceeds in a gene-by-gene fashion over long periods of time—that is, a pest organism's resistance is a single gene change in response to the effect of the pesticide on the organism's physiology. How will organisms respond to crop plants that are engineered to produce several substances? How will they react to what appears to be a very sudden and drastic evolutionary change?

Some ecologists are looking at an even bigger picture. If organisms can be designed to grow in the presence of chemical pesticides, will we increase our use of such chemicals, possibly threatening other species not given protection? Similarly, if organisms can be biologically altered to withstand pollution, will efforts to clean up the environment be slowed or halted? Engineering more hardy species, some argue, is treating the symptoms of environmental problems rather than getting to the roots of the problems. These concerns perhaps summarize the potential impact of agricultural biotechnology on our society—in the long run it can provide valuable new variants; but in the short term, we must understand how altering these organisms will affect the living world.

a.

b.

Figure 1
The very first field tests of genetically engineered microbes applied to plants were conducted by researchers clad in spacesuits and protective gear (*a*). *b*. Today, only a jumpsuit, gloves, goggles, and mask are required.

Table 1 Government Regulation of Transgenic Plants	
Agency	*Role*
Department of Agriculture (USDA)	Approves field tests
Environmental Protection Agency (EPA)	Sets tolerable levels of pesticides in food crops
Food and Drug Administration (FDA)	Ask these questions: Is a genetically engineered food "generally recognized as safe?" and "Does the genetic engineering produce a food additive?"

DNA approach to herbicide resistance focuses on identifying the part of the crop plant's physiology that is damaged by the herbicide and then finding a gene (from any organism) that might enable the plant to prevent or undo whatever damage the herbicide does.

Beyond the Laboratory—Release of Altered Plants to the Environment

A genetically altered plant is more controllable on a laboratory shelf or in a greenhouse than when growing in a field, where it can interact with other species. However, field testing is an essential step in using the agricultural products of biotechnology.

The first "deliberately released" genetically engineered organisms were bacteria that prevent ice crystal formation on crop plants (Reading 32.1). Today, regulatory agencies approving field tests of genetically altered organisms are overrun with requests. The concerns of the agencies as well as the public are several. How long will the altered plant or bacterium survive? How quickly does it multiply? Has it been altered in a way not seen in nature? How far can the organism travel? Most important, what are the effects of the genetically engineered organism on the living and nonliving environment? Can the altered organism pass on its new characteristic—such as herbicide or disease resistance—to a weed?

KEY CONCEPTS

Sometimes an agricultural problem can be solved by several biotechnologies, including traditional breeding, creating a transgenic plant or a somatic hybrid, selecting a resistant mutant, or altering bacteria that live on a plant's roots. After an altered and fertile plant is grown to maturity, it must be field tested. Scientists must determine how a genetically engineered plant affects its environment.

SUMMARY

Plant biotechnology manipulates plants at the level of cells or DNA, producing either new variants or uniformity. Once an individual cell is altered or selected, it must be regenerated into a mature plant, that plant must express the sought-after trait and pass it on to the next generation, and the new variety must be able to withstand conditions in the greenhouse and in the field. Biotechnology differs from traditional plant breeding in that it is asexual, more precise, and faster.

In *protoplast fusion*, two cells are stripped of their cell walls and fused. If a plant can be regenerated from protoplast fusions of two species, an interesting hybrid may result. Culture of an explant produces a *callus*, an undifferentiated mass that can yield embryos or plantlets, which are genetically identical or occasionally variant, depending upon the particular mix of nutrients and plant hormones in which they are cultured. Identical embryos are used to fashion *artificial seeds*. Identical plantlets offer uniform crops. Unusual plantlets, derived from *somaclonal* or *gametoclonal* variation, offer new plant types. Plant cells with novel combinations of cytoplasms, nuclei, and chloroplasts and mitochondria are derived by *cybridization*, in which protoplast fusion is followed by radiation treatment to inactivate one nucleus. Chloroplasts and mitochondria can also be introduced in *liposomes*.

In *recombinant DNA technology*, foreign genes are introduced into bacteria and these bacteria coated onto vulnerable plant parts; or *transgenic* plants are created by transferring foreign genes into fertilized ova and regenerating plants. Recombinant DNA technology in plants can offer resistances to temperature extremes, pests, and disease and alter nutritional qualities. Several environmental concerns are raised by introducing plant varieties derived from altering cells or DNA into the environment.

QUESTIONS

1. List the steps necessary for plants altered at the cell or DNA level to become useful agricultural varieties.

2. How are modern biotechnologies such as cell culture and recombinant DNA technology similar to and different from traditional plant breeding?

3. Name a plant biotechnology that is very precise and one that is very imprecise.

4. If you want to breed identical-looking flowers that are easy to pick, what biotechnology might you choose? Why?

5. List the steps involved in using somaclonal variation to develop an extra-sweet carrot plant.

6. What is an advantage and a drawback of gametoclonal variation?

7. How must dicots and monocots be handled differently in protoplast fusion? In recombinant DNA technology?

TO THINK ABOUT

1. What do you think is the major benefit of plant biotechnology? The major risk?

2. One explanation for the mechanism behind somaclonal variation is that the cell culturing process uncovers preexisting somatic mutations. What is an alternative explanation?

3. Suggest three biotechnology approaches to developing a potato plant that is resistant to the potato beetle, which eats its leaves.

4. On June 18, 1987, a plant pathologist at Montana State University, Gary Strobel, infected 14 elm trees, outdoors, with *Pseudomonas syringae* bacteria. In 1981 he had altered the bacteria to overproduce a protein that they normally manufacture and that kills the fungus that causes Dutch elm disease. As part of the experiment, Strobel injected the 14 trees as well as 14 unprotected control trees with the Dutch elm disease fungus. Strobel had already tried his bacteria on Dutch elm tree sap in the laboratory and greenhouse and found that they indeed killed the disease-causing fungus.

Because it was June, Strobel knew that he needed to conduct the field test immediately or wait another year. So he infected the trees, without clearance from university officials or the EPA. The officials as well as much of the scientific community were enraged by Strobel's disregard of regulations. Strobel, however, defended his view: "We can sit and talk about Dutch elm disease, or we can do something about it. I chose to do something about it."

a. What additional pieces of information do you need to fairly judge whether Gary Strobel was justified in carrying out his field test without standard approvals?

b. Do you think that Strobel acted wisely? Why or why not?

5. Considering what you now know about plant biotechnology, how would you feel about a crop plant that is genetically engineered to tolerate extreme heat being field tested next to your home? If you were the researcher, what measures would you take to ensure the safety of the experiment?

6. Studies have shown that fewer than 20% of Americans know what DNA is and what it does. Many genetic researchers are frustrated by people without knowledge of genetics attempting to regulate their experiments. What do you think can be done so that the public can become better informed about plant biotechnology—or about science in general?

SUGGESTED READINGS

Gould, Fred. January 1988. Evolutionary biology and genetically engineered crops. *BioScience*. Can we learn from evolution about how unaltered plants will respond to altered plants?

Hall, Stephen S. September 1987. One potato patch that is making genetic history. *Smithsonian*. The first release of a genetically altered bacterium onto plants was traumatic for residents of the town that housed the testing site. Today we take such tests for granted.

Hoffman, Carol A. June 1990. Ecological risks of genetic engineering of crop plants. *BioScience*. Genetic engineering of crop plants must proceed with caution because other species may be affected.

Lewis, Ricki. May 1986. Building a better tomato. *High Technology*. A survey of different biotechnological approaches to devising new plant varieties.

Miller, Henry I., and Stephen J. Ackerman. March 1990. Perspectives on food biotechnology. *FDA Consumer*. Biotechnology of plant foods is the latest approach to agriculture.

Potrykus, I. June 1990. Gene transfer to cereals: an assessment. *Biotechnology*. Altering cereals has been biotechnology's greatest challenge.

Ratner, Mark. May 1990. Identifying quantitative traits in plants. *Biotechnology*. How do plant biotechnologists know which combinations of traits are valuable.

Strobel, Gary. October 19, 1987. Strobel: "I have acted in good faith." *The Scientist*. Plant pathologist Gary Strobel released genetically modified bacteria that protect against Dutch elm disease—without appropriate regulatory approval. Was he justified in his action?

UNIT

7

Evolution

C H A P T E R

33

Darwin's View of Evolution

Chapter Outline

Learning Objectives

By the chapter's end, you should be able to answer these questions:

1. What can be learned about the evolution of species when cataloging the types of organisms that populate islands over known periods of time?
2. What natural phenomena does the science of evolution consider?
3. How do geology and geography influence evolution?
4. How did Charles Darwin's observations of the distribution of organisms throughout the world contribute to his theory of evolution?
5. What is natural selection, and how does it explain and underlie evolutionary change?

About 8 million years ago, the island of Mauritius rose in volcanic fury from the depths of the Indian Ocean. By the sixteenth century, the island's surface was covered with low-lying plains and dense, tall forests teeming with colorful birds, scurrying insects, and basking reptiles. The organisms that made the island their home were well adapted to the climate, terrain, and other organisms of Mauritius. Some of the organisms interacted in interesting ways.

Consider the flightless, 30-pound (14-kilogram) relative of the pigeon, the dodo bird, which stood 3 feet (1 meter) tall and nibbled the hard fruits of the abundant Calvaria tree. Were it not for the destruction of the outer hard layer of the Calvaria fruits by the dodo's digestive tract, the seeds of the majestic trees could not germinate. This mutual dependence of two types of organisms is termed **coevolution.**

Food was plentiful on Mauritius, and the populations of its resident species flourished. Life progressed in this splendor and diversity on the little island until the 1500s, when Spanish and Portuguese sailors arrived. It was the beginning of the end for many longtime island inhabitants. The men feasted on dodo meat, and their pet monkeys and pigs ate dodo eggs. The Calvaria trees gradually died out as their route for seed dispersal and germination vanished with the dodo. Today only a few ancient Calvarias remain. Rats and mice swam ashore from the ships to attack native insects and reptiles. Exotic animals accompanying the human invasion also evicted island species from their homes. Indian myna birds took over nesting holes of the native echo parakeet, while Brazilian purple guava and oriental privet plants crowded seedlings of native trees that had been undisturbed for centuries. Soon, the forest that had previously been home mostly to birds, insects, and reptiles had become a mammalian haven, populated largely by newcomers.

By the mid-1600s, only 11 of the original 33 species of native birds remained. The dodo was exterminated by 1681, earning a dubious distinction as the first recorded extinction due to direct human intervention (fig. 33.1). The dodo is known today solely from bones. For years the bird was thought to be mythical. Others of Mauritius's reptile, plant, and insect species may face similar bleak futures.

About 3 million years after Mauritius erupted from the Indian Ocean, the first of a chain of other volcanic islands was born, rising in the central Pacific, 2,000 miles (3,218 kilometers) from the closest continents. After the new land cooled, the invasion of life began. Seeds floated in on mats of vegetation or drifted in on the wind or on migrating birds. Insects and a lone type of mammal, a bat, flew in too. From this original gathering of plants and animals evolved, over the millennia, a diverse collection of new varieties. The original 15 species of land birds diversified into more than 80 species; the 20 founding land molluscs were ancestral to more than 1,000 species living there today; one or two colonizing fruit fly varieties led to hundreds of modern species (fig. 33.2).

Islands continued to erupt from the volcanic earth, and the main island of Hawaii was born a mere half million years ago. Its five volcanoes, two exceeding

Figure 33.1
The dodo—a modern extinction caused by humans. The great, flightless dodo and the Calvaria tree coevolved, the bird dependent upon the tree for food, and the tree dependent on the bird for seed germination.

Figure 33.2
The hundreds of species of fruit flies that inhabit the Hawaiian Islands descended from only one or two colonizing species. This fly hails from the island of Oahu.

13,780 feet (4,200 meters) above sea level, provide a wide range of different habitats, which are frequently altered by fresh lava flows. Local populations confined to isolated habitats accumulate their own distinct sets of adaptations, so that the island is home to hundreds of species seen nowhere else on the planet.

The Hawaiian Islands, like Mauritius, have seen many species vanish. Some species just cannot compete with others for food and shelter. The arrival of humans to the Hawaiian Islands in A.D. 400 transformed the lush lowland forests into fields of crops, destroying the homes of many resident species. The large, flightless birds were eaten, and the vibrantly colored birds were killed and their feathers used to adorn the clothing of Polynesian settlers. Europeans who arrived a dozen centuries later destroyed the land further, hastening the extinctions of even more species.

About 40 million years ago, a group of small volcanoes arose from the sea just off the coast of Ecuador. Over the years the resulting volcanic islands cooled sufficiently so that organisms living on the mainland of South America were able to colonize them, flying or riding natural rafts of floating vegetation to a new habitat. Gradually, these first islands were covered over by the sea, as newer islands arose from volcanic eruptions farther away from the mainland. Perhaps the first group of islands and the second group coexisted for a time, and some of the organisms on the islands closer to the mainland moved to the newer island, which was farther from the mainland.

Today's Galápagos archipelago is a group of volcanic islands 600 miles (965 kilometers) west of Ecuador. The particular species that chanced to inhabit these islands were but a subset of all those present on the mainland. They survived only if they, and their descendants, were adapted to the particular environmental challenges posed by the island habitat. The modern Galápagos Islands are home to an unusual assortment of organisms. The name Galápagos comes from the Spanish for "tortoise," one of the island's most noticeable residents (fig. 33.3).

Figure 33.3
Galápagos tortoises. When he visited the Galápagos Islands, Charles Darwin found that the natives could tell which island a particular tortoise came from by the length of its neck and other obvious adaptations. This tortoise has evolved a long neck, which enables it to reach its vegetable food.

Islands Reveal Evolution in Action

These panoramas of island life are constructed from geological and historical evidence, with the details filled in from present-day ideas about evolution. Islands provide a microcosm for viewing evolution on an accelerated scale, something that life on the mainland does not provide as readily. **Evolution** is the process by which the genetic composition of populations of organisms changes over time. In sexually reproducing organisms, when genetic changes accumulate that prevent members of one population from successfully breeding with members of another population, two **species** have diverged from the ancestral group.

If we are lucky enough to know when an island formed and have evidence indicating which species were the first to colonize it, and we know which species live there today, the events that took place in between can be reconstructed. It is likely that over time, certain individuals were better able to survive and leave offspring than other individuals, because they had inherited variations of traits that were helpful, or **adaptive,** in that particular environment. The island populations accumulated adaptive traits to the point that they eventually became so unlike their distant relatives that originally settled the island that they theoretically could no longer mate with them. Thus did new species arise. The isolation of the island environment prevented members of island populations from breeding with individuals on other landmasses, keeping their collections of traits separate. One way that new species are born is from the isolation afforded by islands.

Macroevolution and Microevolution

Evolution is a continuing process that explains the history of life on earth, as well as the diversity of life forms on the planet today. Evolution includes large-scale events, such as the appearance of new species **(speciation)** and the disappearance of species **(extinction).** These large changes are termed **macroevolution.** Evolution also includes more subtle, incremental single-trait changes that accumulate to the point that two groups of organisms can no longer interbreed. These changes are termed **microevolution.**

Whereas macroevolutionary events tend to span very long periods of time, microevolutionary changes can happen so rapidly that they can be seen experimentally. In one experiment, two populations of a species of fruit fly were bred, each from an original pair of flies. One population was reared at 61°F (16°C) and the other at 80°F (27°C). After 12 years, the flies raised at the colder temperature were on the average 10% larger than the flies raised at the higher temperature.

Microevolutionary change can also be seen yearly as influenza viruses change their surface proteins, reflecting underlying genetic mutation. This is why we must constantly develop new flu vaccines, and it is also why it has been difficult to devise a vaccine against the virus that causes AIDS, which also evolves rapidly.

Evolutionary biologists debate whether macroevolutionary changes reflect the buildup of step-by-step microevolutionary changes or if the more sweeping events involve some other, as yet unknown, genetic mechanism. Although clearly both macroevolutionary and microevolutionary processes are at work, the precise relationship between them is not well understood.

Early in the nineteenth century, biologists realized that inherited traits (rather than acquired traits) are important in evolution. Although later the discovery of genes explained the mechanics of evolutionary change, the idea of evolution as a gradual change in life forms actually grew out of many observations made by several individuals (Reading 33.1). These concepts were crystallized into a coherent theory by Charles Darwin in the nineteenth century. Darwin's theory was radical for its time, but it was so compelling, and so beautifully argued and illustrated, that it achieved scientific acceptance rather easily. Darwin's famous theory of evolution has its roots in the evolution of the earth.

KEY CONCEPTS

The isolation afforded by islands offers views of evolution in action. Evolution is genetic change in a population. Large-scale events such as speciation and extinction illustrate macroevolution. Single-trait changes demonstrate microevolution. The precise relationship between microevolutionary and macroevolutionary events is not known.

The Influence of Geology—Clues to Evolution in Rock Layers

In a sense, it can be said that the science of evolution got its start in ditch digging. The field of geology arose in the latter half of the nineteenth century, as people puzzled over the layers (strata) in the earth that were revealed when ditches were dug or mountains climbed. It had been recognized since the 1600s that lower rock layers were older than those above them, a phenomenon called the **principle of superposition.** Several interesting theories attempted to explain how rock layers came to be (fig. 33.4).

Neptunism, named after the Roman god of the sea, held that a single great flood organized the earth's surface, with the waters receding to reveal the mountains, valleys, plains, and rock strata present to this day. According to Neptunism, the earth has not changed since this initial catastrophic event. A variation of Neptunism, called **mosaic catastrophism,** held that a series of great floods molded the earth's features and were responsible for extinctions of some life forms and creations of others. Mosaic catastrophism explained why preserved remains of sea-dwelling organisms were sometimes found in different rock strata. The single flood of Neptunism could not account for this observation.

At odds with Neptunism and mosaic catastrophism was **uniformitarianism,** the continual remolding of the earth's surface. This idea was proposed by a physician and farmer named James Hutton, who liked to take long walks in the Scottish countryside.

Reading 33.1 *Some Thoughts on Evolution Through History*

THROUGHOUT RECORDED HISTORY AND UNDOUBTEDLY BEFORE THAT AS WELL, HUNDREDS OF THINKERS HAVE PONDERED HOW THE GREAT VARIETY OF LIVING THINGS CAME TO BE, AND HOW THEY ARE RELATED TO ONE ANOTHER, IF AT ALL. Here are a few hypotheses to explain the evolution of life.

One of the earliest thinkers to place human beginnings into a scheme with the origin of other organisms was a fourth-century B.C. Greek physician, Empedocles. He proposed that the appearance of animals was the most recent part of the evolution of the universe. First, parts of animal bodies appeared, and then the parts mixed together randomly to form animals. The first combinations were poorly adapted monsters that did not survive or reproduce and hence did not perpetuate. Eventually, the animals with which we are familiar today formed, and these survived.

Like Empedocles, Aristotle, who lived a century later, considered the appearance of animals to follow the evolution of nonliving matter. He thought that inorganic matter evolved into organic matter, which led to soft living matter, which gave rise to perfect life forms. But he thought each species arose independently and did not change.

Aristotle envisioned a "Great Chain of Being," a detailed ordering of many life forms arranged into a single line of descent. The Chain of Being was uneven—it did not represent all types of organisms in the same degree of detail. Different human races were considered to be distinct types of organisms, and their "order," from less advanced to more advanced, often reflected the ethnic background of the orderer. Yet many species and entire groups of organisms were missing from the Great Chain of Being. Like many plans before it, the Chain of Being traced the evolution of the supposedly imperfect to the perfect. This differs markedly from the modern view of evolutionary change as fostered by adaptations to a particular environment, rather than attainment of a subjective perfection.

The term "evolution" in reference to the appearance of life forms over time was coined by Swiss naturalist Charles Bonnet in the eighteenth century. He envisioned earth's natural history as a series of major catastrophes. After each event, the organisms of the past period became fossils, and new types of organisms appeared, moved up a notch on a scale of increasing biological complexity. The most recent event, according to Bonnet, catapulted apes to human status.

The first thinker to extract a theory from these sometimes arbitrary orderings of organisms was French taxonomist Jean Baptiste Lamarck (1744–1829). He originated some of the ideas found in Darwin's work, such as the evolution of species from preexisting ones and the ability of animals to adapt to a changing environment. He is most remembered, however, for his theory of the inheritance of acquired characteristics, in which evolutionary change results from the appearance of new body parts or functions in response to want or need. According to this view, characteristics acquired during an individual's lifetime are passed to its offspring. An often-quoted example of the inheritance of acquired characteristics is the long neck of the giraffe. The animal "needed" a long neck to reach its treetop food, so the neck grew, a little each generation. Once acquired, a trait was passed on to the next generation. Lamarck is also noted for his law of use and disuse. A used organ would be maintained from generation to generation; an unused organ would gradually disappear.

Darwin was not exactly sure how natural variations arose. The field of evolution had to wait until the early twentieth century, when knowledge of genetics supplied the missing information.

Hutton possessed a great scientific gift—the ability to connect observations to explain phenomena. What set Hutton's mind to contemplating the history of the earth were similarities of rocks in different parts of a particular locale. Why did pebbles brought in by the tide resemble rocks in the hills looking out over the ocean? Why did the rock layers in mountains contain evidence of organisms that clearly had lived in the sea? Hutton noted the weathering of exposed rock on the land and the gradual deposition of sediments in bodies of water. He proposed that the earth's surface features were not formed in just one event, or even several, but were continually being built up and broken down. According to Hutton, the seas receded to reveal new, uplifted sediments, as ancient mountains were eroded down, slowly contributing future sediments to the seas.

These geological changes must have taken a very long time—certainly longer than the 6,000 to 10,000 years of earth's existence according to the Bible. The long time of earth history was inferred because the forces of geological change—erosion and deposition of sediments—occur very slowly, yet vast geological change was clearly evident in many places.

Hutton published his ideas in a book, *Theory of the Earth*, in 1785, earning him long after his death the title "father of geology." His contemplation of the vast stretches of time necessary to account for rock formations, plus the existence of fossils, led Hutton to propose a theory of evolution startlingly like the one that would be described by Charles Darwin in the next century. However, Hutton died before this book could be completed, and the manuscript was discovered in 1947, long after Charles Darwin was recognized as the father of evolution.

Although Hutton did not live long enough to pursue his ideas on the evolution of life, a contemporary was having thoughts along the same lines. William Smith was a grade-school-educated surveyor who dug canals in the English countryside in the late 1700s. Like others before him, Smith was fascinated by the precise rock layers revealed by his excavating, so much so that his friends nicknamed him Strata Smith. Smith noticed that each stratum seemed to have its own characteristic assortment of **fossils** (evidence of past life). Even if the layer meandered, or was sharply disrupted for a distance, when it straightened out or reappeared it always contained fossils of the same types of organisms.

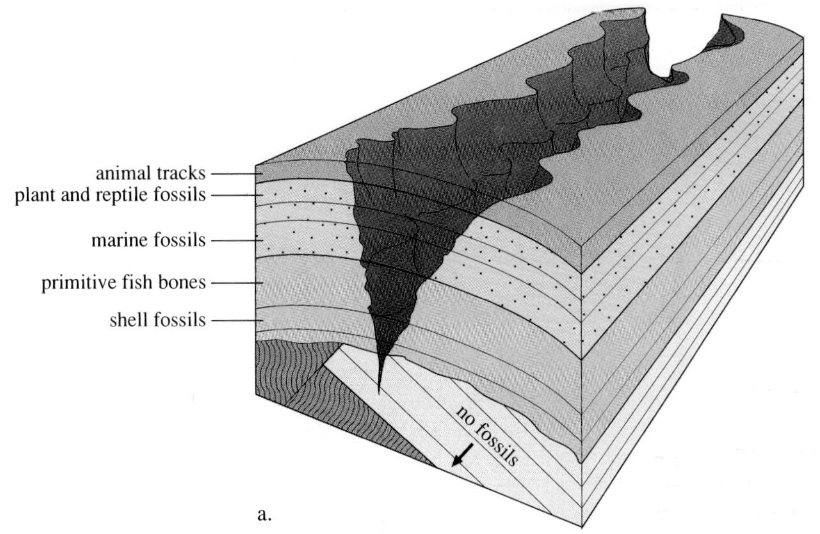

b.

Figure 33.4
Rock layers reveal earth history and sometimes life history. Layers of sedimentary rock form from sand, mud, and gravel deposited in ancient seas. The rock layers on the bottom are older than those on top. Rock strata sometimes contain evidence of organisms that lived when the layer was formed. The position of such fossils within rock layers provides clues about when the organism they represent lived. Sediments are visible along the Grand Canyon. Hiking here is like taking a journey through time. Although the rim now rises over 6,500 feet (2,000 meters) above sea level, when the rock formed it was beneath an ancient ocean.

Because lower strata were older than strata closer to the surface, the positions of certain fossils within rock layers indicated the relative times of existence of the organisms they represent. Fossils could now be put into a relative time frame.

Early in the nineteenth century, another geologist, Charles Lyell, continued Hutton's approach of envisioning long-ago earth features suggested by present-day rock formations. Lyell suggested that sandstone was once sand, shale rock was once mud, and islands were once active volcanoes. He described these ideas in his three-volume *Principles of Geology,* a set of which were to accompany a young man named Charles Darwin on a fascinating voyage. So impressed was Darwin with Lyell's presentation of uniformitarianism that, years later, he would write: "Therefore a man should examine for himself the great piles of superimposed strata, and watch the rivulets bringing down mud, and the waves wearing away the sea-cliffs, in order to comprehend something about the duration of past time, the monuments of which we see all around us."

KEY CONCEPTS

The organization of remains of ancient life in rock layers reveals information about when the organisms lived in time and with respect to each other.

The Voyage of the HMS *Beagle*

It was against the backdrop of the birth of geology and the principle of uniformitarianism that Charles Darwin was born in 1809, the son of a physician and grandson of noted physician and poet Erasmus Darwin. Young Charles did not do particularly well in school, preferring to wander about the countryside examining rock outcroppings, collecting shells, and observing birds and insects. Under family pressure he began to study medicine but abandoned it because he could not stand to watch children undergoing surgery without anesthesia. Next he tried studying for the clergy, but could not maintain interest, leading his father to comment, "You're good for nothing but shooting guns, and rat catching, and you'll be a disgrace to yourself and all of your family."

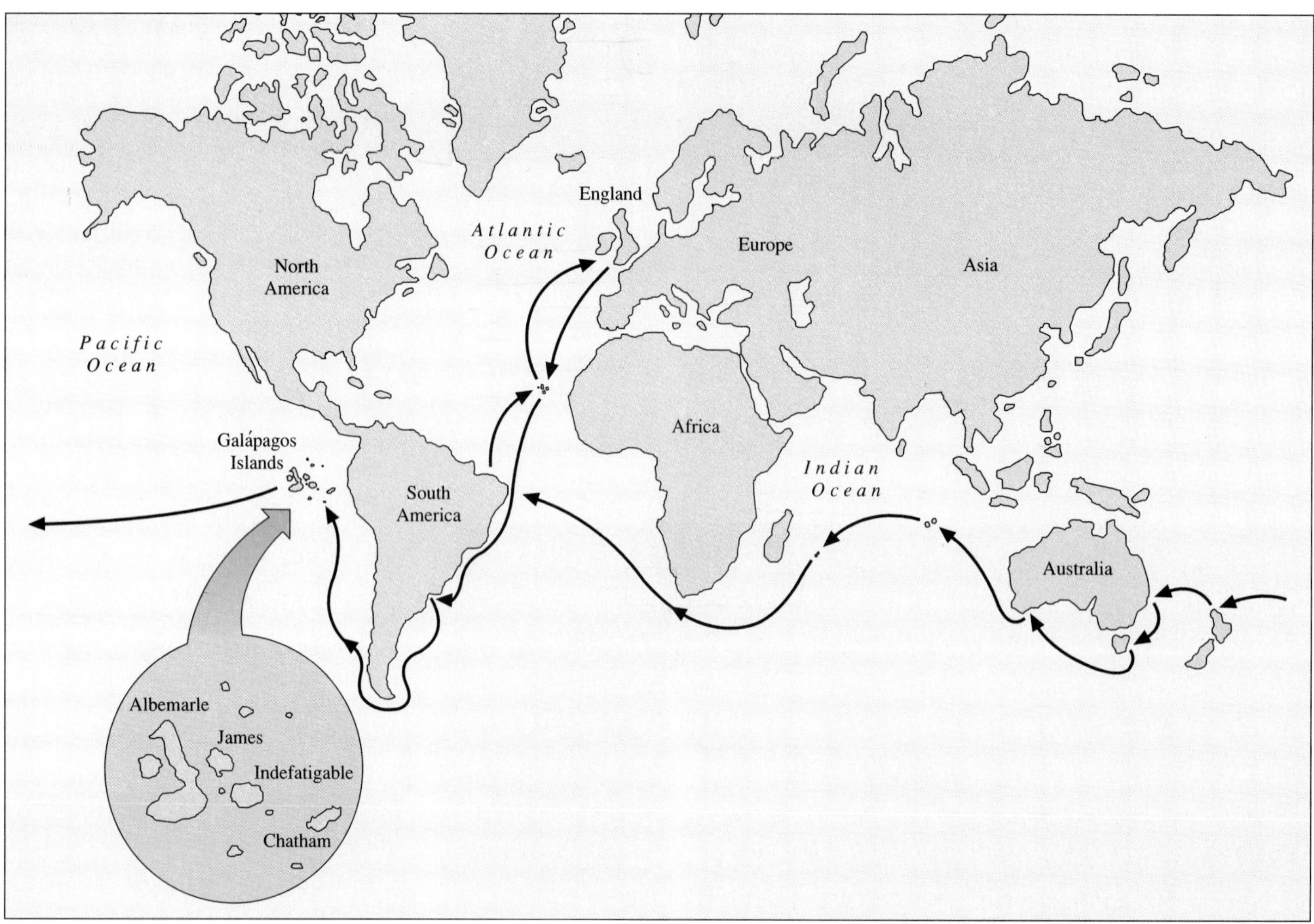

Figure 33.5
The 5-year journey of the HMS *Beagle*. During his voyage, Charles Darwin recorded his observations of geological formations and the distribution of living things. His voluminous notes would serve as the basis for his theory of evolution, synthesized years later.

While Darwin reluctantly explored the subject areas his family thought right for him, he also followed his own interests. He became an avid and valued participant on several geological field trips and met professors in the new science of geology. Darwin's investigations of rock layer formations meshed well with his boyhood love of nature, and so when he was offered a position as captain's companion aboard the HMS *Beagle*, he accepted, despite his father's objections. He would later become the ship's naturalist. Before the ship set sail for its 5-year voyage 2 days after Christmas in 1831, botany professor John Henslow, who had arranged Darwin's new job, gave the young man the first volume of Charles Lyell's *Principles of Geology*. Darwin picked up the second two volumes later in South America. The trilogy was to influence Darwin's thinking profoundly (fig. 33.5).

Darwin read Lyell's geology text as he battled continual seasickness, and he became utterly convinced that uniformitarianism—the gradual changing of the earth's features over long periods of time—was correct. Lyell's ideas fit nicely with what Darwin had observed on his excursions prior to boarding the HMS *Beagle*. Yet Darwin came from a religious family, and he did not set sail with any particular goal of developing a comprehensive theory of evolution. He was actually in search of proof of the biblical version of creation, which he was raised to believe. However, the combination of this unusually perceptive "trained observer," as Darwin has been called, and a voyage to some of the most unusual and undisturbed places on the planet, set the stage for the collection of bountiful evidence upon which the theory of evolution would be built.

Darwin's Observations

Charles Darwin recorded his observations as the ship journeyed down the coast of South America. He paid close attention to forces that uplifted new land, such as earthquakes and volcanoes, and the constant erosion that wore it down. Over and over, he saw what Lyell had written about. Darwin marveled at the intermingling of sea and land in the earth's layers; of fossils of forest plants interspersed with sea sediments; of shell fossils in a mountain cave. The gradual changes evi-

a.

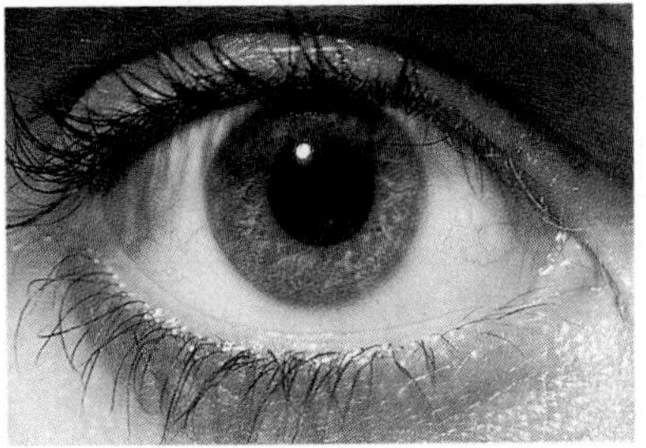

b.

c.

Figure 33.6
Convergent evolution provides common solutions to common challenges. Darwin was puzzled by organisms that lived in similar habitats in different parts of the world but differed in some ways. *a.* For example, the North American wolf and the Tasmanian wolf look remarkably alike, but the North American variety is a placental mammal (which nourishes the unborn through a maternal organ, the placenta) whereas the Tasmanian wolf is a marsupial (which nourishes immature young in a maternal pouch). *b.* Eyes have evolved independently 40 times to meet the need for a light-sensing organ. The eyes of humans and octopuses are coincidentally very similar. Otherwise the two species have little in common. *c.* These three animals look similar. Each has a back fin, flippers, a bilobed tail, and a streamlined body, all of which are adaptations to an aquatic existence. Yet the top animal is an ancient reptile (an ichthyosaur); the middle one is a modern fish; and the bottom animal is a dolphin, a mammal.

denced in the layers of earth were echoed in the fossils found within them. Darwin suggested that fossils, like sediments, were arranged in a chronological sequence, with those most like present forms occupying the highest, or most recent, layers. Like a geologist, he tried to extrapolate past occurrences from contemporary observations and wondered how a particular type of organism got to be where he saw it.

Darwin was particularly aware of differences between organisms, both on a global scale and in localized environments. He noted that different species seen at different latitudes paralleled the different species seen at different elevations on a mountain. Why was this so? If there was a one-time period of special creation, as the Bible held, then why was one sort of animal or plant created for a mountaintop or frigid plain and another for a warmer habitat? Even more puzzling was the fact that similar habitats in different parts of the world were populated by similar-looking types of organisms that could nonetheless have important differences between them. For example, Africa, Australia, and South America all have large, flightless birds, but the ostrich, emu, and rhea are clearly not the same type of animal. The North American wolf and Tasmanian wolf look almost identical, and live in similar environments, yet their methods of nurturing their unborn young are vastly different (fig. 33.6). Today we know that organisms of different species can evolve similar adaptations to similar environments, a phenomenon called **convergent evolution.**

Darwin kept detailed notes on the living things he saw and also on geological formations. However, there is some debate as to whether he realized the influence of geography on the distribution of organisms as he observed it or whether he did so after the journey. He wrote after the voyage that wherever he found a barrier—a desert, river, or mountain—different sorts of organisms populated the two sides. Islands were the ultimate barriers and often lacked the types of organisms that could not get to them. Why would a Creator have put only a few types of organisms on islands, Darwin wondered?

In the fourth year of the voyage, the HMS *Beagle* spent a month in the Galápagos Islands. Although Darwin spent half of the time shipbound due to illness and visited only 4 of the 11 islands, the notes and samples that he brought back would form the backbone of his theory of evolution.

> **KEY CONCEPTS**
>
> *As a young man, Charles Darwin journeyed around the world for 5 years aboard the HMS Beagle, and his observations formed the basis of his theory of evolution. Darwin noticed the constant upheaval in geologic processes, the distribution of fossils in rock strata, differences and similarities between organisms, and the correlation of organisms' characteristics to their environments.*

After the Voyage

Darwin returned to England in 1836 and published the first account of the *Beagle's* voyage, *A Naturalist's Voyage on the Beagle*, 3 years later. In that time he spoke with several other naturalists and geologists whose observations and interpretations helped mold Darwin's thoughts on the evolutionary process.

In March 1837, Darwin spoke with an ornithologist (an expert on birds), who was very excited by the finches brought back from different islands of the Galápagos. By examining beak structure, the ornithologist could tell that some of the birds ate small seeds, some ate large seeds, others ate leaves or fruits, and some ate insects (fig. 33.7). In all, Darwin had brought back or described 14 distinct types of finch, each different from the finches seen on the mainland. A special creation of slightly different finches on each island did not make much sense. Darwin thought it more likely that the different varieties of finch on the Galápagos were descended from a single ancestral type of bird that dwelled on the mainland. Some birds flew over to the islands and, finding a relatively unoccupied new habitat, flourished. Gradually, the finch population branched into several directions, with some groups now eating insects, some eating fruits and leaves, and others consuming different-sized seeds. Darwin called this gradual change from an ancestral type "descent with modification." The different neck lengths of the islands' giant tortoises might also reflect descent with modification. Darwin wondered how and why these variations might have arisen.

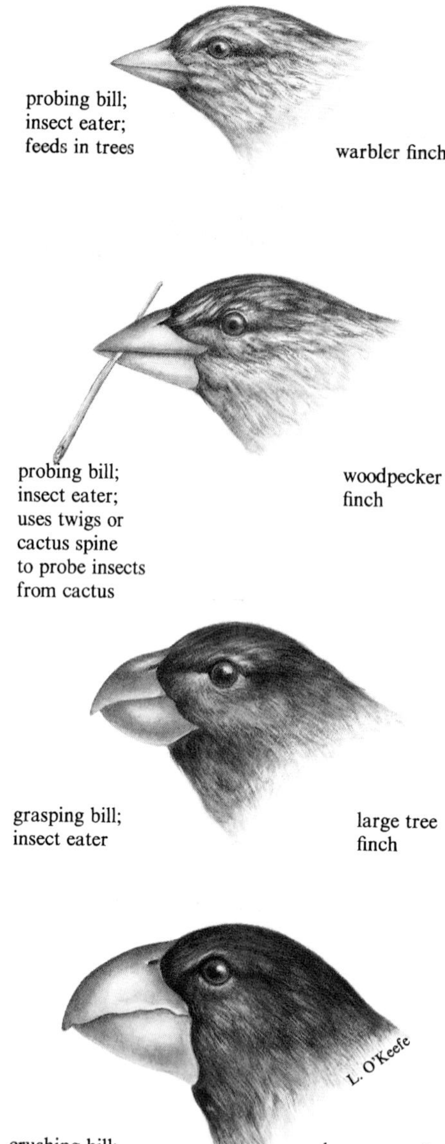

Figure 33.7
The differences in beak structure of the 14 species of finch seen on the Galápagos Islands reflect the birds' different food sources.

In September 1838, Darwin read a book that enabled him to make some sense of the diversity of finches on the Galápagos Islands and of other examples. Economist and theologian Thomas Malthus's "Essay on the Principle of Population," written 40 years earlier, stated that the size of a human population is limited by food availability, disease, and war. Wouldn't populations of other organisms, in the wild, face similar limitations of resources? If so, then individuals who were better able to obtain those resources would survive to reproduce, contributing some individuals like themselves to the next generation (fig. 33.8). This would explain the observation that more individuals were produced in a generation than survived. Over time, the challenges posed by the environment would "select" these better-equipped variants, and gradually the population would change. Darwin compared the selection of adaptive traits by the environment to the **artificial selection** by breeders of certain traits in domesticated pigeons. He called this differential survival and reproduction of better-adapted variants **natural selection.**

Natural selection beautifully explained the diversity of finches on the Galápagos. Originally, some finches flew over from the mainland to populate one island. When that first island population grew too large for all individuals to obtain enough small seeds, those who could eat other things, perhaps because of a quirk in beak structure, began to eat other foods. Finding the new food plentiful, these once-unusual birds gradually made up more of the population. Because each of the islands had slightly different habitats, different varieties of finches were selected on each one. The divergence of several new types from a single ancestral type is termed **adaptive radiation.**

Figure 33.8
Maple saplings compete for natural resources. Near a forest floor, young trees of various sizes attempt to obtain sufficient space, sun, and water to grow and survive. Few of them will succeed.

Darwin's theory of natural selection as the process behind evolutionary change was unveiled in a 35-page sketch published in 1842, and 2 years after that as a 230-page analysis. (Other forces behind evolution are discussed in the next chapter.) In 1856, his treatise expanded but not yet published, Darwin was disturbed to receive a 4,000-word manuscript from British naturalist Alfred Russell Wallace with the rather long-winded title, *On the Tendency of Varieties to Depart Indefinitely From the Original Type*. It was as if Wallace had peered into Darwin's mind. Actually Wallace, like Darwin, had seen the principles of evolution demonstrated among the diverse life forms of South America.

Darwin and Wallace presented their ideas at a scientific meeting later that year. In 1859, Darwin finally published 490 pages of the even longer-titled *On the Origin of Species by Means of Natural Selection, or Preservation of Favoured Races in the Struggle for Life*. It became an overnight success. The direction of biological thought had been changed forever.

KEY CONCEPTS

Darwin noted "descent with modification" in different varieties of finches from the Galápagos. He deduced that natural populations would be affected by the availability of resources and environmental conditions, such that only the better adapted of each generation would survive to reproduce. Darwin termed the differential survival and reproduction of better-adapted variants natural selection. Divergence of several species from one is called adaptive radiation.

Natural Selection—The Mechanism Behind Darwinian Evolution

On his 5-year voyage on the HMS *Beagle*, Darwin saw evidence of the change in life forms through time—evolution. Perhaps his most important observation was not the specifics of beak structure in finches or neck length in giant tortoises but simply that a great deal of variation exists among members of a species. (You can confirm this just by glancing around a classroom and noting the differences among students.) Darwin envisioned natural variation as the raw material of evolution. Depending upon environmental conditions and competition for resources, certain variants would be more fit than others and therefore more likely to have offspring healthy enough to have offspring of their own. The traits that make these individuals healthier in the particular environment are passed on. An individual who is successful in mating is considered to be "fit," and reproductive success is what is meant by the term **survival of the fittest.**

The traits that contribute to reproductive success need not directly affect reproduction. Any trait that ensures an individual's survival can make reproduction more likely. Wrote Darwin, "When we see leaf-eating insects green, and bark-feeders mottled-grey; the alpine ptarmigan white in winter, the red-grouse the colour of heather, we must believe that these tints are of service to these birds and insects in preserving them from danger."

Traits that directly boost reproductive success increase in a population due to a type of natural selection called **sexual selection** (fig. 33.9). Such traits include elaborate feathers in male birds, horns in species ranging from beetles to giant elk, and the courtship songs sung by various insects and birds that attract a mate.

Natural selection does not produce perfection, but it reflects transient adaptation to a prevailing environmental condition. Selective forces act upon preexisting characteristics; they do not create new, perfectly adapted variants. The furry fox that survives a frigid winter to bear pups in the spring is not "perfect" and would certainly not seem so in an unusually mild winter or sweltering summer. Similarly, bighorn sheep whose dwindling numbers grace the crags of the northern Rockies today were far more plentiful at the height of the ice age, when they flourished in the cooler climate.

Natural selection can mold the same population in different directions under different environmental circumstances. Consider the finches on the tiny Galápagos Island of Daphne Major. In a very dry season in the early 1980s, birds with large beaks were more likely to survive because most of the seeds that they eat are large and dry. In 1983, 8 months of extremely heavy rainfall allowed many small seeds to accumulate. Over the next 2 years, finches with small beaks, who could easily eat the tiny seeds, came to predominate.

Although Darwin was quite convinced that the interaction between naturally occurring variants and environmental conditions provides the force behind evolution, he did not know the source of those important variants. He recognized that the natural variations that play a part in the accumulation of new traits and the origin of species are those that are passed on to future generations, but he did not know how they arose or were transmitted.

A predecessor of Darwin, French taxonomist Jean-Baptiste Lamarck, thought that inherited variations were acquired during an individual's lifetime through want or need and then passed on to the next generation. According to this viewpoint, the fox with the heavy coat would have

go to pg 635

a.

b.

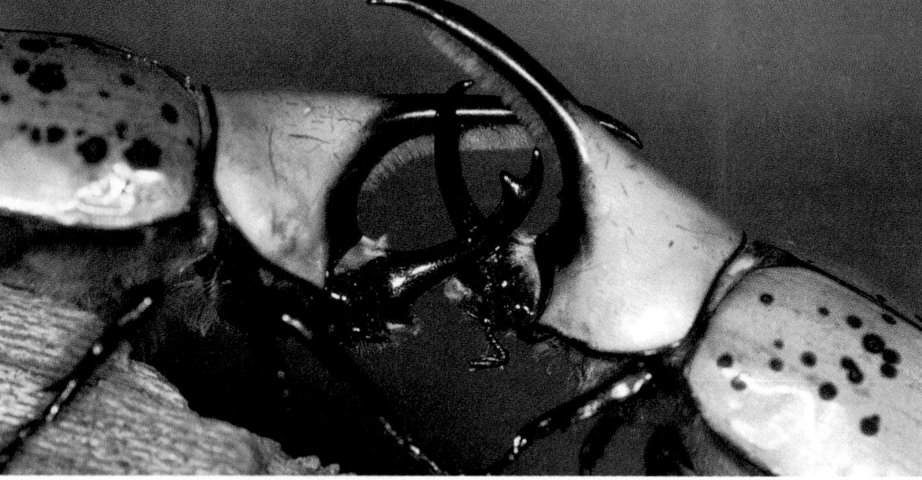

c.

d.

Figure 33.9
Sexual selection. The individual who is most successful in attracting a mate is the most fit, according to Darwin. *a*. The male bowerbird builds intricate towers of sticks and grass to attract a mate. *b*. The male bird of paradise displays bright plumes and capes in his quest for sexual success. Diverse organisms use horns to battle rivals for access to the hornless sex, from the battling Hercules beetles (*c*) to the extinct Irish elk, with its 80-pound antlers (*d*).

Table 33.1
Darwin's Main Ideas

1. Living things are varied. Within a species, no two individuals (except identical siblings) are exactly alike.

2. More individuals are born than survive to reproduce.

3. Individuals compete with one another for the resources that enable them to survive.

4. Within populations, the characteristics of some individuals make them more able to survive and reproduce in the face of certain environmental conditions than others.

5. As a result of this environmentally selected "survival of the fittest," only those individuals with adaptive traits live long enough to pass these traits on. Over time, this natural selection can change the characteristics of populations, even molding new species.

consciously grown it to help him survive the frigid winter, rather than the Darwinian interpretation of the animal's inheriting genes conferring a heavy coat that just happened to be adaptive in the fox's environment. Darwin thought that new characteristics were acquired by an individual's somatic cells, and then transmitted to the sex cells, so that they could be passed to subsequent generations. Table 33.1 presents Darwin's main ideas.

It is ironic that as Darwin spent his later years pondering the source of inherited variation, a contemporary thinker, a breeder of peas in an Austrian monastery, Gregor Mendel, was formulating the answer. But it would be another half a century until the science of genetics would arise to explain the science of evolution.

KEY CONCEPTS

Natural selection acts on preexisting variants. Fitness refers to reproductive success. In a given environment, some individuals will be more fit than others, as a result of particular inherited variations, some of which directly affect sexual behavior and some of which affect fitness indirectly. Lamarck thought that evolution was driven by traits acquired during an individual's lifetime to suit a particular need and were then passed on.

SUMMARY

The biological science of *evolution* examines changing characteristics of populations of organisms that can lead to adaptation and to the formation and demise of species. Islands provide vivid examples of evolutionary change because their geographical isolation separates individuals with particular combinations of traits from ancestral organisms and selects those traits that are adaptive in the particular environment. Evolution consists of both the large-scale, species-level changes of *macroevolution* and the trait-by-trait changes of *microevolution*.

The field of geology laid the groundwork for evolutionary thought. Some people explained the distribution of rock strata with the idea of *Neptunism* (a single flood) or *mosaic catastrophism* (a series of floods), but it was *uniformitarianism* (the continual remolding of the earth's surface) that was most consistent with the planet's appearance. Uniformitarianism suggests that geological change took a very long time. The sequence of rock strata suggests a time frame, with the lowest layers older than layers closer to the earth's surface. It follows that fossils found within rock strata are also organized according to their time of deposition, which demonstrates the *principle of superposition*. Charles Darwin was greatly influenced by geology.

Darwin was an ordinary student given an extraordinary opportunity to observe the distribution of organisms in many diverse habitats as naturalist aboard the HMS *Beagle* on its 5-year journey. In the years following the voyage, Darwin thought over his observations, conferred with biologists and geologists, and was greatly influenced by Thomas Malthus's "Essay on the Principle of Population." From these sources, Darwin eventually synthesized his theory of the origin of species by means of *natural selection*. His theory states that those individuals who are best adapted to their environments are more likely to produce fertile offspring and will pass on their adaptive traits to the next generation. Eventually, enough changes would accumulate so that its members could no longer mate with members of the ancestral population, and a new species would result. The different beak shapes among the several species of finches found on the Galápagos Islands illustrate natural selection. Natural selection does not produce perfection but rather successful adaptation to the environment.

QUESTIONS

1. What characteristic of islands is responsible for the impact they have had on the evolution of species?

2. What is the difference between microevolutionary and macroevolutionary change?

3. How did James Hutton, William Smith, Charles Lyell, and Thomas Malthus influence the thinking of Charles Darwin?

4. What is natural selection? How can natural selection explain evolutionary change?

5. What sorts of traits lead to "fitness" in a Darwinian sense?

6. The term "highly evolved" is often interpreted to mean "perfect." Why is this definition inaccurate if applied to Darwinian evolution?

TO THINK ABOUT

1. Many islands have populations of large, flightless birds found nowhere else in the world. How might they have gotten to the islands?

2. Which theory best explains Darwinian evolution—Neptunism, mosaic catastrophism, or uniformitarianism? Give a reason for your answer.

3. An early and major objection to Darwin's concept of natural selection as the mechanism behind evolution was that it refutes the pleasant idea that we humans are advanced and special when compared to other species. How does Darwinian evolution do this?

4. You have a pet cocker spaniel and believe that if you snip off the end of his tail, then breed him, his puppies will be born with snipped-off tails. Is this idea consistent with Darwinian evolution? What principle is it an example of, and who suggested it?

5. Herbert Spencer was an eighteenth-century English sociologist who believed in applying Darwin's concepts to human society. He popularized the phrase "survival of the fittest" and suggested that the unemployed, the sick, and other "burdens on society" be allowed to die rather than be objects of public assistance and charity. What do you think of this idea and of the validity of his interpretation of the concept of natural selection?

SUGGESTED READINGS

Darwin, Charles. 1958 (1859). *On the origin of species by means of natural selection*. New York: Mentor Books. Darwin's writing is very detailed, but offers fascinating examples and interpretations of the natural world.

Diamond, Jared. October 1990. Bob Dylan and moa's ghosts. *Natural History*. Can extinct, large birds explain the unique distribution of species in New Zealand?

Eldredge, Niles. 1982. *The monkey business*. New York: Washington Square Press. Is evolution really a fact?

Gould, Stephen Jay. *Wonderful life* (1989). *The flamingo's smile* (1986). *Hen's teeth and horse's toes* (1983). *The panda's thumb* (1980). *Ever since Darwin* (1977). New York: W.W. Norton. Humorous and fact-packed evolutionary essays. Gould also writes a monthly column on evolution for *Natural History* magazine.

Monastersky, R. June 13, 1987. Natural selection: Bird seeds of change. *Science News*, vol. 131. Natural selection of beak size among finches on Daphne Major in the Galápagos follows fluctuations in yearly rainfall.

Stone, Irving. 1980. *The origin*. Garden City, N.Y.: Doubleday. A fictionalized but highly informative account of the adventures of a brilliant thinker, Charles Darwin.

C H A P T E R

34

The Forces of Evolutionary Change

Chapter Outline

Learning Objectives

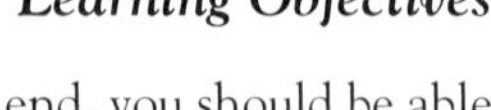

By the chapter's end, you should be able to answer these questions:

1. What is a population?
2. What is a gene pool?
3. Under what conditions does evolutionary change not occur?
4. How can algebra be used to describe whether or not microevolution is occurring?
5. How can migration, nonrandom mating, mutation, and natural selection alter gene frequencies?
6. How do some genetic diseases remain in populations, even though they decrease reproductive success?
7. How do species arise?
8. How do species become extinct?

Bullwinkle J. Moose first appeared on dairy farms in upstate New York in 1980. A refugee from the dense moose populations in nearby New England and Canada, Bullwinkle was one of only 20 moose in all of New York state. Most were males, who were more likely to survive the rigors of migration than females. Bullwinkle developed an attraction to the next best thing to the cows of his own species—domesticated dairy cows (fig. 34.1). The moose's vision was poor, and from a distance the heifers did not look much different from moose cows. Their low-pitched moos were similar to the moans that female moose utter to lure mates. The scent of the dairy cows was enough like that of moose cows that Bullwinkle dug his hooves into the ground and urinated around a small area, marking a territory as a prelude to mating. He then approached a plump heifer.

Once Bullwinkle mounted the heifer, though, the courtship ceased. For in the end, a moose is a moose and a cow is a cow—members of two different species. The moose's genitalia did not fit the cow's, and the activity ended without completion. Bullwinkle continued to frequent upstate New York farms in search of dairy cows until 1987, when he finally found a mate of his own kind, near Tupper Lake.

Not all species are as different appearing as the moose and the dairy cow. For example, two species of fruit fly, *Drosophila persimilis* and *Drosophila pseudoobscura*, look identical, down to the tiniest bristle. The flies will not mate with each other, however, perhaps because of a lack of communication. If the antennae of a female of one species are snipped off, she will indeed mate with a male of the other species.

How does a group of organisms become able to mate successfully only among themselves? In other words, how do species form? Charles Darwin noted that evolution of species was the consequence of different traits being selected by particular environments. Eventually, such accumulated adaptations in a group of organisms changes them to the point that they cannot mate with others not of the group. On a small scale, then, evolution reflects the changing representations of particular traits among groups of individuals. On a large scale, evolution is the formation and the extinction of species.

Figure 34.1
Members of two species cannot successfully mate with one another. In some areas of New England and upstate New York, male moose are rare and will occasionally approach a dairy cow with romance in mind.

Evolution After Darwin—The Genetics of Populations

The "raw material" of evolution is inherited natural variation. The role of genes in evolution is different from genetic functions considered so far—as biochemical instructions that form the biological basis of traits passed through families. Genes influence evolution at the population level. A **population** is a broad term for any group of interbreeding organisms. Human populations include the students in a class or a school or people in a town, a nation, or the entire world.

All the genes in a population constitute its **gene pool**. The proportion of different alleles of each gene determines the characteristics of that population. A Swedish population, for example, would have a preponderance of "blond" hair color alleles; a population of black Africans would have very few, if any, such alleles but would have many alleles conferring darker hair shades.

Microevolution occurs when the frequency of an allele in a population changes. A gene's frequency can be altered when new alleles are introduced into a population by mutation; when individuals migrate between populations; when an environmental condition is more easily tolerated by those who have a particular phenotype (natural selection); and when genes are eliminated because individuals with certain genotypes do not reproduce. Hence, many conditions must be met for evolution *not* to occur.

When Gene Frequencies Stay Constant—Hardy-Weinberg Equilibrium

Genotypes and phenotypes can be predicted when a gene is passed from parents to offspring by using probability or Punnett squares (chapter 11). Changes in gene frequencies in populations from generation to generation can also be calculated, and from these values phenotypes and genotypes can be deduced.

In 1908, a mathematician, H.H. Hardy, and a physician interested in genetics, W. Weinberg, independently proposed that phenotype and genotype frequencies could be determined by applying a simple algebraic expression, the binomial expansion $p^2 + 2pq + q^2 = 1.0$, where p^2 represents homozygous dominant individuals, q^2 represents homozygous recessive individuals, and $2pq$ represents heterozygotes (fig. 34.2). The Hardy-Weinberg equation is analogous to a monohybrid cross (*Aa* x *Aa*), which results in a single way to generate each homozygote (*AA* or *aa*) and two ways to get a heterozygote (*Aa* or *aA*). An example of how phenotype and genotype frequencies are calculated is presented in Reading 34.1.

The Hardy-Weinberg equation can reveal single-gene frequency changes that underlie evolution. If the proportion of genotypes remains the same from generation to generation, as indicated by the equation, then evolution is not occurring for that gene. This situation is called **Hardy-Weinberg equilibrium.** It is an idealized state, only possible if the population is large, if it mates randomly, and there is no mutation, migration, or natural selection.

Algebraic Expression	What It Means
$p + q = 1$	All dominant alleles plus all recessive alleles add up to all of the alleles for a particular gene in a population.
$p^2 + 2pq + q^2 = 1$	For a particular gene, all homozygous dominant individuals (p^2) plus all heterozygotes ($2pq$) plus all homozygous recessives (q^2) add up to all of the individuals in the population.

Figure 34.2
Using algebra to follow gene frequencies.

KEY CONCEPTS

The genes in a population comprise its gene pool. Microevolution reflects changes in gene frequencies in populations. These changes can be traced using the Hardy-Weinberg equation. Microevolution is not occurring if gene frequencies stay constant from generation to generation, a condition called Hardy-Weinberg equilibrium, which is only met if mating is random and the population large, with no migration, mutation, or natural selection.

When Gene Frequencies Change

Migration and Nonrandom Mating

It is easy to see how rarely, if ever, Hardy-Weinberg equilibrium exists by considering the most familiar population—our own. The population of New York City, for example, has been built by waves of immigration. The original Dutch settlers of the 1600s lacked many of the genes present in today's metropolis, contributed by the English, Irish, Slavic peoples, blacks, hispanics, Italians, Asians, and many others.

Although New York City is often described as a "melting pot," many people, there and elsewhere, choose mates who have particular characteristics, and therefore mating is not random. Mates are chosen for any number of reasons, such as physical appearance, ethnic background, or intelligence. Even the fact that women often tend to seek mates who are taller than they are removes the randomness from mating.

Nonrandom mating is also quite common in populations of other species. It is especially pronounced in agriculture, where semen from one prize bull may be used to artificially inseminate thousands of cows. Occasionally such an extreme situation arises in a human population, when a particular male fathers many, many children. In the Cape population of South Africa, for example, a Chinese immigrant known as Arnold had a very rare dominant genetic disease that causes the teeth to fall out before age 20. Arnold was extremely fertile and had seven wives. Of his 356 living descendants, 70 have the dental disorder. The frequency of this allele in the Cape population is exceptionally high, compared to elsewhere in the world, thanks to Arnold.

Genetic Drift

Gene frequencies can change when a small group is separated from a larger population, a phenomenon called **genetic drift.** By chance, the small group may not represent the whole. Could the average academic ability of 200 students in a biology class, for

Reading 34.1 *Using Algebra to Track Gene Frequencies*

ALGEBRA CAN BE USED TO DESCRIBE THE ALLELE, GENOTYPE, AND PHENOTYPE FREQUENCIES IN A POPULATION. When these values change—as they nearly always do—evolution is occurring.

Consider a population of 100 foxes, in whom coat thickness is determined by a single gene. Allele C confers a thick coat, and allele *c* a thin coat. The frequency of dominant alleles is symbolized by the letter p and recessive alleles by q. The sum of p and q is one, so that if the frequency of one allele is known, the other is calculated by subtracting from one.

Among the foxes, the frequency of C alleles is 0.70, and therefore the frequency of *c* alleles is 0.30. The number of thin-coated foxes is calculated first, because this phenotype has only one corresponding genotype, *cc*. The proportion of thin-coated foxes is q^2, or the chance of having two *c* alleles, which equals (0.30) x (0.30), or 0.09, or 9%. Nine percent of 100 foxes equals 9 thin-coated animals.

The number of foxes with thick coats can be determined by subtracting 9 from 100, but applying the appropriate parts of the Hardy-Weinberg equation gives a more precise answer by indicating genotype—that is, the number of thick-coated foxes that are homozygous dominant (CC) and the number that are heterozygous (C*c*). The number of CC foxes equals p^2, or (0.70) x (0.70), or 49% of 100, which is 49 foxes. The number of heterozygotes is $2pq$, or (2) x (0.70) x (0.30), or 42% of the total, or 42 foxes. Adding p^2 and $2pq$ gives the expected 91 foxes with thick coats. Note that although only 9 foxes out of 100 have thin coats, 42 foxes carry the allele for a thin coat.

If none of the foxes dies or leaves the group, if no new foxes of reproductive age enter, if the fur genes do not mutate, if natural selection does not favor one phenotype over another, and if each fox contributes equally to the next generation—then allele frequencies and the proportions of each genotype will remain unchanged in the next generation. Evolution is not occurring. But it is highly unlikely that all of these conditions will be met. Many different scenarios could alter the genetic makeup of this population, with respect to coat thickness. Here are a few:

Twenty foxes wander away and become separated from the main group. If, by chance, 8 of the animals have thin coats, the allele frequency of *c* in the new, small population will be quite higher than it was in the ancestral population. (migration and genetic drift)

A very harsh winter kills 5 of the 9 thin-coated foxes. The allele frequency of *c* is decreased in the new population of 91 foxes. (natural selection)

If thick-coated foxes preferentially seek thick-coated foxes as mates, the frequency of the *c* allele will decline somewhat, but thin-coated individuals will still be born approximately 25% of the time when heterozygotes mate. (nonrandom mating)

If in each generation three C alleles mutate to *c*, the allele frequencies will change accordingly. (mutation)

Can you think of other scenarios that would alter allele frequencies in this population?

example, be determined by considering only those students who earned a grade of F on the final exam—or an A?

An example of genetic drift in human populations is the **founder effect**, which occurs when small groups of people leave their homes to found new settlements. The new colony may have different genotype frequencies than the original population. The fact that Native Americans in North America do not have type B blood illustrates the founder effect. Type B blood is seen in the Asian population from whom the Indians are descended. Perhaps the founding band of Asian settlers who crossed the Bering Strait to America many thousands of years ago did not include a person with type B or AB blood, or if it did, he or she may not have had children survive to pass on the trait.

The founder effect can also increase the proportion of an allele in a population. Consider the 2,500,000 people of the Afrikaner population of South Africa, who are descended from a small group of Dutch immigrants. Today, 30,000 Afrikaners have porphyria variegata, a dominantly inherited enzyme defect whose only symptom is a severe reaction to barbiturate anesthetics. (The disorder was not recognized until these drugs entered medical practice.) All of the affected people are descended from one couple who came from Holland in the 1680s. Today's gene frequency in South Africa is far higher than that in Holland because this couple contributed significantly to the early Afrikaner population.

Genetic drift can occur when members of a small community choose to mate only among themselves, resulting in inbreeding. This happened in the Dunker community of Germantown, Pennsylvania. The frequencies of some genotypes are different among the Dunkers than among their neighbors who are not part of their society and different from people living today in the part of Germany that the Dunkers left between 1719 and 1729 to settle in the New World (table 34.1 and fig. 34.3).

Genetic drift also results from a **population bottleneck**, a situation in which many members of a population die, and the numbers are then restored by mating among a few individuals. The new population has a much more restricted gene pool than the larger, ancestral population. The current world population of cheetahs provides evidence of severe population bottlenecks (fig. 34.4). Until 10,000 years ago, these cats were prevalent in many areas. Today, only two isolated populations live in South and East Africa, numbering only a few thousand animals. Examination of protein sequences from 55 cheetahs found them to be quite uniform in genetic makeup—evidence of a population bottleneck. The cheetahs of the South African population are so alike genetically that even unrelated animals can accept skin grafts from each other. Researchers attribute the genetic uniformity of cheetahs to two bottlenecks—one occurring at the end of the most recent ice age, when habitats were altered, and mass slaughter by nineteenth-century cattle poachers.

Mutation

A major source of genetic variation is mutation, the changing of one allele into another. (Genetic variability also arises from crossing over and independent assortment during meiosis, but these events recombine existing traits rather than introducing new ones.) Spontaneous mutation occurs when a DNA base is in a transient, unusual form at the precise instant that the DNA is replicated, inserting a mismatched DNA base into the new DNA strand (fig. 34.5). If the base change occurs in a part of a gene corresponding to a portion of a protein necessary for its function, then an altered trait may result. If the mutation is in a gamete, then the change can be passed to future generations and therefore ultimately affect an allele's frequency in the population.

The spontaneous mutation rate varies for different genes and in different organisms. Based on prevalence of certain disease-causing genes, it is estimated that each human gene has about a 1 in 100,000 chance of mutating. If there are about 50,000 human genes, then 1 in every 2 gametes (50,000 genes multiplied by a mutation rate per gene of 1/100,000) theoretically carries a new mutation. But because most mutations are recessive and will be masked by a normal allele in the next generation, the mutation situation is not as ominous as it may appear. Dominant mutations, however, are expressed immediately. An achondroplastic dwarf born to normal parents is an example of a spontaneous, dominant mutation.

The frequency of a mutant allele is maintained in a population in heterozygotes and by reintroduction by further mutation. It is removed when homozygous individuals arise who cannot reproduce successfully due to their double dose of the mutant allele. Because of heterozygosity and mutation, all populations have some alleles that would be harmful if homozygous. The collection of deleterious alleles in a population is called the **genetic load**.

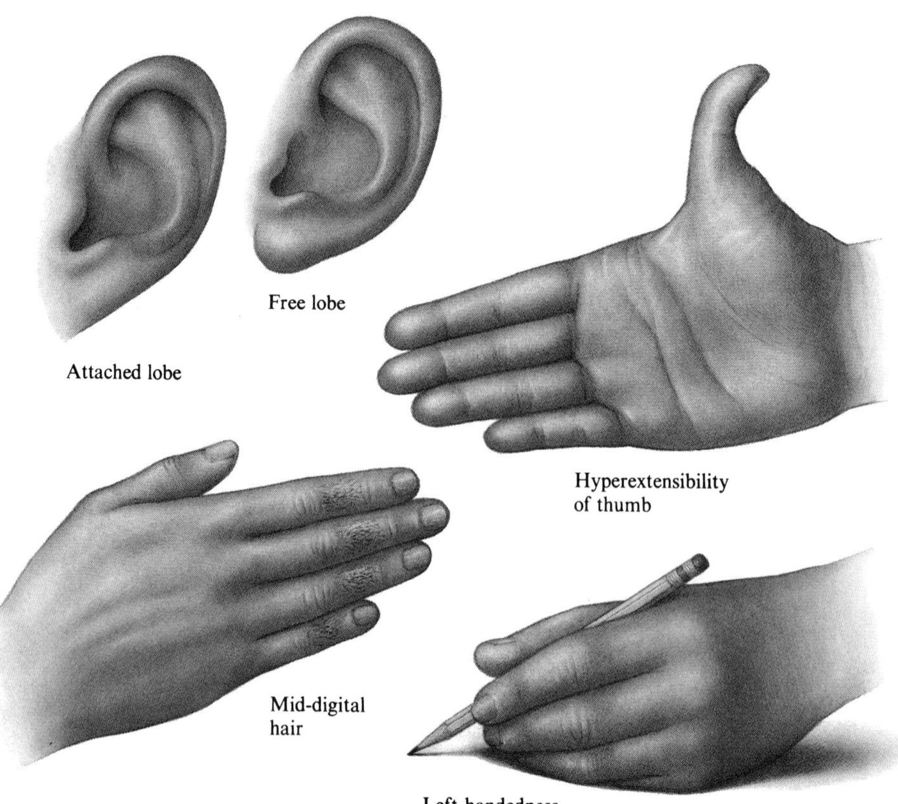

Figure 34.3
Three interesting traits more common among the Dunkers and that occur at unique frequencies are attached earlobes, the ability to bend the thumb backwards, hair on the middle of the fingers, and left-handedness.

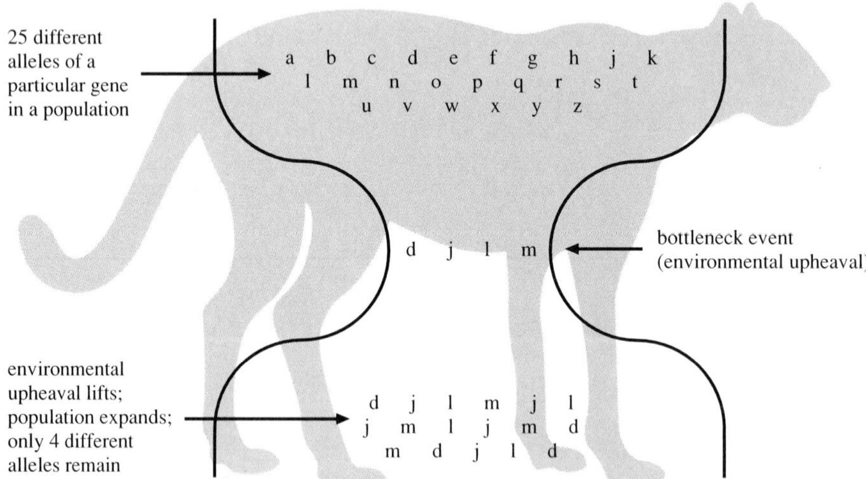

Figure 34.4
A population bottleneck occurs when the size of a genetically diverse population drastically falls, is maintained for a time at this level, and then expands again. The rebuilt population loses some genetic diversity if different alleles are lost in the bottleneck event. The two dwindling cheetah populations in South and East Africa vividly illustrate the result of a population bottleneck. The cheetahs are more genetically alike than are mice bred specifically for genetic uniformity. Cheetahs are difficult to breed in zoos because sperm quality is poor and many newborns die—both due to the lack of genetic diversity.

Table 34.1
Genetic Drift and the Dunkers

Blood Type	*Population*		
	U.S.	*Dunker*	*European*
ABO system			
A	40%	60%	45%
B+AB	15%	5%	15%
Rh-	15%	11%	15%
MN system			
M	30%	44.5%	30%
MN	50%	42%	50%
N	20%	13.5%	20%

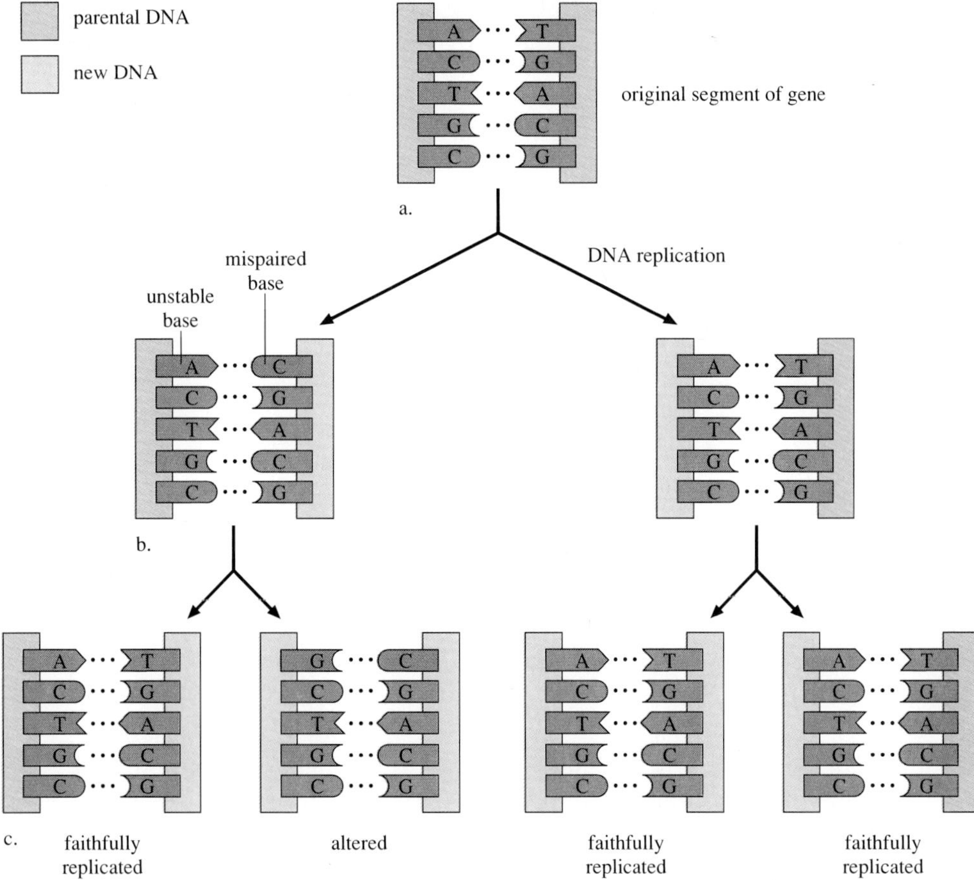

Figure 34.5
Mutation can alter traits. *a.* In DNA, nearly all of the time, base A pairs with base T (and vice versa) and C pairs with G (and vice versa). However, DNA bases are very slightly unstable chemically, and for fleeting moments they exist in altered forms. *b.* If a DNA replication fork encounters a base in its unstable form, a mismatched base pair can result. *c.* After another round of replication, one of the daughter cells has a different base pair than the one in the corresponding position in the original DNA segment. Such a substituted base pair can alter the structure of a gene. If the gene's function is affected, the individual's phenotype may be changed. Because natural selection acts upon phenotypes (a gene's expression), mutation can contribute to evolutionary change.

Each person probably has four or five such deleterious recessive alleles. It may take a very long time, however, for mutant alleles to reveal themselves, because two individuals who have children together must carry the same mutant allele for each child to have a 25% chance of being affected by the particular condition. Many such recessive mutations were induced by the radiation released from the atomic bombs dropped over Japan in 1945. Geneticists estimate that it will take 30 generations for those mutations to start showing up in homozygous form. By 1990, genetic problems were not any more prevalent among Japanese offspring of parents exposed to the bomb than among people of the same age elsewhere. (This may not be true for non-genetic problems.)

When mating occurs among blood relatives, the chance of conceiving a homozygous recessive individual occurs as soon as two people mate who carry the same mutant allele, inherited from a shared ancestor. This is why in some states it is illegal for first cousins to wed. They have one-eighth of their genes in common, and it becomes likely that a pair of harmful recessive alleles could match up.

KEY CONCEPTS

Hardy-Weinberg equilibrium rarely, if ever, exists in nature. In genetic drift, such as the founder effect and population bottlenecks, gene frequencies change when small populations form from larger ones. Spontaneous mutations in germ cells alter allele frequencies. Harmful alleles constitute a population's genetic load. They are maintained in heterozygotes, introduced by mutations, but eliminated by natural selection.

Natural Selection

Gene frequencies that are altered in response to environmental change are vividly illustrated by the 100 or so insect species that have undergone color changes enabling them to blend into their backgrounds as their environments become polluted. This adaptive response is termed **industrial melanism.** Figure 3.4 illustrates the selection of dark pigmentation in the peppered moth *Biston betularia* in rural England.

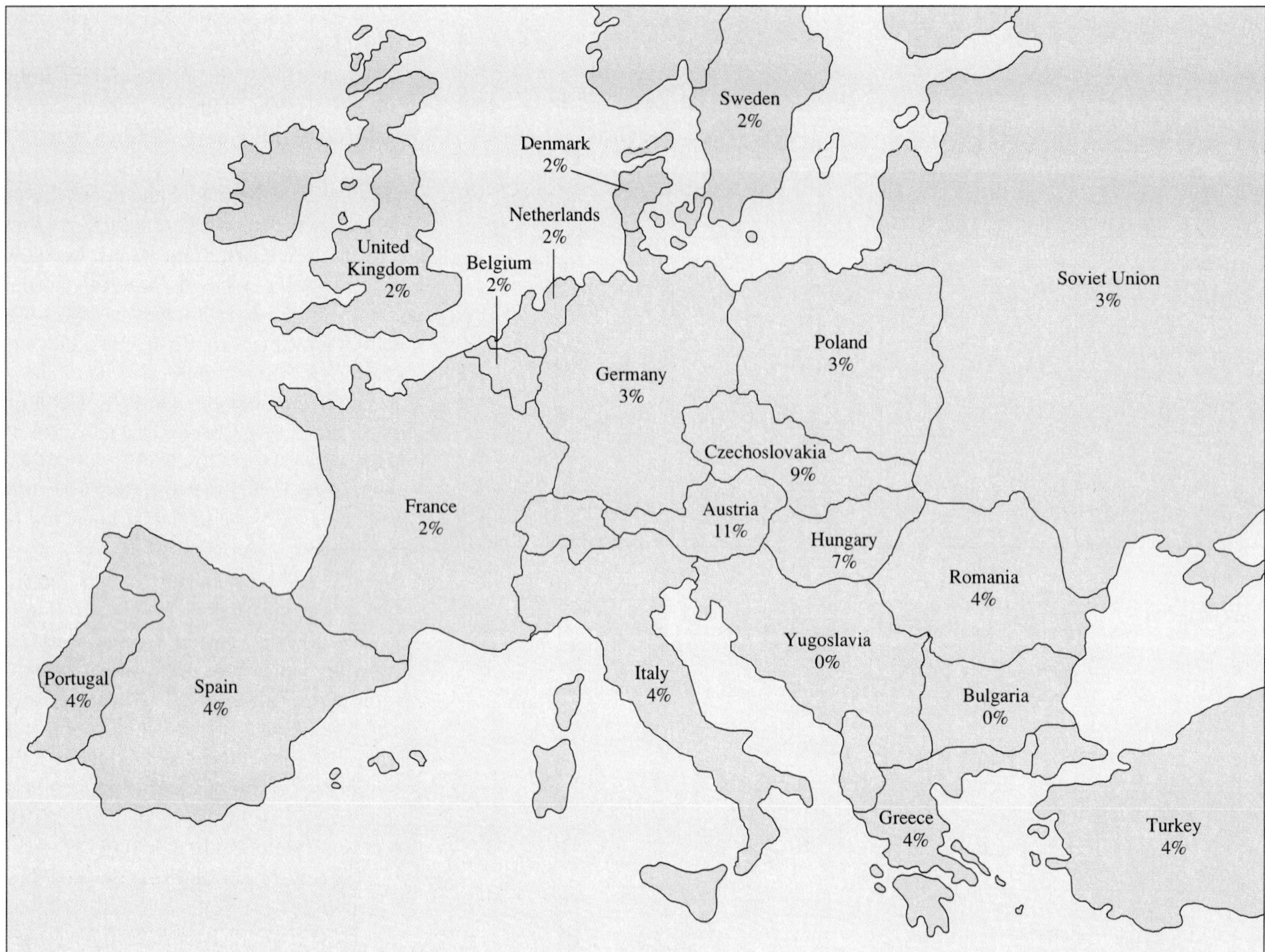

Figure 34.6

Tay-Sachs disease and balanced polymorphism. Just as being a carrier for sickle cell disease protects against malaria, being a carrier for the neurodegenerative Tay-Sachs disease may protect against tuberculosis. Tay-Sachs is quite common in Jewish people whose families came to the United States from Eastern Europe. In these populations, up to 11% of the Jewish people are carriers. Tuberculosis ran rampant in the Eastern European Jewish settlements during World War II, but often healthy relatives in families with Tay-Sachs resisted the infection. The protection against tuberculosis offered by Tay-Sachs disease heterozygosity remained among the Jewish population in Eastern Europe because they could not leave the crowded ghettos. The allele for Tay-Sachs increased in frequency as tuberculosis selectively felled those who did not carry the allele and as the people had children with each other. The map displays the percentage of Tay-Sachs carriers among the Jewish populations in different countries, based on carrier frequencies of American Jews whose ancestors came from these countries.

Before the Industrial Revolution, dark moths were rare because they were easy prey for birds, who spotted them against the background of tree bark turned white from the growth of lichens (see chapter 2, Reading 2.1, "Fungus + Algae = Lichen"). After industrialization, air pollution killed the lichens, darkening the tree bark, and the dark moths were now more easily hidden than the white moths and better able to survive and reproduce. Changes in the coloration of moth populations began in the 1700s and became more pronounced in the 1950s. By 1956, 95% of the moth population was dark—just about the reverse of the situation before the environmental change.

The predominant color of moth populations also varies with the degree of industrialization in a particular geographic region. Dark moths were most common in Liverpool, a sooty, factory town, and were rarely seen in pristine, rural northern Wales. Today, only a small portion of northern England has populations of mostly dark moths, a trend attributed to 20 years of efforts to control smoke pollution.

Natural selection is also evident in the declining virulence of certain human infectious diseases over many years. The spread of tuberculosis in the Plains Indians of the Qu'Appelle Valley Reservation in Saskatchewan, Canada, illustrates the forces of evolution at work. When tuberculosis first appeared on the reservation in the mid-1880s, it struck people swiftly and lethally, infecting many organs and killing 10% of the population. By 1921, tuberculosis in the Indians tended to affect only the lungs, and 7% of the population died annually from it. By 1950, mortality was down to 0.2%.

Outbreaks ran similar courses in other human populations, appearing in crowded settlements where the bacteria were easily spread from person to person in exhaled droplets. In the 1700s, tuberculosis raged through the cities of Europe and was brought to the United States with immigrants in the early 1800s, where it likewise swept through the cities. As in the Plains Indians population, tuberculosis incidence and virulence fell dramatically in the cities of the industrialized world in the first half of the twentieth century—before widespread use of antibiotics.

What tamed tuberculosis? It may have been natural selection, operating on both the bacterial and human populations. Some people inherited resistance to the infection, enabling them to produce protected children. Plus, the most virulent bacteria killed their hosts so quickly that there was no time to pass on the germs. As the deadliest bacteria were selected out of the population, and as people with inherited resistance contributed disproportionately to the next generation, the effect of tuberculosis on human health gradually became more benign, evolving from an acute, systemic infection to an increasingly rare chronic lung infection.

Types of Natural Selection

Industrial melanism and the decline in severity of tuberculosis illustrate **directional selection**, in which a previously "normal" characteristic of a population alters in response to a changing environment as the number of better-adapted individuals increases. Natural selection can have other effects. In **disruptive selection**, two extreme expressions of a trait are the most fit. A population of marine snails, for example, lives among tan rocks encrusted with white barnacles. The snails are either white and camouflaged while near the barnacles or tan and hidden while on the bare rock. Only animals that are not white or tan, or that lie against the opposite-colored background, will be seen, and eaten, by predatory shore birds.

In **stabilizing selection**, extreme phenotypes are less adaptive, and an intermediate phenotype has greater survival and reproductive success. Consider human birthweight. Newborns who are under 5 pounds (2.27 kilograms) or over 10 pounds (4.54 kilograms) are less likely to survive than babies weighing between 5 and 10 pounds.

Balanced Polymorphism—The Sickle Cell Story

A form of stabilizing selection called **balanced polymorphism** allows a genetic disease to remain in a population even though the illness clearly diminishes the fitness of affected individuals. The disease persists because carriers have some advantage over those who have two copies of the normal allele.

Balanced polymorphism maintains sickle cell disease, an autosomal recessive disorder that causes anemia, joint pain, a swollen spleen, and frequent, severe infections (chapter 14). Carriers of sickle cell disease are resistant to malaria, an agonizing cycle of chills and fever that is caused by a protistan parasite, *Plasmodium falciparum*, which spends the first stage of its life cycle in the salivary glands of the mosquito *Anopheles gambiae* (fig. 28.15). When an infected mosquito bites a human, the malaria parasite enters the red blood cells and is taken to the liver. The infected red blood cells eventually burst, releasing the parasite all over the body.

In 1949, British geneticist Anthony Allison found that the frequency of sickle cell carriers in tropical Africa was quite high in regions where malaria was present all year long. Blood tests from children hospitalized with malaria found that nearly all were homozygous for the normal allele. The few sickle cell carriers among them had the mildest cases of malaria. Was the presence of malaria somehow selecting for the sickle cell allele by felling those people who did not inherit it? The fact that sickle cell disease is far less common in the United States, where malaria is not present, supports a protective effect of sickle cell heterozygosity.

The rise of sickle cell disease parallels the cultivation of crops that provide breeding grounds for the *Anopheles* mosquito. About 1,000 B.C., Malayo-Polynesian sailors from southeast Asia traveled in canoes to East Africa, bringing new crops of banana, yam, taro, and coconut. When the jungle was cleared to grow these crops, the open space enticed the mosquitos to flourish. The insects, in turn, provided a habitat for part of the life cycle of the malaria parasite.

The sickle cell gene is believed to have been brought to Africa by people migrating from southern Arabia and India, or it may have arisen by mutation directly in East Africa. However it happened, those people who inherited one copy of the sickle cell allele were better able to fight off the fever and chills of malaria. Something about their red blood cells is inhospitable to the parasite. The healthier carriers had more children, passing on the protective allele to half of them. Gradually, the frequency of the sickle cell allele in East Africa rose, from 0.1% to a spectacular 45% in 35 generations. The price was paid for this genetic protection whenever two carriers produced a child who suffered from sickle cell disease.

A cycle set in. Settlements with large numbers of sickle cell carriers escaped debilitating malaria and were therefore strong enough to clear even more land to grow food—and support the disease-bearing mosquitos. Even today, sickle cell disease is more prevalent in agricultural societies than among people who hunt and gather their food. Figure 34.6 shows another example of balanced polymorphism.

KEY CONCEPTS

In directional selection, better-adapted individuals are selected in a changing environment. In disruptive selection, two extreme expressions of a trait have a survival advantage. In stabilizing selection, the intermediate between two extreme phenotypes is more likely to survive. Balanced polymorphism is a form of stabilizing selection in which a harmful allele is maintained in a population because the heterozygote has some advantage.

How Species Arise

Charles Darwin saw the evolution of species as a gradual series of adaptations molded by a changing environment selecting certain naturally occurring variants (Reading 34.2). He called this process of one species slowly becoming another "descent with modification." When genetic differences

Figure 34.7
Geographical isolation leads to allopatric speciation. The bluish gray pupfish inhabits a warm spring called Devil's Hole at the base of a mountain near Death Valley, Nevada. The summer pupfish population numbers about 600 individuals and the winter residents only 200. What makes these minnow-sized fish of evolutionary note is that their home has been isolated from other bodies of water for 50,000 years. In that time, the genetic makeup of the fish has shifted so greatly that they can no longer mate successfully with pupfish from populations beyond the spring. An entire new species has evolved in a mere flicker of geological time.

Figure 34.8
Hybrids between two species are usually infertile. The liger is the result of mating between a tiger and a lion, but this animal is not fertile. Only four such animals are known in the United States. The first liger we know of was bred by accident, in 1950, at a circus that had undoubtedly kept tigers and lions together. In the wild, these great cats will not normally mate.

prevent members of one population from producing fertile offspring with members of another, a new species is born.

Speciation depends upon natural selection and the physical distribution of organisms, or **biogeography**. These factors contribute to speciation in two stages. First, a population becomes geographically separated, perhaps by an earthquake, flood, or volcanic eruption, so that members of the two newly formed groups cannot interact. Over time, the populations may evolve different allele frequencies because the forces of natural selection, nonrandom mating, migration, and mutation act independently in each group. In addition, the original geographic isolation may have, by chance, created two subgroups with different allele frequencies. Two populations that are geographically isolated from one another are said to be **allopatric**, and formation of new species initiated by geographic isolation is called **allopatric speciation** (fig. 34.7).

For a new species to form, geographic isolation must be followed by reproductive isolation. That is, members of the two separated populations should not be able to reproduce successfully with each other if the geographic barrier is lifted. Two closely related groups of organisms that occupy the same region but cannot reproduce successfully with each other are called **sympatric species.** Sympatric speciation may arise following geographical isolation or by itself.

Reproductive isolating mechanisms prevent production of fertile offspring. The point in development when two species fail to produce viable offspring can occur before or after mating.

Premating Reproductive Isolation

Premating reproductive isolation is not always as obvious as the mismatched genitals of the male moose and the dairy cow. In **ecological isolation**, members of two populations prefer to mate in different habitats. In **temporal isolation**, they have different mating seasons. In **behavioral isolation**, the organisms perform different repertoires of courtship cues as the male and female approach one another.

These mechanisms are like a man and a woman seeking mates, perhaps going to different parties (ecological) on different nights (temporal) and speaking different languages (behavioral). Two species of toad descended from a single ancestral type provide a vivid example of premating reproductive isolation. *Bufo americanus* breeds in the early spring in small, shallow puddles or dry creeks, whereas *Bufo fowleri* breeds in the late spring in large pools and streams. Each type of toad also has a unique mating call.

Postmating Reproductive Isolation

Sometimes organisms mate, but fertile offspring do not result. The incompatibility can occur as early as fertilization, if the genetic material of the two species is packaged into chromosomes differently. A dog's gamete, with 39 chromosomes, could not form a healthy zygote with a cat's gamete, which has 19 chromosomes. Rarely, hybrid offspring result when individuals of different species mate, but they tend to be infertile and so have no evolutionary impact. A mule, for example, is a hybrid between a horse and a donkey. Sometimes two animals that would not mate in the wild do so in captivity. This happens with tigers and lions, but their hybrid offspring, seen only in zoos, are not fertile (fig. 34.8).

Reading 34.2 The Pace of Evolution—Like a Tortoise or a Hare?

Charles Darwin believed that evolution took a very long time, with one life form transforming into another through a series of intermediate stages. He noted, however, that such gradualistic evolution was not well supported by the fossil record because of a lack of intermediate, or transitional, forms. Darwin and others attribute the seeming incompleteness of the fossil record to several factors:

Lack of preservation of animals that did not have hard body parts;
Some species did not exist long enough to leave behind much evidence;
Many species were not widespread;
Geological upheavals that destroyed evidence.

Another explanation for some of the many gaps in the fossil record is that evolution may not always be gradual.

In 1944, paleontologist George Gaylord Simpson suggested that some of the gaps in the fossil record may actually be what they seem to be—true gaps, representing the sudden appearance of a species. Simpson estimated that perhaps 10% of all speciation falls into this "quantum evolution" category. In 1944, Simpson's concept of evolution by "leaps and starts" was not accepted by strict Darwinian gradualists.

In 1972, the idea of fast speciation interspersed with long periods when species changed little or not at all (stasis) was again raised by two young paleontologists, Stephen Jay Gould of Harvard University and Niles Eldredge of the American Museum of Natural History. This time, the concept was termed **punctuated equilibrium** to reflect the long periods of stasis interrupted by times of fast evolutionary change. The fossil record, they claim, lacks transitional forms simply because in some cases of speciation, they do not exist.

Figure 1
Land-dwelling tree frogs of genus *Eleutherodactylus* skip the tadpole stage and hatch directly from an egg as a small frog. This mature body form so early in life widens the range of habitats that the animal can occupy, thus giving it a selective advantage over frogs that spend the earlier parts of their lives as tadpoles.

Instances of very rapid evolution support the operation of punctuated equilibrium. Consider the several species of cichlid fish that live only in Lake Nabugaboo, Uganda, a body of water separated from Lake Victoria by a sand spit. These species have evolved over just the past 4,000 years. Even more recently, banana-eating moths have evolved in Hawaii since Polynesian settlers introduced bananas a few centuries ago. The appearance of new infectious diseases in our lifetime, such as toxic shock syndrome and Lyme disease, may reflect rapid evolution in the microorganisms that cause these illnesses.

How might changes drastic enough to lead to speciation occur rapidly? The answer may lie in the genes, for a single gene can have a profound effect on the appearance or functioning of an organism. A gene that alters the timing of early developmental events may cause obvious changes in the adult. An inherited delay in pigmentation in the embryo, for example, could greatly change the adult's appearance, which could in turn have great selective consequences in a plant or animal whose survival depends upon protective coloration. Frogs of genus *Eleutherodactylus* (fig. 1) skip the tadpole stage, hatching from an egg as a small frog. This characteristic could have evolved by an alteration in a single gene—a drastic and sudden change that would have increased survival. A single genetic "switch" altering timing of cell division could have produced the prolonged brain growth that is characteristic of our own species.

Reproductive isolation due to chromosome incompatibility can occur among individuals of the same species, if something creates subgroups with different organizations of genetic material. This happened, with the help of geographic isolation, in the plant *Clarkia rubicunda*, common along the coast of central California. A severe drought in the region of the Golden Gate Bridge in San Francisco nearly decimated the local population of *C. rubicunda*. By chance, the only survivors had several chromosomal abnormalities. These plants cross-fertilized among themselves, establishing a new population where the chromosomal aberrations were the norm. When the drought ended, *C. rubicunda* came in from surrounding regions, but they could not produce offspring with the Golden Gate group. The gametes of the two groups could no longer unite, although both types of plants were descended from the same ancestors. A new species, *C. franciscana*, had arisen.

Instantaneous reproductive isolation also results from **polyploidy**, when the number of chromosome sets increases (fig. 34.9). Polyploidy can occur when meiosis fails, producing, for example, diploid sex cells in a diploid individual. If diploid sex cells in a plant self-fertilize, a **tetraploid** individual results, having four sets of chromosomes.

When chromosome sets derive from the same species, the organism is called an **autopolyploid**. Autopolyploids of the rose, for example, have 14, 21, 28, 35, 42, or 56 chromosomes—presumably descended from an ancestral species having 7 chromosomes. Polyploids can also form when gametes from two different species fuse, creating a hybrid from which an **allopolyploid** may develop.

a.

b.

Figure 34.9
Polyploidy in plants. A triploid plant is unlikely to give rise to a new species because the three copies of each chromosome cannot segregate during meiosis into daughter cells having a consistent number of chromosomes. As a result, a triploid seed grows into a plant that has few seeds—not very valuable for evolution but a favorite of gardeners who prefer seedless varieties of plants, such as the triploid "seedless" watermelon (*b*). A watermelon is normally diploid.

For example, an "old world" species of cotton has 26 large chromosomes, whereas a species found in Central America and South America has 26 small chromosomes. The type of cotton that is commonly cultivated for cloth is an allopolyploid of the old world and American types. It has 52 chromosomes.

Polyploidy may occur in the wild as a response to unusually low temperatures. Agriculturalists often induce polyploidy, however, because such plants generally have larger leaves, flowers, and fruits. This is done using the drug colchicine, which is an extract of the autumn crocus plant. Colchicine dismantles the spindle apparatus, which normally aligns and then separates the chromosomes in dividing cells. Lacking a spindle, the replicated chromosomes cannot be distributed into two daughter cells, and a polyploid results. Many new varieties of crops are induced polyploids, including alfalfa, apples, bananas, barley, potatoes, and peanuts.

So great is the genetic difference between a polyploid and the plant from which it arises that geographic isolation is not even necessary for speciation. The fact that nearly half of all flowering plant species are natural polyploids indicates that this form of reproductive isolation has been fairly important in plant evolution.

Polyploidy is rarely seen in animals, because the disruption in sex chromosome constitution usually leads to sterility. One exception is the grey tree frog *Hyla versicolor*, which is a tetraploid probably derived from the identical-appearing *Hyla chrysoscelis*, a diploid. Once in a great while a human infant is born who is triploid, but massive birth defects usually end life within days. Certain cells in the healthy human body, however, are normally polyploid. Some liver cells, for example, are octoploid, with eight sets of chromosomes.

KEY CONCEPTS

New species arise when genetic changes prevent the members of one population from mating with the members of another. This may occur rapidly or over a long period of time. Speciation depends upon natural selection and biogeography. Allopatric speciation occurs when two populations are physically separated and develop different gene frequencies because of differences in natural selection, nonrandom mating, mutation, and migration. In sympatric speciation, which may follow allopatric speciation, two closely related groups in the same area cannot reproduce with each other.

Premating reproductive isolation can be ecological, temporal, or behavioral. Postmating reproductive isolation occurs when the genetic material is different, or packaged into chromosomes differently, in two groups of organisms. Mating between members of two animal species may produce an offspring, but it is not fertile. Polyploidy creates new species because the number of chromosome sets increases. An autopolyploid has chromosomes from one ancestral species; an allopolyploid has chromosomes from two different species.

How Species Become Extinct

Not all species emerge in healthy numbers from population bottlenecks or evolve into new species following geographical isolation. Just as speciation does not produce "perfect" organisms, but well-adapted ones, extinction is not a badge of biological failure. Extinction reflects the inability of organisms to adapt to an environmental challenge.

What Causes Mass Extinctions?

Earth history has been marked by at least a dozen periods of mass extinctions (table 34.2). (Geological time periods are discussed in chapter 35.) **Paleontologists**, who study evidence of past life, can find clues in the earth's sediments to the catastrophic events that heralded the disappearances of many species over relatively short expanses of time. Although explanations of the causes of mass extinctions are highly controversial among scientists, two general hypotheses have emerged in recent years.

The **impact theory** suggests that a meteor or comet crashed to earth. If it hit land, then dust, soot, and other debris sent skyward would block the sun, setting into motion a deadly chain reaction. Without sunlight, plants would not be able to photosynthesize, and they would die. The animals that ate plants, and the animals that

Table 34.2
Mass Extinctions

Time	*Species Affected*	*Suggested Cause*
3 bya* (Precambrian)	Anaerobic bacteria	Oxygen in atmosphere
545 mya+ (end of Cambrian)	Trilobites, other marine invertebrates	Meteor impact
440 mya (Ordovician)	Marine invertebrates	Gondwana formed
370 mya (end of Devonian)	Most fish and invertebrates	Meteor impact; Gondwana moved; asteroid shower
240 mya (Permian)	96% of all species	Pangeae formed
200 mya (end of Triassic)	75% marine invertebrates	Meteor impact
140 mya (end of Jurassic)	Marine species	Not known
90 mya (mid-Cretaceous)	Dinosaurs	Flowering plants
60 mya (Cretaceous/Tertiary boundary)	Dinosaurs, marine species	One or more meteor impacts
11,000 years ago	Large mammals	Drought, hunting

*billion years ago
+million years ago

ate those animals, would die too. An extraterrestrial object landing in the ocean would be equally devastating by mixing water layers. Oxygen-poor deeper waters would be shot upward in the turbulence, and upper-dwelling organisms adapted to the oxygen carried in their watery surroundings would die of oxygen starvation.

Evidence for the impact theory includes centimeter-thin layers of earth that are rich in iridium, an element rare here but common in meteors (fig. 34.10). Quartz crystals found in iridium deposits are cracked at angles that suggest an explosion. Where layers of rock unusually devoid of fossils are found near an iridium layer, the impact theory of mass extinction is suggested.

Alternatively, the restlessness of the planet's rocks may explain some mass extinctions. The geological theory of **plate tectonics** views the earth's surface as several rigid plates that can move, like layers of ice on a lake. These plates continually drift away from oceanic ridges, where new molten rock bubbles forth. Older regions of tectonic plates sink back into the earth's interior at huge trenches.

When continents drifted, coalesced, or broke apart, the environmental changes thrust upon organisms must have been profound. Suddenly organisms that had survived well in their particular habitats found themselves among unfamiliar species, whose members possibly competed for limited resources. Weather conditions changed, with ice ages and droughts killing off many species. The shifting of continents often altered shorelines, diminishing shallow sea areas packed with life.

Either a meteor crash or moving continents could explain the chaotic conditions associated with mass extinctions—or perhaps these changes occurred alone. Whatever the initial cause, at various times in earth history, sea levels have dropped, bodies of fresh water flooded their banks, temperatures fluctuated, volcanoes blew their tops, vegetation patterns changed, and tracts of land bridging continents ebbed and flowed. In India, for example, a lava flow the size of France is a reminder of a huge volcanic eruption that may have coincided with the extinction of dinosaurs. Might a long-ago volcano have spewed enough debris into the atmosphere to block the sun, lowering temperatures and triggering mass starvation? A volcano would also have released sulfur emissions, causing global acid rain and threatening life in the waters.

Mass Extinctions Through Geological Time

Extinctions probably occurred even before life as it is known today evolved, with the periodic wiping out of primordial cell-like structures as more efficient ones formed. After each mass extinction, the survivors repopulated the earth. It is interesting to ponder our own fate had different subsets of species vanished or survived.

An early mass extinction reflected the drastic change in the atmosphere—oxygenation—that happened as cyanobacteria evolved. The oxygen they released was toxic to the reigning anaerobic species, many of which died out as the more efficient energy users flourished.

About 440 million years ago, life in the seas was severely disrupted as a huge continent, called Gondwana, formed and covered the South Pole, causing an ice age. The glaciers drew water from the oceans, robbing many species of their habitats. Then 370 million years ago, geological upheaval struck again. Sulfur-containing minerals found in disturbed rock layers in the Canadian Rockies indicate a mixing of ancient ocean layers. Iridium is also here, as is evidence of glaciation. At this time, many fish species and nearly 75% of all marine invertebrates perished.

The Permian period of 240 million years ago saw the greatest mass extinction of them all, when 96% of species died in several waves of death spanning 8 million years. The cause of the Permian extinction is thought to be the fusion of landmasses into a single gigantic continent, Pangaea. The present-day location of the continents suggests that they may once have been connected, with the continental shelf of South America fitting like a puzzle piece along that of western Africa. Remains of the same species of now-extinct organisms are found along the coasts of both South America and West Africa, supporting the hypothesis that the great landmasses were once contiguous.

Just 40 million years after the Permian extinction, earth's inhabitants faced yet another changing environment. Once again, 75% of marine invertebrate species vanished. A crater in Quebec, Canada, roughly half the size of Connecticut, may be evidence of a meteor impact that had devastating repercussions.

a.

b.

Figure 34.10

Evidence for an impact. *a.* A pencil-thin rock layer found everywhere on earth may hold clues to a global meteorite impact about 60 million years ago. The band of rock is rich in iridium, an element rare on earth but prevalent in meteors. On this seaside cliff in Zumaya, Spain, paleontologists and geologists probe the exposed iridium layer. *b.* Further evidence in support of the impact theory is that quartz grains from the iridium-rich region are cracked in a way only possible in a nuclear explosion or meteor impact. Opponents of the impact theory, however, maintain that the cracks could have arisen in a volcanic explosion.

Another impact is hypothesized for the mass extinctions that peaked about 65 million years ago, when 75% of all plant and animal species, including the great dinosaurs, disappeared. In the seas, plankton, microscopic food for many larger marine dwellers, died as well. These extinctions did not all happen at once. Ocean chemistry changes span 2 million years around this time, suggesting a series of impacts rather than a single cataclysmic event.

The most recent round of mass extinctions may have been at least partially caused by our own species. The last ice age occurred during the Pleistocene epoch, from about 1.6 million to 10,000 years ago. In the last 2,000 years of this time period, 35 classes of mammals became extinct in North America. Unlike previous mass extinctions, when many types of organisms perished, the species that disappeared 11,000 years ago were mostly large, plant-eating mammals.

Were the Pleistocene herbivores hunted to extinction? According to the **Pleistocene overkill hypothesis,** at this time humans from Asia crossed a land bridge over the Bering Strait to a corridor east of the Canadian Rockies, traveling down through North America. Along the way, they encountered 50 to 100 million large herbivores, which they hunted to extinction. Flint spears found in remains of mammoths, mastodonts, horses, tapirs, and camels support the overkill hypothesis. The arrival of humans could have been devastating on this continent because it was rather sudden. In contrast, humans and their immediate ancestors had lived in Europe, Asia, and Africa far longer, and herbivore species there may have had the time to adapt to human predation.

Figure 34.11
A Pleistocene scene. The selective extinction of large herbivores (such as the saber-toothed tiger, mastodonts and mammoths, and giant sloths, peccaries, beavers, bears, deer, antelope, and lions) at a time when humans were hunting their way from Alaska to Mexico suggests that the two events may be related. This last round of extinction may have been caused by our own species.

Figure 34.12
A zoo of extinct organisms. This curator of England's Tring Zoological Museum cradles the only organism in this assemblage that is not extinct—an aye-aye from Madagascar. But it may be next.

The disappearance of large mammals may have led to extinctions of smaller animals. The **keystone herbivore hypothesis** suggests that the demise of the large herbivores led to overgrowth of vegetation that these animals ate, and this vegetation change threatened the existence of smaller herbivores (fig. 34.11).

The earth has not seen the last mass extinctions. Largely because of human intervention in the environment, species are vanishing with disturbing rapidity. Among the list of the recently extinct include the 10-foot (3-meter) moa bird, the zebralike quagga, the passenger pigeon, and the dodo bird (fig. 34.12). Within the next 25 years, a million more species may be lost.

Although humans have hastened the extinctions of some species, we have also made efforts to save endangered species (fig. 34.13). Consider the flightless Guam rail bird. In 1968, 80,000 of them lived on the island of Guam. Today, only 50 birds remain, occupying an area of dense brush that runs along an air strip used by B52 bombers at Anderson Air Force Base. The birds' habitat was scheduled to be destroyed because the heavy vegetation could provide cover for terrorists. But preservation overcame politics, and not only has the Guam rail's last refuge been spared, but birds are being relocated in an attempt to increase their numbers.

The whooping crane, the tallest bird in North America, has also benefited from human intervention. In 1941, only 21 birds were known to exist. In 1975, conservationists in northern Alberta placed whooping crane eggs in the nests of closely related species such as the sandhill crane. Soon, the unrelated birds nurtured the whooping crane eggs. By 1985, 140 whooping cranes made up the population. With efforts such as these, perhaps we can make a positive difference in the biological destiny of our planet.

KEY CONCEPTS

Extinction results when species cannot adapt to a changing environment. The earth's dozen or more mass extinctions may have been triggered by meteor impacts or continent movements, which disrupted climatic conditions. Disturbance to the sea would have mixed the oxygen distribution, and disturbance to the land would have raised huge dust clouds. Specific mass extinctions have been linked to oxygenation of the atmosphere, formation of large continents, meteor impacts, and human predation.

Reading 34.3 *Dogs and Cats—Products of Artificial Selection*

THE PAMPERED POODLE AND GRACEFUL GREYHOUND MAY WIN IN THE SHOW RING, BUT IN TERMS OF GENETICS AND EVOLUTION, THEY ARE POOR SPECIMENS. Human notions of attractiveness can lead to bizarre breeds that may not have evolved naturally. Beneath carefully bred quirks lurk small gene pools and extensive inbreeding—all of which spell disaster to the health of many highly prized and highly priced show animals.

The sad eyes of the basset hound make him a favorite in advertisements, but his runny eyes can be quite painful (fig. 1). His short legs make him prone to arthritis, his long abdomen encourages back injuries, and his characteristic floppy ears often hide ear infections. The eyeballs of the Pekingese protrude so much that a mild bump can pop them right out of their sockets. The tiny jaws and massive teeth of pugdogs and bulldogs lead to dental and breathing problems, plus sinusitis, bad colds, and their notorious "dog breath." Folds of skin on their abdomens easily become infected. Larger breeds, such as the Saint Bernard, are plagued by bone problems and short life spans.

We artificially select natural oddities in cats too, our choices of feline traveling companions throughout history has left a legacy of human civilization in the form of cat populations. One of every 10 New England cats has six or seven toes on each paw, thanks to a multitoed ancestor in colonial Boston (fig. 2). Elsewhere, these cats are quite rare. The sizes of the blotched tabby populations in New England, Canada, Australia, and New Zealand reflect the time of colonization by cat-loving Britons. The Vikings brought the orange tabby to the islands off the coast of Scotland, rural Iceland, and the Isle of Man, where these feline favorites flourish today.

A more modern entrant among cat fanciers is the American curl cat, whose origin is traced to a stray female who wandered into the home of a cat-loving family in Lakewood, California, in 1981. She had unusual, curled-up ears, and several of her litters made it obvious that the trait is inherited (fig. 3). The cause—a dominant gene that leads to formation of extra cartilage lining the outer ear. Cat breeders attempting to fashion this natural peculiarity into an official show animal are hoping that the gene does not have other, less loveable effects. Cats with floppy ears, for example, are known to have large feet, stubbed tails, and lazy natures.

Figure 2
Multitoed cats are common in New England but rare elsewhere.

Figure 3
American curl cat

Figure 1
Dogs and cats.

Figure 34.13
The Przewalski's horse, from Russia and China, was nearly extinct when breeders intervened. Twelve of the animals were bred to the related Mongolian domesticated horse to produce 409 "reconstructed" Przewalski's horses that today live in zoos. The animal in the wild is extinct.

SUMMARY

Evolution begins with changes in allele frequencies within the *gene pools* of populations. If allele frequencies are not changing from generation to generation, *Hardy-Weinberg equilibrium* exists and evolution is not occurring. If allele frequencies are known, the binomial expansion $p^2 + 2pq + q^2$ can be used to calculate the proportion of genotypes and phenotypes in a population. The equation reveals allele frequency changes when migration, nonrandom mating, mutation, or natural selection is operating.

In *genetic drift*, allele frequencies change in small populations split off from larger ancestral populations because of chance sampling of alleles. In human populations, a form of genetic drift is the *founder effect*, where a few individuals start a new colony. Genetic drift is also encountered in *population bottlenecks*, when a group of organisms is nearly killed off and the few survivors replenish the population with a restricted gene pool. Mutation can alter gene frequencies by introducing new alleles. Harmful recessive alleles are selected against in homozygotes, and are maintained by heterozygotes and reintroduced by mutation. The deleterious alleles in a population constitute its *genetic load*.

Natural selection is a major driving force behind evolution. In *directional selection*, a trait shifts in one direction. In *disruptive selection*, extreme expressions are selected at the expense of intermediate forms. In *stabilizing selection*, an intermediate phenotype has an advantage. In *balanced polymorphism*, alleles that damage fitness in the homozygote are selected because the heterozygote has an advantage over those lacking the allele.

Speciation is usually caused by a combination of changing allele frequencies and geographical isolation of two populations, which results in two groups whose members can no longer reproduce successfully with each other. *Premating reproductive isolation* prevents two individuals from mating due to ecological, temporal, or behavioral differences. *Postmating reproductive isolation* results from incompatible chromosomes. Hybrid offspring of two species are infertile. *Polyploidy* causes rapid speciation in plants by introducing extra sets of chromosomes and immediate reproductive isolation.

Extinction results from the inability of a species to adapt to a changing environment. The earth has had several periods of mass extinctions, which may be related to meteor impacts and geological upheavals such as continental shifts. Either of these forces can lead to drastic environmental changes, such as mixing ocean layers or blocking the sun, that ultimately make survival impossible for many species.

QUESTIONS

1. The fraggles are a population of mythical, mouselike creatures that live in underground tunnels and chambers beneath a large vegetable garden, which supplies their food. Of the 100 fraggles in this population, 84 have green fur and 16 have gray fur. Green fur is controlled by a dominant allele *F* and gray fur by a recessive allele *f*. Assuming Hardy-Weinberg equilibrium is operating, answer the following questions:

- **a.** What is the frequency of the gray allele *f*?
- **b.** What is the frequency of the green allele *F*?
- **c.** How many fraggles are heterozygotes (*Ff*)?
- **d.** How many fraggles are homozygous recessive (*ff*)?
- **e.** How many fraggles are homozygous dominant (*FF*)?

2. One spring, a dust storm blankets the usually green garden of the fraggles in gray. Under these conditions, the green fraggles become very visible to the Gorgs, monstrous beasts who tend the gardens and try to kill the fraggles underfoot to protect their crops. The gray fraggles, however, blend into the dusty background and find that they can easily steal radishes from the garden. How might this event affect microevolution in this population of fraggles?

3. How did natural selection in this century render tuberculosis a less dangerous disease than it had been in the past century?

4. Women who are extremely thin cease to have menstrual periods. It is thought that this happens because they have too little body fat to support a pregnancy. Women who are extremely obese have fewer children than women of normal weight, because they are also more likely to suffer from heart disease or diabetes, which makes pregnancy difficult. Assuming that extreme thinness and obesity are at least partially genetically determined, what type of natural selection is at work here?

5. Reading 10.1 shows a chimeric animal that is derived from an embryo built of sheep cells and goat cells. How does the reproductive ability, and therefore the potential impact of evolution, of this goat/sheep differ from that of a liger or mule?

6. What evidence would suggest an extraterrestrial rather than an earth-based cause of a mass extinction?

TO THINK ABOUT

1. Conflicting hypotheses are often proposed to account for evolutionary events, such as the pace of evolution and the cause of mass extinctions. Those who do not believe that evolution has taken place sometimes cite these differences of opinion as evidence that even scientists doubt that evolution occurs. Scientists, however, maintain that conflicting hypotheses, or theories that could have been possible simultaneously, do not argue against evolution at all but demonstrate the process of scientific thinking. What is your opinion on the matter? Why are hypotheses particularly important in understanding evolutionary processes compared to other fields of biology?

2. What are three characteristics of human populations in the United States that violate the conditions of Hardy-Weinberg equilibrium?

3. According to the definition of speciation, do you think that the Great Dane and miniature poodle should be considered separate species?

4. In gene therapy, a defective gene is replaced with a functioning one. In what part of an organism would this have to be performed to have the potential to influence evolution? Why?

5. Which evolutionary principles are illustrated by the following scenarios taken from science fiction films?

- **a.** An environmental catastrophe forces one group of humans underground and a smaller group to remain above. One group evolves into grotesque "mutants" and the other into a sleek, handsome race.
- **b.** We cause our own numbers to dwindle dangerously close to extinction following a nuclear holocaust. Only a handful of individuals survive to reestablish the population.
- **c.** An approaching comet renders the earth uninhabitable. Certain humans are chosen to leave the doomed earth to settle another planet.

SUGGESTED READINGS

Cohn, Jeffery P. October 1990. Endangered wolf population increases. *BioScience*. Can humans help to restore wolf populations?

Diamond, Jared M., and Jerome I. Rotter. September 10, 1987. Observing the founder effect in human evolution. What maintains the frequencies of human genetic diseases? *Nature*, vol. 329. Genetic quirks in populations can sometimes be traced to the individuals in whom they arose. Some fascinating examples.

Gore, Rick. June 1989. Extinctions. *National Geographic*. A panorama of earth's dozen largest mass extinctions, including the spectacular evidence.

Kettlewell, B. 1973. *The evolution of melanism*. Oxford: Clarendon. Natural selection is vividly displayed by protective coloration in moths.

Phillips, Kathryn. June 1990. Where have all the frogs and toads gone? *BioScience*, vol. 40 no. 6. Are some of these animals headed for extinction?

Stanley, Steven M. 1981. *The new evolutionary timetable*. New York: Basic Books. Is the pace of evolution like a tortoise or a hare?

Stebbins, G. Ledyard, and Francisco J. Ayala. August 28, 1981. Is a new evolutionary synthesis necessary? *Science*, vol. 213. A classic paper placing rapid evolution in the framework of traditional Darwinian theory.

C H A P T E R

35

Evidence for Evolution

Chapter Outline

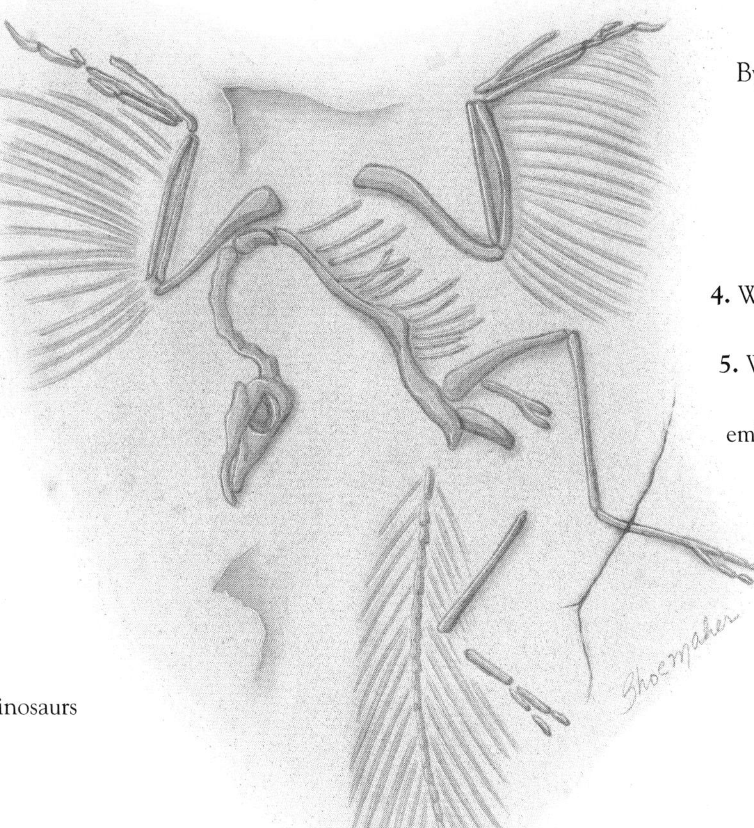

Learning Objectives

By the chapter's end, you should be able to answer these questions:

1. What is a fossil?
2. Why are fossils important to the study of evolution?
3. How do fossils form?
4. What methods are used to determine the age of a fossil?
5. Why is the fossil record incomplete?
6. How can comparative anatomy, embryology, and biochemistry provide clues to evolutionary relationships between species?
7. What were some of the major events in life history that we know about?
8. What sorts of animals living within the past 40 million years were probably our direct ancestors?

Large vultures lived about 12,000 years ago in caves in the walls of the Grand Canyon, so high that only other birds could reach them. The vultures were ancestors of the California condors, which today live only in zoos. The condors of 12,000 years ago had a plentiful food supply of diverse, large mammals. Adult birds would soar about searching for carcasses, bringing back remains of horses, bison, mammoths, camels, and mountain goats for the fledgling birds back in the cave, who were too young to scavenge on their own (fig. 35.1).

To reconstruct the lives of the ancient condors, caves in the Grand Canyon wall were excavated. The great birds left traces of their existence in preserved feather fragments and bones. Some of the vultures were quite young, as evidenced by preserved egg shell fragments and the size of holes in some of the smaller bones. The caves were too high for the large mammals of the time to reach, yet piles of their bones lay about. These mammals must have been brought to the caves by the vultures, as modern condors did, until the last wild bird perished in 1987. By noting the layer of rock within which the remains were found, and by chemically analyzing dried protein in a preserved beak, an approximate date of 120 centuries ago was assigned to the find.

Preserved remains of once-living organisms provide a major tool for deciphering the biological past. Similarities between modern species also offer compelling evidence for evolution. **Paleontologists** use these clues to reconstruct evolutionary relationships between species, which are called lineages, or **phylogenies.** For example, the close evolutionary relationship between humans and chimpanzees is suggested by the ways in which the bones and muscles are arranged and function, by similar chromosome organization, and by nearly identical amino acid sequences of many proteins.

In the biological science of evolution, the pictures painted by these different types of evidence replace the experimental portion of the scientific method, because the best possible experiment—going back in time to actually witness what happened—is impossible. Nonetheless, the evidence for evolution is abundant and rich.

a.

b.

Figure 35.1

a. The California condor is a species recently gone extinct in the wild. *b.* Condors once dwelled in caves high in the walls of the Grand Canyon.

Fossils

How Fossils Form

A **fossil** is evidence of prehistoric life. Many fossils are hard parts of organisms that have been replaced by minerals. An inch-long horn coral dies on an ancient sea bottom in what is now Indiana and is covered, over the eons, with layers of sand and mud. Gradually, the mud hardens into rock. Meanwhile, the horn coral decomposes, leaving an impression of its shell. Millions of years later, a person walking along land that was once that sea bottom sees the impression. Perhaps the mold was filled in with more mud, which also hardened into rock. The explorer may then find a cast of the horn coral, a rocky replica of the ancient animal.

The living matter of trees can be replaced with minerals, producing petrified wood that reveals cellular structures when sliced thin and viewed under a microscope. Evidence of our recent relatives often consists of teeth, which are even harder than bone and the likeliest anatomical parts to be preserved. Bones are fossilized when minerals replace cells.

The most striking fossils form when sudden catastrophes rapidly bury organisms in an oxygen-poor environment. Without oxygen, tissue damage is minimal, and the area is too hostile for scavengers to feed on the dead. Under these unusual conditions, even soft-bodied life forms can leave exquisitely detailed portraits of their anatomies. A block-long section of the Canadian Rockies called the Burgess shale, for example, houses an incredibly varied collection of 530-million-year-old soft invertebrates, preserved, the evidence suggests, in a sudden mudslide. Similarly, a family of rhinoceroslike styracosaurs living along the Milk River in Montana about 75 million years ago was preserved in time by a sudden flood. Today the rock strata in the hills are littered with styracosaur skull and limb fragments, the original bone tissue replaced with mineral.

Muds and floods are not the only agents of fossilization. A gold prospector found a perfectly preserved baby mammoth in the ice of the Arctic circle in Siberia. "Dima" was about 9 months old when she perished 40,000 years ago and stood 3 feet (1 meter) tall and weighed about 140 pounds (64 kilograms) (fig. 35.2). In Los Angeles, mammals were preserved in the La Brea tar pits. Insects and frogs were trapped in sticky tree resin, which then hardened around them, forming translucent tombs known as amber.

Microscopes can aid in identifying fossils whose sources are not as obvious as horn corals, dinosaur bones, and insects in amber. Fossilized grasses, for example, are recognized with the help of a scanning electron microscope. Colonies of microscopic organisms are sometimes preserved. Fossilized coatings of cyanobacteria that lined ancient sea floors 2.5 billion years ago are found in western Australia. In the United States, similar fossils date from 500 million years ago. Fossils need not even be remains themselves. Dinosaur footprints and worm borings reveal how the animals that made them traveled. The pigments in fossilized dinosaur dung offer clues to what the great reptiles ate.

Fossils offer only fleeting glimpses into life's history. Much evidence was destroyed by the formidable forces of nature—erosion, tides, volcanoes, continental shifts, earthquakes, storms, and even meteor impacts. Plus, some species may have lived for such short periods of time, or dwelled in such restricted areas, that their fossils, if they left

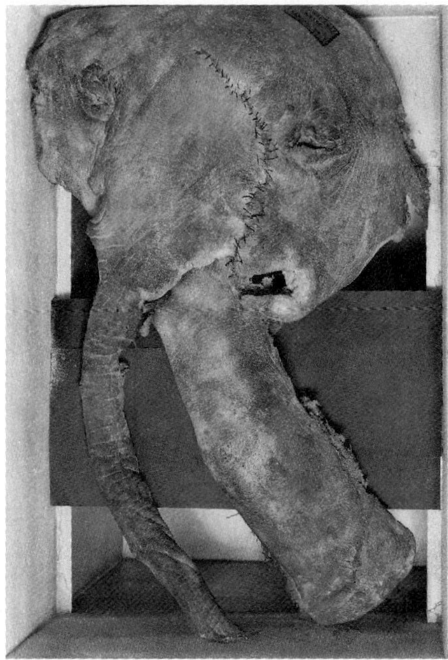

a.

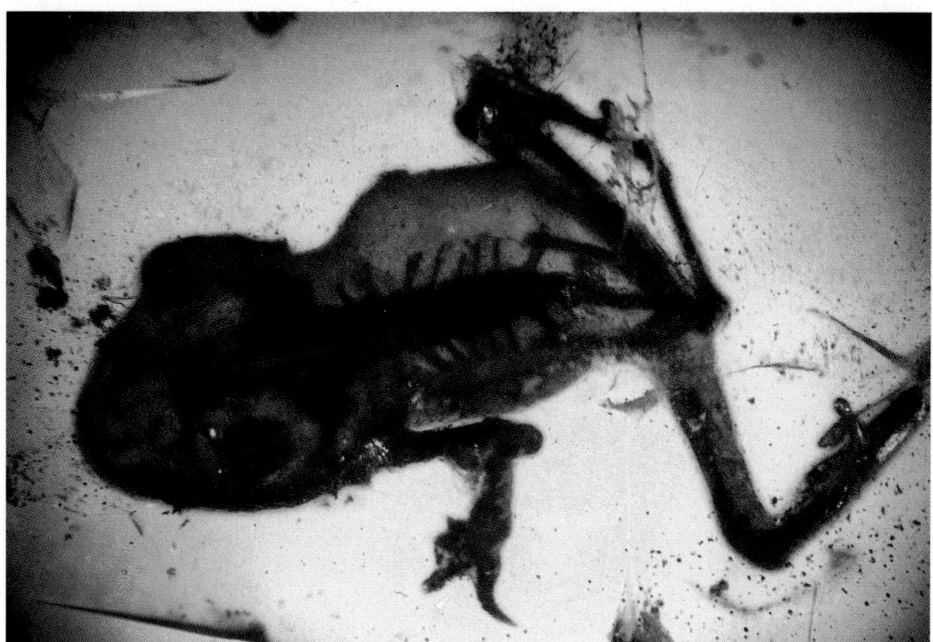

b.

Figure 35.2

Fossils. *a.* This frozen mammoth was found in Alaska and is preserved in a refrigerator at the American Museum of Natural History. *b.* About 40 million years ago, this *Eleutherodactylus* frog was trapped in tree resin. The frog has two broken limbs and other signs of struggle and injury. Perhaps a bird captured the animal, depositing it in the tree that became its tomb.

any, have never been discovered. Still, over the past two decades many fascinating fossils have been found.

Determining the Age of a Fossil

An organism represented by a fossil can be localized in the history of life either relatively or absolutely. A **relative date** is based on how far beneath the surface the rock layer lies in which the fossil is found. This is an application of the principle of superposition. However, because different rock strata are not formed at the same rate, assigning fossils to particular layers cannot offer a precise date.

Arriving at a more meaningful date requires tracking a phenomenon that has occurred at a constant and measurable rate for a very long time, so that fossils hundreds of millions of years old can be dated. With the discovery of radioactive elements in the late 1800s, a method of more accurately dating fossils was eventually developed.

Isotopes of certain elements are naturally unstable, causing them to emit radiation. As the isotopes release radiation (or "radioactively decay"), they change into different isotopes of the same element, or isotopes of a different element. (Recall that an isotope is an alternate form of an element distinguished by a different number of neutrons.) Each radioactive isotope decays to its alternate form at a characteristic and unalterable rate, called its **half-life.** The half-life is the time it takes for half of the isotopes in a sample of the original element to decay into the second form. If the half-life of a radioactive isotope and the amounts of its "before" and "after" forms in a rock or tissue sample are known, the time of formation of the sample can be deduced.

Using natural radioactivity as a "clock" is called **radiometric dating.** A date obtained in this way is absolute because it is expressed in a number of years, although it is not exact. Two radioactive isotopes are often used to assign dates to fossils—potassium 40 and carbon 14. Potassium 40 radioactively decays to argon 40 with a half-life of 1.3 billion years, making it valuable in dating old rocks containing traces of both isotopes (fig. 35.3). Chemical methods can detect amounts of argon 40 small enough to correspond to fossils that are about 300,000 years old. However, many sedimentary (layered) rocks, the source of most fossils, contain argon 40 that was actually part of much older rock. A potassium 40 date for these rocks would not correspond to the age of the fossils within them.

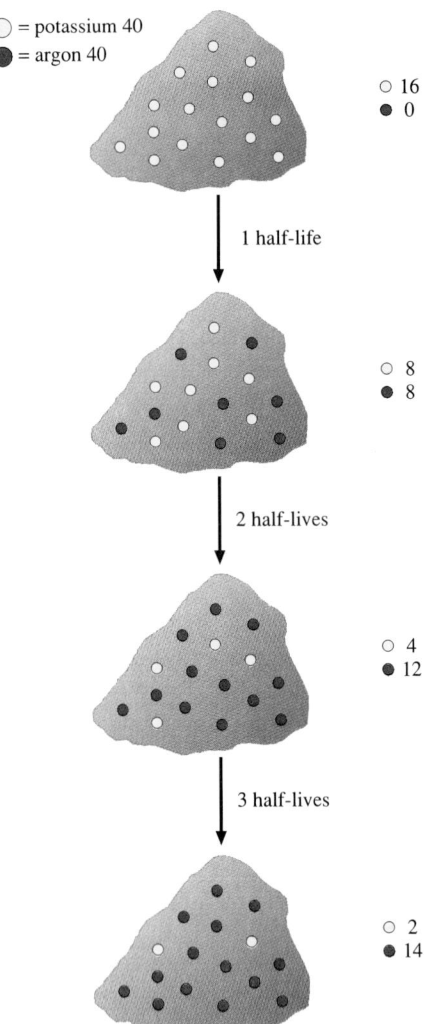

Figure 35.3
Radiometric dating. An approximate age in years can be assigned to some fossils by determining the proportion of potassium 40 to argon 40 in the rock. After a half-life of 1.3 billion years, half of the potassium 40 in a sample has decayed to argon 40. After two half-lives, a quarter of the potassium 40 remains, and after three half-lives, one-eighth of the original isotope is left.

Fossils up to 40,000 years old are radiometrically dated by measuring the proportion of carbon 12 to the rarer carbon 14. Carbon 14 is a radioactive isotope that forms naturally in the atmosphere when the nonradioactive form, carbon 12, is bombarded with cosmic rays (radiation from space). Organisms accumulate a certain proportion of carbon 14 as they assimilate carbon during photosynthesis or by eating organic matter. When an organism dies, however, its intake of carbon 14 stops, and from then on, carbon 14 decays to the more stable carbon 12 with a half-life of 5,710 years. The Grand Canyon vultures have about one-fourth the ratio of carbon 14 to carbon 12 seen in a living organism. Therefore, about two half-lives, or about 11,420 years, have passed since the animals died.

A limitation of potassium-argon and carbon 14 dating is that they leave a gap. That is, carbon dates extend back to 40,000 years ago, but potassium-argon dates do not begin reliably until 300,000 years ago. Several new techniques cover the missing years, which include the time of human origin.

Electron-spin resonance and **thermoluminescence** are physical techniques that measure tiny holes made in crystals over time due to exposure to ionizing radiation. Each method counts the holes differently and works on samples up to 1 million years old. Another approach, **amino acid racemization,** chronicles the rate at which amino acids in biological matter alter to mirror-image chemical structures called isomers. It is used for dating eggs and shells up to 100,000 years old.

The fact that electron-spin resonance and thermoluminescence values are perturbed by natural deposits of radioactive substances, and that amino acid racemization is affected by temperature and moisture, means that results must be corrected for these influences. Still, relative dating, radiometric dating, and these newer methods can be used together to pinpoint more accurately when organisms lived.

KEY CONCEPTS

A fossil is evidence of past life in which biological matter is slowly replaced with minerals or in which organisms are buried rapidly and preserved in a low-oxygen environment. The fossil record is incomplete because not all species had hard parts, were buried rapidly, were widespread, or lived long enough to be sampled. A fossil is relatively dated by the rock layer it is in or absolutely dated by isotope ratios, altered crystal structures, and amino acid isomer ratios.

Comparing Structures in Modern Species

Comparative Anatomy

In general, the more similar two living species are, the more recently they diverged in their lineage from a common ancestor. The structures and organization of the vertebrate skeleton are often cited as evidence of a common ancestor to this familiar group of animals. All vertebrate skeletons are built of the same component parts and provide support (fig. 35.4). An ancestor to modern vertebrates must have originated this skeletal organization, which was then modified in more recent organisms (amphibians, reptiles, birds, fishes, and mammals) for their particular modes of life.

Similarly built structures with the same general function that are inherited from a common ancestor, such as the vertebrate skeleton, are termed **homologous.** However, unrelated organisms can evolve structures in response to the same environmental challenge. Such structures that are similar in function, but not in architecture, are termed **analogous.** Birds and insects, for example, each have wings that enable them to fly, but the bird's wing is a modification of vertebrate limb bones, whereas the insect's wing is an outgrowth of the cuticle that covers its body in an exoskeleton. Analogous structures do not offer evidence for evolution (see fig. 33.6*b*).

Determining whether particular body parts are homologous or analogous can be difficult. In general, though, analogous structures tend to resemble one another only superficially, whereas the similarities in body parts between two species related by common descent—homologies—tend to be complex and numerous.

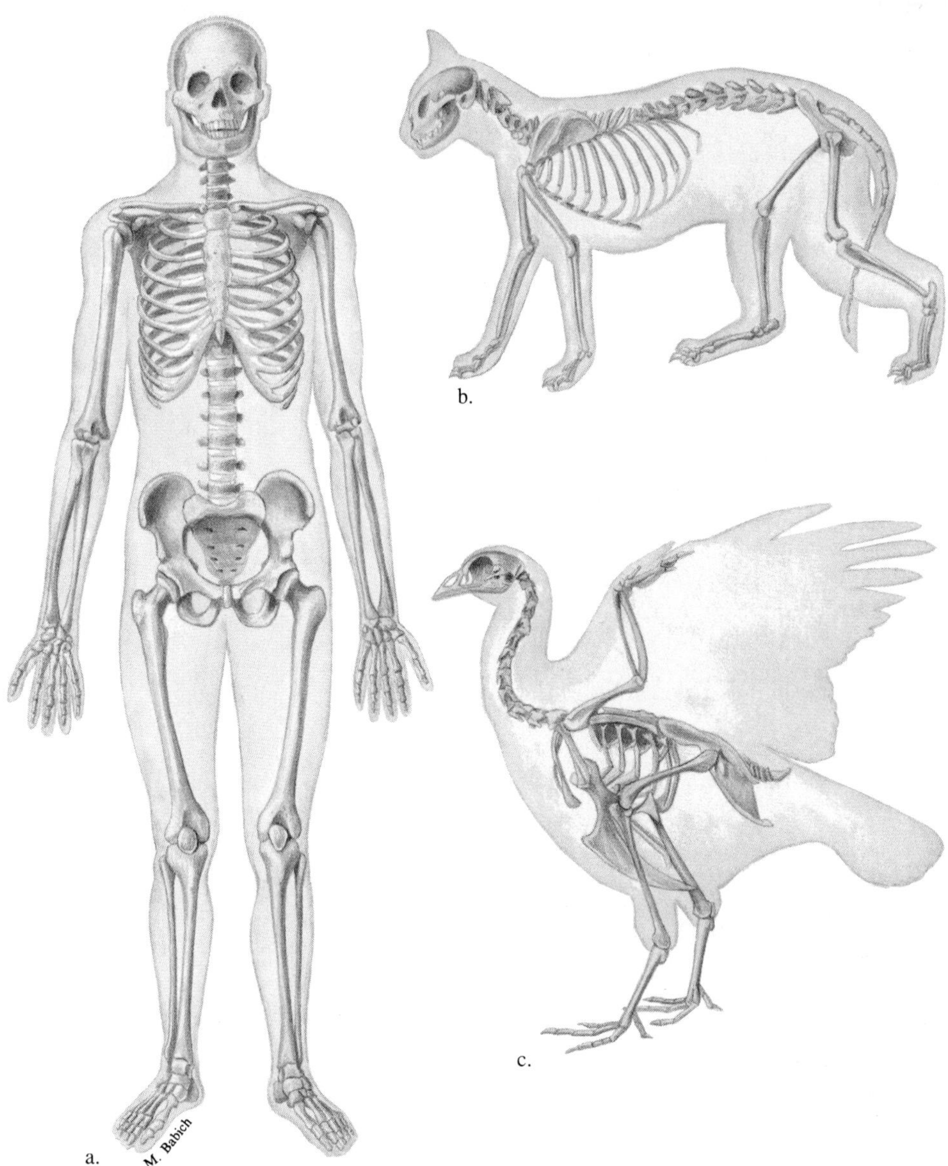

Figure 35.4
Vertebrate skeletal organization reflects common ancestry. Although a human walks erect (*a*), a cat walks on all fours (*b*), and a bird flies (*c*), all have a similar skeletal organization of a skull, shoulder, spine, hips, and limbs. An examination of the forelimbs of these animals accentuates the skeletal similarities.

Vestigial Organs

Evolution is not a perfect process, and as new structures are selected by environmental change, older ones are sometimes retained. A structure that seems not to have a function in an organism, yet resembles a functional organ in another type of organism, is termed **vestigial.** Darwin compared vestigial organs to letters in a word that are not pronounced but offer clues to the word's origin.

Human vestigial organs include wisdom teeth and the appendix, both of which may have been useful to distant ancestors who ate different foods than we do today. Our lowest and smallest vertebrae, the "tailbone," may be a vestige of a monkeylike ancestor who had a tail. In snakes and whales, leg bones are vestigial, perhaps retained from vertebrate ancestors who used their legs (fig. 35.5). A horse has two tiny leg bones that it apparently does not use, but these bones were probably useful to the horse's ancestor, which had a different gait because it had three toes on each foot instead of the modern-day one.

Comparative Embryology

Embryos of a human, monkey, dog, cat, rabbit, and mouse look very much alike. Darwin suggested that the striking similarity between vertebrate embryos reflects adaptations to their similar environments—

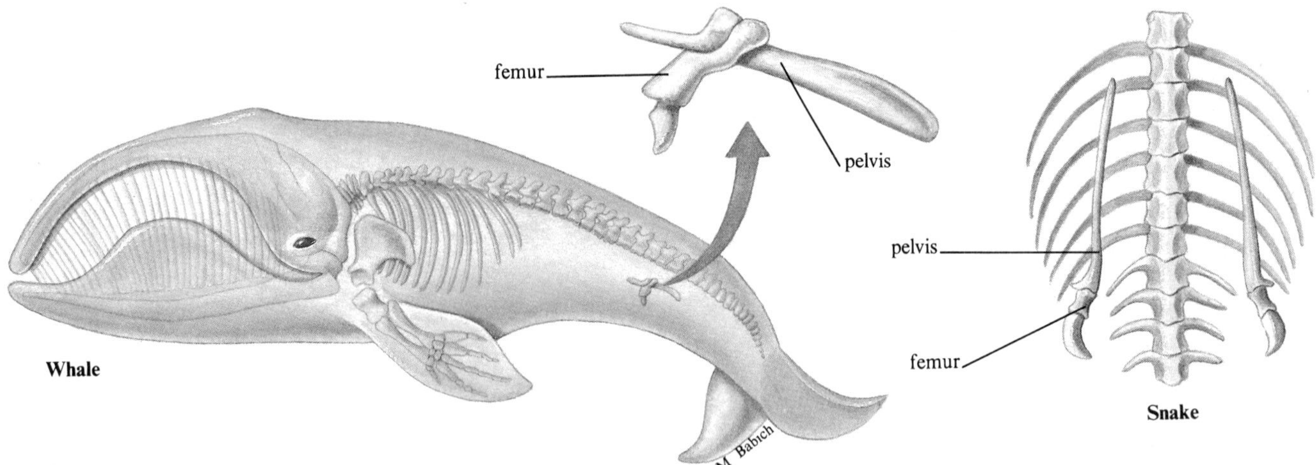

Figure 35.5
Vestigial organs. The whale and the snake have tiny femurs (leg bones), which are apparently of no use to them. These femurs are vestigial organs.

floating in a watery bubble, either in the mother's uterus or in an egg. These embryos may express the same sets of ancestral genes, and as the embryos mature into fetuses, they begin to express other genes and take on the characteristics of their particular species. Shared early genes might explain prenatal similarities, such as the gill slits and rudimentary tail seen in a 7-week-old human embryo.

KEY CONCEPTS

Homologous structures are similar in structure and function, reflecting shared ancestry. Analogous structures are similar in function but not architecture and do not reflect common ancestry. Vestigial organs have no obvious function, resemble structures in other species, and may have had a function in an ancestor. Vertebrate embryos appear similar because of shared, early acting genes.

Molecular Evolution

Large-scale evidence for evolution, such as fossils and similar body parts, can be difficult to interpret. Less ambiguous evidence comes from within the cell, in the molecules of life.

The reasoning behind **molecular evolution** is that genes and proteins are composed of many bits of information. It is highly unlikely that two unrelated species would happen to evolve precisely the same order of chromosome bands, nucleotides, or amino acids simply by chance. As with larger structures, the greater the molecular similarities between two modern species, the closer their evolutionary relationship. Even more direct evidence is found in preserved DNA from extinct species, which can be amplified and sequenced and compared to DNA from its descendants (fig. 35.6).

Several types of molecular evidence for evolution are used to confirm or challenge lineages proposed on the basis of fossil or comparative anatomy evidence.

Comparing Chromosomes

The number, shape, and banding patterns of stained chromosomes can be compared as a measure of relatedness. The chromosomes of humans, chimpanzees, gorillas, and orangutans are very similarly organized. If human chromosome 2 is broken in half, we would have 48 chromosomes instead of 46, as do the three species of apes. Human chromosome banding patterns match closest those of chimpanzees, then gorillas, and then orangutans. Some differences are inversions of band sequences, rather than missing or extra genetic material. Chromosome patterns can be compared between less closely related species as well. All mammals, for example, have identically banded X chromosomes.

Chromosome evidence has settled a long-standing debate—is the giant panda a bear or a raccoon? Although most children would answer "bear," many biologists have placed the panda with raccoons, based on its resemblance to a relative of the raccoon called the red lesser panda. Bears have 74 short chromosomes, with the constrictions near the tips. Pandas have 42 chromosomes, most of them long and with centrally located constrictions. Several pairs of the small bear chromosomes, if fused, would very closely resemble particular chromosomes of the giant panda. As for the red lesser panda, only 2 of its chromosomes match those in giant pandas or bears, but 14 of them are found in raccoons (fig. 35.7). According to its chromosomes, the panda is a bear after all.

Figure 35.6
Probing the molecules of extinct organisms. The last quagga, a relative of the horse and zebra, died in captivity in Amsterdam in 1883. DNA extracted from this preserved quagga is now being deciphered to see how closely related the extinct animal was to its descendants.

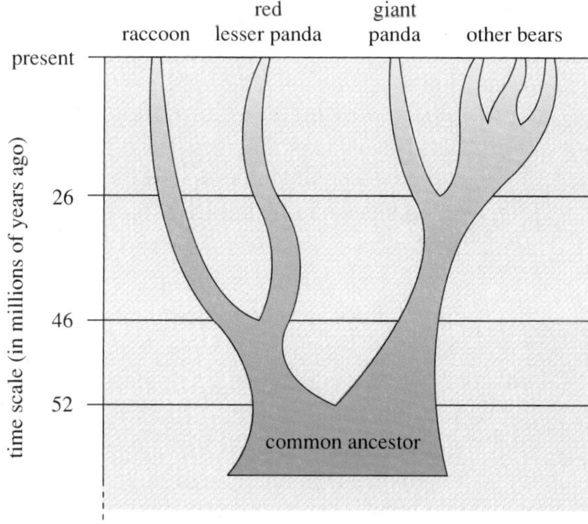

Figure 35.7
The giant panda finds its place in life history. Is the giant panda a bear or a raccoon? It looks most like a bear, but the similar-looking red lesser panda is in the raccoon family. Molecular studies place the giant panda closer to the bears.

Comparing Protein Sequences

The fact that all species utilize the same genetic code to build proteins argues for a common ancestry to all life on earth. In addition, many different types of organisms use the same proteins, with only slight variations in amino acid sequence. One of the most ancient and well-studied proteins is cytochrome C, which is involved in cellular respiration. Twenty of its 104 amino acids occupy identical positions in the cytochrome C of all eukaryotes. The closer related two species are, the more alike their cytochrome C amino acid sequence is (fig. 35.8). Human cytochrome C and chimpanzee cytochrome C are identical.

The amino acid sequences of dozens of proteins have now been deciphered in many species to track lineages, and quite often the results are consistent with fossil or anatomical evidence. The similarities in amino acid sequences between the proteins of humans and chimpanzees are astounding—many proteins are alike in 99% of their amino acids. Several are virtually identical. Chapter 3 discusses two highly conserved proteins—hemoglobin and albumin.

Organism	Number of Amino Acid Differences from Humans
chimpanzee	0
rhesus monkey	1
rabbit	9
cow	10
pigeon	12
bullfrog	20
fruit fly	24
wheat germ	37
yeast	42

a.

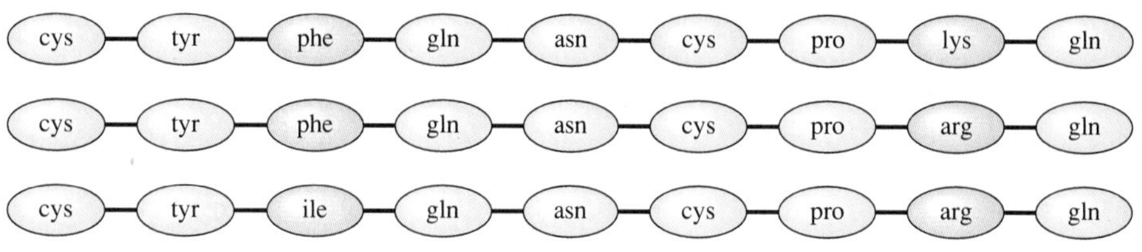

Figure 35.8
a. Amino acid sequence similarities are a measure of evolutionary relatedness. Similarities in amino acid sequence for the respiratory protein cytochrome C between humans and other species parallel our degree of relatedness to them. *b.* Antidiuretic hormone is a peptide hormone eight amino acids long that signals the kidneys to conserve water. Its sequence differs only slightly between major groups of vertebrates.

Comparing DNA Sequences

Similarities in DNA sequences between two species can be assessed for the total genome or for a single gene or piece of DNA. Examining genomic similarity relies on complementary base pairing. DNA double helices from two species are unwound and mixed together. The rate at which hybrid DNA double helices reform—that is, DNA molecules containing one helix from each species—is a direct measure of how similar they are in sequence. The faster the DNA from two species forms hybrids, the more closely related the two species are. Using this approach of **DNA hybridization,** researchers have shown that human DNA differs in 1.8% of its base pairs from chimpanzee DNA; by 2.3% from gorilla DNA; and by 3.7% from orangutan DNA.

DNA sequence similarity can be used to estimate the time when two species diverged from a common ancestor if the rate of base substitution mutation is known. For example, if the DNA from two species differs in 5% of its bases, and substitutions occur at a rate of 1% per 1 million years, then 5 million years must have passed since the species diverged. The rate of base change in DNA is used as a "molecular clock," just as the rate of radioactive decay of an isotope is used as a natural clock in radiometric dating. The information from DNA sequencing is represented in **evolutionary tree diagrams,** which indicate on a time scale the "branching points" when two species diverged from a common ancestor (fig. 35.9). Mitochondrial DNA provides an even better molecular clock, because it mutates about 5 to 10 times faster than nuclear DNA (Reading 35.1).

KEY CONCEPTS

Evolutionary trees are constructed by considering how alike chromosome bands or gene or protein sequences are between species. This evidence indicates that humans are most closely related to chimpanzees.

The History of Life on Earth

The tantalizing clues to past life offered by fossils, comparative anatomy, and biochemistry have made possible one of the most fascinating pursuits a biologist can follow—piecing together the sequence in which species arose, flourished, and died out on the earth. Of course, we can only glimpse brief moments of life's history.

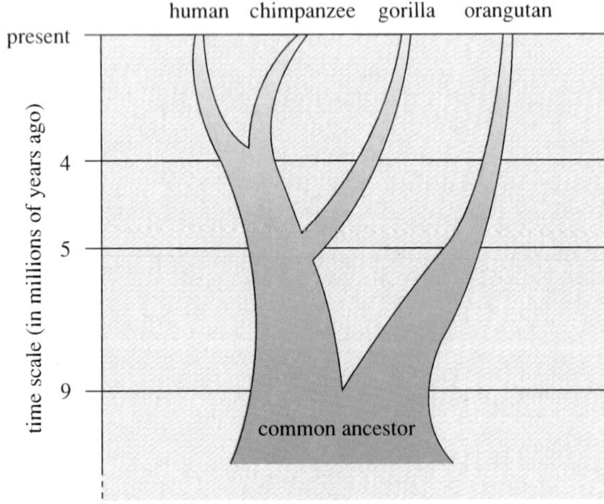

Figure 35.9
Identifying our closest relatives. One way to decipher the evolutionary relationship between humans and apes is to compare the number of chemical building block differences in the sequences of the same gene between pairs of species. This information is then superimposed upon a time scale to estimate when the species diverged from common ancestors by considering the number of years it takes for mutations to accumulate. The resulting "molecular clock" shows the most recent ancestor we have in common with chimpanzees lived about 4 million years ago. According to this theory, the common ancestor of humans and chimps evolved from a common ancestor with gorillas about 5 million years ago, and a common ancestor of all four species lived about 9 million years ago.

The Geological Time Scale

The backdrop for life history is geological history, which began before life debuted yet continues today. The **geological time scale** was developed in the late nineteenth century. It is divided into major **eras** of biological and geological activity lasting vast stretches of time, then **periods** within these eras, and finally **epochs** within some of the periods (table 35.1).

Because evidence of evolution tends to represent accidents, such as mudslides or glaciers, and the more spectacular biological events, the history of life reads like a dramatic series of extinctions followed by adaptive radiation (species entering new habitats and diversifying) of the survivors. Some of the "main events" of evolution include the origin of life (chapter 3), the debut of eukaryotic cells (chapter 4), and mass extinctions (chapter 34). In actuality, the parade of life has probably been continuous since the earliest cell-like collections of organic chemicals formed more than 3 billion years ago.

Following is a probable sequence of events that led from those first cells to the dawn of modern humans. This is a very narrow peek. If the history of life is represented as a vast bush, with each branch representing a type of organism, then those branches leading to humans are but the tiniest of twigs. That is, for each branch leading toward humans, many others led in other directions.

Precambrian Life

Fossil evidence before 600 to 700 million years ago is sparse. This earliest part of earth history, called the **Precambrian era,** accounts for five-sixths of earth's existence so far (fig. 35.10). The oldest rocks to be radiometrically dated come from western Greenland, and they are about 3.8 billion years old. Unfortunately, these rocks were exposed to such high temperatures and pressures that any fossils that might have been in them were undoubtedly destroyed.

Some of the earliest and most intriguing fossils are from the Fig Tree sediments of South Africa, which date back 3.4 billion years. The size and intricate folds of these structures resemble more recent fossils and modern prokaryotes (bacteria and cyanobacteria) (fig. 35.11). Fossils from Rhodesia from 2.8 billion years ago contain

Table 35.1
The Geological Time Scale

Era	*Period*	*Epoch*	*Millions of Years Ago*	*Important Events*
Cenozoic (Age of Mammals)	Quaternary	Recent	0.01	Modern humans
		Pleistocene	2	Early humans
	Tertiary	Pliocene	6	Radiation of apes
		Miocene	23	Abundant grazing mammals
		Oligocene	38	Angiosperms dominant
		Eocene	54	Mammalian radiation
		Paleocene	65	First placental mammals
Mesozoic (Age of Reptiles)	Cretaceous		135	Climax of reptiles; first angiosperms; extinction of ammonoids
	Jurassic		180	Reptiles dominant; first birds; first mammals
	Triassic		225	First dinosaurs; cycads and conifers dominant
Paleozoic	Permian		275	Widespread extinction of marine invertebrates; expansion of primitive reptiles
	Carboniferous*		345	Great swamp trees (coal forests); amphibians prominent
	Devonian		395	Age of fishes; first amphibians
	Silurian		435	First land plants; eurypterids prominent
	Ordovician		500	Earliest known fishes
	Cambrian		600	Abundant marine invertebrates; trilobites and brachiopods dominant; algae prominent
Precambrian			>3000	Soft-bodied primitive life

*The early Carboniferous is often referred to as the "Mississippian" and the late Carboniferous as the "Pennsylvanian."

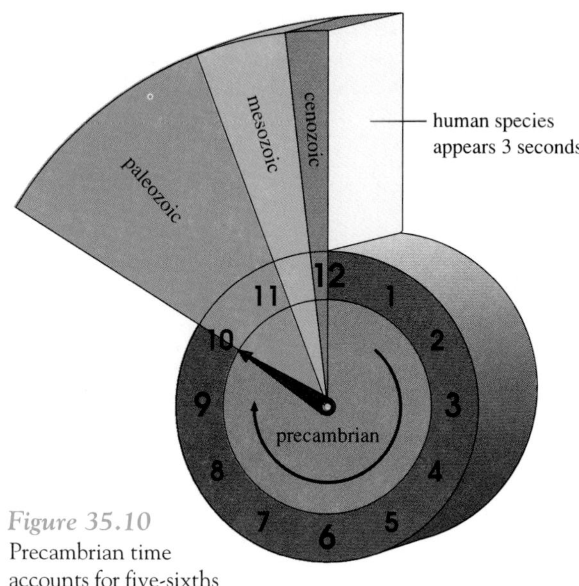

Figure 35.10
Precambrian time accounts for five-sixths of earth's history. The time during which we think life on earth has been abundant—the Paleozoic, Mesozoic, and Cenozoic eras—accounts for only one-sixth of the planet's history.

Figure 35.11
Stromatolites represent very ancient life. Between 2.5 billion years ago and 600 million years ago, bacteria flourished in the shallow seas. Cyanobacteria aggregated into huge colonies, forming great submerged mats that became infiltrated with sediments, which hardened to preserve replicas of these most ancient organisms. These fossils, called stromatolites, are found today in only a few places, such as Shark Bay in Western Australia, where these came from.

Reading 35.1 The Search for Mitochondrial "Eve"

A CLOCK MEASURES THE PASSAGE OF TIME BY MOVING ITS HANDS THROUGH A CERTAIN DEGREE OF A CIRCLE IN A SPECIFIC AND CONSTANT INTERVAL OF TIME—A SECOND OR MINUTE. Similarly, a polymeric molecule can be used as a clock if its building blocks are replaced at a known and constant rate. DNA from a cell's nucleus is used as a molecular clock because the rate of base change mutation is known, and a ratio can be set up using the number of base differences in the nuclear DNA of two organisms to estimate passage of time.

DNA in mitochondria (mtDNA) can also be used as a molecular clock. Recall that mitochondria are organelles that house the reactions of cellular respiration. Mitochondria in human cells contain 16,569 base pairs. This genetic material is sometimes referred to as our 24th chromosome. DNA in the mitochondria is valuable to the charting of evolutionary time because this clock ticks 5 to 10 times faster than the nuclear DNA clock—that is, mtDNA mutates this much faster than nuclear DNA. The mtDNA clock can be used to chart more recent evolutionary events.

Like other measures of evolutionary relatedness, the more similar the mtDNA sequence is between two individuals, the more recently they are assumed to be descended from a common ancestor. But following the inheritance of mtDNA presents an interesting quirk, because mtDNA is maternally inherited. That is, mitochondria are passed from mothers only and received by egg cells when they form in meiosis. The part of the sperm cell that enters the egg at fertilization, in contrast, usually does not contain any mitochondria. Theoretically, if a particular sequence of mtDNA can be identified that could have given rise, by mutation, to mtDNA sequences in modern humans, then that ancestral sequence may represent a very early human or humanlike female—a mitochondrial "Eve," or first woman.

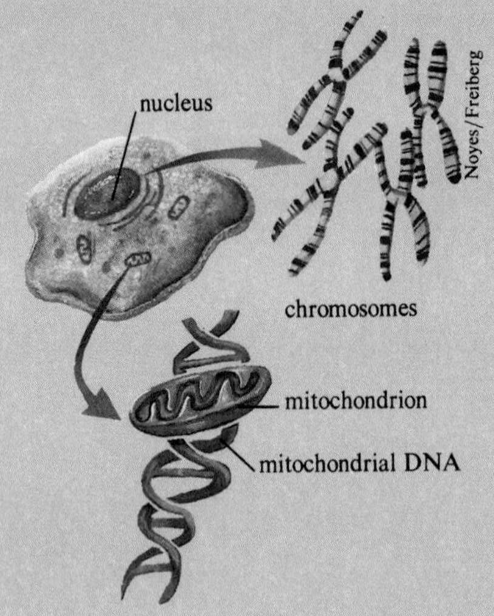

Figure 1
The human mitochondrial genome consists of about 17,000 base pairs.

The search for mitochondrial Eve began in 1986, when Wesley Brown, a graduate student at the University of California at Berkeley, cut mtDNA from 21 living people of diverse ethnic backgrounds with a variety of restriction enzymes. A particular restriction enzyme cuts DNA at a specific base sequence. If the DNA of two individuals differs at a restriction site, and the DNA is cut with an enzyme that cuts at that site, then their DNA will be snipped into a different number of pieces. The similarity of restriction fragment patterns, then, is a measure of relatedness (time of descent from a shared ancestor).

Wesley Brown found that his 21 subjects had very similar mtDNA patterns. The next question—how long would it have taken the differences he did see to accumulate, assuming all the mtDNA types were descended from a single ancestral sequence? By multiplying the rate of mtDNA mutation by the average number of sequence differences between the people's mtDNA, Brown estimated that the ancestral mtDNA came from a woman (or small group of women) who lived about 200,000 years ago.

Where might this figurative first woman have dwelled? Most fossil evidence points to Africa, but some anthropologists maintain that modern peoples came from Asia as well. To answer this question, Berkeley researchers led by Allan Wilson cut mtDNA from 147 people from Africa, Asia, Australia, New Guinea, and Europe with a dozen restriction enzymes. The 133 people who had different mtDNA patterns fell into two groups of similar sequences. One group consisted of all Africans. Because this group had greater sequence diversity within it, the African group is the older of the two. More time must have elapsed from when the ancestral mtDNA existed to produce the present African population than the non-African group. The prevailing view of an African origin is therefore supported by the mtDNA clock.

breakdown products of chlorophyll, suggesting that microbes of the time may have photosynthesized. The Gunflint rock formation from the north shore of Lake Superior in Ontario, Canada, has yielded a diverse assortment of Precambrian life forms, including cyanobacteria, early eukaryotes, red algae, and fungi. A billion years ago algae and fungi left their mark in the Bitter Springs formation in Australia.

Between 700 and 600 million years ago, a profound biological change took place—multicellular life appeared. Signs of worms, jellyfish, and soft corals are evident in the rocks of southern Australia and the Canadian Arctic. Exactly how life proceeded from the single-celled to the many-celled is a mystery. Unicellular organisms may have formed colonies and then undergone a division of labor, with certain cells specializing in certain functions, to eventually form a coordinated multicellular organism. Alternatively, an ancestral, large, single-celled organism may have developed partitions and then undergone division of labor.

The Paleozoic Era

Abundant evidence of life appears quite suddenly in the fossil record of the **Cambrian period** about 600 million years ago—a time quite aptly called the Cambrian explosion. Ancestors of all modern animal phyla debuted in the Cambrian period, plus many now-extinct life forms. Although confined to the seas, life was diverse and plentiful. The ancient oceans were home to abundant soft-bodied organisms, including algae, sponges, jellyfish, and worms, and the earliest known

a.

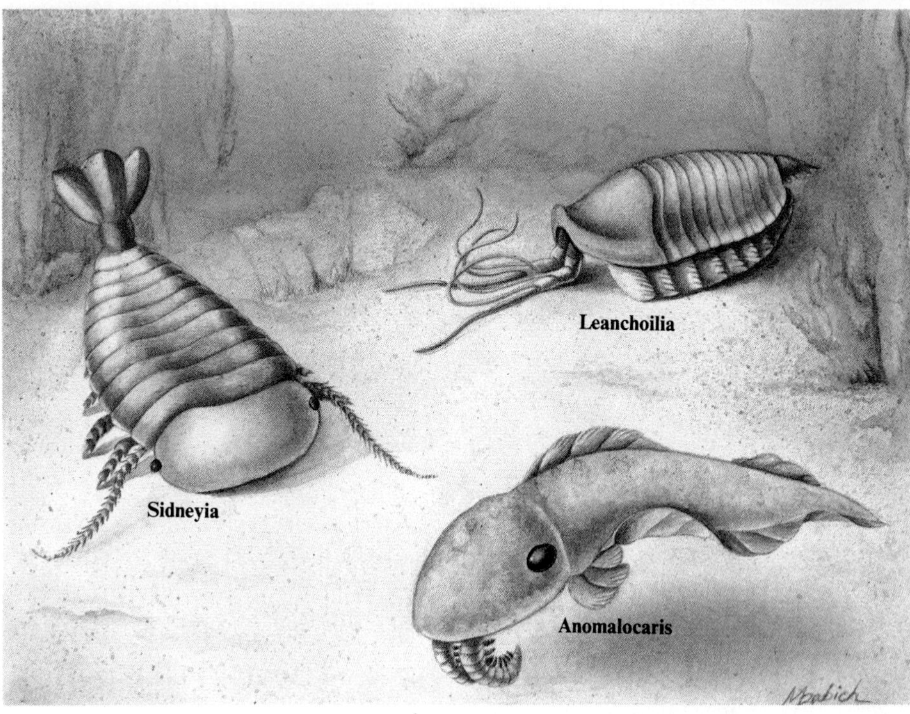

b.

Figure 35.12
The Cambrian seas were home to an explosion of life forms. *a.* This piece of shale was once part of the bottom of a Cambrian period sea. These trilobites were distant forerunners of arthropods, and they ranged in size from that of a pea to an automobile. They left behind many fossils because of their abundance, wide geographical distribution, and hard parts. *b.* The Burgess shale in British Columbia revealed fossils of many soft-bodied invertebrates not seen anywhere else in the world. They were apparently fossilized when they were buried by a mudslide.

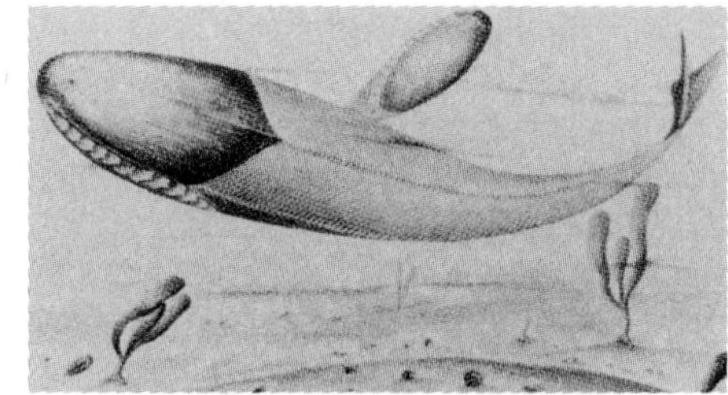

Figure 35.13
The oldest vertebrate fossils are from the Ordovician period. Several complete fossils of jawless fishes dating back to this period were recently found in Bolivia. Shown here is the reconstruction.

organisms with hard parts, such as shellfish, the insectlike **trilobites, nautiloids,** scorpionlike **eurypterids,** and **brachiopods,** which looked like clams and left many fossils (fig. 35.12). The biological blooming at this time is a major puzzlement in the study of life history.

The **Ordovician period** followed the Cambrian 500 to 435 million years ago. The seas continued to support vast communities of algae and invertebrates. Organisms abundant at this time were the **graptolites,** so-named because their fossilized remains resemble pencil markings. The first vertebrates to leave fossils date from the Ordovician. In 1988, complete fossils of 10 jawless fishes dating from 470 million years ago were found in Bolivia, which was then under an ocean (fig. 35.13). These very early animals with backbones were about 18 inches (46 centimeters) long and 6 inches (15 centimeters) wide and were covered by small scales and had bony plates on their round heads.

At the end of the Ordovician period and during the **Silurian period** (435 to 395 million years ago), organisms first ventured onto the land. These pioneers were odd-looking plants called **psilophytes.** They had bare stems from which spores were scattered and underground branches but no leaves or roots. The first animals to leave fossils on the land resembled scorpions.

The **Devonian period** of 395 to 345 million years ago was the "Age of Fishes." The seas continued to support most life, including the now well-established invertebrates, as well as fishes with skeletons built of cartilage or bone. Corals were abundant, as well as animals called crinoids that resembled flowers. The land was still relatively barren, home to scorpions and millipedes.

The **crossopterygians,** or lobe-finned fishes, lived in the Devonian period and were probably ancestral to amphibians. They could gulp air and use its oxygen, and fleshy fins allowed them to haul themselves along the land that separated shallow pools (fig. 35.14). These adaptations probably appeared by chance, but once organisms could spend extended periods of time on land, they thrived in the presence of vast, unused resources. Gradually amphibians evolved as those transitional fishes with the strongest fleshy fins and best ability to breathe air were naturally selected.

By 345 million years ago, the beginning of the **Carboniferous period,** or "Age of Amphibians," the descendants of the lobe-finned fishes were prominent on the land. Like modern amphibians, these animals were

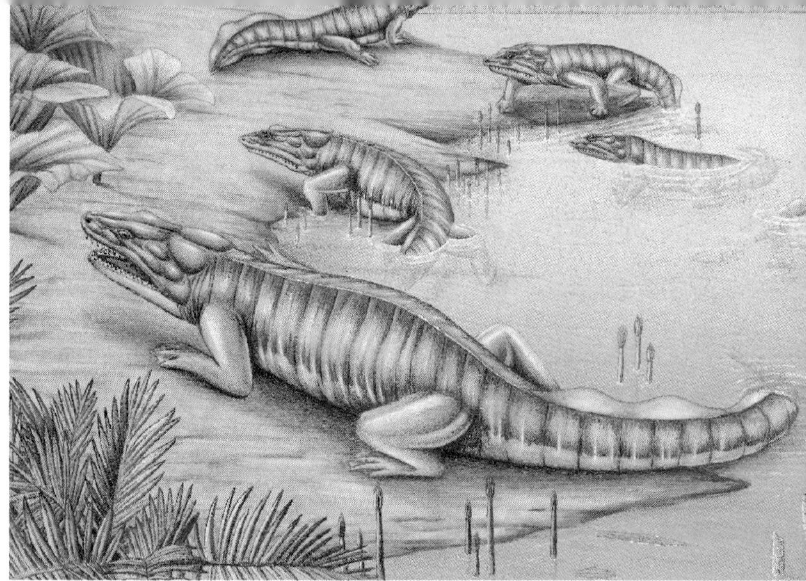

Figure 35.14
From fish to amphibian—how it might have happened. The lobe-finned fish called crossopterygians lived in the Devonian period and had fleshy, powerful fins and the ability to breathe air. These adaptations allowed them to survive on land for short periods of time. By the late Carboniferous period, primitive amphibians appeared that may have descended from the lobe-finned fishes.

a.

b.

Figure 35.15
a. The great forests of the Coal Age. The forests of the Carboniferous period were spectacular and so lush that their remains were preserved in massive coal beds. The fernlike plants in the foreground are very ancient seed-bearing gymnosperms. The thin tree on the right is an extinct ancestor of the modern horsetail, and the thicker trunks on the left, also extinct, gave rise to modern club mosses and ground pines. *b.* In Indiana and Illinois, a piece of coal will occasionally contain the imprint of a fern.

not entirely land-dwelling. Their legs were not strong enough to support full-time walking, their skin needed frequent moisturizing, and their eggs had to be laid in the water in order for the developing embryos to receive sufficient moisture and nutrients. Toward the end of the Carboniferous period about 275 million years ago, some amphibians had evolved eggs that contained water and nutrients and could be laid on land. Fossil evidence of these early reptiles has not been found, but their existence is inferred from organisms appearing over the following several million years that had traits of both amphibians and reptiles.

When the amphibians of the Carboniferous period crawled onto the swampy land, they must have seen fernlike plants at eye level and majestic conifers towering 130 feet (40 meters) above. By the end of the period, many of these plants had died, buried beneath the swamps to form, over the coming millennia, coal beds (fig. 35.15). Because of these bountiful reminders of the ancient first forests, the Carboniferous period is also called the "Coal Age." Today, a piece of coal split in half will sometimes reveal a perfect impression of an ancient fern.

The Paleozoic era ended rather dramatically with the **Permian period** (275 to 225 million years ago), when half the families of vertebrates and more than 90% of species dwelling in the shallow seas became extinct. Permian lands still supported amphibians, but reptiles were becoming more prevalent.

The reptile introduced a new biological structure, the **amniote egg,** in which an embryo could develop completely, without the need to be laid in water. An amniote egg contains nutrient-rich yolk and extraembryonic membranes, which protect and help to nourish the embryo. The amnion encloses the embryo in a sac of fluid that is chemically very similar to seawater, which nurtured so many early life forms. The amniote egg's prepackaged nutrient stores and protection proved so successful that it persists today in the eggs of reptiles, birds, and mammals.

Cotylosaurs were early Permian reptiles that gave rise to the dinosaurs, as well as to modern reptiles, birds, and mammals. They coexisted with their immediate descendants, the **pelycosaurs,** or sailed lizards. The pelycosaurs were distant ancestors of mammals.

KEY CONCEPTS

The geological time scale is divided into eras, periods, and epochs. Most fossils date since the Cambrian period of 600 to 700 million years ago, although the Precambrian period accounts for five-sixths of the planet's existence. The earliest Precambrian fossils are remains of vast bacterial colonies. Life exploded in the Cambrian period as multicellular forms, including many invertebrates, appeared, flourished, and diversified in the seas. During the following Ordovician period, vertebrates joined the invertebrates and algae. Life came to the land during the Silurian period. Bony fishes evolved during the Devonian period, including a forerunner of amphibians. During the Carboniferous period amphibians and the first reptiles dwelled in forests of ferns and conifers. Reptiles were well established by the Permian period.

The Mesozoic Era

When the Age of Reptiles dawned during the **Triassic period** 225 million years ago, small ancestors of the great dinosaurs flourished. These were the **thecodonts,** descendants of the Permian cotylosaurs. Fossil evidence of thecodonts is scant. One of the earliest known is Staurikosaurus, which stood about 6.5 feet (2 meters) tall and weighed about 70 pounds (32 kilograms). It stood upright, with short forearms and strong hind limbs, and held its tail out, much like the carnivorous dinosaurs that would soon appear.

The thecodonts shared the forest of seed-forming gymnosperms (cycads, ginkgos, and conifers) with other animals called **therapsids.** These were reptiles, but they held their limbs and heads in a position more like those of mammals, and their teeth were more mammalian than reptilian (fig. 35.16). By the end of the Mesozoic era, the therapsids would evolve into small, hair-covered animals, most of whom lived on the forest floor—the first mammals. At the close of the Triassic period, about 190 to 185 million years ago, the numbers of both thecodonts and therapsids were dwindling as much larger animals began to infiltrate a wide range of habitats. These new, well-adapted animals were the dinosaurs, and they would dominate for the next 120 million years. Although the dinosaurs have been traditionally classified as reptiles, evidence is accumulating that they may actually be most closely related to birds (Reading 35.2).

By the **Jurassic period** of 185 to 135 million years ago, the dinosaurs had invaded nearly all habitats, from the **ichthyosaurs** in the seas, to **archaeopteryx** in the air, to the familiar **apatosaurs** (brontosaurs), **stegosaurs, diplodocus,** and **allosaurs** that dwelled in the second half of the period. The first flowering plants (angiosperms) appeared, but forests still consisted mostly of tall ferns and conifers, club mosses, and horsetails.

The **Cretaceous period** (135 to 65 million years ago) was a time of great biological change. Angiosperms spread in spectacular diversity, and the number of dinosaur species declined. Although this was the beginning of the end for the dinosaurs, many species soared to quite healthy numbers. Duck-billed **maiasaurs** traveled in herds of thousands in what is now Wyoming. The plains of Alberta supported huge herds of apatosaurs that migrated to the Arctic, northern Europe, and Asia, landmasses that were then connected. By the end of the period, triceratops were so widespread that some paleontologists call them the "cockroaches of the Cretaceous."

The reign of the great dinosaurs ended about 65 million years ago, and each paleontologist seems to have a favorite theory as to how it happened (chapter 34). Whatever sparked the mass extinctions that marked the end of the Mesozoic era, it is clear that the demise of the dinosaurs opened up habitats for many other species, including the primates that eventually gave rise to our own species.

Figure 35.16
The dawn of the Mesozoic era saw many small animals that had characteristics of reptiles and mammals.

KEY CONCEPTS

By 225 million years ago, during the Triassic period, thecodonts shared the gymnosperm forests with therapsids, ancestors of mammals. During the Jurassic period, dinosaurs reigned and angiosperms debuted. During the Cretaceous period, some dinosaur species declined while others rose, and the number of angiosperm species continued to increase. At the end of the Mesozoic era, many species became extinct.

Reading 35.2 A New View of Dinosaurs

Figure 1
Dinosaur families on egg mountain.

About 80 million years ago, a seaway stretching from the Beaufort Sea in northern Canada to the Gulf of Mexico divided North America. The Rockies jutted up from western America; the Appalachians formed the highlands of the east. Between the Rockies and the mid-American sea lay a coastal plain, rich in sediments washed down from the mountains.

Egg Mountain was a hill about 150 feet (46 meters) high in the middle of a seasonal lake in what is now Choteau, Montana. In 1979, paleontologist John Horner, curator of the Museum of the Rockies in Bozeman, Montana, found in this hill a most spectacular fossil assemblage—nine dinosaur nests, complete with eggs and bones from both adult and juvenile dinosaurs. They were part of a huge dinosaur community preserved in time by a sudden volcanic eruption.

Many of the residents of Egg Mountain were **maiasaurs,** duck-billed dinosaurs that stood about 20 feet (6 meters) high and resembled ostriches. The eggs seemed to have been carefully placed in the nests so that they did not touch one another, deposited pointed-end down. The exposed ends were covered with vegetation. The tops were missing from some of the eggs, and nearby were found bones of baby dinosaurs. The presence of beetle egg cases, fossilized cocoons, and what Horner believes is "dinosaur upchuck" suggests that a parent dinosaur fed the young. Grooves in the teeth of the young indicate that they ate seeds and berries. Horner X-rayed the unbroken eggs and found dinosaur embryos, "the flesh rotted around prepackaged bundles of bone." The scene looked very much like groupings of families, with carefully guarded nests and adults who nurtured their young. It was a lifestyle more like that of birds than reptiles.

The dinosaurs resembled birds in their anatomy and physiology as well. The dinosaur egg shells look like those of birds when viewed under a scanning electron microscope. The fast rate of growth and the blood vessel pattern of maiasaurs are more like those of birds than reptiles, findings consistent with the theory of Robert Bakker, curator of the University of Colorado Museum, that dinosaurs were warm-blooded (endothermic), like modern birds and mammals. Bakker examined the spacing between dinosaur footprints, the length of their leg bones, and the structure of the feet and hypothesized that dinosaurs were very active and agile migrants.

This new view of the dinosaur as a swift, warm-blooded social animal led to a reexamination of the evidence that had labeled the dinosaurs as reptiles. The first dinosaur fossils were footprints found in England in 1802, but they were dismissed as the tracks of giant turkeys. Then in 1822, a woman accompanying her physician and fossil-collector husband on a house call in rural England spied a large tooth jutting out of a rock. The tooth reminded the doctor of a tooth from an iguana lizard, so he named the animal that once possessed it Iguanodon. He thought it was from a giant lizard.

By the 1840s, other large fossils had been found. French anatomist George Cuvier concluded that the fossils were from animals that were about the size of rhinoceroses but resembled reptiles. British anatomist Richard Owen coined the term "Dinosauria," meaning "terrible lizards." In museums, curators arranged fossils into typical reptilian poses, the front ends spread out and bent at the elbows like lizards, the hind legs together, and the tails dragging behind.

In the 1960s, Yale University paleontologist John Ostrom ssuggested that fossilized dinosaur foot bones more closely resembled feet of an agile, predatory bird than a plodding lizard. Ostrom's work sent Bakker in search of additional evidence among fossils, and Bakker's controversial view of the dinosaur helped send Horner to the ancient coastal plain of Montana, and the tribute to the dinosaurian way of life revealed at Egg Mountain.

The Cenozoic Era

At the start of the **Cenozoic era** 65 million years ago, a great variety of hoofed mammals grazed upon the grassy Americas. Many were **marsupials** (pouched mammals) or egg-laying **monotremes,** ancestors of the platypus. Both marsupials and monotremes were more prevalent than they are today. Young marsupials and monotremes were quite helpless. The tiny, hairless, and blind offspring of marsupials, for example, crawled from the mother's reproductive tract along her fur to her chest, where they drank from tiny nipples leading to sweat glands modified to secrete milk.

Gradually a new type of mammal evolved whose young were better protected and therefore more likely to survive, reproduce, and perpetuate their species. These newcomers were the **placental mammals.** The young remain within the female's body for a relatively long time, where they are nurtured by a specialized organ, the placenta. Placental mammals have well-developed mammary glands.

The placental mammals eventually replaced most of the marsupials in North America. Changing geography had much to do with their invasion of South America. Until 2 to 3 million years ago, South America was an island continent. Throughout the Tertiary period, several orders of large marsupials thrived there because the placental mammals, which had originated in North America, could not reach the enormous southern island. (Today placental mammals brought to Australia by human settlers threaten several marsupial species.)

Then about 2 to 3 million years ago, the Bering land bridge rose, connecting Asia to North America, and many mammals, including our ancestors, probably journeyed from what is now the Soviet Union to Alaska and southward through what is now Canada and the continental United States. The isthmus of Panama formed, providing a passageway for the placental mammals of the north to invade the marsupial communities of the south. The first animals to arrive, rodents and ground sloths, crossed when the land bridge was only a string of islands. The overwhelming majority of South American marsupials (23 of 25 orders) were driven to extinction by the arrival of the placental mammals, who successfully competed for the same resources. Many of the species found in South America today, including peccaries, llamas, alpacas, deer, tapirs, and jaguars, are descendants of immigrants from the north. A few species, including the opossum, armadillo, and porcupine, traveled in the opposite direction (fig. 35.17).

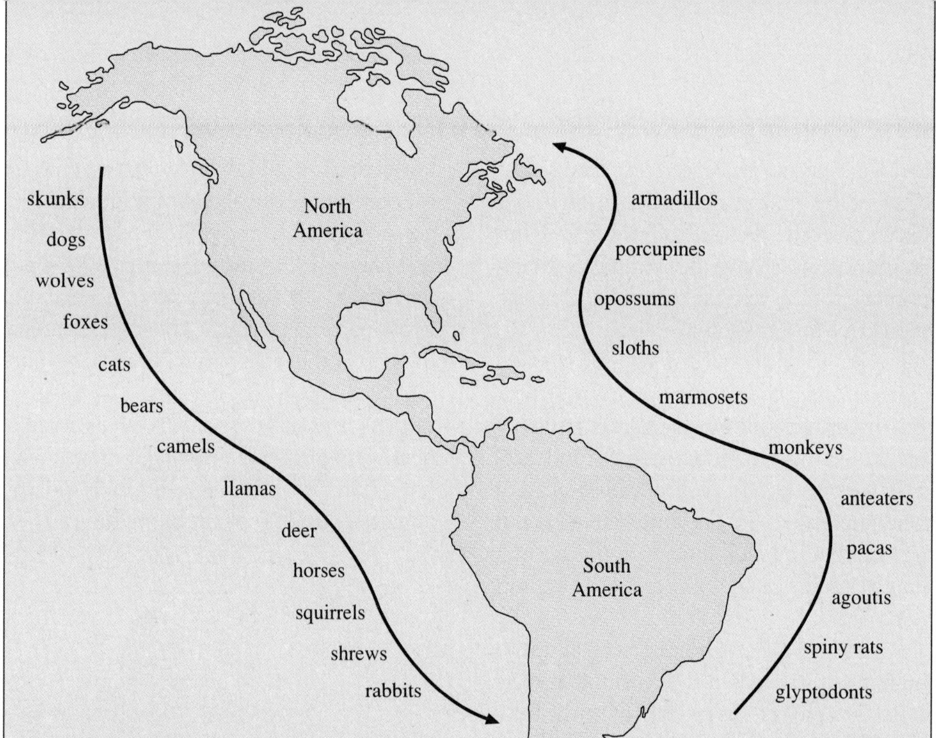

Figure 35.17

Geographical changes alter species distributions. About 3 million years ago, the appearance of the Panama land bridge between the separate continents of North and South America provided a route for the migration of animals between the two huge landmasses. Many species wandered in both directions. The southward invasion of placental mammals drove to extinction many marsupial species living in the former island continent.

KEY CONCEPTS

The Cenozoic era has been dominated by mammals, first by marsupials and monotremes and then by placental mammals. Migrations as land bridges formed altered the geographical distribution of species and led to extinctions as arriving species outcompeted resident ones.

The Evolution of Humans

Our species, *Homo sapiens* (the wise human), probably first appeared during the Pleistocene epoch, about 200,000 years ago. Our ancestry reaches farther back within the order Primates, to the early Tertiary period, about 60 million years ago. The ancestral primates were rodentlike insectivores. Like many mammals, these first primates underwent extensive adaptive radiation. Their ability to grasp and to perceive depth provided the flexibility and coordination needed to dominate the treetops.

About 30 to 40 million years ago, a monkeylike animal about the size of a cat, *Aegyptopithecus,* dwelled in the lush tropical forests of Africa. Although most of the animal's time was probably spent in the trees, fossilized remains of limb bones indicate that it could run on the ground as well. Fossils of different individuals have been found together, indicating that they were social animals. *Aegyptopithecus* had fangs that might have been used for defense, and the large canine teeth seen only in males suggest that males might have provided food for their smaller female mates. *Propliopithecus* was a monkeylike contemporary of *Aegyptopithecus*. Both animals are possible ancestors of gibbons, apes, and humans.

About 20 to 30 million years ago, Africa was inhabited by the first **hominoids,** animals ancestral to only apes and humans. This primate was called *Dryopithecus,* meaning "oak ape," because fossilized bones were found with oak leaves (fig. 35.18). The way that the bones fit together suggests that this animal lived in the trees but could swing and walk farther than *Aegyptopithecus*. Because of the large primate population in the forests, selective pressure to venture onto the grasslands must have been intense. Many primate species probably vanished as the protective forests shrank.

By 15 million years ago, an apelike animal, *Ramapithecus*, had evolved in Africa. Fossilized jawbones and teeth are remarkably humanlike. The small canine teeth and massive molars suggest that *Ramapithecus* ate hard nuts and seeds. The animal lived in the treetops but traveled and ate on the ground. *Ramapithecus* became extinct about 7 million years ago.

Hominoid and **hominid** (ancestral to humans only) fossils from 4 to 8 million years ago are scarce. This is thought to be the time when the stooped, large-brained ape gradually became the upright, smaller-brained ape-human. By 4 million years ago, an animal known as *Australopithecus* had ventured forth from the treetops of the shrinking African forests to walk on his knuckles and use tools. The angle of preserved pelvic bones, plus the fact that *Australopithecus* fossils have been found with those of animals that grazed, indicate that this ape-human had left the forest. *Australopithecus* stood about 4 or 5 feet (1.2 to 1.5 meters) tall and had a brain about the size of a gorilla's but teeth that were very much like those of humans (fig. 35.19*a*).

Four species of *Australopithecus* existed by 3.6 million years ago, and paleontologists are still not certain how they are related to one another. A form called *A. afarensis*, represented by the famous fossil "Lucy" (fig. 35.19*b*), may have been our direct ancestor. Her skull was shaped more like that of a human, with a less prominent face and larger brain than her predecessors. These animals probably lived a hunter-gatherer life-style during the Pleistocene ice age.

By 2 million years ago, *Australopithecus* coexisted with *Homo habilis*, a more humanlike primate who lived communally in caves and cared for young intensely. "Habilis" means handy, and this primate is the first for whom we have evidence of extensive tool use. *H. habilis* may have descended from a group of *Australopithecines*, who ate a greater variety of foods than other ape-humans, allowing them to adapt to a wider range of habitats.

H. habilis coexisted with and was followed by *H. erectus*, who left fossil evidence of cooperation, social organization, and tool use, including the use of fire. Fossilized teeth and jaws of *H. erectus* suggest that he was a meat eater. The fossils are widespread, found in China, tropical Africa, and southeast Asia, indicating that these animals could migrate farther than earlier primates. The distribution of fossils suggests that *H. erectus* lived in families of male-female pairs (most primates have harems). The male hunted and the female nurtured young for long times.

H. erectus lived from the end of the Australopithecine reign some 1.6 million years ago to as recently as 200,000 years ago and probably coexisted for a time with the very first of our own species. Some paleontologists believe that a species intermediate between *H. erectus* and *H. sapiens*, with a big brain and robust build, lived from 50,000 to 30,000 years ago. Several pockets of ancient peoples may have been dispersed throughout the world at this time. Such fossils have been found in Swanscombe, England, Steinheim, Germany, and sites in the Middle East.

a.

b.

c.

d.

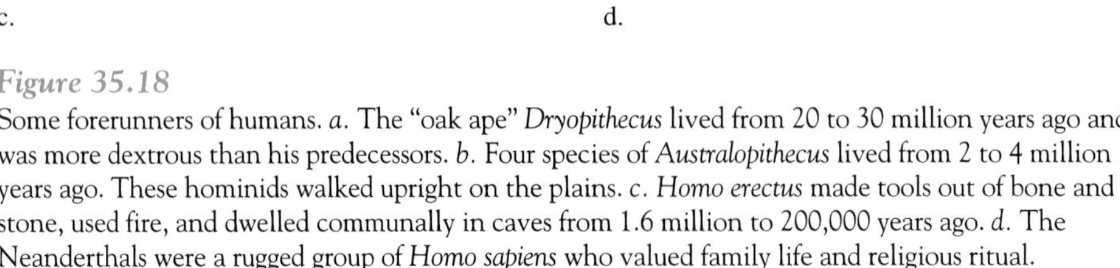

Figure 35.18

Some forerunners of humans. *a*. The "oak ape" *Dryopithecus* lived from 20 to 30 million years ago and was more dextrous than his predecessors. *b*. Four species of *Australopithecus* lived from 2 to 4 million years ago. These hominids walked upright on the plains. *c*. *Homo erectus* made tools out of bone and stone, used fire, and dwelled communally in caves from 1.6 million to 200,000 years ago. *d*. The Neanderthals were a rugged group of *Homo sapiens* who valued family life and religious ritual.

About 75,000 years ago, Europe and Asia were home to the **Neanderthals,** members of *Homo sapiens*. A fossilized crippled skeleton buried with flowers in Shanidar Cave in Iraq reveals that the Neanderthals may have been religious hunter-gatherers clever enough (or lucky enough) to have survived a brutal ice age. The Neanderthals had once been thought to be hunched brutes because the first fossil was from an individual stooped from arthritis. So closely related are the Neanderthals to us that some of them may have been physically indistinguishable from ourselves.

The Neanderthals mysteriously vanished about 40,000 years ago, just as the lighter-weight, finer-boned, and less hairy **Cro-Magnons,** known for their intricate cave art, appeared. The Cro-Magnons may have arisen from a small group of individuals who happened, perhaps by chance mutations, to have high foreheads over well-developed frontal brain regions. It was this group that ultimately flourished and led, over the millennia, to our own kind. The first fossils of modern humans date from about 40,000 years ago.

Telling the story of life has traditionally depended upon the scattered and woefully inadequate clues in rocks. With the new views being offered by physical dating techniques and molecular measurements, scientists will be increasingly able to fill in more of the missing pages of natural history. We will learn more about where we came from, as well as what other organisms have dwelled on this planet in ages past. But perhaps the most compelling questions lie ahead—where will the course of evolution take life on earth in the future?

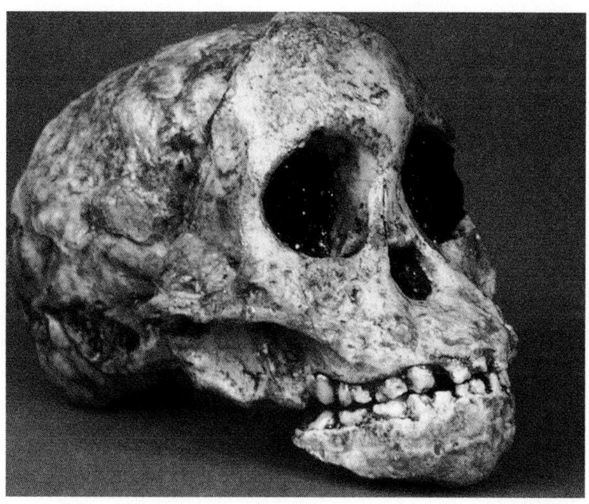

a.

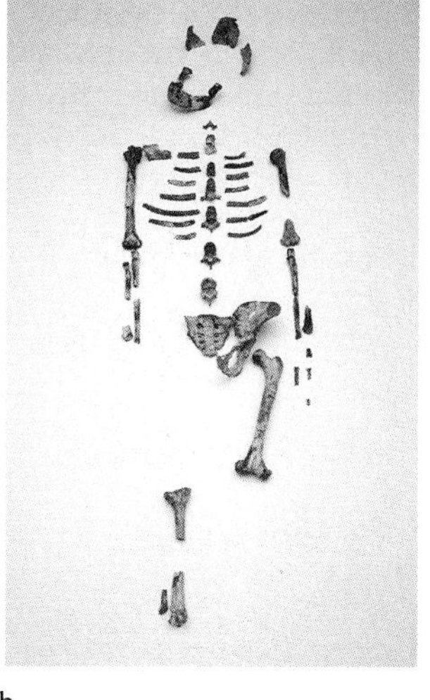

b.

Figure 35.19

Fossils of two *Australopithecines*—the Taung child and Lucy. In 1925 Australian anatomist Raymond Dart discovered the remarkably well-preserved skull of an ancient child in an outcropping of limestone in the South African village of Taung. The fossil was about 2 million years old; the child it came from was about 6 years old when she died. The skull had several humanlike characteristics—lack of brow ridges, an upright vertebral column suggesting the child may have walked erect, and teeth shaped like our own. However, in 1987 researchers at the Washington University in St. Louis used a CT scan (a multidirectional X-ray) on the skull, which revealed the order in which the different teeth had erupted from the jaw. The dental pattern was more like that of apes than humans. Tooth eruption is an important clue to evolutionary relationships because the slower the teeth appear, the longer the period of infancy. Humans have much longer infancies than apes, and many anthropologists attribute our intelligence to this longer nurturing period. The conclusion about the Taung child? Like many early hominids, she is a little of both—ape and human. *b.* About 3.6 million years ago, a small-brained ancestor of the human walked upright in the grasses along a lake in the Afar region of Ethiopia. She skimmed the shores for crabs, turtles, and crocodile eggs to eat. She died at the age of 20, with severe arthritis in the backbone. Nearly 40% of her skeleton was discovered in 1974, and she was named "Lucy," from the Beatles song "Lucy in the Sky with Diamonds," which Donald Johanson of the Cleveland Museum of Natural History and Timothy White of the University of California at Berkeley were listening to when they found her.

KEY CONCEPTS

Aegyptopithecus lived about 30 to 40 million years ago and was a monkeylike animal ancestral to gibbons, apes, and humans. The first hominoid, Dryopithecus, dwelled 20 to 30 million years ago and may have walked onto grasslands. Ramapithecus lived 15 million years ago and had humanlike teeth and jaws. He lived in the trees but ate on the ground. Australopithecus lived about 4 million years ago, walked on his knuckles, and used tools. By 2 million years ago, Australopithecus coexisted with Homo habilis, who was more humanlike in appearance and social structure. Later on, H. Habilis coexisted with H. erectus, who used tools more extensively and was more social. H. sapiens coexisted with or arose from H. erectus about 75,000 years ago. The Neanderthals preceded the Cro-Magnons, and, about 40,000 years ago, modern humans appeared.

SUMMARY

Evidence for evolution comes directly from extinct organisms and from observation of similarities among modern species. A *fossil* may be once-living matter that has been replaced by mineral, formed gradually or after a sudden catastrophe, or indirect evidence such as footprints. Rarely, actual remains are preserved in ice. Microscopes reveal cellular details in fossils. The rock layer in which a fossil is found provides a *relative date*. The proportion of a stable radioactive isotope to its breakdown product and calculating from this how many *half-lives* have passed provides an *absolute date*. Potassium 40 is used to date fossils older than 300,000 years, and carbon 14 dating is used on fossils younger than 40,000 years. New dating techniques are being developed. The fossil record is incomplete. Geographical upheavals destroy much evidence, and some species existed for too short a time or in too restricted an area to leave many fossils.

Homologous structures are similar in architecture and function in two species because they are inherited from a common ancestor.

Analogous structures reflect a common response by unrelated species to the same environmental challenge. *Vestigial* organs and similar prenatal structures may reflect genes retained from an ancestor.

Techniques of *molecular evolution* include comparing chromosome band patterns, protein, and gene sequences. Each of these consists of many bits of information, and it is far more likely that two species are alike by common descent than by chance.

The *geological time scale* charts earth history, divided into eras, periods, and epochs. According to the fossil record, life debuted in *Precambrian* time and exploded in diversity in the *Cambrian period,* about 600 million years ago. The *Paleozoic* oceans were home to diverse algae, invertebrates, and fishes. Plants appeared on land during the *Ordovician* and *Silurian* periods, and the lobe-finned fishes, forerunners of amphibians, conquered the land for parts of their life cycles during the *Devonian* period. By the *Carboniferous* period, the land was lush with conifers and fernlike plants, many now extinct. Terrestrial animal life was still scarce. The evolution of the amniote egg allowed animals to live on land for the entire life cycle. By the *Permian* period, the first reptiles had appeared.

The *Mesozoic era* was the age of reptiles. *Cotylosaurs* gave rise to dinosaurs, modern reptiles, and *therapsids,* forerunners of mammals. The dinosaurs radiated during the *Triassic* and *Jurassic* periods, then began to decline in numbers during the *Cretaceous* period, when flowering plants flourished. The mass extinctions of dinosaurs at the end of the Cretaceous period, about 65 million years ago, opened up many habitats for mammals, which flourished during the *Tertiary period* of the *Cenozoic era.* South America was then home to many marsupial species that were driven to extinction when the isthmus of Panama formed, which allowed the placental mammals of the north to invade the south.

The first primate ancestral to humans, an insectivore, lived about 60 million years ago in Africa. By 40 million years ago, *Aegyptopithecus* had evolved, a monkeylike social animal. By 30 million years ago, the *hominoid Dryopithecus* had appeared. *Ramapithecus* fossils date from 15 million years ago. This apelike animal had humanlike teeth and jaws. *Australopithecus,* the first *hominid,* lived from about 4 million to 2 million years ago. *Homo habilis* coexisted with *Australopithecus.* Later, *Homo erectus* displayed a more complex social structure. Between 500,000 and 75,000 years ago, the first *Homo sapiens* appeared. Groups of early humans included the *Neanderthals* and *Cro-Magnons.* Fossils of modern humans date from about 40,000 years ago.

QUESTIONS

1. Why are a preserved dinosaur bone, a ripple in a rock made by an ancient worm's movements, limestone built from the shells of crustaceans, and insects preserved in amber all fossils?

2. Why don't we have evidence of all extinct forms of life?

3. What type of absolute dating would be useful in dating a fossil of a lobe-finned fish?

4. An elephant uses its trunk to bring food to its mouth; a human uses his hand to do this. Are an elephant's trunk and a human's hand analogous or homologous structures?

TO THINK ABOUT

1. Cytochrome C is a protein that is very similar in all species. Twenty of its 104 amino acids are identical in all eukaryotes. How does the fact that differences among these 20 amino acids are not seen demonstrate that natural selection is operating at the molecular level?

2. What information in this chapter supports the idea, expressed in chapter 1, that the scientific method is a cycle of inquiry?

3. What assumptions underlie the following:
   a. Relative dating by placement in rock strata
   b. Absolute dating by radiometric data
   c. The mitochondrial clock

4. Fossils of horn corals, brachiopods, and trilobites are abundant in Ohio. How can fossils of sea life be found in the middle of a continent?

5. What may have happened in human evolution between 4 and 8 million years ago?

6. How were the development of the following structures or abilities adaptive, and how did they affect evolution?
   a. Strong, fleshy fins in the lobe-finned fish
   b. The amniote egg
   c. The placenta
   d. The ability of primates to walk on the ground

7. What environmental conditions of today might preserve organisms to form the fossils of tomorrow?

SUGGESTED READINGS

Cohn, Jeffrey P. March 1990. Genetics for wildlife conservation. *BioScience*. Molecular techniques reveal the likely lineage of the giant panda.

Gibbons, Ann. February 22, 1991. Systematics goes molecular. *Science*, vol. 251. Molecular analysis is increasingly used to study evolution.

Gould, Stephen Jay. 1989. *Wonderful life*. New York: W.W. Norton. The Burgess shale reveals a slice of Cambrian life.

Horner, John R., and David B. Weishampel. December 1989. Dinosaur eggs: The inside story. *Natural History*. Were dinosaurs birdlike, caring parents?

Johanson, Donald, and James Shreeve. 1989. *Lucy's child*. New York: William Morrow. How paleontologists pieced together our beginnings from scattered 2-million-year-old preserved bones.

Lewis, Ricki. April 1991. An intron in cyanobacteria stirs evolutionary debate. *The Journal of NIH Research*. Clues in modern gene sequences suggest a world based on RNA led to life.

Marshall, Eliot. February 16, 1990. Paleoanthropology gets physical. *Science*, vol. 247. Three new methods may help assign dates to fossils.

Monastersky, Richard. September 22, 1990. Swamped by climate change? *Science News*. The earth was covered in wetlands 315 million years ago.

Palca, Joseph. September 7, 1990. The human genome. *Science*, vol. 249. Mitochondrial genes may hold clues to human disease and origins.

Stolzenberg, William. August 25, 1990. When life got hard. *Science News*. Diverse exoskeletons evolved rapidly in the Cambrian period.

UNIT

8

Behavior and Ecology

C H A P T E R

36

The Behavior of Individuals

Chapter Outline

Learning Objectives

By the chapter's end, you should be able to answer these questions:

1. How do biologists study behavior?
2. How do we know the extent to which a behavior is controlled by heredity or the environment?
3. Why is behavior adaptive?
4. What are the different forms of learning?
5. How do biological rhythms influence behavior?
6. What environmental cues do animals use to navigate?

The ground squirrel sensed danger as soon as she neared her burrow and saw the frenzied tail-flagging and darting about of her neighbors. She had been out foraging for food, leaving behind three pups. The squirrels' behavior sent an unmistakable message—rattlesnake!

A newborn squirrel could be trapped and eaten by a snake commandeering a burrow to keep itself warm. But this squirrel's pups were standing their ground well. Although they were too young to have encountered a rattlesnake before, or to have had their mother teach them how to defend themselves, the pups instinctively knew just what to do. They flagged their tails, pushed out their snouts, jumped back and forth as if anticipating a snake's pounce, and vigorously kicked dirt and sand at the reptile. The arrival of the mother, adding to the ruckus, sent the snake slithering into the temporarily abandoned burrow. Mother and pups immediately kicked sand and dirt into the burrow. They were safe—for a while (fig. 36.1).

Figure 36.1
Ground squirrels mount quite a defense against rattlesnakes and are rarely bitten. When they are, blood proteins detoxify the venom. Many of the squirrels' protective behaviors are innate. Pups who have never encountered rattlers move, kick sand, and flag their tails as if they were old hands at evading snakes.

Behavior Is Shaped By Genes and Experience

Although squirrels instinctively, or innately, know how to deter a snake, newborns are often cornered by the predator and eaten. Experience teaches a pup to refine inborn skills to evade snakes, as well as other predators. In fact, all behaviors are determined by heredity (nature) as well as the environment (nurture), each contributing to different degrees for particular responses. Behavior is limited by the anatomical systems for detecting stimuli, integrating the input, and responding. The sophistication of such systems is set by the genes. On the other hand, learning can be very important in mastering certain very complex behaviors, such as reading, speaking a language, and following the rules of a game.

Closed and Open Behavior Programs

Genetic control of a behavior may be strict, with little room for environmental influence, or more flexible. A **closed behavior program** is rigid and not easily modified by experience. In contrast, an **open behavior program** is flexible and easily altered by learning. The resulting behavior is usually adaptive, whether it is shaped by natural selection perpetuating the appropriate genes to conquer a particular environmental challenge or by learning the best way to deal with a situation.

A primarily **innate** (and therefore mostly genetically determined) behavior occurs when it is critical that the action be performed correctly on the very first trial. Escape behaviors, such as the squirrel's sand-kicking stance, tend to be under fairly precise genetic control. If an individual fails to respond appropriately in its first encounter with a predator, it becomes the predator's dinner and never gets a second chance to refine the response. Animals that respond correctly to predators live to pass the genes enabling them to do so to future generations. In contrast, a predator who fails to capture a prey on the first try will be hungry but lives to try again.

The opportunity to learn the correct behaviors affects whether a behavior is open or closed. Animals with prolonged parental care, such as primates, learn skills from adults, such as how to gather food and find shelter. Species with little parental care, such as fish, usually have closed behavior programs, enabling them to handle life's problems on their own.

Genes May Influence Learning

Genetic constraints on learning are many, as animal trainers know well. One group of trainers attempted to teach raccoons, roosters, and pigs to put coins into a piggy bank for a food reward. The animals were not always cooperative. Roosters would scratch their feet on the floor, pigs rooted on the bare floor, and raccoons became miserly with their coins and fondled them rather than depositing them in the bank. The animals were behaving instinctively to food deprivation—roosters scratch the ground to uncover grain, pigs root around for food, and raccoons wash their food in water by rubbing it in their fingers.

A bird's song also reveals the influence of genes on learning (Reading 36.1). Young, male, white-crowned sparrows learn to sing by hearing the melodies of their fathers. If isolated from older males of their species, they sing abnormal songs as adults. However, the behavior is not solely determined by learning, because the sparrow can only learn the song of its own species. If a young isolated bird hears recordings of a variety of bird songs, including a normal white-crowned sparrow song, it learns to sing properly. The bird innately recognizes its species' song, perfecting the performance through experience.

Brain changes may parallel song learning. In baby male zebra finches, one part of the brain loses half of its neurons, and in another region, the number of neurons increases by 50% during the 70-day period when songs are learned. Perhaps a bird is born with a vast potential for singing specific note combinations, and the repertoire narrows as brain cells die.

KEY CONCEPTS

Heredity and the environment influence behavior to differing degrees. A closed behavior program is rigid and mostly genetically controlled. An open behavior program is flexible and amenable to environmental influences. Innate behaviors are important when immediate success is necessary for survival. Learning can be constrained by genetics.

Innate Behavior

Ethology examines how natural selection shapes behavior to enable an animal to survive. Animals are studied in their natural surroundings where the survival value is most apparent. Natural selection can only work on traits with a genetic basis, so ethologists focus on innate behaviors.

Fixed Action Patterns

A **fixed action pattern (FAP)** is an innate, stereotyped behavior, such as a dog digging on the floor as if trying to bury a bone or a kitten pouncing on a rustling leaf as if it were a mouse. A FAP is performed nearly identically by members of a species and is modified very little, if at all, by the environment.

The egg rolling response of a female greylag goose is a FAP (fig. 36.2). When a brooding female sees an egg outside her nest, she retrieves it in a characteristic manner. She stretches her neck toward the egg so that her bill is just beyond it, and then she scoops it toward her, repeating this motion until the egg is back in the nest. She adjusts her movements as the egg wobbles over uneven terrain. This action is adaptive, because an unincubated egg will not hatch.

Reading 36.1 *Why Does the Mockingbird Sing?*

THE NAME OF *MIMUS POLYGLOTTOS* IS WELL EARNED—THE MALE "MANY-TONGUED MIMIC," OR MOCKINGBIRD, CAN CROON THE SONGS OF SOME 200 SPECIES, ALTERING THE LOUDNESS OF A SONG AND ITS DIRECTION. Just why the bird sings has puzzled ethologists (scientists studying innate animal behavior) for many years, Charles Darwin among them. Are the songs attempts to fool birds of other species into leaving a territory? Does the male mockingbird sing to attract a mate? Is the behavior adaptive at all? An ethology class at the University of Miami, led by Randall Breitwisch, used the scientific method to investigate the matter.

Observations: "Bachelor" males (those without mates) sing much more than mated males. A mated male sings certain songs at different stages of his life cycle—sexual intercourse, egg laying, nurturing nestlings.

Figure 1
A male mockingbird with a family directs his singing in one direction, whereas a bachelor mockingbird sings far and wide.

Hypothesis: If the male mockingbird's songs are strictly to keep other bird species away, mated and unmated birds should sing alike. If the songs are meant to attract a mate, bachelor and paired males should sing differently.

Experiment: Pairs of students were assigned to follow one bird, noting the direction of the animal's bill throughout a half-hour period whenever he sang. A single song could last up to 10 minutes. This measurement is a "song vector."

Results: The researchers recorded song vectors for 19 hours each of unmated and mated male birds' performances and for 2 hours for a lucky male who changed his status while being monitored. The two types of males differed markedly. Unmated males sung in all directions; mated males sang in one specific direction.

Conclusion: The direction of singing in the male mockingbird is consistent with the idea that the songs are used to attract a mate. Unmated males search far and wide; mated males narrow their search.

Further Study: The researchers noted that all males go to an edge of their territory, turn so their backs are to the outside, and sing inward. Why might they do this?

The goose will retrieve a beer bottle as readily as an egg. Apparently she has only a vague sense of what an egg is and automatically responds to all small objects outside the nest. If the egg is removed after she has begun rolling it back to the nest, she continues her retrieval motions until the imaginary egg is back in the nest. Once a FAP is initiated, the action continues until completion, even in the absence of appropriate feedback.

Releasers

The specific factor that triggers a FAP is called a **releaser** or sign stimulus. An animal's sensory systems are bombarded constantly with many more stimuli than could possibly be responded to. The animal must select and respond to only the few key stimuli that are reliable cues to the situation requiring response. Even a single object or organism has many aspects to it—color, shape, odor, taste, sound, and movement are but a few.

To identify the specific releaser for a behavior, an ethologist can build models that isolate a single stimulus, such as color or scent. An organism in an appropriate physiological state is shown the model and observed to see whether it responds as it would to the natural stimulus. For example, a male stickleback fish ignores a model that looks exactly like a rival male except that it lacks a red underbelly. However, a male will attack a model of almost any shape with a bottom half painted red. The color red, then, releases aggressive behavior in the male stickleback.

Releasers are important in human parent-infant interactions. When an infant gazes and smiles at her exhausted parents, the adults respond with tender affection. It is rather ego deflating to realize that an infant up to 2 months old will smile at anything with two dots resembling eyes! A model of a face lacking eyes will not elicit a smile. Doll manufacturers, advertisers, and cartoonists are well aware of the visual releasing power of the juvenile face, with its large eyes, small nose, and chubby cheeks. Many species retain juvenile characteristics of their ancestors, a phenomenon called **neoteny** that is often adaptive (fig. 36.3). An adult human, for example, resembles a chimpanzee fetus.

Figure 36.2
Like all fixed action patterns, egg retrieval in the greylag goose is always the same. She extends her neck so that it is just beyond the egg, then scoops the egg toward her, rolling it back to the nest.

Figure 36.3
Juvenile features release nurturing feelings and behaviors in adults. Note the similarities in facial features between the boy and his puppy.

Many releasers are auditory. A mosquito is attracted to his mate by the buzz of her wings. A model of the stimulus, a tuning fork that vibrates at the same frequency as a female's wings, will also attract him. The pulsing chirps of crickets, the clatter of katydids, and the repetitious droning of a bull frog heard on a summer's eve are auditory signals that release mating behavior.

Other releasers are tactile. During the mating ritual of the stickleback fish, the female enters the nest, stimulating the male to thrust his snout against her rump in a series of quick, rhythmic trembling movements. These thrusts release her spawning behavior. The releaser can be mimicked by prodding the female with a glass rod.

Pheromones are airborne chemical releasers secreted by animals that influence the behavior of other members of the same species. The sex attractants of some insects, crustaceans, fishes, and salamanders are pheromones (see fig. 19.21). To show that a pheromone is a releaser of mating behavior in the Canadian red-sided garter snake, researchers extracted lipids from the skins of sexually mature female and male snakes. Lipids from the females were wetted onto paper towels, and the towels were placed near males anxious to mate. The snakes became excited, some even attempting to mate with the pheromone-soaked paper. Interestingly, when male lipids were added to the towel, the male snakes backed off.

The male silk moth is so exquisitely sensitive to a sex attractant that the minuscule amount released by a female may attract him from as far as 2 miles (3.2 kilometers) away. Sensing her scent, he flies upwind, traveling so that the odor intensifies, until he locates her. Scientists working with such pheromones have been dismayed to find that they themselves serve as models. The pheromone is active in such small amounts that even after a shower, one worker was besieged at a football game by male moths dancing about in amorous frenzy.

A **supernormal releaser** is a model that exaggerates a releaser and elicits a stronger response than the natural object. Birds such as the oystercatcher and the herring gull, for example, prefer to sit on large eggs (fig. 36.4). The female will try diligently to sit on an egg that is more immense than she could possibly have laid, even though her own egg is present. Such a behavior may have evolved to ensure that all sizes of eggs for the species would be protected, but it did not take into account a human ethologist switching eggs!

Figure 36.4
If offered a huge egg, this herring gull will abandon her own eggs to sit on it. A big egg is a supernormal releaser.

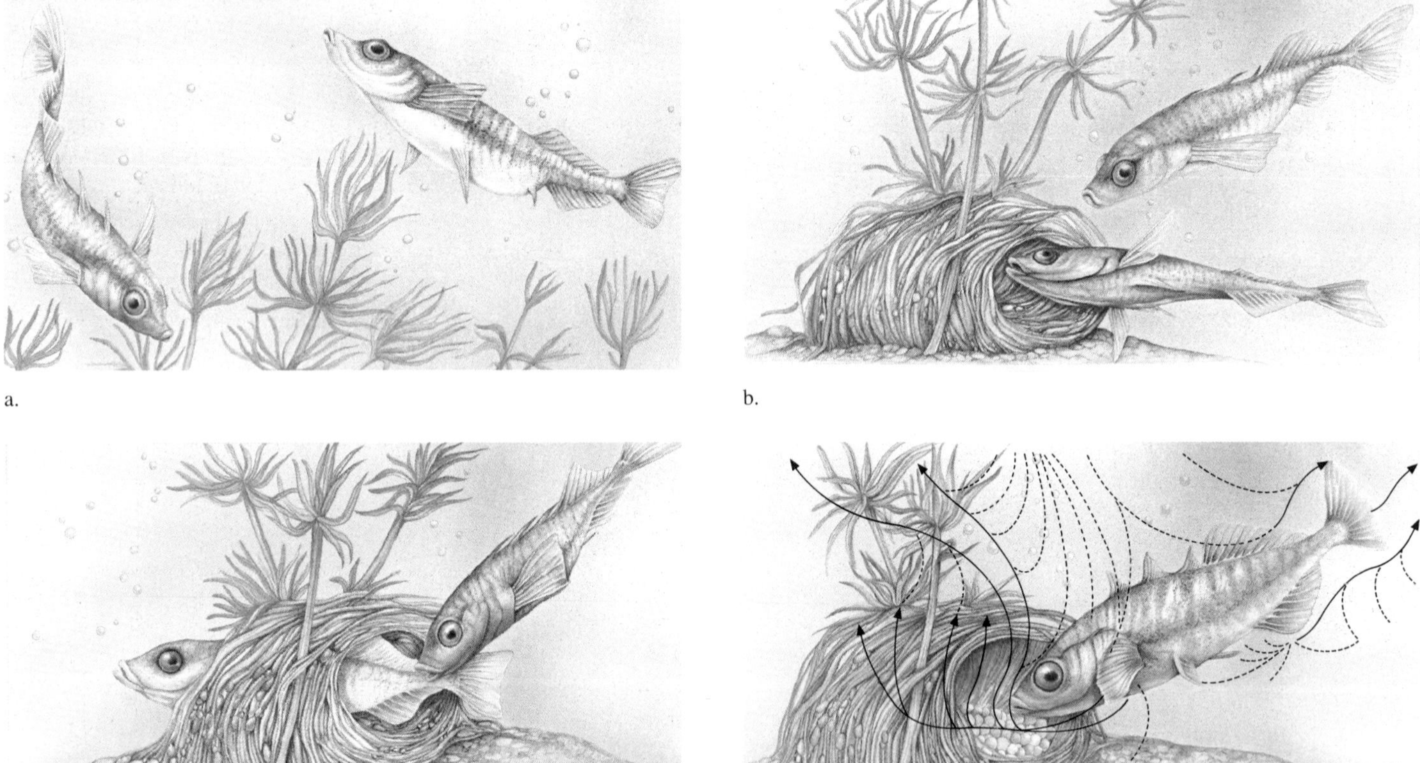

Figure 36.5
a. Mating in the three-spined stickleback is mostly innate. In early spring, in shallow ponds and lakes, male sticklebacks prepare to mate. A male leaves the group, stakes out a territory, and builds a nest of algal weeds, coating the structure with a gooey material from his kidneys. He digs a hole through the mass, which is the nest. Then he changes color, his chin turning red and his back a bluish white. Spying a female, he zigzags toward her. Head lifted, she swims toward him. *b.* The male swims to the nest and pushes his red snout into it. He flips on his side and presents the fins across his back to the female (viewed from above). *c.* The female swims into the burrow in the nest, followed by her mate, who touches the base of her tail. In response, she lays 50 to 100 eggs. *d.* The male fertilizes the eggs, then moves water over them, bringing in oxygen.

Chain Reactions

A complex behavior, sometimes called a chain reaction, is often built from simpler units joined by a sequence of releasers. The female digger wasp, for example, displays parental behavior in an orderly sequence. She constructs a burrow, seals it with rubble, and then captures and stings a caterpillar. She pulls the immobilized caterpillar into the burrow and lays her eggs on it, furnishing her larvae with an ample food supply. This sequence of behaviors is always performed in the same order because the completion of each step is the releaser for the next. Chain reactions can also result from an exchange of releasers between two individuals, such as the mating behavior of stickleback fish (fig. 36.5).

Behavior Is Adaptive

Because behavior has a genetic basis, it is shaped by natural selection and is therefore adaptive. Nobel Prize winning ethologist Niko Tinbergen's studies of parental care in two bird species illustrate this point.

The blackheaded gull lives in colonies in the sand dunes of northwestern England. One parent stays at the nest, incubating the eggs or protecting the chicks, while the other gathers food. The markings on the eggshells blend in with the surroundings. For an hour or two after a chick hatches, the parent stays by its side. Then the parent flies away with pieces of the broken eggshell and deposits them away from the nest. Another ground nesting bird, the oystercatcher, nests far from neighbors and removes eggshell pieces immediately after a chick hatches.

How are the parental behaviors of these birds adaptive? Although the outsides of the eggshells of both birds are camouflaged, the insides are white and very obvious against the grass and twigs around the nests. Tinbergen hypothesized that white pieces of shell make the cryptic intact eggs more noticeable to predators such as crows. He tested this idea by painting hens' eggs to resemble gull eggs. He placed the painted eggs at regular intervals amidst a gull colony, with broken pieces of white shells near some of them. The camouflaged eggs close to white pieces were indeed discovered by predators more often than those not near broken pieces. The parent gull decreased the chance of a predator noticing unhatched eggs by removing broken eggshells. The fastidious parents leave more surviving offspring to perpetuate their genes.

Slight differences in nesting habits may explain why the two bird species differ in how quickly the broken shells are removed. Whereas blackheaded gulls nest in colonies, oystercatchers nest alone. For an hour or so after a chick hatches, its feathers are wet. A wet chick would easily slide down the throat of a cannibalistic adult neighbor. A dry, fluffy chick is not as easy to swallow. It is adaptive for a blackheaded gull, with neighbors, to stay with its chicks until they dry. For an oystercatcher who has no neighbors, it is better to remove the eggshells immediately so that predators are less likely to notice the unhatched eggs.

A behavior amenable to a more precise "cost-benefit" analysis is foraging, because calories used to gather food can be compared to calories gained in eating it. Consider crows on the west coast of Canada. The birds gather whelks (large marine snails) at low tide. A crow flies over a rock and drops a whelk. The bird then can eat the meat inside the smashed shell. The crows take only the largest whelks, and they always drop them from a height of about 16.5 feet (5 meters). The largest whelks contain more meat and break open more easily than smaller ones. Still, a crow usually has to drop a whelk at least twice to break it open.

The greater the height from which the whelk is dropped, the better the chances of breaking it. But flying upward costs energy. If the whelk is dropped from a lower height, the crow expends less energy in flying, but the drop may have to be repeated more times before the shell breaks, so the total energy used may be greater. Also, if the whelk is dropped from a higher height, another crow can zoom in to steal the meal.

In an experiment to determine the lowest dropping height most likely to break open a shell, average-sized whelks were dropped from various heights (fig. 36.6). The smallest total upward flight (the vertical distance times the number of trips to smash the shell) was 17 feet (5.2 meters)—very close to the minimum height commonly used by crows.

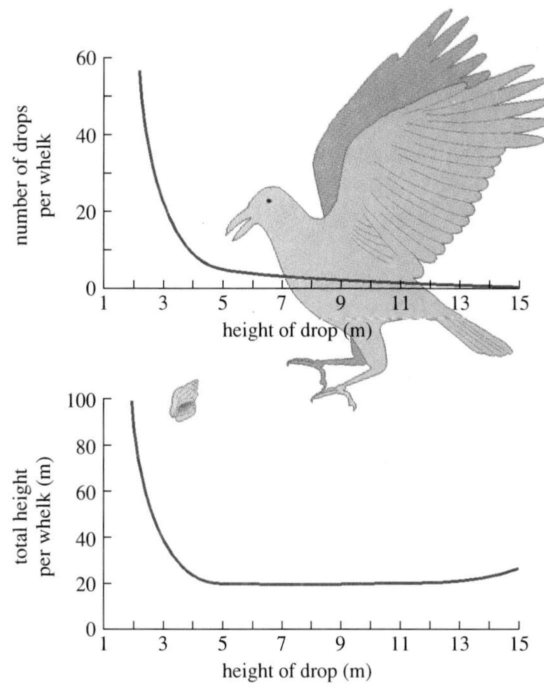

Figure 36.6
The results of a whelk dropping experiment. The greater the height from which the whelk is dropped, the fewer drops are required to crack the shell. The total upward flight needed to break open the whelk is close to the minimum height from which crows commonly drop whelks. This match indicates that the crow is minimizing the energetic cost of obtaining food.

How Genes Can Determine Behavior

One way to demonstrate that genes influence behavior is to mate individuals that perform a particular behavior in distinctly different ways. Then the behavior of the hybrid offspring is observed. Depending on the number of genes influencing the behavior, the offspring may perform the action in the same manner as one of the parents, or they may combine certain aspects of each.

The genetic basis of nest-building behavior in two closely related species of lovebirds was shown by such hybridization studies. When offered sheets of paper, both species tear them into strips. Then, the peach-faced lovebird tucks the materials in its rump feathers, while Fisher's lovebird transports them in its beak. Hybrids between the two species are befuddled. They attempt to tuck the strips into their tail feathers like the peach-faced lovebirds but are unable to raise and lower their feathers correctly. As a result, every piece is dropped during the flight to the nest. The birds gradually learn to use their bills to transport the materials.

Some behaviors can be associated with the protein product of a particular gene. For example, normal fruit flies given a shock in the presence of a certain odor learn to avoid that odor. Mutant fruit flies called dunce forget the association so quickly that they continually receive shocks. The dunce gene reduces the amount of activity of an enzyme that breaks down cyclic AMP, causing levels of this biochemical to rise. Cyclic AMP influences the ability of neurons to communicate with one another, resulting in the inability to associate odors with sensations.

Another genetically controlled behavior easy to observe is egg laying in the sea hare *Aplysia*. A long string of eggs is expelled when the reproductive ducts contract. As the string appears, the snail grasps it in the upper lip and, waving its head in a stereotyped pattern, pulls on the string, coats it with sticky mucus, and winds it into a tangled mass. With a characteristic wave of the head, the snail attaches the knotted string to a rock or other firm surface (fig. 36.7). The egg-laying behavior and physiological changes in *Aplysia* are brought about by proteins in the bag cells, two clusters of nerve cells in the abdomen. A single gene codes for one large protein, which is cut to yield the several smaller proteins that oversee egg laying.

Figure 36.7
Egg laying in the sea hare *Aplysia* is innate—and impressive. The record egg string was 575 feet (175 meters) long and was laid at a rate of 41,000 eggs per minute!

KEY CONCEPTS

Ethology considers the adaptiveness of behavior. A fixed action pattern is an innate, stereotyped behavior triggered by a specific releaser, which may be visual, auditory, tactile, or chemical. A complex behavior may result from sequences of releasers and involve one or more animals. Behavior is shaped by natural selection and is therefore adaptive. Hybridization studies demonstrate the inherited component of behavior. Genes specify proteins that can in turn produce a behavioral phenotype.

Learning

A change in behavior as a result of experience is called **learning.** A young bee innately recognizes a flower pattern of light petals and dark center. An older bee, with experience, recognizes specific flowers. The several types of learning overlap somewhat.

Habituation

In the simplest form of learning, **habituation,** an animal learns not to respond to certain irrelevant stimuli. When a stimulus is presented many times without any consequence, the animal usually decreases its response, perhaps to the point where it is eliminated completely.

Habituation provides protection. When a snail is tapped, it withdraws into its shell. If tapped repeatedly, it stops withdrawing. It is adaptive for young chicks to learn not to escape from innocuous, common stimuli, such as blowing leaves and other nonpredatory birds. Responses to predators are retained because they are rarer.

Habituation also helps to ensure that aggression is adaptive. If seabirds that often nest within a few feet of one another did not habituate to the presence of neighbors, much time would be wasted on aggressive encounters. Instead, they learn not to show aggression toward their neighbors as long as they remain outside a certain territorial limit. They retain aggression toward other birds that are encountered less frequently.

Classical Conditioning

In **classical conditioning,** an animal learns to respond in a familiar way to a new stimulus. A new association between stimulus and response is formed when the new, or **conditioned, stimulus** is presented repeatedly immediately before a familiar, or **unconditioned, stimulus** that normally triggers the response. After the stimuli have been paired in a series of trials, the new stimulus presented first is enough to elicit the response.

For example, a rabbit hears a tone, and then a puff of air is directed at its eye. The rabbit eventually learns to blink upon hearing the tone, because it expects the air puff to follow. If the tone is presented many times without the air puff, the rabbit no longer blinks at the sound. The loss of a conditioned response is called **extinction.** As the rabbit is learning, the activity of certain neurons increases in the hippocampus region of its brain, suggesting a physical cause of the behavior. The most familiar example of classical conditioning is that of dogs associating the sound of a bell with the presence of food. As a result of classical conditioning, the dogs salivate at the sound of a bell.

Operant Conditioning

Operant conditioning is a trial-and-error type of learning in which an animal voluntarily repeats any behavior that meets with success, such as a reward **(positive reinforcement)** or avoiding a painful stimulus **(negative reinforcement).** Reinforcement increases the probability that the animal will repeat the behavior. Trial-and-error learning can refine natural behaviors, such as predatory skills. A grizzly bear, for example, learns that splashing about in a stream does not yield him a salmon dinner. Staying still and quiet in one place, with his head underwater, is much more effective.

Ethologist B. F. Skinner designed an apparatus to demonstrate operant conditioning. When a hungry rat is placed in a Skinner box, it begins to explore its surroundings and eventually presses a lever accidentally, which releases a food pellet. This food reward increases the probability that the rat will press the lever again (fig. 36.8).

In animal training, operant conditioning first reinforces any behavior that vaguely resembles what the trainer wishes and then restricts the reward to better and better approximations. Pigeons learn to play table tennis in this way. Hungry pigeons are taught to use their beaks to hit a ball toward one another. When the ball falls in the trough on either side of the table, a food pellet is delivered to the pigeon that won the point.

Operant conditioning shapes human behavior in many ways. A youngster learns the motor skills required to use a swing for the reward of performing the action more accurately than before. Much adult human behavior is conditioned to rewards of money or social approval. Operant conditioning is also the basis of a technique called **biofeedback,** which gives a person information on physiological processes he or she wishes to control, such as heart rate, blood pressure, muscle activity, or brain-wave pattern.

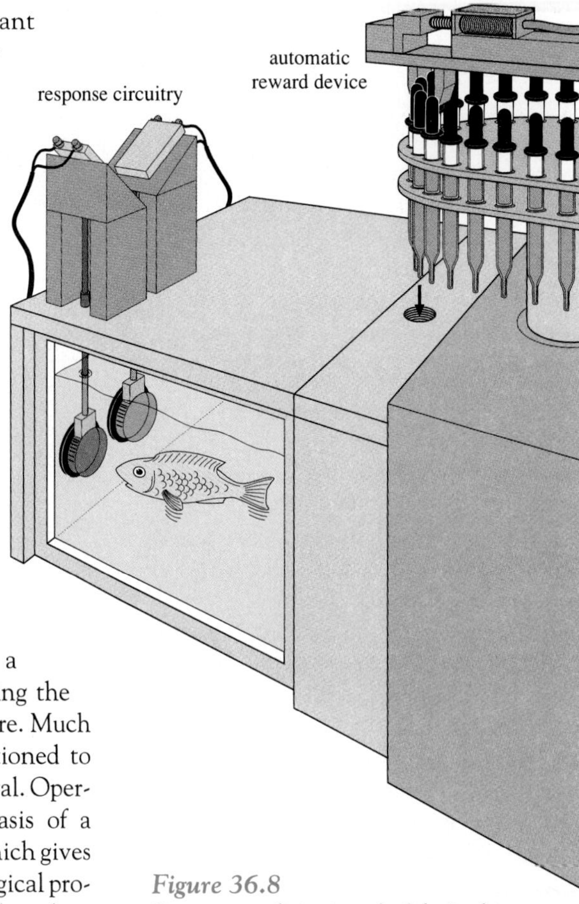

Figure 36.8
Operant conditioning of a fish. In this variation of a Skinner box, called a discrimination tank, a fish is shown two colors. When it taps the one desired by the experimenter, it receives a treat—a worm tidbit.

Imprinting

Imprinting is a type of learning that occurs quickly, during a limited time (called the **critical period**) in an animal's life, and is performed usually without obvious reinforcement. Young chicks, goslings, or ducklings are imprinted to follow the first moving object they see, which in the wild is usually their mother. This "following" response is adaptive, because the mother usually leads them to safe places where they are likely to find food.

Ethologist Konrad Lorenz is famous for his studies of imprinting. Lorenz hatched eggs in an incubator and then had the baby birds waddle after him. Thereafter they showed the same responses to him as they would have to their mother (fig. 36.9). Grade school classes often adopt ducklings or goslings hatched from eggs in the springtime. These class mascots follow the lined-up students through the halls of the school—they look like obedient students but actually are just demonstrating imprinting behavior.

Imprinting is also important in the development of "mother love" in several species, such as goats, sheep, and the Alaska fur seal. In the first few critical minutes after a goat kid's birth, the mother learns to identify her own offspring by their odor. She will accept and nurse any young that she smelled during the critical period and reject any youngsters she does not recognize. Some child development experts maintain that similar bonding occurs between human mother and infant in the 2 hours after birth—but loving fathers and adoptive parents might argue otherwise.

Insight Learning

Insight learning, or reasoning, is the ability to apply prior learning to a new situation without observable trial-and-error activity. For example, in one experiment, captive chimpanzees stacked boxes to climb and reach a banana hanging from the ceiling. This may have involved insight.

Latent Learning

Latent learning takes place without any obvious reward or punishment and is not apparent until sometime after the learning experience. Even without reinforcement, a rat that has been allowed to run freely through a maze will master its twists and turns more quickly than a rat with no previous experience in the maze.

In the wild, animals may learn the details of their surroundings during daily explorations. Although this information may not be of immediate use, knowledge of hiding places may make the difference between life and death when a predator strikes. Reading 36.2 illustrates different types of learning in a human child.

Figure 36.9
Many baby birds imprint on the first moving object they see. This is usually the mother but may be an ethologist, such as Konrad Lorenz.

KEY CONCEPTS

An animal can become habituated to unimportant stimuli. In classical conditioning, an animal associates a known response to an unconditioned stimulus to a new response, the conditioned stimulus. In operant conditioning, an animal learns voluntarily to behave to receive a reward. Imprinting is a rapid form of learning that occurs at a specific time in development and does not entail a reward. In insight learning, prior learning is applied to a new situation. Latent learning uses prior experience.

Biological Clocks

Like fluctuations in such environmental factors as light, temperature, relative humidity, and barometric pressure, many behaviors and physiological processes cycle. It is often adaptive for an organism to perform a specific behavior at a particular time relative to predictable environmental changes. Often the behavior must begin before the environmental change, without a reliable cue to trigger the start. For example, for fruit flies the optimal time to emerge from their pupal cases is at dawn, while the humidity is high and they can expand their wings without danger of excessive water loss through the still permeable cuticle. But this process takes hours and must start long before daybreak. Animals use inner timekeepers called **biological clocks** to sense when it is time to act, without reliable environmental cues. In plants, biological clocks are crucial to photosynthesis, growth, and flowering (chapter 35).

Physiologically, a person in the morning is different than at night. Growth, cell division, respiration, enzyme activity, ability to learn, and sensitivity to drugs are all rhythmic. Heart rate, body temperature, blood composition, hormone secretion, and excretion of electrolytes and water from the kidneys all vary over a day. Eye-hand coordination, the ability to add and count, memory, mood, energy level, and the ability to estimate time also vary during a day. If a student learns better in the morning than in the evening, there might be a sound biological reason.

A Biological Clock Runs Without Environmental Cues

The most extraordinary feature of biological rhythms is that they are not caused by the regular day/night cycles in light and temperature. If plants or animals are maintained in a laboratory at a constant level of illumination and temperature, their rhythms persist. However, the period length of a rhythm (the interval from one peak to the next) usually deviates slightly from the precise 24-hour frequency it displays in nature. For example, a mouse housed in constant light or dark in a laboratory may awaken 30 minutes earlier each day. By week's end, it wakes up 3.5 hours earlier than it did on the first day of the study. Rhythms that are no longer precisely 24 hours under constant conditions are termed **circadian** (about a day).

Disrupted Clocks Affect How We Feel

The fact that biological rhythms take several days to adjust to a new light-dark schedule is the basis for the modern problem of jet lag. When travel across time zones is rapid, a change in the light-dark cycle is encountered. Until circadian rhythms adapt to, or are **entrained** to, the new lighting regime, physiological cycles are no longer synchronous with one another or with the environment. This internal disharmony causes a range of physical and psychological disturbances, including insomnia, constipation, depression, irritability, and decreased intellectual function. Although jet lag may be annoying to a vacationer, it can be terribly disruptive to athletes, diplomats, musicians, and others whose

Reading 36.2 A Child Learns

IT HAS BEEN ONLY A FEW MINUTES SINCE THE CHILD ENTERED THE WORLD, YET ALREADY HE LOOKS ABOUT AT THE LIGHT, TURNS HIS HEAD TOWARD THE BIG PERSON MAKING COOING SOUNDS, AND MOVES HIS LIMBS TO THE RHYTHMIC NOISES IN THE BRIGHT ROOM. Even at this tender age, the infant's hand instinctively closes when the brand new mother places her finger across his tiny palm. The infant also grasps with his toes, perhaps a holdover from primate ancestors in whom grasping was necessary for survival in the treetops.

A human newborn also shows adaptive responses for finding food. When touched on the corner of the mouth or cheek, the baby turns her head toward the stimulus, a motion that would bring a nipple to her mouth. The newborn, with a little prodding, also knows how to suck milk from the nipple, a complex motion involving many muscles. The baby comes equipped with other protective behaviors. If her nose and mouth become covered as she lies in her crib, the baby will use all her strength to lift her comparatively large head and turn it so she can take a deep breath. Some researchers maintain that the infant bonds to whomever she stares at repeatedly during these first few hours. This may be a form of imprinting.

By the age of 1 week, the child easily picks out mom's voice from the babble around the crib and maybe dad's too. By 2 weeks, he recognizes the faces that go with the two familiar voices. By 1 month, he has learned to expect mom or dad to come running whenever he cries. At night, every time he whimpers, one of them appears. Soon, though, they habituate to his nightly noises, only answering the more vigorous wails. Despite some nightly screaming, the baby seems to sleep a lot, through almost any ruckus. Parents may not know that tiptoeing around is not necessary.

Two months is not too early to begin learning language. Mom and dad sing high-pitched words and rhymes as they smile at their infant, not even aware of how their voices instinctively change. Soon the child makes his own noises. One morning, as an exhausted parent comes in to change a diaper, the baby greets him with a sudden, wide grin. It seems to take up his entire face. A wonderful thing happens. The new father is so thrilled that he laughs and cuddles the child, over and over. The child quickly realizes that smiling earns affection—a first experience with operant conditioning.

By 6 months of age, the child starts to eat solid foods, learning by trial and error which foods taste good and which she would prefer to spit out or smear on her shirt. She may become classically conditioned to the sound of the refrigerator door opening or perhaps the sight of a parent going to the cupboard—she knows food is coming.

By 14 months, the child uses trial-and-error learning to fit shaped blocks into the right holes in a box. As a toddler, he begins to show insight learning. It would be great fun, he thinks, to take a big box of detergent sitting on the sink and pour the pink stuff that he's seen mom and dad take from it all over the floor. But he cannot reach the sink. He looks from a chair to the sink, then pushes the chair over to the sink and reaches for the detergent box. Soon the floor is covered in pink dust.

By 4 years of age, the child is an old hand at learning. When a relative visits from France, she is astounded at the ease with which the child picks up her language. The observation that preschool-aged children are remarkably adept at learning a second language may be another demonstration of imprinting.

All of the forms of learning continue throughout life but are perhaps not quite as exciting as these first ventures.

a.

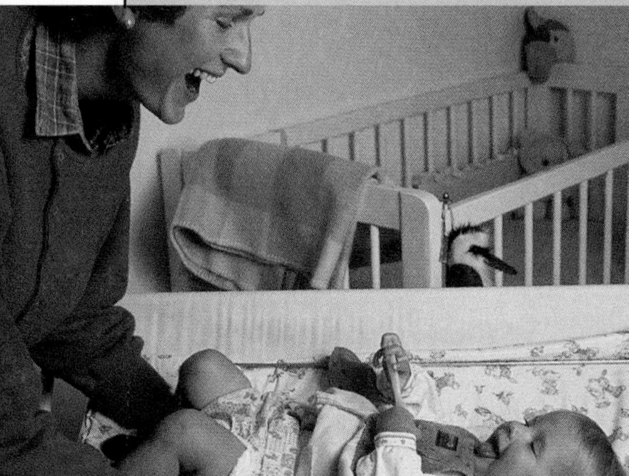

b.

c.

d.

Figure 1
(*a*) When Helen, 6 months, grabs her father's nose, she is showing off her newly developed hand-eye coordination—and possibly the beginnings of a sense of humor. (*b*) A parent makes silly faces and sounds in response to an infant's toothless grin. (*c*) Getting the piece in the puzzle involves trial-and-error learning (operant conditioning) and perhaps latent learning. (*d*) What learning skills is this 4-year-old using?

Reading 36.3 *Sleep/Wake Cycles*

THE BEST-STUDIED CIRCADIAN RHYTHM IN HUMANS IS THE SLEEP-WAKE CYCLE. It is controlled by the suprachiasmatic nuclei (SCN) in the brain's hypothalamus. If the SCN are destroyed, periods of sleep and wakefulness occur at random.

The duration of a night's sleep depends upon body temperature at the time the person falls asleep—the higher the temperature, the longer sleep lasts. Although sleep is not really an active behavior, we spend nearly a third of our lives in this state, and the daily sleep-wake cycle has great influences on all of our behaviors.

Dreams

During a dream, the body is paralyzed, with only the fingers and toes twitching and the eyes darting rapidly beneath closed lids. Head, neck, and rib muscles relax, and heartbeat and respiration become irregular. Blood flow to the brain increases by 50%, and the pressure on this vital organ doubles. Dreams occur during rapid eye movement (REM) sleep, and last from 5 to 15 minutes, repeated every 90 minutes or so, for a total of 1 or 2 hours each night (see Reading 17.2). While the body is in this protected, restrained state, the mind travels to a fantasy that might embarrass the conscious mind or perhaps to a hellish nightmare.

Dreams begin during fetal existence and continue until death, although they decline in frequency with age. Why do we dream? One hypothesis is that this time allows the brain to process what is learned during wakefulness. Or perhaps sleep is a natural debugging system that breaks down faulty neural connections. Lack of dream sleep may lead to expression of these errors during waking hours as fantasy or hallucinations. Still another idea is that the brain dreams because it has nothing else to do while the rest of the body refreshes itself.

Sleep Disorders

About 30% of American adults have sleep problems. The most obvious form of sleep disruption is simply lack of it. A person

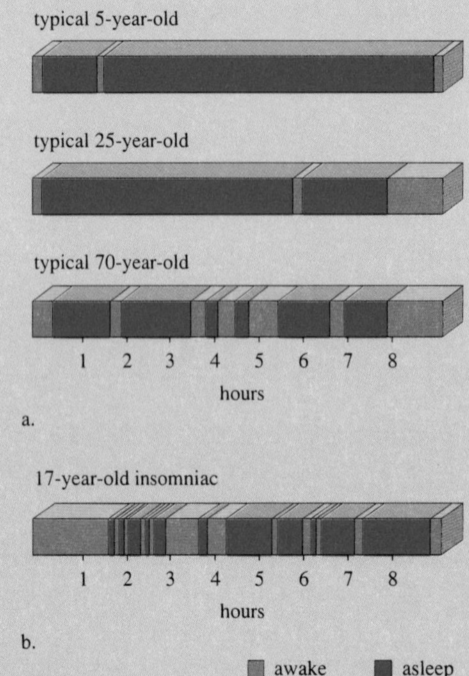

Figure 1
As we age, we sleep less. *a.* An elderly person awakens often. *b.* A 17-year-old insomniac slept very little during the first half of the night indicated and only for an hour at a time during the second half.

deprived of sleep retains normal memory, balance, logic, and reasoning skills and can perform interesting tasks. However, boring tasks rapidly become impossible to do. A sleep-deprived truck driver is far safer pulling off the road for a quick snooze than continuing down a hypnotically dull stretch of highway. If sleep deprivation continues, irritability, paranoia, and even hallucinations may set in.

Disruptions of the sleep-wake cycle are more serious. Manic depressives enter and leave dream sleep too soon and awaken too early, much like experiencing continual jet lag. **Narcolepsy** is a bizarre sleep disorder in which a person suddenly falls asleep during the day, often right in the middle of an activity. The narcoleptic enters dream sleep in 5 to 10 seconds, compared to the normal 60 to 90 minutes. The episodes may be immediately preceded by a hallucination, and the paralysis and muscle weakness characteristic of normal REM sleep occurs just before and after the attack. Episodes last from 2 to 15 minutes and may occur several times a day.

Sometimes night sleep is interrupted by a blocked trachea, and a person jolts to a start. If infrequent, this results in snoring. In **sleep apnea,** breathing stops hundreds of times a night, for 20 to 60 seconds each time. The danger is that the person will not resume breathing. Apnea is especially frightening in newborns, who can die in their sleep. Infants with sleep apnea are hooked up to home monitors that beep when breathing stops. The parent can then stimulate the baby into taking a breath.

A familiar sleep disorder is **insomnia.** In the common "delayed sleep phase" form, the person indeed sleeps for 8 hours but takes at least 45 minutes to fall asleep, compared to the normal 12 minutes. Some insomniacs actually fall asleep faster than they think they do and may even dream that they are awake! Exercise and dietary patterns may contribute to insomnia. Drugs to induce sleep are useful only temporarily, because they lose their effectiveness and can induce dependence. Many insomniacs give in to their "night person" tendencies and do something interesting until fatigue sets in. Insomnia can also be a sign of disease.

Insomnia is common among the 15% to 25% of United States industrial workers who follow nonday shifts. Changing from one shift to another, as often happens among pilots and nurses, can lead to severe insomnia, ulcers, and accident proneness. In fact, the nuclear accident at Three Mile Island in the 1970s may have been due to such an unwound biological clock. The crew on duty at the time, 4 A.M., had been on that shift only a few days and had changed shifts every week for the preceding 6 weeks.

unimpaired performance at the destination is critical. The effects of jet lag may be minimized by exercising on arrival.

Staying up all night to cram for an exam, or a weekend of late nights, similarly offsets the body's natural rhythms. Insomnia also may result from a disturbed biological clock. Elderly people often cannot sleep in the early morning hours due, perhaps, to an accelerated internal clock. Studies in sleep laboratories have shown that sufferers of delayed sleep phase insomnia go through an alert phase 3.6 hours later than do good sleepers (Reading 36.3).

Jet lag and insomnia may be helped by exposure to light at certain intervals, which can "reset" the biological clock. A form of depression called **seasonal affective disorder (SAD)** is also responsive to light therapy. This depression affects people only in the winter (causing decreased energy, excessive sleeping, and overeating) or in the summer (causing agitation, insomnia, and appetite loss). SAD seems to increase in prevalence as people live farther from the equator.

Physical Basis of Biological Clocks

Biological clocks were noted at least as far back as Aristotle, who observed that changes in the sizes of sea urchin ova are cyclic. Today, it is known that multicellular animals seem to have several clocks operating in synch. Clock structures are identified by destroying certain anatomical structures and observing whether biological rhythms are disrupted. Another approach is to culture parts of organs and observe whether their electrical or biochemical activity persists in characteristically cyclic fashion.

Three "master clocks" have been identified—the retina or other light-sensing structure; the **pineal gland,** which sits atop the brain stem; and two clusters of 10,000 neurons each in a region of the brain's hypothalamus called the **suprachiasmatic nuclei (SCN).** These three structures derive from the same part of the embryonic brain. The clock that sets the pace varies among species. The optic lobes are the predominant pacesetter in cockroaches, whereas in sparrows and starlings and other birds that perch, it is the pineal gland. In mammals the SCN is the major clock.

The hormone **melatonin** is thought to be involved in biological timing. In pineal glands of perching birds, photoreceptor cells translate light into rhythmic synthesis of melatonin. Mammalian (including human) pineal glands lack melatonin receptors, but they are found in the SCN.

The hamster is a popular model organism for biological clock studies because its behavior is easy to predict and observe. Fifteen minutes after laboratory lights are turned off, a hamster will run on its exercise wheel for precisely 3 hours. If kept totally in the dark or light, it begins its 3-hour workout every 24.2 hours. Recently, a mutant hamster was found that repeats its exercise period every 20 hours. Crosses between this "supershort" mutant and a normal hamster produced heterozygote offspring who repeated the exercise session at an intermediate interval of 22 hours (fig. 36.10). Transplants of the SCN between wild-type, heterozygote, and supershort hamsters support the genetic evidence. A wild-type hamster with its SCN replaced by the brain region from a supershort mutant will begin to train every 20 hours. If a hamster's SCN is removed, its intervals on the wheel become haphazard.

Rhythms With Other Period Lengths

The earth rotates on its axis in relation to the moon once every 24 hours and 51 minutes. This interval, from moonrise to moonrise, is a lunar day. On most of the coasts of the world, the tide comes in twice during this interval, once each 12.4 hours. Many of the activities of life match these tidal rhythms. For example, some photosynthetic diatoms and flatworms rise to the surface of the sand to bathe their photosynthetic pigments in the sunlight during low tide. Before the tide returns and washes them out to sea, they descend through the sand grains. Tidal rhythms persist in the laboratory with a period length that is only approximately the same as it was in nature and are therefore termed **circalunadian** (about a lunar day).

The interval from full moon to full moon is 29.5 days and is caused by the rotation of the moon around the earth. When the earth, moon, and sun are in line, at new moon and full moon, the tides are at their most extreme. The interval between successive spring tides is a fortnight. The emergence rhythm of the marine insect *Clunio* parallels the tides. It lives very low in the intertidal zone so is only uncovered during the low tides of spring. During the short time when the habitat is uncovered, the adults emerge from their pupae, mate, deposit eggs, and die. If a population of these insects is maintained in the laboratory, a group emerges approximately every 14 days.

Other rhythms have a monthly frequency, such as the reproductive activity of certain marine annelids. The Samoan palolo worm lives in crevices of coral reefs. During October and November, at sunrise on the day of the last quarter of the moon, the long reproductive tail segments break off and rise to the surface, churning like boiling noodles. Synchronously, the tail segments burst open, releasing eggs and sperm to be fertilized in the sea. The rhythm is adaptive because the synchronous release of gametes increases the chances of fertilization.

Many rhythms have a period length of a year. Some of these are controlled by the regular changes in photoperiod that occur over a year, but others seem to be controlled by an annual clock and persist in the laboratory under a constant photoperiod. In the laboratory the rhythms have a period length that is only approximately a year and are termed **circannual** (like a year). Hibernation of the ground squirrel is circannual. Normally these animals gain weight in anticipation of the winter hibernation, when the weight is consumed in metabolism. Squirrels kept up to 4 years in constant light and temperature continue to hibernate at approximately yearly intervals. Human sperm count may also fluctuate circannually. The observations that birth rates are lowest in the spring and that heat kills sperm led researchers to record sperm counts at different times of the year among a large group of men who work outdoors in San Antonio, Texas. Indeed, sperm count drops by 24% in the summer months, presumably due to the warm temperature.

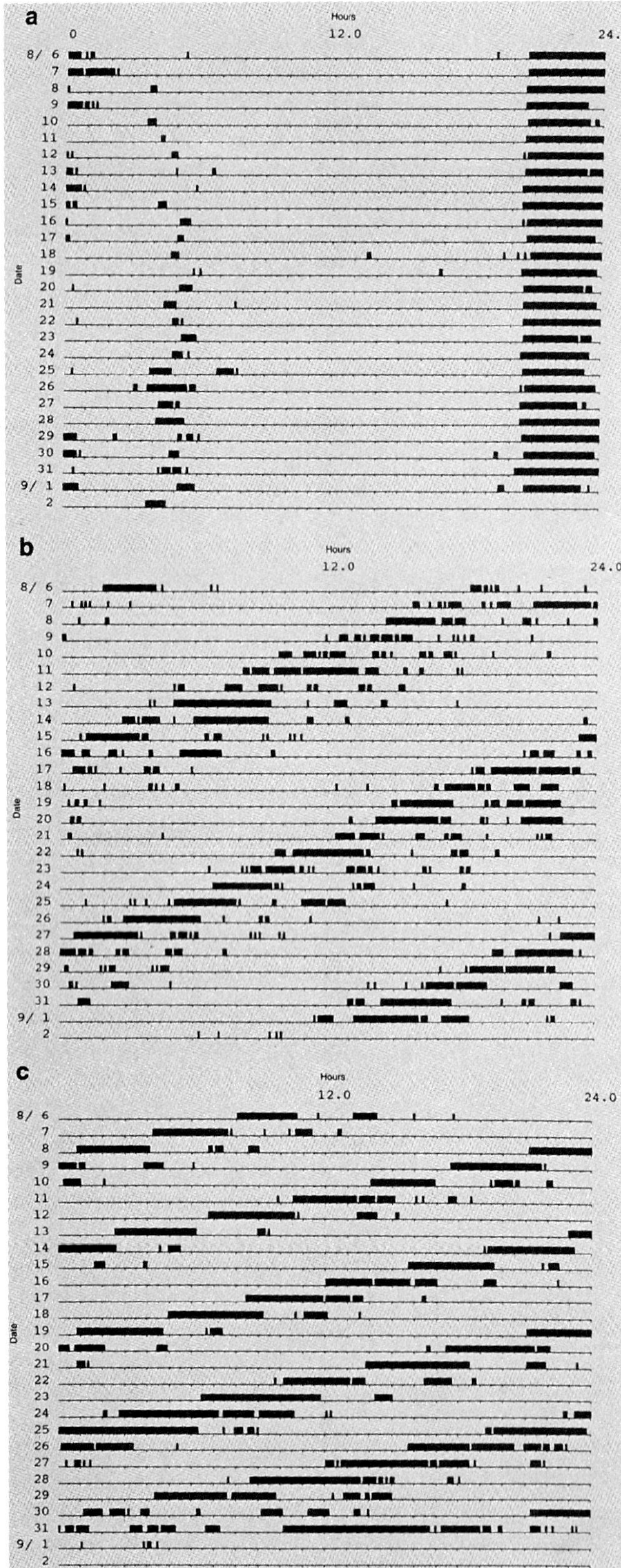

Figure 36.10
Supershort hamsters have abnormal biological clocks. In a biological clock experiment, a circadian rhythm is recorded over many days. Here, the dark bars indicate the times of a hamster's 3-hour run on its exercise wheel. *a.* This hamster has the normal circadian rhythm, working out every 24.2 hours. *b.* This hamster is a heterozygote between a normal hamster and a supershort hamster and exercises every 22 hours. *c.* This hamster, who is supershort, exercises every 20 hours.

KEY CONCEPTS

Biological clocks cause certain behaviors or physiological states to occur before they are actually advantageous. Circadian rhythms deviate slightly from a precise 24-hour cycle if allowed to occur under constant conditions. Jet lag, insomnia, and depression may result from disrupted circadian rhythms. Master clocks are found in the retina, pineal gland, and SCN and may involve melatonin. Rhythms that approximate the tides are called circalunadian, and those that occur yearly are circannual.

Orientation and Navigation

Each spring, the skies of North America are alive with the fluttering of millions of pairs of wings, as birds migrate north from the tropics. Using the position of the sun or stars, subtle shifts in the earth's magnetic field, sights, sounds, and the direction of winds and the aromas they carry, these animals can cover astounding distances and arrive with equally astounding accuracy. Their lives depend on it.

The journeys of birds are remarkable (fig. 36.11). Most notable is the Arctic tern, which flies from pole to pole, some 11,000 miles (17,700 kilometers). A Swainson's hawk can cover 4,000 miles (6,450 kilometers) at a clip as it flies from Alaska to Argentina. Birds are not the only migrants. Atlantic salmon that begin life in New England rivers migrate in the sea to Greenland to feed, then return to the river or stream of their birth to spawn. Insects, crustaceans, and salamanders also orient and navigate.

The simplest type of navigation relies on recognizing landmarks in a random search. When bank swallows are moved from their nesting burrows, they find home by looking around until they recognize a familiar landmark. The further they are displaced, the less likely they are to succeed.

The Compass Sense

Most migrating species use environmental cues as a compass to orient them in one direction. In the fall, for example, garden warblers leave Europe and fly southwest to Spain and then southeast to Africa. Caged

Figure 36.11
These long-eared owls are congregating before they start their annual migration across Lake Ontario.

birds denied the opportunity to migrate still orient in these directions for approximately the right flight times. Likewise, many of the small birds that migrate from North to South America each fall do so by flying to the East Coast and then flying southeast. The flight across the ocean is very long. If birds time their flight so that the winds are with them, they will make it. When the winds fail, they die at sea.

The sun is a compass. It rises in the east and moves across the sky at an average rate of 15° an hour, setting in the west. By knowing one compass direction, all others can be determined. An animal's biological clock tells it the time of day. If it can also tell the position of the sun it can orient in any compass direction.

To demonstrate use of the sun as a compass, birds were placed in a cage with food in boxes positioned along the periphery. The birds were trained to look for food in the box at a particular compass direction. Once trained, they could find food as long as the sun was visible but became disoriented when their view of the sun was blocked. If the apparent position of the sun was shifted with mirrors, the birds appropriately altered their orientation. The birds' ability to compensate for the movement of the sun across the sky was demonstrated by showing them a stationary light source. The birds shifted their orientation to the light by approximately 15° an hour—a change that would keep them right on course if the light were actually the sun.

Some birds migrate at night, using the position of the stars as a compass. In darkness, the indigo bunting navigates northward during its spring trek from the Bahamas, southern Mexico, and Central America to the eastern United States. It returns by flying roughly southward in the fall.

To show that nocturnal migrants use the stars as a compass, a bird was caged in a structure made of blotting paper rolled into a funnel with an ink pad on the floor and a ceiling of wire mesh. At night, when free birds migrate, the caged bird hopped around its cage. When it touched bottom, its feet were stained with the ink, so it left telltale footprints wherever else it alighted. In the morning, the number of footprints at each compass direction were counted. If the birds could see the starry sky, most of the footprints were in the half of the funnel

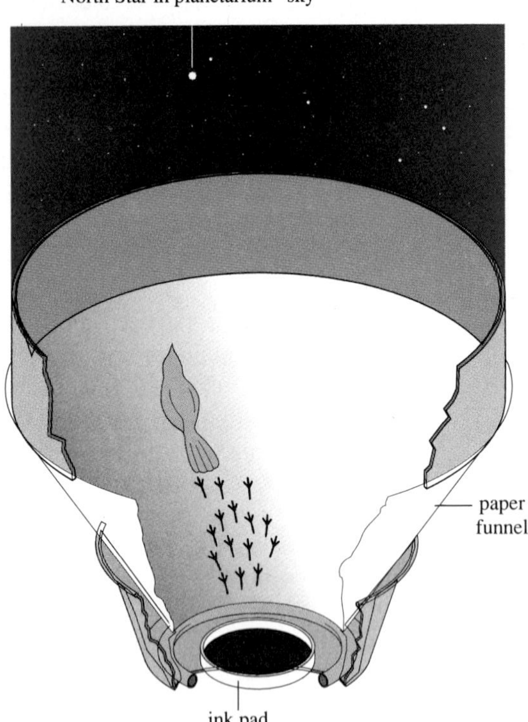

Figure 36.12
Migrating by the stars. The cage used to test star compass orientation consists of a paper funnel with an ink pad on the floor. The cage can be placed outside or in a planetarium, where the position of the stars can be manipulated. At night the bird hops in the direction it would fly if it were free. When it does, it leaves inky footprints on the funnel.

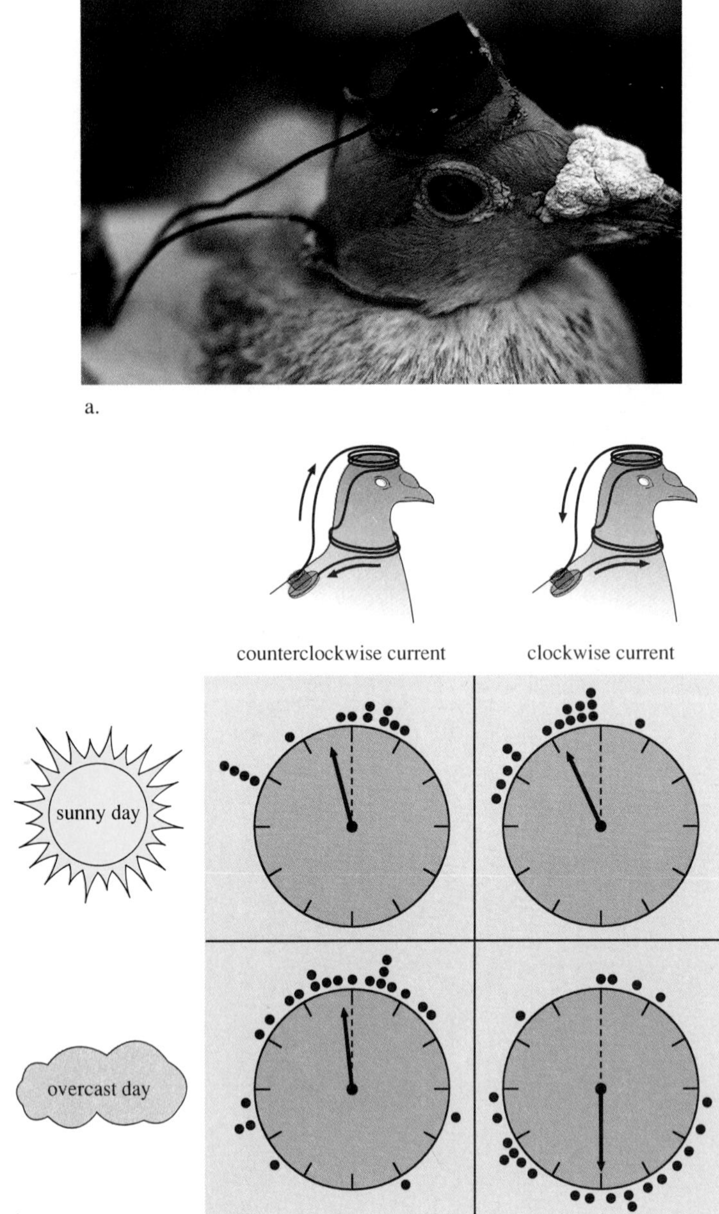

Figure 36.13
Pigeons go home. Pigeon orientation may be altered by magnetic fields. *a.* Pigeon hats were built of Helmholtz coils. When a current runs through the coils, a magnetic field is created. The direction of the field is determined by the direction of current. *b.* On sunny days, pigeons homed normally in spite of their magnetic hats. However, on cloudy days, when the sun could not be used as a compass, the pigeons experiencing a magnetic field with the north pole up flew in the wrong direction.

corresponding to the direction in which the birds would migrate (fig. 36.12). The birds could orient as long as they could see the stars. By placing the cage in a planetarium where the position of the stars could be altered or removed, it was shown that the buntings rely on constellations in the northern part of the sky, within 35° of the North Star, whether flying north or south.

Evidence from pigeons suggests that the earth's magnetic field, which runs in a generally north-south direction, can serve as a compass. A pigeon can find its way home if released up to 1,000 miles (1,609 kilometers) away. Bar magnets placed on a pigeon's wings impair its ability to find its way home on cloudy days. On sunny days, when the sun can be used for cues, the magnets have no effect. Sham magnets (nonmagnetic bars of equal size and weight) never disrupt orientation, indicating that the magnetism, and not the bar itself, serves as the navigational cue.

Further evidence of the role of magnetic cues in orientation is shown by giving pigeons hats made of small Helmholtz coils. When an electric current runs through the coil, a magnetic field is created. The direction of the field depends on the direction of the current. In some birds, nicknamed NUPS, magnetic north was up. In SUPS birds, magnetic south was up. Although there was only a slight difference in the orientation of these pigeons on a sunny day, on a cloudy day the SUPS birds headed home, while the NUPS birds flew directly away from home (fig. 36.13).

Figure 36.14
Sniffing a way home. Bears provide much information on animal navigation, because many are removed from towns as "nuisances," tagged, and transported beyond the 40-mile radius of their home turf. Odors may help bears find familiar surroundings. This bear is sniffing the breeze.

Homing

The most complex navigational skill is the ability to home—that is, to return to a given spot after being displaced to an unfamiliar location using no environmental cues. Homing requires both a compass sense for direction and a **map sense** telling the animal where it is relative to home. Very little is known about the map sense. It may depend on regular variations in the strength of the earth's magnetic field, which is about twice as strong at the poles as it is at the equator. An animal very sensitive to the strength of a magnetic field would know how far north or south it is. Disruptions in the magnetic field, such as is caused by sun spots, disorient pigeons.

A role for familiar senses in fostering a map sense is suggested by mammals who make seemingly miraculous journeys home. A wolf that had spent her whole life in a pen in Barrow, Alaska, was taken 175 miles (282 kilometers) away and released. She found her way home. A deer taken 350 miles (563 kilometers) from its wildlife refuge on the Gulf of Mexico found his way back. Closer scrutiny revealed how the animals may have navigated. The wandering wolf had grown up next to an airport and was accustomed to the sounds of jets taking off and landing. Such loud sounds travel far on the bleak Alaskan landscape, and the animal may have traveled so that the jets became louder. (Another wolf taken on the same trip returned to the wrong airport!) The deer seeking his wildlife refuge may have homed in on scents wafting in on gulf breezes.

Many tales of mammal navigation remain inexplicable. Such is the story of Big Mac, a 450-pound (204-kilogram) black bear tracked by a radio collar by bear expert Lynn Rogers. A bear learns landmarks in a 40-mile (64-kilometer) radius from home. Yet Mac found himself far from home, in search of berries to eat. With the cold weather imminent, he stopped foraging and tirelessly tramped homeward, through forests, backyards, farms, and roads, without stars or sun for guidance. Snow began to fall. His pace started to meander only when he was within 40 miles (64 kilometers) of home. Finally, he arrived—to a tiny crevice in a hillside, 126 miles (203 kilometers) from where his berry hunt had taken him (fig. 36.14).

KEY CONCEPTS

Simple navigation requires recognition of landmarks. A compass sense relies on the sun, stars, or earth's magnetic field to enable an animal to orient in a particular direction. Homing requires a compass sense to show direction, plus a map sense to lead to a specific site. Map sense may rely on magnetic cues or familiar senses.

SUMMARY

Both genes and learning shape behavior. Behaviors are usually adaptive. In a *closed behavior program*, genes are most important and the behavior is stereotyped. In an *open behavior program*, learning is more predominant. *Innate* behaviors may spontaneously appear instead of those that have been learned.

Ethology considers innate behavior in the wild. A *fixed action pattern* is an innate, stereotyped response that, once begun, continues without feedback. A FAP is triggered by a *releaser*, which can be identified using models. Releasers can be visual, auditory, chemical, or tactile. A model more effective than the natural stimulus in triggering the response is called a *supernormal releaser*. *Chain reactions* can be built from a sequence of FAPs. Releasers can be exchanged between individuals.

A *hybridization experiment* can show that a behavior is innate. Genes code for proteins that may underlie a behavior. Behavior is adaptive. Cost and benefit of a behavior can be evaluated in terms of fitness or calories spent and consumed.

Types of learning overlap. In *habituation*, an animal learns to ignore a stimulus that is repeated many times. In *classical conditioning*, an animal learns to give an old response to an *unconditioned stimulus* to a new, *conditioned stimulus*. In *operant conditioning*, frequency of behavior increases because it is positively or negatively reinforced.

Many behaviors and physiological processes fluctuate at regular intervals of a day, fortnight, lunar month, or year. Such biological rhythms can be set by environmental cues, but they are not simply responses to changing intensities of environmental stimuli. The rhythms continue when the organism is kept in constant conditions in the laboratory because they are driven by a *biological clock*. Under constant conditions, the period lengths of biological rhythms become slightly longer or shorter than they are in nature. Disrupted biological clocks affect how we feel. Master clocks are located in light-sensing organs, the pineal gland, or the *suprachiasmatic nuclei*.

The simplest form of navigation involves recognizing landmarks. Birds use the sun or stars as a compass, with the earth's magnetic field possibly serving as a backup. Homing depends on a *compass sense* to set direction and a *map sense* to direct the navigator to a particular place. The map sense may detect magnetic fluctuations or odors.

QUESTIONS

How can the scientific method be used to explain the following behaviors? Identify an initial observation, hypothesis, the experimental design, the results, and the conclusions.

1. The mockingbird's songs
2. Egg-rolling by the greylag goose
3. Aggression in the male stickleback
4. Mating behavior in the male mosquito
5. Nesting behaviors of the oystercatcher and blackheaded gull
6. Crows dropping whelks
7. Nest building in lovebirds
8. A person feels hungry when passing a donut store
9. Ducklings follow the first thing they see after birth
10. Maze running by rats demonstrates latent learning
11. A mouse's sleep-wake cycle is not exactly 24 hours
12. The suprachiasmatic nuclei are the master biological clocks in hamsters
13. Pigeons use the earth's magnetic field as a compass

TO THINK ABOUT

1. The modern feral (wild) horse is unusual in that it was domesticated 5,000 years ago but has returned to the wild behavior of its ancestors. Even a modern broken horse will immediately revert to wild behavior if it is allowed to range freely, with stallions stampeding mares and mock fighting, as if those 5,000 years of domestication had never existed. What type of behavior might explain the primitive actions of the free-ranging modern horse?

2. Two psychologists found that Americans and people of the Minangkabau culture in western Sumatra, Indonesia, have the same facial expressions for disgust, anger, sadness, and fear and have the same physiological changes accompanying these feelings. Does this suggest that nature (genetics) or nurture (culture) plays a more dominant role in determining facial expressions?

3. A blind infant smiles at her mother. Why is this evidence that smiling is an innate behavior?

4. A dog adopted between 4 and 8 weeks of age often develops closer ties to its owner than one adopted at an older age. Suggest an explanation for this.

5. Why is it foolish to stay up all night studying for an exam the next day?

6. How might human settlements disrupt nocturnal bird migration?

7. Pigeons are extremely sensitive animals and are able to detect polarized light, ultraviolet light, very low frequency sounds, and subtle shifts in atmospheric pressure. How might these abilities confound the design of experiments to test the hypothesis that pigeons can find their way home using magnetic cues?

8. Biologists are attempting to reintroduce salmon to the Connecticut River in Massachusetts. Once plentiful, salmon populations have plummeted due to industrialization, pollution, overconstruction, and especially the building of dams, which prevent their returning to home waters to spawn. What information about salmon behavior would help the researchers in their efforts?

SUGGESTED READINGS

Bogin, Barry. January 1990. The evolution of human childhood. *BioScience*, vol. 40, no. 1. Is our extensive childhood responsible for our complex learning abilities?

Coss, Richard G., and Donald H. Owings. May 1989. Rattler battlers. *Natural History*. Ground squirrels inherit behaviors to help them evade snakes.

Dybas, Cheryl Lin. October–November 1990. Secret creatures of the night. *National Wildlife*. American eels are born at sea, migrate to rivers inland, then return to the ocean to die.

Fair, Jeff. 1980. *The great American bear*. Minocqua: Northword Press. Bears may provide answers to our questions about animal navigation.

Levine, Richard J., et al. July 5, 1990. Differences in the quality of semen in outdoor workers during summer and winter. *The New England Journal of Medicine*. Sperm count—and birth rate—may follow a circannual biological clock.

Lewis, Ricki. January 1991. Can salmon make a comeback? *BioScience*. The migration of salmon spans hundreds of kilometers.

Miller, Julie Ann. February 1989. Clockwork in the brain. *BioScience*, vol. 39, no. 2. Mutant hamsters are used to study biological clocks.

Walcott, Charles. November 1989. Show me the way you go home. *Natural History*. The precise mechanism of homing is still a mystery.

Wiley, John P., Jr. March/April 1989. Night Travelers. *The Nature Conservancy Magazine*. Birds can migrate incredible distances.

Willis, Judith. July-August 1990. Keeping time to circadian rhythms. *FDA Consumer*. Much of how we feel is linked to our daily biological rhythms.

C H A P T E R

37

Social Behavior

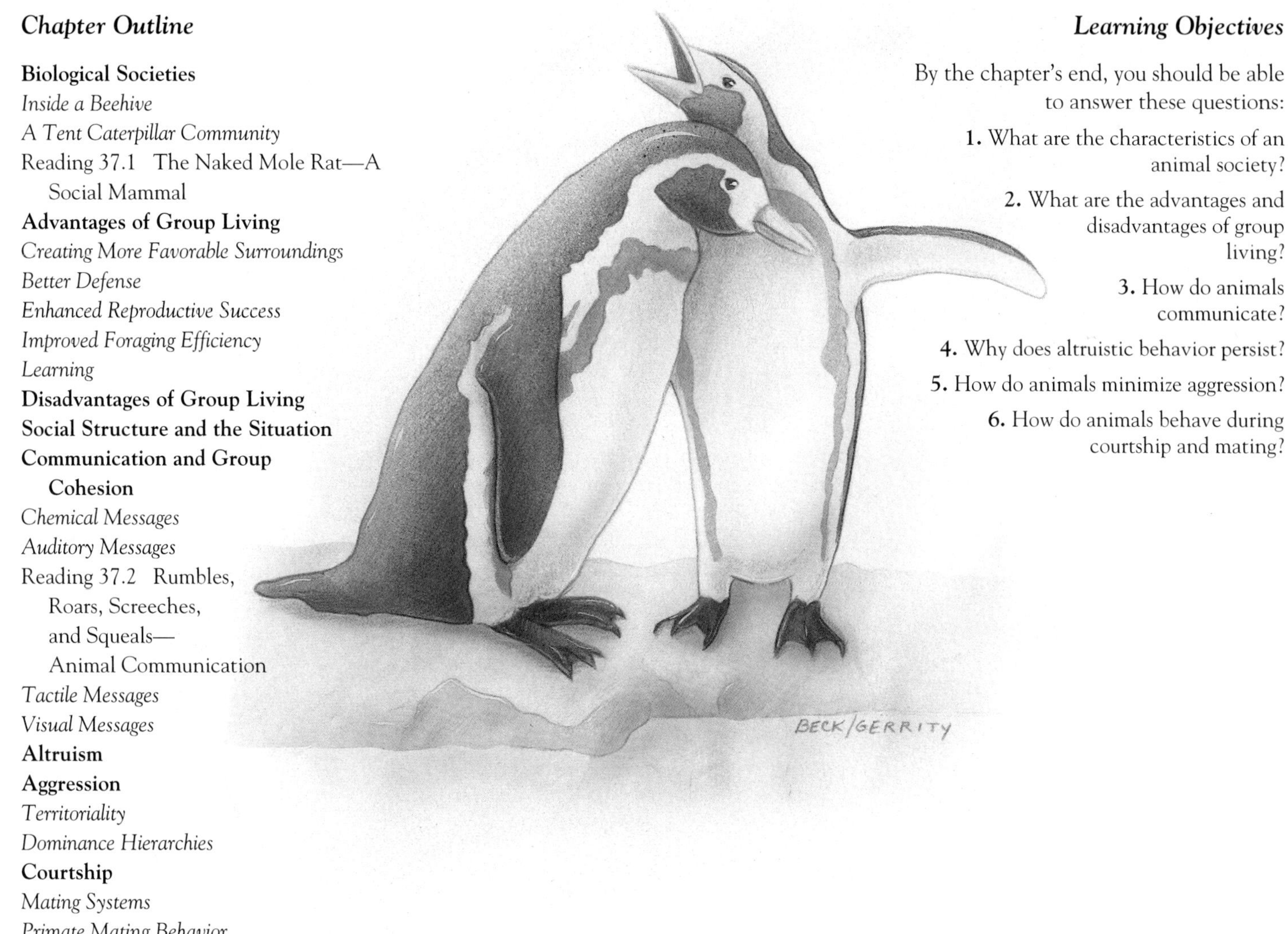

Chapter Outline

Biological Societies
Inside a Beehive
A Tent Caterpillar Community
Reading 37.1 The Naked Mole Rat—A Social Mammal
Advantages of Group Living
Creating More Favorable Surroundings
Better Defense
Enhanced Reproductive Success
Improved Foraging Efficiency
Learning
Disadvantages of Group Living
Social Structure and the Situation
Communication and Group Cohesion
Chemical Messages
Auditory Messages
Reading 37.2 Rumbles, Roars, Screeches, and Squeals—Animal Communication
Tactile Messages
Visual Messages
Altruism
Aggression
Territoriality
Dominance Hierarchies
Courtship
Mating Systems
Primate Mating Behavior

Learning Objectives

By the chapter's end, you should be able to answer these questions:

1. What are the characteristics of an animal society?
2. What are the advantages and disadvantages of group living?
3. How do animals communicate?
4. Why does altruistic behavior persist?
5. How do animals minimize aggression?
6. How do animals behave during courtship and mating?

On a damp, spring morning, all is still in the field. Suddenly, a swarm of red rises from one of the foot-tall mounds that dot the land. The red cloud hovers, as a second, smaller splash of red rises. The second cloud consists of female red fire ants, and as they pass through the upper haze of tiny males, they copulate, each female picking up a packet of 12 million sperm (fig. 37.1). The "lucky" males immediately drop to the ground, dead, their function to the colony served. Worker ants on the ground may pick up their deceased brethren as food.

The females drop to the ground too, shedding their wings. In groups of up to 20, these newly minted queens dig burrows together, into which they deposit their eggs. The first 15 or so eggs from each queen will not develop—they provide a meal for the others. The next 15 to 20 eggs from each female hatch into larvae. They are the workers of the next generation, whose job it will be, once grown, to feed their queen. By the time the larvae are thriving, all but one queen has died. She feeds the larvae a rich broth made of her own degraded flight muscles, which she regurgitates to them.

The red fire ant colony with its single queen reveals social behavior in the process of evolving. Such colonies are being replaced rapidly by societies of up to 500 very active queens. Whereas single-queen colonies inhabit one mound and stay distinctly separate from other colonies, the multi-queen mound is in chemical communication with nearby multi-queen mounds, forming a supercolony. Up to 500 such mounds, housing millions of ants, can occupy a single acre. So successful is this fire ant way of life that the insects demolish nearly every other organism in their midst.

Figure 37.1
A fire ant.

Biological Societies

The fire ants that ravage the southern United States offer a superb illustration of a biological society. In its strictest sense, such a **eusocial** grouping exhibits four characteristics:

1. communication among its members;
2. cooperative care of the young;
3. overlapping generations;
4. division of labor.

Biological societies are so efficient and predictable that they are often called "superorganisms." The evolutionary success of social insect colonies is thought to be due to division of labor. The colony is populated by coordinated groups of individuals, each with a specific "job" to do. One group might locate a new food source, while another assesses the sugar content of a prospective food. Other groups care for the young or rid the colony of dead members. When large groups of individuals perform only a few tasks, errors are minimized. The result is a maximization of the efficiency of the colony as a whole.

Several factors influence a social insect's fate, including nutrition, the temperature of the nest, and pheromones. An important determinant of an insect's job is age. For many species, the younger members dwell in the interior of the hive or nest, where they care for the queen and eggs. As the insects age, they venture outward, first helping to build the nest and finally, at the most dangerous post, guarding the colony. The insect society is a series of temporal castes—groups whose roles change with time.

The proportion of an insect society's population that carries out a specific task is constant, even though individual members proceed through the temporal castes. In laboratory studies, if some members of a caste are removed, the remaining members compensate by adding the tasks of the missing individuals to their own job lists. When the removed insects are returned, everyone resumes the original task assignments.

The best-studied eusocieties are those of ants, termites, and some species of bees and wasps, although other organisms may approach the social insects' level of organization (Reading 37.1). Fossil evidence indicates that insect societies were present 200 million years ago. However, the march of the fire ants, with the multi-queen phenomenon discovered in 1972, indicates that biological social structure is ever-evolving.

Inside a Beehive

The inside of a honeybee hive is a highly regimented, efficient living machine, with each individual knowing his or her place. At the summit of the organizational ladder is the *queen*, a large, specialized female whose major function is to keep the hive populated. She lays about 1,000 eggs each day. Those fertilized by a male *drone* develop through the characteristic insect larval and pupal stages into female *workers*. Unfertilized eggs develop into males. A very few eggs are treated to a substance secreted by the queen called *royal jelly*, which places them on a developmental pathway toward eventual queendom.

The 20,000 to 80,000 workers far outnumber the hundred or so drones. Only one queen reigns per hive. If the queen leaves, workers chemically detect her absence and hasten the development of the young potential queens. The first new queen to emerge from her pupa case kills the others, then takes a "nuptial flight" to attract drones to fertilize her eggs. A drone's life ends when he is stung to death by the workers who have fed him during his short life.

A worker's existence is more complex, including several stages and specializations (fig. 37.2). Newly hatched workers scurry about feeding everyone. The 1-week-old female makes and maintains the wax cells of the hive, where larvae and pupae develop. Some females are undertakers, ridding the hive of dead bees. Older bees, specialized as scouts or recruits, collect food.

A Tent Caterpillar Community

Other animal societies are not as strict as insect colonies, and its members may interact for only one or a few activities. Consider eastern tent caterpillars. Their

Figure 37.2
Honeybee workers build a chamber called a cell for a potential queen to develop in.

Reading 37.1 *The Naked Mole Rat—A Social Mammal*

THEY LIVE IN VAST, SUBTERRANEAN CITIES, DIGGING CONNECTING TUNNELS AND CHAMBERS IN THE SOIL OF KENYA, ETHIOPIA, AND SOMALIA. Roads built above collapse; crops that get in the way, such as yams, are eaten. The naked mole rats of Africa live in societies remarkably like those of eusocial paper wasps and sweat bees, with colonies of fewer than 300 individuals ruled by a powerful queen. Unlike most mammalian social groups, in which many females reproduce, among the naked mole rats the queen is the sole sexually active female.

Other roles in the naked mole rat community are well defined. The "janitors," the smallest and youngest males and females, patrol the colony's tunnels. They keep the walls smooth by rubbing them with their soft, hairless bodies. Janitors eat away roots hanging down from the ceilings and keep excrement confined to the widened dead ends that function as communal bathrooms.

The largest mole rats serve sentinel duty, their senses sharpened to detect intruders. A sentinel will stop and feel the air currents on its nearly sightless eyeballs. A change in current indicates an invader. The sentinel senses the low-frequency sound of footsteps overhead and detects many odors. If these signs, for example, indicate that a snake has poked its head into a tunnel, the sentinels somehow send a message telling the others to descend to the deeper, wider tunnels.

Figure 1
Naked mole rats from different colonies attack each other viciously, each trying to drag the enemy into its own territory to sink its teeth into the other animal's flesh.

In early morning and late afternoon, new tunnels are dug. "Dirt carriers" form groups, with the head animal a "digger" and a "kicker" at the back to get rid of the dirt. Food carriers use their long incisors to break off bits of roots and tubers but often sneak snacks before returning to the group.

The queen is truly in charge. She patrols the tunnels several times an hour. Every 3 months, she gives birth to a litter of up to 27 pups, whom she nurses for a month while three male workers bring her food. For the next 2 months, these workers care for the young, who graduate to worker status by their third month. Apparently the queen's presence is crucial to the integrity of this social colony. In the laboratory, when the queen is removed, chaos ensues. Jobs vanish, and each individual fares for himself or herself.

treetop silken communities are centers for exchanging information about food locations—a culinary clearinghouse of sorts. Several times a day, contingents of 200 caterpillars leave to explore new food sources (fig. 37.3). On the way out, they leave a trail of silk strands permeated with a steroid secreted from the insects' abdomens. This is the "exploratory trail." Should they find food, such as an apple tree, they return to the tent along the trail, dragging their hind ends through the original markings. The path is now a "recruitment trail" that the others back at the tent will know leads to food. The caterpillar's communication and cooperation skills seem to be restricted to foraging behavior.

Social behaviors such as communication and cooperation may be shown by groupings as small as an individual family. For example, a mother beetle of genus *Necrophorus* makes a home for her offspring in a ball of rotting flesh. The young communicate their hunger by rearing up and tapping the mother's mouth parts. In response, she regurgitates food to them.

Figure 37.3
Eastern tent caterpillars cooperate to locate and advertise a food source.

KEY CONCEPTS

A eusocial colony exhibits communication, cooperative care of the young, overlapping generations, and division of labor. Some insect societies are eusocial. Other animal societies may fulfill only some of these requirements.

Advantages of Group Living

Groups of animals, from the dozen or so lions in a pride to the intricate insect societies of thousands, enjoy food, protection, and care of the young more easily than more solitary types. So successful have these cooperative animal societies been to both individual and species survival that they have evolved independently many times among insects and vertebrates.

Creating More Favorable Surroundings

By forming groups, animals may change their environment to their advantage. The simplest sorts of physiological effects are seen in animals that group together to conserve moisture or heat. During the winter, pigs generally sleep in heaps, continuously moving and squealing as those on the perimeter seek the warmth of body heat inside the pile (fig. 37.4).

A group can physically alter its surroundings. Water fleas cannot survive in alkaline water, but when large numbers of them congregate, the carbon dioxide produced by their respiration decreases the alkalinity of the water to acceptable levels. Fruit flies fare badly when too few eggs are laid in an area, because there are not enough larvae to break up the medium and make it soft enough for them to feed. Maintaining prairie dogs' burrows depends upon the animals' constant chewing of the vegetation—otherwise their homes would easily be overgrown.

Better Defense

There is safety in numbers. A predator can more easily fell a lone prey than pick one from a group where a few alert individuals warn the others. Birds yell their warnings. Vervet monkeys have different calls to warn of the approach of an eagle, a leopard, or a poisonous snake. These different signals are adaptive because the route of escape differs depending upon whether the predator strikes from the sky or the land.

Figure 37.4
A pile of pigs conserves heat.

Figure 37.5
Ostriches protect their eggs. A major hen will sit patiently on eggs not her own. But she places her own eggs directly beneath her body, leaving the eggs of others more available to predators.

Group defense can be passive. For example, some small fishes swim at specific distances and angles from each other, forming a **school.** When faced with a school, a predator often suffers a **confusion effect** and is unable to decide which fish to attack. In the presence of a faster-swimming enemy, the school uses the **fountain effect,** splitting in two and regrouping behind the baffled predator.

Ostriches use the **dilution effect** of passive grouping to protect their young. Their societies consist of bonded pairs of a cock and a major hen, as well as minor hens, who mate with many males. The minor hens, however, must deposit their eggs in the nest of a bonded pair, who sit on them, with their own eggs, for 6 weeks. Why would a bonded pair take in eggs not its own? Possibly because if not all of the eggs are theirs, chances are that some of those lost to predators will not be theirs (fig. 37.5). The dilution approach is also practiced by caribou and wildebeests, who round up their young into large herds.

Sometimes the best defense is a good offense, with help from others. Many prey species engage in mobbing behavior, where adults harry a predator that is frequently larger than any individual of the group. Redwing blackbirds make hit-and-run attacks against owls, uttering shrill calls. Baboons and chimpanzees use a similar strategy against leopards. Screaming, they charge and retreat and maybe even throw sticks at the leopard.

Forming a circle with the most formidable parts of the body facing outwards is a common defense strategy. When young catfish are disturbed, they mass together with their large pectoral fins projecting in all directions, resembling the thorns on a cactus. Likewise, adult musk oxen form a ring with their heads pointed outward toward the attacker. The older animals face the threat, while the young are sheltered within the woolly wall.

Enhanced Reproductive Success

Failure to reproduce is evolutionary suicide. An increase in the number of animals in an area may not only raise the chance of finding a mate, but it may also trigger physiological changes in individuals that are necessary for successful reproduction. For example, a mouse's egg may be fertilized, but in the absence of other breeding pairs, the fetus may be reabsorbed into the mother's body. A pig's ovaries may not develop normally unless she has a chance to hear and smell a boar.

The sights, sounds, and scents of other courting individuals appear to enhance and synchronize breeding. Spotted salamanders provide an extreme example of explicitly timed group mating. Each spring in the Midwest, for two to seven consecutive nights, the salamanders perform their courtship dances on pond bottoms, the males depositing bundles of sperm and coaxing females to take them into their bodies. Besides these few evenings when the pond bottom is aswarm with the yellow-dotted black amphibians, the animals do not seem to interact socially at all.

Reproduction occurring simultaneously throughout a population may have additional benefits. By effectively glutting the predators with potential food for a short time, the total number of offspring eaten is minimized. Synchronous breeding may be an advantage if it coincides with the abundance of seasonally available food.

Improved Foraging Efficiency

If food is plentiful in some places and scarce elsewhere, it helps enormously to have others around to locate good feeding sites. When one member of a flock of starlings finds a food source, for example, other birds rapidly change their searching strategies and concentrate both on the general area and on the type of food located. Some birds living in groups establish food stores available to all members to tide them over during periods of food scarcity.

Spiderwebs vividly illustrate the advantage of group living in securing a meal. The community spiders of genus *Stegodyphus* that live in South and East Africa build vast silken empires, 43 feet (4 meters) square, housing hundreds of spiders. At night during the wet season in early autumn, the spiders busily repair their netting, torn in the daytime by visitors, rain, and wind. Maintaining the webs is essential, for these structures very efficiently trap the spiders' food—gnats, termites, ants, stinkbugs, and even huge beetles and grasshoppers (fig. 37.6). By cooperating to erect this food trap, the spiders enjoy a rich and varied menu that an individual could not subdue on its own. An added benefit—the web traps predators.

Figure 37.6
Prey easily becomes trapped in the sticky webs of community spiders.

Learning

Sociality may enhance learning and the passage of tradition. Many animals acquire information by watching others (fig. 37.7). For example, in England around 1921, certain birds learned to break into milk bottles to steal sips of cream. As other birds watched the thieves, they learned this trick and the technique spread throughout the continent. Milk boxes were invented to outsmart the birds. Similarly, in Japan a monkey washing sweet potatoes and wheat in the sea was imitated by others in her troop. Her activity cleaned and seasoned the food. The capacity for rapid transmission and assimilation of new adaptive behaviors has been prominent in human evolution.

Disadvantages of Group Living

Individuals may suffer as a result of group living, although the group is not endangered. The ostrich egg carried off by a vulture or the young wildebeest felled by a predator as it straggles at the back of its herd certainly do not benefit from their species' social behavior. Group living is disadvantageous when members have to compete for scarce resources.

Colonial living also makes it easy for infections to spread. Consider honeybees, whose numbers are dwindling due to mite infections. Microscopic tracheal mites collect in the breathing tubes of the bees, killing them by suffocation or sucking blood. Mated female mites crawl from the tubes and lie in wait on the tips of the bees' hairs, where they are easily picked up by another bee (fig. 37.8). Beehives infested with tracheal mites are morgues by a winter's end.

Figure 37.7
Learning is a benefit of sociality. These cats learned how to raid trash cans for a meal by observing others.

Social Structure and the Situation

The social structure of a species is influenced both by the animals' physiology and their surroundings. When related species occupy similar habitats, yet behave differently, the cause of the diversity may be difficult to identify.

Great differences in social behavior are seen, for example, between two related groups of monkeys that live among the treetops of Peru (fig. 37.9*a*). Brown capuchin monkeys are the larger and stronger of the two species. They live in groups led by a dominant male. One female is dominant and sticks close by the male. When in estrous, she aggressively pursues him, until he mates with her. The other two or three females are submissive. Other males are submissive to the head male, and the females seek only the head male for mating.

Figure 37.8
Parasitic mites spread quickly in beehives. In crowded beehives, both microscopic mites and these large varroa mites are easily passed from insect to insect.

a.

b.

Figure 37.9
a. The dominant male brown capuchin monkey allows an infant near his food—probably because he is the father. *b*. White-fronted capuchin youngsters stick together. Adult males show little, if any, interest in infants and juveniles, possibly because each male does not know which offspring is his.

When plentiful food is found, such as a fig tree, the brown capuchins live peacefully. If food is scarce, the dominant male locates tough, fibrous fruits and opens them. But he is very particular about with whom he shares his dinner. He readily allows infants to eat but screeches at any males or nondominant females who approach. When he is close to eating his fill, he allows the dominant female and youngsters to feed. Next come the nondominant females. Only when the head male is satiated and moves away from the food can the submissive males safely approach. If anyone threatens an infant, the dominant male becomes violent.

Life among the white-fronted capuchins is considerably more tranquil. Their social groups are slightly larger, with up to 20 individuals, and consist of three dominant males, plus females and offspring. All of the adults mate with each other. Although little aggression is shown among adults, the dominant males ignore or are hostile to the juveniles, who are very submissive to them (fig. 37.9*b*). The white-fronted capuchins feed farther up in the forest canopy than their brown cousins, on trees abundant with soft fruits.

Food availability and mating practices are key to the differences in aggression between the two species of monkey. The spotty food supply fosters aggression among the brown capuchins. The dominant male protects his offspring because he is their father. With abundant, easy-to-eat food, aggression is rare among the white-fronted capuchins. However, the males do not bother with the juveniles. Perhaps this is because each adult cannot know which offspring he has fathered.

KEY CONCEPTS

Organisms in groups can keep warm or actually alter their environment. Individuals can cooperate to defend against predation or to find food. Some organisms can only mate if others nearby are mating. Synchronization of reproduction can minimize the number of young lost to predators. Learned adaptive behaviors are passed among members of a group. A species' social structure is influenced by the animals' physiology and by their physical surroundings.

Communication and Group Cohesion

Communication is at the root of behaviors requiring cooperation among individuals, such as foraging, defense, aggression, and mating. Animal signals may be chemical (taste or smell), auditory, tactile, or visual.

Chemical Messages

The best way to send a message depends upon the environment. Beneath the sea, for example, light does not travel far, so visual communication is not very useful. Sound, however, travels far under water. The songs of whales can be heard hundreds of miles away (Reading 37.2).

Lobsters are practically blind, but their bodies are covered with millions of tiny hairs that function as chemoreceptors, providing a sense of smell 1,000 to 1 million times more sensitive than our own. Lobster mating is very dependent upon chemical cues. When a female enters a crevice and faces its male occupant, the two stand face to face, flicking their antennules to sense whether the other is ready to mate (fig. 37.10). After various stereotyped motions, the female ejects a fluid from her pores containing heavily scented urine. The stream hits the male in his antennules, exciting him, and he fans the odor outward. Then the female shrinks and sheds her shell, he flips her onto her back, and they mate.

Auditory Messages

Elephants are masters of auditory communication. These gentle giants live in matriarchal family herds consisting of sisters, cousins, and their offspring (fig. 37.11). Males roam alone or in other groups. Related families greet each other with a cacophony of rumbles, trumpets, and screams.

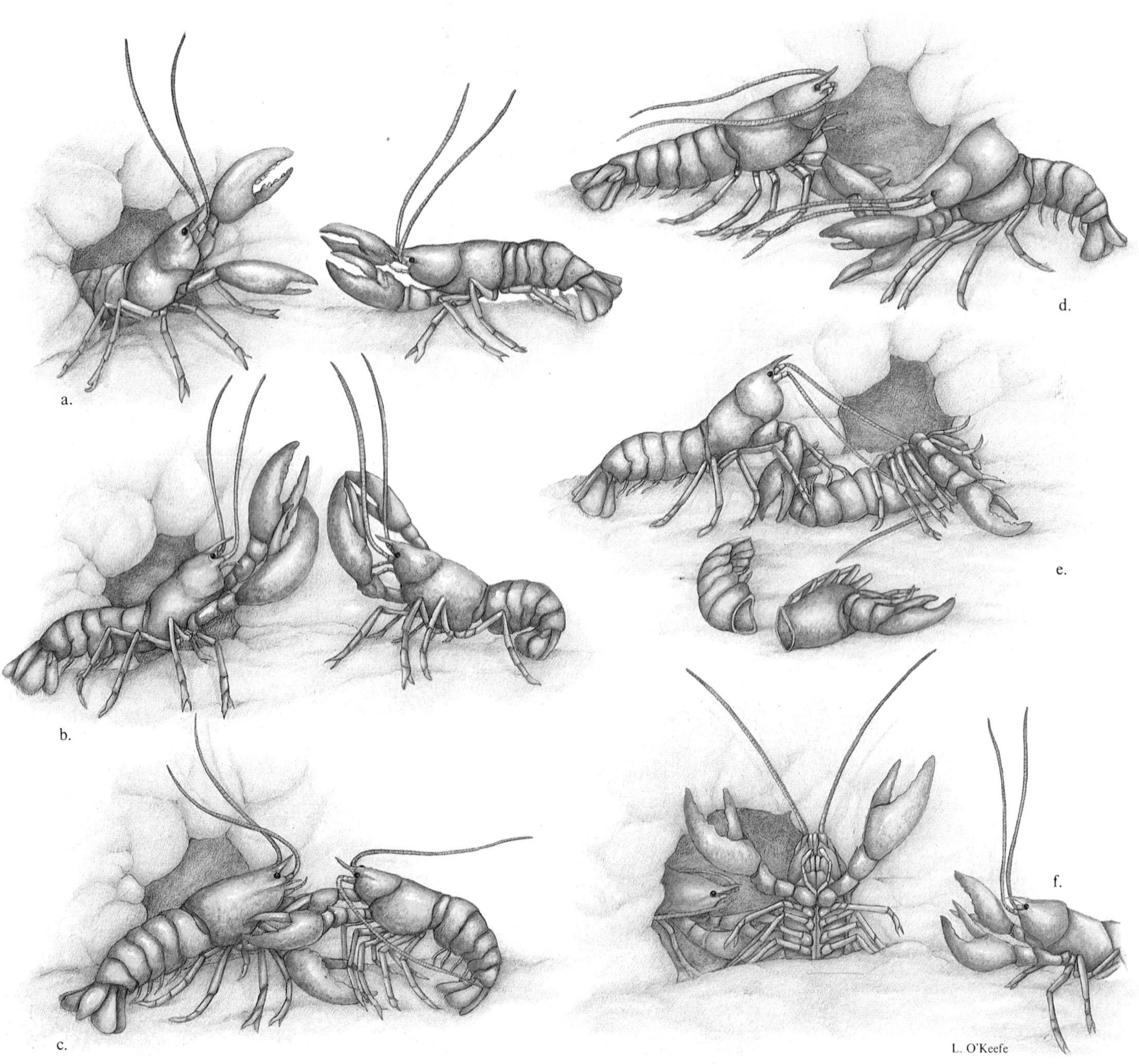

Figure 37.10
Chemical cues are important in lobster mating. *a*. Courtship starts when a female approaches a male in his crevice. *b*. She "knights" him by raising her claws, announcing her intent. *c*. The female's body shrinks, and (*d*) she sheds her shell. *e*. The male turns her gently on her back, and they mate. *f*. For the next week, he shelters her as her new shell hardens. At certain points in the courtship, the female emits a fluid rich in urine, which excites the male—and possibly others.

Reading 37.2 Rumbles, Roars, Screeches, and Squeals—Animal Communication

THE RESIDENTS OF THE HOUSEBOAT COMMUNITY IN SAUSALITO, CALIFORNIA, WERE PERPLEXED AND WORRIED. What was that nightly, hour-long rumbling of a single note? They feared it was a secret military experiment, until Cornell University bioacoustics specialist Andrew Bass brought cameras and computers to the area and located the source—a scaleless, flatheaded, bulgy-eyed fish known now in Sausalito circles as the "humming toadfish" and to ethologists as the plainfin midshipman, or *Porichthys notatus*.

The annoying, foghornlike sound was the love song of male midshipman trying to lure females to lay eggs in their nests. The song is produced by muscles contracting along the sides of a pouch called a swim bladder, which inflates and deflates as the fish rises or submerges in the water (fig. 1).

The midshipman is but one animal whose language is being deciphered by ethologists specializing in bioacoustics—the sounds of life. Another Cornell researcher, Katharine Payne, has monitored the complex songs of humpback whales.

Figure 1
Midshipman nest under rocks near the shore, and their sounds keep nearby houseboat residents awake.

She is also well known for noticing, while observing three newborn elephants in a zoo, that the animals emit obviously meaningful low rumbles barely beyond the range of the human ear. This discovery solved a long-standing mystery among elephant-watching ethologists at Kenya's Amboseli National Park. They had wondered how herds could expect the approach of another herd miles away. The low rumbles were the answer.

Bioacoustics strives to show how different species use sounds to convey a message. A device called a sonograph converts sound waves into electrical signals, providing a common means of expression for the various grunts, squeals, and yelps of the animal kingdom. For example, a dog's growl and a bird's call produce similar patterns of dark bands on a sonogram, indicating that they both send the same message—WARNING! In general, warnings tend to be low and menacing, and sounds of submission are higher and whiny.

When a female approaches a male to mate, she bellows her intent. Her relatives watch, trumpeting loudly.

Tactile Messages

Social bonds are also cemented by physical contact. Touching can reduce tensions. Greeting ceremonies, common in social animals, often involve touching and sometimes even embracing. Members of a wolf pack may surround the dominant male and lick his face and poke his mouth with their muzzles. This ceremony occurs at times when it is useful to reinforce social ties, such as upon awakening, after separations, and just before hunting.

Touch is common among nonhuman primates, who often sit with their arms about one another. These animals groom each other's fur, using their hands, teeth, and tongue. Grooming removes parasitic insects and prevents infection and also promotes social acceptance. Another function is the most obvious of all—it feels good (fig. 37.12).

Figure 37.11
Elephants have strong family bonds.

Figure 37.12
A female brown capuchin monkey grooms her mate.

Visual Messages

The importance of vision in the foraging behavior of honeybees was described by ethologist Karl von Frisch. In 1910, he read a paper stating that bees are color blind. Why, he wondered, are flowers so brightly colored if not to attract the bees that pollinate them? To demonstrate color vision in bees, von Frisch placed sugar water on a blue cardboard disc near a hive. The bees drank the water. Next, he placed blue and red discs near the hive, and the bees went only to the blue disc. Similar trials with different colors and shades showed that bees not only see color (except red), but that they can also detect ultraviolet and polarized light, which humans cannot see. Von Frisch eventually deciphered the "dances" that bees use to communicate the location of food sources.

As a forager or scout bee dances, recruit bees cluster around her, sensing her movements with their antennae. If the food is close to the hive, the scout dances in tight little circles on the face of the comb. This **round dance** incites the others to search for food close to the hive. The longer the dance, the sweeter the food (fig. 37.13*a*). A **waggle dance** signifies food that is farther from the hive. A path similar to a figure eight is traced on the honeycomb. During the straight part of the scout bee's run, she vigorously shakes her abdomen. The speed of the dance, the number of waggles on the straightaway, and the duration of the buzzing sound correlates well with distance to the food source (fig. 37.13*b*).

The orientation of the straight part of the run indicates the direction of the food source. A dance straight up on the comb tells recruits to fly directly toward the sun. A dance straight downward indicates food directly away from the sun. To specify other directions, the dancer changes the angle of the waggle run relative to gravity so it matches the angle between the sun and the food source. Bee dancing is a pretty complex behavior for an animal whose brain contains only 800,000 neurons!

For years, researchers have attempted to decipher the finer nuances of the dances of bees by inventing models to entice them into performing. Until recently, all efforts have failed. The first robot bee to be accepted in the hive revealed what its unsuccessful forerunners lacked—sound (fig. 37.14). The computerized bee dances, vibrates, and delivers sugar water from its front end, taking into account the angle of the sun. Unlike humans, who sense sound as pressure changes, bees hear by sensing the movement of air particles.

KEY CONCEPTS

Communication is important for social behavior, and the type of communication depends upon the environment. Information can be sent by taste and smell, sound, touch, and vision.

Altruism

Altruism is a behavior that harms the individual performing it but helps another organism. Examples of altruism abound. A mother bird risks her life by staying in her nest to protect her young. A worker bee defends its hive by stinging an intruder, even though it will rip out its insides when trying to leave the scene.

Altruistic individuals are less likely to survive and reproduce than are those they assist. This seems to run counter to the evolutionary demand that an individual behave to increase fitness. How could the gene(s) behind altruistic behavior persist in a population if those possessing them are more likely to die or not produce offspring than the more selfish members of the group?

Kin selection is one explanation for altruism. This model was derived from theoretical work by British geneticist William Hamilton in 1964. It proposes a broader definition of fitness called **inclusive fitness,** which considers both personal reproductive success as well as that of relatives that share some of one's genes.

The genes of each diploid offspring of sexually reproducing parents represent one-half of each parent's genotype. When a parent protects its offspring, it is ensuring that copies of half its genes will enter the gene pool of the next generation. Other relatives also share some proportion of the individual's genes. Siblings share half of their genes, the same proportion as is shared between parents and offspring. First cousins share one-eighth of their genes. The more genes in common, the greater the chance that the gene(s) favoring altruism will be present in the altruist's relative.

Natural selection works through differing reproductive successes of individuals. Another way to look at reproductive success is as competition among genes. Since closer relatives share more genes, the degree of altruism should be proportional

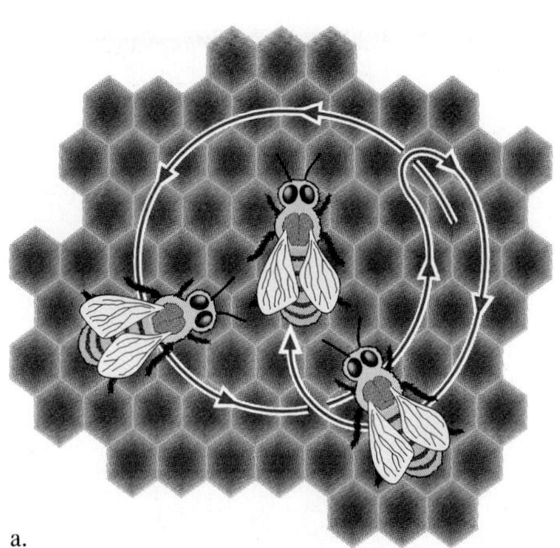

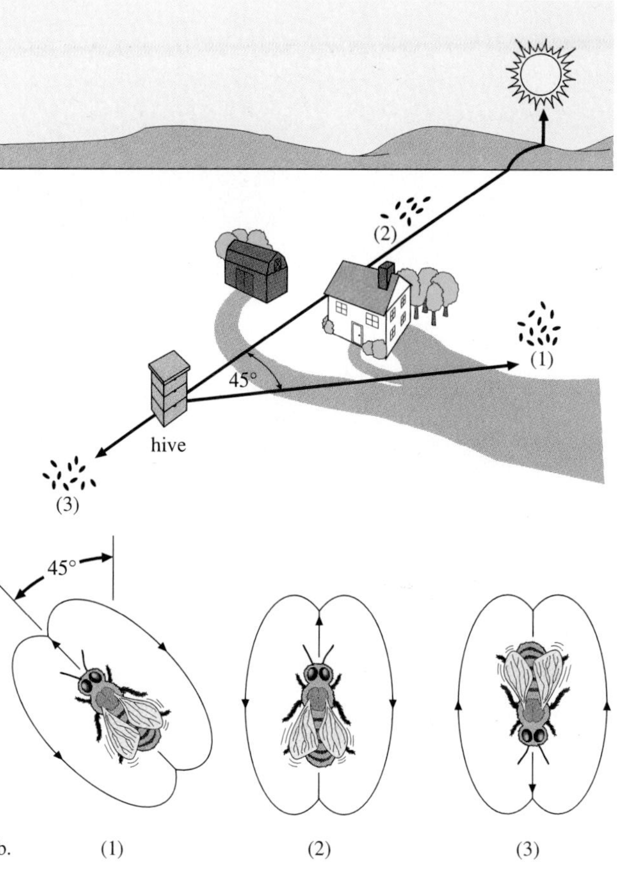

Figure 37.13
a. A round dance informs recruits that food is close by but does not point them in the right direction. *b.* A waggle dance tells recruits the distance and direction to the food source. The dancer loops once to the right, waggles her abdomen along a straight path, and then loops to the left. Distance is indicated by the number of waggles and the duration of buzzing during the waggle run and the speed of the dance. Direction is indicated by the angle of the waggle run relative to gravity. This angle corresponds to the angle that a worker must assume with the sun as she flies to the food source. If the angle formed by the sun, the hive, and the food source is 45° to the left of the sun, then the angle of the waggle run relative to gravity is 45° to the left of vertical (1). If the worker should fly directly toward the sun, the scout dances so that the waggle run is straight up (2). When the worker should fly directly away from the sun to reach the food, the waggle run points down (3).

to the number of shared genes. Therefore, in parent-offspring or sibling relationships, with the individuals sharing 50% of their genes, an altruistic act should at least double the reproductive success of the surviving relative. Surrendering one's life for a first cousin should increase the cousin's reproductive success eightfold. Said population geneticist J.B.S. Haldane, "I would lay down my life for two brothers or eight cousins."

Kin selection cannot, however, explain all instances of altruism. Consider "helper" African bee-eater birds, who forego or postpone raising a family of their own in order to help others raise offspring. Helpers feed and protect the young against predators, enabling the parents to raise twice as many chicks as they could alone. When the interactions of 174 helpers were followed, it was found that 44.8% of them were adult siblings of the chicks, and 9% were half-siblings (fig. 37.15).

Helpers who are unrelated to the young they help raise may derive some benefit from their unselfish behavior, such as a home. A young Florida scrub jay, for

Figure 37.14
A robot bee unravels dance steps. To us it is a fake bee on a rod attached to a computer. A piece of razor blade on its "back" vibrates, the computer moves it, and sugar water sprays from its front. Bees are attracted to this model and display their dance steps to human observers.

Figure 37.15
African bee eaters help their relatives—most of the time. Using leg bands and color-coded wing tags, researchers document the roles of African bee eaters in a multigenerational clan. Most of the helpers who assist in chick-raising are close relatives of the parents.

Figure 37.16
A territory may attract a mate. This male long-horned wood-boring beetle, (*top*), has established his territory on a succulent saguaro fruit, which is likely to attract a female in search of food and a mate.

example, stays with its parents because all available space is occupied. It must wait to inherit territory from its father.

KEY CONCEPTS

Although altruistic behavior appears to be contrary to natural selection, it may enhance the probability that an individual's genes are transmitted to the next generation.

Aggression

Aggressive behavior is often displayed when members of the same species compete for resources, such as mates, food, shelter, and nesting sites. Competition can be fierce, for the stakes are high. Winners survive and leave more offspring than do losers.

Territoriality

One way that animals distribute resources with a minimum of aggression is to defend an area of their habitat against invasion. This defended area, or **territory,** guarantees them access to the resources on that land. The territory may be defended against all members of the species or only those of the same sex. A territory may be fixed, such as a system of prairie dog burrows, or moving, such as baboon settlements. Some territories are held by an individual; others are defended by a group.

Territoriality is widespread in the animal kingdom. Male birds often occupy territories, repelling other males while enticing females to enter. Male insects sometimes take over a food resource to which females are attracted (fig. 37.16). Territories may serve groups of animals, who defend it against strangers from other groups of their species. Group territories are maintained by wolf packs, hamadryas baboons, and pack rats.

Some animals defend territories only during the breeding season, but others hold them year round. Territory size also may vary tremendously. A flightless cormorant defends only the area it can reach while sitting on its nest, and many birds that breed in colonies nest just out of reach of one another. Other territories are large. Tiny African weaver ants defend territories up to 2,153 square yards (1,800 square meters).

Territorial Markers

If territorial boundaries are to be respected, ownership must be indicated in some way. Although fighting may accompany establishment of territories, once boundaries are set, they are generally respected without contest. Actual combat is usually avoided.

An animal becomes less aggressive when it crosses the boundary to leave its territory. A male bird, for example, is most self-assured near the center of his territory. As he approaches the boundary, his confidence falters. The point where the male is as likely to attack an intruder as to flee from it marks the borderline of the territory.

Sometimes a territory is marked by visual advertisement, involving display of **threat postures.** With as much bluff and bluster as can be mustered, each individual tries to appear as intimidating as possible. Two male fish meeting at a territorial boundary puff themselves up. If neither contender submits, they then begin a lateral display. Each propels water against the other, as they assess each other's size and strength. Threat displays often emphasize natural weapons of the opponents. The site of a male baboon's fangs, for example, is enough to encourage respect from any potential opponent.

Territories may be indicated by a strong scent produced in the resident's glands. Badgers, martins, and mongeese have scent glands at the base of the tail. Ungulates may have hoof glands, and the rabbit has both chin and anal glands. Sometimes urine is a territorial marker. The giant galago, a primate, urinates on his hands, rubbing it on his feet and walking around on its territory. Hippopotamuses define their homes with solid waste.

Figure 37.17
A red deer stag roars when another competes for a harem. Roaring requires a great deal of strength and stamina. Threat displays such as this one allow the rivals to assess one another without actually fighting. The weaker male will usually leave without contest.

Some animals vocally proclaim ownership of a territory. A toadfish growls. A bull alligator roars. A male bird sings his presence to opponents. A wolf howling at the moon proclaims his possession of an area.

Traversing Territories

When animals ignore territorial markers, threat behavior follows. A cat arches its back, fluffing its fur to appear larger. A horse prances about, tossing its head and kicking. The sea anemone withdraws its tentacles and inflates a collarlike structure. If one animal is clearly superior, the other may leave without a challenge.

A threat display indicates an animal's strength or endurance. Red deer stags have vocal duels over harems, each taking a turn roaring at the other in an escalating bellowing battle (fig. 37.17). Because roaring is so exhausting, the ability to continue indicates strength. Usually, one male gives up, leaving the other victorious.

At any time during a battle, one combatant can call off the fight with a display of submission, without gloating retaliation from the victor. In wolves, the appeasement gesture is exposure of the vulnerable throat area, which also removes the primary sources of threat—the eyes and mouth. Because males are often inhibited against attacking females, males of some species, such as baboons, indicate submission with the characteristic female behavior of presenting his genitals to the dominant male. In contrast to a threat, in which many organisms attempt to appear larger, appeasing animals try to seem smaller.

Another response to a threat is diversion. The "loser" behaves in a way unrelated to the situation to distract the aggressor. A dog might roll over, paws up, eliciting parental feelings from the other dog rather than a fight. In displacement behavior, both parties cannot decide what to do, so they pursue another activity. Two dogs at a standstill might investigate a fire hydrant instead of tearing each other apart.

When competition does lead to combat, it is more like a tournament with a set of rules than a bloody fight to the death. Although roe deer have antlers that could easily gore an opponent, they are not used against members of the same species. Two males face each other, antler to antler, and push. One buck readied for antler clashing waited for his opponent to finish his lateral display and protect his vulnerable flank.

Dominance Hierarchies

Another way to distribute resources with a minimal amount of aggression is by forming a **dominance hierarchy,** a social ranking of each group member relative to every other adult of the same sex. One of the simplest dominance relationships is a linear hierarchy, in which one animal is dominant over all the rest. A second animal is dominant over all but the first, and a third is dominant over all but the first two, and so on. This type of relationship

Table 37.1
Social Groups

Organism	*Location*	*Mating*	*Social group*	*Territory*
Elephant	African plain	Roving males	Extended family	
Honeybee	North America	Only queen	20,000-80,000 females, 100s males , 1 queen	Hive
Horse (feral)	Bird Shoal Island, east coast, U.S.	Polygyny	2-8 females/males, 6 groups	4 mile strip of land
Lion	Namibia	Polygyny	Pride = 17-19 members, rovings males	Dry, salty lake bed
Magpie	Australia	1 male/2-3 females	3-24 members, roving males	2-18 hectares
Peccary	Argentina	Monogamy	12-15 members	1 square mile

was first studied as the pecking order of domestic hens. Each hen knows who it can peck and who it may be pecked by. Hierarchies in other organisms are often more complex. Female Amboseli baboons, for example, are ranked according to the age of their mothers and the age differences between them and their sisters.

The dominance arrangement is set up during the initial encounters of the members of the group, when there may be many threats and an occasional fight. Once the status of each individual is determined, life within the group is generally peaceful. At these times, it is often possible to determine the degree of dominance of an individual by subtleties of posture. Subordinates indicate submission by lowering their heads and tails and flattening their ears.

KEY CONCEPTS

Animal behavior minimizes aggression. By establishing and marking territories, animals set aside resources for themselves. If boundaries are not respected, threats precede actual fighting, and battles are not as deadly as they might be. Dominance hierarchies minimize aggression.

Courtship

With dominance relationships and territorial boundaries keeping them apart, how do animals come close enough to mate? **Courtship rituals** are stereotyped, elaborate, and conspicuous behaviors that overcome aggressive tendencies long enough for mating to occur. The colorful display of a male peacock's feathers is a vivid example of a courtship ritual (see Unit 8 opening page).

The zigzag dance performed by a male stickleback fish when a female enters his territory, described in chapter 36, illustrates how courtship rituals quell aggression. The zig portion of the maneuver, when the male swims directly towards the female, indicates his aggressive reaction to her invasion of his territory. She tempers his response by revealing her silver belly, swollen with eggs. He swims away from her to lead her to the nest during the zag portion of the dance. As he moves away, though, the female's belly is hidden from his view, so he turns back to attack her, zagging again. She counters by displaying her swollen abdomen. This zigzagging eventually ends at the nest.

Males also behave to reduce aggression in the female. A male orb weaving spider may stand at the edge of a female's web and tug on the suspension cord to announce his presence before venturing in. A male wolf spider cautiously approaches a female and, keeping a safe distance, signals his intent to mate by waving specially adorned appendages.

Courtship displays are specific, preventing the costly error of mating with the wrong species. Male fiddler crabs, for example, lure females into their burrows by waving their large claws, but the exact pattern of movement is unique to different species. Many courtship behaviors hold attractions that we humans simply cannot appreciate. A female jackdaw bird, for example, stuffs regurgitated worms into the ear of her mate.

Mating Systems

Various mating systems are seen among animals. In **monogamy,** a permanent male-female pair is established. In **polygamy,** a member of one sex associates with several members of the opposite sex. The most common polygamous arrangement is **polygyny,** in which one male mates with several females. The sexes of polygynous species such as lions are usually very different in appearance and are said to be highly **sexually dimorphic.** Competition among the males is often intense. In contrast, in monogamous species the sexes usually look more alike. Still other species, including frogs, fruit flies, and antelopes, engage in communal mating displays. Group sex is also associated with decreased sexual dimorphism, perhaps reflecting the lack of competition in a sharing situation.

Related species can have vastly different mating systems, as the brown and white-fronted capuchin monkeys illustrate. Differences may reflect habitats. The white-tailed deer is rather antisocial and monogamous and lives in densely wooded forest. The mule deer is social and polygynous and dwells in the open. It is unclear, however, whether preference for a mating system leads an organism to seek a certain environment or if life in a particular environment makes certain mating systems more practical. Whatever the causes, mating systems are quite diverse.

Primate Mating Behavior

Nonhuman primate societies are a favorite among ethologists because of the evolutionary closeness of these animals to ourselves. The primates are indeed a varied lot. They form male-female pairs, one-sex bands, mixed-sex troops, harems, loners, and social groups that have frequently changing members. Devoted monogamous gibbon pairs sing together in the early morning to defend their territory. Orangutans are polygynous, with a female mating only once every few years.

Primate sexual behavior varies even within species. An aggressive male mountain gorilla savagely rapes and kills an older female who cannot keep up with the troop; yet other mountain gorillas form long-lasting bonds (fig. 37.18). Some chimpanzee males prefer to mate with an unfamiliar female. Other males take turns mating with a single female. Still other chimpanzees spend their lives in monogamy. Every mating system observed in nonhuman primates has also been practiced by humans.

It can be difficult and disturbing to scrutinize human behavior as we would that of other members of the animal kingdom. However, the fact that behavior in many other species is partially genetically determined—from simple fixed action patterns to complex social interactions—suggests that the same is true for us. Such information is frightening when applied to us because it has been the basis of much prejudice and persecution, by one group of people presuming that another is genetically inferior or incapable of some skill. It may help to realize that human behavior, perhaps even more so than that of other animals, is molded by the environment, as well as directed by the genes, leaving much room for modification.

Figure 37.18
Mountain gorillas form deep and lasting bonds to one another.

KEY CONCEPTS

Courtship rituals are stereotyped, conspicuous, and elaborate. They temper aggression so that individuals can mate with an available member of the same species. Mating systems include monogamy, polygamy (particularly polygyny), and variations.

SUMMARY

A *eusocial* group communicates, cooperates in caring for young, has overlapping generations, and divides labor. Other social groups have some of these characteristics. Advantages of group living include ability to alter the environment, defense, improved reproductive success, more effective foraging, and the opportunity to learn. Group living is disadvantageous by increasing competition among its members for resources and by enhancing the spread of infection. A species' social structure reflects physiology as well as the environment. Many behaviors are based on communication, which may be chemical, auditory, tactile, or visual.

Altruistic behavior helps an individual other than the one performing the good deed, who suffers. *Altruism* may have evolved by *kin selection*. *Inclusive fitness*, the basis of kin selection, is a measure of an individual's fitness as the sum of personal reproductive success and that of relatives who share the same genes. Altruism toward a relative increases one's inclusive fitness.

Animals are most likely to fight with members of their own species because they compete for resources. *Territoriality* and *dominance hierarchies* minimize aggression. Threats and appeasement, diversion, or displacement responses often prevent actual combat, which, if it ensues, is restrained.

Courtship rituals calm aggression so that individuals of the same species and physiological readiness can mate. Mating systems include *monogamy*, *polygamy*, and group sex.

QUESTIONS

1. How does a honeybee colony meet the criteria of a eusocial group?

2. What are the advantages of group living?

3. How does the behavior of the brown and white-fronted capuchin monkeys support the theory of kin selection? Does the egg-sitting behavior of the major ostrich hen support or contradict the theory of kin selection? Cite a reason for your answer.

4. Give examples of humans engaged in a threat, appeasement, diversion, and displacement behavior.

5. A pet dog runs to the edge of his owner's property, barking and running to and fro. He looks fierce, but he will not venture past the property lines. What type of behavior is he demonstrating?

6. How is communication important in minimizing aggression and in promoting courtship and mating behaviors?

TO THINK ABOUT

1. A taxonomist working in a South American tropical rain forest uncovers a new species of ant. What signs would she look for to classify the species as eusocial?

2. In laboratories, naked mole rats are studied in elaborate plexiglass colonies; bees establish hives in glass-encased honeycombs built into a window; and spotted salamander mating is observed in huge bathtubs. Do you think that these setups can accurately assess animal behavior? Why or why not? Can you suggest alternative approaches for observing these organisms?

3. For many years it was thought that bees communicated the location of food sources solely by visual cues in their dances. Only recently has the role of sound been discovered. Similarly, the role of sound too low in frequency for us to hear has been described in elephant communication. What do these two observations indicate about our abilities to understand the behaviors of other animals?

4. Geneticists are systematically deciphering the DNA sequence of every human gene. Eventually, we will understand the function of each gene. Within the next decade, it may become possible for a blood test at birth to tell individuals which genetic tendencies they have inherited. Do you think that knowledge of inherited behavioral tendencies—such as violence or nurturing behavior—is valuable or not? Cite a reason for your answer.

5. Suggest ways to limit human aggression, based on the ways that other animals minimize aggression.

SUGGESTED READINGS

Bass, Andrew H. April 1990. Sounds from the intertidal zone: Vocalizing fish. *BioScience*. Midshipmen use their swim bladders to change position and to vocalize.

Blaustein, A.R., and R.K. O'Hara. January 1986. Kin recognition in tadpoles. *Scientific American*. Altruistic behavior may not be completely unselfish.

Coniff, Richard. July 1990. Fire ants: Too hot to handle? *Smithsonian*. Fire ant societies are expanding rapidly.

Fellman, Bruce. April-May 1990. A case of spotted fervor. *National Wildlife*. Spotted salamanders court and mate on only a few nights each year.

Foster, Susan A. November 1990. Courting disaster in cannibal territory. *Natural History*. The three spine stickleback is a favorite of ethologists.

Griswold, Charles E., and Teresa C. Meikle. March 1990. Social life in a web. *BioScience*. Spider webs provide food and protection.

Hurxthal, Lewis M. December 1986. Our gang, ostrich style. *Natural History*. Mating behavior, territoriality, and defense are all intertwined in this enormous bird.

Levine, Joseph S. July 1990. For these fish, school never lets out. *Smithsonian*. Swimming in schools protects fish.

Weiss, Rick. October 28, 1989. New dancer in the hive. *Science News*. A computerized bee enters the hive.

Wilson, Edward O. 1990. *The ants*. Harvard University Press. The author has devoted his life to studying ants and their societies.

C H A P T E R

38

Populations

Chapter Outline

Learning Objectives

By the chapter's end, you should be able to answer these questions:

1. What physical and behavioral responses do organisms have to overpopulation?
2. What factors influence the rate of population growth?
3. Under what conditions would a population grow exponentially?
4. What factors prevent organisms from producing as many offspring as they theoretically could?
5. How do populations respond to environmental constraints on population growth?
6. How can populations of different species coexist?
7. How do populations of predators and their prey interact?
8. Can we—and should we—control human population growth?

Crammed on only 25,000 acres of wildlife preserve in the San Luis Valley in southern Colorado are many millions of birds. Unlike a century ago, before the spread of farms and ranches, today's sparrows, ducks, cranes, owls, wrens, grebes, snowy egrets, and hawks have nowhere to go. As this avian ghetto becomes more crowded, the habitat is destroyed. Biological wastes accumulate and food grows scarce. The dead lie about, decomposing. An infectious disease carried in on a new arrival spreads with astonishing speed. Each year, 5 million birds here die of botulism, cholera, plague, pox, or tuberculosis. Ringed in by mountains and human settlements, the birds' refuge has become their prison.

Grey squirrels inhabiting a city park live very differently than their country cousins. Their nests house more individuals, and they bear more scars from fights. The urban squirrels often die from a poxlike illness not seen in rural squirrels. Reproduction is delayed. A 2-year-old female city squirrel may not have yet had a litter; her country counterpart would almost certainly have been a mother by her first birthday.

"Universe 133" is a mouse colony created by researcher John Calhoun at the National Institute of Mental Health. A typical experiment begins by placing eight pairs of mice into the colony and allowing them to breed freely. Two hundred weeks later, the population reaches a maximum of 1,600 animals. At this point, with food, water, and especially space becoming scarce, behavior radically changes. Interactions between the animals, including mating, gradually cease. Juvenile behaviors persist into adulthood. The males huddle in groups along the periphery of the enclave, some of them biting and swinging others about. Females dart from one section of the colony to another and will follow any strange object introduced into their midst. The animals are clearly disturbed by their overpopulation.

Human Population Growth

Humans, too, are distressed by crowding. On a small scale, violence erupts at crowded rock concerts, sporting events, and in traffic jams (fig. 38.1). This is one reason why people are concerned about the increasing human population.

Figure 38.1
What started out as an exciting soccer semi-final between Liverpool and Nottingham Forest turned to tragedy, when several fans were crushed in the crowd.

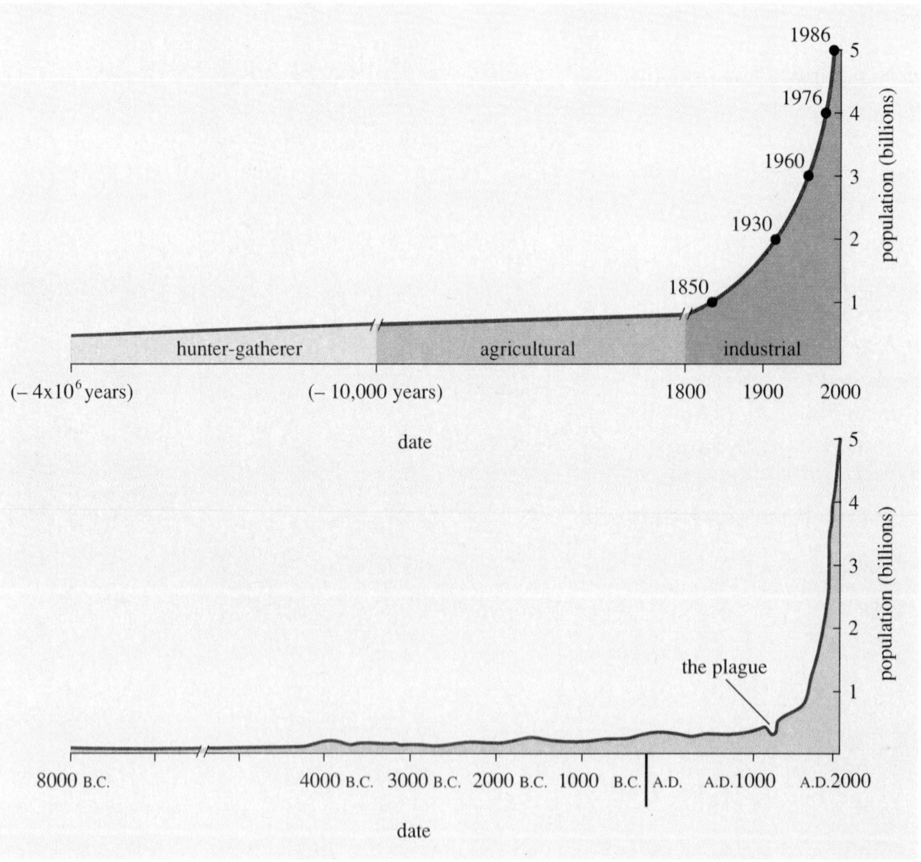

Figure 38.2
Human population growth is on the rise, as indicated by the J-shaped curve. Will we be able to control it

Table 38.1
Population Parameters

Parameter	*Meaning*
Crude birth and death rates	Annual number of births and deaths per 1,000 individuals (reflects population's age distribution)
Rate of natural increase	Birth rate minus death rate (annual population growth ignoring migration)
Population doubling time	Number of years until population doubles, assuming constant rate of natural increase; equals 70 divided by rate of natural increase; indicates potential growth; it is not a prediction
Infant mortality	Annual number of deaths under age 1 per 1,000 births
Total fertility rate	Average number of children per woman aged 15 to 49. A value of 2.1 to 2.5 holds the population constant

Each hour, 11,000 babies are born. As a result, 264,000 more newborns are alive today than were yesterday. The growth of the human population has not been a steady march. It has accelerated in recent years thanks to medical advances and our ability to manipulate the environment. It took from the beginning of humankind some 40,000 years ago until 1850 for the human population to reach 1 billion. Yet only 80 years later, in 1930, we numbered 2 billion. In another 30 years, by 1960, there were 3 billion. In only 15 more years, the world population hit 4 billion, and in July 1986, the 5-billion mark was reached (fig. 38.2).

How long can this explosive growth rate continue? What, if anything, will slow us down? Is there a limit to the number of people this planet can support? What kind of life can our grandchildren expect? We can attempt to answer these questions by understanding the influences and interactions that underlie population growth, in our own species and others. (Reading 38.1 presents some ways that nonhuman populations are studied.)

Population Dynamics

A population is a group of interbreeding organisms. Like anything else, a population grows when more is added to it than is taken away. Births and immigration increase population size; deaths and emigration decrease it. On a global scale, births and deaths are the most important determinants of population growth. However, considering the size or growth of a local population, organisms moving into or out of the area may have a substantial effect. For example, immigration was responsible for 43% of the growth of the human population in the United States during 1981.

Population growth is measured by rates as well as by absolute numbers. A rate is the number of events (such as births or deaths) divided by the average number of individuals in the population during a specific time period. Population growth is usually expressed as the change in the number of individuals during a set interval of time. However, the fact that a population increased by 500 individuals does not predict how large the population will be next week or next year. How long it took to add those 500 individuals, and the number of individuals in the population before the 500 are added, places the significance of the growth into perspective. Five hundred individuals added to a population of 1,000, have a much greater impact than 500 individuals added to a population of 100,000.

The larger a population is, the faster it grows. Since the additional individuals will reproduce, a population grows much the way compound interest accumulates in the bank—the more you have, the more you get. Even if the rate of growth remains the same, an increase in the size of a population means that the number of future members will be greater. Table 38.1 lists terms commonly used to describe population growth.

KEY CONCEPTS

Human population growth has increased rapidly over the past 150 years. Population size reflects births, deaths, and migration. Population growth is measured by rates and considers the size of a population and how many individuals are added in a specified time.

Biotic Potential and Exponential Growth

Imagine a world without hunger, war, disease, or pollution, with adequate space and shelter for all. Under such ideal conditions, populations would grow explosively. Growth would be slow at first, but the increase in numbers would accelerate continuously. Growth resulting from repeated doubling of numbers (1, 2, 4, 8, 16 . . .) is mathematically described as exponential. Plotting the number of individuals in a population over time as it grows exponentially yields a characteristic **J-shaped curve** not unlike that of figure 38.2. Unrestricted by the environment and with each member producing as many offspring as possible, the population would be growing at its **maximum intrinsic rate of increase.**

Under ideal conditions, a population's growth would be determined by how quickly individuals can reproduce. The maximum number of offspring an individual is physiologically capable of producing is called the **biotic potential.** This limit is influenced by such factors as the number of offspring produced at one time, the length of time an individual remains fertile, and how early in life reproduction begins.

The age when reproduction begins is very important in determining how fast the population will grow. Even if each female produces the same total number of offspring, the population will grow faster when reproduction begins at an earlier age, simply because the rate at which new individuals are added is accelerated. A woman who bears five children and has her first baby when she is 30 years old contributes no more to the growth rate of the human population than does a woman who has only three children, one each year, but has her first child when she is 13! Therefore, if one wants to limit population growth but still be a parent, delaying pregnancy may be a compromise.

The biotic potential for even the slowest breeding species is incredibly high, and few organisms ever attain it. Although the gestation period for elephants is 22 months, Charles Darwin calculated that if elephants met their reproductive potential, a single pair could leave 19 million descendants in 750 years! The biotic potential for rapidly breeding organisms is astounding. Recall from chapter 7 that a single *E. coli* bacterium, under ideal conditions, would produce 5 trillion billion offspring in a day! The biotic potential of humans is also impressive. Imagine a couple who starts a family as soon as it is physiologically possible and continues having babies as frequently as possible until menopause. If all their children reproduced in the same manner, within five generations, about a century, the couple would have more than 200,000 descendants.

Reading 38.1 *Conducting a Wildlife Census*

If counting the number of people in the United States, as the Census Bureau attempted to do in 1990, sounds daunting, try estimating the size of a migrating goose population! Wildlife biologists have devised some quite creative ways of tracking natural populations.

In Washington D.C.'s Lafayette Park, biologist Vagn Flyger mans the kiosk that usually supplies information to tourists. Flyger and his helpers distribute nest boxes in trees and then collect them when grey squirrels wander in. The animals are anesthetized and their vital statistics—weight, sex, age, and health status—recorded. Each is tattooed, so its whereabouts can be monitored, and then released.

Elsewhere, a variation of this "capture-mark-count" approach is used on deer. A small number of deer are rounded up and bright orange collars are put on them. Then, when a group of deer is spotted from a plane, an observer takes an aerial photograph. The easily-seen percentage of collared animals in the photo is used to extrapolate the number of animals in the group.

Another way to take a wildlife census is to count deer droppings or moose pellets, as is done in the Adirondack Mountains of upstate New York. A wildlife refuge on the Gulf Coast combines the capture-mark-count approach with detection of droppings. Otters are captured and injected with a radioactive, harmless chemical, which is eliminated in feces. Then droppings are collected. The proportion of radioactively labeled droppings indicates the size of the population.

Wildlife surveys are taken by people in the air counting migrating herds below, and by people on the ground counting migrating birds above. Some biologists estimate population sizes by listening for coos, cackles, gobbles, and howls. Perhaps most inventive is a contraption used in Montana. Rotting meat or fresh sardines are draped in a tree, along with an infrared heat detector attached to a camera. When a bear saunters over to collect the treat, its body heat is detected, and the animal snaps its own picture.

Describing wildlife populations is useful beyond providing information on population dynamics and habitats to biologists. The effects of environmental catastrophes—such as fires, volcanic eruptions, or oil spills—can be better assessed if population figures are available from before the disaster. Hunting, trapping, and fishing regulations and schedules are based on these data, as well as decisions on where to build—or where not to build—houses, dams, bridges, and pipelines.

Figure 1
Each spring United States and Canadian waterfowl spotters traverse 1.3 million square miles (3.4 million square kilometers) in the air, trying to count the number of animals in flocks, such as these snowgeese.

Figure 2
A male grizzly bear takes his own picture while stealing food from a booby-trapped tree.

A Population's Age Structure

The proportion of a population of reproductive age influences the rate of population growth (fig. 38.3). If most members are young, the population will grow as the females enter their reproductive years. Even if pairs produce only two offspring to replace themselves, this mostly young population would continue to grow because the proportion of individuals producing offspring is great. Two-thirds of China's 1 billion people, for example, are under the age of 24, and the population will continue to grow even if the rate of growth slows. By the year 2000, the Chinese will have 200 million more mouths to feed. The concept of population growth reflecting age distribution is also seen in populations of cells, discussed in chapter 7.

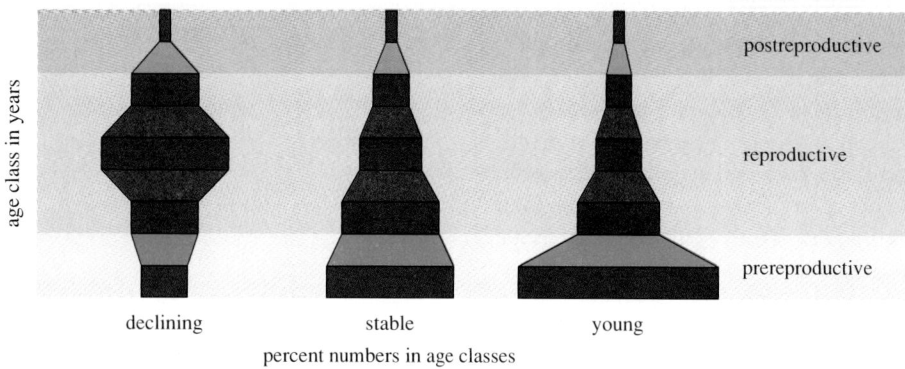

Figure 38.3
The age structure of a population indicates the proportion of individuals who are younger than reproductive age (prereproductive), of reproductive age, or past reproductive age (postreproductive). Theoretically, a population will decline if most of its members are past reproductive age. A population will remain approximately constant in size if the proportion of members of reproductive age stays constant. A population will grow as a larger percentage of its members reach reproductive age.

Table 38.2
Parameters for Selected Populations

	1988 Population (millions)	*Annual Natural Rate of Increase %*	*Projected Population Year 2000 (millions)*	*Total Fertility Rate*	*% Population <15 Years*	*% Population >65 Years*
Africa						
Egypt	54.8	2.8	71.2	5.3	40	4
Nigeria	115.3	2.9	160.9	6.6	45	2
South Africa	38.5	2.6	51.5	4.5	35	5
Asia						
Israel	4.5	1.6	5.4	3.1	32	9
Pakistan	110.4	2.9	145.3	6.5	43	4
China	1,103.9	1.4	1,291.6	2.4	29	6
Europe						
Finland	5.0	0.3	5.0	1.6	19	13
United Kingdom	37.3	0.2	57.3	1.8	19	15
Poland	38.2	0.6	39.9	2.2	26	9
Americas						
United States	248.8	0.7	268.3	1.9	21	12
Canada	26.3	0.8	28.4	1.7	21	11
Mexico	86.7	2.4	107.2	3.8	42	4
Brazil	147.4	2.0	179.5	3.4	36	4
Chile	13.0	1.6	15.3	2.4	31	6
Soviet Union	289.0	1.0	312.0	2.5	25	9

Source: Data from Population Reference Bureau, Inc., Washington, DC.

Assuming that the birth rate remains the same, a population consisting of mostly elderly individuals will decline as the members die. In such a population, the number of females able to bear young is small. Therefore, if each female bears the same number of offspring as in the past, the total young added to the population will be less. Table 38.2 shows proportions of the populations of selected countries under age 15 and over age 65.

The United States and the Soviet Union are experiencing population age trends that will eventually slow population growth. The proportion of 18-year-olds began falling in the United States in the late 1980s and in the Soviet Union since 1978. Not until the year 2002 will the proportion of 18-year-olds match the late 1970s peak in either nation. At the other end of the life span, numbers are swelling rapidly. The boom of seniors in the United States began in the 1970s and is just beginning to peak in the Soviet Union (figs. 9.22 and 38.4).

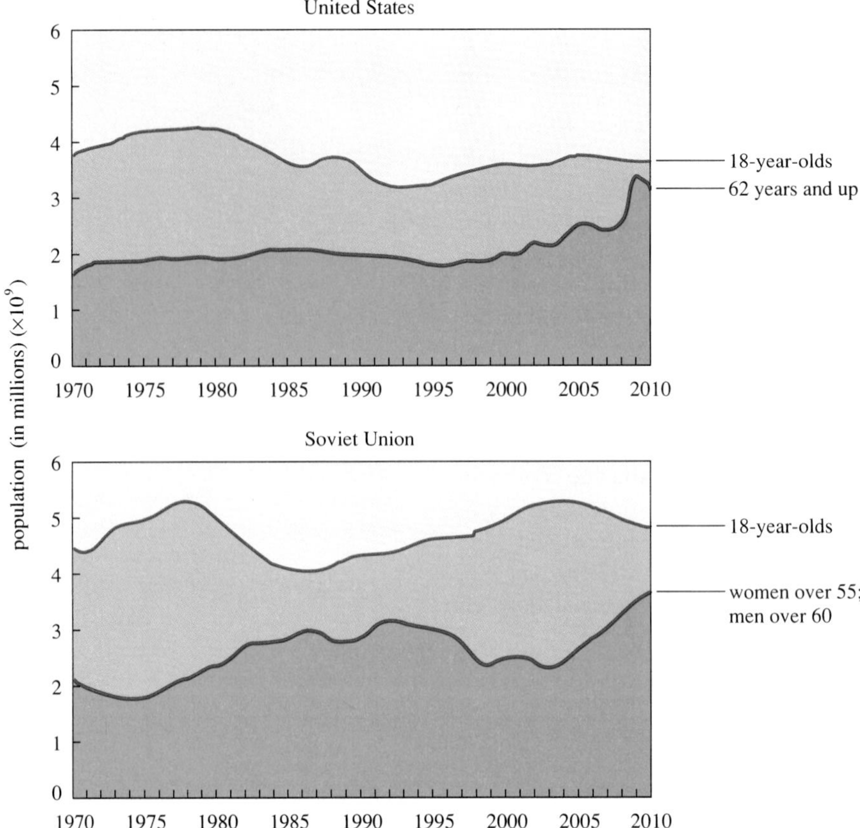

Figure 38.4
In the United States and Soviet Union, the number of 18-year-olds is falling and the number of those reaching pension age is rising. This trend should slow population growth.

KEY CONCEPTS

Under ideal conditions, a population would increase exponentially, producing a J-shaped curve as the maximum intrinsic rate of increase is reached. Growth rate depends upon how quickly individuals reproduce, when they begin reproducing, how long they are fertile, and the proportion of the population of reproductive age. The maximum number of offspring an individual can theoretically produce is its biotic potential.

Regulation of Population Size

Fortunately, populations do not grow as rapidly as is theoretically possible. **Environmental resistance** refers to all the factors that reduce birth rate or increase death rate.

Environmental Resistance and the Carrying Capacity

Environmental resistance may include limited food, pollution that impairs health and ability to reproduce, natural or human-caused environmental disasters, and predation. Figure 38.5 contrasts the exponential rate of

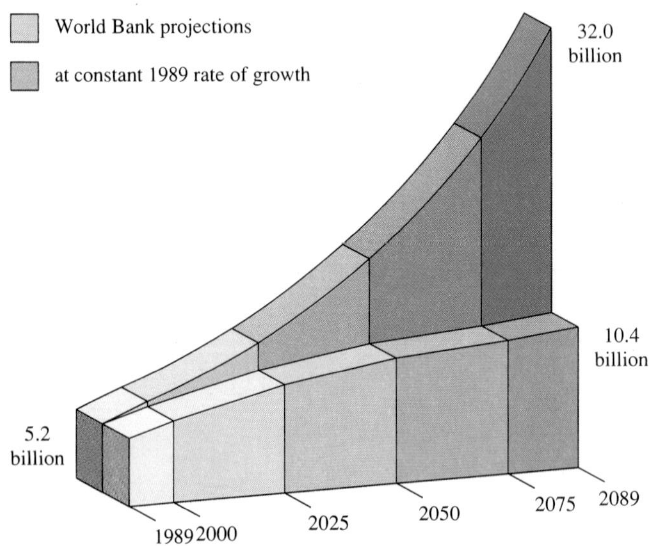

Figure 38.5
Doubling time alone cannot be used to predict population growth. When other real-life factors, such as changes in birth and death rates, age structures, and migration patterns, are considered, the forecast is not quite so gloomy.

human population growth that would occur without considering environmental resistance to the growth predicted when these factors are taken into account. The maximum number of individuals that can be supported by the environment for an indefinite time period is called the **carrying capacity.** A population at its carrying capacity, although no longer growing, may still have very high birth and death rates.

"Boom or Bust" Cycles

Because of environmental resistance, exponential growth can be maintained only for a limited time before the population overshoots the carrying capacity. Then numbers plummet or the population size stabilizes at the carrying capacity.

A population crash, sometimes called a "boom or bust" cycle, is commonly seen in species that colonize a new area with abundant food. A potato beetle population encountering a garden full of potatoes will grow exponentially as the insects incessantly chomp through the greenery until the denuded plants, unable to photosynthesize, die. Without this abundant food, the beetle population crashes. Stanford University population biologist Paul Ehrlich bases his predictions of the dangerous effects of human overpopulation on 3 decades of observing boom or bust cycles in checkerspot butterflies. Ehrlich captured butterflies, marked them, released them, and recaptured them in order to monitor population sizes.

Human population booms and busts often reflect history. The large proportion of middle-aged United States citizens today reflects a "baby boom" after servicemen returned from World War II. From an all-time low in population growth in the depression years of the early 1930s, fertility peaked in 1957 at 3.7 children per woman of childbearing age. The Soviet Union experienced human population busts during the World Wars. In 1940, 40 million Soviet citizens, or 21% of the population, died either directly or indirectly due to the war.

Population crashes, however, are unusual. More often population growth slows as the carrying capacity is approached, in response to environmental resistance. This leveling off produces a characteristic S-shaped (sigmoidal) curve or **logistic growth curve.** Figure 7.8 shows such a growth curve for a bacterial population.

> **KEY CONCEPTS**
>
> *Environmental resistance factors reduce birth rate or increase death rate, preventing populations from growing exponentially. The maximum number of individuals in a population that a particular environment can support is the carrying capacity. When the carrying capacity is reached, a population may crash or level off. A population stable at its carrying capacity has an S-shaped growth curve.*

Density Independent Factors

Density independent factors kill a certain percentage of the population regardless of its size. Natural disasters and severe weather conditions are typical density independent regulating factors, such as the fires that ravaged Yellowstone National Park in the summer of 1988 (fig. 1.1*b*) and hurricane Hugo in the fall of that year, which devastated the rain forest in Nicaragua. Human-caused environmental disasters also operate as density independent factors. The recent oil spills in Alaska and the Persian Gulf, for example, killed whatever became entrapped in the suffocating black mess (chapter 40).

A spectacular density independent event occurred on May 18, 1980, when the north face of Mount St. Helens, in Washington state, began to bulge and emit low rumbling sounds. Underground, water superheated by molten rocks was turning to steam. At 8:32 A.M., Mount St. Helens erupted. The 900° F (477° C) blast pulverized rocks and trees, killing nearly everything within a 6-mile (about 10-kilometer) radius and triggering earthquakes and avalanches. Ten miles (about 16 kilometers) away, trees were blown down; 16 miles (about 26 kilometers) away, trees were scorched. Millions of dead insects rained from the ash-ridden sky. A 25-foot (nearly 8-meter) high tide of mud and molten rock flowed out of control, and it was joined by melting snow and ice in the higher elevations, burying everything in its path under a sticky coat. As the top of the mountain caved in, jets of gas and steam rocketed 65,000 feet (nearly 20 kilometers) into the air, spraying a fine gray dust over nearly half the state.

A forest of 5-century-old fir trees was gone instantly. Hundreds of thousands of tons of new plant growth were suddenly buried or torn away. The animal death toll included some 5,200 elk, 6,000 black-tailed deer, 200 black bears, 11,000 hares, 15 mountain lions, 300 bobcats, 1,400 coyotes, 27,000 grouse, 11 million fish, and uncountable numbers of insects and microorganisms. For those outside the deadly "blast zone," food and shelter would be scarce for some time (fig. 38.6).

Density Dependent Factors

Density dependent factors have a greater effect as the size of a population increases. Behavioral responses to crowding include cessation of mating, poor parental care, and increased aggressive behavior. Physical responses to crowding include increased rate of spontaneous abortion, delayed maturation, and hormonal changes. All of these responses ultimately slow population growth.

Competition for food and space is a commonly encountered density dependent factor in population growth. Members of the same species have the most similar needs and are most likely to compete for limited resources. In **scramble competition,** the individuals compete directly for the limited resource, which may suffice for the individual but may be counterproductive to population growth. When many individuals share a limited food, each may not eat enough to sustain growth and reproduction. As a result, population growth slows. The larvae of many insect species compete in scramble fashion. If several hundred fruit fly larvae hatch on a very small apple already riddled with larvae, the food supply will soon be depleted. Larvae that do not eat enough starve to death or, metamorphose into small adults too weak to reproduce. Population growth slows.

Alternatively, members of a population may compete for a resource indirectly. In **contest competition,** animals compete for social dominance or possession of a territory, factors that guarantee the winners an adequate supply of the limited resource.

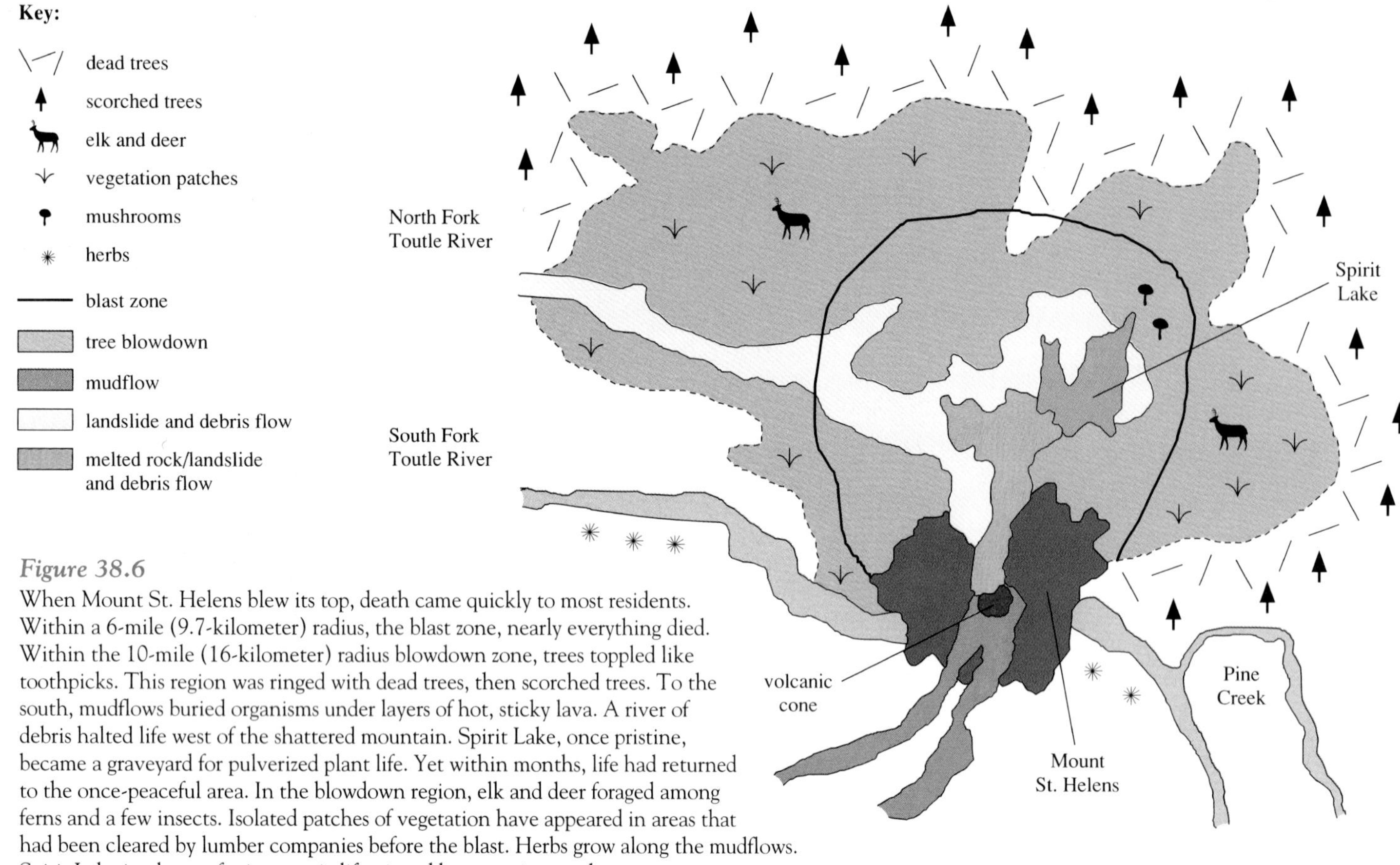

Figure 38.6

When Mount St. Helens blew its top, death came quickly to most residents. Within a 6-mile (9.7-kilometer) radius, the blast zone, nearly everything died. Within the 10-mile (16-kilometer) radius blowdown zone, trees toppled like toothpicks. This region was ringed with dead trees, then scorched trees. To the south, mudflows buried organisms under layers of hot, sticky lava. A river of debris halted life west of the shattered mountain. Spirit Lake, once pristine, became a graveyard for pulverized plant life. Yet within months, life had returned to the once-peaceful area. In the blowdown region, elk and deer foraged among ferns and a few insects. Isolated patches of vegetation have appeared in areas that had been cleared by lumber companies before the blast. Herbs grow along the mudflows. Spirit Lake is a brew of microscopic life, ringed by returning mushrooms.

The losers get less of the resource or none at all. Population growth slows because only the winners of the competition breed successfully.

Members of different species living in the same area may compete for the same resource. The yellow nutsedge, for example, controls the size of soybean populations by crowding the young bean plants and depleting the soil of nutrients.

According to the **principle of competitive exclusion,** coexistence of two competing species will not continue indefinitely; one species will replace the other in the area of overlap. This may take time, however. On Africa's Serengeti Plain, for example, wild dog populations tend to plummet because hyenas are better at hunting the gazelles, antelopes, and wildebeests that both species eat. The wild dogs have not yet gone extinct, but they are endangered.

Laboratory experiments confirm that competition may indeed lead to local extinctions. Two species of the protozoan *Paramecium* thrive when cultured separately. However, if these species are grown together, only one survives. Similarly, when a new species is introduced to a natural area, competition may cause the local extinction of a native species. This happened when starlings were released in New York City's Central Park in 1891, gradually eliminating the native bluebirds, which compete for the same nest sites. Chapter 33 describes how newly arrived species on islands can drive out native residents and come to dominate.

Habitat and Niche

Species with similar or even identical requirements can peacefully coexist when they interact with the environment in different ways. They do not get in each other's way. In understanding how species can do this, it helps to distinguish between a species' **habitat,** the place where it lives, and the ways in which it interacts with the environment, or its **niche.** Habitats include the edge of a pond, the rocky intertidal zone, a desert, or a rain forest. A niche includes physical factors such as temperature, humidity, and lighting; biological factors such as rate of growth, metabolic requirements, or food sources; and behavioral characteristics.

Consider how species of Antarctic seals share their sea, each using the environment differently. The crabeater seal lives on the fringes of ice sheets. The elephant seal breeds on open beaches, whereas the fur seal prefers small, hidden beaches and rocky shores. The Weddell seal winters beneath the ice, in an underground cavern, breathing through holes it cuts with its teeth. Safe beneath the ice, Weddell seal pups escape predation by yet another seal, the leopard seal, as well as by the killer whale. Similarly, in New England, five species of warblers, all small insect eaters, coexist in spruce trees (fig. 38.7). Even more impressive is a single fallen Douglas fir, which may house more than 100 species of beetle, each in a different part of the wood or bark (fig. 39.3).

Although division of a habitat into a variety of niches can allow different species to live together, competition may still narrow their niche so that the population size is restricted. A species' **fundamental niche** includes all the places and ways in which it could possibly live. If a species with a similar but not identical fundamental niche lives in the same area, the competition between them may partition the niche, restricting a species to only part of its fundamental niche. Therefore, a species' **realized niche,** the one it actually fills, is smaller than its fundamental niche due to competition.

Two species of barnacles, for example, live in the intertidal zone along Scotland's shoreline. The fundamental niche of both species is the same, the entire intertidal zone. When present alone, either type of barnacle would be found through the whole range. However, when they coexist, the realized niche of both is reduced. One species grows faster than the other and crowds its competitor out of the lower region of the intertidal zone, which is moister. The other species, however, better resists drying out during exposure to the air while the tide is out and so dominates in the upper part of the intertidal zone.

Figure 38.7

Populations coexist by reducing competition. These five species of North American warbler coexist in the same spruce trees because each concentrates its foraging efforts in a different region of the tree. The feeding zones are indicated by the shaded areas of the tree.

KEY CONCEPTS

Density independent factors, such as weather and natural disasters, kill irrespective of population size. Density dependent factors have a greater effect as the population grows, including the behavioral and physical changes seen with crowding. Competition is a density dependent factor that may be direct (scramble) or indirect (contest). One competing species may win, or competing species can coexist if each interacts with the environment differently. A species' fundamental niche can be restricted to a realized niche by competition.

Predator-Prey Interactions

The size of a population is very dependent upon which species kill and eat which other species. Effects on the prey population are obvious—their numbers diminish as they are eaten. Prey populations are often maintained by high reproduction rates that compensate for the loss to predation. Behaviors such as hiding or escaping allow some individuals to survive and breed. Prey populations are also maintained by natural selection, because predators tend to eliminate the weakest prey.

From the opposite vantage point, predators die if they cannot find enough food. The primary cause of death for the short-tailed weasel, for example, is starvation. These north-dwelling animals must consume their prey of small rodents and birds at least every 4 hours to fuel their very high metabolisms.

Reproduction among predators is intimately tied to the density of their prey populations. Many animals that cannot find sufficient food become temporarily infertile or have small litters. When a bobcat has trouble finding food, she has only one or two kittens, instead of the usual three, and even they may not survive their first 10 months, when the mother alone must care for and feed them. In southeast Idaho, declining litter size among bobcats parallels declines in the populations of cottontail and jackrabbits, their prey. Impaired reproduction in times of famine is seen in our own species. Female athletes who have very low percentages of body fat sometimes

a.

b.

Figure 38.8
a. This lion is bringing home lunch—a baby warthog. *b.* A pack of wild dogs downs a wildebeest. With some dogs grabbing its rear and others its nose and lips, the prey is held still. Next it will be disemboweled.

cease menstruating, rendering them temporarily infertile, possibly because their bodies could not support a pregnancy.

Predators may hunt alone or in packs (fig. 38.8). The type of prey usually depends simply on what is available. Bobcats range throughout the United States, and their menu varies accordingly. In Washington state they eat mountain beaver; in Minnesota they dine on deer and snowshoe hare; in the central and southern states, they eat rabbits, muskrats, voles, and birds.

Predation is not restricted to areas devoid of humans. A study of house cats in an English village found that the animals, although well-fed at home, killed a significant number of local wildlife (fig. 38.9). To participate in the study, the cat owners saved every tidbit their pets brought home for one year. Other interesting ways that researchers investigate predator-prey interactions include probing gut contents from road kills and examining animal droppings.

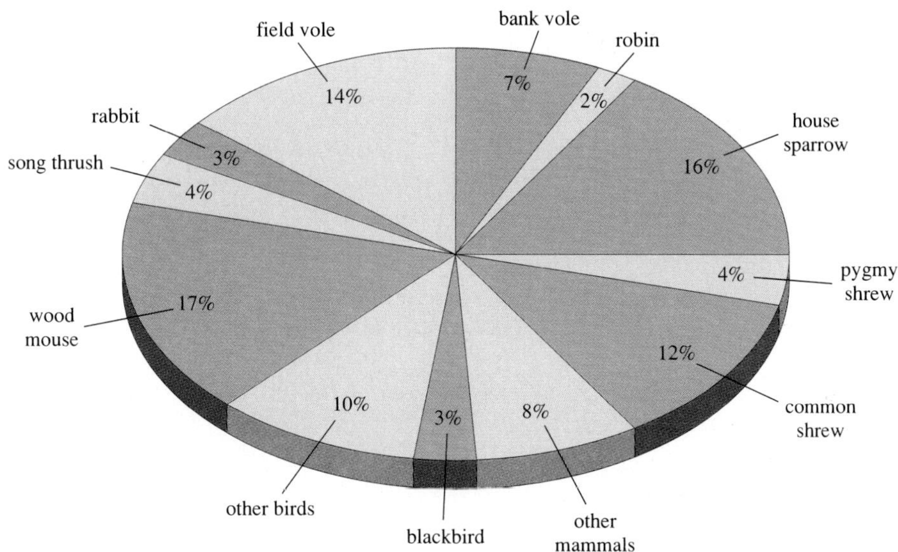

Figure 38.9
Thanks to the carrion-collecting efforts of 78 cat-owning families in a small English village, the effect of predation by house cats has been described.

Figure 38.10
The Tommie gazelle's "stott" communicates information about its ability to escape predation.

Animals evolve adaptations that help them more readily obtain prey or avoid becoming prey. A vivid adaptation is the leaping gait—called "stotting"—of Thomson's gazelles, or "Tommies" (fig. 38.10). The frequency of leaps seems to communicate the endurance of the gazelle. Cheetah and wild dog predators appear able to read this signal, ignoring the frequent leapers and concentrating on those who leap less often—and are inevitably the first to falter. Tommies also zigzag, make U-turns and even do an occasional backflip to avoid being eaten.

Sometimes adaptations seem to be an evolutionary arms race between hunted and hunter. Consider the snouted termites. "Soldier" termites, individuals who flank more vulnerable (and more tasty) members of the colony, have glands near their noses that spray foul-smelling terpene chemicals at would-be predators. Because of the protection afforded by the snouted termites' trademark stench, they, unlike most other termites, venture from their underground nests. The aardvarks, bat-eared foxes, and white-tailed mongeese that eagerly devour less noxious termites stay far away.

The smell of the snouted termites would seem an ideal adaptation, were it not for the aardwolf, an animal that dines almost exclusively on what to everyone else is a most stinky meal. From September through April, in only two known populations in North and South Africa, aardwolves roam areas of bare soil spotted with clumps of grass. An animal will stop at a clump, flick its ears forward and then back, lower its snout, and, using its broad tongue, in less than a minute lap up a snack of a few thousand of the odoriferous, orange termites. The aardwolf either tolerates the stench, or perhaps its huge salivary glands produce a biochemical that neutralizes the offensive chemical. No human has gotten close enough yet to the aardwolf to learn its secrets.

KEY CONCEPTS

Prey populations are maintained by a compensatory reproductive rate and natural selection. Reproduction of predators is dependent upon the abundance of prey. Predator-prey interactions involve many adaptations.

Reading 38.2 Controlling Human Population Growth

CHANGZOU IS A TOWN OF HALF A MILLION PEOPLE THAT IS VERY CLOSE TO ACHIEVING THE CHINESE GOVERNMENT'S GOAL OF ONE CHILD PER FAMILY. In the town's factories, large charts track each woman's reproductive status—"applied for marriage," "married," "approved to try for pregnancy," "waiting to become pregnant," and, finally, "one child." Each factory is allowed a quota of births each year. One government informer oversees 16 female workers, reporting signs of pregnancy. At home, "granny police" continue the surveillance, observing 16 families for second pregnancies and distributing birth-control information and supplies. Prizes are awarded to work and home units achieving 100% success at preventing second pregnancies.

When a first child is born, the parents register him or her at the local police department. They can take out a "one child contract," giving the mother an extra 3 months of maternity leave at full pay. The child receives a yearly sum from the government until age 14, free education, and priority consideration at universities and jobs—in exchange for the parents' agreement not to have another child. If a sibling does come along, the benefits to the first child are revoked.

So far only about 20% of the Chinese population has taken part in the one-child program, begun in 1979. Although China's runaway population growth rate has indeed been cut in half, many rural families, in need of more helping hands, choose not to participate. Abortion and infanticide have increased because some families will consent to having one child only if it is a male.

Figure 1
This class in Beijing teaches Chinese citizens about birth control.

Other countries try to control population growth as well. In Bangladesh, a program instituted in 1975 to encourage women to start small businesses had the side effect of sidetracking them from having babies. Use of birth control among these working women is 75%, compared to 35% among women who do not work. Thailand's rate of population growth has been slashed from 3.2% to 1.6% in 15 years, thanks largely to easy availability of contraceptives.

Curiously, some nations face the opposite population problem—plummeting numbers—and have likewise taken action. In France and the Canadian province of Quebec, the government offers financial incentives and extended new parent leaves from work to encourage citizens to have babies. In Singapore, where the total fertility rate has fallen from a high of 4.7 offspring per female in 1965 to 1.6 in 1987, government matchmakers coach career-oriented people with little social experience in the art of romance.

Human Population Growth Revisited—The Future

Globally, the human population is increasing, but this situation may soon change. The United Nations predicts that by the end of the twenty-first century, the human population will peak at 10.5 billion. From there our numbers will fall, and by 2263, the human population should fall to 2 billion, the level in the early 1900s.

However, population predictions are notorious for being changed. Paul Ehrlich's 1968 book *The Population Bomb* predicted worldwide starvation for the 1970s, which he revised in his 1990 sequel *The Population Explosion*. A United Nations prediction made in the late 1960s placed world population in the year 2000 at 7.5 billion; it was recently revised down to 6.1 billion.

A reason for the downgrades in predictions and lessening of panic concerning human overpopulation is the idea that human populations are not subject to the same growth patterns as other species because we alone can intentionally alter our environment to enhance our survival. We can conserve or find alternate resources, use agriculture to grow crops on barren land and aquaculture to farm the waters, and even control our reproduction (Reading 38.2).

However, many of these activities, while allowing us and our descendants to live in comfort, may have devastating effects on the physical environment and on populations of other organisms with whom we share our increasingly crowded planet. The next two chapters will explore many of the effects that *Homo sapiens* have had on the environment.

SUMMARY

A population grows when more individuals are added through birth or immigration than leave due to death or emigration. Population growth depends upon the initial size of the population, how many individuals are added and at what rate, the age at which individuals begin reproducing, and the age structure of the population.

Growth unrestrained by the environment is exponential and is characterized by a J-shaped curve. An individual reproducing at the maximum rate physiologically possible is meeting its *biotic potential*. Few organisms meet their biotic potentials because of *environmental resistance*. The number of individuals that can be supported by an environment indefinitely is the *carrying capacity* of the environment. After a period of exponential growth, population growth either overshoots the carrying capacity and the population crashes, or the growth slows and levels off at the carrying capacity, producing an S-shaped growth curve.

Population size is regulated by *density independent factors*, such as natural disasters, which kill a proportion of the population regardless of its size, or by *density dependent factors*, which have a greater effect when the population is large. Competition for a limited resource is density dependent when members of the same species compete directly for it *(scramble competition)* and everyone usually gets less of the resource and reproduction slows. Alternatively, territories or social dominance may be contested, with the winner receiving the resource. When members of different species compete, one species may increase its numbers at the expense of the other, or both species may coexist, both at reduced numbers.

A *habitat*, the place where an organism lives, is only one component of the organism's *niche*, the functional role of the organism in the environment. Slight differences in niche allow different species to share their surroundings. Because of competition, a species' *realized niche* is often smaller than its *fundamental niche*.

Predator-prey interactions can influence population size. Reproduction rates of predators often fall when prey becomes scarce. In a natural environment, both predator and prey usually survive because of coevolved adaptations. The human population is predicted to peak towards the end of the next century.

QUESTIONS

1. How can crowding and food scarcity impair reproduction?

2. In the 1960s, folksinger Pete Seeger sang about overpopulation with these lyrics, "We'll all be a-doubling in 32 years." Assuming that the human generation time is 32 years, as the song suggests, what factors could prevent Seeger's prophecy from coming true?

3. Frogs of species *Eleutherodactylus coqui* living in the tropical rain forest compete for hiding places. Those that can hide escape predation more than those frogs left out in the open. Is ability to hide a density dependent, or density independent, factor in population growth for these frogs? Cite a reason for your answer.

4. In Pakistan, 43% of the population is under the age of 15 and 4% is over 65. In the United Kingdom, 19% of the people are under 15 and 15% over 65. Which population will increase faster in the future? Why?

5. In what two ways is the growth of a population of organisms similar to the growth of a population of cells, as discussed in chapter 7?

6. How do a habitat and niche differ? How does a fundamental niche differ from a realized niche? Describe your habitat and niche.

TO THINK ABOUT

1. Each of 13 species of coral that live in the northern Red Sea in Israel reproduces at a different time of year. Although the corals have similar nutritional requirements, they coexist in large populations. How do they all obtain enough resources?

2. In some nations, well-educated, financially better-off families are having fewer children, and poorer families tend to have more children. What short- and long-term problems may result from such a trend, if any? Cite some examples of this phenomenon from table 38.2.

3. Many children growing up in China today will have no brothers or sisters, and their children will have few aunts, uncles, or cousins. The alternative, statistics predict, is mass starvation in the next century. What do you think of China's "one-child" policy? Can you suggest alternative approaches to controlling population growth?

4. How can the human population curve be J-shaped (fig. 38.2) if there is evidence of environmental resistance, such as hurricanes, tornadoes, earthquakes, and oil spills?

5. Suggest an experiment to test the hypothesis that crowding makes humans more aggressive and violent.

SUGGESTED READINGS

Berreby, David. April 1990. The numbers games. *Discover*. Are humans subject to the same population forces as other animals—or can we avert overpopulation?

Churcher, Peter B., and John H. Lawton. July 1989. Beware of well-fed felines. *Natural History*. A fascinating study of the effect of predation by domesticated cats on local wildlife populations in an English village.

Ehrlich, Paul, et al. April 18, 1975. Checkerspot butterflies. *Science*, vol. 188. Ehrlich based his predictions on human population growth on dynamics of wild butterfly populations.

Richardson, Philip R.K. April 1990. The lick of the aardwolf. *Natural History*. This little-known animal competes for food by eating prey unsavory to others.

Steinhart, Peter. January 1991. Essay: beyond pills and condoms. *Audobon*. What will we do to halt Africa's burgeoning human popluation?

Torrey, Barbara Boyle, and W. Ward Kingkade. March 30, 1990. Population dynamics of the United States and the Soviet Union. *Science*, vol. 247. Human population ups and downs mirror history.

Turbak, Gary. April-May 1990. The great American wildlife census. *National Wildlife*. Sizing up nonhuman populations is difficult.

C H A P T E R

39

Ecosystems

Chapter Outline

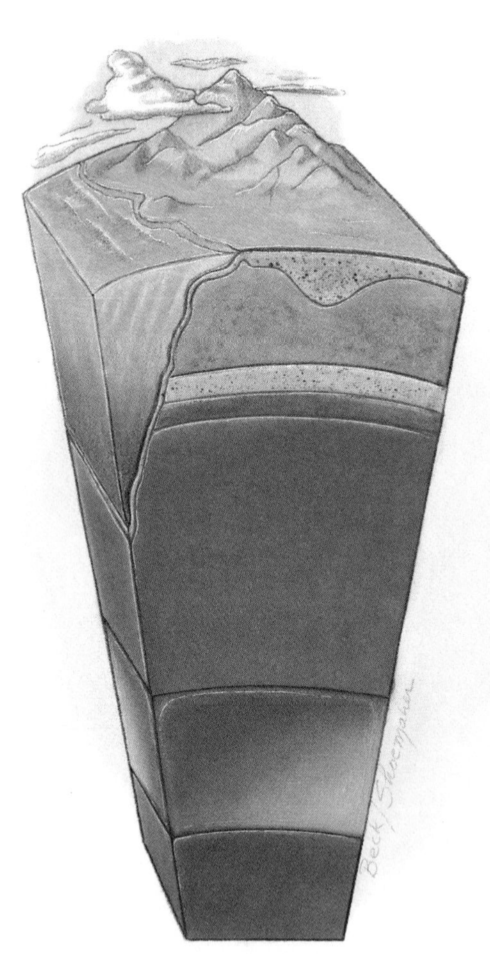

Learning Objectives

By the chapter's end, you should be able to answer these questions:

1. At what levels can an organism's place in the world be described?
2. How does energy flow through an ecosystem?
3. How do organisms interact to form food webs?
4. How are some chemicals concentrated as they proceed up food chains?
5. How do carbon, nitrogen, and phosphorus cycle through an ecosystem?
6. How do ecosystems naturally recover from damage?
7. What are the types of large-scale environments (biomes)?
8. How are plants and animals adapted to their environments?
9. What are the different challenges posed to organisms in a stream, lake, estuary, and ocean?

An adult blue whale is by far the largest animal ever to live on earth. It is 25 times as heavy as an elephant and 98 feet (30 meters) long. The blue whale lives part-time in the southern ocean of Antarctica, and its existence is intimately tied to that of another organism, a 2.4-inch- (6-centimeter-) long shrimplike animal called krill.

The Antarctic krill *Euphausia superba* is a crustacean that provides the diet of the great blue whale. Krill travel in great underwater clouds, circling the South Pole and concentrating in eddies where north-flowing cold surface water meets south-flowing, warmer water. Krill lay their eggs at the ocean's surface from January through March. The eggs sink, hatching at a depth of about 820 yards (750 meters). The larvae rise as they develop, producing a constant stream of organic material as they periodically shed their outer coverings.

Blue whales cruise the Antarctic Ocean every summer. They glide through clouds of krill, their giant jaws partially opened, capturing the abundant animals. The whale pushes water back out of its mouth with its tongue but passes the krill over sheets of feathery, hornlike material, called baleen, that descend from the roof of its mouth like a shower curtain. If a concentrated cloud of krill is not available, the hungry whale stirs one up. By diving beneath a dispersed cloud, and then swimming upward in a spiral pattern, the whale initiates an underwater minivortex of krill. The great animal swims through its whirling meal.

Krill directly or indirectly support nearly all life in Antarctica (fig. 39.1). They feed 6 species of baleen whales, 20 species of squid, 120 species of fishes, 35 species of birds (mostly penguins), and 7 species of seals. The krill themselves eat plankton, which consists of algae, protozoa, small crustaceans, and fish eggs and larvae. Krill feed above and below ice floes and even within tiny channels in the ice that support rich microbial communities. Some krill have pigmented intestines, indicating a recent meal of a brightly colored diatom, or phytoplankton, which contain the photosynthetic green pigment chlorophyll. Phytoplankton "bloom" between October and February, providing an annual feast for the krill, which feed nearly everyone else.

An Organism's Place in the World

The interconnections between the species of Antarctica illustrate the concepts of **ecology,** which is the study of the relationships between organisms and their environments. The plankton, bacteria, crustaceans, squid, fishes, birds, seals, and whales of Antarctica comprise an **ecosystem,** which is a unit of interaction among organisms and between organisms and their physical environment. An ecosystem consists of all living things within a defined area and may be as large as the whole earth or only a small part of it—a forest, or even a single rotting log on the forest floor (fig. 39.2).

Ecosystems always include living (biotic) and nonliving (abiotic) components, and they are relatively self-contained, with more materials cycling within than entering or leaving. An ecosystem is terrestrial if it is on land and aquatic if it is in water (table 39.1). Ecosystems are supported by energy, which is ultimately derived from the sun (solar energy) or from the earth's interior (geothermal energy).

The word *ecology* is Greek for "home," and the field of ecology deals with the homes of organisms. An ecological home is a niche, which includes both the species' relationship to other organisms and its physical habitat. A species is successful in an ecosystem if its niche supplies food and protection from predators and the abiotic environment.

Species coexist within ecosystems by a precise balancing of their requirements, as the warblers occupying the same tree in figure 38.7 demonstrate. Consider a downed Douglas fir (fig. 39.3). The entire log is an ecosystem whose residents differ with location and time. Some species prefer dry, shaded branches, others the soggy, rotting roots, and still others, the deepest heartwood. Soon after the tree falls, insects invade the inner bark, then other species enter the sapwood. Still later, additional species attack the heartwood. Deep within this decomposing tree, bark beetles whittle a labyrinthlike "egg gallery" in which to rear the next generation, as the mother beetle guards the entrance. The chamber is also home to other species, including more than 100 types of beetles. Tiny wasps drill in from the outside and lay eggs on the bark beetle larvae. When wasp larvae hatch from the eggs, they eat the beetle larvae. A scavenging species of beetle eats any dead bark beetle larvae left over from the wasp's meal. Other insect species eat the fungus that grows in indentations in the bark beetle's exoskeleton.

This world-within-a-world of the downed Douglas fir can interact with other ecosystems. When another falling log, or perhaps a push from an impatient bear, moves it into a swamp, the habitats in its various nooks and crannies are drastically altered.

Just as a person's home can be described on various levels—a room in a house on a street in a neighborhood—so can other organisms' places in the world. A population consists of members of a single species in a particular area, such as the bark beetles in a single egg gallery of the fallen tree. An ecological **community** includes all of the organisms, possibly several species, in a given area. Right Whale Bay in the South Georgia region of Antarctica, for example, is a shoreline community populated by fur seals, king penguins, macaroni penguins, dominican gulls, and elephant seals.

Living communities are not distributed evenly on the planet. Biological diversity is highest at and around the equator, possibly due to the relatively unchanging climate there. Toward the poles, biological diversity declines as climatic fluctuations increase, creating harsher, seasonal conditions. In general, similar types of biological communities appear at corresponding latitudes, because they have similar climates. The steamy heat of a South American and African jungle is likely to be home to similarly adapted species. Likewise, there are similarities among organisms that live in a sea of grass, no matter how distant their homes. Differences between communities at the same latitude in different parts of the world reflect regional climatic characteristics or geographical influences.

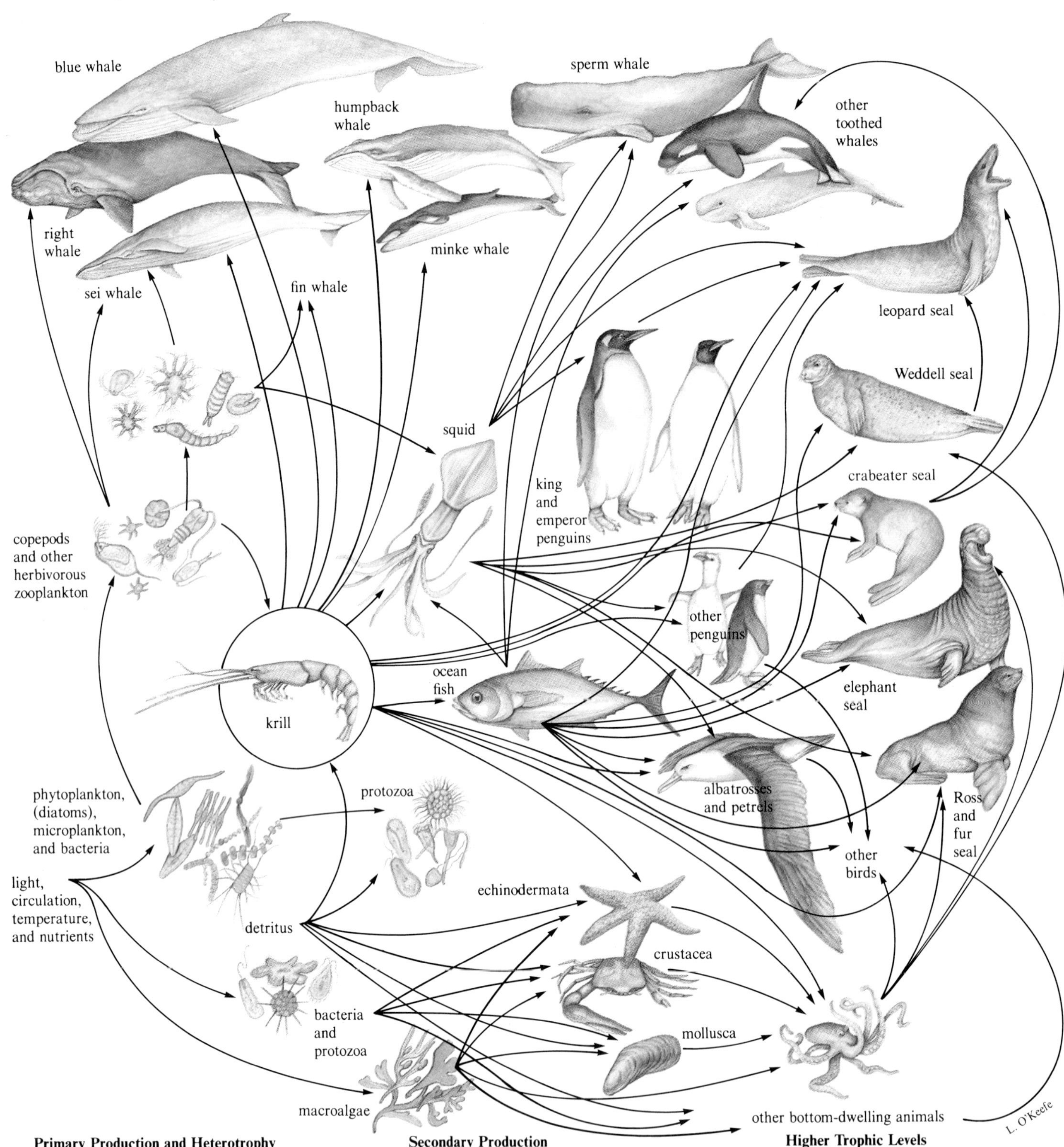

Figure 39.1
The Antarctic web of life. Nearly all species in Antarctica are dependent—either directly or indirectly—upon krill.

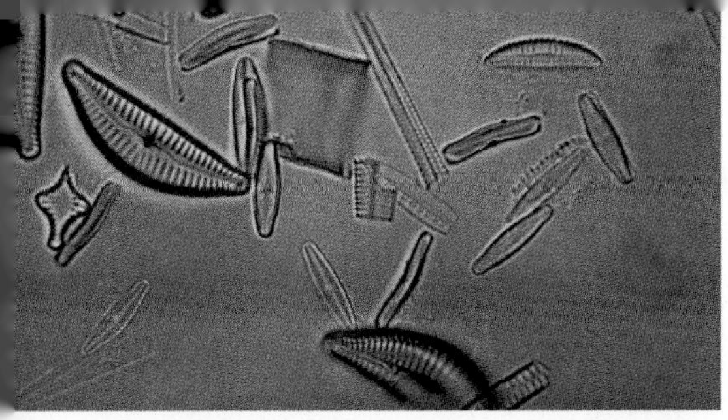

a.

b.

c.

d.

e.

f.

Figure 39.2
An ecosystem can be studied at several levels, as this sequence of photos of the Sycamore Creek drainage basin near Phoenix, Arizona, shows. Fourteen streams feed into this body of water. *a*. The water abounds with microscopic diatoms. *b*. Fallen leaves provide food for fishes. *c*. Snails and insects feed on carpets of algae. *d*. A flash flood is not necessarily a catastrophe—insects move out of the way of the torrent, fish swim into calmer branches, and decimated mayfly populations recover by increasing their reproductive rate. Aerial photographs (*e* and *f*) reveal geological features of the stream system.

Table 39.1
Examples of Ecosystems

Ecosystem	*Location*	*Physical Environment*	*Biotic Environment*	
			Plants	*Animals*
Alpine	Cascade Mountains, northwest Washington state	Strong winds, low soil moisture, frequent freezing and thawing	Twelve species of flowering plants	Mountain goats, cats, eagles, rodents
Tundra (subarctic)	Mt. McKinley National Park, Alaska	Harsh winters, permafrost	Willow, aspen; green in June, brown by September	Moose, wolves, caribou, grizzly bears
Arctic	Antarctica	Two-mile (3.2-kilometer) thick ice, low temperature, high winds	Plankton, algae, lichens	Plankton, krill, fish, squid, seals, whales, birds, mammals
Desert	United States Southwest	Low water, high heat in day	Algae, mosses, lichens, ferns, shrubs, cacti, sagebrush	Snakes, lizards, rodents, deer, fox, puma, peccaries
Lake	United States Northeast	Aquatic	Algae, rushes, water lilies, weeds	Invertebrates, insects, fish, amphibians, waterfowl

Figure 39.3
A fallen tree is a thriving ecosystem. This rotting log is home to hundreds of insect species.

Figure 39.4
Moose foraging changes an ecosystem. When moose preferentially eat foliage of aspen, balsam poplar, and birch, these trees cannot compete for sunlight with the unpalatable spruce and firs. This was demonstrated by sectioning off parts of the forest on Isle Royale in Michigan. Where moose were free to munch, conifers predominated. In areas without moose, the aspen, balsam poplar, and birch thrived.

Ecosystems can and do change. The Antarctic krill population increases by 13-fold in the summer, when some of the ice cover melts, giving the organisms more space in which to breed. A moose's finicky tastes can gradually change a forest of mostly deciduous trees to one where conifers dominate (fig. 39.4).

Ecosystems also interact with one another. The organisms living on the surface of the downed tree interact with residents of the neighboring soil ecosystem. A terrestrial ecosystem may contain an aquatic one, such as a pond or marsh, or an aquatic ecosystem, such as a bay, may be dotted with islands or sandbars. Ecosystems receive seeds in the air and dissolved nutrients in the rivers and streams that link land to water. Animals continually ferry nutrients and organisms between ecosystems. Even physically distant ecosystems interact through air and water conduits. The smokestacks of Cleveland, Ohio, for example, have a very noticeable effect on lakes in the Northeast by polluting them and altering their acidity. Such acid precipitation is discussed in the next chapter.

A group of interacting terrestrial ecosystems characterized by a dominant collection of plant species, or a group of interacting aquatic ecosystems with similar salinities, forms a larger ecological unit, the **biome.** Biomes include grasslands, deserts, and forests.

So connected are ecosystems to each other that the entire planet can be viewed as one huge, interacting ecosystem, the **biosphere.** This is the part of the planet where there is life, reaching about 6.2 miles (10 kilometers) into the atmosphere and extending down about the same distance, into the deepest ocean trenches. The biosphere, then, is only a thin film around the earth, like the outer paint layer on a basketball. Plus, the "paint job" is uneven, with globs in some places and missed spots in others, because the biosphere is not of uniform thickness and does not cover the entire globe. This flimsy veil of life coating the earth is our home (fig. 2.2).

KEY CONCEPTS

Ecology is the study of the interrelationships between organisms and their biotic and abiotic environments. An individual has a niche and is part of a population (organisms of one species), community (groups of organisms), ecosystem (all life within a certain area), biome (a group of interacting ecosystems), and the biosphere (the part of the earth that supports life).

Energy Flow Through an Ecosystem

Energy flows through an ecosystem in one direction, beginning with solar energy that is converted to chemical energy by photosynthetic organisms. This stored energy is then passed through a **food chain,** a series of organisms in which one eats another. The **trophic level,** or feeding level, to which an organism belongs describes its position in the food chain.

Very little of the energy from an organism's body is actually used for growth and development by the organism that eats it. Energy is lost to the environment as heat, or used to power metabolism, including cellular respiration. Some parts of organisms are not eaten (such as thorns and bones), and some parts are consumed but eliminated in the feces (such as cellulose eaten by humans). The total amount of energy converted to chemical energy by photosynthesis in a certain amount of time in a given area is called **gross primary production. Net primary production** is what is left over for growth and reproduction.

From each trophic level to the next, only 10% of the potential energy in the bonds of its molecules fuels growth and development of organisms at the next trophic level. If a food chain consists of four organisms—such as an owl eating a snake that has eaten a bird, which ate seeds—the owl can only use for growth and development 1/1,000 (1/10 x 1/10 x 1/10) of the energy originally available in the seeds. This inefficiency of energy transfer is why food chains rarely extend beyond four trophic levels. One way to maximize energy obtained from food is to lower the number of trophic levels. A person can do this by eating vegetables instead of meat (fig. 39.5).

Trophic Levels

The first link in the food chain and the base of the trophic structure is formed by **primary producers,** which are organisms able to use inorganic materials and energy to produce all the organic material they require. Almost all primary producers photosynthesize and include plants on land and algae in aquatic ecosystems. A few primary pro-

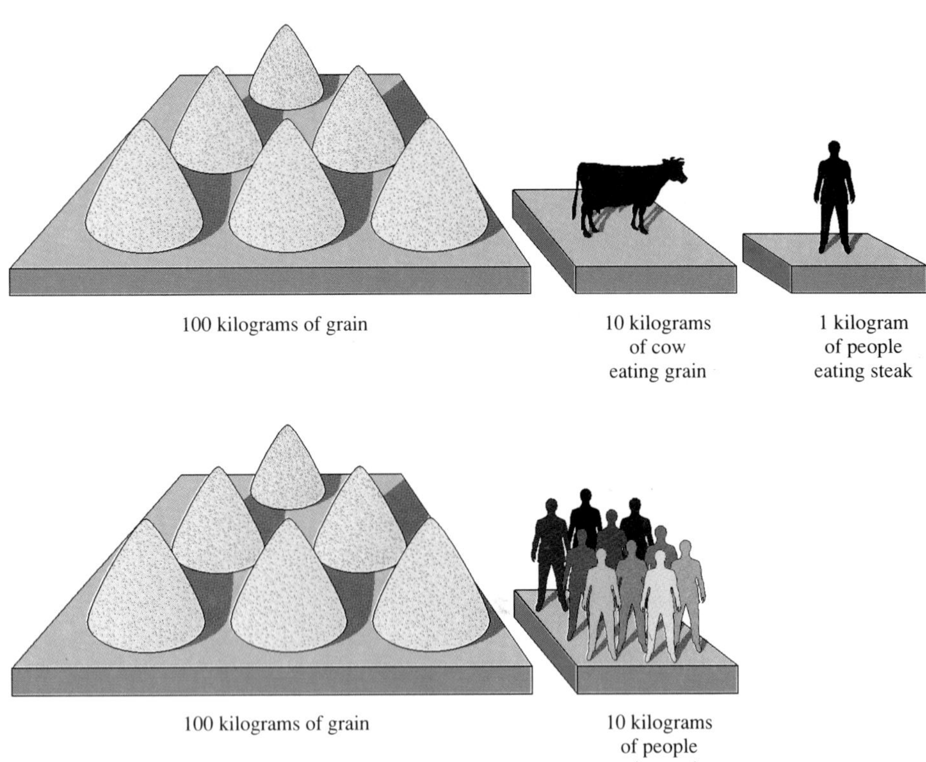

Figure 39.5
On a population level, it is more efficient for consumers to be herbivores than carnivores.

ducers are chemosynthetic organisms, which are bacteria that derive energy from the earth. Only 1/1,000 of the solar energy falling on the earth is converted to chemical energy in photosynthesis. Still, global photosynthesis produces 150 to 200 billion metric tons of organic matter each year.

The energy stored in the tissues of herbivores (plant eaters) and carnivores (meat eaters) is called **secondary production.** At the second trophic level are the **primary consumers,** or herbivores. On the third trophic level are **secondary consumers,** animals that eat herbivores. Carnivores that eat other carnivores form a fourth trophic level and are called **tertiary consumers.** These animals expend a great deal of energy in capturing their prey. **Scavengers** such as vultures eat the leftovers of another's meal, and **decomposers,** such as certain fungi, bacteria, insects, and worms break down dead organisms and feces, liberating minerals held in organic molecules for use by other organisms.

Although energy passes through a series of organisms, chains of "who-eats-whom" often interconnect, forming complex **food webs,** such as the one for Antarctica shown in figure 39.1. Webs form

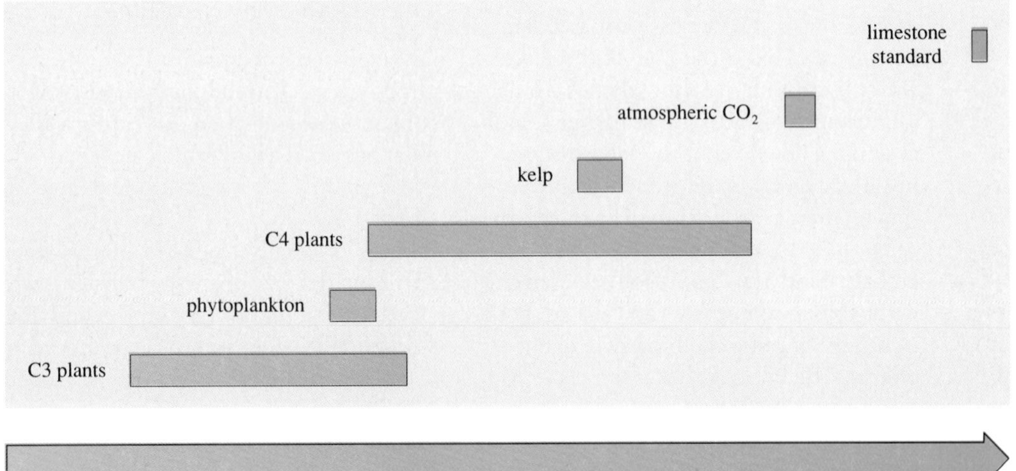

Figure 39.6
a. Stable isotope tracing reveals the ratio of carbon-13 to carbon-12 in a burned tissue sample. Because each type of primary producer has a unique carbon signature, the first trophic level of the food chain that an organism was part of can be identified. The $^{13}C/^{12}C$ ratio is compared to a known standard of limestone from a formation in South Carolina. (C_3 and C_4 plants differ in some of their metabolic reactions.) *b.* Ecologist Steven C. Amstrup sedates and radio collars polar bears. Parings from the bears' claws reveal a $^{13}C/^{12}C$ ratio that identifies the primary producer in its food web.

when an organism functions at more than one trophic level. A person eating a tuna salad sandwich on wheat bread, for example, is simultaneously a primary consumer (an herbivore) and a tertiary consumer (a carnivore eating another carnivore, the tuna).

Stable Isotope Tracing—Deciphering Food Webs

Until recently, ecologists were limited in describing food webs by what they could directly observe. The method of **stable isotope tracing,** borrowed from geology, provides a way to decipher who eats whom without actually being on the scene of a meal. The technique analyzes the proportions of certain isotopes in tissue samples. The basis of the approach is that primary producers each have a characteristic ratio of carbon-13 to carbon-12, called its "carbon signature" (fig. 39.6*a*). (Recall from chapter 3 that isotopes of the same element are distinguished by different numbers of neutrons.)

By determining the $^{13}C/^{12}C$ ratio in a tissue sample, such as a nail paring from a polar bear, the organisms at the base of the food chain can be identified (fig. 39.6*b*). The tissue is thoroughly cleaned and then burned until all that remains is carbon dioxide gas. A device called a mass spectrometer detects the relative amounts of ^{13}C and ^{12}C in the gas.

Stable isotope tracing has shown that the primary producers in many coastal ecosystems are not phytoplankton, as was previously thought, but vast underwater forests of kelp (fig. 2.11*b*). One fascinating use of the technique is to map the migration paths of whales. This is done by examining how the $^{13}C/^{12}C$ ratio varies along their baleen, which are constantly growing and hence accumulating carbon. Because the $^{13}C/^{12}C$ ratio of their plankton food varies according to latitude, changes in the ratio along the baleen reveal where the whale has been to eat. Stable isotope tracing can even reach back in time to reconstruct past food webs. Carbon signatures from 10,000-year-old bison bones show, for example, that the type of grass in the North American prairie has changed.

Ecological Pyramids

If each trophic level is represented by a bar whose length is directly proportional to the number of kilocalories (kcal) available from food for growth and development, the bars representing the different trophic levels, stacked in order, form a steep-sided **energy pyramid** (fig. 39.7). For example, in Cayuga Lake in New York state, 1,000 kcal of energy are stored in the cells of algae, which feed small aquatic animals. The aquatic animals derive 150 kcal from this food for use in growth and development. When a smelt fish eats 150 kcal worth of small aquatic animals, it derives 30 kcal for growth and development. If a trout eats the smelt, only 6 kcal are used. A human eating the trout gets only 1.2 kcal of the original 1,000 kcal contributed by the algae.

Other types of pyramids are used to describe ecosystems. A **pyramid of numbers** shows the number of organisms at each trophic level. The shape of this type of pyramid depends largely on the size of the producers. The pyramid of numbers in a grassland community has a broad base because the producers—grass—are small and numerous. In a forest, however, the pyramid of numbers is propped on a very narrow base, because a single tree can feed many herbivores.

Biomass is the total dry weight of organisms in an area. A **pyramid of biomass** takes into account size or weight. Many pyramids of biomass are wide at the bottom and narrow at the top, because energy that can be converted to biomass is lost with each transfer between trophic levels. However, in some aquatic ecosystems, the pyramid is inverted, because the biomass of the producers (phytoplankton) may be smaller than that of the herbivores (zooplankton). This is because biomass is measured at one time. Phytoplankton reproduce quickly but are eaten almost immediately by zooplankton. In contrast, zooplankton live longer so more of them may be present at one time. If the biomass of all the phytoplankton eaten by the zooplankton during their entire life span was considered, the pyramid would be turned upright.

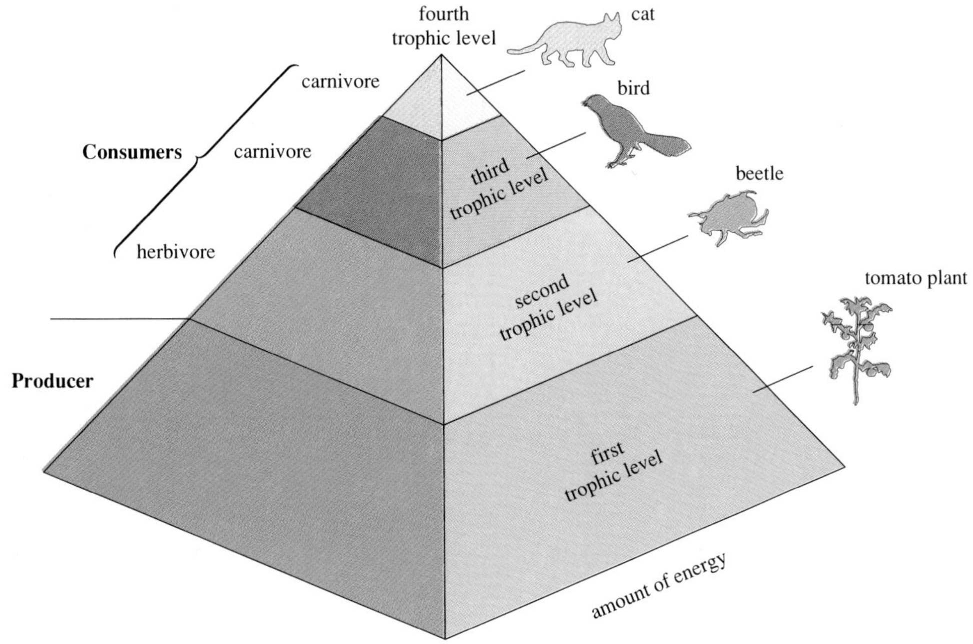

Figure 39.7

Energy flow through an ecosystem—an energy pyramid and a food chain. The cat eats the bird that eats the beetle that eats the tomato plant. On a population level, with each higher trophic level, only 10% of the energy is transferred and used to power growth and development.

KEY CONCEPTS

Primary producers use solar or geothermal energy to synthesize nutrients, which are eaten by secondary producers, including primary consumers at the second trophic level, secondary consumers at the third trophic level, and tertiary consumers at the fourth trophic level. At each level, 90% of the energy is lost. Food webs trace who eats whom. Stable isotope tracing identifies primary producers by analyzing $^{13}C/^{12}C$ ratios in tissues.

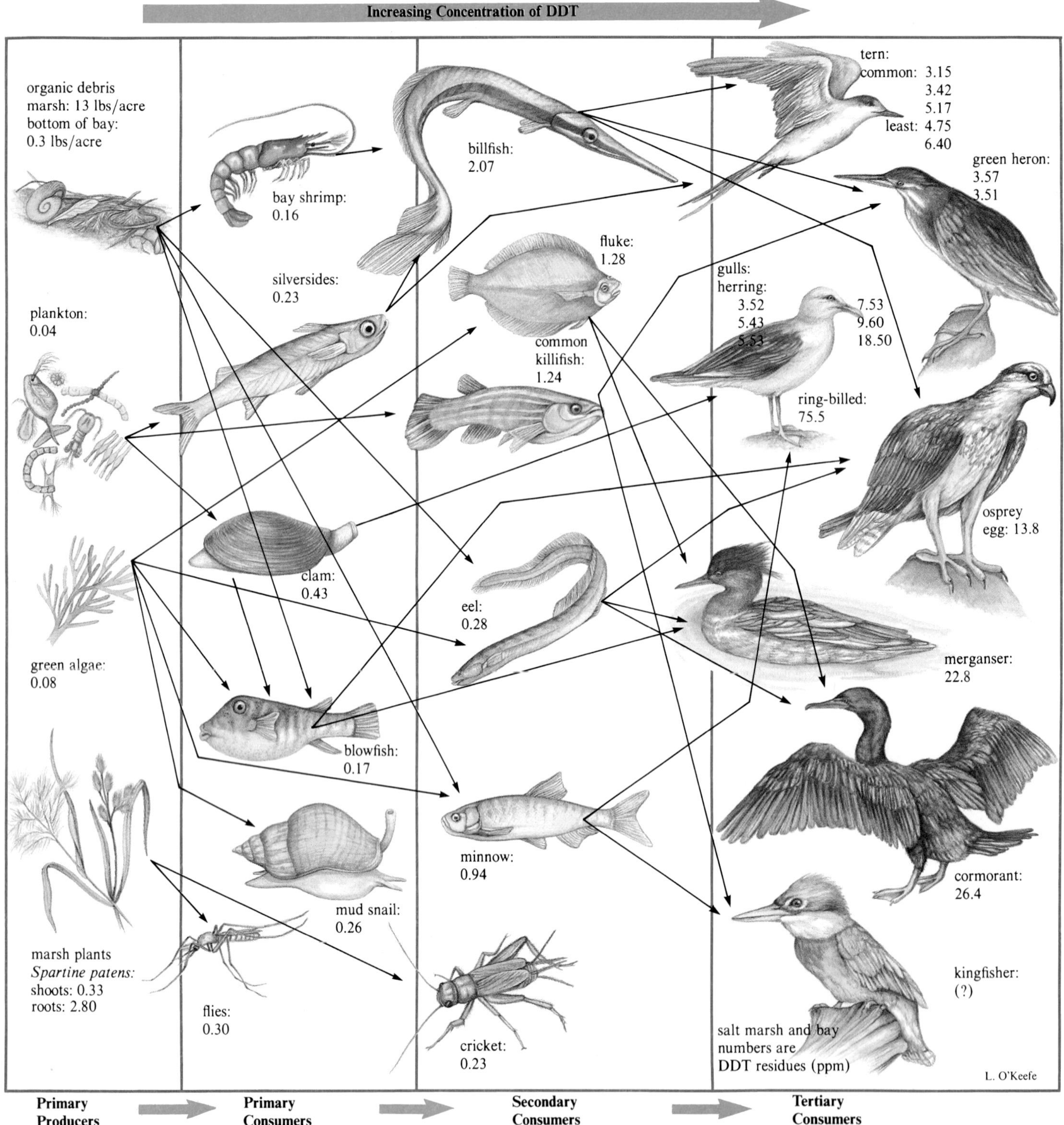

Figure 39.8
Biomagnification. The concentration of the pesticide DDT in organisms' bodies increases as the food web is ascended.

Concentration of Chemicals—Biomagnification

Cells admit some chemicals but not others, resulting in substances that are more concentrated within cells than in the environment. Some substances, particularly those not normally found in ecosystems, become more concentrated in organisms at higher trophic levels. Such **biomagnification** occurs when the chemical is passed along to the next consumer rather than being metabolized and excreted.

DDT

DDT is a pesticide originally used because it damages the nervous systems of insects, such as disease-carrying mosquitoes. However, it soon proved to affect many organisms, including many birds' ability to manufacture eggshell materials. DDT was banned in 1972, but it has crept into nearly all life on earth. The pesticide is even found in the fat of Antarctic penguins, who live where DDT was never used.

Almost all of the DDT taken in by an organism remains in its body, lingering in fatty tissue because it is soluble in fat. The chemical is passed on to those who eat the organism, and at the next trophic level, the process is repeated. Concentration of the pesticide rises with each trophic level, with the organisms at the top of the energy pyramid the most severely affected. By the fourth trophic level, concentration of DDT may be 2,000 times greater than in the organisms at the base of the food web (fig. 39.8).

Mercury

A tragic example of biomagnification occurred in Minimata, Japan, when mercury was discharged into a water supply from an industrial plant. Fish consumed the mercury, storing it in fatty flesh, and when pregnant women ate the fish, the mercury became even more concentrated in their adipose cells, in their milk, and even more so in their unborn children. Nearly 200 infants were born, as a result, with "Minamata disease," grossly deformed and mentally retarded. The tragedy of Minamata may be repeated, because today chemical companies are dumping mercury-laden waste in the province of Natal, South Africa.

Smoking marijuana also demonstrates biomagnification of mercury. When these plants are grown near volcanoes or thermal springs, such as in parts of California, Hawaii, and Mexico, they assimilate mercury from the air through their leaves and from the soil through their roots. Eating marijuana allows only 7% of the mercury to enter a human's circulation, but by smoking it, 75% to 85% of the mercury is absorbed into the bloodstream. From here, it crosses the blood-brain barrier, causing the classic symptoms of mercury toxicity—irritability, paranoia, forgetfulness, tremors, and narrowed vision. (The Mad Hatter in *Alice in Wonderland* suffered from mercury poisoning.) Smoking tobacco is not likely to introduce mercury into the body because the claylike soil in which tobacco is grown is low in natural mercury.

KEY CONCEPTS

Some chemicals, particularly those that are soluble in fat, are concentrated as they ascend a food chain, becoming more and more toxic.

Biogeochemical Cycles

All life, through all time, must use the elements that were here when the earth formed. These elements are continuously recycled through the interactions of organisms and their environment. If not for this constant recycling, these elements would have been depleted as they became bound in the bodies of organisms that lived eons ago. Because the recycling of chemicals essential to life involves both geological and biological processes, the pathways involved are called **biogeochemical cycles.** Water, carbon, hydrogen, nitrogen, oxygen, phosphorus, sulfur, potassium, sodium, calcium, chlorine, iron, and cobalt all pass between the atmosphere, the earth's crust, water, and organisms.

Each type of chemical in an ecosystem has its own characteristic biogeochemical cycle, but all cycles have steps in common. Generally, the nutrient is first taken from the environment and incorporated into the tissues of photosynthetic organisms, such as plants. If the plant is eaten by an animal, the nutrient may become part of animal tissue. If that animal is eaten by another, the elements may be incorporated into the second animal's body. Eventually, all organisms die and their bodies are broken down by other organisms. This decomposition releases the elements into the environment for reuse by photosynthesizers.

The Carbon Cycle

In the carbon cycle, photosynthetic organisms capture the sun's energy and use it to produce carbohydrates, using atmospheric carbon dioxide (CO_2) (fig. 39.9). These carbohydrates are used by the plant and often eaten by animals. Both plants and animals release carbon to the atmosphere through respiration. Animals return carbon to the soil in excrement. When an organism dies, bacterial, insect, and fungal decomposers break down its organic compounds to release carbon to the soil, atmosphere, or aquatic surroundings.

Certain nonliving parts of the environment contain carbon from past life. For instance, limestone was once a colony of shelled sea inhabitants, and fossil fuels, such as coal and petroleum, formed from the remains of plants and animals. When these fuels are burned, carbon is returned to the atmosphere as carbon dioxide.

The Nitrogen Cycle

Nitrogen is an essential component of proteins and nucleic acids, as well as other parts of living cells. Although the atmosphere is 79% nitrogen, most organisms cannot use nitrogen gas to manufacture their biochemicals. They depend on nitrogen-fixing bacteria that convert atmospheric nitrogen into ammonia (NH_3). Lightning also fixes a small amount of nitrogen, and decomposers release ammonia. **Nitrifying bacteria** convert ammonia from dead organisms to nitrites, and then other bacteria convert nitrites to nitrates (NO_3^-), which are compounds that can be used by plants and are then passed on to animals. Finally, nitrogen is returned to the atmosphere by **denitrifying bacteria** that convert ammonia, nitrites, and nitrates to nitrogen gas (fig. 39.10).

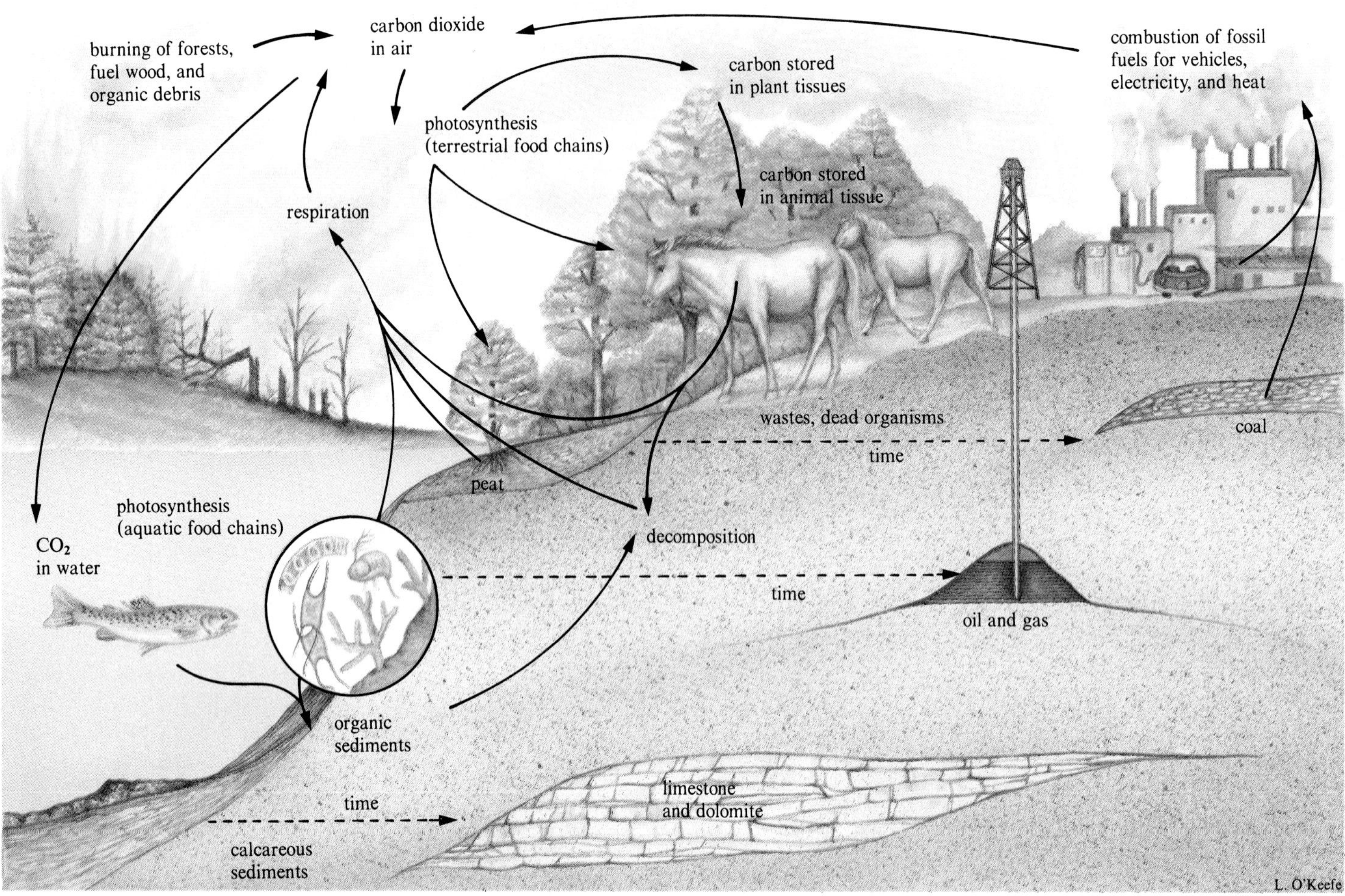

Figure 39.9

The carbon cycle. Carbon dioxide in the atmosphere enters ecosystems through photosynthesis and passes up food chains. Respiration, decomposition, and combustion return carbon to the abiotic environment. Carbon is retained in geological formations for long periods of time.

The enzyme nitrogenase enables nitrogen-fixing bacteria to convert nitrogen to ammonia. Because this enzyme is inactivated when exposed to oxygen, nitrogen-fixing microbes are typically shielded from oxygen. For example, *Rhizobium* bacteria live in swellings (nodules) on the roots of legumes such as beans, peas, and clover (fig. 2.3). Farmers often take advantage of nitrogen-fixing properties of legumes by rotating them with nonleguminous crops, such as corn, so the soil is continually enriched with nitrates.

The Phosphorus Cycle

Phosphorus is a vital component of genetic material, ATP, and the phospholipids of membranes. When phosphorus is scarce, growth is limited. The earth's supply of phosphorus is in certain rocks, excrement, bones, and fossils. As rain falls over these reservoirs, phosphorus is released as phosphate that can be used by organisms to form living tissues. Decomposers eventually return phosphorus to the soil (fig. 39.11).

KEY CONCEPTS

Elements are continually cycled between organisms and the physical environment. Plants use solar energy and atmospheric CO_2 to manufacture carbohydrates, which pass through food webs and are recycled in respiration, excretion, and decay. Carbon is also stored in limestone and fossil fuels. Nitrogen is fixed, converted to usable nitrates, and returned to the atmosphere by bacteria. Phosphorus cycles from nonliving reservoirs to the living.

Succession

Just as the physical components of an ecosystem can change, so can its biological communities, with some species replacing others. The process of change in the community is called **ecological succession,** and it is gradual and directional. Eventually, a **climax community,** one that remains fairly constant if the land and climate are undisturbed, is established.

Primary Succession

Primary succession occurs in an area where no community previously existed. On a patch of bare rock in New England, for example, the first species to invade, called **pioneer species,** are hardy organisms, such as lichens and mosses, that are able to get a

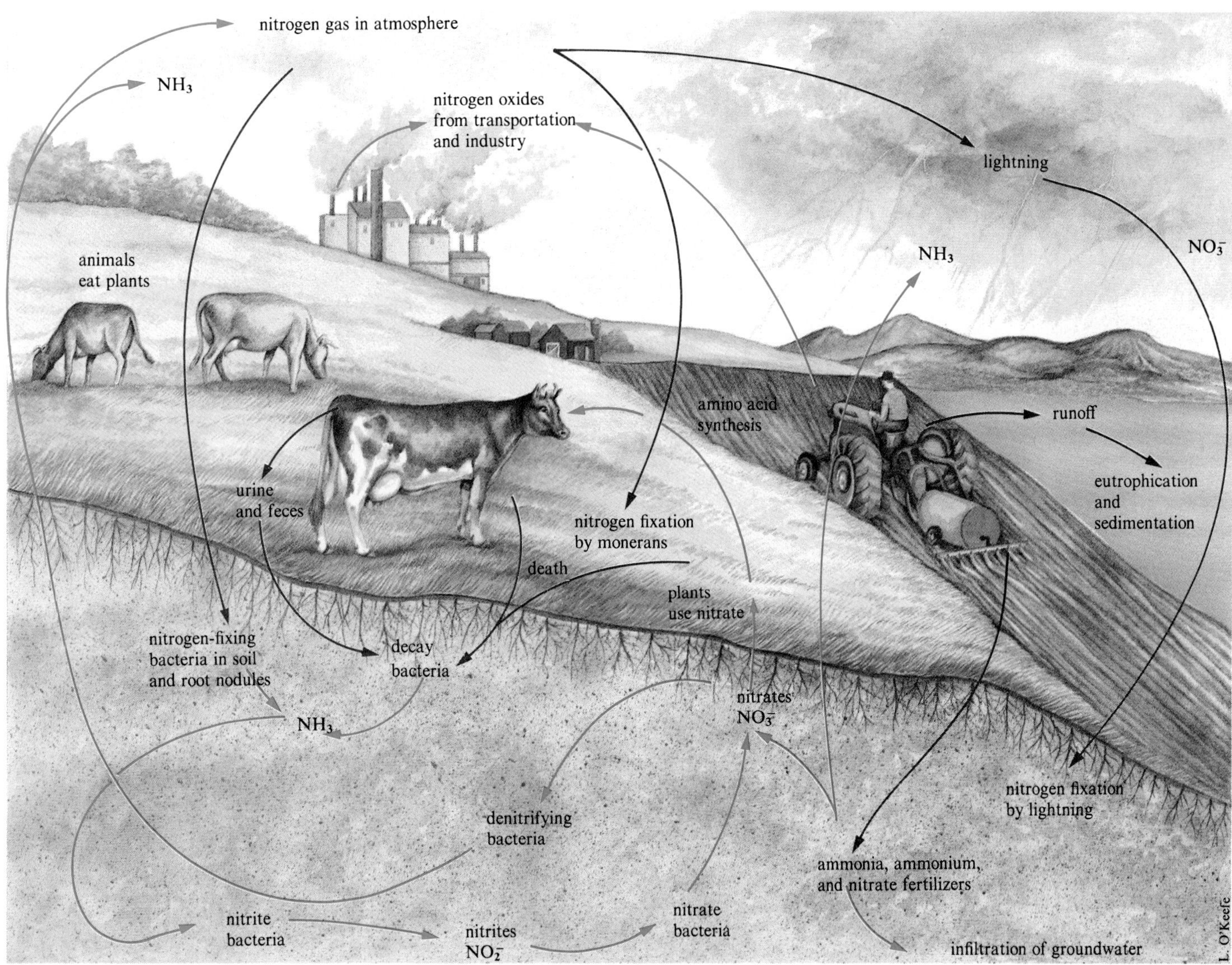

Figure 39.10

The nitrogen cycle. Plants and bacteria use nitrogen to build amino acids and nucleic acids and pass these biochemicals up food chains. Nitrogen is returned to the abiotic environment in urine and feces and by the decaying action of bacteria on dead animals.

foothold on smooth rock. The fungal component of a lichen holds onto the rock and obtains water, while the algal part photosynthesizes and provides food (Reading 2.1). As lichens produce organic acids that erode the rock, sand accumulates in crevices. Decomposition adds organic material to the sand, and eventually a thin covering of soil forms. Then rooted plants, such as herbs and grasses, invade. Soil continues to form, and as it does larger plants, such as shrubs, appear. Next on the scene are aspens and conifers, such as jack pines or black spruces. The soil finally becomes rich enough to support deciduous trees. Eventually, after many years, a climax community of an oak-hickory forest reigns (fig. 39.12*a*).

Secondary Succession

Secondary succession occurs in areas where a community is disturbed, such as by a river changing course and flooding an area or fire destroying a forest. Although the sequence of changes is often similar to that of primary succession, the rate of change is often faster in secondary succession because the soil may already be formed (fig. 39.12*b*).

An abandoned farm illustrates secondary succession, called "old field" succession. It begins when the original deciduous forest is cut down for farmland. As long as crops are cultivated, natural succession is halted. When the land is no longer farmed, fast-growing pioneer grass and weed species, such as black mustard, wild carrot, and dandelion, move in, followed by slower-growing, taller goldenrod and perennial grasses. In a few years, pioneer trees such as pin cherries and aspens arrive, but these are eventually replaced by pine and oak, and finally, a century or so later, the climax community of beech and maple is again well developed.

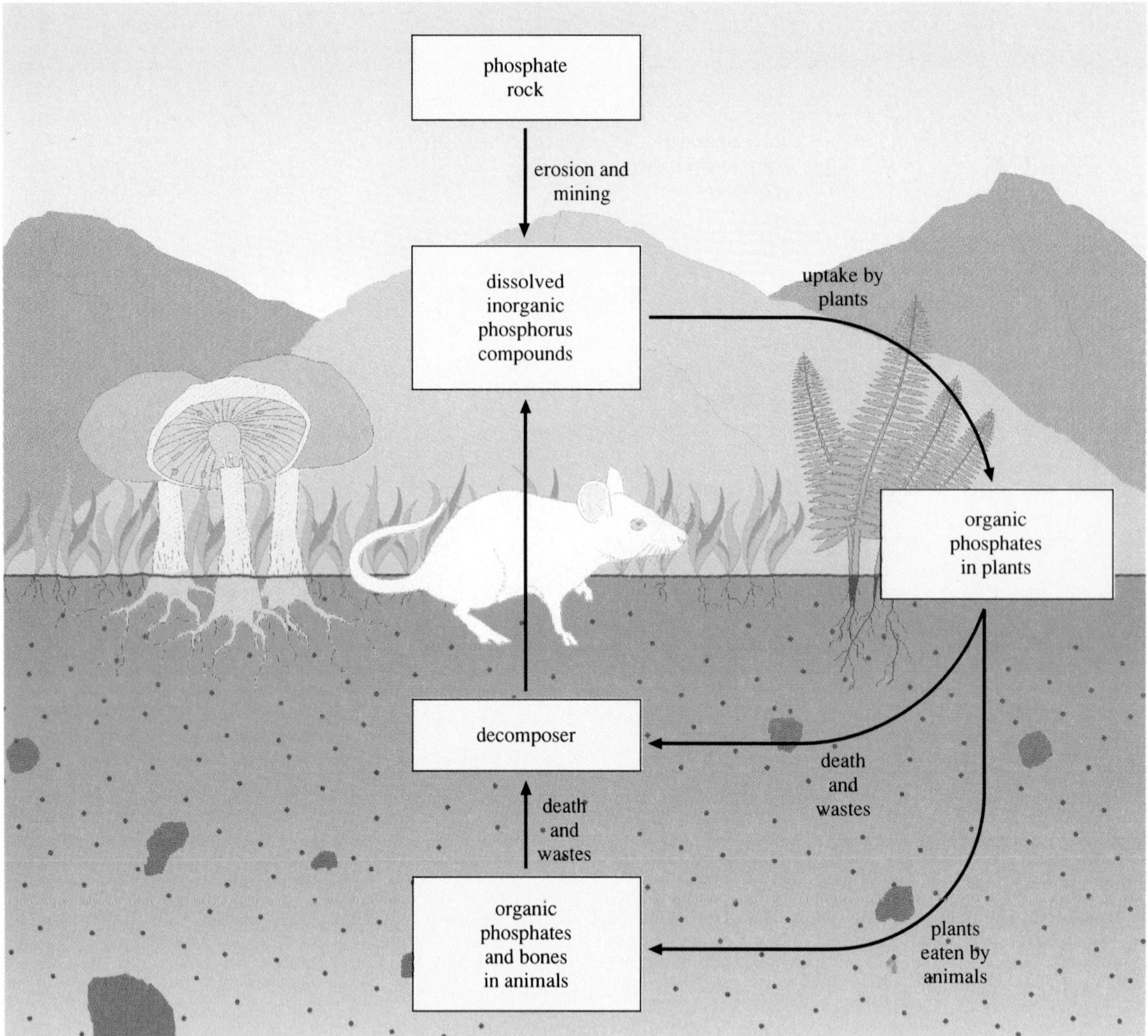

Figure 39.11
The phosphorus cycle. Phosphorus ultimately comes from rock, which is taken up by plants and passed up food chains. Decomposers return phosphorus to the abiotic environment.

Secondary succession is dramatic in the Mount St. Helens area of Washington state, where a volcano erupted in 1980 (fig. 38.6). Dead animals and animal droppings provided the nutrients to fuel new plant growth, and these nutrients were mixed into the soil by gophers. In fact, gopher mounds were the first areas to sprout new plant life. Fallen trees protected areas where flowers and ferns could take hold. Just 1 month after the eruption, the moonlike landscape was dotted with purple and yellow aster blossoms whose buds had been saved by snow cover; green bunches of thimbleberry and fireweed sprouts peeking out from their pumice coats; and a spreading pink carpet of fungus. Four months after the blast, insects were back. Bluebirds and woodpeckers built nests in dead trees, and geologists exploring the area would often be attacked by hummingbirds mistaking their bright clothing for flowers. These pioneer species could live in the destroyed region because they were already adapted to the bare areas resulting from lumber activity.

In the years to come, animals from surrounding areas will continue to wander in, perhaps bringing seeds on their pelts. Other seeds will blow in and eventually germinate. The new inhabitants will return nutrients to the depleted soil in their excrement and when they die. Rain will mix the soil. Over the next 4 decades, a new

Figure 39.12 (facing page)
a. Primary succession. Bare rock becomes an ecosystem when lichens arrive and produce acids that begin to break down the rock into a thin layer of soil. Next, mosses, fungi, small worms, insects, bacteria, and protozoans colonize the scanty soil. Tiny plants appear, and then herbs and weeds settle in. Grasses and shrubs move in and are followed, after many years, by trees. *b*. Secondary succession. Last summer it was a farmer's field, with row after row of corn, beans, and sunflowers, in the southeastern United States. This year it was plowed and then abandoned. But the land does not stay bare for long. Weeds and herbs appear first, then grasses and shrubs. Then small versions of the trees that ring the area will grow. Over many years, tree species will appear in a certain sequence, until the area stabilizes as a climax community.

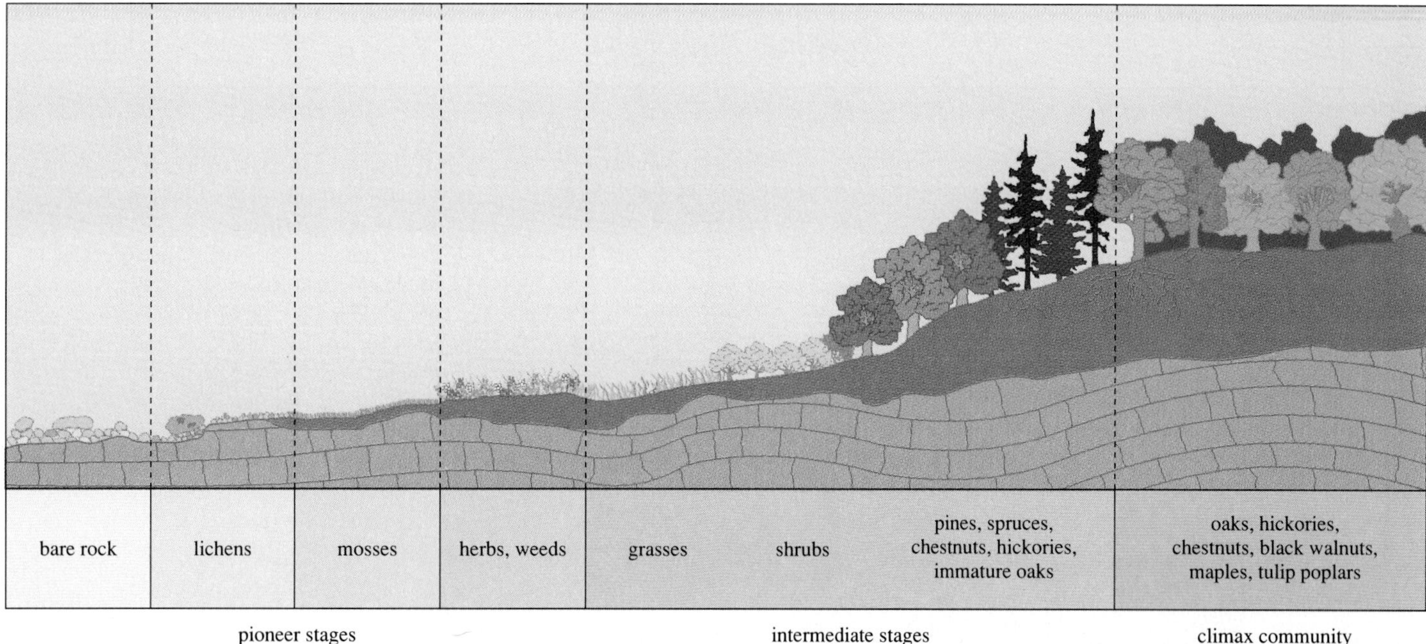

bare rock
lichens
mosses
herbs, weeds
grasses
shrubs
pines, spruces, chestnuts, hickories, immature oaks
oaks, hickories, chestnuts, black walnuts, maples, tulip poplars
pioneer stages
intermediate stages
climax community
hundreds of years

a.

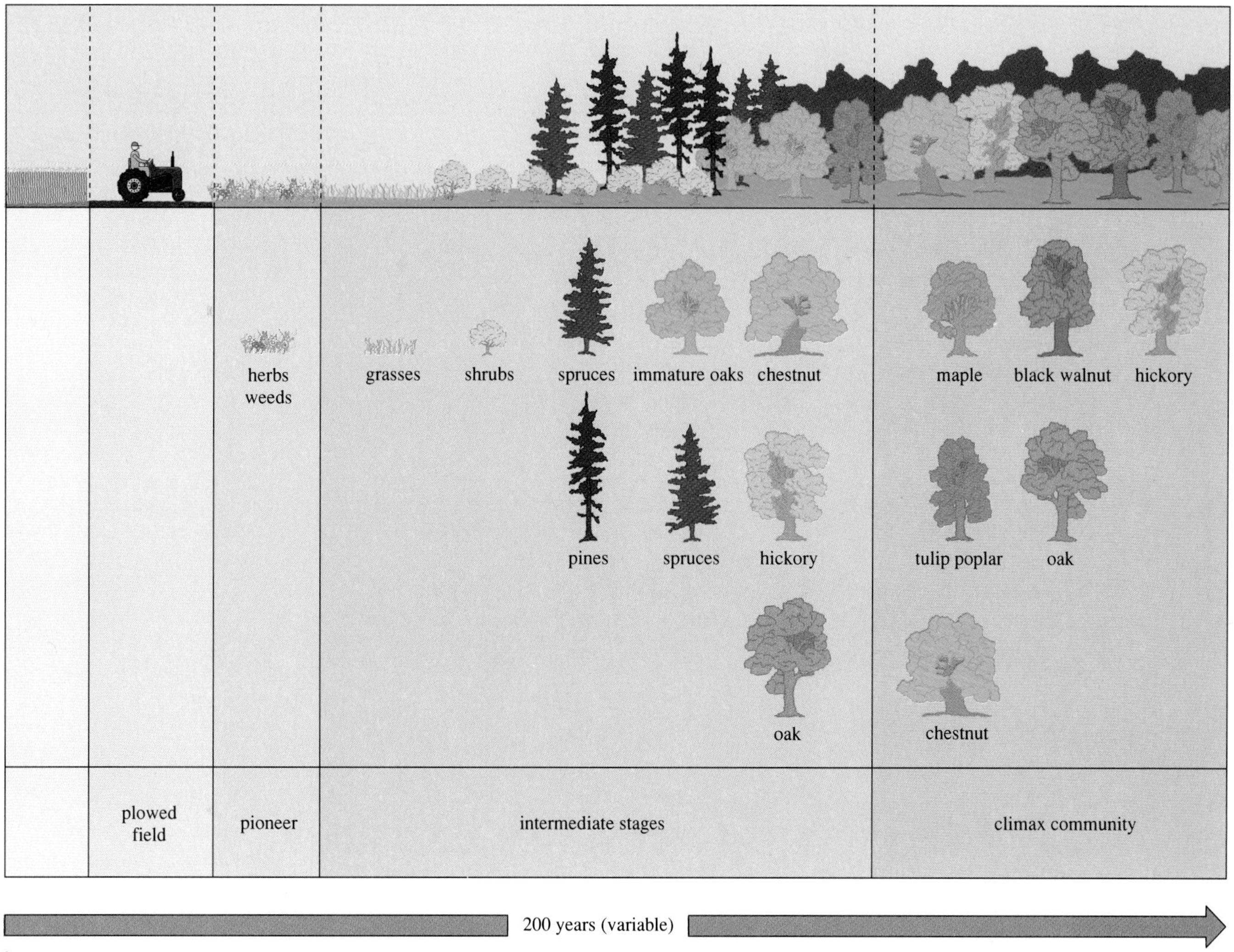

herbs weeds
grasses
shrubs
spruces
immature oaks
chestnut
maple
black walnut
hickory
pines
spruces
hickory
tulip poplar
oak
oak
chestnut
plowed field
pioneer
intermediate stages
climax community
200 years (variable)

b.

Reading 39.1 *FIRE!*

THE SPRING OF 1988 BEGAN NORMALLY ENOUGH IN THE GREATER YELLOWSTONE AREA, WHICH ENCOMPASSES 4.8 MILLION HECTARES IN WYOMING, MONTANA, AND IDAHO. The spring had been rainy, and by early June, there had been 20 fires—typical for the heavily forested area. For the first 3 weeks in July, weather was monitored closely, because a dry spell at this critical time predicts a bad fire season ahead. By the key third week that July, the Yellowstone Park managers had a lot to worry about. By the July 21 order to fight all fires actively (policy is to let many natural fires burn), it was too late.

From mid-July through August, no rain fell. Relative humidity plummeted to 6% as high, dry winds raced through the area. With the deadly combination of weather conditions, accumulated deadwood on the forest floor, lightning, and careless campers, fire was all but inevitable. But no one expected the extent or intensity of the fiery holocaust of the summer of 1988.

Before September rains quenched the flames, 25,000 fire fighters would battle the 248 fires—7 of which caused 95% of the destruction. Some of the fires moved far faster than they could be controlled (fig. 1). Advancing 10 miles (16 kilometers) a day, sparks flew ahead of the front, like scouts, igniting smaller flames ahead of the sweeping tide of fire.

Unlike most fires, these did not halt at nightfall. A year later, aerial photographs would show that nearly two-thirds of the burn occurred in the forest canopy of tree trunks, limbs, leaves, and needles, and about a third raced along on the ground, charring the underbrush but leaving live trees in its wake. The landscape was scarred with a mosaic of "bull's-eye" patches of burn ringed by brown and green (fig. 2).

Yet fire is nothing new to Yellowstone. A similar outbreak in 1910 burned twice as much area as the 1988 inferno, and the early 1700s saw a severe fire season as well. Ecologists now think that periodic fires are a natural component of forest ecosystems, occurring approximately every two to three centuries, coinciding with droughts. Such hellstorms ultimately increase the diversity of life in forest ecosystems, as new species colonize burned areas.

With 45% of the greater Yellowstone area aflame at some time that summer, all ecosystems were affected. Immediate effects included rise in water temperatures and changes in water chemistry, which reverberated up aquatic and ultimately terrestrial food webs. Plant communities were drastically affected, as 250- to 300-year-old lodgepole pines perished.

Wildlife was affected by fire-fighting efforts almost as much as by the fire itself. Vehicles left tracks across meadows, and more than 100 animals were run over. Helicopters clearly disturbed many animals. Nearly 40 million liters of water were removed from lakes and streams to battle the blazes, and fire retardant chemicals, sprayed onto buildings, entered streams, killing fishes.

Just a year after the blazes, evidence of succession abounded. Interspersed with the bull's-eyes that were testimony to the flames were patches of forest floor carpeted in grass and spectacular wildflowers. The new lodgepole pine forest had taken root, its seeds released only by fire. The new "trees" are now about as tall as paper clips. In 30 years, they will stand as tall as a person, perhaps taller.

What lies ahead? Yellowstone waters will continue to grow murkier over the next few years, but as the waters clear and sediments settle, fishes, invertebrates, and algae populations will increase. The first plants to recolonize the newly cleared areas will be herbs. Sagebrush will be noticeably absent for at least a decade, because it cannot resprout following fire. Gradually, meadows and young forests of aspen will grow.

Although it will be another century before the 1.4 million burned acres resemble a forest, the Yellowstone fires of 1988, at first mistakenly viewed as a tragedy, will bring abundant and diverse new life to this beautiful and resilient area.

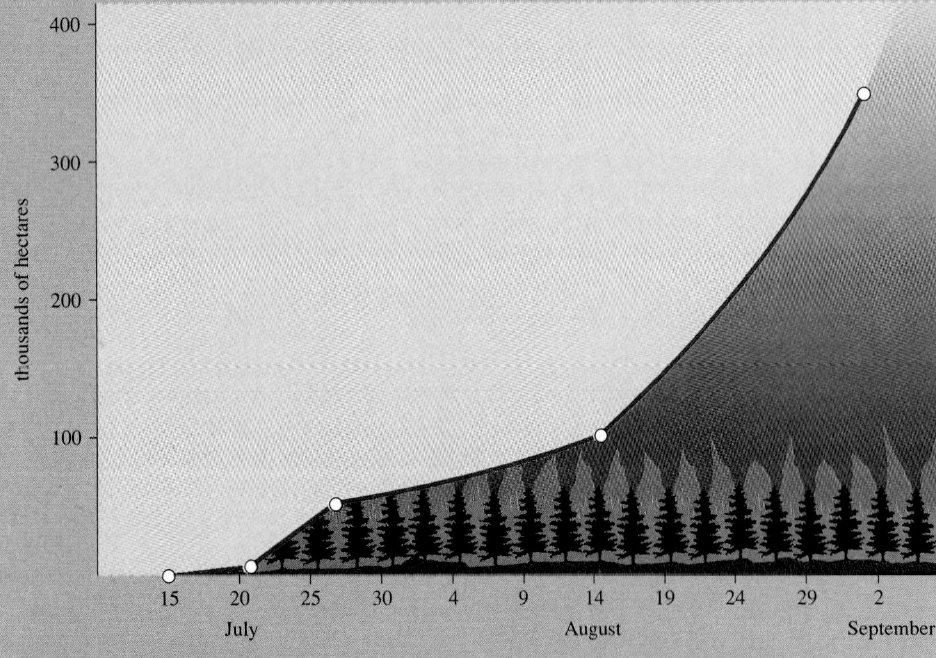

Figure 1
Fires spread rapidly throughout the greater Yellowstone area in the summer of 1988.

Figure 2
The Yellowstone fires left a characteristic mosaic landscape.

forest will grow. Fifty years from now, scattered trees will stand, each 40 to 50 feet (12 to 15 meters) tall. By the century mark, a dense, new forest will flourish—barring further eruptions in the geologically unstable area. Ecosystem recovery is also vivid in parts of the greater Yellowstone area damaged by the fires of 1988 (Reading 39.1).

KEY CONCEPTS

Succession is a gradual, directional change in the communities of an ecosystem. Primary succession occurs on bare rock, beginning with pioneer species of lichens and mosses that form soil, followed by increasingly larger plants. Secondary succession occurs in a damaged, previously inhabited area. Its steps are similar to those of primary succession, but may proceed faster if soil is there.

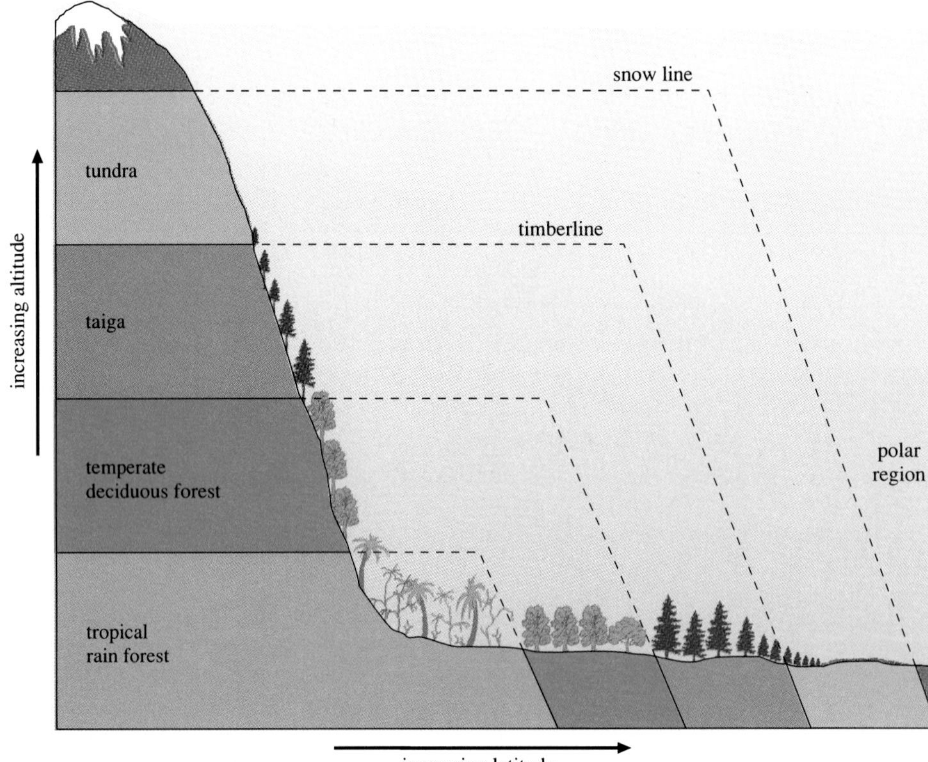

Figure 39.13
Latitude and altitude influence plant communities. As a mountain rises, the types of plants parallel the types seen in the different biomes as distance from the equator (latitude) increases.

Terrestrial Biomes

Plants define terrestrial biomes because they determine which animal and microbe species are present. The types of plants in a biome depend upon altitude and latitude (fig. 39.13). The major terrestrial biomes include the tropics, the temperate zone, the subarctic, and the arctic. Terrestrial biomes within these categories are distinguished by their degrees of moisture. For example, tropical biomes include the very wet tropical rain forest, the drier seasonal forest, the still drier savanna, the semidesert, and the hot but dry desert (fig. 39.14).

Tropical Rain Forest

Tropical rain forests are found where the climate is almost constantly warm and moist. Rainfall is typically between 79 and 157 inches (200 and 400 centimeters) per year, and the ground is perpetually soggy. The tropical rain forest of the Amazon basin in South America is incredibly vast, its meandering rivers and dense foliage covering an area 90% of that of the continental United States. Within the lush maze of intertwined branches and moss-covered vines live a staggering diversity of species. A patch of jungle 4 miles square (10.4 square kilometers) houses, among its 750 tree species, 60 types of amphibians, 100 types of reptiles, 125 varieties of mammals, and 400 kinds of birds. One tree alone may be inhabited by 400 types of insects. A given area of tropical rain forest probably contains 10 times the diversity of species as the same-sized patch of forest in the United States.

From the air, a tropical rain forest is a solid, endless canopy of green, built of treetops 50 to 200 feet (15 to 66 meters) above the forest floor. Plants beneath the canopy compete for sunlight, forming layers of different types of organisms. This arrangement is called **vertical stratification.** Different strategies are used to capture sunlight. Very tall trees poke through the canopy. Others have crowns that are broad and flattened, maximizing sun exposure in this equatorial region where the sun's rays are almost perpendicular to the earth's surface. Vines and epiphytes (small plants that grow on the branches, bark or leaves of another plant) grow piggyback fashion on tall trees. Epiphytes lack roots but are adapted for holding moisture. The ground of the tropical rain forest looks rather bare, shaded as it is by the overwhelming canopy, but it too is home to countless shade-adapted species.

The lush vegetation feeds a variety of herbivores. Insects, sloths, and tapirs devour leaves. Small deer and peccaries, monkeys, rodents, bats, and birds eat primarily fruit. The herbivores are consumed by many carnivores, such as large cats, birds of prey, and snakes. Some animals here are enormous. The pirarvu fish is more than 7 feet (2.1 meters) long, and the wingspans of many insects are several inches.

Cycling of nutrients, including minerals, is quite rapid in the tropical rain forest, where the daily torrential rains rapidly wash away nutrients released in the soil from dead and decaying organisms.

Termites recycle nutrients with the help of protozoans in their guts, which break down the cellulose in plants and release its atoms back to the environment. The soil is not very fertile because nutrients are deposited and reused so quickly. The roots of giant trees recycle so efficiently that almost all nutrients from decaying plant material are saved. Essential in a tree's recycling program are fine roots that permeate the upper 3 inches (7.6 centimeters) of ground. These rootlets grow up to and attach to dead leaves. Fungus on rootlets speeds decomposition, and nutrients released from the dead leaves are quickly absorbed. These extra roots even absorb nutrients from falling dust and rain. Animals help to recycle nutrients. Certain leaf-cutting ants, for example, farm gardens of fungus, which hasten decomposition. Reading 2.2 describes how biologists study the great diversity of life in the tropical rain forest; chapter 40 explores the consequences of destroying this biologically rich biome.

Temperate Deciduous Forest

Deciduous hardwood trees that lose their leaves during the winter thrive where the growing season lasts at least 120 days, annual rainfall ranges from 28 to 59 inches (70 to 150 centimeters), and the soil is rich. Vertical stratification to compete for light is seen here too, although diversity is less than in the tropical rain forest. Usually one or two types of trees predominate, such as oak-hickory or beech-maple forests. The dominant trees, victors in the competition for light, are also best suited for the average rainfall. Trees in the deciduous forest often have ball-shaped tops, which maximize light absorption for the angle of the sun's rays at these latitudes. Shrubs grow beneath the towering trees. Below them are grasses and annual flowers, which grow mostly in early spring when light peeks through the defoliated trees. Mosses and liverworts coat the forest floor.

The decomposers in temperate forests—nematodes, earthworms, and fungi—break down the leaf litter to create rich soil. Herbivores include the cottontail rabbit, whitetail deer, and grouse. The red fox and raccoon are common carnivores. All residents of the deciduous forest must cope with seasonal changes. Some animals hibernate through the cold winter. Others migrate to warmer climes. Some insects enter an inactive state called diapause.

Temperate Coniferous Forest

Coniferous trees are commonly called evergreens. Although they do lose their leaves, or needles, the loss is usually not synchronous, so most conifers remain green all year. In temperate areas, conifers are found in regions where the soil is too poor and water too scarce to support deciduous trees, which need these resources to replace their leaves. Conifers are also suited to areas that experience recurring fires. Fire recycles nutrients and selects resistant species (see Reading 39.1). The thick bark on certain pine trees resists flames, and some pinecones liberate their seeds only after exposure to the extreme temperature of a forest fire.

Taiga

North of the temperate zone is the **taiga,** also known as the northern coniferous forest or boreal forest. The cold, snowy mountainous taiga is home to a few hardy, well-adapted types. Forests of spruce, pine, fir, and hemlock trees dominate, with a few deciduous trees near lakes, streams, and rivers. The long winters and short summers of the region hamper the ability of deciduous trees to shed leaves and grow new ones. Little grows beneath the tall conifers. The forest floor is inhospitable, with decaying needles making the soil too acidic for many plants to grow and the needles above blocking the sun. In addition, frigid temperatures slow decomposition.

Conifers are well adapted to an environment so cold that water is usually frozen throughout the winter. With a scant amount of sap, a conifer has little liquid to freeze. Conifers retain their leaves but have adaptations to reduce water loss through the leaves. The shape of needles minimizes surface area on which water is lost, as does their thick, waxy outer surfaces. The stomatal openings in the needles, through which water is usually lost, are few in number and can close when water is scarce. Photosynthesis slows to 30% to 50% its rate at other times. The conical shape of the trees helps capture the oblique rays of light found at these latitudes and prevents damaging snow and ice buildup.

Common decomposers in the taiga are fungi. Familiar herbivores include squirrels, grouse, warblers, and moose, and carnivores include lynx and wolves. As is true nearly everywhere else on the planet, insect populations can at times grow very large.

Grasslands

Grasslands have 10 to 30 inches (25 to 75 centimeters) of rainfall annually, which is often not sufficient to support trees. These biomes have a variety of names, including **savanna** in the tropics, **prairies** in the temperate zone, and the arctic **tundra.**

The African savanna is a tropical grassland, populated by isolated clumps of trees that are interspersed with grasses rooted deeply in the fine, sandy soil. Grazing the land are zebras, gazelles, antelopes, impalas, giraffes, and wildebeests, and stalking the grazers are lions, cheetahs, tigers, and hyenas (fig. 38.8). A leftover meal is soon scouted by vultures. The particular vegetation pattern of the savanna depends upon the amount of rainfall. In a rainy year, the trees flourish. Their extensive leaves block the sun from the grasses beneath, and the grasses die out. In times of drought, however, the grasses predominate. Their roots store moisture, even when the plant above the soil has dried up.

The North American prairie is a temperate grassland. The height of the grass reflects local moisture. Grass reaches 4 to 8 feet (1.2 to 2.4 meters) around the Mississippi Valley, where moisture from the Great Lakes and Gulf of Mexico contributes to an annual rainfall of about 39 inches (100 centimeters). Westward, toward the Rocky Mountains, the annual rainfall decreases, and shorter grass species dominate the landscape. However, grasses can do quite well with little water. The shape of the blades decreases surface area and, therefore, water loss. Root systems are extensive, with some species sending roots 6 feet (1.8 meters) underground to reach water. The mat of roots holds soil together, preventing it from blowing away during a drought. Growth response to rain is rapid.

Unlike trees, grasses survive damage by fire and herbivores rather easily. Grass roots are so deep that they are protected from flames. Because these plants grow

a.

b.

c.

d.

polar ice cap
tundra
taiga
temperate grassland
temperate deciduous forest
temperate rain forest
desert
tropical rain forest
tropical deciduous forest
tropical scrub forest
tropical grassland
mountain

Figure 39.14
Major terrestrial biomes. *a.* Polar ice cap. *b.* Tundra. *c.* Tropical rain forest. *d.* Temperate deciduous forest. *e.* Temperate rain forest. *f.* Desert. *g.* Grassland. *h.* Taiga.

e.

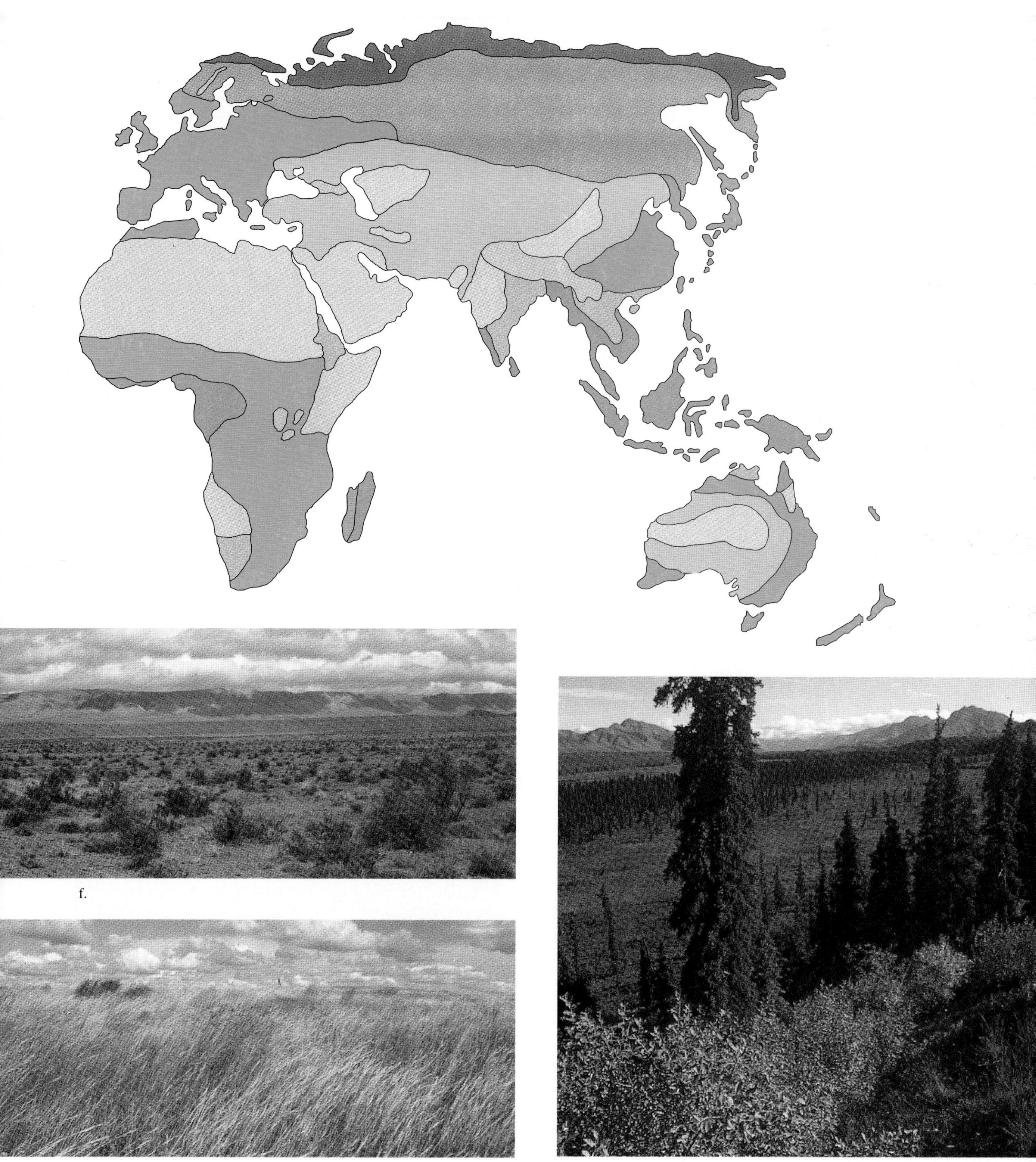

f.

g.

h.

Reading 39.2 *Biosphere II*

IN AN AREA THE SIZE OF TWO FOOTBALL FIELDS, ONE CAN STEP THROUGH THE UNDERGROWTH OF A TROPICAL RAIN FOREST AND LOOK UP AT A CANOPY 85 FEET (26 METERS) TALL, WANDER ONTO A SAVANNA PERCHED ON ROCKY CLIFFS, AND LOOK BELOW TO THREE OTHER MINIBIOMES—AN OCEAN 26 FEET (8 METERS) DEEP, COMPLETE WITH CORAL REEF; A MARSH LEADING FROM SALTY TO FRESH WATER; AND A DESERT SIMILAR TO ONE ON MEXICO'S BAJA PENINSULA.

This is Biosphere II, an attempt to systematically study biome and ecosystem components and interactions—something that is vastly complex in the wild but may be more controllable in this high-tech elaborate greenhouse. Biosphere II, near Tucson, Arizona is 196,200 cubic yards (150,000 cubic meters) in volume and contains 1,452,000 gallons (6.6 million liters) of water, 4,000 species, and 8 people.

The human residents of Biosphere II live in apartments in a wing containing laboratories, contacting the outside world by phone, video, and computer. Half of each day is spent in the lab, and the other half is spent farming the sixth biome, where they grow crops and milk goats. The mostly vegetarian diet is supplemented with an occasional fish or lizard caught on another biome. Pest control is *au naturel*—ladybugs kill aphids, wasps eat white flies, beetles get rid of spider mites, and fungi feed on nematode worms.

Stocking Biosphere II was a far more challenging task than Noah faced when he simply welcomed two of each species onto the ark. The scientists behind the project fell into two camps. The "species packing" approach would overpopulate each biome, releasing many species and letting them, via natural selection, determine which would survive. The "no extinction" approach aimed to stock each biome with precisely those species that would inhabit it in Biosphere I, the real world. In the end, both approaches were used.

The designers quickly found themselves literally entangled in the webs of life, as they tried to simulate "microniches." Consider soil. To support plants, soil would have to harbor communities of mites, bacteria, nematodes, and fungi. Termites would be required to aerate the soil, and they would need dead wood supplied, because the trees of Biosphere II would not be there long enough to die. Ants could be added to back up the termites' aerating duties, but ants and termites fight. Ants eat certain seeds. Building an ecosystem, trying instantly to create what nature has taken eons to direct, is no small task.

What will Biosphere II tell us? Its human residents hope to learn how ecosystem structure and how ecological processes are disrupted by such interventions as acid rain and the greenhouse effect. The more imaginative see Biosphere II as a prototype for eventual colonies on Mars.

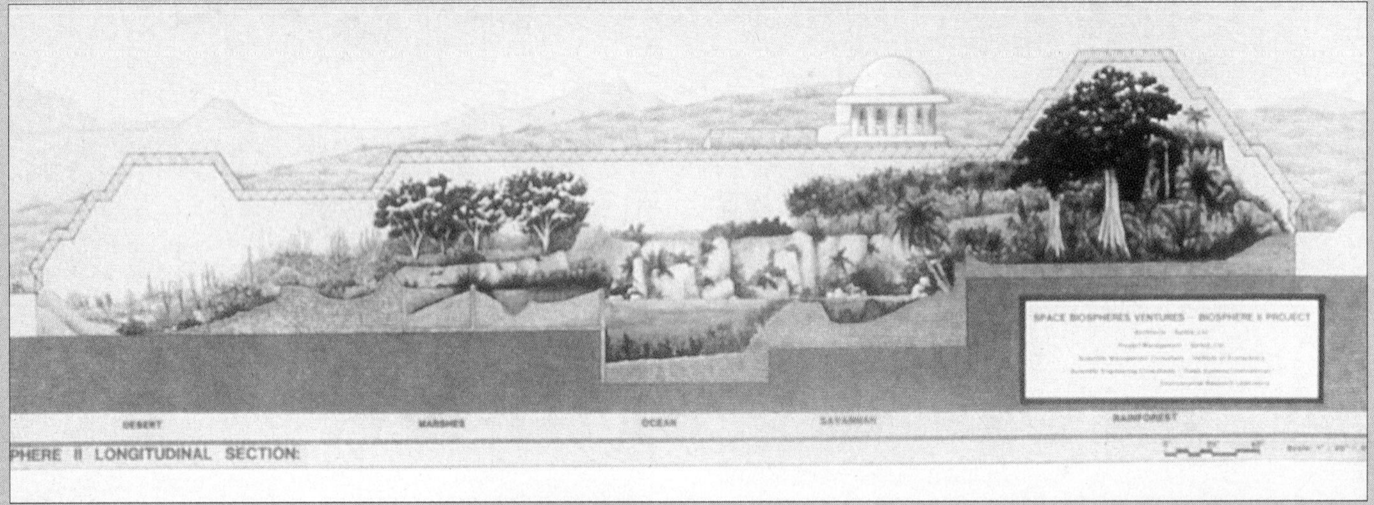

Figure 1
Biosphere II houses biomes and representatives of 4,000 species.

from below, removal of the blade by a hungry animal (or lawn mower) does not hinder growth. Even chunks of grass kicked up by a grazing animal can easily reroot. In contrast, the growing regions of trees—the tips of branches—are the first to be destroyed by fire or herbivores.

Like the grassland community, its corresponding animal community is also simple and uniform compared to the diversity found in the tropical and temperate forests. Many of the herbivores are insects or rodents. Some, such as the prairie dog, retreat into burrows to escape predators. Larger grazers, such as the antelopes, have few places to hide and rely on their speed (fig. 38.10). The bison's massive size and power were sufficient protection against natural predators such as the wolf but were no match for the guns of humans.

Tundra

A band of **tundra** runs across the northern parts of Asia, Europe, and North America. Winter there is bitterly cold, with typical temperatures about -26°F (-3°C). Despite the above-freezing summer temperatures, which range from 40°F to 70°F (4.5°C to 21°C), part of the ground, called the **permafrost,** is frozen all year. Because the permafrost blocks absorption of water, spring runoff from ice and snow drains rapidly into rivers, or it accumulates, forming bogs or small stagnant ponds.

The colonizers of the tundra are lichens, which break down rocks and promote soil formation. The shallow soil supports woody shrubs and short grasses. Plants are stunted because the cycle of freezing and thawing damages roots, hampering nutrition. It is difficult for annual plants to germinate, grow, and flower in the short growing season, which may be less than 60 days. Most flowers are perennials that stay underground, protected by a blanket of snow during the winter.

Tundra plants tend to be low and flat, a shape that lets the wind blow over them. These plants often clump together, which helps to conserve moisture and warmth, and are often buried amid rocks, which provide shelter from the wind. Some plants have protective hairs that insulate them and help break the wind. The dark green color of many plants allows them to absorb more light for photosynthesis.

Animal inhabitants of the tundra include caribou, musk ox, reindeer, lemmings, snowy owls, foxes, and wolverines. Polar bears sometimes wander onto the tundra to feed. In the summer, insect populations expand in the many stagnant bogs and ponds. Migratory birds visit to raise their young.

Like plants, the animals of the tundra have many adaptations to the harsh climate. Both the hunters and the hunted benefit from camouflage. White winter colors of the fox, ptarmigan ermine, and hare are as inconspicuous against snow as are their brown colors of summer against the snow-free landscape. These animals are often round with short extremities, a form that helps to conserve heat. The snowshoe hare's big feet function as snowshoes. The shallow soil, short growing season, and slow decomposition of the tundra make it a very fragile environment.

Desert

Deserts receive fewer than 8 inches (20 centimeters) of rainfall per year. The days are searingly hot because there are few clouds to block or filter the sun's strongest rays. The nights are cool, sometimes 86°F (30°C) below the daytime temperature, because heat is easily dissipated in the air. This harsh, dry climate is home to fewer species than the other tropical biomes, but those that do live here are superbly adapted to the stressful conditions.

Annual plants grow quickly, squeezing their entire life cycles in between droughts. Most of their seeds germinate only after a soaking rain, which rinses a growth inhibitor from their coats. The thick-skinned, fleshy leaves of succulent perennials help hold precious water. In rainy times, perennial flowers bloom magnificently.

Lack of leaves in a cactus minimizes water loss. After a rainfall, the stem expands and stores water, which is guarded from thirsty animals by spines. The root system of a cactus is shallow but widespread, extending 55 to 65 yards (50 to 60 meters). Many desert plants produce chemicals that inhibit the growth of neighboring plants, so they grow evenly spaced.

Desert animals also cope with water scarcity. Body coverings, such as the exoskeleton of a scorpion or the leathery skin of a reptile, minimize water loss. Some small mammals, such as the kangaroo rat, have very concentrated urine, which saves water. Few animals are exposed to the midday sun. Most burrow or seek shelter during the day, becoming active when the sun goes down and the risk of water loss is reduced.

The driest and largest desert in the world is Africa's great Sahara, where rainfall is less than 1 inch (2 centimeters) per year. Because of its extreme dryness and heat, many areas are devoid of life. Chapter 40 takes a closer look at this most desolate of biomes.

KEY CONCEPTS

A biome's characteristics are determined by the types of plants (dependent on altitude and latitude) and moisture. Life is most diverse in the tropical rain forest, where nutrients cycle rapidly. Organisms are vertically stratified to capture sunlight. Temperate deciduous forests have growing seasons more than 120 days long, rainfall of 28 to 59 inches (70 to 150 centimeters), and rich soil. Beneath the trees are shrubs, grasses, flowers, mosses, and liverworts. Temperate areas with poor soil and less rain support conifers. Farther north is the taiga, a rugged, cold area with conifers and a few types of animals. Grasslands have 10 to 30 inches (25 to 75 centimeters) of annual rainfall and many large mammals. Grasses are well adapted to dry conditions. The tundra is cold, with little life besides lichens, shrubs, and short grasses. Deserts receive less than 8 inches (20 centimeters) of rainfall per year and are populated by plants and animals adapted for conserving water.

Freshwater Biomes

The earth's waters house diverse life forms that are adapted to the temperature, light, current, and nutrient availability of their surroundings. Aquatic biomes are distinguished by their degree of salinity. Two types of freshwater biomes are recognized—standing water, such as lakes and ponds **(lentic systems),** and running water, such as rivers and streams **(lotic systems).**

Lakes and Ponds

Light penetrates to differing degrees in the regions of a lake, and these differences determine the types of plants that live in particular areas. The **littoral zone** is the shallow region along the shore where light can penetrate to the bottom with sufficient intensity to allow photosynthesis. Although some photosynthetic organisms in this region are free-floating, most species are rooted to the bottom yet stick out of the water (fig. 39.15). This is the richest area of a lake or pond. Typical producers include cyanobacteria, green algae, and diatoms, as well as emergent vegetation such as cattails. Animal life is also diverse, including damselflies, dragonfly nymphs, crayfish, rotifers, flatworms, hydra, snails, snakes, turtles, frogs, and the young of some species of deep-water fish.

The **limnetic zone** is the layer of open water that is penetrated by light. It is inhabited by phytoplankton, zooplankton, and fish. The **profundal zone** is the deep region beneath the limnetic zone where light does not penetrate. The organisms here rely on the rain of organic material from above and include mostly scavengers and decomposers such as bacteria, fungi, and insect larvae.

Oxygen and mineral nutrients in a lake are distributed unevenly. Oxygen concentration is usually greater in the upper layers, where it comes from the atmosphere and photosynthesis. Decaying organisms bring phosphates and nitrates to the lower layers of the lake. In a shallow lake, wind blowing across the surface mixes the water. Deeper lakes in the temperate regions that have seasons often develop layers with very different water temperatures. The layers prevent complete circulation in the lake. As a result of this **thermal stratification,** the redistribution of oxygen and nutrients occurs seasonally.

In the summer, the upper layer of a lake is heated by the sun, but the lower layers remain cold. Between these two layers is a third region, called the **thermocline,** where water temperature drops quickly. In the fall, the water temperature in the upper layer of the lake drops with the air temperature. Gradually, the water temperature becomes the same throughout the lake. Wind then mixes the upper and lower layers, creating a **fall turnover** of nutrients and oxygen. During the winter, surface water becomes colder. When water cools to 39°F (4°C), the temperature at which it is most dense, it sinks. Water colder than this floats above the 39°F layer. With the warmth of the spring sun, when the temperature of the upper layer rises to 39°F, water begins to sink, bringing oxygen with it. In turn, the nutrient-rich lower layers circulate to the top. This **spring turnover** often results in great algal growth called blooms.

Lakes age. Younger lakes are often deep and steep-sided, making it difficult to retrieve nutrients that settle to the bottom. Because of the low nutrient availability in these **oligotrophic** (few foods) lakes, little phytoplankton colors the water. The lake is sparkling blue. As a lake ages, organic material from decaying organisms and sediment begins to fill it in. The shallow water mixes more easily and light reaches more of it. These nutrient-rich lakes are termed **eutrophic** (good foods). The rich algal growth makes the water green and murky. Weeds abound, and fish and plankton communities change. Gradually, over millions of years, the lake becomes a bog or marsh and eventually, dry land. Eutrophication is hastened by certain types of pollution, such as phosphate fertilizer and dish detergent. Chapter 40 examines Lake Tahoe, which is in the early stages of eutrophication due to rapid urban development along its shores.

Rivers and Streams

Rivers and streams depend heavily on the land, both for the water that fills their banks and the materials that nourish the life within. Dead leaves and other organic material that fall into a river triple the nutrients produced by the organisms that live there. But rivers restore nutrients to land as well as import nutrients from it. When a river approaches the ocean, its current diminishes, depositing fine, rich soil. Many rivers flood each year, swelling with melt water and spring runoff and spreading nutrient-rich silt onto the land.

Oxygen in flowing water is usually abundant because air and water mix in turbulent areas. Many organisms that live in streams require a high oxygen level. Dumping organic material, such as sewage, into a river or stream can create a problem because its decomposition uses oxygen needed by river residents. Nutrients are less abundant where water flows rapidly, because decaying organisms are quickly washed away.

Different organisms are found in different parts of a river, each species adapted to local conditions. In a swift current, some organisms hold onto any available unmoving surface of rocks or logs. This is where to find algae, diatoms, and mosses, as well as the snails that graze on them. Many larval and adult insects hold onto the underside of rocks with hooks or suckers. The bodies of organisms in flowing water are generally streamlined for gliding through the current or flat for squeezing under rocks. As a river or stream is fed by tributaries, it becomes larger and its current slows. Such slower-moving rivers and streams support more diverse life, including crayfish, snails, bass, and catfish. Worms burrow in the murky bottom, and plants line the banks.

KEY CONCEPTS

Lentic systems stand still, such as lakes and ponds. The littoral zone is shallow, where light penetrates. Life is diverse and abundant here. The limnetic zone is lit open water, and it supports fish and plankton. The region where light does not penetrate is the profundal zone, and here scavengers and decomposers dwell. Oxygen and minerals in a lake are distributed unevenly and are moved by winds and seasonal temperature changes. The thermocline lies between a warm and cool layer. Aging lakes accumulate nutrients, becoming eutrophic, and are murky. Rivers and streams are lotic systems. Life within them is adapted to current.

Marine Biomes

Covering 70% of the earth's surface and 7 miles (11.2 kilometers) deep in some places, the ocean is the largest and most stable of biomes. Here, boundaries between habitats are classified by their proximity to land.

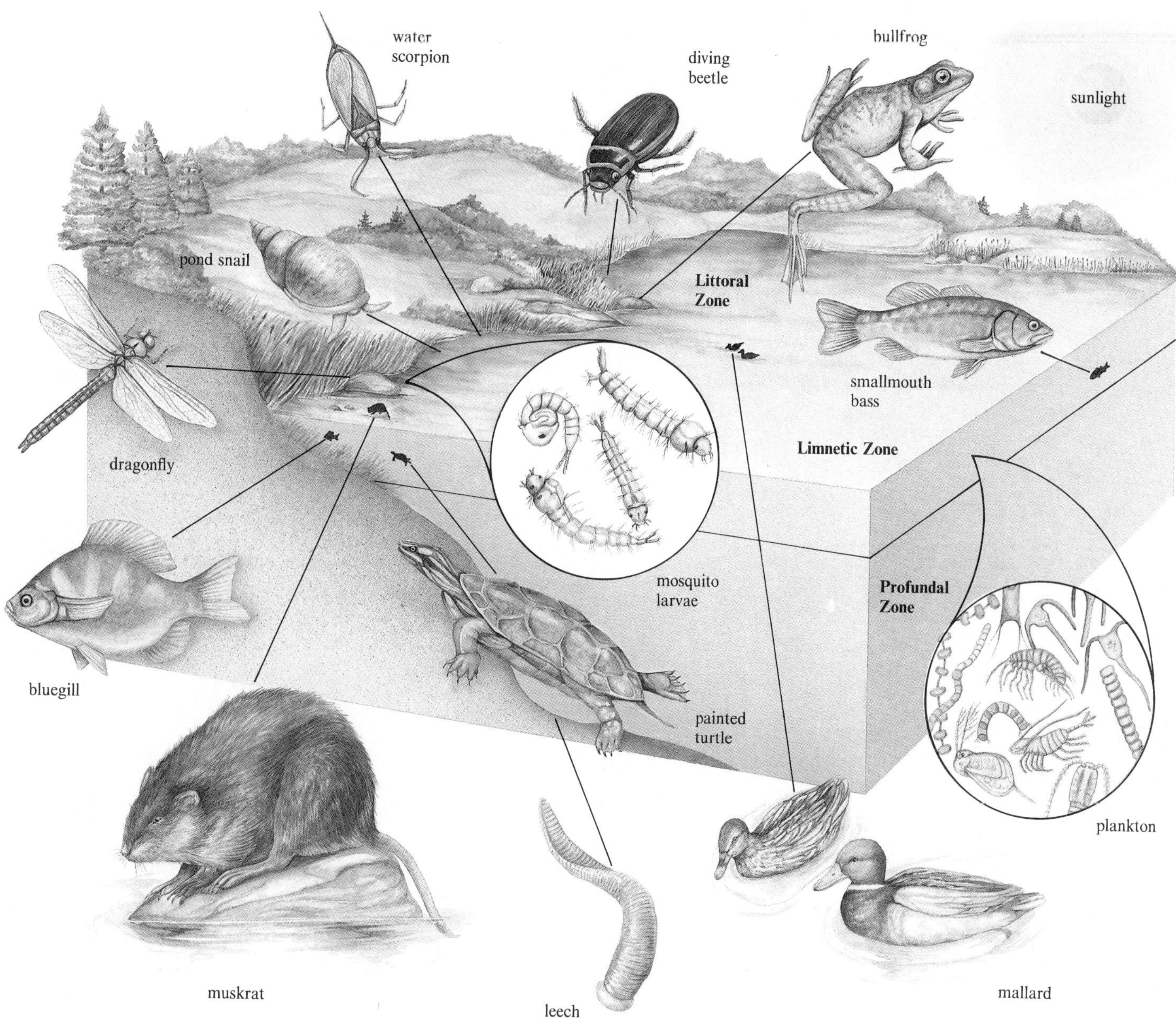

Figure 39.15
Life in a lake.

The Coast

At the margin of the land, where the fresh water of a river meets the salty ocean, is an **estuary.** Life in an estuary must be able to cope with a range of chemical and physical conditions. The salinity of the water fluctuates dramatically. When the tide is out, the water is not much saltier than that of the river that ends there. However, the returning tide may make the water as salty as the sea. As the tide ebbs and flows, much of the land of the estuary is alternately exposed to drying air and then flooded.

Organisms able to withstand these environmental extremes enjoy daily deliveries of nutrients, from the slowing river as well as by the tides. Shallow water encourages photosynthesis. Estuaries house very productive ecosystems, its rocks slippery with algae, its shores lush with vegetation, and the water abounding with plankton. Almost half of an estuary's photosynthetic products go out with the tide to nourish coastal communities. Estuaries are nurseries for many sea animals. More than half of the animals that are commercially important as seafood spend some part of their life cycle in an estuary. Migratory waterfowl feed and nest here as well. Chapter 40 explores a terribly polluted estuary, the Chesapeake.

Bordering estuaries are rocky or sandy areas of the **intertidal zone,** the region that is alternately exposed and covered with water as the tide recedes and returns. The organisms in a rocky intertidal zone often attach to a rock so they are not dragged out to sea by the tide (fig. 39.16). Large marine algae (seaweeds) attach to rocks by holdfasts.

Figure 39.16
Life in the intertidal zone. The places where the sea laps at the land house an assemblage of organisms that can stay in place, such as seaweeds, crabs, mussels, sea urchins, sea anemones, snails, and starfish.

Mussels fasten by threads or by suction. Sea anemones, sea urchins, snails, and starfish live in the pools of water formed amid the rocks as the tide recedes. The organisms of the sandy beach burrow to escape the pounding waves that would wash them away. The sandy beach is not a very productive environment.

The Ocean

Beyond the continental shelf are the deep, open seas of the **oceanic zone.** While the depth of the coastal, or **neritic zone,** averages about .33 mile (200 meters), that of the oceanic zone may dip below 1.8 miles (3 kilometers). The bottom of the ocean is the **benthic zone,** and it is home to crabs, starfish, and the colorful great coral reefs and their diverse tropical fish inhabitants (fig. 39.17). The part of the benthic zone that light never reaches, below 6,500 feet (2,000 meters), is called the **abyssal zone.** The water above the ocean floor is the **pelagic zone** (fig. 39.18). Organisms in each zone are adapted to its conditions.

In the sunlit waters of the pelagic zone, the most abundant producers are algae. The herbivores are zooplankton such as copepods. These are consumed by fish, which may be eaten by other fish. The remains of organisms sink to the bottom, where decomposers release their nutrients.

Very productive ocean environments are created when the cooler, nutrient-rich bottom layers are moved upward, a process called **upwelling.** The resulting sudden influx of nutrients causes phytoplankton to "bloom," and with this widening of the food web base, many ocean populations of species grow. Upwelling occurs on the western side of a continent, where wind blows off the land and out to sea, such as along the coasts of southern California, parts of Africa, and the Antarctic. The importance of upwelling to the communities of the sea is vividly displayed in El Niño, a periodic slack in the trade winds that prevents the upwelling that normally occurs along the coast of Peru. An ecological chain reaction occurs, with populations plummeting at all levels of food webs. Chapter 2 explores El Niño in more depth.

a.

b.

Figure 39.17

a. Coral reefs are vast, underwater structures of calcium carbonate whose nooks and crannies provide habitats for hundreds of species of sea life, including many vibrantly colored tropical fish. *b.* Coral reefs are built by coral animals. The individual organisms, called polyps, house unicellular algae called zooxanthellae, which inhabit the cells lining the coral's digestive tract, where they photosynthesize. Here, the algae can be seen in the tentacles of a Pocillopora coral polyp. Disease, unusual heat, pollution, and storms can kill the algae. This leads to "bleaching" of the corals—they develop white patches. Without their algal comrades, corals must rely on whatever zooplankton food drifts their way. Many die.

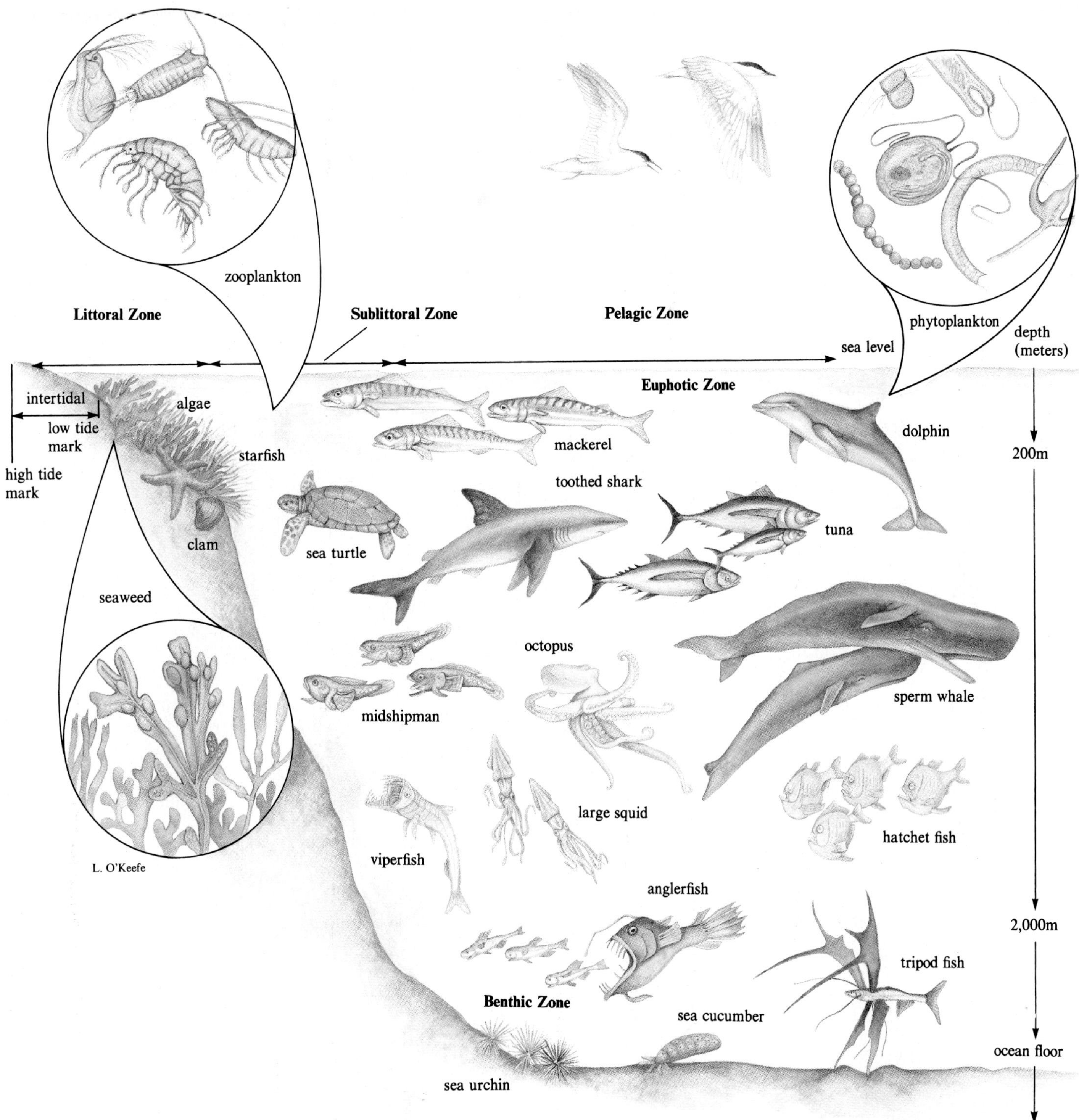

Figure 39.18
The marine biome. Primary producers in the ocean include phytoplankton and algae in the surface waters. Zooplankton are minute animals that feed on phytoplankton. Consumers include many varieties of fishes in the pelagic zone and bottom dwellers such as crabs and starfish in the benthic zone.

The study of ecology shows us how beautiful our planet is, how even the most seemingly forbidding of places is home to some sort of organism. More so than other species, *Homo sapiens* has the ability to alter the associations between organisms and their natural surroundings. The next and final chapter explores many of the impacts that modern human society has had on the planet Earth.

KEY CONCEPTS

In an estuary, a river meets an ocean, life is rich, and salinity and water level fluctuate greatly. Estuaries flow into intertidal zones, home to organisms that can hold fast to rocks, burrow in the sand, or resist damage by the tide. The ocean's bottommost benthic zone includes the abyssal zone, where light does not reach, and a lit pelagic zone, where life abounds. Upwelling of lower, nutrient-rich water layers is important in building food web bases.

SUMMARY

Ecology is the study of the relationships of organisms to the abiotic environment and to each other. An *ecosystem* is a unit of such interaction and can range from a very small area to the entire planet. An organism has its own niche and is also part of a population and the living *community*. Ecosystems change and interact. Related ecosystems form *biomes*, which can be terrestrial or aquatic.

A *food chain* begins with energy from the sun or earth that is harnessed by *primary producers*—photosynthetic or chemosynthetic organisms, which constitute the first *trophic level*. The total amount of energy converted to chemical energy in food is *gross primary production*. The energy remaining after metabolism is *net primary production*.

At the next level are the *primary consumers*, or herbivores, which eat the primary producers. A *secondary consumer* may eat the primary consumer, and perhaps a *tertiary consumer* will eat the secondary one. The energy stored as biochemicals of animal tissues is called *secondary production*. *Decomposers* break down nonliving organic material. The released nutrients can be recycled.

Food chains rarely extend beyond four trophic levels because at each level, 90% of the energy is lost as heat or respiration, or is not eaten or is eliminated. *Ecological pyramids* measure energy, numbers of organisms, or biomass. *Stable isotope tracing* measures $^{13}C/^{12}C$ ratios in primary producers. Some chemicals are *biomagnified* as they ascend food chains.

The carbon in atmospheric CO_2 is used to manufacture carbohydrates in photosynthesis. CO_2 is released by respiration and by burning fossil fuels. Carbon in living tissues is released by decomposers. Atmospheric nitrogen is converted to ammonia and then to usable nitrates by bacteria. Decomposers convert the nitrogen in the tissues of dead organisms to ammonia. *Nitrifying bacteria* convert the ammonia to nitrites, and other bacteria convert nitrites to nitrates. *Denitrifying bacteria* convert ammonia to nitrogen gas. As rain falls over the land, some phosphorus is released as phosphate that can be used by organisms. Decomposers return phosphorus to the soil.

Primary succession is the directional, gradual appearance of life where it did not exist before. *Secondary succession* occurs in disturbed areas. In both types of succession, the first organisms to appear are fast-growing *pioneer species*, which are replaced by slower-growing species, and then more permanent residents, continuing until a stable, *climax community* forms.

Abundance and diversity of life peaks at the equator. The *tropical rain forest* is hot and wet, with diverse life. Competition for light leads to *vertical stratification*. The forest floor have a few shade-adapted species. *Deciduous forests* require a growing season of at least 4 months, are vertically stratified, and have less diverse life than a tropical rain forest. Decomposers form soil from leaf litter. *Coniferous forests* are found in temperate areas with poor soil and little rain. The *taiga* is a very cold northern coniferous forest. Adaptations include the shapes of needles, retention of leaves year round, and the conical shape of the trees.

The *tundra* has very cold and long winters and is covered by *permafrost*. During the spring and summer, melt water forms rivers or pools. Lichens are common in the treeless tundra, and animals include caribou, reindeer, lemmings, and snowy owls. *Grasslands* are found in temperate areas with less water than in the deciduous forest and more water than desert. The more moisture, the taller the grasses. Deserts have less than 8 inches (20 centimeters) of rainfall a year. Plant life is well adapted for obtaining and storing water. Animals cope with water shortage by minimizing water loss and being active at night.

Freshwater ecosystems include standing water *(lentic systems)* and running water *(lotic systems)*. The *littoral zone* of a lake is reached by light; the *limnetic zone* is the lighted upper layer; the *profundal zone* is the deeper dark layer. In the littoral zone most producers are rooted plants. In the limnetic zone, phytoplankton predominate. Nutrients supporting life in the profundal zone fall from the upper layers.

Deep lakes in the temperate zone rely on thermal overturn to mix the oxygen-rich upper layer with the nutrient-rich lower layer. Young, deep, *oligotrophic* lakes are clear blue, with few nutrients to support algae. As a lake *eutrophies*, nutrients accumulate and algae tints the water green.

Shoreline ecosystems include *estuaries*, *rocky intertidal zones*, and sandy beaches. Estuaries occur where a river empties into the sea. The region of ocean near the shore is the *neritic zone*, while that of open water is the *oceanic zone*. The ocean includes the *benthic zone* (the bottom), the *abyssal zone* (the bottom region where light does not reach), and the *pelagic zone* (open water above the ocean floor). The most productive areas are those neritic zones where upwelling occurs.

QUESTIONS

1. The 1959 Antarctic Treaty preserves all land and ice below 60° south latitude for scientific investigation by all nations. However, because of increased populations of krill resulting from whaling, and because hunger is a serious human concern, some people have suggested that we "farm" the krill, taking it from Antarctica to feed people who are starving elsewhere. What effects might krill farming have on the Antarctic ecosystem?

2. A hawk swoops from the sky to eat a water snake that has just consumed a bullfrog that ate a dragonfly that ate a butterfly while it sipped nectar from a flower. Draw a food chain to accommodate these eating activities.

3. Why does it make greater ecological sense for a nation with hungry citizens to use its land to grow crops rather than for animals to graze?

4. Why do we say that energy flow through an ecosystem is one way that nutrient minerals cycle?

5. How can the tropical rain forest support diverse and abundant life with such poor soil?

6. How can grasslands recover from fire?

7. What are the challenges faced by organisms in estuaries? In the intertidal zone?

TO THINK ABOUT

1. The buffalo-bur is a weed found in barnyards and overgrazed areas of Colorado. Although it is poisonous to livestock, the buffalo-bur is consumed by larvae of the striped Colorado potato beetle. In 1859 potatoes were introduced to Colorado by European settlers. As the popular potato crop flourished, so did the voracious beetles. The insect's natural predators—toads, stinkbugs, birds, and snakes—could not control the beetle population. By 1874, the insects had spread across the United States and to Europe. By 1930, potatoes all over the world were infested. The potato beetle became the first insect to be controlled by pesticides. However, many of the pesticides killed other insects as well, and the populations that these insects controlled rose to pest levels. When DDT was used, resistant strains of potato beetles arose.
   - **a.** Draw a food web that includes all of the organisms mentioned.
   - **b.** What evolutionary process is illustrated by the potato beetle?
   - **c.** Suggest an intervention that might benefit agriculture other than the use of a chemical pesticide.

2. After the Yellowstone fires of 1988, forest managers suggested that humans try to help the areas recover by feeding deer, bringing in plants, and planting trees. What might be some of the advantages and disadvantages of intervening in recovery from a natural disaster?

3. How is natural selection apparent in succession? In vertical stratification? In the types of organisms adapted to particular biomes?

4. Is fire a destructive force to ecosystems or a constructive force—or both? Cite reasons for your answer.

5. Do you think that Biosphere II will answer questions raised by ecologists that could not be answered from observing natural biomes? Why or why not?

6. Nowadays it is fashionable for people to call themselves "environmentalists." What does this title mean to you?

SUGGESTED READINGS

BioScience, vol. 39, no. 10. November 1989. The entire issue is devoted to the Yellowstone fires of 1988.

BioScience, vol. 38, no. 11. December 1988. The issue covers the effects of animal species on their ecosystems.

Boucher, Douglas. March 1990. Growing back after hurricanes. *BioScience*, vol. 40, no. 3. Just as forest fires may be a normal component of temperate forest ecosystems, so may be hurricanes in the tropical rain forest.

Bunkley-Williams, Lucy, and Ernest H. Williams Jr. April 1990. Global assault on coral reefs. *Natural History*. Bleached corals indicate something amiss in the ecosystem.

Hart, Stephen. February 1990. Stable carbon isotopes in the study of food chains. *Biology Digest*, vol. 16, no. 6. Ecologists need no longer be on the scene of a meal to trace food chains.

Mohlenbrock, Robert H. June 1990. Mount St. Helens, Washington. *Natural History*. Life returns to the devastated area.

Monastersky, Richard. October 27, 1990. Burning questions. *Science News*. Looking back at the Yellowstone fires.

Siegel, Z., et al. October 1988. Mercury in marijuana. *BioScience*, vol. 38, no. 9. Mercury is biomagnified when marijuana is smoked.

Reingold, Edwin M. September 24, 1990. Noah's Ark—the sequel. *Time*. Constructing and populating six biomes in an area the size of two football fields is a daunting task—as is living there for two years.

Winston, Judith E. September 1990. Life in Antarctic depths. *Natural History*. Life is diverse, even at the murky bottom of the Antarctic sea.

CHAPTER 40

Environmental Concerns

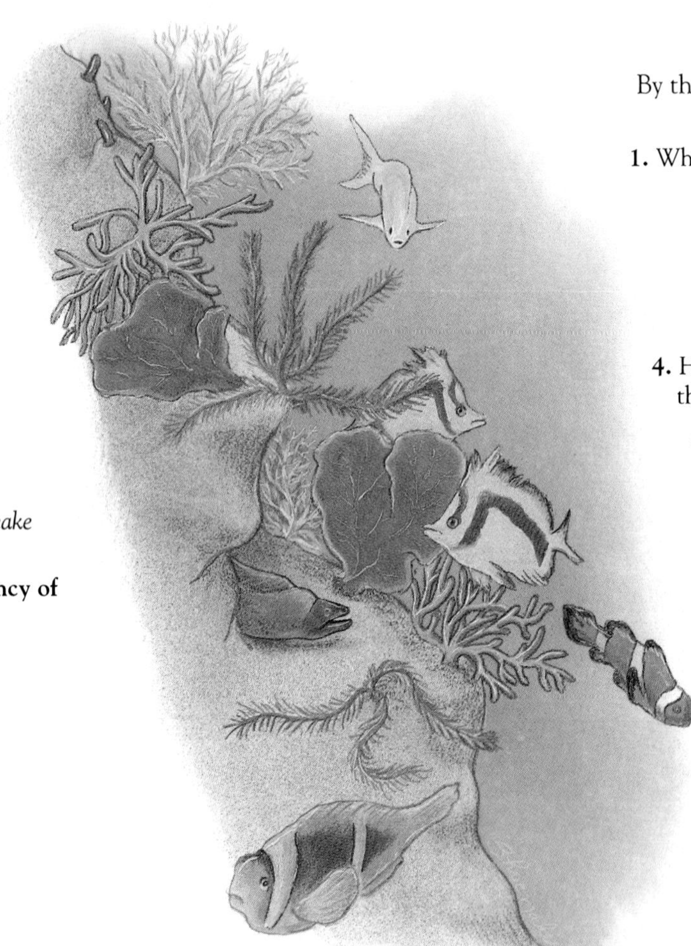

Learning Objectives

By the chapter's end, you should be able to answer these questions:

1. What are some environmental problems caused by humans?
2. How does air pollution cause acid precipitation?
3. How can acid precipitation affect lakes and forests?
4. How can buildup of carbon dioxide in the atmosphere cause global warming?
5. How do certain chemicals deplete the ozone layer?
6. Why is destroying the tropical rain forest ecologically unwise?
7. Why is the African desert expanding?
8. How do human activities pollute lakes, estuaries, and oceans?

Prince William Sound, 1989: When Captain James Cook ventured into Alaska's Prince William Sound in 1778, he discovered a pristine wilderness rich with life. The thousands of square miles of open water, islands, rocky outcroppings, shoreline, and mainland were home to millions of birds, representing more than 200 species; 30 species of land mammals (black bears, lynx, deer, beaver, and porcupines among them); and 10 species of marine mammals (including sea lions, whales, seals, dolphins, and otters). More than 200 types of marine invertebrates dwelt in the oceans, and the 300 streams running into the sea swarmed with salmon and other fishes.

In 1968, another natural resource was discovered in Prince William Sound that would be its undoing—oil. By 1977, the trans-Alaskan pipeline had been built, and tankers became as common a sight in the many inlets of the sound as seals and otters had once been. Some ecologists think that what happened in Prince William Sound on March 24, 1989, was inevitable. On that day, an oil company tanker, the *Valdez*, ran aground on Bligh Reef, spilling 11 million gallons (over 50 million liters) of oil. The spreading stain of sticky blackness would snuff out much of the bountiful life of the once pure area.

Even as thousands of people tried to rescue and clean the dying wildlife, the tragic results of the oil spill began to wash up on the shores. The lungs of otters killed directly by the fumes were reduced to tatters by emphysema. Others suffocated in the muck (fig. 40.1*a*). About 1,200 otters ended up at the Valdez Otter Rescue Center (fig. 40.1*b*), a rapidly converted school gymnasium. Here charcoal was pumped into their stomachs to absorb toxic hydrocarbons. After 2 days of rest, they were tranquilized and scrubbed. Only 200 otters survived, the cleanup adding trauma to the oil drenching. These animals were released 2 months later into a still-polluted inlet.

Birds faced a similar fate in the hands of well-intended human rescuers. They were often chased in boats, netted, and then stuffed into cardboard boxes. Many died of exhaustion trying to escape the unfamiliar humans reaching out to them. The birds

a.

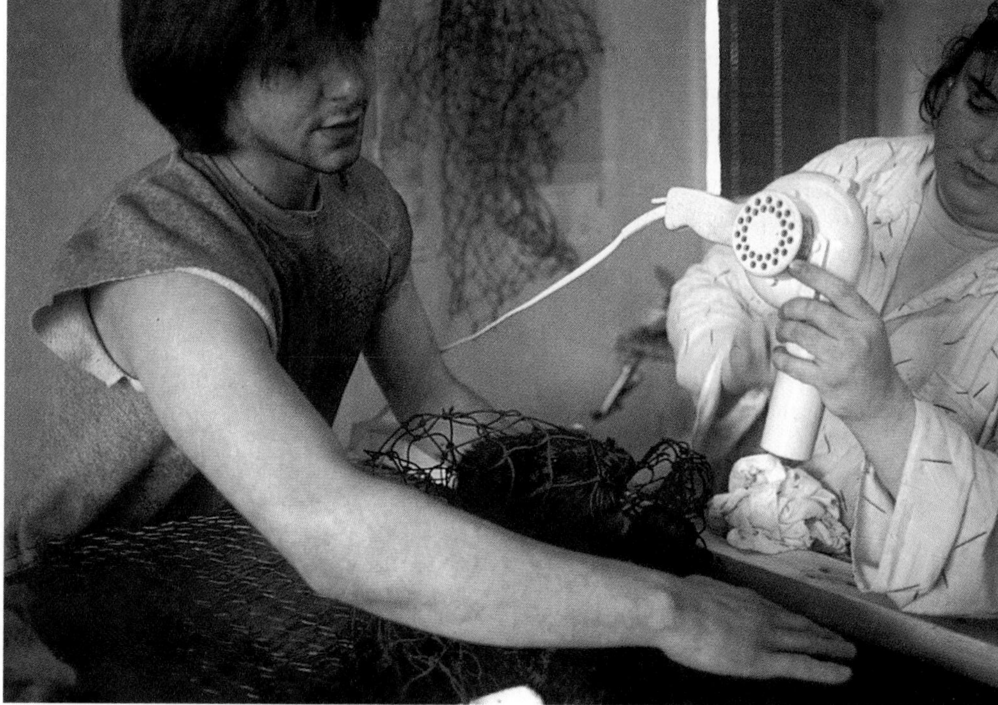

b.

Figure 40.1
a. An oil-drenched, dying otter was a common sight following the spilling of 11 million gallons (500 million liters) of oil by the *Valdez* in the Prince William Sound in March 1989. *b*. Efforts were made to clean many otters and birds, but few survived the terrifying, but well-meant, laundering.

Figure 40.2
The aftermath of a nuclear explosion. This eyeless pig is one of many farm animals born with severe defects following the April 26, 1986, meltdown of the Chernobyl nuclear reactor in the Soviet Union.

wound up in a "triage" room at the Valdez Bird Rehabilitation Center. The worst ones were euthanized immediately. The others were left alone for a few days, except for twice-daily feedings through stomach tubes inserted down their throats. Finally, the birds were held under showers and rinsed repeatedly. Like the otters, few of the birds survived.

Chernobyl, 1986: Some pollutants are not as obvious as a tide of sticky oil. Radiation cannot be seen at all, but its effects on life are devastating, as the residents of a small city in the Soviet Union learned on April 25, 1986.

Early that morning, a series of unwise operating decisions and human error led to a rapid buildup of power in a nuclear reactor, followed within seconds by several explosions and fires. The disaster killed 31 people immediately, sent 500 to the hospital (203 with acute radiation sickness), exposed 24,000 others to dangerously high levels of radiation, and, in the weeks following, exposed another 400 million people to a radioactive cloud drifting across Europe.

Effects linger. Four years later, at a site 37 miles (60 kilometers) from the damaged reactor, radiation is still nine times higher than the acceptable level. Residents complain of great fatigue, anemia, poor appetite and vision, and increased rates of cancer and thyroid disease—all signs of radiation sickness. The immunity of the populace seems to have declined, with common illnesses rampant.

The most chilling reminder of that April night, however, comes from the farm animals that have had a chance to produce a new generation since the disaster. One large collective farm has documented 197 "freak" calves, lacking eyes or with deformed skulls or faces. Another farm notes similar birth defects in piglets (fig. 40.2), and many a farmer has delivered a colt with eight limbs or two heads. It is frightening indeed to think of how life will continue to be affected by the Chernobyl nuclear explosion.

Anywhere, U.S.A.: The sight of a mountain of garbage is not quite as dramatic as an oil spill or a nuclear meltdown,

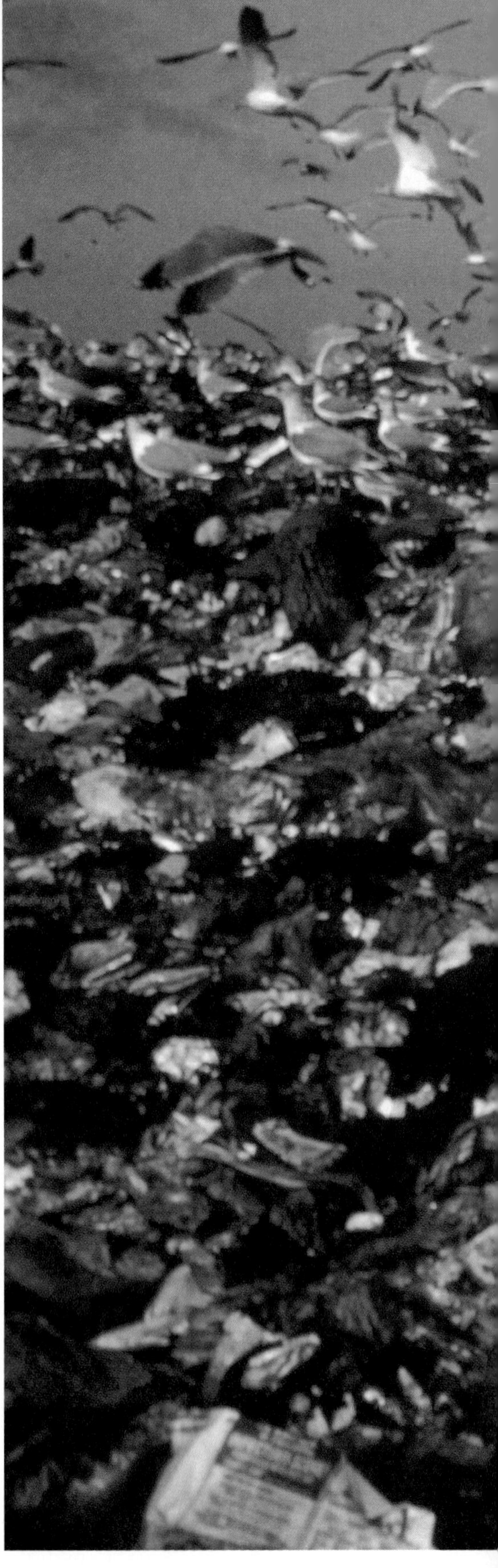

Figure 40.3
A "sanitary landfill" (a.k.a. garbage dump) offers a wide and plentiful, if unpredictable, menu for birds, drastically altering the living communities and population dynamics of the bird species that would normally live in the area. This landfill is in New York City.

but it is equally disturbing, in an ecological sense, precisely because it is so familiar (fig. 40.3). Not only do such "sanitary landfills" introduce toxins into the soil, but they attract many birds, disrupting natural food webs and, ultimately, ecosystems.

The Human Influence Is Everywhere

The influence of one species—*Homo sapiens*—on planet earth is profound. It is most obvious in the cities, where concrete, steel, and glass so extensively replace native vegetation that these areas no longer resemble any natural biome (Reading 40.1). Yet even in the busiest, dirtiest city, species other than ourselves persist.

Homo sapiens has left marks on even the most seemingly remote places. Several years ago on Alaska's North Slope, a bulldozer carved an oil company's initials into the permafrost, leaving a scar that is a testament to human intervention. Hardy lichens that survive unimaginably frigid winds, forming the bases of food webs, cannot weather the sulfur emissions from our machines. The genetic variability underlying adaptation somehow is simply not prepared for all that humans can do to the environment.

At the other end of the earth, at the South Pole, is a gaping hole in the atmosphere's protective ozone layer, opened up by chemical pollutants. In Antarctica, tourists dressed in bright colors and toting video cameras crouch so close to the wildlife that many birds, not recognizing the strange, two-footed creatures, become too frightened to eat or mate (fig. 40.4).

Nearly anywhere on earth can be found the signs of our intervention. We may cause more harm than good to the ecosystems that interact to mold the biosphere. This final chapter examines what we have done to the air, land, and waters of the earth.

The Air

Acid Precipitation

The patterns of air circulation and geography sometimes interact so that the effects of air pollution are experienced far from

Reading 40.1 Urban Ecology

An extraterrestrial visitor looking down a major avenue in Manhattan at noon on a weekday would conclude that Earth is inhabited only by the curious two-footed creatures that hurry past each other. Closer inspection, though, reveals other forms of life—pigeons and sparrows roosting in the facades of buildings, squirrels and chipmunks scurrying across the few patches of green, a tree here and there, and possibly a poodle on a leash. A garbage can in an alley behind a restaurant is banged about by hungry rats and cats—and maybe people. Ants are everywhere. Over in Central Park, the visitor could see a few ducks on the pond and a collection of interesting animals in the children's zoo.

Although urban areas lack the diversity seen in natural ecosystems, they too demonstrate basic ecological phenomena—niches, nutrient and energy flow, trophic levels, carrying capacities, and the various interrelationships between species and their surroundings that define the ecosystem. Yet urban ecosystems present organisms with unique, and sometimes confusing, environmental cues.

The heat rising from a sweltering city drives insects upward, when, in natural surroundings, they would tend to spread out horizontally. Air pollution and poor water quality limit reproduction of some species. Population dynamics are controlled by the whims of *Homo sapiens*—plants deemed "weeds" are doomed, while others with pretty flowers or tasty fruits are artificially selected. Most insects are subjected to nasty chemicals. Buildings and roads interrupt some habitats, causing birds to crash into reflective skyscraper windows, falling to litter the concrete below. Landscaping extends other habitats. In southern California, for example, rats easily travel from house to house because each residence is planted with trees bearing fruits or nuts.

As is true nearly everywhere on earth, life adapts. A tree pokes its roots through the spaces where slabs of concrete meet, and fish and amphibians inhabit filthy makeshift city ponds (fig. 1.1*c*). Cockroaches live and reproduce as well in a snug, city apartment as their larger relatives do in tropical rain forests. Mice thrive inside walls. In Pennsylvania backyards, bears beg for food and prowl among the trash cans; in a western Massachusetts park, wild turkeys run about; in Anchorage, Alaska, once a kindly human feeds a moose, it comes back daily for more. Animals adapt to almost anything—including the peculiarities of our own species.

Figure 1
A raccoon looks for food among garbage on a rooftop.

Figure 40.4
Homo sapiens on South Georgia Island. Although they mean well, these tourists often frighten native birds so much that they cannot eat or mate.

the source of the problem. In 1852, British scientist Angus Smith coined the phrase "acid rain" to describe one effect of the Industrial Revolution on the clean British countryside. Today, the smokestacks of the Midwest that burn coal spew sulfur and nitrogen oxides (SO_2 and NO_2) high into the atmosphere. Gasoline and diesel fuel in internal combustion engines, and waste from heavy metal smelters, add to the problem. In the atmosphere, these oxides join with moisture to fall, at some distance, as sulfuric and nitric acids (H_2SO_4 and HNO_3), forming acid rain, snow, fog, and dew.

Because the atmosphere contains carbon dioxide and water, all rainfall includes some carbonic acid and is therefore slightly acidic. However, burning fossil fuels has made the average rainfall in the eastern United States 25 to 60 times more acidic than normal. Acid precipitation is also a problem in the upper Midwest, the Pacific Northwest, the Rockies, Canada, Scandinavia, Europe, the Orient, and the Soviet Union.

A typical small, mountain lake is clear and blue, with a pH of around 8.0. A lake polluted by acid rain has a pH in the 3.0 to 5.0 range. In this abnormally acidic environment, fish eggs die or hatch to yield deformed offspring. Amphibian eggs do not hatch at all. The shells of crustaceans do not harden. Bacteria and plankton die, and plants are replaced with lake-clogging mosses, fungi, and algae. A type of algae called "mermaid's hair" chokes out phytoplankton and coats the shores where trout usually spawn, preventing them from doing so. Organisms that feed on the doomed species must seek alternate food sources or starve. Over time, succession leads to the establishment of species that can live in increasingly acidic environments.

The effects of acid rain on lakes are difficult to study because of the complexity of these ecosystems, and because detailed studies of a lake's condition prior to acidification are often lacking. To supplement existing anecdotal observations of the biotic changes induced by acid rain, researchers from the Department of Fisheries and Oceans in Winnipeg, Manitoba, Canada, have intentionally added sulfuric acid to a lake 200 miles (322 kilometers) southeast of the city. Since 1968, they have catalogued changes in the biotic, abiotic, and energy components of this experimental ecosystem. The lake's pH has changed from 6.8 to 5.0.

Canada's experimentally acidified lake shows that life continues, but diversity—the backbone of evolutionary success—plummets. Before acidification, the lake was home to 80 species of midge, a type of insect, in healthy numbers. Today, 95% of the midge community is of one previously unknown species. Apparently a formerly rare midge variant, with a quirk enabling it to thrive under low pH conditions, got lucky. The insect found itself in an environment to which it was well suited and flourished, at the expense of less well-adapted species.

a.

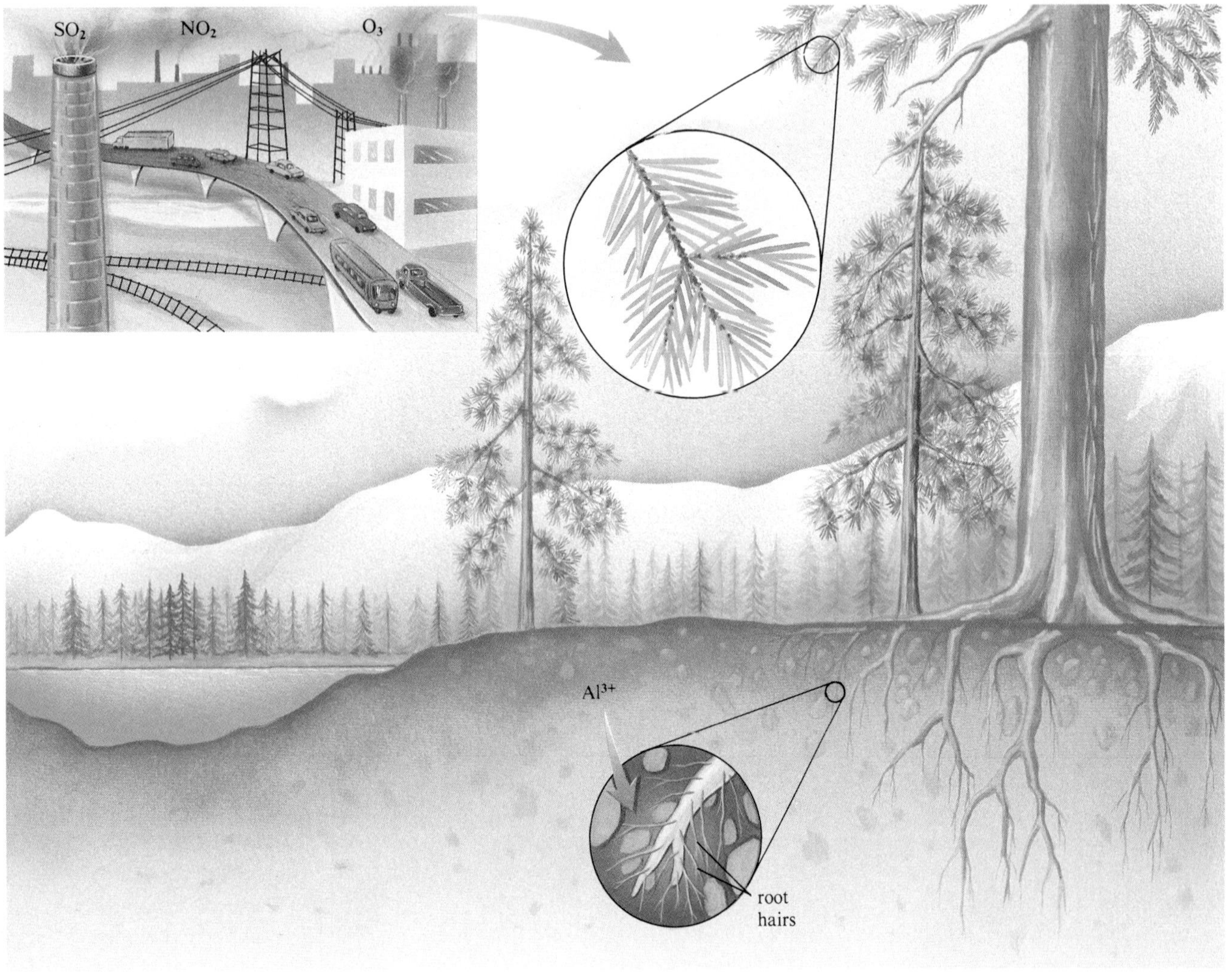

SO_2 NO_2 O_3

Al^{3+}

root hairs

b.

Figure 40.5

a. Contrast among the green needles of a conifer may actually be the first sign of damage by acidified soil. *b.* Acid fog, clouds, mist, and rain harm forests. Red spruce are directly killed as acid precipitation destroys their needles. Other trees are harmed indirectly, as acid soil releases aluminum ions, which block uptake of calcium and magnesium by root hairs. As nitrogen in pollutants acts as fertilizer, plants attempt to grow yet are stressed by malnutrition. The result—increased susceptibility to other problems.

The experimental lake shows that acid precipitation alters food webs and biogeochemical cycles, but it does not seem to affect primary production, nutrient levels, and decomposition rates, at least in the years that the area has been monitored. Most encouraging is the demonstration of natural defenses against acid precipitation. Calcium compounds in the earth show the same buffering effect as similar compounds do in stomach antacids. Acidic rain adhering to the needles of jack pine trees is neutralized by bacteria there. Sphagnum peat mosses ringing lakes soak up acidic rain and buffer it. Bacteria deep within lakes break down nitrogen and sulfur compounds. Researchers at the Canadian lake hypothesize that if we stop polluting lakes, they would recover, thanks to these natural defenses, in just 5 to 10 years.

The optimism of the Canadian researchers is echoed in the United States, where a team of paleolimnologists (biologists or geologists who study ancient lakes) is evaluating deep sediments from 36 of the Adirondack lakes in upstate New York. These high-elevation lakes receive much of the sulfur and nitrogen pollution from the Midwest. Chemical analysis of fossilized diatoms in the lake bottom confirms that acidification has increased recently, but that the problem is indeed being tempered by other natural processes.

Lakes are not the only ecosystems to be affected by acid precipitation. In a coniferous forest high in the mountains of Fichtelgebirge, Germany, the thinning trees and yellowed needles attest to the effects of acid precipitation (fig. 40.5*a*). Here, the nitric and sulfuric acids spark a cascade of effects as they rain down on the soil. As the pH of the soil drops, aluminum ions percolate toward tree roots, where they displace nutrients such as calcium ions (needed for growth of twigs, leaves, stems, and trunks) and magnesium ions (part of chlorophyll). Robbed of nutrients, tree growth is hampered.

Yet the constant input of nitrogen, in nitric acid, is a fertilizer, and the tree attempts to grow in response. These mixed environmental signals apparently stress the tree, which then becomes more likely to succumb to an infection or not survive harsh weather (fig. 40.5*b*). All of this may take up to 30 years. Acid precipitation is thinning high-elevation forests throughout Europe and on the eastern seaboard of the United States from New England to South Carolina.

KEY CONCEPTS

High in the atmosphere, sulfur and nitrogen oxides join with water to form sulfuric and nitric acids, which fall as acid precipitation. Acidic water adversely affects many aquatic species. In coniferous forests, acid rain releases aluminum into soil, which displaces calcium and magnesium, stunting growth. Meanwhile, the excess nitrogen spurs growth. The stressed tree becomes very vulnerable.

The Greenhouse Effect

Carbon dioxide is a colorless, odorless gas that is present in the atmosphere at 350 parts per million. This makes it only a minor atmospheric constituent, but it can have major effects on life by warming the earth.

CO_2 warms the air near the earth's surface by allowing solar radiation of short wavelengths in but not releasing the longer wavelength (infrared) heat to which the energy is converted. The resulting elevation in surface temperature is called the **greenhouse effect,** because the CO_2 blocks the escape of heat much in the same way that the glass panes of a greenhouse retain heat. Other gases that contribute to the greenhouse effect include carbon monoxide, nitric oxide, nitrous oxide, and chlorofluoromethanes, but these chemicals together contribute only half as much to surface warming as does CO_2 alone.

The greenhouse effect was first noted in 1896, after industrialization came to the Western world. It was and is largely caused by burning fossil fuels such as coal, oil, and gas. The total amount of atmospheric CO_2 has increased by 15% to 30% since then. Today, fossil fuels, crops, and to a lesser extent the burning of tropical rain forests, nuclear detonations, and supersonic jets send 5 million tons of CO_2 into the atmosphere each year (fig. 40.6).

What will be the consequences of the greenhouse effect on earth over the next few centuries? The more dramatic predictions foresee sea levels rising and lowland regions flooding as polar ice caps melt. Many changes will be more subtle, tempered by the ability of organisms to adapt. Simulations using greenhouses and artificial atmospheres show, for example, that with increased atmospheric CO_2 and a 6°F to 10°F (3°C to 5°C) increase in temperature, plants use less CO_2 from soil—and that caterpillars prefer to eat plants that have been grown in atmospheres enriched in CO_2. In nature, this may mean that any increase in plant productivity that might occur as a response to increasing atmospheric CO_2 may be offset by extravoracious caterpillars.

Plants will probably be more drastically affected by global warming than animals because they cannot move to find more suitable conditions. Plant communities, over time, will change. Computer simulations predict that the beech forests now stretching from southern Canada to Florida will ultimately respond to a doubling of atmospheric CO_2 by persisting only in the northeastern United States and the southeastern portion of Canada up to Hudson Bay. Similarly, the northern Minnesota coniferous forest of spruce and cedar will gradually be dominated by deciduous species, such as red oak, sugar maple, and yellow birch. As trees vanish, the animals that live in them will have to adapt to the new tree species—or disappear too.

Global warming could reverberate up food chains. Melting of the microbe-lined labyrinths in the Antarctic ice could ultimately lead to starvation of large mammals. Warming will alter when animals mate and produce young, affecting the organisms that are adapted to feeding on them according to a familiar schedule. For example, each May, Arctic shore birds dine on horseshoe crab eggs in Delaware Bay, a refueling stop on their flight south. If the water temperature rises, the crabs will lay their eggs earlier—and the migrating birds will not have enough energy to complete their journey.

Despite dire scenarios of the eastern seaboard under water, it is not clear that the predicted increase in global temperature is actually occurring. Perhaps the rise in CO_2, which would increase temperature, is offset by an increase in dust from pollution, which would reflect more of the sun's energy and lower temperature. Oceans may absorb much of the heat. Clouds formed by increased evaporation could block further absorption of solar energy.

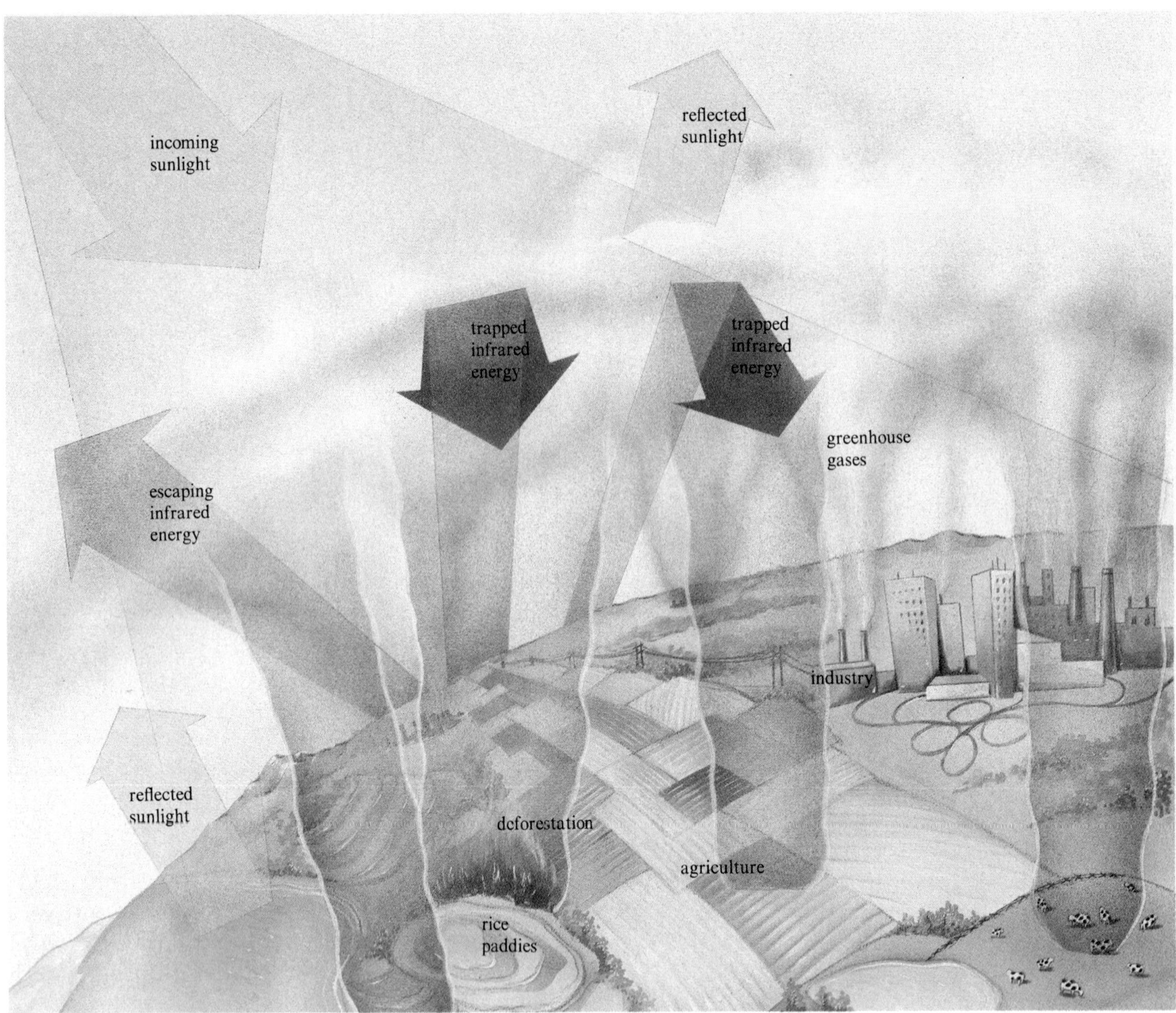

Figure 40.6
The greenhouse effect. The earth's surface is heated by solar radiation. Some energy is returned as infrared radiation, and some is trapped by clouds. This trapping of heat near the surface may be caused by increased levels of CO_2 and other gases in the atmosphere or from industry, farming, and deforestation. Other natural forces, however, may counteract global warming.

Average temperatures did climb from the mid-1800s to the mid-1900s. This trend was reversed for awhile. However, since the mid-1960s, temperatures have risen again. By 1988, worldwide temperatures were 1.2°F (about 0.6°C) above the average temperature of the previous century. Natural variation in temperature would be expected to be only 0.4°F (about 0.2°C). The temperature rise is due primarily to the warmer winters experienced in temperate regions such as New York and Paris. However, some scientists suggest that the hot summers and mild winters of recent years have been caused by the greenhouse effect.

KEY CONCEPTS

CO_2 and other gases may cause global warming by trapping infrared radiation near the earth's surface. Much CO_2 is generated from agriculture, deforestation, jets, and burning fossil fuels. A warmer climate would alter plant distributions, which would alter animal distributions.

Destruction of the Ozone Layer

Ultraviolet (UV) radiation can damage or kill cells. A person's skin peels after a painful sunburn because the UV radiation kills the outer layers. UV radiation also induces mutations, the effects of which are not noticed immediately. Mutations in somatic cells could develop into cancer, and mutations in germ cells could lead to infertility, spontaneous abortion, or birth defects. If the amount of UV light reaching the earth increases it could kill plant cells, and this would disturb food webs at their bases.

Figure 40.7
Tropical rain forest to farm land? It will not work. In South America, local ranchers are tempted by money to burn down the forest so that more land can be used by cattle as pasture. But destroying this delicate ecosystem leads to desert—not arable land.

Fortunately, life is partially shielded from UV radiation by a layer of ozone (O_3), which forms high in the atmosphere when oxygen (O_2) reacts with high-energy ultraviolet light. However, ozone formation is slowed or stopped by nitrogen oxides or chlorine—which we send into the atmosphere in the form of **chlorofluorocarbon (CFC)** compounds, which contain carbon, chlorine, and fluorine.

The colorless and odorless CFCs have found wide applications over the past 50 years because they are nontoxic and do not easily explode, inflame, or corrode. They are used in refrigerants such as freon, and until 1978 were used as propellants in aerosol cans. They are still used as fire retardants, to clean electronic and computer parts, to soften pillows, and to produce foamed plastics. At fast-food restaurants, CFCs keep burgers hot and shakes cold. With so many uses, it is not surprising that millions of tons of CFCs are produced each year.

CFCs do have a drawback with dire ecological consequences, and recognition of this fact has led to efforts to curtail their use. The chemicals are incredibly stable, and once released into the environment—from a leaking refrigerator or a decaying container of foam—they persist. Eventually, they rise in the upper atmosphere, where they react with UV light to produce the chlorine compounds that slow the rate of ozone formation.

Slowly, but noticeably, the earth's protective ozone layer is being destroyed. It is most obvious at the poles because of the pattern of air currents there. In 1983 scientists noticed a "hole" in the ozone layer above the South Pole. The hole appears during the winter, and each year it seems to be larger and it lasts longer into the spring and summer. In 1986, the hole was approximately twice the size of the United States. By mid-1988, a similar hole in the ozone layer was discovered over the North Pole.

KEY CONCEPTS

Ultraviolet radiation can harm or kill cells. Much UV radiation from the sun is shielded from the earth's surface by ozone, but the ozone layer is being depleted by CFC compounds. These widely used chemicals react with UV radiation in the upper atmosphere to produce chlorine compounds that hamper ozone formation. Ozone holes have formed over the poles.

The Land

The Shrinking Tropical Rain Forest

The tropical rain forest, home to more than 3 million species, is gravely endangered by human activity. "Slash and burn" agriculture is rapidly destroying the South American and African tropical rain forests to provide food for increasing human populations (fig. 40.7). Once this bountiful land is

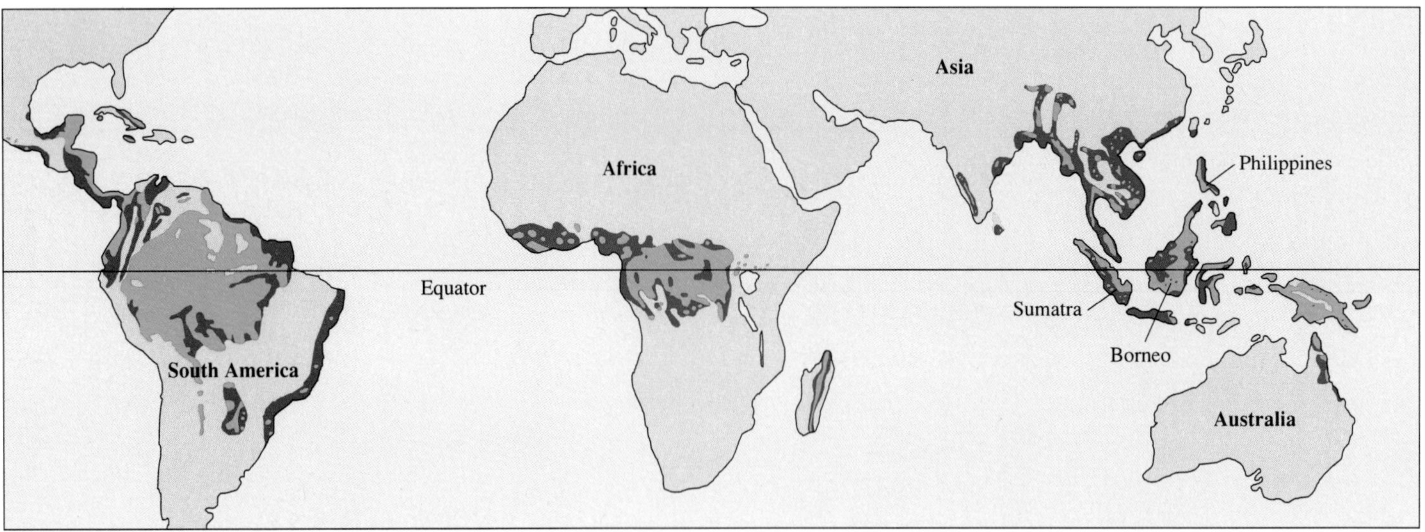

Figure 40.8
The shrinking tropical rain forest. Once, all of the colored area was covered by lush foliage. Now only the areas in green remain in their natural state.
From "Deforestation in the Tropics" by Robert Repetto.

used for crops, timber, or grazing, the nutrients quickly wash away and the soil hardens into a cementlike crust, incapable of supporting plants. It takes at least 60 years for the land to begin to grow into a forest again.

When trees die, food webs topple. Destroying the *Casearia corym bosa* tree in the tropical rain forest of Costa Rica spells famine to many of the 22 bird species that eat its fruits. Other trees dependent upon these birds for seed dispersal are threatened, as are the animals that feed on these trees.

Many attempts to alter the natural diversity of the tropical rain forest have failed. In the 1920s and again in the 1930s, for example, Henry Ford destroyed areas of canopy to grow rubber trees—which quickly died in their denuded environment. Species that do not naturally live in the tropical rain forest are not adapted to use the nutrients rapidly.

The current rate of destruction is appalling (fig. 40.8). A century ago, the area of the lowland tropical forests of South America was twice the area of Europe; today, it is less than one-half Europe's size, and it is still shrinking. Each week, an area the size of Delaware vanishes; each year, an area the size of England disappears. If this rate of destruction continues, it is estimated that the tropical rain forest, with the millions of species in and beneath its lush canopy, may be gone by the year 2000.

KEY CONCEPTS

The lushness of the tropical rain forest has fooled many who try to farm the land. The poor soil quickly turns to dust when species not adapted to the rapid nutrient cycling of the area are introduced.

The Encroaching Desert

With less than 1 inch (2 centimeters) per year of rainfall and nearly as large as the continental United States, Africa's great Sahara is the driest and largest desert in the world. The Sahara is spreading to the 45 nations south of it, an area about twice the size of the United States. Each year the desert moves southward by nearly 4 miles (6 kilometers) encompassing an area the size of Maine. Its dust can be felt as far away as Florida.

The most severely affected nations are those immediately south of the Sahara in an impoverished area called the Sahel. The topsoil in the Sahel is so dry that it cannot hold seeds long enough for them to germinate. Wells are dry. The leaves on the few remaining trees have been eaten by starving cattle and their bark consumed by starving horses. Starving humans eat whatever they can find, including seeds that would in moister times be used for the next year's crop. There are no birds, no basking reptiles, and no succulent cacti. Other regions of the Sahel are completely lifeless. These conditions also prevail in Senegal in West Africa and in Ethiopia, Somalia, and Djibouti in East Africa. Countries farther south, such as Kenya, Uganda, and Tanzania, are threatened as well. Botswana has been in a drought for years, Zimbabwe has experienced crop failures, and the people of Mozambique are malnourished.

The African drought began in 1968, following 10 years of plentiful rain. As the grasses shriveled and trees died in the Sahel, farmers let their cattle overgraze, browsing farther than usual to find food. The hills of Ethiopia were planted with crops, but they were not terraced. Whatever rain did come washed the plants away. Wars destroyed more life. Natural bodies of water were diverted to irrigate cities. Despite relief efforts from many nations, millions of people have already starved to death in the African desert.

Drought, however, is not unusual in this part of the world. The present drought is the third such episode this century, and it has not yet endured for as long as the one that reigned from 1820 through 1840. Some ecologists fear that the present African drought is different from past ones because of its intensity and size. Although the current drought probably started naturally, short-sighted human agricultural practices may be sustaining it, transforming what was once semidesert to the bleakest, most lifeless land on the planet.

KEY CONCEPTS

The Sahara desert is spreading southward. The present drought probably began naturally but is being perpetuated by poor agricultural practices.

The Vanishing Temperate Forest

The current growing desert and shrinking tropical rain forests cap a long history of our altering the land. The carving of the United States from a vast wilderness disrupted many natural ecosystems. In the first half of the eighteenth century, European settlers began clearing the land from east to west to provide fuel and space to build. By 1800, the lush Tidewater region of Virginia and the Carolinas had been cleared for the rapidly expanding towns, as had the banks of the Mississippi River for sugarcane fields. By 1840, the path of destruction had reached Louisiana and Arkansas.

The Industrial Revolution brought new machinery to cut down trees. Forests were sacrificed to build railroads and plantations, and the majestic magnolia and beech trees of the Old South were soon replaced by cotton fields. By 1880, virgin forest remained in less than 35% of the South. Still, the ravaging of the land spread. Ten years later, trees of the southern Appalachians and Gulf coastal plain were felled for timber and turpentine. Meanwhile, farms destroyed during the Civil War displayed new successional growth—nature's attempt to replace the forest. In the next century, the tanning industry in the Smoky Mountains claimed hemlock trees, and soybean fields replaced cypress trees and gum trees in what was left of the Mississippi Valley. Today, less than 1% of the original temperate forest of the southeastern United States prevails.

KEY CONCEPTS

As the vast United States was settled, the native temperate forest was replaced with crops, railroads, and buildings.

The Waters

A Lake in Danger—Tahoe

Lakes age, their populations coming and going, as happens in any dynamic ecosystem. Human activity along a lake's shores can hasten these changes. Lake Tahoe, although only in the early stages of human-induced eutrophication, has already been profoundly altered (fig. 40.9).

Lake Tahoe is the 10th deepest lake in the world, lying along the California-Nevada state line. Measurements of mineral abundancies indicate that the lake formed about 11,000 years ago, making it relatively young by geological standards. It was known only to the Paiute Indians before white settlers discovered it in 1844. Since then, human interference has taken a stiff toll.

Between 1870 and 1900, forests surrounding the lake were cut down to build mine shafts. Since the late 1950s, the shores ringing the crystal-clear lake began to be transformed into a resort. As roads, houses, hotels, and casinos rose, nutrients flowed into the lake, causing the algae and weeds to bloom into a murky tangle. Pollution came from sewage, from runoff from fertilized golf courses and lawns, and from the air. Blackwood Creek was mined for gravel, sending debris to the lake's bottom. In the North End region, 100 acres of marsh were destroyed to make room for two housing developments. Altogether, 75% of the marshes and 50% of the meadows surrounding Lake Tahoe have been disturbed. Marshes and meadows are bridges between ecosystems called **ecotones,** and they are important in maintaining interactions between ecosystems.

What lies ahead for Tahoe and other polluted lakes? If the poisoning continues, the algal population will peak, and bacteria will decompose them, depleting the oxygen supply further. Thanks to this bacterial activity, the lake will soon smell like rotten eggs, and dead fish will litter its shores.

Lake Tahoe is not doomed, however, for natural protections may help. The wetlands ringing the lake block sediments and nutrients from reaching the water, which may retard eutrophication. Artificial marshes are being built to bridge the lake and the condominiums on its shores. If construction companies, waste-dumping industries, and visitors would leave the lake alone, its natural communities of organisms might gradually reestablish themselves.

KEY CONCEPTS

Human destruction of a lake's shores hastens eutrophication by dumping sediments and debris, destroying ecotones, and adding nutrients that encourage algal growth.

An Endangered Estuary—The Chesapeake

As links between fresh water and salt water, estuaries play pivotal roles in ecology. Yet humans have mistreated estuaries, draining them and filling them in to build houses and dumping garbage and other pollutants in these waters. Laws now protect estuaries in some parts of the nation, but in other areas the damage has already gone too far.

Consider the Chesapeake Bay, whose 64,000 square miles (166,000 square kilometers) touch six states and house more than 200 species of fishes and 75 species of birds. The ecological changes in this largest estuary in the nation are ominous. Beneath the once-beautiful waters, the food web base is in upheaval, triggered by sewage, silt, heavy metals, pesticides, and oil pouring in over the past 35 years. Each day 400 million gallons of waste flow into the bay. The seaway is interrupted by 17-miles (27-kilometers) of artificial tunnels, islands, and pilings that can kill whales. In the middle of the estuary, military planes dive toward the water, as young fowl stagger about, disoriented.

The pollution has sharply reduced the oxygen content of the water, which has affected the composition of phytoplankton communities. Dinoflagellate and green algae populations are on the rise, while the numbers of diatoms are falling. As a result of the change in available food, the fish and crustacean populations are declining, much to the distress of those in the seafood industry.

The famed Chesapeake Bay oyster is one victim of pollution. In colonial times, this animal lived up to 10 years and grew to

a.

b.

Figure 40.9
The eutrophication of Lake Tahoe stems from overdevelopment along its shores.

a foot or more in length. Today, an oyster barely lives past 3 or 4 years, and is about the size of a golf ball. In the 1890s, 15 to 18 million bushels of oysters a year were harvested; today it is less than a million. The blue crabs that were once so plentiful that at certain times of the year a wader could not avoid stepping on them, are fast losing their feeding and molting grounds.

However, cleanup efforts may help return the Chesapeake's lost diversity. New sewage facilities are being built, and old ones repaired. Floating plants introduced in some areas use the excess nitrogen. A goal is to reduce the phosphorus and nitrogen flowing into the estuary by 40%.

A model for efforts to revitalize the Chesapeake could be the program to salvage the Delaware River estuary. Pollution of this waterway began in colonial times, and by World War II, the area reeked of oily fumes. After several decades of cleaning up, 33 species of fishes that had virtually vanished have returned.

KEY CONCEPTS

Dumping of sewage into the Chesapeake Bay depleted the oxygen, which encouraged growth of some algae while killing phytoplankton. As food webs toppled, many animal species declined or vanished from the area.

Table 40.1
Are the Oceans Trying to Tell Us Something?

Location	*Event*	*Effect*
Long Island Sound, Chesapeake Bay	Brown algal bloom, lack of oxygen	Many scallops dead
St. Lawrence River	Toxins (PCBs, insecticides)	Dead beluga whales wash ashore
Raritan Bay, N.J.	"Dead zones" lacking oxygen	One million fluke and flounder dead
N.J., N.Y., Mass., R.I.	Medical waste in water	Many beaches closed
East and West Coasts	Toxins	Dolphins, fish, and crustaceans with holes and sores
North Sea	Polluted waters	Many dead seals, cause unknown
Baltic Sea	Polluted waters	Deformed seals
Oregon Coast	Polluted waters	Salmon with tumors
Boston Harbor	0.5 billion gallons of sewage dumped daily	Shellfish with high levels of toxins
Cape Cod	Overgrowth of toxic algae	Dead whales washed ashore

The Oceans

The summer of 1988 was brutally and relentlessly hot in much of the United States. Along the eastern seaboard, beachgoers faced a shock on the shores—medical waste. Soiled bandages, sutures, lumps of solid waste, syringes, and even vials of blood that tested positive for hepatitis B and AIDS washed up on beaches. Elsewhere, dead and dying sea animals littered shores, victims of lack of oxygen or unusual new infectious diseases. Some had sores or tumors, caused by chemical pollutants in the water (table 40.1).

The bottom layers of many watery habitats reveal when certain industrial chemicals were used. Elliott Bay in Seattle, for example, contains cadmium, zinc, lead, copper, and arsenic. Polychlorinated biphenyls (PCBs) appear in the layers approximately when they began being used in 1929 as fire retardants, and the layers tell when lead was removed from gasoline. San Francisco Bay is similarly polluted.

Like other ecological stresses, ocean fouling is a cascade of events. Pollutants pour in from several sources—animal droppings, human sewage, motor oil, fertilizer from large farms and millions of lawns, plus industrial waste, oil spills, and garbage (fig. 40.10). Rivers pick up pollutants, primarily those containing nitrogen and phosphorus, and deliver them to the marshes, wetlands, and estuaries. In saltwater, chemical reactions wrap many of the pollutants in particles, making them heavy enough to sink to the bottom. Here they remain, unless disturbed—which happens often, either by human intervention or by extreme weather.

Once contaminated sediments are disturbed, a deadly chain reaction ensues. The nitrogen and phosphorus trigger algae to bloom, and they rapidly displace other plants. When the algae die and decompose, they deplete the oxygen that is so vital to animals. Some algae produce potent poisons, causing outbreaks of paralytic shellfish poisoning, that ascend the marine food webs, killing sea mammals.

The top layer of the ocean is also essential for supporting the vast webs of marine life, for this is where phytoplankton concentrate—where sunlight is the strongest. The top millimeters of water are also where petroleum-based pollutants concentrate, choking out life from above just as disturbed sediments kill from below.

Another petroleum product with devastating effects on marine life is plastic. More than 2 million seabirds and 100,000 marine mammals, including 50,000 Alaskan fur seals, have been entrapped and choked or starved by carelessly dumped plastic, particularly the rings that hold six-packs of beverages together (fig. 40.11). Many birds build their nests with plastic, and the nestlings eat it. Sea turtles mistake floating blobs of plastic for their natural food, jellyfish, and swallow them. The plastic lodges in their intestines, and they die what must be very painful deaths. The sea turtle is, ironically, a very ancient animal that has weathered many extremes of oceanic existence—but it has nothing in its adaptive repertoire to deal with plastic.

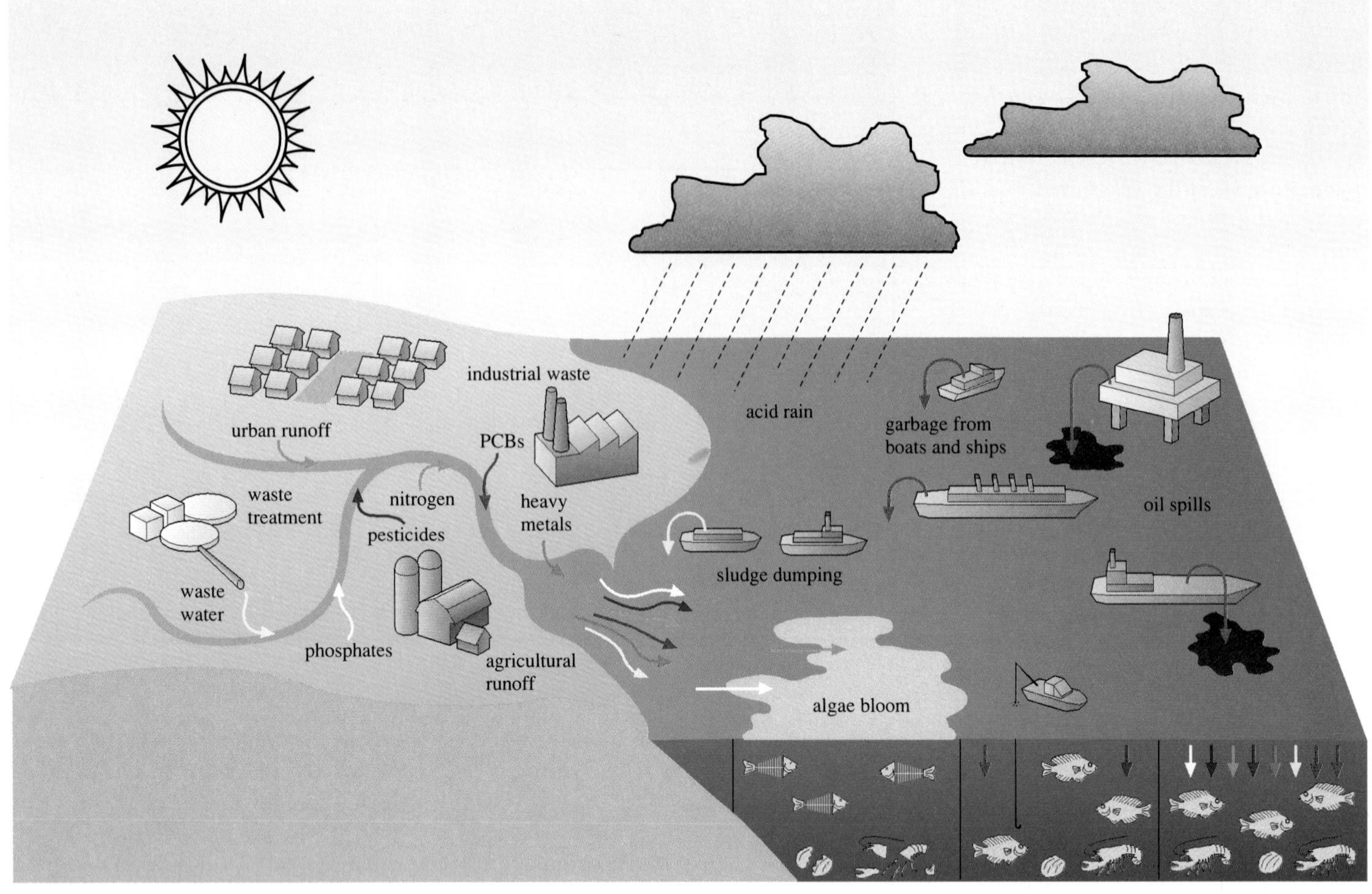

Figure 40.10
Our polluted oceans.

Like other biomes, oceans have natural defenses. Many toxins will remain sequestered if undisturbed. The sheer volume of the oceans dilutes some pollutants, and some may naturally be degraded by microorganisms. For example, hydrocarbons released in a 1979 oil spill in the Gulf of Mexico have broken down, naturally, much faster than was expected.

Meanwhile, powerful images of the befouled beaches and plastic-ringed sea mammals of the 1980s have galvanized many people from the complacent belief that the oceans are so vast that nothing can harm them. The 1990s are a time of action to save our troubled seas. Already plastic beverage carriers are being replaced with cardboard devices, ships can no longer dump waste at sea, cattle are being kept from rivers, and nearly everyone is more aware of what is discarded and poured down the drain.

KEY CONCEPTS

The vastness of the oceans does not protect them from pollution. Sewage, fertilizer, motor oil, garbage, and toxic chemicals dumped or released from sediments upset food webs. Nitrogen and phosphorus trigger algal blooms, which deplete oxygen and release toxins. Infectious diseases spread. Plastic refuse kills many animals.

Epilogue—Thoughts on the Resiliency of Life

Life on earth has had many millions of years to adapt, diversify, and occupy nearly every part of the planet's surface, from tropical rain forest treetops to minute crevices in Antarctic ice. It would be very difficult to halt the parade of life, short of a global catastrophe such as a meteor collision or a nuclear holocaust (Reading 40.2). The biosphere has survived mass extinctions in the past.

Life has prevailed through all manner of localized challenges—from natural events such as the torrential rains and high winds of hurricanes, the eruption of Mount St. Helens

Figure 40.11
A plastic noose.

and the great Yellowstone fires, to the garbage heaps, oil-slicked beaches, and tarnished countrysides caused by our own species. Individual organisms may perish—many of them—but the plasticity built into genes ensures that, in most cases, some individuals will inherit what in one environment is a quirk, but in another, salvation. Life does not end; it changes.

But can we continue to rely on adaptation and diversity to maintain the parade of life? For life may be resilient in an overall sense, yet it is fragile in its interrelationships. Intricate food webs easily collapse if a single member species declines or vanishes—such as the Antarctic krill that feed so many.

Maybe we have just been lucky, so far, that no single event has decimated so many key players in the game of life that other species could not replace them. But could multiple stresses combine to make the earth too inhospitable for life as we know it? Will the deoxygenated, lifeless patches of water in the Gulf of Mexico expand and coalesce, as the ozone holes over the poles and stretches of dead desert do likewise? A single eutrophied lake, an acid or fire ravaged forest, an oil-soaked shore, a cloud of radiation—alone they are terrible; together they could begin to unravel the tangled threads that tie all life together.

The answers to these compelling questions may lie with you, and how you choose to live. This book has shown you the wonder that is life, from its constituent chemicals, to its cells, tissues, and organs, all the way up to the biosphere. Do nothing to harm life—and do whatever you can to preserve its precious diversity. For in diversity lies resiliency, and the future of life on earth.

Reading 40.2 *The End?*

AT FIRST, IT REALLY DOES NOT SEEM MUCH DIFFERENT FROM A TORNADO WARNING, EXCEPT PERHAPS FOR THE CANS OF FOOD STOCKPILED AROUND THE TINY BASEMENT ROOM. You are one of the "lucky" ones, the 2 billion or so humans living far enough from one of the thousand bombs to survive, at least for awhile, this "limited" nuclear war. About half of the earth's human population do not fare as well. Those who did not perish immediately in the blast or fireball either succumb quickly to their injuries or die gradually, in agony. One by one their body systems fail due to radiation sickness acquired in the hail of radiation continuing for the first few days after the detonations. You are at the mercy of the environment. If you do not fall victim to violence committed by another person, you will probably starve to death, as natural food webs collapse at their bases and agriculture vanishes in a single season.

The world above your shelter is dusty, dark, and cold. Because it is, or was, summer, the plummeting temperature kills most plants. The drop of only 4°F (about 2°C) during the growing season decimates half of the Canadian and Russian wheat crops, and if it gets any colder, corn will be the next to die. Just 1 or 2 days near the freezing point, and ice crystals will form in the cells of rice plants, completely ruining the crop. The cereals are particularly sensitive to the vast climatic changes ushered in by the nuclear confrontation because they were cultivated from tropical species having little tolerance for deficits of warmth, light, water, and nutrients—all conditions now present.

In those parts of the world where winter should reign, the following spring planting will not happen, because people, fuel, pesticides, herbicides, and fertilizer will be scarce. But whether an ecosystem was in summer or winter when the bombs fell, the hardy plants that survive the initial blasts will not be able to tolerate the foul chemical smog of nitrogen oxides, sulfur dioxide, carbon monoxide, and the poisons released from destroyed chemical manufacturing facilities.

The great fireballs from the bombs spew soot into the upper atmosphere. Temperature differences at the shores create vast winds that spread this stifling curtain around the globe. The soot absorbs heat and light from the sun, preventing it from reaching the planet's surface. The first dark and cold "nuclear winter" sets in, with the light at 10% its normal level. Plants die as the carbohydrates produced in photosynthesis cannot keep up with those that are metabolized in respiration. Perhaps in a year or two, life-giving light will again trickle through the atmospheric haze, but by then it may be too late. The lethal chain reaction is already in motion, destroying food webs at their photosynthetic bases.

Atmospheric disturbances from the nuclear blasts drastically deplete the ozone layer that normally blocks some of the hazardous ultraviolet B radiation. This electromagnetic radiation of wavelength 280 to 320 nanometers has many effects on life. It can damage DNA beyond the point that enzymes can repair it and disrupts protein structure, and therefore function. The radiation is particularly harmful to plants, decreasing rates of photosynthesis and leaf expansion, disrupting carbohydrate metabolism, and interfering with fruit growth and pollen development.

A statement made by a committee appointed by the International Council of Scientific Unions sums up the bleak situation: "The mechanism most likely to lead to the greatest consequences to humans from a nuclear war is not the blast wave, not the thermal pulse, not direct radiation, nor even fallout; rather, it is mass starvation."

Let's not ever let it happen.

SUMMARY

Human carelessness can cause environmental problems, such as oil spills and nuclear explosions. *Acid precipitation* forms when sulfur and nitrogen oxides from pollution react with water in the upper atmosphere to form hydrochloric and sulfuric acids. Acidification of lakes harms many organisms and changes aquatic communities by selecting organisms that can withstand the altered conditions. In coniferous forests, acid precipitation kills some leaves and acidified soil releases aluminum, which robs roots of calcium and magnesium. Excess nitrogen prompts trees to grow. The conflicting signals stress the tree, and it becomes more vulnerable.

The *greenhouse effect* results from CO_2 and other gases trapping infrared radiation near the earth's surface. The greenhouse gases are generated from burning fossil fuels, agriculture, industry, and destruction of tropical rain forest. Use of CFCs is depleting the ozone layer, which protects life from harmful effects of ultraviolet radiation. Poor agricultural practices are destroying the tropical rain forest while allowing the African desert to grow. Transformation of the American temperate forest has paralleled history. Lake Tahoe is undergoing eutrophication as a result of increased sediments and nutrients from development along its shores. Pollution endangers communities in the Chesapeake Bay estuary. Preserving estuaries is especially important because they are a breeding ground for many species. Oceans are gravely endangered by pollution. Waste washes onto beaches, many organisms die and ocean sediments are contaminated with heavy metals and other industrial waste. When the sediments are disrupted, algae bloom, rob the water of oxygen, and produce toxins. Plastic and oil contaminate surface waters, where much photosynthesis takes place.

QUESTIONS

1. How have experiments added to our understanding of the following:
- **a.** The effects of acid rain on lakes
- **b.** The greenhouse effect

Suggest experiments that could have helped us deal with the *Valdez* oil spill and Chernobyl nuclear plant disaster.

2. What natural protections do ecosystems have against the destructive effects of the following:
- **a.** Acid precipitation
- **b.** Eutrophication of lakes
- **c.** Global warming from the greenhouse effect
- **d.** Ocean pollution

3. How can some types of algae form the bases of aquatic food webs, yet others, when too abundant, choke the life from lakes and oceans?

4. In the Philippines, divers spray live coral with sodium cyanide, which they call "magic." This toxic chemical flushes fishes out of the corals' nooks and crannies, and the diver captures those that he can sell, leaving the others behind to twitch to death from the poison. The coral dies. How might this type of pollution alter the ecosystem?

5. Why is it important that estuaries not be polluted?

TO THINK ABOUT

1. What information from the effects of pollutants on the oceans suggests that it might be foolish to dredge the Hudson River to remove PCBs that have settled on the bottom?

2. Were human efforts to rescue wildlife harmed by the *Valdez* oil spill worthwhile? Cite a reason for your answer.

3. Do you think that humans could destroy the biosphere, or do natural resiliencies make this impossible?

4. The Amazon Basin of South America has 90 million human residents, one-third of whom are malnourished. The population is expected to double in 20 years. Should the tropical rain forest be cultivated to feed them? Why or why not? Suggest a possible solution to the hunger problem in this area.

5. Statistical records show that warmer than normal ocean currents off the Pacific coast of South America often follow volcanic eruptions. The debris spewed into the atmosphere from these eruptions upsets the interaction between the atmosphere and the ocean so that upwelling is prevented. What effect might a nuclear war, which would send dust into the atmosphere equivalent to 1,000 large volcanoes erupting simultaneously, have on oceanic life?

6. A representative of an eastern African nation claims that his country need not join global efforts to halt use of CFCs, and he cited several reasons:
- **a.** His country is already warm. A rise in temperature of a few degrees would hardly be noticeable.
- **b.** The ozone holes do not extend over Africa.
- **c.** His people do not use many CFCs.

Do you agree with this assessment? In what way might his reasoning be short-sighted?

SUGGESTED READINGS

Cohn, Jeffrey P. March 1989. Gauging the biological impacts of the greenhouse effect. *BioScience*, vol. 139, no. 3. Effects of global warming are like biological cascades.

Drew, Lisa. June-July 1990. Truth and consequences along oiled shores. *National Wildlife*. Was the great cleanup effort worth it?

Goldman, Charles R. September 1989. Preserving a fragile ecosystem. *Environment*. Can Lake Tahoe be saved?

Kunzig, Robert. April 1990. Invisible garden. *Discover*. Much of life on earth depends on phytoplankton.

Lawren, Bill. October-November 1990. Plastic rapt. *National Wildlife*. What will we do with all of our plastic?

Lewis, Ricki. January 1991. Can salmon make a comeback? *BioScience*. Salmon populations plummeted, thanks to human intervention. Can we help them return?

Loupe, Diane E. October 6, 1990. To rot or not. *Science News*. Should garbage be buried wet or dry?

McFarland, Mack. October 1989. Chlorofluorocarbons and ozone. *Environmental Science Technology*, vol. 23, no. 10. Chemicals of convenience are destroying the ozone layer.

Monastersky, Richard. July 21, 1990. The fall of the forest. *Science News*. Tropical deforestation has reached dangerous levels.

Repetto, Robert. April 1990. Deforestation in the tropics. *Scientific American*. Burning the tropical rain forest is a political as well as biological problem.

Sullivan, T.J. May 3, 1990. Quantification of changes in lakewater chemistry in response to acidic deposition. *Nature*, vol. 345. The Adirondack lakes show evidence of acid precipitation—and recovery.

The November 1990 issue of *Natural History* is devoted to the effects of war on ecosystems.

Appendix A
Microscopy

As the study of life has progressed steadily from observing organisms to probing the molecules of life, the technology to view living things has grown accordingly. Today's biologist has a range of microscope types to choose from, and the instrument used depends upon the nature of the biological material being observed (table 1). An *ultraviolet microscope* might be used to highlight stained chromosomes; a *polarizing microscope* to focus in on protein arrays of a cytoskeleton; a *phase contrast microscope* to view cells while they are still alive. A *scanning electron microscope* reveals the topography of cell and organelle surfaces. A *confocal microscope* presents startlingly clear peeks at biological structures in action, and a *scanning probe microscope* reveals surfaces of individual atoms.

All microscopes provide two types of power—*magnification* and *resolution* (also called resolving power). A microscope produces an enlarged, or magnified, image of an object. Magnification is defined as the ratio between the image size and the object size. Resolution refers to the smallest degree of separation at which two objects are viewed as distinct from one another, rather than as a blurry, single image. Resolution is important in distinguishing structures from one another. A *compound microscope* commonly used in college biology teaching laboratories can resolve objects that are 0.1 to 0.2 micrometers (4 to 8 millionths of an inch) apart. The resolving power of an electron microscope is 10,000 times greater than this.

Table 1
Compound Microscopes

Method	***Basis***	***Advantages***	***Disadvantages***
Phase contrast microscopy	Converts differences in the velocity of light through different parts of specimen into observable contrasts	Can be used on live cells	Not all subcellular structures are visible; halos seen around structures
Interference microscopy	Two beams of light hit specimen and join in image plane	No halos on structures; fine detail	Cumbersome to use; expensive
Differential-interference (Nomarski-optics) microscopy	Detects localized differences in velocities at which light passes through specimen	Fine transparent detail visible	
Polarizing microscopy	Ray of plane-polarized (i.e., unidirectional) light hits specimen, splits into two directions, at two different velocities, creating image of ordered molecular detail	Highlights detail at molecular level	Works best on highly oriented, crystalline or fibrous structures
Fluorescence microscopy	Light of one wavelength is selectively absorbed by certain molecules that reemit light of a longer wavelength		Only creates image of structures that absorb the wavelength of light used
Ultraviolet (uv) microscopy	Ultraviolet light used with lens made of quartz	High resolving power; excellent for viewing proteins and nucleic acids	

The Light Microscope

The compound light microscope focuses visible light through a specimen. Different regions of the object scatter the light differently, producing an image. In modern microscopes, three sets of lenses contribute to the generation of an image (fig. 1). The *condenser lens* focuses light through the specimen. The *objective lens* receives light that has passed through the specimen, generating an enlarged image. The *ocular lens*, or eyepiece, magnifies the image further. Total magnification is calculated by multiplying the magnification of the objective lens by that of the ocular lens. The coarse and fine adjustment knobs are manipulated to bring the magnified image into sharp focus. The mirror directs the light into the condenser lens, and the *diaphragm* controls the amount of light to which the specimen is exposed.

A limitation of a light microscope is that only one two-dimensional plane of the specimen can be observed at a time. Thus, when a light microscope is focused on the top of a specimen, different structures are visible than when it is focused at a deeper level. It can be difficult to envision the three-dimensional nature of the specimen from the two-dimensional views afforded by the light microscope. The problem is like focusing on particular parts of a scene with a camera. If the photographer focuses on his children in the foreground of a shot, he may miss entirely the antics of a cat and mouse that are several feet behind the children. Similarly, light microscope views at different depths within a cell can reveal different structures.

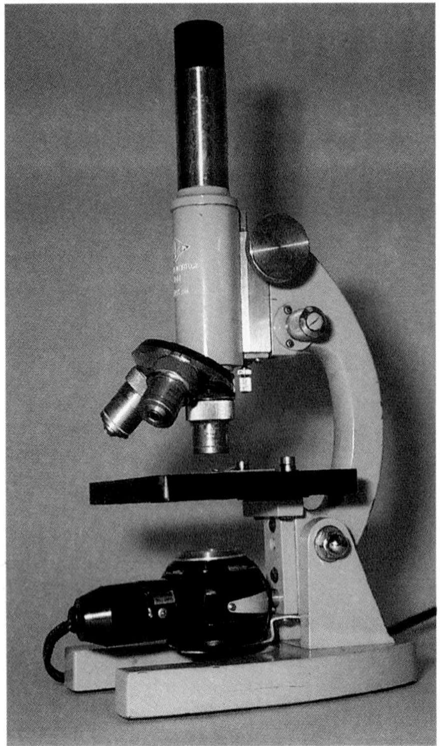

Figure 1
Light Microscope

Electron Microscopes

Electron microscopes provide greater magnification, better resolution, and a better sense of depth than light microscopes. Instead of visible light, the *transmission electron microscope* (TEM) sends a beam of electrons through the specimen, using a magnetic field to focus the beam rather than a glass lens (fig. 2). Different parts of the specimen absorb different numbers of electrons. These contrasts are rendered visible to the human eye by a fluorescent screen coated with a chemical that gives off visible light rays when excited by electrons from the specimen.

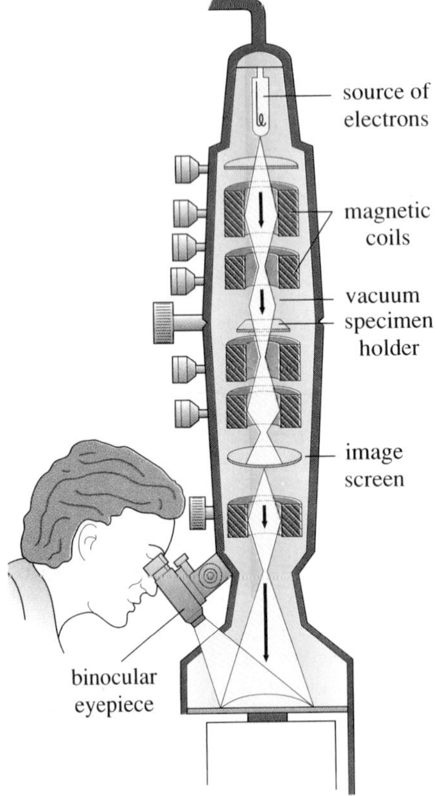

Figure 2
Electron Microscope

Although the TEM has provided some spectacular glimpses into the microscopic structures of life, it does have limitations. For the TEM, the specimen must be killed, treated with chemicals, cut into very thin sections, and placed in a vacuum. This treatment can distort natural structures. A close cousin of the TEM eliminates these drawbacks. The *scanning electron microscope* (SEM) bounces electrons off of a three-dimensional specimen, generating a three-dimensional image on a device similar to a television screen. The resulting depth of field highlights crevices and textures. Although many SEM specimens are coated with a heavy metal to highlight their surfaces, some specimens (such as fruit flies) can be examined while alive, with no apparent harm.

A variation of the electron microscope is the *photoelectron microscope* (PEM), originally used to probe metal surfaces but now used to examine cells as well. The PEM bombards a specimen with ultraviolet light, ejecting the valence shell electrons of molecules on a cell or organelle surface. These electrons are accelerated and focused by an electron lens system. The excited electrons are quite sensitive to the surface detail of the specimen, and their deflection pattern provides a high-resolution view of minute surface details. The PEM is especially useful to zero in on specific molecules that have been labeled with fluorescent antibodies. PEM is an electron-based version of fluorescence microscopy.

While the SEM highlights large surface features, the PEM provides a closer look. It is like comparing a topographic map of a mountain (SEM) to a picture of a bump in the terrain of the mountain (PEM). A light microscope and all three electron microscopes can be used in conjunction to paint a detailed portrait of biological structures, which can clarify functions at the organelle or even the molecular level.

The Confocal Microscope

A limitation of light microscopy is that light reflected from regions of the sample near the object of interest interferes with the image, making it blurry or hazy. A *confocal microscope* avoids interference and enhances resolution by passing white or laser light through a pinhole and a lens to the object (fig. 3). The light is then reflected through a beam splitter and then through another pinhole, a detector, and finally a photomultiplier. The result is a scan of highly focused light on one tiny part of the specimen at a time, usually an area 0.25 μm in diameter and 0.5 μm deep. The microscope is called "confocal" because the objective and the condenser lenses are both focused on the same small area.

The idea of a confocal microscope was patented by Marvin Minsky in 1961, but it was not developed until the mid 1980s, when computers enabled many scans of different sites and at different depths to be integrated and translated into a dynamic image. By using fluorescent dyes that label specific cell parts and are activated by the incoming light, different structures can be distinguished. The first division of a fertilized sea urchin egg, for example, can be captured: the spindle fibers appear green, and the chromosomes being pulled in opposite directions are a vibrant blue. Confocal microscopy has also revealed changing concentrations of calcium ions in a neuron receiving a biochemical message; the cytoskeleton in action; platelets aggregating at the scene of an injury to form a clot; sperm fertilizing an ovum; and nerve cells infiltrating the developing brain of an embryo. When teamed with a tool borrowed from the physical sciences called Raman spectroscopy, confocal microscopy reveals details of chromosome structure, and can distinguish between the chemical bonds of a protein or nucleic acid.

Scanning Probe Microscopes

The world of microscopy was again revolutionized in 1981, with the invention of the *scanning tunneling microscope* (STM) by Gerd K. Binnig and Heinrich Rohrer. This device reveals detail at the atomic level. A very sharp metal needle, its tip as small as an atom, is scanned over a molecule's surface. Electrons "tunnel" across the space between the sample and the needle, thereby creating an electrical current. The closer the needle, the greater the current. An image is generated as the scanner continually adjusts the space between needle and sample, keeping the current constant over the topography of the molecular surface. The needle's movements over the microscopic hills and valleys are expressed as contour lines, which in turn are converted and enhanced by computer into a colored image of the surface.

Electrons do not pass readily from many biological samples, limiting use of STM. However, the same principle of adjusting a probe over a changing surface is used in *scanning ion-conductance microscopy* (SICM), developed by Paul K. Hansma and Calvin Quate. It uses ions instead of electrons—useful in the many biological situations where ions travel between cells. The probe is made of hollow glass filled with a conductive salt solution, which is also applied to the sample. When voltage is passed through the sample and the probe, ions flow to the probe. The rate of ion flow is kept constant, and a portrait is painted by the compensatory movements of the probe. SICM is useful in studying cell membrane surfaces and muscle and nerve function.

Another type of scanning probe microscope, the *atomic force microscope* (AFM), was developed in 1986 by the inventors of the SICM. It uses a diamond-tipped probe that resembles the stylus on a phonograph but that presses a molecule's surface with a force millions of times gentler. As the force is kept constant, the probe moves, generating an image. AFM is especially useful for recording molecular movements, such as those involved in blood clotting and cell division.

New and improved microscopes do not always replace existing models but complement the information that they provide. Many researchers today create their own versions of microscopes to suit their particular experiments. All modern microscopes though, some of them quite technologically sophisticated, support the cell theory advanced by the early microscopists, who had only very crude light microscopes with which to work.

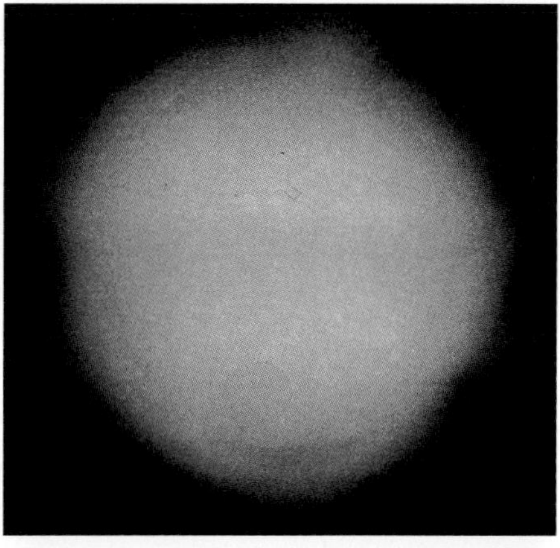

a.

b.

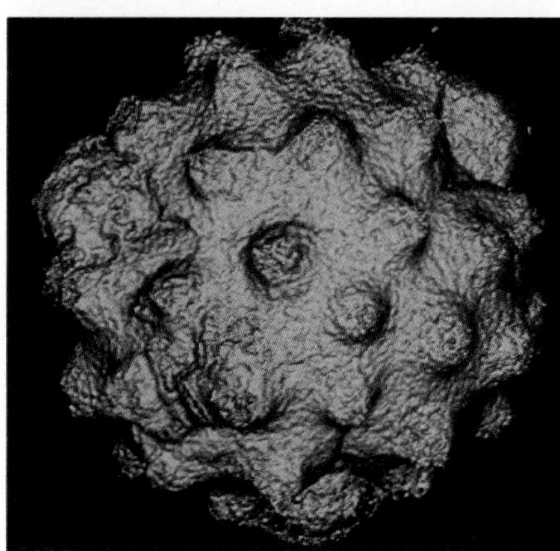

c.

Figure 3
a. Sea urchin embryo at first division stained with fluorescently labeled anti-tubulin antibody, taken with a conventional fluorescence microscope. *b*. Sea urchin embryo at first division stained with fluorescently labeled anti-tubulin antibody, taken with a confocal laser scanning microscope. *c*. Sea urchin embryo at first division, double stained to show tubulin in green and DNA in blue, taken with a tandem-scanning microscope.

Appendix B
Units of Measurement Metric/English Conversions

Length

1 meter = 39.4 inches = 3.28 feet = 1.09 yard
1 foot = 0.305 meters = 12 inches = 0.33 yard
1 inch = 2.54 centimeters
1 centimeter = 10 millimeter = 0.394 inch
1 millimeter = 0.001 meter = 0.01 centimeter = 0.039 inch
1 kilometer = 1,000 meters = 0.621 miles = 0.54 nautical miles
1 mile = 5,280 feet = 1.61 kilometers
1 nautical mile = 1.15 mile

Area

1 square centimeter = 0.155 square inch
1 square foot = 144 square inches = 929 square centimeters
1 square yard = 9 square feet = 0.836 square meters
1 square meter = 10.76 square feet = 1.196 square yards = 1 million square millimeters
1 hectare = 10,000 square meters = 0.01 square kilometers = 2.47 acres
1 acre = 43,560 square feet = 0.405 hectares
1 square kilometer = 100 hectares = 1 million square meters = 0.386 square miles = 247 acres
1 square mile = 640 acres = 2.59 square kilometers

Volume

1 cubic centimeter = 1 milliliter = 0.001 liter
1 cubic meter = 1 million cubic centimeters = 1,000 liters
1 cubic meter = 35.3 cubic feet = 1.307 cubic yards = 264 U.S. gallons
1 cubic yard = 27 cubic feet = 0.765 cubic meters = 202 U.S. gallons
1 cubic kilometer = 1 million cubic meters = 0.24 cubic mile = 264 billion gallons
1 cubic mile = 4.166 cubic kilometers
1 liter = 1,000 milliliters = 1.06 quarts = 0.265 U.S. gallons = 0.035 cubic feet
1 U.S. gallon = 4 quarts = 3.79 liters = 231 cubic inches
1 quart = 2 pints = 4 cups = 0.94 liters

Mass

1 microgram = 0.001 milligram = 0.000001 gram
1 gram = 1,000 milligrams = 0.035 ounce
1 kilogram = 1,000 grams = 2.205 pound
1 pound = 16 ounces = 454 grams
1 short ton = 2,000 pounds = 909 kilograms
1 metric ton = 1,000 kilograms = 2,200 pounds

Temperature

Celsius to Fahrenheit °F = (°C x 1.8) + 32
Fahrenheit to Celsius °C = (°F – 32) ÷ 1.8

Energy and Power

1 kilocalorie = 1,000 calories

Appendix C
Metric Conversion

	Metric Quantities	*Metric to English Conversion*	*English to Metric Conversion*
Length	1 kilometer (km) = 1,000 (10^3) meters 1 meter (m) = 100 centimeters 1 centimeter (cm) = 0.01 (10^{-2}) meter 1 millimeter (mm) = 0.001 (10^{-3}) meter 1 micrometer* (μm) = 0.000001 (10^{-6}) meter 1 nanometer (nm) = 0.000000001 (10^{-9}) meter *formerly called micron	1 km = 0.62 mile 1 m = 1.09 yards = 39.37 inches 1 cm = 0.394 inch 1 mm = 0.039 inch	1 mile = 1.609 km 1 yard = 0.914 m 1 foot = 0.305 m = 30.5 cm 1 inch = 2.54 cm
Area	1 square kilometer (km^2) = 100 hectares 1 hectare (ha) = 10,000 square meters 1 square meter (m^2) = 10,000 square centimeters 1 square centimeter (cm^2) = 100 square millimeters	1 km^2 = 0.3861 square mile 1 ha = 2.471 acres 1 m^2 = 1.1960 square yards = 10.764 square feet 1 cm^2 = 0.155 square inch	1 square mile = 2.590 km^2 1 acre = 0.4047 ha 1 square yard = 0.8361 m^2 1 square foot = 0.0929 m^2 1 square inch = 6.4516 cm^2
Mass	1 metric ton (t) = 1,000 kilograms 1 metric ton (t) = 1,000,000 grams 1 kilogram (kg) = 1,000 grams 1 gram (g) = 1,000 milligrams 1 milligram (mg) = 0.001 gram 1 microgram (μg) = 0.000001 gram	1 t = 1.1025 ton (U.S.) 1 kg = 2.205 pounds 1 g = 0.0353 ounce	1 ton (U.S.) = 0.907 t 1 pound = 0.4536 kg 1 ounce = 28.35 g
Volume (solids)	1 cubic meter (m^3) = 1,000,000 cubic centimeters 1 cubic centimeter (cm^3) = 1,000 cubic millimeters	1 m^3 = 1.3080 cubic yards = 35.315 cubic feet 1 cm^3 = 0.0610 cubic inch	1 cubic yard = 0.7646 m^3 1 cubic foot = 0.0283 m^3 1 cubic inch = 16.387 cm^3
Volume (liquids)	1 liter (l) = 1,000 milliliters 1 milliliter (ml) = 0.001 liter 1 microliter (μl) = 0.000001 liter	1 l = 1.06 quarts (U.S.) 1 ml = 0.034 fluid ounce	1 quart (U.S.) = 0.94 l 1 pint (U.S.) = 0.47 l 1 fluid ounce = 29.57 ml
Time	1 second (sec) = 1,000 milliseconds 1 millisecond (msec) = 0.001 second 1 microsecond (μsec) = 0.000001 second		

Appendix D
Taxonomy

The millions of living and extinct species that have dwelled on earth can be grouped according to many schemes. Taxonomists group organisms to reflect both anatomical similarities and descent from a common ancestor. Two, three, four, five, and most recently, six kingdom classifications have been proposed. The five-kingdom scheme is outlined here with all phyla described briefly. Short statements explaining the rationale behind groupings of phyla are given wherever possible, and indented subheadings reflect these groupings. Scientific names are followed by more familiar names of organisms. Figures from the text accompany the listing to help you visualize and recall the wide diversity of life forms discussed in chapter 2.

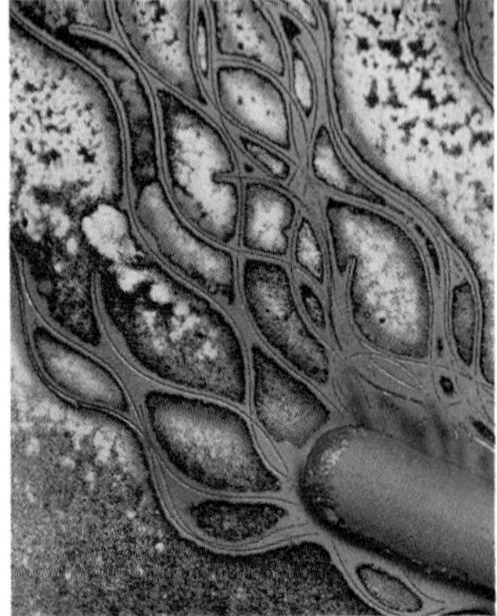

Salmonella

Stromatolites

Kingdom Monera The monerans are unicellular prokaryotes that obtain nutrients by direct absorption or by photosynthesis or chemosynthesis. Most monerans reproduce asexually, but some can exchange genetic material in a primitive form of sexual reproduction. (The six-kingdom classification system divides the monerans into two kingdoms, whose members are distinguished by genetic differences and by whether or not they produce methane as a metabolic by-product. The methane-producing bacteria are thought to be the most primitive organisms, having evolved before the atmosphere contained oxygen.)

- **Phylum Schizophyta** The bacteria.
- **Phylum Cyanobacteria** Photosynthetic bacteria, formerly called blue-green algae.

Kingdom Protista Protists are the structurally simplest eukaryotes, and they can be unicellular or multicellular. They can absorb or ingest nutrients or photosynthesize. Reproduction is asexual or sexual. Some forms move by ciliary or flagellar motion, and others are nonmotile. The protists' early development differs from that of the fungi, plants, and animals. The kingdom includes the protozoans, algae, and the water molds and slime molds.

Protozoans Unicellular, nonphotosynthetic, lack cell walls.

- **Phylum Sarcomastigophora** Locomote by flagella and/or pseudopoda and includes the familiar *Amoeba proteus*.
- **Phylum Labyrinthomorpha** Aquatic, live on algae.
- **Phylum Apicomplexa** Parasitic, with characteristic twisted structure on anterior end at some point in life cycle.
- **Phylum Myxozoa** Parasitic on fish and invertebrates.
- **Phylum Microspora** Parasitic on invertebrates and primitive vertebrates.
- **Phylum Ciliophora** Cilia present at some stage of the life cycle.

Algae Unicellular or multicellular, photosynthetic, some have cell walls. Distinguished by pigments.

- **Phylum Euglenophyta** Unicellular and photosynthetic, with a single flagellum and contractile vacuole.
- **Phylum Chrysophyta** Diatoms, golden-brown algae, and yellow-green algae. Unicellular and photosynthetic.
- **Phylum Pyrrophyta** Dinoflagellates. Unicellular and photosynthetic.
- **Phylum Chlorophyta** Green algae. Unicellular or multicellular, photosynthetic.
- **Phylum Phaeophyta** Brown algae (kelps). Multicellular and photosynthetic.
- **Phylum Rhodophyta** Red algae. Multicellular and photosynthetic.

Water and Slime Molds

- **Phylum Oomycota** The water molds. Unicellular or multinucleate, with cellulose cell walls. Live in fresh water.
- **Phylum Chytridiomycota** The chytrids. Multicellular, with chitinous cell walls. Aquatic.
- **Phylum Myxomycota** Multinucleated, "acellular" slime molds.
- **Phylum Acrasiomycota** Multicellular "cellular" slime molds.

Morel mushroom

Fossil horsetails

Hydra

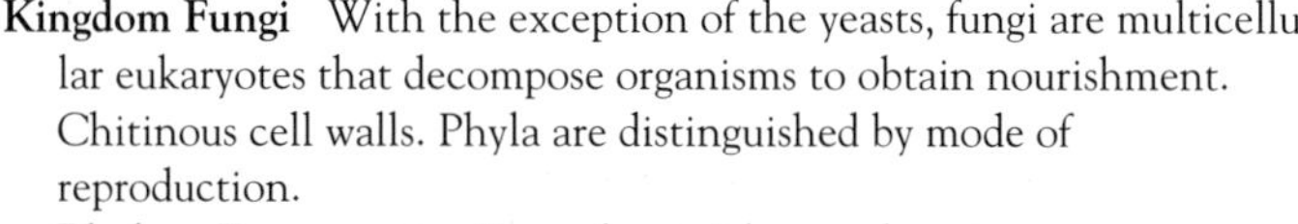

Kingdom Fungi With the exception of the yeasts, fungi are multicellular eukaryotes that decompose organisms to obtain nourishment. Chitinous cell walls. Phyla are distinguished by mode of reproduction.

- **Phylum Zygomycota** Reproduce with sexual resting spores.
- **Phylum Ascomycota** Yeasts, morels, truffles, molds, lichens. Reproduce with sexual spores carried in asci. Some ascomycetes cause food spoilage and some plant diseases; others are used in the production of certain foods, beverages, and antibiotic drugs.
- **Phylum Basidiomycota** Mushrooms, toadstools, puffballs, stinkhorns, shelf fungi, rusts, and smuts. Reproduce by spore-containing basidia.

Kingdom Plantae Plants are multicellular, land dwelling, photosynthetic, and reproduce both asexually and sexually in an alternation of generations. Cellulose cell walls. Plants have specialized tissues and organs but lack nervous and muscular systems.

- **Nonvascular Plants (Bryophytes)** Lack specialized conducting tissues and true roots, stems, and leaves. The gamete-producing reproductive phase predominates.
  - **Division Bryophyta** Liverworts, hornworts, mosses.
- **Vascular Plants (Tracheophytes)** Xylem and phloem transport water and nutrients, respectively, throughout the plant body of roots, stems, and leaves. The spore-producing reproductive phase predominates.
  - **Primitive Plants** Sperm cells travel in water to meet egg cells.
    - **Division Pterophyta** Ferns.
    - **Division Psilophyta** Whisk ferns.
    - **Division Lycophyta** Club mosses and others.
    - **Division Spenophyta** Horsetails.
  - **Seed Plants** Sperm cells and egg cells enclosed in protective structures.
    - **Gymnosperms (naked seed plants)** Male and female cones produce pollen grains and ovules.
      - **Division Coniferophyta** Conifers.
      - **Division Cycadophyta** Cycads.
      - **Division Ginkgophyta** Ginkgos.
      - **Division Gnetophyta** Gnetophytes.
    - **Angiosperms (seeds in a vessel)** The flowering plants.
      - **Division Anthophyta** Flowering plants.

Kingdom Animalia The animals are multicellular with specialized tissues and organs, including nervous and locomotive systems. No cell walls. Animals obtain nutrients from food. Phyla are distinguished largely on the basis of body form and symmetry, characteristics that are generally established in the early embryo.

Mesozoa Simplest animals.

- **Phylum Mesozoa** Very simple, wormlike parasites of marine invertebrates. Consist of only 20 to 30 cells.

Parazoa A separate branch from the evolution of protozoa to metazoa.

- **Phylum Placozoa** A single species, *Trichoplax adhaerens*, characterized by two cell layers with fluid in between them.
- **Phylum Porifera** The sponges. Specialized cell types organized into canal system to transport nutrients in and wastes out.

Eumetazoa Animal phyla descended from protozoa.

- **Radiata** Radially symmetric body plan. Sedentary, saclike bodies with two or three cell layers and a diffuse nerve net.
  - **Phylum Cnidaria** Hydroids, sea anemones, jellyfish, horny corals, hard corals.
  - **Phylum Ctenophora** Sea walnuts, comb jellies.
- **Bilateria** Bilaterally symmetric body plan.
  - **Protostomia (first mouth)** Embryonic characteristics:
    1. Mouth forms close to area of initial folding inward in very early embryo.
    2. Spiral cleavage: At third cell division, second group of four cells sits atop first group of four cells but rotated by 45°.
    3. Determinate cleavage: Cell fates determined very early in development. If a cell from a four-celled embryo is isolated, it will divide and differentiate to form only one-quarter of an embryo.
    4. The protostomes are further grouped by the way in which the body cavity (coelom) forms. A true coelom is a body cavity that develops within mesoderm, the middle layer of the embryo.
    - **Acoelomates** No coelom.
      - **Phylum Platyhelminthes** Flatworms.
      - **Phylum Nemertina** Ribbonworms.
      - **Phylum Gnathostomulida** Jawworms.
    - **Pseudocoelomates** Body cavity derived from a space in the embryo between the mesoderm and endoderm. The body cavity is called "pseudo" because it does not form within mesoderm. In the adult, the pseudocoelom is a cavity but it is not lined with mesoderm-derived peritoneum (seen in more advanced forms).
      - **Phylum Rotifera** The rotifers. Small (40μm–3mm), intricately shaped organisms that have a structure on their anterior ends that resembles rotating wheels. The rotifers occupy a variety of habitats.

Octopus

Sea urchin

Boobies and people

Phylum Gastrotricha Aquatic, microscopic, flattened organisms with a scaly outer covering.

Phylum Kinorhyncha Marine worms less than 1 mm long.

Phylum Nematoda Roundworms. Found everywhere, many parasitic.

Phylum Nematomorpha Horsehair worms. Juveniles are parasitic in arthropods; adults are free-living.

Phylum Acanthocephala Spiny-headed worms. Spiny projection from anterior end used to attach to intestine of host vertebrate. Range in size from 2 mm to more than a meter.

Phylum Entoprocta Nonmotile, sessile, mostly marine animals that look like stalks that are anchored to rocks, shells, algae, or vegetation on one end, with a tufted growth on the other.

Eucoelomates Coelom forms in a schizocoelous fashion, in which the body cavity forms when mesodermal cells invade the space between ectoderm and endoderm, and then proliferate so that a cavity forms within the mesoderm.

Major Eucoelomate Protostomes Three phyla, with many species.

Phylum Mollusca Snails, clams, oysters, squids, octopuses.

Phylum Annelida Segmented worms.

Phylum Arthropoda Spiders, scorpions, ticks, mites, crustaceans, millipedes, centipedes, insects.

The Lesser Protostomes Seven phyla, including many extinct species. Little-understood offshoots of annelid-arthropod line.

Phylum Pripulida Bottom-dwelling marine worms.

Phylum Echiurida Marine worms.

Phylum Sipunculida Bottom-dwelling marine worms.

Phylum Tardigrada "Water bears." Less than 1 mm, live in water film on mosses and lichens.

Phylum Pentastomida Tongue worms. Parasitic on respiratory system of vertebrates, mostly reptiles.

Phylum Onychophora Velvet worms. Live in tropical rain forest and resemble caterpillars, with 14 to 43 pairs of unjointed legs and a velvety skin.

Phylum Pogonophora Beard worms. Live in mud on ocean bottom.

Lophophorates Three phyla distinguished by a ciliary feeding structure called a lophophore.

Phylum Phoronida Small, wormlike bottom-dwellers of shallow, coastal temperate seas. Live in a tube that they secrete.

Phylum Ectoprocta Bryozoa, or "moss animals." Aquatic, less than 1/2 mm long, live in colonies but each individual lives within a chamber secreted by the epidermis. Bryozoa look like crust on rocks, shells, and seaweeds.

Phylum Brachiopoda Lampshells. Attached, bottom-dwelling marine animals that have two shells and resemble mollusks, about 5 to 8 mm long.

Deuterostomia (second mouth) Embryonic characteristics:

1. Mouth forms far from area of initial folding inward in very early embryo.
2. Radial cleavage: At third cell division, second group of four cells sits directly atop first group.
3. Indeterminate cleavage: Cell fates of very early embryo not detmined. If a cell from a four-cell embryo is isolated, it will develop into a complete embryo.
4. Coelom formation is enterocoelous. Body cavity forms from outpouchings of endoderm that become lined with mesoderm.

Phylum Echinodermata Sea stars, brittle stars, sea urchins, sea cucumbers, sea lilies. Radial symmetry in adult but larvae are bilaterally symmetric. Complex organ systems, but no distinct head region.

Phylum Chaetognatha Arrow worms. Marine-dwelling with bristles surrounding mouth.

Phylum Hemichordata Acorn worms and others. Aquatic, bottom-dwelling, nonmotile, wormlike animals.

Phylum Chordata Tunicates, lancelets, hagfishes, lampreys, sharks, bony fishes, amphibians, reptiles, birds, mammals. Chordates have a notochord, dorsal nerve cord, gill slits, and a tail. Some of these characteristics may only be present in embryos.

Glossary

A

abscisic acid *ab-SIS-ik AS-id* A plant hormone that inhibits growth. 588

abscission zone *ab-SCISZ-on ZONE* A region at the base of the petiole from which leaves are shed. 562

abyssal zone *ah-BIS-el ZONE* The part of the bottom of the ocean where light does not reach. 737

acclimation *AK-klah-MA-shun* Changes in a plant in preparation for winter. 599

accommodation *ah-KOM-o-DAY-shun* Changes in the shape of the lens to suit the distance of the object being viewed. 356

acetyl CoA formation *AS-eh-til FOR-MAY-shun* The first step in aerobic respiration, occurring in the mitochondrion. Pyruvic acid loses a carbon dioxide and bonds to coenzyme A to form acetyl CoA. 122

acid *AS-id* A molecule that releases hydrogen ions into water. 46

acquired immune deficiency syndrome (AIDS) *ak-KWY-erd im-MUNE dah-FISH-en-see SIN-drome* Infection by the human immunodeficiency virus (HIV), which kills a certain class of helper T cells, causing profound immune suppression and resulting in opportunistic infections and cancer. 211

acromegaly *AK-ro-MEG-eh-lee* Abnormal thickening of bones in an adult due to excess growth hormone. 374

acrosome *AK-ro-som* A protrusion on the anterior end of a sperm cell containing digestive enzymes that enable the sperm to penetrate the protective layers around the oocyte. 163

actin *AK-tin* A type of protein in the thin myofilaments of skeletal muscle cells. 407

action potential *AK-shun po-TEN-shel* The measurement of an electrochemical change caused by ion movement across the cell membrane of a neuron. The message formed by this change is the nerve impulse. 317

active immunity *AK-tiv im-MUNE-eh-tee* Immunity generated by an organism's production of antibodies. 522

active site *AK-tiv SITE* The portion of an enzyme's conformation that directly participates in catalysis. 56

active transport *AK-tiv TRANZ-port* Movement of a molecule through a membrane against its concentration gradient, using a carrier protein and energy from ATP. 102

adaptation *AD-ap-TAY-shun* An inherited trait that enables an organism to survive a particular environmental challenge. 40

adaptive radiation *ah-DAP-tiv RAID-ee-AY-shun* The divergence of several new types of organisms from a single ancestral type. 626

adenine *AD-eh-neen* One of two purine nitrogenous bases in DNA and RNA. 57, 253

adenosine triphosphate *(ATP) ah-DEN-o-seen tri-FOS-fate* A molecule whose three high-energy phosphate bonds power many biological processes. 101, 109

adipose cell *ADD-eh-pos SEL* A cell filled almost entirely with lipid. 51

adrenal cortex *ad-REE-nal KOR-tex* The outer part of the adrenal glands. 377

adrenal glands *ad-REE-nal GLANZ* Paired, two-part glands that sit atop the kidneys and produce catecholamines, mineralocorticoids, glucocorticoids, and sex hormones. 377

adrenal medulla *ad-REE-nal mah-DUEL-ah* The inner part of the adrenal glands. 377

adrenocorticotropic hormone (ACTH) *ah-DREEN-o-KOR-tah-ko-TROP-ik HOR-moan* A hormone made in the anterior pituitary that stimulates secretion of hormones from the adrenal cortex. 375

adventitious roots *AD-ven-TISH-shus ROOTZ* Roots that form on stems or leaves, replacing the first root (the radicle). 563

aerial roots *AIR-ee-al ROOTZ* Adventitious roots that form and grow in the air. 564

afferent arterioles *AF-fer-ent are-TEAR-ee-olz* Branches of the renal artery approaching the proximal portion of a nephron. 506

agonist *AG-o-nist* A drug that activates a receptor, triggering an action potential, or helps a neurotransmitter to bind to the receptor. 325

agriculture *AG-rah-CUL-tur* The domestication of animals and the planting and cultivation of plants used as crops. 540

AIDS-related complex (ARC) *AIDS re-LAY-tid KOM-plex* The early stages of AIDS, characterized by weakness, swollen glands in the neck, and frequent fever. 523

alcoholic fermentation *AL-ko-HALL-ik FER-men-TAY-shun* An anaerobic step following glycolysis utilized by yeast. Pyruvic acid is converted to ethanol and carbon dioxide. 121

aldosterone *al-DOS-ter-own* The major mineralocorticoid hormone produced by the adrenal cortex. It maintains the level of Na^+ in the blood by altering the amount reabsorbed in the kidneys. 378, 510

aleurone *AL-ah-roan* A protective layer of a seed. 541

algae *AL-gee* Photosynthetic eukaryotes, including the unicellular diatoms, euglenoids, and dinoflagellates and the multicellular red, brown, and green algae. 25

alkaloids *AL-kah-loids* Plant biochemicals that are used to treat cancer and to relieve pain. 547

allele *ah-LEEL* An alternate form of a gene. 159, 220

allergens *AL-er-gens* Substances that provoke an allergic response. 525

allergy *AL-er-gee* An inappropriate response of the immune system against a nonthreatening substance, caused by IgE antibodies binding to mast cells and releasing their allergy mediators. 525

allergy mediators *AL-er-gee MEED-ee-A-terz* Biochemicals, such as histamine and heparin, that are released from mast cells when an allergen is encountered and that cause the symptoms of an allergy. 525

allopatric *AL-o-PAT-rik* Two populations that are geographically isolated from one another. 638

allopatric speciation *AL-o-PAT-rik SPE-she-A-shun* The formation of new species initiated by geographic isolation. 638

allopolyploid *AL-lo-POL-ee-ploid* An organism with multiple chromosome sets resulting from fertilization of an individual of one species by an individual of a different species. 639

allosaurs *AL-lo-SORZ* Carnivorous dinosaurs that stood upright. 659

alternation of generations *ALL-ter-NAY-shun JEN-er-AY-shunz* The existence of a gamete-producing and a spore-producing phase in the life cycle of a plant. 28, 573

altruism *AL-tru-iz-um* A behavior that harms the individual performing it but helps another organism. 692

alveolar ducts *AL-vee-O-ler DUCTS* The narrowed ending of bronchioles, opening into clusters of alveoli. 451

alveoli *AL-vee-O-li* A microscopic air sac in the lung. 451

amaranth *AM-ah-RANTH* A tall plant that can supply many types of food. 547

amino acid *ah-MEEN-o AS-id* An organic molecule built of a central carbon atom bonded to a hydrogen atom, an amino group, a carboxylic acid, and an R group. A polymer of amino acids is a peptide. 19, 52

amino acid racemization *ah-MEEN-o AS-id RACE-eh-mah-ZA-shun* A technique that measures the rate at which amino acids in biological matter alter to isomeric forms. This measurement is used in absolute dating of fossils and remains up to 100,000 years old. 650

ammonia *ah-MOAN-ee-ah* The chemical compound NH_3, which is a nitrogenous waste generated by the deamination of protein. 502

amniocentesis *AM-nee-o-cen-TEE-sis* A prenatal diagnostic procedure, performed during the fourth month of pregnancy. Fetal cells and biochemicals are sampled and then examined to reveal certain abnormalities. 180, 300

amniote egg *AM-nee-oat EGG* An egg in which an embryo could develop completely, without the requirement of being laid in water. 659

ampullae *AM-pew-li* The enlarged bases of the semicircular canals in the inner ear, lined with hair cells that detect fluid movement and convert it into action potentials. 362

amygdala *ah-MIG-dah-lah* A part of the cerebrum involved in encoding factual memory. 341

anabolism *eh-NAB-o-liz-um* Synthetic metabolic reactions, using energy. 110

anal canal *AAN-al kah-NAL* The final section of the digestive tract. 477

analgesic *AN-al-JEE-sik* A pain-relieving treatment. 459

analogous *ah-NAL-eh-ges* Structures similar in function but not in structure that have evolved in unrelated organisms in response to a similar environmental challenge. 651

anaphase *AN-ah-faze* The stage of mitosis when centromeres split and the two sets of chromosomes move to opposite ends of the cell. In anaphase of meiosis I, homologs separate. 134

anaphylactic shock *AN-ah-fah-LAK-tik SHOCK* A potentially life-threatening allergic reaction in which mast cells release allergy mediators throughout the body, causing apprehension, rash, and a closing of the throat. 528

androecium *an-DREE-see-um* The innermost whorl of a flower's corolla, consisting of male reproductive structures. 574

aneuploid *AN-you-ploid* A cell with one or more extra or missing chromosomes. 280

angina pectoris *an-GINE-ah pek-TORE-is* A gripping, viselike pain in the chest. 436

angiosperms *AN-gee-o-spermz* The flowering plants. 29

Animalia *AN-ah-MAIL-ee-ah* Kingdom including eukaryotes that derive energy from food and have nervous systems. 22

anorexia nervosa *AN-eh-REX-ee-ah ner-VO-sah* An eating disorder characterized by self-imposed starvation due to a psychological problem involving self-image. 490

antagonist *an-TAG-o-nist* A drug that binds to a receptor, blocking the docking of a neurotransmitter. 325

antagonistic muscles *an-TAG-o-NIS-tik MUS-selz* The two muscles or muscle groups that flank a movable bone and move it in opposite directions. 414

anthers *an-THERZ* Oval bodies at the tips of stamens that produce pollen. 574

anthocyanins *AN-tho-CY-ah-ninz* Pigments produced in senescent plant cells. 599

antibodies *AN-tah-BOD-eez* Proteins secreted by B cells that recognize and bind to foreign antigens, disabling them or signaling other cells to do so. 518

anticodon *AN-ti-ko-don* A three-base sequence on one loop of a transfer RNA molecule that is complementary to an mRNA codon and therefore serves to bring together the appropriate amino acid and its mRNA instructions. 258

antidiuretic hormone (ADH) (vasopressin) *AN-ti-DI-yur-RET-ik HOR-moan* A hormone made in the hypothalamus and released from the posterior pituitary that acts on the kidneys and smooth muscle cells of blood vessels to maintain the composition of body fluids. 375, 510

antigen *AN-tah-gen* The specific parts of cells or chemicals that elicit an immune response. 517

antigen binding site *AN-tah-gen BIND-ing SITE* Specialized ends of antibodies that bind specific antigens. 519

antihistamines *AN-ti-HIS-tah-meens* Drugs that decrease mucus secretion and alleviate watery eyes and sneezing. 459

antisense strand *AN-ti-sense strand* The side of the DNA double helix for a particular gene that is not transcribed into mRNA. 257

anus *A-nus* The opening to the anal canal. 477

anvil *AN-vil* One of the bones in the middle ear. 359

aorta *a-OR-tah* The largest artery that leaves the heart. 432, 435

aortic semilunar valve *a-OR-tik SEM-i-LOON-er VALVE* The valve between the left ventricle and the aorta. 435

apatosaurs *ah-PAT-o-SORZ* Huge, land-dwelling, herbivorous dinosaurs, also called brontosaurs. 659

apical meristems *A-pik-el MER-eh-STEMZ* Unspecialized cells that divide; found in plants near the tips of roots and shoots. 552

apneustic center *ap-NUS-tik CEN-ter* The part of the brain controlling the ability to take a deep breath. 458

appendicular skeleton *AP-en-DIK-u-lar SKEL-eh-ten* In a vertebrate skeleton, the limb bones and the bones that support them. 397

appendix *ap-PEN-diks* A thin tube from the cecum. 475

aqueous humor *AWK-kwee-es U-mer* A nutritive, watery fluid between the cornea and the lens of the eye that focuses incoming light rays and maintains the shape of the eyeball. 356

archaeopteryx *AR-kee-OP-ter-iks* A type of dinosaur that could fly. 659

arteries *ARE-teh-reez* Large, elastic blood vessels that leave the heart and branch into arterioles. 428

arterioles *are-TER-ee-olz* Small, elastic blood vessels that arise from arteries and lead into capillaries. 428

arthritis *arth-RI-tis* Inflammation of the joints. 401

artificial insemination *AR-teh-FISH-el in-SEM-eh-NAY-shun* Placing donated sperm in a woman's reproductive tract to start a pregnancy. 205

artificial seed *ARE-tah-FISH-al SEED* A somatic embryo placed in a transparent polysaccharide gel containing nutrients and hormones, with an outer, biodegradable polymer coat. 606

artificial selection *AR-tah-FISH-al sah-LEK-shun* Influencing the genetic makeup of a population, as occurs in agriculture and selective breeding of domesticated animals. 540, 626

ascending limb *as-SEN-ding LIM* The distal portion of the loop of Henle, which ascends from the kidney's medulla. Cells here are impermeable to water, and they actively transport Na^+ into the medullary space. 508

ascomycete *AS-ko-my-seat* A fungus with asci as sexual structures, such as the organism that causes athlete's foot. 27

asexual reproduction *A-sex-yu-al re-pro-DUK-shun* A cell's doubling its contents and then splitting in two to yield two identical cells. 42, 158

association areas *ah-SOC-ee-A-shun AIR-ee-ahs* Little-understood parts of the cerebral cortex that control learning and creativity. 337

asthma *AS-mah* Spasm of the bronchial muscles. 451

atherosclerosis *ATH-ee-ro-skle-RO-sis* The accumulation of fatty plaques inside coronary arteries. 436

atom *AT-um* A chemical unit, composed of protons, neutrons, and electrons, that cannot be further broken down by chemical means. 44

atria *A-tree-ah* The paired uppermost chambers of the heart, which receive blood returning to the heart and pump it to the ventricles below. 434

atrial natriuretic factor (ANF) *A-tree-al NAY-tre-yu-RET-ik FAK-ter* A hormone produced in the heart atria of mammals that regulates blood pressure and volume and the excretion of K^+, Na^+, and water. 383

atrioventricular node (AV node) *A-tre-o-ven-TRIK-yu-lar NOOD* Specialized muscle cells that branch into a network of Purkinje fibers, which conduct electrical stimulation six times faster than other parts of the heart. 438

atrioventricular valves *A-tree-o-ven-TRIK-ku-ler VALVZ* Flaps of tissue between each atrium and ventricle that move in response to the pressure changes accompanying the contraction of the ventricle. 434

atrophy *AH-tro-fee* Muscle degeneration resulting from lack of use or immobilization. 417

auditory nerve *AWD-eh-tore-ee NERV* Nerve fibers carrying action potentials from the cochlea in the inner ear to the cortex of the brain. 360

autoantibodies *AW-to-AN-tah-BOD-eez* Antibodies produced by an organism that attack tissue of the body, resulting in an autoimmune disease. 525

autonomic nervous system *AW-toe-NOM-ik NER-ves SIS-tum* Part of the motor pathways of the peripheral nervous system that leads to smooth muscle, cardiac muscle, and glands. 344, 439

autopolyploid *AW-toe-POL-ee-ploid* An organism with multiple chromosome sets derived from the same species. 639

autosome *AW-toe-soam* A non-sex chromosome. 225

autotroph(ic) *AW-toe-trof* An organism that manufactures nutrient molecules using energy harnessed from the environment. 22, 114

auxin *AWK-zin* A type of plant hormone that causes cell elongation in seedlings, shoot tips, embryos, and leaves. 580, 588

axial skeleton *AX-ee-al SKEL-eh-ten* In a vertebrate skeleton, the skull, vertebral column, ribs, and breastbone. 397

axil *AX-el* The regions between a leaf stalk and stem. 557

axon *AX-on* An extension from a neuron that conducts information away from the cell body toward a receiving cell. 315

B

B cells *B SELZ* A class of lymphocytes that produce antibodies. 518

bacteria *BACK-TEAR-ee-ah* Single-celled prokaryotic organisms. 22

balanced polymorphism *BAL-anced POL-ee-MORF-iz-um* A form of stabilizing selection that allows a genetic disease to remain in a population because heterozygotes enjoy a selective advantage. 637

bark *BARK* All of the tissues outside of the vascular cambium. 568

Barr body *bar BOD-ee* The dark-staining body seen in the nucleus of a cell from a female mammal, corresponding to the inactivated X chromosome. 239

basal ganglia *BASE-el GANG-lee-ah* Masses of nerve cell bodies in the cerebrum involved in forming memories required to perform certain skills. 341

basal metabolic rate *BA-sal MET-ah-BALL-ik RATE* The energy required by an organism to stay alive. 112

base *BASE* A molecule that releases hydroxide ions into water. 46

basidiomycete *bah-SID-ee-o-my-SEAT* A fungus with spore-containing basidia, including hallucinogenic mushrooms. 28

basilar membrane *BAY-seh-ler MEM-brane* The membrane beneath the hair cells in the cochlea of the inner ear that vibrates in response to sound. 360

behavioral isolation *be-HAAV-yu-ral I-so-LAY-shun* When members of two populations do not crossbreed because they perform different courtship rituals. 638

benthic zone *BEN-thik ZONE* The bottom of the ocean. 737

bicuspid valve *BI-kus-pid VALVE* The valve between the left atrium and the left ventricle. 435

bile *BILE* A substance produced by the liver and stored in the gallbladder that emulsifies fats. 472

binary fission *BI-nair-ee FISH-en* A type of asexual reproduction in which a cell divides into two identical cells. 22, 131

biofeedback *BI-o-FEED-bak* A technology giving people information on physiological processes they wish to control. 673

biogeochemical cycles *BI-o-GEE-o-KEM-ik-el SI-kelz* The pathways of chemicals between the atmosphere, the earth's crust, water, and organisms. 721

biogeography *BI-o-gee-OG-grah-fee* The physical distribution of organisms. 638

biological clock *BI-o-LOG-ik-kal CLOCK* An internal timing mechanism in an organism that controls its circadian rhythms. 600, 674

biomagnification *BI-o-MAG-nah-fah-KA-shun* The increasing concentration of a substance at higher trophic levels in a food chain. 721

biomass *BI-o-mass* The total dry weight of organisms in an area. 719

biome *BI-oam* A group of interacting terrestrial ecosystems characterized by a dominant collection of plant species, or a group of interacting aquatic ecosystems with similar salinities. 717

biosphere *BI-o-sfer* All of the parts of the earth that support life. 717

biotechnology *BI-o-tek-NAL-eh-gee* The alteration of cells or biological molecules with a specific application, including monoclonal antibody technology, genetic engineering, and cell culture. 268, 603

biotherapy *BI-o-THER-ah-pee* Use of body chemicals as pharmaceuticals. 534

biotic potential *bi-OT-ik po-TEN-shal* The maximum number of offspring an individual is physiologically capable of producing. 701

blade *BLADE* The flattened region of a leaf. 559

blastocyst *BLAS-toe-cyst* The preembryonic stage of human development when the organism is a hollow, fluid-filled ball of cells. 173

blastomere *BLAS-toe-mere* A cell in a preembryonic organism resulting from cleavage divisions. 172

blood *BLOOD* A complex fluid consisting of formed elements (blood cells and platelets) suspended in a watery plasma, in which are dissolved a variety of proteins and other biochemicals. 421

blood-brain barrier *BLOOD BRANE BARR-ee-er* Capillaries in the brain whose endothelial cells are so closely packed that many substances cannot cross from the blood to the brain tissue. 342

blood pressure *BLOOD PRESH-yur* The force exerted outward on blood vessel walls by the blood. 432

blood vessels *BLOOD VES-selz* Conduits that conduct blood throughout the body, including arteries, arterioles, capillaries, venules, and veins. 421

body *BOD-ee* The midsection of the stomach. 470

bolus *BO-lus* Food rolled into a lump by the tongue. 469

bomb calorimeter *BOMB KAL-or-IM-ah-ter* A chamber surrounded by water that is used to measure the caloric content of a food. 484

bone *BONE* A connective tissue consisting of bone-building osteoblasts, stationary osteocytes, and bone-destroying osteoclasts, embedded in a mineralized matrix infused with spaces and canals (lacunae, canaliculi, and Haversian canals). 86

Bowman's capsule *BOW-manz KAP-sul* The cup-shaped proximal end of the renal tubule which surrounds the glomerulus. 506

braced framework *BRACED FRAME-work* A skeleton built of solid structural components that are strong enough to resist pressure without collapsing. 390

brachiopods *BRAK-ee-o-PODZ* Clamlike organisms that appeared in the seas of the Cambrian period. 657

bracts *BRAKS* Floral leaves that protect developing flowers. 561

bronchi *BRON-ki* Two tubules that branch from the trachea as it reaches the lungs. 450

bronchioles *BRON-ki-olz* Microscopic branches of the bronchi within the lungs. 450

bronchitis *bron-KI-tis* Inflammation of the mucous membrane of the bronchi. 459

bryophytes *BRY-o-FIGHTS* Primitive plants that lack specialized tissues to conduct water and nutrients. 29, 571

bulimia *bu-LEEM-ee-ah* An eating disorder characterized by binging and purging. 492

bursae *BUR-si* Small packets within synovial joints that store synovial fluid, which helps to reduce friction between bones and nearby structures. 401

bursitis *bur-SI-tis* Inflammation of the bursae, possibly due to calcium deposits. 401

C

calcitonin *KAL-sah-TOE-nin* A thyroid hormone that decreases blood calcium levels. 375, 396

callus *KAL-lus* An undifferentiated white lump that grows from a cultured plant explant. 605

Calorie *CAL-o-ree* The amount of energy needed to raise the temperature of 1 kilogram of water by 1°C. 112

calyx *KA-liks* One of two outermost whorls of a flower, with no direct role in sexual reproduction. 574

Cambrian period *KAB-ree-an PER-ee-od* The time in earth history about 600 million years ago when many new types of organisms appeared. 656

canaliculi *kah-NAL-ku-LI* Narrow passageways in bone that connect spaces housing osteocytes. 393

cancer *CAN-sir* A group of disorders resulting from the loss of normal control over mitotic rate and number of divisions. 141

capacitation *cah-PASS-eh-TAY-shun* Activation of sperm cells in the human female reproductive tract. 171

capillaries *KAP-ah-lair-eez* The smallest blood vessels, with a lining one cell thick. 428

carbohydrases *KAR-bo-HI-dra-sez* Enzymes that chemically break down certain disaccharides into monosaccharides. 472

carbohydrate loading *KAR-bo-HI-drat LOAD-ing* A regimen of following a high-carbohydrate diet in the week before an endurance athletic event in an attempt to maximize muscle glycogen.

carbohydrates *CAR-bo-HIGH-drates* Compounds containing carbon, hydrogen, and oxygen, with twice as many hydrogens as oxygens. Carbohydrates include the sugars and starches. 49

carbonic anhydrase *kar-BON-ik an-HI-draze* An enzyme in red blood cells that catalyzes the conversion of carbon dioxide to carbonic acid. 457

Carboniferous period *KAR-bah-NIF-er-es PER-ee-od* The time from 345 to 275 million years ago, when the first amphibians and reptiles appeared on the land. 657

cardia *KAR-dee-ah* The neck of the stomach. 470

cardiac cycle *KAR-dee-ak SI-kel* The sequence of a contraction and relaxation that comprises the heartbeat. 438

cardiac muscle *CAR-dee-ak MUS-sel* Striated, involuntary, single-nucleated contractile cells found in the mammalian heart. 87, 405

cardioaccelerator area *KAR-dee-o-ak-SEL-er-ay-ter AIR-ee-ah* Part of the brain's vasomotor center that stimulates circulation by speeding the heart. 439

cardioinhibition area *KAR-de-o-IN-hah-BISH-un AIR-ee-ah* Part of the brain's vasomotor center that slows the heart. 439

carotenoid pigments *KAIR-et-teh-noid PIG-mentz* Yellow and orange plant pigments that become visible in autumn when chlorophyll production declines. 599

carpel *KAR-pel* Leaflike structures in a flower that enclose ovules. 574

carpels *KAR-pelz* The wrist bones. 400

carrying capacity *KARR-e-ing kah-PAS-eh-tee* The maximum number of individuals that can be supported by the environment for an indefinite time period. 705

cartilage *CAR-teh-lij* A supportive connective tissue consisting of chondrocytes embedded in collagen and proteoglycans. 84

Casparian strip *kas-PAHR-ee-an STRIP* The single layer of tightly packed cells comprising the endodermis of a plant. 564

catabolism *cah-TAB-o-liz-um* Metabolic reactions of degradation, releasing energy. 110

catecholamines *KAT-eh-KOL-ah-meenz* A class of hormones, including epinephrine and norepinephrine. 377

cecum *SEE-cum* A pouch at the entrance to the large intestine. 475

cell *SEL* The structural and functional unit of life. 40, 67

cell body *SEL BOD-ee* The central, rounded portion of a neuron from which an axon and dendrites extend. 315

cell cycle *SEL CY-kel* The life of a cell, in terms of whether it is dividing or in interphase. 131

cell membrane (plasmalemma) *SEL MEM-brane* An oily structure built of proteins embedded in a lipid bilayer, which forms the boundary of cells. 92

cell population *SEL POP-u-LAY-shun* A group of cells with characteristic proportions in particular stages of the cell cycle. 139

cell theory *SEL THER-ee* The ideas that all living matter is built of cells, cells are the structural and functional units of life, and all cells come from preexisting cells. 71

cellular respiration *SEL-u-ler RES-pir-AY-shun* Biochemical reactions involved in energy extraction in the mitochondrion. 119

cell wall *SEL WALL* A rigid boundary built of peptidoglycans in prokaryotic cells and cellulose in plant cells. 72

cementum *sah-MEN-tum* An outer layer of the tooth, anchoring it to the gum and jawbone. 469

Cenozoic era *CEN-o-ZO-ik ER-ah* The time from 65 million years ago, including the present. 661
central nervous system (CNS) *SEN-tral NER-vous SIS-tum* The brain and the spinal cord. 331
centrioles *CEN-tre-olz* Paired, oblong structures built of microtubules and found in animal cells, where they organize the mitotic spindle. 81
centromere *CEN-tro-mere* A characteristically located constriction in a chromosome. 133
cereals *SER-ee-alz* Members of the grass family Poaceae, which have seeds that can be stored for long periods of time. 541
cerebellum *SER-eh-BELL-um* A grooved area behind the brain stem and connected to the cerebrum above that receives impulses from the cerebral cortex and the peripheral nervous system and then unconsciously adjusts muscular responses so that they are smooth and coordinated. 335
cerebral cortex *sah-REE-bral KOR-tex* Gray matter comprising the outer layer of the cerebrum that integrates incoming information. 350, 458
cerebrospinal fluid *sah-REE-bro-SPI-nal FLU-id* Fluid similar to blood plasma that bathes and cushions the central nervous system. 342
cerebrum *seh-REE-brum* The higher region of the brain, controlling intelligence, learning, perception, and emotion. 337
cervical vertebrae *SER-vah-kel VER-tah-bray* Seven vertebrae in the neck. 398
cervix *SIR-viks* In the female human, the opening to the uterus. 157
chlorenchyma *klor-REN-kah-mah* Chloroplast-containing parenchyma cells. 553
chlorofluorocarbons (CFCs) *KLOR-o-FLOR-o-KAR-bunz* Compounds containing carbon, chlorine, and fluorine that destroy atmospheric ozone, which filters out ultraviolet radiation. 751
chlorophyll *KLOR-eh-fill* A green pigment used by plants to harness the energy in sunlight. 28, 81, 115
chloroplast *KLOR-o-plast* A plant cell organelle housing the reactions of photosynthesis. 81, 116
chlybrid *KLI-brid* A cell into which a chloroplast from another cell is introduced. 610
cholecystokinin (CCK) *KOL-e-sis-TOE-kah-nin* A hormone produced in the small intestine that signals the release of substances needed for fat digestion. 473
chorionic villi *KOR-ee-ON-ik VIL-i* Fingerlike projections extending from the chorion to the uterine lining. 179
choroid coat *KOR-oid KOAT* The middle layer of the human eyeball, containing many blood vessels. 354
chromatid *CRO-mah-tid* A continuous strand of DNA comprising an unreplicated chromosome or one-half of a replicated chromosome. 133
chromosome *KRO-mo-soam* A dark-staining, rod-shaped structure in the nucleus of a eukaryotic cell built of a continuous molecule of DNA, wrapped in protein. 76, 218
chyme *KIME* Semisolid food in the stomach. 470
chymotrypsin *KEE-mo-TRIP-sin* A pancreatic enzyme that participates in protein digestion in the small intestine. 472
cilia *SIL-ee-ah* Protein projections from cells. Cilia beat in unison, moving substances. 83, 104
ciliary body *SIL-ee-AIR-ee BOD-ee* A highly folded, specialized structure in the center of the choroid coat of the human eye that houses the ciliary muscle, which controls the shape of the lens. 354
ciliary muscle *SIL-e-AIR-ee MUS-sel* A muscle at the center of the choroid coat in the human eye that alters the shape of the lens. 354
circadian rhythms *sir-KA-dee-en RITH-umz* Regular, daily rhythms of particular biological functions. 600
circalunadian *SIR-kah-lu-NAY-di-an* A biological rhythm that repeats approximately every day. 677
circannual *SIR-kah-AN-u-al* A biological rhythm that repeats approximately every year. 677
classical conditioning *KLAS-ik-kal kon-DISH-on-ing* A form of learning in which an animal responds in a familiar way to a new stimulus. 673
cleavage *KLEV-ij* A period of rapid cell division following fertilization but before embryogenesis. 172
climax community *KLI-max kom-MUUN-eh-te* A community that remains fairly constant if the land and climate are undisturbed. 722
clonal propagation *KLO-nel PROP-ah-GAY-shun* Uniform plants grown from cells or protoplasts cultured in the laboratory. 607
clones *KLONZ* Genetically identical individuals. 584
closed behavior program *CLOZED bee-HAIV-yur PRO-gram* A behavior that is largely genetically determined and rigid and not easily influenced by the environment. 667
closed circulatory system *CLOZED SIR-ku-lah-TORE-ee SIS-tum* A circulatory system in which the blood is contained in blood vessels. 422
coagulation *ko-AG-u-LAY-shun* Clotting of blood. 427
coccyx *COK-six* The final four vertebrae, which are fused to form the tailbone. 399
cochlea *COKE-lee-ah* The spiral-shaped, hindmost portion of the inner ear, where vibrations are translated into nerve impulses. 359
cochlear implant *COKE-lee-ar IM-plant* A device that delivers an electronic stimulus directly to the auditory nerve, bypassing the function of hair cells to provide an awareness of sound. 361
codominant *KO-DOM-eh-nent* Alleles that are both expressed in the heterozygote. 225
codon *KO-don* A continuous triplet of mRNA that specifies a particular amino acid. 258
coelom *SEE-loam* A central body cavity in an animal. 31
coevolution *KO-ev-eh-LU-shun* The interdependence of two types of organisms for survival. 619
coleoptile *KOL-ee-OP-tile* A sheathlike structure covering the plumule in monocots. 578
collateral circulation *ko-LAT-er-al SIR-ku-LAY-shun* Rerouting of blood in the heart into different arteries following damage to the heart. 436
collecting duct *ko-LEK-ting DUCT* A structure in the kidney into which nephrons drain urine. 509
collenchyma *kol-LEN-kah-mah* Elongated, living cells that differentiate from parenchyma and support the growing regions of shoots. 553
colon *KOL-en* The large intestine. 475
color blind *KUL-er BLIND* A condition in which one or more types of cone cells in the retina are missing. The individual cannot distinguish among all colors. 358
colorectal cancer *KOL-ah-REK-tal KAN-cer* Cancer of the large intestine and rectum. 477
colostomy *ko-LOS-toe-mee* Surgery that attaches the large intestine to an opening leading to a bag worn outside the body, where fecal matter collects. 477
commensalism *kom-MEN-sah-liz-um* A symbiotic relationship where one partner benefits and the other is unaffected. 14
community *kom-MUN-nah-tee* All of the organisms in a given area. 713
compact bone *KOM-pact BONE* A layer of solid, hard bone covering spongy bone. 393
complement system *KOM-plah-ment SIS-tum* A group of proteins that assist other immune defenses. 515
complementary *kom-ple-MENT-ah-ree* The tendency of adenine to hydrogen bond to thymine and guanine to cytosine in the DNA double helix. 253
complex carbohydrates *KOM-plex kar-bo-HI-drates* The polysaccharides, which are chains of sugars. Polysaccharides include starch, glycogen, cellulose, and chitin. 50
compound *KOM-pound* A molecule consisting of different atoms. 45
compound microscope *KOM-pound MI-kro-scope* A microscope built of two lenses. 70
concentration gradient *KON-sen-TRA-shun GRAY-dee-ent* The phenomenon of ions passively diffusing from an area in which they are highly concentrated to an area where they are less concentrated. 318
concordance *KON-KOR-dance* A measure of the inherited component of a trait, consisting of the number of pairs of either monozygotic or dizygotic twins in which both members express a trait, divided by the number of pairs in which at least one twin expresses the trait. 292
conditioned stimulus *kon-DISH-ond STIM-u-lus* A new stimulus that is coupled to a familiar, or unconditioned, stimulus, so that an animal can learn an association between the two. 673
conductive deafness *kon-DUK-tiv DEF-nes* Hearing loss resulting from blocked transmission of sound through the middle ear. 360
cones *KONZ* Specialized neurons found in the central portion of the retina that detect colors. 355
cones *KONZ* The reproductive structures of pines. 583
conformation *KON-for-MAY-shun* The three-dimensional shape of a protein. 53
confusion effect *kon-FUZ-yun E-fekt* The confusion faced by a predator in the presence of a school of fish. 687
congestive heart failure *kon-JES-tiv HART FAIL-yur* A weakening of the heart, impairing circulation. 436
connective tissue *kon-NECK-tiv TISH-u* Tissues consisting of cells embedded or suspended in a matrix, including loose and fibrous connective tissues, cartilage, bone, and blood. 81
constant regions *KON-stant REE-genz* The sequence of amino acids comprising the lower portions of heavy and light antibody chains, which is very similar in different antibody types. 518
constipation *KON-stah-PAY-shun* The infrequent passage of hard feces, caused by abnormally slow movement of fecal matter through the large intestine. 476
contact inhibition *KON-tact IN-heh-BISH-un* The tendency of a cell to cease dividing once it touches another cell. 138
contest competition *KON-test KOM-pah-TISH-un* Indirect competition of individuals in a population for a limited resource, such as acquiring a territory that provides access to resources. 705
contractile vacuole *KON-tract-till VAK-u-ol* An organelle in paramecium that pumps water out of the cell. 100
convergent evolution *KON-ver-gent EV-o-LU-shun* Organisms that have evolved similar adaptations to a similar environmental challenge but are not related by descent. 625
cork cambium *KORK KAM-bee-um* The lateral meristem that produces the periderm, the outer protective covering on mature stems and roots. 568
cork cells *KORK SELZ* Waxy, densely packed cells covering the surfaces of mature stems and roots. 568

cornea *KOR-nee-ah* A modified portion of the human eye's sclera that forms a transparent curved window admitting light. 353

corolla *kah-ROLE-ah* One of two outermost whorls of a flower, with no direct role in sexual reproduction. 574

corona radiata *kah-ROAN-ah RAID-ee-AH-tah* Cells surrounding the secondary oocyte and the zona pellucida. 171

coronary arteries *KOR-eh-nair-ee AR-ter-eez* Paired arteries that diverge from the aorta and surround and enter the heart. 439

coronary circulation *KOR-eh-nair-ee SIR-ku-LA-shun* The network of blood vessels that supplies blood to the heart. 439

coronary heart disease *KOR-eh-nair-ee HART DIS-eez* Disease of the coronary arteries. 436

corpus callosum *KOR-pes ka-LAWS-um* Tracts of myelinated nerve fibers that form a bridge between the cerebral hemispheres. 340

corpus luteum *KOR-pis LU-te-um* A gland formed from an ovarian follicle from which an oocyte has recently been ovulated that produces estrogen and progesterone. 381

cortex *KOR-teks* In plants, the ground tissue that fills the area between the epidermis and vascular tissue in stems. 558

cortex *KOR-teks* The outermost part of the kidney, consisting of glomeruli, Bowman's capsules, and proximal and distal convoluted tubules of nephrons. 505

cortisol *KOR-teh-sol* A major glucocorticoid hormone produced in the adrenal cortex that helps enable the body to cope with prolonged stress. 378

cotyledons *KOT-ah-LEE-donz* Embryonic leaves in flowering plants that store energy used for germination. 562, 578

cotylosaurs *KOT-el-o-SORZ* Animals living in the early Permian period that were ancestors of the dinosaurs and modern reptiles, birds, and mammals. 659

countercurrent flow *COUNT-er-CURR-ent FLO* A system in which fluid flows in a continuous tubule in opposite directions, which maximizes the amount of a particular substance that diffuses out of the tubule. 447

countercurrent multiplier system *KAUN-ter-CUR-ent MUL-tah-PLI er SIS-tum* The movement of Na^+ and water between the limbs of the loop of Henle and the medullary space in the kidney. The concentration of Na^+ in the medullary space forces water to leave the descending limb, and it then reenters the bloodstream. 509

courtship ritual *KOURT-ship RIT-u-al* A stereotyped, elaborate, and conspicuous behavior that overcomes aggressive tendencies long enough for mating to occur. 696

covalent bond *KO-va-lent bond* The sharing of electrons between atoms. 47

cranial nerves *CRANE-e-al NERVZ* Twelve pairs of somatic nerves that arise from the brain. 344

creatine phosphate *KRE-ah-tin FOS-fate* A molecule stored in muscle fibers that can donate its high-energy phosphate to ADP to regenerate ATP.

Cretaceous period *kra-TAY-shus PER-ee-od* The time from 135 to 65 million years ago when angiosperms were abundant and the number of dinosaur species declined. 659

cretinism *KRE-tin-iz-um* A child who is physically and mentally retarded due to a thyroid gland underactive since birth.

cristae *KRIS-ty* The folds of the inner membrane of a mitochondrion along which many of the reactions of cellular respiration occur. 79, 122

critical period *KRIT-eh-kel PER-ee-od* The time during prenatal development when a specific structure can be altered by a gene or an external influence. 202

critical period *KRIT-eh-kel PER-ee-od* The time in an animal's life when it performs a particular imprinting behavior. 673

Cro-Magnons *kro-MAG-nonz* Lightweight, fine-boned, less hairy members of *Homo sapiens* who lived about 40,000 years ago in Europe and Asia. 663

crossing over *KROS-ing O-ver* The exchange of genetic material between homologous chromosomes during prophase of meiosis I. 159, 237

crossopterygians *cros-SOP-ter-REEG-ee-anz* The lobe-finned fishes, which first appeared in the Devonian period and were probably ancestral to amphibians. 657

cuticle *KU-tah-kal* A covering tissue over all of a plant except the roots. 554

cutin *KU-tin* A fatty material produced by a plant's epidermal cells that forms the cuticle. 554

cyanobacteria *si-AN-o-bak-TEAR-ee-ah* Prokaryotic organisms that contain pigments and can photosynthesize. Also called blue-green algae. 22

cybridization *SI-brid-di-ZAY-shun* The production of a cell having cytoplasm derived from two cells but containing a single nucleus. 610

cyclic ovulator *SI-klik OV-u-LAY-ter* A female mammal that undergoes a monthly cycle of fertility. 382

cyclosporin *SI-klo-SPOR-in* A fungus-derived drug that suppresses immunity and is of great value in assisting transplant recipients in accepting a new organ. 529

cystic fibrosis *SIS-tik fi-BRO-sis* An inherited condition in which excess mucus plugs up the lungs and pancreas. 460

cytokinesis *SI-toe-kin-E-sis* Distribution of cytoplasm, organelles, and macromolecules into two daughter cells in cell division. 131

cytokinins *SI-toe-KI-ninz* A class of plant hormones that promote cytokinesis (division of a cell following division of the genetic material) in seeds, roots, young leaves, and fruits. 588

cytoplasm *SI-toe-PLAZ-um* The jellylike fluid in which organelles are suspended in eukaryotic cells. 75

cytosine *SI-toe-seen* One of the two pyrimidine nitrogenous bases in DNA and RNA. 57, 253

cytoskeleton *SI-toe-SKEL-eh-ten* A framework built of arrays of protein rods and tubules found in animal cells. 75, 92

D

dark reactions *DARK re-AK-shuns* Reactions of photosynthesis that do not require light and that use the products of the light reactions (NADPH and ATP) to synthesize organic molecules. 117

day-neutral plants *DAY NU-trel PLANTZ* Plants that do not rely on photoperiod to flower. 596

dead space *DEAD SPACE* The air in the pharynx, trachea, and the upper third of the lungs, which is not used in gas exchange. 455

deciduous trees *dah-SID-u-us TREEZ* Trees that shed their leaves at the end of a growing season. 562

decomposers DEE-kom-POZ-erz Organisms that consume dead organisms and feces. 718

decongestants *DE-kon-JES-tentz* Drugs that shrink nasal membranes, easing breathing. 459

defibrillator *de-FIB-rah-LAY-ter* A device that sends an electric shock to the heart to restore a normal heartbeat. 436

degenerate *de-JEN-er-at* Different codons specifying the same amino acid. 261

dehydration synthesis *DE-hi-DRA-shun SYN-theh-sis* Formation of a covalent bond between two molecules by the loss of water. 50

dendrites *DEN-dritz* Short, branched, numerous extensions from a neuron that usually receive information from other neurons and transmit it toward the neuron cell body. 315

denitrifying bacteria *DE-ni-trah-FI-ing bak-TER-ee-ah* Bacteria that convert ammonia, nitrite, and nitrate to nitrogen gas. 721

density dependent factors *DEN-seh-tee DE-pen-dent FAK-terz* Factors that kill a greater percentage of a population as population size increases. 705

density independent factors *DEN-seh-tee IN-deh-PEN-dent FAK-terz* Factors that kill a certain percentage of a population regardless of population size, such as natural disasters. 705

dentine *DEN-tin* The bonelike material beneath a tooth's enamel. 469

deoxyhemoglobin *DE-OX-ee-HEEM-o-GLO-bin* Hemoglobin that is deep red after releasing its oxygen to tissues. 426

deoxyribonucleic acid (DNA) *de-OX-ee-RI-bo-nu-KLAY-ic AS-id* A double-stranded nucleic acid built of nucleotides containing a phosphate group, a nitrogenous base (A, T, G, or C), and the sugar deoxyribose. 57, 217

depolarized *DE-pol-er-ized* When the charge of the interior of a neuron at rest becomes less negative by the influx of Na^+. 319

dermal tissue *DER-mal TISH-u* Tissue covering a plant's body. 554

descending limb *de-SEN-ding LIM* The proximal portion of the loop of Henle, which descends into the kidney's medulla. Cells here are permeable to water, which passively diffuses into the kidney's medulla in response to the Na^+ that collects there after leaving the ascending limb. 508

desensitization *de-SEN-sah-teh-ZA-shun* Periodic injection of allergens under the skin in an attempt to plug receptors on mast cells with IgG so that allergy mediators are not released upon encountering the allergen. 526

Devonian period *deh-VOAN-ee-an PER-ee-od* The time following the Silurian period, 395 to 345 million years ago, when fishes first became abundant. 657

diabetes insipidus *DI-ah-BEE-teez IN-sip-eh-dis* A disruption of the synthesis or release of antidiuretic hormone, producing intense thirst and copious, watery urine. 375

diabetes mellitus *DI-ah-BEAT-es MEL-eh-tis* A medical condition in which the body does not produce sufficient insulin or cannot react to the insulin present. 379

diaphragm *DI-ah-fram* A broad sheet of muscle separating the thoracic cavity from the abdominal cavity. 453

diarrhea *DI-ah-REE-ah* The frequent and too-rapid passage of loose feces, caused by abnormally fast movement of fecal matter through the large intestine. 476

diastole *di-AS-toll-ee* The heart's relaxation. 438

diastolic pressure *DI-ah-stol-ik PRESH-yur* The blood pressure at its lowest, when the ventricles relax. 432

dicots *DI-kotz* Flowering plants that have two seed leaves. 557, 578

dihybrid *DI-HI-brid* An individual heterozygous for two particular genes. 222

dilution effect *dah-LU-shun E-fekt* A behavior in which ostriches with mates sit on eggs of females lacking permanent mates, which decreases the chances of their own eggs being eaten. 687

diplodocus *DIP-lo-DOE-kus* Huge, land-dwelling, herbivorous dinosaurs. 659

diploid *DIP-loid* A cell with two copies of each chromosome. 157

directional selection *dah-REK-shun-al sah-LEK-shun* When a previously prevalent characteristic of the individuals of a population is altered in response to a changing environment as the number of better-adapted individuals increases. 637

disaccharide *DI-SAK-eh-ride* A sugar built of two bonded monosaccharides, including sucrose, maltose, and lactose. 50

disruptive selection *dis-RUP-tiv sah-LEK-shun* When either of two extreme expressions of a trait are the most fit. 637

distal convoluted tubule *DIS-tel KON-vo-LU-tid TU-bule* The region of the kidney distal to the loop of Henle and proximal to a collecting duct where Na^+ is reabsorbed into the peritubular capillaries by active transport, blood pH is maintained, and wastes are secreted. 509

diverticulosis *DI-ver-TIK-ku-LO-sis* A weakening of parts of the large intestinal wall. 476

DNA hybridization *HI-brid-i-ZAY-shun* Determining the relatedness of two types of organisms by observing how rapidly separated strands of their DNA form hybrids. 654

DNA polymerase *po-LIM-er-ase* A type of enzyme that participates in DNA replication by inserting new bases and correcting mismatched base pairs. 257

DNA replication *REP-leh-KAY-shun* Construction of a new DNA double helix using the information in parental strands as a template. 253-54

dominance hierarchy *DOM-eh-nance HI-er-AR-kee* A social ranking of members of a group of the same sex, which distributes resources with a minimum of aggression. 695

dominant *DOM-eh-nent* An allele that masks the expression of another allele. 220

dormant *DOR-mant* A period of decreased metabolism that often enables an organism to survive harsh climatic conditions. 599

double-blind *DUB-el BLIND* An experimental protocol where neither the participants nor the researchers know which subjects have received a placebo and which have received the treatment being evaluated. 5

double fertilization *DUB-el FER-til-i-ZAY-shun* The fertilization of both the egg and the polar nuclei in a flowering plant. 578

duodenum *DO-o-DEE-num* The first section of the small intestine. 472

dystrophin *DIS-tro-fin* A protein comprising only 0.002% of the total protein in skeletal muscle but vital for this tissue's function. Lack of dystrophin leads to muscular dystrophy. 412

E

ecological isolation *E-ko-LOG-eh-kel I-so-LAY-shun* When members of two populations do not crossbreed because they prefer to mate in different habitats. 638

ecological succession *E-ko-LODG-ik-el suk-SESH-un* The process of change in an ecological community. 722

ecology *e-KOL-o-gee* The study of the relationships between organisms and their environments. 713

ecosystem *E-ko-SIS-tum* A unit of interaction among organisms and between organisms and their physical environments, including all living things within a defined area. 713

ecotones *E-ko-tonz* Bridges between ecosystems, such as marshes and meadows. 753

ectoderm *EK-TOE-derm* The outermost embryonic germ layer, whose cells become part of the nervous system, sense organs, outer skin layer, and its specializations. 175

ectopic pregnancy *ek-TOP-ik PREG-nan-see* The implantation of a zygote in the wall of a fallopian tube rather than in the uterus. 201

ectotherms *EK-toe-THERMZ* Animals that lose or gain heat to their surroundings by moving into areas where the temperature is suitable. 499

edema *eh-DEEM-ah* Swelling of a body part due to fluid buildup. 441

effector *E-fek-ter* A muscle or gland that receives input from a neuron. 315

efferent arterioles *EF-fer-ent are-TEAR-ee-olz* Branches of the renal artery leaving the proximal portion of a nephron. 506

electrical gradient *e-LEK-trik-el GRAY-dee-ent* The phenomenon of like charges repelling one another and opposite charges attracting one another. 318

electric discharge particle acceleration *e e-LEK-trik DIS-charge PAR-te-kel ak-SEL-er-AY-shun* A gunlike device that shoots tiny metal particles coated with DNA into cells. 612

electron *e-LEK-tron* A subatomic particle carrying a negative electrical charge, and a negligible mass, that orbits the atomic nucleus. 44

electron-spin resonance *e-LEK-tron SPIN REZ-o-nence* A technique that measures the formation of tiny holes in crystals over time, caused by exposure to ionizing radiation. This measurement is used in absolute dating of fossils up to 1 million years old. 650

electron-transport chain *ee-LEK-tron TRANZ-port CHANE* Linked oxidation-reduction reactions. 116

electroporation *ee-LEK-tro-por-AY-shun* Applying a brief jolt of electricity to open up transient holes in cell membranes, allowing foreign DNA to be introduced. 612

element *EL-eh-ment* A pure substance, consisting of atoms containing a characteristic number of protons. 44

embolus *EM-bo-lis* A blood clot that travels in the bloodstream to another location. 428

embryo *EM-bree-o* The stage of prenatal development when organs develop from a three-layered organization. 155

embryonic induction *EM-bree-ON-ik in-DUK-shun* The ability of a group of specialized cells to stimulate neighboring cells to specialize. 180

embryo sac *EM-bree-o SAK* A mature megagametophyte, containing an egg. 574

emphysema *EM-fah-ZEE-ma* Impaired breathing caused by an inherited enzyme deficiency or smoking. The lung's alveoli become overinflated and burst. 459

enamel *ee-NAM-el* The hard covering of a tooth.

endocrine glands *EN-do-crin GLANZ* Structures that secrete hormones directly into the circulatory system. 368

endocytosis *EN-doe-si-TOE-sis* The engulfing of an extracellular substance by the cell membrane. 81, 102

endoderm *EN-doe-derm* The innermost embryonic germ layer, whose cells become the organs and linings of the digestive, respiratory, and urinary systems. 175

endodermis *EN-do-DER-mis* The innermost region of a root's cortex. 564

endogenous pyrogen *en-DODGE-eh-nes PIR-o-gen* A protein secreted by some white blood cells that stimulates the hypothalamus to spike a fever. 516

endometrium *EN-doe-MEE-tree-um* The inner uterine lining. 173

endoplasmic reticulum *EN-doe-PLAZ-mik reh-TIK-u-lum* A maze of interconnected membranous tubules and sacs, winding from the nuclear envelope to the cell membrane, along which proteins are synthesized (in the rough ER) and lipids synthesized (in the smooth ER). 77

endorphins en-DORF-inz Peptides produced in the human body that influence mood and the perception of pain. 326

endoskeleton *EN-do-SKEL-eh-ten* An internal scaffolding type of skeleton in vertebrates. 390

endosperm *EN-do-sperm* A triploid tissue that provides nutrients to the embryo in a seed. 541, 578

endosymbiont theory *EN-doe-SYM-ee-ont THER-ee* The idea that eukaryotic cells evolved from large prokaryotic cells that engulfed once free-living bacteria. 87

endotherms *EN-doe-THERMZ* Animals that regulate their temperatures by using metabolic heat. 499

energy nutrients *EN-er-gee NU-tre-entz* Dietary fats, carbohydrates, and proteins. 482

energy pyramid *EN-er-gee PIR-ah-mid* A depiction of trophic levels, with each level represented by a bar whose length is proportional to the number of kilocalories available from that food for growth and development. 719

energy RDA *EN-er-gee* The recommended dietary allowance for the energy nutrients, which are carbohydrates, proteins, and fats. 485

entrainment *en-TRANE-ment* The resynchronization of a biological clock by the environment. 600, 674

environmental resistance *en-VIR-on-MEN-tal ree-SIS-tance* All factors that reduce birth rate or increase death rate in a population. 704

enzyme *EN-zime* A protein that catalyzes a specific type of chemical reaction. 54

epicotyl *EP-eh-KOT-el* The stemlike region above the cotyledons. 578

epidemiology *EP-eh-dee-mee-OL-o-gee* The analysis of data derived from real-life, nonexperimental situations. 7

epidermis *EP-eh-DER-mis* The covering on the primary plant body. 554

epididymis *EP-eh-DID-eh-mis* In the human male, a tightly coiled tube leading from each testis, where sperm mature and are stored. 155

epiglottis *EP-eh-GLOT-is* A piece of cartilage that covers the glottis, routing food to the digestive tract and air to the respiratory tract. 449

epinephrine (adrenaline) *EP-eh-NEF-rin (ah-DREN-ah-lin)* A catecholamine hormone produced in the adrenal medulla and sent into the bloodstream, where it raises blood pressure, constricts blood vessels, and slows digestion, as part of the "fight or flight" response to a threat. 377

epiphyseal plates *EP-eh-FEEZ-ee-al PLATZ* In children, thin disks of cartilage at the ends of long bones from which new growth occurs. 395

epistasis *EP-eh-STAY-sis* A gene masking another gene's expression. 231

epithelial tissue *EP-eh-THEL-e-al TISH-u* Tissue built of cells that are packed close together to form linings and boundaries. 81

epithelium *EP-eh-THEL-e-um* Cells that form linings and coverings. 81

epochs *EP-okz* Time periods within periods, which are within eras. 654

equational division *ee-QUAY-shun-el deh-VISZ-un* The second meiotic division, when four haploid cells are generated from the two haploid cells that are the products of meiosis I by a mitosislike division. 157

eras *ER-ahs* Very long periods of time of biological or geological activity. 654

erythropoietin *eh-RITH-ro-PO-eh-tin* A hormone produced in the kidneys when the oxygen supply is insufficient that stimulates red blood cell production in the red bone marrow. 426

esophagus *ee-SOF-eh-gus* A muscular tube leading from the pharynx to the stomach. 469

essential nutrients *e-SEN-shal NU-tree-entz* Nutrients that must be obtained from the diet because the body cannot synthesize them. 484

estrogen *ES-tro-gen* A hormone made in the ovaries that increases the rate of mitosis in the uterine lining. 381

estuary *ES-tu-AIR-ee* The point where the fresh water of a river meets the salty water of an ocean. 735

ethology *ee-THOL-o-gee* The study of how natural selection shapes behavior to enable an animal to survive. 668

ethylene *ETH-eh-leen* A simple organic molecule that functions as a hormone in plants, produced in large amounts by a stigma when a pollen tube begins growing. It hastens fruit ripening. 579, 588

etiolated *E-ti-o-LAY-tid* Seedlings that have abnormally elongated stems, small roots, and leaves and a pale color, because they were grown in the dark. 598

eukaryotic cell *u-CARE-ee-OT-ik SEL* A complex cell containing organelles, which carry out a variety of specific functions. 17

eurypterids *yu-RIP-ter-idz* Scorpionlike organisms that appeared in the seas of the Cambrian period. 657

eusocial *YU-sosh-al* A population of animals that communicate with each other, cooperate in caring for young, has overlapping generations, and demonstrates division of labor. 684

eutrophic *yu-TRO-fik* An aging lake, containing many nutrients and decaying organisms, often tinted green with algae. 734

evolution *Ev-eh-LU-shun* The process by which the genetic composition of a population of organisms changes over time. 621

evolutionary tree diagrams *EV-o-LU-shun-air-ee TREE DI-ah-gramz* A depiction of DNA sequence differences indicating evolutionary relationships between different types of organisms. 654

exchange system *ex-CHANGE SIS-tum* A diet planning system based on classifying foods according to their percentages of carbohydrate, protein, and fat. 487

excitatory synapse *ex-SI-TAH-tore-ee SIN-apse* A synapse across which a particular neurotransmitter travels and depolarizes the postsynaptic membrane. 324

exocytosis *EX-o-si-TOE-sis* The fusing of secretion-containing organelles, which travel to the inside surface of the cell membrane, where they transport a substance out of the cell. 79, 102

exon *EX-on* The bases of a gene that code for amino acids. 265

exoskeleton *EX-o-SKEL-eh-ten* A braced framework skeleton on the outside of an organism. 390

experimental control *ex-PEAR-eh-MEN-tel KON-trol* An extra test that does not directly address the hypothesis but can rule out causes other than the one being investigated. 5

expiration *EX-spir-AY-shun* Exhalation. 455

explants *EX-plantz* Small pieces of plant tissue grown in a laboratory dish with nutrients and plant hormones. 605

expressivity *EX-pres-SIV-eh-tee* The degree of expression of a phenotype. 227

extinction *ex-TINK-shun* The disappearance of a type of organism. 621

extinction *ex-TINK-shun* The loss of a conditioned response. 673

extracellular digestion *EX-tra-SEL-yu-lar di-JEST-shun* Dismantling of food by hydrolytic enzymes in a cavity within an organism's body. 465

extraembryonic membranes *EX-tra-EM-bree-on-ik MEM-BRANZ* Structures that support and nourish the mammalian embryo and fetus, including the yolk sac, allantois, and amnion. 168

F

facilitated diffusion *fah-SIL-eh-tay-tid dif-FU-shun* Movement of a substance down its concentration gradient with the aid of a carrier protein. 101

facultative photoperiodism *FAK-kel-TAY-tiv FO-toe-PER-ee-o-DIZ-um* Plants for which an inductive photoperiod speeds flowering. 596

fallopian tubes *fah-LO-pee-an TUBES* In the human female, paired tubes leading from near the ovaries to the uterus, where oocytes can be fertilized. 157

fall turnover *FALL TURN-o-ver* The mixing of upper and lower layers in a lake by wind, which mixes nutrients and oxygen. 734

fast twitch-fatigable fibers *FAST TWITCH fah-TEEG-ab-bel FI-berz* Skeletal muscle fibers that contract rapidly but tire easily due to scarce oxygen. 414

fast twitch-fatigue resistant fibers *FAST TWITCH fah-TEEG re-ZIS-tent FI-berz* Skeletal muscle fibers that contract rapidly, do not tire easily, and have abundant oxygen. 413

fats *FATS* Organic compounds containing carbon, hydrogen, and oxygen but with less oxygen than carbohydrates. 51

feedback loop *FEED-bak LOOP* A complex interaction between the product of a biochemical reaction and the starting material. 370

femur *FE-mer* The thigh bone. 400

fibrillation *FIB-rah-LAY-shun* Wild twitching of the heart muscle. 436

fibroblast *FI-bro-blast* A cell of connective tissue that secretes the proteins collagen and elastin. 83

fibrous root system *FI-bres ROOT SIS-tum* A plant in which the first root (the radicle) is short-lived. 563

fibula *FIB-u-lah* The smaller of the two bones of the lower leg. 400

fixed action pattern (FAP) *FIXED AK-shun PAT-ern* An innate, stereotyped behavior. 668

flagella *fla-GEL-ah* Taillike appendages on prokaryotic cells. 72, 104

flavin *FLA-vin* A yellow pigment in plants that is probably the photoreceptor for phototropism. 592

flowers *FLAU-erz* The reproductive structures of angiosperms. 574

fluid mosaic *FLU-id mo-ZAY-ik* Description of a biological membrane, referring to the arrangement of proteins embedded in the oily lipid bilayer. 97

follicle cells *FOL-ik-kel SELZ* Nourishing cells surrounding oocytes. 155

follicle stimulating hormone (FSH) *FOL-eh-kul STIM-u-la-ting HOR-moan* A hormone made in the anterior pituitary that controls oocyte maturation, the development of ovarian follicles, and their release of estrogen. 374

food chain *FOOD CHANE* A series of organisms in which one eats another. 717

food group plan *FOOD GROUP PLAN* A diet plan based on classifying foods into four groups—meat and meat substitutes, milk and milk products, fruits and vegetables, and grains. 486

food webs *FOOD WEBZ* The interconnection of food chains to form webs. 718

formed elements *FORMED EL-eh-mentz* Blood cells and platelets.

fossils *FOS-silz* Evidence of past life. 622

founder effect *FAUN-der ah-FEKT* A type of genetic drift occurring when small groups of people leave their homes to found new settlements, taking with them a subset of the original population's genes. 633

fountain effect *FOUN-ten E-fekt* The splitting in two and regrouping of a school of fish, which confuses a predator. 687

fovea centralis *FO-ve-ah cen-TRAL-is* An indentation in the retina directly opposite the lens containing only cones and important to the acuity of an animal's sight. 355

free nerve ending *FREE NERV EN-ding* A type of receptor in the skin that responds to touch. 363

free radicals *FREE RAD-eh-kelz* Highly reactive by-products of metabolism that can damage tissue. 188

fruit *FROOT* A ripened plant ovary enclosing a seed. 580

fundamental niche *FUN-dah-MEN-tel NITCH* All the places and ways in which members of a species can live. 707

fundus *FUN-dus* The domelike top of the stomach. 470

Fungi *FUN-ji* The taxonomic kingdom of eukaryotes with chitin cell walls and no nervous systems and distinguished by their reproductive structures. 22, 26

G

G_1 phase *FAZE* The stage of interphase when proteins, lipids, and carbohydrates are synthesized. 133

G_2 phase *FAZE* The stage of interphase when membrane components are synthesized and stored. 133

gallbladder *GALL-blad-er* A structure leading from the liver and toward the small intestine that stores bile. 477

gamete *GAM-eet* A sex cell. The sperm and ovum. 28

gametoclonal variation *gah-ME-toe-KLON-al VAR-ee-AY-shun* Genetically variant plantlets grown from callus initiated by sex cells. 609

gametophyte *gah-MEE-toe-fight* The part of a plant's life cycle when sex cells are manufactured. 28, 573

ganglia *GANG-lee-ah* Cell bodies of neurons. 331

ganglion cells *GANG-lee-on SELZ* Cells comprising the third layer of the retina in the human eye. 356

gastric juice *GAS-trik JUICE* The fluid secreted by stomach cells; responsible for chemical digestion there. 470

gastric lipase *GAS-trik LI-pace* A stomach enzyme that chemically digests certain lipid molecules. 470

gastrin *GAS-trin* A hormone secreted by stomach cells that stimulates more gastric juice to be secreted. 470

gastrointestinal tract *GAS-tro-in-TES-ti-nal TRAKT* A continuous tube along which food is physically and chemically digested into its constituent nutrients. 467

gene *JEAN* A sequence of DNA that specifies the sequence of amino acids in a particular polypeptide. 217

gene library *JEAN LI-brair-ee* The genome of an organism, cut into pieces that are each cultured in recombinant bacteria. 274

gene pool *JEAN PUL* All the genes in a population. 631

generative nucleus *GEN-er-rah-tiv NU-klee-us* A haploid cell resulting from the mitotic division of a microspore, in male plant reproduction. 574

genetic code *jeh-NET-ik KODE* The correspondence between specific DNA base sequences and the amino acids that they specify. 259

genetic drift *jah-NET-ik DRIFT* Changes in gene frequencies caused by the separation of a small group from a larger population. 632
genetic heterogeneity *jeh-NET-ik HET-er-o-jeh-NE-eh-tee* Different genotypes that have identical phenotypes. 231
genetic load *jah-NET-ik LOAD* The collection of deleterious alleles in a population. 634
genetic marker *jeh-NET-ik MAR-ker* A detectable piece of DNA that is closely linked to a gene of interest, whose precise location is not known. 237, 303
genome *jeh-NOME* All of the DNA in a cell of an organism. 253
genotype *JEAN-o-type* The genetic constitution of an individual. 220
geological time scale *GE-o-LODG-ik-kel TIME SKAL* A division of time into major eras of biological and geological activity, then periods within eras, and finally epochs within some periods. 654
geotropism *GEE-o-TRO-piz-um* A plant's growth response toward gravity. 592
gerontology *JER-on-TOL-o-gee* Study of the biological changes of aging at the molecular, cellular, organismal, and population levels. 187
gibberellins *JIB-ah-REL-linz* A class of plant hormones that promote cell elongation and division in seeds, roots, shoots, and young leaves. 588
globin *GLO-bin* A polypeptide chain that binds iron and forms part of a molecule of hemoglobin. 426
glomerular filtrate *glo-MER-u-ler FIL-trate* In a nephron in the kidney, the material that diffuses from the glomerulus to the Bowman's capsule. 508
glomerulus *glo-MER-u-lus* A ball of capillaries lying between the afferent arterioles and efferent arterioles in the proximal region of a nephron. 506
glottis *GLOT-is* The opening from the pharynx to the larynx. 449
glucagon *GLU-ka-gon* A pancreatic hormone that breaks down glycogen into glucose, raising blood-sugar levels. 378
glucocorticoids *GLU-ko-KOR-tah-koidz* Hormones secreted by the adrenal cortex that enable the body to survive prolonged stress. 378
glycolysis *gli-KOL-eh-sis* A catabolic pathway occurring in the cytoplasm of all cells. One molecule of glucose is split and rearranged into two molecules of pyruvic acid. 119
glycoprotein *GLY-ko PRO-teen* A molecule built of a protein and a sugar. 97
goiter GOI-ter A lump in the neck caused by a thyroid gland that swells due to lack of iodine in the diet. 376
Golgi apparatus *GOL-gee AP-ah-rah-tis* A system of flat, stacked, membrane-bound sacs where sugars are polymerized to starches or bonded to proteins or lipids. 77
gonadotropic hormones *go-NAD-o-TRO-pik HOR-moan* Hormones made in the anterior pituitary that affect the ovaries or testes. 374
gout *GOUT* An inborn error of purine metabolism in which uric acid accumulates in the joints. 502
grana *GRAN-ah* Stacks of flattened thylakoid discs comprising the inner membrane of a chloroplast. 81, 116
graptolites *GRAP-toe-litz* Organisms that lived in the Ordovician period, whose fossilized remains resemble pencil markings. 657
gray matter *GRAY MAT-ter* Cell bodies and interneurons that are not myelinated and are often involved in integration. 321
greenhouse effect *GREEN-haus E-fekt* Elevation in surface temperature of the earth by accumulation of carbon dioxide, which allows solar radiation of short wavelengths in but does not release the longer wavelength, infrared heat that the energy is converted to. 749
gross primary production *GROSS PRI-mar-ee pro-DUK-shun* The total amount of energy converted to chemical energy by photosynthesis in a certain amount of time in a given area. 717
ground tissue *GROUND TISH-u* The tissue comprising most of the primary body of a plant, filling much of the interior of roots, stems, and leaves. 552
growth factor *GROWTH FAK-ter* Locally acting proteins that assist in would healing. 137
growth rings *GROWTH RINGZ* Demarcations seen in cross sections of wood, indicating yearly growth. 566
guanine *GWAN-een* One of the two purine nitrogenous bases in DNA and RNA. 57, 253
guard cells *GUARD SELZ* Cells that control the opening and closing of stomata in plants. 554
gymnosperms *JIM-no-spermz* Plants whose sex cells are on cones. 29
gynoecium *gin-NEE-see-um* The second innermost whorl of a flower, consisting of female reproductive structures. 574

H

habitat *HAB-eh-tat* The place where an organism lives. 706
habituation *ha-BIT-ju-AY-shun* The simplest form of learning, in which an animal learns not to respond to certain irrelevant stimuli. 672
hair cells *HAIR SELZ* Mechanoreceptors in the inner ear that lie between the basilar membrane and the tectorial membrane and trigger action potentials in fibers of the auditory nerve. 360
half-life *HAF-life* The time it takes for half of the isotopes in a sample of an element to decay into the second isotopic form. This measurement is used in absolute dating of fossils. 650
hammer *HAM-er* One of the bones in the middle ear. 359
haploid *HAP-loid* A cell with one copy of each chromosome. 157
hardwoods *HARD-woodz* Woods of dicots, such as oak, maple, and ash. 566
Hardy-Weinberg equilibrium *HAR-dee WINE-berg EE-kwah-LEE-BREE-um* Maintenance of the proportion of genotypes from one generation to the next, signifying that for a particular gene, evolution is not occurring. 632
Haversian canal *hah-VER-shun kah-NAL* In bone, a central portal housing blood vessels. 393
heart *HART* A muscular pump that forces blood through conduits throughout the body. 421
heart murmur *HART MUR-mer* A sound heard in the chest when heart valves do not function normally. 434
heartburn *HART-burn* A burning sensation in the upper chest caused by acidic chyme squeezing into the esophagus. 471
heartwood *HART-wood* Wood in the center of a tree, where wastes collect. 566
heavy chain *HEV-ee CHANE* The two larger polypeptide chains comprising a Y-shaped subunit of an antibody. 518
Heimlich maneuver *HEIM-lik mah-NU-ver* A motion of pushing up and in under a choking person's rib cage, which can dislodge the item caught in the throat. 450
helper T cells *HEL-per T SELZ* Lymphocytes that produce lymphokines and stimulate the activities of other T cells and other cell types. 521
heme *HEEM* An iron-containing complex that is the oxygen-binding part of the hemoglobin molecule. 426
hemizygous *HEM-ee-ZY-gus* A gene carried on the Y chromosome in humans. 238
hemoccult test *HEM-ok-kult TEST* An examination of fecal matter for blood, which can indicate a disorder. 477
hepatic portal system *heh-PAH-tik POR-tel SIS-tum* A special division of the circulatory system that enables the liver to harness the chemical energy in digested food rapidly. 432
hepatic portal vein *heh-PAH-tik POR-tel VANE* The vein that leads to the liver. 432
hepatitis *HEP-ah-TI-tis* Inflammation of the liver, usually caused by a viral infection. 478
heterogametic sex *HET-er-o-gah-MEE-tik SEX* The sex with two different sex chromosomes, such as the human male. 245
heterotroph(ic) *HET-er-o-TROF* An organism that obtains nourishment from another organism. 22, 114
heterozygous *HET-er-o-ZI-gus* Possessing two different alleles for a particular gene. 220
hippocampus *HI-po-KAM-pes* A part of the cerebral cortex thought to be involved in forming memories. 341
histamine *HIS-tah-meen* An allergy mediator that widens blood vessels and causes certain allergy symptoms. 515
homeostasis *HOME-ee-o-STA-sis* The ability of an organism to maintain constancy of body temperature, fluid balance, and chemistry. 498
hominid *HAWM-eh-nid* Animals ancestral to humans only. 662
hominoids *HAWM-eh-noidz* Animals ancestral to apes and humans that dwelled in Africa about 20 to 30 million years ago. 661
homogametic sex *HO-mo-gah-MEE-tik SEX* The sex with two identical sex chromosomes, such as the human female. 245
homologous *ho-MOL-eh-gus* Similarly built structures in different organisms that have the same general function, indicating that they are inherited from a common ancestor. 651
homologous pairs *ho-MOL-eh-gus PAIRZ* Chromosome pairs that have the same sequence of genes. 159
homozygous *HO-mo-ZI-gus* Possessing two identical alleles for a particular gene. 220
hormone *HOR-moan* A biochemical manufactured in a gland and transported in the blood to a target organ, where it exerts a characteristic effect. 137, 368
human chorionic gonadotropin *YU-man KOR-ee-on-ik go-NAD-o-TRO-pin* A hormone secreted by the preembryo and embryo that prevents menstruation. 173, 381
human lung surfactant *HU-man LUNG sir-FAK-TANT* A mixture of phospholipid molecules that inflates alveoli in the lungs. 453
humerus *YOOM-eh-ris* The upper arm bone. 400
humoral immune response *HUME-er-al IM-mune ree-SPONZ* The secretion of antibodies by B cells in response to detecting a foreign antigen. 518
hunter-gatherer *HUN-ter GATH-er-er* A person who collects and eats native vegetation. 539
hybridoma *HI-bra-DOE-mah* An artificial cell created by fusing a B cell with a cancer cell that secretes a particular antibody indefinitely. 531
hydrocarbon *HI-dro-kar-bon* A molecule containing carbon and hydrogen. 47
hydrochloric acid *HI-dro-KLOR-ik AS-id* A strong acid found in the stomach, where it provides the pH needed to activate pepsin, which chemically digests protein. 470
hydrogen bond *HI-dro-gen bond* A weak chemical bond between negatively charged portions of molecules and hydrogen ions. 48
hydrolysis *hi-DROL-eh-sis* Splitting of a molecule in two by adding water. 50

hydrophilic *HI-dro-FILL-ik* Attraction of part of a molecule to water. 95
hydrophobic *HI-dro-FOOB-ik* Repulsion of part of a molecule from water. 95
hydrostatic skeleton *HI-dro-STAT-ik SKEL-eh-ten* The simplest type of skeleton, built of a liquid surrounded by a layer of flexible tissue. 390
hyperpolarize *HI-per-POLE-er-ize* The action of an inhibitory neurotransmitter, which causes the postsynaptic neuron's interior to become more negative than the resting potential by admitting Cl^-. 324
hypertension *HI-per-TEN-shun* Higher than normal blood pressure. 434
hyperthyroidism *HI-per-THY-roid-iz-um* A swelling in the neck (toxic goiter) caused by an overactive thyroid gland, which accelerates metabolism. 377
hypertrophy *hi-PER-tro-fee* An increase in muscle mass, possibly due to exercise. 417
hyperventilation *HI-per-VEN-tah-LAY-shun* An increased breathing rate, which decreases the level of carbon dioxide in the blood. 458
hyphae *HI-fee* Threadlike filaments that are part of the bodies of multicellular fungi. 27
hypocotyl *HI-po-KOT-el* The stemlike region below the cotyledons. 578
hypodermis *HI-po-DER-mis* The outermost, protective layer of the cortex of a plant. 564
hypoglycemia *HI-po-gly-SEEM-ee-ah* A low level of glucose in the blood, producing weakness, anxiety, and shakiness. 379
hypotension *HI-po-TEN-shun* Lower than normal blood pressure. 434
hypothalamus *HI-po-THAL-eh-mus* A small area beneath the thalamus that controls many aspects of homeostasis, including hunger, thirst, body temperature, heartbeat, water balance, blood pressure, sexual arousal, and feelings of pain, pleasure, anger, and fear. The hypothalamus links the nervous and endocrine systems. 335
hypothesis *hy-POTH-eh-sis* An educated guess, based on prior knowledge. 3
hypothyroidism *HI-po-THY-ro-diz-um* A slowing of metabolism and heartbeat and lowering of blood pressure and body temperature resulting from an underactive thyroid gland. 376

I

ichthyosaurs *IK-thee-o-SORZ* Dinosaurs that lived in the seas. 659
idiotype *ID-ee-o-TYPE* The particular parts of an antibody's antigen binding site that are complementary in conformation to the conformation of a particular antigen. 519
ileum *IL-ee-um* The last section of the small intestine. 472
imbibition *IM-bah-BISH-un* The absorption of water by a seed. 580
impact theory *IM-pakt THER-ee* The idea that a meteor or comet crashed to earth, throwing soot into the atmosphere, which blocked the sun and hampered photosynthesis, leading to the extinctions of many types of organisms. 640
imprinting *IM-print-ing* A type of learning that occurs for a limited time, usually early in an animal's life, and is performed usually without obvious reinforcement. Chicks following a parent illustrates imprinting. 673
inborn error of metabolism *IN-born ER-er Mah-TAB-o-liz-um* A disorder caused by a missing or inactive enzyme. 287
inclusive fitness *in-KLU-siv FIT-nes* A definition of fitness including personal reproductive success as well as that of relatives sharing an individual's genes. 692
incomplete dominance *IN-kim-plete DOM-eh-nance* A heterozygote whose phenotype is intermediate between the phenotypes of the two homozygotes. 225
independent assortment *IN-deh-PEN-dent ah-SORT-ment* The random arrangement of homologs during metaphase of meiosis I. 160, 223
indoleacetic acid (IAA) *IN-doe-ah-SEE-tik AS-id* The most active auxin, a type of plant hormone that stimulates growth. 589
induced ovulator *in-DEUCED OV-u-LAY-ter* Inducement of ovulation by the presence of a male. 382
industrial melanism *in-DUS-tree-al MEL-an-iz-um* An adaptive response of insects to pollution, in which coloration that is protective against a sooty background is selected. 635
inferior vena cava *in-FEAR-ee-er VE-nah kah-vah* The lower branch of the largest vein that leads to the heart. 432, 434
infertility *IN-fer-TIL-eh-tee* The inability to conceive a child after a year of trying. 196
inflammation *IN-fla-MA-shun* Increased blood flow and accumulation of fluid and phagocytes at the site of an injury, rendering the area inhospitable to bacteria. 515
inhibitory synapse *in-HIB-eh-tore-ee SIN-apse* A synapse across which a particular neurotransmitter has difficulty depolarizing the postsynaptic membrane. 324
innate *in-ATE* Instinctive. 667
inner cell mass *IN-er SEL MASS* The cells in the blastocyst that develop into the embryo. 173
inner ear *IN-ner EAR* A fluid-filled chamber that houses structures important in providing hearing and maintaining balance. 359
insect-trapping leaves *IN-sect TRAP-ing LEEVZ* Leaves of carnivorous plants that attract, capture, and digest prey. 562
insertion *in-SER-shun* The end of a muscle on a movable bone. 414
insight learning *IN-site LEARN-ing* The ability to apply prior learning to a new situation without observable trial-and-error activity. This is reasoning. 674
insomnia *in-SAWM-nee-ah* A sleep disorder in which a person has difficulty falling or remaining asleep. 676
inspiration *IN-spir-AY-shun* Inhalation. 453
insulin *IN-sel-in* A pancreatic hormone that lowers blood-sugar level by stimulating body cells to take up glucose from the blood and to metabolize or store it. 378
insulin-dependent diabetes *IN-sel-in de-PEN-dent DI-ah-BEAT-es* Diabetes mellitus resulting from insufficient insulin, usually beginning in childhood. Without injecting insulin, this condition causes extreme thirst, blurred vision, weakness, fatigue, nausea, and weight loss. 379
insulin-independent diabetes *IN-sel-in IN-de-PEN-dent DI-ah-BEAT-es* Diabetes mellitus resulting from the body's inability to utilize insulin. It usually begins in adulthood, produces fatigue, itchy skin, blurred vision, slow wound healing, and poor circulation and can often be controlled by diet, exercise, and drugs. 379
intercalary meristems *in-TER-kah-LER-ee MER-eh-stemz* Dividing tissues in grasses between mature regions of stem. 552
intercalated disks *in-TER-kah-LAY-tid DISKS* Tight foldings in cardiac muscle cell membranes that join adjacent cells. 405
interferon *IN-ter-FEAR-on* A polypeptide produced by a T cell infected with a virus that diffuses to surrounding cells and stimulates them to manufacture biochemicals that halt viral replication. 515
intermediate-day plants *IN-ter-MEED-ee-at DAY PLANTZ* Plants that flower only when exposed to days of intermediate length; growing vegetatively at other times. 596
interneuron *IN-ter-neur-on* A neuron that connects one neuron to another to integrate information from many sources and to coordinate responses. 315
internodes *IN-ter-noodz* Portions of stem between nodes. 557
interphase *IN-ter-FAZE* The period when the cell synthesizes proteins, lipids, carbohydrates, and nucleic acids. 131
interstitial cell stimulating hormone (ICSH) *IN-ter-STISH-el STIM-u-la-ting HOR-moan* A hormone made in the anterior pituitary that stimulates late development of sperm cells and the synthesis of testosterone. 375
intertidal zone *IN-ter-TI-dal ZONE* The region bordering an estuary where the tide recedes and returns. 735
intracellular digestion *IN-tra-SEL-yu-lar di-JEST-shun* Digestion within food vacuoles in cells. 465
intron *IN-tron* Bases of a gene that are transcribed but are excised from the mRNA before translation into protein. 265
inversion *in-VER-shun* A chromosome with part of its gene sequence inverted. 280
ion *I-on* An atom that has lost or gained electrons, giving it an electrical charge. 46
ionic bond *i-ON-ik bond* Attraction between oppositely charged ions. 46
iris *I-rus* The thin, opaque, colored region of the choroid coat in the human eye. 354
irritability *IR-eh-tah-BIL-eh-tee* An immediate response to a stimulus. 40
islets of Langerhans *I-lets LANG-er-hanz* Clusters of cells in the pancreas that secrete hormones controlling the body's utilization of nutrients. 378
isotope I-so-tope A differently weighted form of an element. 45

J

J-shaped curve *J SHAPED KURV* The mathematical curve resulting from plotting exponential population growth over time. 701
jaundice *JAWN-dis* An overproduction of the bile pigment bilirubin, which causes the skin to turn yellow. 478
jejunum *jah-JU-num* The middle section of the small intestine. 472
Jurassic period *jur-AS-ik PER-ee-od* The time from 185 to 135 million years ago, when dinosaurs were abundant. 659

K

karyokinesis *KAR-ee-o-kah-NEE-sus* Division of the genetic material. 131
karyotype *KAR-ee-o-type* A size-order chart of chromosomes. 298
keystone herbivore hypothesis *Ke-stone ER-bah-vor hi-POTH-eh-sis* The theory that the demise of large herbivores 11,000 years ago led to overgrowth of vegetation, which changed the environment sufficiently to kill many small herbivores. 643
kidney failure *KID-nee FAIL-yur* Damaged renal tubules, which eventually hamper the function of nephrons, resulting in the buildup of toxins in the blood. 511
kidney stones *KID-nee STONZ* Salts that precipitate out of newly formed urine and collect as solid masses in the kidney tubules or the renal pelvis. 511
kidneys *KID-neez* Paired organs built of millions of tubules responsible for excretion of nitrogenous waste and osmoregulation. 504

kilocalories *KIL-o-KAL-o-reez* The energy needed to raise one gram of water one degree Celsius. 484

kwashiorkor *KWASH-ee-OR-ker* Starvation resulting from a switch from breast milk to food deficient in nutrients. 490

L

lactic acid formation *LAK-tik AS-id for-MAY-shun* The conversion of pyruvic acid from glycolysis into lactic acid, occurring in some anaerobic bacteria and tired mammalian muscle cells. 121

lactose intolerance *LAK-tos in-TOLL-eh-rence* Digestive difficulties caused by a deficiency of the enzyme lactase. 472

larynx *LAR-inks* The "voice box" and a conduit for air. 449

latent learning *LA-tent LEARN-ing* Learning without any obvious reward or punishment; not apparent until sometime after the learning experience. 674

lateral meristems *LAT-er-al MER-eh-STEMZ* Actively dividing plant cells that grow outward, thickening the plant. 552

leaf abscission *LEAF ab-SCISZ-on* The shedding of a tree's leaves as a normal part of its life cycle. 562

lean tissue *LEEN-TISH-u* Body weight consisting of muscle, bone, connective tissue, and water. 493

learning *LEARN-ing* A change in behavior as a result of experience. 672

lens *LENZ* The structure in the eye through which light passes and is focused.

lentic system *LEN-tik SIS-tum* Fresh water biomes that have standing water, such as lakes and ponds. 733

lichen *LI-ken* An organism formed by the union of a fungus and a green alga. 16

ligaments *LIG-ah-mentz* Tough bands of fibrous connective tissue that form the joint capsule. 401

light chain *LITE CHANE* The two smaller polypeptide chains comprising a Y-shaped subunit of an antibody. 518

light reactions *LITE re-AK-shunz* The light-requiring reactions of photosynthesis that harness photon energy and use it to convert ADP to ATP. 117

limnetic zone *lim-NET-ik ZONE* The layer of open water in a lake or pond that is penetrated by light. 734

linkage *LINK-ege* The location of genes on the same chromosome. 237

lipases *LI-pay-ses* Enzymes that chemically digest fats. 472

lipid bilayer *LIP-id BI-lay-er* A two-layered structure formed by the alignment of phospholipids, reflecting their hydrophobic and hydrophilic tendencies. 95

lipids *LIP-idz* Organic molecules that are insoluble in water, including the fats. 51

littoral zone *LIT-or-al ZONE* The shallow region along the shore of a lake or pond where light can penetrate to the bottom with sufficient intensity to allow photosynthesis. 734

liver *LIV-er* The largest solid organ in the body, which detoxifies the blood, stores glycogen and fat-soluble vitamins, synthesizes blood proteins, and monitors blood-sugar level, plus other functions. 477

logistic growth curve *lo-JIS-tik GROWTH KURV* An S-shaped mathematical curve reflecting the slowing of population growth as the carrying capacity is reached, in response to environmental resistance. 705

long-day plants *LONG-day PLANTZ* Plants that require light periods longer than some critical length to flower. 596

long-term synaptic potentiation *LONG TERM sin-AP-tik PO-ten-she-A-shun* The hypothesis that long-term memory results from repeated and frequent stimulation of the same neurons, which strengthens their synaptic connections. 341

loop of Henle *LUP HEN-lee* A loop of a nephron, lying between the proximal and distal convoluted tubules, where water is conserved and the urine becomes concentrated by a countercurrent multiplier system that returns water to the blood. 508

lordosis *lor-DOE-sis* An abnormal forward curvature of the spine. 399

lotic system *LO-tik SIS-tum* Fresh water biomes with running water, such as rivers and streams. 733

lumbar vertebrae *LUM-bar VER-tah-bray* The five vertebrae in the small of the back.

lungs *LUNGZ* Paired structures that house the bronchial tree and the alveoli, the sites of gas exchange. 453

luteinizing hormone (LH) *LU-tah-ni-zing HOR-moan* A hormone made in the anterior pituitary that promotes ovulation. 375

luteinizing hormone releasing hormone (LHRH) *LU-ten-I-zing HOR-moan ree-LEAS-ing HOR-moan* A hormone sent from the hypothalamus to the anterior pituitary, where it stimulates release of follicle stimulating hormone and luteinizing hormone. 381

lymph *LIMF* Blood plasma minus some large proteins, which flows through lymph capillaries and lymph vessels. 441

lymphatic system *lim-FAH-tik SIS-tum* A circulatory system consisting of lymph capillaries and lymph vessels that transports lymph, which consists of blood plasma minus large proteins. 441

lymph capillaries *LIMF CAP-eh-LAIR-eez* Dead-ended, microscopic vessels that transport lymph. 441

lymph nodes *LIMF NOODZ* Structures in the lymphatic system that contain white blood cells and protect against infection. 441

lymphocytes *LIM-fo-SITZ* White blood cells that provide immune protection, including the B cells, which secrete antibodies, and the T cells, which secrete lymphokines and control the activities of each other and other cell types. 517

lymphokines *LIM-fo-KINES* Biochemicals secreted by lymphocytes that attack cancer cells or cells coated with viruses. 517

lymph vessels *LIMF VES-selz* Vessels that transport lymph and eventually empty into veins. 441

lysosome *LI-so-soam* A sac in a eukaryotic cell in which molecules and worn-out organelles are enzymatically dismantled. 80

M

macroevolution *MAK-ro-ev-eh-LU-shun* Large-scale evolutionary changes, such as speciation and extinction. 621

macronutrients *MAK-ro-NU-tri-entz* The carbohydrates, fats, and proteins that are obtained from food and required in large amounts. 482

macrophages *MAK-ro-FAH-ges* Very large, wandering phagocytic cells. 515

maiasaurs *MI-ah-SORZ* Duck-billed dinosaurs that lived in Montana about 80 million years ago. 659

major histocompatibility complex (MHC) *MA-jer HIS-toe-kum-PAT-ah-BIL-eh-tee KOM-plex* A family of genes in mammals that specifies cell surface proteins involved in cell-cell recognition. 518

map sense *MAP SENS* A sense that tells an organism its location relative to its home. 681

marasmus *mah-RAS-mus* Starvation due to profound nutrient deficiency. 490

marrow cavity *MAR-o KAV-eh-tee* A space in the shaft of a long bone housing fatty yellow marrow. 393

marsupials *mar-SU-pee-alz* Pouched mammals. 661

mast cells *MAST SELZ* Large cells that are burst by allergens binding to IgE on their surfaces, releasing allergy mediators that cause allergy symptoms. 515

maximum intrinsic rate of increase *MAX-ah-mum in-TRIN-sik RATE IN-krees* The rate of growth of a population when each member produces as many offspring as is possible and the environment does not restrict reproduction. 701

medulla *mah-DUEL-ah* The middle portion of the kidney, consisting of loops of Henle and collecting ducts of nephrons. 505

medulla *mah-DULE-ah* The part of the brain stem closest to the spinal cord; regulates such vital functions as breathing, heartbeat, blood pressure, and certain reflexes. 334

megagametophytes *MEG-ah-gah-MEE-toe-fightz* The large, female egg-producing gametophytes in a plant. 573

megakaryocyte *MEG-ah-KAR-ee-o-site* A huge bone marrow cell that breaks apart to yield platelets. 427

megasporangia *MEG-ah-spor-AN-gee-ah* Structures in which megaspores form. 573

megaspore mother cell *MEG-ah-spor MOTH-er SEL* A cell within an ovule that divides meiotically to produce four haploid cells, three of which degenerate. 574

megaspores *MEG-ah-sporz* Structures in plants that give rise to the female gametophytes. 573

meiosis *Mi-O-sis* Cell division resulting in a halving of the genetic material. 131

melanocyte stimulating hormone (MSH) *mah-LAN-o-site STIM-u-lat-ing HOR-moan* A hormone produced in between the anterior and posterior lobes of the pituitary in some vertebrate species that controls skin pigmentation. 375

melatonin *MEL-ah-TOE-nin* A hormone produced by the pineal gland that may control other hormones by a sensitivity to lightness and darkness. 382, 677

memory cells *MEM-or-ee SELZ* Mature B cells that are specific to an antigen already encountered and respond quickly by secreting antibodies when that antigen is encountered subsequently. 518

meninges *MEN-in-gees* A triple layer of membranes that covers and protects the central nervous system. 342

mesentery *MEZ-en-tear-ee* An epithelial sheet that supports digestive structures in the human.

mesoderm *MEZ-o-derm* The middle embryonic germ layer, whose cells become bone, muscle, blood, dermis, and reproductive organs. 175

messenger RNA (mRNA) *MESS-en-ger* A molecule of ribonucleic acid that is complementary in sequence to the sense strand of a gene. 76, 258

metabolism *meh-TAB-o-liz-um* The biochemical reactions that acquire and utilize energy. 109

metacarpals *MET-ah-KAR-pelz* Bones of the hand. 400

metaphase *MET-ah-faze* The second stage of cell division, when chromosomes align down the center of a cell. In mitosis the chromosomes form a single line. In meiosis I, the chromosomes line up in homologous paris. 134

metastasis *meh-TAH-STAH-sis* The spreading of cancer from its site of origin to other parts of the body. 145

metazoan *MET-ah-ZO-en* An ancient, simple multicellular animal ancestral to modern animals. 31

mibrid *MI-brid* A cell into which a mitochondrion from another cell is introduced. 610

microevolution *MIKE-ro-ev-eh-LU-shun* The more subtle, incremental single-trait changes that underlie speciation. 621
microfilaments *MI-kro-FILL-ah-ment* Tiny rods built of actin found within cells, especially contractile cells. 104
microgametophytes *MIKE-ro-gah-MEE-toe-fightz* The small, male sperm-producing gametophytes in a plant. 573
micronutrients *MIKE-ro-NU-tree-entz* The vitamins and minerals that are obtained from food and required in small amounts. 482
microsporangia *MI-kro-spor-AN-gee-ah* Structures in which microspores form. 573
microspore mother cells *MI-kro-spor MOTH-er SELZ* Cells in pollen sacs that divide meiotically to produce four haploid microspores. 574
microspores *MI-kro-sporz* Structures in plants that give rise to the male gametophytes. 573
microtubules *MI-kro-TO-bules* Long, hollow tubules, built of the protein tubulin, that provide movement within cells. 81, 104
microvilli *MI-kro-VIL-i* Tiny projections on the surfaces of epithelial cells, which comprise intestinal villi. 473
midbrain *MID-brane* The part of the brain stem above the pons where white matter connects with higher brain structures and gray matter contributes to sight and hearing. 335
middle ear *MID-el EAR* The part of the ear consisting of three bones, the hammer, anvil, and stirrup, that transmit and amplify sound. 359
mineralocorticoids *MIN-er-rel-KOR-tah-KOIDZ* Hormones produced in the adrenal cortex that maintain blood volume and electrolyte balance by stimulating the kidneys to return Na^+ and water to the blood and to excrete K^+. 378
mitochondria *MI-toe-KON-dree-ah* Organelles within which the reactions of cellular metabolism occur. 79
mitosis *mi-TOE-sis* A form of cell division in which two genetically identical cells are generated from one. 131
molecular evolution *mo-LEK-yu-ler EV-o-LU-shun* The tracing of sequence differences in proteins and nucleic acids between living species to establish degrees of evolutionary relatedness. 652
molecule *MOLL-eh-kuel* A structure resulting from the combination of atoms. 45
Monera *mo-NER-ah* The taxonomic kingdom including the bacteria and the cyanobacteria. 22
monoclonal antibodies *MON-o-KLON-al AN-tah-BOD-eez* Antibodies descended from a single B cell and therefore are identical. B cells are fused with cancer cells to create hybridomas, which are artificial cells that secrete a particular antibody indefinitely. 531
monocots *MON-o-kotz* Flowering plants that have one seed leaf. 557, 578
monogamy *mah-NAUG-o-mee* The formation of a permanent male-female pair. 696
monohybrid *MON-o-HI-brid* An individual heterozygous for a particular gene. 221
monosaccharide *MON-o-SAK-eh-ride* A sugar built of one 5- or 6-carbon unit, including glucose, galactose, and fructose. 49
monosomy *MON-o-soam-ee* A cell missing one chromosome. 280
monotremes *MON-o-tremz* Egg-laying mammals. 661
morula *MORE-u-lah* The preembryonic stage of a solid ball of cells. 172
mosaic catastrophism *mo-ZAY-ik CAT-as-TROF-iz-um* A variation of Neptunism that holds that a series of great floods molded the earth's features and caused various extinctions and speciations. 621
motor areas *MO-ter AIR-ee-ahs* Parts of the cerebral cortex that send impulses to skeletal muscles. 337
motor (efferent) neuron *MO-ter (EF-fer-ent) NEUR-on* A neuron that transmits a message from the central nervous system toward a muscle or gland. It has a long axon and short dendrites. 315
motor pathways *MO-ter PATH-wayz* Nerve tracts in the peripheral nervous system that carry impulses from the central nervous system to muscles or glands. 342
motor unit *MOW-ter U-nit* A nerve cell and all of the muscle fibers it contacts. 411
mucigel MUUS-eh-gel A slimy, lubricating substance produced by cells of a root cap. 563
muscle fasciculi MUS-sel fah-SIK-u-li Bundles of muscle fibers. 406
muscle spindles MUS-sel SPIN-delz Receptors in skeletal muscle that monitor the degree of muscle tone or how many fibers are contracted at a given time. 416
muscle tone MUS-sel TONE The contraction of some fibers in skeletal muscle at any given time. 416
muscular tissue *MUS-ku-lar TISH-u* Tissue built of contractile cells, providing motion. 81
mutant MU-tent A phenotype or allele that is not the most common for a certain gene in a population. 220
mutant selection *MU-tant sah-LEK-shun* Searching for genetic variants that offer a desired characteristic. 609
mutation *mu-TAY-shun* A change in a gene or chromosome. 232, 261
mutualism *MU-tu-al-iz-um* A symbiotic relationship in which both partners benefit. 14
mycelium *MI-seal-ee-um* An assemblage of hyphae in a fungus. 27
mycorrhizae *MI-ko-RI-zee* A mutualistic association between a plant's roots and fungi that absorb nutrients from soil. 564
myelin sheath *MI-eh-lin SHEATH* A fatty material that insulates some nerve fibers in vertebrates, allowing rapid transmission of nerve impulses. 320
myofibrils *MI-o-FI-brilz* Cylindrical subunits of a muscle fiber. 407
myofilaments *MI-o-FILL-eh-mentz* Actin or myosin "strings" that comprise myofibrils. 407
myosin *MI-o-sin* A type of protein comprising the thick myofilaments of skeletal muscle cells. 407
myxedema *MIX-eh-DEEM-ah* Lethargy, dry sparse hair, and a puffy face, caused by an underactive thyroid beginning in adulthood. 376

N

naked-gene hypothesis *NA-kid JEAN hi-POTH-eh-sis* The theory that nucleic acids evolved before proteins. 62
narcolepsy *NAR-co-LEP-see* A sleep disorder in which a person suddenly falls asleep during the day. 676
nasal conchae *NAZ-al KON-chi* Three shelflike bones in each nasal cavity that partition it into channels. 448
nastic movements *NAS-tic MOVE-mentz* Nondirectional plant motions. 594
natural killer cells *NAT-chu-ral KILL-er SELZ* Lymphocytes that cause infected cells and possibly cancer cells to burst.
natural products chemists *NAT-u-ral PROD-uks KEM-ists* Chemists who examine biochemicals for substances with therapeutic or other value. 547
natural selection *NAT-rul sah-LEK-shun* The differential survival and reproduction of organisms whose genetic traits better adapt them to a particular environment. 626
nautiloids *NAWT-eh-loidz* Organisms that appeared in the seas of the Cambrian period. 657
Neanderthals *nee-AN-der-thalz* Members of *Homo sapiens* who lived about 75,000 years ago in Europe and Asia. 663
negative feedback *NEG-ah-tiv FEED-bak* The turning off of an enzyme's synthesis or activity caused by accumulation of the product of the reaction that the enzyme catalyzes. 111
negative feedback loop *NEG-ah-tiv FEED-bak LOOP* A biochemical pathway in which accumulation of a product inhibits earlier reactions. 370
negative reinforcement *NEG-eh-tiv REE-in-FORC-ment* A painful stimulus given to encourage an animal to avoid a particular behavior. 673
neonatology *NE-o-nah-TOL-eh-gee* The study of the newborn. 204
neoteny *ne-OT-eh-nee* Retaining juvenile features of an ancestral species. 669
nephron *NEF-ron* A microscopic, tubular subunit of a kidney, built of a renal tubule and peritubular capillaries. 505
Neptunism *NEP-tune-iz-um* The idea that a single great flood organized the features of the earth's surface present today. 621
neritic zone *NER-it-ik ZONE* The coastal region of an ocean. 737
nerve fiber *NERVE FI-ber* An axon. 315
nervous tissue *NER-vis TISH-u* A tissue whose cells (neutrons and neuroglia) form a communication network. 81
net primary production *NET PRI-mar-ee pro-DUK-shun* The energy left over for growth and reproduction after an animal has eaten. 717
neural (synaptic) integration *NEUR-el (sin-AP-tik) IN-tah-GRAY-shun* The summing of incoming inhibitory and excitatory messages by a neuron. This information determines whether or not an action potential will occur. 324
neural tube *NEUR-el TUUB* The embryonic precursor of the central nervous system. 181
neuron *NEUR-on* A nerve cell, consisting of a cell body, a long "sending" projection called an axon, and numerous "receiving" projections called dendrites. 86
neurosecretory cells *NUR-o-SEK-rah-tore-ee SELZ* Cells in the hypothalamus that function as neurons at one end but like endocrine cells at the other by receiving neural messages and secreting the hormones ADH and oxytocin. 372
neurotransmitter *NEUR-o-TRANZ-mit-er* A chemical passed from a nerve cell to another nerve cell or to a muscle or gland cell, relaying an electrochemical message. 86, 321
neutron *NEW-tron* A particle in an atom's nucleus that is electrically neutral and has one mass unit. 44
neutrophils *NU-tro-FILLZ* Short-lived phagocytic white blood cells that help combat the initial stages of infection. 515
niche *NITCH* The ways in which an organism interacts with the living and nonliving environment. 706
nitrifying bacteria *NI-trah-FI-ing bak-TER-ee-ah* Bacteria that convert ammonia from dead organisms to nitrite. 721
nitrite bacteria *NI-trit bak-TER-ee-ah* Bacteria that convert nitrite to nitrate.
node of Ranvier *NODE RON-vee-ay* A short region of exposed axon between adjacent Schwann cells on neurons of the peripheral nervous systems of vertebrates. 321
nodes *NOODZ* Areas of leaf attachment. 557

nondisjunction *NON-dis-JUNK-shun* The unequal partition of chromosomes into gametes during meiosis. 280

nonessential nutrients *NON-ee-SEN-shal NU-tree-entz* Nutrients found in food but that can also be synthesized in the body. 484

norepinephrine (noradrenaline) *NOR-EP-eh-NEF-rin (NOR-ah-DREN-ah-lin)* A catecholamine hormone produced in the adrenal medulla and sent into the bloodstream, where it raises blood pressure, constricts blood vessels, and slows digestion, as part of the "fight or flight" response to a threat. 377

notochord *NO-toe-kord* A semirigid rod running down the length of an animal's body, 33, 181

nucleoid *NEW-klee-oid* The part of a prokaryotic cell where the DNA is located. 72

nucleolus *new-KLEE-o-lis* A structure within the nucleus where RNA nucleotides are stored. 76, 134

nucleotide *NEW-klee-o-tide* The building block of a nucleic acid, consisting of a phosphate group, a nitrogenous base, and a five-carbon sugar. 57, 252

nucleus (atomic) *NEW-klee-is* The central region of an atom, consisting of protons and neutrons. 44

nucleus (cellular) *NEW-klee-is* A membrane-bound sac in a eukaryotic cell that contains the genetic material. 22

nutrient *NU-tri-ent* A substance that is obtained from food and used in an organism to promote growth, maintenance, and repair of tissues. 482

nutrient dense *NU-tri-ent DENSE* Foods that offer a maximum amount of nutrients with a minimum number of kilocalories. 487

nyctinasty *NIK-tah-NAS-tee* A nastic (nondirectional) movement caused by daily rhythms of light and dark. 595

O

obese *o-BESE* A person who is 20% above "ideal" weight based on population statistics considering age, sex, and build. 492

obligate photoperiodism *OB-lah-get FO-toe-PER-ee-o-DIZ-um* Plants that will not flower unless they are exposed to the correct photoperiod. 596

oceanic zone *O-she-AN-ik ZONE* The deep, open part of an ocean. 737

olfactory bulb *ol-FAK-tore-ee BULB* A part of the brain that relays a message from olfactory receptor cells. 350

olfactory receptor cells (epithelium) *ol-FAK-tore-ee re-CEP-ter SELZ* Neurons specialized to detect odors, found in a small patch of tissue high in the nostrils. 350

oligotrophic *OL-ah-go-TRO-fik* A lake with few nutrients, usually a sparkling blue. 734

ommatidia *O-mah-TID-ee-ah* The visual units comprising a compound eye. 352

oncogene *ON-ko-jean* A gene that normally controls cell division but when overexpressed leads to cancer. 145

oocyte *O-o-site* The female sex cell before it is fertilized. 155

oogenesis *O-o-GEN-eh-sis* The differentiation of an egg cell, from a diploid oogonium, to a primary oocyte, to two haploid secondary oocytes, to ootids, and finally, after fertilization, to a mature ovum. 165

open behavior program *O-pen bee-HAIV-yur PRO-gram* A behavior that is flexible and easily altered by learning. 667

open circulatory system *O-pen SIR-qu-lah-TORE-ee SIS-tum* A circulatory system in which blood is not always contained in blood vessels. 422

operant conditioning *OP-er-aunt kon-DISH-on-ing* Trial-and-error learning, in which an animal voluntarily repeats any behavior that brings success. 673

operculum *o-PER-ku-lum* A flap of tissue protecting gills. 447

optic nerve *OP-tik NERV* Nerve fibers leading from the retina to the visual cortex of the brain. 356

Ordovician period *OR-do-VEESH-ee-an PER-ee-od* The period following the Cambrian period, 500 to 435 million years ago. 657

organ *OR-gan* A structure built of two or more tissues that functions as an integrated unit. 40

organelles *OR-gan-nellz* Specialized structures in eukaryotic cells that carry out specific functions. 17, 22

organelle transfer *OR-gan-el TRANZ-fer* Engineering combinations of nuclei and organelles in plant cells not seen in nature. 610

origin *ARE-eh-jen* The end of a muscle on an immobile bone. 414

osmoreceptors *OZ-mo-ree-CEP-terz* Specialized cells in the hypothalamus that sense the concentration of water in the blood. 375

osmosis *oz-MO-sis* Passive diffusion of water. 100

ossification *OS-eh-feh-KAY-shun* The process by which cartilage in embryonic bones is replaced with bone tissue. 394

osteocytes *OS-tee-o-sitz* Mature bone cells. 393

osteoporosis *OS-tee-o-por-O-sis* Bones that break easily because calcium is removed from them faster than it is replaced, possibly caused by hyperthyroidism. 377

otolith *O-toe-lith* Calcium carbonate granules in the vestibule of the inner ear whose movements provide information on changes in velocity. 363

outcrossing *OUT-cross-ing* The transfer of pollen grains from one flower to the stigma of another flower. 574

oval window *O-vel WIN-dow* A membrane between the middle ear and inner ear. 359

ovaries *O-var-ees* The paired, female gonads, which house developing oocytes. 155, 374

ovary *O-var-ree* In a flowering plant, the carpels and the ovules they enclose. 574

over dominant *O-ver DOM-en-nant* A heterozygote that is more vigorous that either homozygote. 226

ovulation *OV-u-LAY-shun* The release of an oocyte from the largest ovarian follicle just after the peak of luteinizing hormone in the blood in the middle of a woman's menstrual cycle. 381

ovules *OV-yulz* In flowering plants, a megasporangium containing a megaspore mother cell, which undergoes meiosis to produce four cells, three of which degenerate. 574

oxidation *OX-en-DAY-shun* A chemical reaction in which electrons are lost. 48

oxygen debt *OX-eh-gen DET* The body's need for oxygen to complete the metabolism of lactic acid following heavy exercise that has temporarily shifted metabolism to the anaerobic pathway. 410

oxyhemoglobin *OX-ee-HEEM-o-GLO-bin* Hemoglobin that is bright red because it has just picked up oxygen in the lungs. 426

oxytocin *OX-eh-TOE-sin* A hormone made in the hypothalamus and released from the posterior pituitary that stimulates muscle contraction in the mammary glands and the uterus. 375

P

pacemaker *PACE-may-ker* Specialized cells in the wall of the right atrium that set the pace of the heartbeat. Also called the sinoatrial node. 438

Pacinian and Meissner's corpuscles *pah-SIN-ee-en MICE-nerz KOR-pus-elz* Receptors in the skin that provide information on touch. 363

paleontologist *PAY-lee-on-TOL-ah gist* A scientist who studies evidence of past life. 640, 648

palisade mesophyll *PAL-eh-sade MEZ-o-fil* Long, columnar cells along the upper side of a leaf specialized for light absorption. 561

pancreas *PAN-kre-as* A structure that has an endocrine component, which produces somatostatin, insulin, and glucagon, and a digestive component, which produces a pancreatic juice containing trypsin, chymotrypsin, pancreatic amylase, pancreatic lipase, and nucleases. 477

parasitism *PAR-eh-sah-TIZ-um* A symbiotic relationship where one partner benefits from the other, while harming it. 15

parasympathetic nervous system *PAR-ah-SIM-pah-THE-tik NER-ves SIS-tum* Part of the autonomic nervous system that controls vital functions such as respiration and heart rate when at rest. 344, 439

parathyroid glands *PAR-ah-THY-roid GLANZ* Four small groups of cells embedded in the thyroid gland that secrete a hormone that releases calcium from bones and enhances calcium absorption through the digestive tract and kidneys, actions that regulate calcium level in blood and tissue fluid. 377, 396

parathyroid hormone *PAR-ah-THY-roid HOR-moan* The hormone produced by the parathyroid glands; regulates calcium level in the blood and tissue fluid by releasing calcium from bones and enhancing calcium absorption through the digestive tract and kidneys. 377

parenchyma *pah-REN-kah-mah* Abundant, unspecialized plant cells that can divide. 553

passive immunity *PAS-siv im-MUNE-eh-tee* Immunity generated by an organism's receiving antibodies manufactured by another organism. 522

patella *pah-TEL-lah* The bony part of the kneecap. 400

pectoral girdle PEC-tor-al GIR-del The two clavicles and two scapulae bones that form the shoulders. 400

pedigree *PED-eh-gree* A chart showing the relationships of relatives and which ones have a particular trait. 297

pelagic zone *pah-LA-gik ZONE* The water above the ocean floor. 737

pelvic girdle *PEL-vik GIR-del* The two hipbones. 400

pelvis *PEL-vis* The innermost portion of the kidney, which stores urine. 505

pelycosaurs *PEL-eh-ko-SORZ* The sailed lizards, which were distant ancestors of mammals. 659

penetrance *PEN-eh-trance* The percentage of individuals inheriting a genotype who express the corresponding phenotype. 227

pepsin *PEP-sin* A stomach enzyme that chemically digests protein. 470

pepsinogen *pep-SIN-o-jen* A precursor molecule that is split to yield pepsin, a stomach enzyme that chemically digests protein. 470

peptidase *PEP-tah-daze* A type of intestinal enzyme that completes protein digestion. 472

peptide bond *PEP-tide BOND* A chemical bond between two amino acids resulting from dehydration synthesis. 53, 262

peptide hormone *PEP-tide HOR-moan* A hormone composed of amino acids. It is water soluble but fat insoluble so cannot traverse a cell's membrane. Instead, it binds to a cell surface receptor and triggers a second messenger. 369

pericardium *PEAR-eh-KAR-dee-um* The tough, connective tissue sac enclosing the heart. 434
pericarp *PEAR-ah-karp* A protective layer of a seed. 541
perichondrium *PEAR-eh-KON-dree-um* A layer of connective tissue surrounding embryonic, cartilaginous bones. 394
pericycle *PEAR-ah-SI-kel* A ring of parenchyma cells in a root's cortex that produces branch roots that burst through the cortex and epidermis and into the soil. 564
periderm *PEAR-ah-derm* The outer protective covering on mature stems and roots. 568
periodontal membrane *PEAR-ah-DON-tal MEM-brane* An outer layer of the tooth, anchoring it to the gum and jawbone. 469
periods *PER-ee-odz* Time periods within eras. 654
peripheral nervous system (PNS) *per-RIF-er-al NER-vous SIS-tum* Nerves and cell bodies (ganglia) that transmit information to and from the central nervous system. 331
peristalsis *pear-eh-STAL-sis* Waves of muscle contraction along the digestive tract that propel food. 468
peritubular capillaries *pear-eh-TUUB-yu-lar CAP-eh-LAIR-eez* The capillaries that surround the renal tubule of the nephron, in the kidney. 505
permafrost *PER-mah-frost* The permanently frozen part of the ground in the tundra. 733
Permian period *PER-mee-an PER-ee-od* The time from 275 to 225 million years ago, when many species became extinct. Amphibians and reptiles were still abundant. 658
peroxisome *PER-ox-eh-soam* A membrane-bound sac budded off of the smooth ER housing enzymes important in oxygen utilization. 81
petals *PET-alz* Large and often colorful parts of flowers that sometimes help to attract pollinators. 574
petiole *PET-ee-ol* The stalklike portion of a leaf. 559
phagocytes *FAG-o-sitez* Scavenger cells that engulf and digest foreign cells. 515
phalanges *fah-LAN-gees* The finger bones. 400
pharynx *FAHR-inks* The throat. 449, 469
phelloderm *FEL-ah-derm* Living parenchyma cells in secondary growth. 568
phenocopy *FEEN-o-KOP-ee* An environmentally caused trait that resembles an inherited trait. 232
phenotype *FEEN-o-type* The observable expression of a genotype in a specific environment. 220
pheromones *FER-eh-moanz* Biochemicals secreted by an organism that stimulate a physiological or behavioral response in another individual of the same species. 385
phlebitis *flah-BI-tis* Inflammation of a vein wall. 438
phloem *FLO-um* Plant tissue that transports water and food materials. 29, 555
phospholipid *FOS-fo-LIP-id* A molecule built of a lipid and a phosphate that is hydrophobic at one end and hydrophilic at the other end. 94
photolysis *fo-TOL-eh-sis* A photosynthetic reaction in which electrons from water replace electrons lost by chlorophyll a. 116
photoperiodism *FO-toe-PER-ee-o-DIZ-um* A plant's ability to measure seasonal changes by the length of day and night. 596
photophosphorylation *FO-toe-FOS-for-eh-LAY-shun* A photosynthetic reaction in which energy released by the electron transport chain linking the two photosystems is stored in the high-energy phosphate bonds of ATP. 116
photoreceptors *FO-toe-ree-CEP-terz* Neurons that detect light by means of pigment molecules in contact with sensitive membranes. 351
photosynthesis *FO-toe-SIN-the-sis* The series of biochemical reactions that enable plants to harness the energy in sunlight to manufacture nutrient molecules. 24, 81, 113
photosystem *FO-toe-SIS-tum* A cluster of pigment molecules that enable green plants to absorb, transport, and harness solar energy. 116
phototropism *FO-to-TRO-piz-um* A plant's growth towards unidirectional light. 592
pH scale *SKALE* A measurement of how acidic or basic a solution is. 47
phylogenies *fi-LAWG-ah-nees* The evolutionary relationships between organisms. 648
phytochrome *FI-toe-krom* A pale blue plant pigment that exists in two interconvertible forms. The active form promotes flowering of long-day plants and inhibits flowering of short-day plants. 597
pica *PI-kah* A compulsive disorder in which people consume huge amounts of nonnutritive substances. 488
piloerection *PIL-o-ee-REK-shun* The raising of fur or feathers, which traps a layer of air near the body surface, serving as insulation. 500
pilus *PILL-us* A projection from a bacterial cell that transfers genetic information. 24
pineal gland *pin-EEL GLAND* A small oval structure in the brain, near the hypothalamus, that produces melatonin, a hormone that regulates the activities of other hormones, possibly by a sensitivity to patterns of lightness and darkness. 382, 677
pioneer species *PI-o-neer SPE-shez* The first species to colonize an area, such as lichens and mosses that begin soil formation. 722
pistil *PIS-til* The female reproductive structures and their covering in a flower. 574
pith *PITH* Ground storage tissue in the center of the stem in plants having concentric cylinders of xylem and phloem. 558
pituitary *pah-TU-eh-TEAR-ee* A pea-sized gland in the head. The anterior lobe releases growth hormone, thyroid stimulating hormone, adrenocorticotropic hormone, prolactin, and the gonadotropic hormones. The hypothalamus sends antidiuretic hormone and oxytocin to the posterior pituitary, from where they are released. Melanocyte stimulating hormone is secreted from the region of the pituitary between the lobes in some vertebrates. 371
pituitary dwarfism *pah-TU-eh-TEAR-ee DWARF-is-um* Short stature caused by a deficiency of growth hormone in childhood. 373
pituitary giant *pah-TU-eh-TEAR-ee GI-ant* A very tall child whose height results from a tumor that produces excess growth hormone. 374
placebo *pla-SEE-bo* A substance similar in taste and appearance to a substance under investigation, whose effects are known. 4
placenta *pla-CEN-tah* A specialized organ that develops in certain mammals, connecting mother to unborn offspring. 33, 168
placental mammals *plah-CEN-tel MAM-malz* Mammals that nurture the young in the female's body for a relatively long time, where they are nurtured by an organ called a placenta. 661
Plantae *PLAN-tye* Kingdom including land-dwelling multicellular organisms that extract energy from sunlight and have cell walls built of cellulose. 22
plant embryo *PLANT EM-bree-o* A plant after only a few cell divisions, forming part of a seed. 578
plasma *PLAZ-ma* A watery, protein-rich fluid that is the matrix of blood. 83, 424
plasma cells *PLAZ-mah SELZ* Mature B cells that secrete vast quantities of a single antibody type. 518
plasmid *PLAZ-mid* A small circle of double-stranded DNA found in some bacteria in addition to their DNA, commonly used as a vector for recombinant DNA. 270
platelet *PLATE-let* A cell fragment that is part of the blood and orchestrates clotting. 83, 424
plate tectonics *PLATE tek-TAWN-iks* A geological theory that views the earth's surface as several rigid plates that can move. 641
pleiotropic *PLY-o-TRO-pik* A genotype with multiple expressions. 231
Pleistocene overkill hypothesis *PLEIS-toe-seen O-ver-kill hi-POTH-eh-sis* The theory that the disappearance of many species of herbivores in North America 11,000 years ago was caused by humans hunting them. 642
plumule *PLU-mule* The epicotyl plus the first leaves of a young plantlet. 578
polar body *POLE-er BOD-ee* A small cell generated during female meiosis, enabling much cytoplasm to be partitioned into just one of the four meiotic products, the ovum. 165
polarized *PO-ler-ized* The state of a biological membrane when the electric charge inside differs from the electric charge outside. 317
polar nuclei *PO-lar NU-klee-i* The two nuclei in a cell of a plant's megagametophyte. 574
pollen grains *POL-en GRAANZ* Male microgametophytes. 574
pollen sacs *POL-en SAKS* The four microsporangia in an anther, the male part of a flower. 574
pollination *POL-eh-NA-shun* The transfer of pollen from an anther to a receptive stigma. 574
polygamy *pol-IG-ah-mee* A mating system in which a member of one sex associates with several members of the opposite sex. 696
polygyny *pol-IJ-ah-nee* A mating system in which one male mates with several females. 696
polymer *POL-eh-mer* A long molecule built of similar subunits. 50
polyploidy *POL-ee-PLOID-ee* A condition in which a cell has one or more extra sets of chromosomes. 278, 639
pons *PONZ* An oval mass in the brain stem where white matter connects the medulla to higher brain structures and gray matter helps control respiration. 335
population *POP-u-LAY-shun* Any group of interbreeding organisms. 631
population bottleneck *POP-u-LAY-shun BOT-el-nek* A type of genetic drift resulting from an event that kills many members of a population, and the numbers are restored by mating among a small number of individuals, restricting the gene pool compared to the original population. 633
population growth curve *POP-u-LAY-shun GROWTH CURVE* Description of the growth of a group of cells, influenced by nutrient and space availability and waste removal. Includes lag, log, stationary, and decline phases. 139
positive feedback loop *PAHZ-eh-tiv FEED-bak LOOP* A biochemical pathway in which accumulation of a product stimulates its production. 370
positive reinforcement *POS-eh-tiv REE-in-FORC-ment* A reward given for performing a particular behavior. 673
positron emission tomography (PET) *POS-eh-tron ee-MISH-on tah-MOG-rah-fee* A scanning technology that reveals biochemical activity in a living organism. 340

postsynaptic cell *POST-sin-AP-tik SEL* One of two adjacent neurons that are receiving a message. 321

Precambrian era *pre-KAB-ee-an ER-ah* The earliest part of earth history, from which few fossils are known. 654

presynaptic cell *PRE-sin-AP-tik SEL* One of two adjacent neurons that are transmitting a message. 321

primary body *PRI-mer-ee BOD-ee* A plant's axis, consisting of a root and a shoot. 552

primary consumers *PRI-mar-ee kon-SU-merz* Herbivores, which consume primary producers. 718

primary growth *PRI-mer-ee GROWTH* Lengthening of a plant due to cell division in the apical meristems. 552

primary immune response *PRI-mar-ee IM-mune ree-SPONZ* The immune system's response to its first encounter with a foreign antigen. 518

primary motor cortex *PRI-mare-ee MO-ter KOR-tex* A band of cerebral cortex extending from ear to ear across the top of the head that controls voluntary muscles. 337

primary nutrient deficiencies *PRI-mar-ee NU-tri-ent dee-FISH-en-seez* Too little of a particular nutrient due to an inadequate diet. 487

primary producers *PRI-mar-ee pro-DUCE-erz* Organisms that can use inorganic materials and energy to produce the organic material they require. These organisms form the first trophic level of food chains. 717

primary sensory cortex *PRI-mare-ee SEN-sore-ee KOR-tex* A band of cerebral cortex extending from ear to ear across the top of the head that receives sensory input from the skin. 337

primary structure *PRI-mer-ee STRUK-sure* The amino acid sequence of a protein. 53

primary succession *PRI-mar-ee suk-SESH-un* The arrival of life in an area where no community previously existed. 722

primary tissues *PRI-mer-ee TISH-yuz* Groups of cells with a common function. 552

primitive streak *PRIM-eh-tiv STREEK* The pigmented band along the back of a 3-week embryo that develops into the notochord. 181

principle of competitive exclusion *PRIN-sah-pul of kom-PET-ah-tiv ex-KLU-shun* The observation that two competing species will not continue to coexist indefinitely. 706

principle of superposition *PRIN-sah-pul SU-per-po-ZISH-un* The fact that lower rock layers are older than those above them. 621

prions PRI-onz Infectious protein particles. 43

profundal zone *pro-FUN-dal ZONE* The deep region of a lake or pond where light does not penetrate. 734

progeria *pro-JER-ee-ah* An inherited, accelerated aging disease. 188

progesterone *pro-JES-ter-own* A hormone produced by the ovaries that controls secretion patterns of other hormones involved in reproduction. 381

prokaryotic cell *pro-CARE-ee-OT-ik SEL* A structurally simple cell, lacking organelles. 22

prolactin *pro-LAK-tin* A hormone made in the anterior pituitary that stimulates milk production. 374

pronuclei *pro-NU-kle-eye* The genetic packages of gametes. 172

prophase *PRO-faze* The first stage of cell division, when chromosomes condense and become visible. During prophase of meiosis I, synapsis and crossing over occur. 134

prostaglandins *PROS-tah-GLAN-dinz* Lipid molecules that are released locally and transiently at the site of a cellular disturbance. They control a variety of body functions and are not well understood. 384

prostate *Pros-STATE* A male gland that produces a milky, alkaline fluid that activates sperm. 164

protein *PRO-teen* A long molecule built of amino acids bonded to each other. 19, 52

proteinoid theory *PRO-teh-noid THER-ee* The idea that proteins evolved before nucleic acids. 60

Protista *pro-TEES-tah* The taxonomic kingdom including the simplest eukaryotes—the protozoans, algae, water molds, and slime molds. 22

proton *PRO-ton* A particle in an atom's nucleus carrying a positive charge and having one mass unit. 44

protoplast fusion *PRO-toe-plast FU-zhun* The creation of new types of plants by combining their cells, from which the cell walls have been removed, and then regenerating a mature plant hybrid from the fused cell. 604

protozoans *PRO-toe-ZO-anz* Single-celled eukaryotes, often classified by their mode of movement, including the familiar amoeba, euglena, and paramecium. 25

proximal convoluted tubule *PROX-eh-mel KON-vo-LU-tid TU-bule* The region of the nephron proximal to Bowman's capsule where selective reabsorption of useful components of the glomerular filtrate occurs. 508

psilophytes *SIL-o-FIGHTS* Plants that first ventured onto the land, during the Silurian period 435 to 395 million years ago. 657

pulmonary artery *PULL-mo-NAIR-ee AR-ter-ee* The artery leading from the right ventricle to the lungs. 435

pulmonary semilunar valve *PULL-mo-NAIR-ee SEM-i-LOON-er VALVE* The valve leading from the right ventricle to the pulmonary artery. 435

pulmonary veins *PULL-mo-NAIR-ee VANEZ* Four veins leading from the lungs to the left atrium of the heart. 435

pulp *PULP* The soft inner portion of a tooth, consisting of connective tissue, blood vessels, and nerves. 469

punctuated equilibrium *PUNK-tu-A-tid EE-kwah-LEE-BREE-um* The view that evolution is characterized by long periods of relatively little change interrupted by bursts of rapid evolutionary change.

pupil *PU-pull* The opening in the iris that admits light into the human eye. 354

purine *PURE-een* A type of organic molecule with a double ring structure, including the nitrogenous bases adenine and guanine. 253

Purkinje fibers *per-KIN-gee FI-berz* Muscle fibers that branch from the atrioventricular node and transmit electrical stimulation rapidly. 438

Purkinje system *per-KIN-gee SIS-tum* The sinoatrial node, atrioventricular node, and Purkinje fibers, constituting a network of muscle fibers that triggers contraction of the ventricles. 439

pyloric stenosis *pi-LOR-ik stah-NO-sis* A birth defect in which the pyloric sphincter, which lies between the stomach and the small intestine, fails to open. 471

pylorus *pi-LOR-is* The bottom of the stomach. 470

pyramid of biomass *PIR-ah-mid BI-o-mass* A depiction of trophic levels indicating weight of organisms. 719

pyramid of numbers *PIR-ah-mid NUM-berz* A depiction of the number of organisms at each trophic level in a food chain. 719

pyrimidine *pie-RIM-eh-deen* A type of organic molecule with a single ring structure, including the nitrogenous bases cytosine, thymine, and uracil. 253

Q

quaternary structure *QUAR-teh-nair-ee STRUK-sure* The number and arrangement of polypeptide chains of a protein. 53

quiescent center *kwi-ES-cent CEN-ter* A reservoir of cells behind the root cap that can replace damaged cells in the adjacent meristem. 563

R

radicle *RAD-eh-sil* The first root to emerge from a seed. 563, 578

radiometric dating *RAD-ee-o MET-rik DA-ting* Using the measurement of natural radioactivity as a clock to date fossils. 650

radius *RAY-dee-is* One of the two lower arm bones. 400

realized niche *RE-ah-lized NITCH* The environment in which a species actually lives, which is restricted from its fundamental niche by competition with other species. 707

receptor *re-CEP-ter* A protein protruding from a cell membrane that forms a dock for other molecules. 321

receptor potential *re-CEP-ter po-TEN-shel* A change in membrane potential in a neuron specialized as a sensory receptor, caused by the redistribution of ions, whose magnitude varies with the strength of the stimulus. 349

recessive *re-CESS-ive* An allele whose expression is masked by the activity of another allele. 220

recombinant DNA technology *re-KOM-bah-nent DNA tek-NAL-eh-gee* Transferring a gene from a cell of a member of one species to the cell of a member of a different species. 270, 610

Recommended Dietary Allowances (RDAs) *REK-o-MEN-ded DI-ah-TEAR-ee ah-LAOW-ance* Guidelines on healthy foods to eat issued by the Unites States government every 5 years. 485

rectum *REC-tum* A storage region leading from the large intestine. 477

red blood cell (erythrocyte) *RED BLOOD SEL* A disc-shaped cell, lacking a nucleus, that is packed with the oxygen-carrying molecule hemoglobin. 83, 424

red marrow *RED MAR-o* Immature blood cells and platelets that reside in cavities in spongy bone. 393

reduction *re-DUK-shun* A chemical reaction in which electrons are gained. 48

reduction division *re-DUK-shun dah-VISH-un* Meiosis I, when the diploid chromosome number is halved. 157

reflex *RE-flex* A rapid, involuntary response to a stimulus from within the body or from the outside environment. 331

reflex arc *RE-flex ARK* A neural pathway linking a sensory receptor and an effector, such as a muscle. 333

relative date *REL-ah-tiv DATE* An estimate of the time at which an organism lived based upon the location of its fossils in sedimentary rock. 650

releaser *rah-LEAS-er* The specific factor that triggers a fixed action pattern, also called a sign stimulus. 669

releasing hormone *re-LEES-ing HOR-moan* A hormone produced by the hypothalamus that influences the secretion of a hormone by another gland. 371

renal tubule *REE-nel TU-bule* The tubule portion of a nephron, along which toxins are added and nutrients recycled to the blood, forming urine. 505

renin *REN-in* A hormone that elevates the level of aldosterone, an adrenal hormone that enhances reabsorption of Na^+ in the kidney, salivary glands, sweat glands, and large intestine. 510

residual air *reh-ZID-u-el AIR* The air in the bottom third of the lungs, which is not exchanged with each breath. 455

respiratory chain *RES-pir-ah-TOR-ee CHANE* A series of electron-accepting enzymes embedded in the inner mitochondrial membrane. 123

resting potential *REST-ing po-TEN-shel* The electrical potential (-65 millivolts) on the inside of a neuron not conducting a nerve impulse. 317

restriction enzyme *re-STRIK-shun EN-zime* A bacterial enzyme that cuts DNA at a specific sequence. 270

restriction fragment length polymorphism *re-STRIK-shun FRAG-ment LENGTH POL-e-MORF-iz-um* Differences in restriction enzyme cutting sites between individuals. 303

reticular activating system (RAS) *rah-TIK-u-lar AK-tah-vay-ting SIS-tem* A diffuse network of cell bodies and nerve tracts extending through the brain stem and into the thalamus; screens sensory input approaching the cerebrum. 335

retina *RET-nah* A sheet of photoreceptors at the back of the human eye. 353, 355

retinal *RET-in-al* The pigment portion of rhodopsin, a molecule involved in black-and-white vision. 357

rhinoviruses *RI-no-vi-rus-ez* Viruses that cause the common cold. 459

rhodopsin *ro-DOP-sin* A pigment molecule stored in the rod cells of the retina in the eye. Light splits rhodopsin, which depolarizes the rod cell, provoking a nerve impulse. 357

rib cage *RIB KAGE* The part of the axial skeleton that protects the heart and lungs, consisting in the human of 10 paris of ribs attached to the sternum, plus two floating ribs. 399

ribonucleic acid (RNA) *RI-bo-nu-KLAY-ik AS-id* A single-stranded nucleic acid built of nucleotides containing a phosphate, ribose, and nitrogenous bases adenine, guanine, cytosine, and uracil. 253

ribosomal RNA (rRNA) *RI-bo-SOAM-el* RNA that, along with proteins, comprises the ribosome. 258

ribosome *RI-bo-soam* A structure built of RNA and protein upon which a gene's message (mRNA) anchors during protein synthesis. 72, 258

rods *RODZ* Specialized neurons clustered around the edges of the retina that provide black-and-white vision and night vision. 355

root apical meristem *ROOT AP-eh-kel MER-eh-stem* Dividing tissue at root tips in a plantlet. 578

root cap *ROOT KAP* A thimble-shaped protective structure covering the tip of a root. 563

root hairs *ROOT HAIRZ* Trichomes that appear near root tips and absorb water and minerals from soil. 554

rosettes *ro-ZETTZ* Nonelongated stems, such as in banana. 557

round dance *ROUND DANC* A dance in which bees communicate to members of the colony that food is nearby and is of a certain sweetness. 692

rugae *RU-guy* Folds in the mucosa of the stomach. 470

S

S phase *FAZE* The synthesis phase of interphase, when DNA is replicated and microtubules are produced from tubulin. 133

saccule *SAK-yul* A pouch in the vestibule of the inner ear filled with a jellylike fluid and lined with hair cells, containing calcium carbonate otolith granules that move in response to changes in velocity, firing action potentials. 363

sacrum *SAY-krum* The five fused pelvic vertebrae. 398

saliva *sah-LI-va* A secretion in the mouth that is produced when food is smelled or tasted. 468

salivary amylase *SAL-eh-var-ee AM-eh-lase* An enzyme produced in the mouth that begins the chemical digestion of starch into sugar. 468

salivary glands *SAL-eh-vare-ee GLANDZ* Three pairs of glands near the mouth that secrete saliva, a fluid containing water, mucus, and salivary amylase. 468

saltatory conduction *SAL-teh-tore-ee kon-DUK-shun* The jumping of an action potential between nodes of Ranvier in myelinated nerve axons. 321

sapwood *SAP-wood* Wood located nearest the vascular cambium, which transports water and dissolved nutrients within a plant. 568

sarcolemma *SAR-ko-LEM-ah* The cell membrane of a skeletal muscle cell. 407

sarcoplasmic reticulum *SAR-ko-PLASZ-mik rah-TIK-u-lum* The endoplasmic reticulum of a skeletal muscle cell. 407

saturated (fat) *SAT-yur-ray-tid* A triglyceride with single bonds between the carbons of its fatty acid tails. 52

savanna *s ah-VAN-ah* A grassland. 729

scales *SKALZ* A subunit of a pinecone bearing two ovules. 583

scavengers *SKAHV-en-gerz* Animals that eat the leftovers of another animal's meal. 718

school *SKUL* A formation of fishes swimming at particular distances and angles from each other. 687

Schwann cells *SCHWAN SELZ* Fatty cells that wrap around neurons in the peripheral nervous system, forming myelin sheaths. 320

scientific method *SI-en-TIF-ik METH-id* A systematic approach to interpreting observations, involving reasoning, predicting, testing, and drawing conclusions, which are put into perspective with existing knowledge. 3

sclera *SKLER-ah* The outermost, white layer of the human eye. 353

sclereids *SKLER-ridz* Plant cells with a gritty texture found in pears and in the hulls of peanuts. 553

sclerenchyma *sklah-REN-kah-mah* Elongated supportive plant cells with thick, nonstretchable secondary cell walls. 553

scoliosis *SKOL-ee-O-sis* An abnormal spinal curvature in which the vertebrae shift sideways. 399

scotopsin *sco-TOP-sin* The protein portion of rhodopsin, a pigment molecule important in black-and-white vision. 357

scramble competition *SKRAM-bel KOM-pah-TISH-un* Direct competition of individuals in a population for a limited resource. 705

seasonal affective disorder (SAD) *SEE-son-al AF-fek-tiv DIS-or-der* A form of depression that occurs mostly in the winter and seems to respond to therapy of exposure to light. 677

seasonal ovulator *SEE-son-al OV-u-LAY-ter* A female mammal that has a period of sexual receptivity and fertility. 382

secondary consumers *SEK-on-DAIR-ee kon-SU-merz* Animals that eat herbivores, forming a third trophic level in a food chain. 718

secondary growth *SEK-on-DER-ee GROWTH* Thickening of a plant due to cell division in lateral meristems. 552

secondary immune response *SEK-on-DAIR-ee IM-mune ree-SPONZ* The immune system's response to subsequent encounters with a foreign antigen. 518

secondary nutrient deficiencies *SEK-on-dare-ee NU-tri-ent dah-FISH-en-seez* Too little of a particular nutrient due to an inborn metabolic condition. 488

secondary production *SEK-on-DAIR-ee pro-DUK-shun* The energy stored in the tissues of herbivores and carnivores. 718

secondary structure *SEK-en-DAIR-ee STRUCK-sure* The shape assumed by a protein caused by chemical attractions between amino acids that are close together in the primary structure. 53

secondary succession *SEK-on-DAIR-ee suk-SESH-un* The arrival of new species in an area that already has life. 723

second messengers *SEK-ond MESS-en-gerz* Biochemicals that modulate neurotransmitter action. 325

secretin *sah-KREE-tin* A hormone produced in the small intestine that triggers the release of bicarbonate from the pancreas, which neutralizes stomach acid. 473

seed *SEED* A temporarily dormant sporophyte individual surrounded by a tough protective coat. 578

seed coat *SEED COAT* A tough outer layer protecting a dormant plant embryo and its food supply in a seed. 579

segmentation *SEG-men-TA-shun* Localized muscle contractions in the small intestine that provide mechanical digestion. 472

segregation *SEG-rah-GAY-shun* The distribution of alleles of a gene into separate gametes during meiosis. 219

seismonasty *SIZ-mo-NAS-tee* A nastic (nondirectional) movement resulting from contact or mechanical disturbance. 594

selectively permeable *sah-LEK-tiv-lee PERM-ee-ah-bul* A biological membrane that admits only some substances. 317

semicircular canals *SEM-ee-SIR-ku-ler kah-NALZ* Fluid-filled structures in the inner ear that provide information on the position of the head. 362

semidwarf rices *SEM-i DWARF RI-ses* Highly productive varieties of rice. 545

semilunar valves *SEM-i-LOON-er VALVZ* A ring of tissue flaps in the arteries just outside each ventricle that maintains unidirectional blood flow. 434

seminal vesicles *SEM-en-el VES-eh-kels* In the human male, the paired structures that add fructose and prostaglandins to the sperm. 155

sense strand *SENSE STRAND* The side of the DNA double helix for a particular gene that is transcribed. 257

sensory adaptation *SEN-sore-ee ah-DAP-TAY-shun* The phenomenon of a sensation becoming less noticeable once it has been recognized. 349

sensory (afferent) neuron *SEN-sore-ee (AF-fer-ent) NEUR-on* A neuron that brings information toward the central nervous system, with long dendrites that transmit the message from the stimulated body part to the cell body near the spinal cord, and a short axon. 315

sensory areas *SEN-sore-ee AIR-ee-ahs* Parts of the cerebral cortex that receive and interpret messages from sense organs concerning temperature, body movement, pain, touch, taste, smell, sight, and sound. 337

sensory (neural) deafness *SEN-sore-ee (NEUR-al) DEF-nes* Hearing loss resulting from an inability of the cochlea to generate action potentials in response to detecting sound. 360

sensory pathways *SEN-sore-ee PATH-wayz* Nerve tracts in the peripheral nervous system that transmit impulses from a stimulus to the central nervous system. 342

sensory receptor *SEN-sore-ee re-CEP-ter* A specialized dendrite of a neuron that is specific to detecting a particular sensation and firing an action potential in response, which is transmitted to the spinal cord. 333

sepals *SEE-pelz* Leaflike structures that enclose and protect inner floral parts. 574

severe combined immune deficiency *sah-VEER kom-BIND im-MUNE dah-FISH-en-see* An inborn deficiency of T and B cells. 525

sex chromosome *SEX KRO-mo-soam* A chromosome that determines sex. 225

sex hormones *SEX HOR-moanz* Hormones that provide secondary sexual characteristics and prepare an animal for sexual reproduction, such as estrogen, progesterone, and testosterone. 378

sex-influenced inheritance *SEX IN-flu-enced in-HAIR-eh-tence* An allele that is dominant in one sex but recessive in the other. 225

sex-limited trait *SEX LIM-eh-tid TRAIT* A trait affecting a structure or function of the body that is present in only one sex. 225

sex-linked *SEX LINKED* A gene located on the X chromosome or a trait that results from the activity of such a gene. 238

sex ratio *SEX RAY-she-o* The ratio of males to females at conception (primary), birth (secondary), and 10-year intervals thereafter (tertiary). 246

sexual dimorphism *SEX-u-al di-MOR-fiz-um* The difference in appearance between males and females of the same species. 696

sexual reproduction *SEX-u-el RE-pro-DUK-shun* The combination of genetic material from two individuals to create a third individual. 42, 158

sexual selection *SEX-u-el sah-LEK-shun* Natural selection of traits that increase an individual's reproductive success. 627

shoot apical meristem *SHOOT AP-eh-kel MER-eh-stem* Dividing tissue at the tip of a shoot in a seedling. 578

short-day plants *SHORT-day PLANTZ* Plants that require light periods shorter than some critical length to flower. 596

sieve cells *SIV SELZ* Less specialized conducting cells in phloem. 555

sieve plate *SIV PLATE* End walls of aligned sieve tubes in a plant's phloem. 555

sieve pores *SIV PORZ* Perforations in phloem, through which solutes move from cell to cell. 555

sieve tube members *SIV TUUB MEM-berz* More complex and specialized conducting cells in phloem that form long sieve tubes. 555

Silurian period *sah-LUR-ee-an PER-ee-od* The period following the Ordovician period, 435 to 395 million years ago, when organisms first ventured onto the land. 657

simple carbohydrates *SIM-pel KAR-bo-HIGH-drates* Monosaccharides and disaccharides. 50

sinoatrial node (SA node) *SI-no-A-tree-al NOOD* Specialized cells in the wall of the right atrium that set the pace of the heartbeat. Also called the pacemaker. 438

skeletal muscle *SKEL-eh-tel MUS-sel* Voluntary, striated muscle consisting of single, multinucleated cells that are contractile due to sliding filaments of actin and myosin. 87, 405

skull *SKULL* A hard, bony structure protecting the brain. 397

sleep apnea *SLEEP AP-nee-ah* A sleep disorder in which breathing stops several hundred times a night, for 20 to 60 seconds each time.

sliding filament model *SLY-ding FILL-eh-ment MOD-el* The movement of protein myofilaments past each other to shorten skeletal muscle cells, leading to muscle contraction. 409

slow twitch-fatigue resistant fibers *SLO TWITCH fah-TEEG re-ZIS-tent FI-berz* Skeletal muscle fibers that contract slowly but are resistant to fatigue because of a plentiful supply of oxygen. 413

smooth muscle *SMOOTH MUS-sel* Involuntary, nonstriated contractile tissue found lining the digestive tract and other organs. 87, 405

sodium-potassium pump *SO-dee-um po-TAS-ee-um PUMP* A mechanism that uses energy released from splitting ATP to transport Na^+ out of cells and K^+ into cells. 317

softwoods *SOFT-woodz* Woods of gymnosperms, such as pine, spruce, and fir. 566

solution *so-LU-shun* A homogenous mixture of a substance (the solute) dissolved in water (the solvent). 98

somaclonal variation *SOAM-ah-KLON-al VAR-ee-AY-shun* Genetically variant embryos or plantlets grown from callus initiated by somatic cells. 608

somatic cell *so-MAT-ik SEL* A body cell; a cell other than the sperm or ovum. 131, 157

somatic embryo *so-MAT-ik EM-bree-o* A plant embryo grown from callus. 605

somatic hybrid *so-MAT-ik HI-brid* A plant regenerated from a protoplast fusion of cells from two types of plants. 605

somatic nervous system *so-MAT-ik NER-ves SIS-tum* Part of the motor pathways of the peripheral nervous system that leads to skeletal muscles. 342

somatostatin *so-MAH-toe-STAH-tin* A pancreatic hormone that controls the rate of nutrient absorption into the bloodstream. 378

Southern blotting *SOU-thern BLOT-ting* Use of DNA probes to identify specific fragments of DNA. 302

speciation *SPE-she-AY-shun* The appearance of a new type of organism. 621

species *SPE-shez* A group of similar individuals that interbreed in nature and are reproductively isolated from all other such groups. 19, 621

sperm *SPERM* The male sex cell. 155

spermatogenesis *sper-MAT-o-JEN-eh-sis* The differentiation of a sperm cell, from a diploid spermatogonium, to primary spermatocyte, to two haploid secondary spermatocytes, to spermatids, and finally to mature spermatozoa. 163

sphincters *SFINK-terz* Muscular rings that control the passage of a substance from one area to another. 470

sphygmomanometer *SFIG-mo-mah-NOM-eh-ter* A gauge that measures blood pressure by the displacement of a column of mercury. 432

spinal cord *SPI-nal KORD* A tube of neural tissue extending from the base of the brain to just below the lowest rib that carries impulses to and from the brain. 331

spinal nerves *SPI-nal NERVZ* Thirty-one pairs of somatic nerves that exit the spinal cord and emerge from between the vertebrae. 344

spinal reflex *SPI-nel RE-flex* A neural connection made entirely within the spinal cord. 333

spindle apparatus *SPIN-del AP-ah-RAH-tis* A structure built of microtubules that aligns and separates chromosomes in cell division. 133

spines *SPINZ* Leaves modified to protect plants from predators and excessive sunlight. 561

spleen *SPLEEN* An organ located in the abdomen that produces and stores lymphocytes and contains reserve supplies of red blood cells. 442

spongy bone *SPON-gee BONE* Flat bones and tips of long bones that have many large spaces between a web of bony struts. 393

spongy mesophyll *SPON-gee MEZ-o-fil* Irregularly shaped chlorenchyma cells separated by large spaces that are found below the palisade layer in leaves. 561

spontaneous generation *spon-TAY-nee-us JEN-er-RAY-shun* The idea, proven untrue, that living things can arise from nonliving matter. 59

sporophyte *SPOR-o-fight* The part of a plant's life cycle when spores are produced. 28, 573

spring turnover *SPRING TURN-o-ver* The rising of nutrient-rich lower layers of a lake and sinking of oxygen-rich layers from the top, often causing algal blooms. 734

stabilizing selection *STA-bil-I-zing sah-LEK-shun* When extreme phenotypes are less adaptive than an intermediate phenotype. 637

stable isotope tracing *STA-bel I-so-toap TRAC-ing* A technique that analyzes the proportions of certain isotopes in tissue samples, providing clues to which types of organisms consume others. 719

stamens *STA-menz* Male reproductive structures in flowers built of stalklike filaments bearing pollen-producing anthers at their tips. 574

stegosaurs *STEG-ah-SORZ* Herbivorous dinosaurs with panels down their backs. 659

stem cell *STEM SEL* A cell that divides often. 139

steroid hormone *STAIR-oid HOR-moan* A hormone composed of lipid that can pass through the target cell's membrane and enter the cell's nucleus. 369

stigma *STIG-mah* A pollen receptacle at the tip of a style in a flower. 574

stirrup *STIR-up* One of the bones in the middle ear. 359

stolons *STOL-onz* Stems that grow along the soil surface; also called runners. 558

stomata *sto-MAH-tah* Pores in a plant's cuticle through which water and gases are exchanged between the plant and the atmosphere. 554

storage leaves *STOR-age LEEVZ* Fleshy leaves that store nutrients. 562

stress test *STRESS TEST* An electrocardiogram taken while the subject is exercising. 440

stroma *STRO-ma* The nonmembranous inner region of the chloroplast. 116

stroma lamellae *STRO-ma la-MEL-i* Loosely packed inner membranes of the chloroplast, containing pigment molecules. 116

style *STILE* A stalk forming from an ovary in a flower. 574

subapical region *sub-APE-eh-kel REE-jen* The region behind the root cap, which is divided into zones of cellular division, cellular elongation, and cellular maturation. 563

subclinical *sub-KLIN-eh-kel* The stage of a nutrient deficiency when abnormalities can be detected with biochemical tests, but symptoms are not yet experienced. 488

suberin *SU-ber-in* A waxy, waterproof biochemical in the interior of a root's cortex. 564

succulent *SUK-ku-lent* Fleshy plant tissue that can store large amounts of water. 558

summation *sum-A-shun* An increase in the strength of contraction of a muscle that is stimulated a second time very soon after an initial stimulation. 413

superior vena cava *su-PER-ee-er VEE-nah KAH-vah* The upper branch of the largest vein that leads to the heart. 434

supernormal releaser *su-per-NOR-mal ree-LEAS-er* In animal behavior, a model that exaggerates a releaser and elicits a stronger response than the natural object. 670

suppressor T cells *su-PRES-ser T SELZ* T cells that inhibit the response of all lymphocytes to foreign antigens, shutting off the immune response when an infection is under control. 521

suprachiasmatic nuclei (SCN) *SU-pra-KI-as-MAT-ik NU-klee-i* Two clusters of 10,000 neurons each in the hypothalamus that control certain biological clocks in some species. 677

survival of the fittest *ser-VI-val of the FIT-tist* The idea that those individuals best able to reproduce healthy offspring contribute the most genes to the next generation. 627

symbiosis *SYM-bee-o-sis* An intimate relationship between two types of organisms. 14

sympathetic nervous system *SIM-pah-THE-tik NER-ves SIS-tum* Part of the autonomic nervous system that mobilizes the body to respond to environmental stimuli. 344, 439
sympatric species *SIM-pat-rik SPE-shez* Two closely related groups of organisms that occupy the same geographic region but cannot reproduce successfully with each other. 638
synapse *SIN-apse* A space between two adjacent neurons. 321
synapsis *SIN-ap-sis* The gene-by-gene alignment of homologous chromosomes during prophase of meiosis I. 159
synaptic knobs *sin-AP-tik NOBZ* The enlarged tips of branches at the ends of axons. 321
synaptic vesicles *sin-AP-tik VES-eh-kelz* Small sacs within synaptic knobs at the ends of axons that contain neurotransmitters. 321
synovial joint *sin-OV-ee-el JOINT* A capsule of fluid-filled fibrous connective tissue between freely movable bones. 401
synovial membrane *sin-OV-ee-el MEM-brane* The lining of the interior of a joint capsule, which secretes lubricating synovial fluid. 401
systole *SIS-toll-ee* The heart's contraction. 438
systolic pressure *SIS-tol-ik PRESH-yur* The blood pressure at its peak, when the ventricles contract. 432

T

taiga *TI-gah* The northern coniferous forest, north of the temperate zone. 729
taproot system *TAP-root SIS-tum* A plant in which the first root (the radicle) enlarges to form a major root that persists through the life of the plant. 563
target cell *TAR-get SEL* A cell that is affected directly by a particular hormone. 368
tarsals *TAR-salz* The ankle bones. 400
taste receptors *TASTE ree-CEP-terz* Specialized neurons that detect taste. 351
taxonomy *tax-ON-o-mee* The branch of biology concerned with classifying organisms on the basis of evolutionary relationships. Taxonomic levels include, in order, kingdom, phylum (or division), class, order, family, genus, and species. 19
tectorial membrane *TEK-TORE-ee-al MEM-brane* The membrane above the hair cells in the cochlea of the inner ear that is pressed by the hair cells responding to the basilar membrane's vibration in the presence of sound waves. 360
telophase *TELL-o-faze* The final stage of cell division, when two cells form from one and the spindle is disassembled. 135
temporal isolation *TEM-por-al I-so-LAY-shun* When members of two populations do not crossbreed because they have different mating seasons. 368
tendon *TEN-din* A heavy band of fibrous connective tissue that attaches a muscle to a bone. 406
tendrils *TEN-drilz* Shoots or modified leaves that support plants by coiling around objects. 558, 561
teosinte *TE-o-SIN-tee* A grass that may have been ancestral to corn. 544
teratogen *teh-RAT-eh-jen* A chemical or other environmental agent that causes a birth defect. 202
territory *TEAR-eh-TOR-ee* A portion of land defended by an individual. 694
tertiary consumers *TER-she-AIR-ee kon-SU-merz* Carnivores that eat other carnivores, forming a fourth trophic level. 718
tertiary structure *TER-she-air-ee STRUK-sure* The shape assumed by a protein caused by chemical attractions between amino acids that are far apart in the primary structure. 53
test cross *TEST CROSS.* Crossing an individual of unknown genotype to a homozygous recessive individual. 221
testes *TES-teez* The paired, male gonads, containing the seminiferous tubules, in which sperm are manufactured. 155, 374
tetanus *TET-nes* A smooth and continuous muscle contraction resulting from repeated strong stimulations that occur before the muscle has time to relax. Also, an infectious disease called "lockjaw."
tetraploid *TET-rah-ploid* An individual with four sets of chromosomes, usually resulting from self-fertilization in a diploid plant. 639
thalamus *THAL-eh-mus* A gray, tight package of nerve cell bodies and glia beneath the cerebrum that relays sensory input to the appropriate part of the cerebrum. 335
thecodonts *THEK-o-dontz* Descendants of the Permian cotylosaurs that were ancestors of the great dinosaurs. 659
therapsids *ther-AP-sidz* Reptiles living in the Mesozoic era, which had some characteristics of mammals. 659
thermal stratification *THER-mal STRAH-tah-fah-KAY-shun* Layers within lakes that have different temperatures. 734
thermocline *THER-mo-kline* A middle layer of a lake where water temperature changes rapidly and drastically. 734
thermoluminescence *THER-mo-LU-mah-NES-ence* A technique that measures the formation of tiny holes in crystals over time, caused by exposure to ionizing radiation. This measurement is used in absolute dating of fossils up to 1 million years old. 650
thigmomorphogenesis *THIG-mo-MOR-pho-GEN-ah-sis* A plant's responses to mechanical disturbances, including inhibition of cellular elongation and production of thick-walled supportive tissue. 596
thigmotropism *THIG-mo-TRO-piz-um* A plant's response toward touch. 592
thoracic vertebrae *thor-AS-ik VER-tah-bray* The 12 vertebrae in the upper back. 398
thorns *THORNZ* Stems modified for protection. 558
threat posture *THRET POS-tur* A visual display marking a territory. 694
thrombophlebitis *THROM-bo-flah-BI-tis* Inflammation of a vein wall complicated by the formation of blood clots. 438
thromboplastin *THROM-bo-plas-tin* A protein released from blood vessel walls following injury that, in the presence of calcium, converts the blood protein prothrombin into thrombin. 428
thrombus *THROM-bus* A blood clot that blocks a blood vessel or the heart. 428
thylakoids *THI-lah-koidz* Membranous discs comprising the inner membrane of a chloroplast. 81, 116
thymine *THI-meen* One of the two pyrimidine bases in DNA. 57, 253
thymus *THY-mis* A lymphatic organ in the upper chest where lymphocytes called T cells learn to distinguish foreign from self antigens. 442
thyroid gland *THI-roid Gland* A gland in the neck that manufactures thyroxine, a hormone that increases energy expenditure. 112
thyroid stimulating hormone (TSH) *THY-roid STIM-u-lat-ing HOR-moan* A hormone made in the anterior pituitary that stimulates the thyroid gland to release its two hormones. 375
thyroxine *thy-ROX-in* A thyroid hormone that increases the rate of cellular metabolism. 375
tibia *TIB-ee-ah* The larger of the two bones of the lower leg. 400
tidal volume *TI-del VOL-yum* The amount of air inhaled or exhaled during a normal breath. 455
tinnitus *tin-I-tus* A condition in which a persistent ringing sound is heard. 361
Ti plasmid *TI PLAZ-mid* A ring of DNA in the microorganism *Agrobacterium tumefaciens* that is used to introduce foreign plant genes in recombinant DNA technology. 611
tissue *TISH-u* In multicellular organisms, groups of cells with related functions. 40
tonsils *TAWN-silz* Collections of lymphatic tissue in the throat. 442
trachae *TRAY-ki* A branching system of tubules that brings the outside environment in close contact with an organism's cells so that gas exchange can occur. 446
trachea *TRAY-kee-ah* The respiratory tube just beneath the larynx, held open by rings of cartilage. Also called the windpipe. 449
tracheids *TRA-kee-idz* Less specialized conducting cells in plants that are elongate, are dead at maturity, and have thick walls. 555
tracheophytes *TRAY-key-o-fights* Plants that have specialized tubes to conduct water and nutrients. 29, 571
transcription *tranz-SKRIP-shun* Manufacturing RNA from DNA. 257
transfer RNA (tRNA) *TRANZ-fer* A small RNA molecule that binds an amino acid at one site and an mRNA codon at another site. 258
transgenic organism *TRANZ-jen-ik OR-gan-niz-um* Genetic engineering of a gamete or fertilized ovum, leading to development of an individual with the alteration in every cell. 146, 610
translation *tranz-LAY-shun* Assembly of an amino acid chain according to the sequence of base triplets in a molecule of mRNA. 257
translocation *TRANZ-lo-KAY-shun* Exchange of genetic material between nonhomologous chromosomes. 280
transverse (T) tubules *TRANZ-verse TU-bules* Portions of the sarcolemma that jut into the sarcoplasmic reticulum of a skeletal muscle cell. 407
Triassic period *tri-AS-ik PER-ee-od* The period from 225 to 185 million years ago, when small ancestors of the great dinosaurs flourished. 659
trichomes *TRI-koamz* Outgrowths of a plant's epidermis that provide protection. 554
triiodothyronine *TRI-i-ode-o-THY-ro-neen* A thyroid hormone that increases the rate of cellular metabolism. 375
trilobites *TRI-lo-bitz* Insectlike organisms that appeared in the seas of the Cambrian period. 657
trisomy *TRI-som-mee* A cell with one extra chromosome. 280
trophic level *TRO-fik LEV-el* A feeding level in a food chain or web. 717
trophoblast *TRO-fo-blast* A layer of cells in the preembryo that develops into the chorion and then the placenta. 173
tropical rain forest *TROP-e-kel RAIN FOR-est* A warm, moist terrestrial region where rainfall is 79 to 157 inches (200 to 400 centimeters) per year; life is diverse and plentiful, and nutrient cycling is rapid. 728
tropic hormone *TRO-pik HOR-moan* A hormone produced by one gland that influences the secretion of a hormone by another gland. 371
tropism *TRO-piz-um* Plant growth toward or away from an environmental stimulus. 591
tropomyosin *TRO-po-MI-o-sin* A type of protein in the thin myofilaments of skeletal muscle cells. 407

troponin *tro-PO-nin* A type of protein in the thin myofilaments of skeletal muscle cells. 407

trypsin *TRIP-sin* A pancreatic enzyme that participates in protein digestion in the small intestine. 472

tube nucleus *TUUB NU-klee-us* A haploid cell resulting from the mitotic division of a microspore, in male plant reproduction. 574

tubercle *to-BER-kel* A section of lung walled off by a fibrous connective tissue capsule as a result of tuberculosis. 461

tuberculosis *to-BER-ku-LO-sis* A bacterial infection of the lungs. 460

tubers *TU-berz* Swollen regions of stems that store nutrients. 558

tundra *TUN-drah* A band of land running across the northern parts of Asia, Europe, and North America, where the climate is harsh and few organisms live. 729, 733

turgor pressure *TER-ger PRESH-yur* Rigidity of a plant cell caused by water pressing against the cell wall. 100

twitch *TWITCH* A rapid contraction and relaxation of a muscle cell following a single stimulation. 413

twitch types *TWITCH TYPEZ* Varieties of skeletal muscle fibers distinguished by how quickly they contract and tire. 413

tympanal organ *TIM-PAN-al OR-gan* A thin part of an insect's cuticle that detects vibrations and therefore sound. 359

tympanic membrane *TIM-PAN-ik MEM-brane* The eardrum, a structure upon which sound waves impinge. 359

U

ulcer *UL-sir* A raw, craterlike sore. 470

ulcerative colitis *UL-sir-AH-tiv koal-I-tis* Inflammation of the inner lining of the colon and rectum, producing pain, bloody diarrhea, and weight loss. 477

ulna *UL-nah* One of the two lower arm bones. 400

umbilical cord *um-BIL-ik-kel KORD* A ropelike structure containing one artery and two veins that connects mother to unborn child. 168

unconditioned stimulus *un-kon-DISH-ond STIM-u-lus* A stimulus that normally triggers a particular response. 673

uniformitarianism *U-nah-FOR-mah-TER-ee-ah-niz-um* The view that the earth's surface is continually remolded. 621

unsaturated (fat) *un-SAT-yur-RAY tid* A triglyceride with double bonds between some of its carbons. 52

upwelling *up-WELL-ing* The movement upward of cooler, nutrient-rich bottom layers of the ocean, causing nutrients to bloom. 737

uracil *YUR-eh-sil* One of the two pyrimidine bases in RNA. 257

urea *u-REE-ah* A nitrogenous waste derived from ammonia. 502

ureter *u-REE-ter* A muscular tube that transports urine from the kidney to the bladder. 504

urethra *u-RETH-rah* The tube leading from the bladder through which urine exits the body. 505

uric acid *YUR-ik AS-id* A nitrogenous waste derived from ammonia. 502

urinary bladder *YUR-eh-NAIR-ee BLAD-er* A muscular sac in which urine collects. 505

urinary tract infection *YUR-eh-NAIR-ee TRACT in-FEK-shun* A bacterial infection of the urethra, with symptoms of frequent, painful urination and sometimes fever and lower abdominal pain. 511

uterus *U-ter-us* The muscular, saclike organ in the human female in which the embryo and fetus develop. 157

utricle *U-trah-kel* A pouch in the vestibule of the inner ear filled with a jellylike fluid and lined with hair cells, containing calcium carbonate otolith granules that move in response to changes in velocity, firing action potentials. 362

V

vaccine *VAK-seen* A killed or weakened form of, or part of, an infectious agent that initiates an immune response so that when the real agent is encountered, antibodies are already available to deactivate it. 529

vagus nerve *VA-ges NERVE* The one cranial nerve that innervates internal organs, rather than the head or neck. 344

variable regions *VAIR-ee-ah-bul REE-genz* The sequence of amino acids comprising the upper portions of heavy and light antibody chains, which varies greatly in different antibody types. 518

varicose veins *VAR-eh-kos VANEZ* Distension of the superficial veins in the legs. 438

vascular bundles *VAS-ku-ler BUN-delz* Organized groups of vascular tissues in stems. 557

vascular cambium *VAS-ku-ler KAM-bee-um* A thin cylinder of meristematic tissue found in roots and stems that produces most of the secondary plant body. 566

vas deferens *VAS DEF-er-enz* In the human male, a tube from the epididymis that continues to become the vas deferens, which joins the urethra in the penis. 155

vasoconstriction *VAZ-o-kon-strik-shun* The narrowing of blood vessels, which raises blood pressure. 433

vasoconstriction area *VAZ-o-kon-STRIK-shun AIR-ee-ah* Part of the brain's vasomotor center that stimulates circulation by constricting blood flow. 439

vasodilation *VAZ-o-di-LAY-shun* The widening of blood vessels, which lowers blood pressure. 433

vasodilation area *VAZ-o-di-LAY-shun AIR-ee-ah* Part of the brain's vasomotor center that dilates blood vessels. 439

vasomotor center *va-ZOM-eh-ter CEN-ter* A part of the brain that controls blood flow to the heart and heart rate by sending nerve impulses through the spinal cord to the sympathetic nervous system. 439

veins *VANEZ* Large blood vessels arising from venules that return blood to the heart. 428

veins *VANEZ* Strands of vascular tissue in leaves. 561

venous valves *VEEN-is VALVES* Flaplike structures in veins that keep blood flow in one direction. 432

ventricles *VEN-tree-kelz* Spaces in the brain into which cerebrospinal fluid is secreted. Also, the two muscular chambers of the heart located beneath the atria. 342

venules *VANE-yules* Vessels that arise from capillaries and drain into veins. 428

vertebral column *VER-teh-bral KOL-um* Bones along the back and neck that protect the spinal cord.

vertical stratification *VER-tah-kel STRAH-tah-fah-KAY-shun* The formation of layers of different types of organisms beneath the canopy of a tropical rain forest, caused by competition of organisms for sunlight. 728

vessel elements *VES-el EL-eh-mentz* More specialized conducting cells in plants that are elongate, are dead at maturity, and have thick walls. 555

vestibule *VES-teh-bule* A structure in the inner ear that provides information on the position of the head with respect to gravity and changes in velocity. 362

vestigial *ves-TEEG-el* A structure that seems not to have a function in an organism but resembles a functional organ in another type of organism. 651

villi *VIL-i* Tiny projections on the inner lining of the small intestine, which greatly increase surface area. 473

viroid *VEAR-oid* Infectious genetic material. 43

virus *VI-rus* An infectious particle consisting of a nucleic acid (DNA or RNA) wrapped in protein. 42

vital capacity *VI-tel kah-PASS-eh-tee* The maximal amount of air that can be moved in and out of the lungs during forceful breathing. 455

vitamin *VI-tah-min* An organic molecule essential in small amounts for the normal growth and function of an organism. 58

vitreous humor *VIT-ree-es U-mer* A jellylike substance behind the lens, comprising most of the volume of the eye. 356

vocal cords *VO-kel KORDZ* Two elastic bands of tissue stretched over the glottis, which vibrate as air passes, producing sounds. 449

W

waggle dance *WAG-gel DANC* A bee's dance signifying that food is farther from the hive than a round dance would indicate. The speed of the dance, the number of waggles during the straight part, and the duration of buzzing signal the distance to the food source. 692

white blood cell (leukocyte) *WHITE BLOOD SEL* A cell that helps fight infection. 83, 424

white matter *WHITE MAT-ter* Myelinated nerve fibers, found in pathways that transmit impulses over long distances. 321

wild type *WILD TYPE* A phenotype or allele that is the most common for a certain gene in a population. 220

X

X inactivation *X IN-ak-tah-VA-shun* The turning off of one X chromosome in each cell of a female mammal at a certain point in prenatal development. 239

xylem *ZI-lum* Tubules in a plant that transport water and minerals from the roots to the leaves. 29, 555

Z

zona pellucida *ZO-nah pel-LU-seh-dah* A thin, clear layer of proteins and sugars surrounding a secondary oocyte. 171

zone of cellular division *ZONE of SEL-yu-ler dah-VISH-on* The meristematic part of the subapical region in a plant's root. 563

zone of cellular elongation *ZONE of SEL-yu-ler e-long-GAY-shun* The middle part of the subapical region of a plant's root, where rapid cellular elongation lengthens the root. 563

zone of cellular maturation *ZONE of SEL-yu-ler MAT-ur-AY-shun* The hindmost region of the subapical region of a plant's root, where tiny root hairs protrude from epidermal cells. 563

zygomycete *ZI-go-my-SEAT* A fungus with sexual spores, such as bread mold. 27

zygote *ZI-goat* In prenatal humans, the organism during the first 2 weeks of development. Also called a preembryo. 172, 573, 578

Credits

Photographs

Title page © Dr. Lloyd M. Beidler/Science Photo Library/Photo Researchers, Inc.; **page vi:** © Jeff Henry/Peter Arnold, Inc.; **page viii:** © Science Photo Library/ Science Source/Photo Researchers, Inc.

Table of Contents:

Page ix (all): © Dr. Lloyd M. Beidler/ Science Photo Library/Photo Researchers, Inc.; **page x:** © William De Grado/ Dupont; **page xii:** © Bruce Iverson; **page xiv:** © Petit Format/Nestle/Photo Researchers, Inc.; **page xv:** © Erika Stone/ Peter Arnold, Inc.

Unit Openers

Unit 1: © Biophoto Associates/Science Source/Photo Researchers, Inc.; **Unit 2:** © Bruce Iverson; **Unit 3:** © Petit Format/ Nestle/Photo Researchers, Inc.; **Unit 4:** © Erika Stone/Peter Arnold, Inc.; **Unit 5:** © Steven C. Kaufman/Peter Arnold, Inc.; **Unit 6:** © Andrew J. Martinez/Photo Researchers, Inc.; **Unit 7:** © W.H. Moller/Peter Arnold, Inc.; **Unit 8:** © A.P. Barnes/Planet Earth Pictures.

Chapter 1

1.1*b*: © Jeff Henry/Peter Arnold, Inc.; **1.4:** © Stranton K. Short/The Jackson Laboratory; **1.5:** Courtesy Biology Digest; **1.6:** © Thomas Novitsky/Cape Cod Associates; **1.7:** National Library of Medicine; **1.8:** © William E. Ferguson

Chapter 2

2.1: © M. I. Walker/Science Source/ Photo Researchers, Inc.; **2.2:** J. Tucker/ NASA, GFSC; **2.3:** Courtesy Dr. Allen R. J. Eaglesham; **page 16 (top):** © S. J. Krasemann/Peter Arnold, Inc.; **(bottom):** © Bruce Iverson; **2.6 (both):** © Barbara Mensch; **2.8:** © USDA/Science Source/ Photo Researchers, Inc.; **2.9*a*:** © CNRI/ Science Photo Library/Photo Researchers, Inc.; **2.9*b*:** © Dr. Tony Brain/Science Photo Library/Photo Researchers, Inc.; **2.9c (insert):** Carolina Biological Supply Company; **2.9*c*:** © Stanley F. Hayes, National Institutes of Health/Science Photo Library/Photo Researchers, Inc.; **2.11*b*:** © William E. Ferguson; **2.11*c*:** © Biophoto Associates/Science Source/ Photo Researchers, Inc.; **2.14:** © Michel Viard/Peter Arnold, Inc.; **2.15 (all):** © Bruce Iverson; **2.16:** © Jeff Rotman/ Peter Arnold, Inc.; **page 32:** © Peter Arnold/Peter Arnold, Inc.; **2.17*a*:** © Fred Bavendam/Peter Arnold, Inc.; **2.17*b*:** © Tom Branch/Photo Researchers, Inc.; **2.17*c*:** © Ken Lucas/Biological Photo Service; **2.17*d*:** © Norm Thomas/Photo Researchers, Inc.; **2.17*e*:** © Fred Bavendam/Peter Arnold, Inc.; **2.17*f*:** © Andrew Martinez/Photo Researchers, Inc.; **2.17*j*:** © S. Nagendra/Photo Researchers, Inc.

Chapter 3

3.1: © Carl Purcell/Photo Researchers, Inc.; **3.3*a*:** © Bruce Iverson; **3.3*b*:** © Raymond A. Mendez; **3.4*a*:** © William E. Ferguson; **3.4*b*:** © Don Riepe/Peter Arnold, Inc.; **3.4*c*:** © B. I. Ullmann/ Taurus Photos; **3.5 (both):** Theodore Diener/USDA Plant Virology Laboratory; **3.9:** From Ruth Kavenoff, Lynn C. Klotz, and Bruno H. Zimm, Symposia on Quantitative Biology (Cold Spring Harbor) 38 (1973): 4; **3.15*b*:** © Hans Pfletschinger/Peter Arnold, Inc.; **3.21:** © William De Grado/Dupont; **3.22:** © Leonard Lessin/Peter Arnold, Inc.

Chapter 4

4.1*a*,*b*: © Manfred Kage/Peter Arnold, Inc.; **4.1*c*:** © Edwin A. Reschke/Peter Arnold, Inc.; **4.3*a*:** © William E. Ferguson; **4.3*b*:** © Mark Cochran; **4.4 (all):** © Dr. Tony Brain/David Parker/ Science Photo Library/Photo Researchers, Inc.; **4.7*b*:** © Dr. S. Kim/Peter Arnold, Inc.; **4.9*b*:** © Barry King, University of California, Davis/Biological Photo Service; **page 84 (top):** © J. & R. Weber/ Peter Arnold, Inc.; **(second to the bottom):** © Edwin A. Reschke/Peter Arnold, Inc.; **4.15*a*:** © OMIKRON/ Science Source/Photo Researchers, Inc.; **4.15*b*:** Carolina Biological Supply Company; **4.15*c*:** © Chuck Brown/Photo Researchers, Inc.; **4.15*d*:** © J. & R. Weber/Peter Arnold, Inc.

Chapter 5

5.1: © Dr. Arnold Brody/Science Photo Library/Photo Researchers, Inc.; **5.2:** © Dr. A. Liepins/Science Photo Library/ Photo Researchers, Inc.; **5.3 (both):** © Edwin A. Reschke/Peter Arnold, Inc.; **5.5:** © D. W. Fawcett/Photo Researchers, Inc.; **page 97:** Lysosome Company; **5.9 (all):** © S. J. Singer; **5.16*a*:** Reproduced with the permission of the American Lung Association; **5.17:** © C. Edelmann/ La Villette/Photo Researchers, Inc.

Chapter 6

6.5: © Pat and Tom Leeson/Photo Researchers, Inc.; **page 119:** © William Ruf Photography; **6.11:** © Biophoto Associates/Science Source/Photo Researchers, Inc.; **6.14:** Bettmann Newsphotos

Chapter 7

7.1*b*: © David M. Phillips/Taurus Photos; **7.1*c*:** © Junebug Clark/Photo Researchers, Inc.; **7.1*d*:** © Lee D. Simon/Photo Researchers, Inc.; **7.4 (all):** © Edwin A. Reschke; **7.5 (all):** © Bruce Iverson; **7.6:** Dr. Martin Carter/Rockefeller University; **7.10:** Courtesy of Dr. Robert Turgeon and Dr. B. Gillian Turgeon, Cornell University; **page 142 (left):** GE Corporate Research and Development; **(right):** Becton Dickinson Labware; **7.11:** Courtesy Dr. Garth L. Nicolson; **7.13:** © Cecil H. Fox/Science Source/Photo Researchers, Inc.

Chapter 8

Page 158 (left) *a-f*: Carolina Biological Supply Company; **(right):** © David Scharf/Peter Arnold, Inc.; **page 159:** E. L. Wollman, F. Jacob, and W. Hayes, Symposia on Quantitative Biology (Cold Spring Harbor), vol. 21, p. 153, 1956; **8.10:** © David M. Phillips/The Population Council/Taurus Photos

Chapter 9

9.1: Carolina Biological Supply Company; **9.2:** © David Scharf/Peter Arnold, Inc.; **9.5 (all):** © R. G. Edwards; **page 177 (top):** Furnished courtesy of C. L. Markert; **(bottom):** © A. Villeneuve and B. Meyer; **9.13*e*:** © Petit Format/ Nestle/Science Source/Photo Researchers, Inc.; **9.17:** © Mike Greeler/The Image Works; **9.19*a-e*, *g*, *h*:** UPI/Bettmann Newsphotos; **9.19*f*, 9.20:** AP/Wide World Photos; **9.21:** © Dr. Dennis Dickson, Albert Einstein School of Medicine/Peter Arnold, Inc.

Chapter 10

10.1*a*: AP/Wide World Photos; **10.1*b*:** Courtesy Cornell University; **page 195:** Courtesy University of California, Davis; **10.2:** © Dr. Tony Brain/Science Photo Library/Photo Researchers, Inc.; **10.4*a*:** © Jim Sulley/The Image Works, **10.4*b*:** © Alan Carey/The Image Works; **10.6:** © Topham/The Image Works; **10.7*b-d*:** Dr. Ann Pytkowicz Streissguth, University of Washington: Science 209 (18 July 1980) 353–61 © 1980 by the AAAS

Chapter 11

11.1*a*: W. Dorsey Stuart, University of Hawaii; **11.1*b*:** Hans Stubbe, Akademie der Wessenchagten der DDR; **11.6:** Courtesy Dr. Myron Neuffer, University of Missouri; **11.7*a*:** Field Museum of Natural History, Negative number 118; **11.7*b*:** R. L. Brinster, Cover of Cell, vol. 27, Nov. 1981. Reprinted with permission of Cell Press; **11.11:** Carolina Biological Supply Company; **11.12 (all):** The Bettmann Archive; **11.14:** © Lester V. Bergman and Associates, Inc.; **page 228:** American Institute of Applied Science, Syracuse, New York; **page 228c (both):** © Bob Coyle; **page 228*d* (both):** © Tom Ballard/EKM-Nepenthe; **11.15*a*:** © Jim Goodwin/Photo Researchers, Inc.; **11.15*b-d*:** © Phil Randazzo and Rudi Turner; **11.17:** © David Scharf/Peter Arnold, Inc.; **11.18*b*:** Courtesy Nobel Foundation/1977 Science/97/943. Unconventional Viruses: The Orgins and Dosaperanes of Kurin by D. Carlton Gajdusck. © 1977 of the AAAS

Chapter 12

12.3*b*: Historical Pictures Service, Chicago; **12.4 (all):** S. D. Sigamony; **12.5*a*,*b*:** Wilson and Foster, Williams Textbook of Endocrinology, 7th ed. © W.B. Saunders 1985; **12.5*c*,*d*:** National Jewish Hospital and Research Center; **12.6*a*:** © Horst Schaefer/Peter Arnold, Inc.; **12.6*b*:** © William E. Ferguson; **page 244 (top):** © Lee D. Simon/Photo Researchers, Inc.; **(bottom):** © Eric V. Grave/Photo Researchers, Inc.; **page 245 (top left):** © James Richardson/Visuals Unlimited; **top right (both):** Gerald M. Rubin, Cover of Cell, vol. 40, April 1985.

Reprinted with permission of Cell Press; **(bottom)**: Courtesy Dr. Ricki Lewis; **12.8*a*:** © James H. Karales/Peter Arnold, Inc.; **12.8*b*:** © Bob Daemmrich/The Image Works; **12.8*c*:** © Patsy Davidson/ The Image Works

Chapter 13

13.1*a*: © Lee D. Simon/Science Source/ Photo Researchers, Inc.; **13.1*b*:** © John Carld' Annibale; **13.1*c*:** © Stuart Lindsay, Ph.D; page 255: The Bettmann Archive; **13.8*a*:** John Cairns, Symposia on Quantitative Biology (Cold Spring Harbor), vol. 28, p. 44, 1963; **13.8*b*:** David Hogness/Stanford University; **13.15:** D. W. Ow, Keith V. Wood, Marlene De Luca, Jeffrey R. Dewet, Donald R. Helsinki, Stephen H. Howell, "Transient and Stable Expression of the Firefly Luciferase Gene in Plant Cells and Trasgenic Plants." Science 234 (Nov. 14, 1986): 856–859, **fig. 1.** Copyright © 1986 by the AAAS; **13.17*a*, 13.18:** Dr. O. L. Miller; **13.22:** © Dr. Paul Englund; **13.23:** From Federoff, N. 1984. Transposable Genetic Elements in Maize, Scientific American, June 84/98. Photos by Fritz W. Goro; **13.25:** © Science Photo Library/Science Source/Photo Researchers, Inc.

Chapter 14

14.1: © Laura Dwight/Peter Arnold, Inc; **14.2*a,b,d*:** Courtesy Dr. Fred Elder, Department of Pediatrics, University of Texas; **14.2*c*:** Courtesy Ann Cork and Dr. J. M. Trujillo, Cytogenetics Laboratory, M. D. Anderson Cancer Center; **14.6*a* (both), *b*:** From the British Medical Journal, vol. 12, Dec. 1886, p. 1189; **14.7:** © Dr. McKusick/John Hopkins Hospital; **page 286:** March of Dimes Birth Defects Foundation; **page 287 (both):** Woody Guthrie Publications, Inc.; **14.11 (both):** © Bill Longcore/Photo Researchers, Inc.; **14.13*a*:** Library of Congress

Chapter 15

15.3: Photograph by permission of Linda Larsen from Human Chromosome Analysis: Methodology and Applications. American Journal of Medical Technology, 10:687 (1983); **15.5:** © Michael Bender; **15.6 (both):** Courtesy of Jason C. Birnholz, M.D., Rush-Presbyterian St. Lukes Medical Center, Chicago; **15.12:** © Dr. Nancy Wexler; **page 306 (left):** March of Dimes Birth Defects Foundation; **(right)**: © Leonard Lessin/Peter Arnold, Inc.; **15.14*a*:** © Dr. R. L. Brinster/Peter Arnold, Inc.; **15.14*b*:** © Frank Constantini, Kiran Chada, Jeanne Magram

Chapter 16

16.1*b*: © SECCHI-LEAGUE/Roussel-UCLAF/CNRI/ Science Photo Library/ Photo Researchers, Inc.; **16.4*b*:** Chikashi Toyoshima and Nigel Unwin; **16.6*d*, *e*:** National Multiple Sclerosis Society; **16.8:** © CNRI/ Science Photo Researchers, Inc.; **16.10:** © Paul Chesley/ Photographer's Aspen

Chapter 17

Page 340: Mallenckrodt Institute of Radiology © 1989 Discover Publications; **17.11*a*:** © Chesher/ Photo Researchers, Inc.; **17.11*b*:** © Dr. Jon W. Jacklet, Professor and Chairman Department of Biology, SUNY, Albany

Chapter 18

18.1 (both): © Thomas Eisner; **18.3*b*:** © Omekron/ Science Source/ Photo Researchers, Inc.; **18.5:** © Thomas Eisner; **18.8*a*:** © Frank S. Werblin; **page 355 (left):** © Peter A Rora/ National Oceanic and Atmosphere Administration/ Woods Hole: Oceangraphic Institute/ MIT, **(right):** © Steve Chamberlain, Syracuse University; **18.12:** © Martin Dohrn/ Science Library/ Photo Researchers, Inc.; **18.15 (both), 18.18 (both):** © Molly Webster © 1982 Discover Publications

Chapter 19

19.1: © Phillip A. Harrington/ Peter Arnold, Inc.; **19.9:** Weidenfeld and Nicolson Ltd., Dept. of Medical Photography Westminister Hospital; **19.10 (all):** Dept. of Illustrations, Washington University School of Medicine 20 (Jan. 1956) p. 133; **19.12 (all):** Joseph Bagnara, Chromatophores and Color Change, J. T. Bagnara and M. E. Hadley, Prentice Hall, 1973; **19.14*b*:** © Lester V. Bergman & Associates; **19.16:** F. A. Davis Company, Philadelphia; and R. H. Kampmeier; **19.17*b*:** © Edwin A. Reschke; **page 380:** © Martin Dohrn/ Photo Researchers, Inc.; **page 382:** © Tom McHugh/ Photo Researchers, Inc.; **page 383:** © G. Gransanti/ Sygma; **19.21a:** © John Bova/ Photo Researchers, Inc.; **19.21*b*:** © Dwight Kuhn; **19.21c:** © Thomas Eisner

Chapter 20

20.1: © Bill Bachman/ Photo Researchers, Inc.; **20.3:** © Bryan Hitchcock/ Photo Researchers, Inc.; **page 392:** © Gerald Lacz/ NNPA; **20.7:** © Biophoto Associates/ Photo Researchers, Inc.; **page 396:** © Science Source Library/ Photo Researchers, Inc.; **20.10:** © World Health Organization; **20.12:** © Ken Rucas/ Biological Photo Services; **20.13:** © Edward Lettau/ Peter Arnold, Inc.; **20.17:** © Alexander Tsiaras/ Science Source/ Photo Researchers, Inc.

Chapter 21

21.1 (all): © Manfred Kage/ Peter Arnold, Inc.; **21.5 (both):** © K. G. Murti/ Visuals Unlimited; **21.6 (both):** © Toni Michaels; **21.12:** © G. W. Willis, Ochsner Institution/ Biological Photo Service; **21.16*b*:** © Biophoto Associates/ Photo Researchers, Inc.

Chapter 22

Page 423 (top): © Dr. Leland Clark, **(bottom):** Organogenesis Inc.; **22.8*c*:** © Eila Kairinen/ Gilette Research Institute, Gaithersburg, MD; **22.9*a*:** © The Royal College of Surgeons of England; **22.11:** © Biophoto Associates/ Photo Researchers, Inc.; **22.15*b*:** © SIU Biomedical Communications/ Photo Researchers, Inc.; **22.17*a*:** Courtesy of Igaku Shoin, LTD; **22.23:** E. K. Markell and M. Voge: Medical Parasitology, 6th ed. © W.B. Saunders Company, 1986; **22.24:** © Dr. Carole Berger/ Peter Arnold, Inc.; **page 437:** © Science Photo Library/ Photo Researchers, Inc.

Chapter 23

23.1: © Douglas M. Munnecke/ Biological Photo Service; **23.2*a*:** © Andrew J. Martinez/ Photo Researchers, Inc.; **23.2*b*:** © R. Andrew Odum/ Peter Arnold, Inc.; **23.2*c*:** © Tom McHugh/ Photo Researchers, Inc.; **23.2*d*:** © Hans Pfletschinger/ Peter Arnold, Inc.; **23.2*e*:** © Fritz Pokling GDT/ Peter Arnold, Inc.; **23.4:** © F. Gohier/ Photo Researchers, Inc.; **23.8:** © John Watney Photo Library; **23.10*b*:** from TISSUES AND ORGANS: A TEXT-ATLAS OF SCANNING ELECTRON MICROSCOPY by R. G. Kessel and R. H. Kardon, © 1979 W.H. Freeman; **page 454 (both):** American Cancer Society; **23.12*e*, *f*:** J. H. Comroe: PHYSIOLOGY OF RESPIRATION © 1974 Yearbook Medical Publishers, Inc.; **23.15:** © Weinstein/ Custom Medical Stock Photo; **23.17 (both):** © Martin Rocker/ Peter Arnold, Inc.; **23.18:** The Bettmann Archive

Chapter 24

24.2*a*: © Jeff Lepore/ Photo Researchers, Inc.; **24.2*b*:** © John R. MacGregor/ Peter Arnold, Inc.; **page 466 (top):** © Arthus Bertrand/ Photo Researchers, Inc., **(bottom):** © Tom McHugh/ Photo Researchers, Inc.; **24.5b, 24.7*b*:** Courtesy of Utah Valley Hospital, Department of Radiology; **24.10:** The Bettmann Archive; **24.11:** © Martin M. Rocker/ Taurus Photos; **24.16*a*:** © David Scharf/ Peter Arnold, Inc.; **24.16*b*:** D. H. Alpers and B. Seetharan, New England Journal of Medicine 296 (1977) 1047; **page 476 (left):** © Susan Leavires/ Photo Researchers, Inc., **(right):** © Dr. Leonard Crowley/ INTRODUCTION TO HUMAN DISEASE, 2nd edition, fig. 23.22, p. 615 Jones and Bartlett, 1988; **24.21*a*:** © Carroll Weiss/ RBP; **24.21*b*:** © Sherril D. Burton; **page 479:** © Kenneth Murray/ Photo Researchers, Inc.

Chapter 25

25.5 (both): © G.W. Willis, Ochsner Medical Institution/ BPS; **25.6*b*:** UNICEF; **25.7:** © National Medical Slide Bank/ Custom Medical Stock Photo

Chapter 26

Page 498: © Fred Bavendam/ Peter Arnold, Inc.; **26.2:** © Scott Camazine/ Photo Researchers, Inc.; **26.3:** © Dr. R. P. Clark and M. Goff/ Photo Researchers, Inc.; **26.4:** © Demi McIntyre/ Photo Researchers, Inc.; **26.9b, 26.11:** © Biophoto Associates/ Science Source/ Photo Researchers, Inc.; **26.14:** © J. & L. Weber/ Peter Arnold, Inc.

Chapter 27

27.2*b*: © NIBSC/ Science Photo Library/ Photo Researchers, Inc.; **27.3:** © Manfred Kage/ Peter Arnold, Inc.; **27.9:** Courtesy of Schering-Plough; **27.11:** Courtesy of Blackwell Scientific, Reprinted from ESSENTIAL IMMUNOLOGY by Ivan Roitt, Fig. 8/7 P. 130; **27.12:** © Zeva Oelbaum/ Peter Arnold, Inc.; **27.13*a*:** © Professor Luc Montagnier/ Institute Pasteur/ CNRI/ Science Photo/ Photo Researchers, Inc.; **27.14:** © Sygma; **27.15*b*:** © Lennart Nilsson/ Boehringer Ingelhaim International Gmbh; **27.16*a*:** © David Scharf/ Peter Arnold, Inc.; **27.16*b*:** © Phil Harrington/ Peter Arnold, Inc.; **page 530:** Courtesy, National Library of Medicine; **27.17:** © English, MD/ Custom Medical Stock Photo; **27.18:** © Phillip Harrington/ Schering-Plough Corporation

Chapter 28

28.1*a*: © Scott Canazine/ Photo Researchers, Inc.; **28.1*b*:** © W. H. Hodge/ Peter Arnold, Inc.; **28.2*a*:** © Noron Thomas/ Photo Researchers, Inc.; **28.2*b*:** © W. H. Hodge/ Peter Arnold, Inc.; **28.3*a*:** © Charlie Oto/ Photo Researchers, Inc.; **28.3*b*:** © Toni Michaels; **28.6:** © Walter H. Hodge/ Peter Arnold, Inc.; **28.9:** © Kellogg's and Kellogg's Corn Flakes Cereal are registered Trademarks of Kellogg Company. All rights reserved; **28.10:** © W. H. Hodge/ Peter Arnold, Inc.; **28.11:** © Douglas Kirkland/ Sygma; **28.12:** © B & H Kunz/ Okapia/ Photo Reseachers, Inc.; **28.13 1,2:** © Steven R. King/ Peter Arnold, Inc.; **28.14:** © Gilbert Grant/ Photo Researchers, Inc.; **28.15*a*:** © J. & L. Weber/ Peter Arnold, Inc.; **28.15*b*:** U.S. Aid Photo; **28.15*c*:** Courtesy of Dr. Daniel L. Klayman/ Science, Fig. 1, Vol. 228, Page 1051, 31 May 1985, "Quinghausu (artemysinin): Am Antimalarial Drug From China." © 1985 by the AAAS

Chapter 29

29.2: © D. E. Akin/ Visuals Unlimited; **29.3:** © David M. Phillips/ Visuals Unlimited; **29.6 1,2:** © P. Gates, University of Durham/ Biological Photo Service; **29.8:** © Kjeil B. Sandved & Coleman; **29.11:** © Michael Ederegger/ Peter Arnold, Inc.; **29.15:** © P. Dayanander/ Photo Researchers, Inc.; **29.16*a*:** © A. W. Ambler/ Photo Researchers, Inc.; **29.16*b*:** © W. H. Hodge/ Peter Arnold, Inc.; **29.16*c*:** © Walter Hodge/ Peter Arnold, Inc.

Chapter 30

30.1: © J. Alcock/ Visuals Unlimted; **30.2:** William E. Ferguson; **30.3*b*:** © Walter H. Hodge/ Peter Arnold, Inc.; **30.6*a*:** © William E. Ferguson; **30.6*b*:** © Francois Gohier/ Photo Researchers, Inc.; **30.7:** © Thomas Eisner; **30.9*a*:** © R. J. Erwin/ Photo Researchers Inc.; **30.9*b*:** © Toni Michaels; **30.9*c*:** © James Welgos/ Photo Researchers, Inc.; **30.10*a*:** © W. H. Hodge/ Peter Arnold, Inc.; **30.10*b*:** © Rod Planck/ Photo Researchers, Inc.; **30.10*c*:** © William E. Ferguson; **30.10*d*:** © Adam Hart-Davis/ Science Photo Library/ Photo Researchers, Inc.; **30.10*e*:** ARS/USDA; **30.12*b*:** © Dr. Jeremy Burgess/ Science Photo Library/ Photo Researchers, Inc.; **30.13*a*:** © Runk/ Schoenberger/ Grant Heilman; **30.13*b*:** © W. H. Hodge/ Peter Arnold, Inc.; **30.13*c*:** © Runk/ Schoenberger/ Grant Heilman

Chapter 31

31.1: © David Whitcomb; **31.3:** © David Newman/ Visuals Unlimited; **31.4:** © R. Lyons/ Visuals Unlimited; **31.5:** © Walter H. Hodge/ Peter Arnold, Inc.; **31.6:** © Runk/ Schoenberger/ Grant Heilman; **page 593 (both):** © Dr. Randy Moore; **31.7:** © John D. Cunningham/ Visuals Unlimited; **31.8:** © William E. Ferguson; **31.9:** © Adrian Davies/ Bruce Coleman, Inc.; **31.10:** © Tom McHugh/ Photo Researchers, Inc.; **31.13:** © Stephanie Ferguson/ William Ferguson; **31.14:** © Runk/ Schoenberger/ Grant Heilman; **31.15:** © Galen Rowell/ Peter Arnold, Inc.

Chapter 32

32.1: © Wilfred G. Oltis/ William E. Ferguson Photography; **32.2:** © Elliot Meyerowitz; **32.3:** © Dr. Jeremy Burgess/ Science Photo Library/ Photo Researchers, Inc.; **32.5a (all):** © Christian T. Harms; **32.7:** Courtesy, Calgene/ SCIENCE AND THE FUTURE YEARBOOK, 1986, p. 105 Encyclopedia Britannica; **32.8a, b, c, d:** © Runk/ Schoenberger/ Grant Heilman; **32.9a:** © Ted Spiegel/ Black Star; **32.9b:** DNA Plant Technology; **32.10 (both):** © Dan McCoy/ Rainbow; **32.11:** © Curt Maas/ Pioneer Hi-Bred International; **32.15:** Courtesy, Calgene; **32.16 (all):** © Dr. Paul Christou; **32.17:** Courtesy, Monsanto; **32.17b:** Courtesy, Calgene/ SCIENCE AND THE FUTURE YEARBOOK, 1986, p. 114 Encyclopedia Britannica; **page 615 (top):** Courtesy, DNA Plant Technology, **(bottom):** © Mike Greenlar/ The Image Works

Chapter 33

33.2: © William E. Ferguson; **33.3:** © Trans Lanting/ Photo Researchers, Inc.; **33.4:** © William E. Ferguson; **33.6a1:** © Lynn Rogers; **33.6a2:** © Michael Viard/ Peter Arnold, Inc.; **33.6b1:** © Ralph Eagle/ Science Source/ Photo Researchers, Inc.; **33.6b2:** © Jeff Rotman/ Peter Arnold, Inc.; **33.8:** © S. J. Krasemann/ Peter Arnold, Inc.; **33.9a:** © Tom McHugh/ Photo Researchers, Inc.; **33.9b:** © Tom McHugh/ Rapho/ Photo Researchers, Inc.; **33.9c:** © Robert and Linda Mitchell; **33.9d:** Field Museum of Natural History, Chicago, IL, Negative Number CKIT

Chapter 34

34.1: © Sandy Macys/ Gamma Liaison; **34.7:** © William E. Ferguson; **34.8:** © Porterfield/ Chickering/ Photo Researchers, Inc.; **page 639:** © Ray Pfortner/ Peter Arnold, Inc.; **34.9b:** Courtesy, W. Atlee Burpee and Company; **34.9d:** © Michael Viard/ Peter Arnold, Inc.; **34.10a:** © Jonathan Blair/ Woodfin Camp and Associates; **34.10b:** © Glenn Izett/ United States Dept. of Interior, Geological Survey; **34.12:** © Jonathan Blair/ Woodfin Camp & Associates; **page 644 (top right):** © Leonard Lee Rue III/ Animals Animals, **(bottom right):** © Barb Zurawski; **34.13:** © William E. Ferguson

Chapter 35

35.1a: © Jonathan Blair/ Woodfin Camp and Associates; **35.1b:** © Albert Copley/ Visuals Unlimited; **35.2a:** Courtesy Department Library Services/ American Museum of Natural History, Neg. No. 320496; **35.2b:** © George O. Poinar, Jr.; **35.6:** Zoological Society of London; **35.7a1:** © Ellan Young/ Photo Researchers, Inc.; **35.7a2:** © Lynn Rogers; **35.7a3:** © Steve Kaufman/ Peter Arnold, Inc.; **35.7a4:** © Tom McHugh/ Photo Researchers, Inc.; **35.11, 35.12a:** © William E. Ferguson; **35.15a:** Field Museum of Natural History; **35.15b:** © Martin Land/ Science Photo Library/ Photo Researchers, Inc.; **35.19a:** © Cabisco/ Visuals Unlimited; **35.19b:** © John Reader/ Science Photo Library/ Photo Researchers, Inc.

Chapter 36

36.1: J. A. L. Cooke; **page 668:** © Laura Riley; **36.3a:** © Doug Vargas/ The Image Works; **36.4:** © Roger Wilmshurst/ Bruce Coleman; **36.7:** © Marty Snyderman/ Visuals Unlimited; **36.9:** Nina Leen/ LIFE Magazine; **36.10:** Courtesy, Dr. Martin Ralph/ "CLOCKWORK IN THE BRAIN," Dr. Martin Ralph, BIO-SCIENCE, Vol. 39, No. 2, p. 76; **36.11:** © Gary Meszaros; **36.13a:** © Charles Walcott; **36.14:** © Lynn Rogers; **page 675a:** © Laura Dwight/ Peter Arnold, Inc.; **page 675b:** © Bill Bachman/ Photo Researchers, Inc.; **page 675c:** © Laura Dwight/ Peter Arnold, Inc.; **page 675d:** © Bob Busby/ Photo Researchers, Inc.

Chapter 37

37.1: © Runk/ Schoenberger/ Grant Heilman; **37.2:** © Stephan Dalton/ Photo Researchers, Inc.; **page 685 (bottom):** © Raymond A. Mendez; **37.3:** © Louis Quitt/ M.A.S./ Photo Researchers, Inc.; **37.4:** © Horst Schafer/ Peter Arnold, Inc.; **37.5:** © Laura Riley/ Bruce Coleman, Inc.; **37.6:** © Oxford Scientific 2 films/ Animals Animals; **37.7:** © J. P. Thomas/ Jacana/ Photo Researchers, Inc.; **37.8:** © 1990 Scott Canazine; **page 691 37.9a, b:** © Charles H. Janson; **page 691 (top):** © M. Marchaterre/ Cornell University; **37.11:** © Jonathan Scott/ Planet Earth Pictures; **37.12:** © Charles H. Jason/ NATURAL HISTORY, February, 1986, p. 444; **37.14:** © Professor Axel Michelsan; **37.15:** © G. Ziesler/ Peter Arnold, Inc.; **37.16:** © John Alcock; **37.17:** © V. Arthus-Bertrand/ Peter Arnold, Inc.; **37.18:** © Bildorchiv Okapia/ Photo Researchers, Inc.

Chapter 38

38.1: © Bob Thomas/ Gamma Liaison; **page 702 (left):** © C. Gable Ray; **page 702 (right):** Montana Department of Fish, Wildlife and Parks; **38.8a:** © Stephen J. Krasemann/ Peter Arnold, Inc.; **38.8b:** © Jonathan Scott/ Planet Earth Pictures; **38.10:** © Gregory G. Dimijian, M.D./ Photo Researchers, Inc.; **page 711:** © Alan Reininger/ Contact Press Images

Chapter 39

39.2a-f: Courtesy, Dr. Stuart Fisher, Department of Zoology, Arizona State University; **39.3a:** © Fritz Polking/ Peter Arnold, Inc.; **39.4:** © Charlie Ott/ Photo Researchers, Inc.; **39.6b:** © Steven C. Amstrup/ U.S. Fish and Wildlife; **page 727:** © Ken Spencer; **39.14a:** © Gordon Wiltsie/ Peter Arnold, Inc.; **39.14b:** © Jim Zippo/ Photo Researchers, Inc.; **39.14c:** © Luiz C. Marigo/ Peter Arnold, Inc.; **39.14d:** © L. West/ Photo Researchers, Inc.; **39.14e:** © Gregory G. Dimijian, M.D./ Photo Researchers, Inc.; **39.14f:** © Toni Michaels; **39.14g:** © John Bova/ Photo Researchers, Inc.; **39.14h:** © Stephen J. Krasemann/ Peter Arnold, Inc.; **page 732 (both):** © Space Biospheres Ventures; **39.16:** © G. Ziesler/ Peter Arnold, Inc.; **39.17a:** © David Hall/ Photo Researchers, Inc.; **39.17b:** © Peter Parks/ Oxford Scientific Films/ Animals Animals

Chapter 40

40.1a, b: © Al Grillo/ Picture Group; **40.2:** © TASS/ Sovofoto; **40.3:** © Ray Pfortner/ Peter Arnold, Inc.; **page 746:** © C. C. Lockwood/ Animals, Animals; **40.4:** © Joel Simon/ Society Expeditions; **40.5a:** © Ray Pfortner/ Peter Arnold, Inc.; **40.7:** © Michael Nichols/ Magnum Photos; **40.9a:** © George H. Harrison/ Grant Heilman; **40.9b:** © Imre De Pozsgay/ Alpine Color Lab; **40.11:** © George Antonelis, National Marine Fisheries Service

Appendix

Page A.2 (top): © Leonard Lessin/Peter Arnold, Inc.; **page A.6 (left):** USDA/ Science Source/Photo Researchers, Inc., **(right):** © William E. Ferguson; **page A.7 (left):** © Michel Viard/Peter Arnold, Inc., **(middle):** © Martin Land/Science Library/ Photo Researchers, Inc., **(right):** © Tom Branch/Photo Researchers, Inc., **page A.8 (left):** © Fred Bavendam/Peter Arnold, Inc., **(middle):** © Andrew Martinez/Photo Researchers, Inc., **(right):** © Carl Purcell/ Photo Researchers, Inc.

Text/Line Art

Chapter 2

2.10c: From Eldon D. Enger, et al., *Concepts in Biology*, 5th ed. Copyright © 1988 Wm. C. Brown Publishers, Dubuque, Iowa. All Rights Reserved. Reprinted by permission.

Chapter 3

3.26: From E. Peter Volpe, *Understanding Evolution*, 5th ed. Copyright © 1985 Wm. C. Brown Publishers, Dubuque, Iowa. All Rights Reserved. Reprinted by permission.

Chapter 4

4.5: From Boston Medical Library in The Francis A. Countway Library of Medicine. Used with permission. **4.7:** Frederick C. Ross, *Introductory Microbiology*, 2nd ed. Reprinted by permission of the author.

Chapter 5

5.13, 5.14: From Bruce Alberts, et al., *Molecular Biology of the Cell*. Copyright © 1983 Garland Publishing, Inc., New York, NY. Reprinted with permission.

Chapter 7

7.2: From Stuart Ira Fox, *Human Physiology*, 3d ed. Copyright © 1990 Wm. C. Brown Publishers, Dubuque, Iowa. All Rights Reserved. Reprinted by permission. **7.7:** From Bruce Alberts, et al., *Molecular Biology of the Cell*. Copyright © 1983 Garland Publishing, Inc., New York, NY. Reprinted with permission. **7.12a-e:** Reprinted with permission from *Chem. Eng. News*, February 25, 1985, 63(8), p. 16. Copyright © 1985 American Chemical Society.

Chapter 8

8.8: From Linda R. Maxson and Charles H. Daugherty, *Genetics: A Human Perspective*, 2d ed. Copyright © 1989 Wm. C. Brown Publishers, Dubuque, Iowa. All Rights Reserved. Reprinted by permission.

Chapter 9

Page 178, fig. 3: Gilbert, *Developmental Biology*, Second Edition, 1988. Reprinted by permission of Sinauer Associates, Inc., Sunderland, MA. **9.4, 9.9 (text):** From Kent M. Van De Graaff, *Human Anatomy*, 2d ed. Copyright © 1988 Wm. C. Brown Publishers, Dubuque, Iowa. All Rights Reserved. Reprinted by permission. **9.9 (line art):** From Kent M. Van De Graaff and Stuart Ira Fox, *Concepts in Human Anatomy and Physiology*, 2d ed. Copyright © 1989 Wm. C. Brown Publishers, Dubuque, Iowa. All Rights Reserved. Reprinted by permission. **9.16:** From K. L. Moore, *Before We Are Born: Basic Embryology and Birth Defects*, 3d ed. Copyright © 1989 W.B. Saunders Company, Philadelphia, PA. Reprinted by permission of the publisher and author.

Chapter 10

10.8: Reprinted with permission of Macmillan Publishing Company from *July 20, 2019; Life in the 21st Century* by Arthur C. Clarke. Copyright © 1986 by Serendib, B. V. and OMNI Publications International, Ltd; and From *July 20, 2019: A Day in the Life of the Future*, edited by Arthur C. Clarke. Reprinted by permission of the author and the author's agents, Scott Meredith Literary Agency, Inc., 845 Third Avenue, New York, New York 10022.

Chapter 11

11.2: From E. Peter Volpe, *Biology and Human Concerns*, 3d ed. Copyright © 1983 Wm. C. Brown Publishers, Dubuque, Iowa. All Rights Reserved. Reprinted by permission. **11.4, 11.9:** Reproduced by permission from Nagle, James J.: *Heredity and human affairs*, ed. 3, St. Louis, 1984, Times Mirror/Mosby College Publishing.

Chapter 12

12.5b: From Linda R. Maxson and Charles H. Daugherty, *Genetics: A Human Perspective*, 2d ed. Copyright © 1989 Wm. C. Brown Publishers, Dubuque, Iowa. All Rights Reserved. Reprinted by permission.

Chapter 13

13.7, 13.8a: From *Genetics*, 2/e by Peter J. Russell. Copyright © 1990 by Peter J. Russell. Reprinted by permission of HarperCollins Publishers. **13.8b:** David Hogness, Department of Biochemistry, Stanford University School of Medicine, Stanford, California as in *Proceedings of the National Academy of Sciences*, U.S., Vol. 71 (1974), p. 135. **13.12:** From A. L. Lehninger, *Principles of Biochemistry*, Worth Publishers, New York, 1982. Reprinted by permission.

Chapter 14

14.4b: From *Human Genetics: An Introduction to the Principles of Heredity* 2/E. By Sam Singer. Copyright © 1978, 1985 by W.H. Freeman and Company. Reprinted by permission. **14.5:** From *Human Genetics* by Elof Carlson. Copyright © 1984 by D.C. Heath and Company. Reprinted by permission of the publisher. **14.9:** From Stuart Ira Fox, *Human Physiology*, 3d ed. Copyright © 1990 Wm. C. Brown Publishers, Dubuque, Iowa. All Rights Reserved. Reprinted by permission. **14.10:** Reprinted by permission of Macmillan Publishing Company from *Genetics* by Monroe W. Strickberger. Copyright © 1985, Monroe W. Strickberger. **14.13b:** Source: Terman and Merrill, *Measuring Intelligence*. Copyright © 1937 Houghton Mifflin Company, Boston, MA. **Table 14.2:** From Heredity, Evolution, and Society 2/E. By I. Michael Lerner and William J. Libby. Copyright © 1968, 1976 by W.H. Freeman and Company. Reprinted with permission.

Chapter 15

15.11: Source: Bob Conrad, *Newsweek*, March 5, 1984.

Chapter 17

17.8: From T. L. Peele, *Neuroanatomical Basis for Clinical Neurology*, 2d ed. Copyright © 1961 McGraw-Hill Book Company, New York, NY.

Chapter 22

22.15*a*: From Wallace, King, and Sanders, *Biosphere, 2d ed.* Copyright © 1986 Scott, Foresman and Company, Glenview, IL.

Chapter 23

23.2 (line art): From Wallace, King, and Sanders, *Biosphere, 2d ed.* Copyright © 1986 Scott, Foresman and Company, Glenview, IL.

Chapter 24

24.13: "Reprinted by permission from page 101 of *Understanding Nutrition*, Fourth Edition by Whitney and Hamilton. Copyright © 1987 by West Publishing Company. All rights reserved."

Chapter 25

25.2: From Y. H. Hui, *Principles and Issues in Nutrition*. Copyright © Wadsworth Publishing Company, Belmont, CA; **25.9:** From Journal of the *American Dietetic Association*, American Dietetic Association, Chicago, IL.

Chapter 27

27.6: From Linda R. Maxson and Charles H. Daugherty, *Genetics: A Human Perspective*, 2d ed. Copyright © 1989 Wm. C. Brown Publishers, Dubuque, Iowa. All Rights Reserved. Reprinted by permission; **27.8:** From Bruce Alberts, et al., *Molecular Biology of the Cell*. Copyright © 1983 Garland Publishing Company, New York, NY. Reprinted by permission.

Chapter 30

30.11*a*: From Street and Opik, *The Physiology of Flowering Plants*. Copyright © 1976 Elsevier Science Publishing Co., Inc., New York, NY; **30.11*b*:** From Devlin and Witham, *Plant Physiology*, 4th ed. Copyright c PWS-Kent Publishing Company, Boston, MA.

Chapter 31

31.12: From Raven, *Biology of Plants*. Copyright © Worth Publishing, New York, NY.

Chapter 35

35.13: As appeared in *Science News*, January 9, 1988, page 21.

Chapter 36

36.6: From Krebs and Davies, *An Introduction to Behavioural Ecology*, 2d ed. Copyright © 1987 Blackwell Scientific Publications, Ltd., Oxford, England.

Chapter 37

37.13*a*: Reproduced with permission from Goodenough, Judith, *Animal Communication*; Carolina Biology Reader Series, No. 143. Copyright © 1984, Carolina Biology Supply Co., Burlington, North Carolina, U.S.A. **37.13*b*:** From Keeton and Gould, *Biological Science*, 4th ed. Copyright © W. W. Norton & Co., Inc., New York, NY.

Chapter 38

38.2 (top): "Reprinted with permission from *The New England Journal of Medicine*, vol. 1321, page 1581"; **38.3:** From Arthur Boughey, *Ecology of Populations*, 2d ed. Copyright © 1973 Macmillan Publishing Company, New York, NY; **38.4:** From Barbara Boyle Torrey and W. Ward Kingkade, "Population Dynamics of the United States and the Soviet Union" in *Science*, page 1551, March 30, 1990. Copyright © 1990 by the American Association for the Advancement of Science, Washington, DC. Reprinted by permission of the publisher and authors; **38.5:** Population Reference Bureau, *1989 World Population Data Sheet* (Washington, D.C.: Population Reference Bureau, Inc., 1989). **38.9:** Adapted from *Natural History Magazine*, July 1989. Used with permission.

Chapter 39

39.6*a*: From "Stable Carbon Isotopes in the Study of Food Chains," by Stephen Hart, *Biology Digest*, February 1990. Used with permission; **39.8:** From Eldon D. Enger, et al., *Environmental Science*, 3d ed. Copyright © 1989 Wm. C. Brown Publishers, Dubuque, Iowa. All Rights Reserved. Reprinted by permission; **39.12*a*, *b*:** From Eldon D. Enger, et al., *Concepts in Biology*, 6th ed. Copyright © 1991 Wm. C. Brown Publishers, Dubuque, Iowa. All Rights Reserved. Reprinted by permission; **39.13** and **39.14 (line art):** From McLaren, et al., *Health Biology*. Copyright © 1991 D. C. Heath and Company, Lexington, MA.

Illustrators

John Hagen: 4.13, 8.1, 8.2, 9.9 (after Diane Nelson), 9.13, 9.16, 9.18, 10.7*a*, 12.2, 15.7, Reading 11.1, figures 1, 2*a*,*b*, **34.3**

Beck Visual Communication: 2.5, 2.11*a*, 2.12, 2.13, 2.18, 3.2, 3.25, 4.7*a*, 4.9*a*, 4.10, 4.11, 4.14, 4.16, 4.17, 5.10, 5.11, 5.15, 5.16*b*, 6.4, 6.8, 6.15, 6.21, 7.1*a*, 7.7, 7.9, 7.12, 9.2, 9.3, 9.4, 9.6, 9.7, 9.8, 9.10, 9.11, 9.12, 9.14, 9.15, 10.3, 10.10, 11.2, 11.3, 11.8, 11.13, 12.7, 15.8, 18.2, 18.3, 18.4, 18.6, 18.7, 18.8*b*, 18.11, 18.13, 18.14, 18.16, 18.19, 18.20, 18.21, 19.2, 19.3, 19.7, 19.8, 19.11, 19.13, 19.15, 19.17, 19.19, 19.20, 20.2, 20.4, 20.5, 20.6, 20.8, 20.9, 20.11, 20.14, 20.15, 20.16, 22.1, 22.5, 22.6, 22.7, 22.9*b*, *c*, 22.10, 22.12, 22.13, 22.14, 22.16, 22.17, 22.19, 22.20, 22.21, 22.22, 23.3, 23.5, 23.6. 23.7, 23.9, 23.10*a*, 23.12, 23.13, 23.14
Reading 4.1, figure 1
Reading 4.2, figure 1
Reading 9.2, figures 2, 3, 4
Reading 13.3, figure 1
Reading 20.3, figure 1
Reading 22.2, figure 1*a-d*
Reading 23.1, figure 1
Table 4.4
Table 10.4

Illustrious, Inc.: 1.1*a*, 1.3, 2.7, 3.7, 3.8, 3.10, 3.11, 3.12, 3.13, 3.14, 3.15, 3.16, 3.17, 3.18, 3.19, 3.20, 3.26, 4.6, 4.8, 5.4, 5.6, 5.7, 5.8, 5.12, 5.13, 5.14, 6.1, 6.2, 6.3, 6.6, 6.7, 6.9, 6.10, 6.12, 6.13, 6.16, 6.17, 6.18, 6.19, 6.20, 6.22, 7.2, 7.3, 7.4, 7.8, 7.14, 7.15, 8.3, 8.4, 8.5, 8.6, 8.7, 8.8, 8.9, 8.11, 9.19, 9.22, 10.5, 10.9, 11.4, 11.5, 11.9, 11.10, 11.16, 11.18*a*, 12.1, 12.3, 12.5, 13.2, 13.3, 13.4, 13.5, 13.6, 13.7, 13.9, 13.10, 13.11, 13.12, 13.14, 13.16, 13.17, 13.19, 13.20, 13.21, 13.24, 13.26, 14.2*e*, *f*, 14.3, 14.4, 14.5, 14.8, 14.9, 14.10, 14.12, 14.13, 15.1, 15.2, 15.9, 15.10, 15.11, 15.13, 16.3, 16.4*a*, 16.5, 16.7, 16.9, 17.12, 18.9, 18.10, 18.17, 19.4, 19.5, 19.6, 19.18, 21.4, 21.5, 21.8, 21.9, 21.11, 22.2, 22.3, 22.8*a*, *b*, 22.15*a*, 23.2, 23.11, 23.16, 24.1, 24.9, 24.13, 24.14, 24.19, 25.2, 25.3, 25.4, 25.9, 25.10, 25.11, 26.5, 26.7, 26.10, 26.13, 27.2*a*, 27.6, 27.7, 27.8, 28.2*c*, 28.4, 28.5, 28.7, 28.8, 30.11, 31.2, 31.11, 31.12, 33.4*a*, 33.5, 34.2, 34.4, 34.5, 34.6, 35.3, 35.7, 35.8, 35.9, 35.10, 35.17, 36.6*a*, 36.8, 36.12, 36.13*b*, 37.13, 38.2, 38.3, 38.4, 38.5, 38.6, 38.9, 39.5, 39.6*a*, 39.7, 39.11, 39.12, 39.13, 39.14, 40.8, 40.10
Reading 3.2, figure 1
Reading 17.2, figure 1
Reading 29.1, figure 1
Reading 36.3, figure 1
Reading 39.1, figure 1
Text art page 275

Todd Buck: 16.1, 16.2, 16.6, 16.8, 17.1, 17.2, 17.3, 17.4, 17.5, 17.6, 17.7, 17.8, 17.9, 17.10, 17.13

Carlyn Iverson: 17.14, 21.2, 21.3, 21.7, 21.10, 21.13, 21.14, 21.15, 21.16, 22.18, 25.1, 25.6*a*, 25.8, 29.1, 29.4, 29.5, 29.7, 29.9, 29.10, 29.12, 29.13, 29.14, 29.17, 30.2, 30.3, 30.4, 30.5, 30.8, 30.12*a*, 32.4, 32.6, 32.12, 32.13, 32.14, 40.5, 40.6
Reading 17.1, figure 1
Reading 21.1, figure 1
Reading 40.2, figure 1

Medical Art Services, Inc.: 24.3, 24.4, 24.5*b*, 24.6, 24.7*a*, 24.8, 24.12, 24.15, 24.17, 24.18, 24.20, 27.1, 27.4, 27.5, 27.10, 27.13*b*, 27.15, 27.18*a*
Reading 24.1, figure 2
Reading 35.1, figure 1

Molly Babich: 26.8, 26.9, 26.12, 35.4, 35.5, 35.12*b*

Phil Sims: 34.11, 35.14, 35.16, 35.18
Reading 35.2, figure 1

Laurie O'Keefe: 26.1, 33.1, 33.6*c*, 33.7, 36.2, 36.5, 37.10, 38.7, 39.1, 39.8, 39.9, 39.10, 39.15, 39.18
Reading 34.3, figure 1

Index

G

P

Q

R